S0-BPW-296

A Cross-Polity Survey

A Cross-Polity Survey

Arthur S. Banks

Robert B. Textor

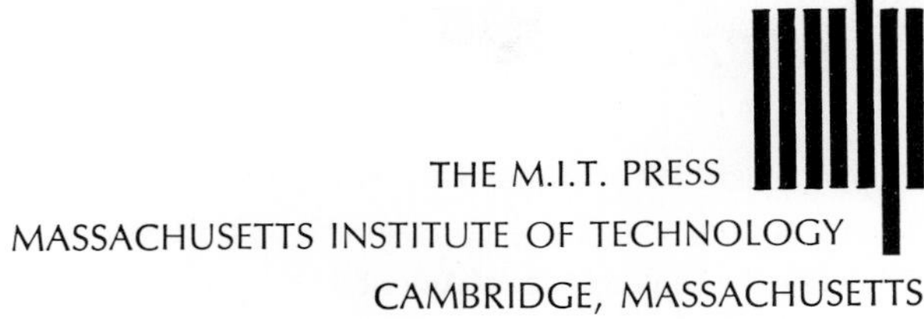

THE M.I.T. PRESS
MASSACHUSETTS INSTITUTE OF TECHNOLOGY
CAMBRIDGE, MASSACHUSETTS

Library of Congress Catalog Card Number: 63-22602
Printed in the United States of America

Preface

THE publication of *A Cross-Polity Survey* is the result of a series of accidents. Banks had for some years intended to produce a book in comparative government that would embrace all of the independent polities of the world. Early in 1962, on a chance re-meeting of two ex-Cornellians, he learned of Textor's Pattern Search and Table Translation Technique. After reflection and experimentation, Banks decided that this technique would lend itself well to cross-polity analysis. Textor welcomed the opportunity to try out the technique on a corpus of data quite different from any to which it had previously been applied. The result, we feel, has been felicitous. The technique has, we believe, proved capable of handling cross-polity data effectively. And, in the process of applying the technique to these data, a number of previously unsuspected weaknesses have been discovered, and improvements made.

The bulk of the present work consists of carefully selected and edited computer printout. It was not our original intention to issue the printout as a separate publication. However, we were surprised by its length, richness, and complexity, which will make it necessary for Banks to devote far more time than had been first anticipated to the preparation of an interpretive monograph. In the meantime, a number of scholars who saw the printout in an earlier form suggested that an edited version be brought out immediately, so that they would not be delayed in proceeding with their own interpretations. Other scholars expressed interest in using the printout as a teaching device, as "raw material" for graduate seminars and research projects. We are thus happy to bring out the printout at this time as a tool for use by the profession.

As will soon become clear, many of the variables included in this study involve coding decisions of a "judgmental" rather than an "objective" nature. While we certainly would prefer wherever possible to use variables of the latter type, we also feel that it would have been improper to exclude important variables simply because social science has, so far at least, been unable to provide more objective criteria for coming to grips with them. Indeed, one might say that one function served by our judgmental codings is simply to make *explicit,* as best we can, a large number of judgments of the type normally left *implicit* in the literature on comparative government. We are quite aware that our decision to be explicit exposes us to criticism from area experts and others. This prospect does not alarm us unduly, provided only that the criticisms are of a nature that will be conducive to improvement of data quality and to progress in the field of comparative politics generally. For cross-national political analysis involves, as Karl W. Deutsch has put it, "a constant interaction between man and data." And it is only by such interaction that there can be improvement both in the data themselves, and in man's understanding of them.

Banks assumes primary responsibility for the conceptualization and coding of the variables that appear in the printout. Textor is primarily responsible for the methodology.

For financial support during a year of intensive work devoted to the project, Banks wishes to thank the Wenner-Gren Foundation for Anthropological Research for two successive research grants. In addition, certain research expenses were met through an award to him from the Permanent Science Fund of the American Academy of Arts and Sciences. Under Public Health Service Research Grant No. M 05478, the National Institute of Mental Health provided support for Textor's methodological research in developing the Pattern Search and Table Translation Technique. This support is gratefully acknowledged.

Textor wishes to express his thanks to the Department of Anthropology and the Southeast Asia Program at Cornell University, to the Southeast Asia Studies Program and Department of Anthropology at Yale University, and to the Department of Statistics at Harvard University for their support of his work over the past several years. At each university, he was generously accorded all facilities, including those of the computation center, and was encouraged and assisted by stimulating and interested colleagues.

For valuable consultation on methodology, we wish to express our sincere gratitude to Professors William G. Cochran, Robert M. Elashoff, J. E. Morton, James N. Mosél, and Frederick Mosteller.

Special thanks are due to Professor Miles Davis of Johns Hopkins University, who served as statistical consultant and programmer during the earlier phases of the development of the Pattern Search and Table Translation Technique. His place as programmer was later ably assumed by Byron A. Marshall, Jr.

Other persons who helped our work in one way or another, and to whom thanks are due, include Professors C. E. Black, Robert B. Dishman, Karl J. Pelzer, Ithiel de Sola Pool, Lucian W. Pye, Fred W. Riggs, Lauriston Sharp, William J. Siffin, and Wayne E. Thompson.

We are indebted to the Computation Center at the Massachusetts Institute of Technology, where much of the computer work was done, and to Mr. Michael V. Solomita, its Operations Manager.

And finally, may we thank those innumerable area experts and other specialists, mainly in the Boston area, who were kind enough to respond with co-operation to our requests for help in coding decisions. We were thus enabled to enter more confident codings on matters as disparate as GNP in Guatemala, polyarchy in Pakistan, and religion in Rwanda.

A. S. B.
R. B. T.

Cambridge, Mass.
August 1, 1963

Contents

1. Introduction

THIS book is intended as a research and reference tool in comparative politics. It summarizes a large amount of current information in a form that should enable the trained reader to discern new relationships among political phenomena.

We depart from tradition in that the bulk of the book consists of computer "printout." This is material printed by a 7090 computer in the form of grammatical English-language sentences, which are grouped by subject into paragraphs. We believe that this printout provides several advantages to the student of comparative politics, which may be summarized as follows:

1. The printout includes in its scope *all* of the world's currently independent polities.[1] This will, we hope, contribute toward making the study of comparative politics less parochial and more nearly general.

2. The sentences and paragraphs employ variables based on concepts considered analytically important by leading scholars in the field of comparative politics. A majority of these variables are of a directly "political" character, but a substantial minority of them are of an "environmental" character. Many more variables were considered for inclusion than actually appear in the present printout. Prominent among those excluded were a number of variables derived from recent literature, which we found impossible to "operationalize" on a world-wide cross-polity basis. (See pages 6 and 22.)

3. The sentences that appear in the printout have been rigorously and systematically selected by means of a new computer technique. The computer automatically "winnows out" the weaker, less important, and less interesting sentences. This "chaff" having been removed, the reader is left free to examine the solid kernels of important information. (See pages 27–28 and 36.)

4. Each sentence constitutes a rigorous "translation" of a statistical table. Such translations appear to be effective in stimulating the creative

[1] As of April 1, 1963.

imagination of the average reader, who is probably more "word-minded" than "number-minded." As an additional convenience to the reader, the statistical table that forms the basis of each sentence also appears in the printout as a "side-note." (See pages 37–40.)

5. The order in which the paragraphs are presented is identical with the order in which sentences appear within a paragraph. This makes it easy for the reader to find his way. Since this ordering is based on general usage, the reader becomes accustomed to it quickly. (See page 25.)

6. Experience to date suggests that when a trained reader scans a paragraph of the printout, he quickly develops a "feel" for the over-all patterning of the data. The degree of internal consistency within most paragraphs is gratifyingly high. (See pages 44–45 and 51–53.)

7. Experience to date also reveals that the trained reader, once having discerned a pattern from the reading of a paragraph, quite naturally finds himself developing more explicit insights and hypotheses as to trend, process, and cause. The reader will discover certain sentences that appear to indicate crucial relationships, and other sentences that, he suspects, indicate intervening variables. The printout is planned in such a way as to facilitate follow-up inquiry that will confirm or disconfirm such suspicions. (See pages 14–17.)

THE PATTERN SEARCH AND TABLE TRANSLATION TECHNIQUE

The Pattern Search and Table Translation Technique is the basic technique used in this book. By means of this technique, the computer performs the following operations: (1) It receives as input punched-card data about each polity, in the form of dichotomous variables. (See pages 25–27.) (2) It crosses each dichotomous variable with each other such variable. (See page 27.) (3) It "winnows out" those resulting two-by-two contingency tables, where the strength of association between the two variables is too low. (See pages 27–28.) (4) It takes the remaining "strong" tables and "translates" them, printing them out as grammatical English sentences. (See page 28.)

UNITS AND UNIVERSE OF ANALYSIS

The unit of analysis is the independent polity. This is the "object" that is classified, cross-classified, and cross-tabulated. The universe of analysis comprises all of the 115 independent polities in the world.[2] We have drawn a "total sample" of this universe. The problem of "sampling error"

[2] Two nations, while technically independent, are excluded: Nationalist China for jurisdictonal reasons and Western Samoa for reasons of self-imposed limitations on the exercise of sovereignty.

is thereby obviated.[3] In cross-tabulations, each of 115 polities in our universe is given equal weighting.[4]

Our decision to include all of the world's independent polities represents a deliberate departure from the older tradition in comparative government, which typically confined itself to the analysis of *Western* political forms. A more recent trend in the field has emphasized the study of *non*-Western polities. Studies resulting from this newer trend have, however, tended to be confined to limited portions of the non-Western world, such as a single geographic region, the Communist bloc, the "underdeveloped" nations, and so forth. The present work builds on this newer trend, but attempts to go further and include all of the world's independent polities within its scope. If it succeeds in nothing else than focusing more interest on the need for a global approach, we shall be satisfied.

THE CONCEPT OF PATTERN

A basic concept used in this book is that of "pattern," which we have borrowed and adapted from cultural anthropology. For present purposes, we define a pattern as "a concatenation of co-occurrences among attributes considered important by the observer."[5] A pattern so defined might be quite simple or highly complex. An example of a fairly complex pattern is the following quotation from a hypothetical book written in 1963 by a political scientist: "The pattern among the independent nations of Africa is that of sparse populations deriving their livelihood primarily from agriculture, suffering from low income and literacy, riven by religious and linguistic diversity, and governed by one-party systems with presidential-type chief executives and unicameral legislatures."

To be of value to the analyst of comparative politics, a pattern must state, or at least imply, some kind of *contrast* between classes of polities. To take a nonsense example, it would be of no analytical value to point out that all nations in Africa are characterized by the presence of some kind of religion, because the same is true of all nations located outside of Africa. In similar fashion, it would be of little value from most analytical points of view to learn that the populations of the countries of Africa are predominantly non-Buddhist, because the same is true of almost all of the non-African countries as well.

The illustrative pattern just quoted supplies contrasts between African

[3] This statement holds when certain assumptions are made with respect to a larger theoretical universe. This matter is discussed on pages 29–30.

[4] Equal weighting is also practiced in cross-cultural research, regardless of whether the number of members of a particular culture is counted in the hundreds or in the millions. For a brief discussion of the relevance of cross-cultural research to the present study, see page 4.

[5] In the present work, we use dichotomous (two-attribute) variables and perform two-way cross-classifications. It would, of course, be possible to use polychotomous variables and to perform cross-classifications that are more than two-way.

and non-African polities in every case. It tells us that African nations are inclined toward sparse populations, dependence on agriculture, and so forth. It also implies that non-African nations are by and large inclined toward denser populations, less dependence on agriculture, and so forth. As will become clear on pages 37–39, each such contrastive element in our illustrative pattern can be derived from a two-by-two table. Indeed, this particular pattern was derived from a number of such tables in the present printout.[6]

The tables, sentences, and paragraphs in the printout are designed to facilitate the *search* for patterns of political phenomena that might not otherwise be apparent to the reader. In no sense, however, does the printout actually *supply* such patterns. Only the intellect of the reader can do that.

THE CROSS-POLITY AND CROSS-CULTURAL METHODS

Our method of cross-polity analysis builds upon, and adapts from, the methodology of cross-cultural analysis. The use of statistical techniques in cross-cultural analysis dates back as far as 1889, when the British anthropologist Sir Edward Burnett Tylor first presented it to a learned audience. Since the 1930's, the cross-cultural method has been vastly improved by a group of cultural anthropologists centering around Yale University and the Human Relations Area Files, and led by George Peter Murdock.[7]

It is somewhat surprising to us that apparently no scholar has heretofore attempted to adapt the cross-cultural method to cross-polity analysis. Indeed, this method would seem to be *more* applicable to polities than to cultures, in at least two senses: (1) The unit that we count in the present study is the independent polity, which is considerably more discrete and rigorously definable than the unit used in cross-cultural research, namely, the individual culture. Moreover, (2) it is possible in cross-polity analysis to take a "complete sample" of the world's independent polities; this ideal cannot be reached in cross-cultural analysis because of the lack of ethnographic data for a large number of cultures.

VARIABLES USED

Each polity in our "total sample" of 115 has been coded on the basis of 57 variables, or, as we call them, "raw characteristics." Approximately

[6] The paragraph that contrasts polities located in Africa with those located elsewhere is Paragraph 10. This is identified in the printout by the number 10 in the upper-left corner. The statements (pairs of contrastive sentences) in Paragraph 10 on which the illustrative pattern is based are numbered 26, 30, 37, 46, 66, 69, 139, 170, and 178.

[7] Readers interested in the original Tylor paper, and in developments in the cross-cultural method since then, can consult Frank W. Moore, *Readings in Cross-Cultural Methodology*. Human Relations Area Files Press, 1961.

two-thirds of these raw characteristics are of direct political significance (for example, No. 27, "Governmental Stability"; No. 29, "Current Electoral System"; No. 54, "Political Participation by the Military"). The remaining one-third of the characteristics are political in a derivative or environmental sense (for example, No. 4, "Population Density"; No. 9, "Per Capita GNP"; No. 15, "Religious Configuration").

Subsumed under each raw characteristic are two or more specific "raw attributes." For some characteristics, the attributes constitute an ordinal continuum; for others, the attributes are in the form of a nominal typology.

A typical ordinal characteristic is Raw Characteristic 4, which deals with Population Density and exhibits the following raw attributes:

A. Very high (600/sq. mi. & above)
B. High (300–599/sq. mi.)
G. Medium (100–299/sq. mi.)
H. Low (below 100/sq. mi.)

A typical nominal-typological characteristic is Raw Characteristic 56 (Legal System) with raw attributes that include the following:

A. Civil law
G. Common law
K. Muslim law
Y. Communist law

After the initial codings had been completed, the raw characteristics, with appropriate reference to the frequency distribution of their subsumed attributes, were dichotomized in a variety of strategically promising ways. For example, one dichotomy contrasts "Polities where interest articulation by associational groups is significant" with "Polities where interest articulation by associational groups is moderate, limited, or negligible."

These dichotomized characteristics—194 in number—are called "finished characteristics." Each finished characteristic is subsequently cross-tabulated with each other such characteristic, by means of the 7090 computer. A considerable proportion of the resulting two-by-two tables, together with their English-language "translations," appear in the computer printout in this book. Details of our methodology are presented in the next chapter.

SELECTING AND REFINING THE RAW CHARACTERISTICS

Some of the raw characteristics used in this study have been established *ad hoc* in accordance with our own interests, though we have made no effort to be original simply for the sake of originality. Most of the raw characteristics, however, have been drawn from published materials in comparative politics.[8] We have conscientiously canvassed the existing literature, and have attempted to select and adapt—experimentally at least—every possible raw characteristic that gave promise of being workable and analytically powerful. As will become increasingly clear in later pages, we found it necessary to reformulate most of the categories taken from the existing literature in order to render them wholly meaningful and operational for purposes of world-wide cross-polity comparison.

Some characteristics that we would like to have used were discarded for conceptual reasons, and others because adequate data for a sufficient number of polities proved unavailable. Ideally, we should have included characteristics on patterns of legitimacy, elite structure, political and social mobility, decision-making, and a host of other subjects of current interest for which conceptualization, on a global basis, at least, is still in its infancy. Tremendous analytical leverage would have resulted from the use of "Social Structure" as an "environmental" characteristic, particularly if this made possible the cross-tabulation of types of stratification along vertical axes with types of differentiation along horizontal axes.[9] But neither adequate data nor appropriate classificatory schemes are yet available, partly because sociologists have not yet attempted to study social structure on a comprehensive cross-national basis, and partly because anthropologists have tended to confine their attention to non-national ethnic groups.

It must be conceded that we have *not* employed as many "quantitative" indices as we might have. This is for a variety of reasons, four of which may be noted.

First, there is the fact that quantitative data for many of the world's polities are simply *unavailable*, especially for many of the newer nations.

Second, there is the problem of *reliability*. Statistics for many variables, especially where the newer nations are concerned, are notoriously suspect.

Third, there is the matter of redundance. In the economic and communications realms particularly, one may be confronted with a variety of indices, several of which yield information that, for purposes of political analysis, covers essentially the same ground. All else being equal, there is, for example, no particular value in supplementing newspaper circulation percentages with figures dealing with magazines, books, and so forth.

Fourth, there is the broader question of validity. Except in the relatively few instances where the quantitative index is political, *per se*, one must ask whether an assumed relationship between a particular measurable quantity

[8] Two particularly useful sources have been: Karl W. Deutsch, "Toward an Inventory of Basic Trends and Patterns in Comparative and International Politics," *American Political Science Review*, Vol. LIV, No. 1, March, 1960; and Gabriel A. Almond and James S. Coleman, eds., *The Politics of the Developing Areas*. Princeton University Press, 1960. The latter is hereafter cited as *PDA*.

[9] For a limited attempt of this type for one country, *see* Leon D. Poullada, "Problems of Social Development," in *An Asia Society Paper: Afghanistan*, January, 1963.

and a given political phenomenon is genuine. It is true, for example, that "number of central government employees as percentage of total population" provides us with valid information as to "degree of political centralization"? The answer must depend, in part, on whether a sizable bureaucracy is required to meet the regime's objectives, or whether it is merely a reflection of pressures on the regime to provide jobs for the politically faithful. In the latter case, the "index" would presumably be far less valid than in the former.

It is our feeling that the stage has not yet been reached where a comprehensive analysis of political systems can fruitfully be undertaken by restricting oneself exclusively to purely quantitative indices. We employ ordinal variables (preferably based on metric scales) wherever we feel that this can validly be done. Otherwise, we rely on nominal variables—often of a "judgmental" type—which we consider to be valid, and we then cope with the reliability problem as best we can. Our position finds support in the writings of John W. Tukey, Professor of Mathematics at Princeton University, who says that "the most important maxim for data analysis to heed, and one which many statisticians seem to have shunned, is this: 'Far better an approximate answer to the *right* question, which is often vague, than an *exact* answer to the wrong question, which can always be made precise.' Data analysis must progress by approximate answers, at best, since its knowledge of what the problem really is will at best be approximate. It would be a mistake not to face up to this fact, for by denying it, we would deny ourselves the use of a great body of approximate knowledge, as well as failing to maintain alertness to the possible importance in each particular instance of particular ways in which our knowledge is incomplete."[10]

Some readers might point out that most of our raw characteristics have a "static" aspect to them—that they describe a situation as of a certain moment or period in time but offer little insight into the direction and rate of ongoing change. To this charge, we must plead largely guilty. Although we are strongly in favor of the proper use of "dynamic" variables, we have nonetheless found that for purposes of global analysis they are, at present, extremely difficult to devise and to code.[11] This applies to the *direction* of change, and even more strongly to the *rate* of change. Thus, despite considerable effort, we have been able to include only one purely "dynamic" variable: "Population Growth Rate" (Raw Characteristic 5).

[10] John W. Tukey, "The Future of Data Analysis," *Annals of Mathematical Statistics,* Vol. 33, 1962, pp. 13–14. Emphasis his.

[11] As we see it, "dynamic" analyses are most easily carried out when the analyst has restricted himself to (1) relatively few *variables* for which highly accurate data are available, or (2) relatively few *polities* (whether dealt with on either a limited-variable or "configurative" basis) for which highly accurate data are likewise available. Neither situation obtains in the present study.

If this study were to be repeated every five years until, say, 1983, then it would be possible in the 1983 edition to analyze the data in all five editions by means of a stochastic process model.

Certain other raw characteristics have a distinctly dynamic coloration, and if a revised edition of the survey were to be prepared some years hence, their potentialities might be further explored. In such an event we would also wish to include any promising dynamic variables that might have been developed in the interim.

CODING CLASSES

It has been our concern to provide reasonably reliable information about particular *classes* of polities rather than, *necessarily,* about any given polity. This is a distinction that in the main causes little difficulty. For example, every polity whose per capita gross national product is coded as "high" will have a higher GNP per capita than every polity coded as "intermediate," "low," or "very low." Now and then, however, the abovementioned distinction can lead to difficulty unless it is clearly understood. Cases of this sort are normally those where the raw characteristic is composite in nature. Take, for example, Raw Characteristic 27, dealing with Governmental Stability. Its structure is as follows:

A. Government generally stable since World War I or major inter-war constitutional change
B. Government generally stable since World War II or major post-war constitutional change
G. Government moderately stable since World War II or major post-war constitutional change
H. Government unstable since World War II or major post-war constitutional change

In coding for this characteristic, Finland and Sa'udi Arabia are listed under "A" and Mongolia under "B." The reader might deduce from this that we regard Finland and Sa'udi Arabia as possessing "more stable" governments than that of Mongolia. This is not the case. Indeed, we might well regard the Mongolian regime as being the most stable of the three. What we *do* contend, however, is that Mongolia belongs to a *class* of governments that, by and large, are probably less stable than the class of governments to which Finland and Sa'udi Arabia belong. Similar examples could be cited with respect to specific codings from a number of other raw characteristics.

RELATIVISM IN CODING

Of necessity, the grouping of polities into classes is often a highly relativistic procedure. This is particularly true of those raw characteristics that are judgmental in nature. Take, for example, the raw characteristic,

"Ideological Orientation." Subsumed under this are the categories, "doctrinal," "developmental," "conventional," "traditional," and "situational." A polity that is "doctrinal" will tend to pursue doctrine, even at the expense of developmental activity, of conventional political values (in the Western sense), and of traditional values (largely in a non-Western sense). Similarly, a polity that is "developmental" will pursue development, even if it means departing from doctrine, or doing violence to conventional or traditional values. Thus in most cases it is possible to render a fairly confident judgment that Polity X tends to give primary emphasis to development *relative to* polities assigned to other classes. Such a coding, however, is clearly more a matter of judgment than of fact. One might almost say that the principal "fact" about such a coding is simply the fact that it has been judgmentally coded that way.

Unless the relativistic nature of many of our codings is clearly understood, we feel that some area specialists might seriously misunderstand the present work. An example will illustrate. Banks, who was primarily responsible for the codings, coded Thailand's ideological orientation as "traditional." Textor, with a background of five years of field research in Thailand, at first objected. He argued that all recent regimes in Thailand have evidenced considerable desire to "develop" that country economically, educationally, and in numerous other ways. He also pointed out that political behavior in Thailand has a strong "situational" emphasis. After some discussion, however, Textor agreed that, of the five available categories, Thailand more properly belongs, relativistically speaking, in the "traditional" category than in any other. Despite its developmental activities, Thailand does not really belong in the same class with such countries as Algeria, Ghana, Guinea, Israel, the United Arab Republic, and Venezuela under Betancourt. Moreover, the present Thai regime's strong emphasis on traditional ritual and behavior would seem properly to bar Thailand from the "situational" class, which embraces such countries as El Salvador and Haiti. Admittedly, Thailand is more developmentally oriented than any of the other polities coded as being "traditional": Afghanistan, Ethiopia, Jordan, and Sa'udi Arabia. Nonetheless, Textor agreed that, viewing the matter in global and relativistic terms, this coding of Thailand was the one that made the most sense.

THE USE OF EXPERTS TO ENHANCE CODING RELIABILITY

We have made use of all readily available published materials by area experts. Indeed, the sheer volume of sources consulted prohibits the specific citation of each for the thousands of individual coding decisions. We have also consulted frequently with area experts in the Boston area. Unfortunately, funds did not permit direct and systematic consultation with area experts elsewhere in the United States or abroad.

Inadequacy of funds also prohibited a formal reliability check. Ideally, it would have been desirable to perform such a check on an areally stratified sample of polities and on a stratified or random sample of our raw characteristics, using area experts as coders. Coefficients of reliability could then have been computed.

Unquestionably, procedures such as the foregoing would have resulted in higher reliability. If a revision of the present survey should ever be undertaken, we hope that funds to support such procedures will be available. Meanwhile, we welcome communications from area experts and other readers concerning ways in which our codings might be improved. (For a discussion of the extent to which suspected coding unreliability might distort the tables and statements in the printout, see pages 42–44.)

COMBINING THE "HOLISTIC" AND "ATOMISTIC" APPROACHES

The Pattern Search Technique, in our opinion, enables the reader to employ both the "holistic" and the "atomistic" approaches to comparative analysis. The holistic approach emphasizes validity at the possible expense of reliability and results in essentially judgmental conclusions. Anthropology and psychoanalysis are among the professions exemplifying the holistic approach. By contrast, the atomistic approach emphasizes reliability at the possible expense of validity and results in relatively rigorous conclusions of limited scope. Sociology and experimental psychology are among the professions most clearly identified with the atomistic approach. The Pattern Search and Table Translation Technique described in Chapter 2 goes some distance toward the holistic approach, in that it provides a very large number of ordered statements generally characterized by a reassuring degree of internal consistency. These statements, when read and analyzed by a trained person, provide a wide-ranging context from which holistic judgments can sometimes be derived. The technique also goes some distance toward the atomistic approach, in that each of the statements in the printout is at least somewhat atomistic in character. We hope that the present technique possesses most of the advantages of both of these approaches, with few of the disadvantages of either. For the reader whose normal habits emphasize the holistic approach, we hope that the tables and statements will be suggestively stimulating, yet also provide a series of moorings that will discourage unduly reckless leaps of imagination. For the reader whose normal habits emphasize the atomistic approach, we hope that the tables and statements will supply rigorous, reliable, and hard information, yet at the same time provide enough over-all context to discourage unduly hasty or narrow conclusions based on co-occurrences among too limited a range of variables.

SOME QUESTIONS THAT MIGHT BE RAISED

Before concluding this chapter, we would like to deal specifically with some questions which many readers would doubtless wish to raise. One such question is: "Does the present survey have any theory?" Here the answer is No. The printout does not present any theory or any hypothesis. It is, instead, designed to stimulate the *generation* of hypotheses on the part of the professionally trained reader. To this end, the raw characteristics reflect, wherever possible, theoretical concepts and analytical categories current in the literature of political and social science.

A second question that might be raised is: "Does the printout permit the testing of hypotheses?" The printout, as we have chosen to use it, does not permit the acceptance or rejection of the null hypothesis in the usual sense, because our 115 polities do not constitute a sample, but rather the total universe itself.[12] However, the printout *does* permit the reader to formulate a hypothesis and then to check whether or not there is a strong association between the variables involved (as distinct from testing the significance of the hypothesized relationship). Suppose one were to hypothesize that polities coded as low in per capita gross national product tend to be those coded as high for the variable "Interest Articulation by Institutional Groups." One need only turn to the appropriate place in the printout, Statement 36/119, in order to determine whether or not there is a strong association between these variables. A precise measure of strength of association is also provided, which the reader is free to evaluate as he pleases.

Finally there is the question: "Does the printout explain or predict political phenomena?" Here, again, the answer is No. Although we share with other social scientists the ultimate goals of explanation and prediction, these are not the immediate goals of the present survey. For the moment, we shall be satisfied if the printout succeeds in stimulating colleagues to discern patterns. Once this occurs, ideas and hypotheses relating to trend, process, and causation are likely to follow. It is in this indirect fashion that we hope the printout will serve the goals of explanation and prediction.

SUMMARY

Our survey is global in scope. The variables used are in general those that most colleagues would consider relevant and important. However, we have avoided variables that proved impossible to operationalize on a

[12] Some readers might wish to make other assumptions under which, it could be argued, significance testing would be possible. Such a set of assumptions would be that our "N" is fixed, and that the row and column totals for any particular table are fixed, but that the cell frequencies are not fixed. This problem is discussed on pages 29–30.

world-wide cross-polity basis. The printout includes *all* the important information produced by cross-tabulation, and nothing *but* important information.[13] All statistical information is "translated" into grammatical English sentences, which are presented in convenient order. From these, the reader is enabled to discern over-all patterns possessing high internal consistency. The final result, we hope, is that the reader is stimulated to generate, and enabled to check, hypotheses concerning process and cause. In essence, the computer has been utilized to do the "dirty work," and the trained reader is stimulated and enabled to engage in "inductive leaps" in the direction of a more rigorous and all-inclusive science of comparative politics.

[13] That is, it includes all information resulting from two-way cross-classification of dichotomous variables where the strength of association is high. See pages 27–28.

2. Methodology

THIS chapter will explain the basic technique used in this study, the "Pattern Search and Table Translation Technique."[1] The purpose of the technique is to assist the reader in discovering patterned co-occurrences and inter-relationships in political institutions, structures, and behavior. To facilitate this search for patterns, statistical tables are first constructed and are then "translated" into grammatical English. We believe that these translations are of some value, in that most social scientists tend to think more creatively in terms of words than of numbers.

THE TWO-BY-TWO MATRIX

The core of the present technique is the two-by-two matrix that results from the cross-classification and cross-tabulation of two dichotomous variables, or what we call "characteristics."

For purposes of illustration, let us take as *column* headings, "Polities that were formerly dependencies of Britain," as contrasted with "Polities that were formerly dependencies of France." Let us take as *row* headings, "Polities where the literacy rate is above ten percent," as contrasted with "Polities where the literacy rate is below ten percent." Initial coding reveals that there are 19 ex-British and 24 ex-French territories in the world today (excluding those that became independent before 1914). A separate coding operation results in our being able to classify 33 of these 43 polities as having literacy rates that are unambiguously either above or below ten percent. (The other 10 polities are excluded from our tabulations because we could not obtain sufficiently reliable data to permit their coding on the basis of the latter dichotomy.)

Thus we know, without the aid of any cross-tabulation, which countries are ex-British and which are ex-French. We also know which have literacy rates above ten percent, and which below. What we do *not* know is

[1] Since we are writing primarily for non-statisticians, we ask the indulgence of those readers who are statistically trained.

which *ex-British* areas are above ten percent and which below, and which *ex-French* areas are above and below the ten-percent figure. To obtain such information, we resort to cross-classification, as exemplified by Matrix 1.

MATRIX 1

	COUNTRIES FORMERLY UNDER BRITISH RULE			COUNTRIES FORMERLY UNDER FRENCH RULE		
LITERACY ABOVE TEN PERCENT	Burma India Jamaica Pakistan (total: 12)	Ceylon Iraq Malaya Trinidad	Cyprus Israel Nigeria UAR	Algeria (total: 3)	Cambodia	Malagasy R
LITERACY BELOW TEN PERCENT	Sierre Leo Uganda (total: 4)	Sudan	Tanganyika	Cameroun Congo (Bra) Guinea Mauritania Togo (total: 14)	Cen Afr Rep Dahomey Ivory Coast Niger Upper Volta	Chad Gabon Mali Senegal

Matrix 1 reveals clearly that three-quarters of the ex-British territories for which codings were available fall into the above-ten-percent category, while an even greater majority of the ex-French territories fall into the below-ten-percent category. Since it is generally assumed that literacy rates in most ex-British and ex-French dependencies were low at the time of colonial take-over, the initial conclusion of some analysts might be that the British exerted more effort in promoting literacy and achieved more success than did the French.

Proper analysis requires, however, that we persist further. We must ask: Is this matrix explainable *simply* in terms of differential skill and effort as between the British and the French? Or are there perhaps intervening variables that help explain the difference? One such variable might be whether religion in the ex-colony is predominantly literate, or whether it is predominantly or partially non-literate. It is reasonable to assume that areas with literate religions at the time of colonial take-over would have had higher literacy rates at that time than would be true of areas where religion was largely or wholly non-literate. Moreover, as linguists would point out, if Country X has a literate religion and Country Y does not, the long-term odds would favor Country X over Country Y, even if the two countries had started from the same level of literacy. This would probably hold even where, as is often the case, the literate language of the religion of Country X is other than the day-to-day vernacular language of that country. In Matrix 2, the ex-British and ex-French areas with which we are concerned are cross-classified with the "literate religion" variable.

MATRIX 2

	COUNTRIES FORMERLY UNDER BRITISH RULE	COUNTRIES FORMERLY UNDER FRENCH RULE
RELIGION LITERATE	Burma, Ceylon, Cyprus, India, Iraq, Israel, Jamaica, Malaya, Pakistan, Trinidad, UAR (total: 11)	Algeria, Cambodia (total: 2)
RELIGION PREDOMINANTLY OR PARTLY NON-LITERATE	Nigeria, Sierre Leo, Sudan, Tanganyika, Uganda (total: 5)	Cameroun, Cen Afr Rep, Chad, Congo (Bra), Dahomey, Gabon, Guinea, Ivory Coast, Malagasy R, Mali, Mauritania, Niger, Senegal, Togo, Upper Volta (total: 15)

Matrix 2 clearly suggests that literate religion is an important intervening variable. In the rough-and-tumble of colonial empire building, the British seem to have gained dominance primarily in areas possessing literate religions and, hence, to have enjoyed a literacy base on which to build. The French seem, by and large, to have inherited areas with non-literate religions and, hence, to have lacked such a base. The three-variable Matrix 3 sums up this phenomenon.

MATRIX 3

		COUNTRIES FORMERLY UNDER BRITISH RULE	COUNTRIES FORMERLY UNDER FRENCH RULE
LITERACY ABOVE TEN PERCENT	RELIGION PREDOMINANTLY LITERATE	Burma Ceylon Cyprus India Iraq Israel Malaya Pakistan UAR (Jamaica) (Trinidad) [total: 11 (or 9)]	Algeria Cambodia [total: 2]
	RELIGION PREDOMINANTLY OR PARTLY NON-LITERATE	Nigeria [total: 1]	Malagasy R [total: 1]
LITERACY BELOW TEN PERCENT	RELIGION PREDOMINANTLY LITERATE	None	None
	RELIGION PREDOMINANTLY OR PARTLY NON-LITERATE	Sierre Leo Sudan Tanganyika Uganda [total: 4]	Cameroun Cen Afr Rep Chad Congo (Bra) Dahomey Gabon Guinea Ivory Coast Mali Mauritania Niger Senegal Togo Upper Volta [total: 14]

We have placed Jamaica and Trinidad in parentheses because Matrix 3 happens to be based on the status of religion in each country as of *now*, rather than as of the time of colonial take-over. In most cases the present status and the status at the time of take-over are more or less the same, but in the cases of Jamaica and Trinidad this is doubtful. Matrix 3, thus interpreted, shows that Jamaica, Trinidad, and Nigeria are the only three countries that Britain took over that at the time of take-over were religiously at least partly non-literate, *and* that at the end of the period of British rule enjoyed a literacy rate above ten percent. This makes three countries out of seven. For France, the record is one—Malagasy—out of fifteen. There are thus four "deviant cases": Jamaica, Trinidad, Nigeria, and Malagasy. In further analysis of these deviant cases, the analyst would normally ponder all available information about them. Upon reflection, he would probably decide that Jamaica and Trinidad are somewhat special. These two polities were British dependencies as early as the seventeenth and eighteenth centuries, respectively. Their populations are concentrated on relatively small islands and, hence, comparatively easy of access. In contrast, the French in Africa by and large enjoyed neither the advantage of time nor that of easy accessibility. On balance, then, the use of the "literate religion" characteristic as an intervening variable, plus deviant case analysis, makes the French record look less inadequate than it did at first glance.

The example just given illustrates, in a preliminary way, the manner in which we use matrices to reach conclusions, or at least to refine problems. A number of two-by-two matrices are included in the computer-produced portion of this study. Because of limitations of space, we were forced to be highly selective in deciding which matrices to include. In making the selection, we gave preference to

1. Matrices that would portray important relationships.
2. Matrices that would illustrate interesting deviant cases.
3. Matrices dealing with subjects that would, we thought, evoke fairly general reader interest.
4. Matrices that cross a "political" with an "environmental" characteristic.

Because only a small number of matrices could be included, the reader must in the vast majority of cases be satisfied with two-by-two tables that merely give the *number* of polities in each quadrant but do not *name* the polities as a matrix does. Readers who wish to construct their own matrices (two-by-two, two-by-N, or M-by-N) may easily do so. (See pages 49–50.)

OUTLINE OF PROCEDURE

The following outline is skeletal. In some respects it is distorted for the sake of simplicity. If the reader should wish to apply this technique to a problem of his own, we suggest that he first consult Textor's forthcoming technical monograph, *The Pattern Search and Table Translation Technique*.

DEFINITION OF TERMS

We have found it desirable to employ the following terms:

"Raw Characteristic": A raw characteristic is any basic variable. For example, "Former Colonial Ruler."

"Raw Attribute": Subsumed under each raw characteristic are two or more raw attributes. For example, "Britain," "France," "Spain," "Other." Raw attributes are *always* mutually exclusive.

"Finished Characteristic": A finished characteristic is always a dichotomous characteristic. An example would be: " 'Polities that were formerly dependencies of Britain,' as contrasted with 'Polities that were formerly dependencies of France.' " Another finished characteristic contrasts former dependencies of Britain *or* France, on the one hand, with those of Spain, on the other. Several finished characteristics can, and in practice usually do, result from the dichotomization of a single raw characteristic in a variety of strategically promising ways. Occasionally a raw characteristic is dichotomous to begin with. In such cases, the raw characteristic and the finished characteristic are identical. The reasons why finished characteristics must always be dichotomous will become clear later (pages 23–25).

"Finished Attribute": Either of the two attributes subsumed under a finished characteristic is called a "finished attribute."

1. TENTATIVE SELECTION OF RAW CHARACTERISTICS

Our initial procedural step was to assemble a tentative group of raw characteristics. Two such tentative raw characteristics were: (1) "Almond-Shils System Characterization" and (2) "Former Colonial Ruler." We wished to include the first because it reflected the highly stimulating heuristic typology developed by Gabriel A. Almond and others in *The Politics of the Developing Areas.* In our opinion, this typology offered promise of proving useful in the building of new and more powerful theory. We wished to include the second raw characteristic because we felt it worth while to ask the question, "What difference does it make whether a particular polity is an ex-dependency of one colonizing power rather than another?" Approximately one hundred economic, demographic, cultural, and political variables were tentatively considered for incorporation into the present study. A majority of these were either completely discarded or else substantially revised because of coding difficulties, inadequate data, redundancy, and so forth.

2. ESTABLISHING RULES FOR CODING

The early stages of this study witnessed a constant interaction between specific coding decisions and decisions about general coding rules. This was particularly true with respect to judgmental raw characteristics, but it was true as well for raw characteristics based primarily on objective fact.

For example, in establishing rules for coding the raw characteristic, "Former Colonial Ruler," we decided to limit the scope of the characteristic to European colonizing powers.[2] Thus Sa'udi Arabia, a former dependency of the Ottoman Empire, and North and South Korea, former dependencies of Japan, were excluded as "irrelevant," along with such countries as Liberia, Thailand, and the USSR. Another rule we found necessary to establish was that in cases where more than one colonial power had successively occupied a particular country, we would base our coding on the most recent occupying power. Thus Trinidad is coded under Britain rather than Spain, Ceylon under Britain rather than the Netherlands or Portugal, Togo under France rather than Germany, and so forth.

Coding rules and the considerations underlying them are spelled out in detail in Chapter 3.

3. ESTABLISHING RAW ATTRIBUTES

In establishing raw attributes, one often has considerable freedom of choice. Take again the example of "Former Colonial Ruler." We discover that the former British territories are 24 in number; ex-French, 25; ex-Spanish, 18; ex-Belgian, 3; ex-Italian, 2; and ex-American, Dutch, and Portuguese, 1 each. The decision to combine the former Belgian, Italian, American, Dutch, and Portuguese territories into the single raw attribute, "Other," was clearly an arbitrary one. If our interests had been different, we would have coded differently. For example, we might conceivably have combined former French, Spanish, Belgian, Italian, and Portuguese territories into one raw attribute, "Polities whose former colonial ruler was a primarily Catholic country." This raw attribute category might have been contrasted with polities formerly dependent on Britain or America, under the label "Polities whose former colonial ruler was a primarily Protestant country." Indonesia, as a former Dutch territory, would then be excluded as "ambiguous," since the Netherlands is approximately half Catholic and half Protestant.

In deciding on the range of attributes for any one raw characteristic, we were guided by two considerations. First, we wanted to provide maximum flexibility for the subsequent combining of raw attributes into a variety of finished attributes on a dichotomous basis. Second, we wished to avoid raw attribute categories with too few members. To take an extreme example, there would be no point in establishing a raw attribute, "Polities where the former colonial ruler was the United States," because only one country, the Philippines, would be so codable. It is easier to describe the Philippines by ordinary descriptive techniques than to put it into the computer and contrast it with the other 114 independent polities

[2] Here taken to include the United States as a "European-derived" colonial power.

in the world. The same consideration would usually apply to any attribute class of five or fewer polities, though for reasons of typological consistency and completeness a limited number of such very small classes do appear in the present study.

4. ASSIGNING EACH POLITY TO A RAW ATTRIBUTE CATEGORY

Once the coding rules and raw attributes were established, the next step was to assign each polity to one or another of the various raw attribute categories. For some raw characteristics, such as "Former Colonial Ruler," the procedure was straightforward and presented little or no difficulty. In other instances, the assignments were largely judgmental, rather than factual. In coding for the raw characteristic "[Degree of] Westernization," for example, it was necessary to employ holistic, relativistic judgment.

In order to minimize error and bias, a number of built-in checks were used. First, we coded alphabetically, from Afghanistan to Yugoslavia, for each characteristic. This, we felt, provided some protection against regional or other bias. Second, we attempted conscientiously to avoid "contamination." For example, in coding for RC 55 ("Role of Police" in politics) we attempted to confine our attention to this characteristic exclusively. Thus, in coding polities such as Lebanon, Malagasy, Malaya, and Nigeria, we made an effort to disregard such irrelevant facts as that the first two are ex-French dependencies and the latter two ex-British dependencies.

A third safeguard is what might be called an "internal homogeneity check." For characteristics requiring largely judgmental coding, each polity was given an initial *tentative* assignment to a raw attribute category. The polities assigned to each category were then re-examined to see if each "really belonged" with the rest. This meant, in effect, that to some extent we adopted the tactic of permitting each coding assignment to help define the category itself. In deciding whether a polity is properly to be regarded as "significantly Westernized," "partially Westernized," or "non-Westernized," one is forced to ask whether Burma "really belongs" in the same class with Bolivia; Ceylon with Cambodia; Honduras with Hungary; and so forth. In attempting to achieve maximum internal homogeneity within raw attribute categories of this sort, we often found it necessary to process all of the polities several times. The assignments would become progressively less tentative, until finally we arrived at a set of assignments which were intuitively and logically satisfying in the sense that each polity had been properly coded *relative to* each other polity.

5. ELIMINATING THE RESIDUE

In assigning polities to appropriate raw attribute classes, we would often find that a residue of polities remained which could not meaningfully and reliably be assigned to any specific attribute class. Nor in most cases was it possible or desirable to set up an additional raw attribute class to which some or all of such polities could be assigned. Such polities were therefore assigned to the "residue." There are four grounds for assigning a polity to the residue.

AMBIGUITY

Suppose that the New Hebrides were today an independent polity and hence a member of our universe of polities. In coding this country for the raw characteristic, "Former Colonial Ruler," we would be forced to place it in the residue, on grounds of ambiguity. The reason, of course, is that the New Hebrides' former colonial ruler would have been a condominium of Britain and France. To place it either in the "Britain" raw attribute category or in that of "France" would violate the basic coding rule of mutual exclusiveness. A theoretical alternative would be to create a new category, "Condominium of Britain and France." But this would be unproductive, because the only polity assignable to such a category would be the New Hebrides itself.

IRRELEVANCE

Of the 115 independent polities in the world, 75 are former European dependencies. The remaining 40 have, by our coding rules, never been under European domination. As indicated earlier, these 40 are coded as "irrelevant," and placed in the residue.

UNASCERTAINEDNESS

A third basis for assigning a polity to the residue is lack of available information. It will be recalled that when we compared the literacy rate in ex-British countries with that in ex-French countries, some countries dropped out of the matrix because their literacy rates had not been reliably ascertained.

UNASCERTAINABILITY

A polity that is "unascertainable" is one that is in such a state of flux that, regardless of how much reliable information we might possess about it, we still could not code it with respect to the raw characteristic in question. For example, countries like Mauritania, Niger, and Chad have so recently gained their independence that they cannot meaningfully be coded for "Governmental Stability." "Unascertainable" differs, therefore, from "unascertained." "Unascertainable" means, in effect, that no data relevant for coding are *possible*. "Unascertained" means that the relevant data are *unavailable* to the coder.

As already indicated, the residue is always excluded from the matrices and tables and is not reflected in the sentences into which the matrices and tables are translated. What this means is that the matrices, tables, and sentences deal only with those polities that they have a "right" to deal with. That is, they deal only with those polities that are ascertainable rather than in flux; that are "ascertained" in the sense that adequate reliable information is at hand; that are relevant to the characteristic in question; and that can be unambiguously classified with respect to that characteristic.

6. *DEFINITIVE SELECTION OF RAW CHARACTERISTICS*

At this point a definite decision is made whether a given raw characteristic, however desirable, can be included or not. The "Former Colonial Ruler" characteristic is straightforward, and there was never any serious doubt about its practicability. However, other desirable characteristics proved impossible to operationalize. An example is the "Almond-Shils System Characterization." In this case, despite persistent effort we were not able to formulate satisfactory rules that would serve to discriminate polities validly and reliably into raw attribute classes on a satisfactory basis, with due regard for the need for mutual exclusiveness and a reasonable degree of exhaustiveness. Illustrative of our difficulties were the two Almond-Shils attributes, "Tutelary Democracy" and "Modernizing Oligarchy." We found it impossible to set up coding rules that would render these two attributes mutually exclusive. The same was true of our efforts to utilize or adapt interesting and stimulating typologies appearing in the published works of a number of other scholars. The only non-ordinal classificatory schemes that we took over *in toto* from an external source were C. E. Black's modernization typologies (RCs 22 and 23).

Banks devoted ten months, full-time, to the coding work underlying the present study. The problems he encountered were comparable in nature and inherent difficulty to those frequently complained about by cross-cultural researchers. There was, first, the serious problem of lack of data on many subjects. Beyond this was the equally serious problem that many of the analytical concepts and categories currently in use by comparative government specialists are not globally relevant and hence do not lend themselves to global comparisons.[3] In short, it was often necessary to resist the blandishments of an otherwise interesting typology because of insoluble reliability problems, while it was also necessary to reject the temptation to fall back on categories of easy reliability where little would result that was fresh, valid, or analytically powerful.

7. *PREPARING THE RAW CHARACTERISTIC DECK*

The Raw Characteristic Deck contains the basic data used in all computations. In key-punching the raw characteristics, one IBM card is used to represent each polity. This card is uniquely identified with the polity's

[3] It is worthy of note that of all the non-ordinal variables considered, only Black's had been previously tested on a *comprehensive* global basis.

name and a serial number based on alphabetical ordering. The code representing the particular polity's raw attribute for the first raw characteristic, "Areal Grouping," is punched in column 1. For example, in column 1 for Card 056 representing Japan, the punch is "I" for East Asia. The appropriate coding for Japan's raw attribute under Raw Characteristic 2 is punched on the same card in column 2. We end up with a deck of 115 cards, each punched in as many columns as there are raw characteristics. Alphabetic codes are used for raw attributes. Numeric codes are used for the residue. Within the residue, punch 3 means "ambiguous"; punch 4 means "irrelevant"; punch 5 means "unascertained"; and punch 6 means "unascertainable."[4]

8. ESTABLISHING FINISHED CHARACTERISTICS

As stated earlier, The Pattern Search and Table Translation Technique, as presently constituted, handles only dichotomous characteristics. The justification for dichotomizing at this particular stage in the analysis is straightforward: it reduces highly complex matters to a format simple enough so that the analyst can readily perceive significant, patterned relationships. The immediate yield is, in effect, a sort of preliminary conceptual distillation on the basis of which more sophisticated, "polychotomous" relationships may later be perceived.

The four principal forms of dichotomization are as follows:

SIMPLE DICHOTOMIZATION WITH ONE POSITIVE CATEGORY

This type of dichotomization contrasts polities occupying a single positive category with those in the "all other" category. An example is FC 1, which contrasts "Polities located in East Europe" with "Polities located elsewhere than in East Europe." The positive attribute is always placed in the left column and the "all other" attribute in the right column. In this kind of dichotomization, no polities are placed in the "irrelevant" part of the residue.

SIMPLE DICHOTOMIZATION WITH TWO POSITIVE CATEGORIES

An example of this type of dichotomization would be "Polities that were formerly dependencies of Britain, rather than France," as contrasted with "Polities that were formerly dependencies of France, rather than Britain." Both finished attributes are equally "positive," and hence the decision concerning which will occupy the left column is purely arbitrary. Except in cases where the original raw characteristic is dichotomous,

[4] For convenience in using the sorting machine on the Raw Characteristic Deck, we have avoided using alphabetic punches C, D, E, F, L, M, N, O, T, U, V, or W, which, for technical reasons, would result in confusion with numeric punches 3, 4, 5, and 6. Raw attributes are thus serialized as A, B, G, H, I, J, K, P, Q, R, S, X, Y, Z.

finished charcteristics of this type will normally fail to exhaust all of the polities in the world; there will usually be some polities in the "irrelevant" part of the residue. In this example, polities become irrelevant if they (1) were formerly ruled by a European colonial power other than either Britain or France, or (2) were never ruled by any European colonial power.

COMPLEX DICHOTOMIZATION WITH ONE POSITIVE CATEGORY

This type of dichotomization contrasts a positive attribute with an "all other" attribute. The positive attribute is "complex" in that it combines raw attributes from one, two, or more raw characteristics. Such combinations may be (1) additive, (2) alternative, or (3) both additive and alternative. An example of the additive type would be "Polities whose date of independence is after 1914, and that were formerly dependencies of Britain," as contrasted with "All others." Here, *both* of two conditions must be satisfied by a polity, or it will fall into the "all other" category. An example of the "alternative" type would be "Polities that were formerly dependencies of either Britain or France," as contrasted with "All others." Here, a polity can qualify for the left column if it has been a dependency of either of these two colonial powers. (It could qualify also if it had been a dependency of both Britain and France conjointly, because the conjunction "or" as we use it is always the "inclusive or," meaning "and/or," unless expressly specified to be otherwise.) An example of the "additive and alternative" type would be "Polities whose date of independence is after 1914, and that were formerly dependencies of Britain or France," as contrasted with "All others."

Finished characteristics formed by complex dichotomization of either this or the following type are called "compound finished characteristics."

COMPLEX DICHOTOMIZATION WITH TWO POSITIVE CATEGORIES

Finished Characteristic 80 exemplifies this type of dichotomization. The left subject is "Polities whose date of independence is after 1914 and that were formerly dependencies of Britain, rather than France." The right subject is "Polities whose date of independence is after 1914 and that were formerly dependencies of France, rather than Britain." In such a case, the following classes of polities would drop into the residue as irrelevant:

a. Polities that were formerly dependencies of Britain or France, but whose date of independence is before 1914, such as the United States and Haiti.

b. Polities that were formerly dependencies of some other European colonial power.

c. Polities that were never dependencies of any European power.

Dichotomization into finished characteristics can, of course, proceed in as many cross-cutting ways as the analyst wishes, in order to extract from the data all contrasts that seem to be strategically promising. Thus, our original 57 raw characteristics have been dichotomized into 194 finished characteristics.

9. ORDERING THE FINISHED CHARACTERISTICS

When all the finished characteristics have been established, it is necessary to place them in whatever order is most likely to maximize clarity of presentation. In the present study we have started with finished characteristics dealing with areal grouping, size, population, and other demographic variables. We have then moved to economic variables. After that, the finished characteristics deal with literacy, religion, language, racial homogeneity, and historical background. The remainder of the printout deals with political subjects. We have thus followed more or less the order that many political analysts would follow if they were writing a political study of a particular country. Our decision to do so was deliberate: We wished to conform as closely as possible to the accustomed intellectual habits of our readers.

Within the group of paragraphs stemming from any one raw characteristic, the order is usually from the general to the specific. Contrasts "within" a particular area of analysis are usually dealt with first and are followed by contrasts "between" two areas of analysis. However, there are no hard-and-fast rules. Our overriding objective throughout has been *clarity* and ease of understanding, and we have sometimes sacrificed pure consistency in an effort to achieve this end.

10. PREPARING THE FINISHED CHARACTERISTIC DECK

At this point the computer enters our procedure for the first time. The Column and Punch Reallocation Program Deck is fed into the computer, followed by the Raw Characteristic Deck. The result is the production of the Finished Characteristic Deck.

An example will suffice. The Raw Characteristic Deck comprises 115 cards. Each card stands for a particular polity and is uniquely identified by means of an identification number. Column 21 of the Raw Characteristic Deck stands for Raw Characteristic 21, "Former Colonial Ruler." In this column, a particular card might be punched A, B, G, H, or 4. The distribution of these punches is as follows:

Punch A: 24 cards, each standing for a former British dependency.
Punch B: 25 cards, each standing for a former French dependency.
Punch G: 18 cards, each standing for a former Spanish dependency.
Punch H: 8 cards, each standing for a former dependency of some other European colonial power.

Punch 4: 40 cards, each standing for a polity that is "irrelevant" in the sense that it has never been a dependency of any European power.

Using Raw Characteristic 21, we wish to create Finished Characteristic 79, which in essence asks the question: "What difference does it make whether a polity is an ex-British or ex-French dependency on the one hand, as contrasted with an ex-Spanish dependency on the other?" In Finished Characteristic 79, the ex-British or ex-French dependencies become the left subject, and the ex-Spanish dependencies become the right subject.
Employing the Column and Punch Reallocation Program Deck, plus appropriate instruction cards, we instruct the computer as follows:

First, examine the Raw Characteristic Deck and find all cards that are punched "A" or "B" in column 21. Now move to the Finished Characteristic Deck and punch all corresponding cards "1" in column 79. (Punch "1" means "left subject.")

Second, examine the Raw Characteristic Deck and find all cards that are punched "G" in column 21. Now move to the Finished Characteristic Deck and punch all corresponding cards "2" in column 79. (Punch "2" means "right subject.")

Third, examine the Raw Characteristic Deck and find all cards that are punched "H" or "4" in column 21. Now move to the Finished Characteristic Deck and punch all corresponding cards "4" in column 79. (Punch "4" means "irrelevant.")

In such a manner, the 49 ex-British and ex-French dependencies become Left Subject 79, entitled "Polities that were formerly dependencies of Britain or France, rather than Spain (49)." The 18 ex-Spanish dependencies become Right Subject 79, entitled "Polities that were formerly dependencies of Spain, rather than Britain or France (18)." The 48 polities that were formerly dependencies of other European colonial powers, or that have never been dependencies of any European colonial power, drop into the residue as "irrelevant."

11. REASSIGNMENT OF RESIDUE

After the Finished Characteristic Deck has been prepared, an additional operation is called for in connection with the residue. Step 5 resulted in the assignment of certain polities to the residue on the basis of coding for *raw* characteristics. In connection with the *finished* characteristics, however, it is possible to "retrieve" some of the earlier residual assignments.

For example, in coding for Raw Characteristic 12, specific literacy figures were unavailable for ten polities, which were therefore placed in the residue on grounds of being "unascertained" (punch 5). But Finished Characteristic 44, which contrasts "Polities where the literacy rate is ninety percent or above" with "Polities where the literacy rate is below ninety percent," is of narrower scope than the raw characteristic from which it is derived. And it so happens that, although we do not have *precise* literacy-level data for the ten polities in question, we *do* know that, whatever the precise figures might be, they are certainly in all ten cases "below ninety percent." Hence, for this *particular* finished characteristic, all ten polities may be removed from the residual category and coded substantively as "punch 2." This is accomplished by key-punching ten new finished characteristic cards, one for each of the polities affected, changing the coding under column 44 from punch 5 to punch 2. These ten cards represent the following polities: China, Guinea, North Korea, Lebanon, Liberia, Mauritania, Mongolia, Morocco, Syria, and North Vietnam.

A similar operation is performed, where possible, for all other finished characteristics. The result is a much smaller residual count for the Finished Characteristic Deck than for the Raw Characteristic Deck.[5] Needless waste of information is thus avoided, and the statements, paragraphs and patterns that ultimately emerge are much more clearly defined.

12. ADMINISTRATION OF THE PATTERN SEARCH PROGRAM

At this point the computer is used for the second time. The Pattern Search Program Deck is fed into the computer, followed by the Finished Characteristic Deck. The result is the production of a large number of automatically punched cards. Each of these represents one two-by-two table, and is called a "table card."

To illustrate, let us take Finished Characteristic 80, which provides a dichotomous contrast between ex-British and ex-French dependencies. The computer crosses this dichotomy with the dichotomies contained in each of the other 193 finished characteristics. In this way the computer generates a very large number of two-by-two tables. For each such table, the computer computes the Fisher Exact calculation. This Fisher value, indicated as "P" in the tables in the printout, is the basic calculation of the Pattern Search Technique. Details are discussed on pages 29–30. Suffice to say here that in the present survey (but not necessarily elsewhere) this value is used as a measure of *strength of association*. It can vary from 1.00 down to almost zero. The lower its value, the stronger is the association between the two finished characteristics entered in the table. The Pattern Search Technique permits each investigator to select the critical level of the Fisher value which is best suited to his research needs. In the present survey, the critical level is set at .14999. As the computer generates tables, it looks at the Fisher value of each table. If this value is too weak—that is, .150 or higher—that table is automatically disre-

[5] For this reason, the raw attribute totals in Chapter 3 often do not tally with the corresponding finished attribute totals.

garded, or "winnowed out." If this value is strong—that is, .14999 or lower—the computer automatically punches for that table a "table card."

The punched information appearing on each table card includes all information needed to carry the analysis forward:

a. The identification number of the finished characteristic that will provide the subjects of the printout statement.

b. The identification number of the finished characteristic that will provide the predicates of the statement.

c. The number of polities to be found in each of the four cells of the two-by-two table.

d. Grammatical instructions for later translating the table into a statement.

e. Other numerical and technical information.

13. ADMINISTRATION OF THE TABLE TRANSLATION PROGRAM

The table cards produced by administration of the Pattern Search program are next arranged in exact numerical order. They are now ready for our third use of the computer, which is designed to translate the tables into English. The Table Translation Program Deck is fed into the computer, followed by the large deck of ordered table cards. The result is the production of printout. The bulk of the present study is simply a carefully edited version of this printout as produced by the computer.

The Table Translation Program translates each table into two English "sentences" which, taken together, comprise one "statement." Each sentence contains the following parts of speech: subject lead, subject clause, verb, and predicate clause. In some cases there is also an adverb. The precise grammatical rules for building a sentence are described on pages 37–39.

14. GENERATING MATRICES

The computer printout will reveal numerous interesting relationships. In some cases, the analyst will be dissatisfied with the mere knowledge of *how many* polities occupy each of the four quadrants of the two-by-two table. He will wish to know, in addition, just *which* polities occupy each quadrant, so that he can, among other things, analyze the deviant cases. To gain this information, he makes use of the computer for the fourth time. The Matrix Generation Program Deck is fed into the computer, followed by appropriate instruction cards and all the table cards representing tables for which matrices are desired. The result is the printout of each of the desired matrices.

We have already seen on page 14 an example of a typical matrix—the one that crossed "Ex-British versus Ex-French" with "Literacy Above

Ten Percent versus Literacy Below Ten Percent." A number of matrices appear in the computer-produced printout in the present study. For each statement for which a matrix is provided, the letter "M" appears alongside the two-by-two table associated with the statement. The matrix appears at the end of the paragraph concerned.

15. RE-RUNS

Careful study of the printout from the first run, and of the printout of the matrices, will probably indicate that a new, revised run is required. Some raw characteristics will probably prove to have discriminated inadequately and, hence, will best be dropped from the Raw Characteristic Deck. The same will probably be true with respect to some of the finished characteristics, which will need to be dropped from the Finished Characteristic Deck. Study of the printout will probably also suggest certain new raw or finished characteristics that ought to be added. If a re-run is decided upon, the procedure outlined in the foregoing pages must, of course, be repeated.

THE FISHER CALCULATION

At this point it would be well to discuss briefly the Fisher calculation, which is the basic calculation employed in the Pattern Search Technique. The Fisher values have been calculated as if for a two-tailed test of significance, in accordance with a procedure suggested by P. Armsen.[6]

The Fisher calculation is usually employed as the Fisher Exact Test of significance. We are instead employing it, in this survey, as a measure of strength of association. We do this because our sample is a "total sample" rather than the usual partial sample. We therefore regard the statements in the printout as summaries of information rather than as statements of significance.

Some statisticians would approach the matter differently. They would assume a theoretical universe consisting of a very large number of collections of all the world's independent polities, of which the present collection is but one. If this assumption is made, then choices must be made concerning what further assumptions are to be made. Illustrative of these choices are the following: First, one might assume that the number (N) of polities in our sample is not fixed. Second, one might assume that the N is fixed but that, for a particular table, the row and column totals are not fixed. Third, one might assume that the N is fixed, and that the

[6] P. Armsen, "Tables for Significance Tests of 2 × 2 Contingency Tables," *Biometrika*, Vol. 42, 1955, pp. 494–511. The decision to use Armsen's procedure was taken on the advice of Professor Miles Davis, a statistician who served as consultant and programmer. Professor Davis is satisfied that the "P" values appearing in the printout are those following from Armsen's Definition D-3.

row and column totals are fixed, but that the cell frequencies are not fixed. Such choices are vexing and complex. Statisticians differ about the way to proceed. Our decision to regard the Fisher value as a measure of strength of association appears to us to be the simplest way out. We assume that the N is fixed, and that for any particular table the row and column totals are fixed, and that the cell frequencies are likewise fixed. However, we are quite aware that this simple way out might not be satisfactory to some readers, or for some purposes. We have no strong reason for arguing against making some other set of assumptions, such as the third one just outlined.

It should be mentioned that the Pattern Search Technique can be, and already has been, applied to data gathered in accordance with random sample designs. In such cases, it would be legitimate to regard the Fisher calculation in the usual manner, as the Fisher Exact Test of significance (making the assumption that row and column totals are fixed).[7]

THE FORMAT OF THE PRINTOUT

The format of the printout is composed of three principal elements: (1) subject listings, (2) polity listings, and (3) predicate listings. Pages 32–33 are two non-consecutive pages of the printout from Paragraph 80, which will be used to illustrate these listings.

THE SUBJECT LISTINGS

Paragraph 80, like all paragraphs in this study, contains two contrasting subjects, namely: "Polities whose date of independence is after 1914, and that were formerly dependencies of Britain, rather than France," versus "Polities whose date of independence is after 1914, and that were formerly dependencies of France, rather than Britain." Each subject is followed by a number in parentheses: 19 in the case of the left subject, and 24 in

[7] We adopted the Fisher calculation rather than some other calculation, because of its extreme versatility. In the first place, the Fisher calculation is, in our view, capable of being used either to measure strength of association or to test significance, depending on the nature of the data being treated and on the assumptions that the analyst wishes to make. In the second place, the Fisher calculation can be used even where N is small and where the expected frequency of a cell is very small. This is an advantage not offered by the chi-square (to test significance) nor by the phi or C coefficients (to measure strength of association).

Nonetheless, the chi-square value appears in the printout in addition to the Fisher value. One reason for this is to afford convenience to readers who are accustomed to thinking in chi-square terms. An added reason for calculating the chi-square value is that, where N becomes too large, the computer exhausts its capacity to compute the Fisher value. In such a case, it is necessary to make a slight change in the Pattern Search Program, keying the winnowing and verb-selecting procedures (see pages 36, 38) to the chi-square value. The revised program would instruct the computer to do this whenever N reaches a size of, say, 200 or greater. Where N is that large, the P associated with the chi-square value is an extremely close approximation to the Fisher value, and hence is satisfactory.

the case of the right subject. The number 19 means that our original count of the number of polities that possess *both* the raw attribute "date of independence after 1914" *and* the raw attribute "former dependency of Britain" produced a total of 19. The number 24 means that the corresponding figure for the right-hand subject was 24.

THE POLITY LISTINGS

Immediately beneath the subject listings are the polity listings. Each polity is listed once. The total number of polities listed is always 115. A polity may be listed as belonging to the finished attribute category of the left-hand subject, the right-hand subject, or the residue. If it is listed as belonging to the residue, then the ground on which it was assigned to the residue is specified, namely, "ambiguous," "irrelevant," "unascertained," or "unascertainable." The number of polities in each listing is specified. These numbers are the result of the computer's count, and thus act as a check to ensure that the computer's count and our own original count are identical. Note, for example, that the computer found 19 polities in the left column, and that this number is identical with the number of polities that we originally counted and stated in parentheses at the end of the left-hand subject. Note also that the polities are listed alphabetically and ten-across, so as to make it more convenient for the reader to check our codings.

PREDICATE LISTINGS, SENTENCES, AND STATEMENTS

Each pair of predicates, like each pair of subjects, has its own identifying number. Thus, Left Subject 80, when joined with Left Predicate 10, becomes Left Sentence 80/10. Left Sentences will always provide some kind of contrast with corresponding Right Sentences. These two sentences, when read together, form a compound sentence, which we call a "statement." Thus Statement 80/10 reads: "Polities whose date of independence is after 1914, and that were formerly dependencies of Britain, rather than France, tend to be located elsewhere than in . . . Africa, [while] polities whose date of independence is after 1914, and that were formerly dependencies of France, rather than Britain, tend to be those located in . . . Africa." This statement, if somewhat ungainly, is nonetheless reasonably clear. A more precise statement would be *more* ungainly, because it would expand the predicate, "tend to be those located in . . . Africa," to read "tend to belong to the category of polities bearing the attribute, 'located in . . . Africa.' " Thus we have compromised somewhat between ungainliness and precision. The reader who might wish to make use of Statement 80/10 could easily simplify it to read something like this: "Leaving out of

80 POLITIES
WHOSE DATE OF INDEPENDENCE IS AFTER 1914, AND THAT WERE FORMERLY DEPENDENCIES OF BRITAIN, RATHER THAN FRANCE (19)

80 POLITIES
WHOSE DATE OF INDEPENDENCE IS AFTER 1914, AND THAT WERE FORMERLY DEPENDENCIES OF FRANCE, RATHER THAN BRITAIN (24)

BOTH SUBJECT AND PREDICATE

[illegible] N LEFT COLUMN

BURMA	CEYLON	CYPRUS	GHANA	INDIA	IRAQ	I[illegible]	ISRAEL	JAMAICA	JORDAN
MALAYA	NIGERIA	PAKISTAN	SIERRE LEO	[illegible]DAN	TANGANYIKA	TR[illegible]	UGANDA	UAR	

24 IN RIGHT COLUMN

ALGERIA	CAMBODIA	CAMEROUN	CEN AFR REP	CHAD	[illegible]NGO(BRA)	DAHOMEY	GABON	[illegible]EA	IVORY COAST
LAOS	LEBANON	MALAGASY R	MALI	MAURITANIA	MOR[illegible]CO	NIGER	SENEGAL	S[illegible]	TOGO
TUNISIA	UPPER VOLTA	VIETNAM, N	VIETNAM REP						

72 EXCLUDED BECAUSE IRRELEVANT

AFGHANISTAN	ALBANIA	ARGENTINA	AUSTRALIA	AUSTRIA	BELGIUM	BOLIVIA	BRAZIL	BULGARI[illegible]	BURUNDI
CANADA	CHILE	CHINA, PR	COLOMBIA	CONGO(LEO)	COSTA RICA	CUBA	[illegible]ZECHOS'KIA	DENMARK	DOMIN REP
ECUADOR	EL SALVADOR	ETHIOPIA	FINLAND	FRANCE	GERMANY, E	GERMAN FR	[illegible]CE	GUATEMALA	[illegible]AITI
HONDURAS	HUNGARY	ICELAND	INDONESIA	IRAN	ITALY	JAPAN	K[illegible]N	KOREA REP	[illegible]BERIA
LIBYA	LUXEMBOURG	MEXICO	MONGOLIA	NEPAL	NETHERLANDS	NEW ZEALAND	NICA[illegible]	NORWAY	[illegible]MA
PARAGUAY	PERU	PHILIPPINES	POLAND	PORTUGAL	RUMANIA	RWANDA	SA'U AR[illegible]	[illegible]MALIA	S[illegible]RICA
SPAIN	SWEDEN	SWITZERLAND	THAILAND	TURKEY	USSR	UK	US	[illegible]AY	VE[illegible]ELA
YEMEN	YUGOSLAVIA								

10	TEND TO BE THOSE LOCATED ELSEWHERE THAN IN NORTH AFRICA, OR CENTRAL AND SOUTH AFRICA (82)	0.63	10	TEND TO BE THOSE LOCATED IN NORTH AFRICA, OR CENTRAL AND SOUTH AFRICA (33)	0.75	7 18 12 6 X SQ= 4.87 P = 0.016 RV YES (YES)
11	LEAN TOWARD BEING THOSE LOCATED ELSEWHERE THAN IN CENTRAL AND SOUTH AFRICA (87)	0.68	11	LEAN TOWARD BEING THOSE LOCATED IN CENTRAL AND SOUTH AFRICA (28)	0.62	6 15 13 9 X SQ= 2.91 P = 0.067 RV YES (YES)
19	TILT LESS TOWARD BEING THOSE LOCATED IN NORTH AFRICA OR CENTRAL AND SOUTH AFRICA, RATHER THAN IN THE CARIBBEAN, CENTRAL AMERICA, OR SOUTH AMERICA (33)	0.78	19	ALWAYS ARE THOSE LOCATED IN NORTH AFRICA OR CENTRAL AND SOUTH AFRICA, RATHER THAN IN THE CARIBBEAN, CENTRAL AMERICA, OR SOUTH AMERICA (33)	1.00	7 18 2 0 X SQ= 1.69 P = 0.103 RV NO (YES)

125 TILT [illegible]SS TOWARD BEING THOSE 0.67
WHE[illegible]REST ARTICULATION
BY [illegible]UPS
IS FR[illegible]CCASIONAL (64)

125 TILT MORE TOWARD BEING THOSE 0.91
WHERE INTEREST ARTICULATION
BY ANOMIC GROUPS
IS FREQUENT OR OCCASIONAL (64)

	12	20
	6	2
X SQ=		2.28
P =		0.110
RV NO		(YES)

128 TEND TO BE THO[illegible] 0.77
WHERE INTEREST AR[illegible]
BY POLITICAL PARTIES
IS SIGNIFICANT OR MODER[illegible] (48)

128 TEND TO BE THOSE
WHERE INTEREST ARTICULATION
BY POLITICAL PARTIES
IS LIMITED OR NEGLIGIBLE

M

	10	5
	3	17
X SQ=		7.71
P =		0.004
RV YES		(YES)

129 TEND TO BE THOSE 0.77
WHERE INTEREST ARTICULATION
BY POLITICAL PARTIES
IS SIGNIFICANT, MODERATE, OR
LIMITED (56)

129 TEND TO BE THOSE
WHERE INTEREST ARTICULATION
BY POLITICAL PARTIES
IS NEGLIGIBLE (37)

	10	7
	3	15
[illegible]		4.97
[illegible]		0.015
RV [illegible]		(YES)

133 TEND TO BE THOSE 0.71
WHERE INTEREST AGGREGATION
BY THE EXECUTIVE
IS MODERATE, LIMITED, OR
NEGLIGIBLE (76)

133 TEND TO BE THOSE 0.80
WHERE INTEREST AGGREGATION
BY THE EXECUTIVE
IS SIGNIFICANT (29)

M

	5	16
	12	4
X SQ=		7.63
P =		0.003
RV YES		(YES)

138 TEND TO BE THOSE 0.64
WHERE INTEREST AGGREGATION
BY THE LEGISLATURE
IS SIGNIFICANT, MODERATE, OR
LIMITED (48)

138 TEND TO BE THOSE 0.90
WHERE INTEREST AGGREGATION
BY THE LEGISLATURE
IS NEGLIGIBLE (49)

	9	2
	5	19
X SQ=		9.29
P =		0.002
RV YES		(YES)

139 TEND TO BE THOSE 0.83
WHERE THE PARTY SYST[illegible]NTITATIVELY
OTHER THAN ONE-PA[illegible]

139 TEND TO BE THOSE [illegible]
WHERE THE PARTY SYSTEM IS QUANTITATI[illegible]
ONE-PARTY (34)

M

	3	15
	15	7
X SQ=		8.64
P =		0.002
RV YES		(YES)

141 TILT LESS TOWARD BEING THOSE 0.86
WHERE THE PARTY SYSTEM IS QUANTITATIVELY
OTHER THAN TWO-PARTY (87)

141 ALWAYS ARE THOS[illegible] 1.00
WHERE THE PARTY[illegible] QUANTITATIVELY
OTHER THAN TWO-[illegible]

	2	0
	12	22
X SQ=		1.16
P =		0.144
RV NO		(YES)

142 TEND LESS TO BE THOSE 0.54
WHERE THE PARTY SYSTEM IS QUANTITATIVELY
OTHER THAN MULTI-PARTY (66)

142 TEND MORE TO BE THOSE 0.90
WHERE THE PARTY SYSTEM IS QUANTITATIVELY
OTHER THAN MULTI-PARTY (66)

	6	2
	7	19
X SQ=		4.12
P =		0.033
RV NO		(YES)

144 LEAN LESS TOWARD BEING THOSE 0.60
WHERE THE PARTY SYSTEM IS QUANTITATIVELY
ONE-PARTY, RATHER THAN
TWO-PARTY (34)

144 ALWAYS ARE THOSE 1.00
WHERE THE PARTY SYSTEM IS QUANTITATIVELY
ONE-PARTY, RATHER THAN
TWO-PARTY (34)

	3	15
	2	0
X SQ=		2.96
P =		0.053
RV NO		(YES)

Total number of polities coded under this attribute

Left majority cell is 64% of column total

Table 80/125 is non-reversal

Table 129/80 is reversal

Matrix provided for this table

Right majority cell is 90% of column total

consideration polities that gained their independence prior to 1914, the ex-French dependencies tend to be located in Africa, while the ex-British dependencies tend to be located elsewhere."

SUBJECT-ONLY AND PREDICATE-ONLY CHARACTERISTICS

As has by now become clear, in the typical case a table appears twice in the printout. For example, the computer has found that the table resulting from the crossing of FCs 10 and 80 bears a sufficiently strong Fisher value to justify its appearing in the printout. This table therefore appears twice, once as Table 10/80, and once as Table 80/10. In the former case, FC 10 is the subject characteristic, and FC 80 is the predicate characteristic. In the latter case, the table is "turned over," and FC 80 provides the two subjects, while FC 10 provides the two predicates. In some cases, however, for the convenience of the reader we have decided to run a particular characteristic as subject only or as predicate only. There are 29 finished characteristics run as "subject only," and 69 run as "predicate only."

SUBJECT-ONLY CHARACTERISTICS

The "subject-only" characteristics are indicated in the serial listing of finished characteristics in Chapter 3 by the notation "[SO]". On the printout, each subject-only characteristic bears the label "Subject Only" in the lower right-hand corner of the subject listing.

The reason for our decision to run a particular finished characteristic as subject only, and not as predicate, was often that the positive attribute category had a relative frequency so low that the characteristic would not be instructive if run as a predicate. An example of such a subject-only characteristic is seen in Paragraph 12. Here the left-hand, or "positive," attribute category is "Polities located in North Africa (5)." The right-hand, or "all-other," category is "Polities located elsewhere than in North Africa (110)." The frequency of the positive category is 1/22 that of the all-other category. In the final computer run, which produced the printout in this book, we retained the left-hand group of polities (Algeria, Libya, Morocco, Tunisia, and the United Arab Republic) as a subject because we wanted to know what predicate attributes showed an inclination to co-occur with the positive attribute "North African-ness." (The "all-other" attribute "located elsewhere than in North Africa," is of no intrinsic interest; its only function is to provide a contrast with the positive attribute.) We did not use Finished Characteristic 12 as a predicate, however. To have done so would have produced sentences most of which would have told us little of value with respect to the various *subjects* to which this predicate would have been attached. The frequency of the "North Africa" category is simply too small, relative to the frequency of the analytically uninteresting "all-other" category. Thus, for example, Statement 12/93 appears in the printout, but potential Statement 93/12 does not. Statement 12/93 tells us something of value about North Africa, namely that polities located there tend to have a system style that is mobilizational or limited mobilizational (in contrast to those located elsewhere, which tend to be non-mobilizational). On the other hand, if potential Statement 93/12 had appeared in the printout, it would have told us relatively little of value about the characteristic, "System Style." It would merely have told us that polities where the system style is mobilizational or limited mobilizational tend less to be located elsewhere than in North Africa, while polities where the system style is non-mobilizational tend more to be located elsewhere than in North Africa.

PREDICATE-ONLY CHARACTERISTICS

The "predicate-only" characteristics are indicated in the serial listing of finished characteristics in Chapter 3 by the notation "[PO]". Although a predicate-only characteristic is not run as a subject, the printout includes polity listings for each such characteristic. This is done so as to enable the reader to check the coding of the finished characteristic in question. The listing bears the label, "Predicate Only." After the polity listing, the printout proceeds immediately to the next paragraph, since there are no predicates to appear.[8]

Finished characteristics that we decided to run as predicate only are usually derived from raw characteristics of an ordinal (as distinct from a nominal) nature. Such ordinal raw characteristics are usually dichotomized into several finished characteristics by means of a moving cutting point. This is done so as to enable the reader to discern a staircase effect among successive predicates. (See pages 45–47.) Thus, in the case of the raw characteristic, "Gross National Product Per Capita," we established five attributes: Very High, High, Medium, Low, and Very Low. We then dichotomized three times by placing the cutting point successively between High and Medium, between Medium and Low, and between Low and Very Low. In setting this characteristic up as a subject variable, however, we dichotomized only once, locating the cutting point where the left and right attribute frequencies would be most nearly equal. This location proved to be between Low and Very Low. We adopted this procedure because, in treating this raw characteristic as a subject variable, we merely wanted to discover what were, in general, the implications of "higher" GNP per capita, as contrasted with "lower" GNP per capita. In other words, the moving cutting-point technique provides valuable results when applied

[8] In order to prevent the subject-only characteristics from appearing as predicates, and the predicate-only characteristics from appearing as subjects, one need only remove the appropriate table cards prior to the printout run.

to predicate variables but often results in confusion and quasi-redundancy when applied to subject variables.

In the foregoing example, we had no theoretical or other reason for wanting to discover, for example, what might be the implications of "Very High" GNP per capita, as contrasted with "All Lower." *If* we had had such reasons, then this dichotomy would have been run as a subject. If a particular reader should have reason for wishing to contrast this (or any other) omitted dichotomy with some other characteristic, he can always do so by building his own matrix (pages 49–50).

We can now briefly summarize with respect to both subject-only and predicate-only characteristics. It should first be pointed out that these two types of characteristics, taken together, comprise only about half of the total number of finished characteristics. The other half are run both as subject and as predicate. A second point to emphasize is that running a finished characteristic as subject-only or as predicate-only does *not* mean that any information is thrown away. What it *does* mean is simply that the information is presented only once, rather than twice, solely for the purpose of easing the burden on the reader. We would be throwing away information only if we were to decide not to run a given finished characteristic *either* as subject *or* as predicate—if, in other words, we were completely to remove the characteristic from our corpus of finished characteristics. For the convenience of the reader, we have listed in Appendix C all tables that have been removed because their predicate characteristics have been designated to appear as "Subject Only."[9]

COMPUTER-WINNOWED STATEMENTS

In Paragraph 80, the first statement is Statement 80/10. Potential Statements 80/1 through 80/9 do not appear. The reason is that the Fisher values of these potential tables were weaker than the critical level of .149 established for this study. All such weak statements have, in effect, been "winnowed out" by the computer, leaving only the solid kernels of interesting information. Thus the reader is spared the chore of poring through a large number of statements that are, in general, of little or no importance. It should be emphasized, however, that under some circumstances the *non*-appearance of a potential statement can be highly important, depending on the analyst's purposes and theoretical interests. The failure to confirm an assumed correlation is often as scientifically important as the confirmation of another.

[9] These removals are indicated by the abbreviation "SO." Each entry in Appendix C is comprised of six numbers. Reading from left to right, these numbers signify: (1) The identification number of the finished characteristic that constitutes the subject, or column, characteristic. (2) The identification number of the finished characteristic that constitutes the predicate, or row, characteristic. (3) The frequency of the upper left cell. (4) The frequency of the upper right cell. (5) The frequency of the lower left cell. (6) The frequency of the lower right cell.

THE GRAMMAR OF THE SENTENCES

The sentences produced by the computer are in every case completely grammatical. The subject invariably contains a "subject lead" and a "subject clause." The predicate invariably contains a "verb" and a "predicate clause." In addition, some predicates also contain "adverbs." The computer is programmed to look for, select, and print the subject lead, subject clause, verb, and predicate clause. It is also programmed to decide whether an adverb is necessary, and, if so, to select and print it. These operations will be illustrated with reference to Statement 80/10.

SUBJECT LEAD

In the present study, the subject lead of every sentence is "Polities. . . ."

SUBJECT CLAUSE

The subject clause is next retrieved from the computer's memory. The left subject clause is simply the left attribute clause for Finished Characteristic 80. Joining the subject lead to the left subject clause, the computer prints as the entire Left Subject 80: "Polities whose date of independence is after 1914, and that were formerly dependencies of Britain, rather than France."

PREDICATE CLAUSE

The next task of the computer is to select the predicate clause from between two alternatives. To do this, the computer examines Table 80/10, which is printed just to the right of Statement 80/10. The left column of the table is keyed to the left subject, and the right column to the right subject. This table reads as follows:

	Ex-British	Ex-French
Located in Africa	7	18
Located elsewhere than in Africa	12	6

The computer notices that a majority of 12 out of 19, or 63 percent of the ex-British dependencies are located outside of Africa. The computer selects the predicate clause for Left Sentence 80/10 from the "majority" row, in this case, the lower row. It therefore selects and prints the predicate clause associated with the lower row, namely, "located elsewhere than in . . . Africa (82)."[10]

The computer then chooses the predicate clause for Right Predicate 80/10 in a like manner.[11] It notices that 18 out of 24, or 75 percent of the ex-French dependencies, are "located in . . . Africa," and it therefore chooses and prints that predicate clause.

[10] The number 82 indicates that of all the 115 independent polities in the world, there are 82 that are located elsewhere than in Africa.

[11] When, in a particular column, there is no "majority" row, the computer selects the row that is other than the majority row in the other column.

As a convenience to the reader, the column ratios—here 63 and 75 percent—are printed just to the right of the Left and Right Predicates, respectively.

VERB

The computer must next decide which left verb to select and print, out of several possibilities. For the most part, the selection of the verb depends on the Fisher value, which is printed below the table. If this value is stronger, that is, lower, than .05, then the computer selects and prints the verb, "tend." This is the case in Left Sentence 80/10, where the Fisher value is .016. In cases where the value is stronger than .10 but not stronger than .05, the computer selects and prints the somewhat "weaker" verb, "lean." This is illustrated in Left Sentence 80/11. If the value is stronger than .150, but not stronger than .10, then the computer selects and prints its weakest verb, "tilt." If the value is .150 or weaker, then, of course, the table and its statement will already have been computer-winnowed out (page 36).

THE ADVERBS "MORE" AND "LESS"

None of the statements cited to exemplify the verbs "tend," "lean," and "tilt" contained an adverb. None was necessary, because in all these statements, the ex-British polities were inclined one way and the ex-French polities were inclined the other way. Thus, for example, the ex-French polities were inclined to be located in Africa, while the ex-British polities were inclined to be located outside of Africa. Such a relationship is called a "reversal." A reversal is characterized by having the two majority cells in diagonally opposed positions. A "non-reversal" is characterized by having both majority cells in the same row.

The adverbs "more" and "less" are used only in cases where there is a *non*-reversal, in other words, where both classes of polities are inclined in the same direction. An example is seen in Statement 80/125. In this instance, ex-French polities tilt toward being those where interest articulation by anomic groups is frequent or occasional—to the extent of 20 out of 22, or 91 percent. Ex-British polities also tilt in the same direction, but only to the extent of 12 out of 18, or 67 percent. Thus, both the ex-French and the ex-British polities are inclined in the same direction, but the ex-French territories show *more* of such inclination, and the ex-British territories show *less*. Therefore, the adverb "more" is added to the right predicate, and the adverb "less" is added to the left predicate.

At the bottom of Table 80/125 is printed the expression "Reversal: No (Yes)." The "Reversal: No . . ." means that Table 80/125 does not contain a reversal. The ". . . (Yes)" simply adds the parenthetical information that when the table is "turned over" and becomes Table 125/80, a reversal does occur.

The adverbs "more" and "less" can be added to any predicate where the verb is "tend," "lean," or "tilt."

THE VERB-PHRASE "ALWAYS ARE"

The verb-phrase "always are" is selected whenever one of the four cells in the table contains a zero. An example is seen in Right Sentence 80/144. If the cell diagonally opposed to the zero cell also contained a zero (instead of a two), then the verb-phrase for Left Sentence 80/144 would also be "always are." Such instances do occur, although rarely. In the actual case, Left Sentence 80/144 takes the verb "lean," reflecting the fact that the Fisher value is between .099 and .050, together with the adverb "less," indicating that the statement is a non-reversal.

SYNTAX

We recommend that in first reading a paragraph the reader devote primary attention to the left sentences, with occasional glances at the right sentences and tables, as necessary. This will afford him a quick preliminary contextual overview of the entire "terrain" of the paragraph. Our syntax is planned to facilitate this quick overview. That is, we have designed the format of our syntax in such a way that the reader, as his eye moves down the left sentences, will usually know what is contained in the corresponding right sentences without having to look at them. Chart 2.1 gives the relevant syntactic rules.

CHART 2.1 SYNTACTIC RULES

If the Left Verb is	Then the Right Verb will be
Tend	Tend — in the opposite direction, or Always — in the opposite direction
Lean	Lean — in the opposite direction, or Always — in the opposite direction
Tilt	Tilt — in the opposite direction, or Always — in the opposite direction
Tend More	Tend Less — in the same direction
Lean More	Lean Less — in the same direction
Tilt More	Tilt Less — in the same direction
Tend Less	Tend More — in the same direction, or Always — in the same direction
Lean Less	Lean More — in the same direction, or Always — in the same direction
Tilt Less	Tilt More — in the same direction, or Always — in the same direction
Always Are	[There are many possibilities. It is best to examine the Right Predicate rather than to attempt to guess.]

There is one other general principle of syntax that should be mentioned. Where the left subject or predicate clause does not specify otherwise, the reader may safely assume that this left clause defines a class of polities that is being contrasted with "all other" polities in the world. To illustrate, when the reader sees a left predicate that involves polities "located in East Europe," he will know immediately that this class of polities is being contrasted with all other substantively codable polities, and that no polities have been classified as "irrelevant." On the other hand, when the reader sees a left predicate that includes the words "rather than" (as in "Polities located in Scandinavia or West Europe, rather than in East Europe"), he knows that two classes of polities are being contrasted, and that others are *excluded* and drop into the residue as "irrelevant."

HAND-WINNOWED STATEMENTS

There are only two means by which statements can be winnowed out of the printout. The first, described on page 36, is computer-winnowing. The second is hand-winnowing. Hand-winnowing is not an automatic process, though it does follow fixed rules. In brief, the next-to-last printout is inspected to decide what statements are redundant. The table cards for each such redundant statement are then removed, so that they do not appear in the final printout. The purpose of hand-winnowing is to minimize the reading burden without sacrificing information and, in some cases, to eliminate confusion that would otherwise result. Each such hand-winnowed statement is summarized in Appendix C at the end of the printout.[12]

Redundant statements take two principal forms: the "pure" redundant and the "quasi"-redundant. Pure redundant statements result from tables where both the "subject" finished characteristic and the "predicate" finished characteristic derive from the same raw characteristic. For example, if it had not been winnowed out, Left Sentence 1/11 would have appeared in the final printout with the information that "Polities located in East Europe always are those located elsewhere than in Central or South Africa."

Quasi-redundant statements are both more common and more complex. Such a statement can occur where there appears in the preliminary printout more than one predicate-pair deriving from the same raw characteristic, and where that raw characteristic is of an *ordinal* nature.[13]

[12] In Appendix C, statements winnowed for reason of redundancy are identified by the abbreviation "RD." (The other abbreviation, "SO," meaning "Subject Only," is explained on pages 34–35.) Since the cell frequencies are provided, the reader is enabled to reassemble the deleted table if he wishes. Appendix C is a convenient means of checking whether a particular table and its accompanying statement have been eliminated on a "redundant" or a "subject-only" basis. If neither, the only other possibility is that the statement was computer-winnowed.

[13] Where only one predicate-pair deriving from a particular ordinal raw characteristic appears, it is never regarded as quasi-redundant.

In such a case, we *retain* all predicate-pairs that contain a reversal (that is, that do not possess the adverbs "more" or "less") and, with one exception, all predicate-pairs that contain the verb-phrase, "always are."[14] We hand-winnowed *out* those pairs of predicates where each predicate contains the adverb "more" or "less." We do this because experience has shown that such statements are the most cumbersome to read and the least informative.[15]

Examples are found in Paragraph 41. Statement 41/38 was deleted. Here, the left predicate was "tend less to be those whose international financial status is medium, low, or very low." The right predicate was "tend more to be those whose international financial status is medium, low, or very low." The essential "thrust" of this deleted statement is more succinctly expressed in Statement 41/39. Statement 41/40 was allowed to remain in the printout because its left sentence adds new information, namely, that it *never* happens that polities whose economic developmental status is developed, have an international financial status that is very low.

Our policy on the hand-winnowing of quasi-redundant statements means that in some cases as few as one statement remains which derives from a particular ordinal raw characteristic. Therefore, if the reader finds even one such statement, and it deals with a subject that interests him, he would be well advised to assemble the two raw characteristics concerned into a 2-by-N or M-by-N matrix, as described on pages 49–50.

HOW TO USE THE PRINTOUT

It should be emphasized, once again, that the printout simply tells us what classes of polities co-occur or overlap with what other classes. It tells us, for example, that the class of polities that possesses the attribute "former British dependency" tends to overlap with the class of polities that possesses the attribute "literate religion." What the computer *cannot* do is to tell us *why* this is so. Indeed, like most other researchers who have worked with the computer, we are impressed with the essential "stupidity" of the machine. The computer, at least as we use it, can perform only those operations that it has been instructed to perform. And while many of these tasks are of a highly complex character, they are all wholly mechanical. The proper interpretation of our computer printout

14 Occasionally *two* "always are" predicate-pairs will appear, one of which contains more precise information than the other. In such a case, the less precise statement will be hand-winnowed out. For example, we deleted Statement 10/53, which tells us that polities located in Africa always are those where newspaper circulation is less than three hundred per thousand. We retained the more precise and informative Statement 10/54, which tells us that African polities always are those where newspaper circulation is less than one hundred per thousand. Both statements are true, but the first one is less precise and, indeed, misleading.

15 Again, however, there is an exception. In the rare case where two or more predicate-pairs appear, *all of which* contain adverbs in both the left and right predicates, *none* of the pairs are deleted.

must, then, always come from the free creative intellect of the analyst himself. The following recommendations are designed to aid the reader in the task of interpreting a portion of the printout.

LISTINGS OF RAW AND FINISHED CHARACTERISTICS

It would be best to begin by briefly scanning the serial listings of raw and finished characteristics in Chapter 3. These provide a general idea of the scope of the present study.[16]

SELECTING A RAW CHARACTERISTIC

Next, it is suggested that the reader select a raw characteristic that is of particular interest to him, and that covers a subject with which he is familiar. It would be well to examine the commentary that follows after the raw characteristic listing. This states what coding rules were used, how borderline decisions were reached, and so on. Immediately following the commentary are listed the finished characteristics derived from this raw characteristic.

PARAGRAPH CODINGS

The reader is now ready to turn to the printout, and to the beginning of the first paragraph representing a finished characteristic derived from the raw characteristic in question. He may wish first to examine the polity listings for this paragraph, to determine whether he would have coded certain polities differently.[17]

The question now arises: To what extent can the reader disagree with our codings and still be justified in proceeding to read the paragraph concerned? The answer to this question depends on the extent and distribution of the cases involved. If the reader disagrees with a substantial proportion of the codings listed at the top of a particular paragraph, then obviously there would be little point in proceeding to read the paragraph. However, if the reader disagrees with only a small proportion of our codings, and especially if one questionable coding would seem to cancel out another, then he can proceed to read the paragraph with a measure of confidence. Of crucial importance is the extent to which the *proportions* of a particular column or row may thus be affected. If, for example, a particular column has a relatively small marginal total to begin with,

[16] The Raw Characteristic Code Sheet appears as Appendix A. The Finished Characteristic Code Sheet appears as Appendix B. These code sheets enable the reader to check all codings for a particular polity.

[17] Owing to the "reassignment of residue," discussed on pages 26–27, discrepancies will be noted between the codings as given in the Raw Characteristic Code Sheet and the codings presented in the polity listings for a particular paragraph. There should, however, be no discrepancies between polity listings and the *Finished* Characteristic Code Sheet.

then considerable disturbance could result from the shift of two or three polities from one cell to the other within the column, or from that column to the other column.

It is possible to illustrate with concrete examples. Paragraph 67 contrasts the 82 polities coded as "racially homogeneous" with the 27 polities coded as "racially heterogeneous." As will be noted in Table 67/2, however, the left column adds up to only 80 instead of 82, while the right column total is 29 instead of 27. This is the result of a clerical error. Two polities—Uganda and Tanganyika—that we had coded as "homogeneous" were inadvertently punched on the input cards as "heterogeneous." Since the cost of correcting this error would have been great, we decided to let the error stand.

The right column of Paragraph 67 should contain 27 polities, but because of the above-mentioned error, it actually contains 29. This column total is relatively low, and under such circumstances the loss of two polities can disturb a column's proportions considerably. While in the great majority of cases no substantial change results, it is worth while to cite a few extreme cases so that the reader will be aware of the upsets that can sometimes result when coding or punching errors occur. Table 67/45 should have cell frequencies of 43, 10, 33, and 17. This renders the table less "lopsided" than appears in the printout, with the result that the true Fisher value is .116 instead of .048. In other words, the verb ought to be a "tilt" instead of the "tend" that appears in the printout. Table 67/50's true cell frequencies are 28, 13, 42, and 10. The true Fisher value, instead of being .098, is larger than the critical level of .149, so that this table actually should not have appeared in the printout at all. In the main, however, these punching errors have not proved serious, and a coding error of equal proportions elsewhere would likewise not be serious.[18]

An added safety feature derives from the fact that many of our raw characteristics are of an ordinal nature, and have been dichotomized into two or more finished characteristics by means of a moving cutting point (page 45). The result of such dichotomization is often a quasi-redundancy among predicates resulting from the same raw characteristic, as explained on pages 40–41. The essential "thrust" of such a set of quasi-redundant predicate-pairs will, of course, be in the same direction. It seems probable to us that in many such instances coding errors might distort particular predicate-pairs, yet still not distort the general thrust of such a set of predicate-pairs found in the printout or in Appendix C.[19]

[18] A few additional changes in Fisher values would seem worth listing. Table 67/41 changes its value from .022 to .039. Table 67/42 goes from .061 to .096. Table 67/53 changes from .107 to .116. Table 67/54 drops from .002 to .004.

[19] However, there will also be cases, such as that of predicate-pair 67/50 mentioned previously, where this will not be true. Here, only two predicate-pairs emerge: 67/50 in the printout, and 67/51 which was hand-winnowed out and appears in Appendix C. Both of these predicate-pairs would have been machine-winnowed out if the punching error had not occurred.

In the examples cited, the correction of punching errors has led to decreases in strength of association. It could just as easily happen, however, that such corrections could lead to increases in strength of association.[20]

QUICK READING

The next recommended step is to obtain a quick over-all picture of the "general terrain" of the section. The reader will thus gain a "holistic" contextual "feel" for the data before embarking on a more atomistic search for crucial relationships. As noted on pages 39–40 in the discussion of syntax, the statements are arranged in a format such that the reader can develop this over-all feel quite easily by concentrating primarily on the left-hand sentences during the first reading.

DETAILED READING

The reader will probably now wish to return to the beginning of the section and this time read the paragraphs more thoroughly. Both left and right predicates should be studied. Attention should also be paid to the tables on which statements are based, and to any matrices that might appear at the end of a paragraph.

DRAFT SUMMARY

At this point it should be possible to prepare a draft summary of the section, listing the major conclusions that seem to emerge. Each such conclusion should indicate the statement numbers from which it is derived.

In reaching these conclusions, we recommend that the reader place most weight on the "tend" statements, and only moderate weight on the "lean" statements. We regard the "tilt" statements as being of value primarily insofar as they give contextual support to the "tend" and "lean"

[20] To our knowledge, there is only one other substantive error in the printout. This is in Paragraph 70. Malagasy is religiously heterogeneous, and hence belongs in the right column. Because of a punching error, Malagasy was placed in the residue. The right column of Table 70/2, therefore, totals only 84 polities, whereas it ought to total 85. This case is different from that in Paragraph 67 in that a change occurs in the total for only one of the two columns. The case is instructive in that this column total is by far the larger one. Adding one polity to 84 alters the table's proportions hardly at all. For example, it changes the Fisher value of Table 70/36 from .024 to .023, while Tables 70/54, 56, 63 retain the same Fisher values as before.

The polity listings for Paragraphs 67 and 70, and Matrices 67/153 and 70/125 have been corrected so as to show the proper codings for Tanganyika, Uganda, and Malagasy.

As far as procedural errors are concerned, we are aware of only two, both of them inconsequential. The first is that we inadvertently neglected to correct a "bug" in the program prior to the final print run, with the result that a predicate-pair where both verb phrases are "always are" can creep into the printout, even though its associated Fisher value is higher than .149. To our knowledge, this occurred only twice, in Statements 191/48, 99.

The second procedural error is of even less consequence. It was almost inevitable that hand-winnowing would be less than 100 percent thorough. We know of one predicate-pair, 67/70, that should have been hand-winnowed out, but was not. There might, of course, be a few more such cases.

statements. "Always" statements should be evaluated somewhat more individually, but certainly the Fisher level would be an important consideration in their case as well. Occasionally the reader might reach a conclusion from a single isolated statement, and such conclusions can deal with quite crucial relationships. However, our experience has been that a conclusion of real value will more frequently assume the form of a pattern based on a whole congeries of inter-related statements. And indeed, the aspect of the present survey that we regard as the most reassuring is the degree of internal consistency among the statements that appear in the printout.

CROSS-CHECKING

One of the advantages of the Pattern Search Technique is that tables are "turned over" so that an interesting predicate can be examined in another paragraph where it becomes the subject. By reading this other paragraph, the analyst can obtain a rich contextual view of the new subject characteristic. This is valuable both for identifying important intervening variables, and for suggesting new research leads involving variables not included in the present study.

STAIRCASING

Most of the characteristics and statements thus far used for illustrative purposes have been supported by data of a qualitative, or nominal, variety. However, many of our raw characteristics are not nominal, but ordinal. An example is the (originally metric) characteristic "Gross National Product Per Capita," under which we have subsumed the five ordinal raw attributes "Very High," "High," "Medium," "Low," and "Very Low." Since the Pattern Search Technique, as presently constituted, requires that finished characteristics be dichotomous, we have dichotomized this raw characteristic into three finished characteristics, by means of a moving cutting point, as follows:

Very High or High	*versus*	Medium, Low, or Very Low
Very High, High, or Medium	*versus*	Low or Very Low
Very High, High, Medium, or Low	*versus*	Very Low

This moving cutting point is designed to help the analyst notice "staircasing" among predicates. For example, suppose we are concerned as to whether there is any relationship between the above-mentioned raw characteristic and the raw characteristic "Role of Police." The latter contains only two attributes, which serve as subjects for Paragraph 187: "Polities where the role of the police is politically significant" versus "Polities where the role of the police is not politically significant." When the analyst confronts Statements 187/35–37, he will immediately note

MATRIX 4

GROSS NATIONAL PRODUCT PER CAPITA

	VERY HIGH	HIGH	MEDIUM	LOW	VERY LOW
POLICE IN POLITICAL ROLE		*Czechos'kia* *Germany, E* *USSR* Venezuela	Argentina *Bulgaria* *Cuba* *Hungary* Lebanon *Poland* *Rumania* So Africa Spain *Yugoslavia*	*Albania* Algeria Colombia Ecuador El Salvador Guatemala Honduras Iraq Nicaragua Portugal Sa'u Arabia Syria Tunisia Turkey UAR	Afghanistan Bolivia Burma Cambodia Cameroun Cen Afr Rep *China, PR* Congo (Bra) Congo (Leo) Dahomey Ethiopia Gabon Ghana Guinea Haiti Indonesia Iran Ivory Coast Jordan *Korea, N* Korea Rep Laos Liberia *Mongolia* Morocco Nepal Nigeria Pakistan Paraguay Senegal Somalia Sudan Thailand Togo *Vietnam, N* Vietnam Rep Yemen
POLICE NOT IN POLITICAL ROLE	AUSTRALIA BELGIUM CANADA FRANCE LUXEMBOURG NEW ZEALAND NORWAY SWEDEN SWITZERLAND UK US	AUSTRIA DENMARK FINLAND GERMAN FR ICELAND IRELAND Israel ITALY NETHERLANDS	Greece Jamaica Japan Malaya Trinidad Uruguay	Mexico Philippines	Ceylon India Malagasy R Mauritania Sierre Leo Tanganyika Uganda

Note: Developed West in capitals
Communist bloc in italics

a staircase phenomenon. These statements clearly suggest that there is an inverse relationship between high GNP per capita and political involvement on the part of the police. This relationship is summarized by Matrix 4.

In watching for the staircase phenomenon, the reader should bear in mind that quasi-redundant statements have been removed. Hence, if even *one* statement appears that suggests staircasing, it might well be that a 2-by-N or M-by-N matrix would bear out the presence of such a phenomenon. (The reader can check to see what statements and tables have been removed on grounds of quasi-redundance by referring to Appendix C.)

MATRICES OF RAW ATTRIBUTES

Matrix 4 is an example of a matrix including all raw attributes from two raw characteristics. It happens to be a "2-by-N" matrix, but one can also construct useful "M-by-N" matrices. Such a matrix might result from the crossing of two nominal characteristics, or two that are ordinal, or one of each. Matrix 5 involves one nominal raw characteristic, "Legislative-Executive Structure." The other raw characteristic, "Party System: Quantitative," is in a sense an ordinal variable, although we do not treat it as such in this matrix.

Matrix 5 in some ways merely confirms the obvious, as when it tells us that if a polity is "Communist" in its legislative-executive structure, then it will have only one party. But it also reveals other relationships that the reader might not have anticipated. For example: If a polity is "one-party presidential," it will invariably be located in Africa and will almost invariably be a former French territory. On the other hand, a "multi-party presidential" system will almost invariably be Latin American.

STUDY OF DEVIANT CASES

One of the main reasons why we have suggested that the reader plot out matrices concerning the crossing of characteristics that interest him is that a matrix clearly reveals which polities are deviant cases. Deviant case analysis can be a powerful technique. It enables the analyst to see beyond the attribute categories immediately involved, and also to perceive the relevance of outside factors not included among the characteristics employed in the present study. In some cases, such factors would perhaps have been readily codable, but we simply happened not to use them. In other cases, however, it will turn out that these outside factors are inherently difficult or impossible to code for on a global basis. It is here that the well-informed reader, particularly the historian and area expert, can bring to bear his broad knowledge of the polities involved and their particular or unique characteristics.

MATRIX 5

LEGISLATIVE-EXECUTIVE STRUCTURE*

PARTY SYSTEM: QUANTITATIVE	Communist	Presidential-Parliamentary	Presidential	Parliamentary-Republican	Parliamentary-Royalist	Monarchical or Monarchical-Parliamentary
One-Party	Albania Bulgaria China, PR Czechos'kia Germany, E Hungary Korea, N Mongolia Poland Rumania USSR Vietnam, N Yugoslavia	Chad	Cen Afr Rep Dahomey Gabon Ghana Guinea Ivory Coast Liberia Mauritania Niger Senegal Tunisia Upper Volta			
One Party Dominant		France	Bolivia Cameroun Congo (Bra) Malagasy R Mexico Paraguay Rwanda Vietnam Rep	India Somalia	Burundi Greece	
Multi-Party		Lebanon	Argentina Brazil Chile Costa Rica Cyprus Domin Rep Ecuador El Salvador Guatemala Panama Venezuela	Congo (Leo) Finland Iceland Ireland Israel Italy	Belgium Ceylon Denmark Laos Luxembourg Netherlands Nigeria Norway Sweden Uganda	
Two-Party			Colombia Honduras Philippines US	Austria	Australia Canada Jamaica New Zealand Sierre Leo UK	
One-and-a-Half Party				German FR	Japan Trinidad	
No Parties			Haiti			Afghanistan Ethiopia Libya

*The following thirty polities do not appear in the matrix for the following reasons:
Because there is no matrix column for "Pure Parliamentary": *Mali* (1).
Because there is no matrix column for "Other": *Switzerland* and *Uruguay* (2).
Because "ambiguous" with respect to legislative-executive structure: *Cambodia, Portugal, Spain, and Tanganyika* (4).
Because "unascertainable" with respect to legislative-executive structure: *Algeria, Cuba, Peru,* and the *United Arab Republic* (4).
Because "irrelevant" with respect to legislative-executive structure: *Sa'udi Arabia* (1).
Because "unascertainable" with respect to *both* legislative-executive structure and party system: *Burma, Iran, Iraq, South Korea, Sudan, Syria, Thailand, Togo,* and *Yemen* (9).
Because "ambiguous" with respect to party system: *Indonesia, Malaya, Morocco, Nicaragua, South Africa,* and *Turkey* (6).
Because "unascertainable" with respect to party system: *Jordan, Nepal,* and *Pakistan* (3).

Skillful analysis of deviant cases can also lead to other fruitful discoveries, such as (1) the existence of coding errors, (2) the need for a refinement of coding rules, (3) the need for redefinition of coding categories, or even (4) the need for a complete redesigning of one's conceptual or analytical approach.

THE RESIDUE

No statement should be regarded as authoritative until the reader has checked the residue applying to the table on which the statement is based. In cross-checking and in constructing matrices, it becomes readily apparent how many polities drop out into the residue. Generally speaking, if such polities drop out for the reason of "irrelevance," there is no cause for concern. A polity that has been coded as "irrelevant" to either of the finished characteristics of the matrix is, by definition, a polity that need not, and should not, enter into the over-all analytic result. If, however, an excessive number of polities drop out on grounds of "ambiguity," "unascertainedness," or "unascertainability," then a serious bias might be introduced.[21] This is particularly true when such dropouts tend to possess some common quality, such as being located mostly in Africa, being mostly Communist nations, being mostly "underdeveloped," and so on.

USE OF THE SORTER AND TABULATOR

If the reader discovers a table for which he desires a matrix, and if a matrix for that table does not appear in the printout, it will be necessary for him to prepare his own. It is possible to do this by copying our codings from the polity listings at the top of each paragraph, or from Appendix A, the Raw Characteristic Code Sheet, as the case may be. However, a much more efficient and accurate method would be to utilize the sorter and tabulator, two extremely simple data processing machines which require no previous experience. The analyst will also need a copy of our Raw Characteristic Deck. Both this and the Finished Characteristic Deck are planned in such a way as to make them maximally convenient to use. For example, each individual card is punched not only with its alphabetically ordered polity identification number but also with the name of that polity, so that listing is simple and error improbable.[22]

For the benefit of those readers who have not used the sorter and tabulator, an example will be given. Suppose that we wished to build Matrix 4 (page 46), which crosses "Role of Police in Politics" with "Per

[21] The loss from a table of polities on any of these three grounds will, of course, affect the Fisher value to an indeterminate degree.

[22] These decks are available at cost plus postage from the Carnegie Seminar, Department of Government, Indiana University, Bloomington, Indiana. Please specify whether only the Raw Characteristic Deck or both Raw and Finished Decks are desired.

Capita Gross National Product." We first sort on Raw Column 55, and the sorter deposits 66 cards into the "A" pocket and 35 cards into the "B" pocket. (It also drops 12 cards into the "5" pocket, and 2 into the "6" pocket.) The 66 cards represent the 66 polities where the role of the police is politically significant, and the 35 cards represent the polities where it is not. (The 12 cards in the "5" pocket are residual because "unascertained," and the 2 cards in the "6" pocket are residual because "unascertainable.") We now sort the 66 cards on Raw Column 9, and the sorter will deposit 4 cards in the "B" pocket, 10 cards in the "G" pocket, 15 cards in the "H" pocket, and 37 cards in the "I" pocket. A similar operation is now performed for the 35 cards that represent polities where the role of the police is not politically significant. The A, B, G, H, and I pockets will contain 11, 9, 6, 2, and 7 cards, respectively. This makes a total of 9 small decks. Each of these is now listed on the tabulator, and copied off onto squared paper in matrix form.

ADDITIONAL ANALYTICAL TECHNIQUES

As has been pointed out, the Pattern Search Technique employs characteristics in dichotomous form because this simplifies the analysis and hence helps to "trigger off" in the mind of the reader a new insight or hypothesis. To follow up such an insight, it is necessary to move once again to the polychotomous raw characteristics, as they afford a more comprehensive view of the inter-relations between the variables under study. In addition to the simple building of matrices as already described, there are numerous techniques of data manipulation and of measurement that might be employed. A few illustrative examples follow.

OTHER MEASURES OF CORRELATION

Multiple and partial correlation matrices could be constructed, using various measures of correlation. The raw characteristics used might be nominal, ordinal, or a mixture of the two.[23]

CLUSTER ANALYSIS

Polities can be clustered in such a way that a polity falling into a given cluster possesses more attributes in common with other polities in that cluster than with polities in any other cluster. This kind of analysis can also be used to produce clusters of attributes.[24]

[23] See, in this regard, Sidney Siegel, *Nonparametric Statistics for the Behavioral Sciences*. McGraw-Hill, 1956; and John E. Walsh, *Handbook of Nonparametric Statistics*. D. Van Nostrand, 1962.

[24] See C. Radhakrishna Rao, *Advanced Statistical Methods in Biometric Research*. John Wiley & Sons, 1952.

SCALING

Raw attributes can possibly be converted into an ordinal scale by means of such techniques as the Guttman Scale. If a number of carefully chosen attributes can be successfully scaled in accordance with Guttman's criteria, the result could be the establishment of one or more fairly general profiles of political evolution. Such profiles would be especially interesting if they contained, in addition to political attributes, at least a few "environmental" attributes from the economic, demographic, and cultural realms.[25]

RECONVERTING DATA TO INTERVAL FORM

A number of our ordinal raw characteristics are derived from interval data, as for example, per capita gross national product. Under some circumstances, it would be well to reconvert such data to interval form. This would afford opportunities for correlational and regression analysis.[26]

A PARTIAL TEST OF THE MEANINGFULNESS OF THE FINDINGS

We are convinced that the degree of meaningfulness of most of the paragraphs in the printout is reasonably high. The primary ground for this conviction is that there is a reassuringly high degree of internal consistency among the statements appearing in almost every paragraph. We decided, nonetheless, to subject our technique and findings to a partial test of meaningfulness.

We asked ourselves the question: "How sure can we be that the co-occurrences stated in the printout are not simply the consequence of chance?" To answer this question, we set up five nonsense variables which appear as Paragraphs 195 to 199. These are the "whiskers" characteristics, which straight-facedly tell us what color whiskers are worn, or are not worn, by particular polities. Paragraph 195 contrasts "Polities that belong to the fifty percent that have purple whiskers" with the "fifty percent that do not have purple whiskers." Paragraph 196 contrasts the "forty percent that have blue whiskers" with the "sixty percent that do not have blue whiskers," and so forth. To determine which fifty percent of the 115 polities were to be regarded as sporting purple whiskers, we simply consulted a table of random numbers. We did the same, independently, with each other "whiskers" characteristic. To gain some

[25] For a suggestive discussion of the use of Guttman scaling as a means of establishing sequences of evolutionary development, see Robert L. Carneiro, "Scale Analysis as an Instrument for the Study of Cultural Evolution," *Southwest Journal of Anthropology*, Vol. 18, 1962, pp. 149–169.

[26] For a discussion of techniques for dealing with an interval and an ordinal variable, or with an interval and a nominal variable, see Robert M. Elashoff, *Multivariate Two-Sample Problems with Discrete and Continuous Variables*. Unpublished doctoral dissertation, Harvard University, 1963.

idea of the quantity and nature of the statements that result from these purely chance co-occurrences, the reader may glance at the paragraphs in question. In doing so, it would be well to keep in mind that the present technique *can* be applied to a random sample of objects, as distinct from a whole universe of objects as in the present survey.

Our *a priori* expectation was that the average "whiskers" paragraph would contain about 25 predicate-pairs, that is, about 15 percent of the total number of potential predicate-pairs that theoretically could appear under a particular paragraph.[27] We also expected that about one-third, or 8, of these predicate-pairs would contain the verb "tend," that there would be another 8 containing the verb "lean," and 8 containing the verb "tilt."[28]

The actual results are somewhat different from the above-mentioned expectations. None of the five "whiskers" paragraphs yields as many as 25 predicate-pairs, although Paragraphs 197 and 199 come close. The mean number of "tends" in the five paragraphs is 6.2, rather than the expected 8.2 The mean number of "leans" is only 4.2, as against the expected 8.2. And the mean number of "tilts" is 5.4, again well short of 8.2. The reason for this "conservative" discrepancy has to do with the manner in which the Fisher calculation was made. Following the procedure suggested by Armsen, this calculation is characterized by a significance level smaller than the nominal level. (See pages 29–30.) This "conservative" discrepancy should, if anything, increase the credibility of the entire survey.

The small number of predicate-pairs appearing in the nonsense paragraphs is in striking contrast to the large average number appearing in the meaningful paragraphs.[29] The median number of statements appearing in the 125 paragraphs is 70. This is almost three times the number that would be expected by chance, and is over 40 percent of the total number, 165, of statements that are theoretically possible. The number of statements ranges from 110 in Paragraphs 30 and 116 down to 7 in Paragraph 150. The semi-interquartile range is from 92 down to 49.

It should be pointed out that our decision to retain or discard a particular paragraph after examining the printout from a preliminary run was not governed primarily by the *number* of predicate-pairs appearing therein, but rather by whether these predicate-pairs revealed information which we considered interesting. In the present printout we have, indeed,

[27] There are 194 finished characteristics. Of these, 29 are "subject-only" characteristics, and hence could not appear as predicates. The remaining 165 finished characteristics could theoretically appear as predicate-pairs in a particular paragraph. Fifteen percent of 165 equals 24.75.

[28] Predicates containing the verb-phrase "always are" were counted as if they were "tends," "leans," or "tilts," depending on their Fisher values.

[29] While there are 194 meaningful finished characteristics, it will be remembered that 69 of these are run as "predicate-only," and hence do not serve as subject-pairs of paragraphs. Of the remaining 125 paragraphs, 29 are those in which the subject characteristic is run as "subject-only," while in the remaining 96 cases the characteristic is run as "both subject and predicate."

retained some paragraphs where the number of predicate-pairs is small, precisely *because* we believe that the *paucity* of such predicate-pairs is an interesting fact in itself.[30] Paragraph 150, for example, brings out the point that it makes little difference, in terms of the present survey, whether a polity is of the "Latin Liberal Conservative" type or of the "Latin Social Revolutionary" type. We might add that none of the figures just given include those predicate-pairs that were hand-winnowed out of the print-out on grounds of redundance.

Aside from the question of the quantity of predicate-pairs appearing in the nonsense paragraphs, as contrasted with the meaningful paragraphs, there is the question of the quality and internal consistency of the statements appearing in each of the two kinds of paragraph. In the case of the meaningful paragraphs, large numbers of redundant predicate-pairs were hand-winnowed out, as is immediately clear from a glance at Appendix C.[31] This can be taken to mean that there is a considerable degree of consistency in the meaningful paragraphs. By contrast, in all of the five "whiskers" paragraphs, it was necessary to remove only one quasi-redundant predicate-pair.

As one glances through the whiskers paragraphs, it becomes readily evident that their internal consistency is generally low. When it does happen that a nonsense paragraph reveals a fair degree of internal consistency, an explanation is usually readily apparent. Paragraph 199 is an example. Here, through the workings of chance, it turns out that 7 of the 12 polities with yellow whiskers are new nations located in Central Africa. It should come as no surprise that, given such a distribution, the statements in Paragraph 199 are not only fairly numerous but reveal a fair degree of internal consistency.[32]

[30] These include eight paragraphs where the number of predicate-pairs is lower than the 24.75 that would be expected on the basis of chance alone, namely, Paragraphs 7, 12, 16, 88, 103, 150, 151, and 191.

[31] In this Appendix, both "pure" redundant and "quasi"-redundant removals are identified by the symbol "RD." However, the great majority of these are quasi-redundant.

[32] In only one instance does a "whiskers" finished attribute co-occur with another such attribute. If we had permitted "whiskers" predicates to appear in the final printout, Statements 197/199 and 199/197 would have appeared. These would have revealed that green-whiskered polities tilt more toward being non-yellow, and that yellow-whiskered polities tilt more toward being non-green, respectively.

3. The Raw and Finished Characteristics

THE format of this chapter is quite simple. Each of the 57 raw characteristics, with its subsumed raw attributes, is presented in serial order, accompanied by a brief commentary on coding procedures and problems. Immediately following the commentary are listed the finished characteristics derived from the raw characteristics. For the sake of convenience, the following abbreviations are used:

Pol = Polities
rt = rather than
v = as contrasted with
AO = all others
CONV = converse
[SO] = Subject Only
[PO] = Predicate Only

---

1. AREAL GROUPING

A. Australasia [2]
B. Caribbean [5]
G. Central America [7]
H. Central and South Africa [28]
I. East Asia [5]
J. East Europe [9]
K. Middle East [11]
P. North Africa [5]
Q. North America [2]
R. Scandinavia [5]
S. South America [10]
X. South Asia [4]
Y. Southeast Asia [9]
Z. West Europe [13]

n = 115

Areal grouping proved to be one of the most discriminating characteristics employed in this study, much more so, in fact, than we had originally anticipated.

The raw characteristic breakdown adheres, insofar as possible, to generally accepted geographic criteria. Some readers will object to the inclusion of Mexico and Panama in "Central America," but with some hesitation we retain this expression rather than the less familiar "Middle America." More serious difficulties concern delimitation of the Middle East and North Africa, respectively. As Dankwart A. Rustow has pointed out,[1] there are "minimal" and "maximal" definitions of the Middle East. The "maximal" definition sometimes embraces all of North Africa, together with portions of the Balkans and Southwest Asia. We incline toward the "minimal" view and take the Middle East to include Afghanistan, Cyprus, Iran, Iraq, Israel, Jordan, Lebanon, Sa'udi Arabia, Syria, Turkey, and Yemen. We have restricted our definition of North Africa to the five Mediterranean states: Algeria, Libya, Morocco, Tunisia, and the United Arab Republic (Egypt). A case can admittedly be made for the inclusion of Sudan, but the criteria employed might just as easily require the inclusion of Mauritania and perhaps Mali, Niger, and Chad. Somewhat arbitrarily, we have included all five states in "Central and South Africa." We have included the Malagasy Republic in Central and South Africa with full awareness that it does not, strictly speaking, consider itself an "African" state. In distinguishing between East and West Europe, we have adhered to the boundary established by the Cold War, though this differs significantly from our demarcation of the "traditional West" in Raw Characteristic 20.

In the preparation of finished characteristics, it was necessary to combine certain areas into larger groupings so as to avoid both self-obvious and statistically unimportant results.[2] In one such case, the larger grouping is not defined in strictly contiguous terms, namely, West Europe, Scandinavia, North America, and Australasia.

- FC 1. Pol located in East Europe (9) v AO (106) [SO]
- FC 2. Pol located in West Europe, Scandinavia, North America, or Australasia (22) v AO (93)
- FC 3. Pol located in West Europe (13) v AO (102)
- FC 4. Pol located in Scandinavia (5) v AO (110) [SO]
- FC 5. Pol located in Scandinavia or West Europe, rt in East Europe (18) v CONV (9)
- FC 6. Pol located in the Caribbean, Central America, or South America (22) v AO (93)

[1] Dankwart A. Rustow, "The Politics of the Near East," in *PDA*, pp. 369 ff.

[2] South Asia v AO and East Asia v AO were employed as finished characteristics in an earlier computer run. The first did not yield results markedly at variance with those of Asia as a whole. The latter produced few statements of any interest, apparently because of the extreme diversity among the relatively few countries involved.

FC 7. Pol located in Central America, rt in South America (7) v CONV (10)
FC 8. Pol located in East Asia, South Asia, or Southeast Asia (18) v AO (97)
FC 9. Pol located in Southeast Asia (9) v AO (106)
FC 10. Pol located in North Africa, or Central and South Africa (33) v AO (82)
FC 11. Pol located in Central and South Africa (28) v AO (87)
FC 12. Pol located in North Africa (5) v AO (110) [SO]
FC 13. Pol located in North Africa or the Middle East (16) v AO (99)
FC 14. Pol located in the Middle East (11) v AO (104)
FC 15. Pol located in the Middle East, rt in North Africa, or Central and South Africa (11) v CONV (33)
FC 16. Pol located in the Middle East, rt in East Asia, South Asia, or Southeast Asia (11) v CONV (18)
FC 17. Pol located in the Middle East, rt in the Caribbean, Central America, or South America (11) v CONV (22)
FC 18. Pol located in North Africa or Central and South Africa, rt in East Asia, South Asia, or Southeast Asia (33) v CONV (18)
FC 19. Pol located in North Africa or Central and South Africa, rt in the Caribbean, Central America, or South America (33) v CONV (22)
FC 20. Pol located in East Asia, South Asia, or Southeast Asia, rt in the Caribbean, Central America, or South America (18) v CONV (22)

---

2. SIZE

A. Very large (2 million sq. mi. & above) [6]
B. Large (300,000–1.9 million sq. mi.) [26]
G. Medium (75,000–299,000 sq. mi.) [36]
H. Small (below 75,000 sq. mi.) [47]

n = 115

This raw characteristic, like a number to follow, is an ordinal characteristic derived from an "interval" continuum. Our first step was to arrange all the countries in precise order from the largest to smallest. We then proceeded to divide the continuum into an appropriate number of segments by establishing cutting points at places that seemed most satisfactory in terms of our research objectives.

In deciding on a procedure by which to establish the cutting points, we considered and rejected two arbitrary procedures. One would have been to space the cutting points at equal intervals in terms of numbers of square miles. The other would have been to space them so that each segment would contain an equal number of countries.

The procedure we actually followed was somewhat more complex. It contains an admittedly "intuitive" element, but has yielded results that appear to be analytically fruitful. We examined the list of countries for gaps or breaks in the continuum, so as to decide whether segmentation on the basis of two, three, or four cutting points would be most "natural." Three such intervals readily appeared, sufficiently near the 75,000, 300,000, and 2,000,000 sq. mi. levels to permit our using these as our cutting points in the construction of a four-attribute characteristic. The resulting distribution is "pyramidal" in shape. United Nations figures were used,[3] with conversion from square kilometers to square miles and rounding to the nearest 1000.

Of the finished characteristics tested during the first computer run, only two were retained. Those based on extreme contrast and on adjacent contrast were found to be relatively unproductive and were discarded. The result is a modified use of the "moving cutting point" approach (page 45).[4]

FC 21. Pol whose territorial size is very large or large (32) v Pol whose territorial size is medium or small (83)

FC 22. Pol whose territorial size is very large, large, or medium (68) v Pol whose territorial size is small (47) [PO]

---

3. POPULATION

A. Very large (100 million & above) [4]
B. Large (17–99.9 million) [23]
G. Medium (6–16.9 million) [34]
H. Small (under 6 million) [54]

n = 115

Despite laudable efforts by the Statistical Office of the United Nations and other agencies, the reliability of basic census data for many countries, particularly the newer ones, ranges from the somewhat questionable to the highly suspect. The situation is complicated by the fact that in several countries—Lebanon and Nigeria, for example—even the *desirability* of an accurate census has, for reasons associated with problems of communal or ethnic representation, become a distinctly political issue. Other countries, for somewhat more obscure reasons, appear to have issued grossly inflated "official" figures. Ethiopia is perhaps the worst offender in this respect.

Our caveat having been entered, we make no apologies for having

[3] United Nations, *Statistical Yearbook: 1961*. New York, 1961. Hereafter cited as *SY: 1961*.

[4] Very large v Large, medium, or small was not used because of the limited number of "Very large" countries.

employed the latest available UN data.[5] These are, in virtually all instances, estimated or provisional data for the years 1960, 1961, or 1962. We have used P. N. Rosenstein-Rodan's figure of 15 million (as contrasted with the official 20 million) for Ethiopia.[6] Figures for Burundi and Rwanda were calculated from density figures provided by the 1961 *Report of the United Nations Commission for Ruanda-Urundi.*[7]

Cutting points for the present raw characteristic were established in the same manner as were those for territorial size. The finished characteristics derived from RC 3 are structurally similar to those for RC 2.

FC 23. Pol whose population is very large or large (27) v Pol whose population is medium or small (88)
FC 24. Pol whose population is very large, large, or medium (61) v Pol whose population is small (54) [PO]

---

4. POPULATION DENSITY

A. Very high (600/sq. mi. & above) [4]
B. High (300–599/sq. mi.) [13]
G. Medium (100–299/sq. mi) [31]
H. Low (below 100/sq. mi.) [67]

n = 115

Population density varies from a scant 2 persons per square mile in Libya, Mauritania, and Mongolia to a teeming 903 in the Netherlands. Belgium, with 769, remains the second most densely populated country, but it is of interest to note that South Korea, rather than Japan, now ranks third with 687.

It need hardly be pointed out that density figures can often be quite deceptive. India, with 375 per square mile, actually has a far greater "population problem" than does the German Federal Republic with 590. However else they may differ, both France and the Philippines, with densities of 217 and 252, respectively, could easily support higher populations, while Israel, with 280, could do so only with difficulty. Indeed, while the "population explosion" is a frightening reality for many countries, *under-population* is almost as great a problem for others. Again, however, caution is in order. Jordan has a relatively low population density of 46 per square

[5] United Nations, Department of Economic and Social Affairs, Statistical Office of the United Nations, "Population and Vital Statistics Report: Data Available as of 1 July 1962," *Statistical Papers, Series A,* Vol. XIV, No. 3.

[6] P. N. Rosenstein-Rodan, "International Aid for Underdeveloped Countries," *The Review of Economics and Statistics,* Vol. XLIII, No. 2, May 1961.

[7] United Nations, General Assembly, Sixteenth Session, "Report of the United Nations Commission for Ruanda-Urundi," *Annexes, Addendum to Agenda Item 49,* Document A/4994, 30 November 1961.

mile, but, as George L. Harris has pointed out, "since . . . the area under cultivation is considerably under 2,000 square miles (4.8 percent of the total), the density of the planted area [is] about 780 persons per square mile—a high density for an agrarian country."[8] A far more extreme case is that of Egypt, with a gross population density of 69, but with nearly 2,000 persons per square mile of arable area.[9]

Cutting points for the present raw characteristic were established in the same manner as were those for size and population. In the case of the finished characteristics, two were retained from the first computer run, while one new (and rather unusual) characteristic was added. Very little is usually gained by constrasting the median component of an ordinal series with *both* extremes. But in examining the first-run printout it became apparent that in some respects polities with high and low population densities had more in common than did either extreme group with those of medium density. For this reason, we added Medium v Very high, high, or low to the present group of finished characteristics.

FC 25. Pol whose population density is very high or high (17) v Pol whose population density is medium or low (98) [PO]
FC 26. Pol whose population density is very high, high, or medium (48) v Pol whose population density is low (67)
FC 27. Pol whose population density is medium (31) v Pol whose population density is very high, high, or low (84) [SO]

5. POPULATION GROWTH RATE

A. High (2% or above) [62]
B. Low (less than 2%) [48]

n = 110
punch 5 = 5

The sophistication of any ordinal segmentation must depend, in part, upon the reliability of its supporting data, and perhaps the least reliable of current demographic data are those dealing with population growth rates. For this reason, we have confined ourselves here to a simple high-low schema, using 2 percent as our cutting point.

Two sets of data have been collated for this characteristic: the latest available "official" United Nations figures[10] and a somewhat shorter list

[8] George L. Harris, *Jordan: Its People, Its Society, Its Culture*. Human Relations Area Files Press, 1958, p. 23. Harris has a gross population density for Jordan of 39 per square mile, but the figure is based on a 1956 population estimate.

[9] The exact figure is 1969, based on the 1961 population estimate of 26,578,000 and an arable area estimate of 13,500 sq. mi.

[10] *SY: 1961*.

for the "underdeveloped" countries compiled by Rosenstein-Rodan.[11] In general, Rosenstein-Rodan's data are regarded as the more accurate and were relied upon in most doubtful cases, except for Ghana and Togo, which were coded as unascertained (punch 5). Neither list contained data for Ethiopia, North Korea, or the USSR.

On the basis of United Nations figures, three countries (East Germany, Ireland, and North Vietnam) would appear to have net declining populations, the rates being —0.7 percent, —0.6 percent, and —0.2 percent respectively.

FC 28. Pol whose population growth rate is high (62) v Pol whose population growth rate is low (48)

---

6. URBANIZATION

A. High (20% or more of population in cities of 20,000 or more and 12.5% or more of population in cities of 100,000 or more) [56]

B. Low (less than 20% of population in cities of 20,000 or more and less than 12.5% of population in cities of 100,000 or more) [49]

n = 105
punch 3 = 7
punch 5 = 3

There is no generally accepted formula for rating "degree of urbanization." United Nations urbanization studies usually consist of data supplied by member states, adorned with extensive footnotes attesting to the wide variety of criteria employed by the reporting governments.

The UN does, however, issue data periodically on percentage of population in cities of specified size, and we have keyed our urbanization characteristic to this material. Again, we have restricted ourselves to a simple high-low schema. We have based this schema on two criteria: (1) 20 percent or more of the population in cities of 20,000 or more *and* (2) 12.5 percent or more of the population in cities of 100,000 or more. Six polities (Malaya, Paraguay, Philippines, Portugal, Trinidad, Republic of Vietnam) satisfy one of these criteria but not the other, and hence have been coded as ambiguous (punch 3). Costa Rica has also been coded as ambiguous because of apparent uncertainty concerning the boundaries of its single large city, San José.[12]

The UN has provided data on percentage of population in cities of

[11] Rosenstein-Rodan, *op. cit.*

[12] Iceland and Luxembourg, neither of which has cities as large as 100,000, have been coded as highly urbanized: 40.5 percent of Iceland's population of 179,000 is located in the capital, Reykjavik; 30.5 percent of Luxembourg's population of 316,000 is located in cities of 20,000 or more.

100,000 or more for 93 countries, and on percentage of population in cities of 20,000 or more for 74 countries.[13] For countries not included in either or both UN lists, individual computations were prepared from two separate sources: *The Statesman's Yearbook* and the *Encyclopaedia Britannica World Atlas*. These computations were then collated and appropriate percentage estimates made. There were only three countries, all Communist, for which reasonably reliable data could not thus be assembled: China, North Korea, and North Vietnam.

It should be observed that while urbanization (however defined) correlates to some extent in an *over-all* way with "economic development" and related variables, there are many *specific* exceptions. The capital of the relatively underpopulated Republic of Congo, Brazzaville, will shortly attain a population of 100,000, representing approximately 12.5 percent of the national total—thus satisfying one of our urbanization criteria. Jordan is already "urbanized" by our definition, with about 14 percent of its population dwelling in the capital city of Amman, and 20 percent in Amman and the Jordanian sector of Jerusalem combined. These illustrations would suggest that a clear distinction should be made between what economists have designated as "push" and "pull" factors, that is, between the character of the various forces *generating* urbanization. Until this is done, it will be difficult to reach conclusions concerning the economic significance of the phenomenon of urbanization in various of the underdeveloped countries.

FC 29. Pol where the degree of urbanization is high (56) v Pol where the degree of urbanization is low (49)

---

7. AGRICULTURAL POPULATION

A. High (over 66%) [56]
B. Medium (34–66%) [33]
G. Low (16–33%) [17]
H. Very low (under 16%) [7]

n = 113
punch 5 = 2

The Food and Agricultural Organization of the United Nations issues statistics periodically on world agricultural population, the most recent of which provide data for about 80 countries.[14] These data are subsumed under two time spans (1930–1944; 1945–1960) for each of two overlapping

13 United Nations, *Demographic Yearbook: 1960*. New York, 1960. Hereafter cited as *DY:1960*.

14 Food and Agricultural Organization of the United Nations, *Production Yearbook: 1961*. Rome, 1962.

groups, "Agricultural Population" and "Population Engaged in Agricultural Occupations." The first, generally speaking, includes those engaged directly in agricultural pursuits, plus their dependents. The second provides indices of work force in agricultural occupations, including those engaged in processing agricultural commodities.

Our coding has been keyed to the "Agricultural Population" group. For a few countries we have had to rely on data provided by the *1950 World Census of Agriculture*,[15] and in some cases have had to extrapolate from the 1930–44 figures.

In the case of the following 20 countries, for which no precise figures are available, we have estimated that the agricultural population (including hunting, fishing, and forestry, where significant) constitutes at least 66 percent of the total: Burundi, Cameroun, Central African Republic, Chad, Congo (Bra), Dahomey, Gabon, Guinea, Ivory Coast, Laos, Mali, Mauritania, Mongolia, Niger, Rwanda, Senegal, Somalia, Upper Volta, Republic of Vietnam, and Yemen. No figures could be obtained, or appropriate estimates made, for Sa'udi Arabia and North Vietnam.

Two finished characteristics are provided. We did not dichotomize on the basis of High, medium, or low v Very low, since the seven polities in the latter category (Australia, Belgium, Canada, German Federal Republic, Israel, United Kingdom, United States) are, with the possible or partial exception of Israel, all members of the "developed West."

FC 30. Pol whose agricultural population is high (56) v Pol whose agricultural population is medium, low, or very low (57)

FC 31. Pol whose agricultural population is high or medium (90) v Pol whose agricultural population is low or very low (24) [PO]

---

8. GROSS NATIONAL PRODUCT

A. Very high ($125 billion & above) [2]
B. High ($25–124.9 billion) [8]
G. Medium ($5–24.9 billion) [20]
H. Low ($1–4.9 billion) [32]
I. Very low (under $1 billion) [53]

n = 115

The United States and the USSR are the only two nations with Gross National Products exceeding $100 billion (515 for the US; 176 for the USSR). At the other extreme is Togo, with an estimated GNP for 1961 of $63 million.

[15] Food and Agricultural Organization of the United Nations, *Report on the 1950 World Census of Agriculture*, Vol. I. Rome, 1955.

The greater part of the data for this characteristic is taken or calculated from Rosenstein-Rodan.[16] Figures for Australia, Austria, Iceland, and New Zealand were calculated from the *Yearbook of National Accounts Statistics.*[17] Figures for Burundi and Rwanda, respectively, were each derived on the basis of 50 percent of the estimate for the former mandated territory of Ruanda-Urundi. Mongolia and 13 newly independent Central and West African states for which specific data could not be obtained were coded as "under $1 billion."

FC 32. Pol whose gross national product is very high or high (10) v Pol whose gross national product is medium, low, or very low (105) [PO]
FC 33. Pol whose gross national product is very high, high, or medium (30) v Pol whose gross national product is low or very low (85)
FC 34. Pol whose gross national product is very high, high, medium, or low (62) v Pol whose gross national product is very low (53) [PO]

---

9. PER CAPITA GROSS NATIONAL PRODUCT

A. Very high ($1200 & above) [11]
B. High ($600–1199) [13]
G. Medium ($300–599) [18]
H. Low ($150–299) [22]
I. Very low (under $150) [51]

n = 115

GNP per capita is, for many purposes, far from being the most useful per capita economic indicator. Ideally, "real" GNP per capita, per capita income, and "real" per capita income afford successively more meaningful indices of individual economic status. But unfortunately this holds in a manner almost inversely proportional to the reliability of the relevant data. For this reason we have confined ourselves to GNP per capita.

Our figures, again, are largely taken or calculated from Rosenstein-Rodan, with exceptions as noted in the commentary for RC 8.

FC 35. Pol whose per capita gross national product is very high or high (24) v Pol whose per capita gross national product is medium, low, or very low (91) [PO]
FC 36. Pol whose per capita gross national product is very high, high, or medium (42) v Pol whose per capita gross national product is low or very low (73)

[16] Rosenstein-Rodan, *op. cit.*
[17] United Nations, *Yearbook of National Accounts Statistics: 1961.* New York, 1962.

FC 37. Pol whose per capita gross national product is very high, high, medium, or low (64) v Pol whose per capita gross national product is very low (51) [PO]

---

10. INTERNATIONAL FINANCIAL STATUS

A. Very high (UN assessment of 10% or above) [2]
B. High (UN assessment of 1.50–9.99%) [8]
G. Medium (UN assessment of 0.25–1.49%) [27]
H. Low (UN assessment of 0.05–0.24%) [32]
I. Very low (minimum UN assessment of 0.04%) [39]

n = 108
punch 5 = 7

Although Article 17 of the United Nations Charter specifies only that "the expenses of the Organization shall be borne by the Members as apportioned by the General Assembly," ability to pay has long been recognized as the principal criterion to be employed in determining the scale of assessments.[18] This being the case, it seemed reasonable to base a scale dealing with each polity's financial status on the UN assessment rates.[19]

Coding for four countries (German Federal Republic, Republic of Korea, Switzerland, Republic of Vietnam) that are not members of the United Nations was possible on the basis of assessments for UN-affiliated agencies. Data could not be obtained for seven countries (Algeria, China, East Germany, Jamaica, North Korea, Trinidad, North Vietnam) that are not members of the UN or affiliated agencies, or for which assessments as members have not yet been announced. Burundi and Rwanda, for which data were also unobtainable, were coded as "Very low."

FC 38. Pol whose international financial status is very high or high (10) v Pol whose international financial status is medium, low, or very low (103) [PO]

FC 39. Pol whose international financial status is very high, high, or medium (38) v Pol whose international financial status is low or very low (76)

[18] Except that "in principle, the maximum contribution of any Member State to the ordinary expenses of the United Nations shall not exceed 30 percent of the total." [United Nations, Public Inquiries Unit, *United Nations Budget,* 9 February 1961, p. 2.] The United States assessment is currently 32.02 percent.

[19] United Nations, General Assembly, *Scale of Assessments for the Apportionment of the Expenses of the United Nations,* A/Res/1870 (XVI), 27 December 1961.

United Nations, General Assembly, *Scale of Assessments for the Apportionment of the Expenses of the United Nations,* A/Res/1870 (XVII), 27 December 1962.

FC 40. Pol whose international financial status is very high, high, medium, or low (71) v Pol whose international financial status is very low (39) [PO]

---

11. ECONOMIC DEVELOPMENTAL STATUS

A. Developed (self-sustaining economic growth; GNP/capita over $600) [19]
B. Intermediate (sustained and near self-sustaining economic growth) [17]
G. Underdeveloped (reasonable prospect of attaining sustained economic growth by the mid-1970's) [17]
H. Very underdeveloped (little or no prospect of attaining sustained economic growth within the foreseeable future) [57]

n = 110
punch 3 = 5

In establishing cutting points for this raw characteristic, we rely primarily upon the distinction between "sustained" and "self-sustaining" growth. As Rosenstein-Rodan puts it, " 'economic factors' are a necessary, but not sufficient condition of sustained growth. . . . Important symptoms of sustained growth are, on the one hand, the ability to imitate and to absorb other countries' methods of production—frequently referred to as 'technological progress'—and, on the other hand, a differentiated structure of production and investment, notably including a minimum quantum and growth of industrial production."[20]

A nation that has attained a capacity for sustained growth is not necessarily lacking in foreign aid requirements. Indeed, an important aspect of sustained growth is usually the capacity to absorb more foreign aid more efficiently than would be possible for a relatively *less* developed economy. Only with self-sustaining growth does a nation ordinarily reach the stage where public foreign aid is no longer required; though normal capital imports may continue.[21]

We have characterized "developed" economic status as requiring self-sustaining growth, *in addition* to a GNP per capita of at least $600. A few countries (for example, Ireland, Venezuela) have GNP per capita rates in excess of $600, but do not otherwise qualify as fully developed. Other states (for example, Japan) have attained a level of self-sustaining growth but rank quite low in GNP per capita.

For a number of countries it is extremely difficult to estimate the

[20] Rosenstein-Rodan, *op. cit.*, p. 113.
[21] *Ibid.*, p. 115.

impact of current political developments on the economy. Codings for countries currently experiencing internal difficulties of the sort faced by Argentina, Burma, and Ceylon must necessarily be regarded as highly tentative.

The developmental status of five countries (Costa Rica, Israel, Lebanon, Sa'udi Arabia, Venezuela) is coded as ambiguous. Costa Rica has a relatively high income level but little apparent growth potential, for which its current high rate of population increase is at least partly responsible. The present oil revenues of Sa'udi Arabia are impressive, but that country otherwise appears to have reached the limit of its developmental potential.

Libya (here coded as "underdeveloped") is an interesting case in that it has formally abandoned all reliance on direct foreign aid. There is, however, considerable question concerning Libya's future yield from petroleum investment.

FC 41. Pol whose economic developmental status is developed (19) v Pol whose economic developmental status is intermediate, underdeveloped, or very underdeveloped (94)
FC 42. Pol whose economic developmental status is developed or intermediate (36) v Pol whose economic developmental status is underdeveloped or very underdeveloped (76) [PO]
FC 43. Pol whose economic developmental status is developed, intermediate, or underdeveloped (55) v Pol whose economic developmental status is very underdeveloped (57)

---

12. LITERACY RATE

A. High (90% or above) [25]
B. Medium (50–89%) [30]
G. Low (10–49%) [24]
H. Very low (under 10%) [26]

n = 105
punch 5 = 10

There is no universal consensus as to the precise minimal proficiency required before a person is considered "literate." An individual classified as "literate" by one nation's standards might well be designated as semiliterate or even as illiterate by those of another.

Another difficulty, quite apart from wide variation in the reliability of reported figures, concerns the age-group covered. Some countries report only on adult literacy (usually 15 years or older), while others report on all persons of school age or older. We have tried, wherever necessary, to weigh relevant data so as to yield estimates covering the latter group.

Our principal source for this characteristic has been United Nations data,[22] though we have not hesitated to depart therefrom in certain cases. For example, the UN *il*literacy rate for Ecuador of 44.2 percent seemed excessively low, even though the Amerindian jungle population was specifically excluded.

Ten polities (China, Guinea, North Korea, Lebanon, Liberia, Mauritania, Mongolia, Morocco, Syria, North Vietnam) are unascertained (punch 5) for this characteristic.

FC 44. Pol where the literacy rate is ninety percent or above (25) v Pol where the literacy rate is below ninety percent (90) [PO]
FC 45. Pol where the literacy rate is fifty percent or above (55) v Pol where the literacy rate is below fifty percent (54)
FC 46. Pol where the literacy rate is ten percent or above (84) v Pol where the literacy rate is below ten percent (26) [PO]
FC 47. Pol where the literacy rate is ninety percent or above, rt between fifty and ninety percent (25) v CONV (30) [PO]
FC 48. Pol where the literacy rate is between fifty and ninety percent, rt between ten and fifty percent (30) v CONV (24) [PO]
FC 49. Pol where the literacy rate is between ten and fifty percent, rt below ten percent (24) v CONV (26) [PO]

---

13. FREEDOM OF THE PRESS

A. Complete (no censorship or government control of either domestic press or foreign correspondents) [43]
B. Intermittent (occasional or selective censorship of either domestic press or foreign correspondents) [17]
G. Internally absent (strict domestic censorship; no restraint on foreign newsgathering, or selective cable-head censorship) [21]
H. Internally and externally absent (strict direct or indirect censorship or control, domestic and foreign) [16]

n = 97
punch 3 = 1
punch 5 = 13
punch 6 = 4

Criteria for designating specific degrees of press freedom are not easily established, and the foregoing is a rather crude attempt to combine elements of two different but related ordinal continua dealing with the

[22] *DY:1960.* A limited number of more recent figures are contained in UNESCO, General Conference, Twelfth Session, *World Campaign for Universal Literacy,* 12C/PRG/3, Paris, 1963. This source should, however, be used with extreme caution. It contains several typographical errors in addition to data that are, in part, suspect on other grounds.

frequency of restrictive action and the *scope* of such action, respectively.

Our principal sources have been the Associated Press year-end censorship reports for 1961[23] and 1962.[24] We have referred also to the 1959 study by the International Press Institute on censorship in authoritarian countries.[25]

The Associated Press identifies three major types of censorship as follows: "formal censorship involves blue-penciling or outright killing of stories that do not fit a government's pattern of propaganda. Then there is 'responsibility censorship,' which in some countries holds a threat of expulsion over correspondents who offend the regime.

"The most widespread sort of censorship is a withholding of information at the source. This often involves a lack of access to news sources, especially public officials.

"Even in the United States, complaints were heard in 1962 of 'management of news.' These were applied especially to officials who leaked information or parceled it out in a manner presumed to suit government objectives."[23]

Taking a somewhat narrower view than that of the Associated Press, we do not regard "managed news" as necessarily constituting censorship. We simply ask the question, "Is the reporter, newspaper, or news-gathering agency restrained, directly or indirectly, from filing or publishing legitimate news already in his or its possession?"

Of the thirteen countries for which information on freedom of the press was unascertained at the time of coding, nine are Central African and four are Central American. These two high residual concentrations should be borne in mind when drawing conclusions from Paragraphs 50, 51, and 52 in the printout.

The four countries coded as currently "unascertainable" for this characteristic are Ceylon, Iraq, Togo, and Yemen. The last three have recently undergone drastic changes of regime. In the case of Ceylon, no censorship was imposed in 1962, but a drastic censorship bill was before the legislature as of April, 1963.

Greece is coded as ambiguous (punch 3) in view of the fact that while no censorship, as such, is imposed, the only domestic news agency is government-owned.

FC 50. Pol where freedom of the press is complete (43) v Pol where freedom of the press is intermittent, internally absent, or internally and externally absent (56)

23 Associated Press, "Annual Report on World-Wide Press Curbs," *New York Times*, January 7, 1962.

24 Associated Press, "Annual Censorship Roundup," dated December 28, 1962 (copy provided by AP, New York).

25 International Press Institute, *The Press in Authoritarian Countries*. Zurich, 1959.

FC 51. Pol where freedom of the press is complete or intermittent (60) v Pol where freedom of the press is internally absent, or internally and externally absent (37) [PO]
FC 52. Pol where freedom of the press is complete, intermittent, or internally absent (82) v Pol where freedom of the press is internally and externally absent (16) [PO]

---

14. NEWSPAPER CIRCULATION PER 1000 POPULATION

A. High (300 & over) [14]
B. Medium (100–299) [23]
G. Low (10–99) [41]
H. Very low (below 10) [35]

n = 113
punch 5 = 2

In the communications realm, several kinds of data of varying comprehensiveness and reliability exist. The present raw characteristic could easily be supplemented by others dealing with such artifacts of modern life as cinema seats, radio receivers, and television sets. However, space limitations have compelled us to be selective. We have therefore confined ourselves to newspaper circulation as a universally applicable and representative communications variable, and as one for which reasonably adequate and reliable data are available.

Codings for 108 countries are based on current UN figures.[26] Of the seven countries for which figures are not available, five (Burundi, Gabon, Mauritania, Rwanda, Yemen) have been estimated to be in the "Very low" category. The remaining two (North Korea and North Vietnam) are coded as unascertained (punch 5).

FC 53. Pol where newspaper circulation is three hundred or more per thousand (14) v Pol where newspaper circulation is less than three hundred per thousand (101) [PO]
FC 54. Pol where newspaper circulation is one hundred or more per thousand (37) v Pol where newspaper circulation is less than one hundred per thousand (76)
FC 55. Pol where newspaper circulation is ten or more per thousand (78) v Pol where newspaper circulation is less than ten per thousand (35) [PO]

[26] *SY: 1961.*

---

15. RELIGIOUS CONFIGURATION

A. Protestant [5]
B. Catholic [25]
G. East Orthodox [3]
H. Mixed Christian [13]
I. Hindu [1]
J. Buddhist [4]
K. Muslim [18]
P. Jewish [1]
Q. Mixed literate non-Christian [4]
R. Mixed: Christian, literate non-Christian [4]
S. Non-literate [1]
X. Mixed: Christian, non-literate [13]
Y. Mixed: literate non-Christian, non-literate [7]
Z. Mixed: Christian, literate non-Christian, non-literate [8]

n = 107
punch 3 = 2
punch 5 = 6

This typology is designed to afford approximate, yet meaningful, religious identification for each polity for which relevant information is presently available. In general, we have adhered to a rule of 80 to 85 percent predominance. This means, for example, that a polity coded as "Muslim" could have a non-Muslim religious minority of as much as 15 to 20 percent, but no greater. The lack of a more precise cutoff point is due to a lack of reliability in religious statistics generally.

A certain ethnocentrism must be conceded in that Christianity is divided into its leading components, while Islam and Buddhism are not. The schism between Sunni and Shi'i Muslim is undoubtedly as significant, for certain analytical purposes, as that between Catholic and Protestant Christian. Nor is a collateral distinction between Theravada and Mahayana Buddhist to be ignored. But limits had to be imposed on the number of attributes subsumed under any one raw characteristic.[27]

Some readers might question our coding Colombia and Mexico under (B), while assigning Ecuador and Guatemala to (X). We can only say that the Amerindian population in the first two countries appears to be far more thoroughly "Catholicized" than in the last two.

Some sources identify Nepal as being "predominantly Hindu." There appears, however, to be a high degree of syncretism between Hinduism and Buddhism in Nepal, and for this reason we have coded it under (Q).

South Korea and South Vietnam are coded as ambiguous (punch 3) in view of the unusually complex intermixture of literate and non-literate

[27] Ethiopia is coded under (Z), the Christian component being, of course, Coptic Christian. "Muslim" is here taken to include *all* Islamic sects including, for example, Ismailis in addition to Sunnis and Shi'is.

components in each case. There is also a question concerning the status of Confucianism as a *bona fide* religion in these two countries. Six polities (Burundi, China, North Korea, Rwanda, USSR, North Vietnam) are unascertained (punch 5) for this characteristic.

In establishing the following finished characteristics, we have not dichotomized on the basis of Protestant v AO, since the five predominantly Protestant countries are identical with the five Scandinavian countries, and the reader may refer to FC 4.

FC 56. Pol where the religion is predominantly literate (79) v Pol where the religion is predominantly or partly non-literate (31)
FC 57. Pol where the religion is Catholic (25) v AO (90)
FC 58. Pol where the religion is Muslim (18) v AO (97)
FC 59. Pol where the religion is mixed: Christian, non-literate (13) v AO (100) [SO]
FC 60. Pol where the religion is mixed: literate non-Christian, non-literate (7) v AO (103) [SO]
FC 61. Pol where the religion is mixed: Christian, literate non-Christian, non-literate (8) v AO (104) [SO]
FC 62. Pol where the religion is Protestant, rt Catholic (5) v CONV (25)
FC 63. Pol where the religion is predominantly some kind of Christian (46) v Pol where the religion is predominantly or partly other than Christian (68)
FC 64. Pol where the religion is Christian, rt Muslim (46) v CONV (18)
FC 65. Pol where the religion is Catholic, rt Muslim (25) v CONV (18) [PO]

---

16. RELIGIOUS HOMOGENEITY

A. Homogeneous (A, B, G, I, J, K, or P, above) [57]
B. Heterogeneous (H, Q, R, X, Y, or Z, above [49]

n = 106
punch 3 = 3
punch 5 = 6

In coding for this characteristic, attribute (S) of RC 15 is regarded as ambiguous (punch 3).

FC 66. Pol that are religiously homogeneous (57) v Pol that are religiously heterogeneous (49)

---

17. RACIAL HOMOGENEITY

A. Homogeneous (90% or more of one race) [82]
B. Heterogeneous (less than 90% of one race) [27]

n = 109
punch 3 = 4
punch 5 = 2

The essential question we ask here is: "In a particular polity, do people generally look alike, in a racial sense, or not?" In coding for this raw characteristic, we confine ourselves to the three most populous racial stocks recognized by most anthropologists: Caucasoid, Mongoloid, and Negroid.[28] Where 90 percent or more of a particular polity's population belong to only one of these three major stocks, we code that polity as "homogeneous." Otherwise, we code the polity as "heterogeneous."

It should be emphasized that we refrain from asking "racist" questions, such as which race has "superior" polity-building abilities. Thus, we have no finished characteristic that contrasts "Caucasoid polities" with "Negroid polities," or the like. We simply contrast "homogeneous" polities with "heterogeneous" polities, because it seems clear that, generally speaking, racial heterogeneity often accompanies political divisiveness.

Chile, the Malagasy Republic, Nepal, and Paraguay are coded as ambiguous (punch 3). The people of the Malagasy Republic appear to be, for all practical purposes, of a single mixed stock of Negroid and Mongoloid. The other three are apparently ambiguous in a similar sense, the mixed stocks being of Mongoloid and Caucasoid composition.

We have coded Sa'udi Arabia and Yemen as unascertained (punch 5). There are many coastal and urban Negroes in both countries, but we are unable to estimate whether they constitute as much as 10 percent of the total population, partly because the governments concerned have discouraged research in such matters.

FC 67. Pol that are racially homogeneous (82) v Pol that are racially heterogeneous (27)

---

18. LINGUISTIC HOMOGENEITY

A. Homogeneous (majority of 85% or more; no significant single minority) [52]
B. Weakly heterogeneous (majority of 85% or more; significant minority of 15% or less) [12]

[28] The Bushmanoid and Pygmoid races are not sufficiently numerous or concentrated to warrant their consideration here.

G. Strongly heterogeneous (no single group of 85% or more) [50]

n = 114
punch 3 = 1

In setting up this characteristic, our central question was: "Are people in a particular country culturally homogeneous or culturally fragmented?" We employ the criterion of language as a convenient index. For each country we have estimated the adult population that speaks each of that country's languages as a *native* tongue. A "native" speaker is defined as a person who has learned the language in childhood or adolescence as an integral part of his over-all enculturation. Thus, for example, a language learned in an elementary or secondary schoolroom but not used "naturally" outside the school by the person as he was growing up would not be considered a language "native" to that person. It is, of course, recognized that some persons might grow up speaking more than one language natively or near-natively. This might take either of two principal forms: (1) diglossia, as found, for example, in Haiti or Sa'udi Arabia; and (2) bilingualism, as found, for example, in Paraguay and among many Sino-Thai persons in Thailand.

In dealing with the vexing question of defining a language, we follow Ferguson and Gumperz: "A language consists of all varieties (whether only a single variety or an indefinitely large number of them) which share a single superposed variety (such as a literary standard) having substantial similarity in phonology and grammar with the included varieties or which are either mutually intelligible or are connected by a series of mutually intelligible varieties."[29]

We have coded a particular country as linguistically "homogeneous" in cases where an estimated 85 percent of the adult population are native speakers of a common language, and where there is no other significant language present. We have coded a country as "weakly heterogeneous" where at least an estimated 85 percent of the adults are native speakers of a common language, but where another significant language is also present. Finally, we have coded as "strongly heterogeneous" those countries where no single language is spoken as a native tongue by as many as 85 percent of the adult population.

Linguistic census data often reveal large gaps. Even where gaps are no problem, the available material usually gives numbers of persons who speak a given language, which can often be quite different from the numbers of persons who learned that language natively in childhood or adolescence. For these reasons, coding for this raw characteristic has been difficult. The most difficult decisions involved cases on the borderline between "homogeneous" and "weakly heterogeneous." Here, the decision hinged on whether to regard any one minority language as

[29] Charles A. Ferguson and John J. Gumperz, "Linguistic Diversity in South Asia," *International Journal of American Linguistics,* Vol. 26, No. 3, July 1960, p. 5.

"significant." Illustrative of our coding problem were our decisions to regard as significant the Turkish language in Bulgaria, English in Panama (not counting U. S. citizens in the Canal Zone), Basque in Spain, and Chinese and Khmer in South Vietnam.

Occasionally there were also difficult borderline decisions between the "weakly heterogeneous" and the "strongly heterogeneous" categories. Illustrative of these difficulties were our decisions to place Algeria, Ecuador, and Iraq in the "strongly heterogeneous" category, because the following languages appeared respectively to comprise a larger than 15 percent minority: Berber, Quechua, and Kurdish.

It would, of course, be desirable to code additionally on the basis of a variety of non-linguistic aspects of culture. However, the cost of such a procedure would be prohibitive. In selecting language as a manageable index, yet one that is reasonably reliable and valid, we are in accord with common practice among ethnologists.

We are aware of the limitations on the validity of this technique. There are deviant cases where the linguistic situation is not an adequately valid index of the total cultural situation. The four "native" languages of Switzerland and the two found in Belgium might suggest to some persons a somewhat greater cultural fragmentation than actually exists. On the other hand, the single "native" language of the United States might obscure, to some observers, the very real cultural or subcultural fragmentation that exists in this country as a result of the "separate caste" structure of the Negro community. However, it should not be overlooked that one reason why the Swiss and Belgians are somewhat more socially cohesive than their linguistic situation might suggest is that they have had relatively long periods of time in which to learn to live together, despite ethnolinguistic differences. Consult, in this regard, RC 19, dealing with the date of independence of each polity. Another reason in the case of Belgium is that persons who profess any religion at all are overwhelmingly of the Catholic faith. Consult here RC 15, dealing with religion. Similarly, in connection with the example of the United States, it should be noted that we have introduced (RC 17) the variable of racial homogeneity. In short, we recognize that the linguistic variable will carry only so much analytical weight, after which it, like all the other characteristics that we have introduced, begins to lose strength.

Following the two finished characteristics dealing with linguistic homogeneity, we have added a composite finished characteristic covering the full range of religious, racial, and linguistic homogeneity.

FC 68. Pol that are linguistically homogeneous (52) v Pol that are linguistically weakly heterogeneous or strongly heterogeneous (62)

FC 69. Pol that are linguistically homogeneous or weakly heterogeneous (64) v Pol that are linguistically strongly heterogeneous (50) [PO]

FC 70. Pol that are religiously, racially, and linguistically homogeneous (21) v Pol that are religiously, racially, or linguistically heterogeneous (85)

---

19. DATE OF INDEPENDENCE

A. Before nineteenth century [21]
B. 1800–1913 [31]
G. 1914–1945 [13]
H. After 1945 [46]

n = 111
punch 3 = 4

Cutting points for a time-continuum dealing with a polity's date of independence were dictated primarily by events in Latin America during the nineteenth century and by the impact of the two major wars of this century.

For states once autonomous but later subjugated (for example, Poland), the date of modern independence has been used. German, Italian, and Japanese aggrandizement during the period 1930–1945 has been disregarded.

For original members of the British Commonwealth, the date used is that of the acquisition of dominion status. In the case of Norway, union with Denmark (1381–1814) and with Sweden (1814–1905) is disregarded. Also disregarded is the British "residency" in Nepal, 1816–1846.

The two Germanies and the two Koreas are considered ambiguous (punch 3).

FC 71. Pol whose date of independence is before 1800 (21) v Pol whose date of independence is after 1800 (90)

FC 72. Pol whose date of independence is before 1914 (52) v Pol whose date of independence is after 1914 (59)

FC 73. Pol whose date of independence is before 1945 (65) v Pol whose date of independence is after 1945 (46)

---

20. WESTERNIZATION

A. Historically Western nation [26]
B. Significantly Westernized (no colonial relationship) [7]
G. Significantly Westernized (colonial relationship) [28]

H. Partially Westernized (no colonial relationship) [8]
I. Partially Westernized (colonial relationship) [41]
J. Non-Westernized (no colonial relationship; little or no visible Westernization) [2]

n = 112
punch 3 = 2
punch 6 = 1

For this characteristic, the demarcation of the "historic West" on the European continent is based on the limits of the Ottoman Empire, with the exception that Greece is considered Western. Czechoslovakia, Hungary, and Poland are thus regarded as historically Western; Rumania, Bulgaria, and Yugoslavia are not.

Australia, Canada, New Zealand, and the United States are included in A; the former Latin American colonies of Spain and Portugal are included in G or I.

Israel and South Africa have been coded as ambiguous (punch 3). Israel was so coded because the greater part of its population immigrated after the Palestinian mandate had been lifted, and such immigrants came from a wide variety of Western and non-Western areas. South Africa was considered ambiguous because of the mixed character of its population and the separateness of its social institutions. Yemen has been coded as unascertainable (punch 6) under current conditions.

The present raw characteristic, like a number of others that follow, is of a hybrid character. Essentially typological in structure, it also exhibits an ordinal component. This has been taken into account in framing the finished characteristics.

FC 74. Pol that are historically Western (26) v AO (87)
FC 75. Pol that are historically Western or significantly Westernized (62) v AO (52)
FC 76. Pol that are historically Western, rt having been significantly or partially Westernized through a colonial relationship (26) v CONV (70) [PO]
FC 77. Pol that have been significantly Westernized, rt partially Westernized, through a colonial relationship (28) v CONV (41) [PO]
FC 78. Pol that have been significantly Westernized through a colonial relationship, rt without such a relationship (28) v CONV (7)

---

21. FORMER COLONIAL RULER

A. Britain [24]
B. France [25]
G. Spain [18]

H. Other [8]

n = 75
punch 4 = 40

In establishing this raw characteristic, we have restricted ourselves to instances of Western colonialism, including mandates and protectorates. Polities with records of other kinds of dependency (for example, as components of the Ottoman Empire), or with no record of dependency, are excluded as irrelevant (punch 4).

In cases of successive dependency, the more recent is utilized for coding purposes. Thus, the Philippines is regarded as a former dependency of the United States, rather than of Spain.

We have coded Cameroun and Morocco under B and Somalia under H. In all three cases, substantially less than half of the present polity's territory was held by a second colonial power: the southern portion of former British Cameroons, Spanish Morocco, British Somaliland.

The eight polities in the "other" category were formerly dependencies of Belgium (Burundi, Congo [Leo], Rwanda), Italy (Libya, Somalia), the Netherlands (Indonesia), Portugal (Brazil), and the United States (Philippines).

In contrasting ex-British with ex-French dependencies, it is somewhat more meaningful if one concentrates on those countries that have become independent since 1914. Hence FC 80 is composite in nature.

It would perhaps have been equally desirable to have made FC 79 a composite characteristic with a cutoff point of, say, 1800. But the only country affected as a result would have been the United States.

FC 79. Pol that were formerly dependencies of Britain or France, rt Spain (49) v CONV (18)

FC 80. Pol whose date of independence is after 1914, and that were formerly dependencies of Britain, rt France (19) v Pol whose date of independence is after 1914, and that were formerly dependencies of France, rt Britain (24)

---

22. POLITICAL MODERNIZATION: HISTORICAL TYPE

A. Early European or early European derived (early modernizing European society or offshoot) [11]

B. Later European or later European derived (later modernizing European society or offshoot) [40]

G. Non-European autochthonous (self-modernizing extra-European society) [9]

H. Developed tutelary (developed society modernizing under tutelage) [31]
I. Undeveloped tutelary (undeveloped society modernizing under tutelage) [24]

n = 115

---

23. POLITICAL MODERNIZATION: PERIODIZATION

A. Advanced (transitional phase completed) [60]
B. Mid-transitional (entered transitional phase prior to 1945) [16]
G. Early transitional (entered transitional phase 1945 or later) [38]
H. Pre-transitional (not yet entered transitional phase) [1]

n = 115

These two raw characteristics are adapted, without substantial change, from current research by C. E. Black on political modernization.[30] Black's material is particularly useful because it focuses sharply on leadership as a factor in modernization. In establishing his historical typology, Black is interested in "the characteristic problems that modernizing leaders have faced in gaining political power and in implementing their programs." As regards periodization, he is concerned with "the principal phases in the transfer of political power from traditional to modernizing leaders."[31]

As Black puts it, "political leaders in the first societies to modernize had to find solution to their problems without the assistance of outside models, but they were favored by an essential continuity of political and territorial structure. The second type embraces those societies where the seizure of power by modernizing leaders involved a fundamental re-organization of political structure and of territory. In these cases the establishment of national identity was a predominant issue for several generations, relegating to a secondary place the intellectual, economic, and social aspects of modernization. The third type comprises those later-modernizing societies which, by virtue of a long tradition of effective central government, never came under the direct rule of more modern societies. In these societies the struggle between traditional and modernizing leaders took place within the established framework of territory and institutions, but under great pressure from the example of the more modern societies. The fourth and fifth types are those in which the

[30] C. E. Black, *Modernizing Societies: Typology and Periodization,* unpublished mimeograph draft, February 2, 1963. See also Black's earlier paper, *Political Modernization in Historical Perspective,* prepared for the Conference on Political Modernization sponsored by the Committee on Comparative Politics, Social Science Research Council, Dobbs Ferry, N. Y., June 8–11, 1959. We are indebted to Professor Black for his courtesy in permitting this use of his material prior to its publication.

[31] *Modernizing Societies,* p. 1.

direct tutelage of a more modern society was the principal factor in political modernization. Where the traditional institutions of a society were well developed, modernization resulted from an interaction between these institutions and those of a tutelary society. Where the societies under tutelage did not have well-developed institutions, the influence of the tutelary society was correspondingly much greater."[32]

With respect to the overseas territories of the European colonial powers, Black distinguishes between "the 'offshoots' of metropolitan societies, and the other societies which came under their tutelage. Offshoots are those societies where immigrants from the metropolitan society form the majority or a dominant element of the population, bringing with them their traditional ideas and institutions. . . . In the case of the societies under tutelage, the influence of the tutelary society may be very great but it is the native ideas and institutions which remain the bases from which the modern society evolves."[33]

Black divides the modernization process into three phases. "The initial phase is that in which modern ideas first begin to have an impact on a traditional society, without as yet causing a profound upheaval."[34]

Black's "transitional phase" is "the phase of active struggle," consisting of "three essential features. . . . The first is the assertion on the part of political leaders of the determination to modernize. This assertion may take the form of a revolution by disaffected members of the traditional oligarchy, or by modernizing leaders representing new political interests. Occasionally the traditional oligarchy itself initiates the process. . . . The second feature is an effective and decisive break with the institutions associated with a predominantly agrarian way of life, permitting the transition to an industrial way of life. . . . Finally, the creation of a politically organized society in those cases where one did not exist in the initial phase is also essential."[35]

Black's "advanced phase" is that in which "all the principal groups of the elite are agreed that modernization is desirable, and the political struggle is now engaged between the supporters of rival programs of modernization rather than between modernizers and traditionalists. Whereas political struggles in the intermediate phase are concerned with a fundamental social revolution, they now (as in the initial phase) take the form rather of rebellions or *coups d'état.* Nationalism and liberalism, which were revolutionary forces in the intermediate phase . . . now become conservative forces in that their main purpose is to preserve the newly-won order."[36]

[32] *Ibid.*, p. 2.
[33] *Ibid.*, pp. 2–3.
[34] *Ibid.*, p. 10.
[35] *Ibid.*, pp. 11–12.
[36] *Ibid.*, p. 12.

The first of Black's raw characteristics is essentially typological but also has an ordinal component. For this reason, A and B are collectively contrasted with H and I, while G is the only attribute contrasted with "all other." We have also provided the *adjacent* contrasts of A v B and H v I.

The periodization characteristic is essentially ordinal in nature, and this is reflected in the structure of the finished characteristics. We have excluded "pre-transitional" (Black's "initial phase"), since Sa'udi Arabia is the only independent polity covered by this attribute.

FC 81. Pol where the type of political modernization is early European or early European derived, rt later European or later European derived (11) v CONV (40)
FC 82. Pol where the type of political modernization is early or later European or European derived, rt developed tutelary or undeveloped tutelary (51) v CONV (55)
FC 83. Pol where the type of political modernization is non-European autochthonous (9) v AO (106)
FC 84. Pol where the type of political modernization is developed tutelary, rt undeveloped tutelary (31) v CONV (24)
FC 85. Pol where the stage of political modernization is advanced, rt mid- or early transitional (60) v CONV (54)
FC 86. Pol where the stage of political modernization is advanced or mid-transitional, rt early transitional (76) v CONV (38)

---

24. IDEOLOGICAL ORIENTATION

A. Doctrinal [13]
B. Developmental [31]
G. Situational [5]
H. Conventional [33]
I. Traditional [5]

n = 87
punch 3 = 25
punch 6 = 3

It is exceedingly difficult to construct a workable ideological typology for coding polities on a global basis. Since only a totalitarian regime is capable of committing the nation as a whole to a single doctrine, it was necessary to establish the present raw characteristic on the basis of over-all "ideological orientation," with specific "doctrinal" commitment as only one of the five attributes.

Formal doctrine, however, is presently the major component of ideological orientation only for the Communist states. For this reason, we have not established a separate finished characteristic, "doctrinal v AO," because the reader need simply consult FC 194 (Communist v Non-Communist).

"Developmental" seems to us to be a more meaningful attribute-class than the much broader category "nationalistic." All modern states exhibit some degree of nationalist fervor. *Not* all states, however, are equally committed to development as a national goal. Yet, as an inspection of the public utterances of virtually any leader of the "underdeveloped" world will quickly reveal, dedication to developmental objectives tends to transcend all other forms of ideological commitment.

A "situational" outlook necessarily entails a rejection of ideology. But the very negation of commitment involves an orientation *toward* ideology that is, we feel, distinctly relevant to the present characteristic.

The term "conventional" is taken from Leonard Binder's recent book on Iran.[37] Binder's highly stimulating schema of "traditional," "conventional," and "rational" system types could not be directly incorporated into the present study because of its "ideal-type" character. Relatively few polities can actually be *coded* on the basis of this trichotomy without the creation of an excessive number of composite or "hybrid" attribute classes.

Unlike Binder, we do not necessarily regard a conventional system as the equivalent of a "working constitutional democracy." Thus we regard South Africa, for example, as relying on more or less "conventionalized procedures for achieving the legitimization of new or changed power relationships,"[38] even though *access* to the conventions in question is effectively denied to a majority of the population.

Three polities (South Korea, Peru, Yemen) are coded as currently unascertainable (punch 6) for this characteristic. We have coded 25 polities as ambiguous (punch 3) on the ground that, in each instance, no single ideological orientation *clearly* predominates over all others. We regard Cuba, for example, as being "doctrinal-developmental"; India and Jamaica as being "developmental-conventional"; Cambodia, Libya, and Uganda as being "developmental-traditional"; and so forth.

FC 87. Pol whose ideological orientation is developmental (31) v AO (58)
FC 88. Pol whose ideological orientation is situational (5) v AO (95) [SO]
FC 89. Pol whose ideological orientation is conventional (33) v AO (62)
FC 90. Pol whose ideological orientation is traditional (5) v AO (102) [SO]

[37] Leonard Binder, *Iran: Political Development in a Changing Society*. University of California Press, 1962, pp. 40, ff.

[38] *Ibid.*, p. 41

FC 91. Pol whose ideological orientation is developmental, rt traditional (31) v CONV (5)

---

25. SYSTEM STYLE

A. Mobilizational [20]
B. Limited mobilizational [12]
C. Non-mobilizational [78]

n = 110
punch 3 = 1
punch 6 = 4

In this raw characteristic we attempt to discriminate between those polities that have mobilized or partially mobilized their resources (human and otherwise) to meet what are perceived to be compelling problems of national urgency, and those that have not. It is, of course, important that "mobilization" be interpreted here in a *political,* rather than in a military sense. Communist nations are, by doctrine, committed to a mobilizational style, but are by no means the only nations so committed. It should also be noted that a "developmental" ideological commitment is by no means invariably supported by a "mobilizational" system style.

We have coded Spain, whose system style might be termed "decayed mobilizational," as ambiguous (punch 3).

South Korea, Nepal, Pakistan, and Yemen have been coded as currently unascertainable (punch 6). Nepal may here appear to be in rather strange company, but recent events seem to suggest that efforts are being made to "mobilize" the socio-political structure of that Himalayan state, albeit along avowedly traditionalist lines.[39]

FC 92. Pol where the system style is mobilizational (20) v Pol where the system style is limited mobilizational, or non-mobilizational (93)

FC 93. Pol where the system style is mobilizational, or limited mobilizational (32) v Pol where the system style is non-mobilizational (78)

---

[39] See, for example, the comments of Leo E. Rose in "The Himalayan Border States: 'Buffers' in Transition," *Asian Survey,* Vol. III, No. 2, February 1963.

26. CONSTITUTIONAL STATUS OF PRESENT REGIME

A. Constitutional (government conducted with reference to recognized constitutional norms) [51]
B. Authoritarian (no effective constitutional limitation, or fairly regular recourse to extra-constitutional powers. Arbitrary exercise of power confined largely to the political sector) [23]
G. Totalitarian (no effective constitutional limitation. Broad exercise of power by the regime in both political and social spheres) [16]

n = 90
punch 3 = 5
punch 5 = 11
punch 6 = 9

The more or less value-laden distinctions of Western political science between constitutional, authoritarian, and totalitarian regimes are premised on the ethical concept of the moral worth of the individual—a concept often located in somewhat uneasy juxtaposition with the principle of majority rule. Constitutionalism is, in part at least, regarded as a defense of individual rights. As a practical matter, the limits of these rights are established on a consensual basis as formalized by some variation of majoritarianism in which the vote of each citizen counts equally. The implicit paradox between "right" and "majoritarianism" is of ancient vintage and need not detain us. It is important to note, however, that in many nations of the contemporary world, the concept of individual rights, *per se,* tends to be downgraded, while the notion of consensus tends to be magnified—but without much emphasis on majoritarianism. This means that it is extremely difficult, particularly with respect to many of the new African states, to differentiate with any degree of confidence between "constitutional" and "authoritarian" regimes. This is one reason, among several, explaining the relatively high residual count for the present raw characteristic.

While Argentina, the Ivory Coast, and Haiti are here coded as ambiguous (punch 3), it should be pointed out that the first two are certainly not totalitarian, and that Haiti is certainly not constitutional.

FC 94. Pol where the status of the regime is constitutional (51) v Pol where the status of the regime is authoritarian or totalitarian (41)
FC 95. Pol where the status of the regime is constitutional or authoritarian (95) v Pol where the status of the regime is totalitarian (16)
FC 96. Pol where the status of the regime is constitutional or totalitarian (67) v Pol where the status of the regime is authoritarian (23)
FC 97. Pol where the status of the regime is authoritarian, rt totalitarian (23) v CONV (16)

27. GOVERNMENTAL STABILITY

A. Government generally stable since World War I or major inter-war constitutional change [22]
B. Government generally stable since World War II or major post-war constitutional change [28]
G. Government moderately stable since World War II or major post-war constitutional change [11]
H. Government unstable since World War II or major post-war constitutional change [22]

n = 83
punch 3 = 3
punch 6 = 29

There are no convenient rules of thumb for evaluating governmental stability. Instability is obviously present when a constitution is suspended or when an elected president is forcibly ousted from office. And yet major changes of this sort may have a purgative effect and be followed by periods of relative stability. Even more difficult to assess is the significance of frequent cabinet changes under a parliamentary system. The coder must rely primarily on his own judgment in determining the point at which cabinet changes become so frequent as to justify coding a particular polity as "unstable." This raw characteristic constitutes an illustration, par excellence, of the importance of the "relativistic" approach to coding (pages 8–9).

We have regarded differentiation between B and G as unascertainable (punch 6) when the major constitutional change in question occurred after 1 January 1959. This entails, of course, the disadvantage that most of the twenty-odd states (primarily African) that achieved independence after this date are excluded. The disadvantage is somewhat mitigated, however, in that those newly independent polities evidencing obvious *in*stability since independence are coded under H.

Haiti, Hungary, and Paraguay are coded as ambiguous (punch 3). Haiti has experienced no change of regime since the advent of President Duvalier in 1957, but the current situation could hardly be described as tranquil. Hungary is rendered ambiguous by the revolutionary events of 1956. Paraguay's apparent stability has been of a peculiarly enforced nature since the installation of General Stroessner's dictatorship in 1954.

FC 98. Pol where governmental stability is generally present and dates at least from the inter-war period (22) v Pol where governmental stability is generally or moderately present and dates from the post-war period, or is absent (93)

FC 99. Pol where governmental stability is generally present and dates from at least the inter-war period, or from the post-war period

(50) v Pol where governmental stability is moderately present and dates from the post-war period, or is absent (36)

FC 100. Pol where governmental stability is generally present and dates from at least the inter-war period, or is generally or moderately present, and dates from the post-war period (64) v Pol where governmental stability is absent (22)

---

28. REPRESENTATIVE CHARACTER OF CURRENT REGIME

A. Polyarchic (broadly representative system) [41]
B. Limited polyarchic (mass-sector representative or broadly oligarchic system) [8]
G. Pseudo-polyarchic (ineffective representative, or disguised oligarchic or autocratic system) [43]
H. Non-polyarchic (non-representative in form as well as content) [6]

n = 98
punch 3 = 7
punch 5 = 2
punch 6 = 8

"Limited polyarchic" is used here to include situations where a relatively broad-based, but nonetheless discriminatory, franchise prevails. This might take the form of a literacy requirement in a nation where illiteracy is widespread, as in the case of Brazil. Or it might take the form of racial or regional disqualification, and so on.

"Pseudo-polyarchic" includes polities where single-list elections prevail, or where elected representatives have only limited or consultative functions to perform.

Seven polities (Argentina, Cameroun, Lebanon, Libya, Nigeria, Somalia, Uganda) are coded as ambiguous (punch 3). Lebanon has an ostensibly polyarchic system, but the communal distribution of seats has been "frozen" on the basis of the 1942 census, with no official census subsequent to that date. Nigeria and Uganda have mixed mass representative and oligarchic systems. Libya's government is only partially representative.

Burundi and Rwanda have been classified as unascertained (punch 5). Congo (Leo), Guatemala, Iraq, South Korea, Peru, Syria, Togo, and Yemen are regarded as unascertainable (punch 6) under present conditions.

FC 101. Pol where the representative character of the regime is polyarchic (41) v Pol where the representative character of the regime is limited polyarchic, pseudo-polyarchic, or non-polyarchic (57) [PO]

FC 102. Pol where the representative character of the regime is polyarchic or limited polyarchic (59) v Pol where the representative character of the regime is pseudo-polyarchic or non-polyarchic (49) [PO]

FC 103. Pol where the representative character of the regime is pseudo-polyarchic, rt non-polyarchic (43) v CONV (6) [SO]

---

29. CURRENT ELECTORAL SYSTEM

A. Competitive (no party ban, or ban on extremist or extra-constitutional parties only) [43]
B. Partially competitive (one party with 85% or more of legislative seats) [9]
G. Non-competitive (single-list voting or no elected opposition) [30]

n = 82
punch 3 = 11
punch 4 = 1
punch 6 = 21

Many of the newly independent African states that exhibited competitive or partially competitive electoral systems shortly after independence have recently been moving, at varying rates, toward the adoption of one-party or fusional single-list systems. Current practices vary all the way from the arresting of particular opposition leaders to the wholesale banning of opposition parties. In only a few African polities does there seem to be much doubt that this process will continue.

Of the countries classified as ambiguous for this characteristic, Afghanistan, Ethiopia, Jordan, Laos, and Libya have only partially elected legislatures. Colombia, Cyprus, and Lebanon each have "parity" formulas for the distribution of legislative seats that act to limit the competitive principle, albeit in the avowed interest of more genuinely "representative" government.

FC 104. Pol where the electoral system is competitive, rt non-competitive (43) v CONV (30) [SO]

FC 105. Pol where the electoral system is competitive (43) v Pol where the electoral system is partially competitive or non-competitive (47) [PO]

FC 106. Pol where the electoral system is competitive or partially competitive (52) v Pol where the electoral system is non-competitive (30) [PO]

---

30. FREEDOM OF GROUP OPPOSITION

A. Autonomous groups free to enter politics and able to oppose government (save for extremist groups, where banned) [46]
B. Autonomous groups free to organize in politics, but limited in capacity to oppose government (includes absorption of actual or potential opposition leadership into government) [18]
G. Autonomous groups tolerated informally and outside politics [24]
H. No genuinely autonomous groups tolerated [11]

n = 99
punch 3 = 3
punch 6 = 13

Virtually any restraint on political activity by opposition or potential opposition groups is likely to be justified by the government concerned as being "in the national interest." For this reason, it is often as difficult to distinguish "freedom" from "license" in the case of groups, as it is in the case of individuals. It is with some hesitancy, therefore, that we code a country such as Argentina, with its "Peronista" problem, as A for this characteristic. Other countries present similar difficulties. Guatemala, for example, is currently a nation of essentially "centrist" political groupings, both extremes of the political spectrum being more or less repressed.

Despite difficulties in coding, only three polities were regarded as genuinely ambiguous (punch 3) for this characteristic: Jordan, Laos, and South Africa. The regime in Jordan has since 1957 vacillated considerably with respect to the degree of freedom it has been willing to accord political groups, particularly political parties.

FC 107. Pol where autonomous groups are fully tolerated in politics (46) v Pol where autonomous groups are partially tolerated in politics, are tolerated only outside politics, or are not tolerated at all (65) [PO]
FC 108. Pol where autonomous groups are fully or partially tolerated in politics (65) v Pol where autonomous groups are tolerated only outside politics or are not tolerated at all (35) [PO]

---

31. POLITICAL ENCULTURATION

A. High (integrated and homogeneous polity with little or no extreme opposition, communalism, fractionalism, disenfranchisement, or political non-assimilation) [15]
B. Medium (less fully integrated polity with significant minority in extreme opposition, communalized, fractionalized, disenfranchised, or politically non-assimilated) [38]

G. Low (relatively non-integrated or restrictive polity with majority or near majority in extreme opposition, communalized, fractionalized, disenfranchised, or politically non-assimilated) [42]

n = 95
punch 3 = 2
punch 5 = 18

In including the raw characteristic, "political enculturation," in the present survey, we are again going a little out of our way to include a concept and a label that are currently the subject of much interest among American political scientists. Unfortunately, much theoretical work remains to be done toward the refining of this concept. We have therefore decided to construe it in the broadest possible sense. A possible weakness of such a freewheeling approach is that it permits no distinction between instances of conflict and of non-assimilation, respectively, within a body politic.

Precise assignment for most Communist states is necessarily unascertained (punch 5) for this characteristic. Switzerland and Uruguay have been coded as ambiguous (punch 3).

FC 109. Pol where political enculturation is high, rt low (15) v CONV (42) [SO]
FC 110. Pol where political enculturation is high (15) v Pol where political enculturation is medium or low (80) [PO]
FC 111. Pol where political enculturation is high or medium (53) v Pol where political enculturation is low (42) [PO]

---

32. SECTIONALISM

A. Extreme (one or more groups with extreme sectional feeling) [27]
B. Moderate (one group with strong sectional feeling or several with moderate sectional feeling) [34]
G. Negligible (no significant sectional feeling) [47]

n = 108
punch 3 = 3
punch 5 = 4

Sectionalism is here defined as the phenomenon in which a significant percentage of the population of a nation lives in a sizable geographic area and indentifies self-consciously and distinctively with that area to a degree that the cohesion of the polity as a whole is appreciably challenged or impaired. Sectionalism may or may not be characterized or reinforced by communalism or tribalism. It is never to be equated with localism in the sense of a group's identifying with a very small locality. Nor should the existence of a sectionally confined non-assimilated group necessarily be regarded as "sectionalism," *per se*. Liberia, for example, is coded under

B because of what may be described as moderate sectionalism on the part of the coastal population, rather than of the less assimilated interior population.

Three countries (Argentina, Mexico, Uruguay) are coded as ambiguous (punch 3) for this characteristic. In both Argentina and Uruguay there is a distinct political cleavage between capital and hinterland, but there is some question whether this may not be simple ruralism versus urbanism, rather than a matter of sectional geography. In the case of Mexico, *"localismo"* is a widespread and often politically articulated phenomenon, but, as stated earlier, this does not in itself constitute "sectionalism" in our sense. It need hardly be noted that a polity lacking sectionalism is not necessarily politically "homogeneous."

FC 112. Pol where sectionalism is extreme, rt negligible (27) v CONV (47) [SO]
FC 113. Pol where sectionalism is extreme (27) v Pol where sectionalism is moderate or negligible (81) [PO]
FC 114. Pol where sectionalism is extreme or moderate (61) v Pol where sectionalism is negligible (47) [PO]

---

33. INTEREST ARTICULATION BY ASSOCIATIONAL GROUPS

A. Significant [20]
B. Moderate [12]
G. Limited [28]
H. Negligible [51]

n = 111
punch 3 = 3
punch 6 = 1

This raw characteristic and the three that follow are based on the interest articulation schema set forth in *The Politics of the Developing Areas.* Almond and Coleman describe associational interest groups as "the specialized structures of interest articulation—trade unions, organizations of businessmen or industrialists, ethnic associations [but not the ethnic groups themselves], associations organized by religious denominations [but not the churches themselves], civic groups, and the like. Their characteristics are explicit representation of the interests of a particular group, orderly procedures for the formulation of interests and demands, and transmission of these demands to other political structures such as political parties, legislatures, bureaucracies."[40]

Portugal, Spain, and Yugoslavia are regarded as ambiguous (punch 3)

[40]*PDA*, p. 34.

for this characteristic. All three have attempted to eliminate "boundary maintenance between the polity and the society"[41] by assigning formal governmental functions to groups which in most societies serve as informal transmission belts from the private sector to the public sector. Spain and Portugal formally embrace "corporativism"; Yugoslavia does so in all but name in its legislative upper house (Council of Producers) and in other official organs.[42]

South Korea is regarded as unascertainable (punch 6) for this characteristic under present circumstances.

FC 115. Pol where interest articulation by associational groups is significant (20) v Pol where interest articulation by associational groups is moderate, limited, or negligible (91) [PO]
FC 116. Pol where interest articulation by associational groups is significant or moderate (32) v Pol where interest articulation by associational groups is limited or negligible (79)
FC 117. Pol where interest articulation by associational groups is significant, moderate, or limited (60) v Pol where interest articulation by associational groups is negligible (51) [PO]

---

34. INTEREST ARTICULATION BY INSTITUTIONAL GROUPS

A. Very significant [40]
B. Significant [34]
G. Moderate [16]
H. Limited [10]

n = 100
punch 6 = 15

In their discussion of institutional interest groups, Almond and Coleman state that "we have in mind phenomena occurring within such organizations as legislatures, political executives, armies, bureaucracies, churches, and the like. These are organizations which perform other social or political functions but which, as corporate bodies or through groups within them (such as legislative blocs, officer cliques, higher or lower clergy or religious orders, departments, skill groups, and ideological cliques in bureaucracies), may articulate their own interests or represent the interests of groups in the society."[43]

[41] *Ibid.*, p. 35.

[42] For an interesting attempt to relate Yugoslavian interest groups to David B. Truman's classificatory scheme (in *The Governmental Process*. Alfred A. Knopf, 1951), see Jovan Djordjevik, "Interest Groups and the Political System of Yugoslavia," in Henry W. Ehrmann, ed., *Interest Groups on Four Continents*. University of Pittsburgh Press, 1958, pp. 197 ff.

[43] *PDA*, p. 33.

For present purposes we go somewhat beyond the apparent scope of the foregoing quotation by regarding totalitarian parties as institutional groups. Thus, all Communist-bloc polities are classified under A.

The term "institutional group" is possibly the most broad-gauged term found in the Almond-Coleman interest-group formulations. More sophisticated use of this concept as a research tool will probably require comparative analysis of interest articulation by *specific kinds* of institutional groups. Indeed, an initial step of this sort has already been taken by a group of scholars (some of whom contributed to the Almond-Coleman volume) investigating the role of the military in certain underdeveloped areas.[44] Our own interest in this subject is reflected in FCs 54 and 55, which examine the role of the military and the police in politics.

It will be noted that the phrasing of the attributes for the present raw characteristic differs somewhat from that used in RC 33, in that the category "negligible" is not here used. The reason for this omission is that *any* polity among the 115 is bound to exhibit a congeries of institutional groups, at least some of which will possess "limited" capacity to articulate interests. By contrast, in some of our 115 polities, associational group structures are virtually absent, and possess at best only "negligible" capacity to articulate interests.

Most sub-Saharan African states are regarded as currently unascertainable (punch 6) for this characteristic.[45]

FC 118. Pol where interest articulation by institutional groups is very significant (40) v Pol where interest articulation by institutional groups is significant, moderate, or limited (60) [PO]

FC 119. Pol where interest articulation by institutional groups is very significant or significant (74) v Pol where interest articulation by institutional groups is moderate or limited (26)

FC 120. Pol where interest articulation by institutional groups is very significant, significant, or moderate (90) v Pol where interest articulation by institutional groups is limited (10) [PO]

---

35. INTEREST ARTICULATION BY NON-ASSOCIATIONAL GROUPS

A. Significant [54]
B. Moderate [29]
G. Limited [24]
H. Negligible [8]

n = 115

[44] John J. Johnson, ed., *The Role of the Military in Underdeveloped Countries*. Princeton University Press, 1962.

[45] See, in this regard, James S. Coleman's comments in *PDA*, pp. 313 ff.

Almond and Coleman describe non-associational groups as embracing "kinship and lineage groups, ethnic, regional, religious, status, and class groups which articulate interests informally and intermittently, through individuals, cliques, family and religious heads, and the like."[46]

In some respects, the term "non-associational group" is an unfortunate one, suggesting to the unwary reader that it includes all groups not classifiable as "associational." The particular set of non-"associational" groups actually referred to might more accurately be designated as "ascriptive," in that one usually becomes a member by birth, rather than by conscious choice or deliberate achievement. However, since the term "non-associational" seems now to have entered the political science lexicon, we retain it here.

As in the case of "institutional group," the "non-associational" category is probably too comprehensive for sophisticated analytic purposes. At the very least, it would seem desirable to be able to differentiate between what are usually called "primary group" structures, as contrasted with other, distinctly different group structures now subsumed under the "non-associational" rubric.

FC 121. Pol where interest articulation by non-associational groups is significant (54) v Pol where interest articulation by non-associational groups is moderate, limited, or negligible (61) [PO]

FC 122. Pol where interest articulation by non-associational groups is significant or moderate (83) v Pol where interest articulation by non-associational groups is limited or negligible (32)

FC 123. Pol where interest articulation by non-associational groups is significant, moderate, or limited (107) v Pol where interest articulation by non-associational groups is negligible (8) [PO]

---

36. INTEREST ARTICULATION BY ANOMIC GROUPS

A. Frequent [14]
B. Occasional [50]
G. Infrequent [19]
H. Very infrequent [16]

n = 99
punch 3 = 6
punch 5 = 6
punch 6 = 4

Almond and Coleman define anomic interest groups as "more or less spontaneous breakthroughs into the political system from the society,

[46] *PDA*, p. 33.

such as riots and demonstrations. Their distinguishing characteristic is their relative structural and functional lability."[47]

We have coded six polities (Argentina, Cyprus, France, Hungary, the United States, Venezuela) as ambiguous (punch 3); six polities (Albania, Bulgaria, North Korea, Mongolia, Rumania, North Vietnam) as unascertained (punch 5); and four polities (Algeria, Cuba, Dominican Republic, Yemen) as currently unascertainable (punch 6).

FC 124. Pol where interest articulation by anomic groups is frequent (14) v Pol where interest articulation by anomic groups is occasional, infrequent, or very infrequent (85) [PO]
FC 125. Pol where interest articulation by anomic groups is frequent or occasional (64) v Pol where interest articulation by anomic groups is infrequent or very infrequent (35)
FC 126. Pol where interest articulation by anomic groups is frequent, occasional, or infrequent (83) v Pol where interest articulation by anomic groups is very infrequent (16) [PO]

---

37. INTEREST ARTICULATION BY POLITICAL PARTIES

A. Significant [21]
B. Moderate [27]
G. Limited [8]
H. Negligible [37]

n = 93
punch 3 = 5
punch 4 = 5
punch 6 = 12

With the present raw characteristic we depart somewhat from the Almond-Coleman schema as strictly interpreted. According to Almond, the "modern" (by which he means the modern-*Western*) party performs an essentially aggregative, rather than an articulative, function. He readily admits circumstances under which the aggregative function, even under modern conditions, may not be performed, particularly where "interest groups control parties."[48] But he does not refer specifically to interest articulation *by* parties. We regard the latter as a meaningful category of analysis, however pathological a condition it may represent to those who prefer thinking in terms of a "pure" party function. Even if no articulation were to occur without the party being the "captive" of an interest group, it would nonetheless be extremely useful to be able to distinguish between

[47] *PDA*, p. 34.

[48] *PDA*, p. 38, quoting Almond, "A Comparative Study of Interest Groups and the Political Process," *American Political Science Review*, Vol. LII, No. 1, March 1958, p. 276.

cases of this type and cases where no "captive" party served as an intervening mechanism.

It may be mentioned in passing that a more or less typical "modern" political system may well have one or more "aggregative" parties in addition to minor parties not only serving interest groups but serving *as* interest groups (for example, the Prohibition Party in the United States).

Five polities (Cyprus, Indonesia, Laos, South Africa, Uruguay) are coded as ambiguous (punch 3) for this characteristic. We have coded as "irrelevant" five polities where no parties currently exist or are likely to exist within the foreseeable future: Afghanistan, Ethiopia, Haiti, Libya, and Sa'udi Arabia. The twelve polities coded as presently unascertainable (punch 6) are Algeria, Burma, Cuba, Iran, Iraq, Jordan, South Korea, Nepal, Pakistan, Sudan, Thailand, and Yemen.

FC 127. Pol where interest articulation by political parties is significant (21) v Pol where interest articulation by political parties is moderate, limited, or negligible (72) [PO]

FC 128. Pol where interest articulation by political parties is significant or moderate (48) v Pol where interest articulation by political parties is limited or negligible (45)

FC 129. Pol where interest articulation by political parties is significant, moderate, or limited (56) v Pol where interest articulation by political parties is negligible (37) [PO]

---

38. INTEREST AGGREGATION BY POLITICAL PARTIES

A. Significant [12]
B. Moderate [18]
G. Limited [19]
H. Negligible [9]

n = 58
punch 3 = 7
punch 4 = 5
punch 5 = 29
punch 6 = 16

In this age of mass parties, it is in many cases extremely difficult to determine, or even to estimate, the extent of interest aggregation by political parties. This is almost universally true with respect to one-party or one party dominant systems. Many of these systems are of quite recent vintage and cannot yet be said to have acquired either internal stability or equilibrium within the broader political complex. All of this has made coding for this raw characteristic particularly difficult.

There are some scholars who maintain that true interest aggregation cannot occur in a one-party context. They argue that the basic incentive for interest aggregation derives from inter-party competition, preferably of a bi-polar character. We cannot agree. There are examples of one-party aggregation. Mexico is perhaps the best contemporary example.[49]

It is our contention that what may be called the breadth of the interest spectrum is of major importance in determining whether a one-party system can successfully aggregate interests. If the spectrum in a particular polity is relatively narrow (by which we mean relatively non-diverse), and particularly if the polity has only recently attained independence, then a one-party system may aggregate just as successfully, or more so, than a two- or multi-party system. In other words, a one-party system for, say, Senegal, may aggregate existing interests with some success, while it could not hope to do so in the United States[50] or in the Soviet Union. There is, of course, no assurance that even under "narrow spectrum" conditions a one-party system *will* perform an aggregative function.

The foregoing speculations are, however, highly tentative. We admit to a complete inability to code 29 polities for the present characteristic. Over half of these are located in sub-Saharan Africa. Also unascertained are the totalitarian systems. The reader is cautioned to bear these biasing factors in mind when reading Paragraphs 130, 131, and 132 in the printout.

Seven polities are coded as ambiguous (punch 3): Argentina, Colombia, Cyprus, France, Greece, Indonesia, and South Africa. The five polities coded as irrelevant for the preceding characteristic are similarly coded here. Sixteen polities are coded as currently unascertainable (punch 6).

FC 130. Pol where interest aggregation by political parties is significant (12) v Pol where interest aggregation by political parties is moderate, limited, or negligible (71) [PO]

FC 131. Pol where interest aggregation by political parties is significant or moderate (30) v Pol where interest aggregation by political parties is limited or negligible (35)

FC 132. Pol where interest aggregation by political parties is significant, moderate, or limited (67) v Pol where interest aggregation by political parties is negligible (9) [PO]

---

49 See, in this connection, Robert E. Scott, *Mexican Government in Transition*. University of Illinois Press, 1959, ch. 6.

50 Even this observation may have to be qualified. There is considerable evidence that interests are successfully and continuously aggregated in many of the traditional "one-party" areas of the American South.

39. INTEREST AGGREGATION BY EXECUTIVE

A. Significant [29]
B. Moderate [28]
G. Limited [18]
H. Negligible [13]

n = 88
punch 3 = 5
punch 5 = 11
punch 6 = 11

There are at least two factors that must be weighed in estimating degree of interest aggregation by the executive: (1) whether alternative mechanisms (party, legislature, bureaucracy, and so forth) are available and functioning components in the aggregative process, and (2) the extent to which the executive facilitates or impedes aggregative activity by these other organs. Needless to say, estimates made on the basis of such intangible criteria must necessarily be highly intuitive, and the most we can claim for codings under this characteristic is that they have been conscientiously entered.

Five polities are coded as ambiguous (punch 3): Cyprus, Laos, Lebanon, South Africa, and Uruguay. All of these except South Africa have distinctly atypical executive structures. In the case of South Africa it is a fair question as to *whose* interests are being aggregated.

All of the polities coded as unascertained are Communist-bloc nations. Coded as unascertainable (punch 6) are Algeria, Burma, Congo (Leo), Iraq, South Korea, Pakistan, Peru, Sudan, Syria, Togo, and Yemen.

FC 133. Pol where interest aggregation by the executive is significant (29) v Pol where interest aggregation by the executive is moderate, limited, or negligible (76) [PO]
FC 134. Pol where interest aggregation by the executive is significant or moderate (57) v Pol where interest aggregation by the executive is limited or negligible (46)
FC 135. Pol where interest aggregation by the executive is significant, moderate, or limited (77) v Pol where interest aggregation by the executive is negligible (13) [PO]

---

40. INTEREST AGGREGATION BY LEGISLATURE

A. Significant [12]
B. Moderate [17]
G. Limited [16]
H. Negligible [49]

n = 94
punch 3 = 6
punch 4 = 1
punch 6 = 14

Interest aggregation by the legislature is most significant in the so-called "working multi-party" context, where parliamentary accommodation by the parties concerned is an indispensible requisite of reasonably stable government.[51] At the other extreme, of course, are the rubber-stamp legislatures of the totalitarian countries and the essentially "consultative" legislatures of the traditionalist regimes.

Six polities are coded as ambiguous (punch 3): Colombia, Cyprus, Indonesia, Portugal, South Africa, and Spain. Sa'udi Arabia, with no legislature and no foreseeable prospect of acquiring one, is coded as irrelevant. Algeria, Burma, Cuba, Dominican Republic, Iran, Iraq, South Korea, Morocco, Pakistan, Peru, Sudan, Syria, UAR, and Yemen are regarded as currently unascertainable (punch 6).

FC 136. Pol where interest aggregation by the legislature is significant (12) v Pol where interest aggregation by the legislature is moderate, limited, or negligible (85) [PO]
FC 137. Pol where interest aggregation by the legislature is significant or moderate (29) v Pol where interest aggregation by the legislature is limited or negligible (68)
FC 138. Pol where interest aggregation by the legislature is significant, moderate, or limited (48) v Pol where interest aggregation by the legislature is negligible (49) [PO]

---

41. PARTY SYSTEM: QUANTITATIVE

A. One-party (all others non-existent, banned, non-participant, or adjuncts of dominant party in electoral activity. Includes "national fronts" and one-party fusional systems) [34]
B. One party dominant (opposition, but numerically ineffective at national level. Includes minority participation in government while retaining party identity for electoral purposes) [13]
G. One-and-a-half-party (opposition significant, but unable to win majority) [3]
H. Two-party or effectively two-party (reasonable expectation of party rotation) [11]
I. Multi-party (coalition or minority party government normally mandatory if parliamentary system) [30]

[51] See, in this connection, Dankwart A. Rustow, "Scandinavia: Working Multiparty Systems" and also Felix E. Oppenheim, "Belgium: Party Cleavage and Compromise," in Sigmund Neumann, ed., *Modern Political Parties*. University of Chicago Press, 1956. For the Netherlands, see Robert C. Bone, "The Dynamics of Dutch Politics," *Journal of Politics*, Vol. 24, No. 1, February 1962.

J. No parties, or all parties illegal or ineffective [5]

n = 96
punch 3 = 7
punch 6 = 12

This raw characteristic is quite orthodox, except, perhaps, for the raw attribute "one-and-a-half-party," a concept apparently originating with Robert A. Scalapino.[52] Although we have coded only three polities (German Federal Republic, Japan, Trinidad) in this category, we feel that the term represents a worthwhile addition to the vocabulary of quantitative party analysis. It would appear that the lot of a doctrinal party competing with a broadly aggregative party within what would normally be a two-party system is not an enviable one. Such a party must either broaden its ideological appeal (that is, become *less* doctrinal) or resign itself to a more or less permanent minority status. The British Labour Party has pursued the former course in recent years. A similar trend is perhaps evident within the German Social Democratic Party, although it remains to be seen whether the latter will emerge as a serious challenge to Christian Democratic hegemony in a post-Adenauer election.

Indonesia, Malaya, Morocco, Nicaragua, South Africa, Turkey, and Uruguay are coded as ambiguous (punch 3) for this characteristic. In Indonesia, parties are not now permitted an electoral function. Nicaragua has two parties, but under present conditions there is virtually no possibility of party rotation. The South African parties cater only to the white minority. Turkey seems currently poised between a one party dominant and a multi-party system. In the case of Uruguay, there is some question as to whether the Blanco and Colorado organizations or their respective *sublemas* should be regarded as the significant party groupings.

Burma, Iran, Iraq, Jordan, South Korea, Nepal, Pakistan, Sudan, Syria, Thailand, Togo, and Yemen are coded as currently unascertainable (punch 6).

FC 139. Pol where the party system is quantitatively one-party (34) v AO (71)
FC 140. Pol where the party system is quantitatively one-party dominant (13) v AO (87)
FC 141. Pol where the party system is quantitatively two-party (11) v AO (87)
FC 142. Pol where the party system is quantitatively multi-party (30) v AO (66)
FC 143. Pol where the party system is quantitatively no-party (5) v AO (100) [SO]
FC 144. Pol where the party system is quantitatively one-party, rt two-party (34) v CONV (11) [PO]

[52] Robert A. Scalapino and Junnosuke Masumi, *Parties and Politics in Contemporary Japan*. University of California Press, 1962.

FC 145. Pol where the party system is quantitatively multi-party, rt two-party (30) v CONV (11) [PO]

---

42. PARTY SYSTEM: QUALITATIVE

A. Communist [13]
B. Mass-based territorial [8]
G. Regional or regional-ethnic [5]
H. Communal [2]
I. Corporative [2]
J. Broadly aggregative [4]
K. Class-oriented or multi-ideological [23]
P. Personalistic, situational, or *ad hoc* [4]
Q. Latin Liberal-Conservative [5]
R. Latin Social Revolutionary [4]
S. African transitional [14]

n = 84
punch 3 = 14
punch 4 = 5
punch 6 = 12

Whatever difficulties may bedevil the construction of a typology of political parties, they are compounded tenfold in attempting to build a typology of party systems. Except in one-party states and in two-party aggregative states, party systems tend to be complex conglomerates, no two of which exhibit identical spectra of specific components. Any attempt, therefore, to conceptualize within manageable limits on a global basis must be a rather rough-and-ready procedure. We are far less satisfied with the structure of this raw characteristic than with any other used in this study. We have nonetheless included it, because of the crucial importance of the subject it covers, and also because we hope it will stimulate other students to typologize more adequately.

"Mass-based territorial" refers to a one-party or one party dominant system with membership theoretically open to all (as contrasted with Communism's quasi-elitism) and without specific ideological content. The eight polities so coded are all African, though there is no inherent reason why this need be so. The "regional or regional-ethnic" systems, of which the Congo (Leopoldville) is perhaps the most flamboyant example, are also exclusively African. Other African systems are in varying stages of movement from multi-party to one-party or one party dominant systems and are coded as "African transitional."

Only two systems (Cyprus and Malaya) are regarded as wholly communal in character, though varying admixtures of communalism are also present in Ceylon, India, and Lebanon.

Spain and Portugal are coded as "corporative." Canada, Mexico, the Philippines, and the United States are coded as "broadly aggregative."

A "class-oriented or multi-ideological" system is always a two- or multi-party system. Each significant party organization in this category will exhibit some form of class or ideological bias, ranging from doctrinal radicalism to the mild class orientation of, say, the British Conservative Party. No party, in such a system, is exclusively "aggregative" in character.

"Personalistic, situational, or *ad hoc*" systems are those in which all significant parties are of the character suggested, namely, "personal" groupings called into being to serve the electoral purposes of a particular candidate or clique. Needless to say, such systems tend to be quite unstable.

A "Latin Liberal-Conservative" system is one in which the "traditional" Hispanic-American complex of Liberal and Conservative parties remains relatively unimpaired. We have coded five polities (Colombia, Ecuador, Honduras, Nicaragua, Paraguay) within this category, though the "opposition" in the latter two cases exists in rather emaciated form.

"Latin Social Revolutionary" systems are those in which the leading parties exhibit what may be called neo-Aprista orientation. The four systems so coded are those of Bolivia, Costa Rica (somewhat hesitantly), the Dominican Republic (perhaps prematurely), and Venezuela.

Communist v AO is not included among the following finished characteristics, since the equivalent may be found under RC 57. It should be noted that "regional or regional-ethnic" has been contrasted with all other sub-Saharan African systems in the form of a composite finished characteristic, FC 151.

FC 146. Pol where the party system is qualitatively mass-based territorial (8) v AO (92)

FC 147. Pol where the party system is qualitatively class-oriented or multi-ideological (23) v AO (67)

FC 148. Pol where the party system is qualitatively African transitional (14) v AO (96)

FC 149. Pol where the party system is qualitatively broadly aggregative, rt personalistic, situational, or *ad hoc* (4) v CONV (4) [SO]

FC 150. Pol where the party system is qualitatively Latin Liberal-Conservative, rt Latin Social Revolutionary (5) v CONV (4) [SO]

FC 151. Pol located in Central and South Africa and where the party system is qualitatively regional or regional-ethnic (5) v Pol located in Central and South Africa and where the party system is qualitatively other than regional or regional-ethnic (22) [SO]

43. STABILITY OF PARTY SYSTEM

A. Stable (all significant parties stable and organizationally non-situational) [42]
B. Moderately stable (relatively infrequent or non-abrupt system changes, or mixed situational-permanent party complex) [13]
G. Unstable (all parties unstable, situational, personalistic, or *ad hoc*) [25]

n = 80
punch 3 = 3
punch 4 = 5
punch 6 = 27

It is impossible to take a "snapshot" of stability, hence it would be meaningless to talk of the "current stability" of any aspect of a political system. But granting that a temporal dimension of *some* magnitude is required, it is nontheless true that coding for "stability" remains a highly relativistic matter. We have not confined ourselves to any hard-and-fast time span in coding for this characteristic. We have, however, regarded most of the recently independent states as being currently unascertainable. Hence, there is a rather large "punch 6" residue of 27 polities, 20 of which are African.

We have coded Indonesia, Colombia, and Uruguay as ambiguous (punch 3), the latter two-party systems being more or less constitutionally "frozen." Once again, we code as irrelevant the five states with no parties and little current likelihood of establishing party systems: Afghanistan, Ethiopia, Haiti, Libya, and Sa'udi Arabia.

In FC 155 we have added a composite finished characteristic dealing with stability within the sub-universe of polities that have multi-party systems.

FC 152. Pol where the party system is stable, rt unstable (42) v CONV (25) [SO]
FC 153. Pol where the party system is stable (42) v Pol where the party system is moderately stable or unstable (38) [PO]
FC 154. Pol where the party system is stable or moderately stable (55) v Pol where the party system is unstable (25) [PO]
FC 155. Pol where the party system is stable multi-party rt unstable multi-party (11) v CONV (13) [SO]

---

44. *PERSONALISMO*

A. Pronounced (all parties highly personalistic or fractionalized along personalistic lines) [14]

B. Moderate (some tendency toward personalism by all parties, or mixed personalistic, non-personalistic party complex) [26]
G. Negligible (no parties with significant personalist tendencies) [56]

n = 96
punch 4 = 5
punch 5 = 3
punch 6 = 11

According to George I. Blanksten, "*Personalismo* may be defined as the tendency of the politically active sectors of the population to follow or oppose a leader for personal, individual, and family reasons rather than because of the influence of a political idea, program, or party."[53] *Personalismo,* as thus defined, is pre-eminently a Latin American phenomenon, but it is by no means restricted to that area.

Personalismo should not be confused with "charismatic" leadership (see RC 46, pages 104–105), though the two phenomena may indeed co-occur, and in such cases each undoubtedly serves to reinforce the other. Contemporary Cambodia provides an illustrative example. However, it is also true that each of these phenomena can exist quite apart from the other, and indeed no statement appears in the printout indicating a strong association between the two. Panamanian politics are steeped in *personalismo,* yet charismatic leaders are extremely rare. Conversely, President Nyerere

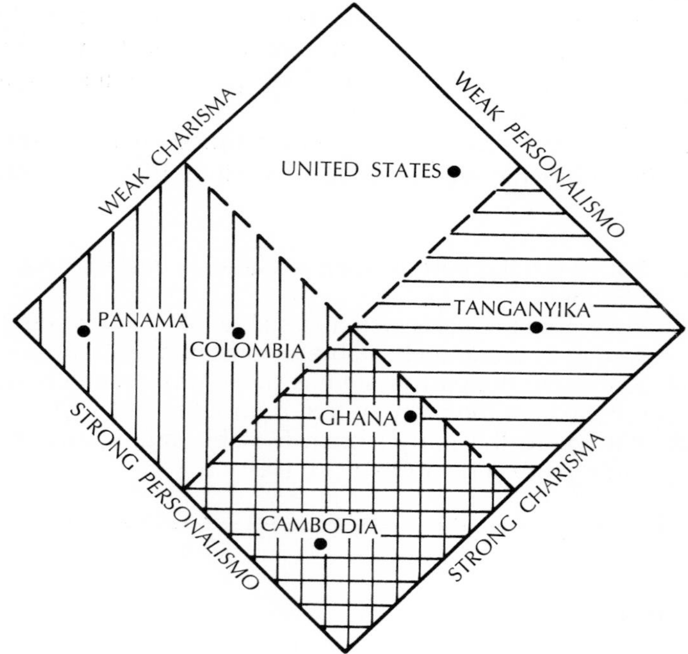

Figure 3.1. Interaction of charisma and *personalismo*

[53]Harold E. Davis, ed., *Government and Politics in Latin America.* Ronald Press, 1958, p. 144

of Tanganyika is a highly charismatic leader, but the Tanganyika African National Union could hardly be described as a personalist organization. Polities such as Colombia and Ghana are ranged between these two poles. The gamut of possibilities is roughly suggested by Figure 3.1.

Burundi, Cyprus, and Rwanda are unascertained (punch 5) for this characteristic. Eleven polities are coded as currently unascertainable (punch 6): Burma, Dominican Republic, Iran, Iraq, Jordan, South Korea, Nepal, Pakistan, Sudan, Togo, and Yemen.

FC 156. Pol where personalismo is pronounced, rt negligible (14) v CONV (56) [SO]
FC 157. Pol where personalismo is pronounced (14) v Pol where personalismo is moderate or negligible (84) [PO]
FC 158. Pol where personalismo is pronounced or moderate (40) v Pol where personalismo is negligible (56) [PO]

---

45. POLITICAL LEADERSHIP

A. Elitist (recruitment confined to a particular racial, social, or ideological stratum) [30]
B. Moderate elitist (recruitment largely, but not wholly confined to a particular racial, social, or ideological stratum) [17]
G. Non-elitist (recruitment largely on the basis of achievement criteria only) [50]

n = 97
punch 3 = 10
punch 6 = 8

In coding for this raw characteristic, two more or less arbitrary decisions were made. First, leadership in a Communist system is regarded as "elitist," since party membership itself is of a cadre character. Second, we assume that for most of the newly independent sub-Saharan African states leadership does *not* come from a particular social class (one sometimes referred to as the "educated elite") but from a leadership pool, membership in which is acquired on the basis of achievement. When, however, social stratification has a manifest bearing on political recruitment (as, for example, in Northern Nigeria), this fact is of course taken into account.

Ten polities are coded as ambiguous (punch 3) for this characteristic: Cambodia, Cyprus, Haiti, Iran, Lebanon, Nigeria, Pakistan, Rwanda, Uganda, and the UAR. It is interesting to note that ambiguity in the cases of Haiti and Rwanda stems from the existence of what may be called "reverse elitism." President Duvalier of Haiti insists upon "all-black" rule. In Rwanda, the formerly privileged Watusi element has been partially barred from political participation.

Eight polities (Algeria, Burma, Cuba, Iraq, South Korea, Sudan, Syria, Yemen) are coded as currently unascertainable (punch 6).

FC 159. Pol where the political leadership is elitist, rt non-elitist (30) v CONV (50) [SO]
FC 160. Pol where the political leadership is elitist (30) v Pol where the political leadership is moderately elitist or non-elitist (67) [PO]
FC 161. Pol where the political leadership is elitist or moderately elitist (47) v Pol where the political leadership is non-elitist (50) [PO]

---

46. LEADERSHIP CHARISMA

A. Pronounced [13]
B. Moderate [21]
G. Negligible [65]

n = 99
punch 3 = 7
punch 5 = 4
punch 6 = 5

Max Weber, at one point, defined "charismatic authority" as a form of rule "to which the governed submit because of their belief in the extraordinary quality of the specific person."[54] We are not here concerned with the question whether the powers of the charismatic leader need necessarily be regarded as "magical" or "supernatural." It suffices, for our purposes, that they be viewed as going (again, in Weber's words) "beyond the normal human qualities"—a phrase which we take to mean "beyond the qualities that most humans usually possess."

It may be argued that such a dilution of Weber's concept results in too great an overlap between the concept of "charisma" and that of "leadership." This does not bother us greatly. Probably *all* leadership necessarily involves some charisma, particularly where the attitudes of the leader's immediate followers are concerned. But some leaders are more charismatically regarded than others. And it is the *relatively more* charismatic that we have here attempted to code under attribute A—Nkrumah of Ghana, for example, as contrasted with Kennedy of the USA.

Seven polities have been coded as ambiguous (punch 3) for this characteristic: Afghanistan, Congo (Leo), Cyprus, Iran, Nigeria, Uganda, and the USSR. Four polities (Burundi, Laos, Nepal, Rwanda) are unascertained (punch 5). Five are regarded as currently unascertainable: Burma, Iraq, Syria, Togo, and Yemen.

[54] Hans Gerth and C. Wright Mills, eds., *From Max Weber: Essays in Sociology*. Oxford University Press, 1946, p. 295.

FC 162. Pol where the regime's leadership charisma is pronounced, rt negligible (13) v CONV (65) [SO]

FC 163. Pol where the regime's leadership charisma is pronounced (13) v Pol where the regime's leadership charisma is moderate or negligible (87) [PO]

FC 164. Pol where the regime's leadership charisma is pronounced or moderate (34) v Pol where the regime's leadership charisma is negligible (65) [PO]

---

47. VERTICAL POWER DISTRIBUTION

A. Effective federalism [8]
B. Limited federalism (federal structure with limited separation or pronounced "centralist" tendencies) [7]
G. Former federalism (formal or limited formal federal structure only) [6]
H. Formal and effective unitarism [93]

n = 114
punch 6 = 1

Federalism has been defined by a leading authority as "the principle of . . . general and regional governments being coordinate and independent in their respective spheres."[55] On the basis of this definition, we have coded Cameroun and Uganda under A, though one may question whether either state has yet attained stability with respect to its internal distribution of power. It may be noted that the Ethiopian-Eritrean federation has recently been abandoned in favor of a unitary Ethiopian state.

By "formal federalism" we mean formal adherence to the federal principle, unaccompanied by substantive adherence. The USSR is a conspicuous example.

The Congo (Leo) has been coded as currently unascertainable (punch 6) for this characteristic.

FC 165. Pol where the vertical power distribution is that of effective federalism (8) v Pol where the vertical power distribution is that of limited federalism, formal federalism, or formal and effective unitarism (106) [SO]

FC 166. Pol where the vertical power distribution is that of effective federalism or limited federalism (15) v Pol where the vertical power distribution is that of formal federalism or formal and effective unitarism (99)

---

[55] K. C. Wheare, *Federal Government*. Oxford University Press, 1946, p. 5.

48. HORIZONTAL POWER DISTRIBUTION

A. Significant (effective allocation of power to functionally autonomous legislative, executive and judicial organs) [34]
B. Limited (one branch of government without genuine functional autonomy, or two branches with limited functional autonomy) [24]
G. Negligible (complete dominance of government by one branch or by extra-governmental agency) [48]

n = 106
punch 5 = 1
punch 6 = 8

We are here concerned with the *functional distribution* of power in the Lockean sense. That is, we regard a "functionally autonomous" legislature, executive, or judiciary as one capable of exercising the functions normally attributable to it, regardless of whether the legislative-executive structure is parliamentary or presidential, or whether the judiciary possesses "judicial review," and so forth.

Uganda is unascertained for this characteristic. Eight polities are regarded as currently unascertainable: Argentina, Burma, Congo (Leo), Iraq, South Korea, Peru, Syria, and Yemen.

FC 167. Pol where the horizontal power distribution is significant, rt negligible (34) v CONV (48) [SO]
FC 168. Pol where the horizontal power distribution is significant (34) v Pol where the horizontal power distribution is limited or negligible (72) [PO]
FC 169. Pol where the horizontal power distribution is significant or limited (58) v Pol where the horizontal power distribution is negligible (48) [PO]

---

49. LEGISLATIVE-EXECUTIVE STRUCTURE

A. Presidential [39]
B. Presidential-Parliamentary [3]
G. Parliamentary-Republican [12]
H. Pure Parliamentary [1]
I. Parliamentary-Royalist [21]
J. Monarchical-Parliamentary [4]
K. Monarchical [2]
P. Communist [13]
Q. Other [2]

n = 97
punch 3 = 4
punch 4 = 1
punch 6 = 13

This characteristic is based on an eightfold classificatory scheme. The scheme in turn reflects three general criteria: (1) whether the Head of State is monarchical or non-monarchical, (2) whether effective executive power is exercised by the Head of State or by the formal executive (where the two are not constitutionally or otherwise conjoined), and (3) whether the formal executive is or is not dependent upon parliamentary support for continuance in office. Figure 3.2 illustrates this classificatory scheme.

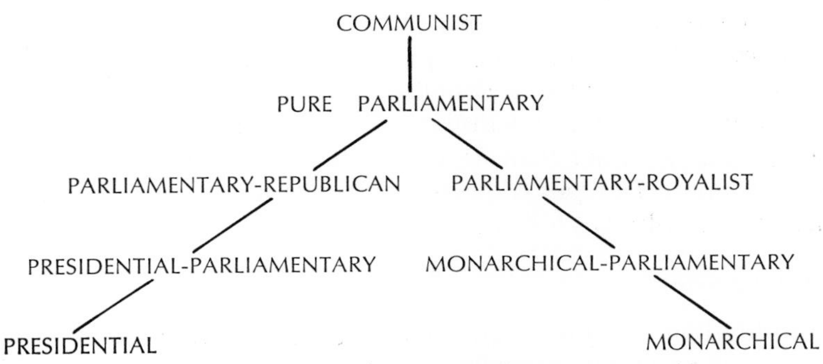

Figure 3.2. Legislative-executive structures

The "pure parliamentary" form, that is, absolute parliamentary control over both Head of State and Executive, is an exceedingly rare and manifestly unstable political phenomenon. Viewed in historical perspective, such "assembly government" would seem to be of an essentially transitional character. The French Convention 1792–1795 is undoubtedly the leading example. And it is of more than passing interest that a large number of France's former African dependencies exhibited essentially "pure parliamentary" forms immediately after independence. Such men as Tombalbaye of Chad, Dacko of the Central African Republic, and M'ba of Gabon were all premiers constitutionally dependent in 1959–1960 upon parliamentary majorities. Then came independence, and each incumbent was simply designated additionally as Chief of State. With the later adoption of new constitutions embodying variations of the presidential system, the pure parliamentary forms were abandoned. Only in the West African state of Mali do the essentials of this type of legislative-executive structure remain (as of April 1, 1963).[56]

[56]In Tanganyika the President is elected by the legislature, and if the legislature "over-rides" him with respect to proposed legislation, he must either assent to the legislation or dissolve parliament, thus putting his own job on the line. This is regarded by the Tanganyikans as a variation of the Ghanaian system. Since it may also perhaps be regarded as exhibiting an element of the "pure parliamentary," we have coded Tanganyika as ambiguous for this characteristic.

The thirteen Communist states are the closest in form, but the most distant in reality, from the pure parliamentary type. In all thirteen states, the legislature has complete *formal* control over the appointment and tenure of both the Head of State and the Premier. Effective control, however, is exercised by the party hierarchy.

The Communist states are, of course, variations on a single theme. There are, however, discernible differences in the form, if not the content, of their executive structures. These may briefly be indentified as (1) *Presidential-Premierial* (China, Czechoslovakia, North Korea, Yugoslavia); (2) *Conciliar-Premierial* (East Germany, Poland, Rumania); and (3) *Praesidial-Premierial* (Albania, Bulgaria, Hungary, Mongolia, USSR). A Communist "Praesidium" is, of course, a conciliar body which differs from a "State Council" only in that it is constitutionally recognized as a superior council of the legislature, rather than as a theoretically autonomous body.

The remaining six types of non-Communist legislative-executive structure may be described as follows:

1. *PRESIDENTIAL*

A presidential structure is characterized by the presence of an elected official (whether directly or indirectly designated) who serves for a stated term as both Head of State and Chief Executive, or as Head of State and *effective* executive, in conjunction with a "weak" premier not dependent upon a parliamentary majority.

Presidential structures tend to occur in conjunction with either one-party or multi-party systems. All of the current *one-party* presidential systems are African. All but one of the *multi-party* presidential systems are Latin American. Cyprus is the lone exception, and in some respects its multi-party configuration is quite atypical.

2. *PRESIDENTIAL-PARLIAMENTARY*

The presidential-parliamentary structure has a "strong" president serving as Head of State and as effective executive, in conjunction with a "weak" premier who is dependent upon parliamentary support. Gaullist France is the leading example. In Chad, where both offices are currently held by the same individual, Tombalbaye can evidently be ousted as Premier, but not as President.

3. *PARLIAMENTARY-REPUBLICAN*

The parliamentary-republican structure differs from the presidential-parliamentary in that the power positions of president and premier, respectively, are reversed. Here a "weak" president serves as Head of State, in conjunction with a "strong" premier dependent upon parliamentary support. India and West Germany are examples.

4. *PARLIAMENTARY-ROYALIST*

The parliamentary-royalist structure differs from the parliamentary-republican in that the Head of State is a "weak" monarch, rather than an elected president. Here again, the premier is "strong" and is dependent upon parliamentary support. The United Kingdom is the classic example.

Parliamentary-Royalist structures tend to occur in *multi-party* or *two-party* contexts. Where the context is two-party the polity is invariably a member of the British Commonwealth: Australia, Canada, Jamaica, New Zealand, Sierre Leone, United Kingdom.

5. *MONARCHICAL-PARLIAMENTARY*

A monarchical-parliamentary structure has a "strong" monarch serving as Head of State and as effective executive, in conjunction with a "weak" premier dependent, at least to some degree, upon parliamentary support. The "constitutional" monarchies of Afghanistan, Jordan, Libya, and Morocco are of this type.

6. *MONARCHICAL*

The monarchical structure has a "strong" monarch serving as Head of State and as Chief Executive, or as Head of State and effective executive, in conjunction with a "weak" premier not dependent upon parliamentary support. Ethiopia and Nepal are currently of this type. Sa'udi Arabia has been classified as irrelevant (punch 4) for this raw characteristic, since it possesses no legislature of any kind.

Switzerland and Uruguay are here coded as "other" (Q), since both appoint annual chiefs of state from conciliar executive bodies.

Cambodia, Portugal, and Spain, in addition to Tanganyika[56] are coded as ambiguous (punch 3). Cambodia is unique in possessing both a "monarch" (Queen Sisowath Kossamak) *and* a Chief of State (Prince Sihanouk), who also serves as Premier.[57]

Thirteen polities are regarded as currently unascertainable for this characteristic: Algeria, Burma, Cuba, Iran, Iraq, South Korea, Peru, Sudan, Syria, Thailand, Togo, UAR, and Yemen.

With respect to the finished characteristics, parliamentary-republican v parliamentary-royalist and presidential-parliamentary v monarchical-parliamentary were included in an earlier computer run. Neither dichotomous contrast discriminated sufficiently, in our opinion, to warrant its inclusion in the final printout.

FN 170. Pol where the legislative-executive structure is presidential (39) v AO (63)

[57] Prince Sihanouk has declared that Cambodia will become a republic after the next election.

FN 171. Pol where the legislative-executive structure is parliamentary-republican (12) v AO (90)
FN 172. Pol where the legislative-executive structure is parliamentary-royalist (21) v AO (88)
FN 173. Pol where the legislative-executive structure is parliamentary-republican or parliamentary-royalist, rt presidential-parliamentary or monarchical-parliamentary (33) v CONV (7) [SO]

---

50. CURRENT STATUS OF LEGISLATURE

A. Fully effective (performs normal legislative function as reasonably "co-equal" branch of national government) [28]
B. Partially effective (tendency toward domination by executive, or otherwise partially limited in effective exercise of legislative function) [23]
G. Largely ineffective (virtually complete domination by executive or by one-party or dominant party organization) [21]
H. Wholly ineffective (restricted to consultative or "rubber-stamp" legislative function) [28]

n = 100
punch 4 = 1
punch 6 = 14

This raw characteristic is largely self-explanatory. As the printout indicates, the "developed West" is virtually the exclusive locale of A, while B tends to occur in Latin American states, and G in African states. All of the Communist states fall in H.

Sa'udi Arabia is coded as irrelevant for this characteristic. Fourteen states are coded as currently unascertainable: Burma, Congo (Leo), Cuba, Dominican Republic, Guatemala, Iran, Iraq, South Korea, Pakistan, Peru, Sudan, Syria, Togo, and Yemen.

FC 174. Pol where the legislature is fully effective (28) v Pol where the legislature is partially effective, largely ineffective, or wholly ineffective (72) [PO]
FC 175. Pol where the legislature is fully effective or partially effective (51) v Pol where the legislature is largely ineffective or wholly ineffective (49)
FC 176. Pol where the legislature is fully effective, partially effective, or largely ineffective (72) v Pol where the legislature is wholly ineffective (28) [PO]
FC 177. Pol where the legislature is partially effective, rt largely ineffective (23) v CONV (21) [SO]

---

51. CHARACTER OF LEGISLATURE

A. Unicameral [53]
B. Bicameral [51]

n = 104
punch 3 = 1
punch 4 = 1
punch 6 = 9

Our original raw characteristic on "character of the legislature" had three attributes: A. Unicameral; B. Bicameral: unequal powers; and G. Bicameral: equal powers. For many polities, however, it proved exceedingly difficult to differentiate between B and G, and for this reason the threefold scheme was abandoned in favor of a simple dichotomy.

Even a simple discrimination between "unicameral" and "bicameral" is not always as easily codable as one might imagine. There is some question, for example, whether the Laotian "King's Council" and its Cambodian counterpart, the "Council of the Kingdom," are genuine upper houses. Both polities are, however, here coded as bicameral.

Upper and lower chambers differ markedly not only in powers but in methods of selection. Most lower houses are elected bodies. Upper house membership may be by appointment, by election, or by a combination of the two. In some cases (for example, the Norwegian *Storting*), designation of upper house membership is by vote of the legislators themselves, all members not so designated constituting the lower chamber.

There is no theoretical reason why a legislature must be *either* unicameral or bicameral. Yet in practice virtually all of the 115 polities can be coded as one or the other.[58]

Nine polities are coded as currently unascertainable (punch 6) for this characteristic: Burma, Cuba, Iraq, South Korea, Sudan, Syria, Thailand, UAR, and Yemen.

FC 178. Pol where the legislature is unicameral (53) v Pol where the legislature is bicameral (51)

---

52. CURRENT STATUS OF EXECUTIVE

A. Dominant [52]
B. Strong [39]
G. Weak [2]

[58] A significant exception is Cyprus, which we have coded as ambiguous. Cyprus has a single House of Representatives, but certain matters are within the jurisdiction of two communal chambers representing the Greek and Turkish ethnic groups, respectively. It may also be noted that Yugoslavia will shortly adopt a new constitution, embodying a unique five-chambered national parliament.

n = 93
punch 3 = 4
punch 5 = 6
punch 6 = 12

In coding for this characteristic, we have assigned all instances of rule by military junta to the residue. In some cases it is not possible to describe the regime as being unambiguously civilian *or* military, and in such cases the assignment has also been to the residue.

We regard a "dominant" executive as one generally capable of imposing his will throughout the government establishment. A "strong" executive, on the other hand, is one whose effective dominance is largely confined to the executive realm. The distinction should not be construed in strictly "constitutional" terms. Thus, we regard Nehru as "dominant" within the effective governing structure of contemporary India, despite the fact that constitutional forms are largely respected.[59]

Only two polities (Switzerland and Uruguay) have been coded as possessing genuinely "weak" executives. Both are of a collegial character.

Four polities (Ceylon, Chile, Greece, Laos) are coded as ambiguous (punch 3) for this characteristic. For six polities, the precise status of the executive has not been ascertained: Burundi, Chad, Niger, Rwanda, Somalia, and Uganda. The twelve polities regarded as currently unascertainable (punch 6) are Argentina, Burma, Congo (Leo), Guatemala, Iraq, South Korea, Peru, Sudan, Syria, Thailand, Togo, and Yemen.

FC 179. Pol where the executive is dominant (52) v Pol where the executive is strong (39)

---

53. CHARACTER OF BUREAUCRACY

A. Modern (generally effective and responsible civil service or equivalent, performing in a functionally specific, non-ascriptive social context) [21]

B. Semi-modern (largely "rationalized" bureaucratic structure of limited efficiency because of shortage of skilled personnel, inadequacy of recruitment or performance criteria, excessive intrusion by non-administrative organs, or partially non-congruent social institutions) [55]

G. Post-colonial transitional (largely rationalized ex-colonial bureaucratic structure in process of personnel "nationalization" and adaptation to the servicing or restructuring of autochthonous social institutions) [25]

[59] The events surrounding Menon's resignation as Defense Minister in November, 1962, while virtually unprecedented, may, of course, presage a severe decline in Nehru's stature.

H. Traditional (largely non-rationalized bureaucractic structure performing in the context of an ascriptive or "deferential" stratification system) [9]

n = 110
punch 3 = 3
punch 6 = 2

The comparative study of public administration is still very much in its infancy. The pioneering 1959 volume edited by William J. Siffin[60] contains useful case studies but needs to be supplemented tenfold before it will be possible to make valid global generalizations in this area.

Because of this dearth of relevant country studies, our typology for this characteristic is less than adequate. It fails to deal with specific bureaucratic structure or with the calibre of performance of specific functions. The result is a somewhat unproductive characteristic—"unproductive" in the sense that much of its yield was fairly predictable from the beginning.

Indonesia, Pakistan, and South Vietnam are regarded as ambiguous (punch 3) for this characteristic. Iraq and Yemen are coded as currently unascertainable (punch 6).

It should be noted that we have dichotomized for the following finished characteristics by contrasting adjacent categories, rather than on the basis of a moving cutting point.

FC 180. Pol where the bureaucracy is modern or semi-modern (76) v Pol where the bureaucracy is post-colonial transitional or traditional (34) [SO]
FC 181. Pol where the bureaucracy is modern, rt semi-modern (21) v CONV (55) [PO]
FC 182. Pol where the bureaucracy is semi-modern, rt post-colonial transitional (55) v CONV (25) [PO]
FC 183. Pol where the bureaucracy is post-colonial transitional, rt traditional (25) v CONV (9) [PO]

---

54. POLITICAL PARTICIPATION BY THE MILITARY

A. Interventive (presently exercises or has recently exercised direct power) [21]
B. Supportive (performs para-political role in support of traditionalist, authoritarian, totalitarian, or modernizing regime) [31]
G. Neutral (apolitical, or of minor political importance) [56]

n = 108
punch 3 = 7

Throughout history, military establishments have enjoyed political power, often much more power than would ordinarily accrue to them

[60] William J. Siffin, ed., *Toward the Comparative Study of Public Administration.* Indiana University Press, 1959.

if they were to confine their concern to defense against outside threats. For nearly a century and a half, the nations of Latin America have suffered from the uneasy alliances between arms and politics in the form of recurrent *caudillismo*. In more recent times, rule by the military has become a more or less typical means of providing "transitional" leadership for fledgling nations as they move toward "development."[61]

The present raw characteristic asks only whether a nation's military establishment performs in a politically "interventive," "supportive," or "neutral" manner. It does not attempt to distinguish between "oppressive" and "constructive" intervention or support. In the case of both B and G, the military is presumed to be under civilian control. But in B, as contrasted with G, the military is used by the regime as a means of consolidating, enlarging, or defending political power.

Colombia is coded G, despite the not-too-distant events of the Rojas era.[62] Coding of the Dominican Republic as G may be somewhat premature in view of the critical role played by the military in the 1961 ouster of Trujillo coupled with uncertainties concerning the viability of the present government of Juan Bosch.

Seven polities are coded as ambiguous for this characteristic: Congo (Leo), Cuba, France, Laos, Nepal, Somalia, and Venezuela. In the case of Somalia there is quite evidently a *faction* of the military in former British Somaliland that would *like* to alter the status quo if they could. In Venezuela, the army, which generally controlled politics until 1959, is still being treated with delicacy by President Betancourt.

FC 184. Pol where participation by the military in politics is interventive, rt neutral (21) v CONV (56) [SO]
FC 185. Pol where participation by the military in politics is interventive, rt supportive (21) v CONV (31) [PO]
FC 186. Pol where participation by the military in politics is supportive, rt neutral (13) v CONV (56) [PO]

---

55. ROLE OF POLICE

A. Politically significant (important continuing or intermittent political function in addition to law enforcement) [66]

B. Not politically significant (role confined primarily to law enforcement) [35]

n = 101
punch 5 = 12
punch 6 = 2

[61] See particularly Johnson, *op. cit.* and Edwin Lieuwen, *Arms and Politics in Latin America*. Praeger, 1961.

[62] Lieuwen regards the Colombian armed forces as "nonpolitical" (Lieuwen, *op. cit.*, p. 168); Herbert L. Mathews calls Colombia a "borderline case" ("When Generals Take Over in Latin America," *The New York Times Magazine*, September 9, 1962).

For many polities, the term "police" embraces a variety of organizations with functions ranging all the way from traffic control to para-military activity. In coding for this characteristic we have taken the term in its broadest possible sense and have then asked the question: "Is the activity of the police in Polity X essentially confined to law enforcement, or does it also significantly impinge upon the political process?" We grant that this is not a wholly satisfactory formula. However, in the absence of any extensive literature on this subject, a simple dichotomous characteristic was the best that we could devise.

Twelve polities are coded as unascertained (punch 5): Brazil, Burundi, Chad, Chile, Costa Rica, Cyprus, Libya, Mali, Niger, Panama, Rwanda, and Upper Volta. The Dominican Republic and Peru are regarded as currently unascertainable (punch 6).

FC 187. Pol where the role of the police is politically significant (66) v Pol where the role of the police is not politically significant (35)

---

56. CHARACTER OF LEGAL SYSTEM

A. Civil law [32]
B. Mixed civil-indigenous [11]
G. Common law [7]
H. Mixed common-indigenous [1]
I. Mixed civil-common [2]
J. Mixed civil-common-indigenous [1]
K. Muslim [2]
P. Mixed civil-Muslim [9]
Q. Mixed common-Muslim [3]
R. Mixed civil-Muslim-indigenous [8]
S. Mixed common-Muslim-indigenous [5]
X. Scandinavian [5]
Y. Communist [13]
Z. Other [12]

n = 111
punch 3 = 2
punch 6 = 2

As far as we can determine, the only existing attempt to provide a comprehensive typology of the world's legal systems is the now somewhat dated survey by John Henry Wigmore.[63] Writing some three decades ago,

[63] John Henry Wigmore, *A Panorama of the World's Legal Systems,* rev. ed., Washington Law Book Company, 1936. The original three-volume edition was published in 1928.

Wigmore identified ten system types: (1) Anglican, (2) Chinese, (3) Germanic, (4) Hindu, (5) Japanese, (6) Mohammedan, (7) Romanesque, (8) Slavic, (9) Soviet Slavic, and (10) Tribal Customary.

We have used the more familiar terms "common law" and "civil law" in place of Wigmore's "Anglican" and "Romanesque," and "Muslim" for his "Mohammedan." Our term "indigenous" may be defined as "unique to a particular locality." As such, it embraces all essentially non-literate law, and hence is of somewhat broader scope than Wigmore's "Tribal Customary." On the other hand, a relatively autochthonous legal system that is no longer unique to a particular area (for example, English Common Law) is not here regarded as "indigenous."

Wigmore's remaining six systems have no counterparts in our basic scheme. His "Chinese System" is now significant only with respect to the Nationalist government of Taiwan, which is not included in this survey. Hindu law does not exist as a national system, though it does comprise a component of several of the systems coded under Z. Japanese law, combined with elements of civil and common law, is to be found only in Japan and South Korea, both of which are also coded as Z. Slavic law has been largely absorbed into Communist law. Wigmore's "Germanic" law appeared only in the composite "Romanesque-Germanic" of the five Scandinavian countries, which we have coded under the single term "Scandinavian."

In Cameroun we have disregarded the common-law component found in the small portion of that country which was once part of the British Cameroons. Mongolia is coded as "Communist," although some Buddhist-tribal elements undoubtedly remain. Scottish civil law is disregarded as a component of the legal system of the United Kingdom.

The twelve polities coded as "Other" (Z) are as follows:

1. Afghanistan (Muslim, indigenous)
2. Cambodia (civil, Hindu)
3. Ceylon (civil, common, Hindu)
4. Cyprus (common, sectarian)
5. India (common, Hindu, Muslim)
6. Iraq (civil, sectarian, indigenous)
7. Israel (civil, common, Ottoman, rabbinical)
8. Japan (civil, common, Japanese)
9. So. Korea (civil, common, Japanese)
10. Lebanon (civil, sectarian, indigenous)
11. Nepal (Hindu, indigenous)
12. Somalia (civil, common, Muslim)

Ethiopia and Guinea are regarded as ambiguous (punch 3). Ethiopian law includes civil, Coptic Christian ecclesiastical, and Muslim components.

A European legal commission has been working in that country for some years on a codification of the law, but without notable implementation to date. Judicial inefficiency and incompetence add to the difficulty of correctly assessing Ethiopia's status.[64] Guinea's legal system is based primarily on the French civil law tradition, but elements have been added that are suggestive of Communist law. For example, crimes against the Guinean state "may be judged outside the judicial framework by people's courts."[65]

Burma and Cuba are regarded as currently unascertainable (punch 6) for this characteristic.

In the following finished characteristics, we have not dichotomized on the basis of Scandinavian v AO or on the basis of Communist v AO, since equivalents can be found by consulting FC 4 and FC 194, respectively.

FC 188. Pol where the character of the legal system is civil law (32) v AO (81)
FC 189. Pol where the character of the legal system is common law (7) v AO (108)
FC 190. Pol where the character of the legal system is civil law, rt common law (32) v CONV (7)
FC 191. Pol where the character of the legal system is mixed civil-Muslim-indigenous, rt mixed common-Muslim-indigenous (8) v CONV (5) [SO]
FC 192. Pol where the character of the legal system is Muslim or partly Muslim (27) v AO (86)
FC 193. Pol where the character of the legal system is partly indigenous (28) v AO (86)

---

57. COMMUNIST BLOC

A. Communist [13]
B. Quasi-Communist [1]
G. Non-Communist [101]

n = 115

Cuba, despite its definite alignment with international Communism, has not (as of April 1, 1963) adopted many of the domestic organizational features of a typical Communist state. It is therefore coded as "quasi-Communist."

FC 194. Pol that are Communist (13) v Pol that are non-Communist (101)

[64]See Ernest W. Luther, *Ethiopia Today*. Stanford University Press, 1958, p. 46.

[65]Gray Cowan, "Guinea," in Gwendolen M. Carter, ed., *African One-Party States*. Cornell University Press, 1962, p. 215.

WHISKERS CHARACTERISTICS (see pages 51–53)

FC 195. Pol belonging to the fifty percent that have purple whiskers (58) v AO (57) [SO]
FC 196. Pol belonging to the forty percent that have blue whiskers (46) v AO (69) [SO]
FC 197. Pol belonging to the thirty percent that have green whiskers (35) v AO (80) [SO]
FC 198. Pol belonging to the twenty percent that have pink whiskers (23) v AO (92) [SO]
FC 199. Pol belonging to the ten percent that have yellow whiskers (12) v AO (103) [SO]

Computer Printout

1 POLITIES LOCATED IN EAST EUROPE (9)

1 POLITIES LOCATED ELSEWHERE THAN IN EAST EUROPE (106)

SUBJECT ONLY

9 IN LEFT COLUMN

ALBANIA, BULGARIA, CZECHOS'KIA, GERMANY, E, HUNGARY, POLAND, RUMANIA, USSR, YUGOSLAVIA

106 IN RIGHT COLUMN

AFGHANISTAN	ALGERIA	ARGENTINA	AUSTRALIA	AUSTRIA	BELGIUM	BOLIVIA	BRAZIL	BURMA	BURUNDI
CAMBODIA	CAMEROUN	CANADA	CEN AFR REP	CEYLON	CHAD	CHILE	CHINA, PR	COLOMBIA	CONGO(BRA)
CONGO(LEO)	COSTA RICA	CUBA	CYPRUS	DAHOMEY	DENMARK	DOMIN REP	ECUADOR	EL SALVADOR	ETHIOPIA
FINLAND	FRANCE	GABON	GERMAN FR	GHANA	GREECE	GUATEMALA	GUINEA	HAITI	HONDURAS
ICELAND	INDIA	INDONESIA	IRAN	IRAQ	IRELAND	ISRAEL	ITALY	IVORY COAST	JAMAICA
JAPAN	JORDAN	KOREA, N	KOREA REP	LAOS	LEBANON	LIBERIA	LIBYA	LUXEMBOURG	MALAGASY R
MALAYA	MALI	MAURITANIA	MEXICO	MONGOLIA	MOROCCO	NEPAL	NETHERLANDS	NEW ZEALAND	NICARAGUA
NIGER	NIGERIA	NORWAY	PAKISTAN	PANAMA	PARAGUAY	PERU	PHILIPPINES	PORTUGAL	RWANDA
SA'U ARABIA	SENEGAL	SIERRE LEO	SOMALIA	SO AFRICA	SPAIN	SUDAN	SWEDEN	SWITZERLAND	SYRIA
TANGANYIKA	THAILAND	TOGO	TRINIDAD	TUNISIA	TURKEY	UGANDA	UAR	UK	US
UPPER VOLTA	URUGUAY	VENEZUELA	VIETNAM, N	VIETNAM REP	YEMEN				

	Left column			Right column		
24	TEND TO BE THOSE WHOSE POPULATION IS VERY LARGE, LARGE, OR MEDIUM (61)	0.89	24	TEND TO BE THOSE WHOSE POPULATION IS SMALL (54)	0.50	8 53 1 53 X SQ= 3.60 P = 0.035 RV YES (NO)
26	TEND TO BE THOSE WHOSE POPULATION DENSITY IS VERY HIGH, HIGH, OR MEDIUM (48)	0.89	26	TEND TO BE THOSE WHOSE POPULATION DENSITY IS LOW (67)	0.62	8 40 1 66 X SQ= 6.95 P = 0.004 RV YES (NO)
28	TEND TO BE THOSE WHOSE POPULATION GROWTH RATE IS LOW (48)	0.87	28	TEND TO BE THOSE WHOSE POPULATION GROWTH RATE IS HIGH (62)	0.60	1 61 7 41 X SQ= 4.96 P = 0.020 RV YES (NO)
33	TEND TO BE THOSE WHOSE GROSS NATIONAL PRODUCT IS VERY HIGH, HIGH, OR MEDIUM (30)	0.67	33	TEND TO BE THOSE WHOSE GROSS NATIONAL PRODUCT IS LOW OR VERY LOW (85)	0.77	6 24 3 82 X SQ= 6.21 P = 0.009 RV YES (NO)

1/

36	TEND TO BE THOSE WHOSE PER CAPITA GROSS NATIONAL PRODUCT IS VERY HIGH, HIGH, OR MEDIUM (42)	0.89	36	TEND TO BE THOSE WHOSE PER CAPITA GROSS NATIONAL PRODUCT IS LOW OR VERY LOW (73)	0.68	8 34 1 72 X SQ= 9.23 P = 0.001 RV YES (NO)
37	ALWAYS ARE THOSE WHOSE PER CAPITA GROSS NATIONAL PRODUCT IS VERY HIGH, HIGH, MEDIUM, OR LOW (64)	1.00	37	TEND LESS TO BE THOSE WHOSE PER CAPITA GROSS NATIONAL PRODUCT IS VERY HIGH, HIGH, MEDIUM, OR LOW (64)	0.52	9 55 0 51 X SQ= 5.95 P = 0.004 RV NO (NO)
39	TEND TO BE THOSE WHOSE INTERNATIONAL FINANCIAL STATUS IS VERY HIGH, HIGH, OR MEDIUM (38)	0.75	39	TEND TO BE THOSE WHOSE INTERNATIONAL FINANCIAL STATUS IS LOW OR VERY LOW (76)	0.70	6 32 2 74 X SQ= 4.86 P = 0.016 RV YES (NO)
42	TEND TO BE THOSE WHOSE ECONOMIC DEVELOPMENTAL STATUS IS DEVELOPED OR INTERMEDIATE (36)	0.89	42	TEND TO BE THOSE WHOSE ECONOMIC DEVELOPMENTAL STATUS IS UNDERDEVELOPED OR VERY UNDERDEVELOPED (76)	0.73	8 28 1 75 X SQ= 11.76 P = 0. RV YES (NO)
44	TEND TO BE THOSE WHERE THE LITERACY RATE IS NINETY PERCENT OR ABOVE (25)	0.67	44	TEND TO BE THOSE WHERE THE LITERACY RATE IS BELOW NINETY PERCENT (90)	0.82	6 19 3 87 X SQ= 8.90 P = 0.003 RV YES (NO)
45	ALWAYS ARE THOSE WHERE THE LITERACY RATE IS FIFTY PERCENT OR ABOVE (55)	1.00	45	TEND TO BE THOSE WHERE THE LITERACY RATE IS BELOW FIFTY PERCENT (54)	0.54	9 46 0 54 X SQ= 7.59 P = 0.003 RV YES (NO)
51	ALWAYS ARE THOSE WHERE FREEDOM OF THE PRESS IS INTERNALLY ABSENT, OR INTERNALLY AND EXTERNALLY ABSENT (37)	1.00	51	TEND TO BE THOSE WHERE FREEDOM OF THE PRESS IS COMPLETE OR INTERMITTENT (60)	0.68	0 60 9 28 X SQ= 13.33 P = 0. RV YES (NO)
52	TEND TO BE THOSE WHERE FREEDOM OF THE PRESS IS INTERNALLY AND EXTERNALLY ABSENT (16)	0.67	52	TEND TO BE THOSE WHERE FREEDOM OF THE PRESS IS COMPLETE, INTERMITTENT, OR INTERNALLY ABSENT (82)	0.89	3 79 6 10 X SQ= 14.55 P = 0. RV YES (NO)
54	TEND TO BE THOSE WHERE NEWSPAPER CIRCULATION IS ONE HUNDRED OR MORE PER THOUSAND (37)	0.78	54	TEND TO BE THOSE WHERE NEWSPAPER CIRCULATION IS LESS THAN ONE HUNDRED PER THOUSAND (76)	0.71	7 30 2 74 X SQ= 6.92 P = 0.005 RV YES (NO)

1/

55	ALWAYS ARE THOSE WHERE NEWSPAPER CIRCULATION IS TEN OR MORE PER THOUSAND (78)	1.00	55	LEAN LESS TOWARD BEING THOSE WHERE NEWSPAPER CIRCULATION IS TEN OR MORE PER THOUSAND (78)	0.66	9 69 0 35 X SQ= 2.96 P = 0.055 RV NO (NO)
56	ALWAYS ARE THOSE WHERE THE RELIGION IS PREDOMINANTLY LITERATE (79)	1.00	56	LEAN LESS TOWARD BEING THOSE WHERE THE RELIGION IS PREDOMINANTLY LITERATE (79)	0.69	9 70 0 31 X SQ= 2.48 P = 0.059 RV NO (NO)
63	TEND TO BE THOSE WHERE THE RELIGION IS PREDOMINANTLY SOME KIND OF CHRISTIAN (46)	0.87	63	TEND TO BE THOSE WHERE THE RELIGION IS PREDOMINANTLY OR PARTLY OTHER THAN CHRISTIAN (68)	0.63	7 39 1 67 X SQ= 5.98 P = 0.007 RV YES (NO)
64	ALWAYS ARE THOSE WHERE THE RELIGION IS CHRISTIAN, RATHER THAN MUSLIM (46)	1.00	64	TILT LESS TOWARD BEING THOSE WHERE THE RELIGION IS CHRISTIAN, RATHER THAN MUSLIM (46)	0.68	7 39 0 18 X SQ= 1.71 P = 0.102 RV NO (NO)
67	ALWAYS ARE THOSE THAT ARE RACIALLY HOMOGENEOUS (82)	1.00	67	LEAN LESS TOWARD BEING THOSE THAT ARE RACIALLY HOMOGENEOUS (82)	0.71	9 71 0 29 X SQ= 2.23 P = 0.065 RV NO (NO)
73	ALWAYS ARE THOSE WHOSE DATE OF INDEPENDENCE IS BEFORE 1945 (65)	1.00	73	TEND LESS TO BE THOSE WHOSE DATE OF INDEPENDENCE IS BEFORE 1945 (65)	0.55	8 57 0 46 X SQ= 4.40 P = 0.020 RV NO (NO)
75	TEND MORE TO BE THOSE THAT ARE HISTORICALLY WESTERN OR SIGNIFICANTLY WESTERNIZED (62)	0.89	75	TEND LESS TO BE THOSE THAT ARE HISTORICALLY WESTERN OR SIGNIFICANTLY WESTERNIZED (62)	0.51	8 54 1 51 X SQ= 3.30 P = 0.038 RV NO (NO)
76	ALWAYS ARE THOSE THAT ARE HISTORICALLY WESTERN, RATHER THAN HAVING BEEN SIGNIFICANTLY OR PARTIALLY WESTERNIZED THROUGH A COLONIAL RELATIONSHIP (26)	1.00	76	TEND TO BE THOSE THAT HAVE BEEN SIGNIFICANTLY OR PARTIALLY WESTERNIZED THROUGH A COLONIAL RELATIONSHIP, RATHER THAN BEING HISTORICALLY WESTERN (70)	0.76	4 22 0 70 X SQ= 7.71 P = 0.005 RV YES (NO)
82	ALWAYS ARE THOSE WHERE THE TYPE OF POLITICAL MODERNIZATION IS EARLY OR LATER EUROPEAN OR EUROPEAN DERIVED, RATHER THAN DEVELOPED TUTELARY OR UNDEVELOPED TUTELARY (51)	1.00	82	TEND TO BE THOSE WHERE THE TYPE OF POLITICAL MODERNIZATION IS DEVELOPED TUTELARY OR UNDEVELOPED TUTELARY, RATHER THAN EARLY OR LATER EUROPEAN OR EUROPEAN DERIVED (55)	0.56	8 43 0 55 X SQ= 7.22 P = 0.002 RV YES (NO)

1/

```
 85  ALWAYS ARE THOSE                          1.00     85  TEND TO BE THOSE                          0.51         9   51
     WHERE THE STAGE OF                                    WHERE THE STAGE OF                                     0   54
     POLITICAL MODERNIZATION IS                            POLITICAL MODERNIZATION IS                        X SQ=    6.85
     ADVANCED, RATHER THAN                                 MID- OR EARLY TRANSITIONAL,                       P =    0.003
     MID- OR EARLY TRANSITIONAL  (60)                      RATHER THAN ADVANCED  (54)                        RV YES (NO )

 87  ALWAYS ARE THOSE                          1.00     87  TEND LESS TO BE THOSE                     0.61         0   31
     WHOSE IDEOLOGICAL ORIENTATION                         WHOSE IDEOLOGICAL ORIENTATION                          9   49
     IS OTHER THAN DEVELOPMENTAL  (58)                     IS OTHER THAN DEVELOPMENTAL  (58)                 X SQ=    3.78
                                                                                                             P =    0.024
                                                                                                             RV NO  (NO )

 89  ALWAYS ARE THOSE                          1.00     89  TEND LESS TO BE THOSE                     0.62         0   33
     WHOSE IDEOLOGICAL ORIENTATION                         WHOSE IDEOLOGICAL ORIENTATION                          9   53
     IS OTHER THAN CONVENTIONAL  (62)                      IS OTHER THAN CONVENTIONAL  (62)                  X SQ=    3.73
                                                                                                             P =    0.025
                                                                                                             RV NO  (NO )

 92  ALWAYS ARE THOSE                          1.00     92  TEND TO BE THOSE                          0.89         9   11
     WHERE THE SYSTEM STYLE                                WHERE THE SYSTEM STYLE                                 0   93
     IS MOBILIZATIONAL  (20)                               IS LIMITED MOBILIZATIONAL, OR                     X SQ=   39.54
                                                           NON-MOBILIZATIONAL  (93)                          P =    0.
                                                                                                             RV YES (NO )

 95  ALWAYS ARE THOSE                          1.00     95  TEND TO BE THOSE                          0.93         0   95
     WHERE THE STATUS OF THE REGIME                        WHERE THE STATUS OF THE REGIME                         9    7
     IS TOTALITARIAN  (16)                                 IS CONSTITUTIONAL OR                              X SQ=   50.85
                                                           AUTHORITARIAN  (95)                               P =    0.
                                                                                                             RV YES (YES)

 99  ALWAYS ARE THOSE                          1.00     99  TEND LESS TO BE THOSE                     0.54         8   42
     WHERE GOVERNMENTAL STABILITY                          WHERE GOVERNMENTAL STABILITY                           0   36
     IS GENERALLY PRESENT AND DATES                        IS GENERALLY PRESENT AND DATES                    X SQ=    4.60
     FROM AT LEAST THE INTER-WAR PERIOD,                   FROM AT LEAST THE INTER-WAR PERIOD,               P =    0.019
     OR FROM THE POST-WAR PERIOD  (50)                     OR FROM THE POST-WAR PERIOD  (50)                 RV NO  (NO )

102  ALWAYS ARE THOSE                          1.00    102  TEND TO BE THOSE                          0.60         0   59
     WHERE THE REPRESENTATIVE CHARACTER                    WHERE THE REPRESENTATIVE CHARACTER                     9   40
     OF THE REGIME IS PSEUDO-POLYARCHIC                    OF THE REGIME IS POLYARCHIC                       X SQ=    9.54
     OR NON-POLYARCHIC  (49)                               OR LIMITED POLYARCHIC  (59)                       P =    0.001
                                                                                                             RV YES (NO )

106  ALWAYS ARE THOSE                          1.00    106  TEND TO BE THOSE                          0.71         0   52
     WHERE THE ELECTORAL SYSTEM IS                         WHERE THE ELECTORAL SYSTEM IS                          9   21
     NON-COMPETITIVE  (30)                                 COMPETITIVE OR                                    X SQ=   14.59
                                                           PARTIALLY COMPETITIVE  (52)                       P =    0.
                                                                                                             RV YES (NO )

108  ALWAYS ARE THOSE                          1.00    108  TEND TO BE THOSE                          0.71         0   65
     WHERE AUTONOMOUS GROUPS                               WHERE AUTONOMOUS GROUPS                                9   26
     ARE TOLERATED ONLY OUTSIDE POLITICS                   ARE FULLY OR PARTIALLY TOLERATED                  X SQ=   15.36
     OR ARE NOT TOLERATED AT ALL  (35)                     IN POLITICS  (65)                                 P =    0.
                                                                                                             RV YES (NO )
```

1/

116	ALWAYS ARE THOSE WHERE INTEREST ARTICULATION BY ASSOCIATIONAL GROUPS IS LIMITED OR NEGLIGIBLE (79)	1.00	116	TILT LESS TOWARD BEING THOSE WHERE INTEREST ARTICULATION BY ASSOCIATIONAL GROUPS IS LIMITED OR NEGLIGIBLE (79)	0.69	0 32 8 71 X SQ= 2.14 P = 0.102 RV NO (NO)
118	ALWAYS ARE THOSE WHERE INTEREST ARTICULATION BY INSTITUTIONAL GROUPS IS VERY SIGNIFICANT (40)	1.00	118	TEND TO BE THOSE WHERE INTEREST ARTICULATION BY INSTITUTIONAL GROUPS IS SIGNIFICANT, MODERATE, OR LIMITED (60)	0.66	9 31 0 60 X SQ= 12.22 P = 0. RV YES (NO)
121	ALWAYS ARE THOSE WHERE INTEREST ARTICULATION BY NON-ASSOCIATIONAL GROUPS IS MODERATE, LIMITED, OR NEGLIGIBLE (61)	1.00	121	TEND TO BE THOSE WHERE INTEREST ARTICULATION BY NON-ASSOCIATIONAL GROUPS IS SIGNIFICANT (54)	0.51	0 54 9 52 X SQ= 6.72 P = 0.003 RV YES (NO)
129	ALWAYS ARE THOSE WHERE INTEREST ARTICULATION BY POLITICAL PARTIES IS NEGLIGIBLE (37)	1.00	129	TEND TO BE THOSE WHERE INTEREST ARTICULATION BY POLITICAL PARTIES IS SIGNIFICANT, MODERATE, OR LIMITED (56)	0.67	0 56 9 28 X SQ= 12.43 P = 0. RV YES (NO)
134	ALWAYS ARE THOSE WHERE INTEREST AGGREGATION BY THE EXECUTIVE IS LIMITED OR NEGLIGIBLE (46)	1.00	134	TEND TO BE THOSE WHERE INTEREST AGGREGATION BY THE EXECUTIVE IS SIGNIFICANT OR MODERATE (57)	0.61	0 57 9 37 X SQ= 9.89 P = 0. RV YES (NO)
138	ALWAYS ARE THOSE WHERE INTEREST AGGREGATION BY THE LEGISLATURE IS NEGLIGIBLE (49)	1.00	138	TEND TO BE THOSE WHERE INTEREST AGGREGATION BY THE LEGISLATURE IS SIGNIFICANT, MODERATE, OR LIMITED (48)	0.55	0 48 9 40 X SQ= 7.66 P = 0.003 RV YES (NO)
139	ALWAYS ARE THOSE WHERE THE PARTY SYSTEM IS QUANTITATIVELY ONE-PARTY (34)	1.00	139	TEND TO BE THOSE WHERE THE PARTY SYSTEM IS QUANTITATIVELY OTHER THAN ONE-PARTY (71)	0.74	9 25 0 71 X SQ= 17.32 P = 0. RV YES (NO)
147	ALWAYS ARE THOSE WHERE THE PARTY SYSTEM IS QUALITATIVELY OTHER THAN CLASS-ORIENTED OR MULTI-IDEOLOGICAL (67)	1.00	147	TILT LESS TOWARD BEING THOSE WHERE THE PARTY SYSTEM IS QUALITATIVELY OTHER THAN CLASS-ORIENTED OR MULTI-IDEOLOGICAL (67)	0.72	0 23 9 58 X SQ= 2.10 P = 0.105 RV NO (NO)
153	ALWAYS ARE THOSE WHERE THE PARTY SYSTEM IS STABLE (42)	1.00	153	TEND TO BE THOSE WHERE THE PARTY SYSTEM IS MODERATELY STABLE OR UNSTABLE (38)	0.54	9 33 0 38 X SQ= 7.15 P = 0.003 RV YES (NO)

1/

158	ALWAYS ARE THOSE WHERE PERSONALISMO IS NEGLIGIBLE (56)	1.00	158	TEND LESS TO BE THOSE WHERE PERSONALISMO IS NEGLIGIBLE (56)	0.54	0 40 9 47 X SQ= 5.33 P = 0.009 RV NO (NO)
160	ALWAYS ARE THOSE WHERE THE POLITICAL LEADERSHIP IS ELITIST (30)	1.00	160	TEND TO BE THOSE WHERE THE POLITICAL LEADERSHIP IS MODERATELY ELITIST OR NON-ELITIST (67)	0.76	9 21 0 67 X SQ= 18.73 P = 0. RV YES (NO)
169	ALWAYS ARE THOSE WHERE THE HORIZONTAL POWER DISTRIBUTION IS NEGLIGIBLE (48)	1.00	169	TEND TO BE THOSE WHERE THE HORIZONTAL POWER DISTRIBUTION IS SIGNIFICANT OR LIMITED (58)	0.60	0 58 9 39 X SQ= 9.59 P = 0.001 RV YES (NO)
170	ALWAYS ARE THOSE WHERE THE LEGISLATIVE-EXECUTIVE STRUCTURE IS OTHER THAN PRESIDENTIAL (63)	1.00	170	TEND LESS TO BE THOSE WHERE THE LEGISLATIVE-EXECUTIVE STRUCTURE IS OTHER THAN PRESIDENTIAL (63)	0.58	0 39 9 54 X SQ= 4.46 P = 0.012 RV NO (NO)
176	ALWAYS ARE THOSE WHERE THE LEGISLATURE IS WHOLLY INEFFECTIVE (28)	1.00	176	TEND TO BE THOSE WHERE THE LEGISLATURE IS FULLY EFFECTIVE, PARTIALLY EFFECTIVE, OR LARGELY INEFFECTIVE (72)	0.79	0 72 9 19 X SQ= 21.66 P = 0. RV YES (NO)
179	ALWAYS ARE THOSE WHERE THE EXECUTIVE IS DOMINANT (52)	1.00	179	TEND LESS TO BE THOSE WHERE THE EXECUTIVE IS DOMINANT (52)	0.52	9 43 0 39 X SQ= 5.67 P = 0.009 RV NO (NO)
181	ALWAYS ARE THOSE WHERE THE BUREAUCRACY IS SEMI-MODERN, RATHER THAN MODERN (55)	1.00	181	LEAN LESS TOWARD BEING THOSE WHERE THE BUREAUCRACY IS SEMI-MODERN, RATHER THAN MODERN (55)	0.69	0 21 9 46 X SQ= 2.49 P = 0.056 RV NO (NO)
182	ALWAYS ARE THOSE WHERE THE BUREAUCRACY IS SEMI-MODERN, RATHER THAN POST-COLONIAL TRANSITIONAL (55)	1.00	182	LEAN LESS TOWARD BEING THOSE WHERE THE BUREAUCRACY IS SEMI-MODERN, RATHER THAN POST-COLONIAL TRANSITIONAL (55)	0.65	9 46 0 25 X SQ= 3.12 P = 0.051 RV NO (NO)
185	ALWAYS ARE THOSE WHERE PARTICIPATION BY THE MILITARY IN POLITICS IS SUPPORTIVE, RATHER THAN INTERVENTIVE (31)	1.00	185	TEND LESS TO BE THOSE WHERE PARTICIPATION BY THE MILITARY IN POLITICS IS SUPPORTIVE, RATHER THAN INTERVENTIVE (31)	0.51	0 21 9 22 X SQ= 5.48 P = 0.007 RV NO (NO)

1/

186	ALWAYS ARE THOSE WHERE PARTICIPATION BY THE MILITARY IN POLITICS IS SUPPORTIVE, RATHER THAN NEUTRAL (31)	1.00	186	TEND TO BE THOSE WHERE PARTICIPATION BY THE MILITARY IN POLITICS IS NEUTRAL, RATHER THAN SUPPORTIVE (56)	0.72	9 22 0 56 X SQ= 15.14 P = 0. RV YES (NO)
187	ALWAYS ARE THOSE WHERE THE ROLE OF THE POLICE IS POLITICALLY SIGNIFICANT (66)	1.00	187	TEND LESS TO BE THOSE WHERE THE ROLE OF THE POLICE IS POLITICALLY SIGNIFICANT (66)	0.62	9 57 0 35 X SQ= 3.69 P = 0.025 RV NO (NO)
194	ALWAYS ARE THOSE THAT ARE COMMUNIST (13)	1.00	194	TEND TO BE THOSE THAT ARE NON-COMMUNIST (101)	0.96	9 4 0 101 X SQ= 66.69 P = 0. RV YES (YES)

```
****************************************************************************************************

  2  POLITIES                                           2  POLITIES
     LOCATED IN WEST EUROPE, SCANDINAVIA,                  LOCATED ELSEWHERE THAN IN
     NORTH AMERICA, OR AUSTRALASIA  (22)                   WEST EUROPE, SCANDINAVIA,
                                                           NORTH AMERICA, OR AUSTRALASIA  (93)

                                                                            BOTH SUBJECT AND PREDICATE
....................................................................................................

      22 IN LEFT COLUMN

AUSTRALIA  AUSTRIA    BELGIUM    CANADA     DENMARK    FINLAND    FRANCE     GERMAN FR  GREECE     ICELAND
IRELAND    ITALY      LUXEMBOURG NETHERLANDS NEW ZEALAND NORWAY   PORTUGAL   SPAIN      SWEDEN     SWITZERLAND
UK         US

      93 IN RIGHT COLUMN

AFGHANISTAN ALBANIA   ALGERIA    ARGENTINA  BOLIVIA    BRAZIL     BULGARIA   BURMA      BURUNDI    CAMBODIA
CAMEROUN   CEN AFR REP CEYLON    CHAD       CHILE      CHINA, PR  COLOMBIA   CONGO(BRA) CONGO(LEO) COSTA RICA
CUBA       CYPRUS     CZECHOS'KIA DAHOMEY   DOMIN REP  ECUADOR    EL SALVADOR ETHIOPIA  GABON      GERMANY, E
GHANA      GUATEMALA  GUINEA     HAITI      HONDURAS   HUNGARY    INDIA      INDONESIA  IRAN       IRAQ
ISRAEL     IVORY COAST JAMAICA   JAPAN      JORDAN     KOREA, N   KOREA REP  LAOS       LEBANON    LIBERIA
LIBYA      MALAGASY R MALAYA     MALI       MAURITANIA MEXICO     MONGOLIA   MOROCCO    NEPAL      NICARAGUA
NIGER      NIGERIA    PAKISTAN   PANAMA     PARAGUAY   PERU       PHILIPPINES POLAND    RUMANIA    RWANDA
SA'U ARABIA SENEGAL   SIERRE LEO SOMALIA    SO AFRICA  SUDAN      SYRIA      TANGANYIKA THAILAND   TOGO
TRINIDAD   TUNISIA    TURKEY     UGANDA     USSR       UAR        UPPER VOLTA URUGUAY   VENEZUELA  VIETNAM, N
VIETNAM REP YEMEN     YUGOSLAVIA
----------------------------------------------------------------------------------------------------

 21  TILT MORE TOWARD BEING THOSE          0.86      21  TILT LESS TOWARD BEING THOSE         0.69         3   29
     WHOSE TERRITORIAL SIZE IS                           WHOSE TERRITORIAL SIZE IS                        19   64
     MEDIUM OR SMALL  (83)                               MEDIUM OR SMALL  (83)                        X SQ=   1.92
                                                                                                      P =    0.118
                                                                                                      RV NO  (NO )

 26  TEND TO BE THOSE                      0.64      26  TEND TO BE THOSE                     0.63        14   34
     WHOSE POPULATION DENSITY IS                         WHOSE POPULATION DENSITY IS                       8   59
     VERY HIGH, HIGH, OR MEDIUM  (48)                    LOW  (67)                                    X SQ=   4.31
                                                                                                      P =    0.030
                                                                                                      RV YES (NO )

 28  TEND TO BE THOSE                      0.82      28  TEND TO BE THOSE                     0.66         4   58
     WHOSE POPULATION GROWTH RATE                        WHOSE POPULATION GROWTH RATE                     18   30
     IS LOW  (48)                                        IS HIGH  (62)                                X SQ=  14.42
                                                                                                      P =    0.
                                                                                                      RV YES (NO )

 29  ALWAYS ARE THOSE                      1.00      29  TEND TO BE THOSE                     0.58        21   35
     WHERE THE DEGREE OF URBANIZATION                    WHERE THE DEGREE OF URBANIZATION                  0   49
     IS HIGH  (56)                                       IS LOW  (49)                                 X SQ=  20.68
                                                                                                      P =    0.
                                                                                                      RV YES (NO )
```

30	ALWAYS ARE THOSE WHOSE AGRICULTURAL POPULATION IS MEDIUM, LOW, OR VERY LOW (57)	1.00	30	TEND TO BE THOSE WHOSE AGRICULTURAL POPULATION IS HIGH (56)	0.62	0 56 22 35 X SQ= 24.43 P = 0. RV YES (NO)
33	TEND TO BE THOSE WHOSE GROSS NATIONAL PRODUCT IS VERY HIGH, HIGH, OR MEDIUM (30)	0.64	33	TEND TO BE THOSE WHOSE GROSS NATIONAL PRODUCT IS LOW OR VERY LOW (85)	0.83	14 16 8 77 X SQ= 17.56 P = 0. RV YES (NO)
34	TEND TO BE THOSE WHOSE GROSS NATIONAL PRODUCT IS VERY HIGH, HIGH, MEDIUM, OR LOW (62)	0.91	34	TEND TO BE THOSE WHOSE GROSS NATIONAL PRODUCT IS VERY LOW (53)	0.55	20 42 2 51 X SQ= 13.20 P = 0. RV YES (NO)
35	TEND TO BE THOSE WHOSE PER CAPITA GROSS NATIONAL PRODUCT IS VERY HIGH OR HIGH (24)	0.86	35	TEND TO BE THOSE WHOSE PER CAPITA GROSS NATIONAL PRODUCT IS MEDIUM, LOW, OR VERY LOW (91)	0.95	19 5 3 88 X SQ= 65.84 P = 0. RV YES (YES)
36	TEND TO BE THOSE WHOSE PER CAPITA GROSS NATIONAL PRODUCT IS VERY HIGH, HIGH, OR MEDIUM (42)	0.95	36	TEND TO BE THOSE WHOSE PER CAPITA GROSS NATIONAL PRODUCT IS LOW OR VERY LOW (73)	0.77	21 21 1 72 X SQ= 37.67 P = 0. RV YES (YES)
37	ALWAYS ARE THOSE WHOSE PER CAPITA GROSS NATIONAL PRODUCT IS VERY HIGH, HIGH, MEDIUM, OR LOW (64)	1.00	37	TEND TO BE THOSE WHOSE PER CAPITA GROSS NATIONAL PRODUCT IS VERY LOW (51)	0.55	22 42 0 51 X SQ= 19.51 P = 0. RV YES (NO)
39	TEND TO BE THOSE WHOSE INTERNATIONAL FINANCIAL STATUS IS VERY HIGH, HIGH, OR MEDIUM (38)	0.77	39	TEND TO BE THOSE WHOSE INTERNATIONAL FINANCIAL STATUS IS LOW OR VERY LOW (76)	0.77	17 21 5 71 X SQ= 21.30 P = 0. RV YES (NO)
41	TEND TO BE THOSE WHOSE ECONOMIC DEVELOPMENTAL STATUS IS DEVELOPED (19)	0.73	41	TEND TO BE THOSE WHOSE ECONOMIC DEVELOPMENTAL STATUS IS INTERMEDIATE, UNDERDEVELOPED, OR VERY UNDERDEVELOPED (94)	0.97	16 3 6 88 X SQ= 56.20 P = 0. RV YES (YES)
42	TEND TO BE THOSE WHOSE ECONOMIC DEVELOPMENTAL STATUS IS DEVELOPED OR INTERMEDIATE (36)	0.91	42	TEND TO BE THOSE WHOSE ECONOMIC DEVELOPMENTAL STATUS IS UNDERDEVELOPED OR VERY UNDERDEVELOPED (76)	0.82	20 16 2 74 X SQ= 40.06 P = 0. RV YES (YES)

No.	Statement		No.	Statement		Statistics
43	ALWAYS ARE THOSE WHOSE ECONOMIC DEVELOPMENTAL STATUS IS DEVELOPED, INTERMEDIATE, OR UNDERDEVELOPED (55)	1.00	43	TEND TO BE THOSE WHOSE ECONOMIC DEVELOPMENTAL STATUS IS VERY UNDERDEVELOPED (57)	0.63	22 33 0 57 X SQ= 25.90 P = 0. RV YES (NO)
44	TEND TO BE THOSE WHERE THE LITERACY RATE IS NINETY PERCENT OR ABOVE (25)	0.82	44	TEND TO BE THOSE WHERE THE LITERACY RATE IS BELOW NINETY PERCENT (90)	0.92	18 7 4 86 X SQ= 53.43 P = 0. RV YES (YES)
45	ALWAYS ARE THOSE WHERE THE LITERACY RATE IS FIFTY PERCENT OR ABOVE (55)	1.00	45	TEND TO BE THOSE WHERE THE LITERACY RATE IS BELOW FIFTY PERCENT (54)	0.62	22 33 0 54 X SQ= 24.64 P = 0. RV YES (NO)
50	TEND TO BE THOSE WHERE FREEDOM OF THE PRESS IS COMPLETE (43)	0.86	50	TEND TO BE THOSE WHERE FREEDOM OF THE PRESS IS INTERMITTENT, INTERNALLY ABSENT, OR INTERNALLY AND EXTERNALLY ABSENT (56)	0.68	18 25 3 53 X SQ= 17.27 P = 0. RV YES (NO)
51	TEND MORE TO BE THOSE WHERE FREEDOM OF THE PRESS IS COMPLETE OR INTERMITTENT (60)	0.90	51	TEND LESS TO BE THOSE WHERE FREEDOM OF THE PRESS IS COMPLETE OR INTERMITTENT (60)	0.54	19 41 2 35 X SQ= 7.82 P = 0.002 RV NO (NO)
52	ALWAYS ARE THOSE WHERE FREEDOM OF THE PRESS IS COMPLETE, INTERMITTENT, OR INTERNALLY ABSENT (82)	1.00	52	TEND LESS TO BE THOSE WHERE FREEDOM OF THE PRESS IS COMPLETE, INTERMITTENT, OR INTERNALLY ABSENT (82)	0.79	22 60 0 16 X SQ= 4.10 P = 0.019 RV NO (NO)
53	TEND TO BE THOSE WHERE NEWSPAPER CIRCULATION IS THREE HUNDRED OR MORE PER THOUSAND (14)	0.55	53	TEND TO BE THOSE WHERE NEWSPAPER CIRCULATION IS LESS THAN THREE HUNDRED PER THOUSAND (101)	0.98	12 2 10 91 X SQ= 40.91 P = 0. RV YES (YES)
54	TEND TO BE THOSE WHERE NEWSPAPER CIRCULATION IS ONE HUNDRED OR MORE PER THOUSAND (37)	0.91	54	TEND TO BE THOSE WHERE NEWSPAPER CIRCULATION IS LESS THAN ONE HUNDRED PER THOUSAND (76)	0.81	20 17 2 74 X SQ= 38.75 P = 0. RV YES (YES)
55	ALWAYS ARE THOSE WHERE NEWSPAPER CIRCULATION IS TEN OR MORE PER THOUSAND (78)	1.00	55	TEND LESS TO BE THOSE WHERE NEWSPAPER CIRCULATION IS TEN OR MORE PER THOUSAND (78)	0.62	22 56 0 35 X SQ= 10.53 P = 0. RV NO (NO)

56	ALWAYS ARE THOSE WHERE THE RELIGION IS PREDOMINANTLY LITERATE (79)	1.00	56	TEND LESS TO BE THOSE WHERE THE RELIGION IS PREDOMINANTLY LITERATE (79)	0.65	22 57 0 31 X SQ= 9.12 P = 0. RV NO (NO)
62	TEND LESS TO BE THOSE WHERE THE RELIGION IS CATHOLIC, RATHER THAN PROTESTANT (25)	0.62	62	ALWAYS ARE THOSE WHERE THE RELIGION IS CATHOLIC, RATHER THAN PROTESTANT (25)	1.00	5 0 8 17 X SQ= 5.32 P = 0.009 RV NO (YES)
63	ALWAYS ARE THOSE WHERE THE RELIGION IS PREDOMINANTLY SOME KIND OF CHRISTIAN (46)	1.00	63	TEND TO BE THOSE WHERE THE RELIGION IS PREDOMINANTLY OR PARTLY OTHER THAN CHRISTIAN (68)	0.74	22 24 0 68 X SQ= 37.29 P = 0. RV YES (NO)
67	TEND MORE TO BE THOSE THAT ARE RACIALLY HOMOGENEOUS (82)	0.95	67	TEND LESS TO BE THOSE THAT ARE RACIALLY HOMOGENEOUS (82)	0.68	21 59 1 28 X SQ= 5.53 P = 0.007 RV NO (NO)
68	TEND TO BE THOSE THAT ARE LINGUISTICALLY HOMOGENEOUS (52)	0.77	68	TEND TO BE THOSE THAT ARE LINGUISTICALLY WEAKLY HETEROGENEOUS OR STRONGLY HETEROGENEOUS (62)	0.62	17 35 5 57 X SQ= 9.49 P = 0.002 RV YES (NO)
69	TEND TO BE THOSE THAT ARE LINGUISTICALLY HOMOGENEOUS OR WEAKLY HETEROGENEOUS (64)	0.86	69	TEND TO BE THOSE THAT ARE LINGUISTICALLY STRONGLY HETEROGENEOUS (50)	0.51	19 45 3 47 X SQ= 8.65 P = 0.002 RV YES (NO)
70	TEND TO BE THOSE THAT ARE RELIGIOUSLY, RACIALLY, AND LINGUISTICALLY HOMOGENEOUS (21)	0.50	70	TEND TO BE THOSE THAT ARE RELIGIOUSLY, RACIALLY, OR LINGUISTICALLY HETEROGENEOUS (85)	0.88	11 10 11 73 X SQ= 13.37 P = 0. RV YES (YES)
71	TEND TO BE THOSE WHOSE DATE OF INDEPENDENCE IS BEFORE 1800 (21)	0.57	71	TEND TO BE THOSE WHOSE DATE OF INDEPENDENCE IS AFTER 1800 (90)	0.90	12 9 9 81 X SQ= 21.69 P = 0. RV YES (YES)
72	TEND TO BE THOSE WHOSE DATE OF INDEPENDENCE IS BEFORE 1914 (52)	0.86	72	TEND TO BE THOSE WHOSE DATE OF INDEPENDENCE IS AFTER 1914 (59)	0.62	18 34 3 56 X SQ= 13.85 P = 0. RV YES (NO)

73	ALWAYS ARE THOSE WHOSE DATE OF INDEPENDENCE IS BEFORE 1945 (65)	1.00	73	TEND TO BE THOSE WHOSE DATE OF INDEPENDENCE IS AFTER 1945 (46)	0.51	21 44 0 46 X SQ= 16.28 P = 0. RV YES (NO)
74	ALWAYS ARE THOSE THAT ARE HISTORICALLY WESTERN (26)	1.00	74	TEND TO BE THOSE THAT ARE NOT HISTORICALLY WESTERN (87)	0.96	22 4 0 87 X SQ= 86.10 P = 0. RV YES (YES)
81	TEND LESS TO BE THOSE WHERE THE TYPE OF POLITICAL MODERNIZATION IS LATER EUROPEAN OR LATER EUROPEAN DERIVED, RATHER THAN EARLY EUROPEAN OR EARLY EUROPEAN DERIVED (40)	0.55	81	TEND MORE TO BE THOSE WHERE THE TYPE OF POLITICAL MODERNIZATION IS LATER EUROPEAN OR LATER EUROPEAN DERIVED, RATHER THAN EARLY EUROPEAN OR EARLY EUROPEAN DERIVED (40)	0.97	10 1 12 28 X SQ= 10.68 P = 0. RV NO (YES)
85	ALWAYS ARE THOSE WHERE THE STAGE OF POLITICAL MODERNIZATION IS ADVANCED, RATHER THAN MID- OR EARLY TRANSITIONAL (60)	1.00	85	TEND TO BE THOSE WHERE THE STAGE OF POLITICAL MODERNIZATION IS MID- OR EARLY TRANSITIONAL, RATHER THAN ADVANCED (54)	0.59	22 38 0 54 X SQ= 22.24 P = 0. RV YES (NO)
89	ALWAYS ARE THOSE WHOSE IDEOLOGICAL ORIENTATION IS CONVENTIONAL (33)	1.00	89	TEND TO BE THOSE WHOSE IDEOLOGICAL ORIENTATION IS OTHER THAN CONVENTIONAL (62)	0.83	20 13 0 62 X SQ= 44.02 P = 0. RV YES (YES)
93	ALWAYS ARE THOSE WHERE THE SYSTEM STYLE IS NON-MOBILIZATIONAL (78)	1.00	93	TEND LESS TO BE THOSE WHERE THE SYSTEM STYLE IS NON-MOBILIZATIONAL (78)	0.64	0 32 21 57 X SQ= 8.98 P = 0. RV NO (NO)
94	TEND TO BE THOSE WHERE THE STATUS OF THE REGIME IS CONSTITUTIONAL (51)	0.91	94	TEND TO BE THOSE WHERE THE STATUS OF THE REGIME IS AUTHORITARIAN OR TOTALITARIAN (41)	0.56	20 31 2 39 X SQ= 12.90 P = 0. RV YES (NO)
96	ALWAYS ARE THOSE WHERE THE STATUS OF THE REGIME IS CONSTITUTIONAL OR TOTALITARIAN (67)	1.00	96	TEND LESS TO BE THOSE WHERE THE STATUS OF THE REGIME IS CONSTITUTIONAL OR TOTALITARIAN (67)	0.66	22 45 0 23 X SQ= 8.30 P = 0.001 RV NO (NO)
98	TEND TO BE THOSE WHERE GOVERNMENTAL STABILITY IS GENERALLY PRESENT AND DATES AT LEAST FROM THE INTERWAR PERIOD (22)	0.77	98	TEND TO BE THOSE WHERE GOVERNMENTAL STABILITY IS GENERALLY OR MODERATELY PRESENT AND DATES FROM THE POST-WAR PERIOD, OR IS ABSENT (93)	0.95	17 5 5 88 X SQ= 54.89 P = 0. RV YES (YES)

99	TEND TO BE THOSE WHERE GOVERNMENTAL STABILITY IS GENERALLY PRESENT AND DATES FROM AT LEAST THE INTER-WAR PERIOD, OR FROM THE POST-WAR PERIOD (50)	0.95	99	TEND TO BE THOSE WHERE GOVERNMENTAL STABILITY IS MODERATELY PRESENT AND DATES FROM THE POST-WAR PERIOD, OR IS ABSENT (36)	0.55	21 29 1 35 X SQ= 14.92 P = 0. RV YES (NO)
100	ALWAYS ARE THOSE WHERE GOVERNMENTAL STABILITY IS GENERALLY PRESENT AND DATES FROM AT LEAST THE INTER-WAR PERIOD, OR IS GENERALLY OR MODERATELY PRESENT, AND DATES FROM THE POST-WAR PERIOD (64)	1.00	100	TEND LESS TO BE THOSE WHERE GOVERNMENTAL STABILITY IS GENERALLY PRESENT AND DATES FROM AT LEAST THE INTER-WAR PERIOD, OR IS GENERALLY OR MODERATELY PRESENT, AND DATES FROM THE POST-WAR PERIOD (64)	0.66	22 42 0 22 X SQ= 8.44 P = 0.001 RV NO (NO)
101	TEND TO BE THOSE WHERE THE REPRESENTATIVE CHARACTER OF THE REGIME IS POLYARCHIC (41)	0.91	101	TEND TO BE THOSE WHERE THE REPRESENTATIVE CHARACTER OF THE REGIME IS LIMITED POLYARCHIC, PSEUDO-POLYARCHIC, OR NON-POLYARCHIC (57)	0.72	20 21 2 55 X SQ= 25.53 P = 0. RV YES (NO)
102	TEND TO BE THOSE WHERE THE REPRESENTATIVE CHARACTER OF THE REGIME IS POLYARCHIC OR LIMITED POLYARCHIC (59)	0.91	102	TEND TO BE THOSE WHERE THE REPRESENTATIVE CHARACTER OF THE REGIME IS PSEUDO-POLYARCHIC OR NON-POLYARCHIC (49)	0.55	20 39 2 47 X SQ= 12.89 P = 0. RV YES (NO)
105	TEND TO BE THOSE WHERE THE ELECTORAL SYSTEM IS COMPETITIVE (43)	0.91	105	TEND TO BE THOSE WHERE THE ELECTORAL SYSTEM IS PARTIALLY COMPETITIVE OR NON-COMPETITIVE (47)	0.66	20 23 2 45 X SQ= 19.48 P = 0. RV YES (NO)
106	TEND MORE TO BE THOSE WHERE THE ELECTORAL SYSTEM IS COMPETITIVE OR PARTIALLY COMPETITIVE (52)	0.91	106	TEND LESS TO BE THOSE WHERE THE ELECTORAL SYSTEM IS COMPETITIVE OR PARTIALLY COMPETITIVE (52)	0.53	20 32 2 28 X SQ= 8.24 P = 0.002 RV NO (NO)
107	TEND TO BE THOSE WHERE AUTONOMOUS GROUPS ARE FULLY TOLERATED IN POLITICS (46)	0.91	107	TEND TO BE THOSE WHERE AUTONOMOUS GROUPS ARE PARTIALLY TOLERATED IN POLITICS, ARE TOLERATED ONLY OUTSIDE POLITICS, OR ARE NOT TOLERATED AT ALL (65)	0.71	20 26 2 63 X SQ= 25.18 P = 0. RV YES (NO)
110	TEND TO BE THOSE WHERE POLITICAL ENCULTURATION IS HIGH (15)	0.50	110	TEND TO BE THOSE WHERE POLITICAL ENCULTURATION IS MEDIUM OR LOW (80)	0.93	10 5 10 70 X SQ= 19.16 P = 0. RV YES (YES)
111	TEND TO BE THOSE WHERE POLITICAL ENCULTURATION IS HIGH OR MEDIUM (53)	0.90	111	TEND TO BE THOSE WHERE POLITICAL ENCULTURATION IS LOW (42)	0.53	18 35 2 40 X SQ= 10.33 P = 0.001 RV YES (NO)

115	TEND TO BE THOSE WHERE INTEREST ARTICULATION BY ASSOCIATIONAL GROUPS IS SIGNIFICANT (20)	0.80	115	TEND TO BE THOSE WHERE INTEREST ARTICULATION BY ASSOCIATIONAL GROUPS IS MODERATE, LIMITED, OR NEGLIGIBLE (91)	0.96	16 4 4 87 X SQ= 58.43 P = 0. RV YES (YES)
116	ALWAYS ARE THOSE WHERE INTEREST ARTICULATION BY ASSOCIATIONAL GROUPS IS SIGNIFICANT OR MODERATE (32)	1.00	116	TEND TO BE THOSE WHERE INTEREST ARTICULATION BY ASSOCIATIONAL GROUPS IS LIMITED OR NEGLIGIBLE (79)	0.87	20 12 0 79 X SQ= 56.07 P = 0. RV YES (YES)
119	TEND TO BE THOSE WHERE INTEREST ARTICULATION BY INSTITUTIONAL GROUPS IS MODERATE OR LIMITED (26)	0.86	119	TEND TO BE THOSE WHERE INTEREST ARTICULATION BY INSTITUTIONAL GROUPS IS VERY SIGNIFICANT OR SIGNIFICANT (74)	0.91	3 71 19 7 X SQ= 49.47 P = 0. RV YES (YES)
120	TEND LESS TO BE THOSE WHERE INTEREST ARTICULATION BY INSTITUTIONAL GROUPS IS VERY SIGNIFICANT, SIGNIFICANT, OR MODERATE (90)	0.55	120	ALWAYS ARE THOSE WHERE INTEREST ARTICULATION BY INSTITUTIONAL GROUPS IS VERY SIGNIFICANT, SIGNIFICANT, OR MODERATE (90)	1.00	12 78 10 0 X SQ= 34.51 P = 0. RV NO (YES)
121	ALWAYS ARE THOSE WHERE INTEREST ARTICULATION BY NON-ASSOCIATIONAL GROUPS IS MODERATE, LIMITED, OR NEGLIGIBLE (61)	1.00	121	TEND TO BE THOSE WHERE INTEREST ARTICULATION BY NON-ASSOCIATIONAL GROUPS IS SIGNIFICANT (54)	0.58	0 54 22 39 X SQ= 21.81 P = 0. RV YES (NO)
122	TEND TO BE THOSE WHERE INTEREST ARTICULATION BY NON-ASSOCIATIONAL GROUPS IS LIMITED OR NEGLIGIBLE (32)	0.82	122	TEND TO BE THOSE WHERE INTEREST ARTICULATION BY NON-ASSOCIATIONAL GROUPS IS SIGNIFICANT OR MODERATE (83)	0.85	4 79 18 14 X SQ= 36.23 P = 0. RV YES (YES)
124	ALWAYS ARE THOSE WHERE INTEREST ARTICULATION BY ANOMIC GROUPS IS OCCASIONAL, INFREQUENT, OR VERY INFREQUENT (85)	1.00	124	TEND LESS TO BE THOSE WHERE INTEREST ARTICULATION BY ANOMIC GROUPS IS OCCASIONAL, INFREQUENT, OR VERY INFREQUENT (85)	0.82	0 14 20 65 X SQ= 2.80 P = 0.040 RV NO (NO)
125	TEND TO BE THOSE WHERE INTEREST ARTICULATION BY ANOMIC GROUPS IS INFREQUENT OR VERY INFREQUENT (35)	0.95	125	TEND TO BE THOSE WHERE INTEREST ARTICULATION BY ANOMIC GROUPS IS FREQUENT OR OCCASIONAL (64)	0.80	1 63 19 16 X SQ= 35.81 P = 0. RV YES (YES)
126	TEND TO BE THOSE WHERE INTEREST ARTICULATION BY ANOMIC GROUPS IS VERY INFREQUENT (16)	0.80	126	ALWAYS ARE THOSE WHERE INTEREST ARTICULATION BY ANOMIC GROUPS IS FREQUENT, OCCASIONAL, OR INFREQUENT (83)	1.00	4 79 16 0 X SQ= 69.59 P = 0. RV YES (YES)

128	TEND TO BE THOSE WHERE INTEREST ARTICULATION BY POLITICAL PARTIES IS SIGNIFICANT OR MODERATE (48)	0.82	128	TEND TO BE THOSE WHERE INTEREST ARTICULATION BY POLITICAL PARTIES IS LIMITED OR NEGLIGIBLE (45)	0.58	18 30 4 41 X SQ= 9.00 P = 0.001 RV YES (NO)
132	ALWAYS ARE THOSE WHERE INTEREST AGGREGATION BY POLITICAL PARTIES IS SIGNIFICANT, MODERATE, OR LIMITED (67)	1.00	132	TILT LESS TOWARD BEING THOSE WHERE INTEREST AGGREGATION BY POLITICAL PARTIES IS SIGNIFICANT, MODERATE, OR LIMITED (67)	0.84	18 49 0 9 X SQ= 1.86 P = 0.105 RV NO (NO)
134	TEND TO BE THOSE WHERE INTEREST AGGREGATION BY THE EXECUTIVE IS SIGNIFICANT OR MODERATE (57)	0.86	134	TEND TO BE THOSE WHERE INTEREST AGGREGATION BY THE EXECUTIVE IS LIMITED OR NEGLIGIBLE (46)	0.53	19 38 3 43 X SQ= 9.36 P = 0.001 RV YES (NO)
137	TEND TO BE THOSE WHERE INTEREST AGGREGATION BY THE LEGISLATURE IS SIGNIFICANT OR MODERATE (29)	0.86	137	TEND TO BE THOSE WHERE INTEREST AGGREGATION BY THE LEGISLATURE IS LIMITED OR NEGLIGIBLE (68)	0.87	19 10 3 65 X SQ= 39.87 P = 0. RV YES (YES)
138	ALWAYS ARE THOSE WHERE INTEREST AGGREGATION BY THE LEGISLATURE IS SIGNIFICANT, MODERATE, OR LIMITED (48)	1.00	138	TEND TO BE THOSE WHERE INTEREST AGGREGATION BY THE LEGISLATURE IS NEGLIGIBLE (49)	0.64	20 28 0 49 X SQ= 23.24 P = 0. RV YES (NO)
144	TEND TO BE THOSE WHERE THE PARTY SYSTEM IS QUANTITATIVELY TWO-PARTY, RATHER THAN ONE-PARTY (11)	0.75	144	TEND TO BE THOSE WHERE THE PARTY SYSTEM IS QUANTITATIVELY ONE-PARTY, RATHER THAN TWO-PARTY (34)	0.86	2 32 6 5 X SQ= 10.34 P = 0.001 RV YES (YES)
147	TEND TO BE THOSE WHERE THE PARTY SYSTEM IS QUALITATIVELY CLASS-ORIENTED OR MULTI-IDEOLOGICAL (23)	0.79	147	TEND TO BE THOSE WHERE THE PARTY SYSTEM IS QUALITATIVELY OTHER THAN CLASS-ORIENTED OR MULTI-IDEOLOGICAL (67)	0.89	15 8 4 63 X SQ= 32.62 P = 0. RV YES (YES)
153	TEND TO BE THOSE WHERE THE PARTY SYSTEM IS STABLE (42)	0.86	153	TEND TO BE THOSE WHERE THE PARTY SYSTEM IS MODERATELY STABLE OR UNSTABLE (38)	0.60	19 23 3 35 X SQ= 12.14 P = 0. RV YES (NO)
158	TEND TO BE THOSE WHERE PERSONALISMO IS NEGLIGIBLE (56)	0.91	158	TEND TO BE THOSE WHERE PERSONALISMO IS PRONOUNCED OR MODERATE (40)	0.51	2 38 20 36 X SQ= 10.78 P = 0. RV YES (NO)

161	TEND TO BE THOSE WHERE THE POLITICAL LEADERSHIP IS NON-ELITIST (50)	0.82	161	TEND TO BE THOSE WHERE THE POLITICAL LEADERSHIP IS ELITIST OR MODERATELY ELITIST (47)	0.57	4 43 18 32 X SQ= 8.93 P = 0.001 RV YES (NO)
164	TEND MORE TO BE THOSE WHERE THE REGIME'S LEADERSHIP CHARISMA IS NEGLIGIBLE (65)	0.95	164	TEND LESS TO BE THOSE WHERE THE REGIME'S LEADERSHIP CHARISMA IS NEGLIGIBLE (65)	0.57	1 33 21 44 X SQ= 9.50 P = 0.001 RV NO (NO)
168	TEND TO BE THOSE WHERE THE HORIZONTAL POWER DISTRIBUTION IS SIGNIFICANT (34)	0.86	168	TEND TO BE THOSE WHERE THE HORIZONTAL POWER DISTRIBUTION IS LIMITED OR NEGLIGIBLE (72)	0.82	19 15 3 69 X SQ= 34.48 P = 0. RV YES (YES)
169	TEND TO BE THOSE WHERE THE HORIZONTAL POWER DISTRIBUTION IS SIGNIFICANT OR LIMITED (58)	0.91	169	TEND TO BE THOSE WHERE THE HORIZONTAL POWER DISTRIBUTION IS NEGLIGIBLE (48)	0.55	20 38 2 46 X SQ= 12.89 P = 0. RV YES (NO)
170	TEND MORE TO BE THOSE WHERE THE LEGISLATIVE-EXECUTIVE STRUCTURE IS OTHER THAN PRESIDENTIAL (63)	0.95	170	TEND LESS TO BE THOSE WHERE THE LEGISLATIVE-EXECUTIVE STRUCTURE IS OTHER THAN PRESIDENTIAL (63)	0.52	1 38 21 42 X SQ= 11.72 P = 0. RV NO (NO)
172	TEND TO BE THOSE WHERE THE LEGISLATIVE-EXECUTIVE STRUCTURE IS PARLIAMENTARY-ROYALIST (21)	0.52	172	TEND TO BE THOSE WHERE THE LEGISLATIVE-EXECUTIVE STRUCTURE IS OTHER THAN PARLIAMENTARY-ROYALIST (88)	0.89	11 10 10 78 X SQ= 15.80 P = 0. RV YES (YES)
174	TEND TO BE THOSE WHERE THE LEGISLATURE IS FULLY EFFECTIVE (28)	0.82	174	TEND TO BE THOSE WHERE THE LEGISLATURE IS PARTIALLY EFFECTIVE, LARGELY INEFFECTIVE, OR WHOLLY INEFFECTIVE (72)	0.87	18 10 4 68 X SQ= 37.17 P = 0. RV YES (YES)
175	TEND TO BE THOSE WHERE THE LEGISLATURE IS FULLY EFFECTIVE OR PARTIALLY EFFECTIVE (51)	0.91	175	TEND TO BE THOSE WHERE THE LEGISLATURE IS LARGELY INEFFECTIVE OR WHOLLY INEFFECTIVE (49)	0.60	20 31 2 47 X SQ= 15.99 P = 0. RV YES (NO)
178	TEND TO BE THOSE WHERE THE LEGISLATURE IS BICAMERAL (51)	0.77	178	TEND TO BE THOSE WHERE THE LEGISLATURE IS UNICAMERAL (53)	0.59	5 48 17 34 X SQ= 7.53 P = 0.004 RV YES (NO)

179	TEND TO BE THOSE WHERE THE EXECUTIVE IS STRONG (39)	0.85	179	TEND TO BE THOSE WHERE THE EXECUTIVE IS DOMINANT (52)	0.69	3 49 17 22 X SQ= 16.45 P = 0. RV YES (NO)
181	TEND TO BE THOSE WHERE THE BUREAUCRACY IS MODERN, RATHER THAN SEMI-MODERN (21)	0.86	181	TEND TO BE THOSE WHERE THE BUREAUCRACY IS SEMI-MODERN, RATHER THAN MODERN (55)	0.96	19 2 3 52 X SQ= 49.36 P = 0. RV YES (YES)
186	TEND MORE TO BE THOSE WHERE PARTICIPATION BY THE MILITARY IN POLITICS IS NEUTRAL, RATHER THAN SUPPORTIVE (56)	0.90	186	TEND LESS TO BE THOSE WHERE PARTICIPATION BY THE MILITARY IN POLITICS IS NEUTRAL, RATHER THAN SUPPORTIVE (56)	0.56	2 29 19 37 X SQ= 6.79 P = 0.004 RV NO (NO)
187	TEND TO BE THOSE WHERE THE ROLE OF THE POLICE IS NOT POLITICALLY SIGNIFICANT (35)	0.91	187	TEND TO BE THOSE WHERE THE ROLE OF THE POLICE IS POLITICALLY SIGNIFICANT (66)	0.81	2 64 20 15 X SQ= 36.20 P = 0. RV YES (YES)
189	TEND LESS TO BE THOSE WHERE THE CHARACTER OF THE LEGAL SYSTEM IS OTHER THAN COMMON LAW (108)	0.77	189	TEND MORE TO BE THOSE WHERE THE CHARACTER OF THE LEGAL SYSTEM IS OTHER THAN COMMON LAW (108)	0.98	5 2 17 91 X SQ= 9.82 P = 0.003 RV NO (YES)
190	TILT LESS TOWARD BEING THOSE WHERE THE CHARACTER OF THE LEGAL SYSTEM IS CIVIL LAW, RATHER THAN COMMON LAW (32)	0.69	190	TILT MORE TOWARD BEING THOSE WHERE THE CHARACTER OF THE LEGAL SYSTEM IS CIVIL LAW, RATHER THAN COMMON LAW (32)	0.91	11 21 5 2 X SQ= 1.91 P = 0.101 RV NO (YES)
193	ALWAYS ARE THOSE WHERE THE CHARACTER OF THE LEGAL SYSTEM IS OTHER THAN PARTLY INDIGENOUS (86)	1.00	193	TEND LESS TO BE THOSE WHERE THE CHARACTER OF THE LEGAL SYSTEM IS OTHER THAN PARTLY INDIGENOUS (86)	0.70	0 28 22 64 X SQ= 7.31 P = 0.002 RV NO (NO)
194	ALWAYS ARE THOSE THAT ARE NON-COMMUNIST (101)	1.00	194	LEAN LESS TOWARD BEING THOSE THAT ARE NON-COMMUNIST (101)	0.86	0 13 22 79 X SQ= 2.25 P = 0.070 RV NO (NO)

3 POLITIES
LOCATED IN WEST EUROPE (13)

3 POLITIES
LOCATED ELSEWHERE THAN IN
WEST EUROPE (102)

BOTH SUBJECT AND PREDICATE

13 IN LEFT COLUMN

AUSTRIA	BELGIUM	FRANCE	GERMAN FR	GREECE	IRELAND	ITALY	LUXEMBOURG	NETHERLANDS	PORTUGAL
SPAIN	SWITZERLAND	UK							

102 IN RIGHT COLUMN

AFGHANISTAN	ALBANIA	ALGERIA	ARGENTINA	AUSTRALIA	BOLIVIA	BRAZIL	BULGARIA	BURMA	BURUNDI
CAMBODIA	CAMEROUN	CANADA	CEN AFR REP	CEYLON	CHAD	CHILE	CHINA, PR	COLOMBIA	CONGO(BRA)
CONGO(LEO)	COSTA RICA	CUBA	CYPRUS	CZECHOS'KIA	DAHOMEY	DENMARK	DOMIN REP	ECUADOR	EL SALVADOR
ETHIOPIA	FINLAND	GABON	GERMANY, E	GHANA	GUATEMALA	GUINEA	HAITI	HONDURAS	HUNGARY
ICELAND	INDIA	INDONESIA	IRAN	IRAQ	ISRAEL	IVORY COAST	JAMAICA	JAPAN	JORDAN
KOREA, N	KOREA REP	LAOS	LEBANON	LIBERIA	LIBYA	MALAGASY R	MALAYA	MALI	MAURITANIA
MEXICO	MONGOLIA	MOROCCO	NEPAL	NEW ZEALAND	NICARAGUA	NIGER	NIGERIA	NORWAY	PAKISTAN
PANAMA	PARAGUAY	PERU	PHILIPPINES	POLAND	RUMANIA	RWANDA	SA'U ARABIA	SENEGAL	SIERRE LEO
SOMALIA	SO AFRICA	SUDAN	SWEDEN	SYRIA	TANGANYIKA	THAILAND	TOGO	TRINIDAD	TUNISIA
TURKEY	UGANDA	USSR	UAR	US	UPPER VOLTA	URUGUAY	VENEZUELA	VIETNAM, N	VIETNAM REP
YEMEN	YUGOSLAVIA								

	Left column			Right column		
21	ALWAYS ARE THOSE WHOSE TERRITORIAL SIZE IS MEDIUM OR SMALL (83)	1.00	21	TEND LESS TO BE THOSE WHOSE TERRITORIAL SIZE IS MEDIUM OR SMALL (83)	0.69	0 32 13 70 X SQ= 4.20 P = 0.018 RV NO (NO)
22	TILT TOWARD BEING THOSE WHOSE TERRITORIAL SIZE IS SMALL (47)	0.62	22	TILT TOWARD BEING THOSE WHOSE TERRITORIAL SIZE IS VERY LARGE, LARGE, OR MEDIUM (68)	0.62	5 63 8 39 X SQ= 1.72 P = 0.138 RV YES (NO)
24	LEAN TOWARD BEING THOSE WHOSE POPULATION IS VERY LARGE, LARGE, OR MEDIUM (61)	0.77	24	LEAN TOWARD BEING THOSE WHOSE POPULATION IS SMALL (54)	0.50	10 51 3 51 X SQ= 2.36 P = 0.082 RV YES (NO)
25	TEND TO BE THOSE WHOSE POPULATION DENSITY IS VERY HIGH OR HIGH (17)	0.54	25	TEND TO BE THOSE WHOSE POPULATION DENSITY IS MEDIUM OR LOW (98)	0.90	7 10 6 92 X SQ= 14.43 P = 0. RV YES (NO)

26	ALWAYS ARE THOSE WHOSE POPULATION DENSITY IS VERY HIGH, HIGH, OR MEDIUM (48)	1.00	26	TEND TO BE THOSE WHOSE POPULATION DENSITY IS LOW (67)	0.66	13 35 0 67 X SQ= 17.85 P = 0. RV YES (NO)
28	ALWAYS ARE THOSE WHOSE POPULATION GROWTH RATE IS LOW (48)	1.00	28	TEND TO BE THOSE WHOSE POPULATION GROWTH RATE IS HIGH (62)	0.64	0 62 13 35 X SQ= 16.53 P = 0. RV YES (NO)
29	ALWAYS ARE THOSE WHERE THE DEGREE OF URBANIZATION IS HIGH (56)	1.00	29	TEND TO BE THOSE WHERE THE DEGREE OF URBANIZATION IS LOW (49)	0.53	12 44 0 49 X SQ= 9.83 P = 0. RV YES (NO)
30	ALWAYS ARE THOSE WHOSE AGRICULTURAL POPULATION IS MEDIUM, LOW, OR VERY LOW (57)	1.00	30	TEND TO BE THOSE WHOSE AGRICULTURAL POPULATION IS HIGH (56)	0.56	0 56 13 44 X SQ= 12.28 P = 0. RV YES (NO)
31	TEND TO BE THOSE WHOSE AGRICULTURAL POPULATION IS LOW OR VERY LOW (24)	0.69	31	TEND TO BE THOSE WHOSE AGRICULTURAL POPULATION IS HIGH OR MEDIUM (90)	0.85	4 86 9 15 X SQ= 17.35 P = 0. RV YES (NO)
33	TEND TO BE THOSE WHOSE GROSS NATIONAL PRODUCT IS VERY HIGH, HIGH, OR MEDIUM (30)	0.69	33	TEND TO BE THOSE WHOSE GROSS NATIONAL PRODUCT IS LOW OR VERY LOW (85)	0.79	9 21 4 81 X SQ= 11.74 P = 0.001 RV YES (NO)
34	TEND TO BE THOSE WHOSE GROSS NATIONAL PRODUCT IS VERY HIGH, HIGH, MEDIUM, OR LOW (62)	0.92	34	TEND TO BE THOSE WHOSE GROSS NATIONAL PRODUCT IS VERY LOW (53)	0.51	12 50 1 52 X SQ= 7.04 P = 0.003 RV YES (NO)
35	TEND TO BE THOSE WHOSE PER CAPITA GROSS NATIONAL PRODUCT IS VERY HIGH OR HIGH (24)	0.77	35	TEND TO BE THOSE WHOSE PER CAPITA GROSS NATIONAL PRODUCT IS MEDIUM, LOW, OR VERY LOW (91)	0.86	10 14 3 88 X SQ= 24.19 P = 0. RV YES (NO)
36	TEND TO BE THOSE WHOSE PER CAPITA GROSS NATIONAL PRODUCT IS VERY HIGH, HIGH, OR MEDIUM (42)	0.92	36	TEND TO BE THOSE WHOSE PER CAPITA GROSS NATIONAL PRODUCT IS LOW OR VERY LOW (73)	0.71	12 30 1 72 X SQ= 17.06 P = 0. RV YES (NO)

```
37  ALWAYS ARE THOSE                        1.00    37  TEND TO BE THOSE                          0.50     13   51
    WHOSE PER CAPITA GROSS NATIONAL PRODUCT             WHOSE PER CAPITA GROSS NATIONAL PRODUCT             0   51
    IS VERY HIGH, HIGH, MEDIUM, OR LOW  (64)            IS VERY LOW  (51)                                X SQ=   9.74
                                                                                                         P =     0.
                                                                                                         RV YES (NO )

39  TEND TO BE THOSE                        0.69    39  TEND TO BE THOSE                          0.71      9   29
    WHOSE INTERNATIONAL FINANCIAL STATUS                WHOSE INTERNATIONAL FINANCIAL STATUS                4   72
    IS VERY HIGH, HIGH, OR MEDIUM  (38)                 IS LOW OR VERY LOW  (76)                         X SQ=   6.78
                                                                                                         P =     0.009
                                                                                                         RV YES (NO )

40  ALWAYS ARE THOSE                        1.00    40  TEND LESS TO BE THOSE                     0.60     13   58
    WHOSE INTERNATIONAL FINANCIAL STATUS                WHOSE INTERNATIONAL FINANCIAL STATUS                0   39
    IS VERY HIGH, HIGH,                                 IS VERY HIGH, HIGH,                              X SQ=   6.44
    MEDIUM, OR LOW  (71)                                MEDIUM, OR LOW  (71)                             P =     0.004
                                                                                                         RV NO  (NO )

41  TEND TO BE THOSE                        0.62    41  TEND TO BE THOSE                          0.89      8   11
    WHOSE ECONOMIC DEVELOPMENTAL STATUS                 WHOSE ECONOMIC DEVELOPMENTAL STATUS                 5   89
    IS DEVELOPED  (19)                                  IS INTERMEDIATE, UNDERDEVELOPED,                 X SQ=  17.55
                                                        OR VERY UNDERDEVELOPED  (94)                     P =     0.
                                                                                                         RV YES (NO )

42  TEND TO BE THOSE                        0.85    42  TEND TO BE THOSE                          0.75     11   25
    WHOSE ECONOMIC DEVELOPMENTAL STATUS                 WHOSE ECONOMIC DEVELOPMENTAL STATUS                 2   74
    IS DEVELOPED OR INTERMEDIATE  (36)                  IS UNDERDEVELOPED OR                             X SQ=  15.94
                                                        VERY UNDERDEVELOPED  (76)                        P =     0.
                                                                                                         RV YES (NO )

43  ALWAYS ARE THOSE                        1.00    43  TEND TO BE THOSE                          0.58     13   42
    WHOSE ECONOMIC DEVELOPMENTAL STATUS                 WHOSE ECONOMIC DEVELOPMENTAL STATUS                 0   57
    IS DEVELOPED, INTERMEDIATE, OR                      IS VERY UNDERDEVELOPED  (57)                     X SQ=  13.03
    UNDERDEVELOPED  (55)                                                                                 P =     0.
                                                                                                         RV YES (NO )

44  TEND TO BE THOSE                        0.69    44  TEND TO BE THOSE                          0.84      9   16
    WHERE THE LITERACY RATE IS                          WHERE THE LITERACY RATE IS                          4   86
    NINETY PERCENT OR ABOVE  (25)                       BELOW NINETY PERCENT  (90)                       X SQ=  16.41
                                                                                                         P =     0.
                                                                                                         RV YES (NO )

45  ALWAYS ARE THOSE                        1.00    45  TEND TO BE THOSE                          0.56     13   42
    WHERE THE LITERACY RATE IS                          WHERE THE LITERACY RATE IS                          0   54
    FIFTY PERCENT OR ABOVE  (55)                        BELOW FIFTY PERCENT  (54)                        X SQ=  12.33
                                                                                                         P =     0.
                                                                                                         RV YES (NO )

50  TEND TO BE THOSE                        0.75    50  TEND TO BE THOSE                          0.61      9   34
    WHERE FREEDOM OF THE PRESS IS                       WHERE FREEDOM OF THE PRESS IS                       3   53
    COMPLETE  (43)                                      INTERMITTENT, INTERNALLY ABSENT, OR              X SQ=   4.17
                                                        INTERNALLY AND EXTERNALLY ABSENT  (56)           P =     0.028
                                                                                                         RV YES (NO )
```

No.	Statement	Prop.	No.	Statement	Prop.	Statistics
51	TILT MORE TOWARD BEING THOSE WHERE FREEDOM OF THE PRESS IS COMPLETE OR INTERMITTENT (60)	0.83	51	TILT LESS TOWARD BEING THOSE WHERE FREEDOM OF THE PRESS IS COMPLETE OR INTERMITTENT (60)	0.59	10 50 2 35 X SQ= 1.74 P = 0.124 RV NO (NO)
52	ALWAYS ARE THOSE WHERE FREEDOM OF THE PRESS IS COMPLETE, INTERMITTENT, OR INTERNALLY ABSENT (82)	1.00	52	TILT LESS TOWARD BEING THOSE WHERE FREEDOM OF THE PRESS IS COMPLETE, INTERMITTENT, OR INTERNALLY ABSENT (82)	0.81	13 69 0 16 X SQ= 1.71 P = 0.118 RV NO (NO)
54	TEND TO BE THOSE WHERE NEWSPAPER CIRCULATION IS ONE HUNDRED OR MORE PER THOUSAND (37)	0.85	54	TEND TO BE THOSE WHERE NEWSPAPER CIRCULATION IS LESS THAN ONE HUNDRED PER THOUSAND (76)	0.74	11 26 2 74 X SQ= 15.39 P = 0. RV YES (NO)
55	ALWAYS ARE THOSE WHERE NEWSPAPER CIRCULATION IS TEN OR MORE PER THOUSAND (78)	1.00	55	TEND LESS TO BE THOSE WHERE NEWSPAPER CIRCULATION IS TEN OR MORE PER THOUSAND (78)	0.65	13 65 0 35 X SQ= 5.06 P = 0.009 RV NO (NO)
56	ALWAYS ARE THOSE WHERE THE RELIGION IS PREDOMINANTLY LITERATE (79)	1.00	56	TEND LESS TO BE THOSE WHERE THE RELIGION IS PREDOMINANTLY LITERATE (79)	0.68	13 66 0 31 X SQ= 4.31 P = 0.018 RV NO (NO)
63	ALWAYS ARE THOSE WHERE THE RELIGION IS PREDOMINANTLY SOME KIND OF CHRISTIAN (46)	1.00	63	TEND TO BE THOSE WHERE THE RELIGION IS PREDOMINANTLY OR PARTLY OTHER THAN CHRISTIAN (68)	0.67	13 33 0 68 X SQ= 18.98 P = 0. RV YES (NO)
67	ALWAYS ARE THOSE THAT ARE RACIALLY HOMOGENEOUS (82)	1.00	67	TEND LESS TO BE THOSE THAT ARE RACIALLY HOMOGENEOUS (82)	0.70	13 67 0 29 X SQ= 3.92 P = 0.019 RV NO (NO)
68	TEND TO BE THOSE THAT ARE LINGUISTICALLY HOMOGENEOUS (52)	0.77	68	TEND TO BE THOSE THAT ARE LINGUISTICALLY WEAKLY HETEROGENEOUS OR STRONGLY HETEROGENEOUS (62)	0.58	10 42 3 59 X SQ= 4.46 P = 0.020 RV YES (NO)
69	TEND MORE TO BE THOSE THAT ARE LINGUISTICALLY HOMOGENEOUS OR WEAKLY HETEROGENEOUS (64)	0.85	69	TEND LESS TO BE THOSE THAT ARE LINGUISTICALLY HOMOGENEOUS OR WEAKLY HETEROGENEOUS (64)	0.52	11 53 2 48 X SQ= 3.61 P = 0.037 RV NO (NO)

3/

70	TEND TO BE THOSE THAT ARE RELIGIOUSLY, RACIALLY, AND LINGUISTICALLY HOMOGENEOUS (21)	0.54	70	TEND TO BE THOSE THAT ARE RELIGIOUSLY, RACIALLY, OR LINGUISTICALLY HETEROGENEOUS (85)	0.85	7 14 6 78 X SQ= 8.35 P = 0.004 RV YES (NO)
71	TEND TO BE THOSE WHOSE DATE OF INDEPENDENCE IS BEFORE 1800 (21)	0.67	71	TEND TO BE THOSE WHOSE DATE OF INDEPENDENCE IS AFTER 1800 (90)	0.87	8 13 4 86 X SQ= 16.66 P = 0. RV YES (NO)
72	TEND TO BE THOSE WHOSE DATE OF INDEPENDENCE IS BEFORE 1914 (52)	0.92	72	TEND TO BE THOSE WHOSE DATE OF INDEPENDENCE IS AFTER 1914 (59)	0.59	11 41 1 58 X SQ= 8.93 P = 0.001 RV YES (NO)
73	ALWAYS ARE THOSE WHOSE DATE OF INDEPENDENCE IS BEFORE 1945 (65)	1.00	73	TEND LESS TO BE THOSE WHOSE DATE OF INDEPENDENCE IS BEFORE 1945 (65)	0.54	12 53 0 46 X SQ= 7.70 P = 0.001 RV NO (NO)
85	ALWAYS ARE THOSE WHERE THE STAGE OF POLITICAL MODERNIZATION IS ADVANCED, RATHER THAN MID- OR EARLY TRANSITIONAL (60)	1.00	85	TEND TO BE THOSE WHERE THE STAGE OF POLITICAL MODERNIZATION IS MID- OR EARLY TRANSITIONAL, RATHER THAN ADVANCED (54)	0.53	13 47 0 54 X SQ= 11.15 P = 0. RV YES (NO)
89	ALWAYS ARE THOSE WHOSE IDEOLOGICAL ORIENTATION IS CONVENTIONAL (33)	1.00	89	TEND TO BE THOSE WHOSE IDEOLOGICAL ORIENTATION IS OTHER THAN CONVENTIONAL (62)	0.74	11 22 0 62 X SQ= 20.23 P = 0. RV YES (NO)
93	ALWAYS ARE THOSE WHERE THE SYSTEM STYLE IS NON-MOBILIZATIONAL (78)	1.00	93	TEND LESS TO BE THOSE WHERE THE SYSTEM STYLE IS NON-MOBILIZATIONAL (78)	0.67	0 32 12 66 X SQ= 4.06 P = 0.017 RV NO (NO)
94	TEND MORE TO BE THOSE WHERE THE STATUS OF THE REGIME IS CONSTITUTIONAL (51)	0.85	94	TEND LESS TO BE THOSE WHERE THE STATUS OF THE REGIME IS CONSTITUTIONAL (51)	0.51	11 40 2 39 X SQ= 3.93 P = 0.033 RV NO (NO)
96	ALWAYS ARE THOSE WHERE THE STATUS OF THE REGIME IS CONSTITUTIONAL OR TOTALITARIAN (67)	1.00	96	TEND LESS TO BE THOSE WHERE THE STATUS OF THE REGIME IS CONSTITUTIONAL OR TOTALITARIAN (67)	0.70	13 54 0 23 X SQ= 3.76 P = 0.019 RV NO (NO)

98	TEND TO BE THOSE WHERE GOVERNMENTAL STABILITY IS GENERALLY PRESENT AND DATES AT LEAST FROM THE INTERWAR PERIOD (22)	0.62	98	TEND TO BE THOSE WHERE GOVERNMENTAL STABILITY IS GENERALLY OR MODERATELY PRESENT AND DATES FROM THE POST-WAR PERIOD, OR IS ABSENT (93)	0.86	8 14 5 88 X SQ= 14.09 P = 0. RV YES (NO)
99	TEND MORE TO BE THOSE WHERE GOVERNMENTAL STABILITY IS GENERALLY PRESENT AND DATES FROM AT LEAST THE INTER-WAR PERIOD, OR FROM THE POST-WAR PERIOD (50)	0.92	99	TEND LESS TO BE THOSE WHERE GOVERNMENTAL STABILITY IS GENERALLY PRESENT AND DATES FROM AT LEAST THE INTER-WAR PERIOD, OR FROM THE POST-WAR PERIOD (50)	0.52	12 38 1 35 X SQ= 5.79 P = 0.006 RV NO (NO)
100	ALWAYS ARE THOSE WHERE GOVERNMENTAL STABILITY IS GENERALLY PRESENT AND DATES FROM AT LEAST THE INTER-WAR PERIOD, OR IS GENERALLY OR MODERATELY PRESENT, AND DATES FROM THE POST-WAR PERIOD (64)	1.00	100	TEND LESS TO BE THOSE WHERE GOVERNMENTAL STABILITY IS GENERALLY PRESENT AND DATES FROM AT LEAST THE INTER-WAR PERIOD, OR IS GENERALLY OR MODERATELY PRESENT, AND DATES FROM THE POST-WAR PERIOD (64)	0.70	13 51 0 22 X SQ= 3.80 P = 0.018 RV NO (NO)
101	TEND TO BE THOSE WHERE THE REPRESENTATIVE CHARACTER OF THE REGIME IS POLYARCHIC (41)	0.85	101	TEND TO BE THOSE WHERE THE REPRESENTATIVE CHARACTER OF THE REGIME IS LIMITED POLYARCHIC, PSEUDO-POLYARCHIC, OR NON-POLYARCHIC (57)	0.65	11 30 2 55 X SQ= 9.34 P = 0.001 RV YES (NO)
105	TEND TO BE THOSE WHERE THE ELECTORAL SYSTEM IS COMPETITIVE (43)	0.85	105	TEND TO BE THOSE WHERE THE ELECTORAL SYSTEM IS PARTIALLY COMPETITIVE OR NON-COMPETITIVE (47)	0.58	11 32 2 45 X SQ= 6.63 P = 0.006 RV YES (NO)
107	TEND TO BE THOSE WHERE AUTONOMOUS GROUPS ARE FULLY TOLERATED IN POLITICS (46)	0.85	107	TEND TO BE THOSE WHERE AUTONOMOUS GROUPS ARE PARTIALLY TOLERATED IN POLITICS, ARE TOLERATED ONLY OUTSIDE POLITICS, OR ARE NOT TOLERATED AT ALL (65)	0.64	11 35 2 63 X SQ= 9.38 P = 0.002 RV YES (NO)
111	TILT MORE TOWARD BEING THOSE WHERE POLITICAL ENCULTURATION IS HIGH OR MEDIUM (53)	0.82	111	TILT LESS TOWARD BEING THOSE WHERE POLITICAL ENCULTURATION IS HIGH OR MEDIUM (53)	0.52	9 44 2 40 X SQ= 2.33 P = 0.105 RV NO (NO)
115	TEND TO BE THOSE WHERE INTEREST ARTICULATION BY ASSOCIATIONAL GROUPS IS SIGNIFICANT (20)	0.73	115	TEND TO BE THOSE WHERE INTEREST ARTICULATION BY ASSOCIATIONAL GROUPS IS MODERATE, LIMITED, OR NEGLIGIBLE (91)	0.88	8 12 3 88 X SQ= 20.80 P = 0. RV YES (NO)
116	ALWAYS ARE THOSE WHERE INTEREST ARTICULATION BY ASSOCIATIONAL GROUPS IS SIGNIFICANT OR MODERATE (32)	1.00	116	TEND TO BE THOSE WHERE INTEREST ARTICULATION BY ASSOCIATIONAL GROUPS IS LIMITED OR NEGLIGIBLE (79)	0.79	11 21 0 79 X SQ= 26.42 P = 0. RV YES (NO)

118	LEAN MORE TOWARD BEING THOSE WHERE INTEREST ARTICULATION BY INSTITUTIONAL GROUPS IS SIGNIFICANT, MODERATE, OR LIMITED (60)	0.85	118	LEAN LESS TOWARD BEING THOSE WHERE INTEREST ARTICULATION BY INSTITUTIONAL GROUPS IS SIGNIFICANT, MODERATE, OR LIMITED (60)	0.56	2 38 11 49 X SQ= 2.69 P = 0.070 RV NO (NO)
119	TEND TO BE THOSE WHERE INTEREST ARTICULATION BY INSTITUTIONAL GROUPS IS MODERATE OR LIMITED (26)	0.77	119	TEND TO BE THOSE WHERE INTEREST ARTICULATION BY INSTITUTIONAL GROUPS IS VERY SIGNIFICANT OR SIGNIFICANT (74)	0.82	3 71 10 16 X SQ= 17.21 P = 0. RV YES (NO)
121	ALWAYS ARE THOSE WHERE INTEREST ARTICULATION BY NON-ASSOCIATIONAL GROUPS IS MODERATE, LIMITED, OR NEGLIGIBLE (61)	1.00	121	TEND TO BE THOSE WHERE INTEREST ARTICULATION BY NON-ASSOCIATIONAL GROUPS IS SIGNIFICANT (54)	0.53	0 54 13 48 X SQ= 10.94 P = 0. RV YES (NO)
122	TEND TO BE THOSE WHERE INTEREST ARTICULATION BY NON-ASSOCIATIONAL GROUPS IS LIMITED OR NEGLIGIBLE (32)	0.77	122	TEND TO BE THOSE WHERE INTEREST ARTICULATION BY NON-ASSOCIATIONAL GROUPS IS SIGNIFICANT OR MODERATE (83)	0.78	3 80 10 22 X SQ= 14.94 P = 0. RV YES (NO)
125	TEND TO BE THOSE WHERE INTEREST ARTICULATION BY ANOMIC GROUPS IS INFREQUENT OR VERY INFREQUENT (35)	0.92	125	TEND TO BE THOSE WHERE INTEREST ARTICULATION BY ANOMIC GROUPS IS FREQUENT OR OCCASIONAL (64)	0.72	1 63 11 24 X SQ= 16.25 P = 0. RV YES (NO)
126	TEND TO BE THOSE WHERE INTEREST ARTICULATION BY ANOMIC GROUPS IS VERY INFREQUENT (16)	0.67	126	TEND TO BE THOSE WHERE INTEREST ARTICULATION BY ANOMIC GROUPS IS FREQUENT, OCCASIONAL, OR INFREQUENT (83)	0.91	4 79 8 8 X SQ= 21.64 P = 0. RV YES (YES)
128	TEND TO BE THOSE WHERE INTEREST ARTICULATION BY POLITICAL PARTIES IS SIGNIFICANT OR MODERATE (48)	0.85	128	TEND TO BE THOSE WHERE INTEREST ARTICULATION BY POLITICAL PARTIES IS LIMITED OR NEGLIGIBLE (45)	0.54	11 37 2 43 X SQ= 5.14 P = 0.015 RV YES (NO)
133	TILT MORE TOWARD BEING THOSE WHERE INTEREST AGGREGATION BY THE EXECUTIVE IS MODERATE, LIMITED, OR NEGLIGIBLE (76)	0.92	133	TILT LESS TOWARD BEING THOSE WHERE INTEREST AGGREGATION BY THE EXECUTIVE IS MODERATE, LIMITED, OR NEGLIGIBLE (76)	0.70	1 28 12 64 X SQ= 1.92 P = 0.107 RV NO (NO)
134	TILT MORE TOWARD BEING THOSE WHERE INTEREST AGGREGATION BY THE EXECUTIVE IS SIGNIFICANT OR MODERATE (57)	0.77	134	TILT LESS TOWARD BEING THOSE WHERE INTEREST AGGREGATION BY THE EXECUTIVE IS SIGNIFICANT OR MODERATE (57)	0.52	10 47 3 43 X SQ= 1.89 P = 0.136 RV NO (NO)

No.	Statement	Prop.	No.	Statement	Prop.	Statistics
137	TEND TO BE THOSE WHERE INTEREST AGGREGATION BY THE LEGISLATURE IS SIGNIFICANT OR MODERATE (29)	0.77	137	TEND TO BE THOSE WHERE INTEREST AGGREGATION BY THE LEGISLATURE IS LIMITED OR NEGLIGIBLE (68)	0.77	10 19 3 65 X SQ= 13.35 P = 0. RV YES (NO)
138	ALWAYS ARE THOSE WHERE INTEREST AGGREGATION BY THE LEGISLATURE IS SIGNIFICANT, MODERATE, OR LIMITED (48)	1.00	138	TEND TO BE THOSE WHERE INTEREST AGGREGATION BY THE LEGISLATURE IS NEGLIGIBLE (49)	0.57	11 37 0 49 X SQ= 10.49 P = 0. RV YES (NO)
147	TEND TO BE THOSE WHERE THE PARTY SYSTEM IS QUALITATIVELY CLASS-ORIENTED OR MULTI-IDEOLOGICAL (23)	0.80	147	TEND TO BE THOSE WHERE THE PARTY SYSTEM IS QUALITATIVELY OTHER THAN CLASS-ORIENTED OR MULTI-IDEOLOGICAL (67)	0.81	8 15 2 65 X SQ= 14.46 P = 0. RV YES (NO)
153	LEAN TOWARD BEING THOSE WHERE THE PARTY SYSTEM IS STABLE (42)	0.77	153	LEAN TOWARD BEING THOSE WHERE THE PARTY SYSTEM IS MODERATELY STABLE OR UNSTABLE (38)	0.52	10 32 3 35 X SQ= 2.64 P = 0.071 RV YES (NO)
158	LEAN MORE TOWARD BEING THOSE WHERE PERSONALISMO IS NEGLIGIBLE (56)	0.85	158	LEAN LESS TOWARD BEING THOSE WHERE PERSONALISMO IS NEGLIGIBLE (56)	0.54	2 38 11 45 X SQ= 3.11 P = 0.066 RV NO (NO)
161	LEAN TOWARD BEING THOSE WHERE THE POLITICAL LEADERSHIP IS NON-ELITIST (50)	0.77	161	LEAN TOWARD BEING THOSE WHERE THE POLITICAL LEADERSHIP IS ELITIST OR MODERATELY ELITIST (47)	0.52	3 44 10 40 X SQ= 2.79 P = 0.073 RV YES (NO)
164	TEND MORE TO BE THOSE WHERE THE REGIME'S LEADERSHIP CHARISMA IS NEGLIGIBLE (65)	0.92	164	TEND LESS TO BE THOSE WHERE THE REGIME'S LEADERSHIP CHARISMA IS NEGLIGIBLE (65)	0.62	1 33 12 53 X SQ= 3.45 P = 0.032 RV NO (NO)
168	TEND TO BE THOSE WHERE THE HORIZONTAL POWER DISTRIBUTION IS SIGNIFICANT (34)	0.77	168	TEND TO BE THOSE WHERE THE HORIZONTAL POWER DISTRIBUTION IS LIMITED OR NEGLIGIBLE (72)	0.74	10 24 3 69 X SQ= 11.43 P = 0.001 RV YES (NO)
170	ALWAYS ARE THOSE WHERE THE LEGISLATIVE-EXECUTIVE STRUCTURE IS OTHER THAN PRESIDENTIAL (63)	1.00	170	TEND LESS TO BE THOSE WHERE THE LEGISLATIVE-EXECUTIVE STRUCTURE IS OTHER THAN PRESIDENTIAL (63)	0.56	0 39 13 50 X SQ= 7.46 P = 0.001 RV NO (NO)

174	TEND TO BE THOSE WHERE THE LEGISLATURE IS FULLY EFFECTIVE (28)	0.69	174	TEND TO BE THOSE WHERE THE LEGISLATURE IS PARTIALLY EFFECTIVE, LARGELY INEFFECTIVE, OR WHOLLY INEFFECTIVE (72)	0.78	9 19 4 68 X SQ= 10.36 P = 0.001 RV YES (NO)
175	TEND TO BE THOSE WHERE THE LEGISLATURE IS FULLY EFFECTIVE OR PARTIALLY EFFECTIVE (51)	0.85	175	TEND TO BE THOSE WHERE THE LEGISLATURE IS LARGELY INEFFECTIVE OR WHOLLY INEFFECTIVE (49)	0.54	11 40 2 47 X SQ= 5.30 P = 0.015 RV YES (NO)
178	TEND TO BE THOSE WHERE THE LEGISLATURE IS BICAMERAL (51)	0.85	178	TEND TO BE THOSE WHERE THE LEGISLATURE IS UNICAMERAL (53)	0.56	2 51 11 40 X SQ= 5.99 P = 0.007 RV YES (NO)
179	TEND TO BE THOSE WHERE THE EXECUTIVE IS STRONG (39)	0.73	179	TEND TO BE THOSE WHERE THE EXECUTIVE IS DOMINANT (52)	0.61	3 49 8 31 X SQ= 3.28 P = 0.050 RV YES (NO)
181	TEND TO BE THOSE WHERE THE BUREAUCRACY IS MODERN, RATHER THAN SEMI-MODERN (21)	0.77	181	TEND TO BE THOSE WHERE THE BUREAUCRACY IS SEMI-MODERN, RATHER THAN MODERN (55)	0.83	10 11 3 52 X SQ= 16.20 P = 0. RV YES (NO)
187	TEND TO BE THOSE WHERE THE ROLE OF THE POLICE IS NOT POLITICALLY SIGNIFICANT (35)	0.85	187	TEND TO BE THOSE WHERE THE ROLE OF THE POLICE IS POLITICALLY SIGNIFICANT (66)	0.73	2 64 11 24 X SQ= 14.01 P = 0. RV YES (NO)
188	TEND TO BE THOSE WHERE THE CHARACTER OF THE LEGAL SYSTEM IS CIVIL LAW (32)	0.85	188	TEND TO BE THOSE WHERE THE CHARACTER OF THE LEGAL SYSTEM IS OTHER THAN CIVIL LAW (81)	0.79	11 21 2 79 X SQ= 19.91 P = 0. RV YES (NO)
193	ALWAYS ARE THOSE WHERE THE CHARACTER OF THE LEGAL SYSTEM IS OTHER THAN PARTLY INDIGENOUS (86)	1.00	193	TEND LESS TO BE THOSE WHERE THE CHARACTER OF THE LEGAL SYSTEM IS OTHER THAN PARTLY INDIGENOUS (86)	0.72	0 28 13 73 X SQ= 3.40 P = 0.036 RV NO (NO)

4 POLITIES
LOCATED IN SCANDINAVIA (5)

4 POLITIES
LOCATED ELSEWHERE THAN IN
SCANDINAVIA (110)

SUBJECT ONLY

5 IN LEFT COLUMN

DENMARK	FINLAND	ICELAND	NORWAY	SWEDEN

110 IN RIGHT COLUMN

AFGHANISTAN	ALBANIA	ALGERIA	ARGENTINA	AUSTRALIA	AUSTRIA	BELGIUM	BOLIVIA	BRAZIL	BULGARIA
BURMA	BURUNDI	CAMBODIA	CAMEROUN	CANADA	CEN AFR REP	CEYLON	CHAD	CHILE	CHINA, PR
COLOMBIA	CONGO(BRA)	CONGO(LEO)	COSTA RICA	CUBA	CYPRUS	CZECHOS'KIA	DAHOMEY	DOMIN REP	ECUADOR
EL SALVADOR	ETHIOPIA	FRANCE	GABON	GERMANY, E	GERMAN FR	GHANA	GREECE	GUATEMALA	GUINEA
HAITI	HONDURAS	HUNGARY	INDIA	INDONESIA	IRAN	IRAQ	IRELAND	ISRAEL	ITALY
IVORY COAST	JAMAICA	JAPAN	JORDAN	KOREA, N	KOREA REP	LAOS	LEBANON	LIBERIA	LIBYA
LUXEMBOURG	MALAGASY R	MALAYA	MALI	MAURITANIA	MEXICO	MONGOLIA	MOROCCO	NEPAL	NETHERLANDS
NEW ZEALAND	NICARAGUA	NIGER	NIGERIA	PAKISTAN	PANAMA	PARAGUAY	PERU	PHILIPPINES	POLAND
PORTUGAL	RUMANIA	RWANDA	SA'U ARABIA	SENEGAL	SIERRE LEO	SOMALIA	SO AFRICA	SPAIN	SUDAN
SWITZERLAND	SYRIA	TANGANYIKA	THAILAND	TOGO	TRINIDAD	TUNISIA	TURKEY	UGANDA	USSR
UAR	UK	US	UPPER VOLTA	URUGUAY	VENEZUELA	VIETNAM, N	VIETNAM REP	YEMEN	YUGOSLAVIA

	Left column			Right column		
29	ALWAYS ARE THOSE WHERE THE DEGREE OF URBANIZATION IS HIGH (56)	1.00	29	LEAN LESS TOWARD BEING THOSE WHERE THE DEGREE OF URBANIZATION IS HIGH (56)	0.51	5 51 0 49 X SQ= 2.84 P = 0.059 RV NO (NO)
30	ALWAYS ARE THOSE WHOSE AGRICULTURAL POPULATION IS MEDIUM, LOW, OR VERY LOW (57)	1.00	30	LEAN TOWARD BEING THOSE WHOSE AGRICULTURAL POPULATION IS HIGH (56)	0.52	0 56 5 52 X SQ= 3.27 P = 0.057 RV YES (NO)
31	TEND TO BE THOSE WHOSE AGRICULTURAL POPULATION IS LOW OR VERY LOW (24)	0.80	31	TEND TO BE THOSE WHOSE AGRICULTURAL POPULATION IS HIGH OR MEDIUM (90)	0.82	1 89 4 20 X SQ= 7.54 P = 0.007 RV YES (NO)
35	ALWAYS ARE THOSE WHOSE PER CAPITA GROSS NATIONAL PRODUCT IS VERY HIGH OR HIGH (24)	1.00	35	TEND TO BE THOSE WHOSE PER CAPITA GROSS NATIONAL PRODUCT IS MEDIUM, LOW, OR VERY LOW (91)	0.83	5 19 0 91 X SQ= 15.13 P = 0. RV YES (NO)

39	TEND TO BE THOSE WHOSE INTERNATIONAL FINANCIAL STATUS IS VERY HIGH, HIGH, OR MEDIUM (38)	0.80	39	TEND TO BE THOSE WHOSE INTERNATIONAL FINANCIAL STATUS IS LOW OR VERY LOW (76)	0.69	4 34 1 75 X SQ= 3.16 P = 0.042 RV YES (NO)
41	TEND TO BE THOSE WHOSE ECONOMIC DEVELOPMENTAL STATUS IS DEVELOPED (19)	0.80	41	TEND TO BE THOSE WHOSE ECONOMIC DEVELOPMENTAL STATUS IS INTERMEDIATE, UNDERDEVELOPED, OR VERY UNDERDEVELOPED (94)	0.86	4 15 1 93 X SQ= 10.58 P = 0.003 RV YES (NO)
42	ALWAYS ARE THOSE WHOSE ECONOMIC DEVELOPMENTAL STATUS IS DEVELOPED OR INTERMEDIATE (36)	1.00	42	TEND TO BE THOSE WHOSE ECONOMIC DEVELOPMENTAL STATUS IS UNDERDEVELOPED OR VERY UNDERDEVELOPED (76)	0.71	5 31 0 76 X SQ= 8.03 P = 0.003 RV YES (NO)
44	ALWAYS ARE THOSE WHERE THE LITERACY RATE IS NINETY PERCENT OR ABOVE (25)	1.00	44	TEND TO BE THOSE WHERE THE LITERACY RATE IS BELOW NINETY PERCENT (90)	0.82	5 20 0 90 X SQ= 14.32 P = 0. RV YES (NO)
50	ALWAYS ARE THOSE WHERE FREEDOM OF THE PRESS IS COMPLETE (43)	1.00	50	TEND TO BE THOSE WHERE FREEDOM OF THE PRESS IS INTERMITTENT, INTERNALLY ABSENT, OR INTERNALLY AND EXTERNALLY ABSENT (56)	0.60	5 38 0 56 X SQ= 4.65 P = 0.013 RV YES (NO)
53	ALWAYS ARE THOSE WHERE NEWSPAPER CIRCULATION IS THREE HUNDRED OR MORE PER THOUSAND (14)	1.00	53	TEND TO BE THOSE WHERE NEWSPAPER CIRCULATION IS LESS THAN THREE HUNDRED PER THOUSAND (101)	0.92	5 9 0 101 X SQ= 29.61 P = 0. RV YES (NO)
62	ALWAYS ARE THOSE WHERE THE RELIGION IS PROTESTANT, RATHER THAN CATHOLIC (5)	1.00	62	ALWAYS ARE THOSE WHERE THE RELIGION IS CATHOLIC, RATHER THAN PROTESTANT (25)	1.00	5 0 0 25 X SQ= 23.23 P = 0. RV YES (YES)
63	ALWAYS ARE THOSE WHERE THE RELIGION IS PREDOMINANTLY SOME KIND OF CHRISTIAN (46)	1.00	63	TEND TO BE THOSE WHERE THE RELIGION IS PREDOMINANTLY OR PARTLY OTHER THAN CHRISTIAN (68)	0.62	5 41 0 68 X SQ= 5.36 P = 0.009 RV YES (NO)
66	ALWAYS ARE THOSE THAT ARE RELIGIOUSLY HOMOGENEOUS (57)	1.00	66	LEAN LESS TOWARD BEING THOSE THAT ARE RELIGIOUSLY HOMOGENEOUS (57)	0.51	5 52 0 49 X SQ= 2.77 P = 0.060 RV NO (NO)

```
69  ALWAYS ARE THOSE                       1.00     69  LEAN LESS TOWARD BEING THOSE          0.54          5   59
    THAT ARE LINGUISTICALLY                             THAT ARE LINGUISTICALLY                             0   50
    HOMOGENEOUS OR                                      HOMOGENEOUS OR                                  X SQ=   2.43
    WEAKLY HETEROGENEOUS  (64)                          WEAKLY HETEROGENEOUS  (64)                      P =     0.066
                                                                                                        RV NO   (NO )

70  TEND TO BE THOSE                       0.80     70  TEND TO BE THOSE                      0.83          4   17
    THAT ARE RELIGIOUSLY, RACIALLY,                     THAT ARE RELIGIOUSLY, RACIALLY,                     1   83
    AND LINGUISTICALLY HOMOGENEOUS  (21)                OR LINGUISTICALLY HETEROGENEOUS  (85)           X SQ=   8.20
                                                                                                        P =     0.005
                                                                                                        RV YES (NO )

71  TEND TO BE THOSE                       0.60     71  TEND TO BE THOSE                      0.83          3   18
    WHOSE DATE OF INDEPENDENCE                          WHOSE DATE OF INDEPENDENCE                          2   88
    IS BEFORE 1800  (21)                                IS AFTER 1800  (90)                             X SQ=   3.30
                                                                                                        P =     0.046
                                                                                                        RV YES (NO )

73  ALWAYS ARE THOSE                       1.00     73  LEAN LESS TOWARD BEING THOSE          0.57          5   60
    WHOSE DATE OF INDEPENDENCE                          WHOSE DATE OF INDEPENDENCE                          0   46
    IS BEFORE 1945  (65)                                IS BEFORE 1945  (65)                            X SQ=   2.13
                                                                                                        P =     0.075
                                                                                                        RV NO   (NO )

85  ALWAYS ARE THOSE                       1.00     85  LEAN LESS TOWARD BEING THOSE          0.50          5   55
    WHERE THE STAGE OF                                  WHERE THE STAGE OF                                  0   54
    POLITICAL MODERNIZATION IS                          POLITICAL MODERNIZATION IS                      X SQ=   2.93
    ADVANCED, RATHER THAN                               ADVANCED, RATHER THAN                           P =     0.059
    MID- OR EARLY TRANSITIONAL  (60)                    MID- OR EARLY TRANSITIONAL  (60)                RV NO   (NO )

89  ALWAYS ARE THOSE                       1.00     89  TEND TO BE THOSE                      0.69          5   28
    WHOSE IDEOLOGICAL ORIENTATION                       WHOSE IDEOLOGICAL ORIENTATION                       0   62
    IS CONVENTIONAL  (33)                               IS OTHER THAN CONVENTIONAL  (62)                X SQ=   7.11
                                                                                                        P =     0.004
                                                                                                        RV YES (NO )

94  ALWAYS ARE THOSE                       1.00     94  LEAN LESS TOWARD BEING THOSE          0.53          5   46
    WHERE THE STATUS OF THE REGIME                      WHERE THE STATUS OF THE REGIME                      0   41
    IS CONSTITUTIONAL  (51)                             IS CONSTITUTIONAL  (51)                         X SQ=   2.56
                                                                                                        P =     0.063
                                                                                                        RV NO   (NO )

98  ALWAYS ARE THOSE                       1.00     98  TEND TO BE THOSE                      0.85          5   17
    WHERE GOVERNMENTAL STABILITY                        WHERE GOVERNMENTAL STABILITY                        0   93
    IS GENERALLY PRESENT AND                            IS GENERALLY OR MODERATELY PRESENT              X SQ=  16.97
    DATES AT LEAST FROM                                 AND DATES FROM THE POST-WAR PERIOD,             P =     0.
    THE INTERWAR PERIOD  (22)                           OR IS ABSENT  (93)                              RV YES (NO )

101 ALWAYS ARE THOSE                       1.00     101 TEND TO BE THOSE                      0.61          5   36
    WHERE THE REPRESENTATIVE CHARACTER                  WHERE THE REPRESENTATIVE CHARACTER                  0   57
    OF THE REGIME IS POLYARCHIC  (41)                   OF THE REGIME IS LIMITED POLYARCHIC,            X SQ=   5.02
                                                        PSEUDO-POLYARCHIC, OR                           P =     0.011
                                                        NON-POLYARCHIC  (57)                            RV YES (NO )
```

105	ALWAYS ARE THOSE WHERE THE ELECTORAL SYSTEM IS COMPETITIVE (43)	1.00	105	TEND TO BE THOSE WHERE THE ELECTORAL SYSTEM IS PARTIALLY COMPETITIVE OR NON-COMPETITIVE (47)	0.55	5 38 0 47 X SQ= 3.78 P = 0.022 RV YES (NO)
107	ALWAYS ARE THOSE WHERE AUTONOMOUS GROUPS ARE FULLY TOLERATED IN POLITICS (46)	1.00	107	TEND TO BE THOSE WHERE AUTONOMOUS GROUPS ARE PARTIALLY TOLERATED IN POLITICS, ARE TOLERATED ONLY OUTSIDE POLITICS, OR ARE NOT TOLERATED AT ALL (65)	0.61	5 41 0 65 X SQ= 5.09 P = 0.011 RV YES (NO)
110	TEND TO BE THOSE WHERE POLITICAL ENCULTURATION IS HIGH (15)	0.80	110	TEND TO BE THOSE WHERE POLITICAL ENCULTURATION IS MEDIUM OR LOW (80)	0.88	4 11 1 79 X SQ= 11.66 P = 0.002 RV YES (NO)
111	ALWAYS ARE THOSE WHERE POLITICAL ENCULTURATION IS HIGH OR MEDIUM (53)	1.00	111	LEAN LESS TOWARD BEING THOSE WHERE POLITICAL ENCULTURATION IS HIGH OR MEDIUM (53)	0.53	5 48 0 42 X SQ= 2.50 P = 0.064 RV NO (NO)
114	ALWAYS ARE THOSE WHERE SECTIONALISM IS NEGLIGIBLE (47)	1.00	114	TEND TO BE THOSE WHERE SECTIONALISM IS EXTREME OR MODERATE (61)	0.59	0 61 5 42 X SQ= 4.61 P = 0.014 RV YES (NO)
115	TEND TO BE THOSE WHERE INTEREST ARTICULATION BY ASSOCIATIONAL GROUPS IS SIGNIFICANT (20)	0.80	115	TEND TO BE THOSE WHERE INTEREST ARTICULATION BY ASSOCIATIONAL GROUPS IS MODERATE, LIMITED, OR NEGLIGIBLE (91)	0.85	4 16 1 90 X SQ= 9.58 P = 0.004 RV YES (NO)
116	ALWAYS ARE THOSE WHERE INTEREST ARTICULATION BY ASSOCIATIONAL GROUPS IS SIGNIFICANT OR MODERATE (32)	1.00	116	TEND TO BE THOSE WHERE INTEREST ARTICULATION BY ASSOCIATIONAL GROUPS IS LIMITED OR NEGLIGIBLE (79)	0.75	5 27 0 79 X SQ= 9.55 P = 0.002 RV YES (NO)
120	ALWAYS ARE THOSE WHERE INTEREST ARITUCLATION BY INSTITUTIONAL GROUPS IS LIMITED (10)	1.00	120	TEND TO BE THOSE WHERE INTEREST ARTICULATION BY INSTITUTIONAL GROUPS IS VERY SIGNIFICANT, SIGNIFICANT, OR MODERATE (90)	0.95	0 90 5 5 X SQ= 37.43 P = 0. RV YES (YES)
122	ALWAYS ARE THOSE WHERE INTEREST ARTICULATION BY NON-ASSOCIATIONAL GROUPS IS LIMITED OR NEGLIGIBLE (32)	1.00	122	TEND TO BE THOSE WHERE INTEREST ARTICULATION BY NON-ASSOCIATIONAL GROUPS IS SIGNIFICANT OR MODERATE (83)	0.75	0 83 5 27 X SQ= 10.06 P = 0.001 RV YES (NO)

123	TEND TO BE THOSE WHERE INTEREST ARTICULATION BY NON-ASSOCIATIONAL GROUPS IS NEGLIGIBLE (8)	0.80	123	TEND TO BE THOSE WHERE INTEREST ARTICULATION BY NON-ASSOCIATIONAL GROUPS IS SIGNIFICANT, MODERATE, OR LIMITED (107)	0.96	1 106 4 4 X SQ= 32.10 P = 0. RV YES (YES)
126	ALWAYS ARE THOSE WHERE INTEREST ARTICULATION BY ANOMIC GROUPS IS VERY INFREQUENT (16)	1.00	126	TEND TO BE THOSE WHERE INTEREST ARTICULATION BY ANOMIC GROUPS IS FREQUENT, OCCASIONAL, OR INFREQUENT (83)	0.88	0 83 5 11 X SQ= 21.19 P = 0. RV YES (NO)
127	ALWAYS ARE THOSE WHERE INTEREST ARTICULATION BY POLITICAL PARTIES IS SIGNIFICANT (21)	1.00	127	TEND TO BE THOSE WHERE INTEREST ARTICULATION BY POLITICAL PARTIES IS MODERATE, LIMITED, OR NEGLIGIBLE (72)	0.82	5 16 0 72 X SQ= 13.74 P = 0. RV YES (NO)
131	ALWAYS ARE THOSE WHERE INTEREST AGGREGATION BY POLITICAL PARTIES IS LIMITED OR NEGLIGIBLE (35)	1.00	131	LEAN TOWARD BEING THOSE WHERE INTEREST AGGREGATION BY POLITICAL PARTIES IS SIGNIFICANT OR MODERATE (30)	0.50	0 30 5 30 X SQ= 2.85 P = 0.057 RV YES (NO)
134	ALWAYS ARE THOSE WHERE INTEREST AGGREGATION BY THE EXECUTIVE IS SIGNIFICANT OR MODERATE (57)	1.00	134	LEAN LESS TOWARD BEING THOSE WHERE INTEREST AGGREGATION BY THE EXECUTIVE IS SIGNIFICANT OR MODERATE (57)	0.53	5 52 0 46 X SQ= 2.55 P = 0.063 RV NO (NO)
136	TEND TO BE THOSE WHERE INTEREST AGGREGATION BY THE LEGISLATURE IS SIGNIFICANT (12)	0.80	136	TEND TO BE THOSE WHERE INTEREST AGGREGATION BY THE LEGISLATURE IS MODERATE, LIMITED, OR NEGLIGIBLE (85)	0.91	4 8 1 84 X SQ= 16.15 P = 0.001 RV YES (NO)
137	ALWAYS ARE THOSE WHERE INTEREST AGGREGATION BY THE LEGISLATURE IS SIGNIFICANT OR MODERATE (29)	1.00	137	TEND TO BE THOSE WHERE INTEREST AGGREGATION BY THE LEGISLATURE IS LIMITED OR NEGLIGIBLE (68)	0.74	5 24 0 68 X SQ= 9.09 P = 0.002 RV YES (NO)
142	ALWAYS ARE THOSE WHERE THE PARTY SYSTEM IS QUANTITATIVELY MULTI-PARTY (30)	1.00	142	TEND TO BE THOSE WHERE THE PARTY SYSTEM IS QUANTITATIVELY OTHER THAN MULTI-PARTY (66)	0.73	5 25 0 66 X SQ= 8.47 P = 0.002 RV YES (NO)
147	ALWAYS ARE THOSE WHERE THE PARTY SYSTEM IS QUALITATIVELY CLASS-ORIENTED OR MULTI-IDEOLOGICAL (23)	1.00	147	TEND TO BE THOSE WHERE THE PARTY SYSTEM IS QUALITATIVELY OTHER THAN CLASS-ORIENTED OR MULTI-IDEOLOGICAL (67)	0.79	5 18 0 67 X SQ= 11.56 P = 0.001 RV YES (NO)

153	ALWAYS ARE THOSE WHERE THE PARTY SYSTEM IS STABLE (42)	1.00	153	LEAN TOWARD BEING THOSE WHERE THE PARTY SYSTEM IS MODERATELY STABLE OR UNSTABLE (38)	0.51	5 37 0 38 X SQ= 3.01 P = 0.056 RV YES (NO)
158	ALWAYS ARE THOSE WHERE PERSONALISMO IS NEGLIGIBLE (56)	1.00	158	LEAN LESS TOWARD BEING THOSE WHERE PERSONALISMO IS NEGLIGIBLE (56)	0.56	0 40 5 51 X SQ= 2.18 P = 0.073 RV NO (NO)
161	ALWAYS ARE THOSE WHERE THE POLITICAL LEADERSHIP IS NON-ELITIST (50)	1.00	161	LEAN TOWARD BEING THOSE WHERE THE POLITICAL LEADERSHIP IS ELITIST OR MODERATELY ELITIST (47)	0.51	0 47 5 45 X SQ= 3.12 P = 0.057 RV YES (NO)
168	ALWAYS ARE THOSE WHERE THE HORIZONTAL POWER DISTRIBUTION IS SIGNIFICANT (34)	1.00	168	TEND TO BE THOSE WHERE THE HORIZONTAL POWER DISTRIBUTION IS LIMITED OR NEGLIGIBLE (72)	0.71	5 29 0 72 X SQ= 8.08 P = 0.003 RV YES (NO)
172	TEND TO BE THOSE WHERE THE LEGISLATIVE-EXECUTIVE STRUCTURE IS PARLIAMENTARY-ROYALIST (21)	0.60	172	TEND TO BE THOSE WHERE THE LEGISLATIVE-EXECUTIVE STRUCTURE IS OTHER THAN PARLIAMENTARY-ROYALIST (88)	0.83	3 18 2 86 X SQ= 3.18 P = 0.048 RV YES (NO)
174	ALWAYS ARE THOSE WHERE THE LEGISLATURE IS FULLY EFFECTIVE (28)	1.00	174	TEND TO BE THOSE WHERE THE LEGISLATURE IS PARTIALLY EFFECTIVE, LARGELY INEFFECTIVE, OR WHOLLY INEFFECTIVE (72)	0.76	5 23 0 72 X SQ= 10.04 P = 0.001 RV YES (NO)
179	ALWAYS ARE THOSE WHERE THE EXECUTIVE IS STRONG (39)	1.00	179	TEND TO BE THOSE WHERE THE EXECUTIVE IS DOMINANT (52)	0.60	0 52 5 34 X SQ= 4.80 P = 0.012 RV YES (NO)
181	ALWAYS ARE THOSE WHERE THE BUREAUCRACY IS MODERN, RATHER THAN SEMI-MODERN (21)	1.00	181	TEND TO BE THOSE WHERE THE BUREAUCRACY IS SEMI-MODERN, RATHER THAN MODERN (55)	0.77	5 16 0 55 X SQ= 10.41 P = 0.001 RV YES (NO)
187	ALWAYS ARE THOSE WHERE THE ROLE OF THE POLICE IS NOT POLITICALLY SIGNIFICANT (35)	1.00	187	TEND TO BE THOSE WHERE THE ROLE OF THE POLICE IS POLITICALLY SIGNIFICANT (66)	0.69	0 66 5 30 X SQ= 7.12 P = 0.004 RV YES (NO)

5 POLITIES
LOCATED IN SCANDINAVIA OR WEST EUROPE, RATHER THAN IN EAST EUROPE (18)

5 POLITIES
LOCATED IN EAST EUROPE, RATHER THAN IN SCANDINAVIA OR WEST EUROPE (9)

BOTH SUBJECT AND PREDICATE

18 IN LEFT COLUMN

AUSTRIA	BELGIUM	DENMARK	FINLAND	FRANCE	GERMAN FR	GREECE	ICELAND	IRELAND	ITALY
LUXEMBOURG	NETHERLANDS	NORWAY	PORTUGAL	SPAIN	SWEDEN	SWITZERLAND	UK		

9 IN RIGHT COLUMN

ALBANIA	BULGARIA	CZECHOS'KIA	GERMANY, E	HUNGARY	POLAND	RUMANIA	USSR	YUGOSLAVIA

88 EXCLUDED BECAUSE IRRELEVANT

AFGHANISTAN	ALGERIA	ARGENTINA	AUSTRALIA	BOLIVIA	BRAZIL	BURMA	BURUNDI	CAMBODIA	CAMEROUN
CANADA	CEN AFR REP	CEYLON	CHAD	CHILE	CHINA, PR	COLOMBIA	CONGO(BRA)	CONGO(LEO)	COSTA RICA
CUBA	CYPRUS	DAHOMEY	DOMIN REP	ECUADOR	EL SALVADOR	ETHIOPIA	GABON	GHANA	GUATEMALA
GUINEA	HAITI	HONDURAS	INDIA	INDONESIA	IRAN	IRAQ	ISRAEL	IVORY COAST	JAMAICA
JAPAN	JORDAN	KOREA, N	KOREA REP	LAOS	LEBANON	LIBERIA	LIBYA	MALAGASY R	MALAYA
MALI	MAURITANIA	MEXICO	MONGOLIA	MOROCCO	NEPAL	NEW ZEALAND	NICARAGUA	NIGER	NIGERIA
PAKISTAN	PANAMA	PARAGUAY	PERU	PHILIPPINES	RWANDA	SA'U ARABIA	SENEGAL	SIERRE LEO	SOMALIA
SO AFRICA	SUDAN	SYRIA	TANGANYIKA	THAILAND	TOGO	TRINIDAD	TUNISIA	TURKEY	UGANDA
UAR	US	UPPER VOLTA	URUGUAY	VENEZUELA	VIETNAM, N	VIETNAM REP	YEMEN		

	Left column			Right column		
29	ALWAYS ARE THOSE WHERE THE DEGREE OF URBANIZATION IS HIGH (56)	1.00	29	TILT LESS TOWARD BEING THOSE WHERE THE DEGREE OF URBANIZATION IS HIGH (56)	0.78	17 7 0 2 X SQ= 1.56 P = 0.111 RV NO (YES)
30	ALWAYS ARE THOSE WHOSE AGRICULTURAL POPULATION IS MEDIUM, LOW, OR VERY LOW (57)	1.00	30	TEND LESS TO BE THOSE WHOSE AGRICULTURAL POPULATION IS MEDIUM, LOW, OR VERY LOW (57)	0.67	0 3 18 6 X SQ= 3.80 P = 0.029 RV NO (YES)
31	TEND TO BE THOSE WHOSE AGRICULTURAL POPULATION IS LOW OR VERY LOW (24)	0.72	31	TEND TO BE THOSE WHOSE AGRICULTURAL POPULATION IS HIGH OR MEDIUM (90)	0.89	5 8 13 1 X SQ= 6.69 P = 0.004 RV YES (YES)
35	TEND TO BE THOSE WHOSE PER CAPITA GROSS NATIONAL PRODUCT IS VERY HIGH OR HIGH (24)	0.83	35	TEND TO BE THOSE WHOSE PER CAPITA GROSS NATIONAL PRODUCT IS MEDIUM, LOW, OR VERY LOW (91)	0.67	15 3 3 6 X SQ= 4.69 P = 0.026 RV YES (YES)

51

50	TEND TO BE THOSE WHERE FREEDOM OF THE PRESS IS COMPLETE (43)	0.82	50	ALWAYS ARE THOSE WHERE FREEDOM OF THE PRESS IS INTERMITTENT, INTERNALLY ABSENT, OR INTERNALLY AND EXTERNALLY ABSENT (56)	1.00	14 0 3 9 X SQ= 12.92 P = 0. RV YES (YES)
51	TEND TO BE THOSE WHERE FREEDOM OF THE PRESS IS COMPLETE OR INTERMITTENT (60)	0.88	51	ALWAYS ARE THOSE WHERE FREEDOM OF THE PRESS IS INTERNALLY ABSENT, OR INTERNALLY AND EXTERNALLY ABSENT (37)	1.00	15 0 2 9 X SQ= 15.33 P = 0. RV YES (YES)
52	ALWAYS ARE THOSE WHERE FREEDOM OF THE PRESS IS COMPLETE, INTERMITTENT, OR INTERNALLY ABSENT (82)	1.00	52	TEND TO BE THOSE WHERE FREEDOM OF THE PRESS IS INTERNALLY AND EXTERNALLY ABSENT (16)	0.67	18 3 0 6 X SQ= 11.81 P = 0. RV YES (YES)
53	LEAN TOWARD BEING THOSE WHERE NEWSPAPER CIRCULATION IS THREE HUNDRED OR MORE PER THOUSAND (14)	0.50	53	LEAN TOWARD BEING THOSE WHERE NEWSPAPER CIRCULATION IS LESS THAN THREE HUNDRED PER THOUSAND (101)	0.89	9 1 9 8 X SQ= 2.40 P = 0.091 RV YES (NO)
57	TILT LESS TOWARD BEING THOSE WHERE THE RELIGION IS OTHER THAN CATHOLIC (90)	0.56	57	TILT MORE TOWARD BEING THOSE WHERE THE RELIGION IS OTHER THAN CATHOLIC (90)	0.89	8 1 10 8 X SQ= 1.69 P = 0.110 RV NO (NO)
66	LEAN TOWARD BEING THOSE THAT ARE RELIGIOUSLY HOMOGENEOUS (57)	0.78	66	LEAN TOWARD BEING THOSE THAT ARE RELIGIOUSLY HETEROGENEOUS (49)	0.62	14 3 4 5 X SQ= 2.39 P = 0.078 RV YES (YES)
70	TEND TO BE THOSE THAT ARE RELIGIOUSLY, RACIALLY, AND LINGUISTICALLY HOMOGENEOUS (21)	0.61	70	TEND TO BE THOSE THAT ARE RELIGIOUSLY, RACIALLY, OR LINGUISTICALLY HETEROGENEOUS (85)	0.89	11 1 7 8 X SQ= 4.22 P = 0.019 RV YES (YES)
71	TEND TO BE THOSE WHOSE DATE OF INDEPENDENCE IS BEFORE 1800 (21)	0.65	71	TEND TO BE THOSE WHOSE DATE OF INDEPENDENCE IS AFTER 1800 (90)	0.87	11 1 6 7 X SQ= 4.03 P = 0.030 RV YES (YES)
74	ALWAYS ARE THOSE THAT ARE HISTORICALLY WESTERN (26)	1.00	74	TEND TO BE THOSE THAT ARE NOT HISTORICALLY WESTERN (87)	0.56	18 4 0 5 X SQ= 8.87 P = 0.002 RV YES (YES)

81	TILT LESS TOWARD BEING THOSE WHERE THE TYPE OF POLITICAL MODERNIZATION IS LATER EUROPEAN OR LATER EUROPEAN DERIVED, RATHER THAN EARLY EUROPEAN OR EARLY EUROPEAN DERIVED (40)	0.67	81	ALWAYS ARE THOSE WHERE THE TYPE OF POLITICAL MODERNIZATION IS LATER EUROPEAN OR LATER EUROPEAN DERIVED, RATHER THAN EARLY EUROPEAN OR EARLY EUROPEAN DERIVED (40)	1.00	6 0 12 8 X SQ= 1.84 P = 0.132 RV NO (NO)
89	ALWAYS ARE THOSE WHOSE IDEOLOGICAL ORIENTATION IS CONVENTIONAL (33)	1.00	89	ALWAYS ARE THOSE WHOSE IDEOLOGICAL ORIENTATION IS OTHER THAN CONVENTIONAL (62)	1.00	16 0 0 9 X SQ= 20.85 P = 0. RV YES (YES)
93	ALWAYS ARE THOSE WHERE THE SYSTEM STYLE IS NON-MOBILIZATIONAL (78)	1.00	93	ALWAYS ARE THOSE WHERE THE SYSTEM STYLE IS MOBILIZATIONAL, OR LIMITED MOBILIZATIONAL (32)	1.00	0 9 17 0 X SQ= 21.77 P = 0. RV YES (YES)
94	TEND TO BE THOSE WHERE THE STATUS OF THE REGIME IS CONSTITUTIONAL (51)	0.89	94	ALWAYS ARE THOSE WHERE THE STATUS OF THE REGIME IS AUTHORITARIAN OR TOTALITARIAN (41)	1.00	16 0 2 9 X SQ= 16.13 P = 0. RV YES (YES)
95	TEND TO BE THOSE WHERE THE STATUS OF THE REGIME IS CONSTITUTIONAL OR AUTHORITARIAN (95)	0.89	95	ALWAYS ARE THOSE WHERE THE STATUS OF THE REGIME IS TOTALITARIAN (16)	1.00	16 0 2 9 X SQ= 16.13 P = 0. RV YES (YES)
98	TEND TO BE THOSE WHERE GOVERNMENTAL STABILITY IS GENERALLY PRESENT AND DATES AT LEAST FROM THE INTERWAR PERIOD (22)	0.72	98	TEND TO BE THOSE WHERE GOVERNMENTAL STABILITY IS GENERALLY OR MODERATELY PRESENT AND DATES FROM THE POST-WAR PERIOD, OR IS ABSENT (93)	0.89	13 1 5 8 X SQ= 6.69 P = 0.004 RV YES (YES)
101	TEND TO BE THOSE WHERE THE REPRESENTATIVE CHARACTER OF THE REGIME IS POLYARCHIC (41)	0.89	101	ALWAYS ARE THOSE WHERE THE REPRESENTATIVE CHARACTER OF THE REGIME IS LIMITED POLYARCHIC, PSEUDO-POLYARCHIC, OR NON-POLYARCHIC (57)	1.00	16 0 2 9 X SQ= 16.13 P = 0. RV YES (YES)
102	TEND TO BE THOSE WHERE THE REPRESENTATIVE CHARACTER OF THE REGIME IS POLYARCHIC OR LIMITED POLYARCHIC (59)	0.89	102	ALWAYS ARE THOSE WHERE THE REPRESENTATIVE CHARACTER OF THE REGIME IS PSEUDO-POLYARCHIC OR NON-POLYARCHIC (49)	1.00	16 0 2 9 X SQ= 16.13 P = 0. RV YES (YES)
105	TEND TO BE THOSE WHERE THE ELECTORAL SYSTEM IS COMPETITIVE (43)	0.89	105	ALWAYS ARE THOSE WHERE THE ELECTORAL SYSTEM IS PARTIALLY COMPETITIVE OR NON-COMPETITIVE (47)	1.00	16 0 2 9 X SQ= 16.13 P = 0. RV YES (YES)

106	TEND TO BE THOSE WHERE THE ELECTORAL SYSTEM IS COMPETITIVE OR PARTIALLY COMPETITIVE (52)	0.89	106	ALWAYS ARE THOSE WHERE THE ELECTORAL SYSTEM IS NON-COMPETITIVE (30)	1.00	16 0 2 9 X SQ= 16.13 P = 0. RV YES (YES)
107	TEND TO BE THOSE WHERE AUTONOMOUS GROUPS ARE FULLY TOLERATED IN POLITICS (46)	0.89	107	ALWAYS ARE THOSE WHERE AUTONOMOUS GROUPS ARE PARTIALLY TOLERATED IN POLITICS, ARE TOLERATED ONLY OUTSIDE POLITICS, OR ARE NOT TOLERATED AT ALL (65)	1.00	16 0 2 9 X SQ= 16.13 P = 0. RV YES (YES)
108	TEND TO BE THOSE WHERE AUTONOMOUS GROUPS ARE FULLY OR PARTIALLY TOLERATED IN POLITICS (65)	0.89	108	ALWAYS ARE THOSE WHERE AUTONOMOUS GROUPS ARE TOLERATED ONLY OUTSIDE POLITICS OR ARE NOT TOLERATED AT ALL (35)	1.00	16 0 2 9 X SQ= 16.13 P = 0. RV YES (YES)
115	TEND TO BE THOSE WHERE INTEREST ARTICULATION BY ASSOCIATIONAL GROUPS IS SIGNIFICANT (20)	0.75	115	ALWAYS ARE THOSE WHERE INTEREST ARTICULATION BY ASSOCIATIONAL GROUPS IS MODERATE, LIMITED, OR NEGLIGIBLE (91)	1.00	12 0 4 8 X SQ= 9.19 P = 0.001 RV YES (YES)
116	ALWAYS ARE THOSE WHERE INTEREST ARTICULATION BY ASSOCIATIONAL GROUPS IS SIGNIFICANT OR MODERATE (32)	1.00	116	ALWAYS ARE THOSE WHERE INTEREST ARTICULATION BY ASSOCIATIONAL GROUPS IS LIMITED OR NEGLIGIBLE (79)	1.00	16 0 0 8 X SQ= 19.71 P = 0. RV YES (YES)
117	ALWAYS ARE THOSE WHERE INTEREST ARTICULATION BY ASSOCIATIONAL GROUPS IS SIGNIFICANT, MODERATE, OR LIMITED (60)	1.00	117	TEND TO BE THOSE WHERE INTEREST ARTICULATION BY ASSOCIATIONAL GROUPS IS NEGLIGIBLE (51)	0.62	16 3 0 5 X SQ= 9.13 P = 0.001 RV YES (YES)
118	TEND TO BE THOSE WHERE INTEREST ARTICULATION BY INSTITUTIONAL GROUPS IS SIGNIFICANT, MODERATE, OR LIMITED (60)	0.89	118	ALWAYS ARE THOSE WHERE INTEREST ARTICULATION BY INSTITUTIONAL GROUPS IS VERY SIGNIFICANT (40)	1.00	2 9 16 0 X SQ= 16.13 P = 0. RV YES (YES)
119	TEND TO BE THOSE WHERE INTEREST ARTICULATION BY INSTITUTIONAL GROUPS IS MODERATE OR LIMITED (26)	0.83	119	ALWAYS ARE THOSE WHERE INTEREST ARTICULATION BY INSTITUTIONAL GROUPS IS VERY SIGNIFICANT OR SIGNIFICANT (74)	1.00	3 9 15 0 X SQ= 13.67 P = 0. RV YES (YES)
122	LEAN TOWARD BEING THOSE WHERE INTEREST ARTICULATION BY NON-ASSOCIATIONAL GROUPS IS LIMITED OR NEGLIGIBLE (32)	0.83	122	LEAN TOWARD BEING THOSE WHERE INTEREST ARTICULATION BY NON-ASSOCIATIONAL GROUPS IS SIGNIFICANT OR MODERATE (83)	0.56	3 5 15 4 X SQ= 2.69 P = 0.072 RV YES (YES)

123	LEAN LESS TOWARD BEING THOSE WHERE INTEREST ARTICULATION BY NON-ASSOCIATIONAL GROUPS IS SIGNIFICANT, MODERATE, OR LIMITED (107)	0.67	123	ALWAYS ARE THOSE WHERE INTEREST ARTICULATION BY NON-ASSOCIATIONAL GROUPS IS SIGNIFICANT, MODERATE, OR LIMITED (107)	1.00	12 9 6 0 X SQ= 2.17 P = 0.071 RV NO (NO)
125	TEND TO BE THOSE WHERE INTEREST ARTICULATION BY ANOMIC GROUPS IS INFREQUENT OR VERY INFREQUENT (35)	0.94	125	TEND TO BE THOSE WHERE INTEREST ARTICULATION BY ANOMIC GROUPS IS FREQUENT OR OCCASIONAL (64)	0.80	1 4 16 1 X SQ= 8.23 P = 0.003 RV YES (YES)
126	TEND TO BE THOSE WHERE INTEREST ARTICULATION BY ANOMIC GROUPS IS VERY INFREQUENT (16)	0.76	126	ALWAYS ARE THOSE WHERE INTEREST ARTICULATION BY ANOMIC GROUPS IS FREQUENT, OCCASIONAL, OR INFREQUENT (83)	1.00	4 5 13 0 X SQ= 6.45 P = 0.005 RV YES (YES)
127	TEND TO BE THOSE WHERE INTEREST ARTICULATION BY POLITICAL PARTIES IS SIGNIFICANT (21)	0.50	127	ALWAYS ARE THOSE WHERE INTEREST ARTICULATION BY POLITICAL PARTIES IS MODERATE, LIMITED, OR NEGLIGIBLE (72)	1.00	9 0 9 9 X SQ= 4.69 P = 0.012 RV YES (YES)
128	TEND TO BE THOSE WHERE INTEREST ARTICULATION BY POLITICAL PARTIES IS SIGNIFICANT OR MODERATE (48)	0.89	128	ALWAYS ARE THOSE WHERE INTEREST ARTICULATION BY POLITICAL PARTIES IS LIMITED OR NEGLIGIBLE (45)	1.00	16 0 2 9 X SQ= 16.13 P = 0. RV YES (YES)
129	TEND TO BE THOSE WHERE INTEREST ARTICULATION BY POLITICAL PARTIES IS SIGNIFICANT, MODERATE, OR LIMITED (56)	0.89	129	ALWAYS ARE THOSE WHERE INTEREST ARTICULATION BY POLITICAL PARTIES IS NEGLIGIBLE (37)	1.00	16 0 2 9 X SQ= 16.13 P = 0. RV YES (YES)
134	TEND TO BE THOSE WHERE INTEREST AGGREGATION BY THE EXECUTIVE IS SIGNIFICANT OR MODERATE (57)	0.83	134	ALWAYS ARE THOSE WHERE INTEREST AGGREGATION BY THE EXECUTIVE IS LIMITED OR NEGLIGIBLE (46)	1.00	15 0 3 9 X SQ= 13.67 P = 0. RV YES (YES)
136	TEND TO BE THOSE WHERE INTEREST AGGREGATION BY THE LEGISLATURE IS SIGNIFICANT (12)	0.50	136	ALWAYS ARE THOSE WHERE INTEREST AGGREGATION BY THE LEGISLATURE IS MODERATE, LIMITED, OR NEGLIGIBLE (85)	1.00	9 0 9 9 X SQ= 4.69 P = 0.012 RV YES (YES)
137	TEND TO BE THOSE WHERE INTEREST AGGREGATION BY THE LEGISLATURE IS SIGNIFICANT OR MODERATE (29)	0.83	137	ALWAYS ARE THOSE WHERE INTEREST AGGREGATION BY THE LEGISLATURE IS LIMITED OR NEGLIGIBLE (68)	1.00	15 0 3 9 X SQ= 13.67 P = 0. RV YES (YES)

5/

138	ALWAYS ARE THOSE WHERE INTEREST AGGREGATION BY THE LEGISLATURE IS SIGNIFICANT, MODERATE, OR LIMITED (48)	1.00	138	ALWAYS ARE THOSE WHERE INTEREST AGGREGATION BY THE LEGISLATURE IS NEGLIGIBLE (49)	1.00	16 0 0 9 X SQ= 20.85 P = 0. RV YES (YES)
139	TEND TO BE THOSE WHERE THE PARTY SYSTEM IS QUANTITATIVELY OTHER THAN ONE-PARTY (71)	0.89	139	ALWAYS ARE THOSE WHERE THE PARTY SYSTEM IS QUANTITATIVELY ONE-PARTY (34)	1.00	2 9 16 0 X SQ= 16.13 P = 0. RV YES (YES)
142	TEND TO BE THOSE WHERE THE PARTY SYSTEM IS QUANTITATIVELY MULTI-PARTY (30)	0.61	142	ALWAYS ARE THOSE WHERE THE PARTY SYSTEM IS QUANTITATIVELY OTHER THAN MULTI-PARTY (66)	1.00	11 0 7 9 X SQ= 6.92 P = 0.003 RV YES (YES)
144	LEAN TOWARD BEING THOSE WHERE THE PARTY SYSTEM IS QUANTITATIVELY TWO-PARTY, RATHER THAN ONE-PARTY (11)	0.50	144	ALWAYS ARE THOSE WHERE THE PARTY SYSTEM IS QUANTITATIVELY ONE-PARTY, RATHER THAN TWO-PARTY (34)	1.00	2 9 2 0 X SQ= 2.17 P = 0.077 RV YES (YES)
147	TEND TO BE THOSE WHERE THE PARTY SYSTEM IS QUALITATIVELY CLASS-ORIENTED OR MULTI-IDEOLOGICAL (23)	0.87	147	ALWAYS ARE THOSE WHERE THE PARTY SYSTEM IS QUALITATIVELY OTHER THAN CLASS-ORIENTED OR MULTI-IDEOLOGICAL (67)	1.00	13 0 2 9 X SQ= 13.71 P = 0. RV YES (YES)
160	TEND TO BE THOSE WHERE THE POLITICAL LEADERSHIP IS MODERATELY ELITIST OR NON-ELITIST (67)	0.89	160	ALWAYS ARE THOSE WHERE THE POLITICAL LEADERSHIP IS ELITIST (30)	1.00	2 9 16 0 X SQ= 16.13 P = 0. RV YES (YES)
161	TEND TO BE THOSE WHERE THE POLITICAL LEADERSHIP IS NON-ELITIST (50)	0.83	161	ALWAYS ARE THOSE WHERE THE POLITICAL LEADERSHIP IS ELITIST OR MODERATELY ELITIST (47)	1.00	3 9 15 0 X SQ= 13.67 P = 0. RV YES (YES)
168	TEND TO BE THOSE WHERE THE HORIZONTAL POWER DISTRIBUTION IS SIGNIFICANT (34)	0.83	168	ALWAYS ARE THOSE WHERE THE HORIZONTAL POWER DISTRIBUTION IS LIMITED OR NEGLIGIBLE (72)	1.00	15 0 3 9 X SQ= 13.67 P = 0. RV YES (YES)
169	TEND TO BE THOSE WHERE THE HORIZONTAL POWER DISTRIBUTION IS SIGNIFICANT OR LIMITED (58)	0.89	169	ALWAYS ARE THOSE WHERE THE HORIZONTAL POWER DISTRIBUTION IS NEGLIGIBLE (48)	1.00	16 0 2 9 X SQ= 16.13 P = 0. RV YES (YES)

171	LEAN LESS TOWARD BEING THOSE WHERE THE LEGISLATIVE-EXECUTIVE STRUCTURE IS OTHER THAN PARLIAMENTARY-REPUBLICAN (90)	0.62	171	ALWAYS ARE THOSE WHERE THE LEGISLATIVE-EXECUTIVE STRUCTURE IS OTHER THAN PARLIAMENTARY-REPUBLICAN (90)	1.00	6 0 10 9 X SQ= 2.62 P = 0.057 RV NO (NO)
172	TEND LESS TO BE THOSE WHERE THE LEGISLATIVE-EXECUTIVE STRUCTURE IS OTHER THAN PARLIAMENTARY-ROYALIST (88)	0.53	172	ALWAYS ARE THOSE WHERE THE LEGISLATIVE-EXECUTIVE STRUCTURE IS OTHER THAN PARLIAMENTARY-ROYALIST (88)	1.00	8 0 9 9 X SQ= 4.11 P = 0.023 RV NO (YES)
174	TEND TO BE THOSE WHERE THE LEGISLATURE IS FULLY EFFECTIVE (28)	0.78	174	ALWAYS ARE THOSE WHERE THE LEGISLATURE IS PARTIALLY EFFECTIVE, LARGELY INEFFECTIVE, OR WHOLLY INEFFECTIVE (72)	1.00	14 0 4 9 X SQ= 11.59 P = 0. RV YES (YES)
175	TEND TO BE THOSE WHERE THE LEGISLATURE IS FULLY EFFECTIVE OR PARTIALLY EFFECTIVE (51)	0.89	175	ALWAYS ARE THOSE WHERE THE LEGISLATURE IS LARGELY INEFFECTIVE OR WHOLLY INEFFECTIVE (49)	1.00	16 0 2 9 X SQ= 16.13 P = 0. RV YES (YES)
176	TEND TO BE THOSE WHERE THE LEGISLATURE IS FULLY EFFECTIVE, PARTIALLY EFFECTIVE, OR LARGELY INEFFECTIVE (72)	0.89	176	ALWAYS ARE THOSE WHERE THE LEGISLATURE IS WHOLLY INEFFECTIVE (28)	1.00	16 0 2 9 X SQ= 16.13 P = 0. RV YES (YES)
178	TEND TO BE THOSE WHERE THE LEGISLATURE IS BICAMERAL (51)	0.78	178	TEND TO BE THOSE WHERE THE LEGISLATURE IS UNICAMERAL (53)	0.78	4 7 14 2 X SQ= 5.54 P = 0.011 RV YES (YES)
179	TEND TO BE THOSE WHERE THE EXECUTIVE IS STRONG (39)	0.81	179	ALWAYS ARE THOSE WHERE THE EXECUTIVE IS DOMINANT (52)	1.00	3 9 13 0 X SQ= 12.15 P = 0. RV YES (YES)
181	TEND TO BE THOSE WHERE THE BUREAUCRACY IS MODERN, RATHER THAN SEMI-MODERN (21)	0.83	181	ALWAYS ARE THOSE WHERE THE BUREAUCRACY IS SEMI-MODERN, RATHER THAN MODERN (55)	1.00	15 0 3 9 X SQ= 13.67 P = 0. RV YES (YES)
186	TEND TO BE THOSE WHERE PARTICIPATION BY THE MILITARY IN POLITICS IS NEUTRAL, RATHER THAN SUPPORTIVE (56)	0.88	186	ALWAYS ARE THOSE WHERE PARTICIPATION BY THE MILITARY IN POLITICS IS SUPPORTIVE, RATHER THAN NEUTRAL (31)	1.00	2 9 15 0 X SQ= 15.33 P = 0. RV YES (YES)

5/

```
187  TEND TO BE THOSE                          0.89      187  ALWAYS ARE THOSE                          1.00        2     9
     WHERE THE ROLE OF THE POLICE                              WHERE THE ROLE OF THE POLICE                            16     0
     IS NOT POLITICALLY SIGNIFICANT  (35)                      IS POLITICALLY SIGNIFICANT  (66)                    X SQ=  16.13
                                                                                                                   P =     0.
                                                                                                                   RV YES (YES)

188  TEND TO BE THOSE                          0.61      188  ALWAYS ARE THOSE                          1.00       11     0
     WHERE THE CHARACTER OF THE LEGAL SYSTEM                   WHERE THE CHARACTER OF THE LEGAL SYSTEM                  7     9
     IS CIVIL LAW  (32)                                        IS OTHER THAN CIVIL LAW  (81)                       X SQ=   6.92
                                                                                                                   P =     0.003
                                                                                                                   RV YES (YES)

194  ALWAYS ARE THOSE                          1.00      194  ALWAYS ARE THOSE                          1.00        0     9
     THAT ARE NON-COMMUNIST  (101)                             THAT ARE COMMUNIST  (13)                                18     0
                                                                                                                   X SQ=  22.69
                                                                                                                   P =     0.
                                                                                                                   RV YES (YES)
```

6 POLITIES
LOCATED IN THE CARIBBEAN,
CENTRAL AMERICA, OR SOUTH AMERICA (22)

6 POLITIES
LOCATED ELSEWHERE THAN IN THE
CARIBBEAN, CENTRAL AMERICA,
OR SOUTH AMERICA (93)

BOTH SUBJECT AND PREDICATE

22 IN LEFT COLUMN

ARGENTINA	BOLIVIA	BRAZIL	CHILE	COLOMBIA	COSTA RICA	CUBA	DOMIN REP	ECUADOR	EL SALVADOR
GUATEMALA	HAITI	HONDURAS	JAMAICA	MEXICO	NICARAGUA	PANAMA	PARAGUAY	PERU	TRINIDAD
URUGUAY	VENEZUELA								

93 IN RIGHT COLUMN

AFGHANISTAN	ALBANIA	ALGERIA	AUSTRALIA	AUSTRIA	BELGIUM	BULGARIA	BURMA	BURUNDI	CAMBODIA
CAMEROUN	CANADA	CEN AFR REP	CEYLON	CHAD	CHINA, PR	CONGO(BRA)	CONGO(LEO)	CYPRUS	CZECHOS'KIA
DAHOMEY	DENMARK	ETHIOPIA	FINLAND	FRANCE	GABON	GERMANY, E	GERMAN FR	GHANA	GREECE
GUINEA	HUNGARY	ICELAND	INDIA	INDONESIA	IRAN	IRAQ	IRELAND	ISRAEL	ITALY
IVORY COAST	JAPAN	JORDAN	KOREA, N	KOREA REP	LAOS	LEBANON	LIBERIA	LIBYA	LUXEMBOURG
MALAGASY R	MALAYA	MALI	MAURITANIA	MONGOLIA	MOROCCO	NEPAL	NETHERLANDS	NEW ZEALAND	NIGER
NIGERIA	NORWAY	PAKISTAN	PHILIPPINES	POLAND	PORTUGAL	RUMANIA	RWANDA	SA'U ARABIA	SENEGAL
SIERRE LEO	SOMALIA	SO AFRICA	SPAIN	SUDAN	SWEDEN	SWITZERLAND	SYRIA	TANGANYIKA	THAILAND
TOGO	TUNISIA	TURKEY	UGANDA	USSR	UAR	UK	US	UPPER VOLTA	VIETNAM, N
VIETNAM REP	YEMEN	YUGOSLAVIA							

---

24 LEAN TOWARD BEING THOSE
WHOSE POPULATION IS
SMALL (54)
0.64

24 LEAN TOWARD BEING THOSE
WHOSE POPULATION IS
VERY LARGE, LARGE, OR MEDIUM (61)
0.57

8 53
14 40
X SQ= 2.27
P = 0.099
RV YES (NO)

28 TEND TO BE THOSE
WHOSE POPULATION GROWTH RATE
IS HIGH (62)
0.86

28 TEND TO BE THOSE
WHOSE POPULATION GROWTH RATE
IS LOW (48)
0.51

19 43
3 45
X SQ= 8.60
P = 0.002
RV YES (NO)

29 LEAN TOWARD BEING THOSE
WHERE THE DEGREE OF URBANIZATION
IS HIGH (56)
0.74

29 LEAN TOWARD BEING THOSE
WHERE THE DEGREE OF URBANIZATION
IS LOW (49)
0.51

14 42
5 44
X SQ= 2.93
P = 0.074
RV YES (NO)

30 TEND TO BE THOSE
WHOSE AGRICULTURAL POPULATION
IS MEDIUM, LOW, OR VERY LOW (57)
0.77

30 TEND TO BE THOSE
WHOSE AGRICULTURAL POPULATION
IS HIGH (56)
0.56

5 51
17 40
X SQ= 6.59
P = 0.008
RV YES (NO)

37	TEND TO BE THOSE WHOSE PER CAPITA GROSS NATIONAL PRODUCT IS VERY HIGH, HIGH, MEDIUM, OR LOW (64)	0.86	37	TEND TO BE THOSE WHOSE PER CAPITA GROSS NATIONAL PRODUCT IS VERY LOW (51)	0.52	19 45 3 48 X SQ= 8.91 P = 0.002 RV YES (NO)
41	ALWAYS ARE THOSE WHOSE ECONOMIC DEVELOPMENTAL STATUS IS INTERMEDIATE, UNDERDEVELOPED, OR VERY UNDERDEVELOPED (94)	1.00	41	TEND LESS TO BE THOSE WHOSE ECONOMIC DEVELOPMENTAL STATUS IS INTERMEDIATE, UNDERDEVELOPED, OR VERY UNDERDEVELOPED (94)	0.79	0 19 21 73 X SQ= 3.84 P = 0.022 RV NO (NO)
44	ALWAYS ARE THOSE WHERE THE LITERACY RATE IS BELOW NINETY PERCENT (90)	1.00	44	TEND LESS TO BE THOSE WHERE THE LITERACY RATE IS BELOW NINETY PERCENT (90)	0.73	0 25 22 68 X SQ= 6.06 P = 0.003 RV NO (NO)
46	ALWAYS ARE THOSE WHERE THE LITERACY RATE IS TEN PERCENT OR ABOVE (84)	1.00	46	TEND LESS TO BE THOSE WHERE THE LITERACY RATE IS TEN PERCENT OR ABOVE (84)	0.70	22 62 0 26 X SQ= 6.95 P = 0.002 RV NO (NO)
53	ALWAYS ARE THOSE WHERE NEWSPAPER CIRCULATION IS LESS THAN THREE HUNDRED PER THOUSAND (101)	1.00	53	LEAN LESS TOWARD BEING THOSE WHERE NEWSPAPER CIRCULATION IS LESS THAN THREE HUNDRED PER THOUSAND (101)	0.85	0 14 22 79 X SQ= 2.49 P = 0.068 RV NO (NO)
55	ALWAYS ARE THOSE WHERE NEWSPAPER CIRCULATION IS TEN OR MORE PER THOUSAND (78)	1.00	55	TEND LESS TO BE THOSE WHERE NEWSPAPER CIRCULATION IS TEN OR MORE PER THOUSAND (78)	0.62	22 56 0 35 X SQ= 10.53 P = 0. RV NO (NO)
57	TEND TO BE THOSE WHERE THE RELIGION IS CATHOLIC (25)	0.68	57	TEND TO BE THOSE WHERE THE RELIGION IS OTHER THAN CATHOLIC (90)	0.89	15 10 7 83 X SQ= 31.20 P = 0. RV YES (YES)
62	ALWAYS ARE THOSE WHERE THE RELIGION IS CATHOLIC, RATHER THAN PROTESTANT (25)	1.00	62	TEND LESS TO BE THOSE WHERE THE RELIGION IS CATHOLIC, RATHER THAN PROTESTANT (25)	0.67	0 5 15 10 X SQ= 3.84 P = 0.042 RV NO (YES)
67	TEND TO BE THOSE THAT ARE RACIALLY HETEROGENEOUS (27)	0.85	67	TEND TO BE THOSE THAT ARE RACIALLY HOMOGENEOUS (82)	0.87	3 77 17 12 X SQ= 39.19 P = 0. RV YES (YES)

68	TEND TO BE THOSE THAT ARE LINGUISTICALLY HOMOGENEOUS (52)	0.73	68	TEND TO BE THOSE THAT ARE LINGUISTICALLY WEAKLY HETEROGENEOUS OR STRONGLY HETEROGENEOUS (62)	0.61	16 36 6 56 X SQ= 6.78 P = 0.008 RV YES (NO)
69	TEND TO BE THOSE THAT ARE LINGUISTICALLY HOMOGENEOUS OR WEAKLY HETEROGENEOUS (64)	0.82	69	TEND TO BE THOSE THAT ARE LINGUISTICALLY STRONGLY HETEROGENEOUS (50)	0.50	18 46 4 46 X SQ= 6.06 P = 0.008 RV YES (NO)
71	ALWAYS ARE THOSE WHOSE DATE OF INDEPENDENCE IS AFTER 1800 (90)	1.00	71	TEND LESS TO BE THOSE WHOSE DATE OF INDEPENDENCE IS AFTER 1800 (90)	0.76	0 21 22 68 X SQ= 4.96 P = 0.007 RV NO (NO)
72	TEND TO BE THOSE WHOSE DATE OF INDEPENDENCE IS BEFORE 1914 (52)	0.91	72	TEND TO BE THOSE WHOSE DATE OF INDEPENDENCE IS AFTER 1914 (59)	0.64	20 32 2 57 X SQ= 19.24 P = 0. RV YES (NO)
75	TEND TO BE THOSE THAT ARE HISTORICALLY WESTERN OR SIGNIFICANTLY WESTERNIZED (62)	0.95	75	TEND TO BE THOSE THAT ARE NOT HISTORICALLY WESTERN AND ARE NOT SIGNIFICANTLY WESTERNIZED (52)	0.55	21 41 1 51 X SQ= 16.54 P = 0. RV YES (NO)
76	ALWAYS ARE THOSE THAT HAVE BEEN SIGNIFICANTLY OR PARTIALLY WESTERNIZED THROUGH A COLONIAL RELATIONSHIP, RATHER THAN BEING HISTORICALLY WESTERN (70)	1.00	76	TEND LESS TO BE THOSE THAT HAVE BEEN SIGNIFICANTLY OR PARTIALLY WESTERNIZED THROUGH A COLONIAL RELATIONSHIP, RATHER THAN BEING HISTORICALLY WESTERN (70)	0.65	0 26 22 48 X SQ= 8.90 P = 0.001 RV NO (NO)
77	TEND TO BE THOSE THAT HAVE BEEN SIGNIFICANTLY WESTERNIZED, RATHER THAN PARTIALLY WESTERNIZED, THROUGH A COLONIAL RELATIONSHIP (28)	0.95	77	TEND TO BE THOSE THAT HAVE BEEN PARTIALLY WESTERNIZED, RATHER THAN SIGNIFICANTLY WESTERNIZED, THROUGH A COLONIAL RELATIONSHIP (41)	0.85	21 7 1 40 X SQ= 37.06 P = 0. RV YES (YES)
78	ALWAYS ARE THOSE THAT HAVE BEEN SIGNIFICANTLY WESTERNIZED THROUGH A COLONIAL RELATIONSHIP, RATHER THAN WITHOUT SUCH A RELATIONSHIP (28)	1.00	78	TEND TO BE THOSE THAT HAVE BEEN SIGNIFICANTLY WESTERNIZED WITHOUT A COLONIAL RELATIONSHIP, RATHER THAN THROUGH SUCH A RELATIONSHIP (7)	0.50	21 7 0 7 X SQ= 10.19 P = 0.001 RV YES (YES)
79	TEND TO BE THOSE THAT WERE FORMERLY DEPENDENCIES OF SPAIN, RATHER THAN BRITAIN OR FRANCE (18)	0.86	79	ALWAYS ARE THOSE THAT WERE FORMERLY DEPENDENCIES OF BRITAIN OR FRANCE, RATHER THAN SPAIN (49)	1.00	3 46 18 0 X SQ= 49.64 P = 0. RV YES (YES)

81 ALWAYS ARE THOSE 1.00
WHERE THE TYPE OF POLITICAL MODERNIZATION
IS LATER EUROPEAN OR
LATER EUROPEAN DERIVED, RATHER THAN
EARLY EUROPEAN OR
EARLY EUROPEAN DERIVED (40)

82 TEND TO BE THOSE 0.91
WHERE THE TYPE OF POLITICAL MODERNIZATION
IS EARLY OR LATER EUROPEAN OR
EUROPEAN DERIVED, RATHER THAN
DEVELOPED TUTELARY OR
UNDEVELOPED TUTELARY (51)

86 TEND MORE TO BE THOSE 0.91
WHERE THE STAGE OF
POLITICAL MODERNIZATION IS
ADVANCED OR MID-TRANSITIONAL,
RATHER THAN EARLY TRANSITIONAL (76)

92 TILT MORE TOWARD BEING THOSE 0.95
WHERE THE SYSTEM STYLE
IS LIMITED MOBILIZATIONAL, OR
NON-MOBILIZATIONAL (93)

93 TILT MORE TOWARD BEING THOSE 0.86
WHERE THE SYSTEM STYLE
IS NON-MOBILIZATIONAL (78)

94 TILT MORE TOWARD BEING THOSE 0.74
WHERE THE STATUS OF THE REGIME
IS CONSTITUTIONAL (51)

98 ALWAYS ARE THOSE 1.00
WHERE GOVERNMENTAL STABILITY
IS GENERALLY OR MODERATELY PRESENT
AND DATES FROM THE POST-WAR PERIOD,
OR IS ABSENT (93)

99 TEND TO BE THOSE 0.87
WHERE GOVERNMENTAL STABILITY
IS MODERATELY PRESENT AND DATES
FROM THE POST-WAR PERIOD,
OR IS ABSENT (36)

100 TEND TO BE THOSE 0.50
WHERE GOVERNMENTAL STABILITY
IS ABSENT (22)

81 TEND LESS TO BE THOSE 0.65
WHERE THE TYPE OF POLITICAL MODERNIZATION
IS LATER EUROPEAN OR
LATER EUROPEAN DERIVED, RATHER THAN
EARLY EUROPEAN OR
EARLY EUROPEAN DERIVED (40)

0 11
20 20
X SQ= 7.07
P = 0.004
RV NO (YES)

82 TEND TO BE THOSE 0.63
WHERE THE TYPE OF POLITICAL MODERNIZATION
IS DEVELOPED TUTELARY OR
UNDEVELOPED TUTELARY, RATHER THAN
EARLY OR LATER EUROPEAN OR
EUROPEAN DERIVED (55)

20 31
2 53
X SQ= 18.26
P = 0.
RV YES (NO)

86 TEND LESS TO BE THOSE 0.61
WHERE THE STAGE OF
POLITICAL MODERNIZATION IS
ADVANCED OR MID-TRANSITIONAL,
RATHER THAN EARLY TRANSITIONAL (76)

20 56
2 36
X SQ= 5.92
P = 0.006
RV NO (NO)

92 TILT LESS TOWARD BEING THOSE 0.79
WHERE THE SYSTEM STYLE
IS LIMITED MOBILIZATIONAL, OR
NON-MOBILIZATIONAL (93)

1 19
21 72
X SQ= 2.22
P = 0.116
RV NO (NO)

93 TILT LESS TOWARD BEING THOSE 0.67
WHERE THE SYSTEM STYLE
IS NON-MOBILIZATIONAL (78)

3 29
19 59
X SQ= 2.32
P = 0.114
RV NO (NO)

94 TILT LESS TOWARD BEING THOSE 0.51
WHERE THE STATUS OF THE REGIME
IS CONSTITUTIONAL (51)

14 37
5 36
X SQ= 2.36
P = 0.119
RV NO (NO)

98 TEND LESS TO BE THOSE 0.76
WHERE GOVERNMENTAL STABILITY
IS GENERALLY OR MODERATELY PRESENT
AND DATES FROM THE POST-WAR PERIOD,
OR IS ABSENT (93)

0 22
22 71
X SQ= 5.00
P = 0.007
RV NO (NO)

99 TEND TO BE THOSE 0.69
WHERE GOVERNMENTAL STABILITY
IS GENERALLY PRESENT AND DATES
FROM AT LEAST THE INTER-WAR PERIOD,
OR FROM THE POST-WAR PERIOD (50)

2 48
14 22
X SQ= 14.60
P = 0.
RV YES (NO)

100 TEND TO BE THOSE 0.81
WHERE GOVERNMENTAL STABILITY
IS GENERALLY PRESENT AND DATES FROM
AT LEAST THE INTER-WAR PERIOD, OR IS
GENERALLY OR MODERATELY PRESENT, AND
DATES FROM THE POST-WAR PERIOD (64)

9 55
9 13
X SQ= 5.60
P = 0.014
RV YES (NO)

102	TEND TO BE THOSE WHERE THE REPRESENTATIVE CHARACTER OF THE REGIME IS POLYARCHIC OR LIMITED POLYARCHIC (59)	0.75	102	TEND TO BE THOSE WHERE THE REPRESENTATIVE CHARACTER OF THE REGIME IS PSEUDO-POLYARCHIC OR NON-POLYARCHIC (49)	0.50	15 44 5 44 X SQ= 3.16 P = 0.050 RV YES (NO)
105	TEND TO BE THOSE WHERE THE ELECTORAL SYSTEM IS COMPETITIVE (43)	0.76	105	TEND TO BE THOSE WHERE THE ELECTORAL SYSTEM IS PARTIALLY COMPETITIVE OR NON-COMPETITIVE (47)	0.59	13 30 4 43 X SQ= 5.57 P = 0.014 RV YES (NO)
107	TEND TO BE THOSE WHERE AUTONOMOUS GROUPS ARE FULLY TOLERATED IN POLITICS (46)	0.67	107	TEND TO BE THOSE WHERE AUTONOMOUS GROUPS ARE PARTIALLY TOLERATED IN POLITICS, ARE TOLERATED ONLY OUTSIDE POLITICS, OR ARE NOT TOLERATED AT ALL (65)	0.64	M 14 32 7 58 X SQ= 5.57 P = 0.013 RV YES (NO)
114	TEND TO BE THOSE WHERE SECTIONALISM IS NEGLIGIBLE (47)	0.78	114	TEND TO BE THOSE WHERE SECTIONALISM IS EXTREME OR MODERATE (61)	0.63	4 57 14 33 X SQ= 8.71 P = 0.002 RV YES (NO)
117	TEND TO BE THOSE WHERE INTEREST ARTICULATION BY ASSOCIATIONAL GROUPS IS SIGNIFICANT, MODERATE, OR LIMITED (60)	0.86	117	TEND TO BE THOSE WHERE INTEREST ARTICULATION BY ASSOCIATIONAL GROUPS IS NEGLIGIBLE (51)	0.54	19 41 3 48 X SQ= 9.97 P = 0.001 RV YES (NO)
118	TEND MORE TO BE THOSE WHERE INTEREST ARTICULATION BY INSTITUTIONAL GROUPS IS SIGNIFICANT, MODERATE, OR LIMITED (60)	0.82	118	TEND LESS TO BE THOSE WHERE INTEREST ARTICULATION BY INSTITUTIONAL GROUPS IS SIGNIFICANT, MODERATE, OR LIMITED (60)	0.54	4 36 18 42 X SQ= 4.49 P = 0.025 RV NO (NO)
120	ALWAYS ARE THOSE WHERE INTEREST ARTICULATION BY INSTITUTIONAL GROUPS IS VERY SIGNIFICANT, SIGNIFICANT, OR MODERATE (90)	1.00	120	TILT LESS TOWARD BEING THOSE WHERE INTEREST ARTICULATION BY INSTITUTIONAL GROUPS IS VERY SIGNIFICANT, SIGNIFICANT, OR MODERATE (90)	0.87	22 68 0 10 X SQ= 1.87 P = 0.112 RV NO (NO)
121	TEND TO BE THOSE WHERE INTEREST ARTICULATION BY NON-ASSOCIATIONAL GROUPS IS MODERATE, LIMITED, OR NEGLIGIBLE (61)	0.95	121	TEND TO BE THOSE WHERE INTEREST ARTICULATION BY NON-ASSOCIATIONAL GROUPS IS SIGNIFICANT (54)	0.57	1 53 21 40 X SQ= 17.60 P = 0. RV YES (NO)
124	ALWAYS ARE THOSE WHERE INTEREST ARTICULATION BY ANOMIC GROUPS IS OCCASIONAL, INFREQUENT, OR VERY INFREQUENT (85)	1.00	124	LEAN LESS TOWARD BEING THOSE WHERE INTEREST ARTICULATION BY ANOMIC GROUPS IS OCCASIONAL, INFREQUENT, OR VERY INFREQUENT (85)	0.83	0 14 18 67 X SQ= 2.34 P = 0.067 RV NO (NO)

126	ALWAYS ARE THOSE WHERE INTEREST ARTICULATION BY ANOMIC GROUPS IS FREQUENT, OCCASIONAL, OR INFREQUENT (83)	1.00	126	TEND LESS TO BE THOSE WHERE INTEREST ARTICULATION BY ANOMIC GROUPS IS FREQUENT, OCCASIONAL, OR INFREQUENT (83)	0.80	18 65 0 16 X SQ= 2.91 P = 0.038 RV NO (NO)
128	TEND TO BE THOSE WHERE INTEREST ARTICULATION BY POLITICAL PARTIES IS SIGNIFICANT OR MODERATE (48)	0.74	128	TEND TO BE THOSE WHERE INTEREST ARTICULATION BY POLITICAL PARTIES IS LIMITED OR NEGLIGIBLE (45)	0.54	14 34 5 40 X SQ= 3.61 P = 0.040 RV YES (NO)
132	TEND LESS TO BE THOSE WHERE INTEREST AGGREGATION BY POLITICAL PARTIES IS SIGNIFICANT, MODERATE, OR LIMITED (67)	0.72	132	TEND MORE TO BE THOSE WHERE INTEREST AGGREGATION BY POLITICAL PARTIES IS SIGNIFICANT, MODERATE, OR LIMITED (67)	0.93	13 54 5 4 X SQ= 3.91 P = 0.030 RV NO (YES)
134	TILT TOWARD BEING THOSE WHERE INTEREST AGGREGATION BY THE EXECUTIVE IS LIMITED OR NEGLIGIBLE (46)	0.60	134	TILT TOWARD BEING THOSE WHERE INTEREST AGGREGATION BY THE EXECUTIVE IS SIGNIFICANT OR MODERATE (57)	0.59	8 49 12 34 X SQ= 1.66 P = 0.140 RV YES (NO)
138	TEND TO BE THOSE WHERE INTEREST AGGREGATION BY THE LEGISLATURE IS SIGNIFICANT, MODERATE, OR LIMITED (48)	0.80	138	TEND TO BE THOSE WHERE INTEREST AGGREGATION BY THE LEGISLATURE IS NEGLIGIBLE (49)	0.58	16 32 4 45 X SQ= 7.91 P = 0.003 RV YES (NO)
142	TEND TO BE THOSE WHERE THE PARTY SYSTEM IS QUANTITATIVELY MULTI-PARTY (30)	0.55	142	TEND TO BE THOSE WHERE THE PARTY SYSTEM IS QUANTITATIVELY OTHER THAN MULTI-PARTY (66)	0.75	11 19 9 57 X SQ= 5.31 P = 0.015 RV YES (NO)
144	TEND TO BE THOSE WHERE THE PARTY SYSTEM IS QUANTITATIVELY TWO-PARTY, RATHER THAN ONE-PARTY (11)	0.75	144	TEND TO BE THOSE WHERE THE PARTY SYSTEM IS QUANTITATIVELY ONE-PARTY, RATHER THAN TWO-PARTY (34)	0.80	1 33 3 8 X SQ= 3.44 P = 0.040 RV YES (NO)
153	TEND TO BE THOSE WHERE THE PARTY SYSTEM IS MODERATELY STABLE OR UNSTABLE (38)	0.93	153	TEND TO BE THOSE WHERE THE PARTY SYSTEM IS STABLE (42)	0.63	1 41 14 24 X SQ= 13.37 P = 0. RV YES (NO)
154	TEND TO BE THOSE WHERE THE PARTY SYSTEM IS UNSTABLE (25)	0.73	154	TEND TO BE THOSE WHERE THE PARTY SYSTEM IS STABLE OR MODERATELY STABLE (55)	0.78	4 51 11 14 X SQ= 12.90 P = 0. RV YES (NO)

158	TEND TO BE THOSE WHERE PERSONALISMO IS PRONOUNCED OR MODERATE (40)	0.70	158	TEND TO BE THOSE WHERE PERSONALISMO IS NEGLIGIBLE (56)	0.66	14 26 6 50 X SQ= 6.94 P = 0.005 RV YES (NO)
164	TEND MORE TO BE THOSE WHERE THE REGIME'S LEADERSHIP CHARISMA IS NEGLIGIBLE (65)	0.91	164	TEND LESS TO BE THOSE WHERE THE REGIME'S LEADERSHIP CHARISMA IS NEGLIGIBLE (65)	0.58	2 32 20 45 X SQ= 6.62 P = 0.005 RV NO (NO)
169	TEND TO BE THOSE WHERE THE HORIZONTAL POWER DISTRIBUTION IS SIGNIFICANT OR LIMITED (58)	0.85	169	TEND TO BE THOSE WHERE THE HORIZONTAL POWER DISTRIBUTION IS NEGLIGIBLE (48)	0.52	17 41 3 45 X SQ= 7.68 P = 0.003 RV YES (NO)
170	TEND TO BE THOSE WHERE THE LEGISLATIVE-EXECUTIVE STRUCTURE IS PRESIDENTIAL (39)	0.85	170	TEND TO BE THOSE WHERE THE LEGISLATIVE-EXECUTIVE STRUCTURE IS OTHER THAN PRESIDENTIAL (63)	0.73	17 22 3 60 X SQ= 20.64 P = 0. RV YES (NO)
171	ALWAYS ARE THOSE WHERE THE LEGISLATIVE-EXECUTIVE STRUCTURE IS OTHER THAN PARLIAMENTARY-REPUBLICAN (90)	1.00	171	TILT LESS TOWARD BEING THOSE WHERE THE LEGISLATIVE-EXECUTIVE STRUCTURE IS OTHER THAN PARLIAMENTARY-REPUBLICAN (90)	0.85	0 12 20 70 X SQ= 2.06 P = 0.117 RV NO (NO)
175	TEND TO BE THOSE WHERE THE LEGISLATURE IS FULLY EFFECTIVE OR PARTIALLY EFFECTIVE (51)	0.78	175	TEND TO BE THOSE WHERE THE LEGISLATURE IS LARGELY INEFFECTIVE OR WHOLLY INEFFECTIVE (49)	0.55	14 37 4 45 X SQ= 5.06 P = 0.018 RV YES (NO)
178	LEAN TOWARD BEING THOSE WHERE THE LEGISLATURE IS BICAMERAL (51)	0.67	178	LEAN TOWARD BEING THOSE WHERE THE LEGISLATURE IS UNICAMERAL (53)	0.55	7 46 14 37 X SQ= 2.45 P = 0.089 RV YES (NO)
181	ALWAYS ARE THOSE WHERE THE BUREAUCRACY IS SEMI-MODERN, RATHER THAN MODERN (55)	1.00	181	TEND LESS TO BE THOSE WHERE THE BUREAUCRACY IS SEMI-MODERN, RATHER THAN MODERN (55)	0.61	0 21 22 33 X SQ= 9.96 P = 0. RV NO (NO)
182	ALWAYS ARE THOSE WHERE THE BUREAUCRACY IS SEMI-MODERN, RATHER THAN POST-COLONIAL TRANSITIONAL (55)	1.00	182	TEND LESS TO BE THOSE WHERE THE BUREAUCRACY IS SEMI-MODERN, RATHER THAN POST-COLONIAL TRANSITIONAL (55)	0.57	22 33 0 25 X SQ= 11.86 P = 0. RV NO (NO)

185	TEND TO BE THOSE WHERE PARTICIPATION BY THE MILITARY IN POLITICS IS INTERVENTIVE, RATHER THAN SUPPORTIVE (21)	0.91	185	TEND TO BE THOSE WHERE PARTICIPATION BY THE MILITARY IN POLITICS IS SUPPORTIVE, RATHER THAN INTERVENTIVE (31)	0.73	10 11 1 30 X SQ= 12.25 P = 0. RV YES (NO)
186	LEAN MORE TOWARD BEING THOSE WHERE PARTICIPATION BY THE MILITARY IN POLITICS IS NEUTRAL, RATHER THAN SUPPORTIVE (56)	0.90	186	LEAN LESS TOWARD BEING THOSE WHERE PARTICIPATION BY THE MILITARY IN POLITICS IS NEUTRAL, RATHER THAN SUPPORTIVE (56)	0.61	1 30 9 47 X SQ= 2.10 P = 0.089 RV NO (NO)
188	TEND TO BE THOSE WHERE THE CHARACTER OF THE LEGAL SYSTEM IS CIVIL LAW (32)	0.86	188	TEND TO BE THOSE WHERE THE CHARACTER OF THE LEGAL SYSTEM IS OTHER THAN CIVIL LAW (81)	0.86	19 13 3 78 X SQ= 41.86 P = 0. RV YES (YES)
193	ALWAYS ARE THOSE WHERE THE CHARACTER OF THE LEGAL SYSTEM IS OTHER THAN PARTLY INDIGENOUS (86)	1.00	193	TEND LESS TO BE THOSE WHERE THE CHARACTER OF THE LEGAL SYSTEM IS OTHER THAN PARTLY INDIGENOUS (86)	0.70	0 28 22 64 X SQ= 7.31 P = 0.002 RV NO (NO)
194	ALWAYS ARE THOSE THAT ARE NON-COMMUNIST (101)	1.00	194	LEAN LESS TOWARD BEING THOSE THAT ARE NON-COMMUNIST (101)	0.86	0 13 21 80 X SQ= 2.07 P = 0.074 RV NO (NO)

```
*************************************************MATRIX*************************************************

       6  POLITIES                                                  6  POLITIES
          LOCATED IN THE CARIBBEAN,                                    LOCATED ELSEWHERE THAN IN THE
          CENTRAL AMERICA, OR SOUTH AMERICA  (22)                      CARIBBEAN, CENTRAL AMERICA,
                                                                       OR SOUTH AMERICA  (93)

                            107  POLITIES
                                 WHERE AUTONOMOUS GROUPS
                                 ARE FULLY TOLERATED IN POLITICS  (46)

ARGENTINA  BOLIVIA    BRAZIL     CHILE      COLOMBIA  .AUSTRALIA   AUSTRIA     BELGIUM     CANADA      CEYLON
COSTA RICA DOMIN REP  ECUADOR    HONDURAS   JAMAICA   .CYPRUS      DENMARK     FINLAND     FRANCE      GERMAN FR
PANAMA     TRINIDAD   URUGUAY    VENEZUELA            .GREECE      ICELAND     INDIA       IRELAND     ISRAEL
                                                      .ITALY       JAPAN       LEBANON     LUXEMBOURG  MALAGASY R
                                                      .MALAYA      NETHERLANDS NEW ZEALAND NORWAY      PHILIPPINES
                                                      .SIERRE LEO  SWEDEN      SWITZERLAND TURKEY      UGANDA
                                                      .UK          US
                                                      .
                                                    14 .  32
. . . . . . . . . . . . . . . . . . . . . . . . . . . . . . . . . . . . . . . . . . . . . . . . . . . . . . . . . .
                                                     7 .  58

                            107  POLITIES
                                 WHERE AUTONOMOUS GROUPS
                                 ARE PARTIALLY TOLERATED IN POLITICS,
                                 ARE TOLERATED ONLY OUTSIDE POLITICS,
                                 OR ARE NOT TOLERATED AT ALL  (65)

CUBA       EL SALVADOR GUATEMALA HAITI      MEXICO    .AFGHANISTAN ALBANIA     ALGERIA     BULGARIA    BURMA
NICARAGUA  PARAGUAY                                   .BURUNDI     CAMBODIA    CAMEROUN    CEN AFR REP CHAD
                                                      .CHINA, PR   CONGO(BRA)  CONGO(LEO)  CZECHOS'KIA DAHOMEY
                                                      .ETHIOPIA    GABON       GERMANY, E  GHANA       GUINEA
                                                      .HUNGARY     INDONESIA   IRAN        IRAQ        IVORY COAST
                                                      .JORDAN      KOREA, N    KOREA REP   LIBERIA     LIBYA
                                                      .MALI        MAURITANIA  MONGOLIA    MOROCCO     NEPAL
                                                      .NIGER       NIGERIA     PAKISTAN    POLAND      PORTUGAL
                                                      .RUMANIA     RWANDA      SA'U ARABIA SENEGAL     SOMALIA
                                                      .SO AFRICA   SPAIN       SUDAN       SYRIA       TANGANYIKA
                                                      .THAILAND    TUNISIA     USSR        UAR         UPPER VOLTA
                                                      .VIETNAM, N  VIETNAM REP YUGOSLAVIA
```

```
***************************************************************************************************

 7  POLITIES                                       7  POLITIES
    LOCATED IN CENTRAL AMERICA, RATHER THAN           LOCATED IN SOUTH AMERICA, RATHER THAN
    IN SOUTH AMERICA  (7)                             IN CENTRAL AMERICA  (10)


                                                                      BOTH SUBJECT AND PREDICATE
...................................................................................................

      7 IN LEFT COLUMN

COSTA RICA  EL SALVADOR GUATEMALA   HONDURAS    MEXICO      NICARAGUA   PANAMA

     10 IN RIGHT COLUMN

ARGENTINA   BOLIVIA     BRAZIL      CHILE       COLOMBIA    ECUADOR     PARAGUAY    PERU        URUGUAY     VENEZUELA

     98 EXCLUDED BECAUSE IRRELEVANT

AFGHANISTAN ALBANIA     ALGERIA     AUSTRALIA   AUSTRIA     BELGIUM     BULGARIA    BURMA       BURUNDI     CAMBODIA
CAMEROUN    CANADA      CEN AFR REP CEYLON      CHAD        CHINA, PR   CONGO(BRA)  CONGO(LEO)  CUBA        CYPRUS
CZECHOS'KIA DAHOMEY     DENMARK     DOMIN REP   ETHIOPIA    FINLAND     FRANCE      GABON       GERMANY, E  GERMAN FR
GHANA       GREECE      GUINEA      HAITI       HUNGARY     ICELAND     INDIA       INDONESIA   IRAN        IRAQ
IRELAND     ISRAEL      ITALY       IVORY COAST JAMAICA     JAPAN       JORDAN      KOREA, N    KOREA REP   LAOS
LEBANON     LIBERIA     LIBYA       LUXEMBOURG  MALAGASY R  MALAYA      MALI        MAURITANIA  MONGOLIA    MOROCCO
NEPAL       NETHERLANDS NEW ZEALAND NIGER       NIGERIA     NORWAY      PAKISTAN    PHILIPPINES POLAND      PORTUGAL
RUMANIA     RWANDA      SA'U ARABIA SENEGAL     SIERRE LEO  SOMALIA     SO AFRICA   SPAIN       SUDAN       SWEDEN
SWITZERLAND SYRIA       TANGANYIKA  THAILAND    TOGO        TRINIDAD    TUNISIA     TURKEY      UGANDA      USSR
UAR         UK          US          UPPER VOLTA VIETNAM, N  VIETNAM REP YEMEN       YUGOSLAVIA
---------------------------------------------------------------------------------------------------

 21  TILT TOWARD BEING THOSE           0.86      21  TILT TOWARD BEING THOSE          0.60        1    6
     WHOSE TERRITORIAL SIZE IS                       WHOSE TERRITORIAL SIZE IS                    6    4
     MEDIUM OR SMALL  (83)                           VERY LARGE OR LARGE  (32)               X SQ=   1.92
                                                                                             P =    0.134
                                                                                             RV YES (YES)

 22  TEND TO BE THOSE                  0.86      22  TEND TO BE THOSE                 0.90        1    9
     WHOSE TERRITORIAL SIZE IS                       WHOSE TERRITORIAL SIZE IS                    6    1
     SMALL  (47)                                     VERY LARGE, LARGE, OR MEDIUM  (68)      X SQ=   6.87
                                                                                             P =    0.004
                                                                                             RV YES (YES)

 24  TILT TOWARD BEING THOSE           0.86      24  TILT TOWARD BEING THOSE          0.60        1    6
     WHOSE POPULATION IS                             WHOSE POPULATION IS                          6    4
     SMALL  (54)                                     VERY LARGE, LARGE, OR MEDIUM  (61)      X SQ=   1.92
                                                                                             P =    0.134
                                                                                             RV YES (YES)

 34  TEND TO BE THOSE                  0.86      34  TEND TO BE THOSE                 0.70        1    7
     WHOSE GROSS NATIONAL PRODUCT                    WHOSE GROSS NATIONAL PRODUCT                 6    3
     IS VERY LOW  (53)                               IS VERY HIGH, HIGH, MEDIUM,             X SQ=   3.14
                                                     OR LOW  (62)                            P =    0.050
                                                                                             RV YES (YES)
```

36	ALWAYS ARE THOSE WHOSE PER CAPITA GROSS NATIONAL PRODUCT IS LOW OR VERY LOW (73)	1.00	36	TILT LESS TOWARD BEING THOSE WHOSE PER CAPITA GROSS NATIONAL PRODUCT IS LOW OR VERY LOW (73)	0.60	0 4 7 6 X SQ= 1.78 P = 0.103 RV NO (YES)
40	LEAN TOWARD BEING THOSE WHOSE INTERNATIONAL FINANCIAL STATUS IS VERY LOW (39)	0.71	40	LEAN TOWARD BEING THOSE WHOSE INTERNATIONAL FINANCIAL STATUS IS VERY HIGH, HIGH, MEDIUM, OR LOW (71)	0.80	2 8 5 2 X SQ= 2.62 P = 0.058 RV YES (YES)
43	TEND TO BE THOSE WHOSE ECONOMIC DEVELOPMENTAL STATUS IS VERY UNDERDEVELOPED (57)	0.83	43	TEND TO BE THOSE WHOSE ECONOMIC DEVELOPMENTAL STATUS IS DEVELOPED, INTERMEDIATE, OR UNDERDEVELOPED (55)	0.80	1 8 5 2 X SQ= 3.81 P = 0.035 RV YES (YES)
107	TEND TO BE THOSE WHERE AUTONOMOUS GROUPS ARE PARTIALLY TOLERATED IN POLITICS, ARE TOLERATED ONLY OUTSIDE POLITICS, OR ARE NOT TOLERATED AT ALL (65)	0.71	107	TEND TO BE THOSE WHERE AUTONOMOUS GROUPS ARE FULLY TOLERATED IN POLITICS (46)	0.89	2 8 5 1 X SQ= 3.81 P = 0.035 RV YES (YES)
114	ALWAYS ARE THOSE WHERE SECTIONALISM IS NEGLIGIBLE (47)	1.00	114	TILT TOWARD BEING THOSE WHERE SECTIONALISM IS EXTREME OR MODERATE (61)	0.50	0 4 5 4 X SQ= 1.65 P = 0.105 RV YES (YES)
147	ALWAYS ARE THOSE WHERE THE PARTY SYSTEM IS QUALITATIVELY OTHER THAN CLASS-ORIENTED OR MULTI-IDEOLOGICAL (67)	1.00	147	TILT LESS TOWARD BEING THOSE WHERE THE PARTY SYSTEM IS QUALITATIVELY OTHER THAN CLASS-ORIENTED OR MULTI-IDEOLOGICAL (67)	0.56	0 4 6 5 X SQ= 1.72 P = 0.103 RV NO (YES)
157	LEAN TOWARD BEING THOSE WHERE PERSONALISMO IS PRONOUNCED (14)	0.71	157	LEAN TOWARD BEING THOSE WHERE PERSONALISMO IS MODERATE OR NEGLIGIBLE (84)	0.80	5 2 2 8 X SQ= 2.62 P = 0.058 RV YES (YES)
178	TEND TO BE THOSE WHERE THE LEGISLATURE IS UNICAMERAL (53)	0.71	178	TEND TO BE THOSE WHERE THE LEGISLATURE IS BICAMERAL (51)	0.90	5 1 2 9 X SQ= 4.38 P = 0.035 RV YES (YES)

```
****************************************************************************************************

 8  POLITIES                                        8  POLITIES
    LOCATED IN EAST ASIA, SOUTH ASIA,                  LOCATED ELSEWHERE THAN IN EAST ASIA
    OR SOUTHEAST ASIA  (18)                            SOUTH ASIA, OR SOUTHEAST ASIA  (97)


                                                                          BOTH SUBJECT AND PREDICATE
....................................................................................................

     18 IN LEFT COLUMN

BURMA       CAMBODIA    CEYLON      CHINA, PR   INDIA       INDONESIA   JAPAN       KOREA, N    KOREA REP   LAOS
MALAYA      MONGOLIA    NEPAL       PAKISTAN    PHILIPPINES THAILAND    VIETNAM, N  VIETNAM REP

     97 IN RIGHT COLUMN

AFGHANISTAN ALBANIA     ALGERIA     ARGENTINA   AUSTRALIA   AUSTRIA     BELGIUM     BOLIVIA     BRAZIL      BULGARIA
BURUNDI     CAMEROUN    CANADA      CEN AFR REP CHAD        CHILE       COLOMBIA    CONGO(BRA)  CONGO(LEO)  COSTA RICA
CUBA        CYPRUS      CZECHOS'KIA DAHOMEY     DENMARK     DOMIN REP   ECUADOR     EL SALVADOR ETHIOPIA    FINLAND
FRANCE      GABON       GERMANY, E  GERMAN FR   GHANA       GREECE      GUATEMALA   GUINEA      HAITI       HONDURAS
HUNGARY     ICELAND     IRAN        IRAQ        IRELAND     ISRAEL      ITALY       IVORY COAST JAMAICA     JORDAN
LEBANON     LIBERIA     LIBYA       LUXEMBOURG  MALAGASY R  MALI        MAURITANIA  MEXICO      MOROCCO     NETHERLANDS
NEW ZEALAND NICARAGUA   NIGER       NIGERIA     NORWAY      PANAMA      PARAGUAY    PERU        POLAND      PORTUGAL
RUMANIA     RWANDA      SA'U ARABIA SENEGAL     SIERRE LEO  SOMALIA     SO AFRICA   SPAIN       SUDAN       SWEDEN
SWITZERLAND SYRIA       TANGANYIKA  TOGO        TRINIDAD    TUNISIA     TURKEY      UGANDA      USSR        UAR
UK          US          UPPER VOLTA URUGUAY     VENEZUELA   YEMEN       YUGOSLAVIA
----------------------------------------------------------------------------------------------------

 24  TEND TO BE THOSE                     0.83     24  TEND TO BE THOSE                     0.53        15   46
     WHOSE POPULATION IS                               WHOSE POPULATION IS                               3   51
     VERY LARGE, LARGE, OR MEDIUM  (61)                SMALL  (54)                                 X SQ=   6.49
                                                                                                   P =    0.009
                                                                                                   RV YES (NO )

 26  TEND TO BE THOSE                     0.78     26  TEND TO BE THOSE                     0.65        14   34
     WHOSE POPULATION DENSITY IS                       WHOSE POPULATION DENSITY IS                       4   63
     VERY HIGH, HIGH, OR MEDIUM  (48)                  LOW  (67)                                   X SQ=   9.71
                                                                                                   P =    0.001
                                                                                                   RV YES (NO )

 28  TILT MORE TOWARD BEING THOSE         0.76     28  TILT LESS TOWARD BEING THOSE         0.53        13   49
     WHOSE POPULATION GROWTH RATE                      WHOSE POPULATION GROWTH RATE                      4   44
     IS HIGH  (62)                                     IS HIGH  (62)                               X SQ=   2.41
                                                                                                   P =    0.109
                                                                                                   RV NO  (NO )

 29  TEND TO BE THOSE                     0.83     29  TEND TO BE THOSE                     0.58         2   54
     WHERE THE DEGREE OF URBANIZATION                  WHERE THE DEGREE OF URBANIZATION                 10   39
     IS LOW  (49)                                      IS HIGH  (56)                               X SQ=   5.75
                                                                                                   P =    0.011
                                                                                                   RV YES (NO )
```

30	LEAN TOWARD BEING THOSE WHOSE AGRICULTURAL POPULATION IS HIGH (56)	0.71	30	LEAN TOWARD BEING THOSE WHOSE AGRICULTURAL POPULATION IS MEDIUM, LOW, OR VERY LOW (57)	0.54	12 44 5 52 X SQ= 2.62 P = 0.070 RV YES (NO)
31	ALWAYS ARE THOSE WHOSE AGRICULTURAL POPULATION IS HIGH OR MEDIUM (90)	1.00	31	TEND LESS TO BE THOSE WHOSE AGRICULTURAL POPULATION IS HIGH OR MEDIUM (90)	0.75	18 72 0 24 X SQ= 4.30 P = 0.012 RV NO (NO)
34	TILT MORE TOWARD BEING THOSE WHOSE GROSS NATIONAL PRODUCT IS VERY HIGH, HIGH, MEDIUM, OR LOW (62)	0.72	34	TILT LESS TOWARD BEING THOSE WHOSE GROSS NATIONAL PRODUCT IS VERY HIGH, HIGH, MEDIUM, OR LOW (62)	0.51	13 49 5 48 X SQ= 2.07 P = 0.123 RV NO (NO)
35	ALWAYS ARE THOSE WHOSE PER CAPITA GROSS NATIONAL PRODUCT IS MEDIUM, LOW, OR VERY LOW (91)	1.00	35	TEND LESS TO BE THOSE WHOSE PER CAPITA GROSS NATIONAL PRODUCT IS MEDIUM, LOW, OR VERY LOW (91)	0.75	0 24 18 73 X SQ= 4.23 P = 0.013 RV NO (NO)
36	TEND MORE TO BE THOSE WHOSE PER CAPITA GROSS NATIONAL PRODUCT IS LOW OR VERY LOW (73)	0.89	36	TEND LESS TO BE THOSE WHOSE PER CAPITA GROSS NATIONAL PRODUCT IS LOW OR VERY LOW (73)	0.59	2 40 16 57 X SQ= 4.72 P = 0.016 RV NO (NO)
37	TEND TO BE THOSE WHOSE PER CAPITA GROSS NATIONAL PRODUCT IS VERY LOW (51)	0.83	37	TEND TO BE THOSE WHOSE PER CAPITA GROSS NATIONAL PRODUCT IS VERY HIGH, HIGH, MEDIUM, OR LOW (64)	0.63	3 61 15 36 X SQ= 11.34 P = 0. RV YES (NO)
41	ALWAYS ARE THOSE WHOSE ECONOMIC DEVELOPMENTAL STATUS IS INTERMEDIATE, UNDERDEVELOPED, OR VERY UNDERDEVELOPED (94)	1.00	41	TEND LESS TO BE THOSE WHOSE ECONOMIC DEVELOPMENTAL STATUS IS INTERMEDIATE, UNDERDEVELOPED, OR VERY UNDERDEVELOPED (94)	0.80	0 19 18 76 X SQ= 3.02 P = 0.039 RV NO (NO)
42	TEND MORE TO BE THOSE WHOSE ECONOMIC DEVELOPMENTAL STATUS IS UNDERDEVELOPED OR VERY UNDERDEVELOPED (76)	0.94	42	TEND LESS TO BE THOSE WHOSE ECONOMIC DEVELOPMENTAL STATUS IS UNDERDEVELOPED OR VERY UNDERDEVELOPED (76)	0.63	1 35 17 59 X SQ= 5.57 P = 0.006 RV NO (NO)
44	LEAN MORE TOWARD BEING THOSE WHERE THE LITERACY RATE IS BELOW NINETY PERCENT (90)	0.94	44	LEAN LESS TOWARD BEING THOSE WHERE THE LITERACY RATE IS BELOW NINETY PERCENT (90)	0.75	1 24 17 73 X SQ= 2.25 P = 0.074 RV NO (NO)

47	LEAN TOWARD BEING THOSE WHERE THE LITERACY RATE IS BETWEEN FIFTY AND NINETY PERCENT, RATHER THAN NINETY PERCENT OR ABOVE (30)	0.87	47	LEAN TOWARD BEING THOSE WHERE THE LITERACY RATE IS NINETY PERCENT OR ABOVE, RATHER THAN BETWEEN FIFTY AND NINETY PERCENT (25)	0.51	1 24 7 23 X SQ= 2.69 P = 0.059 RV YES (NO)
51	TEND TO BE THOSE WHERE FREEDOM OF THE PRESS IS INTERNALLY ABSENT, OR INTERNALLY AND EXTERNALLY ABSENT (37)	0.65	51	TEND TO BE THOSE WHERE FREEDOM OF THE PRESS IS COMPLETE OR INTERMITTENT (60)	0.67	6 54 11 26 X SQ= 4.87 P = 0.025 RV YES (NO)
54	LEAN MORE TOWARD BEING THOSE WHERE NEWSPAPER CIRCULATION IS LESS THAN ONE HUNDRED PER THOUSAND (76)	0.87	54	LEAN LESS TOWARD BEING THOSE WHERE NEWSPAPER CIRCULATION IS LESS THAN ONE HUNDRED PER THOUSAND (76)	0.64	2 35 14 62 X SQ= 2.48 P = 0.085 RV NO (NO)
56	TILT MORE TOWARD BEING THOSE WHERE THE RELIGION IS PREDOMINANTLY LITERATE (79)	0.92	56	TILT LESS TOWARD BEING THOSE WHERE THE RELIGION IS PREDOMINANTLY LITERATE (79)	0.69	12 67 1 30 X SQ= 2.02 P = 0.105 RV NO (NO)
63	TEND MORE TO BE THOSE WHERE THE RELIGION IS PREDOMINANTLY OR PARTLY OTHER THAN CHRISTIAN (68)	0.94	63	TEND LESS TO BE THOSE WHERE THE RELIGION IS PREDOMINANTLY OR PARTLY OTHER THAN CHRISTIAN (68)	0.53	1 45 17 51 X SQ= 9.10 P = 0.001 RV NO (NO)
67	TEND MORE TO BE THOSE THAT ARE RACIALLY HOMOGENEOUS (82)	0.94	67	TEND LESS TO BE THOSE THAT ARE RACIALLY HOMOGENEOUS (82)	0.70	16 64 1 28 X SQ= 3.26 P = 0.038 RV NO (NO)
68	LEAN MORE TOWARD BEING THOSE THAT ARE LINGUISTICALLY WEAKLY HETEROGENEOUS OR STRONGLY HETEROGENEOUS (62)	0.76	68	LEAN LESS TOWARD BEING THOSE THAT ARE LINGUISTICALLY WEAKLY HETEROGENEOUS OR STRONGLY HETEROGENEOUS (62)	0.51	4 48 13 49 X SQ= 2.95 P = 0.065 RV NO (NO)
72	LEAN TOWARD BEING THOSE WHOSE DATE OF INDEPENDENCE IS AFTER 1914 (59)	0.75	72	LEAN TOWARD BEING THOSE WHOSE DATE OF INDEPENDENCE IS BEFORE 1914 (52)	0.51	4 48 12 47 X SQ= 2.63 P = 0.065 RV YES (NO)
73	TEND TO BE THOSE WHOSE DATE OF INDEPENDENCE IS AFTER 1945 (46)	0.69	73	TEND TO BE THOSE WHOSE DATE OF INDEPENDENCE IS BEFORE 1945 (65)	0.63	5 60 11 35 X SQ= 4.51 P = 0.026 RV YES (NO)

75	TEND TO BE THOSE THAT ARE NOT HISTORICALLY WESTERN AND ARE NOT SIGNIFICANTLY WESTERNIZED (52)	0.72	75	TEND TO BE THOSE THAT ARE HISTORICALLY WESTERN OR SIGNIFICANTLY WESTERNIZED (62)	0.59	5 57 13 39 X SQ= 4.89 P = 0.019 RV YES (NO)
79	ALWAYS ARE THOSE THAT WERE FORMERLY DEPENDENCIES OF BRITAIN OR FRANCE, RATHER THAN SPAIN (49)	1.00	79	LEAN LESS TOWARD BEING THOSE THAT WERE FORMERLY DEPENDENCIES OF BRITAIN OR FRANCE, RATHER THAN SPAIN (49)	0.69	9 40 0 18 X SQ= 2.40 P = 0.057 RV NO (NO)
82	ALWAYS ARE THOSE WHERE THE TYPE OF POLITICAL MODERNIZATION IS DEVELOPED TUTELARY OR UNDEVELOPED TUTELARY, RATHER THAN EARLY OR LATER EUROPEAN OR EUROPEAN DERIVED (55)	1.00	82	TEND TO BE THOSE WHERE THE TYPE OF POLITICAL MODERNIZATION IS EARLY OR LATER EUROPEAN OR EUROPEAN DERIVED, RATHER THAN DEVELOPED TUTELARY OR UNDEVELOPED TUTELARY (51)	0.55	0 51 14 41 X SQ= 12.82 P = 0. RV YES (NO)
84	ALWAYS ARE THOSE WHERE THE TYPE OF POLITICAL MODERNIZATION IS DEVELOPED TUTELARY, RATHER THAN UNDEVELOPED TUTELARY (31)	1.00	84	TEND TO BE THOSE WHERE THE TYPE OF POLITICAL MODERNIZATION IS UNDEVELOPED TUTELARY, RATHER THAN DEVELOPED TUTELARY (24)	0.59	M 14 17 0 24 X SQ= 12.26 P = 0. RV YES (NO)
89	LEAN MORE TOWARD BEING THOSE WHOSE IDEOLOGICAL ORIENTATION IS OTHER THAN CONVENTIONAL (62)	0.87	89	LEAN LESS TOWARD BEING THOSE WHOSE IDEOLOGICAL ORIENTATION IS OTHER THAN CONVENTIONAL (62)	0.61	2 31 13 49 X SQ= 2.57 P = 0.077 RV NO (NO)
93	TEND TO BE THOSE WHERE THE SYSTEM STYLE IS MOBILIZATIONAL, OR LIMITED MOBILIZATIONAL (32)	0.53	93	TEND TO BE THOSE WHERE THE SYSTEM STYLE IS NON-MOBILIZATIONAL (78)	0.75	8 24 7 71 X SQ= 3.68 P = 0.035 RV YES (NO)
94	TEND TO BE THOSE WHERE THE STATUS OF THE REGIME IS AUTHORITARIAN OR TOTALITARIAN (41)	0.72	94	TEND TO BE THOSE WHERE THE STATUS OF THE REGIME IS CONSTITUTIONAL (51)	0.62	5 46 13 28 X SQ= 5.61 P = 0.016 RV YES (NO)
98	ALWAYS ARE THOSE WHERE GOVERNMENTAL STABILITY IS GENERALLY OR MODERATELY PRESENT AND DATES FROM THE POST-WAR PERIOD, OR IS ABSENT (93)	1.00	98	TEND LESS TO BE THOSE WHERE GOVERNMENTAL STABILITY IS GENERALLY OR MODERATELY PRESENT AND DATES FROM THE POST-WAR PERIOD, OR IS ABSENT (93)	0.77	0 22 18 75 X SQ= 3.69 P = 0.022 RV NO (NO)
102	TEND TO BE THOSE WHERE THE REPRESENTATIVE CHARACTER OF THE REGIME IS PSEUDO-POLYARCHIC OR NON-POLYARCHIC (49)	0.71	102	TEND TO BE THOSE WHERE THE REPRESENTATIVE CHARACTER OF THE REGIME IS POLYARCHIC OR LIMITED POLYARCHIC (59)	0.59	5 54 12 37 X SQ= 4.04 P = 0.033 RV YES (NO)

108	LEAN TOWARD BEING THOSE WHERE AUTONOMOUS GROUPS ARE TOLERATED ONLY OUTSIDE POLITICS OR ARE NOT TOLERATED AT ALL (35)	0.57	108	LEAN TOWARD BEING THOSE WHERE AUTONOMOUS GROUPS ARE FULLY OR PARTIALLY TOLERATED IN POLITICS (65)	0.69	6 59 8 27 X SQ= 2.47 P = 0.075 RV YES (NO)
116	TILT MORE TOWARD BEING THOSE WHERE INTEREST ARTICULATION BY ASSOCIATIONAL GROUPS IS LIMITED OR NEGLIGIBLE (79)	0.88	116	TILT LESS TOWARD BEING THOSE WHERE INTEREST ARTICULATION BY ASSOCIATIONAL GROUPS IS LIMITED OR NEGLIGIBLE (79)	0.68	2 30 15 64 X SQ= 1.95 P = 0.144 RV NO (NO)
118	LEAN TOWARD BEING THOSE WHERE INTEREST ARTICULATION BY INSTITUTIONAL GROUPS IS VERY SIGNIFICANT (40)	0.61	118	LEAN TOWARD BEING THOSE WHERE INTEREST ARTICULATION BY INSTITUTIONAL GROUPS IS SIGNIFICANT, MODERATE, OR LIMITED (60)	0.65	11 29 7 53 X SQ= 3.07 P = 0.062 RV YES (NO)
121	TEND TO BE THOSE WHERE INTEREST ARTICULATION BY NON-ASSOCIATIONAL GROUPS IS SIGNIFICANT (54)	0.94	121	TEND TO BE THOSE WHERE INTEREST ARTICULATION BY NON-ASSOCIATIONAL GROUPS IS MODERATE, LIMITED, OR NEGLIGIBLE (61)	0.62	17 37 1 60 X SQ= 17.13 P = 0. RV YES (NO)
122	ALWAYS ARE THOSE WHERE INTEREST ARTICULATION BY NON-ASSOCIATIONAL GROUPS IS SIGNIFICANT OR MODERATE (83)	1.00	122	TEND LESS TO BE THOSE WHERE INTEREST ARTICULATION BY NON-ASSOCIATIONAL GROUPS IS SIGNIFICANT OR MODERATE (83)	0.67	18 65 0 32 X SQ= 6.67 P = 0.003 RV NO (NO)
126	ALWAYS ARE THOSE WHERE INTEREST ARTICULATION BY ANOMIC GROUPS IS FREQUENT, OCCASIONAL, OR INFREQUENT (83)	1.00	126	LEAN LESS TOWARD BEING THOSE WHERE INTEREST ARTICULATION BY ANOMIC GROUPS IS FREQUENT, OCCASIONAL, OR INFREQUENT (83)	0.81	15 68 0 16 X SQ= 2.15 P = 0.071 RV NO (NO)
129	TILT TOWARD BEING THOSE WHERE INTEREST ARTICULATION BY POLITICAL PARTIES IS NEGLIGIBLE (37)	0.64	129	TILT TOWARD BEING THOSE WHERE INTEREST ARTICULATION BY POLITICAL PARTIES IS SIGNIFICANT, MODERATE, OR LIMITED (56)	0.63	4 52 7 30 X SQ= 1.94 P = 0.107 RV YES (NO)
138	TEND TO BE THOSE WHERE INTEREST AGGREGATION BY THE LEGISLATURE IS NEGLIGIBLE (49)	0.79	138	TEND TO BE THOSE WHERE INTEREST AGGREGATION BY THE LEGISLATURE IS SIGNIFICANT, MODERATE, OR LIMITED (48)	0.54	3 45 11 38 X SQ= 3.92 P = 0.040 RV YES (NO)
161	LEAN TOWARD BEING THOSE WHERE THE POLITICAL LEADERSHIP IS ELITIST OR MODERATELY ELITIST (47)	0.71	161	LEAN TOWARD BEING THOSE WHERE THE POLITICAL LEADERSHIP IS NON-ELITIST (50)	0.55	10 37 4 46 X SQ= 2.47 P = 0.084 RV YES (NO)

No.	Statement	Value	No.	Statement	Value	Statistics
169	LEAN TOWARD BEING THOSE WHERE THE HORIZONTAL POWER DISTRIBUTION IS NEGLIGIBLE (48)	0.69	169	LEAN TOWARD BEING THOSE WHERE THE HORIZONTAL POWER DISTRIBUTION IS SIGNIFICANT OR LIMITED (58)	0.59	5 53 11 37 X SQ= 3.15 P = 0.056 RV YES (NO)
176	TEND TO BE THOSE WHERE THE LEGISLATURE IS WHOLLY INEFFECTIVE (28)	0.60	176	TEND TO BE THOSE WHERE THE LEGISLATURE IS FULLY EFFECTIVE, PARTIALLY EFFECTIVE, OR LARGELY INEFFECTIVE (72)	0.78	6 66 9 19 X SQ= 7.19 P = 0.010 RV YES (NO)
179	TILT MORE TOWARD BEING THOSE WHERE THE EXECUTIVE IS DOMINANT (52)	0.77	179	TILT LESS TOWARD BEING THOSE WHERE THE EXECUTIVE IS DOMINANT (52)	0.54	10 42 3 36 X SQ= 1.57 P = 0.142 RV NO (NO)
181	ALWAYS ARE THOSE WHERE THE BUREAUCRACY IS SEMI-MODERN, RATHER THAN MODERN (55)	1.00	181	LEAN LESS TOWARD BEING THOSE WHERE THE BUREAUCRACY IS SEMI-MODERN, RATHER THAN MODERN (55)	0.68	0 21 10 45 X SQ= 2.95 P = 0.054 RV NO (NO)
183	TEND TO BE THOSE WHERE THE BUREAUCRACY IS TRADITIONAL, RATHER THAN POST-COLONIAL TRANSITIONAL (9)	0.80	183	TEND TO BE THOSE WHERE THE BUREAUCRACY IS POST-COLONIAL TRANSITIONAL, RATHER THAN TRADITIONAL (25)	0.83	1 24 4 5 X SQ= 5.71 P = 0.012 RV YES (NO)
186	TILT TOWARD BEING THOSE WHERE PARTICIPATION BY THE MILITARY IN POLITICS IS SUPPORTIVE, RATHER THAN NEUTRAL (31)	0.58	186	TILT TOWARD BEING THOSE WHERE PARTICIPATION BY THE MILITARY IN POLITICS IS NEUTRAL, RATHER THAN SUPPORTIVE (56)	0.68	7 24 5 51 X SQ= 2.08 P = 0.106 RV YES (NO)
188	TEND MORE TO BE THOSE WHERE THE CHARACTER OF THE LEGAL SYSTEM IS OTHER THAN CIVIL LAW (81)	0.94	188	TEND LESS TO BE THOSE WHERE THE CHARACTER OF THE LEGAL SYSTEM IS OTHER THAN CIVIL LAW (81)	0.67	1 31 17 64 X SQ= 4.21 P = 0.021 RV NO (NO)

M

MATRIX

8 POLITIES
LOCATED IN EAST ASIA, SOUTH ASIA, OR SOUTHEAST ASIA (18)

8 POLITIES
LOCATED ELSEWHERE THAN IN EAST ASIA SOUTH ASIA, OR SOUTHEAST ASIA (97)

84 POLITIES
WHERE THE TYPE OF POLITICAL MODERNIZATION IS DEVELOPED TUTELARY, RATHER THAN UNDEVELOPED TUTELARY (31)

BURMA
KOREA, N
PAKISTAN
CAMBODIA
KOREA REP
PHILIPPINES
CEYLON
LAOS
VIETNAM, N
INDIA
MALAYA
VIETNAM REP
INDONESIA
MONGOLIA

ALGERIA
JORDAN
SA'U ARABIA
UAR
CYPRUS
LEBANON
SUDAN
YEMEN
IRAQ
LIBERIA
SYRIA
ISRAEL
LIBYA
TRINIDAD
JAMAICA
MOROCCO
TUNISIA

14	17
0	24

84 POLITIES
WHERE THE TYPE OF POLITICAL MODERNIZATION IS UNDEVELOPED TUTELARY, RATHER THAN DEVELOPED TUTELARY (24)

BURUNDI
CONGO(LEO)
IVORY COAST
NIGERIA
TANGANYIKA
CAMEROUN
DAHOMEY
MALAGASY R
RWANDA
TOGO
CEN AFR REP
GABON
MALI
SENEGAL
UGANDA
CHAD
GHANA
MAURITANIA
SIERRE LEO
UPPER VOLTA
CONGO(BRA)
GUINEA
NIGER
SOMALIA

9 POLITIES
LOCATED IN SOUTHEAST ASIA (9)

9 POLITIES
LOCATED ELSEWHERE THAN IN
SOUTHEAST ASIA (106)

BOTH SUBJECT AND PREDICATE

9 IN LEFT COLUMN

BURMA, CAMBODIA, INDONESIA, LAOS, MALAYA, PHILIPPINES, THAILAND, VIETNAM, N, VIETNAM REP

106 IN RIGHT COLUMN

AFGHANISTAN	ALBANIA	ALGERIA	ARGENTINA	AUSTRALIA	AUSTRIA	BELGIUM	BOLIVIA	BRAZIL	BULGARIA
BURUNDI	CAMEROUN	CANADA	CEN AFR REP	CEYLON	CHAD	CHILE	CHINA, PR	COLOMBIA	CONGO(BRA)
CONGO(LEO)	COSTA RICA	CUBA	CYPRUS	CZECHOS'KIA	DAHOMEY	DENMARK	DOMIN REP	ECUADOR	EL SALVADOR
ETHIOPIA	FINLAND	FRANCE	GABON	GERMANY, E	GERMAN FR	GHANA	GREECE	GUATEMALA	GUINEA
HAITI	HONDURAS	HUNGARY	ICELAND	INDIA	IRAN	IRAQ	IRELAND	ISRAEL	ITALY
IVORY COAST	JAMAICA	JAPAN	JORDAN	KOREA, N	KOREA REP	LEBANON	LIBERIA	LIBYA	LUXEMBOURG
MALAGASY R	MALI	MAURITANIA	MEXICO	MONGOLIA	MOROCCO	NEPAL	NETHERLANDS	NEW ZEALAND	NICARAGUA
NIGER	NIGERIA	NORWAY	PAKISTAN	PANAMA	PARAGUAY	PERU	POLAND	PORTUGAL	RUMANIA
RWANDA	SA'U ARABIA	SENEGAL	SIERRE LEO	SOMALIA	SO AFRICA	SPAIN	SUDAN	SWEDEN	SWITZERLAND
SYRIA	TANGANYIKA	TOGO	TRINIDAD	TUNISIA	TURKEY	UGANDA	USSR	UAR	UK
US	UPPER VOLTA	URUGUAY	VENEZUELA	YEMEN	YUGOSLAVIA				

	Left column			Right column		
29	ALWAYS ARE THOSE WHERE THE DEGREE OF URBANIZATION IS LOW (49)	1.00	29	TEND TO BE THOSE WHERE THE DEGREE OF URBANIZATION IS HIGH (56)	0.56	0 56 5 44 X SQ= 3.96 P = 0.020 RV YES (NO)
37	LEAN TOWARD BEING THOSE WHOSE PER CAPITA GROSS NATIONAL PRODUCT IS VERY LOW (51)	0.78	37	LEAN TOWARD BEING THOSE WHOSE PER CAPITA GROSS NATIONAL PRODUCT IS VERY HIGH, HIGH, MEDIUM, OR LOW (64)	0.58	2 62 7 44 X SQ= 3.07 P = 0.075 RV YES (NO)
42	ALWAYS ARE THOSE WHOSE ECONOMIC DEVELOPMENTAL STATUS IS UNDERDEVELOPED OR VERY UNDERDEVELOPED (76)	1.00	42	TEND LESS TO BE THOSE WHOSE ECONOMIC DEVELOPMENTAL STATUS IS UNDERDEVELOPED OR VERY UNDERDEVELOPED (76)	0.65	0 36 9 67 X SQ= 3.17 P = 0.031 RV NO (NO)
44	ALWAYS ARE THOSE WHERE THE LITERACY RATE IS BELOW NINETY PERCENT (90)	1.00	44	TILT LESS TOWARD BEING THOSE WHERE THE LITERACY RATE IS BELOW NINETY PERCENT (90)	0.76	0 25 9 81 X SQ= 1.50 P = 0.123 RV NO (NO)

54	ALWAYS ARE THOSE WHERE NEWSPAPER CIRCULATION IS LESS THAN ONE HUNDRED PER THOUSAND (76)	1.00	54	LEAN LESS TOWARD BEING THOSE WHERE NEWSPAPER CIRCULATION IS LESS THAN ONE HUNDRED PER THOUSAND (76)	0.65	0 37 8 68 X SQ= 2.74 P = 0.051 RV NO (NO)
63	LEAN MORE TOWARD BEING THOSE WHERE THE RELIGION IS PREDOMINANTLY OR PARTLY OTHER THAN CHRISTIAN (68)	0.89	63	LEAN LESS TOWARD BEING THOSE WHERE THE RELIGION IS PREDOMINANTLY OR PARTLY OTHER THAN CHRISTIAN (68)	0.57	1 45 8 60 X SQ= 2.28 P = 0.082 RV NO (NO)
68	ALWAYS ARE THOSE THAT ARE LINGUISTICALLY WEAKLY HETEROGENEOUS OR STRONGLY HETEROGENEOUS (62)	1.00	68	TEND LESS TO BE THOSE THAT ARE LINGUISTICALLY WEAKLY HETEROGENEOUS OR STRONGLY HETEROGENEOUS (62)	0.50	0 52 9 53 X SQ= 6.32 P = 0.004 RV NO (NO)
72	TEND TO BE THOSE WHOSE DATE OF INDEPENDENCE IS AFTER 1914 (59)	0.89	72	TEND TO BE THOSE WHOSE DATE OF INDEPENDENCE IS BEFORE 1914 (52)	0.50	1 51 8 51 X SQ= 3.58 P = 0.035 RV YES (NO)
73	TEND TO BE THOSE WHOSE DATE OF INDEPENDENCE IS AFTER 1945 (46)	0.89	73	TEND TO BE THOSE WHOSE DATE OF INDEPENDENCE IS BEFORE 1945 (65)	0.63	1 64 8 38 X SQ= 7.08 P = 0.004 RV YES (NO)
75	TEND TO BE THOSE THAT ARE NOT HISTORICALLY WESTERN AND ARE NOT SIGNIFICANTLY WESTERNIZED (52)	0.89	75	TEND TO BE THOSE THAT ARE HISTORICALLY WESTERN OR SIGNIFICANTLY WESTERNIZED (62)	0.58	1 61 8 44 X SQ= 5.60 P = 0.011 RV YES (NO)
76	ALWAYS ARE THOSE THAT HAVE BEEN SIGNIFICANTLY OR PARTIALLY WESTERNIZED THROUGH A COLONIAL RELATIONSHIP, RATHER THAN BEING HISTORICALLY WESTERN (70)	1.00	76	TILT LESS TOWARD BEING THOSE THAT HAVE BEEN SIGNIFICANTLY OR PARTIALLY WESTERNIZED THROUGH A COLONIAL RELATIONSHIP, RATHER THAN BEING HISTORICALLY WESTERN (70)	0.70	0 26 8 62 X SQ= 1.92 P = 0.103 RV NO (NO)
82	ALWAYS ARE THOSE WHERE THE TYPE OF POLITICAL MODERNIZATION IS DEVELOPED TUTELARY OR UNDEVELOPED TUTELARY, RATHER THAN EARLY OR LATER EUROPEAN OR EUROPEAN DERIVED (55)	1.00	82	TEND TO BE THOSE WHERE THE TYPE OF POLITICAL MODERNIZATION IS EARLY OR LATER EUROPEAN OR EUROPEAN DERIVED, RATHER THAN DEVELOPED TUTELARY OR UNDEVELOPED TUTELARY (51)	0.52	0 51 8 47 X SQ= 6.07 P = 0.006 RV YES (NO)
84	ALWAYS ARE THOSE WHERE THE TYPE OF POLITICAL MODERNIZATION IS DEVELOPED TUTELARY, RATHER THAN UNDEVELOPED TUTELARY (31)	1.00	84	TEND TO BE THOSE WHERE THE TYPE OF POLITICAL MODERNIZATION IS UNDEVELOPED TUTELARY, RATHER THAN DEVELOPED TUTELARY (24)	0.51	8 23 0 24 X SQ= 5.32 P = 0.007 RV YES (NO)

No.	Left subject		No.	Right subject		Statistics
85	TEND TO BE THOSE WHERE THE STAGE OF POLITICAL MODERNIZATION IS MID- OR EARLY TRANSITIONAL, RATHER THAN ADVANCED (54)	0.89	85	TEND TO BE THOSE WHERE THE STAGE OF POLITICAL MODERNIZATION IS ADVANCED, RATHER THAN MID- OR EARLY TRANSITIONAL (60)	0.56	1 59 8 46 X SQ= 5.07 P = 0.013 RV YES (NO)
86	TEND TO BE THOSE WHERE THE STAGE OF POLITICAL MODERNIZATION IS EARLY TRANSITIONAL, RATHER THAN ADVANCED OR MID-TRANSITIONAL (38)	0.78	86	TEND TO BE THOSE WHERE THE STAGE OF POLITICAL MODERNIZATION IS ADVANCED OR MID-TRANSITIONAL, RATHER THAN EARLY TRANSITIONAL (76)	0.70	2 74 7 31 X SQ= 6.65 P = 0.006 RV YES (NO)
93	TILT TOWARD BEING THOSE WHERE THE SYSTEM STYLE IS MOBILIZATIONAL, OR LIMITED MOBILIZATIONAL (32)	0.56	93	TILT TOWARD BEING THOSE WHERE THE SYSTEM STYLE IS NON-MOBILIZATIONAL (78)	0.73	5 27 4 74 X SQ= 2.08 P = 0.118 RV YES (NO)
94	LEAN TOWARD BEING THOSE WHERE THE STATUS OF THE REGIME IS AUTHORITARIAN OR TOTALITARIAN (41)	0.78	94	LEAN TOWARD BEING THOSE WHERE THE STATUS OF THE REGIME IS CONSTITUTIONAL (51)	0.59	2 49 7 34 X SQ= 3.09 P = 0.073 RV YES (NO)
96	TEND TO BE THOSE WHERE THE STATUS OF THE REGIME IS AUTHORITARIAN (23)	0.62	96	TEND TO BE THOSE WHERE THE STATUS OF THE REGIME IS CONSTITUTIONAL OR TOTALITARIAN (67)	0.78	3 64 5 18 X SQ= 4.35 P = 0.024 RV YES (NO)
102	LEAN TOWARD BEING THOSE WHERE THE REPRESENTATIVE CHARACTER OF THE REGIME IS PSEUDO-POLYARCHIC OR NON-POLYARCHIC (49)	0.78	102	LEAN TOWARD BEING THOSE WHERE THE REPRESENTATIVE CHARACTER OF THE REGIME IS POLYARCHIC OR LIMITED POLYARCHIC (59)	0.58	2 57 7 42 X SQ= 2.86 P = 0.076 RV YES (NO)
121	ALWAYS ARE THOSE WHERE INTEREST ARTICULATION BY NON-ASSOCIATIONAL GROUPS IS SIGNIFICANT (54)	1.00	121	TEND TO BE THOSE WHERE INTEREST ARTICULATION BY NON-ASSOCIATIONAL GROUPS IS MODERATE, LIMITED, OR NEGLIGIBLE (61)	0.58	9 45 0 61 X SQ= 8.84 P = 0.001 RV YES (NO)
122	ALWAYS ARE THOSE WHERE INTEREST ARTICULATION BY NON-ASSOCIATIONAL GROUPS IS SIGNIFICANT OR MODERATE (83)	1.00	122	LEAN LESS TOWARD BEING THOSE WHERE INTEREST ARTICULATION BY NON-ASSOCIATIONAL GROUPS IS SIGNIFICANT OR MODERATE (83)	0.70	9 74 0 32 X SQ= 2.41 P = 0.060 RV NO (NO)
138	TILT TOWARD BEING THOSE WHERE INTEREST AGGREGATION BY THE LEGISLATURE IS NEGLIGIBLE (49)	0.86	138	TILT TOWARD BEING THOSE WHERE INTEREST AGGREGATION BY THE LEGISLATURE IS SIGNIFICANT, MODERATE, OR LIMITED (48)	0.52	1 47 6 43 X SQ= 2.38 P = 0.111 RV YES (NO)

169	TILT TOWARD BEING THOSE WHERE THE HORIZONTAL POWER DISTRIBUTION IS NEGLIGIBLE (48)	0.75	169	TILT TOWARD BEING THOSE WHERE THE HORIZONTAL POWER DISTRIBUTION IS SIGNIFICANT OR LIMITED (58)	0.57	2 56 6 42 X SQ= 1.92 P = 0.137 RV YES (NO)
176	TEND TO BE THOSE WHERE THE LEGISLATURE IS WHOLLY INEFFECTIVE (28)	0.62	176	TEND TO BE THOSE WHERE THE LEGISLATURE IS FULLY EFFECTIVE, PARTIALLY EFFECTIVE, OR LARGELY INEFFECTIVE (72)	0.75	3 69 5 23 X SQ= 3.44 P = 0.037 RV YES (NO)
183	TEND TO BE THOSE WHERE THE BUREAUCRACY IS TRADITIONAL, RATHER THAN POST-COLONIAL TRANSITIONAL (9)	0.75	183	TEND TO BE THOSE WHERE THE BUREAUCRACY IS POST-COLONIAL TRANSITIONAL, RATHER THAN TRADITIONAL (25)	0.80	1 24 3 6 X SQ= 3.02 P = 0.048 RV YES (NO)

```
****************************************************************************************************

10  POLITIES                                              10  POLITIES
    LOCATED IN NORTH AFRICA, OR                               LOCATED ELSEWHERE THAN IN NORTH AFRICA,
    CENTRAL AND SOUTH AFRICA  (33)                            OR CENTRAL AND SOUTH AFRICA  (82)


                                                                                BOTH SUBJECT AND PREDICATE
....................................................................................................

      33 IN LEFT COLUMN

ALGERIA     BURUNDI     CAMEROUN    CEN AFR REP CHAD        CONGO(BRA)  CONGO(LEO)  DAHOMEY     ETHIOPIA    GABON
GHANA       GUINEA      IVORY COAST LIBERIA     LIBYA       MALAGASY R  MALI        MAURITANIA  MOROCCO     NIGER
NIGERIA     RWANDA      SENEGAL     SIERRE LEO  SOMALIA     SO AFRICA   SUDAN       TANGANYIKA  TOGO        TUNISIA
UGANDA      UAR         UPPER VOLTA

      82 IN RIGHT COLUMN

AFGHANISTAN ALBANIA     ARGENTINA   AUSTRALIA   AUSTRIA     BELGIUM     BOLIVIA     BRAZIL      BULGARIA    BURMA
CAMBODIA    CANADA      CEYLON      CHILE       CHINA, PR   COLOMBIA    COSTA RICA  CUBA        CYPRUS      CZECHOS'KIA
DENMARK     DOMIN REP   ECUADOR     EL SALVADOR FINLAND     FRANCE      GERMANY, E  GERMAN FR   GREECE      GUATEMALA
HAITI       HONDURAS    HUNGARY     ICELAND     INDIA       INDONESIA   IRAN        IRAQ        IRELAND     ISRAEL
ITALY       JAMAICA     JAPAN       JORDAN      KOREA, N    KOREA REP   LAOS        LEBANON     LUXEMBOURG  MALAYA
MEXICO      MONGOLIA    NEPAL       NETHERLANDS NEW ZEALAND NICARAGUA   NORWAY      PAKISTAN    PANAMA      PARAGUAY
PERU        PHILIPPINES POLAND      PORTUGAL    RUMANIA     SA'U ARABIA SPAIN       SWEDEN      SWITZERLAND SYRIA
THAILAND    TRINIDAD    TURKEY      USSR        UK          US          URUGUAY     VENEZUELA   VIETNAM, N  VIETNAM REP
YEMEN       YUGOSLAVIA
----------------------------------------------------------------------------------------------------

 22  TEND MORE TO BE THOSE                    0.79        22  TEND LESS TO BE THOSE                    0.51       26   42
     WHOSE TERRITORIAL SIZE IS                                WHOSE TERRITORIAL SIZE IS                                7   40
     VERY LARGE, LARGE, OR MEDIUM  (68)                       VERY LARGE, LARGE, OR MEDIUM  (68)                 X SQ=   6.30
                                                                                                                 P =    0.007
                                                                                                                 RV NO  (NO )

 24  TEND TO BE THOSE                         0.67        24  TEND TO BE THOSE                         0.61       11   50
     WHOSE POPULATION IS                                      WHOSE POPULATION IS                                     22   32
     SMALL  (54)                                              VERY LARGE, LARGE, OR MEDIUM  (61)                 X SQ=   6.15
                                                                                                                 P =    0.013
                                                                                                                 RV YES (NO )

 25  ALWAYS ARE THOSE                         1.00        25  TEND LESS TO BE THOSE                    0.79        0   17
     WHOSE POPULATION DENSITY IS                              WHOSE POPULATION DENSITY IS                             33   65
     MEDIUM OR LOW  (98)                                      MEDIUM OR LOW  (98)                                X SQ=   6.47
                                                                                                                 P =    0.003
                                                                                                                 RV NO  (NO )

 26  TEND TO BE THOSE                         0.91        26  TEND TO BE THOSE                         0.55        3   45
     WHOSE POPULATION DENSITY IS                              WHOSE POPULATION DENSITY IS                             30   37
     LOW  (67)                                                VERY HIGH, HIGH, OR MEDIUM  (48)                   X SQ=  18.45
                                                                                                                 P =    0.
                                                                                                                 RV YES (NO )
```

29	TEND TO BE THOSE WHERE THE DEGREE OF URBANIZATION IS LOW (49)	0.85	29	TEND TO BE THOSE WHERE THE DEGREE OF URBANIZATION IS HIGH (56)	0.71	5 51 28 21 X SQ= 26.00 P = 0. RV YES (YES)
30	TEND TO BE THOSE WHOSE AGRICULTURAL POPULATION IS HIGH (56)	0.91	30	TEND TO BE THOSE WHOSE AGRICULTURAL POPULATION IS MEDIUM, LOW, OR VERY LOW (57)	0.67	30 26 3 54 X SQ= 29.59 P = 0. RV YES (YES)
31	TEND MORE TO BE THOSE WHOSE AGRICULTURAL POPULATION IS HIGH OR MEDIUM (90)	0.97	31	TEND LESS TO BE THOSE WHOSE AGRICULTURAL POPULATION IS HIGH OR MEDIUM (90)	0.72	32 58 1 23 X SQ= 7.61 P = 0.002 RV NO (NO)
34	TEND TO BE THOSE WHOSE GROSS NATIONAL PRODUCT IS VERY LOW (53)	0.79	34	TEND TO BE THOSE WHOSE GROSS NATIONAL PRODUCT IS VERY HIGH, HIGH, MEDIUM, OR LOW (62)	0.67	7 55 26 27 X SQ= 18.12 P = 0. RV YES (NO)
35	ALWAYS ARE THOSE WHOSE PER CAPITA GROSS NATIONAL PRODUCT IS MEDIUM, LOW, OR VERY LOW (91)	1.00	35	TEND LESS TO BE THOSE WHOSE PER CAPITA GROSS NATIONAL PRODUCT IS MEDIUM, LOW, OR VERY LOW (91)	0.71	0 24 33 58 X SQ= 10.50 P = 0. RV NO (NO)
36	TEND TO BE THOSE WHOSE PER CAPITA GROSS NATIONAL PRODUCT IS LOW OR VERY LOW (73)	0.97	36	TEND TO BE THOSE WHOSE PER CAPITA GROSS NATIONAL PRODUCT IS VERY HIGH, HIGH, OR MEDIUM (42)	0.50	1 41 32 41 X SQ= 20.41 P = 0. RV YES (NO)
37	TEND TO BE THOSE WHOSE PER CAPITA GROSS NATIONAL PRODUCT IS VERY LOW (51)	0.88	37	TEND TO BE THOSE WHOSE PER CAPITA GROSS NATIONAL PRODUCT IS VERY HIGH, HIGH, MEDIUM, OR LOW (64)	0.73	4 60 29 22 X SQ= 33.10 P = 0. RV YES (YES)
40	TEND TO BE THOSE WHOSE INTERNATIONAL FINANCIAL STATUS IS VERY LOW (39)	0.69	40	TEND TO BE THOSE WHOSE INTERNATIONAL FINANCIAL STATUS IS VERY HIGH, HIGH, MEDIUM, OR LOW (71)	0.78	10 61 22 17 X SQ= 19.86 P = 0. RV YES (YES)
43	TEND TO BE THOSE WHOSE ECONOMIC DEVELOPMENTAL STATUS IS VERY UNDERDEVELOPED (57)	0.94	43	TEND TO BE THOSE WHOSE ECONOMIC DEVELOPMENTAL STATUS IS DEVELOPED, INTERMEDIATE, OR UNDERDEVELOPED (55)	0.67	2 53 31 26 X SQ= 32.29 P = 0. RV YES (YES)

45	ALWAYS ARE THOSE WHERE THE LITERACY RATE IS BELOW FIFTY PERCENT (54)	1.00	45	TEND TO BE THOSE WHERE THE LITERACY RATE IS FIFTY PERCENT OR ABOVE (55)	0.71	0 55 32 22 X SQ= 43.32 P = 0. RV YES (YES)
46	TEND TO BE THOSE WHERE THE LITERACY RATE IS BELOW TEN PERCENT (26)	0.70	46	TEND TO BE THOSE WHERE THE LITERACY RATE IS TEN PERCENT OR ABOVE (84)	0.94	9 75 21 5 X SQ= 45.66 P = 0. RV YES (YES)
52	TILT MORE TOWARD BEING THOSE WHERE FREEDOM OF THE PRESS IS COMPLETE, INTERMITTENT, OR INTERNALLY ABSENT (82)	0.96	52	TILT LESS TOWARD BEING THOSE WHERE FREEDOM OF THE PRESS IS COMPLETE, INTERMITTENT, OR INTERNALLY ABSENT (82)	0.80	22 60 1 15 X SQ= 2.11 P = 0.108 RV NO (NO)
54	ALWAYS ARE THOSE WHERE NEWSPAPER CIRCULATION IS LESS THAN ONE HUNDRED PER THOUSAND (76)	1.00	54	TEND LESS TO BE THOSE WHERE NEWSPAPER CIRCULATION IS LESS THAN ONE HUNDRED PER THOUSAND (76)	0.54	0 37 33 43 X SQ= 20.64 P = 0. RV NO (NO)
55	TEND TO BE THOSE WHERE NEWSPAPER CIRCULATION IS LESS THAN TEN PER THOUSAND (35)	0.82	55	TEND TO BE THOSE WHERE NEWSPAPER CIRCULATION IS TEN OR MORE PER THOUSAND (78)	0.90	6 72 27 8 X SQ= 53.05 P = 0. RV YES (YES)
56	TEND TO BE THOSE WHERE THE RELIGION IS PREDOMINANTLY OR PARTLY NON-LITERATE (31)	0.76	56	TEND TO BE THOSE WHERE THE RELIGION IS PREDOMINANTLY LITERATE (79)	0.92	8 71 25 6 X SQ= 49.42 P = 0. RV YES (YES)
64	ALWAYS ARE THOSE WHERE THE RELIGION IS MUSLIM, RATHER THAN CHRISTIAN (18)	1.00	64	TEND TO BE THOSE WHERE THE RELIGION IS CHRISTIAN, RATHER THAN MUSLIM (46)	0.82	0 46 8 10 X SQ= 19.48 P = 0. RV YES (NO)
66	TEND TO BE THOSE THAT ARE RELIGIOUSLY HETEROGENEOUS (49)	0.73	66	TEND TO BE THOSE THAT ARE RELIGIOUSLY HOMOGENEOUS (57)	0.64	8 49 22 27 X SQ= 10.89 P = 0.001 RV YES (NO)
68	TEND TO BE THOSE THAT ARE LINGUISTICALLY WEAKLY HETEROGENEOUS OR STRONGLY HETEROGENEOUS (62)	0.79	68	TEND TO BE THOSE THAT ARE LINGUISTICALLY HOMOGENEOUS (52)	0.56	7 45 26 36 X SQ= 9.81 P = 0.001 RV YES (NO)

69	TEND TO BE THOSE THAT ARE LINGUISTICALLY STRONGLY HETEROGENEOUS (50)	0.79	69	TEND TO BE THOSE THAT ARE LINGUISTICALLY HOMOGENEOUS OR WEAKLY HETEROGENEOUS (64)	0.70	7 57 26 24 X SQ= 21.06 P = 0. RV YES (YES)
72	TEND TO BE THOSE WHOSE DATE OF INDEPENDENCE IS AFTER 1914 (59)	0.91	72	TEND TO BE THOSE WHOSE DATE OF INDEPENDENCE IS BEFORE 1914 (52)	0.63	3 49 30 29 X SQ= 24.77 P = 0. RV YES (YES)
73	TEND TO BE THOSE WHOSE DATE OF INDEPENDENCE IS AFTER 1945 (46)	0.91	73	TEND TO BE THOSE WHOSE DATE OF INDEPENDENCE IS BEFORE 1945 (65)	0.79	3 62 30 16 X SQ= 44.50 P = 0. RV YES (YES)
76	ALWAYS ARE THOSE THAT HAVE BEEN SIGNIFICANTLY OR PARTIALLY WESTERNIZED THROUGH A COLONIAL RELATIONSHIP, RATHER THAN BEING HISTORICALLY WESTERN (70)	1.00	76	TEND LESS TO BE THOSE THAT HAVE BEEN SIGNIFICANTLY OR PARTIALLY WESTERNIZED THROUGH A COLONIAL RELATIONSHIP, RATHER THAN BEING HISTORICALLY WESTERN (70)	0.60	0 26 31 39 X SQ= 15.04 P = 0. RV NO (NO)
77	TEND TO BE THOSE THAT HAVE BEEN PARTIALLY WESTERNIZED, RATHER THAN SIGNIFICANTLY WESTERNIZED, THROUGH A COLONIAL RELATIONSHIP (41)	0.93	77	TEND TO BE THOSE THAT HAVE BEEN SIGNIFICANTLY WESTERNIZED, RATHER THAN PARTIALLY WESTERNIZED, THROUGH A COLONIAL RELATIONSHIP (28)	0.67	2 26 28 13 X SQ= 22.89 P = 0. RV YES (YES)
79	ALWAYS ARE THOSE THAT WERE FORMERLY DEPENDENCIES OF BRITAIN OR FRANCE, RATHER THAN SPAIN (49)	1.00	79	TEND LESS TO BE THOSE THAT WERE FORMERLY DEPENDENCIES OF BRITAIN OR FRANCE, RATHER THAN SPAIN (49)	0.56	26 23 0 18 X SQ= 13.45 P = 0. RV NO (YES)
80	TEND TO BE THOSE WHOSE DATE OF INDEPENDENCE IS AFTER 1914, AND THAT WERE FORMERLY DEPENDENCIES OF FRANCE, RATHER THAN BRITAIN (24)	0.72	80	TEND TO BE THOSE WHOSE DATE OF INDEPENDENCE IS AFTER 1914, AND THAT WERE FORMERLY DEPENDENCIES OF BRITAIN, RATHER THAN FRANCE (19)	0.67	7 12 18 6 X SQ= 4.87 P = 0.016 RV YES (YES)
82	TEND TO BE THOSE WHERE THE TYPE OF POLITICAL MODERNIZATION IS DEVELOPED TUTELARY OR UNDEVELOPED TUTELARY, RATHER THAN EARLY OR LATER EUROPEAN OR EUROPEAN DERIVED (55)	0.97	82	TEND TO BE THOSE WHERE THE TYPE OF POLITICAL MODERNIZATION IS EARLY OR LATER EUROPEAN OR EUROPEAN DERIVED, RATHER THAN DEVELOPED TUTELARY OR UNDEVELOPED TUTELARY (51)	0.68	1 50 31 24 X SQ= 34.63 P = 0. RV YES (YES)
84	TEND TO BE THOSE WHERE THE TYPE OF POLITICAL MODERNIZATION IS UNDEVELOPED TUTELARY, RATHER THAN DEVELOPED TUTELARY (24)	0.77	84	ALWAYS ARE THOSE WHERE THE TYPE OF POLITICAL MODERNIZATION IS DEVELOPED TUTELARY, RATHER THAN UNDEVELOPED TUTELARY (31)	1.00	7 24 24 0 X SQ= 29.89 P = 0. RV YES (YES)

No.	Left statement		No.	Right statement		Statistics
85	TEND TO BE THOSE WHERE THE STAGE OF POLITICAL MODERNIZATION IS MID- OR EARLY TRANSITIONAL, RATHER THAN ADVANCED (54)	0.85	85	TEND TO BE THOSE WHERE THE STAGE OF POLITICAL MODERNIZATION IS ADVANCED, RATHER THAN MID- OR EARLY TRANSITIONAL (60)	0.68	5 55 28 26 X SQ= 24.10 P = 0. RV YES (YES)
86	TEND TO BE THOSE WHERE THE STAGE OF POLITICAL MODERNIZATION IS EARLY TRANSITIONAL, RATHER THAN ADVANCED OR MID-TRANSITIONAL (38)	0.79	86	TEND TO BE THOSE WHERE THE STAGE OF POLITICAL MODERNIZATION IS ADVANCED OR MID-TRANSITIONAL, RATHER THAN EARLY TRANSITIONAL (76)	0.85	7 69 26 12 X SQ= 40.35 P = 0. RV YES (YES)
87	TEND TO BE THOSE WHOSE IDEOLOGICAL ORIENTATION IS DEVELOPMENTAL (31)	0.93	87	TEND TO BE THOSE WHOSE IDEOLOGICAL ORIENTATION IS OTHER THAN DEVELOPMENTAL (58)	0.90	25 6 2 56 X SQ= 53.37 P = 0. RV YES (YES)
95	ALWAYS ARE THOSE WHERE THE STATUS OF THE REGIME IS CONSTITUTIONAL OR AUTHORITARIAN (95)	1.00	95	TEND LESS TO BE THOSE WHERE THE STATUS OF THE REGIME IS CONSTITUTIONAL OR AUTHORITARIAN (95)	0.80	32 63 0 16 X SQ= 6.02 P = 0.005 RV NO (NO)
96	TEND TO BE THOSE WHERE THE STATUS OF THE REGIME IS AUTHORITARIAN (23)	0.50	96	TEND TO BE THOSE WHERE THE STATUS OF THE REGIME IS CONSTITUTIONAL OR TOTALITARIAN (67)	0.80	8 59 8 15 X SQ= 4.65 P = 0.024 RV YES (NO)
97	ALWAYS ARE THOSE WHERE THE STATUS OF THE REGIME IS AUTHORITARIAN, RATHER THAN TOTALITARIAN (23)	1.00	97	TEND TO BE THOSE WHERE THE STATUS OF THE REGIME IS TOTALITARIAN, RATHER THAN AUTHORITARIAN (16)	0.52	8 15 0 16 X SQ= 5.03 P = 0.012 RV YES (NO)
98	TEND MORE TO BE THOSE WHERE GOVERNMENTAL STABILITY IS GENERALLY OR MODERATELY PRESENT AND DATES FROM THE POST-WAR PERIOD, OR IS ABSENT (93)	0.94	98	TEND LESS TO BE THOSE WHERE GOVERNMENTAL STABILITY IS GENERALLY OR MODERATELY PRESENT AND DATES FROM THE POST-WAR PERIOD, OR IS ABSENT (93)	0.76	2 20 31 62 X SQ= 3.99 P = 0.034 RV NO (NO)
101	TEND TO BE THOSE WHERE THE REPRESENTATIVE CHARACTER OF THE REGIME IS LIMITED POLYARCHIC, PSEUDO-POLYARCHIC, OR NON-POLYARCHIC (57)	0.83	101	TEND TO BE THOSE WHERE THE REPRESENTATIVE CHARACTER OF THE REGIME IS POLYARCHIC (41)	0.50	4 37 20 37 X SQ= 6.96 P = 0.004 RV YES (NO)
105	TEND TO BE THOSE WHERE THE ELECTORAL SYSTEM IS PARTIALLY COMPETITIVE OR NON-COMPETITIVE (47)	0.88	105	TEND TO BE THOSE WHERE THE ELECTORAL SYSTEM IS COMPETITIVE (43)	0.62	3 40 22 25 X SQ= 15.83 P = 0. RV YES (NO)

106	TILT TOWARD BEING THOSE WHERE THE ELECTORAL SYSTEM IS NON-COMPETITIVE (30)	0.52	106	TILT TOWARD BEING THOSE WHERE THE ELECTORAL SYSTEM IS COMPETITIVE OR PARTIALLY COMPETITIVE (52)	0.69	10 42 11 19 X SQ= 2.19 P = 0.115 RV YES (NO)
107	TEND TO BE THOSE WHERE AUTONOMOUS GROUPS ARE PARTIALLY TOLERATED IN POLITICS, ARE TOLERATED ONLY OUTSIDE POLITICS, OR ARE NOT TOLERATED AT ALL (65)	0.91	107	TEND TO BE THOSE WHERE AUTONOMOUS GROUPS ARE FULLY TOLERATED IN POLITICS (46)	0.54	3 43 29 36 X SQ= 17.24 P = 0. RV YES (NO)
110	ALWAYS ARE THOSE WHERE POLITICAL ENCULTURATION IS MEDIUM OR LOW (80)	1.00	110	TEND LESS TO BE THOSE WHERE POLITICAL ENCULTURATION IS MEDIUM OR LOW (80)	0.77	0 15 29 51 X SQ= 6.21 P = 0.004 RV NO (NO)
114	TEND TO BE THOSE WHERE SECTIONALISM IS EXTREME OR MODERATE (61)	0.74	114	TEND TO BE THOSE WHERE SECTIONALISM IS NEGLIGIBLE (47)	0.51	23 38 8 39 X SQ= 4.58 P = 0.020 RV YES (NO)
115	ALWAYS ARE THOSE WHERE INTEREST ARTICULATION BY ASSOCIATIONAL GROUPS IS MODERATE, LIMITED, OR NEGLIGIBLE (91)	1.00	115	TEND LESS TO BE THOSE WHERE INTEREST ARTICULATION BY ASSOCIATIONAL GROUPS IS MODERATE, LIMITED, OR NEGLIGIBLE (91)	0.74	0 20 33 58 X SQ= 8.66 P = 0.001 RV NO (NO)
116	TEND MORE TO BE THOSE WHERE INTEREST ARTICULATION BY ASSOCIATIONAL GROUPS IS LIMITED OR NEGLIGIBLE (79)	0.97	116	TEND LESS TO BE THOSE WHERE INTEREST ARTICULATION BY ASSOCIATIONAL GROUPS IS LIMITED OR NEGLIGIBLE (79)	0.60	1 31 32 47 X SQ= 13.50 P = 0. RV NO (NO)
117	TEND TO BE THOSE WHERE INTEREST ARTICULATION BY ASSOCIATIONAL GROUPS IS NEGLIGIBLE (51)	0.85	117	TEND TO BE THOSE WHERE INTEREST ARTICULATION BY ASSOCIATIONAL GROUPS IS SIGNIFICANT, MODERATE, OR LIMITED (60)	0.71	5 55 28 23 X SQ= 26.43 P = 0. RV YES (YES)
119	TEND MORE TO BE THOSE WHERE INTEREST ARTICULATION BY INSTITUTIONAL GROUPS IS VERY SIGNIFICANT OR SIGNIFICANT (74)	0.95	119	TEND LESS TO BE THOSE WHERE INTEREST ARTICULATION BY INSTITUTIONAL GROUPS IS VERY SIGNIFICANT OR SIGNIFICANT (74)	0.69	18 56 1 25 X SQ= 4.00 P = 0.021 RV NO (NO)
121	TEND TO BE THOSE WHERE INTEREST ARTICULATION BY NON-ASSOCIATIONAL GROUPS IS SIGNIFICANT (54)	0.82	121	TEND TO BE THOSE WHERE INTEREST ARTICULATION BY NON-ASSOCIATIONAL GROUPS IS MODERATE, LIMITED, OR NEGLIGIBLE (61)	0.67	27 27 6 55 X SQ= 20.66 P = 0. RV YES (YES)

122	ALWAYS ARE THOSE WHERE INTEREST ARTICULATION BY NON-ASSOCIATIONAL GROUPS IS SIGNIFICANT OR MODERATE (83)	1.00	122	TEND LESS TO BE THOSE WHERE INTEREST ARTICULATION BY NON-ASSOCIATIONAL GROUPS IS SIGNIFICANT OR MODERATE (83)	0.61	33 50 0 32 X SQ= 15.95 P = 0. RV NO (NO)
125	LEAN MORE TOWARD BEING THOSE WHERE INTEREST ARTICULATION BY ANOMIC GROUPS IS FREQUENT OR OCCASIONAL (64)	0.78	125	LEAN LESS TOWARD BEING THOSE WHERE INTEREST ARTICULATION BY ANOMIC GROUPS IS FREQUENT OR OCCASIONAL (64)	0.58	25 39 7 28 X SQ= 2.94 P = 0.072 RV NO (NO)
126	ALWAYS ARE THOSE WHERE INTEREST ARTICULATION BY ANOMIC GROUPS IS FREQUENT, OCCASIONAL, OR INFREQUENT (83)	1.00	126	TEND LESS TO BE THOSE WHERE INTEREST ARTICULATION BY ANOMIC GROUPS IS FREQUENT, OCCASIONAL, OR INFREQUENT (83)	0.76	32 51 0 16 X SQ= 7.44 P = 0.001 RV NO (NO)
128	TEND TO BE THOSE WHERE INTEREST ARTICULATION BY POLITICAL PARTIES IS LIMITED OR NEGLIGIBLE (45)	0.68	128	TEND TO BE THOSE WHERE INTEREST ARTICULATION BY POLITICAL PARTIES IS SIGNIFICANT OR MODERATE (48)	0.60	M 9 39 19 26 X SQ= 5.02 P = 0.023 RV YES (NO)
129	TEND TO BE THOSE WHERE INTEREST ARTICULATION BY POLITICAL PARTIES IS NEGLIGIBLE (37)	0.57	129	TEND TO BE THOSE WHERE INTEREST ARTICULATION BY POLITICAL PARTIES IS SIGNIFICANT, MODERATE, OR LIMITED (56)	0.68	12 44 16 21 X SQ= 4.06 P = 0.037 RV YES (NO)
131	TEND TO BE THOSE WHERE INTEREST AGGREGATION BY POLITICAL PARTIES IS SIGNIFICANT OR MODERATE (30)	0.83	131	TEND TO BE THOSE WHERE INTEREST AGGREGATION BY POLITICAL PARTIES IS LIMITED OR NEGLIGIBLE (35)	0.62	10 20 2 33 X SQ= 6.45 P = 0.008 RV YES (NO)
132	ALWAYS ARE THOSE WHERE INTEREST AGGREGATION BY POLITICAL PARTIES IS SIGNIFICANT, MODERATE, OR LIMITED (67)	1.00	132	TEND LESS TO BE THOSE WHERE INTEREST AGGREGATION BY POLITICAL PARTIES IS SIGNIFICANT, MODERATE, OR LIMITED (67)	0.81	29 38 0 9 X SQ= 4.60 P = 0.011 RV NO (NO)
133	TEND TO BE THOSE WHERE INTEREST AGGREGATION BY THE EXECUTIVE IS SIGNIFICANT (29)	0.61	133	TEND TO BE THOSE WHERE INTEREST AGGREGATION BY THE EXECUTIVE IS MODERATE, LIMITED, OR NEGLIGIBLE (76)	0.86	19 10 12 64 X SQ= 22.61 P = 0. RV YES (YES)
134	TEND TO BE THOSE WHERE INTEREST AGGREGATION BY THE EXECUTIVE IS SIGNIFICANT OR MODERATE (57)	0.73	134	TEND TO BE THOSE WHERE INTEREST AGGREGATION BY THE EXECUTIVE IS LIMITED OR NEGLIGIBLE (46)	0.52	22 35 8 38 X SQ= 4.57 P = 0.028 RV YES (NO)

137	ALWAYS ARE THOSE WHERE INTEREST AGGREGATION BY THE LEGISLATURE IS LIMITED OR NEGLIGIBLE (68)	1.00	137	TEND LESS TO BE THOSE WHERE INTEREST AGGREGATION BY THE LEGISLATURE IS LIMITED OR NEGLIGIBLE (68)	0.58	0 29 28 40 X SQ= 14.84 P = 0. RV NO (NO)
138	TEND TO BE THOSE WHERE INTEREST AGGREGATION BY THE LEGISLATURE IS NEGLIGIBLE (49)	0.82	138	TEND TO BE THOSE WHERE INTEREST AGGREGATION BY THE LEGISLATURE IS SIGNIFICANT, MODERATE, OR LIMITED (48)	0.62	5 43 23 26 X SQ= 14.02 P = 0. RV YES (NO)
139	TEND TO BE THOSE WHERE THE PARTY SYSTEM IS QUANTITATIVELY ONE-PARTY (34)	0.53	139	TEND TO BE THOSE WHERE THE PARTY SYSTEM IS QUANTITATIVELY OTHER THAN ONE-PARTY (71)	0.77	M 17 17 15 56 X SQ= 7.73 P = 0.006 RV YES (YES)
146	TEND LESS TO BE THOSE WHERE THE PARTY SYSTEM IS QUALITATIVELY OTHER THAN MASS-BASED TERRITORIAL (92)	0.73	146	ALWAYS ARE THOSE WHERE THE PARTY SYSTEM IS QUALITATIVELY OTHER THAN MASS-BASED TERRITORIAL (92)	1.00	8 0 22 70 X SQ= 16.83 P = 0. RV NO (YES)
147	ALWAYS ARE THOSE WHERE THE PARTY SYSTEM IS QUALITATIVELY OTHER THAN CLASS-ORIENTED OR MULTI-IDEOLOGICAL (67)	1.00	147	TEND LESS TO BE THOSE WHERE THE PARTY SYSTEM IS QUALITATIVELY OTHER THAN CLASS-ORIENTED OR MULTI-IDEOLOGICAL (67)	0.63	0 23 28 39 X SQ= 12.07 P = 0. RV NO (NO)
148	TEND LESS TO BE THOSE WHERE THE PARTY SYSTEM IS QUALITATIVELY OTHER THAN AFRICAN TRANSITIONAL (96)	0.55	148	ALWAYS ARE THOSE WHERE THE PARTY SYSTEM IS QUALITATIVELY OTHER THAN AFRICAN TRANSITIONAL (96)	1.00	14 0 17 79 X SQ= 36.92 P = 0. RV NO (YES)
158	TILT TOWARD BEING THOSE WHERE PERSONALISMO IS PRONOUNCED OR MODERATE (40)	0.56	158	TILT TOWARD BEING THOSE WHERE PERSONALISMO IS NEGLIGIBLE (56)	0.64	15 25 12 44 X SQ= 2.24 P = 0.108 RV YES (NO)
161	LEAN TOWARD BEING THOSE WHERE THE POLITICAL LEADERSHIP IS NON-ELITIST (50)	0.67	161	LEAN TOWARD BEING THOSE WHERE THE POLITICAL LEADERSHIP IS ELITIST OR MODERATELY ELITIST (47)	0.54	9 38 18 32 X SQ= 2.64 P = 0.074 RV YES (NO)
164	TEND TO BE THOSE WHERE THE REGIME'S LEADERSHIP CHARISMA IS PRONOUNCED OR MODERATE (34)	0.78	164	TEND TO BE THOSE WHERE THE REGIME'S LEADERSHIP CHARISMA IS NEGLIGIBLE (65)	0.82	M 21 13 6 59 X SQ= 28.47 P = 0. RV YES (YES)

169	LEAN TOWARD BEING THOSE WHERE THE HORIZONTAL POWER DISTRIBUTION IS NEGLIGIBLE (48)	0.61	169	LEAN TOWARD BEING THOSE WHERE THE HORIZONTAL POWER DISTRIBUTION IS SIGNIFICANT OR LIMITED (58)	0.61	12 46 19 29 X SQ= 3.66 P = 0.053 RV YES (NO)
170	TEND TO BE THOSE WHERE THE LEGISLATIVE-EXECUTIVE STRUCTURE IS PRESIDENTIAL (39)	0.57	170	TEND TO BE THOSE WHERE THE LEGISLATIVE-EXECUTIVE STRUCTURE IS OTHER THAN PRESIDENTIAL (63)	0.69	16 23 12 51 X SQ= 4.79 P = 0.022 RV YES (NO)
174	TEND MORE TO BE THOSE WHERE THE LEGISLATURE IS PARTIALLY EFFECTIVE, LARGELY INEFFECTIVE, OR WHOLLY INEFFECTIVE (72)	0.97	174	TEND LESS TO BE THOSE WHERE THE LEGISLATURE IS PARTIALLY EFFECTIVE, LARGELY INEFFECTIVE, OR WHOLLY INEFFECTIVE (72)	0.61	1 27 29 43 X SQ= 11.25 P = 0. RV NO (NO)
175	TEND TO BE THOSE WHERE THE LEGISLATURE IS LARGELY INEFFECTIVE OR WHOLLY INEFFECTIVE (49)	0.73	175	TEND TO BE THOSE WHERE THE LEGISLATURE IS FULLY EFFECTIVE OR PARTIALLY EFFECTIVE (51)	0.61	8 43 22 27 X SQ= 8.81 P = 0.002 RV YES (NO)
176	LEAN MORE TOWARD BEING THOSE WHERE THE LEGISLATURE IS FULLY EFFECTIVE, PARTIALLY EFFECTIVE, OR LARGELY INEFFECTIVE (72)	0.87	176	LEAN LESS TOWARD BEING THOSE WHERE THE LEGISLATURE IS FULLY EFFECTIVE, PARTIALLY EFFECTIVE, OR LARGELY INEFFECTIVE (72)	0.66	26 46 4 24 X SQ= 3.59 P = 0.050 RV NO (NO)
178	TEND TO BE THOSE WHERE THE LEGISLATURE IS UNICAMERAL (53)	0.77	178	TEND TO BE THOSE WHERE THE LEGISLATURE IS BICAMERAL (51)	0.60	24 29 7 44 X SQ= 10.91 P = 0.001 RV YES (NO)
179	TEND TO BE THOSE WHERE THE EXECUTIVE IS DOMINANT (52)	0.79	179	TEND TO BE THOSE WHERE THE EXECUTIVE IS STRONG (39)	0.51	19 33 5 34 X SQ= 5.29 P = 0.016 RV YES (NO)
182	TEND TO BE THOSE WHERE THE BUREAUCRACY IS POST-COLONIAL TRANSITIONAL, RATHER THAN SEMI-MODERN (25)	0.77	182	TEND TO BE THOSE WHERE THE BUREAUCRACY IS SEMI-MODERN, RATHER THAN POST-COLONIAL TRANSITIONAL (55)	0.98	7 48 24 1 X SQ= 46.77 P = 0. RV YES (YES)
183	TEND TO BE THOSE WHERE THE BUREAUCRACY IS POST-COLONIAL TRANSITIONAL, RATHER THAN TRADITIONAL (25)	0.96	183	TEND TO BE THOSE WHERE THE BUREAUCRACY IS TRADITIONAL, RATHER THAN POST-COLONIAL TRANSITIONAL (9)	0.89	24 1 1 8 X SQ= 20.33 P = 0. RV YES (YES)

187	LEAN MORE TOWARD BEING THOSE WHERE THE ROLE OF THE POLICE IS POLITICALLY SIGNIFICANT (66)	0.81	187	LEAN LESS TOWARD BEING THOSE WHERE THE ROLE OF THE POLICE IS POLITICALLY SIGNIFICANT (66)	0.60	21 45 5 30 X SQ= 2.82 P = 0.061 RV NO (NO)
188	ALWAYS ARE THOSE WHERE THE CHARACTER OF THE LEGAL SYSTEM IS OTHER THAN CIVIL LAW (81)	1.00	188	TEND LESS TO BE THOSE WHERE THE CHARACTER OF THE LEGAL SYSTEM IS OTHER THAN CIVIL LAW (81)	0.61	0 32 31 50 X SQ= 15.01 P = 0. RV NO (NO)
189	ALWAYS ARE THOSE WHERE THE CHARACTER OF THE LEGAL SYSTEM IS OTHER THAN COMMON LAW (108)	1.00	189	TILT LESS TOWARD BEING THOSE WHERE THE CHARACTER OF THE LEGAL SYSTEM IS OTHER THAN COMMON LAW (108)	0.91	0 7 33 75 X SQ= 1.69 P = 0.107 RV NO (NO)
192	TEND TO BE THOSE WHERE THE CHARACTER OF THE LEGAL SYSTEM IS MUSLIM OR PARTLY MUSLIM (27)	0.61	192	TEND TO BE THOSE WHERE THE CHARACTER OF THE LEGAL SYSTEM IS OTHER THAN MUSLIM OR PARTLY MUSLIM (86)	0.90	19 8 12 74 X SQ= 30.08 P = 0. RV YES (YES)
193	TEND TO BE THOSE WHERE THE CHARACTER OF THE LEGAL SYSTEM IS PARTLY INDIGENOUS (28)	0.76	193	TEND TO BE THOSE WHERE THE CHARACTER OF THE LEGAL SYSTEM IS OTHER THAN PARTLY INDIGENOUS (86)	0.96	25 3 8 78 X SQ= 61.87 P = 0. RV YES (YES)
194	ALWAYS ARE THOSE THAT ARE NON-COMMUNIST (101)	1.00	194	TEND LESS TO BE THOSE THAT ARE NON-COMMUNIST (101)	0.84	0 13 33 68 X SQ= 4.49 P = 0.010 RV NO (NO)

M

MATRIX

10 POLITIES LOCATED IN NORTH AFRICA, OR CENTRAL AND SOUTH AFRICA (33)	10 POLITIES LOCATED ELSEWHERE THAN IN NORTH AFRICA, OR CENTRAL AND SOUTH AFRICA (82)
84 POLITIES WHERE THE TYPE OF POLITICAL MODERNIZATION IS DEVELOPED TUTELARY, RATHER THAN UNDEVELOPED TUTELARY (31)	
ALGERIA, TUNISIA, LIBERIA, UAR, LIBYA, MOROCCO, SUDAN	BURMA, INDONESIA, KOREA, N, MONGOLIA, TRINIDAD, CAMBODIA, IRAQ, KOREA REP, PAKISTAN, VIETNAM, N, CEYLON, ISRAEL, LAOS, PHILIPPINES, VIETNAM REP, CYPRUS, JAMAICA, LEBANON, SA'U ARABIA, YEMEN, INDIA, JORDAN, MALAYA, SYRIA
7	24
24	0
84 POLITIES WHERE THE TYPE OF POLITICAL MODERNIZATION IS UNDEVELOPED TUTELARY, RATHER THAN DEVELOPED TUTELARY (24)	
BURUNDI, CONGO(LEO), IVORY COAST, NIGERIA, TANGANYIKA, CAMEROUN, DAHOMEY, MALAGASY R, RWANDA, TOGO, CEN AFR REP, GABON, MALI, SENEGAL, UGANDA, CHAD, GHANA, MAURITANIA, SIERRE LEO, UPPER VOLTA, CONGO(BRA), GUINEA, NIGER, SOMALIA	

MATRIX

10 POLITIES
LOCATED IN NORTH AFRICA, OR
CENTRAL AND SOUTH AFRICA (33)

10 POLITIES
LOCATED ELSEWHERE THAN IN NORTH AFRICA,
OR CENTRAL AND SOUTH AFRICA (82)

128 POLITIES
WHERE INTEREST ARTICULATION
BY POLITICAL PARTIES
IS SIGNIFICANT OR MODERATE (48)

CAMEROUN, CONGO(LEO), LIBERIA, MALAGASY R, MOROCCO, NIGERIA, SIERRE LEO, SOMALIA, UGANDA

ARGENTINA, AUSTRALIA, AUSTRIA, BELGIUM, BOLIVIA, BRAZIL, CEYLON, CHILE, COLOMBIA, DENMARK, ECUADOR, EL SALVADOR, FINLAND, FRANCE, GERMAN FR, GREECE, GUATEMALA, HONDURAS, ICELAND, INDIA, IRELAND, ISRAEL, ITALY, JAMAICA, JAPAN, LEBANON, LUXEMBOURG, MALAYA, NETHERLANDS, NEW ZEALAND, NORWAY, PANAMA, PERU, SWEDEN, SWITZERLAND, SYRIA, TRINIDAD, UK, VENEZUELA

9 . 39

19 . 26

128 POLITIES
WHERE INTEREST ARTICULATION
BY POLITICAL PARTIES
IS LIMITED OR NEGLIGIBLE (45)

BURUNDI, CEN AFR REP, CHAD, CONGO(BRA), DAHOMEY, GABON, GHANA, GUINEA, IVORY COAST, MALI, MAURITANIA, NIGER, RWANDA, SENEGAL, TANGANYIKA, TOGO, TUNISIA, UAR, UPPER VOLTA

ALBANIA, BULGARIA, CAMBODIA, CANADA, CHINA, PR, COSTA RICA, CZECHOS'KIA, DOMIN REP, GERMANY, E, HUNGARY, KOREA, N, MEXICO, MONGOLIA, NICARAGUA, PARAGUAY, PHILIPPINES, POLAND, PORTUGAL, RUMANIA, SPAIN, TURKEY, USSR, US, VIETNAM, N, VIETNAM REP, YUGOSLAVIA

MATRIX

10 POLITIES LOCATED IN NORTH AFRICA, OR CENTRAL AND SOUTH AFRICA (33)	10 POLITIES LOCATED ELSEWHERE THAN IN NORTH AFRICA, OR CENTRAL AND SOUTH AFRICA (82)

139 POLITIES
WHERE THE PARTY SYSTEM IS QUANTITATIVELY
ONE-PARTY (34)

```
ALGERIA      CEN AFR REP CHAD        DAHOMEY     GABON       .ALBANIA     BULGARIA    CAMBODIA    CHINA, PR   CUBA
GHANA        GUINEA      IVORY COAST LIBERIA     MALI        .CZECHOS'KIA GERMANY, E  HUNGARY     KOREA, N    MONGOLIA
MAURITANIA   NIGER       SENEGAL     TANGANYIKA  TUNISIA     .POLAND      PORTUGAL    RUMANIA     SPAIN       USSR
UAR          UPPER VOLTA                                     .VIETNAM, N  YUGOSLAVIA
                                                             .
                                                          17 . 17
. . . . . . . . . . . . . . . . . . . . . . . . . . . . . . . . . . . . . . . . . . . . . . . . . . . . . . . . . . . .
                                                          15 . 56
```

139 POLITIES
WHERE THE PARTY SYSTEM IS QUANTITATIVELY
OTHER THAN ONE-PARTY (71)

```
BURUNDI      CAMEROUN    CONGO(BRA)  CONGO(LEO)  ETHIOPIA    .AFGHANISTAN ARGENTINA   AUSTRALIA   AUSTRIA     BELGIUM
LIBYA        MALAGASY R  MOROCCO     NIGERIA     RWANDA      .BOLIVIA     BRAZIL      BURMA       CANADA      CEYLON
SIERRE LEO   SOMALIA     SO AFRICA   SUDAN       UGANDA      .CHILE       COLOMBIA    COSTA RICA  CYPRUS      DENMARK
                                                             .DOMIN REP   ECUADOR     EL SALVADOR FINLAND     FRANCE
                                                             .GERMAN FR   GREECE      GUATEMALA   HAITI       HONDURAS
                                                             .ICELAND     INDIA       IRELAND     ISRAEL      ITALY
                                                             .JAMAICA     JAPAN       JORDAN      LAOS        LEBANON
                                                             .LUXEMBOURG  MALAYA      MEXICO      NETHERLANDS NEW ZEALAND
                                                             .NORWAY      PAKISTAN    PANAMA      PARAGUAY    PERU
                                                             .PHILIPPINES SA'U ARABIA SWEDEN      SWITZERLAND TRINIDAD
                                                             .TURKEY      UK          US          URUGUAY     VENEZUELA
                                                             .VIETNAM REP
```

M

MATRIX

	10 POLITIES LOCATED IN NORTH AFRICA, OR CENTRAL AND SOUTH AFRICA (33)	10 POLITIES LOCATED ELSEWHERE THAN IN NORTH AFRICA, OR CENTRAL AND SOUTH AFRICA (82)
164 POLITIES WHERE THE REGIME'S LEADERSHIP CHARISMA IS PRONOUNCED OR MODERATE (34)	ALGERIA, CAMEROUN, CEN AFR REP, CHAD, CONGO(BRA), DAHOMEY, ETHIOPIA, GABON, GHANA, GUINEA, IVORY COAST, LIBYA, MALI, MAURITANIA, MOROCCO, NIGER, SENEGAL, TANGANYIKA, TUNISIA, UAR, UPPER VOLTA 21	CAMBODIA, CHINA, PR, CUBA, FRANCE, INDIA, INDONESIA, JAMAICA, JORDAN, KOREA, N, PAKISTAN, SA'U ARABIA, VIETNAM, N, YUGOSLAVIA 13
164 POLITIES WHERE THE REGIME'S LEADERSHIP CHARISMA IS NEGLIGIBLE (65)	6 LIBERIA, MALAGASY R, SIERRE LEO, SOMALIA, SO AFRICA, SUDAN	59 ALBANIA, ARGENTINA, AUSTRALIA, AUSTRIA, BELGIUM, BOLIVIA, BRAZIL, BULGARIA, CANADA, CEYLON, CHILE, COLOMBIA, COSTA RICA, CZECHOS'KIA, DENMARK, DOMIN REP, ECUADOR, EL SALVADOR, FINLAND, GERMANY, E, GERMAN FR, GREECE, GUATEMALA, HAITI, HONDURAS, HUNGARY, ICELAND, IRELAND, ISRAEL, ITALY, JAPAN, KOREA REP, LEBANON, LUXEMBOURG, MALAYA, MEXICO, MONGOLIA, NETHERLANDS, NEW ZEALAND, NICARAGUA, NORWAY, PANAMA, PARAGUAY, PERU, PHILIPPINES, POLAND, PORTUGAL, RUMANIA, SPAIN, SWEDEN, SWITZERLAND, THAILAND, TRINIDAD, TURKEY, UK, US, URUGUAY, VENEZUELA, VIETNAM REP

11 POLITIES
LOCATED IN CENTRAL AND
SOUTH AFRICA (28)

11 POLITIES
LOCATED ELSEWHERE THAN IN CENTRAL
AND SOUTH AFRICA (87)

BOTH SUBJECT AND PREDICATE

28 IN LEFT COLUMN

BURUNDI	CAMEROUN	CEN AFR REP	CHAD	CONGO(BRA)	CONGO(LEO)	DAHOMEY	ETHIOPIA	GABON	GHANA
GUINEA	IVORY COAST	LIBERIA	MALAGASY R	MALI	MAURITANIA	NIGER	NIGERIA	RWANDA	SENEGAL
SIERRE LEO	SOMALIA	SO AFRICA	SUDAN	TANGANYIKA	TOGO	UGANDA	UPPER VOLTA		

87 IN RIGHT COLUMN

AFGHANISTAN	ALBANIA	ALGERIA	ARGENTINA	AUSTRALIA	AUSTRIA	BELGIUM	BOLIVIA	BRAZIL	BULGARIA
BURMA	CAMBODIA	CANADA	CEYLON	CHILE	CHINA, PR	COLOMBIA	COSTA RICA	CUBA	CYPRUS
CZECHOS'KIA	DENMARK	DOMIN REP	ECUADOR	EL SALVADOR	FINLAND	FRANCE	GERMANY, E	GERMAN FR	GREECE
GUATEMALA	HAITI	HONDURAS	HUNGARY	ICELAND	INDIA	INDONESIA	IRAN	IRAQ	IRELAND
ISRAEL	ITALY	JAMAICA	JAPAN	JORDAN	KOREA, N	KOREA REP	LAOS	LEBANON	LIBYA
LUXEMBOURG	MALAYA	MEXICO	MONGOLIA	MOROCCO	NEPAL	NETHERLANDS	NEW ZEALAND	NICARAGUA	NORWAY
PAKISTAN	PANAMA	PARAGUAY	PERU	PHILIPPINES	POLAND	PORTUGAL	RUMANIA	SA'U ARABIA	SPAIN
SWEDEN	SWITZERLAND	SYRIA	THAILAND	TRINIDAD	TUNISIA	TURKEY	USSR	UAR	UK
US	URUGUAY	VENEZUELA	VIETNAM, N	VIETNAM REP	YEMEN	YUGOSLAVIA			

22	TEND MORE TO BE THOSE WHOSE TERRITORIAL SIZE IS VERY LARGE, LARGE, OR MEDIUM (68)	0.79	22	TEND LESS TO BE THOSE WHOSE TERRITORIAL SIZE IS VERY LARGE, LARGE, OR MEDIUM (68)	0.53	22 46 6 41 X SQ= 4.77 P = 0.017 RV NO (NO)
24	TEND TO BE THOSE WHOSE POPULATION IS SMALL (54)	0.71	24	TEND TO BE THOSE WHOSE POPULATION IS VERY LARGE, LARGE, OR MEDIUM (61)	0.61	8 53 20 34 X SQ= 7.65 P = 0.004 RV YES (NO)
25	ALWAYS ARE THOSE WHOSE POPULATION DENSITY IS MEDIUM OR LOW (98)	1.00	25	TEND LESS TO BE THOSE WHOSE POPULATION DENSITY IS MEDIUM OR LOW (98)	0.80	0 17 28 70 X SQ= 4.96 P = 0.011 RV NO (NO)
26	TEND TO BE THOSE WHOSE POPULATION DENSITY IS LOW (67)	0.89	26	TEND TO BE THOSE WHOSE POPULATION DENSITY IS VERY HIGH, HIGH, OR MEDIUM (48)	0.52	3 45 25 42 X SQ= 13.01 P = 0. RV YES (NO

```
29  TEND TO BE THOSE                        0.96     29  TEND TO BE THOSE                       0.71        1    55
    WHERE THE DEGREE OF URBANIZATION                     WHERE THE DEGREE OF URBANIZATION                  27    22
    IS LOW  (49)                                         IS HIGH  (56)                                X SQ=  35.31
                                                                                                      P =     0.
                                                                                                      RV YES (YES)

30  TEND TO BE THOSE                        0.93     30  TEND TO BE THOSE                       0.65       26    30
    WHOSE AGRICULTURAL POPULATION                        WHOSE AGRICULTURAL POPULATION                      2    55
    IS HIGH  (56)                                        IS MEDIUM, LOW, OR VERY LOW  (57)            X SQ=  25.66
                                                                                                      P =     0.
                                                                                                      RV YES (NO )

34  TEND TO BE THOSE                        0.86     34  TEND TO BE THOSE                       0.67        4    58
    WHOSE GROSS NATIONAL PRODUCT                         WHOSE GROSS NATIONAL PRODUCT                      24    29
    IS VERY LOW  (53)                                    IS VERY HIGH, HIGH, MEDIUM,                  X SQ=  21.33
                                                         OR LOW  (62)                                 P =     0.
                                                                                                      RV YES (NO )

37  TEND TO BE THOSE                        0.96     37  TEND TO BE THOSE                       0.72        1    63
    WHOSE PER CAPITA GROSS NATIONAL PRODUCT              WHOSE PER CAPITA GROSS NATIONAL PRODUCT           27    24
    IS VERY LOW  (51)                                    IS VERY HIGH, HIGH, MEDIUM, OR LOW  (64)     X SQ=  37.93
                                                                                                      P =     0.
                                                                                                      RV YES (YES)

40  TEND TO BE THOSE                        0.75     40  TEND TO BE THOSE                       0.78        7    64
    WHOSE INTERNATIONAL FINANCIAL STATUS                 WHOSE INTERNATIONAL FINANCIAL STATUS              21    18
    IS VERY LOW  (39)                                    IS VERY HIGH, HIGH,                          X SQ=  23.40
                                                         MEDIUM, OR LOW  (71)                         P =     0.
                                                                                                      RV YES (YES)

43  TEND TO BE THOSE                        0.96     43  TEND TO BE THOSE                       0.64        1    54
    WHOSE ECONOMIC DEVELOPMENTAL STATUS                  WHOSE ECONOMIC DEVELOPMENTAL STATUS               27    30
    IS VERY UNDERDEVELOPED  (57)                         IS DEVELOPED, INTERMEDIATE, OR               X SQ=  28.59
                                                         UNDERDEVELOPED  (55)                         P =     0.
                                                                                                      RV YES (NO )

45  ALWAYS ARE THOSE                        1.00     45  TEND TO BE THOSE                       0.67        0    55
    WHERE THE LITERACY RATE IS                           WHERE THE LITERACY RATE IS                        27    27
    BELOW FIFTY PERCENT  (54)                            FIFTY PERCENT OR ABOVE  (55)                 X SQ=  33.92
                                                                                                      P =     0.
                                                                                                      RV YES (YES)

46  TEND TO BE THOSE                        0.81     46  TEND TO BE THOSE                       0.94        5    79
    WHERE THE LITERACY RATE IS                           WHERE THE LITERACY RATE IS                        21     5
    BELOW TEN PERCENT  (26)                              TEN PERCENT OR ABOVE  (84)                   X SQ=  57.50
                                                                                                      P =     0.
                                                                                                      RV YES (YES)

52  ALWAYS ARE THOSE                        1.00     52  TEND LESS TO BE THOSE                  0.80       18    64
    WHERE FREEDOM OF THE PRESS IS                        WHERE FREEDOM OF THE PRESS IS                      0    16
    COMPLETE, INTERMITTENT, OR                           COMPLETE, INTERMITTENT, OR                   X SQ=   2.96
    INTERNALLY ABSENT  (82)                              INTERNALLY ABSENT  (82)                      P =     0.038
                                                                                                      RV NO  (NO )
```

```
54  ALWAYS ARE THOSE                       1.00    54  TEND LESS TO BE THOSE                  0.56      0   37
    WHERE NEWSPAPER CIRCULATION IS                     WHERE NEWSPAPER CIRCULATION IS                  28   48
    LESS THAN ONE HUNDRED                              LESS THAN ONE HUNDRED                         X SQ=  16.20
    PER THOUSAND  (76)                                 PER THOUSAND  (76)                            P =     0.
                                                                                                     RV NO  (NO )

55  TEND TO BE THOSE                       0.93    55  TEND TO BE THOSE                       0.89      2   76
    WHERE NEWSPAPER CIRCULATION IS                     WHERE NEWSPAPER CIRCULATION IS                  26    9
    LESS THAN TEN                                      TEN OR MORE                                   X SQ=  62.88
    PER THOUSAND  (35)                                 PER THOUSAND  (78)                            P =     0.
                                                                                                     RV YES (YES)

56  TEND TO BE THOSE                       0.89    56  TEND TO BE THOSE                       0.93      3   76
    WHERE THE RELIGION IS                              WHERE THE RELIGION IS                           25    6
    PREDOMINANTLY OR PARTLY                            PREDOMINANTLY LITERATE  (79)                  X SQ=  65.30
    NON-LITERATE  (31)                                                                               P =     0.
                                                                                                     RV YES (YES)

64  ALWAYS ARE THOSE                       1.00    64  TEND TO BE THOSE                       0.75      0   46
    WHERE THE RELIGION IS                              WHERE THE RELIGION IS                            3   15
    MUSLIM, RATHER THAN                                CHRISTIAN, RATHER THAN                        X SQ=   4.75
    CHRISTIAN  (18)                                    MUSLIM  (46)                                  P =     0.020
                                                                                                     RV YES (NO )

66  TEND TO BE THOSE                       0.88    66  TEND TO BE THOSE                       0.67      3   54
    THAT ARE RELIGIOUSLY HETEROGENEOUS  (49)           THAT ARE RELIGIOUSLY HOMOGENEOUS  (57)          22   27
                                                                                                     X SQ=  20.82
                                                                                                     P =     0.
                                                                                                     RV YES (NO )

68  TEND TO BE THOSE                       0.86    68  TEND TO BE THOSE                       0.56      4   48
    THAT ARE LINGUISTICALLY                            THAT ARE LINGUISTICALLY                         24   38
    WEAKLY HETEROGENEOUS OR                            HOMOGENEOUS  (52)                             X SQ=  13.06
    STRONGLY HETEROGENEOUS  (62)                                                                     P =     0.
                                                                                                     RV YES (NO )

69  TEND TO BE THOSE                       0.86    69  TEND TO BE THOSE                       0.70      4   60
    THAT ARE LINGUISTICALLY                            THAT ARE LINGUISTICALLY                         24   26
    STRONGLY HETEROGENEOUS  (50)                       HOMOGENEOUS OR                                X SQ=  24.20
                                                       WEAKLY HETEROGENEOUS  (64)                    P =     0.
                                                                                                     RV YES (NO )

70  TEND MORE TO BE THOSE                  0.96    70  TEND LESS TO BE THOSE                  0.75      1   20
    THAT ARE RELIGIOUSLY, RACIALLY,                    THAT ARE RELIGIOUSLY, RACIALLY,                 24   60
    OR LINGUISTICALLY HETEROGENEOUS  (85)              OR LINGUISTICALLY HETEROGENEOUS  (85)         X SQ=   4.02
                                                                                                     P =     0.022
                                                                                                     RV NO  (NO )

72  TEND TO BE THOSE                       0.89    72  TEND TO BE THOSE                       0.59      3   49
    WHOSE DATE OF INDEPENDENCE                         WHOSE DATE OF INDEPENDENCE                      25   34
    IS AFTER 1914  (59)                                IS BEFORE 1914  (52)                          X SQ=  17.74
                                                                                                     P =     0.
                                                                                                     RV YES (NO )
```

73	TEND TO BE THOSE WHOSE DATE OF INDEPENDENCE IS AFTER 1945 (46)	0.89	73	TEND TO BE THOSE WHOSE DATE OF INDEPENDENCE IS BEFORE 1945 (65)	0.75	3 62 25 21 X SQ= 32.73 P = 0. RV YES (YES)
76	ALWAYS ARE THOSE THAT HAVE BEEN SIGNIFICANTLY OR PARTIALLY WESTERNIZED THROUGH A COLONIAL RELATIONSHIP, RATHER THAN BEING HISTORICALLY WESTERN (70)	1.00	76	TEND LESS TO BE THOSE THAT HAVE BEEN SIGNIFICANTLY OR PARTIALLY WESTERNIZED THROUGH A COLONIAL RELATIONSHIP, RATHER THAN BEING HISTORICALLY WESTERN (70)	0.63	0 26 26 44 X SQ= 11.43 P = 0. RV NO (NO)
77	ALWAYS ARE THOSE THAT HAVE BEEN PARTIALLY WESTERNIZED, RATHER THAN SIGNIFICANTLY WESTERNIZED, THROUGH A COLONIAL RELATIONSHIP (41)	1.00	77	TEND TO BE THOSE THAT HAVE BEEN SIGNIFICANTLY WESTERNIZED, RATHER THAN PARTIALLY WESTERNIZED, THROUGH A COLONIAL RELATIONSHIP (28)	0.64	0 28 25 16 X SQ= 24.20 P = 0. RV YES (YES)
79	ALWAYS ARE THOSE THAT WERE FORMERLY DEPENDENCIES OF BRITAIN OR FRANCE, RATHER THAN SPAIN (49)	1.00	79	TEND LESS TO BE THOSE THAT WERE FORMERLY DEPENDENCIES OF BRITAIN OR FRANCE, RATHER THAN SPAIN (49)	0.60	22 27 0 18 X SQ= 10.08 P = 0. RV NO (NO)
80	LEAN TOWARD BEING THOSE WHOSE DATE OF INDEPENDENCE IS AFTER 1914, AND THAT WERE FORMERLY DEPENDENCIES OF FRANCE, RATHER THAN BRITAIN (24)	0.71	80	LEAN TOWARD BEING THOSE WHOSE DATE OF INDEPENDENCE IS AFTER 1914, AND THAT WERE FORMERLY DEPENDENCIES OF BRITAIN, RATHER THAN FRANCE (19)	0.59	6 13 15 9 X SQ= 2.91 P = 0.067 RV YES (YES)
82	TEND TO BE THOSE WHERE THE TYPE OF POLITICAL MODERNIZATION IS DEVELOPED TUTELARY OR UNDEVELOPED TUTELARY, RATHER THAN EARLY OR LATER EUROPEAN OR EUROPEAN DERIVED (55)	0.96	82	TEND TO BE THOSE WHERE THE TYPE OF POLITICAL MODERNIZATION IS EARLY OR LATER EUROPEAN OR EUROPEAN DERIVED, RATHER THAN DEVELOPED TUTELARY OR UNDEVELOPED TUTELARY (51)	0.63	1 50 26 29 X SQ= 26.28 P = 0. RV YES (NO)
84	TEND TO BE THOSE WHERE THE TYPE OF POLITICAL MODERNIZATION IS UNDEVELOPED TUTELARY, RATHER THAN DEVELOPED TUTELARY (24)	0.92	84	ALWAYS ARE THOSE WHERE THE TYPE OF POLITICAL MODERNIZATION IS DEVELOPED TUTELARY, RATHER THAN UNDEVELOPED TUTELARY (31)	1.00	2 29 24 0 X SQ= 43.81 P = 0. RV YES (YES)
85	TEND TO BE THOSE WHERE THE STAGE OF POLITICAL MODERNIZATION IS MID- OR EARLY TRANSITIONAL, RATHER THAN ADVANCED (54)	0.96	85	TEND TO BE THOSE WHERE THE STAGE OF POLITICAL MODERNIZATION IS ADVANCED, RATHER THAN MID- OR EARLY TRANSITIONAL (60)	0.69	1 59 27 27 X SQ= 33.27 P = 0. RV YES (YES)
86	TEND TO BE THOSE WHERE THE STAGE OF POLITICAL MODERNIZATION IS EARLY TRANSITIONAL, RATHER THAN ADVANCED OR MID-TRANSITIONAL (38)	0.89	86	TEND TO BE THOSE WHERE THE STAGE OF POLITICAL MODERNIZATION IS ADVANCED OR MID-TRANSITIONAL, RATHER THAN EARLY TRANSITIONAL (76)	0.85	3 73 25 13 X SQ= 49.01 P = 0. RV YES (YES)

No.	Statement	Value	No.	Statement	Value	Statistics
87	TEND TO BE THOSE WHOSE IDEOLOGICAL ORIENTATION IS DEVELOPMENTAL (31)	0.92	87	TEND TO BE THOSE WHOSE IDEOLOGICAL ORIENTATION IS OTHER THAN DEVELOPMENTAL (58)	0.86	22 9 2 56 X SQ= 43.40 P = 0. RV YES (YES)
92	LEAN MORE TOWARD BEING THOSE WHERE THE SYSTEM STYLE IS LIMITED MOBILIZATIONAL, OR NON-MOBILIZATIONAL (93)	0.93	92	LEAN LESS TOWARD BEING THOSE WHERE THE SYSTEM STYLE IS LIMITED MOBILIZATIONAL, OR NON-MOBILIZATIONAL (93)	0.79	2 18 26 67 X SQ= 1.97 P = 0.099 RV NO (NO)
95	ALWAYS ARE THOSE WHERE THE STATUS OF THE REGIME IS CONSTITUTIONAL OR AUTHORITARIAN (95)	1.00	95	TEND LESS TO BE THOSE WHERE THE STATUS OF THE REGIME IS CONSTITUTIONAL OR AUTHORITARIAN (95)	0.81	27 68 0 16 X SQ= 4.56 P = 0.011 RV NO (NO)
96	TEND TO BE THOSE WHERE THE STATUS OF THE REGIME IS AUTHORITARIAN (23)	0.55	96	TEND TO BE THOSE WHERE THE STATUS OF THE REGIME IS CONSTITUTIONAL OR TOTALITARIAN (67)	0.78	5 62 6 17 X SQ= 3.94 P = 0.029 RV YES (NO)
97	ALWAYS ARE THOSE WHERE THE STATUS OF THE REGIME IS AUTHORITARIAN, RATHER THAN TOTALITARIAN (23)	1.00	97	TEND LESS TO BE THOSE WHERE THE STATUS OF THE REGIME IS AUTHORITARIAN, RATHER THAN TOTALITARIAN (23)	0.52	6 17 0 16 X SQ= 3.13 P = 0.033 RV NO (NO)
98	LEAN MORE TOWARD BEING THOSE WHERE GOVERNMENTAL STABILITY IS GENERALLY OR MODERATELY PRESENT AND DATES FROM THE POST-WAR PERIOD, OR IS ABSENT (93)	0.93	98	LEAN LESS TOWARD BEING THOSE WHERE GOVERNMENTAL STABILITY IS GENERALLY OR MODERATELY PRESENT AND DATES FROM THE POST-WAR PERIOD, OR IS ABSENT (93)	0.77	2 20 26 67 X SQ= 2.49 P = 0.095 RV NO (NO)
101	TEND TO BE THOSE WHERE THE REPRESENTATIVE CHARACTER OF THE REGIME IS LIMITED POLYARCHIC, PSEUDO-POLYARCHIC, OR NON-POLYARCHIC (57)	0.90	101	TEND TO BE THOSE WHERE THE REPRESENTATIVE CHARACTER OF THE REGIME IS POLYARCHIC (41)	0.50	2 39 18 39 X SQ= 8.89 P = 0.001 RV YES (NO)
105	TEND TO BE THOSE WHERE THE ELECTORAL SYSTEM IS PARTIALLY COMPETITIVE OR NON-COMPETITIVE (47)	0.86	105	TEND TO BE THOSE WHERE THE ELECTORAL SYSTEM IS COMPETITIVE (43)	0.59	3 40 19 28 X SQ= 11.85 P = 0. RV YES (NO)
107	TEND TO BE THOSE WHERE AUTONOMOUS GROUPS ARE PARTIALLY TOLERATED IN POLITICS, ARE TOLERATED ONLY OUTSIDE POLITICS, OR ARE NOT TOLERATED AT ALL (65)	0.89	107	TEND TO BE THOSE WHERE AUTONOMOUS GROUPS ARE FULLY TOLERATED IN POLITICS (46)	0.51	3 43 24 41 X SQ= 11.92 P = 0. RV YES (NO)

110	ALWAYS ARE THOSE WHERE POLITICAL ENCULTURATION IS MEDIUM OR LOW (80)	1.00	110	TEND LESS TO BE THOSE WHERE POLITICAL ENCULTURATION IS MEDIUM OR LOW (80)	0.79	0 15 25 55 X SQ= 4.85 P = 0.009 RV NO (NO)
114	TEND TO BE THOSE WHERE SECTIONALISM IS EXTREME OR MODERATE (61)	0.77	114	TEND TO BE THOSE WHERE SECTIONALISM IS NEGLIGIBLE (47)	0.50	20 41 6 41 X SQ= 4.78 P = 0.022 RV YES (NO)
117	TEND TO BE THOSE WHERE INTEREST ARTICULATION BY ASSOCIATIONAL GROUPS IS NEGLIGIBLE (51)	0.96	117	TEND TO BE THOSE WHERE INTEREST ARTICULATION BY ASSOCIATIONAL GROUPS IS SIGNIFICANT, MODERATE, OR LIMITED (60)	0.71	1 59 27 24 X SQ= 35.75 P = 0. RV YES (YES)
119	TILT MORE TOWARD BEING THOSE WHERE INTEREST ARTICULATION BY INSTITUTIONAL GROUPS IS VERY SIGNIFICANT OR SIGNIFICANT (74)	0.93	119	TILT LESS TOWARD BEING THOSE WHERE INTEREST ARTICULATION BY INSTITUTIONAL GROUPS IS VERY SIGNIFICANT OR SIGNIFICANT (74)	0.71	13 61 1 25 X SQ= 1.98 P = 0.107 RV NO (NO)
121	TEND TO BE THOSE WHERE INTEREST ARTICULATION BY NON-ASSOCIATIONAL GROUPS IS SIGNIFICANT (54)	0.96	121	TEND TO BE THOSE WHERE INTEREST ARTICULATION BY NON-ASSOCIATIONAL GROUPS IS MODERATE, LIMITED, OR NEGLIGIBLE (61)	0.69	27 27 1 60 X SQ= 33.79 P = 0. RV YES (YES)
122	ALWAYS ARE THOSE WHERE INTEREST ARTICULATION BY NON-ASSOCIATIONAL GROUPS IS SIGNIFICANT OR MODERATE (83)	1.00	122	TEND LESS TO BE THOSE WHERE INTEREST ARTICULATION BY NON-ASSOCIATIONAL GROUPS IS SIGNIFICANT OR MODERATE (83)	0.63	28 55 0 32 X SQ= 12.50 P = 0. RV NO (NO)
125	TEND MORE TO BE THOSE WHERE INTEREST ARTICULATION BY ANOMIC GROUPS IS FREQUENT OR OCCASIONAL (64)	0.86	125	TEND LESS TO BE THOSE WHERE INTEREST ARTICULATION BY ANOMIC GROUPS IS FREQUENT OR OCCASIONAL (64)	0.56	24 40 4 31 X SQ= 6.35 P = 0.006 RV NO (NO)
126	ALWAYS ARE THOSE WHERE INTEREST ARTICULATION BY ANOMIC GROUPS IS FREQUENT, OCCASIONAL, OR INFREQUENT (83)	1.00	126	TEND LESS TO BE THOSE WHERE INTEREST ARTICULATION BY ANOMIC GROUPS IS FREQUENT, OCCASIONAL, OR INFREQUENT (83)	0.77	28 55 0 16 X SQ= 5.95 P = 0.005 RV NO (NO)
128	TEND TO BE THOSE WHERE INTEREST ARTICULATION BY POLITICAL PARTIES IS LIMITED OR NEGLIGIBLE (45)	0.68	128	TEND TO BE THOSE WHERE INTEREST ARTICULATION BY POLITICAL PARTIES IS SIGNIFICANT OR MODERATE (48)	0.59	8 40 17 28 X SQ= 4.25 P = 0.034 RV YES (NO)

129	LEAN TOWARD BEING THOSE WHERE INTEREST ARTICULATION BY POLITICAL PARTIES IS NEGLIGIBLE (37)	0.56	129	LEAN TOWARD BEING THOSE WHERE INTEREST ARTICULATION BY POLITICAL PARTIES IS SIGNIFICANT, MODERATE, OR LIMITED (56)	0.66	11 45 14 23 X SQ= 2.88 P = 0.060 RV YES (NO)
131	TEND TO BE THOSE WHERE INTEREST AGGREGATION BY POLITICAL PARTIES IS SIGNIFICANT OR MODERATE (30)	0.80	131	TEND TO BE THOSE WHERE INTEREST AGGREGATION BY POLITICAL PARTIES IS LIMITED OR NEGLIGIBLE (35)	0.60	8 22 2 33 X SQ= 3.96 P = 0.035 RV YES (NO)
132	ALWAYS ARE THOSE WHERE INTEREST AGGREGATION BY POLITICAL PARTIES IS SIGNIFICANT, MODERATE, OR LIMITED (67)	1.00	132	TEND LESS TO BE THOSE WHERE INTEREST AGGREGATION BY POLITICAL PARTIES IS SIGNIFICANT, MODERATE, OR LIMITED (67)	0.82	25 42 0 9 X SQ= 3.46 P = 0.026 RV NO (NO)
133	TEND TO BE THOSE WHERE INTEREST AGGREGATION BY THE EXECUTIVE IS SIGNIFICANT (29)	0.59	133	TEND TO BE THOSE WHERE INTEREST AGGREGATION BY THE EXECUTIVE IS MODERATE, LIMITED, OR NEGLIGIBLE (76)	0.83	16 13 11 65 X SQ= 16.13 P = 0. RV YES (YES)
134	TEND TO BE THOSE WHERE INTEREST AGGREGATION BY THE EXECUTIVE IS SIGNIFICANT OR MODERATE (57)	0.73	134	TEND TO BE THOSE WHERE INTEREST AGGREGATION BY THE EXECUTIVE IS LIMITED OR NEGLIGIBLE (46)	0.51	19 38 7 39 X SQ= 3.52 P = 0.042 RV YES (NO)
137	ALWAYS ARE THOSE WHERE INTEREST AGGREGATION BY THE LEGISLATURE IS LIMITED OR NEGLIGIBLE (68)	1.00	137	TEND LESS TO BE THOSE WHERE INTEREST AGGREGATION BY THE LEGISLATURE IS LIMITED OR NEGLIGIBLE (68)	0.59	0 29 26 42 X SQ= 13.26 P = 0. RV NO (NO)
138	TEND TO BE THOSE WHERE INTEREST AGGREGATION BY THE LEGISLATURE IS NEGLIGIBLE (49)	0.85	138	TEND TO BE THOSE WHERE INTEREST AGGREGATION BY THE LEGISLATURE IS SIGNIFICANT, MODERATE, OR LIMITED (48)	0.62	4 44 22 27 X SQ= 14.71 P = 0. RV YES (NO)
139	TEND TO BE THOSE WHERE THE PARTY SYSTEM IS QUANTITATIVELY ONE-PARTY (34)	0.52	139	TEND TO BE THOSE WHERE THE PARTY SYSTEM IS QUANTITATIVELY OTHER THAN ONE-PARTY (71)	0.74	14 20 13 58 X SQ= 5.15 P = 0.017 RV YES (NO)
146	TEND LESS TO BE THOSE WHERE THE PARTY SYSTEM IS QUALITATIVELY OTHER THAN MASS-BASED TERRITORIAL (92)	0.81	146	TEND MORE TO BE THOSE WHERE THE PARTY SYSTEM IS QUALITATIVELY OTHER THAN MASS-BASED TERRITORIAL (92)	0.96	5 3 22 70 X SQ= 3.77 P = 0.032 RV NO (YES)

147	ALWAYS ARE THOSE WHERE THE PARTY SYSTEM IS QUALITATIVELY OTHER THAN CLASS-ORIENTED OR MULTI-IDEOLOGICAL (67)	1.00	147	TEND LESS TO BE THOSE WHERE THE PARTY SYSTEM IS QUALITATIVELY OTHER THAN CLASS-ORIENTED OR MULTI-IDEOLOGICAL (67)	0.65	0 23 25 42 X SQ= 10.10 P = 0. RV NO (NO)
148	TEND TO BE THOSE WHERE THE PARTY SYSTEM IS QUALITATIVELY AFRICAN TRANSITIONAL (14)	0.52	148	ALWAYS ARE THOSE WHERE THE PARTY SYSTEM IS QUALITATIVELY OTHER THAN AFRICAN TRANSITIONAL (96)	1.00	14 0 13 83 X SQ= 44.76 P = 0. RV YES (YES)
157	ALWAYS ARE THOSE WHERE PERSONALISMO IS MODERATE OR NEGLIGIBLE (84)	1.00	157	TEND LESS TO BE THOSE WHERE PERSONALISMO IS MODERATE OR NEGLIGIBLE (84)	0.81	0 14 25 59 X SQ= 4.14 P = 0.018 RV NO (NO)
158	LEAN TOWARD BEING THOSE WHERE PERSONALISMO IS PRONOUNCED OR MODERATE (40)	0.61	158	LEAN TOWARD BEING THOSE WHERE PERSONALISMO IS NEGLIGIBLE (56)	0.64	14 26 9 47 X SQ= 3.61 P = 0.051 RV YES (NO)
161	TILT TOWARD BEING THOSE WHERE THE POLITICAL LEADERSHIP IS NON-ELITIST (50)	0.67	161	TILT TOWARD BEING THOSE WHERE THE POLITICAL LEADERSHIP IS ELITIST OR MODERATELY ELITIST (47)	0.53	8 39 16 34 X SQ= 2.17 P = 0.103 RV YES (NO)
164	TEND TO BE THOSE WHERE THE REGIME'S LEADERSHIP CHARISMA IS PRONOUNCED OR MODERATE (34)	0.73	164	TEND TO BE THOSE WHERE THE REGIME'S LEADERSHIP CHARISMA IS NEGLIGIBLE (65)	0.77	16 18 6 59 X SQ= 16.36 P = 0. RV YES (NO)
169	LEAN TOWARD BEING THOSE WHERE THE HORIZONTAL POWER DISTRIBUTION IS NEGLIGIBLE (48)	0.62	169	LEAN TOWARD BEING THOSE WHERE THE HORIZONTAL POWER DISTRIBUTION IS SIGNIFICANT OR LIMITED (58)	0.60	10 48 16 32 X SQ= 2.86 P = 0.071 RV YES (NO)
170	TEND TO BE THOSE WHERE THE LEGISLATIVE-EXECUTIVE STRUCTURE IS PRESIDENTIAL (39)	0.60	170	TEND TO BE THOSE WHERE THE LEGISLATIVE-EXECUTIVE STRUCTURE IS OTHER THAN PRESIDENTIAL (63)	0.69	15 24 10 53 X SQ= 5.48 P = 0.017 RV YES (NO)
174	TEND MORE TO BE THOSE WHERE THE LEGISLATURE IS PARTIALLY EFFECTIVE, LARGELY INEFFECTIVE, OR WHOLLY INEFFECTIVE (72)	0.96	174	TEND LESS TO BE THOSE WHERE THE LEGISLATURE IS PARTIALLY EFFECTIVE, LARGELY INEFFECTIVE, OR WHOLLY INEFFECTIVE (72)	0.64	1 27 24 48 X SQ= 8.00 P = 0.002 RV NO (NO)

175	TEND TO BE THOSE WHERE THE LEGISLATURE IS LARGELY INEFFECTIVE OR WHOLLY INEFFECTIVE (49)	0.76	175	TEND TO BE THOSE WHERE THE LEGISLATURE IS FULLY EFFECTIVE OR PARTIALLY EFFECTIVE (51)	0.60	6 45 19 30 X SQ= 8.34 P = 0.002, RV YES (NO)
176	TEND MORE TO BE THOSE WHERE THE LEGISLATURE IS FULLY EFFECTIVE, PARTIALLY EFFECTIVE, OR LARGELY INEFFECTIVE (72)	0.88	176	TEND LESS TO BE THOSE WHERE THE LEGISLATURE IS FULLY EFFECTIVE, PARTIALLY EFFECTIVE, OR LARGELY INEFFECTIVE (72)	0.67	22 50 3 25 X SQ= 3.24 P = 0.043 RV NO (NO)
178	TEND TO BE THOSE WHERE THE LEGISLATURE IS UNICAMERAL (53)	0.81	178	TEND TO BE THOSE WHERE THE LEGISLATURE IS BICAMERAL (51)	0.60	22 31 5 46 X SQ= 11.99 P = 0. RV YES (NO)
179	TEND MORE TO BE THOSE WHERE THE EXECUTIVE IS DOMINANT (52)	0.79	179	TEND LESS TO BE THOSE WHERE THE EXECUTIVE IS DOMINANT (52)	0.51	15 37 4 35 X SQ= 3.60 P = 0.038 RV NO (NO)
182	TEND TO BE THOSE WHERE THE BUREAUCRACY IS POST-COLONIAL TRANSITIONAL, RATHER THAN SEMI-MODERN (25)	0.88	182	TEND TO BE THOSE WHERE THE BUREAUCRACY IS SEMI-MODERN, RATHER THAN POST-COLONIAL TRANSITIONAL (55)	0.96	3 52 23 2 X SQ= 54.80 P = 0. RV YES (YES)
183	TEND TO BE THOSE WHERE THE BUREAUCRACY IS POST-COLONIAL TRANSITIONAL, RATHER THAN TRADITIONAL (25)	0.96	183	TEND TO BE THOSE WHERE THE BUREAUCRACY IS TRADITIONAL, RATHER THAN POST-COLONIAL TRANSITIONAL (9)	0.80	23 2 1 8 X SQ= 17.14 P = 0. RV YES (YES)
188	ALWAYS ARE THOSE WHERE THE CHARACTER OF THE LEGAL SYSTEM IS OTHER THAN CIVIL LAW (81)	1.00	188	TEND LESS TO BE THOSE WHERE THE CHARACTER OF THE LEGAL SYSTEM IS OTHER THAN CIVIL LAW (81)	0.63	0 32 26 55 X SQ= 11.59 P = 0. RV NO (NO)
192	TEND TO BE THOSE WHERE THE CHARACTER OF THE LEGAL SYSTEM IS MUSLIM OR PARTLY MUSLIM (27)	0.54	192	TEND TO BE THOSE WHERE THE CHARACTER OF THE LEGAL SYSTEM IS OTHER THAN MUSLIM OR PARTLY MUSLIM (86)	0.85	14 13 12 74 X SQ= 14.59 P = 0. RV YES (YES)
193	TEND TO BE THOSE WHERE THE CHARACTER OF THE LEGAL SYSTEM IS PARTLY INDIGENOUS (28)	0.89	193	TEND TO BE THOSE WHERE THE CHARACTER OF THE LEGAL SYSTEM IS OTHER THAN PARTLY INDIGENOUS (86)	0.97	25 3 3 83 X SQ= 79.35 P = 0. RV YES (YES)

11/

```
194  ALWAYS ARE THOSE                        1.00      194  TEND LESS TO BE THOSE                0.85        0   13
     THAT ARE NON-COMMUNIST  (101)                          THAT ARE NON-COMMUNIST  (101)                  28   73
                                                                                                        X SQ=   3.40
                                                                                                        P =    0.036
                                                                                                        RV NO  (NO )
```

12 POLITIES LOCATED IN NORTH AFRICA (5)

12 POLITIES LOCATED ELSEWHERE THAN IN NORTH AFRICA (110)

SUBJECT ONLY

5 IN LEFT COLUMN

ALGERIA	LIBYA	MOROCCO	TUNISIA	UAR

110 IN RIGHT COLUMN

AFGHANISTAN	ALBANIA	ARGENTINA	AUSTRALIA	AUSTRIA	BELGIUM	BOLIVIA	BRAZIL	BULGARIA	BURMA
BURUNDI	CAMBODIA	CAMEROUN	CANADA	CEN AFR REP	CEYLON	CHAD	CHILE	CHINA, PR	COLOMBIA
CONGO(BRA)	CONGO(LEO)	COSTA RICA	CUBA	CYPRUS	CZECHOS'KIA	DAHOMEY	DENMARK	DOMIN REP	ECUADOR
EL SALVADOR	ETHIOPIA	FINLAND	FRANCE	GABON	GERMANY, E	GERMAN FR	GHANA	GREECE	GUATEMALA
GUINEA	HAITI	HONDURAS	HUNGARY	ICELAND	INDIA	INDONESIA	IRAN	IRAQ	IRELAND
ISRAEL	ITALY	IVORY COAST	JAMAICA	JAPAN	JORDAN	KOREA, N	KOREA REP	LAOS	LEBANON
LIBERIA	LUXEMBOURG	MALAGASY R	MALAYA	MALI	MAURITANIA	MEXICO	MONGOLIA	NEPAL	NETHERLANDS
NEW ZEALAND	NICARAGUA	NIGER	NIGERIA	NORWAY	PAKISTAN	PANAMA	PARAGUAY	PERU	PHILIPPINES
POLAND	PORTUGAL	RUMANIA	RWANDA	SA'U ARABIA	SENEGAL	SIERRE LEO	SOMALIA	SO AFRICA	SPAIN
SUDAN	SWEDEN	SWITZERLAND	SYRIA	TANGANYIKA	THAILAND	TOGO	TRINIDAD	TURKEY	UGANDA
USSR	UK	US	UPPER VOLTA	URUGUAY	VENEZUELA	VIETNAM, N	VIETNAM REP	YEMEN	YUGOSLAVIA

	Left column			Right column		
21	TILT TOWARD BEING THOSE WHOSE TERRITORIAL SIZE IS VERY LARGE OR LARGE (32)	0.60	21	TILT TOWARD BEING THOSE WHOSE TERRITORIAL SIZE IS MEDIUM OR SMALL (83)	0.74	3 29 2 81 X SQ= 1.28 P = 0.131 RV YES (NO)
26	ALWAYS ARE THOSE WHOSE POPULATION DENSITY IS LOW (67)	1.00	26	LEAN LESS TOWARD BEING THOSE WHOSE POPULATION DENSITY IS LOW (67)	0.56	0 48 5 62 X SQ= 2.17 P = 0.074 RV NO (NO)
28	ALWAYS ARE THOSE WHOSE POPULATION GROWTH RATE IS HIGH (62)	1.00	28	LEAN LESS TOWARD BEING THOSE WHOSE POPULATION GROWTH RATE IS HIGH (62)	0.54	5 57 0 48 X SQ= 2.41 P = 0.067 RV NO (NO)
45	ALWAYS ARE THOSE WHERE THE LITERACY RATE IS BELOW FIFTY PERCENT (54)	1.00	45	TEND TO BE THOSE WHERE THE LITERACY RATE IS FIFTY PERCENT OR ABOVE (55)	0.53	0 55 5 49 X SQ= 3.43 P = 0.027 RV YES (NO)

```
49  ALWAYS ARE THOSE                        1.00      49  TEND TO BE THOSE                         0.57        4   20
    WHERE THE LITERACY RATE IS                            WHERE THE LITERACY RATE IS                           0   26
    BETWEEN TEN AND FIFTY PERCENT,                        BELOW TEN PERCENT, RATHER THAN               X SQ=   2.72
    RATHER THAN BELOW TEN PERCENT  (24)                   BETWEEN TEN AND FIFTY PERCENT  (26)          P =    0.046
                                                                                                       RV YES (NO )

50  ALWAYS ARE THOSE                        1.00      50  LEAN LESS TOWARD BEING THOSE             0.54        0   43
    WHERE FREEDOM OF THE PRESS IS                         WHERE FREEDOM OF THE PRESS IS                        5   51
    INTERMITTENT, INTERNALLY ABSENT, OR                   INTERMITTENT, INTERNALLY ABSENT, OR          X SQ=   2.40
    INTERNALLY AND EXTERNALLY ABSENT  (56)                INTERNALLY AND EXTERNALLY ABSENT  (56)       P =    0.067
                                                                                                       RV NO  (NO )

58  ALWAYS ARE THOSE                        1.00      58  TEND TO BE THOSE                         0.88        5   13
    WHERE THE RELIGION IS                                 WHERE THE RELIGION IS OTHER THAN                     0   97
    MUSLIM  (18)                                          MUSLIM  (97)                                 X SQ=  21.89
                                                                                                       P =    0.
                                                                                                       RV YES (NO )

66  ALWAYS ARE THOSE                        1.00      66  LEAN LESS TOWARD BEING THOSE             0.51        5   52
    THAT ARE RELIGIOUSLY HOMOGENEOUS  (57)                THAT ARE RELIGIOUSLY HOMOGENEOUS  (57)               0   49
                                                                                                       X SQ=   2.77
                                                                                                       P =    0.060
                                                                                                       RV NO  (NO )

70  LEAN TOWARD BEING THOSE                 0.60      70  LEAN TOWARD BEING THOSE                  0.82        3   18
    THAT ARE RELIGIOUSLY, RACIALLY,                       THAT ARE RELIGIOUSLY, RACIALLY,                      2   82
    AND LINGUISTICALLY HOMOGENEOUS  (21)                  OR LINGUISTICALLY HETEROGENEOUS  (85)        X SQ=   2.95
                                                                                                       P =    0.053
                                                                                                       RV YES (NO )

73  ALWAYS ARE THOSE                        1.00      73  TEND TO BE THOSE                         0.61        0   65
    WHOSE DATE OF INDEPENDENCE                            WHOSE DATE OF INDEPENDENCE                           5   41
    IS AFTER 1945  (46)                                   IS BEFORE 1945  (65)                         X SQ=   5.09
                                                                                                       P =    0.011
                                                                                                       RV YES (NO )

82  ALWAYS ARE THOSE                        1.00      82  LEAN TOWARD BEING THOSE                  0.50        0   51
    WHERE THE TYPE OF POLITICAL MODERNIZATION             WHERE THE TYPE OF POLITICAL MODERNIZATION            5   50
    IS DEVELOPED TUTELARY OR                              IS EARLY OR LATER EUROPEAN OR                X SQ=   3.05
    UNDEVELOPED TUTELARY, RATHER THAN                     EUROPEAN DERIVED, RATHER THAN                P =    0.058
    EARLY OR LATER EUROPEAN OR                            DEVELOPED TUTELARY OR                        RV YES (NO )
    EUROPEAN DERIVED  (55)                                UNDEVELOPED TUTELARY  (51)

84  ALWAYS ARE THOSE                        1.00      84  LEAN LESS TOWARD BEING THOSE             0.52        5   26
    WHERE THE TYPE OF POLITICAL MODERNIZATION             WHERE THE TYPE OF POLITICAL MODERNIZATION            0   24
    IS DEVELOPED TUTELARY, RATHER THAN                    IS DEVELOPED TUTELARY, RATHER THAN           X SQ=   2.53
    UNDEVELOPED TUTELARY  (31)                            UNDEVELOPED TUTELARY  (31)                   P =    0.061
                                                                                                       RV NO  (NO )

87  ALWAYS ARE THOSE                        1.00      87  TEND TO BE THOSE                         0.67        3   28
    WHOSE IDEOLOGICAL ORIENTATION                         WHOSE IDEOLOGICAL ORIENTATION                        0   58
    IS DEVELOPMENTAL  (31)                                IS OTHER THAN DEVELOPMENTAL  (58)            X SQ=   3.22
                                                                                                       P =    0.040
                                                                                                       RV YES (NO )
```

93	TEND TO BE THOSE WHERE THE SYSTEM STYLE IS MOBILIZATIONAL, OR LIMITED MOBILIZATIONAL (32)	0.80	93	TEND TO BE THOSE WHERE THE SYSTEM STYLE IS NON-MOBILIZATIONAL (78)	0.73	4 28 1 77 X SQ= 4.25 P = 0.025 RV YES (NO)
99	ALWAYS ARE THOSE WHERE GOVERNMENTAL STABILITY IS GENERALLY PRESENT AND DATES FROM AT LEAST THE INTER-WAR PERIOD, OR FROM THE POST-WAR PERIOD (50)	1.00	99	TILT LESS TOWARD BEING THOSE WHERE GOVERNMENTAL STABILITY IS GENERALLY PRESENT AND DATES FROM AT LEAST THE INTER-WAR PERIOD, OR FROM THE POST-WAR PERIOD (50)	0.56	4 46 0 36 X SQ= 1.49 P = 0.136 RV NO (NO)
106	ALWAYS ARE THOSE WHERE THE ELECTORAL SYSTEM IS NON-COMPETITIVE (30)	1.00	106	TILT TOWARD BEING THOSE WHERE THE ELECTORAL SYSTEM IS COMPETITIVE OR PARTIALLY COMPETITIVE (52)	0.65	0 52 2 28 X SQ= 1.30 P = 0.131 RV YES (NO)
107	ALWAYS ARE THOSE WHERE AUTONOMOUS GROUPS ARE PARTIALLY TOLERATED IN POLITICS, ARE TOLERATED ONLY OUTSIDE POLITICS, OR ARE NOT TOLERATED AT ALL (65)	1.00	107	LEAN LESS TOWARD BEING THOSE WHERE AUTONOMOUS GROUPS ARE PARTIALLY TOLERATED IN POLITICS, ARE TOLERATED ONLY OUTSIDE POLITICS, OR ARE NOT TOLERATED AT ALL (65)	0.57	0 46 5 60 X SQ= 2.13 P = 0.075 RV NO (NO)
111	ALWAYS ARE THOSE WHERE POLITICAL ENCULTURATION IS HIGH OR MEDIUM (53)	1.00	111	TILT LESS TOWARD BEING THOSE WHERE POLITICAL ENCULTURATION IS HIGH OR MEDIUM (53)	0.54	4 49 0 42 X SQ= 1.70 P = 0.127 RV NO (NO)
121	ALWAYS ARE THOSE WHERE INTEREST ARTICULATION BY NON-ASSOCIATIONAL GROUPS IS MODERATE, LIMITED, OR NEGLIGIBLE (61)	1.00	121	LEAN LESS TOWARD BEING THOSE WHERE INTEREST ARTICULATION BY NON-ASSOCIATIONAL GROUPS IS MODERATE, LIMITED, OR NEGLIGIBLE (61)	0.51	0 54 5 56 X SQ= 2.87 P = 0.059 RV NO (NO)
125	TILT TOWARD BEING THOSE WHERE INTEREST ARTICULATION BY ANOMIC GROUPS IS INFREQUENT OR VERY INFREQUENT (35)	0.75	125	TILT TOWARD BEING THOSE WHERE INTEREST ARTICULATION BY ANOMIC GROUPS IS FREQUENT OR OCCASIONAL (64)	0.66	1 63 3 32 X SQ= 1.34 P = 0.125 RV YES (NO)
133	LEAN TOWARD BEING THOSE WHERE INTEREST AGGREGATION BY THE EXECUTIVE IS SIGNIFICANT (29)	0.75	133	LEAN TOWARD BEING THOSE WHERE INTEREST AGGREGATION BY THE EXECUTIVE IS MODERATE, LIMITED, OR NEGLIGIBLE (76)	0.74	3 26 1 75 X SQ= 2.53 P = 0.063 RV YES (NO)
146	ALWAYS ARE THOSE WHERE THE PARTY SYSTEM IS QUALITATIVELY MASS-BASED TERRITORIAL (8)	1.00	146	TEND TO BE THOSE WHERE THE PARTY SYSTEM IS QUALITATIVELY OTHER THAN MASS-BASED TERRITORIAL (92)	0.95	3 5 0 92 X SQ= 23.85 P = 0. RV YES (NO)

164	ALWAYS ARE THOSE WHERE THE REGIME'S LEADERSHIP CHARISMA IS PRONOUNCED OR MODERATE (34)	1.00	164	TEND TO BE THOSE WHERE THE REGIME'S LEADERSHIP CHARISMA IS NEGLIGIBLE (65)	0.69	5 29 0 65 X SQ= 7.23 P = 0.004 RV YES (NO)
192	ALWAYS ARE THOSE WHERE THE CHARACTER OF THE LEGAL SYSTEM IS MUSLIM OR PARTLY MUSLIM (27)	1.00	192	TEND TO BE THOSE WHERE THE CHARACTER OF THE LEGAL SYSTEM IS OTHER THAN MUSLIM OR PARTLY MUSLIM (86)	0.80	5 22 0 86 X SQ= 12.57 P = 0.001 RV YES (NO)

```
************************************************************************************************************

13  POLITIES                                          13  POLITIES
    LOCATED IN NORTH AFRICA OR                            LOCATED ELSEWHERE THAN IN NORTH AFRICA OR
    THE MIDDLE EAST  (16)                                 THE MIDDLE EAST  (99)


                                                                                BOTH SUBJECT AND PREDICATE
............................................................................................................

     16 IN LEFT COLUMN

AFGHANISTAN ALGERIA     CYPRUS      IRAN        IRAQ        ISRAEL      JORDAN      LEBANON     LIBYA       MOROCCO
SA'U ARABIA SYRIA       TUNISIA     TURKEY      UAR         YEMEN

     99 IN RIGHT COLUMN

ALBANIA     ARGENTINA   AUSTRALIA   AUSTRIA     BELGIUM     BOLIVIA     BRAZIL      BULGARIA    BURMA       BURUNDI
CAMBODIA    CAMEROUN    CANADA      CEN AFR REP CEYLON      CHAD        CHILE       CHINA, PR   COLOMBIA    CONGO(BRA)
CONGO(LEO)  COSTA RICA  CUBA        CZECHOS'KIA DAHOMEY     DENMARK     DOMIN REP   ECUADOR     EL SALVADOR ETHIOPIA
FINLAND     FRANCE      GABON       GERMANY, E  GERMAN FR   GHANA       GREECE      GUATEMALA   GUINEA      HAITI
HONDURAS    HUNGARY     ICELAND     INDIA       INDONESIA   IRELAND     ITALY       IVORY COAST JAMAICA     JAPAN
KOREA, N    KOREA REP   LAOS        LIBERIA     LUXEMBOURG  MALAGASY R  MALAYA      MALI        MAURITANIA  MEXICO
MONGOLIA    NEPAL       NETHERLANDS NEW ZEALAND NICARAGUA   NIGER       NIGERIA     NORWAY      PAKISTAN    PANAMA
PARAGUAY    PERU        PHILIPPINES POLAND      PORTUGAL    RUMANIA     RWANDA      SENEGAL     SIERRE LEO  SOMALIA
SO AFRICA   SPAIN       SUDAN       SWEDEN      SWITZERLAND TANGANYIKA  THAILAND    TOGO        TRINIDAD    UGANDA
USSR        UK          US          UPPER VOLTA URUGUAY     VENEZUELA   VIETNAM, N  VIETNAM REP YUGOSLAVIA
------------------------------------------------------------------------------------------------------------

 26  LEAN MORE TOWARD BEING THOSE           0.81        26  LEAN LESS TOWARD BEING THOSE          0.55          3   45
     WHOSE POPULATION DENSITY IS                            WHOSE POPULATION DENSITY IS                        13   54
     LOW  (67)                                              LOW  (67)                                      X SQ=   3.02
                                                                                                           P =    0.057
                                                                                                           RV NO  (NO )

 28  TEND MORE TO BE THOSE                  0.81        28  TEND LESS TO BE THOSE                 0.52         13   49
     WHOSE POPULATION GROWTH RATE                           WHOSE POPULATION GROWTH RATE                        3   45
     IS HIGH  (62)                                          IS HIGH  (62)                                  X SQ=   3.61
                                                                                                           P =    0.033
                                                                                                           RV NO  (NO )

 33  LEAN MORE TOWARD BEING THOSE           0.94        33  LEAN LESS TOWARD BEING THOSE          0.71          1   29
     WHOSE GROSS NATIONAL PRODUCT                           WHOSE GROSS NATIONAL PRODUCT                       15   70
     IS LOW OR VERY LOW  (85)                               IS LOW OR VERY LOW  (85)                       X SQ=   2.69
                                                                                                           P =    0.066
                                                                                                           RV NO  (NO )

 39  LEAN MORE TOWARD BEING THOSE           0.87        39  LEAN LESS TOWARD BEING THOSE          0.63          2   36
     WHOSE INTERNATIONAL FINANCIAL STATUS                   WHOSE INTERNATIONAL FINANCIAL STATUS               14   62
     IS LOW OR VERY LOW  (76)                               IS LOW OR VERY LOW  (76)                       X SQ=   2.63
                                                                                                           P =    0.084
                                                                                                           RV NO  (NO )
```

41	ALWAYS ARE THOSE WHOSE ECONOMIC DEVELOPMENTAL STATUS IS INTERMEDIATE, UNDERDEVELOPED, OR VERY UNDERDEVELOPED (94)	1.00	41	LEAN LESS TOWARD BEING THOSE WHOSE ECONOMIC DEVELOPMENTAL STATUS IS INTERMEDIATE, UNDERDEVELOPED, OR VERY UNDERDEVELOPED (94)	0.81	0 19 15 79 X SQ= 2.25 P = 0.071 RV NO (NO)
42	TEND MORE TO BE THOSE WHOSE ECONOMIC DEVELOPMENTAL STATUS IS UNDERDEVELOPED OR VERY UNDERDEVELOPED (76)	0.93	42	TEND LESS TO BE THOSE WHOSE ECONOMIC DEVELOPMENTAL STATUS IS UNDERDEVELOPED OR VERY UNDERDEVELOPED (76)	0.64	1 35 13 63 X SQ= 3.37 P = 0.035 RV NO (NO)
44	ALWAYS ARE THOSE WHERE THE LITERACY RATE IS BELOW NINETY PERCENT (90)	1.00	44	TEND LESS TO BE THOSE WHERE THE LITERACY RATE IS BELOW NINETY PERCENT (90)	0.75	0 25 16 74 X SQ= 3.79 P = 0.021 RV NO (NO)
45	TEND TO BE THOSE WHERE THE LITERACY RATE IS BELOW FIFTY PERCENT (54)	0.87	45	TEND TO BE THOSE WHERE THE LITERACY RATE IS FIFTY PERCENT OR ABOVE (55)	0.56	2 53 13 41 X SQ= 7.95 P = 0.002 RV YES (NO)
50	TEND TO BE THOSE WHERE FREEDOM OF THE PRESS IS INTERMITTENT, INTERNALLY ABSENT, OR INTERNALLY AND EXTERNALLY ABSENT (56)	0.94	50	TEND TO BE THOSE WHERE FREEDOM OF THE PRESS IS COMPLETE (43)	0.51	1 42 15 41 X SQ= 9.01 P = 0.001 RV YES (NO)
54	LEAN MORE TOWARD BEING THOSE WHERE NEWSPAPER CIRCULATION IS LESS THAN ONE HUNDRED PER THOUSAND (76)	0.87	54	LEAN LESS TOWARD BEING THOSE WHERE NEWSPAPER CIRCULATION IS LESS THAN ONE HUNDRED PER THOUSAND (76)	0.64	2 35 14 62 X SQ= 2.48 P = 0.085 RV NO (NO)
56	ALWAYS ARE THOSE WHERE THE RELIGION IS PREDOMINANTLY LITERATE (79)	1.00	56	TEND LESS TO BE THOSE WHERE THE RELIGION IS PREDOMINANTLY LITERATE (79)	0.67	16 63 0 31 X SQ= 5.81 P = 0.005 RV NO (NO)
58	TEND TO BE THOSE WHERE THE RELIGION IS MUSLIM (18)	0.81	58	TEND TO BE THOSE WHERE THE RELIGION IS OTHER THAN MUSLIM (97)	0.95	13 5 3 94 X SQ= 54.94 P = 0. RV YES (YES)
63	ALWAYS ARE THOSE WHERE THE RELIGION IS PREDOMINANTLY OR PARTLY OTHER THAN CHRISTIAN (68)	1.00	63	TEND LESS TO BE THOSE WHERE THE RELIGION IS PREDOMINANTLY OR PARTLY OTHER THAN CHRISTIAN (68)	0.53	0 46 16 52 X SQ= 10.72 P = 0. RV NO (NO)

64	ALWAYS ARE THOSE WHERE THE RELIGION IS MUSLIM, RATHER THAN CHRISTIAN (18)	1.00	64	TEND TO BE THOSE WHERE THE RELIGION IS CHRISTIAN, RATHER THAN MUSLIM (46)	0.90	0 46 13 5 X SQ= 37.35 P = 0. RV YES (YES)
66	TEND TO BE THOSE THAT ARE RELIGIOUSLY HOMOGENEOUS (57)	0.87	66	TEND TO BE THOSE THAT ARE RELIGIOUSLY HETEROGENEOUS (49)	0.52	14 43 2 47 X SQ= 7.10 P = 0.005 RV YES (NO)
67	TILT MORE TOWARD BEING THOSE THAT ARE RACIALLY HOMOGENEOUS (82)	0.93	67	TILT LESS TOWARD BEING THOSE THAT ARE RACIALLY HOMOGENEOUS (82)	0.71	13 67 1 28 X SQ= 2.08 P = 0.107 RV NO (NO)
72	TEND TO BE THOSE WHOSE DATE OF INDEPENDENCE IS AFTER 1914 (59)	0.81	72	TEND TO BE THOSE WHOSE DATE OF INDEPENDENCE IS BEFORE 1914 (52)	0.52	3 49 13 46 X SQ= 4.68 P = 0.016 RV YES (NO)
76	ALWAYS ARE THOSE THAT HAVE BEEN SIGNIFICANTLY OR PARTIALLY WESTERNIZED THROUGH A COLONIAL RELATIONSHIP, RATHER THAN BEING HISTORICALLY WESTERN (70)	1.00	76	TEND LESS TO BE THOSE THAT HAVE BEEN SIGNIFICANTLY OR PARTIALLY WESTERNIZED THROUGH A COLONIAL RELATIONSHIP, RATHER THAN BEING HISTORICALLY WESTERN (70)	0.69	0 26 11 59 X SQ= 3.20 P = 0.032 RV NO (NO)
79	ALWAYS ARE THOSE THAT WERE FORMERLY DEPENDENCIES OF BRITAIN OR FRANCE, RATHER THAN SPAIN (49)	1.00	79	LEAN LESS TOWARD BEING THOSE THAT WERE FORMERLY DEPENDENCIES OF BRITAIN OR FRANCE, RATHER THAN SPAIN (49)	0.68	10 39 0 18 X SQ= 2.86 P = 0.052 RV NO (NO)
82	ALWAYS ARE THOSE WHERE THE TYPE OF POLITICAL MODERNIZATION IS DEVELOPED TUTELARY OR UNDEVELOPED TUTELARY, RATHER THAN EARLY OR LATER EUROPEAN OR EUROPEAN DERIVED (55)	1.00	82	TEND TO BE THOSE WHERE THE TYPE OF POLITICAL MODERNIZATION IS EARLY OR LATER EUROPEAN OR EUROPEAN DERIVED, RATHER THAN DEVELOPED TUTELARY OR UNDEVELOPED TUTELARY (51)	0.55	0 51 13 42 X SQ= 11.63 P = 0. RV YES (NO)
84	ALWAYS ARE THOSE WHERE THE TYPE OF POLITICAL MODERNIZATION IS DEVELOPED TUTELARY, RATHER THAN UNDEVELOPED TUTELARY (31)	1.00	84	TEND TO BE THOSE WHERE THE TYPE OF POLITICAL MODERNIZATION IS UNDEVELOPED TUTELARY, RATHER THAN DEVELOPED TUTELARY (24)	0.57	13 18 0 24 X SQ= 10.96 P = 0. RV YES (NO)
85	TILT TOWARD BEING THOSE WHERE THE STAGE OF POLITICAL MODERNIZATION IS ADVANCED, RATHER THAN MID- OR EARLY TRANSITIONAL (60)	0.73	85	TILT TOWARD BEING THOSE WHERE THE STAGE OF POLITICAL MODERNIZATION IS MID- OR EARLY TRANSITIONAL, RATHER THAN ADVANCED (54)	0.51	11 49 4 50 X SQ= 2.09 P = 0.102 RV YES (NO)

91	TEND LESS TO BE THOSE WHOSE IDEOLOGICAL ORIENTATION IS DEVELOPMENTAL, RATHER THAN TRADITIONAL (31)	0.57	91	TEND MORE TO BE THOSE WHOSE IDEOLOGICAL ORIENTATION IS DEVELOPMENTAL, RATHER THAN TRADITIONAL (31)	0.93	4 27 3 2 X SQ= 3.46 P = 0.040 RV NO (YES)
95	ALWAYS ARE THOSE WHERE THE STATUS OF THE REGIME IS CONSTITUTIONAL OR AUTHORITARIAN (95)	1.00	95	TILT LESS TOWARD BEING THOSE WHERE THE STATUS OF THE REGIME IS CONSTITUTIONAL OR AUTHORITARIAN (95)	0.83	15 80 0 16 X SQ= 1.73 P = 0.122 RV NO (NO)
97	ALWAYS ARE THOSE WHERE THE STATUS OF THE REGIME IS AUTHORITARIAN, RATHER THAN TOTALITARIAN (23)	1.00	97	TEND LESS TO BE THOSE WHERE THE STATUS OF THE REGIME IS AUTHORITARIAN, RATHER THAN TOTALITARIAN (23)	0.52	6 17 0 16 X SQ= 3.13 P = 0.033 RV NO (NO)
115	ALWAYS ARE THOSE WHERE INTEREST ARTICULATION BY ASSOCIATIONAL GROUPS IS MODERATE, LIMITED, OR NEGLIGIBLE (91)	1.00	115	TEND LESS TO BE THOSE WHERE INTEREST ARTICULATION BY ASSOCIATIONAL GROUPS IS MODERATE, LIMITED, OR NEGLIGIBLE (91)	0.79	0 20 16 75 X SQ= 2.81 P = 0.041 RV NO (NO)
119	ALWAYS ARE THOSE WHERE INTEREST ARTICULATION BY INSTITUTIONAL GROUPS IS VERY SIGNIFICANT OR SIGNIFICANT (74)	1.00	119	TEND LESS TO BE THOSE WHERE INTEREST ARTICULATION BY INSTITUTIONAL GROUPS IS VERY SIGNIFICANT OR SIGNIFICANT (74)	0.69	15 59 0 26 X SQ= 4.71 P = 0.010 RV NO (NO)
122	ALWAYS ARE THOSE WHERE INTEREST ARTICULATION BY NON-ASSOCIATIONAL GROUPS IS SIGNIFICANT OR MODERATE (83)	1.00	122	TEND LESS TO BE THOSE WHERE INTEREST ARTICULATION BY NON-ASSOCIATIONAL GROUPS IS SIGNIFICANT OR MODERATE (83)	0.68	16 67 0 32 X SQ= 5.65 P = 0.005 RV NO (NO)
126	ALWAYS ARE THOSE WHERE INTEREST ARTICULATION BY ANOMIC GROUPS IS FREQUENT, OCCASIONAL, OR INFREQUENT (83)	1.00	126	TILT LESS TOWARD BEING THOSE WHERE INTEREST ARTICULATION BY ANOMIC GROUPS IS FREQUENT, OCCASIONAL, OR INFREQUENT (83)	0.81	13 70 0 16 X SQ= 1.68 P = 0.119 RV NO (NO)
164	TEND TO BE THOSE WHERE THE REGIME'S LEADERSHIP CHARISMA IS PRONOUNCED OR MODERATE (34)	0.70	164	TEND TO BE THOSE WHERE THE REGIME'S LEADERSHIP CHARISMA IS NEGLIGIBLE (65)	0.70	7 27 3 62 X SQ= 4.64 P = 0.029 RV YES (NO)
172	ALWAYS ARE THOSE WHERE THE LEGISLATIVE-EXECUTIVE STRUCTURE IS OTHER THAN PARLIAMENTARY-ROYALIST (88)	1.00	172	LEAN LESS TOWARD BEING THOSE WHERE THE LEGISLATIVE-EXECUTIVE STRUCTURE IS OTHER THAN PARLIAMENTARY-ROYALIST (88)	0.78	0 21 13 75 X SQ= 2.26 P = 0.069 RV NO (NO)

183	TEND TO BE THOSE WHERE THE BUREAUCRACY IS TRADITIONAL, RATHER THAN POST-COLONIAL TRANSITIONAL (9)	0.80	183 TEND TO BE THOSE WHERE THE BUREAUCRACY IS POST-COLONIAL TRANSITIONAL, RATHER THAN TRADITIONAL (25)	0.83	1 24 4 5 X SQ= 5.71 P = 0.012 RV YES (NO)
187	TEND MORE TO BE THOSE WHERE THE ROLE OF THE POLICE IS POLITICALLY SIGNIFICANT (66)	0.93	187 TEND LESS TO BE THOSE WHERE THE ROLE OF THE POLICE IS POLITICALLY SIGNIFICANT (66)	0.61	13 53 1 34 X SQ= 4.11 P = 0.031 RV NO (NO)
192	TEND TO BE THOSE WHERE THE CHARACTER OF THE LEGAL SYSTEM IS MUSLIM OR PARTLY MUSLIM (27)	0.62	192 TEND TO BE THOSE WHERE THE CHARACTER OF THE LEGAL SYSTEM IS OTHER THAN MUSLIM OR PARTLY MUSLIM (86)	0.82	10 17 6 80 X SQ= 12.90 P = 0. RV YES (NO)
193	ALWAYS ARE THOSE WHERE THE CHARACTER OF THE LEGAL SYSTEM IS OTHER THAN PARTLY INDIGENOUS (86)	1.00	193 TEND LESS TO BE THOSE WHERE THE CHARACTER OF THE LEGAL SYSTEM IS OTHER THAN PARTLY INDIGENOUS (86)	0.71	0 28 16 70 X SQ= 4.62 P = 0.011 RV NO (NO)

14 POLITIES
LOCATED IN THE MIDDLE EAST (11)

14 POLITIES
LOCATED ELSEWHERE THAN IN
THE MIDDLE EAST (104)

BOTH SUBJECT AND PREDICATE

11 IN LEFT COLUMN

AFGHANISTAN	CYPRUS	IRAN	IRAQ	ISRAEL	JORDAN	LEBANON	SA'U ARABIA	SYRIA	TURKEY
YEMEN									

104 IN RIGHT COLUMN

ALBANIA	ALGERIA	ARGENTINA	AUSTRALIA	AUSTRIA	BELGIUM	BOLIVIA	BRAZIL	BULGARIA	BURMA
BURUNDI	CAMBODIA	CAMEROUN	CANADA	CEN AFR REP	CEYLON	CHAD	CHILE	CHINA, PR	COLOMBIA
CONGO(BRA)	CONGO(LEO)	COSTA RICA	CUBA	CZECHOS'KIA	DAHOMEY	DENMARK	DOMIN REP	ECUADOR	EL SALVADOR
ETHIOPIA	FINLAND	FRANCE	GABON	GERMANY, E	GERMAN FR	GHANA	GREECE	GUATEMALA	GUINEA
HAITI	HONDURAS	HUNGARY	ICELAND	INDIA	INDONESIA	IRELAND	ITALY	IVORY COAST	JAMAICA
JAPAN	KOREA, N	KOREA REP	LAOS	LIBERIA	LIBYA	LUXEMBOURG	MALAGASY R	MALAYA	MALI
MAURITANIA	MEXICO	MONGOLIA	MOROCCO	NEPAL	NETHERLANDS	NEW ZEALAND	NICARAGUA	NIGER	NIGERIA
NORWAY	PAKISTAN	PANAMA	PARAGUAY	PERU	PHILIPPINES	POLAND	PORTUGAL	RUMANIA	RWANDA
SENEGAL	SIERRE LEO	SOMALIA	SO AFRICA	SPAIN	SUDAN	SWEDEN	SWITZERLAND	TANGANYIKA	THAILAND
TOGO	TRINIDAD	TUNISIA	UGANDA	USSR	UAR	UK	US	UPPER VOLTA	URUGUAY
VENEZUELA	VIETNAM, N	VIETNAM REP	YUGOSLAVIA						

	Left column			Right column		Statistics
39	LEAN MORE TOWARD BEING THOSE WHOSE INTERNATIONAL FINANCIAL STATUS IS LOW OR VERY LOW (76)	0.91	39	LEAN LESS TOWARD BEING THOSE WHOSE INTERNATIONAL FINANCIAL STATUS IS LOW OR VERY LOW (76)	0.64	1 37 10 66 X SQ= 2.13 P = 0.097 RV NO (NO)
44	ALWAYS ARE THOSE WHERE THE LITERACY RATE IS BELOW NINETY PERCENT (90)	1.00	44	LEAN LESS TOWARD BEING THOSE WHERE THE LITERACY RATE IS BELOW NINETY PERCENT (90)	0.76	0 25 11 79 X SQ= 2.11 P = 0.071 RV NO (NO)
45	LEAN TOWARD BEING THOSE WHERE THE LITERACY RATE IS BELOW FIFTY PERCENT (54)	0.80	45	LEAN TOWARD BEING THOSE WHERE THE LITERACY RATE IS FIFTY PERCENT OR ABOVE (55)	0.54	2 53 8 46 X SQ= 2.85 P = 0.052 RV YES (NO)
50	TEND MORE TO BE THOSE WHERE FREEDOM OF THE PRESS IS INTERMITTENT, INTERNALLY ABSENT, OR INTERNALLY AND EXTERNALLY ABSENT (56)	0.91	50	TEND LESS TO BE THOSE WHERE FREEDOM OF THE PRESS IS INTERMITTENT, INTERNALLY ABSENT, OR INTERNALLY AND EXTERNALLY ABSENT (56)	0.52	1 42 10 46 X SQ= 4.47 P = 0.021 RV NO (NO)

56	ALWAYS ARE THOSE WHERE THE RELIGION IS PREDOMINANTLY LITERATE (79)	1.00	56	TEND LESS TO BE THOSE WHERE THE RELIGION IS PREDOMINANTLY LITERATE (79)	0.69	11 68 0 31 X SQ= 3.37 P = 0.032 RV NO (NO)
58	TEND TO BE THOSE WHERE THE RELIGION IS MUSLIM (18)	0.73	58	TEND TO BE THOSE WHERE THE RELIGION IS OTHER THAN MUSLIM (97)	0.90	8 10 3 94 X SQ= 25.42 P = 0. RV YES (NO)
63	ALWAYS ARE THOSE WHERE THE RELIGION IS PREDOMINANTLY OR PARTLY OTHER THAN CHRISTIAN (68)	1.00	63	TEND LESS TO BE THOSE WHERE THE RELIGION IS PREDOMINANTLY OR PARTLY OTHER THAN CHRISTIAN (68)	0.55	0 46 11 57 X SQ= 6.48 P = 0.003 RV NO (NO)
66	LEAN MORE TOWARD BEING THOSE THAT ARE RELIGIOUSLY HOMOGENEOUS (57)	0.82	66	LEAN LESS TOWARD BEING THOSE THAT ARE RELIGIOUSLY HOMOGENEOUS (57)	0.51	9 48 2 47 X SQ= 2.73 P = 0.060 RV NO (NO)
82	ALWAYS ARE THOSE WHERE THE TYPE OF POLITICAL MODERNIZATION IS DEVELOPED TUTELARY OR UNDEVELOPED TUTELARY, RATHER THAN EARLY OR LATER EUROPEAN OR EUROPEAN DERIVED (55)	1.00	82	TEND TO BE THOSE WHERE THE TYPE OF POLITICAL MODERNIZATION IS EARLY OR LATER EUROPEAN OR EUROPEAN DERIVED, RATHER THAN DEVELOPED TUTELARY OR UNDEVELOPED TUTELARY (51)	0.52	0 51 8 47 X SQ= 6.07 P = 0.006 RV YES (NO)
84	ALWAYS ARE THOSE WHERE THE TYPE OF POLITICAL MODERNIZATION IS DEVELOPED TUTELARY, RATHER THAN UNDEVELOPED TUTELARY (31)	1.00	84	TEND TO BE THOSE WHERE THE TYPE OF POLITICAL MODERNIZATION IS UNDEVELOPED TUTELARY, RATHER THAN DEVELOPED TUTELARY (24)	0.51	8 23 0 24 X SQ= 5.32 P = 0.007 RV YES (NO)
91	TEND TO BE THOSE WHOSE IDEOLOGICAL ORIENTATION IS TRADITIONAL, RATHER THAN DEVELOPMENTAL (5)	0.75	91	TEND TO BE THOSE WHOSE IDEOLOGICAL ORIENTATION IS DEVELOPMENTAL, RATHER THAN TRADITIONAL (31)	0.94	1 30 3 2 X SQ= 8.89 P = 0.005 RV YES (YES)
97	ALWAYS ARE THOSE WHERE THE STATUS OF THE REGIME IS AUTHORITARIAN, RATHER THAN TOTALITARIAN (23)	1.00	97	TILT LESS TOWARD BEING THOSE WHERE THE STATUS OF THE REGIME IS AUTHORITARIAN, RATHER THAN TOTALITARIAN (23)	0.54	4 19 0 16 X SQ= 1.50 P = 0.130 RV NO (NO)
100	TEND TO BE THOSE WHERE GOVERNMENTAL STABILITY IS ABSENT (22)	0.62	100	TEND TO BE THOSE WHERE GOVERNMENTAL STABILITY IS GENERALLY PRESENT AND DATES FROM AT LEAST THE INTER-WAR PERIOD, OR IS GENERALLY OR MODERATELY PRESENT, AND DATES FROM THE POST-WAR PERIOD (64)	0.78	3 61 5 17 X SQ= 4.36 P = 0.024 RV YES (NO)

111	LEAN TOWARD BEING THOSE WHERE POLITICAL ENCULTURATION IS LOW (42)	0.73	111	LEAN TOWARD BEING THOSE WHERE POLITICAL ENCULTURATION IS HIGH OR MEDIUM (53)	0.60	3 50 8 34 X SQ= 2.90 P = 0.056 RV YES (NO)
115	ALWAYS ARE THOSE WHERE INTEREST ARTICULATION BY ASSOCIATIONAL GROUPS IS MODERATE, LIMITED, OR NEGLIGIBLE (91)	1.00	115	TILT LESS TOWARD BEING THOSE WHERE INTEREST ARTICULATION BY ASSOCIATIONAL GROUPS IS MODERATE, LIMITED, OR NEGLIGIBLE (91)	0.80	0 20 11 80 X SQ= 1.50 P = 0.126 RV NO (NO)
118	LEAN TOWARD BEING THOSE WHERE INTEREST ARTICULATION BY INSTITUTIONAL GROUPS IS VERY SIGNIFICANT (40)	0.70	118	LEAN TOWARD BEING THOSE WHERE INTEREST ARTICULATION BY INSTITUTIONAL GROUPS IS SIGNIFICANT, MODERATE, OR LIMITED (60)	0.63	7 33 3 57 X SQ= 2.89 P = 0.084 RV YES (NO)
119	ALWAYS ARE THOSE WHERE INTEREST ARTICULATION BY INSTITUTIONAL GROUPS IS VERY SIGNIFICANT OR SIGNIFICANT (74)	1.00	119	LEAN LESS TOWARD BEING THOSE WHERE INTEREST ARTICULATION BY INSTITUTIONAL GROUPS IS VERY SIGNIFICANT OR SIGNIFICANT (74)	0.71	10 64 0 26 X SQ= 2.55 P = 0.060 RV NO (NO)
121	TEND TO BE THOSE WHERE INTEREST ARTICULATION BY NON-ASSOCIATIONAL GROUPS IS SIGNIFICANT (54)	0.82	121	TEND TO BE THOSE WHERE INTEREST ARTICULATION BY NON-ASSOCIATIONAL GROUPS IS MODERATE, LIMITED, OR NEGLIGIBLE (61)	0.57	9 45 2 59 X SQ= 4.49 P = 0.023 RV YES (NO)
122	ALWAYS ARE THOSE WHERE INTEREST ARTICULATION BY NON-ASSOCIATIONAL GROUPS IS SIGNIFICANT OR MODERATE (83)	1.00	122	TEND LESS TO BE THOSE WHERE INTEREST ARTICULATION BY NON-ASSOCIATIONAL GROUPS IS SIGNIFICANT OR MODERATE (83)	0.69	11 72 0 32 X SQ= 3.28 P = 0.033 RV NO (NO)
127	TEND TO BE THOSE WHERE INTEREST ARTICULATION BY POLITICAL PARTIES IS SIGNIFICANT (21)	0.75	127	TEND TO BE THOSE WHERE INTEREST ARTICULATION BY POLITICAL PARTIES IS MODERATE, LIMITED, OR NEGLIGIBLE (72)	0.80	3 18 1 71 X SQ= 3.81 P = 0.035 RV YES (NO)
129	ALWAYS ARE THOSE WHERE INTEREST ARTICULATION BY POLITICAL PARTIES IS SIGNIFICANT, MODERATE, OR LIMITED (56)	1.00	129	TILT LESS TOWARD BEING THOSE WHERE INTEREST ARTICULATION BY POLITICAL PARTIES IS SIGNIFICANT, MODERATE, OR LIMITED (56)	0.58	4 52 0 37 X SQ= 1.30 P = 0.148 RV NO (NO)
132	TEND TO BE THOSE WHERE INTEREST AGGREGATION BY POLITICAL PARTIES IS NEGLIGIBLE (9)	0.67	132	TEND TO BE THOSE WHERE INTEREST AGGREGATION BY POLITICAL PARTIES IS SIGNIFICANT, MODERATE, OR LIMITED (67)	0.90	1 66 2 7 X SQ= 4.36 P = 0.036 RV YES (NO)

139	ALWAYS ARE THOSE WHERE THE PARTY SYSTEM IS QUANTITATIVELY OTHER THAN ONE-PARTY (71)	1.00	139	LEAN LESS TOWARD BEING THOSE WHERE THE PARTY SYSTEM IS QUANTITATIVELY OTHER THAN ONE-PARTY (71)	0.65	0 34 7 64 X SQ= 2.18 P = 0.093 RV NO (NO)
153	LEAN TOWARD BEING THOSE WHERE THE PARTY SYSTEM IS MODERATELY STABLE OR UNSTABLE (38)	0.83	153	LEAN TOWARD BEING THOSE WHERE THE PARTY SYSTEM IS STABLE (42)	0.55	1 41 5 33 X SQ= 1.97 P = 0.097 RV YES (NO)
154	LEAN TOWARD BEING THOSE WHERE THE PARTY SYSTEM IS UNSTABLE (25)	0.67	154	LEAN TOWARD BEING THOSE WHERE THE PARTY SYSTEM IS STABLE OR MODERATELY STABLE (55)	0.72	2 53 4 21 X SQ= 2.21 P = 0.073 RV YES (NO)
183	ALWAYS ARE THOSE WHERE THE BUREAUCRACY IS TRADITIONAL, RATHER THAN POST-COLONIAL TRANSITIONAL (9)	1.00	183	TEND TO BE THOSE WHERE THE BUREAUCRACY IS POST-COLONIAL TRANSITIONAL, RATHER THAN TRADITIONAL (25)	0.83	0 25 4 5 X SQ= 8.68 P = 0.003 RV YES (NO)
186	LEAN TOWARD BEING THOSE WHERE PARTICIPATION BY THE MILITARY IN POLITICS IS SUPPORTIVE, RATHER THAN NEUTRAL (31)	0.71	186	LEAN TOWARD BEING THOSE WHERE PARTICIPATION BY THE MILITARY IN POLITICS IS NEUTRAL, RATHER THAN SUPPORTIVE (56)	0.67	5 26 2 54 X SQ= 2.73 P = 0.092 RV YES (NO)
187	LEAN MORE TOWARD BEING THOSE WHERE THE ROLE OF THE POLICE IS POLITICALLY SIGNIFICANT (66)	0.90	187	LEAN LESS TOWARD BEING THOSE WHERE THE ROLE OF THE POLICE IS POLITICALLY SIGNIFICANT (66)	0.63	9 57 1 34 X SQ= 1.89 P = 0.097 RV NO (NO)
193	ALWAYS ARE THOSE WHERE THE CHARACTER OF THE LEGAL SYSTEM IS OTHER THAN PARTLY INDIGENOUS (86)	1.00	193	LEAN LESS TOWARD BEING THOSE WHERE THE CHARACTER OF THE LEGAL SYSTEM IS OTHER THAN PARTLY INDIGENOUS (86)	0.73	0 28 11 75 X SQ= 2.63 P = 0.063 RV NO (NO)

```
****************************************************************************************************

15  POLITIES                                          15  POLITIES
    LOCATED IN THE MIDDLE EAST, RATHER THAN               LOCATED IN NORTH AFRICA OR
    IN NORTH AFRICA OR                                    CENTRAL AND SOUTH AFRICA, RATHER THAN
    CENTRAL AND SOUTH AFRICA  (11)                        IN THE MIDDLE EAST  (33)

                                                                        BOTH SUBJECT AND PREDICATE
....................................................................................................

      11 IN LEFT COLUMN

AFGHANISTAN CYPRUS      IRAN        IRAQ        ISRAEL      JORDAN      LEBANON     SA'U ARABIA SYRIA       TURKEY
YEMEN

      33 IN RIGHT COLUMN

ALGERIA     BURUNDI     CAMEROUN    CEN AFR REP CHAD        CONGO(BRA)  CONGO(LEO)  DAHOMEY     ETHIOPIA    GABON
GHANA       GUINEA      IVORY COAST LIBERIA     LIBYA       MALAGASY R  MALI        MAURITANIA  MOROCCO     NIGER
NIGERIA     RWANDA      SENEGAL     SIERRE LEO  SOMALIA     SO AFRICA   SUDAN       TANGANYIKA  TOGO        TUNISIA
UGANDA      UAR         UPPER VOLTA

      71 EXCLUDED BECAUSE IRRELEVANT

ALBANIA     ARGENTINA   AUSTRALIA   AUSTRIA     BELGIUM     BOLIVIA     BRAZIL      BULGARIA    BURMA       CAMBODIA
CANADA      CEYLON      CHILE       CHINA, PR   COLOMBIA    COSTA RICA  CUBA        CZECHOS'KIA DENMARK     DOMIN REP
ECUADOR     EL SALVADOR FINLAND     FRANCE      GERMANY, E  GERMAN FR   GREECE      GUATEMALA   HAITI       HONDURAS
HUNGARY     ICELAND     INDIA       INDONESIA   IRELAND     ITALY       JAMAICA     JAPAN       KOREA, N    KOREA REP
LAOS        LUXEMBOURG  MALAYA      MEXICO      MONGOLIA    NEPAL       NETHERLANDS NEW ZEALAND NICARAGUA   NORWAY
PAKISTAN    PANAMA      PARAGUAY    PERU        PHILIPPINES POLAND      PORTUGAL    RUMANIA     SPAIN       SWEDEN
SWITZERLAND THAILAND    TRINIDAD    USSR        UK          US          URUGUAY     VENEZUELA   VIETNAM, N  VIETNAM REP
YUGOSLAVIA

----------------------------------------------------------------------------------------------------

29  TEND TO BE THOSE                    0.64      29  TEND TO BE THOSE                    0.85         7    5
    WHERE THE DEGREE OF URBANIZATION                  WHERE THE DEGREE OF URBANIZATION                 4   28
    IS HIGH  (56)                                     IS LOW  (49)                                 X SQ=  7.49
                                                                                                   P =   0.004
                                                                                                   RV YES (YES)

30  TEND LESS TO BE THOSE               0.60      30  TEND MORE TO BE THOSE               0.91         6   30
    WHOSE AGRICULTURAL POPULATION                     WHOSE AGRICULTURAL POPULATION                    4    3
    IS HIGH  (56)                                     IS HIGH  (56)                                X SQ=  3.35
                                                                                                   P =   0.040
                                                                                                   RV NO  (YES)

37  TEND TO BE THOSE                    0.64      37  TEND TO BE THOSE                    0.88         7    4
    WHOSE PER CAPITA GROSS NATIONAL PRODUCT           WHOSE PER CAPITA GROSS NATIONAL PRODUCT          4   29
    IS VERY HIGH, HIGH, MEDIUM, OR LOW  (64)          IS VERY LOW  (51)                            X SQ=  9.09
                                                                                                   P =   0.002
                                                                                                   RV YES (YES)
```

40	TEND TO BE THOSE WHOSE INTERNATIONAL FINANCIAL STATUS IS VERY HIGH, HIGH, MEDIUM, OR LOW (71)	0.73	40	TEND TO BE THOSE WHOSE INTERNATIONAL FINANCIAL STATUS IS VERY LOW (39)	0.69	8 10 3 22 X SQ= 4.21 P = 0.031 RV YES (NO)
43	TEND LESS TO BE THOSE WHOSE ECONOMIC DEVELOPMENTAL STATUS IS VERY UNDERDEVELOPED (57)	0.56	43	TEND MORE TO BE THOSE WHOSE ECONOMIC DEVELOPMENTAL STATUS IS VERY UNDERDEVELOPED (57)	0.94	4 2 5 31 X SQ= 5.66 P = 0.013 RV NO (YES)
45	LEAN LESS TOWARD BEING THOSE WHERE THE LITERACY RATE IS BELOW FIFTY PERCENT (54)	0.80	45	ALWAYS ARE THOSE WHERE THE LITERACY RATE IS BELOW FIFTY PERCENT (54)	1.00	2 0 8 32 X SQ= 3.03 P = 0.052 RV NO (YES)
46	TEND TO BE THOSE WHERE THE LITERACY RATE IS TEN PERCENT OR ABOVE (84)	0.73	46	TEND TO BE THOSE WHERE THE LITERACY RATE IS BELOW TEN PERCENT (26)	0.70	8 9 3 21 X SQ= 4.42 P = 0.029 RV YES (NO)
50	LEAN MORE TOWARD BEING THOSE WHERE FREEDOM OF THE PRESS IS INTERMITTENT, INTERNALLY ABSENT, OR INTERNALLY AND EXTERNALLY ABSENT (56)	0.91	50	LEAN LESS TOWARD BEING THOSE WHERE FREEDOM OF THE PRESS IS INTERMITTENT, INTERNALLY ABSENT, OR INTERNALLY AND EXTERNALLY ABSENT (56)	0.57	1 10 10 13 X SQ= 2.60 P = 0.060 RV NO (NO)
52	LEAN LESS TOWARD BEING THOSE WHERE FREEDOM OF THE PRESS IS COMPLETE, INTERMITTENT, OR INTERNALLY ABSENT (82)	0.67	52	LEAN MORE TOWARD BEING THOSE WHERE FREEDOM OF THE PRESS IS COMPLETE, INTERMITTENT, OR INTERNALLY ABSENT (82)	0.96	6 22 3 1 X SQ= 2.67 P = 0.057 RV NO (YES)
54	LEAN LESS TOWARD BEING THOSE WHERE NEWSPAPER CIRCULATION IS LESS THAN ONE HUNDRED PER THOUSAND (76)	0.82	54	ALWAYS ARE THOSE WHERE NEWSPAPER CIRCULATION IS LESS THAN ONE HUNDRED PER THOUSAND (76)	1.00	2 0 9 33 X SQ= 2.79 P = 0.058 RV NO (YES)
55	TEND TO BE THOSE WHERE NEWSPAPER CIRCULATION IS TEN OR MORE PER THOUSAND (78)	0.64	55	TEND TO BE THOSE WHERE NEWSPAPER CIRCULATION IS LESS THAN TEN PER THOUSAND (35)	0.82	7 6 4 27 X SQ= 6.15 P = 0.008 RV YES (YES)
56	ALWAYS ARE THOSE WHERE THE RELIGION IS PREDOMINANTLY LITERATE (79)	1.00	56	TEND TO BE THOSE WHERE THE RELIGION IS PREDOMINANTLY OR PARTLY NON-LITERATE (31)	0.76	11 8 0 25 X SQ= 16.33 P = 0. RV YES (YES)

58	TEND TO BE THOSE WHERE THE RELIGION IS MUSLIM (18)	0.73	58	TEND TO BE THOSE WHERE THE RELIGION IS OTHER THAN MUSLIM (97)	0.76	8 8 3 25 X SQ= 6.42 P = 0.009 RV YES (YES)
66	TEND TO BE THOSE THAT ARE RELIGIOUSLY HOMOGENEOUS (57)	0.82	66	TEND TO BE THOSE THAT ARE RELIGIOUSLY HETEROGENEOUS (49)	0.73	9 8 2 22 X SQ= 7.94 P = 0.003 RV YES (YES)
69	LEAN TOWARD BEING THOSE THAT ARE LINGUISTICALLY HOMOGENEOUS OR WEAKLY HETEROGENEOUS (64)	0.55	69	LEAN TOWARD BEING THOSE THAT ARE LINGUISTICALLY STRONGLY HETEROGENEOUS (50)	0.79	6 7 5 26 X SQ= 2.95 P = 0.057 RV YES (NO)
73	TEND TO BE THOSE WHOSE DATE OF INDEPENDENCE IS BEFORE 1945 (65)	0.73	73	TEND TO BE THOSE WHOSE DATE OF INDEPENDENCE IS AFTER 1945 (46)	0.91	8 3 3 30 X SQ= 14.59 P = 0. RV YES (YES)
75	TEND LESS TO BE THOSE THAT ARE NOT HISTORICALLY WESTERN AND ARE NOT SIGNIFICANTLY WESTERNIZED (52)	0.64	75	TEND MORE TO BE THOSE THAT ARE NOT HISTORICALLY WESTERN AND ARE NOT SIGNIFICANTLY WESTERNIZED (52)	0.94	4 2 7 30 X SQ= 3.93 P = 0.029 RV NO (YES)
83	TEND LESS TO BE THOSE WHERE THE TYPE OF POLITICAL MODERNIZATION IS OTHER THAN NON-EUROPEAN AUTOCHTHONOUS (106)	0.73	83	TEND MORE TO BE THOSE WHERE THE TYPE OF POLITICAL MODERNIZATION IS OTHER THAN NON-EUROPEAN AUTOCHTHONOUS (106)	0.97	3 1 8 32 X SQ= 3.30 P = 0.043 RV NO (YES)
84	ALWAYS ARE THOSE WHERE THE TYPE OF POLITICAL MODERNIZATION IS DEVELOPED TUTELARY, RATHER THAN UNDEVELOPED TUTELARY (31)	1.00	84	TEND TO BE THOSE WHERE THE TYPE OF POLITICAL MODERNIZATION IS UNDEVELOPED TUTELARY, RATHER THAN DEVELOPED TUTELARY (24)	0.77	8 7 0 24 X SQ= 13.00 P = 0. RV YES (YES)
85	TEND TO BE THOSE WHERE THE STAGE OF POLITICAL MODERNIZATION IS ADVANCED, RATHER THAN MID- OR EARLY TRANSITIONAL (60)	0.70	85	TEND TO BE THOSE WHERE THE STAGE OF POLITICAL MODERNIZATION IS MID- OR EARLY TRANSITIONAL, RATHER THAN ADVANCED (54)	0.85	7 5 3 28 X SQ= 8.91 P = 0.002 RV YES (YES)
86	TEND TO BE THOSE WHERE THE STAGE OF POLITICAL MODERNIZATION IS ADVANCED OR MID-TRANSITIONAL, RATHER THAN EARLY TRANSITIONAL (76)	0.90	86	TEND TO BE THOSE WHERE THE STAGE OF POLITICAL MODERNIZATION IS EARLY TRANSITIONAL, RATHER THAN ADVANCED OR MID-TRANSITIONAL (38)	0.79	9 7 1 26 X SQ= 12.74 P = 0. RV YES (YES)

87	TEND TO BE THOSE WHOSE IDEOLOGICAL ORIENTATION IS OTHER THAN DEVELOPMENTAL (58)	0.89	87	TEND TO BE THOSE WHOSE IDEOLOGICAL ORIENTATION IS DEVELOPMENTAL (31)	0.93	1 25 8 2 X SQ= 18.46 P = 0. RV YES (YES)
89	LEAN LESS TOWARD BEING THOSE WHOSE IDEOLOGICAL ORIENTATION IS OTHER THAN CONVENTIONAL (62)	0.73	89	LEAN MORE TOWARD BEING THOSE WHOSE IDEOLOGICAL ORIENTATION IS OTHER THAN CONVENTIONAL (62)	0.96	3 1 8 26 X SQ= 2.45 P = 0.065 RV NO (YES)
91	TEND TO BE THOSE WHOSE IDEOLOGICAL ORIENTATION IS TRADITIONAL, RATHER THAN DEVELOPMENTAL (5)	0.75	91	TEND TO BE THOSE WHOSE IDEOLOGICAL ORIENTATION IS DEVELOPMENTAL, RATHER THAN TRADITIONAL (31)	0.96	1 25 3 1 X SQ= 9.66 P = 0.004 RV YES (YES)
100	LEAN TOWARD BEING THOSE WHERE GOVERNMENTAL STABILITY IS ABSENT (22)	0.62	100	LEAN TOWARD BEING THOSE WHERE GOVERNMENTAL STABILITY IS GENERALLY PRESENT AND DATES FROM AT LEAST THE INTER-WAR PERIOD, OR IS GENERALLY OR MODERATELY PRESENT, AND DATES FROM THE POST-WAR PERIOD (64)	0.82	3 9 5 2 X SQ= 2.24 P = 0.074 RV YES (YES)
107	TEND LESS TO BE THOSE WHERE AUTONOMOUS GROUPS ARE PARTIALLY TOLERATED IN POLITICS, ARE TOLERATED ONLY OUTSIDE POLITICS, OR ARE NOT TOLERATED AT ALL (65)	0.60	107	TEND MORE TO BE THOSE WHERE AUTONOMOUS GROUPS ARE PARTIALLY TOLERATED IN POLITICS, ARE TOLERATED ONLY OUTSIDE POLITICS, OR ARE NOT TOLERATED AT ALL (65)	0.91	4 3 6 29 X SQ= 3.18 P = 0.043 RV NO (YES)
117	TEND TO BE THOSE WHERE INTEREST ARTICULATION BY ASSOCIATIONAL GROUPS IS SIGNIFICANT, MODERATE, OR LIMITED (60)	0.55	117	TEND TO BE THOSE WHERE INTEREST ARTICULATION BY ASSOCIATIONAL GROUPS IS NEGLIGIBLE (51)	0.85	6 5 5 28 X SQ= 4.89 P = 0.016 RV YES (YES)
118	TILT TOWARD BEING THOSE WHERE INTEREST ARTICULATION BY INSTITUTIONAL GROUPS IS VERY SIGNIFICANT (40)	0.70	118	TILT TOWARD BEING THOSE WHERE INTEREST ARTICULATION BY INSTITUTIONAL GROUPS IS SIGNIFICANT, MODERATE, OR LIMITED (60)	0.63	7 7 3 12 X SQ= 1.71 P = 0.128 RV YES (YES)
127	TEND TO BE THOSE WHERE INTEREST ARTICULATION BY POLITICAL PARTIES IS SIGNIFICANT (21)	0.75	127	TEND TO BE THOSE WHERE INTEREST ARTICULATION BY POLITICAL PARTIES IS MODERATE, LIMITED, OR NEGLIGIBLE (72)	0.89	3 3 1 25 X SQ= 5.74 P = 0.015 RV YES (YES)
129	ALWAYS ARE THOSE WHERE INTEREST ARTICULATION BY POLITICAL PARTIES IS SIGNIFICANT, MODERATE, OR LIMITED (56)	1.00	129	TILT TOWARD BEING THOSE WHERE INTEREST ARTICULATION BY POLITICAL PARTIES IS NEGLIGIBLE (37)	0.57	4 12 0 16 X SQ= 2.57 P = 0.101 RV YES (NO)

131	TEND TO BE THOSE WHERE INTEREST AGGREGATION BY POLITICAL PARTIES IS LIMITED OR NEGLIGIBLE (35)	0.80	131	TEND TO BE THOSE WHERE INTEREST AGGREGATION BY POLITICAL PARTIES IS SIGNIFICANT OR MODERATE (30)	0.83	1 10 4 2 X SQ= 3.74 P = 0.028 RV YES (YES)
132	TEND TO BE THOSE WHERE INTEREST AGGREGATION BY POLITICAL PARTIES IS NEGLIGIBLE (9)	0.67	132	ALWAYS ARE THOSE WHERE INTEREST AGGREGATION BY POLITICAL PARTIES IS SIGNIFICANT, MODERATE, OR LIMITED (67)	1.00	1 29 2 0 X SQ= 10.81 P = 0.006 RV YES (YES)
133	ALWAYS ARE THOSE WHERE INTEREST AGGREGATION BY THE EXECUTIVE IS MODERATE, LIMITED, OR NEGLIGIBLE (76)	1.00	133	TEND TO BE THOSE WHERE INTEREST AGGREGATION BY THE EXECUTIVE IS SIGNIFICANT (29)	0.61	0 19 6 12 X SQ= 5.30 P = 0.008 RV YES (NO)
134	TILT TOWARD BEING THOSE WHERE INTEREST AGGREGATION BY THE EXECUTIVE IS LIMITED OR NEGLIGIBLE (46)	0.67	134	TILT TOWARD BEING THOSE WHERE INTEREST AGGREGATION BY THE EXECUTIVE IS SIGNIFICANT OR MODERATE (57)	0.73	2 22 4 8 X SQ= 2.02 P = 0.149 RV YES (NO)
137	TEND TO BE THOSE WHERE INTEREST AGGREGATION BY THE LEGISLATURE IS SIGNIFICANT OR MODERATE (29)	0.60	137	ALWAYS ARE THOSE WHERE INTEREST AGGREGATION BY THE LEGISLATURE IS LIMITED OR NEGLIGIBLE (68)	1.00	3 0 2 28 X SQ= 11.93 P = 0.002 RV YES (YES)
138	TEND TO BE THOSE WHERE INTEREST AGGREGATION BY THE LEGISLATURE IS SIGNIFICANT, MODERATE, OR LIMITED (48)	0.67	138	TEND TO BE THOSE WHERE INTEREST AGGREGATION BY THE LEGISLATURE IS NEGLIGIBLE (49)	0.82	4 5 2 23 X SQ= 3.80 P = 0.031 RV YES (NO)
139	ALWAYS ARE THOSE WHERE THE PARTY SYSTEM IS QUANTITATIVELY OTHER THAN ONE-PARTY (71)	1.00	139	TEND TO BE THOSE WHERE THE PARTY SYSTEM IS QUANTITATIVELY ONE-PARTY (34)	0.53	0 17 7 15 X SQ= 4.61 P = 0.012 RV YES (NO)
142	TEND TO BE THOSE WHERE THE PARTY SYSTEM IS QUANTITATIVELY MULTI-PARTY (30)	0.60	142	TEND TO BE THOSE WHERE THE PARTY SYSTEM IS QUANTITATIVELY OTHER THAN MULTI-PARTY (66)	0.90	3 3 2 26 X SQ= 4.22 P = 0.029 RV YES (YES)
147	LEAN LESS TOWARD BEING THOSE WHERE THE PARTY SYSTEM IS QUALITATIVELY OTHER THAN CLASS-ORIENTED OR MULTI-IDEOLOGICAL (67)	0.67	147	ALWAYS ARE THOSE WHERE THE PARTY SYSTEM IS QUALITATIVELY OTHER THAN CLASS-ORIENTED OR MULTI-IDEOLOGICAL (67)	1.00	1 0 2 28 X SQ= 1.92 P = 0.097 RV NO (YES)

148	ALWAYS ARE THOSE WHERE THE PARTY SYSTEM IS QUALITATIVELY OTHER THAN AFRICAN TRANSITIONAL (96)	1.00	148	TEND LESS TO BE THOSE WHERE THE PARTY SYSTEM IS QUALITATIVELY OTHER THAN AFRICAN TRANSITIONAL (96)	0.55	0 14 9 17 X SQ= 4.43 P = 0.016 RV NO (NO)
154	TILT TOWARD BEING THOSE WHERE THE PARTY SYSTEM IS UNSTABLE (25)	0.67	154	TILT TOWARD BEING THOSE WHERE THE PARTY SYSTEM IS STABLE OR MODERATELY STABLE (55)	0.75	2 9 4 3 X SQ= 1.43 P = 0.141 RV YES (YES)
157	TEND TO BE THOSE WHERE PERSONALISMO IS PRONOUNCED (14)	0.50	157	TEND TO BE THOSE WHERE PERSONALISMO IS MODERATE OR NEGLIGIBLE (84)	0.97	2 1 2 28 X SQ= 4.45 P = 0.033 RV YES (YES)
160	TEND TO BE THOSE WHERE THE POLITICAL LEADERSHIP IS ELITIST (30)	0.60	160	TEND TO BE THOSE WHERE THE POLITICAL LEADERSHIP IS MODERATELY ELITIST OR NON-ELITIST (67)	0.89	3 3 2 24 X SQ= 3.80 P = 0.034 RV YES (YES)
164	TILT TOWARD BEING THOSE WHERE THE REGIME'S LEADERSHIP CHARISMA IS NEGLIGIBLE (65)	0.60	164	TILT TOWARD BEING THOSE WHERE THE REGIME'S LEADERSHIP CHARISMA IS PRONOUNCED OR MODERATE (34)	0.78	2 21 3 6 X SQ= 1.40 P = 0.121 RV YES (NO)
168	TEND LESS TO BE THOSE WHERE THE HORIZONTAL POWER DISTRIBUTION IS LIMITED OR NEGLIGIBLE (72)	0.62	168	TEND MORE TO BE THOSE WHERE THE HORIZONTAL POWER DISTRIBUTION IS LIMITED OR NEGLIGIBLE (72)	0.97	3 1 5 30 X SQ= 4.82 P = 0.022 RV NO (YES)
170	LEAN TOWARD BEING THOSE WHERE THE LEGISLATIVE-EXECUTIVE STRUCTURE IS OTHER THAN PRESIDENTIAL (63)	0.86	170	LEAN TOWARD BEING THOSE WHERE THE LEGISLATIVE-EXECUTIVE STRUCTURE IS PRESIDENTIAL (39)	0.57	1 16 6 12 X SQ= 2.58 P = 0.088 RV YES (NO)
175	TILT TOWARD BEING THOSE WHERE THE LEGISLATURE IS FULLY EFFECTIVE OR PARTIALLY EFFECTIVE (51)	0.67	175	TILT TOWARD BEING THOSE WHERE THE LEGISLATURE IS LARGELY INEFFECTIVE OR WHOLLY INEFFECTIVE (49)	0.73	4 8 2 22 X SQ= 2.02 P = 0.149 RV YES (NO)
178	LEAN TOWARD BEING THOSE WHERE THE LEGISLATURE IS BICAMERAL (51)	0.67	178	LEAN TOWARD BEING THOSE WHERE THE LEGISLATURE IS UNICAMERAL (53)	0.77	2 24 4 7 X SQ= 2.80 P = 0.051 RV YES (NO)

182	ALWAYS ARE THOSE WHERE THE BUREAUCRACY IS SEMI-MODERN, RATHER THAN POST-COLONIAL TRANSITIONAL (55)	1.00	182	TEND TO BE THOSE WHERE THE BUREAUCRACY IS POST-COLONIAL TRANSITIONAL, RATHER THAN SEMI-MODERN (25)	0.77	4 7 0 24 X SQ= 6.59 P = 0.006 RV YES (NO)
183	ALWAYS ARE THOSE WHERE THE BUREAUCRACY IS TRADITIONAL, RATHER THAN POST-COLONIAL TRANSITIONAL (9)	1.00	183	TEND TO BE THOSE WHERE THE BUREAUCRACY IS POST-COLONIAL TRANSITIONAL, RATHER THAN TRADITIONAL (25)	0.96	0 24 4 1 X SQ= 16.05 P = 0. RV YES (YES)
186	TEND TO BE THOSE WHERE PARTICIPATION BY THE MILITARY IN POLITICS IS SUPPORTIVE, RATHER THAN NEUTRAL (31)	0.71	186	TEND TO BE THOSE WHERE PARTICIPATION BY THE MILITARY IN POLITICS IS NEUTRAL, RATHER THAN SUPPORTIVE (56)	0.75	5 7 2 21 X SQ= 3.50 P = 0.033 RV YES (NO)
193	ALWAYS ARE THOSE WHERE THE CHARACTER OF THE LEGAL SYSTEM IS OTHER THAN PARTLY INDIGENOUS (86)	1.00	193	TEND TO BE THOSE WHERE THE CHARACTER OF THE LEGAL SYSTEM IS PARTLY INDIGENOUS (28)	0.76	0 25 11 8 X SQ= 16.33 P = 0. RV YES (YES)

16	POLITIES LOCATED IN THE MIDDLE EAST, RATHER THAN IN EAST ASIA, SOUTH ASIA, OR SOUTHEAST ASIA (11)	16	POLITIES LOCATED IN EAST ASIA, SOUTH ASIA, OR SOUTHEAST ASIA, RATHER THAN IN THE MIDDLE EAST (18)

BOTH SUBJECT AND PREDICATE

11 IN LEFT COLUMN

AFGHANISTAN	CYPRUS	IRAN	IRAQ	ISRAEL	JORDAN	LEBANON	SA'U ARABIA	SYRIA	TURKEY
YEMEN									

18 IN RIGHT COLUMN

BURMA	CAMBODIA	CEYLON	CHINA, PR	INDIA	INDONESIA	JAPAN	KOREA, N	KOREA REP	LAOS
MALAYA	MONGOLIA	NEPAL	PAKISTAN	PHILIPPINES	THAILAND	VIETNAM, N	VIETNAM REP		

86 EXCLUDED BECAUSE IRRELEVANT

ALBANIA	ALGERIA	ARGENTINA	AUSTRALIA	AUSTRIA	BELGIUM	BOLIVIA	BRAZIL	BULGARIA	BURUNDI
CAMEROUN	CANADA	CEN AFR REP	CHAD	CHILE	COLOMBIA	CONGO(BRA)	CONGO(LEO)	COSTA RICA	CUBA
CZECHOS'KIA	DAHOMEY	DENMARK	DOMIN REP	ECUADOR	EL SALVADOR	ETHIOPIA	FINLAND	FRANCE	GABON
GERMANY, E	GERMAN FR	GHANA	GREECE	GUATEMALA	GUINEA	HAITI	HONDURAS	HUNGARY	ICELAND
IRELAND	ITALY	IVORY COAST	JAMAICA	LIBERIA	LIBYA	LUXEMBOURG	MALAGASY R	MALI	MAURITANIA
MEXICO	MOROCCO	NETHERLANDS	NEW ZEALAND	NICARAGUA	NIGER	NIGERIA	NORWAY	PANAMA	PARAGUAY
PERU	POLAND	PORTUGAL	RUMANIA	RWANDA	SENEGAL	SIERRE LEO	SOMALIA	SO AFRICA	SPAIN
SUDAN	SWEDEN	SWITZERLAND	TANGANYIKA	TOGO	TRINIDAD	TUNISIA	UGANDA	USSR	UAR
UK	US	UPPER VOLTA	URUGUAY	VENEZUELA	YUGOSLAVIA				

23	TILT TOWARD BEING THOSE WHOSE POPULATION IS MEDIUM OR SMALL (88)	0.82	23	TILT TOWARD BEING THOSE WHOSE POPULATION IS VERY LARGE OR LARGE (27)	0.50	2 9 9 9 X SQ= 1.74 P = 0.125 RV YES (YES)
24	TEND TO BE THOSE WHOSE POPULATION IS SMALL (54)	0.55	24	TEND TO BE THOSE WHOSE POPULATION IS VERY LARGE, LARGE, OR MEDIUM (61)	0.83	5 15 6 3 X SQ= 2.98 P = 0.048 RV YES (YES)
26	TEND TO BE THOSE WHOSE POPULATION DENSITY IS LOW (67)	0.73	26	TEND TO BE THOSE WHOSE POPULATION DENSITY IS VERY HIGH, HIGH, OR MEDIUM (48)	0.78	3 14 8 4 X SQ= 5.25 P = 0.018 RV YES (YES)

29	TEND TO BE THOSE WHERE THE DEGREE OF URBANIZATION IS HIGH (56)	0.64	29	TEND TO BE THOSE WHERE THE DEGREE OF URBANIZATION IS LOW (49)	0.83	7 2 4 10 X SQ= 3.53 P = 0.036 RV YES (YES)
37	TEND TO BE THOSE WHOSE PER CAPITA GROSS NATIONAL PRODUCT IS VERY HIGH, HIGH, MEDIUM, OR LOW (64)	0.64	37	TEND TO BE THOSE WHOSE PER CAPITA GROSS NATIONAL PRODUCT IS VERY LOW (51)	0.83	7 3 4 15 X SQ= 4.75 P = 0.017 RV YES (YES)
45	TILT TOWARD BEING THOSE WHERE THE LITERACY RATE IS BELOW FIFTY PERCENT (54)	0.80	45	TILT TOWARD BEING THOSE WHERE THE LITERACY RATE IS FIFTY PERCENT OR ABOVE (55)	0.57	2 8 8 6 X SQ= 1.96 P = 0.104 RV YES (YES)
58	TEND TO BE THOSE WHERE THE RELIGION IS MUSLIM (18)	0.73	58	TEND TO BE THOSE WHERE THE RELIGION IS OTHER THAN MUSLIM (97)	0.89	8 2 3 16 X SQ= 8.91 P = 0.001 RV YES (YES)
73	LEAN TOWARD BEING THOSE WHOSE DATE OF INDEPENDENCE IS BEFORE 1945 (65)	0.73	73	LEAN TOWARD BEING THOSE WHOSE DATE OF INDEPENDENCE IS AFTER 1945 (46)	0.69	8 5 3 11 X SQ= 2.98 P = 0.054 RV YES (YES)
86	TEND TO BE THOSE WHERE THE STAGE OF POLITICAL MODERNIZATION IS ADVANCED OR MID-TRANSITIONAL, RATHER THAN EARLY TRANSITIONAL (76)	0.90	86	TEND TO BE THOSE WHERE THE STAGE OF POLITICAL MODERNIZATION IS EARLY TRANSITIONAL, RATHER THAN ADVANCED OR MID-TRANSITIONAL (38)	0.50	9 9 1 9 X SQ= 2.91 P = 0.048 RV YES (YES)
92	ALWAYS ARE THOSE WHERE THE SYSTEM STYLE IS LIMITED MOBILIZATIONAL, OR NON-MOBILIZATIONAL (93)	1.00	92	LEAN LESS TOWARD BEING THOSE WHERE THE SYSTEM STYLE IS LIMITED MOBILIZATIONAL, OR NON-MOBILIZATIONAL (93)	0.67	0 6 10 12 X SQ= 2.49 P = 0.062 RV NO (NO)
93	TEND TO BE THOSE WHERE THE SYSTEM STYLE IS NON-MOBILIZATIONAL (78)	0.90	93	TEND TO BE THOSE WHERE THE SYSTEM STYLE IS MOBILIZATIONAL, OR LIMITED MOBILIZATIONAL (32)	0.53	1 8 9 7 X SQ= 3.19 P = 0.040 RV YES (YES)
98	TILT LESS TOWARD BEING THOSE WHERE GOVERNMENTAL STABILITY IS GENERALLY OR MODERATELY PRESENT AND DATES FROM THE POST-WAR PERIOD, OR IS ABSENT (93)	0.82	98	ALWAYS ARE THOSE WHERE GOVERNMENTAL STABILITY IS GENERALLY OR MODERATELY PRESENT AND DATES FROM THE POST-WAR PERIOD, OR IS ABSENT (93)	1.00	2 0 9 18 X SQ= 1.25 P = 0.135 RV NO (YES)

127	TEND TO BE THOSE WHERE INTEREST ARTICULATION BY POLITICAL PARTIES IS SIGNIFICANT (21)	0.75	127	TEND TO BE THOSE WHERE INTEREST ARTICULATION BY POLITICAL PARTIES IS MODERATE, LIMITED, OR NEGLIGIBLE (72)	0.91	3 1 1 10 X SQ= 3.58 P = 0.033 RV YES (YES)
129	ALWAYS ARE THOSE WHERE INTEREST ARTICULATION BY POLITICAL PARTIES IS SIGNIFICANT, MODERATE, OR LIMITED (56)	1.00	129	LEAN TOWARD BEING THOSE WHERE INTEREST ARTICULATION BY POLITICAL PARTIES IS NEGLIGIBLE (37)	0.64	4 4 0 7 X SQ= 2.56 P = 0.077 RV YES (YES)
138	TILT TOWARD BEING THOSE WHERE INTEREST AGGREGATION BY THE LEGISLATURE IS SIGNIFICANT, MODERATE, OR LIMITED (48)	0.67	138	TILT TOWARD BEING THOSE WHERE INTEREST AGGREGATION BY THE LEGISLATURE IS NEGLIGIBLE (49)	0.79	4 3 2 11 X SQ= 2.05 P = 0.122 RV YES (YES)
139	ALWAYS ARE THOSE WHERE THE PARTY SYSTEM IS QUANTITATIVELY OTHER THAN ONE-PARTY (71)	1.00	139	TILT LESS TOWARD BEING THOSE WHERE THE PARTY SYSTEM IS QUANTITATIVELY OTHER THAN ONE-PARTY (71)	0.64	0 5 7 9 X SQ= 1.61 P = 0.123 RV NO (NO)
194	ALWAYS ARE THOSE THAT ARE NON-COMMUNIST (101)	1.00	194	TILT LESS TOWARD BEING THOSE THAT ARE NON-COMMUNIST (101)	0.78	0 4 11 14 X SQ= 1.27 P = 0.143 RV NO (NO)

```
************************************************************************************************************

17  POLITIES                                          17  POLITIES
    LOCATED IN THE MIDDLE EAST, RATHER THAN               LOCATED IN THE CARIBBEAN,
    IN THE CARIBBEAN, CENTRAL AMERICA, OR                 CENTRAL AMERICA, OR SOUTH AMERICA,
    SOUTH AMERICA  (11)                                   RATHER THAN IN THE MIDDLE EAST  (22)

                                                                             BOTH SUBJECT AND PREDICATE
............................................................................................................

       11 IN LEFT COLUMN

AFGHANISTAN CYPRUS      IRAN        IRAQ        ISRAEL      JORDAN      LEBANON     SA'U ARABIA SYRIA       TURKEY
YEMEN

       22 IN RIGHT COLUMN

ARGENTINA   BOLIVIA     BRAZIL      CHILE       COLOMBIA    COSTA RICA  CUBA        DOMIN REP   ECUADOR     EL SALVADOR
GUATEMALA   HAITI       HONDURAS    JAMAICA     MEXICO      NICARAGUA   PANAMA      PARAGUAY    PERU        TRINIDAD
URUGUAY     VENEZUELA

       82 EXCLUDED BECAUSE IRRELEVANT

ALBANIA     ALGERIA     AUSTRALIA   AUSTRIA     BELGIUM     BULGARIA    BURMA       BURUNDI     CAMBODIA    CAMEROUN
CANADA      CEN AFR REP CEYLON      CHAD        CHINA, PR   CONGO(BRA)  CONGO(LEO)  CZECHOS'KIA DAHOMEY     DENMARK
ETHIOPIA    FINLAND     FRANCE      GABON       GERMANY, E  GERMAN FR   GHANA       GREECE      GUINEA      HUNGARY
ICELAND     INDIA       INDONESIA   IRELAND     ITALY       IVORY COAST JAPAN       KOREA, N    KOREA REP   LAOS
LIBERIA     LIBYA       LUXEMBOURG  MALAGASY R  MALAYA      MALI        MAURITANIA  MONGOLIA    MOROCCO     NEPAL
NETHERLANDS NEW ZEALAND NIGER       NIGERIA     NORWAY      PAKISTAN    PHILIPPINES POLAND      PORTUGAL    RUMANIA
RWANDA      SENEGAL     SIERRE LEO  SOMALIA     SO AFRICA   SPAIN       SUDAN       SWEDEN      SWITZERLAND TANGANYIKA
THAILAND    TOGO        TUNISIA     UGANDA      USSR        UAR         UK          US          UPPER VOLTA VIETNAM, N
VIETNAM REP YUGOSLAVIA
------------------------------------------------------------------------------------------------------------

30  LEAN TOWARD BEING THOSE             0.60      30  LEAN TOWARD BEING THOSE             0.77          6    5
    WHOSE AGRICULTURAL POPULATION                     WHOSE AGRICULTURAL POPULATION                     4   17
    IS HIGH  (56)                                     IS MEDIUM, LOW, OR VERY LOW  (57)          X SQ=    2.74
                                                                                                 P =     0.056
                                                                                                 RV YES (YES)

45  LEAN TOWARD BEING THOSE             0.80      45  LEAN TOWARD BEING THOSE             0.64          2   14
    WHERE THE LITERACY RATE IS                        WHERE THE LITERACY RATE IS                        8    8
    BELOW FIFTY PERCENT  (54)                         FIFTY PERCENT OR ABOVE  (55)               X SQ=    3.64
                                                                                                 P =     0.054
                                                                                                 RV YES (YES)

46  TEND LESS TO BE THOSE               0.73      46  ALWAYS ARE THOSE                    1.00          8   22
    WHERE THE LITERACY RATE IS                        WHERE THE LITERACY RATE IS                        3    0
    TEN PERCENT OR ABOVE  (84)                        TEN PERCENT OR ABOVE  (84)                 X SQ=    3.71
                                                                                                 P =     0.030
                                                                                                 RV NO  (YES)
```

50	TEND TO BE THOSE WHERE FREEDOM OF THE PRESS IS INTERMITTENT, INTERNALLY ABSENT, OR INTERNALLY AND EXTERNALLY ABSENT (56)	0.91	50	TEND TO BE THOSE WHERE FREEDOM OF THE PRESS IS COMPLETE (43)	0.56	1 10 10 8 X SQ= 4.44 P = 0.019 RV YES (YES)
55	TEND LESS TO BE THOSE WHERE NEWSPAPER CIRCULATION IS TEN OR MORE PER THOUSAND (78)	0.64	55	ALWAYS ARE THOSE WHERE NEWSPAPER CIRCULATION IS TEN OR MORE PER THOUSAND (78)	1.00	7 22 4 0 X SQ= 6.01 P = 0.008 RV NO (YES)
56	ALWAYS ARE THOSE WHERE THE RELIGION IS PREDOMINANTLY LITERATE (79)	1.00	56	TILT LESS TOWARD BEING THOSE WHERE THE RELIGION IS PREDOMINANTLY LITERATE (79)	0.77	11 17 0 5 X SQ= 1.44 P = 0.143 RV NO (NO)
57	ALWAYS ARE THOSE WHERE THE RELIGION IS OTHER THAN CATHOLIC (90)	1.00	57	TEND TO BE THOSE WHERE THE RELIGION IS CATHOLIC (25)	0.68	0 15 11 7 X SQ= 11.14 P = 0. RV YES (YES)
58	TEND TO BE THOSE WHERE THE RELIGION IS MUSLIM (18)	0.73	58	ALWAYS ARE THOSE WHERE THE RELIGION IS OTHER THAN MUSLIM (97)	1.00	8 0 3 22 X SQ= 17.35 P = 0. RV YES (YES)
63	ALWAYS ARE THOSE WHERE THE RELIGION IS PREDOMINANTLY OR PARTLY OTHER THAN CHRISTIAN (68)	1.00	63	TEND TO BE THOSE WHERE THE RELIGION IS PREDOMINANTLY SOME KIND OF CHRISTIAN (46)	0.73	0 16 11 6 X SQ= 12.75 P = 0. RV YES (YES)
67	TEND TO BE THOSE THAT ARE RACIALLY HOMOGENEOUS (82)	0.89	67	TEND TO BE THOSE THAT ARE RACIALLY HETEROGENEOUS (27)	0.85	8 3 1 17 X SQ= 11.43 P = 0. RV YES (YES)
68	LEAN TOWARD BEING THOSE THAT ARE LINGUISTICALLY WEAKLY HETEROGENEOUS OR STRONGLY HETEROGENEOUS (62)	0.64	68	LEAN TOWARD BEING THOSE THAT ARE LINGUISTICALLY HOMOGENEOUS (52)	0.73	4 16 7 6 X SQ= 2.68 P = 0.065 RV YES (YES)
72	TEND TO BE THOSE WHOSE DATE OF INDEPENDENCE IS AFTER 1914 (59)	0.73	72	TEND TO BE THOSE WHOSE DATE OF INDEPENDENCE IS BEFORE 1914 (52)	0.91	3 20 8 2 X SQ= 11.21 P = 0. RV YES (YES)

75	TEND TO BE THOSE THAT ARE NOT HISTORICALLY WESTERN AND ARE NOT SIGNIFICANTLY WESTERNIZED (52)	0.64	75	TEND TO BE THOSE THAT ARE HISTORICALLY WESTERN OR SIGNIFICANTLY WESTERNIZED (62)	0.95	4 21 7 1 X SQ= 10.91 P = 0.001 RV YES (YES)
77	TEND TO BE THOSE THAT HAVE BEEN PARTIALLY WESTERNIZED, RATHER THAN SIGNIFICANTLY WESTERNIZED, THROUGH A COLONIAL RELATIONSHIP (41)	0.67	77	TEND TO BE THOSE THAT HAVE BEEN SIGNIFICANTLY WESTERNIZED, RATHER THAN PARTIALLY WESTERNIZED, THROUGH A COLONIAL RELATIONSHIP (28)	0.95	2 21 4 1 X SQ= 8.53 P = 0.003 RV YES (YES)
78	TILT LESS TOWARD BEING THOSE THAT HAVE BEEN SIGNIFICANTLY WESTERNIZED THROUGH A COLONIAL RELATIONSHIP, RATHER THAN WITHOUT SUCH A RELATIONSHIP (28)	0.67	78	ALWAYS ARE THOSE THAT HAVE BEEN SIGNIFICANTLY WESTERNIZED THROUGH A COLONIAL RELATIONSHIP, RATHER THAN WITHOUT SUCH A RELATIONSHIP (28)	1.00	2 21 1 0 X SQ= 1.34 P = 0.125 RV NO (YES)
79	ALWAYS ARE THOSE THAT WERE FORMERLY DEPENDENCIES OF BRITAIN OR FRANCE, RATHER THAN SPAIN (49)	1.00	79	TEND TO BE THOSE THAT WERE FORMERLY DEPENDENCIES OF SPAIN, RATHER THAN BRITAIN OR FRANCE (18)	0.86	6 3 0 18 X SQ= 11.81 P = 0. RV YES (YES)
82	ALWAYS ARE THOSE WHERE THE TYPE OF POLITICAL MODERNIZATION IS DEVELOPED TUTELARY OR UNDEVELOPED TUTELARY, RATHER THAN EARLY OR LATER EUROPEAN OR EUROPEAN DERIVED (55)	1.00	82	TEND TO BE THOSE WHERE THE TYPE OF POLITICAL MODERNIZATION IS EARLY OR LATER EUROPEAN OR EUROPEAN DERIVED, RATHER THAN DEVELOPED TUTELARY OR UNDEVELOPED TUTELARY (51)	0.91	0 20 8 2 X SQ= 17.92 P = 0. RV YES (YES)
83	TEND LESS TO BE THOSE WHERE THE TYPE OF POLITICAL MODERNIZATION IS OTHER THAN NON-EUROPEAN AUTOCHTHONOUS (106)	0.73	83	ALWAYS ARE THOSE WHERE THE TYPE OF POLITICAL MODERNIZATION IS OTHER THAN NON-EUROPEAN AUTOCHTHONOUS (106)	1.00	3 0 8 22 X SQ= 3.71 P = 0.030 RV NO (YES)
91	TILT TOWARD BEING THOSE WHOSE IDEOLOGICAL ORIENTATION IS TRADITIONAL, RATHER THAN DEVELOPMENTAL (5)	0.75	91	ALWAYS ARE THOSE WHOSE IDEOLOGICAL ORIENTATION IS DEVELOPMENTAL, RATHER THAN TRADITIONAL (31)	1.00	1 3 3 0 X SQ= 1.47 P = 0.143 RV YES (YES)
96	TILT TOWARD BEING THOSE WHERE THE STATUS OF THE REGIME IS AUTHORITARIAN (23)	0.50	96	TILT TOWARD BEING THOSE WHERE THE STATUS OF THE REGIME IS CONSTITUTIONAL OR TOTALITARIAN (67)	0.83	4 15 4 3 X SQ= 1.66 P = 0.149 RV YES (YES)
98	TILT LESS TOWARD BEING THOSE WHERE GOVERNMENTAL STABILITY IS GENERALLY OR MODERATELY PRESENT AND DATES FROM THE POST-WAR PERIOD, OR IS ABSENT (93)	0.82	98	ALWAYS ARE THOSE WHERE GOVERNMENTAL STABILITY IS GENERALLY OR MODERATELY PRESENT AND DATES FROM THE POST-WAR PERIOD, OR IS ABSENT (93)	1.00	2 0 9 22 X SQ= 1.66 P = 0.104 RV NO (YES)

114	TEND TO BE THOSE WHERE SECTIONALISM IS EXTREME OR MODERATE (61)	0.64	114	TEND TO BE THOSE WHERE SECTIONALISM IS NEGLIGIBLE (47)	0.78	7 4 4 14 X SQ= 3.37 P = 0.048 RV YES (YES)
117	LEAN LESS TOWARD BEING THOSE WHERE INTEREST ARTICULATION BY ASSOCIATIONAL GROUPS IS SIGNIFICANT, MODERATE, OR LIMITED (60)	0.55	117	LEAN MORE TOWARD BEING THOSE WHERE INTEREST ARTICULATION BY ASSOCIATIONAL GROUPS IS SIGNIFICANT, MODERATE, OR LIMITED (60)	0.86	6 19 5 3 X SQ= 2.50 P = 0.082 RV NO (YES)
118	TEND TO BE THOSE WHERE INTEREST ARTICULATION BY INSTITUTIONAL GROUPS IS VERY SIGNIFICANT (40)	0.70	118	TEND TO BE THOSE WHERE INTEREST ARTICULATION BY INSTITUTIONAL GROUPS IS SIGNIFICANT, MODERATE, OR LIMITED (60)	0.82	7 4 3 18 X SQ= 6.05 P = 0.013 RV YES (YES)
121	TEND TO BE THOSE WHERE INTEREST ARTICULATION BY NON-ASSOCIATIONAL GROUPS IS SIGNIFICANT (54)	0.82	121	TEND TO BE THOSE WHERE INTEREST ARTICULATION BY NON-ASSOCIATIONAL GROUPS IS MODERATE, LIMITED, OR NEGLIGIBLE (61)	0.95	9 1 2 21 X SQ= 17.24 P = 0. RV YES (YES)
122	ALWAYS ARE THOSE WHERE INTEREST ARTICULATION BY NON-ASSOCIATIONAL GROUPS IS SIGNIFICANT OR MODERATE (83)	1.00	122	TEND LESS TO BE THOSE WHERE INTEREST ARTICULATION BY NON-ASSOCIATIONAL GROUPS IS SIGNIFICANT OR MODERATE (83)	0.55	11 12 0 10 X SQ= 5.18 P = 0.013 RV NO (NO)
124	TEND LESS TO BE THOSE WHERE INTEREST ARTICULATION BY ANOMIC GROUPS IS OCCASIONAL, INFREQUENT, OR VERY INFREQUENT (85)	0.56	124	ALWAYS ARE THOSE WHERE INTEREST ARTICULATION BY ANOMIC GROUPS IS OCCASIONAL, INFREQUENT, OR VERY INFREQUENT (85)	1.00	4 0 5 18 X SQ= 6.20 P = 0.007 RV NO (YES)
127	TILT TOWARD BEING THOSE WHERE INTEREST ARTICULATION BY POLITICAL PARTIES IS SIGNIFICANT (21)	0.75	127	TILT TOWARD BEING THOSE WHERE INTEREST ARTICULATION BY POLITICAL PARTIES IS MODERATE, LIMITED, OR NEGLIGIBLE (72)	0.74	3 5 1 14 X SQ= 1.64 P = 0.103 RV YES (NO)
169	TILT TOWARD BEING THOSE WHERE THE HORIZONTAL POWER DISTRIBUTION IS NEGLIGIBLE (48)	0.50	169	TILT TOWARD BEING THOSE WHERE THE HORIZONTAL POWER DISTRIBUTION IS SIGNIFICANT OR LIMITED (58)	0.85	4 17 4 3 X SQ= 2.10 P = 0.142 RV YES (YES)
170	TEND TO BE THOSE WHERE THE LEGISLATIVE-EXECUTIVE STRUCTURE IS OTHER THAN PRESIDENTIAL (63)	0.86	170	TEND TO BE THOSE WHERE THE LEGISLATIVE-EXECUTIVE STRUCTURE IS PRESIDENTIAL (39)	0.85	1 17 6 3 X SQ= 8.70 P = 0.002 RV YES (YES)

```
171  LEAN LESS TOWARD BEING THOSE               0.71      171  ALWAYS ARE THOSE                            1.00           2     0
     WHERE THE LEGISLATIVE-EXECUTIVE STRUCTURE                 WHERE THE LEGISLATIVE-EXECUTIVE STRUCTURE                5    20
     IS OTHER THAN                                             IS OTHER THAN                                        X SQ=   2.71
     PARLIAMENTARY-REPUBLICAN  (90)                            PARLIAMENTARY-REPUBLICAN  (90)                       P =    0.060
                                                                                                                    RV NO  (YES)

185  TEND TO BE THOSE                           0.56      185  TEND TO BE THOSE                            0.91           4    10
     WHERE PARTICIPATION BY THE MILITARY                       WHERE PARTICIPATION BY THE MILITARY                      5     1
     IN POLITICS IS                                            IN POLITICS IS                                       X SQ=   3.12
     SUPPORTIVE, RATHER THAN                                   INTERVENTIVE, RATHER THAN                            P =    0.050
     INTERVENTIVE  (31)                                        SUPPORTIVE  (21)                                     RV YES (YES)

186  TEND TO BE THOSE                           0.71      186  TEND TO BE THOSE                            0.90           5     1
     WHERE PARTICIPATION BY THE MILITARY                       WHERE PARTICIPATION BY THE MILITARY                      2     9
     IN POLITICS IS                                            IN POLITICS IS                                       X SQ=   4.38
     SUPPORTIVE, RATHER THAN                                   NEUTRAL, RATHER THAN                                 P =    0.035
     NEUTRAL  (31)                                             SUPPORTIVE  (56)                                     RV YES (YES)

188  TEND TO BE THOSE                           0.91      188  TEND TO BE THOSE                            0.86           1    19
     WHERE THE CHARACTER OF THE LEGAL SYSTEM                   WHERE THE CHARACTER OF THE LEGAL SYSTEM                 10     3
     IS OTHER THAN CIVIL LAW  (81)                             IS CIVIL LAW  (32)                                   X SQ=  15.25
                                                                                                                    P =    0.
                                                                                                                    RV YES (YES)

192  TEND LESS TO BE THOSE                      0.55      192  ALWAYS ARE THOSE                            1.00           5     0
     WHERE THE CHARACTER OF THE LEGAL SYSTEM                   WHERE THE CHARACTER OF THE LEGAL SYSTEM                  6    22
     IS OTHER THAN                                             IS OTHER THAN                                        X SQ=   8.52
     MUSLIM OR PARTLY MUSLIM  (86)                             MUSLIM OR PARTLY MUSLIM  (86)                        P =    0.002
                                                                                                                    RV NO  (YES)
```

```
************************************************************************************************************

 18  POLITIES                                            18  POLITIES
     LOCATED IN NORTH AFRICA OR                              LOCATED IN EAST ASIA, SOUTH ASIA, OR
     CENTRAL AND SOUTH AFRICA, RATHER THAN                   SOUTHEAST ASIA, RATHER THAN IN
     IN EAST ASIA, SOUTH ASIA, OR                            NORTH AFRICA OR
     SOUTHEAST ASIA  (33)                                    CENTRAL AND SOUTH AFRICA  (18)

                                                                            BOTH SUBJECT AND PREDICATE
............................................................................................................

     33 IN LEFT COLUMN

ALGERIA     BURUNDI     CAMEROUN    CEN AFR REP CHAD        CONGO(BRA)  CONGO(LEO)  DAHOMEY     ETHIOPIA    GABON
GHANA       GUINEA      IVORY COAST LIBERIA     LIBYA       MALAGASY R  MALI        MAURITANIA  MOROCCO     NIGER
NIGERIA     RWANDA      SENEGAL     SIERRE LEO  SOMALIA     SO AFRICA   SUDAN       TANGANYIKA  TOGO        TUNISIA
UGANDA      UAR         UPPER VOLTA

     18 IN RIGHT COLUMN

BURMA       CAMBODIA    CEYLON      CHINA, PR   INDIA       INDONESIA   JAPAN       KOREA, N    KOREA REP   LAOS
MALAYA      MONGOLIA    NEPAL       PAKISTAN    PHILIPPINES THAILAND    VIETNAM, N  VIETNAM REP

     64 EXCLUDED BECAUSE IRRELEVANT

AFGHANISTAN ALBANIA     ARGENTINA   AUSTRALIA   AUSTRIA     BELGIUM     BOLIVIA     BRAZIL      BULGARIA    CANADA
CHILE       COLOMBIA    COSTA RICA  CUBA        CYPRUS      CZECHOS'KIA DENMARK     DOMIN REP   ECUADOR     EL SALVADOR
FINLAND     FRANCE      GERMANY, E  GERMAN FR   GREECE      GUATEMALA   HAITI       HONDURAS    HUNGARY     ICELAND
IRAN        IRAQ        IRELAND     ISRAEL      ITALY       JAMAICA     JORDAN      LEBANON     LUXEMBOURG  MEXICO
NETHERLANDS NEW ZEALAND NICARAGUA   NORWAY      PANAMA      PARAGUAY    PERU        POLAND      PORTUGAL    RUMANIA
SA'U ARABIA SPAIN       SWEDEN      SWITZERLAND SYRIA       TRINIDAD    TURKEY      USSR        UK          US
URUGUAY     VENEZUELA   YEMEN       YUGOSLAVIA
------------------------------------------------------------------------------------------------------------

 22  TILT MORE TOWARD BEING THOSE           0.79     22  TILT LESS TOWARD BEING THOSE          0.56        26    10
     WHOSE TERRITORIAL SIZE IS                           WHOSE TERRITORIAL SIZE IS                          7     8
     VERY LARGE, LARGE, OR MEDIUM  (68)                  VERY LARGE, LARGE, OR MEDIUM  (68)             X SQ=   2.01
                                                                                                        P =    0.112
                                                                                                        RV NO  (YES)

 23  TEND TO BE THOSE                       0.94     23  TEND TO BE THOSE                      0.50         2     9
     WHOSE POPULATION IS                                 WHOSE POPULATION IS                               31     9
     MEDIUM OR SMALL  (88)                               VERY LARGE OR LARGE  (27)                      X SQ=  10.82
                                                                                                        P =    0.001
                                                                                                        RV YES (YES)

 24  TEND TO BE THOSE                       0.67     24  TEND TO BE THOSE                      0.83        11    15
     WHOSE POPULATION IS                                 WHOSE POPULATION IS                               22     3
     SMALL  (54)                                         VERY LARGE, LARGE, OR MEDIUM  (61)             X SQ=   9.74
                                                                                                        P =    0.001
                                                                                                        RV YES (YES)
```

25	ALWAYS ARE THOSE WHOSE POPULATION DENSITY IS MEDIUM OR LOW (98)	1.00	25	TEND LESS TO BE THOSE WHOSE POPULATION DENSITY IS MEDIUM OR LOW (98)	0.78	0 4 33 14 X SQ= 5.18 P = 0.012 RV NO (YES)
26	TEND TO BE THOSE WHOSE POPULATION DENSITY IS LOW (67)	0.91	26	TEND TO BE THOSE WHOSE POPULATION DENSITY IS VERY HIGH, HIGH, OR MEDIUM (48)	0.78	3 14 30 4 X SQ= 21.73 P = 0. RV YES (YES)
30	TILT MORE TOWARD BEING THOSE WHOSE AGRICULTURAL POPULATION IS HIGH (56)	0.91	30	TILT LESS TOWARD BEING THOSE WHOSE AGRICULTURAL POPULATION IS HIGH (56)	0.71	30 12 3 5 X SQ= 2.10 P = 0.102 RV NO (YES)
34	TEND TO BE THOSE WHOSE GROSS NATIONAL PRODUCT IS VERY LOW (53)	0.79	34	TEND TO BE THOSE WHOSE GROSS NATIONAL PRODUCT IS VERY HIGH, HIGH, MEDIUM, OR LOW (62)	0.72	7 13 26 5 X SQ= 10.66 P = 0.001 RV YES (YES)
40	TEND TO BE THOSE WHOSE INTERNATIONAL FINANCIAL STATUS IS VERY LOW (39)	0.69	40	TEND TO BE THOSE WHOSE INTERNATIONAL FINANCIAL STATUS IS VERY HIGH, HIGH, MEDIUM, OR LOW (71)	0.75	10 12 22 4 X SQ= 6.56 P = 0.006 RV YES (YES)
43	TEND MORE TO BE THOSE WHOSE ECONOMIC DEVELOPMENTAL STATUS IS VERY UNDERDEVELOPED (57)	0.94	43	TEND LESS TO BE THOSE WHOSE ECONOMIC DEVELOPMENTAL STATUS IS VERY UNDERDEVELOPED (57)	0.61	2 7 31 11 X SQ= 6.53 P = 0.006 RV NO (YES)
45	ALWAYS ARE THOSE WHERE THE LITERACY RATE IS BELOW FIFTY PERCENT (54)	1.00	45	TEND TO BE THOSE WHERE THE LITERACY RATE IS FIFTY PERCENT OR ABOVE (55)	0.57	0 8 32 6 X SQ= 18.34 P = 0. RV YES (YES)
46	TEND TO BE THOSE WHERE THE LITERACY RATE IS BELOW TEN PERCENT (26)	0.70	46	TEND TO BE THOSE WHERE THE LITERACY RATE IS TEN PERCENT OR ABOVE (84)	0.87	9 14 21 2 X SQ= 11.60 P = 0. RV YES (YES)
51	LEAN TOWARD BEING THOSE WHERE FREEDOM OF THE PRESS IS COMPLETE OR INTERMITTENT (60)	0.70	51	LEAN TOWARD BEING THOSE WHERE FREEDOM OF THE PRESS IS INTERNALLY ABSENT, OR INTERNALLY AND EXTERNALLY ABSENT (37)	0.65	16 6 7 11 X SQ= 3.36 P = 0.053 RV YES (YES)

54	ALWAYS ARE THOSE WHERE NEWSPAPER CIRCULATION IS LESS THAN ONE HUNDRED PER THOUSAND (76)	1.00	54	TILT LESS TOWARD BEING THOSE WHERE NEWSPAPER CIRCULATION IS LESS THAN ONE HUNDRED PER THOUSAND (76)	0.87	0 2 33 14 X SQ= 1.70 P = 0.102 RV NO (YES)
55	TEND TO BE THOSE WHERE NEWSPAPER CIRCULATION IS LESS THAN TEN PER THOUSAND (35)	0.82	55	TEND TO BE THOSE WHERE NEWSPAPER CIRCULATION IS TEN OR MORE PER THOUSAND (78)	0.75	6 12 27 4 X SQ= 12.62 P = 0. RV YES (YES)
56	TEND TO BE THOSE WHERE THE RELIGION IS PREDOMINANTLY OR PARTLY NON-LITERATE (31)	0.76	56	TEND TO BE THOSE WHERE THE RELIGION IS PREDOMINANTLY LITERATE (79)	0.92	8 12 25 1 X SQ= 14.92 P = 0. RV YES (YES)
66	TEND TO BE THOSE THAT ARE RELIGIOUSLY HETEROGENEOUS (49)	0.73	66	TEND TO BE THOSE THAT ARE RELIGIOUSLY HOMOGENEOUS (57)	0.62	8 8 22 5 X SQ= 3.35 P = 0.043 RV YES (YES)
67	LEAN LESS TOWARD BEING THOSE THAT ARE RACIALLY HOMOGENEOUS (82)	0.72	67	LEAN MORE TOWARD BEING THOSE THAT ARE RACIALLY HOMOGENEOUS (82)	0.94	23 16 9 1 X SQ= 2.15 P = 0.079 RV NO (NO)
69	TILT MORE TOWARD BEING THOSE THAT ARE LINGUISTICALLY STRONGLY HETEROGENEOUS (50)	0.79	69	TILT LESS TOWARD BEING THOSE THAT ARE LINGUISTICALLY STRONGLY HETEROGENEOUS (50)	0.53	7 8 26 9 X SQ= 2.44 P = 0.102 RV NO (YES)
71	TEND MORE TO BE THOSE WHOSE DATE OF INDEPENDENCE IS AFTER 1800 (90)	0.97	71	TEND LESS TO BE THOSE WHOSE DATE OF INDEPENDENCE IS AFTER 1800 (90)	0.75	1 4 32 12 X SQ= 3.53 P = 0.034 RV NO (YES)
73	LEAN MORE TOWARD BEING THOSE WHOSE DATE OF INDEPENDENCE IS AFTER 1945 (46)	0.91	73	LEAN LESS TOWARD BEING THOSE WHOSE DATE OF INDEPENDENCE IS AFTER 1945 (46)	0.69	3 5 30 11 X SQ= 2.42 P = 0.094 RV NO (YES)
75	LEAN MORE TOWARD BEING THOSE THAT ARE NOT HISTORICALLY WESTERN AND ARE NOT SIGNIFICANTLY WESTERNIZED (52)	0.94	75	LEAN LESS TOWARD BEING THOSE THAT ARE NOT HISTORICALLY WESTERN AND ARE NOT SIGNIFICANTLY WESTERNIZED (52)	0.72	2 5 30 13 X SQ= 2.83 P = 0.083 RV NO (YES)

77	TILT MORE TOWARD BEING THOSE THAT HAVE BEEN PARTIALLY WESTERNIZED, RATHER THAN SIGNIFICANTLY WESTERNIZED, THROUGH A COLONIAL RELATIONSHIP (41)	0.93	77	TILT LESS TOWARD BEING THOSE THAT HAVE BEEN PARTIALLY WESTERNIZED, RATHER THAN SIGNIFICANTLY WESTERNIZED, THROUGH A COLONIAL RELATIONSHIP (41)	0.73	2 3 28 8 X SQ= 1.56 P = 0.110 RV NO (YES)
83	TEND MORE TO BE THOSE WHERE THE TYPE OF POLITICAL MODERNIZATION IS OTHER THAN NON-EUROPEAN AUTOCHTHONOUS (106)	0.97	83	TEND LESS TO BE THOSE WHERE THE TYPE OF POLITICAL MODERNIZATION IS OTHER THAN NON-EUROPEAN AUTOCHTHONOUS (106)	0.78	1 4 32 14 X SQ= 2.92 P = 0.047 RV NO (YES)
84	TEND TO BE THOSE WHERE THE TYPE OF POLITICAL MODERNIZATION IS UNDEVELOPED TUTELARY, RATHER THAN DEVELOPED TUTELARY (24)	0.77	84	ALWAYS ARE THOSE WHERE THE TYPE OF POLITICAL MODERNIZATION IS DEVELOPED TUTELARY, RATHER THAN UNDEVELOPED TUTELARY (31)	1.00	7 14 24 0 X SQ= 20.22 P = 0. RV YES (YES)
85	TEND MORE TO BE THOSE WHERE THE STAGE OF POLITICAL MODERNIZATION IS MID- OR EARLY TRANSITIONAL, RATHER THAN ADVANCED (54)	0.85	85	TEND LESS TO BE THOSE WHERE THE STAGE OF POLITICAL MODERNIZATION IS MID- OR EARLY TRANSITIONAL, RATHER THAN ADVANCED (54)	0.56	5 8 28 10 X SQ= 3.83 P = 0.041 RV NO (YES)
86	LEAN TOWARD BEING THOSE WHERE THE STAGE OF POLITICAL MODERNIZATION IS EARLY TRANSITIONAL, RATHER THAN ADVANCED OR MID-TRANSITIONAL (38)	0.79	86	LEAN TOWARD BEING THOSE WHERE THE STAGE OF POLITICAL MODERNIZATION IS ADVANCED OR MID-TRANSITIONAL, RATHER THAN EARLY TRANSITIONAL (76)	0.50	7 9 26 9 X SQ= 3.25 P = 0.057 RV YES (YES)
87	TEND TO BE THOSE WHOSE IDEOLOGICAL ORIENTATION IS DEVELOPMENTAL (31)	0.93	87	TEND TO BE THOSE WHOSE IDEOLOGICAL ORIENTATION IS OTHER THAN DEVELOPMENTAL (58)	0.80	25 2 2 8 X SQ= 15.99 P = 0. RV YES (YES)
92	TILT MORE TOWARD BEING THOSE WHERE THE SYSTEM STYLE IS LIMITED MOBILIZATIONAL, OR NON-MOBILIZATIONAL (93)	0.88	92	TILT LESS TOWARD BEING THOSE WHERE THE SYSTEM STYLE IS LIMITED MOBILIZATIONAL, OR NON-MOBILIZATIONAL (93)	0.67	4 6 29 12 X SQ= 2.12 P = 0.137 RV NO (YES)
95	ALWAYS ARE THOSE WHERE THE STATUS OF THE REGIME IS CONSTITUTIONAL OR AUTHORITARIAN (95)	1.00	95	TEND LESS TO BE THOSE WHERE THE STATUS OF THE REGIME IS CONSTITUTIONAL OR AUTHORITARIAN (95)	0.76	32 13 0 4 X SQ= 5.36 P = 0.011 RV NO (YES)
97	ALWAYS ARE THOSE WHERE THE STATUS OF THE REGIME IS AUTHORITARIAN, RATHER THAN TOTALITARIAN (23)	1.00	97	TILT LESS TOWARD BEING THOSE WHERE THE STATUS OF THE REGIME IS AUTHORITARIAN, RATHER THAN TOTALITARIAN (23)	0.67	8 8 0 4 X SQ= 1.58 P = 0.117 RV NO (YES)

No.	Statement	Value	No.	Statement	Value	Statistics
105	LEAN MORE TOWARD BEING THOSE WHERE THE ELECTORAL SYSTEM IS PARTIALLY COMPETITIVE OR NON-COMPETITIVE (47)	0.88	105	LEAN LESS TOWARD BEING THOSE WHERE THE ELECTORAL SYSTEM IS PARTIALLY COMPETITIVE OR NON-COMPETITIVE (47)	0.62	3 5 22 8 X SQ= 2.19 P = 0.094 RV NO (YES)
107	TILT MORE TOWARD BEING THOSE WHERE AUTONOMOUS GROUPS ARE PARTIALLY TOLERATED IN POLITICS, ARE TOLERATED ONLY OUTSIDE POLITICS, OR ARE NOT TOLERATED AT ALL (65)	0.91	107	TILT LESS TOWARD BEING THOSE WHERE AUTONOMOUS GROUPS ARE PARTIALLY TOLERATED IN POLITICS, ARE TOLERATED ONLY OUTSIDE POLITICS, OR ARE NOT TOLERATED AT ALL (65)	0.71	3 5 29 12 X SQ= 1.96 P = 0.106 RV NO (YES)
110	ALWAYS ARE THOSE WHERE POLITICAL ENCULTURATION IS MEDIUM OR LOW (80)	1.00	110	TILT LESS TOWARD BEING THOSE WHERE POLITICAL ENCULTURATION IS MEDIUM OR LOW (80)	0.86	0 2 29 12 X SQ= 1.72 P = 0.101 RV NO (YES)
115	ALWAYS ARE THOSE WHERE INTEREST ARTICULATION BY ASSOCIATIONAL GROUPS IS MODERATE, LIMITED, OR NEGLIGIBLE (91)	1.00	115	TILT LESS TOWARD BEING THOSE WHERE INTEREST ARTICULATION BY ASSOCIATIONAL GROUPS IS MODERATE, LIMITED, OR NEGLIGIBLE (91)	0.88	0 2 33 15 X SQ= 1.56 P = 0.111 RV NO (YES)
117	LEAN MORE TOWARD BEING THOSE WHERE INTEREST ARTICULATION BY ASSOCIATIONAL GROUPS IS NEGLIGIBLE (51)	0.85	117	LEAN LESS TOWARD BEING THOSE WHERE INTEREST ARTICULATION BY ASSOCIATIONAL GROUPS IS NEGLIGIBLE (51)	0.59	5 7 28 10 X SQ= 2.86 P = 0.077 RV NO (YES)
124	TILT MORE TOWARD BEING THOSE WHERE INTEREST ARTICULATION BY ANOMIC GROUPS IS OCCASIONAL, INFREQUENT, OR VERY INFREQUENT (85)	0.87	124	TILT LESS TOWARD BEING THOSE WHERE INTEREST ARTICULATION BY ANOMIC GROUPS IS OCCASIONAL, INFREQUENT, OR VERY INFREQUENT (85)	0.67	4 5 28 10 X SQ= 1.68 P = 0.121 RV NO (YES)
131	LEAN TOWARD BEING THOSE WHERE INTEREST AGGREGATION BY POLITICAL PARTIES IS SIGNIFICANT OR MODERATE (30)	0.83	131	LEAN TOWARD BEING THOSE WHERE INTEREST AGGREGATION BY POLITICAL PARTIES IS LIMITED OR NEGLIGIBLE (35)	0.58	10 5 2 7 X SQ= 2.84 P = 0.089 RV YES (YES)
132	ALWAYS ARE THOSE WHERE INTEREST AGGREGATION BY POLITICAL PARTIES IS SIGNIFICANT, MODERATE, OR LIMITED (67)	1.00	132	TEND LESS TO BE THOSE WHERE INTEREST AGGREGATION BY POLITICAL PARTIES IS SIGNIFICANT, MODERATE, OR LIMITED (67)	0.75	29 6 0 2 X SQ= 3.55 P = 0.042 RV NO (YES)
133	LEAN TOWARD BEING THOSE WHERE INTEREST AGGREGATION BY THE EXECUTIVE IS SIGNIFICANT (29)	0.61	133	LEAN TOWARD BEING THOSE WHERE INTEREST AGGREGATION BY THE EXECUTIVE IS MODERATE, LIMITED, OR NEGLIGIBLE (76)	0.71	19 5 12 12 X SQ= 3.28 P = 0.069 RV YES (YES)

```
134  TEND TO BE THOSE                     0.73     134  TEND TO BE THOSE                     0.62        22    6
     WHERE INTEREST AGGREGATION                         WHERE INTEREST AGGREGATION                       8   10
     BY THE EXECUTIVE                                   BY THE EXECUTIVE                             X SQ=   4.22
     IS SIGNIFICANT OR MODERATE  (57)                   IS LIMITED OR NEGLIGIBLE  (46)               P =    0.027
                                                                                                     RV YES (YES)

137  ALWAYS ARE THOSE                     1.00     137  TEND LESS TO BE THOSE                0.80         0    3
     WHERE INTEREST AGGREGATION                         WHERE INTEREST AGGREGATION                      28   12
     BY THE LEGISLATURE                                 BY THE LEGISLATURE                           X SQ=   3.33
     IS LIMITED OR NEGLIGIBLE  (68)                     IS LIMITED OR NEGLIGIBLE  (68)               P =    0.037
                                                                                                     RV NO  (YES)

146  TEND LESS TO BE THOSE                0.73     146  ALWAYS ARE THOSE                     1.00         8    0
     WHERE THE PARTY SYSTEM IS QUALITATIVELY            WHERE THE PARTY SYSTEM IS QUALITATIVELY         22   15
     OTHER THAN                                         OTHER THAN                                   X SQ=   3.21
     MASS-BASED TERRITORIAL  (92)                       MASS-BASED TERRITORIAL  (92)                 P =    0.038
                                                                                                     RV NO  (NO )

157  LEAN MORE TOWARD BEING THOSE         0.97     157  LEAN LESS TOWARD BEING THOSE         0.79         1    3
     WHERE PERSONALISMO IS                              WHERE PERSONALISMO IS                           28   11
     MODERATE OR NEGLIGIBLE  (84)                       MODERATE OR NEGLIGIBLE  (84)                 X SQ=   1.80
                                                                                                     P =    0.094
                                                                                                     RV NO  (YES)

160  TEND TO BE THOSE                     0.89     160  TEND TO BE THOSE                     0.50         3    7
     WHERE THE POLITICAL LEADERSHIP IS                  WHERE THE POLITICAL LEADERSHIP IS               24    7
     MODERATELY ELITIST OR NON-ELITIST  (67)            ELITIST  (30)                                X SQ=   5.60
                                                                                                     P =    0.017
                                                                                                     RV YES (YES)

161  TEND TO BE THOSE                     0.67     161  TEND TO BE THOSE                     0.71         9   10
     WHERE THE POLITICAL LEADERSHIP IS                  WHERE THE POLITICAL LEADERSHIP IS               18    4
     NON-ELITIST  (50)                                  ELITIST OR MODERATELY ELITIST  (47)          X SQ=   3.96
                                                                                                     P =    0.046
                                                                                                     RV YES (YES)

164  LEAN TOWARD BEING THOSE              0.78     164  LEAN TOWARD BEING THOSE              0.53        21    7
     WHERE THE REGIME'S LEADERSHIP CHARISMA             WHERE THE REGIME'S LEADERSHIP CHARISMA           6    8
     IS PRONOUNCED OR MODERATE  (34)                    IS NEGLIGIBLE  (65)                          X SQ=   2.92
                                                                                                     P =    0.085
                                                                                                     RV YES (YES)

168  TEND MORE TO BE THOSE                0.97     168  TEND LESS TO BE THOSE                0.75         1    4
     WHERE THE HORIZONTAL POWER DISTRIBUTION            WHERE THE HORIZONTAL POWER DISTRIBUTION         30   12
     IS LIMITED OR NEGLIGIBLE  (72)                     IS LIMITED OR NEGLIGIBLE  (72)               X SQ=   3.22
                                                                                                     P =    0.040
                                                                                                     RV NO  (YES)

170  LEAN TOWARD BEING THOSE              0.57     170  LEAN TOWARD BEING THOSE              0.75        16    4
     WHERE THE LEGISLATIVE-EXECUTIVE STRUCTURE          WHERE THE LEGISLATIVE-EXECUTIVE STRUCTURE       12   12
     IS PRESIDENTIAL  (39)                              IS OTHER THAN PRESIDENTIAL  (63)             X SQ=   3.05
                                                                                                     P =    0.060
                                                                                                     RV YES (YES)
```

174	TEND MORE TO BE THOSE WHERE THE LEGISLATURE IS PARTIALLY EFFECTIVE, LARGELY INEFFECTIVE, OR WHOLLY INEFFECTIVE (72)	0.97	174	TEND LESS TO BE THOSE WHERE THE LEGISLATURE IS PARTIALLY EFFECTIVE, LARGELY INEFFECTIVE, OR WHOLLY INEFFECTIVE (72)	0.73	1 4 29 11 X SQ= 3.40 P = 0.036 RV NO (YES)
176	TEND TO BE THOSE WHERE THE LEGISLATURE IS FULLY EFFECTIVE, PARTIALLY EFFECTIVE, OR LARGELY INEFFECTIVE (72)	0.87	176	TEND TO BE THOSE WHERE THE LEGISLATURE IS WHOLLY INEFFECTIVE (28)	0.60	26 6 4 9 X SQ= 8.45 P = 0.004 RV YES (YES)
182	TEND TO BE THOSE WHERE THE BUREAUCRACY IS POST-COLONIAL TRANSITIONAL, RATHER THAN SEMI-MODERN (25)	0.77	182	TEND TO BE THOSE WHERE THE BUREAUCRACY IS SEMI-MODERN, RATHER THAN POST-COLONIAL TRANSITIONAL (55)	0.91	7 10 24 1 X SQ= 13.02 P = 0. RV YES (YES)
183	TEND TO BE THOSE WHERE THE BUREAUCRACY IS POST-COLONIAL TRANSITIONAL, RATHER THAN TRADITIONAL (25)	0.96	183	TEND TO BE THOSE WHERE THE BUREAUCRACY IS TRADITIONAL, RATHER THAN POST-COLONIAL TRANSITIONAL (9)	0.80	24 1 1 4 X SQ= 12.29 P = 0.001 RV YES (YES)
186	LEAN TOWARD BEING THOSE WHERE PARTICIPATION BY THE MILITARY IN POLITICS IS NEUTRAL, RATHER THAN SUPPORTIVE (56)	0.75	186	LEAN TOWARD BEING THOSE WHERE PARTICIPATION BY THE MILITARY IN POLITICS IS SUPPORTIVE, RATHER THAN NEUTRAL (31)	0.58	7 7 21 5 X SQ= 2.77 P = 0.071 RV YES (YES)
192	TEND TO BE THOSE WHERE THE CHARACTER OF THE LEGAL SYSTEM IS MUSLIM OR PARTLY MUSLIM (27)	0.61	192	TEND TO BE THOSE WHERE THE CHARACTER OF THE LEGAL SYSTEM IS OTHER THAN MUSLIM OR PARTLY MUSLIM (86)	0.83	19 3 12 15 X SQ= 7.45 P = 0.003 RV YES (YES)
193	TEND TO BE THOSE WHERE THE CHARACTER OF THE LEGAL SYSTEM IS PARTLY INDIGENOUS (28)	0.76	193	TEND TO BE THOSE WHERE THE CHARACTER OF THE LEGAL SYSTEM IS OTHER THAN PARTLY INDIGENOUS (86)	0.82	25 3 8 14 X SQ= 13.11 P = 0. RV YES (YES)
194	ALWAYS ARE THOSE THAT ARE NON-COMMUNIST (101)	1.00	194	TEND LESS TO BE THOSE THAT ARE NON-COMMUNIST (101)	0.78	0 4 33 14 X SQ= 5.18 P = 0.012 RV NO (YES)

```
************************************************************************************************************

19  POLITIES                                          19  POLITIES
    LOCATED IN NORTH AFRICA OR                            LOCATED IN THE CARIBBEAN,
    CENTRAL AND SOUTH AFRICA, RATHER THAN                 CENTRAL AMERICA, OR SOUTH AMERICA,
    IN THE CARIBBEAN, CENTRAL AMERICA, OR                 RATHER THAN IN NORTH AFRICA OR
    SOUTH AMERICA  (33)                                   CENTRAL AND SOUTH AFRICA  (22)

                                                                              BOTH SUBJECT AND PREDICATE
............................................................................................................

     33 IN LEFT COLUMN

ALGERIA     BURUNDI     CAMEROUN    CEN AFR REP CHAD        CONGO(BRA)  CONGO(LEO)  DAHOMEY     ETHIOPIA    GABON
GHANA       GUINEA      IVORY COAST LIBERIA     LIBYA       MALAGASY R  MALI        MAURITANIA  MOROCCO     NIGER
NIGERIA     RWANDA      SENEGAL     SIERRE LEO  SOMALIA     SO AFRICA   SUDAN       TANGANYIKA  TOGO        TUNISIA
UGANDA      UAR         UPPER VOLTA

     22 IN RIGHT COLUMN

ARGENTINA   BOLIVIA     BRAZIL      CHILE       COLOMBIA    COSTA RICA  CUBA        DOMIN REP   ECUADOR     EL SALVADOR
GUATEMALA   HAITI       HONDURAS    JAMAICA     MEXICO      NICARAGUA   PANAMA      PARAGUAY    PERU        TRINIDAD
URUGUAY     VENEZUELA

     60 EXCLUDED BECAUSE IRRELEVANT

AFGHANISTAN ALBANIA     AUSTRALIA   AUSTRIA     BELGIUM     BULGARIA    BURMA       CAMBODIA    CANADA      CEYLON
CHINA, PR   CYPRUS      CZECHOS'KIA DENMARK     FINLAND     FRANCE      GERMANY, E  GERMAN FR   GREECE      HUNGARY
ICELAND     INDIA       INDONESIA   IRAN        IRAQ        IRELAND     ISRAEL      ITALY       JAPAN       JORDAN
KOREA, N    KOREA REP   LAOS        LEBANON     LUXEMBOURG  MALAYA      MONGOLIA    NEPAL       NETHERLANDS NEW ZEALAND
NORWAY      PAKISTAN    PHILIPPINES POLAND      PORTUGAL    RUMANIA     SA'U ARABIA SPAIN       SWEDEN      SWITZERLAND
SYRIA       THAILAND    TURKEY      USSR        UK          US          VIETNAM, N  VIETNAM REP YEMEN       YUGOSLAVIA
------------------------------------------------------------------------------------------------------------

22  TEND TO BE THOSE                        0.79      22  TEND TO BE THOSE                   0.55       26   10
    WHOSE TERRITORIAL SIZE IS                             WHOSE TERRITORIAL SIZE IS                       7   12
    VERY LARGE, LARGE, OR MEDIUM  (68)                    SMALL  (47)                               X SQ=   5.10
                                                                                                    P =    0.020
                                                                                                    RV YES (YES)

25  ALWAYS ARE THOSE                        1.00      25  TEND LESS TO BE THOSE              0.82        0    4
    WHOSE POPULATION DENSITY IS                           WHOSE POPULATION DENSITY IS                    33   18
    MEDIUM OR LOW  (98)                                   MEDIUM OR LOW  (98)                       X SQ=   4.06
                                                                                                    P =    0.021
                                                                                                    RV NO  (YES)

26  TILT MORE TOWARD BEING THOSE            0.91      26  TILT LESS TOWARD BEING THOSE       0.73        3    6
    WHOSE POPULATION DENSITY IS                           WHOSE POPULATION DENSITY IS                    30   16
    LOW  (67)                                             LOW  (67)                                 X SQ=   2.00
                                                                                                    P =    0.134
                                                                                                    RV NO  (YES)
```

28	TEND LESS TO BE THOSE WHOSE POPULATION GROWTH RATE IS HIGH (62)	0.57	28	TEND MORE TO BE THOSE WHOSE POPULATION GROWTH RATE IS HIGH (62)	0.86	17 19 13 3 X SQ= 3.95 P = 0.033 RV NO (YES)
29	TEND TO BE THOSE WHERE THE DEGREE OF URBANIZATION IS LOW (49)	0.85	29	TEND TO BE THOSE WHERE THE DEGREE OF URBANIZATION IS HIGH (56)	0.74	5 14 28 5 X SQ= 15.38 P = 0. RV YES (YES)
30	TEND TO BE THOSE WHOSE AGRICULTURAL POPULATION IS HIGH (56)	0.91	30	TEND TO BE THOSE WHOSE AGRICULTURAL POPULATION IS MEDIUM, LOW, OR VERY LOW (57)	0.77	30 5 3 17 X SQ= 23.65 P = 0. RV YES (YES)
34	TILT MORE TOWARD BEING THOSE WHOSE GROSS NATIONAL PRODUCT IS VERY LOW (53)	0.79	34	TILT LESS TOWARD BEING THOSE WHOSE GROSS NATIONAL PRODUCT IS VERY LOW (53)	0.59	7 9 26 13 X SQ= 1.62 P = 0.139 RV NO (YES)
37	TEND TO BE THOSE WHOSE PER CAPITA GROSS NATIONAL PRODUCT IS VERY LOW (51)	0.88	37	TEND TO BE THOSE WHOSE PER CAPITA GROSS NATIONAL PRODUCT IS VERY HIGH, HIGH, MEDIUM, OR LOW (64)	0.86	4 19 29 3 X SQ= 26.93 P = 0. RV YES (YES)
40	TEND TO BE THOSE WHOSE INTERNATIONAL FINANCIAL STATUS IS VERY LOW (39)	0.69	40	TEND TO BE THOSE WHOSE INTERNATIONAL FINANCIAL STATUS IS VERY HIGH, HIGH, MEDIUM, OR LOW (71)	0.60	10 12 22 8 X SQ= 3.07 P = 0.050 RV YES (YES)
43	TEND TO BE THOSE WHOSE ECONOMIC DEVELOPMENTAL STATUS IS VERY UNDERDEVELOPED (57)	0.94	43	TEND TO BE THOSE WHOSE ECONOMIC DEVELOPMENTAL STATUS IS DEVELOPED, INTERMEDIATE, OR UNDERDEVELOPED (55)	0.57	2 12 31 9 X SQ= 14.88 P = 0. RV YES (YES)
45	ALWAYS ARE THOSE WHERE THE LITERACY RATE IS BELOW FIFTY PERCENT (54)	1.00	45	TEND TO BE THOSE WHERE THE LITERACY RATE IS FIFTY PERCENT OR ABOVE (55)	0.64	0 14 32 8 X SQ= 24.28 P = 0. RV YES (YES)
46	TEND TO BE THOSE WHERE THE LITERACY RATE IS BELOW TEN PERCENT (26)	0.70	46	ALWAYS ARE THOSE WHERE THE LITERACY RATE IS TEN PERCENT OR ABOVE (84)	1.00	9 22 21 0 X SQ= 23.01 P = 0. RV YES (YES)

```
54  ALWAYS ARE THOSE                        1.00     54  TEND LESS TO BE THOSE                   0.73       0    6
    WHERE NEWSPAPER CIRCULATION IS                       WHERE NEWSPAPER CIRCULATION IS                    33   16
    LESS THAN ONE HUNDRED                                LESS THAN ONE HUNDRED                          X SQ=   7.49
    PER THOUSAND  (76)                                   PER THOUSAND  (76)                             P =     0.003
                                                                                                        RV NO  (YES)

55  TEND TO BE THOSE                        0.82     55  ALWAYS ARE THOSE                        1.00       6   22
    WHERE NEWSPAPER CIRCULATION IS                       WHERE NEWSPAPER CIRCULATION IS                    27    0
    LESS THAN TEN                                        TEN OR MORE                                    X SQ=  32.16
    PER THOUSAND  (35)                                   PER THOUSAND  (78)                             P =     0.
                                                                                                        RV YES (YES)

56  TEND TO BE THOSE                        0.76     56  TEND TO BE THOSE                        0.77       8   17
    WHERE THE RELIGION IS                                WHERE THE RELIGION IS                             25    5
    PREDOMINANTLY OR PARTLY                              PREDOMINANTLY LITERATE  (79)                   X SQ=  12.91
    NON-LITERATE  (31)                                                                                  P =     0.
                                                                                                        RV YES (YES)

57  ALWAYS ARE THOSE                        1.00     57  TEND TO BE THOSE                        0.68       0   15
    WHERE THE RELIGION IS OTHER THAN                     WHERE THE RELIGION IS                             33    7
    CATHOLIC  (90)                                       CATHOLIC  (25)                                 X SQ=  27.60
                                                                                                        P =     0.
                                                                                                        RV YES (YES)

58  TEND LESS TO BE THOSE                   0.76     58  ALWAYS ARE THOSE                        1.00       8    0
    WHERE THE RELIGION IS OTHER THAN                     WHERE THE RELIGION IS OTHER THAN                  25   22
    MUSLIM  (97)                                         MUSLIM  (97)                                   X SQ=   4.44
                                                                                                        P =     0.016
                                                                                                        RV NO  (NO )

63  ALWAYS ARE THOSE                        1.00     63  TEND TO BE THOSE                        0.73       0   16
    WHERE THE RELIGION IS                                WHERE THE RELIGION IS                             33    6
    PREDOMINANTLY OR PARTLY                              PREDOMINANTLY                                  X SQ=  30.41
    OTHER THAN CHRISTIAN  (68)                           SOME KIND OF CHRISTIAN  (46)                   P =     0.
                                                                                                        RV YES (YES)

66  TEND TO BE THOSE                        0.73     66  TEND TO BE THOSE                        0.68       8   15
    THAT ARE RELIGIOUSLY HETEROGENEOUS  (49)             THAT ARE RELIGIOUSLY HOMOGENEOUS  (57)            22    7
                                                                                                        X SQ=   7.27
                                                                                                        P =     0.005
                                                                                                        RV YES (YES)

67  TEND TO BE THOSE                        0.72     67  TEND TO BE THOSE                        0.85      23    3
    THAT ARE RACIALLY HOMOGENEOUS  (82)                  THAT ARE RACIALLY HETEROGENEOUS  (27)              9   17
                                                                                                        X SQ=  13.73
                                                                                                        P =     0.
                                                                                                        RV YES (YES)

68  TEND TO BE THOSE                        0.79     68  TEND TO BE THOSE                        0.73       7   16
    THAT ARE LINGUISTICALLY                              THAT ARE LINGUISTICALLY                           26    6
    WEAKLY HETEROGENEOUS OR                              HOMOGENEOUS  (52)                              X SQ=  12.36
    STRONGLY HETEROGENEOUS  (62)                                                                        P =     0.
                                                                                                        RV YES (YES)
```

69	TEND TO BE THOSE THAT ARE LINGUISTICALLY STRONGLY HETEROGENEOUS (50)	0.79	69	TEND TO BE THOSE THAT ARE LINGUISTICALLY HOMOGENEOUS OR WEAKLY HETEROGENEOUS (64)	0.82	7 18 26 4 X SQ= 17.19 P = 0. RV YES (YES)
72	TEND TO BE THOSE WHOSE DATE OF INDEPENDENCE IS AFTER 1914 (59)	0.91	72	TEND TO BE THOSE WHOSE DATE OF INDEPENDENCE IS BEFORE 1914 (52)	0.91	3 20 30 2 X SQ= 33.03 P = 0. RV YES (YES)
73	TEND TO BE THOSE WHOSE DATE OF INDEPENDENCE IS AFTER 1945 (46)	0.91	73	TEND TO BE THOSE WHOSE DATE OF INDEPENDENCE IS BEFORE 1945 (65)	0.91	3 20 30 2 X SQ= 33.03 P = 0. RV YES (YES)
75	TEND TO BE THOSE THAT ARE NOT HISTORICALLY WESTERN AND ARE NOT SIGNIFICANTLY WESTERNIZED (52)	0.94	75	TEND TO BE THOSE THAT ARE HISTORICALLY WESTERN OR SIGNIFICANTLY WESTERNIZED (62)	0.95	2 21 30 1 X SQ= 38.86 P = 0. RV YES (YES)
77	TEND TO BE THOSE THAT HAVE BEEN PARTIALLY WESTERNIZED, RATHER THAN SIGNIFICANTLY WESTERNIZED, THROUGH A COLONIAL RELATIONSHIP (41)	0.93	77	TEND TO BE THOSE THAT HAVE BEEN SIGNIFICANTLY WESTERNIZED, RATHER THAN PARTIALLY WESTERNIZED, THROUGH A COLONIAL RELATIONSHIP (28)	0.95	2 21 28 1 X SQ= 37.04 P = 0. RV YES (YES)
79	ALWAYS ARE THOSE THAT WERE FORMERLY DEPENDENCIES OF BRITAIN OR FRANCE, RATHER THAN SPAIN (49)	1.00	79	TEND TO BE THOSE THAT WERE FORMERLY DEPENDENCIES OF SPAIN, RATHER THAN BRITAIN OR FRANCE (18)	0.86	26 3 0 18 X SQ= 32.58 P = 0. RV YES (YES)
80	TILT TOWARD BEING THOSE WHOSE DATE OF INDEPENDENCE IS AFTER 1914, AND THAT WERE FORMERLY DEPENDENCIES OF FRANCE, RATHER THAN BRITAIN (24)	0.72	80	ALWAYS ARE THOSE WHOSE DATE OF INDEPENDENCE IS AFTER 1914, AND THAT WERE FORMERLY DEPENDENCIES OF BRITAIN, RATHER THAN FRANCE (19)	1.00	7 2 18 0 X SQ= 1.69 P = 0.103 RV YES (NO)
81	ALWAYS ARE THOSE WHERE THE TYPE OF POLITICAL MODERNIZATION IS EARLY EUROPEAN OR EARLY EUROPEAN DERIVED, RATHER THAN LATER EUROPEAN OR LATER EUROPEAN DERIVED (11)	1.00	81	ALWAYS ARE THOSE WHERE THE TYPE OF POLITICAL MODERNIZATION IS LATER EUROPEAN OR LATER EUROPEAN DERIVED, RATHER THAN EARLY EUROPEAN OR EARLY EUROPEAN DERIVED (40)	1.00	1 0 0 20 X SQ= 4.74 P = 0.048 RV YES (YES)
82	TEND TO BE THOSE WHERE THE TYPE OF POLITICAL MODERNIZATION IS DEVELOPED TUTELARY OR UNDEVELOPED TUTELARY, RATHER THAN EARLY OR LATER EUROPEAN OR EUROPEAN DERIVED (55)	0.97	82	TEND TO BE THOSE WHERE THE TYPE OF POLITICAL MODERNIZATION IS EARLY OR LATER EUROPEAN OR EUROPEAN DERIVED, RATHER THAN DEVELOPED TUTELARY OR UNDEVELOPED TUTELARY (51)	0.91	1 20 31 2 X SQ= 38.66 P = 0. RV YES (YES)

84	LEAN TOWARD BEING THOSE WHERE THE TYPE OF POLITICAL MODERNIZATION IS UNDEVELOPED TUTELARY, RATHER THAN DEVELOPED TUTELARY (24)	0.77	84	ALWAYS ARE THOSE WHERE THE TYPE OF POLITICAL MODERNIZATION IS DEVELOPED TUTELARY, RATHER THAN UNDEVELOPED TUTELARY (31)	1.00	7 2 24 0 X SQ= 2.45 P = 0.068 RV YES (NO)
86	TEND TO BE THOSE WHERE THE STAGE OF POLITICAL MODERNIZATION IS EARLY TRANSITIONAL, RATHER THAN ADVANCED OR MID-TRANSITIONAL (38)	0.79	86	TEND TO BE THOSE WHERE THE STAGE OF POLITICAL MODERNIZATION IS ADVANCED OR MID-TRANSITIONAL, RATHER THAN EARLY TRANSITIONAL (76)	0.91	7 20 26 2 X SQ= 22.94 P = 0. RV YES (YES)
87	TEND TO BE THOSE WHOSE IDEOLOGICAL ORIENTATION IS DEVELOPMENTAL (31)	0.93	87	TEND TO BE THOSE WHOSE IDEOLOGICAL ORIENTATION IS OTHER THAN DEVELOPMENTAL (58)	0.75	25 3 2 9 X SQ= 15.55 P = 0. RV YES (YES)
89	TEND TO BE THOSE WHOSE IDEOLOGICAL ORIENTATION IS OTHER THAN CONVENTIONAL (62)	0.96	89	TEND TO BE THOSE WHOSE IDEOLOGICAL ORIENTATION IS CONVENTIONAL (33)	0.54	1 7 26 6 X SQ= 10.83 P = 0.001 RV YES (YES)
93	TILT LESS TOWARD BEING THOSE WHERE THE SYSTEM STYLE IS NON-MOBILIZATIONAL (78)	0.67	93	TILT MORE TOWARD BEING THOSE WHERE THE SYSTEM STYLE IS NON-MOBILIZATIONAL (78)	0.86	11 3 22 19 X SQ= 1.76 P = 0.125 RV NO (NO)
96	LEAN TOWARD BEING THOSE WHERE THE STATUS OF THE REGIME IS AUTHORITARIAN (23)	0.50	96	LEAN TOWARD BEING THOSE WHERE THE STATUS OF THE REGIME IS CONSTITUTIONAL OR TOTALITARIAN (67)	0.83	8 15 8 3 X SQ= 2.91 P = 0.066 RV YES (YES)
99	TEND TO BE THOSE WHERE GOVERNMENTAL STABILITY IS GENERALLY PRESENT AND DATES FROM AT LEAST THE INTER-WAR PERIOD, OR FROM THE POST-WAR PERIOD (50)	0.62	99	TEND TO BE THOSE WHERE GOVERNMENTAL STABILITY IS MODERATELY PRESENT AND DATES FROM THE POST-WAR PERIOD, OR IS ABSENT (36)	0.87	8 2 5 14 X SQ= 5.62 P = 0.016 RV YES (YES)
100	TILT TOWARD BEING THOSE WHERE GOVERNMENTAL STABILITY IS GENERALLY PRESENT AND DATES FROM AT LEAST THE INTER-WAR PERIOD, OR IS GENERALLY OR MODERATELY PRESENT, AND DATES FROM THE POST-WAR PERIOD (64)	0.82	100	TILT TOWARD BEING THOSE WHERE GOVERNMENTAL STABILITY IS ABSENT (22)	0.50	9 9 2 9 X SQ= 1.74 P = 0.125 RV YES (YES)
102	LEAN TOWARD BEING THOSE WHERE THE REPRESENTATIVE CHARACTER OF THE REGIME IS PSEUDO-POLYARCHIC OR NON-POLYARCHIC (49)	0.53	102	LEAN TOWARD BEING THOSE WHERE THE REPRESENTATIVE CHARACTER OF THE REGIME IS POLYARCHIC OR LIMITED POLYARCHIC (59)	0.75	15 15 17 5 X SQ= 2.92 P = 0.082 RV YES (YES)

105	TEND TO BE THOSE WHERE THE ELECTORAL SYSTEM IS PARTIALLY COMPETITIVE OR NON-COMPETITIVE (47)	0.88	105	TEND TO BE THOSE WHERE THE ELECTORAL SYSTEM IS COMPETITIVE (43)	0.76	M 3 13 22 4 X SQ= 15.21 P = 0. RV YES (YES)
106	TEND TO BE THOSE WHERE THE ELECTORAL SYSTEM IS NON-COMPETITIVE (30)	0.52	106	TEND TO BE THOSE WHERE THE ELECTORAL SYSTEM IS COMPETITIVE OR PARTIALLY COMPETITIVE (52)	0.88	10 15 11 2 X SQ= 5.20 P = 0.015 RV YES (YES)
107	TEND TO BE THOSE WHERE AUTONOMOUS GROUPS ARE PARTIALLY TOLERATED IN POLITICS, ARE TOLERATED ONLY OUTSIDE POLITICS, OR ARE NOT TOLERATED AT ALL (65)	0.91	107	TEND TO BE THOSE WHERE AUTONOMOUS GROUPS ARE FULLY TOLERATED IN POLITICS (46)	0.67	3 14 29 7 X SQ= 16.56 P = 0. RV YES (YES)
114	TEND TO BE THOSE WHERE SECTIONALISM IS EXTREME OR MODERATE (61)	0.74	114	TEND TO BE THOSE WHERE SECTIONALISM IS NEGLIGIBLE (47)	0.78	23 4 8 14 X SQ= 10.42 P = 0.001 RV YES (YES)
117	TEND TO BE THOSE WHERE INTEREST ARTICULATION BY ASSOCIATIONAL GROUPS IS NEGLIGIBLE (51)	0.85	117	TEND TO BE THOSE WHERE INTEREST ARTICULATION BY ASSOCIATIONAL GROUPS IS SIGNIFICANT, MODERATE, OR LIMITED (60)	0.86	5 19 28 3 X SQ= 24.40 P = 0. RV YES (YES)
121	TEND TO BE THOSE WHERE INTEREST ARTICULATION BY NON-ASSOCIATIONAL GROUPS IS SIGNIFICANT (54)	0.82	121	TEND TO BE THOSE WHERE INTEREST ARTICULATION BY NON-ASSOCIATIONAL GROUPS IS MODERATE, LIMITED, OR NEGLIGIBLE (61)	0.95	27 1 6 21 X SQ= 28.52 P = 0. RV YES (YES)
122	ALWAYS ARE THOSE WHERE INTEREST ARTICULATION BY NON-ASSOCIATIONAL GROUPS IS SIGNIFICANT OR MODERATE (83)	1.00	122	TEND LESS TO BE THOSE WHERE INTEREST ARTICULATION BY NON-ASSOCIATIONAL GROUPS IS SIGNIFICANT OR MODERATE (83)	0.55	33 12 0 10 X SQ= 15.41 P = 0. RV NO (YES)
128	TEND TO BE THOSE WHERE INTEREST ARTICULATION BY POLITICAL PARTIES IS LIMITED OR NEGLIGIBLE (45)	0.68	128	TEND TO BE THOSE WHERE INTEREST ARTICULATION BY POLITICAL PARTIES IS SIGNIFICANT OR MODERATE (48)	0.74	9 14 19 5 X SQ= 6.24 P = 0.008 RV YES (YES)
129	TEND TO BE THOSE WHERE INTEREST ARTICULATION BY POLITICAL PARTIES IS NEGLIGIBLE (37)	0.57	129	TEND TO BE THOSE WHERE INTEREST ARTICULATION BY POLITICAL PARTIES IS SIGNIFICANT, MODERATE, OR LIMITED (56)	0.95	12 18 16 1 X SQ= 11.04 P = 0. RV YES (YES)

131	TEND TO BE THOSE WHERE INTEREST AGGREGATION BY POLITICAL PARTIES IS SIGNIFICANT OR MODERATE (30)	0.83	131	TEND TO BE THOSE WHERE INTEREST AGGREGATION BY POLITICAL PARTIES IS LIMITED OR NEGLIGIBLE (35)	0.67	10 6 2 12 X SQ= 5.36 P = 0.011 RV YES (YES)
132	ALWAYS ARE THOSE WHERE INTEREST AGGREGATION BY POLITICAL PARTIES IS SIGNIFICANT, MODERATE, OR LIMITED (67)	1.00	132	TEND LESS TO BE THOSE WHERE INTEREST AGGREGATION BY POLITICAL PARTIES IS SIGNIFICANT, MODERATE, OR LIMITED (67)	0.72	29 13 0 5 X SQ= 6.33 P = 0.006 RV NO (YES)
133	TEND TO BE THOSE WHERE INTEREST AGGREGATION BY THE EXECUTIVE IS SIGNIFICANT (29)	0.61	133	TEND TO BE THOSE WHERE INTEREST AGGREGATION BY THE EXECUTIVE IS MODERATE, LIMITED, OR NEGLIGIBLE (76)	0.85	19 3 12 17 X SQ= 8.82 P = 0.001 RV YES (YES)
134	TEND TO BE THOSE WHERE INTEREST AGGREGATION BY THE EXECUTIVE IS SIGNIFICANT OR MODERATE (57)	0.73	134	TEND TO BE THOSE WHERE INTEREST AGGREGATION BY THE EXECUTIVE IS LIMITED OR NEGLIGIBLE (46)	0.60	22 8 8 12 X SQ= 4.25 P = 0.038 RV YES (YES)
137	ALWAYS ARE THOSE WHERE INTEREST AGGREGATION BY THE LEGISLATURE IS LIMITED OR NEGLIGIBLE (68)	1.00	137	TEND LESS TO BE THOSE WHERE INTEREST AGGREGATION BY THE LEGISLATURE IS LIMITED OR NEGLIGIBLE (68)	0.78	0 4 28 14 X SQ= 4.30 P = 0.019 RV NO (YES)
138	TEND TO BE THOSE WHERE INTEREST AGGREGATION BY THE LEGISLATURE IS NEGLIGIBLE (49)	0.82	138	TEND TO BE THOSE WHERE INTEREST AGGREGATION BY THE LEGISLATURE IS SIGNIFICANT, MODERATE, OR LIMITED (48)	0.80	5 16 23 4 X SQ= 15.87 P = 0. RV YES (YES)
139	TEND TO BE THOSE WHERE THE PARTY SYSTEM IS QUANTITATIVELY ONE-PARTY (34)	0.53	139	TEND TO BE THOSE WHERE THE PARTY SYSTEM IS QUANTITATIVELY OTHER THAN ONE-PARTY (71)	0.95	17 1 15 20 X SQ= 11.15 P = 0. RV YES (YES)
142	TEND TO BE THOSE WHERE THE PARTY SYSTEM IS QUANTITATIVELY OTHER THAN MULTI-PARTY (66)	0.90	142	TEND TO BE THOSE WHERE THE PARTY SYSTEM IS QUANTITATIVELY MULTI-PARTY (30)	0.55	3 11 26 9 X SQ= 9.48 P = 0.001 RV YES (YES)
144	TEND TO BE THOSE WHERE THE PARTY SYSTEM IS QUANTITATIVELY ONE-PARTY, RATHER THAN TWO-PARTY (34)	0.94	144	TEND TO BE THOSE WHERE THE PARTY SYSTEM IS QUANTITATIVELY TWO-PARTY, RATHER THAN ONE-PARTY (11)	0.75	17 1 1 3 X SQ= 6.45 P = 0.010 RV YES (YES)

```
146  TEND LESS TO BE THOSE                        0.73     146  ALWAYS ARE THOSE                              1.00        8    0
     WHERE THE PARTY SYSTEM IS QUALITATIVELY                    WHERE THE PARTY SYSTEM IS QUALITATIVELY                   22   20
     OTHER THAN                                                 OTHER THAN                                           X SQ=   4.52
     MASS-BASED TERRITORIAL  (92)                               MASS-BASED TERRITORIAL  (92)                         P =    0.015
                                                                                                                     RV NO  (NO )

147  ALWAYS ARE THOSE                             1.00     147  TEND LESS TO BE THOSE                         0.68        0    6
     WHERE THE PARTY SYSTEM IS QUALITATIVELY                    WHERE THE PARTY SYSTEM IS QUALITATIVELY                   28   13
     OTHER THAN                                                 OTHER THAN                                           X SQ=   7.50
     CLASS-ORIENTED OR MULTI-IDEOLOGICAL  (67)                  CLASS-ORIENTED OR MULTI-IDEOLOGICAL  (67)            P =    0.003
                                                                                                                     RV NO  (YES)

154  TEND TO BE THOSE                             0.75     154  TEND TO BE THOSE                              0.73        9    4
     WHERE THE PARTY SYSTEM IS                                  WHERE THE PARTY SYSTEM IS                                  3   11
     STABLE OR MODERATELY STABLE  (55)                          UNSTABLE  (25)                                       X SQ=   4.45
                                                                                                                     P =    0.021
                                                                                                                     RV YES (YES)

157  TEND MORE TO BE THOSE                        0.97     157  TEND LESS TO BE THOSE                         0.65        1    7
     WHERE PERSONALISMO IS                                      WHERE PERSONALISMO IS                                     28   13
     MODERATE OR NEGLIGIBLE  (84)                               MODERATE OR NEGLIGIBLE  (84)                         X SQ=   6.47
                                                                                                                     P =    0.005
                                                                                                                     RV NO  (YES)

161  LEAN TOWARD BEING THOSE                      0.67     161  LEAN TOWARD BEING THOSE                       0.60        9   12
     WHERE THE POLITICAL LEADERSHIP IS                          WHERE THE POLITICAL LEADERSHIP IS                         18    8
     NON-ELITIST  (50)                                          ELITIST OR MODERATELY ELITIST  (47)                  X SQ=   2.31
                                                                                                                     P =    0.084
                                                                                                                     RV YES (YES)

164  TEND TO BE THOSE                             0.78     164  TEND TO BE THOSE                              0.91       21    2
     WHERE THE REGIME'S LEADERSHIP CHARISMA                     WHERE THE REGIME'S LEADERSHIP CHARISMA                     6   20
     IS PRONOUNCED OR MODERATE  (34)                            IS NEGLIGIBLE  (65)                                  X SQ=  20.29
                                                                                                                     P =    0.
                                                                                                                     RV YES (YES)

169  TEND TO BE THOSE                             0.61     169  TEND TO BE THOSE                              0.85       12   17
     WHERE THE HORIZONTAL POWER DISTRIBUTION                    WHERE THE HORIZONTAL POWER DISTRIBUTION                   19    3
     IS NEGLIGIBLE  (48)                                        IS SIGNIFICANT OR LIMITED  (58)                      X SQ=   8.82
                                                                                                                     P =    0.001
                                                                                                                     RV YES (YES)

170  LEAN LESS TOWARD BEING THOSE                 0.57     170  LEAN MORE TOWARD BEING THOSE                  0.85       16   17
     WHERE THE LEGISLATIVE-EXECUTIVE STRUCTURE                  WHERE THE LEGISLATIVE-EXECUTIVE STRUCTURE                 12    3
     IS PRESIDENTIAL  (39)                                      IS PRESIDENTIAL  (39)                                X SQ=   3.02
                                                                                                                     P =    0.059
                                                                                                                     RV NO  (YES)

175  TEND TO BE THOSE                             0.73     175  TEND TO BE THOSE                              0.78        8   14
     WHERE THE LEGISLATURE IS                                   WHERE THE LEGISLATURE IS                                  22    4
     LARGELY INEFFECTIVE OR                                     FULLY EFFECTIVE OR                                   X SQ=   9.87
     WHOLLY INEFFECTIVE  (49)                                   PARTIALLY EFFECTIVE  (51)                            P =    0.001
                                                                                                                     RV YES (YES)
```

178	TEND TO BE THOSE WHERE THE LEGISLATURE IS UNICAMERAL (53)	0.77	178	TEND TO BE THOSE WHERE THE LEGISLATURE IS BICAMERAL (51)	0.67	24 7 7 14 X SQ= 8.36 P = 0.003 RV YES (YES)
179	TEND TO BE THOSE WHERE THE EXECUTIVE IS DOMINANT (52)	0.79	179	TEND TO BE THOSE WHERE THE EXECUTIVE IS STRONG (39)	0.59	19 7 5 10 X SQ= 4.66 P = 0.021 RV YES (YES)
182	TEND TO BE THOSE WHERE THE BUREAUCRACY IS POST-COLONIAL TRANSITIONAL, RATHER THAN SEMI-MODERN (25)	0.77	182	ALWAYS ARE THOSE WHERE THE BUREAUCRACY IS SEMI-MODERN, RATHER THAN POST-COLONIAL TRANSITIONAL (55)	1.00	7 22 24 0 X SQ= 28.08 P = 0. RV YES (YES)
185	TEND TO BE THOSE WHERE PARTICIPATION BY THE MILITARY IN POLITICS IS SUPPORTIVE, RATHER THAN INTERVENTIVE (31)	0.70	185	TEND TO BE THOSE WHERE PARTICIPATION BY THE MILITARY IN POLITICS IS INTERVENTIVE, RATHER THAN SUPPORTIVE (21)	0.91	3 10 7 1 X SQ= 5.86 P = 0.008 RV YES (YES)
188	ALWAYS ARE THOSE WHERE THE CHARACTER OF THE LEGAL SYSTEM IS OTHER THAN CIVIL LAW (81)	1.00	188	TEND TO BE THOSE WHERE THE CHARACTER OF THE LEGAL SYSTEM IS CIVIL LAW (32)	0.86	0 19 31 3 X SQ= 38.06 P = 0. RV YES (YES)
192	TEND TO BE THOSE WHERE THE CHARACTER OF THE LEGAL SYSTEM IS MUSLIM OR PARTLY MUSLIM (27)	0.61	192	ALWAYS ARE THOSE WHERE THE CHARACTER OF THE LEGAL SYSTEM IS OTHER THAN MUSLIM OR PARTLY MUSLIM (86)	1.00	19 0 12 22 X SQ= 18.44 P = 0. RV YES (YES)
193	TEND TO BE THOSE WHERE THE CHARACTER OF THE LEGAL SYSTEM IS PARTLY INDIGENOUS (28)	0.76	193	ALWAYS ARE THOSE WHERE THE CHARACTER OF THE LEGAL SYSTEM IS OTHER THAN PARTLY INDIGENOUS (86)	1.00	25 0 8 22 X SQ= 27.58 P = 0. RV YES (YES)

MATRIX

	19 POLITIES LOCATED IN NORTH AFRICA OR CENTRAL AND SOUTH AFRICA, RATHER THAN IN THE CARIBBEAN, CENTRAL AMERICA, OR SOUTH AMERICA (33)	19 POLITIES LOCATED IN THE CARIBBEAN, CENTRAL AMERICA, OR SOUTH AMERICA, RATHER THAN IN NORTH AFRICA OR CENTRAL AND SOUTH AFRICA (22)
105 POLITIES WHERE THE ELECTORAL SYSTEM IS COMPETITIVE (43)	3 MALAGASY R SIERRE LEO UGANDA	13 ARGENTINA BOLIVIA BRAZIL CHILE COSTA RICA DOMIN REP ECUADOR HONDURAS JAMAICA PANAMA TRINIDAD URUGUAY VENEZUELA
105 POLITIES WHERE THE ELECTORAL SYSTEM IS PARTIALLY COMPETITIVE OR NON-COMPETITIVE (47)	22 ALGERIA BURUNDI CEN AFR REP CHAD CONGO(BRA) DAHOMEY ETHIOPIA GABON GHANA GUINEA IVORY COAST LIBERIA LIBYA MALI MAURITANIA NIGER RWANDA SENEGAL SOMALIA TANGANYIKA TUNISIA UPPER VOLTA	4 EL SALVADOR HAITI NICARAGUA PARAGUAY

```
**********************************************************************************************

 20  POLITIES                                         20  POLITIES
     LOCATED IN EAST ASIA, SOUTH ASIA, OR                 LOCATED IN THE CARIBBEAN,
     SOUTHEAST ASIA, RATHER THAN IN                       CENTRAL AMERICA, OR SOUTH AMERICA,
     THE CARIBBEAN, CENTRAL AMERICA,                      RATHER THAN IN EAST ASIA, SOUTH ASIA,
     OR SOUTH AMERICA  (18)                               OR SOUTHEAST ASIA  (22)

                                                                         BOTH SUBJECT AND PREDICATE
..............................................................................................

      18 IN LEFT COLUMN

BURMA       CAMBODIA    CEYLON      CHINA, PR   INDIA       INDONESIA   JAPAN       KOREA, N    KOREA REP   LAOS
MALAYA      MONGOLIA    NEPAL       PAKISTAN    PHILIPPINES THAILAND    VIETNAM, N  VIETNAM REP

      22 IN RIGHT COLUMN

ARGENTINA   BOLIVIA     BRAZIL      CHILE       COLOMBIA    COSTA RICA  CUBA        DOMIN REP   ECUADOR     EL SALVADOR
GUATEMALA   HAITI       HONDURAS    JAMAICA     MEXICO      NICARAGUA   PANAMA      PARAGUAY    PERU        TRINIDAD
URUGUAY     VENEZUELA

      75 EXCLUDED BECAUSE IRRELEVANT

AFGHANISTAN ALBANIA     ALGERIA     AUSTRALIA   AUSTRIA     BELGIUM     BULGARIA    BURUNDI     CAMEROUN    CANADA
CEN AFR REP CHAD        CONGO(BRA)  CONGO(LEO)  CYPRUS      CZECHOS'KIA DAHOMEY     DENMARK     ETHIOPIA    FINLAND
FRANCE      GABON       GERMANY, E  GERMAN FR   GHANA       GREECE      GUINEA      HUNGARY     ICELAND     IRAN
IRAQ        IRELAND     ISRAEL      ITALY       IVORY COAST JORDAN      LEBANON     LIBERIA     LIBYA       LUXEMBOURG
MALAGASY R  MALI        MAURITANIA  MOROCCO     NETHERLANDS NEW ZEALAND NIGER       NIGERIA     NORWAY      POLAND
PORTUGAL    RUMANIA     RWANDA      SA'U ARABIA SENEGAL     SIERRE LEO  SOMALIA     SO AFRICA   SPAIN       SUDAN
SWEDEN      SWITZERLAND SYRIA       TANGANYIKA  TOGO        TUNISIA     TURKEY      UGANDA      USSR        UAR
UK          US          UPPER VOLTA YEMEN       YUGOSLAVIA
----------------------------------------------------------------------------------------------

 23  TEND TO BE THOSE                     0.50        23  TEND TO BE THOSE                 0.86        9    3
     WHOSE POPULATION IS                                  WHOSE POPULATION IS                          9   19
     VERY LARGE OR LARGE  (27)                            MEDIUM OR SMALL  (88)                  X SQ=    4.62
                                                                                                 P =     0.018
                                                                                                 RV YES (YES)

 24  TEND TO BE THOSE                     0.83        24  TEND TO BE THOSE                 0.64       15    8
     WHOSE POPULATION IS                                  WHOSE POPULATION IS                          3   14
     VERY LARGE, LARGE, OR MEDIUM  (61)                   SMALL  (54)                            X SQ=    7.12
                                                                                                 P =     0.004
                                                                                                 RV YES (YES)

 26  TEND TO BE THOSE                     0.78        26  TEND TO BE THOSE                 0.73       14    6
     WHOSE POPULATION DENSITY IS                          WHOSE POPULATION DENSITY IS                  4   16
     VERY HIGH, HIGH, OR MEDIUM  (48)                     LOW  (67)                              X SQ=    8.18
                                                                                                 P =     0.004
                                                                                                 RV YES (YES)
```

No.	Statement	Value	No.	Statement	Value	Test
29	TEND TO BE THOSE WHERE THE DEGREE OF URBANIZATION IS LOW (49)	0.83	29	TEND TO BE THOSE WHERE THE DEGREE OF URBANIZATION IS HIGH (56)	0.74	2 14 10 5 X SQ= 7.43 P = 0.003 RV YES (YES)
30	TEND TO BE THOSE WHOSE AGRICULTURAL POPULATION IS HIGH (56)	0.71	30	TEND TO BE THOSE WHOSE AGRICULTURAL POPULATION IS MEDIUM, LOW, OR VERY LOW (57)	0.77	12 5 5 17 X SQ= 7.09 P = 0.004 RV YES (YES)
31	ALWAYS ARE THOSE WHOSE AGRICULTURAL POPULATION IS HIGH OR MEDIUM (90)	1.00	31	TILT LESS TOWARD BEING THOSE WHOSE AGRICULTURAL POPULATION IS HIGH OR MEDIUM (90)	0.82	18 18 0 4 X SQ= 1.90 P = 0.114 RV NO (YES)
34	LEAN TOWARD BEING THOSE WHOSE GROSS NATIONAL PRODUCT IS VERY HIGH, HIGH, MEDIUM, OR LOW (62)	0.72	34	LEAN TOWARD BEING THOSE WHOSE GROSS NATIONAL PRODUCT IS VERY LOW (53)	0.59	13 9 5 13 X SQ= 2.76 P = 0.062 RV YES (YES)
37	TEND TO BE THOSE WHOSE PER CAPITA GROSS NATIONAL PRODUCT IS VERY LOW (51)	0.83	37	TEND TO BE THOSE WHOSE PER CAPITA GROSS NATIONAL PRODUCT IS VERY HIGH, HIGH, MEDIUM, OR LOW (64)	0.86	3 19 15 3 X SQ= 16.72 P = 0. RV YES (YES)
50	LEAN TOWARD BEING THOSE WHERE FREEDOM OF THE PRESS IS INTERMITTENT, INTERNALLY ABSENT, OR INTERNALLY AND EXTERNALLY ABSENT (56)	0.76	50	LEAN TOWARD BEING THOSE WHERE FREEDOM OF THE PRESS IS COMPLETE (43)	0.56	4 10 13 8 X SQ= 2.52 P = 0.086 RV YES (YES)
51	TEND TO BE THOSE WHERE FREEDOM OF THE PRESS IS INTERNALLY ABSENT, OR INTERNALLY AND EXTERNALLY ABSENT (37)	0.65	51	TEND TO BE THOSE WHERE FREEDOM OF THE PRESS IS COMPLETE OR INTERMITTENT (60)	0.78	6 14 11 4 X SQ= 4.83 P = 0.018 RV YES (YES)
55	TEND LESS TO BE THOSE WHERE NEWSPAPER CIRCULATION IS TEN OR MORE PER THOUSAND (78)	0.75	55	ALWAYS ARE THOSE WHERE NEWSPAPER CIRCULATION IS TEN OR MORE PER THOUSAND (78)	1.00	12 22 4 0 X SQ= 3.78 P = 0.025 RV NO (YES)
57	TEND TO BE THOSE WHERE THE RELIGION IS OTHER THAN CATHOLIC (90)	0.94	57	TEND TO BE THOSE WHERE THE RELIGION IS CATHOLIC (25)	0.68	1 15 17 7 X SQ= 13.67 P = 0. RV YES (YES)

63	TEND TO BE THOSE WHERE THE RELIGION IS PREDOMINANTLY OR PARTLY OTHER THAN CHRISTIAN (68)	0.94	63	TEND TO BE THOSE WHERE THE RELIGION IS PREDOMINANTLY SOME KIND OF CHRISTIAN (46)	0.73	1 16 17 6 X SQ= 15.63 P = 0. RV YES (YES)
65	TEND TO BE THOSE WHERE THE RELIGION IS MUSLIM, RATHER THAN CATHOLIC (18)	0.67	65	ALWAYS ARE THOSE WHERE THE RELIGION IS CATHOLIC, RATHER THAN MUSLIM (25)	1.00	1 15 2 0 X SQ= 5.51 P = 0.020 RV YES (YES)
67	TEND TO BE THOSE THAT ARE RACIALLY HOMOGENEOUS (82)	0.94	67	TEND TO BE THOSE THAT ARE RACIALLY HETEROGENEOUS (27)	0.85	16 3 1 17 X SQ= 19.97 P = 0. RV YES (YES)
68	TEND TO BE THOSE THAT ARE LINGUISTICALLY WEAKLY HETEROGENEOUS OR STRONGLY HETEROGENEOUS (62)	0.76	68	TEND TO BE THOSE THAT ARE LINGUISTICALLY HOMOGENEOUS (52)	0.73	4 16 13 6 X SQ= 7.43 P = 0.004 RV YES (YES)
69	TEND TO BE THOSE THAT ARE LINGUISTICALLY STRONGLY HETEROGENEOUS (50)	0.53	69	TEND TO BE THOSE THAT ARE LINGUISTICALLY HOMOGENEOUS OR WEAKLY HETEROGENEOUS (64)	0.82	8 18 9 4 X SQ= 3.77 P = 0.039 RV YES (YES)
72	TEND TO BE THOSE WHOSE DATE OF INDEPENDENCE IS AFTER 1914 (59)	0.75	72	TEND TO BE THOSE WHOSE DATE OF INDEPENDENCE IS BEFORE 1914 (52)	0.91	4 20 12 2 X SQ= 14.58 P = 0. RV YES (YES)
73	TEND TO BE THOSE WHOSE DATE OF INDEPENDENCE IS AFTER 1945 (46)	0.69	73	TEND TO BE THOSE WHOSE DATE OF INDEPENDENCE IS BEFORE 1945 (65)	0.91	5 20 11 2 X SQ= 12.12 P = 0. RV YES (YES)
75	TEND TO BE THOSE THAT ARE NOT HISTORICALLY WESTERN AND ARE NOT SIGNIFICANTLY WESTERNIZED (52)	0.72	75	TEND TO BE THOSE THAT ARE HISTORICALLY WESTERN OR SIGNIFICANTLY WESTERNIZED (62)	0.95	5 21 13 1 X SQ= 17.07 P = 0. RV YES (YES)
77	TEND TO BE THOSE THAT HAVE BEEN PARTIALLY WESTERNIZED, RATHER THAN SIGNIFICANTLY WESTERNIZED, THROUGH A COLONIAL RELATIONSHIP (41)	0.73	77	TEND TO BE THOSE THAT HAVE BEEN SIGNIFICANTLY WESTERNIZED, RATHER THAN PARTIALLY WESTERNIZED, THROUGH A COLONIAL RELATIONSHIP (28)	0.95	3 21 8 1 X SQ= 13.92 P = 0. RV YES (YES)

78	TEND LESS TO BE THOSE THAT HAVE BEEN SIGNIFICANTLY WESTERNIZED THROUGH A COLONIAL RELATIONSHIP, RATHER THAN WITHOUT SUCH A RELATIONSHIP (28)	0.60	78	ALWAYS ARE THOSE THAT HAVE BEEN SIGNIFICANTLY WESTERNIZED THROUGH A COLONIAL RELATIONSHIP, RATHER THAN WITHOUT SUCH A RELATIONSHIP (28)	1.00	3 21 2 0 X SQ= 4.34 P = 0.031 RV NO (YES)
79	ALWAYS ARE THOSE THAT WERE FORMERLY DEPENDENCIES OF BRITAIN OR FRANCE, RATHER THAN SPAIN (49)	1.00	79	TEND TO BE THOSE THAT WERE FORMERLY DEPENDENCIES OF SPAIN, RATHER THAN BRITAIN OR FRANCE (18)	0.86	9 3 0 18 X SQ= 15.88 P = 0. RV YES (YES)
82	ALWAYS ARE THOSE WHERE THE TYPE OF POLITICAL MODERNIZATION IS DEVELOPED TUTELARY OR UNDEVELOPED TUTELARY, RATHER THAN EARLY OR LATER EUROPEAN OR EUROPEAN DERIVED (55)	1.00	82	TEND TO BE THOSE WHERE THE TYPE OF POLITICAL MODERNIZATION IS EARLY OR LATER EUROPEAN OR EUROPEAN DERIVED, RATHER THAN DEVELOPED TUTELARY OR UNDEVELOPED TUTELARY (51)	0.91	0 20 14 2 X SQ= 25.07 P = 0. RV YES (YES)
83	TEND LESS TO BE THOSE WHERE THE TYPE OF POLITICAL MODERNIZATION IS OTHER THAN NON-EUROPEAN AUTOCHTHONOUS (106)	0.78	83	ALWAYS ARE THOSE WHERE THE TYPE OF POLITICAL MODERNIZATION IS OTHER THAN NON-EUROPEAN AUTOCHTHONOUS (106)	1.00	4 0 14 22 X SQ= 3.24 P = 0.033 RV NO (YES)
86	TEND TO BE THOSE WHERE THE STAGE OF POLITICAL MODERNIZATION IS EARLY TRANSITIONAL, RATHER THAN ADVANCED OR MID-TRANSITIONAL (38)	0.50	86	TEND TO BE THOSE WHERE THE STAGE OF POLITICAL MODERNIZATION IS ADVANCED OR MID-TRANSITIONAL, RATHER THAN EARLY TRANSITIONAL (76)	0.91	9 20 9 2 X SQ= 6.38 P = 0.006 RV YES (YES)
89	TEND TO BE THOSE WHOSE IDEOLOGICAL ORIENTATION IS OTHER THAN CONVENTIONAL (62)	0.87	89	TEND TO BE THOSE WHOSE IDEOLOGICAL ORIENTATION IS CONVENTIONAL (33)	0.54	2 7 13 6 X SQ= 3.55 P = 0.042 RV YES (YES)
93	TEND TO BE THOSE WHERE THE SYSTEM STYLE IS MOBILIZATIONAL, OR LIMITED MOBILIZATIONAL (32)	0.53	93	TEND TO BE THOSE WHERE THE SYSTEM STYLE IS NON-MOBILIZATIONAL (78)	0.86	8 3 7 19 X SQ= 4.96 P = 0.025 RV YES (YES)
94	TEND TO BE THOSE WHERE THE STATUS OF THE REGIME IS AUTHORITARIAN OR TOTALITARIAN (41)	0.72	94	TEND TO BE THOSE WHERE THE STATUS OF THE REGIME IS CONSTITUTIONAL (51)	0.74	5 14 13 5 X SQ= 6.07 P = 0.009 RV YES (YES)
96	LEAN LESS TOWARD BEING THOSE WHERE THE STATUS OF THE REGIME IS CONSTITUTIONAL OR TOTALITARIAN (67)	0.53	96	LEAN MORE TOWARD BEING THOSE WHERE THE STATUS OF THE REGIME IS CONSTITUTIONAL OR TOTALITARIAN (67)	0.83	9 15 8 3 X SQ= 2.47 P = 0.075 RV NO (YES)

99	LEAN LESS TOWARD BEING THOSE WHERE GOVERNMENTAL STABILITY IS MODERATELY PRESENT AND DATES FROM THE POST-WAR PERIOD, OR IS ABSENT (36)	0.56	99	LEAN MORE TOWARD BEING THOSE WHERE GOVERNMENTAL STABILITY IS MODERATELY PRESENT AND DATES FROM THE POST-WAR PERIOD, OR IS ABSENT (36)	0.87	8 2 10 14 X SQ= 2.77 P = 0.063 RV NO (YES)
102	TEND TO BE THOSE WHERE THE REPRESENTATIVE CHARACTER OF THE REGIME IS PSEUDO-POLYARCHIC OR NON-POLYARCHIC (49)	0.71	102	TEND TO BE THOSE WHERE THE REPRESENTATIVE CHARACTER OF THE REGIME IS POLYARCHIC OR LIMITED POLYARCHIC (59)	0.75	5 15 12 5 X SQ= 5.96 P = 0.009 RV YES (YES)
105	LEAN TOWARD BEING THOSE WHERE THE ELECTORAL SYSTEM IS PARTIALLY COMPETITIVE OR NON-COMPETITIVE (47)	0.62	105	LEAN TOWARD BEING THOSE WHERE THE ELECTORAL SYSTEM IS COMPETITIVE (43)	0.76	5 13 8 4 X SQ= 2.99 P = 0.061 RV YES (YES)
106	TEND TO BE THOSE WHERE THE ELECTORAL SYSTEM IS NON-COMPETITIVE (30)	0.55	106	TEND TO BE THOSE WHERE THE ELECTORAL SYSTEM IS COMPETITIVE OR PARTIALLY COMPETITIVE (52)	0.88	5 15 6 2 X SQ= 4.08 P = 0.030 RV YES (YES)
107	TEND TO BE THOSE WHERE AUTONOMOUS GROUPS ARE PARTIALLY TOLERATED IN POLITICS, ARE TOLERATED ONLY OUTSIDE POLITICS, OR ARE NOT TOLERATED AT ALL (65)	0.71	107	TEND TO BE THOSE WHERE AUTONOMOUS GROUPS ARE FULLY TOLERATED IN POLITICS (46)	0.67	5 14 12 7 X SQ= 3.83 P = 0.049 RV YES (YES)
108	TEND TO BE THOSE WHERE AUTONOMOUS GROUPS ARE TOLERATED ONLY OUTSIDE POLITICS OR ARE NOT TOLERATED AT ALL (35)	0.57	108	TEND TO BE THOSE WHERE AUTONOMOUS GROUPS ARE FULLY OR PARTIALLY TOLERATED IN POLITICS (65)	0.90	6 19 8 2 X SQ= 7.15 P = 0.006 RV YES (YES)
114	TEND TO BE THOSE WHERE SECTIONALISM IS EXTREME OR MODERATE (61)	0.65	114	TEND TO BE THOSE WHERE SECTIONALISM IS NEGLIGIBLE (47)	0.78	11 4 6 14 X SQ= 4.83 P = 0.018 RV YES (YES)
117	TEND TO BE THOSE WHERE INTEREST ARTICULATION BY ASSOCIATIONAL GROUPS IS NEGLIGIBLE (51)	0.59	117	TEND TO BE THOSE WHERE INTEREST ARTICULATION BY ASSOCIATIONAL GROUPS IS SIGNIFICANT, MODERATE, OR LIMITED (60)	0.86	7 19 10 3 X SQ= 6.90 P = 0.005 RV YES (YES)
118	TEND TO BE THOSE WHERE INTEREST ARTICULATION BY INSTITUTIONAL GROUPS IS VERY SIGNIFICANT (40)	0.61	118	TEND TO BE THOSE WHERE INTEREST ARTICULATION BY INSTITUTIONAL GROUPS IS SIGNIFICANT, MODERATE, OR LIMITED (60)	0.82	11 4 7 18 X SQ= 6.06 P = 0.009 RV YES (YES)

121	TEND TO BE THOSE WHERE INTEREST ARTICULATION BY NON-ASSOCIATIONAL GROUPS IS SIGNIFICANT (54)	0.94	121	TEND TO BE THOSE WHERE INTEREST ARTICULATION BY NON-ASSOCIATIONAL GROUPS IS MODERATE, LIMITED, OR NEGLIGIBLE (61)	0.95	17 1 1 21 X SQ= 28.80 P = 0. RV YES (YES)
122	ALWAYS ARE THOSE WHERE INTEREST ARTICULATION BY NON-ASSOCIATIONAL GROUPS IS SIGNIFICANT OR MODERATE (83)	1.00	122	TEND LESS TO BE THOSE WHERE INTEREST ARTICULATION BY NON-ASSOCIATIONAL GROUPS IS SIGNIFICANT OR MODERATE (83)	0.55	18 12 0 10 X SQ= 8.62 P = 0.001 RV NO (YES)
124	TEND LESS TO BE THOSE WHERE INTEREST ARTICULATION BY ANOMIC GROUPS IS OCCASIONAL, INFREQUENT, OR VERY INFREQUENT (85)	0.67	124	ALWAYS ARE THOSE WHERE INTEREST ARTICULATION BY ANOMIC GROUPS IS OCCASIONAL, INFREQUENT, OR VERY INFREQUENT (85)	1.00	5 0 10 18 X SQ= 4.72 P = 0.013 RV NO (YES)
125	LEAN MORE TOWARD BEING THOSE WHERE INTEREST ARTICULATION BY ANOMIC GROUPS IS FREQUENT OR OCCASIONAL (64)	0.93	125	LEAN LESS TOWARD BEING THOSE WHERE INTEREST ARTICULATION BY ANOMIC GROUPS IS FREQUENT OR OCCASIONAL (64)	0.67	14 12 1 6 X SQ= 2.07 P = 0.095 RV NO (YES)
128	LEAN TOWARD BEING THOSE WHERE INTEREST ARTICULATION BY POLITICAL PARTIES IS LIMITED OR NEGLIGIBLE (45)	0.64	128	LEAN TOWARD BEING THOSE WHERE INTEREST ARTICULATION BY POLITICAL PARTIES IS SIGNIFICANT OR MODERATE (48)	0.74	4 14 7 5 X SQ= 2.64 P = 0.063 RV YES (YES)
129	TEND TO BE THOSE WHERE INTEREST ARTICULATION BY POLITICAL PARTIES IS NEGLIGIBLE (37)	0.64	129	TEND TO BE THOSE WHERE INTEREST ARTICULATION BY POLITICAL PARTIES IS SIGNIFICANT, MODERATE, OR LIMITED (56)	0.95	4 18 7 1 X SQ= 9.34 P = 0.001 RV YES (YES)
138	TEND TO BE THOSE WHERE INTEREST AGGREGATION BY THE LEGISLATURE IS NEGLIGIBLE (49)	0.79	138	TEND TO BE THOSE WHERE INTEREST AGGREGATION BY THE LEGISLATURE IS SIGNIFICANT, MODERATE, OR LIMITED (48)	0.80	3 16 11 4 X SQ= 9.21 P = 0.001 RV YES (YES)
139	TEND LESS TO BE THOSE WHERE THE PARTY SYSTEM IS QUANTITATIVELY OTHER THAN ONE-PARTY (71)	0.64	139	TEND MORE TO BE THOSE WHERE THE PARTY SYSTEM IS QUANTITATIVELY OTHER THAN ONE-PARTY (71)	0.95	5 1 9 20 X SQ= 3.70 P = 0.028 RV NO (YES)
142	LEAN TOWARD BEING THOSE WHERE THE PARTY SYSTEM IS QUANTITATIVELY OTHER THAN MULTI-PARTY (66)	0.82	142	LEAN TOWARD BEING THOSE WHERE THE PARTY SYSTEM IS QUANTITATIVELY MULTI-PARTY (30)	0.55	2 11 9 9 X SQ= 2.58 P = 0.066 RV YES (YES)

154	LEAN TOWARD BEING THOSE WHERE THE PARTY SYSTEM IS STABLE OR MODERATELY STABLE (55)	0.62	154	LEAN TOWARD BEING THOSE WHERE THE PARTY SYSTEM IS UNSTABLE (25)	0.73	10 4 6 11 X SQ= 2.70 P = 0.073 RV YES (YES)
163	LEAN LESS TOWARD BEING THOSE WHERE THE REGIME'S LEADERSHIP CHARISMA IS MODERATE OR NEGLIGIBLE (87)	0.69	163	LEAN MORE TOWARD BEING THOSE WHERE THE REGIME'S LEADERSHIP CHARISMA IS MODERATE OR NEGLIGIBLE (87)	0.95	5 1 11 21 X SQ= 3.16 P = 0.065 RV NO (YES)
164	TEND LESS TO BE THOSE WHERE THE REGIME'S LEADERSHIP CHARISMA IS NEGLIGIBLE (65)	0.53	164	TEND MORE TO BE THOSE WHERE THE REGIME'S LEADERSHIP CHARISMA IS NEGLIGIBLE (65)	0.91	7 2 8 20 X SQ= 4.95 P = 0.017 RV NO (YES)
169	TEND TO BE THOSE WHERE THE HORIZONTAL POWER DISTRIBUTION IS NEGLIGIBLE (48)	0.69	169	TEND TO BE THOSE WHERE THE HORIZONTAL POWER DISTRIBUTION IS SIGNIFICANT OR LIMITED (58)	0.85	5 17 11 3 X SQ= 8.66 P = 0.002 RV YES (YES)
170	TEND TO BE THOSE WHERE THE LEGISLATIVE-EXECUTIVE STRUCTURE IS OTHER THAN PRESIDENTIAL (63)	0.75	170	TEND TO BE THOSE WHERE THE LEGISLATIVE-EXECUTIVE STRUCTURE IS PRESIDENTIAL (39)	0.85	4 17 12 3 X SQ= 10.81 P = 0.001 RV YES (YES)
175	TEND TO BE THOSE WHERE THE LEGISLATURE IS LARGELY INEFFECTIVE OR WHOLLY INEFFECTIVE (49)	0.67	175	TEND TO BE THOSE WHERE THE LEGISLATURE IS FULLY EFFECTIVE OR PARTIALLY EFFECTIVE (51)	0.78	5 14 10 4 X SQ= 4.92 P = 0.015 RV YES (YES)
176	TEND TO BE THOSE WHERE THE LEGISLATURE IS WHOLLY INEFFECTIVE (28)	0.60	176	TEND TO BE THOSE WHERE THE LEGISLATURE IS FULLY EFFECTIVE, PARTIALLY EFFECTIVE, OR LARGELY INEFFECTIVE (72)	0.89	6 16 9 2 X SQ= 6.74 P = 0.008 RV YES (YES)
179	LEAN TOWARD BEING THOSE WHERE THE EXECUTIVE IS DOMINANT (52)	0.77	179	LEAN TOWARD BEING THOSE WHERE THE EXECUTIVE IS STRONG (39)	0.59	10 7 3 10 X SQ= 2.52 P = 0.071 RV YES (YES)
185	TEND TO BE THOSE WHERE PARTICIPATION BY THE MILITARY IN POLITICS IS SUPPORTIVE, RATHER THAN INTERVENTIVE (31)	0.64	185	TEND TO BE THOSE WHERE PARTICIPATION BY THE MILITARY IN POLITICS IS INTERVENTIVE, RATHER THAN SUPPORTIVE (21)	0.91	4 10 7 1 X SQ= 4.91 P = 0.024 RV YES (YES)

186	TEND TO BE THOSE WHERE PARTICIPATION BY THE MILITARY IN POLITICS IS SUPPORTIVE, RATHER THAN NEUTRAL (31)	0.58	186	TEND TO BE THOSE WHERE PARTICIPATION BY THE MILITARY IN POLITICS IS NEUTRAL, RATHER THAN SUPPORTIVE (56)	0.90	7 1 5 9 X SQ= 3.62 P = 0.031 RV YES (YES)
188	TEND TO BE THOSE WHERE THE CHARACTER OF THE LEGAL SYSTEM IS OTHER THAN CIVIL LAW (81)	0.94	188	TEND TO BE THOSE WHERE THE CHARACTER OF THE LEGAL SYSTEM IS CIVIL LAW (32)	0.86	1 19 17 3 X SQ= 22.73 P = 0. RV YES (YES)
192	LEAN LESS TOWARD BEING THOSE WHERE THE CHARACTER OF THE LEGAL SYSTEM IS OTHER THAN MUSLIM OR PARTLY MUSLIM (86)	0.83	192	ALWAYS ARE THOSE WHERE THE CHARACTER OF THE LEGAL SYSTEM IS OTHER THAN MUSLIM OR PARTLY MUSLIM (86)	1.00	3 0 15 22 X SQ= 1.93 P = 0.083 RV NO (YES)
193	LEAN LESS TOWARD BEING THOSE WHERE THE CHARACTER OF THE LEGAL SYSTEM IS OTHER THAN PARTLY INDIGENOUS (86)	0.82	193	ALWAYS ARE THOSE WHERE THE CHARACTER OF THE LEGAL SYSTEM IS OTHER THAN PARTLY INDIGENOUS (86)	1.00	3 0 14 22 X SQ= 2.09 P = 0.074 RV NO (YES)
194	TEND LESS TO BE THOSE THAT ARE NON-COMMUNIST (101)	0.78	194	ALWAYS ARE THOSE THAT ARE NON-COMMUNIST (101)	1.00	4 0 14 21 X SQ= 3.07 P = 0.037 RV NO (YES)

****

21 POLITIES WHOSE TERRITORIAL SIZE IS VERY LARGE OR LARGE (32)	21 POLITIES WHOSE TERRITORIAL SIZE IS MEDIUM OR SMALL (83)

BOTH SUBJECT AND PREDICATE

....

32 IN LEFT COLUMN

ALGERIA	ARGENTINA	AUSTRALIA	BOLIVIA	BRAZIL	CANADA	CHAD	CHINA, PR	COLOMBIA	CONGO(LEO)
ETHIOPIA	INDIA	INDONESIA	IRAN	LIBYA	MALI	MAURITANIA	MEXICO	MONGOLIA	NIGER
NIGERIA	PAKISTAN	PERU	SA'U ARABIA	SO AFRICA	SUDAN	TANGANYIKA	TURKEY	USSR	UAR
US	VENEZUELA								

83 IN RIGHT COLUMN

AFGHANISTAN	ALBANIA	AUSTRIA	BELGIUM	BULGARIA	BURMA	BURUNDI	CAMBODIA	CAMEROUN	CEN AFR REP
CEYLON	CHILE	CONGO(BRA)	COSTA RICA	CUBA	CYPRUS	CZECHOS'KIA	DAHOMEY	DENMARK	DOMIN REP
ECUADOR	EL SALVADOR	FINLAND	FRANCE	GABON	GERMANY, E	GERMAN FR	GHANA	GREECE	GUATEMALA
GUINEA	HAITI	HONDURAS	HUNGARY	ICELAND	IRAQ	IRELAND	ISRAEL	ITALY	IVORY COAST
JAMAICA	JAPAN	JORDAN	KOREA, N	KOREA REP	LAOS	LEBANON	LIBERIA	LUXEMBOURG	MALAGASY R
MALAYA	MOROCCO	NEPAL	NETHERLANDS	NEW ZEALAND	NICARAGUA	NORWAY	PANAMA	PARAGUAY	PHILIPPINES
POLAND	PORTUGAL	RUMANIA	RWANDA	SENEGAL	SIERRE LEO	SOMALIA	SPAIN	SWEDEN	SWITZERLAND
SYRIA	THAILAND	TOGO	TRINIDAD	TUNISIA	UGANDA	UK	UPPER VOLTA	URUGUAY	VIETNAM, N
VIETNAM REP	YEMEN	YUGOSLAVIA							

2	TILT MORE TOWARD BEING THOSE LOCATED ELSEWHERE THAN IN WEST EUROPE, SCANDINAVIA, NORTH AMERICA, OR AUSTRALASIA (93)	0.91	2	TILT LESS TOWARD BEING THOSE LOCATED ELSEWHERE THAN IN WEST EUROPE, SCANDINAVIA, NORTH AMERICA, OR AUSTRALASIA (93)	0.77	3 19 29 64 X SQ= 1.92 P = 0.118 RV NO (NO)
3	ALWAYS ARE THOSE LOCATED ELSEWHERE THAN IN WEST EUROPE (102)	1.00	3	TEND LESS TO BE THOSE LOCATED ELSEWHERE THAN IN WEST EUROPE (102)	0.84	0 13 32 70 X SQ= 4.20 P = 0.018 RV NO (NO)
7	TILT TOWARD BEING THOSE LOCATED IN SOUTH AMERICA, RATHER THAN IN CENTRAL AMERICA (10)	0.86	7	TILT TOWARD BEING THOSE LOCATED IN CENTRAL AMERICA, RATHER THAN IN SOUTH AMERICA (7)	0.60	1 6 6 4 X SQ= 1.92 P = 0.134 RV YES (YES)
24	TEND TO BE THOSE WHOSE POPULATION IS VERY LARGE, LARGE, OR MEDIUM (61)	0.78	24	TEND TO BE THOSE WHOSE POPULATION IS SMALL (54)	0.57	25 36 7 47 X SQ= 9.85 P = 0.001 RV YES (NO)

26	TEND TO BE THOSE WHOSE POPULATION DENSITY IS LOW (67)	0.84	26	TEND TO BE THOSE WHOSE POPULATION DENSITY IS VERY HIGH, HIGH, OR MEDIUM (48)	0.52	5 43 27 40 X SQ= 10.99 P = 0. RV YES (NO)
28	TEND TO BE THOSE WHOSE POPULATION GROWTH RATE IS HIGH (62)	0.80	28	TEND TO BE THOSE WHOSE POPULATION GROWTH RATE IS LOW (48)	0.52	24 38 6 42 X SQ= 8.10 P = 0.002 RV YES (NO)
34	TEND TO BE THOSE WHOSE GROSS NATIONAL PRODUCT IS VERY HIGH, HIGH, MEDIUM, OR LOW (62)	0.72	34	TEND TO BE THOSE WHOSE GROSS NATIONAL PRODUCT IS VERY LOW (53)	0.53	23 39 9 44 X SQ= 4.80 P = 0.022 RV YES (NO)
36	LEAN MORE TOWARD BEING THOSE WHOSE PER CAPITA GROSS NATIONAL PRODUCT IS LOW OR VERY LOW (73)	0.78	36	LEAN LESS TOWARD BEING THOSE WHOSE PER CAPITA GROSS NATIONAL PRODUCT IS LOW OR VERY LOW (73)	0.58	7 35 25 48 X SQ= 3.27 P = 0.053 RV NO (NO)
45	TEND TO BE THOSE WHERE THE LITERACY RATE IS BELOW FIFTY PERCENT (54)	0.67	45	TEND TO BE THOSE WHERE THE LITERACY RATE IS FIFTY PERCENT OR ABOVE (55)	0.57	10 45 20 34 X SQ= 3.96 P = 0.033 RV YES (NO)
54	TEND MORE TO BE THOSE WHERE NEWSPAPER CIRCULATION IS LESS THAN ONE HUNDRED PER THOUSAND (76)	0.81	54	TEND LESS TO BE THOSE WHERE NEWSPAPER CIRCULATION IS LESS THAN ONE HUNDRED PER THOUSAND (76)	0.62	6 31 26 50 X SQ= 3.13 P = 0.049 RV NO (NO)
63	LEAN MORE TOWARD BEING THOSE WHERE THE RELIGION IS PREDOMINANTLY OR PARTLY OTHER THAN CHRISTIAN (68)	0.74	63	LEAN LESS TOWARD BEING THOSE WHERE THE RELIGION IS PREDOMINANTLY OR PARTLY OTHER THAN CHRISTIAN (68)	0.54	8 38 23 45 X SQ= 2.96 P = 0.058 RV NO (NO)
64	TEND TO BE THOSE WHERE THE RELIGION IS MUSLIM, RATHER THAN CHRISTIAN (18)	0.53	64	TEND TO BE THOSE WHERE THE RELIGION IS CHRISTIAN, RATHER THAN MUSLIM (46)	0.81	8 38 9 9 X SQ= 5.48 P = 0.012 RV YES (YES)
65	LEAN TOWARD BEING THOSE WHERE THE RELIGION IS MUSLIM, RATHER THAN CATHOLIC (18)	0.64	65	LEAN TOWARD BEING THOSE WHERE THE RELIGION IS CATHOLIC, RATHER THAN MUSLIM (25)	0.69	5 20 9 9 X SQ= 3.03 P = 0.052 RV YES (YES)

No.	Statement	Value	No.	Statement	Value	Statistics
67	TEND LESS TO BE THOSE THAT ARE RACIALLY HOMOGENEOUS (82)	0.52	67	TEND MORE TO BE THOSE THAT ARE RACIALLY HOMOGENEOUS (82)	0.82	16 64 15 14 X SQ= 9.02 P = 0.003 RV NO (YES)
69	TEND TO BE THOSE THAT ARE LINGUISTICALLY STRONGLY HETEROGENEOUS (50)	0.61	69	TEND TO BE THOSE THAT ARE LINGUISTICALLY HOMOGENEOUS OR WEAKLY HETEROGENEOUS (64)	0.63	12 52 19 31 X SQ= 4.33 P = 0.033 RV YES (NO)
74	TEND MORE TO BE THOSE THAT ARE NOT HISTORICALLY WESTERN (87)	0.90	74	TEND LESS TO BE THOSE THAT ARE NOT HISTORICALLY WESTERN (87)	0.72	3 23 28 59 X SQ= 3.31 P = 0.046 RV NO (NO)
76	TEND MORE TO BE THOSE THAT HAVE BEEN SIGNIFICANTLY OR PARTIALLY WESTERNIZED THROUGH A COLONIAL RELATIONSHIP, RATHER THAN BEING HISTORICALLY WESTERN (70)	0.88	76	TEND LESS TO BE THOSE THAT HAVE BEEN SIGNIFICANTLY OR PARTIALLY WESTERNIZED THROUGH A COLONIAL RELATIONSHIP, RATHER THAN BEING HISTORICALLY WESTERN (70)	0.67	3 23 23 47 X SQ= 3.35 P = 0.041 RV NO (NO)
83	TILT LESS TOWARD BEING THOSE WHERE THE TYPE OF POLITICAL MODERNIZATION IS OTHER THAN NON-EUROPEAN AUTOCHTHONOUS (106)	0.84	83	TILT MORE TOWARD BEING THOSE WHERE THE TYPE OF POLITICAL MODERNIZATION IS OTHER THAN NON-EUROPEAN AUTOCHTHONOUS (106)	0.95	5 4 27 79 X SQ= 2.39 P = 0.113 RV NO (YES)
114	TEND TO BE THOSE WHERE SECTIONALISM IS EXTREME OR MODERATE (61)	0.75	114	TEND TO BE THOSE WHERE SECTIONALISM IS NEGLIGIBLE (47)	0.50	21 40 7 40 X SQ= 4.31 P = 0.027 RV YES (NO)
119	TEND MORE TO BE THOSE WHERE INTEREST ARTICULATION BY INSTITUTIONAL GROUPS IS VERY SIGNIFICANT OR SIGNIFICANT (74)	0.90	119	TEND LESS TO BE THOSE WHERE INTEREST ARTICULATION BY INSTITUTIONAL GROUPS IS VERY SIGNIFICANT OR SIGNIFICANT (74)	0.68	26 48 3 23 X SQ= 4.12 P = 0.025 RV NO (NO)
122	TILT MORE TOWARD BEING THOSE WHERE INTEREST ARTICULATION BY NON-ASSOCIATIONAL GROUPS IS SIGNIFICANT OR MODERATE (83)	0.84	122	TILT LESS TOWARD BEING THOSE WHERE INTEREST ARTICULATION BY NON-ASSOCIATIONAL GROUPS IS SIGNIFICANT OR MODERATE (83)	0.67	27 56 5 27 X SQ= 2.50 P = 0.103 RV NO (NO)
125	TEND MORE TO BE THOSE WHERE INTEREST ARTICULATION BY ANOMIC GROUPS IS FREQUENT OR OCCASIONAL (64)	0.81	125	TEND LESS TO BE THOSE WHERE INTEREST ARTICULATION BY ANOMIC GROUPS IS FREQUENT OR OCCASIONAL (64)	0.58	22 42 5 30 X SQ= 3.65 P = 0.036 RV NO (NO)

131	TEND TO BE THOSE WHERE INTEREST AGGREGATION BY POLITICAL PARTIES IS SIGNIFICANT OR MODERATE (30)	0.73	131	TEND TO BE THOSE WHERE INTEREST AGGREGATION BY POLITICAL PARTIES IS LIMITED OR NEGLIGIBLE (35)	0.62	11 19 4 31 X SQ= 4.46 P = 0.020 RV YES (NO)
132	ALWAYS ARE THOSE WHERE INTEREST AGGREGATION BY POLITICAL PARTIES IS SIGNIFICANT, MODERATE, OR LIMITED (67)	1.00	132	TILT LESS TOWARD BEING THOSE WHERE INTEREST AGGREGATION BY POLITICAL PARTIES IS SIGNIFICANT, MODERATE, OR LIMITED (67)	0.84	19 48 0 9 X SQ= 2.06 P = 0.102 RV NO (NO)
157	TILT MORE TOWARD BEING THOSE WHERE PERSONALISMO IS MODERATE OR NEGLIGIBLE (84)	0.96	157	TILT LESS TOWARD BEING THOSE WHERE PERSONALISMO IS MODERATE OR NEGLIGIBLE (84)	0.82	1 13 25 59 X SQ= 2.10 P = 0.104 RV NO (NO)
166	TEND LESS TO BE THOSE WHERE THE VERTICAL POWER DISTRIBUTION IS THAT OF FORMAL FEDERALISM OR FORMAL AND EFFECTIVE UNITARISM (99)	0.71	166	TEND MORE TO BE THOSE WHERE THE VERTICAL POWER DISTRIBUTION IS THAT OF FORMAL FEDERALISM OR FORMAL AND EFFECTIVE UNITARISM (99)	0.93	9 6 22 77 X SQ= 7.58 P = 0.004 RV NO (YES)
178	TILT TOWARD BEING THOSE WHERE THE LEGISLATURE IS BICAMERAL (51)	0.62	178	TILT TOWARD BEING THOSE WHERE THE LEGISLATURE IS UNICAMERAL (53)	0.56	11 42 18 33 X SQ= 2.06 P = 0.127 RV YES (NO)
192	TEND LESS TO BE THOSE WHERE THE CHARACTER OF THE LEGAL SYSTEM IS OTHER THAN MUSLIM OR PARTLY MUSLIM (86)	0.55	192	TEND MORE TO BE THOSE WHERE THE CHARACTER OF THE LEGAL SYSTEM IS OTHER THAN MUSLIM OR PARTLY MUSLIM (86)	0.84	14 13 17 69 X SQ= 9.08 P = 0.002 RV NO (YES)

```
****************************************************************************************************

 22  POLITIES                                              22  POLITIES
     WHOSE TERRITORIAL SIZE IS                                 WHOSE TERRITORIAL SIZE IS
     VERY LARGE, LARGE, OR MEDIUM  (68)                        SMALL  (47)


                                                                                      PREDICATE ONLY
....................................................................................................

      68 IN LEFT COLUMN

AFGHANISTAN ALGERIA     ARGENTINA   AUSTRALIA   BOLIVIA     BRAZIL      BURMA       CAMEROUN    CANADA      CEN AFR REP
CHAD        CHILE       CHINA, PR   COLOMBIA    CONGO(BRA)  CONGO(LEO)  ECUADOR     ETHIOPIA    FINLAND     FRANCE
GABON       GERMAN FR   GHANA       GUINEA      INDIA       INDONESIA   IRAN        IRAQ        ITALY       IVORY COAST
JAPAN       LAOS        LIBYA       MALAGASY R  MALI        MAURITANIA  MEXICO      MONGOLIA    MOROCCO     NEW ZEALAND
NIGER       NIGERIA     NORWAY      PAKISTAN    PARAGUAY    PERU        PHILIPPINES POLAND      RUMANIA     SA'U ARABIA
SENEGAL     SOMALIA     SO AFRICA   SPAIN       SUDAN       SWEDEN      TANGANYIKA  THAILAND    TURKEY      UGANDA
USSR        UAR         UK          US          UPPER VOLTA VENEZUELA   YEMEN       YUGOSLAVIA

      47 IN RIGHT COLUMN

ALBANIA     AUSTRIA     BELGIUM     BULGARIA    BURUNDI     CAMBODIA    CEYLON      COSTA RICA  CUBA        CYPRUS
CZECHOS'KIA DAHOMEY     DENMARK     DOMIN REP   EL SALVADOR GERMANY, E  GREECE      GUATEMALA   HAITI       HONDURAS
HUNGARY     ICELAND     IRELAND     ISRAEL      JAMAICA     JORDAN      KOREA, N    KOREA REP   LEBANON     LIBERIA
LUXEMBOURG  MALAYA      NEPAL       NETHERLANDS NICARAGUA   PANAMA      PORTUGAL    RWANDA      SIERRE LEO  SWITZERLAND
SYRIA       TOGO        TRINIDAD    TUNISIA     URUGUAY     VIETNAM, N  VIETNAM REP
----------------------------------------------------------------------------------------------------
```

23 POLITIES
WHOSE POPULATION IS
VERY LARGE OR LARGE (27)

23 POLITIES
WHOSE POPULATION IS
MEDIUM OR SMALL (88)

BOTH SUBJECT AND PREDICATE

27 IN LEFT COLUMN

ARGENTINA	BRAZIL	BURMA	CANADA	CHINA, PR	FRANCE	GERMAN FR	INDIA	INDONESIA	IRAN
ITALY	JAPAN	KOREA REP	MEXICO	NIGERIA	PAKISTAN	PHILIPPINES	POLAND	RUMANIA	SPAIN
THAILAND	TURKEY	USSR	UAR	UK	US	YUGOSLAVIA			

88 IN RIGHT COLUMN

AFGHANISTAN	ALBANIA	ALGERIA	AUSTRALIA	AUSTRIA	BELGIUM	BOLIVIA	BULGARIA	BURUNDI	CAMBODIA
CAMEROUN	CEN AFR REP	CEYLON	CHAD	CHILE	COLOMBIA	CONGO(BRA)	CONGO(LEO)	COSTA RICA	CUBA
CYPRUS	CZECHOS'KIA	DAHOMEY	DENMARK	DOMIN REP	ECUADOR	EL SALVADOR	ETHIOPIA	FINLAND	GABON
GERMANY, E	GHANA	GREECE	GUATEMALA	GUINEA	HAITI	HONDURAS	HUNGARY	ICELAND	IRAQ
IRELAND	ISRAEL	IVORY COAST	JAMAICA	JORDAN	KOREA, N	LAOS	LEBANON	LIBERIA	LIBYA
LUXEMBOURG	MALAGASY R	MALAYA	MALI	MAURITANIA	MONGOLIA	MOROCCO	NEPAL	NETHERLANDS	NEW ZEALAND
NICARAGUA	NIGER	NORWAY	PANAMA	PARAGUAY	PERU	PORTUGAL	RWANDA	SA'U ARABIA	SENEGAL
SIERRE LEO	SOMALIA	SO AFRICA	SUDAN	SWEDEN	SWITZERLAND	SYRIA	TANGANYIKA	TOGO	TRINIDAD
TUNISIA	UGANDA	UPPER VOLTA	URUGUAY	VENEZUELA	VIETNAM, N	VIETNAM REP	YEMEN		

	Left column			Right column		Statistics
10	TEND MORE TO BE THOSE LOCATED ELSEWHERE THAN IN NORTH AFRICA, OR CENTRAL AND SOUTH AFRICA (82)	0.93	10	TEND LESS TO BE THOSE LOCATED ELSEWHERE THAN IN NORTH AFRICA, OR CENTRAL AND SOUTH AFRICA (82)	0.65	2 31 25 57 X SQ= 6.51 P = 0.004 RV NO (NO)
16	TILT TOWARD BEING THOSE LOCATED IN EAST ASIA, SOUTH ASIA, OR SOUTHEAST ASIA, RATHER THAN IN THE MIDDLE EAST (18)	0.82	16	TILT TOWARD BEING THOSE LOCATED IN THE MIDDLE EAST, RATHER THAN IN EAST ASIA, SOUTH ASIA, OR SOUTHEAST ASIA (11)	0.50	2 9 9 9 X SQ= 1.74 P = 0.125 RV YES (YES)
18	TEND TO BE THOSE LOCATED IN EAST ASIA, SOUTH ASIA, OR SOUTHEAST ASIA, RATHER THAN IN NORTH AFRICA OR CENTRAL AND SOUTH AFRICA (18)	0.82	18	TEND TO BE THOSE LOCATED IN NORTH AFRICA OR CENTRAL AND SOUTH AFRICA, RATHER THAN IN EAST ASIA, SOUTH ASIA, OR SOUTHEAST ASIA (33)	0.77	2 31 9 9 X SQ= 10.82 P = 0.001 RV YES (YES)
20	TEND TO BE THOSE LOCATED IN EAST ASIA, SOUTH ASIA, OR SOUTHEAST ASIA, RATHER THAN IN THE CARIBBEAN, CENTRAL AMERICA, OR SOUTH AMERICA (18)	0.75	20	TEND TO BE THOSE LOCATED IN THE CARIBBEAN, CENTRAL AMERICA, OR SOUTH AMERICA, RATHER THAN IN EAST ASIA, SOUTH ASIA, OR SOUTHEAST ASIA (22)	0.68	9 9 3 19 X SQ= 4.62 P = 0.018 RV YES (YES)

21	TEND TO BE THOSE WHOSE TERRITORIAL SIZE IS VERY LARGE OR LARGE (32)	0.52	21	TEND TO BE THOSE WHOSE TERRITORIAL SIZE IS MEDIUM OR SMALL (83)	0.80	14 18 13 70 X SQ= 8.64 P = 0.003 RV YES (NO)
22	TEND TO BE THOSE WHOSE TERRITORIAL SIZE IS VERY LARGE, LARGE, OR MEDIUM (68)	0.96	22	TEND TO BE THOSE WHOSE TERRITORIAL SIZE IS SMALL (47)	0.52	26 42 1 46 X SQ= 18.21 P = 0. RV YES (NO)
26	TEND TO BE THOSE WHOSE POPULATION DENSITY IS VERY HIGH, HIGH, OR MEDIUM (48)	0.63	26	TEND TO BE THOSE WHOSE POPULATION DENSITY IS LOW (67)	0.65	17 31 10 57 X SQ= 5.45 P = 0.014 RV YES (NO)
29	TEND TO BE THOSE WHERE THE DEGREE OF URBANIZATION IS HIGH (56)	0.72	29	TEND TO BE THOSE WHERE THE DEGREE OF URBANIZATION IS LOW (49)	0.52	18 38 7 42 X SQ= 3.66 P = 0.040 RV YES (NO)
33	TEND TO BE THOSE WHOSE GROSS NATIONAL PRODUCT IS VERY HIGH, HIGH, OR MEDIUM (30)	0.74	33	TEND TO BE THOSE WHOSE GROSS NATIONAL PRODUCT IS LOW OR VERY LOW (85)	0.89	20 10 7 78 X SQ= 38.95 P = 0. RV YES (YES)
34	ALWAYS ARE THOSE WHOSE GROSS NATIONAL PRODUCT IS VERY HIGH, HIGH, MEDIUM, OR LOW (62)	1.00	34	TEND TO BE THOSE WHOSE GROSS NATIONAL PRODUCT IS VERY LOW (53)	0.60	27 35 0 53 X SQ= 27.79 P = 0. RV YES (NO)
39	TEND TO BE THOSE WHOSE INTERNATIONAL FINANCIAL STATUS IS VERY HIGH, HIGH, OR MEDIUM (38)	0.81	39	TEND TO BE THOSE WHOSE INTERNATIONAL FINANCIAL STATUS IS LOW OR VERY LOW (76)	0.82	22 16 5 71 X SQ= 34.12 P = 0. RV YES (YES)
40	ALWAYS ARE THOSE WHOSE INTERNATIONAL FINANCIAL STATUS IS VERY HIGH, HIGH, MEDIUM, OR LOW (71)	1.00	40	TEND LESS TO BE THOSE WHOSE INTERNATIONAL FINANCIAL STATUS IS VERY HIGH, HIGH, MEDIUM, OR LOW (71)	0.53	27 44 0 39 X SQ= 17.66 P = 0. RV NO (NO)
42	TEND TO BE THOSE WHOSE ECONOMIC DEVELOPMENTAL STATUS IS DEVELOPED OR INTERMEDIATE (36)	0.52	42	TEND TO BE THOSE WHOSE ECONOMIC DEVELOPMENTAL STATUS IS UNDERDEVELOPED OR VERY UNDERDEVELOPED (76)	0.74	14 22 13 63 X SQ= 5.20 P = 0.018 RV YES (NO)

43	TEND TO BE THOSE WHOSE ECONOMIC DEVELOPMENTAL STATUS IS DEVELOPED, INTERMEDIATE, OR UNDERDEVELOPED (55)	0.78	43	TEND TO BE THOSE WHOSE ECONOMIC DEVELOPMENTAL STATUS IS VERY UNDERDEVELOPED (57)	0.60	21 34 6 51 X SQ= 10.24 P = 0.001 RV YES (NO)
45	TEND TO BE THOSE WHERE THE LITERACY RATE IS FIFTY PERCENT OR ABOVE (55)	0.77	45	TEND TO BE THOSE WHERE THE LITERACY RATE IS BELOW FIFTY PERCENT (54)	0.58	20 35 6 48 X SQ= 8.23 P = 0.003 RV YES (NO)
46	ALWAYS ARE THOSE WHERE THE LITERACY RATE IS TEN PERCENT OR ABOVE (84)	1.00	46	TEND LESS TO BE THOSE WHERE THE LITERACY RATE IS TEN PERCENT OR ABOVE (84)	0.69	27 57 0 26 X SQ= 9.41 P = 0. RV NO (NO)
55	TEND MORE TO BE THOSE WHERE NEWSPAPER CIRCULATION IS TEN OR MORE PER THOUSAND (78)	0.89	55	TEND LESS TO BE THOSE WHERE NEWSPAPER CIRCULATION IS TEN OR MORE PER THOUSAND (78)	0.63	24 54 3 32 X SQ= 5.38 P = 0.010 RV NO (NO)
56	TEND MORE TO BE THOSE WHERE THE RELIGION IS PREDOMINANTLY LITERATE (79)	0.96	56	TEND LESS TO BE THOSE WHERE THE RELIGION IS PREDOMINANTLY LITERATE (79)	0.65	24 55 1 30 X SQ= 7.87 P = 0.002 RV NO (NO)
66	LEAN TOWARD BEING THOSE THAT ARE RELIGIOUSLY HOMOGENEOUS (57)	0.71	66	LEAN TOWARD BEING THOSE THAT ARE RELIGIOUSLY HETEROGENEOUS (49)	0.51	17 40 7 42 X SQ= 2.80 P = 0.066 RV YES (NO)
67	TEND MORE TO BE THOSE THAT ARE RACIALLY HOMOGENEOUS (82)	0.89	67	TEND LESS TO BE THOSE THAT ARE RACIALLY HOMOGENEOUS (82)	0.68	24 56 3 26 X SQ= 3.42 P = 0.044 RV NO (NO)
72	LEAN TOWARD BEING THOSE WHOSE DATE OF INDEPENDENCE IS BEFORE 1914 (52)	0.64	72	LEAN TOWARD BEING THOSE WHOSE DATE OF INDEPENDENCE IS AFTER 1914 (59)	0.58	16 36 9 50 X SQ= 2.98 P = 0.069 RV YES (NO)
75	TEND TO BE THOSE THAT ARE HISTORICALLY WESTERN OR SIGNIFICANTLY WESTERNIZED (62)	0.74	75	TEND TO BE THOSE THAT ARE NOT HISTORICALLY WESTERN AND ARE NOT SIGNIFICANTLY WESTERNIZED (52)	0.52	20 42 7 45 X SQ= 4.54 P = 0.026 RV YES (NO)

78	TEND TO BE THOSE THAT HAVE BEEN SIGNIFICANTLY WESTERNIZED WITHOUT A COLONIAL RELATIONSHIP, RATHER THAN THROUGH SUCH A RELATIONSHIP (7)	0.50	78	TEND TO BE THOSE THAT HAVE BEEN SIGNIFICANTLY WESTERNIZED THROUGH A COLONIAL RELATIONSHIP, RATHER THAN WITHOUT SUCH A RELATIONSHIP (28)	0.96	6 22 6 1 X SQ= 7.62 P = 0.003 RV YES (YES)
80	ALWAYS ARE THOSE WHOSE DATE OF INDEPENDENCE IS AFTER 1914, AND THAT WERE FORMERLY DEPENDENCIES OF BRITAIN, RATHER THAN FRANCE (19)	1.00	80	TEND TO BE THOSE WHOSE DATE OF INDEPENDENCE IS AFTER 1914, AND THAT WERE FORMERLY DEPENDENCIES OF FRANCE, RATHER THAN BRITAIN (24)	0.63	5 14 0 24 X SQ= 4.82 P = 0.012 RV YES (NO)
83	TEND LESS TO BE THOSE WHERE THE TYPE OF POLITICAL MODERNIZATION IS OTHER THAN NON-EUROPEAN AUTOCHTHONOUS (106)	0.78	83	TEND MORE TO BE THOSE WHERE THE TYPE OF POLITICAL MODERNIZATION IS OTHER THAN NON-EUROPEAN AUTOCHTHONOUS (106)	0.97	6 3 21 85 X SQ= 7.70 P = 0.005 RV NO (YES)
84	LEAN MORE TOWARD BEING THOSE WHERE THE TYPE OF POLITICAL MODERNIZATION IS DEVELOPED TUTELARY, RATHER THAN UNDEVELOPED TUTELARY (31)	0.87	84	LEAN LESS TOWARD BEING THOSE WHERE THE TYPE OF POLITICAL MODERNIZATION IS DEVELOPED TUTELARY, RATHER THAN UNDEVELOPED TUTELARY (31)	0.51	7 24 1 23 X SQ= 2.36 P = 0.068 RV NO (NO)
85	TEND TO BE THOSE WHERE THE STAGE OF POLITICAL MODERNIZATION IS ADVANCED, RATHER THAN MID- OR EARLY TRANSITIONAL (60)	0.81	85	TEND TO BE THOSE WHERE THE STAGE OF POLITICAL MODERNIZATION IS MID- OR EARLY TRANSITIONAL, RATHER THAN ADVANCED (54)	0.56	22 38 5 49 X SQ= 10.34 P = 0.001 RV YES (NO)
87	TEND MORE TO BE THOSE WHOSE IDEOLOGICAL ORIENTATION IS OTHER THAN DEVELOPMENTAL (58)	0.89	87	TEND LESS TO BE THOSE WHOSE IDEOLOGICAL ORIENTATION IS OTHER THAN DEVELOPMENTAL (58)	0.59	2 29 17 41 X SQ= 5.00 P = 0.014 RV NO (NO)
114	TEND MORE TO BE THOSE WHERE SECTIONALISM IS EXTREME OR MODERATE (61)	0.76	114	TEND LESS TO BE THOSE WHERE SECTIONALISM IS EXTREME OR MODERATE (61)	0.51	19 42 6 41 X SQ= 4.06 P = 0.037 RV NO (NO)
117	TEND TO BE THOSE WHERE INTEREST ARTICULATION BY ASSOCIATIONAL GROUPS IS SIGNIFICANT, MODERATE, OR LIMITED (60)	0.79	117	TEND TO BE THOSE WHERE INTEREST ARTICULATION BY ASSOCIATIONAL GROUPS IS NEGLIGIBLE (51)	0.53	19 41 5 46 X SQ= 6.54 P = 0.006 RV YES (NO)
118	LEAN TOWARD BEING THOSE WHERE INTEREST ARTICULATION BY INSTITUTIONAL GROUPS IS VERY SIGNIFICANT (40)	0.56	118	LEAN TOWARD BEING THOSE WHERE INTEREST ARTICULATION BY INSTITUTIONAL GROUPS IS SIGNIFICANT, MODERATE, OR LIMITED (60)	0.66	15 25 12 48 X SQ= 2.89 P = 0.067 RV YES (NO)

127	TEND MORE TO BE THOSE WHERE INTEREST ARTICULATION BY POLITICAL PARTIES IS MODERATE, LIMITED, OR NEGLIGIBLE (72)	0.95	127	TEND LESS TO BE THOSE WHERE INTEREST ARTICULATION BY POLITICAL PARTIES IS MODERATE, LIMITED, OR NEGLIGIBLE (72)	0.72	1 20 20 52 X SQ= 3.70 P = 0.035 RV NO (NO)
138	TILT TOWARD BEING THOSE WHERE INTEREST AGGREGATION BY THE LEGISLATURE IS SIGNIFICANT, MODERATE, OR LIMITED (48)	0.65	138	TILT TOWARD BEING THOSE WHERE INTEREST AGGREGATION BY THE LEGISLATURE IS NEGLIGIBLE (49)	0.55	13 35 7 42 X SQ= 1.71 P = 0.139 RV YES (NO)
148	ALWAYS ARE THOSE WHERE THE PARTY SYSTEM IS QUALITATIVELY OTHER THAN AFRICAN TRANSITIONAL (96)	1.00	148	TEND LESS TO BE THOSE WHERE THE PARTY SYSTEM IS QUALITATIVELY OTHER THAN AFRICAN TRANSITIONAL (96)	0.83	0 14 27 69 X SQ= 3.81 P = 0.020 RV NO (NO)
178	TEND TO BE THOSE WHERE THE LEGISLATURE IS BICAMERAL (51)	0.74	178	TEND TO BE THOSE WHERE THE LEGISLATURE IS UNICAMERAL (53)	0.58	6 47 17 34 X SQ= 6.09 P = 0.009 RV YES (NO)
182	LEAN MORE TOWARD BEING THOSE WHERE THE BUREAUCRACY IS SEMI-MODERN, RATHER THAN POST-COLONIAL TRANSITIONAL (55)	0.88	182	LEAN LESS TOWARD BEING THOSE WHERE THE BUREAUCRACY IS SEMI-MODERN, RATHER THAN POST-COLONIAL TRANSITIONAL (55)	0.63	15 40 2 23 X SQ= 2.75 P = 0.076 RV NO (NO)
193	TEND MORE TO BE THOSE WHERE THE CHARACTER OF THE LEGAL SYSTEM IS OTHER THAN PARTLY INDIGENOUS (86)	0.92	193	TEND LESS TO BE THOSE WHERE THE CHARACTER OF THE LEGAL SYSTEM IS OTHER THAN PARTLY INDIGENOUS (86)	0.70	2 26 24 62 X SQ= 4.06 P = 0.022 RV NO (NO)

24 POLITIES WHOSE POPULATION IS VERY LARGE, LARGE, OR MEDIUM (61)	24 POLITIES WHOSE POPULATION IS SMALL (54)

PREDICATE ONLY

61 IN LEFT COLUMN

AFGHANISTAN	ALGERIA	ARGENTINA	AUSTRALIA	AUSTRIA	BELGIUM	BRAZIL	BULGARIA	BURMA	CANADA
CEYLON	CHILE	CHINA, PR	COLOMBIA	CONGO(LEO)	CUBA	CZECHOS'KIA	ETHIOPIA	FRANCE	GERMANY, E
GERMAN FR	GHANA	GREECE	HUNGARY	INDIA	INDONESIA	IRAN	IRAQ	ITALY	JAPAN
KOREA, N	KOREA REP	MALAYA	MEXICO	MOROCCO	NEPAL	NETHERLANDS	NIGERIA	PAKISTAN	PERU
PHILIPPINES	POLAND	PORTUGAL	RUMANIA	SA'U ARABIA	SO AFRICA	SPAIN	SUDAN	SWEDEN	TANGANYIKA
THAILAND	TURKEY	UGANDA	USSR	UAR	UK	US	VENEZUELA	VIETNAM, N	VIETNAM REP
YUGOSLAVIA									

54 IN RIGHT COLUMN

ALBANIA	BOLIVIA	BURUNDI	CAMBODIA	CAMEROUN	CEN AFR REP	CHAD	CONGO(BRA)	COSTA RICA	CYPRUS
DAHOMEY	DENMARK	DOMIN REP	ECUADOR	EL SALVADOR	FINLAND	GABON	GUATEMALA	GUINEA	HAITI
HONDURAS	ICELAND	IRELAND	ISRAEL	IVORY COAST	JAMAICA	JORDAN	LAOS	LEBANON	LIBERIA
LIBYA	LUXEMBOURG	MALAGASY R	MALI	MAURITANIA	MONGOLIA	NEW ZEALAND	NICARAGUA	NIGER	NORWAY
PANAMA	PARAGUAY	RWANDA	SENEGAL	SIERRE LEO	SOMALIA	SWITZERLAND	SYRIA	TOGO	TRINIDAD
TUNISIA	UPPER VOLTA	URUGUAY	YEMEN						

25 POLITIES
WHOSE POPULATION DENSITY IS
VERY HIGH OR HIGH (17)

25 POLITIES
WHOSE POPULATION DENSITY IS
MEDIUM OR LOW (98)

PREDICATE ONLY

17 IN LEFT COLUMN

BELGIUM	CEYLON	EL SALVADOR	GERMANY, E	GERMAN FR	HAITI	INDIA	ITALY	JAMAICA	JAPAN
KOREA REP	LEBANON	LUXEMBOURG	NETHERLANDS	SWITZERLAND	TRINIDAD	UK			

98 IN RIGHT COLUMN

AFGHANISTAN	ALBANIA	ALGERIA	ARGENTINA	AUSTRALIA	AUSTRIA	BOLIVIA	BRAZIL	BULGARIA	BURMA
BURUNDI	CAMBODIA	CAMEROUN	CANADA	CEN AFR REP	CHAD	CHILE	CHINA, PR	COLOMBIA	CONGO(BRA)
CONGO(LEO)	COSTA RICA	CUBA	CYPRUS	CZECHOS'KIA	DAHOMEY	DENMARK	DOMIN REP	ECUADOR	ETHIOPIA
FINLAND	FRANCE	GABON	GHANA	GREECE	GUATEMALA	GUINEA	HONDURAS	HUNGARY	ICELAND
INDONESIA	IRAN	IRAQ	IRELAND	ISRAEL	IVORY COAST	JORDAN	KOREA, N	LAOS	LIBERIA
LIBYA	MALAGASY R	MALAYA	MALI	MAURITANIA	MEXICO	MONGOLIA	MOROCCO	NEPAL	NEW ZEALAND
NICARAGUA	NIGER	NIGERIA	NORWAY	PAKISTAN	PANAMA	PARAGUAY	PERU	PHILIPPINES	POLAND
PORTUGAL	RUMANIA	RWANDA	SA'U ARABIA	SENEGAL	SIERRE LEO	SOMALIA	SO AFRICA	SPAIN	SUDAN
SWEDEN	SYRIA	TANGANYIKA	THAILAND	TOGO	TUNISIA	TURKEY	UGANDA	USSR	UAR
US	UPPER VOLTA	URUGUAY	VENEZUELA	VIETNAM, N	VIETNAM REP	YEMEN	YUGOSLAVIA		

26 POLITIES
WHOSE POPULATION DENSITY IS
VERY HIGH, HIGH, OR MEDIUM (48)

26 POLITIES
WHOSE POPULATION DENSITY IS
LOW (67)

BOTH SUBJECT AND PREDICATE

48 IN LEFT COLUMN

ALBANIA	AUSTRIA	BELGIUM	BULGARIA	BURUNDI	CEYLON	CHINA, PR	CUBA	CYPRUS	CZECHOS'KIA
DENMARK	DOMIN REP	EL SALVADOR	FRANCE	GERMANY, E	GERMAN FR	GREECE	HAITI	HUNGARY	INDIA
INDONESIA	IRELAND	ISRAEL	ITALY	JAMAICA	JAPAN	KOREA, N	KOREA REP	LEBANON	LUXEMBOURG
MALAYA	NEPAL	NETHERLANDS	NIGERIA	PAKISTAN	PHILIPPINES	POLAND	PORTUGAL	RUMANIA	RWANDA
SPAIN	SWITZERLAND	THAILAND	TRINIDAD	UK	VIETNAM, N	VIETNAM REP	YUGOSLAVIA		

67 IN RIGHT COLUMN

AFGHANISTAN	ALGERIA	ARGENTINA	AUSTRALIA	BOLIVIA	BRAZIL	BURMA	CAMBODIA	CAMEROUN	CANADA
CEN AFR REP	CHAD	CHILE	COLOMBIA	CONGO(BRA)	CONGO(LEO)	COSTA RICA	DAHOMEY	ECUADOR	ETHIOPIA
FINLAND	GABON	GHANA	GUATEMALA	GUINEA	HONDURAS	ICELAND	IRAN	IRAQ	IVORY COAST
JORDAN	LAOS	LIBERIA	LIBYA	MALAGASY R	MALI	MAURITANIA	MEXICO	MONGOLIA	MOROCCO
NEW ZEALAND	NICARAGUA	NIGER	NORWAY	PANAMA	PARAGUAY	PERU	SA'U ARABIA	SENEGAL	SIERRE LEO
SOMALIA	SO AFRICA	SUDAN	SWEDEN	SYRIA	TANGANYIKA	TOGO	TUNISIA	TURKEY	UGANDA
USSR	UAR	US	UPPER VOLTA	URUGUAY	VENEZUELA	YEMEN			

	Left column			Right column		
2	TEND LESS TO BE THOSE LOCATED ELSEWHERE THAN IN WEST EUROPE, SCANDINAVIA, NORTH AMERICA, OR AUSTRALASIA (93)	0.71	2	TEND MORE TO BE THOSE LOCATED ELSEWHERE THAN IN WEST EUROPE, SCANDINAVIA, NORTH AMERICA, OR AUSTRALASIA (93)	0.88	14 8 34 59 X SQ= 4.31 P = 0.030 RV NO (YES)
3	TEND LESS TO BE THOSE LOCATED ELSEWHERE THAN IN WEST EUROPE (102)	0.73	3	ALWAYS ARE THOSE LOCATED ELSEWHERE THAN IN WEST EUROPE (102)	1.00	13 0 35 67 X SQ= 17.85 P = 0. RV NO (YES)
8	TEND LESS TO BE THOSE LOCATED ELSEWHERE THAN IN EAST ASIA SOUTH ASIA, OR SOUTHEAST ASIA (97)	0.71	8	TEND MORE TO BE THOSE LOCATED ELSEWHERE THAN IN EAST ASIA SOUTH ASIA, OR SOUTHEAST ASIA (97)	0.94	14 4 34 63 X SQ= 9.71 P = 0.001 RV NO (YES)
10	TEND MORE TO BE THOSE LOCATED ELSEWHERE THAN IN NORTH AFRICA, OR CENTRAL AND SOUTH AFRICA (82)	0.94	10	TEND LESS TO BE THOSE LOCATED ELSEWHERE THAN IN NORTH AFRICA, OR CENTRAL AND SOUTH AFRICA (82)	0.55	3 30 45 37 X SQ= 18.45 P = 0. RV NO (YES)

13	LEAN MORE TOWARD BEING THOSE LOCATED ELSEWHERE THAN IN NORTH AFRICA OR THE MIDDLE EAST (99)	0.94	13	LEAN LESS TOWARD BEING THOSE LOCATED ELSEWHERE THAN IN NORTH AFRICA OR THE MIDDLE EAST (99)	0.81	3 13 45 54 X SQ= 3.02 P = 0.057 RV NO (NO)
16	TEND TO BE THOSE LOCATED IN EAST ASIA, SOUTH ASIA, OR SOUTHEAST ASIA, RATHER THAN IN THE MIDDLE EAST (18)	0.82	16	TEND TO BE THOSE LOCATED IN THE MIDDLE EAST, RATHER THAN IN EAST ASIA, SOUTH ASIA, OR SOUTHEAST ASIA (11)	0.67	3 8 14 4 X SQ= 5.25 P = 0.018 RV YES (YES)
18	TEND TO BE THOSE LOCATED IN EAST ASIA, SOUTH ASIA, OR SOUTHEAST ASIA, RATHER THAN IN NORTH AFRICA OR CENTRAL AND SOUTH AFRICA (18)	0.82	18	TEND TO BE THOSE LOCATED IN NORTH AFRICA OR CENTRAL AND SOUTH AFRICA, RATHER THAN IN EAST ASIA, SOUTH ASIA, OR SOUTHEAST ASIA (33)	0.88	3 30 14 4 X SQ= 21.73 P = 0. RV YES (YES)
19	TILT TOWARD BEING THOSE LOCATED IN THE CARIBBEAN, CENTRAL AMERICA, OR SOUTH AMERICA, RATHER THAN IN NORTH AFRICA OR CENTRAL AND SOUTH AFRICA (22)	0.67	19	TILT TOWARD BEING THOSE LOCATED IN NORTH AFRICA OR CENTRAL AND SOUTH AFRICA, RATHER THAN IN THE CARIBBEAN, CENTRAL AMERICA, OR SOUTH AMERICA (33)	0.65	3 30 6 16 X SQ= 2.00 P = 0.134 RV YES (NO)
20	TEND TO BE THOSE LOCATED IN EAST ASIA, SOUTH ASIA, OR SOUTHEAST ASIA, RATHER THAN IN THE CARIBBEAN, CENTRAL AMERICA, OR SOUTH AMERICA (18)	0.70	20	TEND TO BE THOSE LOCATED IN THE CARIBBEAN, CENTRAL AMERICA, OR SOUTH AMERICA, RATHER THAN IN EAST ASIA, SOUTH ASIA, OR SOUTHEAST ASIA (22)	0.80	14 4 6 16 X SQ= 8.18 P = 0.004 RV YES (YES)
22	TEND TO BE THOSE WHOSE TERRITORIAL SIZE IS SMALL (47)	0.67	22	TEND TO BE THOSE WHOSE TERRITORIAL SIZE IS VERY LARGE, LARGE, OR MEDIUM (68)	0.78	16 52 32 15 X SQ= 20.89 P = 0. RV YES (YES)
24	TEND TO BE THOSE WHOSE POPULATION IS VERY LARGE, LARGE, OR MEDIUM (61)	0.69	24	TEND TO BE THOSE WHOSE POPULATION IS SMALL (54)	0.58	33 28 15 39 X SQ= 7.11 P = 0.005 RV YES (YES)
28	TEND TO BE THOSE WHOSE POPULATION GROWTH RATE IS LOW (48)	0.60	28	TEND TO BE THOSE WHOSE POPULATION GROWTH RATE IS HIGH (62)	0.68	19 43 28 20 X SQ= 7.38 P = 0.006 RV YES (YES)
29	LEAN TOWARD BEING THOSE WHERE THE DEGREE OF URBANIZATION IS HIGH (56)	0.65	29	LEAN TOWARD BEING THOSE WHERE THE DEGREE OF URBANIZATION IS LOW (49)	0.54	26 30 14 35 X SQ= 2.82 P = 0.072 RV YES (NO)

30	TEND TO BE THOSE WHOSE AGRICULTURAL POPULATION IS MEDIUM, LOW, OR VERY LOW (57)	0.70	30	TEND TO BE THOSE WHOSE AGRICULTURAL POPULATION IS HIGH (56)	0.64	14 42 33 24 X SQ= 11.26 P = 0.001 RV YES (YES)
34	TEND TO BE THOSE WHOSE GROSS NATIONAL PRODUCT IS VERY HIGH, HIGH, MEDIUM, OR LOW (62)	0.73	34	TEND TO BE THOSE WHOSE GROSS NATIONAL PRODUCT IS VERY LOW (53)	0.60	35 27 13 40 X SQ= 10.70 P = 0.001 RV YES (YES)
36	TEND TO BE THOSE WHOSE PER CAPITA GROSS NATIONAL PRODUCT IS VERY HIGH, HIGH, OR MEDIUM (42)	0.58	36	TEND TO BE THOSE WHOSE PER CAPITA GROSS NATIONAL PRODUCT IS LOW OR VERY LOW (73)	0.79	28 14 20 53 X SQ= 15.33 P = 0. RV YES (YES)
37	TEND TO BE THOSE WHOSE PER CAPITA GROSS NATIONAL PRODUCT IS VERY HIGH, HIGH, MEDIUM, OR LOW (64)	0.69	37	TEND TO BE THOSE WHOSE PER CAPITA GROSS NATIONAL PRODUCT IS VERY LOW (51)	0.54	33 31 15 36 X SQ= 4.85 P = 0.022 RV YES (YES)
39	TEND LESS TO BE THOSE WHOSE INTERNATIONAL FINANCIAL STATUS IS LOW OR VERY LOW (76)	0.55	39	TEND MORE TO BE THOSE WHOSE INTERNATIONAL FINANCIAL STATUS IS LOW OR VERY LOW (76)	0.75	21 17 26 50 X SQ= 3.81 P = 0.043 RV NO (YES)
40	TEND MORE TO BE THOSE WHOSE INTERNATIONAL FINANCIAL STATUS IS VERY HIGH, HIGH, MEDIUM, OR LOW (71)	0.84	40	TEND LESS TO BE THOSE WHOSE INTERNATIONAL FINANCIAL STATUS IS VERY HIGH, HIGH, MEDIUM, OR LOW (71)	0.52	37 34 7 32 X SQ= 10.86 P = 0.001 RV NO (YES)
43	TEND TO BE THOSE WHOSE ECONOMIC DEVELOPMENTAL STATUS IS DEVELOPED, INTERMEDIATE, OR UNDERDEVELOPED (55)	0.68	43	TEND TO BE THOSE WHOSE ECONOMIC DEVELOPMENTAL STATUS IS VERY UNDERDEVELOPED (57)	0.65	32 23 15 42 X SQ= 10.40 P = 0.001 RV YES (YES)
45	TEND TO BE THOSE WHERE THE LITERACY RATE IS FIFTY PERCENT OR ABOVE (55)	0.80	45	TEND TO BE THOSE WHERE THE LITERACY RATE IS BELOW FIFTY PERCENT (54)	0.69	35 20 9 45 X SQ= 23.06 P = 0. RV YES (YES)
51	LEAN LESS TOWARD BEING THOSE WHERE FREEDOM OF THE PRESS IS COMPLETE OR INTERMITTENT (60)	0.51	51	LEAN MORE TOWARD BEING THOSE WHERE FREEDOM OF THE PRESS IS COMPLETE OR INTERMITTENT (60)	0.70	22 38 21 16 X SQ= 2.97 P = 0.061 RV NO (YES)

54	TEND LESS TO BE THOSE WHERE NEWSPAPER CIRCULATION IS LESS THAN ONE HUNDRED PER THOUSAND (76)	0.52	54	TEND MORE TO BE THOSE WHERE NEWSPAPER CIRCULATION IS LESS THAN ONE HUNDRED PER THOUSAND (76)	0.78	22 15 24 52 X SQ= 6.90 P = 0.008 RV NO (YES)
55	TEND MORE TO BE THOSE WHERE NEWSPAPER CIRCULATION IS TEN OR MORE PER THOUSAND (78)	0.89	55	TEND LESS TO BE THOSE WHERE NEWSPAPER CIRCULATION IS TEN OR MORE PER THOUSAND (78)	0.55	41 37 5 30 X SQ= 13.12 P = 0. RV NO (YES)
56	TEND MORE TO BE THOSE WHERE THE RELIGION IS PREDOMINANTLY LITERATE (79)	0.91	56	TEND LESS TO BE THOSE WHERE THE RELIGION IS PREDOMINANTLY LITERATE (79)	0.60	39 40 4 27 X SQ= 10.95 P = 0. RV NO (NO)
58	TEND MORE TO BE THOSE WHERE THE RELIGION IS OTHER THAN MUSLIM (97)	0.96	58	TEND LESS TO BE THOSE WHERE THE RELIGION IS OTHER THAN MUSLIM (97)	0.76	2 16 46 51 X SQ= 6.81 P = 0.004 RV NO (NO)
63	TEND TO BE THOSE WHERE THE RELIGION IS PREDOMINANTLY SOME KIND OF CHRISTIAN (46)	0.54	63	TEND TO BE THOSE WHERE THE RELIGION IS PREDOMINANTLY OR PARTLY OTHER THAN CHRISTIAN (68)	0.70	26 20 22 46 X SQ= 5.62 P = 0.012 RV YES (YES)
67	TEND MORE TO BE THOSE THAT ARE RACIALLY HOMOGENEOUS (82)	0.85	67	TEND LESS TO BE THOSE THAT ARE RACIALLY HOMOGENEOUS (82)	0.65	40 40 7 22 X SQ= 4.80 P = 0.017 RV NO (YES)
68	LEAN TOWARD BEING THOSE THAT ARE LINGUISTICALLY HOMOGENEOUS (52)	0.55	68	LEAN TOWARD BEING THOSE THAT ARE LINGUISTICALLY WEAKLY HETEROGENEOUS OR STRONGLY HETEROGENEOUS (62)	0.61	26 26 21 41 X SQ= 2.41 P = 0.090 RV YES (YES)
69	TEND TO BE THOSE THAT ARE LINGUISTICALLY HOMOGENEOUS OR WEAKLY HETEROGENEOUS (64)	0.70	69	TEND TO BE THOSE THAT ARE LINGUISTICALLY STRONGLY HETEROGENEOUS (50)	0.54	33 31 14 36 X SQ= 5.50 P = 0.013 RV YES (YES)
71	TEND LESS TO BE THOSE WHOSE DATE OF INDEPENDENCE IS AFTER 1800 (90)	0.70	71	TEND MORE TO BE THOSE WHOSE DATE OF INDEPENDENCE IS AFTER 1800 (90)	0.88	13 8 31 59 X SQ= 4.28 P = 0.027 RV NO (YES)

74	TEND LESS TO BE THOSE THAT ARE NOT HISTORICALLY WESTERN (87)	0.62	74	TEND MORE TO BE THOSE THAT ARE NOT HISTORICALLY WESTERN (87)	0.88	18 8 29 58 X SQ= 9.19 P = 0.001 RV NO (YES)
75	TEND TO BE THOSE THAT ARE HISTORICALLY WESTERN OR SIGNIFICANTLY WESTERNIZED (62)	0.71	75	TEND TO BE THOSE THAT ARE NOT HISTORICALLY WESTERN AND ARE NOT SIGNIFICANTLY WESTERNIZED (52)	0.58	34 28 14 38 X SQ= 7.93 P = 0.004 RV YES (YES)
76	TEND LESS TO BE THOSE THAT HAVE BEEN SIGNIFICANTLY OR PARTIALLY WESTERNIZED THROUGH A COLONIAL RELATIONSHIP, RATHER THAN BEING HISTORICALLY WESTERN (70)	0.51	76	TEND MORE TO BE THOSE THAT HAVE BEEN SIGNIFICANTLY OR PARTIALLY WESTERNIZED THROUGH A COLONIAL RELATIONSHIP, RATHER THAN BEING HISTORICALLY WESTERN (70)	0.86	18 8 19 51 X SQ= 12.46 P = 0. RV NO (YES)
80	TEND TO BE THOSE WHOSE DATE OF INDEPENDENCE IS AFTER 1914, AND THAT WERE FORMERLY DEPENDENCIES OF BRITAIN, RATHER THAN FRANCE (19)	0.77	80	TEND TO BE THOSE WHOSE DATE OF INDEPENDENCE IS AFTER 1914, AND THAT WERE FORMERLY DEPENDENCIES OF FRANCE, RATHER THAN BRITAIN (24)	0.70	10 9 3 21 X SQ= 6.31 P = 0.007 RV YES (YES)
82	LEAN TOWARD BEING THOSE WHERE THE TYPE OF POLITICAL MODERNIZATION IS EARLY OR LATER EUROPEAN OR EUROPEAN DERIVED, RATHER THAN DEVELOPED TUTELARY OR UNDEVELOPED TUTELARY (51)	0.59	82	LEAN TOWARD BEING THOSE WHERE THE TYPE OF POLITICAL MODERNIZATION IS DEVELOPED TUTELARY OR UNDEVELOPED TUTELARY, RATHER THAN EARLY OR LATER EUROPEAN OR EUROPEAN DERIVED (55)	0.60	26 25 18 37 X SQ= 2.92 P = 0.076 RV YES (YES)
84	TEND TO BE THOSE WHERE THE TYPE OF POLITICAL MODERNIZATION IS DEVELOPED TUTELARY, RATHER THAN UNDEVELOPED TUTELARY (31)	0.83	84	TEND TO BE THOSE WHERE THE TYPE OF POLITICAL MODERNIZATION IS UNDEVELOPED TUTELARY, RATHER THAN DEVELOPED TUTELARY (24)	0.57	15 16 3 21 X SQ= 6.37 P = 0.008 RV YES (NO)
85	TEND TO BE THOSE WHERE THE STAGE OF POLITICAL MODERNIZATION IS ADVANCED, RATHER THAN MID- OR EARLY TRANSITIONAL (60)	0.69	85	TEND TO BE THOSE WHERE THE STAGE OF POLITICAL MODERNIZATION IS MID- OR EARLY TRANSITIONAL, RATHER THAN ADVANCED (54)	0.59	33 27 15 39 X SQ= 7.56 P = 0.004 RV YES (YES)
87	TEND MORE TO BE THOSE WHOSE IDEOLOGICAL ORIENTATION IS OTHER THAN DEVELOPMENTAL (58)	0.84	87	TEND LESS TO BE THOSE WHOSE IDEOLOGICAL ORIENTATION IS OTHER THAN DEVELOPMENTAL (58)	0.51	6 25 32 26 X SQ= 9.18 P = 0.001 RV NO (YES)
92	TEND LESS TO BE THOSE WHERE THE SYSTEM STYLE IS LIMITED MOBILIZATIONAL, OR NON-MOBILIZATIONAL (93)	0.72	92	TEND MORE TO BE THOSE WHERE THE SYSTEM STYLE IS LIMITED MOBILIZATIONAL, OR NON-MOBILIZATIONAL (93)	0.89	13 7 34 59 X SQ= 4.37 P = 0.025 RV NO (YES)

95	TEND LESS TO BE THOSE WHERE THE STATUS OF THE REGIME IS CONSTITUTIONAL OR AUTHORITARIAN (95)	0.70	95	TEND MORE TO BE THOSE WHERE THE STATUS OF THE REGIME IS CONSTITUTIONAL OR AUTHORITARIAN (95)	0.97	32 63 14 2 X SQ= 14.20 P = 0. RV NO (YES)
96	TEND MORE TO BE THOSE WHERE THE STATUS OF THE REGIME IS CONSTITUTIONAL OR TOTALITARIAN (67)	0.86	96	TEND LESS TO BE THOSE WHERE THE STATUS OF THE REGIME IS CONSTITUTIONAL OR TOTALITARIAN (67)	0.63	38 29 6 17 X SQ= 5.26 P = 0.015 RV NO (YES)
97	TEND TO BE THOSE WHERE THE STATUS OF THE REGIME IS TOTALITARIAN, RATHER THAN AUTHORITARIAN (16)	0.70	97	TEND TO BE THOSE WHERE THE STATUS OF THE REGIME IS AUTHORITARIAN, RATHER THAN TOTALITARIAN (23)	0.89	6 17 14 2 X SQ= 11.89 P = 0. RV YES (YES)
99	TEND TO BE THOSE WHERE GOVERNMENTAL STABILITY IS GENERALLY PRESENT AND DATES FROM AT LEAST THE INTER-WAR PERIOD, OR FROM THE POST-WAR PERIOD (50)	0.74	99	TEND TO BE THOSE WHERE GOVERNMENTAL STABILITY IS MODERATELY PRESENT AND DATES FROM THE POST-WAR PERIOD, OR IS ABSENT (36)	0.54	28 22 10 26 X SQ= 5.66 P = 0.015 RV YES (YES)
101	TILT TOWARD BEING THOSE WHERE THE REPRESENTATIVE CHARACTER OF THE REGIME IS POLYARCHIC (41)	0.51	101	TILT TOWARD BEING THOSE WHERE THE REPRESENTATIVE CHARACTER OF THE REGIME IS LIMITED POLYARCHIC, PSEUDO-POLYARCHIC, OR NON-POLYARCHIC (57)	0.65	22 19 21 36 X SQ= 2.10 P = 0.105 RV YES (YES)
107	TILT TOWARD BEING THOSE WHERE AUTONOMOUS GROUPS ARE FULLY TOLERATED IN POLITICS (46)	0.50	107	TILT TOWARD BEING THOSE WHERE AUTONOMOUS GROUPS ARE PARTIALLY TOLERATED IN POLITICS, ARE TOLERATED ONLY OUTSIDE POLITICS, OR ARE NOT TOLERATED AT ALL (65)	0.65	24 22 24 41 X SQ= 1.97 P = 0.124 RV YES (YES)
110	LEAN LESS TOWARD BEING THOSE WHERE POLITICAL ENCULTURATION IS MEDIUM OR LOW (80)	0.74	110	LEAN MORE TOWARD BEING THOSE WHERE POLITICAL ENCULTURATION IS MEDIUM OR LOW (80)	0.90	9 6 26 54 X SQ= 3.01 P = 0.077 RV NO (YES)
114	LEAN TOWARD BEING THOSE WHERE SECTIONALISM IS NEGLIGIBLE (47)	0.53	114	LEAN TOWARD BEING THOSE WHERE SECTIONALISM IS EXTREME OR MODERATE (61)	0.64	22 39 25 22 X SQ= 2.51 P = 0.082 RV YES (YES)
117	TILT TOWARD BEING THOSE WHERE INTEREST ARTICULATION BY ASSOCIATIONAL GROUPS IS SIGNIFICANT, MODERATE, OR LIMITED (60)	0.64	117	TILT TOWARD BEING THOSE WHERE INTEREST ARTICULATION BY ASSOCIATIONAL GROUPS IS NEGLIGIBLE (51)	0.52	28 32 16 35 X SQ= 2.09 P = 0.121 RV YES (NO)

```
122  TILT LESS TOWARD BEING THOSE          0.65     122  TILT MORE TOWARD BEING THOSE          0.78        31    52
     WHERE INTEREST ARTICULATION                         WHERE INTEREST ARTICULATION                        17    15
     BY NON-ASSOCIATIONAL GROUPS                         BY NON-ASSOCIATIONAL GROUPS                   X SQ=   1.76
     IS SIGNIFICANT OR MODERATE  (83)                    IS SIGNIFICANT OR MODERATE  (83)              P =    0.143
                                                                                                        RV NO  (YES)

125  TEND LESS TO BE THOSE                 0.51     125  TEND MORE TO BE THOSE                 0.73        20    44
     WHERE INTEREST ARTICULATION                         WHERE INTEREST ARTICULATION                        19    16
     BY ANOMIC GROUPS                                    BY ANOMIC GROUPS                              X SQ=   4.11
     IS FREQUENT OR OCCASIONAL  (64)                     IS FREQUENT OR OCCASIONAL  (64)               P =    0.032
                                                                                                        RV NO  (YES)

130  TILT MORE TOWARD BEING THOSE          0.93     130  TILT LESS TOWARD BEING THOSE          0.79         3     9
     WHERE INTEREST AGGREGATION                          WHERE INTEREST AGGREGATION                         38    33
     BY POLITICAL PARTIES                                BY POLITICAL PARTIES                          X SQ=   2.30
     IS MODERATE, LIMITED, OR                            IS MODERATE, LIMITED, OR                      P =    0.116
     NEGLIGIBLE  (71)                                    NEGLIGIBLE  (71)                              RV NO  (YES)

133  TEND MORE TO BE THOSE                 0.87     133  TEND LESS TO BE THOSE                 0.61         6    23
     WHERE INTEREST AGGREGATION                          WHERE INTEREST AGGREGATION                         40    36
     BY THE EXECUTIVE                                    BY THE EXECUTIVE                              X SQ=   7.45
     IS MODERATE, LIMITED, OR                            IS MODERATE, LIMITED, OR                      P =    0.004
     NEGLIGIBLE  (76)                                    NEGLIGIBLE  (76)                              RV NO  (YES)

135  TILT LESS TOWARD BEING THOSE          0.78     135  TILT MORE TOWARD BEING THOSE          0.91        29    48
     WHERE INTEREST AGGREGATION                          WHERE INTEREST AGGREGATION                          8     5
     BY THE EXECUTIVE                                    BY THE EXECUTIVE                              X SQ=   1.73
     IS SIGNIFICANT, MODERATE, OR                        IS SIGNIFICANT, MODERATE, OR                  P =    0.133
     LIMITED  (77)                                       LIMITED  (77)                                 RV NO  (YES)

137  TEND LESS TO BE THOSE                 0.59     137  TEND MORE TO BE THOSE                 0.79        18    11
     WHERE INTEREST AGGREGATION                          WHERE INTEREST AGGREGATION                         26    42
     BY THE LEGISLATURE                                  BY THE LEGISLATURE                            X SQ=   3.75
     IS LIMITED OR NEGLIGIBLE  (68)                      IS LIMITED OR NEGLIGIBLE  (68)                P =    0.045
                                                                                                        RV NO  (YES)

146  ALWAYS ARE THOSE                      1.00     146  TEND LESS TO BE THOSE                 0.86         0     8
     WHERE THE PARTY SYSTEM IS QUALITATIVELY             WHERE THE PARTY SYSTEM IS QUALITATIVELY            42    50
     OTHER THAN                                          OTHER THAN                                    X SQ=   4.56
     MASS-BASED TERRITORIAL  (92)                        MASS-BASED TERRITORIAL  (92)                  P =    0.019
                                                                                                        RV NO  (NO )

148  TEND MORE TO BE THOSE                 0.96     148  TEND LESS TO BE THOSE                 0.81         2    12
     WHERE THE PARTY SYSTEM IS QUALITATIVELY             WHERE THE PARTY SYSTEM IS QUALITATIVELY            45    51
     OTHER THAN                                          OTHER THAN                                    X SQ=   4.05
     AFRICAN TRANSITIONAL  (96)                          AFRICAN TRANSITIONAL  (96)                    P =    0.023
                                                                                                        RV NO  (NO )

153  TEND TO BE THOSE                      0.68     153  TEND TO BE THOSE                      0.62        26    16
     WHERE THE PARTY SYSTEM IS                           WHERE THE PARTY SYSTEM IS                          12    26
     STABLE  (42)                                        MODERATELY STABLE OR UNSTABLE  (38)           X SQ=   6.19
                                                                                                        P =    0.008
                                                                                                        RV YES (YES)
```

158	TEND TO BE THOSE WHERE PERSONALISMO IS NEGLIGIBLE (56)	0.77	158	TEND TO BE THOSE WHERE PERSONALISMO IS PRONOUNCED OR MODERATE (40)	0.55	9 31 31 25 X SQ= 9.06 P = 0.002 RV YES (YES)
160	TILT LESS TOWARD BEING THOSE WHERE THE POLITICAL LEADERSHIP IS MODERATELY ELITIST OR NON-ELITIST (67)	0.60	160	TILT MORE TOWARD BEING THOSE WHERE THE POLITICAL LEADERSHIP IS MODERATELY ELITIST OR NON-ELITIST (67)	0.75	16 14 24 43 X SQ= 1.95 P = 0.122 RV NO (YES)
164	LEAN MORE TOWARD BEING THOSE WHERE THE REGIME'S LEADERSHIP CHARISMA IS NEGLIGIBLE (65)	0.77	164	LEAN LESS TOWARD BEING THOSE WHERE THE REGIME'S LEADERSHIP CHARISMA IS NEGLIGIBLE (65)	0.57	10 24 33 32 X SQ= 3.32 P = 0.055 RV NO (YES)
168	LEAN LESS TOWARD BEING THOSE WHERE THE HORIZONTAL POWER DISTRIBUTION IS LIMITED OR NEGLIGIBLE (72)	0.57	168	LEAN MORE TOWARD BEING THOSE WHERE THE HORIZONTAL POWER DISTRIBUTION IS LIMITED OR NEGLIGIBLE (72)	0.76	20 14 27 45 X SQ= 3.43 P = 0.059 RV NO (YES)
170	TEND TO BE THOSE WHERE THE LEGISLATIVE-EXECUTIVE STRUCTURE IS OTHER THAN PRESIDENTIAL (63)	0.80	170	TEND TO BE THOSE WHERE THE LEGISLATIVE-EXECUTIVE STRUCTURE IS PRESIDENTIAL (39)	0.54	9 30 37 26 X SQ= 10.97 P = 0. RV YES (YES)
172	LEAN LESS TOWARD BEING THOSE WHERE THE LEGISLATIVE-EXECUTIVE STRUCTURE IS OTHER THAN PARLIAMENTARY-ROYALIST (88)	0.72	172	LEAN MORE TOWARD BEING THOSE WHERE THE LEGISLATIVE-EXECUTIVE STRUCTURE IS OTHER THAN PARLIAMENTARY-ROYALIST (88)	0.87	13 8 33 55 X SQ= 3.20 P = 0.051 RV NO (YES)
174	TEND LESS TO BE THOSE WHERE THE LEGISLATURE IS PARTIALLY EFFECTIVE, LARGELY INEFFECTIVE, OR WHOLLY INEFFECTIVE (72)	0.59	174	TEND MORE TO BE THOSE WHERE THE LEGISLATURE IS PARTIALLY EFFECTIVE, LARGELY INEFFECTIVE, OR WHOLLY INEFFECTIVE (72)	0.82	18 10 26 46 X SQ= 5.40 P = 0.014 RV NO (YES)
176	TEND LESS TO BE THOSE WHERE THE LEGISLATURE IS FULLY EFFECTIVE, PARTIALLY EFFECTIVE, OR LARGELY INEFFECTIVE (72)	0.61	176	TEND MORE TO BE THOSE WHERE THE LEGISLATURE IS FULLY EFFECTIVE, PARTIALLY EFFECTIVE, OR LARGELY INEFFECTIVE (72)	0.80	27 45 17 11 X SQ= 3.52 P = 0.045 RV NO (YES)
182	TEND MORE TO BE THOSE WHERE THE BUREAUCRACY IS SEMI-MODERN, RATHER THAN POST-COLONIAL TRANSITIONAL (55)	0.90	182	TEND LESS TO BE THOSE WHERE THE BUREAUCRACY IS SEMI-MODERN, RATHER THAN POST-COLONIAL TRANSITIONAL (55)	0.55	28 27 3 22 X SQ= 9.39 P = 0.001 RV NO (YES)

185	LEAN TOWARD BEING THOSE WHERE PARTICIPATION BY THE MILITARY IN POLITICS IS SUPPORTIVE, RATHER THAN INTERVENTIVE (31)	0.76	185	LEAN TOWARD BEING THOSE WHERE PARTICIPATION BY THE MILITARY IN POLITICS IS INTERVENTIVE, RATHER THAN SUPPORTIVE (21)	0.52	M 5 16 16 15 X SQ= 2.95 P = 0.083 RV YES (YES)
187	LEAN LESS TOWARD BEING THOSE WHERE THE ROLE OF THE POLICE IS POLITICALLY SIGNIFICANT (66)	0.55	187	LEAN MORE TOWARD BEING THOSE WHERE THE ROLE OF THE POLICE IS POLITICALLY SIGNIFICANT (66)	0.74	24 42 20 15 X SQ= 3.22 P = 0.058 RV NO (YES)
192	TEND MORE TO BE THOSE WHERE THE CHARACTER OF THE LEGAL SYSTEM IS OTHER THAN MUSLIM OR PARTLY MUSLIM (86)	0.92	192	TEND LESS TO BE THOSE WHERE THE CHARACTER OF THE LEGAL SYSTEM IS OTHER THAN MUSLIM OR PARTLY MUSLIM (86)	0.65	4 23 44 42 X SQ= 9.67 P = 0.001 RV NO (YES)
193	TEND MORE TO BE THOSE WHERE THE CHARACTER OF THE LEGAL SYSTEM IS OTHER THAN PARTLY INDIGENOUS (86)	0.90	193	TEND LESS TO BE THOSE WHERE THE CHARACTER OF THE LEGAL SYSTEM IS OTHER THAN PARTLY INDIGENOUS (86)	0.65	5 23 43 43 X SQ= 7.68 P = 0.004 RV NO (YES)
194	TEND LESS TO BE THOSE THAT ARE NON-COMMUNIST (101)	0.77	194	TEND MORE TO BE THOSE THAT ARE NON-COMMUNIST (101)	0.97	11 2 36 65 X SQ= 9.47 P = 0.002 RV NO (YES)

M

MATRIX

	26 POLITIES WHOSE POPULATION DENSITY IS VERY HIGH, HIGH, OR MEDIUM (48)	26 POLITIES WHOSE POPULATION DENSITY IS LOW (67)
185 POLITIES WHERE PARTICIPATION BY THE MILITARY IN POLITICS IS INTERVENTIVE, RATHER THAN SUPPORTIVE (21)	5	16
	EL SALVADOR HAITI KOREA REP PAKISTAN THAILAND	ARGENTINA IRAQ SUDAN YEMEN BRAZIL NICARAGUA SYRIA BURMA PANAMA TOGO GUATEMALA PARAGUAY TURKEY HONDURAS PERU UAR
185 POLITIES WHERE PARTICIPATION BY THE MILITARY IN POLITICS IS SUPPORTIVE, RATHER THAN INTERVENTIVE (31)	16	15
	ALBANIA HUNGARY PORTUGAL YUGOSLAVIA BULGARIA INDONESIA RUMANIA CHINA, PR KOREA, N SPAIN CZECHOS'KIA LEBANON VIETNAM, N GERMANY, E POLAND VIETNAM REP	AFGHANISTAN GHANA MONGOLIA ALGERIA GUINEA SA'U ARABIA CAMBODIA IRAN SENEGAL ECUADOR JORDAN SO AFRICA ETHIOPIA LIBERIA USSR

```
***********************************************************************************************************************

 27  POLITIES                                                 27  POLITIES
     WHOSE POPULATION DENSITY IS                                  WHOSE POPULATION DENSITY IS
     MEDIUM  (31)                                                 VERY HIGH, HIGH, OR LOW  (84)


                                                                                                          SUBJECT ONLY
.......................................................................................................................

      31 IN LEFT COLUMN

ALBANIA     AUSTRIA     BULGARIA    BURUNDI     CHINA, PR   CUBA        CYPRUS      CZECHOS'KIA DENMARK     DOMIN REP
FRANCE      GREECE      HUNGARY     INDONESIA   IRELAND     ISRAEL      KOREA, N    MALAYA      NEPAL       NIGERIA
PAKISTAN    PHILIPPINES POLAND      PORTUGAL    RUMANIA     RWANDA      SPAIN       THAILAND    VIETNAM, N  VIETNAM REP
YUGOSLAVIA

      84 IN RIGHT COLUMN

AFGHANISTAN ALGERIA     ARGENTINA   AUSTRALIA   BELGIUM     BOLIVIA     BRAZIL      BURMA       CAMBODIA    CAMEROUN
CANADA      CEN AFR REP CEYLON      CHAD        CHILE       COLOMBIA    CONGO(BRA)  CONGO(LEO)  COSTA RICA  DAHOMEY
ECUADOR     EL SALVADOR ETHIOPIA    FINLAND     GABON       GERMANY, E  GERMAN FR   GHANA       GUATEMALA   GUINEA
HAITI       HONDURAS    ICELAND     INDIA       IRAN        IRAQ        ITALY       IVORY COAST JAMAICA     JAPAN
JORDAN      KOREA REP   LAOS        LEBANON     LIBERIA     LIBYA       LUXEMBOURG  MALAGASY R  MALI        MAURITANIA
MEXICO      MONGOLIA    MOROCCO     NETHERLANDS NEW ZEALAND NICARAGUA   NIGER       NORWAY      PANAMA      PARAGUAY
PERU        SA'U ARABIA SENEGAL     SIERRE LEO  SOMALIA     SO AFRICA   SUDAN       SWEDEN      SWITZERLAND SYRIA
TANGANYIKA  TOGO        TRINIDAD    TUNISIA     TURKEY      UGANDA      USSR        UAR         UK          US
UPPER VOLTA URUGUAY     VENEZUELA   YEMEN
-----------------------------------------------------------------------------------------------------------------------

  5  TILT TOWARD BEING THOSE                    0.50       5  TILT TOWARD BEING THOSE                  0.85          7   11
     LOCATED IN EAST EUROPE, RATHER THAN                      LOCATED IN SCANDINAVIA OR WEST EUROPE,                 7    2
     IN SCANDINAVIA OR WEST EUROPE  (9)                       RATHER THAN IN EAST EUROPE  (18)                X SQ=   2.24
                                                                                                              P =    0.103
                                                                                                              RV YES (YES)

  6  TEND MORE TO BE THOSE                      0.94       6  TEND LESS TO BE THOSE                    0.76          2   20
     LOCATED ELSEWHERE THAN IN THE                            LOCATED ELSEWHERE THAN IN THE                         29   64
     CARIBBEAN, CENTRAL AMERICA,                              CARIBBEAN, CENTRAL AMERICA,                     X SQ=   3.36
     OR SOUTH AMERICA  (93)                                   OR SOUTH AMERICA  (93)                          P =    0.036
                                                                                                              RV NO  (NO )

  8  TEND LESS TO BE THOSE                      0.68       8  TEND MORE TO BE THOSE                    0.90         10    8
     LOCATED ELSEWHERE THAN IN EAST ASIA                      LOCATED ELSEWHERE THAN IN EAST ASIA                   21   76
     SOUTH ASIA, OR SOUTHEAST ASIA  (97)                      SOUTH ASIA, OR SOUTHEAST ASIA  (97)             X SQ=   7.23
                                                                                                              P =    0.007
                                                                                                              RV NO  (YES)

 10  TEND MORE TO BE THOSE                      0.90      10  TEND LESS TO BE THOSE                    0.64          3   30
     LOCATED ELSEWHERE THAN IN NORTH AFRICA,                  LOCATED ELSEWHERE THAN IN NORTH AFRICA,               28   54
     OR CENTRAL AND SOUTH AFRICA  (82)                        OR CENTRAL AND SOUTH AFRICA  (82)               X SQ=   6.28
                                                                                                              P =    0.005
                                                                                                              RV NO  (NO )
```

16	LEAN TOWARD BEING THOSE LOCATED IN EAST ASIA, SOUTH ASIA, OR SOUTHEAST ASIA, RATHER THAN IN THE MIDDLE EAST (18)	0.83	16	LEAN TOWARD BEING THOSE LOCATED IN THE MIDDLE EAST, RATHER THAN IN EAST ASIA, SOUTH ASIA, OR SOUTHEAST ASIA (11)	0.53	2 9 10 8 X SQ= 2.54 P = 0.064 RV YES (YES)
18	TEND TO BE THOSE LOCATED IN EAST ASIA, SOUTH ASIA, OR SOUTHEAST ASIA, RATHER THAN IN NORTH AFRICA OR CENTRAL AND SOUTH AFRICA (18)	0.77	18	TEND TO BE THOSE LOCATED IN NORTH AFRICA OR CENTRAL AND SOUTH AFRICA, RATHER THAN IN EAST ASIA, SOUTH ASIA, OR SOUTHEAST ASIA (33)	0.79	3 30 10 8 X SQ= 10.91 P = 0.001 RV YES (YES)
20	TEND TO BE THOSE LOCATED IN EAST ASIA, SOUTH ASIA, OR SOUTHEAST ASIA, RATHER THAN IN THE CARIBBEAN, CENTRAL AMERICA, OR SOUTH AMERICA (18)	0.83	20	TEND TO BE THOSE LOCATED IN THE CARIBBEAN, CENTRAL AMERICA, OR SOUTH AMERICA, RATHER THAN IN EAST ASIA, SOUTH ASIA, OR SOUTHEAST ASIA (22)	0.71	10 8 2 20 X SQ= 8.09 P = 0.002 RV YES (YES)
22	TEND TO BE THOSE WHOSE TERRITORIAL SIZE IS SMALL (47)	0.65	22	TEND TO BE THOSE WHOSE TERRITORIAL SIZE IS VERY LARGE, LARGE, OR MEDIUM (68)	0.68	11 57 20 27 X SQ= 8.53 P = 0.003 RV YES (NO)
24	TEND TO BE THOSE WHOSE POPULATION IS VERY LARGE, LARGE, OR MEDIUM (61)	0.74	24	TEND TO BE THOSE WHOSE POPULATION IS SMALL (54)	0.55	23 38 8 46 X SQ= 6.50 P = 0.006 RV YES (NO)
25	ALWAYS ARE THOSE WHOSE POPULATION DENSITY IS MEDIUM OR LOW (98)	1.00	25	TEND LESS TO BE THOSE WHOSE POPULATION DENSITY IS MEDIUM OR LOW (98)	0.80	0 17 31 67 X SQ= 5.84 P = 0.006 RV NO (NO)
28	LEAN TOWARD BEING THOSE WHOSE POPULATION GROWTH RATE IS LOW (48)	0.60	28	LEAN TOWARD BEING THOSE WHOSE POPULATION GROWTH RATE IS HIGH (62)	0.62	12 50 18 30 X SQ= 3.62 P = 0.051 RV YES (NO)
30	TILT TOWARD BEING THOSE WHOSE AGRICULTURAL POPULATION IS MEDIUM, LOW, OR VERY LOW (57)	0.63	30	TILT TOWARD BEING THOSE WHOSE AGRICULTURAL POPULATION IS HIGH (56)	0.54	11 45 19 38 X SQ= 2.06 P = 0.136 RV YES (NO)
34	TEND TO BE THOSE WHOSE GROSS NATIONAL PRODUCT IS VERY HIGH, HIGH, MEDIUM, OR LOW (62)	0.77	34	TEND TO BE THOSE WHOSE GROSS NATIONAL PRODUCT IS VERY LOW (53)	0.55	24 38 7 46 X SQ= 8.19 P = 0.003 RV YES (NO)

```
36  LEAN TOWARD BEING THOSE                   0.52     36  LEAN TOWARD BEING THOSE                   0.69          16    26
    WHOSE PER CAPITA GROSS NATIONAL PRODUCT                WHOSE PER CAPITA GROSS NATIONAL PRODUCT                15    58
    IS VERY HIGH, HIGH, OR MEDIUM  (42)                    IS LOW OR VERY LOW  (73)                          X SQ=    3.33
                                                                                                             P =     0.051
                                                                                                             RV YES (NO )

40  TEND MORE TO BE THOSE                     0.83     40  TEND LESS TO BE THOSE                     0.58          24    47
    WHOSE INTERNATIONAL FINANCIAL STATUS                   WHOSE INTERNATIONAL FINANCIAL STATUS                    5    34
    IS VERY HIGH, HIGH,                                    IS VERY HIGH, HIGH,                               X SQ=    4.68
    MEDIUM, OR LOW  (71)                                   MEDIUM, OR LOW  (71)                              P =     0.023
                                                                                                             RV NO  (NO )

43  LEAN TOWARD BEING THOSE                   0.65     43  LEAN TOWARD BEING THOSE                   0.57          20    35
    WHOSE ECONOMIC DEVELOPMENTAL STATUS                    WHOSE ECONOMIC DEVELOPMENTAL STATUS                    11    46
    IS DEVELOPED, INTERMEDIATE, OR                         IS VERY UNDERDEVELOPED  (57)                      X SQ=    3.26
    UNDERDEVELOPED  (55)                                                                                     P =     0.057
                                                                                                             RV YES (NO )

45  TEND TO BE THOSE                          0.79     45  TEND TO BE THOSE                          0.59          22    33
    WHERE THE LITERACY RATE IS                             WHERE THE LITERACY RATE IS                              6    48
    FIFTY PERCENT OR ABOVE  (55)                           BELOW FIFTY PERCENT  (54)                         X SQ=   10.45
                                                                                                             P =     0.001
                                                                                                             RV YES (NO )

50  TEND TO BE THOSE                          0.75     50  TEND TO BE THOSE                          0.51           7    36
    WHERE FREEDOM OF THE PRESS IS                          WHERE FREEDOM OF THE PRESS IS                          21    35
    INTERMITTENT, INTERNALLY ABSENT, OR                    COMPLETE  (43)                                    X SQ=    4.40
    INTERNALLY AND EXTERNALLY ABSENT  (56)                                                                   P =     0.025
                                                                                                             RV YES (NO )

51  TEND TO BE THOSE                          0.64     51  TEND TO BE THOSE                          0.72          10    50
    WHERE FREEDOM OF THE PRESS IS                          WHERE FREEDOM OF THE PRESS IS                          18    19
    INTERNALLY ABSENT, OR                                  COMPLETE OR INTERMITTENT  (60)                    X SQ=    9.90
    INTERNALLY AND EXTERNALLY ABSENT  (37)                                                                   P =     0.001
                                                                                                             RV YES (NO )

53  TILT MORE TOWARD BEING THOSE              0.97     53  TILT LESS TOWARD BEING THOSE              0.85           1    13
    WHERE NEWSPAPER CIRCULATION IS                         WHERE NEWSPAPER CIRCULATION IS                         30    71
    LESS THAN THREE HUNDRED                                LESS THAN THREE HUNDRED                           X SQ=    2.14
    PER THOUSAND  (101)                                    PER THOUSAND  (101)                               P =     0.108
                                                                                                             RV NO  (NO )

55  LEAN MORE TOWARD BEING THOSE              0.83     55  LEAN LESS TOWARD BEING THOSE              0.64          24    54
    WHERE NEWSPAPER CIRCULATION IS                         WHERE NEWSPAPER CIRCULATION IS                          5    30
    TEN OR MORE                                            TEN OR MORE                                       X SQ=    2.63
    PER THOUSAND  (78)                                     PER THOUSAND  (78)                                P =     0.068
                                                                                                             RV NO  (NO )

56  TEND MORE TO BE THOSE                     0.89     56  TEND LESS TO BE THOSE                     0.66          24    55
    WHERE THE RELIGION IS                                  WHERE THE RELIGION IS                                   3    28
    PREDOMINANTLY LITERATE  (79)                           PREDOMINANTLY LITERATE  (79)                      X SQ=    4.09
                                                                                                             P =     0.027
                                                                                                             RV NO  (NO )
```

58	TILT MORE TOWARD BEING THOSE WHERE THE RELIGION IS OTHER THAN MUSLIM (97)	0.94	58	TILT LESS TOWARD BEING THOSE WHERE THE RELIGION IS OTHER THAN MUSLIM (97)	0.81	2 16 29 68 X SQ= 1.85 P = 0.148 RV NO (NO)
64	LEAN MORE TOWARD BEING THOSE WHERE THE RELIGION IS CHRISTIAN, RATHER THAN MUSLIM (46)	0.89	64	LEAN LESS TOWARD BEING THOSE WHERE THE RELIGION IS CHRISTIAN, RATHER THAN MUSLIM (46)	0.65	16 30 2 16 X SQ= 2.51 P = 0.070 RV NO (NO)
65	LEAN TOWARD BEING THOSE WHERE THE RELIGION IS CATHOLIC, RATHER THAN MUSLIM (25)	0.82	65	LEAN TOWARD BEING THOSE WHERE THE RELIGION IS MUSLIM, RATHER THAN CATHOLIC (18)	0.50	9 16 2 16 X SQ= 2.22 P = 0.086 RV YES (NO)
66	TILT TOWARD BEING THOSE THAT ARE RELIGIOUSLY HOMOGENEOUS (57)	0.68	66	TILT TOWARD BEING THOSE THAT ARE RELIGIOUSLY HETEROGENEOUS (49)	0.51	17 40 8 41 X SQ= 1.97 P = 0.115 RV YES (NO)
67	TEND MORE TO BE THOSE THAT ARE RACIALLY HOMOGENEOUS (82)	0.90	67	TEND LESS TO BE THOSE THAT ARE RACIALLY HOMOGENEOUS (82)	0.67	27 53 3 26 X SQ= 4.73 P = 0.016 RV NO (NO)
78	TEND TO BE THOSE THAT HAVE BEEN SIGNIFICANTLY WESTERNIZED WITHOUT A COLONIAL RELATIONSHIP, RATHER THAN THROUGH SUCH A RELATIONSHIP (7)	0.50	78	TEND TO BE THOSE THAT HAVE BEEN SIGNIFICANTLY WESTERNIZED THROUGH A COLONIAL RELATIONSHIP, RATHER THAN WITHOUT SUCH A RELATIONSHIP (28)	0.89	4 24 4 3 X SQ= 3.66 P = 0.033 RV YES (YES)
80	TILT TOWARD BEING THOSE WHOSE DATE OF INDEPENDENCE IS AFTER 1914, AND THAT WERE FORMERLY DEPENDENCIES OF BRITAIN, RATHER THAN FRANCE (19)	0.75	80	TILT TOWARD BEING THOSE WHOSE DATE OF INDEPENDENCE IS AFTER 1914, AND THAT WERE FORMERLY DEPENDENCIES OF FRANCE, RATHER THAN BRITAIN (24)	0.63	6 13 2 22 X SQ= 2.40 P = 0.111 RV YES (NO)
81	TILT MORE TOWARD BEING THOSE WHERE THE TYPE OF POLITICAL MODERNIZATION IS LATER EUROPEAN OR LATER EUROPEAN DERIVED, RATHER THAN EARLY EUROPEAN OR EARLY EUROPEAN DERIVED (40)	0.94	81	TILT LESS TOWARD BEING THOSE WHERE THE TYPE OF POLITICAL MODERNIZATION IS LATER EUROPEAN OR LATER EUROPEAN DERIVED, RATHER THAN EARLY EUROPEAN OR EARLY EUROPEAN DERIVED (40)	0.71	1 10 15 25 X SQ= 2.05 P = 0.140 RV NO (NO)
85	LEAN TOWARD BEING THOSE WHERE THE STAGE OF POLITICAL MODERNIZATION IS ADVANCED, RATHER THAN MID- OR EARLY TRANSITIONAL (60)	0.68	85	LEAN TOWARD BEING THOSE WHERE THE STAGE OF POLITICAL MODERNIZATION IS MID- OR EARLY TRANSITIONAL, RATHER THAN ADVANCED (54)	0.53	21 39 10 44 X SQ= 3.11 P = 0.059 RV YES (NO)

92	TEND LESS TO BE THOSE WHERE THE SYSTEM STYLE IS LIMITED MOBILIZATIONAL, OR NON-MOBILIZATIONAL (93)	0.60	92	TEND MORE TO BE THOSE WHERE THE SYSTEM STYLE IS LIMITED MOBILIZATIONAL, OR NON-MOBILIZATIONAL (93)	0.90	12 8 18 75 X SQ= 11.94 P = 0.001 RV NO (YES)
94	TEND TO BE THOSE WHERE THE STATUS OF THE REGIME IS AUTHORITARIAN OR TOTALITARIAN (41)	0.62	94	TEND TO BE THOSE WHERE THE STATUS OF THE REGIME IS CONSTITUTIONAL (51)	0.63	11 40 18 23 X SQ= 4.27 P = 0.026 RV YES (NO)
97	TEND TO BE THOSE WHERE THE STATUS OF THE REGIME IS TOTALITARIAN, RATHER THAN AUTHORITARIAN (16)	0.76	97	TEND TO BE THOSE WHERE THE STATUS OF THE REGIME IS AUTHORITARIAN, RATHER THAN TOTALITARIAN (23)	0.86	4 19 13 3 X SQ= 13.16 P = 0. RV YES (YES)
99	LEAN MORE TOWARD BEING THOSE WHERE GOVERNMENTAL STABILITY IS GENERALLY PRESENT AND DATES FROM AT LEAST THE INTER-WAR PERIOD, OR FROM THE POST-WAR PERIOD (50)	0.75	99	LEAN LESS TOWARD BEING THOSE WHERE GOVERNMENTAL STABILITY IS GENERALLY PRESENT AND DATES FROM AT LEAST THE INTER-WAR PERIOD, OR FROM THE POST-WAR PERIOD (50)	0.52	18 32 6 30 X SQ= 2.99 P = 0.055 RV NO (NO)
100	TEND MORE TO BE THOSE WHERE GOVERNMENTAL STABILITY IS GENERALLY PRESENT AND DATES FROM AT LEAST THE INTER-WAR PERIOD, OR IS GENERALLY OR MODERATELY PRESENT, AND DATES FROM THE POST-WAR PERIOD (64)	0.92	100	TEND LESS TO BE THOSE WHERE GOVERNMENTAL STABILITY IS GENERALLY PRESENT AND DATES FROM AT LEAST THE INTER-WAR PERIOD, OR IS GENERALLY OR MODERATELY PRESENT, AND DATES FROM THE POST-WAR PERIOD (64)	0.67	23 41 2 20 X SQ= 4.49 P = 0.016 RV NO (NO)
102	TILT TOWARD BEING THOSE WHERE THE REPRESENTATIVE CHARACTER OF THE REGIME IS PSEUDO-POLYARCHIC OR NON-POLYARCHIC (49)	0.58	102	TILT TOWARD BEING THOSE WHERE THE REPRESENTATIVE CHARACTER OF THE REGIME IS POLYARCHIC OR LIMITED POLYARCHIC (59)	0.60	13 46 18 31 X SQ= 2.15 P = 0.134 RV YES (NO)
106	TEND TO BE THOSE WHERE THE ELECTORAL SYSTEM IS NON-COMPETITIVE (30)	0.54	106	TEND TO BE THOSE WHERE THE ELECTORAL SYSTEM IS COMPETITIVE OR PARTIALLY COMPETITIVE (52)	0.71	11 41 13 17 X SQ= 3.51 P = 0.045 RV YES (NO)
108	TEND TO BE THOSE WHERE AUTONOMOUS GROUPS ARE TOLERATED ONLY OUTSIDE POLITICS OR ARE NOT TOLERATED AT ALL (35)	0.52	108	TEND TO BE THOSE WHERE AUTONOMOUS GROUPS ARE FULLY OR PARTIALLY TOLERATED IN POLITICS (65)	0.72	14 51 15 20 X SQ= 4.04 P = 0.037 RV YES (NO)
114	TILT TOWARD BEING THOSE WHERE SECTIONALISM IS NEGLIGIBLE (47)	0.57	114	TILT TOWARD BEING THOSE WHERE SECTIONALISM IS EXTREME OR MODERATE (61)	0.62	13 48 17 30 X SQ= 2.23 P = 0.129 RV YES (NO)

118	TEND TO BE THOSE WHERE INTEREST ARTICULATION BY INSTITUTIONAL GROUPS IS VERY SIGNIFICANT (40)	0.59	118	TEND TO BE THOSE WHERE INTEREST ARTICULATION BY INSTITUTIONAL GROUPS IS SIGNIFICANT, MODERATE, OR LIMITED (60)	0.68	17 23 12 48 X SQ= 4.86 P = 0.024 RV YES (NO)
128	TEND TO BE THOSE WHERE INTEREST ARTICULATION BY POLITICAL PARTIES IS LIMITED OR NEGLIGIBLE (45)	0.69	128	TEND TO BE THOSE WHERE INTEREST ARTICULATION BY POLITICAL PARTIES IS SIGNIFICANT OR MODERATE (48)	0.60	8 40 18 27 X SQ= 5.17 P = 0.020 RV YES (NO)
129	TEND TO BE THOSE WHERE INTEREST ARTICULATION BY POLITICAL PARTIES IS NEGLIGIBLE (37)	0.58	129	TEND TO BE THOSE WHERE INTEREST ARTICULATION BY POLITICAL PARTIES IS SIGNIFICANT, MODERATE, OR LIMITED (56)	0.67	11 45 15 22 X SQ= 3.85 P = 0.035 RV YES (NO)
133	TILT MORE TOWARD BEING THOSE WHERE INTEREST AGGREGATION BY THE EXECUTIVE IS MODERATE, LIMITED, OR NEGLIGIBLE (76)	0.83	133	TILT LESS TOWARD BEING THOSE WHERE INTEREST AGGREGATION BY THE EXECUTIVE IS MODERATE, LIMITED, OR NEGLIGIBLE (76)	0.68	5 24 25 51 X SQ= 1.81 P = 0.149 RV NO (NO)
146	ALWAYS ARE THOSE WHERE THE PARTY SYSTEM IS QUALITATIVELY OTHER THAN MASS-BASED TERRITORIAL (92)	1.00	146	TILT LESS TOWARD BEING THOSE WHERE THE PARTY SYSTEM IS QUALITATIVELY OTHER THAN MASS-BASED TERRITORIAL (92)	0.89	0 8 27 65 X SQ= 1.90 P = 0.104 RV NO (NO)
147	LEAN MORE TOWARD BEING THOSE WHERE THE PARTY SYSTEM IS QUALITATIVELY OTHER THAN CLASS-ORIENTED OR MULTI-IDEOLOGICAL (67)	0.88	147	LEAN LESS TOWARD BEING THOSE WHERE THE PARTY SYSTEM IS QUALITATIVELY OTHER THAN CLASS-ORIENTED OR MULTI-IDEOLOGICAL (67)	0.69	3 20 23 44 X SQ= 2.81 P = 0.064 RV NO (NO)
153	TEND TO BE THOSE WHERE THE PARTY SYSTEM IS STABLE (42)	0.71	153	TEND TO BE THOSE WHERE THE PARTY SYSTEM IS MODERATELY STABLE OR UNSTABLE (38)	0.55	17 25 7 31 X SQ= 3.63 P = 0.050 RV YES (NO)
158	TEND MORE TO BE THOSE WHERE PERSONALISMO IS NEGLIGIBLE (56)	0.80	158	TEND LESS TO BE THOSE WHERE PERSONALISMO IS NEGLIGIBLE (56)	0.51	5 35 20 36 X SQ= 5.38 P = 0.017 RV NO (NO)
160	TEND TO BE THOSE WHERE THE POLITICAL LEADERSHIP IS ELITIST (30)	0.54	160	TEND TO BE THOSE WHERE THE POLITICAL LEADERSHIP IS MODERATELY ELITIST OR NON-ELITIST (67)	0.77	14 16 12 55 X SQ= 7.33 P = 0.006 RV YES (NO)

169	TILT TOWARD BEING THOSE WHERE THE HORIZONTAL POWER DISTRIBUTION IS NEGLIGIBLE (48) 0.58	169	TILT TOWARD BEING THOSE WHERE THE HORIZONTAL POWER DISTRIBUTION IS SIGNIFICANT OR LIMITED (58) 0.60	13 45 18 30 X SQ= 2.21 P = 0.133 RV YES (NO)
170	TEND MORE TO BE THOSE WHERE THE LEGISLATIVE-EXECUTIVE STRUCTURE IS OTHER THAN PRESIDENTIAL (63) 0.77	170	TEND LESS TO BE THOSE WHERE THE LEGISLATIVE-EXECUTIVE STRUCTURE IS OTHER THAN PRESIDENTIAL (63) 0.56	7 32 23 40 X SQ= 3.15 P = 0.049 RV NO (NO)
175	LEAN TOWARD BEING THOSE WHERE THE LEGISLATURE IS LARGELY INEFFECTIVE OR WHOLLY INEFFECTIVE (49) 0.64	175	LEAN TOWARD BEING THOSE WHERE THE LEGISLATURE IS FULLY EFFECTIVE OR PARTIALLY EFFECTIVE (51) 0.57	10 41 18 31 X SQ= 2.84 P = 0.075 RV YES (NO)
176	TEND TO BE THOSE WHERE THE LEGISLATURE IS WHOLLY INEFFECTIVE (28) 0.54	176	TEND TO BE THOSE WHERE THE LEGISLATURE IS FULLY EFFECTIVE, PARTIALLY EFFECTIVE, OR LARGELY INEFFECTIVE (72) 0.82	13 59 15 13 X SQ= 10.91 P = 0.001 RV YES (YES)
178	TEND TO BE THOSE WHERE THE LEGISLATURE IS UNICAMERAL (53) 0.68	178	TEND TO BE THOSE WHERE THE LEGISLATURE IS BICAMERAL (51) 0.55	19 34 9 42 X SQ= 3.50 P = 0.047 RV YES (NO)
182	LEAN MORE TOWARD BEING THOSE WHERE THE BUREAUCRACY IS SEMI-MODERN, RATHER THAN POST-COLONIAL TRANSITIONAL (55) 0.86	182	LEAN LESS TOWARD BEING THOSE WHERE THE BUREAUCRACY IS SEMI-MODERN, RATHER THAN POST-COLONIAL TRANSITIONAL (55) 0.63	18 37 3 22 X SQ= 2.82 P = 0.059 RV NO (NO)
185	TEND TO BE THOSE WHERE PARTICIPATION BY THE MILITARY IN POLITICS IS SUPPORTIVE, RATHER THAN INTERVENTIVE (31) 0.87	185	TEND TO BE THOSE WHERE PARTICIPATION BY THE MILITARY IN POLITICS IS INTERVENTIVE, RATHER THAN SUPPORTIVE (21) 0.53	2 19 14 17 X SQ= 5.88 P = 0.007 RV YES (NO)
186	TEND TO BE THOSE WHERE PARTICIPATION BY THE MILITARY IN POLITICS IS SUPPORTIVE, RATHER THAN NEUTRAL (31) 0.54	186	TEND TO BE THOSE WHERE PARTICIPATION BY THE MILITARY IN POLITICS IS NEUTRAL, RATHER THAN SUPPORTIVE (56) 0.72	14 17 12 44 X SQ= 4.29 P = 0.028 RV YES (NO)
192	TILT MORE TOWARD BEING THOSE WHERE THE CHARACTER OF THE LEGAL SYSTEM IS OTHER THAN MUSLIM OR PARTLY MUSLIM (86) 0.87	192	TILT LESS TOWARD BEING THOSE WHERE THE CHARACTER OF THE LEGAL SYSTEM IS OTHER THAN MUSLIM OR PARTLY MUSLIM (86) 0.72	4 23 27 59 X SQ= 2.07 P = 0.137 RV NO (NO)

```
194  TEND LESS TO BE THOSE                 0.67     194  TEND MORE TO BE THOSE                 0.96       10    3
     THAT ARE NON-COMMUNIST  (101)                       THAT ARE NON-COMMUNIST  (101)                    20   81
                                                                                                      X SQ=  16.55
                                                                                                      P =     0.
                                                                                                      RV NO  (YES)
```

```
*******************************************************************************************************

 28  POLITIES                                             28  POLITIES
     WHOSE POPULATION GROWTH RATE                             WHOSE POPULATION GROWTH RATE
     IS HIGH  (62)                                            IS LOW  (48)


                                                                                   BOTH SUBJECT AND PREDICATE
.......................................................................................................

     62 IN LEFT COLUMN

ALBANIA     ALGERIA     AUSTRALIA   BOLIVIA     BRAZIL      CANADA      CEYLON      CHAD        CHILE       CHINA, PR
COLOMBIA    CONGO(BRA)  COSTA RICA  CUBA        DOMIN REP   ECUADOR     EL SALVADOR GUATEMALA   GUINEA      HONDURAS
ICELAND     INDIA       INDONESIA   IRAN        IRAQ        ISRAEL      IVORY COAST JAMAICA     JORDAN      KOREA REP
LAOS        LEBANON     LIBYA       MALAGASY R  MALAYA      MALI        MAURITANIA  MEXICO      MONGOLIA    MOROCCO
NEPAL       NEW ZEALAND NICARAGUA   NIGER       PAKISTAN    PANAMA      PARAGUAY    PERU        PHILIPPINES SA'U ARABIA
SENEGAL     SIERRE LEO  SO AFRICA   SYRIA       THAILAND    TRINIDAD    TUNISIA     TURKEY      UAR         UPPER VOLTA
VENEZUELA   VIETNAM REP

     48 IN RIGHT COLUMN

AFGHANISTAN ARGENTINA   AUSTRIA     BELGIUM     BULGARIA    BURMA       BURUNDI     CAMBODIA    CAMEROUN    CEN AFR REP
CONGO(LEO)  CYPRUS      CZECHOS'KIA DAHOMEY     DENMARK     FINLAND     FRANCE      GABON       GERMANY, E  GERMAN FR
GREECE      HAITI       HUNGARY     IRELAND     ITALY       JAPAN       LIBERIA     LUXEMBOURG  NETHERLANDS NIGERIA
NORWAY      POLAND      PORTUGAL    RUMANIA     RWANDA      SOMALIA     SPAIN       SUDAN       SWEDEN      SWITZERLAND
TANGANYIKA  UGANDA      UK          US          URUGUAY     VIETNAM, N  YEMEN       YUGOSLAVIA

     5 EXCLUDED BECAUSE UNASCERTAINED

ETHIOPIA    GHANA       KOREA, N    TOGO        USSR

-------------------------------------------------------------------------------------------------------

  2  TEND MORE TO BE THOSE                   0.94         2  TEND LESS TO BE THOSE                  0.62         4   18
     LOCATED ELSEWHERE THAN IN                               LOCATED ELSEWHERE THAN IN                          58   30
     WEST EUROPE, SCANDINAVIA,                               WEST EUROPE, SCANDINAVIA,                    X SQ=  14.42
     NORTH AMERICA, OR AUSTRALASIA  (93)                     NORTH AMERICA, OR AUSTRALASIA  (93)          P =    0.
                                                                                                          RV NO  (YES)

  3  ALWAYS ARE THOSE                        1.00         3  TEND LESS TO BE THOSE                  0.73         0   13
     LOCATED ELSEWHERE THAN IN                               LOCATED ELSEWHERE THAN IN                          62   35
     WEST EUROPE  (102)                                      WEST EUROPE  (102)                           X SQ=  16.53
                                                                                                          P =    0.
                                                                                                          RV NO  (YES)

  6  TEND LESS TO BE THOSE                   0.69         6  TEND MORE TO BE THOSE                  0.94        19    3
     LOCATED ELSEWHERE THAN IN THE                           LOCATED ELSEWHERE THAN IN THE                      43   45
     CARIBBEAN, CENTRAL AMERICA,                             CARIBBEAN, CENTRAL AMERICA,                  X SQ=   8.60
     OR SOUTH AMERICA  (93)                                  OR SOUTH AMERICA  (93)                       P =    0.002
                                                                                                          RV NO  (YES)
```

8	TILT LESS TOWARD BEING THOSE LOCATED ELSEWHERE THAN IN EAST ASIA SOUTH ASIA, OR SOUTHEAST ASIA (97)	0.79	8	TILT MORE TOWARD BEING THOSE LOCATED ELSEWHERE THAN IN EAST ASIA SOUTH ASIA, OR SOUTHEAST ASIA (97)	0.92	13 4 49 44 X SQ= 2.41 P = 0.109 RV NO (NO)
13	TEND LESS TO BE THOSE LOCATED ELSEWHERE THAN IN NORTH AFRICA OR THE MIDDLE EAST (99)	0.79	13	TEND MORE TO BE THOSE LOCATED ELSEWHERE THAN IN NORTH AFRICA OR THE MIDDLE EAST (99)	0.94	13 3 49 45 X SQ= 3.61 P = 0.033 RV NO (NO)
19	TEND TO BE THOSE LOCATED IN THE CARIBBEAN, CENTRAL AMERICA, OR SOUTH AMERICA, RATHER THAN IN NORTH AFRICA OR CENTRAL AND SOUTH AFRICA (22)	0.53	19	TEND TO BE THOSE LOCATED IN NORTH AFRICA OR CENTRAL AND SOUTH AFRICA, RATHER THAN IN THE CARIBBEAN, CENTRAL AMERICA, OR SOUTH AMERICA (33)	0.81	17 13 19 3 X SQ= 3.95 P = 0.033 RV YES (NO)
21	TEND LESS TO BE THOSE WHOSE TERRITORIAL SIZE IS MEDIUM OR SMALL (83)	0.61	21	TEND MORE TO BE THOSE WHOSE TERRITORIAL SIZE IS MEDIUM OR SMALL (83)	0.87	24 6 38 42 X SQ= 8.10 P = 0.002 RV NO (YES)
24	TILT TOWARD BEING THOSE WHOSE POPULATION IS SMALL (54)	0.55	24	TILT TOWARD BEING THOSE WHOSE POPULATION IS VERY LARGE, LARGE, OR MEDIUM (61)	0.60	28 29 34 19 X SQ= 1.95 P = 0.127 RV YES (YES)
26	TEND TO BE THOSE WHOSE POPULATION DENSITY IS LOW (67)	0.69	26	TEND TO BE THOSE WHOSE POPULATION DENSITY IS VERY HIGH, HIGH, OR MEDIUM (48)	0.58	19 28 43 20 X SQ= 7.38 P = 0.006 RV YES (YES)
31	TEND MORE TO BE THOSE WHOSE AGRICULTURAL POPULATION IS HIGH OR MEDIUM (90)	0.87	31	TEND LESS TO BE THOSE WHOSE AGRICULTURAL POPULATION IS HIGH OR MEDIUM (90)	0.67	53 32 8 16 X SQ= 5.27 P = 0.019 RV NO (YES)
34	LEAN TOWARD BEING THOSE WHOSE GROSS NATIONAL PRODUCT IS VERY LOW (53)	0.53	34	LEAN TOWARD BEING THOSE WHOSE GROSS NATIONAL PRODUCT IS VERY HIGH, HIGH, MEDIUM, OR LOW (62)	0.65	29 31 33 17 X SQ= 2.78 P = 0.083 RV YES (YES)
36	TEND TO BE THOSE WHOSE PER CAPITA GROSS NATIONAL PRODUCT IS LOW OR VERY LOW (73)	0.79	36	TEND TO BE THOSE WHOSE PER CAPITA GROSS NATIONAL PRODUCT IS VERY HIGH, HIGH, OR MEDIUM (42)	0.58	M 13 28 49 20 X SQ= 14.60 P = 0. RV YES (YES)

39	TEND MORE TO BE THOSE WHOSE INTERNATIONAL FINANCIAL STATUS IS LOW OR VERY LOW (76)	0.74	39	TEND LESS TO BE THOSE WHOSE INTERNATIONAL FINANCIAL STATUS IS LOW OR VERY LOW (76)	0.55	16 21 46 26 X SQ= 3.45 P = 0.044 RV NO (YES)
42	TEND TO BE THOSE WHOSE ECONOMIC DEVELOPMENTAL STATUS IS UNDERDEVELOPED OR VERY UNDERDEVELOPED (76)	0.86	42	TEND TO BE THOSE WHOSE ECONOMIC DEVELOPMENTAL STATUS IS DEVELOPED OR INTERMEDIATE (36)	0.56	8 27 51 21 X SQ= 20.02 P = 0. RV YES (YES)
43	TEND TO BE THOSE WHOSE ECONOMIC DEVELOPMENTAL STATUS IS VERY UNDERDEVELOPED (57)	0.59	43	TEND TO BE THOSE WHOSE ECONOMIC DEVELOPMENTAL STATUS IS DEVELOPED, INTERMEDIATE, OR UNDERDEVELOPED (55)	0.62	M 24 30 35 18 X SQ= 4.21 P = 0.033 RV YES (YES)
45	TEND TO BE THOSE WHERE THE LITERACY RATE IS BELOW FIFTY PERCENT (54)	0.59	45	TEND TO BE THOSE WHERE THE LITERACY RATE IS FIFTY PERCENT OR ABOVE (55)	0.65	24 30 35 16 X SQ= 5.29 P = 0.018 RV YES (YES)
48	TILT TOWARD BEING THOSE WHERE THE LITERACY RATE IS BETWEEN TEN AND FIFTY PERCENT, RATHER THAN BETWEEN FIFTY AND NINETY PERCENT (24)	0.50	48	TILT TOWARD BEING THOSE WHERE THE LITERACY RATE IS BETWEEN FIFTY AND NINETY PERCENT, RATHER THAN BETWEEN TEN AND FIFTY PERCENT (30)	0.77	20 10 20 3 X SQ= 1.90 P = 0.115 RV YES (NO)
49	TEND TO BE THOSE WHERE THE LITERACY RATE IS BETWEEN TEN AND FIFTY PERCENT, RATHER THAN BELOW TEN PERCENT (24)	0.65	49	TEND TO BE THOSE WHERE THE LITERACY RATE IS BELOW TEN PERCENT, RATHER THAN BETWEEN TEN AND FIFTY PERCENT (26)	0.81	20 3 11 13 X SQ= 7.11 P = 0.005 RV YES (YES)
54	TEND TO BE THOSE WHERE NEWSPAPER CIRCULATION IS LESS THAN ONE HUNDRED PER THOUSAND (76)	0.84	54	TEND TO BE THOSE WHERE NEWSPAPER CIRCULATION IS ONE HUNDRED OR MORE PER THOUSAND (37)	0.55	10 26 52 21 X SQ= 16.83 P = 0. RV YES (YES)
58	TEND LESS TO BE THOSE WHERE THE RELIGION IS OTHER THAN MUSLIM (97)	0.76	58	TEND MORE TO BE THOSE WHERE THE RELIGION IS OTHER THAN MUSLIM (97)	0.94	15 3 47 45 X SQ= 5.12 P = 0.018 RV NO (NO)
63	TEND TO BE THOSE WHERE THE RELIGION IS PREDOMINANTLY OR PARTLY OTHER THAN CHRISTIAN (68)	0.69	63	TEND TO BE THOSE WHERE THE RELIGION IS PREDOMINANTLY SOME KIND OF CHRISTIAN (46)	0.56	19 27 43 21 X SQ= 6.28 P = 0.011 RV YES (YES)

65	LEAN TOWARD BEING THOSE WHERE THE RELIGION IS MUSLIM, RATHER THAN CATHOLIC (18)	0.52	65	LEAN TOWARD BEING THOSE WHERE THE RELIGION IS CATHOLIC, RATHER THAN MUSLIM (25)	0.79	14 11 15 3 X SQ= 2.42 P = 0.099 RV YES (NO)
67	TEND LESS TO BE THOSE THAT ARE RACIALLY HOMOGENEOUS (82)	0.61	67	TEND MORE TO BE THOSE THAT ARE RACIALLY HOMOGENEOUS (82)	0.87	35 41 22 6 X SQ= 7.47 P = 0.004 RV NO (YES)
70	TEND MORE TO BE THOSE THAT ARE RELIGIOUSLY, RACIALLY, OR LINGUISTICALLY HETEROGENEOUS (85)	0.87	70	TEND LESS TO BE THOSE THAT ARE RELIGIOUSLY, RACIALLY, OR LINGUISTICALLY HETEROGENEOUS (85)	0.69	7 14 49 31 X SQ= 4.18 P = 0.028 RV NO (YES)
71	TEND MORE TO BE THOSE WHOSE DATE OF INDEPENDENCE IS AFTER 1800 (90)	0.92	71	TEND LESS TO BE THOSE WHOSE DATE OF INDEPENDENCE IS AFTER 1800 (90)	0.70	5 14 56 32 X SQ= 7.42 P = 0.004 RV NO (YES)
74	TEND MORE TO BE THOSE THAT ARE NOT HISTORICALLY WESTERN (87)	0.93	74	TEND LESS TO BE THOSE THAT ARE NOT HISTORICALLY WESTERN (87)	0.54	4 22 56 26 X SQ= 20.29 P = 0. RV NO (YES)
76	TEND TO BE THOSE THAT HAVE BEEN SIGNIFICANTLY OR PARTIALLY WESTERNIZED THROUGH A COLONIAL RELATIONSHIP, RATHER THAN BEING HISTORICALLY WESTERN (70)	0.92	76	TEND TO BE THOSE THAT ARE HISTORICALLY WESTERN, RATHER THAN HAVING BEEN SIGNIFICANTLY OR PARTIALLY WESTERNIZED THROUGH A COLONIAL RELATIONSHIP (26)	0.54	4 22 49 19 X SQ= 22.31 P = 0. RV YES (YES)
77	TEND TO BE THOSE THAT HAVE BEEN SIGNIFICANTLY WESTERNIZED, RATHER THAN PARTIALLY WESTERNIZED, THROUGH A COLONIAL RELATIONSHIP (28)	0.52	77	TEND TO BE THOSE THAT HAVE BEEN PARTIALLY WESTERNIZED, RATHER THAN SIGNIFICANTLY WESTERNIZED, THROUGH A COLONIAL RELATIONSHIP (41)	0.84	25 3 23 16 X SQ= 5.95 P = 0.007 RV YES (NO)
78	TEND TO BE THOSE THAT HAVE BEEN SIGNIFICANTLY WESTERNIZED THROUGH A COLONIAL RELATIONSHIP, RATHER THAN WITHOUT SUCH A RELATIONSHIP (28)	0.93	78	TEND TO BE THOSE THAT HAVE BEEN SIGNIFICANTLY WESTERNIZED WITHOUT A COLONIAL RELATIONSHIP, RATHER THAN THROUGH SUCH A RELATIONSHIP (7)	0.57	25 3 2 4 X SQ= 6.35 P = 0.010 RV YES (YES)
79	TILT LESS TOWARD BEING THOSE THAT WERE FORMERLY DEPENDENCIES OF BRITAIN OR FRANCE, RATHER THAN SPAIN (49)	0.67	79	TILT MORE TOWARD BEING THOSE THAT WERE FORMERLY DEPENDENCIES OF BRITAIN OR FRANCE, RATHER THAN SPAIN (49)	0.88	32 15 16 2 X SQ= 1.94 P = 0.119 RV NO (NO)

82	TEND TO BE THOSE WHERE THE TYPE OF POLITICAL MODERNIZATION IS DEVELOPED TUTELARY OR UNDEVELOPED TUTELARY, RATHER THAN EARLY OR LATER EUROPEAN OR EUROPEAN DERIVED (55)	0.60	82	TEND TO BE THOSE WHERE THE TYPE OF POLITICAL MODERNIZATION IS EARLY OR LATER EUROPEAN OR EUROPEAN DERIVED, RATHER THAN DEVELOPED TUTELARY OR UNDEVELOPED TUTELARY (51)	0.61	23 28 34 18 X SQ= 3.51 P = 0.048 RV YES (YES)
84	LEAN TOWARD BEING THOSE WHERE THE TYPE OF POLITICAL MODERNIZATION IS DEVELOPED TUTELARY, RATHER THAN UNDEVELOPED TUTELARY (31)	0.68	84	LEAN TOWARD BEING THOSE WHERE THE TYPE OF POLITICAL MODERNIZATION IS UNDEVELOPED TUTELARY, RATHER THAN DEVELOPED TUTELARY (24)	0.61	23 7 11 11 X SQ= 2.90 P = 0.076 RV YES (YES)
85	TILT TOWARD BEING THOSE WHERE THE STAGE OF POLITICAL MODERNIZATION IS MID- OR EARLY TRANSITIONAL, RATHER THAN ADVANCED (54)	0.54	85	TILT TOWARD BEING THOSE WHERE THE STAGE OF POLITICAL MODERNIZATION IS ADVANCED, RATHER THAN MID- OR EARLY TRANSITIONAL (60)	0.62	28 30 33 18 X SQ= 2.34 P = 0.122 RV YES (YES)
87	LEAN LESS TOWARD BEING THOSE WHOSE IDEOLOGICAL ORIENTATION IS OTHER THAN DEVELOPMENTAL (58)	0.56	87	LEAN MORE TOWARD BEING THOSE WHOSE IDEOLOGICAL ORIENTATION IS OTHER THAN DEVELOPMENTAL (58)	0.76	19 10 24 31 X SQ= 2.82 P = 0.069 RV NO (YES)
89	TILT MORE TOWARD BEING THOSE WHOSE IDEOLOGICAL ORIENTATION IS OTHER THAN CONVENTIONAL (62)	0.71	89	TILT LESS TOWARD BEING THOSE WHOSE IDEOLOGICAL ORIENTATION IS OTHER THAN CONVENTIONAL (62)	0.55	14 19 34 23 X SQ= 1.85 P = 0.130 RV NO (YES)
95	TEND MORE TO BE THOSE WHERE THE STATUS OF THE REGIME IS CONSTITUTIONAL OR AUTHORITARIAN (95)	0.93	95	TEND LESS TO BE THOSE WHERE THE STATUS OF THE REGIME IS CONSTITUTIONAL OR AUTHORITARIAN (95)	0.78	56 36 4 10 X SQ= 3.93 P = 0.040 RV NO (YES)
96	TILT LESS TOWARD BEING THOSE WHERE THE STATUS OF THE REGIME IS CONSTITUTIONAL OR TOTALITARIAN (67)	0.69	96	TILT MORE TOWARD BEING THOSE WHERE THE STATUS OF THE REGIME IS CONSTITUTIONAL OR TOTALITARIAN (67)	0.84	34 31 15 6 X SQ= 1.65 P = 0.138 RV NO (NO)
97	TEND TO BE THOSE WHERE THE STATUS OF THE REGIME IS AUTHORITARIAN, RATHER THAN TOTALITARIAN (23)	0.79	97	TEND TO BE THOSE WHERE THE STATUS OF THE REGIME IS TOTALITARIAN, RATHER THAN AUTHORITARIAN (16)	0.62	15 6 4 10 X SQ= 4.61 P = 0.018 RV YES (YES)
99	TEND TO BE THOSE WHERE GOVERNMENTAL STABILITY IS MODERATELY PRESENT AND DATES FROM THE POST-WAR PERIOD, OR IS ABSENT (36)	0.60	99	TEND TO BE THOSE WHERE GOVERNMENTAL STABILITY IS GENERALLY PRESENT AND DATES FROM AT LEAST THE INTER-WAR PERIOD, OR FROM THE POST-WAR PERIOD (50)	0.82	19 28 28 6 X SQ= 12.57 P = 0. RV YES (YES)

110	TILT MORE TOWARD BEING THOSE WHERE POLITICAL ENCULTURATION IS MEDIUM OR LOW (80)	0.89	110	TILT LESS TOWARD BEING THOSE WHERE POLITICAL ENCULTURATION IS MEDIUM OR LOW (80)	0.76	6 9 49 28 X SQ= 2.02 P = 0.148 RV NO (YES)
115	TEND MORE TO BE THOSE WHERE INTEREST ARTICULATION BY ASSOCIATIONAL GROUPS IS MODERATE, LIMITED, OR NEGLIGIBLE (91)	0.92	115	TEND LESS TO BE THOSE WHERE INTEREST ARTICULATION BY ASSOCIATIONAL GROUPS IS MODERATE, LIMITED, OR NEGLIGIBLE (91)	0.67	5 15 56 30 X SQ= 9.11 P = 0.002 RV NO (YES)
116	TEND MORE TO BE THOSE WHERE INTEREST ARTICULATION BY ASSOCIATIONAL GROUPS IS LIMITED OR NEGLIGIBLE (79)	0.79	116	TEND LESS TO BE THOSE WHERE INTEREST ARTICULATION BY ASSOCIATIONAL GROUPS IS LIMITED OR NEGLIGIBLE (79)	0.58	13 19 48 26 X SQ= 4.43 P = 0.032 RV NO (YES)
119	TEND MORE TO BE THOSE WHERE INTEREST ARTICULATION BY INSTITUTIONAL GROUPS IS VERY SIGNIFICANT OR SIGNIFICANT (74)	0.82	119	TEND LESS TO BE THOSE WHERE INTEREST ARTICULATION BY INSTITUTIONAL GROUPS IS VERY SIGNIFICANT OR SIGNIFICANT (74)	0.60	46 24 10 16 X SQ= 4.73 P = 0.021 RV NO (YES)
120	LEAN MORE TOWARD BEING THOSE WHERE INTEREST ARTICULATION BY INSTITUTIONAL GROUPS IS VERY SIGNIFICANT, SIGNIFICANT, OR MODERATE (90)	0.95	120	LEAN LESS TOWARD BEING THOSE WHERE INTEREST ARTICULATION BY INSTITUTIONAL GROUPS IS VERY SIGNIFICANT, SIGNIFICANT, OR MODERATE (90)	0.82	53 33 3 7 X SQ= 2.50 P = 0.088 RV NO (YES)
122	TEND MORE TO BE THOSE WHERE INTEREST ARTICULATION BY NON-ASSOCIATIONAL GROUPS IS SIGNIFICANT OR MODERATE (83)	0.82	122	TEND LESS TO BE THOSE WHERE INTEREST ARTICULATION BY NON-ASSOCIATIONAL GROUPS IS SIGNIFICANT OR MODERATE (83)	0.56	51 27 11 21 X SQ= 7.66 P = 0.005 RV NO (YES)
125	TEND TO BE THOSE WHERE INTEREST ARTICULATION BY ANOMIC GROUPS IS FREQUENT OR OCCASIONAL (64)	0.73	125	TEND TO BE THOSE WHERE INTEREST ARTICULATION BY ANOMIC GROUPS IS INFREQUENT OR VERY INFREQUENT (35)	0.51	41 19 15 20 X SQ= 4.92 P = 0.018 RV YES (YES)
132	TEND LESS TO BE THOSE WHERE INTEREST AGGREGATION BY POLITICAL PARTIES IS SIGNIFICANT, MODERATE, OR LIMITED (67)	0.80	132	ALWAYS ARE THOSE WHERE INTEREST AGGREGATION BY POLITICAL PARTIES IS SIGNIFICANT, MODERATE, OR LIMITED (67)	1.00	37 29 9 0 X SQ= 4.73 P = 0.010 RV NO (NO)
133	TILT LESS TOWARD BEING THOSE WHERE INTEREST AGGREGATION BY THE EXECUTIVE IS MODERATE, LIMITED, OR NEGLIGIBLE (76)	0.65	133	TILT MORE TOWARD BEING THOSE WHERE INTEREST AGGREGATION BY THE EXECUTIVE IS MODERATE, LIMITED, OR NEGLIGIBLE (76)	0.80	19 9 36 36 X SQ= 1.93 P = 0.122 RV NO (YES)

136	TEND MORE TO BE THOSE WHERE INTEREST AGGREGATION BY THE LEGISLATURE IS MODERATE, LIMITED, OR NEGLIGIBLE (85)	0.98	136	TEND LESS TO BE THOSE WHERE INTEREST AGGREGATION BY THE LEGISLATURE IS MODERATE, LIMITED, OR NEGLIGIBLE (85)	0.75	1 11 47 33 X SQ= 8.71 P = 0.001 RV NO (YES)
146	TILT LESS TOWARD BEING THOSE WHERE THE PARTY SYSTEM IS QUALITATIVELY OTHER THAN MASS-BASED TERRITORIAL (92)	0.89	146	TILT MORE TOWARD BEING THOSE WHERE THE PARTY SYSTEM IS QUALITATIVELY OTHER THAN MASS-BASED TERRITORIAL (92)	0.98	6 1 48 41 X SQ= 1.53 P = 0.132 RV NO (NO)
147	LEAN MORE TOWARD BEING THOSE WHERE THE PARTY SYSTEM IS QUALITATIVELY OTHER THAN CLASS-ORIENTED OR MULTI-IDEOLOGICAL (67)	0.81	147	LEAN LESS TOWARD BEING THOSE WHERE THE PARTY SYSTEM IS QUALITATIVELY OTHER THAN CLASS-ORIENTED OR MULTI-IDEOLOGICAL (67)	0.64	9 14 38 25 X SQ= 2.26 P = 0.093 RV NO (YES)
153	TEND TO BE THOSE WHERE THE PARTY SYSTEM IS MODERATELY STABLE OR UNSTABLE (38)	0.68	153	TEND TO BE THOSE WHERE THE PARTY SYSTEM IS STABLE (42)	0.79	M 14 26 30 7 X SQ= 14.84 P = 0. RV YES (YES)
158	LEAN TOWARD BEING THOSE WHERE PERSONALISMO IS PRONOUNCED OR MODERATE (40)	0.51	158	LEAN TOWARD BEING THOSE WHERE PERSONALISMO IS NEGLIGIBLE (56)	0.70	27 12 26 28 X SQ= 3.29 P = 0.057 RV YES (YES)
164	TILT LESS TOWARD BEING THOSE WHERE THE REGIME'S LEADERSHIP CHARISMA IS NEGLIGIBLE (65)	0.61	164	TILT MORE TOWARD BEING THOSE WHERE THE REGIME'S LEADERSHIP CHARISMA IS NEGLIGIBLE (65)	0.77	22 9 35 30 X SQ= 1.89 P = 0.125 RV NO (NO)
168	TILT MORE TOWARD BEING THOSE WHERE THE HORIZONTAL POWER DISTRIBUTION IS LIMITED OR NEGLIGIBLE (72)	0.72	168	TILT LESS TOWARD BEING THOSE WHERE THE HORIZONTAL POWER DISTRIBUTION IS LIMITED OR NEGLIGIBLE (72)	0.58	16 18 42 25 X SQ= 1.66 P = 0.143 RV NO (YES)
170	TEND TO BE THOSE WHERE THE LEGISLATIVE-EXECUTIVE STRUCTURE IS PRESIDENTIAL (39)	0.52	170	TEND TO BE THOSE WHERE THE LEGISLATIVE-EXECUTIVE STRUCTURE IS OTHER THAN PRESIDENTIAL (63)	0.77	28 10 26 34 X SQ= 7.48 P = 0.004 RV YES (YES)
174	LEAN MORE TOWARD BEING THOSE WHERE THE LEGISLATURE IS PARTIALLY EFFECTIVE, LARGELY INEFFECTIVE, OR WHOLLY INEFFECTIVE (72)	0.79	174	LEAN LESS TOWARD BEING THOSE WHERE THE LEGISLATURE IS PARTIALLY EFFECTIVE, LARGELY INEFFECTIVE, OR WHOLLY INEFFECTIVE (72)	0.61	11 17 41 27 X SQ= 2.73 P = 0.074 RV NO (YES)

181	TEND MORE TO BE THOSE WHERE THE BUREAUCRACY IS SEMI-MODERN, RATHER THAN MODERN (55)	0.85	181	TEND LESS TO BE THOSE WHERE THE BUREAUCRACY IS SEMI-MODERN, RATHER THAN MODERN (55)	0.53	6 15 35 17 X SQ= 7.61 P = 0.004 RV NO (YES)
182	LEAN MORE TOWARD BEING THOSE WHERE THE BUREAUCRACY IS SEMI-MODERN, RATHER THAN POST-COLONIAL TRANSITIONAL (55)	0.76	182	LEAN LESS TOWARD BEING THOSE WHERE THE BUREAUCRACY IS SEMI-MODERN, RATHER THAN POST-COLONIAL TRANSITIONAL (55)	0.57	35 17 11 13 X SQ= 2.33 P = 0.085 RV NO (YES)
185	TILT TOWARD BEING THOSE WHERE PARTICIPATION BY THE MILITARY IN POLITICS IS INTERVENTIVE, RATHER THAN SUPPORTIVE (21)	0.52	185	TILT TOWARD BEING THOSE WHERE PARTICIPATION BY THE MILITARY IN POLITICS IS SUPPORTIVE, RATHER THAN INTERVENTIVE (31)	0.72	15 5 14 13 X SQ= 1.72 P = 0.137 RV YES (NO)
187	TILT MORE TOWARD BEING THOSE WHERE THE ROLE OF THE POLICE IS POLITICALLY SIGNIFICANT (66)	0.71	187	TILT LESS TOWARD BEING THOSE WHERE THE ROLE OF THE POLICE IS POLITICALLY SIGNIFICANT (66)	0.56	36 25 15 20 X SQ= 1.73 P = 0.142 RV NO (YES)
192	TEND LESS TO BE THOSE WHERE THE CHARACTER OF THE LEGAL SYSTEM IS OTHER THAN MUSLIM OR PARTLY MUSLIM (86)	0.67	192	TEND MORE TO BE THOSE WHERE THE CHARACTER OF THE LEGAL SYSTEM IS OTHER THAN MUSLIM OR PARTLY MUSLIM (86)	0.87	20 6 41 42 X SQ= 5.02 P = 0.022 RV NO (YES)
194	LEAN MORE TOWARD BEING THOSE THAT ARE NON-COMMUNIST (101)	0.95	194	LEAN LESS TOWARD BEING THOSE THAT ARE NON-COMMUNIST (101)	0.83	3 8 58 40 X SQ= 2.89 P = 0.057 RV NO (YES)

M

```
*************************************************MATRIX*************************************************

          28  POLITIES                                              28  POLITIES
              WHOSE POPULATION GROWTH RATE                              WHOSE POPULATION GROWTH RATE
              IS HIGH  (62)                                             IS LOW  (48)

                                36  POLITIES
                                    WHOSE PER CAPITA GROSS NATIONAL PRODUCT
                                    IS VERY HIGH, HIGH, OR MEDIUM  (42)

AUSTRALIA   CANADA      CHILE       CUBA        ICELAND     .ARGENTINA   AUSTRIA     BELGIUM     BULGARIA    CYPRUS
ISRAEL      JAMAICA     LEBANON     MALAYA      NEW ZEALAND.CZECHOS'KIA DENMARK     FINLAND     FRANCE      GERMANY, E
SO AFRICA   TRINIDAD    VENEZUELA                           .GERMAN FR   GREECE      HUNGARY     IRELAND     ITALY
                                                            .JAPAN       LUXEMBOURG  NETHERLANDS NORWAY      POLAND
                                                            .RUMANIA     SPAIN       SWEDEN      SWITZERLAND UK
                                                            .US          URUGUAY     YUGOSLAVIA

                                                         13 .  28
. . . . . . . . . . . . . . . . . . . . . . . . . . . . . . . . . . . . . . . . . . . . . . . . . . . . . . . .
                                                         49 .  20

                                36  POLITIES
                                    WHOSE PER CAPITA GROSS NATIONAL PRODUCT
                                    IS LOW OR VERY LOW  (73)

ALBANIA     ALGERIA     BOLIVIA     BRAZIL      CEYLON      .AFGHANISTAN BURMA       BURUNDI     CAMBODIA    CAMEROUN
CHAD        CHINA, PR   COLOMBIA    CONGO(BRA)  COSTA RICA  .CEN AFR REP CONGO(LEO)  DAHOMEY     GABON       HAITI
DOMIN REP   ECUADOR     EL SALVADOR GUATEMALA   GUINEA      .LIBERIA     NIGERIA     PORTUGAL    RWANDA      SOMALIA
HONDURAS    INDIA       INDONESIA   IRAN        IRAQ        .SUDAN       TANGANYIKA  UGANDA      VIETNAM, N  YEMEN
IVORY COAST JORDAN      KOREA REP   LAOS        LIBYA       .
MALAGASY R  MALI        MAURITANIA  MEXICO      MONGOLIA    .
MOROCCO     NEPAL       NICARAGUA   NIGER       PAKISTAN    .
PANAMA      PARAGUAY    PERU        PHILIPPINES SA'U ARABIA.
SENEGAL     SIERRE LEO  SYRIA       THAILAND    TUNISIA     .
TURKEY      UAR         UPPER VOLTA VIETNAM REP             .
```

```
M
*****************************************************MATRIX*****************************************************

          28  POLITIES                                              28  POLITIES
              WHOSE POPULATION GROWTH RATE                              WHOSE POPULATION GROWTH RATE
              IS HIGH  (62)                                             IS LOW  (48)

                              43  POLITIES
                                  WHOSE ECONOMIC DEVELOPMENTAL STATUS
                                  IS DEVELOPED, INTERMEDIATE, OR
                                  UNDERDEVELOPED  (55)

AUSTRALIA   BRAZIL      CANADA      CHILE       CHINA, PR   .ARGENTINA   AUSTRIA     BELGIUM     BULGARIA    BURMA
COLOMBIA    CUBA        ECUADOR     ICELAND     INDIA       .CYPRUS      CZECHOS'KIA DENMARK     FINLAND     FRANCE
IRAQ        ISRAEL      JAMAICA     LIBYA       MALAYA      .GERMANY, E  GERMAN FR   GREECE      HUNGARY     IRELAND
MEXICO      NEW ZEALAND PAKISTAN    PERU        PHILIPPINES.ITALY        JAPAN       LUXEMBOURG  NETHERLANDS NORWAY
SO AFRICA   TRINIDAD    TURKEY      VENEZUELA               .POLAND      PORTUGAL    RUMANIA     SPAIN       SWEDEN
                                                            .SWITZERLAND UK          US          URUGUAY     YUGOSLAVIA

                                                         24 .  30
. . . . . . . . . . . . . . . . . . . . . . . . . . . . . . . . . . . . . . . . . . . . . . . . . . . . . . . . . . . .
                                                         35 .  18

                              43  POLITIES
                                  WHOSE ECONOMIC DEVELOPMENTAL STATUS
                                  IS VERY UNDERDEVELOPED  (57)

ALBANIA     ALGERIA     BOLIVIA     CEYLON      CHAD        .AFGHANISTAN BURUNDI     CAMBODIA    CAMEROUN    CEN AFR REP
CONGO(BRA)  DOMIN REP   EL SALVADOR GUATEMALA   GUINEA      .CONGO(LEO)  DAHOMEY     GABON       HAITI       LIBERIA
HONDURAS    INDONESIA   IRAN        IVORY COAST JORDAN      .NIGERIA     RWANDA      SOMALIA     SUDAN       TANGANYIKA
KOREA REP   LAOS        MALAGASY R  MALI        MAURITANIA  .UGANDA      VIETNAM, N  YEMEN
MONGOLIA    MOROCCO     NEPAL       NICARAGUA   NIGER       .
PANAMA      PARAGUAY    SENEGAL     SIERRE LEO  SYRIA       .
THAILAND    TUNISIA     UAR         UPPER VOLTA VIETNAM REP.
```

M

MATRIX

28 POLITIES WHOSE POPULATION GROWTH RATE IS HIGH (62)	28 POLITIES WHOSE POPULATION GROWTH RATE IS LOW (48)
78 POLITIES THAT HAVE BEEN SIGNIFICANTLY WESTERNIZED THROUGH A COLONIAL RELATIONSHIP, RATHER THAN WITHOUT SUCH A RELATIONSHIP (28)	
ALGERIA, BOLIVIA, BRAZIL, CEYLON, CHILE, COLOMBIA, COSTA RICA, CUBA, DOMIN REP, ECUADOR, EL SALVADOR, GUATEMALA, HONDURAS, INDIA, JAMAICA, LEBANON, MEXICO, NICARAGUA, PANAMA, PARAGUAY, PERU, PHILIPPINES, TRINIDAD, UAR, VENEZUELA	ARGENTINA, CYPRUS, URUGUAY
25	3
2	4
78 POLITIES THAT HAVE BEEN SIGNIFICANTLY WESTERNIZED WITHOUT A COLONIAL RELATIONSHIP, RATHER THAN THROUGH SUCH A RELATIONSHIP (7)	
CHINA, PR, TURKEY	BULGARIA, JAPAN, RUMANIA, YUGOSLAVIA

MATRIX

28 POLITIES WHOSE POPULATION GROWTH RATE IS HIGH (62)	28 POLITIES WHOSE POPULATION GROWTH RATE IS LOW (48)
97 POLITIES WHERE THE STATUS OF THE REGIME IS AUTHORITARIAN, RATHER THAN TOTALITARIAN (23)	
ALGERIA, JORDAN, PAKISTAN, EL SALVADOR, KOREA REP, PARAGUAY, GUINEA, LAOS, SA'U ARABIA, INDONESIA, NEPAL, THAILAND, IRAN, NICARAGUA, UAR	AFGHANISTAN, SUDAN, BURMA, CAMBODIA, CEN AFR REP, LIBERIA
15	6
4	10
97 POLITIES WHERE THE STATUS OF THE REGIME IS TOTALITARIAN, RATHER THAN AUTHORITARIAN (16)	
ALBANIA, CHINA, PR, CUBA, MONGOLIA	BULGARIA, PORTUGAL, CZECHOS'KIA, RUMANIA, GERMANY, E, SPAIN, HUNGARY, VIETNAM, N, POLAND, YUGOSLAVIA

MATRIX

	28 POLITIES WHOSE POPULATION GROWTH RATE IS HIGH (62)	28 POLITIES WHOSE POPULATION GROWTH RATE IS LOW (48)
153 POLITIES WHERE THE PARTY SYSTEM IS STABLE (42)	14: ALBANIA, ICELAND, NEW ZEALAND, AUSTRALIA, INDIA, PHILIPPINES, CANADA, ISRAEL, SO AFRICA, CHINA, PR, MEXICO, TUNISIA, GUINEA, MONGOLIA	26: AUSTRIA, FINLAND, JAPAN, POLAND, SWITZERLAND, YUGOSLAVIA, BELGIUM, GERMANY, E, LIBERIA, PORTUGAL, TANGANYIKA, BULGARIA, GERMAN FR, LUXEMBOURG, RUMANIA, UK, CZECHOS'KIA, HUNGARY, NETHERLANDS, SPAIN, US, DENMARK, IRELAND, NORWAY, SWEDEN, VIETNAM, N
153 POLITIES WHERE THE PARTY SYSTEM IS MODERATELY STABLE OR UNSTABLE (38)	30: BOLIVIA, ECUADOR, IVORY COAST, MALAGASY R, PANAMA, SYRIA, BRAZIL, EL SALVADOR, JORDAN, MALAYA, PARAGUAY, THAILAND, CEYLON, GUATEMALA, KOREA REP, NEPAL, PERU, TURKEY, CHILE, HONDURAS, LAOS, NICARAGUA, SENEGAL, VENEZUELA, COSTA RICA, IRAQ, LEBANON, PAKISTAN, SIERRE LEO, VIETNAM REP	7: ARGENTINA, ITALY, CAMBODIA, SOMALIA, CONGO(LEO), FRANCE, GREECE

```
********************************************************************************************************

29  POLITIES                                                  29  POLITIES
    WHERE THE DEGREE OF URBANIZATION                              WHERE THE DEGREE OF URBANIZATION
    IS HIGH  (56)                                                 IS LOW  (49)


                                                                            BOTH SUBJECT AND PREDICATE
........................................................................................................

      56 IN LEFT COLUMN

ARGENTINA    AUSTRALIA    AUSTRIA      BELGIUM      BRAZIL       BULGARIA     CANADA       CHILE        COLOMBIA     CUBA
CZECHOS'KIA  DENMARK      DOMIN REP    ECUADOR      FINLAND      FRANCE       GERMANY, E   GERMAN FR    GREECE       HUNGARY
ICELAND      IRAN         IRAQ         IRELAND      ISRAEL       ITALY        JAMAICA      JAPAN        JORDAN       KOREA REP
LEBANON      LIBYA        LUXEMBOURG   MEXICO       MOROCCO      NETHERLANDS  NEW ZEALAND  NICARAGUA    NORWAY       PANAMA
PERU         POLAND       RUMANIA      SO AFRICA    SPAIN        SWEDEN       SWITZERLAND  SYRIA        TUNISIA      TURKEY
USSR         UAR          UK           US           URUGUAY      VENEZUELA

      49 IN RIGHT COLUMN

AFGHANISTAN  ALBANIA      ALGERIA      BOLIVIA      BURMA        BURUNDI      CAMBODIA     CAMEROUN     CEN AFR REP  CEYLON
CHAD         CONGO(BRA)   CONGO(LEO)   CYPRUS       DAHOMEY      EL SALVADOR  ETHIOPIA     GABON        GHANA        GUATEMALA
GUINEA       HAITI        HONDURAS     INDIA        INDONESIA    IVORY COAST  LAOS         LIBERIA      MALAGASY R   MALI
MAURITANIA   MONGOLIA     NEPAL        NIGER        NIGERIA      PAKISTAN     RWANDA       SA'U ARABIA  SENEGAL      SIERRE LEO
SOMALIA      SUDAN        TANGANYIKA   THAILAND     TOGO         UGANDA       UPPER VOLTA  YEMEN        YUGOSLAVIA

       7 EXCLUDED BECAUSE AMBIGUOUS

COSTA RICA   MALAYA       PARAGUAY     PHILIPPINES  PORTUGAL     TRINIDAD     VIETNAM REP

       3 EXCLUDED BECAUSE UNASCERTAINED

CHINA, PR    KOREA, N     VIETNAM, N
--------------------------------------------------------------------------------------------------------

  2  TEND LESS TO BE THOSE                 0.62      2  ALWAYS ARE THOSE                      1.00        21    0
     LOCATED ELSEWHERE THAN IN                          LOCATED ELSEWHERE THAN IN                         35   49
     WEST EUROPE, SCANDINAVIA,                          WEST EUROPE, SCANDINAVIA,                    X SQ=  20.68
     NORTH AMERICA, OR AUSTRALASIA  (93)                NORTH AMERICA, OR AUSTRALASIA  (93)          P =     0.
                                                                                                     RV NO  (YES)

  5  TILT TOWARD BEING THOSE               0.71      5  ALWAYS ARE THOSE                      1.00        17    0
     LOCATED IN SCANDINAVIA OR WEST EUROPE,             LOCATED IN EAST EUROPE, RATHER THAN                7    2
     RATHER THAN IN EAST EUROPE  (18)                   IN SCANDINAVIA OR WEST EUROPE  (9)           X SQ=   1.56
                                                                                                     P =    0.111
                                                                                                     RV YES (NO )

  6  LEAN LESS TOWARD BEING THOSE          0.75      6  LEAN MORE TOWARD BEING THOSE          0.90        14    5
     LOCATED ELSEWHERE THAN IN THE                      LOCATED ELSEWHERE THAN IN THE                     42   44
     CARIBBEAN, CENTRAL AMERICA,                        CARIBBEAN, CENTRAL AMERICA,                  X SQ=   2.93
     OR SOUTH AMERICA  (93)                             OR SOUTH AMERICA  (93)                       P =    0.074
                                                                                                     RV NO  (YES)
```

```
 8  TEND MORE TO BE THOSE                      0.96      8  TEND LESS TO BE THOSE                     0.80        2    10
    LOCATED ELSEWHERE THAN IN EAST ASIA                     LOCATED ELSEWHERE THAN IN EAST ASIA                  54    39
    SOUTH ASIA, OR SOUTHEAST ASIA  (97)                     SOUTH ASIA, OR SOUTHEAST ASIA  (97)            X SQ=    5.75
                                                                                                           P =     0.011
                                                                                                           RV NO   (YES)

 9  ALWAYS ARE THOSE                           1.00      9  TEND LESS TO BE THOSE                     0.90        0     5
    LOCATED ELSEWHERE THAN IN                               LOCATED ELSEWHERE THAN IN                            56    44
    SOUTHEAST ASIA  (106)                                   SOUTHEAST ASIA  (106)                          X SQ=    3.96
                                                                                                           P =     0.020
                                                                                                           RV NO   (YES)

10  TEND TO BE THOSE                           0.91     10  TEND TO BE THOSE                          0.57        5    28
    LOCATED ELSEWHERE THAN IN NORTH AFRICA,                 LOCATED IN NORTH AFRICA, OR                          51    21
    OR CENTRAL AND SOUTH AFRICA  (82)                       CENTRAL AND SOUTH AFRICA  (33)                 X SQ=   26.00
                                                                                                           P =     0.
                                                                                                           RV YES  (YES)

15  TEND TO BE THOSE                           0.58     15  TEND TO BE THOSE                          0.87        7     4
    LOCATED IN THE MIDDLE EAST, RATHER THAN                 LOCATED IN NORTH AFRICA OR                            5    28
    IN NORTH AFRICA OR                                      CENTRAL AND SOUTH AFRICA, RATHER THAN          X SQ=    7.49
    CENTRAL AND SOUTH AFRICA  (11)                          IN THE MIDDLE EAST  (33)                       P =     0.004
                                                                                                           RV YES  (YES)

16  TEND TO BE THOSE                           0.78     16  TEND TO BE THOSE                          0.71        7     4
    LOCATED IN THE MIDDLE EAST, RATHER THAN                 LOCATED IN EAST ASIA, SOUTH ASIA, OR                  2    10
    IN EAST ASIA, SOUTH ASIA, OR                            SOUTHEAST ASIA, RATHER THAN IN                 X SQ=    3.53
    SOUTHEAST ASIA  (11)                                    THE MIDDLE EAST  (18)                          P =     0.036
                                                                                                           RV YES  (YES)

19  TEND TO BE THOSE                           0.74     19  TEND TO BE THOSE                          0.85        5    28
    LOCATED IN THE CARIBBEAN,                               LOCATED IN NORTH AFRICA OR                           14     5
    CENTRAL AMERICA, OR SOUTH AMERICA,                      CENTRAL AND SOUTH AFRICA, RATHER THAN          X SQ=   15.38
    RATHER THAN IN NORTH AFRICA OR                          IN THE CARIBBEAN, CENTRAL AMERICA, OR          P =     0.
    CENTRAL AND SOUTH AFRICA  (22)                          SOUTH AMERICA  (33)                            RV YES  (YES)

20  TEND TO BE THOSE                           0.87     20  TEND TO BE THOSE                          0.67        2    10
    LOCATED IN THE CARIBBEAN,                               LOCATED IN EAST ASIA, SOUTH ASIA, OR                 14     5
    CENTRAL AMERICA, OR SOUTH AMERICA,                      SOUTHEAST ASIA, RATHER THAN IN                 X SQ=    7.43
    RATHER THAN IN EAST ASIA, SOUTH ASIA,                   THE CARIBBEAN, CENTRAL AMERICA,                P =     0.003
    OR SOUTHEAST ASIA  (22)                                 OR SOUTH AMERICA  (18)                         RV YES  (YES)

24  TEND TO BE THOSE                           0.64     24  TEND TO BE THOSE                          0.63       36    18
    WHOSE POPULATION IS                                     WHOSE POPULATION IS                                  20    31
    VERY LARGE, LARGE, OR MEDIUM  (61)                      SMALL  (54)                                    X SQ=    6.88
                                                                                                           P =     0.006
                                                                                                           RV YES  (YES)

25  TILT LESS TOWARD BEING THOSE               0.79     25  TILT MORE TOWARD BEING THOSE              0.92       12     4
    WHOSE POPULATION DENSITY IS                             WHOSE POPULATION DENSITY IS                          44    45
    MEDIUM OR LOW  (98)                                     MEDIUM OR LOW  (98)                            X SQ=    2.61
                                                                                                           P =     0.100
                                                                                                           RV NO   (YES)
```

26	LEAN LESS TOWARD BEING THOSE WHOSE POPULATION DENSITY IS LOW (67)	0.54	26	LEAN MORE TOWARD BEING THOSE WHOSE POPULATION DENSITY IS LOW (67)	0.71	26 14 30 35 X SQ= 2.82 P = 0.072 RV NO (YES)
30	TEND TO BE THOSE WHOSE AGRICULTURAL POPULATION IS MEDIUM, LOW, OR VERY LOW (57)	0.82	30	TEND TO BE THOSE WHOSE AGRICULTURAL POPULATION IS HIGH (56)	0.90	10 43 46 5 X SQ= 50.38 P = 0. RV YES (YES)
31	TEND LESS TO BE THOSE WHOSE AGRICULTURAL POPULATION IS HIGH OR MEDIUM (90)	0.59	31	ALWAYS ARE THOSE WHOSE AGRICULTURAL POPULATION IS HIGH OR MEDIUM (90)	1.00	33 48 23 0 X SQ= 22.98 P = 0. RV NO (YES)
34	TEND TO BE THOSE WHOSE GROSS NATIONAL PRODUCT IS VERY HIGH, HIGH, MEDIUM, OR LOW (62)	0.79	34	TEND TO BE THOSE WHOSE GROSS NATIONAL PRODUCT IS VERY LOW (53)	0.76	44 12 12 37 X SQ= 28.58 P = 0. RV YES (YES)
35	TEND LESS TO BE THOSE WHOSE PER CAPITA GROSS NATIONAL PRODUCT IS MEDIUM, LOW, OR VERY LOW (91)	0.57	35	ALWAYS ARE THOSE WHOSE PER CAPITA GROSS NATIONAL PRODUCT IS MEDIUM, LOW, OR VERY LOW (91)	1.00	24 0 32 49 X SQ= 24.85 P = 0. RV NO (YES)
36	TEND TO BE THOSE WHOSE PER CAPITA GROSS NATIONAL PRODUCT IS VERY HIGH, HIGH, OR MEDIUM (42)	0.68	36	TEND TO BE THOSE WHOSE PER CAPITA GROSS NATIONAL PRODUCT IS LOW OR VERY LOW (73)	0.96	38 2 18 47 X SQ= 42.41 P = 0. RV YES (YES)
37	TEND TO BE THOSE WHOSE PER CAPITA GROSS NATIONAL PRODUCT IS VERY HIGH, HIGH, MEDIUM, OR LOW (64)	0.91	37	TEND TO BE THOSE WHOSE PER CAPITA GROSS NATIONAL PRODUCT IS VERY LOW (51)	0.84	51 8 5 41 X SQ= 56.31 P = 0. RV YES (YES)
39	TEND TO BE THOSE WHOSE INTERNATIONAL FINANCIAL STATUS IS VERY HIGH, HIGH, OR MEDIUM (38)	0.58	39	TEND TO BE THOSE WHOSE INTERNATIONAL FINANCIAL STATUS IS LOW OR VERY LOW (76)	0.92	32 4 23 45 X SQ= 26.48 P = 0. RV YES (YES)
40	TEND TO BE THOSE WHOSE INTERNATIONAL FINANCIAL STATUS IS VERY HIGH, HIGH, MEDIUM, OR LOW (71)	0.91	40	TEND TO BE THOSE WHOSE INTERNATIONAL FINANCIAL STATUS IS VERY LOW (39)	0.67	50 16 5 32 X SQ= 34.45 P = 0. RV YES (YES)

42	TEND TO BE THOSE WHOSE ECONOMIC DEVELOPMENTAL STATUS IS DEVELOPED OR INTERMEDIATE (36)	0.63	42	TEND TO BE THOSE WHOSE ECONOMIC DEVELOPMENTAL STATUS IS UNDERDEVELOPED OR VERY UNDERDEVELOPED (76)	0.96	34 2 20 46 X SQ= 35.94 P = 0. RV YES (YES)
43	TEND TO BE THOSE WHOSE ECONOMIC DEVELOPMENTAL STATUS IS DEVELOPED, INTERMEDIATE, OR UNDERDEVELOPED (55)	0.82	43	TEND TO BE THOSE WHOSE ECONOMIC DEVELOPMENTAL STATUS IS VERY UNDERDEVELOPED (57)	0.90	45 5 10 43 X SQ= 49.49 P = 0. RV YES (YES)
44	TEND LESS TO BE THOSE WHERE THE LITERACY RATE IS BELOW NINETY PERCENT (90)	0.55	44	ALWAYS ARE THOSE WHERE THE LITERACY RATE IS BELOW NINETY PERCENT (90)	1.00	25 0 31 49 X SQ= 26.30 P = 0. RV NO (YES)
45	TEND TO BE THOSE WHERE THE LITERACY RATE IS FIFTY PERCENT OR ABOVE (55)	0.76	45	TEND TO BE THOSE WHERE THE LITERACY RATE IS BELOW FIFTY PERCENT (54)	0.85	42 7 13 40 X SQ= 35.94 P = 0. RV YES (YES)
46	ALWAYS ARE THOSE WHERE THE LITERACY RATE IS TEN PERCENT OR ABOVE (84)	1.00	46	TEND TO BE THOSE WHERE THE LITERACY RATE IS BELOW TEN PERCENT (26)	0.57	55 20 0 26 X SQ= 38.96 P = 0. RV YES (YES)
51	TILT MORE TOWARD BEING THOSE WHERE FREEDOM OF THE PRESS IS COMPLETE OR INTERMITTENT (60)	0.72	51	TILT LESS TOWARD BEING THOSE WHERE FREEDOM OF THE PRESS IS COMPLETE OR INTERMITTENT (60)	0.53	38 18 15 16 X SQ= 2.41 P = 0.108 RV NO (YES)
53	TEND LESS TO BE THOSE WHERE NEWSPAPER CIRCULATION IS LESS THAN THREE HUNDRED PER THOUSAND (101)	0.75	53	ALWAYS ARE THOSE WHERE NEWSPAPER CIRCULATION IS LESS THAN THREE HUNDRED PER THOUSAND (101)	1.00	14 0 42 49 X SQ= 12.05 P = 0. RV NO (YES)
54	TEND TO BE THOSE WHERE NEWSPAPER CIRCULATION IS ONE HUNDRED OR MORE PER THOUSAND (37)	0.61	54	TEND TO BE THOSE WHERE NEWSPAPER CIRCULATION IS LESS THAN ONE HUNDRED PER THOUSAND (76)	0.96	34 2 22 47 X SQ= 34.73 P = 0. RV YES (YES)
55	TEND TO BE THOSE WHERE NEWSPAPER CIRCULATION IS TEN OR MORE PER THOUSAND (78)	0.96	55	TEND TO BE THOSE WHERE NEWSPAPER CIRCULATION IS LESS THAN TEN PER THOUSAND (35)	0.67	54 16 2 33 X SQ= 45.00 P = 0. RV YES (YES)

56	TEND TO BE THOSE WHERE THE RELIGION IS PREDOMINANTLY LITERATE (79)	0.95	56	TEND TO BE THOSE WHERE THE RELIGION IS PREDOMINANTLY OR PARTLY NON-LITERATE (31)	0.57	52 21 3 28 X SQ= 30.67 P = 0. RV YES (YES)
63	TEND TO BE THOSE WHERE THE RELIGION IS PREDOMINANTLY SOME KIND OF CHRISTIAN (46)	0.71	63	TEND TO BE THOSE WHERE THE RELIGION IS PREDOMINANTLY OR PARTLY OTHER THAN CHRISTIAN (68)	0.94	M 39 3 16 46 X SQ= 42.53 P = 0. RV YES (YES)
66	TEND TO BE THOSE THAT ARE RELIGIOUSLY HOMOGENEOUS (57)	0.69	66	TEND TO BE THOSE THAT ARE RELIGIOUSLY HETEROGENEOUS (49)	0.65	37 16 17 30 X SQ= 10.04 P = 0.001 RV YES (YES)
68	TEND TO BE THOSE THAT ARE LINGUISTICALLY HOMOGENEOUS (52)	0.66	68	TEND TO BE THOSE THAT ARE LINGUISTICALLY WEAKLY HETEROGENEOUS OR STRONGLY HETEROGENEOUS (62)	0.78	37 11 19 38 X SQ= 18.32 P = 0. RV YES (YES)
69	TEND TO BE THOSE THAT ARE LINGUISTICALLY HOMOGENEOUS OR WEAKLY HETEROGENEOUS (64)	0.79	69	TEND TO BE THOSE THAT ARE LINGUISTICALLY STRONGLY HETEROGENEOUS (50)	0.73	44 13 12 36 X SQ= 26.46 P = 0. RV YES (YES)
70	TEND LESS TO BE THOSE THAT ARE RELIGIOUSLY, RACIALLY, OR LINGUISTICALLY HETEROGENEOUS (85)	0.69	70	TEND MORE TO BE THOSE THAT ARE RELIGIOUSLY, RACIALLY, OR LINGUISTICALLY HETEROGENEOUS (85)	0.95	17 2 37 42 X SQ= 9.60 P = 0.001 RV NO (YES)
72	TEND TO BE THOSE WHOSE DATE OF INDEPENDENCE IS BEFORE 1914 (52)	0.70	72	TEND TO BE THOSE WHOSE DATE OF INDEPENDENCE IS AFTER 1914 (59)	0.78	37 11 16 38 X SQ= 21.06 P = 0. RV YES (YES)
73	TEND TO BE THOSE WHOSE DATE OF INDEPENDENCE IS BEFORE 1945 (65)	0.87	73	TEND TO BE THOSE WHOSE DATE OF INDEPENDENCE IS AFTER 1945 (46)	0.69	46 15 7 34 X SQ= 31.13 P = 0. RV YES (YES)
74	TEND LESS TO BE THOSE THAT ARE NOT HISTORICALLY WESTERN (87)	0.54	74	ALWAYS ARE THOSE THAT ARE NOT HISTORICALLY WESTERN (87)	1.00	25 0 29 49 X SQ= 27.49 P = 0. RV NO (YES)

No.	Statement	Value	No.	Statement	Value	Statistics
75	TEND TO BE THOSE THAT ARE HISTORICALLY WESTERN OR SIGNIFICANTLY WESTERNIZED (62)	0.85	75	TEND TO BE THOSE THAT ARE NOT HISTORICALLY WESTERN AND ARE NOT SIGNIFICANTLY WESTERNIZED (52)	0.82	47 9 8 40 X SQ= 44.27 P = 0. RV YES (YES)
76	TEND TO BE THOSE THAT ARE HISTORICALLY WESTERN, RATHER THAN HAVING BEEN SIGNIFICANTLY OR PARTIALLY WESTERNIZED THROUGH A COLONIAL RELATIONSHIP (26)	0.51	76	ALWAYS ARE THOSE THAT HAVE BEEN SIGNIFICANTLY OR PARTIALLY WESTERNIZED THROUGH A COLONIAL RELATIONSHIP, RATHER THAN BEING HISTORICALLY WESTERN (70)	1.00	25 0 24 39 X SQ= 25.34 P = 0. RV YES (YES)
77	TEND TO BE THOSE THAT HAVE BEEN SIGNIFICANTLY WESTERNIZED, RATHER THAN PARTIALLY WESTERNIZED, THROUGH A COLONIAL RELATIONSHIP (28)	0.70	77	TEND TO BE THOSE THAT HAVE BEEN PARTIALLY WESTERNIZED, RATHER THAN SIGNIFICANTLY WESTERNIZED, THROUGH A COLONIAL RELATIONSHIP (41)	0.79	16 8 7 31 X SQ= 12.68 P = 0. RV YES (YES)
79	TEND LESS TO BE THOSE THAT WERE FORMERLY DEPENDENCIES OF BRITAIN OR FRANCE, RATHER THAN SPAIN (49)	0.56	79	TEND MORE TO BE THOSE THAT WERE FORMERLY DEPENDENCIES OF BRITAIN OR FRANCE, RATHER THAN SPAIN (49)	0.88	15 30 12 4 X SQ= 6.70 P = 0.007 RV NO (YES)
82	TEND TO BE THOSE WHERE THE TYPE OF POLITICAL MODERNIZATION IS EARLY OR LATER EUROPEAN OR EUROPEAN DERIVED, RATHER THAN DEVELOPED TUTELARY OR UNDEVELOPED TUTELARY (51)	0.79	82	TEND TO BE THOSE WHERE THE TYPE OF POLITICAL MODERNIZATION IS DEVELOPED TUTELARY OR UNDEVELOPED TUTELARY, RATHER THAN EARLY OR LATER EUROPEAN OR EUROPEAN DERIVED (55)	0.84	41 7 11 38 X SQ= 36.17 P = 0. RV YES (YES)
84	ALWAYS ARE THOSE WHERE THE TYPE OF POLITICAL MODERNIZATION IS DEVELOPED TUTELARY, RATHER THAN UNDEVELOPED TUTELARY (31)	1.00	84	TEND TO BE THOSE WHERE THE TYPE OF POLITICAL MODERNIZATION IS UNDEVELOPED TUTELARY, RATHER THAN DEVELOPED TUTELARY (24)	0.63	11 14 0 24 X SQ= 11.21 P = 0. RV YES (NO)
85	TEND TO BE THOSE WHERE THE STAGE OF POLITICAL MODERNIZATION IS ADVANCED, RATHER THAN MID- OR EARLY TRANSITIONAL (60)	0.82	85	TEND TO BE THOSE WHERE THE STAGE OF POLITICAL MODERNIZATION IS MID- OR EARLY TRANSITIONAL, RATHER THAN ADVANCED (54)	0.81	46 9 10 39 X SQ= 39.18 P = 0. RV YES (YES)
86	TEND TO BE THOSE WHERE THE STAGE OF POLITICAL MODERNIZATION IS ADVANCED OR MID-TRANSITIONAL, RATHER THAN EARLY TRANSITIONAL (76)	0.96	86	TEND TO BE THOSE WHERE THE STAGE OF POLITICAL MODERNIZATION IS EARLY TRANSITIONAL, RATHER THAN ADVANCED OR MID-TRANSITIONAL (38)	0.67	54 16 2 32 X SQ= 43.94 P = 0. RV YES (YES)
87	TEND TO BE THOSE WHOSE IDEOLOGICAL ORIENTATION IS OTHER THAN DEVELOPMENTAL (58)	0.89	87	TEND TO BE THOSE WHOSE IDEOLOGICAL ORIENTATION IS DEVELOPMENTAL (31)	0.68	5 25 40 12 X SQ= 25.51 P = 0. RV YES (YES)

89	TEND TO BE THOSE WHOSE IDEOLOGICAL ORIENTATION IS CONVENTIONAL (33)	0.63	89	TEND TO BE THOSE WHOSE IDEOLOGICAL ORIENTATION IS OTHER THAN CONVENTIONAL (62)	0.95	29 2 17 40 X SQ= 30.18 P = 0. RV YES (YES)
94	TEND TO BE THOSE WHERE THE STATUS OF THE REGIME IS CONSTITUTIONAL (51)	0.73	94	TEND TO BE THOSE WHERE THE STATUS OF THE REGIME IS AUTHORITARIAN OR TOTALITARIAN (41)	0.68	37 10 14 21 X SQ= 11.20 P = 0.001 RV YES (YES)
96	TEND TO BE THOSE WHERE THE STATUS OF THE REGIME IS CONSTITUTIONAL OR TOTALITARIAN (67)	0.90	96	TEND TO BE THOSE WHERE THE STATUS OF THE REGIME IS AUTHORITARIAN (23)	0.57	46 13 5 17 X SQ= 18.67 P = 0. RV YES (YES)
97	TEND TO BE THOSE WHERE THE STATUS OF THE REGIME IS TOTALITARIAN, RATHER THAN AUTHORITARIAN (16)	0.64	97	TEND TO BE THOSE WHERE THE STATUS OF THE REGIME IS AUTHORITARIAN, RATHER THAN TOTALITARIAN (23)	0.85	5 17 9 3 X SQ= 6.73 P = 0.009 RV YES (YES)
99	TEND TO BE THOSE WHERE GOVERNMENTAL STABILITY IS GENERALLY PRESENT AND DATES FROM AT LEAST THE INTER-WAR PERIOD, OR FROM THE POST-WAR PERIOD (50)	0.67	99	TEND TO BE THOSE WHERE GOVERNMENTAL STABILITY IS MODERATELY PRESENT AND DATES FROM THE POST-WAR PERIOD, OR IS ABSENT (36)	0.65	35 9 17 17 X SQ= 6.26 P = 0.008 RV YES (YES)
101	TEND TO BE THOSE WHERE THE REPRESENTATIVE CHARACTER OF THE REGIME IS POLYARCHIC (41)	0.63	101	TEND TO BE THOSE WHERE THE REPRESENTATIVE CHARACTER OF THE REGIME IS LIMITED POLYARCHIC, PSEUDO-POLYARCHIC, OR NON-POLYARCHIC (57)	0.85	31 6 18 33 X SQ= 18.51 P = 0. RV YES (YES)
102	TEND TO BE THOSE WHERE THE REPRESENTATIVE CHARACTER OF THE REGIME IS POLYARCHIC OR LIMITED POLYARCHIC (59)	0.75	102	TEND TO BE THOSE WHERE THE REPRESENTATIVE CHARACTER OF THE REGIME IS PSEUDO-POLYARCHIC OR NON-POLYARCHIC (49)	0.65	39 16 13 30 X SQ= 14.44 P = 0. RV YES (YES)
105	TEND TO BE THOSE WHERE THE ELECTORAL SYSTEM IS COMPETITIVE (43)	0.73	105	TEND TO BE THOSE WHERE THE ELECTORAL SYSTEM IS PARTIALLY COMPETITIVE OR NON-COMPETITIVE (47)	0.81	32 7 12 29 X SQ= 20.42 P = 0. RV YES (YES)
106	TEND TO BE THOSE WHERE THE ELECTORAL SYSTEM IS COMPETITIVE OR PARTIALLY COMPETITIVE (52)	0.79	106	TEND TO BE THOSE WHERE THE ELECTORAL SYSTEM IS NON-COMPETITIVE (30)	0.53	34 14 9 16 X SQ= 6.86 P = 0.006 RV YES (YES)

107	TEND TO BE THOSE WHERE AUTONOMOUS GROUPS ARE FULLY TOLERATED IN POLITICS (46)	0.64	107	TEND TO BE THOSE WHERE AUTONOMOUS GROUPS ARE PARTIALLY TOLERATED IN POLITICS, ARE TOLERATED ONLY OUTSIDE POLITICS, OR ARE NOT TOLERATED AT ALL (65)	0.83	35 8 20 38 X SQ= 20.06 P = 0. RV YES (YES)
108	TEND TO BE THOSE WHERE AUTONOMOUS GROUPS ARE FULLY OR PARTIALLY TOLERATED IN POLITICS (65)	0.82	108	TEND TO BE THOSE WHERE AUTONOMOUS GROUPS ARE TOLERATED ONLY OUTSIDE POLITICS OR ARE NOT TOLERATED AT ALL (35)	0.51	42 19 9 20 X SQ= 9.96 P = 0.001 RV YES (YES)
111	LEAN TOWARD BEING THOSE WHERE POLITICAL ENCULTURATION IS HIGH OR MEDIUM (53)	0.66	111	LEAN TOWARD BEING THOSE WHERE POLITICAL ENCULTURATION IS LOW (42)	0.55	31 20 16 24 X SQ= 3.09 P = 0.059 RV YES (YES)
114	LEAN TOWARD BEING THOSE WHERE SECTIONALISM IS NEGLIGIBLE (47)	0.50	114	LEAN TOWARD BEING THOSE WHERE SECTIONALISM IS EXTREME OR MODERATE (61)	0.70	26 32 26 14 X SQ= 3.10 P = 0.064 RV YES (YES)
116	TEND TO BE THOSE WHERE INTEREST ARTICULATION BY ASSOCIATIONAL GROUPS IS SIGNIFICANT OR MODERATE (32)	0.57	116	ALWAYS ARE THOSE WHERE INTEREST ARTICULATION BY ASSOCIATIONAL GROUPS IS LIMITED OR NEGLIGIBLE (79)	1.00	31 0 23 48 X SQ= 36.92 P = 0. RV YES (YES)
117	TEND TO BE THOSE WHERE INTEREST ARTICULATION BY ASSOCIATIONAL GROUPS IS SIGNIFICANT, MODERATE, OR LIMITED (60)	0.87	117	TEND TO BE THOSE WHERE INTEREST ARTICULATION BY ASSOCIATIONAL GROUPS IS NEGLIGIBLE (51)	0.79	47 10 7 38 X SQ= 42.53 P = 0. RV YES (YES)
119	TEND LESS TO BE THOSE WHERE INTEREST ARTICULATION BY INSTITUTIONAL GROUPS IS VERY SIGNIFICANT OR SIGNIFICANT (74)	0.59	119	TEND MORE TO BE THOSE WHERE INTEREST ARTICULATION BY INSTITUTIONAL GROUPS IS VERY SIGNIFICANT OR SIGNIFICANT (74)	0.97	33 33 23 1 X SQ= 13.84 P = 0. RV NO (YES)
120	TEND LESS TO BE THOSE WHERE INTEREST ARTICULATION BY INSTITUTIONAL GROUPS IS VERY SIGNIFICANT, SIGNIFICANT, OR MODERATE (90)	0.82	120	ALWAYS ARE THOSE WHERE INTEREST ARTICULATION BY INSTITUTIONAL GROUPS IS VERY SIGNIFICANT, SIGNIFICANT, OR MODERATE (90)	1.00	46 34 10 0 X SQ= 5.14 P = 0.012 RV NO (NO)
121	TEND TO BE THOSE WHERE INTEREST ARTICULATION BY NON-ASSOCIATIONAL GROUPS IS MODERATE, LIMITED, OR NEGLIGIBLE (61)	0.87	121	TEND TO BE THOSE WHERE INTEREST ARTICULATION BY NON-ASSOCIATIONAL GROUPS IS SIGNIFICANT (54)	0.84	7 41 49 8 X SQ= 50.52 P = 0. RV YES (YES)

122	TEND TO BE THOSE WHERE INTEREST ARTICULATION BY NON-ASSOCIATIONAL GROUPS IS LIMITED OR NEGLIGIBLE (32)	0.52	122	ALWAYS ARE THOSE WHERE INTEREST ARTICULATION BY NON-ASSOCIATIONAL GROUPS IS SIGNIFICANT OR MODERATE (83)	1.00	27 49 29 0 X SQ= 32.51 P = 0. RV YES (YES)
125	TEND TO BE THOSE WHERE INTEREST ARTICULATION BY ANOMIC GROUPS IS INFREQUENT OR VERY INFREQUENT (35)	0.58	125	TEND TO BE THOSE WHERE INTEREST ARTICULATION BY ANOMIC GROUPS IS FREQUENT OR OCCASIONAL (64)	0.91	20 40 28 4 X SQ= 22.42 P = 0. RV YES (YES)
128	TEND TO BE THOSE WHERE INTEREST ARTICULATION BY POLITICAL PARTIES IS SIGNIFICANT OR MODERATE (48)	0.65	128	TEND TO BE THOSE WHERE INTEREST ARTICULATION BY POLITICAL PARTIES IS LIMITED OR NEGLIGIBLE (45)	0.60	32 14 17 21 X SQ= 4.31 P = 0.027 RV YES (YES)
129	TEND TO BE THOSE WHERE INTEREST ARTICULATION BY POLITICAL PARTIES IS SIGNIFICANT, MODERATE, OR LIMITED (56)	0.73	129	TEND TO BE THOSE WHERE INTEREST ARTICULATION BY POLITICAL PARTIES IS NEGLIGIBLE (37)	0.51	36 17 13 18 X SQ= 4.42 P = 0.024 RV YES (YES)
133	TEND MORE TO BE THOSE WHERE INTEREST AGGREGATION BY THE EXECUTIVE IS MODERATE, LIMITED, OR NEGLIGIBLE (76)	0.86	133	TEND LESS TO BE THOSE WHERE INTEREST AGGREGATION BY THE EXECUTIVE IS MODERATE, LIMITED, OR NEGLIGIBLE (76)	0.56	7 20 43 25 X SQ= 9.35 P = 0.001 RV NO (YES)
135	TEND MORE TO BE THOSE WHERE INTEREST AGGREGATION BY THE EXECUTIVE IS SIGNIFICANT, MODERATE, OR LIMITED (77)	0.95	135	TEND LESS TO BE THOSE WHERE INTEREST AGGREGATION BY THE EXECUTIVE IS SIGNIFICANT, MODERATE, OR LIMITED (77)	0.80	42 32 2 8 X SQ= 3.41 P = 0.042 RV NO (YES)
137	TEND TO BE THOSE WHERE INTEREST AGGREGATION BY THE LEGISLATURE IS SIGNIFICANT OR MODERATE (29)	0.57	137	TEND TO BE THOSE WHERE INTEREST AGGREGATION BY THE LEGISLATURE IS LIMITED OR NEGLIGIBLE (68)	0.98	26 1 20 41 X SQ= 27.77 P = 0. RV YES (YES)
138	TEND TO BE THOSE WHERE INTEREST AGGREGATION BY THE LEGISLATURE IS SIGNIFICANT, MODERATE, OR LIMITED (48)	0.79	138	TEND TO BE THOSE WHERE INTEREST AGGREGATION BY THE LEGISLATURE IS NEGLIGIBLE (49)	0.79	37 9 10 33 X SQ= 26.91 P = 0. RV YES (YES)
139	TEND MORE TO BE THOSE WHERE THE PARTY SYSTEM IS QUANTITATIVELY OTHER THAN ONE-PARTY (71)	0.80	139	TEND LESS TO BE THOSE WHERE THE PARTY SYSTEM IS QUANTITATIVELY OTHER THAN ONE-PARTY (71)	0.57	10 19 41 25 X SQ= 5.13 P = 0.015 RV NO (YES)

No.	Statement		No.	Statement		Statistics
140	TILT MORE TOWARD BEING THOSE WHERE THE PARTY SYSTEM IS QUANTITATIVELY OTHER THAN ONE PARTY DOMINANT (87)	0.94	140	TILT LESS TOWARD BEING THOSE WHERE THE PARTY SYSTEM IS QUANTITATIVELY OTHER THAN ONE PARTY DOMINANT (87)	0.81	3 8 45 35 X SQ= 2.20 P = 0.107 RV NO (YES)
141	TILT LESS TOWARD BEING THOSE WHERE THE PARTY SYSTEM IS QUANTITATIVELY OTHER THAN TWO-PARTY (87)	0.83	141	TILT MORE TOWARD BEING THOSE WHERE THE PARTY SYSTEM IS QUANTITATIVELY OTHER THAN TWO-PARTY (87)	0.95	8 2 40 39 X SQ= 2.01 P = 0.100 RV NO (NO)
142	TEND LESS TO BE THOSE WHERE THE PARTY SYSTEM IS QUANTITATIVELY OTHER THAN MULTI-PARTY (66)	0.54	142	TEND MORE TO BE THOSE WHERE THE PARTY SYSTEM IS QUANTITATIVELY OTHER THAN MULTI-PARTY (66)	0.80	21 8 25 33 X SQ= 5.54 P = 0.012 RV NO (YES)
144	TEND LESS TO BE THOSE WHERE THE PARTY SYSTEM IS QUANTITATIVELY ONE-PARTY, RATHER THAN TWO-PARTY (34)	0.58	144	TEND MORE TO BE THOSE WHERE THE PARTY SYSTEM IS QUANTITATIVELY ONE-PARTY, RATHER THAN TWO-PARTY (34)	0.90	11 19 8 2 X SQ= 4.04 P = 0.028 RV NO (YES)
146	TILT MORE TOWARD BEING THOSE WHERE THE PARTY SYSTEM IS QUALITATIVELY OTHER THAN MASS-BASED TERRITORIAL (92)	0.96	146	TILT LESS TOWARD BEING THOSE WHERE THE PARTY SYSTEM IS QUALITATIVELY OTHER THAN MASS-BASED TERRITORIAL (92)	0.86	2 6 47 36 X SQ= 1.80 P = 0.137 RV NO (YES)
147	TEND TO BE THOSE WHERE THE PARTY SYSTEM IS QUALITATIVELY CLASS-ORIENTED OR MULTI-IDEOLOGICAL (23)	0.51	147	ALWAYS ARE THOSE WHERE THE PARTY SYSTEM IS QUALITATIVELY OTHER THAN CLASS-ORIENTED OR MULTI-IDEOLOGICAL (67)	1.00	22 0 21 37 X SQ= 23.61 P = 0. RV YES (YES)
148	ALWAYS ARE THOSE WHERE THE PARTY SYSTEM IS QUALITATIVELY OTHER THAN AFRICAN TRANSITIONAL (96)	1.00	148	TEND LESS TO BE THOSE WHERE THE PARTY SYSTEM IS QUALITATIVELY OTHER THAN AFRICAN TRANSITIONAL (96)	0.69	0 14 55 31 X SQ= 17.40 P = 0. RV NO (YES)
153	TEND TO BE THOSE WHERE THE PARTY SYSTEM IS STABLE (42)	0.64	153	TEND TO BE THOSE WHERE THE PARTY SYSTEM IS MODERATELY STABLE OR UNSTABLE (38)	0.71	M 30 7 17 17 X SQ= 6.32 P = 0.011 RV YES (YES)
154	LEAN TOWARD BEING THOSE WHERE THE PARTY SYSTEM IS STABLE OR MODERATELY STABLE (55)	0.72	154	LEAN TOWARD BEING THOSE WHERE THE PARTY SYSTEM IS UNSTABLE (25)	0.50	34 12 13 12 X SQ= 2.57 P = 0.072 RV YES (NO)

158	TEND TO BE THOSE WHERE PERSONALISMO IS NEGLIGIBLE (56)	0.70	158	TEND TO BE THOSE WHERE PERSONALISMO IS PRONOUNCED OR MODERATE (40)	0.64	15 23 35 13 X SQ= 8.42 P = 0.002 RV YES (YES)
164	TEND TO BE THOSE WHERE THE REGIME'S LEADERSHIP CHARISMA IS NEGLIGIBLE (65)	0.85	164	TEND TO BE THOSE WHERE THE REGIME'S LEADERSHIP CHARISMA IS PRONOUNCED OR MODERATE (34)	0.62	8 23 44 14 X SQ= 18.83 P = 0. RV YES (YES)
168	TEND TO BE THOSE WHERE THE HORIZONTAL POWER DISTRIBUTION IS SIGNIFICANT (34)	0.55	168	TEND TO BE THOSE WHERE THE HORIZONTAL POWER DISTRIBUTION IS LIMITED OR NEGLIGIBLE (72)	0.96	28 2 23 43 X SQ= 26.03 P = 0. RV YES (YES)
169	TEND TO BE THOSE WHERE THE HORIZONTAL POWER DISTRIBUTION IS SIGNIFICANT OR LIMITED (58)	0.75	169	TEND TO BE THOSE WHERE THE HORIZONTAL POWER DISTRIBUTION IS NEGLIGIBLE (48)	0.64	38 16 13 29 X SQ= 13.20 P = 0. RV YES (YES)
170	TEND TO BE THOSE WHERE THE LEGISLATIVE-EXECUTIVE STRUCTURE IS OTHER THAN PRESIDENTIAL (63)	0.76	170	TEND TO BE THOSE WHERE THE LEGISLATIVE-EXECUTIVE STRUCTURE IS PRESIDENTIAL (39)	0.55	M 12 23 38 19 X SQ= 7.91 P = 0.005 RV YES (YES)
171	TILT LESS TOWARD BEING THOSE WHERE THE LEGISLATIVE-EXECUTIVE STRUCTURE IS OTHER THAN PARLIAMENTARY-REPUBLICAN (90)	0.82	171	TILT MORE TOWARD BEING THOSE WHERE THE LEGISLATIVE-EXECUTIVE STRUCTURE IS OTHER THAN PARLIAMENTARY-REPUBLICAN (90)	0.93	9 3 40 41 X SQ= 1.82 P = 0.127 RV NO (YES)
175	TEND TO BE THOSE WHERE THE LEGISLATURE IS FULLY EFFECTIVE OR PARTIALLY EFFECTIVE (51)	0.76	175	TEND TO BE THOSE WHERE THE LEGISLATURE IS LARGELY INEFFECTIVE OR WHOLLY INEFFECTIVE (49)	0.76	37 10 12 31 X SQ= 21.38 P = 0. RV YES (YES)
178	TEND TO BE THOSE WHERE THE LEGISLATURE IS BICAMERAL (51)	0.71	178	TEND TO BE THOSE WHERE THE LEGISLATURE IS UNICAMERAL (53)	0.74	M 15 32 36 11 X SQ= 17.15 P = 0. RV YES (YES)
179	TEND TO BE THOSE WHERE THE EXECUTIVE IS STRONG (39)	0.62	179	TEND TO BE THOSE WHERE THE EXECUTIVE IS DOMINANT (52)	0.82	18 28 29 6 X SQ= 13.86 P = 0. RV YES (YES)

181	TEND LESS TO BE THOSE WHERE THE BUREAUCRACY IS SEMI-MODERN, RATHER THAN MODERN (55)	0.60	181	ALWAYS ARE THOSE WHERE THE BUREAUCRACY IS SEMI-MODERN, RATHER THAN MODERN (55)	1.00	21 0 32 14 X SQ= 6.34 P = 0.003 RV NO (NO)
182	ALWAYS ARE THOSE WHERE THE BUREAUCRACY IS SEMI-MODERN, RATHER THAN POST-COLONIAL TRANSITIONAL (55)	1.00	182	TEND TO BE THOSE WHERE THE BUREAUCRACY IS POST-COLONIAL TRANSITIONAL, RATHER THAN SEMI-MODERN (25)	0.64	32 14 0 25 X SQ= 28.91 P = 0. RV YES (YES)
183	ALWAYS ARE THOSE WHERE THE BUREAUCRACY IS TRADITIONAL, RATHER THAN POST-COLONIAL TRANSITIONAL (9)	1.00	183	LEAN TOWARD BEING THOSE WHERE THE BUREAUCRACY IS POST-COLONIAL TRANSITIONAL, RATHER THAN TRADITIONAL (25)	0.78	0 25 2 7 X SQ= 2.57 P = 0.064 RV YES (NO)
187	TEND TO BE THOSE WHERE THE ROLE OF THE POLICE IS NOT POLITICALLY SIGNIFICANT (35)	0.50	187	TEND TO BE THOSE WHERE THE ROLE OF THE POLICE IS POLITICALLY SIGNIFICANT (66)	0.83	M 25 35 25 7 X SQ= 9.76 P = 0.001 RV YES (YES)
188	TEND LESS TO BE THOSE WHERE THE CHARACTER OF THE LEGAL SYSTEM IS OTHER THAN CIVIL LAW (81)	0.59	188	TEND MORE TO BE THOSE WHERE THE CHARACTER OF THE LEGAL SYSTEM IS OTHER THAN CIVIL LAW (81)	0.87	23 6 33 41 X SQ= 8.77 P = 0.002 RV NO (YES)
189	TEND LESS TO BE THOSE WHERE THE CHARACTER OF THE LEGAL SYSTEM IS OTHER THAN COMMON LAW (108)	0.89	189	ALWAYS ARE THOSE WHERE THE CHARACTER OF THE LEGAL SYSTEM IS OTHER THAN COMMON LAW (108)	1.00	6 0 50 49 X SQ= 3.76 P = 0.029 RV NO (NO)
192	TEND MORE TO BE THOSE WHERE THE CHARACTER OF THE LEGAL SYSTEM IS OTHER THAN MUSLIM OR PARTLY MUSLIM (86)	0.87	192	TEND LESS TO BE THOSE WHERE THE CHARACTER OF THE LEGAL SYSTEM IS OTHER THAN MUSLIM OR PARTLY MUSLIM (86)	0.60	7 19 49 28 X SQ= 9.13 P = 0.001 RV NO (YES)
193	TEND TO BE THOSE WHERE THE CHARACTER OF THE LEGAL SYSTEM IS OTHER THAN PARTLY INDIGENOUS (86)	0.98	193	TEND TO BE THOSE WHERE THE CHARACTER OF THE LEGAL SYSTEM IS PARTLY INDIGENOUS (28)	0.54	1 26 55 22 X SQ= 34.22 P = 0. RV YES (YES)

M

MATRIX

	29 POLITIES WHERE THE DEGREE OF URBANIZATION IS HIGH (56)	29 POLITIES WHERE THE DEGREE OF URBANIZATION IS LOW (49)
63 POLITIES WHERE THE RELIGION IS PREDOMINANTLY SOME KIND OF CHRISTIAN (46)	39: ARGENTINA, BULGARIA, CZECHOS'KIA, GERMANY, E, IRELAND, NETHERLANDS, POLAND, UK, AUSTRALIA, CANADA, DENMARK, GERMAN FR, ITALY, NEW ZEALAND, RUMANIA, US, AUSTRIA, CHILE, DOMIN REP, GREECE, JAMAICA, NICARAGUA, SPAIN, URUGUAY, BELGIUM, COLOMBIA, FINLAND, HUNGARY, LUXEMBOURG, NORWAY, SWEDEN, VENEZUELA, BRAZIL, CUBA, FRANCE, ICELAND, MEXICO, PANAMA, SWITZERLAND	3: EL SALVADOR, HONDURAS, YUGOSLAVIA
63 POLITIES WHERE THE RELIGION IS PREDOMINANTLY OR PARTLY OTHER THAN CHRISTIAN (68)	16: ECUADOR, JORDAN, PERU, UAR, IRAN, KOREA REP, SO AFRICA, IRAQ, LEBANON, SYRIA, ISRAEL, LIBYA, TUNISIA, JAPAN, MOROCCO, TURKEY	46: AFGHANISTAN, BURUNDI, CHAD, ETHIOPIA, HAITI, LIBERIA, NEPAL, SA'U ARABIA, TANGANYIKA, YEMEN, ALBANIA, CAMBODIA, CONGO(BRA), GABON, INDIA, MALAGASY R, NIGER, SENEGAL, THAILAND, ALGERIA, CAMEROUN, CONGO(LEO), GHANA, INDONESIA, MALI, NIGERIA, SIERRE LEO, TOGO, BOLIVIA, CEN AFR REP, CYPRUS, GUATEMALA, IVORY COAST, MAURITANIA, PAKISTAN, SOMALIA, UGANDA, BURMA, CEYLON, DAHOMEY, GUINEA, LAOS, MONGOLIA, RWANDA, SUDAN, UPPER VOLTA

M

MATRIX

	29 POLITIES WHERE THE DEGREE OF URBANIZATION IS HIGH (56)	29 POLITIES WHERE THE DEGREE OF URBANIZATION IS LOW (49)
97 POLITIES WHERE THE STATUS OF THE REGIME IS AUTHORITARIAN, RATHER THAN TOTALITARIAN (23)	5 IRAN, JORDAN, KOREA REP, NICARAGUA, UAR	17 AFGHANISTAN, ALGERIA, BURMA, CAMBODIA, CEN AFR REP, EL SALVADOR, ETHIOPIA, GHANA, GUINEA, INDONESIA, LAOS, LIBERIA, NEPAL, PAKISTAN, SA'U ARABIA, SUDAN, THAILAND
97 POLITIES WHERE THE STATUS OF THE REGIME IS TOTALITARIAN, RATHER THAN AUTHORITARIAN (16)	9 BULGARIA, CUBA, CZECHOS'KIA, GERMANY, E, HUNGARY, POLAND, RUMANIA, SPAIN, USSR	3 ALBANIA, MONGOLIA, YUGOSLAVIA

MATRIX

29 POLITIES WHERE THE DEGREE OF URBANIZATION IS HIGH (56)	29 POLITIES WHERE THE DEGREE OF URBANIZATION IS LOW (49)
153 POLITIES WHERE THE PARTY SYSTEM IS STABLE (42)	
AUSTRALIA, CZECHOS'KIA, HUNGARY, LUXEMBOURG, POLAND, SWITZERLAND, AUSTRIA, DENMARK, ICELAND, MEXICO, RUMANIA, TUNISIA, BELGIUM, FINLAND, IRELAND, NETHERLANDS, SO AFRICA, USSR, BULGARIA, GERMANY, E, ISRAEL, NEW ZEALAND, SPAIN, UK, CANADA, GERMAN FR, JAPAN, NORWAY, SWEDEN, US	ALBANIA, TANGANYIKA, GUINEA, YUGOSLAVIA, INDIA, LIBERIA, MONGOLIA
30	7
17	17
153 POLITIES WHERE THE PARTY SYSTEM IS MODERATELY STABLE OR UNSTABLE (38)	
ARGENTINA, GREECE, LEBANON, TURKEY, BRAZIL, IRAQ, NICARAGUA, VENEZUELA, CHILE, ITALY, PANAMA, ECUADOR, JORDAN, PERU, FRANCE, KOREA REP, SYRIA	BOLIVIA, GHANA, MALAGASY R, SOMALIA, CAMBODIA, GUATEMALA, NEPAL, THAILAND, CEYLON, HONDURAS, PAKISTAN, CONGO(LEO), IVORY COAST, SENEGAL, EL SALVADOR, LAOS, SIERRE LEO

```
***************************************************MATRIX***************************************************

        29  POLITIES                                              29  POLITIES
            WHERE THE DEGREE OF URBANIZATION                          WHERE THE DEGREE OF URBANIZATION
            IS HIGH  (56)                                             IS LOW  (49)

                              170  POLITIES
                                   WHERE THE LEGISLATIVE-EXECUTIVE STRUCTURE
                                   IS PRESIDENTIAL  (39)

ARGENTINA  BRAZIL     CHILE      COLOMBIA  DOMIN REP .BOLIVIA   CAMEROUN    CEN AFR REP CONGO(BRA) CYPRUS
ECUADOR    MEXICO     NICARAGUA  PANAMA    TUNISIA   .DAHOMEY   EL SALVADOR GABON       GHANA      GUATEMALA
US         VENEZUELA                                 .GUINEA    HAITI       HONDURAS    INDONESIA  IVORY COAST
                                                     .LIBERIA   MALAGASY R  MAURITANIA  NIGER      PAKISTAN
                                                     .RWANDA    SENEGAL     UPPER VOLTA

                                                  12 .  23
. . . . . . . . . . . . . . . . . . . . . . . . . . . . . . . . . . . . . . . . . . . . . . . . . . . . . . .
                                                  38 .  19

                              170  POLITIES
                                   WHERE THE LEGISLATIVE-EXECUTIVE STRUCTURE
                                   IS OTHER THAN PRESIDENTIAL  (63)

AUSTRALIA   AUSTRIA     BELGIUM  BULGARIA    CANADA      .AFGHANISTAN ALBANIA    BURUNDI  CAMBODIA   CEYLON
CZECHOS'KIA DENMARK     FINLAND  FRANCE      GERMANY, E  .CHAD        CONGO(LEO) ETHIOPIA INDIA      LAOS
GERMAN FR   GREECE      HUNGARY  ICELAND     IRAN        .MALI        MONGOLIA   NEPAL    NIGERIA    SIERRE LEO
IRELAND     ISRAEL      ITALY    JAMAICA     JAPAN       .SOMALIA     THAILAND   UGANDA   YUGOSLAVIA
JORDAN      LEBANON     LIBYA    LUXEMBOURG  MOROCCO
NETHERLANDS NEW ZEALAND NORWAY   POLAND      RUMANIA
SO AFRICA   SPAIN       SWEDEN   SWITZERLAND TURKEY
USSR        UK          URUGUAY
```

M

MATRIX

	29 POLITIES WHERE THE DEGREE OF URBANIZATION IS HIGH (56)	29 POLITIES WHERE THE DEGREE OF URBANIZATION IS LOW (49)
178 POLITIES WHERE THE LEGISLATURE IS UNICAMERAL (53)	15	32
178 POLITIES WHERE THE LEGISLATURE IS BICAMERAL (51)	36	11

178 POLITIES WHERE THE LEGISLATURE IS UNICAMERAL (53)

Degree of urbanization high (15):
BULGARIA, GREECE, PANAMA, CZECHOS'KIA, HUNGARY, POLAND, DENMARK, ISRAEL, RUMANIA, FINLAND, LEBANON, SPAIN, GERMANY, E, NEW ZEALAND, TUNISIA

Degree of urbanization low (32):
ALBANIA, CHAD, GABON, HONDURAS, MONGOLIA, SENEGAL, UGANDA, ALGERIA, CONGO(BRA), GHANA, INDONESIA, NEPAL, SIERRE LEO, UPPER VOLTA, BURUNDI, CONGO(LEO), GUATEMALA, IVORY COAST, NIGER, SOMALIA, CAMEROUN, DAHOMEY, GUINEA, MALI, PAKISTAN, TANGANYIKA, CEN AFR REP, EL SALVADOR, HAITI, MAURITANIA, RWANDA, TOGO

178 POLITIES WHERE THE LEGISLATURE IS BICAMERAL (51)

Degree of urbanization high (36):
ARGENTINA, CANADA, FRANCE, ITALY, LUXEMBOURG, NORWAY, TURKEY, VENEZUELA, AUSTRALIA, CHILE, GERMAN FR, JAMAICA, MEXICO, PERU, USSR, AUSTRIA, COLOMBIA, ICELAND, JAPAN, MOROCCO, SO AFRICA, UK, BELGIUM, DOMIN REP, IRAN, JORDAN, NETHERLANDS, SWEDEN, US, BRAZIL, ECUADOR, IRELAND, LIBYA, NICARAGUA, SWITZERLAND, URUGUAY

Degree of urbanization low (11):
AFGHANISTAN, INDIA, YUGOSLAVIA, BOLIVIA, LAOS, CAMBODIA, LIBERIA, CEYLON, MALAGASY R, ETHIOPIA, NIGERIA

```
*******************************************************MATRIX*******************************************************
        29  POLITIES                                                 29  POLITIES
            WHERE THE DEGREE OF URBANIZATION          .                  WHERE THE DEGREE OF URBANIZATION
            IS HIGH  (56)                              .                  IS LOW  (49)

                                 187  POLITIES
                                      WHERE THE ROLE OF THE POLICE
                                      IS POLITICALLY SIGNIFICANT  (66)

ARGENTINA  BULGARIA    COLOMBIA   CUBA        CZECHCS'KIA.AFGHANISTAN ALBANIA     ALGERIA     BOLIVIA     BURMA
ECUADOR    GERMANY, E  HUNGARY    IRAN        IRAQ       .CAMBODIA    CAMERCUN    CEN AFR REP CONGO(BRA)  CONGO(LEO)
JORDAN     KOREA REP   LEBANON    MOROCCO     NICARAGUA  .DAHOMEY     EL SALVADOR ETHIOPIA    GABON       GHANA
POLAND     RUMANIA     SO AFRICA  SPAIN       SYRIA      .GUATEMALA   GUINEA      HAITI       HONDURAS    INDONESIA
TUNISIA    TURKEY      USSR       UAR         VENEZUELA  .IVORY COAST LAOS        LIBERIA     MONGOLIA    NEPAL
                                                         .NIGERIA     PAKISTAN    SA'U ARABIA SENEGAL     SOMALIA
                                                         .SUDAN       THAILAND    TOGO        YEMEN       YUGOSLAVIA
                                                      25 .  35
. . . . . . . . . . . . . . . . . . . . . . . . . . . . . . . . . . . . . . . . . . . . . . . . . . . . . . . . . .
                                                      25 .   7

                                 187  POLITIES
                                      WHERE THE ROLE OF THE POLICE
                                      IS NOT POLITICALLY SIGNIFICANT  (35)

AUSTRALIA  AUSTRIA     BELGIUM     CANADA      DENMARK   .CEYLON      INDIA       MALAGASY R  MAURITANIA  SIERRE LEO
FINLAND    FRANCE      GERMAN FR   GREECE      ICELAND   .TANGANYIKA  UGANDA
IRELAND    ISRAEL      ITALY       JAMAICA     JAPAN     .
LUXEMBOURG MEXICO      NETHERLANDS NEW ZEALAND NORWAY    .
SWEDEN     SWITZERLAND UK          US          URUGUAY   .
```

30 POLITIES
WHOSE AGRICULTURAL POPULATION
IS HIGH (56)

30 POLITIES
WHOSE AGRICULTURAL POPULATION
IS MEDIUM, LOW, OR VERY LOW (57)

BOTH SUBJECT AND PREDICATE

56 IN LEFT COLUMN

AFGHANISTAN	ALBANIA	ALGERIA	BOLIVIA	BURMA	BURUNDI	CAMBODIA	CAMEROUN	CEN AFR REP	CHAD
CHINA, PR	CONGO(BRA)	CONGO(LEO)	DAHOMEY	ETHIOPIA	GABON	GHANA	GUATEMALA	GUINEA	HAITI
HONDURAS	INDIA	INDONESIA	IRAN	IRAQ	IVORY COAST	KOREA, N	KOREA REP	LAOS	LIBERIA
LIBYA	MALAGASY R	MALI	MAURITANIA	MONGOLIA	MOROCCO	NEPAL	NICARAGUA	NIGER	RUMANIA
RWANDA	SENEGAL	SIERRE LEO	SOMALIA	SUDAN	SYRIA	TANGANYIKA	THAILAND	TOGO	TUNISIA
TURKEY	UGANDA	UPPER VOLTA	VIETNAM REP	YEMEN	YUGOSLAVIA				

57 IN RIGHT COLUMN

ARGENTINA	AUSTRALIA	AUSTRIA	BELGIUM	BRAZIL	BULGARIA	CANADA	CEYLON	CHILE	COLOMBIA
COSTA RICA	CUBA	CYPRUS	CZECHOS'KIA	DENMARK	DOMIN REP	ECUADOR	EL SALVADOR	FINLAND	FRANCE
GERMANY, E	GERMAN FR	GREECE	HUNGARY	ICELAND	IRELAND	ISRAEL	ITALY	JAMAICA	JAPAN
JORDAN	LEBANON	LUXEMBOURG	MALAYA	MEXICO	NETHERLANDS	NEW ZEALAND	NIGERIA	NORWAY	PAKISTAN
PANAMA	PARAGUAY	PERU	PHILIPPINES	POLAND	PORTUGAL	SO AFRICA	SPAIN	SWEDEN	SWITZERLAND
TRINIDAD	USSR	UAR	UK	US	URUGUAY	VENEZUELA			

2 EXCLUDED BECAUSE UNASCERTAINED

SA'U ARABIA VIETNAM, N

---

2 ALWAYS ARE THOSE LOCATED ELSEWHERE THAN IN WEST EUROPE, SCANDINAVIA, NORTH AMERICA, OR AUSTRALASIA (93) — 1.00

2 TEND LESS TO BE THOSE LOCATED ELSEWHERE THAN IN WEST EUROPE, SCANDINAVIA, NORTH AMERICA, OR AUSTRALASIA (93) — 0.61

0	22
56	35

X SQ= 24.43
P = 0.
RV NO (YES)

5 ALWAYS ARE THOSE LOCATED IN EAST EUROPE, RATHER THAN IN SCANDINAVIA OR WEST EUROPE (9) — 1.00

5 TEND TO BE THOSE LOCATED IN SCANDINAVIA OR WEST EUROPE, RATHER THAN IN EAST EUROPE (18) — 0.75

0	18
3	6

X SQ= 3.80
P = 0.029
RV YES (NO)

6 TEND MORE TO BE THOSE LOCATED ELSEWHERE THAN IN THE CARIBBEAN, CENTRAL AMERICA, OR SOUTH AMERICA (93) — 0.91

6 TEND LESS TO BE THOSE LOCATED ELSEWHERE THAN IN THE CARIBBEAN, CENTRAL AMERICA, OR SOUTH AMERICA (93) — 0.70

5	17
51	40

X SQ= 6.59
P = 0.008
RV NO (YES)

8	LEAN LESS TOWARD BEING THOSE LOCATED ELSEWHERE THAN IN EAST ASIA SOUTH ASIA, OR SOUTHEAST ASIA (97)	0.79	8	LEAN MORE TOWARD BEING THOSE LOCATED ELSEWHERE THAN IN EAST ASIA SOUTH ASIA, OR SOUTHEAST ASIA (97)	0.91	12 5 44 52 X SQ= 2.62 P = 0.070 RV NO (YES)
10	TEND TO BE THOSE LOCATED IN NORTH AFRICA, OR CENTRAL AND SOUTH AFRICA (33)	0.54	10	TEND TO BE THOSE LOCATED ELSEWHERE THAN IN NORTH AFRICA, OR CENTRAL AND SOUTH AFRICA (82)	0.95	30 3 26 54 X SQ= 29.59 P = 0. RV YES (YES)
11	TEND LESS TO BE THOSE LOCATED ELSEWHERE THAN IN CENTRAL AND SOUTH AFRICA (87)	0.54	11	TEND MORE TO BE THOSE LOCATED ELSEWHERE THAN IN CENTRAL AND SOUTH AFRICA (87)	0.96	26 2 30 55 X SQ= 25.66 P = 0. RV NO (YES)
15	TEND TO BE THOSE LOCATED IN NORTH AFRICA OR CENTRAL AND SOUTH AFRICA, RATHER THAN IN THE MIDDLE EAST (33)	0.83	15	TEND TO BE THOSE LOCATED IN THE MIDDLE EAST, RATHER THAN IN NORTH AFRICA OR CENTRAL AND SOUTH AFRICA (11)	0.57	6 4 30 3 X SQ= 3.35 P = 0.040 RV YES (NO)
17	LEAN TOWARD BEING THOSE LOCATED IN THE MIDDLE EAST, RATHER THAN IN THE CARIBBEAN, CENTRAL AMERICA, OR SOUTH AMERICA (11)	0.55	17	LEAN TOWARD BEING THOSE LOCATED IN THE CARIBBEAN, CENTRAL AMERICA, OR SOUTH AMERICA, RATHER THAN IN THE MIDDLE EAST (22)	0.81	6 4 5 17 X SQ= 2.74 P = 0.056 RV YES (YES)
18	TILT TOWARD BEING THOSE LOCATED IN NORTH AFRICA OR CENTRAL AND SOUTH AFRICA, RATHER THAN IN EAST ASIA, SOUTH ASIA, OR SOUTHEAST ASIA (33)	0.71	18	TILT TOWARD BEING THOSE LOCATED IN EAST ASIA, SOUTH ASIA, OR SOUTHEAST ASIA, RATHER THAN IN NORTH AFRICA OR CENTRAL AND SOUTH AFRICA (18)	0.62	30 3 12 5 X SQ= 2.10 P = 0.102 RV YES (NO)
19	TEND TO BE THOSE LOCATED IN NORTH AFRICA OR CENTRAL AND SOUTH AFRICA, RATHER THAN IN THE CARIBBEAN, CENTRAL AMERICA, OR SOUTH AMERICA (33)	0.86	19	TEND TO BE THOSE LOCATED IN THE CARIBBEAN, CENTRAL AMERICA, OR SOUTH AMERICA, RATHER THAN IN NORTH AFRICA OR CENTRAL AND SOUTH AFRICA (22)	0.85	30 3 5 17 X SQ= 23.65 P = 0. RV YES (YES)
20	TEND TO BE THOSE LOCATED IN EAST ASIA, SOUTH ASIA, OR SOUTHEAST ASIA, RATHER THAN IN THE CARIBBEAN, CENTRAL AMERICA, OR SOUTH AMERICA (18)	0.71	20	TEND TO BE THOSE LOCATED IN THE CARIBBEAN, CENTRAL AMERICA, OR SOUTH AMERICA, RATHER THAN IN EAST ASIA, SOUTH ASIA, OR SOUTHEAST ASIA (22)	0.77	12 5 5 17 X SQ= 7.09 P = 0.004 RV YES (YES)
22	LEAN MORE TOWARD BEING THOSE WHOSE TERRITORIAL SIZE IS VERY LARGE, LARGE, OR MEDIUM (68)	0.68	22	LEAN LESS TOWARD BEING THOSE WHOSE TERRITORIAL SIZE IS VERY LARGE, LARGE, OR MEDIUM (68)	0.51	38 29 18 28 X SQ= 2.71 P = 0.085 RV NO (YES)

24	TEND TO BE THOSE WHOSE POPULATION IS SMALL (54)	0.59	24	TEND TO BE THOSE WHOSE POPULATION IS VERY LARGE, LARGE, OR MEDIUM (61)	0.63	23 36 33 21 X SQ= 4.67 P = 0.024 RV YES (YES)
26	TEND TO BE THOSE WHOSE POPULATION DENSITY IS LOW (67)	0.75	26	TEND TO BE THOSE WHOSE POPULATION DENSITY IS VERY HIGH, HIGH, OR MEDIUM (48)	0.58	14 33 42 24 X SQ= 11.26 P = 0.001 RV YES (YES)
29	TEND TO BE THOSE WHERE THE DEGREE OF URBANIZATION IS LOW (49)	0.81	29	TEND TO BE THOSE WHERE THE DEGREE OF URBANIZATION IS HIGH (56)	0.90	10 46 43 5 X SQ= 50.38 P = 0. RV YES (YES)
34	TEND TO BE THOSE WHOSE GROSS NATIONAL PRODUCT IS VERY LOW (53)	0.71	34	TEND TO BE THOSE WHOSE GROSS NATIONAL PRODUCT IS VERY HIGH, HIGH, MEDIUM, OR LOW (62)	0.77	16 44 40 13 X SQ= 24.90 P = 0. RV YES (YES)
35	ALWAYS ARE THOSE WHOSE PER CAPITA GROSS NATIONAL PRODUCT IS MEDIUM, LOW, OR VERY LOW (91)	1.00	35	TEND LESS TO BE THOSE WHOSE PER CAPITA GROSS NATIONAL PRODUCT IS MEDIUM, LOW, OR VERY LOW (91)	0.58	0 24 56 33 X SQ= 27.47 P = 0. RV NO (YES)
36	TEND TO BE THOSE WHOSE PER CAPITA GROSS NATIONAL PRODUCT IS LOW OR VERY LOW (73)	0.96	36	TEND TO BE THOSE WHOSE PER CAPITA GROSS NATIONAL PRODUCT IS VERY HIGH, HIGH, OR MEDIUM (42)	0.70	2 40 54 17 X SQ= 50.84 P = 0. RV YES (YES)
37	TEND TO BE THOSE WHOSE PER CAPITA GROSS NATIONAL PRODUCT IS VERY LOW (51)	0.80	37	TEND TO BE THOSE WHOSE PER CAPITA GROSS NATIONAL PRODUCT IS VERY HIGH, HIGH, MEDIUM, OR LOW (64)	0.91	11 52 45 5 X SQ= 55.81 P = 0. RV YES (YES)
39	TEND TO BE THOSE WHOSE INTERNATIONAL FINANCIAL STATUS IS LOW OR VERY LOW (76)	0.89	39	TEND TO BE THOSE WHOSE INTERNATIONAL FINANCIAL STATUS IS VERY HIGH, HIGH, OR MEDIUM (38)	0.57	6 32 50 24 X SQ= 24.89 P = 0. RV YES (YES)
40	TEND TO BE THOSE WHOSE INTERNATIONAL FINANCIAL STATUS IS VERY LOW (39)	0.59	40	TEND TO BE THOSE WHOSE INTERNATIONAL FINANCIAL STATUS IS VERY HIGH, HIGH, MEDIUM, OR LOW (71)	0.87	22 48 32 7 X SQ= 23.69 P = 0. RV YES (YES)

41	ALWAYS ARE THOSE WHOSE ECONOMIC DEVELOPMENTAL STATUS IS INTERMEDIATE, UNDERDEVELOPED, OR VERY UNDERDEVELOPED (94)	1.00	41	TEND LESS TO BE THOSE WHOSE ECONOMIC DEVELOPMENTAL STATUS IS INTERMEDIATE, UNDERDEVELOPED, OR VERY UNDERDEVELOPED (94)	0.65	0 19 56 36 X SQ= 20.97 P = 0. RV NO (YES)
42	TEND TO BE THOSE WHOSE ECONOMIC DEVELOPMENTAL STATUS IS UNDERDEVELOPED OR VERY UNDERDEVELOPED (76)	0.96	42	TEND TO BE THOSE WHOSE ECONOMIC DEVELOPMENTAL STATUS IS DEVELOPED OR INTERMEDIATE (36)	0.62	2 34 54 21 X SQ= 40.34 P = 0. RV YES (YES)
43	TEND TO BE THOSE WHOSE ECONOMIC DEVELOPMENTAL STATUS IS VERY UNDERDEVELOPED (57)	0.86	43	TEND TO BE THOSE WHOSE ECONOMIC DEVELOPMENTAL STATUS IS DEVELOPED, INTERMEDIATE, OR UNDERDEVELOPED (55)	0.85	8 47 48 8 X SQ= 53.41 P = 0. RV YES (YES)
45	TEND TO BE THOSE WHERE THE LITERACY RATE IS BELOW FIFTY PERCENT (54)	0.85	45	TEND TO BE THOSE WHERE THE LITERACY RATE IS FIFTY PERCENT OR ABOVE (55)	0.84	8 47 44 9 X SQ= 47.98 P = 0. RV YES (YES)
47	LEAN TOWARD BEING THOSE WHERE THE LITERACY RATE IS BETWEEN FIFTY AND NINETY PERCENT, RATHER THAN NINETY PERCENT OR ABOVE (30)	0.87	47	LEAN TOWARD BEING THOSE WHERE THE LITERACY RATE IS NINETY PERCENT OR ABOVE, RATHER THAN BETWEEN FIFTY AND NINETY PERCENT (25)	0.51	1 24 7 23 X SQ= 2.69 P = 0.059 RV YES (NO)
48	TEND TO BE THOSE WHERE THE LITERACY RATE IS BETWEEN TEN AND FIFTY PERCENT, RATHER THAN BETWEEN FIFTY AND NINETY PERCENT (24)	0.68	48	TEND TO BE THOSE WHERE THE LITERACY RATE IS BETWEEN FIFTY AND NINETY PERCENT, RATHER THAN BETWEEN TEN AND FIFTY PERCENT (30)	0.72	7 23 15 9 X SQ= 6.93 P = 0.005 RV YES (YES)
49	TEND TO BE THOSE WHERE THE LITERACY RATE IS BELOW TEN PERCENT, RATHER THAN BETWEEN TEN AND FIFTY PERCENT (26)	0.62	49	ALWAYS ARE THOSE WHERE THE LITERACY RATE IS BETWEEN TEN AND FIFTY PERCENT, RATHER THAN BELOW TEN PERCENT (24)	1.00	15 9 25 0 X SQ= 9.12 P = 0.001 RV YES (NO)
50	TEND TO BE THOSE WHERE FREEDOM OF THE PRESS IS INTERMITTENT, INTERNALLY ABSENT, OR INTERNALLY AND EXTERNALLY ABSENT (56)	0.74	50	TEND TO BE THOSE WHERE FREEDOM OF THE PRESS IS COMPLETE (43)	0.59	11 32 32 22 X SQ= 9.68 P = 0.001 RV YES (YES)
51	TEND TO BE THOSE WHERE FREEDOM OF THE PRESS IS INTERNALLY ABSENT, OR INTERNALLY AND EXTERNALLY ABSENT (37)	0.51	51	TEND TO BE THOSE WHERE FREEDOM OF THE PRESS IS COMPLETE OR INTERMITTENT (60)	0.74	20 40 21 14 X SQ= 5.37 P = 0.018 RV YES (YES)

53	ALWAYS ARE THOSE WHERE NEWSPAPER CIRCULATION IS LESS THAN THREE HUNDRED PER THOUSAND (101)	1.00	53	TEND LESS TO BE THOSE WHERE NEWSPAPER CIRCULATION IS LESS THAN THREE HUNDRED PER THOUSAND (101)	0.75	0 14 56 43 X SQ= 13.52 P = 0. RV NO (YES)
54	TEND TO BE THOSE WHERE NEWSPAPER CIRCULATION IS LESS THAN ONE HUNDRED PER THOUSAND (76)	0.96	54	TEND TO BE THOSE WHERE NEWSPAPER CIRCULATION IS ONE HUNDRED OR MORE PER THOUSAND (37)	0.61	2 35 53 22 X SQ= 39.65 P = 0. RV YES (YES)
55	TEND TO BE THOSE WHERE NEWSPAPER CIRCULATION IS LESS THAN TEN PER THOUSAND (35)	0.58	55	TEND TO BE THOSE WHERE NEWSPAPER CIRCULATION IS TEN OR MORE PER THOUSAND (78)	0.96	23 55 32 2 X SQ= 37.03 P = 0. RV YES (YES)
56	TEND TO BE THOSE WHERE THE RELIGION IS PREDOMINANTLY OR PARTLY NON-LITERATE (31)	0.52	56	TEND TO BE THOSE WHERE THE RELIGION IS PREDOMINANTLY LITERATE (79)	0.93	25 53 27 4 X SQ= 24.78 P = 0. RV YES (YES)
57	TEND MORE TO BE THOSE WHERE THE RELIGION IS OTHER THAN CATHOLIC (90)	0.96	57	TEND LESS TO BE THOSE WHERE THE RELIGION IS OTHER THAN CATHOLIC (90)	0.60	2 23 54 34 X SQ= 20.09 P = 0. RV NO (YES)
58	TEND LESS TO BE THOSE WHERE THE RELIGION IS OTHER THAN MUSLIM (97)	0.75	58	TEND MORE TO BE THOSE WHERE THE RELIGION IS OTHER THAN MUSLIM (97)	0.95	14 3 42 54 X SQ= 7.13 P = 0.004 RV NO (YES)
63	TEND TO BE THOSE WHERE THE RELIGION IS PREDOMINANTLY OR PARTLY OTHER THAN CHRISTIAN (68)	0.93	63	TEND TO BE THOSE WHERE THE RELIGION IS PREDOMINANTLY SOME KIND OF CHRISTIAN (46)	0.75	4 42 52 14 X SQ= 50.50 P = 0. RV YES (YES)
66	TILT TOWARD BEING THOSE THAT ARE RELIGIOUSLY HETEROGENEOUS (49)	0.55	66	TILT TOWARD BEING THOSE THAT ARE RELIGIOUSLY HOMOGENEOUS (57)	0.61	M 22 34 27 22 X SQ= 2.03 P = 0.120 RV YES (YES)
68	TEND TO BE THOSE THAT ARE LINGUISTICALLY WEAKLY HETEROGENEOUS OR STRONGLY HETEROGENEOUS (62)	0.75	68	TEND TO BE THOSE THAT ARE LINGUISTICALLY HOMOGENEOUS (52)	0.65	14 37 41 20 X SQ= 16.02 P = 0. RV YES (YES)

```
69  TEND TO BE THOSE                          0.64     69  TEND TO BE THOSE                          0.74          20   42
    THAT ARE LINGUISTICALLY                                THAT ARE LINGUISTICALLY                                35   15
    STRONGLY HETEROGENEOUS  (50)                           HOMOGENEOUS OR                                  X SQ=  14.30
                                                           WEAKLY HETEROGENEOUS  (64)                      P =     0.
                                                                                                           RV YES (YES)

70  TEND MORE TO BE THOSE                     0.92     70  TEND LESS TO BE THOSE                     0.69           4   17
    THAT ARE RELIGIOUSLY, RACIALLY,                        THAT ARE RELIGIOUSLY, RACIALLY,                        45   38
    OR LINGUISTICALLY HETEROGENEOUS  (85)                  OR LINGUISTICALLY HETEROGENEOUS  (85)           X SQ=   6.97
                                                                                                           P =     0.006
                                                                                                           RV NO  (YES)

72  TEND TO BE THOSE                          0.72     72  TEND TO BE THOSE                          0.67          15   37
    WHOSE DATE OF INDEPENDENCE                             WHOSE DATE OF INDEPENDENCE                             39   18
    IS AFTER 1914  (59)                                    IS BEFORE 1914  (52)                            X SQ=  15.49
                                                                                                           P =     0.
                                                                                                           RV YES (YES)

73  TEND TO BE THOSE                          0.63     73  TEND TO BE THOSE                          0.80          20   44
    WHOSE DATE OF INDEPENDENCE                             WHOSE DATE OF INDEPENDENCE                             34   11
    IS AFTER 1945  (46)                                    IS BEFORE 1945  (65)                            X SQ=  19.01
                                                                                                           P =     0.
                                                                                                           RV YES (YES)

74  ALWAYS ARE THOSE                          1.00     74  TEND LESS TO BE THOSE                     0.53           0   26
    THAT ARE NOT HISTORICALLY WESTERN  (87)                THAT ARE NOT HISTORICALLY WESTERN  (87)                56   29
                                                                                                           X SQ=  31.98
                                                                                                           P =     0.
                                                                                                           RV NO  (YES)

75  TEND TO BE THOSE                          0.82     75  TEND TO BE THOSE                          0.93          10   52
    THAT ARE NOT HISTORICALLY WESTERN AND                  THAT ARE HISTORICALLY WESTERN OR                       46    4
    ARE NOT SIGNIFICANTLY WESTERNIZED  (52)                SIGNIFICANTLY WESTERNIZED  (62)                 X SQ=  60.73
                                                                                                           P =     0.
                                                                                                           RV YES (YES)

76  ALWAYS ARE THOSE                          1.00     76  TEND LESS TO BE THOSE                     0.51           0   26
    THAT HAVE BEEN SIGNIFICANTLY OR                        THAT HAVE BEEN SIGNIFICANTLY OR                        42   27
    PARTIALLY WESTERNIZED THROUGH A                        PARTIALLY WESTERNIZED THROUGH A                 X SQ=  25.95
    COLONIAL RELATIONSHIP, RATHER THAN                     COLONIAL RELATIONSHIP, RATHER THAN              P =     0.
    BEING HISTORICALLY WESTERN  (70)                       BEING HISTORICALLY WESTERN  (70)                RV NO  (YES)

77  TEND TO BE THOSE                          0.86     77  TEND TO BE THOSE                          0.85           6   22
    THAT HAVE BEEN PARTIALLY WESTERNIZED,                  THAT HAVE BEEN SIGNIFICANTLY WESTERNIZED,              36    4
    RATHER THAN SIGNIFICANTLY WESTERNIZED,                 RATHER THAN PARTIALLY WESTERNIZED,              X SQ=  29.95
    THROUGH A COLONIAL RELATIONSHIP  (41)                  THROUGH A COLONIAL RELATIONSHIP  (28)           P =     0.
                                                                                                           RV YES (YES)

79  TEND MORE TO BE THOSE                     0.89     79  TEND LESS TO BE THOSE                     0.55          31   17
    THAT WERE FORMERLY DEPENDENCIES                        THAT WERE FORMERLY DEPENDENCIES                         4   14
    OF BRITAIN OR FRANCE,                                  OF BRITAIN OR FRANCE,                           X SQ=   7.81
    RATHER THAN SPAIN  (49)                                RATHER THAN SPAIN  (49)                         P =     0.003
                                                                                                           RV NO  (YES)
```

80	TEND TO BE THOSE WHOSE DATE OF INDEPENDENCE IS AFTER 1914, AND THAT WERE FORMERLY DEPENDENCIES OF FRANCE, RATHER THAN BRITAIN (24)	0.73	80	TEND TO BE THOSE WHOSE DATE OF INDEPENDENCE IS AFTER 1914, AND THAT WERE FORMERLY DEPENDENCIES OF BRITAIN, RATHER THAN FRANCE (19)	0.92	8 11 22 1 X SQ= 12.11 P = 0. RV YES (YES)
82	TEND TO BE THOSE WHERE THE TYPE OF POLITICAL MODERNIZATION IS DEVELOPED TUTELARY OR UNDEVELOPED TUTELARY, RATHER THAN EARLY OR LATER EUROPEAN OR EUROPEAN DERIVED (55)	0.84	82	TEND TO BE THOSE WHERE THE TYPE OF POLITICAL MODERNIZATION IS EARLY OR LATER EUROPEAN OR EUROPEAN DERIVED, RATHER THAN DEVELOPED TUTELARY OR UNDEVELOPED TUTELARY (51)	0.78	8 43 41 12 X SQ= 37.24 P = 0. RV YES (YES)
83	LEAN LESS TOWARD BEING THOSE WHERE THE TYPE OF POLITICAL MODERNIZATION IS OTHER THAN NON-EUROPEAN AUTOCHTHONOUS (106)	0.87	83	LEAN MORE TOWARD BEING THOSE WHERE THE TYPE OF POLITICAL MODERNIZATION IS OTHER THAN NON-EUROPEAN AUTOCHTHONOUS (106)	0.96	7 2 49 55 X SQ= 2.01 P = 0.094 RV NO (YES)
84	TEND TO BE THOSE WHERE THE TYPE OF POLITICAL MODERNIZATION IS UNDEVELOPED TUTELARY, RATHER THAN DEVELOPED TUTELARY (24)	0.56	84	TEND TO BE THOSE WHERE THE TYPE OF POLITICAL MODERNIZATION IS DEVELOPED TUTELARY, RATHER THAN UNDEVELOPED TUTELARY (31)	0.92	18 11 23 1 X SQ= 6.73 P = 0.007 RV YES (NO)
85	TEND TO BE THOSE WHERE THE STAGE OF POLITICAL MODERNIZATION IS MID- OR EARLY TRANSITIONAL, RATHER THAN ADVANCED (54)	0.71	85	TEND TO BE THOSE WHERE THE STAGE OF POLITICAL MODERNIZATION IS ADVANCED, RATHER THAN MID- OR EARLY TRANSITIONAL (60)	0.77	16 44 40 13 X SQ= 24.90 P = 0. RV YES (YES)
86	TEND TO BE THOSE WHERE THE STAGE OF POLITICAL MODERNIZATION IS EARLY TRANSITIONAL, RATHER THAN ADVANCED OR MID-TRANSITIONAL (38)	0.57	86	TEND TO BE THOSE WHERE THE STAGE OF POLITICAL MODERNIZATION IS ADVANCED OR MID-TRANSITIONAL, RATHER THAN EARLY TRANSITIONAL (76)	0.91	24 52 32 5 X SQ= 27.86 P = 0. RV YES (YES)
87	TEND TO BE THOSE WHOSE IDEOLOGICAL ORIENTATION IS DEVELOPMENTAL (31)	0.63	87	TEND TO BE THOSE WHOSE IDEOLOGICAL ORIENTATION IS OTHER THAN DEVELOPMENTAL (58)	0.89	26 5 15 41 X SQ= 23.86 P = 0. RV YES (YES)
89	TEND TO BE THOSE WHOSE IDEOLOGICAL ORIENTATION IS OTHER THAN CONVENTIONAL (62)	0.96	89	TEND TO BE THOSE WHOSE IDEOLOGICAL ORIENTATION IS CONVENTIONAL (33)	0.67	2 31 45 15 X SQ= 37.77 P = 0. RV YES (YES)
93	TEND LESS TO BE THOSE WHERE THE SYSTEM STYLE IS NON-MOBILIZATIONAL (78)	0.60	93	TEND MORE TO BE THOSE WHERE THE SYSTEM STYLE IS NON-MOBILIZATIONAL (78)	0.82	21 10 32 45 X SQ= 5.06 P = 0.019 RV NO (YES)

94	TEND TO BE THOSE WHERE THE STATUS OF THE REGIME IS AUTHORITARIAN OR TOTALITARIAN (41)	0.69	94	TEND TO BE THOSE WHERE THE STATUS OF THE REGIME IS CONSTITUTIONAL (51)	0.74	11 40 25 14 X SQ= 14.93 P = 0. RV YES (YES)
96	TEND TO BE THOSE WHERE THE STATUS OF THE REGIME IS AUTHORITARIAN (23)	0.50	96	TEND TO BE THOSE WHERE THE STATUS OF THE REGIME IS CONSTITUTIONAL OR TOTALITARIAN (67)	0.91	17 49 17 5 X SQ= 16.36 P = 0. RV YES (YES)
97	TEND TO BE THOSE WHERE THE STATUS OF THE REGIME IS AUTHORITARIAN, RATHER THAN TOTALITARIAN (23)	0.74	97	TEND TO BE THOSE WHERE THE STATUS OF THE REGIME IS TOTALITARIAN, RATHER THAN AUTHORITARIAN (16)	0.64	17 5 6 9 X SQ= 3.80 P = 0.038 RV YES (YES)
99	TEND TO BE THOSE WHERE GOVERNMENTAL STABILITY IS MODERATELY PRESENT AND DATES FROM THE POST-WAR PERIOD, OR IS ABSENT (36)	0.60	99	TEND TO BE THOSE WHERE GOVERNMENTAL STABILITY IS GENERALLY PRESENT AND DATES FROM AT LEAST THE INTER-WAR PERIOD, OR FROM THE POST-WAR PERIOD (50)	0.69	14 34 21 15 X SQ= 6.05 P = 0.013 RV YES (YES)
101	TEND TO BE THOSE WHERE THE REPRESENTATIVE CHARACTER OF THE REGIME IS LIMITED POLYARCHIC, PSEUDO-POLYARCHIC, OR NON-POLYARCHIC (57)	0.84	101	TEND TO BE THOSE WHERE THE REPRESENTATIVE CHARACTER OF THE REGIME IS POLYARCHIC (41)	0.64	7 34 36 19 X SQ= 20.32 P = 0. RV YES (YES)
102	TEND TO BE THOSE WHERE THE REPRESENTATIVE CHARACTER OF THE REGIME IS PSEUDO-POLYARCHIC OR NON-POLYARCHIC (49)	0.66	102	TEND TO BE THOSE WHERE THE REPRESENTATIVE CHARACTER OF THE REGIME IS POLYARCHIC OR LIMITED POLYARCHIC (59)	0.75	17 42 33 14 X SQ= 16.37 P = 0. RV YES (YES)
105	TEND TO BE THOSE WHERE THE ELECTORAL SYSTEM IS PARTIALLY COMPETITIVE OR NON-COMPETITIVE (47)	0.83	105	TEND TO BE THOSE WHERE THE ELECTORAL SYSTEM IS COMPETITIVE (43)	0.75	7 36 34 12 X SQ= 27.44 P = 0. RV YES (YES)
106	TEND TO BE THOSE WHERE THE ELECTORAL SYSTEM IS NON-COMPETITIVE (30)	0.57	106	TEND TO BE THOSE WHERE THE ELECTORAL SYSTEM IS COMPETITIVE OR PARTIALLY COMPETITIVE (52)	0.80	15 37 20 9 X SQ= 10.63 P = 0.001 RV YES (YES)
107	TEND TO BE THOSE WHERE AUTONOMOUS GROUPS ARE PARTIALLY TOLERATED IN POLITICS, ARE TOLERATED ONLY OUTSIDE POLITICS, OR ARE NOT TOLERATED AT ALL (65)	0.87	107	TEND TO BE THOSE WHERE AUTONOMOUS GROUPS ARE FULLY TOLERATED IN POLITICS (46)	0.70	7 39 46 17 X SQ= 33.28 P = 0. RV YES (YES)

108	TEND TO BE THOSE WHERE AUTONOMOUS GROUPS ARE TOLERATED ONLY OUTSIDE POLITICS OR ARE NOT TOLERATED AT ALL (35)	0.51	108	TEND TO BE THOSE WHERE AUTONOMOUS GROUPS ARE FULLY OR PARTIALLY TOLERATED IN POLITICS (65)	0.81	22 43 23 10 X SQ= 9.93 P = 0.001 RV YES (YES)
111	TILT TOWARD BEING THOSE WHERE POLITICAL ENCULTURATION IS LOW (42)	0.52	111	TILT TOWARD BEING THOSE WHERE POLITICAL ENCULTURATION IS HIGH OR MEDIUM (53)	0.65	23 30 25 16 X SQ= 2.20 P = 0.101 RV YES (YES)
114	LEAN TOWARD BEING THOSE WHERE SECTIONALISM IS EXTREME OR MODERATE (61)	0.67	114	LEAN TOWARD BEING THOSE WHERE SECTIONALISM IS NEGLIGIBLE (47)	0.52	35 26 17 28 X SQ= 3.23 P = 0.052 RV YES (YES)
115	ALWAYS ARE THOSE WHERE INTEREST ARTICULATION BY ASSOCIATIONAL GROUPS IS MODERATE, LIMITED, OR NEGLIGIBLE (91)	1.00	115	TEND LESS TO BE THOSE WHERE INTEREST ARTICULATION BY ASSOCIATIONAL GROUPS IS MODERATE, LIMITED, OR NEGLIGIBLE (91)	0.64	0 20 54 35 X SQ= 21.68 P = 0. RV NO (YES)
116	TEND TO BE THOSE WHERE INTEREST ARTICULATION BY ASSOCIATIONAL GROUPS IS LIMITED OR NEGLIGIBLE (79)	0.98	116	TEND TO BE THOSE WHERE INTEREST ARTICULATION BY ASSOCIATIONAL GROUPS IS SIGNIFICANT OR MODERATE (32)	0.56	M 1 31 53 24 X SQ= 36.46 P = 0. RV YES (YES)
117	TEND TO BE THOSE WHERE INTEREST ARTICULATION BY ASSOCIATIONAL GROUPS IS NEGLIGIBLE (51)	0.76	117	TEND TO BE THOSE WHERE INTEREST ARTICULATION BY ASSOCIATIONAL GROUPS IS SIGNIFICANT, MODERATE, OR LIMITED (60)	0.85	13 47 41 8 X SQ= 39.04 P = 0. RV YES (YES)
118	TEND TO BE THOSE WHERE INTEREST ARTICULATION BY INSTITUTIONAL GROUPS IS VERY SIGNIFICANT (40)	0.54	118	TEND TO BE THOSE WHERE INTEREST ARTICULATION BY INSTITUTIONAL GROUPS IS SIGNIFICANT, MODERATE, OR LIMITED (60)	0.70	M 22 17 19 40 X SQ= 4.70 P = 0.022 RV YES (YES)
121	TEND TO BE THOSE WHERE INTEREST ARTICULATION BY NON-ASSOCIATIONAL GROUPS IS SIGNIFICANT (54)	0.77	121	TEND TO BE THOSE WHERE INTEREST ARTICULATION BY NON-ASSOCIATIONAL GROUPS IS MODERATE, LIMITED, OR NEGLIGIBLE (61)	0.84	M 43 9 13 48 X SQ= 39.89 P = 0. RV YES (YES)
122	ALWAYS ARE THOSE WHERE INTEREST ARTICULATION BY NON-ASSOCIATIONAL GROUPS IS SIGNIFICANT OR MODERATE (83)	1.00	122	TEND TO BE THOSE WHERE INTEREST ARTICULATION BY NON-ASSOCIATIONAL GROUPS IS LIMITED OR NEGLIGIBLE (32)	0.56	56 25 0 32 X SQ= 41.14 P = 0. RV YES (YES)

No.	Statement	Value	No.	Statement	Value	Statistics
125	TEND TO BE THOSE WHERE INTEREST ARTICULATION BY ANOMIC GROUPS IS FREQUENT OR OCCASIONAL (64)	0.86	125	TEND TO BE THOSE WHERE INTEREST ARTICULATION BY ANOMIC GROUPS IS INFREQUENT OR VERY INFREQUENT (35)	0.58	**M** 43 20 7 28 X SQ= 19.08 P = 0. RV YES (YES)
126	ALWAYS ARE THOSE WHERE INTEREST ARTICULATION BY ANOMIC GROUPS IS FREQUENT, OCCASIONAL, OR INFREQUENT (83)	1.00	126	TEND LESS TO BE THOSE WHERE INTEREST ARTICULATION BY ANOMIC GROUPS IS FREQUENT, OCCASIONAL, OR INFREQUENT (83)	0.67	50 32 0 16 X SQ= 17.55 P = 0. RV NO (YES)
128	TEND TO BE THOSE WHERE INTEREST ARTICULATION BY POLITICAL PARTIES IS LIMITED OR NEGLIGIBLE (45)	0.68	128	TEND TO BE THOSE WHERE INTEREST ARTICULATION BY POLITICAL PARTIES IS SIGNIFICANT OR MODERATE (48)	0.69	13 35 28 16 X SQ= 10.98 P = 0.001 RV YES (YES)
129	TEND TO BE THOSE WHERE INTEREST ARTICULATION BY POLITICAL PARTIES IS NEGLIGIBLE (37)	0.56	129	TEND TO BE THOSE WHERE INTEREST ARTICULATION BY POLITICAL PARTIES IS SIGNIFICANT, MODERATE, OR LIMITED (56)	0.75	**M** 18 38 23 13 X SQ= 7.70 P = 0.005 RV YES (YES)
133	TEND LESS TO BE THOSE WHERE INTEREST AGGREGATION BY THE EXECUTIVE IS MODERATE, LIMITED, OR NEGLIGIBLE (76)	0.57	133	TEND MORE TO BE THOSE WHERE INTEREST AGGREGATION BY THE EXECUTIVE IS MODERATE, LIMITED, OR NEGLIGIBLE (76)	0.87	22 7 29 45 X SQ= 9.79 P = 0.001 RV NO (YES)
136	ALWAYS ARE THOSE WHERE INTEREST AGGREGATION BY THE LEGISLATURE IS MODERATE, LIMITED, OR NEGLIGIBLE (85)	1.00	136	TEND LESS TO BE THOSE WHERE INTEREST AGGREGATION BY THE LEGISLATURE IS MODERATE, LIMITED, OR NEGLIGIBLE (85)	0.76	0 12 46 38 X SQ= 10.52 P = 0. RV NO (YES)
137	TEND TO BE THOSE WHERE INTEREST AGGREGATION BY THE LEGISLATURE IS LIMITED OR NEGLIGIBLE (68)	0.96	137	TEND TO BE THOSE WHERE INTEREST AGGREGATION BY THE LEGISLATURE IS SIGNIFICANT OR MODERATE (29)	0.54	2 27 44 23 X SQ= 25.71 P = 0. RV YES (YES)
138	TEND TO BE THOSE WHERE INTEREST AGGREGATION BY THE LEGISLATURE IS NEGLIGIBLE (49)	0.78	138	TEND TO BE THOSE WHERE INTEREST AGGREGATION BY THE LEGISLATURE IS SIGNIFICANT, MODERATE, OR LIMITED (48)	0.76	10 38 36 12 X SQ= 26.09 P = 0. RV YES (YES)
139	TEND TO BE THOSE WHERE THE PARTY SYSTEM IS QUANTITATIVELY ONE-PARTY (34)	0.50	139	TEND TO BE THOSE WHERE THE PARTY SYSTEM IS QUANTITATIVELY OTHER THAN ONE-PARTY (71)	0.82	23 10 23 47 X SQ= 10.87 P = 0.001 RV YES (YES)

140	LEAN LESS TOWARD BEING THOSE WHERE THE PARTY SYSTEM IS QUANTITATIVELY OTHER THAN ONE PARTY DOMINANT (87)	0.80	140	LEAN MORE TOWARD BEING THOSE WHERE THE PARTY SYSTEM IS QUANTITATIVELY OTHER THAN ONE PARTY DOMINANT (87)	0.93	9 4 35 50 X SQ= 2.54 P = 0.075 RV NO (YES)
141	TILT MORE TOWARD BEING THOSE WHERE THE PARTY SYSTEM IS QUANTITATIVELY OTHER THAN TWO-PARTY (87)	0.95	141	TILT LESS TOWARD BEING THOSE WHERE THE PARTY SYSTEM IS QUANTITATIVELY OTHER THAN TWO-PARTY (87)	0.83	2 9 41 44 X SQ= 2.45 P = 0.104 RV NO (NO)
142	TEND TO BE THOSE WHERE THE PARTY SYSTEM IS QUANTITATIVELY OTHER THAN MULTI-PARTY (66)	0.90	142	TEND TO BE THOSE WHERE THE PARTY SYSTEM IS QUANTITATIVELY MULTI-PARTY (30)	0.50	4 26 38 26 X SQ= 15.70 P = 0. RV YES (YES)
146	TEND LESS TO BE THOSE WHERE THE PARTY SYSTEM IS QUALITATIVELY OTHER THAN MASS-BASED TERRITORIAL (92)	0.84	146	TEND MORE TO BE THOSE WHERE THE PARTY SYSTEM IS QUALITATIVELY OTHER THAN MASS-BASED TERRITORIAL (92)	0.98	7 1 38 53 X SQ= 4.50 P = 0.022 RV NO (YES)
147	ALWAYS ARE THOSE WHERE THE PARTY SYSTEM IS QUALITATIVELY OTHER THAN CLASS-ORIENTED OR MULTI-IDEOLOGICAL (67)	1.00	147	TEND LESS TO BE THOSE WHERE THE PARTY SYSTEM IS QUALITATIVELY OTHER THAN CLASS-ORIENTED OR MULTI-IDEOLOGICAL (67)	0.53	0 23 40 26 X SQ= 22.93 P = 0. RV NO (YES)
148	TEND LESS TO BE THOSE WHERE THE PARTY SYSTEM IS QUALITATIVELY OTHER THAN AFRICAN TRANSITIONAL (96)	0.73	148	ALWAYS ARE THOSE WHERE THE PARTY SYSTEM IS QUALITATIVELY OTHER THAN AFRICAN TRANSITIONAL (96)	1.00	14 0 38 57 X SQ= 15.28 P = 0. RV NO (YES)
153	TEND TO BE THOSE WHERE THE PARTY SYSTEM IS MODERATELY STABLE OR UNSTABLE (38)	0.65	153	TEND TO BE THOSE WHERE THE PARTY SYSTEM IS STABLE (42)	0.62	11 30 20 18 X SQ= 4.48 P = 0.023 RV YES (YES)
158	TEND TO BE THOSE WHERE PERSONALISMO IS PRONOUNCED OR MODERATE (40)	0.57	158	TEND TO BE THOSE WHERE PERSONALISMO IS NEGLIGIBLE (56)	0.70	24 16 18 37 X SQ= 5.92 P = 0.012 RV YES (YES)
164	TEND TO BE THOSE WHERE THE REGIME'S LEADERSHIP CHARISMA IS PRONOUNCED OR MODERATE (34)	0.60	164	TEND TO BE THOSE WHERE THE REGIME'S LEADERSHIP CHARISMA IS NEGLIGIBLE (65)	0.89	26 6 17 48 X SQ= 24.19 P = 0. RV YES (YES)

166	TEND MORE TO BE THOSE WHERE THE VERTICAL POWER DISTRIBUTION IS THAT OF FORMAL FEDERALISM OR FORMAL AND EFFECTIVE UNITARISM (99)	0.95	166	TEND LESS TO BE THOSE WHERE THE VERTICAL POWER DISTRIBUTION IS THAT OF FORMAL FEDERALISM OR FORMAL AND EFFECTIVE UNITARISM (99)	0.79	3 12 52 45 X SQ= 4.60 P = 0.024 RV NO (YES)
168	TEND TO BE THOSE WHERE THE HORIZONTAL POWER DISTRIBUTION IS LIMITED OR NEGLIGIBLE (72)	0.96	168	TEND TO BE THOSE WHERE THE HORIZONTAL POWER DISTRIBUTION IS SIGNIFICANT (34)	0.58	2 32 47 23 X SQ= 32.05 P = 0. RV YES (YES)
169	TEND TO BE THOSE WHERE THE HORIZONTAL POWER DISTRIBUTION IS NEGLIGIBLE (48)	0.67	169	TEND TO BE THOSE WHERE THE HORIZONTAL POWER DISTRIBUTION IS SIGNIFICANT OR LIMITED (58)	0.76	16 42 33 13 X SQ= 18.34 P = 0. RV YES (YES)
170	LEAN LESS TOWARD BEING THOSE WHERE THE LEGISLATIVE-EXECUTIVE STRUCTURE IS OTHER THAN PRESIDENTIAL (63)	0.51	170	LEAN MORE TOWARD BEING THOSE WHERE THE LEGISLATIVE-EXECUTIVE STRUCTURE IS OTHER THAN PRESIDENTIAL (63)	0.70	23 16 24 38 X SQ= 3.18 P = 0.065 RV NO (YES)
172	TEND MORE TO BE THOSE WHERE THE LEGISLATIVE-EXECUTIVE STRUCTURE IS OTHER THAN PARLIAMENTARY-ROYALIST (88)	0.92	172	TEND LESS TO BE THOSE WHERE THE LEGISLATIVE-EXECUTIVE STRUCTURE IS OTHER THAN PARLIAMENTARY-ROYALIST (88)	0.70	4 17 48 39 X SQ= 7.45 P = 0.003 RV NO (YES)
174	TEND TO BE THOSE WHERE THE LEGISLATURE IS PARTIALLY EFFECTIVE, LARGELY INEFFECTIVE, OR WHOLLY INEFFECTIVE (72)	0.98	174	TEND TO BE THOSE WHERE THE LEGISLATURE IS FULLY EFFECTIVE (28)	0.51	1 27 45 26 X SQ= 26.52 P = 0. RV YES (YES)
175	TEND TO BE THOSE WHERE THE LEGISLATURE IS LARGELY INEFFECTIVE OR WHOLLY INEFFECTIVE (49)	0.78	175	TEND TO BE THOSE WHERE THE LEGISLATURE IS FULLY EFFECTIVE OR PARTIALLY EFFECTIVE (51)	0.77	10 41 36 12 X SQ= 28.31 P = 0. RV YES (YES)
178	TEND TO BE THOSE WHERE THE LEGISLATURE IS UNICAMERAL (53)	0.71	178	TEND TO BE THOSE WHERE THE LEGISLATURE IS BICAMERAL (51)	0.69	35 17 14 37 X SQ= 14.84 P = 0. RV YES (YES)
179	TEND TO BE THOSE WHERE THE EXECUTIVE IS DOMINANT (52)	0.85	179	TEND TO BE THOSE WHERE THE EXECUTIVE IS STRONG (39)	0.66	33 17 6 33 X SQ= 20.79 P = 0. RV YES (YES)

181	ALWAYS ARE THOSE WHERE THE BUREAUCRACY IS SEMI-MODERN, RATHER THAN MODERN (55)	1.00	181	TEND LESS TO BE THOSE WHERE THE BUREAUCRACY IS SEMI-MODERN, RATHER THAN MODERN (55)	0.61	0 21 21 33 X SQ= 9.50 P = 0. RV NO (NO)
182	TEND TO BE THOSE WHERE THE BUREAUCRACY IS POST-COLONIAL TRANSITIONAL, RATHER THAN SEMI-MODERN (25)	0.53	182	TEND TO BE THOSE WHERE THE BUREAUCRACY IS SEMI-MODERN, RATHER THAN POST-COLONIAL TRANSITIONAL (55)	0.97	21 33 24 1 X SQ= 20.47 P = 0. RV YES (YES)
186	TILT LESS TOWARD BEING THOSE WHERE PARTICIPATION BY THE MILITARY IN POLITICS IS NEUTRAL, RATHER THAN SUPPORTIVE (56)	0.56	186	TILT MORE TOWARD BEING THOSE WHERE PARTICIPATION BY THE MILITARY IN POLITICS IS NEUTRAL, RATHER THAN SUPPORTIVE (56)	0.74	17 12 22 34 X SQ= 2.15 P = 0.111 RV NO (YES)
187	TEND TO BE THOSE WHERE THE ROLE OF THE POLICE IS POLITICALLY SIGNIFICANT (66)	0.88	187	TEND TO BE THOSE WHERE THE ROLE OF THE POLICE IS NOT POLITICALLY SIGNIFICANT (35)	0.58	43 21 6 29 X SQ= 20.71 P = 0. RV YES (YES)
188	TEND MORE TO BE THOSE WHERE THE CHARACTER OF THE LEGAL SYSTEM IS OTHER THAN CIVIL LAW (81)	0.87	188	TEND LESS TO BE THOSE WHERE THE CHARACTER OF THE LEGAL SYSTEM IS OTHER THAN CIVIL LAW (81)	0.56	7 25 47 32 X SQ= 11.44 P = 0. RV NO (YES)
189	ALWAYS ARE THOSE WHERE THE CHARACTER OF THE LEGAL SYSTEM IS OTHER THAN COMMON LAW (108)	1.00	189	TEND LESS TO BE THOSE WHERE THE CHARACTER OF THE LEGAL SYSTEM IS OTHER THAN COMMON LAW (108)	0.88	0 7 56 50 X SQ= 5.37 P = 0.013 RV NO (YES)
192	TEND LESS TO BE THOSE WHERE THE CHARACTER OF THE LEGAL SYSTEM IS OTHER THAN MUSLIM OR PARTLY MUSLIM (86)	0.61	192	TEND MORE TO BE THOSE WHERE THE CHARACTER OF THE LEGAL SYSTEM IS OTHER THAN MUSLIM OR PARTLY MUSLIM (86)	0.91	21 5 33 52 X SQ= 12.39 P = 0. RV NO (YES)
193	TEND LESS TO BE THOSE WHERE THE CHARACTER OF THE LEGAL SYSTEM IS OTHER THAN PARTLY INDIGENOUS (86)	0.53	193	TEND MORE TO BE THOSE WHERE THE CHARACTER OF THE LEGAL SYSTEM IS OTHER THAN PARTLY INDIGENOUS (86)	0.96	26 2 29 55 X SQ= 26.31 P = 0. RV NO (YES)

**********MATRIX**********

	30 POLITIES WHOSE AGRICULTURAL POPULATION IS HIGH (56)	30 POLITIES WHOSE AGRICULTURAL POPULATION IS MEDIUM, LOW, OR VERY LOW (57)
66 POLITIES THAT ARE RELIGIOUSLY HOMOGENEOUS (57)	AFGHANISTAN, ALGERIA, BURMA, CAMBODIA, HONDURAS, INDIA, INDONESIA, IRAN, IRAQ, LIBYA, MAURITANIA, MONGOLIA, MOROCCO, NICARAGUA, RUMANIA, SENEGAL, SOMALIA, SYRIA, THAILAND, TUNISIA, TURKEY, YEMEN	ARGENTINA, AUSTRIA, BELGIUM, BRAZIL, BULGARIA, CHILE, COLOMBIA, COSTA RICA, CUBA, DENMARK, DOMIN REP, EL SALVADOR, FINLAND, FRANCE, GREECE, ICELAND, IRELAND, ISRAEL, ITALY, JORDAN, LUXEMBOURG, MEXICO, NORWAY, PAKISTAN, PANAMA, PARAGUAY, PHILIPPINES, POLAND, PORTUGAL, SPAIN, SWEDEN, UAR, URUGUAY, VENEZUELA
	22	34
	27	22
66 POLITIES THAT ARE RELIGIOUSLY HETEROGENEOUS (49)	ALBANIA, BOLIVIA, CAMEROUN, CHAD, CONGO(BRA), CONGO(LEO), DAHOMEY, ETHIOPIA, GABON, GHANA, GUATEMALA, GUINEA, HAITI, IVORY COAST, LAOS, LIBERIA, MALAGASY R, MALI, NEPAL, NIGER, SIERRE LEO, SUDAN, TANGANYIKA, TOGO, UGANDA, UPPER VOLTA, YUGOSLAVIA	AUSTRALIA, CANADA, CEYLON, CYPRUS, CZECHOS'KIA, ECUADOR, GERMANY, E, GERMAN FR, HUNGARY, JAMAICA, JAPAN, LEBANON, MALAYA, NETHERLANDS, NEW ZEALAND, NIGERIA, PERU, SO AFRICA, SWITZERLAND, TRINIDAD, UK, US

M

```
*****************************************************MATRIX*****************************************************

          30  POLITIES                                                   30  POLITIES
              WHOSE AGRICULTURAL POPULATION                       .          WHOSE AGRICULTURAL POPULATION
              IS HIGH  (56)                                       .          IS MEDIUM, LOW, OR VERY LOW  (57)

                                   101  POLITIES
                                        WHERE THE REPRESENTATIVE CHARACTER
                                        OF THE REGIME IS POLYARCHIC  (41)

BOLIVIA     INDIA       MALAGASY R  MOROCCO     SIERRE LEO .AUSTRALIA  AUSTRIA     BELGIUM     CANADA      CEYLON
TUNISIA     TURKEY                                         .COSTA RICA CYPRUS      DENMARK     DOMIN REP   FINLAND
                                                           .FRANCE     GERMAN FR   GREECE      ICELAND     IRELAND
                                                           .ISRAEL     ITALY       JAMAICA     JAPAN       LUXEMBOURG
                                                           .MALAYA     MEXICO      NETHERLANDS NEW ZEALAND NORWAY
                                                           .PANAMA     PHILIPPINES SWEDEN      SWITZERLAND TRINIDAD
                                                           .UK         US          URUGUAY     VENEZUELA
                                                           .
                                                          7 .  34
. . . . . . . . . . . . . . . . . . . . . . . . . . . . . . . . . . . . . . . . . . . . . . . . . . . . . . . . . . .
                                                         36 .  19

                                   101  POLITIES
                                        WHERE THE REPRESENTATIVE CHARACTER
                                        OF THE REGIME IS LIMITED POLYARCHIC,
                                        PSEUDO-POLYARCHIC, OR
                                        NON-POLYARCHIC  (57)

AFGHANISTAN ALBANIA     ALGERIA     BURMA       CAMBODIA   .BRAZIL     BULGARIA    CHILE       COLOMBIA    CUBA
CEN AFR REP CHAD        CHINA, PR   CONGO(BRA)  DAHOMEY    .CZECHOS'KIA ECUADOR    EL SALVADOR GERMANY, E  HUNGARY
ETHIOPIA    GABON       GHANA       GUINEA      HAITI      .JORDAN     PAKISTAN    PARAGUAY    POLAND      PORTUGAL
HONDURAS    INDONESIA   IRAN        IVORY COAST KOREA, N   .SO AFRICA  SPAIN       USSR        UAR
LAOS        LIBERIA     MALI        MAURITANIA  MONGOLIA   .
NEPAL       NICARAGUA   NIGER       RUMANIA     SENEGAL    .
SUDAN       TANGANYIKA  THAILAND    UPPER VOLTA VIETNAM REP.
YUGOSLAVIA                                                 .
```

M

MATRIX

30 POLITIES WHOSE AGRICULTURAL POPULATION IS HIGH (56)	30 POLITIES WHOSE AGRICULTURAL POPULATION IS MEDIUM, LOW, OR VERY LOW (57)
116 POLITIES WHERE INTEREST ARTICULATION BY ASSOCIATIONAL GROUPS IS SIGNIFICANT OR MODERATE (32)	
TURKEY	ARGENTINA, CANADA, GERMAN FR, ITALY, NETHERLANDS, SWEDEN, VENEZUELA, AUSTRALIA, CHILE, GREECE, JAPAN, NEW ZEALAND, SWITZERLAND, AUSTRIA, DENMARK, ICELAND, LEBANON, NORWAY, UK, BELGIUM, FINLAND, IRELAND, LUXEMBOURG, PHILIPPINES, US, BRAZIL, FRANCE, ISRAEL, MEXICO, SO AFRICA, URUGUAY
1	31
53	24
116 POLITIES WHERE INTEREST ARTICULATION BY ASSOCIATIONAL GROUPS IS LIMITED OR NEGLIGIBLE (79)	
AFGHANISTAN, BURUNDI, CHINA, PR, GABON, HONDURAS, IVORY COAST, MALAGASY R, NEPAL, SENEGAL, TANGANYIKA, UPPER VOLTA, ALBANIA, CAMBODIA, CONGO(BRA), GHANA, INDIA, KOREA, N, MALI, NICARAGUA, SIERRE LEO, THAILAND, VIETNAM REP, ALGERIA, CAMEROUN, CONGO(LEO), GUATEMALA, INDONESIA, LAOS, MAURITANIA, NIGER, SOMALIA, TOGO, YEMEN, BOLIVIA, CEN AFR REP, DAHOMEY, GUINEA, IRAN, LIBERIA, MONGOLIA, RUMANIA, SUDAN, TUNISIA, BURMA, CHAD, ETHIOPIA, HAITI, IRAQ, LIBYA, MOROCCO, RWANDA, SYRIA, UGANDA	BULGARIA, CYPRUS, GERMANY, E, NIGERIA, POLAND, CEYLON, CZECHOS'KIA, HUNGARY, PAKISTAN, TRINIDAD, COLOMBIA, DOMIN REP, JAMAICA, PANAMA, USSR, CUBA, ECUADOR, JORDAN, PARAGUAY, UAR, COSTA RICA, EL SALVADOR, MALAYA, PERU

MATRIX

30 POLITIES WHOSE AGRICULTURAL POPULATION IS HIGH (56)	30 POLITIES WHOSE AGRICULTURAL POPULATION IS MEDIUM, LOW, OR VERY LOW (57)
118 POLITIES WHERE INTEREST ARTICULATION BY INSTITUTIONAL GROUPS IS VERY SIGNIFICANT (40)	
AFGHANISTAN, ALBANIA, ALGERIA, BURMA, CHINA, PR, ETHIOPIA, GHANA, GUINEA, INDONESIA, IRAN, IRAQ, KOREA, N, KOREA REP, LAOS, LIBERIA, MONGOLIA, RUMANIA, SUDAN, SYRIA, THAILAND, VIETNAM REP, YUGOSLAVIA	ARGENTINA, BRAZIL, BULGARIA, CUBA, CYPRUS, CZECHOS'KIA, GERMANY, E, HUNGARY, JORDAN, LEBANON, PAKISTAN, PARAGUAY, POLAND, PORTUGAL, SPAIN, USSR, UAR
22	17
19	40
118 POLITIES WHERE INTEREST ARTICULATION BY INSTITUTIONAL GROUPS IS SIGNIFICANT, MODERATE, OR LIMITED (60)	
BOLIVIA, CAMBODIA, CONGO(LEO), GUATEMALA, HAITI, HONDURAS, INDIA, IVORY COAST, LIBYA, MALAGASY R, MAURITANIA, MOROCCO, NEPAL, NICARAGUA, SENEGAL, TANGANYIKA, TUNISIA, TURKEY, UGANDA	AUSTRALIA, AUSTRIA, BELGIUM, CANADA, CEYLON, CHILE, COLOMBIA, COSTA RICA, DENMARK, DOMIN REP, ECUADOR, EL SALVADOR, FINLAND, FRANCE, GERMAN FR, GREECE, ICELAND, IRELAND, ISRAEL, ITALY, JAMAICA, JAPAN, LUXEMBOURG, MALAYA, MEXICO, NETHERLANDS, NEW ZEALAND, NIGERIA, NORWAY, PANAMA, PERU, PHILIPPINES, SO AFRICA, SWEDEN, SWITZERLAND, TRINIDAD, UK, US, URUGUAY, VENEZUELA

M

```
*****************************************************MATRIX*****************************************************

          30  POLITIES                                              30  POLITIES
              WHOSE AGRICULTURAL POPULATION                             WHOSE AGRICULTURAL POPULATION
              IS HIGH  (56)                                             IS MEDIUM, LOW, OR VERY LOW  (57)

                              121  POLITIES
                                   WHERE INTEREST ARTICULATION
                                   BY NON-ASSOCIATIONAL GROUPS
                                   IS SIGNIFICANT  (54)

AFGHANISTAN BURMA       BURUNDI     CAMBODIA    CAMEROUN    .CEYLON      CYPRUS      JORDAN      LEBANON     MALAYA
CEN AFR REP CHAD        CHINA, PR   CONGO(BRA)  CONGO(LEO)  .NIGERIA     PAKISTAN    PHILIPPINES SO AFRICA
DAHOMEY     ETHIOPIA    GABON       GHANA       GUINEA      .
HAITI       INDIA       INDONESIA   IRAN        IRAQ        .
IVORY COAST KOREA, N    KOREA REP   LAOS        MALAGASY R  .
MALI        MAURITANIA  MONGOLIA    NEPAL       NIGER       .
RWANDA      SENEGAL     SIERRE LEO  SOMALIA     SUDAN       .
SYRIA       TANGANYIKA  THAILAND    TOGO        UGANDA      .
UPPER VOLTA VIETNAM REP YEMEN                               .
                                                         43 .  9
. . . . . . . . . . . . . . . . . . . . . . . . . . . . . . . . . . . . . . . . . . . . . . . . . . . . . . . . .
                                                         13 .  48

                              121  POLITIES
                                   WHERE INTEREST ARTICULATION
                                   BY NON-ASSOCIATIONAL GROUPS
                                   IS MODERATE, LIMITED, OR
                                   NEGLIGIBLE  (61)

ALBANIA     ALGERIA     BOLIVIA     GUATEMALA   HONDURAS    .ARGENTINA   AUSTRALIA   AUSTRIA     BELGIUM     BRAZIL
LIBERIA     LIBYA       MOROCCO     NICARAGUA   RUMANIA     .BULGARIA    CANADA      CHILE       COLOMBIA    CUBA
TUNISIA     TURKEY      YUGOSLAVIA                          .COSTA RICA  CZECHOS'KIA DENMARK     DOMIN REP   ECUADOR
                                                            .EL SALVADOR FINLAND     FRANCE      GERMANY, E  GERMAN FR
                                                            .GREECE      HUNGARY     ICELAND     IRELAND     ISRAEL
                                                            .ITALY       JAMAICA     JAPAN       LUXEMBOURG  MEXICO
                                                            .NETHERLANDS NEW ZEALAND NORWAY      PANAMA      PARAGUAY
                                                            .PERU        POLAND      PORTUGAL    SPAIN       SWEDEN
                                                            .SWITZERLAND TRINIDAD    USSR        UAR         UK
```

MATRIX

	30 POLITIES WHOSE AGRICULTURAL POPULATION IS HIGH (56)	30 POLITIES WHOSE AGRICULTURAL POPULATION IS MEDIUM, LOW, OR VERY LOW (57)
125 POLITIES WHERE INTEREST ARTICULATION BY ANOMIC GROUPS IS FREQUENT OR OCCASIONAL (64)	AFGHANISTAN, BOLIVIA, BURMA, CAMBODIA, CAMEROUN, CEN AFR REP, CHAD, CHINA, PR, CONGO(BRA), CONGO(LEO), DAHOMEY, ETHIOPIA, GABON, GHANA, GUATEMALA, GUINEA, HAITI, HONDURAS, INDIA, INDONESIA, IRAN, IRAQ, IVORY COAST, KOREA REP, LAOS, MALI, MAURITANIA, MOROCCO, NEPAL, NIGER, RWANDA, SENEGAL, SIERRE LEO, SOMALIA, SUDAN, SYRIA, THAILAND, TOGO, TURKEY, UGANDA, UPPER VOLTA, VIETNAM REP, YUGOSLAVIA 43	BRAZIL, CEYLON, COLOMBIA, ECUADOR, EL SALVADOR, GERMANY, E, JAPAN, JORDAN, LEBANON, MALAYA, MEXICO, NIGERIA, PAKISTAN, PANAMA, PARAGUAY, PERU, POLAND, SO AFRICA, SPAIN, USSR 20
125 POLITIES WHERE INTEREST ARTICULATION BY ANOMIC GROUPS IS INFREQUENT OR VERY INFREQUENT (35)	7 BURUNDI, LIBERIA, LIBYA, MALAGASY R, NICARAGUA, TANGANYIKA, TUNISIA	28 AUSTRALIA, AUSTRIA, BELGIUM, CANADA, CHILE, COSTA RICA, CZECHOS'KIA, DENMARK, FINLAND, GERMAN FR, GREECE, ICELAND, IRELAND, ISRAEL, ITALY, JAMAICA, LUXEMBOURG, NETHERLANDS, NEW ZEALAND, NORWAY, PHILIPPINES, PORTUGAL, SWEDEN, SWITZERLAND, TRINIDAD, UAR, UK, URUGUAY

MATRIX

	30 POLITIES WHOSE AGRICULTURAL POPULATION IS HIGH (56)	30 POLITIES WHOSE AGRICULTURAL POPULATION IS MEDIUM, LOW, OR VERY LOW (57)
129 POLITIES WHERE INTEREST ARTICULATION BY POLITICAL PARTIES IS SIGNIFICANT, MODERATE, OR LIMITED (56)	18	38
129 POLITIES WHERE INTEREST ARTICULATION BY POLITICAL PARTIES IS NEGLIGIBLE (37)	23	13

Agricultural population high; interest articulation by parties significant, moderate, or limited (18):

BOLIVIA, CAMEROUN, CONGO(LEO), GUATEMALA, HONDURAS, INDIA, LIBERIA, MALAGASY R, MAURITANIA, MOROCCO, NICARAGUA, RWANDA, SIERRE LEO, SOMALIA, SYRIA, TOGO, TURKEY, UGANDA

Agricultural population medium, low, or very low; interest articulation by parties significant, moderate, or limited (38):

ARGENTINA, AUSTRALIA, AUSTRIA, BELGIUM, BRAZIL, CEYLON, CHILE, COLOMBIA, COSTA RICA, DENMARK, DOMIN REP, ECUADOR, EL SALVADOR, FINLAND, FRANCE, GERMAN FR, GREECE, ICELAND, IRELAND, ISRAEL, ITALY, JAMAICA, JAPAN, LEBANON, LUXEMBOURG, MALAYA, NETHERLANDS, NEW ZEALAND, NIGERIA, NORWAY, PANAMA, PARAGUAY, PERU, SWEDEN, SWITZERLAND, TRINIDAD, UK, VENEZUELA

Agricultural population high; interest articulation by parties negligible (23):

ALBANIA, BURUNDI, CAMBODIA, CEN AFR REP, CHAD, CHINA, PR, CONGO(BRA), DAHOMEY, GABON, GHANA, GUINEA, IVORY COAST, KOREA, N, MALI, MONGOLIA, NIGER, RUMANIA, SENEGAL, TANGANYIKA, TUNISIA, UPPER VOLTA, VIETNAM REP, YUGOSLAVIA

Agricultural population medium, low, or very low; interest articulation by parties negligible (13):

BULGARIA, CANADA, CZECHOS'KIA, GERMANY, E, HUNGARY, MEXICO, PHILIPPINES, POLAND, PORTUGAL, SPAIN, USSR, UAR, US

M MATRIX

	30 POLITIES WHOSE AGRICULTURAL POPULATION IS HIGH (56)	30 POLITIES WHOSE AGRICULTURAL POPULATION IS MEDIUM, LOW, OR VERY LOW (57)
158 POLITIES WHERE PERSONALISMO IS PRONOUNCED OR MODERATE (40)	24	16
158 POLITIES WHERE PERSONALISMO IS NEGLIGIBLE (56)	18	37

Agricultural population high / personalismo pronounced or moderate (24):
BOLIVIA, CONGO(BRA), GUATEMALA, LIBERIA, SYRIA, CAMBODIA, CONGO(LEO), HONDURAS, MALI, THAILAND, CAMEROUN, DAHOMEY, INDONESIA, MAURITANIA, UPPER VOLTA, CEN AFR REP, GABON, IVORY COAST, NICARAGUA, VIETNAM REP, CHAD, GHANA, LAOS, NIGER

Agricultural population medium, low, or very low / personalismo pronounced or moderate (16):
ARGENTINA, COSTA RICA, JAPAN, URUGUAY, BRAZIL, ECUADOR, LEBANON, CEYLON, EL SALVADOR, PANAMA, CHILE, FRANCE, PERU, COLOMBIA, GREECE, UAR

Agricultural population high / personalismo negligible (18):
ALBANIA, KOREA, N, SENEGAL, TURKEY, ALGERIA, MALAGASY R, SIERRE LEO, UGANDA, CHINA, PR, MONGOLIA, SOMALIA, YUGOSLAVIA, GUINEA, MOROCCO, TANGANYIKA, INDIA, RUMANIA, TUNISIA

Agricultural population medium, low, or very low / personalismo negligible (37):
AUSTRALIA, CUBA, GERMAN FR, ITALY, NETHERLANDS, PHILIPPINES, SWEDEN, US, AUSTRIA, CZECHOS'KIA, HUNGARY, JAMAICA, NEW ZEALAND, POLAND, SWITZERLAND, VENEZUELA, BELGIUM, DENMARK, ICELAND, LUXEMBOURG, NIGERIA, PORTUGAL, TRINIDAD, BULGARIA, FINLAND, IRELAND, MALAYA, NORWAY, SO AFRICA, USSR, CANADA, GERMANY, E, ISRAEL, MEXICO, PARAGUAY, SPAIN, UK

MATRIX

	30 POLITIES WHOSE AGRICULTURAL POPULATION IS HIGH (56)	30 POLITIES WHOSE AGRICULTURAL POPULATION IS MEDIUM, LOW, OR VERY LOW (57)
175 POLITIES WHERE THE LEGISLATURE IS FULLY EFFECTIVE OR PARTIALLY EFFECTIVE (51)	10	41
175 POLITIES WHERE THE LEGISLATURE IS LARGELY INEFFECTIVE OR WHOLLY INEFFECTIVE (49)	36	12

High agricultural population; legislature fully or partially effective (10):

BOLIVIA, CAMEROUN, HONDURAS, INDIA, LIBYA, MALAGASY R, MORCCCO, SIERRE LEO, TURKEY, UGANDA

Medium, low, or very low agricultural population; legislature fully or partially effective (41):

ARGENTINA, AUSTRALIA, AUSTRIA, BELGIUM, BRAZIL, CANADA, CEYLON, CHILE, COLOMBIA, COSTA RICA, CYPRUS, DENMARK, ECUADOR, FINLAND, FRANCE, GERMAN FR, GREECE, ICELAND, IRELAND, ISRAEL, ITALY, JAMAICA, JAPAN, LEBANON, LUXEMBOURG, MALAYA, MEXICO, NETHERLANDS, NEW ZEALAND, NIGERIA, NORWAY, PANAMA, PHILIPPINES, SO AFRICA, SWEDEN, SWITZERLAND, TRINIDAD, UK, US, URUGUAY, VENEZUELA

High agricultural population; legislature largely or wholly ineffective (36):

AFGHANISTAN, ALBANIA, ALGERIA, BURUNDI, CAMBODIA, CEN AFR REP, CHAD, CHINA, PR, CONGO(BRA), DAHOMEY, ETHIOPIA, GABON, GHANA, GUINEA, HAITI, INDONESIA, IVORY COAST, KOREA, N, LAOS, LIBERIA, MALI, MAURITANIA, MONGOLIA, NEPAL, NICARAGUA, NIGER, RUMANIA, RWANDA, SENEGAL, SOMALIA, TANGANYIKA, THAILAND, TUNISIA, UPPER VOLTA, VIETNAM REP, YUGOSLAVIA

Medium, low, or very low agricultural population; legislature largely or wholly ineffective (12):

BULGARIA, CZECHOS'KIA, EL SALVADOR, GERMANY, E, HUNGARY, JORDAN, PARAGUAY, POLAND, PORTUGAL, SPAIN, USSR, UAR

31 POLITIES
WHOSE AGRICULTURAL POPULATION
IS HIGH OR MEDIUM (90)

31 POLITIES
WHOSE AGRICULTURAL POPULATION
IS LOW OR VERY LOW (24)

PREDICATE ONLY

90 IN LEFT COLUMN

AFGHANISTAN	ALBANIA	ALGERIA	BOLIVIA	BRAZIL	BULGARIA	BURMA	BURUNDI	CAMBODIA	CAMEROUN
CEN AFR REP	CEYLON	CHAD	CHINA, PR	COLOMBIA	CONGO(BRA)	CONGO(LEO)	COSTA RICA	CUBA	CYPRUS
CZECHOS'KIA	DAHOMEY	DOMIN REP	ECUADOR	EL SALVADOR	ETHIOPIA	FINLAND	GABON	GHANA	GREECE
GUATEMALA	GUINEA	HAITI	HONDURAS	HUNGARY	INDIA	INDONESIA	IRAN	IRAQ	IRELAND
IVORY COAST	JAMAICA	JAPAN	JORDAN	KOREA, N	KOREA REP	LAOS	LEBANON	LIBERIA	LIBYA
MALAGASY R	MALAYA	MALI	MAURITANIA	MEXICO	MONGOLIA	MOROCCO	NEPAL	NICARAGUA	NIGER
NIGERIA	PAKISTAN	PANAMA	PARAGUAY	PERU	PHILIPPINES	POLAND	PORTUGAL	RUMANIA	RWANDA
SENEGAL	SIERRE LEO	SOMALIA	SPAIN	SUDAN	SYRIA	TANGANYIKA	THAILAND	TOGO	TUNISIA
TURKEY	UGANDA	USSR	UAR	UPPER VOLTA	VENEZUELA	VIETNAM, N	VIETNAM REP	YEMEN	YUGOSLAVIA

24 IN RIGHT COLUMN

ARGENTINA	AUSTRALIA	AUSTRIA	BELGIUM	CANADA	CHILE	DENMARK	FRANCE	GERMANY, E	GERMAN FR
ICELAND	ISRAEL	ITALY	LUXEMBOURG	NETHERLANDS	NEW ZEALAND	NORWAY	SO AFRICA	SWEDEN	SWITZERLAND
TRINIDAD	UK	US	URUGUAY						

1 EXCLUDED BECAUSE UNASCERTAINED

SA'U ARABIA

32 POLITIES
WHOSE GROSS NATIONAL PRODUCT
IS VERY HIGH OR HIGH (10)

32 POLITIES
WHOSE GROSS NATIONAL PRODUCT
IS MEDIUM, LOW, OR VERY LOW (105)

PREDICATE ONLY

10 IN LEFT COLUMN

CANADA	CHINA, PR	FRANCE	GERMAN FR	INDIA	ITALY	JAPAN	USSR	UK	US

105 IN RIGHT COLUMN

AFGHANISTAN	ALBANIA	ALGERIA	ARGENTINA	AUSTRALIA	AUSTRIA	BELGIUM	BOLIVIA	BRAZIL	BULGARIA
BURMA	BURUNDI	CAMBODIA	CAMEROUN	CEN AFR REP	CEYLON	CHAD	CHILE	COLOMBIA	CONGO(BRA)
CONGO(LEO)	COSTA RICA	CUBA	CYPRUS	CZECHOS'KIA	DAHOMEY	DENMARK	DOMIN REP	ECUADOR	EL SALVADOR
ETHIOPIA	FINLAND	GABON	GERMANY, E	GHANA	GREECE	GUATEMALA	GUINEA	HAITI	HONDURAS
HUNGARY	ICELAND	INDONESIA	IRAN	IRAQ	IRELAND	ISRAEL	IVORY COAST	JAMAICA	JORDAN
KOREA, N	KOREA REP	LAOS	LEBANON	LIBERIA	LIBYA	LUXEMBOURG	MALAGASY R	MALAYA	MALI
MAURITANIA	MEXICO	MONGOLIA	MOROCCO	NEPAL	NETHERLANDS	NEW ZEALAND	NICARAGUA	NIGER	NIGERIA
NORWAY	PAKISTAN	PANAMA	PARAGUAY	PERU	PHILIPPINES	POLAND	PORTUGAL	RUMANIA	RWANDA
SA'U ARABIA	SENEGAL	SIERRE LEO	SOMALIA	SO AFRICA	SPAIN	SUDAN	SWEDEN	SWITZERLAND	SYRIA
TANGANYIKA	THAILAND	TOGO	TRINIDAD	TUNISIA	TURKEY	UGANDA	UAR	UPPER VOLTA	URUGUAY
VENEZUELA	VIETNAM, N	VIETNAM REP	YEMEN	YUGOSLAVIA					

33 POLITIES
WHOSE GROSS NATIONAL PRODUCT
IS VERY HIGH, HIGH, OR MEDIUM (30)

33 POLITIES
WHOSE GROSS NATIONAL PRODUCT
IS LOW OR VERY LOW (85)

BOTH SUBJECT AND PREDICATE

30 IN LEFT COLUMN

ARGENTINA	AUSTRALIA	AUSTRIA	BELGIUM	BRAZIL	CANADA	CHINA, PR	CZECHOS'KIA	DENMARK	FRANCE
GERMANY, E	GERMAN FR	INDIA	INDONESIA	ITALY	JAPAN	MEXICO	NETHERLANDS	PAKISTAN	POLAND
RUMANIA	SO AFRICA	SPAIN	SWEDEN	SWITZERLAND	TURKEY	USSR	UK	US	YUGOSLAVIA

85 IN RIGHT COLUMN

AFGHANISTAN	ALBANIA	ALGERIA	BOLIVIA	BULGARIA	BURMA	BURUNDI	CAMBODIA	CAMEROUN	CEN AFR REP
CEYLON	CHAD	CHILE	COLOMBIA	CONGO(BRA)	CONGO(LEO)	COSTA RICA	CUBA	CYPRUS	DAHOMEY
DOMIN REP	ECUADOR	EL SALVADOR	ETHIOPIA	FINLAND	GABON	GHANA	GREECE	GUATEMALA	GUINEA
HAITI	HONDURAS	HUNGARY	ICELAND	IRAN	IRAQ	IRELAND	ISRAEL	IVORY COAST	JAMAICA
JORDAN	KOREA, N	KOREA REP	LAOS	LEBANON	LIBERIA	LIBYA	LUXEMBOURG	MALAGASY R	MALAYA
MALI	MAURITANIA	MONGOLIA	MOROCCO	NEPAL	NEW ZEALAND	NICARAGUA	NIGER	NIGERIA	NORWAY
PANAMA	PARAGUAY	PERU	PHILIPPINES	PORTUGAL	RWANDA	SA'U ARABIA	SENEGAL	SIERRE LEO	SOMALIA
SUDAN	SYRIA	TANGANYIKA	THAILAND	TOGO	TRINIDAD	TUNISIA	UGANDA	UAR	UPPER VOLTA
URUGUAY	VENEZUELA	VIETNAM, N	VIETNAM REP	YEMEN					

---

2 TEND LESS TO BE THOSE LOCATED ELSEWHERE THAN IN WEST EUROPE, SCANDINAVIA, NORTH AMERICA, OR AUSTRALASIA (93) 0.53

2 TEND MORE TO BE THOSE LOCATED ELSEWHERE THAN IN WEST EUROPE, SCANDINAVIA, NORTH AMERICA, OR AUSTRALASIA (93) 0.91

14 8
16 77
X SQ= 17.56
P = 0.
RV NO (YES)

10 TEND MORE TO BE THOSE LOCATED ELSEWHERE THAN IN NORTH AFRICA, OR CENTRAL AND SOUTH AFRICA (82) 0.97

10 TEND LESS TO BE THOSE LOCATED ELSEWHERE THAN IN NORTH AFRICA, OR CENTRAL AND SOUTH AFRICA (82) 0.62

1 32
29 53
X SQ= 11.14
P = 0.
RV NO (NO)

13 LEAN MORE TOWARD BEING THOSE LOCATED ELSEWHERE THAN IN NORTH AFRICA OR THE MIDDLE EAST (99) 0.97

13 LEAN LESS TOWARD BEING THOSE LOCATED ELSEWHERE THAN IN NORTH AFRICA OR THE MIDDLE EAST (99) 0.82

1 15
29 70
X SQ= 2.69
P = 0.066
RV NO (NO)

18 TEND TO BE THOSE LOCATED IN EAST ASIA, SOUTH ASIA, OR SOUTHEAST ASIA, RATHER THAN IN NORTH AFRICA OR CENTRAL AND SOUTH AFRICA (18) 0.83

18 TEND TO BE THOSE LOCATED IN NORTH AFRICA OR CENTRAL AND SOUTH AFRICA, RATHER THAN IN EAST ASIA, SOUTH ASIA, OR SOUTHEAST ASIA (33) 0.71

1 32
5 13
X SQ= 4.69
P = 0.017
RV YES (NO)

22	TEND MORE TO BE THOSE WHOSE TERRITORIAL SIZE IS VERY LARGE, LARGE, OR MEDIUM (68)	0.77	22	TEND LESS TO BE THOSE WHOSE TERRITORIAL SIZE IS VERY LARGE, LARGE, OR MEDIUM (68)	0.53	23 45 7 40 X SQ= 4.23 P = 0.030 RV NO (NO)
23	TEND TO BE THOSE WHOSE POPULATION IS VERY LARGE OR LARGE (27)	0.67	23	TEND TO BE THOSE WHOSE POPULATION IS MEDIUM OR SMALL (88)	0.92	20 7 10 78 X SQ= 38.95 P = 0. RV YES (YES)
24	TEND TO BE THOSE WHOSE POPULATION IS VERY LARGE, LARGE, OR MEDIUM (61)	0.93	24	TEND TO BE THOSE WHOSE POPULATION IS SMALL (54)	0.61	28 33 2 52 X SQ= 24.31 P = 0. RV YES (NO)
26	TEND TO BE THOSE WHOSE POPULATION DENSITY IS VERY HIGH, HIGH, OR MEDIUM (48)	0.67	26	TEND TO BE THOSE WHOSE POPULATION DENSITY IS LOW (67)	0.67	20 28 10 57 X SQ= 9.03 P = 0.002 RV YES (NO)
28	TEND TO BE THOSE WHOSE POPULATION GROWTH RATE IS LOW (48)	0.66	28	TEND TO BE THOSE WHOSE POPULATION GROWTH RATE IS HIGH (62)	0.64	10 52 19 29 X SQ= 6.51 P = 0.008 RV YES (NO)
29	TEND TO BE THOSE WHERE THE DEGREE OF URBANIZATION IS HIGH (56)	0.86	29	TEND TO BE THOSE WHERE THE DEGREE OF URBANIZATION IS LOW (49)	0.59	25 31 4 45 X SQ= 15.62 P = 0. RV YES (NO)
30	TEND TO BE THOSE WHOSE AGRICULTURAL POPULATION IS MEDIUM, LOW, OR VERY LOW (57)	0.80	30	TEND TO BE THOSE WHOSE AGRICULTURAL POPULATION IS HIGH (56)	0.60	6 50 24 33 X SQ= 12.71 P = 0. RV YES (NO)
31	TEND TO BE THOSE WHOSE AGRICULTURAL POPULATION IS LOW OR VERY LOW (24)	0.53	31	TEND TO BE THOSE WHOSE AGRICULTURAL POPULATION IS HIGH OR MEDIUM (90)	0.90	14 76 16 8 X SQ= 22.96 P = 0. RV YES (YES)
35	TEND TO BE THOSE WHOSE PER CAPITA GROSS NATIONAL PRODUCT IS VERY HIGH OR HIGH (24)	0.53	35	TEND TO BE THOSE WHOSE PER CAPITA GROSS NATIONAL PRODUCT IS MEDIUM, LOW, OR VERY LOW (91)	0.91	16 8 14 77 X SQ= 23.31 P = 0. RV YES (YES)

36	TEND TO BE THOSE WHOSE PER CAPITA GROSS NATIONAL PRODUCT IS VERY HIGH, HIGH, OR MEDIUM (42)	0.77	36	TEND TO BE THOSE WHOSE PER CAPITA GROSS NATIONAL PRODUCT IS LOW OR VERY LOW (73)	0.78	23 19 7 66 X SQ= 25.92 P = 0. RV YES (YES)
37	TEND TO BE THOSE WHOSE PER CAPITA GROSS NATIONAL PRODUCT IS VERY HIGH, HIGH, MEDIUM, OR LOW (64)	0.87	37	TEND TO BE THOSE WHOSE PER CAPITA GROSS NATIONAL PRODUCT IS VERY LOW (51)	0.55	26 38 4 47 X SQ= 14.16 P = 0. RV YES (NO)
38	TEND LESS TO BE THOSE WHOSE INTERNATIONAL FINANCIAL STATUS IS MEDIUM, LOW, OR VERY LOW (103)	0.64	38	ALWAYS ARE THOSE WHOSE INTERNATIONAL FINANCIAL STATUS IS MEDIUM, LOW, OR VERY LOW (103)	1.00	10 0 18 85 X SQ= 29.02 P = 0. RV NO (YES)
39	ALWAYS ARE THOSE WHOSE INTERNATIONAL FINANCIAL STATUS IS VERY HIGH, HIGH, OR MEDIUM (38)	1.00	39	TEND TO BE THOSE WHOSE INTERNATIONAL FINANCIAL STATUS IS LOW OR VERY LOW (76)	0.89	29 9 0 76 X SQ= 73.82 P = 0. RV YES (YES)
40	ALWAYS ARE THOSE WHOSE INTERNATIONAL FINANCIAL STATUS IS VERY HIGH, HIGH, MEDIUM, OR LOW (71)	1.00	40	TEND LESS TO BE THOSE WHOSE INTERNATIONAL FINANCIAL STATUS IS VERY HIGH, HIGH, MEDIUM, OR LOW (71)	0.51	30 41 0 39 X SQ= 20.58 P = 0. RV NO (NO)
41	TEND TO BE THOSE WHOSE ECONOMIC DEVELOPMENTAL STATUS IS DEVELOPED (19)	0.50	41	TEND TO BE THOSE WHOSE ECONOMIC DEVELOPMENTAL STATUS IS INTERMEDIATE, UNDERDEVELOPED, OR VERY UNDERDEVELOPED (94)	0.95	15 4 15 79 X SQ= 29.01 P = 0. RV YES (YES)
42	TEND TO BE THOSE WHOSE ECONOMIC DEVELOPMENTAL STATUS IS DEVELOPED OR INTERMEDIATE (36)	0.80	42	TEND TO BE THOSE WHOSE ECONOMIC DEVELOPMENTAL STATUS IS UNDERDEVELOPED OR VERY UNDERDEVELOPED (76)	0.85	24 12 6 70 X SQ= 40.08 P = 0. RV YES (YES)
43	TEND TO BE THOSE WHOSE ECONOMIC DEVELOPMENTAL STATUS IS DEVELOPED, INTERMEDIATE, OR UNDERDEVELOPED (55)	0.97	43	TEND TO BE THOSE WHOSE ECONOMIC DEVELOPMENTAL STATUS IS VERY UNDERDEVELOPED (57)	0.68	29 26 1 56 X SQ= 34.53 P = 0. RV YES (YES)
44	TEND TO BE THOSE WHERE THE LITERACY RATE IS NINETY PERCENT OR ABOVE (25)	0.60	44	TEND TO BE THOSE WHERE THE LITERACY RATE IS BELOW NINETY PERCENT (90)	0.92	18 7 12 78 X SQ= 31.95 P = 0. RV YES (YES)

45	TEND TO BE THOSE WHERE THE LITERACY RATE IS FIFTY PERCENT OR ABOVE (55)	0.86	45	TEND TO BE THOSE WHERE THE LITERACY RATE IS BELOW FIFTY PERCENT (54)	0.62	25 30 4 50 X SQ= 18.30 P = 0. RV YES (NO)
46	ALWAYS ARE THOSE WHERE THE LITERACY RATE IS TEN PERCENT OR ABOVE (84)	1.00	46	TEND LESS TO BE THOSE WHERE THE LITERACY RATE IS TEN PERCENT OR ABOVE (84)	0.67	30 54 0 26 X SQ= 11.03 P = 0. RV NO (NO)
54	TEND TO BE THOSE WHERE NEWSPAPER CIRCULATION IS ONE HUNDRED OR MORE PER THOUSAND (37)	0.67	54	TEND TO BE THOSE WHERE NEWSPAPER CIRCULATION IS LESS THAN ONE HUNDRED PER THOUSAND (76)	0.80	20 17 10 66 X SQ= 19.30 P = 0. RV YES (YES)
56	TEND MORE TO BE THOSE WHERE THE RELIGION IS PREDOMINANTLY LITERATE (79)	0.97	56	TEND LESS TO BE THOSE WHERE THE RELIGION IS PREDOMINANTLY LITERATE (79)	0.63	28 51 1 30 X SQ= 10.30 P = 0. RV NO (NO)
63	TEND TO BE THOSE WHERE THE RELIGION IS PREDOMINANTLY SOME KIND OF CHRISTIAN (46)	0.76	63	TEND TO BE THOSE WHERE THE RELIGION IS PREDOMINANTLY OR PARTLY OTHER THAN CHRISTIAN (68)	0.72	22 24 7 61 X SQ= 18.45 P = 0. RV YES (NO)
67	LEAN MORE TOWARD BEING THOSE THAT ARE RACIALLY HOMOGENEOUS (82)	0.87	67	LEAN LESS TOWARD BEING THOSE THAT ARE RACIALLY HOMOGENEOUS (82)	0.68	26 54 4 25 X SQ= 2.86 P = 0.057 RV NO (NO)
72	TEND TO BE THOSE WHOSE DATE OF INDEPENDENCE IS BEFORE 1914 (52)	0.79	72	TEND TO BE THOSE WHOSE DATE OF INDEPENDENCE IS AFTER 1914 (59)	0.64	22 30 6 53 X SQ= 13.48 P = 0. RV YES (NO)
73	TEND TO BE THOSE WHOSE DATE OF INDEPENDENCE IS BEFORE 1945 (65)	0.89	73	TEND TO BE THOSE WHOSE DATE OF INDEPENDENCE IS AFTER 1945 (46)	0.52	25 40 3 43 X SQ= 12.92 P = 0. RV YES (NO)
74	TEND TO BE THOSE THAT ARE HISTORICALLY WESTERN (26)	0.59	74	TEND TO BE THOSE THAT ARE NOT HISTORICALLY WESTERN (87)	0.89	17 9 12 75 X SQ= 25.29 P = 0. RV YES (YES)

75	TEND TO BE THOSE THAT ARE HISTORICALLY WESTERN OR SIGNIFICANTLY WESTERNIZED (62)	0.93	75	TEND TO BE THOSE THAT ARE NOT HISTORICALLY WESTERN AND ARE NOT SIGNIFICANTLY WESTERNIZED (52)	0.59	27 35 2 50 X SQ= 21.46 P = 0. RV YES (NO)
76	TEND TO BE THOSE THAT ARE HISTORICALLY WESTERN, RATHER THAN HAVING BEEN SIGNIFICANTLY OR PARTIALLY WESTERNIZED THROUGH A COLONIAL RELATIONSHIP (26)	0.71	76	TEND TO BE THOSE THAT HAVE BEEN SIGNIFICANTLY OR PARTIALLY WESTERNIZED THROUGH A COLONIAL RELATIONSHIP, RATHER THAN BEING HISTORICALLY WESTERN (70)	0.87	17 9 7 63 X SQ= 28.13 P = 0. RV YES (YES)
78	TEND TO BE THOSE THAT HAVE BEEN SIGNIFICANTLY WESTERNIZED WITHOUT A COLONIAL RELATIONSHIP, RATHER THAN THROUGH SUCH A RELATIONSHIP (7)	0.60	78	TEND TO BE THOSE THAT HAVE BEEN SIGNIFICANTLY WESTERNIZED THROUGH A COLONIAL RELATIONSHIP, RATHER THAN WITHOUT SUCH A RELATIONSHIP (28)	0.96	4 24 6 1 X SQ= 10.72 P = 0.001 RV YES (YES)
81	TEND LESS TO BE THOSE WHERE THE TYPE OF POLITICAL MODERNIZATION IS LATER EUROPEAN OR LATER EUROPEAN DERIVED, RATHER THAN EARLY EUROPEAN OR EARLY EUROPEAN DERIVED (40)	0.61	81	TEND MORE TO BE THOSE WHERE THE TYPE OF POLITICAL MODERNIZATION IS LATER EUROPEAN OR LATER EUROPEAN DERIVED, RATHER THAN EARLY EUROPEAN OR EARLY EUROPEAN DERIVED (40)	0.93	9 2 14 26 X SQ= 5.86 P = 0.014 RV NO (YES)
82	TEND TO BE THOSE WHERE THE TYPE OF POLITICAL MODERNIZATION IS EARLY OR LATER EUROPEAN OR EUROPEAN DERIVED, RATHER THAN DEVELOPED TUTELARY OR UNDEVELOPED TUTELARY (51)	0.88	82	TEND TO BE THOSE WHERE THE TYPE OF POLITICAL MODERNIZATION IS DEVELOPED TUTELARY OR UNDEVELOPED TUTELARY, RATHER THAN EARLY OR LATER EUROPEAN OR EUROPEAN DERIVED (55)	0.65	23 28 3 52 X SQ= 20.38 P = 0. RV YES (NO)
85	TEND TO BE THOSE WHERE THE STAGE OF POLITICAL MODERNIZATION IS ADVANCED, RATHER THAN MID- OR EARLY TRANSITIONAL (60)	0.93	85	TEND TO BE THOSE WHERE THE STAGE OF POLITICAL MODERNIZATION IS MID- OR EARLY TRANSITIONAL, RATHER THAN ADVANCED (54)	0.62	28 32 2 52 X SQ= 24.88 P = 0. RV YES (NO)
87	TEND MORE TO BE THOSE WHOSE IDEOLOGICAL ORIENTATION IS OTHER THAN DEVELOPMENTAL (58)	0.96	87	TEND LESS TO BE THOSE WHOSE IDEOLOGICAL ORIENTATION IS OTHER THAN DEVELOPMENTAL (58)	0.52	1 30 25 33 X SQ= 13.67 P = 0. RV NO (NO)
89	TEND TO BE THOSE WHOSE IDEOLOGICAL ORIENTATION IS CONVENTIONAL (33)	0.65	89	TEND TO BE THOSE WHOSE IDEOLOGICAL ORIENTATION IS OTHER THAN CONVENTIONAL (62)	0.77	17 16 9 53 X SQ= 13.03 P = 0. RV YES (YES)
96	TEND MORE TO BE THOSE WHERE THE STATUS OF THE REGIME IS CONSTITUTIONAL OR TOTALITARIAN (67)	0.93	96	TEND LESS TO BE THOSE WHERE THE STATUS OF THE REGIME IS CONSTITUTIONAL OR TOTALITARIAN (67)	0.66	26 41 2 21 X SQ= 5.91 P = 0.008 RV NO (NO)

97	TEND TO BE THOSE WHERE THE STATUS OF THE REGIME IS TOTALITARIAN, RATHER THAN AUTHORITARIAN (16)	0.80	97	TEND TO BE THOSE WHERE THE STATUS OF THE REGIME IS AUTHORITARIAN, RATHER THAN TOTALITARIAN (23)	0.72	2 21 8 8 X SQ= 6.42 P = 0.007 RV YES (YES)
99	TEND TO BE THOSE WHERE GOVERNMENTAL STABILITY IS GENERALLY PRESENT AND DATES FROM AT LEAST THE INTER-WAR PERIOD, OR FROM THE POST-WAR PERIOD (50)	0.83	99	TEND TO BE THOSE WHERE GOVERNMENTAL STABILITY IS MODERATELY PRESENT AND DATES FROM THE POST-WAR PERIOD, OR IS ABSENT (36)	0.55	25 25 5 31 X SQ= 10.48 P = 0.001 RV YES (YES)
101	TEND TO BE THOSE WHERE THE REPRESENTATIVE CHARACTER OF THE REGIME IS POLYARCHIC (41)	0.59	101	TEND TO BE THOSE WHERE THE REPRESENTATIVE CHARACTER OF THE REGIME IS LIMITED POLYARCHIC, PSEUDO-POLYARCHIC, OR NON-POLYARCHIC (57)	0.65	17 24 12 45 X SQ= 3.84 P = 0.043 RV YES (NO)
102	TILT TOWARD BEING THOSE WHERE THE REPRESENTATIVE CHARACTER OF THE REGIME IS POLYARCHIC OR LIMITED POLYARCHIC (59)	0.67	102	TILT TOWARD BEING THOSE WHERE THE REPRESENTATIVE CHARACTER OF THE REGIME IS PSEUDO-POLYARCHIC OR NON-POLYARCHIC (49)	0.50	20 39 10 39 X SQ= 1.80 P = 0.136 RV YES (NO)
105	TEND TO BE THOSE WHERE THE ELECTORAL SYSTEM IS COMPETITIVE (43)	0.67	105	TEND TO BE THOSE WHERE THE ELECTORAL SYSTEM IS PARTIALLY COMPETITIVE OR NON-COMPETITIVE (47)	0.60	18 25 9 38 X SQ= 4.49 P = 0.023 RV YES (NO)
107	TEND TO BE THOSE WHERE AUTONOMOUS GROUPS ARE FULLY TOLERATED IN POLITICS (46)	0.60	107	TEND TO BE THOSE WHERE AUTONOMOUS GROUPS ARE PARTIALLY TOLERATED IN POLITICS, ARE TOLERATED ONLY OUTSIDE POLITICS, OR ARE NOT TOLERATED AT ALL (65)	0.65	18 28 12 53 X SQ= 4.83 P = 0.018 RV YES (NO)
110	TEND LESS TO BE THOSE WHERE POLITICAL ENCULTURATION IS MEDIUM OR LOW (80)	0.67	110	TEND MORE TO BE THOSE WHERE POLITICAL ENCULTURATION IS MEDIUM OR LOW (80)	0.90	8 7 16 64 X SQ= 5.77 P = 0.019 RV NO (YES)
114	LEAN MORE TOWARD BEING THOSE WHERE SECTIONALISM IS EXTREME OR MODERATE (61)	0.71	114	LEAN LESS TOWARD BEING THOSE WHERE SECTIONALISM IS EXTREME OR MODERATE (61)	0.51	20 41 8 39 X SQ= 2.66 P = 0.078 RV NO (NO)
115	TEND TO BE THOSE WHERE INTEREST ARTICULATION BY ASSOCIATIONAL GROUPS IS SIGNIFICANT (20)	0.54	115	TEND TO BE THOSE WHERE INTEREST ARTICULATION BY ASSOCIATIONAL GROUPS IS MODERATE, LIMITED, OR NEGLIGIBLE (91)	0.94	15 5 13 78 X SQ= 28.91 P = 0. RV YES (YES)

116	TEND TO BE THOSE WHERE INTEREST ARTICULATION BY ASSOCIATIONAL GROUPS IS SIGNIFICANT OR MODERATE (32)	0.68	116	TEND TO BE THOSE WHERE INTEREST ARTICULATION BY ASSOCIATIONAL GROUPS IS LIMITED OR NEGLIGIBLE (79)	0.84	19 13 9 70 X SQ= 25.31 P = 0. RV YES (YES)
117	TEND TO BE THOSE WHERE INTEREST ARTICULATION BY ASSOCIATIONAL GROUPS IS SIGNIFICANT, MODERATE, OR LIMITED (60)	0.82	117	TEND TO BE THOSE WHERE INTEREST ARTICULATION BY ASSOCIATIONAL GROUPS IS NEGLIGIBLE (51)	0.55	23 37 5 46 X SQ= 10.43 P = 0.001 RV YES (NO)
121	TEND TO BE THOSE WHERE INTEREST ARTICULATION BY NON-ASSOCIATIONAL GROUPS IS MODERATE, LIMITED, OR NEGLIGIBLE (61)	0.83	121	TEND TO BE THOSE WHERE INTEREST ARTICULATION BY NON-ASSOCIATIONAL GROUPS IS SIGNIFICANT (54)	0.58	5 49 25 36 X SQ= 13.35 P = 0. RV YES (NO)
126	TEND LESS TO BE THOSE WHERE INTEREST ARTICULATION BY ANOMIC GROUPS IS FREQUENT, OCCASIONAL, OR INFREQUENT (83)	0.62	126	TEND MORE TO BE THOSE WHERE INTEREST ARTICULATION BY ANOMIC GROUPS IS FREQUENT, OCCASIONAL, OR INFREQUENT (83)	0.92	16 67 10 6 X SQ= 10.81 P = 0.001 RV NO (YES)
132	ALWAYS ARE THOSE WHERE INTEREST AGGREGATION BY POLITICAL PARTIES IS SIGNIFICANT, MODERATE, OR LIMITED (67)	1.00	132	TILT LESS TOWARD BEING THOSE WHERE INTEREST AGGREGATION BY POLITICAL PARTIES IS SIGNIFICANT, MODERATE, OR LIMITED (67)	0.84	18 49 0 9 X SQ= 1.86 P = 0.105 RV NO (NO)
135	TILT MORE TOWARD BEING THOSE WHERE INTEREST AGGREGATION BY THE EXECUTIVE IS SIGNIFICANT, MODERATE, OR LIMITED (77)	0.96	135	TILT LESS TOWARD BEING THOSE WHERE INTEREST AGGREGATION BY THE EXECUTIVE IS SIGNIFICANT, MODERATE, OR LIMITED (77)	0.82	24 53 1 12 X SQ= 2.00 P = 0.102 RV NO (NO)
137	TEND TO BE THOSE WHERE INTEREST AGGREGATION BY THE LEGISLATURE IS SIGNIFICANT OR MODERATE (29)	0.54	137	TEND TO BE THOSE WHERE INTEREST AGGREGATION BY THE LEGISLATURE IS LIMITED OR NEGLIGIBLE (68)	0.80	15 14 13 55 X SQ= 9.00 P = 0.003 RV YES (YES)
138	TEND TO BE THOSE WHERE INTEREST AGGREGATION BY THE LEGISLATURE IS SIGNIFICANT, MODERATE, OR LIMITED (48)	0.69	138	TEND TO BE THOSE WHERE INTEREST AGGREGATION BY THE LEGISLATURE IS NEGLIGIBLE (49)	0.58	18 30 8 41 X SQ= 4.51 P = 0.023 RV YES (NO)
146	ALWAYS ARE THOSE WHERE THE PARTY SYSTEM IS QUALITATIVELY OTHER THAN MASS-BASED TERRITORIAL (92)	1.00	146	TILT LESS TOWARD BEING THOSE WHERE THE PARTY SYSTEM IS QUALITATIVELY OTHER THAN MASS-BASED TERRITORIAL (92)	0.89	0 8 28 64 X SQ= 2.04 P = 0.101 RV NO (NO)

147	TEND TO BE THOSE WHERE THE PARTY SYSTEM IS QUALITATIVELY CLASS-ORIENTED OR MULTI-IDEOLOGICAL (23)	0.54	147	TEND TO BE THOSE WHERE THE PARTY SYSTEM IS QUALITATIVELY OTHER THAN CLASS-ORIENTED OR MULTI-IDEOLOGICAL (67)	0.85	13 10 11 56 X SQ= 12.11 P = 0.001 RV YES (YES)
148	ALWAYS ARE THOSE WHERE THE PARTY SYSTEM IS QUALITATIVELY OTHER THAN AFRICAN TRANSITIONAL (96)	1.00	148	TEND LESS TO BE THOSE WHERE THE PARTY SYSTEM IS QUALITATIVELY OTHER THAN AFRICAN TRANSITIONAL (96)	0.82	0 14 30 66 X SQ= 4.54 P = 0.010 RV NO (NO)
153	TEND TO BE THOSE WHERE THE PARTY SYSTEM IS STABLE (42)	0.79	153	TEND TO BE THOSE WHERE THE PARTY SYSTEM IS MODERATELY STABLE OR UNSTABLE (38)	0.63	23 19 6 32 X SQ= 11.48 P = 0. RV YES (YES)
157	ALWAYS ARE THOSE WHERE PERSONALISMO IS MODERATE OR NEGLIGIBLE (84)	1.00	157	TEND LESS TO BE THOSE WHERE PERSONALISMO IS MODERATE OR NEGLIGIBLE (84)	0.80	0 14 29 55 X SQ= 5.31 P = 0.009 RV NO (NO)
158	TEND TO BE THOSE WHERE PERSONALISMO IS NEGLIGIBLE (56)	0.83	158	TEND TO BE THOSE WHERE PERSONALISMO IS PRONOUNCED OR MODERATE (40)	0.52	5 35 24 32 X SQ= 8.81 P = 0.002 RV YES (NO)
164	TILT MORE TOWARD BEING THOSE WHERE THE REGIME'S LEADERSHIP CHARISMA IS NEGLIGIBLE (65)	0.79	164	TILT LESS TOWARD BEING THOSE WHERE THE REGIME'S LEADERSHIP CHARISMA IS NEGLIGIBLE (65)	0.60	6 28 23 42 X SQ= 2.59 P = 0.102 RV NO (NO)
166	TEND LESS TO BE THOSE WHERE THE VERTICAL POWER DISTRIBUTION IS THAT OF FORMAL FEDERALISM OR FORMAL AND EFFECTIVE UNITARISM (99)	0.70	166	TEND MORE TO BE THOSE WHERE THE VERTICAL POWER DISTRIBUTION IS THAT OF FORMAL FEDERALISM OR FORMAL AND EFFECTIVE UNITARISM (99)	0.93	9 6 21 78 X SQ= 8.21 P = 0.003 RV NO (YES)
168	TEND TO BE THOSE WHERE THE HORIZONTAL POWER DISTRIBUTION IS SIGNIFICANT (34)	0.59	168	TEND TO BE THOSE WHERE THE HORIZONTAL POWER DISTRIBUTION IS LIMITED OR NEGLIGIBLE (72)	0.78	17 17 12 60 X SQ= 11.29 P = 0.001 RV YES (YES)
170	TEND MORE TO BE THOSE WHERE THE LEGISLATIVE-EXECUTIVE STRUCTURE IS OTHER THAN PRESIDENTIAL (63)	0.80	170	TEND LESS TO BE THOSE WHERE THE LEGISLATIVE-EXECUTIVE STRUCTURE IS OTHER THAN PRESIDENTIAL (63)	0.54	6 33 24 39 X SQ= 4.94 P = 0.015 RV NO (NO)

174	TEND TO BE THOSE WHERE THE LEGISLATURE IS FULLY EFFECTIVE (28)	0.52	174	TEND TO BE THOSE WHERE THE LEGISLATURE IS PARTIALLY EFFECTIVE, LARGELY INEFFECTIVE, OR WHOLLY INEFFECTIVE (72)	0.82	15 13 14 58 X SQ= 9.81 P = 0.001 RV YES (YES)
175	TEND TO BE THOSE WHERE THE LEGISLATURE IS FULLY EFFECTIVE OR PARTIALLY EFFECTIVE (51)	0.69	175	TEND TO BE THOSE WHERE THE LEGISLATURE IS LARGELY INEFFECTIVE OR WHOLLY INEFFECTIVE (49)	0.56	20 31 9 40 X SQ= 4.31 P = 0.028 RV YES (NO)
178	TEND TO BE THOSE WHERE THE LEGISLATURE IS BICAMERAL (51)	0.70	178	TEND TO BE THOSE WHERE THE LEGISLATURE IS UNICAMERAL (53)	0.59	9 44 21 30 X SQ= 6.28 P = 0.009 RV YES (NO)
181	TEND TO BE THOSE WHERE THE BUREAUCRACY IS MODERN, RATHER THAN SEMI-MODERN (21)	0.50	181	TEND TO BE THOSE WHERE THE BUREAUCRACY IS SEMI-MODERN, RATHER THAN MODERN (55)	0.85	14 7 14 41 X SQ= 9.39 P = 0.001 RV YES (YES)
182	ALWAYS ARE THOSE WHERE THE BUREAUCRACY IS SEMI-MODERN, RATHER THAN POST-COLONIAL TRANSITIONAL (55)	1.00	182	TEND LESS TO BE THOSE WHERE THE BUREAUCRACY IS SEMI-MODERN, RATHER THAN POST-COLONIAL TRANSITIONAL (55)	0.62	14 41 0 25 X SQ= 6.05 P = 0.004 RV NO (NO)
187	TEND TO BE THOSE WHERE THE ROLE OF THE POLICE IS NOT POLITICALLY SIGNIFICANT (35)	0.55	187	TEND TO BE THOSE WHERE THE ROLE OF THE POLICE IS POLITICALLY SIGNIFICANT (66)	0.74	13 53 16 19 X SQ= 6.35 P = 0.010 RV YES (NO)
192	TEND MORE TO BE THOSE WHERE THE CHARACTER OF THE LEGAL SYSTEM IS OTHER THAN MUSLIM OR PARTLY MUSLIM (86)	0.93	192	TEND LESS TO BE THOSE WHERE THE CHARACTER OF THE LEGAL SYSTEM IS OTHER THAN MUSLIM OR PARTLY MUSLIM (86)	0.70	2 25 28 58 X SQ= 5.44 P = 0.011 RV NO (NO)
193	TEND MORE TO BE THOSE WHERE THE CHARACTER OF THE LEGAL SYSTEM IS OTHER THAN PARTLY INDIGENOUS (86)	0.93	193	TEND LESS TO BE THOSE WHERE THE CHARACTER OF THE LEGAL SYSTEM IS OTHER THAN PARTLY INDIGENOUS (86)	0.69	2 26 28 58 X SQ= 5.79 P = 0.007 RV NO (NO)
194	TEND LESS TO BE THOSE THAT ARE NON-COMMUNIST (101)	0.77	194	TEND MORE TO BE THOSE THAT ARE NON-COMMUNIST (101)	0.93	7 6 23 78 X SQ= 4.24 P = 0.038 RV NO (YES)

34 POLITIES WHOSE GROSS NATIONAL PRODUCT IS VERY HIGH, HIGH, MEDIUM, OR LOW (62)

34 POLITIES WHOSE GROSS NATIONAL PRODUCT IS VERY LOW (53)

PREDICATE ONLY

62 IN LEFT COLUMN

ALGERIA	ARGENTINA	AUSTRALIA	AUSTRIA	BELGIUM	BRAZIL	BULGARIA	BURMA	CANADA	CEYLON
CHILE	CHINA, PR	COLOMBIA	CONGO(LEO)	CUBA	CZECHOS'KIA	DENMARK	ETHIOPIA	FINLAND	FRANCE
GERMANY, E	GERMAN FR	GREECE	HUNGARY	INDIA	INDONESIA	IRAN	IRAQ	IRELAND	ISRAEL
ITALY	JAPAN	KOREA REP	MALAYA	MEXICO	MOROCCO	NETHERLANDS	NEW ZEALAND	NIGERIA	NORWAY
PAKISTAN	PERU	PHILIPPINES	POLAND	PORTUGAL	RUMANIA	SA'U ARABIA	SO AFRICA	SPAIN	SWEDEN
SWITZERLAND	THAILAND	TURKEY	USSR	UAR	UK	US	URUGUAY	VENEZUELA	VIETNAM, N
VIETNAM REP	YUGOSLAVIA								

53 IN RIGHT COLUMN

AFGHANISTAN	ALBANIA	BOLIVIA	BURUNDI	CAMBODIA	CAMEROUN	CEN AFR REP	CHAD	CONGO(BRA)	COSTA RICA
CYPRUS	DAHOMEY	DOMIN REP	ECUADOR	EL SALVADOR	GABON	GHANA	GUATEMALA	GUINEA	HAITI
HONDURAS	ICELAND	IVORY COAST	JAMAICA	JORDAN	KOREA, N	LAOS	LEBANON	LIBERIA	LIBYA
LUXEMBOURG	MALAGASY R	MALI	MAURITANIA	MONGOLIA	NEPAL	NICARAGUA	NIGER	PANAMA	PARAGUAY
RWANDA	SENEGAL	SIERRE LEO	SOMALIA	SUDAN	SYRIA	TANGANYIKA	TOGO	TRINIDAD	TUNISIA
UGANDA	UPPER VOLTA	YEMEN							

35 POLITIES
WHOSE PER CAPITA GROSS NATIONAL PRODUCT
IS VERY HIGH OR HIGH (24)

35 POLITIES
WHOSE PER CAPITA GROSS NATIONAL PRODUCT
IS MEDIUM, LOW, OR VERY LOW (91)

PREDICATE ONLY

24 IN LEFT COLUMN

AUSTRALIA	AUSTRIA	BELGIUM	CANADA	CZECHOS'KIA	DENMARK	FINLAND	FRANCE	GERMANY, E	GERMAN FR
ICELAND	IRELAND	ISRAEL	ITALY	LUXEMBOURG	NETHERLANDS	NEW ZEALAND	NORWAY	SWEDEN	SWITZERLAND
USSR	UK	US	VENEZUELA						

91 IN RIGHT COLUMN

AFGHANISTAN	ALBANIA	ALGERIA	ARGENTINA	BOLIVIA	BRAZIL	BULGARIA	BURMA	BURUNDI	CAMBODIA
CAMEROUN	CEN AFR REP	CEYLON	CHAD	CHILE	CHINA, PR	COLOMBIA	CONGO(BRA)	CONGO(LEO)	COSTA RICA
CUBA	CYPRUS	DAHOMEY	DOMIN REP	ECUADOR	EL SALVADOR	ETHIOPIA	GABON	GHANA	GREECE
GUATEMALA	GUINEA	HAITI	HONDURAS	HUNGARY	INDIA	INDONESIA	IRAN	IRAQ	IVORY COAST
JAMAICA	JAPAN	JORDAN	KOREA, N	KOREA REP	LAOS	LEBANON	LIBERIA	LIBYA	MALAGASY R
MALAYA	MALI	MAURITANIA	MEXICO	MONGOLIA	MOROCCO	NEPAL	NICARAGUA	NIGER	NIGERIA
PAKISTAN	PANAMA	PARAGUAY	PERU	PHILIPPINES	POLAND	PORTUGAL	RUMANIA	RWANDA	SA'U ARABIA
SENEGAL	SIERRE LEO	SOMALIA	SO AFRICA	SPAIN	SUDAN	SYRIA	TANGANYIKA	THAILAND	TOGO
TRINIDAD	TUNISIA	TURKEY	UGANDA	UAR	UPPER VOLTA	URUGUAY	VIETNAM, N	VIETNAM REP	YEMEN
YUGOSLAVIA									

36 POLITIES
WHOSE PER CAPITA GROSS NATIONAL PRODUCT
IS VERY HIGH, HIGH, OR MEDIUM (42)

36 POLITIES
WHOSE PER CAPITA GROSS NATIONAL PRODUCT
IS LOW OR VERY LOW (73)

BOTH SUBJECT AND PREDICATE

42 IN LEFT COLUMN

ARGENTINA	AUSTRALIA	AUSTRIA	BELGIUM	BULGARIA	CANADA	CHILE	CUBA	CYPRUS	CZECHOS'KIA
DENMARK	FINLAND	FRANCE	GERMANY, E	GERMAN FR	GREECE	HUNGARY	ICELAND	IRELAND	ISRAEL
ITALY	JAMAICA	JAPAN	LEBANON	LUXEMBOURG	MALAYA	NETHERLANDS	NEW ZEALAND	NORWAY	POLAND
RUMANIA	SO AFRICA	SPAIN	SWEDEN	SWITZERLAND	TRINIDAD	USSR	UK	US	URUGUAY
VENEZUELA	YUGOSLAVIA								

73 IN RIGHT COLUMN

AFGHANISTAN	ALBANIA	ALGERIA	BOLIVIA	BRAZIL	BURMA	BURUNDI	CAMBODIA	CAMEROUN	CEN AFR REP
CEYLON	CHAD	CHINA, PR	COLOMBIA	CONGO(BRA)	CONGO(LEO)	COSTA RICA	DAHOMEY	DOMIN REP	ECUADOR
EL SALVADOR	ETHIOPIA	GABON	GHANA	GUATEMALA	GUINEA	HAITI	HONDURAS	INDIA	INDONESIA
IRAN	IRAQ	IVORY COAST	JORDAN	KOREA, N	KOREA REP	LAOS	LIBERIA	LIBYA	MALAGASY R
MALI	MAURITANIA	MEXICO	MONGOLIA	MOROCCO	NEPAL	NICARAGUA	NIGER	NIGERIA	PAKISTAN
PANAMA	PARAGUAY	PERU	PHILIPPINES	PORTUGAL	RWANDA	SA'U ARABIA	SENEGAL	SIERRE LEO	SOMALIA
SUDAN	SYRIA	TANGANYIKA	THAILAND	TOGO	TUNISIA	TURKEY	UGANDA	UAR	UPPER VOLTA
VIETNAM, N	VIETNAM REP	YEMEN							

Left		Right		
2 TEND TO BE THOSE LOCATED IN WEST EUROPE, SCANDINAVIA, NORTH AMERICA, OR AUSTRALASIA (22)	0.50	2 TEND TO BE THOSE LOCATED ELSEWHERE THAN IN WEST EUROPE, SCANDINAVIA, NORTH AMERICA, OR AUSTRALASIA (93)	0.99	21 1 21 72 X SQ= 37.67 P = 0. RV YES (YES)
7 ALWAYS ARE THOSE LOCATED IN SOUTH AMERICA, RATHER THAN IN CENTRAL AMERICA (10)	1.00	7 TILT TOWARD BEING THOSE LOCATED IN CENTRAL AMERICA, RATHER THAN IN SOUTH AMERICA (7)	0.54	0 7 4 6 X SQ= 1.78 P = 0.103 RV YES (NO)
8 TEND MORE TO BE THOSE LOCATED ELSEWHERE THAN IN EAST ASIA SOUTH ASIA, OR SOUTHEAST ASIA (97)	0.95	8 TEND LESS TO BE THOSE LOCATED ELSEWHERE THAN IN EAST ASIA SOUTH ASIA, OR SOUTHEAST ASIA (97)	0.78	2 16 40 57 X SQ= 4.72 P = 0.016 RV NO (NO)
10 TEND MORE TO BE THOSE LOCATED ELSEWHERE THAN IN NORTH AFRICA, OR CENTRAL AND SOUTH AFRICA (82)	0.98	10 TEND LESS TO BE THOSE LOCATED ELSEWHERE THAN IN NORTH AFRICA, OR CENTRAL AND SOUTH AFRICA (82)	0.56	1 32 41 41 X SQ= 20.41 P = 0. RV NO (YES)

15	TEND TO BE THOSE LOCATED IN THE MIDDLE EAST, RATHER THAN IN NORTH AFRICA OR CENTRAL AND SOUTH AFRICA (11)	0.75	15	TEND TO BE THOSE LOCATED IN NORTH AFRICA OR CENTRAL AND SOUTH AFRICA, RATHER THAN IN THE MIDDLE EAST (33)	0.80	3 8 1 32 X SQ= 3.30 P = 0.043 RV YES (NO)
19	TEND TO BE THOSE LOCATED IN THE CARIBBEAN, CENTRAL AMERICA, OR SOUTH AMERICA, RATHER THAN IN NORTH AFRICA OR CENTRAL AND SOUTH AFRICA (22)	0.87	19	TEND TO BE THOSE LOCATED IN NORTH AFRICA OR CENTRAL AND SOUTH AFRICA, RATHER THAN IN THE CARIBBEAN, CENTRAL AMERICA, OR SOUTH AMERICA (33)	0.68	1 32 7 15 X SQ= 6.64 P = 0.005 RV YES (NO)
20	TILT TOWARD BEING THOSE LOCATED IN THE CARIBBEAN, CENTRAL AMERICA, OR SOUTH AMERICA, RATHER THAN IN EAST ASIA, SOUTH ASIA, OR SOUTHEAST ASIA (22)	0.78	20	TILT TOWARD BEING THOSE LOCATED IN EAST ASIA, SOUTH ASIA, OR SOUTHEAST ASIA, RATHER THAN IN THE CARIBBEAN, CENTRAL AMERICA, OR SOUTH AMERICA (18)	0.52	2 16 7 15 X SQ= 1.39 P = 0.149 RV YES (NO)
21	LEAN MORE TOWARD BEING THOSE WHOSE TERRITORIAL SIZE IS MEDIUM OR SMALL (83)	0.83	21	LEAN LESS TOWARD BEING THOSE WHOSE TERRITORIAL SIZE IS MEDIUM OR SMALL (83)	0.66	7 25 35 48 X SQ= 3.27 P = 0.053 RV NO (NO)
24	TEND TO BE THOSE WHOSE POPULATION IS VERY LARGE, LARGE, OR MEDIUM (61)	0.67	24	TEND TO BE THOSE WHOSE POPULATION IS SMALL (54)	0.55	28 33 14 40 X SQ= 4.11 P = 0.033 RV YES (NO)
26	TEND TO BE THOSE WHOSE POPULATION DENSITY IS VERY HIGH, HIGH, OR MEDIUM (48)	0.67	26	TEND TO BE THOSE WHOSE POPULATION DENSITY IS LOW (67)	0.73	28 20 14 53 X SQ= 15.33 P = 0. RV YES (YES)
28	TEND TO BE THOSE WHOSE POPULATION GROWTH RATE IS LOW (48)	0.68	28	TEND TO BE THOSE WHOSE POPULATION GROWTH RATE IS HIGH (62)	0.71	13 49 28 20 X SQ= 14.60 P = 0. RV YES (YES)
29	TEND TO BE THOSE WHERE THE DEGREE OF URBANIZATION IS HIGH (56)	0.95	29	TEND TO BE THOSE WHERE THE DEGREE OF URBANIZATION IS LOW (49)	0.72	38 18 2 47 X SQ= 42.41 P = 0. RV YES (YES)
30	TEND TO BE THOSE WHOSE AGRICULTURAL POPULATION IS MEDIUM, LOW, OR VERY LOW (57)	0.95	30	TEND TO BE THOSE WHOSE AGRICULTURAL POPULATION IS HIGH (56)	0.76	2 54 40 17 X SQ= 50.84 P = 0. RV YES (YES)

31	TEND TC BE THOSE WHOSE AGRICULTURAL POPULATION IS LOW OR VERY LOW (24)	0.57	31	ALWAYS ARE THOSE WHOSE AGRICULTURAL POPULATION IS HIGH OR MEDIUM (90)	1.00	18 72 24 0 X SQ= 48.73 P = 0. RV YES (YES)
33	TEND TO BE THOSE WHOSE GROSS NATIONAL PRODUCT IS VERY HIGH, HIGH, OR MEDIUM (30)	0.55	33	TEND TO BE THOSE WHOSE GROSS NATIONAL PRODUCT IS LOW OR VERY LOW (85)	0.90	23 7 19 66 X SQ= 25.92 P = 0. RV YES (YES)
34	TEND TO BE THOSE WHOSE GROSS NATIONAL PRODUCT IS VERY HIGH, HIGH, MEDIUM, OR LOW (62)	0.86	34	TEND TO BE THOSE WHOSE GROSS NATIONAL PRODUCT IS VERY LOW (53)	0.64	36 26 6 47 X SQ= 24.95 P = 0. RV YES (YES)
39	TEND TO BE THOSE WHOSE INTERNATIONAL FINANCIAL STATUS IS VERY HIGH, HIGH, OR MEDIUM (38)	0.68	39	TEND TO BE THOSE WHOSE INTERNATIONAL FINANCIAL STATUS IS LOW OR VERY LOW (76)	0.86	28 10 13 63 X SQ= 32.80 P = 0. RV YES (YES)
40	TEND TO BE THOSE WHOSE INTERNATIONAL FINANCIAL STATUS IS VERY HIGH, HIGH, MEDIUM, OR LOW (71)	0.95	40	TEND TO BE THOSE WHOSE INTERNATIONAL FINANCIAL STATUS IS VERY LOW (39)	0.53	38 33 2 37 X SQ= 23.43 P = 0. RV YES (YES)
41	TEND LESS TO BE THOSE WHOSE ECONOMIC DEVELOPMENTAL STATUS IS INTERMEDIATE, UNDERDEVELOPED, OR VERY UNDERDEVELOPED (94)	0.52	41	ALWAYS ARE THOSE WHOSE ECONOMIC DEVELOPMENTAL STATUS IS INTERMEDIATE, UNDERDEVELOPED, OR VERY UNDERDEVELOPED (94)	1.00	19 0 21 73 X SQ= 38.36 P = 0. RV NO (YES)
42	TEND TO BE THOSE WHOSE ECONOMIC DEVELOPMENTAL STATUS IS DEVELOPED OR INTERMEDIATE (36)	0.82	42	TEND TO BE THOSE WHOSE ECONOMIC DEVELOPMENTAL STATUS IS UNDERDEVELOPED OR VERY UNDERDEVELOPED (76)	0.96	33 3 7 69 X SQ= 68.79 P = 0. RV YES (YES)
43	ALWAYS ARE THOSE WHOSE ECONOMIC DEVELOPMENTAL STATUS IS DEVELOPED, INTERMEDIATE, OR UNDERDEVELOPED (55)	1.00	43	TEND TO BE THOSE WHOSE ECONOMIC DEVELOPMENTAL STATUS IS VERY UNDERDEVELOPED (57)	0.80	41 14 0 57 X SQ= 63.85 P = 0. RV YES (YES)
44	TEND TO BE THOSE WHERE THE LITERACY RATE IS NINETY PERCENT OR ABOVE (25)	0.60	44	ALWAYS ARE THOSE WHERE THE LITERACY RATE IS BELOW NINETY PERCENT (90)	1.00	25 0 17 73 X SQ= 52.08 P = 0. RV YES (YES)

45	TEND TO BE THOSE WHERE THE LITERACY RATE IS FIFTY PERCENT OR ABOVE (55)	0.95	45	TEND TO BE THOSE WHERE THE LITERACY RATE IS BELOW FIFTY PERCENT (54)	0.76	39 16 2 52 X SQ= 49.62 P = 0. RV YES (YES)
46	ALWAYS ARE THOSE WHERE THE LITERACY RATE IS TEN PERCENT OR ABOVE (84)	1.00	46	TEND LESS TO BE THOSE WHERE THE LITERACY RATE IS TEN PERCENT OR ABOVE (84)	0.62	42 42 0 26 X SQ= 18.96 P = 0. RV NO (YES)
50	TEND TO BE THOSE WHERE FREEDOM OF THE PRESS IS COMPLETE (43)	0.61	50	TEND TO BE THOSE WHERE FREEDOM OF THE PRESS IS INTERMITTENT, INTERNALLY ABSENT, OR INTERNALLY AND EXTERNALLY ABSENT (56)	0.69	25 18 16 40 X SQ= 7.59 P = 0.004 RV YES (YES)
51	LEAN MORE TOWARD BEING THOSE WHERE FREEDOM OF THE PRESS IS COMPLETE OR INTERMITTENT (60)	0.73	51	LEAN LESS TOWARD BEING THOSE WHERE FREEDOM OF THE PRESS IS COMPLETE OR INTERMITTENT (60)	0.54	30 30 11 26 X SQ= 3.07 P = 0.059 RV NO (YES)
53	TEND LESS TO BE THOSE WHERE NEWSPAPER CIRCULATION IS LESS THAN THREE HUNDRED PER THOUSAND (101)	0.67	53	ALWAYS ARE THOSE WHERE NEWSPAPER CIRCULATION IS LESS THAN THREE HUNDRED PER THOUSAND (101)	1.00	14 0 28 73 X SQ= 24.68 P = 0. RV NO (YES)
54	TEND TO BE THOSE WHERE NEWSPAPER CIRCULATION IS ONE HUNDRED OR MORE PER THOUSAND (37)	0.81	54	TEND TO BE THOSE WHERE NEWSPAPER CIRCULATION IS LESS THAN ONE HUNDRED PER THOUSAND (76)	0.96	34 3 8 68 X SQ= 67.10 P = 0. RV YES (YES)
55	ALWAYS ARE THOSE WHERE NEWSPAPER CIRCULATION IS TEN OR MORE PER THOUSAND (78)	1.00	55	TEND LESS TO BE THOSE WHERE NEWSPAPER CIRCULATION IS TEN OR MORE PER THOUSAND (78)	0.51	42 36 0 35 X SQ= 27.73 P = 0. RV NO (YES)
56	TEND MORE TO BE THOSE WHERE THE RELIGION IS PREDOMINANTLY LITERATE (79)	0.98	56	TEND LESS TO BE THOSE WHERE THE RELIGION IS PREDOMINANTLY LITERATE (79)	0.56	41 38 1 30 X SQ= 20.33 P = 0. RV NO (YES)
57	LEAN LESS TOWARD BEING THOSE WHERE THE RELIGION IS OTHER THAN CATHOLIC (90)	0.69	57	LEAN MORE TOWARD BEING THOSE WHERE THE RELIGION IS OTHER THAN CATHOLIC (90)	0.84	13 12 29 61 X SQ= 2.50 P = 0.099 RV NO (YES)

No.	Description	Value	No.	Description	Value	Statistics
58	ALWAYS ARE THOSE WHERE THE RELIGION IS OTHER THAN MUSLIM (97)	1.00	58	TEND LESS TO BE THOSE WHERE THE RELIGION IS OTHER THAN MUSLIM (97)	0.75	0 18 42 55 X SQ= 10.48 P = 0. RV NO (NO)
62	LEAN LESS TOWARD BEING THOSE WHERE THE RELIGION IS CATHOLIC, RATHER THAN PROTESTANT (25)	0.72	62	ALWAYS ARE THOSE WHERE THE RELIGION IS CATHOLIC, RATHER THAN PROTESTANT (25)	1.00	5 0 13 12 X SQ= 2.25 P = 0.066 RV NO (NO)
63	TEND TO BE THOSE WHERE THE RELIGION IS PREDOMINANTLY SOME KIND OF CHRISTIAN (46)	0.83	63	TEND TO BE THOSE WHERE THE RELIGION IS PREDOMINANTLY OR PARTLY OTHER THAN CHRISTIAN (68)	0.84	M 34 12 7 61 X SQ= 45.50 P = 0. RV YES (YES)
67	TILT MORE TOWARD BEING THOSE THAT ARE RACIALLY HOMOGENEOUS (82)	0.83	67	TILT LESS TOWARD BEING THOSE THAT ARE RACIALLY HOMOGENEOUS (82)	0.68	34 46 7 22 X SQ= 2.33 P = 0.117 RV NO (NO)
68	TEND TO BE THOSE THAT ARE LINGUISTICALLY HOMOGENEOUS (52)	0.64	68	TEND TO BE THOSE THAT ARE LINGUISTICALLY WEAKLY HETEROGENEOUS OR STRONGLY HETEROGENEOUS (62)	0.65	M 27 25 15 47 X SQ= 8.19 P = 0.003 RV YES (YES)
69	TEND TO BE THOSE THAT ARE LINGUISTICALLY HOMOGENEOUS OR WEAKLY HETEROGENEOUS (64)	0.76	69	TEND TO BE THOSE THAT ARE LINGUISTICALLY STRONGLY HETEROGENEOUS (50)	0.56	32 32 10 40 X SQ= 9.61 P = 0.002 RV YES (YES)
70	TEND LESS TO BE THOSE THAT ARE RELIGIOUSLY, RACIALLY, OR LINGUISTICALLY HETEROGENEOUS (85)	0.68	70	TEND MORE TO BE THOSE THAT ARE RELIGIOUSLY, RACIALLY, OR LINGUISTICALLY HETEROGENEOUS (85)	0.87	13 8 28 56 X SQ= 4.62 P = 0.024 RV NO (YES)
72	TEND TO BE THOSE WHOSE DATE OF INDEPENDENCE IS BEFORE 1914 (52)	0.67	72	TEND TO BE THOSE WHOSE DATE OF INDEPENDENCE IS AFTER 1914 (59)	0.65	27 25 13 46 X SQ= 9.45 P = 0.001 RV YES (YES)
73	TEND TO BE THOSE WHOSE DATE OF INDEPENDENCE IS BEFORE 1945 (65)	0.87	73	TEND TO BE THOSE WHOSE DATE OF INDEPENDENCE IS AFTER 1945 (46)	0.58	35 30 5 41 X SQ= 19.76 P = 0. RV YES (YES)

74	TEND TO BE THOSE THAT ARE HISTORICALLY WESTERN (26)	0.62	74	TEND TO BE THOSE THAT ARE NOT HISTORICALLY WESTERN (87)	0.99	25 1 15 72 X SQ= 51.11 P = 0. RV YES (YES)
75	TEND TO BE THOSE THAT ARE HISTORICALLY WESTERN OR SIGNIFICANTLY WESTERNIZED (62)	0.98	75	TEND TO BE THOSE THAT ARE NOT HISTORICALLY WESTERN AND ARE NOT SIGNIFICANTLY WESTERNIZED (52)	0.70	40 22 1 51 X SQ= 45.43 P = 0. RV YES (YES)
76	TEND TO BE THOSE THAT ARE HISTORICALLY WESTERN, RATHER THAN HAVING BEEN SIGNIFICANTLY OR PARTIALLY WESTERNIZED THROUGH A COLONIAL RELATIONSHIP (26)	0.69	76	TEND TO BE THOSE THAT HAVE BEEN SIGNIFICANTLY OR PARTIALLY WESTERNIZED THROUGH A COLONIAL RELATIONSHIP, RATHER THAN BEING HISTORICALLY WESTERN (70)	0.98	25 1 11 59 X SQ= 48.96 P = 0. RV YES (YES)
77	TEND TO BE THOSE THAT HAVE BEEN SIGNIFICANTLY WESTERNIZED, RATHER THAN PARTIALLY WESTERNIZED, THROUGH A COLONIAL RELATIONSHIP (28)	0.90	77	TEND TO BE THOSE THAT HAVE BEEN PARTIALLY WESTERNIZED, RATHER THAN SIGNIFICANTLY WESTERNIZED, THROUGH A COLONIAL RELATIONSHIP (41)	0.68	9 19 1 40 X SQ= 9.57 P = 0.001 RV YES (NO)
78	LEAN LESS TOWARD BEING THOSE THAT HAVE BEEN SIGNIFICANTLY WESTERNIZED THROUGH A COLONIAL RELATIONSHIP, RATHER THAN WITHOUT SUCH A RELATIONSHIP (28)	0.64	78	LEAN MORE TOWARD BEING THOSE THAT HAVE BEEN SIGNIFICANTLY WESTERNIZED THROUGH A COLONIAL RELATIONSHIP, RATHER THAN WITHOUT SUCH A RELATIONSHIP (28)	0.90	9 19 5 2 X SQ= 2.15 P = 0.090 RV NO (YES)
80	TEND TO BE THOSE WHOSE DATE OF INDEPENDENCE IS AFTER 1914, AND THAT WERE FORMERLY DEPENDENCIES OF BRITAIN, RATHER THAN FRANCE (19)	0.86	80	TEND TO BE THOSE WHOSE DATE OF INDEPENDENCE IS AFTER 1914, AND THAT WERE FORMERLY DEPENDENCIES OF FRANCE, RATHER THAN BRITAIN (24)	0.64	6 13 1 23 X SQ= 4.01 P = 0.033 RV YES (NO)
81	TEND LESS TO BE THOSE WHERE THE TYPE OF POLITICAL MODERNIZATION IS LATER EUROPEAN OR LATER EUROPEAN DERIVED, RATHER THAN EARLY EUROPEAN OR EARLY EUROPEAN DERIVED (40)	0.68	81	ALWAYS ARE THOSE WHERE THE TYPE OF POLITICAL MODERNIZATION IS LATER EUROPEAN OR LATER EUROPEAN DERIVED, RATHER THAN EARLY EUROPEAN OR EARLY EUROPEAN DERIVED (40)	1.00	11 0 23 17 X SQ= 5.23 P = 0.009 RV NO (NO)
82	TEND TO BE THOSE WHERE THE TYPE OF POLITICAL MODERNIZATION IS EARLY OR LATER EUROPEAN OR EUROPEAN DERIVED, RATHER THAN DEVELOPED TUTELARY OR UNDEVELOPED TUTELARY (51)	0.85	82	TEND TO BE THOSE WHERE THE TYPE OF POLITICAL MODERNIZATION IS DEVELOPED TUTELARY OR UNDEVELOPED TUTELARY, RATHER THAN EARLY OR LATER EUROPEAN OR EUROPEAN DERIVED (55)	0.74	34 17 6 49 X SQ= 32.68 P = 0. RV YES (YES)
84	ALWAYS ARE THOSE WHERE THE TYPE OF POLITICAL MODERNIZATION IS DEVELOPED TUTELARY, RATHER THAN UNDEVELOPED TUTELARY (31)	1.00	84	TEND LESS TO BE THOSE WHERE THE TYPE OF POLITICAL MODERNIZATION IS DEVELOPED TUTELARY, RATHER THAN UNDEVELOPED TUTELARY (31)	0.51	6 25 0 24 X SQ= 3.41 P = 0.030 RV NO (NO)

85	TEND TO BE THOSE WHERE THE STAGE OF POLITICAL MODERNIZATION IS ADVANCED, RATHER THAN MID- OR EARLY TRANSITIONAL (60)	0.90	TEND TO BE THOSE WHERE THE STAGE OF POLITICAL MODERNIZATION IS MID- OR EARLY TRANSITIONAL, RATHER THAN ADVANCED (54)	0.69	38 22 4 50 X SQ= 35.84 P = 0. RV YES (YES)
87	TEND TO BE THOSE WHOSE IDEOLOGICAL ORIENTATION IS OTHER THAN DEVELOPMENTAL (58)	0.92	TEND TO BE THOSE WHOSE IDEOLOGICAL ORIENTATION IS DEVELOPMENTAL (31)	0.55	M 3 28 35 23 X SQ= 19.18 P = 0. RV YES (YES)
89	TEND TO BE THOSE WHOSE IDEOLOGICAL ORIENTATION IS CONVENTIONAL (33)	0.68	TEND TO BE THOSE WHOSE IDEOLOGICAL ORIENTATION IS OTHER THAN CONVENTIONAL (62)	0.88	26 7 12 50 X SQ= 29.27 P = 0. RV YES (YES)
94	TEND TO BE THOSE WHERE THE STATUS OF THE REGIME IS CONSTITUTIONAL (51)	0.75	TEND TO BE THOSE WHERE THE STATUS OF THE REGIME IS AUTHORITARIAN OR TOTALITARIAN (41)	0.60	30 21 10 31 X SQ= 9.61 P = 0.001 RV YES (YES)
95	TEND LESS TO BE THOSE WHERE THE STATUS OF THE REGIME IS CONSTITUTIONAL OR AUTHORITARIAN (95)	0.76	TEND MORE TO BE THOSE WHERE THE STATUS OF THE REGIME IS CONSTITUTIONAL OR AUTHORITARIAN (95)	0.91	31 64 10 6 X SQ= 4.04 P = 0.047 RV NO (YES)
96	ALWAYS ARE THOSE WHERE THE STATUS OF THE REGIME IS CONSTITUTIONAL OR TOTALITARIAN (67)	1.00	TEND LESS TO BE THOSE WHERE THE STATUS OF THE REGIME IS CONSTITUTIONAL OR TOTALITARIAN (67)	0.54	40 27 0 23 X SQ= 22.36 P = 0. RV NO (YES)
97	ALWAYS ARE THOSE WHERE THE STATUS OF THE REGIME IS TOTALITARIAN, RATHER THAN AUTHORITARIAN (16)	1.00	TEND TO BE THOSE WHERE THE STATUS OF THE REGIME IS AUTHORITARIAN, RATHER THAN TOTALITARIAN (23)	0.79	M 0 23 10 6 X SQ= 16.19 P = 0. RV YES (YES)
99	TEND TO BE THOSE WHERE GOVERNMENTAL STABILITY IS GENERALLY PRESENT AND DATES FROM AT LEAST THE INTER-WAR PERIOD, OR FROM THE POST-WAR PERIOD (50)	0.86	TEND TO BE THOSE WHERE GOVERNMENTAL STABILITY IS MODERATELY PRESENT AND DATES FROM THE POST-WAR PERIOD, OR IS ABSENT (36)	0.63	32 18 5 31 X SQ= 19.45 P = 0. RV YES (YES)
101	TEND TO BE THOSE WHERE THE REPRESENTATIVE CHARACTER OF THE REGIME IS POLYARCHIC (41)	0.70	TEND TO BE THOSE WHERE THE REPRESENTATIVE CHARACTER OF THE REGIME IS LIMITED POLYARCHIC, PSEUDO-POLYARCHIC, OR NON-POLYARCHIC (57)	0.78	28 13 12 45 X SQ= 20.12 P = 0. RV YES (YES)

102	TEND TO BE THOSE WHERE THE REPRESENTATIVE CHARACTER OF THE REGIME IS POLYARCHIC OR LIMITED POLYARCHIC (59)	0.76	102	TEND TO BE THOSE WHERE THE REPRESENTATIVE CHARACTER OF THE REGIME IS PSEUDO-POLYARCHIC OR NON-POLYARCHIC (49)	0.59	32 27 10 39 X SQ= 11.51 P = 0. RV YES (YES)
105	TEND TO BE THOSE WHERE THE ELECTORAL SYSTEM IS COMPETITIVE (43)	0.76	105	TEND TO BE THOSE WHERE THE ELECTORAL SYSTEM IS PARTIALLY COMPETITIVE OR NON-COMPETITIVE (47)	0.73	29 14 9 38 X SQ= 19.53 P = 0. RV YES (YES)
107	TEND TO BE THOSE WHERE AUTONOMOUS GROUPS ARE FULLY TOLERATED IN POLITICS (46)	0.76	107	TEND TO BE THOSE WHERE AUTONOMOUS GROUPS ARE PARTIALLY TOLERATED IN POLITICS, ARE TOLERATED ONLY OUTSIDE POLITICS, OR ARE NOT TOLERATED AT ALL (65)	0.80	32 14 10 55 X SQ= 31.35 P = 0. RV YES (YES)
111	TEND TO BE THOSE WHERE POLITICAL ENCULTURATION IS HIGH OR MEDIUM (53)	0.74	111	TEND TO BE THOSE WHERE POLITICAL ENCULTURATION IS LOW (42)	0.54	25 28 9 33 X SQ= 5.68 P = 0.011 RV YES (NO)
114	LEAN TOWARD BEING THOSE WHERE SECTIONALISM IS NEGLIGIBLE (47)	0.55	114	LEAN TOWARD BEING THOSE WHERE SECTIONALISM IS EXTREME OR MODERATE (61)	0.63	18 43 22 25 X SQ= 2.71 P = 0.074 RV YES (NO)
116	TEND TO BE THOSE WHERE INTEREST ARTICULATION BY ASSOCIATIONAL GROUPS IS SIGNIFICANT OR MODERATE (32)	0.70	116	TEND TO BE THOSE WHERE INTEREST ARTICULATION BY ASSOCIATIONAL GROUPS IS LIMITED OR NEGLIGIBLE (79)	0.94	28 4 12 67 X SQ= 48.57 P = 0. RV YES (YES)
117	TEND TO BE THOSE WHERE INTEREST ARTICULATION BY ASSOCIATIONAL GROUPS IS SIGNIFICANT, MODERATE, OR LIMITED (60)	0.90	117	TEND TO BE THOSE WHERE INTEREST ARTICULATION BY ASSOCIATIONAL GROUPS IS NEGLIGIBLE (51)	0.66	36 24 4 47 X SQ= 30.31 P = 0. RV YES (YES)
119	TEND TO BE THOSE WHERE INTEREST ARTICULATION BY INSTITUTIONAL GROUPS IS MODERATE OR LIMITED (26)	0.57	119	TEND TO BE THOSE WHERE INTEREST ARTICULATION BY INSTITUTIONAL GROUPS IS VERY SIGNIFICANT OR SIGNIFICANT (74)	0.97	18 56 24 2 X SQ= 33.77 P = 0. RV YES (YES)
120	TEND LESS TO BE THOSE WHERE INTEREST ARTICULATION BY INSTITUTIONAL GROUPS IS VERY SIGNIFICANT, SIGNIFICANT, OR MODERATE (90)	0.76	120	ALWAYS ARE THOSE WHERE INTEREST ARTICULATION BY INSTITUTIONAL GROUPS IS VERY SIGNIFICANT, SIGNIFICANT, OR MODERATE (90)	1.00	32 58 10 0 X SQ= 12.81 P = 0. RV NO (YES)

No.	Left statement	Value	No.	Right statement	Value	Statistics
121	TEND TO BE THOSE WHERE INTEREST ARTICULATION BY NON-ASSOCIATIONAL GROUPS IS MODERATE, LIMITED, OR NEGLIGIBLE (61)	0.90	121	TEND TO BE THOSE WHERE INTEREST ARTICULATION BY NON-ASSOCIATIONAL GROUPS IS SIGNIFICANT (54)	0.68	M 4 50 38 23 X SQ= 34.89 P = 0. RV YES (YES)
122	TEND TO BE THOSE WHERE INTEREST ARTICULATION BY NON-ASSOCIATIONAL GROUPS IS LIMITED OR NEGLIGIBLE (32)	0.62	122	TEND TO BE THOSE WHERE INTEREST ARTICULATION BY NON-ASSOCIATIONAL GROUPS IS SIGNIFICANT OR MODERATE (83)	0.92	16 67 26 6 X SQ= 35.63 P = 0. RV YES (YES)
123	TEND LESS TO BE THOSE WHERE INTEREST ARTICULATION BY NON-ASSOCIATIONAL GROUPS IS SIGNIFICANT, MODERATE, OR LIMITED (107)	0.81	123	ALWAYS ARE THOSE WHERE INTEREST ARTICULATION BY NON-ASSOCIATIONAL GROUPS IS SIGNIFICANT, MODERATE, OR LIMITED (107)	1.00	34 73 8 0 X SQ= 12.15 P = 0. RV NO (YES)
125	TEND TO BE THOSE WHERE INTEREST ARTICULATION BY ANOMIC GROUPS IS INFREQUENT OR VERY INFREQUENT (35)	0.74	125	TEND TO BE THOSE WHERE INTEREST ARTICULATION BY ANOMIC GROUPS IS FREQUENT OR OCCASIONAL (64)	0.85	M 9 55 25 10 X SQ= 30.53 P = 0. RV YES (YES)
126	TEND LESS TO BE THOSE WHERE INTEREST ARTICULATION BY ANOMIC GROUPS IS FREQUENT, OCCASIONAL, OR INFREQUENT (83)	0.53	126	ALWAYS ARE THOSE WHERE INTEREST ARTICULATION BY ANOMIC GROUPS IS FREQUENT, OCCASIONAL, OR INFREQUENT (83)	1.00	18 65 16 0 X SQ= 33.09 P = 0. RV NO (YES)
128	TEND TO BE THOSE WHERE INTEREST ARTICULATION BY POLITICAL PARTIES IS SIGNIFICANT OR MODERATE (48)	0.69	128	TEND TO BE THOSE WHERE INTEREST ARTICULATION BY POLITICAL PARTIES IS LIMITED OR NEGLIGIBLE (45)	0.61	M 27 21 12 33 X SQ= 7.18 P = 0.006 RV YES (YES)
133	TEND MORE TO BE THOSE WHERE INTEREST AGGREGATION BY THE EXECUTIVE IS MODERATE, LIMITED, OR NEGLIGIBLE (76)	0.89	133	TEND LESS TO BE THOSE WHERE INTEREST AGGREGATION BY THE EXECUTIVE IS MODERATE, LIMITED, OR NEGLIGIBLE (76)	0.63	4 25 34 42 X SQ= 7.41 P = 0.003 RV NO (NO)
134	LEAN TOWARD BEING THOSE WHERE INTEREST AGGREGATION BY THE EXECUTIVE IS SIGNIFICANT OR MODERATE (57)	0.68	134	LEAN TOWARD BEING THOSE WHERE INTEREST AGGREGATION BY THE EXECUTIVE IS LIMITED OR NEGLIGIBLE (46)	0.52	26 31 12 34 X SQ= 3.37 P = 0.064 RV YES (NO)
135	TEND MORE TO BE THOSE WHERE INTEREST AGGREGATION BY THE EXECUTIVE IS SIGNIFICANT, MODERATE, OR LIMITED (77)	0.97	135	TEND LESS TO BE THOSE WHERE INTEREST AGGREGATION BY THE EXECUTIVE IS SIGNIFICANT, MODERATE, OR LIMITED (77)	0.79	32 45 1 12 X SQ= 4.13 P = 0.027 RV NO (NO)

136	TEND LESS TO BE THOSE WHERE INTEREST AGGREGATION BY THE LEGISLATURE IS MODERATE, LIMITED, OR NEGLIGIBLE (85)	0.70	136	ALWAYS ARE THOSE WHERE INTEREST AGGREGATION BY THE LEGISLATURE IS MODERATE, LIMITED, OR NEGLIGIBLE (85)	1.00	12 0 28 57 X SQ= 16.84 P = 0. RV NO (YES)
137	TEND TO BE THOSE WHERE INTEREST AGGREGATION BY THE LEGISLATURE IS SIGNIFICANT OR MODERATE (29)	0.65	137	TEND TO BE THOSE WHERE INTEREST AGGREGATION BY THE LEGISLATURE IS LIMITED OR NEGLIGIBLE (68)	0.95	26 3 14 54 X SQ= 37.22 P = 0. RV YES (YES)
138	TEND TO BE THOSE WHERE INTEREST AGGREGATION BY THE LEGISLATURE IS SIGNIFICANT, MODERATE, OR LIMITED (48)	0.77	138	TEND TO BE THOSE WHERE INTEREST AGGREGATION BY THE LEGISLATURE IS NEGLIGIBLE (49)	0.70	31 17 9 40 X SQ= 19.51 P = 0. RV YES (YES)
139	LEAN MORE TOWARD BEING THOSE WHERE THE PARTY SYSTEM IS QUANTITATIVELY OTHER THAN ONE-PARTY (71)	0.79	139	LEAN LESS TOWARD BEING THOSE WHERE THE PARTY SYSTEM IS QUANTITATIVELY OTHER THAN ONE-PARTY (71)	0.60	9 25 33 38 X SQ= 3.05 P = 0.058 RV NO (NO)
140	LEAN MORE TOWARD BEING THOSE WHERE THE PARTY SYSTEM IS QUANTITATIVELY OTHER THAN ONE PARTY DOMINANT (87)	0.95	140	LEAN LESS TOWARD BEING THOSE WHERE THE PARTY SYSTEM IS QUANTITATIVELY OTHER THAN ONE PARTY DOMINANT (87)	0.82	2 11 38 49 X SQ= 2.69 P = 0.070 RV NO (NO)
141	TILT LESS TOWARD BEING THOSE WHERE THE PARTY SYSTEM IS QUANTITATIVELY OTHER THAN TWO-PARTY (87)	0.82	141	TILT MORE TOWARD BEING THOSE WHERE THE PARTY SYSTEM IS QUANTITATIVELY OTHER THAN TWO-PARTY (87)	0.93	7 4 32 55 X SQ= 1.93 P = 0.108 RV NO (YES)
142	TEND LESS TO BE THOSE WHERE THE PARTY SYSTEM IS QUANTITATIVELY OTHER THAN MULTI-PARTY (66)	0.56	142	TEND MORE TO BE THOSE WHERE THE PARTY SYSTEM IS QUANTITATIVELY OTHER THAN MULTI-PARTY (66)	0.77	17 13 22 44 X SQ= 3.74 P = 0.043 RV NO (YES)
146	ALWAYS ARE THOSE WHERE THE PARTY SYSTEM IS QUALITATIVELY OTHER THAN MASS-BASED TERRITORIAL (92)	1.00	146	TEND LESS TO BE THOSE WHERE THE PARTY SYSTEM IS QUALITATIVELY OTHER THAN MASS-BASED TERRITORIAL (92)	0.87	0 8 39 53 X SQ= 3.92 P = 0.021 RV NO (NO)
147	TEND TO BE THOSE WHERE THE PARTY SYSTEM IS QUALITATIVELY CLASS-ORIENTED OR MULTI-IDEOLOGICAL (23)	0.57	147	TEND TO BE THOSE WHERE THE PARTY SYSTEM IS QUALITATIVELY OTHER THAN CLASS-ORIENTED OR MULTI-IDEOLOGICAL (67)	0.96	21 2 16 51 X SQ= 29.43 P = 0. RV YES (YES)

148	ALWAYS ARE THOSE 1.00 WHERE THE PARTY SYSTEM IS QUALITATIVELY OTHER THAN AFRICAN TRANSITIONAL (96)	148	TEND LESS TO BE THOSE 0.79 WHERE THE PARTY SYSTEM IS QUALITATIVELY OTHER THAN AFRICAN TRANSITIONAL (96)	0 14 42 54 X SQ= 8.14 P = 0.001 RV NO (NO)
153	TEND TO BE THOSE 0.78 WHERE THE PARTY SYSTEM IS STABLE (42)	153	TEND TO BE THOSE 0.70 WHERE THE PARTY SYSTEM IS MODERATELY STABLE OR UNSTABLE (38)	M 29 13 8 30 X SQ= 16.61 P = 0. RV YES (YES)
158	TEND TO BE THOSE 0.83 WHERE PERSONALISMO IS NEGLIGIBLE (56)	158	TEND TO BE THOSE 0.60 WHERE PERSONALISMO IS PRONOUNCED OR MODERATE (40)	7 33 34 22 X SQ= 16.09 P = 0. RV YES (YES)
161	TEND TO BE THOSE 0.67 WHERE THE POLITICAL LEADERSHIP IS NON-ELITIST (50)	161	TEND TO BE THOSE 0.59 WHERE THE POLITICAL LEADERSHIP IS ELITIST OR MODERATELY ELITIST (47)	13 34 26 24 X SQ= 5.00 P = 0.022 RV YES (YES)
164	TEND TO BE THOSE 0.90 WHERE THE REGIME'S LEADERSHIP CHARISMA IS NEGLIGIBLE (65)	164	TEND TO BE THOSE 0.51 WHERE THE REGIME'S LEADERSHIP CHARISMA IS PRONOUNCED OR MODERATE (34)	M 4 30 36 29 X SQ= 15.87 P = 0. RV YES (YES)
166	LEAN LESS TOWARD BEING THOSE 0.79 WHERE THE VERTICAL POWER DISTRIBUTION IS THAT OF FORMAL FEDERALISM OR FORMAL AND EFFECTIVE UNITARISM (99)	166	LEAN MORE TOWARD BEING THOSE 0.92 WHERE THE VERTICAL POWER DISTRIBUTION IS THAT OF FORMAL FEDERALISM OR FORMAL AND EFFECTIVE UNITARISM (99)	9 6 33 66 X SQ= 2.92 P = 0.082 RV NO (YES)
168	TEND TO BE THOSE 0.68 WHERE THE HORIZONTAL POWER DISTRIBUTION IS SIGNIFICANT (34)	168	TEND TO BE THOSE 0.91 WHERE THE HORIZONTAL POWER DISTRIBUTION IS LIMITED OR NEGLIGIBLE (72)	28 6 13 59 X SQ= 37.59 P = 0. RV YES (YES)
169	TEND TO BE THOSE 0.76 WHERE THE HORIZONTAL POWER DISTRIBUTION IS SIGNIFICANT OR LIMITED (58)	169	TEND TO BE THOSE 0.58 WHERE THE HORIZONTAL POWER DISTRIBUTION IS NEGLIGIBLE (48)	M 31 27 10 38 X SQ= 10.44 P = 0.001 RV YES (YES)
170	TEND TO BE THOSE 0.88 WHERE THE LEGISLATIVE-EXECUTIVE STRUCTURE IS OTHER THAN PRESIDENTIAL (63)	170	TEND TO BE THOSE 0.56 WHERE THE LEGISLATIVE-EXECUTIVE STRUCTURE IS PRESIDENTIAL (39)	5 34 36 27 X SQ= 17.88 P = 0. RV YES (YES)

171	LEAN LESS TOWARD BEING THOSE WHERE THE LEGISLATIVE-EXECUTIVE STRUCTURE IS OTHER THAN PARLIAMENTARY-REPUBLICAN (90)	0.80	171	LEAN MORE TOWARD BEING THOSE WHERE THE LEGISLATIVE-EXECUTIVE STRUCTURE IS OTHER THAN PARLIAMENTARY-REPUBLICAN (90)	0.94	8 4 32 58 X SQ= 3.09 P = 0.057 RV NO (YES)
172	TEND LESS TO BE THOSE WHERE THE LEGISLATIVE-EXECUTIVE STRUCTURE IS OTHER THAN PARLIAMENTARY-ROYALIST (88)	0.63	172	TEND MORE TO BE THOSE WHERE THE LEGISLATIVE-EXECUTIVE STRUCTURE IS OTHER THAN PARLIAMENTARY-ROYALIST (88)	0.91	15 6 26 62 X SQ= 10.95 P = 0.001 RV NO (YES)
174	TEND TO BE THOSE WHERE THE LEGISLATURE IS FULLY EFFECTIVE (28)	0.63	174	TEND TO BE THOSE WHERE THE LEGISLATURE IS PARTIALLY EFFECTIVE, LARGELY INEFFECTIVE, OR WHOLLY INEFFECTIVE (72)	0.97	26 2 15 57 X SQ= 40.31 P = 0. RV YES (YES)
175	TEND TO BE THOSE WHERE THE LEGISLATURE IS FULLY EFFECTIVE OR PARTIALLY EFFECTIVE (51)	0.78	175	TEND TO BE THOSE WHERE THE LEGISLATURE IS LARGELY INEFFECTIVE OR WHOLLY INEFFECTIVE (49)	0.68	32 19 9 40 X SQ= 18.55 P = 0. RV YES (YES)
178	TEND TO BE THOSE WHERE THE LEGISLATURE IS BICAMERAL (51)	0.67	178	TEND TO BE THOSE WHERE THE LEGISLATURE IS UNICAMERAL (53)	0.62	13 40 27 24 X SQ= 7.71 P = 0.005 RV YES (YES)
179	TEND TO BE THOSE WHERE THE EXECUTIVE IS STRONG (39)	0.68	179	TEND TO BE THOSE WHERE THE EXECUTIVE IS DOMINANT (52)	0.74	12 40 25 14 X SQ= 13.89 P = 0. RV YES (YES)
181	TEND TO BE THOSE WHERE THE BUREAUCRACY IS MODERN, RATHER THAN SEMI-MODERN (21)	0.50	181	ALWAYS ARE THOSE WHERE THE BUREAUCRACY IS SEMI-MODERN, RATHER THAN MODERN (55)	1.00	21 0 21 34 X SQ= 21.06 P = 0. RV YES (YES)
182	ALWAYS ARE THOSE WHERE THE BUREAUCRACY IS SEMI-MODERN, RATHER THAN POST-COLONIAL TRANSITIONAL (55)	1.00	182	TEND LESS TO BE THOSE WHERE THE BUREAUCRACY IS SEMI-MODERN, RATHER THAN POST-COLONIAL TRANSITIONAL (55)	0.58	21 34 0 25 X SQ= 11.05 P = 0. RV NO (NO)
185	TEND TO BE THOSE WHERE PARTICIPATION BY THE MILITARY IN POLITICS IS SUPPORTIVE, RATHER THAN INTERVENTIVE (31)	0.92	185	TEND TO BE THOSE WHERE PARTICIPATION BY THE MILITARY IN POLITICS IS INTERVENTIVE, RATHER THAN SUPPORTIVE (21)	0.50	1 20 11 20 X SQ= 5.04 P = 0.017 RV YES (NO)

No.	Left	Left value	No.	Right	Right value	Statistics
187	TEND TO BE THOSE WHERE THE ROLE OF THE POLICE IS NOT POLITICALLY SIGNIFICANT (35)	0.65	187	TEND TO BE THOSE WHERE THE ROLE OF THE POLICE IS POLITICALLY SIGNIFICANT (66)	0.85	14 52 26 9 X SQ= 24.76 P = 0. RV YES (YES)
189	TEND LESS TO BE THOSE WHERE THE CHARACTER OF THE LEGAL SYSTEM IS OTHER THAN COMMON LAW (108)	0.83	189	ALWAYS ARE THOSE WHERE THE CHARACTER OF THE LEGAL SYSTEM IS OTHER THAN COMMON LAW (108)	1.00	7 0 35 73 X SQ= 10.20 P = 0.001 RV NO (YES)
190	TEND LESS TO BE THOSE WHERE THE CHARACTER OF THE LEGAL SYSTEM IS CIVIL LAW, RATHER THAN COMMON LAW (32)	0.67	190	ALWAYS ARE THOSE WHERE THE CHARACTER OF THE LEGAL SYSTEM IS CIVIL LAW, RATHER THAN COMMON LAW (32)	1.00	14 18 7 0 X SQ= 5.22 P = 0.010 RV NO (YES)
192	TEND MORE TO BE THOSE WHERE THE CHARACTER OF THE LEGAL SYSTEM IS OTHER THAN MUSLIM OR PARTLY MUSLIM (86)	0.98	192	TEND LESS TO BE THOSE WHERE THE CHARACTER OF THE LEGAL SYSTEM IS OTHER THAN MUSLIM OR PARTLY MUSLIM (86)	0.63	1 26 41 45 X SQ= 15.18 P = 0. RV NO (NO)
193	TEND MORE TO BE THOSE WHERE THE CHARACTER OF THE LEGAL SYSTEM IS OTHER THAN PARTLY INDIGENOUS (86)	0.98	193	TEND LESS TO BE THOSE WHERE THE CHARACTER OF THE LEGAL SYSTEM IS OTHER THAN PARTLY INDIGENOUS (86)	0.62	1 27 41 45 X SQ= 15.81 P = 0. RV NO (NO)
194	LEAN LESS TOWARD BEING THOSE THAT ARE NON-COMMUNIST (101)	0.80	194	LEAN MORE TOWARD BEING THOSE THAT ARE NON-COMMUNIST (101)	0.93	8 5 33 68 X SQ= 3.01 P = 0.063 RV NO (YES)

M

MATRIX

36 POLITIES
WHOSE PER CAPITA GROSS NATIONAL PRODUCT
IS VERY HIGH, HIGH, OR MEDIUM (42)

36 POLITIES
WHOSE PER CAPITA GROSS NATIONAL PRODUCT
IS LOW OR VERY LOW (73)

26 POLITIES
WHOSE POPULATION DENSITY IS
VERY HIGH, HIGH, OR MEDIUM (48)

AUSTRIA	BELGIUM	BULGARIA	CUBA	CYPRUS
CZECHOS'KIA	DENMARK	FRANCE	GERMANY, E	GERMAN FR
GREECE	HUNGARY	IRELAND	ISRAEL	ITALY
JAMAICA	JAPAN	LEBANON	LUXEMBOURG	MALAYA
NETHERLANDS	POLAND	RUMANIA	SPAIN	SWITZERLAND
TRINIDAD	UK	YUGOSLAVIA		

ALBANIA	BURUNDI	CEYLON	CHINA, PR	DOMIN REP
EL SALVADOR	HAITI	INDIA	INDONESIA	KOREA, N
KOREA REP	NEPAL	NIGERIA	PAKISTAN	PHILIPPINES
PORTUGAL	RWANDA	THAILAND	VIETNAM, N	VIETNAM REP

28 . 20

14 . 53

26 POLITIES
WHOSE POPULATION DENSITY IS
LOW (67)

ARGENTINA	AUSTRALIA	CANADA	CHILE	FINLAND
ICELAND	NEW ZEALAND	NORWAY	SO AFRICA	SWEDEN
USSR	US	URUGUAY	VENEZUELA	

AFGHANISTAN	ALGERIA	BOLIVIA	BRAZIL	BURMA
CAMBODIA	CAMERCUN	CEN AFR REP	CHAD	COLOMBIA
CONGO(BRA)	CONGO(LEO)	COSTA RICA	DAHOMEY	ECUADOR
ETHIOPIA	GABON	GHANA	GUATEMALA	GUINEA
HONDURAS	IRAN	IRAQ	IVORY COAST	JORDAN
LAOS	LIBERIA	LIBYA	MALAGASY R	MALI
MAURITANIA	MEXICO	MONGOLIA	MOROCCO	NICARAGUA
NIGER	PANAMA	PARAGUAY	PERU	SA'U ARABIA
SENEGAL	SIERRE LEO	SOMALIA	SUDAN	SYRIA
TANGANYIKA	TOGO	TUNISIA	TURKEY	UGANDA
UAR	UPPER VOLTA	YEMEN		

MATRIX

	36 POLITIES WHOSE PER CAPITA GROSS NATIONAL PRODUCT IS VERY HIGH, HIGH, OR MEDIUM (42)	36 POLITIES WHOSE PER CAPITA GROSS NATIONAL PRODUCT IS LOW OR VERY LOW (73)
63 POLITIES WHERE THE RELIGION IS PREDOMINANTLY SOME KIND OF CHRISTIAN (46)	34	12
63 POLITIES WHERE THE RELIGION IS PREDOMINANTLY OR PARTLY OTHER THAN CHRISTIAN (68)	7	61

High GNP, Christian (34):
ARGENTINA, AUSTRALIA, AUSTRIA, BELGIUM, BULGARIA, CANADA, CHILE, CUBA, CZECHOS'KIA, DENMARK, FINLAND, FRANCE, GERMANY, E, GERMAN FR, GREECE, HUNGARY, ICELAND, IRELAND, ITALY, JAMAICA, LUXEMBOURG, NETHERLANDS, NEW ZEALAND, NORWAY, POLAND, RUMANIA, SPAIN, SWEDEN, SWITZERLAND, UK, US, URUGUAY, VENEZUELA, YUGOSLAVIA

Low GNP, Christian (12):
BRAZIL, COLOMBIA, COSTA RICA, DOMIN REP, EL SALVADOR, HONDURAS, MEXICO, NICARAGUA, PANAMA, PARAGUAY, PHILIPPINES, PORTUGAL

High GNP, other than Christian (7):
CYPRUS, ISRAEL, JAPAN, LEBANON, MALAYA, SO AFRICA, TRINIDAD

Low GNP, other than Christian (61):
AFGHANISTAN, ALBANIA, ALGERIA, BOLIVIA, BURMA, BURUNDI, CAMBODIA, CAMEROUN, CEN AFR REP, CEYLON, CHAD, CHINA, PR, CONGO(BRA), CONGO(LEO), DAHOMEY, ECUADOR, ETHIOPIA, GABON, GHANA, GUATEMALA, GUINEA, HAITI, INDIA, INDONESIA, IRAN, IRAQ, IVORY COAST, JORDAN, KOREA, N, KOREA REP, LAOS, LIBERIA, LIBYA, MALAGASY R, MALI, MAURITANIA, MONGOLIA, MOROCCO, NEPAL, NIGER, NIGERIA, PAKISTAN, PERU, RWANDA, SA'U ARABIA, SENEGAL, SIERRE LEO, SOMALIA, SUDAN, SYRIA, TANGANYIKA, THAILAND, TOGO, TUNISIA, TURKEY

M

```
*************************************************MATRIX*************************************************

         36  POLITIES                                                36  POLITIES
             WHOSE PER CAPITA GROSS NATIONAL PRODUCT                     WHOSE PER CAPITA GROSS NATIONAL PRODUCT
             IS VERY HIGH, HIGH, OR MEDIUM  (42)                         IS LOW OR VERY LOW  (73)

                          68  POLITIES
                              THAT ARE LINGUISTICALLY
                              HOMOGENEOUS  (52)

ARGENTINA  AUSTRALIA  AUSTRIA      CHILE        CUBA        .ALBANIA      BRAZIL       BURUNDI   COLOMBIA     COSTA RICA
DENMARK    FRANCE     GERMANY, E   GERMAN FR    GREECE      .DOMIN REP    EL SALVADOR  HAITI     HONDURAS     JORDAN
HUNGARY    ICELAND    IRELAND      ITALY        JAMAICA     .KOREA, N     KOREA REP    LIBYA     MALAGASY R   MEXICO
JAPAN      LEBANON    LUXEMBOURG   NETHERLANDS  NEW ZEALAND.MONGOLIA      NICARAGUA    PARAGUAY  PORTUGAL     RWANDA
NORWAY     POLAND     SWEDEN       UK           US          .SA'U ARABIA  SOMALIA      TUNISIA   UAR          YEMEN
URUGUAY    VENEZUELA

                                                        27 .  25
. . . . . . . . . . . . . . . . . . . . . . . . . . . . . . . . . . . . . . . . . . . . . . . . . . . .
                                                        15 .  47

                          68  POLITIES
                              THAT ARE LINGUISTICALLY
                              WEAKLY HETEROGENEOUS OR
                              STRONGLY HETEROGENEOUS  (62)

BELGIUM    BULGARIA     CANADA     CYPRUS     CZECHOS'KIA.AFGHANISTAN  ALGERIA       BOLIVIA       BURMA        CAMBODIA
FINLAND    ISRAEL       MALAYA     RUMANIA    SO AFRICA  .CAMEROUN     CEN AFR REP   CEYLON        CHAD         CONGO(BRA)
SPAIN      SWITZERLAND  TRINIDAD   USSR       YUGOSLAVIA .CONGO(LEO)   DAHOMEY       ECUADOR       ETHIOPIA     GABON
                                                         .GHANA        GUATEMALA     GUINEA        INDIA        INDONESIA
                                                         .IRAN         IRAQ          IVORY COAST   LAOS         LIBERIA
                                                         .MALI         MAURITANIA    MOROCCO       NEPAL        NIGER
                                                         .NIGERIA      PAKISTAN      PANAMA        PERU         PHILIPPINES
                                                         .SENEGAL      SIERRE LEO    SUDAN         SYRIA        TANGANYIKA
                                                         .THAILAND     TOGO          TURKEY        UGANDA       UPPER VOLTA
                                                         .VIETNAM, N   VIETNAM REP
                                                         .UGANDA       UAR           UPPER VOLTA   VIETNAM, N   VIETNAM REP
                                                         .YEMEN
```

MATRIX

	36 POLITIES WHOSE PER CAPITA GROSS NATIONAL PRODUCT IS VERY HIGH, HIGH, OR MEDIUM (42)	36 POLITIES WHOSE PER CAPITA GROSS NATIONAL PRODUCT IS LOW OR VERY LOW (73)
87 POLITIES WHOSE IDEOLOGICAL ORIENTATION IS DEVELOPMENTAL (31)	ISRAEL, MALAYA, VENEZUELA (3)	ALGERIA, BOLIVIA, BURUNDI, CAMEROUN, CEN AFR REP, CHAD, CONGO(BRA), CONGO(LEO), DAHOMEY, DOMIN REP, GABON, GHANA, GUINEA, INDONESIA, IVORY COAST, MALI, MAURITANIA, NIGER, RWANDA, SENEGAL, SIERRE LEO, SOMALIA, SUDAN, TANGANYIKA, TOGO, TUNISIA, UAR, UPPER VCLTA (28)
87 POLITIES WHOSE IDEOLOGICAL ORIENTATION IS OTHER THAN DEVELOPMENTAL (58)	AUSTRALIA, AUSTRIA, BELGIUM, BULGARIA, CANADA, CHILE, CYPRUS, CZECHOS'KIA, DENMARK, FINLAND, FRANCE, GERMANY, E, GERMAN FR, GREECE, HUNGARY, ICELAND, IRELAND, ITALY, JAPAN, LEBANON, LUXEMBOURG, NETHERLANDS, NEW ZEALAND, NORWAY, POLAND, RUMANIA, SO AFRICA, SPAIN, SWEDEN, SWITZERLAND, USSR, UK, US, URUGUAY, YUGOSLAVIA (35)	AFGHANISTAN, ALBANIA, CHINA, PR, COLOMBIA, COSTA RICA, ECUADOR, EL SALVADOR, ETHIOPIA, HAITI, HONDURAS, IRAQ, JORDAN, KOREA, N, LAOS, MEXICO, MONGOLIA, PHILIPPINES, PORTUGAL, SA'U ARABIA, SYRIA, THAILAND, TURKEY, VIETNAM, N (23)

M

MATRIX

36 POLITIES WHOSE PER CAPITA GROSS NATIONAL PRODUCT IS VERY HIGH, HIGH, OR MEDIUM (42)	36 POLITIES WHOSE PER CAPITA GROSS NATIONAL PRODUCT IS LOW OR VERY LOW (73)
97 POLITIES WHERE THE STATUS OF THE REGIME IS AUTHORITARIAN, RATHER THAN TOTALITARIAN (23)	
0	23 AFGHANISTAN, ALGERIA, BURMA, CAMBODIA, CEN AFR REP, EL SALVADOR, ETHIOPIA, GHANA, GUINEA, INDONESIA, IRAN, JORDAN, KOREA REP, LAOS, LIBERIA, NEPAL, NICARAGUA, PAKISTAN, PARAGUAY, SA'U ARABIA, SUDAN, THAILAND, UAR
97 POLITIES WHERE THE STATUS OF THE REGIME IS TOTALITARIAN, RATHER THAN AUTHORITARIAN (16)	
10 BULGARIA, CUBA, CZECHOS'KIA, GERMANY, E, HUNGARY, POLAND, RUMANIA, SPAIN, USSR, YUGOSLAVIA	6 ALBANIA, CHINA, PR, KOREA, N, MONGOLIA, PORTUGAL, VIETNAM, N

M

```
*********************************************MATRIX*********************************************

        36  POLITIES                                                     36  POLITIES
            WHOSE PER CAPITA GROSS NATIONAL PRODUCT      .                   WHOSE PER CAPITA GROSS NATIONAL PRODUCT
            IS VERY HIGH, HIGH, OR MEDIUM  (42)          .                   IS LOW OR VERY LOW  (73)

                          119  POLITIES
                               WHERE INTEREST ARTICULATION
                               BY INSTITUTIONAL GROUPS
                               IS VERY SIGNIFICANT OR SIGNIFICANT  (74)

ARGENTINA   BULGARIA   CUBA       CYPRUS      CZECHOS'KIA.AFGHANISTAN ALBANIA     ALGERIA     BOLIVIA     BRAZIL
GERMANY, E  HUNGARY    ISRAEL     ITALY       JAPAN      .BURMA       CAMBODIA    CEYLON      CHINA, PR   COLOMBIA
LEBANON     MALAYA     POLAND     RUMANIA     SO AFRICA  .CONGO(LEO)  DOMIN REP   ECUADOR     EL SALVADOR ETHIOPIA
SPAIN       USSR       VENEZUELA  YUGOSLAVIA             .GHANA       GUATEMALA   GUINEA      HAITI       HONDURAS
                                                         .INDIA       INDONESIA   IRAN        IRAQ        IVORY COAST
                                                         .JORDAN      KOREA, N    KOREA REP   LAOS        LIBERIA
                                                         .LIBYA       MAURITANIA  MEXICO      MONGOLIA    MOROCCO
                                                         .NEPAL       NICARAGUA   NIGERIA     PAKISTAN    PANAMA
                                                         .PARAGUAY    PERU        PORTUGAL    SA'U ARABIA SENEGAL
                                                         .SUDAN       SYRIA       TANGANYIKA  THAILAND    TUNISIA
                                                         .TURKEY      UGANDA      UAR         VIETNAM, N  VIETNAM REP

                                                      19 .  55
. . . . . . . . . . . . . . . . . . . . . . . . . . . . . . . . . . . . . . . . . . . . . . . . . . . . . . . . .
                                                      23 .   3

                          119  POLITIES
                               WHERE INTEREST ARTICULATION
                               BY INSTITUTIONAL GROUPS
                               IS MODERATE OR LIMITED  (26)

AUSTRALIA   AUSTRIA    BELGIUM    CANADA      CHILE      .COSTA RICA  MALAGASY R  PHILIPPINES
DENMARK     FINLAND    FRANCE     GERMAN FR   GREECE     .
ICELAND     IRELAND    JAMAICA    LUXEMBOURG  NETHERLANDS.
NEW ZEALAND NORWAY     SWEDEN     SWITZERLAND TRINIDAD   .
UK          US         URUGUAY                           .
```

M

MATRIX

	36 POLITIES WHOSE PER CAPITA GROSS NATIONAL PRODUCT IS VERY HIGH, HIGH, OR MEDIUM (42)	36 POLITIES WHOSE PER CAPITA GROSS NATIONAL PRODUCT IS LOW OR VERY LOW (73)
121 POLITIES WHERE INTEREST ARTICULATION BY NON-ASSOCIATIONAL GROUPS IS SIGNIFICANT (54)	CYPRUS, LEBANON, MALAYA, SO AFRICA 4	AFGHANISTAN, BURMA, BURUNDI, CAMBODIA, CAMEROUN, CEN AFR REP, CEYLON, CHAD, CHINA, PR, CONGO(BRA), CONGO(LEO), DAHOMEY, ETHIOPIA, GABON, GHANA, GUINEA, HAITI, INDIA, INDONESIA, IRAN, IRAQ, IVCRY COAST, JORDAN, KOREA, N, KOREA REP, LAOS, MALAGASY R, MALI, MAURITANIA, MONGOLIA, NEPAL, NIGER, NIGERIA, PAKISTAN, PHILIPPINES, RWANDA, SA'U ARABIA, SENEGAL, SIERRE LEO, SOMALIA, SUDAN, SYRIA, TANGANYIKA, THAILAND, TOGO, UGANDA, UPPER VOLTA, VIETNAM, N, VIETNAM REP, YEMEN 50
121 POLITIES WHERE INTEREST ARTICULATION BY NON-ASSOCIATIONAL GROUPS IS MODERATE, LIMITED, OR NEGLIGIBLE (61)	38 ARGENTINA, AUSTRALIA, AUSTRIA, BELGIUM, BULGARIA, CANADA, CHILE, CUBA, CZECHOS'KIA, DENMARK, FINLAND, FRANCE, GERMANY, E, GERMAN FR, GREECE, HUNGARY, ICELAND, IRELAND, ISRAEL, ITALY, JAMAICA, JAPAN, LUXEMBOURG, NETHERLANDS, NEW ZEALAND, NORWAY, POLAND, RUMANIA, SPAIN, SWEDEN, SWITZERLAND, TRINIDAD, USSR, UK, US, URUGUAY, VENEZUELA, YUGOSLAVIA	23 ALBANIA, ALGERIA, BOLIVIA, BRAZIL, COLOMBIA, COSTA RICA, DOMIN REP, ECUADOR, EL SALVADOR, GUATEMALA, HONDURAS, LIBERIA, LIBYA, MEXICO, MOROCCO, NICARAGUA, PANAMA, PARAGUAY, PERU, PORTUGAL, TUNISIA, TURKEY, UAR

MATRIX

	36 POLITIES WHOSE PER CAPITA GROSS NATIONAL PRODUCT IS VERY HIGH, HIGH, OR MEDIUM (42)	36 POLITIES WHOSE PER CAPITA GROSS NATIONAL PRODUCT IS LOW OR VERY LOW (73)
125 POLITIES WHERE INTEREST ARTICULATION BY ANOMIC GROUPS IS FREQUENT OR OCCASIONAL (64)	9	55
125 POLITIES WHERE INTEREST ARTICULATION BY ANOMIC GROUPS IS INFREQUENT OR VERY INFREQUENT (35)	24	11

125 POLITIES WHERE INTEREST ARTICULATION BY ANOMIC GROUPS IS FREQUENT OR OCCASIONAL (64)

High/medium GNP (9):

GERMANY, E   JAPAN   LEBANON   MALAYA   POLAND
SO AFRICA   SPAIN   USSR   YUGOSLAVIA

Low/very low GNP (55):

AFGHANISTAN   BOLIVIA   BRAZIL   BURMA   CAMBODIA
CAMEROUN   CEN AFR REP   CEYLON   CHAD   CHINA, PR
COLOMBIA   CONGO(BRA)   CONGO(LEO)   DAHOMEY   ECUADOR
EL SALVADOR   ETHIOPIA   GABON   GHANA   GUATEMALA
GUINEA   HAITI   HONDURAS   INDIA   INDONESIA
IRAN   IRAQ   IVORY COAST   JORDAN   KOREA REP
LAOS   MALI   MAURITANIA   MEXICO   MOROCCO
NEPAL   NIGER   NIGERIA   PAKISTAN   PANAMA
PARAGUAY   PERU   RWANDA   SA'U ARABIA   SENEGAL
SIERRE LEO   SOMALIA   SUDAN   SYRIA   THAILAND
TOGO   TURKEY   UGANDA   UPPER VOLTA   VIETNAM REP

125 POLITIES WHERE INTEREST ARTICULATION BY ANOMIC GROUPS IS INFREQUENT OR VERY INFREQUENT (35)

High/medium GNP (24):

AUSTRALIA   AUSTRIA   BELGIUM   CANADA   CHILE
CZECHOS'KIA   DENMARK   FINLAND   GERMAN FR   GREECE
ICELAND   IRELAND   ISRAEL   ITALY   JAMAICA
LUXEMBOURG   NETHERLANDS   NEW ZEALAND   NORWAY   SWEDEN
SWITZERLAND   TRINIDAD   UK   URUGUAY

Low/very low GNP (11):

BURUNDI   COSTA RICA   LIBERIA   LIBYA   MALAGASY R
NICARAGUA   PHILIPPINES   PORTUGAL   TANGANYIKA   TUNISIA
UAR

M

MATRIX

	36 POLITIES WHOSE PER CAPITA GROSS NATIONAL PRODUCT IS VERY HIGH, HIGH, OR MEDIUM (42)	36 POLITIES WHOSE PER CAPITA GROSS NATIONAL PRODUCT IS LOW OR VERY LOW (73)
128 POLITIES WHERE INTEREST ARTICULATION BY POLITICAL PARTIES IS SIGNIFICANT OR MODERATE (48)	ARGENTINA, DENMARK, ICELAND, JAPAN, NEW ZEALAND, UK, AUSTRALIA, FINLAND, IRELAND, LEBANON, NORWAY, VENEZUELA, AUSTRIA, FRANCE, ISRAEL, LUXEMBOURG, SWEDEN, BELGIUM, GERMAN FR, ITALY, MALAYA, SWITZERLAND, CHILE, GREECE, JAMAICA, NETHERLANDS, TRINIDAD 27	BOLIVIA, CONGO(LEO), INDIA, PANAMA, UGANDA, BRAZIL, ECUADOR, LIBERIA, PERU, CAMEROUN, EL SALVADOR, MALAGASY R, SIERRE LEO, CEYLON, GUATEMALA, MOROCCO, SOMALIA, COLOMBIA, HONDURAS, NIGERIA, SYRIA 21
128 POLITIES WHERE INTEREST ARTICULATION BY POLITICAL PARTIES IS LIMITED OR NEGLIGIBLE (45)	11 BULGARIA, POLAND, YUGOSLAVIA, CANADA, RUMANIA, CZECHOS'KIA, SPAIN, GERMANY, E, USSR, HUNGARY, US	34 ALBANIA, CHINA, PR, GABON, MALI, NIGER, SENEGAL, UAR, BURUNDI, CONGO(BRA), GHANA, MAURITANIA, PARAGUAY, TANGANYIKA, UPPER VOLTA, CAMBODIA, COSTA RICA, GUINEA, MEXICO, PHILIPPINES, TOGO, VIETNAM, N, CEN AFR REP, DAHOMEY, IVORY COAST, MONGOLIA, PORTUGAL, TUNISIA, VIETNAM REP, CHAD, DOMIN REP, KOREA, N, NICARAGUA, RWANDA, TURKEY

M

MATRIX

	36 POLITIES WHOSE PER CAPITA GROSS NATIONAL PRODUCT IS VERY HIGH, HIGH, OR MEDIUM (42)	36 POLITIES WHOSE PER CAPITA GROSS NATIONAL PRODUCT IS LOW OR VERY LOW (73)
153 POLITIES WHERE THE PARTY SYSTEM IS STABLE (42)	AUSTRALIA, CZECHCS'KIA, HUNGARY, LUXEMBOURG, RUMANIA, USSR, AUSTRIA, DENMARK, ICELAND, NETHERLANDS, SO AFRICA, UK, BELGIUM, FINLAND, IRELAND, NEW ZEALAND, SPAIN, US, BULGARIA, GERMANY, E, ISRAEL, NORWAY, SWEDEN, YUGOSLAVIA, CANADA, GERMAN FR, JAPAN, POLAND, SWITZERLAND	ALBANIA, LIBERIA, TANGANYIKA, CHINA, PR, MEXICO, TUNISIA, GUINEA, MONGOLIA, VIETNAM, N, INDIA, PHILIPPINES, KOREA, N, PORTUGAL
	29	13
	8	30
153 POLITIES WHERE THE PARTY SYSTEM IS MODERATELY STABLE OR UNSTABLE (38)	ARGENTINA, LEBANON, CHILE, MALAYA, FRANCE, VENEZUELA, GREECE, ITALY	BOLIVIA, COSTA RICA, HONDURAS, LAOS, PANAMA, SOMALIA, BRAZIL, ECUADOR, IRAQ, MALAGASY R, PARAGUAY, SYRIA, CAMBODIA, EL SALVADOR, IVORY COAST, NEPAL, PERU, THAILAND, CEYLON, GHANA, JORDAN, NICARAGUA, SENEGAL, TURKEY, CONGO(LEO), GUATEMALA, KOREA REP, PAKISTAN, SIERRE LEO, VIETNAM REP

MATRIX

36 POLITIES
WHOSE PER CAPITA GROSS NATIONAL PRODUCT
IS VERY HIGH, HIGH, OR MEDIUM (42)

36 POLITIES
WHOSE PER CAPITA GROSS NATIONAL PRODUCT
IS LOW OR VERY LOW (73)

164 POLITIES
WHERE THE REGIME'S LEADERSHIP CHARISMA
IS PRONOUNCED OR MODERATE (34)

CUBA FRANCE JAMAICA YUGOSLAVIA

ALGERIA, CAMBODIA, CAMEROUN, CEN AFR REP, CHAD, CHINA, PR, CONGO(BRA), DAHOMEY, ETHIOPIA, GABON, GHANA, GUINEA, INDIA, INDONESIA, IVORY COAST, JORDAN, KOREA, N, LIBYA, MALI, MAURITANIA, MOROCCO, NIGER, PAKISTAN, SA'U ARABIA, SENEGAL, TANGANYIKA, TUNISIA, UAR, UPPER VOLTA, VIETNAM, N

4	30
36	29

164 POLITIES
WHERE THE REGIME'S LEADERSHIP CHARISMA
IS NEGLIGIBLE (65)

ARGENTINA, AUSTRALIA, AUSTRIA, BELGIUM, BULGARIA, CANADA, CHILE, CZECHOS'KIA, DENMARK, FINLAND, GERMANY, E, GERMAN FR, GREECE, HUNGARY, ICELAND, IRELAND, ISRAEL, ITALY, JAPAN, LEBANON, LUXEMBOURG, MALAYA, NETHERLANDS, NEW ZEALAND, NORWAY, POLAND, RUMANIA, SO AFRICA, SPAIN, SWEDEN, SWITZERLAND, TRINIDAD, UK, US, URUGUAY, VENEZUELA

ALBANIA, BOLIVIA, BRAZIL, CEYLON, COLOMBIA, COSTA RICA, DOMIN REP, ECUADOR, EL SALVADOR, GUATEMALA, HAITI, HONDURAS, KOREA REP, LIBERIA, MALAGASY R, MEXICO, MONGOLIA, NICARAGUA, PANAMA, PARAGUAY, PERU, PHILIPPINES, PORTUGAL, SIERRE LEO, SOMALIA, SUDAN, THAILAND, TURKEY, VIETNAM REP

M

MATRIX

36 POLITIES WHOSE PER CAPITA GROSS NATIONAL PRODUCT IS VERY HIGH, HIGH, OR MEDIUM (42)	36 POLITIES WHOSE PER CAPITA GROSS NATIONAL PRODUCT IS LOW OR VERY LOW (73)
169 POLITIES WHERE THE HORIZONTAL POWER DISTRIBUTION IS SIGNIFICANT OR LIMITED (58)	
AUSTRALIA, AUSTRIA, BELGIUM, CANADA, CHILE, CYPRUS, DENMARK, FINLAND, FRANCE, GERMAN FR, GREECE, ICELAND, IRELAND, ISRAEL, ITALY, JAMAICA, JAPAN, LEBANON, LUXEMBOURG, MALAYA, NETHERLANDS, NEW ZEALAND, NORWAY, SO AFRICA, SWEDEN, SWITZERLAND, TRINIDAD, UK, US, URUGUAY, VENEZUELA	BOLIVIA, BRAZIL, BURUNDI, CAMEROUN, CEYLON, COLOMBIA, COSTA RICA, DOMIN REP, ECUADOR, EL SALVADOR, GUATEMALA, HONDURAS, INDIA, LIBYA, MALAGASY R, MALI, MAURITANIA, MEXICO, MOROCCO, NICARAGUA, NIGERIA, PANAMA, PHILIPPINES, RWANDA, SIERRE LEO, SOMALIA, TURKEY
31	27
10	38
169 POLITIES WHERE THE HORIZONTAL POWER DISTRIBUTION IS NEGLIGIBLE (48)	
BULGARIA, CUBA, CZECHOS'KIA, GERMANY, E, HUNGARY, POLAND, RUMANIA, SPAIN, USSR, YUGOSLAVIA	AFGHANISTAN, ALBANIA, ALGERIA, CAMBODIA, CEN AFR REP, CHAD, CHINA, PR, CONGO(BRA), DAHOMEY, ETHIOPIA, GABON, GHANA, GUINEA, HAITI, INDONESIA, IRAN, IVORY COAST, JORDAN, KOREA, N, LAOS, LIBERIA, MONGOLIA, NEPAL, NIGER, PAKISTAN, PARAGUAY, PORTUGAL, SA'U ARABIA, SENEGAL, SUDAN, TANGANYIKA, THAILAND, TOGO, TUNISIA, UAR, UPPER VOLTA, VIETNAM, N, VIETNAM REP

37 POLITIES
WHOSE PER CAPITA GROSS NATIONAL PRODUCT
IS VERY HIGH, HIGH, MEDIUM, OR LOW (64)

37 POLITIES
WHOSE PER CAPITA GROSS NATIONAL PRODUCT
IS VERY LOW (51)

PREDICATE ONLY

64 IN LEFT COLUMN

ALBANIA	ALGERIA	ARGENTINA	AUSTRALIA	AUSTRIA	BELGIUM	BRAZIL	BULGARIA	CANADA	CHILE
COLOMBIA	COSTA RICA	CUBA	CYPRUS	CZECHOS'KIA	DENMARK	DOMIN REP	ECUADOR	EL SALVADOR	FINLAND
FRANCE	GERMANY, E	GERMAN FR	GREECE	GUATEMALA	HONDURAS	HUNGARY	ICELAND	IRAQ	IRELAND
ISRAEL	ITALY	JAMAICA	JAPAN	LEBANON	LUXEMBOURG	MALAYA	MEXICO	NETHERLANDS	NEW ZEALAND
NICARAGUA	NORWAY	PANAMA	PERU	PHILIPPINES	POLAND	PORTUGAL	RUMANIA	SA'U ARABIA	SO AFRICA
SPAIN	SWEDEN	SWITZERLAND	SYRIA	TRINIDAD	TUNISIA	TURKEY	USSR	UAR	UK
US	URUGUAY	VENEZUELA	YUGOSLAVIA						

51 IN RIGHT COLUMN

AFGHANISTAN	BOLIVIA	BURMA	BURUNDI	CAMBODIA	CAMEROUN	CEN AFR REP	CEYLON	CHAD	CHINA, PR
CONGO(BRA)	CONGO(LEO)	DAHOMEY	ETHIOPIA	GABON	GHANA	GUINEA	HAITI	INDIA	INDONESIA
IRAN	IVORY COAST	JORDAN	KOREA, N	KOREA REP	LAOS	LIBERIA	LIBYA	MALAGASY R	MALI
MAURITANIA	MONGOLIA	MOROCCO	NEPAL	NIGER	NIGERIA	PAKISTAN	PARAGUAY	RWANDA	SENEGAL
SIERRE LEO	SOMALIA	SUDAN	TANGANYIKA	THAILAND	TOGO	UGANDA	UPPER VOLTA	VIETNAM, N	VIETNAM REP
YEMEN									

38 POLITIES
WHOSE INTERNATIONAL FINANCIAL STATUS
IS VERY HIGH OR HIGH (10)

38 POLITIES
WHOSE INTERNATIONAL FINANCIAL STATUS
IS MEDIUM, LOW, OR VERY LOW (103)

PREDICATE ONLY

10 IN LEFT COLUMN

AUSTRALIA CANADA FRANCE GERMAN FR INDIA ITALY JAPAN USSR UK US

103 IN RIGHT COLUMN

AFGHANISTAN ALBANIA ALGERIA ARGENTINA AUSTRIA BELGIUM BOLIVIA BRAZIL BULGARIA BURMA
BURUNDI CAMBODIA CAMEROUN CEN AFR REP CEYLON CHAD CHILE COLOMPIA CONGO(BRA) CONGO(LEO)
COSTA RICA CUBA CYPRUS CZECHOS'KIA DAHOMEY DENMARK DOMIN REP ECUADOR EL SALVADOR ETHIOPIA
FINLAND GABON GHANA GREECE GUATEMALA GUINEA HAITI HONDURAS HUNGARY ICELAND
INDONESIA IRAN IRAQ IRELAND ISRAEL IVORY COAST JAMAICA JORDAN KOREA, N KOREA REP
LAOS LEBANON LIBERIA LIBYA LUXEMBOURG MALAGASY R MALAYA MALI MAURITANIA MEXICO
MONGOLIA MOROCCO NEPAL NETHERLANDS NEW ZEALAND NICARAGUA NIGER NIGERIA NORWAY PAKISTAN
PANAMA PARAGUAY PERU PHILIPPINES POLAND PORTUGAL RUMANIA RWANDA SA'U ARABIA SENEGAL
SIERRE LEO SOMALIA SO AFRICA SPAIN SUDAN SWEDEN SWITZERLAND SYRIA TANGANYIKA THAILAND
TOGO TRINIDAD TUNISIA TURKEY UGANDA UAR UPPER VOLTA URUGUAY VENEZUELA VIETNAM, N
VIETNAM REP YEMEN YUGOSLAVIA

2 FXCLUDED BECAUSE UNASCERTAINED

CHINA, PR GERMANY, E

```
************************************************************************************************************

39  POLITIES                                          39  POLITIES
    WHOSE INTERNATIONAL FINANCIAL STATUS                  WHOSE INTERNATIONAL FINANCIAL STATUS
    IS VERY HIGH, HIGH, OR MEDIUM  (38)                   IS LOW OR VERY LOW  (76)


                                                                                   BOTH SUBJECT AND PREDICATE
............................................................................................................

      38 IN LEFT COLUMN

ARGENTINA    AUSTRALIA   AUSTRIA     BELGIUM     BRAZIL      CANADA      CHILE       CHINA, PR   COLOMBIA    CZECHOS'KIA
DENMARK      FINLAND     FRANCE      GERMAN FR   HUNGARY     INDIA       INDONESIA   ITALY       JAPAN       MEXICO
NETHERLANDS  NEW ZEALAND NORWAY      PAKISTAN    PHILIPPINES POLAND      RUMANIA     SO AFRICA   SPAIN       SWEDEN
SWITZERLAND  TURKEY      USSR        UAR         UK          US          VENEZUELA   YUGOSLAVIA

      76 IN RIGHT COLUMN

AFGHANISTAN  ALBANIA     ALGERIA     BOLIVIA     BULGARIA    BURMA       BURUNDI     CAMBODIA    CAMEROUN    CEN AFR REP
CEYLON       CHAD        CONGO(BRA)  CONGO(LEO)  COSTA RICA  CUBA        CYPRUS      DAHOMEY     DOMIN REP   ECUADOR
EL SALVADOR  ETHIOPIA    GABON       GHANA       GREECE      GUATEMALA   GUINEA      HAITI       HONDURAS    ICELAND
IRAN         IRAQ        IRELAND     ISRAEL      IVORY COAST JAMAICA     JORDAN      KOREA, N    KOREA REP   LAOS
LEBANON      LIBERIA     LIBYA       LUXEMBOURG  MALAGASY R  MALAYA      MALI        MAURITANIA  MONGOLIA    MOROCCO
NEPAL        NICARAGUA   NIGER       NIGERIA     PANAMA      PARAGUAY    PERU        PORTUGAL    RWANDA      SA'U ARABIA
SENEGAL      SIERRE LEO  SOMALIA     SUDAN       SYRIA       TANGANYIKA  THAILAND    TOGO        TRINIDAD    TUNISIA
UGANDA       UPPER VOLTA URUGUAY     VIETNAM, N  VIETNAM REP YEMEN

      1 EXCLUDED BECAUSE UNASCERTAINED

GERMANY, E
------------------------------------------------------------------------------------------------------------

  2  TEND LESS TO BE THOSE                   0.55     2  TEND MORE TO BE THOSE                   0.93        17    5
     LOCATED ELSEWHERE THAN IN                           LOCATED ELSEWHERE THAN IN                           21   71
     WEST EUROPE, SCANDINAVIA,                           WEST EUROPE, SCANDINAVIA,                       X SQ=  21.30
     NORTH AMERICA, OR AUSTRALASIA  (93)                 NORTH AMERICA, OR AUSTRALASIA  (93)             P =    0.
                                                                                                         RV NO  (YES)

 10  TEND MORE TO BE THOSE                   0.95    10  TEND LESS TO BE THOSE                   0.59         2   31
     LOCATED ELSEWHERE THAN IN NORTH AFRICA,             LOCATED ELSEWHERE THAN IN NORTH AFRICA,             36   45
     OR CENTRAL AND SOUTH AFRICA  (82)                   OR CENTRAL AND SOUTH AFRICA  (82)               X SQ=  13.87
                                                                                                         P =    0.
                                                                                                         RV NO  (NO )

 13  LEAN MORE TOWARD BEING THOSE            0.95    13  LEAN LESS TOWARD BEING THOSE            0.82         2   14
     LOCATED ELSEWHERE THAN IN NORTH AFRICA OR           LOCATED ELSEWHERE THAN IN NORTH AFRICA OR           36   62
     THE MIDDLE EAST  (99)                               THE MIDDLE EAST  (99)                           X SQ=   2.63
                                                                                                         P =    0.084
                                                                                                         RV NO  (NO )
```

18	TEND TO BE THOSE LOCATED IN EAST ASIA, SOUTH ASIA, OR SOUTHEAST ASIA, RATHER THAN IN NORTH AFRICA OR CENTRAL AND SOUTH AFRICA (18)	0.75	18	TEND TO BE THOSE LOCATED IN NORTH AFRICA OR CENTRAL AND SOUTH AFRICA, RATHER THAN IN EAST ASIA, SOUTH ASIA, OR SOUTHEAST ASIA (33)	0.72	2 31 6 12 X SQ= 4.65 P = 0.017 RV YES (NO)
19	TEND TO BE THOSE LOCATED IN THE CARIBBEAN, CENTRAL AMERICA, OR SOUTH AMERICA, RATHER THAN IN NORTH AFRICA OR CENTRAL AND SOUTH AFRICA (22)	0.75	19	TEND TO BE THOSE LOCATED IN NORTH AFRICA OR CENTRAL AND SOUTH AFRICA, RATHER THAN IN THE CARIBBEAN, CENTRAL AMERICA, OR SOUTH AMERICA (33)	0.66	2 31 6 16 X SQ= 3.22 P = 0.049 RV YES (NO)
22	TEND TO BE THOSE WHOSE TERRITORIAL SIZE IS VERY LARGE, LARGE, OR MEDIUM (68)	0.82	22	TEND TO BE THOSE WHOSE TERRITORIAL SIZE IS SMALL (47)	0.51	31 37 7 39 X SQ= 10.06 P = 0.001 RV YES (NO)
23	TEND TO BE THOSE WHOSE POPULATION IS VERY LARGE OR LARGE (27)	0.58	23	TEND TO BE THOSE WHOSE POPULATION IS MEDIUM OR SMALL (88)	0.93	22 5 16 71 X SQ= 34.12 P = 0. RV YES (YES)
24	TEND TO BE THOSE WHOSE POPULATION IS VERY LARGE, LARGE, OR MEDIUM (61)	0.87	24	TEND TO BE THOSE WHOSE POPULATION IS SMALL (54)	0.64	33 27 5 49 X SQ= 24.74 P = 0. RV YES (YES)
26	TEND TO BE THOSE WHOSE POPULATION DENSITY IS VERY HIGH, HIGH, OR MEDIUM (48)	0.55	26	TEND TO BE THOSE WHOSE POPULATION DENSITY IS LOW (67)	0.66	21 26 17 50 X SQ= 3.81 P = 0.043 RV YES (NO)
28	TEND TO BE THOSE WHOSE POPULATION GROWTH RATE IS LOW (48)	0.57	28	TEND TO BE THOSE WHOSE POPULATION GROWTH RATE IS HIGH (62)	0.64	16 46 21 26 X SQ= 3.45 P = 0.044 RV YES (NO)
29	TEND TO BE THOSE WHERE THE DEGREE OF URBANIZATION IS HIGH (56)	0.89	29	TEND TO BE THOSE WHERE THE DEGREE OF URBANIZATION IS LOW (49)	0.66	32 23 4 45 X SQ= 26.48 P = 0. RV YES (YES)
30	TEND TO BE THOSE WHOSE AGRICULTURAL POPULATION IS MEDIUM, LOW, OR VERY LOW (57)	0.84	30	TEND TO BE THOSE WHOSE AGRICULTURAL POPULATION IS HIGH (56)	0.68	6 50 32 24 X SQ= 24.89 P = 0. RV YES (YES)

```
32  TEND LESS TO BE THOSE                  0.74     32  ALWAYS ARE THOSE                          1.00       10    0
    WHOSE GROSS NATIONAL PRODUCT                         WHOSE GROSS NATIONAL PRODUCT                         28   76
    IS MEDIUM, LOW, OR VERY LOW  (105)                   IS MEDIUM, LOW, OR VERY LOW  (105)              X SQ=  18.76
                                                                                                         P =     0.
                                                                                                         RV NO  (YES)

33  TEND TO BE THOSE                       0.76     33  ALWAYS ARE THOSE                          1.00       29    0
    WHOSE GROSS NATIONAL PRODUCT                         WHOSE GROSS NATIONAL PRODUCT                          9   76
    IS VERY HIGH, HIGH, OR MEDIUM  (30)                  IS LOW OR VERY LOW  (85)                        X SQ=  73.82
                                                                                                         P =     0.
                                                                                                         RV YES (YES)

34  ALWAYS ARE THOSE                       1.00     34  TEND TO BE THOSE                          0.70       38   23
    WHOSE GROSS NATIONAL PRODUCT                         WHOSE GROSS NATIONAL PRODUCT                          0   53
    IS VERY HIGH, HIGH, MEDIUM,                          IS VERY LOW  (53)                               X SQ=  46.76
    OR LOW  (62)                                                                                         P =     0.
                                                                                                         RV YES (YES)

35  TEND TO BE THOSE                       0.50     35  TEND TO BE THOSE                          0.95       19    4
    WHOSE PER CAPITA GROSS NATIONAL PRODUCT              WHOSE PER CAPITA GROSS NATIONAL PRODUCT              19   72
    IS VERY HIGH OR HIGH  (24)                           IS MEDIUM, LOW, OR VERY LOW  (91)               X SQ=  28.77
                                                                                                         P =     0.
                                                                                                         RV YES (YES)

36  TEND TO BE THOSE                       0.74     36  TEND TO BE THOSE                          0.83       28   13
    WHOSE PER CAPITA GROSS NATIONAL PRODUCT              WHOSE PER CAPITA GROSS NATIONAL PRODUCT              10   63
    IS VERY HIGH, HIGH, OR MEDIUM  (42)                  IS LOW OR VERY LOW  (73)                        X SQ=  32.80
                                                                                                         P =     0.
                                                                                                         RV YES (YES)

37  TEND TO BE THOSE                       0.89     37  TEND TO BE THOSE                          0.62       34   29
    WHOSE PER CAPITA GROSS NATIONAL PRODUCT              WHOSE PER CAPITA GROSS NATIONAL PRODUCT               4   47
    IS VERY HIGH, HIGH, MEDIUM, OR LOW  (64)             IS VERY LOW  (51)                               X SQ=  24.95
                                                                                                         P =     0.
                                                                                                         RV YES (YES)

42  TEND TO BE THOSE                       0.76     42  TEND TO BE THOSE                          0.91       28    7
    WHOSE ECONOMIC DEVELOPMENTAL STATUS                  WHOSE ECONOMIC DEVELOPMENTAL STATUS                   9   67
    IS DEVELOPED OR INTERMEDIATE  (36)                   IS UNDERDEVELOPED OR                            X SQ=  47.08
                                                         VERY UNDERDEVELOPED  (76)                       P =     0.
                                                                                                         RV YES (YES)

43  TEND TO BE THOSE                       0.95     43  TEND TO BE THOSE                          0.75       36   18
    WHOSE ECONOMIC DEVELOPMENTAL STATUS                  WHOSE ECONOMIC DEVELOPMENTAL STATUS                   2   55
    IS DEVELOPED, INTERMEDIATE, OR                       IS VERY UNDERDEVELOPED  (57)                    X SQ=  46.36
    UNDERDEVELOPED  (55)                                                                                 P =     0.
                                                                                                         RV YES (YES)

44  TEND TO BE THOSE                       0.55     44  TEND TO BE THOSE                          0.96       21    3
    WHERE THE LITERACY RATE IS                           WHERE THE LITERACY RATE IS                           17   73
    NINETY PERCENT OR ABOVE  (25)                        BELOW NINETY PERCENT  (90)                      X SQ=  37.11
                                                                                                         P =     0.
                                                                                                         RV YES (YES)
```

45	TEND TO BE THOSE WHERE THE LITERACY RATE IS FIFTY PERCENT OR ABOVE (55)	0.86	45	TEND TO BE THOSE WHERE THE LITERACY RATE IS BELOW FIFTY PERCENT (54)	0.69	32 22 5 49 X SQ= 27.79 P = 0. RV YES (YES)
46	ALWAYS ARE THOSE WHERE THE LITERACY RATE IS TEN PERCENT OR ABOVE (84)	1.00	46	TEND LESS TO BE THOSE WHERE THE LITERACY RATE IS TEN PERCENT OR ABOVE (84)	0.63	38 45 0 26 X SQ= 16.31 P = 0. RV NO (NO)
50	LEAN TOWARD BEING THOSE WHERE FREEDOM OF THE PRESS IS COMPLETE (43)	0.55	50	LEAN TOWARD BEING THOSE WHERE FREEDOM OF THE PRESS IS INTERMITTENT, INTERNALLY ABSENT, OR INTERNALLY AND EXTERNALLY ABSENT (56)	0.63	21 22 17 38 X SQ= 2.56 P = 0.095 RV YES (NO)
54	TEND TO BE THOSE WHERE NEWSPAPER CIRCULATION IS ONE HUNDRED OR MORE PER THOUSAND (37)	0.63	54	TEND TO BE THOSE WHERE NEWSPAPER CIRCULATION IS LESS THAN ONE HUNDRED PER THOUSAND (76)	0.84	24 12 14 62 X SQ= 23.26 P = 0. RV YES (YES)
56	TEND MORE TO BE THOSE WHERE THE RELIGION IS PREDOMINANTLY LITERATE (79)	0.97	56	TEND LESS TO BE THOSE WHERE THE RELIGION IS PREDOMINANTLY LITERATE (79)	0.58	36 42 1 30 X SQ= 16.37 P = 0. RV NO (NO)
63	TEND TO BE THOSE WHERE THE RELIGION IS PREDOMINANTLY SOME KIND OF CHRISTIAN (46)	0.78	63	TEND TO BE THOSE WHERE THE RELIGION IS PREDOMINANTLY OR PARTLY OTHER THAN CHRISTIAN (68)	0.79	29 16 8 60 X SQ= 31.77 P = 0. RV YES (YES)
67	TILT MORE TOWARD BEING THOSE THAT ARE RACIALLY HOMOGENEOUS (82)	0.84	67	TILT LESS TOWARD BEING THOSE THAT ARE RACIALLY HOMOGENEOUS (82)	0.68	31 48 6 23 X SQ= 2.47 P = 0.108 RV NO (NO)
68	TEND TO BE THOSE THAT ARE LINGUISTICALLY HOMOGENEOUS (52)	0.59	68	TEND TO BE THOSE THAT ARE LINGUISTICALLY WEAKLY HETEROGENEOUS OR STRONGLY HETEROGENEOUS (62)	0.62	22 29 15 47 X SQ= 3.74 P = 0.044 RV YES (NO)
69	TEND TO BE THOSE THAT ARE LINGUISTICALLY HOMOGENEOUS OR WEAKLY HETEROGENEOUS (64)	0.70	69	TEND TO BE THOSE THAT ARE LINGUISTICALLY STRONGLY HETEROGENEOUS (50)	0.51	26 37 11 39 X SQ= 3.87 P = 0.043 RV YES (NO)

72	TEND TO BE THOSE WHOSE DATE OF INDEPENDENCE IS BEFORE 1914 (52)	0.73	TEND TO BE THOSE WHOSE DATE OF INDEPENDENCE IS AFTER 1914 (59)	0.66	27 25 10 49 X SQ= 13.68 P = 0. RV YES (YES)
73	TEND TO BE THOSE WHOSE DATE OF INDEPENDENCE IS BEFORE 1945 (65)	0.86	TEND TO BE THOSE WHOSE DATE OF INDEPENDENCE IS AFTER 1945 (46)	0.55	32 33 5 41 X SQ= 16.15 P = 0. RV YES (NO)
74	TEND TO BE THOSE THAT ARE HISTORICALLY WESTERN (26)	0.54	TEND TO BE THOSE THAT ARE NOT HISTORICALLY WESTERN (87)	0.93	20 5 17 70 X SQ= 29.41 P = 0. RV YES (YES)
75	TEND TO BE THOSE THAT ARE HISTORICALLY WESTERN OR SIGNIFICANTLY WESTERNIZED (62)	0.95	TEND TO BE THOSE THAT ARE NOT HISTORICALLY WESTERN AND ARE NOT SIGNIFICANTLY WESTERNIZED (52)	0.66	35 26 2 50 X SQ= 34.14 P = 0. RV YES (YES)
76	TEND TO BE THOSE THAT ARE HISTORICALLY WESTERN, RATHER THAN HAVING BEEN SIGNIFICANTLY OR PARTIALLY WESTERNIZED THROUGH A COLONIAL RELATIONSHIP (26)	0.62	TEND TO BE THOSE THAT HAVE BEEN SIGNIFICANTLY OR PARTIALLY WESTERNIZED THROUGH A COLONIAL RELATIONSHIP, RATHER THAN BEING HISTORICALLY WESTERN (70)	0.92	20 5 12 58 X SQ= 29.83 P = 0. RV YES (YES)
77	TEND TO BE THOSE THAT HAVE BEEN SIGNIFICANTLY WESTERNIZED, RATHER THAN PARTIALLY WESTERNIZED, THROUGH A COLONIAL RELATIONSHIP (28)	0.82	TEND TO BE THOSE THAT HAVE BEEN PARTIALLY WESTERNIZED, RATHER THAN SIGNIFICANTLY WESTERNIZED, THROUGH A COLONIAL RELATIONSHIP (41)	0.67	9 19 2 39 X SQ= 7.31 P = 0.005 RV YES (NO)
78	TEND LESS TO BE THOSE THAT HAVE BEEN SIGNIFICANTLY WESTERNIZED THROUGH A COLONIAL RELATIONSHIP, RATHER THAN WITHOUT SUCH A RELATIONSHIP (28)	0.60	TEND MORE TO BE THOSE THAT HAVE BEEN SIGNIFICANTLY WESTERNIZED THROUGH A COLONIAL RELATIONSHIP, RATHER THAN WITHOUT SUCH A RELATIONSHIP (28)	0.95	9 19 6 1 X SQ= 4.56 P = 0.027 RV NO (YES)
80	ALWAYS ARE THOSE WHOSE DATE OF INDEPENDENCE IS AFTER 1914, AND THAT WERE FORMERLY DEPENDENCIES OF BRITAIN, RATHER THAN FRANCE (19)	1.00	LEAN TOWARD BEING THOSE WHOSE DATE OF INDEPENDENCE IS AFTER 1914, AND THAT WERE FORMERLY DEPENDENCIES OF FRANCE, RATHER THAN BRITAIN (24)	0.60	3 16 0 24 X SQ= 2.00 P = 0.079 RV YES (NO)
81	TEND LESS TO BE THOSE WHERE THE TYPE OF POLITICAL MODERNIZATION IS LATER EUROPEAN OR LATER EUROPEAN DERIVED, RATHER THAN EARLY EUROPEAN OR EARLY EUROPEAN DERIVED (40)	0.66	TEND MORE TO BE THOSE WHERE THE TYPE OF POLITICAL MODERNIZATION IS LATER EUROPEAN OR LATER EUROPEAN DERIVED, RATHER THAN EARLY EUROPEAN OR EARLY EUROPEAN DERIVED (40)	0.95	10 1 19 20 X SQ= 4.66 P = 0.016 RV NO (YES)

82	TEND TO BE THOSE WHERE THE TYPE OF POLITICAL MODERNIZATION IS EARLY OR LATER EUROPEAN OR EUROPEAN DERIVED, RATHER THAN DEVELOPED TUTELARY OR UNDEVELOPED TUTELARY (51)	0.85	82	TEND TO BE THOSE WHERE THE TYPE OF POLITICAL MODERNIZATION IS DEVELOPED TUTELARY OR UNDEVELOPED TUTELARY, RATHER THAN EARLY OR LATER EUROPEAN OR EUROPEAN DERIVED (55)	0.70	29 21 5 50 X SQ= 26.42 P = 0. RV YES (YES)
84	ALWAYS ARE THOSE WHERE THE TYPE OF POLITICAL MODERNIZATION IS DEVELOPED TUTELARY, RATHER THAN UNDEVELOPED TUTELARY (31)	1.00	84	LEAN LESS TOWARD BEING THOSE WHERE THE TYPE OF POLITICAL MODERNIZATION IS DEVELOPED TUTELARY, RATHER THAN UNDEVELOPED TUTELARY (31)	0.52	5 26 0 24 X SQ= 2.53 P = 0.061 RV NO (NO)
85	TEND TO BE THOSE WHERE THE STAGE OF POLITICAL MODERNIZATION IS ADVANCED, RATHER THAN MID- OR EARLY TRANSITIONAL (60)	0.92	85	TEND TO BE THOSE WHERE THE STAGE OF POLITICAL MODERNIZATION IS MID- OR EARLY TRANSITIONAL, RATHER THAN ADVANCED (54)	0.68	35 24 3 51 X SQ= 34.15 P = 0. RV YES (YES)
87	TEND TO BE THOSE WHOSE IDEOLOGICAL ORIENTATION IS OTHER THAN DEVELOPMENTAL (58)	0.91	87	TEND TO BE THOSE WHOSE IDEOLOGICAL ORIENTATION IS DEVELOPMENTAL (31)	0.52	3 28 31 26 X SQ= 15.10 P = 0. RV YES (YES)
89	TEND TO BE THOSE WHOSE IDEOLOGICAL ORIENTATION IS CONVENTIONAL (33)	0.68	89	TEND TO BE THOSE WHOSE IDEOLOGICAL ORIENTATION IS OTHER THAN CONVENTIONAL (62)	0.83	23 10 11 50 X SQ= 22.57 P = 0. RV YES (YES)
94	LEAN TOWARD BEING THOSE WHERE THE STATUS OF THE REGIME IS CONSTITUTIONAL (51)	0.69	94	LEAN TOWARD BEING THOSE WHERE THE STATUS OF THE REGIME IS AUTHORITARIAN OR TOTALITARIAN (41)	0.53	25 26 11 29 X SQ= 3.49 P = 0.052 RV YES (NO)
95	TILT LESS TOWARD BEING THOSE WHERE THE STATUS OF THE REGIME IS CONSTITUTIONAL OR AUTHORITARIAN (95)	0.78	95	TILT MORE TOWARD BEING THOSE WHERE THE STATUS OF THE REGIME IS CONSTITUTIONAL OR AUTHORITARIAN (95)	0.90	29 66 8 7 X SQ= 2.08 P = 0.139 RV NO (YES)
96	TEND MORE TO BE THOSE WHERE THE STATUS OF THE REGIME IS CONSTITUTIONAL OR TOTALITARIAN (67)	0.92	96	TEND LESS TO BE THOSE WHERE THE STATUS OF THE REGIME IS CONSTITUTIONAL OR TOTALITARIAN (67)	0.62	33 33 3 20 X SQ= 8.20 P = 0.003 RV NO (YES)
97	TEND TO BE THOSE WHERE THE STATUS OF THE REGIME IS TOTALITARIAN, RATHER THAN AUTHORITARIAN (16)	0.73	97	TEND TO BE THOSE WHERE THE STATUS OF THE REGIME IS AUTHORITARIAN, RATHER THAN TOTALITARIAN (23)	0.74	3 20 8 7 X SQ= 5.34 P = 0.012 RV YES (YES)

99	TEND TO BE THOSE WHERE GOVERNMENTAL STABILITY IS GENERALLY PRESENT AND DATES FROM AT LEAST THE INTER-WAR PERIOD, OR FROM THE POST-WAR PERIOD (50)	0.78	99	TEND TO BE THOSE WHERE GOVERNMENTAL STABILITY IS MODERATELY PRESENT AND DATES FROM THE POST-WAR PERIOD, OR IS ABSENT (36)	0.58	29 20 8 28 X SQ= 10.08 P = 0.001 RV YES (YES)
101	TEND TO BE THOSE WHERE THE REPRESENTATIVE CHARACTER OF THE REGIME IS POLYARCHIC (41)	0.59	101	TEND TO BE THOSE WHERE THE REPRESENTATIVE CHARACTER OF THE REGIME IS LIMITED POLYARCHIC, PSEUDO-POLYARCHIC, OR NON-POLYARCHIC (57)	0.68	22 19 15 41 X SQ= 6.15 P = 0.011 RV YES (YES)
102	TEND TO BE THOSE WHERE THE REPRESENTATIVE CHARACTER OF THE REGIME IS POLYARCHIC OR LIMITED POLYARCHIC (59)	0.71	102	TEND TO BE THOSE WHERE THE REPRESENTATIVE CHARACTER OF THE REGIME IS PSEUDO-POLYARCHIC OR NON-POLYARCHIC (49)	0.54	27 32 11 37 X SQ= 5.08 P = 0.016 RV YES (NO)
105	TEND TO BE THOSE WHERE THE ELECTORAL SYSTEM IS COMPETITIVE (43)	0.73	105	TEND TO BE THOSE WHERE THE ELECTORAL SYSTEM IS PARTIALLY COMPETITIVE OR NON-COMPETITIVE (47)	0.66	24 19 9 37 X SQ= 11.01 P = 0. RV YES (YES)
107	TEND TO BE THOSE WHERE AUTONOMOUS GROUPS ARE FULLY TOLERATED IN POLITICS (46)	0.66	107	TEND TO BE THOSE WHERE AUTONOMOUS GROUPS ARE PARTIALLY TOLERATED IN POLITICS, ARE TOLERATED ONLY OUTSIDE POLITICS, OR ARE NOT TOLERATED AT ALL (65)	0.71	25 21 13 51 X SQ= 12.25 P = 0. RV YES (YES)
111	TEND TO BE THOSE WHERE POLITICAL ENCULTURATION IS HIGH OR MEDIUM (53)	0.71	111	TEND TO BE THOSE WHERE POLITICAL ENCULTURATION IS LOW (42)	0.52	22 31 9 33 X SQ= 3.43 P = 0.048 RV YES (NO)
115	TEND TO BE THOSE WHERE INTEREST ARTICULATION BY ASSOCIATIONAL GROUPS IS SIGNIFICANT (20)	0.53	115	TEND TO BE THOSE WHERE INTEREST ARTICULATION BY ASSOCIATIONAL GROUPS IS MODERATE, LIMITED, OR NEGLIGIBLE (91)	0.99	19 1 17 73 X SQ= 39.67 P = 0. RV YES (YES)
116	TEND TO BE THOSE WHERE INTEREST ARTICULATION BY ASSOCIATIONAL GROUPS IS SIGNIFICANT OR MODERATE (32)	0.69	116	TEND TO BE THOSE WHERE INTEREST ARTICULATION BY ASSOCIATIONAL GROUPS IS LIMITED OR NEGLIGIBLE (79)	0.91	25 7 11 67 X SQ= 39.39 P = 0. RV YES (YES)
117	TEND TO BE THOSE WHERE INTEREST ARTICULATION BY ASSOCIATIONAL GROUPS IS SIGNIFICANT, MODERATE, OR LIMITED (60)	0.89	117	TEND TO BE THOSE WHERE INTEREST ARTICULATION BY ASSOCIATIONAL GROUPS IS NEGLIGIBLE (51)	0.62	32 28 4 46 X SQ= 23.44 P = 0. RV YES (YES)

119	TEND LESS TO BE THOSE WHERE INTEREST ARTICULATION BY INSTITUTIONAL GROUPS IS VERY SIGNIFICANT OR SIGNIFICANT (74)	0.55	TEND MORE TO BE THOSE WHERE INTEREST ARTICULATION BY INSTITUTIONAL GROUPS IS VERY SIGNIFICANT OR SIGNIFICANT (74)	0.85	21 52 17 9 X SQ= 9.38 P = 0.002 RV NO (YES)
120	TEND LESS TO BE THOSE WHERE INTEREST ARTICULATION BY INSTITUTIONAL GROUPS IS VERY SIGNIFICANT, SIGNIFICANT, OR MODERATE (90)	0.79	TEND MORE TO BE THOSE WHERE INTEREST ARTICULATION BY INSTITUTIONAL GROUPS IS VERY SIGNIFICANT, SIGNIFICANT, OR MODERATE (90)	0.97	30 59 8 2 X SQ= 6.31 P = 0.012 RV NO (YES)
121	TEND TO BE THOSE WHERE INTEREST ARTICULATION BY NON-ASSOCIATIONAL GROUPS IS MODERATE, LIMITED, OR NEGLIGIBLE (61)	0.84	TEND TO BE THOSE WHERE INTEREST ARTICULATION BY NON-ASSOCIATIONAL GROUPS IS SIGNIFICANT (54)	0.63	6 48 32 28 X SQ= 20.94 P = 0. RV YES (YES)
122	TEND TO BE THOSE WHERE INTEREST ARTICULATION BY NON-ASSOCIATIONAL GROUPS IS LIMITED OR NEGLIGIBLE (32)	0.50	TEND TO BE THOSE WHERE INTEREST ARTICULATION BY NON-ASSOCIATIONAL GROUPS IS SIGNIFICANT OR MODERATE (83)	0.84	19 64 19 12 X SQ= 13.30 P = 0. RV YES (YES)
125	TEND TO BE THOSE WHERE INTEREST ARTICULATION BY ANOMIC GROUPS IS INFREQUENT OR VERY INFREQUENT (35)	0.56	TEND TO BE THOSE WHERE INTEREST ARTICULATION BY ANOMIC GROUPS IS FREQUENT OR OCCASIONAL (64)	0.74	14 49 18 17 X SQ= 7.45 P = 0.006 RV YES (YES)
130	LEAN LESS TOWARD BEING THOSE WHERE INTEREST AGGREGATION BY POLITICAL PARTIES IS MODERATE, LIMITED, OR NEGLIGIBLE (71)	0.76	LEAN MORE TOWARD BEING THOSE WHERE INTEREST AGGREGATION BY POLITICAL PARTIES IS MODERATE, LIMITED, OR NEGLIGIBLE (71)	0.92	8 4 25 45 X SQ= 2.90 P = 0.058 RV NO (YES)
132	ALWAYS ARE THOSE WHERE INTEREST AGGREGATION BY POLITICAL PARTIES IS SIGNIFICANT, MODERATE, OR LIMITED (67)	1.00	TEND LESS TO BE THOSE WHERE INTEREST AGGREGATION BY POLITICAL PARTIES IS SIGNIFICANT, MODERATE, OR LIMITED (67)	0.82	25 42 0 9 X SQ= 3.46 P = 0.026 RV NO (NO)
134	TILT TOWARD BEING THOSE WHERE INTEREST AGGREGATION BY THE EXECUTIVE IS SIGNIFICANT OR MODERATE (57)	0.67	TILT TOWARD BEING THOSE WHERE INTEREST AGGREGATION BY THE EXECUTIVE IS LIMITED OR NEGLIGIBLE (46)	0.50	24 33 12 33 X SQ= 1.99 P = 0.144 RV YES (NO)
135	ALWAYS ARE THOSE WHERE INTEREST AGGREGATION BY THE EXECUTIVE IS SIGNIFICANT, MODERATE, OR LIMITED (77)	1.00	TEND LESS TO BE THOSE WHERE INTEREST AGGREGATION BY THE EXECUTIVE IS SIGNIFICANT, MODERATE, OR LIMITED (77)	0.79	32 45 0 12 X SQ= 6.09 P = 0.003 RV NO (NO)

```
137  TEND TO BE THOSE                          0.59    137  TEND TO BE THOSE                          0.85         20     9
     WHERE INTEREST AGGREGATION                             WHERE INTEREST AGGREGATION                              14    53
     BY THE LEGISLATURE                                     BY THE LEGISLATURE                                 X SQ=  18.40
     IS SIGNIFICANT OR MODERATE  (29)                       IS LIMITED OR NEGLIGIBLE  (68)                     P =     0.
                                                                                                               RV YES (YES)

138  TEND TO BE THOSE                          0.76    138  TEND TO BE THOSE                          0.63         25    23
     WHERE INTEREST AGGREGATION                             WHERE INTEREST AGGREGATION                               8    40
     BY THE LEGISLATURE                                     BY THE LEGISLATURE                                 X SQ=  11.82
     IS SIGNIFICANT, MODERATE, OR                           IS NEGLIGIBLE  (49)                                P =     0.
     LIMITED  (48)                                                                                             RV YES (YES)

141  TEND LESS TO BE THOSE                     0.76    141  TEND MORE TO BE THOSE                     0.95          8     3
     WHERE THE PARTY SYSTEM IS QUANTITATIVELY               WHERE THE PARTY SYSTEM IS QUANTITATIVELY                26    60
     OTHER THAN TWO-PARTY  (87)                             OTHER THAN TWO-PARTY  (87)                         X SQ=   5.98
                                                                                                               P =     0.015
                                                                                                               RV NO  (YES)

144  TEND LESS TO BE THOSE                     0.53    144  TEND MORE TO BE THOSE                     0.89          9    24
     WHERE THE PARTY SYSTEM IS QUANTITATIVELY               WHERE THE PARTY SYSTEM IS QUANTITATIVELY                 8     3
     ONE-PARTY, RATHER THAN                                 ONE-PARTY, RATHER THAN                             X SQ=   5.40
     TWO-PARTY  (34)                                        TWO-PARTY  (34)                                    P =     0.012
                                                                                                               RV NO  (YES)

145  LEAN LESS TOWARD BEING THOSE              0.60    145  LEAN MORE TOWARD BEING THOSE              0.86         12    18
     WHERE THE PARTY SYSTEM IS QUANTITATIVELY               WHERE THE PARTY SYSTEM IS QUANTITATIVELY                 8     3
     MULTI-PARTY, RATHER THAN                               MULTI-PARTY, RATHER THAN                           X SQ=   2.26
     TWO-PARTY  (30)                                        TWO-PARTY  (30)                                    P =     0.085
                                                                                                               RV NO  (YES)

147  TEND TO BE THOSE                          0.53    147  TEND TO BE THOSE                          0.89         17     6
     WHERE THE PARTY SYSTEM IS QUALITATIVELY                WHERE THE PARTY SYSTEM IS QUALITATIVELY                 15    51
     CLASS-ORIENTED OR MULTI-IDEOLOGICAL  (23)              OTHER THAN                                         X SQ=  17.25
                                                            CLASS-ORIENTED OR MULTI-IDEOLOGICAL  (67)          P =     0.
                                                                                                               RV YES (YES)

148  ALWAYS ARE THOSE                          1.00    148  TEND LESS TO BE THOSE                     0.80          0    14
     WHERE THE PARTY SYSTEM IS QUALITATIVELY                WHERE THE PARTY SYSTEM IS QUALITATIVELY                 38    57
     OTHER THAN                                             OTHER THAN                                         X SQ=   6.93
     AFRICAN TRANSITIONAL  (96)                             AFRICAN TRANSITIONAL  (96)                         P =     0.002
                                                                                                               RV NO  (NO )

153  TEND TO BE THOSE                          0.77    153  TEND TO BE THOSE                          0.68         27    14
     WHERE THE PARTY SYSTEM IS                              WHERE THE PARTY SYSTEM IS                                8    30
     STABLE  (42)                                           MODERATELY STABLE OR UNSTABLE  (38)                X SQ=  14.28
                                                                                                               P =     0.
                                                                                                               RV YES (YES)

158  TEND TO BE THOSE                          0.78    158  TEND TO BE THOSE                          0.55          8    32
     WHERE PERSONALISMO IS                                  WHERE PERSONALISMO IS                                   29    26
     NEGLIGIBLE  (56)                                       PRONOUNCED OR MODERATE  (40)                       X SQ=   9.10
                                                                                                               P =     0.001
                                                                                                               RV YES (YES)
```

164	TEND MORE TO BE THOSE WHERE THE REGIME'S LEADERSHIP CHARISMA IS NEGLIGIBLE (65)	0.81	164	TEND LESS TO BE THOSE WHERE THE REGIME'S LEADERSHIP CHARISMA IS NEGLIGIBLE (65)	0.56	7 27 30 34 X SQ= 5.46 P = 0.015 RV NO (NO)
166	TEND LESS TO BE THOSE WHERE THE VERTICAL POWER DISTRIBUTION IS THAT OF FORMAL FEDERALISM OR FORMAL AND EFFECTIVE UNITARISM (99)	0.74	166	TEND MORE TO BE THOSE WHERE THE VERTICAL POWER DISTRIBUTION IS THAT OF FORMAL FEDERALISM OR FORMAL AND EFFECTIVE UNITARISM (99)	0.93	10 5 28 70 X SQ= 6.84 P = 0.007 RV NO (YES)
168	TEND TO BE THOSE WHERE THE HORIZONTAL POWER DISTRIBUTION IS SIGNIFICANT (34)	0.59	168	TEND TO BE THOSE WHERE THE HORIZONTAL POWER DISTRIBUTION IS LIMITED OR NEGLIGIBLE (72)	0.82	22 12 15 56 X SQ= 17.27 P = 0. RV YES (YES)
169	TEND TO BE THOSE WHERE THE HORIZONTAL POWER DISTRIBUTION IS SIGNIFICANT OR LIMITED (58)	0.70	169	TEND TO BE THOSE WHERE THE HORIZONTAL POWER DISTRIBUTION IS NEGLIGIBLE (48)	0.53	26 32 11 36 X SQ= 4.32 P = 0.025 RV YES (NO)
170	LEAN MORE TOWARD BEING THOSE WHERE THE LEGISLATIVE-EXECUTIVE STRUCTURE IS OTHER THAN PRESIDENTIAL (63)	0.73	170	LEAN LESS TOWARD BEING THOSE WHERE THE LEGISLATIVE-EXECUTIVE STRUCTURE IS OTHER THAN PRESIDENTIAL (63)	0.55	10 29 27 35 X SQ= 2.58 P = 0.090 RV NO (NO)
171	TILT LESS TOWARD BEING THOSE WHERE THE LEGISLATIVE-EXECUTIVE STRUCTURE IS OTHER THAN PARLIAMENTARY-REPUBLICAN (90)	0.81	171	TILT MORE TOWARD BEING THOSE WHERE THE LEGISLATIVE-EXECUTIVE STRUCTURE IS OTHER THAN PARLIAMENTARY-REPUBLICAN (90)	0.92	7 5 29 60 X SQ= 2.04 P = 0.109 RV NO (YES)
174	TEND TO BE THOSE WHERE THE LEGISLATURE IS FULLY EFFECTIVE (28)	0.51	174	TEND TO BE THOSE WHERE THE LEGISLATURE IS PARTIALLY EFFECTIVE, LARGELY INEFFECTIVE, OR WHOLLY INEFFECTIVE (72)	0.85	19 9 18 53 X SQ= 13.74 P = 0. RV YES (YES)
175	TEND TO BE THOSE WHERE THE LEGISLATURE IS FULLY EFFECTIVE OR PARTIALLY EFFECTIVE (51)	0.73	175	TEND TO BE THOSE WHERE THE LEGISLATURE IS LARGELY INEFFECTIVE OR WHOLLY INEFFECTIVE (49)	0.61	27 24 10 38 X SQ= 9.56 P = 0.002 RV YES (YES)
178	TEND TO BE THOSE WHERE THE LEGISLATURE IS BICAMERAL (51)	0.70	178	TEND TO BE THOSE WHERE THE LEGISLATURE IS UNICAMERAL (53)	0.62	11 41 26 25 X SQ= 8.70 P = 0.002 RV YES (YES)

179	TEND TO BE THOSE WHERE THE EXECUTIVE IS STRONG (39)	0.57	179	TEND TO BE THOSE WHERE THE EXECUTIVE IS DOMINANT (52)	0.65	15 36 20 19 X SQ= 3.58 P = 0.049 RV YES (YES)
181	TEND LESS TO BE THOSE WHERE THE BUREAUCRACY IS SEMI-MODERN, RATHER THAN MODERN (55)	0.53	181	TEND MORE TO BE THOSE WHERE THE BUREAUCRACY IS SEMI-MODERN, RATHER THAN MODERN (55)	0.90	17 4 19 35 X SQ= 10.92 P = 0.001 RV NO (YES)
182	ALWAYS ARE THOSE WHERE THE BUREAUCRACY IS SEMI-MODERN, RATHER THAN POST-COLONIAL TRANSITIONAL (55)	1.00	182	TEND LESS TO BE THOSE WHERE THE BUREAUCRACY IS SEMI-MODERN, RATHER THAN POST-COLONIAL TRANSITIONAL (55)	0.58	19 35 0 25 X SQ= 9.74 P = 0. RV NO (NO)
187	TEND TO BE THOSE WHERE THE ROLE OF THE POLICE IS NOT POLITICALLY SIGNIFICANT (35)	0.56	187	TEND TO BE THOSE WHERE THE ROLE OF THE POLICE IS POLITICALLY SIGNIFICANT (66)	0.77	16 49 20 15 X SQ= 9.08 P = 0.002 RV YES (YES)
192	TEND MORE TO BE THOSE WHERE THE CHARACTER OF THE LEGAL SYSTEM IS OTHER THAN MUSLIM OR PARTLY MUSLIM (86)	0.92	192	TEND LESS TO BE THOSE WHERE THE CHARACTER OF THE LEGAL SYSTEM IS OTHER THAN MUSLIM OR PARTLY MUSLIM (86)	0.68	3 24 35 50 X SQ= 6.98 P = 0.005 RV NO (NO)
193	TEND MORE TO BE THOSE WHERE THE CHARACTER OF THE LEGAL SYSTEM IS OTHER THAN PARTLY INDIGENOUS (86)	0.95	193	TEND LESS TO BE THOSE WHERE THE CHARACTER OF THE LEGAL SYSTEM IS OTHER THAN PARTLY INDIGENOUS (86)	0.65	2 26 36 49 X SQ= 10.17 P = 0. RV NO (NO)
194	TILT LESS TOWARD BEING THOSE THAT ARE NON-COMMUNIST (101)	0.82	194	TILT MORE TOWARD BEING THOSE THAT ARE NON-COMMUNIST (101)	0.93	7 5 31 70 X SQ= 2.54 P = 0.102 RV NO (YES)

40 POLITIES
WHOSE INTERNATIONAL FINANCIAL STATUS
IS VERY HIGH, HIGH,
MEDIUM, OR LOW (71)

40 POLITIES
WHOSE INTERNATIONAL FINANCIAL STATUS
IS VERY LOW (39)

PREDICATE ONLY

71 IN LEFT COLUMN

AFGHANISTAN	ARGENTINA	AUSTRALIA	AUSTRIA	BELGIUM	BRAZIL	BULGARIA	BURMA	CANADA	CEYLON
CHILE	CHINA, PR	COLOMBIA	CONGO(LEO)	CUBA	CZECHOS'KIA	DENMARK	DOMIN REP	ECUADOR	ETHIOPIA
FINLAND	FRANCE	GERMANY, E	GERMAN FR	GHANA	GREECE	GUATEMALA	HUNGARY	INDIA	INDONESIA
IRAN	IRAQ	IRELAND	ISRAEL	ITALY	JAPAN	KOREA REP	LEBANON	LUXEMBOURG	MALAYA
MEXICO	MOROCCO	NETHERLANDS	NEW ZEALAND	NIGERIA	NORWAY	PAKISTAN	PERU	PHILIPPINES	POLAND
PORTUGAL	RUMANIA	SA'U ARABIA	SENEGAL	SO AFRICA	SPAIN	SUDAN	SWEDEN	SWITZERLAND	SYRIA
THAILAND	TUNISIA	TURKEY	USSR	UAR	UK	US	URUGUAY	VENEZUELA	VIETNAM REP
YUGOSLAVIA									

39 IN RIGHT COLUMN

ALBANIA	BOLIVIA	BURUNDI	CAMBODIA	CAMEROUN	CEN AFR REP	CHAD	CONGO(BRA)	COSTA RICA	CYPRUS
DAHOMEY	EL SALVADOR	GABON	GUINEA	HAITI	HONDURAS	ICELAND	IVORY COAST	JORDAN	LAOS
LIBERIA	LIBYA	MALAGASY R	MALI	MAURITANIA	MONGOLIA	NEPAL	NICARAGUA	NIGER	PANAMA
PARAGUAY	RWANDA	SIERRE LEO	SOMALIA	TANGANYIKA	TOGO	UGANDA	UPPER VOLTA	YEMEN	

5 EXCLUDED BECAUSE UNASCERTAINED

ALGERIA	JAMAICA	KOREA, N	TRINIDAD	VIETNAM, N

41 POLITIES
WHOSE ECONOMIC DEVELOPMENTAL STATUS
IS DEVELOPED (19)

41 POLITIES
WHOSE ECONOMIC DEVELOPMENTAL STATUS
IS INTERMEDIATE, UNDERDEVELOPED,
OR VERY UNDERDEVELOPED (94)

BOTH SUBJECT AND PREDICATE

19 IN LEFT COLUMN

AUSTRALIA	BELGIUM	CANADA	CZECHOS'KIA	DENMARK	FINLAND	FRANCE	GERMANY, E	GERMAN FR	ITALY
LUXEMBOURG	NETHERLANDS	NEW ZEALAND	NORWAY	SWEDEN	SWITZERLAND	USSR	UK	US	

94 IN RIGHT COLUMN

AFGHANISTAN	ALBANIA	ALGERIA	ARGENTINA	AUSTRIA	BOLIVIA	BRAZIL	BULGARIA	BURMA	BURUNDI
CAMBODIA	CAMEROUN	CEN AFR REP	CEYLON	CHAD	CHILE	CHINA, PR	COLOMBIA	CONGO(BRA)	CONGO(LEO)
COSTA RICA	CUBA	CYPRUS	DAHOMEY	DOMIN REP	ECUADOR	EL SALVADOR	ETHIOPIA	GABON	GHANA
GREECE	GUATEMALA	GUINEA	HAITI	HONDURAS	HUNGARY	ICELAND	INDIA	INDONESIA	IRAN
IRAQ	IRELAND	IVORY COAST	JAMAICA	JAPAN	JORDAN	KOREA, N	KOREA REP	LAOS	LEBANON
LIBERIA	LIBYA	MALAGASY R	MALAYA	MALI	MAURITANIA	MEXICO	MONGOLIA	MOROCCO	NEPAL
NICARAGUA	NIGER	NIGERIA	PAKISTAN	PANAMA	PARAGUAY	PERU	PHILIPPINES	POLAND	PORTUGAL
RUMANIA	RWANDA	SA'U ARABIA	SENEGAL	SIERRE LEO	SOMALIA	SO AFRICA	SPAIN	SUDAN	SYRIA
TANGANYIKA	THAILAND	TOGO	TRINIDAD	TUNISIA	TURKEY	UGANDA	UAR	UPPER VOLTA	URUGUAY
VIETNAM, N	VIETNAM REP	YEMEN	YUGOSLAVIA						

2 EXCLUDED BECAUSE AMBIGUOUS

ISRAEL VENEZUELA

	Subject			Predicate		
2	TEND TO BE THOSE LOCATED IN WEST EUROPE, SCANDINAVIA, NORTH AMERICA, OR AUSTRALASIA (22)	0.84	2	TEND TO BE THOSE LOCATED ELSEWHERE THAN IN WEST EUROPE, SCANDINAVIA, NORTH AMERICA, OR AUSTRALASIA (93)	0.94	16 6 3 88 X SQ= 56.20 P = 0. RV YES (YES)
3	TEND LESS TO BE THOSE LOCATED ELSEWHERE THAN IN WEST EUROPE (102)	0.58	3	TEND MORE TO BE THOSE LOCATED ELSEWHERE THAN IN WEST EUROPE (102)	0.95	8 5 11 89 X SQ= 17.55 P = 0. RV NO (YES)
6	ALWAYS ARE THOSE LOCATED ELSEWHERE THAN IN THE CARIBBEAN, CENTRAL AMERICA, OR SOUTH AMERICA (93)	1.00	6	TEND LESS TO BE THOSE LOCATED ELSEWHERE THAN IN THE CARIBBEAN, CENTRAL AMERICA, OR SOUTH AMERICA (93)	0.78	0 21 19 73 X SQ= 3.84 P = 0.022 RV NO (NO)

8	ALWAYS ARE THOSE LOCATED ELSEWHERE THAN IN EAST ASIA SOUTH ASIA, OR SOUTHEAST ASIA (97)	1.00	8	TEND LESS TO BE THOSE LOCATED ELSEWHERE THAN IN EAST ASIA SOUTH ASIA, OR SOUTHEAST ASIA (97)	0.81	0 18 19 76 X SQ= 3.02 P = 0.039 RV NO (NO)
10	ALWAYS ARE THOSE LOCATED ELSEWHERE THAN IN NORTH AFRICA, OR CENTRAL AND SOUTH AFRICA (82)	1.00	10	TEND LESS TO BE THOSE LOCATED ELSEWHERE THAN IN NORTH AFRICA, OR CENTRAL AND SOUTH AFRICA (82)	0.65	0 33 19 61 X SQ= 7.80 P = 0.001 RV NO (NO)
13	ALWAYS ARE THOSE LOCATED ELSEWHERE THAN IN NORTH AFRICA OR THE MIDDLE EAST (99)	1.00	13	LEAN LESS TOWARD BEING THOSE LOCATED ELSEWHERE THAN IN NORTH AFRICA OR THE MIDDLE EAST (99)	0.84	0 15 19 79 X SQ= 2.25 P = 0.071 RV NO (NO)
26	TILT TOWARD BEING THOSE WHOSE POPULATION DENSITY IS VERY HIGH, HIGH, OR MEDIUM (48)	0.58	26	TILT TOWARD BEING THOSE WHOSE POPULATION DENSITY IS LOW (67)	0.62	11 36 8 58 X SQ= 1.76 P = 0.132 RV YES (NO)
28	TEND TO BE THOSE WHOSE POPULATION GROWTH RATE IS LOW (48)	0.83	28	TEND TO BE THOSE WHOSE POPULATION GROWTH RATE IS HIGH (62)	0.63	3 57 15 33 X SQ= 11.41 P = 0. RV YES (NO)
29	ALWAYS ARE THOSE WHERE THE DEGREE OF URBANIZATION IS HIGH (56)	1.00	29	TEND TO BE THOSE WHERE THE DEGREE OF URBANIZATION IS LOW (49)	0.58	19 35 0 49 X SQ= 18.87 P = 0. RV YES (NO)
30	ALWAYS ARE THOSE WHOSE AGRICULTURAL POPULATION IS MEDIUM, LOW, OR VERY LOW (57)	1.00	30	TEND TO BE THOSE WHOSE AGRICULTURAL POPULATION IS HIGH (56)	0.61	0 56 19 36 X SQ= 20.97 P = 0. RV YES (NO)
31	TEND TO BE THOSE WHOSE AGRICULTURAL POPULATION IS LOW OR VERY LOW (24)	0.84	31	TEND TO BE THOSE WHOSE AGRICULTURAL POPULATION IS HIGH OR MEDIUM (90)	0.92	3 86 16 7 X SQ= 52.25 P = 0. RV YES (YES)
33	TEND TO BE THOSE WHOSE GROSS NATIONAL PRODUCT IS VERY HIGH, HIGH, OR MEDIUM (30)	0.79	33	TEND TO BE THOSE WHOSE GROSS NATIONAL PRODUCT IS LOW OR VERY LOW (85)	0.84	15 15 4 79 X SQ= 29.01 P = 0. RV YES (YES)

34	TEND TO BE THOSE WHOSE GROSS NATIONAL PRODUCT IS VERY HIGH, HIGH, MEDIUM, OR LOW (62)	0.95	34	TEND TO BE THOSE WHOSE GROSS NATIONAL PRODUCT IS VERY LOW (53)	0.55	18 42 1 52 X SQ= 13.96 P = 0. RV YES (NO)
35	ALWAYS ARE THOSE WHOSE PER CAPITA GROSS NATIONAL PRODUCT IS VERY HIGH OR HIGH (24)	1.00	35	TEND TO BE THOSE WHOSE PER CAPITA GROSS NATIONAL PRODUCT IS MEDIUM, LOW, OR VERY LOW (91)	0.97	19 3 0 91 X SQ= 88.40 P = 0. RV YES (YES)
36	ALWAYS ARE THOSE WHOSE PER CAPITA GROSS NATIONAL PRODUCT IS VERY HIGH, HIGH, OR MEDIUM (42)	1.00	36	TEND TO BE THOSE WHOSE PER CAPITA GROSS NATIONAL PRODUCT IS LOW OR VERY LOW (73)	0.78	19 21 0 73 X SQ= 38.36 P = 0. RV YES (NO)
37	ALWAYS ARE THOSE WHOSE PER CAPITA GROSS NATIONAL PRODUCT IS VERY HIGH, HIGH, MEDIUM, OR LOW (64)	1.00	37	TEND TO BE THOSE WHOSE PER CAPITA GROSS NATIONAL PRODUCT IS VERY LOW (51)	0.54	19 43 0 51 X SQ= 16.66 P = 0. RV YES (NO)
39	TEND TO BE THOSE WHOSE INTERNATIONAL FINANCIAL STATUS IS VERY HIGH, HIGH, OR MEDIUM (38)	0.94	39	TEND TO BE THOSE WHOSE INTERNATIONAL FINANCIAL STATUS IS LOW OR VERY LOW (76)	0.79	17 20 1 74 X SQ= 33.33 P = 0. RV YES (NO)
40	ALWAYS ARE THOSE WHOSE INTERNATIONAL FINANCIAL STATUS IS VERY HIGH, HIGH, MEDIUM, OR LOW (71)	1.00	40	TEND LESS TO BE THOSE WHOSE INTERNATIONAL FINANCIAL STATUS IS VERY HIGH, HIGH, MEDIUM, OR LOW (71)	0.56	19 50 0 39 X SQ= 11.20 P = 0. RV NO (NO)
44	TEND TO BE THOSE WHERE THE LITERACY RATE IS NINETY PERCENT OR ABOVE (25)	0.95	44	TEND TO BE THOSE WHERE THE LITERACY RATE IS BELOW NINETY PERCENT (90)	0.93	18 7 1 87 X SQ= 64.92 P = 0. RV YES (YES)
45	ALWAYS ARE THOSE WHERE THE LITERACY RATE IS FIFTY PERCENT OR ABOVE (55)	1.00	45	TEND TO BE THOSE WHERE THE LITERACY RATE IS BELOW FIFTY PERCENT (54)	0.61	19 34 0 54 X SQ= 21.15 P = 0. RV YES (NO)
50	TEND TO BE THOSE WHERE FREEDOM OF THE PRESS IS COMPLETE (43)	0.79	50	TEND TO BE THOSE WHERE FREEDOM OF THE PRESS IS INTERMITTENT, INTERNALLY ABSENT, OR INTERNALLY AND EXTERNALLY ABSENT (56)	0.64	15 28 4 50 X SQ= 9.80 P = 0.001 RV YES (NO)

53	TEND TO BE THOSE WHERE NEWSPAPER CIRCULATION IS THREE HUNDRED OR MORE PER THOUSAND (14)	0.63	53	TEND TO BE THOSE WHERE NEWSPAPER CIRCULATION IS LESS THAN THREE HUNDRED PER THOUSAND (101)	0.98	12 2 7 92 X SQ= 48.76 P = 0. RV YES (YES)
54	ALWAYS ARE THOSE WHERE NEWSPAPER CIRCULATION IS ONE HUNDRED OR MORE PER THOUSAND (37)	1.00	54	TEND TO BE THOSE WHERE NEWSPAPER CIRCULATION IS LESS THAN ONE HUNDRED PER THOUSAND (76)	0.82	19 17 0 75 X SQ= 44.11 P = 0. RV YES (YES)
56	ALWAYS ARE THOSE WHERE THE RELIGION IS PREDOMINANTLY LITERATE (79)	1.00	56	TEND LESS TO BE THOSE WHERE THE RELIGION IS PREDOMINANTLY LITERATE (79)	0.65	19 58 0 31 X SQ= 7.66 P = 0.001 RV NO (NO)
58	ALWAYS ARE THOSE WHERE THE RELIGION IS OTHER THAN MUSLIM (97)	1.00	58	TEND LESS TO BE THOSE WHERE THE RELIGION IS OTHER THAN MUSLIM (97)	0.81	0 18 19 76 X SQ= 3.02 P = 0.039 RV NO (NO)
62	TEND TO BE THOSE WHERE THE RELIGION IS PROTESTANT, RATHER THAN CATHOLIC (5)	0.50	62	TEND TO BE THOSE WHERE THE RELIGION IS CATHOLIC, RATHER THAN PROTESTANT (25)	0.95	4 1 4 20 X SQ= 5.44 P = 0.013 RV YES (YES)
63	ALWAYS ARE THOSE WHERE THE RELIGION IS PREDOMINANTLY SOME KIND OF CHRISTIAN (46)	1.00	63	TEND TO BE THOSE WHERE THE RELIGION IS PREDOMINANTLY OR PARTLY OTHER THAN CHRISTIAN (68)	0.71	18 27 0 67 X SQ= 29.04 P = 0. RV YES (NO)
67	TEND MORE TO BE THOSE THAT ARE RACIALLY HOMOGENEOUS (82)	0.95	67	TEND LESS TO BE THOSE THAT ARE RACIALLY HOMOGENEOUS (82)	0.69	18 61 1 27 X SQ= 3.99 P = 0.022 RV NO (NO)
68	TEND TO BE THOSE THAT ARE LINGUISTICALLY HOMOGENEOUS (52)	0.68	68	TEND TO BE THOSE THAT ARE LINGUISTICALLY WEAKLY HETEROGENEOUS OR STRONGLY HETEROGENEOUS (62)	0.59	13 38 6 55 X SQ= 3.78 P = 0.042 RV YES (NO)
71	TEND TO BE THOSE WHOSE DATE OF INDEPENDENCE IS BEFORE 1800 (21)	0.59	71	TEND TO BE THOSE WHOSE DATE OF INDEPENDENCE IS AFTER 1800 (90)	0.88	10 11 7 81 X SQ= 17.36 P = 0. RV YES (NO)

72	TEND TO BE THOSE WHOSE DATE OF INDEPENDENCE IS BEFORE 1914 (52)	0.88	72	TEND TO BE THOSE WHOSE DATE OF INDEPENDENCE IS AFTER 1914 (59)	0.61	15 36 2 56 X SQ= 11.99 P = 0. RV YES (NO)
73	ALWAYS ARE THOSE WHOSE DATE OF INDEPENDENCE IS BEFORE 1945 (65)	1.00	73	TEND LESS TO BE THOSE WHOSE DATE OF INDEPENDENCE IS BEFORE 1945 (65)	0.51	17 47 0 45 X SQ= 12.22 P = 0. RV NO (NO)
74	TEND TO BE THOSE THAT ARE HISTORICALLY WESTERN (26)	0.95	74	TEND TO BE THOSE THAT ARE NOT HISTORICALLY WESTERN (87)	0.91	18 8 1 85 X SQ= 60.92 P = 0. RV YES (YES)
75	ALWAYS ARE THOSE THAT ARE HISTORICALLY WESTERN OR SIGNIFICANTLY WESTERNIZED (62)	1.00	75	TEND TO BE THOSE THAT ARE NOT HISTORICALLY WESTERN AND ARE NOT SIGNIFICANTLY WESTERNIZED (52)	0.56	19 41 0 52 X SQ= 17.65 P = 0. RV YES (NO)
81	TEND TO BE THOSE WHERE THE TYPE OF POLITICAL MODERNIZATION IS EARLY EUROPEAN OR EARLY EUROPEAN DERIVED, RATHER THAN LATER EUROPEAN OR LATER EUROPEAN DERIVED (11)	0.56	81	TEND TO BE THOSE WHERE THE TYPE OF POLITICAL MODERNIZATION IS LATER EUROPEAN OR LATER EUROPEAN DERIVED, RATHER THAN EARLY EUROPEAN OR EARLY EUROPEAN DERIVED (40)	0.97	10 1 8 31 X SQ= 15.53 P = 0. RV YES (YES)
82	ALWAYS ARE THOSE WHERE THE TYPE OF POLITICAL MODERNIZATION IS EARLY OR LATER EUROPEAN OR EUROPEAN DERIVED, RATHER THAN DEVELOPED TUTELARY OR UNDEVELOPED TUTELARY (51)	1.00	82	TEND TO BE THOSE WHERE THE TYPE OF POLITICAL MODERNIZATION IS DEVELOPED TUTELARY OR UNDEVELOPED TUTELARY, RATHER THAN EARLY OR LATER EUROPEAN OR EUROPEAN DERIVED (55)	0.63	18 32 0 54 X SQ= 21.06 P = 0. RV YES (NO)
85	ALWAYS ARE THOSE WHERE THE STAGE OF POLITICAL MODERNIZATION IS ADVANCED, RATHER THAN MID- OR EARLY TRANSITIONAL (60)	1.00	85	TEND TO BE THOSE WHERE THE STAGE OF POLITICAL MODERNIZATION IS MID- OR EARLY TRANSITIONAL, RATHER THAN ADVANCED (54)	0.58	19 39 0 54 X SQ= 19.04 P = 0. RV YES (NO)
87	ALWAYS ARE THOSE WHOSE IDEOLOGICAL ORIENTATION IS OTHER THAN DEVELOPMENTAL (58)	1.00	87	TEND LESS TO BE THOSE WHOSE IDEOLOGICAL ORIENTATION IS OTHER THAN DEVELOPMENTAL (58)	0.57	0 29 19 39 X SQ= 10.31 P = 0. RV NO (NO)
89	TEND TO BE THOSE WHOSE IDEOLOGICAL ORIENTATION IS CONVENTIONAL (33)	0.84	89	TEND TO BE THOSE WHOSE IDEOLOGICAL ORIENTATION IS OTHER THAN CONVENTIONAL (62)	0.77	16 17 3 57 X SQ= 22.16 P = 0. RV YES (NO)

94	TEND TO BE THOSE WHERE THE STATUS OF THE REGIME IS CONSTITUTIONAL (51)	0.84	94	TEND TO BE THOSE WHERE THE STATUS OF THE REGIME IS AUTHORITARIAN OR TOTALITARIAN (41)	0.54	16 33 3 38 X SQ= 7.15 P = 0.004 RV YES (NO)
96	ALWAYS ARE THOSE WHERE THE STATUS OF THE REGIME IS CONSTITUTIONAL OR TOTALITARIAN (67)	1.00	96	TEND LESS TO BE THOSE WHERE THE STATUS OF THE REGIME IS CONSTITUTIONAL OR TOTALITARIAN (67)	0.67	19 46 0 23 X SQ= 6.93 P = 0.002 RV NO (NO)
97	ALWAYS ARE THOSE WHERE THE STATUS OF THE REGIME IS TOTALITARIAN, RATHER THAN AUTHORITARIAN (16)	1.00	97	LEAN TOWARD BEING THOSE WHERE THE STATUS OF THE REGIME IS AUTHORITARIAN, RATHER THAN TOTALITARIAN (23)	0.64	0 23 3 13 X SQ= 2.40 P = 0.061 RV YES (NO)
98	TEND TO BE THOSE WHERE GOVERNMENTAL STABILITY IS GENERALLY PRESENT AND DATES AT LEAST FROM THE INTERWAR PERIOD (22)	0.74	98	TEND TO BE THOSE WHERE GOVERNMENTAL STABILITY IS GENERALLY OR MODERATELY PRESENT AND DATES FROM THE POST-WAR PERIOD, OR IS ABSENT (93)	0.91	14 8 5 86 X SQ= 38.76 P = 0. RV YES (YES)
99	ALWAYS ARE THOSE WHERE GOVERNMENTAL STABILITY IS GENERALLY PRESENT AND DATES FROM AT LEAST THE INTER-WAR PERIOD, OR FROM THE POST-WAR PERIOD (50)	1.00	99	TEND TO BE THOSE WHERE GOVERNMENTAL STABILITY IS MODERATELY PRESENT AND DATES FROM THE POST-WAR PERIOD, OR IS ABSENT (36)	0.54	19 30 0 35 X SQ= 15.39 P = 0. RV YES (NO)
101	TEND TO BE THOSE WHERE THE REPRESENTATIVE CHARACTER OF THE REGIME IS POLYARCHIC (41)	0.84	101	TEND TO BE THOSE WHERE THE REPRESENTATIVE CHARACTER OF THE REGIME IS LIMITED POLYARCHIC, PSEUDO-POLYARCHIC, OR NON-POLYARCHIC (57)	0.70	16 23 3 54 X SQ= 16.47 P = 0. RV YES (NO)
102	TEND TO BE THOSE WHERE THE REPRESENTATIVE CHARACTER OF THE REGIME IS POLYARCHIC OR LIMITED POLYARCHIC (59)	0.84	102	TEND TO BE THOSE WHERE THE REPRESENTATIVE CHARACTER OF THE REGIME IS PSEUDO-POLYARCHIC OR NON-POLYARCHIC (49)	0.53	16 41 3 46 X SQ= 7.20 P = 0.004 RV YES (NO)
105	TEND TO BE THOSE WHERE THE ELECTORAL SYSTEM IS COMPETITIVE (43)	0.84	105	TEND TO BE THOSE WHERE THE ELECTORAL SYSTEM IS PARTIALLY COMPETITIVE OR NON-COMPETITIVE (47)	0.64	16 25 3 44 X SQ= 11.92 P = 0. RV YES (NO)
107	TEND TO BE THOSE WHERE AUTONOMOUS GROUPS ARE FULLY TOLERATED IN POLITICS (46)	0.84	107	TEND TO BE THOSE WHERE AUTONOMOUS GROUPS ARE PARTIALLY TOLERATED IN POLITICS, ARE TOLERATED ONLY OUTSIDE POLITICS, OR ARE NOT TOLERATED AT ALL (65)	0.69	16 28 3 62 X SQ= 16.24 P = 0. RV YES (NO)

111	TEND TO BE THOSE WHERE POLITICAL ENCULTURATION IS HIGH OR MEDIUM (53)	0.93	111	TEND TO BE THOSE WHERE POLITICAL ENCULTURATION IS LOW (42)	0.53	14 37 1 41 X SQ= 8.93 P = 0.001 RV YES (NO)
115	TEND TO BE THOSE WHERE INTEREST ARTICULATION BY ASSOCIATIONAL GROUPS IS SIGNIFICANT (20)	0.84	115	TEND TO BE THOSE WHERE INTEREST ARTICULATION BY ASSOCIATIONAL GROUPS IS MODERATE, LIMITED, OR NEGLIGIBLE (91)	0.96	16 4 3 86 X SQ= 61.41 P = 0. RV YES (YES)
116	TEND TO BE THOSE WHERE INTEREST ARTICULATION BY ASSOCIATIONAL GROUPS IS SIGNIFICANT OR MODERATE (32)	0.84	116	TEND TO BE THOSE WHERE INTEREST ARTICULATION BY ASSOCIATIONAL GROUPS IS LIMITED OR NEGLIGIBLE (79)	0.84	16 14 3 76 X SQ= 33.71 P = 0. RV YES (YES)
117	TEND TO BE THOSE WHERE INTEREST ARTICULATION BY ASSOCIATIONAL GROUPS IS SIGNIFICANT, MODERATE, OR LIMITED (60)	0.89	117	TEND TO BE THOSE WHERE INTEREST ARTICULATION BY ASSOCIATIONAL GROUPS IS NEGLIGIBLE (51)	0.54	17 41 2 49 X SQ= 10.45 P = 0.001 RV YES (NO)
119	TEND TO BE THOSE WHERE INTEREST ARTICULATION BY INSTITUTIONAL GROUPS IS MODERATE OR LIMITED (26)	0.79	119	TEND TO BE THOSE WHERE INTEREST ARTICULATION BY INSTITUTIONAL GROUPS IS VERY SIGNIFICANT OR SIGNIFICANT (74)	0.86	4 68 15 11 X SQ= 29.97 P = 0. RV YES (YES)
121	ALWAYS ARE THOSE WHERE INTEREST ARTICULATION BY NON-ASSOCIATIONAL GROUPS IS MODERATE, LIMITED, OR NEGLIGIBLE (61)	1.00	121	TEND TO BE THOSE WHERE INTEREST ARTICULATION BY NON-ASSOCIATIONAL GROUPS IS SIGNIFICANT (54)	0.57	0 54 19 40 X SQ= 18.67 P = 0. RV YES (NO)
122	TEND TO BE THOSE WHERE INTEREST ARTICULATION BY NON-ASSOCIATIONAL GROUPS IS LIMITED OR NEGLIGIBLE (32)	0.79	122	TEND TO BE THOSE WHERE INTEREST ARTICULATION BY NON-ASSOCIATIONAL GROUPS IS SIGNIFICANT OR MODERATE (83)	0.83	4 78 15 16 X SQ= 27.41 P = 0. RV YES (NO)
125	TEND TO BE THOSE WHERE INTEREST ARTICULATION BY ANOMIC GROUPS IS INFREQUENT OR VERY INFREQUENT (35)	0.88	125	TEND TO BE THOSE WHERE INTEREST ARTICULATION BY ANOMIC GROUPS IS FREQUENT OR OCCASIONAL (64)	0.77	2 62 15 19 X SQ= 23.24 P = 0. RV YES (NO)
126	TEND TO BE THOSE WHERE INTEREST ARTICULATION BY ANOMIC GROUPS IS VERY INFREQUENT (16)	0.76	126	TEND TO BE THOSE WHERE INTEREST ARTICULATION BY ANOMIC GROUPS IS FREQUENT, OCCASIONAL, OR INFREQUENT (83)	0.96	4 78 13 3 X SQ= 49.27 P = 0. RV YES (YES)

No.			No.			
128	TEND TO BE THOSE WHERE INTEREST ARTICULATION BY POLITICAL PARTIES IS SIGNIFICANT OR MODERATE (48)	0.74	128	TEND TO BE THOSE WHERE INTEREST ARTICULATION BY POLITICAL PARTIES IS LIMITED OR NEGLIGIBLE (45)	0.56	14 32 5 40 X SQ= 4.04 P = 0.038 RV YES (NO)
133	LEAN MORE TOWARD BEING THOSE WHERE INTEREST AGGREGATION BY THE EXECUTIVE IS MODERATE, LIMITED, OR NEGLIGIBLE (76)	0.89	133	LEAN LESS TOWARD BEING THOSE WHERE INTEREST AGGREGATION BY THE EXECUTIVE IS MODERATE, LIMITED, OR NEGLIGIBLE (76)	0.69	2 26 17 58 X SQ= 2.32 P = 0.090 RV NO (NO)
134	TEND TO BE THOSE WHERE INTEREST AGGREGATION BY THE EXECUTIVE IS SIGNIFICANT OR MODERATE (57)	0.79	134	TEND TO BE THOSE WHERE INTEREST AGGREGATION BY THE EXECUTIVE IS LIMITED OR NEGLIGIBLE (46)	0.51	15 40 4 42 X SQ= 4.51 P = 0.022 RV YES (NO)
137	TEND TO BE THOSE WHERE INTEREST AGGREGATION BY THE LEGISLATURE IS SIGNIFICANT OR MODERATE (29)	0.79	137	TEND TO BE THOSE WHERE INTEREST AGGREGATION BY THE LEGISLATURE IS LIMITED OR NEGLIGIBLE (68)	0.83	15 13 4 63 X SQ= 25.07 P = 0. RV YES (YES)
138	TEND TO BE THOSE WHERE INTEREST AGGREGATION BY THE LEGISLATURE IS SIGNIFICANT, MODERATE, OR LIMITED (48)	0.84	138	TEND TO BE THOSE WHERE INTEREST AGGREGATION BY THE LEGISLATURE IS NEGLIGIBLE (49)	0.61	16 30 3 46 X SQ= 10.46 P = 0.001 RV YES (NO)
139	TILT MORE TOWARD BEING THOSE WHERE THE PARTY SYSTEM IS QUANTITATIVELY OTHER THAN ONE-PARTY (71)	0.84	139	TILT LESS TOWARD BEING THOSE WHERE THE PARTY SYSTEM IS QUANTITATIVELY OTHER THAN ONE-PARTY (71)	0.63	3 31 16 53 X SQ= 2.24 P = 0.106 RV NO (NO)
144	TEND TO BE THOSE WHERE THE PARTY SYSTEM IS QUANTITATIVELY TWO-PARTY, RATHER THAN ONE-PARTY (11)	0.62	144	TEND TO BE THOSE WHERE THE PARTY SYSTEM IS QUANTITATIVELY ONE-PARTY, RATHER THAN TWO-PARTY (34)	0.84	3 31 5 6 X SQ= 5.33 P = 0.014 RV YES (NO)
147	TEND TO BE THOSE WHERE THE PARTY SYSTEM IS QUALITATIVELY CLASS-ORIENTED OR MULTI-IDEOLOGICAL (23)	0.72	147	TEND TO BE THOSE WHERE THE PARTY SYSTEM IS QUALITATIVELY OTHER THAN CLASS-ORIENTED OR MULTI-IDEOLOGICAL (67)	0.87	13 9 5 61 X SQ= 23.84 P = 0. RV YES (YES)
148	ALWAYS ARE THOSE WHERE THE PARTY SYSTEM IS QUALITATIVELY OTHER THAN AFRICAN TRANSITIONAL (96)	1.00	148	LEAN LESS TOWARD BEING THOSE WHERE THE PARTY SYSTEM IS QUALITATIVELY OTHER THAN AFRICAN TRANSITIONAL (96)	0.84	0 14 19 75 X SQ= 2.18 P = 0.071 RV NO (NO)

153	TEND TO BE THOSE WHERE THE PARTY SYSTEM IS STABLE (42)	0.89	153	TEND TO BE THOSE WHERE THE PARTY SYSTEM IS MODERATELY STABLE OR UNSTABLE (38)	0.59	17 24 2 35 X SQ= 11.84 P = 0. RV YES (NO)
154	ALWAYS ARE THOSE WHERE THE PARTY SYSTEM IS STABLE OR MODERATELY STABLE (55)	1.00	154	TEND LESS TO BE THOSE WHERE THE PARTY SYSTEM IS STABLE OR MODERATELY STABLE (55)	0.59	19 35 0 24 X SQ= 9.34 P = 0. RV NO (NO)
157	ALWAYS ARE THOSE WHERE PERSONALISMO IS MODERATE OR NEGLIGIBLE (84)	1.00	157	LEAN LESS TOWARD BEING THOSE WHERE PERSONALISMO IS MODERATE OR NEGLIGIBLE (84)	0.82	0 14 19 63 X SQ= 2.72 P = 0.065 RV NO (NO)
158	TEND TO BE THOSE WHERE PERSONALISMO IS NEGLIGIBLE (56)	0.95	158	TEND TO BE THOSE WHERE PERSONALISMO IS PRONOUNCED OR MODERATE (40)	0.52	1 39 18 36 X SQ= 11.70 P = 0. RV YES (NO)
161	TEND TO BE THOSE WHERE THE POLITICAL LEADERSHIP IS NON-ELITIST (50)	0.74	161	TEND TO BE THOSE WHERE THE POLITICAL LEADERSHIP IS ELITIST OR MODERATELY ELITIST (47)	0.55	5 42 14 34 X SQ= 4.00 P = 0.039 RV YES (NO)
164	TEND MORE TO BE THOSE WHERE THE REGIME'S LEADERSHIP CHARISMA IS NEGLIGIBLE (65)	0.94	164	TEND LESS TO BE THOSE WHERE THE REGIME'S LEADERSHIP CHARISMA IS NEGLIGIBLE (65)	0.58	1 33 17 46 X SQ= 6.93 P = 0.003 RV NO (NO)
168	TEND TO BE THOSE WHERE THE HORIZONTAL POWER DISTRIBUTION IS SIGNIFICANT (34)	0.79	168	TEND TO BE THOSE WHERE THE HORIZONTAL POWER DISTRIBUTION IS LIMITED OR NEGLIGIBLE (72)	0.79	15 18 4 67 X SQ= 21.33 P = 0. RV YES (NO)
169	TEND TO BE THOSE WHERE THE HORIZONTAL POWER DISTRIBUTION IS SIGNIFICANT OR LIMITED (58)	0.84	169	TEND TO BE THOSE WHERE THE HORIZONTAL POWER DISTRIBUTION IS NEGLIGIBLE (48)	0.53	16 40 3 45 X SQ= 7.19 P = 0.004 RV YES (NO)
170	TEND MORE TO BE THOSE WHERE THE LEGISLATIVE-EXECUTIVE STRUCTURE IS OTHER THAN PRESIDENTIAL (63)	0.95	170	TEND LESS TO BE THOSE WHERE THE LEGISLATIVE-EXECUTIVE STRUCTURE IS OTHER THAN PRESIDENTIAL (63)	0.54	1 37 18 44 X SQ= 9.02 P = 0.001 RV NO (NO)

172	TEND TO BE THOSE WHERE THE LEGISLATIVE-EXECUTIVE STRUCTURE IS PARLIAMENTARY-ROYALIST (21)	0.53	172	TEND TO BE THOSE WHERE THE LEGISLATIVE-EXECUTIVE STRUCTURE IS OTHER THAN PARLIAMENTARY-ROYALIST (88)	0.87	10 11 9 77 X SQ= 13.51 P = 0. RV YES (NO)
174	TEND TO BE THOSE WHERE THE LEGISLATURE IS FULLY EFFECTIVE (28)	0.79	174	TEND TO BE THOSE WHERE THE LEGISLATURE IS PARTIALLY EFFECTIVE, LARGELY INEFFECTIVE, OR WHOLLY INEFFECTIVE (72)	0.85	15 12 4 67 X SQ= 28.08 P = 0. RV YES (YES)
175	TEND TO BE THOSE WHERE THE LEGISLATURE IS FULLY EFFECTIVE OR PARTIALLY EFFECTIVE (51)	0.84	175	TEND TO BE THOSE WHERE THE LEGISLATURE IS LARGELY INEFFECTIVE OR WHOLLY INEFFECTIVE (49)	0.58	16 33 3 46 X SQ= 9.40 P = 0.002 RV YES (NO)
178	TEND TO BE THOSE WHERE THE LEGISLATURE IS BICAMERAL (51)	0.74	178	TEND TO BE THOSE WHERE THE LEGISLATURE IS UNICAMERAL (53)	0.57	5 47 14 36 X SQ= 4.54 P = 0.022 RV YES (NO)
179	TEND TO BE THOSE WHERE THE EXECUTIVE IS STRONG (39)	0.78	179	TEND TO BE THOSE WHERE THE EXECUTIVE IS DOMINANT (52)	0.68	4 48 14 23 X SQ= 10.38 P = 0.001 RV YES (NO)
181	TEND TO BE THOSE WHERE THE BUREAUCRACY IS MODERN, RATHER THAN SEMI-MODERN (21)	0.84	181	TEND TO BE THOSE WHERE THE BUREAUCRACY IS SEMI-MODERN, RATHER THAN MODERN (55)	0.93	16 4 3 51 X SQ= 38.57 P = 0. RV YES (YES)
186	LEAN MORE TOWARD BEING THOSE WHERE PARTICIPATION BY THE MILITARY IN POLITICS IS NEUTRAL, RATHER THAN SUPPORTIVE (56)	0.83	186	LEAN LESS TOWARD BEING THOSE WHERE PARTICIPATION BY THE MILITARY IN POLITICS IS NEUTRAL, RATHER THAN SUPPORTIVE (56)	0.59	3 28 15 40 X SQ= 2.72 P = 0.060 RV NO (NO)
187	TEND TO BE THOSE WHERE THE ROLE OF THE POLICE IS NOT POLITICALLY SIGNIFICANT (35)	0.84	187	TEND TO BE THOSE WHERE THE ROLE OF THE POLICE IS POLITICALLY SIGNIFICANT (66)	0.77	3 62 16 18 X SQ= 23.27 P = 0. RV YES (NO)
189	TEND LESS TO BE THOSE WHERE THE CHARACTER OF THE LEGAL SYSTEM IS OTHER THAN COMMON LAW (108)	0.79	189	TEND MORE TO BE THOSE WHERE THE CHARACTER OF THE LEGAL SYSTEM IS OTHER THAN COMMON LAW (108)	0.97	4 3 15 91 X SQ= 5.88 P = 0.015 RV NO (YES)

192	ALWAYS ARE THOSE WHERE THE CHARACTER OF THE LEGAL SYSTEM IS OTHER THAN MUSLIM OR PARTLY MUSLIM (86) 1.00	192	TEND LESS TO BE THOSE WHERE THE CHARACTER OF THE LEGAL SYSTEM IS OTHER THAN MUSLIM OR PARTLY MUSLIM (86) 0.71	0 27 19 65 X SQ= 5.86 P = 0.003 RV NO (NO)
193	ALWAYS ARE THOSE WHERE THE CHARACTER OF THE LEGAL SYSTEM IS OTHER THAN PARTLY INDIGENOUS (86) 1.00	193	TEND LESS TO BE THOSE WHERE THE CHARACTER OF THE LEGAL SYSTEM IS OTHER THAN PARTLY INDIGENOUS (86) 0.70	0 28 19 65 X SQ= 6.11 P = 0.003 RV NO (NO)

```
*******************************************************************************************************

42  POLITIES                                            42  POLITIES
    WHOSE ECONOMIC DEVELOPMENTAL STATUS                     WHOSE ECONOMIC DEVELOPMENTAL STATUS
    IS DEVELOPED OR INTERMEDIATE  (36)                      IS UNDERDEVELOPED OR
                                                            VERY UNDERDEVELOPED  (76)


                                                                                        PREDICATE ONLY
.......................................................................................................

    36 IN LEFT COLUMN

ARGENTINA   AUSTRALIA   AUSTRIA     BELGIUM     BRAZIL      BULGARIA    CANADA      COLOMBIA    CYPRUS      CZECHOS'KIA
DENMARK     FINLAND     FRANCE      GERMANY, E  GERMAN FR   GREECE      HUNGARY     ICELAND     IRELAND     ITALY
JAPAN       LUXEMBOURG  MEXICO      NETHERLANDS NEW ZEALAND NORWAY      POLAND      RUMANIA     SO AFRICA   SWEDEN
SWITZERLAND USSR        UK          US          URUGUAY     YUGOSLAVIA

    76 IN RIGHT COLUMN

AFGHANISTAN ALBANIA     ALGERIA     BOLIVIA     BURMA       BURUNDI     CAMBODIA    CAMEROUN    CEN AFR REP CEYLON
CHAD        CHILE       CHINA, PR   CONGO(BRA)  CONGO(LEO)  COSTA RICA  CUBA        DAHOMEY     DOMIN REP   ECUADOR
EL SALVADOR ETHIOPIA    GABON       GHANA       GUATEMALA   GUINEA      HAITI       HONDURAS    INDIA       INDONESIA
IRAN        IRAQ        IVORY COAST JAMAICA     JORDAN      KOREA, N    KOREA REP   LAOS        LEBANON     LIBERIA
LIBYA       MALAGASY R  MALAYA      MALI        MAURITANIA  MONGOLIA    MOROCCO     NEPAL       NICARAGUA   NIGER
NIGERIA     PAKISTAN    PANAMA      PARAGUAY    PERU        PHILIPPINES PORTUGAL    RWANDA      SENEGAL     SIERRE LEO
SOMALIA     SPAIN       SUDAN       SYRIA       TANGANYIKA  THAILAND    TOGO        TRINIDAD    TUNISIA     TURKEY
UGANDA      UAR         UPPER VOLTA VIETNAM, N  VIETNAM REP YEMEN

    3 EXCLUDED BECAUSE AMBIGUOUS

ISRAEL      SA'U ARABIA VENEZUELA

-------------------------------------------------------------------------------------------------------
```

```
****************************************************************************************************

43  POLITIES                                          43  POLITIES
    WHOSE ECONOMIC DEVELOPMENTAL STATUS                   WHOSE ECONOMIC DEVELOPMENTAL STATUS
    IS DEVELOPED, INTERMEDIATE, OR                        IS VERY UNDERDEVELOPED  (57)
    UNDERDEVELOPED  (55)

                                                                       BOTH SUBJECT AND PREDICATE
....................................................................................................

      55 IN LEFT COLUMN

ARGENTINA   AUSTRALIA   AUSTRIA     BELGIUM     BRAZIL      BULGARIA    BURMA       CANADA      CHILE       CHINA, PR
COLOMBIA    CUBA        CYPRUS      CZECHOS'KIA DENMARK     ECUADOR     FINLAND     FRANCE      GERMANY, E  GERMAN FR
GREECE      HUNGARY     ICELAND     INDIA       IRAQ        IRELAND     ISRAEL      ITALY       JAMAICA     JAPAN
LIBYA       LUXEMBOURG  MALAYA      MEXICO      NETHERLANDS NEW ZEALAND NORWAY      PAKISTAN    PERU        PHILIPPINES
POLAND      PORTUGAL    RUMANIA     SO AFRICA   SPAIN       SWEDEN      SWITZERLAND TRINIDAD    TURKEY      USSR
UK          US          URUGUAY     VENEZUELA   YUGOSLAVIA

      57 IN RIGHT COLUMN

AFGHANISTAN ALBANIA     ALGERIA     BOLIVIA     BURUNDI     CAMBODIA    CAMEROUN    CEN AFR REP CEYLON      CHAD
CONGO(BRA)  CONGO(LEO)  DAHOMEY     DOMIN REP   EL SALVADOR ETHIOPIA    GABON       GHANA       GUATEMALA   GUINEA
HAITI       HONDURAS    INDONESIA   IRAN        IVORY COAST JORDAN      KOREA, N    KOREA REP   LAOS        LIBERIA
MALAGASY R  MALI        MAURITANIA  MONGOLIA    MOROCCO     NEPAL       NICARAGUA   NIGER       NIGERIA     PANAMA
PARAGUAY    RWANDA      SENEGAL     SIERRE LEO  SOMALIA     SUDAN       SYRIA       TANGANYIKA  THAILAND    TOGO
TUNISIA     UGANDA      UAR         UPPER VOLTA VIETNAM, N  VIETNAM REP YEMEN

       3 EXCLUDED BECAUSE AMBIGUOUS

COSTA RICA  LEBANON     SA'U ARABIA

----------------------------------------------------------------------------------------------------

  2  TEND LESS TO BE THOSE                  0.60     2  ALWAYS ARE THOSE                      1.00        22     0
     LOCATED ELSEWHERE THAN IN                          LOCATED ELSEWHERE THAN IN                         33    57
     WEST EUROPE, SCANDINAVIA,                          WEST EUROPE, SCANDINAVIA,                     X SQ=  25.90
     NORTH AMERICA, OR AUSTRALASIA  (93)                NORTH AMERICA, OR AUSTRALASIA  (93)           P =     0.
                                                                                                      RV NO  (YES)

  7  TEND TO BE THOSE                       0.89     7  TEND TO BE THOSE                      0.71         1     5
     LOCATED IN SOUTH AMERICA, RATHER THAN              LOCATED IN CENTRAL AMERICA, RATHER THAN            8     2
     IN CENTRAL AMERICA  (10)                           IN SOUTH AMERICA  (7)                         X SQ=   3.81
                                                                                                      P =     0.035
                                                                                                      RV YES (YES)

 10  TEND TO BE THOSE                       0.96    10  TEND TO BE THOSE                      0.54         2    31
     LOCATED ELSEWHERE THAN IN NORTH AFRICA,            LOCATED IN NORTH AFRICA, OR                       53    26
     OR CENTRAL AND SOUTH AFRICA  (82)                  CENTRAL AND SOUTH AFRICA  (33)                X SQ=  32.29
                                                                                                      P =     0.
                                                                                                      RV YES (YES)
```

11	TEND MORE TO BE THOSE LOCATED ELSEWHERE THAN IN CENTRAL AND SOUTH AFRICA (87)	0.98	11	TEND LESS TO BE THOSE LOCATED ELSEWHERE THAN IN CENTRAL AND SOUTH AFRICA (87)	0.53	1 27 54 30 X SQ= 28.59 P = 0. RV NO (YES)
15	TEND TO BE THOSE LOCATED IN THE MIDDLE EAST, RATHER THAN IN NORTH AFRICA OR CENTRAL AND SOUTH AFRICA (11)	0.67	15	TEND TO BE THOSE LOCATED IN NORTH AFRICA OR CENTRAL AND SOUTH AFRICA, RATHER THAN IN THE MIDDLE EAST (33)	0.86	4 5 2 31 X SQ= 5.66 P = 0.013 RV YES (NO)
18	TEND TO BE THOSE LOCATED IN EAST ASIA, SOUTH ASIA, OR SOUTHEAST ASIA, RATHER THAN IN NORTH AFRICA OR CENTRAL AND SOUTH AFRICA (18)	0.78	18	TEND TO BE THOSE LOCATED IN NORTH AFRICA OR CENTRAL AND SOUTH AFRICA, RATHER THAN IN EAST ASIA, SOUTH ASIA, OR SOUTHEAST ASIA (33)	0.74	2 31 7 11 X SQ= 6.53 P = 0.006 RV YES (NO)
19	TEND TO BE THOSE LOCATED IN THE CARIBBEAN, CENTRAL AMERICA, OR SOUTH AMERICA, RATHER THAN IN NORTH AFRICA OR CENTRAL AND SOUTH AFRICA (22)	0.86	19	TEND TO BE THOSE LOCATED IN NORTH AFRICA OR CENTRAL AND SOUTH AFRICA, RATHER THAN IN THE CARIBBEAN, CENTRAL AMERICA, OR SOUTH AMERICA (33)	0.77	2 31 12 9 X SQ= 14.88 P = 0. RV YES (YES)
24	TEND TO BE THOSE WHOSE POPULATION IS VERY LARGE, LARGE, OR MEDIUM (61)	0.73	24	TEND TO BE THOSE WHOSE POPULATION IS SMALL (54)	0.65	40 20 15 37 X SQ= 14.47 P = 0. RV YES (YES)
26	TEND TO BE THOSE WHOSE POPULATION DENSITY IS VERY HIGH, HIGH, OR MEDIUM (48)	0.58	26	TEND TO BE THOSE WHOSE POPULATION DENSITY IS LOW (67)	0.74	32 15 23 42 X SQ= 10.40 P = 0.001 RV YES (YES)
28	TEND TO BE THOSE WHOSE POPULATION GROWTH RATE IS LOW (48)	0.56	28	TEND TO BE THOSE WHOSE POPULATION GROWTH RATE IS HIGH (62)	0.66	24 35 30 18 X SQ= 4.21 P = 0.033 RV YES (YES)
29	TEND TO BE THOSE WHERE THE DEGREE OF URBANIZATION IS HIGH (56)	0.90	29	TEND TO BE THOSE WHERE THE DEGREE OF URBANIZATION IS LOW (49)	0.81	45 10 5 43 X SQ= 49.49 P = 0. RV YES (YES)
30	TEND TO BE THOSE WHOSE AGRICULTURAL POPULATION IS MEDIUM, LOW, OR VERY LOW (57)	0.85	30	TEND TO BE THOSE WHOSE AGRICULTURAL POPULATION IS HIGH (56)	0.86	8 48 47 8 X SQ= 53.41 P = 0. RV YES (YES)

No.	Statement	Value	No.	Statement	Value	Statistics
31	TEND LESS TO BE THOSE WHOSE AGRICULTURAL POPULATION IS HIGH OR MEDIUM (90)	0.56	31	ALWAYS ARE THOSE WHOSE AGRICULTURAL POPULATION IS HIGH OR MEDIUM (90)	1.00	31 57 24 0 X SQ= 29.12 P = 0. RV NO (YES)
32	TEND LESS TO BE THOSE WHOSE GROSS NATIONAL PRODUCT IS MEDIUM, LOW, OR VERY LOW (105)	0.82	32	ALWAYS ARE THOSE WHOSE GROSS NATIONAL PRODUCT IS MEDIUM, LOW, OR VERY LOW (105)	1.00	10 0 45 57 X SQ= 9.25 P = 0.001 RV NO (YES)
33	TEND TO BE THOSE WHOSE GROSS NATIONAL PRODUCT IS VERY HIGH, HIGH, OR MEDIUM (30)	0.53	33	TEND TO BE THOSE WHOSE GROSS NATIONAL PRODUCT IS LOW OR VERY LOW (85)	0.98	29 1 26 56 X SQ= 34.53 P = 0. RV YES (YES)
34	TEND TO BE THOSE WHOSE GROSS NATIONAL PRODUCT IS VERY HIGH, HIGH, MEDIUM, OR LOW (62)	0.87	34	TEND TO BE THOSE WHOSE GROSS NATIONAL PRODUCT IS VERY LOW (53)	0.77	48 13 7 44 X SQ= 44.34 P = 0. RV YES (YES)
36	TEND TO BE THOSE WHOSE PER CAPITA GROSS NATIONAL PRODUCT IS VERY HIGH, HIGH, OR MEDIUM (42)	0.75	36	ALWAYS ARE THOSE WHOSE PER CAPITA GROSS NATIONAL PRODUCT IS LOW OR VERY LOW (73)	1.00	41 0 14 57 X SQ= 63.85 P = 0. RV YES (YES)
37	TEND TO BE THOSE WHOSE PER CAPITA GROSS NATIONAL PRODUCT IS VERY HIGH, HIGH, MEDIUM, OR LOW (64)	0.91	37	TEND TO BE THOSE WHOSE PER CAPITA GROSS NATIONAL PRODUCT IS VERY LOW (51)	0.81	50 11 5 46 X SQ= 55.03 P = 0. RV YES (YES)
38	TEND LESS TO BE THOSE WHOSE INTERNATIONAL FINANCIAL STATUS IS MEDIUM, LOW, OR VERY LOW (103)	0.81	38	ALWAYS ARE THOSE WHOSE INTERNATIONAL FINANCIAL STATUS IS MEDIUM, LOW, OR VERY LOW (103)	1.00	10 0 43 57 X SQ= 9.66 P = 0. RV NO (YES)
39	TEND TO BE THOSE WHOSE INTERNATIONAL FINANCIAL STATUS IS VERY HIGH, HIGH, OR MEDIUM (38)	0.67	39	TEND TO BE THOSE WHOSE INTERNATIONAL FINANCIAL STATUS IS LOW OR VERY LOW (76)	0.96	36 2 18 55 X SQ= 46.36 P = 0. RV YES (YES)
40	TEND TO BE THOSE WHOSE INTERNATIONAL FINANCIAL STATUS IS VERY HIGH, HIGH, MEDIUM, OR LOW (71)	0.94	40	TEND TO BE THOSE WHOSE INTERNATIONAL FINANCIAL STATUS IS VERY LOW (39)	0.65	50 19 3 35 X SQ= 38.33 P = 0. RV YES (YES)

44	TEND LESS TO BE THOSE WHERE THE LITERACY RATE IS BELOW NINETY PERCENT (90)	0.55	44	ALWAYS ARE THOSE WHERE THE LITERACY RATE IS BELOW NINETY PERCENT (90)	1.00	25 0 30 57 X SQ= 30.78 P = 0. RV NO (YES)
45	TEND TO BE THOSE WHERE THE LITERACY RATE IS FIFTY PERCENT OR ABOVE (55)	0.83	45	TEND TO BE THOSE WHERE THE LITERACY RATE IS BELOW FIFTY PERCENT (54)	0.83	45 9 9 44 X SQ= 44.49 P = 0. RV YES (YES)
46	ALWAYS ARE THOSE WHERE THE LITERACY RATE IS TEN PERCENT OR ABOVE (84)	1.00	46	TEND LESS TO BE THOSE WHERE THE LITERACY RATE IS TEN PERCENT OR ABOVE (84)	0.52	55 27 0 25 X SQ= 31.87 P = 0. RV NO (YES)
50	TEND TO BE THOSE WHERE FREEDOM OF THE PRESS IS COMPLETE (43)	0.56	50	TEND TO BE THOSE WHERE FREEDOM OF THE PRESS IS INTERMITTENT, INTERNALLY ABSENT, OR INTERNALLY AND EXTERNALLY ABSENT (56)	0.71	M 30 12 24 30 X SQ= 5.94 P = 0.012 RV YES (YES)
51	TEND TO BE THOSE WHERE FREEDOM OF THE PRESS IS COMPLETE OR INTERMITTENT (60)	0.74	51	TEND TO BE THOSE WHERE FREEDOM OF THE PRESS IS INTERNALLY ABSENT, OR INTERNALLY AND EXTERNALLY ABSENT (37)	0.54	39 19 14 22 X SQ= 6.15 P = 0.010 RV YES (YES)
53	TEND LESS TO BE THOSE WHERE NEWSPAPER CIRCULATION IS LESS THAN THREE HUNDRED PER THOUSAND (101)	0.75	53	ALWAYS ARE THOSE WHERE NEWSPAPER CIRCULATION IS LESS THAN THREE HUNDRED PER THOUSAND (101)	1.00	14 0 41 57 X SQ= 14.34 P = 0. RV NO (YES)
54	TEND TO BE THOSE WHERE NEWSPAPER CIRCULATION IS ONE HUNDRED OR MORE PER THOUSAND (37)	0.62	54	TEND TO BE THOSE WHERE NEWSPAPER CIRCULATION IS LESS THAN ONE HUNDRED PER THOUSAND (76)	0.96	34 2 21 53 X SQ= 39.68 P = 0. RV YES (YES)
55	TEND TO BE THOSE WHERE NEWSPAPER CIRCULATION IS TEN OR MORE PER THOUSAND (78)	0.96	55	TEND TO BE THOSE WHERE NEWSPAPER CIRCULATION IS LESS THAN TEN PER THOUSAND (35)	0.58	53 23 2 32 X SQ= 35.80 P = 0. RV YES (YES)
56	TEND TO BE THOSE WHERE THE RELIGION IS PREDOMINANTLY LITERATE (79)	0.94	56	TEND TO BE THOSE WHERE THE RELIGION IS PREDOMINANTLY OR PARTLY NON-LITERATE (31)	0.53	51 25 3 28 X SQ= 26.80 P = 0. RV YES (YES)

63	TEND TO BE THOSE WHERE THE RELIGION IS PREDOMINANTLY SOME KIND OF CHRISTIAN (46)	0.72	63	TEND TO BE THOSE WHERE THE RELIGION IS PREDOMINANTLY OR PARTLY OTHER THAN CHRISTIAN (68)	0.89	M 39 6 15 51 X SQ= 41.27 P = 0. RV YES (YES)
64	TEND TO BE THOSE WHERE THE RELIGION IS CHRISTIAN, RATHER THAN MUSLIM (46)	0.91	64	TEND TO BE THOSE WHERE THE RELIGION IS MUSLIM, RATHER THAN CHRISTIAN (18)	0.68	39 6 4 13 X SQ= 20.27 P = 0. RV YES (YES)
66	LEAN TOWARD BEING THOSE THAT ARE RELIGIOUSLY HOMOGENEOUS (57)	0.62	66	LEAN TOWARD BEING THOSE THAT ARE RELIGIOUSLY HETEROGENEOUS (49)	0.56	33 22 20 28 X SQ= 2.75 P = 0.077 RV YES (YES)
68	TEND TO BE THOSE THAT ARE LINGUISTICALLY HOMOGENEOUS (52)	0.57	68	TEND TO BE THOSE THAT ARE LINGUISTICALLY WEAKLY HETEROGENEOUS OR STRONGLY HETEROGENEOUS (62)	0.68	M 31 18 23 39 X SQ= 6.49 P = 0.008 RV YES (YES)
69	TEND TO BE THOSE THAT ARE LINGUISTICALLY HOMOGENEOUS OR WEAKLY HETEROGENEOUS (64)	0.69	69	TEND TO BE THOSE THAT ARE LINGUISTICALLY STRONGLY HETEROGENEOUS (50)	0.58	37 24 17 33 X SQ= 6.78 P = 0.007 RV YES (YES)
70	TEND LESS TO BE THOSE THAT ARE RELIGIOUSLY, RACIALLY, OR LINGUISTICALLY HETEROGENEOUS (85)	0.72	70	TEND MORE TO BE THOSE THAT ARE RELIGIOUSLY, RACIALLY, OR LINGUISTICALLY HETEROGENEOUS (85)	0.90	15 5 38 45 X SQ= 4.40 P = 0.025 RV NO (YES)
72	TEND TO BE THOSE WHOSE DATE OF INDEPENDENCE IS BEFORE 1914 (52)	0.66	72	TEND TO BE THOSE WHOSE DATE OF INDEPENDENCE IS AFTER 1914 (59)	0.71	35 16 18 39 X SQ= 13.34 P = 0. RV YES (YES)
73	TEND TO BE THOSE WHOSE DATE OF INDEPENDENCE IS BEFORE 1945 (65)	0.81	73	TEND TO BE THOSE WHOSE DATE OF INDEPENDENCE IS AFTER 1945 (46)	0.65	43 19 10 36 X SQ= 22.09 P = 0. RV YES (YES)
74	TEND LESS TO BE THOSE THAT ARE NOT HISTORICALLY WESTERN (87)	0.51	74	ALWAYS ARE THOSE THAT ARE NOT HISTORICALLY WESTERN (87)	1.00	26 0 27 57 X SQ= 33.95 P = 0. RV NO (YES)

75	TEND TO BE THOSE 0.91 THAT ARE HISTORICALLY WESTERN OR SIGNIFICANTLY WESTERNIZED (62)	75	TEND TO BE THOSE 0.81 THAT ARE NOT HISTORICALLY WESTERN AND ARE NOT SIGNIFICANTLY WESTERNIZED (52)	49 11 5 46 X SQ= 54.15 P = 0. RV YES (YES)
76	TEND TO BE THOSE 0.55 THAT ARE HISTORICALLY WESTERN, RATHER THAN HAVING BEEN SIGNIFICANTLY OR PARTIALLY WESTERNIZED THROUGH A COLONIAL RELATIONSHIP (26)	76	ALWAYS ARE THOSE 1.00 THAT HAVE BEEN SIGNIFICANTLY OR PARTIALLY WESTERNIZED THROUGH A COLONIAL RELATIONSHIP, RATHER THAN BEING HISTORICALLY WESTERN (70)	26 0 21 47 X SQ= 33.23 P = 0. RV YES (YES)
77	TEND TO BE THOSE 0.75 THAT HAVE BEEN SIGNIFICANTLY WESTERNIZED, RATHER THAN PARTIALLY WESTERNIZED, THROUGH A COLONIAL RELATIONSHIP (28)	77	TEND TO BE THOSE 0.77 THAT HAVE BEEN PARTIALLY WESTERNIZED, RATHER THAN SIGNIFICANTLY WESTERNIZED, THROUGH A COLONIAL RELATIONSHIP (41)	15 11 5 36 X SQ= 13.63 P = 0. RV YES (YES)
78	LEAN LESS TOWARD BEING THOSE 0.68 THAT HAVE BEEN SIGNIFICANTLY WESTERNIZED THROUGH A COLONIAL RELATIONSHIP, RATHER THAN WITHOUT SUCH A RELATIONSHIP (28)	78	ALWAYS ARE THOSE 1.00 THAT HAVE BEEN SIGNIFICANTLY WESTERNIZED THROUGH A COLONIAL RELATIONSHIP, RATHER THAN WITHOUT SUCH A RELATIONSHIP (28)	15 11 7 0 X SQ= 2.74 P = 0.067 RV NO (NO)
79	TILT LESS TOWARD BEING THOSE 0.62 THAT WERE FORMERLY DEPENDENCIES OF BRITAIN OR FRANCE, RATHER THAN SPAIN (49)	79	TILT MORE TOWARD BEING THOSE 0.80 THAT WERE FORMERLY DEPENDENCIES OF BRITAIN OR FRANCE, RATHER THAN SPAIN (49)	15 33 9 8 X SQ= 1.69 P = 0.147 RV NO (YES)
80	ALWAYS ARE THOSE 1.00 WHOSE DATE OF INDEPENDENCE IS AFTER 1914, AND THAT WERE FORMERLY DEPENDENCIES OF BRITAIN, RATHER THAN FRANCE (19)	80	TEND TO BE THOSE 0.72 WHOSE DATE OF INDEPENDENCE IS AFTER 1914, AND THAT WERE FORMERLY DEPENDENCIES OF FRANCE, RATHER THAN BRITAIN (24)	10 9 0 23 X SQ= 13.12 P = 0. RV YES (YES)
81	LEAN LESS TOWARD BEING THOSE 0.72 WHERE THE TYPE OF POLITICAL MODERNIZATION IS LATER EUROPEAN OR LATER EUROPEAN DERIVED, RATHER THAN EARLY EUROPEAN OR EARLY EUROPEAN DERIVED (40)	81	ALWAYS ARE THOSE 1.00 WHERE THE TYPE OF POLITICAL MODERNIZATION IS LATER EUROPEAN OR LATER EUROPEAN DERIVED, RATHER THAN EARLY EUROPEAN OR EARLY EUROPEAN DERIVED (40)	11 0 29 10 X SQ= 2.11 P = 0.092 RV NO (NO)
82	TEND TO BE THOSE 0.78 WHERE THE TYPE OF POLITICAL MODERNIZATION IS EARLY OR LATER EUROPEAN OR EUROPEAN DERIVED, RATHER THAN DEVELOPED TUTELARY OR UNDEVELOPED TUTELARY (51)	82	TEND TO BE THOSE 0.81 WHERE THE TYPE OF POLITICAL MODERNIZATION IS DEVELOPED TUTELARY OR UNDEVELOPED TUTELARY, RATHER THAN EARLY OR LATER EUROPEAN OR EUROPEAN DERIVED (55)	40 10 11 42 X SQ= 33.79 P = 0. RV YES (YES)
84	ALWAYS ARE THOSE 1.00 WHERE THE TYPE OF POLITICAL MODERNIZATION IS DEVELOPED TUTELARY, RATHER THAN UNDEVELOPED TUTELARY (31)	84	TEND TO BE THOSE 0.57 WHERE THE TYPE OF POLITICAL MODERNIZATION IS UNDEVELOPED TUTELARY, RATHER THAN DEVELOPED TUTELARY (24)	11 18 0 24 X SQ= 9.30 P = 0. RV YES (NO)

85	TEND TO BE THOSE WHERE THE STAGE OF POLITICAL MODERNIZATION IS ADVANCED, RATHER THAN MID- OR EARLY TRANSITIONAL (60)	0.84	85	TEND TO BE THOSE WHERE THE STAGE OF POLITICAL MODERNIZATION IS MID- OR EARLY TRANSITIONAL, RATHER THAN ADVANCED (54)	0.79	46 12 9 45 X SQ= 41.44 P = 0. RV YES (YES)
86	TEND TO BE THOSE WHERE THE STAGE OF POLITICAL MODERNIZATION IS ADVANCED OR MID-TRANSITIONAL, RATHER THAN EARLY TRANSITIONAL (76)	0.91	86	TEND TO BE THOSE WHERE THE STAGE OF POLITICAL MODERNIZATION IS EARLY TRANSITIONAL, RATHER THAN ADVANCED OR MID-TRANSITIONAL (38)	0.58	50 24 5 33 X SQ= 27.60 P = 0. RV YES (YES)
87	TEND TO BE THOSE WHOSE IDEOLOGICAL ORIENTATION IS OTHER THAN DEVELOPMENTAL (58)	0.93	87	TEND TO BE THOSE WHOSE IDEOLOGICAL ORIENTATION IS DEVELOPMENTAL (31)	0.68	3 28 42 13 X SQ= 32.72 P = 0. RV YES (YES)
89	TEND TO BE THOSE WHOSE IDEOLOGICAL ORIENTATION IS CONVENTIONAL (33)	0.65	89	TEND TO BE THOSE WHOSE IDEOLOGICAL ORIENTATION IS OTHER THAN CONVENTIONAL (62)	0.98	30 1 16 45 X SQ= 38.14 P = 0. RV YES (YES)
94	TEND TO BE THOSE WHERE THE STATUS OF THE REGIME IS CONSTITUTIONAL (51)	0.73	94	TEND TO BE THOSE WHERE THE STATUS OF THE REGIME IS AUTHORITARIAN OR TOTALITARIAN (41)	0.68	37 12 14 26 X SQ= 13.16 P = 0. RV YES (YES)
96	TEND TO BE THOSE WHERE THE STATUS OF THE REGIME IS CONSTITUTIONAL OR TOTALITARIAN (67)	0.96	96	TEND TO BE THOSE WHERE THE STATUS OF THE REGIME IS AUTHORITARIAN (23)	0.56	49 16 2 20 X SQ= 27.11 P = 0. RV YES (YES)
97	TEND TO BE THOSE WHERE THE STATUS OF THE REGIME IS TOTALITARIAN, RATHER THAN AUTHORITARIAN (16)	0.86	97	TEND TO BE THOSE WHERE THE STATUS OF THE REGIME IS AUTHORITARIAN, RATHER THAN TOTALITARIAN (23)	0.83	2 20 12 4 X SQ= 14.58 P = 0. RV YES (YES)
99	TEND TO BE THOSE WHERE GOVERNMENTAL STABILITY IS GENERALLY PRESENT AND DATES FROM AT LEAST THE INTER-WAR PERIOD, OR FROM THE POST-WAR PERIOD (50)	0.76	99	TEND TO BE THOSE WHERE GOVERNMENTAL STABILITY IS MODERATELY PRESENT AND DATES FROM THE POST-WAR PERIOD, OR IS ABSENT (36)	0.67	38 11 12 22 X SQ= 13.25 P = 0. RV YES (YES)
101	TEND TO BE THOSE WHERE THE REPRESENTATIVE CHARACTER OF THE REGIME IS POLYARCHIC (41)	0.63	101	TEND TO BE THOSE WHERE THE REPRESENTATIVE CHARACTER OF THE REGIME IS LIMITED POLYARCHIC, PSEUDO-POLYARCHIC, OR NON-POLYARCHIC (57)	0.82	32 8 19 37 X SQ= 18.08 P = 0. RV YES (YES)

102	TEND TO BE THOSE WHERE THE REPRESENTATIVE CHARACTER OF THE REGIME IS POLYARCHIC OR LIMITED POLYARCHIC (59)	0.74	102	TEND TO BE THOSE WHERE THE REPRESENTATIVE CHARACTER OF THE REGIME IS PSEUDO-POLYARCHIC OR NON-POLYARCHIC (49)	0.65	39 18 14 34 X SQ= 14.53 P = 0. RV YES (YES)
105	TEND TO BE THOSE WHERE THE ELECTORAL SYSTEM IS COMPETITIVE (43)	0.72	105	TEND TO BE THOSE WHERE THE ELECTORAL SYSTEM IS PARTIALLY COMPETITIVE OR NON-COMPETITIVE (47)	0.81	M 34 8 13 34 X SQ= 23.18 P = 0. RV YES (YES)
106	TEND TO BE THOSE WHERE THE ELECTORAL SYSTEM IS COMPETITIVE OR PARTIALLY COMPETITIVE (52)	0.76	106	TEND TO BE THOSE WHERE THE ELECTORAL SYSTEM IS NON-COMPETITIVE (30)	0.53	35 17 11 19 X SQ= 6.06 P = 0.011 RV YES (YES)
107	TEND TO BE THOSE WHERE AUTONOMOUS GROUPS ARE FULLY TOLERATED IN POLITICS (46)	0.69	107	TEND TO BE THOSE WHERE AUTONOMOUS GROUPS ARE PARTIALLY TOLERATED IN POLITICS, ARE TOLERATED ONLY OUTSIDE POLITICS, OR ARE NOT TOLERATED AT ALL (65)	0.85	37 8 17 46 X SQ= 29.87 P = 0. RV YES (YES)
111	TEND TO BE THOSE WHERE POLITICAL ENCULTURATION IS HIGH OR MEDIUM (53)	0.71	111	TEND TO BE THOSE WHERE POLITICAL ENCULTURATION IS LOW (42)	0.56	32 21 13 27 X SQ= 6.02 P = 0.012 RV YES (YES)
115	TEND LESS TO BE THOSE WHERE INTEREST ARTICULATION BY ASSOCIATIONAL GROUPS IS MODERATE, LIMITED, OR NEGLIGIBLE (91)	0.62	115	ALWAYS ARE THOSE WHERE INTEREST ARTICULATION BY ASSOCIATIONAL GROUPS IS MODERATE, LIMITED, OR NEGLIGIBLE (91)	1.00	20 0 32 56 X SQ= 23.95 P = 0. RV NO (YES)
116	TEND TO BE THOSE WHERE INTEREST ARTICULATION BY ASSOCIATIONAL GROUPS IS SIGNIFICANT OR MODERATE (32)	0.60	116	ALWAYS ARE THOSE WHERE INTEREST ARTICULATION BY ASSOCIATIONAL GROUPS IS LIMITED OR NEGLIGIBLE (79)	1.00	M 31 0 21 56 X SQ= 43.96 P = 0. RV YES (YES)
117	TEND TO BE THOSE WHERE INTEREST ARTICULATION BY ASSOCIATIONAL GROUPS IS SIGNIFICANT, MODERATE, OR LIMITED (60)	0.85	117	TEND TO BE THOSE WHERE INTEREST ARTICULATION BY ASSOCIATIONAL GROUPS IS NEGLIGIBLE (51)	0.73	44 15 8 41 X SQ= 34.08 P = 0. RV YES (YES)
118	LEAN TOWARD BEING THOSE WHERE INTEREST ARTICULATION BY INSTITUTIONAL GROUPS IS SIGNIFICANT, MODERATE, OR LIMITED (60)	0.69	118	LEAN TOWARD BEING THOSE WHERE INTEREST ARTICULATION BY INSTITUTIONAL GROUPS IS VERY SIGNIFICANT (40)	0.50	17 21 38 21 X SQ= 2.89 P = 0.063 RV YES (YES)

No.	Left statement		No.	Right statement		Statistics
119	TEND LESS TO BE THOSE WHERE INTEREST ARTICULATION BY INSTITUTIONAL GROUPS IS VERY SIGNIFICANT OR SIGNIFICANT (74)	0.55	119	TEND MORE TO BE THOSE WHERE INTEREST ARTICULATION BY INSTITUTIONAL GROUPS IS VERY SIGNIFICANT OR SIGNIFICANT (74)	0.98	30 41 25 1 X SQ= 20.38 P = 0. RV NO (YES)
120	TEND LESS TO BE THOSE WHERE INTEREST ARTICULATION BY INSTITUTIONAL GROUPS IS VERY SIGNIFICANT, SIGNIFICANT, OR MODERATE (90)	0.82	120	ALWAYS ARE THOSE WHERE INTEREST ARTICULATION BY INSTITUTIONAL GROUPS IS VERY SIGNIFICANT, SIGNIFICANT, OR MODERATE (90)	1.00	45 42 10 0 X SQ= 6.66 P = 0.004 RV NO (NO)
121	TEND TO BE THOSE WHERE INTEREST ARTICULATION BY NON-ASSOCIATIONAL GROUPS IS MODERATE, LIMITED, OR NEGLIGIBLE (61)	0.84	121	TEND TO BE THOSE WHERE INTEREST ARTICULATION BY NON-ASSOCIATIONAL GROUPS IS SIGNIFICANT (54)	0.75	9 43 46 14 X SQ= 36.94 P = 0. RV YES (YES)
122	TEND TO BE THOSE WHERE INTEREST ARTICULATION BY NON-ASSOCIATIONAL GROUPS IS LIMITED OR NEGLIGIBLE (32)	0.51	122	TEND TO BE THOSE WHERE INTEREST ARTICULATION BY NON-ASSOCIATIONAL GROUPS IS SIGNIFICANT OR MODERATE (83)	0.95	27 54 28 3 X SQ= 26.90 P = 0. RV YES (YES)
123	TEND LESS TO BE THOSE WHERE INTEREST ARTICULATION BY NON-ASSOCIATIONAL GROUPS IS SIGNIFICANT, MODERATE, OR LIMITED (107)	0.85	123	ALWAYS ARE THOSE WHERE INTEREST ARTICULATION BY NON-ASSOCIATIONAL GROUPS IS SIGNIFICANT, MODERATE, OR LIMITED (107)	1.00	47 57 8 0 X SQ= 6.87 P = 0.003 RV NO (YES)
125	TEND TO BE THOSE WHERE INTEREST ARTICULATION BY ANOMIC GROUPS IS INFREQUENT OR VERY INFREQUENT (35)	0.60	125	TEND TO BE THOSE WHERE INTEREST ARTICULATION BY ANOMIC GROUPS IS FREQUENT OR OCCASIONAL (64)	0.86	19 43 28 7 X SQ= 19.89 P = 0. RV YES (YES)
126	TEND LESS TO BE THOSE WHERE INTEREST ARTICULATION BY ANOMIC GROUPS IS FREQUENT, OCCASIONAL, OR INFREQUENT (83)	0.66	126	ALWAYS ARE THOSE WHERE INTEREST ARTICULATION BY ANOMIC GROUPS IS FREQUENT, OCCASIONAL, OR INFREQUENT (83)	1.00	31 50 16 0 X SQ= 17.99 P = 0. RV NO (YES)
128	TEND TO BE THOSE WHERE INTEREST ARTICULATION BY POLITICAL PARTIES IS SIGNIFICANT OR MODERATE (48)	0.65	128	TEND TO BE THOSE WHERE INTEREST ARTICULATION BY POLITICAL PARTIES IS LIMITED OR NEGLIGIBLE (45)	0.64	31 16 17 28 X SQ= 6.23 P = 0.012 RV YES (YES)
129	LEAN TOWARD BEING THOSE WHERE INTEREST ARTICULATION BY POLITICAL PARTIES IS SIGNIFICANT, MODERATE, OR LIMITED (56)	0.69	129	LEAN TOWARD BEING THOSE WHERE INTEREST ARTICULATION BY POLITICAL PARTIES IS NEGLIGIBLE (37)	0.50	33 22 15 22 X SQ= 2.62 P = 0.089 RV YES (YES)

```
132  LEAN MORE TOWARD BEING THOSE            0.97     132  LEAN LESS TOWARD BEING THOSE            0.82        33   33
     WHERE INTEREST AGGREGATION                            WHERE INTEREST AGGREGATION                             1    7
     BY POLITICAL PARTIES                                  BY POLITICAL PARTIES                            X SQ=   2.67
     IS SIGNIFICANT, MODERATE, OR                          IS SIGNIFICANT, MODERATE, OR                    P =    0.063
     LIMITED  (67)                                         LIMITED  (67)                                   RV NO  (YES)

133  TEND MORE TO BE THOSE                   0.86     133  TEND LESS TO BE THOSE                   0.58         7   22
     WHERE INTEREST AGGREGATION                            WHERE INTEREST AGGREGATION                           43   31
     BY THE EXECUTIVE                                      BY THE EXECUTIVE                                X SQ=   8.31
     IS MODERATE, LIMITED, OR                              IS MODERATE, LIMITED, OR                        P =    0.002
     NEGLIGIBLE  (76)                                      NEGLIGIBLE  (76)                                RV NO  (YES)

135  TEND MORE TO BE THOSE                   0.95     135  TEND LESS TO BE THOSE                   0.78        41   36
     WHERE INTEREST AGGREGATION                            WHERE INTEREST AGGREGATION                            2   10
     BY THE EXECUTIVE                                      BY THE EXECUTIVE                                X SQ=   4.19
     IS SIGNIFICANT, MODERATE, OR                          IS SIGNIFICANT, MODERATE, OR                    P =    0.028
     LIMITED  (77)                                         LIMITED  (77)                                   RV NO  (YES)

137  TEND TO BE THOSE                        0.57     137  ALWAYS ARE THOSE                        1.00        28    0
     WHERE INTEREST AGGREGATION                            WHERE INTEREST AGGREGATION                           21   47
     BY THE LEGISLATURE                                    BY THE LEGISLATURE                              X SQ=  35.20
     IS SIGNIFICANT OR MODERATE  (29)                      IS LIMITED OR NEGLIGIBLE  (68)                  P =    0.
                                                                                                           RV YES (YES)

138  TEND TO BE THOSE                        0.77     138  TEND TO BE THOSE                        0.79        37   10
     WHERE INTEREST AGGREGATION                            WHERE INTEREST AGGREGATION                           11   38
     BY THE LEGISLATURE                                    BY THE LEGISLATURE                              X SQ=  28.18
     IS SIGNIFICANT, MODERATE, OR                          IS NEGLIGIBLE  (49)                             P =    0.
     LIMITED  (48)                                                                                         RV YES (YES)

139  TEND MORE TO BE THOSE                   0.80     139  TEND LESS TO BE THOSE                   0.54        11   22
     WHERE THE PARTY SYSTEM IS QUANTITATIVELY              WHERE THE PARTY SYSTEM IS QUANTITATIVELY             43   26
     OTHER THAN ONE-PARTY  (71)                            OTHER THAN ONE-PARTY  (71)                      X SQ=   6.41
                                                                                                           P =    0.010
                                                                                                           RV NO  (YES)

140  TILT MORE TOWARD BEING THOSE            0.92     140  TILT LESS TOWARD BEING THOSE            0.81         4    9
     WHERE THE PARTY SYSTEM IS QUANTITATIVELY              WHERE THE PARTY SYSTEM IS QUANTITATIVELY             46   38
     OTHER THAN ONE PARTY DOMINANT  (87)                   OTHER THAN ONE PARTY DOMINANT  (87)             X SQ=   1.72
                                                                                                           P =    0.140
                                                                                                           RV NO  (YES)

141  LEAN LESS TOWARD BEING THOSE            0.81     141  LEAN MORE TOWARD BEING THOSE            0.96         9    2
     WHERE THE PARTY SYSTEM IS QUANTITATIVELY              WHERE THE PARTY SYSTEM IS QUANTITATIVELY             39   45
     OTHER THAN TWO-PARTY  (87)                            OTHER THAN TWO-PARTY  (87)                      X SQ=   3.56
                                                                                                           P =    0.051
                                                                                                           RV NO  (YES)

142  TEND LESS TO BE THOSE                   0.60     142  TEND MORE TO BE THOSE                   0.80        19    9
     WHERE THE PARTY SYSTEM IS QUANTITATIVELY              WHERE THE PARTY SYSTEM IS QUANTITATIVELY             29   36
     OTHER THAN MULTI-PARTY  (66)                          OTHER THAN MULTI-PARTY  (66)                    X SQ=   3.35
                                                                                                           P =    0.045
                                                                                                           RV NO  (YES)
```

146	ALWAYS ARE THOSE WHERE THE PARTY SYSTEM IS QUALITATIVELY OTHER THAN MASS-BASED TERRITORIAL (92)	1.00	146	TEND LESS TO BE THOSE WHERE THE PARTY SYSTEM IS QUALITATIVELY OTHER THAN MASS-BASED TERRITORIAL (92)	0.84	0 8 49 41 X SQ= 6.67 P = 0.006 RV NO (YES)
147	TEND TO BE THOSE WHERE THE PARTY SYSTEM IS QUALITATIVELY CLASS-ORIENTED OR MULTI-IDEOLOGICAL (23)	0.52	147	ALWAYS ARE THOSE WHERE THE PARTY SYSTEM IS QUALITATIVELY OTHER THAN CLASS-ORIENTED OR MULTI-IDEOLOGICAL (67)	1.00	23 0 21 44 X SQ= 28.49 P = 0. RV YES (YES)
148	ALWAYS ARE THOSE WHERE THE PARTY SYSTEM IS QUALITATIVELY OTHER THAN AFRICAN TRANSITIONAL (96)	1.00	148	TEND LESS TO BE THOSE WHERE THE PARTY SYSTEM IS QUALITATIVELY OTHER THAN AFRICAN TRANSITIONAL (96)	0.74	0 14 54 40 X SQ= 13.87 P = 0. RV NO (YES)
153	TEND TO BE THOSE WHERE THE PARTY SYSTEM IS STABLE (42)	0.72	153	TEND TO BE THOSE WHERE THE PARTY SYSTEM IS MODERATELY STABLE OR UNSTABLE (38)	0.74	34 8 13 23 X SQ= 14.46 P = 0. RV YES (YES)
158	TEND TO BE THOSE WHERE PERSONALISMO IS NEGLIGIBLE (56)	0.80	158	TEND TO BE THOSE WHERE PERSONALISMO IS PRONOUNCED OR MODERATE (40)	0.64	10 28 40 16 X SQ= 16.74 P = 0. RV YES (YES)
164	TEND TO BE THOSE WHERE THE REGIME'S LEADERSHIP CHARISMA IS NEGLIGIBLE (65)	0.84	164	TEND TO BE THOSE WHERE THE REGIME'S LEADERSHIP CHARISMA IS PRONOUNCED OR MODERATE (34)	0.56	8 25 43 20 X SQ= 15.12 P = 0. RV YES (YES)
166	TEND LESS TO BE THOSE WHERE THE VERTICAL POWER DISTRIBUTION IS THAT OF FORMAL FEDERALISM OR FORMAL AND EFFECTIVE UNITARISM (99)	0.78	166	TEND MORE TO BE THOSE WHERE THE VERTICAL POWER DISTRIBUTION IS THAT OF FORMAL FEDERALISM OR FORMAL AND EFFECTIVE UNITARISM (99)	0.95	12 3 43 53 X SQ= 5.10 P = 0.013 RV NO (YES)
168	TEND TO BE THOSE WHERE THE HORIZONTAL POWER DISTRIBUTION IS SIGNIFICANT (34)	0.63	168	TEND TO BE THOSE WHERE THE HORIZONTAL POWER DISTRIBUTION IS LIMITED OR NEGLIGIBLE (72)	0.98	32 1 19 51 X SQ= 41.00 P = 0. RV YES (YES)
169	TEND TO BE THOSE WHERE THE HORIZONTAL POWER DISTRIBUTION IS SIGNIFICANT OR LIMITED (58)	0.75	169	TEND TO BE THOSE WHERE THE HORIZONTAL POWER DISTRIBUTION IS NEGLIGIBLE (48)	0.65	38 18 13 34 X SQ= 14.95 P = 0. RV YES (YES)

170	TEND TO BE THOSE WHERE THE LEGISLATIVE-EXECUTIVE STRUCTURE IS OTHER THAN PRESIDENTIAL (63)	0.78	170	TEND TO BE THOSE WHERE THE LEGISLATIVE-EXECUTIVE STRUCTURE IS PRESIDENTIAL (39)	0.55	11 27 40 22 X SQ= 10.55 P = 0.001 RV YES (YES)
171	TEND LESS TO BE THOSE WHERE THE LEGISLATIVE-EXECUTIVE STRUCTURE IS OTHER THAN PARLIAMENTARY-REPUBLICAN (90)	0.80	171	TEND MORE TO BE THOSE WHERE THE LEGISLATIVE-EXECUTIVE STRUCTURE IS OTHER THAN PARLIAMENTARY-REPUBLICAN (90)	0.96	10 2 39 49 X SQ= 4.97 P = 0.014 RV NO (YES)
172	LEAN LESS TOWARD BEING THOSE WHERE THE LEGISLATIVE-EXECUTIVE STRUCTURE IS OTHER THAN PARLIAMENTARY-ROYALIST (88)	0.72	172	LEAN MORE TOWARD BEING THOSE WHERE THE LEGISLATIVE-EXECUTIVE STRUCTURE IS OTHER THAN PARLIAMENTARY-ROYALIST (88)	0.89	15 6 39 47 X SQ= 3.61 P = 0.050 RV NO (YES)
174	TEND TO BE THOSE WHERE THE LEGISLATURE IS FULLY EFFECTIVE (28)	0.56	174	ALWAYS ARE THOSE WHERE THE LEGISLATURE IS PARTIALLY EFFECTIVE, LARGELY INEFFECTIVE, OR WHOLLY INEFFECTIVE (72)	1.00	28 0 22 48 X SQ= 34.94 P = 0. RV YES (YES)
175	TEND TO BE THOSE WHERE THE LEGISLATURE IS FULLY EFFECTIVE OR PARTIALLY EFFECTIVE (51)	0.78	175	TEND TO BE THOSE WHERE THE LEGISLATURE IS LARGELY INEFFECTIVE OR WHOLLY INEFFECTIVE (49)	0.79	39 10 11 38 X SQ= 29.77 P = 0. RV YES (YES)
178	TEND TO BE THOSE WHERE THE LEGISLATURE IS BICAMERAL (51)	0.73	178	TEND TO BE THOSE WHERE THE LEGISLATURE IS UNICAMERAL (53)	0.73	14 37 37 14 X SQ= 18.98 P = 0. RV YES (YES)
179	TEND TO BE THOSE WHERE THE EXECUTIVE IS STRONG (39)	0.64	179	TEND TO BE THOSE WHERE THE EXECUTIVE IS DOMINANT (52)	0.83	17 34 30 7 X SQ= 17.77 P = 0. RV YES (YES)
181	TEND LESS TO BE THOSE WHERE THE BUREAUCRACY IS SEMI-MODERN, RATHER THAN MODERN (55)	0.60	181	ALWAYS ARE THOSE WHERE THE BUREAUCRACY IS SEMI-MODERN, RATHER THAN MODERN (55)	1.00	21 0 31 22 X SQ= 10.50 P = 0. RV NO (NO)
182	TEND TO BE THOSE WHERE THE BUREAUCRACY IS SEMI-MODERN, RATHER THAN POST-COLONIAL TRANSITIONAL (55)	0.97	182	TEND TO BE THOSE WHERE THE BUREAUCRACY IS POST-COLONIAL TRANSITIONAL, RATHER THAN SEMI-MODERN (25)	0.52	31 22 1 24 X SQ= 18.66 P = 0. RV YES (YES)

187	TEND TO BE THOSE WHERE THE ROLE OF THE POLICE IS NOT POLITICALLY SIGNIFICANT (35)	0.58	187	TEND TO BE THOSE WHERE THE ROLE OF THE POLICE IS POLITICALLY SIGNIFICANT (66)	0.88	21 43 29 6 X SQ= 20.71 P = 0. RV YES (YES)
188	TEND LESS TO BE THOSE WHERE THE CHARACTER OF THE LEGAL SYSTEM IS OTHER THAN CIVIL LAW (81)	0.62	188	TEND MORE TO BE THOSE WHERE THE CHARACTER OF THE LEGAL SYSTEM IS OTHER THAN CIVIL LAW (81)	0.82	21 10 34 45 X SQ= 4.49 P = 0.033 RV NO (YES)
189	TEND LESS TO BE THOSE WHERE THE CHARACTER OF THE LEGAL SYSTEM IS OTHER THAN COMMON LAW (108)	0.87	189	ALWAYS ARE THOSE WHERE THE CHARACTER OF THE LEGAL SYSTEM IS OTHER THAN COMMON LAW (108)	1.00	7 0 48 57 X SQ= 5.72 P = 0.006 RV NO (YES)
192	TEND MORE TO BE THOSE WHERE THE CHARACTER OF THE LEGAL SYSTEM IS OTHER THAN MUSLIM OR PARTLY MUSLIM (86)	0.95	192	TEND LESS TO BE THOSE WHERE THE CHARACTER OF THE LEGAL SYSTEM IS OTHER THAN MUSLIM OR PARTLY MUSLIM (86)	0.58	3 23 52 32 X SQ= 18.18 P = 0. RV NO (YES)
193	TEND MORE TO BE THOSE WHERE THE CHARACTER OF THE LEGAL SYSTEM IS OTHER THAN PARTLY INDIGENOUS (86)	0.98	193	TEND LESS TO BE THOSE WHERE THE CHARACTER OF THE LEGAL SYSTEM IS OTHER THAN PARTLY INDIGENOUS (86)	0.53	1 27 53 30 X SQ= 28.09 P = 0. RV NO (YES)
194	TILT LESS TOWARD BEING THOSE THAT ARE NON-COMMUNIST (101)	0.83	194	TILT MORE TOWARD BEING THOSE THAT ARE NON-COMMUNIST (101)	0.93	9 4 45 53 X SQ= 1.65 P = 0.145 RV NO (YES)

M

MATRIX

43 POLITIES
WHOSE ECONOMIC DEVELOPMENTAL STATUS
IS DEVELOPED, INTERMEDIATE, OR
UNDERDEVELOPED (55)

43 POLITIES
WHOSE ECONOMIC DEVELOPMENTAL STATUS
IS VERY UNDERDEVELOPED (57)

50 POLITIES
WHERE FREEDOM OF THE PRESS IS
COMPLETE (43)

AUSTRALIA	AUSTRIA	BELGIUM	CANADA	CHILE	CAMEROUN	CHAD	DOMIN REP	MALAGASY R	MALI
COLOMBIA	CYPRUS	DENMARK	ECUADOR	FINLAND	MAURITANIA	NIGER	PANAMA	SIERRE LEO	SOMALIA
GERMAN FR	ICELAND	INDIA	IRELAND	ITALY	TANGANYIKA	UGANDA			
JAMAICA	JAPAN	LUXEMBOURG	MALAYA	MEXICO					
NETHERLANDS	NEW ZEALAND	NORWAY	PHILIPPINES	SWEDEN					
SWITZERLAND	TRINIDAD	UK	US	URUGUAY					

30 . 12

24 . 30

50 POLITIES
WHERE FREEDOM OF THE PRESS IS
INTERMITTENT, INTERNALLY ABSENT, OR
INTERNALLY AND EXTERNALLY ABSENT (56)

ARGENTINA	BRAZIL	BULGARIA	BURMA	CHINA, PR	AFGHANISTAN	ALBANIA	ALGERIA	BOLIVIA	CAMBODIA
CUBA	CZECHOS'KIA	FRANCE	GERMANY, E	HUNGARY	CONGO(LEO)	ETHIOPIA	GHANA	GUINEA	HAITI
IRAQ	ISRAEL	LIBYA	PAKISTAN	PERU	INDONESIA	IRAN	JORDAN	KOREA, N	KOREA REP
POLAND	PORTUGAL	RUMANIA	SO AFRICA	SPAIN	LAOS	LIBERIA	MONGOLIA	MOROCCO	NEPAL
TURKEY	USSR	VENEZUELA	YUGOSLAVIA		NIGERIA	PARAGUAY	SUDAN	SYRIA	THAILAND
					TUNISIA	UAR	VIETNAM, N	VIETNAM REP	YEMEN

M

MATRIX

	43 POLITIES WHOSE ECONOMIC DEVELOPMENTAL STATUS IS DEVELOPED, INTERMEDIATE, OR UNDERDEVELOPED (55)	43 POLITIES WHOSE ECONOMIC DEVELOPMENTAL STATUS IS VERY UNDERDEVELOPED (57)
63 POLITIES WHERE THE RELIGION IS PREDOMINANTLY SOME KIND OF CHRISTIAN (46)	39	6
63 POLITIES WHERE THE RELIGION IS PREDOMINANTLY OR PARTLY OTHER THAN CHRISTIAN (68)	15	51

Developed, intermediate, or underdeveloped / Christian (39):

ARGENTINA, AUSTRALIA, AUSTRIA, BELGIUM, BRAZIL, BULGARIA, CANADA, CHILE, COLOMBIA, CUBA, CZECHOS'KIA, DENMARK, FINLAND, FRANCE, GERMANY, E, GERMAN FR, GREECE, HUNGARY, ICELAND, IRELAND, ITALY, JAMAICA, LUXEMBOURG, MEXICO, NETHERLANDS, NEW ZEALAND, NORWAY, PHILIPPINES, POLAND, PORTUGAL, RUMANIA, SPAIN, SWEDEN, SWITZERLAND, UK, US, URUGUAY, VENEZUELA, YUGOSLAVIA

Very underdeveloped / Christian (6):

DOMIN REP, EL SALVADOR, HONDURAS, NICARAGUA, PANAMA, PARAGUAY

Developed, intermediate, or underdeveloped / Other than Christian (15):

BURMA, CHINA, PR, CYPRUS, ECUADOR, INDIA, IRAQ, ISRAEL, JAPAN, LIBYA, MALAYA, PAKISTAN, PERU, SO AFRICA, TRINIDAD, TURKEY

Very underdeveloped / Other than Christian (51):

AFGHANISTAN, ALBANIA, ALGERIA, BOLIVIA, BURUNDI, CAMBODIA, CAMEROUN, CEN_AFR REP, CEYLON, CHAD, CONGO(BRA), CONGO(LEO), DAHOMEY, ETHIOPIA, GABON, GHANA, GUATEMALA, GUINEA, HAITI, INDONESIA, IRAN, IVORY COAST, JORDAN, KOREA, N, KOREA REP, LAOS, LIBERIA, MALAGASY R, MALI, MAURITANIA, MONGOLIA, MOROCCO, NEPAL, NIGER, NIGERIA, RWANDA, SENEGAL, SIERRE LEO, SOMALIA, SUDAN, SYRIA, TANGANYIKA, THAILAND, TOGO, TUNISIA, UGANDA, UAR, UPPER VOLTA, VIETNAM, N, VIETNAM REP, YEMEN

M

MATRIX

	43 POLITIES WHOSE ECONOMIC DEVELOPMENTAL STATUS IS DEVELOPED, INTERMEDIATE, OR UNDERDEVELOPED (55)	43 POLITIES WHOSE ECONOMIC DEVELOPMENTAL STATUS IS VERY UNDERDEVELOPED (57)
68 POLITIES THAT ARE LINGUISTICALLY HOMOGENEOUS (52)	31	18
68 POLITIES THAT ARE LINGUISTICALLY WEAKLY HETEROGENEOUS OR STRONGLY HETEROGENEOUS (62)	23	39

43 developed, intermediate, or underdeveloped / 68 linguistically homogeneous (31):

ARGENTINA, COLOMBIA, GERMAN FR, ITALY, MEXICO, PORTUGAL, VENEZUELA, AUSTRALIA, CUBA, GREECE, JAMAICA, NETHERLANDS, SWEDEN, AUSTRIA, DENMARK, HUNGARY, JAPAN, NEW ZEALAND, UK, BRAZIL, FRANCE, ICELAND, LIBYA, NORWAY, US, CHILE, GERMANY, E, IRELAND, LUXEMBOURG, POLAND, URUGUAY

43 very underdeveloped / 68 linguistically homogeneous (18):

ALBANIA, HONDURAS, MONGOLIA, TUNISIA, BURUNDI, JORDAN, NICARAGUA, UAR, DOMIN REP, KOREA, N, PARAGUAY, YEMEN, EL SALVADOR, KOREA REP, RWANDA, HAITI, MALAGASY R, SOMALIA

43 developed, intermediate, or underdeveloped / 68 weakly or strongly heterogeneous (23):

BELGIUM, CZECHOS'KIA, ISRAEL, RUMANIA, TURKEY, BULGARIA, ECUADOR, MALAYA, SO AFRICA, USSR, BURMA, FINLAND, PAKISTAN, SPAIN, YUGOSLAVIA, CANADA, INDIA, PERU, SWITZERLAND, CYPRUS, IRAQ, PHILIPPINES, TRINIDAD

43 very underdeveloped / 68 weakly or strongly heterogeneous (39):

AFGHANISTAN, CEN AFR REP, DAHOMEY, GUINEA, LIBERIA, NIGER, SUDAN, UGANDA, ALGERIA, CEYLON, ETHIOPIA, INDONESIA, MALI, NIGERIA, SYRIA, UPPER VOLTA, BOLIVIA, CHAD, GABON, IRAN, MAURITANIA, PANAMA, TANGANYIKA, VIETNAM, N, CAMBODIA, CONGO(BRA), GHANA, IVORY COAST, MOROCCO, SENEGAL, THAILAND, VIETNAM REP, CAMEROUN, CONGO(LEO), GUATEMALA, LAOS, NEPAL, SIERRE LEO, TOGO

MATRIX

	43 POLITIES WHOSE ECONOMIC DEVELOPMENTAL STATUS IS DEVELOPED, INTERMEDIATE, OR UNDERDEVELOPED (55)	43 POLITIES WHOSE ECONOMIC DEVELOPMENTAL STATUS IS VERY UNDERDEVELOPED (57)
105 POLITIES WHERE THE ELECTORAL SYSTEM IS COMPETITIVE (43)	ARGENTINA, CANADA, FRANCE, IRELAND, LUXEMBOURG, PHILIPPINES, UK, AUSTRALIA, CHILE, GERMAN FR, ISRAEL, MALAYA, SWEDEN, US, AUSTRIA, DENMARK, GREECE, ITALY, NETHERLANDS, SWITZERLAND, URUGUAY, BELGIUM, ECUADOR, ICELAND, JAMAICA, NEW ZEALAND, TRINIDAD, VENEZUELA, BRAZIL, FINLAND, INDIA, JAPAN, NORWAY, TURKEY (34)	BOLIVIA, PANAMA, CEYLON, SIERRE LEO, DOMIN REP, UGANDA, HONDURAS, MALAGASY R (8)
105 POLITIES WHERE THE ELECTORAL SYSTEM IS PARTIALLY COMPETITIVE OR NON-COMPETITIVE (47)	BULGARIA, LIBYA, SPAIN, CHINA, PR, PAKISTAN, USSR, CZECHOS'KIA, POLAND, YUGOSLAVIA, GERMANY, E, PORTUGAL, HUNGARY, RUMANIA (13)	AFGHANISTAN, CEN AFR REP, ETHIOPIA, IVORY COAST, MALI, PARAGUAY, TUNISIA, ALBANIA, CHAD, GABON, JORDAN, MAURITANIA, RWANDA, UPPER VOLTA, ALGERIA, CONGO(BRA), GHANA, KOREA, N, MONGOLIA, SENEGAL, VIETNAM, N, BURUNDI, DAHOMEY, GUINEA, LAOS, NICARAGUA, SOMALIA, VIETNAM REP, CAMBODIA, EL SALVADOR, HAITI, LIBERIA, NIGER, TANGANYIKA (34)

MATRIX

43 POLITIES
WHOSE ECONOMIC DEVELOPMENTAL STATUS
IS DEVELOPED, INTERMEDIATE, OR
UNDERDEVELOPED (55)

43 POLITIES
WHOSE ECONOMIC DEVELOPMENTAL STATUS
IS VERY UNDERDEVELOPED (57)

116 POLITIES
WHERE INTEREST ARTICULATION
BY ASSOCIATIONAL GROUPS
IS SIGNIFICANT OR MODERATE (32)

ARGENTINA, AUSTRALIA, AUSTRIA, BELGIUM, BRAZIL, CANADA, CHILE, DENMARK, FINLAND, FRANCE, GERMAN FR, GREECE, ICELAND, IRELAND, ISRAEL, ITALY, JAPAN, LUXEMBOURG, MEXICO, NETHERLANDS, NEW ZEALAND, NORWAY, PHILIPPINES, SO AFRICA, SWEDEN, SWITZERLAND, TURKEY, UK, US, URUGUAY, VENEZUELA

31	0
21	56

116 POLITIES
WHERE INTEREST ARTICULATION
BY ASSOCIATIONAL GROUPS
IS LIMITED OR NEGLIGIBLE (79)

BULGARIA, BURMA, CHINA, PR, COLOMBIA, CUBA, CYPRUS, CZECHOS'KIA, ECUADOR, GERMANY, E, HUNGARY, INDIA, IRAQ, JAMAICA, LIBYA, MALAYA, PAKISTAN, PERU, POLAND, RUMANIA, TRINICAD, USSR

AFGHANISTAN, ALBANIA, ALGERIA, BOLIVIA, BURUNDI, CAMBODIA, CAMEROUN, CEN AFR REP, CEYLON, CHAD, CONGO(BRA), CONGO(LEO), DAHOMEY, DOMIN REP, EL SALVADOR, ETHIOPIA, GABON, GHANA, GUATEMALA, GUINEA, HAITI, HONDURAS, INDONESIA, IRAN, IVORY COAST, JORDAN, KOREA, N, LAOS, LIBERIA, MALAGASY R, MALI, MAURITANIA, MONGOLIA, MOROCCO, NEPAL, NICARAGUA, NIGER, NIGERIA, PANAMA, PARAGUAY, RWANDA, SENEGAL, SIERRE LEO, SOMALIA, SUDAN, SYRIA, TANGANYIKA, THAILAND, TOGO, TUNISIA, UGANDA, UAR, UPPER VOLTA, VIETNAM, N, VIETNAM REP, YEMEN

44 POLITIES
WHERE THE LITERACY RATE IS
NINETY PERCENT OR ABOVE (25)

44 POLITIES
WHERE THE LITERACY RATE IS
BELOW NINETY PERCENT (90)

PREDICATE ONLY

25 IN LEFT COLUMN

AUSTRALIA	AUSTRIA	BELGIUM	CANADA	CZECHOS'KIA	DENMARK	FINLAND	FRANCE	GERMANY, E	GERMAN FR
HUNGARY	ICELAND	IRELAND	JAPAN	LUXEMBOURG	NETHERLANDS	NEW ZEALAND	NORWAY	POLAND	RUMANIA
SWEDEN	SWITZERLAND	USSR	UK	US					

90 IN RIGHT COLUMN

AFGHANISTAN	ALBANIA	ALGERIA	ARGENTINA	BOLIVIA	BRAZIL	BULGARIA	BURMA	BURUNDI	CAMBODIA
CAMEROUN	CEN AFR REP	CEYLON	CHAD	CHILE	CHINA, PR	COLOMBIA	CONGO(BRA)	CONGO(LEO)	COSTA RICA
CUBA	CYPRUS	DAHOMEY	DOMIN REP	ECUADOR	EL SALVADOR	ETHIOPIA	GABON	GHANA	GREECE
GUATEMALA	GUINEA	HAITI	HONDURAS	INDIA	INDONESIA	IRAN	IRAQ	ISRAEL	ITALY
IVORY COAST	JAMAICA	JORDAN	KOREA, N	KOREA REP	LAOS	LEBANON	LIBERIA	LIBYA	MALAGASY R
MALAYA	MALI	MAURITANIA	MEXICO	MONGOLIA	MOROCCO	NEPAL	NICARAGUA	NIGER	NIGERIA
PAKISTAN	PANAMA	PARAGUAY	PERU	PHILIPPINES	PORTUGAL	RWANDA	SA'U ARABIA	SENEGAL	SIERRE LEO
SOMALIA	SO AFRICA	SPAIN	SUDAN	SYRIA	TANGANYIKA	THAILAND	TOGO	TRINIDAD	TUNISIA
TURKEY	UGANDA	UAR	UPPER VOLTA	URUGUAY	VENEZUELA	VIETNAM, N	VIETNAM REP	YEMEN	YUGOSLAVIA

```
****************************************************************************************************

45  POLITIES                                          45  POLITIES
    WHERE THE LITERACY RATE IS                            WHERE THE LITERACY RATE IS
    FIFTY PERCENT OR ABOVE  (55)                          BELOW FIFTY PERCENT  (54)


                                                                         BOTH SUBJECT AND PREDICATE
....................................................................................................

      55 IN LEFT COLUMN

ALBANIA     ARGENTINA   AUSTRALIA   AUSTRIA     BELGIUM     BRAZIL      BULGARIA    BURMA       CANADA      CEYLON
CHILE       COLOMBIA    COSTA RICA  CUBA        CYPRUS      CZECHOS'KIA DENMARK     DOMIN REP   FINLAND     FRANCE
GERMANY, E  GERMAN FR   GREECE      HUNGARY     ICELAND     INDONESIA   IRELAND     ISRAEL      ITALY       JAMAICA
JAPAN       KOREA REP   LUXEMBOURG  MEXICO      NETHERLANDS NEW ZEALAND NORWAY      PANAMA      PARAGUAY    PHILIPPINES
POLAND      PORTUGAL    RUMANIA     SPAIN       SWEDEN      SWITZERLAND THAILAND    TRINIDAD    USSR        UK
US          URUGUAY     VENEZUELA   VIETNAM REP YUGOSLAVIA

      54 IN RIGHT COLUMN

AFGHANISTAN ALGERIA     BOLIVIA     BURUNDI     CAMBODIA    CAMEROUN    CEN AFR REP CHAD        CONGO(BRA)  CONGO(LEO)
DAHOMEY     ECUADOR     EL SALVADOR ETHIOPIA    GABON       GHANA       GUATEMALA   GUINEA      HAITI       HONDURAS
INDIA       IRAN        IRAQ        IVORY COAST JORDAN      LAOS        LIBYA       MALAGASY R  MALAYA      MALI
MAURITANIA  MOROCCO     NEPAL       NICARAGUA   NIGER       NIGERIA     PAKISTAN    PERU        RWANDA      SA'U ARABIA
SENEGAL     SIERRE LEO  SOMALIA     SO AFRICA   SUDAN       SYRIA       TANGANYIKA  TOGO        TUNISIA     TURKEY
UGANDA      UAR         UPPER VOLTA YEMEN

       6 EXCLUDED BECAUSE UNASCERTAINED

CHINA, PR   KOREA, N    LEBANON     LIBERIA     MONGOLIA    VIETNAM, N

----------------------------------------------------------------------------------------------------

  2  TEND LESS TO BE THOSE                  0.60       2  ALWAYS ARE THOSE                      1.00      22    0
     LOCATED ELSEWHERE THAN IN                            LOCATED ELSEWHERE THAN IN                       33   54
     WEST EUROPE, SCANDINAVIA,                            WEST EUROPE, SCANDINAVIA,                  X SQ=  24.64
     NORTH AMERICA, OR AUSTRALASIA  (93)                  NORTH AMERICA, OR AUSTRALASIA  (93)        P =     0.
                                                                                                     RV NO  (YES)

 10  ALWAYS ARE THOSE                       1.00      10  TEND TO BE THOSE                      0.59       0   32
     LOCATED ELSEWHERE THAN IN NORTH AFRICA,              LOCATED IN NORTH AFRICA, OR                     55   22
     OR CENTRAL AND SOUTH AFRICA  (82)                    CENTRAL AND SOUTH AFRICA  (33)             X SQ=  43.32
                                                                                                     P =     0.
                                                                                                     RV YES (YES)

 13  TEND MORE TO BE THOSE                  0.96      13  TEND LESS TO BE THOSE                 0.76       2   13
     LOCATED ELSEWHERE THAN IN NORTH AFRICA OR            LOCATED ELSEWHERE THAN IN NORTH AFRICA OR       53   41
     THE MIDDLE EAST  (99)                                THE MIDDLE EAST  (99)                      X SQ=   7.95
                                                                                                     P =     0.002
                                                                                                     RV NO  (YES)
```

14	LEAN MORE TOWARD BEING THOSE LOCATED ELSEWHERE THAN IN THE MIDDLE EAST (104)	0.96	14	LEAN LESS TOWARD BEING THOSE LOCATED ELSEWHERE THAN IN THE MIDDLE EAST (104)	0.85	2 8 53 46 X SQ= 2.85 P = 0.052 RV NO (YES)
16	TILT TOWARD BEING THOSE LOCATED IN EAST ASIA, SOUTH ASIA, OR SOUTHEAST ASIA, RATHER THAN IN THE MIDDLE EAST (18)	0.80	16	TILT TOWARD BEING THOSE LOCATED IN THE MIDDLE EAST, RATHER THAN IN EAST ASIA, SOUTH ASIA, OR SOUTHEAST ASIA (11)	0.57	2 8 8 6 X SQ= 1.96 P = 0.104 RV YES (YES)
17	LEAN TOWARD BEING THOSE LOCATED IN THE CARIBBEAN, CENTRAL AMERICA, OR SOUTH AMERICA, RATHER THAN IN THE MIDDLE EAST (22)	0.87	17	LEAN TOWARD BEING THOSE LOCATED IN THE MIDDLE EAST, RATHER THAN IN THE CARIBBEAN, CENTRAL AMERICA, OR SOUTH AMERICA (11)	0.50	2 8 14 8 X SQ= 3.64 P = 0.054 RV YES (YES)
21	TEND MORE TO BE THOSE WHOSE TERRITORIAL SIZE IS MEDIUM OR SMALL (83)	0.82	21	TEND LESS TO BE THOSE WHOSE TERRITORIAL SIZE IS MEDIUM OR SMALL (83)	0.63	10 20 45 34 X SQ= 3.96 P = 0.033 RV NO (YES)
22	LEAN LESS TOWARD BEING THOSE WHOSE TERRITORIAL SIZE IS VERY LARGE, LARGE, OR MEDIUM (68)	0.51	22	LEAN MORE TOWARD BEING THOSE WHOSE TERRITORIAL SIZE IS VERY LARGE, LARGE, OR MEDIUM (68)	0.70	28 38 27 16 X SQ= 3.54 P = 0.050 RV NO (YES)
24	TEND TO BE THOSE WHOSE POPULATION IS VERY LARGE, LARGE, OR MEDIUM (61)	0.67	24	TEND TO BE THOSE WHOSE POPULATION IS SMALL (54)	0.61	37 21 18 33 X SQ= 7.71 P = 0.004 RV YES (YES)
26	TEND TO BE THOSE WHOSE POPULATION DENSITY IS VERY HIGH, HIGH, OR MEDIUM (48)	0.64	26	TEND TO BE THOSE WHOSE POPULATION DENSITY IS LOW (67)	0.83	35 9 20 45 X SQ= 23.06 P = 0. RV YES (YES)
28	TEND TO BE THOSE WHOSE POPULATION GROWTH RATE IS LOW (48)	0.56	28	TEND TO BE THOSE WHOSE POPULATION GROWTH RATE IS HIGH (62)	0.69	24 35 30 16 X SQ= 5.29 P = 0.018 RV YES (YES)
29	TEND TO BE THOSE WHERE THE DEGREE OF URBANIZATION IS HIGH (56)	0.86	29	TEND TO BE THOSE WHERE THE DEGREE OF URBANIZATION IS LOW (49)	0.75	M 42 13 7 40 X SQ= 35.94 P = 0. RV YES (YES)

```
30  TEND TO BE THOSE                      0.85     30  TEND TO BE THOSE                      0.83          8   44
    WHOSE AGRICULTURAL POPULATION                      WHOSE AGRICULTURAL POPULATION                      47    9
    IS MEDIUM, LOW, OR VERY LOW  (57)                  IS HIGH  (56)                                 X SQ=  47.98
                                                                                                     P =     0.
                                                                                                     RV YES (YES)

34  TEND TO BE THOSE                      0.82     34  TEND TO BE THOSE                      0.72         45   15
    WHOSE GROSS NATIONAL PRODUCT                       WHOSE GROSS NATIONAL PRODUCT                       10   39
    IS VERY HIGH, HIGH, MEDIUM,                        IS VERY LOW  (53)                             X SQ=  30.01
    OR LOW  (62)                                                                                     P =     0.
                                                                                                     RV YES (YES)

35  TEND LESS TO BE THOSE                 0.56     35  ALWAYS ARE THOSE                      1.00         24    0
    WHOSE PER CAPITA GROSS NATIONAL PRODUCT            WHOSE PER CAPITA GROSS NATIONAL PRODUCT            31   54
    IS MEDIUM, LOW, OR VERY LOW  (91)                  IS MEDIUM, LOW, OR VERY LOW  (91)             X SQ=  27.73
                                                                                                     P =     0.
                                                                                                     RV NO  (YES)

36  TEND TO BE THOSE                      0.71     36  TEND TO BE THOSE                      0.96         39    2
    WHOSE PER CAPITA GROSS NATIONAL PRODUCT            WHOSE PER CAPITA GROSS NATIONAL PRODUCT            16   52
    IS VERY HIGH, HIGH, OR MEDIUM  (42)                IS LOW OR VERY LOW  (73)                      X SQ=  49.62
                                                                                                     P =     0.
                                                                                                     RV YES (YES)

37  TEND TO BE THOSE                      0.87     37  TEND TO BE THOSE                      0.72         48   15
    WHOSE PER CAPITA GROSS NATIONAL PRODUCT            WHOSE PER CAPITA GROSS NATIONAL PRODUCT             7   39
    IS VERY HIGH, HIGH, MEDIUM, OR LOW  (64)           IS VERY LOW  (51)                             X SQ=  37.14
                                                                                                     P =     0.
                                                                                                     RV YES (YES)

39  TEND TO BE THOSE                      0.59     39  TEND TO BE THOSE                      0.91         32    5
    WHOSE INTERNATIONAL FINANCIAL STATUS               WHOSE INTERNATIONAL FINANCIAL STATUS               22   49
    IS VERY HIGH, HIGH, OR MEDIUM  (38)                IS LOW OR VERY LOW  (76)                      X SQ=  27.79
                                                                                                     P =     0.
                                                                                                     RV YES (YES)

40  TEND TO BE THOSE                      0.89     40  TEND TO BE THOSE                      0.58         47   22
    WHOSE INTERNATIONAL FINANCIAL STATUS               WHOSE INTERNATIONAL FINANCIAL STATUS                6   31
    IS VERY HIGH, HIGH,                                IS VERY LOW  (39)                             X SQ=  23.92
    MEDIUM, OR LOW  (71)                                                                             P =     0.
                                                                                                     RV YES (YES)

41  TEND LESS TO BE THOSE                 0.64     41  ALWAYS ARE THOSE                      1.00         19    0
    WHOSE ECONOMIC DEVELOPMENTAL STATUS                WHOSE ECONOMIC DEVELOPMENTAL STATUS                34   54
    IS INTERMEDIATE, UNDERDEVELOPED,                   IS INTERMEDIATE, UNDERDEVELOPED,              X SQ=  21.15
    OR VERY UNDERDEVELOPED  (94)                       OR VERY UNDERDEVELOPED  (94)                  P =     0.
                                                                                                     RV NO  (YES)

42  TEND TO BE THOSE                      0.66     42  TEND TO BE THOSE                      0.98         35    1
    WHOSE ECONOMIC DEVELOPMENTAL STATUS                WHOSE ECONOMIC DEVELOPMENTAL STATUS                18   52
    IS DEVELOPED OR INTERMEDIATE  (36)                 IS UNDERDEVELOPED OR                          X SQ=  45.81
                                                       VERY UNDERDEVELOPED  (76)                     P =     0.
                                                                                                     RV YES (YES)
```

43	TEND TO BE THOSE WHOSE ECONOMIC DEVELOPMENTAL STATUS IS DEVELOPED, INTERMEDIATE, OR UNDERDEVELOPED (55)	0.83	43	TEND TO BE THOSE WHOSE ECONOMIC DEVELOPMENTAL STATUS IS VERY UNDERDEVELOPED (57)	0.83	45 9 9 44 X SQ= 44.49 P = 0. RV YES (YES)
50	TEND TO BE THOSE WHERE FREEDOM OF THE PRESS IS COMPLETE (43)	0.57	50	TEND TO BE THOSE WHERE FREEDOM OF THE PRESS IS INTERMITTENT, INTERNALLY ABSENT, OR INTERNALLY AND EXTERNALLY ABSENT (56)	0.67	30 13 23 27 X SQ= 4.40 P = 0.023 RV YES (YES)
54	TEND TO BE THOSE WHERE NEWSPAPER CIRCULATION IS ONE HUNDRED OR MORE PER THOUSAND (37)	0.65	54	ALWAYS ARE THOSE WHERE NEWSPAPER CIRCULATION IS LESS THAN ONE HUNDRED PER THOUSAND (76)	1.00	36 0 19 54 X SQ= 49.86 P = 0. RV YES (YES)
55	ALWAYS ARE THOSE WHERE NEWSPAPER CIRCULATION IS TEN OR MORE PER THOUSAND (78)	1.00	55	TEND TO BE THOSE WHERE NEWSPAPER CIRCULATION IS LESS THAN TEN PER THOUSAND (35)	0.63	55 20 0 34 X SQ= 47.44 P = 0. RV YES (YES)
56	ALWAYS ARE THOSE WHERE THE RELIGION IS PREDOMINANTLY LITERATE (79)	1.00	56	TEND TO BE THOSE WHERE THE RELIGION IS PREDOMINANTLY OR PARTLY NON-LITERATE (31)	0.56	53 24 0 30 X SQ= 38.21 P = 0. RV YES (YES)
63	TEND TO BE THOSE WHERE THE RELIGION IS PREDOMINANTLY SOME KIND OF CHRISTIAN (46)	0.80	63	TEND TO BE THOSE WHERE THE RELIGION IS PREDOMINANTLY OR PARTLY OTHER THAN CHRISTIAN (68)	0.94	43 3 11 51 X SQ= 57.60 P = 0. RV YES (YES)
64	TEND TO BE THOSE WHERE THE RELIGION IS CHRISTIAN, RATHER THAN MUSLIM (46)	0.98	64	TEND TO BE THOSE WHERE THE RELIGION IS MUSLIM, RATHER THAN CHRISTIAN (18)	0.85	43 3 1 17 X SQ= 42.55 P = 0. RV YES (YES)
66	TEND TO BE THOSE THAT ARE RELIGIOUSLY HOMOGENEOUS (57)	0.65	66	TEND TO BE THOSE THAT ARE RELIGIOUSLY HETEROGENEOUS (49)	0.57	34 22 18 29 X SQ= 4.28 P = 0.030 RV YES (YES)
67	TEND MORE TO BE THOSE THAT ARE RACIALLY HOMOGENEOUS (82)	0.81	67	TEND LESS TO BE THOSE THAT ARE RACIALLY HOMOGENEOUS (82)	0.62	43 31 10 19 X SQ= 3.76 P = 0.048 RV NO (YES)

68	TEND TO BE THOSE THAT ARE LINGUISTICALLY HOMOGENEOUS (52)	0.64	68	TEND TO BE THOSE THAT ARE LINGUISTICALLY WEAKLY HETEROGENEOUS OR STRONGLY HETEROGENEOUS (62)	0.74	35 14 20 40 X SQ= 14.17 P = 0. RV YES (YES)
69	TEND TO BE THOSE THAT ARE LINGUISTICALLY HOMOGENEOUS OR WEAKLY HETEROGENEOUS (64)	0.78	69	TEND TO BE THOSE THAT ARE LINGUISTICALLY STRONGLY HETEROGENEOUS (50)	0.69	43 17 12 37 X SQ= 22.16 P = 0. RV YES (YES)
70	TEND LESS TO BE THOSE THAT ARE RELIGIOUSLY, RACIALLY, OR LINGUISTICALLY HETEROGENEOUS (85)	0.71	70	TEND MORE TO BE THOSE THAT ARE RELIGIOUSLY, RACIALLY, OR LINGUISTICALLY HETEROGENEOUS (85)	0.90	15 5 37 44 X SQ= 4.41 P = 0.024 RV NO (YES)
72	TEND TO BE THOSE WHOSE DATE OF INDEPENDENCE IS BEFORE 1914 (52)	0.69	72	TEND TO BE THOSE WHOSE DATE OF INDEPENDENCE IS AFTER 1914 (59)	0.74	36 14 16 40 X SQ= 18.24 P = 0. RV YES (YES)
73	TEND TO BE THOSE WHOSE DATE OF INDEPENDENCE IS BEFORE 1945 (65)	0.83	73	TEND TO BE THOSE WHOSE DATE OF INDEPENDENCE IS AFTER 1945 (46)	0.67	43 18 9 36 X SQ= 24.44 P = 0. RV YES (YES)
74	TEND LESS TO BE THOSE THAT ARE NOT HISTORICALLY WESTERN (87)	0.52	74	ALWAYS ARE THOSE THAT ARE NOT HISTORICALLY WESTERN (87)	1.00	26 0 28 53 X SQ= 31.14 P = 0. RV NO (YES)
75	TEND TO BE THOSE THAT ARE HISTORICALLY WESTERN OR SIGNIFICANTLY WESTERNIZED (62)	0.89	75	TEND TO BE THOSE THAT ARE NOT HISTORICALLY WESTERN AND ARE NOT SIGNIFICANTLY WESTERNIZED (52)	0.79	49 11 6 42 X SQ= 48.32 P = 0. RV YES (YES)
76	TEND TO BE THOSE THAT ARE HISTORICALLY WESTERN, RATHER THAN HAVING BEEN SIGNIFICANTLY OR PARTIALLY WESTERNIZED THROUGH A COLONIAL RELATIONSHIP (26)	0.57	76	ALWAYS ARE THOSE THAT HAVE BEEN SIGNIFICANTLY OR PARTIALLY WESTERNIZED THROUGH A COLONIAL RELATIONSHIP, RATHER THAN BEING HISTORICALLY WESTERN (70)	1.00	26 0 20 48 X SQ= 34.73 P = 0. RV YES (YES)
77	TEND TO BE THOSE THAT HAVE BEEN SIGNIFICANTLY WESTERNIZED, RATHER THAN PARTIALLY WESTERNIZED, THROUGH A COLONIAL RELATIONSHIP (28)	0.85	77	TEND TO BE THOSE THAT HAVE BEEN PARTIALLY WESTERNIZED, RATHER THAN SIGNIFICANTLY WESTERNIZED, THROUGH A COLONIAL RELATIONSHIP (41)	0.79	17 10 3 37 X SQ= 21.11 P = 0. RV YES (YES)

79	TEND LESS TO BE THOSE THAT WERE FORMERLY DEPENDENCIES OF BRITAIN OR FRANCE, RATHER THAN SPAIN (49)	0.52	79	TEND MORE TO BE THOSE THAT WERE FORMERLY DEPENDENCIES OF BRITAIN OR FRANCE, RATHER THAN SPAIN (49)	0.83	12 35 11 7 X SQ= 5.73 P = 0.010 RV NO (YES)
80	TEND TO BE THOSE WHOSE DATE OF INDEPENDENCE IS AFTER 1914, AND THAT WERE FORMERLY DEPENDENCIES OF BRITAIN, RATHER THAN FRANCE (19)	0.87	80	TEND TO BE THOSE WHOSE DATE OF INDEPENDENCE IS AFTER 1914, AND THAT WERE FORMERLY DEPENDENCIES OF FRANCE, RATHER THAN BRITAIN (24)	0.64	7 12 1 21 X SQ= 4.87 P = 0.016 RV YES (NO)
82	TEND TO BE THOSE WHERE THE TYPE OF POLITICAL MODERNIZATION IS EARLY OR LATER EUROPEAN OR EUROPEAN DERIVED, RATHER THAN DEVELOPED TUTELARY OR UNDEVELOPED TUTELARY (51)	0.81	82	TEND TO BE THOSE WHERE THE TYPE OF POLITICAL MODERNIZATION IS DEVELOPED TUTELARY OR UNDEVELOPED TUTELARY, RATHER THAN EARLY OR LATER EUROPEAN OR EUROPEAN DERIVED (55)	0.82	42 9 10 40 X SQ= 36.84 P = 0. RV YES (YES)
84	ALWAYS ARE THOSE WHERE THE TYPE OF POLITICAL MODERNIZATION IS DEVELOPED TUTELARY, RATHER THAN UNDEVELOPED TUTELARY (31)	1.00	84	TEND TO BE THOSE WHERE THE TYPE OF POLITICAL MODERNIZATION IS UNDEVELOPED TUTELARY, RATHER THAN DEVELOPED TUTELARY (24)	0.60	10 16 0 24 X SQ= 9.26 P = 0.001 RV YES (NO)
85	TEND TO BE THOSE WHERE THE STAGE OF POLITICAL MODERNIZATION IS ADVANCED, RATHER THAN MID- OR EARLY TRANSITIONAL (60)	0.80	85	TEND TO BE THOSE WHERE THE STAGE OF POLITICAL MODERNIZATION IS MID- OR EARLY TRANSITIONAL, RATHER THAN ADVANCED (54)	0.77	44 12 11 41 X SQ= 33.31 P = 0. RV YES (YES)
86	TEND TO BE THOSE WHERE THE STAGE OF POLITICAL MODERNIZATION IS ADVANCED OR MID-TRANSITIONAL, RATHER THAN EARLY TRANSITIONAL (76)	0.89	86	TEND TO BE THOSE WHERE THE STAGE OF POLITICAL MODERNIZATION IS EARLY TRANSITIONAL, RATHER THAN ADVANCED OR MID-TRANSITIONAL (38)	0.57	49 23 6 30 X SQ= 23.35 P = 0. RV YES (YES)
87	TEND TO BE THOSE WHOSE IDEOLOGICAL ORIENTATION IS OTHER THAN DEVELOPMENTAL (58)	0.91	87	TEND TO BE THOSE WHOSE IDEOLOGICAL ORIENTATION IS DEVELOPMENTAL (31)	0.67	4 27 40 13 X SQ= 28.24 P = 0. RV YES (YES)
89	TEND TO BE THOSE WHOSE IDEOLOGICAL ORIENTATION IS CONVENTIONAL (33)	0.62	89	TEND TO BE THOSE WHOSE IDEOLOGICAL ORIENTATION IS OTHER THAN CONVENTIONAL (62)	0.91	28 4 17 41 X SQ= 25.65 P = 0. RV YES (YES)
92	LEAN LESS TOWARD BEING THOSE WHERE THE SYSTEM STYLE IS LIMITED MOBILIZATIONAL, OR NON-MOBILIZATIONAL (93)	0.78	92	LEAN MORE TOWARD BEING THOSE WHERE THE SYSTEM STYLE IS LIMITED MOBILIZATIONAL, OR NON-MOBILIZATIONAL (93)	0.92	12 4 42 49 X SQ= 3.45 P = 0.055 RV NO (YES)

94	TEND TO BE THOSE WHERE THE STATUS OF THE REGIME IS CONSTITUTIONAL (51)	0.67	94	TEND TO BE THOSE WHERE THE STATUS OF THE REGIME IS AUTHORITARIAN OR TOTALITARIAN (41)	0.56	36 14 18 18 X SQ= 3.45 P = 0.044 RV YES (YES)
95	TEND LESS TO BE THOSE WHERE THE STATUS OF THE REGIME IS CONSTITUTIONAL OR AUTHORITARIAN (95)	0.78	95	ALWAYS ARE THOSE WHERE THE STATUS OF THE REGIME IS CONSTITUTIONAL OR AUTHORITARIAN (95)	1.00	42 51 12 0 X SQ= 10.69 P = 0. RV NO (YES)
96	TEND TO BE THOSE WHERE THE STATUS OF THE REGIME IS CONSTITUTIONAL OR TOTALITARIAN (67)	0.91	96	TEND TO BE THOSE WHERE THE STATUS OF THE REGIME IS AUTHORITARIAN (23)	0.55	48 14 5 17 X SQ= 18.58 P = 0. RV YES (YES)
97	TEND TO BE THOSE WHERE THE STATUS OF THE REGIME IS TOTALITARIAN, RATHER THAN AUTHORITARIAN (16)	0.71	97	ALWAYS ARE THOSE WHERE THE STATUS OF THE REGIME IS AUTHORITARIAN, RATHER THAN TOTALITARIAN (23)	1.00	5 17 12 0 X SQ= 15.58 P = 0. RV YES (YES)
99	TEND TO BE THOSE WHERE GOVERNMENTAL STABILITY IS GENERALLY PRESENT AND DATES FROM AT LEAST THE INTER-WAR PERIOD, OR FROM THE POST-WAR PERIOD (50)	0.71	99	TEND TO BE THOSE WHERE GOVERNMENTAL STABILITY IS MODERATELY PRESENT AND DATES FROM THE POST-WAR PERIOD, OR IS ABSENT (36)	0.66	34 11 14 21 X SQ= 8.94 P = 0.003 RV YES (YES)
101	TEND TO BE THOSE WHERE THE REPRESENTATIVE CHARACTER OF THE REGIME IS POLYARCHIC (41)	0.62	101	TEND TO BE THOSE WHERE THE REPRESENTATIVE CHARACTER OF THE REGIME IS LIMITED POLYARCHIC, PSEUDO-POLYARCHIC, OR NON-POLYARCHIC (57)	0.80	33 8 20 32 X SQ= 14.85 P = 0. RV YES (YES)
102	TEND TO BE THOSE WHERE THE REPRESENTATIVE CHARACTER OF THE REGIME IS POLYARCHIC OR LIMITED POLYARCHIC (59)	0.69	102	TEND TO BE THOSE WHERE THE REPRESENTATIVE CHARACTER OF THE REGIME IS PSEUDO-POLYARCHIC OR NON-POLYARCHIC (49)	0.58	37 20 17 28 X SQ= 6.38 P = 0.009 RV YES (YES)
105	TEND TO BE THOSE WHERE THE ELECTORAL SYSTEM IS COMPETITIVE (43)	0.72	105	TEND TO BE THOSE WHERE THE ELECTORAL SYSTEM IS PARTIALLY COMPETITIVE OR NON-COMPETITIVE (47)	0.76	34 9 13 29 X SQ= 18.00 P = 0. RV YES (YES)
107	TEND TO BE THOSE WHERE AUTONOMOUS GROUPS ARE FULLY TOLERATED IN POLITICS (46)	0.65	107	TEND TO BE THOSE WHERE AUTONOMOUS GROUPS ARE PARTIALLY TOLERATED IN POLITICS, ARE TOLERATED ONLY OUTSIDE POLITICS, OR ARE NOT TOLERATED AT ALL (65)	0.82	36 9 19 41 X SQ= 22.18 P = 0. RV YES (YES)

111	TEND TO BE THOSE WHERE POLITICAL ENCULTURATION IS HIGH OR MEDIUM (53)	0.74	111	TEND TO BE THOSE WHERE POLITICAL ENCULTURATION IS LOW (42)	0.58	M 32 21 11 29 X SQ= 8.63 P = 0.002 RV YES (YES)
114	TEND TO BE THOSE WHERE SECTIONALISM IS NEGLIGIBLE (47)	0.56	114	TEND TO BE THOSE WHERE SECTIONALISM IS EXTREME OR MODERATE (61)	0.70	23 35 29 15 X SQ= 5.89 P = 0.010 RV YES (YES)
115	TEND LESS TO BE THOSE WHERE INTEREST ARTICULATION BY ASSOCIATIONAL GROUPS IS MODERATE, LIMITED, OR NEGLIGIBLE (91)	0.61	115	ALWAYS ARE THOSE WHERE INTEREST ARTICULATION BY ASSOCIATIONAL GROUPS IS MODERATE, LIMITED, OR NEGLIGIBLE (91)	1.00	20 0 31 54 X SQ= 23.68 P = 0. RV NO (YES)
116	TEND TO BE THOSE WHERE INTEREST ARTICULATION BY ASSOCIATIONAL GROUPS IS SIGNIFICANT OR MODERATE (32)	0.57	116	TEND TO BE THOSE WHERE INTEREST ARTICULATION BY ASSOCIATIONAL GROUPS IS LIMITED OR NEGLIGIBLE (79)	0.96	29 2 22 52 X SQ= 33.11 P = 0. RV YES (YES)
117	TEND TO BE THOSE WHERE INTEREST ARTICULATION BY ASSOCIATIONAL GROUPS IS SIGNIFICANT, MODERATE, OR LIMITED (60)	0.82	117	TEND TO BE THOSE WHERE INTEREST ARTICULATION BY ASSOCIATIONAL GROUPS IS NEGLIGIBLE (51)	0.69	M 42 17 9 37 X SQ= 25.55 P = 0. RV YES (YES)
119	TEND LESS TO BE THOSE WHERE INTEREST ARTICULATION BY INSTITUTIONAL GROUPS IS VERY SIGNIFICANT OR SIGNIFICANT (74)	0.55	119	TEND MORE TO BE THOSE WHERE INTEREST ARTICULATION BY INSTITUTIONAL GROUPS IS VERY SIGNIFICANT OR SIGNIFICANT (74)	0.97	30 38 25 1 X SQ= 18.89 P = 0. RV NO (YES)
120	TEND LESS TO BE THOSE WHERE INTEREST ARTICULATION BY INSTITUTIONAL GROUPS IS VERY SIGNIFICANT, SIGNIFICANT, OR MODERATE (90)	0.82	120	ALWAYS ARE THOSE WHERE INTEREST ARTICULATION BY INSTITUTIONAL GROUPS IS VERY SIGNIFICANT, SIGNIFICANT, OR MODERATE (90)	1.00	45 39 10 0 X SQ= 6.14 P = 0.005 RV NO (NO)
121	TEND TO BE THOSE WHERE INTEREST ARTICULATION BY NON-ASSOCIATIONAL GROUPS IS MODERATE, LIMITED, OR NEGLIGIBLE (61)	0.85	121	TEND TO BE THOSE WHERE INTEREST ARTICULATION BY NON-ASSOCIATIONAL GROUPS IS SIGNIFICANT (54)	0.76	M 8 41 47 13 X SQ= 39.04 P = 0. RV YES (YES)
122	TEND TO BE THOSE WHERE INTEREST ARTICULATION BY NON-ASSOCIATIONAL GROUPS IS LIMITED OR NEGLIGIBLE (32)	0.58	122	ALWAYS ARE THOSE WHERE INTEREST ARTICULATION BY NON-ASSOCIATIONAL GROUPS IS SIGNIFICANT OR MODERATE (83)	1.00	23 54 32 0 X SQ= 41.71 P = 0. RV YES (YES)

125	TEND TO BE THOSE WHERE INTEREST ARTICULATION BY ANOMIC GROUPS IS INFREQUENT OR VERY INFREQUENT (35)	0.61	125	TEND TO BE THOSE WHERE INTEREST ARTICULATION BY ANOMIC GROUPS IS FREQUENT OR OCCASIONAL (64)	0.87	17 45 27 7 X SQ= 21.86 P = 0. RV YES (YES)
128	TILT TOWARD BEING THOSE WHERE INTEREST ARTICULATION BY POLITICAL PARTIES IS SIGNIFICANT OR MODERATE (48)	0.60	128	TILT TOWARD BEING THOSE WHERE INTEREST ARTICULATION BY POLITICAL PARTIES IS LIMITED OR NEGLIGIBLE (45)	0.56	29 17 19 22 X SQ= 1.82 P = 0.135 RV YES (YES)
133	TEND MORE TO BE THOSE WHERE INTEREST AGGREGATION BY THE EXECUTIVE IS MODERATE, LIMITED, OR NEGLIGIBLE (76)	0.89	133	TEND LESS TO BE THOSE WHERE INTEREST AGGREGATION BY THE EXECUTIVE IS MODERATE, LIMITED, OR NEGLIGIBLE (76)	0.51	6 23 47 24 X SQ= 15.34 P = 0. RV NO (YES)
136	TEND LESS TO BE THOSE WHERE INTEREST AGGREGATION BY THE LEGISLATURE IS MODERATE, LIMITED, OR NEGLIGIBLE (85)	0.75	136	ALWAYS ARE THOSE WHERE INTEREST AGGREGATION BY THE LEGISLATURE IS MODERATE, LIMITED, OR NEGLIGIBLE (85)	1.00	12 0 36 43 X SQ= 10.30 P = 0. RV NO (YES)
137	TEND TO BE THOSE WHERE INTEREST AGGREGATION BY THE LEGISLATURE IS SIGNIFICANT OR MODERATE (29)	0.54	137	TEND TO BE THOSE WHERE INTEREST AGGREGATION BY THE LEGISLATURE IS LIMITED OR NEGLIGIBLE (68)	0.95	26 2 22 41 X SQ= 23.83 P = 0. RV YES (YES)
138	TEND TO BE THOSE WHERE INTEREST AGGREGATION BY THE LEGISLATURE IS SIGNIFICANT, MODERATE, OR LIMITED (48)	0.71	138	TEND TO BE THOSE WHERE INTEREST AGGREGATION BY THE LEGISLATURE IS NEGLIGIBLE (49)	0.71	35 12 14 30 X SQ= 14.96 P = 0. RV YES (YES)
141	LEAN LESS TOWARD BEING THOSE WHERE THE PARTY SYSTEM IS QUANTITATIVELY OTHER THAN TWO-PARTY (87)	0.82	141	LEAN MORE TOWARD BEING THOSE WHERE THE PARTY SYSTEM IS QUANTITATIVELY OTHER THAN TWO-PARTY (87)	0.95	9 2 41 40 X SQ= 2.65 P = 0.060 RV NO (NO)
142	TEND LESS TO BE THOSE WHERE THE PARTY SYSTEM IS QUANTITATIVELY OTHER THAN MULTI-PARTY (66)	0.58	142	TEND MORE TO BE THOSE WHERE THE PARTY SYSTEM IS QUANTITATIVELY OTHER THAN MULTI-PARTY (66)	0.80	21 8 29 32 X SQ= 3.97 P = 0.040 RV NO (YES)
144	TEND LESS TO BE THOSE WHERE THE PARTY SYSTEM IS QUANTITATIVELY ONE-PARTY, RATHER THAN TWO-PARTY (34)	0.57	144	TEND MORE TO BE THOSE WHERE THE PARTY SYSTEM IS QUANTITATIVELY ONE-PARTY, RATHER THAN TWO-PARTY (34)	0.89	12 17 9 2 X SQ= 3.73 P = 0.034 RV NO (YES)

```
146  ALWAYS ARE THOSE                                   1.00   146  TEND LESS TO BE THOSE                              0.81       0    8
     WHERE THE PARTY SYSTEM IS QUALITATIVELY                        WHERE THE PARTY SYSTEM IS QUALITATIVELY                     51   35
     OTHER THAN                                                     OTHER THAN                                            X SQ=   8.12
     MASS-BASED TERRITORIAL  (92)                                   MASS-BASED TERRITORIAL  (92)                          P =    0.001
                                                                                                                          RV NO  (YES)

147  TEND LESS TO BE THOSE                              0.53   147  TEND MORE TO BE THOSE                              0.97      22    1
     WHERE THE PARTY SYSTEM IS QUALITATIVELY                        WHERE THE PARTY SYSTEM IS QUALITATIVELY                     25   37
     OTHER THAN                                                     OTHER THAN                                            X SQ=  18.60
     CLASS-ORIENTED OR MULTI-IDEOLOGICAL  (67)                      CLASS-ORIENTED OR MULTI-IDEOLOGICAL  (67)             P =    0.
                                                                                                                          RV NO  (YES)

148  ALWAYS ARE THOSE                                   1.00   148  TEND LESS TO BE THOSE                              0.71       0   14
     WHERE THE PARTY SYSTEM IS QUALITATIVELY                        WHERE THE PARTY SYSTEM IS QUALITATIVELY                     55   35
     OTHER THAN                                                     OTHER THAN                                            X SQ=  15.79
     AFRICAN TRANSITIONAL  (96)                                     AFRICAN TRANSITIONAL  (96)                            P =    0.
                                                                                                                          RV NO  (YES)

153  TEND TO BE THOSE                                   0.70   153  TEND TO BE THOSE                                   0.82      32    5
     WHERE THE PARTY SYSTEM IS                                      WHERE THE PARTY SYSTEM IS                                   14   23
     STABLE  (42)                                                   MODERATELY STABLE OR UNSTABLE  (38)                   X SQ=  16.60
                                                                                                                          P =    0.
                                                                                                                          RV YES (YES)

154  TEND TO BE THOSE                                   0.80   154  TEND TO BE THOSE                                   0.54  M   37   13
     WHERE THE PARTY SYSTEM IS                                      WHERE THE PARTY SYSTEM IS                                    9   15
     STABLE OR MODERATELY STABLE  (55)                              UNSTABLE  (25)                                        X SQ=   7.70
                                                                                                                          P =    0.004
                                                                                                                          RV YES (YES)

158  TEND TO BE THOSE                                   0.73   158  TEND TO BE THOSE                                   0.62  M   14   24
     WHERE PERSONALISMO IS                                          WHERE PERSONALISMO IS                                       37   15
     NEGLIGIBLE  (56)                                               PRONOUNCED OR MODERATE  (40)                          X SQ=   9.18
                                                                                                                          P =    0.001
                                                                                                                          RV YES (YES)

164  TEND TO BE THOSE                                   0.90   164  TEND TO BE THOSE                                   0.63  M    5   26
     WHERE THE REGIME'S LEADERSHIP CHARISMA                         WHERE THE REGIME'S LEADERSHIP CHARISMA                      47   15
     IS NEGLIGIBLE  (65)                                            IS PRONOUNCED OR MODERATE  (34)                       X SQ=  27.49
                                                                                                                          P =    0.
                                                                                                                          RV YES (YES)

168  TEND TO BE THOSE                                   0.58   168  TEND TO BE THOSE                                   0.92      30    4
     WHERE THE HORIZONTAL POWER DISTRIBUTION                        WHERE THE HORIZONTAL POWER DISTRIBUTION                     22   44
     IS SIGNIFICANT  (34)                                           IS LIMITED OR NEGLIGIBLE  (72)                        X SQ=  24.94
                                                                                                                          P =    0.
                                                                                                                          RV YES (YES)

169  TEND TO BE THOSE                                   0.69   169  TEND TO BE THOSE                                   0.56      36   21
     WHERE THE HORIZONTAL POWER DISTRIBUTION                        WHERE THE HORIZONTAL POWER DISTRIBUTION                     16   27
     IS SIGNIFICANT OR LIMITED  (58)                                IS NEGLIGIBLE  (48)                                   X SQ=   5.61
                                                                                                                          P =    0.015
                                                                                                                          RV YES (YES)
```

170	TEND TO BE THOSE WHERE THE LEGISLATIVE-EXECUTIVE STRUCTURE IS OTHER THAN PRESIDENTIAL (63)	0.71	TEND TO BE THOSE WHERE THE LEGISLATIVE-EXECUTIVE STRUCTURE IS PRESIDENTIAL (39)	0.52	15 23 37 21 X SQ= 4.53 P = 0.023 RV YES (YES)
172	LEAN LESS TOWARD BEING THOSE WHERE THE LEGISLATIVE-EXECUTIVE STRUCTURE IS OTHER THAN PARLIAMENTARY-ROYALIST (88)	0.72	LEAN MORE TOWARD BEING THOSE WHERE THE LEGISLATIVE-EXECUTIVE STRUCTURE IS OTHER THAN PARLIAMENTARY-ROYALIST (88)	0.88	15 6 38 44 X SQ= 3.27 P = 0.051 RV NO (YES)
175	TEND TO BE THOSE WHERE THE LEGISLATURE IS FULLY EFFECTIVE OR PARTIALLY EFFECTIVE (51)	0.71	TEND TO BE THOSE WHERE THE LEGISLATURE IS LARGELY INEFFECTIVE OR WHOLLY INEFFECTIVE (49)	0.67	36 14 15 29 X SQ= 12.07 P = 0. RV YES (YES)
178	TEND TO BE THOSE WHERE THE LEGISLATURE IS BICAMERAL (51)	0.64	TEND TO BE THOSE WHERE THE LEGISLATURE IS UNICAMERAL (53)	0.62	18 30 32 18 X SQ= 5.86 P = 0.015 RV YES (YES)
179	TEND TO BE THOSE WHERE THE EXECUTIVE IS STRONG (39)	0.63	TEND TO BE THOSE WHERE THE EXECUTIVE IS DOMINANT (52)	0.77	17 30 29 9 X SQ= 12.07 P = 0. RV YES (YES)
181	TEND LESS TO BE THOSE WHERE THE BUREAUCRACY IS SEMI-MODERN, RATHER THAN MODERN (55)	0.61	TEND MORE TO BE THOSE WHERE THE BUREAUCRACY IS SEMI-MODERN, RATHER THAN MODERN (55)	0.95	20 1 31 18 X SQ= 6.07 P = 0.007 RV NO (NO)
182	TEND TO BE THOSE WHERE THE BUREAUCRACY IS SEMI-MODERN, RATHER THAN POST-COLONIAL TRANSITIONAL (55)	0.97	TEND TO BE THOSE WHERE THE BUREAUCRACY IS POST-COLONIAL TRANSITIONAL, RATHER THAN SEMI-MODERN (25)	0.57	31 18 1 24 X SQ= 21.34 P = 0. RV YES (YES)
187	TEND TO BE THOSE WHERE THE ROLE OF THE POLICE IS NOT POLITICALLY SIGNIFICANT (35)	0.57	TEND TO BE THOSE WHERE THE ROLE OF THE POLICE IS POLITICALLY SIGNIFICANT (66)	0.85	21 39 28 7 X SQ= 16.17 P = 0. RV YES (YES)
188	TEND LESS TO BE THOSE WHERE THE CHARACTER OF THE LEGAL SYSTEM IS OTHER THAN CIVIL LAW (81)	0.58	TEND MORE TO BE THOSE WHERE THE CHARACTER OF THE LEGAL SYSTEM IS OTHER THAN CIVIL LAW (81)	0.83	23 9 32 43 X SQ= 6.54 P = 0.006 RV NO (YES)

189	TEND LESS TO BE THOSE WHERE THE CHARACTER OF THE LEGAL SYSTEM IS OTHER THAN COMMON LAW (108)	0.87	189	ALWAYS ARE THOSE WHERE THE CHARACTER OF THE LEGAL SYSTEM IS OTHER THAN COMMON LAW (108)	1.00	7 0 48 54 X SQ= 5.38 P = 0.013 RV NO (YES)
192	TEND TO BE THOSE WHERE THE CHARACTER OF THE LEGAL SYSTEM IS OTHER THAN MUSLIM OR PARTLY MUSLIM (86)	0.98	192	TEND TO BE THOSE WHERE THE CHARACTER OF THE LEGAL SYSTEM IS MUSLIM OR PARTLY MUSLIM (27)	0.50	1 26 54 26 X SQ= 30.39 P = 0. RV YES (YES)
193	TEND MORE TO BE THOSE WHERE THE CHARACTER OF THE LEGAL SYSTEM IS OTHER THAN PARTLY INDIGENOUS (86)	0.96	193	TEND LESS TO BE THOSE WHERE THE CHARACTER OF THE LEGAL SYSTEM IS OTHER THAN PARTLY INDIGENOUS (86)	0.54	2 25 52 29 X SQ= 23.90 P = 0. RV NO (YES)
194	TEND LESS TO BE THOSE THAT ARE NON-COMMUNIST (101)	0.83	194	ALWAYS ARE THOSE THAT ARE NON-COMMUNIST (101)	1.00	9 0 45 54 X SQ= 7.76 P = 0.003 RV NO (YES)

46 POLITIES
WHERE THE LITERACY RATE IS
TEN PERCENT OR ABOVE (84)

46 POLITIES
WHERE THE LITERACY RATE IS
BELOW TEN PERCENT (26)

PREDICATE ONLY

84 IN LEFT COLUMN

ALBANIA	ALGERIA	ARGENTINA	AUSTRALIA	AUSTRIA	BELGIUM	BOLIVIA	BRAZIL	BULGARIA	BURMA
CAMBODIA	CANADA	CEYLON	CHILE	CHINA, PR	COLOMBIA	COSTA RICA	CUBA	CYPRUS	CZECHOS'KIA
DENMARK	DOMIN REP	ECUADOR	EL SALVADOR	FINLAND	FRANCE	GERMANY, E	GERMAN FR	GHANA	GREECE
GUATEMALA	HAITI	HONDURAS	HUNGARY	ICELAND	INDIA	INDONESIA	IRAN	IRAQ	IRELAND
ISRAEL	ITALY	JAMAICA	JAPAN	JORDAN	KOREA, N	KOREA REP	LEBANON	LIBERIA	LIBYA
LUXEMBOURG	MALAGASY R	MALAYA	MEXICO	NETHERLANDS	NEW ZEALAND	NICARAGUA	NIGERIA	NORWAY	PAKISTAN
PANAMA	PARAGUAY	PERU	PHILIPPINES	POLAND	PORTUGAL	RUMANIA	SO AFRICA	SPAIN	SWEDEN
SWITZERLAND	SYRIA	THAILAND	TRINIDAD	TUNISIA	TURKEY	USSR	UAR	UK	US
URUGUAY	VENEZUELA	VIETNAM REP	YUGOSLAVIA						

26 IN RIGHT COLUMN

AFGHANISTAN	BURUNDI	CAMEROUN	CEN AFR REP	CHAD	CONGO(BRA)	CONGO(LEO)	DAHOMEY	ETHIOPIA	GABON
IVORY COAST	LAOS	MALI	NEPAL	NIGER	RWANDA	SA'U ARABIA	SENEGAL	SIERRE LEO	SOMALIA
SUDAN	TANGANYIKA	TOGO	UGANDA	UPPER VOLTA	YEMEN				

5 EXCLUDED BECAUSE UNASCERTAINED

GUINEA	MAURITANIA	MONGOLIA	MOROCCO	VIETNAM, N

45/ 29

M

MATRIX

	45 POLITIES WHERE THE LITERACY RATE IS FIFTY PERCENT OR ABOVE (55)	45 POLITIES WHERE THE LITERACY RATE IS BELOW FIFTY PERCENT (54)
29 POLITIES WHERE THE DEGREE OF URBANIZATION IS HIGH (56)	42	13
	ARGENTINA, BULGARIA, CZECHOS'KIA, GERMANY, E, IRELAND, KOREA REP, NORWAY, SWEDEN, URUGUAY, AUSTRALIA, CANADA, DENMARK, GERMAN FR, ISRAEL, LUXEMBOURG, PANAMA, SWITZERLAND, VENEZUELA, AUSTRIA, CHILE, DOMIN REP, GREECE, ITALY, MEXICO, POLAND, USSR, BELGIUM, COLOMBIA, FINLAND, HUNGARY, JAMAICA, NETHERLANDS, RUMANIA, UK, BRAZIL, CUBA, FRANCE, ICELAND, JAPAN, NEW ZEALAND, SPAIN, US	ECUADOR, MOROCCO, TUNISIA, IRAN, NICARAGUA, TURKEY, IRAQ, PERU, UAR, JORDAN, SO AFRICA, LIBYA, SYRIA
29 POLITIES WHERE THE DEGREE OF URBANIZATION IS LOW (49)	7	40
	ALBANIA, THAILAND, BURMA, YUGOSLAVIA, CEYLON, CYPRUS, INDONESIA	AFGHANISTAN, CAMEROUN, DAHOMEY, GUATEMALA, IVORY COAST, NEPAL, SA'U ARABIA, TANGANYIKA, ALGERIA, CEN AFR REP, EL SALVADOR, GUINEA, LAOS, NIGER, SENEGAL, TOGO, BOLIVIA, CHAD, ETHIOPIA, HAITI, MALAGASY R, NIGERIA, SIERRE LEO, UGANDA, BURUNDI, CONGO(BRA), GABON, HONDURAS, MALI, PAKISTAN, SOMALIA, UPPER VOLTA, CAMBODIA, CONGO(LEO), GHANA, INDIA, MAURITANIA, RWANDA, SUDAN, YEMEN

M

```
****************************************************MATRIX****************************************************

        45  POLITIES                                                  45  POLITIES
            WHERE THE LITERACY RATE IS                                    WHERE THE LITERACY RATE IS
            FIFTY PERCENT OR ABOVE  (55)                                  BELOW FIFTY PERCENT  (54)

                          63  POLITIES
                              WHERE THE RELIGION IS
                              PREDOMINANTLY
                              SOME KIND OF CHRISTIAN  (46)

ARGENTINA  AUSTRALIA   AUSTRIA     BELGIUM    BRAZIL     .EL SALVADOR HONDURAS    NICARAGUA
BULGARIA   CANADA      CHILE       COLOMBIA   COSTA RICA .
CUBA       CZECHOS'KIA DENMARK     DOMIN REP  FINLAND    .
FRANCE     GERMANY, E  GERMAN FR   GREECE     HUNGARY    .
ICELAND    IRELAND     ITALY       JAMAICA    LUXEMBOURG .
MEXICO     NETHERLANDS NEW ZEALAND NORWAY     PANAMA     .
PARAGUAY   PHILIPPINES POLAND      PORTUGAL   RUMANIA    .
SPAIN      SWEDEN      SWITZERLAND UK         US         .
URUGUAY    VENEZUELA   YUGOSLAVIA                        .

                                                      43 .  3
. . . . . . . . . . . . . . . . . . . . . . . . . . . . . . . . . . . . . . . . . . . . . . . . . .
                                                      11 .  51

                          63  POLITIES
                              WHERE THE RELIGION IS
                              PREDOMINANTLY OR PARTLY
                              OTHER THAN CHRISTIAN  (68)

ALBANIA    BURMA       CEYLON      CYPRUS     INDONESIA  .AFGHANISTAN ALGERIA     BOLIVIA     BURUNDI     CAMBODIA
ISRAEL     JAPAN       KOREA REP   THAILAND   TRINIDAD   .CAMEROUN    CEN AFR REP CHAD        CONGO(BRA)  CONGO(LEO)
VIETNAM REP                                              .DAHOMEY     ECUADOR     ETHIOPIA    GABON       GHANA
                                                         .GUATEMALA   GUINEA      HAITI       INDIA       IRAN
                                                         .IRAQ        IVORY COAST JORDAN      LAOS        LIBYA
                                                         .MALAGASY R  MALAYA      MALI        MAURITANIA  MOROCCO
                                                         .NEPAL       NIGER       NIGERIA     PAKISTAN    PERU
                                                         .RWANDA      SA'U ARABIA SENEGAL     SIERRE LEO  SOMALIA
                                                         .SO AFRICA   SUDAN       SYRIA       TANGANYIKA  TOGO
                                                         .TUNISIA     TURKEY      UGANDA      UAR         UPPER VOLTA
                                                         .YEMEN
```

MATRIX

	45 POLITIES WHERE THE LITERACY RATE IS FIFTY PERCENT OR ABOVE (55)	45 POLITIES WHERE THE LITERACY RATE IS BELOW FIFTY PERCENT (54)
97 POLITIES WHERE THE STATUS OF THE REGIME IS AUTHORITARIAN, RATHER THAN TOTALITARIAN (23)	BURMA, INDONESIA, KOREA REP, PARAGUAY, THAILAND	AFGHANISTAN, ETHIOPIA, LAOS, SUDAN, ALGERIA, GHANA, NEPAL, UAR, CAMBODIA, GUINEA, NICARAGUA, CEN AFR REP, IRAN, PAKISTAN, EL SALVADOR, JORDAN, SA'U ARABIA
	5	17
97 POLITIES WHERE THE STATUS OF THE REGIME IS TOTALITARIAN, RATHER THAN AUTHORITARIAN (16)	ALBANIA, HUNGARY, USSR, BULGARIA, POLAND, YUGOSLAVIA, CUBA, PORTUGAL, CZECHOS'KIA, RUMANIA, GERMANY, E, SPAIN	
	12	0

********MATRIX********

45 POLITIES
WHERE THE LITERACY RATE IS
FIFTY PERCENT OR ABOVE (55)

45 POLITIES
WHERE THE LITERACY RATE IS
BELOW FIFTY PERCENT (54)

111 POLITIES
WHERE POLITICAL ENCULTURATION
IS HIGH OR MEDIUM (53)

AUSTRALIA AUSTRIA BURMA CANADA CEYLON
COLOMBIA COSTA RICA DENMARK DOMIN REP FINLAND
FRANCE GERMAN FR GREECE ICELAND IRELAND
ISRAEL ITALY JAMAICA JAPAN LUXEMBOURG
MEXICO NETHERLANDS NEW ZEALAND NORWAY PHILIPPINES
POLAND SWEDEN THAILAND TRINIDAD UK
US VENEZUELA

AFGHANISTAN ALGERIA CAMBODIA CAMEROUN GHANA
GUINEA HONDURAS INDIA IVORY COAST LIBYA
MALAGASY R MAURITANIA MOROCCO SENEGAL SIERRE LEO
SOMALIA SUDAN TANGANYIKA TOGO TUNISIA
TURKEY

32	21
11	29

111 POLITIES
WHERE POLITICAL ENCULTURATION
IS LOW (42)

ARGENTINA BELGIUM BRAZIL CHILE CYPRUS

BOLIVIA CEN AFR REP CHAD CONGO(BRA) CONGO(LEO)

M

MATRIX

45 POLITIES WHERE THE LITERACY RATE IS FIFTY PERCENT OR ABOVE (55)	45 POLITIES WHERE THE LITERACY RATE IS BELOW FIFTY PERCENT (54)
117 POLITIES WHERE INTEREST ARTICULATION BY ASSOCIATIONAL GROUPS IS SIGNIFICANT, MODERATE, OR LIMITED (60)	
42	17
ARGENTINA, BURMA, COSTA RICA, FRANCE, INDONESIA, JAPAN, NORWAY, SWITZERLAND, URUGUAY, AUSTRALIA, CANADA, CYPRUS, GERMAN FR, IRELAND, LUXEMBOURG, PANAMA, TRINIDAD, VENEZUELA, AUSTRIA, CEYLON, DENMARK, GREECE, ISRAEL, MEXICO, PHILIPPINES, USSR, BELGIUM, CHILE, DOMIN REP, HUNGARY, ITALY, NETHERLANDS, POLAND, UK, BRAZIL, COLOMBIA, FINLAND, ICELAND, JAMAICA, NEW ZEALAND, SWEDEN, US	ALGERIA, HONDURAS, NICARAGUA, TURKEY, BOLIVIA, INDIA, PERU, UAR, ECUADOR, IRAN, SO AFRICA, EL SALVADOR, MALAYA, SYRIA, GUATEMALA, MOROCCO, TUNISIA
117 POLITIES WHERE INTEREST ARTICULATION BY ASSOCIATIONAL GROUPS IS NEGLIGIBLE (51)	
9	37
ALBANIA, PARAGUAY, BULGARIA, RUMANIA, CUBA, THAILAND, CZECHOS'KIA, VIETNAM REP, GERMANY, E	AFGHANISTAN, CHAD, GABON, IVORY COAST, MALI, PAKISTAN, SOMALIA, UPPER VOLTA, BURUNDI, CONGO(BRA), GHANA, JORDAN, MAURITANIA, RWANDA, SUDAN, YEMEN, CAMBODIA, CONGO(LEO), GUINEA, LAOS, NEPAL, SA'U ARABIA, TANGANYIKA, CAMEROUN, DAHOMEY, HAITI, LIBYA, NIGER, SENEGAL, TOGO, CEN AFR REP, ETHIOPIA, IRAQ, MALAGASY R, NIGERIA, SIERRE LEO, UGANDA

MATRIX

45 POLITIES WHERE THE LITERACY RATE IS FIFTY PERCENT OR ABOVE (55)	45 POLITIES WHERE THE LITERACY RATE IS BELOW FIFTY PERCENT (54)

121 POLITIES WHERE INTEREST ARTICULATION BY NON-ASSOCIATIONAL GROUPS IS SIGNIFICANT (54)

BURMA	CEYLON	CYPRUS	INDONESIA	KOREA REP	AFGHANISTAN	BURUNDI	CAMBODIA	CAMEROUN	CEN AFR REP
PHILIPPINES	THAILAND	VIETNAM REP			CHAD	CONGO(BRA)	CONGO(LEO)	DAHOMEY	ETHIOPIA
					GABON	GHANA	GUINEA	HAITI	INDIA
					IRAN	IRAQ	IVORY COAST	JORDAN	LAOS
					MALAGASY R	MALAYA	MALI	MAURITANIA	NEPAL
					NIGER	NIGERIA	PAKISTAN	RWANDA	SA'U ARABIA
					SENEGAL	SIERRE LEO	SOMALIA	SO AFRICA	SUDAN
					SYRIA	TANGANYIKA	TOGO	UGANDA	UPPER VOLTA
					YEMEN				

8	41
47	13

121 POLITIES WHERE INTEREST ARTICULATION BY NON-ASSOCIATIONAL GROUPS IS MODERATE, LIMITED, OR NEGLIGIBLE (61)

ALBANIA	ARGENTINA	AUSTRALIA	AUSTRIA	BELGIUM	ALGERIA	BOLIVIA	ECUADOR	EL SALVADOR	GUATEMALA
BRAZIL	BULGARIA	CANADA	CHILE	COLOMBIA	HONDURAS	LIBYA	MOROCCO	NICARAGUA	PERU
CUBA	COSTA RICA	CZECHOS'KIA	DENMARK	DOMIN REP	TUNISIA	TURKEY	UAR		
FINLAND	FRANCE	GERMANY, E	GERMAN FR	GREECE					
HUNGARY	ICELAND	IRELAND	ISRAEL	ITALY					
JAMAICA	JAPAN	LUXEMBOURG	MEXICO	NETHERLANDS					
NEW ZEALAND	NORWAY	PANAMA	PARAGUAY	POLAND					
PORTUGAL	RUMANIA	SPAIN	SWEDEN	SWITZERLAND					
TRINIDAD	USSR	UK	US	URUGUAY					
VENEZUELA	YUGOSLAVIA								

M

MATRIX

	45 POLITIES WHERE THE LITERACY RATE IS FIFTY PERCENT OR ABOVE (55)	45 POLITIES WHERE THE LITERACY RATE IS BELOW FIFTY PERCENT (54)
154 POLITIES WHERE THE PARTY SYSTEM IS STABLE OR MODERATELY STABLE (55)	ALBANIA, CANADA, FRANCE, IRELAND, MEXICO, PHILIPPINES, SWEDEN, VIETNAM REP, AUSTRALIA, COSTA RICA, GERMANY, E, ISRAEL, NETHERLANDS, POLAND, SWITZERLAND, YUGOSLAVIA, AUSTRIA, CZECHOS'KIA, GERMAN FR, ITALY, NEW ZEALAND, PORTUGAL, USSR, BELGIUM, DENMARK, HUNGARY, JAPAN, NORWAY, RUMANIA, UK, BULGARIA, FINLAND, ICELAND, LUXEMBOURG, PARAGUAY, SPAIN, US 37	CAMBODIA, MALAGASY R, TANGANYIKA, GHANA, MALAYA, TUNISIA, GUINEA, NICARAGUA, TURKEY, INDIA, SENEGAL, IVORY COAST, SO AFRICA 13
154 POLITIES WHERE THE PARTY SYSTEM IS UNSTABLE (25)	9 ARGENTINA, KOREA REP, BRAZIL, PANAMA, CEYLON, THAILAND, CHILE, VENEZUELA, GREECE	15 BOLIVIA, HONDURAS, PAKISTAN, CONGO(LEO), IRAQ, PERU, ECUADOR, JORDAN, SIERRE LEO, EL SALVADOR, LAOS, SOMALIA, GUATEMALA, NEPAL, SYRIA

MATRIX

45 POLITIES WHERE THE LITERACY RATE IS FIFTY PERCENT OR ABOVE (55)	45 POLITIES WHERE THE LITERACY RATE IS BELOW FIFTY PERCENT (54)
158 POLITIES WHERE PERSONALISMO IS PRONOUNCED OR MODERATE (40)	
ARGENTINA, COSTA RICA, PANAMA, BRAZIL, FRANCE, THAILAND, CEYLON, GREECE, URUGUAY, CHILE, INDONESIA, VIETNAM REP, COLOMBIA, JAPAN	BOLIVIA, CONGO(BRA), GABON, LAOS, PERU, CAMBODIA, CONGO(LEO), GHANA, MALI, SYRIA, CAMEROUN, DAHOMEY, GUATEMALA, MAURITANIA, UAR, CEN AFR REP, ECUADOR, HONDURAS, NICARAGUA, UPPER VOLTA, CHAD, EL SALVADOR, IVORY COAST, NIGER
14	24
37	15
158 POLITIES WHERE PERSONALISMO IS NEGLIGIBLE (56)	
ALBANIA, CANADA, GERMANY, E, ISRAEL, NETHERLANDS, POLAND, SWITZERLAND, VENEZUELA, AUSTRALIA, CUBA, GERMAN FR, ITALY, NEW ZEALAND, PORTUGAL, TRINIDAD, YUGOSLAVIA, AUSTRIA, CZECHOS'KIA, HUNGARY, JAMAICA, NORWAY, RUMANIA, USSR, BELGIUM, DENMARK, ICELAND, LUXEMBOURG, PARAGUAY, SPAIN, UK, BULGARIA, FINLAND, IRELAND, MEXICO, PHILIPPINES, SWEDEN, US	ALGERIA, MOROCCO, SO AFRICA, GUINEA, NIGERIA, TANGANYIKA, INDIA, SENEGAL, TUNISIA, MALAGASY R, SIERRE LEO, TURKEY, MALAYA, SOMALIA, UGANDA

MATRIX

45 POLITIES WHERE THE LITERACY RATE IS FIFTY PERCENT OR ABOVE (55)	45 POLITIES WHERE THE LITERACY RATE IS BELOW FIFTY PERCENT (54)
164 POLITIES WHERE THE REGIME'S LEADERSHIP CHARISMA IS PRONOUNCED OR MODERATE (34)	
CUBA FRANCE INDONESIA JAMAICA YUGOSLAVIA	ALGERIA CONGO(BRA) GUINEA MALI SA'U ARABIA UPPER VOLTA CAMBODIA DAHOMEY INDIA MAURITANIA SENEGAL CAMEROUN ETHIOPIA IVORY COAST MOROCCO TANGANYIKA CEN AFR REP GABON JORDAN NIGER TUNISIA CHAD GHANA LIBYA PAKISTAN UAR
5	26
47	15
164 POLITIES WHERE THE REGIME'S LEADERSHIP CHARISMA IS NEGLIGIBLE (65)	
ALBANIA BRAZIL COLOMBIA FINLAND ICELAND KOREA REP NORWAY PORTUGAL THAILAND VENEZUELA ARGENTINA BULGARIA COSTA RICA GERMANY, E IRELAND LUXEMBOURG PANAMA RUMANIA TRINIDAD VIETNAM REP AUSTRALIA CANADA CZECHOS'KIA GERMAN FR ISRAEL MEXICO PARAGUAY SPAIN UK AUSTRIA CEYLON DENMARK GREECE ITALY NETHERLANDS PHILIPPINES SWEDEN US BELGIUM CHILE DOMIN REP HUNGARY JAPAN NEW ZEALAND POLAND SWITZERLAND URUGUAY	BOLIVIA HONDURAS SIERRE LEO ECUADOR MALAGASY R SOMALIA EL SALVADOR MALAYA SO AFRICA GUATEMALA NICARAGUA SUDAN HAITI PERU TURKEY

47 POLITIES
WHERE THE LITERACY RATE IS
NINETY PERCENT OR ABOVE, RATHER THAN
BETWEEN FIFTY AND NINETY PERCENT (25)

47 POLITIES
WHERE THE LITERACY RATE IS
BETWEEN FIFTY AND NINETY PERCENT,
RATHER THAN NINETY PERCENT OR ABOVE (30)

PREDICATE ONLY

25 IN LEFT COLUMN

AUSTRALIA	AUSTRIA	BELGIUM	CANADA	CZECHOS'KIA	DENMARK	FINLAND	FRANCE	GERMANY, E	GERMAN FR
HUNGARY	ICELAND	IRELAND	JAPAN	LUXEMBOURG	NETHERLANDS	NEW ZEALAND	NORWAY	POLAND	RUMANIA
SWEDEN	SWITZERLAND	USSR	UK	US					

30 IN RIGHT COLUMN

ALBANIA	ARGENTINA	BRAZIL	BULGARIA	BURMA	CEYLON	CHILE	COLOMBIA	COSTA RICA	CUBA
CYPRUS	DOMIN REP	GREECE	INDONESIA	ISRAEL	ITALY	JAMAICA	KOREA REP	MEXICO	PANAMA
PARAGUAY	PHILIPPINES	PORTUGAL	SPAIN	THAILAND	TRINIDAD	URUGUAY	VENEZUELA	VIETNAM REP	YUGOSLAVIA

54 EXCLUDED BECAUSE IRRELEVANT

AFGHANISTAN	ALGERIA	BOLIVIA	BURUNDI	CAMBODIA	CAMEROUN	CEN AFR REP	CHAD	CONGO(BRA)	CONGO(LEO)
DAHOMEY	ECUADOR	EL SALVADOR	ETHIOPIA	GABON	GHANA	GUATEMALA	GUINEA	HAITI	HONDURAS
INDIA	IRAN	IRAQ	IVORY COAST	JORDAN	LAOS	LIBYA	MALAGASY R	MALAYA	MALI
MAURITANIA	MOROCCO	NEPAL	NICARAGUA	NIGER	NIGERIA	PAKISTAN	PERU	RWANDA	SA'U ARABIA
SENEGAL	SIERRE LEO	SOMALIA	SO AFRICA	SUDAN	SYRIA	TANGANYIKA	TOGO	TUNISIA	TURKEY
UGANDA	UAR	UPPER VOLTA	YEMEN						

6 EXCLUDED BECAUSE UNASCERTAINED

CHINA, PR	KOREA, N	LEBANON	LIBERIA	MONGOLIA	VIETNAM, N

48 POLITIES
WHERE THE LITERACY RATE IS
BETWEEN FIFTY AND NINETY PERCENT,
RATHER THAN BETWEEN
TEN AND FIFTY PERCENT (30)

48 POLITIES
WHERE THE LITERACY RATE IS
BETWEEN TEN AND FIFTY PERCENT,
RATHER THAN BETWEEN
FIFTY AND NINETY PERCENT (24)

PREDICATE ONLY

30 IN LEFT COLUMN

ALBANIA	ARGENTINA	BRAZIL	BULGARIA	BURMA	CEYLON	CHILE	COLOMBIA	COSTA RICA	CUBA
CYPRUS	DOMIN REP	GREECE	INDONESIA	ISRAEL	ITALY	JAMAICA	KOREA REP	MEXICO	PANAMA
PARAGUAY	PHILIPPINES	PORTUGAL	SPAIN	THAILAND	TRINIDAD	URUGUAY	VENEZUELA	VIETNAM REP	YUGOSLAVIA

24 IN RIGHT COLUMN

ALGERIA	BOLIVIA	CAMBODIA	ECUADOR	EL SALVADOR	GHANA	GUATEMALA	HAITI	HONDURAS	INDIA
IRAN	IRAQ	JORDAN	LIBYA	MALAGASY R	MALAYA	NICARAGUA	NIGERIA	PAKISTAN	PERU
SO AFRICA	TUNISIA	TURKEY	UAR						

51 EXCLUDED BECAUSE IRRELEVANT

AFGHANISTAN	AUSTRALIA	AUSTRIA	BELGIUM	BURUNDI	CAMEROUN	CANADA	CEN AFR REP	CHAD	CONGO(BRA)
CONGO(LEO)	CZECHOS'KIA	DAHOMEY	DENMARK	ETHIOPIA	FINLAND	FRANCE	GABON	GERMANY, E	GERMAN FR
HUNGARY	ICELAND	IRELAND	IVORY COAST	JAPAN	LAOS	LUXEMBOURG	MALI	NEPAL	NETHERLANDS
NEW ZEALAND	NIGER	NORWAY	POLAND	RUMANIA	RWANDA	SA'U ARABIA	SENEGAL	SIERRE LEO	SOMALIA
SUDAN	SWEDEN	SWITZERLAND	TANGANYIKA	TOGO	UGANDA	USSR	UK	US	UPPER VOLTA
YEMEN									

10 EXCLUDED BECAUSE UNASCERTAINED

CHINA, PR	GUINEA	KOREA, N	LEBANON	LIBERIA	MAURITANIA	MONGOLIA	MOROCCO	SYRIA	VIETNAM, N

49 POLITIES
WHERE THE LITERACY RATE IS
BETWEEN TEN AND FIFTY PERCENT,
RATHER THAN BELOW TEN PERCENT (24)

49 POLITIES
WHERE THE LITERACY RATE IS
BELOW TEN PERCENT, RATHER THAN
BETWEEN TEN AND FIFTY PERCENT (26)

PREDICATE ONLY

24 IN LEFT COLUMN

ALGERIA	BOLIVIA	CAMBODIA	ECUADOR	EL SALVADOR	GHANA	GUATEMALA	HAITI	HONDURAS	INDIA
IRAN	IRAQ	JORDAN	LIBYA	MALAGASY R	MALAYA	NICARAGUA	NIGERIA	PAKISTAN	PERU
SO AFRICA	TUNISIA	TURKEY	UAR						

26 IN RIGHT COLUMN

AFGHANISTAN	BURUNDI	CAMEROUN	CEN AFR REP	CHAD	CONGO(BRA)	CONGO(LEO)	DAHOMEY	ETHIOPIA	GABON
IVORY COAST	LAOS	MALI	NEPAL	NIGER	RWANDA	SA'U ARABIA	SENEGAL	SIERRE LEO	SOMALIA
SUDAN	TANGANYIKA	TOGO	UGANDA	UPPER VOLTA	YEMEN				

55 EXCLUDED BECAUSE IRRELEVANT

ALBANIA	ARGENTINA	AUSTRALIA	AUSTRIA	BELGIUM	BRAZIL	BULGARIA	BURMA	CANADA	CEYLON
CHILE	COLOMBIA	COSTA RICA	CUBA	CYPRUS	CZECHOS'KIA	DENMARK	DOMIN REP	FINLAND	FRANCE
GERMANY, E	GERMAN FR	GREECE	HUNGARY	ICELAND	INDONESIA	IRELAND	ISRAEL	ITALY	JAMAICA
JAPAN	KOREA REP	LUXEMBOURG	MEXICO	NETHERLANDS	NEW ZEALAND	NORWAY	PANAMA	PARAGUAY	PHILIPPINES
POLAND	PORTUGAL	RUMANIA	SPAIN	SWEDEN	SWITZERLAND	THAILAND	TRINIDAD	USSR	UK
US	URUGUAY	VENEZUELA	VIETNAM REP	YUGOSLAVIA					

10 EXCLUDED BECAUSE UNASCERTAINED

CHINA, PR	GUINEA	KOREA, N	LEBANON	LIBERIA	MAURITANIA	MONGOLIA	MOROCCO	SYRIA	VIETNAM, N

```
******************************************************************************************

50  POLITIES                                          50  POLITIES
    WHERE FREEDOM OF THE PRESS IS                         WHERE FREEDOM OF THE PRESS IS
    COMPLETE  (43)                                        INTERMITTENT, INTERNALLY ABSENT, OR
                                                          INTERNALLY AND EXTERNALLY ABSENT  (56)

                                                                      BOTH SUBJECT AND PREDICATE
..........................................................................................

      43 IN LEFT COLUMN

AUSTRALIA   AUSTRIA     BELGIUM     CAMEROUN    CANADA      CHAD        CHILE       COLOMBIA    COSTA RICA  CYPRUS
DENMARK     DOMIN REP   ECUADOR     FINLAND     GERMAN FR   ICELAND     INDIA       IRELAND     ITALY       JAMAICA
JAPAN       LUXEMBOURG  MALAGASY R  MALAYA      MALI        MAURITANIA  MEXICO      NETHERLANDS NEW ZEALAND NIGER
NORWAY      PANAMA      PHILIPPINES SIERRE LEO  SOMALIA     SWEDEN      SWITZERLAND TANGANYIKA  TRINIDAD    UGANDA
UK          US          URUGUAY

      56 IN RIGHT COLUMN

AFGHANISTAN ALBANIA     ALGERIA     ARGENTINA   BOLIVIA     BRAZIL      BULGARIA    BURMA       CAMBODIA    CHINA, PR
CONGO(LEO)  CUBA        CZECHOS'KIA ETHIOPIA    FRANCE      GERMANY, E  GHANA       GUINEA      HAITI       HUNGARY
INDONESIA   IRAN        IRAQ        ISRAEL      JORDAN      KOREA, N    KOREA REP   LAOS        LEBANON     LIBERIA
LIBYA       MONGOLIA    MOROCCO     NEPAL       NIGERIA     PAKISTAN    PARAGUAY    PERU        POLAND      PORTUGAL
RUMANIA     SA'U ARABIA SO AFRICA   SPAIN       SUDAN       SYRIA       THAILAND    TUNISIA     TURKEY      USSR
UAR         VENEZUELA   VIETNAM, N  VIETNAM REP YEMEN       YUGOSLAVIA

       1 EXCLUDED BECAUSE AMBIGUOUS

GREECE

      13 EXCLUDED BECAUSE UNASCERTAINED

BURUNDI     CEN AFR REP CONGO(BRA)  DAHOMEY     EL SALVADOR GABON       GUATEMALA   HONDURAS    IVORY COAST NICARAGUA
RWANDA      SENEGAL     UPPER VOLTA

       2 EXCLUDED BECAUSE UNASCERTAINABLE

CEYLON      TOGO

------------------------------------------------------------------------------------------

  2  TEND LESS TO BE THOSE                      0.58    2  TEND MORE TO BE THOSE                   0.95        18    3
     LOCATED ELSEWHERE THAN IN                             LOCATED ELSEWHERE THAN IN                           25   53
     WEST EUROPE, SCANDINAVIA,                             WEST EUROPE, SCANDINAVIA,                       X SQ=  17.27
     NORTH AMERICA, OR AUSTRALASIA  (93)                   NORTH AMERICA, OR AUSTRALASIA  (93)             P =     0.
                                                                                                           RV NO  (YES)

  5  ALWAYS ARE THOSE                           1.00    5  TEND TO BE THOSE                        0.75        14    3
     LOCATED IN SCANDINAVIA OR WEST EUROPE,                LOCATED IN EAST EUROPE, RATHER THAN                  0    9
     RATHER THAN IN EAST EUROPE  (18)                      IN SCANDINAVIA OR WEST EUROPE  (9)              X SQ=  12.92
                                                                                                           P =     0.
                                                                                                           RV YES (YES)
```

8	TILT MORE TOWARD BEING THOSE LOCATED ELSEWHERE THAN IN EAST ASIA SOUTH ASIA, OR SOUTHEAST ASIA (97)	0.91	8	TILT LESS TOWARD BEING THOSE LOCATED ELSEWHERE THAN IN EAST ASIA SOUTH ASIA, OR SOUTHEAST ASIA (97)	0.77	4 13 39 43 X SQ= 2.40 P = 0.106 RV NO (NO)
13	TEND MORE TO BE THOSE LOCATED ELSEWHERE THAN IN NORTH AFRICA OR THE MIDDLE EAST (99)	0.98	13	TEND LESS TO BE THOSE LOCATED ELSEWHERE THAN IN NORTH AFRICA OR THE MIDDLE EAST (99)	0.73	1 15 42 41 X SQ= 9.01 P = 0.001 RV NO (YES)
15	LEAN MORE TOWARD BEING THOSE LOCATED IN NORTH AFRICA OR CENTRAL AND SOUTH AFRICA, RATHER THAN IN THE MIDDLE EAST (33)	0.91	15	LEAN LESS TOWARD BEING THOSE LOCATED IN NORTH AFRICA OR CENTRAL AND SOUTH AFRICA, RATHER THAN IN THE MIDDLE EAST (33)	0.57	1 10 10 13 X SQ= 2.60 P = 0.060 RV NO (NO)
17	TEND TO BE THOSE LOCATED IN THE CARIBBEAN, CENTRAL AMERICA, OR SOUTH AMERICA, RATHER THAN IN THE MIDDLE EAST (22)	0.91	17	TEND TO BE THOSE LOCATED IN THE MIDDLE EAST, RATHER THAN IN THE CARIBBEAN, CENTRAL AMERICA, OR SOUTH AMERICA (11)	0.56	1 10 10 8 X SQ= 4.44 P = 0.019 RV YES (YES)
20	LEAN TOWARD BEING THOSE LOCATED IN THE CARIBBEAN, CENTRAL AMERICA, OR SOUTH AMERICA, RATHER THAN IN EAST ASIA, SOUTH ASIA, OR SOUTHEAST ASIA (22)	0.71	20	LEAN TOWARD BEING THOSE LOCATED IN EAST ASIA, SOUTH ASIA, OR SOUTHEAST ASIA, RATHER THAN IN THE CARIBBEAN, CENTRAL AMERICA, OR SOUTH AMERICA (18)	0.62	4 13 10 8 X SQ= 2.52 P = 0.086 RV YES (YES)
24	TEND TO BE THOSE WHOSE POPULATION IS SMALL (54)	0.56	24	TEND TO BE THOSE WHOSE POPULATION IS VERY LARGE, LARGE, OR MEDIUM (61)	0.71	19 40 24 16 X SQ= 6.41 P = 0.008 RV YES (YES)
25	TEND LESS TO BE THOSE WHOSE POPULATION DENSITY IS MEDIUM OR LOW (98)	0.74	25	TEND MORE TO BE THOSE WHOSE POPULATION DENSITY IS MEDIUM OR LOW (98)	0.93	11 4 32 52 X SQ= 5.08 P = 0.021 RV NO (YES)
30	TEND TO BE THOSE WHOSE AGRICULTURAL POPULATION IS MEDIUM, LOW, OR VERY LOW (57)	0.74	30	TEND TO BE THOSE WHOSE AGRICULTURAL POPULATION IS HIGH (56)	0.59	11 32 32 22 X SQ= 9.68 P = 0.001 RV YES (YES)
32	LEAN LESS TOWARD BEING THOSE WHOSE GROSS NATIONAL PRODUCT IS MEDIUM, LOW, OR VERY LOW (105)	0.84	32	LEAN MORE TOWARD BEING THOSE WHOSE GROSS NATIONAL PRODUCT IS MEDIUM, LOW, OR VERY LOW (105)	0.95	7 3 36 53 X SQ= 2.11 P = 0.097 RV NO (YES)

36	TEND TO BE THOSE WHOSE PER CAPITA GROSS NATIONAL PRODUCT IS VERY HIGH, HIGH, OR MEDIUM (42)	0.58	36	TEND TO BE THOSE WHOSE PER CAPITA GROSS NATIONAL PRODUCT IS LOW OR VERY LOW (73)	0.71	25 16 18 40 X SQ= 7.59 P = 0.004 RV YES (YES)
37	TEND TO BE THOSE WHOSE PER CAPITA GROSS NATIONAL PRODUCT IS VERY HIGH, HIGH, MEDIUM, OR LOW (64)	0.74	37	TEND TO BE THOSE WHOSE PER CAPITA GROSS NATIONAL PRODUCT IS VERY LOW (51)	0.52	32 27 11 29 X SQ= 5.89 P = 0.013 RV YES (YES)
38	TEND LESS TO BE THOSE WHOSE INTERNATIONAL FINANCIAL STATUS IS MEDIUM, LOW, OR VERY LOW (103)	0.81	38	TEND MORE TO BE THOSE WHOSE INTERNATIONAL FINANCIAL STATUS IS MEDIUM, LOW, OR VERY LOW (103)	0.96	8 2 35 52 X SQ= 4.25 P = 0.021 RV NO (YES)
39	LEAN LESS TOWARD BEING THOSE WHOSE INTERNATIONAL FINANCIAL STATUS IS LOW OR VERY LOW (76)	0.51	39	LEAN MORE TOWARD BEING THOSE WHOSE INTERNATIONAL FINANCIAL STATUS IS LOW OR VERY LOW (76)	0.69	21 17 22 38 X SQ= 2.56 P = 0.095 RV NO (YES)
42	TEND TO BE THOSE WHOSE ECONOMIC DEVELOPMENTAL STATUS IS DEVELOPED OR INTERMEDIATE (36)	0.53	42	TEND TO BE THOSE WHOSE ECONOMIC DEVELOPMENTAL STATUS IS UNDERDEVELOPED OR VERY UNDERDEVELOPED (76)	0.77	23 12 20 41 X SQ= 8.46 P = 0.003 RV YES (YES)
43	TEND TO BE THOSE WHOSE ECONOMIC DEVELOPMENTAL STATUS IS DEVELOPED, INTERMEDIATE, OR UNDERDEVELOPED (55)	0.71	43	TEND TO BE THOSE WHOSE ECONOMIC DEVELOPMENTAL STATUS IS VERY UNDERDEVELOPED (57)	0.56	30 24 12 30 X SQ= 5.94 P = 0.012 RV YES (YES)
45	TEND TO BE THOSE WHERE THE LITERACY RATE IS FIFTY PERCENT OR ABOVE (55)	0.70	45	TEND TO BE THOSE WHERE THE LITERACY RATE IS BELOW FIFTY PERCENT (54)	0.54	30 23 13 27 X SQ= 4.40 P = 0.023 RV YES (YES)
47	LEAN TOWARD BEING THOSE WHERE THE LITERACY RATE IS NINETY PERCENT OR ABOVE, RATHER THAN BETWEEN FIFTY AND NINETY PERCENT (25)	0.60	47	LEAN TOWARD BEING THOSE WHERE THE LITERACY RATE IS BETWEEN FIFTY AND NINETY PERCENT, RATHER THAN NINETY PERCENT OR ABOVE (30)	0.70	18 7 12 16 X SQ= 3.46 P = 0.052 RV YES (YES)
48	TILT TOWARD BEING THOSE WHERE THE LITERACY RATE IS BETWEEN FIFTY AND NINETY PERCENT, RATHER THAN BETWEEN TEN AND FIFTY PERCENT (30)	0.75	48	TILT TOWARD BEING THOSE WHERE THE LITERACY RATE IS BETWEEN TEN AND FIFTY PERCENT, RATHER THAN BETWEEN FIFTY AND NINETY PERCENT (24)	0.50	12 16 4 16 X SQ= 1.81 P = 0.127 RV YES (NO)

49	LEAN TOWARD BEING THOSE WHERE THE LITERACY RATE IS BELOW TEN PERCENT, RATHER THAN BETWEEN TEN AND FIFTY PERCENT (26)	0.67	49	LEAN TOWARD BEING THOSE WHERE THE LITERACY RATE IS BETWEEN TEN AND FIFTY PERCENT, RATHER THAN BELOW TEN PERCENT (24)	0.67	4 16 8 8 X SQ= 2.38 P = 0.081 RV YES (YES)
54	TEND TO BE THOSE WHERE NEWSPAPER CIRCULATION IS ONE HUNDRED OR MORE PER THOUSAND (37)	0.56	54	TEND TO BE THOSE WHERE NEWSPAPER CIRCULATION IS LESS THAN ONE HUNDRED PER THOUSAND (76)	0.78	24 12 19 42 X SQ= 10.18 P = 0.001 RV YES (YES)
57	TILT LESS TOWARD BEING THOSE WHERE THE RELIGION IS OTHER THAN CATHOLIC (90)	0.70	57	TILT MORE TOWARD BEING THOSE WHERE THE RELIGION IS OTHER THAN CATHOLIC (90)	0.84	13 9 30 47 X SQ= 2.06 P = 0.142 RV NO (YES)
58	TEND MORE TO BE THOSE WHERE THE RELIGION IS OTHER THAN MUSLIM (97)	0.95	58	TEND LESS TO BE THOSE WHERE THE RELIGION IS OTHER THAN MUSLIM (97)	0.73	2 15 41 41 X SQ= 6.89 P = 0.006 RV NO (YES)
62	TILT LESS TOWARD BEING THOSE WHERE THE RELIGION IS CATHOLIC, RATHER THAN PROTESTANT (25)	0.72	62	ALWAYS ARE THOSE WHERE THE RELIGION IS CATHOLIC, RATHER THAN PROTESTANT (25)	1.00	5 0 13 9 X SQ= 1.50 P = 0.136 RV NO (NO)
63	TEND TO BE THOSE WHERE THE RELIGION IS PREDOMINANTLY SOME KIND OF CHRISTIAN (46)	0.63	63	TEND TO BE THOSE WHERE THE RELIGION IS PREDOMINANTLY OR PARTLY OTHER THAN CHRISTIAN (68)	0.73	M 27 15 16 40 X SQ= 11.02 P = 0.001 RV YES (YES)
67	LEAN LESS TOWARD BEING THOSE THAT ARE RACIALLY HOMOGENEOUS (82)	0.63	67	LEAN MORE TOWARD BEING THOSE THAT ARE RACIALLY HOMOGENEOUS (82)	0.81	26 42 15 10 X SQ= 2.69 P = 0.098 RV NO (YES)
75	TEND TO BE THOSE THAT ARE HISTORICALLY WESTERN OR SIGNIFICANTLY WESTERNIZED (62)	0.74	75	TEND TO BE THOSE THAT ARE NOT HISTORICALLY WESTERN AND ARE NOT SIGNIFICANTLY WESTERNIZED (52)	0.56	32 24 11 31 X SQ= 8.12 P = 0.004 RV YES (YES)
76	TEND LESS TO BE THOSE THAT HAVE BEEN SIGNIFICANTLY OR PARTIALLY WESTERNIZED THROUGH A COLONIAL RELATIONSHIP, RATHER THAN BEING HISTORICALLY WESTERN (70)	0.57	76	TEND MORE TO BE THOSE THAT HAVE BEEN SIGNIFICANTLY OR PARTIALLY WESTERNIZED THROUGH A COLONIAL RELATIONSHIP, RATHER THAN BEING HISTORICALLY WESTERN (70)	0.82	18 7 24 31 X SQ= 4.47 P = 0.029 RV NO (YES)

78	LEAN MORE TOWARD BEING THOSE THAT HAVE BEEN SIGNIFICANTLY WESTERNIZED THROUGH A COLONIAL RELATIONSHIP, RATHER THAN WITHOUT SUCH A RELATIONSHIP (28)	0.93	78	LEAN LESS TOWARD BEING THOSE THAT HAVE BEEN SIGNIFICANTLY WESTERNIZED THROUGH A COLONIAL RELATIONSHIP, RATHER THAN WITHOUT SUCH A RELATIONSHIP (28)	0.62	13 10 1 6 X SQ= 2.34 P = 0.086 RV NO (YES)
81	LEAN LESS TOWARD BEING THOSE WHERE THE TYPE OF POLITICAL MODERNIZATION IS LATER EUROPEAN OR LATER EUROPEAN DERIVED, RATHER THAN EARLY EUROPEAN OR EARLY EUROPEAN DERIVED (40)	0.65	81	LEAN MORE TOWARD BEING THOSE WHERE THE TYPE OF POLITICAL MODERNIZATION IS LATER EUROPEAN OR LATER EUROPEAN DERIVED, RATHER THAN EARLY EUROPEAN OR EARLY EUROPEAN DERIVED (40)	0.90	9 2 17 18 X SQ= 2.53 P = 0.082 RV NO (YES)
82	LEAN TOWARD BEING THOSE WHERE THE TYPE OF POLITICAL MODERNIZATION IS EARLY OR LATER EUROPEAN OR EUROPEAN DERIVED, RATHER THAN DEVELOPED TUTELARY OR UNDEVELOPED TUTELARY (51)	0.62	82	LEAN TOWARD BEING THOSE WHERE THE TYPE OF POLITICAL MODERNIZATION IS DEVELOPED TUTELARY OR UNDEVELOPED TUTELARY, RATHER THAN EARLY OR LATER EUROPEAN OR EUROPEAN DERIVED (55)	0.58	26 20 16 28 X SQ= 2.91 P = 0.061 RV YES (YES)
83	LEAN MORE TOWARD BEING THOSE WHERE THE TYPE OF POLITICAL MODERNIZATION IS OTHER THAN NON-EUROPEAN AUTOCHTHONOUS (106)	0.98	83	LEAN LESS TOWARD BEING THOSE WHERE THE TYPE OF POLITICAL MODERNIZATION IS OTHER THAN NON-EUROPEAN AUTOCHTHONOUS (106)	0.86	1 8 42 48 X SQ= 2.89 P = 0.073 RV NO (NO)
84	TEND TO BE THOSE WHERE THE TYPE OF POLITICAL MODERNIZATION IS UNDEVELOPED TUTELARY, RATHER THAN DEVELOPED TUTELARY (24)	0.62	84	TEND TO BE THOSE WHERE THE TYPE OF POLITICAL MODERNIZATION IS DEVELOPED TUTELARY, RATHER THAN UNDEVELOPED TUTELARY (31)	0.86	6 24 10 4 X SQ= 8.80 P = 0.002 RV YES (YES)
89	TEND TO BE THOSE WHOSE IDEOLOGICAL ORIENTATION IS CONVENTIONAL (33)	0.73	89	TEND TO BE THOSE WHOSE IDEOLOGICAL ORIENTATION IS OTHER THAN CONVENTIONAL (62)	0.91	27 4 10 41 X SQ= 32.79 P = 0. RV YES (YES)
91	ALWAYS ARE THOSE WHOSE IDEOLOGICAL ORIENTATION IS DEVELOPMENTAL, RATHER THAN TRADITIONAL (31)	1.00	91	TILT LESS TOWARD BEING THOSE WHOSE IDEOLOGICAL ORIENTATION IS DEVELOPMENTAL, RATHER THAN TRADITIONAL (31)	0.69	10 11 0 5 X SQ= 2.12 P = 0.121 RV NO (NO)
92	ALWAYS ARE THOSE WHERE THE SYSTEM STYLE IS LIMITED MOBILIZATIONAL, OR NON-MOBILIZATIONAL (93)	1.00	92	TEND LESS TO BE THOSE WHERE THE SYSTEM STYLE IS LIMITED MOBILIZATIONAL, OR NON-MOBILIZATIONAL (93)	0.63	0 20 43 34 X SQ= 17.86 P = 0. RV NO (YES)
93	TEND TO BE THOSE WHERE THE SYSTEM STYLE IS NON-MOBILIZATIONAL (78)	0.95	93	TEND TO BE THOSE WHERE THE SYSTEM STYLE IS MOBILIZATIONAL, OR LIMITED MOBILIZATIONAL (32)	0.55	2 28 41 23 X SQ= 24.85 P = 0. RV YES (YES)

94	ALWAYS ARE THOSE WHERE THE STATUS OF THE REGIME IS CONSTITUTIONAL (51)	1.00	94	TEND TO BE THOSE WHERE THE STATUS OF THE REGIME IS AUTHORITARIAN OR TOTALITARIAN (41)	0.78	37 11 0 38 X SQ= 48.31 P = 0. RV YES (YES)
98	TEND LESS TO BE THOSE WHERE GOVERNMENTAL STABILITY IS GENERALLY OR MODERATELY PRESENT AND DATES FROM THE POST-WAR PERIOD, OR IS ABSENT (93)	0.65	98	TEND MORE TO BE THOSE WHERE GOVERNMENTAL STABILITY IS GENERALLY OR MODERATELY PRESENT AND DATES FROM THE POST-WAR PERIOD, OR IS ABSENT (93)	0.87	15 7 28 49 X SQ= 5.82 P = 0.014 RV NO (YES)
99	TEND MORE TO BE THOSE WHERE GOVERNMENTAL STABILITY IS GENERALLY PRESENT AND DATES FROM AT LEAST THE INTER-WAR PERIOD, OR FROM THE POST-WAR PERIOD (50)	0.83	99	TEND LESS TO BE THOSE WHERE GOVERNMENTAL STABILITY IS GENERALLY PRESENT AND DATES FROM AT LEAST THE INTER-WAR PERIOD, OR FROM THE POST-WAR PERIOD (50)	0.53	24 26 5 23 X SQ= 5.75 P = 0.014 RV NO (NO)
100	TEND MORE TO BE THOSE WHERE GOVERNMENTAL STABILITY IS GENERALLY PRESENT AND DATES FROM AT LEAST THE INTER-WAR PERIOD, OR IS GENERALLY OR MODERATELY PRESENT, AND DATES FROM THE POST-WAR PERIOD (64)	0.93	100	TEND LESS TO BE THOSE WHERE GOVERNMENTAL STABILITY IS GENERALLY PRESENT AND DATES FROM AT LEAST THE INTER-WAR PERIOD, OR IS GENERALLY OR MODERATELY PRESENT, AND DATES FROM THE POST-WAR PERIOD (64)	0.69	27 35 2 16 X SQ= 5.03 P = 0.013 RV NO (NO)
101	TEND TO BE THOSE WHERE THE REPRESENTATIVE CHARACTER OF THE REGIME IS POLYARCHIC (41)	0.80	101	TEND TO BE THOSE WHERE THE REPRESENTATIVE CHARACTER OF THE REGIME IS LIMITED POLYARCHIC, PSEUDO-POLYARCHIC, OR NON-POLYARCHIC (57)	0.85	32 7 8 39 X SQ= 33.66 P = 0. RV YES (YES)
102	TEND TO BE THOSE WHERE THE REPRESENTATIVE CHARACTER OF THE REGIME IS POLYARCHIC OR LIMITED POLYARCHIC (59)	0.91	102	TEND TO BE THOSE WHERE THE REPRESENTATIVE CHARACTER OF THE REGIME IS PSEUDO-POLYARCHIC OR NON-POLYARCHIC (49)	0.71	39 15 4 36 X SQ= 33.38 P = 0. RV YES (YES)
105	TEND TO BE THOSE WHERE THE ELECTORAL SYSTEM IS COMPETITIVE (43)	0.85	105	TEND TO BE THOSE WHERE THE ELECTORAL SYSTEM IS PARTIALLY COMPETITIVE OR NON-COMPETITIVE (47)	0.81	M 33 7 6 30 X SQ= 30.29 P = 0. RV YES (YES)
106	TEND TO BE THOSE WHERE THE ELECTORAL SYSTEM IS COMPETITIVE OR PARTIALLY COMPETITIVE (52)	0.95	106	TEND TO BE THOSE WHERE THE ELECTORAL SYSTEM IS NON-COMPETITIVE (30)	0.72	35 9 2 23 X SQ= 30.00 P = 0. RV YES (YES)
107	TEND TO BE THOSE WHERE AUTONOMOUS GROUPS ARE FULLY TOLERATED IN POLITICS (46)	0.79	107	TEND TO BE THOSE WHERE AUTONOMOUS GROUPS ARE PARTIALLY TOLERATED IN POLITICS, ARE TOLERATED ONLY OUTSIDE POLITICS, OR ARE NOT TOLERATED AT ALL (65)	0.83	34 9 9 44 X SQ= 34.54 P = 0. RV YES (YES)

108	TEND TO BE THOSE WHERE AUTONOMOUS GROUPS ARE FULLY OR PARTIALLY TOLERATED IN POLITICS (65)	0.93	108	TEND TO BE THOSE WHERE AUTONOMOUS GROUPS ARE TOLERATED ONLY OUTSIDE POLITICS OR ARE NOT TOLERATED AT ALL (35)	0.64	40 16 3 28 X SQ= 28.02 P = 0. RV YES (YES)
111	TEND TO BE THOSE WHERE POLITICAL ENCULTURATION IS HIGH OR MEDIUM (53)	0.77	111	TEND TO BE THOSE WHERE POLITICAL ENCULTURATION IS LOW (42)	0.60	30 17 9 25 X SQ= 9.58 P = 0.001 RV YES (YES)
116	TEND TO BE THOSE WHERE INTEREST ARTICULATION BY ASSOCIATIONAL GROUPS IS SIGNIFICANT OR MODERATE (32)	0.53	116	TEND TO BE THOSE WHERE INTEREST ARTICULATION BY ASSOCIATIONAL GROUPS IS LIMITED OR NEGLIGIBLE (79)	0.85	23 8 20 44 X SQ= 13.86 P = 0. RV YES (YES)
117	TEND TO BE THOSE WHERE INTEREST ARTICULATION BY ASSOCIATIONAL GROUPS IS SIGNIFICANT, MODERATE, OR LIMITED (60)	0.74	117	TEND TO BE THOSE WHERE INTEREST ARTICULATION BY ASSOCIATIONAL GROUPS IS NEGLIGIBLE (51)	0.58	32 22 11 30 X SQ= 8.63 P = 0.002 RV YES (YES)
118	TEND TO BE THOSE WHERE INTEREST ARTICULATION BY INSTITUTIONAL GROUPS IS SIGNIFICANT, MODERATE, OR LIMITED (60)	0.95	118	TEND TO BE THOSE WHERE INTEREST ARTICULATION BY INSTITUTIONAL GROUPS IS VERY SIGNIFICANT (40)	0.69	2 38 35 17 X SQ= 33.96 P = 0. RV YES (YES)
119	TEND TO BE THOSE WHERE INTEREST ARTICULATION BY INSTITUTIONAL GROUPS IS MODERATE OR LIMITED (26)	0.62	119	TEND TO BE THOSE WHERE INTEREST ARTICULATION BY INSTITUTIONAL GROUPS IS VERY SIGNIFICANT OR SIGNIFICANT (74)	0.96	14 53 23 2 X SQ= 35.39 P = 0. RV YES (YES)
121	LEAN TOWARD BEING THOSE WHERE INTEREST ARTICULATION BY NON-ASSOCIATIONAL GROUPS IS MODERATE, LIMITED, OR NEGLIGIBLE (61)	0.67	121	LEAN TOWARD BEING THOSE WHERE INTEREST ARTICULATION BY NON-ASSOCIATIONAL GROUPS IS SIGNIFICANT (54)	0.52	14 29 29 27 X SQ= 2.92 P = 0.067 RV YES (YES)
122	TEND LESS TO BE THOSE WHERE INTEREST ARTICULATION BY NON-ASSOCIATIONAL GROUPS IS SIGNIFICANT OR MODERATE (83)	0.51	122	TEND MORE TO BE THOSE WHERE INTEREST ARTICULATION BY NON-ASSOCIATIONAL GROUPS IS SIGNIFICANT OR MODERATE (83)	0.82	22 46 21 10 X SQ= 9.46 P = 0.002 RV NO (YES)
123	TEND LESS TO BE THOSE WHERE INTEREST ARTICULATION BY NON-ASSOCIATIONAL GROUPS IS SIGNIFICANT, MODERATE, OR LIMITED (107)	0.81	123	ALWAYS ARE THOSE WHERE INTEREST ARTICULATION BY NON-ASSOCIATIONAL GROUPS IS SIGNIFICANT, MODERATE, OR LIMITED (107)	1.00	35 56 8 0 X SQ= 8.97 P = 0.001 RV NO (YES)

124	ALWAYS ARE THOSE WHERE INTEREST ARTICULATION BY ANOMIC GROUPS IS OCCASIONAL, INFREQUENT, OR VERY INFREQUENT (85)	1.00	124	TEND LESS TO BE THOSE WHERE INTEREST ARTICULATION BY ANOMIC GROUPS IS OCCASIONAL, INFREQUENT, OR VERY INFREQUENT (85)	0.73	0 12 39 32 X SQ= 10.33 P = 0. RV NO (YES)
125	TEND TO BE THOSE WHERE INTEREST ARTICULATION BY ANOMIC GROUPS IS INFREQUENT OR VERY INFREQUENT (35)	0.62	125	TEND TO BE THOSE WHERE INTEREST ARTICULATION BY ANOMIC GROUPS IS FREQUENT OR OCCASIONAL (64)	0.82	15 36 24 8 X SQ= 14.63 P = 0. RV YES (YES)
126	TEND LESS TO BE THOSE WHERE INTEREST ARTICULATION BY ANOMIC GROUPS IS FREQUENT, OCCASIONAL, OR INFREQUENT (83)	0.59	126	ALWAYS ARE THOSE WHERE INTEREST ARTICULATION BY ANOMIC GROUPS IS FREQUENT, OCCASIONAL, OR INFREQUENT (83)	1.00	23 44 16 0 X SQ= 19.80 P = 0. RV NO (YES)
128	TEND TO BE THOSE WHERE INTEREST ARTICULATION BY POLITICAL PARTIES IS SIGNIFICANT OR MODERATE (48)	0.75	128	TEND TO BE THOSE WHERE INTEREST ARTICULATION BY POLITICAL PARTIES IS LIMITED OR NEGLIGIBLE (45)	0.65	30 13 10 24 X SQ= 10.82 P = 0.001 RV YES (YES)
129	TEND TO BE THOSE WHERE INTEREST ARTICULATION BY POLITICAL PARTIES IS SIGNIFICANT, MODERATE, OR LIMITED (56)	0.80	129	TEND TO BE THOSE WHERE INTEREST ARTICULATION BY POLITICAL PARTIES IS NEGLIGIBLE (37)	0.57	32 16 8 21 X SQ= 9.55 P = 0.001 RV YES (YES)
130	TEND LESS TO BE THOSE WHERE INTEREST AGGREGATION BY POLITICAL PARTIES IS MODERATE, LIMITED, OR NEGLIGIBLE (71)	0.73	130	TEND MORE TO BE THOSE WHERE INTEREST AGGREGATION BY POLITICAL PARTIES IS MODERATE, LIMITED, OR NEGLIGIBLE (71)	0.97	10 1 27 39 X SQ= 7.55 P = 0.003 RV NO (YES)
132	TEND MORE TO BE THOSE WHERE INTEREST AGGREGATION BY POLITICAL PARTIES IS SIGNIFICANT, MODERATE, OR LIMITED (67)	0.98	132	TEND LESS TO BE THOSE WHERE INTEREST AGGREGATION BY POLITICAL PARTIES IS SIGNIFICANT, MODERATE, OR LIMITED (67)	0.76	40 16 1 5 X SQ= 5.02 P = 0.014 RV NO (YES)
134	TEND TO BE THOSE WHERE INTEREST AGGREGATION BY THE EXECUTIVE IS SIGNIFICANT OR MODERATE (57)	0.83	134	TEND TO BE THOSE WHERE INTEREST AGGREGATION BY THE EXECUTIVE IS LIMITED OR NEGLIGIBLE (46)	0.72	34 13 7 33 X SQ= 23.93 P = 0. RV YES (YES)
135	ALWAYS ARE THOSE WHERE INTEREST AGGREGATION BY THE EXECUTIVE IS SIGNIFICANT, MODERATE, OR LIMITED (77)	1.00	135	TEND LESS TO BE THOSE WHERE INTEREST AGGREGATION BY THE EXECUTIVE IS SIGNIFICANT, MODERATE, OR LIMITED (77)	0.71	40 25 0 10 X SQ= 10.83 P = 0. RV NO (YES)

137	TEND TO BE THOSE WHERE INTEREST AGGREGATION BY THE LEGISLATURE IS SIGNIFICANT OR MODERATE (29)	0.64	137	TEND TO BE THOSE WHERE INTEREST AGGREGATION BY THE LEGISLATURE IS LIMITED OR NEGLIGIBLE (68)	0.93	25 3 14 39 X SQ= 26.54 P = 0. RV YES (YES)
138	TEND TO BE THOSE WHERE INTEREST AGGREGATION BY THE LEGISLATURE IS SIGNIFICANT, MODERATE, OR LIMITED (48)	0.79	138	TEND TO BE THOSE WHERE INTEREST AGGREGATION BY THE LEGISLATURE IS NEGLIGIBLE (49)	0.72	33 11 9 28 X SQ= 18.69 P = 0. RV YES (YES)
139	TEND MORE TO BE THOSE WHERE THE PARTY SYSTEM IS QUANTITATIVELY OTHER THAN ONE-PARTY (71)	0.86	139	TEND LESS TO BE THOSE WHERE THE PARTY SYSTEM IS QUANTITATIVELY OTHER THAN ONE-PARTY (71)	0.54	6 22 37 26 X SQ= 9.38 P = 0.001 RV NO (YES)
141	TEND LESS TO BE THOSE WHERE THE PARTY SYSTEM IS QUANTITATIVELY OTHER THAN TWO-PARTY (87)	0.76	141	ALWAYS ARE THOSE WHERE THE PARTY SYSTEM IS QUANTITATIVELY OTHER THAN TWO-PARTY (87)	1.00	10 0 31 43 X SQ= 9.69 P = 0. RV NO (YES)
142	TEND LESS TO BE THOSE WHERE THE PARTY SYSTEM IS QUANTITATIVELY OTHER THAN MULTI-PARTY (66)	0.56	142	TEND MORE TO BE THOSE WHERE THE PARTY SYSTEM IS QUANTITATIVELY OTHER THAN MULTI-PARTY (66)	0.78	18 9 23 32 X SQ= 3.53 P = 0.039 RV NO (YES)
147	TEND LESS TO BE THOSE WHERE THE PARTY SYSTEM IS QUALITATIVELY OTHER THAN CLASS-ORIENTED OR MULTI-IDEOLOGICAL (67)	0.52	147	TEND MORE TO BE THOSE WHERE THE PARTY SYSTEM IS QUALITATIVELY OTHER THAN CLASS-ORIENTED OR MULTI-IDEOLOGICAL (67)	0.89	19 4 21 33 X SQ= 10.66 P = 0. RV NO (YES)
148	TEND LESS TO BE THOSE WHERE THE PARTY SYSTEM IS QUALITATIVELY OTHER THAN AFRICAN TRANSITIONAL (96)	0.86	148	ALWAYS ARE THOSE WHERE THE PARTY SYSTEM IS QUALITATIVELY OTHER THAN AFRICAN TRANSITIONAL (96)	1.00	6 0 37 51 X SQ= 5.45 P = 0.007 RV NO (YES)
153	LEAN TOWARD BEING THOSE WHERE THE PARTY SYSTEM IS STABLE (42)	0.71	153	LEAN TOWARD BEING THOSE WHERE THE PARTY SYSTEM IS MODERATELY STABLE OR UNSTABLE (38)	0.51	22 20 9 21 X SQ= 2.72 P = 0.091 RV YES (YES)
160	TEND TO BE THOSE WHERE THE POLITICAL LEADERSHIP IS MODERATELY ELITIST OR NON-ELITIST (67)	0.98	160	TEND TO BE THOSE WHERE THE POLITICAL LEADERSHIP IS ELITIST (30)	0.61	1 25 40 16 X SQ= 29.79 P = 0. RV YES (YES)

161	TEND TO BE THOSE WHERE THE POLITICAL LEADERSHIP IS NON-ELITIST (50)	0.73	161	TEND TO BE THOSE WHERE THE POLITICAL LEADERSHIP IS ELITIST OR MODERATELY ELITIST (47)	0.76	11 31 30 10 X SQ= 17.62 P = 0. RV YES (YES)
163	TEND MORE TO BE THOSE WHERE THE REGIME'S LEADERSHIP CHARISMA IS MODERATE OR NEGLIGIBLE (87)	0.95	163	TEND LESS TO BE THOSE WHERE THE REGIME'S LEADERSHIP CHARISMA IS MODERATE OR NEGLIGIBLE (87)	0.78	2 10 39 36 X SQ= 3.86 P = 0.030 RV NO (YES)
164	TEND MORE TO BE THOSE WHERE THE REGIME'S LEADERSHIP CHARISMA IS NEGLIGIBLE (65)	0.80	164	TEND LESS TO BE THOSE WHERE THE REGIME'S LEADERSHIP CHARISMA IS NEGLIGIBLE (65)	0.58	8 19 33 26 X SQ= 4.14 P = 0.036 RV NO (YES)
168	TEND TO BE THOSE WHERE THE HORIZONTAL POWER DISTRIBUTION IS SIGNIFICANT (34)	0.69	168	TEND TO BE THOSE WHERE THE HORIZONTAL POWER DISTRIBUTION IS LIMITED OR NEGLIGIBLE (72)	0.92	29 4 13 44 X SQ= 32.99 P = 0. RV YES (YES)
169	TEND TO BE THOSE WHERE THE HORIZONTAL POWER DISTRIBUTION IS SIGNIFICANT OR LIMITED (58)	0.93	169	TEND TO BE THOSE WHERE THE HORIZONTAL POWER DISTRIBUTION IS NEGLIGIBLE (48)	0.77	39 11 3 37 X SQ= 41.59 P = 0. RV YES (YES)
172	TEND LESS TO BE THOSE WHERE THE LEGISLATIVE-EXECUTIVE STRUCTURE IS OTHER THAN PARLIAMENTARY-ROYALIST (88)	0.63	172	TEND MORE TO BE THOSE WHERE THE LEGISLATIVE-EXECUTIVE STRUCTURE IS OTHER THAN PARLIAMENTARY-ROYALIST (88)	0.96	16 2 27 48 X SQ= 14.28 P = 0. RV NO (YES)
174	TEND TO BE THOSE WHERE THE LEGISLATURE IS FULLY EFFECTIVE (28)	0.62	174	TEND TO BE THOSE WHERE THE LEGISLATURE IS PARTIALLY EFFECTIVE, LARGELY INEFFECTIVE, OR WHOLLY INEFFECTIVE (72)	0.95	26 2 16 42 X SQ= 29.64 P = 0. RV YES (YES)
175	TEND TO BE THOSE WHERE THE LEGISLATURE IS FULLY EFFECTIVE OR PARTIALLY EFFECTIVE (51)	0.86	175	TEND TO BE THOSE WHERE THE LEGISLATURE IS LARGELY INEFFECTIVE OR WHOLLY INEFFECTIVE (49)	0.73	36 12 6 32 X SQ= 27.44 P = 0. RV YES (YES)
176	ALWAYS ARE THOSE WHERE THE LEGISLATURE IS FULLY EFFECTIVE, PARTIALLY EFFECTIVE, OR LARGELY INEFFECTIVE (72)	1.00	176	TEND TO BE THOSE WHERE THE LEGISLATURE IS WHOLLY INEFFECTIVE (28)	0.64	42 16 0 28 X SQ= 36.78 P = 0. RV YES (YES)

No.	Statement	Value	No.	Statement	Value	Statistics
178	LEAN TOWARD BEING THOSE WHERE THE LEGISLATURE IS BICAMERAL (51)	0.67	178	LEAN TOWARD BEING THOSE WHERE THE LEGISLATURE IS UNICAMERAL (53)	0.54	14 25 28 21 X SQ= 3.12 P = 0.056 RV YES (YES)
179	TEND TO BE THOSE WHERE THE EXECUTIVE IS STRONG (39)	0.86	179	TEND TO BE THOSE WHERE THE EXECUTIVE IS DOMINANT (52)	0.82	5 37 31 8 X SQ= 34.72 P = 0. RV YES (YES)
181	TEND TO BE THOSE WHERE THE BUREAUCRACY IS MODERN, RATHER THAN SEMI-MODERN (21)	0.55	181	TEND TO BE THOSE WHERE THE BUREAUCRACY IS SEMI-MODERN, RATHER THAN MODERN (55)	0.92	18 3 15 34 X SQ= 15.77 P = 0. RV YES (YES)
182	TEND LESS TO BE THOSE WHERE THE BUREAUCRACY IS SEMI-MODERN, RATHER THAN POST-COLONIAL TRANSITIONAL (55)	0.60	182	TEND MORE TO BE THOSE WHERE THE BUREAUCRACY IS SEMI-MODERN, RATHER THAN POST-COLONIAL TRANSITIONAL (55)	0.87	15 34 10 5 X SQ= 4.85 P = 0.017 RV NO (YES)
183	ALWAYS ARE THOSE WHERE THE BUREAUCRACY IS POST-COLONIAL TRANSITIONAL, RATHER THAN TRADITIONAL (25)	1.00	183	TEND TO BE THOSE WHERE THE BUREAUCRACY IS TRADITIONAL, RATHER THAN POST-COLONIAL TRANSITIONAL (9)	0.64	10 5 0 9 X SQ= 7.73 P = 0.002 RV YES (YES)
186	TEND TO BE THOSE WHERE PARTICIPATION BY THE MILITARY IN POLITICS IS NEUTRAL, RATHER THAN SUPPORTIVE (56)	0.98	186	TEND TO BE THOSE WHERE PARTICIPATION BY THE MILITARY IN POLITICS IS SUPPORTIVE, RATHER THAN NEUTRAL (31)	0.83	1 29 40 6 X SQ= 47.80 P = 0. RV YES (YES)
187	TEND TO BE THOSE WHERE THE ROLE OF THE POLICE IS NOT POLITICALLY SIGNIFICANT (35)	0.89	187	TEND TO BE THOSE WHERE THE ROLE OF THE POLICE IS POLITICALLY SIGNIFICANT (66)	0.96	4 51 31 2 X SQ= 61.11 P = 0. RV YES (YES)
190	TEND LESS TO BE THOSE WHERE THE CHARACTER OF THE LEGAL SYSTEM IS CIVIL LAW, RATHER THAN COMMON LAW (32)	0.68	190	ALWAYS ARE THOSE WHERE THE CHARACTER OF THE LEGAL SYSTEM IS CIVIL LAW, RATHER THAN COMMON LAW (32)	1.00	15 12 7 0 X SQ= 3.06 P = 0.036 RV NO (NO)
194	ALWAYS ARE THOSE THAT ARE NON-COMMUNIST (101)	1.00	194	TEND LESS TO BE THOSE THAT ARE NON-COMMUNIST (101)	0.76	0 13 43 42 X SQ= 9.75 P = 0. RV NO (YES)

M

MATRIX

	50 POLITIES WHERE FREEDOM OF THE PRESS IS COMPLETE (43)	50 POLITIES WHERE FREEDOM OF THE PRESS IS INTERMITTENT, INTERNALLY ABSENT, OR INTERNALLY AND EXTERNALLY ABSENT (56)
63 POLITIES WHERE THE RELIGION IS PREDOMINANTLY SOME KIND OF CHRISTIAN (46)	27: AUSTRALIA, COLOMBIA, GERMAN FR, LUXEMBOURG, PANAMA, US, AUSTRIA, COSTA RICA, ICELAND, MEXICO, PHILIPPINES, URUGUAY, BELGIUM, DENMARK, IRELAND, NETHERLANDS, SWEDEN, CANADA, DOMIN REP, ITALY, NEW ZEALAND, SWITZERLAND, CHILE, FINLAND, JAMAICA, NORWAY, UK	15: ARGENTINA, FRANCE, PORTUGAL, BRAZIL, GERMANY, E, RUMANIA, BULGARIA, HUNGARY, SPAIN, CUBA, PARAGUAY, VENEZUELA, CZECHOS'KIA, POLAND, YUGOSLAVIA
63 POLITIES WHERE THE RELIGION IS PREDOMINANTLY OR PARTLY OTHER THAN CHRISTIAN (68)	16: CAMEROUN, JAPAN, NIGER, UGANDA, CHAD, MALAGASY R, SIERRE LEO, CYPRUS, MALAYA, SOMALIA, ECUADOR, MALI, TANGANYIKA, INDIA, MAURITANIA, TRINIDAD	40: AFGHANISTAN, CAMBODIA, GUINEA, ISRAEL, LEBANON, NEPAL, SO AFRICA, TURKEY, ALBANIA, CHINA, PR, HAITI, JORDAN, LIBERIA, NIGERIA, SUDAN, UAR, ALGERIA, CONGO(LEO), INDONESIA, KOREA, N, LIBYA, PAKISTAN, SYRIA, VIETNAM, N, BOLIVIA, ETHIOPIA, IRAN, KOREA REP, MONGOLIA, PERU, THAILAND, VIETNAM REP, BURMA, GHANA, IRAQ, LAOS, MOROCCO, SA'U ARABIA, TUNISIA, YEMEN

M

MATRIX

	50 POLITIES WHERE FREEDOM OF THE PRESS IS COMPLETE (43)	50 POLITIES WHERE FREEDOM OF THE PRESS IS INTERMITTENT, INTERNALLY ABSENT, OR INTERNALLY AND EXTERNALLY ABSENT (56)
105 POLITIES WHERE THE ELECTORAL SYSTEM IS COMPETITIVE (43)	33	7
105 POLITIES WHERE THE ELECTORAL SYSTEM IS PARTIALLY COMPETITIVE OR NON-COMPETITIVE (47)	6	30

Electoral system competitive, freedom of the press complete (33):

AUSTRALIA, AUSTRIA, BELGIUM, CANADA, CHILE, COSTA RICA, DENMARK, DOMIN REP, ECUADOR, FINLAND, GERMAN FR, ICELAND, INDIA, IRELAND, ITALY, JAMAICA, JAPAN, LUXEMBOURG, MALAGASY R, MALAYA, NETHERLANDS, NEW ZEALAND, NORWAY, PANAMA, PHILIPPINES, SIERRE LEO, SWEDEN, SWITZERLAND, TRINIDAD, UGANDA, UK, US, URUGUAY

Electoral system competitive, freedom of the press intermittent or absent (7):

ARGENTINA, BOLIVIA, BRAZIL, FRANCE, ISRAEL, TURKEY, VENEZUELA

Electoral system partially competitive or non-competitive, freedom of the press complete (6):

CHAD, MALI, MAURITANIA, NIGER, SOMALIA, TANGANYIKA

Electoral system partially competitive or non-competitive, freedom of the press intermittent or absent (30):

AFGHANISTAN, ALBANIA, ALGERIA, BULGARIA, CAMBODIA, CHINA, PR, CZECHOS'KIA, ETHIOPIA, GERMANY, E, GHANA, GUINEA, HAITI, HUNGARY, JORDAN, KOREA, N, LAOS, LIBERIA, LIBYA, MONGOLIA, PAKISTAN, PARAGUAY, POLAND, PORTUGAL, RUMANIA, SPAIN, TUNISIA, USSR, VIETNAM, N, VIETNAM REP, YUGOSLAVIA

****

51 POLITIES
WHERE FREEDOM OF THE PRESS IS
COMPLETE OR INTERMITTENT (60)

51 POLITIES
WHERE FREEDOM OF THE PRESS IS
INTERNALLY ABSENT, OR
INTERNALLY AND EXTERNALLY ABSENT (37)

PREDICATE ONLY

....

60 IN LEFT COLUMN

ARGENTINA	AUSTRALIA	AUSTRIA	BELGIUM	BOLIVIA	BRAZIL	BURMA	CAMEROUN	CANADA	CHAD
CHILE	COLOMBIA	CONGO(LEO)	COSTA RICA	CYPRUS	DENMARK	DOMIN REP	ECUADOR	FINLAND	FRANCE
GERMAN FR	ICELAND	INDIA	IRELAND	ISRAEL	ITALY	JAMAICA	JAPAN	LAOS	LEBANON
LIBERIA	LIBYA	LUXEMBOURG	MALAGASY R	MALAYA	MALI	MAURITANIA	MEXICO	MOROCCO	NETHERLANDS
NEW ZEALAND	NIGER	NIGERIA	NORWAY	PANAMA	PERU	PHILIPPINES	SIERRE LEO	SOMALIA	SO AFRICA
SWEDEN	SWITZERLAND	SYRIA	TANGANYIKA	TRINIDAD	TURKEY	UGANDA	UK	US	URUGUAY

37 IN RIGHT COLUMN

AFGHANISTAN	ALBANIA	ALGERIA	BULGARIA	CAMBODIA	CHINA, PR	CUBA	CZECHOS'KIA	ETHIOPIA	GERMANY, E
GHANA	GUINEA	HAITI	HUNGARY	INDONESIA	IRAN	JORDAN	KOREA, N	KOREA REP	MONGOLIA
NEPAL	PAKISTAN	PARAGUAY	POLAND	PORTUGAL	RUMANIA	SA'U ARABIA	SPAIN	SUDAN	THAILAND
TUNISIA	USSR	UAR	VENEZUELA	VIETNAM, N	VIETNAM REP	YUGOSLAVIA			

1 EXCLUDED BECAUSE AMBIGUOUS

GREECE

13 EXCLUDED BECAUSE UNASCERTAINED

BURUNDI	CEN AFR REP	CONGO(BRA)	DAHOMEY	EL SALVADOR	GABON	GUATEMALA	HONDURAS	IVORY COAST	NICARAGUA
RWANDA	SENEGAL	UPPER VOLTA							

4 EXCLUDED BECAUSE UNASCERTAINABLE

CEYLON	IRAQ	TOGO	YEMEN

----

52 POLITIES
WHERE FREEDOM OF THE PRESS IS
COMPLETE, INTERMITTENT, OR
INTERNALLY ABSENT (82)

52 POLITIES
WHERE FREEDOM OF THE PRESS IS
INTERNALLY AND EXTERNALLY ABSENT (16)

PREDICATE ONLY

82 IN LEFT COLUMN

ALGERIA	ARGENTINA	AUSTRALIA	AUSTRIA	BELGIUM	BOLIVIA	BRAZIL	BURMA	CAMBODIA	CAMEROUN
CANADA	CHAD	CHILE	COLOMBIA	CONGO(LEO)	COSTA RICA	CYPRUS	DENMARK	DOMIN REP	ECUADOR
ETHIOPIA	FINLAND	FRANCE	GERMAN FR	GHANA	GREECE	GUINEA	HAITI	ICELAND	INDIA
INDONESIA	IRAN	IRELAND	ISRAEL	ITALY	JAMAICA	JAPAN	KOREA REP	LAOS	LEBANON
LIBERIA	LIBYA	LUXEMBOURG	MALAGASY R	MALAYA	MALI	MAURITANIA	MEXICO	MOROCCO	NEPAL
NETHERLANDS	NEW ZEALAND	NIGER	NIGERIA	NORWAY	PAKISTAN	PANAMA	PERU	PHILIPPINES	POLAND
PORTUGAL	SIERRE LEO	SOMALIA	SO AFRICA	SPAIN	SUDAN	SWEDEN	SWITZERLAND	SYRIA	TANGANYIKA
THAILAND	TRINIDAD	TUNISIA	TURKEY	UGANDA	USSR	UK	US	URUGUAY	VENEZUELA
VIETNAM REP	YUGOSLAVIA								

16 IN RIGHT COLUMN

AFGHANISTAN	ALBANIA	BULGARIA	CHINA, PR	CUBA	CZECHOS'KIA	GERMANY, E	HUNGARY	JORDAN	KOREA, N
MONGOLIA	PARAGUAY	RUMANIA	SA'U ARABIA	UAR	VIETNAM, N				

13 EXCLUDED BECAUSE UNASCERTAINED

BURUNDI	CEN AFR REP	CONGO(BRA)	DAHOMEY	EL SALVADOR	GABON	GUATEMALA	HONDURAS	IVORY COAST	NICARAGUA
RWANDA	SENEGAL	UPPER VOLTA							

4 EXCLUDED BECAUSE UNASCERTAINABLE

CEYLON	IRAQ	TOGO	YEMEN

****

```
53  POLITIES                                        53  POLITIES
    WHERE NEWSPAPER CIRCULATION IS                      WHERE NEWSPAPER CIRCULATION IS
    THREE HUNDRED OR MORE                               LESS THAN THREE HUNDRED
    PER THOUSAND  (14)                                  PER THOUSAND  (101)

                                                                                   PREDICATE ONLY
.............................................................................................

     14 IN LEFT COLUMN

AUSTRALIA   DENMARK     FINLAND     GERMANY, E  GERMAN FR   ICELAND     JAPAN       LUXEMBOURG  NEW ZEALAND NORWAY
SWEDEN      SWITZERLAND UK          US

     101 IN RIGHT COLUMN

AFGHANISTAN ALBANIA     ALGERIA     ARGENTINA   AUSTRIA     BELGIUM     BOLIVIA     BRAZIL      BULGARIA    BURMA
BURUNDI     CAMBODIA    CAMEROUN    CANADA      CEN AFR REP CEYLON      CHAD        CHILE       CHINA, PR   COLOMBIA
CONGO(BRA)  CONGO(LEO)  COSTA RICA  CUBA        CYPRUS      CZECHOS'KIA DAHOMEY     DOMIN REP   ECUADOR     EL SALVADOR
ETHIOPIA    FRANCE      GABON       GHANA       GREECE      GUATEMALA   GUINEA      HAITI       HONDURAS    HUNGARY
INDIA       INDONESIA   IRAN        IRAQ        IRELAND     ISRAEL      ITALY       IVORY COAST JAMAICA     JORDAN
KOREA, N    KOREA REP   LAOS        LEBANON     LIBERIA     LIBYA       MALAGASY R  MALAYA      MALI        MAURITANIA
MEXICO      MONGOLIA    MOROCCO     NEPAL       NETHERLANDS NICARAGUA   NIGER       NIGERIA     PAKISTAN    PANAMA
PARAGUAY    PERU        PHILIPPINES POLAND      PORTUGAL    RUMANIA     RWANDA      SA'U ARABIA SENEGAL     SIERRE LEO
SOMALIA     SO AFRICA   SPAIN       SUDAN       SYRIA       TANGANYIKA  THAILAND    TOGO        TRINIDAD    TUNISIA
TURKEY      UGANDA      USSR        UAR         UPPER VOLTA URUGUAY     VENEZUELA   VIETNAM, N  VIETNAM REP YEMEN
YUGOSLAVIA
---------------------------------------------------------------------------------------------
```

54 POLITIES WHERE NEWSPAPER CIRCULATION IS ONE HUNDRED OR MORE PER THOUSAND (37)	54 POLITIES WHERE NEWSPAPER CIRCULATION IS LESS THAN ONE HUNDRED PER THOUSAND (76)

BOTH SUBJECT AND PREDICATE

37 IN LEFT COLUMN

ARGENTINA	AUSTRALIA	AUSTRIA	BELGIUM	BULGARIA	CANADA	CHILE	COSTA RICA	CUBA	CYPRUS
CZECHOS'KIA	DENMARK	FINLAND	FRANCE	GERMANY, E	GERMAN FR	GREECE	HUNGARY	ICELAND	IRELAND
ISRAEL	ITALY	JAPAN	LUXEMBOURG	MONGOLIA	NETHERLANDS	NEW ZEALAND	NORWAY	PANAMA	POLAND
RUMANIA	SWEDEN	SWITZERLAND	USSR	UK	US	URUGUAY			

76 IN RIGHT COLUMN

AFGHANISTAN	ALBANIA	ALGERIA	BOLIVIA	BRAZIL	BURMA	BURUNDI	CAMBODIA	CAMEROUN	CEN AFR REP
CEYLON	CHAD	CHINA, PR	COLOMBIA	CONGO(BRA)	CONGO(LEO)	DAHOMEY	DOMIN REP	ECUADOR	EL SALVADOR
ETHIOPIA	GABON	GHANA	GUATEMALA	GUINEA	HAITI	HONDURAS	INDIA	INDONESIA	IRAN
IRAQ	IVORY COAST	JAMAICA	JORDAN	KOREA REP	LAOS	LEBANON	LIBERIA	LIBYA	MALAGASY R
MALAYA	MALI	MAURITANIA	MEXICO	MOROCCO	NEPAL	NICARAGUA	NIGER	NIGERIA	PAKISTAN
PARAGUAY	PERU	PHILIPPINES	PORTUGAL	RWANDA	SA'U ARABIA	SENEGAL	SIERRE LEO	SOMALIA	SO AFRICA
SPAIN	SUDAN	SYRIA	TANGANYIKA	THAILAND	TOGO	TRINIDAD	TUNISIA	TURKEY	UGANDA
UAR	UPPER VOLTA	VENEZUELA	VIETNAM REP	YEMEN	YUGOSLAVIA				

2 EXCLUDED BECAUSE UNASCERTAINED

KOREA, N   VIETNAM, N

	Left column			Right column		
2	TEND TO BE THOSE LOCATED IN WEST EUROPE, SCANDINAVIA, NORTH AMERICA, OR AUSTRALASIA (22)	0.54	2	TEND TO BE THOSE LOCATED ELSEWHERE THAN IN WEST EUROPE, SCANDINAVIA, NORTH AMERICA, OR AUSTRALASIA (93)	0.97	20 2 17 74 X SQ= 38.75 P = 0. RV YES (YES)
3	TEND LESS TO BE THOSE LOCATED ELSEWHERE THAN IN WEST EUROPE (102)	0.70	3	TEND MORE TO BE THOSE LOCATED ELSEWHERE THAN IN WEST EUROPE (102)	0.97	11 2 26 74 X SQ= 15.39 P = 0. RV NO (YES)
8	LEAN MORE TOWARD BEING THOSE LOCATED ELSEWHERE THAN IN EAST ASIA SOUTH ASIA, OR SOUTHEAST ASIA (97)	0.95	8	LEAN LESS TOWARD BEING THOSE LOCATED ELSEWHERE THAN IN EAST ASIA SOUTH ASIA, OR SOUTHEAST ASIA (97)	0.82	2 14 35 62 X SQ= 2.48 P = 0.085 RV NO (NO)

9	ALWAYS ARE THOSE LOCATED ELSEWHERE THAN IN SOUTHEAST ASIA (106)	1.00	9	LEAN LESS TOWARD BEING THOSE LOCATED ELSEWHERE THAN IN SOUTHEAST ASIA (106)	0.89	0 8 37 68 X SQ= 2.74 P = 0.051 RV NO (NO)
10	ALWAYS ARE THOSE LOCATED ELSEWHERE THAN IN NORTH AFRICA, OR CENTRAL AND SOUTH AFRICA (82)	1.00	10	TEND LESS TO BE THOSE LOCATED ELSEWHERE THAN IN NORTH AFRICA, OR CENTRAL AND SOUTH AFRICA (82)	0.57	0 33 37 43 X SQ= 20.64 P = 0. RV NO (NO)
11	ALWAYS ARE THOSE LOCATED ELSEWHERE THAN IN CENTRAL AND SOUTH AFRICA (87)	1.00	11	TEND LESS TO BE THOSE LOCATED ELSEWHERE THAN IN CENTRAL AND SOUTH AFRICA (87)	0.63	0 28 37 48 X SQ= 16.20 P = 0. RV NO (NO)
13	LEAN MORE TOWARD BEING THOSE LOCATED ELSEWHERE THAN IN NORTH AFRICA OR THE MIDDLE EAST (99)	0.95	13	LEAN LESS TOWARD BEING THOSE LOCATED ELSEWHERE THAN IN NORTH AFRICA OR THE MIDDLE EAST (99)	0.82	2 14 35 62 X SQ= 2.48 P = 0.085 RV NO (NO)
22	TILT TOWARD BEING THOSE WHOSE TERRITORIAL SIZE IS SMALL (47)	0.51	22	TILT TOWARD BEING THOSE WHOSE TERRITORIAL SIZE IS VERY LARGE, LARGE, OR MEDIUM (68)	0.66	18 50 19 26 X SQ= 2.38 P = 0.102 RV YES (NO)
26	TEND TO BE THOSE WHOSE POPULATION DENSITY IS VERY HIGH, HIGH, OR MEDIUM (48)	0.59	26	TEND TO BE THOSE WHOSE POPULATION DENSITY IS LOW (67)	0.68	22 24 15 52 X SQ= 6.90 P = 0.008 RV YES (NO)
28	TEND TO BE THOSE WHOSE POPULATION GROWTH RATE IS LOW (48)	0.72	28	TEND TO BE THOSE WHOSE POPULATION GROWTH RATE IS HIGH (62)	0.71	10 52 26 21 X SQ= 16.83 P = 0. RV YES (YES)
29	TEND TO BE THOSE WHERE THE DEGREE OF URBANIZATION IS HIGH (56)	0.94	29	TEND TO BE THOSE WHERE THE DEGREE OF URBANIZATION IS LOW (49)	0.68	34 22 2 47 X SQ= 34.73 P = 0. RV YES (YES)
30	TEND TO BE THOSE WHOSE AGRICULTURAL POPULATION IS MEDIUM, LOW, OR VERY LOW (57)	0.95	30	TEND TO BE THOSE WHOSE AGRICULTURAL POPULATION IS HIGH (56)	0.71	2 53 35 22 X SQ= 39.65 P = 0. RV YES (YES)

31	TEND TO BE THOSE WHOSE AGRICULTURAL POPULATION IS LOW OR VERY LOW (24)	0.59	31	TEND TO BE THOSE WHOSE AGRICULTURAL POPULATION IS HIGH OR MEDIUM (90)	0.97	15 73 22 2 X SQ= 44.15 P = 0. RV YES (YES)
33	TEND TO BE THOSE WHOSE GROSS NATIONAL PRODUCT IS VERY HIGH, HIGH, OR MEDIUM (30)	0.54	33	TEND TO BE THOSE WHOSE GROSS NATIONAL PRODUCT IS LOW OR VERY LOW (85)	0.87	20 10 17 66 X SQ= 19.30 P = 0. RV YES (YES)
34	TEND TO BE THOSE WHOSE GROSS NATIONAL PRODUCT IS VERY HIGH, HIGH, MEDIUM, OR LOW (62)	0.84	34	TEND TO BE THOSE WHOSE GROSS NATIONAL PRODUCT IS VERY LOW (53)	0.61	31 30 6 46 X SQ= 17.92 P = 0. RV YES (YES)
35	TEND TO BE THOSE WHOSE PER CAPITA GROSS NATIONAL PRODUCT IS VERY HIGH OR HIGH (24)	0.62	35	TEND TO BE THOSE WHOSE PER CAPITA GROSS NATIONAL PRODUCT IS MEDIUM, LOW, OR VERY LOW (91)	0.99	23 1 14 75 X SQ= 51.50 P = 0. RV YES (YES)
36	TEND TO BE THOSE WHOSE PER CAPITA GROSS NATIONAL PRODUCT IS VERY HIGH, HIGH, OR MEDIUM (42)	0.92	36	TEND TO BE THOSE WHOSE PER CAPITA GROSS NATIONAL PRODUCT IS LOW OR VERY LOW (73)	0.89	34 8 3 68 X SQ= 67.10 P = 0. RV YES (YES)
37	TEND TO BE THOSE WHOSE PER CAPITA GROSS NATIONAL PRODUCT IS VERY HIGH, HIGH, MEDIUM, OR LOW (64)	0.97	37	TEND TO BE THOSE WHOSE PER CAPITA GROSS NATIONAL PRODUCT IS VERY LOW (51)	0.63	36 28 1 48 X SQ= 34.61 P = 0. RV YES (YES)
39	TEND TO BE THOSE WHOSE INTERNATIONAL FINANCIAL STATUS IS VERY HIGH, HIGH, OR MEDIUM (38)	0.67	39	TEND TO BE THOSE WHOSE INTERNATIONAL FINANCIAL STATUS IS LOW OR VERY LOW (76)	0.82	24 14 12 62 X SQ= 23.26 P = 0. RV YES (YES)
41	TEND TO BE THOSE WHOSE ECONOMIC DEVELOPMENTAL STATUS IS DEVELOPED (19)	0.53	41	ALWAYS ARE THOSE WHOSE ECONOMIC DEVELOPMENTAL STATUS IS INTERMEDIATE, UNDERDEVELOPED, OR VERY UNDERDEVELOPED (94)	1.00	19 0 17 75 X SQ= 44.11 P = 0. RV YES (YES)
42	TEND TO BE THOSE WHOSE ECONOMIC DEVELOPMENTAL STATUS IS DEVELOPED OR INTERMEDIATE (36)	0.86	42	TEND TO BE THOSE WHOSE ECONOMIC DEVELOPMENTAL STATUS IS UNDERDEVELOPED OR VERY UNDERDEVELOPED (76)	0.93	31 5 5 69 X SQ= 65.71 P = 0. RV YES (YES)

43	TEND TO BE THOSE WHOSE ECONOMIC DEVELOPMENTAL STATUS IS DEVELOPED, INTERMEDIATE, OR UNDERDEVELOPED (55)	0.94	43	TEND TO BE THOSE WHOSE ECONOMIC DEVELOPMENTAL STATUS IS VERY UNDERDEVELOPED (57)	0.72	34 21 2 53 X SQ= 39.68 P = 0. RV YES (YES)
44	TEND TO BE THOSE WHERE THE LITERACY RATE IS NINETY PERCENT OR ABOVE (25)	0.68	44	ALWAYS ARE THOSE WHERE THE LITERACY RATE IS BELOW NINETY PERCENT (90)	1.00	25 0 12 76 X SQ= 62.08 P = 0. RV YES (YES)
45	ALWAYS ARE THOSE WHERE THE LITERACY RATE IS FIFTY PERCENT OR ABOVE (55)	1.00	45	TEND TO BE THOSE WHERE THE LITERACY RATE IS BELOW FIFTY PERCENT (54)	0.74	36 19 0 54 X SQ= 49.86 P = 0. RV YES (YES)
50	TEND TO BE THOSE WHERE FREEDOM OF THE PRESS IS COMPLETE (43)	0.67	50	TEND TO BE THOSE WHERE FREEDOM OF THE PRESS IS INTERMITTENT, INTERNALLY ABSENT, OR INTERNALLY AND EXTERNALLY ABSENT (56)	0.69	24 19 12 42 X SQ= 10.18 P = 0.001 RV YES (YES)
56	ALWAYS ARE THOSE WHERE THE RELIGION IS PREDOMINANTLY LITERATE (79)	1.00	56	TEND LESS TO BE THOSE WHERE THE RELIGION IS PREDOMINANTLY LITERATE (79)	0.58	37 42 0 31 X SQ= 19.83 P = 0. RV NO (NO)
57	TEND LESS TO BE THOSE WHERE THE RELIGION IS OTHER THAN CATHOLIC (90)	0.65	57	TEND MORE TO BE THOSE WHERE THE RELIGION IS OTHER THAN CATHOLIC (90)	0.84	13 12 24 64 X SQ= 4.34 P = 0.029 RV NO (YES)
58	ALWAYS ARE THOSE WHERE THE RELIGION IS OTHER THAN MUSLIM (97)	1.00	58	TEND LESS TO BE THOSE WHERE THE RELIGION IS OTHER THAN MUSLIM (97)	0.76	0 18 37 58 X SQ= 8.73 P = 0.001 RV NO (NO)
62	LEAN LESS TOWARD BEING THOSE WHERE THE RELIGION IS CATHOLIC, RATHER THAN PROTESTANT (25)	0.72	62	ALWAYS ARE THOSE WHERE THE RELIGION IS CATHOLIC, RATHER THAN PROTESTANT (25)	1.00	5 0 13 12 X SQ= 2.25 P = 0.066 RV NO (NO)
63	TEND TO BE THOSE WHERE THE RELIGION IS PREDOMINANTLY SOME KIND OF CHRISTIAN (46)	0.89	63	TEND TO BE THOSE WHERE THE RELIGION IS PREDOMINANTLY OR PARTLY OTHER THAN CHRISTIAN (68)	0.82	32 14 4 62 X SQ= 47.25 P = 0. RV YES (YES)

No.			No.			
67	TEND MORE TO BE THOSE THAT ARE RACIALLY HOMOGENEOUS (82)	0.92	67	TEND LESS TO BE THOSE THAT ARE RACIALLY HOMOGENEOUS (82)	0.63	33 45 3 26 X SQ= 8.30 P = 0.002 RV NO (NO)
68	TEND TO BE THOSE THAT ARE LINGUISTICALLY HOMOGENEOUS (52)	0.70	68	TEND TO BE THOSE THAT ARE LINGUISTICALLY WEAKLY HETEROGENEOUS OR STRONGLY HETEROGENEOUS (62)	0.67	26 25 11 50 X SQ= 12.18 P = 0. RV YES (YES)
69	TEND TO BE THOSE THAT ARE LINGUISTICALLY HOMOGENEOUS OR WEAKLY HETEROGENEOUS (64)	0.81	69	TEND TO BE THOSE THAT ARE LINGUISTICALLY STRONGLY HETEROGENEOUS (50)	0.57	30 32 7 43 X SQ= 13.28 P = 0. RV YES (NO)
70	TEND LESS TO BE THOSE THAT ARE RELIGIOUSLY, RACIALLY, OR LINGUISTICALLY HETEROGENEOUS (85)	0.58	70	TEND MORE TO BE THOSE THAT ARE RELIGIOUSLY, RACIALLY, OR LINGUISTICALLY HETEROGENEOUS (85)	0.91	15 6 21 62 X SQ= 13.78 P = 0. RV NO (YES)
72	TEND TO BE THOSE WHOSE DATE OF INDEPENDENCE IS BEFORE 1914 (52)	0.74	72	TEND TO BE THOSE WHOSE DATE OF INDEPENDENCE IS AFTER 1914 (59)	0.65	26 26 9 49 X SQ= 13.48 P = 0. RV YES (YES)
73	TEND TO BE THOSE WHOSE DATE OF INDEPENDENCE IS BEFORE 1945 (65)	0.94	73	TEND TO BE THOSE WHOSE DATE OF INDEPENDENCE IS AFTER 1945 (46)	0.57	33 32 2 43 X SQ= 24.21 P = 0. RV YES (YES)
74	TEND TO BE THOSE THAT ARE HISTORICALLY WESTERN (26)	0.67	74	TEND TO BE THOSE THAT ARE NOT HISTORICALLY WESTERN (87)	0.97	24 2 12 73 X SQ= 52.04 P = 0. RV YES (YES)
75	TEND TO BE THOSE THAT ARE HISTORICALLY WESTERN OR SIGNIFICANTLY WESTERNIZED (62)	0.97	75	TEND TO BE THOSE THAT ARE NOT HISTORICALLY WESTERN AND ARE NOT SIGNIFICANTLY WESTERNIZED (52)	0.65	36 26 1 49 X SQ= 36.83 P = 0. RV YES (YES)
76	TEND TO BE THOSE THAT ARE HISTORICALLY WESTERN, RATHER THAN HAVING BEEN SIGNIFICANTLY OR PARTIALLY WESTERNIZED THROUGH A COLONIAL RELATIONSHIP (26)	0.77	76	TEND TO BE THOSE THAT HAVE BEEN SIGNIFICANTLY OR PARTIALLY WESTERNIZED THROUGH A COLONIAL RELATIONSHIP, RATHER THAN BEING HISTORICALLY WESTERN (70)	0.97	24 2 7 62 X SQ= 54.31 P = 0. RV YES (YES)

77	ALWAYS ARE THOSE THAT HAVE BEEN SIGNIFICANTLY WESTERNIZED, RATHER THAN PARTIALLY WESTERNIZED, THROUGH A COLONIAL RELATIONSHIP (28)	1.00	77	TEND TO BE THOSE THAT HAVE BEEN PARTIALLY WESTERNIZED, RATHER THAN SIGNIFICANTLY WESTERNIZED, THROUGH A COLONIAL RELATIONSHIP (41)	0.66	7 21 0 40 X SQ= 8.60 P = 0.001 RV YES (NO)
80	ALWAYS ARE THOSE WHOSE DATE OF INDEPENDENCE IS AFTER 1914, AND THAT WERE FORMERLY DEPENDENCIES OF BRITAIN, RATHER THAN FRANCE (19)	1.00	80	LEAN TOWARD BEING THOSE WHOSE DATE OF INDEPENDENCE IS AFTER 1914, AND THAT WERE FORMERLY DEPENDENCIES OF FRANCE, RATHER THAN BRITAIN (24)	0.59	3 16 0 23 X SQ= 1.89 P = 0.084 RV YES (NO)
81	TEND LESS TO BE THOSE WHERE THE TYPE OF POLITICAL MODERNIZATION IS LATER EUROPEAN OR LATER EUROPEAN DERIVED, RATHER THAN EARLY EUROPEAN OR EARLY EUROPEAN DERIVED (40)	0.69	81	TEND MORE TO BE THOSE WHERE THE TYPE OF POLITICAL MODERNIZATION IS LATER EUROPEAN OR LATER EUROPEAN DERIVED, RATHER THAN EARLY EUROPEAN OR EARLY EUROPEAN DERIVED (40)	0.95	10 1 22 18 X SQ= 3.35 P = 0.037 RV NO (NO)
82	TEND TO BE THOSE WHERE THE TYPE OF POLITICAL MODERNIZATION IS EARLY OR LATER EUROPEAN OR EUROPEAN DERIVED, RATHER THAN DEVELOPED TUTELARY OR UNDEVELOPED TUTELARY (51)	0.91	82	TEND TO BE THOSE WHERE THE TYPE OF POLITICAL MODERNIZATION IS DEVELOPED TUTELARY OR UNDEVELOPED TUTELARY, RATHER THAN EARLY OR LATER EUROPEAN OR EUROPEAN DERIVED (55)	0.72	32 19 3 50 X SQ= 35.42 P = 0. RV YES (YES)
85	TEND TO BE THOSE WHERE THE STAGE OF POLITICAL MODERNIZATION IS ADVANCED, RATHER THAN MID- OR EARLY TRANSITIONAL (60)	0.97	85	TEND TO BE THOSE WHERE THE STAGE OF POLITICAL MODERNIZATION IS MID- OR EARLY TRANSITIONAL, RATHER THAN ADVANCED (54)	0.69	36 23 1 52 X SQ= 41.49 P = 0. RV YES (YES)
86	ALWAYS ARE THOSE WHERE THE STAGE OF POLITICAL MODERNIZATION IS ADVANCED OR MID-TRANSITIONAL, RATHER THAN EARLY TRANSITIONAL (76)	1.00	86	TEND LESS TO BE THOSE WHERE THE STAGE OF POLITICAL MODERNIZATION IS ADVANCED OR MID-TRANSITIONAL, RATHER THAN EARLY TRANSITIONAL (76)	0.51	37 38 0 37 X SQ= 25.07 P = 0. RV NO (NO)
87	TEND TO BE THOSE WHOSE IDEOLOGICAL ORIENTATION IS OTHER THAN DEVELOPMENTAL (58)	0.97	87	TEND TO BE THOSE WHOSE IDEOLOGICAL ORIENTATION IS DEVELOPMENTAL (31)	0.57	1 30 33 23 X SQ= 23.72 P = 0. RV YES (YES)
89	TEND TO BE THOSE WHOSE IDEOLOGICAL ORIENTATION IS CONVENTIONAL (33)	0.71	89	TEND TO BE THOSE WHOSE IDEOLOGICAL ORIENTATION IS OTHER THAN CONVENTIONAL (62)	0.86	25 8 10 50 X SQ= 29.21 P = 0. RV YES (YES)
94	TEND TO BE THOSE WHERE THE STATUS OF THE REGIME IS CONSTITUTIONAL (51)	0.75	94	TEND TO BE THOSE WHERE THE STATUS OF THE REGIME IS AUTHORITARIAN OR TOTALITARIAN (41)	0.56	27 24 9 30 X SQ= 7.02 P = 0.005 RV YES (YES)

95	TEND LESS TO BE THOSE WHERE THE STATUS OF THE REGIME IS CONSTITUTIONAL OR AUTHORITARIAN (95)	0.76	95	TEND MORE TO BE THOSE WHERE THE STATUS OF THE REGIME IS CONSTITUTIONAL OR AUTHORITARIAN (95)	0.93	28 67 9 5 X SQ= 5.13 P = 0.015 RV NO (YES)
96	ALWAYS ARE THOSE WHERE THE STATUS OF THE REGIME IS CONSTITUTIONAL OR TOTALITARIAN (67)	1.00	96	TEND LESS TO BE THOSE WHERE THE STATUS OF THE REGIME IS CONSTITUTIONAL OR TOTALITARIAN (67)	0.56	36 29 0 23 X SQ= 19.33 P = 0. RV NO (YES)
97	ALWAYS ARE THOSE WHERE THE STATUS OF THE REGIME IS TOTALITARIAN, RATHER THAN AUTHORITARIAN (16)	1.00	97	TEND TO BE THOSE WHERE THE STATUS OF THE REGIME IS AUTHORITARIAN, RATHER THAN TOTALITARIAN (23)	0.82	0 23 9 5 X SQ= 16.20 P = 0. RV YES (YES)
99	TEND TO BE THOSE WHERE GOVERNMENTAL STABILITY IS GENERALLY PRESENT AND DATES FROM AT LEAST THE INTER-WAR PERIOD, OR FROM THE POST-WAR PERIOD (50)	0.85	99	TEND TO BE THOSE WHERE GOVERNMENTAL STABILITY IS MODERATELY PRESENT AND DATES FROM THE POST-WAR PERIOD, OR IS ABSENT (36)	0.62	29 19 5 31 X SQ= 16.60 P = 0. RV YES (YES)
101	TEND TO BE THOSE WHERE THE REPRESENTATIVE CHARACTER OF THE REGIME IS POLYARCHIC (41)	0.72	101	TEND TO BE THOSE WHERE THE REPRESENTATIVE CHARACTER OF THE REGIME IS LIMITED POLYARCHIC, PSEUDO-POLYARCHIC, OR NON-POLYARCHIC (57)	0.75	26 15 10 45 X SQ= 18.62 P = 0. RV YES (YES)
102	TEND TO BE THOSE WHERE THE REPRESENTATIVE CHARACTER OF THE REGIME IS POLYARCHIC OR LIMITED POLYARCHIC (59)	0.76	102	TEND TO BE THOSE WHERE THE REPRESENTATIVE CHARACTER OF THE REGIME IS PSEUDO-POLYARCHIC OR NON-POLYARCHIC (49)	0.55	28 31 9 38 X SQ= 8.02 P = 0.004 RV YES (NO)
105	TEND TO BE THOSE WHERE THE ELECTORAL SYSTEM IS COMPETITIVE (43)	0.77	105	TEND TO BE THOSE WHERE THE ELECTORAL SYSTEM IS PARTIALLY COMPETITIVE OR NON-COMPETITIVE (47)	0.70	27 16 8 37 X SQ= 16.77 P = 0. RV YES (YES)
107	TEND TO BE THOSE WHERE AUTONOMOUS GROUPS ARE FULLY TOLERATED IN POLITICS (46)	0.76	107	TEND TO BE THOSE WHERE AUTONOMOUS GROUPS ARE PARTIALLY TOLERATED IN POLITICS, ARE TOLERATED ONLY OUTSIDE POLITICS, OR ARE NOT TOLERATED AT ALL (65)	0.75	28 18 9 54 X SQ= 23.70 P = 0. RV YES (YES)
111	TEND TO BE THOSE WHERE POLITICAL ENCULTURATION IS HIGH OR MEDIUM (53)	0.81	111	TEND TO BE THOSE WHERE POLITICAL ENCULTURATION IS LOW (42)	0.54	M 22 31 5 37 X SQ= 8.69 P = 0.001 RV YES (NO)

114	TEND TO BE THOSE WHERE SECTIONALISM IS NEGLIGIBLE (47)	0.63	114	TEND TO BE THOSE WHERE SECTIONALISM IS EXTREME OR MODERATE (61)	0.68	13 48 22 23 X SQ= 7.70 P = 0.004 RV YES (NO)
116	TEND TO BE THOSE WHERE INTEREST ARTICULATION BY ASSOCIATIONAL GROUPS IS SIGNIFICANT OR MODERATE (32)	0.68	116	TEND TO BE THOSE WHERE INTEREST ARTICULATION BY ASSOCIATIONAL GROUPS IS LIMITED OR NEGLIGIBLE (79)	0.90	25 7 12 65 X SQ= 36.69 P = 0. RV YES (YES)
117	TEND TO BE THOSE WHERE INTEREST ARTICULATION BY ASSOCIATIONAL GROUPS IS SIGNIFICANT, MODERATE, OR LIMITED (60)	0.84	117	TEND TO BE THOSE WHERE INTEREST ARTICULATION BY ASSOCIATIONAL GROUPS IS NEGLIGIBLE (51)	0.60	31 29 6 43 X SQ= 16.98 P = 0. RV YES (YES)
119	TEND TO BE THOSE WHERE INTEREST ARTICULATION BY INSTITUTIONAL GROUPS IS MODERATE OR LIMITED (26)	0.59	119	TEND TO BE THOSE WHERE INTEREST ARTICULATION BY INSTITUTIONAL GROUPS IS VERY SIGNIFICANT OR SIGNIFICANT (74)	0.93	15 57 22 4 X SQ= 30.41 P = 0. RV YES (YES)
120	TEND LESS TO BE THOSE WHERE INTEREST ARTICULATION BY INSTITUTIONAL GROUPS IS VERY SIGNIFICANT, SIGNIFICANT, OR MODERATE (90)	0.73	120	ALWAYS ARE THOSE WHERE INTEREST ARTICULATION BY INSTITUTIONAL GROUPS IS VERY SIGNIFICANT, SIGNIFICANT, OR MODERATE (90)	1.00	27 61 10 0 X SQ= 15.53 P = 0. RV NO (YES)
121	TEND TO BE THOSE WHERE INTEREST ARTICULATION BY NON-ASSOCIATIONAL GROUPS IS MODERATE, LIMITED, OR NEGLIGIBLE (61)	0.95	121	TEND TO BE THOSE WHERE INTEREST ARTICULATION BY NON-ASSOCIATIONAL GROUPS IS SIGNIFICANT (54)	0.66	2 50 35 26 X SQ= 34.14 P = 0. RV YES (YES)
122	TEND TO BE THOSE WHERE INTEREST ARTICULATION BY NON-ASSOCIATIONAL GROUPS IS LIMITED OR NEGLIGIBLE (32)	0.73	122	TEND TO BE THOSE WHERE INTEREST ARTICULATION BY NON-ASSOCIATIONAL GROUPS IS SIGNIFICANT OR MODERATE (83)	0.93	10 71 27 5 X SQ= 50.82 P = 0. RV YES (YES)
123	TEND LESS TO BE THOSE WHERE INTEREST ARTICULATION BY NON-ASSOCIATIONAL GROUPS IS SIGNIFICANT, MODERATE, OR LIMITED (107)	0.78	123	ALWAYS ARE THOSE WHERE INTEREST ARTICULATION BY NON-ASSOCIATIONAL GROUPS IS SIGNIFICANT, MODERATE, OR LIMITED (107)	1.00	29 76 8 0 X SQ= 14.55 P = 0. RV NO (YES)
125	TEND TO BE THOSE WHERE INTEREST ARTICULATION BY ANOMIC GROUPS IS INFREQUENT OR VERY INFREQUENT (35)	0.82	125	TEND TO BE THOSE WHERE INTEREST ARTICULATION BY ANOMIC GROUPS IS FREQUENT OR OCCASIONAL (64)	0.83	5 59 23 12 X SQ= 34.60 P = 0. RV YES (YES)

No.	Statement	Value	No.	Statement	Value	Statistics
126	TEND TO BE THOSE WHERE INTEREST ARTICULATION BY ANOMIC GROUPS IS VERY INFREQUENT (16)	0.57	126	ALWAYS ARE THOSE WHERE INTEREST ARTICULATION BY ANOMIC GROUPS IS FREQUENT, OCCASIONAL, OR INFREQUENT (83)	1.00	12 71 16 0 X SQ= 44.27 P = 0. RV YES (YES)
128	TEND TO BE THOSE WHERE INTEREST ARTICULATION BY POLITICAL PARTIES IS SIGNIFICANT OR MODERATE (48)	0.68	128	TEND TO BE THOSE WHERE INTEREST ARTICULATION BY POLITICAL PARTIES IS LIMITED OR NEGLIGIBLE (45)	0.56	23 25 11 32 X SQ= 3.93 P = 0.032 RV YES (NO)
133	TEND MORE TO BE THOSE WHERE INTEREST AGGREGATION BY THE EXECUTIVE IS MODERATE, LIMITED, OR NEGLIGIBLE (76)	0.94	133	TEND LESS TO BE THOSE WHERE INTEREST AGGREGATION BY THE EXECUTIVE IS MODERATE, LIMITED, OR NEGLIGIBLE (76)	0.60	2 27 33 41 X SQ= 11.57 P = 0. RV NO (NO)
135	LEAN MORE TOWARD BEING THOSE WHERE INTEREST AGGREGATION BY THE EXECUTIVE IS SIGNIFICANT, MODERATE, OR LIMITED (77)	0.96	135	LEAN LESS TOWARD BEING THOSE WHERE INTEREST AGGREGATION BY THE EXECUTIVE IS SIGNIFICANT, MODERATE, OR LIMITED (77)	0.81	27 50 1 12 X SQ= 2.72 P = 0.057 RV NO (NO)
136	TEND LESS TO BE THOSE WHERE INTEREST AGGREGATION BY THE LEGISLATURE IS MODERATE, LIMITED, OR NEGLIGIBLE (85)	0.66	136	ALWAYS ARE THOSE WHERE INTEREST AGGREGATION BY THE LEGISLATURE IS MODERATE, LIMITED, OR NEGLIGIBLE (85)	1.00	12 0 23 60 X SQ= 20.54 P = 0. RV NO (YES)
137	TEND TO BE THOSE WHERE INTEREST AGGREGATION BY THE LEGISLATURE IS SIGNIFICANT OR MODERATE (29)	0.66	137	TEND TO BE THOSE WHERE INTEREST AGGREGATION BY THE LEGISLATURE IS LIMITED OR NEGLIGIBLE (68)	0.90	23 6 12 54 X SQ= 29.78 P = 0. RV YES (YES)
138	TEND TO BE THOSE WHERE INTEREST AGGREGATION BY THE LEGISLATURE IS SIGNIFICANT, MODERATE, OR LIMITED (48)	0.78	138	TEND TO BE THOSE WHERE INTEREST AGGREGATION BY THE LEGISLATURE IS NEGLIGIBLE (49)	0.66	28 20 8 39 X SQ= 15.51 P = 0. RV YES (YES)
140	TILT MORE TOWARD BEING THOSE WHERE THE PARTY SYSTEM IS QUANTITATIVELY OTHER THAN ONE PARTY DOMINANT (87)	0.95	140	TILT LESS TOWARD BEING THOSE WHERE THE PARTY SYSTEM IS QUANTITATIVELY OTHER THAN ONE PARTY DOMINANT (87)	0.82	2 11 35 50 X SQ= 2.19 P = 0.122 RV NO (NO)
142	TEND LESS TO BE THOSE WHERE THE PARTY SYSTEM IS QUANTITATIVELY OTHER THAN MULTI-PARTY (66)	0.53	142	TEND MORE TO BE THOSE WHERE THE PARTY SYSTEM IS QUANTITATIVELY OTHER THAN MULTI-PARTY (66)	0.78	17 13 19 45 X SQ= 5.20 P = 0.022 RV NO (YES)

146	ALWAYS ARE THOSE WHERE THE PARTY SYSTEM IS QUALITATIVELY OTHER THAN MASS-BASED TERRITORIAL (92)	1.00	146	TEND LESS TO BE THOSE WHERE THE PARTY SYSTEM IS QUALITATIVELY OTHER THAN MASS-BASED TERRITORIAL (92)	0.87	0 8 34 56 X SQ= 3.11 P = 0.048 RV NO (NO)
147	TEND TO BE THOSE WHERE THE PARTY SYSTEM IS QUALITATIVELY CLASS-ORIENTED OR MULTI-IDEOLOGICAL (23)	0.58	147	TEND TO BE THOSE WHERE THE PARTY SYSTEM IS QUALITATIVELY OTHER THAN CLASS-ORIENTED OR MULTI-IDEOLOGICAL (67)	0.93	M 19 4 14 51 X SQ= 24.49 P = 0. RV YES (YES)
148	ALWAYS ARE THOSE WHERE THE PARTY SYSTEM IS QUALITATIVELY OTHER THAN AFRICAN TRANSITIONAL (96)	1.00	148	TEND LESS TO BE THOSE WHERE THE PARTY SYSTEM IS QUALITATIVELY OTHER THAN AFRICAN TRANSITIONAL (96)	0.80	0 14 37 57 X SQ= 6.73 P = 0.002 RV NO (NO)
153	TEND TO BE THOSE WHERE THE PARTY SYSTEM IS STABLE (42)	0.79	153	TEND TO BE THOSE WHERE THE PARTY SYSTEM IS MODERATELY STABLE OR UNSTABLE (38)	0.70	27 13 7 31 X SQ= 17.15 P = 0. RV YES (YES)
158	TEND TO BE THOSE WHERE PERSONALISMO IS NEGLIGIBLE (56)	0.78	158	TEND TO BE THOSE WHERE PERSONALISMO IS PRONOUNCED OR MODERATE (40)	0.55	8 32 28 26 X SQ= 8.56 P = 0.002 RV YES (YES)
161	LEAN TOWARD BEING THOSE WHERE THE POLITICAL LEADERSHIP IS NON-ELITIST (50)	0.66	161	LEAN TOWARD BEING THOSE WHERE THE POLITICAL LEADERSHIP IS ELITIST OR MODERATELY ELITIST (47)	0.55	12 33 23 27 X SQ= 3.02 P = 0.058 RV YES (NO)
164	TEND MORE TO BE THOSE WHERE THE REGIME'S LEADERSHIP CHARISMA IS NEGLIGIBLE (65)	0.94	164	TEND LESS TO BE THOSE WHERE THE REGIME'S LEADERSHIP CHARISMA IS NEGLIGIBLE (65)	0.52	2 30 33 32 X SQ= 16.55 P = 0. RV NO (YES)
168	TEND TO BE THOSE WHERE THE HORIZONTAL POWER DISTRIBUTION IS SIGNIFICANT (34)	0.69	168	TEND TO BE THOSE WHERE THE HORIZONTAL POWER DISTRIBUTION IS LIMITED OR NEGLIGIBLE (72)	0.87	25 9 11 59 X SQ= 31.29 P = 0. RV YES (YES)
169	TEND TO BE THOSE WHERE THE HORIZONTAL POWER DISTRIBUTION IS SIGNIFICANT OR LIMITED (58)	0.75	169	TEND TO BE THOSE WHERE THE HORIZONTAL POWER DISTRIBUTION IS NEGLIGIBLE (48)	0.54	27 31 9 37 X SQ= 7.11 P = 0.007 RV YES (NO)

170	TEND TO BE THOSE WHERE THE LEGISLATIVE-EXECUTIVE STRUCTURE IS OTHER THAN PRESIDENTIAL (63)	0.83	170	TEND TO BE THOSE WHERE THE LEGISLATIVE-EXECUTIVE STRUCTURE IS PRESIDENTIAL (39)	0.52	6 33 30 31 X SQ= 10.37 P = 0.001 RV YES (NO)
171	TILT LESS TOWARD BEING THOSE WHERE THE LEGISLATIVE-EXECUTIVE STRUCTURE IS OTHER THAN PARLIAMENTARY-REPUBLICAN (90)	0.81	171	TILT MORE TOWARD BEING THOSE WHERE THE LEGISLATIVE-EXECUTIVE STRUCTURE IS OTHER THAN PARLIAMENTARY-REPUBLICAN (90)	0.92	7 5 29 59 X SQ= 1.95 P = 0.112 RV NO (YES)
172	TEND LESS TO BE THOSE WHERE THE LEGISLATIVE-EXECUTIVE STRUCTURE IS OTHER THAN PARLIAMENTARY-ROYALIST (88)	0.68	172	TEND MORE TO BE THOSE WHERE THE LEGISLATIVE-EXECUTIVE STRUCTURE IS OTHER THAN PARLIAMENTARY-ROYALIST (88)	0.87	12 9 25 61 X SQ= 4.70 P = 0.021 RV NO (YES)
174	TEND TO BE THOSE WHERE THE LEGISLATURE IS FULLY EFFECTIVE (28)	0.61	174	TEND TO BE THOSE WHERE THE LEGISLATURE IS PARTIALLY EFFECTIVE, LARGELY INEFFECTIVE, OR WHOLLY INEFFECTIVE (72)	0.90	22 6 14 56 X SQ= 27.06 P = 0. RV YES (YES)
175	TEND TO BE THOSE WHERE THE LEGISLATURE IS FULLY EFFECTIVE OR PARTIALLY EFFECTIVE (51)	0.78	175	TEND TO BE THOSE WHERE THE LEGISLATURE IS LARGELY INEFFECTIVE OR WHOLLY INEFFECTIVE (49)	0.63	28 23 8 39 X SQ= 13.52 P = 0. RV YES (YES)
179	TEND TO BE THOSE WHERE THE EXECUTIVE IS STRONG (39)	0.69	179	TEND TO BE THOSE WHERE THE EXECUTIVE IS DOMINANT (52)	0.70	10 40 22 17 X SQ= 11.08 P = 0.001 RV YES (YES)
181	TEND TO BE THOSE WHERE THE BUREAUCRACY IS MODERN, RATHER THAN SEMI-MODERN (21)	0.54	181	TEND TO BE THOSE WHERE THE BUREAUCRACY IS SEMI-MODERN, RATHER THAN MODERN (55)	0.97	20 1 17 36 X SQ= 21.54 P = 0. RV YES (YES)
182	ALWAYS ARE THOSE WHERE THE BUREAUCRACY IS SEMI-MODERN, RATHER THAN POST-COLONIAL TRANSITIONAL (55)	1.00	182	TEND LESS TO BE THOSE WHERE THE BUREAUCRACY IS SEMI-MODERN, RATHER THAN POST-COLONIAL TRANSITIONAL (55)	0.59	17 36 0 25 X SQ= 8.46 P = 0.001 RV NO (NO)
187	TEND TO BE THOSE WHERE THE ROLE OF THE POLICE IS NOT POLITICALLY SIGNIFICANT (35)	0.70	187	TEND TO BE THOSE WHERE THE ROLE OF THE POLICE IS POLITICALLY SIGNIFICANT (66)	0.82	10 54 23 12 X SQ= 23.34 P = 0. RV YES (YES)

M

MATRIX

	54 POLITIES WHERE NEWSPAPER CIRCULATION IS ONE HUNDRED OR MORE PER THOUSAND (37)	54 POLITIES WHERE NEWSPAPER CIRCULATION IS LESS THAN ONE HUNDRED PER THOUSAND (76)
111 POLITIES WHERE POLITICAL ENCULTURATION IS HIGH OR MEDIUM (53)	AUSTRALIA, AUSTRIA, CANADA, COSTA RICA, DENMARK, FINLAND, FRANCE, GERMAN FR, GREECE, ICELAND, IRELAND, ISRAEL, ITALY, JAPAN, LUXEMBOURG, NETHERLANDS, NEW ZEALAND, NORWAY, POLAND, SWEDEN, UK, US (22)	AFGHANISTAN, ALGERIA, BURMA, CAMBODIA, CAMEROUN, CEYLON, COLOMBIA, DOMIN REP, GHANA, GUINEA, HONDURAS, INDIA, IVORY COAST, JAMAICA, LIBYA, MALAGASY R, MAURITANIA, MEXICO, MOROCCO, PHILIPPINES, SENEGAL, SIERRE LEO, SOMALIA, SUDAN, TANGANYIKA, THAILAND, TOGO, TRINIDAD, TUNISIA, TURKEY, VENEZUELA (31)
111 POLITIES WHERE POLITICAL ENCULTURATION IS LOW (42)	ARGENTINA, BELGIUM, CHILE, CYPRUS, PANAMA (5)	BOLIVIA, BRAZIL, CEN AFR REP, CHAD, CONGO(BRA), CONGO(LEO), DAHOMEY, ECUADOR, EL SALVADOR, ETHIOPIA, GABON, GUATEMALA, HAITI, INDONESIA, IRAN, IRAQ, JORDAN, KOREA REP, LAOS, LEBANON, LIBERIA, MALAYA, NEPAL, NICARAGUA, NIGER, NIGERIA, PAKISTAN, PERU, SA'U ARABIA, SO AFRICA, SPAIN, SYRIA, UGANDA, UPPER VOLTA, VIETNAM REP, YEMEN, YUGOSLAVIA (37)

MATRIX

54 POLITIES WHERE NEWSPAPER CIRCULATION IS ONE HUNDRED OR MORE PER THOUSAND (37)	54 POLITIES WHERE NEWSPAPER CIRCULATION IS LESS THAN ONE HUNDRED PER THOUSAND (76)
119 POLITIES WHERE INTEREST ARTICULATION BY INSTITUTIONAL GROUPS IS VERY SIGNIFICANT OR SIGNIFICANT (74)	
ARGENTINA, GERMANY, E, MONGOLIA, BULGARIA, HUNGARY, PANAMA, CUBA, ISRAEL, POLAND, CYPRUS, ITALY, RUMANIA, CZECHOS'KIA, JAPAN, USSR	AFGHANISTAN, BURMA, CONGO(LEO), GHANA, INDIA, JORDAN, LIBYA, NEPAL, PERU, SPAIN, TUNISIA, VIETNAM REP, ALBANIA, CAMBODIA, DOMIN REP, GUATEMALA, INDONESIA, KOREA REP, MALAYA, NICARAGUA, PORTUGAL, SUDAN, TURKEY, YUGOSLAVIA, ALGERIA, CEYLON, ECUADOR, GUINEA, IRAN, LAOS, MAURITANIA, NIGERIA, SA'U ARABIA, SYRIA, UGANDA, BOLIVIA, CHINA, PR, EL SALVADOR, HAITI, IRAQ, LEBANON, MEXICO, PAKISTAN, SENEGAL, TANGANYIKA, UAR, BRAZIL, COLOMBIA, ETHIOPIA, HONDURAS, IVORY COAST, LIBERIA, MOROCCO, PARAGUAY, SO AFRICA, THAILAND, VENEZUELA
15	57
22	4
119 POLITIES WHERE INTEREST ARTICULATION BY INSTITUTIONAL GROUPS IS MODERATE OR LIMITED (26)	
AUSTRALIA, COSTA RICA, GREECE, NEW ZEALAND, US, AUSTRIA, DENMARK, ICELAND, NORWAY, URUGUAY, BELGIUM, FINLAND, IRELAND, SWEDEN, CANADA, FRANCE, LUXEMBOURG, SWITZERLAND, CHILE, GERMAN FR, NETHERLANDS, UK	JAMAICA, MALAGASY R, PHILIPPINES, TRINIDAD

MATRIX

54 POLITIES
WHERE NEWSPAPER CIRCULATION IS
ONE HUNDRED OR MORE
PER THOUSAND (37)

54 POLITIES
WHERE NEWSPAPER CIRCULATION IS
LESS THAN ONE HUNDRED
PER THOUSAND (76)

147 POLITIES
WHERE THE PARTY SYSTEM IS QUALITATIVELY
CLASS-ORIENTED OR MULTI-IDEOLOGICAL (23)

ARGENTINA, AUSTRALIA, AUSTRIA, BELGIUM, CHILE, DENMARK, FINLAND, GERMAN FR, ICELAND, ISRAEL, ITALY, JAPAN, LUXEMBOURG, NETHERLANDS, NEW ZEALAND, NORWAY, SWEDEN, SWITZERLAND, UK

BRAZIL, JAMAICA, PERU, TRINIDAD

	≥100 per thousand	<100 per thousand
Class-oriented or multi-ideological	19	4
Other	14	51

147 POLITIES
WHERE THE PARTY SYSTEM IS QUALITATIVELY
OTHER THAN
CLASS-ORIENTED OR MULTI-IDEOLOGICAL (67)

BULGARIA, CANADA, COSTA RICA, CUBA, CYPRUS, CZECHOS'KIA, GERMANY, E, HUNGARY, MONGOLIA, PANAMA, POLAND, RUMANIA, USSR, US

ALBANIA, ALGERIA, BOLIVIA, BURUNDI, CAMBODIA, CAMEROUN, CEN AFR REP, CHAD, CHINA, PR, COLOMBIA, CONGO(BRA), CONGO(LEO), DAHOMEY, DOMIN REP, ECUADOR, EL SALVADOR, GABON, GHANA, GUINEA, HONDURAS, IVORY COAST, LAOS, LEBANON, MALAGASY R, MALAYA, MALI, MAURITANIA, MEXICO, NEPAL, NICARAGUA, NIGER, NIGERIA, PARAGUAY, PHILIPPINES, PORTUGAL, RWANDA, SENEGAL, SIERRE LEO, SOMALIA, SPAIN, SUDAN, TANGANYIKA, THAILAND, TOGO, TUNISIA, UGANDA, UAR, UPPER VOLTA, VENEZUELA, VIETNAM REP, YUGOSLAVIA

MATRIX

54 POLITIES WHERE NEWSPAPER CIRCULATION IS ONE HUNDRED OR MORE PER THOUSAND (37)

54 POLITIES WHERE NEWSPAPER CIRCULATION IS LESS THAN ONE HUNDRED PER THOUSAND (76)

187 POLITIES WHERE THE ROLE OF THE POLICE IS POLITICALLY SIGNIFICANT (66)

ARGENTINA HUNGARY BULGARIA MONGOLIA CUBA POLAND CZECHOS'KIA RUMANIA GERMANY, E USSR	AFGHANISTAN CAMBODIA CONGO(BRA) ETHIOPIA HAITI IVORY COAST LIBERIA PAKISTAN SOMALIA THAILAND VENEZUELA ALBANIA CAMEROUN CONGO(LEO) GABON HONDURAS JORDAN MOROCCO PARAGUAY SO AFRICA TOGO VIETNAM REP ALGERIA CEN AFR REP DAHOMEY GHANA INDONESIA KOREA REP NEPAL PORTUGAL SPAIN TUNISIA YEMEN BOLIVIA CHINA, PR ECUADOR GUATEMALA IRAN LAOS NICARAGUA SA'U ARABIA SUDAN TURKEY YUGOSLAVIA BURMA COLOMBIA EL SALVADOR GUINEA IRAQ LEBANON NIGERIA SENEGAL SYRIA UAR
10	54
23	12
AUSTRALIA FINLAND IRELAND NETHERLANDS UK AUSTRIA FRANCE ISRAEL NEW ZEALAND US BELGIUM GERMAN FR ITALY NORWAY URUGUAY CANADA GREECE JAPAN SWEDEN DENMARK ICELAND LUXEMBOURG SWITZERLAND	CEYLON MAURITANIA TRINIDAD INDIA MEXICO UGANDA JAMAICA PHILIPPINES MALAGASY R SIERRE LEO MALAYA TANGANYIKA

187 POLITIES WHERE THE ROLE OF THE POLICE IS NOT POLITICALLY SIGNIFICANT (35)

189 TEND LESS TO BE THOSE 0.86
WHERE THE CHARACTER OF THE LEGAL SYSTEM
IS OTHER THAN COMMON LAW (108)

192 ALWAYS ARE THOSE 1.00
WHERE THE CHARACTER OF THE LEGAL SYSTEM
IS OTHER THAN
MUSLIM OR PARTLY MUSLIM (86)

193 ALWAYS ARE THOSE 1.00
WHERE THE CHARACTER OF THE LEGAL SYSTEM
IS OTHER THAN PARTLY INDIGENOUS (86)

194 TEND LESS TO BE THOSE 0.78
THAT ARE NON-COMMUNIST (101)

189 TEND MORE TO BE THOSE 0.97
WHERE THE CHARACTER OF THE LEGAL SYSTEM
IS OTHER THAN COMMON LAW (108)

5	2
32	74
X SQ=	3.37
P =	0.037
RV NO	(YES)

192 TEND LESS TO BE THOSE 0.64
WHERE THE CHARACTER OF THE LEGAL SYSTEM
IS OTHER THAN
MUSLIM OR PARTLY MUSLIM (86)

0	27
37	47
X SQ=	15.91
P =	0.
RV NO	(NO)

193 TEND LESS TO BE THOSE 0.63
WHERE THE CHARACTER OF THE LEGAL SYSTEM
IS OTHER THAN PARTLY INDIGENOUS (86)

0	28
37	47
X SQ=	16.48
P =	0.
RV NO	(NO)

194 TEND MORE TO BE THOSE 0.96
THAT ARE NON-COMMUNIST (101)

8	3
28	73
X SQ=	7.26
P =	0.005
RV NO	(YES)

55 POLITIES
WHERE NEWSPAPER CIRCULATION IS
TEN OR MORE
PER THOUSAND (78)

55 POLITIES
WHERE NEWSPAPER CIRCULATION IS
LESS THAN TEN
PER THOUSAND (35)

PREDICATE ONLY

78 IN LEFT COLUMN

ALBANIA	ALGERIA	ARGENTINA	AUSTRALIA	AUSTRIA	BELGIUM	BOLIVIA	BRAZIL	BULGARIA	BURMA
CANADA	CEYLON	CHILE	CHINA, PR	COLOMBIA	COSTA RICA	CUBA	CYPRUS	CZECHOS'KIA	DENMARK
DOMIN REP	ECUADOR	EL SALVADOR	FINLAND	FRANCE	GERMANY, E	GERMAN FR	GHANA	GREECE	GUATEMALA
HAITI	HONDURAS	HUNGARY	ICELAND	INDIA	INDONESIA	IRAQ	IRELAND	ISRAEL	ITALY
JAMAICA	JAPAN	JORDAN	KOREA REP	LEBANON	LUXEMBOURG	MALAYA	MEXICO	MONGOLIA	MOROCCO
NETHERLANDS	NEW ZEALAND	NICARAGUA	NORWAY	PANAMA	PARAGUAY	PERU	PHILIPPINES	POLAND	PORTUGAL
RUMANIA	SO AFRICA	SPAIN	SWEDEN	SWITZERLAND	SYRIA	THAILAND	TRINIDAD	TUNISIA	TURKEY
USSR	UAR	UK	US	URUGUAY	VENEZUELA	VIETNAM REP	YUGOSLAVIA		

35 IN RIGHT COLUMN

AFGHANISTAN	BURUNDI	CAMBODIA	CAMEROUN	CEN AFR REP	CHAD	CONGO(BRA)	CONGO(LEO)	DAHOMEY	ETHIOPIA
GABON	GUINEA	IRAN	IVORY COAST	LAOS	LIBERIA	LIBYA	MALAGASY R	MALI	MAURITANIA
NEPAL	NIGER	NIGERIA	PAKISTAN	RWANDA	SA'U ARABIA	SENEGAL	SIERRE LEO	SOMALIA	SUDAN
TANGANYIKA	TOGO	UGANDA	UPPER VOLTA	YEMEN					

2 EXCLUDED BECAUSE UNASCERTAINED

KOREA, N   VIETNAM, N

****

56 POLITIES WHERE THE RELIGION IS PREDOMINANTLY LITERATE (79)	56 POLITIES WHERE THE RELIGION IS PREDOMINANTLY OR PARTLY NON-LITERATE (31)

BOTH SUBJECT AND PREDICATE

....

79 IN LEFT COLUMN

AFGHANISTAN	ALBANIA	ALGERIA	ARGENTINA	AUSTRALIA	AUSTRIA	BELGIUM	BRAZIL	BULGARIA	BURMA
CAMBODIA	CANADA	CEYLON	CHILE	COLOMBIA	COSTA RICA	CUBA	CYPRUS	CZECHOS'KIA	DENMARK
DOMIN REP	EL SALVADOR	FINLAND	FRANCE	GERMANY, E	GERMAN FR	GREECE	HONDURAS	HUNGARY	ICELAND
INDIA	INDONESIA	IRAN	IRAQ	IRELAND	ISRAEL	ITALY	JAMAICA	JAPAN	JORDAN
LEBANON	LIBYA	LUXEMBOURG	MALAYA	MAURITANIA	MEXICO	MONGOLIA	MOROCCO	NEPAL	NETHERLANDS
NEW ZEALAND	NICARAGUA	NORWAY	PAKISTAN	PANAMA	PARAGUAY	PHILIPPINES	POLAND	PORTUGAL	RUMANIA
SA'U ARABIA	SENEGAL	SOMALIA	SPAIN	SWEDEN	SWITZERLAND	SYRIA	THAILAND	TRINIDAD	TUNISIA
TURKEY	USSR	UAR	UK	US	URUGUAY	VENEZUELA	YEMEN	YUGOSLAVIA	

31 IN RIGHT COLUMN

BOLIVIA	BURUNDI	CAMEROUN	CEN AFR REP	CHAD	CONGO(BRA)	CONGO(LEO)	DAHOMEY	ECUADOR	ETHIOPIA
GABON	GHANA	GUATEMALA	GUINEA	HAITI	IVORY COAST	LAOS	LIBERIA	MALAGASY R	MALI
NIGER	NIGERIA	PERU	RWANDA	SIERRE LEO	SO AFRICA	SUDAN	TANGANYIKA	TOGO	UGANDA
UPPER VOLTA									

5 EXCLUDED BECAUSE AMBIGUOUS

CHINA, PR   KOREA, N   KOREA REP   VIETNAM, N   VIETNAM REP

	Left column			Right column		
2	TEND LESS TO BE THOSE LOCATED ELSEWHERE THAN IN WEST EUROPE, SCANDINAVIA, NORTH AMERICA, OR AUSTRALASIA (93)	0.72	2	ALWAYS ARE THOSE LOCATED ELSEWHERE THAN IN WEST EUROPE, SCANDINAVIA, NORTH AMERICA, OR AUSTRALASIA (93)	1.00	22 0 57 31 X SQ= 9.12 P = 0. RV NO (NO)
8	TILT LESS TOWARD BEING THOSE LOCATED ELSEWHERE THAN IN EAST ASIA SOUTH ASIA, OR SOUTHEAST ASIA (97)	0.85	8	TILT MORE TOWARD BEING THOSE LOCATED ELSEWHERE THAN IN EAST ASIA SOUTH ASIA, OR SOUTHEAST ASIA (97)	0.97	12 1 67 30 X SQ= 2.02 P = 0.105 RV NO (NO)
10	TEND TO BE THOSE LOCATED ELSEWHERE THAN IN NORTH AFRICA, OR CENTRAL AND SOUTH AFRICA (82)	0.90	10	TEND TO BE THOSE LOCATED IN NORTH AFRICA, OR CENTRAL AND SOUTH AFRICA (33)	0.81	8 25 71 6 X SQ= 49.42 P = 0. RV YES (YES)

13	TEND LESS TO BE THOSE LOCATED ELSEWHERE THAN IN NORTH AFRICA OR THE MIDDLE EAST (99)	0.80	13	ALWAYS ARE THOSE LOCATED ELSEWHERE THAN IN NORTH AFRICA OR THE MIDDLE EAST (99)	1.00	16 0 63 31 X SQ= 5.81 P = 0.005 RV NO (NO)
15	TEND TO BE THOSE LOCATED IN THE MIDDLE EAST, RATHER THAN IN NORTH AFRICA OR CENTRAL AND SOUTH AFRICA (11)	0.58	15	ALWAYS ARE THOSE LOCATED IN NORTH AFRICA OR CENTRAL AND SOUTH AFRICA, RATHER THAN IN THE MIDDLE EAST (33)	1.00	11 0 8 25 X SQ= 16.33 P = 0. RV YES (YES)
17	TILT LESS TOWARD BEING THOSE LOCATED IN THE CARIBBEAN, CENTRAL AMERICA, OR SOUTH AMERICA, RATHER THAN IN THE MIDDLE EAST (22)	0.61	17	ALWAYS ARE THOSE LOCATED IN THE CARIBBEAN, CENTRAL AMERICA, OR SOUTH AMERICA, RATHER THAN IN THE MIDDLE EAST (22)	1.00	11 0 17 5 X SQ= 1.44 P = 0.143 RV NO (NO)
18	TEND TO BE THOSE LOCATED IN EAST ASIA, SOUTH ASIA, OR SOUTHEAST ASIA, RATHER THAN IN NORTH AFRICA OR CENTRAL AND SOUTH AFRICA (18)	0.60	18	TEND TO BE THOSE LOCATED IN NORTH AFRICA OR CENTRAL AND SOUTH AFRICA, RATHER THAN IN EAST ASIA, SOUTH ASIA, OR SOUTHEAST ASIA (33)	0.96	8 25 12 1 X SQ= 14.92 P = 0. RV YES (YES)
19	TEND TO BE THOSE LOCATED IN THE CARIBBEAN, CENTRAL AMERICA, OR SOUTH AMERICA, RATHER THAN IN NORTH AFRICA OR CENTRAL AND SOUTH AFRICA (22)	0.68	19	TEND TO BE THOSE LOCATED IN NORTH AFRICA OR CENTRAL AND SOUTH AFRICA, RATHER THAN IN THE CARIBBEAN, CENTRAL AMERICA, OR SOUTH AMERICA (33)	0.83	8 25 17 5 X SQ= 12.91 P = 0. RV YES (YES)
22	LEAN LESS TOWARD BEING THOSE WHOSE TERRITORIAL SIZE IS VERY LARGE, LARGE, OR MEDIUM (68)	0.56	22	LEAN MORE TOWARD BEING THOSE WHOSE TERRITORIAL SIZE IS VERY LARGE, LARGE, OR MEDIUM (68)	0.74	44 23 35 8 X SQ= 2.47 P = 0.086 RV NO (NO)
24	TEND TO BE THOSE WHOSE POPULATION IS VERY LARGE, LARGE, OR MEDIUM (61)	0.59	24	TEND TO BE THOSE WHOSE POPULATION IS SMALL (54)	0.71	47 9 32 22 X SQ= 7.09 P = 0.006 RV YES (NO)
25	TEND LESS TO BE THOSE WHOSE POPULATION DENSITY IS MEDIUM OR LOW (98)	0.81	25	TEND MORE TO BE THOSE WHOSE POPULATION DENSITY IS MEDIUM OR LOW (98)	0.97	15 1 64 30 X SQ= 3.27 P = 0.037 RV NO (NO)
26	TEND LESS TO BE THOSE WHOSE POPULATION DENSITY IS LOW (67)	0.51	26	TEND MORE TO BE THOSE WHOSE POPULATION DENSITY IS LOW (67)	0.87	39 4 40 27 X SQ= 10.95 P = 0. RV NO (NO)

29	TEND TO BE THOSE WHERE THE DEGREE OF URBANIZATION IS HIGH (56)	0.71	29	TEND TO BE THOSE WHERE THE DEGREE OF URBANIZATION IS LOW (49)	0.90	52 3 21 28 X SQ= 30.67 P = 0. RV YES (YES)
30	TEND TO BE THOSE WHOSE AGRICULTURAL POPULATION IS MEDIUM, LOW, OR VERY LOW (57)	0.68	30	TEND TO BE THOSE WHOSE AGRICULTURAL POPULATION IS HIGH (56)	0.87	25 27 53 4 X SQ= 24.78 P = 0. RV YES (YES)
32	LEAN LESS TOWARD BEING THOSE WHOSE GROSS NATIONAL PRODUCT IS MEDIUM, LOW, OR VERY LOW (105)	0.89	32	ALWAYS ARE THOSE WHOSE GROSS NATIONAL PRODUCT IS MEDIUM, LOW, OR VERY LOW (105)	1.00	9 0 70 31 X SQ= 2.48 P = 0.059 RV NO (NO)
33	TEND LESS TO BE THOSE WHOSE GROSS NATIONAL PRODUCT IS LOW OR VERY LOW (85)	0.65	33	TEND MORE TO BE THOSE WHOSE GROSS NATIONAL PRODUCT IS LOW OR VERY LOW (85)	0.97	28 1 51 30 X SQ= 10.30 P = 0. RV NO (NO)
34	TEND TO BE THOSE WHOSE GROSS NATIONAL PRODUCT IS VERY HIGH, HIGH, MEDIUM, OR LOW (62)	0.67	34	TEND TO BE THOSE WHOSE GROSS NATIONAL PRODUCT IS VERY LOW (53)	0.84	53 5 26 26 X SQ= 21.20 P = 0. RV YES (YES)
35	TEND LESS TO BE THOSE WHOSE PER CAPITA GROSS NATIONAL PRODUCT IS MEDIUM, LOW, OR VERY LOW (91)	0.70	35	ALWAYS ARE THOSE WHOSE PER CAPITA GROSS NATIONAL PRODUCT IS MEDIUM, LOW, OR VERY LOW (91)	1.00	24 0 55 31 X SQ= 10.33 P = 0. RV NO (NO)
36	TEND TO BE THOSE WHOSE PER CAPITA GROSS NATIONAL PRODUCT IS VERY HIGH, HIGH, OR MEDIUM (42)	0.52	36	TEND TO BE THOSE WHOSE PER CAPITA GROSS NATIONAL PRODUCT IS LOW OR VERY LOW (73)	0.97	41 1 38 30 X SQ= 20.33 P = 0. RV YES (NO)
37	TEND TO BE THOSE WHOSE PER CAPITA GROSS NATIONAL PRODUCT IS VERY HIGH, HIGH, MEDIUM, OR LOW (64)	0.76	37	TEND TO BE THOSE WHOSE PER CAPITA GROSS NATIONAL PRODUCT IS VERY LOW (51)	0.87	60 4 19 27 X SQ= 33.83 P = 0. RV YES (YES)
38	TEND LESS TO BE THOSE WHOSE INTERNATIONAL FINANCIAL STATUS IS MEDIUM, LOW, OR VERY LOW (103)	0.87	38	ALWAYS ARE THOSE WHOSE INTERNATIONAL FINANCIAL STATUS IS MEDIUM, LOW, OR VERY LOW (103)	1.00	10 0 68 31 X SQ= 2.97 P = 0.035 RV NO (NO)

```
39  TEND LESS TO BE THOSE                  0.54     39  TEND MORE TO BE THOSE                  0.97        36    1
    WHOSE INTERNATIONAL FINANCIAL STATUS                WHOSE INTERNATIONAL FINANCIAL STATUS               42   30
    IS LOW OR VERY LOW  (76)                            IS LOW OR VERY LOW  (76)                       X SQ=  16.37
                                                                                                       P =    0.
                                                                                                       RV NO  (NO )

40  TEND TO BE THOSE                       0.78     40  TEND TO BE THOSE                       0.71        59    9
    WHOSE INTERNATIONAL FINANCIAL STATUS                WHOSE INTERNATIONAL FINANCIAL STATUS               17   22
    IS VERY HIGH, HIGH,                                 IS VERY LOW  (39)                              X SQ=  20.40
    MEDIUM, OR LOW  (71)                                                                               P =    0.
                                                                                                       RV YES (YES)

41  TEND LESS TO BE THOSE                  0.75     41  ALWAYS ARE THOSE                       1.00        19    0
    WHOSE ECONOMIC DEVELOPMENTAL STATUS                 WHOSE ECONOMIC DEVELOPMENTAL STATUS                58   31
    IS INTERMEDIATE, UNDERDEVELOPED,                    IS INTERMEDIATE, UNDERDEVELOPED,               X SQ=   7.66
    OR VERY UNDERDEVELOPED  (94)                        OR VERY UNDERDEVELOPED  (94)                   P =    0.001
                                                                                                       RV NO  (NO )

42  TEND LESS TO BE THOSE                  0.54     42  TEND MORE TO BE THOSE                  0.97        35    1
    WHOSE ECONOMIC DEVELOPMENTAL STATUS                 WHOSE ECONOMIC DEVELOPMENTAL STATUS                41   30
    IS UNDERDEVELOPED OR                                IS UNDERDEVELOPED OR                           X SQ=  16.22
    VERY UNDERDEVELOPED  (76)                           VERY UNDERDEVELOPED  (76)                      P =    0.
                                                                                                       RV NO  (NO )

43  TEND TO BE THOSE                       0.67     43  TEND TO BE THOSE                       0.90        51    3
    WHOSE ECONOMIC DEVELOPMENTAL STATUS                 WHOSE ECONOMIC DEVELOPMENTAL STATUS                25   28
    IS DEVELOPED, INTERMEDIATE, OR                      IS VERY UNDERDEVELOPED  (57)                   X SQ=  26.80
    UNDERDEVELOPED  (55)                                                                               P =    0.
                                                                                                       RV YES (YES)

45  TEND TO BE THOSE                       0.69     45  ALWAYS ARE THOSE                       1.00        53    0
    WHERE THE LITERACY RATE IS                          WHERE THE LITERACY RATE IS                         24   30
    FIFTY PERCENT OR ABOVE  (55)                        BELOW FIFTY PERCENT  (54)                      X SQ=  38.21
                                                                                                       P =    0.
                                                                                                       RV YES (YES)

46  TEND TO BE THOSE                       0.92     46  TEND TO BE THOSE                       0.67        70   10
    WHERE THE LITERACY RATE IS                          WHERE THE LITERACY RATE IS                          6   20
    TEN PERCENT OR ABOVE  (84)                          BELOW TEN PERCENT  (26)                        X SQ=  37.02
                                                                                                       P =    0.
                                                                                                       RV YES (YES)

52  TEND LESS TO BE THOSE                  0.82     52  ALWAYS ARE THOSE                       1.00        59   21
    WHERE FREEDOM OF THE PRESS IS                       WHERE FREEDOM OF THE PRESS IS                      13    0
    COMPLETE, INTERMITTENT, OR                          COMPLETE, INTERMITTENT, OR                     X SQ=   3.03
    INTERNALLY ABSENT  (82)                             INTERNALLY ABSENT  (82)                        P =    0.036
                                                                                                       RV NO  (NO )

54  TEND LESS TO BE THOSE                  0.53     54  ALWAYS ARE THOSE                       1.00        37    0
    WHERE NEWSPAPER CIRCULATION IS                      WHERE NEWSPAPER CIRCULATION IS                     42   31
    LESS THAN ONE HUNDRED                               LESS THAN ONE HUNDRED                          X SQ=  19.83
    PER THOUSAND  (76)                                  PER THOUSAND  (76)                             P =    0.
                                                                                                       RV NO  (NO )
```

55	TEND TO BE THOSE WHERE NEWSPAPER CIRCULATION IS TEN OR MORE PER THOUSAND (78)	0.86	55	TEND TO BE THOSE WHERE NEWSPAPER CIRCULATION IS LESS THAN TEN PER THOUSAND (35)	0.77	68 7 11 24 X SQ= 38.50 P = 0. RV YES (YES)
57	TEND LESS TO BE THOSE WHERE THE RELIGION IS OTHER THAN CATHOLIC (90)	0.68	57	ALWAYS ARE THOSE WHERE THE RELIGION IS OTHER THAN CATHOLIC (90)	1.00	25 0 54 31 X SQ= 10.96 P = 0. RV NO (NO)
58	TEND LESS TO BE THOSE WHERE THE RELIGION IS OTHER THAN MUSLIM (97)	0.77	58	ALWAYS ARE THOSE WHERE THE RELIGION IS OTHER THAN MUSLIM (97)	1.00	18 0 61 31 X SQ= 6.86 P = 0.002 RV NO (NO)
63	TEND TO BE THOSE WHERE THE RELIGION IS PREDOMINANTLY SOME KIND OF CHRISTIAN (46)	0.59	63	ALWAYS ARE THOSE WHERE THE RELIGION IS PREDOMINANTLY OR PARTLY OTHER THAN CHRISTIAN (68)	1.00	46 0 32 31 X SQ= 29.26 P = 0. RV YES (NO)
66	TEND TO BE THOSE THAT ARE RELIGIOUSLY HOMOGENEOUS (57)	0.73	66	ALWAYS ARE THOSE THAT ARE RELIGIOUSLY HETEROGENEOUS (49)	1.00	57 0 21 28 X SQ= 41.37 P = 0. RV YES (YES)
68	TEND TO BE THOSE THAT ARE LINGUISTICALLY HOMOGENEOUS (52)	0.58	68	TEND TO BE THOSE THAT ARE LINGUISTICALLY WEAKLY HETEROGENEOUS OR STRONGLY HETEROGENEOUS (62)	0.87	46 4 33 27 X SQ= 16.66 P = 0. RV YES (NO)
69	TEND TO BE THOSE THAT ARE LINGUISTICALLY HOMOGENEOUS OR WEAKLY HETEROGENEOUS (64)	0.71	69	TEND TO BE THOSE THAT ARE LINGUISTICALLY STRONGLY HETEROGENEOUS (50)	0.87	56 4 23 27 X SQ= 27.90 P = 0. RV YES (YES)
70	TEND LESS TO BE THOSE THAT ARE RELIGIOUSLY, RACIALLY, OR LINGUISTICALLY HETEROGENEOUS (85)	0.72	70	ALWAYS ARE THOSE THAT ARE RELIGIOUSLY, RACIALLY, OR LINGUISTICALLY HETEROGENEOUS (85)	1.00	21 0 54 28 X SQ= 8.20 P = 0.001 RV NO (NO)
72	TEND TO BE THOSE WHOSE DATE OF INDEPENDENCE IS BEFORE 1914 (52)	0.56	72	TEND TO BE THOSE WHOSE DATE OF INDEPENDENCE IS AFTER 1914 (59)	0.74	43 8 34 23 X SQ= 6.84 P = 0.006 RV YES (NO)

	Subject			Predicate		Statistics
73	TEND TO BE THOSE WHOSE DATE OF INDEPENDENCE IS BEFORE 1945 (65)	0.73	73	TEND TO BE THOSE WHOSE DATE OF INDEPENDENCE IS AFTER 1945 (46)	0.74	56 8 21 23 X SQ= 18.26 P = 0. RV YES (YES)
74	TEND LESS TO BE THOSE THAT ARE NOT HISTORICALLY WESTERN (87)	0.67	74	ALWAYS ARE THOSE THAT ARE NOT HISTORICALLY WESTERN (87)	1.00	26 0 52 30 X SQ= 11.41 P = 0. RV NO (NO)
75	TEND TO BE THOSE THAT ARE HISTORICALLY WESTERN OR SIGNIFICANTLY WESTERNIZED (62)	0.72	75	TEND TO BE THOSE THAT ARE NOT HISTORICALLY WESTERN AND ARE NOT SIGNIFICANTLY WESTERNIZED (52)	0.87	57 4 22 26 X SQ= 28.18 P = 0. RV YES (YES)
76	TEND LESS TO BE THOSE THAT HAVE BEEN SIGNIFICANTLY OR PARTIALLY WESTERNIZED THROUGH A COLONIAL RELATIONSHIP, RATHER THAN BEING HISTORICALLY WESTERN (70)	0.60	76	ALWAYS ARE THOSE THAT HAVE BEEN SIGNIFICANTLY OR PARTIALLY WESTERNIZED THROUGH A COLONIAL RELATIONSHIP, RATHER THAN BEING HISTORICALLY WESTERN (70)	1.00	26 0 39 29 X SQ= 14.10 P = 0. RV NO (NO)
77	TEND TO BE THOSE THAT HAVE BEEN SIGNIFICANTLY WESTERNIZED, RATHER THAN PARTIALLY WESTERNIZED, THROUGH A COLONIAL RELATIONSHIP (28)	0.62	77	TEND TO BE THOSE THAT HAVE BEEN PARTIALLY WESTERNIZED, RATHER THAN SIGNIFICANTLY WESTERNIZED, THROUGH A COLONIAL RELATIONSHIP (41)	0.86	M 24 4 15 24 X SQ= 13.08 P = 0. RV YES (YES)
79	LEAN LESS TOWARD BEING THOSE THAT WERE FORMERLY DEPENDENCIES OF BRITAIN OR FRANCE, RATHER THAN SPAIN (49)	0.64	79	LEAN MORE TOWARD BEING THOSE THAT WERE FORMERLY DEPENDENCIES OF BRITAIN OR FRANCE, RATHER THAN SPAIN (49)	0.85	25 22 14 4 X SQ= 2.33 P = 0.093 RV NO (NO)
80	LEAN TOWARD BEING THOSE WHOSE DATE OF INDEPENDENCE IS AFTER 1914, AND THAT WERE FORMERLY DEPENDENCIES OF BRITAIN, RATHER THAN FRANCE (19)	0.62	80	LEAN TOWARD BEING THOSE WHOSE DATE OF INDEPENDENCE IS AFTER 1914, AND THAT WERE FORMERLY DEPENDENCIES OF FRANCE, RATHER THAN BRITAIN (24)	0.70	13 6 8 14 X SQ= 3.01 P = 0.062 RV YES (YES)
82	TEND TO BE THOSE WHERE THE TYPE OF POLITICAL MODERNIZATION IS EARLY OR LATER EUROPEAN OR EUROPEAN DERIVED, RATHER THAN DEVELOPED TUTELARY OR UNDEVELOPED TUTELARY (51)	0.62	82	TEND TO BE THOSE WHERE THE TYPE OF POLITICAL MODERNIZATION IS DEVELOPED TUTELARY OR UNDEVELOPED TUTELARY, RATHER THAN EARLY OR LATER EUROPEAN OR EUROPEAN DERIVED (55)	0.80	45 6 27 24 X SQ= 13.65 P = 0. RV YES (NO)
84	TEND TO BE THOSE WHERE THE TYPE OF POLITICAL MODERNIZATION IS DEVELOPED TUTELARY, RATHER THAN UNDEVELOPED TUTELARY (31)	0.89	84	TEND TO BE THOSE WHERE THE TYPE OF POLITICAL MODERNIZATION IS UNDEVELOPED TUTELARY, RATHER THAN DEVELOPED TUTELARY (24)	0.87	M 24 3 3 21 X SQ= 26.77 P = 0. RV YES (YES)

85	TEND TO BE THOSE WHERE THE STAGE OF POLITICAL MODERNIZATION IS ADVANCED, RATHER THAN MID- OR EARLY TRANSITIONAL (60)	0.71	85	TEND TO BE THOSE WHERE THE STAGE OF POLITICAL MODERNIZATION IS MID- OR EARLY TRANSITIONAL, RATHER THAN ADVANCED (54)	0.94	55 2 23 29 X SQ= 33.97 P = 0. RV YES (YES)
86	TEND TO BE THOSE WHERE THE STAGE OF POLITICAL MODERNIZATION IS ADVANCED OR MID-TRANSITIONAL, RATHER THAN EARLY TRANSITIONAL (76)	0.83	86	TEND TO BE THOSE WHERE THE STAGE OF POLITICAL MODERNIZATION IS EARLY TRANSITIONAL, RATHER THAN ADVANCED OR MID-TRANSITIONAL (38)	0.74	65 8 13 23 X SQ= 30.64 P = 0. RV YES (YES)
87	TEND TO BE THOSE WHOSE IDEOLOGICAL ORIENTATION IS OTHER THAN DEVELOPMENTAL (58)	0.82	87	TEND TO BE THOSE WHOSE IDEOLOGICAL ORIENTATION IS DEVELOPMENTAL (31)	0.80	11 20 50 5 X SQ= 26.91 P = 0. RV YES (YES)
89	TEND LESS TO BE THOSE WHOSE IDEOLOGICAL ORIENTATION IS OTHER THAN CONVENTIONAL (62)	0.53	89	TEND MORE TO BE THOSE WHOSE IDEOLOGICAL ORIENTATION IS OTHER THAN CONVENTIONAL (62)	0.92	31 2 35 23 X SQ= 10.29 P = 0. RV NO (NO)
91	TILT LESS TOWARD BEING THOSE WHOSE IDEOLOGICAL ORIENTATION IS DEVELOPMENTAL, RATHER THAN TRADITIONAL (31)	0.73	91	TILT MORE TOWARD BEING THOSE WHOSE IDEOLOGICAL ORIENTATION IS DEVELOPMENTAL, RATHER THAN TRADITIONAL (31)	0.95	11 20 4 1 X SQ= 1.92 P = 0.138 RV NO (YES)
92	TILT LESS TOWARD BEING THOSE WHERE THE SYSTEM STYLE IS LIMITED MOBILIZATIONAL, OR NON-MOBILIZATIONAL (93)	0.82	92	TILT MORE TOWARD BEING THOSE WHERE THE SYSTEM STYLE IS LIMITED MOBILIZATIONAL, OR NON-MOBILIZATIONAL (93)	0.94	14 2 63 29 X SQ= 1.57 P = 0.145 RV NO (NO)
95	TEND LESS TO BE THOSE WHERE THE STATUS OF THE REGIME IS CONSTITUTIONAL OR AUTHORITARIAN (95)	0.83	95	ALWAYS ARE THOSE WHERE THE STATUS OF THE REGIME IS CONSTITUTIONAL OR AUTHORITARIAN (95)	1.00	65 29 13 0 X SQ= 4.05 P = 0.018 RV NO (NO)
96	TEND TO BE THOSE WHERE THE STATUS OF THE REGIME IS CONSTITUTIONAL OR TOTALITARIAN (67)	0.79	96	TEND TO BE THOSE WHERE THE STATUS OF THE REGIME IS AUTHORITARIAN (23)	0.54	58 6 15 7 X SQ= 4.80 P = 0.033 RV YES (NO)
97	TEND LESS TO BE THOSE WHERE THE STATUS OF THE REGIME IS AUTHORITARIAN, RATHER THAN TOTALITARIAN (23)	0.54	97	ALWAYS ARE THOSE WHERE THE STATUS OF THE REGIME IS AUTHORITARIAN, RATHER THAN TOTALITARIAN (23)	1.00	15 7 13 0 X SQ= 3.37 P = 0.031 RV NO (NO)

99	LEAN TOWARD BEING THOSE WHERE GOVERNMENTAL STABILITY IS GENERALLY PRESENT AND DATES FROM AT LEAST THE INTER-WAR PERIOD, OR FROM THE POST-WAR PERIOD (50)	0.63	LEAN TOWARD BEING THOSE WHERE GOVERNMENTAL STABILITY IS MODERATELY PRESENT AND DATES FROM THE POST-WAR PERIOD, OR IS ABSENT (36)	0.69	43 4 25 9 X SQ= 3.48 P = 0.063 RV YES (NO)
101	TEND TO BE THOSE WHERE THE REPRESENTATIVE CHARACTER OF THE REGIME IS POLYARCHIC (41)	0.53	TEND TO BE THOSE WHERE THE REPRESENTATIVE CHARACTER OF THE REGIME IS LIMITED POLYARCHIC, PSEUDO-POLYARCHIC, OR NON-POLYARCHIC (57)	0.86	38 3 34 19 X SQ= 8.97 P = 0.001 RV YES (NO)
105	TEND TO BE THOSE WHERE THE ELECTORAL SYSTEM IS COMPETITIVE (43)	0.60	TEND TO BE THOSE WHERE THE ELECTORAL SYSTEM IS PARTIALLY COMPETITIVE OR NON-COMPETITIVE (47)	0.78	38 5 25 18 X SQ= 8.55 P = 0.003 RV YES (NO)
107	TEND TO BE THOSE WHERE AUTONOMOUS GROUPS ARE FULLY TOLERATED IN POLITICS (46)	0.53	TEND TO BE THOSE WHERE AUTONOMOUS GROUPS ARE PARTIALLY TOLERATED IN POLITICS, ARE TOLERATED ONLY OUTSIDE POLITICS, OR ARE NOT TOLERATED AT ALL (65)	0.82	41 5 37 23 X SQ= 8.74 P = 0.002 RV YES (NO)
111	TEND TO BE THOSE WHERE POLITICAL ENCULTURATION IS HIGH OR MEDIUM (53)	0.68	TEND TO BE THOSE WHERE POLITICAL ENCULTURATION IS LOW (42)	0.68	44 9 21 19 X SQ= 8.69 P = 0.003 RV YES (NO)
115	TEND LESS TO BE THOSE WHERE INTEREST ARTICULATION BY ASSOCIATIONAL GROUPS IS MODERATE, LIMITED, OR NEGLIGIBLE (91)	0.74	ALWAYS ARE THOSE WHERE INTEREST ARTICULATION BY ASSOCIATIONAL GROUPS IS MODERATE, LIMITED, OR NEGLIGIBLE (91)	1.00	20 0 56 31 X SQ= 8.38 P = 0.001 RV NO (NO)
116	TEND LESS TO BE THOSE WHERE INTEREST ARTICULATION BY ASSOCIATIONAL GROUPS IS LIMITED OR NEGLIGIBLE (79)	0.59	TEND MORE TO BE THOSE WHERE INTEREST ARTICULATION BY ASSOCIATIONAL GROUPS IS LIMITED OR NEGLIGIBLE (79)	0.97	31 1 45 30 X SQ= 13.08 P = 0. RV NO (NO)
117	TEND TO BE THOSE WHERE INTEREST ARTICULATION BY ASSOCIATIONAL GROUPS IS SIGNIFICANT, MODERATE, OR LIMITED (60)	0.72	TEND TO BE THOSE WHERE INTEREST ARTICULATION BY ASSOCIATIONAL GROUPS IS NEGLIGIBLE (51)	0.84	M 55 5 21 26 X SQ= 26.04 P = 0. RV YES (YES)
119	TEND LESS TO BE THOSE WHERE INTEREST ARTICULATION BY INSTITUTIONAL GROUPS IS VERY SIGNIFICANT OR SIGNIFICANT (74)	0.68	TEND MORE TO BE THOSE WHERE INTEREST ARTICULATION BY INSTITUTIONAL GROUPS IS VERY SIGNIFICANT OR SIGNIFICANT (74)	0.94	52 17 25 1 X SQ= 4.05 P = 0.020 RV NO (NO)

121	TEND TO BE THOSE WHERE INTEREST ARTICULATION BY NON-ASSOCIATIONAL GROUPS IS MODERATE, LIMITED, OR NEGLIGIBLE (61)	0.71	121	TEND TO BE THOSE WHERE INTEREST ARTICULATION BY NON-ASSOCIATIONAL GROUPS IS SIGNIFICANT (54)	0.84	23 26 56 5 X SQ= 24.85 P = 0. RV YES (YES)
122	TEND LESS TO BE THOSE WHERE INTEREST ARTICULATION BY NON-ASSOCIATIONAL GROUPS IS SIGNIFICANT OR MODERATE (83)	0.59	122	ALWAYS ARE THOSE WHERE INTEREST ARTICULATION BY NON-ASSOCIATIONAL GROUPS IS SIGNIFICANT OR MODERATE (83)	1.00	47 31 32 0 X SQ= 15.80 P = 0. RV NO (NO)
125	TEND LESS TO BE THOSE WHERE INTEREST ARTICULATION BY ANOMIC GROUPS IS FREQUENT OR OCCASIONAL (64)	0.52	125	TEND MORE TO BE THOSE WHERE INTEREST ARTICULATION BY ANOMIC GROUPS IS FREQUENT OR OCCASIONAL (64)	0.87	34 27 31 4 X SQ= 9.52 P = 0.001 RV NO (NO)
126	TEND LESS TO BE THOSE WHERE INTEREST ARTICULATION BY ANOMIC GROUPS IS FREQUENT, OCCASIONAL, OR INFREQUENT (83)	0.75	126	ALWAYS ARE THOSE WHERE INTEREST ARTICULATION BY ANOMIC GROUPS IS FREQUENT, OCCASIONAL, OR INFREQUENT (83)	1.00	49 31 16 0 X SQ= 7.47 P = 0.001 RV NO (NO)
129	TILT TOWARD BEING THOSE WHERE INTEREST ARTICULATION BY POLITICAL PARTIES IS SIGNIFICANT, MODERATE, OR LIMITED (56)	0.68	129	TILT TOWARD BEING THOSE WHERE INTEREST ARTICULATION BY POLITICAL PARTIES IS NEGLIGIBLE (37)	0.50	43 13 20 13 X SQ= 1.90 P = 0.147 RV YES (NO)
133	TEND TO BE THOSE WHERE INTEREST AGGREGATION BY THE EXECUTIVE IS MODERATE, LIMITED, OR NEGLIGIBLE (76)	0.81	133	TEND TO BE THOSE WHERE INTEREST AGGREGATION BY THE EXECUTIVE IS SIGNIFICANT (29)	0.54	14 15 58 13 X SQ= 9.81 P = 0.001 RV YES (YES)
137	TEND LESS TO BE THOSE WHERE INTEREST AGGREGATION BY THE LEGISLATURE IS LIMITED OR NEGLIGIBLE (68)	0.55	137	ALWAYS ARE THOSE WHERE INTEREST AGGREGATION BY THE LEGISLATURE IS LIMITED OR NEGLIGIBLE (68)	1.00	29 0 36 28 X SQ= 16.13 P = 0. RV NO (NO)
138	TEND TO BE THOSE WHERE INTEREST AGGREGATION BY THE LEGISLATURE IS SIGNIFICANT, MODERATE, OR LIMITED (48)	0.63	138	TEND TO BE THOSE WHERE INTEREST AGGREGATION BY THE LEGISLATURE IS NEGLIGIBLE (49)	0.75	41 7 24 21 X SQ= 9.89 P = 0.001 RV YES (NO)
144	TILT LESS TOWARD BEING THOSE WHERE THE PARTY SYSTEM IS QUANTITATIVELY ONE-PARTY, RATHER THAN TWO-PARTY (34)	0.66	144	TILT MORE TOWARD BEING THOSE WHERE THE PARTY SYSTEM IS QUANTITATIVELY ONE-PARTY, RATHER THAN TWO-PARTY (34)	0.92	19 12 10 1 X SQ= 2.09 P = 0.127 RV NO (NO)

147	TEND LESS TO BE THOSE WHERE THE PARTY SYSTEM IS QUALITATIVELY OTHER THAN CLASS-ORIENTED OR MULTI-IDEOLOGICAL (67)	0.63	147	TEND MORE TO BE THOSE WHERE THE PARTY SYSTEM IS QUALITATIVELY OTHER THAN CLASS-ORIENTED OR MULTI-IDEOLOGICAL (67)	0.96	22 1 38 25 X SQ= 8.37 P = 0.001 RV NO (NO)
148	TEND MORE TO BE THOSE WHERE THE PARTY SYSTEM IS QUALITATIVELY OTHER THAN AFRICAN TRANSITIONAL (96)	0.99	148	TEND LESS TO BE THOSE WHERE THE PARTY SYSTEM IS QUALITATIVELY OTHER THAN AFRICAN TRANSITIONAL (96)	0.55	1 13 75 16 X SQ= 30.73 P = 0. RV NO (YES)
153	LEAN TOWARD BEING THOSE WHERE THE PARTY SYSTEM IS STABLE (42)	0.57	153	LEAN TOWARD BEING THOSE WHERE THE PARTY SYSTEM IS MODERATELY STABLE OR UNSTABLE (38)	0.71	35 4 26 10 X SQ= 2.72 P = 0.075 RV YES (NO)
158	TEND TO BE THOSE WHERE PERSONALISMO IS NEGLIGIBLE (56)	0.69	158	TEND TO BE THOSE WHERE PERSONALISMO IS PRONOUNCED OR MODERATE (40)	0.72	21 18 46 7 X SQ= 10.71 P = 0.001 RV YES (NO)
164	TEND TO BE THOSE WHERE THE REGIME'S LEADERSHIP CHARISMA IS NEGLIGIBLE (65)	0.76	164	TEND TO BE THOSE WHERE THE REGIME'S LEADERSHIP CHARISMA IS PRONOUNCED OR MODERATE (34)	0.58	17 14 53 10 X SQ= 7.90 P = 0.005 RV YES (NO)
169	TEND TO BE THOSE WHERE THE HORIZONTAL POWER DISTRIBUTION IS SIGNIFICANT OR LIMITED (58)	0.64	169	TEND TO BE THOSE WHERE THE HORIZONTAL POWER DISTRIBUTION IS NEGLIGIBLE (48)	0.61	47 11 27 17 X SQ= 3.92 P = 0.043 RV YES (NO)
170	TEND TO BE THOSE WHERE THE LEGISLATIVE-EXECUTIVE STRUCTURE IS OTHER THAN PRESIDENTIAL (63)	0.70	170	TEND TO BE THOSE WHERE THE LEGISLATIVE-EXECUTIVE STRUCTURE IS PRESIDENTIAL (39)	0.63	21 17 50 10 X SQ= 7.83 P = 0.005 RV YES (NO)
175	TEND TO BE THOSE WHERE THE LEGISLATURE IS FULLY EFFECTIVE OR PARTIALLY EFFECTIVE (51)	0.61	175	TEND TO BE THOSE WHERE THE LEGISLATURE IS LARGELY INEFFECTIVE OR WHOLLY INEFFECTIVE (49)	0.69	43 8 27 18 X SQ= 5.98 P = 0.011 RV YES (NO)
178	TEND TO BE THOSE WHERE THE LEGISLATURE IS BICAMERAL (51)	0.60	178	TEND TO BE THOSE WHERE THE LEGISLATURE IS UNICAMERAL (53)	0.70	28 21 42 9 X SQ= 6.41 P = 0.008 RV YES (NO)

182	TEND TO BE THOSE WHERE THE BUREAUCRACY IS SEMI-MODERN, RATHER THAN POST-COLONIAL TRANSITIONAL (55)	0.90	182	TEND TO BE THOSE WHERE THE BUREAUCRACY IS POST-COLONIAL TRANSITIONAL, RATHER THAN SEMI-MODERN (25)	0.71	43 8 5 20 X SQ= 27.12 P = 0. RV YES (YES)
183	TEND TO BE THOSE WHERE THE BUREAUCRACY IS TRADITIONAL, RATHER THAN POST-COLONIAL TRANSITIONAL (9)	0.58	183	TEND TO BE THOSE WHERE THE BUREAUCRACY IS POST-COLONIAL TRANSITIONAL, RATHER THAN TRADITIONAL (25)	0.91	M 5 20 7 2 X SQ= 7.31 P = 0.004 RV YES (YES)
187	TEND LESS TO BE THOSE WHERE THE ROLE OF THE POLICE IS POLITICALLY SIGNIFICANT (66)	0.57	187	TEND MORE TO BE THOSE WHERE THE ROLE OF THE POLICE IS POLITICALLY SIGNIFICANT (66)	0.83	41 20 31 4 X SQ= 4.33 P = 0.027 RV NO (NO)
188	TILT LESS TOWARD BEING THOSE WHERE THE CHARACTER OF THE LEGAL SYSTEM IS OTHER THAN CIVIL LAW (81)	0.66	188	TILT MORE TOWARD BEING THOSE WHERE THE CHARACTER OF THE LEGAL SYSTEM IS OTHER THAN CIVIL LAW (81)	0.83	27 5 52 24 X SQ= 2.16 P = 0.101 RV NO (NO)
189	TILT LESS TOWARD BEING THOSE WHERE THE CHARACTER OF THE LEGAL SYSTEM IS OTHER THAN COMMON LAW (108)	0.91	189	ALWAYS ARE THOSE WHERE THE CHARACTER OF THE LEGAL SYSTEM IS OTHER THAN COMMON LAW (108)	1.00	7 0 72 31 X SQ= 1.63 P = 0.109 RV NO (NO)
193	TEND TO BE THOSE WHERE THE CHARACTER OF THE LEGAL SYSTEM IS OTHER THAN PARTLY INDIGENOUS (86)	0.97	193	TEND TO BE THOSE WHERE THE CHARACTER OF THE LEGAL SYSTEM IS PARTLY INDIGENOUS (28)	0.81	2 25 76 6 X SQ= 68.45 P = 0. RV YES (YES)
194	TEND LESS TO BE THOSE THAT ARE NON-COMMUNIST (101)	0.87	194	ALWAYS ARE THOSE THAT ARE NON-COMMUNIST (101)	1.00	10 0 68 31 X SQ= 2.97 P = 0.035 RV NO (NO)

M

MATRIX

	56 POLITIES WHERE THE RELIGION IS PREDOMINANTLY LITERATE (79)	56 POLITIES WHERE THE RELIGION IS PREDOMINANTLY OR PARTLY NON-LITERATE (31)
77 POLITIES THAT HAVE BEEN SIGNIFICANTLY WESTERNIZED, RATHER THAN PARTIALLY WESTERNIZED, THROUGH A COLONIAL RELATIONSHIP (28)	24 ALGERIA, ARGENTINA, BRAZIL, CEYLON, CHILE, COLOMBIA, COSTA RICA, CUBA, CYPRUS, DOMIN REP, EL SALVADOR, HONDURAS, INDIA, JAMAICA, LEBANON, MEXICO, NICARAGUA, PANAMA, PARAGUAY, PHILIPPINES, TRINIDAD, UAR, URUGUAY, VENEZUELA	4 BOLIVIA, ECUADOR, GUATEMALA, PERU
77 POLITIES THAT HAVE BEEN PARTIALLY WESTERNIZED, RATHER THAN SIGNIFICANTLY WESTERNIZED, THROUGH A COLONIAL RELATIONSHIP (41)	15 BURMA, CAMBODIA, INDONESIA, IRAN, IRAQ, JORDAN, LIBYA, MALAYA, MAURITANIA, MOROCCO, PAKISTAN, SENEGAL, SOMALIA, SYRIA, TUNISIA	24 BURUNDI, CAMEROUN, CEN AFR REP, CHAD, CONGO(BRA), CONGO(LEO), DAHOMEY, GABON, GHANA, GUINEA, HAITI, IVORY COAST, LAOS, MALAGASY R, MALI, NIGER, NIGERIA, RWANDA, SIERRE LEO, SUDAN, TANGANYIKA, TOGO, UGANDA, UPPER VOLTA

MATRIX

	56 POLITIES WHERE THE RELIGION IS PREDOMINANTLY LITERATE (79)	56 POLITIES WHERE THE RELIGION IS PREDOMINANTLY OR PARTLY NON-LITERATE (31)
84 POLITIES WHERE THE TYPE OF POLITICAL MODERNIZATION IS DEVELOPED TUTELARY, RATHER THAN UNDEVELOPED TUTELARY (31)	24	3
84 POLITIES WHERE THE TYPE OF POLITICAL MODERNIZATION IS UNDEVELOPED TUTELARY, RATHER THAN DEVELOPED TUTELARY (24)	3	21

Literate / Developed tutelary (24):
ALGERIA, INDIA, JORDAN, MOROCCO, TRINIDAD, BURMA, INDONESIA, LEBANON, PAKISTAN, TUNISIA, CAMBODIA, IRAQ, LIBYA, PHILIPPINES, UAR, CEYLON, ISRAEL, MALAYA, SA'U ARABIA, YEMEN, CYPRUS, JAMAICA, MONGOLIA, SYRIA

Non-literate / Developed tutelary (3):
LAOS, LIBERIA, SUDAN

Literate / Undeveloped tutelary (3):
MAURITANIA, SENEGAL, SOMALIA

Non-literate / Undeveloped tutelary (21):
BURUNDI, CONGO(LEO), IVORY COAST, RWANDA, UPPER VOLTA, CAMEROUN, DAHOMEY, MALAGASY R, SIERRE LEO, CEN AFR REP, GABON, MALI, TANGANYIKA, CHAD, GHANA, NIGER, TOGO, CONGO(BRA), GUINEA, NIGERIA, UGANDA

MATRIX

56 POLITIES WHERE THE RELIGION IS PREDOMINANTLY LITERATE (79)	56 POLITIES WHERE THE RELIGION IS PREDOMINANTLY OR PARTLY NON-LITERATE (31)
117 POLITIES WHERE INTEREST ARTICULATION BY ASSOCIATIONAL GROUPS IS SIGNIFICANT, MODERATE, OR LIMITED (60)	
55	5
ALGERIA, BRAZIL, COLOMBIA, EL SALVADOR, HONDURAS, IRAN, JAPAN, MOROCCO, PANAMA, SYRIA, UAR, ARGENTINA, BURMA, COSTA RICA, FINLAND, HUNGARY, IRELAND, LEBANON, NETHERLANDS, PHILIPPINES, TRINIDAD, UK, AUSTRALIA, CANADA, CYPRUS, FRANCE, ICELAND, ISRAEL, LUXEMBOURG, NEW ZEALAND, POLAND, TUNISIA, US, AUSTRIA, CEYLON, DENMARK, GERMAN FR, INDIA, ITALY, MALAYA, NICARAGUA, SWEDEN, TURKEY, URUGUAY, BELGIUM, CHILE, DOMIN REP, GREECE, INDONESIA, JAMAICA, MEXICO, NORWAY, SWITZERLAND, USSR, VENEZUELA	BOLIVIA, ECUADOR, GUATEMALA, PERU, SO AFRICA
117 POLITIES WHERE INTEREST ARTICULATION BY ASSOCIATIONAL GROUPS IS NEGLIGIBLE (51)	
21	26
AFGHANISTAN, CZECHOS'KIA, MAURITANIA, RUMANIA, YEMEN, ALBANIA, GERMANY, E, MONGOLIA, SA'U ARABIA, BULGARIA, IRAQ, NEPAL, SENEGAL, CAMBODIA, JORDAN, PAKISTAN, SOMALIA, CUBA, LIBYA, PARAGUAY, THAILAND	BURUNDI, CONGO(LEO), GUINEA, MALAGASY R, SIERRE LEO, UPPER VOLTA, CAMEROUN, DAHOMEY, HAITI, MALI, SUDAN, CEN AFR REP, ETHIOPIA, IVORY COAST, NIGER, TANGANYIKA, CHAD, GABON, LAOS, NIGERIA, TOGO, CONGO(BRA), GHANA, LIBERIA, RWANDA, UGANDA

56/183

M

MATRIX

	56 POLITIES WHERE THE RELIGION IS PREDOMINANTLY LITERATE (79)	56 POLITIES WHERE THE RELIGION IS PREDOMINANTLY OR PARTLY NON-LITERATE (31)
183 POLITIES WHERE THE BUREAUCRACY IS POST-COLONIAL TRANSITIONAL, RATHER THAN TRADITIONAL (25)	ALGERIA BURMA MAURITANIA SENEGAL SOMALIA 5	BURUNDI CAMEROUN CEN AFR REP CHAD CONGO(BRA) CONGO(LEO) DAHOMEY GABON IVORY COAST MALAGASY R MALI NIGER NIGERIA RWANDA SIERRE LEO SUDAN TANGANYIKA TOGO UGANDA UPPER VOLTA 20
183 POLITIES WHERE THE BUREAUCRACY IS TRADITIONAL, RATHER THAN POST-COLONIAL TRANSITIONAL (9)	7 AFGHANISTAN CAMBODIA IRAN JORDAN NEPAL SA'U ARABIA THAILAND	2 ETHIOPIA LAOS

57 POLITIES WHERE THE RELIGION IS CATHOLIC (25)	57 POLITIES WHERE THE RELIGION IS OTHER THAN CATHOLIC (90)

BOTH SUBJECT AND PREDICATE

25 IN LEFT COLUMN

ARGENTINA	AUSTRIA	BELGIUM	BRAZIL	CHILE	COLOMBIA	COSTA RICA	CUBA	DOMIN REP	EL SALVADOR
FRANCE	HONDURAS	IRELAND	ITALY	LUXEMBOURG	MEXICO	NICARAGUA	PANAMA	PARAGUAY	PHILIPPINES
POLAND	PORTUGAL	SPAIN	URUGUAY	VENEZUELA					

90 IN RIGHT COLUMN

AFGHANISTAN	ALBANIA	ALGERIA	AUSTRALIA	BOLIVIA	BULGARIA	BURMA	BURUNDI	CAMBODIA	CAMEROUN
CANADA	CEN AFR REP	CEYLON	CHAD	CHINA, PR	CONGO(BRA)	CONGO(LEO)	CYPRUS	CZECHOS'KIA	DAHOMEY
DENMARK	ECUADOR	ETHIOPIA	FINLAND	GABON	GERMANY, E	GERMAN FR	GHANA	GREECE	GUATEMALA
GUINEA	HAITI	HUNGARY	ICELAND	INDIA	INDONESIA	IRAN	IRAQ	ISRAEL	IVORY COAST
JAMAICA	JAPAN	JORDAN	KOREA, N	KOREA REP	LAOS	LEBANON	LIBERIA	LIBYA	MALAGASY R
MALAYA	MALI	MAURITANIA	MONGOLIA	MOROCCO	NEPAL	NETHERLANDS	NEW ZEALAND	NIGER	NIGERIA
NORWAY	PAKISTAN	PERU	RUMANIA	RWANDA	SA'U ARABIA	SENEGAL	SIERRE LEO	SOMALIA	SO AFRICA
SUDAN	SWEDEN	SWITZERLAND	SYRIA	TANGANYIKA	THAILAND	TOGO	TRINIDAD	TUNISIA	TURKEY
UGANDA	USSR	UAR	UK	US	UPPER VOLTA	VIETNAM, N	VIETNAM REP	YEMEN	YUGOSLAVIA

3	TEND LESS TO BE THOSE LOCATED ELSEWHERE THAN IN WEST EUROPE (102)	0.68	TEND MORE TO BE THOSE LOCATED ELSEWHERE THAN IN WEST EUROPE (102)	0.94	8 5 17 85 X SQ= 11.14 P = 0.001 RV NO (YES)
5	TILT MORE TOWARD BEING THOSE LOCATED IN SCANDINAVIA OR WEST EUROPE, RATHER THAN IN EAST EUROPE (18)	0.89	TILT LESS TOWARD BEING THOSE LOCATED IN SCANDINAVIA OR WEST EUROPE, RATHER THAN IN EAST EUROPE (18)	0.56	8 10 1 8 X SQ= 1.69 P = 0.110 RV NO (NO)
6	TEND TO BE THOSE LOCATED IN THE CARIBBEAN, CENTRAL AMERICA, OR SOUTH AMERICA (22)	0.60	TEND TO BE THOSE LOCATED ELSEWHERE THAN IN THE CARIBBEAN, CENTRAL AMERICA, OR SOUTH AMERICA (93)	0.92	15 7 10 83 X SQ= 31.20 P = 0. RV YES (YES)
8	LEAN MORE TOWARD BEING THOSE LOCATED ELSEWHERE THAN IN EAST ASIA SOUTH ASIA, OR SOUTHEAST ASIA (97)	0.96	LEAN LESS TOWARD BEING THOSE LOCATED ELSEWHERE THAN IN EAST ASIA SOUTH ASIA, OR SOUTHEAST ASIA (97)	0.81	1 17 24 73 X SQ= 2.25 P = 0.074 RV NO (NO)

No.	Statement	Value	No.	Statement	Value	Statistics
10	ALWAYS ARE THOSE LOCATED ELSEWHERE THAN IN NORTH AFRICA, OR CENTRAL AND SOUTH AFRICA (82)	1.00	10	TEND LESS TO BE THOSE LOCATED ELSEWHERE THAN IN NORTH AFRICA, OR CENTRAL AND SOUTH AFRICA (82)	0.63	0 33 25 57 X SQ= 11.13 P = 0. RV NO (NO)
13	ALWAYS ARE THOSE LOCATED ELSEWHERE THAN IN NORTH AFRICA OR THE MIDDLE EAST (99)	1.00	13	TEND LESS TO BE THOSE LOCATED ELSEWHERE THAN IN NORTH AFRICA OR THE MIDDLE EAST (99)	0.82	0 16 25 74 X SQ= 3.79 P = 0.021 RV NO (NO)
14	ALWAYS ARE THOSE LOCATED ELSEWHERE THAN IN THE MIDDLE EAST (104)	1.00	14	LEAN LESS TOWARD BEING THOSE LOCATED ELSEWHERE THAN IN THE MIDDLE EAST (104)	0.88	0 11 25 79 X SQ= 2.11 P = 0.071 RV NO (NO)
20	TEND TO BE THOSE LOCATED IN THE CARIBBEAN, CENTRAL AMERICA, OR SOUTH AMERICA, RATHER THAN IN EAST ASIA, SOUTH ASIA, OR SOUTHEAST ASIA (22)	0.94	20	TEND TO BE THOSE LOCATED IN EAST ASIA, SOUTH ASIA, OR SOUTHEAST ASIA, RATHER THAN IN THE CARIBBEAN, CENTRAL AMERICA, OR SOUTH AMERICA (18)	0.71	1 17 15 7 X SQ= 13.67 P = 0. RV YES (YES)
29	TEND TO BE THOSE WHERE THE DEGREE OF URBANIZATION IS HIGH (56)	0.90	29	TEND TO BE THOSE WHERE THE DEGREE OF URBANIZATION IS LOW (49)	0.56	19 37 2 47 X SQ= 12.74 P = 0. RV YES (NO)
30	TEND TO BE THOSE WHOSE AGRICULTURAL POPULATION IS MEDIUM, LOW, OR VERY LOW (57)	0.92	30	TEND TO BE THOSE WHOSE AGRICULTURAL POPULATION IS HIGH (56)	0.61	2 54 23 34 X SQ= 20.09 P = 0. RV YES (NO)
34	TILT TOWARD BEING THOSE WHOSE GROSS NATIONAL PRODUCT IS VERY HIGH, HIGH, MEDIUM, OR LOW (62)	0.68	34	TILT TOWARD BEING THOSE WHOSE GROSS NATIONAL PRODUCT IS VERY LOW (53)	0.50	17 45 8 45 X SQ= 1.88 P = 0.120 RV YES (NO)
36	LEAN TOWARD BEING THOSE WHOSE PER CAPITA GROSS NATIONAL PRODUCT IS VERY HIGH, HIGH, OR MEDIUM (42)	0.52	36	LEAN TOWARD BEING THOSE WHOSE PER CAPITA GROSS NATIONAL PRODUCT IS LOW OR VERY LOW (73)	0.68	13 29 12 61 X SQ= 2.50 P = 0.099 RV YES (NO)
37	TEND TO BE THOSE WHOSE PER CAPITA GROSS NATIONAL PRODUCT IS VERY HIGH, HIGH, MEDIUM, OR LOW (64)	0.96	37	TEND TO BE THOSE WHOSE PER CAPITA GROSS NATIONAL PRODUCT IS VERY LOW (51)	0.56	24 40 1 50 X SQ= 19.03 P = 0. RV YES (NO)

39	TEND TO BE THOSE WHOSE INTERNATIONAL FINANCIAL STATUS IS VERY HIGH, HIGH, OR MEDIUM (38)	0.52	39	TEND TO BE THOSE WHOSE INTERNATIONAL FINANCIAL STATUS IS LOW OR VERY LOW (76)	0.72	13 25 12 64 X SQ= 4.00 P = 0.032 RV YES (NO)
43	TEND TO BE THOSE WHOSE ECONOMIC DEVELOPMENTAL STATUS IS DEVELOPED, INTERMEDIATE, OR UNDERDEVELOPED (55)	0.75	43	TEND TO BE THOSE WHOSE ECONOMIC DEVELOPMENTAL STATUS IS VERY UNDERDEVELOPED (57)	0.58	18 37 6 51 X SQ= 6.93 P = 0.005 RV YES (NO)
45	TEND TO BE THOSE WHERE THE LITERACY RATE IS FIFTY PERCENT OR ABOVE (55)	0.88	45	TEND TO BE THOSE WHERE THE LITERACY RATE IS BELOW FIFTY PERCENT (54)	0.61	22 33 3 51 X SQ= 16.39 P = 0. RV YES (NO)
46	ALWAYS ARE THOSE WHERE THE LITERACY RATE IS TEN PERCENT OR ABOVE (84)	1.00	46	TEND LESS TO BE THOSE WHERE THE LITERACY RATE IS TEN PERCENT OR ABOVE (84)	0.69	25 59 0 26 X SQ= 8.39 P = 0.001 RV NO (NO)
47	TEND TO BE THOSE WHERE THE LITERACY RATE IS BETWEEN FIFTY AND NINETY PERCENT, RATHER THAN NINETY PERCENT OR ABOVE (30)	0.73	47	TEND TO BE THOSE WHERE THE LITERACY RATE IS NINETY PERCENT OR ABOVE, RATHER THAN BETWEEN FIFTY AND NINETY PERCENT (25)	0.58	6 19 16 14 X SQ= 3.74 P = 0.032 RV YES (YES)
48	TEND TO BE THOSE WHERE THE LITERACY RATE IS BETWEEN FIFTY AND NINETY PERCENT, RATHER THAN BETWEEN TEN AND FIFTY PERCENT (30)	0.84	48	TEND TO BE THOSE WHERE THE LITERACY RATE IS BETWEEN TEN AND FIFTY PERCENT, RATHER THAN BETWEEN FIFTY AND NINETY PERCENT (24)	0.60	16 14 3 21 X SQ= 8.04 P = 0.004 RV YES (YES)
50	TILT TOWARD BEING THOSE WHERE FREEDOM OF THE PRESS IS COMPLETE (43)	0.59	50	TILT TOWARD BEING THOSE WHERE FREEDOM OF THE PRESS IS INTERMITTENT, INTERNALLY ABSENT, OR INTERNALLY AND EXTERNALLY ABSENT (56)	0.61	13 30 9 47 X SQ= 2.06 P = 0.142 RV YES (NO)
54	TEND TO BE THOSE WHERE NEWSPAPER CIRCULATION IS ONE HUNDRED OR MORE PER THOUSAND (37)	0.52	54	TEND TO BE THOSE WHERE NEWSPAPER CIRCULATION IS LESS THAN ONE HUNDRED PER THOUSAND (76)	0.73	13 24 12 64 X SQ= 4.34 P = 0.029 RV YES (NO)
55	ALWAYS ARE THOSE WHERE NEWSPAPER CIRCULATION IS TEN OR MORE PER THOUSAND (78)	1.00	55	TEND LESS TO BE THOSE WHERE NEWSPAPER CIRCULATION IS TEN OR MORE PER THOUSAND (78)	0.60	25 53 0 35 X SQ= 12.60 P = 0. RV NO (NO)

68	TEND TO BE THOSE THAT ARE LINGUISTICALLY HOMOGENEOUS (52)	0.84	68	TEND TO BE THOSE THAT ARE LINGUISTICALLY WEAKLY HETEROGENEOUS OR STRONGLY HETEROGENEOUS (62)	0.65	21 31 4 58 X SQ= 17.09 P = 0. RV YES (NO)
69	TEND TO BE THOSE THAT ARE LINGUISTICALLY HOMOGENEOUS OR WEAKLY HETEROGENEOUS (64)	0.92	69	TEND TO BE THOSE THAT ARE LINGUISTICALLY STRONGLY HETEROGENEOUS (50)	0.54	23 41 2 48 X SQ= 14.91 P = 0. RV YES (NO)
72	TEND TO BE THOSE WHOSE DATE OF INDEPENDENCE IS BEFORE 1914 (52)	0.88	72	TEND TO BE THOSE WHOSE DATE OF INDEPENDENCE IS AFTER 1914 (59)	0.65	22 30 3 56 X SQ= 19.87 P = 0. RV YES (NO)
73	TEND TO BE THOSE WHOSE DATE OF INDEPENDENCE IS BEFORE 1945 (65)	0.96	73	TEND TO BE THOSE WHOSE DATE OF INDEPENDENCE IS AFTER 1945 (46)	0.52	24 41 1 45 X SQ= 16.70 P = 0. RV YES (NO)
75	ALWAYS ARE THOSE THAT ARE HISTORICALLY WESTERN OR SIGNIFICANTLY WESTERNIZED (62)	1.00	75	TEND TO BE THOSE THAT ARE NOT HISTORICALLY WESTERN AND ARE NOT SIGNIFICANTLY WESTERNIZED (52)	0.58	25 37 0 52 X SQ= 24.55 P = 0. RV YES (NO)
78	ALWAYS ARE THOSE THAT HAVE BEEN SIGNIFICANTLY WESTERNIZED THROUGH A COLONIAL RELATIONSHIP, RATHER THAN WITHOUT SUCH A RELATIONSHIP (28)	1.00	78	TEND LESS TO BE THOSE THAT HAVE BEEN SIGNIFICANTLY WESTERNIZED THROUGH A COLONIAL RELATIONSHIP, RATHER THAN WITHOUT SUCH A RELATIONSHIP (28)	0.63	16 12 0 7 X SQ= 5.25 P = 0.009 RV NO (YES)
79	TEND TO BE THOSE THAT WERE FORMERLY DEPENDENCIES OF SPAIN, RATHER THAN BRITAIN OR FRANCE (18)	0.93	79	TEND TO BE THOSE THAT WERE FORMERLY DEPENDENCIES OF BRITAIN OR FRANCE, RATHER THAN SPAIN (49)	0.92	1 48 14 4 X SQ= 39.21 P = 0. RV YES (YES)
82	TEND TO BE THOSE WHERE THE TYPE OF POLITICAL MODERNIZATION IS EARLY OR LATER EUROPEAN OR EUROPEAN DERIVED, RATHER THAN DEVELOPED TUTELARY OR UNDEVELOPED TUTELARY (51)	0.96	82	TEND TO BE THOSE WHERE THE TYPE OF POLITICAL MODERNIZATION IS DEVELOPED TUTELARY OR UNDEVELOPED TUTELARY, RATHER THAN EARLY OR LATER EUROPEAN OR EUROPEAN DERIVED (55)	0.67	24 27 1 54 X SQ= 27.59 P = 0. RV YES (NO)
83	ALWAYS ARE THOSE WHERE THE TYPE OF POLITICAL MODERNIZATION IS OTHER THAN NON-EUROPEAN AUTOCHTHONOUS (106)	1.00	83	TILT LESS TOWARD BEING THOSE WHERE THE TYPE OF POLITICAL MODERNIZATION IS OTHER THAN NON-EUROPEAN AUTOCHTHONOUS (106)	0.90	0 9 25 81 X SQ= 1.50 P = 0.123 RV NO (NO)

85	TEND TO BE THOSE WHERE THE STAGE OF POLITICAL MODERNIZATION IS ADVANCED, RATHER THAN MID- OR EARLY TRANSITIONAL (60)	0.72	85	TEND TO BE THOSE WHERE THE STAGE OF POLITICAL MODERNIZATION IS MID- OR EARLY TRANSITIONAL, RATHER THAN ADVANCED (54)	0.53	18 42 7 47 X SQ= 3.87 P = 0.040 RV YES (NO)
86	ALWAYS ARE THOSE WHERE THE STAGE OF POLITICAL MODERNIZATION IS ADVANCED OR MID-TRANSITIONAL, RATHER THAN EARLY TRANSITIONAL (76)	1.00	86	TEND LESS TO BE THOSE WHERE THE STAGE OF POLITICAL MODERNIZATION IS ADVANCED OR MID-TRANSITIONAL, RATHER THAN EARLY TRANSITIONAL (76)	0.57	25 51 0 38 X SQ= 14.15 P = 0. RV NO (NO)
87	TEND MORE TO BE THOSE WHOSE IDEOLOGICAL ORIENTATION IS OTHER THAN DEVELOPMENTAL (58)	0.89	87	TEND LESS TO BE THOSE WHOSE IDEOLOGICAL ORIENTATION IS OTHER THAN DEVELOPMENTAL (58)	0.59	2 29 17 41 X SQ= 5.00 P = 0.014 RV NO (NO)
89	TEND TO BE THOSE WHOSE IDEOLOGICAL ORIENTATION IS CONVENTIONAL (33)	0.72	89	TEND TO BE THOSE WHOSE IDEOLOGICAL ORIENTATION IS OTHER THAN CONVENTIONAL (62)	0.74	13 20 5 57 X SQ= 11.80 P = 0.001 RV YES (NO)
93	TEND MORE TO BE THOSE WHERE THE SYSTEM STYLE IS NON-MOBILIZATIONAL (78)	0.87	93	TEND LESS TO BE THOSE WHERE THE SYSTEM STYLE IS NON-MOBILIZATIONAL (78)	0.66	3 29 21 57 X SQ= 3.13 P = 0.046 RV NO (NO)
94	LEAN TOWARD BEING THOSE WHERE THE STATUS OF THE REGIME IS CONSTITUTIONAL (51)	0.71	94	LEAN TOWARD BEING THOSE WHERE THE STATUS OF THE REGIME IS AUTHORITARIAN OR TOTALITARIAN (41)	0.50	17 34 7 34 X SQ= 2.33 P = 0.097 RV YES (NO)
96	TILT MORE TOWARD BEING THOSE WHERE THE STATUS OF THE REGIME IS CONSTITUTIONAL OR TOTALITARIAN (67)	0.87	96	TILT LESS TOWARD BEING THOSE WHERE THE STATUS OF THE REGIME IS CONSTITUTIONAL OR TOTALITARIAN (67)	0.70	21 46 3 20 X SQ= 2.07 P = 0.106 RV NO (NO)
102	LEAN TOWARD BEING THOSE WHERE THE REPRESENTATIVE CHARACTER OF THE REGIME IS POLYARCHIC OR LIMITED POLYARCHIC (59)	0.72	102	LEAN TOWARD BEING THOSE WHERE THE REPRESENTATIVE CHARACTER OF THE REGIME IS PSEUDO-POLYARCHIC OR NON-POLYARCHIC (49)	0.51	18 41 7 42 X SQ= 3.10 P = 0.066 RV YES (NO)
105	TEND TO BE THOSE WHERE THE ELECTORAL SYSTEM IS COMPETITIVE (43)	0.73	105	TEND TO BE THOSE WHERE THE ELECTORAL SYSTEM IS PARTIALLY COMPETITIVE OR NON-COMPETITIVE (47)	0.60	16 27 6 41 X SQ= 6.00 P = 0.013 RV YES (NO)

107	TEND TO BE THOSE WHERE AUTONOMOUS GROUPS ARE FULLY TOLERATED IN POLITICS (46)	0.68	107	TEND TO BE THOSE WHERE AUTONOMOUS GROUPS ARE PARTIALLY TOLERATED IN POLITICS, ARE TOLERATED ONLY OUTSIDE POLITICS, OR ARE NOT TOLERATED AT ALL (65)	0.66	M 17 29 8 57 X SQ= 8.02 P = 0.003 RV YES (NO)
114	TEND TO BE THOSE WHERE SECTIONALISM IS NEGLIGIBLE (47)	0.67	114	TEND TO BE THOSE WHERE SECTIONALISM IS EXTREME OR MODERATE (61)	0.62	7 54 14 33 X SQ= 4.57 P = 0.026 RV YES (NO)
116	TEND TO BE THOSE WHERE INTEREST ARTICULATION BY ASSOCIATIONAL GROUPS IS SIGNIFICANT OR MODERATE (32)	0.57	116	TEND TO BE THOSE WHERE INTEREST ARTICULATION BY ASSOCIATIONAL GROUPS IS LIMITED OR NEGLIGIBLE (79)	0.78	13 19 10 69 X SQ= 9.21 P = 0.002 RV YES (NO)
117	TEND TO BE THOSE WHERE INTEREST ARTICULATION BY ASSOCIATIONAL GROUPS IS SIGNIFICANT, MODERATE, OR LIMITED (60)	0.91	117	TEND TO BE THOSE WHERE INTEREST ARTICULATION BY ASSOCIATIONAL GROUPS IS NEGLIGIBLE (51)	0.56	21 39 2 49 X SQ= 14.37 P = 0. RV YES (NO)
121	TEND TO BE THOSE WHERE INTEREST ARTICULATION BY NON-ASSOCIATIONAL GROUPS IS MODERATE, LIMITED, OR NEGLIGIBLE (61)	0.96	121	TEND TO BE THOSE WHERE INTEREST ARTICULATION BY NON-ASSOCIATIONAL GROUPS IS SIGNIFICANT (54)	0.59	1 53 24 37 X SQ= 21.51 P = 0. RV YES (NO)
122	TEND TO BE THOSE WHERE INTEREST ARTICULATION BY NON-ASSOCIATIONAL GROUPS IS LIMITED OR NEGLIGIBLE (32)	0.64	122	TEND TO BE THOSE WHERE INTEREST ARTICULATION BY NON-ASSOCIATIONAL GROUPS IS SIGNIFICANT OR MODERATE (83)	0.82	9 74 16 16 X SQ= 18.58 P = 0. RV YES (YES)
124	ALWAYS ARE THOSE WHERE INTEREST ARTICULATION BY ANOMIC GROUPS IS OCCASIONAL, INFREQUENT, OR VERY INFREQUENT (85)	1.00	124	TEND LESS TO BE THOSE WHERE INTEREST ARTICULATION BY ANOMIC GROUPS IS OCCASIONAL, INFREQUENT, OR VERY INFREQUENT (85)	0.82	0 14 20 65 X SQ= 2.80 P = 0.040 RV NO (NO)
125	LEAN TOWARD BEING THOSE WHERE INTEREST ARTICULATION BY ANOMIC GROUPS IS INFREQUENT OR VERY INFREQUENT (35)	0.55	125	LEAN TOWARD BEING THOSE WHERE INTEREST ARTICULATION BY ANOMIC GROUPS IS FREQUENT OR OCCASIONAL (64)	0.70	9 55 11 24 X SQ= 3.22 P = 0.065 RV YES (NO)
129	LEAN MORE TOWARD BEING THOSE WHERE INTEREST ARTICULATION BY POLITICAL PARTIES IS SIGNIFICANT, MODERATE, OR LIMITED (56)	0.78	129	LEAN LESS TOWARD BEING THOSE WHERE INTEREST ARTICULATION BY POLITICAL PARTIES IS SIGNIFICANT, MODERATE, OR LIMITED (56)	0.54	18 38 5 32 X SQ= 3.21 P = 0.051 RV NO (NO)

138	TEND TO BE THOSE WHERE INTEREST AGGREGATION BY THE LEGISLATURE IS SIGNIFICANT, MODERATE, OR LIMITED (48)	0.82	138	TEND TO BE THOSE WHERE INTEREST AGGREGATION BY THE LEGISLATURE IS NEGLIGIBLE (49)	0.60	18 30 4 45 X SQ= 10.29 P = 0.001 RV YES (NO)
139	LEAN MORE TOWARD BEING THOSE WHERE THE PARTY SYSTEM IS QUANTITATIVELY OTHER THAN ONE-PARTY (71)	0.83	139	LEAN LESS TOWARD BEING THOSE WHERE THE PARTY SYSTEM IS QUANTITATIVELY OTHER THAN ONE-PARTY (71)	0.63	4 30 20 51 X SQ= 2.64 P = 0.082 RV NO (NO)
142	TEND TO BE THOSE WHERE THE PARTY SYSTEM IS QUANTITATIVELY MULTI-PARTY (30)	0.52	142	TEND TO BE THOSE WHERE THE PARTY SYSTEM IS QUANTITATIVELY OTHER THAN MULTI-PARTY (66)	0.75	12 18 11 55 X SQ= 4.95 P = 0.020 RV YES (NO)
148	ALWAYS ARE THOSE WHERE THE PARTY SYSTEM IS QUALITATIVELY OTHER THAN AFRICAN TRANSITIONAL (96)	1.00	148	TEND LESS TO BE THOSE WHERE THE PARTY SYSTEM IS QUALITATIVELY OTHER THAN AFRICAN TRANSITIONAL (96)	0.84	0 14 25 71 X SQ= 3.35 P = 0.037 RV NO (NO)
164	TEND MORE TO BE THOSE WHERE THE REGIME'S LEADERSHIP CHARISMA IS NEGLIGIBLE (65)	0.92	164	TEND LESS TO BE THOSE WHERE THE REGIME'S LEADERSHIP CHARISMA IS NEGLIGIBLE (65)	0.57	2 32 23 42 X SQ= 8.79 P = 0.001 RV NO (NO)
169	TEND TO BE THOSE WHERE THE HORIZONTAL POWER DISTRIBUTION IS SIGNIFICANT OR LIMITED (58)	0.79	169	TEND TO BE THOSE WHERE THE HORIZONTAL POWER DISTRIBUTION IS NEGLIGIBLE (48)	0.52	19 39 5 43 X SQ= 6.26 P = 0.009 RV YES (NO)
170	TEND TO BE THOSE WHERE THE LEGISLATIVE-EXECUTIVE STRUCTURE IS PRESIDENTIAL (39)	0.58	170	TEND TO BE THOSE WHERE THE LEGISLATIVE-EXECUTIVE STRUCTURE IS OTHER THAN PRESIDENTIAL (63)	0.68	14 25 10 53 X SQ= 4.31 P = 0.030 RV YES (NO)
175	TEND TO BE THOSE WHERE THE LEGISLATURE IS FULLY EFFECTIVE OR PARTIALLY EFFECTIVE (51)	0.74	175	TEND TO BE THOSE WHERE THE LEGISLATURE IS LARGELY INEFFECTIVE OR WHOLLY INEFFECTIVE (49)	0.56	17 34 6 43 X SQ= 5.14 P = 0.017 RV YES (NO)
178	TEND TO BE THOSE WHERE THE LEGISLATURE IS BICAMERAL (51)	0.71	178	TEND TO BE THOSE WHERE THE LEGISLATURE IS UNICAMERAL (53)	0.57	7 46 17 34 X SQ= 4.85 P = 0.020 RV YES (NO)

```
182  ALWAYS ARE THOSE                          1.00      182  TEND LESS TO BE THOSE                      0.59         19    36
     WHERE THE BUREAUCRACY                                    WHERE THE BUREAUCRACY                                    0    25
     IS SEMI-MODERN, RATHER THAN                              IS SEMI-MODERN, RATHER THAN                          X SQ=   9.50
     POST-COLONIAL TRANSITIONAL  (55)                         POST-COLONIAL TRANSITIONAL  (55)                     P =     0.
                                                                                                                   RV NO  (NO )

185  LEAN TOWARD BEING THOSE                   0.70      185  LEAN TOWARD BEING THOSE                    0.67          7    14
     WHERE PARTICIPATION BY THE MILITARY                      WHERE PARTICIPATION BY THE MILITARY                      3    28
     IN POLITICS IS                                           IN POLITICS IS                                       X SQ=   3.12
     INTERVENTIVE, RATHER THAN                                SUPPORTIVE, RATHER THAN                              P =     0.069
     SUPPORTIVE  (21)                                         INTERVENTIVE  (31)                                   RV YES (NO )

188  TEND TO BE THOSE                          0.84      188  TEND TO BE THOSE                           0.87         21    11
     WHERE THE CHARACTER OF THE LEGAL SYSTEM                  WHERE THE CHARACTER OF THE LEGAL SYSTEM                  4    77
     IS CIVIL LAW  (32)                                       IS OTHER THAN CIVIL LAW  (81)                        X SQ=  45.57
                                                                                                                   P =     0.
                                                                                                                   RV YES (YES)

190  TEND MORE TO BE THOSE                     0.95      190  TEND LESS TO BE THOSE                      0.65         21    11
     WHERE THE CHARACTER OF THE LEGAL SYSTEM                  WHERE THE CHARACTER OF THE LEGAL SYSTEM                  1     6
     IS CIVIL LAW, RATHER THAN                                IS CIVIL LAW, RATHER THAN                            X SQ=   4.25
     COMMON LAW  (32)                                         COMMON LAW  (32)                                     P =     0.030
                                                                                                                   RV NO  (YES)

192  ALWAYS ARE THOSE                          1.00      192  TEND LESS TO BE THOSE                      0.69          0    27
     WHERE THE CHARACTER OF THE LEGAL SYSTEM                  WHERE THE CHARACTER OF THE LEGAL SYSTEM                 25    61
     IS OTHER THAN                                            IS OTHER THAN                                        X SQ=   8.46
     MUSLIM OR PARTLY MUSLIM  (86)                            MUSLIM OR PARTLY MUSLIM  (86)                        P =     0.
                                                                                                                   RV NO  (NO )

193  ALWAYS ARE THOSE                          1.00      193  TEND LESS TO BE THOSE                      0.69          0    28
     WHERE THE CHARACTER OF THE LEGAL SYSTEM                  WHERE THE CHARACTER OF THE LEGAL SYSTEM                 25    61
     IS OTHER THAN PARTLY INDIGENOUS  (86)                    IS OTHER THAN PARTLY INDIGENOUS  (86)                X SQ=   8.80
                                                                                                                   P =     0.
                                                                                                                   RV NO  (NO )
```

MATRIX

57 POLITIES WHERE THE RELIGION IS CATHOLIC (25)	57 POLITIES WHERE THE RELIGION IS OTHER THAN CATHOLIC (90)
36 POLITIES WHOSE PER CAPITA GROSS NATIONAL PRODUCT IS VERY HIGH, HIGH, OR MEDIUM (42)	
13	29
ARGENTINA, FRANCE, SPAIN, AUSTRIA, IRELAND, URUGUAY, BELGIUM, ITALY, VENEZUELA, CHILE, LUXEMBOURG, CUBA, POLAND	AUSTRALIA, DENMARK, HUNGARY, LEBANON, RUMANIA, USSR, BULGARIA, FINLAND, ICELAND, MALAYA, SO AFRICA, UK, CANADA, GERMANY, E, ISRAEL, NETHERLANDS, SWEDEN, US, CYPRUS, GERMAN FR, JAMAICA, NEW ZEALAND, SWITZERLAND, YUGOSLAVIA, CZECHOS'KIA, GREECE, JAPAN, NORWAY, TRINIDAD
36 POLITIES WHOSE PER CAPITA GROSS NATIONAL PRODUCT IS LOW OR VERY LOW (73)	
12	61
BRAZIL, HONDURAS, PHILIPPINES, COLOMBIA, MEXICO, PORTUGAL, COSTA RICA, NICARAGUA, DOMIN REP, PANAMA, EL SALVADOR, PARAGUAY	AFGHANISTAN, BURUNDI, CHAD, ECUADOR, GUINEA, IRAQ, LAOS, MAURITANIA, NIGERIA, SENEGAL, TANGANYIKA, UGANDA, YEMEN, ALBANIA, CAMBODIA, CHINA, PR, ETHIOPIA, HAITI, IVORY COAST, LIBERIA, MONGOLIA, PAKISTAN, SIERRE LEO, THAILAND, UAR, ALGERIA, CAMEROUN, CONGO(BRA), GABON, INDIA, JORDAN, LIBYA, MOROCCO, PERU, SOMALIA, TOGO, UPPER VOLTA, BOLIVIA, CEN AFR REP, CONGO(LEO), GHANA, INDONESIA, KOREA, N, MALAGASY R, NEPAL, RWANDA, SUDAN, TUNISIA, VIETNAM, N, BURMA, CEYLON, DAHOMEY, GUATEMALA, IRAN, KOREA REP, MALI, NIGER, SA'U ARABIA, SYRIA, TURKEY, VIETNAM REP

M

```
*****************************************************MATRIX*****************************************************

         57  POLITIES                                           57  POLITIES
             WHERE THE RELIGION IS                                  WHERE THE RELIGION IS OTHER THAN
             CATHOLIC  (25)                            .            CATHOLIC  (90)
                                                       .
                                                       .
                                  107  POLITIES
                                       WHERE AUTONOMOUS GROUPS
                                       ARE FULLY TOLERATED IN POLITICS  (46)
                                                       .
ARGENTINA  AUSTRIA     BELGIUM     BRAZIL     CHILE       .AUSTRALIA  BOLIVIA     CANADA     CEYLON      CYPRUS
COLOMBIA   COSTA RICA  DOMIN REP   FRANCE     HONDURAS    .DENMARK    ECUADOR     FINLAND    GERMAN FR   GREECE
IRELAND    ITALY       LUXEMBOURG  PANAMA     PHILIPPINES.ICELAND     INDIA       ISRAEL     JAMAICA     JAPAN
URUGUAY    VENEZUELA                                      .LEBANON    MALAGASY R  MALAYA     NETHERLANDS NEW ZEALAND
                                                          .NORWAY     SIERRE LEO  SWEDEN     SWITZERLAND TRINIDAD
                                                          .TURKEY     UGANDA      UK         US
                                                       .
                                                   17  .  29
. . . . . . . . . . . . . . . . . . . . . . . . . . . . . . . . . . . . . . . . . . . . . . . . . . . . . . . . . .
                                                    8  .  57
                                                       .
                                  107  POLITIES
                                       WHERE AUTONOMOUS GROUPS
                                       ARE PARTIALLY TOLERATED IN POLITICS,
                                       ARE TOLERATED ONLY OUTSIDE POLITICS,
                                       OR ARE NOT TOLERATED AT ALL  (65)
                                                       .
CUBA       EL SALVADOR MEXICO      NICARAGUA  PARAGUAY    .AFGHANISTAN ALBANIA    ALGERIA     BULGARIA    BURMA
POLAND     PORTUGAL    SPAIN                              .BURUNDI    CAMBODIA    CAMEROUN    CEN AFR REP CHAD
                                                          .CHINA, PR  CONGO(BRA)  CONGO(LEO)  CZECHOS'KIA DAHOMEY
                                                          .ETHIOPIA   GABON       GERMANY, E  GHANA       GUATEMALA
                                                          .GUINEA     HAITI       HUNGARY     INDONESIA   IRAN
                                                          .IRAQ       IVORY COAST JORDAN      KOREA, N    KOREA REP
                                                          .LIBERIA    LIBYA       MALI        MAURITANIA  MONGOLIA
                                                          .MOROCCO    NEPAL       NIGER       NIGERIA     PAKISTAN
                                                          .RUMANIA    RWANDA      SA'U ARABIA SENEGAL     SOMALIA
                                                          .SO AFRICA  SUDAN       SYRIA       TANGANYIKA  THAILAND
                                                          .TUNISIA    USSR        UAR         UPPER VOLTA VIETNAM, N
                                                          .VIETNAM REP YUGOSLAVIA
```

58 POLITIES
WHERE THE RELIGION IS
MUSLIM (18)

58 POLITIES
WHERE THE RELIGION IS OTHER THAN
MUSLIM (97)

BOTH SUBJECT AND PREDICATE

18 IN LEFT COLUMN

AFGHANISTAN	ALGERIA	INDONESIA	IRAN	IRAQ	JORDAN	LIBYA	MAURITANIA	MOROCCO	PAKISTAN
SA'U ARABIA	SENEGAL	SOMALIA	SYRIA	TUNISIA	TURKEY	UAR	YEMEN		

97 IN RIGHT COLUMN

ALBANIA	ARGENTINA	AUSTRALIA	AUSTRIA	BELGIUM	BOLIVIA	BRAZIL	BULGARIA	BURMA	BURUNDI
CAMBODIA	CAMEROUN	CANADA	CEN AFR REP	CEYLON	CHAD	CHILE	CHINA, PR	COLOMBIA	CONGO(BRA)
CONGO(LEO)	COSTA RICA	CUBA	CYPRUS	CZECHOS'KIA	DAHOMEY	DENMARK	DOMIN REP	ECUADOR	EL SALVADOR
ETHIOPIA	FINLAND	FRANCE	GABON	GERMANY, E	GERMAN FR	GHANA	GREECE	GUATEMALA	GUINEA
HAITI	HONDURAS	HUNGARY	ICELAND	INDIA	IRELAND	ISRAEL	ITALY	IVORY COAST	JAMAICA
JAPAN	KOREA, N	KOREA REP	LAOS	LEBANON	LIBERIA	LUXEMBOURG	MALAGASY R	MALAYA	MALI
MEXICO	MONGOLIA	NEPAL	NETHERLANDS	NEW ZEALAND	NICARAGUA	NIGER	NIGERIA	NORWAY	PANAMA
PARAGUAY	PERU	PHILIPPINES	POLAND	PORTUGAL	RUMANIA	RWANDA	SIERRE LEO	SO AFRICA	SPAIN
SUDAN	SWEDEN	SWITZERLAND	TANGANYIKA	THAILAND	TOGO	TRINIDAD	UGANDA	USSR	UK
US	UPPER VOLTA	URUGUAY	VENEZUELA	VIETNAM, N	VIETNAM REP	YUGOSLAVIA			

	Left column			Right column		
2	ALWAYS ARE THOSE LOCATED ELSEWHERE THAN IN WEST EUROPE, SCANDINAVIA, NORTH AMERICA, OR AUSTRALASIA (93)	1.00	2	TEND LESS TO BE THOSE LOCATED ELSEWHERE THAN IN WEST EUROPE, SCANDINAVIA, NORTH AMERICA, OR AUSTRALASIA (93)	0.77	0 22 18 75 X SQ= 3.69 P = 0.022 RV NO (NO)
6	ALWAYS ARE THOSE LOCATED ELSEWHERE THAN IN THE CARIBBEAN, CENTRAL AMERICA, OR SOUTH AMERICA (93)	1.00	6	TEND LESS TO BE THOSE LOCATED ELSEWHERE THAN IN THE CARIBBEAN, CENTRAL AMERICA, OR SOUTH AMERICA (93)	0.77	0 22 18 75 X SQ= 3.69 P = 0.022 RV NO (NO)
13	TEND TO BE THOSE LOCATED IN NORTH AFRICA OR THE MIDDLE EAST (16)	0.72	13	TEND TO BE THOSE LOCATED ELSEWHERE THAN IN NORTH AFRICA OR THE MIDDLE EAST (99)	0.97	13 3 5 94 X SQ= 54.94 P = 0. RV YES (YES)
14	TEND LESS TO BE THOSE LOCATED ELSEWHERE THAN IN THE MIDDLE EAST (104)	0.56	14	TEND MORE TO BE THOSE LOCATED ELSEWHERE THAN IN THE MIDDLE EAST (104)	0.97	8 3 10 94 X SQ= 25.42 P = 0. RV NO (YES)

15	TEND TO BE THOSE LOCATED IN THE MIDDLE EAST, RATHER THAN IN NORTH AFRICA OR CENTRAL AND SOUTH AFRICA (11)	0.50	15	TEND TO BE THOSE LOCATED IN NORTH AFRICA OR CENTRAL AND SOUTH AFRICA, RATHER THAN IN THE MIDDLE EAST (33)	0.89	8 3 8 25 X SQ= 6.42 P = 0.009 RV YES (YES)
16	TEND TO BE THOSE LOCATED IN THE MIDDLE EAST, RATHER THAN IN EAST ASIA, SOUTH ASIA, OR SOUTHEAST ASIA (11)	0.80	16	TEND TO BE THOSE LOCATED IN EAST ASIA, SOUTH ASIA, OR SOUTHEAST ASIA, RATHER THAN IN THE MIDDLE EAST (18)	0.84	8 3 2 16 X SQ= 8.91 P = 0.001 RV YES (YES)
22	TEND MORE TO BE THOSE WHOSE TERRITORIAL SIZE IS VERY LARGE, LARGE, OR MEDIUM (68)	0.83	22	TEND LESS TO BE THOSE WHOSE TERRITORIAL SIZE IS VERY LARGE, LARGE, OR MEDIUM (68)	0.55	15 53 3 44 X SQ= 4.05 P = 0.035 RV NO (NO)
25	ALWAYS ARE THOSE WHOSE POPULATION DENSITY IS MEDIUM OR LOW (98)	1.00	25	LEAN LESS TOWARD BEING THOSE WHOSE POPULATION DENSITY IS MEDIUM OR LOW (98)	0.82	0 17 18 80 X SQ= 2.44 P = 0.070 RV NO (NO)
26	TEND MORE TO BE THOSE WHOSE POPULATION DENSITY IS LOW (67)	0.89	26	TEND LESS TO BE THOSE WHOSE POPULATION DENSITY IS LOW (67)	0.53	2 46 16 51 X SQ= 6.81 P = 0.004 RV NO (NO)
28	TEND MORE TO BE THOSE WHOSE POPULATION GROWTH RATE IS HIGH (62)	0.83	28	TEND LESS TO BE THOSE WHOSE POPULATION GROWTH RATE IS HIGH (62)	0.51	15 47 3 45 X SQ= 5.12 P = 0.018 RV NO (NO)
30	TEND TO BE THOSE WHOSE AGRICULTURAL POPULATION IS HIGH (56)	0.82	30	TEND TO BE THOSE WHOSE AGRICULTURAL POPULATION IS MEDIUM, LOW, OR VERY LOW (57)	0.56	14 42 3 54 X SQ= 7.13 P = 0.004 RV YES (NO)
31	ALWAYS ARE THOSE WHOSE AGRICULTURAL POPULATION IS HIGH OR MEDIUM (90)	1.00	31	TEND LESS TO BE THOSE WHOSE AGRICULTURAL POPULATION IS HIGH OR MEDIUM (90)	0.75	17 73 0 24 X SQ= 3.94 P = 0.021 RV NO (NO)
36	ALWAYS ARE THOSE WHOSE PER CAPITA GROSS NATIONAL PRODUCT IS LOW OR VERY LOW (73)	1.00	36	TEND LESS TO BE THOSE WHOSE PER CAPITA GROSS NATIONAL PRODUCT IS LOW OR VERY LOW (73)	0.57	0 42 18 55 X SQ= 10.48 P = 0. RV NO (NO)

37	TILT TOWARD BEING THOSE WHOSE PER CAPITA GROSS NATIONAL PRODUCT IS VERY LOW (51)	0.61	37	TILT TOWARD BEING THOSE WHOSE PER CAPITA GROSS NATIONAL PRODUCT IS VERY HIGH, HIGH, MEDIUM, OR LOW (64)	0.59	7 57 11 40 X SQ= 1.69 P = 0.131 RV YES (NO)
42	ALWAYS ARE THOSE WHOSE ECONOMIC DEVELOPMENTAL STATUS IS UNDERDEVELOPED OR VERY UNDERDEVELOPED (76)	1.00	42	TEND LESS TO BE THOSE WHOSE ECONOMIC DEVELOPMENTAL STATUS IS UNDERDEVELOPED OR VERY UNDERDEVELOPED (76)	0.62	0 36 17 59 X SQ= 7.84 P = 0.001 RV NO (NO)
43	TEND TO BE THOSE WHOSE ECONOMIC DEVELOPMENTAL STATUS IS VERY UNDERDEVELOPED (57)	0.76	43	TEND TO BE THOSE WHOSE ECONOMIC DEVELOPMENTAL STATUS IS DEVELOPED, INTERMEDIATE, OR UNDERDEVELOPED (55)	0.54	4 51 13 44 X SQ= 4.11 P = 0.033 RV YES (NO)
44	ALWAYS ARE THOSE WHERE THE LITERACY RATE IS BELOW NINETY PERCENT (90)	1.00	44	TEND LESS TO BE THOSE WHERE THE LITERACY RATE IS BELOW NINETY PERCENT (90)	0.74	0 25 18 72 X SQ= 4.51 P = 0.012 RV NO (NO)
45	TEND TO BE THOSE WHERE THE LITERACY RATE IS BELOW FIFTY PERCENT (54)	0.94	45	TEND TO BE THOSE WHERE THE LITERACY RATE IS FIFTY PERCENT OR ABOVE (55)	0.59	1 54 17 37 X SQ= 15.31 P = 0. RV YES (NO)
50	TEND TO BE THOSE WHERE FREEDOM OF THE PRESS IS INTERMITTENT, INTERNALLY ABSENT, OR INTERNALLY AND EXTERNALLY ABSENT (56)	0.88	50	TEND TO BE THOSE WHERE FREEDOM OF THE PRESS IS COMPLETE (43)	0.50	2 41 15 41 X SQ= 6.89 P = 0.006 RV YES (NO)
51	LEAN TOWARD BEING THOSE WHERE FREEDOM OF THE PRESS IS INTERNALLY ABSENT, OR INTERNALLY AND EXTERNALLY ABSENT (37)	0.60	51	LEAN TOWARD BEING THOSE WHERE FREEDOM OF THE PRESS IS COMPLETE OR INTERMITTENT (60)	0.66	6 54 9 28 X SQ= 2.58 P = 0.082 RV YES (NO)
54	ALWAYS ARE THOSE WHERE NEWSPAPER CIRCULATION IS LESS THAN ONE HUNDRED PER THOUSAND (76)	1.00	54	TEND LESS TO BE THOSE WHERE NEWSPAPER CIRCULATION IS LESS THAN ONE HUNDRED PER THOUSAND (76)	0.61	0 37 18 58 X SQ= 8.73 P = 0.001 RV NO (NO)
55	LEAN TOWARD BEING THOSE WHERE NEWSPAPER CIRCULATION IS LESS THAN TEN PER THOUSAND (35)	0.50	55	LEAN TOWARD BEING THOSE WHERE NEWSPAPER CIRCULATION IS TEN OR MORE PER THOUSAND (78)	0.73	9 69 9 26 X SQ= 2.64 P = 0.092 RV YES (NO)

72	TEND TO BE THOSE WHOSE DATE OF INDEPENDENCE IS AFTER 1914 (59)	0.83	72	TEND TO BE THOSE WHOSE DATE OF INDEPENDENCE IS BEFORE 1914 (52)	0.53	3 49 15 44 X SQ= 6.48 P = 0.008 RV YES (NO)
73	LEAN TOWARD BEING THOSE WHOSE DATE OF INDEPENDENCE IS AFTER 1945 (46)	0.61	73	LEAN TOWARD BEING THOSE WHOSE DATE OF INDEPENDENCE IS BEFORE 1945 (65)	0.62	7 58 11 35 X SQ= 2.53 P = 0.073 RV YES (NO)
74	ALWAYS ARE THOSE THAT ARE NOT HISTORICALLY WESTERN (87)	1.00	74	TEND LESS TO BE THOSE THAT ARE NOT HISTORICALLY WESTERN (87)	0.73	0 26 18 69 X SQ= 4.95 P = 0.007 RV NO (NO)
75	TEND TO BE THOSE THAT ARE NOT HISTORICALLY WESTERN AND ARE NOT SIGNIFICANTLY WESTERNIZED (52)	0.83	75	TEND TO BE THOSE THAT ARE HISTORICALLY WESTERN OR SIGNIFICANTLY WESTERNIZED (62)	0.61	3 59 15 37 X SQ= 10.52 P = 0.001 RV YES (NO)
76	ALWAYS ARE THOSE THAT HAVE BEEN SIGNIFICANTLY OR PARTIALLY WESTERNIZED THROUGH A COLONIAL RELATIONSHIP, RATHER THAN BEING HISTORICALLY WESTERN (70)	1.00	76	TEND LESS TO BE THOSE THAT HAVE BEEN SIGNIFICANTLY OR PARTIALLY WESTERNIZED THROUGH A COLONIAL RELATIONSHIP, RATHER THAN BEING HISTORICALLY WESTERN (70)	0.68	0 26 14 56 X SQ= 4.59 P = 0.010 RV NO (NO)
77	TEND MORE TO BE THOSE THAT HAVE BEEN PARTIALLY WESTERNIZED, RATHER THAN SIGNIFICANTLY WESTERNIZED, THROUGH A COLONIAL RELATIONSHIP (41)	0.86	77	TEND LESS TO BE THOSE THAT HAVE BEEN PARTIALLY WESTERNIZED, RATHER THAN SIGNIFICANTLY WESTERNIZED, THROUGH A COLONIAL RELATIONSHIP (41)	0.53	2 26 12 29 X SQ= 3.76 P = 0.033 RV NO (NO)
79	ALWAYS ARE THOSE THAT WERE FORMERLY DEPENDENCIES OF BRITAIN OR FRANCE, RATHER THAN SPAIN (49)	1.00	79	LEAN LESS TOWARD BEING THOSE THAT WERE FORMERLY DEPENDENCIES OF BRITAIN OR FRANCE, RATHER THAN SPAIN (49)	0.68	10 39 0 18 X SQ= 2.86 P = 0.052 RV NO (NO)
82	ALWAYS ARE THOSE WHERE THE TYPE OF POLITICAL MODERNIZATION IS DEVELOPED TUTELARY OR UNDEVELOPED TUTELARY, RATHER THAN EARLY OR LATER EUROPEAN OR EUROPEAN DERIVED (55)	1.00	82	TEND TO BE THOSE WHERE THE TYPE OF POLITICAL MODERNIZATION IS EARLY OR LATER EUROPEAN OR EUROPEAN DERIVED, RATHER THAN DEVELOPED TUTELARY OR UNDEVELOPED TUTELARY (51)	0.56	0 51 15 40 X SQ= 14.03 P = 0. RV YES (NO)
84	TEND TO BE THOSE WHERE THE TYPE OF POLITICAL MODERNIZATION IS DEVELOPED TUTELARY, RATHER THAN UNDEVELOPED TUTELARY (31)	0.80	84	TEND TO BE THOSE WHERE THE TYPE OF POLITICAL MODERNIZATION IS UNDEVELOPED TUTELARY, RATHER THAN DEVELOPED TUTELARY (24)	0.52	12 19 3 21 X SQ= 3.46 P = 0.037 RV YES (NO)

89	TEND MORE TO BE THOSE WHOSE IDEOLOGICAL ORIENTATION IS OTHER THAN CONVENTIONAL (62)	0.94	89	TEND LESS TO BE THOSE WHOSE IDEOLOGICAL ORIENTATION IS OTHER THAN CONVENTIONAL (62)	0.59	1 32 15 47 X SQ= 5.46 P = 0.009 RV NO (NO)
91	TILT LESS TOWARD BEING THOSE WHOSE IDEOLOGICAL ORIENTATION IS DEVELOPMENTAL, RATHER THAN TRADITIONAL (31)	0.70	91	TILT MORE TOWARD BEING THOSE WHOSE IDEOLOGICAL ORIENTATION IS DEVELOPMENTAL, RATHER THAN TRADITIONAL (31)	0.92	7 24 3 2 X SQ= 1.43 P = 0.119 RV NO (YES)
95	ALWAYS ARE THOSE WHERE THE STATUS OF THE REGIME IS CONSTITUTIONAL OR AUTHORITARIAN (95)	1.00	95	LEAN LESS TOWARD BEING THOSE WHERE THE STATUS OF THE REGIME IS CONSTITUTIONAL OR AUTHORITARIAN (95)	0.83	17 78 0 16 X SQ= 2.14 P = 0.073 RV NO (NO)
96	TEND TO BE THOSE WHERE THE STATUS OF THE REGIME IS AUTHORITARIAN (23)	0.62	96	TEND TO BE THOSE WHERE THE STATUS OF THE REGIME IS CONSTITUTIONAL OR TOTALITARIAN (67)	0.81	5 62 8 15 X SQ= 8.25 P = 0.003 RV YES (NO)
97	ALWAYS ARE THOSE WHERE THE STATUS OF THE REGIME IS AUTHORITARIAN, RATHER THAN TOTALITARIAN (23)	1.00	97	TEND TO BE THOSE WHERE THE STATUS OF THE REGIME IS TOTALITARIAN, RATHER THAN AUTHORITARIAN (16)	0.52	8 15 0 16 X SQ= 5.03 P = 0.012 RV YES (NO)
102	LEAN TOWARD BEING THOSE WHERE THE REPRESENTATIVE CHARACTER OF THE REGIME IS PSEUDO-POLYARCHIC OR NON-POLYARCHIC (49)	0.67	102	LEAN TOWARD BEING THOSE WHERE THE REPRESENTATIVE CHARACTER OF THE REGIME IS POLYARCHIC OR LIMITED POLYARCHIC (59)	0.58	5 54 10 39 X SQ= 2.27 P = 0.096 RV YES (NO)
105	TEND TO BE THOSE WHERE THE ELECTORAL SYSTEM IS PARTIALLY COMPETITIVE OR NON-COMPETITIVE (47)	0.90	105	TEND TO BE THOSE WHERE THE ELECTORAL SYSTEM IS COMPETITIVE (43)	0.52	1 42 9 38 X SQ= 4.84 P = 0.016 RV YES (NO)
107	TEND MORE TO BE THOSE WHERE AUTONOMOUS GROUPS ARE PARTIALLY TOLERATED IN POLITICS, ARE TOLERATED ONLY OUTSIDE POLITICS, OR ARE NOT TOLERATED AT ALL (65)	0.94	107	TEND LESS TO BE THOSE WHERE AUTONOMOUS GROUPS ARE PARTIALLY TOLERATED IN POLITICS, ARE TOLERATED ONLY OUTSIDE POLITICS, OR ARE NOT TOLERATED AT ALL (65)	0.52	1 45 16 49 X SQ= 8.80 P = 0.001 RV NO (NO)
114	LEAN MORE TOWARD BEING THOSE WHERE SECTIONALISM IS EXTREME OR MODERATE (61)	0.78	114	LEAN LESS TOWARD BEING THOSE WHERE SECTIONALISM IS EXTREME OR MODERATE (61)	0.52	14 47 4 43 X SQ= 3.01 P = 0.067 RV NO (NO)

115	ALWAYS ARE THOSE WHERE INTEREST ARTICULATION BY ASSOCIATIONAL GROUPS IS MODERATE, LIMITED, OR NEGLIGIBLE (91)	1.00	115	TEND LESS TO BE THOSE WHERE INTEREST ARTICULATION BY ASSOCIATIONAL GROUPS IS MODERATE, LIMITED, OR NEGLIGIBLE (91)	0.78	0 20 18 73 X SQ= 3.38 P = 0.039 RV NO (NO)
116	TEND MORE TO BE THOSE WHERE INTEREST ARTICULATION BY ASSOCIATIONAL GROUPS IS LIMITED OR NEGLIGIBLE (79)	0.94	116	TEND LESS TO BE THOSE WHERE INTEREST ARTICULATION BY ASSOCIATIONAL GROUPS IS LIMITED OR NEGLIGIBLE (79)	0.67	1 31 17 62 X SQ= 4.40 P = 0.021 RV NO (NO)
119	ALWAYS ARE THOSE WHERE INTEREST ARTICULATION BY INSTITUTIONAL GROUPS IS VERY SIGNIFICANT OR SIGNIFICANT (74)	1.00	119	TEND LESS TO BE THOSE WHERE INTEREST ARTICULATION BY INSTITUTIONAL GROUPS IS VERY SIGNIFICANT OR SIGNIFICANT (74)	0.69	16 58 0 26 X SQ= 5.18 P = 0.006 RV NO (NO)
121	LEAN TOWARD BEING THOSE WHERE INTEREST ARTICULATION BY NON-ASSOCIATIONAL GROUPS IS SIGNIFICANT (54)	0.67	121	LEAN TOWARD BEING THOSE WHERE INTEREST ARTICULATION BY NON-ASSOCIATIONAL GROUPS IS MODERATE, LIMITED, OR NEGLIGIBLE (61)	0.57	12 42 6 55 X SQ= 2.46 P = 0.078 RV YES (NO)
122	ALWAYS ARE THOSE WHERE INTEREST ARTICULATION BY NON-ASSOCIATIONAL GROUPS IS SIGNIFICANT OR MODERATE (83)	1.00	122	TEND LESS TO BE THOSE WHERE INTEREST ARTICULATION BY NON-ASSOCIATIONAL GROUPS IS SIGNIFICANT OR MODERATE (83)	0.67	18 65 0 32 X SQ= 6.67 P = 0.003 RV NO (NO)
126	ALWAYS ARE THOSE WHERE INTEREST ARTICULATION BY ANOMIC GROUPS IS FREQUENT, OCCASIONAL, OR INFREQUENT (83)	1.00	126	LEAN LESS TOWARD BEING THOSE WHERE INTEREST ARTICULATION BY ANOMIC GROUPS IS FREQUENT, OCCASIONAL, OR INFREQUENT (83)	0.81	16 67 0 16 X SQ= 2.39 P = 0.067 RV NO (NO)
142	ALWAYS ARE THOSE WHERE THE PARTY SYSTEM IS QUANTITATIVELY OTHER THAN MULTI-PARTY (66)	1.00	142	LEAN LESS TOWARD BEING THOSE WHERE THE PARTY SYSTEM IS QUANTITATIVELY OTHER THAN MULTI-PARTY (66)	0.66	0 30 9 57 X SQ= 3.05 P = 0.053 RV NO (NO)
153	TEND TO BE THOSE WHERE THE PARTY SYSTEM IS MODERATELY STABLE OR UNSTABLE (38)	0.87	153	TEND TO BE THOSE WHERE THE PARTY SYSTEM IS STABLE (42)	0.57	1 41 7 31 X SQ= 4.06 P = 0.024 RV YES (NO)
154	TILT TOWARD BEING THOSE WHERE THE PARTY SYSTEM IS UNSTABLE (25)	0.62	154	TILT TOWARD BEING THOSE WHERE THE PARTY SYSTEM IS STABLE OR MODERATELY STABLE (55)	0.72	3 52 5 20 X SQ= 2.59 P = 0.100 RV YES (NO)

164	TEND TO BE THOSE WHERE THE REGIME'S LEADERSHIP CHARISMA IS PRONOUNCED OR MODERATE (34)	0.85	164	TEND TO BE THOSE WHERE THE REGIME'S LEADERSHIP CHARISMA IS NEGLIGIBLE (65)	0.73	M 11 23 2 63 X SQ= 14.30 P = 0. RV YES (NO)
169	LEAN TOWARD BEING THOSE WHERE THE HORIZONTAL POWER DISTRIBUTION IS NEGLIGIBLE (48)	0.67	169	LEAN TOWARD BEING THOSE WHERE THE HORIZONTAL POWER DISTRIBUTION IS SIGNIFICANT OR LIMITED (58)	0.58	5 53 10 38 X SQ= 2.30 P = 0.095 RV YES (NO)
172	ALWAYS ARE THOSE WHERE THE LEGISLATIVE-EXECUTIVE STRUCTURE IS OTHER THAN PARLIAMENTARY-ROYALIST (88)	1.00	172	TEND LESS TO BE THOSE WHERE THE LEGISLATIVE-EXECUTIVE STRUCTURE IS OTHER THAN PARLIAMENTARY-ROYALIST (88)	0.78	0 21 15 73 X SQ= 2.84 P = 0.040 RV NO (NO)
174	ALWAYS ARE THOSE WHERE THE LEGISLATURE IS PARTIALLY EFFECTIVE, LARGELY INEFFECTIVE, OR WHOLLY INEFFECTIVE (72)	1.00	174	TEND LESS TO BE THOSE WHERE THE LEGISLATURE IS PARTIALLY EFFECTIVE, LARGELY INEFFECTIVE, OR WHOLLY INEFFECTIVE (72)	0.68	0 28 12 60 X SQ= 3.84 P = 0.018 RV NO (NO)
175	LEAN TOWARD BEING THOSE WHERE THE LEGISLATURE IS LARGELY INEFFECTIVE OR WHOLLY INEFFECTIVE (49)	0.75	175	LEAN TOWARD BEING THOSE WHERE THE LEGISLATURE IS FULLY EFFECTIVE OR PARTIALLY EFFECTIVE (51)	0.55	3 48 9 40 X SQ= 2.60 P = 0.069 RV YES (NO)
179	TEND MORE TO BE THOSE WHERE THE EXECUTIVE IS DOMINANT (52)	0.86	179	TEND LESS TO BE THOSE WHERE THE EXECUTIVE IS DOMINANT (52)	0.52	12 40 2 37 X SQ= 4.22 P = 0.021 RV NO (NO)
186	TEND TO BE THOSE WHERE PARTICIPATION BY THE MILITARY IN POLITICS IS SUPPORTIVE, RATHER THAN NEUTRAL (31)	0.64	186	TEND TO BE THOSE WHERE PARTICIPATION BY THE MILITARY IN POLITICS IS NEUTRAL, RATHER THAN SUPPORTIVE (56)	0.68	7 24 4 52 X SQ= 3.02 P = 0.049 RV YES (NO)
187	TEND MORE TO BE THOSE WHERE THE ROLE OF THE POLICE IS POLITICALLY SIGNIFICANT (66)	0.94	187	TEND LESS TO BE THOSE WHERE THE ROLE OF THE POLICE IS POLITICALLY SIGNIFICANT (66)	0.60	16 50 1 34 X SQ= 6.02 P = 0.005 RV NO (NO)
188	TEND MORE TO BE THOSE WHERE THE CHARACTER OF THE LEGAL SYSTEM IS OTHER THAN CIVIL LAW (81)	0.94	188	TEND LESS TO BE THOSE WHERE THE CHARACTER OF THE LEGAL SYSTEM IS OTHER THAN CIVIL LAW (81)	0.67	1 31 17 64 X SQ= 4.21 P = 0.021 RV NO (NO)

192	TEND TO BE THOSE WHERE THE CHARACTER OF THE LEGAL SYSTEM IS MUSLIM OR PARTLY MUSLIM (27)	0.78	192	TEND TO BE THOSE WHERE THE CHARACTER OF THE LEGAL SYSTEM IS OTHER THAN MUSLIM OR PARTLY MUSLIM (86)	0.86	14 13 4 82 X SQ= 30.75 P = 0. RV YES (YES)
194	ALWAYS ARE THOSE THAT ARE NON-COMMUNIST (101)	1.00	194	TILT LESS TOWARD BEING THOSE THAT ARE NON-COMMUNIST (101)	0.86	0 13 18 83 X SQ= 1.57 P = 0.125 RV NO (NO)

M

MATRIX

	58 POLITIES WHERE THE RELIGION IS MUSLIM (18)	58 POLITIES WHERE THE RELIGION IS OTHER THAN MUSLIM (97)
164 POLITIES WHERE THE REGIME'S LEADERSHIP CHARISMA IS PRONOUNCED OR MODERATE (34)	11 ALGERIA, MOROCCO, UAR, INDONESIA, PAKISTAN, JORDAN, SA'U ARABIA, LIBYA, SENEGAL, MAURITANIA, TUNISIA	23 CAMBODIA, CONGO(BRA), GABON, JAMAICA, UPPER VOLTA, CAMEROUN, CUBA, GHANA, KOREA, N, VIETNAM, N, CEN AFR REP, DAHOMEY, GUINEA, MALI, YUGOSLAVIA, CHAD, ETHIOPIA, INDIA, NIGER, CHINA, PR, FRANCE, IVORY COAST, TANGANYIKA
164 POLITIES WHERE THE REGIME'S LEADERSHIP CHARISMA IS NEGLIGIBLE (65)	2 SOMALIA, TURKEY	63 ALBANIA, BOLIVIA, CHILE, DOMIN REP, GERMAN FR, HUNGARY, JAPAN, MALAGASY R, NEW ZEALAND, PERU, SIERRE LEO, SWITZERLAND, URUGUAY, ARGENTINA, BRAZIL, COLOMBIA, ECUADOR, GREECE, ICELAND, KOREA REP, MALAYA, NICARAGUA, PHILIPPINES, SO AFRICA, THAILAND, VENEZUELA, AUSTRALIA, BULGARIA, COSTA RICA, EL SALVADOR, GUATEMALA, IRELAND, LEBANON, MEXICO, NORWAY, POLAND, SPAIN, TRINIDAD, VIETNAM REP, AUSTRIA, CANADA, CZECHOS'KIA, FINLAND, HAITI, ISRAEL, LIBERIA, MONGOLIA, PANAMA, PORTUGAL, SUDAN, UK, BELGIUM, CEYLON, DENMARK, GERMANY, E, HONDURAS, ITALY, LUXEMBOURG, NETHERLANDS, PARAGUAY, RUMANIA, SWEDEN, US

59 POLITIES
WHERE THE RELIGION IS
MIXED - CHRISTIAN, NON-LITERATE (13)

59 POLITIES
WHERE THE RELIGION IS OTHER THAN
MIXED - CHRISTIAN, NON-LITERATE (100)

SUBJECT ONLY

13 IN LEFT COLUMN

BOLIVIA	CONGO(BRA)	CONGO(LEO)	ECUADOR	GABON	GUATEMALA	HAITI	LIBERIA	MALAGASY R	PERU
SO AFRICA	TOGO	UGANDA							

100 IN RIGHT COLUMN

AFGHANISTAN	ALBANIA	ALGERIA	ARGENTINA	AUSTRALIA	AUSTRIA	BELGIUM	BRAZIL	BULGARIA	BURMA
CAMBODIA	CAMEROUN	CANADA	CEN AFR REP	CEYLON	CHAD	CHILE	CHINA, PR	COLOMBIA	COSTA RICA
CUBA	CYPRUS	CZECHOS'KIA	DAHOMEY	DENMARK	DOMIN REP	EL SALVADOR	ETHIOPIA	FINLAND	FRANCE
GERMANY, E	GERMAN FR	GHANA	GREECE	GUINEA	HONDURAS	HUNGARY	ICELAND	INDIA	INDONESIA
IRAN	IRAQ	IRELAND	ISRAEL	ITALY	IVORY COAST	JAMAICA	JAPAN	JORDAN	KOREA, N
KOREA REP	LAOS	LEBANON	LIBYA	LUXEMBOURG	MALAYA	MALI	MAURITANIA	MEXICO	MONGOLIA
MOROCCO	NEPAL	NETHERLANDS	NEW ZEALAND	NICARAGUA	NIGER	NIGERIA	NORWAY	PAKISTAN	PANAMA
PARAGUAY	PHILIPPINES	POLAND	PORTUGAL	RUMANIA	SA'U ARABIA	SENEGAL	SIERRE LEO	SOMALIA	SPAIN
SUDAN	SWEDEN	SWITZERLAND	SYRIA	TANGANYIKA	THAILAND	TRINIDAD	TUNISIA	TURKEY	USSR
UAR	UK	US	UPPER VOLTA	URUGUAY	VENEZUELA	VIETNAM, N	VIETNAM REP	YEMEN	YUGOSLAVIA

2 EXCLUDED BECAUSE UNASCERTAINED

BURUNDI RWANDA

	Left column			Right column		
2	ALWAYS ARE THOSE LOCATED ELSEWHERE THAN IN WEST EUROPE, SCANDINAVIA, NORTH AMERICA, OR AUSTRALASIA (93)	1.00	2	LEAN LESS TOWARD BEING THOSE LOCATED ELSEWHERE THAN IN WEST EUROPE, SCANDINAVIA, NORTH AMERICA, OR AUSTRALASIA (93)	0.78	0 22 13 78 X SQ= 2.29 P = 0.069 RV NO (NO)
8	ALWAYS ARE THOSE LOCATED ELSEWHERE THAN IN EAST ASIA SOUTH ASIA, OR SOUTHEAST ASIA (97)	1.00	8	TILT LESS TOWARD BEING THOSE LOCATED ELSEWHERE THAN IN EAST ASIA SOUTH ASIA, OR SOUTHEAST ASIA (97)	0.82	0 18 13 82 X SQ= 1.60 P = 0.124 RV NO (NO)
10	TEND TO BE THOSE LOCATED IN NORTH AFRICA, OR CENTRAL AND SOUTH AFRICA (33)	0.62	10	TEND TO BE THOSE LOCATED ELSEWHERE THAN IN NORTH AFRICA, OR CENTRAL AND SOUTH AFRICA (82)	0.77	8 23 5 77 X SQ= 6.76 P = 0.007 RV YES (NO)

15	ALWAYS ARE THOSE LOCATED IN NORTH AFRICA OR CENTRAL AND SOUTH AFRICA, RATHER THAN IN THE MIDDLE EAST (33)	1.00	15	LEAN LESS TOWARD BEING THOSE LOCATED IN NORTH AFRICA OR CENTRAL AND SOUTH AFRICA, RATHER THAN IN THE MIDDLE EAST (33)	0.68	0 11 8 23 X SQ= 2.03 P = 0.086 RV NO (NO)
17	ALWAYS ARE THOSE LOCATED IN THE CARIBBEAN, CENTRAL AMERICA, OR SOUTH AMERICA, RATHER THAN IN THE MIDDLE EAST (22)	1.00	17	TILT LESS TOWARD BEING THOSE LOCATED IN THE CARIBBEAN, CENTRAL AMERICA, OR SOUTH AMERICA, RATHER THAN IN THE MIDDLE EAST (22)	0.61	0 11 5 17 X SQ= 1.44 P = 0.143 RV NO (NO)
23	ALWAYS ARE THOSE WHOSE POPULATION IS MEDIUM OR SMALL (88)	1.00	23	TEND LESS TO BE THOSE WHOSE POPULATION IS MEDIUM OR SMALL (88)	0.73	0 27 13 73 X SQ= 3.25 P = 0.036 RV NO (NO)
24	LEAN TOWARD BEING THOSE WHOSE POPULATION IS SMALL (54)	0.69	24	LEAN TOWARD BEING THOSE WHOSE POPULATION IS VERY LARGE, LARGE, OR MEDIUM (61)	0.57	4 57 9 43 X SQ= 2.22 P = 0.085 RV YES (NO)
26	TEND MORE TO BE THOSE WHOSE POPULATION DENSITY IS LOW (67)	0.92	26	TEND LESS TO BE THOSE WHOSE POPULATION DENSITY IS LOW (67)	0.55	1 45 12 55 X SQ= 5.18 P = 0.014 RV NO (NO)
29	TEND TO BE THOSE WHERE THE DEGREE OF URBANIZATION IS LOW (49)	0.77	29	TEND TO BE THOSE WHERE THE DEGREE OF URBANIZATION IS HIGH (56)	0.59	3 53 10 37 X SQ= 4.52 P = 0.019 RV YES (NO)
30	TEND TO BE THOSE WHOSE AGRICULTURAL POPULATION IS HIGH (56)	0.77	30	TEND TO BE THOSE WHOSE AGRICULTURAL POPULATION IS MEDIUM, LOW, OR VERY LOW (57)	0.55	10 44 3 54 X SQ= 3.52 P = 0.039 RV YES (NO)
34	TEND TO BE THOSE WHOSE GROSS NATIONAL PRODUCT IS VERY LOW (53)	0.77	34	TEND TO BE THOSE WHOSE GROSS NATIONAL PRODUCT IS VERY HIGH, HIGH, MEDIUM, OR LOW (62)	0.59	3 59 10 41 X SQ= 4.63 P = 0.018 RV YES (NO)
35	ALWAYS ARE THOSE WHOSE PER CAPITA GROSS NATIONAL PRODUCT IS MEDIUM, LOW, OR VERY LOW (91)	1.00	35	LEAN LESS TOWARD BEING THOSE WHOSE PER CAPITA GROSS NATIONAL PRODUCT IS MEDIUM, LOW, OR VERY LOW (91)	0.76	0 24 13 76 X SQ= 2.66 P = 0.067 RV NO (NO)

36	TEND MORE TO BE THOSE WHOSE PER CAPITA GROSS NATIONAL PRODUCT IS LOW OR VERY LOW (73)	0.92	36	TEND LESS TO BE THOSE WHOSE PER CAPITA GROSS NATIONAL PRODUCT IS LOW OR VERY LOW (73)	0.59	1 41 12 59 X SQ= 4.13 P = 0.029 RV NO (NO)
37	LEAN TOWARD BEING THOSE WHOSE PER CAPITA GROSS NATIONAL PRODUCT IS VERY LOW (51)	0.69	37	LEAN TOWARD BEING THOSE WHOSE PER CAPITA GROSS NATIONAL PRODUCT IS VERY HIGH, HIGH, MEDIUM, OR LOW (64)	0.60	4 60 9 40 X SQ= 2.90 P = 0.072 RV YES (NO)
40	LEAN TOWARD BEING THOSE WHOSE INTERNATIONAL FINANCIAL STATUS IS VERY LOW (39)	0.62	40	LEAN TOWARD BEING THOSE WHOSE INTERNATIONAL FINANCIAL STATUS IS VERY HIGH, HIGH, MEDIUM, OR LOW (71)	0.69	5 66 8 29 X SQ= 3.60 P = 0.057 RV YES (NO)
41	ALWAYS ARE THOSE WHOSE ECONOMIC DEVELOPMENTAL STATUS IS INTERMEDIATE, UNDERDEVELOPED, OR VERY UNDERDEVELOPED (94)	1.00	41	TILT LESS TOWARD BEING THOSE WHOSE ECONOMIC DEVELOPMENTAL STATUS IS INTERMEDIATE, UNDERDEVELOPED, OR VERY UNDERDEVELOPED (94)	0.81	0 19 13 79 X SQ= 1.83 P = 0.120 RV NO (NO)
42	LEAN MORE TOWARD BEING THOSE WHOSE ECONOMIC DEVELOPMENTAL STATUS IS UNDERDEVELOPED OR VERY UNDERDEVELOPED (76)	0.92	42	LEAN LESS TOWARD BEING THOSE WHOSE ECONOMIC DEVELOPMENTAL STATUS IS UNDERDEVELOPED OR VERY UNDERDEVELOPED (76)	0.64	1 35 12 62 X SQ= 3.01 P = 0.057 RV NO (NO)
43	TEND TO BE THOSE WHOSE ECONOMIC DEVELOPMENTAL STATUS IS VERY UNDERDEVELOPED (57)	0.77	43	TEND TO BE THOSE WHOSE ECONOMIC DEVELOPMENTAL STATUS IS DEVELOPED, INTERMEDIATE, OR UNDERDEVELOPED (55)	0.54	3 52 10 45 X SQ= 3.14 P = 0.044 RV YES (NO)
45	ALWAYS ARE THOSE WHERE THE LITERACY RATE IS BELOW FIFTY PERCENT (54)	1.00	45	TEND TO BE THOSE WHERE THE LITERACY RATE IS FIFTY PERCENT OR ABOVE (55)	0.58	0 55 12 40 X SQ= 12.07 P = 0. RV YES (NO)
51	TILT MORE TOWARD BEING THOSE WHERE FREEDOM OF THE PRESS IS COMPLETE OR INTERMITTENT (60)	0.89	51	TILT LESS TOWARD BEING THOSE WHERE FREEDOM OF THE PRESS IS COMPLETE OR INTERMITTENT (60)	0.59	8 52 1 36 X SQ= 1.94 P = 0.147 RV NO (NO)
54	ALWAYS ARE THOSE WHERE NEWSPAPER CIRCULATION IS LESS THAN ONE HUNDRED PER THOUSAND (76)	1.00	54	TEND LESS TO BE THOSE WHERE NEWSPAPER CIRCULATION IS LESS THAN ONE HUNDRED PER THOUSAND (76)	0.62	0 37 13 61 X SQ= 5.76 P = 0.004 RV NO (NO)

55	LEAN TOWARD BEING THOSE WHERE NEWSPAPER CIRCULATION IS LESS THAN TEN PER THOUSAND (35)	0.54	55	LEAN TOWARD BEING THOSE WHERE NEWSPAPER CIRCULATION IS TEN OR MORE PER THOUSAND (78)	0.73	6 72 7 26 X SQ= 2.90 P = 0.056 RV YES (NO)
67	TEND TO BE THOSE THAT ARE RACIALLY HETEROGENEOUS (27)	0.58	67	TEND TO BE THOSE THAT ARE RACIALLY HOMOGENEOUS (82)	0.77	5 73 7 22 X SQ= 5.01 P = 0.016 RV YES (NO)
69	TEND TO BE THOSE THAT ARE LINGUISTICALLY STRONGLY HETEROGENEOUS (50)	0.85	69	TEND TO BE THOSE THAT ARE LINGUISTICALLY HOMOGENEOUS OR WEAKLY HETEROGENEOUS (64)	0.61	2 60 11 39 X SQ= 7.77 P = 0.003 RV YES (NO)
70	ALWAYS ARE THOSE THAT ARE RELIGIOUSLY, RACIALLY, OR LINGUISTICALLY HETEROGENEOUS (85)	1.00	70	LEAN LESS TOWARD BEING THOSE THAT ARE RELIGIOUSLY, RACIALLY, OR LINGUISTICALLY HETEROGENEOUS (85)	0.77	0 21 12 72 X SQ= 2.12 P = 0.071 RV NO (NO)
71	ALWAYS ARE THOSE WHOSE DATE OF INDEPENDENCE IS AFTER 1800 (90)	1.00	71	LEAN LESS TOWARD BEING THOSE WHOSE DATE OF INDEPENDENCE IS AFTER 1800 (90)	0.78	0 21 13 75 X SQ= 2.26 P = 0.069 RV NO (NO)
74	ALWAYS ARE THOSE THAT ARE NOT HISTORICALLY WESTERN (87)	1.00	74	LEAN LESS TOWARD BEING THOSE THAT ARE NOT HISTORICALLY WESTERN (87)	0.74	0 26 12 73 X SQ= 2.78 P = 0.065 RV NO (NO)
75	TILT TOWARD BEING THOSE THAT ARE NOT HISTORICALLY WESTERN AND ARE NOT SIGNIFICANTLY WESTERNIZED (52)	0.67	75	TILT TOWARD BEING THOSE THAT ARE HISTORICALLY WESTERN OR SIGNIFICANTLY WESTERNIZED (62)	0.58	4 58 8 42 X SQ= 1.73 P = 0.130 RV YES (NO)
76	ALWAYS ARE THOSE THAT HAVE BEEN SIGNIFICANTLY OR PARTIALLY WESTERNIZED THROUGH A COLONIAL RELATIONSHIP, RATHER THAN BEING HISTORICALLY WESTERN (70)	1.00	76	TEND LESS TO BE THOSE THAT HAVE BEEN SIGNIFICANTLY OR PARTIALLY WESTERNIZED THROUGH A COLONIAL RELATIONSHIP, RATHER THAN BEING HISTORICALLY WESTERN (70)	0.68	0 26 12 56 X SQ= 3.79 P = 0.018 RV NO (NO)
84	TEND TO BE THOSE WHERE THE TYPE OF POLITICAL MODERNIZATION IS UNDEVELOPED TUTELARY, RATHER THAN DEVELOPED TUTELARY (24)	0.86	84	TEND TO BE THOSE WHERE THE TYPE OF POLITICAL MODERNIZATION IS DEVELOPED TUTELARY, RATHER THAN UNDEVELOPED TUTELARY (31)	0.65	1 30 6 16 X SQ= 4.56 P = 0.016 RV YES (NO)

85	TEND TO BE THOSE WHERE THE STAGE OF POLITICAL MODERNIZATION IS MID- OR EARLY TRANSITIONAL, RATHER THAN ADVANCED (54)	0.92	85	TEND TO BE THOSE WHERE THE STAGE OF POLITICAL MODERNIZATION IS ADVANCED, RATHER THAN MID- OR EARLY TRANSITIONAL (60)	0.60	1 59 12 40 X SQ= 10.45 P = 0.001 RV YES (NO)
86	TILT TOWARD BEING THOSE WHERE THE STAGE OF POLITICAL MODERNIZATION IS EARLY TRANSITIONAL, RATHER THAN ADVANCED OR MID-TRANSITIONAL (38)	0.54	86	TILT TOWARD BEING THOSE WHERE THE STAGE OF POLITICAL MODERNIZATION IS ADVANCED OR MID-TRANSITIONAL, RATHER THAN EARLY TRANSITIONAL (76)	0.71	6 70 7 29 X SQ= 2.15 P = 0.112 RV YES (NO)
87	TILT TOWARD BEING THOSE WHOSE IDEOLOGICAL ORIENTATION IS DEVELOPMENTAL (31)	0.62	87	TILT TOWARD BEING THOSE WHOSE IDEOLOGICAL ORIENTATION IS OTHER THAN DEVELOPMENTAL (58)	0.70	5 24 3 55 X SQ= 2.08 P = 0.111 RV YES (NO)
92	ALWAYS ARE THOSE WHERE THE SYSTEM STYLE IS LIMITED MOBILIZATIONAL, OR NON-MOBILIZATIONAL (93)	1.00	92	TILT LESS TOWARD BEING THOSE WHERE THE SYSTEM STYLE IS LIMITED MOBILIZATIONAL, OR NON-MOBILIZATIONAL (93)	0.80	0 20 13 78 X SQ= 2.00 P = 0.120 RV NO (NO)
93	TILT MORE TOWARD BEING THOSE WHERE THE SYSTEM STYLE IS NON-MOBILIZATIONAL (78)	0.92	93	TILT LESS TOWARD BEING THOSE WHERE THE SYSTEM STYLE IS NON-MOBILIZATIONAL (78)	0.67	1 31 12 64 X SQ= 2.32 P = 0.103 RV NO (NO)
99	LEAN TOWARD BEING THOSE WHERE GOVERNMENTAL STABILITY IS MODERATELY PRESENT AND DATES FROM THE POST-WAR PERIOD, OR IS ABSENT (36)	0.75	99	LEAN TOWARD BEING THOSE WHERE GOVERNMENTAL STABILITY IS GENERALLY PRESENT AND DATES FROM AT LEAST THE INTER-WAR PERIOD, OR FROM THE POST-WAR PERIOD (50)	0.62	2 48 6 30 X SQ= 2.62 P = 0.064 RV YES (NO)
110	ALWAYS ARE THOSE WHERE POLITICAL ENCULTURATION IS MEDIUM OR LOW (80)	1.00	110	TILT LESS TOWARD BEING THOSE WHERE POLITICAL ENCULTURATION IS MEDIUM OR LOW (80)	0.82	0 15 13 67 X SQ= 1.62 P = 0.120 RV NO (NO)
111	TEND TO BE THOSE WHERE POLITICAL ENCULTURATION IS LOW (42)	0.85	111	TEND TO BE THOSE WHERE POLITICAL ENCULTURATION IS HIGH OR MEDIUM (53)	0.62	2 51 11 31 X SQ= 8.16 P = 0.002 RV YES (NO)
114	TILT MORE TOWARD BEING THOSE WHERE SECTIONALISM IS EXTREME OR MODERATE (61)	0.77	114	TILT LESS TOWARD BEING THOSE WHERE SECTIONALISM IS EXTREME OR MODERATE (61)	0.55	10 51 3 42 X SQ= 1.46 P = 0.149 RV NO (NO)

115	ALWAYS ARE THOSE WHERE INTEREST ARTICULATION BY ASSOCIATIONAL GROUPS IS MODERATE, LIMITED, OR NEGLIGIBLE (91)	1.00	115	LEAN LESS TOWARD BEING THOSE WHERE INTEREST ARTICULATION BY ASSOCIATIONAL GROUPS IS MODERATE, LIMITED, OR NEGLIGIBLE (91)	0.79	0 20 13 76 X SQ= 2.07 P = 0.074 RV NO (NO)
116	TILT MORE TOWARD BEING THOSE WHERE INTEREST ARTICULATION BY ASSOCIATIONAL GROUPS IS LIMITED OR NEGLIGIBLE (79)	0.92	116	TILT LESS TOWARD BEING THOSE WHERE INTEREST ARTICULATION BY ASSOCIATIONAL GROUPS IS LIMITED OR NEGLIGIBLE (79)	0.68	1 31 12 65 X SQ= 2.26 P = 0.103 RV NO (NO)
118	TEND MORE TO BE THOSE WHERE INTEREST ARTICULATION BY INSTITUTIONAL GROUPS IS SIGNIFICANT, MODERATE, OR LIMITED (60)	0.90	118	TEND LESS TO BE THOSE WHERE INTEREST ARTICULATION BY INSTITUTIONAL GROUPS IS SIGNIFICANT, MODERATE, OR LIMITED (60)	0.57	1 39 9 51 X SQ= 2.89 P = 0.047 RV NO (NO)
122	ALWAYS ARE THOSE WHERE INTEREST ARTICULATION BY NON-ASSOCIATIONAL GROUPS IS SIGNIFICANT OR MODERATE (83)	1.00	122	TEND LESS TO BE THOSE WHERE INTEREST ARTICULATION BY NON-ASSOCIATIONAL GROUPS IS SIGNIFICANT OR MODERATE (83)	0.68	13 68 0 32 X SQ= 4.33 P = 0.018 RV NO (NO)
125	TILT MORE TOWARD BEING THOSE WHERE INTEREST ARTICULATION BY ANOMIC GROUPS IS FREQUENT OR OCCASIONAL (64)	0.85	125	TILT LESS TOWARD BEING THOSE WHERE INTEREST ARTICULATION BY ANOMIC GROUPS IS FREQUENT OR OCCASIONAL (64)	0.62	11 52 2 32 X SQ= 1.65 P = 0.131 RV NO (NO)
126	ALWAYS ARE THOSE WHERE INTEREST ARTICULATION BY ANOMIC GROUPS IS FREQUENT, OCCASIONAL, OR INFREQUENT (83)	1.00	126	TILT LESS TOWARD BEING THOSE WHERE INTEREST ARTICULATION BY ANOMIC GROUPS IS FREQUENT, OCCASIONAL, OR INFREQUENT (83)	0.81	13 68 0 16 X SQ= 1.74 P = 0.118 RV NO (NO)
137	ALWAYS ARE THOSE WHERE INTEREST AGGREGATION BY THE LEGISLATURE IS LIMITED OR NEGLIGIBLE (68)	1.00	137	TEND LESS TO BE THOSE WHERE INTEREST AGGREGATION BY THE LEGISLATURE IS LIMITED OR NEGLIGIBLE (68)	0.65	0 29 11 55 X SQ= 3.96 P = 0.017 RV NO (NO)
153	TILT TOWARD BEING THOSE WHERE THE PARTY SYSTEM IS MODERATELY STABLE OR UNSTABLE (38)	0.75	153	TILT TOWARD BEING THOSE WHERE THE PARTY SYSTEM IS STABLE (42)	0.56	2 40 6 32 X SQ= 1.61 P = 0.141 RV YES (NO)
154	TILT TOWARD BEING THOSE WHERE THE PARTY SYSTEM IS UNSTABLE (25)	0.62	154	TILT TOWARD BEING THOSE WHERE THE PARTY SYSTEM IS STABLE OR MODERATELY STABLE (55)	0.72	3 52 5 20 X SQ= 2.59 P = 0.100 RV YES (NO)

158	TEND TO BE THOSE WHERE PERSONALISMO IS PRONOUNCED OR MODERATE (40)	0.73	158	TEND TO BE THOSE WHERE PERSONALISMO IS NEGLIGIBLE (56)	0.62	8 32 3 53 X SQ= 3.59 P = 0.047 RV YES (NO)
170	TEND TO BE THOSE WHERE THE LEGISLATIVE-EXECUTIVE STRUCTURE IS PRESIDENTIAL (39)	0.73	170	TEND TO BE THOSE WHERE THE LEGISLATIVE-EXECUTIVE STRUCTURE IS OTHER THAN PRESIDENTIAL (63)	0.66	8 30 3 59 X SQ= 4.78 P = 0.019 RV YES (NO)
193	TEND TO BE THOSE WHERE THE CHARACTER OF THE LEGAL SYSTEM IS PARTLY INDIGENOUS (28)	0.62	193	TEND TO BE THOSE WHERE THE CHARACTER OF THE LEGAL SYSTEM IS OTHER THAN PARTLY INDIGENOUS (86)	0.82	8 18 5 81 X SQ= 9.81 P = 0.002 RV YES (NO)

60 POLITIES
WHERE THE RELIGION IS
MIXED - LITERATE NON-CHRISTIAN,
NON-LITERATE (7)

60 POLITIES
WHERE THE RELIGION IS OTHER THAN
MIXED - LITERATE NON-CHRISTIAN,
NON-LITERATE (103)

SUBJECT ONLY

7 IN LEFT COLUMN

CHAD GUINEA LAOS MALI NIGER SUDAN UPPER VOLTA

103 IN RIGHT COLUMN

AFGHANISTAN	ALBANIA	ALGERIA	ARGENTINA	AUSTRALIA	AUSTRIA	BELGIUM	BOLIVIA	BRAZIL	BULGARIA
BURMA	BURUNDI	CAMBODIA	CAMEROUN	CANADA	CEN AFR REP	CEYLON	CHILE	COLOMBIA	CONGO(BRA)
CONGO(LEO)	COSTA RICA	CUBA	CYPRUS	CZECHOS'KIA	DAHOMEY	DENMARK	DOMIN REP	ECUADOR	EL SALVADOR
ETHIOPIA	FINLAND	FRANCE	GABON	GERMANY, E	GERMAN FR	GHANA	GREECE	GUATEMALA	HAITI
HONDURAS	HUNGARY	ICELAND	INDIA	INDONESIA	IRAN	IRAQ	IRELAND	ISRAEL	ITALY
IVORY COAST	JAMAICA	JAPAN	JORDAN	LEBANON	LIBERIA	LIBYA	LUXEMBOURG	MALAGASY R	MALAYA
MAURITANIA	MEXICO	MONGOLIA	MOROCCO	NEPAL	NETHERLANDS	NEW ZEALAND	NICARAGUA	NIGERIA	NORWAY
PAKISTAN	PANAMA	PARAGUAY	PERU	PHILIPPINES	POLAND	PORTUGAL	RUMANIA	RWANDA	SA'U ARABIA
SENEGAL	SIERRE LEO	SOMALIA	SO AFRICA	SPAIN	SWEDEN	SWITZERLAND	SYRIA	TANGANYIKA	THAILAND
TOGO	TRINIDAD	TUNISIA	TURKEY	UGANDA	USSR	UAR	UK	US	URUGUAY
VENEZUELA	YEMEN	YUGOSLAVIA							

2 EXCLUDED BECAUSE AMBIGUOUS

KOREA REP VIETNAM REP

3 EXCLUDED BECAUSE UNASCERTAINED

CHINA, PR KOREA, N VIETNAM, N

	Left column			Right column		
10	TEND TO BE THOSE LOCATED IN NORTH AFRICA, OR CENTRAL AND SOUTH AFRICA (33)	0.86	10	TEND TO BE THOSE LOCATED ELSEWHERE THAN IN NORTH AFRICA, OR CENTRAL AND SOUTH AFRICA (82)	0.74	6 27 1 76 X SQ= 8.40 P = 0.003 RV YES (NO)
19	ALWAYS ARE THOSE LOCATED IN NORTH AFRICA OR CENTRAL AND SOUTH AFRICA, RATHER THAN IN THE CARIBBEAN, CENTRAL AMERICA, OR SOUTH AMERICA (33)	1.00	19	LEAN LESS TOWARD BEING THOSE LOCATED IN NORTH AFRICA OR CENTRAL AND SOUTH AFRICA, RATHER THAN IN THE CARIBBEAN, CENTRAL AMERICA, OR SOUTH AMERICA (33)	0.55	6 27 0 22 X SQ= 2.81 P = 0.071 RV NO (NO)
22	ALWAYS ARE THOSE WHOSE TERRITORIAL SIZE IS VERY LARGE, LARGE, OR MEDIUM (68)	1.00	22	TEND LESS TO BE THOSE WHOSE TERRITORIAL SIZE IS VERY LARGE, LARGE, OR MEDIUM (68)	0.58	7 60 0 43 X SQ= 3.20 P = 0.041 RV NO (NO)

24	LEAN TOWARD BEING THOSE WHOSE POPULATION IS SMALL (54)	0.86	24	LEAN TOWARD BEING THOSE WHOSE POPULATION IS VERY LARGE, LARGE, OR MEDIUM (61)	0.53	1 55 6 48 X SQ= 2.60 P = 0.058 RV YES (NO)
26	ALWAYS ARE THOSE WHOSE POPULATION DENSITY IS LOW (67)	1.00	26	TEND LESS TO BE THOSE WHOSE POPULATION DENSITY IS LOW (67)	0.58	0 43 7 60 X SQ= 3.20 P = 0.041 RV NO (NO)
28	TILT MORE TOWARD BEING THOSE WHOSE POPULATION GROWTH RATE IS HIGH (62)	0.86	28	TILT LESS TOWARD BEING THOSE WHOSE POPULATION GROWTH RATE IS HIGH (62)	0.54	6 53 1 46 X SQ= 1.59 P = 0.129 RV NO (NO)
29	ALWAYS ARE THOSE WHERE THE DEGREE OF URBANIZATION IS LOW (49)	1.00	29	TEND TO BE THOSE WHERE THE DEGREE OF URBANIZATION IS HIGH (56)	0.57	0 55 7 42 X SQ= 6.30 P = 0.004 RV YES (NO)
30	ALWAYS ARE THOSE WHOSE AGRICULTURAL POPULATION IS HIGH (56)	1.00	30	TEND TO BE THOSE WHOSE AGRICULTURAL POPULATION IS MEDIUM, LOW, OR VERY LOW (57)	0.56	7 45 0 57 X SQ= 6.11 P = 0.004 RV YES (NO)
34	ALWAYS ARE THOSE WHOSE GROSS NATIONAL PRODUCT IS VERY LOW (53)	1.00	34	TEND TO BE THOSE WHOSE GROSS NATIONAL PRODUCT IS VERY HIGH, HIGH, MEDIUM, OR LOW (62)	0.56	0 58 7 45 X SQ= 6.23 P = 0.004 RV YES (NO)
37	ALWAYS ARE THOSE WHOSE PER CAPITA GROSS NATIONAL PRODUCT IS VERY LOW (51)	1.00	37	TEND TO BE THOSE WHOSE PER CAPITA GROSS NATIONAL PRODUCT IS VERY HIGH, HIGH, MEDIUM, OR LOW (64)	0.62	0 64 7 39 X SQ= 8.00 P = 0.002 RV YES (NO)
39	ALWAYS ARE THOSE WHOSE INTERNATIONAL FINANCIAL STATUS IS LOW OR VERY LOW (76)	1.00	39	LEAN LESS TOWARD BEING THOSE WHOSE INTERNATIONAL FINANCIAL STATUS IS LOW OR VERY LOW (76)	0.64	0 37 7 65 X SQ= 2.40 P = 0.093 RV NO (NO)
40	TEND TO BE THOSE WHOSE INTERNATIONAL FINANCIAL STATUS IS VERY LOW (39)	0.86	40	TEND TO BE THOSE WHOSE INTERNATIONAL FINANCIAL STATUS IS VERY HIGH, HIGH, MEDIUM, OR LOW (71)	0.67	1 67 6 33 X SQ= 5.74 P = 0.009 RV YES (NO)

No.	Left		No.	Right		Statistics
43	ALWAYS ARE THOSE WHOSE ECONOMIC DEVELOPMENTAL STATUS IS VERY UNDERDEVELOPED (57)	1.00	43	TEND TO BE THOSE WHOSE ECONOMIC DEVELOPMENTAL STATUS IS DEVELOPED, INTERMEDIATE, OR UNDERDEVELOPED (55)	0.54	0 54 7 46 X SQ= 5.62 P = 0.006 RV YES (NO)
46	ALWAYS ARE THOSE WHERE THE LITERACY RATE IS BELOW TEN PERCENT (26)	1.00	46	TEND TO BE THOSE WHERE THE LITERACY RATE IS TEN PERCENT OR ABOVE (84)	0.80	0 80 6 20 X SQ= 15.49 P = 0. RV YES (NO)
55	ALWAYS ARE THOSE WHERE NEWSPAPER CIRCULATION IS LESS THAN TEN PER THOUSAND (35)	1.00	55	TEND TO BE THOSE WHERE NEWSPAPER CIRCULATION IS TEN OR MORE PER THOUSAND (78)	0.73	0 75 7 28 X SQ= 12.84 P = 0. RV YES (NO)
67	LEAN TOWARD BEING THOSE THAT ARE RACIALLY HETEROGENEOUS (27)	0.57	67	LEAN TOWARD BEING THOSE THAT ARE RACIALLY HOMOGENEOUS (82)	0.74	3 72 4 25 X SQ= 1.83 P = 0.093 RV YES (NO)
69	ALWAYS ARE THOSE THAT ARE LINGUISTICALLY STRONGLY HETEROGENEOUS (50)	1.00	69	TEND TO BE THOSE THAT ARE LINGUISTICALLY HOMOGENEOUS OR WEAKLY HETEROGENEOUS (64)	0.58	0 60 7 43 X SQ= 6.78 P = 0.003 RV YES (NO)
73	ALWAYS ARE THOSE WHOSE DATE OF INDEPENDENCE IS AFTER 1945 (46)	1.00	73	TEND TO BE THOSE WHOSE DATE OF INDEPENDENCE IS BEFORE 1945 (65)	0.63	0 64 7 37 X SQ= 8.42 P = 0.001 RV YES (NO)
75	ALWAYS ARE THOSE THAT ARE NOT HISTORICALLY WESTERN AND ARE NOT SIGNIFICANTLY WESTERNIZED (52)	1.00	75	TEND TO BE THOSE THAT ARE HISTORICALLY WESTERN OR SIGNIFICANTLY WESTERNIZED (62)	0.60	0 61 7 41 X SQ= 7.23 P = 0.002 RV YES (NO)
77	ALWAYS ARE THOSE THAT HAVE BEEN PARTIALLY WESTERNIZED, RATHER THAN SIGNIFICANTLY WESTERNIZED, THROUGH A COLONIAL RELATIONSHIP (41)	1.00	77	TEND LESS TO BE THOSE THAT HAVE BEEN PARTIALLY WESTERNIZED, RATHER THAN SIGNIFICANTLY WESTERNIZED, THROUGH A COLONIAL RELATIONSHIP (41)	0.53	0 28 7 32 X SQ= 3.86 P = 0.019 RV NO (NO)
80	LEAN TOWARD BEING THOSE WHOSE DATE OF INDEPENDENCE IS AFTER 1914, AND THAT WERE FORMERLY DEPENDENCIES OF FRANCE, RATHER THAN BRITAIN (24)	0.86	80	LEAN TOWARD BEING THOSE WHOSE DATE OF INDEPENDENCE IS AFTER 1914, AND THAT WERE FORMERLY DEPENDENCIES OF BRITAIN, RATHER THAN FRANCE (19)	0.53	1 18 6 16 X SQ= 2.11 P = 0.099 RV YES (NO)

82	ALWAYS ARE THOSE WHERE THE TYPE OF POLITICAL MODERNIZATION IS DEVELOPED TUTELARY OR UNDEVELOPED TUTELARY, RATHER THAN EARLY OR LATER EUROPEAN OR EUROPEAN DERIVED (55)	1.00	82	TEND TO BE THOSE WHERE THE TYPE OF POLITICAL MODERNIZATION IS EARLY OR LATER EUROPEAN OR EUROPEAN DERIVED, RATHER THAN DEVELOPED TUTELARY OR UNDEVELOPED TUTELARY (51)	0.54	0 51 7 44 X SQ= 5.52 P = 0.013 RV YES (NO)
85	LEAN TOWARD BEING THOSE WHERE THE STAGE OF POLITICAL MODERNIZATION IS MID- OR EARLY TRANSITIONAL, RATHER THAN ADVANCED (54)	0.86	85	LEAN TOWARD BEING THOSE WHERE THE STAGE OF POLITICAL MODERNIZATION IS ADVANCED, RATHER THAN MID- OR EARLY TRANSITIONAL (60)	0.55	1 56 6 46 X SQ= 2.86 P = 0.052 RV YES (NO)
86	TEND TO BE THOSE WHERE THE STAGE OF POLITICAL MODERNIZATION IS EARLY TRANSITIONAL, RATHER THAN ADVANCED OR MID-TRANSITIONAL (38)	0.86	86	TEND TO BE THOSE WHERE THE STAGE OF POLITICAL MODERNIZATION IS ADVANCED OR MID-TRANSITIONAL, RATHER THAN EARLY TRANSITIONAL (76)	0.71	1 72 6 30 X SQ= 7.01 P = 0.005 RV YES (NO)
87	TEND TO BE THOSE WHOSE IDEOLOGICAL ORIENTATION IS DEVELOPMENTAL (31)	0.86	87	TEND TO BE THOSE WHOSE IDEOLOGICAL ORIENTATION IS OTHER THAN DEVELOPMENTAL (58)	0.68	6 25 1 54 X SQ= 5.98 P = 0.008 RV YES (NO)
89	ALWAYS ARE THOSE WHOSE IDEOLOGICAL ORIENTATION IS OTHER THAN CONVENTIONAL (62)	1.00	89	TEND LESS TO BE THOSE WHOSE IDEOLOGICAL ORIENTATION IS OTHER THAN CONVENTIONAL (62)	0.61	0 33 7 51 X SQ= 2.78 P = 0.046 RV NO (NO)
96	ALWAYS ARE THOSE WHERE THE STATUS OF THE REGIME IS AUTHORITARIAN (23)	1.00	96	TEND TO BE THOSE WHERE THE STATUS OF THE REGIME IS CONSTITUTIONAL OR TOTALITARIAN (67)	0.77	0 64 3 19 X SQ= 5.45 P = 0.015 RV YES (NO)
102	ALWAYS ARE THOSE WHERE THE REPRESENTATIVE CHARACTER OF THE REGIME IS PSEUDO-POLYARCHIC OR NON-POLYARCHIC (49)	1.00	102	TEND TO BE THOSE WHERE THE REPRESENTATIVE CHARACTER OF THE REGIME IS POLYARCHIC OR LIMITED POLYARCHIC (59)	0.61	0 59 7 38 X SQ= 7.52 P = 0.002 RV YES (NO)
105	ALWAYS ARE THOSE WHERE THE ELECTORAL SYSTEM IS PARTIALLY COMPETITIVE OR NON-COMPETITIVE (47)	1.00	105	TEND TO BE THOSE WHERE THE ELECTORAL SYSTEM IS COMPETITIVE (43)	0.54	0 43 6 37 X SQ= 4.48 P = 0.026 RV YES (NO)
107	ALWAYS ARE THOSE WHERE AUTONOMOUS GROUPS ARE PARTIALLY TOLERATED IN POLITICS, ARE TOLERATED ONLY OUTSIDE POLITICS, OR ARE NOT TOLERATED AT ALL (65)	1.00	107	TEND LESS TO BE THOSE WHERE AUTONOMOUS GROUPS ARE PARTIALLY TOLERATED IN POLITICS, ARE TOLERATED ONLY OUTSIDE POLITICS, OR ARE NOT TOLERATED AT ALL (65)	0.54	0 46 6 54 X SQ= 3.18 P = 0.035 RV NO (NO)

108	TEND TO BE THOSE WHERE AUTONOMOUS GROUPS ARE TOLERATED ONLY OUTSIDE POLITICS OR ARE NOT TOLERATED AT ALL (35)	0.83	108	TEND TO BE THOSE WHERE AUTONOMOUS GROUPS ARE FULLY OR PARTIALLY TOLERATED IN POLITICS (65)	0.71	1 64 5 26 X SQ= 5.34 P = 0.013 RV YES (NO)
117	ALWAYS ARE THOSE WHERE INTEREST ARTICULATION BY ASSOCIATIONAL GROUPS IS NEGLIGIBLE (51)	1.00	117	TEND TO BE THOSE WHERE INTEREST ARTICULATION BY ASSOCIATIONAL GROUPS IS SIGNIFICANT, MODERATE, OR LIMITED (60)	0.60	0 60 7 40 X SQ= 7.28 P = 0.002 RV YES (NO)
118	ALWAYS ARE THOSE WHERE INTEREST ARTICULATION BY INSTITUTIONAL GROUPS IS VERY SIGNIFICANT (40)	1.00	118	TEND TO BE THOSE WHERE INTEREST ARTICULATION BY INSTITUTIONAL GROUPS IS SIGNIFICANT, MODERATE, OR LIMITED (60)	0.65	3 32 0 60 X SQ= 2.88 P = 0.047 RV YES (NO)
121	ALWAYS ARE THOSE WHERE INTEREST ARTICULATION BY NON-ASSOCIATIONAL GROUPS IS SIGNIFICANT (54)	1.00	121	TEND TO BE THOSE WHERE INTEREST ARTICULATION BY NON-ASSOCIATIONAL GROUPS IS MODERATE, LIMITED, OR NEGLIGIBLE (61)	0.59	7 42 0 61 X SQ= 7.06 P = 0.003 RV YES (NO)
125	ALWAYS ARE THOSE WHERE INTEREST ARTICULATION BY ANOMIC GROUPS IS FREQUENT OR OCCASIONAL (64)	1.00	125	TEND LESS TO BE THOSE WHERE INTEREST ARTICULATION BY ANOMIC GROUPS IS FREQUENT OR OCCASIONAL (64)	0.61	7 54 0 35 X SQ= 2.80 P = 0.045 RV NO (NO)
129	ALWAYS ARE THOSE WHERE INTEREST ARTICULATION BY POLITICAL PARTIES IS NEGLIGIBLE (37)	1.00	129	TEND TO BE THOSE WHERE INTEREST ARTICULATION BY POLITICAL PARTIES IS SIGNIFICANT, MODERATE, OR LIMITED (56)	0.67	0 56 5 28 X SQ= 6.36 P = 0.006 RV YES (NO)
133	TEND TO BE THOSE WHERE INTEREST AGGREGATION BY THE EXECUTIVE IS SIGNIFICANT (29)	0.83	133	TEND TO BE THOSE WHERE INTEREST AGGREGATION BY THE EXECUTIVE IS MODERATE, LIMITED, OR NEGLIGIBLE (76)	0.74	5 24 1 70 X SQ= 6.56 P = 0.007 RV YES (NO)
138	ALWAYS ARE THOSE WHERE INTEREST AGGREGATION BY THE LEGISLATURE IS NEGLIGIBLE (49)	1.00	138	TEND TO BE THOSE WHERE INTEREST AGGREGATION BY THE LEGISLATURE IS SIGNIFICANT, MODERATE, OR LIMITED (48)	0.55	0 48 6 39 X SQ= 4.81 P = 0.011 RV YES (NO)
139	TEND TO BE THOSE WHERE THE PARTY SYSTEM IS QUANTITATIVELY ONE-PARTY (34)	0.71	139	TEND TO BE THOSE WHERE THE PARTY SYSTEM IS QUANTITATIVELY OTHER THAN ONE-PARTY (71)	0.72	5 26 2 68 X SQ= 3.99 P = 0.027 RV YES (NO)

148	TEND TO BE THOSE WHERE THE PARTY SYSTEM IS QUALITATIVELY AFRICAN TRANSITIONAL (14)	0.57	148	TEND TO BE THOSE WHERE THE PARTY SYSTEM IS QUALITATIVELY OTHER THAN AFRICAN TRANSITIONAL (96)	0.90	4 10 3 88 X SQ= 8.73 P = 0.006 RV YES (NO)
158	LEAN TOWARD BEING THOSE WHERE PERSONALISMO IS PRONOUNCED OR MODERATE (40)	0.83	158	LEAN TOWARD BEING THOSE WHERE PERSONALISMO IS NEGLIGIBLE (56)	0.60	5 34 1 52 X SQ= 2.79 P = 0.080 RV YES (NO)
164	TEND TO BE THOSE WHERE THE REGIME'S LEADERSHIP CHARISMA IS PRONOUNCED OR MODERATE (34)	0.83	164	TEND TO BE THOSE WHERE THE REGIME'S LEADERSHIP CHARISMA IS NEGLIGIBLE (65)	0.70	5 26 1 62 X SQ= 5.12 P = 0.014 RV YES (NO)
168	ALWAYS ARE THOSE WHERE THE HORIZONTAL POWER DISTRIBUTION IS LIMITED OR NEGLIGIBLE (72)	1.00	168	LEAN LESS TOWARD BEING THOSE WHERE THE HORIZONTAL POWER DISTRIBUTION IS LIMITED OR NEGLIGIBLE (72)	0.64	0 34 7 61 X SQ= 2.32 P = 0.092 RV NO (NO)
169	TEND TO BE THOSE WHERE THE HORIZONTAL POWER DISTRIBUTION IS NEGLIGIBLE (48)	0.86	169	TEND TO BE THOSE WHERE THE HORIZONTAL POWER DISTRIBUTION IS SIGNIFICANT OR LIMITED (58)	0.60	1 57 6 38 X SQ= 3.85 P = 0.041 RV YES (NO)
175	ALWAYS ARE THOSE WHERE THE LEGISLATURE IS LARGELY INEFFECTIVE OR WHOLLY INEFFECTIVE (49)	1.00	175	TEND TO BE THOSE WHERE THE LEGISLATURE IS FULLY EFFECTIVE OR PARTIALLY EFFECTIVE (51)	0.57	0 51 6 39 X SQ= 5.16 P = 0.009 RV YES (NO)
178	TILT TOWARD BEING THOSE WHERE THE LEGISLATURE IS UNICAMERAL (53)	0.83	178	TILT TOWARD BEING THOSE WHERE THE LEGISLATURE IS BICAMERAL (51)	0.53	5 44 1 50 X SQ= 1.73 P = 0.108 RV YES (NO)
182	TEND TO BE THOSE WHERE THE BUREAUCRACY IS POST-COLONIAL TRANSITIONAL, RATHER THAN SEMI-MODERN (25)	0.83	182	TEND TO BE THOSE WHERE THE BUREAUCRACY IS SEMI-MODERN, RATHER THAN POST-COLONIAL TRANSITIONAL (55)	0.71	1 50 5 20 X SQ= 5.23 P = 0.013 RV YES (NO)
192	TEND TO BE THOSE WHERE THE CHARACTER OF THE LEGAL SYSTEM IS MUSLIM OR PARTLY MUSLIM (27)	0.83	192	TEND TO BE THOSE WHERE THE CHARACTER OF THE LEGAL SYSTEM IS OTHER THAN MUSLIM OR PARTLY MUSLIM (86)	0.78	5 22 1 80 X SQ= 8.47 P = 0.004 RV YES (NO)

193 TEND TO BE THOSE 0.86
WHERE THE CHARACTER OF THE LEGAL SYSTEM
IS PARTLY INDIGENOUS (28)

193 TEND TO BE THOSE 0.79
WHERE THE CHARACTER OF THE LEGAL SYSTEM
IS OTHER THAN PARTLY INDIGENOUS (86)

```
       6   21
       1   81
X SQ=  11.62
P =    0.001
RV YES (NO )
```

```
**************************************************************************************************

 61  POLITIES                                                  61  POLITIES
     WHERE THE RELIGION IS                                         WHERE THE RELIGION IS OTHER THAN
     MIXED - CHRISTIAN,                                            MIXED - CHRISTIAN,
     LITERATE NON-CHRISTIAN,                                       LITERATE NON-CHRISTIAN,
     NON-LITERATE  (8)                                             NON-LITERATE  (104)

                                                                                              SUBJECT ONLY
..................................................................................................

      8 IN LEFT COLUMN

CAMEROUN     DAHOMEY     ETHIOPIA    GHANA       IVORY COAST NIGERIA     SIERRE LEO  TANGANYIKA

     104 IN RIGHT COLUMN

AFGHANISTAN ALBANIA     ALGERIA     ARGENTINA   AUSTRALIA   AUSTRIA     BELGIUM     BOLIVIA     BRAZIL      BULGARIA
BURMA       BURUNDI     CAMBODIA    CANADA      CEN AFR REP CEYLON      CHAD        CHILE       CHINA, PR   COLOMBIA
CONGO(BRA)  CONGO(LEO)  COSTA RICA  CUBA        CYPRUS      CZECHOS'KIA DENMARK     DOMIN REP   ECUADOR     EL SALVADOR
FINLAND     FRANCE      GABON       GERMANY, E  GERMAN FR   GREECE      GUATEMALA   GUINEA      HAITI       HONDURAS
HUNGARY     ICELAND     INDIA       INDONESIA   IRAN        IRAQ        IRELAND     ISRAEL      ITALY       JAMAICA
JAPAN       JORDAN      KOREA, N    LAOS        LEBANON     LIBERIA     LIBYA       LUXEMBOURG  MALAGASY R  MALAYA
MALI        MAURITANIA  MEXICO      MONGOLIA    MOROCCO     NEPAL       NETHERLANDS NEW ZEALAND NICARAGUA   NIGER
NORWAY      PAKISTAN    PANAMA      PARAGUAY    PERU        PHILIPPINES POLAND      PORTUGAL    RUMANIA     RWANDA
SA'U ARABIA SENEGAL     SOMALIA     SO AFRICA   SPAIN       SUDAN       SWEDEN      SWITZERLAND SYRIA       THAILAND
TOGO        TRINIDAD    TUNISIA     TURKEY      UGANDA      USSR        UAR         UK          US          UPPER VOLTA
URUGUAY     VENEZUELA   YEMEN       YUGOSLAVIA

      2 EXCLUDED BECAUSE AMBIGUOUS

KOREA REP   VIETNAM REP

      1 EXCLUDED BECAUSE UNASCERTAINED

VIETNAM, N

--------------------------------------------------------------------------------------------------

 10  ALWAYS ARE THOSE                        1.00        10  TEND TO BE THOSE                      0.76          8   25
     LOCATED IN NORTH AFRICA, OR                             LOCATED ELSEWHERE THAN IN NORTH AFRICA,            0   79
     CENTRAL AND SOUTH AFRICA  (33)                          OR CENTRAL AND SOUTH AFRICA  (82)              X SQ=  17.13
                                                                                                            P =     0.
                                                                                                            RV YES (NO )

 26  TILT MORE TOWARD BEING THOSE            0.87        26  TILT LESS TOWARD BEING THOSE          0.58          1   44
     WHOSE POPULATION DENSITY IS                             WHOSE POPULATION DENSITY IS                        7   60
     LOW  (67)                                               LOW  (67)                                      X SQ=   1.65
                                                                                                            P =    0.141
                                                                                                            RV NO  (NO )

 29  ALWAYS ARE THOSE                        1.00        29  TEND TO BE THOSE                      0.57          0   55
     WHERE THE DEGREE OF URBANIZATION                        WHERE THE DEGREE OF URBANIZATION                   8   41
     IS LOW  (49)                                            IS HIGH  (56)                                  X SQ=   7.56
                                                                                                            P =    0.002
                                                                                                            RV YES (NO )
```

30	TEND TO BE THOSE WHOSE AGRICULTURAL POPULATION IS HIGH (56)	0.87	30	TEND TO BE THOSE WHOSE AGRICULTURAL POPULATION IS MEDIUM, LOW, OR VERY LOW (57)	0.54	7 47 1 56 X SQ= 3.67 P = 0.029 RV YES (NO)
33	ALWAYS ARE THOSE WHOSE GROSS NATIONAL PRODUCT IS LOW OR VERY LOW (85)	1.00	33	TILT LESS TOWARD BEING THOSE WHOSE GROSS NATIONAL PRODUCT IS LOW OR VERY LOW (85)	0.71	0 30 8 74 X SQ= 1.85 P = 0.106 RV NO (NO)
34	TILT TOWARD BEING THOSE WHOSE GROSS NATIONAL PRODUCT IS VERY LOW (53)	0.75	34	TILT TOWARD BEING THOSE WHOSE GROSS NATIONAL PRODUCT IS VERY HIGH, HIGH, MEDIUM, OR LOW (62)	0.55	2 57 6 47 X SQ= 1.59 P = 0.146 RV YES (NO)
37	ALWAYS ARE THOSE WHOSE PER CAPITA GROSS NATIONAL PRODUCT IS VERY LOW (51)	1.00	37	TEND TO BE THOSE WHOSE PER CAPITA GROSS NATIONAL PRODUCT IS VERY HIGH, HIGH, MEDIUM, OR LOW (64)	0.62	0 64 8 40 X SQ= 9.11 P = 0.001 RV YES (NO)
39	ALWAYS ARE THOSE WHOSE INTERNATIONAL FINANCIAL STATUS IS LOW OR VERY LOW (76)	1.00	39	TEND LESS TO BE THOSE WHOSE INTERNATIONAL FINANCIAL STATUS IS LOW OR VERY LOW (76)	0.63	0 38 8 65 X SQ= 3.00 P = 0.049 RV NO (NO)
40	TILT TOWARD BEING THOSE WHOSE INTERNATIONAL FINANCIAL STATUS IS VERY LOW (39)	0.62	40	TILT TOWARD BEING THOSE WHOSE INTERNATIONAL FINANCIAL STATUS IS VERY HIGH, HIGH, MEDIUM, OR LOW (71)	0.66	3 66 5 34 X SQ= 1.52 P = 0.134 RV YES (NO)
43	ALWAYS ARE THOSE WHOSE ECONOMIC DEVELOPMENTAL STATUS IS VERY UNDERDEVELOPED (57)	1.00	43	TEND TO BE THOSE WHOSE ECONOMIC DEVELOPMENTAL STATUS IS DEVELOPED, INTERMEDIATE, OR UNDERDEVELOPED (55)	0.54	0 55 8 46 X SQ= 6.75 P = 0.003 RV YES (NO)
45	ALWAYS ARE THOSE WHERE THE LITERACY RATE IS BELOW FIFTY PERCENT (54)	1.00	45	TEND TO BE THOSE WHERE THE LITERACY RATE IS FIFTY PERCENT OR ABOVE (55)	0.54	0 53 8 46 X SQ= 6.48 P = 0.006 RV YES (NO)
46	TEND TO BE THOSE WHERE THE LITERACY RATE IS BELOW TEN PERCENT (26)	0.75	46	TEND TO BE THOSE WHERE THE LITERACY RATE IS TEN PERCENT OR ABOVE (84)	0.80	2 80 6 20 X SQ= 9.43 P = 0.002 RV YES (NO)

54	ALWAYS ARE THOSE WHERE NEWSPAPER CIRCULATION IS LESS THAN ONE HUNDRED PER THOUSAND (76)	1.00	54	LEAN LESS TOWARD BEING THOSE WHERE NEWSPAPER CIRCULATION IS LESS THAN ONE HUNDRED PER THOUSAND (76)	0.64	0 37 8 66 X SQ= 2.85 P = 0.050 RV NO (NO)
55	TEND TO BE THOSE WHERE NEWSPAPER CIRCULATION IS LESS THAN TEN PER THOUSAND (35)	0.87	55	TEND TO BE THOSE WHERE NEWSPAPER CIRCULATION IS TEN OR MORE PER THOUSAND (78)	0.73	1 75 7 28 X SQ= 9.87 P = 0.001 RV YES (NO)
69	ALWAYS ARE THOSE THAT ARE LINGUISTICALLY STRONGLY HETEROGENEOUS (50)	1.00	69	TEND TO BE THOSE THAT ARE LINGUISTICALLY HOMOGENEOUS OR WEAKLY HETEROGENEOUS (64)	0.59	0 61 8 42 X SQ= 8.26 P = 0.001 RV YES (NO)
72	LEAN TOWARD BEING THOSE WHOSE DATE OF INDEPENDENCE IS AFTER 1914 (59)	0.87	72	LEAN TOWARD BEING THOSE WHOSE DATE OF INDEPENDENCE IS BEFORE 1914 (52)	0.50	1 51 7 50 X SQ= 2.90 P = 0.063 RV YES (NO)
73	TEND TO BE THOSE WHOSE DATE OF INDEPENDENCE IS AFTER 1945 (46)	0.87	73	TEND TO BE THOSE WHOSE DATE OF INDEPENDENCE IS BEFORE 1945 (65)	0.63	1 64 7 37 X SQ= 5.99 P = 0.007 RV YES (NO)
75	ALWAYS ARE THOSE THAT ARE NOT HISTORICALLY WESTERN AND ARE NOT SIGNIFICANTLY WESTERNIZED (52)	1.00	75	TEND TO BE THOSE THAT ARE HISTORICALLY WESTERN OR SIGNIFICANTLY WESTERNIZED (62)	0.60	0 62 8 41 X SQ= 8.60 P = 0.001 RV YES (NO)
77	ALWAYS ARE THOSE THAT HAVE BEEN PARTIALLY WESTERNIZED, RATHER THAN SIGNIFICANTLY WESTERNIZED, THROUGH A COLONIAL RELATIONSHIP (41)	1.00	77	TEND LESS TO BE THOSE THAT HAVE BEEN PARTIALLY WESTERNIZED, RATHER THAN SIGNIFICANTLY WESTERNIZED, THROUGH A COLONIAL RELATIONSHIP (41)	0.53	0 28 7 32 X SQ= 3.86 P = 0.019 RV NO (NO)
82	ALWAYS ARE THOSE WHERE THE TYPE OF POLITICAL MODERNIZATION IS DEVELOPED TUTELARY OR UNDEVELOPED TUTELARY, RATHER THAN EARLY OR LATER EUROPEAN OR EUROPEAN DERIVED (55)	1.00	82	TEND TO BE THOSE WHERE THE TYPE OF POLITICAL MODERNIZATION IS EARLY OR LATER EUROPEAN OR EUROPEAN DERIVED, RATHER THAN DEVELOPED TUTELARY OR UNDEVELOPED TUTELARY (51)	0.53	0 51 7 45 X SQ= 5.39 P = 0.013 RV YES (NO)
84	ALWAYS ARE THOSE WHERE THE TYPE OF POLITICAL MODERNIZATION IS UNDEVELOPED TUTELARY, RATHER THAN DEVELOPED TUTELARY (24)	1.00	84	TEND TO BE THOSE WHERE THE TYPE OF POLITICAL MODERNIZATION IS DEVELOPED TUTELARY, RATHER THAN UNDEVELOPED TUTELARY (31)	0.62	0 28 7 17 X SQ= 7.10 P = 0.003 RV YES (NO)

85	ALWAYS ARE THOSE WHERE THE STAGE OF POLITICAL MODERNIZATION IS MID- OR EARLY TRANSITIONAL, RATHER THAN ADVANCED (54)	1.00	85	TEND TO BE THOSE WHERE THE STAGE OF POLITICAL MODERNIZATION IS ADVANCED, RATHER THAN MID- OR EARLY TRANSITIONAL (60)	0.57	0 59 8 44 X SQ= 7.62 P = 0.002 RV YES (NO)
86	TEND TO BE THOSE WHERE THE STAGE OF POLITICAL MODERNIZATION IS EARLY TRANSITIONAL, RATHER THAN ADVANCED OR MID-TRANSITIONAL (38)	0.87	86	TEND TO BE THOSE WHERE THE STAGE OF POLITICAL MODERNIZATION IS ADVANCED OR MID-TRANSITIONAL, RATHER THAN EARLY TRANSITIONAL (76)	0.72	1 74 7 29 X SQ= 9.38 P = 0.001 RV YES (NO)
87	TEND TO BE THOSE WHOSE IDEOLOGICAL ORIENTATION IS DEVELOPMENTAL (31)	0.86	87	TEND TO BE THOSE WHOSE IDEOLOGICAL ORIENTATION IS OTHER THAN DEVELOPMENTAL (58)	0.69	6 25 1 56 X SQ= 6.26 P = 0.007 RV YES (NO)
89	ALWAYS ARE THOSE WHOSE IDEOLOGICAL ORIENTATION IS OTHER THAN CONVENTIONAL (62)	1.00	89	TEND LESS TO BE THOSE WHOSE IDEOLOGICAL ORIENTATION IS OTHER THAN CONVENTIONAL (62)	0.62	0 33 7 53 X SQ= 2.66 P = 0.048 RV NO (NO)
107	TILT MORE TOWARD BEING THOSE WHERE AUTONOMOUS GROUPS ARE PARTIALLY TOLERATED IN POLITICS, ARE TOLERATED ONLY OUTSIDE POLITICS, OR ARE NOT TOLERATED AT ALL (65)	0.87	107	TILT LESS TOWARD BEING THOSE WHERE AUTONOMOUS GROUPS ARE PARTIALLY TOLERATED IN POLITICS, ARE TOLERATED ONLY OUTSIDE POLITICS, OR ARE NOT TOLERATED AT ALL (65)	0.55	1 45 7 55 X SQ= 2.01 P = 0.134 RV NO (NO)
117	ALWAYS ARE THOSE WHERE INTEREST ARTICULATION BY ASSOCIATIONAL GROUPS IS NEGLIGIBLE (51)	1.00	117	TEND TO BE THOSE WHERE INTEREST ARTICULATION BY ASSOCIATIONAL GROUPS IS SIGNIFICANT, MODERATE, OR LIMITED (60)	0.59	0 60 8 41 X SQ= 8.31 P = 0.001 RV YES (NO)
121	ALWAYS ARE THOSE WHERE INTEREST ARTICULATION BY NON-ASSOCIATIONAL GROUPS IS SIGNIFICANT (54)	1.00	121	TEND TO BE THOSE WHERE INTEREST ARTICULATION BY NON-ASSOCIATIONAL GROUPS IS MODERATE, LIMITED, OR NEGLIGIBLE (61)	0.59	8 43 0 61 X SQ= 8.08 P = 0.001 RV YES (NO)
133	TEND TO BE THOSE WHERE INTEREST AGGREGATION BY THE EXECUTIVE IS SIGNIFICANT (29)	0.62	133	TEND TO BE THOSE WHERE INTEREST AGGREGATION BY THE EXECUTIVE IS MODERATE, LIMITED, OR NEGLIGIBLE (76)	0.74	5 24 3 70 X SQ= 3.30 P = 0.040 RV YES (NO)
137	ALWAYS ARE THOSE WHERE INTEREST AGGREGATION BY THE LEGISLATURE IS LIMITED OR NEGLIGIBLE (68)	1.00	137	LEAN LESS TOWARD BEING THOSE WHERE INTEREST AGGREGATION BY THE LEGISLATURE IS LIMITED OR NEGLIGIBLE (68)	0.67	0 29 8 58 X SQ= 2.43 P = 0.057 RV NO (NO)

163	TEND TO BE THOSE WHERE THE REGIME'S LEADERSHIP CHARISMA IS PRONOUNCED (13)	0.57	163	TEND TO BE THOSE WHERE THE REGIME'S LEADERSHIP CHARISMA IS MODERATE OR NEGLIGIBLE (87)	0.91	4 8 3 82 X SQ= 9.85 P = 0.004 RV YES (NO)
164	TEND TO BE THOSE WHERE THE REGIME'S LEADERSHIP CHARISMA IS PRONOUNCED OR MODERATE (34)	0.86	164	TEND TO BE THOSE WHERE THE REGIME'S LEADERSHIP CHARISMA IS NEGLIGIBLE (65)	0.70	6 27 1 62 X SQ= 6.54 P = 0.006 RV YES (NO)
168	ALWAYS ARE THOSE WHERE THE HORIZONTAL POWER DISTRIBUTION IS LIMITED OR NEGLIGIBLE (72)	1.00	168	LEAN LESS TOWARD BEING THOSE WHERE THE HORIZONTAL POWER DISTRIBUTION IS LIMITED OR NEGLIGIBLE (72)	0.65	0 34 8 62 X SQ= 2.75 P = 0.051 RV NO (NO)
174	ALWAYS ARE THOSE WHERE THE LEGISLATURE IS PARTIALLY EFFECTIVE, LARGELY INEFFECTIVE, OR WHOLLY INEFFECTIVE (72)	1.00	174	TILT LESS TOWARD BEING THOSE WHERE THE LEGISLATURE IS PARTIALLY EFFECTIVE, LARGELY INEFFECTIVE, OR WHOLLY INEFFECTIVE (72)	0.69	0 28 8 62 X SQ= 2.13 P = 0.101 RV NO (NO)
182	TEND TO BE THOSE WHERE THE BUREAUCRACY IS POST-COLONIAL TRANSITIONAL, RATHER THAN SEMI-MODERN (25)	0.86	182	TEND TO BE THOSE WHERE THE BUREAUCRACY IS SEMI-MODERN, RATHER THAN POST-COLONIAL TRANSITIONAL (55)	0.73	1 52 6 19 X SQ= 7.64 P = 0.004 RV YES (NO)
188	ALWAYS ARE THOSE WHERE THE CHARACTER OF THE LEGAL SYSTEM IS OTHER THAN CIVIL LAW (81)	1.00	188	TILT LESS TOWARD BEING THOSE WHERE THE CHARACTER OF THE LEGAL SYSTEM IS OTHER THAN CIVIL LAW (81)	0.69	0 32 7 71 X SQ= 1.75 P = 0.104 RV NO (NO)
192	TEND TO BE THOSE WHERE THE CHARACTER OF THE LEGAL SYSTEM IS MUSLIM OR PARTLY MUSLIM (27)	0.86	192	TEND TO BE THOSE WHERE THE CHARACTER OF THE LEGAL SYSTEM IS OTHER THAN MUSLIM OR PARTLY MUSLIM (86)	0.80	6 21 1 82 X SQ= 11.78 P = 0.001 RV YES (NO)
193	ALWAYS ARE THOSE WHERE THE CHARACTER OF THE LEGAL SYSTEM IS PARTLY INDIGENOUS (28)	1.00	193	TEND TO BE THOSE WHERE THE CHARACTER OF THE LEGAL SYSTEM IS OTHER THAN PARTLY INDIGENOUS (86)	0.82	8 19 0 84 X SQ= 22.57 P = 0. RV YES (NO)

62 POLITIES
WHERE THE RELIGION IS
PROTESTANT, RATHER THAN
CATHOLIC (5)

62 POLITIES
WHERE THE RELIGION IS
CATHOLIC, RATHER THAN
PROTESTANT (25)

BOTH SUBJECT AND PREDICATE

5 IN LEFT COLUMN

DENMARK FINLAND ICELAND NORWAY SWEDEN

25 IN RIGHT COLUMN

ARGENTINA	AUSTRIA	BELGIUM	BRAZIL	CHILE	COLOMBIA	COSTA RICA	CUBA	DOMIN REP	EL SALVADOR
FRANCE	HONDURAS	IRELAND	ITALY	LUXEMBOURG	MEXICO	NICARAGUA	PANAMA	PARAGUAY	PHILIPPINES
POLAND	PORTUGAL	SPAIN	URUGUAY	VENEZUELA					

85 EXCLUDED BECAUSE IRRELEVANT

AFGHANISTAN	ALBANIA	ALGERIA	AUSTRALIA	BOLIVIA	BULGARIA	BURMA	BURUNDI	CAMBODIA	CAMEROUN
CANADA	CEN AFR REP	CEYLON	CHAD	CHINA, PR	CONGO(BRA)	CONGO(LEO)	CYPRUS	CZECHOS'KIA	DAHOMEY
ECUADOR	ETHIOPIA	GABON	GERMANY, E	GERMAN FR	GHANA	GREECE	GUATEMALA	GUINEA	HAITI
HUNGARY	INDIA	INDONESIA	IRAN	IRAQ	ISRAEL	IVORY COAST	JAMAICA	JAPAN	JORDAN
KOREA, N	KOREA REP	LAOS	LEBANON	LIBERIA	LIBYA	MALAGASY R	MALAYA	MALI	MAURITANIA
MONGOLIA	MOROCCO	NEPAL	NETHERLANDS	NEW ZEALAND	NIGER	NIGERIA	PAKISTAN	PERU	RUMANIA
RWANDA	SA'U ARABIA	SENEGAL	SIERRE LEO	SOMALIA	SO AFRICA	SUDAN	SWITZERLAND	SYRIA	TANGANYIKA
THAILAND	TOGO	TRINIDAD	TUNISIA	TURKEY	UGANDA	USSR	UAR	UK	US
UPPER VOLTA	VIETNAM, N	VIETNAM REP	YEMEN	YUGOSLAVIA					

2	ALWAYS ARE THOSE LOCATED IN WEST EUROPE, SCANDINAVIA, NORTH AMERICA, OR AUSTRALASIA (22)	1.00	2 TEND TO BE THOSE LOCATED ELSEWHERE THAN IN WEST EUROPE, SCANDINAVIA, NORTH AMERICA, OR AUSTRALASIA (93)	0.68	5 8 0 17 X SQ= 5.32 P = 0.009 RV YES (NO)
6	ALWAYS ARE THOSE LOCATED ELSEWHERE THAN IN THE CARIBBEAN, CENTRAL AMERICA, OR SOUTH AMERICA (93)	1.00	6 TEND TO BE THOSE LOCATED IN THE CARIBBEAN, CENTRAL AMERICA, OR SOUTH AMERICA (22)	0.60	0 15 5 10 X SQ= 3.84 P = 0.042 RV YES (NO)
31	TILT TOWARD BEING THOSE WHOSE AGRICULTURAL POPULATION IS LOW OR VERY LOW (24)	0.80	31 TILT TOWARD BEING THOSE WHOSE AGRICULTURAL POPULATION IS HIGH OR MEDIUM (90)	0.68	1 17 4 8 X SQ= 2.25 P = 0.128 RV YES (NO)

35	ALWAYS ARE THOSE WHOSE PER CAPITA GROSS NATIONAL PRODUCT IS VERY HIGH OR HIGH (24)	1.00	35	TEND TO BE THOSE WHOSE PER CAPITA GROSS NATIONAL PRODUCT IS MEDIUM, LOW, OR VERY LOW (91)	0.72	5 7 0 18 X SQ= 6.25 P = 0.006 RV YES (NO)
41	TEND TO BE THOSE WHOSE ECONOMIC DEVELOPMENTAL STATUS IS DEVELOPED (19)	0.80	41	TEND TO BE THOSE WHOSE ECONOMIC DEVELOPMENTAL STATUS IS INTERMEDIATE, UNDERDEVELOPED, OR VERY UNDERDEVELOPED (94)	0.83	4 4 1 20 X SQ= 5.44 P = 0.013 RV YES (YES)
42	ALWAYS ARE THOSE WHOSE ECONOMIC DEVELOPMENTAL STATUS IS DEVELOPED OR INTERMEDIATE (36)	1.00	42	LEAN TOWARD BEING THOSE WHOSE ECONOMIC DEVELOPMENTAL STATUS IS UNDERDEVELOPED OR VERY UNDERDEVELOPED (76)	0.50	5 12 0 12 X SQ= 2.45 P = 0.059 RV YES (NO)
44	ALWAYS ARE THOSE WHERE THE LITERACY RATE IS NINETY PERCENT OR ABOVE (25)	1.00	44	TEND TO BE THOSE WHERE THE LITERACY RATE IS BELOW NINETY PERCENT (90)	0.76	5 6 0 19 X SQ= 7.35 P = 0.003 RV YES (NO)
50	ALWAYS ARE THOSE WHERE FREEDOM OF THE PRESS IS COMPLETE (43)	1.00	50	TILT LESS TOWARD BEING THOSE WHERE FREEDOM OF THE PRESS IS COMPLETE (43)	0.59	5 13 0 9 X SQ= 1.50 P = 0.136 RV NO (NO)
53	ALWAYS ARE THOSE WHERE NEWSPAPER CIRCULATION IS THREE HUNDRED OR MORE PER THOUSAND (14)	1.00	53	TEND TO BE THOSE WHERE NEWSPAPER CIRCULATION IS LESS THAN THREE HUNDRED PER THOUSAND (101)	0.96	5 1 0 24 X SQ= 18.37 P = 0. RV YES (YES)
67	ALWAYS ARE THOSE THAT ARE RACIALLY HOMOGENEOUS (82)	1.00	67	TILT LESS TOWARD BEING THOSE THAT ARE RACIALLY HOMOGENEOUS (82)	0.57	5 13 0 10 X SQ= 1.75 P = 0.128 RV NO (NO)
71	TILT TOWARD BEING THOSE WHOSE DATE OF INDEPENDENCE IS BEFORE 1800 (21)	0.60	71	TILT TOWARD BEING THOSE WHOSE DATE OF INDEPENDENCE IS AFTER 1800 (90)	0.80	3 5 2 20 X SQ= 1.67 P = 0.102 RV YES (NO)
74	ALWAYS ARE THOSE THAT ARE HISTORICALLY WESTERN (26)	1.00	74	TEND TO BE THOSE THAT ARE NOT HISTORICALLY WESTERN (87)	0.64	5 9 0 16 X SQ= 4.53 P = 0.014 RV YES (NO)

76	ALWAYS ARE THOSE THAT ARE HISTORICALLY WESTERN, RATHER THAN HAVING BEEN SIGNIFICANTLY OR PARTIALLY WESTERNIZED THROUGH A COLONIAL RELATIONSHIP (26)	1.00	76	TEND TO BE THOSE THAT HAVE BEEN SIGNIFICANTLY OR PARTIALLY WESTERNIZED THROUGH A COLONIAL RELATIONSHIP, RATHER THAN BEING HISTORICALLY WESTERN (70)	0.64	5 9 0 16 X SQ= 4.53 P = 0.014 RV YES (NO)
98	ALWAYS ARE THOSE WHERE GOVERNMENTAL STABILITY IS GENERALLY PRESENT AND DATES AT LEAST FROM THE INTERWAR PERIOD (22)	1.00	98	TEND TO BE THOSE WHERE GOVERNMENTAL STABILITY IS GENERALLY OR MODERATELY PRESENT AND DATES FROM THE POST-WAR PERIOD, OR IS ABSENT (93)	0.80	5 5 0 20 X SQ= 8.67 P = 0.002 RV YES (YES)
101	ALWAYS ARE THOSE WHERE THE REPRESENTATIVE CHARACTER OF THE REGIME IS POLYARCHIC (41)	1.00	101	TILT LESS TOWARD BEING THOSE WHERE THE REPRESENTATIVE CHARACTER OF THE REGIME IS POLYARCHIC (41)	0.54	5 13 0 11 X SQ= 2.00 P = 0.126 RV NO (NO)
110	TEND TO BE THOSE WHERE POLITICAL ENCULTURATION IS HIGH (15)	0.80	110	TEND TO BE THOSE WHERE POLITICAL ENCULTURATION IS MEDIUM OR LOW (80)	0.76	4 5 1 16 X SQ= 3.42 P = 0.034 RV YES (NO)
115	LEAN TOWARD BEING THOSE WHERE INTEREST ARTICULATION BY ASSOCIATIONAL GROUPS IS SIGNIFICANT (20)	0.80	115	LEAN TOWARD BEING THOSE WHERE INTEREST ARTICULATION BY ASSOCIATIONAL GROUPS IS MODERATE, LIMITED, OR NEGLIGIBLE (91)	0.70	4 7 1 16 X SQ= 2.41 P = 0.062 RV YES (NO)
116	ALWAYS ARE THOSE WHERE INTEREST ARTICULATION BY ASSOCIATIONAL GROUPS IS SIGNIFICANT OR MODERATE (32)	1.00	116	TILT LESS TOWARD BEING THOSE WHERE INTEREST ARTICULATION BY ASSOCIATIONAL GROUPS IS SIGNIFICANT OR MODERATE (32)	0.57	5 13 0 10 X SQ= 1.75 P = 0.128 RV NO (NO)
120	ALWAYS ARE THOSE WHERE INTEREST ARITUCLATION BY INSTITUTIONAL GROUPS IS LIMITED (10)	1.00	120	TEND TO BE THOSE WHERE INTEREST ARTICULATION BY INSTITUTIONAL GROUPS IS VERY SIGNIFICANT, SIGNIFICANT, OR MODERATE (90)	0.96	0 24 5 1 X SQ= 18.37 P = 0. RV YES (YES)
123	TEND TO BE THOSE WHERE INTEREST ARTICULATION BY NON-ASSOCIATIONAL GROUPS IS NEGLIGIBLE (8)	0.80	123	TEND TO BE THOSE WHERE INTEREST ARTICULATION BY NON-ASSOCIATIONAL GROUPS IS SIGNIFICANT, MODERATE, OR LIMITED (107)	0.92	1 23 4 2 X SQ= 9.37 P = 0.003 RV YES (YES)
126	ALWAYS ARE THOSE WHERE INTEREST ARTICULATION BY ANOMIC GROUPS IS VERY INFREQUENT (16)	1.00	126	TEND TO BE THOSE WHERE INTEREST ARTICULATION BY ANOMIC GROUPS IS FREQUENT, OCCASIONAL, OR INFREQUENT (83)	0.80	M 0 16 5 4 X SQ= 7.91 P = 0.002 RV YES (YES)

127	ALWAYS ARE THOSE WHERE INTEREST ARTICULATION BY POLITICAL PARTIES IS SIGNIFICANT (21)	1.00	127	TEND TO BE THOSE WHERE INTEREST ARTICULATION BY POLITICAL PARTIES IS MODERATE, LIMITED, OR NEGLIGIBLE (72)	0.83	5 4 0 19 X SQ= 9.34 P = 0.001 RV YES (YES)
134	ALWAYS ARE THOSE WHERE INTEREST AGGREGATION BY THE EXECUTIVE IS SIGNIFICANT OR MODERATE (57)	1.00	134	LEAN TOWARD BEING THOSE WHERE INTEREST AGGREGATION BY THE EXECUTIVE IS LIMITED OR NEGLIGIBLE (46)	0.50	5 12 0 12 X SQ= 2.45 P = 0.059 RV YES (NO)
136	TEND TO BE THOSE WHERE INTEREST AGGREGATION BY THE LEGISLATURE IS SIGNIFICANT (12)	0.80	136	TEND TO BE THOSE WHERE INTEREST AGGREGATION BY THE LEGISLATURE IS MODERATE, LIMITED, OR NEGLIGIBLE (85)	0.82	4 4 1 18 X SQ= 4.80 P = 0.017 RV YES (YES)
137	ALWAYS ARE THOSE WHERE INTEREST AGGREGATION BY THE LEGISLATURE IS SIGNIFICANT OR MODERATE (29)	1.00	137	TEND TO BE THOSE WHERE INTEREST AGGREGATION BY THE LEGISLATURE IS LIMITED OR NEGLIGIBLE (68)	0.64	5 8 0 14 X SQ= 4.31 P = 0.016 RV YES (NO)
142	ALWAYS ARE THOSE WHERE THE PARTY SYSTEM IS QUANTITATIVELY MULTI-PARTY (30)	1.00	142	TILT LESS TOWARD BEING THOSE WHERE THE PARTY SYSTEM IS QUANTITATIVELY MULTI-PARTY (30)	0.52	5 12 0 11 X SQ= 2.19 P = 0.125 RV NO (NO)
147	ALWAYS ARE THOSE WHERE THE PARTY SYSTEM IS QUALITATIVELY CLASS-ORIENTED OR MULTI-IDEOLOGICAL (23)	1.00	147	TEND TO BE THOSE WHERE THE PARTY SYSTEM IS QUALITATIVELY OTHER THAN CLASS-ORIENTED OR MULTI-IDEOLOGICAL (67)	0.68	5 7 0 15 X SQ= 5.16 P = 0.010 RV YES (NO)
153	ALWAYS ARE THOSE WHERE THE PARTY SYSTEM IS STABLE (42)	1.00	153	TEND TO BE THOSE WHERE THE PARTY SYSTEM IS MODERATELY STABLE OR UNSTABLE (38)	0.57	5 9 0 12 X SQ= 3.26 P = 0.042 RV YES (NO)
158	ALWAYS ARE THOSE WHERE PERSONALISMO IS NEGLIGIBLE (56)	1.00	158	TILT LESS TOWARD BEING THOSE WHERE PERSONALISMO IS NEGLIGIBLE (56)	0.54	0 11 5 13 X SQ= 2.00 P = 0.126 RV NO (NO)
161	ALWAYS ARE THOSE WHERE THE POLITICAL LEADERSHIP IS NON-ELITIST (50)	1.00	161	TEND TO BE THOSE WHERE THE POLITICAL LEADERSHIP IS ELITIST OR MODERATELY ELITIST (47)	0.54	0 13 5 11 X SQ= 2.96 P = 0.048 RV YES (NO)

168	ALWAYS ARE THOSE WHERE THE HORIZONTAL POWER DISTRIBUTION IS SIGNIFICANT (34)	1.00	168	TEND TO BE THOSE WHERE THE HORIZONTAL POWER DISTRIBUTION IS LIMITED OR NEGLIGIBLE (72)	0.54	5 11 0 13 X SQ= 2.96 P = 0.048 RV YES (NO)
170	ALWAYS ARE THOSE WHERE THE LEGISLATIVE-EXECUTIVE STRUCTURE IS OTHER THAN PRESIDENTIAL (63)	1.00	170	TEND TO BE THOSE WHERE THE LEGISLATIVE-EXECUTIVE STRUCTURE IS PRESIDENTIAL (39)	0.58	0 14 5 10 X SQ= 3.54 P = 0.042 RV YES (NO)
172	TEND TO BE THOSE WHERE THE LEGISLATIVE-EXECUTIVE STRUCTURE IS PARLIAMENTARY-ROYALIST (21)	0.60	172	TEND TO BE THOSE WHERE THE LEGISLATIVE-EXECUTIVE STRUCTURE IS OTHER THAN PARLIAMENTARY-ROYALIST (88)	0.92	3 2 2 22 X SQ= 4.54 P = 0.024 RV YES (YES)
174	ALWAYS ARE THOSE WHERE THE LEGISLATURE IS FULLY EFFECTIVE (28)	1.00	174	TEND TO BE THOSE WHERE THE LEGISLATURE IS PARTIALLY EFFECTIVE, LARGELY INEFFECTIVE, OR WHOLLY INEFFECTIVE (72)	0.70	5 7 0 16 X SQ= 5.52 P = 0.008 RV YES (NO)
179	ALWAYS ARE THOSE WHERE THE EXECUTIVE IS STRONG (39)	1.00	179	TILT LESS TOWARD BEING THOSE WHERE THE EXECUTIVE IS STRONG (39)	0.55	0 10 5 12 X SQ= 1.92 P = 0.124 RV NO (NO)
181	ALWAYS ARE THOSE WHERE THE BUREAUCRACY IS MODERN, RATHER THAN SEMI-MODERN (21)	1.00	181	TEND TO BE THOSE WHERE THE BUREAUCRACY IS SEMI-MODERN, RATHER THAN MODERN (55)	0.76	5 6 0 19 X SQ= 7.35 P = 0.003 RV YES (NO)
187	ALWAYS ARE THOSE WHERE THE ROLE OF THE POLICE IS NOT POLITICALLY SIGNIFICANT (35)	1.00	187	TEND TO BE THOSE WHERE THE ROLE OF THE POLICE IS POLITICALLY SIGNIFICANT (66)	0.55	0 11 5 9 X SQ= 2.93 P = 0.046 RV YES (NO)

MATRIX

62 POLITIES
WHERE THE RELIGION IS
PROTESTANT, RATHER THAN
CATHOLIC (5)

62 POLITIES
WHERE THE RELIGION IS
CATHOLIC, RATHER THAN
PROTESTANT (25)

126 POLITIES
WHERE INTEREST ARTICULATION
BY ANOMIC GROUPS
IS FREQUENT, OCCASIONAL, OR
INFREQUENT (83)

BRAZIL CHILE COLOMBIA COSTA RICA EL SALVADOR
HONDURAS ITALY MEXICO NICARAGUA PANAMA
PARAGUAY PHILIPPINES POLAND PORTUGAL SPAIN
URUGUAY

0	16
5	4

126 POLITIES
WHERE INTEREST ARTICULATION
BY ANOMIC GROUPS
IS VERY INFREQUENT (16)

DENMARK FINLAND ICELAND NORWAY SWEDEN

AUSTRIA BELGIUM IRELAND LUXEMBOURG

63 POLITIES
WHERE THE RELIGION IS
PREDOMINANTLY
SOME KIND OF CHRISTIAN (46)

63 POLITIES
WHERE THE RELIGION IS
PREDOMINANTLY OR PARTLY
OTHER THAN CHRISTIAN (68)

BOTH SUBJECT AND PREDICATE

46 IN LEFT COLUMN

ARGENTINA	AUSTRALIA	AUSTRIA	BELGIUM	BRAZIL	BULGARIA	CANADA	CHILE	COLOMBIA	COSTA RICA
CUBA	CZECHOS'KIA	DENMARK	DOMIN REP	EL SALVADOR	FINLAND	FRANCE	GERMANY, E	GERMAN FR	GREECE
HONDURAS	HUNGARY	ICELAND	IRELAND	ITALY	JAMAICA	LUXEMBOURG	MEXICO	NETHERLANDS	NEW ZEALAND
NICARAGUA	NORWAY	PANAMA	PARAGUAY	PHILIPPINES	POLAND	PORTUGAL	RUMANIA	SPAIN	SWEDEN
SWITZERLAND	UK	US	URUGUAY	VENEZUELA	YUGOSLAVIA				

68 IN RIGHT COLUMN

AFGHANISTAN	ALBANIA	ALGERIA	BOLIVIA	BURMA	BURUNDI	CAMBODIA	CAMEROUN	CEN AFR REP	CEYLON
CHAD	CHINA, PR	CONGO(BRA)	CONGO(LEO)	CYPRUS	DAHOMEY	ECUADOR	ETHIOPIA	GABON	GHANA
GUATEMALA	GUINEA	HAITI	INDIA	INDONESIA	IRAN	IRAQ	ISRAEL	IVORY COAST	JAPAN
JORDAN	KOREA, N	KOREA REP	LAOS	LEBANON	LIBERIA	LIBYA	MALAGASY R	MALAYA	MALI
MAURITANIA	MONGOLIA	MOROCCO	NEPAL	NIGER	NIGERIA	PAKISTAN	PERU	RWANDA	SA'U ARABIA
SENEGAL	SIERRE LEO	SOMALIA	SO AFRICA	SUDAN	SYRIA	TANGANYIKA	THAILAND	TOGO	TRINIDAD
TUNISIA	TURKEY	UGANDA	UAR	UPPER VOLTA	VIETNAM, N	VIETNAM REP	YEMEN		

1 EXCLUDED BECAUSE UNASCERTAINED

USSR

2	TEND LESS TO BE THOSE LOCATED ELSEWHERE THAN IN WEST EUROPE, SCANDINAVIA, NORTH AMERICA, OR AUSTRALASIA (93)	0.52	2	ALWAYS ARE THOSE LOCATED ELSEWHERE THAN IN WEST EUROPE, SCANDINAVIA, NORTH AMERICA, OR AUSTRALASIA (93)	1.00	22 0 24 68 X SQ= 37.29 P = 0. RV NO (YES)
6	TEND LESS TO BE THOSE LOCATED ELSEWHERE THAN IN THE CARIBBEAN, CENTRAL AMERICA, OR SOUTH AMERICA (93)	0.65	6	TEND MORE TO BE THOSE LOCATED ELSEWHERE THAN IN THE CARIBBEAN, CENTRAL AMERICA, OR SOUTH AMERICA (93)	0.91	16 6 30 62 X SQ= 10.26 P = 0.001 RV NO (YES)
8	TEND MORE TO BE THOSE LOCATED ELSEWHERE THAN IN EAST ASIA SOUTH ASIA, OR SOUTHEAST ASIA (97)	0.98	8	TEND LESS TO BE THOSE LOCATED ELSEWHERE THAN IN EAST ASIA SOUTH ASIA, OR SOUTHEAST ASIA (97)	0.75	1 17 45 51 X SQ= 9.10 P = 0.001 RV NO (NO)

9	LEAN MORE TOWARD BEING THOSE LOCATED ELSEWHERE THAN IN SOUTHEAST ASIA (106)	0.98	9	LEAN LESS TOWARD BEING THOSE LOCATED ELSEWHERE THAN IN SOUTHEAST ASIA (106)	0.88	1 8 45 60 X SQ= 2.28 P = 0.082 RV NO (NO)
10	ALWAYS ARE THOSE LOCATED ELSEWHERE THAN IN NORTH AFRICA, OR CENTRAL AND SOUTH AFRICA (82)	1.00	10	TEND LESS TO BE THOSE LOCATED ELSEWHERE THAN IN NORTH AFRICA, OR CENTRAL AND SOUTH AFRICA (82)	0.51	0 33 46 35 X SQ= 29.10 P = 0. RV NO (YES)
13	ALWAYS ARE THOSE LOCATED ELSEWHERE THAN IN NORTH AFRICA OR THE MIDDLE EAST (99)	1.00	13	TEND LESS TO BE THOSE LOCATED ELSEWHERE THAN IN NORTH AFRICA OR THE MIDDLE EAST (99)	0.76	0 16 46 52 X SQ= 10.72 P = 0. RV NO (NO)
22	TILT TOWARD BEING THOSE WHOSE TERRITORIAL SIZE IS SMALL (47)	0.50	22	TILT TOWARD BEING THOSE WHOSE TERRITORIAL SIZE IS VERY LARGE, LARGE, OR MEDIUM (68)	0.65	23 44 23 24 X SQ= 1.88 P = 0.126 RV YES (NO)
24	LEAN TOWARD BEING THOSE WHOSE POPULATION IS VERY LARGE, LARGE, OR MEDIUM (61)	0.63	24	LEAN TOWARD BEING THOSE WHOSE POPULATION IS SMALL (54)	0.54	29 31 17 37 X SQ= 2.69 P = 0.086 RV YES (NO)
26	TEND TO BE THOSE WHOSE POPULATION DENSITY IS VERY HIGH, HIGH, OR MEDIUM (48)	0.57	26	TEND TO BE THOSE WHOSE POPULATION DENSITY IS LOW (67)	0.68	26 22 20 46 X SQ= 5.62 P = 0.012 RV YES (YES)
28	TEND TO BE THOSE WHOSE POPULATION GROWTH RATE IS LOW (48)	0.59	28	TEND TO BE THOSE WHOSE POPULATION GROWTH RATE IS HIGH (62)	0.67	19 43 27 21 X SQ= 6.28 P = 0.011 RV YES (YES)
29	TEND TO BE THOSE WHERE THE DEGREE OF URBANIZATION IS HIGH (56)	0.93	29	TEND TO BE THOSE WHERE THE DEGREE OF URBANIZATION IS LOW (49)	0.74	39 16 3 46 X SQ= 42.53 P = 0. RV YES (YES)
30	TEND TO BE THOSE WHOSE AGRICULTURAL POPULATION IS MEDIUM, LOW, OR VERY LOW (57)	0.91	30	TEND TO BE THOSE WHOSE AGRICULTURAL POPULATION IS HIGH (56)	0.79	4 52 42 14 X SQ= 50.50 P = 0. RV YES (YES)

34	TEND TO BE THOSE WHOSE GROSS NATIONAL PRODUCT IS VERY HIGH, HIGH, MEDIUM, OR LOW (62)	0.78	34	TEND TO BE THOSE WHOSE GROSS NATIONAL PRODUCT IS VERY LOW (53)	0.63	36 25 10 43 X SQ= 17.36 P = 0. RV YES (YES)
36	TEND TO BE THOSE WHOSE PER CAPITA GROSS NATIONAL PRODUCT IS VERY HIGH, HIGH, OR MEDIUM (42)	0.74	36	TEND TO BE THOSE WHOSE PER CAPITA GROSS NATIONAL PRODUCT IS LOW OR VERY LOW (73)	0.90	34 7 12 61 X SQ= 45.50 P = 0. RV YES (YES)
37	TEND TO BE THOSE WHOSE PER CAPITA GROSS NATIONAL PRODUCT IS VERY HIGH, HIGH, MEDIUM, OR LOW (64)	0.98	37	TEND TO BE THOSE WHOSE PER CAPITA GROSS NATIONAL PRODUCT IS VERY LOW (51)	0.74	45 18 1 50 X SQ= 53.66 P = 0. RV YES (YES)
39	TEND TO BE THOSE WHOSE INTERNATIONAL FINANCIAL STATUS IS VERY HIGH, HIGH, OR MEDIUM (38)	0.64	39	TEND TO BE THOSE WHOSE INTERNATIONAL FINANCIAL STATUS IS LOW OR VERY LOW (76)	0.88	29 8 16 60 X SQ= 31.77 P = 0. RV YES (YES)
40	TEND TO BE THOSE WHOSE INTERNATIONAL FINANCIAL STATUS IS VERY HIGH, HIGH, MEDIUM, OR LOW (71)	0.84	40	TEND TO BE THOSE WHOSE INTERNATIONAL FINANCIAL STATUS IS VERY LOW (39)	0.50	38 32 7 32 X SQ= 12.18 P = 0. RV YES (YES)
41	TEND LESS TO BE THOSE WHOSE ECONOMIC DEVELOPMENTAL STATUS IS INTERMEDIATE, UNDERDEVELOPED, OR VERY UNDERDEVELOPED (94)	0.60	41	ALWAYS ARE THOSE WHOSE ECONOMIC DEVELOPMENTAL STATUS IS INTERMEDIATE, UNDERDEVELOPED, OR VERY UNDERDEVELOPED (94)	1.00	18 0 27 67 X SQ= 29.04 P = 0. RV NO (YES)
42	TEND TO BE THOSE WHOSE ECONOMIC DEVELOPMENTAL STATUS IS DEVELOPED OR INTERMEDIATE (36)	0.71	42	TEND TO BE THOSE WHOSE ECONOMIC DEVELOPMENTAL STATUS IS UNDERDEVELOPED OR VERY UNDERDEVELOPED (76)	0.95	32 3 13 63 X SQ= 51.88 P = 0. RV YES (YES)
43	TEND TO BE THOSE WHOSE ECONOMIC DEVELOPMENTAL STATUS IS DEVELOPED, INTERMEDIATE, OR UNDERDEVELOPED (55)	0.87	43	TEND TO BE THOSE WHOSE ECONOMIC DEVELOPMENTAL STATUS IS VERY UNDERDEVELOPED (57)	0.77	39 15 6 51 X SQ= 41.27 P = 0. RV YES (YES)
44	TEND TO BE THOSE WHERE THE LITERACY RATE IS NINETY PERCENT OR ABOVE (25)	0.50	44	TEND TO BE THOSE WHERE THE LITERACY RATE IS BELOW NINETY PERCENT (90)	0.99	23 1 23 67 X SQ= 36.02 P = 0. RV YES (YES)

45	TEND TO BE THOSE WHERE THE LITERACY RATE IS FIFTY PERCENT OR ABOVE (55)	0.93	45	TEND TO BE THOSE WHERE THE LITERACY RATE IS BELOW FIFTY PERCENT (54)	0.82	43 11 3 51 X SQ= 57.60 P = 0. RV YES (YES)
46	ALWAYS ARE THOSE WHERE THE LITERACY RATE IS TEN PERCENT OR ABOVE (84)	1.00	46	TEND LESS TO BE THOSE WHERE THE LITERACY RATE IS TEN PERCENT OR ABOVE (84)	0.59	46 37 0 26 X SQ= 22.71 P = 0. RV NO (YES)
50	TEND TO BE THOSE WHERE FREEDOM OF THE PRESS IS COMPLETE (43)	0.64	50	TEND TO BE THOSE WHERE FREEDOM OF THE PRESS IS INTERMITTENT, INTERNALLY ABSENT, OR INTERNALLY AND EXTERNALLY ABSENT (56)	0.71	27 16 15 40 X SQ= 11.02 P = 0.001 RV YES (YES)
54	TEND TO BE THOSE WHERE NEWSPAPER CIRCULATION IS ONE HUNDRED OR MORE PER THOUSAND (37)	0.70	54	TEND TO BE THOSE WHERE NEWSPAPER CIRCULATION IS LESS THAN ONE HUNDRED PER THOUSAND (76)	0.94	32 4 14 62 X SQ= 47.25 P = 0. RV YES (YES)
55	ALWAYS ARE THOSE WHERE NEWSPAPER CIRCULATION IS TEN OR MORE PER THOUSAND (78)	1.00	55	TEND TO BE THOSE WHERE NEWSPAPER CIRCULATION IS LESS THAN TEN PER THOUSAND (35)	0.53	46 31 0 35 X SQ= 33.06 P = 0. RV YES (YES)
57	TEND TO BE THOSE WHERE THE RELIGION IS CATHOLIC (25)	0.54	57	ALWAYS ARE THOSE WHERE THE RELIGION IS OTHER THAN CATHOLIC (90)	1.00	25 0 21 68 X SQ= 44.22 P = 0. RV YES (YES)
66	TEND TO BE THOSE THAT ARE RELIGIOUSLY HOMOGENEOUS (57)	0.72	66	TEND TO BE THOSE THAT ARE RELIGIOUSLY HETEROGENEOUS (49)	0.60	33 24 13 36 X SQ= 9.31 P = 0.002 RV YES (YES)
68	TEND TO BE THOSE THAT ARE LINGUISTICALLY HOMOGENEOUS (52)	0.76	68	TEND TO BE THOSE THAT ARE LINGUISTICALLY WEAKLY HETEROGENEOUS OR STRONGLY HETEROGENEOUS (62)	0.75	35 17 11 50 X SQ= 26.23 P = 0. RV YES (YES)
69	TEND TO BE THOSE THAT ARE LINGUISTICALLY HOMOGENEOUS OR WEAKLY HETEROGENEOUS (64)	0.87	69	TEND TO BE THOSE THAT ARE LINGUISTICALLY STRONGLY HETEROGENEOUS (50)	0.64	40 24 6 43 X SQ= 26.99 P = 0. RV YES (YES)

No.	Left	Prop.	No.	Right	Prop.	Statistics
70	TEND LESS TO BE THOSE THAT ARE RELIGIOUSLY, RACIALLY, OR LINGUISTICALLY HETEROGENEOUS (85)	0.66	70	TEND MORE TO BE THOSE THAT ARE RELIGIOUSLY, RACIALLY, OR LINGUISTICALLY HETEROGENEOUS (85)	0.90	15 6 29 54 X SQ= 7.71 P = 0.003 RV NO (YES)
72	TEND TO BE THOSE WHOSE DATE OF INDEPENDENCE IS BEFORE 1914 (52)	0.80	72	TEND TO BE THOSE WHOSE DATE OF INDEPENDENCE IS AFTER 1914 (59)	0.76	35 16 9 50 X SQ= 30.28 P = 0. RV YES (YES)
73	TEND TO BE THOSE WHOSE DATE OF INDEPENDENCE IS BEFORE 1945 (65)	0.95	73	TEND TO BE THOSE WHOSE DATE OF INDEPENDENCE IS AFTER 1945 (46)	0.67	42 22 2 44 X SQ= 39.36 P = 0. RV YES (YES)
74	TEND TO BE THOSE THAT ARE HISTORICALLY WESTERN (26)	0.57	74	ALWAYS ARE THOSE THAT ARE NOT HISTORICALLY WESTERN (87)	1.00	26 0 20 66 X SQ= 45.46 P = 0. RV YES (YES)
75	ALWAYS ARE THOSE THAT ARE HISTORICALLY WESTERN OR SIGNIFICANTLY WESTERNIZED (62)	1.00	75	TEND TO BE THOSE THAT ARE NOT HISTORICALLY WESTERN AND ARE NOT SIGNIFICANTLY WESTERNIZED (52)	0.78	46 15 0 52 X SQ= 63.05 P = 0. RV YES (YES)
76	TEND TO BE THOSE THAT ARE HISTORICALLY WESTERN, RATHER THAN HAVING BEEN SIGNIFICANTLY OR PARTIALLY WESTERNIZED THROUGH A COLONIAL RELATIONSHIP (26)	0.60	76	ALWAYS ARE THOSE THAT HAVE BEEN SIGNIFICANTLY OR PARTIALLY WESTERNIZED THROUGH A COLONIAL RELATIONSHIP, RATHER THAN BEING HISTORICALLY WESTERN (70)	1.00	26 0 17 53 X SQ= 40.94 P = 0. RV YES (YES)
77	ALWAYS ARE THOSE THAT HAVE BEEN SIGNIFICANTLY WESTERNIZED, RATHER THAN PARTIALLY WESTERNIZED, THROUGH A COLONIAL RELATIONSHIP (28)	1.00	77	TEND TO BE THOSE THAT HAVE BEEN PARTIALLY WESTERNIZED, RATHER THAN SIGNIFICANTLY WESTERNIZED, THROUGH A COLONIAL RELATIONSHIP (41)	0.79	17 11 0 41 X SQ= 29.84 P = 0. RV YES (YES)
79	TEND TO BE THOSE THAT WERE FORMERLY DEPENDENCIES OF SPAIN, RATHER THAN BRITAIN OR FRANCE (18)	0.70	79	TEND TO BE THOSE THAT WERE FORMERLY DEPENDENCIES OF BRITAIN OR FRANCE, RATHER THAN SPAIN (49)	0.91	6 43 14 4 X SQ= 23.96 P = 0. RV YES (YES)
82	TEND TO BE THOSE WHERE THE TYPE OF POLITICAL MODERNIZATION IS EARLY OR LATER EUROPEAN OR EUROPEAN DERIVED, RATHER THAN DEVELOPED TUTELARY OR UNDEVELOPED TUTELARY (51)	0.96	82	TEND TO BE THOSE WHERE THE TYPE OF POLITICAL MODERNIZATION IS DEVELOPED TUTELARY OR UNDEVELOPED TUTELARY, RATHER THAN EARLY OR LATER EUROPEAN OR EUROPEAN DERIVED (55)	0.88	44 7 2 53 X SQ= 70.24 P = 0. RV YES (YES)

83	ALWAYS ARE THOSE WHERE THE TYPE OF POLITICAL MODERNIZATION IS OTHER THAN NON-EUROPEAN AUTOCHTHONOUS (106)	1.00	83	TEND LESS TO BE THOSE WHERE THE TYPE OF POLITICAL MODERNIZATION IS OTHER THAN NON-EUROPEAN AUTOCHTHONOUS (106)	0.88	0 8 46 60 X SQ= 4.16 P = 0.021 RV NO (NO)
85	TEND TO BE THOSE WHERE THE STAGE OF POLITICAL MODERNIZATION IS ADVANCED, RATHER THAN MID- OR EARLY TRANSITIONAL (60)	0.83	85	TEND TO BE THOSE WHERE THE STAGE OF POLITICAL MODERNIZATION IS MID- OR EARLY TRANSITIONAL, RATHER THAN ADVANCED (54)	0.69	38 21 8 46 X SQ= 26.71 P = 0. RV YES (YES)
86	TEND TO BE THOSE WHERE THE STAGE OF POLITICAL MODERNIZATION IS ADVANCED OR MID-TRANSITIONAL, RATHER THAN EARLY TRANSITIONAL (76)	0.98	86	TEND TO BE THOSE WHERE THE STAGE OF POLITICAL MODERNIZATION IS EARLY TRANSITIONAL, RATHER THAN ADVANCED OR MID-TRANSITIONAL (38)	0.55	45 30 1 . 37 X SQ= 32.05 P = 0. RV YES (YES)
87	TEND TO BE THOSE WHOSE IDEOLOGICAL ORIENTATION IS OTHER THAN DEVELOPMENTAL (58)	0.95	87	TEND TO BE THOSE WHOSE IDEOLOGICAL ORIENTATION IS DEVELOPMENTAL (31)	0.59	2 29 37 20 X SQ= 25.49 P = 0. RV YES (YES)
89	TEND TO BE THOSE WHOSE IDEOLOGICAL ORIENTATION IS CONVENTIONAL (33)	0.71	89	TEND TO BE THOSE WHOSE IDEOLOGICAL ORIENTATION IS OTHER THAN CONVENTIONAL (62)	0.89	27 6 11 50 X SQ= 33.58 P = 0. RV YES (YES)
93	TILT MORE TOWARD BEING THOSE WHERE THE SYSTEM STYLE IS NON-MOBILIZATIONAL (78)	0.80	93	TILT LESS TOWARD BEING THOSE WHERE THE SYSTEM STYLE IS NON-MOBILIZATIONAL (78)	0.66	9 22 36 42 X SQ= 2.02 P = 0.132 RV NO (NO)
94	TEND TO BE THOSE WHERE THE STATUS OF THE REGIME IS CONSTITUTIONAL (51)	0.71	94	TEND TO BE THOSE WHERE THE STATUS OF THE REGIME IS AUTHORITARIAN OR TOTALITARIAN (41)	0.59	32 19 13 27 X SQ= 7.04 P = 0.006 RV YES (YES)
95	TEND LESS TO BE THOSE WHERE THE STATUS OF THE REGIME IS CONSTITUTIONAL OR AUTHORITARIAN (95)	0.78	95	TEND MORE TO BE THOSE WHERE THE STATUS OF THE REGIME IS CONSTITUTIONAL OR AUTHORITARIAN (95)	0.92	36 59 10 5 X SQ= 3.30 P = 0.049 RV NO (YES)
96	TEND MORE TO BE THOSE WHERE THE STATUS OF THE REGIME IS CONSTITUTIONAL OR TOTALITARIAN (67)	0.93	96	TEND LESS TO BE THOSE WHERE THE STATUS OF THE REGIME IS CONSTITUTIONAL OR TOTALITARIAN (67)	0.55	42 24 3 20 X SQ= 15.50 P = 0. RV NO (YES)

97	TEND TO BE THOSE WHERE THE STATUS OF THE REGIME IS TOTALITARIAN, RATHER THAN AUTHORITARIAN (16)	0.77	97	TEND TO BE THOSE WHERE THE STATUS OF THE REGIME IS AUTHORITARIAN, RATHER THAN TOTALITARIAN (23)	0.80	M 3 20 10 5 X SQ= 9.34 P = 0.001 RV YES (YES)
99	TEND TO BE THOSE WHERE GOVERNMENTAL STABILITY IS GENERALLY PRESENT AND DATES FROM AT LEAST THE INTER-WAR PERIOD, OR FROM THE POST-WAR PERIOD (50)	0.73	99	TEND TO BE THOSE WHERE GOVERNMENTAL STABILITY IS MODERATELY PRESENT AND DATES FROM THE POST-WAR PERIOD, OR IS ABSENT (36)	0.57	30 19 11 25 X SQ= 6.64 P = 0.008 RV YES (YES)
101	TEND TO BE THOSE WHERE THE REPRESENTATIVE CHARACTER OF THE REGIME IS POLYARCHIC (41)	0.62	101	TEND TO BE THOSE WHERE THE REPRESENTATIVE CHARACTER OF THE REGIME IS LIMITED POLYARCHIC, PSEUDO-POLYARCHIC, OR NON-POLYARCHIC (57)	0.75	28 13 17 39 X SQ= 12.21 P = 0. RV YES (YES)
102	TEND TO BE THOSE WHERE THE REPRESENTATIVE CHARACTER OF THE REGIME IS POLYARCHIC OR LIMITED POLYARCHIC (59)	0.72	102	TEND TO BE THOSE WHERE THE REPRESENTATIVE CHARACTER OF THE REGIME IS PSEUDO-POLYARCHIC OR NON-POLYARCHIC (49)	0.57	33 26 13 35 X SQ= 7.85 P = 0.003 RV YES (YES)
105	TEND TO BE THOSE WHERE THE ELECTORAL SYSTEM IS COMPETITIVE (43)	0.72	105	TEND TO BE THOSE WHERE THE ELECTORAL SYSTEM IS PARTIALLY COMPETITIVE OR NON-COMPETITIVE (47)	0.74	31 12 12 34 X SQ= 17.04 P = 0. RV YES (YES)
106	TEND TO BE THOSE WHERE THE ELECTORAL SYSTEM IS COMPETITIVE OR PARTIALLY COMPETITIVE (52)	0.77	106	TEND TO BE THOSE WHERE THE ELECTORAL SYSTEM IS NON-COMPETITIVE (30)	0.50	33 19 10 19 X SQ= 5.17 P = 0.020 RV YES (YES)
107	TEND TO BE THOSE WHERE AUTONOMOUS GROUPS ARE FULLY TOLERATED IN POLITICS (46)	0.70	107	TEND TO BE THOSE WHERE AUTONOMOUS GROUPS ARE PARTIALLY TOLERATED IN POLITICS, ARE TOLERATED ONLY OUTSIDE POLITICS, OR ARE NOT TOLERATED AT ALL (65)	0.78	32 14 14 50 X SQ= 23.10 P = 0. RV YES (YES)
111	TEND TO BE THOSE WHERE POLITICAL ENCULTURATION IS HIGH OR MEDIUM (53)	0.75	111	TEND TO BE THOSE WHERE POLITICAL ENCULTURATION IS LOW (42)	0.56	27 26 9 33 X SQ= 7.46 P = 0.005 RV YES (YES)
114	TEND TO BE THOSE WHERE SECTIONALISM IS NEGLIGIBLE (47)	0.60	114	TEND TO BE THOSE WHERE SECTIONALISM IS EXTREME OR MODERATE (61)	0.66	17 43 25 22 X SQ= 5.83 P = 0.010 RV YES (YES)

116	TEND TO BE THOSE WHERE INTEREST ARTICULATION BY ASSOCIATIONAL GROUPS IS SIGNIFICANT OR MODERATE (32)	0.63	116	TEND TO BE THOSE WHERE INTEREST ARTICULATION BY ASSOCIATIONAL GROUPS IS LIMITED OR NEGLIGIBLE (79)	0.93	27 5 16 62 X SQ= 36.23 P = 0. RV YES (YES)
117	TEND TO BE THOSE WHERE INTEREST ARTICULATION BY ASSOCIATIONAL GROUPS IS SIGNIFICANT, MODERATE, OR LIMITED (60)	0.86	117	TEND TO BE THOSE WHERE INTEREST ARTICULATION BY ASSOCIATIONAL GROUPS IS NEGLIGIBLE (51)	0.67	37 22 6 45 X SQ= 27.72 P = 0. RV YES (YES)
119	TEND TO BE THOSE WHERE INTEREST ARTICULATION BY INSTITUTIONAL GROUPS IS MODERATE OR LIMITED (26)	0.52	119	TEND TO BE THOSE WHERE INTEREST ARTICULATION BY INSTITUTIONAL GROUPS IS VERY SIGNIFICANT OR SIGNIFICANT (74)	0.96	M 22 51 24 2 X SQ= 27.34 P = 0. RV YES (YES)
121	TEND TO BE THOSE WHERE INTEREST ARTICULATION BY NON-ASSOCIATIONAL GROUPS IS MODERATE, LIMITED, OR NEGLIGIBLE (61)	0.98	121	TEND TO BE THOSE WHERE INTEREST ARTICULATION BY NON-ASSOCIATIONAL GROUPS IS SIGNIFICANT (54)	0.78	1 53 45 15 X SQ= 60.18 P = 0. RV YES (YES)
122	TEND TO BE THOSE WHERE INTEREST ARTICULATION BY NON-ASSOCIATIONAL GROUPS IS LIMITED OR NEGLIGIBLE (32)	0.70	122	ALWAYS ARE THOSE WHERE INTEREST ARTICULATION BY NON-ASSOCIATIONAL GROUPS IS SIGNIFICANT OR MODERATE (83)	1.00	14 68 32 0 X SQ= 62.36 P = 0. RV YES (YES)
125	TEND TO BE THOSE WHERE INTEREST ARTICULATION BY ANOMIC GROUPS IS INFREQUENT OR VERY INFREQUENT (35)	0.70	125	TEND TO BE THOSE WHERE INTEREST ARTICULATION BY ANOMIC GROUPS IS FREQUENT OR OCCASIONAL (64)	0.85	M 11 52 26 9 X SQ= 28.55 P = 0. RV YES (YES)
128	TILT TOWARD BEING THOSE WHERE INTEREST ARTICULATION BY POLITICAL PARTIES IS SIGNIFICANT OR MODERATE (48)	0.61	128	TILT TOWARD BEING THOSE WHERE INTEREST ARTICULATION BY POLITICAL PARTIES IS LIMITED OR NEGLIGIBLE (45)	0.56	27 21 17 27 X SQ= 2.19 P = 0.100 RV YES (YES)
130	TILT LESS TOWARD BEING THOSE WHERE INTEREST AGGREGATION BY POLITICAL PARTIES IS MODERATE, LIMITED, OR NEGLIGIBLE (71)	0.79	130	TILT MORE TOWARD BEING THOSE WHERE INTEREST AGGREGATION BY POLITICAL PARTIES IS MODERATE, LIMITED, OR NEGLIGIBLE (71)	0.92	9 3 33 37 X SQ= 2.16 P = 0.117 RV NO (YES)
133	TEND MORE TO BE THOSE WHERE INTEREST AGGREGATION BY THE EXECUTIVE IS MODERATE, LIMITED, OR NEGLIGIBLE (76)	0.89	133	TEND LESS TO BE THOSE WHERE INTEREST AGGREGATION BY THE EXECUTIVE IS MODERATE, LIMITED, OR NEGLIGIBLE (76)	0.59	5 24 40 35 X SQ= 9.68 P = 0.001 RV NO (YES)

No.	Left statement		No.	Right statement		Statistics
137	TEND TO BE THOSE WHERE INTEREST AGGREGATION BY THE LEGISLATURE IS SIGNIFICANT OR MODERATE (29)	0.53	137	TEND TO BE THOSE WHERE INTEREST AGGREGATION BY THE LEGISLATURE IS LIMITED OR NEGLIGIBLE (68)	0.89	23 6 20 47 X SQ= 18.07 P = 0. RV YES (YES)
138	TEND TO BE THOSE WHERE INTEREST AGGREGATION BY THE LEGISLATURE IS SIGNIFICANT, MODERATE, OR LIMITED (48)	0.77	138	TEND TO BE THOSE WHERE INTEREST AGGREGATION BY THE LEGISLATURE IS NEGLIGIBLE (49)	0.72	33 15 10 38 X SQ= 20.39 P = 0. RV YES (YES)
139	LEAN MORE TOWARD BEING THOSE WHERE THE PARTY SYSTEM IS QUANTITATIVELY OTHER THAN ONE-PARTY (71)	0.78	139	LEAN LESS TOWARD BEING THOSE WHERE THE PARTY SYSTEM IS QUANTITATIVELY OTHER THAN ONE-PARTY (71)	0.61	10 23 35 36 X SQ= 2.58 P = 0.090 RV NO (NO)
141	TEND LESS TO BE THOSE WHERE THE PARTY SYSTEM IS QUANTITATIVELY OTHER THAN TWO-PARTY (87)	0.77	141	TEND MORE TO BE THOSE WHERE THE PARTY SYSTEM IS QUANTITATIVELY OTHER THAN TWO-PARTY (87)	0.98	10 1 34 52 X SQ= 8.42 P = 0.002 RV NO (YES)
142	TEND LESS TO BE THOSE WHERE THE PARTY SYSTEM IS QUANTITATIVELY OTHER THAN MULTI-PARTY (66)	0.57	142	TEND MORE TO BE THOSE WHERE THE PARTY SYSTEM IS QUANTITATIVELY OTHER THAN MULTI-PARTY (66)	0.78	19 11 25 40 X SQ= 4.16 P = 0.028 RV NO (YES)
145	TILT LESS TOWARD BEING THOSE WHERE THE PARTY SYSTEM IS QUANTITATIVELY MULTI-PARTY, RATHER THAN TWO-PARTY (30)	0.66	145	TILT MORE TOWARD BEING THOSE WHERE THE PARTY SYSTEM IS QUANTITATIVELY MULTI-PARTY, RATHER THAN TWO-PARTY (30)	0.92	19 11 10 1 X SQ= 1.77 P = 0.128 RV NO (NO)
146	ALWAYS ARE THOSE WHERE THE PARTY SYSTEM IS QUALITATIVELY OTHER THAN MASS-BASED TERRITORIAL (92)	1.00	146	TEND LESS TO BE THOSE WHERE THE PARTY SYSTEM IS QUALITATIVELY OTHER THAN MASS-BASED TERRITORIAL (92)	0.86	0 8 43 48 X SQ= 4.90 P = 0.009 RV NO (NO)
147	TEND LESS TO BE THOSE WHERE THE PARTY SYSTEM IS QUALITATIVELY OTHER THAN CLASS-ORIENTED OR MULTI-IDEOLOGICAL (67)	0.55	147	TEND MORE TO BE THOSE WHERE THE PARTY SYSTEM IS QUALITATIVELY OTHER THAN CLASS-ORIENTED OR MULTI-IDEOLOGICAL (67)	0.91	19 4 23 43 X SQ= 13.75 P = 0. RV NO (YES)
148	ALWAYS ARE THOSE WHERE THE PARTY SYSTEM IS QUALITATIVELY OTHER THAN AFRICAN TRANSITIONAL (96)	1.00	148	TEND LESS TO BE THOSE WHERE THE PARTY SYSTEM IS QUALITATIVELY OTHER THAN AFRICAN TRANSITIONAL (96)	0.78	0 14 46 49 X SQ= 9.83 P = 0. RV NO (NO)

153	TEND TO BE THOSE WHERE THE PARTY SYSTEM IS STABLE (42)	0.68	153	TEND TO BE THOSE WHERE THE PARTY SYSTEM IS MODERATELY STABLE OR UNSTABLE (38)	0.66	28 13 13 25 X SQ= 7.86 P = 0.003 RV YES (YES)
158	TEND TO BE THOSE WHERE PERSONALISMO IS NEGLIGIBLE (56)	0.73	158	TEND TO BE THOSE WHERE PERSONALISMO IS PRONOUNCED OR MODERATE (40)	0.56	M 12 28 33 22 X SQ= 7.20 P = 0.006 RV YES (YES)
164	TEND TO BE THOSE WHERE THE REGIME'S LEADERSHIP CHARISMA IS NEGLIGIBLE (65)	0.91	164	TEND TO BE THOSE WHERE THE REGIME'S LEADERSHIP CHARISMA IS PRONOUNCED OR MODERATE (34)	0.57	4 30 42 23 X SQ= 22.99 P = 0. RV YES (YES)
168	TEND TO BE THOSE WHERE THE HORIZONTAL POWER DISTRIBUTION IS SIGNIFICANT (34)	0.58	168	TEND TO BE THOSE WHERE THE HORIZONTAL POWER DISTRIBUTION IS LIMITED OR NEGLIGIBLE (72)	0.87	26 8 19 52 X SQ= 21.21 P = 0. RV YES (YES)
169	TEND TO BE THOSE WHERE THE HORIZONTAL POWER DISTRIBUTION IS SIGNIFICANT OR LIMITED (58)	0.76	169	TEND TO BE THOSE WHERE THE HORIZONTAL POWER DISTRIBUTION IS NEGLIGIBLE (48)	0.60	34 24 11 36 X SQ= 11.75 P = 0. RV YES (YES)
172	TILT LESS TOWARD BEING THOSE WHERE THE LEGISLATIVE-EXECUTIVE STRUCTURE IS OTHER THAN PARLIAMENTARY-ROYALIST (88)	0.73	172	TILT MORE TOWARD BEING THOSE WHERE THE LEGISLATIVE-EXECUTIVE STRUCTURE IS OTHER THAN PARLIAMENTARY-ROYALIST (88)	0.86	12 9 33 54 X SQ= 1.84 P = 0.140 RV NO (YES)
175	TEND TO BE THOSE WHERE THE LEGISLATURE IS FULLY EFFECTIVE OR PARTIALLY EFFECTIVE (51)	0.73	175	TEND TO BE THOSE WHERE THE LEGISLATURE IS LARGELY INEFFECTIVE OR WHOLLY INEFFECTIVE (49)	0.65	32 19 12 36 X SQ= 12.78 P = 0. RV YES (YES)
178	TEND TO BE THOSE WHERE THE LEGISLATURE IS BICAMERAL (51)	0.64	178	TEND TO BE THOSE WHERE THE LEGISLATURE IS UNICAMERAL (53)	0.64	16 37 29 21 X SQ= 7.00 P = 0.006 RV YES (YES)
179	TEND TO BE THOSE WHERE THE EXECUTIVE IS STRONG (39)	0.61	179	TEND TO BE THOSE WHERE THE EXECUTIVE IS DOMINANT (52)	0.71	16 35 25 14 X SQ= 8.27 P = 0.003 RV YES (YES)

181	TEND LESS TO BE THOSE WHERE THE BUREAUCRACY IS SEMI-MODERN, RATHER THAN MODERN (55)	0.59	181	TEND MORE TO BE THOSE WHERE THE BUREAUCRACY IS SEMI-MODERN, RATHER THAN MODERN (55)	0.93	19 2 27 27 X SQ= 8.81 P = 0.001 RV NO (YES)
182	ALWAYS ARE THOSE WHERE THE BUREAUCRACY IS SEMI-MODERN, RATHER THAN POST-COLONIAL TRANSITIONAL (55)	1.00	182	TEND LESS TO BE THOSE WHERE THE BUREAUCRACY IS SEMI-MODERN, RATHER THAN POST-COLONIAL TRANSITIONAL (55)	0.52	27 27 0 25 X SQ= 16.83 P = 0. RV NO (YES)
186	TILT MORE TOWARD BEING THOSE WHERE PARTICIPATION BY THE MILITARY IN POLITICS IS NEUTRAL, RATHER THAN SUPPORTIVE (56)	0.75	186	TILT LESS TOWARD BEING THOSE WHERE PARTICIPATION BY THE MILITARY IN POLITICS IS NEUTRAL, RATHER THAN SUPPORTIVE (56)	0.58	9 21 27 29 X SQ= 1.97 P = 0.115 RV NO (NO)
187	TEND TO BE THOSE WHERE THE ROLE OF THE POLICE IS NOT POLITICALLY SIGNIFICANT (35)	0.59	187	TEND TO BE THOSE WHERE THE ROLE OF THE POLICE IS POLITICALLY SIGNIFICANT (66)	0.81	17 48 24 11 X SQ= 15.21 P = 0. RV YES (YES)
188	TEND TO BE THOSE WHERE THE CHARACTER OF THE LEGAL SYSTEM IS CIVIL LAW (32)	0.54	188	TEND TO BE THOSE WHERE THE CHARACTER OF THE LEGAL SYSTEM IS OTHER THAN CIVIL LAW (81)	0.89	25 7 21 59 X SQ= 23.32 P = 0. RV YES (YES)
189	TEND LESS TO BE THOSE WHERE THE CHARACTER OF THE LEGAL SYSTEM IS OTHER THAN COMMON LAW (108)	0.87	189	TEND MORE TO BE THOSE WHERE THE CHARACTER OF THE LEGAL SYSTEM IS OTHER THAN COMMON LAW (108)	0.99	6 1 40 67 X SQ= 4.53 P = 0.017 RV NO (YES)
192	ALWAYS ARE THOSE WHERE THE CHARACTER OF THE LEGAL SYSTEM IS OTHER THAN MUSLIM OR PARTLY MUSLIM (86)	1.00	192	TEND LESS TO BE THOSE WHERE THE CHARACTER OF THE LEGAL SYSTEM IS OTHER THAN MUSLIM OR PARTLY MUSLIM (86)	0.59	0 27 46 39 X SQ= 22.61 P = 0. RV NO (YES)
193	ALWAYS ARE THOSE WHERE THE CHARACTER OF THE LEGAL SYSTEM IS OTHER THAN PARTLY INDIGENOUS (86)	1.00	193	TEND LESS TO BE THOSE WHERE THE CHARACTER OF THE LEGAL SYSTEM IS OTHER THAN PARTLY INDIGENOUS (86)	0.58	0 28 46 39 X SQ= 23.36 P = 0. RV NO (YES)

MATRIX

63 POLITIES WHERE THE RELIGION IS PREDOMINANTLY SOME KIND OF CHRISTIAN (46)	63 POLITIES WHERE THE RELIGION IS PREDOMINANTLY OR PARTLY OTHER THAN CHRISTIAN (68)

42 POLITIES WHOSE ECONOMIC DEVELOPMENTAL STATUS IS DEVELOPED OR INTERMEDIATE (36)

ARGENTINA	AUSTRALIA	AUSTRIA	BELGIUM	BRAZIL	CYPRUS	JAPAN	SO AFRICA
BULGARIA	CANADA	COLOMBIA	CZECHOS'KIA	DENMARK			
FINLAND	FRANCE	GERMANY, E	GERMAN FR	GREECE			
HUNGARY	ICELAND	IRELAND	ITALY	LUXEMBOURG			
MEXICO	NETHERLANDS	NEW ZEALAND	NORWAY	POLAND			
RUMANIA	SWEDEN	SWITZERLAND	UK	US			
URUGUAY	YUGOSLAVIA						

32	3
13	63

42 POLITIES WHOSE ECONOMIC DEVELOPMENTAL STATUS IS UNDERDEVELOPED OR VERY UNDERDEVELOPED (76)

CHILE	COSTA RICA	CUBA	DOMIN REP	EL SALVADOR	AFGHANISTAN	ALBANIA	ALGERIA	BOLIVIA	BURMA
HONDURAS	JAMAICA	NICARAGUA	PANAMA	PARAGUAY	BURUNDI	CAMBODIA	CAMEROUN	CEN AFR REP	CEYLON
PHILIPPINES	PORTUGAL	SPAIN			CHAD	CHINA, PR	CONGO(BRA)	CONGO(LEO)	DAHOMEY
					ECUADOR	ETHIOPIA	GABON	GHANA	GUATEMALA
					GUINEA	HAITI	INDIA	INDONESIA	IRAN
					IRAQ	IVORY COAST	JORDAN	KOREA, N	KOREA REP
					LAOS	LEBANON	LIBERIA	LIBYA	MALAGASY R
					MALAYA	MALI	MAURITANIA	MONGOLIA	MOROCCO
					NEPAL	NIGER	NIGERIA	PAKISTAN	PERU
					RWANDA	SENEGAL	SIERRE LEO	SOMALIA	SUDAN
					SYRIA	TANGANYIKA	THAILAND	TOGO	TRINIDAD
					TUNISIA	TURKEY	UGANDA	UAR	UPPER VOLTA
					VIETNAM, N	VIETNAM REP	YEMEN		

MATRIX

	63 POLITIES WHERE THE RELIGION IS PREDOMINANTLY SOME KIND OF CHRISTIAN (46)	63 POLITIES WHERE THE RELIGION IS PREDOMINANTLY OR PARTLY OTHER THAN CHRISTIAN (68)
97 POLITIES WHERE THE STATUS OF THE REGIME IS AUTHORITARIAN, RATHER THAN TOTALITARIAN (23)	3 EL SALVADOR NICARAGUA PARAGUAY	20 AFGHANISTAN ETHIOPIA JORDAN PAKISTAN ALGERIA GHANA KOREA REP SA'U ARABIA BURMA GUINEA LAOS SUDAN CAMBODIA INDONESIA LIBERIA THAILAND CEN AFR REP IRAN NEPAL UAR
97 POLITIES WHERE THE STATUS OF THE REGIME IS TOTALITARIAN, RATHER THAN AUTHORITARIAN (16)	10 BULGARIA POLAND CUBA PORTUGAL CZECHOS'KIA RUMANIA GERMANY, E SPAIN HUNGARY YUGOSLAVIA	5 ALBANIA CHINA, PR KOREA, N MONGOLIA VIETNAM, N

M

MATRIX

	63 POLITIES WHERE THE RELIGION IS PREDOMINANTLY SOME KIND OF CHRISTIAN (46)	63 POLITIES WHERE THE RELIGION IS PREDOMINANTLY OR PARTLY OTHER THAN CHRISTIAN (68)
119 POLITIES WHERE INTEREST ARTICULATION BY INSTITUTIONAL GROUPS IS VERY SIGNIFICANT OR SIGNIFICANT (74)	ARGENTINA, BRAZIL, BULGARIA, COLOMBIA, CUBA, CZECHOS'KIA, DOMIN REP, EL SALVADOR, GERMANY, E, HONDURAS, HUNGARY, ITALY, MEXICO, NICARAGUA, PANAMA, PARAGUAY, POLAND, PORTUGAL, RUMANIA, SPAIN, VENEZUELA, YUGOSLAVIA 22	AFGHANISTAN, ALBANIA, ALGERIA, BOLIVIA, BURMA, CAMBODIA, CEYLON, CHINA, PR, CONGO(LEO), CYPRUS, ECUADOR, ETHIOPIA, GHANA, GUATEMALA, GUINEA, HAITI, INDIA, INDONESIA, IRAN, IRAQ, ISRAEL, IVORY COAST, JAPAN, JORDAN, KOREA, N, KOREA REP, LAOS, LEBANON, LIBERIA, LIBYA, MALAYA, MAURITANIA, MONGOLIA, MOROCCO, NEPAL, NIGERIA, PAKISTAN, PERU, SA'U ARABIA, SENEGAL, SO AFRICA, SUDAN, SYRIA, TANGANYIKA, THAILAND, TUNISIA, TURKEY, UGANDA, UAR, VIETNAM, N, VIETNAM REP 51
119 POLITIES WHERE INTEREST ARTICULATION BY INSTITUTIONAL GROUPS IS MODERATE OR LIMITED (26)	24 AUSTRALIA, AUSTRIA, BELGIUM, CANADA, CHILE, COSTA RICA, DENMARK, FINLAND, FRANCE, GERMAN FR, GREECE, ICELAND, IRELAND, JAMAICA, LUXEMBOURG, NETHERLANDS, NEW ZEALAND, NORWAY, PHILIPPINES, SWEDEN, SWITZERLAND, UK, US, URUGUAY	2 MALAGASY R, TRINIDAD

M

MATRIX

63 POLITIES WHERE THE RELIGION IS PREDOMINANTLY SOME KIND OF CHRISTIAN (46)	63 POLITIES WHERE THE RELIGION IS PREDOMINANTLY OR PARTLY OTHER THAN CHRISTIAN (68)
125 POLITIES WHERE INTEREST ARTICULATION BY ANOMIC GROUPS IS FREQUENT OR OCCASIONAL (64)	
BRAZIL, COLOMBIA, EL SALVADOR, GERMANY, E, HONDURAS, MEXICO, PANAMA, PARAGUAY, POLAND, SPAIN, YUGOSLAVIA	AFGHANISTAN, BOLIVIA, BURMA, CAMBODIA, CAMEROUN, CEN AFR REP, CEYLON, CHAD, CHINA, PR, CONGO(BRA), CONGO(LEO), DAHOMEY, ECUADOR, ETHIOPIA, GABON, GHANA, GUATEMALA, GUINEA, HAITI, INDIA, INDONESIA, IRAN, IRAQ, IVORY COAST, JAPAN, JORDAN, KOREA REP, LAOS, LEBANON, MALAYA, MALI, MAURITANIA, MOROCCO, NEPAL, NIGER, NIGERIA, PAKISTAN, PERU, RWANDA, SA'U ARABIA, SENEGAL, SIERRE LEO, SOMALIA, SO AFRICA, SUDAN, SYRIA, THAILAND, TOGO, TURKEY, UGANDA, UPPER VOLTA, VIETNAM REP
11	52
26	9
125 POLITIES WHERE INTEREST ARTICULATION BY ANOMIC GROUPS IS INFREQUENT OR VERY INFREQUENT (35)	
AUSTRALIA, AUSTRIA, BELGIUM, CANADA, CHILE, COSTA RICA, CZECHOS'KIA, DENMARK, FINLAND, GERMAN FR, GREECE, ICELAND, IRELAND, ITALY, JAMAICA, LUXEMBOURG, NETHERLANDS, NEW ZEALAND, NICARAGUA, NORWAY, PHILIPPINES, PORTUGAL, SWEDEN, SWITZERLAND, UK, URUGUAY	BURUNDI, ISRAEL, LIBERIA, LIBYA, MALAGASY R, TANGANYIKA, TRINIDAD, TUNISIA, UAR

MATRIX

63 POLITIES WHERE THE RELIGION IS PREDOMINANTLY SOME KIND OF CHRISTIAN (46)	63 POLITIES WHERE THE RELIGION IS PREDOMINANTLY OR PARTLY OTHER THAN CHRISTIAN (68)
158 POLITIES WHERE PERSONALISMO IS PRONOUNCED OR MODERATE (40)	
ARGENTINA, BRAZIL, CHILE, COLOMBIA, COSTA RICA, EL SALVADOR, FRANCE, GREECE, HONDURAS, NICARAGUA, PANAMA, URUGUAY	BOLIVIA, CAMBODIA, CAMEROUN, CEN AFR REP, CEYLON, CHAD, CONGO(BRA), CONGO(LEO), DAHOMEY, ECUADOR, GABON, GHANA, GUATEMALA, INDONESIA, IVORY COAST, JAPAN, LAOS, LEBANON, LIBERIA, MALI, MAURITANIA, NIGER, PERU, SYRIA, THAILAND, UAR, UPPER VOLTA, VIETNAM REP
12	28
33	22
158 POLITIES WHERE PERSONALISMO IS NEGLIGIBLE (56)	
AUSTRALIA, AUSTRIA, BELGIUM, BULGARIA, CANADA, CUBA, CZECHOS'KIA, DENMARK, FINLAND, GERMANY, E, GERMAN FR, HUNGARY, ICELAND, IRELAND, ITALY, JAMAICA, LUXEMBOURG, MEXICO, NETHERLANDS, NEW ZEALAND, NORWAY, PARAGUAY, PHILIPPINES, POLAND, PORTUGAL, RUMANIA, SPAIN, SWEDEN, SWITZERLAND, UK, US, VENEZUELA, YUGOSLAVIA	ALBANIA, ALGERIA, CHINA, PR, GUINEA, INDIA, ISRAEL, KOREA, N, MALAGASY R, MALAYA, MONGOLIA, MOROCCO, NIGERIA, SENEGAL, SIERRE LEO, SOMALIA, SO AFRICA, TANGANYIKA, TRINIDAD, TUNISIA, TURKEY, UGANDA, VIETNAM, N

64 POLITIES
WHERE THE RELIGION IS
CHRISTIAN, RATHER THAN
MUSLIM (46)

64 POLITIES
WHERE THE RELIGION IS
MUSLIM, RATHER THAN
CHRISTIAN (18)

BOTH SUBJECT AND PREDICATE

46 IN LEFT COLUMN

ARGENTINA, AUSTRALIA, AUSTRIA, BELGIUM, BRAZIL, BULGARIA, CANADA, CHILE, COLOMBIA, COSTA RICA, CUBA, CZECHOS'KIA, DENMARK, DOMIN REP, EL SALVADOR, FINLAND, FRANCE, GERMANY, E, GERMAN FR, GREECE, HONDURAS, HUNGARY, ICELAND, IRELAND, ITALY, JAMAICA, LUXEMBOURG, MEXICO, NETHERLANDS, NEW ZEALAND, NICARAGUA, NORWAY, PANAMA, PARAGUAY, PHILIPPINES, POLAND, PORTUGAL, RUMANIA, SPAIN, SWEDEN, SWITZERLAND, UK, US, URUGUAY, VENEZUELA, YUGOSLAVIA

18 IN RIGHT COLUMN

AFGHANISTAN, ALGERIA, INDONESIA, IRAN, IRAQ, JORDAN, LIBYA, MAURITANIA, MOROCCO, PAKISTAN, SA'U ARABIA, SENEGAL, SOMALIA, SYRIA, TUNISIA, TURKEY, UAR, YEMEN

51 EXCLUDED BECAUSE IRRELEVANT

ALBANIA, BOLIVIA, BURMA, BURUNDI, CAMBODIA, CAMEROUN, CEN AFR REP, CEYLON, CHAD, CHINA, PR, CONGO(BRA), CONGO(LEO), CYPRUS, DAHOMEY, ECUADOR, ETHIOPIA, GABON, GHANA, GUATEMALA, GUINEA, HAITI, INDIA, ISRAEL, IVORY COAST, JAPAN, KOREA, N, KOREA REP, LAOS, LEBANON, LIBERIA, MALAGASY R, MALAYA, MALI, MONGOLIA, NEPAL, NIGER, NIGERIA, PERU, RWANDA, SIERRE LEO, SO AFRICA, SUDAN, TANGANYIKA, THAILAND, TOGO, TRINIDAD, UGANDA, USSR, UPPER VOLTA, VIETNAM, N, VIETNAM REP

	Left column (Christian)			Right column (Muslim)		
2	TEND LESS TO BE THOSE LOCATED ELSEWHERE THAN IN WEST EUROPE, SCANDINAVIA, NORTH AMERICA, OR AUSTRALASIA (93)	0.52	2	ALWAYS ARE THOSE LOCATED ELSEWHERE THAN IN WEST EUROPE, SCANDINAVIA, NORTH AMERICA, OR AUSTRALASIA (93)	1.00	22 0 24 18 X SQ= 11.08 P = 0. RV NO (NO)
6	TEND LESS TO BE THOSE LOCATED ELSEWHERE THAN IN THE CARIBBEAN, CENTRAL AMERICA, OR SOUTH AMERICA (93)	0.65	6	ALWAYS ARE THOSE LOCATED ELSEWHERE THAN IN THE CARIBBEAN, CENTRAL AMERICA, OR SOUTH AMERICA (93)	1.00	16 0 30 18 X SQ= 6.60 P = 0.003 RV NO (NO)
10	ALWAYS ARE THOSE LOCATED ELSEWHERE THAN IN NORTH AFRICA, OR CENTRAL AND SOUTH AFRICA (82)	1.00	10	TEND LESS TO BE THOSE LOCATED ELSEWHERE THAN IN NORTH AFRICA, OR CENTRAL AND SOUTH AFRICA (82)	0.56	0 8 46 10 X SQ= 19.48 P = 0. RV NO (YES)

13	ALWAYS ARE THOSE LOCATED ELSEWHERE THAN IN NORTH AFRICA OR THE MIDDLE EAST (99)	1.00	13	TEND TO BE THOSE LOCATED IN NORTH AFRICA OR THE MIDDLE EAST (16)	0.72	0 13 46 5 X SQ= 37.35 P = 0. RV YES (YES)
14	ALWAYS ARE THOSE LOCATED ELSEWHERE THAN IN THE MIDDLE EAST (104)	1.00	14	TEND LESS TO BE THOSE LOCATED ELSEWHERE THAN IN THE MIDDLE EAST (104)	0.56	0 8 46 10 X SQ= 19.48 P = 0. RV NO (YES)
20	TEND TO BE THOSE LOCATED IN THE CARIBBEAN, CENTRAL AMERICA, OR SOUTH AMERICA, RATHER THAN IN EAST ASIA, SOUTH ASIA, OR SOUTHEAST ASIA (22)	0.94	20	ALWAYS ARE THOSE LOCATED IN EAST ASIA, SOUTH ASIA, OR SOUTHEAST ASIA, RATHER THAN IN THE CARIBBEAN, CENTRAL AMERICA, OR SOUTH AMERICA (18)	1.00	1 2 16 0 X SQ= 5.89 P = 0.018 RV YES (YES)
21	TEND TO BE THOSE WHOSE TERRITORIAL SIZE IS MEDIUM OR SMALL (83)	0.83	21	TEND TO BE THOSE WHOSE TERRITORIAL SIZE IS VERY LARGE OR LARGE (32)	0.50	8 9 38 9 X SQ= 5.48 P = 0.012 RV YES (YES)
22	TEND TO BE THOSE WHOSE TERRITORIAL SIZE IS SMALL (47)	0.50	22	TEND TO BE THOSE WHOSE TERRITORIAL SIZE IS VERY LARGE, LARGE, OR MEDIUM (68)	0.83	23 15 23 3 X SQ= 4.66 P = 0.022 RV YES (NO)
25	LEAN LESS TOWARD BEING THOSE WHOSE POPULATION DENSITY IS MEDIUM OR LOW (98)	0.78	25	ALWAYS ARE THOSE WHOSE POPULATION DENSITY IS MEDIUM OR LOW (98)	1.00	10 0 36 18 X SQ= 3.14 P = 0.050 RV NO (NO)
26	TEND TO BE THOSE WHOSE POPULATION DENSITY IS VERY HIGH, HIGH, OR MEDIUM (48)	0.57	26	TEND TO BE THOSE WHOSE POPULATION DENSITY IS LOW (67)	0.89	26 2 20 16 X SQ= 9.07 P = 0.002 RV YES (NO)
28	TEND TO BE THOSE WHOSE POPULATION GROWTH RATE IS LOW (48)	0.59	28	TEND TO BE THOSE WHOSE POPULATION GROWTH RATE IS HIGH (62)	0.83	19 15 27 3 X SQ= 7.57 P = 0.005 RV YES (NO)
29	TEND TO BE THOSE WHERE THE DEGREE OF URBANIZATION IS HIGH (56)	0.93	29	TEND TO BE THOSE WHERE THE DEGREE OF URBANIZATION IS LOW (49)	0.50	39 9 3 9 X SQ= 11.91 P = 0. RV YES (YES)

30	TEND TO BE THOSE WHOSE AGRICULTURAL POPULATION IS MEDIUM, LOW, OR VERY LOW (57)	0.91	30	TEND TO BE THOSE WHOSE AGRICULTURAL POPULATION IS HIGH (56)	0.82	4 14 42 3 X SQ= 29.49 P = 0. RV YES (YES)
34	TEND TO BE THOSE WHOSE GROSS NATIONAL PRODUCT IS VERY HIGH, HIGH, MEDIUM, OR LOW (62)	0.78	34	TEND TO BE THOSE WHOSE GROSS NATIONAL PRODUCT IS VERY LOW (53)	0.50	36 9 10 9 X SQ= 3.69 P = 0.036 RV YES (NO)
36	TEND TO BE THOSE WHOSE PER CAPITA GROSS NATIONAL PRODUCT IS VERY HIGH, HIGH, OR MEDIUM (42)	0.74	36	ALWAYS ARE THOSE WHOSE PER CAPITA GROSS NATIONAL PRODUCT IS LOW OR VERY LOW (73)	1.00	34 0 12 18 X SQ= 25.49 P = 0. RV YES (YES)
37	TEND TO BE THOSE WHOSE PER CAPITA GROSS NATIONAL PRODUCT IS VERY HIGH, HIGH, MEDIUM, OR LOW (64)	0.98	37	TEND TO BE THOSE WHOSE PER CAPITA GROSS NATIONAL PRODUCT IS VERY LOW (51)	0.61	45 7 1 11 X SQ= 25.76 P = 0. RV YES (YES)
38	LEAN LESS TOWARD BEING THOSE WHOSE INTERNATIONAL FINANCIAL STATUS IS MEDIUM, LOW, OR VERY LOW (103)	0.84	38	ALWAYS ARE THOSE WHOSE INTERNATIONAL FINANCIAL STATUS IS MEDIUM, LOW, OR VERY LOW (103)	1.00	7 0 38 18 X SQ= 1.77 P = 0.099 RV NO (NO)
39	TEND TO BE THOSE WHOSE INTERNATIONAL FINANCIAL STATUS IS VERY HIGH, HIGH, OR MEDIUM (38)	0.64	39	TEND TO BE THOSE WHOSE INTERNATIONAL FINANCIAL STATUS IS LOW OR VERY LOW (76)	0.78	29 4 16 14 X SQ= 7.57 P = 0.004 RV YES (NO)
42	TEND TO BE THOSE WHOSE ECONOMIC DEVELOPMENTAL STATUS IS DEVELOPED OR INTERMEDIATE (36)	0.71	42	ALWAYS ARE THOSE WHOSE ECONOMIC DEVELOPMENTAL STATUS IS UNDERDEVELOPED OR VERY UNDERDEVELOPED (76)	1.00	32 0 13 17 X SQ= 22.22 P = 0. RV YES (YES)
43	TEND TO BE THOSE WHOSE ECONOMIC DEVELOPMENTAL STATUS IS DEVELOPED, INTERMEDIATE, OR UNDERDEVELOPED (55)	0.87	43	TEND TO BE THOSE WHOSE ECONOMIC DEVELOPMENTAL STATUS IS VERY UNDERDEVELOPED (57)	0.76	39 4 6 13 X SQ= 20.27 P = 0. RV YES (YES)
44	TEND TO BE THOSE WHERE THE LITERACY RATE IS NINETY PERCENT OR ABOVE (25)	0.50	44	ALWAYS ARE THOSE WHERE THE LITERACY RATE IS BELOW NINETY PERCENT (90)	1.00	23 0 23 18 X SQ= 11.96 P = 0. RV YES (NO)

45	TEND TO BE THOSE WHERE THE LITERACY RATE IS FIFTY PERCENT OR ABOVE (55)	0.93	45	TEND TO BE THOSE WHERE THE LITERACY RATE IS BELOW FIFTY PERCENT (54)	0.94	43 1 3 17 X SQ= 42.55 P = 0. RV YES (YES)
46	ALWAYS ARE THOSE WHERE THE LITERACY RATE IS TEN PERCENT OR ABOVE (84)	1.00	46	TEND LESS TO BE THOSE WHERE THE LITERACY RATE IS TEN PERCENT OR ABOVE (84)	0.69	46 11 0 5 X SQ= 11.71 P = 0.001 RV NO (YES)
50	TEND TO BE THOSE WHERE FREEDOM OF THE PRESS IS COMPLETE (43)	0.64	50	TEND TO BE THOSE WHERE FREEDOM OF THE PRESS IS INTERMITTENT, INTERNALLY ABSENT, OR INTERNALLY AND EXTERNALLY ABSENT (56)	0.88	27 2 15 15 X SQ= 11.34 P = 0. RV YES (YES)
51	LEAN TOWARD BEING THOSE WHERE FREEDOM OF THE PRESS IS COMPLETE OR INTERMITTENT (60)	0.71	51	LEAN TOWARD BEING THOSE WHERE FREEDOM OF THE PRESS IS INTERNALLY ABSENT, OR INTERNALLY AND EXTERNALLY ABSENT (37)	0.60	30 6 12 9 X SQ= 3.44 P = 0.059 RV YES (NO)
54	TEND TO BE THOSE WHERE NEWSPAPER CIRCULATION IS ONE HUNDRED OR MORE PER THOUSAND (37)	0.70	54	ALWAYS ARE THOSE WHERE NEWSPAPER CIRCULATION IS LESS THAN ONE HUNDRED PER THOUSAND (76)	1.00	32 0 14 18 X SQ= 22.34 P = 0. RV YES (YES)
55	ALWAYS ARE THOSE WHERE NEWSPAPER CIRCULATION IS TEN OR MORE PER THOUSAND (78)	1.00	55	TEND TO BE THOSE WHERE NEWSPAPER CIRCULATION IS LESS THAN TEN PER THOUSAND (35)	0.50	46 9 0 9 X SQ= 22.79 P = 0. RV YES (YES)
68	TEND TO BE THOSE THAT ARE LINGUISTICALLY HOMOGENEOUS (52)	0.76	68	TEND TO BE THOSE THAT ARE LINGUISTICALLY WEAKLY HETEROGENEOUS OR STRONGLY HETEROGENEOUS (62)	0.61	35 7 11 11 X SQ= 6.37 P = 0.008 RV YES (YES)
69	TEND TO BE THOSE THAT ARE LINGUISTICALLY HOMOGENEOUS OR WEAKLY HETEROGENEOUS (64)	0.87	69	TEND TO BE THOSE THAT ARE LINGUISTICALLY STRONGLY HETEROGENEOUS (50)	0.50	40 9 6 9 X SQ= 7.90 P = 0.006 RV YES (YES)
72	TEND TO BE THOSE WHOSE DATE OF INDEPENDENCE IS BEFORE 1914 (52)	0.80	72	TEND TO BE THOSE WHOSE DATE OF INDEPENDENCE IS AFTER 1914 (59)	0.83	35 3 9 15 X SQ= 18.72 P = 0. RV YES (YES)

73	TEND TO BE THOSE WHOSE DATE OF INDEPENDENCE IS BEFORE 1945 (65)	0.95	73	TEND TO BE THOSE WHOSE DATE OF INDEPENDENCE IS AFTER 1945 (46)	0.61	42 7 2 11 X SQ= 21.37 P = 0. RV YES (YES)
74	TEND TO BE THOSE THAT ARE HISTORICALLY WESTERN (26)	0.57	74	ALWAYS ARE THOSE THAT ARE NOT HISTORICALLY WESTERN (87)	1.00	26 0 20 18 X SQ= 14.87 P = 0. RV YES (NO)
75	ALWAYS ARE THOSE THAT ARE HISTORICALLY WESTERN OR SIGNIFICANTLY WESTERNIZED (62)	1.00	75	TEND TO BE THOSE THAT ARE NOT HISTORICALLY WESTERN AND ARE NOT SIGNIFICANTLY WESTERNIZED (52)	0.83	46 3 0 15 X SQ= 45.53 P = 0. RV YES (YES)
76	TEND TO BE THOSE THAT ARE HISTORICALLY WESTERN, RATHER THAN HAVING BEEN SIGNIFICANTLY OR PARTIALLY WESTERNIZED THROUGH A COLONIAL RELATIONSHIP (26)	0.60	76	ALWAYS ARE THOSE THAT HAVE BEEN SIGNIFICANTLY OR PARTIALLY WESTERNIZED THROUGH A COLONIAL RELATIONSHIP, RATHER THAN BEING HISTORICALLY WESTERN (70)	1.00	26 0 17 14 X SQ= 13.22 P = 0. RV YES (NO)
77	ALWAYS ARE THOSE THAT HAVE BEEN SIGNIFICANTLY WESTERNIZED, RATHER THAN PARTIALLY WESTERNIZED, THROUGH A COLONIAL RELATIONSHIP (28)	1.00	77	TEND TO BE THOSE THAT HAVE BEEN PARTIALLY WESTERNIZED, RATHER THAN SIGNIFICANTLY WESTERNIZED, THROUGH A COLONIAL RELATIONSHIP (41)	0.86	17 2 0 12 X SQ= 20.30 P = 0. RV YES (YES)
79	TEND TO BE THOSE THAT WERE FORMERLY DEPENDENCIES OF SPAIN, RATHER THAN BRITAIN OR FRANCE (18)	0.70	79	ALWAYS ARE THOSE THAT WERE FORMERLY DEPENDENCIES OF BRITAIN OR FRANCE, RATHER THAN SPAIN (49)	1.00	6 10 14 0 X SQ= 10.46 P = 0. RV YES (YES)
82	TEND TO BE THOSE WHERE THE TYPE OF POLITICAL MODERNIZATION IS EARLY OR LATER EUROPEAN OR EUROPEAN DERIVED, RATHER THAN DEVELOPED TUTELARY OR UNDEVELOPED TUTELARY (51)	0.96	82	ALWAYS ARE THOSE WHERE THE TYPE OF POLITICAL MODERNIZATION IS DEVELOPED TUTELARY OR UNDEVELOPED TUTELARY, RATHER THAN EARLY OR LATER EUROPEAN OR EUROPEAN DERIVED (55)	1.00	44 0 2 15 X SQ= 46.84 P = 0. RV YES (YES)
83	ALWAYS ARE THOSE WHERE THE TYPE OF POLITICAL MODERNIZATION IS OTHER THAN NON-EUROPEAN AUTOCHTHONOUS (106)	1.00	83	TEND LESS TO BE THOSE WHERE THE TYPE OF POLITICAL MODERNIZATION IS OTHER THAN NON-EUROPEAN AUTOCHTHONOUS (106)	0.83	0 3 46 15 X SQ= 4.75 P = 0.020 RV NO (YES)
85	TEND MORE TO BE THOSE WHERE THE STAGE OF POLITICAL MODERNIZATION IS ADVANCED, RATHER THAN MID- OR EARLY TRANSITIONAL (60)	0.83	85	TEND LESS TO BE THOSE WHERE THE STAGE OF POLITICAL MODERNIZATION IS ADVANCED, RATHER THAN MID- OR EARLY TRANSITIONAL (60)	0.53	38 9 8 8 X SQ= 4.31 P = 0.024 RV NO (YES)

86	TEND MORE TO BE THOSE WHERE THE STAGE OF POLITICAL MODERNIZATION IS ADVANCED OR MID-TRANSITIONAL, RATHER THAN EARLY TRANSITIONAL (76)	0.98	86	TEND LESS TO BE THOSE WHERE THE STAGE OF POLITICAL MODERNIZATION IS ADVANCED OR MID-TRANSITIONAL, RATHER THAN EARLY TRANSITIONAL (76)	0.65	45 11 1 6 X SQ= 10.64 P = 0.001 RV NO (YES)
87	TEND TO BE THOSE WHOSE IDEOLOGICAL ORIENTATION IS OTHER THAN DEVELOPMENTAL (58)	0.95	87	TEND TO BE THOSE WHOSE IDEOLOGICAL ORIENTATION IS DEVELOPMENTAL (31)	0.54	2 7 37 6 X SQ= 12.94 P = 0. RV YES (YES)
89	TEND TO BE THOSE WHOSE IDEOLOGICAL ORIENTATION IS CONVENTIONAL (33)	0.71	89	TEND TO BE THOSE WHOSE IDEOLOGICAL ORIENTATION IS OTHER THAN CONVENTIONAL (62)	0.94	27 1 11 15 X SQ= 16.43 P = 0. RV YES (YES)
94	TEND TO BE THOSE WHERE THE STATUS OF THE REGIME IS CONSTITUTIONAL (51)	0.71	94	TEND TO BE THOSE WHERE THE STATUS OF THE REGIME IS AUTHORITARIAN OR TOTALITARIAN (41)	0.62	32 5 13 8 X SQ= 3.35 P = 0.049 RV YES (NO)
95	LEAN LESS TOWARD BEING THOSE WHERE THE STATUS OF THE REGIME IS CONSTITUTIONAL OR AUTHORITARIAN (95)	0.78	95	ALWAYS ARE THOSE WHERE THE STATUS OF THE REGIME IS CONSTITUTIONAL OR AUTHORITARIAN (95)	1.00	36 17 10 0 X SQ= 2.92 P = 0.050 RV NO (NO)
96	TEND TO BE THOSE WHERE THE STATUS OF THE REGIME IS CONSTITUTIONAL OR TOTALITARIAN (67)	0.93	96	TEND TO BE THOSE WHERE THE STATUS OF THE REGIME IS AUTHORITARIAN (23)	0.62	42 5 3 8 X SQ= 16.35 P = 0. RV YES (YES)
97	TEND TO BE THOSE WHERE THE STATUS OF THE REGIME IS TOTALITARIAN, RATHER THAN AUTHORITARIAN (16)	0.77	97	ALWAYS ARE THOSE WHERE THE STATUS OF THE REGIME IS AUTHORITARIAN, RATHER THAN TOTALITARIAN (23)	1.00	3 8 10 0 X SQ= 8.87 P = 0.001 RV YES (YES)
99	LEAN TOWARD BEING THOSE WHERE GOVERNMENTAL STABILITY IS GENERALLY PRESENT AND DATES FROM AT LEAST THE INTER-WAR PERIOD, OR FROM THE POST-WAR PERIOD (50)	0.73	99	LEAN TOWARD BEING THOSE WHERE GOVERNMENTAL STABILITY IS MODERATELY PRESENT AND DATES FROM THE POST-WAR PERIOD, OR IS ABSENT (36)	0.57	30 6 11 8 X SQ= 3.01 P = 0.054 RV YES (NO)
101	TEND TO BE THOSE WHERE THE REPRESENTATIVE CHARACTER OF THE REGIME IS POLYARCHIC (41)	0.62	101	TEND TO BE THOSE WHERE THE REPRESENTATIVE CHARACTER OF THE REGIME IS LIMITED POLYARCHIC, PSEUDO-POLYARCHIC, OR NON-POLYARCHIC (57)	0.77	28 3 17 10 X SQ= 4.74 P = 0.025 RV YES (NO)

102	TEND TO BE THOSE WHERE THE REPRESENTATIVE CHARACTER OF THE REGIME IS POLYARCHIC OR LIMITED POLYARCHIC (59)	0.72	102	TEND TO BE THOSE WHERE THE REPRESENTATIVE CHARACTER OF THE REGIME IS PSEUDO-POLYARCHIC OR NON-POLYARCHIC (49)	0.67	33 5 13 10 X SQ= 5.56 P = 0.013 RV YES (NO)
105	TEND TO BE THOSE WHERE THE ELECTORAL SYSTEM IS COMPETITIVE (43)	0.72	105	TEND TO BE THOSE WHERE THE ELECTORAL SYSTEM IS PARTIALLY COMPETITIVE OR NON-COMPETITIVE (47)	0.90	31 1 12 9 X SQ= 10.61 P = 0.001 RV YES (NO)
107	TEND TO BE THOSE WHERE AUTONOMOUS GROUPS ARE FULLY TOLERATED IN POLITICS (46)	0.70	107	TEND TO BE THOSE WHERE AUTONOMOUS GROUPS ARE PARTIALLY TOLERATED IN POLITICS, ARE TOLERATED ONLY OUTSIDE POLITICS, OR ARE NOT TOLERATED AT ALL (65)	0.94	32 1 14 16 X SQ= 17.71 P = 0. RV YES (YES)
110	TEND LESS TO BE THOSE WHERE POLITICAL ENCULTURATION IS MEDIUM OR LOW (80)	0.67	110	TEND MORE TO BE THOSE WHERE POLITICAL ENCULTURATION IS MEDIUM OR LOW (80)	0.94	12 1 24 16 X SQ= 3.33 P = 0.041 RV NO (NO)
113	TEND TO BE THOSE WHERE SECTIONALISM IS MODERATE OR NEGLIGIBLE (81)	0.81	113	TEND TO BE THOSE WHERE SECTIONALISM IS EXTREME (27)	0.50	8 9 34 9 X SQ= 4.52 P = 0.027 RV YES (YES)
114	TEND TO BE THOSE WHERE SECTIONALISM IS NEGLIGIBLE (47)	0.60	114	TEND TO BE THOSE WHERE SECTIONALISM IS EXTREME OR MODERATE (61)	0.78	17 14 25 4 X SQ= 5.61 P = 0.011 RV YES (NO)
115	TEND LESS TO BE THOSE WHERE INTEREST ARTICULATION BY ASSOCIATIONAL GROUPS IS MODERATE, LIMITED, OR NEGLIGIBLE (91)	0.56	115	ALWAYS ARE THOSE WHERE INTEREST ARTICULATION BY ASSOCIATIONAL GROUPS IS MODERATE, LIMITED, OR NEGLIGIBLE (91)	1.00	19 0 24 18 X SQ= 9.58 P = 0. RV NO (NO)
116	TEND TO BE THOSE WHERE INTEREST ARTICULATION BY ASSOCIATIONAL GROUPS IS SIGNIFICANT OR MODERATE (32)	0.63	116	TEND TO BE THOSE WHERE INTEREST ARTICULATION BY ASSOCIATIONAL GROUPS IS LIMITED OR NEGLIGIBLE (79)	0.94	27 1 16 17 X SQ= 14.51 P = 0. RV YES (YES)
117	TEND TO BE THOSE WHERE INTEREST ARTICULATION BY ASSOCIATIONAL GROUPS IS SIGNIFICANT, MODERATE, OR LIMITED (60)	0.86	117	TEND TO BE THOSE WHERE INTEREST ARTICULATION BY ASSOCIATIONAL GROUPS IS NEGLIGIBLE (51)	0.56	37 8 6 10 X SQ= 9.30 P = 0.003 RV YES (YES)

118	LEAN TOWARD BEING THOSE WHERE INTEREST ARTICULATION BY INSTITUTIONAL GROUPS IS SIGNIFICANT, MODERATE, OR LIMITED (60)	0.72	118	LEAN TOWARD BEING THOSE WHERE INTEREST ARTICULATION BY INSTITUTIONAL GROUPS IS VERY SIGNIFICANT (40)	0.56	13 9 33 7 X SQ= 2.93 P = 0.068 RV YES (NO)
119	TEND TO BE THOSE WHERE INTEREST ARTICULATION BY INSTITUTIONAL GROUPS IS MODERATE OR LIMITED (26)	0.52	119	ALWAYS ARE THOSE WHERE INTEREST ARTICULATION BY INSTITUTIONAL GROUPS IS VERY SIGNIFICANT OR SIGNIFICANT (74)	1.00	22 16 24 0 X SQ= 11.51 P = 0. RV YES (NO)
121	TEND TO BE THOSE WHERE INTEREST ARTICULATION BY NON-ASSOCIATIONAL GROUPS IS MODERATE, LIMITED, OR NEGLIGIBLE (61)	0.98	121	TEND TO BE THOSE WHERE INTEREST ARTICULATION BY NON-ASSOCIATIONAL GROUPS IS SIGNIFICANT (54)	0.67	1 12 45 6 X SQ= 29.38 P = 0. RV YES (YES)
122	TEND TO BE THOSE WHERE INTEREST ARTICULATION BY NON-ASSOCIATIONAL GROUPS IS LIMITED OR NEGLIGIBLE (32)	0.70	122	ALWAYS ARE THOSE WHERE INTEREST ARTICULATION BY NON-ASSOCIATIONAL GROUPS IS SIGNIFICANT OR MODERATE (83)	1.00	14 18 32 0 X SQ= 22.34 P = 0. RV YES (YES)
125	TEND TO BE THOSE WHERE INTEREST ARTICULATION BY ANOMIC GROUPS IS INFREQUENT OR VERY INFREQUENT (35)	0.70	125	TEND TO BE THOSE WHERE INTEREST ARTICULATION BY ANOMIC GROUPS IS FREQUENT OR OCCASIONAL (64)	0.81	11 13 26 3 X SQ= 9.98 P = 0.001 RV YES (YES)
126	TEND LESS TO BE THOSE WHERE INTEREST ARTICULATION BY ANOMIC GROUPS IS FREQUENT, OCCASIONAL, OR INFREQUENT (83)	0.57	126	ALWAYS ARE THOSE WHERE INTEREST ARTICULATION BY ANOMIC GROUPS IS FREQUENT, OCCASIONAL, OR INFREQUENT (83)	1.00	21 16 16 0 X SQ= 7.97 P = 0.001 RV NO (NO)
133	TEND MORE TO BE THOSE WHERE INTEREST AGGREGATION BY THE EXECUTIVE IS MODERATE, LIMITED, OR NEGLIGIBLE (76)	0.89	133	TEND LESS TO BE THOSE WHERE INTEREST AGGREGATION BY THE EXECUTIVE IS MODERATE, LIMITED, OR NEGLIGIBLE (76)	0.57	5 6 40 8 X SQ= 5.16 P = 0.015 RV NO (YES)
137	TEND TO BE THOSE WHERE INTEREST AGGREGATION BY THE LEGISLATURE IS SIGNIFICANT OR MODERATE (29)	0.53	137	TEND TO BE THOSE WHERE INTEREST AGGREGATION BY THE LEGISLATURE IS LIMITED OR NEGLIGIBLE (68)	0.89	23 1 20 8 X SQ= 3.81 P = 0.028 RV YES (NO)
138	TEND TO BE THOSE WHERE INTEREST AGGREGATION BY THE LEGISLATURE IS SIGNIFICANT, MODERATE, OR LIMITED (48)	0.77	138	TEND TO BE THOSE WHERE INTEREST AGGREGATION BY THE LEGISLATURE IS NEGLIGIBLE (49)	0.75	33 2 10 6 X SQ= 6.16 P = 0.008 RV YES (NO)

141	TILT LESS TOWARD BEING THOSE 0.77 WHERE THE PARTY SYSTEM IS QUANTITATIVELY OTHER THAN TWO-PARTY (87)	141	ALWAYS ARE THOSE 1.00 WHERE THE PARTY SYSTEM IS QUANTITATIVELY OTHER THAN TWO-PARTY (87)	10 0 34 11 X SQ= 1.72 P = 0.104 RV NO (NO)
142	TEND LESS TO BE THOSE 0.57 WHERE THE PARTY SYSTEM IS QUANTITATIVELY OTHER THAN MULTI-PARTY (66)	142	ALWAYS ARE THOSE 1.00 WHERE THE PARTY SYSTEM IS QUANTITATIVELY OTHER THAN MULTI-PARTY (66)	19 0 25 9 X SQ= 4.33 P = 0.019 RV NO (NO)
144	LEAN TOWARD BEING THOSE 0.50 WHERE THE PARTY SYSTEM IS QUANTITATIVELY TWO-PARTY, RATHER THAN ONE-PARTY (11)	144	ALWAYS ARE THOSE 1.00 WHERE THE PARTY SYSTEM IS QUANTITATIVELY ONE-PARTY, RATHER THAN TWO-PARTY (34)	10 5 10 0 X SQ= 2.34 P = 0.061 RV YES (NO)
146	ALWAYS ARE THOSE 1.00 WHERE THE PARTY SYSTEM IS QUALITATIVELY OTHER THAN MASS-BASED TERRITORIAL (92)	146	TEND LESS TO BE THOSE 0.64 WHERE THE PARTY SYSTEM IS QUALITATIVELY OTHER THAN MASS-BASED TERRITORIAL (92)	0 4 43 7 X SQ= 12.00 P = 0.001 RV NO (YES)
147	LEAN LESS TOWARD BEING THOSE 0.55 WHERE THE PARTY SYSTEM IS QUALITATIVELY OTHER THAN CLASS-ORIENTED OR MULTI-IDEOLOGICAL (67)	147	ALWAYS ARE THOSE 1.00 WHERE THE PARTY SYSTEM IS QUALITATIVELY OTHER THAN CLASS-ORIENTED OR MULTI-IDEOLOGICAL (67)	19 0 23 6 X SQ= 2.80 P = 0.068 RV NO (NO)
153	TEND TO BE THOSE 0.68 WHERE THE PARTY SYSTEM IS STABLE (42)	153	TEND TO BE THOSE 0.87 WHERE THE PARTY SYSTEM IS MODERATELY STABLE OR UNSTABLE (38)	28 1 13 7 X SQ= 6.47 P = 0.005 RV YES (NO)
154	TEND TO BE THOSE 0.80 WHERE THE PARTY SYSTEM IS STABLE OR MODERATELY STABLE (55)	154	TEND TO BE THOSE 0.62 WHERE THE PARTY SYSTEM IS UNSTABLE (25)	33 3 8 5 X SQ= 4.33 P = 0.023 RV YES (NO)
164	TEND TO BE THOSE 0.91 WHERE THE REGIME'S LEADERSHIP CHARISMA IS NEGLIGIBLE (65)	164	TEND TO BE THOSE 0.85 WHERE THE REGIME'S LEADERSHIP CHARISMA IS PRONOUNCED OR MODERATE (34)	4 11 42 2 X SQ= 26.94 P = 0. RV YES (YES)
168	TEND TO BE THOSE 0.58 WHERE THE HORIZONTAL POWER DISTRIBUTION IS SIGNIFICANT (34)	168	TEND TO BE THOSE 0.93 WHERE THE HORIZONTAL POWER DISTRIBUTION IS LIMITED OR NEGLIGIBLE (72)	26 1 19 14 X SQ= 9.90 P = 0.001 RV YES (NO)

No.			No.			
169	TEND TO BE THOSE WHERE THE HORIZONTAL POWER DISTRIBUTION IS SIGNIFICANT OR LIMITED (58)	0.76	169	TEND TO BE THOSE WHERE THE HORIZONTAL POWER DISTRIBUTION IS NEGLIGIBLE (48)	0.67	34 5 11 10 X SQ= 7.06 P = 0.005 RV YES (NO)
172	TEND LESS TO BE THOSE WHERE THE LEGISLATIVE-EXECUTIVE STRUCTURE IS OTHER THAN PARLIAMENTARY-ROYALIST (88)	0.73	172	ALWAYS ARE THOSE WHERE THE LEGISLATIVE-EXECUTIVE STRUCTURE IS OTHER THAN PARLIAMENTARY-ROYALIST (88)	1.00	12 0 33 15 X SQ= 3.47 P = 0.027 RV NO (NO)
174	TEND LESS TO BE THOSE WHERE THE LEGISLATURE IS PARTIALLY EFFECTIVE, LARGELY INEFFECTIVE, OR WHOLLY INEFFECTIVE (72)	0.52	174	ALWAYS ARE THOSE WHERE THE LEGISLATURE IS PARTIALLY EFFECTIVE, LARGELY INEFFECTIVE, OR WHOLLY INEFFECTIVE (72)	1.00	21 0 23 12 X SQ= 7.24 P = 0.002 RV NO (NO)
175	TEND TO BE THOSE WHERE THE LEGISLATURE IS FULLY EFFECTIVE OR PARTIALLY EFFECTIVE (51)	0.73	175	TEND TO BE THOSE WHERE THE LEGISLATURE IS LARGELY INEFFECTIVE OR WHOLLY INEFFECTIVE (49)	0.75	32 3 12 9 X SQ= 7.24 P = 0.005 RV YES (NO)
179	TEND TO BE THOSE WHERE THE EXECUTIVE IS STRONG (39)	0.61	179	TEND TO BE THOSE WHERE THE EXECUTIVE IS DOMINANT (52)	0.86	16 12 25 2 X SQ= 7.33 P = 0.004 RV YES (NO)
181	LEAN LESS TOWARD BEING THOSE WHERE THE BUREAUCRACY IS SEMI-MODERN, RATHER THAN MODERN (55)	0.59	181	ALWAYS ARE THOSE WHERE THE BUREAUCRACY IS SEMI-MODERN, RATHER THAN MODERN (55)	1.00	19 0 27 6 X SQ= 2.33 P = 0.075 RV NO (NO)
182	ALWAYS ARE THOSE WHERE THE BUREAUCRACY IS SEMI-MODERN, RATHER THAN POST-COLONIAL TRANSITIONAL (55)	1.00	182	TEND LESS TO BE THOSE WHERE THE BUREAUCRACY IS SEMI-MODERN, RATHER THAN POST-COLONIAL TRANSITIONAL (55)	0.60	27 6 0 4 X SQ= 8.32 P = 0.003 RV NO (YES)
186	TEND TO BE THOSE WHERE PARTICIPATION BY THE MILITARY IN POLITICS IS NEUTRAL, RATHER THAN SUPPORTIVE (56)	0.75	186	TEND TO BE THOSE WHERE PARTICIPATION BY THE MILITARY IN POLITICS IS SUPPORTIVE, RATHER THAN NEUTRAL (31)	0.64	9 7 27 4 X SQ= 4.01 P = 0.029 RV YES (NO)
187	TEND TO BE THOSE WHERE THE ROLE OF THE POLICE IS NOT POLITICALLY SIGNIFICANT (35)	0.59	187	TEND TO BE THOSE WHERE THE ROLE OF THE POLICE IS POLITICALLY SIGNIFICANT (66)	0.94	17 16 24 1 X SQ= 11.52 P = 0. RV YES (NO)

188	TEND TO BE THOSE WHERE THE CHARACTER OF THE LEGAL SYSTEM IS CIVIL LAW (32)	0.54	188	TEND TO BE THOSE WHERE THE CHARACTER OF THE LEGAL SYSTEM IS OTHER THAN CIVIL LAW (81)	0.94	25 1 21 17 X SQ= 10.83 P = 0. RV YES (NO)
192	ALWAYS ARE THOSE WHERE THE CHARACTER OF THE LEGAL SYSTEM IS OTHER THAN MUSLIM OR PARTLY MUSLIM (86)	1.00	192	TEND TO BE THOSE WHERE THE CHARACTER OF THE LEGAL SYSTEM IS MUSLIM OR PARTLY MUSLIM (27)	0.78	0 14 46 4 X SQ= 41.36 P = 0. RV YES (YES)
193	ALWAYS ARE THOSE WHERE THE CHARACTER OF THE LEGAL SYSTEM IS OTHER THAN PARTLY INDIGENOUS (86)	1.00	93	LEAN LESS TOWARD BEING THOSE WHERE THE CHARACTER OF THE LEGAL SYSTEM IS OTHER THAN PARTLY INDIGENOUS (86)	0.89	0 2 46 16 X SQ= 2.24 P = 0.076 RV NO (YES)
194	LEAN LESS TOWARD BEING THOSE THAT ARE NON-COMMUNIST (101)	0.84	194	ALWAYS ARE THOSE THAT ARE NON-COMMUNIST (101)	1.00	7 0 38 18 X SQ= 1.77 P = 0.099 RV NO (NO)

65 POLITIES
WHERE THE RELIGION IS
CATHOLIC, RATHER THAN
MUSLIM (25)

65 POLITIES
WHERE THE RELIGION IS
MUSLIM, RATHER THAN
CATHOLIC (18)

PREDICATE ONLY

25 IN LEFT COLUMN

ARGENTINA	AUSTRIA	BELGIUM	BRAZIL	CHILE	COLOMBIA	COSTA RICA	CUBA	DOMIN REP	EL SALVADOR
FRANCE	HONDURAS	IRELAND	ITALY	LUXEMBOURG	MEXICO	NICARAGUA	PANAMA	PARAGUAY	PHILIPPINES
POLAND	PORTUGAL	SPAIN	URUGUAY	VENEZUELA					

18 IN RIGHT COLUMN

AFGHANISTAN	ALGERIA	INDONESIA	IRAN	IRAQ	JORDAN	LIBYA	MAURITANIA	MOROCCO	PAKISTAN
SA'U ARABIA	SENEGAL	SOMALIA	SYRIA	TUNISIA	TURKEY	UAR	YEMEN		

72 EXCLUDED BECAUSE IRRELEVANT

ALBANIA	AUSTRALIA	BOLIVIA	BULGARIA	BURMA	BURUNDI	CAMBODIA	CAMEROUN	CANADA	CEN AFR REP
CEYLON	CHAD	CHINA, PR	CONGO(BRA)	CONGO(LEO)	CYPRUS	CZECHOS'KIA	DAHOMEY	DENMARK	ECUADOR
ETHIOPIA	FINLAND	GABON	GERMANY, E	GERMAN FR	GHANA	GREECE	GUATEMALA	GUINEA	HAITI
HUNGARY	ICELAND	INDIA	ISRAEL	IVORY COAST	JAMAICA	JAPAN	KOREA, N	KOREA REP	LAOS
LEBANON	LIBERIA	MALAGASY R	MALAYA	MALI	MONGOLIA	NEPAL	NETHERLANDS	NEW ZEALAND	NIGER
NIGERIA	NORWAY	PERU	RUMANIA	RWANDA	SIERRE LEO	SO AFRICA	SUDAN	SWEDEN	SWITZERLAND
TANGANYIKA	THAILAND	TOGO	TRINIDAD	UGANDA	USSR	UK	US	UPPER VOLTA	VIETNAM, N
VIETNAM REP	YUGOSLAVIA								

66 POLITIES
THAT ARE RELIGIOUSLY HOMOGENEOUS (57)

66 POLITIES
THAT ARE RELIGIOUSLY HETEROGENEOUS (49)

BOTH SUBJECT AND PREDICATE

57 IN LEFT COLUMN

AFGHANISTAN	ALGERIA	ARGENTINA	AUSTRIA	BELGIUM	BRAZIL	BULGARIA	BURMA	CAMBODIA	CHILE
COLOMBIA	COSTA RICA	CUBA	DENMARK	DOMIN REP	EL SALVADOR	FINLAND	FRANCE	GREECE	HONDURAS
ICELAND	INDIA	INDONESIA	IRAN	IRAQ	IRELAND	ISRAEL	ITALY	JORDAN	LIBYA
LUXEMBOURG	MAURITANIA	MEXICO	MONGOLIA	MOROCCO	NICARAGUA	NORWAY	PAKISTAN	PANAMA	PARAGUAY
PHILIPPINES	POLAND	PORTUGAL	RUMANIA	SA'U ARABIA	SENEGAL	SOMALIA	SPAIN	SWEDEN	SYRIA
THAILAND	TUNISIA	TURKEY	UAR	URUGUAY	VENEZUELA	YEMEN			

49 IN RIGHT COLUMN

ALBANIA	AUSTRALIA	BOLIVIA	CAMEROUN	CANADA	CEYLON	CHAD	CONGO(BRA)	CONGO(LEO)	CYPRUS
CZECHOS'KIA	DAHOMEY	ECUADOR	ETHIOPIA	GABON	GERMANY, E	GERMAN FR	GHANA	GUATEMALA	GUINEA
HAITI	HUNGARY	IVORY COAST	JAMAICA	JAPAN	LAOS	LEBANON	LIBERIA	MALAGASY R	MALAYA
MALI	NEPAL	NETHERLANDS	NEW ZEALAND	NIGER	NIGERIA	PERU	SIERRE LEO	SO AFRICA	SUDAN
SWITZERLAND	TANGANYIKA	TOGO	TRINIDAD	UGANDA	UK	US	UPPER VOLTA	YUGOSLAVIA	

3 EXCLUDED BECAUSE AMBIGUOUS

CEN AFR REP KOREA REP VIETNAM REP

6 EXCLUDED BECAUSE UNASCERTAINED

BURUNDI CHINA, PR KOREA, N RWANDA USSR VIETNAM, N

5	LEAN TOWARD BEING THOSE LOCATED IN SCANDINAVIA OR WEST EUROPE, RATHER THAN IN EAST EUROPE (18)	0.82	5 LEAN TOWARD BEING THOSE LOCATED IN EAST EUROPE, RATHER THAN IN SCANDINAVIA OR WEST EUROPE (9)	0.56	14 4 3 5 X SQ= 2.39 P = 0.078 RV YES (YES)
10	TEND MORE TO BE THOSE LOCATED ELSEWHERE THAN IN NORTH AFRICA, OR CENTRAL AND SOUTH AFRICA (82)	0.86	10 TEND LESS TO BE THOSE LOCATED ELSEWHERE THAN IN NORTH AFRICA, OR CENTRAL AND SOUTH AFRICA (82)	0.55	8 22 49 27 X SQ= 10.89 P = 0.001 RV NO (YES)
13	TEND LESS TO BE THOSE LOCATED ELSEWHERE THAN IN NORTH AFRICA OR THE MIDDLE EAST (99)	0.75	13 TEND MORE TO BE THOSE LOCATED ELSEWHERE THAN IN NORTH AFRICA OR THE MIDDLE EAST (99)	0.96	14 2 43 47 X SQ= 7.10 P = 0.005 RV NO (YES)

14	LEAN LESS TOWARD BEING THOSE LOCATED ELSEWHERE THAN IN THE MIDDLE EAST (104)	0.84	14	LEAN MORE TOWARD BEING THOSE LOCATED ELSEWHERE THAN IN THE MIDDLE EAST (104)	0.96	9 2 48 47 X SQ= 2.73 P = 0.060 RV NO (NO)
15	TEND TO BE THOSE LOCATED IN THE MIDDLE EAST, RATHER THAN IN NORTH AFRICA OR CENTRAL AND SOUTH AFRICA (11)	0.53	15	TEND TO BE THOSE LOCATED IN NORTH AFRICA OR CENTRAL AND SOUTH AFRICA, RATHER THAN IN THE MIDDLE EAST (33)	0.92	9 2 8 22 X SQ= 7.94 P = 0.003 RV YES (YES)
18	TEND TO BE THOSE LOCATED IN EAST ASIA, SOUTH ASIA, OR SOUTHEAST ASIA, RATHER THAN IN NORTH AFRICA OR CENTRAL AND SOUTH AFRICA (18)	0.50	18	TEND TO BE THOSE LOCATED IN NORTH AFRICA OR CENTRAL AND SOUTH AFRICA, RATHER THAN IN EAST ASIA, SOUTH ASIA, OR SOUTHEAST ASIA (33)	0.81	8 22 8 5 X SQ= 3.35 P = 0.043 RV YES (YES)
19	TEND TO BE THOSE LOCATED IN THE CARIBBEAN, CENTRAL AMERICA, OR SOUTH AMERICA, RATHER THAN IN NORTH AFRICA OR CENTRAL AND SOUTH AFRICA (22)	0.65	19	TEND TO BE THOSE LOCATED IN NORTH AFRICA OR CENTRAL AND SOUTH AFRICA, RATHER THAN IN THE CARIBBEAN, CENTRAL AMERICA, OR SOUTH AMERICA (33)	0.76	8 22 15 7 X SQ= 7.27 P = 0.005 RV YES (YES)
23	LEAN LESS TOWARD BEING THOSE WHOSE POPULATION IS MEDIUM OR SMALL (88)	0.70	23	LEAN MORE TOWARD BEING THOSE WHOSE POPULATION IS MEDIUM OR SMALL (88)	0.86	17 7 40 42 X SQ= 2.80 P = 0.066 RV NO (YES)
25	LEAN MORE TOWARD BEING THOSE WHOSE POPULATION DENSITY IS MEDIUM OR LOW (98)	0.91	25	LEAN LESS TOWARD BEING THOSE WHOSE POPULATION DENSITY IS MEDIUM OR LOW (98)	0.78	5 11 52 38 X SQ= 2.85 P = 0.060 RV NO (YES)
29	TEND TO BE THOSE WHERE THE DEGREE OF URBANIZATION IS HIGH (56)	0.70	29	TEND TO BE THOSE WHERE THE DEGREE OF URBANIZATION IS LOW (49)	0.64	37 17 16 30 X SQ= 10.04 P = 0.001 RV YES (YES)
30	TILT TOWARD BEING THOSE WHOSE AGRICULTURAL POPULATION IS MEDIUM, LOW, OR VERY LOW (57)	0.61	30	TILT TOWARD BEING THOSE WHOSE AGRICULTURAL POPULATION IS HIGH (56)	0.55	22 27 34 22 X SQ= 2.03 P = 0.120 RV YES (YES)
34	TEND TO BE THOSE WHOSE GROSS NATIONAL PRODUCT IS VERY HIGH, HIGH, MEDIUM, OR LOW (62)	0.65	34	TEND TO BE THOSE WHOSE GROSS NATIONAL PRODUCT IS VERY LOW (53)	0.59	37 20 20 29 X SQ= 5.22 P = 0.019 RV YES (YES)

No.	Left		No.	Right		Statistics
37	TEND TO BE THOSE WHOSE PER CAPITA GROSS NATIONAL PRODUCT IS VERY HIGH, HIGH, MEDIUM, OR LOW (64)	0.70	37	TEND TO BE THOSE WHOSE PER CAPITA GROSS NATIONAL PRODUCT IS VERY LOW (51)	0.53	40 23 17 26 X SQ= 4.98 P = 0.018 RV YES (YES)
40	TEND MORE TO BE THOSE WHOSE INTERNATIONAL FINANCIAL STATUS IS VERY HIGH, HIGH, MEDIUM, OR LOW (71)	0.75	40	TEND LESS TO BE THOSE WHOSE INTERNATIONAL FINANCIAL STATUS IS VERY HIGH, HIGH, MEDIUM, OR LOW (71)	0.53	42 25 14 22 X SQ= 4.43 P = 0.024 RV NO (YES)
43	LEAN TOWARD BEING THOSE WHOSE ECONOMIC DEVELOPMENTAL STATUS IS DEVELOPED, INTERMEDIATE, OR UNDERDEVELOPED (55)	0.60	43	LEAN TOWARD BEING THOSE WHOSE ECONOMIC DEVELOPMENTAL STATUS IS VERY UNDERDEVELOPED (57)	0.58	33 20 22 28 X SQ= 2.75 P = 0.077 RV YES (YES)
45	TEND TO BE THOSE WHERE THE LITERACY RATE IS FIFTY PERCENT OR ABOVE (55)	0.61	45	TEND TO BE THOSE WHERE THE LITERACY RATE IS BELOW FIFTY PERCENT (54)	0.62	34 18 22 29 X SQ= 4.28 P = 0.030 RV YES (YES)
47	TEND TO BE THOSE WHERE THE LITERACY RATE IS BETWEEN FIFTY AND NINETY PERCENT, RATHER THAN NINETY PERCENT OR ABOVE (30)	0.65	47	TEND TO BE THOSE WHERE THE LITERACY RATE IS NINETY PERCENT OR ABOVE, RATHER THAN BETWEEN FIFTY AND NINETY PERCENT (25)	0.67	12 12 22 6 X SQ= 3.48 P = 0.043 RV YES (YES)
48	TILT TOWARD BEING THOSE WHERE THE LITERACY RATE IS BETWEEN FIFTY AND NINETY PERCENT, RATHER THAN BETWEEN TEN AND FIFTY PERCENT (30)	0.61	48	TILT TOWARD BEING THOSE WHERE THE LITERACY RATE IS BETWEEN TEN AND FIFTY PERCENT, RATHER THAN BETWEEN FIFTY AND NINETY PERCENT (24)	0.62	22 6 14 10 X SQ= 1.63 P = 0.141 RV YES (NO)
49	TEND TO BE THOSE WHERE THE LITERACY RATE IS BETWEEN TEN AND FIFTY PERCENT, RATHER THAN BELOW TEN PERCENT (24)	0.74	49	TEND TO BE THOSE WHERE THE LITERACY RATE IS BELOW TEN PERCENT, RATHER THAN BETWEEN TEN AND FIFTY PERCENT (26)	0.64	14 10 5 18 X SQ= 5.10 P = 0.017 RV YES (YES)
55	TEND MORE TO BE THOSE WHERE NEWSPAPER CIRCULATION IS TEN OR MORE PER THOUSAND (78)	0.82	55	TEND LESS TO BE THOSE WHERE NEWSPAPER CIRCULATION IS TEN OR MORE PER THOUSAND (78)	0.55	47 27 10 22 X SQ= 8.10 P = 0.003 RV NO (YES)
56	ALWAYS ARE THOSE WHERE THE RELIGION IS PREDOMINANTLY LITERATE (79)	1.00	56	TEND TO BE THOSE WHERE THE RELIGION IS PREDOMINANTLY OR PARTLY NON-LITERATE (31)	0.57	57 21 0 28 X SQ= 41.37 P = 0. RV YES (YES)

57	TEND LESS TO BE THOSE WHERE THE RELIGION IS OTHER THAN CATHOLIC (90)	0.56	57	ALWAYS ARE THOSE WHERE THE RELIGION IS OTHER THAN CATHOLIC (90)	1.00	25 0 32 49 X SQ= 25.74 P = 0. RV NO (YES)
58	TEND LESS TO BE THOSE WHERE THE RELIGION IS OTHER THAN MUSLIM (97)	0.68	58	ALWAYS ARE THOSE WHERE THE RELIGION IS OTHER THAN MUSLIM (97)	1.00	18 0 39 49 X SQ= 16.47 P = 0. RV NO (YES)
63	TEND TO BE THOSE WHERE THE RELIGION IS PREDOMINANTLY SOME KIND OF CHRISTIAN (46)	0.58	63	TEND TO BE THOSE WHERE THE RELIGION IS PREDOMINANTLY OR PARTLY OTHER THAN CHRISTIAN (68)	0.73	33 13 24 36 X SQ= 9.31 P = 0.002 RV YES (YES)
64	TEND LESS TO BE THOSE WHERE THE RELIGION IS CHRISTIAN, RATHER THAN MUSLIM (46)	0.65	64	ALWAYS ARE THOSE WHERE THE RELIGION IS CHRISTIAN, RATHER THAN MUSLIM (46)	1.00	33 13 18 0 X SQ= 4.76 P = 0.013 RV NO (NO)
68	TEND TO BE THOSE THAT ARE LINGUISTICALLY HOMOGENEOUS (52)	0.60	68	TEND TO BE THOSE THAT ARE LINGUISTICALLY WEAKLY HETEROGENEOUS OR STRONGLY HETEROGENEOUS (62)	0.71	34 14 23 35 X SQ= 9.05 P = 0.002 RV YES (YES)
69	TEND TO BE THOSE THAT ARE LINGUISTICALLY HOMOGENEOUS OR WEAKLY HETEROGENEOUS (64)	0.75	69	TEND TO BE THOSE THAT ARE LINGUISTICALLY STRONGLY HETEROGENEOUS (50)	0.69	43 15 14 34 X SQ= 19.60 P = 0. RV YES (YES)
72	LEAN TOWARD BEING THOSE WHOSE DATE OF INDEPENDENCE IS BEFORE 1914 (52)	0.56	72	LEAN TOWARD BEING THOSE WHOSE DATE OF INDEPENDENCE IS AFTER 1914 (59)	0.62	32 18 25 29 X SQ= 2.61 P = 0.079 RV YES (YES)
73	TEND TO BE THOSE WHOSE DATE OF INDEPENDENCE IS BEFORE 1945 (65)	0.72	73	TEND TO BE THOSE WHOSE DATE OF INDEPENDENCE IS AFTER 1945 (46)	0.53	41 22 16 25 X SQ= 5.80 P = 0.015 RV YES (YES)
75	TEND TO BE THOSE THAT ARE HISTORICALLY WESTERN OR SIGNIFICANTLY WESTERNIZED (62)	0.67	75	TEND TO BE THOSE THAT ARE NOT HISTORICALLY WESTERN AND ARE NOT SIGNIFICANTLY WESTERNIZED (52)	0.54	38 22 19 26 X SQ= 3.81 P = 0.047 RV YES (YES)

77	TEND TO BE THOSE THAT HAVE BEEN SIGNIFICANTLY WESTERNIZED, RATHER THAN PARTIALLY WESTERNIZED, THROUGH A COLONIAL RELATIONSHIP (28)	0.58	77	TEND TO BE THOSE THAT HAVE BEEN PARTIALLY WESTERNIZED, RATHER THAN SIGNIFICANTLY WESTERNIZED, THROUGH A COLONIAL RELATIONSHIP (41)	0.71	19 9 14 22 X SQ= 4.20 P = 0.026 RV YES (YES)
79	TEND LESS TO BE THOSE THAT WERE FORMERLY DEPENDENCIES OF BRITAIN OR FRANCE, RATHER THAN SPAIN (49)	0.52	79	TEND MORE TO BE THOSE THAT WERE FORMERLY DEPENDENCIES OF BRITAIN OR FRANCE, RATHER THAN SPAIN (49)	0.89	15 31 14 4 X SQ= 8.91 P = 0.002 RV NO (YES)
81	TEND MORE TO BE THOSE WHERE THE TYPE OF POLITICAL MODERNIZATION IS LATER EUROPEAN OR LATER EUROPEAN DERIVED, RATHER THAN EARLY EUROPEAN OR EARLY EUROPEAN DERIVED (40)	0.91	81	TEND LESS TO BE THOSE WHERE THE TYPE OF POLITICAL MODERNIZATION IS LATER EUROPEAN OR LATER EUROPEAN DERIVED, RATHER THAN EARLY EUROPEAN OR EARLY EUROPEAN DERIVED (40)	0.58	3 8 29 11 X SQ= 5.74 P = 0.012 RV NO (YES)
82	LEAN TOWARD BEING THOSE WHERE THE TYPE OF POLITICAL MODERNIZATION IS EARLY OR LATER EUROPEAN OR EUROPEAN DERIVED, RATHER THAN DEVELOPED TUTELARY OR UNDEVELOPED TUTELARY (51)	0.60	82	LEAN TOWARD BEING THOSE WHERE THE TYPE OF POLITICAL MODERNIZATION IS DEVELOPED TUTELARY OR UNDEVELOPED TUTELARY, RATHER THAN EARLY OR LATER EUROPEAN OR EUROPEAN DERIVED (55)	0.59	32 19 21 27 X SQ= 2.86 P = 0.071 RV YES (YES)
84	TEND TO BE THOSE WHERE THE TYPE OF POLITICAL MODERNIZATION IS DEVELOPED TUTELARY, RATHER THAN UNDEVELOPED TUTELARY (31)	0.86	84	TEND TO BE THOSE WHERE THE TYPE OF POLITICAL MODERNIZATION IS UNDEVELOPED TUTELARY, RATHER THAN DEVELOPED TUTELARY (24)	0.67	M 18 9 3 18 X SQ= 11.13 P = 0. RV YES (YES)
85	TEND TO BE THOSE WHERE THE STAGE OF POLITICAL MODERNIZATION IS ADVANCED, RATHER THAN MID- OR EARLY TRANSITIONAL (60)	0.68	85	TEND TO BE THOSE WHERE THE STAGE OF POLITICAL MODERNIZATION IS MID- OR EARLY TRANSITIONAL, RATHER THAN ADVANCED (54)	0.63	38 18 18 31 X SQ= 8.96 P = 0.002 RV YES (YES)
86	TEND TO BE THOSE WHERE THE STAGE OF POLITICAL MODERNIZATION IS ADVANCED OR MID-TRANSITIONAL, RATHER THAN EARLY TRANSITIONAL (76)	0.86	86	TEND TO BE THOSE WHERE THE STAGE OF POLITICAL MODERNIZATION IS EARLY TRANSITIONAL, RATHER THAN ADVANCED OR MID-TRANSITIONAL (38)	0.51	48 24 8 25 X SQ= 14.70 P = 0. RV YES (YES)
87	TEND MORE TO BE THOSE WHOSE IDEOLOGICAL ORIENTATION IS OTHER THAN DEVELOPMENTAL (58)	0.77	87	TEND LESS TO BE THOSE WHOSE IDEOLOGICAL ORIENTATION IS OTHER THAN DEVELOPMENTAL (58)	0.54	10 18 33 21 X SQ= 3.80 P = 0.037 RV NO (YES)
91	TILT LESS TOWARD BEING THOSE WHOSE IDEOLOGICAL ORIENTATION IS DEVELOPMENTAL, RATHER THAN TRADITIONAL (31)	0.71	91	TILT MORE TOWARD BEING THOSE WHOSE IDEOLOGICAL ORIENTATION IS DEVELOPMENTAL, RATHER THAN TRADITIONAL (31)	0.95	10 18 4 1 X SQ= 1.83 P = 0.138 RV NO (YES)

111	LEAN TOWARD BEING THOSE WHERE POLITICAL ENCULTURATION IS HIGH OR MEDIUM (53)	0.67	111	LEAN TOWARD BEING THOSE WHERE POLITICAL ENCULTURATION IS LOW (42)	0.53	33 20 16 23 X SQ= 3.26 P = 0.058 RV YES (YES)
114	TEND TO BE THOSE WHERE SECTIONALISM IS NEGLIGIBLE (47)	0.51	114	TEND TO BE THOSE WHERE SECTIONALISM IS EXTREME OR MODERATE (61)	0.72	26 33 27 13 X SQ= 4.36 P = 0.025 RV YES (YES)
117	TEND TO BE THOSE WHERE INTEREST ARTICULATION BY ASSOCIATIONAL GROUPS IS SIGNIFICANT, MODERATE, OR LIMITED (60)	0.69	117	TEND TO BE THOSE WHERE INTEREST ARTICULATION BY ASSOCIATIONAL GROUPS IS NEGLIGIBLE (51)	0.56	38 21 17 27 X SQ= 5.73 P = 0.016 RV YES (YES)
121	TEND TO BE THOSE WHERE INTEREST ARTICULATION BY NON-ASSOCIATIONAL GROUPS IS MODERATE, LIMITED, OR NEGLIGIBLE (61)	0.68	121	TEND TO BE THOSE WHERE INTEREST ARTICULATION BY NON-ASSOCIATIONAL GROUPS IS SIGNIFICANT (54)	0.57	18 28 39 21 X SQ= 6.01 P = 0.011 RV YES (YES)
125	TILT LESS TOWARD BEING THOSE WHERE INTEREST ARTICULATION BY ANOMIC GROUPS IS FREQUENT OR OCCASIONAL (64)	0.55	125	TILT MORE TOWARD BEING THOSE WHERE INTEREST ARTICULATION BY ANOMIC GROUPS IS FREQUENT OR OCCASIONAL (64)	0.71	26 32 21 13 X SQ= 1.83 P = 0.135 RV NO (YES)
131	TEND TO BE THOSE WHERE INTEREST AGGREGATION BY POLITICAL PARTIES IS LIMITED OR NEGLIGIBLE (35)	0.64	131	TEND TO BE THOSE WHERE INTEREST AGGREGATION BY POLITICAL PARTIES IS SIGNIFICANT OR MODERATE (30)	0.63	13 17 23 10 X SQ= 3.45 P = 0.044 RV YES (YES)
138	LEAN TOWARD BEING THOSE WHERE INTEREST AGGREGATION BY THE LEGISLATURE IS SIGNIFICANT, MODERATE, OR LIMITED (48)	0.65	138	LEAN TOWARD BEING THOSE WHERE INTEREST AGGREGATION BY THE LEGISLATURE IS NEGLIGIBLE (49)	0.57	28 20 15 26 X SQ= 3.36 P = 0.056 RV YES (YES)
148	TEND MORE TO BE THOSE WHERE THE PARTY SYSTEM IS QUALITATIVELY OTHER THAN AFRICAN TRANSITIONAL (96)	0.98	148	TEND LESS TO BE THOSE WHERE THE PARTY SYSTEM IS QUALITATIVELY OTHER THAN AFRICAN TRANSITIONAL (96)	0.79	1 10 53 37 X SQ= 7.87 P = 0.002 RV NO (YES)
171	TILT LESS TOWARD BEING THOSE WHERE THE LEGISLATIVE-EXECUTIVE STRUCTURE IS OTHER THAN PARLIAMENTARY-REPUBLICAN (90)	0.81	171	TILT MORE TOWARD BEING THOSE WHERE THE LEGISLATIVE-EXECUTIVE STRUCTURE IS OTHER THAN PARLIAMENTARY-REPUBLICAN (90)	0.94	9 3 38 44 X SQ= 2.39 P = 0.120 RV NO (YES)

172	TEND MORE TO BE THOSE WHERE THE LEGISLATIVE-EXECUTIVE STRUCTURE IS OTHER THAN PARLIAMENTARY-ROYALIST (88)	0.88	172	TEND LESS TO BE THOSE WHERE THE LEGISLATIVE-EXECUTIVE STRUCTURE IS OTHER THAN PARLIAMENTARY-ROYALIST (88)	0.71	6 14 45 35 X SQ= 3.42 P = 0.046 RV NO (YES)
182	TEND MORE TO BE THOSE WHERE THE BUREAUCRACY IS SEMI-MODERN, RATHER THAN POST-COLONIAL TRANSITIONAL (55)	0.86	182	TEND LESS TO BE THOSE WHERE THE BUREAUCRACY IS SEMI-MODERN, RATHER THAN POST-COLONIAL TRANSITIONAL (55)	0.54	30 20 5 17 X SQ= 7.07 P = 0.005 RV NO (YES)
183	TEND TO BE THOSE WHERE THE BUREAUCRACY IS TRADITIONAL, RATHER THAN POST-COLONIAL TRANSITIONAL (9)	0.55	183	TEND TO BE THOSE WHERE THE BUREAUCRACY IS POST-COLONIAL TRANSITIONAL, RATHER THAN TRADITIONAL (25)	0.85	5 17 6 3 X SQ= 3.64 P = 0.038 RV YES (YES)
188	TEND LESS TO BE THOSE WHERE THE CHARACTER OF THE LEGAL SYSTEM IS OTHER THAN CIVIL LAW (81)	0.58	188	TEND MORE TO BE THOSE WHERE THE CHARACTER OF THE LEGAL SYSTEM IS OTHER THAN CIVIL LAW (81)	0.83	24 8 33 39 X SQ= 6.48 P = 0.010 RV NO (YES)
189	TEND MORE TO BE THOSE WHERE THE CHARACTER OF THE LEGAL SYSTEM IS OTHER THAN COMMON LAW (108)	0.98	189	TEND LESS TO BE THOSE WHERE THE CHARACTER OF THE LEGAL SYSTEM IS OTHER THAN COMMON LAW (108)	0.88	1 6 56 43 X SQ= 3.15 P = 0.047 RV NO (YES)
190	TEND MORE TO BE THOSE WHERE THE CHARACTER OF THE LEGAL SYSTEM IS CIVIL LAW, RATHER THAN COMMON LAW (32)	0.96	190	TEND LESS TO BE THOSE WHERE THE CHARACTER OF THE LEGAL SYSTEM IS CIVIL LAW, RATHER THAN COMMON LAW (32)	0.57	24 8 1 6 X SQ= 6.75 P = 0.005 RV NO (YES)
193	TEND MORE TO BE THOSE WHERE THE CHARACTER OF THE LEGAL SYSTEM IS OTHER THAN PARTLY INDIGENOUS (86)	0.96	193	TEND LESS TO BE THOSE WHERE THE CHARACTER OF THE LEGAL SYSTEM IS OTHER THAN PARTLY INDIGENOUS (86)	0.55	2 22 54 27 X SQ= 23.02 P = 0. RV NO (YES)

M

MATRIX

	66 POLITIES THAT ARE RELIGIOUSLY HOMOGENEOUS (57)	66 POLITIES THAT ARE RELIGIOUSLY HETEROGENEOUS (49)
84 POLITIES WHERE THE TYPE OF POLITICAL MODERNIZATION IS DEVELOPED TUTELARY, RATHER THAN UNDEVELOPED TUTELARY (31)	ALGERIA, IRAQ, MOROCCO, TUNISIA, BURMA, ISRAEL, PAKISTAN, UAR, CAMBODIA, JORDAN, PHILIPPINES, YEMEN, INDIA, LIBYA, SA'U ARABIA, INDONESIA, MONGOLIA, SYRIA 18	CEYLON, LIBERIA, CYPRUS, MALAYA, JAMAICA, SUDAN, LAOS, TRINIDAD, LEBANON 9
84 POLITIES WHERE THE TYPE OF POLITICAL MODERNIZATION IS UNDEVELOPED TUTELARY, RATHER THAN DEVELOPED TUTELARY (24)	MAURITANIA, SENEGAL, SOMALIA 3	CAMEROUN, GABON, MALI, TOGO, CHAD, GHANA, NIGER, UGANDA, CONGO(BRA), GUINEA, NIGERIA, UPPER VOLTA, CONGO(LEO), IVORY COAST, SIERRE LEO, DAHOMEY, MALAGASY R, TANGANYIKA 18

67 POLITIES
THAT ARE RACIALLY HOMOGENEOUS (82)

67 POLITIES
THAT ARE RACIALLY HETEROGENEOUS (27)

BOTH SUBJECT AND PREDICATE

82 IN LEFT COLUMN

ALBANIA	ALGERIA	ARGENTINA	AUSTRALIA	AUSTRIA	BELGIUM	BULGARIA	BURMA	BURUNDI	CAMBODIA
CAMEROUN	CANADA	CEN AFR REP	CEYLON	CHINA, PR	CONGO(BRA)	CONGO(LEO)	COSTA RICA	CYPRUS	CZECHOS'KIA
DAHOMEY	DENMARK	FINLAND	FRANCE	GABON	GERMANY, E	GERMAN FR	GHANA	GREECE	GUINEA
HUNGARY	ICELAND	INDIA	INDONESIA	IRAN	IRAQ	IRELAND	ISRAEL	ITALY	IVORY COAST
JAPAN	JORDAN	KOREA, N	KOREA REP	LAOS	LEBANON	LIBERIA	LIBYA	LUXEMBOURG	MONGOLIA
MOROCCO	NETHERLANDS	NEW ZEALAND	NIGERIA	NORWAY	PAKISTAN	PHILIPPINES	POLAND	PORTUGAL	RUMANIA
RWANDA	SENEGAL	SIERRE LEO	SOMALIA	SPAIN	SWEDEN	SWITZERLAND	SYRIA	TANGANYIKA	THAILAND
TOGO	TUNISIA	TURKEY	UGANDA	USSR	UAR	UK	UPPER VOLTA	URUGUAY	VIETNAM, N
VIETNAM REP	YUGOSLAVIA								

27 IN RIGHT COLUMN

AFGHANISTAN	BOLIVIA	BRAZIL	CHAD	COLOMBIA	CUBA	DOMIN REP	ECUADOR	EL SALVADOR	ETHIOPIA
GUATEMALA	HAITI	HONDURAS	JAMAICA	MALAYA	MALI	MAURITANIA	MEXICO	NICARAGUA	NIGER
PANAMA	PERU	SO AFRICA	SUDAN	TRINIDAD	US	VENEZUELA			

4 EXCLUDED BECAUSE AMBIGUOUS

CHILE MALAGASY R NEPAL PARAGUAY

2 EXCLUDED BECAUSE UNASCERTAINED

SA'U ARABIA YEMEN

2	TEND LESS TO BE THOSE LOCATED ELSEWHERE THAN IN WEST EUROPE, SCANDINAVIA, NORTH AMERICA, OR AUSTRALASIA (93)	0.74	2 TEND MORE TO BE THOSE LOCATED ELSEWHERE THAN IN WEST EUROPE, SCANDINAVIA, NORTH AMERICA, OR AUSTRALASIA (93)	0.97	21 1 59 28 X SQ= 5.53 P = 0.007 RV NO (NO)
3	TEND LESS TO BE THOSE LOCATED ELSEWHERE THAN IN WEST EUROPE (102)	0.84	3 ALWAYS ARE THOSE LOCATED ELSEWHERE THAN IN WEST EUROPE (102)	1.00	13 0 67 29 X SQ= 3.92 P = 0.019 RV NO (NO)
6	TEND TO BE THOSE LOCATED ELSEWHERE THAN IN THE CARIBBEAN, CENTRAL AMERICA, OR SOUTH AMERICA (93)	0.96	6 TEND TO BE THOSE LOCATED IN THE CARIBBEAN, CENTRAL AMERICA, OR SOUTH AMERICA (22)	0.59	3 17 77 12 X SQ= 39.19 P = 0. RV YES (YES)

8	TEND LESS TO BE THOSE LOCATED ELSEWHERE THAN IN EAST ASIA SOUTH ASIA, OR SOUTHEAST ASIA (97)	0.80	8	TEND MORE TO BE THOSE LOCATED ELSEWHERE THAN IN EAST ASIA SOUTH ASIA, OR SOUTHEAST ASIA (97)	0.97	16 1 64 28 X SQ= 3.26 P = 0.038 RV NO (NO)
13	TILT LESS TOWARD BEING THOSE LOCATED ELSEWHERE THAN IN NORTH AFRICA OR THE MIDDLE EAST (99)	0.84	13	TILT MORE TOWARD BEING THOSE LOCATED ELSEWHERE THAN IN NORTH AFRICA OR THE MIDDLE EAST (99)	0.97	13 1 67 28 X SQ= 2.08 P = 0.107 RV NO (NO)
17	TEND TO BE THOSE LOCATED IN THE MIDDLE EAST, RATHER THAN IN THE CARIBBEAN, CENTRAL AMERICA, OR SOUTH AMERICA (11)	0.73	17	TEND TO BE THOSE LOCATED IN THE CARIBBEAN, CENTRAL AMERICA, OR SOUTH AMERICA, RATHER THAN IN THE MIDDLE EAST (22)	0.94	8 1 3 17 X SQ= 11.43 P = 0. RV YES (YES)
18	LEAN LESS TOWARD BEING THOSE LOCATED IN NORTH AFRICA OR CENTRAL AND SOUTH AFRICA, RATHER THAN IN EAST ASIA, SOUTH ASIA, OR SOUTHEAST ASIA (33)	0.59	18	LEAN MORE TOWARD BEING THOSE LOCATED IN NORTH AFRICA OR CENTRAL AND SOUTH AFRICA, RATHER THAN IN EAST ASIA, SOUTH ASIA, OR SOUTHEAST ASIA (33)	0.90	23 9 16 1 X SQ= 2.15 P = 0.079 RV NO (NO)
19	TEND TO BE THOSE LOCATED IN NORTH AFRICA OR CENTRAL AND SOUTH AFRICA, RATHER THAN IN THE CARIBBEAN, CENTRAL AMERICA, OR SOUTH AMERICA (33)	0.88	19	TEND TO BE THOSE LOCATED IN THE CARIBBEAN, CENTRAL AMERICA, OR SOUTH AMERICA, RATHER THAN IN NORTH AFRICA OR CENTRAL AND SOUTH AFRICA (22)	0.65	23 9 3 17 X SQ= 13.73 P = 0. RV YES (YES)
20	TEND TO BE THOSE LOCATED IN EAST ASIA, SOUTH ASIA, OR SOUTHEAST ASIA, RATHER THAN IN THE CARIBBEAN, CENTRAL AMERICA, OR SOUTH AMERICA (18)	0.84	20	TEND TO BE THOSE LOCATED IN THE CARIBBEAN, CENTRAL AMERICA, OR SOUTH AMERICA, RATHER THAN IN EAST ASIA, SOUTH ASIA, OR SOUTHEAST ASIA (22)	0.94	16 1 3 17 X SQ= 19.97 P = 0. RV YES (YES)
21	TEND TO BE THOSE WHOSE TERRITORIAL SIZE IS MEDIUM OR SMALL (83)	0.80	21	TEND TO BE THOSE WHOSE TERRITORIAL SIZE IS VERY LARGE OR LARGE (32)	0.52	16 15 64 14 X SQ= 9.02 P = 0.003 RV YES (NO)
23	TEND LESS TO BE THOSE WHOSE POPULATION IS MEDIUM OR SMALL (88)	0.70	23	TEND MORE TO BE THOSE WHOSE POPULATION IS MEDIUM OR SMALL (88)	0.90	24 3 56 26 X SQ= 3.42 P = 0.044 RV NO (NO)
26	TEND TO BE THOSE WHOSE POPULATION DENSITY IS VERY HIGH, HIGH, OR MEDIUM (48)	0.50	26	TEND TO BE THOSE WHOSE POPULATION DENSITY IS LOW (67)	0.76	40 7 40 22 X SQ= 4.80 P = 0.017 RV YES (NO)

28	TEND TO BE THOSE WHOSE POPULATION GROWTH RATE IS LOW (48)	0.54	28	TEND TO BE THOSE WHOSE POPULATION GROWTH RATE IS HIGH (62)	0.79	35 22 41 6 X SQ= 7.47 P = 0.004 RV YES (NO)
31	TILT LESS TOWARD BEING THOSE WHOSE AGRICULTURAL POPULATION IS HIGH OR MEDIUM (90)	0.75	31	TILT MORE TOWARD BEING THOSE WHOSE AGRICULTURAL POPULATION IS HIGH OR MEDIUM (90)	0.90	60 26 20 3 X SQ= 1.94 P = 0.117 RV NO (NO)
34	TEND TO BE THOSE WHOSE GROSS NATIONAL PRODUCT IS VERY HIGH, HIGH, MEDIUM, OR LOW (62)	0.62	34	TEND TO BE THOSE WHOSE GROSS NATIONAL PRODUCT IS VERY LOW (53)	0.66	50 10 30 19 X SQ= 5.67 P = 0.016 RV YES (NO)
35	TEND LESS TO BE THOSE WHOSE PER CAPITA GROSS NATIONAL PRODUCT IS MEDIUM, LOW, OR VERY LOW (91)	0.72	35	TEND MORE TO BE THOSE WHOSE PER CAPITA GROSS NATIONAL PRODUCT IS MEDIUM, LOW, OR VERY LOW (91)	0.93	22 2 58 27 X SQ= 4.13 P = 0.021 RV NO (NO)
36	TILT LESS TOWARD BEING THOSE WHOSE PER CAPITA GROSS NATIONAL PRODUCT IS LOW OR VERY LOW (73)	0.57	36	TILT MORE TOWARD BEING THOSE WHOSE PER CAPITA GROSS NATIONAL PRODUCT IS LOW OR VERY LOW (73)	0.76	34 7 46 22 X SQ= 2.33 P = 0.117 RV NO (NO)
39	TILT LESS TOWARD BEING THOSE WHOSE INTERNATIONAL FINANCIAL STATUS IS LOW OR VERY LOW (76)	0.61	39	TILT MORE TOWARD BEING THOSE WHOSE INTERNATIONAL FINANCIAL STATUS IS LOW OR VERY LOW (76)	0.79	31 6 48 23 X SQ= 2.47 P = 0.108 RV NO (NO)
41	TEND LESS TO BE THOSE WHOSE ECONOMIC DEVELOPMENTAL STATUS IS INTERMEDIATE, UNDERDEVELOPED, OR VERY UNDERDEVELOPED (94)	0.77	41	TEND MORE TO BE THOSE WHOSE ECONOMIC DEVELOPMENTAL STATUS IS INTERMEDIATE, UNDERDEVELOPED, OR VERY UNDERDEVELOPED (94)	0.96	18 1 61 27 X SQ= 3.99 P = 0.022 RV NO (NO)
42	LEAN LESS TOWARD BEING THOSE WHOSE ECONOMIC DEVELOPMENTAL STATUS IS UNDERDEVELOPED OR VERY UNDERDEVELOPED (76)	0.61	42	LEAN MORE TOWARD BEING THOSE WHOSE ECONOMIC DEVELOPMENTAL STATUS IS UNDERDEVELOPED OR VERY UNDERDEVELOPED (76)	0.82	31 5 48 23 X SQ= 3.33 P = 0.061 RV NO (NO)
45	TEND TO BE THOSE WHERE THE LITERACY RATE IS FIFTY PERCENT OR ABOVE (55)	0.58	45	TEND TO BE THOSE WHERE THE LITERACY RATE IS BELOW FIFTY PERCENT (54)	0.66	43 10 31 19 X SQ= 3.76 P = 0.048 RV YES (NO)

47	TEND TO BE THOSE WHERE THE LITERACY RATE IS NINETY PERCENT OR ABOVE, RATHER THAN BETWEEN FIFTY AND NINETY PERCENT (25)	0.56	47	TEND TO BE THOSE WHERE THE LITERACY RATE IS BETWEEN FIFTY AND NINETY PERCENT, RATHER THAN NINETY PERCENT OR ABOVE (30)	0.90	24 1 19 9 X SQ= 5.12 P = 0.013 RV YES (NO)
50	LEAN TOWARD BEING THOSE WHERE FREEDOM OF THE PRESS IS INTERMITTENT, INTERNALLY ABSENT, OR INTERNALLY AND EXTERNALLY ABSENT (56)	0.62	50	LEAN TOWARD BEING THOSE WHERE FREEDOM OF THE PRESS IS COMPLETE (43)	0.60	26 15 42 10 X SQ= 2.69 P = 0.098 RV YES (NO)
51	TILT LESS TOWARD BEING THOSE WHERE FREEDOM OF THE PRESS IS COMPLETE OR INTERMITTENT (60)	0.58	51	TILT MORE TOWARD BEING THOSE WHERE FREEDOM OF THE PRESS IS COMPLETE OR INTERMITTENT (60)	0.76	39 19 28 6 X SQ= 1.77 P = 0.148 RV NO (NO)
53	TILT LESS TOWARD BEING THOSE WHERE NEWSPAPER CIRCULATION IS LESS THAN THREE HUNDRED PER THOUSAND (101)	0.84	53	TILT MORE TOWARD BEING THOSE WHERE NEWSPAPER CIRCULATION IS LESS THAN THREE HUNDRED PER THOUSAND (101)	0.97	13 1 67 28 X SQ= 2.08 P = 0.107 RV NO (NO)
54	TEND LESS TO BE THOSE WHERE NEWSPAPER CIRCULATION IS LESS THAN ONE HUNDRED PER THOUSAND (76)	0.58	54	TEND MORE TO BE THOSE WHERE NEWSPAPER CIRCULATION IS LESS THAN ONE HUNDRED PER THOUSAND (76)	0.90	33 3 45 26 X SQ= 8.30 P = 0.002 RV NO (NO)
62	TILT LESS TOWARD BEING THOSE WHERE THE RELIGION IS CATHOLIC, RATHER THAN PROTESTANT (25)	0.72	62	ALWAYS ARE THOSE WHERE THE RELIGION IS CATHOLIC, RATHER THAN PROTESTANT (25)	1.00	5 0 13 10 X SQ= 1.75 P = 0.128 RV NO (NO)
65	LEAN TOWARD BEING THOSE WHERE THE RELIGION IS MUSLIM, RATHER THAN CATHOLIC (18)	0.52	65	LEAN TOWARD BEING THOSE WHERE THE RELIGION IS CATHOLIC, RATHER THAN MUSLIM (25)	0.83	13 10 14 2 X SQ= 2.92 P = 0.076 RV YES (NO)
70	TEND LESS TO BE THOSE THAT ARE RELIGIOUSLY, RACIALLY, OR LINGUISTICALLY HETEROGENEOUS (85)	0.72	70	ALWAYS ARE THOSE THAT ARE RELIGIOUSLY, RACIALLY, OR LINGUISTICALLY HETEROGENEOUS (85)	1.00	21 0 54 29 X SQ= 8.51 P = 0.001 RV NO (NO)
72	TEND TO BE THOSE WHOSE DATE OF INDEPENDENCE IS AFTER 1914 (59)	0.61	72	TEND TO BE THOSE WHOSE DATE OF INDEPENDENCE IS BEFORE 1914 (52)	0.66	30 19 46 10 X SQ= 4.72 P = 0.028 RV YES (NO)

74	TEND LESS TO BE THOSE THAT ARE NOT HISTORICALLY WESTERN (87)	0.68	74	TEND MORE TO BE THOSE THAT ARE NOT HISTORICALLY WESTERN (87)	0.96	25 1 54 27 X SQ= 7.40 P = 0.002 RV NO (NO)
76	TEND LESS TO BE THOSE THAT HAVE BEEN SIGNIFICANTLY OR PARTIALLY WESTERNIZED THROUGH A COLONIAL RELATIONSHIP, RATHER THAN BEING HISTORICALLY WESTERN (70)	0.62	76	TEND MORE TO BE THOSE THAT HAVE BEEN SIGNIFICANTLY OR PARTIALLY WESTERNIZED THROUGH A COLONIAL RELATIONSHIP, RATHER THAN BEING HISTORICALLY WESTERN (70)	0.96	25 1 41 26 X SQ= 9.48 P = 0.001 RV NO (NO)
77	TEND TO BE THOSE THAT HAVE BEEN PARTIALLY WESTERNIZED, RATHER THAN SIGNIFICANTLY WESTERNIZED, THROUGH A COLONIAL RELATIONSHIP (41)	0.76	77	TEND TO BE THOSE THAT HAVE BEEN SIGNIFICANTLY WESTERNIZED, RATHER THAN PARTIALLY WESTERNIZED, THROUGH A COLONIAL RELATIONSHIP (28)	0.64	10 16 31 9 X SQ= 8.61 P = 0.002 RV YES (YES)
78	TEND LESS TO BE THOSE THAT HAVE BEEN SIGNIFICANTLY WESTERNIZED THROUGH A COLONIAL RELATIONSHIP, RATHER THAN WITHOUT SUCH A RELATIONSHIP (28)	0.59	78	ALWAYS ARE THOSE THAT HAVE BEEN SIGNIFICANTLY WESTERNIZED THROUGH A COLONIAL RELATIONSHIP, RATHER THAN WITHOUT SUCH A RELATIONSHIP (28)	1.00	10 16 7 0 X SQ= 6.08 P = 0.007 RV NO (YES)
79	TEND TO BE THOSE THAT WERE FORMERLY DEPENDENCIES OF BRITAIN OR FRANCE, RATHER THAN SPAIN (49)	0.92	79	TEND TO BE THOSE THAT WERE FORMERLY DEPENDENCIES OF SPAIN, RATHER THAN BRITAIN OR FRANCE (18)	0.50	35 13 3 13 X SQ= 12.44 P = 0. RV YES (YES)
82	TILT TOWARD BEING THOSE WHERE THE TYPE OF POLITICAL MODERNIZATION IS DEVELOPED TUTELARY OR UNDEVELOPED TUTELARY, RATHER THAN EARLY OR LATER EUROPEAN OR EUROPEAN DERIVED (55)	0.57	82	TILT TOWARD BEING THOSE WHERE THE TYPE OF POLITICAL MODERNIZATION IS EARLY OR LATER EUROPEAN OR EUROPEAN DERIVED, RATHER THAN DEVELOPED TUTELARY OR UNDEVELOPED TUTELARY (51)	0.63	32 17 42 10 X SQ= 2.34 P = 0.115 RV YES (NO)
85	TEND TO BE THOSE WHERE THE STAGE OF POLITICAL MODERNIZATION IS ADVANCED, RATHER THAN MID- OR EARLY TRANSITIONAL (60)	0.65	85	TEND TO BE THOSE WHERE THE STAGE OF POLITICAL MODERNIZATION IS MID- OR EARLY TRANSITIONAL, RATHER THAN ADVANCED (54)	0.76	52 7 28 22 X SQ= 12.71 P = 0. RV YES (NO)
92	TEND LESS TO BE THOSE WHERE THE SYSTEM STYLE IS LIMITED MOBILIZATIONAL, OR NON-MOBILIZATIONAL (93)	0.76	92	TEND MORE TO BE THOSE WHERE THE SYSTEM STYLE IS LIMITED MOBILIZATIONAL, OR NON-MOBILIZATIONAL (93)	0.97	19 1 60 28 X SQ= 4.68 P = 0.013 RV NO (NO)
95	LEAN LESS TOWARD BEING THOSE WHERE THE STATUS OF THE REGIME IS CONSTITUTIONAL OR AUTHORITARIAN (95)	0.81	95	LEAN MORE TOWARD BEING THOSE WHERE THE STATUS OF THE REGIME IS CONSTITUTIONAL OR AUTHORITARIAN (95)	0.96	64 26 15 1 X SQ= 2.57 P = 0.065 RV NO (NO)

99	TEND TO BE THOSE WHERE GOVERNMENTAL STABILITY IS GENERALLY PRESENT AND DATES FROM AT LEAST THE INTER-WAR PERIOD, OR FROM THE POST-WAR PERIOD (50)	0.68	99	TEND TO BE THOSE WHERE GOVERNMENTAL STABILITY IS MODERATELY PRESENT AND DATES FROM THE POST-WAR PERIOD, OR IS ABSENT (36)	0.72	44 5 21 13 X SQ= 7.71 P = 0.003 RV YES (NO)
106	LEAN LESS TOWARD BEING THOSE WHERE THE ELECTORAL SYSTEM IS COMPETITIVE OR PARTIALLY COMPETITIVE (52)	0.56	106	LEAN MORE TOWARD BEING THOSE WHERE THE ELECTORAL SYSTEM IS COMPETITIVE OR PARTIALLY COMPETITIVE (52)	0.80	33 16 26 4 X SQ= 2.72 P = 0.066 RV NO (NO)
110	TEND LESS TO BE THOSE WHERE POLITICAL ENCULTURATION IS MEDIUM OR LOW (80)	0.77	110	TEND MORE TO BE THOSE WHERE POLITICAL ENCULTURATION IS MEDIUM OR LOW (80)	0.96	14 1 48 27 X SQ= 3.74 P = 0.031 RV NO (NO)
115	LEAN LESS TOWARD BEING THOSE WHERE INTEREST ARTICULATION BY ASSOCIATIONAL GROUPS IS MODERATE, LIMITED, OR NEGLIGIBLE (91)	0.76	115	LEAN MORE TOWARD BEING THOSE WHERE INTEREST ARTICULATION BY ASSOCIATIONAL GROUPS IS MODERATE, LIMITED, OR NEGLIGIBLE (91)	0.93	18 2 58 27 X SQ= 2.82 P = 0.056 RV NO (NO)
116	LEAN LESS TOWARD BEING THOSE WHERE INTEREST ARTICULATION BY ASSOCIATIONAL GROUPS IS LIMITED OR NEGLIGIBLE (79)	0.66	116	LEAN MORE TOWARD BEING THOSE WHERE INTEREST ARTICULATION BY ASSOCIATIONAL GROUPS IS LIMITED OR NEGLIGIBLE (79)	0.83	26 5 50 24 X SQ= 2.15 P = 0.100 RV NO (NO)
118	TEND TO BE THOSE WHERE INTEREST ARTICULATION BY INSTITUTIONAL GROUPS IS VERY SIGNIFICANT (40)	0.51	118	TEND TO BE THOSE WHERE INTEREST ARTICULATION BY INSTITUTIONAL GROUPS IS SIGNIFICANT, MODERATE, OR LIMITED (60)	0.85	35 4 34 22 X SQ= 8.34 P = 0.002 RV YES (NO)
123	TILT LESS TOWARD BEING THOSE WHERE INTEREST ARTICULATION BY NON-ASSOCIATIONAL GROUPS IS SIGNIFICANT, MODERATE, OR LIMITED (107)	0.90	123	ALWAYS ARE THOSE WHERE INTEREST ARTICULATION BY NON-ASSOCIATIONAL GROUPS IS SIGNIFICANT, MODERATE, OR LIMITED (107)	1.00	72 29 8 0 X SQ= 1.83 P = 0.106 RV NO (NO)
125	LEAN LESS TOWARD BEING THOSE WHERE INTEREST ARTICULATION BY ANOMIC GROUPS IS FREQUENT OR OCCASIONAL (64)	0.59	125	LEAN MORE TOWARD BEING THOSE WHERE INTEREST ARTICULATION BY ANOMIC GROUPS IS FREQUENT OR OCCASIONAL (64)	0.81	40 21 28 5 X SQ= 3.07 P = 0.055 RV NO (NO)
126	TEND LESS TO BE THOSE WHERE INTEREST ARTICULATION BY ANOMIC GROUPS IS FREQUENT, OCCASIONAL, OR INFREQUENT (83)	0.76	126	ALWAYS ARE THOSE WHERE INTEREST ARTICULATION BY ANOMIC GROUPS IS FREQUENT, OCCASIONAL, OR INFREQUENT (83)	1.00	52 26 16 0 X SQ= 5.80 P = 0.005 RV NO (NO)

129	LEAN LESS TOWARD BEING THOSE WHERE INTEREST ARTICULATION BY POLITICAL PARTIES IS SIGNIFICANT, MODERATE, OR LIMITED (56)	0.53	129	LEAN MORE TOWARD BEING THOSE WHERE INTEREST ARTICULATION BY POLITICAL PARTIES IS SIGNIFICANT, MODERATE, OR LIMITED (56)	0.75	35 18 31 6 X SQ= 2.66 P = 0.089 RV NO (NO)
137	TEND LESS TO BE THOSE WHERE INTEREST AGGREGATION BY THE LEGISLATURE IS LIMITED OR NEGLIGIBLE (68)	0.64	137	TEND MORE TO BE THOSE WHERE INTEREST AGGREGATION BY THE LEGISLATURE IS LIMITED OR NEGLIGIBLE (68)	0.87	25 3 44 21 X SQ= 3.70 P = 0.038 RV NO (NO)
139	LEAN LESS TOWARD BEING THOSE WHERE THE PARTY SYSTEM IS QUANTITATIVELY OTHER THAN ONE-PARTY (71)	0.60	139	LEAN MORE TOWARD BEING THOSE WHERE THE PARTY SYSTEM IS QUANTITATIVELY OTHER THAN ONE-PARTY (71)	0.82	29 5 44 23 X SQ= 3.41 P = 0.059 RV NO (NO)
153	TEND TO BE THOSE WHERE THE PARTY SYSTEM IS STABLE (42)	0.62	153	TEND TO BE THOSE WHERE THE PARTY SYSTEM IS MODERATELY STABLE OR UNSTABLE (38)	0.73	M 38 4 23 11 X SQ= 4.82 P = 0.019 RV YES (NO)
154	TEND TO BE THOSE WHERE THE PARTY SYSTEM IS STABLE OR MODERATELY STABLE (55)	0.77	154	TEND TO BE THOSE WHERE THE PARTY SYSTEM IS UNSTABLE (25)	0.60	47 6 14 9 X SQ= 6.17 P = 0.010 RV YES (NO)
158	LEAN TOWARD BEING THOSE WHERE PERSONALISMO IS NEGLIGIBLE (56)	0.64	158	LEAN TOWARD BEING THOSE WHERE PERSONALISMO IS PRONOUNCED OR MODERATE (40)	0.58	25 14 44 10 X SQ= 2.72 P = 0.092 RV YES (NO)
169	LEAN TOWARD BEING THOSE WHERE THE HORIZONTAL POWER DISTRIBUTION IS NEGLIGIBLE (48)	0.50	169	LEAN TOWARD BEING THOSE WHERE THE HORIZONTAL POWER DISTRIBUTION IS SIGNIFICANT OR LIMITED (58)	0.70	37 19 37 8 X SQ= 2.55 P = 0.076 RV YES (NO)
170	TEND TO BE THOSE WHERE THE LEGISLATIVE-EXECUTIVE STRUCTURE IS OTHER THAN PRESIDENTIAL (63)	0.73	170	TEND TO BE THOSE WHERE THE LEGISLATIVE-EXECUTIVE STRUCTURE IS PRESIDENTIAL (39)	0.64	20 16 53 9 X SQ= 9.22 P = 0.002 RV YES (NO)
176	LEAN LESS TOWARD BEING THOSE WHERE THE LEGISLATURE IS FULLY EFFECTIVE, PARTIALLY EFFECTIVE, OR LARGELY INEFFECTIVE (72)	0.68	176	LEAN MORE TOWARD BEING THOSE WHERE THE LEGISLATURE IS FULLY EFFECTIVE, PARTIALLY EFFECTIVE, OR LARGELY INEFFECTIVE (72)	0.87	49 21 23 3 X SQ= 2.53 P = 0.070 RV NO (NO)

181	TEND LESS TO BE THOSE WHERE THE BUREAUCRACY IS SEMI-MODERN, RATHER THAN MODERN (55)	0.65	181	TEND MORE TO BE THOSE WHERE THE BUREAUCRACY IS SEMI-MODERN, RATHER THAN MODERN (55)	0.90	19 2 35 18 X SQ= 3.40 P = 0.042 RV NO (NO)
185	TEND TO BE THOSE WHERE PARTICIPATION BY THE MILITARY IN POLITICS IS SUPPORTIVE, RATHER THAN INTERVENTIVE (31)	0.72	185	TEND TO BE THOSE WHERE PARTICIPATION BY THE MILITARY IN POLITICS IS INTERVENTIVE, RATHER THAN SUPPORTIVE (21)	0.69	10 9 26 4 X SQ= 5.28 P = 0.018 RV YES (NO)
194	TEND LESS TO BE THOSE THAT ARE NON-COMMUNIST (101)	0.84	194	ALWAYS ARE THOSE THAT ARE NON-COMMUNIST (101)	1.00	13 0 67 28 X SQ= 3.75 P = 0.020 RV NO (NO)

M

MATRIX

	67 POLITIES THAT ARE RACIALLY HOMOGENEOUS (82)	67 POLITIES THAT ARE RACIALLY HETEROGENEOUS (27)
153 POLITIES WHERE THE PARTY SYSTEM IS STABLE (42)	ALBANIA, AUSTRALIA, AUSTRIA, BELGIUM, BULGARIA, CANADA, CHINA, PR, CZECHOS'KIA, DENMARK, FINLAND, GERMANY, E, GERMAN FR, GUINEA, HUNGARY, ICELAND, INDIA, IRELAND, ISRAEL, JAPAN, KOREA, N, LIBERIA, LUXEMBOURG, MONGOLIA, NETHERLANDS, NEW ZEALAND, NORWAY, PHILIPPINES, POLAND, PORTUGAL, RUMANIA, SPAIN, SWEDEN, SWITZERLAND, TANGANYIKA, TUNISIA, USSR, UK, VIETNAM, N, YUGOSLAVIA	MEXICO, SO AFRICA, US
	39	3
	23	11
153 POLITIES WHERE THE PARTY SYSTEM IS MODERATELY STABLE OR UNSTABLE (38)	ARGENTINA, CAMBODIA, CEYLON, CONGO(LEO), COSTA RICA, FRANCE, GHANA, GREECE, IRAQ, ITALY, IVORY COAST, JORDAN, KOREA REP, LAOS, LEBANON, PAKISTAN, SENEGAL, SIERRE LEO, SOMALIA, SYRIA, THAILAND, TURKEY, VIETNAM REP	BOLIVIA, BRAZIL, ECUADOR, EL SALVADOR, GUATEMALA, HONDURAS, MALAYA, NICARAGUA, PANAMA, PERU, VENEZUELA

```
****************************************************************************************************

68  POLITIES                                          68  POLITIES
    THAT ARE LINGUISTICALLY                               THAT ARE LINGUISTICALLY
    HOMOGENEOUS  (52)                                     WEAKLY HETEROGENEOUS OR
                                                          STRONGLY HETEROGENEOUS  (62)

                                                                        BOTH SUBJECT AND PREDICATE
....................................................................................................

     52 IN LEFT COLUMN

ALBANIA     ARGENTINA   AUSTRALIA   AUSTRIA     BRAZIL      BURUNDI     CHILE       COLOMBIA    COSTA RICA  CUBA
DENMARK     DOMIN REP   EL SALVADOR FRANCE      GERMANY, E  GERMAN FR   GREECE      HAITI       HONDURAS    HUNGARY
ICELAND     IRELAND     ITALY       JAMAICA     JAPAN       JORDAN      KOREA, N    KOREA REP   LEBANON     LIBYA
LUXEMBOURG  MALAGASY R  MEXICO      MONGOLIA    NETHERLANDS NEW ZEALAND NICARAGUA   NORWAY      PARAGUAY    POLAND
PORTUGAL    RWANDA      SA'U ARABIA SOMALIA     SWEDEN      TUNISIA     UAR         UK          US          URUGUAY
VENEZUELA   YEMEN

     62 IN RIGHT COLUMN

AFGHANISTAN ALGERIA     BELGIUM     BOLIVIA     BULGARIA    BURMA       CAMBODIA    CAMEROUN    CANADA      CEN AFR REP
CEYLON      CHAD        CONGO(BRA)  CONGO(LEO)  CYPRUS      CZECHOS'KIA DAHOMEY     ECUADOR     ETHIOPIA    FINLAND
GABON       GHANA       GUATEMALA   GUINEA      INDIA       INDONESIA   IRAN        IRAQ        ISRAEL      IVORY COAST
LAOS        LIBERIA     MALAYA      MALI        MAURITANIA  MOROCCO     NEPAL       NIGER       NIGERIA     PAKISTAN
PANAMA      PERU        PHILIPPINES RUMANIA     SENEGAL     SIERRE LEO  SO AFRICA   SPAIN       SUDAN       SWITZERLAND
SYRIA       TANGANYIKA  THAILAND    TOGO        TRINIDAD    TURKEY      UGANDA      USSR        UPPER VOLTA VIETNAM, N
VIETNAM REP YUGOSLAVIA

      1 EXCLUDED BECAUSE AMBIGUOUS

CHINA, PR
----------------------------------------------------------------------------------------------------

  2  TEND LESS TO BE THOSE                    0.67      2  TEND MORE TO BE THOSE                   0.92        17    5
     LOCATED ELSEWHERE THAN IN                             LOCATED ELSEWHERE THAN IN                            35   57
     WEST EUROPE, SCANDINAVIA,                             WEST EUROPE, SCANDINAVIA,                       X SQ=   9.49
     NORTH AMERICA, OR AUSTRALASIA  (93)                   NORTH AMERICA, OR AUSTRALASIA  (93)             P =    0.002
                                                                                                           RV NO  (YES)

  6  TEND LESS TO BE THOSE                    0.69      6  TEND MORE TO BE THOSE                   0.90        16    6
     LOCATED ELSEWHERE THAN IN THE                         LOCATED ELSEWHERE THAN IN THE                        36   56
     CARIBBEAN, CENTRAL AMERICA,                           CARIBBEAN, CENTRAL AMERICA,                     X SQ=   6.78
     OR SOUTH AMERICA  (93)                                OR SOUTH AMERICA  (93)                          P =    0.008
                                                                                                           RV NO  (YES)

  8  LEAN MORE TOWARD BEING THOSE             0.92      8  LEAN LESS TOWARD BEING THOSE            0.79         4   13
     LOCATED ELSEWHERE THAN IN EAST ASIA                   LOCATED ELSEWHERE THAN IN EAST ASIA                  48   49
     SOUTH ASIA, OR SOUTHEAST ASIA  (97)                   SOUTH ASIA, OR SOUTHEAST ASIA  (97)             X SQ=   2.95
                                                                                                           P =    0.065
                                                                                                           RV NO  (NO )
```

9	ALWAYS ARE THOSE LOCATED ELSEWHERE THAN IN SOUTHEAST ASIA (106)	1.00	9	TEND LESS TO BE THOSE LOCATED ELSEWHERE THAN IN SOUTHEAST ASIA (106)	0.85	0 9 52 53 X SQ= 6.32 P = 0.004 RV NO (NO)
10	TEND MORE TO BE THOSE LOCATED ELSEWHERE THAN IN NORTH AFRICA, OR CENTRAL AND SOUTH AFRICA (82)	0.87	10	TEND LESS TO BE THOSE LOCATED ELSEWHERE THAN IN NORTH AFRICA, OR CENTRAL AND SOUTH AFRICA (82)	0.58	7 26 45 36 X SQ= 9.81 P = 0.001 RV NO (YES)
17	LEAN TOWARD BEING THOSE LOCATED IN THE CARIBBEAN, CENTRAL AMERICA, OR SOUTH AMERICA, RATHER THAN IN THE MIDDLE EAST (22)	0.80	17	LEAN TOWARD BEING THOSE LOCATED IN THE MIDDLE EAST, RATHER THAN IN THE CARIBBEAN, CENTRAL AMERICA, OR SOUTH AMERICA (11)	0.54	4 7 16 6 X SQ= 2.68 P = 0.065 RV YES (YES)
19	TEND TO BE THOSE LOCATED IN THE CARIBBEAN, CENTRAL AMERICA, OR SOUTH AMERICA, RATHER THAN IN NORTH AFRICA OR CENTRAL AND SOUTH AFRICA (22)	0.70	19	TEND TO BE THOSE LOCATED IN NORTH AFRICA OR CENTRAL AND SOUTH AFRICA, RATHER THAN IN THE CARIBBEAN, CENTRAL AMERICA, OR SOUTH AMERICA (33)	0.81	7 26 16 6 X SQ= 12.36 P = 0. RV YES (YES)
20	TEND TO BE THOSE LOCATED IN THE CARIBBEAN, CENTRAL AMERICA, OR SOUTH AMERICA, RATHER THAN IN EAST ASIA, SOUTH ASIA, OR SOUTHEAST ASIA (22)	0.80	20	TEND TO BE THOSE LOCATED IN EAST ASIA, SOUTH ASIA, OR SOUTHEAST ASIA, RATHER THAN IN THE CARIBBEAN, CENTRAL AMERICA, OR SOUTH AMERICA (18)	0.68	4 13 16 6 X SQ= 7.43 P = 0.004 RV YES (YES)
22	TEND TO BE THOSE WHOSE TERRITORIAL SIZE IS SMALL (47)	0.52	22	TEND TO BE THOSE WHOSE TERRITORIAL SIZE IS VERY LARGE, LARGE, OR MEDIUM (68)	0.68	25 42 27 20 X SQ= 3.74 P = 0.038 RV YES (YES)
26	LEAN TOWARD BEING THOSE WHOSE POPULATION DENSITY IS VERY HIGH, HIGH, OR MEDIUM (48)	0.50	26	LEAN TOWARD BEING THOSE WHOSE POPULATION DENSITY IS LOW (67)	0.66	26 21 26 41 X SQ= 2.41 P = 0.090 RV YES (YES)
29	TEND TO BE THOSE WHERE THE DEGREE OF URBANIZATION IS HIGH (56)	0.77	29	TEND TO BE THOSE WHERE THE DEGREE OF URBANIZATION IS LOW (49)	0.67	37 19 11 38 X SQ= 18.32 P = 0. RV YES (YES)
30	TEND TO BE THOSE WHOSE AGRICULTURAL POPULATION IS MEDIUM, LOW, OR VERY LOW (57)	0.73	30	TEND TO BE THOSE WHOSE AGRICULTURAL POPULATION IS HIGH (56)	0.67	14 41 37 20 X SQ= 16.02 P = 0. RV YES (YES)

36	TEND TO BE THOSE WHOSE PER CAPITA GROSS NATIONAL PRODUCT IS VERY HIGH, HIGH, OR MEDIUM (42)	0.52	36	TEND TO BE THOSE WHOSE PER CAPITA GROSS NATIONAL PRODUCT IS LOW OR VERY LOW (73)	0.76	27 15 25 47 X SQ= 8.19 P = 0.003 RV YES (YES)
37	TEND TO BE THOSE WHOSE PER CAPITA GROSS NATIONAL PRODUCT IS VERY HIGH, HIGH, MEDIUM, OR LOW (64)	0.77	37	TEND TO BE THOSE WHOSE PER CAPITA GROSS NATIONAL PRODUCT IS VERY LOW (51)	0.61	40 24 12 38 X SQ= 15.26 P = 0. RV YES (YES)
39	TEND LESS TO BE THOSE WHOSE INTERNATIONAL FINANCIAL STATUS IS LOW OR VERY LOW (76)	0.57	39	TEND MORE TO BE THOSE WHOSE INTERNATIONAL FINANCIAL STATUS IS LOW OR VERY LOW (76)	0.76	22 15 29 47 X SQ= 3.74 P = 0.044 RV NO (YES)
42	TEND TO BE THOSE WHOSE ECONOMIC DEVELOPMENTAL STATUS IS DEVELOPED OR INTERMEDIATE (36)	0.50	42	TEND TO BE THOSE WHOSE ECONOMIC DEVELOPMENTAL STATUS IS UNDERDEVELOPED OR VERY UNDERDEVELOPED (76)	0.82	M 25 11 25 50 X SQ= 11.40 P = 0. RV YES (YES)
43	TEND TO BE THOSE WHOSE ECONOMIC DEVELOPMENTAL STATUS IS DEVELOPED, INTERMEDIATE, OR UNDERDEVELOPED (55)	0.63	43	TEND TO BE THOSE WHOSE ECONOMIC DEVELOPMENTAL STATUS IS VERY UNDERDEVELOPED (57)	0.63	31 23 18 39 X SQ= 6.49 P = 0.008 RV YES (YES)
45	TEND TO BE THOSE WHERE THE LITERACY RATE IS FIFTY PERCENT OR ABOVE (55)	0.71	45	TEND TO BE THOSE WHERE THE LITERACY RATE IS BELOW FIFTY PERCENT (54)	0.67	35 20 14 40 X SQ= 14.17 P = 0. RV YES (YES)
54	TEND TO BE THOSE WHERE NEWSPAPER CIRCULATION IS ONE HUNDRED OR MORE PER THOUSAND (37)	0.51	54	TEND TO BE THOSE WHERE NEWSPAPER CIRCULATION IS LESS THAN ONE HUNDRED PER THOUSAND (76)	0.82	26 11 25 50 X SQ= 12.18 P = 0. RV YES (YES)
56	TEND MORE TO BE THOSE WHERE THE RELIGION IS PREDOMINANTLY LITERATE (79)	0.92	56	TEND LESS TO BE THOSE WHERE THE RELIGION IS PREDOMINANTLY LITERATE (79)	0.55	46 33 4 27 X SQ= 16.66 P = 0. RV NO (YES)
57	TEND LESS TO BE THOSE WHERE THE RELIGION IS OTHER THAN CATHOLIC (90)	0.60	57	TEND MORE TO BE THOSE WHERE THE RELIGION IS OTHER THAN CATHOLIC (90)	0.94	21 4 31 58 X SQ= 17.09 P = 0. RV NO (YES)

63	TEND TO BE THOSE WHERE THE RELIGION IS PREDOMINANTLY SOME KIND OF CHRISTIAN (46)	0.67	63	TEND TO BE THOSE WHERE THE RELIGION IS PREDOMINANTLY OR PARTLY OTHER THAN CHRISTIAN (68)	0.82	35 11 17 50 X SQ= 26.23 P = 0. RV YES (YES)
66	TEND TO BE THOSE THAT ARE RELIGIOUSLY HOMOGENEOUS (57)	0.71	66	TEND TO BE THOSE THAT ARE RELIGIOUSLY HETEROGENEOUS (49)	0.60	34 23 14 35 X SQ= 9.05 P = 0.002 RV YES (YES)
70	TEND LESS TO BE THOSE THAT ARE RELIGIOUSLY, RACIALLY, OR LINGUISTICALLY HETEROGENEOUS (85)	0.51	70	ALWAYS ARE THOSE THAT ARE RELIGIOUSLY, RACIALLY, OR LINGUISTICALLY HETEROGENEOUS (85)	1.00	21 0 22 62 X SQ= 34.86 P = 0. RV NO (YES)
72	TEND TO BE THOSE WHOSE DATE OF INDEPENDENCE IS BEFORE 1914 (52)	0.65	72	TEND TO BE THOSE WHOSE DATE OF INDEPENDENCE IS AFTER 1914 (59)	0.68	31 20 17 42 X SQ= 10.11 P = 0.001 RV YES (YES)
73	TEND TO BE THOSE WHOSE DATE OF INDEPENDENCE IS BEFORE 1945 (65)	0.81	73	TEND TO BE THOSE WHOSE DATE OF INDEPENDENCE IS AFTER 1945 (46)	0.60	39 25 9 37 X SQ= 16.98 P = 0. RV YES (YES)
74	TEND LESS TO BE THOSE THAT ARE NOT HISTORICALLY WESTERN (87)	0.62	74	TEND MORE TO BE THOSE THAT ARE NOT HISTORICALLY WESTERN (87)	0.90	20 6 32 54 X SQ= 11.11 P = 0.001 RV NO (YES)
75	TEND TO BE THOSE THAT ARE HISTORICALLY WESTERN OR SIGNIFICANTLY WESTERNIZED (62)	0.73	75	TEND TO BE THOSE THAT ARE NOT HISTORICALLY WESTERN AND ARE NOT SIGNIFICANTLY WESTERNIZED (52)	0.62	38 23 14 38 X SQ= 12.75 P = 0. RV YES (YES)
76	TEND LESS TO BE THOSE THAT HAVE BEEN SIGNIFICANTLY OR PARTIALLY WESTERNIZED THROUGH A COLONIAL RELATIONSHIP, RATHER THAN BEING HISTORICALLY WESTERN (70)	0.56	76	TEND MORE TO BE THOSE THAT HAVE BEEN SIGNIFICANTLY OR PARTIALLY WESTERNIZED THROUGH A COLONIAL RELATIONSHIP, RATHER THAN BEING HISTORICALLY WESTERN (70)	0.88	20 6 25 45 X SQ= 11.33 P = 0. RV NO (YES)
77	TEND TO BE THOSE THAT HAVE BEEN SIGNIFICANTLY WESTERNIZED, RATHER THAN PARTIALLY WESTERNIZED, THROUGH A COLONIAL RELATIONSHIP (28)	0.68	77	TEND TO BE THOSE THAT HAVE BEEN PARTIALLY WESTERNIZED, RATHER THAN SIGNIFICANTLY WESTERNIZED, THROUGH A COLONIAL RELATIONSHIP (41)	0.75	17 11 8 33 X SQ= 10.51 P = 0.001 RV YES (YES)

78	LEAN MORE TOWARD BEING THOSE THAT HAVE BEEN SIGNIFICANTLY WESTERNIZED THROUGH A COLONIAL RELATIONSHIP, RATHER THAN WITHOUT SUCH A RELATIONSHIP (28)	0.94	78	LEAN LESS TOWARD BEING THOSE THAT HAVE BEEN SIGNIFICANTLY WESTERNIZED THROUGH A COLONIAL RELATIONSHIP, RATHER THAN WITHOUT SUCH A RELATIONSHIP (28)	0.69	17 11 1 5 X SQ= 2.28 P = 0.078 RV NO (YES)
79	TEND TO BE THOSE THAT WERE FORMERLY DEPENDENCIES OF SPAIN, RATHER THAN BRITAIN OR FRANCE (18)	0.54	79	TEND TO BE THOSE THAT WERE FORMERLY DEPENDENCIES OF BRITAIN OR FRANCE, RATHER THAN SPAIN (49)	0.88	11 38 13 5 X SQ= 12.10 P = 0. RV YES (YES)
82	TEND TO BE THOSE WHERE THE TYPE OF POLITICAL MODERNIZATION IS EARLY OR LATER EUROPEAN OR EUROPEAN DERIVED, RATHER THAN DEVELOPED TUTELARY OR UNDEVELOPED TUTELARY (51)	0.71	82	TEND TO BE THOSE WHERE THE TYPE OF POLITICAL MODERNIZATION IS DEVELOPED TUTELARY OR UNDEVELOPED TUTELARY, RATHER THAN EARLY OR LATER EUROPEAN OR EUROPEAN DERIVED (55)	0.73	36 15 15 40 X SQ= 18.19 P = 0. RV YES (YES)
83	LEAN MORE TOWARD BEING THOSE WHERE THE TYPE OF POLITICAL MODERNIZATION IS OTHER THAN NON-EUROPEAN AUTOCHTHONOUS (106)	0.98	83	LEAN LESS TOWARD BEING THOSE WHERE THE TYPE OF POLITICAL MODERNIZATION IS OTHER THAN NON-EUROPEAN AUTOCHTHONOUS (106)	0.89	1 7 51 55 X SQ= 2.50 P = 0.069 RV NO (NO)
84	TILT TOWARD BEING THOSE WHERE THE TYPE OF POLITICAL MODERNIZATION IS DEVELOPED TUTELARY, RATHER THAN UNDEVELOPED TUTELARY (31)	0.73	84	TILT TOWARD BEING THOSE WHERE THE TYPE OF POLITICAL MODERNIZATION IS UNDEVELOPED TUTELARY, RATHER THAN DEVELOPED TUTELARY (24)	0.50	11 20 4 20 X SQ= 1.56 P = 0.141 RV YES (NO)
85	TEND TO BE THOSE WHERE THE STAGE OF POLITICAL MODERNIZATION IS ADVANCED, RATHER THAN MID- OR EARLY TRANSITIONAL (60)	0.73	85	TEND TO BE THOSE WHERE THE STAGE OF POLITICAL MODERNIZATION IS MID- OR EARLY TRANSITIONAL, RATHER THAN ADVANCED (54)	0.65	37 22 14 40 X SQ= 13.96 P = 0. RV YES (YES)
86	TEND TO BE THOSE WHERE THE STAGE OF POLITICAL MODERNIZATION IS ADVANCED OR MID-TRANSITIONAL, RATHER THAN EARLY TRANSITIONAL (76)	0.86	86	TEND TO BE THOSE WHERE THE STAGE OF POLITICAL MODERNIZATION IS EARLY TRANSITIONAL, RATHER THAN ADVANCED OR MID-TRANSITIONAL (38)	0.50	44 31 7 31 X SQ= 14.91 P = 0. RV YES (YES)
87	TEND TO BE THOSE WHOSE IDEOLOGICAL ORIENTATION IS OTHER THAN DEVELOPMENTAL (58)	0.83	87	TEND TO BE THOSE WHOSE IDEOLOGICAL ORIENTATION IS DEVELOPMENTAL (31)	0.52	7 24 35 22 X SQ= 10.62 P = 0.001 RV YES (YES)
89	TEND TO BE THOSE WHOSE IDEOLOGICAL ORIENTATION IS CONVENTIONAL (33)	0.56	89	TEND TO BE THOSE WHOSE IDEOLOGICAL ORIENTATION IS OTHER THAN CONVENTIONAL (62)	0.82	24 9 19 42 X SQ= 13.29 P = 0. RV YES (YES)

93	LEAN MORE TOWARD BEING THOSE WHERE THE SYSTEM STYLE IS NON-MOBILIZATIONAL (78)	0.80	93	LEAN LESS TOWARD BEING THOSE WHERE THE SYSTEM STYLE IS NON-MOBILIZATIONAL (78)	0.64	10 21 40 38 X SQ= 2.51 P = 0.090 RV NO (YES)
94	LEAN TOWARD BEING THOSE WHERE THE STATUS OF THE REGIME IS CONSTITUTIONAL (51)	0.66	94	LEAN TOWARD BEING THOSE WHERE THE STATUS OF THE REGIME IS AUTHORITARIAN OR TOTALITARIAN (41)	0.55	31 20 16 24 X SQ= 3.09 P = 0.059 RV YES (YES)
96	TEND MORE TO BE THOSE WHERE THE STATUS OF THE REGIME IS CONSTITUTIONAL OR TOTALITARIAN (67)	0.85	96	TEND LESS TO BE THOSE WHERE THE STATUS OF THE REGIME IS CONSTITUTIONAL OR TOTALITARIAN (67)	0.63	39 27 7 16 X SQ= 4.52 P = 0.028 RV NO (YES)
99	LEAN TOWARD BEING THOSE WHERE GOVERNMENTAL STABILITY IS GENERALLY PRESENT AND DATES FROM AT LEAST THE INTER-WAR PERIOD, OR FROM THE POST-WAR PERIOD (50)	0.68	99	LEAN TOWARD BEING THOSE WHERE GOVERNMENTAL STABILITY IS MODERATELY PRESENT AND DATES FROM THE POST-WAR PERIOD, OR IS ABSENT (36)	0.52	28 21 13 23 X SQ= 2.88 P = 0.079 RV YES (YES)
101	TEND TO BE THOSE WHERE THE REPRESENTATIVE CHARACTER OF THE REGIME IS POLYARCHIC (41)	0.57	101	TEND TO BE THOSE WHERE THE REPRESENTATIVE CHARACTER OF THE REGIME IS LIMITED POLYARCHIC, PSEUDO-POLYARCHIC, OR NON-POLYARCHIC (57)	0.70	25 16 19 37 X SQ= 5.94 P = 0.013 RV YES (YES)
102	TEND TO BE THOSE WHERE THE REPRESENTATIVE CHARACTER OF THE REGIME IS POLYARCHIC OR LIMITED POLYARCHIC (59)	0.70	102	TEND TO BE THOSE WHERE THE REPRESENTATIVE CHARACTER OF THE REGIME IS PSEUDO-POLYARCHIC OR NON-POLYARCHIC (49)	0.58	35 24 15 33 X SQ= 7.29 P = 0.006 RV YES (YES)
105	TEND TO BE THOSE WHERE THE ELECTORAL SYSTEM IS COMPETITIVE (43)	0.61	105	TEND TO BE THOSE WHERE THE ELECTORAL SYSTEM IS PARTIALLY COMPETITIVE OR NON-COMPETITIVE (47)	0.64	27 16 17 29 X SQ= 4.95 P = 0.020 RV YES (YES)
107	TEND TO BE THOSE WHERE AUTONOMOUS GROUPS ARE FULLY TOLERATED IN POLITICS (46)	0.57	107	TEND TO BE THOSE WHERE AUTONOMOUS GROUPS ARE PARTIALLY TOLERATED IN POLITICS, ARE TOLERATED ONLY OUTSIDE POLITICS, OR ARE NOT TOLERATED AT ALL (65)	0.71	M 29 17 22 42 X SQ= 7.73 P = 0.004 RV YES (YES)
111	TEND TO BE THOSE WHERE POLITICAL ENCULTURATION IS HIGH OR MEDIUM (53)	0.72	111	TEND TO BE THOSE WHERE POLITICAL ENCULTURATION IS LOW (42)	0.56	29 24 11 31 X SQ= 6.70 P = 0.007 RV YES (YES)

114	TEND TO BE THOSE WHERE SECTIONALISM IS NEGLIGIBLE (47)	0.67	114	TEND TO BE THOSE WHERE SECTIONALISM IS EXTREME OR MODERATE (61)	0.75	16 44 32 15 X SQ= 16.64 P = 0. RV YES (YES)
117	TEND TO BE THOSE WHERE INTEREST ARTICULATION BY ASSOCIATIONAL GROUPS IS SIGNIFICANT, MODERATE, OR LIMITED (60)	0.70	117	TEND TO BE THOSE WHERE INTEREST ARTICULATION BY ASSOCIATIONAL GROUPS IS NEGLIGIBLE (51)	0.58	M 35 25 15 35 X SQ= 7.72 P = 0.004 RV YES (YES)
119	TEND LESS TO BE THOSE WHERE INTEREST ARTICULATION BY INSTITUTIONAL GROUPS IS VERY SIGNIFICANT OR SIGNIFICANT (74)	0.58	119	TEND MORE TO BE THOSE WHERE INTEREST ARTICULATION BY INSTITUTIONAL GROUPS IS VERY SIGNIFICANT OR SIGNIFICANT (74)	0.88	28 45 20 6 X SQ= 9.93 P = 0.001 RV NO (YES)
120	TEND LESS TO BE THOSE WHERE INTEREST ARTICULATION BY INSTITUTIONAL GROUPS IS VERY SIGNIFICANT, SIGNIFICANT, OR MODERATE (90)	0.81	120	TEND MORE TO BE THOSE WHERE INTEREST ARTICULATION BY INSTITUTIONAL GROUPS IS VERY SIGNIFICANT, SIGNIFICANT, OR MODERATE (90)	0.98	39 50 9 1 X SQ= 5.94 P = 0.007 RV NO (YES)
121	TEND TO BE THOSE WHERE INTEREST ARTICULATION BY NON-ASSOCIATIONAL GROUPS IS MODERATE, LIMITED, OR NEGLIGIBLE (61)	0.77	121	TEND TO BE THOSE WHERE INTEREST ARTICULATION BY NON-ASSOCIATIONAL GROUPS IS SIGNIFICANT (54)	0.66	12 41 40 21 X SQ= 19.38 P = 0. RV YES (YES)
122	TEND TO BE THOSE WHERE INTEREST ARTICULATION BY NON-ASSOCIATIONAL GROUPS IS LIMITED OR NEGLIGIBLE (32)	0.54	122	TEND TO BE THOSE WHERE INTEREST ARTICULATION BY NON-ASSOCIATIONAL GROUPS IS SIGNIFICANT OR MODERATE (83)	0.94	M 24 58 28 4 X SQ= 29.16 P = 0. RV YES (YES)
123	TEND LESS TO BE THOSE WHERE INTEREST ARTICULATION BY NON-ASSOCIATIONAL GROUPS IS SIGNIFICANT, MODERATE, OR LIMITED (107)	0.85	123	ALWAYS ARE THOSE WHERE INTEREST ARTICULATION BY NON-ASSOCIATIONAL GROUPS IS SIGNIFICANT, MODERATE, OR LIMITED (107)	1.00	44 62 8 0 X SQ= 8.04 P = 0.001 RV NO (YES)
125	TEND TO BE THOSE WHERE INTEREST ARTICULATION BY ANOMIC GROUPS IS INFREQUENT OR VERY INFREQUENT (35)	0.61	125	TEND TO BE THOSE WHERE INTEREST ARTICULATION BY ANOMIC GROUPS IS FREQUENT OR OCCASIONAL (64)	0.82	M 16 47 25 10 X SQ= 17.75 P = 0. RV YES (YES)
128	TILT TOWARD BEING THOSE WHERE INTEREST ARTICULATION BY POLITICAL PARTIES IS SIGNIFICANT OR MODERATE (48)	0.61	128	TILT TOWARD BEING THOSE WHERE INTEREST ARTICULATION BY POLITICAL PARTIES IS LIMITED OR NEGLIGIBLE (45)	0.56	27 21 17 27 X SQ= 2.19 P = 0.100 RV YES (YES)

129	TEND TO BE THOSE WHERE INTEREST ARTICULATION BY POLITICAL PARTIES IS SIGNIFICANT, MODERATE, OR LIMITED (56)	0.73	129	TEND TO BE THOSE WHERE INTEREST ARTICULATION BY POLITICAL PARTIES IS NEGLIGIBLE (37)	0.50	32 24 12 24 X SQ= 4.07 P = 0.033 RV YES (YES)
133	TEND MORE TO BE THOSE WHERE INTEREST AGGREGATION BY THE EXECUTIVE IS MODERATE, LIMITED, OR NEGLIGIBLE (76)	0.86	133	TEND LESS TO BE THOSE WHERE INTEREST AGGREGATION BY THE EXECUTIVE IS MODERATE, LIMITED, OR NEGLIGIBLE (76)	0.60	7 22 42 33 X SQ= 7.29 P = 0.004 RV NO (YES)
138	TEND TO BE THOSE WHERE INTEREST AGGREGATION BY THE LEGISLATURE IS SIGNIFICANT, MODERATE, OR LIMITED (48)	0.65	138	TEND TO BE THOSE WHERE INTEREST AGGREGATION BY THE LEGISLATURE IS NEGLIGIBLE (49)	0.64	30 18 16 32 X SQ= 7.05 P = 0.005 RV YES (YES)
139	TEND MORE TO BE THOSE WHERE THE PARTY SYSTEM IS QUANTITATIVELY OTHER THAN ONE-PARTY (71)	0.80	139	TEND LESS TO BE THOSE WHERE THE PARTY SYSTEM IS QUANTITATIVELY OTHER THAN ONE-PARTY (71)	0.58	10 23 39 32 X SQ= 4.54 P = 0.022 RV NO (YES)
141	TILT LESS TOWARD BEING THOSE WHERE THE PARTY SYSTEM IS QUANTITATIVELY OTHER THAN TWO-PARTY (87)	0.83	141	TILT MORE TOWARD BEING THOSE WHERE THE PARTY SYSTEM IS QUANTITATIVELY OTHER THAN TWO-PARTY (87)	0.94	8 3 40 46 X SQ= 1.74 P = 0.121 RV NO (YES)
147	TEND LESS TO BE THOSE WHERE THE PARTY SYSTEM IS QUALITATIVELY OTHER THAN CLASS-ORIENTED OR MULTI-IDEOLOGICAL (67)	0.60	147	TEND MORE TO BE THOSE WHERE THE PARTY SYSTEM IS QUALITATIVELY OTHER THAN CLASS-ORIENTED OR MULTI-IDEOLOGICAL (67)	0.87	17 6 25 41 X SQ= 7.50 P = 0.004 RV NO (YES)
148	LEAN MORE TOWARD BEING THOSE WHERE THE PARTY SYSTEM IS QUALITATIVELY OTHER THAN AFRICAN TRANSITIONAL (96)	0.94	148	LEAN LESS TOWARD BEING THOSE WHERE THE PARTY SYSTEM IS QUALITATIVELY OTHER THAN AFRICAN TRANSITIONAL (96)	0.82	3 11 46 49 X SQ= 2.58 P = 0.084 RV NO (NO)
158	LEAN TOWARD BEING THOSE WHERE PERSONALISMO IS NEGLIGIBLE (56)	0.67	158	LEAN TOWARD BEING THOSE WHERE PERSONALISMO IS PRONOUNCED OR MODERATE (40)	0.50	14 26 29 26 X SQ= 2.27 P = 0.099 RV YES (YES)
163	TILT MORE TOWARD BEING THOSE WHERE THE REGIME'S LEADERSHIP CHARISMA IS MODERATE OR NEGLIGIBLE (87)	0.94	163	TILT LESS TOWARD BEING THOSE WHERE THE REGIME'S LEADERSHIP CHARISMA IS MODERATE OR NEGLIGIBLE (87)	0.82	3 9 46 41 X SQ= 2.26 P = 0.121 RV NO (YES)

164	TEND MORE TO BE THOSE WHERE THE REGIME'S LEADERSHIP CHARISMA IS NEGLIGIBLE (65)	0.82	164	TEND LESS TO BE THOSE WHERE THE REGIME'S LEADERSHIP CHARISMA IS NEGLIGIBLE (65)	0.51	9 24 40 25 X SQ= 8.95 P = 0.002 RV NO (YES)
169	TEND TO BE THOSE WHERE THE HORIZONTAL POWER DISTRIBUTION IS SIGNIFICANT OR LIMITED (58)	0.71	169	TEND TO BE THOSE WHERE THE HORIZONTAL POWER DISTRIBUTION IS NEGLIGIBLE (48)	0.59	35 23 14 33 X SQ= 8.55 P = 0.003 RV YES (YES)
175	TEND TO BE THOSE WHERE THE LEGISLATURE IS FULLY EFFECTIVE OR PARTIALLY EFFECTIVE (51)	0.64	175	TEND TO BE THOSE WHERE THE LEGISLATURE IS LARGELY INEFFECTIVE OR WHOLLY INEFFECTIVE (49)	0.60	30 21 17 31 X SQ= 4.53 P = 0.027 RV YES (YES)
178	LEAN TOWARD BEING THOSE WHERE THE LEGISLATURE IS BICAMERAL (51)	0.60	178	LEAN TOWARD BEING THOSE WHERE THE LEGISLATURE IS UNICAMERAL (53)	0.59	19 33 28 23 X SQ= 2.80 P = 0.076 RV YES (YES)
179	TEND TO BE THOSE WHERE THE EXECUTIVE IS STRONG (39)	0.56	179	TEND TO BE THOSE WHERE THE EXECUTIVE IS DOMINANT (52)	0.68	M 19 32 24 15 X SQ= 4.30 P = 0.033 RV YES (YES)
182	TEND MORE TO BE THOSE WHERE THE BUREAUCRACY IS SEMI-MODERN, RATHER THAN POST-COLONIAL TRANSITIONAL (55)	0.88	182	TEND LESS TO BE THOSE WHERE THE BUREAUCRACY IS SEMI-MODERN, RATHER THAN POST-COLONIAL TRANSITIONAL (55)	0.53	30 24 4 21 X SQ= 9.35 P = 0.001 RV NO (YES)
187	TEND LESS TO BE THOSE WHERE THE ROLE OF THE POLICE IS POLITICALLY SIGNIFICANT (66)	0.53	187	TEND MORE TO BE THOSE WHERE THE ROLE OF THE POLICE IS POLITICALLY SIGNIFICANT (66)	0.75	24 41 21 14 X SQ= 4.01 P = 0.035 RV NO (YES)
188	TEND LESS TO BE THOSE WHERE THE CHARACTER OF THE LEGAL SYSTEM IS OTHER THAN CIVIL LAW (81)	0.58	188	TEND MORE TO BE THOSE WHERE THE CHARACTER OF THE LEGAL SYSTEM IS OTHER THAN CIVIL LAW (81)	0.83	22 10 30 50 X SQ= 7.76 P = 0.003 RV NO (YES)
189	TEND LESS TO BE THOSE WHERE THE CHARACTER OF THE LEGAL SYSTEM IS OTHER THAN COMMON LAW (108)	0.88	189	TEND MORE TO BE THOSE WHERE THE CHARACTER OF THE LEGAL SYSTEM IS OTHER THAN COMMON LAW (108)	0.98	6 1 46 61 X SQ= 3.27 P = 0.046 RV NO (YES)

```
192  TEND MORE TO BE THOSE                     0.88      192  TEND LESS TO BE THOSE                     0.65          6    21
     WHERE THE CHARACTER OF THE LEGAL SYSTEM                  WHERE THE CHARACTER OF THE LEGAL SYSTEM                46    39
     IS OTHER THAN                                            IS OTHER THAN                                     X SQ=    7.15
     MUSLIM OR PARTLY MUSLIM  (86)                            MUSLIM OR PARTLY MUSLIM  (86)                     P =      0.004
                                                                                                                RV NO   (YES)

193  TEND MORE TO BE THOSE                     0.94      193  TEND LESS TO BE THOSE                     0.59          3    25
     WHERE THE CHARACTER OF THE LEGAL SYSTEM                  WHERE THE CHARACTER OF THE LEGAL SYSTEM                49    36
     IS OTHER THAN PARTLY INDIGENOUS  (86)                    IS OTHER THAN PARTLY INDIGENOUS  (86)             X SQ=   16.83
                                                                                                                P =      0.
                                                                                                                RV NO   (YES)
```

MATRIX

68 POLITIES THAT ARE LINGUISTICALLY HOMOGENEOUS (52)	68 POLITIES THAT ARE LINGUISTICALLY WEAKLY HETEROGENEOUS OR STRONGLY HETEROGENEOUS (62)
42 POLITIES WHOSE ECONOMIC DEVELOPMENTAL STATUS IS DEVELOPED OR INTERMEDIATE (36)	
ARGENTINA, DENMARK, HUNGARY, LUXEMBOURG, POLAND, AUSTRALIA, FRANCE, ICELAND, MEXICO, SWEDEN, AUSTRIA, GERMANY, E, IRELAND, NETHERLANDS, UK, BRAZIL, GERMAN FR, ITALY, NEW ZEALAND, US, COLOMBIA, GREECE, JAPAN, NORWAY, URUGUAY	BELGIUM, FINLAND, YUGOSLAVIA, BULGARIA, RUMANIA, CANADA, SO AFRICA, CYPRUS, SWITZERLAND, CZECHOS'KIA, USSR
25	11
25	50
42 POLITIES WHOSE ECONOMIC DEVELOPMENTAL STATUS IS UNDERDEVELOPED OR VERY UNDERDEVELOPED (76)	
ALBANIA, DOMIN REP, JORDAN, MALAGASY R, RWANDA, BURUNDI, EL SALVADOR, KOREA, N, MONGOLIA, SOMALIA, CHILE, HAITI, KOREA REP, NICARAGUA, TUNISIA, COSTA RICA, HONDURAS, LEBANON, PARAGUAY, UAR, CUBA, JAMAICA, LIBYA, PORTUGAL, YEMEN	AFGHANISTAN, CAMEROUN, CONGO(LEO), GHANA, IRAN, MALAYA, NIGER, PHILIPPINES, SYRIA, TURKEY, ALGERIA, CEN AFR REP, DAHOMEY, GUATEMALA, IRAQ, MALI, NIGERIA, SENEGAL, TANGANYIKA, UGANDA, BOLIVIA, CEYLON, ECUADOR, GUINEA, IVORY COAST, MAURITANIA, PAKISTAN, SIERRE LEO, THAILAND, UPPER VOLTA, BURMA, CHAD, ETHIOPIA, INDIA, LAOS, MOROCCO, PANAMA, SPAIN, TOGO, VIETNAM, N, CAMBODIA, CONGO(BRA), GABON, INDONESIA, LIBERIA, NEPAL, PERU, SUDAN, TRINIDAD, VIETNAM REP

68/ 77

M

MATRIX

	68 POLITIES THAT ARE LINGUISTICALLY HOMOGENEOUS (52)	68 POLITIES THAT ARE LINGUISTICALLY WEAKLY HETEROGENEOUS OR STRONGLY HETEROGENEOUS (62)
77 POLITIES THAT HAVE BEEN SIGNIFICANTLY WESTERNIZED, RATHER THAN PARTIALLY WESTERNIZED, THROUGH A COLONIAL RELATIONSHIP (28)	ARGENTINA, CUBA, LEBANON, URUGUAY, BRAZIL, DOMIN REP, MEXICO, VENEZUELA, CHILE, EL SALVADOR, NICARAGUA, COLOMBIA, HONDURAS, PARAGUAY, COSTA RICA, JAMAICA, UAR 17	ALGERIA, GUATEMALA, TRINIDAD, BOLIVIA, INDIA, CEYLON, PANAMA, CYPRUS, PERU, ECUADOR, PHILIPPINES 11
77 POLITIES THAT HAVE BEEN PARTIALLY WESTERNIZED, RATHER THAN SIGNIFICANTLY WESTERNIZED, THROUGH A COLONIAL RELATIONSHIP (41)	8 BURUNDI, RWANDA, HAITI, SOMALIA, JORDAN, TUNISIA, LIBYA, MALAGASY R	33 BURMA, CONGO(BRA), GUINEA, LAOS, NIGER, SUDAN, UPPER VOLTA, CAMBODIA, CONGO(LEO), INDONESIA, MALAYA, NIGERIA, SYRIA, VIETNAM, N, CAMEROUN, DAHOMEY, IRAN, MALI, PAKISTAN, TANGANYIKA, VIETNAM REP, CEN AFR REP, GABON, IRAQ, MAURITANIA, SENEGAL, TOGO, CHAD, GHANA, IVORY COAST, MOROCCO, SIERRE LEO, UGANDA

68/ 85

M

MATRIX

	68 POLITIES THAT ARE LINGUISTICALLY HOMOGENEOUS (52)	68 POLITIES THAT ARE LINGUISTICALLY WEAKLY HETEROGENEOUS OR STRONGLY HETEROGENEOUS (62)
85 POLITIES WHERE THE STAGE OF POLITICAL MODERNIZATION IS ADVANCED, RATHER THAN MID- OR EARLY TRANSITIONAL (60)	37	22
85 POLITIES WHERE THE STAGE OF POLITICAL MODERNIZATION IS MID- OR EARLY TRANSITIONAL, RATHER THAN ADVANCED (54)	14	40

Linguistically homogeneous; advanced (37):

ALBANIA, CHILE, GERMANY, E, IRELAND, KOREA REP, NETHERLANDS, SWEDEN, URUGUAY, ARGENTINA, COSTA RICA, GERMAN FR, ITALY, LEBANON, NEW ZEALAND, TUNISIA, VENEZUELA, AUSTRALIA, CUBA, GREECE, JAPAN, LUXEMBOURG, NORWAY, UAR, AUSTRIA, DENMARK, HUNGARY, JORDAN, MEXICO, POLAND, UK, BRAZIL, FRANCE, ICELAND, KOREA, N, MONGOLIA, PORTUGAL, US

Linguistically heterogeneous; advanced (22):

.ALGERIA, .CZECHOS'KIA, .ISRAEL, .SPAIN, .USSR, BELGIUM, FINLAND, MOROCCO, SUDAN, YUGOSLAVIA, BULGARIA, GUATEMALA, PAKISTAN, SWITZERLAND, CANADA, INDIA, PHILIPPINES, SYRIA, CYPRUS, IRAQ, RUMANIA, TURKEY

Linguistically homogeneous; mid- or early transitional (14):

BURUNDI, HONDURAS, PARAGUAY, COLOMBIA, JAMAICA, RWANDA, DOMIN REP, LIBYA, SOMALIA, EL SALVADOR, MALAGASY R, YEMEN, HAITI, NICARAGUA

Linguistically heterogeneous; mid- or early transitional (40):

.AFGHANISTAN, .CEN AFR REP, .DAHOMEY, .GUINEA, .LIBERIA, .NIGER, .SIERRE LEO, .TRINIDAD, BOLIVIA, CEYLON, ECUADOR, INDONESIA, MALAYA, NIGERIA, SO AFRICA, UGANDA, BURMA, CHAD, ETHIOPIA, IRAN, MALI, PANAMA, TANGANYIKA, UPPER VOLTA, CAMBODIA, CONGO(BRA), GABON, IVORY COAST, MAURITANIA, PERU, THAILAND, VIETNAM, N, CAMEROUN, CONGO(LEO), GHANA, LAOS, NEPAL, SENEGAL, TOGO, VIETNAM REP

M

MATRIX

	68 POLITIES THAT ARE LINGUISTICALLY HOMOGENEOUS (52)	68 POLITIES THAT ARE LINGUISTICALLY WEAKLY HETEROGENEOUS OR STRONGLY HETEROGENEOUS (62)
107 POLITIES WHERE AUTONOMOUS GROUPS ARE FULLY TOLERATED IN POLITICS (46)	ARGENTINA, COLOMBIA, GERMAN FR, ITALY, MALAGASY R, UK, AUSTRALIA, COSTA RICA, GREECE, JAMAICA, NETHERLANDS, US, AUSTRIA, DENMARK, HONDURAS, JAPAN, NEW ZEALAND, URUGUAY, BRAZIL, DOMIN REP, ICELAND, LEBANON, NORWAY, VENEZUELA, CHILE, FRANCE, IRELAND, LUXEMBOURG, SWEDEN 29	BELGIUM, ECUADOR, PANAMA, TURKEY, BOLIVIA, FINLAND, PHILIPPINES, UGANDA, CANADA, INDIA, SIERRE LEO, CEYLON, ISRAEL, SWITZERLAND, CYPRUS, MALAYA, TRINIDAD 17
107 POLITIES WHERE AUTONOMOUS GROUPS ARE PARTIALLY TOLERATED IN POLITICS, ARE TOLERATED ONLY OUTSIDE POLITICS, OR ARE NOT TOLERATED AT ALL (65)	22 ALBANIA, HAITI, LIBYA, POLAND, TUNISIA, BURUNDI, HUNGARY, MEXICO, PORTUGAL, UAR, CUBA, JORDAN, MONGOLIA, RWANDA, EL SALVADOR, KOREA, N, NICARAGUA, SA'U ARABIA, GERMANY, E, KOREA REP, PARAGUAY, SOMALIA	42 AFGHANISTAN, CAMEROUN, CZECHOS'KIA, GUATEMALA, IVORY COAST, NEPAL, SENEGAL, TANGANYIKA, VIETNAM REP, ALGERIA, CEN AFR REP, DAHOMEY, GUINEA, LIBERIA, NIGER, SO AFRICA, THAILAND, YUGOSLAVIA, BULGARIA, CHAD, ETHIOPIA, INDONESIA, MALI, NIGERIA, SPAIN, USSR, BURMA, CONGO(BRA), GABON, IRAN, MAURITANIA, PAKISTAN, SUDAN, UPPER VOLTA, CAMBODIA, CONGO(LEO), GHANA, IRAQ, MOROCCO, RUMANIA, SYRIA, VIETNAM, N

MATRIX

	68 POLITIES THAT ARE LINGUISTICALLY HOMOGENEOUS (52)	68 POLITIES THAT ARE LINGUISTICALLY WEAKLY HETEROGENEOUS OR STRONGLY HETEROGENEOUS (62)
117 POLITIES WHERE INTEREST ARTICULATION BY ASSOCIATIONAL GROUPS IS SIGNIFICANT, MODERATE, OR LIMITED (60)	35	25
117 POLITIES WHERE INTEREST ARTICULATION BY ASSOCIATIONAL GROUPS IS NEGLIGIBLE (51)	15	35

Linguistically homogeneous; interest articulation significant, moderate, or limited (35):

ARGENTINA, COLOMBIA, FRANCE, ICELAND, LEBANON, NICARAGUA, UAR, AUSTRALIA, COSTA RICA, GERMAN FR, IRELAND, LUXEMBOURG, NORWAY, UK, AUSTRIA, DENMARK, GREECE, ITALY, MEXICO, POLAND, US, BRAZIL, DOMIN REP, HONDURAS, JAMAICA, NETHERLANDS, SWEDEN, URUGUAY, CHILE, EL SALVADOR, HUNGARY, JAPAN, NEW ZEALAND, TUNISIA, VENEZUELA

Linguistically heterogeneous; interest articulation significant, moderate, or limited (25):

ALGERIA, CEYLON, INDIA, MOROCCO, SWITZERLAND, BELGIUM, CYPRUS, INDONESIA, PANAMA, SYRIA, BOLIVIA, ECUADOR, IRAN, PERU, TRINIDAD, BURMA, FINLAND, ISRAEL, PHILIPPINES, TURKEY, CANADA, GUATEMALA, MALAYA, SO AFRICA, USSR

Linguistically homogeneous; interest articulation negligible (15):

ALBANIA, JORDAN, PARAGUAY, BURUNDI, KOREA, N, RWANDA, CUBA, LIBYA, SA'U ARABIA, GERMANY, E, MALAGASY R, SOMALIA, HAITI, MONGOLIA, YEMEN

Linguistically heterogeneous; interest articulation negligible (35):

AFGHANISTAN, CHAD, ETHIOPIA, IVORY COAST, NEPAL, SENEGAL, TOGO, BULGARIA, CONGO(BRA), GABON, LAOS, NIGER, SIERRE LEO, UGANDA, CAMBODIA, CONGO(LEO), GHANA, LIBERIA, NIGERIA, SUDAN, UPPER VOLTA, CAMEROUN, CZECHOS'KIA, GUINEA, MALI, PAKISTAN, TANGANYIKA, VIETNAM, N, CEN AFR REP, DAHOMEY, IRAQ, MAURITANIA, RUMANIA, THAILAND, VIETNAM REP

M

```
*****************************************************MATRIX*****************************************************

        68  POLITIES                                                      68  POLITIES
            THAT ARE LINGUISTICALLY                   .                       THAT ARE LINGUISTICALLY
            HOMOGENEOUS  (52)                         .                       WEAKLY HETEROGENEOUS OR
                                                      .                       STRONGLY HETEROGENEOUS  (62)

                          122  POLITIES
                               WHERE INTEREST ARTICULATION
                               BY NON-ASSOCIATIONAL GROUPS
                               IS SIGNIFICANT OR MODERATE  (83)

ALBANIA     BRAZIL      BURUNDI     EL SALVADOR HAITI       .AFGHANISTAN ALGERIA     BELGIUM     BOLIVIA     BURMA
HONDURAS    JAMAICA     JAPAN       JORDAN      KOREA, N    .CAMBODIA    CAMEROUN    CANADA      CEN AFR REP CEYLON
KOREA REP   LEBANON     LIBYA       MALAGASY R  MEXICO      .CHAD        CONGO(BRA)  CONGO(LEO)  CYPRUS      DAHOMEY
MONGOLIA    NICARAGUA   POLAND      RWANDA      SA'U ARABIA.ECUADOR      ETHIOPIA    GABON       GHANA       GUATEMALA
SOMALIA     TUNISIA     UAR         YEMEN                   .GUINEA      INDIA       INDONESIA   IRAN        IRAQ
                                                            .ISRAEL      IVORY COAST LAOS        LIBERIA     MALAYA
                                                            .MALI        MAURITANIA  MOROCCO     NEPAL       NIGER
                                                            .NIGERIA     PAKISTAN    PERU        PHILIPPINES RUMANIA
                                                            .SENEGAL     SIERRE LEO  SO AFRICA   SPAIN       SUDAN
                                                            .SWITZERLAND SYRIA       TANGANYIKA  THAILAND    TOGO
                                                            .TRINIDAD    TURKEY      UGANDA      USSR        UPPER VOLTA
                                                            .VIETNAM, N  VIETNAM REP YUGOSLAVIA
                                                          24 .  58
. . . . . . . . . . . . . . . . . . . . . . . . . . . . . . .:. . . . . . . . . . . . . . . . . . . . . . . . . . . .
                                                          28 .   4

                          122  POLITIES
                               WHERE INTEREST ARTICULATION
                               BY NON-ASSOCIATIONAL GROUPS
                               IS LIMITED OR NEGLIGIBLE  (32)

ARGENTINA   AUSTRALIA   AUSTRIA     CHILE       COLOMBIA    .BULGARIA    CZECHOS'KIA FINLAND     PANAMA
CUBA        COSTA RICA  DENMARK     DOMIN REP   FRANCE      .
GERMANY, E  GERMAN FR   GREECE      HUNGARY     ICELAND     .
IRELAND     ITALY       LUXEMBOURG  NETHERLANDS NEW ZEALAND.
NORWAY      PARAGUAY    PORTUGAL    SWEDEN      UK          .
US          URUGUAY     VENEZUELA                           .
```

M

MATRIX

	68 POLITIES THAT ARE LINGUISTICALLY HOMOGENEOUS (52)	68 POLITIES THAT ARE LINGUISTICALLY WEAKLY HETEROGENEOUS OR STRONGLY HETEROGENEOUS (62)
125 POLITIES WHERE INTEREST ARTICULATION BY ANOMIC GROUPS IS FREQUENT OR OCCASIONAL (64)	16	47
125 POLITIES WHERE INTEREST ARTICULATION BY ANOMIC GROUPS IS INFREQUENT OR VERY INFREQUENT (35)	25	10

Linguistically homogeneous; anomic interest articulation frequent or occasional (16):

BRAZIL, COLOMBIA, EL SALVADOR, GERMANY, E, HAITI, HONDURAS, JAPAN, JORDAN, KOREA REP, LEBANON, MEXICO, PARAGUAY, POLAND, RWANDA, SA'U ARABIA, SOMALIA

Linguistically heterogeneous; anomic interest articulation frequent or occasional (47):

AFGHANISTAN, BOLIVIA, BURMA, CAMBODIA, CAMEROUN, CEN AFR REP, CEYLON, CHAD, CONGO(BRA), CONGO(LEO), DAHOMEY, ECUADOR, ETHIOPIA, GABON, GHANA, GUATEMALA, GUINEA, INDIA, INDONESIA, IRAN, IRAQ, IVORY COAST, LAOS, MALAYA, MALI, MAURITANIA, MOROCCO, NEPAL, NIGER, NIGERIA, PAKISTAN, PANAMA, PERU, SENEGAL, SIERRE LEO, SO AFRICA, SPAIN, SUDAN, SYRIA, THAILAND, TOGO, TURKEY, UGANDA, USSR, UPPER VOLTA, VIETNAM REP, YUGOSLAVIA

Linguistically homogeneous; anomic interest articulation infrequent or very infrequent (25):

AUSTRALIA, AUSTRIA, BURUNDI, CHILE, COSTA RICA, DENMARK, GERMAN FR, GREECE, ICELAND, IRELAND, ITALY, JAMAICA, LIBYA, LUXEMBOURG, MALAGASY R, NETHERLANDS, NEW ZEALAND, NICARAGUA, NORWAY, PORTUGAL, SWEDEN, TUNISIA, UAR, UK, URUGUAY

Linguistically heterogeneous; anomic interest articulation infrequent or very infrequent (10):

BELGIUM, CANADA, CZECHOS'KIA, FINLAND, ISRAEL, LIBERIA, PHILIPPINES, SWITZERLAND, TANGANYIKA, TRINIDAD

M

MATRIX

	68 POLITIES THAT ARE LINGUISTICALLY HOMOGENEOUS (52)	68 POLITIES THAT ARE LINGUISTICALLY WEAKLY HETEROGENEOUS OR STRONGLY HETEROGENEOUS (62)
179 POLITIES WHERE THE EXECUTIVE IS DOMINANT (52)	ALBANIA, HAITI, MEXICO, PORTUGAL, CUBA, HONDURAS, MONGOLIA, SA'U ARABIA, EL SALVADOR, HUNGARY, NICARAGUA, TUNISIA, FRANCE, JORDAN, PARAGUAY, UAR, GERMANY, E, KOREA, N, POLAND	AFGHANISTAN, CONGO(BRA), GHANA, IVORY COAST, NEPAL, SPAIN, VIETNAM REP, ALGERIA, CZECHOS'KIA, GUINEA, LIBERIA, PAKISTAN, TANGANYIKA, YUGOSLAVIA, BULGARIA, DAHOMEY, INDIA, MALI, RUMANIA, USSR, CAMBODIA, ETHIOPIA, INDONESIA, MAURITANIA, SENEGAL, UPPER VOLTA, CEN AFR REP, GABON, IRAN, MOROCCO, SO AFRICA, VIETNAM, N
	19	32
	24	15
179 POLITIES WHERE THE EXECUTIVE IS STRONG (39)	AUSTRALIA, DENMARK, ITALY, LUXEMBOURG, SWEDEN, AUSTRIA, DOMIN REP, JAMAICA, MALAGASY R, UK, BRAZIL, GERMAN FR, JAPAN, NETHERLANDS, US, COLOMBIA, ICELAND, LEBANON, NEW ZEALAND, VENEZUELA, COSTA RICA, IRELAND, LIBYA, NORWAY	BELGIUM, ECUADOR, PANAMA, BOLIVIA, FINLAND, PHILIPPINES, CAMEROUN, ISRAEL, SIERRE LEO, CANADA, MALAYA, TRINIDAD, CYPRUS, NIGERIA, TURKEY

69 POLITIES
THAT ARE LINGUISTICALLY
HOMOGENEOUS OR
WEAKLY HETEROGENEOUS (64)

69 POLITIES
THAT ARE LINGUISTICALLY
STRONGLY HETEROGENEOUS (50)

PREDICATE ONLY

64 IN LEFT COLUMN

ALBANIA	ARGENTINA	AUSTRALIA	AUSTRIA	BRAZIL	BULGARIA	BURUNDI	CAMBODIA	CHILE	COLOMBIA
COSTA RICA	CUBA	DENMARK	DOMIN REP	EL SALVADOR	FINLAND	FRANCE	GERMANY, E	GERMAN FR	GREECE
HAITI	HONDURAS	HUNGARY	ICELAND	IRELAND	ITALY	JAMAICA	JAPAN	JORDAN	KOREA, N
KOREA REP	LEBANON	LIBYA	LUXEMBOURG	MALAGASY R	MEXICO	MONGOLIA	NETHERLANDS	NEW ZEALAND	NICARAGUA
NORWAY	PANAMA	PARAGUAY	POLAND	PORTUGAL	RUMANIA	RWANDA	SA'U ARABIA	SOMALIA	SPAIN
SWEDEN	SYRIA	THAILAND	TRINIDAD	TUNISIA	TURKEY	UAR	UK	US	URUGUAY
VENEZUELA	VIETNAM, N	VIETNAM REP	YEMEN						

50 IN RIGHT COLUMN

AFGHANISTAN	ALGERIA	BELGIUM	BOLIVIA	BURMA	CAMEROUN	CANADA	CEN AFR REP	CEYLON	CHAD
CONGO(BRA)	CONGO(LEO)	CYPRUS	CZECHOS'KIA	DAHOMEY	ECUADOR	ETHIOPIA	GABON	GHANA	GUATEMALA
GUINEA	INDIA	INDONESIA	IRAN	IRAQ	ISRAEL	IVORY COAST	LAOS	LIBERIA	MALAYA
MALI	MAURITANIA	MOROCCO	NEPAL	NIGER	NIGERIA	PAKISTAN	PERU	PHILIPPINES	SENEGAL
SIERRE LEO	SO AFRICA	SUDAN	SWITZERLAND	TANGANYIKA	TOGO	UGANDA	USSR	UPPER VOLTA	YUGOSLAVIA

1 EXCLUDED BECAUSE AMBIGUOUS

CHINA, PR

```
**************************************************************************************************

70  POLITIES                                               70  POLITIES
    THAT ARE RELIGIOUSLY, RACIALLY,                            THAT ARE RELIGIOUSLY, RACIALLY,
    AND LINGUISTICALLY HOMOGENEOUS  (21)                       OR LINGUISTICALLY HETEROGENEOUS  (85)


                                                                          BOTH SUBJECT AND PREDICATE
..................................................................................................
```

```
      21 IN LEFT COLUMN

ARGENTINA   AUSTRIA     COSTA RICA  DENMARK     FRANCE      GREECE      ICELAND     IRELAND     ITALY       JORDAN
LIBYA       LUXEMBOURG  MONGOLIA    NORWAY      POLAND      PORTUGAL    SOMALIA     SWEDEN      TUNISIA     UAR
URUGUAY

      85 IN RIGHT COLUMN

AFGHANISTAN ALBANIA     ALGERIA     AUSTRALIA   BELGIUM     BOLIVIA     BRAZIL      BULGARIA    BURMA       CAMBODIA
CAMEROUN    CANADA      CEN AFR REP CEYLON      CHAD        COLOMBIA    CONGO(BRA)  CONGO(LEO)  CUBA        CYPRUS
CZECHOS'KIA DAHOMEY     DOMIN REP   ECUADOR     EL SALVADOR ETHIOPIA    FINLAND     GABON       GERMANY, E  GERMAN FR
GHANA       GUATEMALA   GUINEA      HAITI       HONDURAS    HUNGARY     INDIA       INDONESIA   IRAN        IRAQ
ISRAEL      IVORY COAST JAMAICA     JAPAN       LAOS        LEBANON     LIBERIA     MALAGASY R  MALAYA      MALI
MAURITANIA  MEXICO      MOROCCO     NEPAL       NETHERLANDS NEW ZEALAND NICARAGUA   NIGER       NIGERIA     PAKISTAN
PANAMA      PERU        PHILIPPINES RUMANIA     SENEGAL     SIERRE LEO  SO AFRICA   SPAIN       SUDAN       SWITZERLAND
SYRIA       TANGANYIKA  THAILAND    TOGO        TRINIDAD    TURKEY      UGANDA      USSR        UK          US
UPPER VOLTA VENEZUELA   VIETNAM, N  VIETNAM REP YUGOSLAVIA

      3 EXCLUDED BECAUSE AMBIGUOUS

CHILE       KOREA REP   PARAGUAY

      6 EXCLUDED BECAUSE INSUFFICIENTLY ASCERTAINED

BURUNDI     CHINA, PR   KOREA, N    RWANDA      SA'U ARABIA YEMEN
```

```
--------------------------------------------------------------------------------------------------

 2  TEND TO BE THOSE                        0.52     2  TEND TO BE THOSE                     0.87       11    11
    LOCATED IN WEST EUROPE, SCANDINAVIA,                LOCATED ELSEWHERE THAN IN                       10    73
    NORTH AMERICA, OR AUSTRALASIA  (22)                 WEST EUROPE, SCANDINAVIA,                  X SQ=  13.37
                                                        NORTH AMERICA, OR AUSTRALASIA  (93)        P =    0.
                                                                                                   RV YES (YES)

 3  TEND LESS TO BE THOSE                   0.67     3  TEND MORE TO BE THOSE                0.93        7     6
    LOCATED ELSEWHERE THAN IN                           LOCATED ELSEWHERE THAN IN                       14    78
    WEST EUROPE  (102)                                  WEST EUROPE  (102)                         X SQ=   8.35
                                                                                                   P =    0.004
                                                                                                   RV NO  (YES)

 5  TEND TO BE THOSE                        0.92     5  TEND TO BE THOSE                     0.53       11     7
    LOCATED IN SCANDINAVIA OR WEST EUROPE,              LOCATED IN EAST EUROPE, RATHER THAN              1     8
    RATHER THAN IN EAST EUROPE  (18)                    IN SCANDINAVIA OR WEST EUROPE  (9)         X SQ=   4.22
                                                                                                   P =    0.019
                                                                                                   RV YES (YES)
```

11	TEND MORE TO BE THOSE LOCATED ELSEWHERE THAN IN CENTRAL AND SOUTH AFRICA (87)	0.95	11	TEND LESS TO BE THOSE LOCATED ELSEWHERE THAN IN CENTRAL AND SOUTH AFRICA (87)	0.71	1 24 20 60 X SQ= 4.02 P = 0.022 RV NO (NO)
28	TEND TO BE THOSE WHOSE POPULATION GROWTH RATE IS LOW (48)	0.67	28	TEND TO BE THOSE WHOSE POPULATION GROWTH RATE IS HIGH (62)	0.61	7 49 14 31 X SQ= 4.18 P = 0.028 RV YES (NO)
29	TEND TO BE THOSE WHERE THE DEGREE OF URBANIZATION IS HIGH (56)	0.89	29	TEND TO BE THOSE WHERE THE DEGREE OF URBANIZATION IS LOW (49)	0.53	17 37 2 42 X SQ= 9.60 P = 0.001 RV YES (NO)
30	TEND TO BE THOSE WHOSE AGRICULTURAL POPULATION IS MEDIUM, LOW, OR VERY LOW (57)	0.81	30	TEND TO BE THOSE WHOSE AGRICULTURAL POPULATION IS HIGH (56)	0.54	4 45 17 38 X SQ= 6.97 P = 0.006 RV YES (NO)
36	TEND TO BE THOSE WHOSE PER CAPITA GROSS NATIONAL PRODUCT IS VERY HIGH, HIGH, OR MEDIUM (42)	0.62	36	TEND TO BE THOSE WHOSE PER CAPITA GROSS NATIONAL PRODUCT IS LOW OR VERY LOW (73)	0.67	13 28 8 56 X SQ= 4.62 P = 0.024 RV YES (NO)
42	TEND TO BE THOSE WHOSE ECONOMIC DEVELOPMENTAL STATUS IS DEVELOPED OR INTERMEDIATE (36)	0.62	42	TEND TO BE THOSE WHOSE ECONOMIC DEVELOPMENTAL STATUS IS UNDERDEVELOPED OR VERY UNDERDEVELOPED (76)	0.72	13 23 8 59 X SQ= 7.01 P = 0.009 RV YES (NO)
43	TEND TO BE THOSE WHOSE ECONOMIC DEVELOPMENTAL STATUS IS DEVELOPED, INTERMEDIATE, OR UNDERDEVELOPED (55)	0.75	43	TEND TO BE THOSE WHOSE ECONOMIC DEVELOPMENTAL STATUS IS VERY UNDERDEVELOPED (57)	0.54	15 38 5 45 X SQ= 4.40 P = 0.025 RV YES (NO)
45	TEND TO BE THOSE WHERE THE LITERACY RATE IS FIFTY PERCENT OR ABOVE (55)	0.75	45	TEND TO BE THOSE WHERE THE LITERACY RATE IS BELOW FIFTY PERCENT (54)	0.54	15 37 5 44 X SQ= 4.41 P = 0.024 RV YES (NO)
54	TEND TO BE THOSE WHERE NEWSPAPER CIRCULATION IS ONE HUNDRED OR MORE PER THOUSAND (37)	0.71	54	TEND TO BE THOSE WHERE NEWSPAPER CIRCULATION IS LESS THAN ONE HUNDRED PER THOUSAND (76)	0.75	15 21 6 62 X SQ= 13.78 P = 0. RV YES (NO)

56	ALWAYS ARE THOSE WHERE THE RELIGION IS PREDOMINANTLY LITERATE (79)	1.00	56	TEND LESS TO BE THOSE WHERE THE RELIGION IS PREDOMINANTLY LITERATE (79)	0.66	21 54 0 28 X SQ= 8.20 P = 0.001 RV NO (NO)
63	TEND TO BE THOSE WHERE THE RELIGION IS PREDOMINANTLY SOME KIND OF CHRISTIAN (46)	0.71	63	TEND TO BE THOSE WHERE THE RELIGION IS PREDOMINANTLY OR PARTLY OTHER THAN CHRISTIAN (68)	0.65	15 29 6 54 X SQ= 7.71 P = 0.003 RV YES (NO)
73	LEAN MORE TOWARD BEING THOSE WHOSE DATE OF INDEPENDENCE IS BEFORE 1945 (65)	0.76	73	LEAN LESS TOWARD BEING THOSE WHOSE DATE OF INDEPENDENCE IS BEFORE 1945 (65)	0.54	16 44 5 38 X SQ= 2.63 P = 0.083 RV NO (NO)
74	TEND TO BE THOSE THAT ARE HISTORICALLY WESTERN (26)	0.57	74	TEND TO BE THOSE THAT ARE NOT HISTORICALLY WESTERN (87)	0.83	12 14 9 68 X SQ= 12.18 P = 0. RV YES (NO)
75	LEAN MORE TOWARD BEING THOSE THAT ARE HISTORICALLY WESTERN OR SIGNIFICANTLY WESTERNIZED (62)	0.76	75	LEAN LESS TOWARD BEING THOSE THAT ARE HISTORICALLY WESTERN OR SIGNIFICANTLY WESTERNIZED (62)	0.52	16 43 5 40 X SQ= 3.13 P = 0.051 RV NO (NO)
76	TEND TO BE THOSE THAT ARE HISTORICALLY WESTERN, RATHER THAN HAVING BEEN SIGNIFICANTLY OR PARTIALLY WESTERNIZED THROUGH A COLONIAL RELATIONSHIP (26)	0.60	76	TEND TO BE THOSE THAT HAVE BEEN SIGNIFICANTLY OR PARTIALLY WESTERNIZED THROUGH A COLONIAL RELATIONSHIP, RATHER THAN BEING HISTORICALLY WESTERN (70)	0.80	12 14 8 57 X SQ= 10.51 P = 0.001 RV YES (NO)
82	TEND TO BE THOSE WHERE THE TYPE OF POLITICAL MODERNIZATION IS EARLY OR LATER EUROPEAN OR EUROPEAN DERIVED, RATHER THAN DEVELOPED TUTELARY OR UNDEVELOPED TUTELARY (51)	0.71	82	TEND TO BE THOSE WHERE THE TYPE OF POLITICAL MODERNIZATION IS DEVELOPED TUTELARY OR UNDEVELOPED TUTELARY, RATHER THAN EARLY OR LATER EUROPEAN OR EUROPEAN DERIVED (55)	0.55	15 34 6 42 X SQ= 3.68 P = 0.047 RV YES (NO)
85	TEND TO BE THOSE WHERE THE STAGE OF POLITICAL MODERNIZATION IS ADVANCED, RATHER THAN MID- OR EARLY TRANSITIONAL (60)	0.90	85	TEND TO BE THOSE WHERE THE STAGE OF POLITICAL MODERNIZATION IS MID- OR EARLY TRANSITIONAL, RATHER THAN ADVANCED (54)	0.56	19 37 2 47 X SQ= 12.74 P = 0. RV YES (NO)
87	LEAN MORE TOWARD BEING THOSE WHOSE IDEOLOGICAL ORIENTATION IS OTHER THAN DEVELOPMENTAL (58)	0.84	87	LEAN LESS TOWARD BEING THOSE WHOSE IDEOLOGICAL ORIENTATION IS OTHER THAN DEVELOPMENTAL (58)	0.59	3 26 16 38 X SQ= 2.96 P = 0.057 RV NO (NO)

89	TEND TO BE THOSE WHOSE IDEOLOGICAL ORIENTATION IS CONVENTIONAL (33)	0.67	89	TEND TO BE THOSE WHOSE IDEOLOGICAL ORIENTATION IS OTHER THAN CONVENTIONAL (62)	0.71	12 20 6 50 X SQ= 7.41 P = 0.005 RV YES (NO)
94	TILT MORE TOWARD BEING THOSE WHERE THE STATUS OF THE REGIME IS CONSTITUTIONAL (51)	0.74	94	TILT LESS TOWARD BEING THOSE WHERE THE STATUS OF THE REGIME IS CONSTITUTIONAL (51)	0.53	14 35 5 31 X SQ= 1.80 P = 0.123 RV NO (NO)
96	TILT MORE TOWARD BEING THOSE WHERE THE STATUS OF THE REGIME IS CONSTITUTIONAL OR TOTALITARIAN (67)	0.89	96	TILT LESS TOWARD BEING THOSE WHERE THE STATUS OF THE REGIME IS CONSTITUTIONAL OR TOTALITARIAN (67)	0.72	17 46 2 18 X SQ= 1.61 P = 0.139 RV NO (NO)
99	TEND MORE TO BE THOSE WHERE GOVERNMENTAL STABILITY IS GENERALLY PRESENT AND DATES FROM AT LEAST THE INTER-WAR PERIOD, OR FROM THE POST-WAR PERIOD (50)	0.80	99	TEND LESS TO BE THOSE WHERE GOVERNMENTAL STABILITY IS GENERALLY PRESENT AND DATES FROM AT LEAST THE INTER-WAR PERIOD, OR FROM THE POST-WAR PERIOD (50)	0.51	16 31 4 30 X SQ= 4.14 P = 0.035 RV NO (NO)
100	LEAN MORE TOWARD BEING THOSE WHERE GOVERNMENTAL STABILITY IS GENERALLY PRESENT AND DATES FROM AT LEAST THE INTER-WAR PERIOD, OR IS GENERALLY OR MODERATELY PRESENT, AND DATES FROM THE POST-WAR PERIOD (64)	0.90	100	LEAN LESS TOWARD BEING THOSE WHERE GOVERNMENTAL STABILITY IS GENERALLY PRESENT AND DATES FROM AT LEAST THE INTER-WAR PERIOD, OR IS GENERALLY OR MODERATELY PRESENT, AND DATES FROM THE POST-WAR PERIOD (64)	0.68	18 41 2 19 X SQ= 2.60 P = 0.079 RV NO (NO)
101	TEND TO BE THOSE WHERE THE REPRESENTATIVE CHARACTER OF THE REGIME IS POLYARCHIC (41)	0.72	101	TEND TO BE THOSE WHERE THE REPRESENTATIVE CHARACTER OF THE REGIME IS LIMITED POLYARCHIC, PSEUDO-POLYARCHIC, OR NON-POLYARCHIC (57)	0.64	13 27 5 47 X SQ= 6.14 P = 0.008 RV YES (NO)
102	TEND TO BE THOSE WHERE THE REPRESENTATIVE CHARACTER OF THE REGIME IS POLYARCHIC OR LIMITED POLYARCHIC (59)	0.76	102	TEND TO BE THOSE WHERE THE REPRESENTATIVE CHARACTER OF THE REGIME IS PSEUDO-POLYARCHIC OR NON-POLYARCHIC (49)	0.51	16 39 5 40 X SQ= 3.80 P = 0.047 RV YES (NO)
105	TILT TOWARD BEING THOSE WHERE THE ELECTORAL SYSTEM IS COMPETITIVE (43)	0.65	105	TILT TOWARD BEING THOSE WHERE THE ELECTORAL SYSTEM IS PARTIALLY COMPETITIVE OR NON-COMPETITIVE (47)	0.56	13 28 7 35 X SQ= 1.81 P = 0.129 RV YES (NO)
110	TEND TO BE THOSE WHERE POLITICAL ENCULTURATION IS HIGH (15)	0.50	110	TEND TO BE THOSE WHERE POLITICAL ENCULTURATION IS MEDIUM OR LOW (80)	0.91 •	8 7 8 67 X SQ= 12.79 P = 0.001 RV YES (YES)

114	TEND TO BE THOSE WHERE SECTIONALISM IS NEGLIGIBLE (47)	0.74	114	TEND TO BE THOSE WHERE SECTIONALISM IS EXTREME OR MODERATE (61)	0.67	5 53 14 26 X SQ= 8.92 P = 0.002 RV YES (NO)
116	TEND TO BE THOSE WHERE INTEREST ARTICULATION BY ASSOCIATIONAL GROUPS IS SIGNIFICANT OR MODERATE (32)	0.60	116	TEND TO BE THOSE WHERE INTEREST ARTICULATION BY ASSOCIATIONAL GROUPS IS LIMITED OR NEGLIGIBLE (79)	0.77	12 19 8 63 X SQ= 8.64 P = 0.002 RV YES (NO)
119	TEND TO BE THOSE WHERE INTEREST ARTICULATION BY INSTITUTIONAL GROUPS IS MODERATE OR LIMITED (26)	0.50	119	TEND TO BE THOSE WHERE INTEREST ARTICULATION BY INSTITUTIONAL GROUPS IS VERY SIGNIFICANT OR SIGNIFICANT (74)	0.81	10 59 10 14 X SQ= 6.26 P = 0.009 RV YES (NO)
121	TEND TO BE THOSE WHERE INTEREST ARTICULATION BY NON-ASSOCIATIONAL GROUPS IS MODERATE, LIMITED, OR NEGLIGIBLE (61)	0.86	121	TEND TO BE THOSE WHERE INTEREST ARTICULATION BY NON-ASSOCIATIONAL GROUPS IS SIGNIFICANT (54)	0.51	3 43 18 41 X SQ= 7.86 P = 0.003 RV YES (NO)
122	TEND TO BE THOSE WHERE INTEREST ARTICULATION BY NON-ASSOCIATIONAL GROUPS IS LIMITED OR NEGLIGIBLE (32)	0.67	122	TEND TO BE THOSE WHERE INTEREST ARTICULATION BY NON-ASSOCIATIONAL GROUPS IS SIGNIFICANT OR MODERATE (83)	0.81	7 68 14 16 X SQ= 16.41 P = 0. RV YES (NO)
125	TEND TO BE THOSE WHERE INTEREST ARTICULATION BY ANOMIC GROUPS IS INFREQUENT OR VERY INFREQUENT (35)	0.82	125	TEND TO BE THOSE WHERE INTEREST ARTICULATION BY ANOMIC GROUPS IS FREQUENT OR OCCASIONAL (64)	0.76	3 56 14 18 X SQ= 17.95 P = 0. RV YES (NO)
137	TEND TO BE THOSE WHERE INTEREST AGGREGATION BY THE LEGISLATURE IS SIGNIFICANT OR MODERATE (29)	0.53	137	TEND TO BE THOSE WHERE INTEREST AGGREGATION BY THE LEGISLATURE IS LIMITED OR NEGLIGIBLE (68)	0.75	10 18 9 53 X SQ= 4.01 P = 0.048 RV YES (NO)
138	LEAN TOWARD BEING THOSE WHERE INTEREST AGGREGATION BY THE LEGISLATURE IS SIGNIFICANT, MODERATE, OR LIMITED (48)	0.72	138	LEAN TOWARD BEING THOSE WHERE INTEREST AGGREGATION BY THE LEGISLATURE IS NEGLIGIBLE (49)	0.53	13 34 5 38 X SQ= 2.67 P = 0.069 RV YES (NO)
147	TEND TO BE THOSE WHERE THE PARTY SYSTEM IS QUALITATIVELY CLASS-ORIENTED OR MULTI-IDEOLOGICAL (23)	0.53	147	TEND TO BE THOSE WHERE THE PARTY SYSTEM IS QUALITATIVELY OTHER THAN CLASS-ORIENTED OR MULTI-IDEOLOGICAL (67)	0.79	8 14 7 54 X SQ= 5.19 P = 0.020 RV YES (NO)

148	ALWAYS ARE THOSE WHERE THE PARTY SYSTEM IS QUALITATIVELY OTHER THAN AFRICAN TRANSITIONAL (96)	1.00	148	TILT LESS TOWARD BEING THOSE WHERE THE PARTY SYSTEM IS QUALITATIVELY OTHER THAN AFRICAN TRANSITIONAL (96)	0.86	0 11 20 70 X SQ= 1.81 P = 0.115 RV NO (NO)
161	TILT TOWARD BEING THOSE WHERE THE POLITICAL LEADERSHIP IS NON-ELITIST (50)	0.70	161	TILT TOWARD BEING THOSE WHERE THE POLITICAL LEADERSHIP IS ELITIST OR MODERATELY ELITIST (47)	0.51	6 36 14 34 X SQ= 2.07 P = 0.128 RV YES (NO)
168	TEND TO BE THOSE WHERE THE HORIZONTAL POWER DISTRIBUTION IS SIGNIFICANT (34)	0.55	168	TEND TO BE THOSE WHERE THE HORIZONTAL POWER DISTRIBUTION IS LIMITED OR NEGLIGIBLE (72)	0.72	11 22 9 56 X SQ= 3.99 P = 0.034 RV YES (NO)
170	TEND MORE TO BE THOSE WHERE THE LEGISLATIVE-EXECUTIVE STRUCTURE IS OTHER THAN PRESIDENTIAL (63)	0.85	170	TEND LESS TO BE THOSE WHERE THE LEGISLATIVE-EXECUTIVE STRUCTURE IS OTHER THAN PRESIDENTIAL (63)	0.57	3 32 17 43 X SQ= 4.07 P = 0.035 RV NO (NO)
187	TEND TO BE THOSE WHERE THE ROLE OF THE POLICE IS NOT POLITICALLY SIGNIFICANT (35)	0.58	187	TEND TO BE THOSE WHERE THE ROLE OF THE POLICE IS POLITICALLY SIGNIFICANT (66)	0.69	8 52 11 23 X SQ= 3.76 P = 0.035 RV YES (NO)
193	ALWAYS ARE THOSE WHERE THE CHARACTER OF THE LEGAL SYSTEM IS OTHER THAN PARTLY INDIGENOUS (86)	1.00	193	TEND LESS TO BE THOSE WHERE THE CHARACTER OF THE LEGAL SYSTEM IS OTHER THAN PARTLY INDIGENOUS (86)	0.70	0 25 21 58 X SQ= 6.76 P = 0.002 RV NO (NO)

M

MATRIX

70 POLITIES THAT ARE RELIGIOUSLY, RACIALLY, AND LINGUISTICALLY HOMOGENEOUS (21)	70 POLITIES THAT ARE RELIGIOUSLY, RACIALLY, OR LINGUISTICALLY HETEROGENEOUS (85)

125 POLITIES WHERE INTEREST ARTICULATION BY ANOMIC GROUPS IS FREQUENT OR OCCASIONAL (64)

JORDAN POLAND SOMALIA

AFGHANISTAN BOLIVIA BRAZIL BURMA CAMBODIA
CAMEROUN CEN AFR REP CEYLON CHAD COLOMBIA
CONGO(BRA) CONGO(LEO) DAHOMEY ECUADOR EL SALVADOR
ETHIOPIA GABON GERMANY, E GHANA GUATEMALA
GUINEA HAITI HONDURAS INDIA INDONESIA
IRAN IRAQ IVORY COAST JAPAN LAOS
LEBANON MALAYA MALI MAURITANIA MEXICO
MOROCCO NEPAL NIGER NIGERIA PAKISTAN
PANAMA PERU SENEGAL SIERRE LEO SO AFRICA
SPAIN SUDAN SYRIA THAILAND TOGO
TURKEY UGANDA USSR UPPER VOLTA VIETNAM REP
YUGOSLAVIA

3	56
15	18

125 POLITIES WHERE INTEREST ARTICULATION BY ANOMIC GROUPS IS INFREQUENT OR VERY INFREQUENT (35)

AUSTRIA COSTA RICA DENMARK GREECE ICELAND
IRELAND ITALY LIBYA LUXEMBOURG NORWAY
PORTUGAL SWEDEN TUNISIA UAR URUGUAY

AUSTRALIA BELGIUM CANADA CZECHOS'KIA FINLAND
GERMAN FR ISRAEL JAMAICA LIBERIA MALAGASY R
NETHERLANDS NEW ZEALAND NICARAGUA PHILIPPINES SWITZERLAND
TANGANYIKA TRINIDAD UK

71 POLITIES
WHOSE DATE OF INDEPENDENCE
IS BEFORE 1800 (21)

71 POLITIES
WHOSE DATE OF INDEPENDENCE
IS AFTER 1800 (90)

BOTH SUBJECT AND PREDICATE

21 IN LEFT COLUMN

AFGHANISTAN AUSTRIA CHINA, PR DENMARK ETHIOPIA FRANCE IRAN JAPAN LUXEMBOURG NEPAL
NETHERLANDS NORWAY PORTUGAL SPAIN SWEDEN SWITZERLAND THAILAND TURKEY USSR UK
US

90 IN RIGHT COLUMN

ALBANIA ALGERIA ARGENTINA AUSTRALIA BELGIUM BOLIVIA BRAZIL BULGARIA BURMA BURUNDI
CAMBODIA CAMEROUN CANADA CEN AFR REP CEYLON CHAD CHILE COLOMBIA CONGO(BRA) CONGO(LEO)
COSTA RICA CUBA CYPRUS CZECHOS'KIA DAHOMEY DOMIN REP ECUADOR EL SALVADOR FINLAND GABON
GHANA GREECE GUATEMALA GUINEA HAITI HONDURAS HUNGARY ICELAND INDIA INDONESIA
IRAQ IRELAND ISRAEL ITALY IVORY COAST JAMAICA JORDAN LAOS LEBANON LIBERIA
LIBYA MALAGASY R MALAYA MALI MAURITANIA MEXICO MONGOLIA MOROCCO NEW ZEALAND NICARAGUA
NIGER NIGERIA PAKISTAN PANAMA PARAGUAY PERU PHILIPPINES POLAND RUMANIA RWANDA
SA'U ARABIA SENEGAL SIERRE LEO SOMALIA SO AFRICA SUDAN SYRIA TANGANYIKA TOGO TRINIDAD
TUNISIA UGANDA UAR UPPER VOLTA URUGUAY VENEZUELA VIETNAM, N VIETNAM REP YEMEN YUGOSLAVIA

4 EXCLUDED BECAUSE AMBIGUOUS

GERMANY, E GERMAN FR KOREA, N KOREA REP

	Left column			Right column		
2	TEND TO BE THOSE LOCATED IN WEST EUROPE, SCANDINAVIA, NORTH AMERICA, OR AUSTRALASIA (22)	0.57	2	TEND TO BE THOSE LOCATED ELSEWHERE THAN IN WEST EUROPE, SCANDINAVIA, NORTH AMERICA, OR AUSTRALASIA (93)	0.90	12 9 9 81 X SQ= 21.69 P = 0. RV YES (YES)
3	TEND LESS TO BE THOSE LOCATED ELSEWHERE THAN IN WEST EUROPE (102)	0.62	3	TEND MORE TO BE THOSE LOCATED ELSEWHERE THAN IN WEST EUROPE (102)	0.96	8 4 13 86 X SQ= 16.66 P = 0. RV NO (YES)
5	TEND TO BE THOSE LOCATED IN SCANDINAVIA OR WEST EUROPE, RATHER THAN IN EAST EUROPE (18)	0.92	5	TEND TO BE THOSE LOCATED IN EAST EUROPE, RATHER THAN IN SCANDINAVIA OR WEST EUROPE (9)	0.54	11 6 1 7 X SQ= 4.03 P = 0.030 RV YES (YES)

6	ALWAYS ARE THOSE LOCATED ELSEWHERE THAN IN THE CARIBBEAN, CENTRAL AMERICA, OR SOUTH AMERICA (93)	1.00	6	TEND LESS TO BE THOSE LOCATED ELSEWHERE THAN IN THE CARIBBEAN, CENTRAL AMERICA, OR SOUTH AMERICA (93)	0.76	0 22 21 68 X SQ= 4.96 P = 0.007 RV NO (NO)
10	TEND MORE TO BE THOSE LOCATED ELSEWHERE THAN IN NORTH AFRICA, OR CENTRAL AND SOUTH AFRICA (82)	0.95	10	TEND LESS TO BE THOSE LOCATED ELSEWHERE THAN IN NORTH AFRICA, OR CENTRAL AND SOUTH AFRICA (82)	0.64	1 32 20 58 X SQ= 6.32 P = 0.004 RV NO (NO)
15	TEND TO BE THOSE LOCATED IN THE MIDDLE EAST, RATHER THAN IN NORTH AFRICA OR CENTRAL AND SOUTH AFRICA (11)	0.75	15	TEND TO BE THOSE LOCATED IN NORTH AFRICA OR CENTRAL AND SOUTH AFRICA, RATHER THAN IN THE MIDDLE EAST (33)	0.80	3 8 1 32 X SQ= 3.30 P = 0.043 RV YES (NO)
18	TEND TO BE THOSE LOCATED IN EAST ASIA, SOUTH ASIA, OR SOUTHEAST ASIA, RATHER THAN IN NORTH AFRICA OR CENTRAL AND SOUTH AFRICA (18)	0.80	18	TEND TO BE THOSE LOCATED IN NORTH AFRICA OR CENTRAL AND SOUTH AFRICA, RATHER THAN IN EAST ASIA, SOUTH ASIA, OR SOUTHEAST ASIA (33)	0.73	1 32 4 12 X SQ= 3.53 P = 0.034 RV YES (NO)
20	ALWAYS ARE THOSE LOCATED IN EAST ASIA, SOUTH ASIA, OR SOUTHEAST ASIA, RATHER THAN IN THE CARIBBEAN, CENTRAL AMERICA, OR SOUTH AMERICA (18)	1.00	20	TEND TO BE THOSE LOCATED IN THE CARIBBEAN, CENTRAL AMERICA, OR SOUTH AMERICA, RATHER THAN IN EAST ASIA, SOUTH ASIA, OR SOUTHEAST ASIA (22)	0.65	4 12 0 22 X SQ= 3.78 P = 0.025 RV YES (NO)
24	TEND TO BE THOSE WHOSE POPULATION IS VERY LARGE, LARGE, OR MEDIUM (61)	0.81	24	TEND TO BE THOSE WHOSE POPULATION IS SMALL (54)	0.56	17 40 4 50 X SQ= 7.68 P = 0.003 RV YES (NO)
26	TEND TO BE THOSE WHOSE POPULATION DENSITY IS VERY HIGH, HIGH, OR MEDIUM (48)	0.62	26	TEND TO BE THOSE WHOSE POPULATION DENSITY IS LOW (67)	0.66	13 31 8 59 X SQ= 4.28 P = 0.027 RV YES (NO)
28	TEND TO BE THOSE WHOSE POPULATION GROWTH RATE IS LOW (48)	0.74	28	TEND TO BE THOSE WHOSE POPULATION GROWTH RATE IS HIGH (62)	0.64	5 56 14 32 X SQ= 7.42 P = 0.004 RV YES (NO)
29	TEND TO BE THOSE WHERE THE DEGREE OF URBANIZATION IS HIGH (56)	0.79	29	TEND TO BE THOSE WHERE THE DEGREE OF URBANIZATION IS LOW (49)	0.54	15 38 4 45 X SQ= 5.55 P = 0.011 RV YES (NO)

30	TILT TOWARD BEING THOSE WHOSE AGRICULTURAL POPULATION IS MEDIUM, LOW, OR VERY LOW (57)	0.67	30	TILT TOWARD BEING THOSE WHOSE AGRICULTURAL POPULATION IS HIGH (56)	0.53	7 47 14 41 X SQ= 1.99 P = 0.144 RV YES (NO)
33	TEND TO BE THOSE WHOSE GROSS NATIONAL PRODUCT IS VERY HIGH, HIGH, OR MEDIUM (30)	0.62	33	TEND TO BE THOSE WHOSE GROSS NATIONAL PRODUCT IS LOW OR VERY LOW (85)	0.83	13 15 8 75 X SQ= 16.15 P = 0. RV YES (NO)
34	TEND TO BE THOSE WHOSE GROSS NATIONAL PRODUCT IS VERY HIGH, HIGH, MEDIUM, OR LOW (62)	0.86	34	TEND TO BE THOSE WHOSE GROSS NATIONAL PRODUCT IS VERY LOW (53)	0.54	18 41 3 49 X SQ= 9.47 P = 0.001 RV YES (NO)
35	TEND TO BE THOSE WHOSE PER CAPITA GROSS NATIONAL PRODUCT IS VERY HIGH OR HIGH (24)	0.52	35	TEND TO BE THOSE WHOSE PER CAPITA GROSS NATIONAL PRODUCT IS MEDIUM, LOW, OR VERY LOW (91)	0.88	11 11 10 79 X SQ= 14.84 P = 0. RV YES (YES)
36	TEND TO BE THOSE WHOSE PER CAPITA GROSS NATIONAL PRODUCT IS VERY HIGH, HIGH, OR MEDIUM (42)	0.62	36	TEND TO BE THOSE WHOSE PER CAPITA GROSS NATIONAL PRODUCT IS LOW OR VERY LOW (73)	0.70	13 27 8 63 X SQ= 6.20 P = 0.010 RV YES (NO)
39	TEND TO BE THOSE WHOSE INTERNATIONAL FINANCIAL STATUS IS VERY HIGH, HIGH, OR MEDIUM (38)	0.67	39	TEND TO BE THOSE WHOSE INTERNATIONAL FINANCIAL STATUS IS LOW OR VERY LOW (76)	0.74	14 23 7 67 X SQ= 11.17 P = 0.001 RV YES (NO)
42	TEND TO BE THOSE WHOSE ECONOMIC DEVELOPMENTAL STATUS IS DEVELOPED OR INTERMEDIATE (36)	0.57	42	TEND TO BE THOSE WHOSE ECONOMIC DEVELOPMENTAL STATUS IS UNDERDEVELOPED OR VERY UNDERDEVELOPED (76)	0.75	12 22 9 65 X SQ= 6.55 P = 0.008 RV YES (NO)
43	TEND TO BE THOSE WHOSE ECONOMIC DEVELOPMENTAL STATUS IS DEVELOPED, INTERMEDIATE, OR UNDERDEVELOPED (55)	0.76	43	TEND TO BE THOSE WHOSE ECONOMIC DEVELOPMENTAL STATUS IS VERY UNDERDEVELOPED (57)	0.57	16 37 5 50 X SQ= 6.38 P = 0.007 RV YES (NO)
44	TEND TO BE THOSE WHERE THE LITERACY RATE IS NINETY PERCENT OR ABOVE (25)	0.57	44	TEND TO BE THOSE WHERE THE LITERACY RATE IS BELOW NINETY PERCENT (90)	0.88	12 11 9 79 X SQ= 18.27 P = 0. RV YES (YES)

45	TEND TO BE THOSE WHERE THE LITERACY RATE IS FIFTY PERCENT OR ABOVE (55)	0.75	45	TEND TO BE THOSE WHERE THE LITERACY RATE IS BELOW FIFTY PERCENT (54)	0.57	15 37 5 49 X SQ= 5.42 P = 0.013 RV YES (NO)
54	TEND TO BE THOSE WHERE NEWSPAPER CIRCULATION IS ONE HUNDRED OR MORE PER THOUSAND (37)	0.57	54	TEND TO BE THOSE WHERE NEWSPAPER CIRCULATION IS LESS THAN ONE HUNDRED PER THOUSAND (76)	0.74	12 23 9 66 X SQ= 6.30 P = 0.009 RV YES (NO)
56	TEND MORE TO BE THOSE WHERE THE RELIGION IS PREDOMINANTLY LITERATE (79)	0.95	56	TEND LESS TO BE THOSE WHERE THE RELIGION IS PREDOMINANTLY LITERATE (79)	0.66	19 58 1 30 X SQ= 5.39 P = 0.012 RV NO (NO)
62	TILT LESS TOWARD BEING THOSE WHERE THE RELIGION IS CATHOLIC, RATHER THAN PROTESTANT (25)	0.62	62	TILT MORE TOWARD BEING THOSE WHERE THE RELIGION IS CATHOLIC, RATHER THAN PROTESTANT (25)	0.91	3 2 5 20 X SQ= 1.67 P = 0.102 RV NO (YES)
63	LEAN TOWARD BEING THOSE WHERE THE RELIGION IS PREDOMINANTLY SOME KIND OF CHRISTIAN (46)	0.60	63	LEAN TOWARD BEING THOSE WHERE THE RELIGION IS PREDOMINANTLY OR PARTLY OTHER THAN CHRISTIAN (68)	0.64	12 32 8 58 X SQ= 3.12 P = 0.075 RV YES (NO)
69	TILT MORE TOWARD BEING THOSE THAT ARE LINGUISTICALLY HOMOGENEOUS OR WEAKLY HETEROGENEOUS (64)	0.70	69	TILT LESS TOWARD BEING THOSE THAT ARE LINGUISTICALLY HOMOGENEOUS OR WEAKLY HETEROGENEOUS (64)	0.51	14 46 6 44 X SQ= 1.65 P = 0.144 RV NO (NO)
74	TEND TO BE THOSE THAT ARE HISTORICALLY WESTERN (26)	0.57	74	TEND TO BE THOSE THAT ARE NOT HISTORICALLY WESTERN (87)	0.86	12 12 9 76 X SQ= 16.24 P = 0. RV YES (YES)
75	TEND TO BE THOSE THAT ARE HISTORICALLY WESTERN OR SIGNIFICANTLY WESTERNIZED (62)	0.76	75	TEND TO BE THOSE THAT ARE NOT HISTORICALLY WESTERN AND ARE NOT SIGNIFICANTLY WESTERNIZED (52)	0.51	16 44 5 45 X SQ= 3.88 P = 0.030 RV YES (NO)
76	TEND TO BE THOSE THAT ARE HISTORICALLY WESTERN, RATHER THAN HAVING BEEN SIGNIFICANTLY OR PARTIALLY WESTERNIZED THROUGH A COLONIAL RELATIONSHIP (26)	0.92	76	TEND TO BE THOSE THAT HAVE BEEN SIGNIFICANTLY OR PARTIALLY WESTERNIZED THROUGH A COLONIAL RELATIONSHIP, RATHER THAN BEING HISTORICALLY WESTERN (70)	0.85	12 12 1 69 X SQ= 31.42 P = 0. RV YES (YES)

78	ALWAYS ARE THOSE THAT HAVE BEEN SIGNIFICANTLY WESTERNIZED WITHOUT A COLONIAL RELATIONSHIP, RATHER THAN THROUGH SUCH A RELATIONSHIP (7)	1.00	78	TEND TO BE THOSE THAT HAVE BEEN SIGNIFICANTLY WESTERNIZED THROUGH A COLONIAL RELATIONSHIP, RATHER THAN WITHOUT SUCH A RELATIONSHIP (28)	0.90	0 28 4 3 X SQ= 12.86 P = 0.001 RV YES (YES)
81	TEND TO BE THOSE WHERE THE TYPE OF POLITICAL MODERNIZATION IS EARLY EUROPEAN OR EARLY EUROPEAN DERIVED, RATHER THAN LATER EUROPEAN OR LATER EUROPEAN DERIVED (11)	0.50	81	TEND TO BE THOSE WHERE THE TYPE OF POLITICAL MODERNIZATION IS LATER EUROPEAN OR LATER EUROPEAN DERIVED, RATHER THAN EARLY EUROPEAN OR EARLY EUROPEAN DERIVED (40)	0.86	6 5 6 32 X SQ= 4.99 P = 0.016 RV YES (YES)
82	ALWAYS ARE THOSE WHERE THE TYPE OF POLITICAL MODERNIZATION IS EARLY OR LATER EUROPEAN OR EUROPEAN DERIVED, RATHER THAN DEVELOPED TUTELARY OR UNDEVELOPED TUTELARY (51)	1.00	82	TEND TO BE THOSE WHERE THE TYPE OF POLITICAL MODERNIZATION IS DEVELOPED TUTELARY OR UNDEVELOPED TUTELARY, RATHER THAN EARLY OR LATER EUROPEAN OR EUROPEAN DERIVED (55)	0.59	12 37 0 53 X SQ= 12.45 P = 0. RV YES (NO)
83	TEND LESS TO BE THOSE WHERE THE TYPE OF POLITICAL MODERNIZATION IS OTHER THAN NON-EUROPEAN AUTOCHTHONOUS (106)	0.57	83	ALWAYS ARE THOSE WHERE THE TYPE OF POLITICAL MODERNIZATION IS OTHER THAN NON-EUROPEAN AUTOCHTHONOUS (106)	1.00	9 0 12 90 X SQ= 36.42 P = 0. RV NO (YES)
85	TEND TO BE THOSE WHERE THE STAGE OF POLITICAL MODERNIZATION IS ADVANCED, RATHER THAN MID- OR EARLY TRANSITIONAL (60)	0.76	85	TEND TO BE THOSE WHERE THE STAGE OF POLITICAL MODERNIZATION IS MID- OR EARLY TRANSITIONAL, RATHER THAN ADVANCED (54)	0.55	16 40 5 49 X SQ= 5.45 P = 0.014 RV YES (NO)
87	ALWAYS ARE THOSE WHOSE IDEOLOGICAL ORIENTATION IS OTHER THAN DEVELOPMENTAL (58)	1.00	87	TEND LESS TO BE THOSE WHOSE IDEOLOGICAL ORIENTATION IS OTHER THAN DEVELOPMENTAL (58)	0.54	0 31 19 36 X SQ= 11.81 P = 0. RV NO (NO)
89	TEND TO BE THOSE WHOSE IDEOLOGICAL ORIENTATION IS CONVENTIONAL (33)	0.63	89	TEND TO BE THOSE WHOSE IDEOLOGICAL ORIENTATION IS OTHER THAN CONVENTIONAL (62)	0.73	12 20 7 53 X SQ= 7.00 P = 0.006 RV YES (NO)
91	ALWAYS ARE THOSE WHOSE IDEOLOGICAL ORIENTATION IS TRADITIONAL, RATHER THAN DEVELOPMENTAL (5)	1.00	91	TEND TO BE THOSE WHOSE IDEOLOGICAL ORIENTATION IS DEVELOPMENTAL, RATHER THAN TRADITIONAL (31)	0.94	0 31 3 2 X SQ= 13.20 P = 0.001 RV YES (YES)
93	LEAN MORE TOWARD BEING THOSE WHERE THE SYSTEM STYLE IS NON-MOBILIZATIONAL (78)	0.89	93	LEAN LESS TOWARD BEING THOSE WHERE THE SYSTEM STYLE IS NON-MOBILIZATIONAL (78)	0.68	2 28 17 60 X SQ= 2.54 P = 0.090 RV NO (NO)

98	TEND TO BE THOSE WHERE GOVERNMENTAL STABILITY IS GENERALLY PRESENT AND DATES AT LEAST FROM THE INTERWAR PERIOD (22)	0.57	98	TEND TO BE THOSE WHERE GOVERNMENTAL STABILITY IS GENERALLY OR MODERATELY PRESENT AND DATES FROM THE POST-WAR PERIOD, OR IS ABSENT (93)	0.89	12 10 9 80 X SQ= 19.90 P = 0. RV YES (YES)
101	TILT TOWARD BEING THOSE WHERE THE REPRESENTATIVE CHARACTER OF THE REGIME IS POLYARCHIC (41)	0.57	101	TILT TOWARD BEING THOSE WHERE THE REPRESENTATIVE CHARACTER OF THE REGIME IS LIMITED POLYARCHIC, PSEUDO-POLYARCHIC, OR NON-POLYARCHIC (57)	0.62	12 28 9 46 X SQ= 1.77 P = 0.137 RV YES (NO)
105	TILT TOWARD BEING THOSE WHERE THE ELECTORAL SYSTEM IS COMPETITIVE (43)	0.67	105	TILT TOWARD BEING THOSE WHERE THE ELECTORAL SYSTEM IS PARTIALLY COMPETITIVE OR NON-COMPETITIVE (47)	0.57	12 30 6 39 X SQ= 2.22 P = 0.112 RV YES (NO)
107	TILT TOWARD BEING THOSE WHERE AUTONOMOUS GROUPS ARE FULLY TOLERATED IN POLITICS (46)	0.57	107	TILT TOWARD BEING THOSE WHERE AUTONOMOUS GROUPS ARE PARTIALLY TOLERATED IN POLITICS, ARE TOLERATED ONLY OUTSIDE POLITICS, OR ARE NOT TOLERATED AT ALL (65)	0.62	12 33 9 53 X SQ= 1.73 P = 0.143 RV YES (NO)
110	TEND TO BE THOSE WHERE POLITICAL ENCULTURATION IS HIGH (15)	0.53	110	TEND TO BE THOSE WHERE POLITICAL ENCULTURATION IS MEDIUM OR LOW (80)	0.92	9 6 8 70 X SQ= 17.64 P = 0. RV YES (YES)
111	LEAN MORE TOWARD BEING THOSE WHERE POLITICAL ENCULTURATION IS HIGH OR MEDIUM (53)	0.76	111	LEAN LESS TOWARD BEING THOSE WHERE POLITICAL ENCULTURATION IS HIGH OR MEDIUM (53)	0.51	13 39 4 37 X SQ= 2.62 P = 0.066 RV NO (NO)
115	TEND TO BE THOSE WHERE INTEREST ARTICULATION BY ASSOCIATIONAL GROUPS IS SIGNIFICANT (20)	0.53	115	TEND TO BE THOSE WHERE INTEREST ARTICULATION BY ASSOCIATIONAL GROUPS IS MODERATE, LIMITED, OR NEGLIGIBLE (91)	0.90	10 9 9 80 X SQ= 16.70 P = 0. RV YES (YES)
116	TEND TO BE THOSE WHERE INTEREST ARTICULATION BY ASSOCIATIONAL GROUPS IS SIGNIFICANT OR MODERATE (32)	0.63	116	TEND TO BE THOSE WHERE INTEREST ARTICULATION BY ASSOCIATIONAL GROUPS IS LIMITED OR NEGLIGIBLE (79)	0.79	12 19 7 70 X SQ= 11.41 P = 0.001 RV YES (NO)
120	TEND LESS TO BE THOSE WHERE INTEREST ARTICULATION BY INSTITUTIONAL GROUPS IS VERY SIGNIFICANT, SIGNIFICANT, OR MODERATE (90)	0.71	120	TEND MORE TO BE THOSE WHERE INTEREST ARTICULATION BY INSTITUTIONAL GROUPS IS VERY SIGNIFICANT, SIGNIFICANT, OR MODERATE (90)	0.95	15 71 6 4 X SQ= 7.17 P = 0.007 RV NO (YES)

121	LEAN TOWARD BEING THOSE WHERE INTEREST ARTICULATION BY NON-ASSOCIATIONAL GROUPS IS MODERATE, LIMITED, OR NEGLIGIBLE (61)	0.71	121	LEAN TOWARD BEING THOSE WHERE INTEREST ARTICULATION BY NON-ASSOCIATIONAL GROUPS IS SIGNIFICANT (54)	0.51	6 46 15 44 X SQ= 2.63 P = 0.089 RV YES (NO)
126	TEND LESS TO BE THOSE WHERE INTEREST ARTICULATION BY ANOMIC GROUPS IS FREQUENT, OCCASIONAL, OR INFREQUENT (83)	0.58	126	TEND MORE TO BE THOSE WHERE INTEREST ARTICULATION BY ANOMIC GROUPS IS FREQUENT, OCCASIONAL, OR INFREQUENT (83)	0.91	11 70 8 7 X SQ= 10.22 P = 0.002 RV NO (YES)
133	TEND MORE TO BE THOSE WHERE INTEREST AGGREGATION BY THE EXECUTIVE IS MODERATE, LIMITED, OR NEGLIGIBLE (76)	0.90	133	TEND LESS TO BE THOSE WHERE INTEREST AGGREGATION BY THE EXECUTIVE IS MODERATE, LIMITED, OR NEGLIGIBLE (76)	0.66	2 27 19 53 X SQ= 3.66 P = 0.032 RV NO (NO)
137	TEND TO BE THOSE WHERE INTEREST AGGREGATION BY THE LEGISLATURE IS SIGNIFICANT OR MODERATE (29)	0.55	137	TEND TO BE THOSE WHERE INTEREST AGGREGATION BY THE LEGISLATURE IS LIMITED OR NEGLIGIBLE (68)	0.77	11 17 9 57 X SQ= 6.27 P = 0.011 RV YES (NO)
138	TILT TOWARD BEING THOSE WHERE INTEREST AGGREGATION BY THE LEGISLATURE IS SIGNIFICANT, MODERATE, OR LIMITED (48)	0.67	138	TILT TOWARD BEING THOSE WHERE INTEREST AGGREGATION BY THE LEGISLATURE IS NEGLIGIBLE (49)	0.54	12 35 6 41 X SQ= 1.72 P = 0.127 RV YES (NO)
147	TEND TO BE THOSE WHERE THE PARTY SYSTEM IS QUALITATIVELY CLASS-ORIENTED OR MULTI-IDEOLOGICAL (23)	0.56	147	TEND TO BE THOSE WHERE THE PARTY SYSTEM IS QUALITATIVELY OTHER THAN CLASS-ORIENTED OR MULTI-IDEOLOGICAL (67)	0.82	9 13 7 58 X SQ= 8.04 P = 0.003 RV YES (NO)
148	ALWAYS ARE THOSE WHERE THE PARTY SYSTEM IS QUALITATIVELY OTHER THAN AFRICAN TRANSITIONAL (96)	1.00	148	LEAN LESS TOWARD BEING THOSE WHERE THE PARTY SYSTEM IS QUALITATIVELY OTHER THAN AFRICAN TRANSITIONAL (96)	0.84	0 14 19 73 X SQ= 2.26 P = 0.069 RV NO (NO)
153	TEND TO BE THOSE WHERE THE PARTY SYSTEM IS STABLE (42)	0.78	153	TEND TO BE THOSE WHERE THE PARTY SYSTEM IS MODERATELY STABLE OR UNSTABLE (38)	0.57	14 25 4 33 X SQ= 5.30 P = 0.015 RV YES (NO)
158	TEND MORE TO BE THOSE WHERE PERSONALISMO IS NEGLIGIBLE (56)	0.82	158	TEND LESS TO BE THOSE WHERE PERSONALISMO IS NEGLIGIBLE (56)	0.51	3 37 14 39 X SQ= 4.27 P = 0.029 RV NO (NO)

168	TEND TO BE THOSE WHERE THE HORIZONTAL POWER DISTRIBUTION IS SIGNIFICANT (34)	0.52	168	TEND TO BE THOSE WHERE THE HORIZONTAL POWER DISTRIBUTION IS LIMITED OR NEGLIGIBLE (72)	0.73	11 22 10 60 X SQ= 3.91 P = 0.036 RV YES (NO)
170	TEND MORE TO BE THOSE WHERE THE LEGISLATIVE-EXECUTIVE STRUCTURE IS OTHER THAN PRESIDENTIAL (63)	0.95	170	TEND LESS TO BE THOSE WHERE THE LEGISLATIVE-EXECUTIVE STRUCTURE IS OTHER THAN PRESIDENTIAL (63)	0.51	1 38 20 40 X SQ= 11.61 P = 0. RV NO (NO)
178	TEND TO BE THOSE WHERE THE LEGISLATURE IS BICAMERAL (51)	0.80	178	TEND TO BE THOSE WHERE THE LEGISLATURE IS UNICAMERAL (53)	0.58	4 47 16 34 X SQ= 7.82 P = 0.003 RV YES (NO)
181	TEND TO BE THOSE WHERE THE BUREAUCRACY IS MODERN, RATHER THAN SEMI-MODERN (21)	0.63	181	TEND TO BE THOSE WHERE THE BUREAUCRACY IS SEMI-MODERN, RATHER THAN MODERN (55)	0.82	10 10 6 46 X SQ= 10.24 P = 0.001 RV YES (YES)
183	ALWAYS ARE THOSE WHERE THE BUREAUCRACY IS TRADITIONAL, RATHER THAN POST-COLONIAL TRANSITIONAL (9)	1.00	183	TEND TO BE THOSE WHERE THE BUREAUCRACY IS POST-COLONIAL TRANSITIONAL, RATHER THAN TRADITIONAL (25)	0.86	0 25 5 4 X SQ= 12.16 P = 0. RV YES (YES)
187	LEAN TOWARD BEING THOSE WHERE THE ROLE OF THE POLICE IS NOT POLITICALLY SIGNIFICANT (35)	0.52	187	LEAN TOWARD BEING THOSE WHERE THE ROLE OF THE POLICE IS POLITICALLY SIGNIFICANT (66)	0.70	10 53 11 23 X SQ= 2.63 P = 0.074 RV YES (NO)
192	TEND MORE TO BE THOSE WHERE THE CHARACTER OF THE LEGAL SYSTEM IS OTHER THAN MUSLIM OR PARTLY MUSLIM (86)	0.95	192	TEND LESS TO BE THOSE WHERE THE CHARACTER OF THE LEGAL SYSTEM IS OTHER THAN MUSLIM OR PARTLY MUSLIM (86)	0.71	1 26 19 63 X SQ= 3.92 P = 0.023 RV NO (NO)
193	TEND MORE TO BE THOSE WHERE THE CHARACTER OF THE LEGAL SYSTEM IS OTHER THAN PARTLY INDIGENOUS (86)	0.95	193	TEND LESS TO BE THOSE WHERE THE CHARACTER OF THE LEGAL SYSTEM IS OTHER THAN PARTLY INDIGENOUS (86)	0.70	1 27 20 62 X SQ= 4.59 P = 0.013 RV NO (NO)

```
**************************************************************************************************

72  POLITIES                                          72  POLITIES
    WHOSE DATE OF INDEPENDENCE                            WHOSE DATE OF INDEPENDENCE
    IS BEFORE 1914  (52)                                  IS AFTER 1914  (59)


                                                                        BOTH SUBJECT AND PREDICATE
..................................................................................................

     52 IN LEFT COLUMN

AFGHANISTAN ALBANIA     ARGENTINA   AUSTRALIA   AUSTRIA     BELGIUM     BOLIVIA     BRAZIL      BULGARIA    CANADA
CHILE       CHINA, PR   COLOMBIA    COSTA RICA  CUBA        DENMARK     DOMIN REP   ECUADOR     EL SALVADOR ETHIOPIA
FRANCE      GREECE      GUATEMALA   HAITI       HONDURAS    IRAN        ITALY       JAPAN       LIBERIA     LUXEMBOURG
MEXICO      NEPAL       NETHERLANDS NEW ZEALAND NICARAGUA   NORWAY      PANAMA      PARAGUAY    PERU        PORTUGAL
RUMANIA     SO AFRICA   SPAIN       SWEDEN      SWITZERLAND THAILAND    TURKEY      USSR        UK          US
URUGUAY     VENEZUELA

     59 IN RIGHT COLUMN

ALGERIA     BURMA       BURUNDI     CAMBODIA    CAMEROUN    CEN AFR REP CEYLON      CHAD        CONGO(BRA)  CONGO(LEO)
CYPRUS      CZECHOS'KIA DAHOMEY     FINLAND     GABON       GHANA       GUINEA      HUNGARY     ICELAND     INDIA
INDONESIA   IRAQ        IRELAND     ISRAEL      IVORY COAST JAMAICA     JORDAN      LAOS        LEBANON     LIBYA
MALAGASY R  MALAYA      MALI        MAURITANIA  MONGOLIA    MOROCCO     NIGER       NIGERIA     PAKISTAN    PHILIPPINES
POLAND      RWANDA      SA'U ARABIA SENEGAL     SIERRE LEO  SOMALIA     SUDAN       SYRIA       TANGANYIKA  TOGO
TRINIDAD    TUNISIA     UGANDA      UAR         UPPER VOLTA VIETNAM, N  VIETNAM REP YEMEN       YUGOSLAVIA

     4 EXCLUDED BECAUSE AMBIGUOUS

GERMANY, E  GERMAN FR   KOREA, N    KOREA REP

--------------------------------------------------------------------------------------------------

 2  TEND LESS TO BE THOSE                  0.65       2  TEND MORE TO BE THOSE                  0.95         18     3
    LOCATED ELSEWHERE THAN IN                            LOCATED ELSEWHERE THAN IN                           34    56
    WEST EUROPE, SCANDINAVIA,                            WEST EUROPE, SCANDINAVIA,                     X SQ=  13.85
    NORTH AMERICA, OR AUSTRALASIA  (93)                  NORTH AMERICA, OR AUSTRALASIA  (93)           P =     0.
                                                                                                       RV NO  (YES)

 6  TEND LESS TO BE THOSE                  0.62       6  TEND MORE TO BE THOSE                  0.97         20     2
    LOCATED ELSEWHERE THAN IN THE                        LOCATED ELSEWHERE THAN IN THE                       32    57
    CARIBBEAN, CENTRAL AMERICA,                          CARIBBEAN, CENTRAL AMERICA,                   X SQ=  19.24
    OR SOUTH AMERICA  (93)                               OR SOUTH AMERICA  (93)                        P =     0.
                                                                                                       RV NO  (YES)

 8  LEAN MORE TOWARD BEING THOSE           0.92       8  LEAN LESS TOWARD BEING THOSE           0.80          4    12
    LOCATED ELSEWHERE THAN IN EAST ASIA                  LOCATED ELSEWHERE THAN IN EAST ASIA                 48    47
    SOUTH ASIA, OR SOUTHEAST ASIA  (97)                  SOUTH ASIA, OR SOUTHEAST ASIA  (97)           X SQ=   2.63
                                                                                                       P =     0.065
                                                                                                       RV NO  (YES)
```

9	TEND MORE TO BE THOSE LOCATED ELSEWHERE THAN IN SOUTHEAST ASIA (106)	0.98	9	TEND LESS TO BE THOSE LOCATED ELSEWHERE THAN IN SOUTHEAST ASIA (106)	0.86	1 8 51 51 X SQ= 3.58 P = 0.035 RV NO (YES)
10	TEND TO BE THOSE LOCATED ELSEWHERE THAN IN NORTH AFRICA, OR CENTRAL AND SOUTH AFRICA (82)	0.94	10	TEND TO BE THOSE LOCATED IN NORTH AFRICA, OR CENTRAL AND SOUTH AFRICA (33)	0.51	3 30 49 29 X SQ= 24.77 P = 0. RV YES (YES)
11	TEND MORE TO BE THOSE LOCATED ELSEWHERE THAN IN CENTRAL AND SOUTH AFRICA (87)	0.94	11	TEND LESS TO BE THOSE LOCATED ELSEWHERE THAN IN CENTRAL AND SOUTH AFRICA (87)	0.58	3 25 49 34 X SQ= 17.74 P = 0. RV NO (YES)
13	TEND MORE TO BE THOSE LOCATED ELSEWHERE THAN IN NORTH AFRICA OR THE MIDDLE EAST (99)	0.94	13	TEND LESS TO BE THOSE LOCATED ELSEWHERE THAN IN NORTH AFRICA OR THE MIDDLE EAST (99)	0.78	3 13 49 46 X SQ= 4.68 P = 0.016 RV NO (YES)
17	TEND TO BE THOSE LOCATED IN THE CARIBBEAN, CENTRAL AMERICA, OR SOUTH AMERICA, RATHER THAN IN THE MIDDLE EAST (22)	0.87	17	TEND TO BE THOSE LOCATED IN THE MIDDLE EAST, RATHER THAN IN THE CARIBBEAN, CENTRAL AMERICA, OR SOUTH AMERICA (11)	0.80	3 8 20 2 X SQ= 11.21 P = 0. RV YES (YES)
19	TEND TO BE THOSE LOCATED IN THE CARIBBEAN, CENTRAL AMERICA, OR SOUTH AMERICA, RATHER THAN IN NORTH AFRICA OR CENTRAL AND SOUTH AFRICA (22)	0.87	19	TEND TO BE THOSE LOCATED IN NORTH AFRICA OR CENTRAL AND SOUTH AFRICA, RATHER THAN IN THE CARIBBEAN, CENTRAL AMERICA, OR SOUTH AMERICA (33)	0.94	3 30 20 2 X SQ= 33.03 P = 0. RV YES (YES)
20	TEND TO BE THOSE LOCATED IN THE CARIBBEAN, CENTRAL AMERICA, OR SOUTH AMERICA, RATHER THAN IN EAST ASIA, SOUTH ASIA, OR SOUTHEAST ASIA (22)	0.83	20	TEND TO BE THOSE LOCATED IN EAST ASIA, SOUTH ASIA, OR SOUTHEAST ASIA, RATHER THAN IN THE CARIBBEAN, CENTRAL AMERICA, OR SOUTH AMERICA (18)	0.86	4 12 20 2 X SQ= 14.58 P = 0. RV YES (YES)
24	TEND TO BE THOSE WHOSE POPULATION IS VERY LARGE, LARGE, OR MEDIUM (61)	0.63	24	TEND TO BE THOSE WHOSE POPULATION IS SMALL (54)	0.59	33 24 19 35 X SQ= 4.87 P = 0.022 RV YES (YES)
29	TEND TO BE THOSE WHERE THE DEGREE OF URBANIZATION IS HIGH (56)	0.77	29	TEND TO BE THOSE WHERE THE DEGREE OF URBANIZATION IS LOW (49)	0.70	37 16 11 38 X SQ= 21.06 P = 0. RV YES (YES)

30	TEND TO BE THOSE WHOSE AGRICULTURAL POPULATION IS MEDIUM, LOW, OR VERY LOW (57)	0.71	30	TEND TO BE THOSE WHOSE AGRICULTURAL POPULATION IS HIGH (56)	0.68	15 39 37 18 X SQ= 15.49 P = 0. RV YES (YES)
36	TEND TO BE THOSE WHOSE PER CAPITA GROSS NATIONAL PRODUCT IS VERY HIGH, HIGH, OR MEDIUM (42)	0.52	36	TEND TO BE THOSE WHOSE PER CAPITA GROSS NATIONAL PRODUCT IS LOW OR VERY LOW (73)	0.78	27 13 25 46 X SQ= 9.45 P = 0.001 RV YES (YES)
37	TEND TO BE THOSE WHOSE PER CAPITA GROSS NATIONAL PRODUCT IS VERY HIGH, HIGH, MEDIUM, OR LOW (64)	0.81	37	TEND TO BE THOSE WHOSE PER CAPITA GROSS NATIONAL PRODUCT IS VERY LOW (51)	0.66	42 20 10 39 X SQ= 22.76 P = 0. RV YES (YES)
39	TEND TO BE THOSE WHOSE INTERNATIONAL FINANCIAL STATUS IS VERY HIGH, HIGH, OR MEDIUM (38)	0.52	39	TEND TO BE THOSE WHOSE INTERNATIONAL FINANCIAL STATUS IS LOW OR VERY LOW (76)	0.83	27 10 25 49 X SQ= 13.68 P = 0. RV YES (YES)
40	TEND TO BE THOSE WHOSE INTERNATIONAL FINANCIAL STATUS IS VERY HIGH, HIGH, MEDIUM, OR LOW (71)	0.79	40	TEND TO BE THOSE WHOSE INTERNATIONAL FINANCIAL STATUS IS VERY LOW (39)	0.51	41 27 11 28 X SQ= 8.97 P = 0.002 RV YES (YES)
42	TEND TO BE THOSE WHOSE ECONOMIC DEVELOPMENTAL STATUS IS DEVELOPED OR INTERMEDIATE (36)	0.51	42	TEND TO BE THOSE WHOSE ECONOMIC DEVELOPMENTAL STATUS IS UNDERDEVELOPED OR VERY UNDERDEVELOPED (76)	0.86	26 8 25 49 X SQ= 15.36 P = 0. RV YES (YES)
43	TEND TO BE THOSE WHOSE ECONOMIC DEVELOPMENTAL STATUS IS DEVELOPED, INTERMEDIATE, OR UNDERDEVELOPED (55)	0.69	43	TEND TO BE THOSE WHOSE ECONOMIC DEVELOPMENTAL STATUS IS VERY UNDERDEVELOPED (57)	0.68	35 18 16 39 X SQ= 13.34 P = 0. RV YES (YES)
45	TEND TO BE THOSE WHERE THE LITERACY RATE IS FIFTY PERCENT OR ABOVE (55)	0.72	45	TEND TO BE THOSE WHERE THE LITERACY RATE IS BELOW FIFTY PERCENT (54)	0.71	36 16 14 40 X SQ= 18.24 P = 0. RV YES (YES)
49	TEND TO BE THOSE WHERE THE LITERACY RATE IS BETWEEN TEN AND FIFTY PERCENT, RATHER THAN BELOW TEN PERCENT (24)	0.79	49	TEND TO BE THOSE WHERE THE LITERACY RATE IS BELOW TEN PERCENT, RATHER THAN BETWEEN TEN AND FIFTY PERCENT (26)	0.64	11 13 3 23 X SQ= 5.68 P = 0.011 RV YES (NO)

54	TEND TO BE THOSE WHERE NEWSPAPER CIRCULATION IS ONE HUNDRED OR MORE PER THOUSAND (37)	0.50	54	TEND TO BE THOSE WHERE NEWSPAPER CIRCULATION IS LESS THAN ONE HUNDRED PER THOUSAND (76)	0.84	26 9 26 49 X SQ= 13.48 P = 0. RV YES (YES)
55	TEND TO BE THOSE WHERE NEWSPAPER CIRCULATION IS TEN OR MORE PER THOUSAND (78)	0.90	55	TEND TO BE THOSE WHERE NEWSPAPER CIRCULATION IS LESS THAN TEN PER THOUSAND (35)	0.52	47 28 5 30 X SQ= 20.51 P = 0. RV YES (YES)
56	TEND MORE TO BE THOSE WHERE THE RELIGION IS PREDOMINANTLY LITERATE (79)	0.84	56	TEND LESS TO BE THOSE WHERE THE RELIGION IS PREDOMINANTLY LITERATE (79)	0.60	43 34 8 23 X SQ= 6.84 P = 0.006 RV NO (YES)
63	TEND TO BE THOSE WHERE THE RELIGION IS PREDOMINANTLY SOME KIND OF CHRISTIAN (46)	0.69	63	TEND TO BE THOSE WHERE THE RELIGION IS PREDOMINANTLY OR PARTLY OTHER THAN CHRISTIAN (68)	0.85	35 9 16 50 X SQ= 30.28 P = 0. RV YES (YES)
66	LEAN TOWARD BEING THOSE THAT ARE RELIGIOUSLY HOMOGENEOUS (57)	0.64	66	LEAN TOWARD BEING THOSE THAT ARE RELIGIOUSLY HETEROGENEOUS (49)	0.54	32 25 18 29 X SQ= 2.61 P = 0.079 RV YES (YES)
67	TEND LESS TO BE THOSE THAT ARE RACIALLY HOMOGENEOUS (82)	0.61	67	TEND MORE TO BE THOSE THAT ARE RACIALLY HOMOGENEOUS (82)	0.82	30 46 19 10 X SQ= 4.72 P = 0.028 RV NO (YES)
68	TEND TO BE THOSE THAT ARE LINGUISTICALLY HOMOGENEOUS (52)	0.61	68	TEND TO BE THOSE THAT ARE LINGUISTICALLY WEAKLY HETEROGENEOUS OR STRONGLY HETEROGENEOUS (62)	0.71	31 17 20 42 X SQ= 10.11 P = 0.001 RV YES (YES)
69	TEND TO BE THOSE THAT ARE LINGUISTICALLY HOMOGENEOUS OR WEAKLY HETEROGENEOUS (64)	0.73	69	TEND TO BE THOSE THAT ARE LINGUISTICALLY STRONGLY HETEROGENEOUS (50)	0.61	37 23 14 36 X SQ= 11.11 P = 0.001 RV YES (YES)
75	TEND TO BE THOSE THAT ARE HISTORICALLY WESTERN OR SIGNIFICANTLY WESTERNIZED (62)	0.84	75	TEND TO BE THOSE THAT ARE NOT HISTORICALLY WESTERN AND ARE NOT SIGNIFICANTLY WESTERNIZED (52)	0.71	43 17 8 42 X SQ= 31.78 P = 0. RV YES (YES)

76	TEND LESS TO BE THOSE THAT HAVE BEEN SIGNIFICANTLY OR PARTIALLY WESTERNIZED THROUGH A COLONIAL RELATIONSHIP, RATHER THAN BEING HISTORICALLY WESTERN (70)	0.55	76	TEND MORE TO BE THOSE THAT HAVE BEEN SIGNIFICANTLY OR PARTIALLY WESTERNIZED THROUGH A COLONIAL RELATIONSHIP, RATHER THAN BEING HISTORICALLY WESTERN (70)	0.89	18 6 22 48 X SQ= 12.15 P = 0. RV NO (YES)
77	TEND TO BE THOSE THAT HAVE BEEN SIGNIFICANTLY WESTERNIZED, RATHER THAN PARTIALLY WESTERNIZED, THROUGH A COLONIAL RELATIONSHIP (28)	0.90	77	TEND TO BE THOSE THAT HAVE BEEN PARTIALLY WESTERNIZED, RATHER THAN SIGNIFICANTLY WESTERNIZED, THROUGH A COLONIAL RELATIONSHIP (41)	0.81	19 9 2 39 X SQ= 28.27 P = 0. RV YES (YES)
79	TEND TO BE THOSE THAT WERE FORMERLY DEPENDENCIES OF SPAIN, RATHER THAN BRITAIN OR FRANCE (18)	0.75	79	ALWAYS ARE THOSE THAT WERE FORMERLY DEPENDENCIES OF BRITAIN OR FRANCE, RATHER THAN SPAIN (49)	1.00	6 43 18 0 X SQ= 40.36 P = 0. RV YES (YES)
82	TEND TO BE THOSE WHERE THE TYPE OF POLITICAL MODERNIZATION IS EARLY OR LATER EUROPEAN OR EUROPEAN DERIVED, RATHER THAN DEVELOPED TUTELARY OR UNDEVELOPED TUTELARY (51)	0.98	82	TEND TO BE THOSE WHERE THE TYPE OF POLITICAL MODERNIZATION IS DEVELOPED TUTELARY OR UNDEVELOPED TUTELARY, RATHER THAN EARLY OR LATER EUROPEAN OR EUROPEAN DERIVED (55)	0.88	42 7 1 52 X SQ= 69.97 P = 0. RV YES (YES)
83	TEND LESS TO BE THOSE WHERE THE TYPE OF POLITICAL MODERNIZATION IS OTHER THAN NON-EUROPEAN AUTOCHTHONOUS (106)	0.83	83	ALWAYS ARE THOSE WHERE THE TYPE OF POLITICAL MODERNIZATION IS OTHER THAN NON-EUROPEAN AUTOCHTHONOUS (106)	1.00	9 0 43 59 X SQ= 8.91 P = 0.001 RV NO (YES)
85	TEND TO BE THOSE WHERE THE STAGE OF POLITICAL MODERNIZATION IS ADVANCED, RATHER THAN MID- OR EARLY TRANSITIONAL (60)	0.65	85	TEND TO BE THOSE WHERE THE STAGE OF POLITICAL MODERNIZATION IS MID- OR EARLY TRANSITIONAL, RATHER THAN ADVANCED (54)	0.62	34 22 18 36 X SQ= 7.21 P = 0.005 RV YES (YES)
86	TEND TO BE THOSE WHERE THE STAGE OF POLITICAL MODERNIZATION IS ADVANCED OR MID-TRANSITIONAL, RATHER THAN EARLY TRANSITIONAL (76)	0.96	86	TEND TO BE THOSE WHERE THE STAGE OF POLITICAL MODERNIZATION IS EARLY TRANSITIONAL, RATHER THAN ADVANCED OR MID-TRANSITIONAL (38)	0.62	50 22 2 36 X SQ= 38.57 P = 0. RV YES (YES)
87	TEND TO BE THOSE WHOSE IDEOLOGICAL ORIENTATION IS OTHER THAN DEVELOPMENTAL (58)	0.93	87	TEND TO BE THOSE WHOSE IDEOLOGICAL ORIENTATION IS DEVELOPMENTAL (31)	0.62	3 28 38 17 X SQ= 25.72 P = 0. RV YES (YES)
89	TEND TO BE THOSE WHOSE IDEOLOGICAL ORIENTATION IS CONVENTIONAL (33)	0.62	89	TEND TO BE THOSE WHOSE IDEOLOGICAL ORIENTATION IS OTHER THAN CONVENTIONAL (62)	0.88	26 6 16 44 X SQ= 22.91 P = 0. RV YES (YES)

No.	Statement	Value	No.	Statement	Value	Statistics
91	TEND TO BE THOSE WHOSE IDEOLOGICAL ORIENTATION IS TRADITIONAL, RATHER THAN DEVELOPMENTAL (5)	0.50	91	TEND TO BE THOSE WHOSE IDEOLOGICAL ORIENTATION IS DEVELOPMENTAL, RATHER THAN TRADITIONAL (31)	0.93	3 28 3 2 X SQ= 4.65 P = 0.024 RV YES (YES)
93	TEND MORE TO BE THOSE WHERE THE SYSTEM STYLE IS NON-MOBILIZATIONAL (78)	0.84	93	TEND LESS TO BE THOSE WHERE THE SYSTEM STYLE IS NON-MOBILIZATIONAL (78)	0.61	8 22 42 35 X SQ= 5.67 P = 0.010 RV NO (YES)
96	TILT MORE TOWARD BEING THOSE WHERE THE STATUS OF THE REGIME IS CONSTITUTIONAL OR TOTALITARIAN (67)	0.81	96	TILT LESS TOWARD BEING THOSE WHERE THE STATUS OF THE REGIME IS CONSTITUTIONAL OR TOTALITARIAN (67)	0.67	38 26 9 13 X SQ= 1.57 P = 0.146 RV NO (YES)
98	TEND LESS TO BE THOSE WHERE GOVERNMENTAL STABILITY IS GENERALLY OR MODERATELY PRESENT AND DATES FROM THE POST-WAR PERIOD, OR IS ABSENT (93)	0.65	98	TEND MORE TO BE THOSE WHERE GOVERNMENTAL STABILITY IS GENERALLY OR MODERATELY PRESENT AND DATES FROM THE POST-WAR PERIOD, OR IS ABSENT (93)	0.93	18 4 34 55 X SQ= 11.78 P = 0. RV NO (YES)
101	LEAN TOWARD BEING THOSE WHERE THE REPRESENTATIVE CHARACTER OF THE REGIME IS POLYARCHIC (41)	0.51	101	LEAN TOWARD BEING THOSE WHERE THE REPRESENTATIVE CHARACTER OF THE REGIME IS LIMITED POLYARCHIC, PSEUDO-POLYARCHIC, OR NON-POLYARCHIC (57)	0.67	25 15 24 31 X SQ= 2.59 P = 0.096 RV YES (YES)
102	TEND TO BE THOSE WHERE THE REPRESENTATIVE CHARACTER OF THE REGIME IS POLYARCHIC OR LIMITED POLYARCHIC (59)	0.66	102	TEND TO BE THOSE WHERE THE REPRESENTATIVE CHARACTER OF THE REGIME IS PSEUDO-POLYARCHIC OR NON-POLYARCHIC (49)	0.55	33 25 17 30 X SQ= 3.68 P = 0.049 RV YES (YES)
105	TEND TO BE THOSE WHERE THE ELECTORAL SYSTEM IS COMPETITIVE (43)	0.67	105	TEND TO BE THOSE WHERE THE ELECTORAL SYSTEM IS PARTIALLY COMPETITIVE OR NON-COMPETITIVE (47)	0.70	29 13 14 31 X SQ= 11.04 P = 0.001 RV YES (YES)
107	TEND TO BE THOSE WHERE AUTONOMOUS GROUPS ARE FULLY TOLERATED IN POLITICS (46)	0.59	107	TEND TO BE THOSE WHERE AUTONOMOUS GROUPS ARE PARTIALLY TOLERATED IN POLITICS, ARE TOLERATED ONLY OUTSIDE POLITICS, OR ARE NOT TOLERATED AT ALL (65)	0.73	30 15 21 41 X SQ= 9.97 P = 0.001 RV YES (YES)
110	TEND LESS TO BE THOSE WHERE POLITICAL ENCULTURATION IS MEDIUM OR LOW (80)	0.71	110	TEND MORE TO BE THOSE WHERE POLITICAL ENCULTURATION IS MEDIUM OR LOW (80)	0.94	12 3 30 48 X SQ= 7.17 P = 0.004 RV NO (YES)

116	TEND TO BE THOSE WHERE INTEREST ARTICULATION BY ASSOCIATIONAL GROUPS IS SIGNIFICANT OR MODERATE (32)	0.50	116	TEND TO BE THOSE WHERE INTEREST ARTICULATION BY ASSOCIATIONAL GROUPS IS LIMITED OR NEGLIGIBLE (79)	0.90	25 6 25 52 X SQ= 18.74 P = 0. RV YES (YES)
117	TEND TO BE THOSE WHERE INTEREST ARTICULATION BY ASSOCIATIONAL GROUPS IS SIGNIFICANT, MODERATE, OR LIMITED (60)	0.76	117	TEND TO BE THOSE WHERE INTEREST ARTICULATION BY ASSOCIATIONAL GROUPS IS NEGLIGIBLE (51)	0.64	38 21 12 37 X SQ= 15.59 P = 0. RV YES (YES)
119	LEAN LESS TOWARD BEING THOSE WHERE INTEREST ARTICULATION BY INSTITUTIONAL GROUPS IS VERY SIGNIFICANT OR SIGNIFICANT (74)	0.65	119	LEAN MORE TOWARD BEING THOSE WHERE INTEREST ARTICULATION BY INSTITUTIONAL GROUPS IS VERY SIGNIFICANT OR SIGNIFICANT (74)	0.84	34 37 18 7 X SQ= 3.41 P = 0.061 RV NO (YES)
120	TILT LESS TOWARD BEING THOSE WHERE INTEREST ARTICULATION BY INSTITUTIONAL GROUPS IS VERY SIGNIFICANT, SIGNIFICANT, OR MODERATE (90)	0.85	120	TILT MORE TOWARD BEING THOSE WHERE INTEREST ARTICULATION BY INSTITUTIONAL GROUPS IS VERY SIGNIFICANT, SIGNIFICANT, OR MODERATE (90)	0.95	44 42 8 2 X SQ= 1.95 P = 0.103 RV NO (NO)
121	TEND TO BE THOSE WHERE INTEREST ARTICULATION BY NON-ASSOCIATIONAL GROUPS IS MODERATE, LIMITED, OR NEGLIGIBLE (61)	0.85	121	TEND TO BE THOSE WHERE INTEREST ARTICULATION BY NON-ASSOCIATIONAL GROUPS IS SIGNIFICANT (54)	0.75	8 44 44 15 X SQ= 36.55 P = 0. RV YES (YES)
124	TILT MORE TOWARD BEING THOSE WHERE INTEREST ARTICULATION BY ANOMIC GROUPS IS OCCASIONAL, INFREQUENT, OR VERY INFREQUENT (85)	0.93	124	TILT LESS TOWARD BEING THOSE WHERE INTEREST ARTICULATION BY ANOMIC GROUPS IS OCCASIONAL, INFREQUENT, OR VERY INFREQUENT (85)	0.81	3 10 40 43 X SQ= 1.94 P = 0.134 RV NO (NO)
125	LEAN LESS TOWARD BEING THOSE WHERE INTEREST ARTICULATION BY ANOMIC GROUPS IS FREQUENT OR OCCASIONAL (64)	0.53	125	LEAN MORE TOWARD BEING THOSE WHERE INTEREST ARTICULATION BY ANOMIC GROUPS IS FREQUENT OR OCCASIONAL (64)	0.74	23 39 20 14 X SQ= 3.36 P = 0.054 RV NO (YES)
126	TEND LESS TO BE THOSE WHERE INTEREST ARTICULATION BY ANOMIC GROUPS IS FREQUENT, OCCASIONAL, OR INFREQUENT (83)	0.72	126	TEND MORE TO BE THOSE WHERE INTEREST ARTICULATION BY ANOMIC GROUPS IS FREQUENT, OCCASIONAL, OR INFREQUENT (83)	0.94	31 50 12 3 X SQ= 7.30 P = 0.004 RV NO (YES)
128	TEND TO BE THOSE WHERE INTEREST ARTICULATION BY POLITICAL PARTIES IS SIGNIFICANT OR MODERATE (48)	0.65	128	TEND TO BE THOSE WHERE INTEREST ARTICULATION BY POLITICAL PARTIES IS LIMITED OR NEGLIGIBLE (45)	0.60	28 19 15 28 X SQ= 4.54 P = 0.022 RV YES (YES)

129	TEND TO BE THOSE WHERE INTEREST ARTICULATION BY POLITICAL PARTIES IS SIGNIFICANT, MODERATE, OR LIMITED (56)	0.77	129	TEND TO BE THOSE WHERE INTEREST ARTICULATION BY POLITICAL PARTIES IS NEGLIGIBLE (37)	0.53	33 22 10 25 X SQ= 7.25 P = 0.005 RV YES (YES)
131	LEAN TOWARD BEING THOSE WHERE INTEREST AGGREGATION BY POLITICAL PARTIES IS LIMITED OR NEGLIGIBLE (35)	0.64	131	LEAN TOWARD BEING THOSE WHERE INTEREST AGGREGATION BY POLITICAL PARTIES IS SIGNIFICANT OR MODERATE (30)	0.59	13 16 23 11 X SQ= 2.46 P = 0.080 RV YES (YES)
134	LEAN TOWARD BEING THOSE WHERE INTEREST AGGREGATION BY THE EXECUTIVE IS LIMITED OR NEGLIGIBLE (46)	0.53	134	LEAN TOWARD BEING THOSE WHERE INTEREST AGGREGATION BY THE EXECUTIVE IS SIGNIFICANT OR MODERATE (57)	0.66	23 33 26 17 X SQ= 2.92 P = 0.069 RV YES (YES)
138	TEND TO BE THOSE WHERE INTEREST AGGREGATION BY THE LEGISLATURE IS SIGNIFICANT, MODERATE, OR LIMITED (48)	0.70	138	TEND TO BE THOSE WHERE INTEREST AGGREGATION BY THE LEGISLATURE IS NEGLIGIBLE (49)	0.69	32 15 14 33 X SQ= 12.30 P = 0. RV YES (YES)
139	TEND MORE TO BE THOSE WHERE THE PARTY SYSTEM IS QUANTITATIVELY OTHER THAN ONE-PARTY (71)	0.81	139	TEND LESS TO BE THOSE WHERE THE PARTY SYSTEM IS QUANTITATIVELY OTHER THAN ONE-PARTY (71)	0.57	9 23 39 31 X SQ= 5.65 P = 0.011 RV NO (YES)
141	TILT LESS TOWARD BEING THOSE WHERE THE PARTY SYSTEM IS QUANTITATIVELY OTHER THAN TWO-PARTY (87)	0.82	141	TILT MORE TOWARD BEING THOSE WHERE THE PARTY SYSTEM IS QUANTITATIVELY OTHER THAN TWO-PARTY (87)	0.94	8 3 37 47 X SQ= 2.16 P = 0.108 RV NO (YES)
142	LEAN LESS TOWARD BEING THOSE WHERE THE PARTY SYSTEM IS QUANTITATIVELY OTHER THAN MULTI-PARTY (66)	0.58	142	LEAN MORE TOWARD BEING THOSE WHERE THE PARTY SYSTEM IS QUANTITATIVELY OTHER THAN MULTI-PARTY (66)	0.77	19 11 26 37 X SQ= 3.13 P = 0.075 RV NO (YES)
146	ALWAYS ARE THOSE WHERE THE PARTY SYSTEM IS QUALITATIVELY OTHER THAN MASS-BASED TERRITORIAL (92)	1.00	146	TEND LESS TO BE THOSE WHERE THE PARTY SYSTEM IS QUALITATIVELY OTHER THAN MASS-BASED TERRITORIAL (92)	0.84	0 8 46 42 X SQ= 6.07 P = 0.006 RV NO (YES)
147	TEND LESS TO BE THOSE WHERE THE PARTY SYSTEM IS QUALITATIVELY OTHER THAN CLASS-ORIENTED OR MULTI-IDEOLOGICAL (67)	0.59	147	TEND MORE TO BE THOSE WHERE THE PARTY SYSTEM IS QUALITATIVELY OTHER THAN CLASS-ORIENTED OR MULTI-IDEOLOGICAL (67)	0.89	17 5 24 41 X SQ= 9.18 P = 0.001 RV NO (YES)

148	ALWAYS ARE THOSE WHERE THE PARTY SYSTEM IS QUALITATIVELY OTHER THAN AFRICAN TRANSITIONAL (96)	1.00	148	TEND LESS TO BE THOSE WHERE THE PARTY SYSTEM IS QUALITATIVELY OTHER THAN AFRICAN TRANSITIONAL (96)	0.75	0 14 49 43 X SQ= 11.81 P = 0. RV NO (YES)
160	LEAN LESS TOWARD BEING THOSE WHERE THE POLITICAL LEADERSHIP IS MODERATELY ELITIST OR NON-ELITIST (67)	0.61	160	LEAN MORE TOWARD BEING THOSE WHERE THE POLITICAL LEADERSHIP IS MODERATELY ELITIST OR NON-ELITIST (67)	0.80	19 9 30 36 X SQ= 3.11 P = 0.070 RV NO (YES)
164	TEND TO BE THOSE WHERE THE REGIME'S LEADERSHIP CHARISMA IS NEGLIGIBLE (65)	0.92	164	TEND TO BE THOSE WHERE THE REGIME'S LEADERSHIP CHARISMA IS PRONOUNCED OR MODERATE (34)	0.62	4 29 44 18 X SQ= 27.53 P = 0. RV YES (YES)
169	TEND TO BE THOSE WHERE THE HORIZONTAL POWER DISTRIBUTION IS SIGNIFICANT OR LIMITED (58)	0.68	169	TEND TO BE THOSE WHERE THE HORIZONTAL POWER DISTRIBUTION IS NEGLIGIBLE (48)	0.57	34 23 16 30 X SQ= 5.35 P = 0.017 RV YES (YES)
175	TEND TO BE THOSE WHERE THE LEGISLATURE IS FULLY EFFECTIVE OR PARTIALLY EFFECTIVE (51)	0.66	175	TEND TO BE THOSE WHERE THE LEGISLATURE IS LARGELY INEFFECTIVE OR WHOLLY INEFFECTIVE (49)	0.62	31 19 16 31 X SQ= 6.50 P = 0.008 RV YES (YES)
178	TEND TO BE THOSE WHERE THE LEGISLATURE IS BICAMERAL (51)	0.68	178	TEND TO BE THOSE WHERE THE LEGISLATURE IS UNICAMERAL (53)	0.69	16 35 34 16 X SQ= 12.12 P = 0. RV YES (YES)
179	LEAN TOWARD BEING THOSE WHERE THE EXECUTIVE IS STRONG (39)	0.52	179	LEAN TOWARD BEING THOSE WHERE THE EXECUTIVE IS DOMINANT (52)	0.66	21 29 23 15 X SQ= 2.27 P = 0.092 RV YES (YES)
182	ALWAYS ARE THOSE WHERE THE BUREAUCRACY IS SEMI-MODERN, RATHER THAN POST-COLONIAL TRANSITIONAL (55)	1.00	182	TEND TO BE THOSE WHERE THE BUREAUCRACY IS POST-COLONIAL TRANSITIONAL, RATHER THAN SEMI-MODERN (25)	0.54	31 21 0 25 X SQ= 22.53 P = 0. RV YES (YES)
183	ALWAYS ARE THOSE WHERE THE BUREAUCRACY IS TRADITIONAL, RATHER THAN POST-COLONIAL TRANSITIONAL (9)	1.00	183	TEND TO BE THOSE WHERE THE BUREAUCRACY IS POST-COLONIAL TRANSITIONAL, RATHER THAN TRADITIONAL (25)	0.86	0 25 5 4 X SQ= 12.16 P = 0. RV YES (YES)

188	TEND TO BE THOSE WHERE THE CHARACTER OF THE LEGAL SYSTEM IS CIVIL LAW (32)	0.61	ALWAYS ARE THOSE WHERE THE CHARACTER OF THE LEGAL SYSTEM IS OTHER THAN CIVIL LAW (81)	1.00	31 0 20 58 X SQ= 46.33 P = 0. RV YES (YES)
190	TEND TO BE THOSE WHERE THE CHARACTER OF THE LEGAL SYSTEM IS CIVIL LAW, RATHER THAN COMMON LAW (32)	0.89	ALWAYS ARE THOSE WHERE THE CHARACTER OF THE LEGAL SYSTEM IS COMMON LAW, RATHER THAN CIVIL LAW (7)	1.00	31 0 4 3 X SQ= 9.13 P = 0.004 RV YES (NO)
192	TEND MORE TO BE THOSE WHERE THE CHARACTER OF THE LEGAL SYSTEM IS OTHER THAN MUSLIM OR PARTLY MUSLIM (86)	0.98	TEND LESS TO BE THOSE WHERE THE CHARACTER OF THE LEGAL SYSTEM IS OTHER THAN MUSLIM OR PARTLY MUSLIM (86)	0.55	1 26 50 32 X SQ= 24.51 P = 0. RV NO (YES)
193	TEND MORE TO BE THOSE WHERE THE CHARACTER OF THE LEGAL SYSTEM IS OTHER THAN PARTLY INDIGENOUS (86)	0.94	TEND LESS TO BE THOSE WHERE THE CHARACTER OF THE LEGAL SYSTEM IS OTHER THAN PARTLY INDIGENOUS (86)	0.57	3 25 49 33 X SQ= 18.22 P = 0. RV NO (YES)

73 POLITIES WHOSE DATE OF INDEPENDENCE IS BEFORE 1945 (65)	73 POLITIES WHOSE DATE OF INDEPENDENCE IS AFTER 1945 (46)

BOTH SUBJECT AND PREDICATE

65 IN LEFT COLUMN

AFGHANISTAN	ALBANIA	ARGENTINA	AUSTRALIA	AUSTRIA	BELGIUM	BOLIVIA	BRAZIL	BULGARIA	CANADA
CHILE	CHINA, PR	COLOMBIA	COSTA RICA	CUBA	CZECHOS'KIA	DENMARK	DOMIN REP	ECUADOR	EL SALVADOR
ETHIOPIA	FINLAND	FRANCE	GREECE	GUATEMALA	HAITI	HONDURAS	HUNGARY	ICELAND	IRAN
IRAQ	IRELAND	ITALY	JAPAN	LEBANON	LIBERIA	LUXEMBOURG	MEXICO	MONGOLIA	NEPAL
NETHERLANDS	NEW ZEALAND	NICARAGUA	NORWAY	PANAMA	PARAGUAY	PERU	POLAND	PORTUGAL	RUMANIA
SA'U ARABIA	SO AFRICA	SPAIN	SWEDEN	SWITZERLAND	SYRIA	THAILAND	TURKEY	USSR	UK
US	URUGUAY	VENEZUELA	YEMEN	YUGOSLAVIA					

46 IN RIGHT COLUMN

ALGERIA	BURMA	BURUNDI	CAMBODIA	CAMEROUN	CEN AFR REP	CEYLON	CHAD	CONGO(BRA)	CONGO(LEO)
CYPRUS	DAHOMEY	GABON	GHANA	GUINEA	INDIA	INDONESIA	ISRAEL	IVORY COAST	JAMAICA
JORDAN	LAOS	LIBYA	MALAGASY R	MALAYA	MALI	MAURITANIA	MOROCCO	NIGER	NIGERIA
PAKISTAN	PHILIPPINES	RWANDA	SENEGAL	SIERRE LEO	SOMALIA	SUDAN	TANGANYIKA	TOGO	TRINIDAD
TUNISIA	UGANDA	UAR	UPPER VOLTA	VIETNAM, N	VIETNAM REP				

4 EXCLUDED BECAUSE AMBIGUOUS

GERMANY, E   GERMAN FR   KOREA, N   KOREA REP

2	TEND LESS TO BE THOSE LOCATED ELSEWHERE THAN IN WEST EUROPE, SCANDINAVIA, NORTH AMERICA, OR AUSTRALASIA (93)	0.68	2	ALWAYS ARE THOSE LOCATED ELSEWHERE THAN IN WEST EUROPE, SCANDINAVIA, NORTH AMERICA, OR AUSTRALASIA (93)	1.00	21 0 44 46 X SQ= 16.28 P = 0. RV NO (YES)
6	TEND LESS TO BE THOSE LOCATED ELSEWHERE THAN IN THE CARIBBEAN, CENTRAL AMERICA, OR SOUTH AMERICA (93)	0.69	6	TEND MORE TO BE THOSE LOCATED ELSEWHERE THAN IN THE CARIBBEAN, CENTRAL AMERICA, OR SOUTH AMERICA (93)	0.96	20 2 45 44 X SQ= 10.23 P = 0.001 RV NO (NO)
8	TEND MORE TO BE THOSE LOCATED ELSEWHERE THAN IN EAST ASIA SOUTH ASIA, OR SOUTHEAST ASIA (97)	0.92	8	TEND LESS TO BE THOSE LOCATED ELSEWHERE THAN IN EAST ASIA SOUTH ASIA, OR SOUTHEAST ASIA (97)	0.76	5 11 60 35 X SQ= 4.51 P = 0.026 RV NO (YES)

10	TEND TO BE THOSE LOCATED ELSEWHERE THAN IN NORTH AFRICA, OR CENTRAL AND SOUTH AFRICA (82)	0.95	10	TEND TO BE THOSE LOCATED IN NORTH AFRICA, OR CENTRAL AND SOUTH AFRICA (33)	0.65	3 30 62 16 X SQ= 44.50 P = 0. RV YES (YES)
15	TEND TO BE THOSE LOCATED IN THE MIDDLE EAST, RATHER THAN IN NORTH AFRICA OR CENTRAL AND SOUTH AFRICA (11)	0.73	15	TEND TO BE THOSE LOCATED IN NORTH AFRICA OR CENTRAL AND SOUTH AFRICA, RATHER THAN IN THE MIDDLE EAST (33)	0.91	8 3 3 30 X SQ= 14.59 P = 0. RV YES (YES)
16	LEAN TOWARD BEING THOSE LOCATED IN THE MIDDLE EAST, RATHER THAN IN EAST ASIA, SOUTH ASIA, OR SOUTHEAST ASIA (11)	0.62	16	LEAN TOWARD BEING THOSE LOCATED IN EAST ASIA, SOUTH ASIA, OR SOUTHEAST ASIA, RATHER THAN IN THE MIDDLE EAST (18)	0.79	8 3 5 11 X SQ= 2.98 P = 0.054 RV YES (YES)
18	LEAN TOWARD BEING THOSE LOCATED IN EAST ASIA, SOUTH ASIA, OR SOUTHEAST ASIA, RATHER THAN IN NORTH AFRICA OR CENTRAL AND SOUTH AFRICA (18)	0.62	18	LEAN TOWARD BEING THOSE LOCATED IN NORTH AFRICA OR CENTRAL AND SOUTH AFRICA, RATHER THAN IN EAST ASIA, SOUTH ASIA, OR SOUTHEAST ASIA (33)	0.73	3 30 5 11 X SQ= 2.42 P = 0.094 RV YES (NO)
19	TEND TO BE THOSE LOCATED IN THE CARIBBEAN, CENTRAL AMERICA, OR SOUTH AMERICA, RATHER THAN IN NORTH AFRICA OR CENTRAL AND SOUTH AFRICA (22)	0.87	19	TEND TO BE THOSE LOCATED IN NORTH AFRICA OR CENTRAL AND SOUTH AFRICA, RATHER THAN IN THE CARIBBEAN, CENTRAL AMERICA, OR SOUTH AMERICA (33)	0.94	3 30 20 2 X SQ= 33.03 P = 0. RV YES (YES)
20	TEND TO BE THOSE LOCATED IN THE CARIBBEAN, CENTRAL AMERICA, OR SOUTH AMERICA, RATHER THAN IN EAST ASIA, SOUTH ASIA, OR SOUTHEAST ASIA (22)	0.80	20	TEND TO BE THOSE LOCATED IN EAST ASIA, SOUTH ASIA, OR SOUTHEAST ASIA, RATHER THAN IN THE CARIBBEAN, CENTRAL AMERICA, OR SOUTH AMERICA (18)	0.85	5 11 20 2 X SQ= 12.12 P = 0. RV YES (YES)
24	TEND TO BE THOSE WHOSE POPULATION IS VERY LARGE, LARGE, OR MEDIUM (61)	0.60	24	TEND TO BE THOSE WHOSE POPULATION IS SMALL (54)	0.61	39 18 26 28 X SQ= 3.90 P = 0.035 RV YES (YES)
29	TEND TO BE THOSE WHERE THE DEGREE OF URBANIZATION IS HIGH (56)	0.75	29	TEND TO BE THOSE WHERE THE DEGREE OF URBANIZATION IS LOW (49)	0.83	46 7 15 34 X SQ= 31.13 P = 0. RV YES (YES)
30	TEND TO BE THOSE WHOSE AGRICULTURAL POPULATION IS MEDIUM, LOW, OR VERY LOW (57)	0.69	30	TEND TO BE THOSE WHOSE AGRICULTURAL POPULATION IS HIGH (56)	0.76	20 34 44 11 X SQ= 19.01 P = 0. RV YES (YES)

```
34  TEND TO BE THOSE                      0.68      34  TEND TO BE THOSE                      0.67        44   15
    WHOSE GROSS NATIONAL PRODUCT                        WHOSE GROSS NATIONAL PRODUCT                   21   31
    IS VERY HIGH, HIGH, MEDIUM,                         IS VERY LOW  (53)                          X SQ=  11.94
    OR LOW  (62)                                                                                   P =     0.
                                                                                                   RV YES (YES)

36  TEND TO BE THOSE                      0.54      36  TEND TO BE THOSE                      0.89        35    5
    WHOSE PER CAPITA GROSS NATIONAL PRODUCT             WHOSE PER CAPITA GROSS NATIONAL PRODUCT        30   41
    IS VERY HIGH, HIGH, OR MEDIUM  (42)                 IS LOW OR VERY LOW  (73)                   X SQ=  19.76
                                                                                                   P =     0.
                                                                                                   RV YES (YES)

37  TEND TO BE THOSE                      0.82      37  TEND TO BE THOSE                      0.80        53    9
    WHOSE PER CAPITA GROSS NATIONAL PRODUCT             WHOSE PER CAPITA GROSS NATIONAL PRODUCT        12   37
    IS VERY HIGH, HIGH, MEDIUM, OR LOW  (64)            IS VERY LOW  (51)                          X SQ=  39.48
                                                                                                   P =     0.
                                                                                                   RV YES (YES)

40  TEND TO BE THOSE                      0.78      40  TEND TO BE THOSE                      0.60        51   17
    WHOSE INTERNATIONAL FINANCIAL STATUS                WHOSE INTERNATIONAL FINANCIAL STATUS           14   25
    IS VERY HIGH, HIGH,                                 IS VERY LOW  (39)                          X SQ=  14.30
    MEDIUM, OR LOW  (71)                                                                           P =     0.
                                                                                                   RV YES (YES)

41  TEND LESS TO BE THOSE                 0.73      41  ALWAYS ARE THOSE                      1.00        17    0
    WHOSE ECONOMIC DEVELOPMENTAL STATUS                 WHOSE ECONOMIC DEVELOPMENTAL STATUS            47   45
    IS INTERMEDIATE, UNDERDEVELOPED,                    IS INTERMEDIATE, UNDERDEVELOPED,           X SQ=  12.22
    OR VERY UNDERDEVELOPED  (94)                        OR VERY UNDERDEVELOPED  (94)               P =     0.
                                                                                                   RV NO  (NO )

42  TEND TO BE THOSE                      0.52      42  TEND TO BE THOSE                      0.98        33    1
    WHOSE ECONOMIC DEVELOPMENTAL STATUS                 WHOSE ECONOMIC DEVELOPMENTAL STATUS            30   44
    IS DEVELOPED OR INTERMEDIATE  (36)                  IS UNDERDEVELOPED OR                       X SQ=  28.34
                                                        VERY UNDERDEVELOPED  (76)                  P =     0.
                                                                                                   RV YES (YES)

43  TEND TO BE THOSE                      0.69      43  TEND TO BE THOSE                      0.78        43   10
    WHOSE ECONOMIC DEVELOPMENTAL STATUS                 WHOSE ECONOMIC DEVELOPMENTAL STATUS            19   36
    IS DEVELOPED, INTERMEDIATE, OR                      IS VERY UNDERDEVELOPED  (57)               X SQ=  22.09
    UNDERDEVELOPED  (55)                                                                           P =     0.
                                                                                                   RV YES (YES)

44  TEND LESS TO BE THOSE                 0.65      44  ALWAYS ARE THOSE                      1.00        23    0
    WHERE THE LITERACY RATE IS                          WHERE THE LITERACY RATE IS                     42   46
    BELOW NINETY PERCENT  (90)                          BELOW NINETY PERCENT  (90)                 X SQ=  18.43
                                                                                                   P =     0.
                                                                                                   RV NO  (YES)

45  TEND TO BE THOSE                      0.70      45  TEND TO BE THOSE                      0.80        43    9
    WHERE THE LITERACY RATE IS                          WHERE THE LITERACY RATE IS                     18   36
    FIFTY PERCENT OR ABOVE  (55)                        BELOW FIFTY PERCENT  (54)                  X SQ=  24.44
                                                                                                   P =     0.
                                                                                                   RV YES (YES)
```

46	TEND TO BE THOSE WHERE THE LITERACY RATE IS TEN PERCENT OR ABOVE (84)	0.92	46	TEND TO BE THOSE WHERE THE LITERACY RATE IS BELOW TEN PERCENT (26)	0.50	59 21 5 21 X SQ= 22.15 P = 0. RV YES (YES)
53	TEND LESS TO BE THOSE WHERE NEWSPAPER CIRCULATION IS LESS THAN THREE HUNDRED PER THOUSAND (101)	0.82	53	ALWAYS ARE THOSE WHERE NEWSPAPER CIRCULATION IS LESS THAN THREE HUNDRED PER THOUSAND (101)	1.00	12 0 53 46 X SQ= 7.70 P = 0.001 RV NO (NO)
54	TEND TO BE THOSE WHERE NEWSPAPER CIRCULATION IS ONE HUNDRED OR MORE PER THOUSAND (37)	0.51	54	TEND TO BE THOSE WHERE NEWSPAPER CIRCULATION IS LESS THAN ONE HUNDRED PER THOUSAND (76)	0.96	33 2 32 43 X SQ= 24.21 P = 0. RV YES (YES)
55	TEND TO BE THOSE WHERE NEWSPAPER CIRCULATION IS TEN OR MORE PER THOUSAND (78)	0.89	55	TEND TO BE THOSE WHERE NEWSPAPER CIRCULATION IS LESS THAN TEN PER THOUSAND (35)	0.62	58 17 7 28 X SQ= 30.12 P = 0. RV YES (YES)
56	TEND TO BE THOSE WHERE THE RELIGION IS PREDOMINANTLY LITERATE (79)	0.87	56	TEND TO BE THOSE WHERE THE RELIGION IS PREDOMINANTLY OR PARTLY NON-LITERATE (31)	0.52	56 21 8 23 X SQ= 18.26 P = 0. RV YES (YES)
57	TEND LESS TO BE THOSE WHERE THE RELIGION IS OTHER THAN CATHOLIC (90)	0.63	57	TEND MORE TO BE THOSE WHERE THE RELIGION IS OTHER THAN CATHOLIC (90)	0.98	24 1 41 45 X SQ= 16.70 P = 0. RV NO (YES)
58	LEAN MORE TOWARD BEING THOSE WHERE THE RELIGION IS OTHER THAN MUSLIM (97)	0.89	58	LEAN LESS TOWARD BEING THOSE WHERE THE RELIGION IS OTHER THAN MUSLIM (97)	0.76	7 11 58 35 X SQ= 2.53 P = 0.073 RV NO (YES)
63	TEND TO BE THOSE WHERE THE RELIGION IS PREDOMINANTLY SOME KIND OF CHRISTIAN (46)	0.66	63	TEND TO BE THOSE WHERE THE RELIGION IS PREDOMINANTLY OR PARTLY OTHER THAN CHRISTIAN (68)	0.96	42 2 22 44 X SQ= 39.36 P = 0. RV YES (YES)
66	TEND TO BE THOSE THAT ARE RELIGIOUSLY HOMOGENEOUS (57)	0.65	66	TEND TO BE THOSE THAT ARE RELIGIOUSLY HETEROGENEOUS (49)	0.61	41 16 22 25 X SQ= 5.80 P = 0.015 RV YES (YES)

68	TEND TO BE THOSE THAT ARE LINGUISTICALLY HOMOGENEOUS (52)	0.61	68	TEND TO BE THOSE THAT ARE LINGUISTICALLY WEAKLY HETEROGENEOUS OR STRONGLY HETEROGENEOUS (62)	0.80	39 9 25 37 X SQ= 16.98 P = 0. RV YES (YES)
69	TEND TO BE THOSE THAT ARE LINGUISTICALLY HOMOGENEOUS OR WEAKLY HETEROGENEOUS (64)	0.73	69	TEND TO BE THOSE THAT ARE LINGUISTICALLY STRONGLY HETEROGENEOUS (50)	0.72	47 13 17 33 X SQ= 20.25 P = 0. RV YES (YES)
70	LEAN LESS TOWARD BEING THOSE THAT ARE RELIGIOUSLY, RACIALLY, OR LINGUISTICALLY HETEROGENEOUS (85)	0.73	70	LEAN MORE TOWARD BEING THOSE THAT ARE RELIGIOUSLY, RACIALLY, OR LINGUISTICALLY HETEROGENEOUS (85)	0.88	16 5 44 38 X SQ= 2.63 P = 0.083 RV NO (NO)
74	TEND LESS TO BE THOSE THAT ARE NOT HISTORICALLY WESTERN (87)	0.63	74	ALWAYS ARE THOSE THAT ARE NOT HISTORICALLY WESTERN (87)	1.00	24 0 40 45 X SQ= 19.51 P = 0. RV NO (YES)
75	TEND TO BE THOSE THAT ARE HISTORICALLY WESTERN OR SIGNIFICANTLY WESTERNIZED (62)	0.80	75	TEND TO BE THOSE THAT ARE NOT HISTORICALLY WESTERN AND ARE NOT SIGNIFICANTLY WESTERNIZED (52)	0.80	51 9 13 37 X SQ= 36.63 P = 0. RV YES (YES)
76	TEND LESS TO BE THOSE THAT HAVE BEEN SIGNIFICANTLY OR PARTIALLY WESTERNIZED THROUGH A COLONIAL RELATIONSHIP, RATHER THAN BEING HISTORICALLY WESTERN (70)	0.51	76	ALWAYS ARE THOSE THAT HAVE BEEN SIGNIFICANTLY OR PARTIALLY WESTERNIZED THROUGH A COLONIAL RELATIONSHIP, RATHER THAN BEING HISTORICALLY WESTERN (70)	1.00	24 0 25 45 X SQ= 27.08 P = 0. RV NO (YES)
77	TEND TO BE THOSE THAT HAVE BEEN SIGNIFICANTLY WESTERNIZED, RATHER THAN PARTIALLY WESTERNIZED, THROUGH A COLONIAL RELATIONSHIP (28)	0.83	77	TEND TO BE THOSE THAT HAVE BEEN PARTIALLY WESTERNIZED, RATHER THAN SIGNIFICANTLY WESTERNIZED, THROUGH A COLONIAL RELATIONSHIP (41)	0.82	20 8 4 37 X SQ= 25.24 P = 0. RV YES (YES)
79	TEND TO BE THOSE THAT WERE FORMERLY DEPENDENCIES OF SPAIN, RATHER THAN BRITAIN OR FRANCE (18)	0.64	79	ALWAYS ARE THOSE THAT WERE FORMERLY DEPENDENCIES OF BRITAIN OR FRANCE, RATHER THAN SPAIN (49)	1.00	10 39 18 0 X SQ= 31.09 P = 0. RV YES (YES)
82	TEND TO BE THOSE WHERE THE TYPE OF POLITICAL MODERNIZATION IS EARLY OR LATER EUROPEAN OR EUROPEAN DERIVED, RATHER THAN DEVELOPED TUTELARY OR UNDEVELOPED TUTELARY (51)	0.88	82	ALWAYS ARE THOSE WHERE THE TYPE OF POLITICAL MODERNIZATION IS DEVELOPED TUTELARY OR UNDEVELOPED TUTELARY, RATHER THAN EARLY OR LATER EUROPEAN OR EUROPEAN DERIVED (55)	1.00	49 0 7 46 X SQ= 74.00 P = 0. RV YES (YES)

83	TEND LESS TO BE THOSE WHERE THE TYPE OF POLITICAL MODERNIZATION IS OTHER THAN NON-EUROPEAN AUTOCHTHONOUS (106)	0.86	83	ALWAYS ARE THOSE WHERE THE TYPE OF POLITICAL MODERNIZATION IS OTHER THAN NON-EUROPEAN AUTOCHTHONOUS (106)	1.00	9 0 56 46 X SQ= 5.20 P = 0.010 RV NO (NO)
84	ALWAYS ARE THOSE WHERE THE TYPE OF POLITICAL MODERNIZATION IS DEVELOPED TUTELARY, RATHER THAN UNDEVELOPED TUTELARY (31)	1.00	84	TEND TO BE THOSE WHERE THE TYPE OF POLITICAL MODERNIZATION IS UNDEVELOPED TUTELARY, RATHER THAN DEVELOPED TUTELARY (24)	0.52	7 22 0 24 X SQ= 4.74 P = 0.012 RV YES (NO)
85	TEND TO BE THOSE WHERE THE STAGE OF POLITICAL MODERNIZATION IS ADVANCED, RATHER THAN MID- OR EARLY TRANSITIONAL (60)	0.70	85	TEND TO BE THOSE WHERE THE STAGE OF POLITICAL MODERNIZATION IS MID- OR EARLY TRANSITIONAL, RATHER THAN ADVANCED (54)	0.76	45 11 19 35 X SQ= 21.24 P = 0. RV YES (YES)
86	TEND TO BE THOSE WHERE THE STAGE OF POLITICAL MODERNIZATION IS ADVANCED OR MID-TRANSITIONAL, RATHER THAN EARLY TRANSITIONAL (76)	0.95	86	TEND TO BE THOSE WHERE THE STAGE OF POLITICAL MODERNIZATION IS EARLY TRANSITIONAL, RATHER THAN ADVANCED OR MID-TRANSITIONAL (38)	0.76	61 11 3 35 X SQ= 57.22 P = 0. RV YES (YES)
87	TEND TO BE THOSE WHOSE IDEOLOGICAL ORIENTATION IS OTHER THAN DEVELOPMENTAL (58)	0.94	87	TEND TO BE THOSE WHOSE IDEOLOGICAL ORIENTATION IS DEVELOPMENTAL (31)	0.85	3 28 50 5 X SQ= 51.94 P = 0. RV YES (YES)
89	TEND TO BE THOSE WHOSE IDEOLOGICAL ORIENTATION IS CONVENTIONAL (33)	0.55	89	TEND TO BE THOSE WHOSE IDEOLOGICAL ORIENTATION IS OTHER THAN CONVENTIONAL (62)	0.95	30 2 25 35 X SQ= 21.43 P = 0. RV YES (YES)
91	TEND TO BE THOSE WHOSE IDEOLOGICAL ORIENTATION IS TRADITIONAL, RATHER THAN DEVELOPMENTAL (5)	0.57	91	TEND TO BE THOSE WHOSE IDEOLOGICAL ORIENTATION IS DEVELOPMENTAL, RATHER THAN TRADITIONAL (31)	0.97	3 28 4 1 X SQ= 9.47 P = 0.003 RV YES (YES)
93	LEAN MORE TOWARD BEING THOSE WHERE THE SYSTEM STYLE IS NON-MOBILIZATIONAL (78)	0.79	93	LEAN LESS TOWARD BEING THOSE WHERE THE SYSTEM STYLE IS NON-MOBILIZATIONAL (78)	0.62	13 17 49 28 X SQ= 2.87 P = 0.081 RV NO (YES)
95	TEND LESS TO BE THOSE WHERE THE STATUS OF THE REGIME IS CONSTITUTIONAL OR AUTHORITARIAN (95)	0.79	95	TEND MORE TO BE THOSE WHERE THE STATUS OF THE REGIME IS CONSTITUTIONAL OR AUTHORITARIAN (95)	0.98	49 44 13 1 X SQ= 6.49 P = 0.004 RV NO (NO)

96	TEND MORE TO BE THOSE WHERE THE STATUS OF THE REGIME IS CONSTITUTIONAL OR TOTALITARIAN (67)	0.82	96	TEND LESS TO BE THOSE WHERE THE STATUS OF THE REGIME IS CONSTITUTIONAL OR TOTALITARIAN (67)	0.59
					47 17 10 12 X SQ= 4.55 P = 0.035 RV NO (YES)
97	TEND TO BE THOSE WHERE THE STATUS OF THE REGIME IS TOTALITARIAN, RATHER THAN AUTHORITARIAN (16)	0.57	97	TEND TO BE THOSE WHERE THE STATUS OF THE REGIME IS AUTHORITARIAN, RATHER THAN TOTALITARIAN (23)	0.92
					10 12 13 1 X SQ= 6.40 P = 0.005 RV YES (YES)
98	TEND LESS TO BE THOSE WHERE GOVERNMENTAL STABILITY IS GENERALLY OR MODERATELY PRESENT AND DATES FROM THE POST-WAR PERIOD, OR IS ABSENT (93)	0.66	98	ALWAYS ARE THOSE WHERE GOVERNMENTAL STABILITY IS GENERALLY OR MODERATELY PRESENT AND DATES FROM THE POST-WAR PERIOD, OR IS ABSENT (93)	1.00
					22 0 43 46 X SQ= 17.35 P = 0. RV NO (YES)
101	TILT LESS TOWARD BEING THOSE WHERE THE REPRESENTATIVE CHARACTER OF THE REGIME IS LIMITED POLYARCHIC, PSEUDO-POLYARCHIC, OR NON-POLYARCHIC (57)	0.52	101	TILT MORE TOWARD BEING THOSE WHERE THE REPRESENTATIVE CHARACTER OF THE REGIME IS LIMITED POLYARCHIC, PSEUDO-POLYARCHIC, OR NON-POLYARCHIC (57)	0.68
					28 12 30 25 X SQ= 1.72 P = 0.142 RV NO (NO)
105	TEND TO BE THOSE WHERE THE ELECTORAL SYSTEM IS COMPETITIVE (43)	0.63	105	TEND TO BE THOSE WHERE THE ELECTORAL SYSTEM IS PARTIALLY COMPETITIVE OR NON-COMPETITIVE (47)	0.72
					32 10 19 26 X SQ= 8.98 P = 0.002 RV YES (YES)
107	TEND TO BE THOSE WHERE AUTONOMOUS GROUPS ARE FULLY TOLERATED IN POLITICS (46)	0.54	107	TEND TO BE THOSE WHERE AUTONOMOUS GROUPS ARE PARTIALLY TOLERATED IN POLITICS, ARE TOLERATED ONLY OUTSIDE POLITICS, OR ARE NOT TOLERATED AT ALL (65)	0.75
					34 11 29 33 X SQ= 7.77 P = 0.003 RV YES (YES)
110	TEND LESS TO BE THOSE WHERE POLITICAL ENCULTURATION IS MEDIUM OR LOW (80)	0.71	110	ALWAYS ARE THOSE WHERE POLITICAL ENCULTURATION IS MEDIUM OR LOW (80)	1.00
					15 0 37 41 X SQ= 12.05 P = 0. RV NO (YES)
114	LEAN TOWARD BEING THOSE WHERE SECTIONALISM IS NEGLIGIBLE (47)	0.50	114	LEAN TOWARD BEING THOSE WHERE SECTIONALISM IS EXTREME OR MODERATE (61)	0.68
					30 30 30 14 X SQ= 2.73 P = 0.073 RV YES (YES)
117	TEND TO BE THOSE WHERE INTEREST ARTICULATION BY ASSOCIATIONAL GROUPS IS SIGNIFICANT, MODERATE, OR LIMITED (60)	0.73	117	TEND TO BE THOSE WHERE INTEREST ARTICULATION BY ASSOCIATIONAL GROUPS IS NEGLIGIBLE (51)	0.70
					45 14 17 32 X SQ= 17.26 P = 0. RV YES (YES)

119	TEND LESS TO BE THOSE WHERE INTEREST ARTICULATION BY INSTITUTIONAL GROUPS IS VERY SIGNIFICANT OR SIGNIFICANT (74)	0.67	119	TEND MORE TO BE THOSE WHERE INTEREST ARTICULATION BY INSTITUTIONAL GROUPS IS VERY SIGNIFICANT OR SIGNIFICANT (74)	0.88	43 28 21 4 X SQ= 3.58 P = 0.047 RV NO (NO)
120	TEND LESS TO BE THOSE WHERE INTEREST ARTICULATION BY INSTITUTIONAL GROUPS IS VERY SIGNIFICANT, SIGNIFICANT, OR MODERATE (90)	0.84	120	ALWAYS ARE THOSE WHERE INTEREST ARTICULATION BY INSTITUTIONAL GROUPS IS VERY SIGNIFICANT, SIGNIFICANT, OR MODERATE (90)	1.00	54 32 10 0 X SQ= 4.03 P = 0.015 RV NO (NO)
121	TEND TO BE THOSE WHERE INTEREST ARTICULATION BY NON-ASSOCIATIONAL GROUPS IS MODERATE, LIMITED, OR NEGLIGIBLE (61)	0.78	121	TEND TO BE THOSE WHERE INTEREST ARTICULATION BY NON-ASSOCIATIONAL GROUPS IS SIGNIFICANT (54)	0.83	14 38 51 8 X SQ= 37.93 P = 0. RV YES (YES)
122	TEND LESS TO BE THOSE WHERE INTEREST ARTICULATION BY NON-ASSOCIATIONAL GROUPS IS SIGNIFICANT OR MODERATE (83)	0.54	122	ALWAYS ARE THOSE WHERE INTEREST ARTICULATION BY NON-ASSOCIATIONAL GROUPS IS SIGNIFICANT OR MODERATE (83)	1.00	35 46 30 0 X SQ= 26.80 P = 0. RV NO (YES)
125	TEND LESS TO BE THOSE WHERE INTEREST ARTICULATION BY ANOMIC GROUPS IS FREQUENT OR OCCASIONAL (64)	0.55	125	TEND MORE TO BE THOSE WHERE INTEREST ARTICULATION BY ANOMIC GROUPS IS FREQUENT OR OCCASIONAL (64)	0.77	29 33 24 10 X SQ= 4.12 P = 0.032 RV NO (YES)
126	TEND LESS TO BE THOSE WHERE INTEREST ARTICULATION BY ANOMIC GROUPS IS FREQUENT, OCCASIONAL, OR INFREQUENT (83)	0.72	126	ALWAYS ARE THOSE WHERE INTEREST ARTICULATION BY ANOMIC GROUPS IS FREQUENT, OCCASIONAL, OR INFREQUENT (83)	1.00	38 43 15 0 X SQ= 12.36 P = 0. RV NO (YES)
128	TEND TO BE THOSE WHERE INTEREST ARTICULATION BY POLITICAL PARTIES IS SIGNIFICANT OR MODERATE (48)	0.62	128	TEND TO BE THOSE WHERE INTEREST ARTICULATION BY POLITICAL PARTIES IS LIMITED OR NEGLIGIBLE (45)	0.62	33 14 20 23 X SQ= 4.28 P = 0.032 RV YES (YES)
129	TEND TO BE THOSE WHERE INTEREST ARTICULATION BY POLITICAL PARTIES IS SIGNIFICANT, MODERATE, OR LIMITED (56)	0.72	129	TEND TO BE THOSE WHERE INTEREST ARTICULATION BY POLITICAL PARTIES IS NEGLIGIBLE (37)	0.54	38 17 15 20 X SQ= 5.04 P = 0.017 RV YES (YES)
131	TEND TO BE THOSE WHERE INTEREST AGGREGATION BY POLITICAL PARTIES IS LIMITED OR NEGLIGIBLE (35)	0.65	131	TEND TO BE THOSE WHERE INTEREST AGGREGATION BY POLITICAL PARTIES IS SIGNIFICANT OR MODERATE (30)	0.65	14 15 26 8 X SQ= 4.22 P = 0.035 RV YES (YES)

133	TEND TO BE THOSE WHERE INTEREST AGGREGATION BY THE EXECUTIVE IS MODERATE, LIMITED, OR NEGLIGIBLE (76)	0.91	133	TEND TO BE THOSE WHERE INTEREST AGGREGATION BY THE EXECUTIVE IS SIGNIFICANT (29)	0.56	5 24 53 19 X SQ= 24.61 P = 0. RV YES (YES)
134	TEND TO BE THOSE WHERE INTEREST AGGREGATION BY THE EXECUTIVE IS LIMITED OR NEGLIGIBLE (46)	0.55	134	TEND TO BE THOSE WHERE INTEREST AGGREGATION BY THE EXECUTIVE IS SIGNIFICANT OR MODERATE (57)	0.73	26 30 32 11 X SQ= 6.74 P = 0.007 RV YES (YES)
138	TEND TO BE THOSE WHERE INTEREST AGGREGATION BY THE LEGISLATURE IS SIGNIFICANT, MODERATE, OR LIMITED (48)	0.65	138	TEND TO BE THOSE WHERE INTEREST AGGREGATION BY THE LEGISLATURE IS NEGLIGIBLE (49)	0.72	36 11 19 28 X SQ= 11.22 P = 0.001 RV YES (YES)
139	LEAN MORE TOWARD BEING THOSE WHERE THE PARTY SYSTEM IS QUANTITATIVELY OTHER THAN ONE-PARTY (71)	0.76	139	LEAN LESS TOWARD BEING THOSE WHERE THE PARTY SYSTEM IS QUANTITATIVELY OTHER THAN ONE-PARTY (71)	0.59	14 18 44 26 X SQ= 2.54 P = 0.087 RV NO (YES)
142	TEND LESS TO BE THOSE WHERE THE PARTY SYSTEM IS QUANTITATIVELY OTHER THAN MULTI-PARTY (66)	0.58	142	TEND MORE TO BE THOSE WHERE THE PARTY SYSTEM IS QUANTITATIVELY OTHER THAN MULTI-PARTY (66)	0.82	23 7 32 31 X SQ= 4.61 P = 0.024 RV NO (NO)
146	ALWAYS ARE THOSE WHERE THE PARTY SYSTEM IS QUALITATIVELY OTHER THAN MASS-BASED TERRITORIAL (92)	1.00	146	TEND LESS TO BE THOSE WHERE THE PARTY SYSTEM IS QUALITATIVELY OTHER THAN MASS-BASED TERRITORIAL (92)	0.80	0 8 55 33 X SQ= 9.29 P = 0.001 RV NO (YES)
147	TEND LESS TO BE THOSE WHERE THE PARTY SYSTEM IS QUALITATIVELY OTHER THAN CLASS-ORIENTED OR MULTI-IDEOLOGICAL (67)	0.61	147	TEND MORE TO BE THOSE WHERE THE PARTY SYSTEM IS QUALITATIVELY OTHER THAN CLASS-ORIENTED OR MULTI-IDEOLOGICAL (67)	0.92	19 3 30 35 X SQ= 9.23 P = 0.001 RV NO (YES)
148	ALWAYS ARE THOSE WHERE THE PARTY SYSTEM IS QUALITATIVELY OTHER THAN AFRICAN TRANSITIONAL (96)	1.00	148	TEND LESS TO BE THOSE WHERE THE PARTY SYSTEM IS QUALITATIVELY OTHER THAN AFRICAN TRANSITIONAL (96)	0.69	0 14 61 31 X SQ= 19.24 P = 0. RV NO (YES)
153	LEAN TOWARD BEING THOSE WHERE THE PARTY SYSTEM IS STABLE (42)	0.58	153	LEAN TOWARD BEING THOSE WHERE THE PARTY SYSTEM IS MODERATELY STABLE OR UNSTABLE (38)	0.67	32 7 23 14 X SQ= 2.83 P = 0.073 RV YES (NO)

157	TILT LESS TOWARD BEING THOSE WHERE PERSONALISMO IS MODERATE OR NEGLIGIBLE (84)	0.80	157	TILT MORE TOWARD BEING THOSE WHERE PERSONALISMO IS MODERATE OR NEGLIGIBLE (84)	0.92	11 3 45 36 X SQ= 1.75 P = 0.144 RV NO (NO)
161	TEND TO BE THOSE WHERE THE POLITICAL LEADERSHIP IS ELITIST OR MODERATELY ELITIST (47)	0.57	161	TEND TO BE THOSE WHERE THE POLITICAL LEADERSHIP IS NON-ELITIST (50)	0.67	33 12 25 24 X SQ= 4.04 P = 0.034 RV YES (NO)
164	TEND TO BE THOSE WHERE THE REGIME'S LEADERSHIP CHARISMA IS NEGLIGIBLE (65)	0.90	164	TEND TO BE THOSE WHERE THE REGIME'S LEADERSHIP CHARISMA IS PRONOUNCED OR MODERATE (34)	0.73	6 27 52 10 X SQ= 36.37 P = 0. RV YES (YES)
169	LEAN TOWARD BEING THOSE WHERE THE HORIZONTAL POWER DISTRIBUTION IS SIGNIFICANT OR LIMITED (58)	0.63	169	LEAN TOWARD BEING THOSE WHERE THE HORIZONTAL POWER DISTRIBUTION IS NEGLIGIBLE (48)	0.56	38 19 22 24 X SQ= 2.98 P = 0.071 RV YES (YES)
170	LEAN TOWARD BEING THOSE WHERE THE LEGISLATIVE-EXECUTIVE STRUCTURE IS OTHER THAN PRESIDENTIAL (63)	0.68	170	LEAN TOWARD BEING THOSE WHERE THE LEGISLATIVE-EXECUTIVE STRUCTURE IS PRESIDENTIAL (39)	0.50	19 20 40 20 X SQ= 2.46 P = 0.095 RV YES (YES)
175	TEND TO BE THOSE WHERE THE LEGISLATURE IS FULLY EFFECTIVE OR PARTIALLY EFFECTIVE (51)	0.63	175	TEND TO BE THOSE WHERE THE LEGISLATURE IS LARGELY INEFFECTIVE OR WHOLLY INEFFECTIVE (49)	0.63	35 15 21 26 X SQ= 5.37 P = 0.014 RV YES (YES)
178	TEND TO BE THOSE WHERE THE LEGISLATURE IS BICAMERAL (51)	0.63	178	TEND TO BE THOSE WHERE THE LEGISLATURE IS UNICAMERAL (53)	0.69	22 29 37 13 X SQ= 8.67 P = 0.002 RV YES (YES)
179	TILT TOWARD BEING THOSE WHERE THE EXECUTIVE IS STRONG (39)	0.50	179	TILT TOWARD BEING THOSE WHERE THE EXECUTIVE IS DOMINANT (52)	0.68	27 23 27 11 X SQ= 1.98 P = 0.125 RV YES (NO)
181	LEAN LESS TOWARD BEING THOSE WHERE THE BUREAUCRACY IS SEMI-MODERN, RATHER THAN MODERN (55)	0.67	181	LEAN MORE TOWARD BEING THOSE WHERE THE BUREAUCRACY IS SEMI-MODERN, RATHER THAN MODERN (55)	0.93	19 1 38 14 X SQ= 2.98 P = 0.053 RV NO (NO)

182	ALWAYS ARE THOSE WHERE THE BUREAUCRACY IS SEMI-MODERN, RATHER THAN POST-COLONIAL TRANSITIONAL (55)	1.00	182	TEND TO BE THOSE WHERE THE BUREAUCRACY IS POST-COLONIAL TRANSITIONAL, RATHER THAN SEMI-MODERN (25)	0.64	38 14 0 25 X SQ= 33.21 P = 0. RV YES (YES)
183	ALWAYS ARE THOSE WHERE THE BUREAUCRACY IS TRADITIONAL, RATHER THAN POST-COLONIAL TRANSITIONAL (9)	1.00	183	TEND TO BE THOSE WHERE THE BUREAUCRACY IS POST-COLONIAL TRANSITIONAL, RATHER THAN TRADITIONAL (25)	0.89	0 25 6 3 X SQ= 15.91 P = 0. RV YES (YES)
186	LEAN LESS TOWARD BEING THOSE WHERE PARTICIPATION BY THE MILITARY IN POLITICS IS NEUTRAL, RATHER THAN SUPPORTIVE (56)	0.57	186	LEAN MORE TOWARD BEING THOSE WHERE PARTICIPATION BY THE MILITARY IN POLITICS IS NEUTRAL, RATHER THAN SUPPORTIVE (56)	0.76	20 9 26 29 X SQ= 2.78 P = 0.068 RV NO (YES)
188	TEND LESS TO BE THOSE WHERE THE CHARACTER OF THE LEGAL SYSTEM IS OTHER THAN CIVIL LAW (81)	0.52	188	ALWAYS ARE THOSE WHERE THE CHARACTER OF THE LEGAL SYSTEM IS OTHER THAN CIVIL LAW (81)	1.00	31 0 33 45 X SQ= 28.13 P = 0. RV NO (YES)
190	TEND TO BE THOSE WHERE THE CHARACTER OF THE LEGAL SYSTEM IS CIVIL LAW, RATHER THAN COMMON LAW (32)	0.86	190	ALWAYS ARE THOSE WHERE THE CHARACTER OF THE LEGAL SYSTEM IS COMMON LAW, RATHER THAN CIVIL LAW (7)	1.00	31 0 5 2 X SQ= 4.50 P = 0.030 RV YES (NO)
192	TEND TO BE THOSE WHERE THE CHARACTER OF THE LEGAL SYSTEM IS OTHER THAN MUSLIM OR PARTLY MUSLIM (86)	0.94	192	TEND TO BE THOSE WHERE THE CHARACTER OF THE LEGAL SYSTEM IS MUSLIM OR PARTLY MUSLIM (27)	0.51	4 23 60 22 X SQ= 26.18 P = 0. RV YES (YES)
193	TEND TO BE THOSE WHERE THE CHARACTER OF THE LEGAL SYSTEM IS OTHER THAN PARTLY INDIGENOUS (86)	0.95	193	TEND TO BE THOSE WHERE THE CHARACTER OF THE LEGAL SYSTEM IS PARTLY INDIGENOUS (28)	0.56	3 25 62 20 X SQ= 33.73 P = 0. RV YES (YES)
194	TEND LESS TO BE THOSE THAT ARE NON-COMMUNIST (101)	0.84	194	TEND MORE TO BE THOSE THAT ARE NON-COMMUNIST (101)	0.98	10 1 54 45 X SQ= 3.99 P = 0.024 RV NO (NO)

74 POLITIES
THAT ARE HISTORICALLY WESTERN (26)

74 POLITIES
THAT ARE NOT HISTORICALLY WESTERN (87)

BOTH SUBJECT AND PREDICATE

26 IN LEFT COLUMN

AUSTRALIA	AUSTRIA	BELGIUM	CANADA	CZECHOS'KIA	DENMARK	FINLAND	FRANCE	GERMANY, E	GERMAN FR
GREECE	HUNGARY	ICELAND	IRELAND	ITALY	LUXEMBOURG	NETHERLANDS	NEW ZEALAND	NORWAY	POLAND
PORTUGAL	SPAIN	SWEDEN	SWITZERLAND	UK	US				

87 IN RIGHT COLUMN

AFGHANISTAN	ALBANIA	ALGERIA	ARGENTINA	BOLIVIA	BRAZIL	BULGARIA	BURMA	BURUNDI	CAMBODIA
CAMEROUN	CEN AFR REP	CEYLON	CHAD	CHILE	CHINA, PR	COLOMBIA	CONGO(BRA)	CONGO(LEO)	COSTA RICA
CUBA	CYPRUS	DAHOMEY	DOMIN REP	ECUADOR	EL SALVADOR	ETHIOPIA	GABON	GHANA	GUATEMALA
GUINEA	HAITI	HONDURAS	INDIA	INDONESIA	IRAN	IRAQ	IVORY COAST	JAMAICA	JAPAN
JORDAN	KOREA, N	KOREA REP	LAOS	LEBANON	LIBERIA	LIBYA	MALAGASY R	MALAYA	MALI
MAURITANIA	MEXICO	MONGOLIA	MOROCCO	NEPAL	NICARAGUA	NIGER	NIGERIA	PAKISTAN	PANAMA
PARAGUAY	PERU	PHILIPPINES	RUMANIA	RWANDA	SA'U ARABIA	SENEGAL	SIERRE LEO	SOMALIA	SUDAN
SYRIA	TANGANYIKA	THAILAND	TOGO	TRINIDAD	TUNISIA	TURKEY	UGANDA	USSR	UAR
UPPER VOLTA	URUGUAY	VENEZUELA	VIETNAM, N	VIETNAM REP	YEMEN	YUGOSLAVIA			

2 EXCLUDED BECAUSE AMBIGUOUS

ISRAEL SO AFRICA

---

2 TEND TO BE THOSE 0.85
LOCATED IN WEST EUROPE, SCANDINAVIA,
NORTH AMERICA, OR AUSTRALASIA (22)

2 ALWAYS ARE THOSE 1.00
LOCATED ELSEWHERE THAN IN
WEST EUROPE, SCANDINAVIA,
NORTH AMERICA, OR AUSTRALASIA (93)

22 0
4 87
X SQ= 86.10
P = 0.
RV YES (YES)

5 TEND TO BE THOSE 0.82
LOCATED IN SCANDINAVIA OR WEST EUROPE,
RATHER THAN IN EAST EUROPE (18)

5 ALWAYS ARE THOSE 1.00
LOCATED IN EAST EUROPE, RATHER THAN
IN SCANDINAVIA OR WEST EUROPE (9)

18 0
4 5
X SQ= 8.87
P = 0.002
RV YES (YES)

6 ALWAYS ARE THOSE 1.00
LOCATED ELSEWHERE THAN IN THE
CARIBBEAN, CENTRAL AMERICA,
OR SOUTH AMERICA (93)

6 TEND LESS TO BE THOSE 0.75
LOCATED ELSEWHERE THAN IN THE
CARIBBEAN, CENTRAL AMERICA,
OR SOUTH AMERICA (93)

0 22
26 65
X SQ= 6.63
P = 0.002
RV NO (NO)

8	ALWAYS ARE THOSE LOCATED ELSEWHERE THAN IN EAST ASIA SOUTH ASIA, OR SOUTHEAST ASIA (97)	1.00	8	TEND LESS TO BE THOSE LOCATED ELSEWHERE THAN IN EAST ASIA SOUTH ASIA, OR SOUTHEAST ASIA (97)	0.79	0 18 26 69 X SQ= 4.95 P = 0.007 RV NO (NO)
10	ALWAYS ARE THOSE LOCATED ELSEWHERE THAN IN NORTH AFRICA, OR CENTRAL AND SOUTH AFRICA (82)	1.00	10	TEND LESS TO BE THOSE LOCATED ELSEWHERE THAN IN NORTH AFRICA, OR CENTRAL AND SOUTH AFRICA (82)	0.63	0 32 26 55 X SQ= 11.59 P = 0. RV NO (NO)
13	ALWAYS ARE THOSE LOCATED ELSEWHERE THAN IN NORTH AFRICA OR THE MIDDLE EAST (99)	1.00	13	TEND LESS TO BE THOSE LOCATED ELSEWHERE THAN IN NORTH AFRICA OR THE MIDDLE EAST (99)	0.83	0 15 26 72 X SQ= 3.78 P = 0.021 RV NO (NO)
21	TEND MORE TO BE THOSE WHOSE TERRITORIAL SIZE IS MEDIUM OR SMALL (83)	0.88	21	TEND LESS TO BE THOSE WHOSE TERRITORIAL SIZE IS MEDIUM OR SMALL (83)	0.68	3 28 23 59 X SQ= 3.31 P = 0.046 RV NO (NO)
24	LEAN TOWARD BEING THOSE WHOSE POPULATION IS VERY LARGE, LARGE, OR MEDIUM (61)	0.69	24	LEAN TOWARD BEING THOSE WHOSE POPULATION IS SMALL (54)	0.52	18 42 8 45 X SQ= 2.74 P = 0.075 RV YES (NO)
26	TEND TO BE THOSE WHOSE POPULATION DENSITY IS VERY HIGH, HIGH, OR MEDIUM (48)	0.69	26	TEND TO BE THOSE WHOSE POPULATION DENSITY IS LOW (67)	0.67	18 29 8 58 X SQ= 9.19 P = 0.001 RV YES (NO)
28	TEND TO BE THOSE WHOSE POPULATION GROWTH RATE IS LOW (48)	0.85	28	TEND TO BE THOSE WHOSE POPULATION GROWTH RATE IS HIGH (62)	0.68	4 56 22 26 X SQ= 20.29 P = 0. RV YES (NO)
29	ALWAYS ARE THOSE WHERE THE DEGREE OF URBANIZATION IS HIGH (56)	1.00	29	TEND TO BE THOSE WHERE THE DEGREE OF URBANIZATION IS LOW (49)	0.63	25 29 0 49 X SQ= 27.49 P = 0. RV YES (NO)
30	ALWAYS ARE THOSE WHOSE AGRICULTURAL POPULATION IS MEDIUM, LOW, OR VERY LOW (57)	1.00	30	TEND TO BE THOSE WHOSE AGRICULTURAL POPULATION IS HIGH (56)	0.66	0 56 26 29 X SQ= 31.98 P = 0. RV YES (NO)

31	TEND TO BE THOSE WHOSE AGRICULTURAL POPULATION IS LOW OR VERY LOW (24)	0.69	31	TEND TO BE THOSE WHOSE AGRICULTURAL POPULATION IS HIGH OR MEDIUM (90)	0.95	8 82 18 4 X SQ= 48.74 P = 0. RV YES (YES)
33	TEND TO BE THOSE WHOSE GROSS NATIONAL PRODUCT IS VERY HIGH, HIGH, OR MEDIUM (30)	0.65	33	TEND TO BE THOSE WHOSE GROSS NATIONAL PRODUCT IS LOW OR VERY LOW (85)	0.86	17 12 9 75 X SQ= 25.29 P = 0. RV YES (YES)
34	TEND TO BE THOSE WHOSE GROSS NATIONAL PRODUCT IS VERY HIGH, HIGH, MEDIUM, OR LOW (62)	0.92	34	TEND TO BE THOSE WHOSE GROSS NATIONAL PRODUCT IS VERY LOW (53)	0.59	24 36 2 51 X SQ= 18.85 P = 0. RV YES (NO)
35	TEND TO BE THOSE WHOSE PER CAPITA GROSS NATIONAL PRODUCT IS VERY HIGH OR HIGH (24)	0.81	35	TEND TO BE THOSE WHOSE PER CAPITA GROSS NATIONAL PRODUCT IS MEDIUM, LOW, OR VERY LOW (91)	0.98	21 2 5 85 X SQ= 71.27 P = 0. RV YES (YES)
36	TEND TO BE THOSE WHOSE PER CAPITA GROSS NATIONAL PRODUCT IS VERY HIGH, HIGH, OR MEDIUM (42)	0.96	36	TEND TO BE THOSE WHOSE PER CAPITA GROSS NATIONAL PRODUCT IS LOW OR VERY LOW (73)	0.83	25 15 1 72 X SQ= 51.11 P = 0. RV YES (YES)
37	ALWAYS ARE THOSE WHOSE PER CAPITA GROSS NATIONAL PRODUCT IS VERY HIGH, HIGH, MEDIUM, OR LOW (64)	1.00	37	TEND TO BE THOSE WHOSE PER CAPITA GROSS NATIONAL PRODUCT IS VERY LOW (51)	0.59	26 36 0 51 X SQ= 25.46 P = 0. RV YES (NO)
39	TEND TO BE THOSE WHOSE INTERNATIONAL FINANCIAL STATUS IS VERY HIGH, HIGH, OR MEDIUM (38)	0.80	39	TEND TO BE THOSE WHOSE INTERNATIONAL FINANCIAL STATUS IS LOW OR VERY LOW (76)	0.80	20 17 5 70 X SQ= 29.41 P = 0. RV YES (YES)
41	TEND TO BE THOSE WHOSE ECONOMIC DEVELOPMENTAL STATUS IS DEVELOPED (19)	0.69	41	TEND TO BE THOSE WHOSE ECONOMIC DEVELOPMENTAL STATUS IS INTERMEDIATE, UNDERDEVELOPED, OR VERY UNDERDEVELOPED (94)	0.99	18 1 8 85 X SQ= 60.92 P = 0. RV YES (YES)
42	TEND TO BE THOSE WHOSE ECONOMIC DEVELOPMENTAL STATUS IS DEVELOPED OR INTERMEDIATE (36)	0.92	42	TEND TO BE THOSE WHOSE ECONOMIC DEVELOPMENTAL STATUS IS UNDERDEVELOPED OR VERY UNDERDEVELOPED (76)	0.87	24 11 2 74 X SQ= 54.47 P = 0. RV YES (YES)

No.	Left column	Value	Right column	Value	Statistics
43	ALWAYS ARE THOSE WHOSE ECONOMIC DEVELOPMENTAL STATUS IS DEVELOPED, INTERMEDIATE, OR UNDERDEVELOPED (55)	1.00	TEND TO BE THOSE WHOSE ECONOMIC DEVELOPMENTAL STATUS IS VERY UNDERDEVELOPED (57)	0.68	26 27 0 57 X SQ= 33.95 P = 0. RV YES (NO)
44	TEND TO BE THOSE WHERE THE LITERACY RATE IS NINETY PERCENT OR ABOVE (25)	0.85	TEND TO BE THOSE WHERE THE LITERACY RATE IS BELOW NINETY PERCENT (90)	0.97	22 3 4 84 X SQ= 71.91 P = 0. RV YES (YES)
45	ALWAYS ARE THOSE WHERE THE LITERACY RATE IS FIFTY PERCENT OR ABOVE (55)	1.00	TEND TO BE THOSE WHERE THE LITERACY RATE IS BELOW FIFTY PERCENT (54)	0.65	26 28 0 53 X SQ= 31.14 P = 0. RV YES (NO)
50	TEND TO BE THOSE WHERE FREEDOM OF THE PRESS IS COMPLETE (43)	0.72	TEND TO BE THOSE WHERE FREEDOM OF THE PRESS IS INTERMITTENT, INTERNALLY ABSENT, OR INTERNALLY AND EXTERNALLY ABSENT (56)	0.65	18 25 7 47 X SQ= 8.99 P = 0.002 RV YES (NO)
53	TEND TO BE THOSE WHERE NEWSPAPER CIRCULATION IS THREE HUNDRED OR MORE PER THOUSAND (14)	0.50	TEND TO BE THOSE WHERE NEWSPAPER CIRCULATION IS LESS THAN THREE HUNDRED PER THOUSAND (101)	0.99	13 1 13 86 X SQ= 39.62 P = 0. RV YES (YES)
54	TEND TO BE THOSE WHERE NEWSPAPER CIRCULATION IS ONE HUNDRED OR MORE PER THOUSAND (37)	0.92	TEND TO BE THOSE WHERE NEWSPAPER CIRCULATION IS LESS THAN ONE HUNDRED PER THOUSAND (76)	0.86	24 12 2 73 X SQ= 52.04 P = 0. RV YES (YES)
55	ALWAYS ARE THOSE WHERE NEWSPAPER CIRCULATION IS TEN OR MORE PER THOUSAND (78)	1.00	TEND LESS TO BE THOSE WHERE NEWSPAPER CIRCULATION IS TEN OR MORE PER THOUSAND (78)	0.59	26 50 0 35 X SQ= 13.79 P = 0. RV NO (NO)
56	ALWAYS ARE THOSE WHERE THE RELIGION IS PREDOMINANTLY LITERATE (79)	1.00	TEND LESS TO BE THOSE WHERE THE RELIGION IS PREDOMINANTLY LITERATE (79)	0.63	26 52 0 30 X SQ= 11.41 P = 0. RV NO (NO)
62	TEND LESS TO BE THOSE WHERE THE RELIGION IS CATHOLIC, RATHER THAN PROTESTANT (25)	0.64	ALWAYS ARE THOSE WHERE THE RELIGION IS CATHOLIC, RATHER THAN PROTESTANT (25)	1.00	5 0 9 16 X SQ= 4.53 P = 0.014 RV NO (YES)

63	ALWAYS ARE THOSE WHERE THE RELIGION IS PREDOMINANTLY SOME KIND OF CHRISTIAN (46)	1.00	63	TEND TO BE THOSE WHERE THE RELIGION IS PREDOMINANTLY OR PARTLY OTHER THAN CHRISTIAN (68)	0.77	26 20 0 66 X SQ= 45.46 P = 0. RV YES (YES)
67	TEND MORE TO BE THOSE THAT ARE RACIALLY HOMOGENEOUS (82)	0.96	67	TEND LESS TO BE THOSE THAT ARE RACIALLY HOMOGENEOUS (82)	0.67	25 54 1 27 X SQ= 7.40 P = 0.002 RV NO (NO)
68	TEND TO BE THOSE THAT ARE LINGUISTICALLY HOMOGENEOUS (52)	0.77	68	TEND TO BE THOSE THAT ARE LINGUISTICALLY WEAKLY HETEROGENEOUS OR STRONGLY HETEROGENEOUS (62)	0.63	20 32 6 54 X SQ= 11.11 P = 0.001 RV YES (NO)
69	TEND TO BE THOSE THAT ARE LINGUISTICALLY HOMOGENEOUS OR WEAKLY HETEROGENEOUS (64)	0.85	69	TEND TO BE THOSE THAT ARE LINGUISTICALLY STRONGLY HETEROGENEOUS (50)	0.51	22 42 4 44 X SQ= 9.03 P = 0.001 RV YES (NO)
70	TEND LESS TO BE THOSE THAT ARE RELIGIOUSLY, RACIALLY, OR LINGUISTICALLY HETEROGENEOUS (85)	0.54	70	TEND MORE TO BE THOSE THAT ARE RELIGIOUSLY, RACIALLY, OR LINGUISTICALLY HETEROGENEOUS (85)	0.88	12 9 14 68 X SQ= 12.18 P = 0. RV NO (YES)
71	TEND TO BE THOSE WHOSE DATE OF INDEPENDENCE IS BEFORE 1800 (21)	0.50	71	TEND TO BE THOSE WHOSE DATE OF INDEPENDENCE IS AFTER 1800 (90)	0.89	12 9 12 76 X SQ= 16.24 P = 0. RV YES (YES)
72	TEND TO BE THOSE WHOSE DATE OF INDEPENDENCE IS BEFORE 1914 (52)	0.75	72	TEND TO BE THOSE WHOSE DATE OF INDEPENDENCE IS AFTER 1914 (59)	0.61	18 33 6 52 X SQ= 8.44 P = 0.002 RV YES (NO)
73	ALWAYS ARE THOSE WHOSE DATE OF INDEPENDENCE IS BEFORE 1945 (65)	1.00	73	TEND TO BE THOSE WHOSE DATE OF INDEPENDENCE IS AFTER 1945 (46)	0.53	24 40 0 45 X SQ= 19.51 P = 0. RV YES (NO)
81	TEND LESS TO BE THOSE WHERE THE TYPE OF POLITICAL MODERNIZATION IS LATER EUROPEAN OR LATER EUROPEAN DERIVED, RATHER THAN EARLY EUROPEAN OR EARLY EUROPEAN DERIVED (40)	0.62	81	ALWAYS ARE THOSE WHERE THE TYPE OF POLITICAL MODERNIZATION IS LATER EUROPEAN OR LATER EUROPEAN DERIVED, RATHER THAN EARLY EUROPEAN OR EARLY EUROPEAN DERIVED (40)	1.00	10 0 16 24 X SQ= 9.26 P = 0.001 RV NO (YES)

```
82  ALWAYS ARE THOSE                              1.00     82  TEND TO BE THOSE                              0.69        26    24
    WHERE THE TYPE OF POLITICAL MODERNIZATION               WHERE THE TYPE OF POLITICAL MODERNIZATION                0    54
    IS EARLY OR LATER EUROPEAN OR                           IS DEVELOPED TUTELARY OR                          X SQ=  34.72
    EUROPEAN DERIVED, RATHER THAN                           UNDEVELOPED TUTELARY, RATHER THAN                 P =     0.
    DEVELOPED TUTELARY OR                                   EARLY OR LATER EUROPEAN OR                        RV YES (YES)
    UNDEVELOPED TUTELARY  (51)                              EUROPEAN DERIVED  (55)

83  ALWAYS ARE THOSE                              1.00     83  TILT LESS TOWARD BEING THOSE                  0.90         0     9
    WHERE THE TYPE OF POLITICAL MODERNIZATION               WHERE THE TYPE OF POLITICAL MODERNIZATION               26    78
    IS OTHER THAN                                           IS OTHER THAN                                     X SQ=   1.68
    NON-EUROPEAN AUTOCHTHONOUS  (106)                       NON-EUROPEAN AUTOCHTHONOUS  (106)                 P =     0.115
                                                                                                              RV NO  (NO )

85  ALWAYS ARE THOSE                              1.00     85  TEND TO BE THOSE                              0.62        26    33
    WHERE THE STAGE OF                                      WHERE THE STAGE OF                                       0    53
    POLITICAL MODERNIZATION IS                              POLITICAL MODERNIZATION IS                        X SQ=  28.00
    ADVANCED, RATHER THAN                                   MID- OR EARLY TRANSITIONAL,                       P =     0.
    MID- OR EARLY TRANSITIONAL  (60)                        RATHER THAN ADVANCED  (54)                        RV YES (NO )

87  ALWAYS ARE THOSE                              1.00     87  TEND LESS TO BE THOSE                         0.51         0    30
    WHOSE IDEOLOGICAL ORIENTATION                           WHOSE IDEOLOGICAL ORIENTATION                           26    31
    IS OTHER THAN DEVELOPMENTAL  (58)                       IS OTHER THAN DEVELOPMENTAL  (58)                 X SQ=  17.40
                                                                                                              P =     0.
                                                                                                              RV NO  (NO )

89  TEND TO BE THOSE                              0.83     89  TEND TO BE THOSE                              0.83        20    12
    WHOSE IDEOLOGICAL ORIENTATION                           WHOSE IDEOLOGICAL ORIENTATION                            4    57
    IS CONVENTIONAL  (33)                                   IS OTHER THAN CONVENTIONAL  (62)                  X SQ=  31.45
                                                                                                              P =     0.
                                                                                                              RV YES (YES)

93  TILT MORE TOWARD BEING THOSE                  0.84     93  TILT LESS TOWARD BEING THOSE                  0.67         4    27
    WHERE THE SYSTEM STYLE                                  WHERE THE SYSTEM STYLE                                  21    56
    IS NON-MOBILIZATIONAL  (78)                             IS NON-MOBILIZATIONAL  (78)                       X SQ=   1.82
                                                                                                              P =     0.135
                                                                                                              RV NO  (NO )

94  TEND TO BE THOSE                              0.77     94  TEND TO BE THOSE                              0.54        20    30
    WHERE THE STATUS OF THE REGIME                          WHERE THE STATUS OF THE REGIME                           6    35
    IS CONSTITUTIONAL  (51)                                 IS AUTHORITARIAN OR                               X SQ=   5.91
                                                            TOTALITARIAN  (41)                                P =     0.010
                                                                                                              RV YES (NO )

96  ALWAYS ARE THOSE                              1.00     96  TEND LESS TO BE THOSE                         0.63        26    40
    WHERE THE STATUS OF THE REGIME                          WHERE THE STATUS OF THE REGIME                           0    23
    IS CONSTITUTIONAL OR                                    IS CONSTITUTIONAL OR                              X SQ=  10.97
    TOTALITARIAN (67)                                       TOTALITARIAN (67)                                 P =     0.
                                                                                                              RV NO  (NO )

97  ALWAYS ARE THOSE                              1.00     97  TEND TO BE THOSE                              0.70         0    23
    WHERE THE STATUS OF THE REGIME                          WHERE THE STATUS OF THE REGIME                           6    10
    IS TOTALITARIAN, RATHER THAN                            IS AUTHORITARIAN, RATHER THAN                     X SQ=   7.52
    AUTHORITARIAN  (16)                                     TOTALITARIAN  (23)                                P =     0.002
                                                                                                              RV YES (NO )
```

98	TEND TO BE THOSE WHERE GOVERNMENTAL STABILITY IS GENERALLY PRESENT AND DATES AT LEAST FROM THE INTERWAR PERIOD (22)	0.65	98	TEND TO BE THOSE WHERE GOVERNMENTAL STABILITY IS GENERALLY OR MODERATELY PRESENT AND DATES FROM THE POST-WAR PERIOD, OR IS ABSENT (93)	0.95	17 4 9 83 X SQ= 44.95 P = 0. RV YES (YES)
99	TEND TO BE THOSE WHERE GOVERNMENTAL STABILITY IS GENERALLY PRESENT AND DATES FROM AT LEAST THE INTER-WAR PERIOD, OR FROM THE POST-WAR PERIOD (50)	0.96	99	TEND TO BE THOSE WHERE GOVERNMENTAL STABILITY IS MODERATELY PRESENT AND DATES FROM THE POST-WAR PERIOD, OR IS ABSENT (36)	0.59	24 24 1 35 X SQ= 19.74 P = 0. RV YES (YES)
100	ALWAYS ARE THOSE WHERE GOVERNMENTAL STABILITY IS GENERALLY PRESENT AND DATES FROM AT LEAST THE INTER-WAR PERIOD, OR IS GENERALLY OR MODERATELY PRESENT, AND DATES FROM THE POST-WAR PERIOD (64)	1.00	100	TEND LESS TO BE THOSE WHERE GOVERNMENTAL STABILITY IS GENERALLY PRESENT AND DATES FROM AT LEAST THE INTER-WAR PERIOD, OR IS GENERALLY OR MODERATELY PRESENT, AND DATES FROM THE POST-WAR PERIOD (64)	0.62	26 36 0 22 X SQ= 11.47 P = 0. RV NO (NO)
101	TEND TO BE THOSE WHERE THE REPRESENTATIVE CHARACTER OF THE REGIME IS POLYARCHIC (41)	0.77	101	TEND TO BE THOSE WHERE THE REPRESENTATIVE CHARACTER OF THE REGIME IS LIMITED POLYARCHIC, PSEUDO-POLYARCHIC, OR NON-POLYARCHIC (57)	0.71	20 20 6 50 X SQ= 16.30 P = 0. RV YES (YES)
102	TEND TO BE THOSE WHERE THE REPRESENTATIVE CHARACTER OF THE REGIME IS POLYARCHIC OR LIMITED POLYARCHIC (59)	0.77	102	TEND TO BE THOSE WHERE THE REPRESENTATIVE CHARACTER OF THE REGIME IS PSEUDO-POLYARCHIC OR NON-POLYARCHIC (49)	0.54	20 37 6 43 X SQ= 6.24 P = 0.007 RV YES (NO)
105	TEND TO BE THOSE WHERE THE ELECTORAL SYSTEM IS COMPETITIVE (43)	0.77	105	TEND TO BE THOSE WHERE THE ELECTORAL SYSTEM IS PARTIALLY COMPETITIVE OR NON-COMPETITIVE (47)	0.65	20 22 6 41 X SQ= 11.40 P = 0. RV YES (NO)
107	TEND TO BE THOSE WHERE AUTONOMOUS GROUPS ARE FULLY TOLERATED IN POLITICS (46)	0.77	107	TEND TO BE THOSE WHERE AUTONOMOUS GROUPS ARE PARTIALLY TOLERATED IN POLITICS, ARE TOLERATED ONLY OUTSIDE POLITICS, OR ARE NOT TOLERATED AT ALL (65)	0.70	20 25 6 58 X SQ= 16.01 P = 0. RV YES (NO)
110	TEND TO BE THOSE WHERE POLITICAL ENCULTURATION IS HIGH (15)	0.52	110	TEND TO BE THOSE WHERE POLITICAL ENCULTURATION IS MEDIUM OR LOW (80)	0.94	11 4 10 68 X SQ= 23.00 P = 0. RV YES (YES)
111	TEND TO BE THOSE WHERE POLITICAL ENCULTURATION IS HIGH OR MEDIUM (53)	0.90	111	TEND TO BE THOSE WHERE POLITICAL ENCULTURATION IS LOW (42)	0.54	19 33 2 39 X SQ= 11.40 P = 0. RV YES (NO)

No.	Left	Value	No.	Right	Value	Statistics
115	TEND TO BE THOSE WHERE INTEREST ARTICULATION BY ASSOCIATIONAL GROUPS IS SIGNIFICANT (20)	0.67	115	TEND TO BE THOSE WHERE INTEREST ARTICULATION BY ASSOCIATIONAL GROUPS IS MODERATE, LIMITED, OR NEGLIGIBLE (91)	0.95	16 4 8 81 X SQ= 43.91 P = 0. RV YES (YES)
116	TEND TO BE THOSE WHERE INTEREST ARTICULATION BY ASSOCIATIONAL GROUPS IS SIGNIFICANT OR MODERATE (32)	0.83	116	TEND TO BE THOSE WHERE INTEREST ARTICULATION BY ASSOCIATIONAL GROUPS IS LIMITED OR NEGLIGIBLE (79)	0.88	20 10 4 75 X SQ= 44.54 P = 0. RV YES (YES)
117	TEND TO BE THOSE WHERE INTEREST ARTICULATION BY ASSOCIATIONAL GROUPS IS SIGNIFICANT, MODERATE, OR LIMITED (60)	0.92	117	TEND TO BE THOSE WHERE INTEREST ARTICULATION BY ASSOCIATIONAL GROUPS IS NEGLIGIBLE (51)	0.58	22 36 2 49 X SQ= 16.35 P = 0. RV YES (NO)
119	TEND TO BE THOSE WHERE INTEREST ARTICULATION BY INSTITUTIONAL GROUPS IS MODERATE OR LIMITED (26)	0.73	119	TEND TO BE THOSE WHERE INTEREST ARTICULATION BY INSTITUTIONAL GROUPS IS VERY SIGNIFICANT OR SIGNIFICANT (74)	0.90	7 65 19 7 X SQ= 36.15 P = 0. RV YES (YES)
120	TEND LESS TO BE THOSE WHERE INTEREST ARTICULATION BY INSTITUTIONAL GROUPS IS VERY SIGNIFICANT, SIGNIFICANT, OR MODERATE (90)	0.62	120	ALWAYS ARE THOSE WHERE INTEREST ARTICULATION BY INSTITUTIONAL GROUPS IS VERY SIGNIFICANT, SIGNIFICANT, OR MODERATE (90)	1.00	16 72 10 0 X SQ= 26.78 P = 0. RV NO (YES)
121	ALWAYS ARE THOSE WHERE INTEREST ARTICULATION BY NON-ASSOCIATIONAL GROUPS IS MODERATE, LIMITED, OR NEGLIGIBLE (61)	1.00	121	TEND TO BE THOSE WHERE INTEREST ARTICULATION BY NON-ASSOCIATIONAL GROUPS IS SIGNIFICANT (54)	0.61	0 53 26 34 X SQ= 27.43 P = 0. RV YES (NO)
122	TEND TO BE THOSE WHERE INTEREST ARTICULATION BY NON-ASSOCIATIONAL GROUPS IS LIMITED OR NEGLIGIBLE (32)	0.81	122	TEND TO BE THOSE WHERE INTEREST ARTICULATION BY NON-ASSOCIATIONAL GROUPS IS SIGNIFICANT OR MODERATE (83)	0.87	5 76 21 11 X SQ= 42.47 P = 0. RV YES (YES)
123	TEND LESS TO BE THOSE WHERE INTEREST ARTICULATION BY NON-ASSOCIATIONAL GROUPS IS SIGNIFICANT, MODERATE, OR LIMITED (107)	0.69	123	ALWAYS ARE THOSE WHERE INTEREST ARTICULATION BY NON-ASSOCIATIONAL GROUPS IS SIGNIFICANT, MODERATE, OR LIMITED (107)	1.00	18 87 8 0 X SQ= 24.32 P = 0. RV NO (YES)
125	TEND TO BE THOSE WHERE INTEREST ARTICULATION BY ANOMIC GROUPS IS INFREQUENT OR VERY INFREQUENT (35)	0.87	125	TEND TO BE THOSE WHERE INTEREST ARTICULATION BY ANOMIC GROUPS IS FREQUENT OR OCCASIONAL (64)	0.81	3 60 20 14 X SQ= 32.75 P = 0. RV YES (YES)

126	TEND TO BE THOSE WHERE INTEREST ARTICULATION BY ANOMIC GROUPS IS VERY INFREQUENT (16)	0.70	126	ALWAYS ARE THOSE WHERE INTEREST ARTICULATION BY ANOMIC GROUPS IS FREQUENT, OCCASIONAL, OR INFREQUENT (83)	1.00	7 74 16 0 X SQ= 56.70 P = 0. RV YES (YES)
128	TEND TO BE THOSE WHERE INTEREST ARTICULATION BY POLITICAL PARTIES IS SIGNIFICANT OR MODERATE (48)	0.69	128	TEND TO BE THOSE WHERE INTEREST ARTICULATION BY POLITICAL PARTIES IS LIMITED OR NEGLIGIBLE (45)	0.56	18 29 8 37 X SQ= 3.82 P = 0.038 RV YES (NO)
134	TEND TO BE THOSE WHERE INTEREST AGGREGATION BY THE EXECUTIVE IS SIGNIFICANT OR MODERATE (57)	0.73	134	TEND TO BE THOSE WHERE INTEREST AGGREGATION BY THE EXECUTIVE IS LIMITED OR NEGLIGIBLE (46)	0.51	19 37 7 39 X SQ= 3.72 P = 0.040 RV YES (NO)
137	TEND TO BE THOSE WHERE INTEREST AGGREGATION BY THE LEGISLATURE IS SIGNIFICANT OR MODERATE (29)	0.73	137	TEND TO BE THOSE WHERE INTEREST AGGREGATION BY THE LEGISLATURE IS LIMITED OR NEGLIGIBLE (68)	0.87	19 9 7 61 X SQ= 30.43 P = 0. RV YES (YES)
138	TEND TO BE THOSE WHERE INTEREST AGGREGATION BY THE LEGISLATURE IS SIGNIFICANT, MODERATE, OR LIMITED (48)	0.83	138	TEND TO BE THOSE WHERE INTEREST AGGREGATION BY THE LEGISLATURE IS NEGLIGIBLE (49)	0.63	20 27 4 45 X SQ= 13.35 P = 0. RV YES (NO)
141	LEAN LESS TOWARD BEING THOSE WHERE THE PARTY SYSTEM IS QUANTITATIVELY OTHER THAN TWO-PARTY (87)	0.77	141	LEAN MORE TOWARD BEING THOSE WHERE THE PARTY SYSTEM IS QUANTITATIVELY OTHER THAN TWO-PARTY (87)	0.93	6 5 20 66 X SQ= 3.40 P = 0.063 RV NO (YES)
144	TEND TO BE THOSE WHERE THE PARTY SYSTEM IS QUANTITATIVELY TWO-PARTY, RATHER THAN ONE-PARTY (11)	0.50	144	TEND TO BE THOSE WHERE THE PARTY SYSTEM IS QUANTITATIVELY ONE-PARTY, RATHER THAN TWO-PARTY (34)	0.85	6 28 6 5 X SQ= 4.05 P = 0.044 RV YES (YES)
146	ALWAYS ARE THOSE WHERE THE PARTY SYSTEM IS QUALITATIVELY OTHER THAN MASS-BASED TERRITORIAL (92)	1.00	146	TILT LESS TOWARD BEING THOSE WHERE THE PARTY SYSTEM IS QUALITATIVELY OTHER THAN MASS-BASED TERRITORIAL (92)	0.89	0 8 24 66 X SQ= 1.57 P = 0.116 RV NO (NO)
147	TEND TO BE THOSE WHERE THE PARTY SYSTEM IS QUALITATIVELY CLASS-ORIENTED OR MULTI-IDEOLOGICAL (23)	0.65	147	TEND TO BE THOSE WHERE THE PARTY SYSTEM IS QUALITATIVELY OTHER THAN CLASS-ORIENTED OR MULTI-IDEOLOGICAL (67)	0.89	15 7 8 59 X SQ= 24.48 P = 0. RV YES (YES)

153	TEND TO BE THOSE WHERE THE PARTY SYSTEM IS STABLE (42)	0.88	153	TEND TO BE THOSE WHERE THE PARTY SYSTEM IS MODERATELY STABLE OR UNSTABLE (38)	0.67	23 17 3 35 X SQ= 19.40 P = 0. RV YES (YES)
158	TEND TO BE THOSE WHERE PERSONALISMO IS NEGLIGIBLE (56)	0.92	158	TEND TO BE THOSE WHERE PERSONALISMO IS PRONOUNCED OR MODERATE (40)	0.56	2 38 24 30 X SQ= 15.95 P = 0. RV YES (NO)
161	TEND TO BE THOSE WHERE THE POLITICAL LEADERSHIP IS NON-ELITIST (50)	0.69	161	TEND TO BE THOSE WHERE THE POLITICAL LEADERSHIP IS ELITIST OR MODERATELY ELITIST (47)	0.55	8 38 18 31 X SQ= 3.55 P = 0.041 RV YES (NO)
164	TEND MORE TO BE THOSE WHERE THE REGIME'S LEADERSHIP CHARISMA IS NEGLIGIBLE (65)	0.96	164	TEND LESS TO BE THOSE WHERE THE REGIME'S LEADERSHIP CHARISMA IS NEGLIGIBLE (65)	0.54	1 33 25 38 X SQ= 13.38 P = 0. RV NO (NO)
168	TEND TO BE THOSE WHERE THE HORIZONTAL POWER DISTRIBUTION IS SIGNIFICANT (34)	0.73	168	TEND TO BE THOSE WHERE THE HORIZONTAL POWER DISTRIBUTION IS LIMITED OR NEGLIGIBLE (72)	0.83	19 13 7 65 X SQ= 26.54 P = 0. RV YES (YES)
169	TEND TO BE THOSE WHERE THE HORIZONTAL POWER DISTRIBUTION IS SIGNIFICANT OR LIMITED (58)	0.77	169	TEND TO BE THOSE WHERE THE HORIZONTAL POWER DISTRIBUTION IS NEGLIGIBLE (48)	0.54	20 36 6 42 X SQ= 6.24 P = 0.007 RV YES (NO)
170	TEND TO BE THOSE WHERE THE LEGISLATIVE-EXECUTIVE STRUCTURE IS OTHER THAN PRESIDENTIAL (63)	0.96	170	TEND TO BE THOSE WHERE THE LEGISLATIVE-EXECUTIVE STRUCTURE IS PRESIDENTIAL (39)	0.51	1 38 25 36 X SQ= 16.31 P = 0. RV YES (NO)
171	TEND LESS TO BE THOSE WHERE THE LEGISLATIVE-EXECUTIVE STRUCTURE IS OTHER THAN PARLIAMENTARY-REPUBLICAN (90)	0.75	171	TEND MORE TO BE THOSE WHERE THE LEGISLATIVE-EXECUTIVE STRUCTURE IS OTHER THAN PARLIAMENTARY-REPUBLICAN (90)	0.95	6 4 18 72 X SQ= 5.85 P = 0.012 RV NO (YES)
172	TEND LESS TO BE THOSE WHERE THE LEGISLATIVE-EXECUTIVE STRUCTURE IS OTHER THAN PARLIAMENTARY-ROYALIST (88)	0.56	172	TEND MORE TO BE THOSE WHERE THE LEGISLATIVE-EXECUTIVE STRUCTURE IS OTHER THAN PARLIAMENTARY-ROYALIST (88)	0.88	11 10 14 72 X SQ= 10.35 P = 0.001 RV NO (YES)

174	TEND TO BE THOSE WHERE THE LEGISLATURE IS FULLY EFFECTIVE (28)	0.69	174	TEND TO BE THOSE WHERE THE LEGISLATURE IS PARTIALLY EFFECTIVE, LARGELY INEFFECTIVE, OR WHOLLY INEFFECTIVE (72)	0.89	18 8 8 64 X SQ= 30.19 P = 0. RV YES (YES)
175	TEND TO BE THOSE WHERE THE LEGISLATURE IS FULLY EFFECTIVE OR PARTIALLY EFFECTIVE (51)	0.77	175	TEND TO BE THOSE WHERE THE LEGISLATURE IS LARGELY INEFFECTIVE OR WHOLLY INEFFECTIVE (49)	0.60	20 29 6 43 X SQ= 8.85 P = 0.001 RV YES (NO)
178	LEAN TOWARD BEING THOSE WHERE THE LEGISLATURE IS BICAMERAL (51)	0.65	178	LEAN TOWARD BEING THOSE WHERE THE LEGISLATURE IS UNICAMERAL (53)	0.57	9 43 17 33 X SQ= 2.91 P = 0.070 RV YES (NO)
179	TEND TO BE THOSE WHERE THE EXECUTIVE IS STRONG (39)	0.71	179	TEND TO BE THOSE WHERE THE EXECUTIVE IS DOMINANT (52)	0.68	7 44 17 21 X SQ= 9.12 P = 0.002 RV YES (NO)
181	TEND TO BE THOSE WHERE THE BUREAUCRACY IS MODERN, RATHER THAN SEMI-MODERN (21)	0.73	181	ALWAYS ARE THOSE WHERE THE BUREAUCRACY IS SEMI-MODERN, RATHER THAN MODERN (55)	1.00	19 0 7 48 X SQ= 43.44 P = 0. RV YES (YES)
182	ALWAYS ARE THOSE WHERE THE BUREAUCRACY IS SEMI-MODERN, RATHER THAN POST-COLONIAL TRANSITIONAL (55)	1.00	182	LEAN LESS TOWARD BEING THOSE WHERE THE BUREAUCRACY IS SEMI-MODERN, RATHER THAN POST-COLONIAL TRANSITIONAL (55)	0.66	7 48 0 25 X SQ= 2.08 P = 0.092 RV NO (NO)
185	ALWAYS ARE THOSE WHERE PARTICIPATION BY THE MILITARY IN POLITICS IS SUPPORTIVE, RATHER THAN INTERVENTIVE (31)	1.00	185	TEND LESS TO BE THOSE WHERE PARTICIPATION BY THE MILITARY IN POLITICS IS SUPPORTIVE, RATHER THAN INTERVENTIVE (31)	0.53	0 21 6 24 X SQ= 3.03 P = 0.036 RV NO (NO)
187	TEND TO BE THOSE WHERE THE ROLE OF THE POLICE IS NOT POLITICALLY SIGNIFICANT (35)	0.77	187	TEND TO BE THOSE WHERE THE ROLE OF THE POLICE IS POLITICALLY SIGNIFICANT (66)	0.81	6 59 20 14 X SQ= 25.85 P = 0. RV YES (YES)
189	TEND LESS TO BE THOSE WHERE THE CHARACTER OF THE LEGAL SYSTEM IS OTHER THAN COMMON LAW (108)	0.81	189	TEND MORE TO BE THOSE WHERE THE CHARACTER OF THE LEGAL SYSTEM IS OTHER THAN COMMON LAW (108)	0.98	5 2 21 85 X SQ= 7.18 P = 0.007 RV NO (YES)

```
190  TILT LESS TOWARD BEING THOSE            0.69      190  TILT MORE TOWARD BEING THOSE            0.91        11    21
     WHERE THE CHARACTER OF THE LEGAL SYSTEM                WHERE THE CHARACTER OF THE LEGAL SYSTEM               5     2
     IS CIVIL LAW, RATHER THAN                              IS CIVIL LAW, RATHER THAN                        X SQ=   1.91
     COMMON LAW  (32)                                       COMMON LAW  (32)                                 P =    0.101
                                                                                                             RV NO  (YES)

192  ALWAYS ARE THOSE                        1.00      192  TEND LESS TO BE THOSE                   0.68         0    27
     WHERE THE CHARACTER OF THE LEGAL SYSTEM                WHERE THE CHARACTER OF THE LEGAL SYSTEM              26    58
     IS OTHER THAN                                          IS OTHER THAN                                    X SQ=   9.26
     MUSLIM OR PARTLY MUSLIM  (86)                          MUSLIM OR PARTLY MUSLIM  (86)                    P =    0.
                                                                                                             RV NO  (NO )

193  ALWAYS ARE THOSE                        1.00      193  TEND LESS TO BE THOSE                   0.69         0    27
     WHERE THE CHARACTER OF THE LEGAL SYSTEM                WHERE THE CHARACTER OF THE LEGAL SYSTEM              26    59
     IS OTHER THAN PARTLY INDIGENOUS  (86)                  IS OTHER THAN PARTLY INDIGENOUS  (86)            X SQ=   9.11
                                                                                                             P =    0.
                                                                                                             RV NO  (NO )
```

75 POLITIES
THAT ARE HISTORICALLY WESTERN OR
SIGNIFICANTLY WESTERNIZED (62)

75 POLITIES
THAT ARE NOT HISTORICALLY WESTERN AND
ARE NOT SIGNIFICANTLY WESTERNIZED (52)

BOTH SUBJECT AND PREDICATE

62 IN LEFT COLUMN

ALGERIA	ARGENTINA	AUSTRALIA	AUSTRIA	BELGIUM	BOLIVIA	BRAZIL	BULGARIA	CANADA	CEYLON
CHILE	CHINA, PR	COLOMBIA	COSTA RICA	CUBA	CYPRUS	CZECHOS'KIA	DENMARK	DOMIN REP	ECUADOR
EL SALVADOR	FINLAND	FRANCE	GERMANY, E	GERMAN FR	GREECE	GUATEMALA	HONDURAS	HUNGARY	ICELAND
INDIA	IRELAND	ISRAEL	ITALY	JAMAICA	JAPAN	LEBANON	LUXEMBOURG	MEXICO	NETHERLANDS
NEW ZEALAND	NICARAGUA	NORWAY	PANAMA	PARAGUAY	PERU	PHILIPPINES	POLAND	PORTUGAL	RUMANIA
SPAIN	SWEDEN	SWITZERLAND	TRINIDAD	TURKEY	USSR	UAR	UK	US	URUGUAY
VENEZUELA	YUGOSLAVIA								

52 IN RIGHT COLUMN

AFGHANISTAN	ALBANIA	BURMA	BURUNDI	CAMBODIA	CAMEROUN	CEN AFR REP	CHAD	CONGO(BRA)	CONGO(LEO)
DAHOMEY	ETHIOPIA	GABON	GHANA	GUINEA	HAITI	INDONESIA	IRAN	IRAQ	IVORY COAST
JORDAN	KOREA, N	KOREA REP	LAOS	LIBERIA	LIBYA	MALAGASY R	MALAYA	MALI	MAURITANIA
MONGOLIA	MOROCCO	NEPAL	NIGER	NIGERIA	PAKISTAN	RWANDA	SA'U ARABIA	SENEGAL	SIERRE LEO
SOMALIA	SUDAN	SYRIA	TANGANYIKA	THAILAND	TOGO	TUNISIA	UGANDA	UPPER VOLTA	VIETNAM, N
VIETNAM REP	YEMEN								

1 EXCLUDED BECAUSE AMBIGUOUS

SO AFRICA

2	TEND LESS TO BE THOSE LOCATED ELSEWHERE THAN IN WEST EUROPE, SCANDINAVIA, NORTH AMERICA, OR AUSTRALASIA (93)	0.65	2	ALWAYS ARE THOSE LOCATED ELSEWHERE THAN IN WEST EUROPE, SCANDINAVIA, NORTH AMERICA, OR AUSTRALASIA (93)	1.00	22 0 40 52 X SQ= 20.64 P = 0. RV NO (YES)
6	TEND LESS TO BE THOSE LOCATED ELSEWHERE THAN IN THE CARIBBEAN, CENTRAL AMERICA, OR SOUTH AMERICA (93)	0.66	6	TEND MORE TO BE THOSE LOCATED ELSEWHERE THAN IN THE CARIBBEAN, CENTRAL AMERICA, OR SOUTH AMERICA (93)	0.98	21 1 41 51 X SQ= 16.54 P = 0. RV NO (YES)
8	TEND MORE TO BE THOSE LOCATED ELSEWHERE THAN IN EAST ASIA SOUTH ASIA, OR SOUTHEAST ASIA (97)	0.92	8	TEND LESS TO BE THOSE LOCATED ELSEWHERE THAN IN EAST ASIA SOUTH ASIA, OR SOUTHEAST ASIA (97)	0.75	5 13 57 39 X SQ= 4.89 P = 0.019 RV NO (YES)

10	TEND TO BE THOSE LOCATED ELSEWHERE THAN IN NORTH AFRICA, OR CENTRAL AND SOUTH AFRICA (82)	0.97	10	TEND TO BE THOSE LOCATED IN NORTH AFRICA, OR CENTRAL AND SOUTH AFRICA (33)	0.58	2 30 60 22 X SQ= 38.90 P = 0. RV YES (YES)
11	ALWAYS ARE THOSE LOCATED ELSEWHERE THAN IN CENTRAL AND SOUTH AFRICA (87)	1.00	11	TEND TO BE THOSE LOCATED IN CENTRAL AND SOUTH AFRICA (28)	0.52	0 27 62 25 X SQ= 39.36 P = 0. RV YES (YES)
15	TEND TO BE THOSE LOCATED IN THE MIDDLE EAST, RATHER THAN IN NORTH AFRICA OR CENTRAL AND SOUTH AFRICA (11)	0.67	15	TEND TO BE THOSE LOCATED IN NORTH AFRICA OR CENTRAL AND SOUTH AFRICA, RATHER THAN IN THE MIDDLE EAST (33)	0.81	4 7 2 30 X SQ= 3.93 P = 0.029 RV YES (NO)
17	TEND TO BE THOSE LOCATED IN THE CARIBBEAN, CENTRAL AMERICA, OR SOUTH AMERICA, RATHER THAN IN THE MIDDLE EAST (22)	0.84	17	TEND TO BE THOSE LOCATED IN THE MIDDLE EAST, RATHER THAN IN THE CARIBBEAN, CENTRAL AMERICA, OR SOUTH AMERICA (11)	0.87	4 7 21 1 X SQ= 10.91 P = 0.001 RV YES (YES)
18	LEAN TOWARD BEING THOSE LOCATED IN EAST ASIA, SOUTH ASIA, OR SOUTHEAST ASIA, RATHER THAN IN NORTH AFRICA OR CENTRAL AND SOUTH AFRICA (18)	0.71	18	LEAN TOWARD BEING THOSE LOCATED IN NORTH AFRICA OR CENTRAL AND SOUTH AFRICA, RATHER THAN IN EAST ASIA, SOUTH ASIA, OR SOUTHEAST ASIA (33)	0.70	2 30 5 13 X SQ= 2.83 P = 0.083 RV YES (NO)
19	TEND TO BE THOSE LOCATED IN THE CARIBBEAN, CENTRAL AMERICA, OR SOUTH AMERICA, RATHER THAN IN NORTH AFRICA OR CENTRAL AND SOUTH AFRICA (22)	0.91	19	TEND TO BE THOSE LOCATED IN NORTH AFRICA OR CENTRAL AND SOUTH AFRICA, RATHER THAN IN THE CARIBBEAN, CENTRAL AMERICA, OR SOUTH AMERICA (33)	0.97	2 30 21 1 X SQ= 38.86 P = 0. RV YES (YES)
20	TEND TO BE THOSE LOCATED IN THE CARIBBEAN, CENTRAL AMERICA, OR SOUTH AMERICA, RATHER THAN IN EAST ASIA, SOUTH ASIA, OR SOUTHEAST ASIA (22)	0.81	20	TEND TO BE THOSE LOCATED IN EAST ASIA, SOUTH ASIA, OR SOUTHEAST ASIA, RATHER THAN IN THE CARIBBEAN, CENTRAL AMERICA, OR SOUTH AMERICA (18)	0.93	5 13 21 1 X SQ= 17.07 P = 0. RV YES (YES)
24	LEAN TOWARD BEING THOSE WHOSE POPULATION IS VERY LARGE, LARGE, OR MEDIUM (61)	0.61	24	LEAN TOWARD BEING THOSE WHOSE POPULATION IS SMALL (54)	0.58	38 22 24 30 X SQ= 3.36 P = 0.060 RV YES (YES)
26	TEND TO BE THOSE WHOSE POPULATION DENSITY IS VERY HIGH, HIGH, OR MEDIUM (48)	0.55	26	TEND TO BE THOSE WHOSE POPULATION DENSITY IS LOW (67)	0.73	34 14 28 38 X SQ= 7.93 P = 0.004 RV YES (YES)

No.	Statement		No.	Statement		Statistics
29	TEND TO BE THOSE WHERE THE DEGREE OF URBANIZATION IS HIGH (56)	0.84	29	TEND TO BE THOSE WHERE THE DEGREE OF URBANIZATION IS LOW (49)	0.83	47 8 9 40 X SQ= 44.27 P = 0. RV YES (YES)
30	TEND TO BE THOSE WHOSE AGRICULTURAL POPULATION IS MEDIUM, LOW, OR VERY LOW (57)	0.84	30	TEND TO BE THOSE WHOSE AGRICULTURAL POPULATION IS HIGH (56)	0.92	10 46 52 4 X SQ= 60.73 P = 0. RV YES (YES)
31	TEND LESS TO BE THOSE WHOSE AGRICULTURAL POPULATION IS HIGH OR MEDIUM (90)	0.63	31	ALWAYS ARE THOSE WHOSE AGRICULTURAL POPULATION IS HIGH OR MEDIUM (90)	1.00	39 51 23 0 X SQ= 21.52 P = 0. RV NO (YES)
32	TEND LESS TO BE THOSE WHOSE GROSS NATIONAL PRODUCT IS MEDIUM, LOW, OR VERY LOW (105)	0.84	32	ALWAYS ARE THOSE WHOSE GROSS NATIONAL PRODUCT IS MEDIUM, LOW, OR VERY LOW (105)	1.00	10 0 52 52 X SQ= 7.29 P = 0.002 RV NO (YES)
33	TEND LESS TO BE THOSE WHOSE GROSS NATIONAL PRODUCT IS LOW OR VERY LOW (85)	0.56	33	TEND MORE TO BE THOSE WHOSE GROSS NATIONAL PRODUCT IS LOW OR VERY LOW (85)	0.96	27 2 35 50 X SQ= 21.46 P = 0. RV NO (YES)
34	TEND TO BE THOSE WHOSE GROSS NATIONAL PRODUCT IS VERY HIGH, HIGH, MEDIUM, OR LOW (62)	0.74	34	TEND TO BE THOSE WHOSE GROSS NATIONAL PRODUCT IS VERY LOW (53)	0.71	46 15 16 37 X SQ= 21.59 P = 0. RV YES (YES)
35	TEND LESS TO BE THOSE WHOSE PER CAPITA GROSS NATIONAL PRODUCT IS MEDIUM, LOW, OR VERY LOW (91)	0.61	35	ALWAYS ARE THOSE WHOSE PER CAPITA GROSS NATIONAL PRODUCT IS MEDIUM, LOW, OR VERY LOW (91)	1.00	24 0 38 52 X SQ= 23.22 P = 0. RV NO (YES)
36	TEND TO BE THOSE WHOSE PER CAPITA GROSS NATIONAL PRODUCT IS VERY HIGH, HIGH, OR MEDIUM (42)	0.65	36	TEND TO BE THOSE WHOSE PER CAPITA GROSS NATIONAL PRODUCT IS LOW OR VERY LOW (73)	0.98	40 1 22 51 X SQ= 45.43 P = 0. RV YES (YES)
37	TEND TO BE THOSE WHOSE PER CAPITA GROSS NATIONAL PRODUCT IS VERY HIGH, HIGH, MEDIUM, OR LOW (64)	0.92	37	TEND TO BE THOSE WHOSE PER CAPITA GROSS NATIONAL PRODUCT IS VERY LOW (51)	0.88	57 6 5 46 X SQ= 70.72 P = 0. RV YES (YES)

38	TEND LESS TO BE THOSE WHOSE INTERNATIONAL FINANCIAL STATUS IS MEDIUM, LOW, OR VERY LOW (103)	0.83	38	ALWAYS ARE THOSE WHOSE INTERNATIONAL FINANCIAL STATUS IS MEDIUM, LOW, OR VERY LOW (103)	1.00	10 0 50 52 X SQ= 7.58 P = 0.002 RV NO (YES)
39	TEND TO BE THOSE WHOSE INTERNATIONAL FINANCIAL STATUS IS VERY HIGH, HIGH, OR MEDIUM (38)	0.57	39	TEND TO BE THOSE WHOSE INTERNATIONAL FINANCIAL STATUS IS LOW OR VERY LOW (76)	0.96	35 2 26 50 X SQ= 34.14 P = 0. RV YES (YES)
40	TEND TO BE THOSE WHOSE INTERNATIONAL FINANCIAL STATUS IS VERY HIGH, HIGH, MEDIUM, OR LOW (71)	0.85	40	TEND TO BE THOSE WHOSE INTERNATIONAL FINANCIAL STATUS IS VERY LOW (39)	0.60	50 20 9 30 X SQ= 21.68 P = 0. RV YES (YES)
42	TEND TO BE THOSE WHOSE ECONOMIC DEVELOPMENTAL STATUS IS DEVELOPED OR INTERMEDIATE (36)	0.58	42	ALWAYS ARE THOSE WHOSE ECONOMIC DEVELOPMENTAL STATUS IS UNDERDEVELOPED OR VERY UNDERDEVELOPED (76)	1.00	35 0 25 51 X SQ= 40.79 P = 0. RV YES (YES)
43	TEND TO BE THOSE WHOSE ECONOMIC DEVELOPMENTAL STATUS IS DEVELOPED, INTERMEDIATE, OR UNDERDEVELOPED (55)	0.82	43	TEND TO BE THOSE WHOSE ECONOMIC DEVELOPMENTAL STATUS IS VERY UNDERDEVELOPED (57)	0.90	49 5 11 46 X SQ= 54.15 P = 0. RV YES (YES)
44	TEND LESS TO BE THOSE WHERE THE LITERACY RATE IS BELOW NINETY PERCENT (90)	0.60	44	ALWAYS ARE THOSE WHERE THE LITERACY RATE IS BELOW NINETY PERCENT (90)	1.00	25 0 37 52 X SQ= 24.55 P = 0. RV NO (YES)
45	TEND TO BE THOSE WHERE THE LITERACY RATE IS FIFTY PERCENT OR ABOVE (55)	0.82	45	TEND TO BE THOSE WHERE THE LITERACY RATE IS BELOW FIFTY PERCENT (54)	0.87	49 6 11 42 X SQ= 48.32 P = 0. RV YES (YES)
46	ALWAYS ARE THOSE WHERE THE LITERACY RATE IS TEN PERCENT OR ABOVE (84)	1.00	46	TEND TO BE THOSE WHERE THE LITERACY RATE IS BELOW TEN PERCENT (26)	0.55	62 21 0 26 X SQ= 42.05 P = 0. RV YES (YES)
50	TEND TO BE THOSE WHERE FREEDOM OF THE PRESS IS COMPLETE (43)	0.57	50	TEND TO BE THOSE WHERE FREEDOM OF THE PRESS IS INTERMITTENT, INTERNALLY ABSENT, OR INTERNALLY AND EXTERNALLY ABSENT (56)	0.74	32 11 24 31 X SQ= 8.12 P = 0.004 RV YES (YES)

51	TEND TO BE THOSE WHERE FREEDOM OF THE PRESS IS COMPLETE OR INTERMITTENT (60)	0.71	51	TEND TO BE THOSE WHERE FREEDOM OF THE PRESS IS INTERNALLY ABSENT, OR INTERNALLY AND EXTERNALLY ABSENT (37)	0.52	40 19 16 21 X SQ= 4.68 P = 0.021 RV YES (YES)
53	TEND LESS TO BE THOSE WHERE NEWSPAPER CIRCULATION IS LESS THAN THREE HUNDRED PER THOUSAND (101)	0.77	53	ALWAYS ARE THOSE WHERE NEWSPAPER CIRCULATION IS LESS THAN THREE HUNDRED PER THOUSAND (101)	1.00	14 0 48 52 X SQ= 11.37 P = 0. RV NO (YES)
54	TEND TO BE THOSE WHERE NEWSPAPER CIRCULATION IS ONE HUNDRED OR MORE PER THOUSAND (37)	0.58	54	TEND TO BE THOSE WHERE NEWSPAPER CIRCULATION IS LESS THAN ONE HUNDRED PER THOUSAND (76)	0.98	36 1 26 49 X SQ= 36.83 P = 0. RV YES (YES)
55	ALWAYS ARE THOSE WHERE NEWSPAPER CIRCULATION IS TEN OR MORE PER THOUSAND (78)	1.00	55	TEND TO BE THOSE WHERE NEWSPAPER CIRCULATION IS LESS THAN TEN PER THOUSAND (35)	0.70	62 15 0 35 X SQ= 59.91 P = 0. RV YES (YES)
56	TEND TO BE THOSE WHERE THE RELIGION IS PREDOMINANTLY LITERATE (79)	0.93	56	TEND TO BE THOSE WHERE THE RELIGION IS PREDOMINANTLY OR PARTLY NON-LITERATE (31)	0.54	57 22 4 26 X SQ= 28.18 P = 0. RV YES (YES)
63	TEND TO BE THOSE WHERE THE RELIGION IS PREDOMINANTLY SOME KIND OF CHRISTIAN (46)	0.75	63	ALWAYS ARE THOSE WHERE THE RELIGION IS PREDOMINANTLY OR PARTLY OTHER THAN CHRISTIAN (68)	1.00	46 0 15 52 X SQ= 63.05 P = 0. RV YES (YES)
66	TEND TO BE THOSE THAT ARE RELIGIOUSLY HOMOGENEOUS (57)	0.63	66	TEND TO BE THOSE THAT ARE RELIGIOUSLY HETEROGENEOUS (49)	0.58	38 19 22 26 X SQ= 3.81 P = 0.047 RV YES (YES)
68	TEND TO BE THOSE THAT ARE LINGUISTICALLY HOMOGENEOUS (52)	0.62	68	TEND TO BE THOSE THAT ARE LINGUISTICALLY WEAKLY HETEROGENEOUS OR STRONGLY HETEROGENEOUS (62)	0.73	38 14 23 38 X SQ= 12.75 P = 0. RV YES (YES)
69	TEND TO BE THOSE THAT ARE LINGUISTICALLY HOMOGENEOUS OR WEAKLY HETEROGENEOUS (64)	0.74	69	TEND TO BE THOSE THAT ARE LINGUISTICALLY STRONGLY HETEROGENEOUS (50)	0.63	45 19 16 33 X SQ= 14.36 P = 0. RV YES (YES)

	Left			Right		Statistics
70	LEAN LESS TOWARD BEING THOSE THAT ARE RELIGIOUSLY, RACIALLY, OR LINGUISTICALLY HETEROGENEOUS (85)	0.73	70	LEAN MORE TOWARD BEING THOSE THAT ARE RELIGIOUSLY, RACIALLY, OR LINGUISTICALLY HETEROGENEOUS (85)	0.89	16 5 43 40 X SQ= 3.13 P = 0.051 RV NO (NO)
72	TEND TO BE THOSE WHOSE DATE OF INDEPENDENCE IS BEFORE 1914 (52)	0.72	72	TEND TO BE THOSE WHOSE DATE OF INDEPENDENCE IS AFTER 1914 (59)	0.84	43 8 17 42 X SQ= 31.78 P = 0. RV YES (YES)
73	TEND TO BE THOSE WHOSE DATE OF INDEPENDENCE IS BEFORE 1945 (65)	0.85	73	TEND TO BE THOSE WHOSE DATE OF INDEPENDENCE IS AFTER 1945 (46)	0.74	51 13 9 37 X SQ= 36.63 P = 0. RV YES (YES)
79	TEND TO BE THOSE THAT WERE FORMERLY DEPENDENCIES OF SPAIN, RATHER THAN BRITAIN OR FRANCE (18)	0.56	79	ALWAYS ARE THOSE THAT WERE FORMERLY DEPENDENCIES OF BRITAIN OR FRANCE, RATHER THAN SPAIN (49)	1.00	14 34 18 0 X SQ= 23.54 P = 0. RV YES (YES)
80	TEND TO BE THOSE WHOSE DATE OF INDEPENDENCE IS AFTER 1914, AND THAT WERE FORMERLY DEPENDENCIES OF BRITAIN, RATHER THAN FRANCE (19)	0.80	80	TEND TO BE THOSE WHOSE DATE OF INDEPENDENCE IS AFTER 1914, AND THAT WERE FORMERLY DEPENDENCIES OF FRANCE, RATHER THAN BRITAIN (24)	0.67	8 11 2 22 X SQ= 5.02 P = 0.013 RV YES (NO)
82	TEND TO BE THOSE WHERE THE TYPE OF POLITICAL MODERNIZATION IS EARLY OR LATER EUROPEAN OR EUROPEAN DERIVED, RATHER THAN DEVELOPED TUTELARY OR UNDEVELOPED TUTELARY (51)	0.83	82	TEND TO BE THOSE WHERE THE TYPE OF POLITICAL MODERNIZATION IS DEVELOPED TUTELARY OR UNDEVELOPED TUTELARY, RATHER THAN EARLY OR LATER EUROPEAN OR EUROPEAN DERIVED (55)	0.96	48 2 10 45 X SQ= 61.04 P = 0. RV YES (YES)
84	ALWAYS ARE THOSE WHERE THE TYPE OF POLITICAL MODERNIZATION IS DEVELOPED TUTELARY, RATHER THAN UNDEVELOPED TUTELARY (31)	1.00	84	TEND TO BE THOSE WHERE THE TYPE OF POLITICAL MODERNIZATION IS UNDEVELOPED TUTELARY, RATHER THAN DEVELOPED TUTELARY (24)	0.53	10 21 0 24 X SQ= 7.42 P = 0.003 RV YES (NO)
85	TEND TO BE THOSE WHERE THE STAGE OF POLITICAL MODERNIZATION IS ADVANCED, RATHER THAN MID- OR EARLY TRANSITIONAL (60)	0.79	85	TEND TO BE THOSE WHERE THE STAGE OF POLITICAL MODERNIZATION IS MID- OR EARLY TRANSITIONAL, RATHER THAN ADVANCED (54)	0.78	49 11 13 40 X SQ= 34.83 P = 0. RV YES (YES)
86	TEND TO BE THOSE WHERE THE STAGE OF POLITICAL MODERNIZATION IS ADVANCED OR MID-TRANSITIONAL, RATHER THAN EARLY TRANSITIONAL (76)	0.95	86	TEND TO BE THOSE WHERE THE STAGE OF POLITICAL MODERNIZATION IS EARLY TRANSITIONAL, RATHER THAN ADVANCED OR MID-TRANSITIONAL (38)	0.69	59 16 3 35 X SQ= 48.20 P = 0. RV YES (YES)

87	TEND TO BE THOSE WHOSE IDEOLOGICAL ORIENTATION IS OTHER THAN DEVELOPMENTAL (58)	0.88	87	TEND TO BE THOSE WHOSE IDEOLOGICAL ORIENTATION IS DEVELOPMENTAL (31)	0.66	6 25 44 13 X SQ= 25.07 P = 0. RV YES (YES)
89	TEND TO BE THOSE WHOSE IDEOLOGICAL ORIENTATION IS CONVENTIONAL (33)	0.65	89	ALWAYS ARE THOSE WHOSE IDEOLOGICAL ORIENTATION IS OTHER THAN CONVENTIONAL (62)	1.00	32 0 17 45 X SQ= 41.69 P = 0. RV YES (YES)
94	TEND TO BE THOSE WHERE THE STATUS OF THE REGIME IS CONSTITUTIONAL (51)	0.71	94	TEND TO BE THOSE WHERE THE STATUS OF THE REGIME IS AUTHORITARIAN OR TOTALITARIAN (41)	0.73	42 9 17 24 X SQ= 14.79 P = 0. RV YES (YES)
95	TILT LESS TOWARD BEING THOSE WHERE THE STATUS OF THE REGIME IS CONSTITUTIONAL OR AUTHORITARIAN (95)	0.81	95	TILT MORE TOWARD BEING THOSE WHERE THE STATUS OF THE REGIME IS CONSTITUTIONAL OR AUTHORITARIAN (95)	0.92	50 45 12 4 X SQ= 1.95 P = 0.110 RV NO (NO)
96	TEND TO BE THOSE WHERE THE STATUS OF THE REGIME IS CONSTITUTIONAL OR TOTALITARIAN (67)	0.92	96	TEND TO BE THOSE WHERE THE STATUS OF THE REGIME IS AUTHORITARIAN (23)	0.58	54 13 5 18 X SQ= 23.73 P = 0. RV YES (YES)
97	TEND TO BE THOSE WHERE THE STATUS OF THE REGIME IS TOTALITARIAN, RATHER THAN AUTHORITARIAN (16)	0.71	97	TEND TO BE THOSE WHERE THE STATUS OF THE REGIME IS AUTHORITARIAN, RATHER THAN TOTALITARIAN (23)	0.82	5 18 12 4 X SQ= 8.83 P = 0.003 RV YES (YES)
99	TEND TO BE THOSE WHERE GOVERNMENTAL STABILITY IS GENERALLY PRESENT AND DATES FROM AT LEAST THE INTER-WAR PERIOD, OR FROM THE POST-WAR PERIOD (50)	0.67	99	TEND TO BE THOSE WHERE GOVERNMENTAL STABILITY IS MODERATELY PRESENT AND DATES FROM THE POST-WAR PERIOD, OR IS ABSENT (36)	0.58	36 13 18 18 X SQ= 3.97 P = 0.040 RV YES (YES)
101	TEND TO BE THOSE WHERE THE REPRESENTATIVE CHARACTER OF THE REGIME IS POLYARCHIC (41)	0.62	101	TEND TO BE THOSE WHERE THE REPRESENTATIVE CHARACTER OF THE REGIME IS LIMITED POLYARCHIC, PSEUDO-POLYARCHIC, OR NON-POLYARCHIC (57)	0.87	36 5 22 34 X SQ= 21.20 P = 0. RV YES (YES)
102	TEND TO BE THOSE WHERE THE REPRESENTATIVE CHARACTER OF THE REGIME IS POLYARCHIC OR LIMITED POLYARCHIC (59)	0.72	102	TEND TO BE THOSE WHERE THE REPRESENTATIVE CHARACTER OF THE REGIME IS PSEUDO-POLYARCHIC OR NON-POLYARCHIC (49)	0.68	43 15 17 32 X SQ= 15.21 P = 0. RV YES (YES)

105	TEND TO BE THOSE WHERE THE ELECTORAL SYSTEM IS COMPETITIVE (43)	0.72	105	TEND TO BE THOSE WHERE THE ELECTORAL SYSTEM IS PARTIALLY COMPETITIVE OR NON-COMPETITIVE (47)	0.89	39 4 15 32 X SQ= 29.93 P = 0. RV YES (YES)
106	TEND TO BE THOSE WHERE THE ELECTORAL SYSTEM IS COMPETITIVE OR PARTIALLY COMPETITIVE (52)	0.76	106	TEND TO BE THOSE WHERE THE ELECTORAL SYSTEM IS NON-COMPETITIVE (30)	0.61	41 11 13 17 X SQ= 9.15 P = 0.002 RV YES (YES)
107	TEND TO BE THOSE WHERE AUTONOMOUS GROUPS ARE FULLY TOLERATED IN POLITICS (46)	0.69	107	TEND TO BE THOSE WHERE AUTONOMOUS GROUPS ARE PARTIALLY TOLERATED IN POLITICS, ARE TOLERATED ONLY OUTSIDE POLITICS, OR ARE NOT TOLERATED AT ALL (65)	0.92	42 4 19 45 X SQ= 38.68 P = 0. RV YES (YES)
108	TEND TO BE THOSE WHERE AUTONOMOUS GROUPS ARE FULLY OR PARTIALLY TOLERATED IN POLITICS (65)	0.78	108	TEND TO BE THOSE WHERE AUTONOMOUS GROUPS ARE TOLERATED ONLY OUTSIDE POLITICS OR ARE NOT TOLERATED AT ALL (35)	0.55	47 18 13 22 X SQ= 10.30 P = 0.001 RV YES (YES)
111	TEND TO BE THOSE WHERE POLITICAL ENCULTURATION IS HIGH OR MEDIUM (53)	0.69	111	TEND TO BE THOSE WHERE POLITICAL ENCULTURATION IS LOW (42)	0.58	34 19 15 26 X SQ= 5.98 P = 0.012 RV YES (YES)
114	TEND TO BE THOSE WHERE SECTIONALISM IS NEGLIGIBLE (47)	0.57	114	TEND TO BE THOSE WHERE SECTIONALISM IS EXTREME OR MODERATE (61)	0.71	25 35 33 14 X SQ= 7.54 P = 0.004 RV YES (YES)
115	TEND LESS TO BE THOSE WHERE INTEREST ARTICULATION BY ASSOCIATIONAL GROUPS IS MODERATE, LIMITED, OR NEGLIGIBLE (91)	0.66	115	ALWAYS ARE THOSE WHERE INTEREST ARTICULATION BY ASSOCIATIONAL GROUPS IS MODERATE, LIMITED, OR NEGLIGIBLE (91)	1.00	20 0 39 51 X SQ= 18.91 P = 0. RV NO (YES)
116	TEND TO BE THOSE WHERE INTEREST ARTICULATION BY ASSOCIATIONAL GROUPS IS SIGNIFICANT OR MODERATE (32)	0.53	116	ALWAYS ARE THOSE WHERE INTEREST ARTICULATION BY ASSOCIATIONAL GROUPS IS LIMITED OR NEGLIGIBLE (79)	1.00	31 0 28 51 X SQ= 34.76 P = 0. RV YES (YES)
117	TEND TO BE THOSE WHERE INTEREST ARTICULATION BY ASSOCIATIONAL GROUPS IS SIGNIFICANT, MODERATE, OR LIMITED (60)	0.88	117	TEND TO BE THOSE WHERE INTEREST ARTICULATION BY ASSOCIATIONAL GROUPS IS NEGLIGIBLE (51)	0.86	52 7 7 44 X SQ= 57.95 P = 0. RV YES (YES)

118	TEND TO BE THOSE WHERE INTEREST ARTICULATION BY INSTITUTIONAL GROUPS IS SIGNIFICANT, MODERATE, OR LIMITED (60)	0.69	118	TEND TO BE THOSE WHERE INTEREST ARTICULATION BY INSTITUTIONAL GROUPS IS VERY SIGNIFICANT (40)	0.57	19 21 43 16 X SQ= 5.52 P = 0.012 RV YES (YES)
119	TEND LESS TO BE THOSE WHERE INTEREST ARTICULATION BY INSTITUTIONAL GROUPS IS VERY SIGNIFICANT OR SIGNIFICANT (74)	0.60	119	TEND MORE TO BE THOSE WHERE INTEREST ARTICULATION BY INSTITUTIONAL GROUPS IS VERY SIGNIFICANT OR SIGNIFICANT (74)	0.97	37 36 25 1 X SQ= 15.05 P = 0. RV NO (NO)
120	TEND LESS TO BE THOSE WHERE INTEREST ARTICULATION BY INSTITUTIONAL GROUPS IS VERY SIGNIFICANT, SIGNIFICANT, OR MODERATE (90)	0.84	120	ALWAYS ARE THOSE WHERE INTEREST ARTICULATION BY INSTITUTIONAL GROUPS IS VERY SIGNIFICANT, SIGNIFICANT, OR MODERATE (90)	1.00	52 37 10 0 X SQ= 4.98 P = 0.012 RV NO (NO)
121	TEND TO BE THOSE WHERE INTEREST ARTICULATION BY NON-ASSOCIATIONAL GROUPS IS MODERATE, LIMITED, OR NEGLIGIBLE (61)	0.90	121	TEND TO BE THOSE WHERE INTEREST ARTICULATION BY NON-ASSOCIATIONAL GROUPS IS SIGNIFICANT (54)	0.90	6 47 56 5 X SQ= 70.84 P = 0. RV YES (YES)
122	TEND TO BE THOSE WHERE INTEREST ARTICULATION BY NON-ASSOCIATIONAL GROUPS IS LIMITED OR NEGLIGIBLE (32)	0.52	122	ALWAYS ARE THOSE WHERE INTEREST ARTICULATION BY NON-ASSOCIATIONAL GROUPS IS SIGNIFICANT OR MODERATE (83)	1.00	30 52 32 0 X SQ= 34.80 P = 0. RV YES (YES)
125	TEND TO BE THOSE WHERE INTEREST ARTICULATION BY ANOMIC GROUPS IS INFREQUENT OR VERY INFREQUENT (35)	0.57	125	TEND TO BE THOSE WHERE INTEREST ARTICULATION BY ANOMIC GROUPS IS FREQUENT OR OCCASIONAL (64)	0.87	22 41 29 6 X SQ= 18.84 P = 0. RV YES (YES)
126	TEND LESS TO BE THOSE WHERE INTEREST ARTICULATION BY ANOMIC GROUPS IS FREQUENT, OCCASIONAL, OR INFREQUENT (83)	0.69	126	ALWAYS ARE THOSE WHERE INTEREST ARTICULATION BY ANOMIC GROUPS IS FREQUENT, OCCASIONAL, OR INFREQUENT (83)	1.00	35 47 16 0 X SQ= 15.40 P = 0. RV NO (YES)
128	TEND TO BE THOSE WHERE INTEREST ARTICULATION BY POLITICAL PARTIES IS SIGNIFICANT OR MODERATE (48)	0.64	128	TEND TO BE THOSE WHERE INTEREST ARTICULATION BY POLITICAL PARTIES IS LIMITED OR NEGLIGIBLE (45)	0.69	37 11 21 24 X SQ= 7.91 P = 0.003 RV YES (YES)
129	TEND TO BE THOSE WHERE INTEREST ARTICULATION BY POLITICAL PARTIES IS SIGNIFICANT, MODERATE, OR LIMITED (56)	0.72	129	TEND TO BE THOSE WHERE INTEREST ARTICULATION BY POLITICAL PARTIES IS NEGLIGIBLE (37)	0.60	42 14 16 21 X SQ= 8.27 P = 0.002 RV YES (YES)

133	TEND MORE TO BE THOSE WHERE INTEREST AGGREGATION BY THE EXECUTIVE IS MODERATE, LIMITED, OR NEGLIGIBLE (76)	0.86	133	TEND LESS TO BE THOSE WHERE INTEREST AGGREGATION BY THE EXECUTIVE IS MODERATE, LIMITED, OR NEGLIGIBLE (76)	0.56	8 21 49 27 X SQ= 10.07 P = 0.001 RV NO (YES)
136	TEND LESS TO BE THOSE WHERE INTEREST AGGREGATION BY THE LEGISLATURE IS MODERATE, LIMITED, OR NEGLIGIBLE (85)	0.78	136	ALWAYS ARE THOSE WHERE INTEREST AGGREGATION BY THE LEGISLATURE IS MODERATE, LIMITED, OR NEGLIGIBLE (85)	1.00	12 0 43 42 X SQ= 8.54 P = 0.001 RV NO (NO)
137	TEND TO BE THOSE WHERE INTEREST AGGREGATION BY THE LEGISLATURE IS SIGNIFICANT OR MODERATE (29)	0.53	137	ALWAYS ARE THOSE WHERE INTEREST AGGREGATION BY THE LEGISLATURE IS LIMITED OR NEGLIGIBLE (68)	1.00	29 0 26 42 X SQ= 29.12 P = 0. RV YES (YES)
138	TEND TO BE THOSE WHERE INTEREST AGGREGATION BY THE LEGISLATURE IS SIGNIFICANT, MODERATE, OR LIMITED (48)	0.77	138	TEND TO BE THOSE WHERE INTEREST AGGREGATION BY THE LEGISLATURE IS NEGLIGIBLE (49)	0.88	43 5 13 36 X SQ= 36.96 P = 0. RV YES (YES)
139	TEND MORE TO BE THOSE WHERE THE PARTY SYSTEM IS QUANTITATIVELY OTHER THAN ONE-PARTY (71)	0.77	139	TEND LESS TO BE THOSE WHERE THE PARTY SYSTEM IS QUANTITATIVELY OTHER THAN ONE-PARTY (71)	0.53	14 20 47 23 X SQ= 5.34 P = 0.019 RV NO (YES)
141	TEND LESS TO BE THOSE WHERE THE PARTY SYSTEM IS QUANTITATIVELY OTHER THAN TWO-PARTY (87)	0.83	141	TEND MORE TO BE THOSE WHERE THE PARTY SYSTEM IS QUANTITATIVELY OTHER THAN TWO-PARTY (87)	0.97	10 1 49 38 X SQ= 3.54 P = 0.046 RV NO (NO)
142	TEND LESS TO BE THOSE WHERE THE PARTY SYSTEM IS QUANTITATIVELY OTHER THAN MULTI-PARTY (66)	0.56	142	TEND MORE TO BE THOSE WHERE THE PARTY SYSTEM IS QUANTITATIVELY OTHER THAN MULTI-PARTY (66)	0.89	26 4 33 33 X SQ= 10.21 P = 0.001 RV NO (YES)
146	LEAN MORE TOWARD BEING THOSE WHERE THE PARTY SYSTEM IS QUALITATIVELY OTHER THAN MASS-BASED TERRITORIAL (92)	0.97	146	LEAN LESS TOWARD BEING THOSE WHERE THE PARTY SYSTEM IS QUALITATIVELY OTHER THAN MASS-BASED TERRITORIAL (92)	0.85	2 6 56 35 X SQ= 2.68 P = 0.063 RV NO (YES)
147	TEND LESS TO BE THOSE WHERE THE PARTY SYSTEM IS QUALITATIVELY OTHER THAN CLASS-ORIENTED OR MULTI-IDEOLOGICAL (67)	0.57	147	ALWAYS ARE THOSE WHERE THE PARTY SYSTEM IS QUALITATIVELY OTHER THAN CLASS-ORIENTED OR MULTI-IDEOLOGICAL (67)	1.00	23 0 31 36 X SQ= 18.42 P = 0. RV NO (YES)

153	TEND TO BE THOSE WHERE THE PARTY SYSTEM IS STABLE (42)	0.62	153	TEND TO BE THOSE WHERE THE PARTY SYSTEM IS MODERATELY STABLE OR UNSTABLE (38)	0.69	33 8 20 18 X SQ= 5.73 P = 0.016 RV YES (NO)
158	TEND TO BE THOSE WHERE PERSONALISMO IS NEGLIGIBLE (56)	0.67	158	TEND TO BE THOSE WHERE PERSONALISMO IS PRONOUNCED OR MODERATE (40)	0.57	20 20 40 15 X SQ= 4.21 P = 0.031 RV YES (YES)
164	TEND TO BE THOSE WHERE THE REGIME'S LEADERSHIP CHARISMA IS NEGLIGIBLE (65)	0.87	164	TEND TO BE THOSE WHERE THE REGIME'S LEADERSHIP CHARISMA IS PRONOUNCED OR MODERATE (34)	0.68	8 26 52 12 X SQ= 28.78 P = 0. RV YES (YES)
168	TEND TO BE THOSE WHERE THE HORIZONTAL POWER DISTRIBUTION IS SIGNIFICANT (34)	0.53	168	TEND TO BE THOSE WHERE THE HORIZONTAL POWER DISTRIBUTION IS LIMITED OR NEGLIGIBLE (72)	0.98	32 1 28 44 X SQ= 28.84 P = 0. RV YES (YES)
169	TEND TO BE THOSE WHERE THE HORIZONTAL POWER DISTRIBUTION IS SIGNIFICANT OR LIMITED (58)	0.75	169	TEND TO BE THOSE WHERE THE HORIZONTAL POWER DISTRIBUTION IS NEGLIGIBLE (48)	0.73	45 12 15 33 X SQ= 22.30 P = 0. RV YES (YES)
171	TILT LESS TOWARD BEING THOSE WHERE THE LEGISLATIVE-EXECUTIVE STRUCTURE IS OTHER THAN PARLIAMENTARY-REPUBLICAN (90)	0.84	171	TILT MORE TOWARD BEING THOSE WHERE THE LEGISLATIVE-EXECUTIVE STRUCTURE IS OTHER THAN PARLIAMENTARY-REPUBLICAN (90)	0.96	9 2 47 43 X SQ= 2.38 P = 0.106 RV NO (NO)
172	TILT LESS TOWARD BEING THOSE WHERE THE LEGISLATIVE-EXECUTIVE STRUCTURE IS OTHER THAN PARLIAMENTARY-ROYALIST (88)	0.75	172	TILT MORE TOWARD BEING THOSE WHERE THE LEGISLATIVE-EXECUTIVE STRUCTURE IS OTHER THAN PARLIAMENTARY-ROYALIST (88)	0.87	15 6 46 41 X SQ= 1.67 P = 0.147 RV NO (NO)
175	TEND TO BE THOSE WHERE THE LEGISLATURE IS FULLY EFFECTIVE OR PARTIALLY EFFECTIVE (51)	0.72	175	TEND TO BE THOSE WHERE THE LEGISLATURE IS LARGELY INEFFECTIVE OR WHOLLY INEFFECTIVE (49)	0.80	42 8 16 33 X SQ= 24.82 P = 0. RV YES (YES)
178	TEND TO BE THOSE WHERE THE LEGISLATURE IS BICAMERAL (51)	0.64	178	TEND TO BE THOSE WHERE THE LEGISLATURE IS UNICAMERAL (53)	0.73	21 32 38 12 X SQ= 12.47 P = 0. RV YES (YES)

179	TEND TO BE THOSE WHERE THE EXECUTIVE IS STRONG (39)	0.61	179	TEND TO BE THOSE WHERE THE EXECUTIVE IS DOMINANT (52)	0.83	21 30 33 6 X SQ= 15.61 P = 0. RV YES (YES)
181	TEND LESS TO BE THOSE WHERE THE BUREAUCRACY IS SEMI-MODERN, RATHER THAN MODERN (55)	0.67	181	ALWAYS ARE THOSE WHERE THE BUREAUCRACY IS SEMI-MODERN, RATHER THAN MODERN (55)	1.00	20 0 41 14 X SQ= 4.69 P = 0.015 RV NO (NO)
182	TEND TO BE THOSE WHERE THE BUREAUCRACY IS SEMI-MODERN, RATHER THAN POST-COLONIAL TRANSITIONAL (55)	0.98	182	TEND TO BE THOSE WHERE THE BUREAUCRACY IS POST-COLONIAL TRANSITIONAL, RATHER THAN SEMI-MODERN (25)	0.63	41 14 1 24 X SQ= 31.53 P = 0. RV YES (YES)
187	TEND TO BE THOSE WHERE THE ROLE OF THE POLICE IS NOT POLITICALLY SIGNIFICANT (35)	0.53	187	TEND TO BE THOSE WHERE THE ROLE OF THE POLICE IS POLITICALLY SIGNIFICANT (66)	0.87	26 39 29 6 X SQ= 15.20 P = 0. RV YES (YES)
188	TEND LESS TO BE THOSE WHERE THE CHARACTER OF THE LEGAL SYSTEM IS OTHER THAN CIVIL LAW (81)	0.52	188	TEND MORE TO BE THOSE WHERE THE CHARACTER OF THE LEGAL SYSTEM IS OTHER THAN CIVIL LAW (81)	0.96	30 2 32 48 X SQ= 24.59 P = 0. RV NO (YES)
189	TEND LESS TO BE THOSE WHERE THE CHARACTER OF THE LEGAL SYSTEM IS OTHER THAN COMMON LAW (108)	0.89	189	ALWAYS ARE THOSE WHERE THE CHARACTER OF THE LEGAL SYSTEM IS OTHER THAN COMMON LAW (108)	1.00	7 0 55 52 X SQ= 4.45 P = 0.015 RV NO (NO)
192	TEND TO BE THOSE WHERE THE CHARACTER OF THE LEGAL SYSTEM IS OTHER THAN MUSLIM OR PARTLY MUSLIM (86)	0.97	192	TEND TO BE THOSE WHERE THE CHARACTER OF THE LEGAL SYSTEM IS MUSLIM OR PARTLY MUSLIM (27)	0.50	2 25 60 25 X SQ= 30.59 P = 0. RV YES (YES)
193	ALWAYS ARE THOSE WHERE THE CHARACTER OF THE LEGAL SYSTEM IS OTHER THAN PARTLY INDIGENOUS (86)	1.00	193	TEND TO BE THOSE WHERE THE CHARACTER OF THE LEGAL SYSTEM IS PARTLY INDIGENOUS (28)	0.53	0 27 62 24 X SQ= 40.27 P = 0. RV YES (YES)

76 POLITIES
THAT ARE HISTORICALLY WESTERN,
RATHER THAN HAVING BEEN
SIGNIFICANTLY OR PARTIALLY WESTERNIZED
THROUGH A COLONIAL RELATIONSHIP (26)

76 POLITIES
THAT HAVE BEEN SIGNIFICANTLY OR
PARTIALLY WESTERNIZED THROUGH A
COLONIAL RELATIONSHIP, RATHER THAN
BEING HISTORICALLY WESTERN (70)

PREDICATE ONLY

26 IN LEFT COLUMN

AUSTRALIA	AUSTRIA	BELGIUM	CANADA	CZECHOS'KIA	DENMARK	FINLAND	FRANCE	GERMANY, E	GERMAN FR
GREECE	HUNGARY	ICELAND	IRELAND	ITALY	LUXEMBOURG	NETHERLANDS	NEW ZEALAND	NORWAY	POLAND
PORTUGAL	SPAIN	SWEDEN	SWITZERLAND	UK	US				

70 IN RIGHT COLUMN

ALGERIA	ARGENTINA	BOLIVIA	BRAZIL	BURMA	BURUNDI	CAMBODIA	CAMEROUN	CEN AFR REP	CEYLON
CHAD	CHILE	COLOMBIA	CONGO(BRA)	CONGO(LEO)	COSTA RICA	CUBA	CYPRUS	DAHOMEY	DOMIN REP
ECUADOR	EL SALVADOR	GABON	GHANA	GUATEMALA	GUINEA	HAITI	HONDURAS	INDIA	INDONESIA
IRAN	IRAQ	IVORY COAST	JAMAICA	JORDAN	LAOS	LEBANON	LIBYA	MALAGASY R	MALAYA
MALI	MAURITANIA	MEXICO	MOROCCO	NICARAGUA	NIGER	NIGERIA	PAKISTAN	PANAMA	PARAGUAY
PERU	PHILIPPINES	RWANDA	SENEGAL	SIERRE LEO	SOMALIA	SO AFRICA	SUDAN	SYRIA	TANGANYIKA
TOGO	TRINIDAD	TUNISIA	UGANDA	UAR	UPPER VOLTA	URUGUAY	VENEZUELA	VIETNAM, N	VIETNAM REP

1 EXCLUDED BECAUSE AMBIGUOUS

ISRAEL

18 EXCLUDED BECAUSE IRRELEVANT

AFGHANISTAN	ALBANIA	BULGARIA	CHINA, PR	ETHIOPIA	JAPAN	KOREA, N	KOREA REP	LIBERIA	MONGOLIA
NEPAL	RUMANIA	SA'U ARABIA	THAILAND	TURKEY	USSR	YEMEN	YUGOSLAVIA		

77 POLITIES
THAT HAVE BEEN SIGNIFICANTLY WESTERNIZED,
RATHER THAN PARTIALLY WESTERNIZED,
THROUGH A COLONIAL RELATIONSHIP (28)

77 POLITIES
THAT HAVE BEEN PARTIALLY WESTERNIZED,
RATHER THAN SIGNIFICANTLY WESTERNIZED,
THROUGH A COLONIAL RELATIONSHIP (41)

PREDICATE ONLY

28 IN LEFT COLUMN

ALGERIA	ARGENTINA	BOLIVIA	BRAZIL	CEYLON	CHILE	COLOMBIA	COSTA RICA	CUBA	CYPRUS
DOMIN REP	ECUADOR	EL SALVADOR	GUATEMALA	HONDURAS	INDIA	JAMAICA	LEBANON	MEXICO	NICARAGUA
PANAMA	PARAGUAY	PERU	PHILIPPINES	TRINIDAD	UAR	URUGUAY	VENEZUELA		

41 IN RIGHT COLUMN

BURMA	BURUNDI	CAMBODIA	CAMEROUN	CEN AFR REP	CHAD	CONGO(BRA)	CONGO(LEO)	DAHOMEY	GABON
GHANA	GUINEA	HAITI	INDONESIA	IRAN	IRAQ	IVORY COAST	JORDAN	LAOS	LIBYA
MALAGASY R	MALAYA	MALI	MAURITANIA	MOROCCO	NIGER	NIGERIA	PAKISTAN	RWANDA	SENEGAL
SIERRE LEO	SOMALIA	SUDAN	SYRIA	TANGANYIKA	TOGO	TUNISIA	UGANDA	UPPER VOLTA	VIETNAM, N
VIETNAM REP									

2 EXCLUDED BECAUSE AMBIGUOUS

ISRAEL	SO AFRICA

44 EXCLUDED BECAUSE IRRELEVANT

AFGHANISTAN	ALBANIA	AUSTRALIA	AUSTRIA	BELGIUM	BULGARIA	CANADA	CHINA, PR	CZECHOS'KIA	DENMARK
ETHIOPIA	FINLAND	FRANCE	GERMANY, E	GERMAN FR	GREECE	HUNGARY	ICELAND	IRELAND	ITALY
JAPAN	KOREA, N	KOREA REP	LIBERIA	LUXEMBOURG	MONGOLIA	NEPAL	NETHERLANDS	NEW ZEALAND	NORWAY
POLAND	PORTUGAL	RUMANIA	SA'U ARABIA	SPAIN	SWEDEN	SWITZERLAND	THAILAND	TURKEY	USSR
UK	US	YEMEN	YUGOSLAVIA						

```
**********************************************************************************************

78  POLITIES                                          78  POLITIES
    THAT HAVE BEEN SIGNIFICANTLY WESTERNIZED              THAT HAVE BEEN SIGNIFICANTLY WESTERNIZED
    THROUGH A COLONIAL RELATIONSHIP, RATHER               WITHOUT A COLONIAL RELATIONSHIP, RATHER
    THAN WITHOUT SUCH A RELATIONSHIP  (28)                THAN THROUGH SUCH A RELATIONSHIP  (7)

                                                                          BOTH SUBJECT AND PREDICATE
..............................................................................................

     28 IN LEFT COLUMN

ALGERIA     ARGENTINA   BOLIVIA     BRAZIL      CEYLON      CHILE       COLOMBIA    COSTA RICA  CUBA        CYPRUS
DOMIN REP   ECUADOR     EL SALVADOR GUATEMALA   HONDURAS    INDIA       JAMAICA     LEBANON     MEXICO      NICARAGUA
PANAMA      PARAGUAY    PERU        PHILIPPINES TRINIDAD    UAR         URUGUAY     VENEZUELA

      7 IN RIGHT COLUMN

BULGARIA    CHINA, PR   JAPAN       RUMANIA     TURKEY      USSR        YUGOSLAVIA

      2 EXCLUDED BECAUSE AMBIGUOUS

ISRAEL      SO AFRICA

     78 EXCLUDED BECAUSE IRRELEVANT

AFGHANISTAN ALBANIA     AUSTRALIA   AUSTRIA     BELGIUM     BURMA       BURUNDI     CAMBODIA    CAMEROUN    CANADA
CEN AFR REP CHAD        CONGO(BRA)  CONGO(LEO)  CZECHOS'KIA DAHOMEY     DENMARK     ETHIOPIA    FINLAND     FRANCE
GABON       GERMANY, E  GERMAN FR   GHANA       GREECE      GUINEA      HAITI       HUNGARY     ICELAND     INDONESIA
IRAN        IRAQ        IRELAND     ITALY       IVORY COAST JORDAN      KOREA, N    KOREA REP   LAOS        LIBERIA
LIBYA       LUXEMBOURG  MALAGASY R  MALAYA      MALI        MAURITANIA  MONGOLIA    MOROCCO     NEPAL       NETHERLANDS
NEW ZEALAND NIGER       NIGERIA     NORWAY      PAKISTAN    POLAND      PORTUGAL    RWANDA      SA'U ARABIA SENEGAL
SIERRE LEO  SOMALIA     SPAIN       SUDAN       SWEDEN      SWITZERLAND SYRIA       TANGANYIKA  THAILAND    TOGO
TUNISIA     UGANDA      UK          US          UPPER VOLTA VIETNAM, N  VIETNAM REP YEMEN
----------------------------------------------------------------------------------------------

  6  TEND TO BE THOSE                        0.75     6  ALWAYS ARE THOSE                       1.00       21    0
     LOCATED IN THE CARIBBEAN,                           LOCATED ELSEWHERE THAN IN THE                      7    7
     CENTRAL AMERICA, OR SOUTH AMERICA  (22)             CARIBBEAN, CENTRAL AMERICA,                X SQ=  10.19
                                                         OR SOUTH AMERICA  (93)                     P =    0.001
                                                                                                    RV YES (YES)

 17  TILT TOWARD BEING THOSE                 0.91    17  ALWAYS ARE THOSE                       1.00        2    1
     LOCATED IN THE CARIBBEAN,                           LOCATED IN THE MIDDLE EAST, RATHER THAN           21    0
     CENTRAL AMERICA, OR SOUTH AMERICA,                  IN THE CARIBBEAN, CENTRAL AMERICA, OR      X SQ=   1.34
     RATHER THAN IN THE MIDDLE EAST  (22)                SOUTH AMERICA  (11)                        P =    0.125
                                                                                                    RV YES (NO )

 20  TEND TO BE THOSE                        0.87    20  ALWAYS ARE THOSE                       1.00        3    2
     LOCATED IN THE CARIBBEAN,                           LOCATED IN EAST ASIA, SOUTH ASIA, OR              21    0
     CENTRAL AMERICA, OR SOUTH AMERICA,                  SOUTHEAST ASIA, RATHER THAN IN             X SQ=   4.34
     RATHER THAN IN EAST ASIA, SOUTH ASIA,               THE CARIBBEAN, CENTRAL AMERICA,            P =    0.031
     OR SOUTHEAST ASIA  (22)                             OR SOUTH AMERICA  (18)                     RV YES (NO )
```

22	TILT TOWARD BEING THOSE WHOSE TERRITORIAL SIZE IS SMALL (47)	0.50	22	TILT TOWARD BEING THOSE WHOSE TERRITORIAL SIZE IS VERY LARGE, LARGE, OR MEDIUM (68)	0.86	14 6 14 1 X SQ= 1.64 P = 0.114 RV YES (NO)
23	TEND TO BE THOSE WHOSE POPULATION IS MEDIUM OR SMALL (88)	0.79	23	TEND TO BE THOSE WHOSE POPULATION IS VERY LARGE OR LARGE (27)	0.86	6 6 22 1 X SQ= 7.62 P = 0.003 RV YES (YES)
24	TEND TO BE THOSE WHOSE POPULATION IS SMALL (54)	0.54	24	ALWAYS ARE THOSE WHOSE POPULATION IS VERY LARGE, LARGE, OR MEDIUM (61)	1.00	13 7 15 0 X SQ= 4.56 P = 0.012 RV YES (NO)
28	TEND TO BE THOSE WHOSE POPULATION GROWTH RATE IS HIGH (62)	0.89	28	TEND TO BE THOSE WHOSE POPULATION GROWTH RATE IS LOW (48)	0.67	25 2 3 4 X SQ= 6.35 P = 0.010 RV YES (YES)
33	TEND TO BE THOSE WHOSE GROSS NATIONAL PRODUCT IS LOW OR VERY LOW (85)	0.86	33	TEND TO BE THOSE WHOSE GROSS NATIONAL PRODUCT IS VERY HIGH, HIGH, OR MEDIUM (30)	0.86	4 6 24 1 X SQ= 10.72 P = 0.001 RV YES (YES)
34	TEND TO BE THOSE WHOSE GROSS NATIONAL PRODUCT IS VERY LOW (53)	0.50	34	ALWAYS ARE THOSE WHOSE GROSS NATIONAL PRODUCT IS VERY HIGH, HIGH, MEDIUM, OR LOW (62)	1.00	14 7 14 0 X SQ= 3.94 P = 0.027 RV YES (NO)
36	LEAN TOWARD BEING THOSE WHOSE PER CAPITA GROSS NATIONAL PRODUCT IS LOW OR VERY LOW (73)	0.68	36	LEAN TOWARD BEING THOSE WHOSE PER CAPITA GROSS NATIONAL PRODUCT IS VERY HIGH, HIGH, OR MEDIUM (42)	0.71	9 5 19 2 X SQ= 2.15 P = 0.090 RV YES (NO)
39	TEND TO BE THOSE WHOSE INTERNATIONAL FINANCIAL STATUS IS LOW OR VERY LOW (76)	0.68	39	TEND TO BE THOSE WHOSE INTERNATIONAL FINANCIAL STATUS IS VERY HIGH, HIGH, OR MEDIUM (38)	0.86	9 6 19 1 X SQ= 4.56 P = 0.027 RV YES (NO)
42	TEND TO BE THOSE WHOSE ECONOMIC DEVELOPMENTAL STATUS IS UNDERDEVELOPED OR VERY UNDERDEVELOPED (76)	0.78	42	TEND TO BE THOSE WHOSE ECONOMIC DEVELOPMENTAL STATUS IS DEVELOPED OR INTERMEDIATE (36)	0.71	6 5 21 2 X SQ= 4.11 P = 0.024 RV YES (NO)

43	LEAN LESS TOWARD BEING THOSE WHOSE ECONOMIC DEVELOPMENTAL STATUS IS DEVELOPED, INTERMEDIATE, OR UNDERDEVELOPED (55)	0.58	43	ALWAYS ARE THOSE WHOSE ECONOMIC DEVELOPMENTAL STATUS IS DEVELOPED, INTERMEDIATE, OR UNDERDEVELOPED (55)	1.00	15 7 11 0 X SQ= 2.74 P = 0.067 RV NO (NO)
44	ALWAYS ARE THOSE WHERE THE LITERACY RATE IS BELOW NINETY PERCENT (90)	1.00	44	TEND LESS TO BE THOSE WHERE THE LITERACY RATE IS BELOW NINETY PERCENT (90)	0.57	0 3 28 4 X SQ= 8.23 P = 0.005 RV NO (YES)
47	ALWAYS ARE THOSE WHERE THE LITERACY RATE IS BETWEEN FIFTY AND NINETY PERCENT, RATHER THAN NINETY PERCENT OR ABOVE (30)	1.00	47	TEND TO BE THOSE WHERE THE LITERACY RATE IS NINETY PERCENT OR ABOVE, RATHER THAN BETWEEN FIFTY AND NINETY PERCENT (25)	0.60	0 3 17 2 X SQ= 7.27 P = 0.006 RV YES (YES)
50	LEAN TOWARD BEING THOSE WHERE FREEDOM OF THE PRESS IS COMPLETE (43)	0.57	50	LEAN TOWARD BEING THOSE WHERE FREEDOM OF THE PRESS IS INTERMITTENT, INTERNALLY ABSENT, OR INTERNALLY AND EXTERNALLY ABSENT (56)	0.86	13 1 10 6 X SQ= 2.34 P = 0.086 RV YES (NO)
52	TILT MORE TOWARD BEING THOSE WHERE FREEDOM OF THE PRESS IS COMPLETE, INTERMITTENT, OR INTERNALLY ABSENT (82)	0.87	52	TILT LESS TOWARD BEING THOSE WHERE FREEDOM OF THE PRESS IS COMPLETE, INTERMITTENT, OR INTERNALLY ABSENT (82)	0.57	20 4 3 3 X SQ= 1.41 P = 0.120 RV NO (YES)
57	TEND TO BE THOSE WHERE THE RELIGION IS CATHOLIC (25)	0.57	57	ALWAYS ARE THOSE WHERE THE RELIGION IS OTHER THAN CATHOLIC (90)	1.00	16 0 12 7 X SQ= 5.25 P = 0.009 RV YES (NO)
67	TEND TO BE THOSE THAT ARE RACIALLY HETEROGENEOUS (27)	0.62	67	ALWAYS ARE THOSE THAT ARE RACIALLY HOMOGENEOUS (82)	1.00	10 7 16 0 X SQ= 6.08 P = 0.007 RV YES (NO)
68	LEAN TOWARD BEING THOSE THAT ARE LINGUISTICALLY HOMOGENEOUS (52)	0.61	68	LEAN TOWARD BEING THOSE THAT ARE LINGUISTICALLY WEAKLY HETEROGENEOUS OR STRONGLY HETEROGENEOUS (62)	0.83	17 1 11 5 X SQ= 2.28 P = 0.078 RV YES (NO)
71	ALWAYS ARE THOSE WHOSE DATE OF INDEPENDENCE IS AFTER 1800 (90)	1.00	71	TEND TO BE THOSE WHOSE DATE OF INDEPENDENCE IS BEFORE 1800 (21)	0.57	0 4 28 3 X SQ= 12.86 P = 0.001 RV YES (YES)

83	ALWAYS ARE THOSE WHERE THE TYPE OF POLITICAL MODERNIZATION IS OTHER THAN NON-EUROPEAN AUTOCHTHONOUS (106)	1.00	83	TEND TO BE THOSE WHERE THE TYPE OF POLITICAL MODERNIZATION IS NON-EUROPEAN AUTOCHTHONOUS (9)	0.57	0 4 28 3 X SQ= 12.86 P = 0.001 RV YES (YES)
85	TEND LESS TO BE THOSE WHERE THE STAGE OF POLITICAL MODERNIZATION IS ADVANCED, RATHER THAN MID- OR EARLY TRANSITIONAL (60)	0.54	85	ALWAYS ARE THOSE WHERE THE STAGE OF POLITICAL MODERNIZATION IS ADVANCED, RATHER THAN MID- OR EARLY TRANSITIONAL (60)	1.00	15 7 13 0 X SQ= 3.37 P = 0.031 RV NO (NO)
87	TILT LESS TOWARD BEING THOSE WHOSE IDEOLOGICAL ORIENTATION IS OTHER THAN DEVELOPMENTAL (58)	0.69	87	ALWAYS ARE THOSE WHOSE IDEOLOGICAL ORIENTATION IS OTHER THAN DEVELOPMENTAL (58)	1.00	5 0 11 7 X SQ= 1.26 P = 0.147 RV NO (NO)
92	TEND TO BE THOSE WHERE THE SYSTEM STYLE IS LIMITED MOBILIZATIONAL, OR NON-MOBILIZATIONAL (93)	0.89	92	TEND TO BE THOSE WHERE THE SYSTEM STYLE IS MOBILIZATIONAL (20)	0.71	3 5 25 2 X SQ= 8.52 P = 0.003 RV YES (YES)
93	TEND TO BE THOSE WHERE THE SYSTEM STYLE IS NON-MOBILIZATIONAL (78)	0.82	93	TEND TO BE THOSE WHERE THE SYSTEM STYLE IS MOBILIZATIONAL, OR LIMITED MOBILIZATIONAL (32)	0.71	5 5 23 2 X SQ= 5.47 P = 0.012 RV YES (YES)
94	TEND TO BE THOSE WHERE THE STATUS OF THE REGIME IS CONSTITUTIONAL (51)	0.76	94	TEND TO BE THOSE WHERE THE STATUS OF THE REGIME IS AUTHORITARIAN OR TOTALITARIAN (41)	0.71	19 2 6 5 X SQ= 3.55 P = 0.032 RV YES (NO)
95	TEND TO BE THOSE WHERE THE STATUS OF THE REGIME IS CONSTITUTIONAL OR AUTHORITARIAN (95)	0.96	95	TEND TO BE THOSE WHERE THE STATUS OF THE REGIME IS TOTALITARIAN (16)	0.71	27 2 1 5 X SQ= 13.69 P = 0. RV YES (YES)
97	TEND TO BE THOSE WHERE THE STATUS OF THE REGIME IS AUTHORITARIAN, RATHER THAN TOTALITARIAN (23)	0.83	97	ALWAYS ARE THOSE WHERE THE STATUS OF THE REGIME IS TOTALITARIAN, RATHER THAN AUTHORITARIAN (16)	1.00	5 0 1 5 X SQ= 4.65 P = 0.015 RV YES (YES)
99	TEND TO BE THOSE WHERE GOVERNMENTAL STABILITY IS MODERATELY PRESENT AND DATES FROM THE POST-WAR PERIOD, OR IS ABSENT (36)	0.76	99	TEND TO BE THOSE WHERE GOVERNMENTAL STABILITY IS GENERALLY PRESENT AND DATES FROM AT LEAST THE INTER-WAR PERIOD, OR FROM THE POST-WAR PERIOD (50)	0.86	5 6 16 1 X SQ= 6.04 P = 0.007 RV YES (YES)

100	LEAN TOWARD BEING THOSE WHERE GOVERNMENTAL STABILITY IS ABSENT (22)	0.50	100	ALWAYS ARE THOSE WHERE GOVERNMENTAL STABILITY IS GENERALLY PRESENT AND DATES FROM AT LEAST THE INTER-WAR PERIOD, OR IS GENERALLY OR MODERATELY PRESENT, AND DATES FROM THE POST-WAR PERIOD (64)	1.00	11 6 11 0 X SQ= 3.07 P = 0.055 RV YES (NO)
102	TEND TO BE THOSE WHERE THE REPRESENTATIVE CHARACTER OF THE REGIME IS POLYARCHIC OR LIMITED POLYARCHIC (59)	0.77	102	TEND TO BE THOSE WHERE THE REPRESENTATIVE CHARACTER OF THE REGIME IS PSEUDO-POLYARCHIC OR NON-POLYARCHIC (49)	0.71	20 2 6 5 X SQ= 3.83 P = 0.027 RV YES (NO)
105	TEND TO BE THOSE WHERE THE ELECTORAL SYSTEM IS COMPETITIVE (43)	0.80	105	TEND TO BE THOSE WHERE THE ELECTORAL SYSTEM IS PARTIALLY COMPETITIVE OR NON-COMPETITIVE (47)	0.71	16 2 4 5 X SQ= 4.07 P = 0.023 RV YES (YES)
106	TEND TO BE THOSE WHERE THE ELECTORAL SYSTEM IS COMPETITIVE OR PARTIALLY COMPETITIVE (52)	0.90	106	TEND TO BE THOSE WHERE THE ELECTORAL SYSTEM IS NON-COMPETITIVE (30)	0.71	18 2 2 5 X SQ= 7.24 P = 0.005 RV YES (YES)
107	LEAN TOWARD BEING THOSE WHERE AUTONOMOUS GROUPS ARE FULLY TOLERATED IN POLITICS (46)	0.70	107	LEAN TOWARD BEING THOSE WHERE AUTONOMOUS GROUPS ARE PARTIALLY TOLERATED IN POLITICS, ARE TOLERATED ONLY OUTSIDE POLITICS, OR ARE NOT TOLERATED AT ALL (65)	0.71	19 2 8 5 X SQ= 2.53 P = 0.079 RV YES (NO)
108	TEND TO BE THOSE WHERE AUTONOMOUS GROUPS ARE FULLY OR PARTIALLY TOLERATED IN POLITICS (65)	0.92	108	TEND TO BE THOSE WHERE AUTONOMOUS GROUPS ARE TOLERATED ONLY OUTSIDE POLITICS OR ARE NOT TOLERATED AT ALL (35)	0.71	24 2 2 5 X SQ= 9.86 P = 0.002 RV YES (YES)
110	TEND TO BE THOSE WHERE POLITICAL ENCULTURATION IS MEDIUM OR LOW (80)	0.96	110	TEND TO BE THOSE WHERE POLITICAL ENCULTURATION IS HIGH (15)	0.67	1 2 23 1 X SQ= 5.17 P = 0.025 RV YES (YES)
117	TEND TO BE THOSE WHERE INTEREST ARTICULATION BY ASSOCIATIONAL GROUPS IS SIGNIFICANT, MODERATE, OR LIMITED (60)	0.93	117	TEND TO BE THOSE WHERE INTEREST ARTICULATION BY ASSOCIATIONAL GROUPS IS NEGLIGIBLE (51)	0.50	26 3 2 3 X SQ= 4.22 P = 0.029 RV YES (YES)
118	LEAN TOWARD BEING THOSE WHERE INTEREST ARTICULATION BY INSTITUTIONAL GROUPS IS SIGNIFICANT, MODERATE, OR LIMITED (60)	0.71	118	LEAN TOWARD BEING THOSE WHERE INTEREST ARTICULATION BY INSTITUTIONAL GROUPS IS VERY SIGNIFICANT (40)	0.71	8 5 20 2 X SQ= 2.76 P = 0.075 RV YES (NO)

128	TEND TO BE THOSE WHERE INTEREST ARTICULATION BY POLITICAL PARTIES IS SIGNIFICANT OR MODERATE (48)	0.71	128	TEND TO BE THOSE WHERE INTEREST ARTICULATION BY POLITICAL PARTIES IS LIMITED OR NEGLIGIBLE (45)	0.86	17 1 7 6 X SQ= 4.98 P = 0.025 RV YES (NO)
129	TEND TO BE THOSE WHERE INTEREST ARTICULATION BY POLITICAL PARTIES IS SIGNIFICANT, MODERATE, OR LIMITED (56)	0.88	129	TEND TO BE THOSE WHERE INTEREST ARTICULATION BY POLITICAL PARTIES IS NEGLIGIBLE (37)	0.71	21 2 3 5 X SQ= 6.99 P = 0.006 RV YES (YES)
138	TEND TO BE THOSE WHERE INTEREST AGGREGATION BY THE LEGISLATURE IS SIGNIFICANT, MODERATE, OR LIMITED (48)	0.83	138	TEND TO BE THOSE WHERE INTEREST AGGREGATION BY THE LEGISLATURE IS NEGLIGIBLE (49)	0.71	20 2 4 5 X SQ= 5.45 P = 0.012 RV YES (YES)
139	TEND TO BE THOSE WHERE THE PARTY SYSTEM IS QUANTITATIVELY OTHER THAN ONE-PARTY (71)	0.89	139	TEND TO BE THOSE WHERE THE PARTY SYSTEM IS QUANTITATIVELY ONE-PARTY (34)	0.71	3 5 24 2 X SQ= 8.14 P = 0.004 RV YES (YES)
142	TEND TO BE THOSE WHERE THE PARTY SYSTEM IS QUANTITATIVELY MULTI-PARTY (30)	0.54	142	ALWAYS ARE THOSE WHERE THE PARTY SYSTEM IS QUANTITATIVELY OTHER THAN MULTI-PARTY (66)	1.00	14 0 12 6 X SQ= 3.76 P = 0.024 RV YES (NO)
144	LEAN TOWARD BEING THOSE WHERE THE PARTY SYSTEM IS QUANTITATIVELY TWO-PARTY, RATHER THAN ONE-PARTY (11)	0.57	144	ALWAYS ARE THOSE WHERE THE PARTY SYSTEM IS QUANTITATIVELY ONE-PARTY, RATHER THAN TWO-PARTY (34)	1.00	3 5 4 0 X SQ= 2.10 P = 0.081 RV YES (YES)
153	TEND TO BE THOSE WHERE THE PARTY SYSTEM IS MODERATELY STABLE OR UNSTABLE (38)	0.84	153	TEND TO BE THOSE WHERE THE PARTY SYSTEM IS STABLE (42)	0.86	3 6 16 1 X SQ= 8.18 P = 0.002 RV YES (YES)
154	TEND TO BE THOSE WHERE THE PARTY SYSTEM IS UNSTABLE (25)	0.68	154	ALWAYS ARE THOSE WHERE THE PARTY SYSTEM IS STABLE OR MODERATELY STABLE (55)	1.00	6 7 13 0 X SQ= 7.04 P = 0.005 RV YES (YES)
157	TILT LESS TOWARD BEING THOSE WHERE PERSONALISMO IS MODERATE OR NEGLIGIBLE (84)	0.65	157	ALWAYS ARE THOSE WHERE PERSONALISMO IS MODERATE OR NEGLIGIBLE (84)	1.00	9 0 17 7 X SQ= 1.82 P = 0.149 RV NO (NO)

158	TEND TO BE THOSE WHERE PERSONALISMO IS PRONOUNCED OR MODERATE (40)	0.65	TEND TO BE THOSE WHERE PERSONALISMO IS NEGLIGIBLE (56)	0.86	17 1 9 6 X SQ= 3.93 P = 0.030 RV YES (NO)
160	LEAN TOWARD BEING THOSE WHERE THE POLITICAL LEADERSHIP IS MODERATELY ELITIST OR NON-ELITIST (67)	0.74	LEAN TOWARD BEING THOSE WHERE THE POLITICAL LEADERSHIP IS ELITIST (30)	0.71	6 5 17 2 X SQ= 3.00 P = 0.068 RV YES (NO)
169	TEND TO BE THOSE WHERE THE HORIZONTAL POWER DISTRIBUTION IS SIGNIFICANT OR LIMITED (58)	0.85	TEND TO BE THOSE WHERE THE HORIZONTAL POWER DISTRIBUTION IS NEGLIGIBLE (48)	0.71	22 2 4 5 X SQ= 6.14 P = 0.009 RV YES (YES)
170	TEND TO BE THOSE WHERE THE LEGISLATIVE-EXECUTIVE STRUCTURE IS PRESIDENTIAL (39)	0.75	ALWAYS ARE THOSE WHERE THE LEGISLATIVE-EXECUTIVE STRUCTURE IS OTHER THAN PRESIDENTIAL (63)	1.00	18 0 6 7 X SQ= 9.63 P = 0.001 RV YES (YES)
175	TEND TO BE THOSE WHERE THE LEGISLATURE IS FULLY EFFECTIVE OR PARTIALLY EFFECTIVE (51)	0.79	TEND TO BE THOSE WHERE THE LEGISLATURE IS LARGELY INEFFECTIVE OR WHOLLY INEFFECTIVE (49)	0.71	19 2 5 5 X SQ= 4.24 P = 0.022 RV YES (YES)
176	TEND TO BE THOSE WHERE THE LEGISLATURE IS FULLY EFFECTIVE, PARTIALLY EFFECTIVE, OR LARGELY INEFFECTIVE (72)	0.92	TEND TO BE THOSE WHERE THE LEGISLATURE IS WHOLLY INEFFECTIVE (28)	0.71	22 2 2 5 X SQ= 9.00 P = 0.002 RV YES (YES)
185	TEND TO BE THOSE WHERE PARTICIPATION BY THE MILITARY IN POLITICS IS INTERVENTIVE, RATHER THAN SUPPORTIVE (21)	0.77	TEND TO BE THOSE WHERE PARTICIPATION BY THE MILITARY IN POLITICS IS SUPPORTIVE, RATHER THAN INTERVENTIVE (31)	0.83	10 1 3 5 X SQ= 3.89 P = 0.041 RV YES (YES)
186	TEND TO BE THOSE WHERE PARTICIPATION BY THE MILITARY IN POLITICS IS NEUTRAL, RATHER THAN SUPPORTIVE (56)	0.81	TEND TO BE THOSE WHERE PARTICIPATION BY THE MILITARY IN POLITICS IS SUPPORTIVE, RATHER THAN NEUTRAL (31)	0.83	3 5 13 1 X SQ= 5.32 P = 0.011 RV YES (YES)
188	TEND TO BE THOSE WHERE THE CHARACTER OF THE LEGAL SYSTEM IS CIVIL LAW (32)	0.64	TEND TO BE THOSE WHERE THE CHARACTER OF THE LEGAL SYSTEM IS OTHER THAN CIVIL LAW (81)	0.86	18 1 10 6 X SQ= 3.81 P = 0.032 RV YES (NO)

```
194  ALWAYS ARE THOSE                    1.00     194  TEND TO BE THOSE              0.71        0    5
     THAT ARE NON-COMMUNIST  (101)                     THAT ARE COMMUNIST  (13)                 27    2
                                                                                            X SQ=  17.27
                                                                                            P =     0.
                                                                                            RV YES (YES)
```

```
*****************************************************************************************************

79  POLITIES                                        79  POLITIES
    THAT WERE FORMERLY DEPENDENCIES                     THAT WERE FORMERLY DEPENDENCIES
    OF BRITAIN OR FRANCE,                               OF SPAIN, RATHER THAN
    RATHER THAN SPAIN  (49)                             BRITAIN OR FRANCE  (18)

                                                                        BOTH SUBJECT AND PREDICATE
.....................................................................................................

     49 IN LEFT COLUMN

ALGERIA      AUSTRALIA    BURMA        CAMBODIA     CAMEROUN     CANADA       CEN AFR REP  CEYLON       CHAD         CONGO(BRA)
CYPRUS       DAHOMEY      GABON        GHANA        GUINEA       HAITI        INDIA        IRAQ         IRELAND      ISRAEL
IVORY COAST  JAMAICA      JORDAN       LAOS         LEBANON      MALAGASY R   MALAYA       MALI         MAURITANIA   MOROCCO
NEW ZEALAND  NIGER        NIGERIA      PAKISTAN     SENEGAL      SIERRE LEO   SO AFRICA    SUDAN        SYRIA        TANGANYIKA
TOGO         TRINIDAD     TUNISIA      UGANDA       UAR          US           UPPER VOLTA  VIETNAM, N   VIETNAM REP

     18 IN RIGHT COLUMN

ARGENTINA    BOLIVIA      CHILE        COLOMBIA     COSTA RICA   CUBA         DOMIN REP    ECUADOR      EL SALVADOR  GUATEMALA
HONDURAS     MEXICO       NICARAGUA    PANAMA       PARAGUAY     PERU         URUGUAY      VENEZUELA

     48 EXCLUDED BECAUSE IRRELEVANT

AFGHANISTAN  ALBANIA      AUSTRIA      BELGIUM      BRAZIL       BULGARIA     BURUNDI      CHINA, PR    CONGO(LEO)   CZECHOS'KIA
DENMARK      ETHIOPIA     FINLAND      FRANCE       GERMANY, E   GERMAN FR    GREECE       HUNGARY      ICELAND      INDONESIA
IRAN         ITALY        JAPAN        KOREA, N     KOREA REP    LIBERIA      LIBYA        LUXEMBOURG   MONGOLIA     NEPAL
NETHERLANDS  NORWAY       PHILIPPINES  POLAND       PORTUGAL     RUMANIA      RWANDA       SA'U ARABIA  SOMALIA      SPAIN
SWEDEN       SWITZERLAND  THAILAND     TURKEY       USSR         UK           YEMEN        YUGOSLAVIA
-----------------------------------------------------------------------------------------------------

 6  TEND TO BE THOSE                   0.94      6  ALWAYS ARE THOSE                   1.00        3   18
    LOCATED ELSEWHERE THAN IN THE                   LOCATED IN THE CARIBBEAN,                     46    0
    CARIBBEAN, CENTRAL AMERICA,                     CENTRAL AMERICA, OR SOUTH AMERICA  (22)   X SQ=  49.64
    OR SOUTH AMERICA  (93)                                                                    P =    0.
                                                                                              RV YES (YES)

 8  LEAN LESS TOWARD BEING THOSE       0.82      8  ALWAYS ARE THOSE                   1.00        9    0
    LOCATED ELSEWHERE THAN IN EAST ASIA             LOCATED ELSEWHERE THAN IN EAST ASIA           40   18
    SOUTH ASIA, OR SOUTHEAST ASIA  (97)             SOUTH ASIA, OR SOUTHEAST ASIA  (97)       X SQ=   2.40
                                                                                              P =    0.057
                                                                                              RV NO  (NO )

10  TEND TO BE THOSE                   0.53     10  ALWAYS ARE THOSE                   1.00       26    0
    LOCATED IN NORTH AFRICA, OR                     LOCATED ELSEWHERE THAN IN NORTH AFRICA,       23   18
    CENTRAL AND SOUTH AFRICA  (33)                  OR CENTRAL AND SOUTH AFRICA  (82)         X SQ=  13.45
                                                                                              P =    0.
                                                                                              RV YES (NO )

11  TEND LESS TO BE THOSE              0.55     11  ALWAYS ARE THOSE                   1.00       22    0
    LOCATED ELSEWHERE THAN IN CENTRAL               LOCATED ELSEWHERE THAN IN CENTRAL             27   18
    AND SOUTH AFRICA  (87)                          AND SOUTH AFRICA  (87)                    X SQ=  10.08
                                                                                              P =    0.
                                                                                              RV NO  (NO )
```

13	LEAN LESS TOWARD BEING THOSE LOCATED ELSEWHERE THAN IN NORTH AFRICA OR THE MIDDLE EAST (99)	0.80	13	ALWAYS ARE THOSE LOCATED ELSEWHERE THAN IN NORTH AFRICA OR THE MIDDLE EAST (99)	1.00	10 0 39 18 X SQ= 2.86 P = 0.052 RV NO (NO)
17	TEND TO BE THOSE LOCATED IN THE MIDDLE EAST, RATHER THAN IN THE CARIBBEAN, CENTRAL AMERICA, OR SOUTH AMERICA (11)	0.67	17	ALWAYS ARE THOSE LOCATED IN THE CARIBBEAN, CENTRAL AMERICA, OR SOUTH AMERICA, RATHER THAN IN THE MIDDLE EAST (22)	1.00	6 0 3 18 X SQ= 11.81 P = 0. RV YES (YES)
28	TILT LESS TOWARD BEING THOSE WHOSE POPULATION GROWTH RATE IS HIGH (62)	0.68	28	TILT MORE TOWARD BEING THOSE WHOSE POPULATION GROWTH RATE IS HIGH (62)	0.89	32 16 15 2 X SQ= 1.94 P = 0.119 RV NO (NO)
29	TEND TO BE THOSE WHERE THE DEGREE OF URBANIZATION IS LOW (49)	0.67	29	TEND TO BE THOSE WHERE THE DEGREE OF URBANIZATION IS HIGH (56)	0.75	15 12 30 4 X SQ= 6.70 P = 0.007 RV YES (NO)
30	TEND TO BE THOSE WHOSE AGRICULTURAL POPULATION IS HIGH (56)	0.65	30	TEND TO BE THOSE WHOSE AGRICULTURAL POPULATION IS MEDIUM, LOW, OR VERY LOW (57)	0.78	31 4 17 14 X SQ= 7.81 P = 0.003 RV YES (NO)
37	TEND TO BE THOSE WHOSE PER CAPITA GROSS NATIONAL PRODUCT IS VERY LOW (51)	0.65	37	TEND TO BE THOSE WHOSE PER CAPITA GROSS NATIONAL PRODUCT IS VERY HIGH, HIGH, MEDIUM, OR LOW (64)	0.89	17 16 32 2 X SQ= 13.38 P = 0. RV YES (NO)
43	TILT TOWARD BEING THOSE WHOSE ECONOMIC DEVELOPMENTAL STATUS IS VERY UNDERDEVELOPED (57)	0.69	43	TILT TOWARD BEING THOSE WHOSE ECONOMIC DEVELOPMENTAL STATUS IS DEVELOPED, INTERMEDIATE, OR UNDERDEVELOPED (55)	0.53	15 9 33 8 X SQ= 1.69 P = 0.147 RV YES (NO)
45	TEND TO BE THOSE WHERE THE LITERACY RATE IS BELOW FIFTY PERCENT (54)	0.74	45	TEND TO BE THOSE WHERE THE LITERACY RATE IS FIFTY PERCENT OR ABOVE (55)	0.61	12 11 35 7 X SQ= 5.73 P = 0.010 RV YES (NO)
46	TEND LESS TO BE THOSE WHERE THE LITERACY RATE IS TEN PERCENT OR ABOVE (84)	0.62	46	ALWAYS ARE THOSE WHERE THE LITERACY RATE IS TEN PERCENT OR ABOVE (84)	1.00	28 18 17 0 X SQ= 7.49 P = 0.001 RV NO (NO)

47	TEND LESS TO BE THOSE WHERE THE LITERACY RATE IS BETWEEN FIFTY AND NINETY PERCENT, RATHER THAN NINETY PERCENT OR ABOVE (30)	0.58	47	ALWAYS ARE THOSE WHERE THE LITERACY RATE IS BETWEEN FIFTY AND NINETY PERCENT, RATHER THAN NINETY PERCENT OR ABOVE (30)	1.00	5 0 7 11 X SQ= 3.66 P = 0.037 RV NO (YES)
49	TEND TO BE THOSE WHERE THE LITERACY RATE IS BELOW TEN PERCENT, RATHER THAN BETWEEN TEN AND FIFTY PERCENT (26)	0.55	49	ALWAYS ARE THOSE WHERE THE LITERACY RATE IS BETWEEN TEN AND FIFTY PERCENT, RATHER THAN BELOW TEN PERCENT (24)	1.00	14 7 17 0 X SQ= 4.91 P = 0.011 RV YES (NO)
55	TEND LESS TO BE THOSE WHERE NEWSPAPER CIRCULATION IS TEN OR MORE PER THOUSAND (78)	0.52	55	ALWAYS ARE THOSE WHERE NEWSPAPER CIRCULATION IS TEN OR MORE PER THOUSAND (78)	1.00	25 18 23 0 X SQ= 11.21 P = 0. RV NO (NO)
56	LEAN LESS TOWARD BEING THOSE WHERE THE RELIGION IS PREDOMINANTLY LITERATE (79)	0.53	56	LEAN MORE TOWARD BEING THOSE WHERE THE RELIGION IS PREDOMINANTLY LITERATE (79)	0.78	25 14 22 4 X SQ= 2.33 P = 0.093 RV NO (NO)
57	TEND TO BE THOSE WHERE THE RELIGION IS OTHER THAN CATHOLIC (90)	0.98	57	TEND TO BE THOSE WHERE THE RELIGION IS CATHOLIC (25)	0.78	1 14 48 4 X SQ= 39.21 P = 0. RV YES (YES)
58	LEAN LESS TOWARD BEING THOSE WHERE THE RELIGION IS OTHER THAN MUSLIM (97)	0.80	58	ALWAYS ARE THOSE WHERE THE RELIGION IS OTHER THAN MUSLIM (97)	1.00	10 0 39 18 X SQ= 2.86 P = 0.052 RV NO (NO)
63	TEND TO BE THOSE WHERE THE RELIGION IS PREDOMINANTLY OR PARTLY OTHER THAN CHRISTIAN (68)	0.88	63	TEND TO BE THOSE WHERE THE RELIGION IS PREDOMINANTLY SOME KIND OF CHRISTIAN (46)	0.78	6 14 43 4 X SQ= 23.96 P = 0. RV YES (YES)
64	TEND TO BE THOSE WHERE THE RELIGION IS MUSLIM, RATHER THAN CHRISTIAN (18)	0.62	64	ALWAYS ARE THOSE WHERE THE RELIGION IS CHRISTIAN, RATHER THAN MUSLIM (46)	1.00	6 14 10 0 X SQ= 10.46 P = 0. RV YES (YES)
66	TEND TO BE THOSE THAT ARE RELIGIOUSLY HETEROGENEOUS (49)	0.67	66	TEND TO BE THOSE THAT ARE RELIGIOUSLY HOMOGENEOUS (57)	0.78	15 14 31 4 X SQ= 8.91 P = 0.002 RV YES (NO)

67	TEND TO BE THOSE THAT ARE RACIALLY HOMOGENEOUS (82)	0.73	67	TEND TO BE THOSE THAT ARE RACIALLY HETEROGENEOUS (27)	0.81	35 3 13 13 X SQ= 12.44 P = 0. RV YES (YES)
68	TEND TO BE THOSE THAT ARE LINGUISTICALLY WEAKLY HETEROGENEOUS OR STRONGLY HETEROGENEOUS (62)	0.78	68	TEND TO BE THOSE THAT ARE LINGUISTICALLY HOMOGENEOUS (52)	0.72	11 13 38 5 X SQ= 12.10 P = 0. RV YES (YES)
69	TEND TO BE THOSE THAT ARE LINGUISTICALLY STRONGLY HETEROGENEOUS (50)	0.67	69	TEND TO BE THOSE THAT ARE LINGUISTICALLY HOMOGENEOUS OR WEAKLY HETEROGENEOUS (64)	0.78	16 14 33 4 X SQ= 9.09 P = 0.002 RV YES (NO)
72	TEND TO BE THOSE WHOSE DATE OF INDEPENDENCE IS AFTER 1914 (59)	0.88	72	ALWAYS ARE THOSE WHOSE DATE OF INDEPENDENCE IS BEFORE 1914 (52)	1.00	6 18 43 0 X SQ= 40.36 P = 0. RV YES (YES)
73	TEND TO BE THOSE WHOSE DATE OF INDEPENDENCE IS AFTER 1945 (46)	0.80	73	ALWAYS ARE THOSE WHOSE DATE OF INDEPENDENCE IS BEFORE 1945 (65)	1.00	10 18 39 0 X SQ= 31.09 P = 0. RV YES (YES)
75	TEND TO BE THOSE THAT ARE NOT HISTORICALLY WESTERN AND ARE NOT SIGNIFICANTLY WESTERNIZED (52)	0.71	75	ALWAYS ARE THOSE THAT ARE HISTORICALLY WESTERN OR SIGNIFICANTLY WESTERNIZED (62)	1.00	14 18 34 0 X SQ= 23.54 P = 0. RV YES (YES)
77	TEND TO BE THOSE THAT HAVE BEEN PARTIALLY WESTERNIZED, RATHER THAN SIGNIFICANTLY WESTERNIZED, THROUGH A COLONIAL RELATIONSHIP (41)	0.81	77	ALWAYS ARE THOSE THAT HAVE BEEN SIGNIFICANTLY WESTERNIZED, RATHER THAN PARTIALLY WESTERNIZED, THROUGH A COLONIAL RELATIONSHIP (28)	1.00	8 18 34 0 X SQ= 30.41 P = 0. RV YES (YES)
81	TEND TO BE THOSE WHERE THE TYPE OF POLITICAL MODERNIZATION IS EARLY EUROPEAN OR EARLY EUROPEAN DERIVED, RATHER THAN LATER EUROPEAN OR LATER EUROPEAN DERIVED (11)	0.71	81	ALWAYS ARE THOSE WHERE THE TYPE OF POLITICAL MODERNIZATION IS LATER EUROPEAN OR LATER EUROPEAN DERIVED, RATHER THAN EARLY EUROPEAN OR EARLY EUROPEAN DERIVED (40)	1.00	5 0 2 18 X SQ= 11.92 P = 0. RV YES (YES)
82	TEND TO BE THOSE WHERE THE TYPE OF POLITICAL MODERNIZATION IS DEVELOPED TUTELARY OR UNDEVELOPED TUTELARY, RATHER THAN EARLY OR LATER EUROPEAN OR EUROPEAN DERIVED (55)	0.86	82	ALWAYS ARE THOSE WHERE THE TYPE OF POLITICAL MODERNIZATION IS EARLY OR LATER EUROPEAN OR EUROPEAN DERIVED, RATHER THAN DEVELOPED TUTELARY OR UNDEVELOPED TUTELARY (51)	1.00	7 18 42 0 X SQ= 37.77 P = 0. RV YES (YES)

86	TEND TO BE THOSE WHERE THE STAGE OF POLITICAL MODERNIZATION IS EARLY TRANSITIONAL, RATHER THAN ADVANCED OR MID-TRANSITIONAL (38)	0.59	86	ALWAYS ARE THOSE WHERE THE STAGE OF POLITICAL MODERNIZATION IS ADVANCED OR MID-TRANSITIONAL, RATHER THAN EARLY TRANSITIONAL (76)	1.00	20 18 29 0 X SQ= 16.45 P = 0. RV YES (NO)
87	LEAN TOWARD BEING THOSE WHOSE IDEOLOGICAL ORIENTATION IS DEVELOPMENTAL (31)	0.62	87	LEAN TOWARD BEING THOSE WHOSE IDEOLOGICAL ORIENTATION IS OTHER THAN DEVELOPMENTAL (58)	0.73	23 3 14 8 X SQ= 2.87 P = 0.082 RV YES (NO)
89	TEND TO BE THOSE WHOSE IDEOLOGICAL ORIENTATION IS OTHER THAN CONVENTIONAL (62)	0.80	89	TEND TO BE THOSE WHOSE IDEOLOGICAL ORIENTATION IS CONVENTIONAL (33)	0.58	8 7 33 5 X SQ= 5.11 P = 0.024 RV YES (NO)
99	TEND TO BE THOSE WHERE GOVERNMENTAL STABILITY IS GENERALLY PRESENT AND DATES FROM AT LEAST THE INTER-WAR PERIOD, OR FROM THE POST-WAR PERIOD (50)	0.54	99	TEND TO BE THOSE WHERE GOVERNMENTAL STABILITY IS MODERATELY PRESENT AND DATES FROM THE POST-WAR PERIOD, OR IS ABSENT (36)	0.87	15 2 13 13 X SQ= 5.04 P = 0.020 RV YES (YES)
102	LEAN TOWARD BEING THOSE WHERE THE REPRESENTATIVE CHARACTER OF THE REGIME IS PSEUDO-POLYARCHIC OR NON-POLYARCHIC (49)	0.52	102	LEAN TOWARD BEING THOSE WHERE THE REPRESENTATIVE CHARACTER OF THE REGIME IS POLYARCHIC OR LIMITED POLYARCHIC (59)	0.75	22 12 24 4 X SQ= 2.53 P = 0.082 RV YES (NO)
105	TEND TO BE THOSE WHERE THE ELECTORAL SYSTEM IS PARTIALLY COMPETITIVE OR NON-COMPETITIVE (47)	0.62	105	TEND TO BE THOSE WHERE THE ELECTORAL SYSTEM IS COMPETITIVE (43)	0.77	14 10 23 3 X SQ= 4.43 P = 0.024 RV YES (NO)
107	TEND TO BE THOSE WHERE AUTONOMOUS GROUPS ARE PARTIALLY TOLERATED IN POLITICS, ARE TOLERATED ONLY OUTSIDE POLITICS, OR ARE NOT TOLERATED AT ALL (65)	0.66	107	TEND TO BE THOSE WHERE AUTONOMOUS GROUPS ARE FULLY TOLERATED IN POLITICS (46)	0.65	16 11 31 6 X SQ= 3.64 P = 0.044 RV YES (NO)
114	TEND TO BE THOSE WHERE SECTIONALISM IS EXTREME OR MODERATE (61)	0.68	114	TEND TO BE THOSE WHERE SECTIONALISM IS NEGLIGIBLE (47)	0.79	32 3 15 11 X SQ= 7.79 P = 0.004 RV YES (NO)
117	TEND TO BE THOSE WHERE INTEREST ARTICULATION BY ASSOCIATIONAL GROUPS IS NEGLIGIBLE (51)	0.59	117	TEND TO BE THOSE WHERE INTEREST ARTICULATION BY ASSOCIATIONAL GROUPS IS SIGNIFICANT, MODERATE, OR LIMITED (60)	0.89	20 16 29 2 X SQ= 10.38 P = 0.001 RV YES (NO)

118	TILT LESS TOWARD BEING THOSE WHERE INTEREST ARTICULATION BY INSTITUTIONAL GROUPS IS SIGNIFICANT, MODERATE, OR LIMITED (60)	0.61	118	TILT MORE TOWARD BEING THOSE WHERE INTEREST ARTICULATION BY INSTITUTIONAL GROUPS IS SIGNIFICANT, MODERATE, OR LIMITED (60)	0.83	15 3 23 15 X SQ= 1.96 P = 0.128 RV NO (NO)
121	TEND TO BE THOSE WHERE INTEREST ARTICULATION BY NON-ASSOCIATIONAL GROUPS IS SIGNIFICANT (54)	0.76	121	ALWAYS ARE THOSE WHERE INTEREST ARTICULATION BY NON-ASSOCIATIONAL GROUPS IS MODERATE, LIMITED, OR NEGLIGIBLE (61)	1.00	37 0 12 18 X SQ= 27.38 P = 0. RV YES (YES)
122	TEND TO BE THOSE WHERE INTEREST ARTICULATION BY NON-ASSOCIATIONAL GROUPS IS SIGNIFICANT OR MODERATE (83)	0.92	122	TEND TO BE THOSE WHERE INTEREST ARTICULATION BY NON-ASSOCIATIONAL GROUPS IS LIMITED OR NEGLIGIBLE (32)	0.56	45 8 4 10 X SQ= 15.14 P = 0. RV YES (YES)
124	LEAN LESS TOWARD BEING THOSE WHERE INTEREST ARTICULATION BY ANOMIC GROUPS IS OCCASIONAL, INFREQUENT, OR VERY INFREQUENT (85)	0.80	124	ALWAYS ARE THOSE WHERE INTEREST ARTICULATION BY ANOMIC GROUPS IS OCCASIONAL, INFREQUENT, OR VERY INFREQUENT (85)	1.00	9 0 36 14 X SQ= 1.94 P = 0.098 RV NO (NO)
128	TILT TOWARD BEING THOSE WHERE INTEREST ARTICULATION BY POLITICAL PARTIES IS LIMITED OR NEGLIGIBLE (45)	0.56	128	TILT TOWARD BEING THOSE WHERE INTEREST ARTICULATION BY POLITICAL PARTIES IS SIGNIFICANT OR MODERATE (48)	0.69	17 11 22 5 X SQ= 1.96 P = 0.138 RV YES (NO)
129	TEND TO BE THOSE WHERE INTEREST ARTICULATION BY POLITICAL PARTIES IS NEGLIGIBLE (37)	0.51	129	TEND TO BE THOSE WHERE INTEREST ARTICULATION BY POLITICAL PARTIES IS SIGNIFICANT, MODERATE, OR LIMITED (56)	0.94	19 15 20 1 X SQ= 7.93 P = 0.002 RV YES (NO)
131	TEND TO BE THOSE WHERE INTEREST AGGREGATION BY POLITICAL PARTIES IS SIGNIFICANT OR MODERATE (30)	0.70	131	TEND TO BE THOSE WHERE INTEREST AGGREGATION BY POLITICAL PARTIES IS LIMITED OR NEGLIGIBLE (35)	0.73	19 4 8 11 X SQ= 5.78 P = 0.010 RV YES (YES)
133	TEND TO BE THOSE WHERE INTEREST AGGREGATION BY THE EXECUTIVE IS SIGNIFICANT (29)	0.52	133	TEND TO BE THOSE WHERE INTEREST AGGREGATION BY THE EXECUTIVE IS MODERATE, LIMITED, OR NEGLIGIBLE (76)	0.81	22 3 20 13 X SQ= 4.06 P = 0.036 RV YES (NO)
134	TEND TO BE THOSE WHERE INTEREST AGGREGATION BY THE EXECUTIVE IS SIGNIFICANT OR MODERATE (57)	0.76	134	TEND TO BE THOSE WHERE INTEREST AGGREGATION BY THE EXECUTIVE IS LIMITED OR NEGLIGIBLE (46)	0.63	31 6 10 10 X SQ= 5.76 P = 0.012 RV YES (YES)

No.			No.			
138	TEND TO BE THOSE WHERE INTEREST AGGREGATION BY THE LEGISLATURE IS NEGLIGIBLE (49)	0.63	138	TEND TO BE THOSE WHERE INTEREST AGGREGATION BY THE LEGISLATURE IS SIGNIFICANT, MODERATE, OR LIMITED (48)	0.81	15 13 25 3 X SQ= 7.09 P = 0.007 RV YES (NO)
139	TEND LESS TO BE THOSE WHERE THE PARTY SYSTEM IS QUANTITATIVELY OTHER THAN ONE-PARTY (71)	0.61	139	TEND MORE TO BE THOSE WHERE THE PARTY SYSTEM IS QUANTITATIVELY OTHER THAN ONE-PARTY (71)	0.94	18 1 28 16 X SQ= 5.03 P = 0.013 RV NO (NO)
142	TEND TO BE THOSE WHERE THE PARTY SYSTEM IS QUANTITATIVELY OTHER THAN MULTI-PARTY (66)	0.79	142	TEND TO BE THOSE WHERE THE PARTY SYSTEM IS QUANTITATIVELY MULTI-PARTY (30)	0.63	8 10 31 6 X SQ= 7.28 P = 0.004 RV YES (YES)
146	LEAN LESS TOWARD BEING THOSE WHERE THE PARTY SYSTEM IS QUALITATIVELY OTHER THAN MASS-BASED TERRITORIAL (92)	0.81	146	ALWAYS ARE THOSE WHERE THE PARTY SYSTEM IS QUALITATIVELY OTHER THAN MASS-BASED TERRITORIAL (92)	1.00	8 0 34 17 X SQ= 2.30 P = 0.090 RV NO (NO)
148	TEND LESS TO BE THOSE WHERE THE PARTY SYSTEM IS QUALITATIVELY OTHER THAN AFRICAN TRANSITIONAL (96)	0.75	148	ALWAYS ARE THOSE WHERE THE PARTY SYSTEM IS QUALITATIVELY OTHER THAN AFRICAN TRANSITIONAL (96)	1.00	12 0 36 18 X SQ= 3.95 P = 0.027 RV NO (NO)
154	TEND TO BE THOSE WHERE THE PARTY SYSTEM IS STABLE OR MODERATELY STABLE (55)	0.70	154	TEND TO BE THOSE WHERE THE PARTY SYSTEM IS UNSTABLE (25)	0.71	19 4 8 10 X SQ= 4.95 P = 0.019 RV YES (YES)
158	TEND TO BE THOSE WHERE PERSONALISMO IS NEGLIGIBLE (56)	0.54	158	TEND TO BE THOSE WHERE PERSONALISMO IS PRONOUNCED OR MODERATE (40)	0.76	19 13 22 4 X SQ= 3.28 P = 0.046 RV YES (NO)
161	TEND TO BE THOSE WHERE THE POLITICAL LEADERSHIP IS NON-ELITIST (50)	0.69	161	TEND TO BE THOSE WHERE THE POLITICAL LEADERSHIP IS ELITIST OR MODERATELY ELITIST (47)	0.65	11 11 25 6 X SQ= 4.23 P = 0.035 RV YES (YES)
164	TEND TO BE THOSE WHERE THE REGIME'S LEADERSHIP CHARISMA IS PRONOUNCED OR MODERATE (34)	0.61	164	TEND TO BE THOSE WHERE THE REGIME'S LEADERSHIP CHARISMA IS NEGLIGIBLE (65)	0.94	25 1 16 17 X SQ= 13.42 P = 0. RV YES (YES)

```
169  TEND TO BE THOSE                          0.53     169  TEND TO BE THOSE                          0.88        21    14
     WHERE THE HORIZONTAL POWER DISTRIBUTION                 WHERE THE HORIZONTAL POWER DISTRIBUTION            24     2
     IS NEGLIGIBLE  (48)                                     IS SIGNIFICANT OR LIMITED  (58)                X SQ=   6.46
                                                                                                            P =    0.007
                                                                                                            RV YES (NO )

170  TEND TO BE THOSE                          0.54     170  TEND TO BE THOSE                          0.94        19    15
     WHERE THE LEGISLATIVE-EXECUTIVE STRUCTURE               WHERE THE LEGISLATIVE-EXECUTIVE STRUCTURE          22     1
     IS OTHER THAN PRESIDENTIAL  (63)                        IS PRESIDENTIAL  (39)                          X SQ=   8.87
                                                                                                            P =    0.001
                                                                                                            RV YES (NO )

172  TEND LESS TO BE THOSE                     0.77     172  ALWAYS ARE THOSE                          1.00        11     0
     WHERE THE LEGISLATIVE-EXECUTIVE STRUCTURE               WHERE THE LEGISLATIVE-EXECUTIVE STRUCTURE          37    18
     IS OTHER THAN PARLIAMENTARY-ROYALIST  (88)              IS OTHER THAN PARLIAMENTARY-ROYALIST  (88)     X SQ=   3.44
                                                                                                            P =    0.028
                                                                                                            RV NO  (NO )

175  LEAN TOWARD BEING THOSE                   0.53     175  LEAN TOWARD BEING THOSE                   0.79        20    11
     WHERE THE LEGISLATURE IS                                WHERE THE LEGISLATURE IS                           23     3
     LARGELY INEFFECTIVE OR                                  FULLY EFFECTIVE OR                             X SQ=   3.18
     WHOLLY INEFFECTIVE  (49)                                PARTIALLY EFFECTIVE  (51)                      P =    0.062
                                                                                                            RV YES (NO )

178  LEAN TOWARD BEING THOSE                   0.63     178  LEAN TOWARD BEING THOSE                   0.65        27     6
     WHERE THE LEGISLATURE IS UNICAMERAL  (53)               WHERE THE LEGISLATURE IS BICAMERAL  (51)           16    11
                                                                                                            X SQ=   2.69
                                                                                                            P =    0.084
                                                                                                            RV YES (NO )

181  TEND LESS TO BE THOSE                     0.68     181  ALWAYS ARE THOSE                          1.00         7     0
     WHERE THE BUREAUCRACY                                   WHERE THE BUREAUCRACY                              15    18
     IS SEMI-MODERN, RATHER THAN                             IS SEMI-MODERN, RATHER THAN                    X SQ=   4.91
     MODERN  (55)                                            MODERN  (55)                                   P =    0.011
                                                                                                            RV NO  (YES)

182  TEND TO BE THOSE                          0.58     182  ALWAYS ARE THOSE                          1.00        15    18
     WHERE THE BUREAUCRACY                                   WHERE THE BUREAUCRACY                              21     0
     IS POST-COLONIAL TRANSITIONAL,                          IS SEMI-MODERN, RATHER THAN                    X SQ=  14.81
     RATHER THAN SEMI-MODERN  (25)                           POST-COLONIAL TRANSITIONAL  (55)               P =    0.
                                                                                                            RV YES (YES)

185  TEND TO BE THOSE                          0.56     185  TEND TO BE THOSE                          0.89         8     8
     WHERE PARTICIPATION BY THE MILITARY                     WHERE PARTICIPATION BY THE MILITARY                10     1
     IN POLITICS IS                                          IN POLITICS IS                                 X SQ=   3.24
     SUPPORTIVE, RATHER THAN                                 INTERVENTIVE, RATHER THAN                      P =    0.042
     INTERVENTIVE  (31)                                      SUPPORTIVE  (21)                               RV YES (YES)

188  TEND TO BE THOSE                          0.98     188  TEND TO BE THOSE                          0.94         1    17
     WHERE THE CHARACTER OF THE LEGAL SYSTEM                 WHERE THE CHARACTER OF THE LEGAL SYSTEM            47     1
     IS OTHER THAN CIVIL LAW  (81)                           IS CIVIL LAW  (32)                             X SQ=  51.74
                                                                                                            P =    0.
                                                                                                            RV YES (YES)
```

```
190  TEND TO BE THOSE                          0.86     190  ALWAYS ARE THOSE                          1.00        1   17
     WHERE THE CHARACTER OF THE LEGAL SYSTEM                 WHERE THE CHARACTER OF THE LEGAL SYSTEM                6    0
     IS COMMON LAW, RATHER THAN                              IS CIVIL LAW, RATHER THAN                         X SQ=  15.13
     CIVIL LAW  (7)                                          COMMON LAW  (32)                                  P =     0.
                                                                                                               RV YES (YES)

192  TEND LESS TO BE THOSE                     0.54     192  ALWAYS ARE THOSE                          1.00       22    0
     WHERE THE CHARACTER OF THE LEGAL SYSTEM                 WHERE THE CHARACTER OF THE LEGAL SYSTEM               26   18
     IS OTHER THAN                                           IS OTHER THAN                                     X SQ=  10.40
     MUSLIM OR PARTLY MUSLIM  (86)                           MUSLIM OR PARTLY MUSLIM  (86)                     P =     0.
                                                                                                               RV NO  (NO )

193  TEND LESS TO BE THOSE                     0.54     193  ALWAYS ARE THOSE                          1.00       22    0
     WHERE THE CHARACTER OF THE LEGAL SYSTEM                 WHERE THE CHARACTER OF THE LEGAL SYSTEM               26   18
     IS OTHER THAN PARTLY INDIGENOUS  (86)                   IS OTHER THAN PARTLY INDIGENOUS  (86)             X SQ=  10.40
                                                                                                               P =     0.
                                                                                                               RV NO  (NO )
```

```
************************************************************************************************************************

80  POLITIES                                          80  POLITIES
    WHOSE DATE OF INDEPENDENCE IS AFTER 1914,            WHOSE DATE OF INDEPENDENCE IS AFTER 1914,
    AND THAT WERE FORMERLY DEPENDENCIES OF               AND THAT WERE FORMERLY DEPENDENCIES OF
    BRITAIN, RATHER THAN FRANCE  (19)                    FRANCE, RATHER THAN BRITAIN  (24)

                                                                                    BOTH SUBJECT AND PREDICATE
........................................................................................................................

     19 IN LEFT COLUMN

BURMA       CEYLON      CYPRUS      GHANA       INDIA       IRAQ        IRELAND     ISRAEL      JAMAICA     JORDAN
MALAYA      NIGERIA     PAKISTAN    SIERRE LEO  SUDAN       TANGANYIKA  TRINIDAD    UGANDA      UAR

     24 IN RIGHT COLUMN

ALGERIA     CAMBODIA    CAMEROUN    CEN AFR REP CHAD        CONGO(BRA)  DAHOMEY     GABON       GUINEA      IVORY COAST
LAOS        LEBANON     MALAGASY R  MALI        MAURITANIA  MOROCCO     NIGER       SENEGAL     SYRIA       TOGO
TUNISIA     UPPER VOLTA VIETNAM, N  VIETNAM REP

     72 EXCLUDED BECAUSE IRRELEVANT

AFGHANISTAN ALBANIA     ARGENTINA   AUSTRALIA   AUSTRIA     BELGIUM     BOLIVIA     BRAZIL      BULGARIA    BURUNDI
CANADA      CHILE       CHINA, PR   COLOMBIA    CONGO(LEO)  COSTA RICA  CUBA        CZECHOS'KIA DENMARK     DOMIN REP
ECUADOR     EL SALVADOR ETHIOPIA    FINLAND     FRANCE      GERMANY, E  GERMAN FR   GREECE      GUATEMALA   HAITI
HONDURAS    HUNGARY     ICELAND     INDONESIA   IRAN        ITALY       JAPAN       KOREA, N    KOREA REP   LIBERIA
LIBYA       LUXEMBOURG  MEXICO      MONGOLIA    NEPAL       NETHERLANDS NEW ZEALAND NICARAGUA   NORWAY      PANAMA
PARAGUAY    PERU        PHILIPPINES POLAND      PORTUGAL    RUMANIA     RWANDA      SA'U ARABIA SOMALIA     SO AFRICA
SPAIN       SWEDEN      SWITZERLAND THAILAND    TURKEY      USSR        UK          US          URUGUAY     VENEZUELA
YEMEN       YUGOSLAVIA
------------------------------------------------------------------------------------------------------------------------

 10  TEND TO BE THOSE                        0.63     10  TEND TO BE THOSE                      0.75         7   18
     LOCATED ELSEWHERE THAN IN NORTH AFRICA,              LOCATED IN NORTH AFRICA, OR                       12    6
     OR CENTRAL AND SOUTH AFRICA  (82)                    CENTRAL AND SOUTH AFRICA  (33)               X SQ=   4.87
                                                                                                       P =    0.016
                                                                                                       RV YES (YES)

 11  LEAN TOWARD BEING THOSE                 0.68     11  LEAN TOWARD BEING THOSE               0.62         6   15
     LOCATED ELSEWHERE THAN IN CENTRAL                    LOCATED IN CENTRAL AND                            13    9
     AND SOUTH AFRICA  (87)                               SOUTH AFRICA  (28)                           X SQ=   2.91
                                                                                                       P =    0.067
                                                                                                       RV YES (YES)

 19  TILT LESS TOWARD BEING THOSE            0.78     19  ALWAYS ARE THOSE                      1.00         7   18
     LOCATED IN NORTH AFRICA OR                           LOCATED IN NORTH AFRICA OR                         2    0
     CENTRAL AND SOUTH AFRICA, RATHER THAN                CENTRAL AND SOUTH AFRICA, RATHER THAN        X SQ=   1.69
     IN THE CARIBBEAN, CENTRAL AMERICA, OR                IN THE CARIBBEAN, CENTRAL AMERICA, OR        P =    0.103
     SOUTH AMERICA  (33)                                  SOUTH AMERICA  (33)                          RV NO  (YES)
```

23	TEND LESS TO BE THOSE WHOSE POPULATION IS MEDIUM OR SMALL (88)	0.74	23	ALWAYS ARE THOSE WHOSE POPULATION IS MEDIUM OR SMALL (88)	1.00	5 0 14 24 X SQ= 4.82 P = 0.012 RV NO (YES)
24	TEND TO BE THOSE WHOSE POPULATION IS VERY LARGE, LARGE, OR MEDIUM (61)	0.63	24	TEND TO BE THOSE WHOSE POPULATION IS SMALL (54)	0.83	12 4 7 20 X SQ= 7.92 P = 0.004 RV YES (YES)
26	TEND TO BE THOSE WHOSE POPULATION DENSITY IS VERY HIGH, HIGH, OR MEDIUM (48)	0.53	26	TEND TO BE THOSE WHOSE POPULATION DENSITY IS LOW (67)	0.87	10 3 9 21 X SQ= 6.31 P = 0.007 RV YES (YES)
30	TEND TO BE THOSE WHOSE AGRICULTURAL POPULATION IS MEDIUM, LOW, OR VERY LOW (57)	0.58	30	TEND TO BE THOSE WHOSE AGRICULTURAL POPULATION IS HIGH (56)	0.96	8 22 11 1 X SQ= 12.11 P = 0. RV YES (YES)
34	TEND TO BE THOSE WHOSE GROSS NATIONAL PRODUCT IS VERY HIGH, HIGH, MEDIUM, OR LOW (62)	0.53	34	TEND TO BE THOSE WHOSE GROSS NATIONAL PRODUCT IS VERY LOW (53)	0.83	M 10 4 9 20 X SQ= 4.72 P = 0.021 RV YES (YES)
36	TEND LESS TO BE THOSE WHOSE PER CAPITA GROSS NATIONAL PRODUCT IS LOW OR VERY LOW (73)	0.68	36	TEND MORE TO BE THOSE WHOSE PER CAPITA GROSS NATIONAL PRODUCT IS LOW OR VERY LOW (73)	0.96	6 1 13 23 X SQ= 4.01 P = 0.033 RV NO (YES)
37	LEAN LESS TOWARD BEING THOSE WHOSE PER CAPITA GROSS NATIONAL PRODUCT IS VERY LOW (51)	0.58	37	LEAN MORE TOWARD BEING THOSE WHOSE PER CAPITA GROSS NATIONAL PRODUCT IS VERY LOW (51)	0.83	8 4 11 20 X SQ= 2.26 P = 0.091 RV NO (YES)
39	LEAN LESS TOWARD BEING THOSE WHOSE INTERNATIONAL FINANCIAL STATUS IS LOW OR VERY LOW (76)	0.84	39	ALWAYS ARE THOSE WHOSE INTERNATIONAL FINANCIAL STATUS IS LOW OR VERY LOW (76)	1.00	3 0 16 24 X SQ= 2.00 P = 0.079 RV NO (YES)
40	TEND TO BE THOSE WHOSE INTERNATIONAL FINANCIAL STATUS IS VERY HIGH, HIGH, MEDIUM, OR LOW (71)	0.71	40	TEND TO BE THOSE WHOSE INTERNATIONAL FINANCIAL STATUS IS VERY LOW (39)	0.73	12 6 5 16 X SQ= 5.60 P = 0.011 RV YES (YES)

43	TEND TO BE THOSE WHOSE ECONOMIC DEVELOPMENTAL STATUS IS DEVELOPED, INTERMEDIATE, OR UNDERDEVELOPED (55)	0.53	43	ALWAYS ARE THOSE WHOSE ECONOMIC DEVELOPMENTAL STATUS IS VERY UNDERDEVELOPED (57)	1.00	10 0 9 23 X SQ= 13.12 P = 0. RV YES (YES)
46	TEND TO BE THOSE WHERE THE LITERACY RATE IS TEN PERCENT OR ABOVE (84)	0.79	46	TEND TO BE THOSE WHERE THE LITERACY RATE IS BELOW TEN PERCENT (26)	0.65	M 15 7 4 13 X SQ= 5.97 P = 0.010 RV YES (YES)
54	LEAN LESS TOWARD BEING THOSE WHERE NEWSPAPER CIRCULATION IS LESS THAN ONE HUNDRED PER THOUSAND (76)	0.84	54	ALWAYS ARE THOSE WHERE NEWSPAPER CIRCULATION IS LESS THAN ONE HUNDRED PER THOUSAND (76)	1.00	3 0 16 23 X SQ= 1.89 P = 0.084 RV NO (YES)
55	TEND TO BE THOSE WHERE NEWSPAPER CIRCULATION IS TEN OR MORE PER THOUSAND (78)	0.68	55	TEND TO BE THOSE WHERE NEWSPAPER CIRCULATION IS LESS THAN TEN PER THOUSAND (35)	0.74	M 13 6 6 17 X SQ= 5.92 P = 0.012 RV YES (YES)
56	LEAN TOWARD BEING THOSE WHERE THE RELIGION IS PREDOMINANTLY LITERATE (79)	0.68	56	LEAN TOWARD BEING THOSE WHERE THE RELIGION IS PREDOMINANTLY OR PARTLY NON-LITERATE (31)	0.64	M 13 8 6 14 X SQ= 3.01 P = 0.062 RV YES (YES)
75	TEND LESS TO BE THOSE THAT ARE NOT HISTORICALLY WESTERN AND ARE NOT SIGNIFICANTLY WESTERNIZED (52)	0.58	75	TEND MORE TO BE THOSE THAT ARE NOT HISTORICALLY WESTERN AND ARE NOT SIGNIFICANTLY WESTERNIZED (52)	0.92	8 2 11 22 X SQ= 5.02 P = 0.013 RV NO (YES)
77	TEND LESS TO BE THOSE THAT HAVE BEEN PARTIALLY WESTERNIZED, RATHER THAN SIGNIFICANTLY WESTERNIZED, THROUGH A COLONIAL RELATIONSHIP (41)	0.65	77	TEND MORE TO BE THOSE THAT HAVE BEEN PARTIALLY WESTERNIZED, RATHER THAN SIGNIFICANTLY WESTERNIZED, THROUGH A COLONIAL RELATIONSHIP (41)	0.92	6 2 11 22 X SQ= 3.05 P = 0.049 RV NO (YES)
84	TEND TO BE THOSE WHERE THE TYPE OF POLITICAL MODERNIZATION IS DEVELOPED TUTELARY, RATHER THAN UNDEVELOPED TUTELARY (31)	0.72	84	TEND TO BE THOSE WHERE THE TYPE OF POLITICAL MODERNIZATION IS UNDEVELOPED TUTELARY, RATHER THAN DEVELOPED TUTELARY (24)	0.63	M 13 9 5 15 X SQ= 3.68 P = 0.033 RV YES (YES)
85	TILT LESS TOWARD BEING THOSE WHERE THE STAGE OF POLITICAL MODERNIZATION IS MID- OR EARLY TRANSITIONAL, RATHER THAN ADVANCED (54)	0.53	85	TILT MORE TOWARD BEING THOSE WHERE THE STAGE OF POLITICAL MODERNIZATION IS MID- OR EARLY TRANSITIONAL, RATHER THAN ADVANCED (54)	0.79	9 5 10 19 X SQ= 2.30 P = 0.102 RV NO (YES)

No.	Left		No.	Right		Stats
86	TILT LESS TOWARD BEING THOSE WHERE THE STAGE OF POLITICAL MODERNIZATION IS EARLY TRANSITIONAL, RATHER THAN ADVANCED OR MID-TRANSITIONAL (38)	0.53	86	TILT MORE TOWARD BEING THOSE WHERE THE STAGE OF POLITICAL MODERNIZATION IS EARLY TRANSITIONAL, RATHER THAN ADVANCED OR MID-TRANSITIONAL (38)	0.79	9 5 10 19 X SQ= 2.30 P = 0.102 RV NO (YES)
101	TEND TO BE THOSE WHERE THE REPRESENTATIVE CHARACTER OF THE REGIME IS POLYARCHIC (41)	0.56	101	TEND TO BE THOSE WHERE THE REPRESENTATIVE CHARACTER OF THE REGIME IS LIMITED POLYARCHIC, PSEUDO-POLYARCHIC, OR NON-POLYARCHIC (57)	0.85	9 3 7 17 X SQ= 5.08 P = 0.014 RV YES (YES)
102	TEND TO BE THOSE WHERE THE REPRESENTATIVE CHARACTER OF THE REGIME IS POLYARCHIC OR LIMITED POLYARCHIC (59)	0.67	102	TEND TO BE THOSE WHERE THE REPRESENTATIVE CHARACTER OF THE REGIME IS PSEUDO-POLYARCHIC OR NON-POLYARCHIC (49)	0.77	M 12 5 6 17 X SQ= 6.13 P = 0.010 RV YES (YES)
105	TEND TO BE THOSE WHERE THE ELECTORAL SYSTEM IS COMPETITIVE (43)	0.69	105	TEND TO BE THOSE WHERE THE ELECTORAL SYSTEM IS PARTIALLY COMPETITIVE OR NON-COMPETITIVE (47)	0.95	9 1 4 18 X SQ= 11.87 P = 0. RV YES (YES)
106	TEND TO BE THOSE WHERE THE ELECTORAL SYSTEM IS COMPETITIVE OR PARTIALLY COMPETITIVE (52)	0.91	106	TEND TO BE THOSE WHERE THE ELECTORAL SYSTEM IS NON-COMPETITIVE (30)	0.71	10 5 1 12 X SQ= 7.83 P = 0.002 RV YES (YES)
107	TEND TO BE THOSE WHERE AUTONOMOUS GROUPS ARE FULLY TOLERATED IN POLITICS (46)	0.53	107	TEND TO BE THOSE WHERE AUTONOMOUS GROUPS ARE PARTIALLY TOLERATED IN POLITICS, ARE TOLERATED ONLY OUTSIDE POLITICS, OR ARE NOT TOLERATED AT ALL (65)	0.91	10 2 9 20 X SQ= 7.35 P = 0.005 RV YES (YES)
108	LEAN TOWARD BEING THOSE WHERE AUTONOMOUS GROUPS ARE FULLY OR PARTIALLY TOLERATED IN POLITICS (65)	0.79	108	LEAN TOWARD BEING THOSE WHERE AUTONOMOUS GROUPS ARE TOLERATED ONLY OUTSIDE POLITICS OR ARE NOT TOLERATED AT ALL (35)	0.53	11 9 3 10 X SQ= 2.11 P = 0.087 RV YES (YES)
113	LEAN LESS TOWARD BEING THOSE WHERE SECTIONALISM IS MODERATE OR NEGLIGIBLE (81)	0.58	113	LEAN MORE TOWARD BEING THOSE WHERE SECTIONALISM IS MODERATE OR NEGLIGIBLE (81)	0.86	8 3 11 19 X SQ= 2.88 P = 0.075 RV NO (YES)
117	LEAN TOWARD BEING THOSE WHERE INTEREST ARTICULATION BY ASSOCIATIONAL GROUPS IS SIGNIFICANT, MODERATE, OR LIMITED (60)	0.53	117	LEAN TOWARD BEING THOSE WHERE INTEREST ARTICULATION BY ASSOCIATIONAL GROUPS IS NEGLIGIBLE (51)	0.79	10 5 9 19 X SQ= 3.42 P = 0.052 RV YES (YES)

125	TILT LESS TOWARD BEING THOSE WHERE INTEREST ARTICULATION BY ANOMIC GROUPS IS FREQUENT OR OCCASIONAL (64)	0.67	125	TILT MORE TOWARD BEING THOSE WHERE INTEREST ARTICULATION BY ANOMIC GROUPS IS FREQUENT OR OCCASIONAL (64)	0.91	12 20 6 2 X SQ= 2.28 P = 0.110 RV NO (YES)
128	TEND TO BE THOSE WHERE INTEREST ARTICULATION BY POLITICAL PARTIES IS SIGNIFICANT OR MODERATE (48)	0.77	128	TEND TO BE THOSE WHERE INTEREST ARTICULATION BY POLITICAL PARTIES IS LIMITED OR NEGLIGIBLE (45)	0.77	M 10 5 3 17 X SQ= 7.71 P = 0.004 RV YES (YES)
129	TEND TO BE THOSE WHERE INTEREST ARTICULATION BY POLITICAL PARTIES IS SIGNIFICANT, MODERATE, OR LIMITED (56)	0.77	129	TEND TO BE THOSE WHERE INTEREST ARTICULATION BY POLITICAL PARTIES IS NEGLIGIBLE (37)	0.68	10 7 3 15 X SQ= 4.97 P = 0.015 RV YES (YES)
133	TEND TO BE THOSE WHERE INTEREST AGGREGATION BY THE EXECUTIVE IS MODERATE, LIMITED, OR NEGLIGIBLE (76)	0.71	133	TEND TO BE THOSE WHERE INTEREST AGGREGATION BY THE EXECUTIVE IS SIGNIFICANT (29)	0.80	M 5 16 12 4 X SQ= 7.63 P = 0.003 RV YES (YES)
138	TEND TO BE THOSE WHERE INTEREST AGGREGATION BY THE LEGISLATURE IS SIGNIFICANT, MODERATE, OR LIMITED (48)	0.64	138	TEND TO BE THOSE WHERE INTEREST AGGREGATION BY THE LEGISLATURE IS NEGLIGIBLE (49)	0.90	9 2 5 19 X SQ= 9.29 P = 0.002 RV YES (YES)
139	TEND TO BE THOSE WHERE THE PARTY SYSTEM IS QUANTITATIVELY OTHER THAN ONE-PARTY (71)	0.83	139	TEND TO BE THOSE WHERE THE PARTY SYSTEM IS QUANTITATIVELY ONE-PARTY (34)	0.68	M 3 15 15 7 X SQ= 8.64 P = 0.002 RV YES (YES)
141	TILT LESS TOWARD BEING THOSE WHERE THE PARTY SYSTEM IS QUANTITATIVELY OTHER THAN TWO-PARTY (87)	0.86	141	ALWAYS ARE THOSE WHERE THE PARTY SYSTEM IS QUANTITATIVELY OTHER THAN TWO-PARTY (87)	1.00	2 0 12 22 X SQ= 1.16 P = 0.144 RV NO (YES)
142	TEND LESS TO BE THOSE WHERE THE PARTY SYSTEM IS QUANTITATIVELY OTHER THAN MULTI-PARTY (66)	0.54	142	TEND MORE TO BE THOSE WHERE THE PARTY SYSTEM IS QUANTITATIVELY OTHER THAN MULTI-PARTY (66)	0.90	6 2 7 19 X SQ= 4.12 P = 0.033 RV NO (YES)
144	LEAN LESS TOWARD BEING THOSE WHERE THE PARTY SYSTEM IS QUANTITATIVELY ONE-PARTY, RATHER THAN TWO-PARTY (34)	0.60	144	ALWAYS ARE THOSE WHERE THE PARTY SYSTEM IS QUANTITATIVELY ONE-PARTY, RATHER THAN TWO-PARTY (34)	1.00	3 15 2 0 X SQ= 2.96 P = 0.053 RV NO (YES)

147	TEND LESS TO BE THOSE WHERE THE PARTY SYSTEM IS QUALITATIVELY OTHER THAN CLASS-ORIENTED OR MULTI-IDEOLOGICAL (67)	0.75	147	ALWAYS ARE THOSE WHERE THE PARTY SYSTEM IS QUALITATIVELY OTHER THAN CLASS-ORIENTED OR MULTI-IDEOLOGICAL (67)	1.00	3 0 9 22 X SQ= 3.32 P = 0.037 RV NO (YES)
148	TEND MORE TO BE THOSE WHERE THE PARTY SYSTEM IS QUALITATIVELY OTHER THAN AFRICAN TRANSITIONAL (96)	0.95	148	TEND LESS TO BE THOSE WHERE THE PARTY SYSTEM IS QUALITATIVELY OTHER THAN AFRICAN TRANSITIONAL (96)	0.54	1 11 18 13 X SQ= 6.78 P = 0.005 RV NO (YES)
158	TEND TO BE THOSE WHERE PERSONALISMO IS NEGLIGIBLE (56)	0.77	158	TEND TO BE THOSE WHERE PERSONALISMO IS PRONOUNCED OR MODERATE (40)	0.70	3 16 10 7 X SQ= 5.46 P = 0.014 RV YES (YES)
164	LEAN TOWARD BEING THOSE WHERE THE REGIME'S LEADERSHIP CHARISMA IS NEGLIGIBLE (65)	0.50	164	LEAN TOWARD BEING THOSE WHERE THE REGIME'S LEADERSHIP CHARISMA IS PRONOUNCED OR MODERATE (34)	0.86	M 7 18 7 3 X SQ= 3.65 P = 0.053 RV YES (YES)
168	TEND LESS TO BE THOSE WHERE THE HORIZONTAL POWER DISTRIBUTION IS LIMITED OR NEGLIGIBLE (72)	0.56	168	ALWAYS ARE THOSE WHERE THE HORIZONTAL POWER DISTRIBUTION IS LIMITED OR NEGLIGIBLE (72)	1.00	7 0 9 23 X SQ= 9.47 P = 0.001 RV NO (YES)
169	TEND TO BE THOSE WHERE THE HORIZONTAL POWER DISTRIBUTION IS SIGNIFICANT OR LIMITED (58)	0.63	169	TEND TO BE THOSE WHERE THE HORIZONTAL POWER DISTRIBUTION IS NEGLIGIBLE (48)	0.74	10 6 6 17 X SQ= 3.78 P = 0.046 RV YES (YES)
170	TEND TO BE THOSE WHERE THE LEGISLATIVE-EXECUTIVE STRUCTURE IS OTHER THAN PRESIDENTIAL (63)	0.79	170	TEND TO BE THOSE WHERE THE LEGISLATIVE-EXECUTIVE STRUCTURE IS PRESIDENTIAL (39)	0.67	3 14 11 7 X SQ= 5.19 P = 0.015 RV YES (YES)
171	LEAN LESS TOWARD BEING THOSE WHERE THE LEGISLATIVE-EXECUTIVE STRUCTURE IS OTHER THAN PARLIAMENTARY-REPUBLICAN (90)	0.80	171	ALWAYS ARE THOSE WHERE THE LEGISLATIVE-EXECUTIVE STRUCTURE IS OTHER THAN PARLIAMENTARY-REPUBLICAN (90)	1.00	3 0 12 22 X SQ= 2.48 P = 0.059 RV NO (YES)
172	TEND LESS TO BE THOSE WHERE THE LEGISLATIVE-EXECUTIVE STRUCTURE IS OTHER THAN PARLIAMENTARY-ROYALIST (88)	0.63	172	TEND MORE TO BE THOSE WHERE THE LEGISLATIVE-EXECUTIVE STRUCTURE IS OTHER THAN PARLIAMENTARY-ROYALIST (88)	0.96	7 1 12 22 X SQ= 5.17 P = 0.015 RV NO (YES)

174	TEND LESS TO BE THOSE WHERE THE LEGISLATURE IS PARTIALLY EFFECTIVE, LARGELY INEFFECTIVE, OR WHOLLY INEFFECTIVE (72)	0.53	174	ALWAYS ARE THOSE WHERE THE LEGISLATURE IS PARTIALLY EFFECTIVE, LARGELY INEFFECTIVE, OR WHOLLY INEFFECTIVE (72)	1.00	7 0 8 22 X SQ= 9.80 P = 0.001 RV NO (YES)
175	TEND TO BE THOSE WHERE THE LEGISLATURE IS FULLY EFFECTIVE OR PARTIALLY EFFECTIVE (51)	0.73	175	TEND TO BE THOSE WHERE THE LEGISLATURE IS LARGELY INEFFECTIVE OR WHOLLY INEFFECTIVE (49)	0.82	11 4 4 18 X SQ= 9.08 P = 0.002 RV YES (YES)
178	TEND TO BE THOSE WHERE THE LEGISLATURE IS BICAMERAL (51)	0.57	178	TEND TO BE THOSE WHERE THE LEGISLATURE IS UNICAMERAL (53)	0.83	6 19 8 4 X SQ= 4.59 P = 0.027 RV YES (YES)
179	TEND TO BE THOSE WHERE THE EXECUTIVE IS STRONG (39)	0.57	179	TEND TO BE THOSE WHERE THE EXECUTIVE IS DOMINANT (52)	0.84	6 16 8 3 X SQ= 4.48 P = 0.024 RV YES (YES)
185	TILT TOWARD BEING THOSE WHERE PARTICIPATION BY THE MILITARY IN POLITICS IS INTERVENTIVE, RATHER THAN SUPPORTIVE (21)	0.71	185	TILT TOWARD BEING THOSE WHERE PARTICIPATION BY THE MILITARY IN POLITICS IS SUPPORTIVE, RATHER THAN INTERVENTIVE (31)	0.78	5 2 2 7 X SQ= 2.13 P = 0.126 RV YES (YES)
187	TEND TO BE THOSE WHERE THE ROLE OF THE POLICE IS NOT POLITICALLY SIGNIFICANT (35)	0.56	187	TEND TO BE THOSE WHERE THE ROLE OF THE POLICE IS POLITICALLY SIGNIFICANT (66)	0.90	8 18 10 2 X SQ= 7.11 P = 0.004 RV YES (YES)
189	LEAN LESS TOWARD BEING THOSE WHERE THE CHARACTER OF THE LEGAL SYSTEM IS OTHER THAN COMMON LAW (108)	0.84	189	ALWAYS ARE THOSE WHERE THE CHARACTER OF THE LEGAL SYSTEM IS OTHER THAN COMMON LAW (108)	1.00	3 0 16 24 X SQ= 2.00 P = 0.079 RV NO (YES)
193	TEND TO BE THOSE WHERE THE CHARACTER OF THE LEGAL SYSTEM IS OTHER THAN PARTLY INDIGENOUS (86)	0.72	193	TEND TO BE THOSE WHERE THE CHARACTER OF THE LEGAL SYSTEM IS PARTLY INDIGENOUS (28)	0.67	5 16 13 8 X SQ= 4.76 P = 0.028 RV YES (YES)

M

MATRIX

	80 POLITIES WHOSE DATE OF INDEPENDENCE IS AFTER 1914, AND THAT WERE FORMERLY DEPENDENCIES OF BRITAIN, RATHER THAN FRANCE (19)	80 POLITIES WHOSE DATE OF INDEPENDENCE IS AFTER 1914, AND THAT WERE FORMERLY DEPENDENCIES OF FRANCE, RATHER THAN BRITAIN (24)
34 POLITIES WHOSE GROSS NATIONAL PRODUCT IS VERY HIGH, HIGH, MEDIUM, OR LOW (62)	10 BURMA, ISRAEL, CEYLON, MALAYA, INDIA, NIGERIA, IRAQ, PAKISTAN, IRELAND, UAR	4 ALGERIA, MOROCCO, VIETNAM, N, VIETNAM REP
34 POLITIES WHOSE GROSS NATIONAL PRODUCT IS VERY LOW (53)	9 CYPRUS, SUDAN, GHAN[illegible], TANG[illegible]NYIKA, JAMAICA, TRINIDAD, JORDAN, UGANDA, SIERRE LEO	20 CAMBODIA, DAHOMEY, LEBANON, SENEGAL, CAMEROUN, GABON, MALAGASY R, SYRIA, CEN AFR REP, GUINEA, MALI, TOGO, CHAD, IVORY COAST, MAURITANIA, TUNISIA, CONGO(BRA), LAOS, NIGER, UPPER VOLTA

MATRIX

80 POLITIES WHOSE DATE OF INDEPENDENCE IS AFTER 1914, AND THAT WERE FORMERLY DEPENDENCIES OF BRITAIN, RATHER THAN FRANCE (19)	8C POLITIES WHOSE DATE OF INDEPENDENCE IS AFTER 1914, AND THAT WERE FORMERLY DEPENDENCIES OF FRANCE, RATHER THAN BRITAIN (24)
46 POLITIES WHERE THE LITERACY RATE IS TEN PERCENT OR ABOVE (84)	
BURMA, IRAQ, MALAYA, CEYLON, IRELAND, NIGERIA, CYPRUS, ISRAEL, PAKISTAN, GHANA, JAMAICA, TRINIDAD, INDIA, JORDAN, UAR	ALGERIA, TUNISIA, CAMBODIA, VIETNAM REP, LEBANON, MALAGASY R, SYRIA
15	7
4	13
46 POLITIES WHERE THE LITERACY RATE IS BELOW TEN PERCENT (26)	
SIERRE LEO, SUDAN, TANGANYIKA, UGANDA	CAMEROUN, GABON, SENEGAL, CEN AFR REP, IVORY COAST, TOGC, CHAD, LAOS, UPPER VOLTA, CONGO(BRA), MALI, DAHOMEY, NIGER

M

MATRIX

	80 POLITIES WHOSE DATE OF INDEPENDENCE IS AFTER 1914, AND THAT WERE FORMERLY DEPENDENCIES OF BRITAIN, RATHER THAN FRANCE (19)	80 POLITIES WHOSE DATE OF INDEPENDENCE IS AFTER 1914, AND THAT WERE FORMERLY DEPENDENCIES OF FRANCE, RATHER THAN BRITAIN (24)
55 POLITIES WHERE NEWSPAPER CIRCULATION IS TEN OR MORE PER THOUSAND (78)	13 BURMA, IRAQ, MALAYA, CEYLON, IRELAND, TRINIDAD, CYPRUS, ISRAEL, UAR, GHANA, JAMAICA, INDIA, JORDAN	6 ALGERIA, VIETNAM REP, LEBANON, MOROCCO, SYRIA, TUNISIA
55 POLITIES WHERE NEWSPAPER CIRCULATION IS LESS THAN TEN PER THOUSAND (35)	6 NIGERIA, UGANDA, PAKISTAN, SIERRE LEO, SUDAN, TANGANYIKA	17 CAMBODIA, DAHOMEY, MALAGASY R, TOGO, CAMEROUN, GABON, MALI, UPPER VOLTA, CEN AFR REP, GUINEA, MAURITANIA, CHAD, IVORY COAST, NIGER, CONGO(BRA), LAOS, SENEGAL

MATRIX

80 POLITIES
WHOSE DATE OF INDEPENDENCE IS AFTER 1914, AND THAT WERE FORMERLY DEPENDENCIES OF BRITAIN, RATHER THAN FRANCE (19)

80 POLITIES
WHOSE DATE OF INDEPENDENCE IS AFTER 1914, AND THAT WERE FORMERLY DEPENDENCIES OF FRANCE, RATHER THAN BRITAIN (24)

56 POLITIES
WHERE THE RELIGION IS PREDOMINANTLY LITERATE (79)

BURMA IRELAND PAKISTAN CEYLON ISRAEL TRINIDAD CYPRUS JAMAICA UAR INDIA JORDAN IRAQ MALAYA

ALGERIA SENEGAL CAMBODIA SYRIA LEBANON TUNISIA MAURITANIA MOROCCO

13	8
6	14

56 POLITIES
WHERE THE RELIGION IS PREDOMINANTLY OR PARTLY NON-LITERATE (31)

GHANA UGANDA NIGERIA SIERRE LEO SUDAN TANGANYIKA

CAMEROUN GABON MALI CEN AFR REP GUINEA NIGER CHAD IVORY COAST TOGO CONGO(BRA) LAOS UPPER VOLTA DAHOMEY MALAGASY R

M

MATRIX

	80 POLITIES WHOSE DATE OF INDEPENDENCE IS AFTER 1914, AND THAT WERE FORMERLY DEPENDENCIES OF BRITAIN, RATHER THAN FRANCE (19)	80 POLITIES WHOSE DATE OF INDEPENDENCE IS AFTER 1914, AND THAT WERE FORMERLY DEPENDENCIES OF FRANCE, RATHER THAN BRITAIN (24)
84 POLITIES WHERE THE TYPE OF POLITICAL MODERNIZATION IS DEVELOPED TUTELARY, RATHER THAN UNDEVELOPED TUTELARY (31)	13 BURMA, ISRAEL, SUDAN, CEYLON, JAMAICA, TRINIDAD, CYPRUS, JORDAN, UAR, INDIA, MALAYA, IRAQ, PAKISTAN	9 ALGERIA, SYRIA, CAMBODIA, TUNISIA, LAOS, VIETNAM, N, LEBANON, VIETNAM REP, MOROCCO
84 POLITIES WHERE THE TYPE OF POLITICAL MODERNIZATION IS UNDEVELOPED TUTELARY, RATHER THAN DEVELOPED TUTELARY (24)	5 GHANA, NIGERIA, SIERRE LEO, TANGANYIKA, UGANDA	15 CAMEROUN, GABON, MAURITANIA, CEN AFR REP, GUINEA, NIGER, CHAD, IVORY COAST, SENEGAL, CONGO(BRA), MALAGASY R, TOGO, DAHOMEY, MALI, UPPER VOLTA

M

MATRIX

80 POLITIES WHOSE DATE OF INDEPENDENCE IS AFTER 1914, AND THAT WERE FORMERLY DEPENDENCIES OF BRITAIN, RATHER THAN FRANCE (19)	80 POLITIES WHOSE DATE OF INDEPENDENCE IS AFTER 1914, AND THAT WERE FORMERLY DEPENDENCIES OF FRANCE, RATHER THAN BRITAIN (24)
102 POLITIES WHERE THE REPRESENTATIVE CHARACTER OF THE REGIME IS POLYARCHIC OR LIMITED POLYARCHIC (59)	
CEYLON, JAMAICA, TRINIDAD, CYPRUS, MALAYA, UGANDA, INDIA, NIGERIA, IRELAND, SIERRE LEO, ISRAEL, TANGANYIKA	CAMEROUN, LEBANON, MALAGASY R, MOROCCO, TUNISIA
12	5
6	17
102 POLITIES WHERE THE REPRESENTATIVE CHARACTER OF THE REGIME IS PSEUDO-POLYARCHIC OR NON-POLYARCHIC (49)	
BURMA, UAR, GHANA, JORDAN, PAKISTAN, SUDAN	ALGERIA, DAHOMEY, MALI, VIETNAM, N, CAMBODIA, GABON, MAURITANIA, VIETNAM REP, CEN AFR REP, GUINEA, NIGER, CHAD, IVORY COAST, SENEGAL, CONGO(BRA), LAOS, UPPER VOLTA

MATRIX

80 POLITIES
WHOSE DATE OF INDEPENDENCE IS AFTER 1914,
AND THAT WERE FORMERLY DEPENDENCIES OF
BRITAIN, RATHER THAN FRANCE (19)

80 POLITIES
WHOSE DATE OF INDEPENDENCE IS AFTER 1914,
AND THAT WERE FORMERLY DEPENDENCIES OF
FRANCE, RATHER THAN BRITAIN (24)

128 POLITIES
WHERE INTEREST ARTICULATION
BY POLITICAL PARTIES
IS SIGNIFICANT OR MODERATE (48)

CEYLON, MALAYA, INDIA, NIGERIA, IRELAND, SIERRE LEO, ISRAEL, TRINIDAD, JAMAICA, UGANDA

CAMEROUN, LEBANON, MALAGASY R, MORROCCO, SYRIA

	Britain (19)	France (24)
Significant or moderate (48)	10	5
Limited or negligible (45)	3	17

128 POLITIES
WHERE INTEREST ARTICULATION
BY POLITICAL PARTIES
IS LIMITED OR NEGLIGIBLE (45)

GHANA, TANGANYIKA, UAR

CAMBODIA, GABON, NIGER, VIETNAM, N, CEN AFR REP, GUINEA, SENEGAL, VIETNAM REP, CHAD, IVORY COAST, TOGO, CONGO(BRA), MALI, TUNISIA, DAHOMEY, MAURITANIA, UPPER VOLTA

80/133

M

```
******************************************************MATRIX******************************************************

        80  POLITIES                                                   80  POLITIES
            WHOSE DATE OF INDEPENDENCE IS AFTER 1914,                      WHOSE DATE OF INDEPENDENCE IS AFTER 1914,
            AND THAT WERE FORMERLY DEPENDENCIES OF                         AND THAT WERE FORMERLY DEPENDENCIES OF
            BRITAIN, RATHER THAN FRANCE  (19)                              FRANCE, RATHER THAN BRITAIN  (24)

                              133  POLITIES
                                   WHERE INTEREST AGGREGATION
                                   BY THE EXECUTIVE
                                   IS SIGNIFICANT  (29)

GHANA      INDIA      MALAYA     TANGANYIKA  UAR    .CAMBODIA     CAMEROUN    CEN AFR REP CHAD        CONGO(BRA)
                                                    .DAHOMEY      GABON       GUINEA      IVORY COAST MALI
                                                    .MAURITANIA   MOROCCO     NIGER       SENEGAL     TUNISIA
                                                    .UPPER VOLTA

                                                  5 .  16
. . . . . . . . . . . . . . . . . . . . . . . . . . . . . . . . . . . . . . . . . . . . . . . . . . . . . . . .
                                                 12 .   4

                              133  POLITIES
                                   WHERE INTEREST AGGREGATION
                                   BY THE EXECUTIVE
                                   IS MODERATE, LIMITED, OR
                                   NEGLIGIBLE  (76)

BURMA      CEYLON     IRELAND    ISRAEL      JAMAICA  .MALAGASY R  TOGO        VIETNAM, N  VIETNAM REP
JORDAN     NIGERIA    PAKISTAN   SIERRE LEO  SUDAN    .
TRINICAD   UGANDA                                     .
```

80 POLITIES WHOSE DATE OF INDEPENDENCE IS AFTER 1914, AND THAT WERE FORMERLY DEPENDENCIES OF BRITAIN, RATHER THAN FRANCE (19)	80 POLITIES WHOSE DATE OF INDEPENDENCE IS AFTER 1914, AND THAT WERE FORMERLY DEPENDENCIES OF FRANCE, RATHER THAN BRITAIN (24)
139 POLITIES WHERE THE PARTY SYSTEM IS QUANTITATIVELY ONE-PARTY (34)	
GHANA TANGANYIKA UAR	ALGERIA GABON NIGER CAMBODIA GUINEA SENEGAL CEN AFR REP IVORY COAST TUNISIA CHAD MALI UPPER VOLTA DAHOMEY MAURITANIA VIETNAM, N
3	15
139 POLITIES WHERE THE PARTY SYSTEM IS QUANTITATIVELY OTHER THAN ONE-PARTY (71)	
15	7
BURMA ISRAEL PAKISTAN CEYLON JAMAICA SIERRE LEO CYPRUS JORDAN SUDAN INDIA MALAYA TRINIDAD IRELAND NIGERIA UGANDA	CAMEROUN MOROCCO CONGO(BRA) VIETNAM REP LAOS LEBANON MALAGASY R

MATRIX

80 POLITIES WHOSE DATE OF INDEPENDENCE IS AFTER 1914, AND THAT WERE FORMERLY DEPENDENCIES OF BRITAIN, RATHER THAN FRANCE (19)	80 POLITIES WHOSE DATE OF INDEPENDENCE IS AFTER 1914, AND THAT WERE FORMERLY DEPENDENCIES OF FRANCE, RATHER THAN BRITAIN (24)

164 POLITIES
WHERE THE REGIME'S LEADERSHIP CHARISMA
IS PRONOUNCED OR MODERATE (34)

Britain	France
GHANA, TANGANYIKA, INDIA, UAR, JAMAICA, JORDAN, PAKISTAN	ALGERIA, CONGO(BRA), MALI, TUNISIA, CAMBODIA, DAHOMEY, MAURITANIA, UPPER VOLTA, CAMEROUN, GABON, MOROCCO, VIETNAM, N, CEN AFR REP, GUINEA, NIGER, CHAD, IVORY COAST, SENEGAL
7	18
7	3
CEYLON, SUDAN, IRELAND, TRINIDAD, ISRAEL, MALAYA, SIERRE LEO	LEBANON, MALAGASY R, VIETNAM REP

164 POLITIES
WHERE THE REGIME'S LEADERSHIP CHARISMA
IS NEGLIGIBLE (65)

M

MATRIX

	80 POLITIES WHOSE DATE OF INDEPENDENCE IS AFTER 1914, AND THAT WERE FORMERLY DEPENDENCIES OF BRITAIN, RATHER THAN FRANCE (19)	80 POLITIES WHOSE DATE OF INDEPENDENCE IS AFTER 1914, AND THAT WERE FORMERLY DEPENDENCIES OF FRANCE, RATHER THAN BRITAIN (24)
185 POLITIES WHERE PARTICIPATION BY THE MILITARY IN POLITICS IS INTERVENTIVE, RATHER THAN SUPPORTIVE (21)	BURMA IRAQ PAKISTAN SUDAN UAR 5	SYRIA TOGO 2
185 POLITIES WHERE PARTICIPATION BY THE MILITARY IN POLITICS IS SUPPORTIVE, RATHER THAN INTERVENTIVE (31)	2 GHANA JORDAN	7 ALGERIA CAMBODIA GUINEA LEBANON SENEGAL VIETNAM, N VIETNAM REP

```
*****************************************************************************************************

81  POLITIES                                          81  POLITIES
    WHERE THE TYPE OF POLITICAL MODERNIZATION             WHERE THE TYPE OF POLITICAL MODERNIZATION
    IS EARLY EUROPEAN OR                                  IS LATER EUROPEAN OR
    EARLY EUROPEAN DERIVED, RATHER THAN                   LATER EUROPEAN DERIVED, RATHER THAN
    LATER EUROPEAN OR                                     EARLY EUROPEAN OR
    LATER EUROPEAN DERIVED  (11)                          EARLY EUROPEAN DERIVED  (40)

                                                                            BOTH SUBJECT AND PREDICATE
.....................................................................................................

     11 IN LEFT COLUMN

AUSTRALIA   BELGIUM     CANADA      FRANCE      LUXEMBOURG  NETHERLANDS NEW ZEALAND SO AFRICA   SWITZERLAND UK
US

     40 IN RIGHT COLUMN

ALBANIA     ARGENTINA   AUSTRIA     BOLIVIA     BRAZIL      BULGARIA    CHILE       COLOMBIA    COSTA RICA  CUBA
CZECHOS'KIA DENMARK     DOMIN REP   ECUADOR     EL SALVADOR FINLAND     GERMANY, E  GERMAN FR   GREECE      GUATEMALA
HAITI       HONDURAS    HUNGARY     ICELAND     IRELAND     ITALY       MEXICO      NICARAGUA   NORWAY      PANAMA
PARAGUAY    PERU        POLAND      PORTUGAL    RUMANIA     SPAIN       SWEDEN      URUGUAY     VENEZUELA   YUGOSLAVIA

     64 EXCLUDED BECAUSE IRRELEVANT

AFGHANISTAN ALGERIA     BURMA       BURUNDI     CAMBODIA    CAMEROUN    CEN AFR REP CEYLON      CHAD        CHINA, PR
CONGO(BRA)  CONGO(LEO)  CYPRUS      DAHOMEY     ETHIOPIA    GABON       GHANA       GUINEA      INDIA       INDONESIA
IRAN        IRAQ        ISRAEL      IVORY COAST JAMAICA     JAPAN       JORDAN      KOREA, N    KOREA REP   LAOS
LEBANON     LIBERIA     LIBYA       MALAGASY R  MALAYA      MALI        MAURITANIA  MONGOLIA    MOROCCO     NEPAL
NIGER       NIGERIA     PAKISTAN    PHILIPPINES RWANDA      SA'U ARABIA SENEGAL     SIERRE LEO  SOMALIA     SUDAN
SYRIA       TANGANYIKA  THAILAND    TOGO        TRINIDAD    TUNISIA     TURKEY      UGANDA      USSR        UAR
UPPER VOLTA VIETNAM, N  VIETNAM REP YEMEN

-----------------------------------------------------------------------------------------------------

 2  TEND TO BE THOSE                      0.91       2  TEND TO BE THOSE                      0.70      10   12
    LOCATED IN WEST EUROPE, SCANDINAVIA,                LOCATED ELSEWHERE THAN IN                          1   28
    NORTH AMERICA, OR AUSTRALASIA  (22)                 WEST EUROPE, SCANDINAVIA,                   X SQ=  10.68
                                                        NORTH AMERICA, OR AUSTRALASIA  (93)         P =    0.
                                                                                                    RV YES (NO )

 5  ALWAYS ARE THOSE                      1.00       5  TILT LESS TOWARD BEING THOSE          0.60       6   12
    LOCATED IN SCANDINAVIA OR WEST EUROPE,              LOCATED IN SCANDINAVIA OR WEST EUROPE,        0    8
    RATHER THAN IN EAST EUROPE  (18)                    RATHER THAN IN EAST EUROPE  (18)            X SQ=   1.84
                                                                                                    P =    0.132
                                                                                                    RV NO  (NO )

 6  ALWAYS ARE THOSE                      1.00       6  TEND TO BE THOSE                      0.50       0   20
    LOCATED ELSEWHERE THAN IN THE                       LOCATED IN THE CARIBBEAN,                     11   20
    CARIBBEAN, CENTRAL AMERICA,                         CENTRAL AMERICA, OR SOUTH AMERICA  (22)     X SQ=   7.07
    OR SOUTH AMERICA  (93)                                                                          P =    0.004
                                                                                                    RV YES (NO )
```

31	ALWAYS ARE THOSE WHOSE AGRICULTURAL POPULATION IS LOW OR VERY LOW (24)	1.00	31	TEND TO BE THOSE WHOSE AGRICULTURAL POPULATION IS HIGH OR MEDIUM (90)	0.72	0 29 11 11 X SQ= 15.65 P = 0. RV YES (YES)
33	TEND TO BE THOSE WHOSE GROSS NATIONAL PRODUCT IS VERY HIGH, HIGH, OR MEDIUM (30)	0.82	33	TEND TO BE THOSE WHOSE GROSS NATIONAL PRODUCT IS LOW OR VERY LOW (85)	0.65	9 14 2 26 X SQ= 5.86 P = 0.014 RV YES (NO)
35	TEND TO BE THOSE WHOSE PER CAPITA GROSS NATIONAL PRODUCT IS VERY HIGH OR HIGH (24)	0.91	35	TEND TO BE THOSE WHOSE PER CAPITA GROSS NATIONAL PRODUCT IS MEDIUM, LOW, OR VERY LOW (91)	0.70	10 12 1 28 X SQ= 10.68 P = 0. RV YES (NO)
36	ALWAYS ARE THOSE WHOSE PER CAPITA GROSS NATIONAL PRODUCT IS VERY HIGH, HIGH, OR MEDIUM (42)	1.00	36	TEND LESS TO BE THOSE WHOSE PER CAPITA GROSS NATIONAL PRODUCT IS VERY HIGH, HIGH, OR MEDIUM (42)	0.57	11 23 0 17 X SQ= 5.23 P = 0.009 RV NO (NO)
39	TEND TO BE THOSE WHOSE INTERNATIONAL FINANCIAL STATUS IS VERY HIGH, HIGH, OR MEDIUM (38)	0.91	39	TEND TO BE THOSE WHOSE INTERNATIONAL FINANCIAL STATUS IS LOW OR VERY LOW (76)	0.51	10 19 1 20 X SQ= 4.66 P = 0.016 RV YES (NO)
40	ALWAYS ARE THOSE WHOSE INTERNATIONAL FINANCIAL STATUS IS VERY HIGH, HIGH, MEDIUM, OR LOW (71)	1.00	40	LEAN LESS TOWARD BEING THOSE WHOSE INTERNATIONAL FINANCIAL STATUS IS VERY HIGH, HIGH, MEDIUM, OR LOW (71)	0.75	11 30 0 10 X SQ= 2.02 P = 0.094 RV NO (NO)
41	TEND TO BE THOSE WHOSE ECONOMIC DEVELOPMENTAL STATUS IS DEVELOPED (19)	0.91	41	TEND TO BE THOSE WHOSE ECONOMIC DEVELOPMENTAL STATUS IS INTERMEDIATE, UNDERDEVELOPED, OR VERY UNDERDEVELOPED (94)	0.79	10 8 1 31 X SQ= 15.53 P = 0. RV YES (YES)
42	ALWAYS ARE THOSE WHOSE ECONOMIC DEVELOPMENTAL STATUS IS DEVELOPED OR INTERMEDIATE (36)	1.00	42	TEND LESS TO BE THOSE WHOSE ECONOMIC DEVELOPMENTAL STATUS IS DEVELOPED OR INTERMEDIATE (36)	0.56	11 22 0 17 X SQ= 5.45 P = 0.009 RV NO (NO)
44	TEND TO BE THOSE WHERE THE LITERACY RATE IS NINETY PERCENT OR ABOVE (25)	0.91	44	TEND TO BE THOSE WHERE THE LITERACY RATE IS BELOW NINETY PERCENT (90)	0.67	10 13 1 27 X SQ= 9.65 P = 0.001 RV YES (NO)

50	LEAN TOWARD BEING THOSE WHERE FREEDOM OF THE PRESS IS COMPLETE (43)	0.82	50	LEAN TOWARD BEING THOSE WHERE FREEDOM OF THE PRESS IS INTERMITTENT, INTERNALLY ABSENT, OR INTERNALLY AND EXTERNALLY ABSENT (56)	0.51	9 17 2 18 X SQ= 2.53 P = 0.082 RV YES (NO)
51	ALWAYS ARE THOSE WHERE FREEDOM OF THE PRESS IS COMPLETE OR INTERMITTENT (60)	1.00	51	TEND LESS TO BE THOSE WHERE FREEDOM OF THE PRESS IS COMPLETE OR INTERMITTENT (60)	0.60	11 21 0 14 X SQ= 4.58 P = 0.011 RV NO (NO)
53	TEND TO BE THOSE WHERE NEWSPAPER CIRCULATION IS THREE HUNDRED OR MORE PER THOUSAND (14)	0.55	53	TEND TO BE THOSE WHERE NEWSPAPER CIRCULATION IS LESS THAN THREE HUNDRED PER THOUSAND (101)	0.82	6 7 5 33 X SQ= 4.44 P = 0.021 RV YES (NO)
66	TEND TO BE THOSE THAT ARE RELIGIOUSLY HETEROGENEOUS (49)	0.73	66	TEND TO BE THOSE THAT ARE RELIGIOUSLY HOMOGENEOUS (57)	0.72	3 29 8 11 X SQ= 5.74 P = 0.012 RV YES (NO)
71	TEND TO BE THOSE WHOSE DATE OF INDEPENDENCE IS BEFORE 1800 (21)	0.55	71	TEND TO BE THOSE WHOSE DATE OF INDEPENDENCE IS AFTER 1800 (90)	0.84	6 6 5 32 X SQ= 4.99 P = 0.016 RV YES (YES)
79	ALWAYS ARE THOSE THAT WERE FORMERLY DEPENDENCIES OF BRITAIN OR FRANCE, RATHER THAN SPAIN (49)	1.00	79	TEND TO BE THOSE THAT WERE FORMERLY DEPENDENCIES OF SPAIN, RATHER THAN BRITAIN OR FRANCE (18)	0.90	5 2 0 18 X SQ= 11.92 P = 0. RV YES (YES)
89	ALWAYS ARE THOSE WHOSE IDEOLOGICAL ORIENTATION IS CONVENTIONAL (33)	1.00	89	TEND LESS TO BE THOSE WHOSE IDEOLOGICAL ORIENTATION IS CONVENTIONAL (33)	0.55	11 17 0 14 X SQ= 5.56 P = 0.007 RV NO (NO)
93	ALWAYS ARE THOSE WHERE THE SYSTEM STYLE IS NON-MOBILIZATIONAL (78)	1.00	93	LEAN LESS TOWARD BEING THOSE WHERE THE SYSTEM STYLE IS NON-MOBILIZATIONAL (78)	0.72	0 11 11 28 X SQ= 2.50 P = 0.053 RV NO (NO)
94	ALWAYS ARE THOSE WHERE THE STATUS OF THE REGIME IS CONSTITUTIONAL (51)	1.00	94	TEND LESS TO BE THOSE WHERE THE STATUS OF THE REGIME IS CONSTITUTIONAL (51)	0.59	10 22 0 15 X SQ= 4.23 P = 0.019 RV NO (NO)

98	TEND TO BE THOSE WHERE GOVERNMENTAL STABILITY IS GENERALLY PRESENT AND DATES AT LEAST FROM THE INTERWAR PERIOD (22)	0.91	98	TEND TO BE THOSE WHERE GOVERNMENTAL STABILITY IS GENERALLY OR MODERATELY PRESENT AND DATES FROM THE POST-WAR PERIOD, OR IS ABSENT (93)	0.80	10 8 1 32 X SQ= 16.02 P = 0. RV YES (YES)
99	ALWAYS ARE THOSE WHERE GOVERNMENTAL STABILITY IS GENERALLY PRESENT AND DATES FROM AT LEAST THE INTER-WAR PERIOD, OR FROM THE POST-WAR PERIOD (50)	1.00	99	TEND LESS TO BE THOSE WHERE GOVERNMENTAL STABILITY IS GENERALLY PRESENT AND DATES FROM AT LEAST THE INTER-WAR PERIOD, OR FROM THE POST-WAR PERIOD (50)	0.57	11 20 0 15 X SQ= 5.18 P = 0.009 RV NO (NO)
101	TEND TO BE THOSE WHERE THE REPRESENTATIVE CHARACTER OF THE REGIME IS POLYARCHIC (41)	0.91	101	TEND TO BE THOSE WHERE THE REPRESENTATIVE CHARACTER OF THE REGIME IS LIMITED POLYARCHIC, PSEUDO-POLYARCHIC, OR NON-POLYARCHIC (57)	0.54	10 17 1 20 X SQ= 5.26 P = 0.013 RV YES (NO)
102	ALWAYS ARE THOSE WHERE THE REPRESENTATIVE CHARACTER OF THE REGIME IS POLYARCHIC OR LIMITED POLYARCHIC (59)	1.00	102	TEND LESS TO BE THOSE WHERE THE REPRESENTATIVE CHARACTER OF THE REGIME IS POLYARCHIC OR LIMITED POLYARCHIC (59)	0.61	11 23 0 15 X SQ= 4.54 P = 0.011 RV NO (NO)
105	ALWAYS ARE THOSE WHERE THE ELECTORAL SYSTEM IS COMPETITIVE (43)	1.00	105	TEND LESS TO BE THOSE WHERE THE ELECTORAL SYSTEM IS COMPETITIVE (43)	0.60	10 21 0 14 X SQ= 4.09 P = 0.019 RV NO (NO)
107	TEND MORE TO BE THOSE WHERE AUTONOMOUS GROUPS ARE FULLY TOLERATED IN POLITICS (46)	0.91	107	TEND LESS TO BE THOSE WHERE AUTONOMOUS GROUPS ARE FULLY TOLERATED IN POLITICS (46)	0.56	10 22 1 17 X SQ= 3.06 P = 0.041 RV NO (NO)
108	ALWAYS ARE THOSE WHERE AUTONOMOUS GROUPS ARE FULLY OR PARTIALLY TOLERATED IN POLITICS (65)	1.00	108	LEAN LESS TOWARD BEING THOSE WHERE AUTONOMOUS GROUPS ARE FULLY OR PARTIALLY TOLERATED IN POLITICS (65)	0.69	10 27 0 12 X SQ= 2.58 P = 0.051 RV NO (NO)
114	TEND TO BE THOSE WHERE SECTIONALISM IS EXTREME OR MODERATE (61)	0.82	114	TEND TO BE THOSE WHERE SECTIONALISM IS NEGLIGIBLE (47)	0.69	9 11 2 25 X SQ= 7.08 P = 0.004 RV YES (NO)
115	TEND TO BE THOSE WHERE INTEREST ARTICULATION BY ASSOCIATIONAL GROUPS IS SIGNIFICANT (20)	0.91	115	TEND TO BE THOSE WHERE INTEREST ARTICULATION BY ASSOCIATIONAL GROUPS IS MODERATE, LIMITED, OR NEGLIGIBLE (91)	0.78	10 8 1 29 X SQ= 14.54 P = 0. RV YES (YES)

116	ALWAYS ARE THOSE WHERE INTEREST ARTICULATION BY ASSOCIATIONAL GROUPS IS SIGNIFICANT OR MODERATE (32)	1.00	116	TEND TO BE THOSE WHERE INTEREST ARTICULATION BY ASSOCIATIONAL GROUPS IS LIMITED OR NEGLIGIBLE (79)	0.57	11 16 0 21 X SQ= 8.91 P = 0.001 RV YES (NO)
118	ALWAYS ARE THOSE WHERE INTEREST ARTICULATION BY INSTITUTIONAL GROUPS IS SIGNIFICANT, MODERATE, OR LIMITED (60)	1.00	118	TEND LESS TO BE THOSE WHERE INTEREST ARTICULATION BY INSTITUTIONAL GROUPS IS SIGNIFICANT, MODERATE, OR LIMITED (60)	0.65	0 14 11 26 X SQ= 3.69 P = 0.023 RV NO (NO)
119	TEND TO BE THOSE WHERE INTEREST ARTICULATION BY INSTITUTIONAL GROUPS IS MODERATE OR LIMITED (26)	0.91	119	TEND TO BE THOSE WHERE INTEREST ARTICULATION BY INSTITUTIONAL GROUPS IS VERY SIGNIFICANT OR SIGNIFICANT (74)	0.70	1 28 10 12 X SQ= 10.68 P = 0. RV YES (NO)
125	LEAN TOWARD BEING THOSE WHERE INTEREST ARTICULATION BY ANOMIC GROUPS IS INFREQUENT OR VERY INFREQUENT (35)	0.89	125	LEAN TOWARD BEING THOSE WHERE INTEREST ARTICULATION BY ANOMIC GROUPS IS FREQUENT OR OCCASIONAL (64)	0.50	1 16 8 16 X SQ= 2.92 P = 0.056 RV YES (NO)
126	TEND TO BE THOSE WHERE INTEREST ARTICULATION BY ANOMIC GROUPS IS VERY INFREQUENT (16)	0.89	126	TEND TO BE THOSE WHERE INTEREST ARTICULATION BY ANOMIC GROUPS IS FREQUENT, OCCASIONAL, OR INFREQUENT (83)	0.75	1 24 8 8 X SQ= 9.51 P = 0.001 RV YES (YES)
130	TEND TO BE THOSE WHERE INTEREST AGGREGATION BY POLITICAL PARTIES IS SIGNIFICANT (12)	0.56	130	TEND TO BE THOSE WHERE INTEREST AGGREGATION BY POLITICAL PARTIES IS MODERATE, LIMITED, OR NEGLIGIBLE (71)	0.92	5 3 4 33 X SQ= 7.99 P = 0.004 RV YES (YES)
134	TEND TO BE THOSE WHERE INTEREST AGGREGATION BY THE EXECUTIVE IS SIGNIFICANT OR MODERATE (57)	0.90	134	TEND TO BE THOSE WHERE INTEREST AGGREGATION BY THE EXECUTIVE IS LIMITED OR NEGLIGIBLE (46)	0.58	9 16 1 22 X SQ= 5.48 P = 0.011 RV YES (NO)
137	TEND TO BE THOSE WHERE INTEREST AGGREGATION BY THE LEGISLATURE IS SIGNIFICANT OR MODERATE (29)	0.90	137	TEND TO BE THOSE WHERE INTEREST AGGREGATION BY THE LEGISLATURE IS LIMITED OR NEGLIGIBLE (68)	0.67	9 12 1 24 X SQ= 7.97 P = 0.003 RV YES (NO)
138	ALWAYS ARE THOSE WHERE INTEREST AGGREGATION BY THE LEGISLATURE IS SIGNIFICANT, MODERATE, OR LIMITED (48)	1.00	138	TEND LESS TO BE THOSE WHERE INTEREST AGGREGATION BY THE LEGISLATURE IS SIGNIFICANT, MODERATE, OR LIMITED (48)	0.67	10 24 0 12 X SQ= 2.95 P = 0.044 RV NO (NO)

```
139  ALWAYS ARE THOSE                       1.00    139  LEAN LESS TOWARD BEING THOSE          0.72        0   11
     WHERE THE PARTY SYSTEM IS QUANTITATIVELY            WHERE THE PARTY SYSTEM IS QUANTITATIVELY         11   28
     OTHER THAN ONE-PARTY  (71)                          OTHER THAN ONE-PARTY  (71)                  X SQ=    2.50
                                                                                                     P =     0.053
                                                                                                     RV NO   (NO )

141  TEND TO BE THOSE                       0.50    141  TEND TO BE THOSE                      0.92        5    3
     WHERE THE PARTY SYSTEM IS QUANTITATIVELY            WHERE THE PARTY SYSTEM IS QUANTITATIVELY          5   35
     TWO-PARTY  (11)                                     OTHER THAN TWO-PARTY  (87)                  X SQ=    7.30
                                                                                                     P =     0.006
                                                                                                     RV YES (YES)

145  TEND TO BE THOSE                       0.56    145  TEND TO BE THOSE                      0.86        4   18
     WHERE THE PARTY SYSTEM IS QUANTITATIVELY            WHERE THE PARTY SYSTEM IS QUANTITATIVELY          5    3
     TWO-PARTY, RATHER THAN                              MULTI-PARTY, RATHER THAN                    X SQ=    3.58
     MULTI-PARTY  (11)                                   TWO-PARTY  (30)                             P =     0.032
                                                                                                     RV YES (YES)

147  TEND TO BE THOSE                       0.78    147  TEND TO BE THOSE                      0.66        7   12
     WHERE THE PARTY SYSTEM IS QUALITATIVELY             WHERE THE PARTY SYSTEM IS QUALITATIVELY           2   23
     CLASS-ORIENTED OR MULTI-IDEOLOGICAL  (23)           OTHER THAN                                  X SQ=    3.89
                                                         CLASS-ORIENTED OR MULTI-IDEOLOGICAL  (67)   P =     0.027
                                                                                                     RV YES (NO )

153  TEND MORE TO BE THOSE                  0.91    153  TEND LESS TO BE THOSE                 0.54       10   19
     WHERE THE PARTY SYSTEM IS                           WHERE THE PARTY SYSTEM IS                         1   16
     STABLE  (42)                                        STABLE  (42)                                X SQ=    3.37
                                                                                                     P =     0.036
                                                                                                     RV NO   (NO )

154  ALWAYS ARE THOSE                       1.00    154  TEND LESS TO BE THOSE                 0.66       11   23
     WHERE THE PARTY SYSTEM IS                           WHERE THE PARTY SYSTEM IS                         0   12
     STABLE OR MODERATELY STABLE  (55)                   STABLE OR MODERATELY STABLE  (55)           X SQ=    3.48
                                                                                                     P =     0.024
                                                                                                     RV NO   (NO )

158  LEAN MORE TOWARD BEING THOSE           0.91    158  LEAN LESS TOWARD BEING THOSE          0.61        1   15
     WHERE PERSONALISMO IS                               WHERE PERSONALISMO IS                            10   23
     NEGLIGIBLE  (56)                                    NEGLIGIBLE  (56)                            X SQ=    2.33
                                                                                                     P =     0.076
                                                                                                     RV NO   (NO )

161  LEAN TOWARD BEING THOSE                0.73    161  LEAN TOWARD BEING THOSE               0.58        3   22
     WHERE THE POLITICAL LEADERSHIP IS                   WHERE THE POLITICAL LEADERSHIP IS                 8   16
     NON-ELITIST  (50)                                   ELITIST OR MODERATELY ELITIST  (47)         X SQ=    2.09
                                                                                                     P =     0.095
                                                                                                     RV YES (NO )

168  TEND TO BE THOSE                       0.91    168  TEND TO BE THOSE                      0.61       10   15
     WHERE THE HORIZONTAL POWER DISTRIBUTION             WHERE THE HORIZONTAL POWER DISTRIBUTION           1   23
     IS SIGNIFICANT  (34)                                IS LIMITED OR NEGLIGIBLE  (72)              X SQ=    7.09
                                                                                                     P =     0.005
                                                                                                     RV YES (NO )
```

169	ALWAYS ARE THOSE WHERE THE HORIZONTAL POWER DISTRIBUTION IS SIGNIFICANT OR LIMITED (58)	1.00	169	TEND LESS TO BE THOSE WHERE THE HORIZONTAL POWER DISTRIBUTION IS SIGNIFICANT OR LIMITED (58)	0.66	11 25 0 13 X SQ= 3.52 P = 0.024 RV NO (NO)
170	TEND MORE TO BE THOSE WHERE THE LEGISLATIVE-EXECUTIVE STRUCTURE IS OTHER THAN PRESIDENTIAL (63)	0.91	170	TEND LESS TO BE THOSE WHERE THE LEGISLATIVE-EXECUTIVE STRUCTURE IS OTHER THAN PRESIDENTIAL (63)	0.55	1 17 10 21 X SQ= 3.26 P = 0.038 RV NO (NO)
172	TEND TO BE THOSE WHERE THE LEGISLATIVE-EXECUTIVE STRUCTURE IS PARLIAMENTARY-ROYALIST (21)	0.64	172	TEND TO BE THOSE WHERE THE LEGISLATIVE-EXECUTIVE STRUCTURE IS OTHER THAN PARLIAMENTARY-ROYALIST (88)	0.90	7 4 4 35 X SQ= 11.31 P = 0.001 RV YES (YES)
174	TEND TO BE THOSE WHERE THE LEGISLATURE IS FULLY EFFECTIVE (28)	0.91	174	TEND TO BE THOSE WHERE THE LEGISLATURE IS PARTIALLY EFFECTIVE, LARGELY INEFFECTIVE, OR WHOLLY INEFFECTIVE (72)	0.72	10 10 1 26 X SQ= 11.28 P = 0. RV YES (YES)
175	ALWAYS ARE THOSE WHERE THE LEGISLATURE IS FULLY EFFECTIVE OR PARTIALLY EFFECTIVE (51)	1.00	175	TEND LESS TO BE THOSE WHERE THE LEGISLATURE IS FULLY EFFECTIVE OR PARTIALLY EFFECTIVE (51)	0.61	11 22 0 14 X SQ= 4.38 P = 0.020 RV NO (NO)
178	TEND MORE TO BE THOSE WHERE THE LEGISLATURE IS BICAMERAL (51)	0.91	178	TEND LESS TO BE THOSE WHERE THE LEGISLATURE IS BICAMERAL (51)	0.54	1 18 10 21 X SQ= 3.55 P = 0.035 RV NO (NO)
179	TILT TOWARD BEING THOSE WHERE THE EXECUTIVE IS STRONG (39)	0.80	179	TILT TOWARD BEING THOSE WHERE THE EXECUTIVE IS DOMINANT (52)	0.50	2 17 8 17 X SQ= 1.74 P = 0.148 RV YES (NO)
181	ALWAYS ARE THOSE WHERE THE BUREAUCRACY IS MODERN, RATHER THAN SEMI-MODERN (21)	1.00	181	TEND TO BE THOSE WHERE THE BUREAUCRACY IS SEMI-MODERN, RATHER THAN MODERN (55)	0.77	11 9 0 31 X SQ= 18.61 P = 0. RV YES (YES)
186	TILT MORE TOWARD BEING THOSE WHERE PARTICIPATION BY THE MILITARY IN POLITICS IS NEUTRAL, RATHER THAN SUPPORTIVE (56)	0.90	186	TILT LESS TOWARD BEING THOSE WHERE PARTICIPATION BY THE MILITARY IN POLITICS IS NEUTRAL, RATHER THAN SUPPORTIVE (56)	0.61	1 11 9 17 X SQ= 1.73 P = 0.124 RV NO (NO)

187	TEND TO BE THOSE WHERE THE ROLE OF THE POLICE IS NOT POLITICALLY SIGNIFICANT (35)	0.91	187	TEND TO BE THOSE WHERE THE ROLE OF THE POLICE IS POLITICALLY SIGNIFICANT (66)	0.65	1 22 10 12 X SQ= 8.18 P = 0.002 RV YES (NO)
189	TEND LESS TO BE THOSE WHERE THE CHARACTER OF THE LEGAL SYSTEM IS OTHER THAN COMMON LAW (108)	0.64	189	TEND MORE TO BE THOSE WHERE THE CHARACTER OF THE LEGAL SYSTEM IS OTHER THAN COMMON LAW (108)	0.97	4 1 7 39 X SQ= 7.69 P = 0.006 RV NO (YES)
190	TEND LESS TO BE THOSE WHERE THE CHARACTER OF THE LEGAL SYSTEM IS CIVIL LAW, RATHER THAN COMMON LAW (32)	0.56	190	TEND MORE TO BE THOSE WHERE THE CHARACTER OF THE LEGAL SYSTEM IS CIVIL LAW, RATHER THAN COMMON LAW (32)	0.96	5 25 4 1 X SQ= 5.99 P = 0.010 RV NO (YES)

82 POLITIES
WHERE THE TYPE OF POLITICAL MODERNIZATION
IS EARLY OR LATER EUROPEAN OR
EUROPEAN DERIVED, RATHER THAN
DEVELOPED TUTELARY OR
UNDEVELOPED TUTELARY (51)

82 POLITIES
WHERE THE TYPE OF POLITICAL MODERNIZATION
IS DEVELOPED TUTELARY OR
UNDEVELOPED TUTELARY, RATHER THAN
EARLY OR LATER EUROPEAN OR
EUROPEAN DERIVED (55)

BOTH SUBJECT AND PREDICATE

51 IN LEFT COLUMN

ALBANIA	ARGENTINA	AUSTRALIA	AUSTRIA	BELGIUM	BOLIVIA	BRAZIL	BULGARIA	CANADA	CHILE
COLOMBIA	COSTA RICA	CUBA	CZECHOS'KIA	DENMARK	DOMIN REP	ECUADOR	EL SALVADOR	FINLAND	FRANCE
GERMANY, E	GERMAN FR	GREECE	GUATEMALA	HAITI	HONDURAS	HUNGARY	ICELAND	IRELAND	ITALY
LUXEMBOURG	MEXICO	NETHERLANDS	NEW ZEALAND	NICARAGUA	NORWAY	PANAMA	PARAGUAY	PERU	POLAND
PORTUGAL	RUMANIA	SO AFRICA	SPAIN	SWEDEN	SWITZERLAND	UK	US	URUGUAY	VENEZUELA
YUGOSLAVIA									

55 IN RIGHT COLUMN

ALGERIA	BURMA	BURUNDI	CAMBODIA	CAMEROUN	CEN AFR REP	CEYLON	CHAD	CONGO(BRA)	CONGO(LEO)
CYPRUS	DAHOMEY	GABON	GHANA	GUINEA	INDIA	INDONESIA	IRAQ	ISRAEL	IVORY COAST
JAMAICA	JORDAN	KOREA, N	KOREA REP	LAOS	LEBANON	LIBERIA	LIBYA	MALAGASY R	MALAYA
MALI	MAURITANIA	MONGOLIA	MOROCCO	NIGER	NIGERIA	PAKISTAN	PHILIPPINES	RWANDA	SA'U ARABIA
SENEGAL	SIERRE LEO	SOMALIA	SUDAN	SYRIA	TANGANYIKA	TOGO	TRINIDAD	TUNISIA	UGANDA
UAR	UPPER VOLTA	VIETNAM, N	VIETNAM REP	YEMEN					

9 EXCLUDED BECAUSE IRRELEVANT

AFGHANISTAN CHINA, PR ETHIOPIA IRAN JAPAN NEPAL THAILAND TURKEY USSR

2	TEND LESS TO BE THOSE LOCATED ELSEWHERE THAN IN WEST EUROPE, SCANDINAVIA, NORTH AMERICA, OR AUSTRALASIA (93)	0.57	2 ALWAYS ARE THOSE LOCATED ELSEWHERE THAN IN WEST EUROPE, SCANDINAVIA, NORTH AMERICA, OR AUSTRALASIA (93)	1.00	22 0 29 55 X SQ= 27.37 P = 0. RV NO (YES)
6	TEND LESS TO BE THOSE LOCATED ELSEWHERE THAN IN THE CARIBBEAN, CENTRAL AMERICA, OR SOUTH AMERICA (93)	0.61	6 TEND MORE TO BE THOSE LOCATED ELSEWHERE THAN IN THE CARIBBEAN, CENTRAL AMERICA, OR SOUTH AMERICA (93)	0.96	20 2 31 53 X SQ= 18.26 P = 0. RV NO (YES)
8	ALWAYS ARE THOSE LOCATED ELSEWHERE THAN IN EAST ASIA SOUTH ASIA, OR SOUTHEAST ASIA (97)	1.00	8 TEND LESS TO BE THOSE LOCATED ELSEWHERE THAN IN EAST ASIA SOUTH ASIA, OR SOUTHEAST ASIA (97)	0.75	0 14 51 41 X SQ= 12.82 P = 0. RV NO (YES)

10	TEND TO BE THOSE LOCATED ELSEWHERE THAN IN NORTH AFRICA, OR CENTRAL AND SOUTH AFRICA (82)	0.98	10	TEND TO BE THOSE LOCATED IN NORTH AFRICA, OR CENTRAL AND SOUTH AFRICA (33)	0.56	1 31 50 24 X SQ= 34.63 P = 0. RV YES (YES)
11	TEND MORE TO BE THOSE LOCATED ELSEWHERE THAN IN CENTRAL AND SOUTH AFRICA (87)	0.98	11	TEND LESS TO BE THOSE LOCATED ELSEWHERE THAN IN CENTRAL AND SOUTH AFRICA (87)	0.53	1 26 50 29 X SQ= 26.28 P = 0. RV NO (YES)
13	ALWAYS ARE THOSE LOCATED ELSEWHERE THAN IN NORTH AFRICA OR THE MIDDLE EAST (99)	1.00	13	TEND LESS TO BE THOSE LOCATED ELSEWHERE THAN IN NORTH AFRICA OR THE MIDDLE EAST (99)	0.76	0 13 51 42 X SQ= 11.63 P = 0. RV NO (YES)
19	TEND TO BE THOSE LOCATED IN THE CARIBBEAN, CENTRAL AMERICA, OR SOUTH AMERICA, RATHER THAN IN NORTH AFRICA OR CENTRAL AND SOUTH AFRICA (22)	0.95	19	TEND TO BE THOSE LOCATED IN NORTH AFRICA OR CENTRAL AND SOUTH AFRICA, RATHER THAN IN THE CARIBBEAN, CENTRAL AMERICA, OR SOUTH AMERICA (33)	0.94	1 31 20 2 X SQ= 38.66 P = 0. RV YES (YES)
24	LEAN TOWARD BEING THOSE WHOSE POPULATION IS VERY LARGE, LARGE, OR MEDIUM (61)	0.59	24	LEAN TOWARD BEING THOSE WHOSE POPULATION IS SMALL (54)	0.60	30 22 21 33 X SQ= 3.04 P = 0.080 RV YES (YES)
26	LEAN TOWARD BEING THOSE WHOSE POPULATION DENSITY IS VERY HIGH, HIGH, OR MEDIUM (48)	0.51	26	LEAN TOWARD BEING THOSE WHOSE POPULATION DENSITY IS LOW (67)	0.67	26 18 25 37 X SQ= 2.92 P = 0.076 RV YES (YES)
28	TEND TO BE THOSE WHOSE POPULATION GROWTH RATE IS LOW (48)	0.55	28	TEND TO BE THOSE WHOSE POPULATION GROWTH RATE IS HIGH (62)	0.65	23 34 28 18 X SQ= 3.51 P = 0.048 RV YES (YES)
29	TEND TO BE THOSE WHERE THE DEGREE OF URBANIZATION IS HIGH (56)	0.85	29	TEND TO BE THOSE WHERE THE DEGREE OF URBANIZATION IS LOW (49)	0.78	41 11 7 38 X SQ= 36.17 P = 0. RV YES (YES)
30	TEND TO BE THOSE WHOSE AGRICULTURAL POPULATION IS MEDIUM, LOW, OR VERY LOW (57)	0.84	30	TEND TO BE THOSE WHOSE AGRICULTURAL POPULATION IS HIGH (56)	0.77	8 41 43 12 X SQ= 37.24 P = 0. RV YES (YES)

34	TEND TO BE THOSE WHOSE GROSS NATIONAL PRODUCT IS VERY HIGH, HIGH, MEDIUM, OR LOW (62)	0.73	34	TEND TO BE THOSE WHOSE GROSS NATIONAL PRODUCT IS VERY LOW (53)	0.67	37 18 14 37 X SQ= 15.25 P = 0. RV YES (YES)
36	TEND TO BE THOSE WHOSE PER CAPITA GROSS NATIONAL PRODUCT IS VERY HIGH, HIGH, OR MEDIUM (42)	0.67	36	TEND TO BE THOSE WHOSE PER CAPITA GROSS NATIONAL PRODUCT IS LOW OR VERY LOW (73)	0.89	34 6 17 49 X SQ= 32.68 P = 0. RV YES (YES)
37	TEND TO BE THOSE WHOSE PER CAPITA GROSS NATIONAL PRODUCT IS VERY HIGH, HIGH, MEDIUM, OR LOW (64)	0.94	37	TEND TO BE THOSE WHOSE PER CAPITA GROSS NATIONAL PRODUCT IS VERY LOW (51)	0.76	48 13 3 42 X SQ= 50.96 P = 0. RV YES (YES)
39	TEND TO BE THOSE WHOSE INTERNATIONAL FINANCIAL STATUS IS VERY HIGH, HIGH, OR MEDIUM (38)	0.58	39	TEND TO BE THOSE WHOSE INTERNATIONAL FINANCIAL STATUS IS LOW OR VERY LOW (76)	0.91	29 5 21 50 X SQ= 26.42 P = 0. RV YES (YES)
40	TEND TO BE THOSE WHOSE INTERNATIONAL FINANCIAL STATUS IS VERY HIGH, HIGH, MEDIUM, OR LOW (71)	0.80	40	TEND TO BE THOSE WHOSE INTERNATIONAL FINANCIAL STATUS IS VERY LOW (39)	0.56	41 22 10 28 X SQ= 12.74 P = 0. RV YES (YES)
41	TEND LESS TO BE THOSE WHOSE ECONOMIC DEVELOPMENTAL STATUS IS INTERMEDIATE, UNDERDEVELOPED, OR VERY UNDERDEVELOPED (94)	0.64	41	ALWAYS ARE THOSE WHOSE ECONOMIC DEVELOPMENTAL STATUS IS INTERMEDIATE, UNDERDEVELOPED, OR VERY UNDERDEVELOPED (94)	1.00	18 0 32 54 X SQ= 21.06 P = 0. RV NO (YES)
42	TEND TO BE THOSE WHOSE ECONOMIC DEVELOPMENTAL STATUS IS DEVELOPED OR INTERMEDIATE (36)	0.66	42	TEND TO BE THOSE WHOSE ECONOMIC DEVELOPMENTAL STATUS IS UNDERDEVELOPED OR VERY UNDERDEVELOPED (76)	0.98	33 1 17 52 X SQ= 44.97 P = 0. RV YES (YES)
43	TEND TO BE THOSE WHOSE ECONOMIC DEVELOPMENTAL STATUS IS DEVELOPED, INTERMEDIATE, OR UNDERDEVELOPED (55)	0.80	43	TEND TO BE THOSE WHOSE ECONOMIC DEVELOPMENTAL STATUS IS VERY UNDERDEVELOPED (57)	0.79	40 11 10 42 X SQ= 33.79 P = 0. RV YES (YES)
44	TEND LESS TO BE THOSE WHERE THE LITERACY RATE IS BELOW NINETY PERCENT (90)	0.55	44	ALWAYS ARE THOSE WHERE THE LITERACY RATE IS BELOW NINETY PERCENT (90)	1.00	23 0 28 55 X SQ= 29.08 P = 0. RV NO (YES)

45	TEND TO BE THOSE WHERE THE LITERACY RATE IS FIFTY PERCENT OR ABOVE (55)	0.82	45	TEND TO BE THOSE WHERE THE LITERACY RATE IS BELOW FIFTY PERCENT (54)	0.80	42 10 9 40 X SQ= 36.84 P = 0. RV YES (YES)
46	ALWAYS ARE THOSE WHERE THE LITERACY RATE IS TEN PERCENT OR ABOVE (84)	1.00	46	TEND LESS TO BE THOSE WHERE THE LITERACY RATE IS TEN PERCENT OR ABOVE (84)	0.54	51 27 0 23 X SQ= 27.82 P = 0. RV NO (YES)
50	LEAN TOWARD BEING THOSE WHERE FREEDOM OF THE PRESS IS COMPLETE (43)	0.57	50	LEAN TOWARD BEING THOSE WHERE FREEDOM OF THE PRESS IS INTERMITTENT, INTERNALLY ABSENT, OR INTERNALLY AND EXTERNALLY ABSENT (56)	0.64	26 16 20 28 X SQ= 2.91 P = 0.061 RV YES (YES)
53	TEND LESS TO BE THOSE WHERE NEWSPAPER CIRCULATION IS LESS THAN THREE HUNDRED PER THOUSAND (101)	0.75	53	ALWAYS ARE THOSE WHERE NEWSPAPER CIRCULATION IS LESS THAN THREE HUNDRED PER THOUSAND (101)	1.00	13 0 38 55 X SQ= 13.70 P = 0. RV NO (YES)
54	TEND TO BE THOSE WHERE NEWSPAPER CIRCULATION IS ONE HUNDRED OR MORE PER THOUSAND (37)	0.63	54	TEND TO BE THOSE WHERE NEWSPAPER CIRCULATION IS LESS THAN ONE HUNDRED PER THOUSAND (76)	0.94	32 3 19 50 X SQ= 35.42 P = 0. RV YES (YES)
55	ALWAYS ARE THOSE WHERE NEWSPAPER CIRCULATION IS TEN OR MORE PER THOUSAND (78)	1.00	55	TEND TO BE THOSE WHERE NEWSPAPER CIRCULATION IS LESS THAN TEN PER THOUSAND (35)	0.58	51 22 0 31 X SQ= 39.75 P = 0. RV YES (YES)
56	TEND MORE TO BE THOSE WHERE THE RELIGION IS PREDOMINANTLY LITERATE (79)	0.88	56	TEND LESS TO BE THOSE WHERE THE RELIGION IS PREDOMINANTLY LITERATE (79)	0.53	45 27 6 24 X SQ= 13.65 P = 0. RV NO (YES)
57	TEND LESS TO BE THOSE WHERE THE RELIGION IS OTHER THAN CATHOLIC (90)	0.53	57	TEND MORE TO BE THOSE WHERE THE RELIGION IS OTHER THAN CATHOLIC (90)	0.98	24 1 27 54 X SQ= 27.59 P = 0. RV NO (YES)
58	ALWAYS ARE THOSE WHERE THE RELIGION IS OTHER THAN MUSLIM (97)	1.00	58	TEND LESS TO BE THOSE WHERE THE RELIGION IS OTHER THAN MUSLIM (97)	0.73	0 15 51 40 X SQ= 14.03 P = 0. RV NO (YES)

63	TEND TO BE THOSE WHERE THE RELIGION IS PREDOMINANTLY SOME KIND OF CHRISTIAN (46)	0.86	63	TEND TO BE THOSE WHERE THE RELIGION IS PREDOMINANTLY OR PARTLY OTHER THAN CHRISTIAN (68)	0.96	44 2 7 53 X SQ= 70.24 P = 0. RV YES (YES)
66	LEAN TOWARD BEING THOSE THAT ARE RELIGIOUSLY HOMOGENEOUS (57)	0.63	66	LEAN TOWARD BEING THOSE THAT ARE RELIGIOUSLY HETEROGENEOUS (49)	0.56	32 21 19 27 X SQ= 2.86 P = 0.071 RV YES (YES)
67	TILT LESS TOWARD BEING THOSE THAT ARE RACIALLY HOMOGENEOUS (82)	0.65	67	TILT MORE TOWARD BEING THOSE THAT ARE RACIALLY HOMOGENEOUS (82)	0.81	32 42 17 10 X SQ= 2.34 P = 0.115 RV NO (YES)
68	TEND TO BE THOSE THAT ARE LINGUISTICALLY HOMOGENEOUS (52)	0.71	68	TEND TO BE THOSE THAT ARE LINGUISTICALLY WEAKLY HETEROGENEOUS OR STRONGLY HETEROGENEOUS (62)	0.73	36 15 15 40 X SQ= 18.19 P = 0. RV YES (YES)
69	TEND TO BE THOSE THAT ARE LINGUISTICALLY HOMOGENEOUS OR WEAKLY HETEROGENEOUS (64)	0.80	69	TEND TO BE THOSE THAT ARE LINGUISTICALLY STRONGLY HETEROGENEOUS (50)	0.64	41 20 10 35 X SQ= 19.23 P = 0. RV YES (YES)
70	TEND LESS TO BE THOSE THAT ARE RELIGIOUSLY, RACIALLY, OR LINGUISTICALLY HETEROGENEOUS (85)	0.69	70	TEND MORE TO BE THOSE THAT ARE RELIGIOUSLY, RACIALLY, OR LINGUISTICALLY HETEROGENEOUS (85)	0.87	15 6 34 42 X SQ= 3.68 P = 0.047 RV NO (YES)
71	TEND LESS TO BE THOSE WHOSE DATE OF INDEPENDENCE IS AFTER 1800 (90)	0.76	71	ALWAYS ARE THOSE WHOSE DATE OF INDEPENDENCE IS AFTER 1800 (90)	1.00	12 0 37 53 X SQ= 12.45 P = 0. RV NO (YES)
72	TEND TO BE THOSE WHOSE DATE OF INDEPENDENCE IS BEFORE 1914 (52)	0.86	72	TEND TO BE THOSE WHOSE DATE OF INDEPENDENCE IS AFTER 1914 (59)	0.98	42 1 7 52 X SQ= 69.97 P = 0. RV YES (YES)
73	ALWAYS ARE THOSE WHOSE DATE OF INDEPENDENCE IS BEFORE 1945 (65)	1.00	73	TEND TO BE THOSE WHOSE DATE OF INDEPENDENCE IS AFTER 1945 (46)	0.87	49 7 0 46 X SQ= 74.00 P = 0. RV YES (YES)

79	TEND TO BE THOSE THAT WERE FORMERLY DEPENDENCIES OF SPAIN, RATHER THAN BRITAIN OR FRANCE (18)	0.72	79	ALWAYS ARE THOSE THAT WERE FORMERLY DEPENDENCIES OF BRITAIN OR FRANCE, RATHER THAN SPAIN (49)	1.00	7 42 18 0 X SQ= 37.77 P = 0. RV YES (YES)
85	TEND TO BE THOSE WHERE THE STAGE OF POLITICAL MODERNIZATION IS ADVANCED, RATHER THAN MID- OR EARLY TRANSITIONAL (60)	0.76	85	TEND TO BE THOSE WHERE THE STAGE OF POLITICAL MODERNIZATION IS MID- OR EARLY TRANSITIONAL, RATHER THAN ADVANCED (54)	0.69	39 17 12 37 X SQ= 19.56 P = 0. RV YES (YES)
86	ALWAYS ARE THOSE WHERE THE STAGE OF POLITICAL MODERNIZATION IS ADVANCED OR MID-TRANSITIONAL, RATHER THAN EARLY TRANSITIONAL (76)	1.00	86	TEND TO BE THOSE WHERE THE STAGE OF POLITICAL MODERNIZATION IS EARLY TRANSITIONAL, RATHER THAN ADVANCED OR MID-TRANSITIONAL (38)	0.69	51 17 0 37 X SQ= 51.00 P = 0. RV YES (YES)
87	TEND TO BE THOSE WHOSE IDEOLOGICAL ORIENTATION IS OTHER THAN DEVELOPMENTAL (58)	0.93	87	TEND TO BE THOSE WHOSE IDEOLOGICAL ORIENTATION IS DEVELOPMENTAL (31)	0.72	3 28 40 11 X SQ= 33.84 P = 0. RV YES (YES)
89	TEND TO BE THOSE WHOSE IDEOLOGICAL ORIENTATION IS CONVENTIONAL (33)	0.67	89	TEND TO BE THOSE WHOSE IDEOLOGICAL ORIENTATION IS OTHER THAN CONVENTIONAL (62)	0.93	28 3 14 41 X SQ= 30.84 P = 0. RV YES (YES)
93	TILT MORE TOWARD BEING THOSE WHERE THE SYSTEM STYLE IS NON-MOBILIZATIONAL (78)	0.78	93	TILT LESS TOWARD BEING THOSE WHERE THE SYSTEM STYLE IS NON-MOBILIZATIONAL (78)	0.63	11 19 39 33 X SQ= 1.94 P = 0.131 RV NO (YES)
94	LEAN TOWARD BEING THOSE WHERE THE STATUS OF THE REGIME IS CONSTITUTIONAL (51)	0.68	94	LEAN TOWARD BEING THOSE WHERE THE STATUS OF THE REGIME IS AUTHORITARIAN OR TOTALITARIAN (41)	0.53	32 17 15 19 X SQ= 2.86 P = 0.073 RV YES (YES)
95	TEND LESS TO BE THOSE WHERE THE STATUS OF THE REGIME IS CONSTITUTIONAL OR AUTHORITARIAN (95)	0.78	95	TEND MORE TO BE THOSE WHERE THE STATUS OF THE REGIME IS CONSTITUTIONAL OR AUTHORITARIAN (95)	0.94	38 50 11 3 X SQ= 4.73 P = 0.020 RV NO (YES)
96	TEND MORE TO BE THOSE WHERE THE STATUS OF THE REGIME IS CONSTITUTIONAL OR TOTALITARIAN (67)	0.93	96	TEND LESS TO BE THOSE WHERE THE STATUS OF THE REGIME IS CONSTITUTIONAL OR TOTALITARIAN (67)	0.57	43 20 3 15 X SQ= 13.15 P = 0. RV NO (YES)

97	TEND TO BE THOSE WHERE THE STATUS OF THE REGIME IS TOTALITARIAN, RATHER THAN AUTHORITARIAN (16)	0.79	97	TEND TO BE THOSE WHERE THE STATUS OF THE REGIME IS AUTHORITARIAN, RATHER THAN TOTALITARIAN (23)	0.83	3 15 11 3 X SQ= 9.88 P = 0.001 RV YES (YES)
99	TILT TOWARD BEING THOSE WHERE GOVERNMENTAL STABILITY IS GENERALLY PRESENT AND DATES FROM AT LEAST THE INTER-WAR PERIOD, OR FROM THE POST-WAR PERIOD (50)	0.67	99	TILT TOWARD BEING THOSE WHERE GOVERNMENTAL STABILITY IS MODERATELY PRESENT AND DATES FROM THE POST-WAR PERIOD, OR IS ABSENT (36)	0.52	31 15 15 16 X SQ= 2.05 P = 0.105 RV YES (YES)
101	TEND TO BE THOSE WHERE THE REPRESENTATIVE CHARACTER OF THE REGIME IS POLYARCHIC (41)	0.56	101	TEND TO BE THOSE WHERE THE REPRESENTATIVE CHARACTER OF THE REGIME IS LIMITED POLYARCHIC, PSEUDO-POLYARCHIC, OR NON-POLYARCHIC (57)	0.71	27 12 21 29 X SQ= 5.49 P = 0.018 RV YES (YES)
102	TEND TO BE THOSE WHERE THE REPRESENTATIVE CHARACTER OF THE REGIME IS POLYARCHIC OR LIMITED POLYARCHIC (59)	0.69	102	TEND TO BE THOSE WHERE THE REPRESENTATIVE CHARACTER OF THE REGIME IS PSEUDO-POLYARCHIC OR NON-POLYARCHIC (49)	0.54	34 23 15 27 X SQ= 4.63 P = 0.025 RV YES (YES)
105	TEND TO BE THOSE WHERE THE ELECTORAL SYSTEM IS COMPETITIVE (43)	0.69	105	TEND TO BE THOSE WHERE THE ELECTORAL SYSTEM IS PARTIALLY COMPETITIVE OR NON-COMPETITIVE (47)	0.74	31 10 14 29 X SQ= 13.96 P = 0. RV YES (YES)
107	TEND TO BE THOSE WHERE AUTONOMOUS GROUPS ARE FULLY TOLERATED IN POLITICS (46)	0.64	107	TEND TO BE THOSE WHERE AUTONOMOUS GROUPS ARE PARTIALLY TOLERATED IN POLITICS, ARE TOLERATED ONLY OUTSIDE POLITICS, OR ARE NOT TOLERATED AT ALL (65)	0.77	32 12 18 40 X SQ= 15.77 P = 0. RV YES (YES)
110	TEND LESS TO BE THOSE WHERE POLITICAL ENCULTURATION IS MEDIUM OR LOW (80)	0.70	110	ALWAYS ARE THOSE WHERE POLITICAL ENCULTURATION IS MEDIUM OR LOW (80)	1.00	12 0 28 48 X SQ= 14.22 P = 0. RV NO (YES)
114	TEND TO BE THOSE WHERE SECTIONALISM IS NEGLIGIBLE (47)	0.57	114	TEND TO BE THOSE WHERE SECTIONALISM IS EXTREME OR MODERATE (61)	0.66	20 35 27 18 X SQ= 4.64 P = 0.027 RV YES (YES)
116	TEND TO BE THOSE WHERE INTEREST ARTICULATION BY ASSOCIATIONAL GROUPS IS SIGNIFICANT OR MODERATE (32)	0.56	116	TEND TO BE THOSE WHERE INTEREST ARTICULATION BY ASSOCIATIONAL GROUPS IS LIMITED OR NEGLIGIBLE (79)	0.94	27 3 21 51 X SQ= 29.06 P = 0. RV YES (YES)

117	TEND TO BE THOSE WHERE INTEREST ARTICULATION BY ASSOCIATIONAL GROUPS IS SIGNIFICANT, MODERATE, OR LIMITED (60)	0.83	117	TEND TO BE THOSE WHERE INTEREST ARTICULATION BY ASSOCIATIONAL GROUPS IS NEGLIGIBLE (51)	0.70	40 16 8 38 X SQ= 27.47 P = 0. RV YES (YES)
118	TEND TO BE THOSE WHERE INTEREST ARTICULATION BY INSTITUTIONAL GROUPS IS SIGNIFICANT, MODERATE, OR LIMITED (60)	0.73	118	TEND TO BE THOSE WHERE INTEREST ARTICULATION BY INSTITUTIONAL GROUPS IS VERY SIGNIFICANT (40)	0.50	14 20 37 20 X SQ= 3.95 P = 0.031 RV YES (YES)
119	TEND LESS TO BE THOSE WHERE INTEREST ARTICULATION BY INSTITUTIONAL GROUPS IS VERY SIGNIFICANT OR SIGNIFICANT (74)	0.57	119	TEND MORE TO BE THOSE WHERE INTEREST ARTICULATION BY INSTITUTIONAL GROUPS IS VERY SIGNIFICANT OR SIGNIFICANT (74)	0.90	29 36 22 4 X SQ= 10.49 P = 0. RV NO (YES)
120	TEND LESS TO BE THOSE WHERE INTEREST ARTICULATION BY INSTITUTIONAL GROUPS IS VERY SIGNIFICANT, SIGNIFICANT, OR MODERATE (90)	0.80	120	ALWAYS ARE THOSE WHERE INTEREST ARTICULATION BY INSTITUTIONAL GROUPS IS VERY SIGNIFICANT, SIGNIFICANT, OR MODERATE (90)	1.00	41 40 10 0 X SQ= 6.92 P = 0.002 RV NO (NO)
121	TEND TO BE THOSE WHERE INTEREST ARTICULATION BY NON-ASSOCIATIONAL GROUPS IS MODERATE, LIMITED, OR NEGLIGIBLE (61)	0.96	121	TEND TO BE THOSE WHERE INTEREST ARTICULATION BY NON-ASSOCIATIONAL GROUPS IS SIGNIFICANT (54)	0.84	2 46 49 9 X SQ= 64.69 P = 0. RV YES (YES)
122	TEND TO BE THOSE WHERE INTEREST ARTICULATION BY NON-ASSOCIATIONAL GROUPS IS LIMITED OR NEGLIGIBLE (32)	0.63	122	ALWAYS ARE THOSE WHERE INTEREST ARTICULATION BY NON-ASSOCIATIONAL GROUPS IS SIGNIFICANT OR MODERATE (83)	1.00	19 55 32 0 X SQ= 46.50 P = 0. RV YES (YES)
125	TEND TO BE THOSE WHERE INTEREST ARTICULATION BY ANOMIC GROUPS IS INFREQUENT OR VERY INFREQUENT (35)	0.59	125	TEND TO BE THOSE WHERE INTEREST ARTICULATION BY ANOMIC GROUPS IS FREQUENT OR OCCASIONAL (64)	0.78	17 38 24 11 X SQ= 10.76 P = 0.001 RV YES (YES)
126	TEND LESS TO BE THOSE WHERE INTEREST ARTICULATION BY ANOMIC GROUPS IS FREQUENT, OCCASIONAL, OR INFREQUENT (83)	0.61	126	ALWAYS ARE THOSE WHERE INTEREST ARTICULATION BY ANOMIC GROUPS IS FREQUENT, OCCASIONAL, OR INFREQUENT (83)	1.00	25 49 16 0 X SQ= 20.66 P = 0. RV NO (YES)
128	TEND TO BE THOSE WHERE INTEREST ARTICULATION BY POLITICAL PARTIES IS SIGNIFICANT OR MODERATE (48)	0.64	128	TEND TO BE THOSE WHERE INTEREST ARTICULATION BY POLITICAL PARTIES IS LIMITED OR NEGLIGIBLE (45)	0.60	30 17 17 25 X SQ= 3.96 P = 0.035 RV YES (YES)

129	TEND TO BE THOSE WHERE INTEREST ARTICULATION BY POLITICAL PARTIES IS SIGNIFICANT, MODERATE, OR LIMITED (56)	0.72	129	TEND TO BE THOSE WHERE INTEREST ARTICULATION BY POLITICAL PARTIES IS NEGLIGIBLE (37)	0.52	34 20 13 22 X SQ= 4.69 P = 0.029 RV YES (YES)
131	LEAN TOWARD BEING THOSE WHERE INTEREST AGGREGATION BY POLITICAL PARTIES IS LIMITED OR NEGLIGIBLE (35)	0.65	131	LEAN TOWARD BEING THOSE WHERE INTEREST AGGREGATION BY POLITICAL PARTIES IS SIGNIFICANT OR MODERATE (30)	0.62	12 16 22 10 X SQ= 3.09 P = 0.067 RV YES (YES)
133	TEND TO BE THOSE WHERE INTEREST AGGREGATION BY THE EXECUTIVE IS MODERATE, LIMITED, OR NEGLIGIBLE (76)	0.90	133	TEND TO BE THOSE WHERE INTEREST AGGREGATION BY THE EXECUTIVE IS SIGNIFICANT (29)	0.50	5 24 43 24 X SQ= 16.01 P = 0. RV YES (YES)
138	TEND TO BE THOSE WHERE INTEREST AGGREGATION BY THE LEGISLATURE IS SIGNIFICANT, MODERATE, OR LIMITED (48)	0.74	138	TEND TO BE THOSE WHERE INTEREST AGGREGATION BY THE LEGISLATURE IS NEGLIGIBLE (49)	0.72	34 12 12 31 X SQ= 17.04 P = 0. RV YES (YES)
139	TEND MORE TO BE THOSE WHERE THE PARTY SYSTEM IS QUANTITATIVELY OTHER THAN ONE-PARTY (71)	0.78	139	TEND LESS TO BE THOSE WHERE THE PARTY SYSTEM IS QUANTITATIVELY OTHER THAN ONE-PARTY (71)	0.57	11 21 39 28 X SQ= 4.01 P = 0.033 RV NO (YES)
142	TEND LESS TO BE THOSE WHERE THE PARTY SYSTEM IS QUANTITATIVELY OTHER THAN MULTI-PARTY (66)	0.54	142	TEND MORE TO BE THOSE WHERE THE PARTY SYSTEM IS QUANTITATIVELY OTHER THAN MULTI-PARTY (66)	0.81	22 8 26 35 X SQ= 6.43 P = 0.007 RV NO (YES)
146	ALWAYS ARE THOSE WHERE THE PARTY SYSTEM IS QUALITATIVELY OTHER THAN MASS-BASED TERRITORIAL (92)	1.00	146	TEND LESS TO BE THOSE WHERE THE PARTY SYSTEM IS QUALITATIVELY OTHER THAN MASS-BASED TERRITORIAL (92)	0.83	0 8 47 38 X SQ= 6.87 P = 0.003 RV NO (YES)
147	TEND LESS TO BE THOSE WHERE THE PARTY SYSTEM IS QUALITATIVELY OTHER THAN CLASS-ORIENTED OR MULTI-IDEOLOGICAL (67)	0.57	147	TEND MORE TO BE THOSE WHERE THE PARTY SYSTEM IS QUALITATIVELY OTHER THAN CLASS-ORIENTED OR MULTI-IDEOLOGICAL (67)	0.93	19 3 25 38 X SQ= 12.42 P = 0. RV NO (YES)
148	ALWAYS ARE THOSE WHERE THE PARTY SYSTEM IS QUALITATIVELY OTHER THAN AFRICAN TRANSITIONAL (96)	1.00	148	TEND LESS TO BE THOSE WHERE THE PARTY SYSTEM IS QUALITATIVELY OTHER THAN AFRICAN TRANSITIONAL (96)	0.74	0 14 50 39 X SQ= 13.12 P = 0. RV NO (YES)

153	TEND TO BE THOSE WHERE THE PARTY SYSTEM IS STABLE (42)	0.63	153	TEND TO BE THOSE WHERE THE PARTY SYSTEM IS MODERATELY STABLE OR UNSTABLE (38)	0.64	29 10 17 18 X SQ= 4.18 P = 0.031 RV YES (YES)
158	LEAN TOWARD BEING THOSE WHERE PERSONALISMO IS NEGLIGIBLE (56)	0.67	158	LEAN TOWARD BEING THOSE WHERE PERSONALISMO IS PRONOUNCED OR MODERATE (40)	0.52	16 22 33 20 X SQ= 2.85 P = 0.088 RV YES (YES)
160	LEAN LESS TOWARD BEING THOSE WHERE THE POLITICAL LEADERSHIP IS MODERATELY ELITIST OR NON-ELITIST (67)	0.65	160	LEAN MORE TOWARD BEING THOSE WHERE THE POLITICAL LEADERSHIP IS MODERATELY ELITIST OR NON-ELITIST (67)	0.82	17 7 32 33 X SQ= 2.49 P = 0.093 RV NO (YES)
164	TEND TO BE THOSE WHERE THE REGIME'S LEADERSHIP CHARISMA IS NEGLIGIBLE (65)	0.94	164	TEND TO BE THOSE WHERE THE REGIME'S LEADERSHIP CHARISMA IS PRONOUNCED OR MODERATE (34)	0.67	3 29 48 14 X SQ= 36.68 P = 0. RV YES (YES)
168	TEND TO BE THOSE WHERE THE HORIZONTAL POWER DISTRIBUTION IS SIGNIFICANT (34)	0.51	168	TEND TO BE THOSE WHERE THE HORIZONTAL POWER DISTRIBUTION IS LIMITED OR NEGLIGIBLE (72)	0.85	25 7 24 41 X SQ= 12.96 P = 0. RV YES (YES)
169	TEND TO BE THOSE WHERE THE HORIZONTAL POWER DISTRIBUTION IS SIGNIFICANT OR LIMITED (58)	0.73	169	TEND TO BE THOSE WHERE THE HORIZONTAL POWER DISTRIBUTION IS NEGLIGIBLE (48)	0.58	36 20 13 28 X SQ= 8.79 P = 0.002 RV YES (YES)
175	TEND TO BE THOSE WHERE THE LEGISLATURE IS FULLY EFFECTIVE OR PARTIALLY EFFECTIVE (51)	0.70	175	TEND TO BE THOSE WHERE THE LEGISLATURE IS LARGELY INEFFECTIVE OR WHOLLY INEFFECTIVE (49)	0.64	33 16 14 29 X SQ= 9.74 P = 0.002 RV YES (YES)
178	TEND TO BE THOSE WHERE THE LEGISLATURE IS BICAMERAL (51)	0.62	178	TEND TO BE THOSE WHERE THE LEGISLATURE IS UNICAMERAL (53)	0.70	19 32 31 14 X SQ= 8.36 P = 0.002 RV YES (YES)
179	TEND TO BE THOSE WHERE THE EXECUTIVE IS STRONG (39)	0.57	179	TEND TO BE THOSE WHERE THE EXECUTIVE IS DOMINANT (52)	0.69	19 27 25 12 X SQ= 4.67 P = 0.027 RV YES (YES)

No.	Statement	Value	No.	Statement	Value	Statistics
181	TEND LESS TO BE THOSE WHERE THE BUREAUCRACY IS SEMI-MODERN, RATHER THAN MODERN (55)	0.61	181	TEND MORE TO BE THOSE WHERE THE BUREAUCRACY IS SEMI-MODERN, RATHER THAN MODERN (55)	0.95	20 1 31 20 X SQ= 6.96 P = 0.004 RV NO (NO)
182	ALWAYS ARE THOSE WHERE THE BUREAUCRACY IS SEMI-MODERN, RATHER THAN POST-COLONIAL TRANSITIONAL (55)	1.00	182	TEND TO BE THOSE WHERE THE BUREAUCRACY IS POST-COLONIAL TRANSITIONAL, RATHER THAN SEMI-MODERN (25)	0.56	31 20 0 25 X SQ= 23.21 P = 0. RV YES (YES)
187	TEND LESS TO BE THOSE WHERE THE ROLE OF THE POLICE IS POLITICALLY SIGNIFICANT (66)	0.51	187	TEND MORE TO BE THOSE WHERE THE ROLE OF THE POLICE IS POLITICALLY SIGNIFICANT (66)	0.74	23 35 22 12 X SQ= 4.43 P = 0.030 RV NO (YES)
188	TEND TO BE THOSE WHERE THE CHARACTER OF THE LEGAL SYSTEM IS CIVIL LAW (32)	0.59	188	ALWAYS ARE THOSE WHERE THE CHARACTER OF THE LEGAL SYSTEM IS OTHER THAN CIVIL LAW (81)	1.00	30 0 21 54 X SQ= 41.63 P = 0. RV YES (YES)
190	TEND TO BE THOSE WHERE THE CHARACTER OF THE LEGAL SYSTEM IS CIVIL LAW, RATHER THAN COMMON LAW (32)	0.86	190	ALWAYS ARE THOSE WHERE THE CHARACTER OF THE LEGAL SYSTEM IS COMMON LAW, RATHER THAN CIVIL LAW (7)	1.00	30 0 5 2 X SQ= 4.33 P = 0.032 RV YES (NO)
192	ALWAYS ARE THOSE WHERE THE CHARACTER OF THE LEGAL SYSTEM IS OTHER THAN MUSLIM OR PARTLY MUSLIM (86)	1.00	192	TEND LESS TO BE THOSE WHERE THE CHARACTER OF THE LEGAL SYSTEM IS OTHER THAN MUSLIM OR PARTLY MUSLIM (86)	0.52	0 26 51 28 X SQ= 30.10 P = 0. RV NO (YES)
193	TEND MORE TO BE THOSE WHERE THE CHARACTER OF THE LEGAL SYSTEM IS OTHER THAN PARTLY INDIGENOUS (86)	0.98	193	TEND LESS TO BE THOSE WHERE THE CHARACTER OF THE LEGAL SYSTEM IS OTHER THAN PARTLY INDIGENOUS (86)	0.52	1 26 50 28 X SQ= 26.92 P = 0. RV NO (YES)
194	TILT LESS TOWARD BEING THOSE THAT ARE NON-COMMUNIST (101)	0.84	194	TILT MORE TOWARD BEING THOSE THAT ARE NON-COMMUNIST (101)	0.95	8 3 42 52 X SQ= 2.08 P = 0.111 RV NO (YES)

83 POLITIES
WHERE THE TYPE OF POLITICAL MODERNIZATION
IS NON-EUROPEAN AUTOCHTHONOUS (9)

83 POLITIES
WHERE THE TYPE OF POLITICAL MODERNIZATION
IS OTHER THAN
NON-EUROPEAN AUTOCHTHONOUS (106)

BOTH SUBJECT AND PREDICATE

9 IN LEFT COLUMN

AFGHANISTAN CHINA, PR ETHIOPIA IRAN JAPAN NEPAL THAILAND TURKEY USSR

106 IN RIGHT COLUMN

ALBANIA	ALGERIA	ARGENTINA	AUSTRALIA	AUSTRIA	BELGIUM	BOLIVIA	BRAZIL	BULGARIA	BURMA
BURUNDI	CAMBODIA	CAMEROUN	CANADA	CEN AFR REP	CEYLON	CHAD	CHILE	COLOMBIA	CONGO(BRA)
CONGO(LEO)	COSTA RICA	CUBA	CYPRUS	CZECHOS'KIA	DAHOMEY	DENMARK	DOMIN REP	ECUADOR	EL SALVADOR
FINLAND	FRANCE	GABON	GERMANY, E	GERMAN FR	GHANA	GREECE	GUATEMALA	GUINEA	HAITI
HONDURAS	HUNGARY	ICELAND	INDIA	INDONESIA	IRAQ	IRELAND	ISRAEL	ITALY	IVORY COAST
JAMAICA	JORDAN	KOREA, N	KOREA REP	LAOS	LEBANON	LIBERIA	LIBYA	LUXEMBOURG	MALAGASY R
MALAYA	MALI	MAURITANIA	MEXICO	MONGOLIA	MOROCCO	NETHERLANDS	NEW ZEALAND	NICARAGUA	NIGER
NIGERIA	NORWAY	PAKISTAN	PANAMA	PARAGUAY	PERU	PHILIPPINES	POLAND	PORTUGAL	RUMANIA
RWANDA	SA'U ARABIA	SENEGAL	SIERRE LEO	SOMALIA	SO AFRICA	SPAIN	SUDAN	SWEDEN	SWITZERLAND
SYRIA	TANGANYIKA	TOGO	TRINIDAD	TUNISIA	UGANDA	UAR	UK	US	UPPER VOLTA
URUGUAY	VENEZUELA	VIETNAM, N	VIETNAM REP	YEMEN	YUGOSLAVIA				

---

15 TEND TO BE THOSE 0.75
LOCATED IN THE MIDDLE EAST, RATHER THAN
IN NORTH AFRICA OR
CENTRAL AND SOUTH AFRICA (11)

15 TEND TO BE THOSE 0.80
LOCATED IN NORTH AFRICA OR
CENTRAL AND SOUTH AFRICA, RATHER THAN
IN THE MIDDLE EAST (33)

3 8
1 32
X SQ= 3.30
P = 0.043
RV YES (NO)

17 ALWAYS ARE THOSE 1.00
LOCATED IN THE MIDDLE EAST, RATHER THAN
IN THE CARIBBEAN, CENTRAL AMERICA, OR
SOUTH AMERICA (11)

17 TEND TO BE THOSE 0.73
LOCATED IN THE CARIBBEAN,
CENTRAL AMERICA, OR SOUTH AMERICA,
RATHER THAN IN THE MIDDLE EAST (22)

3 8
0 22
X SQ= 3.71
P = 0.030
RV YES (NO)

18 TEND TO BE THOSE 0.80
LOCATED IN EAST ASIA, SOUTH ASIA, OR
SOUTHEAST ASIA, RATHER THAN IN
NORTH AFRICA OR
CENTRAL AND SOUTH AFRICA (18)

18 TEND TO BE THOSE 0.70
LOCATED IN NORTH AFRICA OR
CENTRAL AND SOUTH AFRICA, RATHER THAN
IN EAST ASIA, SOUTH ASIA, OR
SOUTHEAST ASIA (33)

1 32
4 14
X SQ= 2.92
P = 0.047
RV YES (NO)

20 ALWAYS ARE THOSE 1.00
LOCATED IN EAST ASIA, SOUTH ASIA, OR
SOUTHEAST ASIA, RATHER THAN IN
THE CARIBBEAN, CENTRAL AMERICA,
OR SOUTH AMERICA (18)

20 TEND TO BE THOSE 0.61
LOCATED IN THE CARIBBEAN,
CENTRAL AMERICA, OR SOUTH AMERICA,
RATHER THAN IN EAST ASIA, SOUTH ASIA,
OR SOUTHEAST ASIA (22)

4 14
0 22
X SQ= 3.24
P = 0.033
RV YES (NO)

83/

```
21  TILT TOWARD BEING THOSE              0.56      21  TILT TOWARD BEING THOSE             0.75        5   27
    WHOSE TERRITORIAL SIZE IS                          WHOSE TERRITORIAL SIZE IS                       4   79
    VERY LARGE OR LARGE  (32)                          MEDIUM OR SMALL  (83)                     X SQ=    2.39
                                                                                                 P =     0.113
                                                                                                 RV YES (NO )

23  TEND TO BE THOSE                     0.67      23  TEND TO BE THOSE                    0.80        6   21
    WHOSE POPULATION IS                                WHOSE POPULATION IS                             3   85
    VERY LARGE OR LARGE  (27)                          MEDIUM OR SMALL  (88)                     X SQ=    7.70
                                                                                                 P =     0.005
                                                                                                 RV YES (NO )

24  ALWAYS ARE THOSE                     1.00      24  TEND TO BE THOSE                    0.51        9   52
    WHOSE POPULATION IS                                WHOSE POPULATION IS                             0   54
    VERY LARGE, LARGE, OR MEDIUM  (61)                 SMALL  (54)                               X SQ=    6.72
                                                                                                 P =     0.003
                                                                                                 RV YES (NO )

30  LEAN TOWARD BEING THOSE              0.78      30  LEAN TOWARD BEING THOSE             0.53        7   49
    WHOSE AGRICULTURAL POPULATION                      WHOSE AGRICULTURAL POPULATION                   2   55
    IS HIGH  (56)                                      IS MEDIUM, LOW, OR VERY LOW  (57)         X SQ=    2.01
                                                                                                 P =     0.094
                                                                                                 RV YES (NO )

51  TEND TO BE THOSE                     0.78      51  TEND TO BE THOSE                    0.66        2   58
    WHERE FREEDOM OF THE PRESS IS                      WHERE FREEDOM OF THE PRESS IS                   7   30
    INTERNALLY ABSENT, OR                              COMPLETE OR INTERMITTENT  (60)            X SQ=    4.88
    INTERNALLY AND EXTERNALLY ABSENT  (37)                                                       P =     0.025
                                                                                                 RV YES (NO )

63  ALWAYS ARE THOSE                     1.00      63  TEND LESS TO BE THOSE               0.57        0   46
    WHERE THE RELIGION IS                              WHERE THE RELIGION IS                           8   60
    PREDOMINANTLY OR PARTLY                            PREDOMINANTLY OR PARTLY                   X SQ=    4.16
    OTHER THAN CHRISTIAN  (68)                         OTHER THAN CHRISTIAN  (68)                P =     0.021
                                                                                                 RV NO  (NO )

68  LEAN MORE TOWARD BEING THOSE         0.87      68  LEAN LESS TOWARD BEING THOSE        0.52        1   51
    THAT ARE LINGUISTICALLY                            THAT ARE LINGUISTICALLY                         7   55
    WEAKLY HETEROGENEOUS OR                            WEAKLY HETEROGENEOUS OR                   X SQ=    2.50
    STRONGLY HETEROGENEOUS  (62)                       STRONGLY HETEROGENEOUS  (62)              P =     0.069
                                                                                                 RV NO  (NO )

71  ALWAYS ARE THOSE                     1.00      71  TEND TO BE THOSE                    0.88        9   12
    WHOSE DATE OF INDEPENDENCE                         WHOSE DATE OF INDEPENDENCE                      0   90
    IS BEFORE 1800  (21)                               IS AFTER 1800  (90)                       X SQ=   36.42
                                                                                                 P =     0.
                                                                                                 RV YES (NO )

87  ALWAYS ARE THOSE                     1.00      87  LEAN LESS TOWARD BEING THOSE        0.62        0   31
    WHOSE IDEOLOGICAL ORIENTATION                      WHOSE IDEOLOGICAL ORIENTATION                   7   51
    IS OTHER THAN DEVELOPMENTAL  (58)                  IS OTHER THAN DEVELOPMENTAL  (58)         X SQ=    2.57
                                                                                                 P =     0.050
                                                                                                 RV NO  (NO )
```

94	LEAN TOWARD BEING THOSE WHERE THE STATUS OF THE REGIME IS AUTHORITARIAN OR TOTALITARIAN (41)	0.78	94	LEAN TOWARD BEING THOSE WHERE THE STATUS OF THE REGIME IS CONSTITUTIONAL (51)	0.59	2 49 7 34 X SQ= 3.09 P = 0.073 RV YES (NO)
96	TEND TO BE THOSE WHERE THE STATUS OF THE REGIME IS AUTHORITARIAN (23)	0.56	96	TEND TO BE THOSE WHERE THE STATUS OF THE REGIME IS CONSTITUTIONAL OR TOTALITARIAN (67)	0.78	4 63 5 18 X SQ= 3.14 P = 0.044 RV YES (NO)
102	LEAN TOWARD BEING THOSE WHERE THE REPRESENTATIVE CHARACTER OF THE REGIME IS PSEUDO-POLYARCHIC OR NON-POLYARCHIC (49)	0.78	102	LEAN TOWARD BEING THOSE WHERE THE REPRESENTATIVE CHARACTER OF THE REGIME IS POLYARCHIC OR LIMITED POLYARCHIC (59)	0.58	2 57 7 42 X SQ= 2.86 P = 0.076 RV YES (NO)
108	LEAN TOWARD BEING THOSE WHERE AUTONOMOUS GROUPS ARE TOLERATED ONLY OUTSIDE POLITICS OR ARE NOT TOLERATED AT ALL (35)	0.67	108	LEAN TOWARD BEING THOSE WHERE AUTONOMOUS GROUPS ARE FULLY OR PARTIALLY TOLERATED IN POLITICS (65)	0.68	3 62 6 29 X SQ= 2.96 P = 0.062 RV YES (NO)
119	ALWAYS ARE THOSE WHERE INTEREST ARTICULATION BY INSTITUTIONAL GROUPS IS VERY SIGNIFICANT OR SIGNIFICANT (74)	1.00	119	TILT LESS TOWARD BEING THOSE WHERE INTEREST ARTICULATION BY INSTITUTIONAL GROUPS IS VERY SIGNIFICANT OR SIGNIFICANT (74)	0.71	9 65 0 26 X SQ= 2.15 P = 0.107 RV NO (NO)
122	ALWAYS ARE THOSE WHERE INTEREST ARTICULATION BY NON-ASSOCIATIONAL GROUPS IS SIGNIFICANT OR MODERATE (83)	1.00	122	LEAN LESS TOWARD BEING THOSE WHERE INTEREST ARTICULATION BY NON-ASSOCIATIONAL GROUPS IS SIGNIFICANT OR MODERATE (83)	0.70	9 74 0 32 X SQ= 2.41 P = 0.060 RV NO (NO)
125	ALWAYS ARE THOSE WHERE INTEREST ARTICULATION BY ANOMIC GROUPS IS FREQUENT OR OCCASIONAL (64)	1.00	125	TEND LESS TO BE THOSE WHERE INTEREST ARTICULATION BY ANOMIC GROUPS IS FREQUENT OR OCCASIONAL (64)	0.61	9 55 0 35 X SQ= 3.85 P = 0.025 RV NO (NO)
133	ALWAYS ARE THOSE WHERE INTEREST AGGREGATION BY THE EXECUTIVE IS MODERATE, LIMITED, OR NEGLIGIBLE (76)	1.00	133	LEAN LESS TOWARD BEING THOSE WHERE INTEREST AGGREGATION BY THE EXECUTIVE IS MODERATE, LIMITED, OR NEGLIGIBLE (76)	0.70	0 29 9 67 X SQ= 2.40 P = 0.060 RV NO (NO)
134	LEAN TOWARD BEING THOSE WHERE INTEREST AGGREGATION BY THE EXECUTIVE IS LIMITED OR NEGLIGIBLE (46)	0.78	134	LEAN TOWARD BEING THOSE WHERE INTEREST AGGREGATION BY THE EXECUTIVE IS SIGNIFICANT OR MODERATE (57)	0.59	2 55 7 39 X SQ= 3.03 P = 0.074 RV YES (NO)

135	TEND TO BE THOSE WHERE INTEREST AGGREGATION BY THE EXECUTIVE IS NEGLIGIBLE (13)	0.57	135	TEND TO BE THOSE WHERE INTEREST AGGREGATION BY THE EXECUTIVE IS SIGNIFICANT, MODERATE, OR LIMITED (77)	0.89	3 74 4 9 X SQ= 7.76 P = 0.008 RV YES (NO)
160	TEND TO BE THOSE WHERE THE POLITICAL LEADERSHIP IS ELITIST (30)	0.75	160	TEND TO BE THOSE WHERE THE POLITICAL LEADERSHIP IS MODERATELY ELITIST OR NON-ELITIST (67)	0.73	6 24 2 65 X SQ= 5.84 P = 0.010 RV YES (NO)
169	LEAN TOWARD BEING THOSE WHERE THE HORIZONTAL POWER DISTRIBUTION IS NEGLIGIBLE (48)	0.78	169	LEAN TOWARD BEING THOSE WHERE THE HORIZONTAL POWER DISTRIBUTION IS SIGNIFICANT OR LIMITED (58)	0.58	2 56 7 41 X SQ= 2.88 P = 0.075 RV YES (NO)
170	ALWAYS ARE THOSE WHERE THE LEGISLATIVE-EXECUTIVE STRUCTURE IS OTHER THAN PRESIDENTIAL (63)	1.00	170	TEND LESS TO BE THOSE WHERE THE LEGISLATIVE-EXECUTIVE STRUCTURE IS OTHER THAN PRESIDENTIAL (63)	0.58	0 39 9 54 X SQ= 4.46 P = 0.012 RV NO (NO)
176	TEND TO BE THOSE WHERE THE LEGISLATURE IS WHOLLY INEFFECTIVE (28)	0.75	176	TEND TO BE THOSE WHERE THE LEGISLATURE IS FULLY EFFECTIVE, PARTIALLY EFFECTIVE, OR LARGELY INEFFECTIVE (72)	0.76	2 70 6 22 X SQ= 7.16 P = 0.006 RV YES (NO)
183	ALWAYS ARE THOSE WHERE THE BUREAUCRACY IS TRADITIONAL, RATHER THAN POST-COLONIAL TRANSITIONAL (9)	1.00	183	TEND TO BE THOSE WHERE THE BUREAUCRACY IS POST-COLONIAL TRANSITIONAL, RATHER THAN TRADITIONAL (25)	0.86	0 25 5 4 X SQ= 12.16 P = 0. RV YES (YES)
186	TEND TO BE THOSE WHERE PARTICIPATION BY THE MILITARY IN POLITICS IS SUPPORTIVE, RATHER THAN NEUTRAL (31)	0.83	186	TEND TO BE THOSE WHERE PARTICIPATION BY THE MILITARY IN POLITICS IS NEUTRAL, RATHER THAN SUPPORTIVE (56)	0.68	5 26 1 55 X SQ= 4.35 P = 0.020 RV YES (NO)

84 POLITIES
WHERE THE TYPE OF POLITICAL MODERNIZATION
IS DEVELOPED TUTELARY, RATHER THAN
UNDEVELOPED TUTELARY (31)

84 POLITIES
WHERE THE TYPE OF POLITICAL MODERNIZATION
IS UNDEVELOPED TUTELARY, RATHER THAN
DEVELOPED TUTELARY (24)

BOTH SUBJECT AND PREDICATE

31 IN LEFT COLUMN

ALGERIA	BURMA	CAMBODIA	CEYLON	CYPRUS	INDIA	INDONESIA	IRAQ	ISRAEL	JAMAICA
JORDAN	KOREA, N	KOREA REP	LAOS	LEBANON	LIBERIA	LIBYA	MALAYA	MONGOLIA	MOROCCO
PAKISTAN	PHILIPPINES	SA'U ARABIA	SUDAN	SYRIA	TRINIDAD	TUNISIA	UAR	VIETNAM, N	VIETNAM REP
YEMEN									

24 IN RIGHT COLUMN

BURUNDI	CAMEROUN	CEN AFR REP	CHAD	CONGO(BRA)	CONGO(LEO)	DAHOMEY	GABON	GHANA	GUINEA
IVORY COAST	MALAGASY R	MALI	MAURITANIA	NIGER	NIGERIA	RWANDA	SENEGAL	SIERRE LEO	SOMALIA
TANGANYIKA	TOGO	UGANDA	UPPER VOLTA						

60 EXCLUDED BECAUSE IRRELEVANT

AFGHANISTAN	ALBANIA	ARGENTINA	AUSTRALIA	AUSTRIA	BELGIUM	BOLIVIA	BRAZIL	BULGARIA	CANADA
CHILE	CHINA, PR	COLOMBIA	COSTA RICA	CUBA	CZECHOS'KIA	DENMARK	DOMIN REP	ECUADOR	EL SALVADOR
ETHIOPIA	FINLAND	FRANCE	GERMANY, E	GERMAN FR	GREECE	GUATEMALA	HAITI	HONDURAS	HUNGARY
ICELAND	IRAN	IRELAND	ITALY	JAPAN	LUXEMBOURG	MEXICO	NEPAL	NETHERLANDS	NEW ZEALAND
NICARAGUA	NORWAY	PANAMA	PARAGUAY	PERU	POLAND	PORTUGAL	RUMANIA	SO AFRICA	SPAIN
SWEDEN	SWITZERLAND	THAILAND	TURKEY	USSR	UK	US	URUGUAY	VENEZUELA	YUGOSLAVIA

8	TEND LESS TO BE THOSE LOCATED ELSEWHERE THAN IN EAST ASIA SOUTH ASIA, OR SOUTHEAST ASIA (97)	0.55	8	ALWAYS ARE THOSE LOCATED ELSEWHERE THAN IN EAST ASIA SOUTH ASIA, OR SOUTHEAST ASIA (97)	1.00	14 0 17 24 X SQ= 12.26 P = 0. RV NO (YES)
10	TEND TO BE THOSE LOCATED ELSEWHERE THAN IN NORTH AFRICA, OR CENTRAL AND SOUTH AFRICA (82)	0.77	10	ALWAYS ARE THOSE LOCATED IN NORTH AFRICA, OR CENTRAL AND SOUTH AFRICA (33)	1.00	7 24 24 0 X SQ= 29.89 P = 0. RV YES (YES)
13	TEND LESS TO BE THOSE LOCATED ELSEWHERE THAN IN NORTH AFRICA OR THE MIDDLE EAST (99)	0.58	13	ALWAYS ARE THOSE LOCATED ELSEWHERE THAN IN NORTH AFRICA OR THE MIDDLE EAST (99)	1.00	13 0 18 24 X SQ= 10.96 P = 0. RV NO (YES)

15	TEND TO BE THOSE LOCATED IN THE MIDDLE EAST, RATHER THAN IN NORTH AFRICA OR CENTRAL AND SOUTH AFRICA (11)	0.53	15	ALWAYS ARE THOSE LOCATED IN NORTH AFRICA OR CENTRAL AND SOUTH AFRICA, RATHER THAN IN THE MIDDLE EAST (33)	1.00	8 0 7 24 X SQ= 13.00 P = 0. RV YES (YES)
18	TEND TO BE THOSE LOCATED IN EAST ASIA, SOUTH ASIA, OR SOUTHEAST ASIA, RATHER THAN IN NORTH AFRICA OR CENTRAL AND SOUTH AFRICA (18)	0.67	18	ALWAYS ARE THOSE LOCATED IN NORTH AFRICA OR CENTRAL AND SOUTH AFRICA, RATHER THAN IN EAST ASIA, SOUTH ASIA, OR SOUTHEAST ASIA (33)	1.00	7 24 14 0 X SQ= 20.22 P = 0. RV YES (YES)
19	LEAN LESS TOWARD BEING THOSE LOCATED IN NORTH AFRICA OR CENTRAL AND SOUTH AFRICA, RATHER THAN IN THE CARIBBEAN, CENTRAL AMERICA, OR SOUTH AMERICA (33)	0.78	19	ALWAYS ARE THOSE LOCATED IN NORTH AFRICA OR CENTRAL AND SOUTH AFRICA, RATHER THAN IN THE CARIBBEAN, CENTRAL AMERICA, OR SOUTH AMERICA (33)	1.00	7 24 2 0 X SQ= 2.45 P = 0.068 RV NO (YES)
22	TEND TO BE THOSE WHOSE TERRITORIAL SIZE IS SMALL (47)	0.52	22	TEND TO BE THOSE WHOSE TERRITORIAL SIZE IS VERY LARGE, LARGE, OR MEDIUM (68)	0.79	15 19 16 5 X SQ= 4.20 P = 0.027 RV YES (YES)
24	TEND TO BE THOSE WHOSE POPULATION IS VERY LARGE, LARGE, OR MEDIUM (61)	0.55	24	TEND TO BE THOSE WHOSE POPULATION IS SMALL (54)	0.79	17 5 14 19 X SQ= 5.18 P = 0.014 RV YES (YES)
25	TEND LESS TO BE THOSE WHOSE POPULATION DENSITY IS MEDIUM OR LOW (98)	0.81	25	ALWAYS ARE THOSE WHOSE POPULATION DENSITY IS MEDIUM OR LOW (98)	1.00	6 0 25 24 X SQ= 3.41 P = 0.030 RV NO (NO)
26	TEND LESS TO BE THOSE WHOSE POPULATION DENSITY IS LOW (67)	0.52	26	TEND MORE TO BE THOSE WHOSE POPULATION DENSITY IS LOW (67)	0.87	15 3 16 21 X SQ= 6.37 P = 0.008 RV NO (YES)
28	LEAN TOWARD BEING THOSE WHOSE POPULATION GROWTH RATE IS HIGH (62)	0.77	28	LEAN TOWARD BEING THOSE WHOSE POPULATION GROWTH RATE IS LOW (48)	0.50	23 11 7 11 X SQ= 2.90 P = 0.076 RV YES (YES)
29	TEND LESS TO BE THOSE WHERE THE DEGREE OF URBANIZATION IS LOW (49)	0.56	29	ALWAYS ARE THOSE WHERE THE DEGREE OF URBANIZATION IS LOW (49)	1.00	11 0 14 24 X SQ= 11.21 P = 0. RV NO (YES)

No.	Statement	Value	No.	Statement	Value	Statistics
30	TEND LESS TO BE THOSE WHOSE AGRICULTURAL POPULATION IS HIGH (56)	0.62	30	TEND MORE TO BE THOSE WHOSE AGRICULTURAL POPULATION IS HIGH (56)	0.96	18 23 11 1 X SQ= 6.73 P = 0.007 RV NO (YES)
34	TEND TO BE THOSE WHOSE GROSS NATIONAL PRODUCT IS VERY HIGH, HIGH, MEDIUM, OR LOW (62)	0.52	34	TEND TO BE THOSE WHOSE GROSS NATIONAL PRODUCT IS VERY LOW (53)	0.92	16 2 15 22 X SQ= 9.63 P = 0.001 RV YES (YES)
36	TEND LESS TO BE THOSE WHOSE PER CAPITA GROSS NATIONAL PRODUCT IS LOW OR VERY LOW (73)	0.81	36	ALWAYS ARE THOSE WHOSE PER CAPITA GROSS NATIONAL PRODUCT IS LOW OR VERY LOW (73)	1.00	6 0 25 24 X SQ= 3.41 P = 0.030 RV NO (NO)
37	TEND LESS TO BE THOSE WHOSE PER CAPITA GROSS NATIONAL PRODUCT IS VERY LOW (51)	0.58	37	ALWAYS ARE THOSE WHOSE PER CAPITA GROSS NATIONAL PRODUCT IS VERY LOW (51)	1.00	13 0 18 24 X SQ= 10.96 P = 0. RV NO (YES)
39	LEAN LESS TOWARD BEING THOSE WHOSE INTERNATIONAL FINANCIAL STATUS IS LOW OR VERY LOW (76)	0.84	39	ALWAYS ARE THOSE WHOSE INTERNATIONAL FINANCIAL STATUS IS LOW OR VERY LOW (76)	1.00	5 0 26 24 X SQ= 2.53 P = 0.061 RV NO (NO)
40	TEND TO BE THOSE WHOSE INTERNATIONAL FINANCIAL STATUS IS VERY HIGH, HIGH, MEDIUM, OR LOW (71)	0.69	40	TEND TO BE THOSE WHOSE INTERNATIONAL FINANCIAL STATUS IS VERY LOW (39)	0.83	18 4 8 20 X SQ= 11.94 P = 0. RV YES (YES)
43	TEND LESS TO BE THOSE WHOSE ECONOMIC DEVELOPMENTAL STATUS IS VERY UNDERDEVELOPED (57)	0.62	43	ALWAYS ARE THOSE WHOSE ECONOMIC DEVELOPMENTAL STATUS IS VERY UNDERDEVELOPED (57)	1.00	11 0 18 24 X SQ= 9.30 P = 0. RV NO (YES)
45	TEND LESS TO BE THOSE WHERE THE LITERACY RATE IS BELOW FIFTY PERCENT (54)	0.62	45	ALWAYS ARE THOSE WHERE THE LITERACY RATE IS BELOW FIFTY PERCENT (54)	1.00	10 0 16 24 X SQ= 9.26 P = 0.001 RV NO (YES)
46	TEND TO BE THOSE WHERE THE LITERACY RATE IS TEN PERCENT OR ABOVE (84)	0.86	46	TEND TO BE THOSE WHERE THE LITERACY RATE IS BELOW TEN PERCENT (26)	0.86	24 3 4 19 X SQ= 22.95 P = 0. RV YES (YES)

50	TEND TO BE THOSE WHERE FREEDOM OF THE PRESS IS INTERMITTENT, INTERNALLY ABSENT, OR INTERNALLY AND EXTERNALLY ABSENT (56)	0.80	50	TEND TO BE THOSE WHERE FREEDOM OF THE PRESS IS COMPLETE (43)	0.71	6 10 24 4 X SQ= 8.80 P = 0.002 RV YES (YES)
51	TEND TO BE THOSE WHERE FREEDOM OF THE PRESS IS INTERNALLY ABSENT, OR INTERNALLY AND EXTERNALLY ABSENT (37)	0.50	51	TEND TO BE THOSE WHERE FREEDOM OF THE PRESS IS COMPLETE OR INTERMITTENT (60)	0.86	14 12 14 2 X SQ= 3.65 P = 0.042 RV YES (NO)
52	LEAN LESS TOWARD BEING THOSE WHERE FREEDOM OF THE PRESS IS COMPLETE, INTERMITTENT, OR INTERNALLY ABSENT (82)	0.79	52	ALWAYS ARE THOSE WHERE FREEDOM OF THE PRESS IS COMPLETE, INTERMITTENT, OR INTERNALLY ABSENT (82)	1.00	22 14 6 0 X SQ= 1.97 P = 0.083 RV NO (NO)
55	TEND TO BE THOSE WHERE NEWSPAPER CIRCULATION IS TEN OR MORE PER THOUSAND (78)	0.72	55	TEND TO BE THOSE WHERE NEWSPAPER CIRCULATION IS LESS THAN TEN PER THOUSAND (35)	0.96	21 1 8 23 X SQ= 22.46 P = 0. RV YES (YES)
56	TEND TO BE THOSE WHERE THE RELIGION IS PREDOMINANTLY LITERATE (79)	0.89	56	TEND TO BE THOSE WHERE THE RELIGION IS PREDOMINANTLY OR PARTLY NON-LITERATE (31)	0.87	24 3 3 21 X SQ= 26.77 P = 0. RV YES (YES)
58	TEND LESS TO BE THOSE WHERE THE RELIGION IS OTHER THAN MUSLIM (97)	0.61	58	TEND MORE TO BE THOSE WHERE THE RELIGION IS OTHER THAN MUSLIM (97)	0.87	12 3 19 21 X SQ= 3.46 P = 0.037 RV NO (YES)
66	TEND TO BE THOSE THAT ARE RELIGIOUSLY HOMOGENEOUS (57)	0.67	66	TEND TO BE THOSE THAT ARE RELIGIOUSLY HETEROGENEOUS (49)	0.86	18 3 9 18 X SQ= 11.13 P = 0. RV YES (YES)
69	TEND TO BE THOSE THAT ARE LINGUISTICALLY HOMOGENEOUS OR WEAKLY HETEROGENEOUS (64)	0.52	69	TEND TO BE THOSE THAT ARE LINGUISTICALLY STRONGLY HETEROGENEOUS (50)	0.83	16 4 15 20 X SQ= 5.71 P = 0.011 RV YES (YES)
73	TEND LESS TO BE THOSE WHOSE DATE OF INDEPENDENCE IS AFTER 1945 (46)	0.76	73	ALWAYS ARE THOSE WHOSE DATE OF INDEPENDENCE IS AFTER 1945 (46)	1.00	7 0 22 24 X SQ= 4.74 P = 0.012 RV NO (YES)

80	TEND TO BE THOSE WHOSE DATE OF INDEPENDENCE IS AFTER 1914, AND THAT WERE FORMERLY DEPENDENCIES OF BRITAIN, RATHER THAN FRANCE (19)	0.59	80	TEND TO BE THOSE WHOSE DATE OF INDEPENDENCE IS AFTER 1914, AND THAT WERE FORMERLY DEPENDENCIES OF FRANCE, RATHER THAN BRITAIN (24)	0.75	13 5 9 15 X SQ= 3.68 P = 0.033 RV YES (YES)
85	TEND TO BE THOSE WHERE THE STAGE OF POLITICAL MODERNIZATION IS ADVANCED, RATHER THAN MID- OR EARLY TRANSITIONAL (60)	0.57	85	ALWAYS ARE THOSE WHERE THE STAGE OF POLITICAL MODERNIZATION IS MID- OR EARLY TRANSITIONAL, RATHER THAN ADVANCED (54)	1.00	17 0 13 24 X SQ= 17.31 P = 0. RV YES (YES)
86	TEND TO BE THOSE WHERE THE STAGE OF POLITICAL MODERNIZATION IS ADVANCED OR MID-TRANSITIONAL, RATHER THAN EARLY TRANSITIONAL (76)	0.57	86	ALWAYS ARE THOSE WHERE THE STAGE OF POLITICAL MODERNIZATION IS EARLY TRANSITIONAL, RATHER THAN ADVANCED OR MID-TRANSITIONAL (38)	1.00	17 0 13 24 X SQ= 17.31 P = 0. RV YES (YES)
87	TEND TO BE THOSE WHOSE IDEOLOGICAL ORIENTATION IS OTHER THAN DEVELOPMENTAL (58)	0.61	87	ALWAYS ARE THOSE WHOSE IDEOLOGICAL ORIENTATION IS DEVELOPMENTAL (31)	1.00	7 21 11 0 X SQ= 14.98 P = 0. RV YES (YES)
91	LEAN LESS TOWARD BEING THOSE WHOSE IDEOLOGICAL ORIENTATION IS DEVELOPMENTAL, RATHER THAN TRADITIONAL (31)	0.78	91	ALWAYS ARE THOSE WHOSE IDEOLOGICAL ORIENTATION IS DEVELOPMENTAL, RATHER THAN TRADITIONAL (31)	1.00	7 21 2 0 X SQ= 2.07 P = 0.083 RV NO (YES)
101	LEAN LESS TOWARD BEING THOSE WHERE THE REPRESENTATIVE CHARACTER OF THE REGIME IS LIMITED POLYARCHIC, PSEUDO-POLYARCHIC, OR NON-POLYARCHIC (57)	0.60	101	LEAN MORE TOWARD BEING THOSE WHERE THE REPRESENTATIVE CHARACTER OF THE REGIME IS LIMITED POLYARCHIC, PSEUDO-POLYARCHIC, OR NON-POLYARCHIC (57)	0.88	10 2 15 14 X SQ= 2.36 P = 0.084 RV NO (NO)
117	TEND TO BE THOSE WHERE INTEREST ARTICULATION BY ASSOCIATIONAL GROUPS IS SIGNIFICANT, MODERATE, OR LIMITED (60)	0.53	117	ALWAYS ARE THOSE WHERE INTEREST ARTICULATION BY ASSOCIATIONAL GROUPS IS NEGLIGIBLE (51)	1.00	16 0 14 24 X SQ= 15.72 P = 0. RV YES (YES)
118	LEAN TOWARD BEING THOSE WHERE INTEREST ARTICULATION BY INSTITUTIONAL GROUPS IS VERY SIGNIFICANT (40)	0.60	118	LEAN TOWARD BEING THOSE WHERE INTEREST ARTICULATION BY INSTITUTIONAL GROUPS IS SIGNIFICANT, MODERATE, OR LIMITED (60)	0.80	18 2 12 8 X SQ= 3.33 P = 0.065 RV YES (NO)
121	TEND LESS TO BE THOSE WHERE INTEREST ARTICULATION BY NON-ASSOCIATIONAL GROUPS IS SIGNIFICANT (54)	0.71	121	ALWAYS ARE THOSE WHERE INTEREST ARTICULATION BY NON-ASSOCIATIONAL GROUPS IS SIGNIFICANT (54)	1.00	22 24 9 0 X SQ= 6.34 P = 0.003 RV NO (YES)

124	LEAN LESS TOWARD BEING THOSE WHERE INTEREST ARTICULATION BY ANOMIC GROUPS IS OCCASIONAL, INFREQUENT, OR VERY INFREQUENT (85)	0.68	124	LEAN MORE TOWARD BEING THOSE WHERE INTEREST ARTICULATION BY ANOMIC GROUPS IS OCCASIONAL, INFREQUENT, OR VERY INFREQUENT (85)	0.92	8 2 17 22 X SQ= 2.89 P = 0.074 RV NO (YES)
128	TILT TOWARD BEING THOSE WHERE INTEREST ARTICULATION BY POLITICAL PARTIES IS SIGNIFICANT OR MODERATE (48)	0.56	128	TILT TOWARD BEING THOSE WHERE INTEREST ARTICULATION BY POLITICAL PARTIES IS LIMITED OR NEGLIGIBLE (45)	0.71	10 7 8 17 X SQ= 1.98 P = 0.117 RV YES (YES)
132	TEND LESS TO BE THOSE WHERE INTEREST AGGREGATION BY POLITICAL PARTIES IS SIGNIFICANT, MODERATE, OR LIMITED (67)	0.75	132	ALWAYS ARE THOSE WHERE INTEREST AGGREGATION BY POLITICAL PARTIES IS SIGNIFICANT, MODERATE, OR LIMITED (67)	1.00	12 23 4 0 X SQ= 3.98 P = 0.022 RV NO (YES)
133	TEND TO BE THOSE WHERE INTEREST AGGREGATION BY THE EXECUTIVE IS MODERATE, LIMITED, OR NEGLIGIBLE (76)	0.67	133	TEND TO BE THOSE WHERE INTEREST AGGREGATION BY THE EXECUTIVE IS SIGNIFICANT (29)	0.67	8 16 16 8 X SQ= 4.08 P = 0.025 RV YES (YES)
134	TEND TO BE THOSE WHERE INTEREST AGGREGATION BY THE EXECUTIVE IS LIMITED OR NEGLIGIBLE (46)	0.52	134	TEND TO BE THOSE WHERE INTEREST AGGREGATION BY THE EXECUTIVE IS SIGNIFICANT OR MODERATE (57)	0.83	11 19 12 4 X SQ= 4.70 P = 0.029 RV YES (YES)
135	LEAN LESS TOWARD BEING THOSE WHERE INTEREST AGGREGATION BY THE EXECUTIVE IS SIGNIFICANT, MODERATE, OR LIMITED (77)	0.84	135	ALWAYS ARE THOSE WHERE INTEREST AGGREGATION BY THE EXECUTIVE IS SIGNIFICANT, MODERATE, OR LIMITED (77)	1.00	16 22 3 0 X SQ= 1.78 P = 0.091 RV NO (YES)
137	TEND LESS TO BE THOSE WHERE INTEREST AGGREGATION BY THE LEGISLATURE IS LIMITED OR NEGLIGIBLE (68)	0.68	137	ALWAYS ARE THOSE WHERE INTEREST AGGREGATION BY THE LEGISLATURE IS LIMITED OR NEGLIGIBLE (68)	1.00	6 0 13 24 X SQ= 6.37 P = 0.004 RV NO (YES)
138	LEAN LESS TOWARD BEING THOSE WHERE INTEREST AGGREGATION BY THE LEGISLATURE IS NEGLIGIBLE (49)	0.58	138	LEAN MORE TOWARD BEING THOSE WHERE INTEREST AGGREGATION BY THE LEGISLATURE IS NEGLIGIBLE (49)	0.83	8 4 11 20 X SQ= 2.26 P = 0.091 RV NO (YES)
139	LEAN TOWARD BEING THOSE WHERE THE PARTY SYSTEM IS QUANTITATIVELY OTHER THAN ONE-PARTY (71)	0.69	139	LEAN TOWARD BEING THOSE WHERE THE PARTY SYSTEM IS QUANTITATIVELY ONE-PARTY (34)	0.57	8 13 18 10 X SQ= 2.34 P = 0.088 RV YES (YES)

147	LEAN LESS TOWARD BEING THOSE WHERE THE PARTY SYSTEM IS QUALITATIVELY OTHER THAN CLASS-ORIENTED OR MULTI-IDEOLOGICAL (67)	0.82	147	ALWAYS ARE THOSE WHERE THE PARTY SYSTEM IS QUALITATIVELY OTHER THAN CLASS-ORIENTED OR MULTI-IDEOLOGICAL (67)	1.00	3 0 14 24 X SQ= 2.34 P = 0.064 RV NO (YES)
148	ALWAYS ARE THOSE WHERE THE PARTY SYSTEM IS QUALITATIVELY OTHER THAN AFRICAN TRANSITIONAL (96)	1.00	148	TEND TO BE THOSE WHERE THE PARTY SYSTEM IS QUALITATIVELY AFRICAN TRANSITIONAL (14)	0.58	0 14 29 10 X SQ= 20.09 P = 0. RV YES (YES)
157	TEND LESS TO BE THOSE WHERE PERSONALISMO IS MODERATE OR NEGLIGIBLE (84)	0.76	157	ALWAYS ARE THOSE WHERE PERSONALISMO IS MODERATE OR NEGLIGIBLE (84)	1.00	5 0 16 23 X SQ= 4.04 P = 0.019 RV NO (YES)
160	TEND LESS TO BE THOSE WHERE THE POLITICAL LEADERSHIP IS MODERATELY ELITIST OR NON-ELITIST (67)	0.63	160	ALWAYS ARE THOSE WHERE THE POLITICAL LEADERSHIP IS MODERATELY ELITIST OR NON-ELITIST (67)	1.00	7 0 12 21 X SQ= 7.00 P = 0.003 RV NO (YES)
161	LEAN TOWARD BEING THOSE WHERE THE POLITICAL LEADERSHIP IS ELITIST OR MODERATELY ELITIST (47)	0.58	161	LEAN TOWARD BEING THOSE WHERE THE POLITICAL LEADERSHIP IS NON-ELITIST (50)	0.76	11 5 8 16 X SQ= 3.51 P = 0.051 RV YES (YES)
164	LEAN LESS TOWARD BEING THOSE WHERE THE REGIME'S LEADERSHIP CHARISMA IS PRONOUNCED OR MODERATE (34)	0.56	164	LEAN MORE TOWARD BEING THOSE WHERE THE REGIME'S LEADERSHIP CHARISMA IS PRONOUNCED OR MODERATE (34)	0.83	14 15 11 3 X SQ= 2.42 P = 0.099 RV NO (YES)
168	TEND LESS TO BE THOSE WHERE THE HORIZONTAL POWER DISTRIBUTION IS LIMITED OR NEGLIGIBLE (72)	0.73	168	ALWAYS ARE THOSE WHERE THE HORIZONTAL POWER DISTRIBUTION IS LIMITED OR NEGLIGIBLE (72)	1.00	7 0 19 22 X SQ= 4.94 P = 0.011 RV NO (YES)
170	TEND TO BE THOSE WHERE THE LEGISLATIVE-EXECUTIVE STRUCTURE IS OTHER THAN PRESIDENTIAL (63)	0.68	170	TEND TO BE THOSE WHERE THE LEGISLATIVE-EXECUTIVE STRUCTURE IS PRESIDENTIAL (39)	0.64	7 14 15 8 X SQ= 3.28 P = 0.042 RV YES (YES)
174	TEND LESS TO BE THOSE WHERE THE LEGISLATURE IS PARTIALLY EFFECTIVE, LARGELY INEFFECTIVE, OR WHOLLY INEFFECTIVE (72)	0.70	174	ALWAYS ARE THOSE WHERE THE LEGISLATURE IS PARTIALLY EFFECTIVE, LARGELY INEFFECTIVE, OR WHOLLY INEFFECTIVE (72)	1.00	7 0 16 22 X SQ= 5.78 P = 0.009 RV NO (YES)

175	TILT LESS TOWARD BEING THOSE WHERE THE LEGISLATURE IS LARGELY INEFFECTIVE OR WHOLLY INEFFECTIVE (49)	0.52	175	TILT MORE TOWARD BEING THOSE WHERE THE LEGISLATURE IS LARGELY INEFFECTIVE OR WHOLLY INEFFECTIVE (49)	0.77	11 5 12 17 X SQ= 2.09 P = 0.120 RV NO (YES)
178	TEND TO BE THOSE WHERE THE LEGISLATURE IS BICAMERAL (51)	0.55	178	TEND TO BE THOSE WHERE THE LEGISLATURE IS UNICAMERAL (53)	0.92	10 22 12 2 X SQ= 9.50 P = 0.001 RV YES (YES)
182	TEND TO BE THOSE WHERE THE BUREAUCRACY IS SEMI-MODERN, RATHER THAN POST-COLONIAL TRANSITIONAL (55)	0.86	182	TEND TO BE THOSE WHERE THE BUREAUCRACY IS POST-COLONIAL TRANSITIONAL, RATHER THAN SEMI-MODERN (25)	0.92	18 2 3 22 X ´SQ= 24.12 P = 0. RV YES (YES)
183	TEND TO BE THOSE WHERE THE BUREAUCRACY IS TRADITIONAL, RATHER THAN POST-COLONIAL TRANSITIONAL (9)	0.57	183	ALWAYS ARE THOSE WHERE THE BUREAUCRACY IS POST-COLONIAL TRANSITIONAL, RATHER THAN TRADITIONAL (25)	1.00	3 22 4 0 X SQ= 10.17 P = 0.001 RV YES (YES)
186	TEND TO BE THOSE WHERE PARTICIPATION BY THE MILITARY IN POLITICS IS SUPPORTIVE, RATHER THAN NEUTRAL (31)	0.50	186	TEND TO BE THOSE WHERE PARTICIPATION BY THE MILITARY IN POLITICS IS NEUTRAL, RATHER THAN SUPPORTIVE (56)	0.86	11 3 11 18 X SQ= 4.72 P = 0.022 RV YES (YES)
193	TEND TO BE THOSE WHERE THE CHARACTER OF THE LEGAL SYSTEM IS OTHER THAN PARTLY INDIGENOUS (86)	0.87	193	TEND TO BE THOSE WHERE THE CHARACTER OF THE LEGAL SYSTEM IS PARTLY INDIGENOUS (28)	0.92	4 22 26 2 X SQ= 29.71 P = 0. RV YES (YES)

```
****************************************************************************************************

85  POLITIES                                          85  POLITIES
    WHERE THE STAGE OF                                    WHERE THE STAGE OF
    POLITICAL MODERNIZATION IS                            POLITICAL MODERNIZATION IS
    ADVANCED, RATHER THAN                                 MID- OR EARLY TRANSITIONAL,
    MID- OR EARLY TRANSITIONAL  (60)                      RATHER THAN ADVANCED  (54)

                                                                      BOTH SUBJECT AND PREDICATE
....................................................................................................

     60 IN LEFT COLUMN

ALBANIA     ALGERIA     ARGENTINA   AUSTRALIA   AUSTRIA     BELGIUM     BRAZIL      BULGARIA    CANADA      CHILE
CHINA, PR   COSTA RICA  CUBA        CYPRUS      CZECHOS'KIA DENMARK     FINLAND     FRANCE      GERMANY, E  GERMAN FR
GREECE      GUATEMALA   HUNGARY     ICELAND     INDIA       IRAQ        IRELAND     ISRAEL      ITALY       JAPAN
JORDAN      KOREA, N    KOREA REP   LEBANON     LUXEMBOURG  MEXICO      MONGOLIA    MOROCCO     NETHERLANDS NEW ZEALAND
NORWAY      PAKISTAN    PHILIPPINES POLAND      PORTUGAL    RUMANIA     SPAIN       SUDAN       SWEDEN      SWITZERLAND
SYRIA       TUNISIA     TURKEY      USSR        UAR         UK          US          URUGUAY     VENEZUELA   YUGOSLAVIA

     54 IN RIGHT COLUMN

AFGHANISTAN BOLIVIA     BURMA       BURUNDI     CAMBODIA    CAMEROUN    CEN AFR REP CEYLON      CHAD        COLOMBIA
CONGO(BRA)  CONGO(LEO)  DAHOMEY     DOMIN REP   ECUADOR     EL SALVADOR ETHIOPIA    GABON       GHANA       GUINEA
HAITI       HONDURAS    INDONESIA   IRAN        IVORY COAST JAMAICA     LAOS        LIBERIA     LIBYA       MALAGASY R
MALAYA      MALI        MAURITANIA  NEPAL       NICARAGUA   NIGER       NIGERIA     PANAMA      PARAGUAY    PERU
RWANDA      SENEGAL     SIERRE LEO  SOMALIA     SO AFRICA   TANGANYIKA  THAILAND    TOGO        TRINIDAD    UGANDA
UPPER VOLTA VIETNAM, N  VIETNAM REP YEMEN

      1 EXCLUDED BECAUSE IRRELEVANT

SA'U ARABIA

----------------------------------------------------------------------------------------------------

  2  TEND LESS TO BE THOSE                    0.63       2  ALWAYS ARE THOSE                      1.00        22     0
     LOCATED ELSEWHERE THAN IN                              LOCATED ELSEWHERE THAN IN                        38    54
     WEST EUROPE, SCANDINAVIA,                              WEST EUROPE, SCANDINAVIA,                    X SQ=  22.24
     NORTH AMERICA, OR AUSTRALASIA  (93)                    NORTH AMERICA, OR AUSTRALASIA  (93)          P =    0.
                                                                                                         RV NO  (YES)

  9  TEND MORE TO BE THOSE                    0.98       9  TEND LESS TO BE THOSE                 0.85         1     8
     LOCATED ELSEWHERE THAN IN                              LOCATED ELSEWHERE THAN IN                        59    46
     SOUTHEAST ASIA  (106)                                  SOUTHEAST ASIA  (106)                        X SQ=   5.07
                                                                                                         P =    0.013
                                                                                                         RV NO  (YES)

 10  TEND TO BE THOSE                         0.92      10  TEND TO BE THOSE                      0.52         5    28
     LOCATED ELSEWHERE THAN IN NORTH AFRICA,                LOCATED IN NORTH AFRICA, OR                      55    26
     OR CENTRAL AND SOUTH AFRICA  (82)                      CENTRAL AND SOUTH AFRICA  (33)               X SQ=  24.10
                                                                                                         P =    0.
                                                                                                         RV YES (YES)
```

13	TILT LESS TOWARD BEING THOSE LOCATED ELSEWHERE THAN IN NORTH AFRICA OR THE MIDDLE EAST (99)	0.82	13	TILT MORE TOWARD BEING THOSE LOCATED ELSEWHERE THAN IN NORTH AFRICA OR THE MIDDLE EAST (99)	0.93	11 4 49 50 X SQ= 2.09 P = 0.102 RV NO (YES)
15	TEND TO BE THOSE LOCATED IN THE MIDDLE EAST, RATHER THAN IN NORTH AFRICA OR CENTRAL AND SOUTH AFRICA (11)	0.58	15	TEND TO BE THOSE LOCATED IN NORTH AFRICA OR CENTRAL AND SOUTH AFRICA, RATHER THAN IN THE MIDDLE EAST (33)	0.90	7 3 5 28 X SQ= 8.91 P = 0.002 RV YES (YES)
18	TEND TO BE THOSE LOCATED IN EAST ASIA, SOUTH ASIA, OR SOUTHEAST ASIA, RATHER THAN IN NORTH AFRICA OR CENTRAL AND SOUTH AFRICA (18)	0.62	18	TEND TO BE THOSE LOCATED IN NORTH AFRICA OR CENTRAL AND SOUTH AFRICA, RATHER THAN IN EAST ASIA, SOUTH ASIA, OR SOUTHEAST ASIA (33)	0.74	5 28 8 10 X SQ= 3.83 P = 0.041 RV YES (NO)
19	LEAN TOWARD BEING THOSE LOCATED IN THE CARIBBEAN, CENTRAL AMERICA, OR SOUTH AMERICA, RATHER THAN IN NORTH AFRICA OR CENTRAL AND SOUTH AFRICA (22)	0.64	19	LEAN TOWARD BEING THOSE LOCATED IN NORTH AFRICA OR CENTRAL AND SOUTH AFRICA, RATHER THAN IN THE CARIBBEAN, CENTRAL AMERICA, OR SOUTH AMERICA (33)	0.68	5 28 9 13 X SQ= 3.36 P = 0.056 RV YES (NO)
24	TEND TO BE THOSE WHOSE POPULATION IS VERY LARGE, LARGE, OR MEDIUM (61)	0.68	24	TEND TO BE THOSE WHOSE POPULATION IS SMALL (54)	0.65	41 19 19 35 X SQ= 11.23 P = 0.001 RV YES (YES)
26	TEND TO BE THOSE WHOSE POPULATION DENSITY IS VERY HIGH, HIGH, OR MEDIUM (48)	0.55	26	TEND TO BE THOSE WHOSE POPULATION DENSITY IS LOW (67)	0.72	33 15 27 39 X SQ= 7.56 P = 0.004 RV YES (YES)
28	TILT TOWARD BEING THOSE WHOSE POPULATION GROWTH RATE IS LOW (48)	0.52	28	TILT TOWARD BEING THOSE WHOSE POPULATION GROWTH RATE IS HIGH (62)	0.65	28 33 30 18 X SQ= 2.34 P = 0.122 RV YES (YES)
29	TEND TO BE THOSE WHERE THE DEGREE OF URBANIZATION IS HIGH (56)	0.84	29	TEND TO BE THOSE WHERE THE DEGREE OF URBANIZATION IS LOW (49)	0.80	46 10 9 39 X SQ= 39.18 P = 0. RV YES (YES)
30	TEND TO BE THOSE WHOSE AGRICULTURAL POPULATION IS MEDIUM, LOW, OR VERY LOW (57)	0.73	30	TEND TO BE THOSE WHOSE AGRICULTURAL POPULATION IS HIGH (56)	0.75	16 40 44 13 X SQ= 24.90 P = 0. RV YES (YES)

32	TEND LESS TO BE THOSE WHOSE GROSS NATIONAL PRODUCT IS MEDIUM, LOW, OR VERY LOW (105)	0.83	32	ALWAYS ARE THOSE WHOSE GROSS NATIONAL PRODUCT IS MEDIUM, LOW, OR VERY LOW (105)	1.00	10 0 50 54 X SQ= 7.89 P = 0.001 RV NO (YES)
33	TEND LESS TO BE THOSE WHOSE GROSS NATIONAL PRODUCT IS LOW OR VERY LOW (85)	0.53	33	TEND MORE TO BE THOSE WHOSE GROSS NATIONAL PRODUCT IS LOW OR VERY LOW (85)	0.96	28 2 32 52 X SQ= 24.88 P = 0. RV NO (YES)
34	TEND TO BE THOSE WHOSE GROSS NATIONAL PRODUCT IS VERY HIGH, HIGH, MEDIUM, OR LOW (62)	0.78	34	TEND TO BE THOSE WHOSE GROSS NATIONAL PRODUCT IS VERY LOW (53)	0.74	47 14 13 40 X SQ= 29.31 P = 0. RV YES (YES)
35	TEND LESS TO BE THOSE WHOSE PER CAPITA GROSS NATIONAL PRODUCT IS MEDIUM, LOW, OR VERY LOW (91)	0.60	35	ALWAYS ARE THOSE WHOSE PER CAPITA GROSS NATIONAL PRODUCT IS MEDIUM, LOW, OR VERY LOW (91)	1.00	24 0 36 54 X SQ= 25.01 P = 0. RV NO (YES)
36	TEND TO BE THOSE WHOSE PER CAPITA GROSS NATIONAL PRODUCT IS VERY HIGH, HIGH, OR MEDIUM (42)	0.63	36	TEND TO BE THOSE WHOSE PER CAPITA GROSS NATIONAL PRODUCT IS LOW OR VERY LOW (73)	0.93	38 4 22 50 X SQ= 35.84 P = 0. RV YES (YES)
37	TEND TO BE THOSE WHOSE PER CAPITA GROSS NATIONAL PRODUCT IS VERY HIGH, HIGH, MEDIUM, OR LOW (64)	0.85	37	TEND TO BE THOSE WHOSE PER CAPITA GROSS NATIONAL PRODUCT IS VERY LOW (51)	0.78	51 12 9 42 X SQ= 42.80 P = 0. RV YES (YES)
38	TEND LESS TO BE THOSE WHOSE INTERNATIONAL FINANCIAL STATUS IS MEDIUM, LOW, OR VERY LOW (103)	0.83	38	ALWAYS ARE THOSE WHOSE INTERNATIONAL FINANCIAL STATUS IS MEDIUM, LOW, OR VERY LOW (103)	1.00	10 0 48 54 X SQ= 8.21 P = 0.001 RV NO (YES)
39	TEND TO BE THOSE WHOSE INTERNATIONAL FINANCIAL STATUS IS VERY HIGH, HIGH, OR MEDIUM (38)	0.59	39	TEND TO BE THOSE WHOSE INTERNATIONAL FINANCIAL STATUS IS LOW OR VERY LOW (76)	0.94	35 3 24 51 X SQ= 34.15 P = 0. RV YES (YES)
40	TEND TO BE THOSE WHOSE INTERNATIONAL FINANCIAL STATUS IS VERY HIGH, HIGH, MEDIUM, OR LOW (71)	0.90	40	TEND TO BE THOSE WHOSE INTERNATIONAL FINANCIAL STATUS IS VERY LOW (39)	0.65	52 18 6 33 X SQ= 32.58 P = 0. RV YES (YES)

41	TEND LESS TO BE THOSE WHOSE ECONOMIC DEVELOPMENTAL STATUS IS INTERMEDIATE, UNDERDEVELOPED, OR VERY UNDERDEVELOPED (94)	0.67	41	ALWAYS ARE THOSE WHOSE ECONOMIC DEVELOPMENTAL STATUS IS INTERMEDIATE, UNDERDEVELOPED, OR VERY UNDERDEVELOPED (94)	1.00	19 0 39 54 X SQ= 19.04 P = 0. RV NO (YES)
42	TEND TO BE THOSE WHOSE ECONOMIC DEVELOPMENTAL STATUS IS DEVELOPED OR INTERMEDIATE (36)	0.59	42	TEND TO BE THOSE WHOSE ECONOMIC DEVELOPMENTAL STATUS IS UNDERDEVELOPED OR VERY UNDERDEVELOPED (76)	0.96	34 2 24 52 X SQ= 36.19 P = 0. RV YES (YES)
43	TEND TO BE THOSE WHOSE ECONOMIC DEVELOPMENTAL STATUS IS DEVELOPED, INTERMEDIATE, OR UNDERDEVELOPED (55)	0.79	43	TEND TO BE THOSE WHOSE ECONOMIC DEVELOPMENTAL STATUS IS VERY UNDERDEVELOPED (57)	0.83	46 9 12 45 X SQ= 41.44 P = 0. RV YES (YES)
44	TEND LESS TO BE THOSE WHERE THE LITERACY RATE IS BELOW NINETY PERCENT (90)	0.58	44	ALWAYS ARE THOSE WHERE THE LITERACY RATE IS BELOW NINETY PERCENT (90)	1.00	25 0 35 54 X SQ= 26.44 P = 0. RV NO (YES)
45	TEND TO BE THOSE WHERE THE LITERACY RATE IS FIFTY PERCENT OR ABOVE (55)	0.79	45	TEND TO BE THOSE WHERE THE LITERACY RATE IS BELOW FIFTY PERCENT (54)	0.79	44 11 12 41 X SQ= 33.31 P = 0. RV YES (YES)
46	TEND MORE TO BE THOSE WHERE THE LITERACY RATE IS TEN PERCENT OR ABOVE (84)	0.98	46	TEND LESS TO BE THOSE WHERE THE LITERACY RATE IS TEN PERCENT OR ABOVE (84)	0.53	57 27 1 24 X SQ= 29.04 P = 0. RV NO (YES)
52	LEAN LESS TOWARD BEING THOSE WHERE FREEDOM OF THE PRESS IS COMPLETE, INTERMITTENT, OR INTERNALLY ABSENT (82)	0.79	52	LEAN MORE TOWARD BEING THOSE WHERE FREEDOM OF THE PRESS IS COMPLETE, INTERMITTENT, OR INTERNALLY ABSENT (82)	0.92	46 36 12 3 X SQ= 2.10 P = 0.095 RV NO (NO)
53	TEND LESS TO BE THOSE WHERE NEWSPAPER CIRCULATION IS LESS THAN THREE HUNDRED PER THOUSAND (101)	0.77	53	ALWAYS ARE THOSE WHERE NEWSPAPER CIRCULATION IS LESS THAN THREE HUNDRED PER THOUSAND (101)	1.00	14 0 46 54 X SQ= 12.28 P = 0. RV NO (YES)
54	TEND TO BE THOSE WHERE NEWSPAPER CIRCULATION IS ONE HUNDRED OR MORE PER THOUSAND (37)	0.61	54	TEND TO BE THOSE WHERE NEWSPAPER CIRCULATION IS LESS THAN ONE HUNDRED PER THOUSAND (76)	0.98	36 1 23 52 X SQ= 41.49 P = 0. RV YES (YES)

55	TEND TO BE THOSE WHERE NEWSPAPER CIRCULATION IS TEN OR MORE PER THOUSAND (78)	0.97	55	TEND TO BE THOSE WHERE NEWSPAPER CIRCULATION IS LESS THAN TEN PER THOUSAND (35)	0.60	57 21 2 32 X SQ= 40.23 P = 0. RV YES (YES)
56	TEND TO BE THOSE WHERE THE RELIGION IS PREDOMINANTLY LITERATE (79)	0.96	56	TEND TO BE THOSE WHERE THE RELIGION IS PREDOMINANTLY OR PARTLY NON-LITERATE (31)	0.56	55 23 2 29 X SQ= 33.97 P = 0. RV YES (YES)
63	TEND TO BE THOSE WHERE THE RELIGION IS PREDOMINANTLY SOME KIND OF CHRISTIAN (46)	0.64	63	TEND TO BE THOSE WHERE THE RELIGION IS PREDOMINANTLY OR PARTLY OTHER THAN CHRISTIAN (68)	0.85	38 8 21 46 X SQ= 26.71 P = 0. RV YES (YES)
66	TEND TO BE THOSE THAT ARE RELIGIOUSLY HOMOGENEOUS (57)	0.68	66	TEND TO BE THOSE THAT ARE RELIGIOUSLY HETEROGENEOUS (49)	0.63	38 18 18 31 X SQ= 8.96 P = 0.002 RV YES (YES)
67	TEND MORE TO BE THOSE THAT ARE RACIALLY HOMOGENEOUS (82)	0.88	67	TEND LESS TO BE THOSE THAT ARE RACIALLY HOMOGENEOUS (82)	0.56	52 28 7 22 X SQ= 12.71 P = 0. RV NO (YES)
68	TEND TO BE THOSE THAT ARE LINGUISTICALLY HOMOGENEOUS (52)	0.63	68	TEND TO BE THOSE THAT ARE LINGUISTICALLY WEAKLY HETEROGENEOUS OR STRONGLY HETEROGENEOUS (62)	0.74	37 14 22 40 X SQ= 13.96 P = 0. RV YES (YES)
69	TEND TO BE THOSE THAT ARE LINGUISTICALLY HOMOGENEOUS OR WEAKLY HETEROGENEOUS (64)	0.73	69	TEND TO BE THOSE THAT ARE LINGUISTICALLY STRONGLY HETEROGENEOUS (50)	0.63	43 20 16 34 X SQ= 13.27 P = 0. RV YES (YES)
70	TEND LESS TO BE THOSE THAT ARE RELIGIOUSLY, RACIALLY, OR LINGUISTICALLY HETEROGENEOUS (85)	0.66	70	TEND MORE TO BE THOSE THAT ARE RELIGIOUSLY, RACIALLY, OR LINGUISTICALLY HETEROGENEOUS (85)	0.96	19 2 37 47 X SQ= 12.74 P = 0. RV NO (YES)
72	TEND TO BE THOSE WHOSE DATE OF INDEPENDENCE IS BEFORE 1914 (52)	0.61	72	TEND TO BE THOSE WHOSE DATE OF INDEPENDENCE IS AFTER 1914 (59)	0.67	34 18 22 36 X SQ= 7.21 P = 0.005 RV YES (YES)

73	TEND TO BE THOSE WHOSE DATE OF INDEPENDENCE IS BEFORE 1945 (65)	0.80	73	TEND TO BE THOSE WHOSE DATE OF INDEPENDENCE IS AFTER 1945 (46)	0.65	45 19 11 35 X SQ= 21.24 P = 0. RV YES (YES)
74	TEND LESS TO BE THOSE THAT ARE NOT HISTORICALLY WESTERN (87)	0.56	74	ALWAYS ARE THOSE THAT ARE NOT HISTORICALLY WESTERN (87)	1.00	26 0 33 53 X SQ= 28.00 P = 0. RV NO (YES)
75	TEND TO BE THOSE THAT ARE HISTORICALLY WESTERN OR SIGNIFICANTLY WESTERNIZED (62)	0.82	75	TEND TO BE THOSE THAT ARE NOT HISTORICALLY WESTERN AND ARE NOT SIGNIFICANTLY WESTERNIZED (52)	0.75	49 13 11 40 X SQ= 34.83 P = 0. RV YES (YES)
76	TEND TO BE THOSE THAT ARE HISTORICALLY WESTERN, RATHER THAN HAVING BEEN SIGNIFICANTLY OR PARTIALLY WESTERNIZED THROUGH A COLONIAL RELATIONSHIP (26)	0.54	76	ALWAYS ARE THOSE THAT HAVE BEEN SIGNIFICANTLY OR PARTIALLY WESTERNIZED THROUGH A COLONIAL RELATIONSHIP, RATHER THAN BEING HISTORICALLY WESTERN (70)	1.00	26 0 22 48 X SQ= 32.97 P = 0. RV YES (YES)
77	TEND TO BE THOSE THAT HAVE BEEN SIGNIFICANTLY WESTERNIZED, RATHER THAN PARTIALLY WESTERNIZED, THROUGH A COLONIAL RELATIONSHIP (28)	0.68	77	TEND TO BE THOSE THAT HAVE BEEN PARTIALLY WESTERNIZED, RATHER THAN SIGNIFICANTLY WESTERNIZED, THROUGH A COLONIAL RELATIONSHIP (41)	0.72	15 13 7 34 X SQ= 8.59 P = 0.003 RV YES (YES)
78	TEND LESS TO BE THOSE THAT HAVE BEEN SIGNIFICANTLY WESTERNIZED THROUGH A COLONIAL RELATIONSHIP, RATHER THAN WITHOUT SUCH A RELATIONSHIP (28)	0.68	78	ALWAYS ARE THOSE THAT HAVE BEEN SIGNIFICANTLY WESTERNIZED THROUGH A COLONIAL RELATIONSHIP, RATHER THAN WITHOUT SUCH A RELATIONSHIP (28)	1.00	15 13 7 0 X SQ= 3.37 P = 0.031 RV NO (NO)
80	TILT TOWARD BEING THOSE WHOSE DATE OF INDEPENDENCE IS AFTER 1914, AND THAT WERE FORMERLY DEPENDENCIES OF BRITAIN, RATHER THAN FRANCE (19)	0.64	80	TILT TOWARD BEING THOSE WHOSE DATE OF INDEPENDENCE IS AFTER 1914, AND THAT WERE FORMERLY DEPENDENCIES OF FRANCE, RATHER THAN BRITAIN (24)	0.66	9 10 5 19 X SQ= 2.30 P = 0.102 RV YES (NO)
82	TEND TO BE THOSE WHERE THE TYPE OF POLITICAL MODERNIZATION IS EARLY OR LATER EUROPEAN OR EUROPEAN DERIVED, RATHER THAN DEVELOPED TUTELARY OR UNDEVELOPED TUTELARY (51)	0.70	82	TEND TO BE THOSE WHERE THE TYPE OF POLITICAL MODERNIZATION IS DEVELOPED TUTELARY OR UNDEVELOPED TUTELARY, RATHER THAN EARLY OR LATER EUROPEAN OR EUROPEAN DERIVED (55)	0.76	39 12 17 37 X SQ= 19.56 P = 0. RV YES (YES)
84	ALWAYS ARE THOSE WHERE THE TYPE OF POLITICAL MODERNIZATION IS DEVELOPED TUTELARY, RATHER THAN UNDEVELOPED TUTELARY (31)	1.00	84	TEND TO BE THOSE WHERE THE TYPE OF POLITICAL MODERNIZATION IS UNDEVELOPED TUTELARY, RATHER THAN DEVELOPED TUTELARY (24)	0.65	17 13 0 24 X SQ= 17.31 P = 0. RV YES (YES)

87	TEND TO BE THOSE WHOSE IDEOLOGICAL ORIENTATION IS OTHER THAN DEVELOPMENTAL (58)	0.88	87	TEND TO BE THOSE WHOSE IDEOLOGICAL ORIENTATION IS DEVELOPMENTAL (31)	0.69	6 25 46 11 X SQ= 28.77 P = 0. RV YES (YES)
89	TEND TO BE THOSE WHOSE IDEOLOGICAL ORIENTATION IS CONVENTIONAL (33)	0.56	89	TEND TO BE THOSE WHOSE IDEOLOGICAL ORIENTATION IS OTHER THAN CONVENTIONAL (62)	0.90	29 4 23 38 X SQ= 19.83 P = 0. RV YES (YES)
92	TEND LESS TO BE THOSE WHERE THE SYSTEM STYLE IS LIMITED MOBILIZATIONAL, OR NON-MOBILIZATIONAL (93)	0.75	92	TEND MORE TO BE THOSE WHERE THE SYSTEM STYLE IS LIMITED MOBILIZATIONAL, OR NON-MOBILIZATIONAL (93)	0.91	15 5 44 48 X SQ= 3.84 P = 0.046 RV NO (YES)
94	TILT TOWARD BEING THOSE WHERE THE STATUS OF THE REGIME IS CONSTITUTIONAL (51)	0.63	94	TILT TOWARD BEING THOSE WHERE THE STATUS OF THE REGIME IS AUTHORITARIAN OR TOTALITARIAN (41)	0.54	35 16 21 19 X SQ= 1.83 P = 0.133 RV YES (NO)
95	TEND LESS TO BE THOSE WHERE THE STATUS OF THE REGIME IS CONSTITUTIONAL OR AUTHORITARIAN (95)	0.75	95	TEND MORE TO BE THOSE WHERE THE STATUS OF THE REGIME IS CONSTITUTIONAL OR AUTHORITARIAN (95)	0.98	45 49 15 1 X SQ= 9.83 P = 0.001 RV NO (YES)
96	TEND MORE TO BE THOSE WHERE THE STATUS OF THE REGIME IS CONSTITUTIONAL OR TOTALITARIAN (67)	0.89	96	TEND LESS TO BE THOSE WHERE THE STATUS OF THE REGIME IS CONSTITUTIONAL OR TOTALITARIAN (67)	0.52	50 17 6 16 X SQ= 13.95 P = 0. RV NO (YES)
97	TEND TO BE THOSE WHERE THE STATUS OF THE REGIME IS TOTALITARIAN, RATHER THAN AUTHORITARIAN (16)	0.71	97	TEND TO BE THOSE WHERE THE STATUS OF THE REGIME IS AUTHORITARIAN, RATHER THAN TOTALITARIAN (23)	0.94	6 16 15 1 X SQ= 13.98 P = 0. RV YES (YES)
99	TEND TO BE THOSE WHERE GOVERNMENTAL STABILITY IS GENERALLY PRESENT AND DATES FROM AT LEAST THE INTER-WAR PERIOD, OR FROM THE POST-WAR PERIOD (50)	0.73	99	TEND TO BE THOSE WHERE GOVERNMENTAL STABILITY IS MODERATELY PRESENT AND DATES FROM THE POST-WAR PERIOD, OR IS ABSENT (36)	0.72	41 8 15 21 X SQ= 14.48 P = 0. RV YES (YES)
101	TEND TO BE THOSE WHERE THE REPRESENTATIVE CHARACTER OF THE REGIME IS POLYARCHIC (41)	0.59	101	TEND TO BE THOSE WHERE THE REPRESENTATIVE CHARACTER OF THE REGIME IS LIMITED POLYARCHIC, PSEUDO-POLYARCHIC, OR NON-POLYARCHIC (57)	0.79	32 9 22 34 X SQ= 12.88 P = 0. RV YES (YES)

102	LEAN TOWARD BEING THOSE WHERE THE REPRESENTATIVE CHARACTER OF THE REGIME IS POLYARCHIC OR LIMITED POLYARCHIC (59)	0.64	102	LEAN TOWARD BEING THOSE WHERE THE REPRESENTATIVE CHARACTER OF THE REGIME IS PSEUDO-POLYARCHIC OR NON-POLYARCHIC (49)	0.55	36 23 20 28 X SQ= 3.23 P = 0.054 RV YES (YES)
105	TEND TO BE THOSE WHERE THE ELECTORAL SYSTEM IS COMPETITIVE (43)	0.63	105	TEND TO BE THOSE WHERE THE ELECTORAL SYSTEM IS PARTIALLY COMPETITIVE OR NON-COMPETITIVE (47)	0.71	31 12 18 29 X SQ= 9.02 P = 0.002 RV YES (YES)
107	TEND TO BE THOSE WHERE AUTONOMOUS GROUPS ARE FULLY TOLERATED IN POLITICS (46)	0.55	107	TEND TO BE THOSE WHERE AUTONOMOUS GROUPS ARE PARTIALLY TOLERATED IN POLITICS, ARE TOLERATED ONLY OUTSIDE POLITICS, OR ARE NOT TOLERATED AT ALL (65)	0.74	33 13 27 37 X SQ= 8.27 P = 0.003 RV YES (YES)
111	TEND TO BE THOSE WHERE POLITICAL ENCULTURATION IS HIGH OR MEDIUM (53)	0.69	111	TEND TO BE THOSE WHERE POLITICAL ENCULTURATION IS LOW (42)	0.55	31 22 14 27 X SQ= 4.56 P = 0.023 RV YES (YES)
114	TILT TOWARD BEING THOSE WHERE SECTIONALISM IS NEGLIGIBLE (47)	0.51	114	TILT TOWARD BEING THOSE WHERE SECTIONALISM IS EXTREME OR MODERATE (61)	0.66	28 33 29 17 X SQ= 2.45 P = 0.117 RV YES (YES)
115	TEND LESS TO BE THOSE WHERE INTEREST ARTICULATION BY ASSOCIATIONAL GROUPS IS MODERATE, LIMITED, OR NEGLIGIBLE (91)	0.64	115	ALWAYS ARE THOSE WHERE INTEREST ARTICULATION BY ASSOCIATIONAL GROUPS IS MODERATE, LIMITED, OR NEGLIGIBLE (91)	1.00	20 0 36 54 X SQ= 21.23 P = 0. RV NO (YES)
116	TEND TO BE THOSE WHERE INTEREST ARTICULATION BY ASSOCIATIONAL GROUPS IS SIGNIFICANT OR MODERATE (32)	0.55	116	TEND TO BE THOSE WHERE INTEREST ARTICULATION BY ASSOCIATIONAL GROUPS IS LIMITED OR NEGLIGIBLE (79)	0.98	31 1 25 53 X SQ= 35.60 P = 0. RV YES (YES)
117	TEND TO BE THOSE WHERE INTEREST ARTICULATION BY ASSOCIATIONAL GROUPS IS SIGNIFICANT, MODERATE, OR LIMITED (60)	0.77	117	TEND TO BE THOSE WHERE INTEREST ARTICULATION BY ASSOCIATIONAL GROUPS IS NEGLIGIBLE (51)	0.69	43 17 13 37 X SQ= 20.97 P = 0. RV YES (YES)
119	TEND LESS TO BE THOSE WHERE INTEREST ARTICULATION BY INSTITUTIONAL GROUPS IS VERY SIGNIFICANT OR SIGNIFICANT (74)	0.62	119	TEND MORE TO BE THOSE WHERE INTEREST ARTICULATION BY INSTITUTIONAL GROUPS IS VERY SIGNIFICANT OR SIGNIFICANT (74)	0.92	37 36 23 3 X SQ= 9.93 P = 0.001 RV NO (NO)

120	TEND LESS TO BE THOSE WHERE INTEREST ARTICULATION BY INSTITUTIONAL GROUPS IS VERY SIGNIFICANT, SIGNIFICANT, OR MODERATE (90)	0.83	120	ALWAYS ARE THOSE WHERE INTEREST ARTICULATION BY INSTITUTIONAL GROUPS IS VERY SIGNIFICANT, SIGNIFICANT, OR MODERATE (90)	1.00	50 39 10 0 X SQ= 5.51 P = 0.006 RV NO (NO)
121	TEND TO BE THOSE WHERE INTEREST ARTICULATION BY NON-ASSOCIATIONAL GROUPS IS MODERATE, LIMITED, OR NEGLIGIBLE (61)	0.78	121	TEND TO BE THOSE WHERE INTEREST ARTICULATION BY NON-ASSOCIATIONAL GROUPS IS SIGNIFICANT (54)	0.74	13 40 47 14 X SQ= 29.31 P = 0. RV YES (YES)
122	TEND LESS TO BE THOSE WHERE INTEREST ARTICULATION BY NON-ASSOCIATIONAL GROUPS IS SIGNIFICANT OR MODERATE (83)	0.53	122	TEND MORE TO BE THOSE WHERE INTEREST ARTICULATION BY NON-ASSOCIATIONAL GROUPS IS SIGNIFICANT OR MODERATE (83)	0.93	32 50 28 4 X SQ= 19.79 P = 0. RV NO (YES)
123	TEND LESS TO BE THOSE WHERE INTEREST ARTICULATION BY NON-ASSOCIATIONAL GROUPS IS SIGNIFICANT, MODERATE, OR LIMITED (107)	0.87	123	ALWAYS ARE THOSE WHERE INTEREST ARTICULATION BY NON-ASSOCIATIONAL GROUPS IS SIGNIFICANT, MODERATE, OR LIMITED (107)	1.00	52 54 8 0 X SQ= 5.83 P = 0.007 RV NO (YES)
125	TEND TO BE THOSE WHERE INTEREST ARTICULATION BY ANOMIC GROUPS IS INFREQUENT OR VERY INFREQUENT (35)	0.57	125	TEND TU BE THOSE WHERE INTEREST ARTICULATION BY ANOMIC GROUPS IS FREQUENT OR OCCASIONAL (64)	0.84	20 43 27 8 X SQ= 16.80 P = 0. RV YES (YES)
126	TEND LESS TO BE THOSE WHERE INTEREST ARTICULATION BY ANOMIC GROUPS IS FREQUENT, OCCASIONAL, OR INFREQUENT (83)	0.66	126	ALWAYS ARE THOSE WHERE INTEREST ARTICULATION BY ANOMIC GROUPS IS FREQUENT, OCCASIONAL, OR INFREQUENT (83)	1.00	31 51 16 0 X SQ= 18.33 P = 0. RV NO (YES)
127	TILT LESS TOWARD BEING THOSE WHERE INTEREST ARTICULATION BY POLITICAL PARTIES IS MODERATE, LIMITED, OR NEGLIGIBLE (72)	0.71	127	TILT MORE TOWARD BEING THOSE WHERE INTEREST ARTICULATION BY POLITICAL PARTIES IS MODERATE, LIMITED, OR NEGLIGIBLE (72)	0.86	15 6 36 36 X SQ= 2.21 P = 0.134 RV NO (YES)
130	TILT LESS TOWARD BEING THOSE WHERE INTEREST AGGREGATION BY POLITICAL PARTIES IS MODERATE, LIMITED, OR NEGLIGIBLE (71)	0.81	130	TILT MORE TOWARD BEING THOSE WHERE INTEREST AGGREGATION BY POLITICAL PARTIES IS MODERATE, LIMITED, OR NEGLIGIBLE (71)	0.94	10 2 42 29 X SQ= 1.64 P = 0.125 RV NO (NO)
133	TEND MORE TO BE THOSE WHERE INTEREST AGGREGATION BY THE EXECUTIVE IS MODERATE, LIMITED, OR NEGLIGIBLE (76)	0.83	133	TEND LESS TO BE THOSE WHERE INTEREST AGGREGATION BY THE EXECUTIVE IS MODERATE, LIMITED, OR NEGLIGIBLE (76)	0.60	9 20 45 30 X SQ= 5.92 P = 0.009 RV NO (YES)

No.	Left		No.	Right		Statistics
135	TEND MORE TO BE THOSE WHERE INTEREST AGGREGATION BY THE EXECUTIVE IS SIGNIFICANT, MODERATE, OR LIMITED (77)	0.95	135	TEND LESS TO BE THOSE WHERE INTEREST AGGREGATION BY THE EXECUTIVE IS SIGNIFICANT, MODERATE, OR LIMITED (77)	0.78	41 36 2 10 X SQ= 4.19 P = 0.028 RV NO (YES)
136	TEND LESS TO BE THOSE WHERE INTEREST AGGREGATION BY THE LEGISLATURE IS MODERATE, LIMITED, OR NEGLIGIBLE (85)	0.76	136	ALWAYS ARE THOSE WHERE INTEREST AGGREGATION BY THE LEGISLATURE IS MODERATE, LIMITED, OR NEGLIGIBLE (85)	1.00	12 0 39 46 X SQ= 10.28 P = 0. RV NO (YES)
137	TEND TO BE THOSE WHERE INTEREST AGGREGATION BY THE LEGISLATURE IS SIGNIFICANT OR MODERATE (29)	0.53	137	TEND TO BE THOSE WHERE INTEREST AGGREGATION BY THE LEGISLATURE IS LIMITED OR NEGLIGIBLE (68)	0.96	27 2 24 44 X SQ= 24.98 P = 0. RV YES (YES)
138	TEND TO BE THOSE WHERE INTEREST AGGREGATION BY THE LEGISLATURE IS SIGNIFICANT, MODERATE, OR LIMITED (48)	0.69	138	TEND TO BE THOSE WHERE INTEREST AGGREGATION BY THE LEGISLATURE IS NEGLIGIBLE (49)	0.71	34 14 15 34 X SQ= 14.12 P = 0. RV YES (YES)
140	LEAN MORE TOWARD BEING THOSE WHERE THE PARTY SYSTEM IS QUANTITATIVELY OTHER THAN ONE PARTY DOMINANT (87)	0.93	140	LEAN LESS TOWARD BEING THOSE WHERE THE PARTY SYSTEM IS QUANTITATIVELY OTHER THAN ONE PARTY DOMINANT (87)	0.80	4 9 50 36 X SQ= 2.40 P = 0.079 RV NO (YES)
142	TILT LESS TOWARD BEING THOSE WHERE THE PARTY SYSTEM IS QUANTITATIVELY OTHER THAN MULTI-PARTY (66)	0.61	142	TILT MORE TOWARD BEING THOSE WHERE THE PARTY SYSTEM IS QUANTITATIVELY OTHER THAN MULTI-PARTY (66)	0.77	20 10 31 34 X SQ= 2.26 P = 0.121 RV NO (YES)
147	TEND LESS TO BE THOSE WHERE THE PARTY SYSTEM IS QUALITATIVELY OTHER THAN CLASS-ORIENTED OR MULTI-IDEOLOGICAL (67)	0.57	147	TEND MORE TO BE THOSE WHERE THE PARTY SYSTEM IS QUALITATIVELY OTHER THAN CLASS-ORIENTED OR MULTI-IDEOLOGICAL (67)	0.93	20 3 27 40 X SQ= 13.13 P = 0. RV NO (YES)
148	ALWAYS ARE THOSE WHERE THE PARTY SYSTEM IS QUALITATIVELY OTHER THAN AFRICAN TRANSITIONAL (96)	1.00	148	TEND LESS TO BE THOSE WHERE THE PARTY SYSTEM IS QUALITATIVELY OTHER THAN AFRICAN TRANSITIONAL (96)	0.72	0 14 60 36 X SQ= 16.81 P = 0. RV NO (YES)
153	TEND TO BE THOSE WHERE THE PARTY SYSTEM IS STABLE (42)	0.70	153	TEND TO BE THOSE WHERE THE PARTY SYSTEM IS MODERATELY STABLE OR UNSTABLE (38)	0.81	37 5 16 22 X SQ= 16.87 P = 0. RV YES (YES)

158	TEND TO BE THOSE WHERE PERSONALISMO IS NEGLIGIBLE (56)	0.78	158	TEND TO BE THOSE WHERE PERSONALISMO IS PRONOUNCED OR MODERATE (40)	0.67	12 28 42 14 X SQ= 17.41 P = 0. RV YES (YES)
164	TEND TO BE THOSE WHERE THE REGIME'S LEADERSHIP CHARISMA IS NEGLIGIBLE (65)	0.79	164	TEND TO BE THOSE WHERE THE REGIME'S LEADERSHIP CHARISMA IS PRONOUNCED OR MODERATE (34)	0.50	12 21 44 21 X SQ= 7.54 P = 0.005 RV YES (YES)
168	TEND TO BE THOSE WHERE THE HORIZONTAL POWER DISTRIBUTION IS SIGNIFICANT (34)	0.52	168	TEND TO BE THOSE WHERE THE HORIZONTAL POWER DISTRIBUTION IS LIMITED OR NEGLIGIBLE (72)	0.90	29 5 27 44 X SQ= 18.78 P = 0. RV YES (YES)
169	TILT TOWARD BEING THOSE WHERE THE HORIZONTAL POWER DISTRIBUTION IS SIGNIFICANT OR LIMITED (58)	0.63	169	TILT TOWARD BEING THOSE WHERE THE HORIZONTAL POWER DISTRIBUTION IS NEGLIGIBLE (48)	0.53	35 23 21 26 X SQ= 1.97 P = 0.120 RV YES (YES)
170	TEND TO BE THOSE WHERE THE LEGISLATIVE-EXECUTIVE STRUCTURE IS OTHER THAN PRESIDENTIAL (63)	0.77	170	TEND TO BE THOSE WHERE THE LEGISLATIVE-EXECUTIVE STRUCTURE IS PRESIDENTIAL (39)	0.55	12 27 41 22 X SQ= 10.03 P = 0.001 RV YES (YES)
171	TILT LESS TOWARD BEING THOSE WHERE THE LEGISLATIVE-EXECUTIVE STRUCTURE IS OTHER THAN PARLIAMENTARY-REPUBLICAN (90)	0.82	171	TILT MORE TOWARD BEING THOSE WHERE THE LEGISLATIVE-EXECUTIVE STRUCTURE IS OTHER THAN PARLIAMENTARY-REPUBLICAN (90)	0.94	9 3 42 48 X SQ= 2.36 P = 0.122 RV NO (YES)
175	TEND TO BE THOSE WHERE THE LEGISLATURE IS FULLY EFFECTIVE OR PARTIALLY EFFECTIVE (51)	0.66	175	TEND TO BE THOSE WHERE THE LEGISLATURE IS LARGELY INEFFECTIVE OR WHOLLY INEFFECTIVE (49)	0.66	35 16 18 31 X SQ= 8.96 P = 0.002 RV YES (YES)
178	LEAN TOWARD BEING THOSE WHERE THE LEGISLATURE IS BICAMERAL (51)	0.58	178	LEAN TOWARD BEING THOSE WHERE THE LEGISLATURE IS UNICAMERAL (53)	0.61	22 31 31 20 X SQ= 3.13 P = 0.054 RV YES (YES)
179	LEAN TOWARD BEING THOSE WHERE THE EXECUTIVE IS STRONG (39)	0.52	179	LEAN TOWARD BEING THOSE WHERE THE EXECUTIVE IS DOMINANT (52)	0.67	24 27 26 13 X SQ= 2.69 P = 0.087 RV YES (YES)

181	TEND LESS TO BE THOSE WHERE THE BUREAUCRACY IS SEMI-MODERN, RATHER THAN MODERN (55)	0.64	181	TEND MORE TO BE THOSE WHERE THE BUREAUCRACY IS SEMI-MODERN, RATHER THAN MODERN (55)	0.95	20 1 35 20 X SQ= 6.09 P = 0.005 RV NO (NO)
182	TEND TO BE THOSE WHERE THE BUREAUCRACY IS SEMI-MODERN, RATHER THAN POST-COLONIAL TRANSITIONAL (55)	0.95	182	TEND TO BE THOSE WHERE THE BUREAUCRACY IS POST-COLONIAL TRANSITIONAL, RATHER THAN SEMI-MODERN (25)	0.53	35 20 2 23 X SQ= 19.22 P = 0. RV YES (YES)
187	TEND LESS TO BE THOSE WHERE THE ROLE OF THE POLICE IS POLITICALLY SIGNIFICANT (66)	0.54	187	TEND MORE TO BE THOSE WHERE THE ROLE OF THE POLICE IS POLITICALLY SIGNIFICANT (66)	0.80	30 35 26 9 X SQ= 6.21 P = 0.011 RV NO (YES)
192	TEND MORE TO BE THOSE WHERE THE CHARACTER OF THE LEGAL SYSTEM IS OTHER THAN MUSLIM OR PARTLY MUSLIM (86)	0.87	192	TEND LESS TO BE THOSE WHERE THE CHARACTER OF THE LEGAL SYSTEM IS OTHER THAN MUSLIM OR PARTLY MUSLIM (86)	0.65	8 18 52 34 X SQ= 5.93 P = 0.013 RV NO (YES)
193	ALWAYS ARE THOSE WHERE THE CHARACTER OF THE LEGAL SYSTEM IS OTHER THAN PARTLY INDIGENOUS (86)	1.00	193	TEND TO BE THOSE WHERE THE CHARACTER OF THE LEGAL SYSTEM IS PARTLY INDIGENOUS (28)	0.53	0 28 60 25 X SQ= 39.35 P = 0. RV YES (YES)
194	TEND LESS TO BE THOSE THAT ARE NON-COMMUNIST (101)	0.80	194	TEND MORE TO BE THOSE THAT ARE NON-COMMUNIST (101)	0.98	12 1 47 53 X SQ= 7.74 P = 0.002 RV NO (YES)

```
86  POLITIES                                       86  POLITIES
    WHERE THE STAGE OF                                 WHERE THE STAGE OF
    POLITICAL MODERNIZATION IS                         POLITICAL MODERNIZATION IS
    ADVANCED OR MID-TRANSITIONAL,                      EARLY TRANSITIONAL, RATHER THAN
    RATHER THAN EARLY TRANSITIONAL  (76)               ADVANCED OR MID-TRANSITIONAL  (38)

                                                                    BOTH SUBJECT AND PREDICATE

      76 IN LEFT COLUMN

AFGHANISTAN ALBANIA     ALGERIA     ARGENTINA   AUSTRALIA   AUSTRIA     BELGIUM     BOLIVIA     BRAZIL      BULGARIA
CANADA      CHILE       CHINA, PR   COLOMBIA    COSTA RICA  CUBA        CYPRUS      CZECHOS'KIA DENMARK     DOMIN REP
ECUADOR     EL SALVADOR ETHIOPIA    FINLAND     FRANCE      GERMANY, E  GERMAN FR   GREECE      GUATEMALA   HAITI
HONDURAS    HUNGARY     ICELAND     INDIA       IRAN        IRAQ        IRELAND     ISRAEL      ITALY       JAPAN
JORDAN      KOREA, N    KOREA REP   LEBANON     LUXEMBOURG  MEXICO      MONGOLIA    MOROCCO     NETHERLANDS NEW ZEALAND
NICARAGUA   NORWAY      PAKISTAN    PANAMA      PARAGUAY    PERU        PHILIPPINES POLAND      PORTUGAL    RUMANIA
SO AFRICA   SPAIN       SUDAN       SWEDEN      SWITZERLAND SYRIA       THAILAND    TUNISIA     TURKEY      USSR
UAR         UK          US          URUGUAY     VENEZUELA   YUGOSLAVIA

      38 IN RIGHT COLUMN

BURMA       BURUNDI     CAMBODIA    CAMEROUN    CEN AFR REP CEYLON      CHAD        CONGO(BRA)  CONGO(LEO)  DAHOMEY
GABON       GHANA       GUINEA      INDONESIA   IVORY COAST JAMAICA     LAOS        LIBERIA     LIBYA       MALAGASY R
MALAYA      MALI        MAURITANIA  NEPAL       NIGER       NIGERIA     RWANDA      SENEGAL     SIERRE LEO  SOMALIA
TANGANYIKA  TOGO        TRINIDAD    UGANDA      UPPER VOLTA VIETNAM, N  VIETNAM REP YEMEN

      1 EXCLUDED BECAUSE IRRELEVANT

SA'U ARABIA

------------------------------------------------------------------------------------------------------

  2 TEND LESS TO BE THOSE                 0.71      2 ALWAYS ARE THOSE                  1.00        22     0
    LOCATED ELSEWHERE THAN IN                         LOCATED ELSEWHERE THAN IN                     54    38
    WEST EUROPE, SCANDINAVIA,                         WEST EUROPE, SCANDINAVIA,               X SQ=  11.84
    NORTH AMERICA, OR AUSTRALASIA  (93)               NORTH AMERICA, OR AUSTRALASIA  (93)     P =    0.
                                                                                              RV NO  (NO )

  6 TEND LESS TO BE THOSE                 0.74      6 TEND MORE TO BE THOSE             0.95        20     2
    LOCATED ELSEWHERE THAN IN THE                     LOCATED ELSEWHERE THAN IN THE                 56    36
    CARIBBEAN, CENTRAL AMERICA,                       CARIBBEAN, CENTRAL AMERICA,             X SQ=   5.92
    OR SOUTH AMERICA  (93)                            OR SOUTH AMERICA  (93)                  P =    0.006
                                                                                              RV NO  (NO )

  9 TEND MORE TO BE THOSE                 0.97      9 TEND LESS TO BE THOSE             0.82         2     7
    LOCATED ELSEWHERE THAN IN                         LOCATED ELSEWHERE THAN IN                     74    31
    SOUTHEAST ASIA  (106)                             SOUTHEAST ASIA  (106)                   X SQ=   6.65
                                                                                              P =    0.006
                                                                                              RV NO  (YES)
```

10	TEND TO BE THOSE LOCATED ELSEWHERE THAN IN NORTH AFRICA, OR CENTRAL AND SOUTH AFRICA (82)	0.91	10	TEND TO BE THOSE LOCATED IN NORTH AFRICA, OR CENTRAL AND SOUTH AFRICA (33)	0.68	7 26 69 12 X SQ= 40.35 P = 0. RV YES (YES)
13	LEAN LESS TOWARD BEING THOSE LOCATED ELSEWHERE THAN IN NORTH AFRICA OR THE MIDDLE EAST (99)	0.83	13	LEAN MORE TOWARD BEING THOSE LOCATED ELSEWHERE THAN IN NORTH AFRICA OR THE MIDDLE EAST (99)	0.95	13 2 63 36 X SQ= 2.16 P = 0.088 RV NO (NO)
15	TEND TO BE THOSE LOCATED IN THE MIDDLE EAST, RATHER THAN IN NORTH AFRICA OR CENTRAL AND SOUTH AFRICA (11)	0.56	15	TEND TO BE THOSE LOCATED IN NORTH AFRICA OR CENTRAL AND SOUTH AFRICA, RATHER THAN IN THE MIDDLE EAST (33)	0.96	9 1 7 26 X SQ= 12.74 P = 0. RV YES (YES)
16	TEND TO BE THOSE LOCATED IN THE MIDDLE EAST, RATHER THAN IN EAST ASIA, SOUTH ASIA, OR SOUTHEAST ASIA (11)	0.50	16	TEND TO BE THOSE LOCATED IN EAST ASIA, SOUTH ASIA, OR SOUTHEAST ASIA, RATHER THAN IN THE MIDDLE EAST (18)	0.90	9 1 9 9 X SQ= 2.91 P = 0.048 RV YES (YES)
18	LEAN TOWARD BEING THOSE LOCATED IN EAST ASIA, SOUTH ASIA, OR SOUTHEAST ASIA, RATHER THAN IN NORTH AFRICA OR CENTRAL AND SOUTH AFRICA (18)	0.56	18	LEAN TOWARD BEING THOSE LOCATED IN NORTH AFRICA OR CENTRAL AND SOUTH AFRICA, RATHER THAN IN EAST ASIA, SOUTH ASIA, OR SOUTHEAST ASIA (33)	0.74	7 26 9 9 X SQ= 3.25 P = 0.057 RV YES (YES)
19	TEND TO BE THOSE LOCATED IN THE CARIBBEAN, CENTRAL AMERICA, OR SOUTH AMERICA, RATHER THAN IN NORTH AFRICA OR CENTRAL AND SOUTH AFRICA (22)	0.74	19	TEND TO BE THOSE LOCATED IN NORTH AFRICA OR CENTRAL AND SOUTH AFRICA, RATHER THAN IN THE CARIBBEAN, CENTRAL AMERICA, OR SOUTH AMERICA (33)	0.93	7 26 20 2 X SQ= 22.94 P = 0. RV YES (YES)
20	TEND TO BE THOSE LOCATED IN THE CARIBBEAN, CENTRAL AMERICA, OR SOUTH AMERICA, RATHER THAN IN EAST ASIA, SOUTH ASIA, OR SOUTHEAST ASIA (22)	0.69	20	TEND TO BE THOSE LOCATED IN EAST ASIA, SOUTH ASIA, OR SOUTHEAST ASIA, RATHER THAN IN THE CARIBBEAN, CENTRAL AMERICA, OR SOUTH AMERICA (18)	0.82	9 9 20 2 X SQ= 6.38 P = 0.006 RV YES (YES)
24	TEND TO BE THOSE WHOSE POPULATION IS VERY LARGE, LARGE, OR MEDIUM (61)	0.63	24	TEND TO BE THOSE WHOSE POPULATION IS SMALL (54)	0.68	48 12 28 26 X SQ= 8.91 P = 0.003 RV YES (NO)
26	TEND LESS TO BE THOSE WHOSE POPULATION DENSITY IS LOW (67)	0.51	26	TEND MORE TO BE THOSE WHOSE POPULATION DENSITY IS LOW (67)	0.71	37 11 39 27 X SQ= 3.28 P = 0.048 RV NO (NO)

29	TEND TO BE THOSE WHERE THE DEGREE OF URBANIZATION IS HIGH (56)	0.77	29	TEND TO BE THOSE WHERE THE DEGREE OF URBANIZATION IS LOW (49)	0.94	54 2 16 32 X SQ= 43.94 P = 0. RV YES (YES)
30	TEND TO BE THOSE WHOSE AGRICULTURAL POPULATION IS MEDIUM, LOW, OR VERY LOW (57)	0.68	30	TEND TO BE THOSE WHOSE AGRICULTURAL POPULATION IS HIGH (56)	0.86	24 32 52 5 X SQ= 27.86 P = 0. RV YES (YES)
32	TEND LESS TO BE THOSE WHOSE GROSS NATIONAL PRODUCT IS MEDIUM, LOW, OR VERY LOW (105)	0.87	32	ALWAYS ARE THOSE WHOSE GROSS NATIONAL PRODUCT IS MEDIUM, LOW, OR VERY LOW (105)	1.00	10 0 66 38 X SQ= 3.96 P = 0.016 RV NO (NO)
33	TEND LESS TO BE THOSE WHOSE GROSS NATIONAL PRODUCT IS LOW OR VERY LOW (85)	0.62	33	TEND MORE TO BE THOSE WHOSE GROSS NATIONAL PRODUCT IS LOW OR VERY LOW (85)	0.97	29 1 47 37 X SQ= 14.71 P = 0. RV NO (NO)
34	TEND TO BE THOSE WHOSE GROSS NATIONAL PRODUCT IS VERY HIGH, HIGH, MEDIUM, OR LOW (62)	0.70	34	TEND TO BE THOSE WHOSE GROSS NATIONAL PRODUCT IS VERY LOW (53)	0.79	53 8 23 30 X SQ= 22.22 P = 0. RV YES (YES)
35	TEND LESS TO BE THOSE WHOSE PER CAPITA GROSS NATIONAL PRODUCT IS MEDIUM, LOW, OR VERY LOW (91)	0.68	35	ALWAYS ARE THOSE WHOSE PER CAPITA GROSS NATIONAL PRODUCT IS MEDIUM, LOW, OR VERY LOW (91)	1.00	24 0 52 38 X SQ= 13.36 P = 0. RV NO (NO)
36	TEND TO BE THOSE WHOSE PER CAPITA GROSS NATIONAL PRODUCT IS VERY HIGH, HIGH, OR MEDIUM (42)	0.51	36	TEND TO BE THOSE WHOSE PER CAPITA GROSS NATIONAL PRODUCT IS LOW OR VERY LOW (73)	0.92	39 3 37 35 X SQ= 18.70 P = 0. RV YES (NO)
37	TEND TO BE THOSE WHOSE PER CAPITA GROSS NATIONAL PRODUCT IS VERY HIGH, HIGH, MEDIUM, OR LOW (64)	0.79	37	TEND TO BE THOSE WHOSE PER CAPITA GROSS NATIONAL PRODUCT IS VERY LOW (51)	0.92	60 3 16 35 X SQ= 48.90 P = 0. RV YES (YES)
38	TEND LESS TO BE THOSE WHOSE INTERNATIONAL FINANCIAL STATUS IS MEDIUM, LOW, OR VERY LOW (103)	0.86	38	ALWAYS ARE THOSE WHOSE INTERNATIONAL FINANCIAL STATUS IS MEDIUM, LOW, OR VERY LOW (103)	1.00	10 0 64 38 X SQ= 4.10 P = 0.015 RV NO (NO)

```
39  TEND LESS TO BE THOSE                  0.51     39  TEND MORE TO BE THOSE                  0.97       37    1
    WHOSE INTERNATIONAL FINANCIAL STATUS                WHOSE INTERNATIONAL FINANCIAL STATUS              38   37
    IS LOW OR VERY LOW  (76)                            IS LOW OR VERY LOW  (76)                     X SQ=  22.60
                                                                                                     P =     0.
                                                                                                     RV NO  (NO )

40  TEND TO BE THOSE                       0.82     40  TEND TO BE THOSE                       0.74       61    9
    WHOSE INTERNATIONAL FINANCIAL STATUS                WHOSE INTERNATIONAL FINANCIAL STATUS              13   26
    IS VERY HIGH, HIGH,                                 IS VERY LOW  (39)                            X SQ=  30.84
    MEDIUM, OR LOW  (71)                                                                             P =     0.
                                                                                                     RV YES (YES)

42  TEND LESS TO BE THOSE                  0.51     42  ALWAYS ARE THOSE                       1.00       36    0
    WHOSE ECONOMIC DEVELOPMENTAL STATUS                 WHOSE ECONOMIC DEVELOPMENTAL STATUS               38   38
    IS UNDERDEVELOPED OR                                IS UNDERDEVELOPED OR                         X SQ=  25.06
    VERY UNDERDEVELOPED  (76)                           VERY UNDERDEVELOPED  (76)                    P =     0.
                                                                                                     RV NO  (YES)

43  TEND TO BE THOSE                       0.68     43  TEND TO BE THOSE                       0.87       50    5
    WHOSE ECONOMIC DEVELOPMENTAL STATUS                 WHOSE ECONOMIC DEVELOPMENTAL STATUS               24   33
    IS DEVELOPED, INTERMEDIATE, OR                      IS VERY UNDERDEVELOPED  (57)                 X SQ=  27.60
    UNDERDEVELOPED  (55)                                                                             P =     0.
                                                                                                     RV YES (YES)

44  TEND LESS TO BE THOSE                  0.67     44  ALWAYS ARE THOSE                       1.00       25    0
    WHERE THE LITERACY RATE IS                          WHERE THE LITERACY RATE IS                        51   38
    BELOW NINETY PERCENT  (90)                          BELOW NINETY PERCENT  (90)                   X SQ=  14.15
                                                                                                     P =     0.
                                                                                                     RV NO  (NO )

45  TEND TO BE THOSE                       0.68     45  TEND TO BE THOSE                       0.83       49    6
    WHERE THE LITERACY RATE IS                          WHERE THE LITERACY RATE IS                        23   30
    FIFTY PERCENT OR ABOVE  (55)                        BELOW FIFTY PERCENT  (54)                    X SQ=  23.35
                                                                                                     P =     0.
                                                                                                     RV YES (YES)

46  TEND TO BE THOSE                       0.96     46  TEND TO BE THOSE                       0.63       71   13
    WHERE THE LITERACY RATE IS                          WHERE THE LITERACY RATE IS                         3   22
    TEN PERCENT OR ABOVE  (84)                          BELOW TEN PERCENT  (26)                      X SQ=  43.22
                                                                                                     P =     0.
                                                                                                     RV YES (YES)

52  LEAN LESS TOWARD BEING THOSE           0.80     52  LEAN MORE TOWARD BEING THOSE           0.96       57   25
    WHERE FREEDOM OF THE PRESS IS                       WHERE FREEDOM OF THE PRESS IS                     14    1
    COMPLETE, INTERMITTENT, OR                          COMPLETE, INTERMITTENT, OR                   X SQ=   2.55
    INTERNALLY ABSENT  (82)                             INTERNALLY ABSENT  (82)                      P =     0.063
                                                                                                     RV NO  (NO )

54  TEND LESS TO BE THOSE                  0.51     54  ALWAYS ARE THOSE                       1.00       37    0
    WHERE NEWSPAPER CIRCULATION IS                      WHERE NEWSPAPER CIRCULATION IS                    38   37
    LESS THAN ONE HUNDRED                               LESS THAN ONE HUNDRED                        X SQ=  25.07
    PER THOUSAND  (76)                                  PER THOUSAND  (76)                           P =     0.
                                                                                                     RV NO  (NO )
```

55	TEND TO BE THOSE WHERE NEWSPAPER CIRCULATION IS TEN OR MORE PER THOUSAND (78)	0.93	55	TEND TO BE THOSE WHERE NEWSPAPER CIRCULATION IS LESS THAN TEN PER THOUSAND (35)	0.78	70 8 5 29 X SQ= 56.92 P = 0. RV YES (YES)
56	TEND TO BE THOSE WHERE THE RELIGION IS PREDOMINANTLY LITERATE (79)	0.89	56	TEND TO BE THOSE WHERE THE RELIGION IS PREDOMINANTLY OR PARTLY NON-LITERATE (31)	0.64	65 13 8 23 X SQ= 30.64 P = 0. RV YES (YES)
57	TEND LESS TO BE THOSE WHERE THE RELIGION IS OTHER THAN CATHOLIC (90)	0.67	57	ALWAYS ARE THOSE WHERE THE RELIGION IS OTHER THAN CATHOLIC (90)	1.00	25 0 51 38 X SQ= 14.15 P = 0. RV NO (NO)
63	TEND TO BE THOSE WHERE THE RELIGION IS PREDOMINANTLY SOME KIND OF CHRISTIAN (46)	0.60	63	TEND TO BE THOSE WHERE THE RELIGION IS PREDOMINANTLY OR PARTLY OTHER THAN CHRISTIAN (68)	0.97	45 1 30 37 X SQ= 32.05 P = 0. RV YES (YES)
66	TEND TO BE THOSE THAT ARE RELIGIOUSLY HOMOGENEOUS (57)	0.67	66	TEND TO BE THOSE THAT ARE RELIGIOUSLY HETEROGENEOUS (49)	0.76	48 8 24 25 X SQ= 14.70 P = 0. RV YES (YES)
68	TEND TO BE THOSE THAT ARE LINGUISTICALLY HOMOGENEOUS (52)	0.59	68	TEND TO BE THOSE THAT ARE LINGUISTICALLY WEAKLY HETEROGENEOUS OR STRONGLY HETEROGENEOUS (62)	0.82	44 7 31 31 X SQ= 14.91 P = 0. RV YES (YES)
69	TEND TO BE THOSE THAT ARE LINGUISTICALLY HOMOGENEOUS OR WEAKLY HETEROGENEOUS (64)	0.69	69	TEND TO BE THOSE THAT ARE LINGUISTICALLY STRONGLY HETEROGENEOUS (50)	0.71	52 11 23 27 X SQ= 15.08 P = 0. RV YES (YES)
70	TEND LESS TO BE THOSE THAT ARE RELIGIOUSLY, RACIALLY, OR LINGUISTICALLY HETEROGENEOUS (85)	0.73	70	TEND MORE TO BE THOSE THAT ARE RELIGIOUSLY, RACIALLY, OR LINGUISTICALLY HETEROGENEOUS (85)	0.94	19 2 52 32 X SQ= 5.03 P = 0.017 RV NO (NO)
72	TEND TO BE THOSE WHOSE DATE OF INDEPENDENCE IS BEFORE 1914 (52)	0.69	72	TEND TO BE THOSE WHOSE DATE OF INDEPENDENCE IS AFTER 1914 (59)	0.95	50 2 22 36 X SQ= 38.57 P = 0. RV YES (YES)

73	TEND TO BE THOSE WHOSE DATE OF INDEPENDENCE IS BEFORE 1945 (65)	0.85	73	TEND TO BE THOSE WHOSE DATE OF INDEPENDENCE IS AFTER 1945 (46)	0.92	61 3 11 35 X SQ= 57.22 P = 0. RV YES (YES)
74	TEND LESS TO BE THOSE THAT ARE NOT HISTORICALLY WESTERN (87)	0.65	74	ALWAYS ARE THOSE THAT ARE NOT HISTORICALLY WESTERN (87)	1.00	26 0 48 38 X SQ= 15.47 P = 0. RV NO (NO)
75	TEND TO BE THOSE THAT ARE HISTORICALLY WESTERN OR SIGNIFICANTLY WESTERNIZED (62)	0.79	75	TEND TO BE THOSE THAT ARE NOT HISTORICALLY WESTERN AND ARE NOT SIGNIFICANTLY WESTERNIZED (52)	0.92	59 3 16 35 X SQ= 48.20 P = 0. RV YES (YES)
76	TEND LESS TO BE THOSE THAT HAVE BEEN SIGNIFICANTLY OR PARTIALLY WESTFRNIZED THROUGH A COLONIAL RELATIONSHIP, RATHER THAN BEING HISTORICALLY WESTERN (70)	0.57	76	ALWAYS ARE THOSE THAT HAVE BEEN SIGNIFICANTLY OR PARTIALLY WESTERNIZED THROUGH A COLONIAL RELATIONSHIP, RATHER THAN BEING HISTORICALLY WESTERN (70)	1.00	26 0 35 35 X SQ= 18.36 P = 0. RV NO (YES)
77	TEND TO BE THOSE THAT HAVE BEEN SIGNIFICANTLY WESTERNIZED, RATHER THAN PARTIALLY WESTERNIZED, THROUGH A COLONIAL RELATIONSHIP (28)	0.74	77	TEND TO BE THOSE THAT HAVE BEEN PARTIALLY WESTERNIZED, RATHER THAN SIGNIFICANTLY WESTERNIZED, THROUGH A COLONIAL RELATIONSHIP (41)	0.91	25 3 9 32 X SQ= 27.55 P = 0. RV YES (YES)
79	TEND LESS TO BE THOSE THAT WERE FORMERLY DEPENDENCIES OF BRITAIN OR FRANCE, RATHER THAN SPAIN (49)	0.53	79	ALWAYS ARE THOSE THAT WERE FORMERLY DEPENDENCIES OF BRITAIN OR FRANCE, RATHER THAN SPAIN (49)	1.00	20 29 18 0 X SQ= 16.45 P = 0. RV NO (YES)
80	TILT TOWARD BEING THOSE WHOSE DATE OF INDEPENDENCE IS AFTER 1914, AND THAT WERE FORMERLY DEPENDENCIES OF BRITAIN, RATHER THAN FRANCE (19)	0.64	80	TILT TOWARD BEING THOSE WHOSE DATE OF INDEPENDENCE IS AFTER 1914, AND THAT WERE FORMERLY DEPENDENCIES OF FRANCE, RATHER THAN BRITAIN (24)	0.66	9 10 5 19 X SQ= 2.30 P = 0.102 RV YES (NO)
82	TEND TO BE THOSE WHERE THE TYPE OF POLITICAL MODERNIZATION IS EARLY OR LATER EUROPEAN OR EUROPEAN DERIVED, RATHER THAN DEVELOPED TUTELARY OR UNDEVELOPED TUTELARY (51)	0.75	82	ALWAYS ARE THOSE WHERE THE TYPE OF POLITICAL MODERNIZATION IS DEVELOPED TUTELARY OR UNDEVELOPED TUTELARY, RATHER THAN EARLY OR LATER EUROPEAN OR EUROPEAN DERIVED (55)	1.00	51 0 17 37 X SQ= 51.00 P = 0. RV YES (YES)
84	ALWAYS ARE THOSE WHERE THE TYPE OF POLITICAL MODERNIZATION IS DEVELOPED TUTELARY, RATHER THAN UNDEVELOPED TUTELARY (31)	1.00	84	TEND TO BE THOSE WHERE THE TYPE OF POLITICAL MODERNIZATION IS UNDEVELOPED TUTELARY, RATHER THAN DEVELOPED TUTELARY (24)	0.65	17 13 0 24 X SQ= 17.31 P = 0. RV YES (YES)

87	TEND TO BE THOSE WHOSE IDEOLOGICAL ORIENTATION IS OTHER THAN DEVELOPMENTAL (58)	0.87	87	TEND TO BE THOSE WHOSE IDEOLOGICAL ORIENTATION IS DEVELOPMENTAL (31)	0.92	8 23 55 2 X SQ= 45.91 P = 0. RV YES (YES)
89	TEND TO BE THOSE WHOSE IDEOLOGICAL ORIENTATION IS CONVENTIONAL (33)	0.52	89	ALWAYS ARE THOSE WHOSE IDEOLOGICAL ORIENTATION IS OTHER THAN CONVENTIONAL (62)	1.00	33 0 31 30 X SQ= 21.63 P = 0. RV YES (NO)
91	TEND LESS TO BE THOSE WHOSE IDEOLOGICAL ORIENTATION IS DEVELOPMENTAL, RATHER THAN TRADITIONAL (31)	0.67	91	ALWAYS ARE THOSE WHOSE IDEOLOGICAL ORIENTATION IS DEVELOPMENTAL, RATHER THAN TRADITIONAL (31)	1.00	8 23 4 0 X SQ= 5.68 P = 0.009 RV NO (YES)
95	TEND LESS TO BE THOSE WHERE THE STATUS OF THE REGIME IS CONSTITUTIONAL OR AUTHORITARIAN (95)	0.80	95	TEND MORE TO BE THOSE WHERE THE STATUS OF THE REGIME IS CONSTITUTIONAL OR AUTHORITARIAN (95)	0.97	59 35 15 1 X SQ= 4.64 P = 0.019 RV NO (NO)
97	TEND TO BE THOSE WHERE THE STATUS OF THE REGIME IS TOTALITARIAN, RATHER THAN AUTHORITARIAN (16)	0.54	97	TEND TO BE THOSE WHERE THE STATUS OF THE REGIME IS AUTHORITARIAN, RATHER THAN TOTALITARIAN (23)	0.90	13 9 15 1 X SQ= 4.09 P = 0.025 RV YES (NO)
99	LEAN TOWARD BEING THOSE WHERE GOVERNMENTAL STABILITY IS GENERALLY PRESENT AND DATES FROM AT LEAST THE INTER-WAR PERIOD, OR FROM THE POST-WAR PERIOD (50)	0.62	99	LEAN TOWARD BEING THOSE WHERE GOVERNMENTAL STABILITY IS MODERATELY PRESENT AND DATES FROM THE POST-WAR PERIOD, OR IS ABSENT (36)	0.63	43 6 26 10 X SQ= 2.34 P = 0.094 RV YES (NO)
101	TEND TO BE THOSE WHERE THE REPRESENTATIVE CHARACTER OF THE REGIME IS POLYARCHIC (41)	0.51	101	TEND TO BE THOSE WHERE THE REPRESENTATIVE CHARACTER OF THE REGIME IS LIMITED POLYARCHIC, PSEUDO-POLYARCHIC, OR NON-POLYARCHIC (57)	0.79	35 6 34 22 X SQ= 5.86 P = 0.012 RV YES (NO)
105	TEND TO BE THOSE WHERE THE ELECTORAL SYSTEM IS COMPETITIVE (43)	0.60	105	TEND TO BE THOSE WHERE THE ELECTORAL SYSTEM IS PARTIALLY COMPETITIVE OR NON-COMPETITIVE (47)	0.77	36 7 24 23 X SQ= 9.36 P = 0.002 RV YES (NO)
107	TEND TO BE THOSE WHERE AUTONOMOUS GROUPS ARE FULLY TOLERATED IN POLITICS (46)	0.52	107	TEND TO BE THOSE WHERE AUTONOMOUS GROUPS ARE PARTIALLY TOLERATED IN POLITICS, ARE TOLERATED ONLY OUTSIDE POLITICS, OR ARE NOT TOLERATED AT ALL (65)	0.80	39 7 36 28 X SQ= 8.77 P = 0.002 RV YES (NO)

110	TEND LESS TO BE THOSE WHERE POLITICAL ENCULTURATION IS MEDIUM OR LOW (80)	0.75	110	ALWAYS ARE THOSE WHERE POLITICAL ENCULTURATION IS MEDIUM OR LOW (80)	1.00	15 0 45 34 X SQ= 8.34 P = 0.001 RV NO (NO)
114	TEND TO BE THOSE WHERE SECTIONALISM IS NEGLIGIBLE (47)	0.50	114	TEND TO BE THOSE WHERE SECTIONALISM IS EXTREME OR MODERATE (61)	0.71	36 25 36 10 X SQ= 3.58 P = 0.040 RV YES (NO)
116	TEND LESS TO BE THOSE WHERE INTEREST ARTICULATION BY ASSOCIATIONAL GROUPS IS LIMITED OR NEGLIGIBLE (79)	0.56	116	ALWAYS ARE THOSE WHERE INTEREST ARTICULATION BY ASSOCIATIONAL GROUPS IS LIMITED OR NEGLIGIBLE (79)	1.00	32 0 40 38 X SQ= 21.71 P = 0. RV NO (NO)
117	TEND TO BE THOSE WHERE INTEREST ARTICULATION BY ASSOCIATIONAL GROUPS IS SIGNIFICANT, MODERATE, OR LIMITED (60)	0.75	117	TEND TO BE THOSE WHERE INTEREST ARTICULATION BY ASSOCIATIONAL GROUPS IS NEGLIGIBLE (51)	0.84	54 6 18 32 X SQ= 32.82 P = 0. RV YES (YES)
119	TILT LESS TOWARD BEING THOSE WHERE INTEREST ARTICULATION BY INSTITUTIONAL GROUPS IS VERY SIGNIFICANT OR SIGNIFICANT (74)	0.70	119	TILT MORE TOWARD BEING THOSE WHERE INTEREST ARTICULATION BY INSTITUTIONAL GROUPS IS VERY SIGNIFICANT OR SIGNIFICANT (74)	0.87	53 20 23 3 X SQ= 1.89 P = 0.114 RV NO (NO)
120	TILT LESS TOWARD BEING THOSE WHERE INTEREST ARTICULATION BY INSTITUTIONAL GROUPS IS VERY SIGNIFICANT, SIGNIFICANT, OR MODERATE (90)	0.87	120	ALWAYS ARE THOSE WHERE INTEREST ARTICULATION BY INSTITUTIONAL GROUPS IS VERY SIGNIFICANT, SIGNIFICANT, OR MODERATE (90)	1.00	66 23 10 0 X SQ= 2.07 P = 0.111 RV NO (NO)
121	TEND TO BE THOSE WHERE INTEREST ARTICULATION BY NON-ASSOCIATIONAL GROUPS IS MODERATE, LIMITED, OR NEGLIGIBLE (61)	0.75	121	TEND TO BE THOSE WHERE INTEREST ARTICULATION BY NON-ASSOCIATIONAL GROUPS IS SIGNIFICANT (54)	0.89	19 34 57 4 X SQ= 39.78 P = 0. RV YES (YES)
122	TEND LESS TO BE THOSE WHERE INTEREST ARTICULATION BY NON-ASSOCIATIONAL GROUPS IS SIGNIFICANT OR MODERATE (83)	0.58	122	ALWAYS ARE THOSE WHERE INTEREST ARTICULATION BY NON-ASSOCIATIONAL GROUPS IS SIGNIFICANT OR MODERATE (83)	1.00	44 38 32 0 X SQ= 20.21 P = 0. RV NO (NO)
125	TEND LESS TO BE THOSE WHERE INTEREST ARTICULATION BY ANOMIC GROUPS IS FREQUENT OR OCCASIONAL (64)	0.55	125	TEND MORE TO BE THOSE WHERE INTEREST ARTICULATION BY ANOMIC GROUPS IS FREQUENT OR OCCASIONAL (64)	0.81	34 29 28 7 X SQ= 5.49 P = 0.016 RV NO (NO)

126	TEND LESS TO BE THOSE WHERE INTEREST ARTICULATION BY ANOMIC GROUPS IS FREQUENT, OCCASIONAL, OR INFREQUENT (83)	0.74	126	ALWAYS ARE THOSE WHERE INTEREST ARTICULATION BY ANOMIC GROUPS IS FREQUENT, OCCASIONAL, OR INFREQUENT (83)	1.00	46 36 16 0 X SQ= 9.29 P = 0. RV NO (NO)
128	LEAN TOWARD BEING THOSE WHERE INTEREST ARTICULATION BY POLITICAL PARTIES IS SIGNIFICANT OR MODERATE (48)	0.59	128	LEAN TOWARD BEING THOSE WHERE INTEREST ARTICULATION BY POLITICAL PARTIES IS LIMITED OR NEGLIGIBLE (45)	0.63	36 12 25 20 X SQ= 3.08 P = 0.054 RV YES (NO)
129	LEAN TOWARD BEING THOSE WHERE INTEREST ARTICULATION BY POLITICAL PARTIES IS SIGNIFICANT, MODERATE, OR LIMITED (56)	0.67	129	LEAN TOWARD BEING THOSE WHERE INTEREST ARTICULATION BY POLITICAL PARTIES IS NEGLIGIBLE (37)	0.53	41 15 20 17 X SQ= 2.82 P = 0.075 RV YES (NO)
131	LEAN TOWARD BEING THOSE WHERE INTEREST AGGREGATION BY POLITICAL PARTIES IS LIMITED OR NEGLIGIBLE (35)	0.62	131	LEAN TOWARD BEING THOSE WHERE INTEREST AGGREGATION BY POLITICAL PARTIES IS SIGNIFICANT OR MODERATE (30)	0.67	18 12 29 6 X SQ= 3.15 P = 0.054 RV YES (NO)
133	TEND TO BE THOSE WHERE INTEREST AGGREGATION BY THE EXECUTIVE IS MODERATE, LIMITED, OR NEGLIGIBLE (76)	0.85	133	TEND TO BE THOSE WHERE INTEREST AGGREGATION BY THE EXECUTIVE IS SIGNIFICANT (29)	0.53	10 19 58 17 X SQ= 15.13 P = 0. RV YES (YES)
134	LEAN TOWARD BEING THOSE WHERE INTEREST AGGREGATION BY THE EXECUTIVE IS LIMITED OR NEGLIGIBLE (46)	0.51	134	LEAN TOWARD BEING THOSE WHERE INTEREST AGGREGATION BY THE EXECUTIVE IS SIGNIFICANT OR MODERATE (57)	0.69	33 24 34 11 X SQ= 2.74 P = 0.092 RV YES (NO)
136	TEND LESS TO BE THOSE WHERE INTEREST AGGREGATION BY THE LEGISLATURE IS MODERATE, LIMITED, OR NEGLIGIBLE (85)	0.81	136	ALWAYS ARE THOSE WHERE INTEREST AGGREGATION BY THE LEGISLATURE IS MODERATE, LIMITED, OR NEGLIGIBLE (85)	1.00	12 0 50 35 X SQ= 6.05 P = 0.004 RV NO (NO)
137	TEND LESS TO BE THOSE WHERE INTEREST AGGREGATION BY THE LEGISLATURE IS LIMITED OR NEGLIGIBLE (68)	0.56	137	TEND MORE TO BE THOSE WHERE INTEREST AGGREGATION BY THE LEGISLATURE IS LIMITED OR NEGLIGIBLE (68)	0.94	27 2 35 33 X SQ= 13.53 P = 0. RV NO (NO)
138	TEND TO BE THOSE WHERE INTEREST AGGREGATION BY THE LEGISLATURE IS SIGNIFICANT, MODERATE, OR LIMITED (48)	0.66	138	TEND TO BE THOSE WHERE INTEREST AGGREGATION BY THE LEGISLATURE IS NEGLIGIBLE (49)	0.80	41 7 21 28 X SQ= 17.24 P = 0. RV YES (YES)

140	TILT MORE TOWARD BEING THOSE 0.91 WHERE THE PARTY SYSTEM IS QUANTITATIVELY OTHER THAN ONE PARTY DOMINANT (87)	140	TILT LESS TOWARD BEING THOSE 0.79 WHERE THE PARTY SYSTEM IS QUANTITATIVELY OTHER THAN ONE PARTY DOMINANT (87)	6 7 60 26 X SQ= 1.87 P = 0.118 RV NO (YES)
142	TEND LESS TO BE THOSE 0.60 WHERE THE PARTY SYSTEM IS QUANTITATIVELY OTHER THAN MULTI-PARTY (66)	142	TEND MORE TO BE THOSE 0.84 WHERE THE PARTY SYSTEM IS QUANTITATIVELY OTHER THAN MULTI-PARTY (66)	25 5 38 27 X SQ= 4.63 P = 0.020 RV NO (NO)
146	TILT MORE TOWARD BEING THOSE 0.95 WHERE THE PARTY SYSTEM IS QUALITATIVELY OTHER THAN MASS-BASED TERRITORIAL (92)	146	TILT LESS TOWARD BEING THOSE 0.85 WHERE THE PARTY SYSTEM IS QUALITATIVELY OTHER THAN MASS-BASED TERRITORIAL (92)	3 5 63 29 X SQ= 1.92 P = 0.117 RV NO (YES)
147	TEND LESS TO BE THOSE 0.64 WHERE THE PARTY SYSTEM IS QUALITATIVELY OTHER THAN CLASS-ORIENTED OR MULTI-IDEOLOGICAL (67)	147	TEND MORE TO BE THOSE 0.94 WHERE THE PARTY SYSTEM IS QUALITATIVELY OTHER THAN CLASS-ORIENTED OR MULTI-IDEOLOGICAL (67)	21 2 37 30 X SQ= 8.22 P = 0.002 RV NO (NO)
148	ALWAYS ARE THOSE 1.00 WHERE THE PARTY SYSTEM IS QUALITATIVELY OTHER THAN AFRICAN TRANSITIONAL (96)	148	TEND LESS TO BE THOSE 0.62 WHERE THE PARTY SYSTEM IS QUALITATIVELY OTHER THAN AFRICAN TRANSITIONAL (96)	0 14 73 23 X SQ= 28.33 P = 0. RV NO (YES)
153	TEND TO BE THOSE 0.60 WHERE THE PARTY SYSTEM IS STABLE (42)	153	TEND TO BE THOSE 0.76 WHERE THE PARTY SYSTEM IS MODERATELY STABLE OR UNSTABLE (38)	38 4 25 13 X SQ= 5.87 P = 0.012 RV YES (NO)
158	TEND TO BE THOSE 0.68 WHERE PERSONALISMO IS NEGLIGIBLE (56)	158	TEND TO BE THOSE 0.61 WHERE PERSONALISMO IS PRONOUNCED OR MODERATE (40)	21 19 44 12 X SQ= 6.11 P = 0.009 RV YES (NO)
160	TEND LESS TO BE THOSE 0.61 WHERE THE POLITICAL LEADERSHIP IS MODERATELY ELITIST OR NON-ELITIST (67)	160	TEND MORE TO BE THOSE 0.88 WHERE THE POLITICAL LEADERSHIP IS MODERATELY ELITIST OR NON-ELITIST (67)	25 4 39 28 X SQ= 5.94 P = 0.009 RV NO (NO)
164	TEND TO BE THOSE 0.81 WHERE THE REGIME'S LEADERSHIP CHARISMA IS NEGLIGIBLE (65)	164	TEND TO BE THOSE 0.71 WHERE THE REGIME'S LEADERSHIP CHARISMA IS PRONOUNCED OR MODERATE (34)	13 20 57 8 X SQ= 22.71 P = 0. RV YES (YES)

169	LEAN TOWARD BEING THOSE WHERE THE HORIZONTAL POWER DISTRIBUTION IS SIGNIFICANT OR LIMITED (58)	0.62	169	LEAN TOWARD BEING THOSE WHERE THE HORIZONTAL POWER DISTRIBUTION IS NEGLIGIBLE (48)	0.59	44 14 27 20 X SQ= 3.22 P = 0.059 RV YES (NO)
175	TEND TO BE THOSE WHERE THE LEGISLATURE IS FULLY EFFECTIVE OR PARTIALLY EFFECTIVE (51)	0.62	175	TEND TO BE THOSE WHERE THE LEGISLATURE IS LARGELY INEFFECTIVE OR WHOLLY INEFFECTIVE (49)	0.71	41 10 25 24 X SQ= 8.34 P = 0.003 RV YES (NO)
178	TEND TO BE THOSE WHERE THE LEGISLATURE IS BICAMERAL (51)	0.60	178	TEND TO BE THOSE WHERE THE LEGISLATURE IS UNICAMERAL (53)	0.72	27 26 41 10 X SQ= 8.70 P = 0.002 RV YES (NO)
181	LEAN LESS TOWARD BEING THOSE WHERE THE BUREAUCRACY IS SEMI-MODERN, RATHER THAN MODERN (55)	0.69	181	ALWAYS ARE THOSE WHERE THE BUREAUCRACY IS SEMI-MODERN, RATHER THAN MODERN (55)	1.00	21 0 46 9 X SQ= 2.49 P = 0.056 RV NO (NO)
182	TEND TO BE THOSE WHERE THE BUREAUCRACY IS SEMI-MODERN, RATHER THAN POST-COLONIAL TRANSITIONAL (55)	0.96	182	TEND TO BE THOSE WHERE THE BUREAUCRACY IS POST-COLONIAL TRANSITIONAL, RATHER THAN SEMI-MODERN (25)	0.72	46 9 2 23 X SQ= 37.88 P = 0. RV YES (YES)
183	TEND TO BE THOSE WHERE THE BUREAUCRACY IS TRADITIONAL, RATHER THAN POST-COLONIAL TRANSITIONAL (9)	0.71	183	TEND TO BE THOSE WHERE THE BUREAUCRACY IS POST-COLONIAL TRANSITIONAL, RATHER THAN TRADITIONAL (25)	0.88	2 23 5 3 X SQ= 7.76 P = 0.004 RV YES (YES)
188	TEND LESS TO BE THOSE WHERE THE CHARACTER OF THE LEGAL SYSTEM IS OTHER THAN CIVIL LAW (81)	0.57	188	ALWAYS ARE THOSE WHERE THE CHARACTER OF THE LEGAL SYSTEM IS OTHER THAN CIVIL LAW (81)	1.00	32 0 43 37 X SQ= 20.06 P = 0. RV NO (NO)
190	TEND TO BE THOSE WHERE THE CHARACTER OF THE LEGAL SYSTEM IS CIVIL LAW, RATHER THAN COMMON LAW (32)	0.86	190	ALWAYS ARE THOSE WHERE THE CHARACTER OF THE LEGAL SYSTEM IS COMMON LAW, RATHER THAN CIVIL LAW (7)	1.00	32 0 5 2 X SQ= 4.66 P = 0.028 RV YES (NO)
192	TEND MORE TO BE THOSE WHERE THE CHARACTER OF THE LEGAL SYSTEM IS OTHER THAN MUSLIM OR PARTLY MUSLIM (86)	0.88	192	TEND LESS TO BE THOSE WHERE THE CHARACTER OF THE LEGAL SYSTEM IS OTHER THAN MUSLIM OR PARTLY MUSLIM (86)	0.54	9 17 66 20 X SQ= 14.17 P = 0. RV NO (YES)

```
193  TEND TO BE THOSE                               0.97        193  TEND TO BE THOSE                              0.70          2    26
     WHERE THE CHARACTER OF THE LEGAL SYSTEM                         WHERE THE CHARACTER OF THE LEGAL SYSTEM                    74    11
     IS OTHER THAN PARTLY INDIGENOUS  (86)                           IS PARTLY INDIGENOUS  (28)                              X SQ=  57.51
                                                                                                                             P =     0.
                                                                                                                             RV YES (YES)

194  LEAN LESS TOWARD BEING THOSE                   0.84        194  LEAN MORE TOWARD BEING THOSE                  0.97         12     1
     THAT ARE NON-COMMUNIST  (101)                                   THAT ARE NON-COMMUNIST  (101)                              63    37
                                                                                                                             X SQ=   3.21
                                                                                                                             P =    0.057
                                                                                                                             RV NO  (NO )
```

87 POLITIES
WHOSE IDEOLOGICAL ORIENTATION
IS DEVELOPMENTAL (31)

87 POLITIES
WHOSE IDEOLOGICAL ORIENTATION
IS OTHER THAN DEVELOPMENTAL (58)

BOTH SUBJECT AND PREDICATE

31 IN LEFT COLUMN

ALGERIA	BOLIVIA	BURUNDI	CAMEROUN	CEN AFR REP	CHAD	CONGO(BRA)	CONGO(LEO)	DAHOMEY	DOMIN REP
GABON	GHANA	GUINEA	INDONESIA	ISRAEL	IVORY COAST	MALAYA	MALI	MAURITANIA	NIGER
RWANDA	SENEGAL	SIERRE LEO	SOMALIA	SUDAN	TANGANYIKA	TOGO	TUNISIA	UAR	UPPER VOLTA
VENEZUELA									

58 IN RIGHT COLUMN

AFGHANISTAN	ALBANIA	AUSTRALIA	AUSTRIA	BELGIUM	BULGARIA	CANADA	CHILE	CHINA, PR	COLOMBIA
COSTA RICA	CYPRUS	CZECHOS'KIA	DENMARK	ECUADOR	EL SALVADOR	ETHIOPIA	FINLAND	FRANCE	GERMANY, E
GERMAN FR	GREECE	HAITI	HONDURAS	HUNGARY	ICELAND	IRAQ	IRELAND	ITALY	JAPAN
JORDAN	KOREA, N	LAOS	LEBANON	LUXEMBOURG	MEXICO	MONGOLIA	NETHERLANDS	NEW ZEALAND	NORWAY
PHILIPPINES	POLAND	PORTUGAL	RUMANIA	SA'U ARABIA	SO AFRICA	SPAIN	SWEDEN	SWITZERLAND	SYRIA
THAILAND	TURKEY	USSR	UK	US	URUGUAY	VIETNAM, N	YUGOSLAVIA		

23 EXCLUDED BECAUSE AMBIGUOUS

ARGENTINA	BRAZIL	BURMA	CAMBODIA	CEYLON	CUBA	GUATEMALA	INDIA	IRAN	JAMAICA
LIBERIA	LIBYA	MALAGASY R	MOROCCO	NEPAL	NICARAGUA	NIGERIA	PAKISTAN	PANAMA	PARAGUAY
TRINIDAD	UGANDA	VIETNAM REP							

3 EXCLUDED BECAUSE UNASCERTAINABLE

KOREA REP PERU YEMEN

2	ALWAYS ARE THOSE LOCATED ELSEWHERE THAN IN WEST EUROPE, SCANDINAVIA, NORTH AMERICA, OR AUSTRALASIA (93)	1.00	2	TEND LESS TO BE THOSE LOCATED ELSEWHERE THAN IN WEST EUROPE, SCANDINAVIA, NORTH AMERICA, OR AUSTRALASIA (93)	0.62	0 22 31 36 X SQ= 13.65 P = 0. RV NO (NO)
10	TEND TO BE THOSE LOCATED IN NORTH AFRICA, OR CENTRAL AND SOUTH AFRICA (33)	0.81	10	TEND TO BE THOSE LOCATED ELSEWHERE THAN IN NORTH AFRICA, OR CENTRAL AND SOUTH AFRICA (82)	0.97	25 2 6 56 X SQ= 53.37 P = 0. RV YES (YES)
15	TEND TO BE THOSE LOCATED IN NORTH AFRICA OR CENTRAL AND SOUTH AFRICA, RATHER THAN IN THE MIDDLE EAST (33)	0.96	15	TEND TO BE THOSE LOCATED IN THE MIDDLE EAST, RATHER THAN IN NORTH AFRICA OR CENTRAL AND SOUTH AFRICA (11)	0.80	1 8 25 2 X SQ= 18.46 P = 0. RV YES (YES)

18	TEND TO BE THOSE LOCATED IN NORTH AFRICA OR CENTRAL AND SOUTH AFRICA, RATHER THAN IN EAST ASIA, SOUTH ASIA, OR SOUTHEAST ASIA (33)	0.93	18	TEND TO BE THOSE LOCATED IN EAST ASIA, SOUTH ASIA, OR SOUTHEAST ASIA, RATHER THAN IN NORTH AFRICA OR CENTRAL AND SOUTH AFRICA (18)	0.80	25 2 2 8 X SQ= 15.99 P = 0. RV YES (YES)
19	TEND TO BE THOSE LOCATED IN NORTH AFRICA OR CENTRAL AND SOUTH AFRICA, RATHER THAN IN THE CARIBBEAN, CENTRAL AMERICA, OR SOUTH AMERICA (33)	0.89	19	TEND TO BE THOSE LOCATED IN THE CARIBBEAN, CENTRAL AMERICA, OR SOUTH AMERICA, RATHER THAN IN NORTH AFRICA OR CENTRAL AND SOUTH AFRICA (22)	0.82	25 2 3 9 X SQ= 15.55 P = 0. RV YES (YES)
24	TEND TO BE THOSE WHOSE POPULATION IS SMALL (54)	0.71	24	TEND TO BE THOSE WHOSE POPULATION IS VERY LARGE, LARGE, OR MEDIUM (61)	0.64	9 37 22 21 X SQ= 8.43 P = 0.002 RV YES (YES)
25	ALWAYS ARE THOSE WHOSE POPULATION DENSITY IS MEDIUM OR LOW (98)	1.00	25	TEND LESS TO BE THOSE WHOSE POPULATION DENSITY IS MEDIUM OR LOW (98)	0.79	0 12 31 46 X SQ= 5.75 P = 0.007 RV NO (NO)
26	TEND TO BE THOSE WHOSE POPULATION DENSITY IS LOW (67)	0.81	26	TEND TO BE THOSE WHOSE POPULATION DENSITY IS VERY HIGH, HIGH, OR MEDIUM (48)	0.55	6 32 25 26 X SQ= 9.18 P = 0.001 RV YES (NO)
28	LEAN TOWARD BEING THOSE WHOSE POPULATION GROWTH RATE IS HIGH (62)	0.66	28	LEAN TOWARD BEING THOSE WHOSE POPULATION GROWTH RATE IS LOW (48)	0.56	19 24 10 31 X SQ= 2.82 P = 0.069 RV YES (NO)
29	TEND TO BE THOSE WHERE THE DEGREE OF URBANIZATION IS LOW (49)	0.83	29	TEND TO BE THOSE WHERE THE DEGREE OF URBANIZATION IS HIGH (56)	0.77	5 40 25 12 X SQ= 25.51 P = 0. RV YES (YES)
30	TEND TO BE THOSE WHOSE AGRICULTURAL POPULATION IS HIGH (56)	0.84	30	TEND TO BE THOSE WHOSE AGRICULTURAL POPULATION IS MEDIUM, LOW, OR VERY LOW (57)	0.73	26 15 5 41 X SQ= 23.86 P = 0. RV YES (YES)
32	ALWAYS ARE THOSE WHOSE GROSS NATIONAL PRODUCT IS MEDIUM, LOW, OR VERY LOW (105)	1.00	32	TEND LESS TO BE THOSE WHOSE GROSS NATIONAL PRODUCT IS MEDIUM, LOW, OR VERY LOW (105)	0.84	0 9 31 49 X SQ= 3.78 P = 0.024 RV NO (NO)

33	TEND MORE TO BE THOSE WHOSE GROSS NATIONAL PRODUCT IS LOW OR VERY LOW (85)	0.97	33	TEND LESS TO BE THOSE WHOSE GROSS NATIONAL PRODUCT IS LOW OR VERY LOW (85)	0.57	1 25 30 33 X SQ= 13.67 P = 0. RV NO (NO)
34	TEND TO BE THOSE WHOSE GROSS NATIONAL PRODUCT IS VERY LOW (53)	0.77	34	TEND TO BE THOSE WHOSE GROSS NATIONAL PRODUCT IS VERY HIGH, HIGH, MEDIUM, OR LOW (62)	0.72	7 42 24 16 X SQ= 18.31 P = 0. RV YES (YES)
36	TEND TO BE THOSE WHOSE PER CAPITA GROSS NATIONAL PRODUCT IS LOW OR VERY LOW (73)	0.90	36	TEND TO BE THOSE WHOSE PER CAPITA GROSS NATIONAL PRODUCT IS VERY HIGH, HIGH, OR MEDIUM (42)	0.60	3 35 28 23 X SQ= 19.18 P = 0. RV YES (YES)
37	TEND TO BE THOSE WHOSE PER CAPITA GROSS NATIONAL PRODUCT IS VERY LOW (51)	0.77	37	TEND TO BE THOSE WHOSE PER CAPITA GROSS NATIONAL PRODUCT IS VERY HIGH, HIGH, MEDIUM, OR LOW (64)	0.83	7 48 24 10 X SQ= 28.49 P = 0. RV YES (YES)
38	ALWAYS ARE THOSE WHOSE INTERNATIONAL FINANCIAL STATUS IS MEDIUM, LOW, OR VERY LOW (103)	1.00	38	TEND LESS TO BE THOSE WHOSE INTERNATIONAL FINANCIAL STATUS IS MEDIUM, LOW, OR VERY LOW (103)	0.84	0 9 31 47 X SQ= 3.96 P = 0.024 RV NO (NO)
39	TEND TO BE THOSE WHOSE INTERNATIONAL FINANCIAL STATUS IS LOW OR VERY LOW (76)	0.90	39	TEND TO BE THOSE WHOSE INTERNATIONAL FINANCIAL STATUS IS VERY HIGH, HIGH, OR MEDIUM (38)	0.54	3 31 28 26 X SQ= 15.10 P = 0. RV YES (YES)
40	TEND TO BE THOSE WHOSE INTERNATIONAL FINANCIAL STATUS IS VERY LOW (39)	0.63	40	TEND TO BE THOSE WHOSE INTERNATIONAL FINANCIAL STATUS IS VERY HIGH, HIGH, MEDIUM, OR LOW (71)	0.82	11 46 19 10 X SQ= 16.10 P = 0. RV YES (YES)
42	ALWAYS ARE THOSE WHOSE ECONOMIC DEVELOPMENTAL STATUS IS UNDERDEVELOPED OR VERY UNDERDEVELOPED (76)	1.00	42	TEND TO BE THOSE WHOSE ECONOMIC DEVELOPMENTAL STATUS IS DEVELOPED OR INTERMEDIATE (36)	0.60	0 34 29 23 X SQ= 26.17 P = 0. RV YES (YES)
43	TEND TO BE THOSE WHOSE ECONOMIC DEVELOPMENTAL STATUS IS VERY UNDERDEVELOPED (57)	0.90	43	TEND TO BE THOSE WHOSE ECONOMIC DEVELOPMENTAL STATUS IS DEVELOPED, INTERMEDIATE, OR UNDERDEVELOPED (55)	0.76	3 42 28 13 X SQ= 32.72 P = 0. RV YES (YES)

44	ALWAYS ARE THOSE WHERE THE LITERACY RATE IS BELOW NINETY PERCENT (90)	1.00	44	TEND LESS TO BE THOSE WHERE THE LITERACY RATE IS BELOW NINETY PERCENT (90)	0.57	0 25 31 33 X SQ= 16.51 P = 0. RV NO (NO)
45	TEND TO BE THOSE WHERE THE LITERACY RATE IS BELOW FIFTY PERCENT (54)	0.87	45	TEND TO BE THOSE WHERE THE LITERACY RATE IS FIFTY PERCENT OR ABOVE (55)	0.75	4 40 27 13 X SQ= 28.24 P = 0. RV YES (YES)
46	TEND TO BE THOSE WHERE THE LITERACY RATE IS BELOW TEN PERCENT (26)	0.66	46	TEND TO BE THOSE WHERE THE LITERACY RATE IS TEN PERCENT OR ABOVE (84)	0.93	10 52 19 4 X SQ= 30.09 P = 0. RV YES (YES)
52	LEAN MORE TOWARD BEING THOSE WHERE FREEDOM OF THE PRESS IS COMPLETE, INTERMITTENT, OR INTERNALLY ABSENT (82)	0.95	52	LEAN LESS TOWARD BEING THOSE WHERE FREEDOM OF THE PRESS IS COMPLETE, INTERMITTENT, OR INTERNALLY ABSENT (82)	0.76	20 42 1 13 X SQ= 2.46 P = 0.095 RV NO (NO)
53	ALWAYS ARE THOSE WHERE NEWSPAPER CIRCULATION IS LESS THAN THREE HUNDRED PER THOUSAND (101)	1.00	53	TEND LESS TO BE THOSE WHERE NEWSPAPER CIRCULATION IS LESS THAN THREE HUNDRED PER THOUSAND (101)	0.76	0 14 31 44 X SQ= 7.15 P = 0.002 RV NO (NO)
54	TEND TO BE THOSE WHERE NEWSPAPER CIRCULATION IS LESS THAN ONE HUNDRED PER THOUSAND (76)	0.97	54	TEND TO BE THOSE WHERE NEWSPAPER CIRCULATION IS ONE HUNDRED OR MORE PER THOUSAND (37)	0.59	1 33 30 23 X SQ= 23.72 P = 0. RV YES (YES)
55	TEND TO BE THOSE WHERE NEWSPAPER CIRCULATION IS LESS THAN TEN PER THOUSAND (35)	0.68	55	TEND TO BE THOSE WHERE NEWSPAPER CIRCULATION IS TEN OR MORE PER THOUSAND (78)	0.93	10 52 21 4 X SQ= 32.88 P = 0. RV YES (YES)
56	TEND TO BE THOSE WHERE THE RELIGION IS PREDOMINANTLY OR PARTLY NON-LITERATE (31)	0.65	56	TEND TO BE THOSE WHERE THE RELIGION IS PREDOMINANTLY LITERATE (79)	0.91	11 50 20 5 X SQ= 26.91 P = 0. RV YES (YES)
63	TEND TO BE THOSE WHERE THE RELIGION IS PREDOMINANTLY OR PARTLY OTHER THAN CHRISTIAN (68)	0.94	63	TEND TO BE THOSE WHERE THE RELIGION IS PREDOMINANTLY SOME KIND OF CHRISTIAN (46)	0.65	2 37 29 20 X SQ= 25.49 P = 0. RV YES (YES)

No.	Statement	Prop.	No.	Statement	Prop.	Statistics
66	TEND TO BE THOSE THAT ARE RELIGIOUSLY HETEROGENEOUS (49)	0.64	66	TEND TO BE THOSE THAT ARE RELIGIOUSLY HOMOGENEOUS (57)	0.61	10 33 18 21 X SQ= 3.80 P = 0.037 RV YES (NO)
68	TEND TO BE THOSE THAT ARE LINGUISTICALLY WEAKLY HETEROGENEOUS OR STRONGLY HETEROGENEOUS (62)	0.77	68	TEND TO BE THOSE THAT ARE LINGUISTICALLY HOMOGENEOUS (52)	0.61	7 35 24 22 X SQ= 10.62 P = 0.001 RV YES (YES)
69	TEND TO BE THOSE THAT ARE LINGUISTICALLY STRONGLY HETEROGENEOUS (50)	0.77	69	TEND TO BE THOSE THAT ARE LINGUISTICALLY HOMOGENEOUS OR WEAKLY HETEROGENEOUS (64)	0.75	7 43 24 14 X SQ= 20.76 P = 0. RV YES (YES)
70	LEAN MORE TOWARD BEING THOSE THAT ARE RELIGIOUSLY, RACIALLY, OR LINGUISTICALLY HETEROGENEOUS (85)	0.90	70	LEAN LESS TOWARD BEING THOSE THAT ARE RELIGIOUSLY, RACIALLY, OR LINGUISTICALLY HETEROGENEOUS (85)	0.70	3 16 26 38 X SQ= 2.96 P = 0.057 RV NO (NO)
71	ALWAYS ARE THOSE WHOSE DATE OF INDEPENDENCE IS AFTER 1800 (90)	1.00	71	TEND LESS TO BE THOSE WHOSE DATE OF INDEPENDENCE IS AFTER 1800 (90)	0.65	0 19 31 36 X SQ= 11.81 P = 0. RV NO (NO)
72	TEND TO BE THOSE WHOSE DATE OF INDEPENDENCE IS AFTER 1914 (59)	0.90	72	TEND TO BE THOSE WHOSE DATE OF INDEPENDENCE IS BEFORE 1914 (52)	0.69	3 38 28 17 X SQ= 25.72 P = 0. RV YES (YES)
73	TEND TO BE THOSE WHOSE DATE OF INDEPENDENCE IS AFTER 1945 (46)	0.90	73	TEND TO BE THOSE WHOSE DATE OF INDEPENDENCE IS BEFORE 1945 (65)	0.91	3 50 28 5 X SQ= 51.94 P = 0. RV YES (YES)
74	ALWAYS ARE THOSE THAT ARE NOT HISTORICALLY WESTERN (87)	1.00	74	TEND LESS TO BE THOSE THAT ARE NOT HISTORICALLY WESTERN (87)	0.54	0 26 30 31 X SQ= 17.40 P = 0. RV NO (NO)
75	TEND TO BE THOSE THAT ARE NOT HISTORICALLY WESTERN AND ARE NOT SIGNIFICANTLY WESTERNIZED (52)	0.81	75	TEND TO BE THOSE THAT ARE HISTORICALLY WESTERN OR SIGNIFICANTLY WESTERNIZED (62)	0.77	6 44 25 13 X SQ= 25.07 P = 0. RV YES (YES)

76	ALWAYS ARE THOSE THAT HAVE BEEN SIGNIFICANTLY OR PARTIALLY WESTERNIZED THROUGH A COLONIAL RELATIONSHIP, RATHER THAN BEING HISTORICALLY WESTERN (70)	1.00	76	TEND TO BE THOSE THAT ARE HISTORICALLY WESTERN, RATHER THAN HAVING BEEN SIGNIFICANTLY OR PARTIALLY WESTERNIZED THROUGH A COLONIAL RELATIONSHIP (26)	0.59	0 26 30 18 X SQ= 24.80 P = 0. RV YES (YES)
77	TEND TO BE THOSE THAT HAVE BEEN PARTIALLY WESTERNIZED, RATHER THAN SIGNIFICANTLY WESTERNIZED, THROUGH A COLONIAL RELATIONSHIP (41)	0.83	77	TEND TO BE THOSE THAT HAVE BEEN SIGNIFICANTLY WESTERNIZED, RATHER THAN PARTIALLY WESTERNIZED, THROUGH A COLONIAL RELATIONSHIP (28)	0.65	5 11 25 6 X SQ= 9.12 P = 0.001 RV YES (YES)
78	ALWAYS ARE THOSE THAT HAVE BEEN SIGNIFICANTLY WESTERNIZED THROUGH A COLONIAL RELATIONSHIP, RATHER THAN WITHOUT SUCH A RELATIONSHIP (28)	1.00	78	TILT LESS TOWARD BEING THOSE THAT HAVE BEEN SIGNIFICANTLY WESTERNIZED THROUGH A COLONIAL RELATIONSHIP, RATHER THAN WITHOUT SUCH A RELATIONSHIP (28)	0.61	5 11 0 7 X SQ= 1.26 P = 0.147 RV NO (NO)
79	LEAN MORE TOWARD BEING THOSE THAT WERE FORMERLY DEPENDENCIES OF BRITAIN OR FRANCE, RATHER THAN SPAIN (49)	0.88	79	LEAN LESS TOWARD BEING THOSE THAT WERE FORMERLY DEPENDENCIES OF BRITAIN OR FRANCE, RATHER THAN SPAIN (49)	0.64	23 14 3 8 X SQ= 2.87 P = 0.082 RV NO (YES)
82	TEND TO BE THOSE WHERE THE TYPE OF POLITICAL MODERNIZATION IS DEVELOPED TUTELARY OR UNDEVELOPED TUTELARY, RATHER THAN EARLY OR LATER EUROPEAN OR EUROPEAN DERIVED (55)	0.90	82	TEND TO BE THOSE WHERE THE TYPE OF POLITICAL MODERNIZATION IS EARLY OR LATER EUROPEAN OR EUROPEAN DERIVED, RATHER THAN DEVELOPED TUTELARY OR UNDEVELOPED TUTELARY (51)	0.78	3 40 28 11 X SQ= 33.84 P = 0. RV YES (YES)
83	ALWAYS ARE THOSE WHERE THE TYPE OF POLITICAL MODERNIZATION IS OTHER THAN NON-EUROPEAN AUTOCHTHONOUS (106)	1.00	83	LEAN LESS TOWARD BEING THOSE WHERE THE TYPE OF POLITICAL MODERNIZATION IS OTHER THAN NON-EUROPEAN AUTOCHTHONOUS (106)	0.88	0 7 31 51 X SQ= 2.57 P = 0.050 RV NO (NO)
84	TEND TO BE THOSE WHERE THE TYPE OF POLITICAL MODERNIZATION IS UNDEVELOPED TUTELARY, RATHER THAN DEVELOPED TUTELARY (24)	0.75	84	ALWAYS ARE THOSE WHERE THE TYPE OF POLITICAL MODERNIZATION IS DEVELOPED TUTELARY, RATHER THAN UNDEVELOPED TUTELARY (31)	1.00	7 11 21 0 X SQ= 14.98 P = 0. RV YES (YES)
85	TEND TO BE THOSE WHERE THE STAGE OF POLITICAL MODERNIZATION IS MID- OR EARLY TRANSITIONAL, RATHER THAN ADVANCED (54)	0.81	85	TEND TO BE THOSE WHERE THE STAGE OF POLITICAL MODERNIZATION IS ADVANCED, RATHER THAN MID- OR EARLY TRANSITIONAL (60)	0.81	6 46 25 11 X SQ= 28.77 P = 0. RV YES (YES)
86	TEND TO BE THOSE WHERE THE STAGE OF POLITICAL MODERNIZATION IS EARLY TRANSITIONAL, RATHER THAN ADVANCED OR MID-TRANSITIONAL (38)	0.74	86	TEND TO BE THOSE WHERE THE STAGE OF POLITICAL MODERNIZATION IS ADVANCED OR MID-TRANSITIONAL, RATHER THAN EARLY TRANSITIONAL (76)	0.96	8 55 23 2 X SQ= 45.91 P = 0. RV YES (YES)

93	LEAN LESS TOWARD BEING THOSE WHERE THE SYSTEM STYLE IS NON-MOBILIZATIONAL (78)	0.55	93	LEAN MORE TOWARD BEING THOSE WHERE THE SYSTEM STYLE IS NON-MOBILIZATIONAL (78)	0.77	14 13 17 44 X SQ= 3.73 P = 0.052 RV NO (YES)
95	ALWAYS ARE THOSE WHERE THE STATUS OF THE REGIME IS CONSTITUTIONAL OR AUTHORITARIAN (95)	1.00	95	TEND LESS TO BE THOSE WHERE THE STATUS OF THE REGIME IS CONSTITUTIONAL OR AUTHORITARIAN (95)	0.73	31 41 0 15 X SQ= 8.24 P = 0.001 RV NO (NO)
97	ALWAYS ARE THOSE WHERE THE STATUS OF THE REGIME IS AUTHORITARIAN, RATHER THAN TOTALITARIAN (23)	1.00	97	TEND TO BE THOSE WHERE THE STATUS OF THE REGIME IS TOTALITARIAN, RATHER THAN AUTHORITARIAN (16)	0.68	7 7 0 15 X SQ= 7.34 P = 0.002 RV YES (YES)
98	ALWAYS ARE THOSE WHERE GOVERNMENTAL STABILITY IS GENERALLY OR MODERATELY PRESENT AND DATES FROM THE POST-WAR PERIOD, OR IS ABSENT (93)	1.00	98	TEND LESS TO BE THOSE WHERE GOVERNMENTAL STABILITY IS GENERALLY OR MODERATELY PRESENT AND DATES FROM THE POST-WAR PERIOD, OR IS ABSENT (93)	0.64	0 21 31 37 X SQ= 12.75 P = 0. RV NO (NO)
99	LEAN TOWARD BEING THOSE WHERE GOVERNMENTAL STABILITY IS MODERATELY PRESENT AND DATES FROM THE POST-WAR PERIOD, OR IS ABSENT (36)	0.54	99	LEAN TOWARD BEING THOSE WHERE GOVERNMENTAL STABILITY IS GENERALLY PRESENT AND DATES FROM AT LEAST THE INTER-WAR PERIOD, OR FROM THE POST-WAR PERIOD (50)	0.73	6 40 7 15 X SQ= 2.29 P = 0.098 RV YES (NO)
101	LEAN MORE TOWARD BEING THOSE WHERE THE REPRESENTATIVE CHARACTER OF THE REGIME IS LIMITED POLYARCHIC, PSEUDO-POLYARCHIC, OR NON-POLYARCHIC (57)	0.72	101	LEAN LESS TOWARD BEING THOSE WHERE THE REPRESENTATIVE CHARACTER OF THE REGIME IS LIMITED POLYARCHIC, PSEUDO-POLYARCHIC, OR NON-POLYARCHIC (57)	0.51	7 27 18 28 X SQ= 2.33 P = 0.092 RV NO (NO)
105	TEND TO BE THOSE WHERE THE ELECTORAL SYSTEM IS PARTIALLY COMPETITIVE OR NON-COMPETITIVE (47)	0.76	105	TEND TO BE THOSE WHERE THE ELECTORAL SYSTEM IS COMPETITIVE (43)	0.57	6 28 19 21 X SQ= 6.05 P = 0.008 RV YES (NO)
107	TEND TO BE THOSE WHERE AUTONOMOUS GROUPS ARE PARTIALLY TOLERATED IN POLITICS, ARE TOLERATED ONLY OUTSIDE POLITICS, OR ARE NOT TOLERATED AT ALL (65)	0.80	107	TEND TO BE THOSE WHERE AUTONOMOUS GROUPS ARE FULLY TOLERATED IN POLITICS (46)	0.53	6 30 24 27 X SQ= 7.34 P = 0.006 RV YES (NO)
110	ALWAYS ARE THOSE WHERE POLITICAL ENCULTURATION IS MEDIUM OR LOW (80)	1.00	110	TEND LESS TO BE THOSE WHERE POLITICAL ENCULTURATION IS MEDIUM OR LOW (80)	0.67	0 14 27 29 X SQ= 9.05 P = 0. RV NO (NO)

115	ALWAYS ARE THOSE WHERE INTEREST ARTICULATION BY ASSOCIATIONAL GROUPS IS MODERATE, LIMITED, OR NEGLIGIBLE (91)	1.00	115	TEND LESS TO BE THOSE WHERE INTEREST ARTICULATION BY ASSOCIATIONAL GROUPS IS MODERATE, LIMITED, OR NEGLIGIBLE (91)	0.67	0 18 31 37 X SQ= 10.93 P = 0. RV NO (NO)
116	TEND TO BE THOSE WHERE INTEREST ARTICULATION BY ASSOCIATIONAL GROUPS IS LIMITED OR NEGLIGIBLE (79)	0.94	116	TEND TO BE THOSE WHERE INTEREST ARTICULATION BY ASSOCIATIONAL GROUPS IS SIGNIFICANT OR MODERATE (32)	0.51	2 28 29 27 X SQ= 15.35 P = 0. RV YES (YES)
117	TEND TO BE THOSE WHERE INTEREST ARTICULATION BY ASSOCIATIONAL GROUPS IS NEGLIGIBLE (51)	0.71	117	TEND TO BE THOSE WHERE INTEREST ARTICULATION BY ASSOCIATIONAL GROUPS IS SIGNIFICANT, MODERATE, OR LIMITED (60)	0.67	9 37 22 18 X SQ= 10.17 P = 0.001 RV YES (YES)
119	ALWAYS ARE THOSE WHERE INTEREST ARTICULATION BY INSTITUTIONAL GROUPS IS VERY SIGNIFICANT OR SIGNIFICANT (74)	1.00	119	TEND LESS TO BE THOSE WHERE INTEREST ARTICULATION BY INSTITUTIONAL GROUPS IS VERY SIGNIFICANT OR SIGNIFICANT (74)	0.62	17 36 0 22 X SQ= 7.39 P = 0.002 RV NO (NO)
121	TEND TO BE THOSE WHERE INTEREST ARTICULATION BY NON-ASSOCIATIONAL GROUPS IS SIGNIFICANT (54)	0.77	121	TEND TO BE THOSE WHERE INTEREST ARTICULATION BY NON-ASSOCIATIONAL GROUPS IS MODERATE, LIMITED, OR NEGLIGIBLE (61)	0.71	24 17 7 41 X SQ= 16.93 P = 0. RV YES (YES)
122	TEND MORE TO BE THOSE WHERE INTEREST ARTICULATION BY NON-ASSOCIATIONAL GROUPS IS SIGNIFICANT OR MODERATE (83)	0.94	122	TEND LESS TO BE THOSE WHERE INTEREST ARTICULATION BY NON-ASSOCIATIONAL GROUPS IS SIGNIFICANT OR MODERATE (83)	0.55	29 32 2 26 X SQ= 12.08 P = 0. RV NO (NO)
123	ALWAYS ARE THOSE WHERE INTEREST ARTICULATION BY NON-ASSOCIATIONAL GROUPS IS SIGNIFICANT, MODERATE, OR LIMITED (107)	1.00	123	TEND LESS TO BE THOSE WHERE INTEREST ARTICULATION BY NON-ASSOCIATIONAL GROUPS IS SIGNIFICANT, MODERATE, OR LIMITED (107)	0.86	31 50 0 8 X SQ= 3.16 P = 0.047 RV NO (NO)
125	TEND MORE TO BE THOSE WHERE INTEREST ARTICULATION BY ANOMIC GROUPS IS FREQUENT OR OCCASIONAL (64)	0.82	125	TEND LESS TO BE THOSE WHERE INTEREST ARTICULATION BY ANOMIC GROUPS IS FREQUENT OR OCCASIONAL (64)	0.51	23 24 5 23 X SQ= 5.98 P = 0.008 RV NO (NO)
126	ALWAYS ARE THOSE WHERE INTEREST ARTICULATION BY ANOMIC GROUPS IS FREQUENT, OCCASIONAL, OR INFREQUENT (83)	1.00	126	TEND LESS TO BE THOSE WHERE INTEREST ARTICULATION BY ANOMIC GROUPS IS FREQUENT, OCCASIONAL, OR INFREQUENT (83)	0.66	28 31 0 16 X SQ= 10.17 P = 0. RV NO (NO)

128	TEND TO BE THOSE WHERE INTEREST ARTICULATION BY POLITICAL PARTIES IS LIMITED OR NEGLIGIBLE (45)	0.71	128	TEND TO BE THOSE WHERE INTEREST ARTICULATION BY POLITICAL PARTIES IS SIGNIFICANT OR MODERATE (48)	0.57	8 26 20 20 X SQ= 4.41 P = 0.030 RV YES (YES)
133	TEND TO BE THOSE WHERE INTEREST AGGREGATION BY THE EXECUTIVE IS SIGNIFICANT (29)	0.73	133	TEND TO BE THOSE WHERE INTEREST AGGREGATION BY THE EXECUTIVE IS MODERATE, LIMITED, OR NEGLIGIBLE (76)	0.92	22 4 8 47 X SQ= 34.23 P = 0. RV YES (YES)
134	TEND TO BE THOSE WHERE INTEREST AGGREGATION BY THE EXECUTIVE IS SIGNIFICANT OR MODERATE (57)	0.90	134	TEND TO BE THOSE WHERE INTEREST AGGREGATION BY THE EXECUTIVE IS LIMITED OR NEGLIGIBLE (46)	0.53	26 24 3 27 X SQ= 12.55 P = 0. RV YES (YES)
135	ALWAYS ARE THOSE WHERE INTEREST AGGREGATION BY THE EXECUTIVE IS SIGNIFICANT, MODERATE, OR LIMITED (77)	1.00	135	TEND LESS TO BE THOSE WHERE INTEREST AGGREGATION BY THE EXECUTIVE IS SIGNIFICANT, MODERATE, OR LIMITED (77)	0.80	28 32 0 8 X SQ= 4.57 P = 0.017 RV NO (NO)
138	TEND TO BE THOSE WHERE INTEREST AGGREGATION BY THE LEGISLATURE IS NEGLIGIBLE (49)	0.78	138	TEND TO BE THOSE WHERE INTEREST AGGREGATION BY THE LEGISLATURE IS SIGNIFICANT, MODERATE, OR LIMITED (48)	0.59	6 30 21 21 X SQ= 8.10 P = 0.004 RV YES (YES)
139	TEND TO BE THOSE WHERE THE PARTY SYSTEM IS QUANTITATIVELY ONE-PARTY (34)	0.55	139	TEND TO BE THOSE WHERE THE PARTY SYSTEM IS QUANTITATIVELY OTHER THAN ONE-PARTY (71)	0.71	16 16 13 39 X SQ= 4.43 P = 0.032 RV YES (YES)
140	LEAN LESS TOWARD BEING THOSE WHERE THE PARTY SYSTEM IS QUANTITATIVELY OTHER THAN ONE PARTY DOMINANT (87)	0.79	140	LEAN MORE TOWARD BEING THOSE WHERE THE PARTY SYSTEM IS QUANTITATIVELY OTHER THAN ONE PARTY DOMINANT (87)	0.94	6 3 22 50 X SQ= 3.15 P = 0.058 RV NO (YES)
142	LEAN MORE TOWARD BEING THOSE WHERE THE PARTY SYSTEM IS QUANTITATIVELY OTHER THAN MULTI-PARTY (66)	0.85	142	LEAN LESS TOWARD BEING THOSE WHERE THE PARTY SYSTEM IS QUANTITATIVELY OTHER THAN MULTI-PARTY (66)	0.65	4 18 23 33 X SQ= 2.71 P = 0.068 RV NO (NO)
144	TEND MORE TO BE THOSE WHERE THE PARTY SYSTEM IS QUANTITATIVELY ONE-PARTY, RATHER THAN TWO-PARTY (34)	0.94	144	TEND LESS TO BE THOSE WHERE THE PARTY SYSTEM IS QUANTITATIVELY ONE-PARTY, RATHER THAN TWO-PARTY (34)	0.63	16 15 1 9 X SQ= 3.82 P = 0.028 RV NO (YES)

146	TEND LESS TO BE THOSE 0.74 WHERE THE PARTY SYSTEM IS QUALITATIVELY OTHER THAN MASS-BASED TERRITORIAL (92)	146	ALWAYS ARE THOSE 1.00 WHERE THE PARTY SYSTEM IS QUALITATIVELY OTHER THAN MASS-BASED TERRITORIAL (92)	8 0 23 50 X SQ= 11.56 P = 0. RV NO (YES)
147	TEND MORE TO BE THOSE 0.97 WHERE THE PARTY SYSTEM IS QUALITATIVELY OTHER THAN CLASS-ORIENTED OR MULTI-IDEOLOGICAL (67)	147	TEND LESS TO BE THOSE 0.62 WHERE THE PARTY SYSTEM IS QUALITATIVELY OTHER THAN CLASS-ORIENTED OR MULTI-IDEOLOGICAL (67)	1 17 29 28 X SQ= 9.90 P = 0.001 RV NO (YES)
148	TEND LESS TO BE THOSE 0.58 WHERE THE PARTY SYSTEM IS QUALITATIVELY OTHER THAN AFRICAN TRANSITIONAL (96)	148	ALWAYS ARE THOSE 1.00 WHERE THE PARTY SYSTEM IS QUALITATIVELY OTHER THAN AFRICAN TRANSITIONAL (96)	13 0 18 54 X SQ= 23.59 P = 0. RV NO (YES)
153	TEND TO BE THOSE 0.69 WHERE THE PARTY SYSTEM IS MODERATELY STABLE OR UNSTABLE (38)	153	TEND TO BE THOSE 0.71 WHERE THE PARTY SYSTEM IS STABLE (42)	4 36 9 15 X SQ= 5.41 P = 0.012 RV YES (NO)
158	TEND TO BE THOSE 0.62 WHERE PERSONALISMO IS PRONOUNCED OR MODERATE (40)	158	TEND TO BE THOSE 0.73 WHERE PERSONALISMO IS NEGLIGIBLE (56)	16 14 10 37 X SQ= 7.04 P = 0.006 RV YES (YES)
160	ALWAYS ARE THOSE 1.00 WHERE THE POLITICAL LEADERSHIP IS MODERATELY ELITIST OR NON-ELITIST (67)	160	TEND LESS TO BE THOSE 0.53 WHERE THE POLITICAL LEADERSHIP IS MODERATELY ELITIST OR NON-ELITIST (67)	0 25 27 28 X SQ= 16.39 P = 0. RV NO (NO)
161	TEND TO BE THOSE 0.81 WHERE THE POLITICAL LEADERSHIP IS NON-ELITIST (50)	161	TEND TO BE THOSE 0.58 WHERE THE POLITICAL LEADERSHIP IS ELITIST OR MODERATELY ELITIST (47)	5 31 22 22 X SQ= 9.99 P = 0.001 RV YES (YES)
164	TEND TO BE THOSE 0.70 WHERE THE REGIME'S LEADERSHIP CHARISMA IS PRONOUNCED OR MODERATE (34)	164	TEND TO BE THOSE 0.85 WHERE THE REGIME'S LEADERSHIP CHARISMA IS NEGLIGIBLE (65)	19 8 8 44 X SQ= 21.50 P = 0. RV YES (YES)
169	LEAN TOWARD BEING THOSE 0.60 WHERE THE HORIZONTAL POWER DISTRIBUTION IS NEGLIGIBLE (48)	169	LEAN TOWARD BEING THOSE 0.61 WHERE THE HORIZONTAL POWER DISTRIBUTION IS SIGNIFICANT OR LIMITED (58)	12 34 18 22 X SQ= 2.59 P = 0.075 RV YES (NO)

No.	Statement		No.	Statement		Statistics
170	TEND TO BE THOSE WHERE THE LEGISLATIVE-EXECUTIVE STRUCTURE IS PRESIDENTIAL (39)	0.69	170	TEND TO BE THOSE WHERE THE LEGISLATIVE-EXECUTIVE STRUCTURE IS OTHER THAN PRESIDENTIAL (63)	0.80	18 11 8 44 X SQ= 16.54 P = 0. RV YES (YES)
174	TEND MORE TO BE THOSE WHERE THE LEGISLATURE IS PARTIALLY EFFECTIVE, LARGELY INEFFECTIVE, OR WHOLLY INEFFECTIVE (72)	0.93	174	TEND LESS TO BE THOSE WHERE THE LEGISLATURE IS PARTIALLY EFFECTIVE, LARGELY INEFFECTIVE, OR WHOLLY INEFFECTIVE (72)	0.58	2 23 25 32 X SQ= 8.56 P = 0.002 RV NO (NO)
175	TEND TO BE THOSE WHERE THE LEGISLATURE IS LARGELY INEFFECTIVE OR WHOLLY INEFFECTIVE (49)	0.78	175	TEND TO BE THOSE WHERE THE LEGISLATURE IS FULLY EFFECTIVE OR PARTIALLY EFFECTIVE (51)	0.60	6 33 21 22 X SQ= 8.90 P = 0.002 RV YES (NO)
176	TEND MORE TO BE THOSE WHERE THE LEGISLATURE IS FULLY EFFECTIVE, PARTIALLY EFFECTIVE, OR LARGELY INEFFECTIVE (72)	0.89	176	TEND LESS TO BE THOSE WHERE THE LEGISLATURE IS FULLY EFFECTIVE, PARTIALLY EFFECTIVE, OR LARGELY INEFFECTIVE (72)	0.62	24 34 3 21 X SQ= 5.17 P = 0.011 RV NO (NO)
178	TEND TO BE THOSE WHERE THE LEGISLATURE IS UNICAMERAL (53)	0.86	178	TEND TO BE THOSE WHERE THE LEGISLATURE IS BICAMERAL (51)	0.60	25 21 4 32 X SQ= 14.68 P = 0. RV YES (YES)
179	TILT TOWARD BEING THOSE WHERE THE EXECUTIVE IS DOMINANT (52)	0.70	179	TILT TOWARD BEING THOSE WHERE THE EXECUTIVE IS STRONG (39)	0.50	16 25 7 25 X SQ= 1.72 P = 0.136 RV YES (NO)
181	TILT MORE TOWARD BEING THOSE WHERE THE BUREAUCRACY IS SEMI-MODERN, RATHER THAN MODERN (55)	0.89	181	TILT LESS TOWARD BEING THOSE WHERE THE BUREAUCRACY IS SEMI-MODERN, RATHER THAN MODERN (55)	0.61	1 20 8 31 X SQ= 1.56 P = 0.142 RV NO (NO)
182	TEND TO BE THOSE WHERE THE BUREAUCRACY IS POST-COLONIAL TRANSITIONAL, RATHER THAN SEMI-MODERN (25)	0.72	182	ALWAYS ARE THOSE WHERE THE BUREAUCRACY IS SEMI-MODERN, RATHER THAN POST-COLONIAL TRANSITIONAL (55)	1.00	8 31 21 0 X SQ= 31.43 P = 0. RV YES (YES)
183	ALWAYS ARE THOSE WHERE THE BUREAUCRACY IS POST-COLONIAL TRANSITIONAL, RATHER THAN TRADITIONAL (25)	1.00	183	ALWAYS ARE THOSE WHERE THE BUREAUCRACY IS TRADITIONAL, RATHER THAN POST-COLONIAL TRANSITIONAL (9)	1.00	21 0 0 6 X SQ= 21.52 P = 0. RV YES (YES)

186	TEND MORE TO BE THOSE WHERE PARTICIPATION BY THE MILITARY IN POLITICS IS NEUTRAL, RATHER THAN SUPPORTIVE (56)	0.80	186	TEND LESS TO BE THOSE WHERE PARTICIPATION BY THE MILITARY IN POLITICS IS NEUTRAL, RATHER THAN SUPPORTIVE (56)	0.55	5 22 20 27 X SQ= 3.42 P = 0.043 RV NO (NO)
187	LEAN MORE TOWARD BEING THOSE WHERE THE ROLE OF THE POLICE IS POLITICALLY SIGNIFICANT (66)	0.79	187	LEAN LESS TOWARD BEING THOSE WHERE THE ROLE OF THE POLICE IS POLITICALLY SIGNIFICANT (66)	0.56	19 31 5 24 X SQ= 2.82 P = 0.076 RV NO (NO)
188	TEND MORE TO BE THOSE WHERE THE CHARACTER OF THE LEGAL SYSTEM IS OTHER THAN CIVIL LAW (81)	0.90	188	TEND LESS TO BE THOSE WHERE THE CHARACTER OF THE LEGAL SYSTEM IS OTHER THAN CIVIL LAW (81)	0.61	3 22 27 35 X SQ= 6.51 P = 0.006 RV NO (NO)
192	TEND TO BE THOSE WHERE THE CHARACTER OF THE LEGAL SYSTEM IS MUSLIM OR PARTLY MUSLIM (27)	0.57	192	TEND TO BE THOSE WHERE THE CHARACTER OF THE LEGAL SYSTEM IS OTHER THAN MUSLIM OR PARTLY MUSLIM (86)	0.95	17 3 13 54 X SQ= 26.50 P = 0. RV YES (YES)
193	TEND TO BE THOSE WHERE THE CHARACTER OF THE LEGAL SYSTEM IS PARTLY INDIGENOUS (28)	0.65	193	TEND TO BE THOSE WHERE THE CHARACTER OF THE LEGAL SYSTEM IS OTHER THAN PARTLY INDIGENOUS (86)	0.95	20 3 11 55 X SQ= 34.09 P = 0. RV YES (YES)
194	ALWAYS ARE THOSE THAT ARE NON-COMMUNIST (101)	1.00	194	TEND LESS TO BE THOSE THAT ARE NON-COMMUNIST (101)	0.78	0 13 31 45 X SQ= 6.44 P = 0.003 RV NO (NO)

```
******************************************************************************************************

88  POLITIES                                               88  POLITIES
    WHOSE IDEOLOGICAL ORIENTATION                              WHOSE IDEOLOGICAL ORIENTATION
    IS SITUATIONAL  (5)                                        IS OTHER THAN SITUATIONAL  (95)


                                                                                         SUBJECT ONLY
......................................................................................................

      5 IN LEFT COLUMN

EL SALVADOR HAITI       IRAQ        LAOS        SYRIA

      95 IN RIGHT COLUMN

AFGHANISTAN ALBANIA     ALGERIA     AUSTRALIA   AUSTRIA     BELGIUM     BOLIVIA     BULGARIA    BURUNDI     CAMBODIA
CAMEROUN    CANADA      CEN AFR REP CHAD        CHILE       CHINA, PR   COLOMBIA    CONGO(BRA)  CONGO(LEO)  COSTA RICA
CUBA        CYPRUS      CZECHOS'KIA DAHOMEY     DENMARK     DOMIN REP   ECUADOR     ETHIOPIA    FINLAND     FRANCE
GABON       GERMANY, E  GERMAN FR   GHANA       GREECE      GUINEA      HONDURAS    HUNGARY     ICELAND     INDIA
INDONESIA   IRAN        IRELAND     ISRAEL      ITALY       IVORY COAST JAMAICA     JAPAN       JORDAN      KOREA, N
LEBANON     LIBERIA     LIBYA       LUXEMBOURG  MALAGASY R  MALAYA      MALI        MAURITANIA  MEXICO      MONGOLIA
MOROCCO     NEPAL       NETHERLANDS NEW ZEALAND NIGER       NIGERIA     NORWAY      PHILIPPINES POLAND      RUMANIA
RWANDA      SA'U ARABIA SENEGAL     SIERRE LEO  SOMALIA     SO AFRICA   SUDAN       SWEDEN      SWITZERLAND TANGANYIKA
THAILAND    TOGO        TRINIDAD    TUNISIA     TURKEY      UGANDA      USSR        UAR         UK          US
UPPER VOLTA URUGUAY     VENEZUELA   VIETNAM, N  YUGOSLAVIA

      12 EXCLUDED BECAUSE AMBIGUOUS

ARGENTINA   BRAZIL      BURMA       CEYLON      GUATEMALA   NICARAGUA   PAKISTAN    PANAMA      PARAGUAY    PORTUGAL
SPAIN       VIETNAM REP

      3 EXCLUDED BECAUSE UNASCERTAINABLE

KOREA REP   PERU        YEMEN

------------------------------------------------------------------------------------------------------

 15  ALWAYS ARE THOSE                          1.00     15  TEND TO BE THOSE                     0.80        2    8
     LOCATED IN THE MIDDLE EAST, RATHER THAN                LOCATED IN NORTH AFRICA OR                       0   33
     IN NORTH AFRICA OR                                     CENTRAL AND SOUTH AFRICA, RATHER THAN      X SQ=   3.15
     CENTRAL AND SOUTH AFRICA  (11)                         IN THE MIDDLE EAST  (33)                   P =    0.050
                                                                                                       RV YES (NO )

 19  ALWAYS ARE THOSE                          1.00     19  LEAN TOWARD BEING THOSE              0.72        0   33
     LOCATED IN THE CARIBBEAN,                              LOCATED IN NORTH AFRICA OR                       2   13
     CENTRAL AMERICA, OR SOUTH AMERICA,                     CENTRAL AND SOUTH AFRICA, RATHER THAN      X SQ=   1.86
     RATHER THAN IN NORTH AFRICA OR                         IN THE CARIBBEAN, CENTRAL AMERICA, OR      P =    0.093
     CENTRAL AND SOUTH AFRICA  (22)                         SOUTH AMERICA  (33)                        RV YES (NO )

 36  ALWAYS ARE THOSE                          1.00     36  LEAN LESS TOWARD BEING THOSE         0.58        0   40
     WHOSE PER CAPITA GROSS NATIONAL PRODUCT                WHOSE PER CAPITA GROSS NATIONAL PRODUCT          5   55
     IS LOW OR VERY LOW  (73)                               IS LOW OR VERY LOW  (73)                   X SQ=   1.97
                                                                                                       P =    0.081
                                                                                                       RV NO  (NO )
```

45	ALWAYS ARE THOSE WHERE THE LITERACY RATE IS BELOW FIFTY PERCENT (54)	1.00	45	LEAN TOWARD BEING THOSE WHERE THE LITERACY RATE IS FIFTY PERCENT OR ABOVE (55)	0.51	0 45 5 44 X SQ= 3.04 P = 0.057 RV YES (NO)
50	ALWAYS ARE THOSE WHERE FREEDOM OF THE PRESS IS INTERMITTENT, INTERNALLY ABSENT, OR INTERNALLY AND EXTERNALLY ABSENT (56)	1.00	50	TILT TOWARD BEING THOSE WHERE FREEDOM OF THE PRESS IS COMPLETE (43)	0.51	0 42 4 41 X SQ= 2.15 P = 0.117 RV YES (NO)
96	ALWAYS ARE THOSE WHERE THE STATUS OF THE REGIME IS AUTHORITARIAN (23)	1.00	96	TEND TO BE THOSE WHERE THE STATUS OF THE REGIME IS CONSTITUTIONAL OR TOTALITARIAN (67)	0.79	0 62 2 16 X SQ= 3.24 P = 0.048 RV YES (NO)
99	ALWAYS ARE THOSE WHERE GOVERNMENTAL STABILITY IS MODERATELY PRESENT AND DATES FROM THE POST-WAR PERIOD, OR IS ABSENT (36)	1.00	99	TEND TO BE THOSE WHERE GOVERNMENTAL STABILITY IS GENERALLY PRESENT AND DATES FROM AT LEAST THE INTER-WAR PERIOD, OR FROM THE POST-WAR PERIOD (50)	0.70	0 48 4 21 X SQ= 5.33 P = 0.012 RV YES (NO)
100	TEND TO BE THOSE WHERE GOVERNMENTAL STABILITY IS ABSENT (22)	0.80	100	TEND TO BE THOSE WHERE GOVERNMENTAL STABILITY IS GENERALLY PRESENT AND DATES FROM AT LEAST THE INTER-WAR PERIOD, OR IS GENERALLY OR MODERATELY PRESENT, AND DATES FROM THE POST-WAR PERIOD (64)	0.87	1 58 4 9 X SQ= 9.80 P = 0.003 RV YES (NO)
102	ALWAYS ARE THOSE WHERE THE REPRESENTATIVE CHARACTER OF THE REGIME IS PSEUDO-POLYARCHIC OR NON-POLYARCHIC (49)	1.00	102	LEAN TOWARD BEING THOSE WHERE THE REPRESENTATIVE CHARACTER OF THE REGIME IS POLYARCHIC OR LIMITED POLYARCHIC (59)	0.59	0 55 3 39 X SQ= 2.02 P = 0.078 RV YES (NO)
106	ALWAYS ARE THOSE WHERE THE ELECTORAL SYSTEM IS NON-COMPETITIVE (30)	1.00	106	TILT TOWARD BEING THOSE WHERE THE ELECTORAL SYSTEM IS COMPETITIVE OR PARTIALLY COMPETITIVE (52)	0.65	0 46 2 25 X SQ= 1.27 P = 0.134 RV YES (NO)
107	ALWAYS ARE THOSE WHERE AUTONOMOUS GROUPS ARE PARTIALLY TOLERATED IN POLITICS, ARE TOLERATED ONLY OUTSIDE POLITICS, OR ARE NOT TOLERATED AT ALL (65)	1.00	107	TILT LESS TOWARD BEING THOSE WHERE AUTONOMOUS GROUPS ARE PARTIALLY TOLERATED IN POLITICS, ARE TOLERATED ONLY OUTSIDE POLITICS, OR ARE NOT TOLERATED AT ALL (65)	0.55	0 42 4 52 X SQ= 1.57 P = 0.133 RV NO (NO)
111	ALWAYS ARE THOSE WHERE POLITICAL ENCULTURATION IS LOW (42)	1.00	111	TEND TO BE THOSE WHERE POLITICAL ENCULTURATION IS HIGH OR MEDIUM (53)	0.66	0 51 5 26 X SQ= 6.17 P = 0.006 RV YES (NO)

124	TEND TO BE THOSE WHERE INTEREST ARTICULATION BY ANOMIC GROUPS IS FREQUENT (14)	0.60	124	TEND TO BE THOSE WHERE INTEREST ARTICULATION BY ANOMIC GROUPS IS OCCASIONAL, INFREQUENT, OR VERY INFREQUENT (85)	0.90	3 8 2 73 X SQ= 6.59 P = 0.014 RV YES (NO)
132	ALWAYS ARE THOSE WHERE INTEREST AGGREGATION BY POLITICAL PARTIES IS NEGLIGIBLE (9)	1.00	132	TEND TO BE THOSE WHERE INTEREST AGGREGATION BY POLITICAL PARTIES IS SIGNIFICANT, MODERATE, OR LIMITED (67)	0.97	0 65 2 2 X SQ= 18.06 P = 0.003 RV YES (YES)
135	ALWAYS ARE THOSE WHERE INTEREST AGGREGATION BY THE EXECUTIVE IS NEGLIGIBLE (13)	1.00	135	TEND TO BE THOSE WHERE INTEREST AGGREGATION BY THE EXECUTIVE IS SIGNIFICANT, MODERATE, OR LIMITED (77)	0.92	0 71 2 6 X SQ= 9.49 P = 0.009 RV YES (NO)
154	ALWAYS ARE THOSE WHERE THE PARTY SYSTEM IS UNSTABLE (25)	1.00	154	TEND TO BE THOSE WHERE THE PARTY SYSTEM IS STABLE OR MODERATELY STABLE (55)	0.79	0 50 4 13 X SQ= 8.67 P = 0.003 RV YES (NO)
157	ALWAYS ARE THOSE WHERE PERSONALISMO IS PRONOUNCED (14)	1.00	157	TEND TO BE THOSE WHERE PERSONALISMO IS MODERATE OR NEGLIGIBLE (84)	0.90	3 8 0 76 X SQ= 14.06 P = 0.002 RV YES (NO)
160	ALWAYS ARE THOSE WHERE THE POLITICAL LEADERSHIP IS ELITIST (30)	1.00	160	LEAN TOWARD BEING THOSE WHERE THE POLITICAL LEADERSHIP IS MODERATELY ELITIST OR NON-ELITIST (67)	0.73	2 23 0 61 X SQ= 2.09 P = 0.082 RV YES (NO)
175	ALWAYS ARE THOSE WHERE THE LEGISLATURE IS LARGELY INEFFECTIVE OR WHOLLY INEFFECTIVE (49)	1.00	175	TILT TOWARD BEING THOSE WHERE THE LEGISLATURE IS FULLY EFFECTIVE OR PARTIALLY EFFECTIVE (51)	0.53	0 47 3 41 X SQ= 1.52 P = 0.109 RV YES (NO)
185	ALWAYS ARE THOSE WHERE PARTICIPATION BY THE MILITARY IN POLITICS IS INTERVENTIVE, RATHER THAN SUPPORTIVE (21)	1.00	185	TEND TO BE THOSE WHERE PARTICIPATION BY THE MILITARY IN POLITICS IS SUPPORTIVE, RATHER THAN INTERVENTIVE (31)	0.82	4 6 0 28 X SQ= 8.63 P = 0.003 RV YES (NO)

```
89  POLITIES                                              89  POLITIES
    WHOSE IDEOLOGICAL ORIENTATION                             WHOSE IDEOLOGICAL ORIENTATION
    IS CONVENTIONAL  (33)                                     IS OTHER THAN CONVENTIONAL  (62)


                                                                            BOTH SUBJECT AND PREDICATE

      33 IN LEFT COLUMN

AUSTRALIA   AUSTRIA     BELGIUM     CANADA      CHILE       COLOMBIA    COSTA RICA  CYPRUS      DENMARK     ECUADOR
FINLAND     FRANCE      GERMAN FR   GREECE      HONDURAS    ICELAND     IRELAND     ITALY       JAPAN       LEBANON
LUXEMBOURG  MEXICO      NETHERLANDS NEW ZEALAND NORWAY      PHILIPPINES SO AFRICA   SWEDEN      SWITZERLAND TURKEY
UK          US          URUGUAY

      62 IN RIGHT COLUMN

AFGHANISTAN ALBANIA     ALGERIA     BOLIVIA     BULGARIA    BURMA       BURUNDI     CAMBODIA    CAMEROUN    CEN AFR REP
CHAD        CHINA, PR   CONGO(BRA)  CONGO(LEO)  CUBA        CZECHOS'KIA DAHOMEY     DOMIN REP   EL SALVADOR ETHIOPIA
GABON       GERMANY, E  GHANA       GUINEA      HAITI       HUNGARY     INDONESIA   IRAN        IRAQ        ISRAEL
IVORY COAST JORDAN      KOREA, N    LAOS        MALAYA      MALI        MAURITANIA  MONGOLIA    NEPAL       NIGER
PAKISTAN    POLAND      RUMANIA     RWANDA      SA'U ARABIA SENEGAL     SIERRE LEO  SOMALIA     SUDAN       SYRIA
TANGANYIKA  THAILAND    TOGO        TUNISIA     USSR        UAR         UPPER VOLTA VENEZUELA   VIETNAM, N  VIETNAM REP
YEMEN       YUGOSLAVIA

      18 EXCLUDED BECAUSE AMBIGUOUS

ARGENTINA   BRAZIL      CEYLON      GUATEMALA   INDIA       JAMAICA     LIBERIA     LIBYA       MALAGASY R  MOROCCO
NICARAGUA   NIGERIA     PANAMA      PARAGUAY    PORTUGAL    SPAIN       TRINIDAD    UGANDA

       2 EXCLUDED BECAUSE UNASCERTAINABLE

KOREA REP   PERU

------------------------------------------------------------------------------------------------------------------------

  2  TEND TO BE THOSE                       0.61      2  ALWAYS ARE THOSE                       1.00          20    0
     LOCATED IN WEST EUROPE, SCANDINAVIA,                LOCATED ELSEWHERE THAN IN                            13   62
     NORTH AMERICA, OR AUSTRALASIA  (22)                 WEST EUROPE, SCANDINAVIA,                     X SQ=  44.02
                                                         NORTH AMERICA, OR AUSTRALASIA  (93)           P =     0.
                                                                                                       RV YES (YES)

  3  TEND LESS TO BE THOSE                  0.67      3  ALWAYS ARE THOSE                       1.00          11    0
     LOCATED ELSEWHERE THAN IN                           LOCATED ELSEWHERE THAN IN                            22   62
     WEST EUROPE  (102)                                  WEST EUROPE  (102)                            X SQ=  20.23
                                                                                                       P =     0.
                                                                                                       RV NO  (YES)

  5  ALWAYS ARE THOSE                       1.00      5  ALWAYS ARE THOSE                       1.00          16    0
     LOCATED IN SCANDINAVIA OR WEST EUROPE,              LOCATED IN EAST EUROPE, RATHER THAN                   0    9
     RATHER THAN IN EAST EUROPE  (18)                    IN SCANDINAVIA OR WEST EUROPE  (9)            X SQ=  20.85
                                                                                                       P =     0.
                                                                                                       RV YES (YES)
```

8	LEAN MORE TOWARD BEING THOSE LOCATED ELSEWHERE THAN IN EAST ASIA SOUTH ASIA, OR SOUTHEAST ASIA (97)	0.94	8	LEAN LESS TOWARD BEING THOSE LOCATED ELSEWHERE THAN IN EAST ASIA SOUTH ASIA, OR SOUTHEAST ASIA (97)	0.79	2 13 31 49 X SQ= 2.57 P = 0.077 RV NO (NO)
10	TEND MORE TO BE THOSE LOCATED ELSEWHERE THAN IN NORTH AFRICA, OR CENTRAL AND SOUTH AFRICA (82)	0.97	10	TEND LESS TO BE THOSE LOCATED ELSEWHERE THAN IN NORTH AFRICA, OR CENTRAL AND SOUTH AFRICA (82)	0.58	1 26 32 36 X SQ= 14.17 P = 0. RV NO (NO)
15	LEAN TOWARD BEING THOSE LOCATED IN THE MIDDLE EAST, RATHER THAN IN NORTH AFRICA OR CENTRAL AND SOUTH AFRICA (11)	0.75	15	LEAN TOWARD BEING THOSE LOCATED IN NORTH AFRICA OR CENTRAL AND SOUTH AFRICA, RATHER THAN IN THE MIDDLE EAST (33)	0.76	3 8 1 26 X SQ= 2.45 P = 0.065 RV YES (NO)
19	TEND TO BE THOSE LOCATED IN THE CARIBBEAN, CENTRAL AMERICA, OR SOUTH AMERICA, RATHER THAN IN NORTH AFRICA OR CENTRAL AND SOUTH AFRICA (22)	0.87	19	TEND TO BE THOSE LOCATED IN NORTH AFRICA OR CENTRAL AND SOUTH AFRICA, RATHER THAN IN THE CARIBBEAN, CENTRAL AMERICA, OR SOUTH AMERICA (33)	0.81	1 26 7 6 X SQ= 10.83 P = 0.001 RV YES (YES)
20	TEND TO BE THOSE LOCATED IN THE CARIBBEAN, CENTRAL AMERICA, OR SOUTH AMERICA, RATHER THAN IN EAST ASIA, SOUTH ASIA, OR SOUTHEAST ASIA (22)	0.78	20	TEND TO BE THOSE LOCATED IN EAST ASIA, SOUTH ASIA, OR SOUTHEAST ASIA, RATHER THAN IN THE CARIBBEAN, CENTRAL AMERICA, OR SOUTH AMERICA (18)	0.68	2 13 7 6 X SQ= 3.55 P = 0.042 RV YES (YES)
25	TEND LESS TO BE THOSE WHOSE POPULATION DENSITY IS MEDIUM OR LOW (98)	0.73	25	TEND MORE TO BE THOSE WHOSE POPULATION DENSITY IS MEDIUM OR LOW (98)	0.95	9 3 24 59 X SQ= 7.89 P = 0.003 RV NO (YES)
28	TILT TOWARD BEING THOSE WHOSE POPULATION GROWTH RATE IS LOW (48)	0.58	28	TILT TOWARD BEING THOSE WHOSE POPULATION GROWTH RATE IS HIGH (62)	0.60	14 34 19 23 X SQ= 1.85 P = 0.130 RV YES (NO)
29	TEND TO BE THOSE WHERE THE DEGREE OF URBANIZATION IS HIGH (56)	0.94	29	TEND TO BE THOSE WHERE THE DEGREE OF URBANIZATION IS LOW (49)	0.70	29 17 2 40 X SQ= 30.18 P = 0. RV YES (YES)
30	TEND TO BE THOSE WHOSE AGRICULTURAL POPULATION IS MEDIUM, LOW, OR VERY LOW (57)	0.94	30	TEND TO BE THOSE WHOSE AGRICULTURAL POPULATION IS HIGH (56)	0.75	2 45 31 15 X SQ= 37.77 P = 0. RV YES (YES)

31	TEND TO BE THOSE WHOSE AGRICULTURAL POPULATION IS LOW OR VERY LOW (24)	0.61	31	TEND TO BE THOSE WHOSE AGRICULTURAL POPULATION IS HIGH OR MEDIUM (90)	0.97	13 59 20 2 X SQ= 36.13 P = 0. RV YES (YES)
33	TEND TO BE THOSE WHOSE GROSS NATIONAL PRODUCT IS VERY HIGH, HIGH, OR MEDIUM (30)	0.52	33	TEND TO BE THOSE WHOSE GROSS NATIONAL PRODUCT IS LOW OR VERY LOW (85)	0.85	17 9 16 53 X SQ= 13.03 P = 0. RV YES (YES)
34	TEND TO BE THOSE WHOSE GROSS NATIONAL PRODUCT IS VERY HIGH, HIGH, MEDIUM, OR LOW (62)	0.79	34	TEND TO BE THOSE WHOSE GROSS NATIONAL PRODUCT IS VERY LOW (53)	0.58	26 26 7 36 X SQ= 10.37 P = 0.001 RV YES (YES)
35	TEND TO BE THOSE WHOSE PER CAPITA GROSS NATIONAL PRODUCT IS VERY HIGH OR HIGH (24)	0.58	35	TEND TO BE THOSE WHOSE PER CAPITA GROSS NATIONAL PRODUCT IS MEDIUM, LOW, OR VERY LOW (91)	0.92	19 5 14 57 X SQ= 25.40 P = 0. RV YES (YES)
36	TEND TO BE THOSE WHOSE PER CAPITA GROSS NATIONAL PRODUCT IS VERY HIGH, HIGH, OR MEDIUM (42)	0.79	36	TEND TO BE THOSE WHOSE PER CAPITA GROSS NATIONAL PRODUCT IS LOW OR VERY LOW (73)	0.81	26 12 7 50 X SQ= 29.27 P = 0. RV YES (YES)
37	ALWAYS ARE THOSE WHOSE PER CAPITA GROSS NATIONAL PRODUCT IS VERY HIGH, HIGH, MEDIUM, OR LOW (64)	1.00	37	TEND TO BE THOSE WHOSE PER CAPITA GROSS NATIONAL PRODUCT IS VERY LOW (51)	0.66	33 21 0 41 X SQ= 35.74 P = 0. RV YES (YES)
39	TEND TO BE THOSE WHOSE INTERNATIONAL FINANCIAL STATUS IS VERY HIGH, HIGH, OR MEDIUM (38)	0.70	39	TEND TO BE THOSE WHOSE INTERNATIONAL FINANCIAL STATUS IS LOW OR VERY LOW (76)	0.82	23 11 10 50 X SQ= 22.57 P = 0. RV YES (YES)
42	TEND TO BE THOSE WHOSE ECONOMIC DEVELOPMENTAL STATUS IS DEVELOPED OR INTERMEDIATE (36)	0.79	42	TEND TO BE THOSE WHOSE ECONOMIC DEVELOPMENTAL STATUS IS UNDERDEVELOPED OR VERY UNDERDEVELOPED (76)	0.86	26 8 7 51 X SQ= 35.90 P = 0. RV YES (YES)
43	TEND TO BE THOSE WHOSE ECONOMIC DEVELOPMENTAL STATUS IS DEVELOPED, INTERMEDIATE, OR UNDERDEVELOPED (55)	0.97	43	TEND TO BE THOSE WHOSE ECONOMIC DEVELOPMENTAL STATUS IS VERY UNDERDEVELOPED (57)	0.74	30 16 1 45 X SQ= 38.14 P = 0. RV YES (YES)

44	TEND TO BE THOSE WHERE THE LITERACY RATE IS NINETY PERCENT OR ABOVE (25)	0.58	44	TEND TO BE THOSE WHERE THE LITERACY RATE IS BELOW NINETY PERCENT (90)	0.90	19 6 14 56 X SQ= 23.07 P = 0. RV YES (YES)
45	TEND TO BE THOSE WHERE THE LITERACY RATE IS FIFTY PERCENT OR ABOVE (55)	0.87	45	TEND TO BE THOSE WHERE THE LITERACY RATE IS BELOW FIFTY PERCENT (54)	0.71	28 17 4 41 X SQ= 25.65 P = 0. RV YES (YES)
46	ALWAYS ARE THOSE WHERE THE LITERACY RATE IS TEN PERCENT OR ABOVE (84)	1.00	46	TEND LESS TO BE THOSE WHERE THE LITERACY RATE IS TEN PERCENT OR ABOVE (84)	0.57	33 33 0 25 X SQ= 17.51 P = 0. RV NO (YES)
50	TEND TO BE THOSE WHERE FREEDOM OF THE PRESS IS COMPLETE (43)	0.87	50	TEND TO BE THOSE WHERE FREEDOM OF THE PRESS IS INTERMITTENT, INTERNALLY ABSENT, OR INTERNALLY AND EXTERNALLY ABSENT (56)	0.80	27 10 4 41 X SQ= 32.79 P = 0. RV YES (YES)
51	ALWAYS ARE THOSE WHERE FREEDOM OF THE PRESS IS COMPLETE OR INTERMITTENT (60)	1.00	51	TEND TO BE THOSE WHERE FREEDOM OF THE PRESS IS INTERNALLY ABSENT, OR INTERNALLY AND EXTERNALLY ABSENT (37)	0.67	31 16 0 33 X SQ= 32.81 P = 0. RV YES (YES)
54	TEND TO BE THOSE WHERE NEWSPAPER CIRCULATION IS ONE HUNDRED OR MORE PER THOUSAND (37)	0.76	54	TEND TO BE THOSE WHERE NEWSPAPER CIRCULATION IS LESS THAN ONE HUNDRED PER THOUSAND (76)	0.83	25 10 8 50 X SQ= 29.21 P = 0. RV YES (YES)
55	ALWAYS ARE THOSE WHERE NEWSPAPER CIRCULATION IS TEN OR MORE PER THOUSAND (78)	1.00	55	TEND TO BE THOSE WHERE NEWSPAPER CIRCULATION IS LESS THAN TEN PER THOUSAND (35)	0.50	33 30 0 30 X SQ= 22.12 P = 0. RV YES (YES)
56	TEND MORE TO BE THOSE WHERE THE RELIGION IS PREDOMINANTLY LITERATE (79)	0.94	56	TEND LESS TO BE THOSE WHERE THE RELIGION IS PREDOMINANTLY LITERATE (79)	0.60	31 35 2 23 X SQ= 10.29 P = 0. RV NO (NO)
63	TEND TO BE THOSE WHERE THE RELIGION IS PREDOMINANTLY SOME KIND OF CHRISTIAN (46)	0.82	63	TEND TO BE THOSE WHERE THE RELIGION IS PREDOMINANTLY OR PARTLY OTHER THAN CHRISTIAN (68)	0.82	27 11 6 50 X SQ= 33.58 P = 0. RV YES (YES)

68	TEND TO BE THOSE THAT ARE LINGUISTICALLY HOMOGENEOUS (52)	0.73	68	TEND TO BE THOSE THAT ARE LINGUISTICALLY WEAKLY HETEROGENEOUS OR STRONGLY HETEROGENEOUS (62)	0.69	24 19 9 42 X SQ= 13.29 P = 0. RV YES (YES)
69	TEND TO BE THOSE THAT ARE LINGUISTICALLY HOMOGENEOUS OR WEAKLY HETEROGENEOUS (64)	0.79	69	TEND TO BE THOSE THAT ARE LINGUISTICALLY STRONGLY HETEROGENEOUS (50)	0.57	26 26 7 35 X SQ= 9.92 P = 0.001 RV YES (YES)
70	TEND LESS TO BE THOSE THAT ARE RELIGIOUSLY, RACIALLY, OR LINGUISTICALLY HETEROGENEOUS (85)	0.62	70	TEND MORE TO BE THOSE THAT ARE RELIGIOUSLY, RACIALLY, OR LINGUISTICALLY HETEROGENEOUS (85)	0.89	12 6 20 50 X SQ= 7.41 P = 0.005 RV NO (YES)
72	TEND TO BE THOSE WHOSE DATE OF INDEPENDENCE IS BEFORE 1914 (52)	0.81	72	TEND TO BE THOSE WHOSE DATE OF INDEPENDENCE IS AFTER 1914 (59)	0.73	26 16 6 44 X SQ= 22.91 P = 0. RV YES (YES)
73	TEND TO BE THOSE WHOSE DATE OF INDEPENDENCE IS BEFORE 1945 (65)	0.94	73	TEND TO BE THOSE WHOSE DATE OF INDEPENDENCE IS AFTER 1945 (46)	0.58	30 25 2 35 X SQ= 21.43 P = 0. RV YES (YES)
74	TEND TO BE THOSE THAT ARE HISTORICALLY WESTERN (26)	0.63	74	TEND TO BE THOSE THAT ARE NOT HISTORICALLY WESTERN (87)	0.93	20 4 12 57 X SQ= 31.45 P = 0. RV YES (YES)
75	ALWAYS ARE THOSE THAT ARE HISTORICALLY WESTERN OR SIGNIFICANTLY WESTERNIZED (62)	1.00	75	TEND TO BE THOSE THAT ARE NOT HISTORICALLY WESTERN AND ARE NOT SIGNIFICANTLY WESTERNIZED (52)	0.73	32 17 0 45 X SQ= 41.69 P = 0. RV YES (YES)
76	TEND TO BE THOSE THAT ARE HISTORICALLY WESTERN, RATHER THAN HAVING BEEN SIGNIFICANTLY OR PARTIALLY WESTERNIZED THROUGH A COLONIAL RELATIONSHIP (26)	0.65	76	TEND TO BE THOSE THAT HAVE BEEN SIGNIFICANTLY OR PARTIALLY WESTERNIZED THROUGH A COLONIAL RELATIONSHIP, RATHER THAN BEING HISTORICALLY WESTERN (70)	0.91	20 4 11 43 X SQ= 24.94 P = 0. RV YES (YES)
77	ALWAYS ARE THOSE THAT HAVE BEEN SIGNIFICANTLY WESTERNIZED, RATHER THAN PARTIALLY WESTERNIZED, THROUGH A COLONIAL RELATIONSHIP (28)	1.00	77	TEND TO BE THOSE THAT HAVE BEEN PARTIALLY WESTERNIZED, RATHER THAN SIGNIFICANTLY WESTERNIZED, THROUGH A COLONIAL RELATIONSHIP (41)	0.84	10 7 0 36 X SQ= 22.40 P = 0. RV YES (YES)

79	TEND LESS TO BE THOSE THAT WERE FORMERLY DEPENDENCIES OF BRITAIN OR FRANCE, RATHER THAN SPAIN (49)	0.53	79	TEND MORE TO BE THOSE THAT WERE FORMERLY DEPENDENCIES OF BRITAIN OR FRANCE, RATHER THAN SPAIN (49)	0.87	8 33 7 5 X SQ= 5.11 P = 0.024 RV NO (YES)
81	TEND LESS TO BE THOSE WHERE THE TYPE OF POLITICAL MODERNIZATION IS LATER EUROPEAN OR LATER EUROPEAN DERIVED, RATHER THAN EARLY EUROPEAN OR EARLY EUROPEAN DERIVED (40)	0.61	81	ALWAYS ARE THOSE WHERE THE TYPE OF POLITICAL MODERNIZATION IS LATER EUROPEAN OR LATER EUROPEAN DERIVED, RATHER THAN EARLY EUROPEAN OR EARLY EUROPEAN DERIVED (40)	1.00	11 0 17 14 X SQ= 5.56 P = 0.007 RV NO (NO)
82	TEND TO BE THOSE WHERE THE TYPE OF POLITICAL MODERNIZATION IS EARLY OR LATER EUROPEAN OR EUROPEAN DERIVED, RATHER THAN DEVELOPED TUTELARY OR UNDEVELOPED TUTELARY (51)	0.90	82	TEND TO BE THOSE WHERE THE TYPE OF POLITICAL MODERNIZATION IS DEVELOPED TUTELARY OR UNDEVELOPED TUTELARY, RATHER THAN EARLY OR LATER EUROPEAN OR EUROPEAN DERIVED (55)	0.75	28 14 3 41 X SQ= 30.84 P = 0. RV YES (YES)
85	TEND TO BE THOSE WHERE THE STAGE OF POLITICAL MODERNIZATION IS ADVANCED, RATHER THAN MID- OR EARLY TRANSITIONAL (60)	0.88	85	TEND TO BE THOSE WHERE THE STAGE OF POLITICAL MODERNIZATION IS MID- OR EARLY TRANSITIONAL, RATHER THAN ADVANCED (54)	0.62	29 23 4 38 X SQ= 19.83 P = 0. RV YES (YES)
86	ALWAYS ARE THOSE WHERE THE STAGE OF POLITICAL MODERNIZATION IS ADVANCED OR MID-TRANSITIONAL, RATHER THAN EARLY TRANSITIONAL (76)	1.00	86	TEND LESS TO BE THOSE WHERE THE STAGE OF POLITICAL MODERNIZATION IS ADVANCED OR MID-TRANSITIONAL, RATHER THAN EARLY TRANSITIONAL (76)	0.51	33 31 0 30 X SQ= 21.63 P = 0. RV NO (YES)
93	ALWAYS ARE THOSE WHERE THE SYSTEM STYLE IS NON-MOBILIZATIONAL (78)	1.00	93	TEND TO BE THOSE WHERE THE SYSTEM STYLE IS MOBILIZATIONAL, OR LIMITED MOBILIZATIONAL (32)	0.53	0 31 33 28 X SQ= 23.85 P = 0. RV YES (YES)
94	ALWAYS ARE THOSE WHERE THE STATUS OF THE REGIME IS CONSTITUTIONAL (51)	1.00	94	TEND TO BE THOSE WHERE THE STATUS OF THE REGIME IS AUTHORITARIAN OR TOTALITARIAN (41)	0.81	32 8 0 35 X SQ= 45.62 P = 0. RV YES (YES)
99	LEAN TOWARD BEING THOSE WHERE GOVERNMENTAL STABILITY IS GENERALLY PRESENT AND DATES FROM AT LEAST THE INTER-WAR PERIOD, OR FROM THE POST-WAR PERIOD (50)	0.75	99	LEAN TOWARD BEING THOSE WHERE GOVERNMENTAL STABILITY IS MODERATELY PRESENT AND DATES FROM THE POST-WAR PERIOD, OR IS ABSENT (36)	0.50	24 20 8 20 X SQ= 3.68 P = 0.051 RV YES (YES)
101	TEND TO BE THOSE WHERE THE REPRESENTATIVE CHARACTER OF THE REGIME IS POLYARCHIC (41)	0.84	101	TEND TO BE THOSE WHERE THE REPRESENTATIVE CHARACTER OF THE REGIME IS LIMITED POLYARCHIC, PSEUDO-POLYARCHIC, OR NON-POLYARCHIC (57)	0.87	27 7 5 46 X SQ= 39.19 P = 0. RV YES (YES)

102	ALWAYS ARE THOSE WHERE THE REPRESENTATIVE CHARACTER OF THE REGIME IS POLYARCHIC OR LIMITED POLYARCHIC (59)	1.00	102	TEND TO BE THOSE WHERE THE REPRESENTATIVE CHARACTER OF THE REGIME IS PSEUDO-POLYARCHIC OR NON-POLYARCHIC (49)	0.78	33 13 0 45 X SQ= 47.59 P = 0. RV YES (YES)
105	ALWAYS ARE THOSE WHERE THE ELECTORAL SYSTEM IS COMPETITIVE (43)	1.00	105	TEND TO BE THOSE WHERE THE ELECTORAL SYSTEM IS PARTIALLY COMPETITIVE OR NON-COMPETITIVE (47)	0.87	28 6 0 41 X SQ= 50.42 P = 0. RV YES (YES)
107	TEND TO BE THOSE WHERE AUTONOMOUS GROUPS ARE FULLY TOLERATED IN POLITICS (46)	0.91	107	TEND TO BE THOSE WHERE AUTONOMOUS GROUPS ARE PARTIALLY TOLERATED IN POLITICS, ARE TOLERATED ONLY OUTSIDE POLITICS, OR ARE NOT TOLERATED AT ALL (65)	0.88	30 7 3 52 X SQ= 51.76 P = 0. RV YES (YES)
108	TEND TO BE THOSE WHERE AUTONOMOUS GROUPS ARE FULLY OR PARTIALLY TOLERATED IN POLITICS (65)	0.97	108	TEND TO BE THOSE WHERE AUTONOMOUS GROUPS ARE TOLERATED ONLY OUTSIDE POLITICS OR ARE NOT TOLERATED AT ALL (35)	0.61	31 20 1 31 X SQ= 25.21 P = 0. RV YES (YES)
111	TEND TO BE THOSE WHERE POLITICAL ENCULTURATION IS HIGH OR MEDIUM (53)	0.80	111	TEND TO BE THOSE WHERE POLITICAL ENCULTURATION IS LOW (42)	0.53	24 22 6 25 X SQ= 7.06 P = 0.004 RV YES (YES)
115	TEND TO BE THOSE WHERE INTEREST ARTICULATION BY ASSOCIATIONAL GROUPS IS SIGNIFICANT (20)	0.55	115	ALWAYS ARE THOSE WHERE INTEREST ARTICULATION BY ASSOCIATIONAL GROUPS IS MODERATE, LIMITED, OR NEGLIGIBLE (91)	1.00	18 0 15 61 X SQ= 37.71 P = 0. RV YES (YES)
116	TEND TO BE THOSE WHERE INTEREST ARTICULATION BY ASSOCIATIONAL GROUPS IS SIGNIFICANT OR MODERATE (32)	0.85	116	TEND TO BE THOSE WHERE INTEREST ARTICULATION BY ASSOCIATIONAL GROUPS IS LIMITED OR NEGLIGIBLE (79)	0.97	28 2 5 59 X SQ= 61.87 P = 0. RV YES (YES)
117	TEND TO BE THOSE WHERE INTEREST ARTICULATION BY ASSOCIATIONAL GROUPS IS SIGNIFICANT, MODERATE, OR LIMITED (60)	0.97	117	TEND TO BE THOSE WHERE INTEREST ARTICULATION BY ASSOCIATIONAL GROUPS IS NEGLIGIBLE (51)	0.72	32 17 1 44 X SQ= 38.25 P = 0. RV YES (YES)
118	TEND TO BE THOSE WHERE INTEREST ARTICULATION BY INSTITUTIONAL GROUPS IS SIGNIFICANT, MODERATE, OR LIMITED (60)	0.91	118	TEND TO BE THOSE WHERE INTEREST ARTICULATION BY INSTITUTIONAL GROUPS IS VERY SIGNIFICANT (40)	0.64	3 30 30 17 X SQ= 21.77 P = 0. RV YES (YES)

119	TEND TO BE THOSE WHERE INTEREST ARTICULATION BY INSTITUTIONAL GROUPS IS MODERATE OR LIMITED (26)	0.67	119	TEND TO BE THOSE WHERE INTEREST ARTICULATION BY INSTITUTIONAL GROUPS IS VERY SIGNIFICANT OR SIGNIFICANT (74)	0.98	11 46 22 1 X SQ= 36.33 P = 0. RV YES (YES)
120	TEND LESS TO BE THOSE WHERE INTEREST ARTICULATION BY INSTITUTIONAL GROUPS IS VERY SIGNIFICANT, SIGNIFICANT, OR MODERATE (90)	0.70	120	ALWAYS ARE THOSE WHERE INTEREST ARTICULATION BY INSTITUTIONAL GROUPS IS VERY SIGNIFICANT, SIGNIFICANT, OR MODERATE (90)	1.00	23 47 10 0 X SQ= 13.62 P = 0. RV NO (YES)
121	TEND TO BE THOSE WHERE INTEREST ARTICULATION BY NON-ASSOCIATIONAL GROUPS IS MODERATE, LIMITED, OR NEGLIGIBLE (61)	0.88	121	TEND TO BE THOSE WHERE INTEREST ARTICULATION BY NON-ASSOCIATIONAL GROUPS IS SIGNIFICANT (54)	0.71	4 44 29 18 X SQ= 27.53 P = 0. RV YES (YES)
122	TEND TO BE THOSE WHERE INTEREST ARTICULATION BY NON-ASSOCIATIONAL GROUPS IS LIMITED OR NEGLIGIBLE (32)	0.64	122	TEND TO BE THOSE WHERE INTEREST ARTICULATION BY NON-ASSOCIATIONAL GROUPS IS SIGNIFICANT OR MODERATE (83)	0.89	12 55 21 7 X SQ= 25.93 P = 0. RV YES (YES)
123	TEND LESS TO BE THOSE WHERE INTEREST ARTICULATION BY NON-ASSOCIATIONAL GROUPS IS SIGNIFICANT, MODERATE, OR LIMITED (107)	0.76	123	ALWAYS ARE THOSE WHERE INTEREST ARTICULATION BY NON-ASSOCIATIONAL GROUPS IS SIGNIFICANT, MODERATE, OR LIMITED (107)	1.00	25 62 8 0 X SQ= 13.42 P = 0. RV NO (YES)
125	TEND TO BE THOSE WHERE INTEREST ARTICULATION BY ANOMIC GROUPS IS INFREQUENT OR VERY INFREQUENT (35)	0.72	125	TEND TO BE THOSE WHERE INTEREST ARTICULATION BY ANOMIC GROUPS IS FREQUENT OR OCCASIONAL (64)	0.86	8 44 21 7 X SQ= 25.47 P = 0. RV YES (YES)
126	TEND TO BE THOSE WHERE INTEREST ARTICULATION BY ANOMIC GROUPS IS VERY INFREQUENT (16)	0.55	126	ALWAYS ARE THOSE WHERE INTEREST ARTICULATION BY ANOMIC GROUPS IS FREQUENT, OCCASIONAL, OR INFREQUENT (83)	1.00	13 51 16 0 X SQ= 31.81 P = 0. RV YES (YES)
128	TEND TO BE THOSE WHERE INTEREST ARTICULATION BY POLITICAL PARTIES IS SIGNIFICANT OR MODERATE (48)	0.83	128	TEND TO BE THOSE WHERE INTEREST ARTICULATION BY POLITICAL PARTIES IS LIMITED OR NEGLIGIBLE (45)	0.78	24 10 5 36 X SQ= 24.32 P = 0. RV YES (YES)
129	TEND TO BE THOSE WHERE INTEREST ARTICULATION BY POLITICAL PARTIES IS SIGNIFICANT, MODERATE, OR LIMITED (56)	0.86	129	TEND TO BE THOSE WHERE INTEREST ARTICULATION BY POLITICAL PARTIES IS NEGLIGIBLE (37)	0.67	25 15 4 31 X SQ= 18.43 P = 0. RV YES (YES)

130	TEND LESS TO BE THOSE WHERE INTEREST AGGREGATION BY POLITICAL PARTIES IS MODERATE, LIMITED, OR NEGLIGIBLE (71)	0.67	130	TEND MORE TO BE THOSE WHERE INTEREST AGGREGATION BY POLITICAL PARTIES IS MODERATE, LIMITED, OR NEGLIGIBLE (71)	0.92	9 3 18 34 X SQ= 4.97 P = 0.021 RV NO (YES)
134	TEND MORE TO BE THOSE WHERE INTEREST AGGREGATION BY THE EXECUTIVE IS SIGNIFICANT OR MODERATE (57)	0.83	134	TEND LESS TO BE THOSE WHERE INTEREST AGGREGATION BY THE EXECUTIVE IS SIGNIFICANT OR MODERATE (57)	0.51	24 28 5 27 X SQ= 6.87 P = 0.005 RV NO (NO)
135	ALWAYS ARE THOSE WHERE INTEREST AGGREGATION BY THE EXECUTIVE IS SIGNIFICANT, MODERATE, OR LIMITED (77)	1.00	135	TEND LESS TO BE THOSE WHERE INTEREST AGGREGATION BY THE EXECUTIVE IS SIGNIFICANT, MODERATE, OR LIMITED (77)	0.80	28 35 0 9 X SQ= 4.81 P'= 0.010 RV NO (NO)
137	TEND TO BE THOSE WHERE INTEREST AGGREGATION BY THE LEGISLATURE IS SIGNIFICANT OR MODERATE (29)	0.86	137	TEND TO BE THOSE WHERE INTEREST AGGREGATION BY THE LEGISLATURE IS LIMITED OR NEGLIGIBLE (68)	0.98	25 1 4 50 X SQ= 56.03 P = 0. RV YES (YES)
138	TEND TO BE THOSE WHERE INTEREST AGGREGATION BY THE LEGISLATURE IS SIGNIFICANT, MODERATE, OR LIMITED (48)	0.97	138	TEND TO BE THOSE WHERE INTEREST AGGREGATION BY THE LEGISLATURE IS NEGLIGIBLE (49)	0.86	30 7 1 44 X SQ= 50.40 P = 0. RV YES (YES)
139	TEND TO BE THOSE WHERE THE PARTY SYSTEM IS QUANTITATIVELY OTHER THAN ONE-PARTY (71)	0.97	139	TEND TO BE THOSE WHERE THE PARTY SYSTEM IS QUANTITATIVELY ONE-PARTY (34)	0.56	1 30 32 24 X SQ= 22.40 P = 0. RV YES (YES)
141	TEND LESS TO BE THOSE WHERE THE PARTY SYSTEM IS QUANTITATIVELY OTHER THAN TWO-PARTY (87)	0.70	141	TEND MORE TO BE THOSE WHERE THE PARTY SYSTEM IS QUANTITATIVELY OTHER THAN TWO-PARTY (87)	0.98	9 1 21 49 X SQ= 11.00 P = 0. RV NO (YES)
142	TEND TO BE THOSE WHERE THE PARTY SYSTEM IS QUANTITATIVELY MULTI-PARTY (30)	0.53	142	TEND TO BE THOSE WHERE THE PARTY SYSTEM IS QUANTITATIVELY OTHER THAN MULTI-PARTY (66)	0.88	16 6 14 43 X SQ= 13.66 P = 0. RV YES (YES)
144	ALWAYS ARE THOSE WHERE THE PARTY SYSTEM IS QUANTITATIVELY TWO-PARTY, RATHER THAN ONE-PARTY (11)	1.00	144	TEND TO BE THOSE WHERE THE PARTY SYSTEM IS QUANTITATIVELY ONE-PARTY, RATHER THAN TWO-PARTY (34)	0.97	0 31 9 1 X SQ= 30.69 P = 0. RV YES (YES)

146	ALWAYS ARE THOSE 1.00 WHERE THE PARTY SYSTEM IS QUALITATIVELY OTHER THAN MASS-BASED TERRITORIAL (92)	146	TEND LESS TO BE THOSE 0.85 WHERE THE PARTY SYSTEM IS QUALITATIVELY OTHER THAN MASS-BASED TERRITORIAL (92)	0 8 31 44 X SQ= 3.66 P = 0.023 RV NO (NO)
147	TEND TO BE THOSE 0.63 WHERE THE PARTY SYSTEM IS QUALITATIVELY CLASS-ORIENTED OR MULTI-IDEOLOGICAL (23)	147	TEND TO BE THOSE 0.98 WHERE THE PARTY SYSTEM IS QUALITATIVELY OTHER THAN CLASS-ORIENTED OR MULTI-IDEOLOGICAL (67)	17 1 10 49 X SQ= 33.05 P = 0. RV YES (YES)
148	ALWAYS ARE THOSE 1.00 WHERE THE PARTY SYSTEM IS QUALITATIVELY OTHER THAN AFRICAN TRANSITIONAL (96)	148	TEND LESS TO BE THOSE 0.78 WHERE THE PARTY SYSTEM IS QUALITATIVELY OTHER THAN AFRICAN TRANSITIONAL (96)	0 13 33 45 X SQ= 6.90 P = 0.002 RV NO (NO)
153	LEAN TOWARD BEING THOSE 0.70 WHERE THE PARTY SYSTEM IS STABLE (42)	153	LEAN TOWARD BEING THOSE 0.53 WHERE THE PARTY SYSTEM IS MODERATELY STABLE OR UNSTABLE (38)	21 17 9 19 X SQ= 2.61 P = 0.082 RV YES (YES)
161	TEND TO BE THOSE 0.71 WHERE THE POLITICAL LEADERSHIP IS NON-ELITIST (50)	161	TEND TO BE THOSE 0.55 WHERE THE POLITICAL LEADERSHIP IS ELITIST OR MODERATELY ELITIST (47)	9 27 22 22 X SQ= 4.21 P = 0.037 RV YES (YES)
164	TEND TO BE THOSE 0.97 WHERE THE REGIME'S LEADERSHIP CHARISMA IS NEGLIGIBLE (65)	164	TEND TO BE THOSE 0.59 WHERE THE REGIME'S LEADERSHIP CHARISMA IS PRONOUNCED OR MODERATE (34)	1 29 31 20 X SQ= 23.74 P = 0. RV YES (YES)
168	TEND TO BE THOSE 0.82 WHERE THE HORIZONTAL POWER DISTRIBUTION IS SIGNIFICANT (34)	168	TEND TO BE THOSE 0.95 WHERE THE HORIZONTAL POWER DISTRIBUTION IS LIMITED OR NEGLIGIBLE (72)	27 3 6 54 X SQ= 51.73 P = 0. RV YES (YES)
169	ALWAYS ARE THOSE 1.00 WHERE THE HORIZONTAL POWER DISTRIBUTION IS SIGNIFICANT OR LIMITED (58)	169	TEND TO BE THOSE 0.77 WHERE THE HORIZONTAL POWER DISTRIBUTION IS NEGLIGIBLE (48)	33 13 0 44 X SQ= 46.80 P = 0. RV YES (YES)
171	TEND LESS TO BE THOSE 0.76 WHERE THE LEGISLATIVE-EXECUTIVE STRUCTURE IS OTHER THAN PARLIAMENTARY-REPUBLICAN (90)	171	TEND MORE TO BE THOSE 0.94 WHERE THE LEGISLATIVE-EXECUTIVE STRUCTURE IS OTHER THAN PARLIAMENTARY-REPUBLICAN (90)	8 3 25 50 X SQ= 4.74 P = 0.019 RV NO (YES)

172	TEND LESS TO BE THOSE WHERE THE LEGISLATIVE-EXECUTIVE STRUCTURE IS OTHER THAN PARLIAMENTARY-ROYALIST (88)	0.64	172	TEND MORE TO BE THOSE WHERE THE LEGISLATIVE-EXECUTIVE STRUCTURE IS OTHER THAN PARLIAMENTARY-ROYALIST (88)	0.93	12 4 21 53 X SQ= 10.39 P = 0.001 RV NO (YES)
174	TEND TO BE THOSE WHERE THE LEGISLATURE IS FULLY EFFECTIVE (28)	0.70	174	TEND TO BE THOSE WHERE THE LEGISLATURE IS PARTIALLY EFFECTIVE, LARGELY INEFFECTIVE, OR WHOLLY INEFFECTIVE (72)	0.96	23 2 10 48 X SQ= 37.70 P = 0. RV YES (YES)
175	ALWAYS ARE THOSE WHERE THE LEGISLATURE IS FULLY EFFECTIVE OR PARTIALLY EFFECTIVE (51)	1.00	175	TEND TO BE THOSE WHERE THE LEGISLATURE IS LARGELY INEFFECTIVE OR WHOLLY INEFFECTIVE (49)	0.88	33 6 0 44 X SQ= 58.32 P = 0. RV YES (YES)
178	TEND TO BE THOSE WHERE THE LEGISLATURE IS BICAMERAL (51)	0.78	178	TEND TO BE THOSE WHERE THE LEGISLATURE IS UNICAMERAL (53)	0.77	7 41 25 12 X SQ= 22.78 P = 0. RV YES (YES)
179	TEND TO BE THOSE WHERE THE EXECUTIVE IS STRONG (39)	0.86	179	TEND TO BE THOSE WHERE THE EXECUTIVE IS DOMINANT (52)	0.85	4 41 25 7 X SQ= 35.29 P = 0. RV YES (YES)
181	TEND TO BE THOSE WHERE THE BUREAUCRACY IS MODERN, RATHER THAN SEMI-MODERN (21)	0.61	181	TEND TO BE THOSE WHERE THE BUREAUCRACY IS SEMI-MODERN, RATHER THAN MODERN (55)	0.96	20 1 13 25 X SQ= 18.04 P = 0. RV YES (YES)
182	ALWAYS ARE THOSE WHERE THE BUREAUCRACY IS SEMI-MODERN, RATHER THAN POST-COLONIAL TRANSITIONAL (55)	1.00	182	TEND LESS TO BE THOSE WHERE THE BUREAUCRACY IS SEMI-MODERN, RATHER THAN POST-COLONIAL TRANSITIONAL (55)	0.53	13 25 0 22 X SQ= 7.70 P = 0.001 RV NO (NO)
186	TEND TO BE THOSE WHERE PARTICIPATION BY THE MILITARY IN POLITICS IS NEUTRAL, RATHER THAN SUPPORTIVE (56)	0.90	186	TEND TO BE THOSE WHERE PARTICIPATION BY THE MILITARY IN POLITICS IS SUPPORTIVE, RATHER THAN NEUTRAL (31)	0.56	3 25 27 20 X SQ= 14.08 P = 0. RV YES (YES)
187	TEND TO BE THOSE WHERE THE ROLE OF THE POLICE IS NOT POLITICALLY SIGNIFICANT (35)	0.80	187	TEND TO BE THOSE WHERE THE ROLE OF THE POLICE IS POLITICALLY SIGNIFICANT (66)	0.91	6 50 24 5 X SQ= 40.33 P = 0. RV YES (YES)

```
188  TEND TO BE THOSE                        0.52     188  TEND TO BE THOSE                        0.90        17     6
     WHERE THE CHARACTER OF THE LEGAL SYSTEM               WHERE THE CHARACTER OF THE LEGAL SYSTEM              16    54
     IS CIVIL LAW  (32)                                    IS OTHER THAN CIVIL LAW  (81)                    X SQ=  17.55
                                                                                                            P =     0.
                                                                                                            RV YES (YES)

189  TEND LESS TO BE THOSE                   0.85     189  ALWAYS ARE THOSE                        1.00         5     0
     WHERE THE CHARACTER OF THE LEGAL SYSTEM               WHERE THE CHARACTER OF THE LEGAL SYSTEM              28    62
     IS OTHER THAN COMMON LAW  (108)                       IS OTHER THAN COMMON LAW  (108)                  X SQ=   7.11
                                                                                                            P =     0.004
                                                                                                            RV NO  (YES)

192  ALWAYS ARE THOSE                        1.00     192  TEND LESS TO BE THOSE                   0.62         0    23
     WHERE THE CHARACTER OF THE LEGAL SYSTEM               WHERE THE CHARACTER OF THE LEGAL SYSTEM              33    37
     IS OTHER THAN                                         IS OTHER THAN                                    X SQ=  14.81
     MUSLIM OR PARTLY MUSLIM  (86)                         MUSLIM OR PARTLY MUSLIM  (86)                    P =     0.
                                                                                                            RV NO  (NO )

193  TEND MORE TO BE THOSE                   0.97     193  TEND LESS TO BE THOSE                   0.62         1    23
     WHERE THE CHARACTER OF THE LEGAL SYSTEM               WHERE THE CHARACTER OF THE LEGAL SYSTEM              32    38
     IS OTHER THAN PARTLY INDIGENOUS  (86)                 IS OTHER THAN PARTLY INDIGENOUS  (86)            X SQ=  11.78
                                                                                                            P =     0.
                                                                                                            RV NO  (NO )

194  ALWAYS ARE THOSE                        1.00     194  TEND LESS TO BE THOSE                   0.79         0    13
     THAT ARE NON-COMMUNIST  (101)                         THAT ARE NON-COMMUNIST  (101)                        33    48
                                                                                                            X SQ=   6.47
                                                                                                            P =     0.003
                                                                                                            RV NO  (NO )
```

90 POLITIES WHOSE IDEOLOGICAL ORIENTATION IS TRADITIONAL (5)	90 POLITIES WHOSE IDEOLOGICAL ORIENTATION IS OTHER THAN TRADITIONAL (102)

SUBJECT ONLY

5 IN LEFT COLUMN

AFGHANISTAN ETHIOPIA JORDAN SA'U ARABIA THAILAND

102 IN RIGHT COLUMN

ALBANIA	ALGERIA	ARGENTINA	AUSTRALIA	AUSTRIA	BELGIUM	BOLIVIA	BRAZIL	BULGARIA	BURMA
BURUNDI	CAMEROUN	CANADA	CEN AFR REP	CEYLON	CHAD	CHILE	CHINA, PR	COLOMBIA	CONGO(BRA)
CONGO(LEO)	COSTA RICA	CUBA	CYPRUS	CZECHOS'KIA	DAHOMEY	DENMARK	DOMIN REP	ECUADOR	EL SALVADOR
FINLAND	FRANCE	GABON	GERMANY, E	GERMAN FR	GHANA	GREECE	GUATEMALA	GUINEA	HAITI
HONDURAS	HUNGARY	ICELAND	INDIA	INDONESIA	IRAQ	IRELAND	ISRAEL	ITALY	IVORY COAST
JAMAICA	JAPAN	KOREA, N	KOREA REP	LAOS	LEBANON	LIBERIA	LUXEMBOURG	MALAGASY R	MALAYA
MALI	MAURITANIA	MEXICO	MONGOLIA	NETHERLANDS	NEW ZEALAND	NICARAGUA	NIGER	NORWAY	PAKISTAN
PANAMA	PARAGUAY	PERU	PHILIPPINES	POLAND	PORTUGAL	RUMANIA	RWANDA	SENEGAL	SIERRE LEO
SOMALIA	SO AFRICA	SPAIN	SUDAN	SWEDEN	SWITZERLAND	SYRIA	TANGANYIKA	TOGO	TRINIDAD
TUNISIA	TURKEY	USSR	UAR	UK	US	UPPER VOLTA	URUGUAY	VENEZUELA	VIETNAM, N
VIETNAM REP	YUGOSLAVIA								

7 EXCLUDED BECAUSE AMBIGUOUS

CAMBODIA IRAN LIBYA MOROCCO NEPAL NIGERIA UGANDA

1 EXCLUDED BECAUSE UNASCERTAINABLE

YEMEN

	Left column			Right column		
13	TEND TO BE THOSE LOCATED IN NORTH AFRICA OR THE MIDDLE EAST (16)	0.60	13	TEND TO BE THOSE LOCATED ELSEWHERE THAN IN NORTH AFRICA OR THE MIDDLE EAST (99)	0.91	3 9 2 93 X SQ= 7.92 P = 0.010 RV YES (NO)
15	TEND TO BE THOSE LOCATED IN THE MIDDLE EAST, RATHER THAN IN NORTH AFRICA OR CENTRAL AND SOUTH AFRICA (11)	0.75	15	TEND TO BE THOSE LOCATED IN NORTH AFRICA OR CENTRAL AND SOUTH AFRICA, RATHER THAN IN THE MIDDLE EAST (33)	0.82	3 6 1 28 X SQ= 3.73 P = 0.035 RV YES (NO)
16	TILT TOWARD BEING THOSE LOCATED IN THE MIDDLE EAST, RATHER THAN IN EAST ASIA, SOUTH ASIA, OR SOUTHEAST ASIA (11)	0.75	16	TILT TOWARD BEING THOSE LOCATED IN EAST ASIA, SOUTH ASIA, OR SOUTHEAST ASIA, RATHER THAN IN THE MIDDLE EAST (18)	0.71	3 6 1 15 X SQ= 1.45 P = 0.116 RV YES (NO)

17	ALWAYS ARE THOSE LOCATED IN THE MIDDLE EAST, RATHER THAN IN THE CARIBBEAN, CENTRAL AMERICA, OR SOUTH AMERICA (11)	1.00	17	TEND TO BE THOSE LOCATED IN THE CARIBBEAN, CENTRAL AMERICA, OR SOUTH AMERICA, RATHER THAN IN THE MIDDLE EAST (22)	0.79	3 6 0 22 X SQ= 4.75 P = 0.019 RV YES (NO)
43	ALWAYS ARE THOSE WHOSE ECONOMIC DEVELOPMENTAL STATUS IS VERY UNDERDEVELOPED (57)	1.00	43	LEAN TOWARD BEING THOSE WHOSE ECONOMIC DEVELOPMENTAL STATUS IS DEVELOPED, INTERMEDIATE, OR UNDERDEVELOPED (55)	0.54	0 54 4 46 X SQ= 2.59 P = 0.050 RV YES (NO)
46	LEAN TOWARD BEING THOSE WHERE THE LITERACY RATE IS BELOW TEN PERCENT (26)	0.60	46	LEAN TOWARD BEING THOSE WHERE THE LITERACY RATE IS TEN PERCENT OR ABOVE (84)	0.80	2 78 3 20 X SQ= 2.32 P = 0.072 RV YES (NO)
51	ALWAYS ARE THOSE WHERE FREEDOM OF THE PRESS IS INTERNALLY ABSENT, OR INTERNALLY AND EXTERNALLY ABSENT (37)	1.00	51	TEND TO BE THOSE WHERE FREEDOM OF THE PRESS IS COMPLETE OR INTERMITTENT (60)	0.66	0 56 5 29 X SQ= 6.14 P = 0.006 RV YES (NO)
52	TEND TO BE THOSE WHERE FREEDOM OF THE PRESS IS INTERNALLY AND EXTERNALLY ABSENT (16)	0.60	52	TEND TO BE THOSE WHERE FREEDOM OF THE PRESS IS COMPLETE, INTERMITTENT, OR INTERNALLY ABSENT (82)	0.85	2 73 3 13 X SQ= 3.84 P = 0.036 RV YES (NO)
55	TILT TOWARD BEING THOSE WHERE NEWSPAPER CIRCULATION IS LESS THAN TEN PER THOUSAND (35)	0.60	55	TILT TOWARD BEING THOSE WHERE NEWSPAPER CIRCULATION IS TEN OR MORE PER THOUSAND (78)	0.75	2 75 3 25 X SQ= 1.46 P = 0.117 RV YES (NO)
58	TEND TO BE THOSE WHERE THE RELIGION IS MUSLIM (18)	0.60	58	TEND TO BE THOSE WHERE THE RELIGION IS OTHER THAN MUSLIM (97)	0.89	3 11 2 91 X SQ= 6.29 P = 0.016 RV YES (NO)
63	ALWAYS ARE THOSE WHERE THE RELIGION IS PREDOMINANTLY OR PARTLY OTHER THAN CHRISTIAN (68)	1.00	63	LEAN LESS TOWARD BEING THOSE WHERE THE RELIGION IS PREDOMINANTLY OR PARTLY OTHER THAN CHRISTIAN (68)	0.54	0 46 5 55 X SQ= 2.38 P = 0.067 RV NO (NO)
71	TEND TO BE THOSE WHOSE DATE OF INDEPENDENCE IS BEFORE 1800 (21)	0.60	71	TEND TO BE THOSE WHOSE DATE OF INDEPENDENCE IS AFTER 1800 (90)	0.84	3 16 2 82 X SQ= 3.48 P = 0.042 RV YES (NO)

75	ALWAYS ARE THOSE THAT ARE NOT HISTORICALLY WESTERN AND ARE NOT SIGNIFICANTLY WESTERNIZED (52)	1.00	75	TEND TO BE THOSE THAT ARE HISTORICALLY WESTERN OR SIGNIFICANTLY WESTERNIZED (62)	0.61	0 62 5 39 X SQ= 5.08 P = 0.011 RV YES (NO)
83	TEND TO BE THOSE WHERE THE TYPE OF POLITICAL MODERNIZATION IS NON-EUROPEAN AUTOCHTHONOUS (9)	0.60	83	TEND TO BE THOSE WHERE THE TYPE OF POLITICAL MODERNIZATION IS OTHER THAN NON-EUROPEAN AUTOCHTHONOUS (106)	0.96	3 4 2 98 X SQ= 16.20 P = 0.002 RV YES (NO)
96	ALWAYS ARE THOSE WHERE THE STATUS OF THE REGIME IS AUTHORITARIAN (23)	1.00	96	TEND TO BE THOSE WHERE THE STATUS OF THE REGIME IS CONSTITUTIONAL OR TOTALITARIAN (67)	0.81	0 63 5 15 X SQ= 12.63 P = 0.001 RV YES (NO)
102	ALWAYS ARE THOSE WHERE THE REPRESENTATIVE CHARACTER OF THE REGIME IS PSEUDO-POLYARCHIC OR NON-POLYARCHIC (49)	1.00	102	TEND TO BE THOSE WHERE THE REPRESENTATIVE CHARACTER OF THE REGIME IS POLYARCHIC OR LIMITED POLYARCHIC (59)	0.57	0 55 5 41 X SQ= 4.19 P = 0.017 RV YES (NO)
108	ALWAYS ARE THOSE WHERE AUTONOMOUS GROUPS ARE TOLERATED ONLY OUTSIDE POLITICS OR ARE NOT TOLERATED AT ALL (35)	1.00	108	TEND TO BE THOSE WHERE AUTONOMOUS GROUPS ARE FULLY OR PARTIALLY TOLERATED IN POLITICS (65)	0.68	0 61 4 29 X SQ= 5.03 P = 0.013 RV YES (NO)
117	ALWAYS ARE THOSE WHERE INTEREST ARTICULATION BY ASSOCIATIONAL GROUPS IS NEGLIGIBLE (51)	1.00	117	TEND TO BE THOSE WHERE INTEREST ARTICULATION BY ASSOCIATIONAL GROUPS IS SIGNIFICANT, MODERATE, OR LIMITED (60)	0.59	0 58 5 40 X SQ= 4.58 P = 0.014 RV YES (NO)
121	ALWAYS ARE THOSE WHERE INTEREST ARTICULATION BY NON-ASSOCIATIONAL GROUPS IS SIGNIFICANT (54)	1.00	121	TEND TO BE THOSE WHERE INTEREST ARTICULATION BY NON-ASSOCIATIONAL GROUPS IS MODERATE, LIMITED, OR NEGLIGIBLE (61)	0.58	5 43 0 59 X SQ= 4.32 P = 0.016 RV YES (NO)
134	ALWAYS ARE THOSE WHERE INTEREST AGGREGATION BY THE EXECUTIVE IS LIMITED OR NEGLIGIBLE (46)	1.00	134	TEND TO BE THOSE WHERE INTEREST AGGREGATION BY THE EXECUTIVE IS SIGNIFICANT OR MODERATE (57)	0.60	0 55 5 36 X SQ= 4.82 P = 0.012 RV YES (NO)
135	TEND TO BE THOSE WHERE INTEREST AGGREGATION BY THE EXECUTIVE IS NEGLIGIBLE (13)	0.80	135	TEND TO BE THOSE WHERE INTEREST AGGREGATION BY THE EXECUTIVE IS SIGNIFICANT, MODERATE, OR LIMITED (77)	0.90	1 70 4 8 X SQ= 13.27 P = 0.001 RV YES (NO)

No.	Statement	Value	No.	Statement	Value	Statistics
138	ALWAYS ARE THOSE WHERE INTEREST AGGREGATION BY THE LEGISLATURE IS NEGLIGIBLE (49)	1.00	138	TILT TOWARD BEING THOSE WHERE INTEREST AGGREGATION BY THE LEGISLATURE IS SIGNIFICANT, MODERATE, OR LIMITED (48)	0.51	0 45 4 43 X SQ= 2.22 P = 0.117 RV YES (NO)
154	ALWAYS ARE THOSE WHERE THE PARTY SYSTEM IS UNSTABLE (25)	1.00	154	LEAN TOWARD BEING THOSE WHERE THE PARTY SYSTEM IS STABLE OR MODERATELY STABLE (55)	0.71	0 54 2 22 X SQ= 1.89 P = 0.092 RV YES (NO)
157	ALWAYS ARE THOSE WHERE PERSONALISMO IS PRONOUNCED (14)	1.00	157	TILT TOWARD BEING THOSE WHERE PERSONALISMO IS MODERATE OR NEGLIGIBLE (84)	0.87	1 12 0 81 X SQ= 1.11 P = 0.138 RV YES (NO)
160	ALWAYS ARE THOSE WHERE THE POLITICAL LEADERSHIP IS ELITIST (30)	1.00	160	TEND TO BE THOSE WHERE THE POLITICAL LEADERSHIP IS MODERATELY ELITIST OR NON-ELITIST (67)	0.73	5 24 0 65 X SQ= 8.66 P = 0.002 RV YES (NO)
164	LEAN TOWARD BEING THOSE WHERE THE REGIME'S LEADERSHIP CHARISMA IS PRONOUNCED OR MODERATE (34)	0.75	164	LEAN TOWARD BEING THOSE WHERE THE REGIME'S LEADERSHIP CHARISMA IS NEGLIGIBLE (65)	0.70	3 28 1 64 X SQ= 1.74 P = 0.097 RV YES (NO)
169	ALWAYS ARE THOSE WHERE THE HORIZONTAL POWER DISTRIBUTION IS NEGLIGIBLE (48)	1.00	169	TEND TO BE THOSE WHERE THE HORIZONTAL POWER DISTRIBUTION IS SIGNIFICANT OR LIMITED (58)	0.58	0 55 5 40 X SQ= 4.31 P = 0.016 RV YES (NO)
170	ALWAYS ARE THOSE WHERE THE LEGISLATIVE-EXECUTIVE STRUCTURE IS OTHER THAN PRESIDENTIAL (63)	1.00	170	TILT LESS TOWARD BEING THOSE WHERE THE LEGISLATIVE-EXECUTIVE STRUCTURE IS OTHER THAN PRESIDENTIAL (63)	0.57	0 39 4 52 X SQ= 1.41 P = 0.141 RV NO (NO)
176	ALWAYS ARE THOSE WHERE THE LEGISLATURE IS WHOLLY INEFFECTIVE (28)	1.00	176	TEND TO BE THOSE WHERE THE LEGISLATURE IS FULLY EFFECTIVE, PARTIALLY EFFECTIVE, OR LARGELY INEFFECTIVE (72)	0.76	0 68 4 22 X SQ= 7.48 P = 0.005 RV YES (NO)
178	ALWAYS ARE THOSE WHERE THE LEGISLATURE IS BICAMERAL (51)	1.00	178	TILT TOWARD BEING THOSE WHERE THE LEGISLATURE IS UNICAMERAL (53)	0.54	0 51 3 43 X SQ= 1.60 P = 0.103 RV YES (NO)

179	ALWAYS ARE THOSE WHERE THE EXECUTIVE IS DOMINANT (52)	1.00	179	TILT LESS TOWARD BEING THOSE WHERE THE EXECUTIVE IS DOMINANT (52)	0.54	4 44 0 37 X SQ= 1.64 P = 0.129 RV NO (NO)
183	ALWAYS ARE THOSE WHERE THE BUREAUCRACY IS TRADITIONAL, RATHER THAN POST-COLONIAL TRANSITIONAL (9)	1.00	183	TEND TO BE THOSE WHERE THE BUREAUCRACY IS POST-COLONIAL TRANSITIONAL, RATHER THAN TRADITIONAL (25)	0.96	0 23 5 1 X SQ= 17.69 P = 0. RV YES (YES)
186	ALWAYS ARE THOSE WHERE PARTICIPATION BY THE MILITARY IN POLITICS IS SUPPORTIVE, RATHER THAN NEUTRAL (31)	1.00	186	TEND TO BE THOSE WHERE PARTICIPATION BY THE MILITARY IN POLITICS IS NEUTRAL, RATHER THAN SUPPORTIVE (56)	0.68	4 25 0 52 X SQ= 4.89 P = 0.014 RV YES (NO)

```
******************************************************************************************

91  POLITIES                                     91  POLITIES
    WHOSE IDEOLOGICAL ORIENTATION IS                 WHOSE IDEOLOGICAL ORIENTATION IS
    DEVELOPMENTAL, RATHER THAN                       TRADITIONAL, RATHER THAN
    TRADITIONAL  (31)                                DEVELOPMENTAL  (5)


                                                                    BOTH SUBJECT AND PREDICATE
..........................................................................................

      31 IN LEFT COLUMN

ALGERIA     BOLIVIA     BURUNDI     CAMEROUN    CEN AFR REP CHAD        CONGO(BRA)  CONGO(LEO)  DAHOMEY     DOMIN REP
GABON       GHANA       GUINEA      INDONESIA   ISRAEL      IVORY COAST MALAYA      MALI        MAURITANIA  NIGER
RWANDA      SENEGAL     SIERRE LEO  SOMALIA     SUDAN       TANGANYIKA  TOGO        TUNISIA     UAR         UPPER VOLTA
VENEZUELA

       5 IN RIGHT COLUMN

AFGHANISTAN ETHIOPIA    JORDAN      SA'U ARABIA THAILAND

      20 EXCLUDED BECAUSE AMBIGUOUS

ARGENTINA   BRAZIL      BURMA       CAMBODIA    CEYLON      CUBA        GUATEMALA   INDIA       IRAN        JAMAICA
LIBERIA     LIBYA       MALAGASY R  MOROCCO     NEPAL       NIGERIA     PAKISTAN    TRINIDAD    UGANDA      VIETNAM REP

      56 EXCLUDED BECAUSE IRRELEVANT

ALBANIA     AUSTRALIA   AUSTRIA     BELGIUM     BULGARIA    CANADA      CHILE       CHINA, PR   COLOMBIA    COSTA RICA
CYPRUS      CZECHOS'KIA DENMARK     ECUADOR     EL SALVADOR FINLAND     FRANCE      GERMANY, E  GERMAN FR   GREECE
HAITI       HONDURAS    HUNGARY     ICELAND     IRAQ        IRELAND     ITALY       JAPAN       KOREA, N    LAOS
LEBANON     LUXEMBOURG  MEXICO      MONGOLIA    NETHERLANDS NEW ZEALAND NICARAGUA   NORWAY      PANAMA      PARAGUAY
PHILIPPINES POLAND      PORTUGAL    RUMANIA     SO AFRICA   SPAIN       SWEDEN      SWITZERLAND SYRIA       TURKEY
USSR        UK          US          URUGUAY     VIETNAM, N  YUGOSLAVIA

       3 EXCLUDED BECAUSE UNASCERTAINABLE

KOREA REP   PERU        YEMEN

------------------------------------------------------------------------------------------

 10  TEND TO BE THOSE                    0.81   10  TEND TO BE THOSE                   0.80      25    1
     LOCATED IN NORTH AFRICA, OR                    LOCATED ELSEWHERE THAN IN NORTH AFRICA,       6    4
     CENTRAL AND SOUTH AFRICA  (33)                 OR CENTRAL AND SOUTH AFRICA  (82)        X SQ=   5.16
                                                                                             P =    0.015
                                                                                             RV YES (NO )

 13  TEND TO BE THOSE                    0.87   13  TEND TO BE THOSE                   0.60       4    3
     LOCATED ELSEWHERE THAN IN NORTH AFRICA OR      LOCATED IN NORTH AFRICA OR                   27    2
     THE MIDDLE EAST  (99)                          THE MIDDLE EAST  (16)                    X SQ=   3.46
                                                                                             P =    0.040
                                                                                             RV YES (NO )
```

No.	Description	Value	No.	Description	Value	Statistics
15	TEND TO BE THOSE LOCATED IN NORTH AFRICA OR CENTRAL AND SOUTH AFRICA, RATHER THAN IN THE MIDDLE EAST (33)	0.96	15	TEND TO BE THOSE LOCATED IN THE MIDDLE EAST, RATHER THAN IN NORTH AFRICA OR CENTRAL AND SOUTH AFRICA (11)	0.75	1 3 25 1 X SQ= 9.66 P = 0.004 RV YES (YES)
17	TILT TOWARD BEING THOSE LOCATED IN THE CARIBBEAN, CENTRAL AMERICA, OR SOUTH AMERICA, RATHER THAN IN THE MIDDLE EAST (22)	0.75	17	ALWAYS ARE THOSE LOCATED IN THE MIDDLE EAST, RATHER THAN IN THE CARIBBEAN, CENTRAL AMERICA, OR SOUTH AMERICA (11)	1.00	1 3 3 0 X SQ= 1.47 P = 0.143 RV YES (YES)
24	TEND TO BE THOSE WHOSE POPULATION IS SMALL (54)	0.71	24	TEND TO BE THOSE WHOSE POPULATION IS VERY LARGE, LARGE, OR MEDIUM (61)	0.80	9 4 22 1 X SQ= 2.89 P = 0.047 RV YES (NO)
34	TILT TOWARD BEING THOSE WHOSE GROSS NATIONAL PRODUCT IS VERY LOW (53)	0.77	34	TILT TOWARD BEING THOSE WHOSE GROSS NATIONAL PRODUCT IS VERY HIGH, HIGH, MEDIUM, OR LOW (62)	0.60	7 3 24 2 X SQ= 1.43 P = 0.119 RV YES (NO)
40	TILT TOWARD BEING THOSE WHOSE INTERNATIONAL FINANCIAL STATUS IS VERY LOW (39)	0.63	40	TILT TOWARD BEING THOSE WHOSE INTERNATIONAL FINANCIAL STATUS IS VERY HIGH, HIGH, MEDIUM, OR LOW (71)	0.80	11 4 19 1 X SQ= 1.75 P = 0.141 RV YES (NO)
51	TEND TO BE THOSE WHERE FREEDOM OF THE PRESS IS COMPLETE OR INTERMITTENT (60)	0.62	51	ALWAYS ARE THOSE WHERE FREEDOM OF THE PRESS IS INTERNALLY ABSENT, OR INTERNALLY AND EXTERNALLY ABSENT (37)	1.00	13 0 8 5 X SQ= 3.96 P = 0.039 RV YES (NO)
52	TEND TO BE THOSE WHERE FREEDOM OF THE PRESS IS COMPLETE, INTERMITTENT, OR INTERNALLY ABSENT (82)	0.95	52	TEND TO BE THOSE WHERE FREEDOM OF THE PRESS IS INTERNALLY AND EXTERNALLY ABSENT (16)	0.60	20 2 1 3 X SQ= 5.70 P = 0.014 RV YES (YES)
56	TILT TOWARD BEING THOSE WHERE THE RELIGION IS PREDOMINANTLY OR PARTLY NON-LITERATE (31)	0.65	56	TILT TOWARD BEING THOSE WHERE THE RELIGION IS PREDOMINANTLY LITERATE (79)	0.80	11 4 20 1 X SQ= 1.92 P = 0.138 RV YES (NO)
58	TILT TOWARD BEING THOSE WHERE THE RELIGION IS OTHER THAN MUSLIM (97)	0.77	58	TILT TOWARD BEING THOSE WHERE THE RELIGION IS MUSLIM (18)	0.60	7 3 24 2 X SQ= 1.43 P = 0.119 RV YES (NO)

No.	Statement	Value	No.	Statement	Value	Statistics
66	TILT TOWARD BEING THOSE THAT ARE RELIGIOUSLY HETEROGENEOUS (49)	0.64	66	TILT TOWARD BEING THOSE THAT ARE RELIGIOUSLY HOMOGENEOUS (57)	0.80	10 4 18 1 X SQ= 1.83 P = 0.138 RV YES (NO)
69	TILT TOWARD BEING THOSE THAT ARE LINGUISTICALLY STRONGLY HETEROGENEOUS (50)	0.77	69	TILT TOWARD BEING THOSE THAT ARE LINGUISTICALLY HOMOGENEOUS OR WEAKLY HETEROGENEOUS (64)	0.60	7 3 24 2 X SQ= 1.43 P = 0.119 RV YES (NO)
71	ALWAYS ARE THOSE WHOSE DATE OF INDEPENDENCE IS AFTER 1800 (90)	1.00	71	TEND TO BE THOSE WHOSE DATE OF INDEPENDENCE IS BEFORE 1800 (21)	0.60	0 3 31 2 X SQ= 13.20 P = 0.001 RV YES (YES)
72	TEND TO BE THOSE WHOSE DATE OF INDEPENDENCE IS AFTER 1914 (59)	0.90	72	TEND TO BE THOSE WHOSE DATE OF INDEPENDENCE IS BEFORE 1914 (52)	0.60	3 3 28 2 X SQ= 4.65 P = 0.024 RV YES (YES)
73	TEND TO BE THOSE WHOSE DATE OF INDEPENDENCE IS AFTER 1945 (46)	0.90	73	TEND TO BE THOSE WHOSE DATE OF INDEPENDENCE IS BEFORE 1945 (65)	0.80	3 4 28 1 X SQ= 9.47 P = 0.003 RV YES (YES)
83	ALWAYS ARE THOSE WHERE THE TYPE OF POLITICAL MODERNIZATION IS OTHER THAN NON-EUROPEAN AUTOCHTHONOUS (106)	1.00	83	TEND TO BE THOSE WHERE THE TYPE OF POLITICAL MODERNIZATION IS NON-EUROPEAN AUTOCHTHONOUS (9)	0.60	0 3 31 2 X SQ= 13.20 P = 0.001 RV YES (YES)
84	LEAN TOWARD BEING THOSE WHERE THE TYPE OF POLITICAL MODERNIZATION IS UNDEVELOPED TUTELARY, RATHER THAN DEVELOPED TUTELARY (24)	0.75	84	ALWAYS ARE THOSE WHERE THE TYPE OF POLITICAL MODERNIZATION IS DEVELOPED TUTELARY, RATHER THAN UNDEVELOPED TUTELARY (31)	1.00	7 2 21 0 X SQ= 2.07 P = 0.083 RV YES (NO)
86	TEND TO BE THOSE WHERE THE STAGE OF POLITICAL MODERNIZATION IS EARLY TRANSITIONAL, RATHER THAN ADVANCED OR MID-TRANSITIONAL (38)	0.74	86	ALWAYS ARE THOSE WHERE THE STAGE OF POLITICAL MODERNIZATION IS ADVANCED OR MID-TRANSITIONAL, RATHER THAN EARLY TRANSITIONAL (76)	1.00	8 4 23 0 X SQ= 5.68 P = 0.009 RV YES (NO)
93	TILT LESS TOWARD BEING THOSE WHERE THE SYSTEM STYLE IS NON-MOBILIZATIONAL (78)	0.55	93	ALWAYS ARE THOSE WHERE THE SYSTEM STYLE IS NON-MOBILIZATIONAL (78)	1.00	14 0 17 5 X SQ= 2.04 P = 0.134 RV NO (NO)

94	LEAN TOWARD BEING THOSE WHERE THE STATUS OF THE REGIME IS CONSTITUTIONAL (51)	0.53	94	ALWAYS ARE THOSE WHERE THE STATUS OF THE REGIME IS AUTHORITARIAN OR TOTALITARIAN (41)	1.00	8 0 7 5 X SQ= 2.50 P = 0.055 RV YES (NO)
96	LEAN TOWARD BEING THOSE WHERE THE STATUS OF THE REGIME IS CONSTITUTIONAL OR TOTALITARIAN (67)	0.53	96	ALWAYS ARE THOSE WHERE THE STATUS OF THE REGIME IS AUTHORITARIAN (23)	1.00	8 0 7 5 X SQ= 2.50 P = 0.055 RV YES (NO)
98	ALWAYS ARE THOSE WHERE GOVERNMENTAL STABILITY IS GENERALLY OR MODERATELY PRESENT AND DATES FROM THE POST-WAR PERIOD, OR IS ABSENT (93)	1.00	98	TEND LESS TO BE THOSE WHERE GOVERNMENTAL STABILITY IS GENERALLY OR MODERATELY PRESENT AND DATES FROM THE POST-WAR PERIOD, OR IS ABSENT (93)	0.60	0 2 31 3 X SQ= 6.61 P = 0.016 RV NO (YES)
102	TILT LESS TOWARD BEING THOSE WHERE THE REPRESENTATIVE CHARACTER OF THE REGIME IS PSEUDO-POLYARCHIC OR NON-POLYARCHIC (49)	0.57	102	ALWAYS ARE THOSE WHERE THE REPRESENTATIVE CHARACTER OF THE REGIME IS PSEUDO-POLYARCHIC OR NON-POLYARCHIC (49)	1.00	13 0 17 5 X SQ= 1.84 P = 0.134 RV NO (NO)
108	TEND TO BE THOSE WHERE AUTONOMOUS GROUPS ARE FULLY OR PARTIALLY TOLERATED IN POLITICS (65)	0.63	108	ALWAYS ARE THOSE WHERE AUTONOMOUS GROUPS ARE TOLERATED ONLY OUTSIDE POLITICS OR ARE NOT TOLERATED AT ALL (35)	1.00	17 0 10 4 X SQ= 3.32 P = 0.032 RV YES (NO)
118	TILT TOWARD BEING THOSE WHERE INTEREST ARTICULATION BY INSTITUTIONAL GROUPS IS SIGNIFICANT, MODERATE, OR LIMITED (60)	0.65	118	TILT TOWARD BEING THOSE WHERE INTEREST ARTICULATION BY INSTITUTIONAL GROUPS IS VERY SIGNIFICANT (40)	0.80	6 4 11 1 X SQ= 1.57 P = 0.135 RV YES (NO)
134	TEND TO BE THOSE WHERE INTEREST AGGREGATION BY THE EXECUTIVE IS SIGNIFICANT OR MODERATE (57)	0.90	134	ALWAYS ARE THOSE WHERE INTEREST AGGREGATION BY THE EXECUTIVE IS LIMITED OR NEGLIGIBLE (46)	1.00	26 0 3 5 X SQ= 14.39 P = 0. RV YES (YES)
135	ALWAYS ARE THOSE WHERE INTEREST AGGREGATION BY THE EXECUTIVE IS SIGNIFICANT, MODERATE, OR LIMITED (77)	1.00	135	TEND TO BE THOSE WHERE INTEREST AGGREGATION BY THE EXECUTIVE IS NEGLIGIBLE (13)	0.80	28 1 0 4 X SQ= 18.53 P = 0. RV YES (YES)
139	TILT TOWARD BEING THOSE WHERE THE PARTY SYSTEM IS QUANTITATIVELY ONE-PARTY (34)	0.55	139	ALWAYS ARE THOSE WHERE THE PARTY SYSTEM IS QUANTITATIVELY OTHER THAN ONE-PARTY (71)	1.00	16 0 13 4 X SQ= 2.36 P = 0.103 RV YES (NO)

157	LEAN TOWARD BEING THOSE WHERE PERSONALISMO IS MODERATE OR NEGLIGIBLE (84)	0.96	157	ALWAYS ARE THOSE WHERE PERSONALISMO IS PRONOUNCED (14)	1.00	1 1 27 0 X SQ= 3.00 P = 0.069 RV YES (YES)
160	ALWAYS ARE THOSE WHERE THE POLITICAL LEADERSHIP IS MODERATELY ELITIST OR NON-ELITIST (67)	1.00	160	ALWAYS ARE THOSE WHERE THE POLITICAL LEADERSHIP IS ELITIST (30)	1.00	0 5 27 0 X SQ= 24.86 P = 0. RV YES (YES)
161	TEND TO BE THOSE WHERE THE POLITICAL LEADERSHIP IS NON-ELITIST (50)	0.81	161	ALWAYS ARE THOSE WHERE THE POLITICAL LEADERSHIP IS ELITIST OR MODERATELY ELITIST (47)	1.00	5 5 22 0 X SQ= 9.52 P = 0.001 RV YES (YES)
169	TILT LESS TOWARD BEING THOSE WHERE THE HORIZONTAL POWER DISTRIBUTION IS NEGLIGIBLE (48)	0.60	169	ALWAYS ARE THOSE WHERE THE HORIZONTAL POWER DISTRIBUTION IS NEGLIGIBLE (48)	1.00	12 0 18 5 X SQ= 1.53 P = 0.141 RV NO (NO)
170	TEND TO BE THOSE WHERE THE LEGISLATIVE-EXECUTIVE STRUCTURE IS PRESIDENTIAL (39)	0.69	170	ALWAYS ARE THOSE WHERE THE LEGISLATIVE-EXECUTIVE STRUCTURE IS OTHER THAN PRESIDENTIAL (63)	1.00	18 0 8 4 X SQ= 4.34 P = 0.018 RV YES (NO)
176	TEND TO BE THOSE WHERE THE LEGISLATURE IS FULLY EFFECTIVE, PARTIALLY EFFECTIVE, OR LARGELY INEFFECTIVE (72)	0.89	176	ALWAYS ARE THOSE WHERE THE LEGISLATURE IS WHOLLY INEFFECTIVE (28)	1.00	24 0 3 4 X SQ= 11.07 P = 0.001 RV YES (YES)
178	TEND TO BE THOSE WHERE THE LEGISLATURE IS UNICAMERAL (53)	0.86	178	ALWAYS ARE THOSE WHERE THE LEGISLATURE IS BICAMERAL (51)	1.00	25 0 4 3 X SQ= 7.32 P = 0.007 RV YES (NO)
183	ALWAYS ARE THOSE WHERE THE BUREAUCRACY IS POST-COLONIAL TRANSITIONAL, RATHER THAN TRADITIONAL (25)	1.00	183	ALWAYS ARE THOSE WHERE THE BUREAUCRACY IS TRADITIONAL, RATHER THAN POST-COLONIAL TRANSITIONAL (9)	1.00	21 0 0 5 X SQ= 19.96 P = 0. RV YES (YES)
186	TEND TO BE THOSE WHERE PARTICIPATION BY THE MILITARY IN POLITICS IS NEUTRAL, RATHER THAN SUPPORTIVE (56)	0.80	186	ALWAYS ARE THOSE WHERE PARTICIPATION BY THE MILITARY IN POLITICS IS SUPPORTIVE, RATHER THAN NEUTRAL (31)	1.00	5 4 20 0 X SQ= 6.91 P = 0.005 RV YES (NO)

193 TILT TOWARD BEING THOSE WHERE THE CHARACTER OF THE LEGAL SYSTEM IS PARTLY INDIGENOUS (28)	0.65	193 TILT TOWARD BEING THOSE WHERE THE CHARACTER OF THE LEGAL SYSTEM IS OTHER THAN PARTLY INDIGENOUS (86)	0.80	20 1 11 4 X SQ= 1.92 P = 0.138 RV YES (NO)

```
**********************************************************************************************

92  POLITIES                                         92  POLITIES
    WHERE THE SYSTEM STYLE                               WHERE THE SYSTEM STYLE
    IS MOBILIZATIONAL  (20)                              IS LIMITED MOBILIZATIONAL, OR
                                                         NON-MOBILIZATIONAL  (93)


                                                                   BOTH SUBJECT AND PREDICATE
..............................................................................................

      20 IN LEFT COLUMN

ALBANIA     ALGERIA     BULGARIA    BURMA       CHINA, PR   CUBA        CZECHOS'KIA GERMANY, E  GHANA       GUINEA
HUNGARY     KOREA, N    MONGOLIA    POLAND      RUMANIA     USSR        UAR         VIETNAM, N  VIETNAM REP YUGOSLAVIA

      93 IN RIGHT COLUMN

AFGHANISTAN ARGENTINA   AUSTRALIA   AUSTRIA     BELGIUM     BOLIVIA     BRAZIL      BURUNDI     CAMBODIA    CAMEROUN
CANADA      CEN AFR REP CEYLON      CHAD        CHILE       COLOMBIA    CONGO(BRA)  CONGO(LEO)  COSTA RICA  CYPRUS
DAHOMEY     DENMARK     DOMIN REP   ECUADOR     EL SALVADOR ETHIOPIA    FINLAND     FRANCE      GABON       GERMAN FR
GREECE      GUATEMALA   HAITI       HONDURAS    ICELAND     INDIA       INDONESIA   IRAN        IRAQ        IRELAND
ISRAEL      ITALY       IVORY COAST JAMAICA     JAPAN       JORDAN      KOREA REP   LAOS        LEBANON     LIBERIA
LIBYA       LUXEMBOURG  MALAGASY R  MALAYA      MALI        MAURITANIA  MEXICO      MOROCCO     NEPAL       NETHERLANDS
NEW ZEALAND NICARAGUA   NIGER       NIGERIA     NORWAY      PAKISTAN    PANAMA      PARAGUAY    PERU        PHILIPPINES
PORTUGAL    RWANDA      SA'U ARABIA SENEGAL     SIERRE LEO  SOMALIA     SO AFRICA   SUDAN       SWEDEN      SWITZERLAND
SYRIA       TANGANYIKA  THAILAND    TOGO        TRINIDAD    TUNISIA     TURKEY      UGANDA      UK          US
UPPER VOLTA URUGUAY     VENEZUELA

       1 EXCLUDED BECAUSE AMBIGUOUS

SPAIN

       1 EXCLUDED BECAUSE UNASCERTAINABLE

YEMEN

----------------------------------------------------------------------------------------------

  2  ALWAYS ARE THOSE                       1.00     2  TEND LESS TO BE THOSE              0.77         0    21
     LOCATED ELSEWHERE THAN IN                          LOCATED ELSEWHERE THAN IN                      20    72
     WEST EUROPE, SCANDINAVIA,                          WEST EUROPE, SCANDINAVIA,                  X SQ=   4.15
     NORTH AMERICA, OR AUSTRALASIA  (93)                NORTH AMERICA, OR AUSTRALASIA  (93)        P =     0.013
                                                                                                   RV NO   (NO )

  5  ALWAYS ARE THOSE                       1.00     5  ALWAYS ARE THOSE                   1.00         0    17
     LOCATED IN EAST EUROPE, RATHER THAN                LOCATED IN SCANDINAVIA OR WEST EUROPE,          9     0
     IN SCANDINAVIA OR WEST EUROPE  (9)                 RATHER THAN IN EAST EUROPE  (18)           X SQ=  21.77
                                                                                                   P =     0.
                                                                                                   RV YES (YES)

  6  TILT MORE TOWARD BEING THOSE           0.95     6  TILT LESS TOWARD BEING THOSE       0.77         1    21
     LOCATED ELSEWHERE THAN IN THE                      LOCATED ELSEWHERE THAN IN THE                  19    72
     CARIBBEAN, CENTRAL AMERICA,                        CARIBBEAN, CENTRAL AMERICA,                X SQ=   2.22
     OR SOUTH AMERICA  (93)                             OR SOUTH AMERICA  (93)                     P =     0.116
                                                                                                   RV NO   (NO )
```

11	LEAN MORE TOWARD BEING THOSE LOCATED ELSEWHERE THAN IN CENTRAL AND SOUTH AFRICA (87)	0.90	11	LEAN LESS TOWARD BEING THOSE LOCATED ELSEWHERE THAN IN CENTRAL AND SOUTH AFRICA (87)	0.72	2 26 18 67 X SQ= 1.97 P = 0.099 RV NO (NO)
16	ALWAYS ARE THOSE LOCATED IN EAST ASIA, SOUTH ASIA, OR SOUTHEAST ASIA, RATHER THAN IN THE MIDDLE EAST (18)	1.00	16	LEAN LESS TOWARD BEING THOSE LOCATED IN EAST ASIA, SOUTH ASIA, OR SOUTHEAST ASIA, RATHER THAN IN THE MIDDLE EAST (18)	0.55	0 10 6 12 X SQ= 2.49 P = 0.062 RV NO (NO)
18	TILT TOWARD BEING THOSE LOCATED IN EAST ASIA, SOUTH ASIA, OR SOUTHEAST ASIA, RATHER THAN IN NORTH AFRICA OR CENTRAL AND SOUTH AFRICA (18)	0.60	18	TILT TOWARD BEING THOSE LOCATED IN NORTH AFRICA OR CENTRAL AND SOUTH AFRICA, RATHER THAN IN EAST ASIA, SOUTH ASIA, OR SOUTHEAST ASIA (33)	0.71	4 29 6 12 X SQ= 2.12 P = 0.137 RV YES (NO)
20	TEND TO BE THOSE LOCATED IN EAST ASIA, SOUTH ASIA, OR SOUTHEAST ASIA, RATHER THAN IN THE CARIBBEAN, CENTRAL AMERICA, OR SOUTH AMERICA (18)	0.86	20	TEND TO BE THOSE LOCATED IN THE CARIBBEAN, CENTRAL AMERICA, OR SOUTH AMERICA, RATHER THAN IN EAST ASIA, SOUTH ASIA, OR SOUTHEAST ASIA (22)	0.64	6 12 1 21 X SQ= 3.86 P = 0.033 RV YES (NO)
24	TEND TO BE THOSE WHOSE POPULATION IS VERY LARGE, LARGE, OR MEDIUM (61)	0.85	24	TEND TO BE THOSE WHOSE POPULATION IS SMALL (54)	0.54	17 43 3 50 X SQ= 8.44 P = 0.002 RV YES (NO)
26	TEND TO BE THOSE WHOSE POPULATION DENSITY IS VERY HIGH, HIGH, OR MEDIUM (48)	0.65	26	TEND TO BE THOSE WHOSE POPULATION DENSITY IS LOW (67)	0.63	13 34 7 59 X SQ= 4.37 P = 0.025 RV YES (NO)
31	LEAN MORE TOWARD BEING THOSE WHOSE AGRICULTURAL POPULATION IS HIGH OR MEDIUM (90)	0.95	31	LEAN LESS TOWARD BEING THOSE WHOSE AGRICULTURAL POPULATION IS HIGH OR MEDIUM (90)	0.75	19 69 1 23 X SQ= 2.81 P = 0.069 RV NO (NO)
34	TEND TO BE THOSE WHOSE GROSS NATIONAL PRODUCT IS VERY HIGH, HIGH, MEDIUM, OR LOW (62)	0.75	34	TEND TO BE THOSE WHOSE GROSS NATIONAL PRODUCT IS VERY LOW (53)	0.51	15 46 5 47 X SQ= 3.35 P = 0.048 RV YES (NO)
40	TILT MORE TOWARD BEING THOSE WHOSE INTERNATIONAL FINANCIAL STATUS IS VERY HIGH, HIGH, MEDIUM, OR LOW (71)	0.82	40	TILT LESS TOWARD BEING THOSE WHOSE INTERNATIONAL FINANCIAL STATUS IS VERY HIGH, HIGH, MEDIUM, OR LOW (71)	0.62	14 56 3 35 X SQ= 1.89 P = 0.109 RV NO (NO)

45	LEAN TOWARD BEING THOSE WHERE THE LITERACY RATE IS FIFTY PERCENT OR ABOVE (55)	0.75	45	LEAN TOWARD BEING THOSE WHERE THE LITERACY RATE IS BELOW FIFTY PERCENT (54)	0.54	12 42 4 49 X SQ= 3.45 P = 0.055 RV YES (NO)
46	ALWAYS ARE THOSE WHERE THE LITERACY RATE IS TEN PERCENT OR ABOVE (84)	1.00	46	TEND LESS TO BE THOSE WHERE THE LITERACY RATE IS TEN PERCENT OR ABOVE (84)	0.73	17 66 0 25 X SQ= 4.63 P = 0.011 RV NO (NO)
50	ALWAYS ARE THOSE WHERE FREEDOM OF THE PRESS IS INTERMITTENT, INTERNALLY ABSENT, OR INTERNALLY AND EXTERNALLY ABSENT (56)	1.00	50	TEND TO BE THOSE WHERE FREEDOM OF THE PRESS IS COMPLETE (43)	0.56	0 43 20 34 X SQ= 17.86 P = 0. RV YES (NO)
51	TEND TO BE THOSE WHERE FREEDOM OF THE PRESS IS INTERNALLY ABSENT, OR INTERNALLY AND EXTERNALLY ABSENT (37)	0.95	51	TEND TO BE THOSE WHERE FREEDOM OF THE PRESS IS COMPLETE OR INTERMITTENT (60)	0.78	1 59 19 17 X SQ= 32.61 P = 0. RV YES (YES)
52	TEND TO BE THOSE WHERE FREEDOM OF THE PRESS IS INTERNALLY AND EXTERNALLY ABSENT (16)	0.60	52	TEND TO BE THOSE WHERE FREEDOM OF THE PRESS IS COMPLETE, INTERMITTENT, OR INTERNALLY ABSENT (82)	0.95	8 73 12 4 X SQ= 30.76 P = 0. RV YES (YES)
55	TEND MORE TO BE THOSE WHERE NEWSPAPER CIRCULATION IS TEN OR MORE PER THOUSAND (78)	0.94	55	TEND LESS TO BE THOSE WHERE NEWSPAPER CIRCULATION IS TEN OR MORE PER THOUSAND (78)	0.65	17 60 1 33 X SQ= 5.03 P = 0.011 RV NO (NO)
56	TILT MORE TOWARD BEING THOSE WHERE THE RELIGION IS PREDOMINANTLY LITERATE (79)	0.87	56	TILT LESS TOWARD BEING THOSE WHERE THE RELIGION IS PREDOMINANTLY LITERATE (79)	0.68	14 63 2 29 X SQ= 1.57 P = 0.145 RV NO (NO)
67	TEND MORE TO BE THOSE THAT ARE RACIALLY HOMOGENEOUS (82)	0.95	67	TEND LESS TO BE THOSE THAT ARE RACIALLY HOMOGENEOUS (82)	0.68	19 60 1 28 X SQ= 4.68 P = 0.013 RV NO (NO)
78	TEND TO BE THOSE THAT HAVE BEEN SIGNIFICANTLY WESTERNIZED WITHOUT A COLONIAL RELATIONSHIP, RATHER THAN THROUGH SUCH A RELATIONSHIP (7)	0.63	78	TEND TO BE THOSE THAT HAVE BEEN SIGNIFICANTLY WESTERNIZED THROUGH A COLONIAL RELATIONSHIP, RATHER THAN WITHOUT SUCH A RELATIONSHIP (28)	0.93	3 25 5 2 X SQ= 8.52 P = 0.003 RV YES (YES)

No.	Statement	Value	No.	Statement	Value	Statistics
81	ALWAYS ARE THOSE WHERE THE TYPE OF POLITICAL MODERNIZATION IS LATER EUROPEAN OR LATER EUROPEAN DERIVED, RATHER THAN EARLY EUROPEAN OR EARLY EUROPEAN DERIVED (40)	1.00	81	TILT LESS TOWARD BEING THOSE WHERE THE TYPE OF POLITICAL MODERNIZATION IS LATER EUROPEAN OR LATER EUROPEAN DERIVED, RATHER THAN EARLY EUROPEAN OR EARLY EUROPEAN DERIVED (40)	0.73	0 11 9 30 X SQ= 1.73 P = 0.102 RV NO (NO)
85	TEND TO BE THOSE WHERE THE STAGE OF POLITICAL MODERNIZATION IS ADVANCED, RATHER THAN MID- OR EARLY TRANSITIONAL (60)	0.75	85	TEND TO BE THOSE WHERE THE STAGE OF POLITICAL MODERNIZATION IS MID- OR EARLY TRANSITIONAL, RATHER THAN ADVANCED (54)	0.52	15 44 5 48 X SQ= 3.84 P = 0.046 RV YES (NO)
89	ALWAYS ARE THOSE WHOSE IDEOLOGICAL ORIENTATION IS OTHER THAN CONVENTIONAL (62)	1.00	89	TEND LESS TO BE THOSE WHOSE IDEOLOGICAL ORIENTATION IS OTHER THAN CONVENTIONAL (62)	0.55	0 33 20 41 X SQ= 11.86 P = 0. RV NO (NO)
94	ALWAYS ARE THOSE WHERE THE STATUS OF THE REGIME IS AUTHORITARIAN OR TOTALITARIAN (41)	1.00	94	TEND TO BE THOSE WHERE THE STATUS OF THE REGIME IS CONSTITUTIONAL (51)	0.72	0 51 20 20 X SQ= 29.83 P = 0. RV YES (YES)
95	TEND TO BE THOSE WHERE THE STATUS OF THE REGIME IS TOTALITARIAN (16)	0.74	95	TEND TO BE THOSE WHERE THE STATUS OF THE REGIME IS CONSTITUTIONAL OR AUTHORITARIAN (95)	0.99	5 90 14 1 X SQ= 64.29 P = 0. RV YES (YES)
97	TEND TO BE THOSE WHERE THE STATUS OF THE REGIME IS TOTALITARIAN, RATHER THAN AUTHORITARIAN (16)	0.74	97	TEND TO BE THOSE WHERE THE STATUS OF THE REGIME IS AUTHORITARIAN, RATHER THAN TOTALITARIAN (23)	0.95	5 18 14 1 X SQ= 15.86 P = 0. RV YES (YES)
99	TEND TO BE THOSE WHERE GOVERNMENTAL STABILITY IS GENERALLY PRESENT AND DATES FROM AT LEAST THE INTER-WAR PERIOD, OR FROM THE POST-WAR PERIOD (50)	0.88	99	TEND TO BE THOSE WHERE GOVERNMENTAL STABILITY IS MODERATELY PRESENT AND DATES FROM THE POST-WAR PERIOD, OR IS ABSENT (36)	0.50	15 34 2 34 X SQ= 6.65 P = 0.005 RV YES (NO)
102	ALWAYS ARE THOSE WHERE THE REPRESENTATIVE CHARACTER OF THE REGIME IS PSEUDO-POLYARCHIC OR NON-POLYARCHIC (49)	1.00	102	TEND TO BE THOSE WHERE THE REPRESENTATIVE CHARACTER OF THE REGIME IS POLYARCHIC OR LIMITED POLYARCHIC (59)	0.68	0 59 20 28 X SQ= 27.56 P = 0. RV YES (NO)
105	ALWAYS ARE THOSE WHERE THE ELECTORAL SYSTEM IS PARTIALLY COMPETITIVE OR NON-COMPETITIVE (47)	1.00	105	TEND TO BE THOSE WHERE THE ELECTORAL SYSTEM IS COMPETITIVE (43)	0.60	0 43 17 29 X SQ= 17.32 P = 0. RV YES (NO)

106	TEND TO BE THOSE WHERE THE ELECTORAL SYSTEM IS NON-COMPETITIVE (30)	0.94	106	TEND TO BE THOSE WHERE THE ELECTORAL SYSTEM IS COMPETITIVE OR PARTIALLY COMPETITIVE (52)	0.81	1 51 17 12 X SQ= 31.42 P = 0. RV YES (YES)
107	TEND TO BE THOSE WHERE AUTONOMOUS GROUPS ARE PARTIALLY TOLERATED IN POLITICS, ARE TOLERATED ONLY OUTSIDE POLITICS, OR ARE NOT TOLERATED AT ALL (65)	0.95	107	TEND TO BE THOSE WHERE AUTONOMOUS GROUPS ARE FULLY TOLERATED IN POLITICS (46)	0.50	1 45 19 45 X SQ= 11.83 P = 0. RV YES (NO)
108	TEND TO BE THOSE WHERE AUTONOMOUS GROUPS ARE TOLERATED ONLY OUTSIDE POLITICS OR ARE NOT TOLERATED AT ALL (35)	0.94	108	TEND TO BE THOSE WHERE AUTONOMOUS GROUPS ARE FULLY OR PARTIALLY TOLERATED IN POLITICS (65)	0.79	1 64 17 17 X SQ= 32.06 P = 0. RV YES (YES)
116	ALWAYS ARE THOSE WHERE INTEREST ARTICULATION BY ASSOCIATIONAL GROUPS IS LIMITED OR NEGLIGIBLE (79)	1.00	116	TEND LESS TO BE THOSE WHERE INTEREST ARTICULATION BY ASSOCIATIONAL GROUPS IS LIMITED OR NEGLIGIBLE (79)	0.65	0 32 19 59 X SQ= 7.79 P = 0.001 RV NO (NO)
117	TILT TOWARD BEING THOSE WHERE INTEREST ARTICULATION BY ASSOCIATIONAL GROUPS IS NEGLIGIBLE (51)	0.63	117	TILT TOWARD BEING THOSE WHERE INTEREST ARTICULATION BY ASSOCIATIONAL GROUPS IS SIGNIFICANT, MODERATE, OR LIMITED (60)	0.58	7 53 12 38 X SQ= 2.10 P = 0.128 RV YES (NO)
118	TEND TO BE THOSE WHERE INTEREST ARTICULATION BY INSTITUTIONAL GROUPS IS VERY SIGNIFICANT (40)	0.95	118	TEND TO BE THOSE WHERE INTEREST ARTICULATION BY INSTITUTIONAL GROUPS IS SIGNIFICANT, MODERATE, OR LIMITED (60)	0.75	19 20 1 59 X SQ= 29.61 P = 0. RV YES (NO)
119	TEND MORE TO BE THOSE WHERE INTEREST ARTICULATION BY INSTITUTIONAL GROUPS IS VERY SIGNIFICANT OR SIGNIFICANT (74)	0.95	119	TEND LESS TO BE THOSE WHERE INTEREST ARTICULATION BY INSTITUTIONAL GROUPS IS VERY SIGNIFICANT OR SIGNIFICANT (74)	0.68	19 54 1 25 X SQ= 4.56 P = 0.020 RV NO (NO)
120	ALWAYS ARE THOSE WHERE INTEREST ARTICULATION BY INSTITUTIONAL GROUPS IS VERY SIGNIFICANT, SIGNIFICANT, OR MODERATE (90)	1.00	120	TILT LESS TOWARD BEING THOSE WHERE INTEREST ARTICULATION BY INSTITUTIONAL GROUPS IS VERY SIGNIFICANT, SIGNIFICANT, OR MODERATE (90)	0.87	20 69 0 10 X SQ= 1.59 P = 0.119 RV NO (NO)
126	ALWAYS ARE THOSE WHERE INTEREST ARTICULATION BY ANOMIC GROUPS IS FREQUENT, OCCASIONAL, OR INFREQUENT (83)	1.00	126	TILT LESS TOWARD BEING THOSE WHERE INTEREST ARTICULATION BY ANOMIC GROUPS IS FREQUENT, OCCASIONAL, OR INFREQUENT (83)	0.81	12 70 0 16 X SQ= 1.48 P = 0.126 RV NO (NO)

128	ALWAYS ARE THOSE WHERE INTEREST ARTICULATION BY POLITICAL PARTIES IS LIMITED OR NEGLIGIBLE (45)	1.00	128	TEND TO BE THOSE WHERE INTEREST ARTICULATION BY POLITICAL PARTIES IS SIGNIFICANT OR MODERATE (48)	0.65	0 48 18 26 X SQ= 21.88 P = 0. RV YES (NO)
129	TEND TO BE THOSE WHERE INTEREST ARTICULATION BY POLITICAL PARTIES IS NEGLIGIBLE (37)	0.94	129	TEND TO BE THOSE WHERE INTEREST ARTICULATION BY POLITICAL PARTIES IS SIGNIFICANT, MODERATE, OR LIMITED (56)	0.74	1 55 17 19 X SQ= 25.93 P = 0. RV YES (NO)
130	ALWAYS ARE THOSE WHERE INTEREST AGGREGATION BY POLITICAL PARTIES IS MODERATE, LIMITED, OR NEGLIGIBLE (71)	1.00	130	TILT LESS TOWARD BEING THOSE WHERE INTEREST AGGREGATION BY POLITICAL PARTIES IS MODERATE, LIMITED, OR NEGLIGIBLE (71)	0.82	0 12 15 55 X SQ= 1.88 P = 0.111 RV NO (NO)
134	TEND TO BE THOSE WHERE INTEREST AGGREGATION BY THE EXECUTIVE IS LIMITED OR NEGLIGIBLE (46)	0.79	134	TEND TO BE THOSE WHERE INTEREST AGGREGATION BY THE EXECUTIVE IS SIGNIFICANT OR MODERATE (57)	0.64	4 53 15 30 X SQ= 9.82 P = 0.002 RV YES (NO)
137	ALWAYS ARE THOSE WHERE INTEREST AGGREGATION BY THE LEGISLATURE IS LIMITED OR NEGLIGIBLE (68)	1.00	137	TEND LESS TO BE THOSE WHERE INTEREST AGGREGATION BY THE LEGISLATURE IS LIMITED OR NEGLIGIBLE (68)	0.63	0 29 17 50 X SQ= 7.29 P = 0.001 RV NO (NO)
138	TEND TO BE THOSE WHERE INTEREST AGGREGATION BY THE LEGISLATURE IS NEGLIGIBLE (49)	0.94	138	TEND TO BE THOSE WHERE INTEREST AGGREGATION BY THE LEGISLATURE IS SIGNIFICANT, MODERATE, OR LIMITED (48)	0.59	1 47 16 33 X SQ= 13.63 P = 0. RV YES (NO)
139	TEND TO BE THOSE WHERE THE PARTY SYSTEM IS QUANTITATIVELY ONE-PARTY (34)	0.85	139	TEND TO BE THOSE WHERE THE PARTY SYSTEM IS QUANTITATIVELY OTHER THAN ONE-PARTY (71)	0.81	17 16 3 68 X SQ= 29.46 P = 0. RV YES (YES)
141	ALWAYS ARE THOSE WHERE THE PARTY SYSTEM IS QUANTITATIVELY OTHER THAN TWO-PARTY (87)	1.00	141	TILT LESS TOWARD BEING THOSE WHERE THE PARTY SYSTEM IS QUANTITATIVELY OTHER THAN TWO-PARTY (87)	0.86	0 11 19 67 X SQ= 1.78 P = 0.115 RV NO (NO)
142	ALWAYS ARE THOSE WHERE THE PARTY SYSTEM IS QUANTITATIVELY OTHER THAN MULTI-PARTY (66)	1.00	142	TEND LESS TO BE THOSE WHERE THE PARTY SYSTEM IS QUANTITATIVELY OTHER THAN MULTI-PARTY (66)	0.61	0 30 19 46 X SQ= 9.21 P = 0.001 RV NO (NO)

147	ALWAYS ARE THOSE WHERE THE PARTY SYSTEM IS QUALITATIVELY OTHER THAN CLASS-ORIENTED OR MULTI-IDEOLOGICAL (67)	1.00	147	TEND LESS TO BE THOSE WHERE THE PARTY SYSTEM IS QUALITATIVELY OTHER THAN CLASS-ORIENTED OR MULTI-IDEOLOGICAL (67)	0.67	0 23 19 47 X SQ= 6.79 P = 0.002 RV NO (NO)
148	ALWAYS ARE THOSE WHERE THE PARTY SYSTEM IS QUALITATIVELY OTHER THAN AFRICAN TRANSITIONAL (96)	1.00	148	LEAN LESS TOWARD BEING THOSE WHERE THE PARTY SYSTEM IS QUALITATIVELY OTHER THAN AFRICAN TRANSITIONAL (96)	0.84	0 14 20 74 X SQ= 2.38 P = 0.068 RV NO (NO)
153	TEND TO BE THOSE WHERE THE PARTY SYSTEM IS STABLE (42)	0.88	153	TEND TO BE THOSE WHERE THE PARTY SYSTEM IS MODERATELY STABLE OR UNSTABLE (38)	0.57	14 27 2 36 X SQ= 8.48 P = 0.002 RV YES (NO)
154	ALWAYS ARE THOSE WHERE THE PARTY SYSTEM IS STABLE OR MODERATELY STABLE (55)	1.00	154	TEND LESS TO BE THOSE WHERE THE PARTY SYSTEM IS STABLE OR MODERATELY STABLE (55)	0.60	16 38 0 25 X SQ= 7.54 P = 0.002 RV NO (NO)
158	TEND MORE TO BE THOSE WHERE PERSONALISMO IS NEGLIGIBLE (56)	0.84	158	TEND LESS TO BE THOSE WHERE PERSONALISMO IS NEGLIGIBLE (56)	0.51	3 37 16 39 X SQ= 5.47 P = 0.010 RV NO (NO)
160	TEND TO BE THOSE WHERE THE POLITICAL LEADERSHIP IS ELITIST (30)	0.81	160	TEND TO BE THOSE WHERE THE POLITICAL LEADERSHIP IS MODERATELY ELITIST OR NON-ELITIST (67)	0.80	13 16 3 64 X SQ= 20.91 P = 0. RV YES (NO)
161	ALWAYS ARE THOSE WHERE THE POLITICAL LEADERSHIP IS ELITIST OR MODERATELY ELITIST (47)	1.00	161	TEND TO BE THOSE WHERE THE POLITICAL LEADERSHIP IS NON-ELITIST (50)	0.63	16 30 0 50 X SQ= 18.44 P = 0. RV YES (NO)
166	ALWAYS ARE THOSE WHERE THE VERTICAL POWER DISTRIBUTION IS THAT OF FORMAL FEDERALISM OR FORMAL AND EFFECTIVE UNITARISM (99)	1.00	166	LEAN LESS TOWARD BEING THOSE WHERE THE VERTICAL POWER DISTRIBUTION IS THAT OF FORMAL FEDERALISM OR FORMAL AND EFFECTIVE UNITARISM (99)	0.84	0 15 20 77 X SQ= 2.49 P = 0.069 RV NO (NO)
169	ALWAYS ARE THOSE WHERE THE HORIZONTAL POWER DISTRIBUTION IS NEGLIGIBLE (48)	1.00	169	TEND TO BE THOSE WHERE THE HORIZONTAL POWER DISTRIBUTION IS SIGNIFICANT OR LIMITED (58)	0.67	0 58 19 28 X SQ= 25.96 P = 0. RV YES (NO)

170	LEAN MORE TOWARD BEING THOSE WHERE THE LEGISLATIVE-EXECUTIVE STRUCTURE IS OTHER THAN PRESIDENTIAL (63)	0.81	170	LEAN LESS TOWARD BEING THOSE WHERE THE LEGISLATIVE-EXECUTIVE STRUCTURE IS OTHER THAN PRESIDENTIAL (63)	0.58	3 36 13 49 X SQ= 2.25 P = 0.096 RV NO (NO)
172	ALWAYS ARE THOSE WHERE THE LEGISLATIVE-EXECUTIVE STRUCTURE IS OTHER THAN PARLIAMENTARY-ROYALIST (88)	1.00	172	TEND LESS TO BE THOSE WHERE THE LEGISLATIVE-EXECUTIVE STRUCTURE IS OTHER THAN PARLIAMENTARY-ROYALIST (88)	0.76	0 21 20 68 X SQ= 4.43 P = 0.012 RV NO (NO)
175	ALWAYS ARE THOSE WHERE THE LEGISLATURE IS LARGELY INEFFECTIVE OR WHOLLY INEFFECTIVE (49)	1.00	175	TEND TO BE THOSE WHERE THE LEGISLATURE IS FULLY EFFECTIVE OR PARTIALLY EFFECTIVE (51)	0.63	0 51 18 30 X SQ= 20.92 P = 0. RV YES (NO)
176	TEND TO BE THOSE WHERE THE LEGISLATURE IS WHOLLY INEFFECTIVE (28)	0.94	176	TEND TO BE THOSE WHERE THE LEGISLATURE IS FULLY EFFECTIVE, PARTIALLY EFFECTIVE, OR LARGELY INEFFECTIVE (72)	0.88	1 71 17 10 X SQ= 45.99 P = 0. RV YES (YES)
178	TEND TO BE THOSE WHERE THE LEGISLATURE IS UNICAMERAL (53)	0.88	178	TEND TO BE THOSE WHERE THE LEGISLATURE IS BICAMERAL (51)	0.57	15 37 2 49 X SQ= 9.87 P = 0.001 RV YES (NO)
179	ALWAYS ARE THOSE WHERE THE EXECUTIVE IS DOMINANT (52)	1.00	179	TEND TO BE THOSE WHERE THE EXECUTIVE IS STRONG (39)	0.55	19 32 0 39 X SQ= 16.25 P = 0. RV YES (NO)
181	ALWAYS ARE THOSE WHERE THE BUREAUCRACY IS SEMI-MODERN, RATHER THAN MODERN (55)	1.00	181	TEND LESS TO BE THOSE WHERE THE BUREAUCRACY IS SEMI-MODERN, RATHER THAN MODERN (55)	0.64	0 21 17 37 X SQ= 6.85 P = 0.002 RV NO (NO)
182	TEND MORE TO BE THOSE WHERE THE BUREAUCRACY IS SEMI-MODERN, RATHER THAN POST-COLONIAL TRANSITIONAL (55)	0.89	182	TEND LESS TO BE THOSE WHERE THE BUREAUCRACY IS SEMI-MODERN, RATHER THAN POST-COLONIAL TRANSITIONAL (55)	0.62	17 37 2 23 X SQ= 3.95 P = 0.025 RV NO (NO)
185	TEND TO BE THOSE WHERE PARTICIPATION BY THE MILITARY IN POLITICS IS SUPPORTIVE, RATHER THAN INTERVENTIVE (31)	0.89	185	TEND TO BE THOSE WHERE PARTICIPATION BY THE MILITARY IN POLITICS IS INTERVENTIVE, RATHER THAN SUPPORTIVE (21)	0.58	2 18 17 13 X SQ= 9.20 P = 0.001 RV YES (YES)

```
186  ALWAYS ARE THOSE                              1.00     186  TEND TO BE THOSE                            0.81          17    13
     WHERE PARTICIPATION BY THE MILITARY                         WHERE PARTICIPATION BY THE MILITARY                        0    56
     IN POLITICS IS                                              IN POLITICS IS                                      X SQ=  36.06
     SUPPORTIVE, RATHER THAN                                     NEUTRAL, RATHER THAN                                P =     0.
     NEUTRAL  (31)                                               SUPPORTIVE  (56)                                    RV YES (YES)

187  ALWAYS ARE THOSE                              1.00     187  TEND LESS TO BE THOSE                       0.56          20    44
     WHERE THE ROLE OF THE POLICE                                WHERE THE ROLE OF THE POLICE                               0    35
     IS POLITICALLY SIGNIFICANT  (66)                            IS POLITICALLY SIGNIFICANT  (66)                    X SQ=  11.84
                                                                                                                     P =     0.
                                                                                                                     RV NO  (NO )

188  ALWAYS ARE THOSE                              1.00     188  TEND LESS TO BE THOSE                       0.66           0    31
     WHERE THE CHARACTER OF THE LEGAL SYSTEM                     WHERE THE CHARACTER OF THE LEGAL SYSTEM                   19    61
     IS OTHER THAN CIVIL LAW  (81)                               IS OTHER THAN CIVIL LAW  (81)                       X SQ=   7.29
                                                                                                                     P =     0.001
                                                                                                                     RV NO  (NO )

194  TEND TO BE THOSE                              0.68     194  ALWAYS ARE THOSE                            1.00          13     0
     THAT ARE COMMUNIST  (13)                                    THAT ARE NON-COMMUNIST  (101)                              6    93
                                                                                                                     X SQ=  65.47
                                                                                                                     P =     0.
                                                                                                                     RV YES (YES)
```

```
****************************************************************************************************

 93  POLITIES                                        93  POLITIES
     WHERE THE SYSTEM STYLE                              WHERE THE SYSTEM STYLE
     IS MOBILIZATIONAL, OR                               IS NON-MOBILIZATIONAL  (78)
     LIMITED MOBILIZATIONAL  (32)


                                                                          BOTH SUBJECT AND PREDICATE
....................................................................................................

      32 IN LEFT COLUMN

ALBANIA     ALGERIA     BOLIVIA     BULGARIA    BURMA       CAMBODIA    CHINA, PR   CUBA        CZECHOS'KIA GERMANY, E
GHANA       GUINEA      HUNGARY     INDONESIA   ISRAEL      IVORY COAST KOREA, N    MAURITANIA  MONGOLIA    MOROCCO
POLAND      RUMANIA     SENEGAL     SUDAN       TANGANYIKA  TUNISIA     USSR        UAR         VENEZUELA   VIETNAM, N
VIETNAM REP YUGOSLAVIA

      78 IN RIGHT COLUMN

AFGHANISTAN ARGENTINA   AUSTRALIA   AUSTRIA     BELGIUM     BRAZIL      BURUNDI     CAMEROUN    CANADA      CEN AFR REP
CEYLON      CHAD        CHILE       COLOMBIA    CONGO(BRA)  CONGO(LEO)  COSTA RICA  CYPRUS      DAHOMEY     DENMARK
DOMIN REP   ECUADOR     EL SALVADOR ETHIOPIA    FINLAND     FRANCE      GABON       GERMAN FR   GREECE      GUATEMALA
HAITI       HONDURAS    ICELAND     INDIA       IRAN        IRAQ        IRELAND     ITALY       JAMAICA     JAPAN
JORDAN      LAOS        LEBANON     LIBERIA     LIBYA       LUXEMBOURG  MALAGASY R  MALAYA      MALI        MEXICO
NETHERLANDS NEW ZEALAND NICARAGUA   NIGER       NIGERIA     NORWAY      PANAMA      PARAGUAY    PERU        PHILIPPINES
PORTUGAL    RWANDA      SA'U ARABIA SIERRE LEO  SOMALIA     SO AFRICA   SWEDEN      SWITZERLAND SYRIA       THAILAND
TOGO        TRINIDAD    TURKEY      UGANDA      UK          US          UPPER VOLTA URUGUAY

       1 EXCLUDED BECAUSE AMBIGUOUS

SPAIN

       4 EXCLUDED BECAUSE UNASCERTAINABLE

KOREA REP   NEPAL       PAKISTAN    YEMEN

----------------------------------------------------------------------------------------------------

  2  ALWAYS ARE THOSE                     1.00        2  TEND LESS TO BE THOSE                0.73         0   21
     LOCATED ELSEWHERE THAN IN                           LOCATED ELSEWHERE THAN IN                        32   57
     WEST EUROPE, SCANDINAVIA,                           WEST EUROPE, SCANDINAVIA,                   X SQ=   8.98
     NORTH AMERICA, OR AUSTRALASIA  (93)                 NORTH AMERICA, OR AUSTRALASIA  (93)         P =     0.
                                                                                                     RV NO  (NO )

  5  ALWAYS ARE THOSE                     1.00        5  ALWAYS ARE THOSE                     1.00         0   17
     LOCATED IN EAST EUROPE, RATHER THAN                 LOCATED IN SCANDINAVIA OR WEST EUROPE,            9    0
     IN SCANDINAVIA OR WEST EUROPE  (9)                  RATHER THAN IN EAST EUROPE  (18)            X SQ=  21.77
                                                                                                     P =     0.
                                                                                                     RV YES (YES)

  6  TILT MORE TOWARD BEING THOSE         0.91        6  TILT LESS TOWARD BEING THOSE         0.76         3   19
     LOCATED ELSEWHERE THAN IN THE                       LOCATED ELSEWHERE THAN IN THE                    29   59
     CARIBBEAN, CENTRAL AMERICA,                         CARIBBEAN, CENTRAL AMERICA,                 X SQ=   2.32
     OR SOUTH AMERICA  (93)                              OR SOUTH AMERICA  (93)                      P =     0.114
                                                                                                     RV NO  (NO )
```

8	TEND LESS TO BE THOSE LOCATED ELSEWHERE THAN IN EAST ASIA SOUTH ASIA, OR SOUTHEAST ASIA (97)	0.75	8	TEND MORE TO BE THOSE LOCATED ELSEWHERE THAN IN EAST ASIA SOUTH ASIA, OR SOUTHEAST ASIA (97)	0.91	8 7 24 71 X SQ= 3.68 P = 0.035 RV NO (YES)
9	TILT LESS TOWARD BEING THOSE LOCATED ELSEWHERE THAN IN SOUTHEAST ASIA (106)	0.84	9	TILT MORE TOWARD BEING THOSE LOCATED ELSEWHERE THAN IN SOUTHEAST ASIA (106)	0.95	5 4 27 74 X SQ= 2.08 P = 0.118 RV NO (YES)
16	TEND TO BE THOSE LOCATED IN EAST ASIA, SOUTH ASIA, OR SOUTHEAST ASIA, RATHER THAN IN THE MIDDLE EAST (18)	0.89	16	TEND TO BE THOSE LOCATED IN THE MIDDLE EAST, RATHER THAN IN EAST ASIA, SOUTH ASIA, OR SOUTHEAST ASIA (11)	0.56	1 9 8 7 X SQ= 3.19 P = 0.040 RV YES (YES)
19	TILT MORE TOWARD BEING THOSE LOCATED IN NORTH AFRICA OR CENTRAL AND SOUTH AFRICA, RATHER THAN IN THE CARIBBEAN, CENTRAL AMERICA, OR SOUTH AMERICA (33)	0.79	19	TILT LESS TOWARD BEING THOSE LOCATED IN NORTH AFRICA OR CENTRAL AND SOUTH AFRICA, RATHER THAN IN THE CARIBBEAN, CENTRAL AMERICA, OR SOUTH AMERICA (33)	0.54	11 22 3 19 X SQ= 1.76 P = 0.125 RV NO (NO)
20	TEND TO BE THOSE LOCATED IN EAST ASIA, SOUTH ASIA, OR SOUTHEAST ASIA, RATHER THAN IN THE CARIBBEAN, CENTRAL AMERICA, OR SOUTH AMERICA (18)	0.73	20	TEND TO BE THOSE LOCATED IN THE CARIBBEAN, CENTRAL AMERICA, OR SOUTH AMERICA, RATHER THAN IN EAST ASIA, SOUTH ASIA, OR SOUTHEAST ASIA (22)	0.73	8 7 3 19 X SQ= 4.96 P = 0.025 RV YES (YES)
24	TEND TO BE THOSE WHOSE POPULATION IS VERY LARGE, LARGE, OR MEDIUM (61)	0.69	24	TEND TO BE THOSE WHOSE POPULATION IS SMALL (54)	0.55	22 35 10 43 X SQ= 4.27 P = 0.035 RV YES (NO)
25	TEND MORE TO BE THOSE WHOSE POPULATION DENSITY IS MEDIUM OR LOW (98)	0.97	25	TEND LESS TO BE THOSE WHOSE POPULATION DENSITY IS MEDIUM OR LOW (98)	0.81	1 15 31 63 X SQ= 3.53 P = 0.036 RV NO (NO)
30	TEND TO BE THOSE WHOSE AGRICULTURAL POPULATION IS HIGH (56)	0.68	30	TEND TO BE THOSE WHOSE AGRICULTURAL POPULATION IS MEDIUM, LOW, OR VERY LOW (57)	0.58	21 32 10 45 X SQ= 5.06 P = 0.019 RV YES (NO)
50	TEND TO BE THOSE WHERE FREEDOM OF THE PRESS IS INTERMITTENT, INTERNALLY ABSENT, OR INTERNALLY AND EXTERNALLY ABSENT (56)	0.93	50	TEND TO BE THOSE WHERE FREEDOM OF THE PRESS IS COMPLETE (43)	0.64	2 41 28 23 X SQ= 24.85 P = 0. RV YES (YES)

51	TEND TO BE THOSE WHERE FREEDOM OF THE PRESS IS INTERNALLY ABSENT, OR INTERNALLY AND EXTERNALLY ABSENT (37)	0.80	51	TEND TO BE THOSE WHERE FREEDOM OF THE PRESS IS COMPLETE OR INTERMITTENT (60)	0.86	6 54 24 9 X SQ= 35.52 P = 0. RV YES (YES)
53	LEAN MORE TOWARD BEING THOSE WHERE NEWSPAPER CIRCULATION IS LESS THAN THREE HUNDRED PER THOUSAND (101)	0.97	53	LEAN LESS TOWARD BEING THOSE WHERE NEWSPAPER CIRCULATION IS LESS THAN THREE HUNDRED PER THOUSAND (101)	0.83	1 13 31 65 X SQ= 2.63 P = 0.062 RV NO (NO)
63	TILT MORE TOWARD BEING THOSE WHERE THE RELIGION IS PREDOMINANTLY OR PARTLY OTHER THAN CHRISTIAN (68)	0.71	63	TILT LESS TOWARD BEING THOSE WHERE THE RELIGION IS PREDOMINANTLY OR PARTLY OTHER THAN CHRISTIAN (68)	0.54	9 36 22 42 X SQ= 2.02 P = 0.132 RV NO (NO)
65	LEAN TOWARD BEING THOSE WHERE THE RELIGION IS MUSLIM, RATHER THAN CATHOLIC (18)	0.70	65	LEAN TOWARD BEING THOSE WHERE THE RELIGION IS CATHOLIC, RATHER THAN MUSLIM (25)	0.70	3 21 7 9 X SQ= 3.47 P = 0.059 RV YES (NO)
68	LEAN TOWARD BEING THOSE THAT ARE LINGUISTICALLY WEAKLY HETEROGENEOUS OR STRONGLY HETEROGENEOUS (62)	0.68	68	LEAN TOWARD BEING THOSE THAT ARE LINGUISTICALLY HOMOGENEOUS (52)	0.51	10 40 21 38 X SQ= 2.51 P = 0.090 RV YES (NO)
72	TEND TO BE THOSE WHOSE DATE OF INDEPENDENCE IS AFTER 1914 (59)	0.73	72	TEND TO BE THOSE WHOSE DATE OF INDEPENDENCE IS BEFORE 1914 (52)	0.55	8 42 22 35 X SQ= 5.67 P = 0.010 RV YES (NO)
73	LEAN TOWARD BEING THOSE WHOSE DATE OF INDEPENDENCE IS AFTER 1945 (46)	0.57	73	LEAN TOWARD BEING THOSE WHOSE DATE OF INDEPENDENCE IS BEFORE 1945 (65)	0.64	13 49 17 28 X SQ= 2.87 P = 0.081 RV YES (NO)
74	TILT MORE TOWARD BEING THOSE THAT ARE NOT HISTORICALLY WESTERN (87)	0.87	74	TILT LESS TOWARD BEING THOSE THAT ARE NOT HISTORICALLY WESTERN (87)	0.73	4 21 27 56 X SQ= 1.82 P = 0.135 RV NO (NO)
78	TEND TO BE THOSE THAT HAVE BEEN SIGNIFICANTLY WESTERNIZED WITHOUT A COLONIAL RELATIONSHIP, RATHER THAN THROUGH SUCH A RELATIONSHIP (7)	0.50	78	TEND TO BE THOSE THAT HAVE BEEN SIGNIFICANTLY WESTERNIZED THROUGH A COLONIAL RELATIONSHIP, RATHER THAN WITHOUT SUCH A RELATIONSHIP (28)	0.92	5 23 5 2 X SQ= 5.47 P = 0.012 RV YES (YES)

81	ALWAYS ARE THOSE WHERE THE TYPE OF POLITICAL MODERNIZATION IS LATER EUROPEAN OR LATER EUROPEAN DERIVED, RATHER THAN EARLY EUROPEAN OR EARLY EUROPEAN DERIVED (40)	1.00	81	LEAN LESS TOWARD BEING THOSE WHERE THE TYPE OF POLITICAL MODERNIZATION IS LATER EUROPEAN OR LATER EUROPEAN DERIVED, RATHER THAN EARLY EUROPEAN OR EARLY EUROPEAN DERIVED (40)	0.72	0 11 11 28 X SQ= 2.50 P = 0.053 RV NO (NO)
82	TILT TOWARD BEING THOSE WHERE THE TYPE OF POLITICAL MODERNIZATION IS DEVELOPED TUTELARY OR UNDEVELOPED TUTELARY, RATHER THAN EARLY OR LATER EUROPEAN OR EUROPEAN DERIVED (55)	0.63	82	TILT TOWARD BEING THOSE WHERE THE TYPE OF POLITICAL MODERNIZATION IS EARLY OR LATER EUROPEAN OR EUROPEAN DERIVED, RATHER THAN DEVELOPED TUTELARY OR UNDEVELOPED TUTELARY (51)	0.54	11 39 19 33 X SQ= 1.94 P = 0.131 RV YES (NO)
87	LEAN TOWARD BEING THOSE WHOSE IDEOLOGICAL ORIENTATION IS DEVELOPMENTAL (31)	0.52	87	LEAN TOWARD BEING THOSE WHOSE IDEOLOGICAL ORIENTATION IS OTHER THAN DEVELOPMENTAL (58)	0.72	14 17 13 44 X SQ= 3.73 P = 0.052 RV YES (NO)
89	ALWAYS ARE THOSE WHOSE IDEOLOGICAL ORIENTATION IS OTHER THAN CONVENTIONAL (62)	1.00	89	TEND TO BE THOSE WHOSE IDEOLOGICAL ORIENTATION IS CONVENTIONAL (33)	0.54	0 33 31 28 X SQ= 23.85 P = 0. RV YES (YES)
91	ALWAYS ARE THOSE WHOSE IDEOLOGICAL ORIENTATION IS DEVELOPMENTAL, RATHER THAN TRADITIONAL (31)	1.00	91	TILT LESS TOWARD BEING THOSE WHOSE IDEOLOGICAL ORIENTATION IS DEVELOPMENTAL, RATHER THAN TRADITIONAL (31)	0.77	14 17 0 5 X SQ= 2.04 P = 0.134 RV NO (NO)
94	TEND TO BE THOSE WHERE THE STATUS OF THE REGIME IS AUTHORITARIAN OR TOTALITARIAN (41)	0.79	94	TEND TO BE THOSE WHERE THE STATUS OF THE REGIME IS CONSTITUTIONAL (51)	0.76	6 45 23 14 X SQ= 22.42 P = 0. RV YES (YES)
95	TEND LESS TO BE THOSE WHERE THE STATUS OF THE REGIME IS CONSTITUTIONAL OR AUTHORITARIAN (95)	0.55	95	TEND MORE TO BE THOSE WHERE THE STATUS OF THE REGIME IS CONSTITUTIONAL OR AUTHORITARIAN (95)	0.99	17 75 14 1 X SQ= 31.57 P = 0. RV NO (YES)
97	TEND TO BE THOSE WHERE THE STATUS OF THE REGIME IS TOTALITARIAN, RATHER THAN AUTHORITARIAN (16)	0.64	97	TEND TO BE THOSE WHERE THE STATUS OF THE REGIME IS AUTHORITARIAN, RATHER THAN TOTALITARIAN (23)	0.92	8 12 14 1 X SQ= 8.28 P = 0.002 RV YES (YES)
98	TEND MORE TO BE THOSE WHERE GOVERNMENTAL STABILITY IS GENERALLY OR MODERATELY PRESENT AND DATES FROM THE POST-WAR PERIOD, OR IS ABSENT (93)	0.97	98	TEND LESS TO BE THOSE WHERE GOVERNMENTAL STABILITY IS GENERALLY OR MODERATELY PRESENT AND DATES FROM THE POST-WAR PERIOD, OR IS ABSENT (93)	0.74	1 20 31 58 X SQ= 6.06 P = 0.006 RV NO (NO)

100	TILT MORE TOWARD BEING THOSE WHERE GOVERNMENTAL STABILITY IS GENERALLY PRESENT AND DATES FROM AT LEAST THE INTER-WAR PERIOD, OR IS GENERALLY OR MODERATELY PRESENT, AND DATES FROM THE POST-WAR PERIOD (64)	0.88	100	TILT LESS TOWARD BEING THOSE WHERE GOVERNMENTAL STABILITY IS GENERALLY PRESENT AND DATES FROM AT LEAST THE INTER-WAR PERIOD, OR IS GENERALLY OR MODERATELY PRESENT, AND DATES FROM THE POST-WAR PERIOD (64)	0.71	23 40 3 16 X SQ= 2.02 P = 0.101 RV NO (NO)
101	TEND TO BE THOSE WHERE THE REPRESENTATIVE CHARACTER OF THE REGIME IS LIMITED POLYARCHIC, PSEUDO-POLYARCHIC, OR NON-POLYARCHIC (57)	0.84	101	TEND TO BE THOSE WHERE THE REPRESENTATIVE CHARACTER OF THE REGIME IS POLYARCHIC (41)	0.57	5 36 27 27 X SQ= 13.27 P = 0. RV YES (YES)
102	TEND TO BE THOSE WHERE THE REPRESENTATIVE CHARACTER OF THE REGIME IS PSEUDO-POLYARCHIC OR NON-POLYARCHIC (49)	0.81	102	TEND TO BE THOSE WHERE THE REPRESENTATIVE CHARACTER OF THE REGIME IS POLYARCHIC OR LIMITED POLYARCHIC (59)	0.73	6 53 26 20 X SQ= 24.07 P = 0. RV YES (YES)
105	TEND TO BE THOSE WHERE THE ELECTORAL SYSTEM IS PARTIALLY COMPETITIVE OR NON-COMPETITIVE (47)	0.88	105	TEND TO BE THOSE WHERE THE ELECTORAL SYSTEM IS COMPETITIVE (43)	0.65	3 40 23 22 X SQ= 18.51 P = 0. RV YES (YES)
106	TEND TO BE THOSE WHERE THE ELECTORAL SYSTEM IS NON-COMPETITIVE (30)	0.78	106	TEND TO BE THOSE WHERE THE ELECTORAL SYSTEM IS COMPETITIVE OR PARTIALLY COMPETITIVE (52)	0.85	6 46 21 8 X SQ= 28.37 P = 0. RV YES (YES)
107	TEND TO BE THOSE WHERE AUTONOMOUS GROUPS ARE PARTIALLY TOLERATED IN POLITICS, ARE TOLERATED ONLY OUTSIDE POLITICS, OR ARE NOT TOLERATED AT ALL (65)	0.88	107	TEND TO BE THOSE WHERE AUTONOMOUS GROUPS ARE FULLY TOLERATED IN POLITICS (46)	0.56	4 42 28 33 X SQ= 15.59 P = 0. RV YES (NO)
108	TEND TO BE THOSE WHERE AUTONOMOUS GROUPS ARE TOLERATED ONLY OUTSIDE POLITICS OR ARE NOT TOLERATED AT ALL (35)	0.69	108	TEND TO BE THOSE WHERE AUTONOMOUS GROUPS ARE FULLY OR PARTIALLY TOLERATED IN POLITICS (65)	0.81	9 56 20 13 X SQ= 20.78 P = 0. RV YES (YES)
111	TEND MORE TO BE THOSE WHERE POLITICAL ENCULTURATION IS HIGH OR MEDIUM (53)	0.80	111	TEND LESS TO BE THOSE WHERE POLITICAL ENCULTURATION IS HIGH OR MEDIUM (53)	0.53	16 37 4 33 X SQ= 3.68 P = 0.039 RV NO (NO)
115	ALWAYS ARE THOSE WHERE INTEREST ARTICULATION BY ASSOCIATIONAL GROUPS IS MODERATE, LIMITED, OR NEGLIGIBLE (91)	1.00	115	TEND LESS TO BE THOSE WHERE INTEREST ARTICULATION BY ASSOCIATIONAL GROUPS IS MODERATE, LIMITED, OR NEGLIGIBLE (91)	0.74	0 20 31 57 X SQ= 8.24 P = 0.001 RV NO (NO)

116	TEND MORE TO BE THOSE WHERE INTEREST ARTICULATION BY ASSOCIATIONAL GROUPS IS LIMITED OR NEGLIGIBLE (79)	0.94	116	TEND LESS TO BE THOSE WHERE INTEREST ARTICULATION BY ASSOCIATIONAL GROUPS IS LIMITED OR NEGLIGIBLE (79)	0.61	2 30 29 47 X SQ= 9.70 P = 0. RV NO (NO)
117	LEAN TOWARD BEING THOSE WHERE INTEREST ARTICULATION BY ASSOCIATIONAL GROUPS IS NEGLIGIBLE (51)	0.58	117	LEAN TOWARD BEING THOSE WHERE INTEREST ARTICULATION BY ASSOCIATIONAL GROUPS IS SIGNIFICANT, MODERATE, OR LIMITED (60)	0.61	13 47 18 30 X SQ= 2.54 P = 0.088 RV YES (NO)
118	TEND TO BE THOSE WHERE INTEREST ARTICULATION BY INSTITUTIONAL GROUPS IS VERY SIGNIFICANT (40)	0.66	118	TEND TO BE THOSE WHERE INTEREST ARTICULATION BY INSTITUTIONAL GROUPS IS SIGNIFICANT, MODERATE, OR LIMITED (60)	0.75	21 16 11 48 X SQ= 13.20 P = 0. RV YES (YES)
119	TEND MORE TO BE THOSE WHERE INTEREST ARTICULATION BY INSTITUTIONAL GROUPS IS VERY SIGNIFICANT OR SIGNIFICANT (74)	0.97	119	TEND LESS TO BE THOSE WHERE INTEREST ARTICULATION BY INSTITUTIONAL GROUPS IS VERY SIGNIFICANT OR SIGNIFICANT (74)	0.61	31 39 1 25 X SQ= 12.19 P = 0. RV NO (NO)
120	ALWAYS ARE THOSE WHERE INTEREST ARTICULATION BY INSTITUTIONAL GROUPS IS VERY SIGNIFICANT, SIGNIFICANT, OR MODERATE (90)	1.00	120	TEND LESS TO BE THOSE WHERE INTEREST ARTICULATION BY INSTITUTIONAL GROUPS IS VERY SIGNIFICANT, SIGNIFICANT, OR MODERATE (90)	0.84	32 54 0 10 X SQ= 4.03 P = 0.015 RV NO (NO)
123	ALWAYS ARE THOSE WHERE INTEREST ARTICULATION BY NON-ASSOCIATIONAL GROUPS IS SIGNIFICANT, MODERATE, OR LIMITED (107)	1.00	123	TILT LESS TOWARD BEING THOSE WHERE INTEREST ARTICULATION BY NON-ASSOCIATIONAL GROUPS IS SIGNIFICANT, MODERATE, OR LIMITED (107)	0.90	32 70 0 8 X SQ= 2.18 P = 0.102 RV NO (NO)
126	ALWAYS ARE THOSE WHERE INTEREST ARTICULATION BY ANOMIC GROUPS IS FREQUENT, OCCASIONAL, OR INFREQUENT (83)	1.00	126	TEND LESS TO BE THOSE WHERE INTEREST ARTICULATION BY ANOMIC GROUPS IS FREQUENT, OCCASIONAL, OR INFREQUENT (83)	0.78	23 56 0 16 X SQ= 4.66 P = 0.010 RV NO (NO)
128	TEND TO BE THOSE WHERE INTEREST ARTICULATION BY POLITICAL PARTIES IS LIMITED OR NEGLIGIBLE (45)	0.86	128	TEND TO BE THOSE WHERE INTEREST ARTICULATION BY POLITICAL PARTIES IS SIGNIFICANT OR MODERATE (48)	0.69	4 44 24 20 X SQ= 21.02 P = 0. RV YES (YES)
129	TEND TO BE THOSE WHERE INTEREST ARTICULATION BY POLITICAL PARTIES IS NEGLIGIBLE (37)	0.79	129	TEND TO BE THOSE WHERE INTEREST ARTICULATION BY POLITICAL PARTIES IS SIGNIFICANT, MODERATE, OR LIMITED (56)	0.78	6 50 22 14 X SQ= 23.96 P = 0. RV YES (YES)

131	TILT TOWARD BEING THOSE WHERE INTEREST AGGREGATION BY POLITICAL PARTIES IS SIGNIFICANT OR MODERATE (30)	0.73	131	TILT TOWARD BEING THOSE WHERE INTEREST AGGREGATION BY POLITICAL PARTIES IS LIMITED OR NEGLIGIBLE (35)	0.57	8 22 3 29 X SQ= 2.10 P = 0.101 RV YES (NO)
138	TEND TO BE THOSE WHERE INTEREST AGGREGATION BY THE LEGISLATURE IS NEGLIGIBLE (49)	0.85	138	TEND TO BE THOSE WHERE INTEREST AGGREGATION BY THE LEGISLATURE IS SIGNIFICANT, MODERATE, OR LIMITED (48)	0.63	4 44 22 26 X SQ= 15.24 P = 0. RV YES (NO)
139	TEND TO BE THOSE WHERE THE PARTY SYSTEM IS QUANTITATIVELY ONE-PARTY (34)	0.74	139	TEND TO BE THOSE WHERE THE PARTY SYSTEM IS QUANTITATIVELY OTHER THAN ONE-PARTY (71)	0.86	23 10 8 62 X SQ= 33.48 P = 0. RV YES (YES)
141	ALWAYS ARE THOSE WHERE THE PARTY SYSTEM IS QUANTITATIVELY OTHER THAN TWO-PARTY (87)	1.00	141	TEND LESS TO BE THOSE WHERE THE PARTY SYSTEM IS QUANTITATIVELY OTHER THAN TWO-PARTY (87)	0.84	0 11 29 57 X SQ= 3.80 P = 0.031 RV NO (NO)
142	TEND MORE TO BE THOSE WHERE THE PARTY SYSTEM IS QUANTITATIVELY OTHER THAN MULTI-PARTY (66)	0.93	142	TEND LESS TO BE THOSE WHERE THE PARTY SYSTEM IS QUANTITATIVELY OTHER THAN MULTI-PARTY (66)	0.58	2 28 26 39 X SQ= 9.43 P = 0.001 RV NO (NO)
146	TEND LESS TO BE THOSE WHERE THE PARTY SYSTEM IS QUALITATIVELY OTHER THAN MASS-BASED TERRITORIAL (92)	0.71	146	ALWAYS ARE THOSE WHERE THE PARTY SYSTEM IS QUALITATIVELY OTHER THAN MASS-BASED TERRITORIAL (92)	1.00	8 0 20 68 X SQ= 17.62 P = 0. RV NO (YES)
147	TEND MORE TO BE THOSE WHERE THE PARTY SYSTEM IS QUALITATIVELY OTHER THAN CLASS-ORIENTED OR MULTI-IDEOLOGICAL (67)	0.97	147	TEND LESS TO BE THOSE WHERE THE PARTY SYSTEM IS QUALITATIVELY OTHER THAN CLASS-ORIENTED OR MULTI-IDEOLOGICAL (67)	0.63	1 22 28 37 X SQ= 9.85 P = 0.001 RV NO (NO)
148	LEAN MORE TOWARD BEING THOSE WHERE THE PARTY SYSTEM IS QUALITATIVELY OTHER THAN AFRICAN TRANSITIONAL (96)	0.97	148	LEAN LESS TOWARD BEING THOSE WHERE THE PARTY SYSTEM IS QUALITATIVELY OTHER THAN AFRICAN TRANSITIONAL (96)	0.82	1 13 31 60 X SQ= 2.98 P = 0.059 RV NO (NO)
153	LEAN TOWARD BEING THOSE WHERE THE PARTY SYSTEM IS STABLE (42)	0.71	153	LEAN TOWARD BEING THOSE WHERE THE PARTY SYSTEM IS MODERATELY STABLE OR UNSTABLE (38)	0.54	17 24 7 28 X SQ= 3.09 P = 0.052 RV YES (NO)

158	TEND MORE TO BE THOSE WHERE PERSONALISMO IS NEGLIGIBLE (56)	0.73	158	TEND LESS TO BE THOSE WHERE PERSONALISMO IS NEGLIGIBLE (56)	0.51	8 32 22 33 X SQ= 3.41 P = 0.046 RV NO (NO)
161	TEND TO BE THOSE WHERE THE POLITICAL LEADERSHIP IS ELITIST OR MODERATELY ELITIST (47)	0.65	161	TEND TO BE THOSE WHERE THE POLITICAL LEADERSHIP IS NON-ELITIST (50)	0.59	17 28 9 41 X SQ= 3.72 P = 0.039 RV YES (NO)
164	TEND TO BE THOSE WHERE THE REGIME'S LEADERSHIP CHARISMA IS PRONOUNCED OR MODERATE (34)	0.57	164	TEND TO BE THOSE WHERE THE REGIME'S LEADERSHIP CHARISMA IS NEGLIGIBLE (65)	0.76	17 16 13 50 X SQ= 8.23 P = 0.003 RV YES (YES)
166	LEAN MORE TOWARD BEING THOSE WHERE THE VERTICAL POWER DISTRIBUTION IS THAT OF FORMAL FEDERALISM OR FORMAL AND EFFECTIVE UNITARISM (99)	0.97	166	LEAN LESS TOWARD BEING THOSE WHERE THE VERTICAL POWER DISTRIBUTION IS THAT OF FORMAL FEDERALISM OR FORMAL AND EFFECTIVE UNITARISM (99)	0.83	1 13 31 64 X SQ= 2.69 P = 0.062 RV NO (NO)
169	TEND TO BE THOSE WHERE THE HORIZONTAL POWER DISTRIBUTION IS NEGLIGIBLE (48)	0.84	169	TEND TO BE THOSE WHERE THE HORIZONTAL POWER DISTRIBUTION IS SIGNIFICANT OR LIMITED (58)	0.74	5 53 26 19 X SQ= 26.81 P = 0. RV YES (YES)
172	ALWAYS ARE THOSE WHERE THE LEGISLATIVE-EXECUTIVE STRUCTURE IS OTHER THAN PARLIAMENTARY-ROYALIST (88)	1.00	172	TEND LESS TO BE THOSE WHERE THE LEGISLATIVE-EXECUTIVE STRUCTURE IS OTHER THAN PARLIAMENTARY-ROYALIST (88)	0.72	0 21 31 54 X SQ= 9.13 P = 0. RV NO (NO)
175	TEND TO BE THOSE WHERE THE LEGISLATURE IS LARGELY INEFFECTIVE OR WHOLLY INEFFECTIVE (49)	0.86	175	TEND TO BE THOSE WHERE THE LEGISLATURE IS FULLY EFFECTIVE OR PARTIALLY EFFECTIVE (51)	0.68	4 47 25 22 X SQ= 22.01 P = 0. RV YES (YES)
176	TEND TO BE THOSE WHERE THE LEGISLATURE IS WHOLLY INEFFECTIVE (28)	0.62	176	TEND TO BE THOSE WHERE THE LEGISLATURE IS FULLY EFFECTIVE, PARTIALLY EFFECTIVE, OR LARGELY INEFFECTIVE (72)	0.88	11 61 18 8 X SQ= 24.16 P = 0. RV YES (YES)
178	TEND TO BE THOSE WHERE THE LEGISLATURE IS UNICAMERAL (53)	0.79	178	TEND TO BE THOSE WHERE THE LEGISLATURE IS BICAMERAL (51)	0.62	22 28 6 45 X SQ= 11.53 P = 0. RV YES (NO)

```
179  TEND TO BE THOSE                          0.90    179  TEND TO BE THOSE                          0.62        27    22
     WHERE THE EXECUTIVE IS DOMINANT   (52)                WHERE THE EXECUTIVE IS STRONG  (39)                      3    36
                                                                                                              X SQ=  19.66
                                                                                                              P =     0.
                                                                                                              RV YES (YES)

181  TEND MORE TO BE THOSE                     0.95    181  TEND LESS TO BE THOSE                     0.62         1    20
     WHERE THE BUREAUCRACY                                  WHERE THE BUREAUCRACY                                  21    32
     IS SEMI-MODERN, RATHER THAN                            IS SEMI-MODERN, RATHER THAN                       X SQ=   7.16
     MODERN  (55)                                           MODERN  (55)                                      P =     0.004
                                                                                                              RV NO  (NO )

185  TEND TO BE THOSE                          0.87    185  TEND TO BE THOSE                          0.60         3    15
     WHERE PARTICIPATION BY THE MILITARY                    WHERE PARTICIPATION BY THE MILITARY                    20    10
     IN POLITICS IS                                         IN POLITICS IS                                    X SQ=   9.36
     SUPPORTIVE, RATHER THAN                                INTERVENTIVE, RATHER THAN                         P =     0.001
     INTERVENTIVE  (31)                                     SUPPORTIVE  (21)                                  RV YES (YES)

186  TEND TO BE THOSE                          0.74    186  TEND TO BE THOSE                          0.83        20    10
     WHERE PARTICIPATION BY THE MILITARY                    WHERE PARTICIPATION BY THE MILITARY                     7    49
     IN POLITICS IS                                         IN POLITICS IS                                    X SQ=  24.16
     SUPPORTIVE, RATHER THAN                                NEUTRAL, RATHER THAN                              P =     0.
     NEUTRAL  (31)                                          SUPPORTIVE  (56)                                  RV YES (YES)

187  TEND TO BE THOSE                          0.91    187  TEND TO BE THOSE                          0.50        29    32
     WHERE THE ROLE OF THE POLICE                           WHERE THE ROLE OF THE POLICE                            3    32
     IS POLITICALLY SIGNIFICANT  (66)                       IS NOT POLITICALLY SIGNIFICANT  (35)              X SQ=  13.50
                                                                                                              P =     0.
                                                                                                              RV YES (NO )

188  TEND MORE TO BE THOSE                     0.94    188  TEND LESS TO BE THOSE                     0.62         2    29
     WHERE THE CHARACTER OF THE LEGAL SYSTEM                WHERE THE CHARACTER OF THE LEGAL SYSTEM                29    48
     IS OTHER THAN CIVIL LAW  (81)                          IS OTHER THAN CIVIL LAW  (81)                     X SQ=   9.05
                                                                                                              P =     0.001
                                                                                                              RV NO  (NO )

189  ALWAYS ARE THOSE                          1.00    189  TILT LESS TOWARD BEING THOSE              0.91         0     7
     WHERE THE CHARACTER OF THE LEGAL SYSTEM                WHERE THE CHARACTER OF THE LEGAL SYSTEM                32    71
     IS OTHER THAN COMMON LAW  (108)                        IS OTHER THAN COMMON LAW  (108)                   X SQ=   1.75
                                                                                                              P =     0.104
                                                                                                              RV NO  (NO )

194  TEND LESS TO BE THOSE                     0.58    194  ALWAYS ARE THOSE                          1.00        13     0
     THAT ARE NON-COMMUNIST  (101)                          THAT ARE NON-COMMUNIST  (101)                          18    78
                                                                                                              X SQ=  33.25
                                                                                                              P =     0.
                                                                                                              RV NO  (YES)
```

```
****************************************************************************************************

94  POLITIES                                              94  POLITIES
    WHERE THE STATUS OF THE REGIME                            WHERE THE STATUS OF THE REGIME
    IS CONSTITUTIONAL  (51)                                   IS AUTHORITARIAN OR
                                                              TOTALITARIAN  (41)

                                                                              BOTH SUBJECT AND PREDICATE
....................................................................................................

      51 IN LEFT COLUMN

AUSTRALIA   AUSTRIA     BELGIUM     BOLIVIA     BRAZIL      CANADA      CEYLON      CHILE       COLOMBIA    COSTA RICA
CYPRUS      DENMARK     DOMIN REP   ECUADOR     FINLAND     FRANCE      GERMAN FR   GREECE      HONDURAS    ICELAND
INDIA       IRELAND     ISRAEL      ITALY       JAMAICA     JAPAN       LEBANON     LIBYA       LUXEMBOURG  MALAGASY R
MALAYA      MAURITANIA  MEXICO      MOROCCO     NETHERLANDS NEW ZEALAND NIGERIA     NORWAY      PANAMA      PHILIPPINES
SIERRE LEO  SWEDEN      SWITZERLAND TRINIDAD    TUNISIA     TURKEY      UGANDA      UK          US          URUGUAY
VENEZUELA

      41 IN RIGHT COLUMN

AFGHANISTAN ALBANIA     ALGERIA     BULGARIA    BURMA       CAMBODIA    CEN AFR REP CHINA, PR   CUBA        CZECHOS'KIA
EL SALVADOR ETHIOPIA    GERMANY, E  GHANA       GUINEA      HAITI       HUNGARY     INDONESIA   IRAN        JORDAN
KOREA, N    KOREA REP   LAOS        LIBERIA     MONGOLIA    NEPAL       NICARAGUA   PAKISTAN    PARAGUAY    POLAND
PORTUGAL    RUMANIA     SA'U ARABIA SPAIN       SUDAN       THAILAND    USSR        UAR         VIETNAM, N  VIETNAM REP
YUGOSLAVIA

       3 EXCLUDED BECAUSE AMBIGUOUS

ARGENTINA   IVORY COAST SO AFRICA

      11 EXCLUDED BECAUSE UNASCERTAINED

BURUNDI     CHAD        CONGO(BRA)  DAHOMEY     GABON       MALI        NIGER       RWANDA      SOMALIA     TANGANYIKA
UPPER VOLTA

       9 EXCLUDED BECAUSE UNASCERTAINABLE

CAMEROUN    CONGO(LEO)  GUATEMALA   IRAQ        PERU        SENEGAL     SYRIA       TOGO        YEMEN
----------------------------------------------------------------------------------------------------

  2  TEND LESS TO BE THOSE                  0.61    2  TEND MORE TO BE THOSE                 0.95      20     2
     LOCATED ELSEWHERE THAN IN                         LOCATED ELSEWHERE THAN IN                       31    39
     WEST EUROPE, SCANDINAVIA,                         WEST EUROPE, SCANDINAVIA,                  X SQ=  12.90
     NORTH AMERICA, OR AUSTRALASIA  (93)               NORTH AMERICA, OR AUSTRALASIA  (93)        P =     0.
                                                                                                  RV NO  (YES)

  5  ALWAYS ARE THOSE                       1.00    5  TEND TO BE THOSE                      0.82      16     2
     LOCATED IN SCANDINAVIA OR WEST EUROPE,            LOCATED IN EAST EUROPE, RATHER THAN              0     9
     RATHER THAN IN EAST EUROPE  (18)                  IN SCANDINAVIA OR WEST EUROPE  (9)         X SQ=  16.13
                                                                                                  P =     0.
                                                                                                  RV YES (YES)
```

6	TILT LESS TOWARD BEING THOSE LOCATED ELSEWHERE THAN IN THE CARIBBEAN, CENTRAL AMERICA, OR SOUTH AMERICA (93)	0.73	6	TILT MORE TOWARD BEING THOSE LOCATED ELSEWHERE THAN IN THE CARIBBEAN, CENTRAL AMERICA, OR SOUTH AMERICA (93)	0.88	14 5 37 36 X SQ= 2.36 P = 0.119 RV NO (NO)
8	TEND MORE TO BE THOSE LOCATED ELSEWHERE THAN IN EAST ASIA SOUTH ASIA, OR SOUTHEAST ASIA (97)	0.90	8	TEND LESS TO BE THOSE LOCATED ELSEWHERE THAN IN EAST ASIA SOUTH ASIA, OR SOUTHEAST ASIA (97)	0.68	5 13 46 28 X SQ= 5.61 P = 0.016 RV NO (YES)
9	LEAN MORE TOWARD BEING THOSE LOCATED ELSEWHERE THAN IN SOUTHEAST ASIA (106)	0.96	9	LEAN LESS TOWARD BEING THOSE LOCATED ELSEWHERE THAN IN SOUTHEAST ASIA (106)	0.83	2 7 49 34 X SQ= 3.09 P = 0.073 RV NO (YES)
20	TEND TO BE THOSE LOCATED IN THE CARIBBEAN, CENTRAL AMERICA, OR SOUTH AMERICA, RATHER THAN IN EAST ASIA, SOUTH ASIA, OR SOUTHEAST ASIA (22)	0.74	20	TEND TO BE THOSE LOCATED IN EAST ASIA, SOUTH ASIA, OR SOUTHEAST ASIA, RATHER THAN IN THE CARIBBEAN, CENTRAL AMERICA, OR SOUTH AMERICA (18)	0.72	5 13 14 5 X SQ= 6.07 P = 0.009 RV YES (YES)
24	LEAN LESS TOWARD BEING THOSE WHOSE POPULATION IS VERY LARGE, LARGE, OR MEDIUM (61)	0.51	24	LEAN MORE TOWARD BEING THOSE WHOSE POPULATION IS VERY LARGE, LARGE, OR MEDIUM (61)	0.71	26 29 25 12 X SQ= 2.91 P = 0.086 RV NO (YES)
25	LEAN LESS TOWARD BEING THOSE WHOSE POPULATION DENSITY IS MEDIUM OR LOW (98)	0.75	25	LEAN MORE TOWARD BEING THOSE WHOSE POPULATION DENSITY IS MEDIUM OR LOW (98)	0.90	13 4 38 37 X SQ= 2.76 P = 0.063 RV NO (NO)
29	TEND TO BE THOSE WHERE THE DEGREE OF URBANIZATION IS HIGH (56)	0.79	29	TEND TO BE THOSE WHERE THE DEGREE OF URBANIZATION IS LOW (49)	0.60	37 14 10 21 X SQ= 11.20 P = 0.001 RV YES (YES)
30	TEND TO BE THOSE WHOSE AGRICULTURAL POPULATION IS MEDIUM, LOW, OR VERY LOW (57)	0.78	30	TEND TO BE THOSE WHOSE AGRICULTURAL POPULATION IS HIGH (56)	0.64	11 25 40 14 X SQ= 14.93 P = 0. RV YES (YES)
36	TEND TO BE THOSE WHOSE PER CAPITA GROSS NATIONAL PRODUCT IS VERY HIGH, HIGH, OR MEDIUM (42)	0.59	36	TEND TO BE THOSE WHOSE PER CAPITA GROSS NATIONAL PRODUCT IS LOW OR VERY LOW (73)	0.76	30 10 21 31 X SQ= 9.61 P = 0.001 RV YES (YES)

37	TEND TO BE THOSE WHOSE PER CAPITA GROSS NATIONAL PRODUCT IS VERY HIGH, HIGH, MEDIUM, OR LOW (64)	0.80	37	TEND TO BE THOSE WHOSE PER CAPITA GROSS NATIONAL PRODUCT IS VERY LOW (51)	0.59	41 17 10 24 X SQ= 13.16 P = 0. RV YES (YES)
38	TEND LESS TO BE THOSE WHOSE INTERNATIONAL FINANCIAL STATUS IS MEDIUM, LOW, OR VERY LOW (103)	0.82	38	TEND MORE TO BE THOSE WHOSE INTERNATIONAL FINANCIAL STATUS IS MEDIUM, LOW, OR VERY LOW (103)	0.97	9 1 42 38 X SQ= 3.68 P = 0.039 RV NO (NO)
39	LEAN LESS TOWARD BEING THOSE WHOSE INTERNATIONAL FINANCIAL STATUS IS LOW OR VERY LOW (76)	0.51	39	LEAN MORE TOWARD BEING THOSE WHOSE INTERNATIONAL FINANCIAL STATUS IS LOW OR VERY LOW (76)	0.72	25 11 26 29 X SQ= 3.49 P = 0.052 RV NO (YES)
42	TEND TO BE THOSE WHOSE ECONOMIC DEVELOPMENTAL STATUS IS DEVELOPED OR INTERMEDIATE (36)	0.53	42	TEND TO BE THOSE WHOSE ECONOMIC DEVELOPMENTAL STATUS IS UNDERDEVELOPED OR VERY UNDERDEVELOPED (76)	0.80	26 8 23 32 X SQ= 8.84 P = 0.002 RV YES (YES)
43	TEND TO BE THOSE WHOSE ECONOMIC DEVELOPMENTAL STATUS IS DEVELOPED, INTERMEDIATE, OR UNDERDEVELOPED (55)	0.76	43	TEND TO BE THOSE WHOSE ECONOMIC DEVELOPMENTAL STATUS IS VERY UNDERDEVELOPED (57)	0.65	37 14 12 26 X SQ= 13.16 P = 0. RV YES (YES)
45	TEND TO BE THOSE WHERE THE LITERACY RATE IS FIFTY PERCENT OR ABOVE (55)	0.72	45	TEND TO BE THOSE WHERE THE LITERACY RATE IS BELOW FIFTY PERCENT (54)	0.50	36 18 14 18 X SQ= 3.45 P = 0.044 RV YES (YES)
46	TEND MORE TO BE THOSE WHERE THE LITERACY RATE IS TEN PERCENT OR ABOVE (84)	0.96	46	TEND LESS TO BE THOSE WHERE THE LITERACY RATE IS TEN PERCENT OR ABOVE (84)	0.82	47 31 2 7 X SQ= 3.32 P = 0.038 RV NO (YES)
50	TEND TO BE THOSE WHERE FREEDOM OF THE PRESS IS COMPLETE (43)	0.77	50	ALWAYS ARE THOSE WHERE FREEDOM OF THE PRESS IS INTERMITTENT, INTERNALLY ABSENT, OR INTERNALLY AND EXTERNALLY ABSENT (56)	1.00	37 0 11 38 X SQ= 48.31 P = 0. RV YES (YES)
51	TEND TO BE THOSE WHERE FREEDOM OF THE PRESS IS COMPLETE OR INTERMITTENT (60)	0.96	51	TEND TO BE THOSE WHERE FREEDOM OF THE PRESS IS INTERNALLY ABSENT, OR INTERNALLY AND EXTERNALLY ABSENT (37)	0.92	46 3 2 35 X SQ= 63.37 P = 0. RV YES (YES)

52	ALWAYS ARE THOSE WHERE FREEDOM OF THE PRESS IS COMPLETE, INTERMITTENT, OR INTERNALLY ABSENT (82)	1.00	52	TEND LESS TO BE THOSE WHERE FREEDOM OF THE PRESS IS COMPLETE, INTERMITTENT, OR INTERNALLY ABSENT (82)	0.58	49 22 0 16 X SQ= 22.55 P = 0. RV NO (YES)
54	TEND TO BE THOSE WHERE NEWSPAPER CIRCULATION IS ONE HUNDRED OR MORE PER THOUSAND (37)	0.53	54	TEND TO BE THOSE WHERE NEWSPAPER CIRCULATION IS LESS THAN ONE HUNDRED PER THOUSAND (76)	0.77	27 9 24 30 X SQ= 7.02 P = 0.005 RV YES (YES)
63	TEND TO BE THOSE WHERE THE RELIGION IS PREDOMINANTLY SOME KIND OF CHRISTIAN (46)	0.63	63	TEND TO BE THOSE WHERE THE RELIGION IS PREDOMINANTLY OR PARTLY OTHER THAN CHRISTIAN (68)	0.67	32 13 19 27 X SQ= 7.04 P = 0.006 RV YES (YES)
64	TEND MORE TO BE THOSE WHERE THE RELIGION IS CHRISTIAN, RATHER THAN MUSLIM (46)	0.86	64	TEND LESS TO BE THOSE WHERE THE RELIGION IS CHRISTIAN, RATHER THAN MUSLIM (46)	0.62	32 13 5 8 X SQ= 3.35 P = 0.049 RV NO (YES)
68	LEAN TOWARD BEING THOSE THAT ARE LINGUISTICALLY HOMOGENEOUS (52)	0.61	68	LEAN TOWARD BEING THOSE THAT ARE LINGUISTICALLY WEAKLY HETEROGENEOUS OR STRONGLY HETEROGENEOUS (62)	0.60	31 16 20 24 X SQ= 3.09 P = 0.059 RV YES (YES)
70	TILT LESS TOWARD BEING THOSE THAT ARE RELIGIOUSLY, RACIALLY, OR LINGUISTICALLY HETEROGENEOUS (85)	0.71	70	TILT MORE TOWARD BEING THOSE THAT ARE RELIGIOUSLY, RACIALLY, OR LINGUISTICALLY HETEROGENEOUS (85)	0.86	14 5 35 31 X SQ= 1.80 P = 0.123 RV NO (NO)
74	TEND LESS TO BE THOSE THAT ARE NOT HISTORICALLY WESTERN (87)	0.60	74	TEND MORE TO BE THOSE THAT ARE NOT HISTORICALLY WESTERN (87)	0.85	20 6 30 35 X SQ= 5.91 P = 0.010 RV NO (YES)
75	TEND TO BE THOSE THAT ARE HISTORICALLY WESTERN OR SIGNIFICANTLY WESTERNIZED (62)	0.82	75	TEND TO BE THOSE THAT ARE NOT HISTORICALLY WESTERN AND ARE NOT SIGNIFICANTLY WESTERNIZED (52)	0.59	42 17 9 24 X SQ= 14.79 P = 0. RV YES (YES)
76	TILT LESS TOWARD BEING THOSE THAT HAVE BEEN SIGNIFICANTLY OR PARTIALLY WESTERNIZED THROUGH A COLONIAL RELATIONSHIP, RATHER THAN BEING HISTORICALLY WESTERN (70)	0.58	76	TILT MORE TOWARD BEING THOSE THAT HAVE BEEN SIGNIFICANTLY OR PARTIALLY WESTERNIZED THROUGH A COLONIAL RELATIONSHIP, RATHER THAN BEING HISTORICALLY WESTERN (70)	0.77	20 6 28 20 X SQ= 1.81 P = 0.132 RV NO (NO)

77	TEND TO BE THOSE THAT HAVE BEEN SIGNIFICANTLY WESTERNIZED, RATHER THAN PARTIALLY WESTERNIZED, THROUGH A COLONIAL RELATIONSHIP (28)	0.68	77	TEND TO BE THOSE THAT HAVE BEEN PARTIALLY WESTERNIZED, RATHER THAN SIGNIFICANTLY WESTERNIZED, THROUGH A COLONIAL RELATIONSHIP (41)	0.70	19 6 9 14 X SQ= 5.27 P = 0.018 RV YES (YES)
78	TEND MORE TO BE THOSE THAT HAVE BEEN SIGNIFICANTLY WESTERNIZED THROUGH A COLONIAL RELATIONSHIP, RATHER THAN WITHOUT SUCH A RELATIONSHIP (28)	0.90	78	TEND LESS TO BE THOSE THAT HAVE BEEN SIGNIFICANTLY WESTERNIZED THROUGH A COLONIAL RELATIONSHIP, RATHER THAN WITHOUT SUCH A RELATIONSHIP (28)	0.55	19 6 2 5 X SQ= 3.55 P = 0.032 RV NO (YES)
81	TEND LESS TO BE THOSE WHERE THE TYPE OF POLITICAL MODERNIZATION IS LATER EUROPEAN OR LATER EUROPEAN DERIVED, RATHER THAN EARLY EUROPEAN OR EARLY EUROPEAN DERIVED (40)	0.69	81	ALWAYS ARE THOSE WHERE THE TYPE OF POLITICAL MODERNIZATION IS LATER EUROPEAN OR LATER EUROPEAN DERIVED, RATHER THAN EARLY EUROPEAN OR EARLY EUROPEAN DERIVED (40)	1.00	10 0 22 15 X SQ= 4.23 P = 0.019 RV NO (NO)
82	LEAN TOWARD BEING THOSE WHERE THE TYPE OF POLITICAL MODERNIZATION IS EARLY OR LATER EUROPEAN OR EUROPEAN DERIVED, RATHER THAN DEVELOPED TUTELARY OR UNDEVELOPED TUTELARY (51)	0.65	82	LEAN TOWARD BEING THOSE WHERE THE TYPE OF POLITICAL MODERNIZATION IS DEVELOPED TUTELARY OR UNDEVELOPED TUTELARY, RATHER THAN EARLY OR LATER EUROPEAN OR EUROPEAN DERIVED (55)	0.56	32 15 17 19 X SQ= 2.86 P = 0.073 RV YES (YES)
83	LEAN MORE TOWARD BEING THOSE WHERE THE TYPE OF POLITICAL MODERNIZATION IS OTHER THAN NON-EUROPEAN AUTOCHTHONOUS (106)	0.96	83	LEAN LESS TOWARD BEING THOSE WHERE THE TYPE OF POLITICAL MODERNIZATION IS OTHER THAN NON-EUROPEAN AUTOCHTHONOUS (106)	0.83	2 7 49 34 X SQ= 3.09 P = 0.073 RV NO (YES)
85	TILT MORE TOWARD BEING THOSE WHERE THE STAGE OF POLITICAL MODERNIZATION IS ADVANCED, RATHER THAN MID- OR EARLY TRANSITIONAL (60)	0.69	85	TILT LESS TOWARD BEING THOSE WHERE THE STAGE OF POLITICAL MODERNIZATION IS ADVANCED, RATHER THAN MID- OR EARLY TRANSITIONAL (60)	0.52	35 21 16 19 X SQ= 1.83 P = 0.133 RV NO (YES)
89	TEND TO BE THOSE WHOSE IDEOLOGICAL ORIENTATION IS CONVENTIONAL (33)	0.80	89	ALWAYS ARE THOSE WHOSE IDEOLOGICAL ORIENTATION IS OTHER THAN CONVENTIONAL (62)	1.00	32 0 8 35 X SQ= 45.62 P = 0. RV YES (YES)
91	ALWAYS ARE THOSE WHOSE IDEOLOGICAL ORIENTATION IS DEVELOPMENTAL, RATHER THAN TRADITIONAL (31)	1.00	91	LEAN LESS TOWARD BEING THOSE WHOSE IDEOLOGICAL ORIENTATION IS DEVELOPMENTAL, RATHER THAN TRADITIONAL (31)	0.58	8 7 0 5 X SQ= 2.50 P = 0.055 RV NO (YES)
92	ALWAYS ARE THOSE WHERE THE SYSTEM STYLE IS LIMITED MOBILIZATIONAL, OR NON-MOBILIZATIONAL (93)	1.00	92	TEND TO BE THOSE WHERE THE SYSTEM STYLE IS MOBILIZATIONAL (20)	0.50	0 20 51 20 X SQ= 29.83 P = 0. RV YES (YES)

93	TEND TO BE THOSE WHERE THE SYSTEM STYLE IS NON-MOBILIZATIONAL (78)	0.88	93	TEND TO BE THOSE WHERE THE SYSTEM STYLE IS MOBILIZATIONAL, OR LIMITED MOBILIZATIONAL (32)	0.62	6 23 45 14 X SQ= 22.42 P = 0. RV YES (YES)
98	TILT LESS TOWARD BEING THOSE WHERE GOVERNMENTAL STABILITY IS GENERALLY OR MODERATELY PRESENT AND DATES FROM THE POST-WAR PERIOD, OR IS ABSENT (93)	0.71	98	TILT MORE TOWARD BEING THOSE WHERE GOVERNMENTAL STABILITY IS GENERALLY OR MODERATELY PRESENT AND DATES FROM THE POST-WAR PERIOD, OR IS ABSENT (93)	0.85	15 6 36 35 X SQ= 2.04 P = 0.134 RV NO (NO)
101	TEND TO BE THOSE WHERE THE REPRESENTATIVE CHARACTER OF THE REGIME IS POLYARCHIC (41)	0.87	101	ALWAYS ARE THOSE WHERE THE REPRESENTATIVE CHARACTER OF THE REGIME IS LIMITED POLYARCHIC, PSEUDO-POLYARCHIC, OR NON-POLYARCHIC (57)	1.00	41 0 6 40 X SQ= 62.54 P = 0. RV YES (YES)
102	TEND TO BE THOSE WHERE THE REPRESENTATIVE CHARACTER OF THE REGIME IS POLYARCHIC OR LIMITED POLYARCHIC (59)	0.98	102	TEND TO BE THOSE WHERE THE REPRESENTATIVE CHARACTER OF THE REGIME IS PSEUDO-POLYARCHIC OR NON-POLYARCHIC (49)	0.97	50 1 1 39 X SQ= 79.23 P = 0. RV YES (YES)
105	TEND TO BE THOSE WHERE THE ELECTORAL SYSTEM IS COMPETITIVE (43)	0.93	105	ALWAYS ARE THOSE WHERE THE ELECTORAL SYSTEM IS PARTIALLY COMPETITIVE OR NON-COMPETITIVE (47)	1.00	42 0 3 31 X SQ= 60.95 P = 0. RV YES (YES)
106	TEND TO BE THOSE WHERE THE ELECTORAL SYSTEM IS COMPETITIVE OR PARTIALLY COMPETITIVE (52)	0.95	106	TEND TO BE THOSE WHERE THE ELECTORAL SYSTEM IS NON-COMPETITIVE (30)	0.89	41 3 2 24 X SQ= 46.87 P = 0. RV YES (YES)
107	TEND TO BE THOSE WHERE AUTONOMOUS GROUPS ARE FULLY TOLERATED IN POLITICS (46)	0.86	107	TEND TO BE THOSE WHERE AUTONOMOUS GROUPS ARE PARTIALLY TOLERATED IN POLITICS, ARE TOLERATED ONLY OUTSIDE POLITICS, OR ARE NOT TOLERATED AT ALL (65)	0.97	44 1 7 39 X SQ= 59.63 P = 0. RV YES (YES)
108	TEND TO BE THOSE WHERE AUTONOMOUS GROUPS ARE FULLY OR PARTIALLY TOLERATED IN POLITICS (65)	0.98	108	TEND TO BE THOSE WHERE AUTONOMOUS GROUPS ARE TOLERATED ONLY OUTSIDE POLITICS OR ARE NOT TOLERATED AT ALL (35)	0.83	49 6 1 29 X SQ= 55.45 P = 0. RV YES (YES)
111	TEND TO BE THOSE WHERE POLITICAL ENCULTURATION IS HIGH OR MEDIUM (53)	0.77	111	TEND TO BE THOSE WHERE POLITICAL ENCULTURATION IS LOW (42)	0.63	37 10 11 17 X SQ= 10.20 P = 0.001 RV YES (YES)

115	TEND LESS TO BE THOSE WHERE INTEREST ARTICULATION BY ASSOCIATIONAL GROUPS IS MODERATE, LIMITED, OR NEGLIGIBLE (91)	0.63	115	ALWAYS ARE THOSE WHERE INTEREST ARTICULATION BY ASSOCIATIONAL GROUPS IS MODERATE, LIMITED, OR NEGLIGIBLE (91)	1.00	19 0 32 37 X SQ= 15.45 P = 0. RV NO (YES)
116	TEND TO BE THOSE WHERE INTEREST ARTICULATION BY ASSOCIATIONAL GROUPS IS SIGNIFICANT OR MODERATE (32)	0.59	116	ALWAYS ARE THOSE WHERE INTEREST ARTICULATION BY ASSOCIATIONAL GROUPS IS LIMITED OR NEGLIGIBLE (79)	1.00	30 0 21 37 X SQ= 30.46 P = 0. RV YES (YES)
117	TEND TO BE THOSE WHERE INTEREST ARTICULATION BY ASSOCIATIONAL GROUPS IS SIGNIFICANT, MODERATE, OR LIMITED (60)	0.86	117	TEND TO BE THOSE WHERE INTEREST ARTICULATION BY ASSOCIATIONAL GROUPS IS NEGLIGIBLE (51)	0.70	44 11 7 26 X SQ= 26.89 P = 0. RV YES (YES)
118	TEND TO BE THOSE WHERE INTEREST ARTICULATION BY INSTITUTIONAL GROUPS IS SIGNIFICANT, MODERATE, OR LIMITED (60)	0.92	118	TEND TO BE THOSE WHERE INTEREST ARTICULATION BY INSTITUTIONAL GROUPS IS VERY SIGNIFICANT (40)	0.82	4 33 46 7 X SQ= 47.91 P = 0. RV YES (YES)
119	TEND TO BE THOSE WHERE INTEREST ARTICULATION BY INSTITUTIONAL GROUPS IS MODERATE OR LIMITED (26)	0.50	119	TEND TO BE THOSE WHERE INTEREST ARTICULATION BY INSTITUTIONAL GROUPS IS VERY SIGNIFICANT OR SIGNIFICANT (74)	0.97	25 39 25 1 X SQ= 22.15 P = 0. RV YES (YES)
120	TEND LESS TO BE THOSE WHERE INTEREST ARTICULATION BY INSTITUTIONAL GROUPS IS VERY SIGNIFICANT, SIGNIFICANT, OR MODERATE (90)	0.80	120	ALWAYS ARE THOSE WHERE INTEREST ARTICULATION BY INSTITUTIONAL GROUPS IS VERY SIGNIFICANT, SIGNIFICANT, OR MODERATE (90)	1.00	40 40 10 0 X SQ= 7.09 P = 0.002 RV NO (YES)
121	TEND TO BE THOSE WHERE INTEREST ARTICULATION BY NON-ASSOCIATIONAL GROUPS IS MODERATE, LIMITED, OR NEGLIGIBLE (61)	0.78	121	TEND TO BE THOSE WHERE INTEREST ARTICULATION BY NON-ASSOCIATIONAL GROUPS IS SIGNIFICANT (54)	0.56	11 23 40 18 X SQ= 10.20 P = 0.001 RV YES (YES)
122	TEND LESS TO BE THOSE WHERE INTEREST ARTICULATION BY NON-ASSOCIATIONAL GROUPS IS SIGNIFICANT OR MODERATE (83)	0.53	122	TEND MORE TO BE THOSE WHERE INTEREST ARTICULATION BY NON-ASSOCIATIONAL GROUPS IS SIGNIFICANT OR MODERATE (83)	0.83	27 34 24 7 X SQ= 7.85 P = 0.004 RV NO (YES)
123	TEND LESS TO BE THOSE WHERE INTEREST ARTICULATION BY NON-ASSOCIATIONAL GROUPS IS SIGNIFICANT, MODERATE, OR LIMITED (107)	0.84	123	ALWAYS ARE THOSE WHERE INTEREST ARTICULATION BY NON-ASSOCIATIONAL GROUPS IS SIGNIFICANT, MODERATE, OR LIMITED (107)	1.00	43 41 8 0 X SQ= 5.21 P = 0.008 RV NO (NO)

125	TEND TO BE THOSE WHERE INTEREST ARTICULATION BY ANOMIC GROUPS IS INFREQUENT OR VERY INFREQUENT (35)	0.60	125	TEND TO BE THOSE WHERE INTEREST ARTICULATION BY ANOMIC GROUPS IS FREQUENT OR OCCASIONAL (64)	0.82	18 27 27 6 X SQ= 11.98 P = 0. RV YES (YES)
126	TEND LESS TO BE THOSE WHERE INTEREST ARTICULATION BY ANOMIC GROUPS IS FREQUENT, OCCASIONAL, OR INFREQUENT (83)	0.64	126	ALWAYS ARE THOSE WHERE INTEREST ARTICULATION BY ANOMIC GROUPS IS FREQUENT, OCCASIONAL, OR INFREQUENT (83)	1.00	29 33 16 0 X SQ= 12.66 P = 0. RV NO (YES)
127	TEND LESS TO BE THOSE WHERE INTEREST ARTICULATION BY POLITICAL PARTIES IS MODERATE, LIMITED, OR NEGLIGIBLE (72)	0.64	127	ALWAYS ARE THOSE WHERE INTEREST ARTICULATION BY POLITICAL PARTIES IS MODERATE, LIMITED, OR NEGLIGIBLE (72)	1.00	17 0 30 26 X SQ= 10.32 P = 0. RV NO (NO)
128	TEND TO BE THOSE WHERE INTEREST ARTICULATION BY POLITICAL PARTIES IS SIGNIFICANT OR MODERATE (48)	0.83	128	TEND TO BE THOSE WHERE INTEREST ARTICULATION BY POLITICAL PARTIES IS LIMITED OR NEGLIGIBLE (45)	0.92	39 2 8 24 X SQ= 35.54 P = 0. RV YES (YES)
129	TEND TO BE THOSE WHERE INTEREST ARTICULATION BY POLITICAL PARTIES IS SIGNIFICANT, MODERATE, OR LIMITED (56)	0.89	129	TEND TO BE THOSE WHERE INTEREST ARTICULATION BY POLITICAL PARTIES IS NEGLIGIBLE (37)	0.81	42 5 5 21 X SQ= 32.91 P = 0. RV YES (YES)
130	TEND LESS TO BE THOSE WHERE INTEREST AGGREGATION BY POLITICAL PARTIES IS MODERATE, LIMITED, OR NEGLIGIBLE (71)	0.78	130	ALWAYS ARE THOSE WHERE INTEREST AGGREGATION BY POLITICAL PARTIES IS MODERATE, LIMITED, OR NEGLIGIBLE (71)	1.00	10 0 35 30 X SQ= 5.89 P = 0.005 RV NO (NO)
131	TEND TO BE THOSE WHERE INTEREST AGGREGATION BY POLITICAL PARTIES IS SIGNIFICANT OR MODERATE (30)	0.56	131	TEND TO BE THOSE WHERE INTEREST AGGREGATION BY POLITICAL PARTIES IS LIMITED OR NEGLIGIBLE (35)	0.86	25 2 20 12 X SQ= 5.76 P = 0.012 RV YES (NO)
134	TEND TO BE THOSE WHERE INTEREST AGGREGATION BY THE EXECUTIVE IS SIGNIFICANT OR MODERATE (57)	0.77	134	TEND TO BE THOSE WHERE INTEREST AGGREGATION BY THE EXECUTIVE IS LIMITED OR NEGLIGIBLE (46)	0.82	37 7 11 31 X SQ= 26.91 P = 0. RV YES (YES)
137	TEND TO BE THOSE WHERE INTEREST AGGREGATION BY THE LEGISLATURE IS SIGNIFICANT OR MODERATE (29)	0.63	137	ALWAYS ARE THOSE WHERE INTEREST AGGREGATION BY THE LEGISLATURE IS LIMITED OR NEGLIGIBLE (68)	1.00	29 0 17 33 X SQ= 30.21 P = 0. RV YES (YES)

138	TEND TO BE THOSE WHERE INTEREST AGGREGATION BY THE LEGISLATURE IS SIGNIFICANT, MODERATE, OR LIMITED (48)	0.88	138	TEND TO BE THOSE WHERE INTEREST AGGREGATION BY THE LEGISLATURE IS NEGLIGIBLE (49)	0.93	43 2 6 28 X SQ= 46.66 P = 0. RV YES (YES)
139	TEND TO BE THOSE WHERE THE PARTY SYSTEM IS QUANTITATIVELY OTHER THAN ONE-PARTY (71)	0.94	139	TEND TO BE THOSE WHERE THE PARTY SYSTEM IS QUANTITATIVELY ONE-PARTY (34)	0.63	3 22 48 13 X SQ= 29.97 P = 0. RV YES (YES)
141	TEND LESS TO BE THOSE WHERE THE PARTY SYSTEM IS QUANTITATIVELY OTHER THAN TWO-PARTY (87)	0.77	141	ALWAYS ARE THOSE WHERE THE PARTY SYSTEM IS QUANTITATIVELY OTHER THAN TWO-PARTY (87)	1.00	11 0 37 32 X SQ= 6.68 P = 0.002 RV NO (NO)
142	TEND TO BE THOSE WHERE THE PARTY SYSTEM IS QUANTITATIVELY MULTI-PARTY (30)	0.51	142	TEND TO BE THOSE WHERE THE PARTY SYSTEM IS QUANTITATIVELY OTHER THAN MULTI-PARTY (66)	0.94	24 2 23 29 X SQ= 14.78 P = 0. RV YES (YES)
147	TEND TO BE THOSE WHERE THE PARTY SYSTEM IS QUALITATIVELY CLASS-ORIENTED OR MULTI-IDEOLOGICAL (23)	0.50	147	ALWAYS ARE THOSE WHERE THE PARTY SYSTEM IS QUALITATIVELY OTHER THAN CLASS-ORIENTED OR MULTI-IDEOLOGICAL (67)	1.00	21 0 21 30 X SQ= 18.83 P = 0. RV YES (YES)
160	TEND TO BE THOSE WHERE THE POLITICAL LEADERSHIP IS MODERATELY ELITIST OR NON-ELITIST (67)	0.96	160	TEND TO BE THOSE WHERE THE POLITICAL LEADERSHIP IS ELITIST (30)	0.84	2 26 45 5 X SQ= 48.05 P = 0. RV YES (YES)
161	TEND TO BE THOSE WHERE THE POLITICAL LEADERSHIP IS NON-ELITIST (50)	0.70	161	TEND TO BE THOSE WHERE THE POLITICAL LEADERSHIP IS ELITIST OR MODERATELY ELITIST (47)	0.94	14 29 33 2 X SQ= 28.18 P = 0. RV YES (YES)
163	TEND MORE TO BE THOSE WHERE THE REGIME'S LEADERSHIP CHARISMA IS MODERATE OR NEGLIGIBLE (87)	0.96	163	TEND LESS TO BE THOSE WHERE THE REGIME'S LEADERSHIP CHARISMA IS MODERATE OR NEGLIGIBLE (87)	0.75	2 9 46 27 X SQ= 6.12 P = 0.008 RV NO (YES)
164	TEND MORE TO BE THOSE WHERE THE REGIME'S LEADERSHIP CHARISMA IS NEGLIGIBLE (65)	0.85	164	TEND LESS TO BE THOSE WHERE THE REGIME'S LEADERSHIP CHARISMA IS NEGLIGIBLE (65)	0.54	7 16 41 19 X SQ= 8.30 P = 0.003 RV NO (YES)

166	TEND LESS TO BE THOSE WHERE THE VERTICAL POWER DISTRIBUTION IS THAT OF FORMAL FEDERALISM OR FORMAL AND EFFECTIVE UNITARISM (99)	0.76	166	TEND MORE TO BE THOSE WHERE THE VERTICAL POWER DISTRIBUTION IS THAT OF FORMAL FEDERALISM OR FORMAL AND EFFECTIVE UNITARISM (99)	0.98	12 1 39 40 X SQ= 6.68 P = 0.005 RV NO (YES)
168	TEND TO BE THOSE WHERE THE HORIZONTAL POWER DISTRIBUTION IS SIGNIFICANT (34)	0.66	168	ALWAYS ARE THOSE WHERE THE HORIZONTAL POWER DISTRIBUTION IS LIMITED OR NEGLIGIBLE (72)	1.00	33 0 17 39 X SQ= 38.13 P = 0. RV YES (YES)
169	TEND TO BE THOSE WHERE THE HORIZONTAL POWER DISTRIBUTION IS SIGNIFICANT OR LIMITED (58)	0.98	169	TEND TO BE THOSE WHERE THE HORIZONTAL POWER DISTRIBUTION IS NEGLIGIBLE (48)	0.95	49 2 1 37 X SQ= 73.49 P = 0. RV YES (YES)
171	TEND LESS TO BE THOSE WHERE THE LEGISLATIVE-EXECUTIVE STRUCTURE IS OTHER THAN PARLIAMENTARY-REPUBLICAN (90)	0.82	171	ALWAYS ARE THOSE WHERE THE LEGISLATIVE-EXECUTIVE STRUCTURE IS OTHER THAN PARLIAMENTARY-REPUBLICAN (90)	1.00	9 0 42 32 X SQ= 4.64 P = 0.011 RV NO (NO)
172	TEND LESS TO BE THOSE WHERE THE LEGISLATIVE-EXECUTIVE STRUCTURE IS OTHER THAN PARLIAMENTARY-ROYALIST (88)	0.63	172	TEND MORE TO BE THOSE WHERE THE LEGISLATIVE-EXECUTIVE STRUCTURE IS OTHER THAN PARLIAMENTARY-ROYALIST (88)	0.97	19 1 32 35 X SQ= 12.29 P = 0. RV NO (YES)
174	TEND TO BE THOSE WHERE THE LEGISLATURE IS FULLY EFFECTIVE (28)	0.54	174	ALWAYS ARE THOSE WHERE THE LEGISLATURE IS PARTIALLY EFFECTIVE, LARGELY INEFFECTIVE, OR WHOLLY INEFFECTIVE (72)	1.00	27 0 23 34 X SQ= 24.64 P = 0. RV YES (YES)
175	TEND TO BE THOSE WHERE THE LEGISLATURE IS FULLY EFFECTIVE OR PARTIALLY EFFECTIVE (51)	0.96	175	ALWAYS ARE THOSE WHERE THE LEGISLATURE IS LARGELY INEFFECTIVE OR WHOLLY INEFFECTIVE (49)	1.00	48 0 2 34 X SQ= 72.29 P = 0. RV YES (YES)
176	ALWAYS ARE THOSE WHERE THE LEGISLATURE IS FULLY EFFECTIVE, PARTIALLY EFFECTIVE, OR LARGELY INEFFECTIVE (72)	1.00	176	TEND TO BE THOSE WHERE THE LEGISLATURE IS WHOLLY INEFFECTIVE (28)	0.82	50 6 0 28 X SQ= 58.11 P = 0. RV YES (YES)
178	TEND TO BE THOSE WHERE THE LEGISLATURE IS BICAMERAL (51)	0.74	178	TEND TO BE THOSE WHERE THE LEGISLATURE IS UNICAMERAL (53)	0.68	13 23 37 11 X SQ= 12.68 P = 0. RV YES (YES)

179	TEND TO BE THOSE WHERE THE EXECUTIVE IS STRONG (39)	0.84	179	ALWAYS ARE THOSE WHERE THE EXECUTIVE IS DOMINANT (52)	1.00	7 36 38 0 X SQ= 53.92 P = 0. RV YES (YES)
181	TEND LESS TO BE THOSE WHERE THE BUREAUCRACY IS SEMI-MODERN, RATHER THAN MODERN (55)	0.57	181	ALWAYS ARE THOSE WHERE THE BUREAUCRACY IS SEMI-MODERN, RATHER THAN MODERN (55)	1.00	20 0 26 25 X SQ= 13.06 P = 0. RV NO (NO)
183	ALWAYS ARE THOSE WHERE THE BUREAUCRACY IS POST-COLONIAL TRANSITIONAL, RATHER THAN TRADITIONAL (25)	1.00	183	TEND TO BE THOSE WHERE THE BUREAUCRACY IS TRADITIONAL, RATHER THAN POST-COLONIAL TRANSITIONAL (9)	0.69	5 4 0 9 X SQ= 4.43 P = 0.029 RV YES (YES)
185	LEAN TOWARD BEING THOSE WHERE PARTICIPATION BY THE MILITARY IN POLITICS IS INTERVENTIVE, RATHER THAN SUPPORTIVE (21)	0.67	185	LEAN TOWARD BEING THOSE WHERE PARTICIPATION BY THE MILITARY IN POLITICS IS SUPPORTIVE, RATHER THAN INTERVENTIVE (31)	0.73	4 10 2 27 X SQ= 2.11 P = 0.077 RV YES (NO)
186	TEND TO BE THOSE WHERE PARTICIPATION BY THE MILITARY IN POLITICS IS NEUTRAL, RATHER THAN SUPPORTIVE (56)	0.96	186	TEND TO BE THOSE WHERE PARTICIPATION BY THE MILITARY IN POLITICS IS SUPPORTIVE, RATHER THAN NEUTRAL (31)	0.96	2 27 43 1 X SQ= 57.21 P = 0. RV YES (YES)
187	TEND TO BE THOSE WHERE THE ROLE OF THE POLICE IS NOT POLITICALLY SIGNIFICANT (35)	0.77	187	ALWAYS ARE THOSE WHERE THE ROLE OF THE POLICE IS POLITICALLY SIGNIFICANT (66)	1.00	10 41 34 0 X SQ= 49.63 P = 0. RV YES (YES)
188	TEND LESS TO BE THOSE WHERE THE CHARACTER OF THE LEGAL SYSTEM IS OTHER THAN CIVIL LAW (81)	0.57	188	TEND MORE TO BE THOSE WHERE THE CHARACTER OF THE LEGAL SYSTEM IS OTHER THAN CIVIL LAW (81)	0.82	22 7 29 32 X SQ= 5.32 P = 0.013 RV NO (YES)
189	TEND LESS TO BE THOSE WHERE THE CHARACTER OF THE LEGAL SYSTEM IS OTHER THAN COMMON LAW (108)	0.86	189	ALWAYS ARE THOSE WHERE THE CHARACTER OF THE LEGAL SYSTEM IS OTHER THAN COMMON LAW (108)	1.00	7 0 44 41 X SQ= 4.29 P = 0.016 RV NO (NO)
193	TILT MORE TOWARD BEING THOSE WHERE THE CHARACTER OF THE LEGAL SYSTEM IS OTHER THAN PARTLY INDIGENOUS (86)	0.92	193	TILT LESS TOWARD BEING THOSE WHERE THE CHARACTER OF THE LEGAL SYSTEM IS OTHER THAN PARTLY INDIGENOUS (86)	0.80	4 8 47 32 X SQ= 1.93 P = 0.121 RV NO (YES)

```
194  ALWAYS ARE THOSE                          1.00     194  TEND LESS TO BE THOSE                   0.67        0   13
     THAT ARE NON-COMMUNIST  (101)                           THAT ARE NON-COMMUNIST  (101)                      51   27
                                                                                                           X SQ=  16.77
                                                                                                           P =    0.
                                                                                                           RV NO  (YES)
```

95 POLITIES
WHERE THE STATUS OF THE REGIME
IS CONSTITUTIONAL OR
AUTHORITARIAN (95)

95 POLITIES
WHERE THE STATUS OF THE REGIME
IS TOTALITARIAN (16)

BOTH SUBJECT AND PREDICATE

95 IN LEFT COLUMN

AFGHANISTAN	ALGERIA	ARGENTINA	AUSTRALIA	AUSTRIA	BELGIUM	BOLIVIA	BRAZIL	BURMA	BURUNDI
CAMBODIA	CAMEROUN	CANADA	CEN AFR REP	CEYLON	CHAD	CHILE	COLOMBIA	CONGO(BRA)	CONGO(LEO)
COSTA RICA	CYPRUS	DAHOMEY	DENMARK	DOMIN REP	ECUADOR	EL SALVADOR	ETHIOPIA	FINLAND	FRANCE
GABON	GERMAN FR	GHANA	GREECE	GUATEMALA	GUINEA	HONDURAS	ICELAND	INDIA	INDONESIA
IRAN	IRAQ	IRELAND	ISRAEL	ITALY	IVORY COAST	JAMAICA	JAPAN	JORDAN	KOREA REP
LAOS	LEBANON	LIBERIA	LIBYA	LUXEMBOURG	MALAGASY R	MALAYA	MALI	MAURITANIA	MEXICO
MOROCCO	NEPAL	NETHERLANDS	NEW ZEALAND	NICARAGUA	NIGER	NIGERIA	NORWAY	PAKISTAN	PANAMA
PARAGUAY	PERU	PHILIPPINES	RWANDA	SA'U ARABIA	SENEGAL	SIERRE LEO	SOMALIA	SUDAN	SWEDEN
SWITZERLAND	SYRIA	TANGANYIKA	THAILAND	TOGO	TRINIDAD	TUNISIA	TURKEY	UGANDA	UAR
UK	US	UPPER VOLTA	URUGUAY	VENEZUELA					

16 IN RIGHT COLUMN

ALBANIA	BULGARIA	CHINA, PR	CUBA	CZECHOS'KIA	GERMANY, E	HUNGARY	KOREA, N	MONGOLIA	POLAND
PORTUGAL	RUMANIA	SPAIN	USSR	VIETNAM, N	YUGOSLAVIA				

3 EXCLUDED BECAUSE AMBIGUOUS

HAITI SO AFRICA VIETNAM REP

1 EXCLUDED BECAUSE UNASCERTAINABLE

YEMEN

---

5 ALWAYS ARE THOSE 1.00
LOCATED IN SCANDINAVIA OR WEST EUROPE,
RATHER THAN IN EAST EUROPE (18)

5 TEND TO BE THOSE 0.82
LOCATED IN EAST EUROPE, RATHER THAN
IN SCANDINAVIA OR WEST EUROPE (9)

16 2
0 9
X SQ= 16.13
P = 0.
RV YES (YES)

10 TEND LESS TO BE THOSE 0.66
LOCATED ELSEWHERE THAN IN NORTH AFRICA,
OR CENTRAL AND SOUTH AFRICA (82)

10 ALWAYS ARE THOSE 1.00
LOCATED ELSEWHERE THAN IN NORTH AFRICA,
OR CENTRAL AND SOUTH AFRICA (82)

32 0
63 16
X SQ= 6.02
P = 0.005
RV NO (NO)

13 TILT LESS TOWARD BEING THOSE 0.84
LOCATED ELSEWHERE THAN IN NORTH AFRICA OR
THE MIDDLE EAST (99)

13 ALWAYS ARE THOSE 1.00
LOCATED ELSEWHERE THAN IN NORTH AFRICA OR
THE MIDDLE EAST (99)

15 0
80 16
X SQ= 1.73
P = 0.122
RV NO (NO)

18	TEND TO BE THOSE LOCATED IN NORTH AFRICA OR CENTRAL AND SOUTH AFRICA, RATHER THAN IN EAST ASIA, SOUTH ASIA, OR SOUTHEAST ASIA (33)	0.71	18	ALWAYS ARE THOSE LOCATED IN EAST ASIA, SOUTH ASIA, OR SOUTHEAST ASIA, RATHER THAN IN NORTH AFRICA OR CENTRAL AND SOUTH AFRICA (18)	1.00	32 0 13 4 X SQ= 5.36 P = 0.011 RV YES (NO)
24	TEND TO BE THOSE WHOSE POPULATION IS SMALL (54)	0.53	24	TEND TO BE THOSE WHOSE POPULATION IS VERY LARGE, LARGE, OR MEDIUM (61)	0.87	45 14 50 2 X SQ= 7.32 P = 0.003 RV YES (NO)
26	TEND TO BE THOSE WHOSE POPULATION DENSITY IS LOW (67)	0.66	26	TEND TO BE THOSE WHOSE POPULATION DENSITY IS VERY HIGH, HIGH, OR MEDIUM (48)	0.87	32 14 63 2 X SQ= 14.20 P = 0. RV YES (NO)
28	TEND TO BE THOSE WHOSE POPULATION GROWTH RATE IS HIGH (62)	0.61	28	TEND TO BE THOSE WHOSE POPULATION GROWTH RATE IS LOW (48)	0.71	56 4 36 10 X SQ= 3.93 P = 0.040 RV YES (NO)
29	TILT LESS TOWARD BEING THOSE WHERE THE DEGREE OF URBANIZATION IS HIGH (56)	0.51	29	TILT MORE TOWARD BEING THOSE WHERE THE DEGREE OF URBANIZATION IS HIGH (56)	C.75	46 9 44 3 X SQ= 1.57 P = 0.137 RV NO (NO)
34	TEND TO BE THOSE WHOSE GROSS NATIONAL PRODUCT IS VERY LOW (53)	0.51	34	TEND TO BE THOSE WHOSE GROSS NATIONAL PRODUCT IS VERY HIGH, HIGH, MEDIUM, OR LOW (62)	0.81	47 13 48 3 X SQ= 4.36 P = 0.028 RV YES (NO)
36	TEND TO BE THOSE WHOSE PER CAPITA GROSS NATIONAL PRODUCT IS LOW OR VERY LOW (73)	0.67	36	TEND TO BE THOSE WHOSE PER CAPITA GROSS NATIONAL PRODUCT IS VERY HIGH, HIGH, OR MEDIUM (42)	0.62	31 10 64 6 X SQ= 4.04 P = 0.047 RV YES (NO)
39	TILT TOWARD BEING THOSE WHOSE INTERNATIONAL FINANCIAL STATUS IS LOW OR VERY LOW (76)	0.69	39	TILT TOWARD BEING THOSE WHOSE INTERNATIONAL FINANCIAL STATUS IS VERY HIGH, HIGH, OR MEDIUM (38)	0.53	29 8 66 7 X SQ= 2.08 P = 0.139 RV YES (NO)
43	LEAN TOWARD BEING THOSE WHOSE ECONOMIC DEVELOPMENTAL STATUS IS VERY UNDERDEVELOPED (57)	0.54	43	LEAN TOWARD BEING THOSE WHOSE ECONOMIC DEVELOPMENTAL STATUS IS DEVELOPED, INTERMEDIATE, OR UNDERDEVELOPED (55)	0.75	42 12 50 4 X SQ= 3.60 P = 0.055 RV YES (NO)

45	TEND TO BE THOSE WHERE THE LITERACY RATE IS BELOW FIFTY PERCENT (54)	0.55	45	ALWAYS ARE THOSE WHERE THE LITERACY RATE IS FIFTY PERCENT OR ABOVE (55)	1.00	42 12 51 0 X SQ= 10.69 P = 0. RV YES (NO)
46	TEND LESS TO BE THOSE WHERE THE LITERACY RATE IS TEN PERCENT OR ABOVE (84)	0.73	46	ALWAYS ARE THOSE WHERE THE LITERACY RATE IS TEN PERCENT OR ABOVE (84)	1.00	67 14 25 0 X SQ= 3.58 P = 0.021 RV NO (NO)
50	TEND TO BE THOSE WHERE FREEDOM OF THE PRESS IS COMPLETE (43)	0.54	50	ALWAYS ARE THOSE WHERE FREEDOM OF THE PRESS IS INTERMITTENT, INTERNALLY ABSENT, OR INTERNALLY AND EXTERNALLY ABSENT (56)	1.00	43 0 36 16 X SQ= 13.79 P = 0. RV YES (NO)
51	TEND TO BE THOSE WHERE FREEDOM OF THE PRESS IS COMPLETE OR INTERMITTENT (60)	0.76	51	ALWAYS ARE THOSE WHERE FREEDOM OF THE PRESS IS INTERNALLY ABSENT, OR INTERNALLY AND EXTERNALLY ABSENT (37)	1.00	59 0 19 16 X SQ= 29.35 P = 0. RV YES (NO)
52	TEND TO BE THOSE WHERE FREEDOM OF THE PRESS IS COMPLETE, INTERMITTENT, OR INTERNALLY ABSENT (82)	0.94	52	TEND TO BE THOSE WHERE FREEDOM OF THE PRESS IS INTERNALLY AND EXTERNALLY ABSENT (16)	0.69	74 5 5 11 X SQ= 32.69 P = 0. RV YES (YES)
54	TEND TO BE THOSE WHERE NEWSPAPER CIRCULATION IS LESS THAN ONE HUNDRED PER THOUSAND (76)	0.71	54	TEND TO BE THOSE WHERE NEWSPAPER CIRCULATION IS ONE HUNDRED OR MORE PER THOUSAND (37)	0.64	28 9 67 5 X SQ= 5.13 P = 0.015 RV YES (NO)
55	TEND LESS TO BE THOSE WHERE NEWSPAPER CIRCULATION IS TEN OR MORE PER THOUSAND (78)	0.64	55	ALWAYS ARE THOSE WHERE NEWSPAPER CIRCULATION IS TEN OR MORE PER THOUSAND (78)	1.00	61 14 34 0 X SQ= 5.71 P = 0.005 RV NO (NO)
56	TEND LESS TO BE THOSE WHERE THE RELIGION IS PREDOMINANTLY LITERATE (79)	0.69	56	ALWAYS ARE THOSE WHERE THE RELIGION IS PREDOMINANTLY LITERATE (79)	1.00	65 13 29 0 X SQ= 4.05 P = 0.018 RV NO (NO)
58	LEAN LESS TOWARD BEING THOSE WHERE THE RELIGION IS OTHER THAN MUSLIM (97)	0.82	58	ALWAYS ARE THOSE WHERE THE RELIGION IS OTHER THAN MUSLIM (97)	1.00	17 0 78 16 X SQ= 2.14 P = 0.073 RV NO (NO)

63	TEND TO BE THOSE WHERE THE RELIGION IS PREDOMINANTLY OR PARTLY OTHER THAN CHRISTIAN (68)	0.62	63	TEND TO BE THOSE WHERE THE RELIGION IS PREDOMINANTLY SOME KIND OF CHRISTIAN (46)	0.67	36 10 59 5 X SQ= 3.30 P = 0.049 RV YES (NO)
64	LEAN LESS TOWARD BEING THOSE WHERE THE RELIGION IS CHRISTIAN, RATHER THAN MUSLIM (46)	0.68	64	ALWAYS ARE THOSE WHERE THE RELIGION IS CHRISTIAN, RATHER THAN MUSLIM (46)	1.00	36 10 17 0 X SQ= 2.92 P = 0.050 RV NO (NO)
67	LEAN LESS TOWARD BEING THOSE THAT ARE RACIALLY HOMOGENEOUS (82)	0.71	67	LEAN MORE TOWARD BEING THOSE THAT ARE RACIALLY HOMOGENEOUS (82)	0.94	64 15 26 1 X SQ= 2.57 P = 0.065 RV NO (NO)
69	LEAN LESS TOWARD BEING THOSE THAT ARE LINGUISTICALLY HOMOGENEOUS OR WEAKLY HETEROGENEOUS (64)	0.52	69	LEAN MORE TOWARD BEING THOSE THAT ARE LINGUISTICALLY HOMOGENEOUS OR WEAKLY HETEROGENEOUS (64)	0.80	49 12 46 3 X SQ= 3.16 P = 0.051 RV NO (NO)
76	TEND TO BE THOSE THAT HAVE BEEN SIGNIFICANTLY OR PARTIALLY WESTERNIZED THROUGH A COLONIAL RELATIONSHIP, RATHER THAN BEING HISTORICALLY WESTERN (70)	0.76	76	TEND TO BE THOSE THAT ARE HISTORICALLY WESTERN, RATHER THAN HAVING BEEN SIGNIFICANTLY OR PARTIALLY WESTERNIZED THROUGH A COLONIAL RELATIONSHIP (26)	0.75	20 6 65 2 X SQ= 7.23 P = 0.005 RV YES (NO)
78	TEND TO BE THOSE THAT HAVE BEEN SIGNIFICANTLY WESTERNIZED THROUGH A COLONIAL RELATIONSHIP, RATHER THAN WITHOUT SUCH A RELATIONSHIP (28)	0.93	78	TEND TO BE THOSE THAT HAVE BEEN SIGNIFICANTLY WESTERNIZED WITHOUT A COLONIAL RELATIONSHIP, RATHER THAN THROUGH SUCH A RELATIONSHIP (7)	0.83	27 1 2 5 X SQ= 13.69 P = 0. RV YES (YES)
81	LEAN LESS TOWARD BEING THOSE WHERE THE TYPE OF POLITICAL MODERNIZATION IS LATER EUROPEAN OR LATER EUROPEAN DERIVED, RATHER THAN EARLY EUROPEAN OR EARLY EUROPEAN DERIVED (40)	0.74	81	ALWAYS ARE THOSE WHERE THE TYPE OF POLITICAL MODERNIZATION IS LATER EUROPEAN OR LATER EUROPEAN DERIVED, RATHER THAN EARLY EUROPEAN OR EARLY EUROPEAN DERIVED (40)	1.00	10 0 28 11 X SQ= 2.20 P = 0.090 RV NO (NO)
82	TEND TO BE THOSE WHERE THE TYPE OF POLITICAL MODERNIZATION IS DEVELOPED TUTELARY OR UNDEVELOPED TUTELARY, RATHER THAN EARLY OR LATER EUROPEAN OR EUROPEAN DERIVED (55)	0.57	82	TEND TO BE THOSE WHERE THE TYPE OF POLITICAL MODERNIZATION IS EARLY OR LATER EUROPEAN OR EUROPEAN DERIVED, RATHER THAN DEVELOPED TUTELARY OR UNDEVELOPED TUTELARY (51)	0.79	38 11 50 3 X SQ= 4.73 P = 0.020 RV YES (NO)
85	TEND TO BE THOSE WHERE THE STAGE OF POLITICAL MODERNIZATION IS MID- OR EARLY TRANSITIONAL, RATHER THAN ADVANCED (54)	0.52	85	TEND TO BE THOSE WHERE THE STAGE OF POLITICAL MODERNIZATION IS ADVANCED, RATHER THAN MID- OR EARLY TRANSITIONAL (60)	0.94	45 15 49 1 X SQ= 9.83 P = 0.001 RV YES (NO)

No.	Statement	Value	No.	Statement	Value	Statistics
87	TEND LESS TO BE THOSE WHOSE IDEOLOGICAL ORIENTATION IS OTHER THAN DEVELOPMENTAL (58)	0.57	87	ALWAYS ARE THOSE WHOSE IDEOLOGICAL ORIENTATION IS OTHER THAN DEVELOPMENTAL (58)	1.00	31 0 41 15 X SQ= 8.24 P = 0.001 RV NO (NO)
89	TEND LESS TO BE THOSE WHOSE IDEOLOGICAL ORIENTATION IS OTHER THAN CONVENTIONAL (62)	0.58	89	ALWAYS ARE THOSE WHOSE IDEOLOGICAL ORIENTATION IS OTHER THAN CONVENTIONAL (62)	1.00	32 0 45 14 X SQ= 7.24 P = 0.002 RV NO (NO)
92	TEND TO BE THOSE WHERE THE SYSTEM STYLE IS LIMITED MOBILIZATIONAL, OR NON-MOBILIZATIONAL (93)	0.95	92	TEND TO BE THOSE WHERE THE SYSTEM STYLE IS MOBILIZATIONAL (20)	0.93	5 14 90 1 X SQ= 64.29 P = 0. RV YES (YES)
93	TEND TO BE THOSE WHERE THE SYSTEM STYLE IS NON-MOBILIZATIONAL (78)	0.82	93	TEND TO BE THOSE WHERE THE SYSTEM STYLE IS MOBILIZATIONAL, OR LIMITED MOBILIZATIONAL (32)	0.93	17 14 75 1 X SQ= 31.57 P = 0. RV YES (NO)
99	TEND TO BE THOSE WHERE GOVERNMENTAL STABILITY IS MODERATELY PRESENT AND DATES FROM THE POST-WAR PERIOD, OR IS ABSENT (36)	0.50	99	ALWAYS ARE THOSE WHERE GOVERNMENTAL STABILITY IS GENERALLY PRESENT AND DATES FROM AT LEAST THE INTER-WAR PERIOD, OR FROM THE POST-WAR PERIOD (50)	1.00	35 14 35 0 X SQ= 10.03 P = 0. RV YES (NO)
101	TEND TO BE THOSE WHERE THE REPRESENTATIVE CHARACTER OF THE REGIME IS POLYARCHIC (41)	0.52	101	ALWAYS ARE THOSE WHERE THE REPRESENTATIVE CHARACTER OF THE REGIME IS LIMITED POLYARCHIC, PSEUDO-POLYARCHIC, OR NON-POLYARCHIC (57)	1.00	41 0 38 16 X SQ= 12.57 P = 0. RV YES (NO)
102	TEND TO BE THOSE WHERE THE REPRESENTATIVE CHARACTER OF THE REGIME IS POLYARCHIC OR LIMITED POLYARCHIC (59)	0.65	102	ALWAYS ARE THOSE WHERE THE REPRESENTATIVE CHARACTER OF THE REGIME IS PSEUDO-POLYARCHIC OR NON-POLYARCHIC (49)	1.00	58 0 31 16 X SQ= 20.73 P = 0. RV YES (NO)
105	TEND TO BE THOSE WHERE THE ELECTORAL SYSTEM IS COMPETITIVE (43)	0.59	105	ALWAYS ARE THOSE WHERE THE ELECTORAL SYSTEM IS PARTIALLY COMPETITIVE OR NON-COMPETITIVE (47)	1.00	43 0 30 15 X SQ= 15.00 P = 0. RV YES (NO)
106	TEND TO BE THOSE WHERE THE ELECTORAL SYSTEM IS COMPETITIVE OR PARTIALLY COMPETITIVE (52)	0.80	106	TEND TO BE THOSE WHERE THE ELECTORAL SYSTEM IS NON-COMPETITIVE (30)	0.94	51 1 13 15 X SQ= 27.20 P = 0. RV YES (YES)

108	TEND TO BE THOSE WHERE AUTONOMOUS GROUPS ARE FULLY OR PARTIALLY TOLERATED IN POLITICS (65)	0.78	108	TEND TO BE THOSE WHERE AUTONOMOUS GROUPS ARE TOLERATED ONLY OUTSIDE POLITICS OR ARE NOT TOLERATED AT ALL (35)	0.94	64 1 18 15 X SQ= 27.77 P = 0. RV YES (NO)
115	LEAN LESS TOWARD BEING THOSE WHERE INTEREST ARTICULATION BY ASSOCIATIONAL GROUPS IS MODERATE, LIMITED, OR NEGLIGIBLE (91)	0.79	115	ALWAYS ARE THOSE WHERE INTEREST ARTICULATION BY ASSOCIATIONAL GROUPS IS MODERATE, LIMITED, OR NEGLIGIBLE (91)	1.00	20 0 74 13 X SQ= 2.15 P = 0.071 RV NO (NO)
116	TEND LESS TO BE THOSE WHERE INTEREST ARTICULATION BY ASSOCIATIONAL GROUPS IS LIMITED OR NEGLIGIBLE (79)	0.67	116	ALWAYS ARE THOSE WHERE INTEREST ARTICULATION BY ASSOCIATIONAL GROUPS IS LIMITED OR NEGLIGIBLE (79)	1.00	31 0 63 13 X SQ= 4.54 P = 0.010 RV NO (NO)
117	LEAN TOWARD BEING THOSE WHERE INTEREST ARTICULATION BY ASSOCIATIONAL GROUPS IS SIGNIFICANT, MODERATE, OR LIMITED (60)	0.59	117	LEAN TOWARD BEING THOSE WHERE INTEREST ARTICULATION BY ASSOCIATIONAL GROUPS IS NEGLIGIBLE (51)	0.69	55 4 39 9 X SQ= 2.52 P = 0.077 RV YES (NO)
118	TEND TO BE THOSE WHERE INTEREST ARTICULATION BY INSTITUTIONAL GROUPS IS SIGNIFICANT, MODERATE, OR LIMITED (60)	0.70	118	TEND TO BE THOSE WHERE INTEREST ARTICULATION BY INSTITUTIONAL GROUPS IS VERY SIGNIFICANT (40)	0.94	24 15 57 1 X SQ= 20.26 P = 0. RV YES (NO)
121	TILT LESS TOWARD BEING THOSE WHERE INTEREST ARTICULATION BY NON-ASSOCIATIONAL GROUPS IS MODERATE, LIMITED, OR NEGLIGIBLE (61)	0.52	121	TILT MORE TOWARD BEING THOSE WHERE INTEREST ARTICULATION BY NON-ASSOCIATIONAL GROUPS IS MODERATE, LIMITED, OR NEGLIGIBLE (61)	0.75	46 4 49 12 X SQ= 2.16 P = 0.106 RV NO (NO)
127	TEND LESS TO BE THOSE WHERE INTEREST ARTICULATION BY POLITICAL PARTIES IS MODERATE, LIMITED, OR NEGLIGIBLE (72)	0.72	127	ALWAYS ARE THOSE WHERE INTEREST ARTICULATION BY POLITICAL PARTIES IS MODERATE, LIMITED, OR NEGLIGIBLE (72)	1.00	21 0 55 16 X SQ= 4.27 P = 0.018 RV NO (NO)
128	TEND TO BE THOSE WHERE INTEREST ARTICULATION BY POLITICAL PARTIES IS SIGNIFICANT OR MODERATE (48)	0.63	128	ALWAYS ARE THOSE WHERE INTEREST ARTICULATION BY POLITICAL PARTIES IS LIMITED OR NEGLIGIBLE (45)	1.00	48 0 28 16 X SQ= 18.67 P = 0. RV YES (NO)
129	TEND TO BE THOSE WHERE INTEREST ARTICULATION BY POLITICAL PARTIES IS SIGNIFICANT, MODERATE, OR LIMITED (56)	0.72	129	TEND TO BE THOSE WHERE INTEREST ARTICULATION BY POLITICAL PARTIES IS NEGLIGIBLE (37)	0.94	55 1 21 15 X SQ= 21.56 P = 0. RV YES (NO)

130	TILT LESS TOWARD BEING THOSE WHERE INTEREST AGGREGATION BY POLITICAL PARTIES IS MODERATE, LIMITED, OR NEGLIGIBLE (71)	0.82	130	ALWAYS ARE THOSE WHERE INTFREST AGGREGATION BY POLITICAL PARTIES IS MODERATE, LIMITED, OR NEGLIGIBLE (71)	1.00	12 0 54 16 X SQ= 2.11 P = 0.111 RV NO (NO)
133	TEND LESS TO BE THOSE WHERE INTEREST AGGREGATION BY THE EXECUTIVE IS MODERATE, LIMITED, OR NEGLIGIBLE (76)	0.67	133	ALWAYS ARE THOSE WHERE INTEREST AGGREGATION BY THE EXECUTIVE IS MODERATE, LIMITED, OR NEGLIGIBLE (76)	1.00	29 0 58 16 X SQ= 5.87 P = 0.005 RV NO (NO)
134	TEND TO BE THOSE WHERE INTEREST AGGREGATION BY THE EXECUTIVE IS SIGNIFICANT OR MODERATE (57)	0.66	134	TEND TO BE THOSE WHERE INTEREST AGGREGATION BY THE EXECUTIVE IS LIMITED OR NEGLIGIBLE (46)	0.94	56 1 29 15 X SQ= 17.13 P = 0. RV YES (NO)
137	TEND LESS TO BE THOSE WHERE INTEREST AGGREGATION BY THE LEGISLATURE IS LIMITED OR NEGLIGIBLE (68)	0.63	137	ALWAYS ARE THOSE WHERE INTEREST AGGREGATION BY THE LEGISLATURE IS LIMITED OR NEGLIGIBLE (68)	1.00	29 0 50 16 X SQ= 6.81 P = 0.002 RV NO (NO)
138	TEND TO BE THOSE WHERE INTEREST AGGREGATION BY THE LEGISLATURE IS SIGNIFICANT, MODERATE, OR LIMITED (48)	0.58	138	TEND TO BE THOSE WHERE INTEREST AGGREGATION BY THE LEGISLATURE IS NEGLIGIBLE (49)	0.93	47 1 34 13 X SQ= 10.41 P = 0. RV YES (NO)
139	TEND TO BE THOSE WHERE THE PARTY SYSTEM IS QUANTITATIVELY OTHER THAN ONE-PARTY (71)	0.78	139	TEND TO BE THOSE WHERE THE PARTY SYSTEM IS QUANTITATIVELY ONE-PARTY (34)	0.94	19 15 67 1 X SQ= 28.03 P = 0. RV YES (NO)
140	TILT LESS TOWARD BEING THOSE WHERE THE PARTY SYSTEM IS QUANTITATIVELY OTHER THAN ONE PARTY DOMINANT (87)	0.85	140	ALWAYS ARE THOSE WHERE THE PARTY SYSTEM IS QUANTITATIVELY OTHER THAN ONE PARTY DOMINANT (87)	1.00	12 0 70 16 X SQ= 1.48 P = 0.126 RV NO (NO)
142	TEND LESS TO BE THOSE WHERE THE PARTY SYSTEM IS QUANTITATIVELY OTHER THAN MULTI-PARTY (66)	0.62	142	ALWAYS ARE THOSE WHERE THE PARTY SYSTEM IS QUANTITATIVELY OTHER THAN MULTI-PARTY (66)	1.00	30 0 48 16 X SQ= 7.36 P = 0.002 RV NO (NO)
144	TEND LESS TO BE THOSE WHERE THE PARTY SYSTEM IS QUANTITATIVELY ONE-PARTY, RATHER THAN TWO-PARTY (34)	0.62	144	ALWAYS ARE THOSE WHERE THE PARTY SYSTEM IS QUANTITATIVELY ONE-PARTY, RATHER THAN TWO-PARTY (34)	1.00	18 16 11 0 X SQ= 6.11 P = 0.004 RV NO (NO)

```
147  TEND LESS TO BE THOSE                     0.68     147  ALWAYS ARE THOSE                        1.00       23    0
     WHERE THE PARTY SYSTEM IS QUALITATIVELY              WHERE THE PARTY SYSTEM IS QUALITATIVELY             50   16
     OTHER THAN                                           OTHER THAN                                       X SQ=   5.25
     CLASS-ORIENTED OR MULTI-IDEOLOGICAL  (67)            CLASS-ORIENTED OR MULTI-IDEOLOGICAL  (67)        P =    0.009
                                                                                                           RV NO  (NO )

148  TILT LESS TOWARD BEING THOSE              0.85     148  ALWAYS ARE THOSE                        1.00       14    0
     WHERE THE PARTY SYSTEM IS QUALITATIVELY              WHERE THE PARTY SYSTEM IS QUALITATIVELY             77   16
     OTHER THAN                                           OTHER THAN                                       X SQ=   1.64
     AFRICAN TRANSITIONAL  (96)                           AFRICAN TRANSITIONAL  (96)                       P =    0.122
                                                                                                           RV NO  (NO )

153  TEND TO BE THOSE                          0.59     153  ALWAYS ARE THOSE                        1.00       26   15
     WHERE THE PARTY SYSTEM IS                            WHERE THE PARTY SYSTEM IS                           37    0
     MODERATELY STABLE OR UNSTABLE  (38)                  STABLE  (42)                                     X SQ=  14.49
                                                                                                           P =    0.
                                                                                                           RV YES (NO )

158  TEND TO BE THOSE                          0.50     158  ALWAYS ARE THOSE                        1.00       39    0
     WHERE PERSONALISMO IS                                WHERE PERSONALISMO IS                               39   16
     PRONOUNCED OR MODERATE  (40)                         NEGLIGIBLE  (56)                                 X SQ=  11.69
                                                                                                           P =    0.
                                                                                                           RV YES (NO )

160  TEND TO BE THOSE                          0.82     160  ALWAYS ARE THOSE                        1.00       14   15
     WHERE THE POLITICAL LEADERSHIP IS                    WHERE THE POLITICAL LEADERSHIP IS                   66    0
     MODERATELY ELITIST OR NON-ELITIST  (67)              ELITIST  (30)                                    X SQ=  36.74
                                                                                                           P =    0.
                                                                                                           RV YES (YES)

161  TEND TO BE THOSE                          0.63     161  ALWAYS ARE THOSE                        1.00       30   15
     WHERE THE POLITICAL LEADERSHIP IS                    WHERE THE POLITICAL LEADERSHIP IS                   50    0
     NON-ELITIST  (50)                                    ELITIST OR MODERATELY ELITIST  (47)              X SQ=  17.36
                                                                                                           P =    0.
                                                                                                           RV YES (NO )

166  TILT LESS TOWARD BEING THOSE              0.84     166  ALWAYS ARE THOSE                        1.00       15    0
     WHERE THE VERTICAL POWER DISTRIBUTION                WHERE THE VERTICAL POWER DISTRIBUTION               79   16
     IS THAT OF FORMAL FEDERALISM OR                      IS THAT OF FORMAL FEDERALISM OR                  X SQ=   1.76
     FORMAL AND EFFECTIVE UNITARISM  (99)                 FORMAL AND EFFECTIVE UNITARISM  (99)             P =    0.121
                                                                                                           RV NO  (NO )

169  TEND TO BE THOSE                          0.66     169  ALWAYS ARE THOSE                        1.00       57    0
     WHERE THE HORIZONTAL POWER DISTRIBUTION              WHERE THE HORIZONTAL POWER DISTRIBUTION             30   16
     IS SIGNIFICANT OR LIMITED  (58)                      IS NEGLIGIBLE  (48)                              X SQ=  20.90
                                                                                                           P =    0.
                                                                                                           RV YES (NO )

170  TEND LESS TO BE THOSE                     0.56     170  ALWAYS ARE THOSE                        1.00       37    0
     WHERE THE LEGISLATIVE-EXECUTIVE STRUCTURE            WHERE THE LEGISLATIVE-EXECUTIVE STRUCTURE           47   15
     IS OTHER THAN PRESIDENTIAL  (63)                     IS OTHER THAN PRESIDENTIAL  (63)                 X SQ=   8.75
                                                                                                           P =    0.001
                                                                                                           RV NO  (NO )
```

172	TEND LESS TO BE THOSE WHERE THE LEGISLATIVE-EXECUTIVE STRUCTURE IS OTHER THAN PARLIAMENTARY-ROYALIST (88)	0.77	172	ALWAYS ARE THOSE WHERE THE LEGISLATIVE-EXECUTIVE STRUCTURE IS OTHER THAN PARLIAMENTARY-ROYALIST (88)	1.00	21 0 70 15 X SQ= 2.99 P = 0.038 RV NO (NO)
175	TEND TO BE THOSE WHERE THE LEGISLATURE IS FULLY EFFECTIVE OR PARTIALLY EFFECTIVE (51)	0.61	175	ALWAYS ARE THOSE WHERE THE LEGISLATURE IS LARGELY INEFFECTIVE OR WHOLLY INEFFECTIVE (49)	1.00	50 0 32 15 X SQ= 16.51 P = 0. RV YES (NO)
176	TEND TO BE THOSE WHERE THE LEGISLATURE IS FULLY EFFECTIVE, PARTIALLY EFFECTIVE, OR LARGELY INEFFECTIVE (72)	0.87	176	ALWAYS ARE THOSE WHERE THE LEGISLATURE IS WHOLLY INEFFECTIVE (28)	1.00	71 0 11 15 X SQ= 44.14 P = 0. RV YES (YES)
178	TEND TO BE THOSE WHERE THE LEGISLATURE IS BICAMERAL (51)	0.55	178	TEND TO BE THOSE WHERE THE LEGISLATURE IS UNICAMERAL (53)	0.80	39 12 47 3 X SQ= 4.83 P = 0.023 RV YES (NO)
179	TEND TO BE THOSE WHERE THE EXECUTIVE IS STRONG (39)	0.54	179	ALWAYS ARE THOSE WHERE THE EXECUTIVE IS DOMINANT (52)	1.00	33 16 39 0 X SQ= 13.45 P = 0. RV YES (NO)
181	TEND LESS TO BE THOSE WHERE THE BUREAUCRACY IS SEMI-MODERN, RATHER THAN MODERN (55)	0.66	181	ALWAYS ARE THOSE WHERE THE BUREAUCRACY IS SEMI-MODERN, RATHER THAN MODERN (55)	1.00	20 0 38 16 X SQ= 5.91 P = 0.004 RV NO (NO)
182	TEND LESS TO BE THOSE WHERE THE BUREAUCRACY IS SEMI-MODERN, RATHER THAN POST-COLONIAL TRANSITIONAL (55)	0.60	182	ALWAYS ARE THOSE WHERE THE BUREAUCRACY IS SEMI-MODERN, RATHER THAN POST-COLONIAL TRANSITIONAL (55)	1.00	38 16 25 0 X SQ= 7.54 P = 0.002 RV NO (NO)
185	TEND TO BE THOSE WHERE PARTICIPATION BY THE MILITARY IN POLITICS IS INTERVENTIVE, RATHER THAN SUPPORTIVE (21)	0.58	185	ALWAYS ARE THOSE WHERE PARTICIPATION BY THE MILITARY IN POLITICS IS SUPPORTIVE, RATHER THAN INTERVENTIVE (31)	1.00	19 0 14 15 X SQ= 11.99 P = 0. RV YES (YES)
186	TEND TO BE THOSE WHERE PARTICIPATION BY THE MILITARY IN POLITICS IS NEUTRAL, RATHER THAN SUPPORTIVE (56)	0.80	186	ALWAYS ARE THOSE WHERE PARTICIPATION BY THE MILITARY IN POLITICS IS SUPPORTIVE, RATHER THAN NEUTRAL (31)	1.00	14 15 56 0 X SQ= 31.70 P = 0. RV YES (YES)

187	TEND LESS TO BE THOSE WHERE THE ROLE OF THE POLICE IS POLITICALLY SIGNIFICANT (66)	0.57	187	ALWAYS ARE THOSE WHERE THE ROLE OF THE POLICE IS POLITICALLY SIGNIFICANT (66)	1.00	46 16 35 0 X SQ= 9.02 P = 0. RV NO (NO)
188	TILT LESS TOWARD BEING THOSE WHERE THE CHARACTER OF THE LEGAL SYSTEM IS OTHER THAN CIVIL LAW (81)	0.69	188	TILT MORE TOWARD BEING THOSE WHERE THE CHARACTER OF THE LEGAL SYSTEM IS OTHER THAN CIVIL LAW (81)	0.88	29 2 64 14 X SQ= 1.51 P = 0.148 RV NO (NO)
192	TEND LESS TO BE THOSE WHERE THE CHARACTER OF THE LEGAL SYSTEM IS OTHER THAN MUSLIM OR PARTLY MUSLIM (86)	0.72	192	ALWAYS ARE THOSE WHERE THE CHARACTER OF THE LEGAL SYSTEM IS OTHER THAN MUSLIM OR PARTLY MUSLIM (86)	1.00	26 0 67 16 X SQ= 4.44 P = 0.011 RV NO (NO)
193	TEND LESS TO BE THOSE WHERE THE CHARACTER OF THE LEGAL SYSTEM IS OTHER THAN PARTLY INDIGENOUS (86)	0.72	193	ALWAYS ARE THOSE WHERE THE CHARACTER OF THE LEGAL SYSTEM IS OTHER THAN PARTLY INDIGENOUS (86)	1.00	26 0 68 16 X SQ= 4.36 P = 0.012 RV NO (NO)
194	ALWAYS ARE THOSE THAT ARE NON-COMMUNIST (101)	1.00	194	TEND TO BE THOSE THAT ARE COMMUNIST (13)	0.87	0 13 95 2 X SQ= 85.24 P = 0. RV YES (YES)

96 POLITIES
WHERE THE STATUS OF THE REGIME
IS CONSTITUTIONAL OR
TOTALITARIAN (67)

96 POLITIES
WHERE THE STATUS OF THE REGIME
IS AUTHORITARIAN (23)

BOTH SUBJECT AND PREDICATE

67 IN LEFT COLUMN

ALBANIA AUSTRALIA AUSTRIA BELGIUM BOLIVIA BRAZIL BULGARIA CANADA CEYLON CHILE
CHINA, PR COLOMBIA COSTA RICA CUBA CYPRUS CZECHOS'KIA DENMARK DOMIN REP ECUADOR FINLAND
FRANCE GERMANY, E GERMAN FR GREECE HONDURAS HUNGARY ICELAND INDIA IRELAND ISRAEL
ITALY JAMAICA JAPAN KOREA, N LEBANON LIBYA LUXEMBOURG MALAGASY R MALAYA MAURITANIA
MEXICO MONGOLIA MOROCCO NETHERLANDS NEW ZEALAND NIGERIA NORWAY PANAMA PHILIPPINES POLAND
PORTUGAL RUMANIA SIERRE LEO SPAIN SWEDEN SWITZERLAND TRINIDAD TUNISIA TURKEY UGANDA
USSR UK US URUGUAY VENEZUELA VIETNAM, N YUGOSLAVIA

23 IN RIGHT COLUMN

AFGHANISTAN ALGERIA BURMA CAMBODIA CEN AFR REP EL SALVADOR ETHIOPIA GHANA GUINEA INDONESIA
IRAN JORDAN KOREA REP LAOS LIBERIA NEPAL NICARAGUA PAKISTAN PARAGUAY SA'U ARABIA
SUDAN THAILAND UAR

5 EXCLUDED BECAUSE AMBIGUOUS

ARGENTINA HAITI IVORY COAST SO AFRICA VIETNAM REP

11 EXCLUDED BECAUSE UNASCERTAINED

BURUNDI CHAD CONGO(BRA) DAHOMEY GABON MALI NIGER RWANDA SOMALIA TANGANYIKA
UPPER VOLTA

9 EXCLUDED BECAUSE UNASCERTAINABLE

CAMEROUN CONGO(LEO) GUATEMALA IRAQ PERU SENEGAL SYRIA TOGO YEMEN

---

2 TEND LESS TO BE THOSE LOCATED ELSEWHERE THAN IN WEST EUROPE, SCANDINAVIA, NORTH AMERICA, OR AUSTRALASIA (93) 0.67

2 ALWAYS ARE THOSE LOCATED ELSEWHERE THAN IN WEST EUROPE, SCANDINAVIA, NORTH AMERICA, OR AUSTRALASIA (93) 1.00

22 0
45 23
X SQ= 8.30
P = 0.001
RV NO (NO)

9 TEND MORE TO BE THOSE LOCATED ELSEWHERE THAN IN SOUTHEAST ASIA (106) 0.96

9 TEND LESS TO BE THOSE LOCATED ELSEWHERE THAN IN SOUTHEAST ASIA (106) 0.78

3 5
64 18
X SQ= 4.35
P = 0.024
RV NO (YES)

10	TEND MORE TO BE THOSE LOCATED ELSEWHERE THAN IN NORTH AFRICA, OR CENTRAL AND SOUTH AFRICA (82)	0.88	10	TEND LESS TO BE THOSE LOCATED ELSEWHERE THAN IN NORTH AFRICA, OR CENTRAL AND SOUTH AFRICA (82)	0.65	8 8 59 15 X SQ= 4.65 P = 0.024 RV NO (YES)
17	TILT TOWARD BEING THOSE LOCATED IN THE CARIBBEAN, CENTRAL AMERICA, OR SOUTH AMERICA, RATHER THAN IN THE MIDDLE EAST (22)	0.79	17	TILT TOWARD BEING THOSE LOCATED IN THE MIDDLE EAST, RATHER THAN IN THE CARIBBEAN, CENTRAL AMERICA, OR SOUTH AMERICA (11)	0.57	4 4 15 3 X SQ= 1.66 P = 0.149 RV YES (YES)
19	LEAN TOWARD BEING THOSE LOCATED IN THE CARIBBEAN, CENTRAL AMERICA, OR SOUTH AMERICA, RATHER THAN IN NORTH AFRICA OR CENTRAL AND SOUTH AFRICA (22)	0.65	19	LEAN TOWARD BEING THOSE LOCATED IN NORTH AFRICA OR CENTRAL AND SOUTH AFRICA, RATHER THAN IN THE CARIBBEAN, CENTRAL AMERICA, OR SOUTH AMERICA (33)	0.73	8 8 15 3 X SQ= 2.91 P = 0.066 RV YES (YES)
20	LEAN TOWARD BEING THOSE LOCATED IN THE CARIBBEAN, CENTRAL AMERICA, OR SOUTH AMERICA, RATHER THAN IN EAST ASIA, SOUTH ASIA, OR SOUTHEAST ASIA (22)	0.62	20	LEAN TOWARD BEING THOSE LOCATED IN EAST ASIA, SOUTH ASIA, OR SOUTHEAST ASIA, RATHER THAN IN THE CARIBBEAN, CENTRAL AMERICA, OR SOUTH AMERICA (18)	0.73	9 8 15 3 X SQ= 2.47 P = 0.075 RV YES (NO)
26	TEND TO BE THOSE WHOSE POPULATION DENSITY IS VERY HIGH, HIGH, OR MEDIUM (48)	0.57	26	TEND TO BE THOSE WHOSE POPULATION DENSITY IS LOW (67)	0.74	38 6 29 17 X SQ= 5.26 P = 0.015 RV YES (NO)
28	TILT LESS TOWARD BEING THOSE WHOSE POPULATION GROWTH RATE IS HIGH (62)	0.52	28	TILT MORE TOWARD BEING THOSE WHOSE POPULATION GROWTH RATE IS HIGH (62)	0.71	34 15 31 6 X SQ= 1.65 P = 0.138 RV NO (NO)
29	TEND TO BE THOSE WHERE THE DEGREE OF URBANIZATION IS HIGH (56)	0.78	29	TEND TO BE THOSE WHERE THE DEGREE OF URBANIZATION IS LOW (49)	0.77	46 5 13 17 X SQ= 18.67 P = 0. RV YES (YES)
30	TEND TO BE THOSE WHOSE AGRICULTURAL POPULATION IS MEDIUM, LOW, OR VERY LOW (57)	0.74	30	TEND TO BE THOSE WHOSE AGRICULTURAL POPULATION IS HIGH (56)	0.77	17 17 49 5 X SQ= 16.36 P = 0. RV YES (YES)
31	TEND LESS TO BE THOSE WHOSE AGRICULTURAL POPULATION IS HIGH OR MEDIUM (90)	0.67	31	ALWAYS ARE THOSE WHOSE AGRICULTURAL POPULATION IS HIGH OR MEDIUM (90)	1.00	45 22 22 0 X SQ= 7.91 P = 0.001 RV NO (NO)

32	LEAN LESS TOWARD BEING THOSE WHOSE GROSS NATIONAL PRODUCT IS MEDIUM, LOW, OR VERY LOW (105)	0.85	32	ALWAYS ARE THOSE WHOSE GROSS NATIONAL PRODUCT IS MEDIUM, LOW, OR VERY LOW (105)	1.00	10 0 57 23 X SQ= 2.50 P = 0.059 RV NO (NO)
34	TEND TO BE THOSE WHOSE GROSS NATIONAL PRODUCT IS VERY HIGH, HIGH, MEDIUM, OR LOW (62)	0.69	34	TEND TO BE THOSE WHOSE GROSS NATIONAL PRODUCT IS VERY LOW (53)	0.57	46 10 21 13 X SQ= 3.61 P = 0.046 RV YES (NO)
36	TEND TO BE THOSE WHOSE PER CAPITA GROSS NATIONAL PRODUCT IS VERY HIGH, HIGH, OR MEDIUM (42)	0.60	36	ALWAYS ARE THOSE WHOSE PER CAPITA GROSS NATIONAL PRODUCT IS LOW OR VERY LOW (73)	1.00	40 0 27 23 X SQ= 22.36 P = 0. RV YES (NO)
37	TEND TO BE THOSE WHOSE PER CAPITA GROSS NATIONAL PRODUCT IS VERY HIGH, HIGH, MEDIUM, OR LOW (64)	0.79	37	TEND TO BE THOSE WHOSE PER CAPITA GROSS NATIONAL PRODUCT IS VERY LOW (51)	0.78	53 5 14 18 X SQ= 22.15 P = 0. RV YES (YES)
38	LEAN LESS TOWARD BEING THOSE WHOSE INTERNATIONAL FINANCIAL STATUS IS MEDIUM, LOW, OR VERY LOW (103)	0.85	38	ALWAYS ARE THOSE WHOSE INTERNATIONAL FINANCIAL STATUS IS MEDIUM, LOW, OR VERY LOW (103)	1.00	10 0 55 23 X SQ= 2.61 P = 0.057 RV NO (NO)
39	TEND TO BE THOSE WHOSE INTERNATIONAL FINANCIAL STATUS IS VERY HIGH, HIGH, OR MEDIUM (38)	0.50	39	TEND TO BE THOSE WHOSE INTERNATIONAL FINANCIAL STATUS IS LOW OR VERY LOW (76)	0.87	33 3 33 20 X SQ= 8.20 P = 0.003 RV YES (NO)
42	TEND TO BE THOSE WHOSE ECONOMIC DEVELOPMENTAL STATUS IS DEVELOPED OR INTERMEDIATE (36)	0.52	42	ALWAYS ARE THOSE WHOSE ECONOMIC DEVELOPMENTAL STATUS IS UNDERDEVELOPED OR VERY UNDERDEVELOPED (76)	1.00	34 0 31 22 X SQ= 16.76 P = 0. RV YES (NO)
43	TEND TO BE THOSE WHOSE ECONOMIC DEVELOPMENTAL STATUS IS DEVELOPED, INTERMEDIATE, OR UNDERDEVELOPED (55)	0.75	43	TEND TO BE THOSE WHOSE ECONOMIC DEVELOPMENTAL STATUS IS VERY UNDERDEVELOPED (57)	0.91	49 2 16 20 X SQ= 27.11 P = 0. RV YES (YES)
44	TEND LESS TO BE THOSE WHERE THE LITERACY RATE IS BELOW NINETY PERCENT (90)	0.63	44	ALWAYS ARE THOSE WHERE THE LITERACY RATE IS BELOW NINETY PERCENT (90)	1.00	25 0 42 23 X SQ= 10.10 P = 0. RV NO (NO)

45	TEND TO BE THOSE WHERE THE LITERACY RATE IS FIFTY PERCENT OR ABOVE (55)	0.77	45	TEND TO BE THOSE WHERE THE LITERACY RATE IS BELOW FIFTY PERCENT (54)	0.77	48 5 14 17 X SQ= 18.58 P = 0. RV YES (YES)
46	TEND MORE TO BE THOSE WHERE THE LITERACY RATE IS TEN PERCENT OR ABOVE (84)	0.97	46	TEND LESS TO BE THOSE WHERE THE LITERACY RATE IS TEN PERCENT OR ABOVE (84)	0.68	61 15 2 7 X SQ= 11.27 P = 0.001 RV NO (YES)
50	TEND TO BE THOSE WHERE FREEDOM OF THE PRESS IS COMPLETE (43)	0.58	50	ALWAYS ARE THOSE WHERE FREEDOM OF THE PRESS IS INTERMITTENT, INTERNALLY ABSENT, OR INTERNALLY AND EXTERNALLY ABSENT (56)	1.00	37 0 27 20 X SQ= 18.39 P = 0. RV YES (NO)
51	TEND TO BE THOSE WHERE FREEDOM OF THE PRESS IS COMPLETE OR INTERMITTENT (60)	0.72	51	TEND TO BE THOSE WHERE FREEDOM OF THE PRESS IS INTERNALLY ABSENT, OR INTERNALLY AND EXTERNALLY ABSENT (37)	0.85	46 3 18 17 X SQ= 18.01 P = 0. RV YES (NO)
54	TEND TO BE THOSE WHERE NEWSPAPER CIRCULATION IS ONE HUNDRED OR MORE PER THOUSAND (37)	0.55	54	ALWAYS ARE THOSE WHERE NEWSPAPER CIRCULATION IS LESS THAN ONE HUNDRED PER THOUSAND (76)	1.00	36 0 29 23 X SQ= 19.33 P = 0. RV YES (NO)
55	TEND TO BE THOSE WHERE NEWSPAPER CIRCULATION IS TEN OR MORE PER THOUSAND (78)	0.91	55	TEND TO BE THOSE WHERE NEWSPAPER CIRCULATION IS LESS THAN TEN PER THOUSAND (35)	0.52	59 11 6 12 X SQ= 16.71 P = 0. RV YES (YES)
56	TEND MORE TO BE THOSE WHERE THE RELIGION IS PREDOMINANTLY LITERATE (79)	0.91	56	TEND LESS TO BE THOSE WHERE THE RELIGION IS PREDOMINANTLY LITERATE (79)	0.68	58 15 6 7 X SQ= 4.80 P = 0.033 RV NO (YES)
57	TILT LESS TOWARD BEING THOSE WHERE THE RELIGION IS OTHER THAN CATHOLIC (90)	0.69	57	TILT MORE TOWARD BEING THOSE WHERE THE RELIGION IS OTHER THAN CATHOLIC (90)	0.87	21 3 46 20 X SQ= 2.07 P = 0.106 RV NO (NO)
58	TEND MORE TO BE THOSE WHERE THE RELIGION IS OTHER THAN MUSLIM (97)	0.93	58	TEND LESS TO BE THOSE WHERE THE RELIGION IS OTHER THAN MUSLIM (97)	0.65	5 8 62 15 X SQ= 8.25 P = 0.003 RV NO (YES)

63	TEND TO BE THOSE WHERE THE RELIGION IS PREDOMINANTLY SOME KIND OF CHRISTIAN (46)	0.64	63	TEND TO BE THOSE WHERE THE RELIGION IS PREDOMINANTLY OR PARTLY OTHER THAN CHRISTIAN (68)	0.87	42 3 24 20 X SQ= 15.50 P = 0. RV YES (NO)
68	TEND TO BE THOSE THAT ARE LINGUISTICALLY HOMOGENEOUS (52)	0.59	68	TEND TO BE THOSE THAT ARE LINGUISTICALLY WEAKLY HETEROGENEOUS OR STRONGLY HETEROGENEOUS (62)	0.70	39 7 27 16 X SQ= 4.52 P = 0.028 RV YES (NO)
69	TEND TO BE THOSE THAT ARE LINGUISTICALLY HOMOGENEOUS OR WEAKLY HETEROGENEOUS (64)	0.71	69	TEND TO BE THOSE THAT ARE LINGUISTICALLY STRONGLY HETEROGENEOUS (50)	0.61	47 9 19 14 X SQ= 6.21 P = 0.011 RV YES (NO)
70	TILT LESS TOWARD BEING THOSE THAT ARE RELIGIOUSLY, RACIALLY, OR LINGUISTICALLY HETEROGENEOUS (85)	0.73	70	TILT MORE TOWARD BEING THOSE THAT ARE RELIGIOUSLY, RACIALLY, OR LINGUISTICALLY HETEROGENEOUS (85)	0.90	17 2 46 18 X SQ= 1.61 P = 0.139 RV NO (NO)
72	TILT TOWARD BEING THOSE WHOSE DATE OF INDEPENDENCE IS BEFORE 1914 (52)	0.59	72	TILT TOWARD BEING THOSE WHOSE DATE OF INDEPENDENCE IS AFTER 1914 (59)	0.59	38 9 26 13 X SQ= 1.57 P = 0.146 RV YES (NO)
73	TEND TO BE THOSE WHOSE DATE OF INDEPENDENCE IS BEFORE 1945 (65)	0.73	73	TEND TO BE THOSE WHOSE DATE OF INDEPENDENCE IS AFTER 1945 (46)	0.55	47 10 17 12 X SQ= 4.55 P = 0.035 RV YES (NO)
74	TEND LESS TO BE THOSE THAT ARE NOT HISTORICALLY WESTERN (87)	0.61	74	ALWAYS ARE THOSE THAT ARE NOT HISTORICALLY WESTERN (87)	1.00	26 0 40 23 X SQ= 10.97 P = 0. RV NO (NO)
75	TEND TO BE THOSE THAT ARE HISTORICALLY WESTERN OR SIGNIFICANTLY WESTERNIZED (62)	0.81	75	TEND TO BE THOSE THAT ARE NOT HISTORICALLY WESTERN AND ARE NOT SIGNIFICANTLY WESTERNIZED (52)	0.78	54 5 13 18 X SQ= 23.73 P = 0. RV YES (YES)
76	TEND LESS TO BE THOSE THAT HAVE BEEN SIGNIFICANTLY OR PARTIALLY WESTERNIZED THROUGH A COLONIAL RELATIONSHIP, RATHER THAN BEING HISTORICALLY WESTERN (70)	0.54	76	ALWAYS ARE THOSE THAT HAVE BEEN SIGNIFICANTLY OR PARTIALLY WESTERNIZED THROUGH A COLONIAL RELATIONSHIP, RATHER THAN BEING HISTORICALLY WESTERN (70)	1.00	26 0 30 16 X SQ= 9.70 P = 0. RV NO (NO)

77	TEND TO BE THOSE THAT HAVE BEEN SIGNIFICANTLY WESTERNIZED, RATHER THAN PARTIALLY WESTERNIZED, THROUGH A COLONIAL RELATIONSHIP (28)	0.67	77	TEND TO BE THOSE THAT HAVE BEEN PARTIALLY WESTERNIZED, RATHER THAN SIGNIFICANTLY WESTERNIZED, THROUGH A COLONIAL RELATIONSHIP (41)	0.69	20 5 10 11 X SQ= 3.94 P = 0.031 RV YES (YES)
82	TEND TO BE THOSE WHERE THE TYPE OF POLITICAL MODERNIZATION IS EARLY OR LATER EUROPEAN OR EUROPEAN DERIVED, RATHER THAN DEVELOPED TUTELARY OR UNDEVELOPED TUTELARY (51)	0.68	82	TEND TO BE THOSE WHERE THE TYPE OF POLITICAL MODERNIZATION IS DEVELOPED TUTELARY OR UNDEVELOPED TUTELARY, RATHER THAN EARLY OR LATER EUROPEAN OR EUROPEAN DERIVED (55)	0.83	43 3 20 15 X SQ= 13.15 P = 0. RV YES (NO)
83	TEND MORE TO BE THOSE WHERE THE TYPE OF POLITICAL MODERNIZATION IS OTHER THAN NON-EUROPEAN AUTOCHTHONOUS (106)	0.94	83	TEND LESS TO BE THOSE WHERE THE TYPE OF POLITICAL MODERNIZATION IS OTHER THAN NON-EUROPEAN AUTOCHTHONOUS (106)	0.78	4 5 63 18 X SQ= 3.14 P = 0.044 RV NO (YES)
85	TEND TO BE THOSE WHERE THE STAGE OF POLITICAL MODERNIZATION IS ADVANCED, RATHER THAN MID- OR EARLY TRANSITIONAL (60)	0.75	85	TEND TO BE THOSE WHERE THE STAGE OF POLITICAL MODERNIZATION IS MID- OR EARLY TRANSITIONAL, RATHER THAN ADVANCED (54)	0.73	50 6 17 16 X SQ= 13.95 P = 0. RV YES (NO)
89	TEND TO BE THOSE WHOSE IDEOLOGICAL ORIENTATION IS CONVENTIONAL (33)	0.59	89	ALWAYS ARE THOSE WHOSE IDEOLOGICAL ORIENTATION IS OTHER THAN CONVENTIONAL (62)	1.00	32 0 22 19 X SQ= 17.71 P = 0. RV YES (NO)
91	ALWAYS ARE THOSE WHOSE IDEOLOGICAL ORIENTATION IS DEVELOPMENTAL, RATHER THAN TRADITIONAL (31)	1.00	91	LEAN LESS TOWARD BEING THOSE WHOSE IDEOLOGICAL ORIENTATION IS DEVELOPMENTAL, RATHER THAN TRADITIONAL (31)	0.58	8 7 0 5 X SQ= 2.50 P = 0.055 RV NO (YES)
99	TEND TO BE THOSE WHERE GOVERNMENTAL STABILITY IS GENERALLY PRESENT AND DATES FROM AT LEAST THE INTER-WAR PERIOD, OR FROM THE POST-WAR PERIOD (50)	0.77	99	TEND TO BE THOSE WHERE GOVERNMENTAL STABILITY IS MODERATELY PRESENT AND DATES FROM THE POST-WAR PERIOD, OR IS ABSENT (36)	0.70	43 6 13 14 X SQ= 12.11 P = 0. RV YES (YES)
101	TEND TO BE THOSE WHERE THE REPRESENTATIVE CHARACTER OF THE REGIME IS POLYARCHIC (41)	0.65	101	ALWAYS ARE THOSE WHERE THE REPRESENTATIVE CHARACTER OF THE REGIME IS LIMITED POLYARCHIC, PSEUDO-POLYARCHIC, OR NON-POLYARCHIC (57)	1.00	41 0 22 22 X SQ= 25.11 P = 0. RV YES (YES)
102	TEND TO BE THOSE WHERE THE REPRESENTATIVE CHARACTER OF THE REGIME IS POLYARCHIC OR LIMITED POLYARCHIC (59)	0.75	102	TEND TO BE THOSE WHERE THE REPRESENTATIVE CHARACTER OF THE REGIME IS PSEUDO-POLYARCHIC OR NON-POLYARCHIC (49)	0.95	50 1 17 21 X SQ= 30.44 P = 0. RV YES (YES)

```
105  TEND TO BE THOSE                        0.70    105  ALWAYS ARE THOSE                              1.00        42     0
     WHERE THE ELECTORAL SYSTEM IS                        WHERE THE ELECTORAL SYSTEM IS                             18    14
     COMPETITIVE  (43)                                    PARTIALLY COMPETITIVE OR                          X SQ=   19.90
                                                          NON-COMPETITIVE  (47)                             P =      0.
                                                                                                            RV YES (NO )

106  TEND TO BE THOSE                        0.71    106  TEND TO BE THOSE                              0.78        42     2
     WHERE THE ELECTORAL SYSTEM IS                        WHERE THE ELECTORAL SYSTEM IS                             17     7
     COMPETITIVE OR                                       NON-COMPETITIVE  (30)                             X SQ=    6.19
     PARTIALLY COMPETITIVE  (52)                                                                            P =      0.007
                                                                                                            RV YES (NO )

107  TEND TO BE THOSE                        0.67    107  ALWAYS ARE THOSE                              1.00        45     0
     WHERE AUTONOMOUS GROUPS                              WHERE AUTONOMOUS GROUPS                                   22    22
     ARE FULLY TOLERATED IN POLITICS  (46)                ARE PARTIALLY TOLERATED IN POLITICS,              X SQ=   27.26
                                                          ARE TOLERATED ONLY OUTSIDE POLITICS,              P =      0.
                                                          OR ARE NOT TOLERATED AT ALL  (65)                 RV YES (YES)

108  TEND TO BE THOSE                        0.76    108  TEND TO BE THOSE                              0.71        50     5
     WHERE AUTONOMOUS GROUPS                              WHERE AUTONOMOUS GROUPS                                   16    12
     ARE FULLY OR PARTIALLY TOLERATED                     ARE TOLERATED ONLY OUTSIDE POLITICS               X SQ=   11.00
     IN POLITICS  (65)                                    OR ARE NOT TOLERATED AT ALL  (35)                 P =      0.001
                                                                                                            RV YES (NO )

111  TEND TO BE THOSE                        0.75    111  TEND TO BE THOSE                              0.62        39     8
     WHERE POLITICAL ENCULTURATION                        WHERE POLITICAL ENCULTURATION                             13    13
     IS HIGH OR MEDIUM  (53)                              IS LOW  (42)                                      X SQ=    7.35
                                                                                                            P =      0.006
                                                                                                            RV YES (YES)

114  LEAN TOWARD BEING THOSE                 0.51    114  LEAN TOWARD BEING THOSE                       0.71        32    15
     WHERE SECTIONALISM IS                                WHERE SECTIONALISM IS                                     33     6
     NEGLIGIBLE  (47)                                     EXTREME OR MODERATE  (61)                         X SQ=    2.32
                                                                                                            P =      0.085
                                                                                                            RV YES (NO )

116  TEND LESS TO BE THOSE                   0.53    116  ALWAYS ARE THOSE                              1.00        30     0
     WHERE INTEREST ARTICULATION                          WHERE INTEREST ARTICULATION                               34    22
     BY ASSOCIATIONAL GROUPS                              BY ASSOCIATIONAL GROUPS                           X SQ=   13.84
     IS LIMITED OR NEGLIGIBLE  (79)                       IS LIMITED OR NEGLIGIBLE  (79)                    P =      0.
                                                                                                            RV NO  (NO )

117  TEND TO BE THOSE                        0.75    117  TEND TO BE THOSE                              0.68        48     7
     WHERE INTEREST ARTICULATION                          WHERE INTEREST ARTICULATION                               16    15
     BY ASSOCIATIONAL GROUPS                              BY ASSOCIATIONAL GROUPS                           X SQ=   11.44
     IS SIGNIFICANT, MODERATE, OR                         IS NEGLIGIBLE  (51)                               P =      0.001
     LIMITED  (60)                                                                                          RV YES (NO )

118  TEND TO BE THOSE                        0.71    118  TEND TO BE THOSE                              0.77        19    17
     WHERE INTEREST ARTICULATION                          WHERE INTEREST ARTICULATION                               47     5
     BY INSTITUTIONAL GROUPS                              BY INSTITUTIONAL GROUPS                           X SQ=   14.10
     IS SIGNIFICANT, MODERATE, OR                         IS VERY SIGNIFICANT  (40)                         P =      0.
     LIMITED  (60)                                                                                          RV YES (NO )
```

119	TEND LESS TO BE THOSE WHERE INTEREST ARTICULATION BY INSTITUTIONAL GROUPS IS VERY SIGNIFICANT OR SIGNIFICANT (74)	0.61	119	ALWAYS ARE THOSE WHERE INTEREST ARTICULATION BY INSTITUTIONAL GROUPS IS VERY SIGNIFICANT OR SIGNIFICANT (74)	1.00	40 22 26 0 X SQ= 10.48 P = 0. RV NO (NO)
121	TEND TO BE THOSE WHERE INTEREST ARTICULATION BY NON-ASSOCIATIONAL GROUPS IS MODERATE, LIMITED, OR NEGLIGIBLE (61)	0.78	121	TEND TO BE THOSE WHERE INTEREST ARTICULATION BY NON-ASSOCIATIONAL GROUPS IS SIGNIFICANT (54)	0.74	15 17 52 6 X SQ= 17.65 P = 0. RV YES (YES)
123	TILT LESS TOWARD BEING THOSE WHERE INTEREST ARTICULATION BY NON-ASSOCIATIONAL GROUPS IS SIGNIFICANT, MODERATE, OR LIMITED (107)	0.88	123	ALWAYS ARE THOSE WHERE INTEREST ARTICULATION BY NON-ASSOCIATIONAL GROUPS IS SIGNIFICANT, MODERATE, OR LIMITED (107)	1.00	59 23 8 0 X SQ= 1.72 P = 0.108 RV NO (NO)
125	TEND TO BE THOSE WHERE INTEREST ARTICULATION BY ANOMIC GROUPS IS INFREQUENT OR VERY INFREQUENT (35)	0.56	125	TEND TO BE THOSE WHERE INTEREST ARTICULATION BY ANOMIC GROUPS IS FREQUENT OR OCCASIONAL (64)	0.86	24 19 30 3 X SQ= 9.54 P = 0.001 RV YES (NO)
126	TEND LESS TO BE THOSE WHERE INTEREST ARTICULATION BY ANOMIC GROUPS IS FREQUENT, OCCASIONAL, OR INFREQUENT (83)	0.70	126	ALWAYS ARE THOSE WHERE INTEREST ARTICULATION BY ANOMIC GROUPS IS FREQUENT, OCCASIONAL, OR INFREQUENT (83)	1.00	38 22 16 0 X SQ= 6.57 P = 0.004 RV NO (NO)
127	TILT LESS TOWARD BEING THOSE WHERE INTEREST ARTICULATION BY POLITICAL PARTIES IS MODERATE, LIMITED, OR NEGLIGIBLE (72)	0.73	127	ALWAYS ARE THOSE WHERE INTEREST ARTICULATION BY POLITICAL PARTIES IS MODERATE, LIMITED, OR NEGLIGIBLE (72)	1.00	17 0 46 9 X SQ= 1.86 P = 0.104 RV NO (NO)
128	TEND TO BE THOSE WHERE INTEREST ARTICULATION BY POLITICAL PARTIES IS SIGNIFICANT OR MODERATE (48)	0.62	128	TEND TO BE THOSE WHERE INTEREST ARTICULATION BY POLITICAL PARTIES IS LIMITED OR NEGLIGIBLE (45)	0.78	39 2 24 7 X SQ= 3.57 P = 0.033 RV YES (NO)
131	TEND TO BE THOSE WHERE INTEREST AGGREGATION BY POLITICAL PARTIES IS SIGNIFICANT OR MODERATE (30)	0.54	131	TEND TO BE THOSE WHERE INTEREST AGGREGATION BY POLITICAL PARTIES IS LIMITED OR NEGLIGIBLE (35)	0.83	25 2 21 10 X SQ= 4.02 P = 0.025 RV YES (NO)
134	TEND TO BE THOSE WHERE INTEREST AGGREGATION BY THE EXECUTIVE IS SIGNIFICANT OR MODERATE (57)	0.59	134	TEND TO BE THOSE WHERE INTEREST AGGREGATION BY THE EXECUTIVE IS LIMITED OR NEGLIGIBLE (46)	0.70	38 6 26 14 X SQ= 4.16 P = 0.039 RV YES (NO)

137	TEND LESS TO BE THOSE WHERE INTEREST AGGREGATION BY THE LEGISLATURE IS LIMITED OR NEGLIGIBLE (68)	0.53	137	ALWAYS ARE THOSE WHERE INTEREST AGGREGATION BY THE LEGISLATURE IS LIMITED OR NEGLIGIBLE (68)	1.00	29 0 33 15 X SQ= 9.35 P = 0.001 RV NO (NO)
138	TEND TO BE THOSE WHERE INTEREST AGGREGATION BY THE LEGISLATURE IS SIGNIFICANT, MODERATE, OR LIMITED (48)	0.70	138	TEND TO BE THOSE WHERE INTEREST AGGREGATION BY THE LEGISLATURE IS NEGLIGIBLE (49)	0.93	44 1 19 13 X SQ= 16.05 P = 0. RV YES (NO)
141	TILT LESS TOWARD BEING THOSE WHERE THE PARTY SYSTEM IS QUANTITATIVELY OTHER THAN TWO-PARTY (87)	0.83	141	ALWAYS ARE THOSE WHERE THE PARTY SYSTEM IS QUANTITATIVELY OTHER THAN TWO-PARTY (87)	1.00	11 0 53 14 X SQ= 1.56 P = 0.118 RV NO (NO)
144	LEAN LESS TOWARD BEING THOSE WHERE THE PARTY SYSTEM IS QUANTITATIVELY ONE-PARTY, RATHER THAN TWO-PARTY (34)	0.62	144	ALWAYS ARE THOSE WHERE THE PARTY SYSTEM IS QUANTITATIVELY ONE-PARTY, RATHER THAN TWO-PARTY (34)	1.00	18 7 11 0 X SQ= 2.24 P = 0.076 RV NO (NO)
146	TEND MORE TO BE THOSE WHERE THE PARTY SYSTEM IS QUALITATIVELY OTHER THAN MASS-BASED TERRITORIAL (92)	0.98	146	TEND LESS TO BE THOSE WHERE THE PARTY SYSTEM IS QUALITATIVELY OTHER THAN MASS-BASED TERRITORIAL (92)	0.79	1 4 60 15 X SQ= 6.30 P = 0.010 RV NO (YES)
147	TEND LESS TO BE THOSE WHERE THE PARTY SYSTEM IS QUALITATIVELY OTHER THAN CLASS-ORIENTED OR MULTI-IDEOLOGICAL (67)	0.64	147	ALWAYS ARE THOSE WHERE THE PARTY SYSTEM IS QUALITATIVELY OTHER THAN CLASS-ORIENTED OR MULTI-IDEOLOGICAL (67)	1.00	21 0 37 13 X SQ= 5.06 P = 0.007 RV NO (NO)
153	TEND TO BE THOSE WHERE THE PARTY SYSTEM IS STABLE (42)	0.69	153	TEND TO BE THOSE WHERE THE PARTY SYSTEM IS MODERATELY STABLE OR UNSTABLE (38)	0.85	38 2 17 11 X SQ= 10.40 P = 0.001 RV YES (NO)
154	TEND TO BE THOSE WHERE THE PARTY SYSTEM IS STABLE OR MODERATELY STABLE (55)	0.80	154	TEND TO BE THOSE WHERE THE PARTY SYSTEM IS UNSTABLE (25)	0.54	44 6 11 7 X SQ= 4.57 P = 0.031 RV YES (NO)
158	TEND TO BE THOSE WHERE PERSONALISMO IS NEGLIGIBLE (56)	0.77	158	TEND TO BE THOSE WHERE PERSONALISMO IS PRONOUNCED OR MODERATE (40)	0.77	15 10 49 3 X SQ= 11.76 P = 0. RV YES (NO)

No.			No.			
160	TEND TO BE THOSE WHERE THE POLITICAL LEADERSHIP IS MODERATELY ELITIST OR NON-ELITIST (67)	0.73	160	TEND TO BE THOSE WHERE THE POLITICAL LEADERSHIP IS ELITIST (30)	0.73	17 11 45 4 X SQ= 9.11 P = 0.002 RV YES (NO)
161	TEND TO BE THOSE WHERE THE POLITICAL LEADERSHIP IS NON-ELITIST (50)	0.53	161	TEND TO BE THOSE WHERE THE POLITICAL LEADERSHIP IS ELITIST OR MODERATELY ELITIST (47)	0.87	29 13 33 2 X SQ= 6.23 P = 0.008 RV YES (NO)
164	TEND TO BE THOSE WHERE THE REGIME'S LEADERSHIP CHARISMA IS NEGLIGIBLE (65)	0.81	164	TEND TO BE THOSE WHERE THE REGIME'S LEADERSHIP CHARISMA IS PRONOUNCED OR MODERATE (34)	0.61	12 11 51 7 X SQ= 10.20 P = 0.002 RV YES (NO)
168	TEND TO BE THOSE WHERE THE HORIZONTAL POWER DISTRIBUTION IS SIGNIFICANT (34)	0.50	168	ALWAYS ARE THOSE WHERE THE HORIZONTAL POWER DISTRIBUTION IS LIMITED OR NEGLIGIBLE (72)	1.00	33 0 33 21 X SQ= 14.86 P = 0. RV YES (NO)
169	TEND TO BE THOSE WHERE THE HORIZONTAL POWER DISTRIBUTION IS SIGNIFICANT OR LIMITED (58)	0.74	169	TEND TO BE THOSE WHERE THE HORIZONTAL POWER DISTRIBUTION IS NEGLIGIBLE (48)	0.90	49 2 17 19 X SQ= 24.91 P = 0. RV YES (YES)
170	TEND TO BE THOSE WHERE THE LEGISLATIVE-EXECUTIVE STRUCTURE IS OTHER THAN PRESIDENTIAL (63)	0.74	170	TEND TO BE THOSE WHERE THE LEGISLATIVE-EXECUTIVE STRUCTURE IS PRESIDENTIAL (39)	0.53	17 9 49 8 X SQ= 3.47 P = 0.042 RV YES (NO)
172	TEND LESS TO BE THOSE WHERE THE LEGISLATIVE-EXECUTIVE STRUCTURE IS OTHER THAN PARLIAMENTARY-ROYALIST (88)	0.71	172	TEND MORE TO BE THOSE WHERE THE LEGISLATIVE-EXECUTIVE STRUCTURE IS OTHER THAN PARLIAMENTARY-ROYALIST (88)	0.95	19 1 47 18 X SQ= 3.32 P = 0.035 RV NO (NO)
175	TEND TO BE THOSE WHERE THE LEGISLATURE IS FULLY EFFECTIVE OR PARTIALLY EFFECTIVE (51)	0.74	175	ALWAYS ARE THOSE WHERE THE LEGISLATURE IS LARGELY INEFFECTIVE OR WHOLLY INEFFECTIVE (49)	1.00	48 0 17 17 X SQ= 27.31 P = 0. RV YES (YES)
176	TEND TO BE THOSE WHERE THE LEGISLATURE IS FULLY EFFECTIVE, PARTIALLY EFFECTIVE, OR LARGELY INEFFECTIVE (72)	0.77	176	TEND TO BE THOSE WHERE THE LEGISLATURE IS WHOLLY INEFFECTIVE (28)	0.65	50 6 15 11 X SQ= 8.95 P = 0.002 RV YES (NO)

179	TEND TO BE THOSE WHERE THE EXECUTIVE IS STRONG (39)	0.62	179	ALWAYS ARE THOSE WHERE THE EXECUTIVE IS DOMINANT (52)	1.00	23 18 38 0 X SQ= 19.18 P = 0. RV YES (NO)
181	LEAN LESS TOWARD BEING THOSE WHERE THE BUREAUCRACY IS SEMI-MODERN, RATHER THAN MODERN (55)	0.68	181	ALWAYS ARE THOSE WHERE THE BUREAUCRACY IS SEMI-MODERN, RATHER THAN MODERN (55)	1.00	20 0 42 8 X SQ= 2.21 P = 0.095 RV NO (NO)
183	ALWAYS ARE THOSE WHERE THE BUREAUCRACY IS POST-COLONIAL TRANSITIONAL, RATHER THAN TRADITIONAL (25)	1.00	183	TEND TO BE THOSE WHERE THE BUREAUCRACY IS TRADITIONAL, RATHER THAN POST-COLONIAL TRANSITIONAL (9)	0.69	5 4 0 9 X SQ= 4.43 P = 0.029 RV YES (YES)
185	TILT MORE TOWARD BEING THOSE WHERE PARTICIPATION BY THE MILITARY IN POLITICS IS SUPPORTIVE, RATHER THAN INTERVENTIVE (31)	0.81	185	TILT LESS TOWARD BEING THOSE WHERE PARTICIPATION BY THE MILITARY IN POLITICS IS SUPPORTIVE, RATHER THAN INTERVENTIVE (31)	0.55	4 9 17 11 X SQ= 2.10 P = 0.100 RV NO (YES)
186	TEND TO BE THOSE WHERE PARTICIPATION BY THE MILITARY IN POLITICS IS NEUTRAL, RATHER THAN SUPPORTIVE (56)	0.72	186	TEND TO BE THOSE WHERE PARTICIPATION BY THE MILITARY IN POLITICS IS SUPPORTIVE, RATHER THAN NEUTRAL (31)	0.92	17 11 43 1 X SQ= 14.32 P = 0. RV YES (NO)
187	TEND TO BE THOSE WHERE THE ROLE OF THE POLICE IS NOT POLITICALLY SIGNIFICANT (35)	0.57	187	ALWAYS ARE THOSE WHERE THE ROLE OF THE POLICE IS POLITICALLY SIGNIFICANT (66)	1.00	26 23 34 0 X SQ= 19.80 P = 0. RV YES (NO)
192	TEND MORE TO BE THOSE WHERE THE CHARACTER OF THE LEGAL SYSTEM IS OTHER THAN MUSLIM OR PARTLY MUSLIM (86)	0.88	192	TEND LESS TO BE THOSE WHERE THE CHARACTER OF THE LEGAL SYSTEM IS OTHER THAN MUSLIM OR PARTLY MUSLIM (86)	0.57	8 9 59 12 X SQ= 7.92 P = 0.004 RV NO (YES)
193	TEND MORE TO BE THOSE WHERE THE CHARACTER OF THE LEGAL SYSTEM IS OTHER THAN PARTLY INDIGENOUS (86)	0.94	193	TEND LESS TO BE THOSE WHERE THE CHARACTER OF THE LEGAL SYSTEM IS OTHER THAN PARTLY INDIGENOUS (86)	0.68	4 7 63 15 X SQ= 7.97 P = 0.004 RV NO (YES)
194	TEND LESS TO BE THOSE THAT ARE NON-COMMUNIST (101)	0.80	194	ALWAYS ARE THOSE THAT ARE NON-COMMUNIST (101)	1.00	13 0 53 23 X SQ= 3.84 P = 0.018 RV NO (NO)

```
97  POLITIES                                            97  POLITIES
    WHERE THE STATUS OF THE REGIME                          WHERE THE STATUS OF THE REGIME
    IS AUTHORITARIAN, RATHER THAN                           IS TOTALITARIAN, RATHER THAN
    TOTALITARIAN  (23)                                      AUTHORITARIAN  (16)

                                                                              BOTH SUBJECT AND PREDICATE

      23 IN LEFT COLUMN

AFGHANISTAN ALGERIA     BURMA       CAMBODIA    CEN AFR REP EL SALVADOR ETHIOPIA    GHANA       GUINEA      INDONESIA
IRAN        JORDAN      KOREA REP   LAOS        LIBERIA     NEPAL       NICARAGUA   PAKISTAN    PARAGUAY    SA'U ARABIA
SUDAN       THAILAND    UAR

      16 IN RIGHT COLUMN

ALBANIA     BULGARIA    CHINA, PR   CUBA        CZECHOS'KIA GERMANY, E  HUNGARY     KOREA, N    MONGOLIA    POLAND
PORTUGAL    RUMANIA     SPAIN       USSR        VIETNAM, N  YUGOSLAVIA

       5 EXCLUDED BECAUSE AMBIGUOUS

ARGENTINA   HAITI       IVORY COAST SO AFRICA   VIETNAM REP

      51 EXCLUDED BECAUSE IRRELEVANT

AUSTRALIA   AUSTRIA     BELGIUM     BOLIVIA     BRAZIL      CANADA      CEYLON      CHILE       COLOMBIA    COSTA RICA
CYPRUS      DENMARK     DOMIN REP   ECUADOR     FINLAND     FRANCE      GERMAN FR   GREECE      HONDURAS    ICELAND
INDIA       IRELAND     ISRAEL      ITALY       JAMAICA     JAPAN       LEBANON     LIBYA       LUXEMBOURG  MALAGASY R
MALAYA      MAURITANIA  MEXICO      MOROCCO     NETHERLANDS NEW ZEALAND NIGERIA     NORWAY      PANAMA      PHILIPPINES
SIERRE LEO  SWEDEN      SWITZERLAND TRINIDAD    TUNISIA     TURKEY      UGANDA      UK          US          URUGUAY
VENEZUELA

      11 EXCLUDED BECAUSE UNASCERTAINED

BURUNDI     CHAD        CONGO(BRA)  DAHOMEY     GABON       MALI        NIGER       RWANDA      SOMALIA     TANGANYIKA
UPPER VOLTA

       9 EXCLUDED BECAUSE UNASCERTAINABLE

CAMEROUN    CONGO(LEO)  GUATEMALA   IRAQ        PERU        SENEGAL     SYRIA       TOGO        YEMEN

--------------------------------------------------------------------------------------------------------------------

 10  TEND LESS TO BE THOSE                  0.65        10  ALWAYS ARE THOSE                        1.00          8    0
     LOCATED ELSEWHERE THAN IN NORTH AFRICA,                LOCATED ELSEWHERE THAN IN NORTH AFRICA,              15   16
     OR CENTRAL AND SOUTH AFRICA  (82)                      OR CENTRAL AND SOUTH AFRICA  (82)               X SQ=    5.03
                                                                                                            P =     0.012
                                                                                                            RV NO   (YES)

 13  TEND LESS TO BE THOSE                  0.74        13  ALWAYS ARE THOSE                        1.00          6    0
     LOCATED ELSEWHERE THAN IN NORTH AFRICA OR              LOCATED ELSEWHERE THAN IN NORTH AFRICA OR            17   16
     THE MIDDLE EAST  (99)                                  THE MIDDLE EAST  (99)                           X SQ=    3.13
                                                                                                            P =     0.033
                                                                                                            RV NO   (NO )
```

14	TILT LESS TOWARD BEING THOSE LOCATED ELSEWHERE THAN IN THE MIDDLE EAST (104)	0.83	14	ALWAYS ARE THOSE LOCATED ELSEWHERE THAN IN THE MIDDLE EAST (104)	1.00	4 0 19 16 X SQ= 1.50 P = 0.130 RV NO (NO)
18	TILT TOWARD BEING THOSE LOCATED IN NORTH AFRICA OR CENTRAL AND SOUTH AFRICA, RATHER THAN IN EAST ASIA, SOUTH ASIA, OR SOUTHEAST ASIA (33)	0.50	18	ALWAYS ARE THOSE LOCATED IN EAST ASIA, SOUTH ASIA, OR SOUTHEAST ASIA, RATHER THAN IN NORTH AFRICA OR CENTRAL AND SOUTH AFRICA (18)	1.00	8 0 8 4 X SQ= 1.58 P = 0.117 RV YES (NO)
24	LEAN LESS TOWARD BEING THOSE WHOSE POPULATION IS VERY LARGE, LARGE, OR MEDIUM (61)	0.61	24	LEAN MORE TOWARD BEING THOSE WHOSE POPULATION IS VERY LARGE, LARGE, OR MEDIUM (61)	0.87	14 14 9 2 X SQ= 2.12 P = 0.086 RV NO (YES)
26	TEND TO BE THOSE WHOSE POPULATION DENSITY IS LOW (67)	0.74	26	TEND TO BE THOSE WHOSE POPULATION DENSITY IS VERY HIGH, HIGH, OR MEDIUM (48)	0.87	6 14 17 2 X SQ= 11.89 P = 0. RV YES (YES)
28	TEND TO BE THOSE WHOSE POPULATION GROWTH RATE IS HIGH (62)	0.71	28	TEND TO BE THOSE WHOSE POPULATION GROWTH RATE IS LOW (48)	0.71	15 4 6 10 X SQ= 4.61 P = 0.018 RV YES (YES)
29	TEND TO BE THOSE WHERE THE DEGREE OF URBANIZATION IS LOW (49)	0.77	29	TEND TO BE THOSE WHERE THE DEGREE OF URBANIZATION IS HIGH (56)	0.75	5 9 17 3 X SQ= 6.73 P = 0.009 RV YES (YES)
30	TEND TO BE THOSE WHOSE AGRICULTURAL POPULATION IS HIGH (56)	0.77	30	TEND TO BE THOSE WHOSE AGRICULTURAL POPULATION IS MEDIUM, LOW, OR VERY LOW (57)	0.60	17 6 5 9 X SQ= 3.80 P = 0.038 RV YES (YES)
33	TEND TO BE THOSE WHOSE GROSS NATIONAL PRODUCT IS LOW OR VERY LOW (85)	0.91	33	TEND TO BE THOSE WHOSE GROSS NATIONAL PRODUCT IS VERY HIGH, HIGH, OR MEDIUM (30)	0.50	2 8 21 8 X SQ= 6.42 P = 0.007 RV YES (YES)
34	TEND TO BE THOSE WHOSE GROSS NATIONAL PRODUCT IS VERY LOW (53)	0.57	34	TEND TO BE THOSE WHOSE GROSS NATIONAL PRODUCT IS VERY HIGH, HIGH, MEDIUM, OR LOW (62)	0.81	10 13 13 3 X SQ= 4.11 P = 0.024 RV YES (YES)

36	ALWAYS ARE THOSE WHOSE PER CAPITA GROSS NATIONAL PRODUCT IS LOW OR VERY LOW (73)	1.00	36	TEND TO BE THOSE WHOSE PER CAPITA GROSS NATIONAL PRODUCT IS VERY HIGH, HIGH, OR MEDIUM (42)	0.62	0 10 23 6 X SQ= 16.19 P = 0. RV YES (YES)
37	TEND TO BE THOSE WHOSE PER CAPITA GROSS NATIONAL PRODUCT IS VERY LOW (51)	0.78	37	TEND TO BE THOSE WHOSE PER CAPITA GROSS NATIONAL PRODUCT IS VERY HIGH, HIGH, MEDIUM, OR LOW (64)	0.75	5 12 18 4 X SQ= 8.83 P = 0.003 RV YES (YES)
39	TEND TO BE THOSE WHOSE INTERNATIONAL FINANCIAL STATUS IS LOW OR VERY LOW (76)	0.87	39	TEND TO BE THOSE WHOSE INTERNATIONAL FINANCIAL STATUS IS VERY HIGH, HIGH, OR MEDIUM (38)	0.53	3 8 20 7 X SQ= 5.34 P = 0.012 RV YES (YES)
42	ALWAYS ARE THOSE WHOSE ECONOMIC DEVELOPMENTAL STATUS IS UNDERDEVELOPED OR VERY UNDERDEVELOPED (76)	1.00	42	TEND TO BE THOSE WHOSE ECONOMIC DEVELOPMENTAL STATUS IS DEVELOPED OR INTERMEDIATE (36)	0.50	0 8 22 8 X SQ= 11.09 P = 0. RV YES (YES)
43	TEND TO BE THOSE WHOSE ECONOMIC DEVELOPMENTAL STATUS IS VERY UNDERDEVELOPED (57)	0.91	43	TEND TO BE THOSE WHOSE ECONOMIC DEVELOPMENTAL STATUS IS DEVELOPED, INTERMEDIATE, OR UNDERDEVELOPED (55)	0.75	2 12 20 4 X SQ= 14.58 P = 0. RV YES (YES)
44	ALWAYS ARE THOSE WHERE THE LITERACY RATE IS BELOW NINETY PERCENT (90)	1.00	44	TEND LESS TO BE THOSE WHERE THE LITERACY RATE IS BELOW NINETY PERCENT (90)	0.62	0 6 23 10 X SQ= 7.52 P = 0.002 RV NO (YES)
45	TEND TO BE THOSE WHERE THE LITERACY RATE IS BELOW FIFTY PERCENT (54)	0.77	45	ALWAYS ARE THOSE WHERE THE LITERACY RATE IS FIFTY PERCENT OR ABOVE (55)	1.00	5 12 17 0 X SQ= 15.58 P = 0. RV YES (YES)
46	TEND LESS TO BE THOSE WHERE THE LITERACY RATE IS TEN PERCENT OR ABOVE (84)	0.68	46	ALWAYS ARE THOSE WHERE THE LITERACY RATE IS TEN PERCENT OR ABOVE (84)	1.00	15 14 7 0 X SQ= 3.68 P = 0.029 RV NO (NO)
52	TEND TO BE THOSE WHERE FREEDOM OF THE PRESS IS COMPLETE, INTERMITTENT, OR INTERNALLY ABSENT (82)	0.75	52	TEND TO BE THOSE WHERE FREEDOM OF THE PRESS IS INTERNALLY AND EXTERNALLY ABSENT (16)	0.69	15 5 5 11 X SQ= 5.23 P = 0.017 RV YES (YES)

54	ALWAYS ARE THOSE WHERE NEWSPAPER CIRCULATION IS LESS THAN ONE HUNDRED PER THOUSAND (76)	1.00	54	TEND TO BE THOSE WHERE NEWSPAPER CIRCULATION IS ONE HUNDRED OR MORE PER THOUSAND (37)	0.64	0 9 23 5 X SQ= 16.20 P = 0. RV YES (YES)
55	TEND TO BE THOSE WHERE NEWSPAPER CIRCULATION IS LESS THAN TEN PER THOUSAND (35)	0.52	55	ALWAYS ARE THOSE WHERE NEWSPAPER CIRCULATION IS TEN OR MORE PER THOUSAND (78)	1.00	11 14 12 0 X SQ= 8.56 P = 0.001 RV YES (YES)
56	TEND LESS TO BE THOSE WHERE THE RELIGION IS PREDOMINANTLY LITERATE (79)	0.68	56	ALWAYS ARE THOSE WHERE THE RELIGION IS PREDOMINANTLY LITERATE (79)	1.00	15 13 7 0 X SQ= 3.37 P = 0.031 RV NO (NO)
63	TEND TO BE THOSE WHERE THE RELIGION IS PREDOMINANTLY OR PARTLY OTHER THAN CHRISTIAN (68)	0.87	63	TEND TO BE THOSE WHERE THE RELIGION IS PREDOMINANTLY SOME KIND OF CHRISTIAN (46)	0.67	3 10 20 5 X SQ= 9.34 P = 0.001 RV YES (YES)
64	TEND TO BE THOSE WHERE THE RELIGION IS MUSLIM, RATHER THAN CHRISTIAN (18)	0.73	64	ALWAYS ARE THOSE WHERE THE RELIGION IS CHRISTIAN, RATHER THAN MUSLIM (46)	1.00	3 10 8 0 X SQ= 8.87 P = 0.001 RV YES (YES)
69	TEND TO BE THOSE THAT ARE LINGUISTICALLY STRONGLY HETEROGENEOUS (50)	0.61	69	TEND TO BE THOSE THAT ARE LINGUISTICALLY HOMOGENEOUS OR WEAKLY HETEROGENEOUS (64)	0.80	9 12 14 3 X SQ= 4.59 P = 0.020 RV YES (YES)
73	TEND TO BE THOSE WHOSE DATE OF INDEPENDENCE IS AFTER 1945 (46)	0.55	73	TEND TO BE THOSE WHOSE DATE OF INDEPENDENCE IS BEFORE 1945 (65)	0.93	10 13 12 1 X SQ= 6.40 P = 0.005 RV YES (YES)
74	ALWAYS ARE THOSE THAT ARE NOT HISTORICALLY WESTERN (87)	1.00	74	TEND LESS TO BE THOSE THAT ARE NOT HISTORICALLY WESTERN (87)	0.63	0 6 23 10 X SQ= 7.52 P = 0.002 RV NO (YES)
75	TEND TO BE THOSE THAT ARE NOT HISTORICALLY WESTERN AND ARE NOT SIGNIFICANTLY WESTERNIZED (52)	0.78	75	TEND TO BE THOSE THAT ARE HISTORICALLY WESTERN OR SIGNIFICANTLY WESTERNIZED (62)	0.75	5 12 18 4 X SQ= 8.83 P = 0.003 RV YES (YES)

76	ALWAYS ARE THOSE THAT HAVE BEEN SIGNIFICANTLY OR PARTIALLY WESTERNIZED THROUGH A COLONIAL RELATIONSHIP, RATHER THAN BEING HISTORICALLY WESTERN (70)	1.00	76	TEND TO BE THOSE THAT ARE HISTORICALLY WESTERN, RATHER THAN HAVING BEEN SIGNIFICANTLY OR PARTIALLY WESTERNIZED THROUGH A COLONIAL RELATIONSHIP (26)	0.75	0 6 16 2 X SQ= 12.25 P = 0. RV YES (YES)
78	ALWAYS ARE THOSE THAT HAVE BEEN SIGNIFICANTLY WESTERNIZED THROUGH A COLONIAL RELATIONSHIP, RATHER THAN WITHOUT SUCH A RELATIONSHIP (28)	1.00	78	TEND TO BE THOSE THAT HAVE BEEN SIGNIFICANTLY WESTERNIZED WITHOUT A COLONIAL RELATIONSHIP, RATHER THAN THROUGH SUCH A RELATIONSHIP (7)	0.83	5 1 0 5 X SQ= 4.65 P = 0.015 RV YES (YES)
82	TEND TO BE THOSE WHERE THE TYPE OF POLITICAL MODERNIZATION IS DEVELOPED TUTELARY OR UNDEVELOPED TUTELARY, RATHER THAN EARLY OR LATER EUROPEAN OR EUROPEAN DERIVED (55)	0.83	82	TEND TO BE THOSE WHERE THE TYPE OF POLITICAL MODERNIZATION IS EARLY OR LATER EUROPEAN OR EUROPEAN DERIVED, RATHER THAN DEVELOPED TUTELARY OR UNDEVELOPED TUTELARY (51)	0.79	3 11 15 3 X SQ= 9.88 P = 0.001 RV YES (YES)
85	TEND TO BE THOSE WHERE THE STAGE OF POLITICAL MODERNIZATION IS MID- OR EARLY TRANSITIONAL, RATHER THAN ADVANCED (54)	0.73	85	TEND TO BE THOSE WHERE THE STAGE OF POLITICAL MODERNIZATION IS ADVANCED, RATHER THAN MID- OR EARLY TRANSITIONAL (60)	0.94	6 15 16 1 X SQ= 13.98 P = 0. RV YES (YES)
87	TEND TO BE THOSE WHOSE IDEOLOGICAL ORIENTATION IS DEVELOPMENTAL (31)	0.50	87	ALWAYS ARE THOSE WHOSE IDEOLOGICAL ORIENTATION IS OTHER THAN DEVELOPMENTAL (58)	1.00	7 0 7 15 X SQ= 7.34 P = 0.002 RV YES (YES)
92	TEND TO BE THOSE WHERE THE SYSTEM STYLE IS LIMITED MOBILIZATIONAL, OR NON-MOBILIZATIONAL (93)	0.78	92	TEND TO BE THOSE WHERE THE SYSTEM STYLE IS MOBILIZATIONAL (20)	0.93	5 14 18 1 X SQ= 15.86 P = 0. RV YES (YES)
93	TEND TO BE THOSE WHERE THE SYSTEM STYLE IS NON-MOBILIZATIONAL (78)	0.60	93	TEND TO BE THOSE WHERE THE SYSTEM STYLE IS MOBILIZATIONAL, OR LIMITED MOBILIZATIONAL (32)	0.93	8 14 12 1 X SQ= 8.28 P = 0.002 RV YES (YES)
99	TEND TO BE THOSE WHERE GOVERNMENTAL STABILITY IS MODERATELY PRESENT AND DATES FROM THE POST-WAR PERIOD, OR IS ABSENT (36)	0.70	99	ALWAYS ARE THOSE WHERE GOVERNMENTAL STABILITY IS GENERALLY PRESENT AND DATES FROM AT LEAST THE INTER-WAR PERIOD, OR FROM THE POST-WAR PERIOD (50)	1.00	6 14 14 0 X SQ= 13.90 P = 0. RV YES (YES)
110	LEAN TOWARD BEING THOSE WHERE POLITICAL ENCULTURATION IS MEDIUM OR LOW (80)	0.95	110	LEAN TOWARD BEING THOSE WHERE POLITICAL ENCULTURATION IS HIGH (15)	0.50	1 2 20 2 X SQ= 2.93 P = 0.057 RV YES (YES)

114	TILT TOWARD BEING THOSE WHERE SECTIONALISM IS EXTREME OR MODERATE (61)	0.71	114	TILT TOWARD BEING THOSE WHERE SECTIONALISM IS NEGLIGIBLE (47)	0.56	15 7 6 9 X SQ= 1.85 P = 0.107 RV YES (YES)
121	TEND TO BE THOSE WHERE INTEREST ARTICULATION BY NON-ASSOCIATIONAL GROUPS IS SIGNIFICANT (54)	0.74	121	TEND TO BE THOSE WHERE INTEREST ARTICULATION BY NON-ASSOCIATIONAL GROUPS IS MODERATE, LIMITED, OR NEGLIGIBLE (61)	0.75	17 4 6 12 X SQ= 7.22 P = 0.004 RV YES (YES)
128	TILT LESS TOWARD BEING THOSE WHERE INTEREST ARTICULATION BY POLITICAL PARTIES IS LIMITED OR NEGLIGIBLE (45)	0.78	128	ALWAYS ARE THOSE WHERE INTEREST ARTICULATION BY POLITICAL PARTIES IS LIMITED OR NEGLIGIBLE (45)	1.00	2 0 7 16 X SQ= 1.44 P = 0.120 RV NO (YES)
129	TEND LESS TO BE THOSE WHERE INTEREST ARTICULATION BY POLITICAL PARTIES IS NEGLIGIBLE (37)	0.56	129	TEND MORE TO BE THOSE WHERE INTEREST ARTICULATION BY POLITICAL PARTIES IS NEGLIGIBLE (37)	0.94	4 1 5 15 X SQ= 3.14 P = 0.040 RV NO (YES)
133	TEND LESS TO BE THOSE WHERE INTEREST AGGREGATION BY THE EXECUTIVE IS MODERATE, LIMITED, OR NEGLIGIBLE (76)	0.71	133	ALWAYS ARE THOSE WHERE INTEREST AGGREGATION BY THE EXECUTIVE IS MODERATE, LIMITED, OR NEGLIGIBLE (76)	1.00	6 0 15 16 X SQ= 3.56 P = 0.027 RV NO (YES)
134	TILT LESS TOWARD BEING THOSE WHERE INTEREST AGGREGATION BY THE EXECUTIVE IS LIMITED OR NEGLIGIBLE (46)	0.70	134	TILT MORE TOWARD BEING THOSE WHERE INTEREST AGGREGATION BY THE EXECUTIVE IS LIMITED OR NEGLIGIBLE (46)	0.94	6 1 14 15 X SQ= 1.86 P = 0.104 RV NO (YES)
139	TEND TO BE THOSE WHERE THE PARTY SYSTEM IS QUANTITATIVELY OTHER THAN ONE-PARTY (71)	0.59	139	TEND TO BE THOSE WHERE THE PARTY SYSTEM IS QUANTITATIVELY ONE-PARTY (34)	0.94	7 15 10 1 X SQ= 8.02 P = 0.002 RV YES (YES)
146	TILT LESS TOWARD BEING THOSE WHERE THE PARTY SYSTEM IS QUALITATIVELY OTHER THAN MASS-BASED TERRITORIAL (92)	0.79	146	ALWAYS ARE THOSE WHERE THE PARTY SYSTEM IS QUALITATIVELY OTHER THAN MASS-BASED TERRITORIAL (92)	1.00	4 0 15 15 X SQ= 1.84 P = 0.113 RV NO (YES)
153	TEND TO BE THOSE WHERE THE PARTY SYSTEM IS MODERATELY STABLE OR UNSTABLE (38)	0.85	153	ALWAYS ARE THOSE WHERE THE PARTY SYSTEM IS STABLE (42)	1.00	2 15 11 0 X SQ= 17.51 P = 0. RV YES (YES)

154	TEND TO BE THOSE WHERE THE PARTY SYSTEM IS UNSTABLE (25)	0.54	154	ALWAYS ARE THOSE WHERE THE PARTY SYSTEM IS STABLE OR MODERATELY STABLE (55)	1.00	6 15 7 0 X SQ= 8.09 P = 0.001 RV YES (YES)
158	TEND TO BE THOSE WHERE PERSONALISMO IS PRONOUNCED OR MODERATE (40)	0.77	158	ALWAYS ARE THOSE WHERE PERSONALISMO IS NEGLIGIBLE (56)	1.00	10 0 3 16 X SQ= 15.53 P = 0. RV YES (YES)
160	LEAN LESS TOWARD BEING THOSE WHERE THE POLITICAL LEADERSHIP IS ELITIST (30)	0.73	160	ALWAYS ARE THOSE WHERE THE POLITICAL LEADERSHIP IS ELITIST (30)	1.00	11 15 4 0 X SQ= 2.60 P = 0.100 RV NO (YES)
170	TEND TO BE THOSE WHERE THE LEGISLATIVE-EXECUTIVE STRUCTURE IS PRESIDENTIAL (39)	0.53	170	ALWAYS ARE THOSE WHERE THE LEGISLATIVE-EXECUTIVE STRUCTURE IS OTHER THAN PRESIDENTIAL (63)	1.00	9 0 8 15 X SQ= 8.58 P = 0.001 RV YES (YES)
176	TEND LESS TO BE THOSE WHERE THE LEGISLATURE IS WHOLLY INEFFECTIVE (28)	0.65	176	ALWAYS ARE THOSE WHERE THE LEGISLATURE IS WHOLLY INEFFECTIVE (28)	1.00	6 0 11 15 X SQ= 4.41 P = 0.019 RV NO (YES)
178	TILT LESS TOWARD BEING THOSE WHERE THE LEGISLATURE IS UNICAMERAL (53)	0.53	178	TILT MORE TOWARD BEING THOSE WHERE THE LEGISLATURE IS UNICAMERAL (53)	0.80	9 12 8 3 X SQ= 1.53 P = 0.147 RV NO (YES)
182	TEND LESS TO BE THOSE WHERE THE BUREAUCRACY IS SEMI-MODERN, RATHER THAN POST-COLONIAL TRANSITIONAL (55)	0.67	182	ALWAYS ARE THOSE WHERE THE BUREAUCRACY IS SEMI-MODERN, RATHER THAN POST-COLONIAL TRANSITIONAL (55)	1.00	8 16 4 0 X SQ= 3.80 P = 0.024 RV NO (YES)
185	TEND LESS TO BE THOSE WHERE PARTICIPATION BY THE MILITARY IN POLITICS IS SUPPORTIVE, RATHER THAN INTERVENTIVE (31)	0.55	185	ALWAYS ARE THOSE WHERE PARTICIPATION BY THE MILITARY IN POLITICS IS SUPPORTIVE, RATHER THAN INTERVENTIVE (31)	1.00	9 0 11 15 X SQ= 6.88 P = 0.004 RV NO (YES)
192	TEND LESS TO BE THOSE WHERE THE CHARACTER OF THE LEGAL SYSTEM IS OTHER THAN MUSLIM OR PARTLY MUSLIM (86)	0.57	192	ALWAYS ARE THOSE WHERE THE CHARACTER OF THE LEGAL SYSTEM IS OTHER THAN MUSLIM OR PARTLY MUSLIM (86)	1.00	9 0 12 16 X SQ= 6.88 P = 0.005 RV NO (YES)

```
193  TEND LESS TO BE THOSE                      0.68     193  ALWAYS ARE THOSE                           1.00        7    0
     WHERE THE CHARACTER OF THE LEGAL SYSTEM                  WHERE THE CHARACTER OF THE LEGAL SYSTEM               15   16
     IS OTHER THAN PARTLY INDIGENOUS  (86)                    IS OTHER THAN PARTLY INDIGENOUS  (86)             X SQ=   4.30
                                                                                                                P =     0.014
                                                                                                                RV NO   (YES)

194  ALWAYS ARE THOSE                           1.00     194  TEND TO BE THOSE                           0.87        0   13
     THAT ARE NON-COMMUNIST  (101)                            THAT ARE COMMUNIST  (13)                              23    2
                                                                                                                X SQ=  26.57
                                                                                                                P =    0.
                                                                                                                RV YES (YES)
```

****

98	POLITIES WHERE GOVERNMENTAL STABILITY IS GENERALLY PRESENT AND DATES AT LEAST FROM THE INTERWAR PERIOD (22)	98	POLITIES WHERE GOVERNMENTAL STABILITY IS GENERALLY OR MODERATELY PRESENT AND DATES FROM THE POST-WAR PERIOD, OR IS ABSENT (93)

BOTH SUBJECT AND PREDICATE

....

22 IN LEFT COLUMN

AFGHANISTAN	AUSTRALIA	BELGIUM	CANADA	DENMARK	FINLAND	ICELAND	IRELAND	LIBERIA	LUXEMBOURG
NETHERLANDS	NEW ZEALAND	NORWAY	PORTUGAL	SA'U ARABIA	SO AFRICA	SPAIN	SWEDEN	SWITZERLAND	USSR
UK	US								

93 IN RIGHT COLUMN

ALBANIA	ALGERIA	ARGENTINA	AUSTRIA	BOLIVIA	BRAZIL	BULGARIA	BURMA	BURUNDI	CAMBODIA
CAMEROUN	CEN AFR REP	CEYLON	CHAD	CHILE	CHINA, PR	COLOMBIA	CONGO(BRA)	CONGO(LEO)	COSTA RICA
CUBA	CYPRUS	CZECHOS'KIA	DAHOMEY	DOMIN REP	ECUADOR	EL SALVADOR	ETHIOPIA	FRANCE	GABON
GERMANY, E	GERMAN FR	GHANA	GREECE	GUATEMALA	GUINEA	HAITI	HONDURAS	HUNGARY	INDIA
INDONESIA	IRAN	IRAQ	ISRAEL	ITALY	IVORY COAST	JAMAICA	JAPAN	JORDAN	KOREA, N
KOREA REP	LAOS	LEBANON	LIBYA	MALAGASY R	MALAYA	MALI	MAURITANIA	MEXICO	MONGOLIA
MOROCCO	NEPAL	NICARAGUA	NIGER	NIGERIA	PAKISTAN	PANAMA	PARAGUAY	PERU	PHILIPPINES
POLAND	RUMANIA	RWANDA	SENEGAL	SIERRE LEO	SOMALIA	SUDAN	SYRIA	TANGANYIKA	THAILAND
TOGO	TRINIDAD	TUNISIA	TURKEY	UGANDA	UAR	UPPER VOLTA	URUGUAY	VENEZUELA	VIETNAM, N
VIETNAM REP	YEMEN	YUGOSLAVIA							

2	TEND TO BE THOSE LOCATED IN WEST EUROPE, SCANDINAVIA, NORTH AMERICA, OR AUSTRALASIA (22)	0.77	2	TEND TO BE THOSE LOCATED ELSEWHERE THAN IN WEST EUROPE, SCANDINAVIA, NORTH AMERICA, OR AUSTRALASIA (93)	0.95	17 5 5 88 X SQ= 54.89 P = 0. RV YES (YES)
3	TEND LESS TO BE THOSE LOCATED ELSEWHERE THAN IN WEST EUROPE (102)	0.64	3	TEND MORE TO BE THOSE LOCATED ELSEWHERE THAN IN WEST EUROPE (102)	0.95	8 5 14 88 X SQ= 14.09 P = 0. RV NO (YES)
5	TEND TO BE THOSE LOCATED IN SCANDINAVIA OR WEST EUROPE, RATHER THAN IN EAST EUROPE (18)	0.93	5	TEND TO BE THOSE LOCATED IN EAST EUROPE, RATHER THAN IN SCANDINAVIA OR WEST EUROPE (9)	0.62	13 5 1 8 X SQ= 6.69 P = 0.004 RV YES (YES)
6	ALWAYS ARE THOSE LOCATED ELSEWHERE THAN IN THE CARIBBEAN, CENTRAL AMERICA, OR SOUTH AMERICA (93)	1.00	6	TEND LESS TO BE THOSE LOCATED ELSEWHERE THAN IN THE CARIBBEAN, CENTRAL AMERICA, OR SOUTH AMERICA (93)	0.76	0 22 22 71 X SQ= 5.00 P = 0.007 RV NO (NO)

8	ALWAYS ARE THOSE LOCATED ELSEWHERE THAN IN EAST ASIA SOUTH ASIA, OR SOUTHEAST ASIA (97)	1.00	8	TEND LESS TO BE THOSE LOCATED ELSEWHERE THAN IN EAST ASIA SOUTH ASIA, OR SOUTHEAST ASIA (97)	0.81	0 18 22 75 X SQ= 3.69 P = 0.022 RV NO (NO)
10	TEND MORE TO BE THOSE LOCATED ELSEWHERE THAN IN NORTH AFRICA, OR CENTRAL AND SOUTH AFRICA (82)	0.91	10	TEND LESS TO BE THOSE LOCATED ELSEWHERE THAN IN NORTH AFRICA, OR CENTRAL AND SOUTH AFRICA (82)	0.67	2 31 20 62 X SQ= 3.99 P = 0.034 RV NO (NO)
11	LEAN MORE TOWARD BEING THOSE LOCATED ELSEWHERE THAN IN CENTRAL AND SOUTH AFRICA (87)	0.91	11	LEAN LESS TOWARD BEING THOSE LOCATED ELSEWHERE THAN IN CENTRAL AND SOUTH AFRICA (87)	0.72	2 26 20 67 X SQ= 2.49 P = 0.095 RV NO (NO)
16	ALWAYS ARE THOSE LOCATED IN THE MIDDLE EAST, RATHER THAN IN EAST ASIA, SOUTH ASIA, OR SOUTHEAST ASIA (11)	1.00	16	TILT TOWARD BEING THOSE LOCATED IN EAST ASIA, SOUTH ASIA, OR SOUTHEAST ASIA, RATHER THAN IN THE MIDDLE EAST (18)	0.67	2 9 0 18 X SQ= 1.25 P = 0.135 RV YES (NO)
17	ALWAYS ARE THOSE LOCATED IN THE MIDDLE EAST, RATHER THAN IN THE CARIBBEAN, CENTRAL AMERICA, OR SOUTH AMERICA (11)	1.00	17	TILT TOWARD BEING THOSE LOCATED IN THE CARIBBEAN, CENTRAL AMERICA, OR SOUTH AMERICA, RATHER THAN IN THE MIDDLE EAST (22)	0.71	2 9 0 22 X SQ= 1.66 P = 0.104 RV YES (NO)
28	TEND TO BE THOSE WHOSE POPULATION GROWTH RATE IS LOW (48)	0.71	28	TEND TO BE THOSE WHOSE POPULATION GROWTH RATE IS HIGH (62)	0.63	6 56 15 33 X SQ= 6.81 P = 0.006 RV YES (NO)
29	TEND TO BE THOSE WHERE THE DEGREE OF URBANIZATION IS HIGH (56)	0.86	29	TEND TO BE THOSE WHERE THE DEGREE OF URBANIZATION IS LOW (49)	0.55	18 38 3 46 X SQ= 9.49 P = 0.001 RV YES (NO)
30	TEND TO BE THOSE WHOSE AGRICULTURAL POPULATION IS MEDIUM, LOW, OR VERY LOW (57)	0.90	30	TEND TO BE THOSE WHOSE AGRICULTURAL POPULATION IS HIGH (56)	0.59	2 54 19 38 X SQ= 14.63 P = 0. RV YES (NO)
31	TEND TO BE THOSE WHOSE AGRICULTURAL POPULATION IS LOW OR VERY LOW (24)	0.67	31	TEND TO BE THOSE WHOSE AGRICULTURAL POPULATION IS HIGH OR MEDIUM (90)	0.89	7 83 14 10 X SQ= 28.95 P = 0. RV YES (YES)

33	TEND TO BE THOSE WHOSE GROSS NATIONAL PRODUCT IS VERY HIGH, HIGH, OR MEDIUM (30)	0.55	33	TEND TO BE THOSE WHOSE GROSS NATIONAL PRODUCT IS LOW OR VERY LOW (85)	0.81	12 18 10 75 X SQ= 9.67 P = 0.002 RV YES (NO)
34	TEND TO BE THOSE WHOSE GROSS NATIONAL PRODUCT IS VERY HIGH, HIGH, MEDIUM, OR LOW (62)	0.82	34	TEND TO BE THOSE WHOSE GROSS NATIONAL PRODUCT IS VERY LOW (53)	0.53	18 44 4 49 X SQ= 7.19 P = 0.004 RV YES (NO)
35	TEND TO BE THOSE WHOSE PER CAPITA GROSS NATIONAL PRODUCT IS VERY HIGH OR HIGH (24)	0.73	35	TEND TO BE THOSE WHOSE PER CAPITA GROSS NATIONAL PRODUCT IS MEDIUM, LOW, OR VERY LOW (91)	0.91	16 8 6 85 X SQ= 40.50 P = 0. RV YES (YES)
36	TEND TO BE THOSE WHOSE PER CAPITA GROSS NATIONAL PRODUCT IS VERY HIGH, HIGH, OR MEDIUM (42)	0.82	36	TEND TO BE THOSE WHOSE PER CAPITA GROSS NATIONAL PRODUCT IS LOW OR VERY LOW (73)	0.74	18 24 4 69 X SQ= 21.72 P = 0. RV YES (NO)
37	TEND TO BE THOSE WHOSE PER CAPITA GROSS NATIONAL PRODUCT IS VERY HIGH, HIGH, MEDIUM, OR LOW (64)	0.91	37	TEND TO BE THOSE WHOSE PER CAPITA GROSS NATIONAL PRODUCT IS VERY LOW (51)	0.53	20 44 2 49 X SQ= 11.99 P = 0. RV YES (NO)
39	TEND TO BE THOSE WHOSE INTERNATIONAL FINANCIAL STATUS IS VERY HIGH, HIGH, OR MEDIUM (38)	0.68	39	TEND TO BE THOSE WHOSE INTERNATIONAL FINANCIAL STATUS IS LOW OR VERY LOW (76)	0.75	15 23 7 69 X SQ= 13.02 P = 0. RV YES (NO)
41	TEND TO BE THOSE WHOSE ECONOMIC DEVELOPMENTAL STATUS IS DEVELOPED (19)	0.64	41	TEND TO BE THOSE WHOSE ECONOMIC DEVELOPMENTAL STATUS IS INTERMEDIATE, UNDERDEVELOPED, OR VERY UNDERDEVELOPED (94)	0.95	14 5 8 86 X SQ= 38.76 P = 0. RV YES (YES)
42	TEND TO BE THOSE WHOSE ECONOMIC DEVELOPMENTAL STATUS IS DEVELOPED OR INTERMEDIATE (36)	0.81	42	TEND TO BE THOSE WHOSE ECONOMIC DEVELOPMENTAL STATUS IS UNDERDEVELOPED OR VERY UNDERDEVELOPED (76)	0.79	17 19 4 72 X SQ= 25.54 P = 0. RV YES (NO)
43	TEND TO BE THOSE WHOSE ECONOMIC DEVELOPMENTAL STATUS IS DEVELOPED, INTERMEDIATE, OR UNDERDEVELOPED (55)	0.90	43	TEND TO BE THOSE WHOSE ECONOMIC DEVELOPMENTAL STATUS IS VERY UNDERDEVELOPED (57)	0.60	19 36 2 55 X SQ= 15.72 P = 0. RV YES (NO)

44	TEND TO BE THOSE WHERE THE LITERACY RATE IS NINETY PERCENT OR ABOVE (25)	0.73	44	TEND TO BE THOSE WHERE THE LITERACY RATE IS BELOW NINETY PERCENT (90)	0.90	16 9 6 84 X SQ= 37.95 P = 0. RV YES (YES)
45	TEND TO BE THOSE WHERE THE LITERACY RATE IS FIFTY PERCENT OR ABOVE (55)	0.86	45	TEND TO BE THOSE WHERE THE LITERACY RATE IS BELOW FIFTY PERCENT (54)	0.58	18 37 3 51 X SQ= 11.25 P = 0. RV YES (NO)
46	LEAN MORE TOWARD BEING THOSE WHERE THE LITERACY RATE IS TEN PERCENT OR ABOVE (84)	0.91	46	LEAN LESS TOWARD BEING THOSE WHERE THE LITERACY RATE IS TEN PERCENT OR ABOVE (84)	0.73	20 64 2 24 X SQ= 2.29 P = 0.094 RV NO (NO)
50	TEND TO BE THOSE WHERE FREEDOM OF THE PRESS IS COMPLETE (43)	0.68	50	TEND TO BE THOSE WHERE FREEDOM OF THE PRESS IS INTERMITTENT, INTERNALLY ABSENT, OR INTERNALLY AND EXTERNALLY ABSENT (56)	0.64	15 28 7 49 X SQ= 5.82 P = 0.014 RV YES (NO)
53	TEND TO BE THOSE WHERE NEWSPAPER CIRCULATION IS THREE HUNDRED OR MORE PER THOUSAND (14)	0.50	53	TEND TO BE THOSE WHERE NEWSPAPER CIRCULATION IS LESS THAN THREE HUNDRED PER THOUSAND (101)	0.97	11 3 11 90 X SQ= 32.16 P = 0. RV YES (YES)
54	TEND TO BE THOSE WHERE NEWSPAPER CIRCULATION IS ONE HUNDRED OR MORE PER THOUSAND (37)	0.73	54	TEND TO BE THOSE WHERE NEWSPAPER CIRCULATION IS LESS THAN ONE HUNDRED PER THOUSAND (76)	0.77	16 21 6 70 X SQ= 17.64 P = 0. RV YES (NO)
56	TEND MORE TO BE THOSE WHERE THE RELIGION IS PREDOMINANTLY LITERATE (79)	0.91	56	TEND LESS TO BE THOSE WHERE THE RELIGION IS PREDOMINANTLY LITERATE (79)	0.67	20 59 2 29 X SQ= 3.84 P = 0.033 RV NO (NO)
62	TEND TO BE THOSE WHERE THE RELIGION IS PROTESTANT, RATHER THAN CATHOLIC (5)	0.50	62	ALWAYS ARE THOSE WHERE THE RELIGION IS CATHOLIC, RATHER THAN PROTESTANT (25)	1.00	5 0 5 20 X SQ= 8.67 P = 0.002 RV YES (YES)
63	TEND TO BE THOSE WHERE THE RELIGION IS PREDOMINANTLY SOME KIND OF CHRISTIAN (46)	0.81	63	TEND TO BE THOSE WHERE THE RELIGION IS PREDOMINANTLY OR PARTLY OTHER THAN CHRISTIAN (68)	0.69	17 29 4 64 X SQ= 15.62 P = 0. RV YES (NO)

No.	Left group	Value	No.	Right group	Value	Statistics
64	LEAN MORE TOWARD BEING THOSE WHERE THE RELIGION IS CHRISTIAN, RATHER THAN MUSLIM (46)	0.89	64	LEAN LESS TOWARD BEING THOSE WHERE THE RELIGION IS CHRISTIAN, RATHER THAN MUSLIM (46)	0.64	17 29 2 16 X SQ= 2.99 P = 0.066 RV NO (NO)
71	TEND TO BE THOSE WHOSE DATE OF INDEPENDENCE IS BEFORE 1800 (21)	0.55	71	TEND TO BE THOSE WHOSE DATE OF INDEPENDENCE IS AFTER 1800 (90)	0.90	12 9 10 80 X SQ= 19.90 P = 0. RV YES (YES)
72	TEND TO BE THOSE WHOSE DATE OF INDEPENDENCE IS BEFORE 1914 (52)	0.82	72	TEND TO BE THOSE WHOSE DATE OF INDEPENDENCE IS AFTER 1914 (59)	0.62	18 34 4 55 X SQ= 11.78 P = 0. RV YES (NO)
73	ALWAYS ARE THOSE WHOSE DATE OF INDEPENDENCE IS BEFORE 1945 (65)	1.00	73	TEND TO BE THOSE WHOSE DATE OF INDEPENDENCE IS AFTER 1945 (46)	0.52	22 43 0 46 X SQ= 17.35 P = 0. RV YES (NO)
74	TEND TO BE THOSE THAT ARE HISTORICALLY WESTERN (26)	0.81	74	TEND TO BE THOSE THAT ARE NOT HISTORICALLY WESTERN (87)	0.90	17 9 4 83 X SQ= 44.95 P = 0. RV YES (YES)
75	TEND TO BE THOSE THAT ARE HISTORICALLY WESTERN OR SIGNIFICANTLY WESTERNIZED (62)	0.86	75	TEND TO BE THOSE THAT ARE NOT HISTORICALLY WESTERN AND ARE NOT SIGNIFICANTLY WESTERNIZED (52)	0.53	18 44 3 49 X SQ= 8.70 P = 0.001 RV YES (NO)
76	TEND TO BE THOSE THAT ARE HISTORICALLY WESTERN, RATHER THAN HAVING BEEN SIGNIFICANTLY OR PARTIALLY WESTERNIZED THROUGH A COLONIAL RELATIONSHIP (26)	0.94	76	TEND TO BE THOSE THAT HAVE BEEN SIGNIFICANTLY OR PARTIALLY WESTERNIZED THROUGH A COLONIAL RELATIONSHIP, RATHER THAN BEING HISTORICALLY WESTERN (70)	0.88	17 9 1 69 X SQ= 46.79 P = 0. RV YES (YES)
81	TEND TO BE THOSE WHERE THE TYPE OF POLITICAL MODERNIZATION IS EARLY EUROPEAN OR EARLY EUROPEAN DERIVED, RATHER THAN LATER EUROPEAN OR LATER EUROPEAN DERIVED (11)	0.56	81	TEND TO BE THOSE WHERE THE TYPE OF POLITICAL MODERNIZATION IS LATER EUROPEAN OR LATER EUROPEAN DERIVED, RATHER THAN EARLY EUROPEAN OR EARLY EUROPEAN DERIVED (40)	0.97	10 1 8 32 X SQ= 16.02 P = 0. RV YES (YES)
82	TEND TO BE THOSE WHERE THE TYPE OF POLITICAL MODERNIZATION IS EARLY OR LATER EUROPEAN OR EUROPEAN DERIVED, RATHER THAN DEVELOPED TUTELARY OR UNDEVELOPED TUTELARY (51)	0.90	82	TEND TO BE THOSE WHERE THE TYPE OF POLITICAL MODERNIZATION IS DEVELOPED TUTELARY OR UNDEVELOPED TUTELARY, RATHER THAN EARLY OR LATER EUROPEAN OR EUROPEAN DERIVED (55)	0.62	18 33 2 53 X SQ= 15.32 P = 0. RV YES (NO)

85	TEND TO BE THOSE WHERE THE STAGE OF POLITICAL MODERNIZATION IS ADVANCED, RATHER THAN MID- OR EARLY TRANSITIONAL (60)	0.86	85	TEND TO BE THOSE WHERE THE STAGE OF POLITICAL MODERNIZATION IS MID- OR EARLY TRANSITIONAL, RATHER THAN ADVANCED (54)	0.55	18 42 3 51 X SQ= 9.73 P = 0.001 RV YES (NO)
87	ALWAYS ARE THOSE WHOSE IDEOLOGICAL ORIENTATION IS OTHER THAN DEVELOPMENTAL (58)	1.00	87	TEND LESS TO BE THOSE WHOSE IDEOLOGICAL ORIENTATION IS OTHER THAN DEVELOPMENTAL (58)	0.54	0 31 21 37 X SQ= 12.75 P = 0. RV NO (NO)
89	TEND TO BE THOSE WHOSE IDEOLOGICAL ORIENTATION IS CONVENTIONAL (33)	0.84	89	TEND TO BE THOSE WHOSE IDEOLOGICAL ORIENTATION IS OTHER THAN CONVENTIONAL (62)	0.78	16 17 3 59 X SQ= 22.99 P = 0. RV YES (NO)
91	ALWAYS ARE THOSE WHOSE IDEOLOGICAL ORIENTATION IS TRADITIONAL, RATHER THAN DEVELOPMENTAL (5)	1.00	91	TEND TO BE THOSE WHOSE IDEOLOGICAL ORIENTATION IS DEVELOPMENTAL, RATHER THAN TRADITIONAL (31)	0.91	0 31 2 3 X SQ= 6.61 P = 0.016 RV YES (NO)
92	TILT MORE TOWARD BEING THOSE WHERE THE SYSTEM STYLE IS LIMITED MOBILIZATIONAL, OR NON-MOBILIZATIONAL (93)	0.95	92	TILT LESS TOWARD BEING THOSE WHERE THE SYSTEM STYLE IS LIMITED MOBILIZATIONAL, OR NON-MOBILIZATIONAL (93)	0.79	1 19 20 73 X SQ= 1.97 P = 0.116 RV NO (NO)
93	TEND MORE TO BE THOSE WHERE THE SYSTEM STYLE IS NON-MOBILIZATIONAL (78)	0.95	93	TEND LESS TO BE THOSE WHERE THE SYSTEM STYLE IS NON-MOBILIZATIONAL (78)	0.65	1 31 20 58 X SQ= 6.06 P = 0.006 RV NO (NO)
94	TILT MORE TOWARD BEING THOSE WHERE THE STATUS OF THE REGIME IS CONSTITUTIONAL (51)	0.71	94	TILT LESS TOWARD BEING THOSE WHERE THE STATUS OF THE REGIME IS CONSTITUTIONAL (51)	0.51	15 36 6 35 X SQ= 2.04 P = 0.134 RV NO (NO)
101	TEND TO BE THOSE WHERE THE REPRESENTATIVE CHARACTER OF THE REGIME IS POLYARCHIC (41)	0.68	101	TEND TO BE THOSE WHERE THE REPRESENTATIVE CHARACTER OF THE REGIME IS LIMITED POLYARCHIC, PSEUDO-POLYARCHIC, OR NON-POLYARCHIC (57)	0.66	15 26 7 50 X SQ= 6.76 P = 0.007 RV YES (NO)
102	TEND TO BE THOSE WHERE THE REPRESENTATIVE CHARACTER OF THE REGIME IS POLYARCHIC OR LIMITED POLYARCHIC (59)	0.77	102	TEND TO BE THOSE WHERE THE REPRESENTATIVE CHARACTER OF THE REGIME IS PSEUDO-POLYARCHIC OR NON-POLYARCHIC (49)	0.51	17 42 5 44 X SQ= 4.63 P = 0.018 RV YES (NO)

105	TEND TO BE THOSE WHERE THE ELECTORAL SYSTEM IS COMPETITIVE (43)	0.75	105	TEND TO BE THOSE WHERE THE ELECTORAL SYSTEM IS PARTIALLY COMPETITIVE OR NON-COMPETITIVE (47)	0.60	15 28 5 42 X SQ= 6.30 P = 0.010 RV YES (NO)
107	TEND TO BE THOSE WHERE AUTONOMOUS GROUPS ARE FULLY TOLERATED IN POLITICS (46)	0.68	107	TEND TO BE THOSE WHERE AUTONOMOUS GROUPS ARE PARTIALLY TOLERATED IN POLITICS, ARE TOLERATED ONLY OUTSIDE POLITICS, OR ARE NOT TOLERATED AT ALL (65)	0.65	15 31 7 58 X SQ= 6.77 P = 0.007 RV YES (NO)
110	TEND LESS TO BE THOSE WHERE POLITICAL ENCULTURATION IS MEDIUM OR LOW (80)	0.53	110	TEND MORE TO BE THOSE WHERE POLITICAL ENCULTURATION IS MEDIUM OR LOW (80)	0.92	9 6 10 70 X SQ= 14.97 P = 0. RV NO (YES)
111	TILT MORE TOWARD BEING THOSE WHERE POLITICAL ENCULTURATION IS HIGH OR MEDIUM (53)	0.74	111	TILT LESS TOWARD BEING THOSE WHERE POLITICAL ENCULTURATION IS HIGH OR MEDIUM (53)	0.51	14 39 5 37 X SQ= 2.24 P = 0.120 RV NO (NO)
115	TEND TO BE THOSE WHERE INTEREST ARTICULATION BY ASSOCIATIONAL GROUPS IS SIGNIFICANT (20)	0.65	115	TEND TO BE THOSE WHERE INTEREST ARTICULATION BY ASSOCIATIONAL GROUPS IS MODERATE, LIMITED, OR NEGLIGIBLE (91)	0.92	13 7 7 84 X SQ= 32.68 P = 0. RV YES (YES)
116	TEND TO BE THOSE WHERE INTEREST ARTICULATION BY ASSOCIATIONAL GROUPS IS SIGNIFICANT OR MODERATE (32)	0.80	116	TEND TO BE THOSE WHERE INTEREST ARTICULATION BY ASSOCIATIONAL GROUPS IS LIMITED OR NEGLIGIBLE (79)	0.82	16 16 4 75 X SQ= 28.17 P = 0. RV YES (YES)
117	TEND TO BE THOSE WHERE INTEREST ARTICULATION BY ASSOCIATIONAL GROUPS IS SIGNIFICANT, MODERATE, OR LIMITED (60)	0.85	117	TEND TO BE THOSE WHERE INTEREST ARTICULATION BY ASSOCIATIONAL GROUPS IS NEGLIGIBLE (51)	0.53	17 43 3 48 X SQ= 7.95 P = 0.003 RV YES (NO)
119	TEND TO BE THOSE WHERE INTEREST ARTICULATION BY INSTITUTIONAL GROUPS IS MODERATE OR LIMITED (26)	0.68	119	TEND TO BE THOSE WHERE INTEREST ARTICULATION BY INSTITUTIONAL GROUPS IS VERY SIGNIFICANT OR SIGNIFICANT (74)	0.86	7 67 15 11 X SQ= 23.35 P = 0. RV YES (YES)
120	TEND LESS TO BE THOSE WHERE INTEREST ARTICULATION BY INSTITUTIONAL GROUPS IS VERY SIGNIFICANT, SIGNIFICANT, OR MODERATE (90)	0.55	120	ALWAYS ARE THOSE WHERE INTEREST ARTICULATION BY INSTITUTIONAL GROUPS IS VERY SIGNIFICANT, SIGNIFICANT, OR MODERATE (90)	1.00	12 78 10 0 X SQ= 34.51 P = 0. RV NO (YES)

121	TEND TO BE THOSE WHERE INTEREST ARTICULATION BY NON-ASSOCIATIONAL GROUPS IS MODERATE, LIMITED, OR NEGLIGIBLE (61)	0.86	121	TEND TO BE THOSE WHERE INTEREST ARTICULATION BY NON-ASSOCIATIONAL GROUPS IS SIGNIFICANT (54)	0.55	3 51 19 42 X SQ= 10.53 P = 0.001 RV YES (NO)
122	TEND TO BE THOSE WHERE INTEREST ARTICULATION BY NON-ASSOCIATIONAL GROUPS IS LIMITED OR NEGLIGIBLE (32)	0.59	122	TEND TO BE THOSE WHERE INTEREST ARTICULATION BY NON-ASSOCIATIONAL GROUPS IS SIGNIFICANT OR MODERATE (83)	0.80	9 74 13 19 X SQ= 11.39 P = 0.001 RV YES (NO)
125	TEND TO BE THOSE WHERE INTEREST ARTICULATION BY ANOMIC GROUPS IS INFREQUENT OR VERY INFREQUENT (35)	0.76	125	TEND TO BE THOSE WHERE INTEREST ARTICULATION BY ANOMIC GROUPS IS FREQUENT OR OCCASIONAL (64)	0.76	5 59 16 19 X SQ= 17.25 P = 0. RV YES (NO)
126	TEND TO BE THOSE WHERE INTEREST ARTICULATION BY ANOMIC GROUPS IS VERY INFREQUENT (16)	0.67	126	TEND TO BE THOSE WHERE INTEREST ARTICULATION BY ANOMIC GROUPS IS FREQUENT, OCCASIONAL, OR INFREQUENT (83)	0.97	7 76 14 2 X SQ= 45.56 P = 0. RV YES (YES)
128	TEND TO BE THOSE WHERE INTEREST ARTICULATION BY POLITICAL PARTIES IS SIGNIFICANT OR MODERATE (48)	0.74	128	TEND TO BE THOSE WHERE INTEREST ARTICULATION BY POLITICAL PARTIES IS LIMITED OR NEGLIGIBLE (45)	0.54	14 34 5 40 X SQ= 3.61 P = 0.040 RV YES (NO)
132	ALWAYS ARE THOSE WHERE INTEREST AGGREGATION BY POLITICAL PARTIES IS SIGNIFICANT, MODERATE, OR LIMITED (67)	1.00	132	TILT LESS TOWARD BEING THOSE WHERE INTEREST AGGREGATION BY POLITICAL PARTIES IS SIGNIFICANT, MODERATE, OR LIMITED (67)	0.85	17 50 0 9 X SQ= 1.66 P = 0.111 RV NO (NO)
133	TEND MORE TO BE THOSE WHERE INTEREST AGGREGATION BY THE EXECUTIVE IS MODERATE, LIMITED, OR NEGLIGIBLE (76)	0.95	133	TEND LESS TO BE THOSE WHERE INTEREST AGGREGATION BY THE EXECUTIVE IS MODERATE, LIMITED, OR NEGLIGIBLE (76)	0.67	1 28 20 56 X SQ= 5.51 P = 0.007 RV NO (NO)
136	TEND TO BE THOSE WHERE INTEREST AGGREGATION BY THE LEGISLATURE IS SIGNIFICANT (12)	0.50	136	TEND TO BE THOSE WHERE INTEREST AGGREGATION BY THE LEGISLATURE IS MODERATE, LIMITED, OR NEGLIGIBLE (85)	0.97	10 2 10 75 X SQ= 28.68 P = 0. RV YES (YES)
137	TEND TO BE THOSE WHERE INTEREST AGGREGATION BY THE LEGISLATURE IS SIGNIFICANT OR MODERATE (29)	0.75	137	TEND TO BE THOSE WHERE INTEREST AGGREGATION BY THE LEGISLATURE IS LIMITED OR NEGLIGIBLE (68)	0.82	15 14 5 63 X SQ= 21.82 P = 0. RV YES (YES)

138	TEND TO BE THOSE WHERE INTEREST AGGREGATION BY THE LEGISLATURE IS SIGNIFICANT, MODERATE, OR LIMITED (48)	0.83	138	TEND TO BE THOSE WHERE INTEREST AGGREGATION BY THE LEGISLATURE IS NEGLIGIBLE (49)	0.58	15 33 3 46 X SQ= 8.54 P = 0.002 RV YES (NO)
139	TILT MORE TOWARD BEING THOSE WHERE THE PARTY SYSTEM IS QUANTITATIVELY OTHER THAN ONE-PARTY (71)	0.82	139	TILT LESS TOWARD BEING THOSE WHERE THE PARTY SYSTEM IS QUANTITATIVELY OTHER THAN ONE-PARTY (71)	0.64	4 30 18 53 X SQ= 1.81 P = 0.130 RV NO (NO)
140	ALWAYS ARE THOSE WHERE THE PARTY SYSTEM IS QUANTITATIVELY OTHER THAN ONE PARTY DOMINANT (87)	1.00	140	LEAN LESS TOWARD BEING THOSE WHERE THE PARTY SYSTEM IS QUANTITATIVELY OTHER THAN ONE PARTY DOMINANT (87)	0.84	0 13 21 66 X SQ= 2.65 P = 0.064 RV NO (NO)
144	TEND TO BE THOSE WHERE THE PARTY SYSTEM IS QUANTITATIVELY TWO-PARTY, RATHER THAN ONE-PARTY (11)	0.56	144	TEND TO BE THOSE WHERE THE PARTY SYSTEM IS QUANTITATIVELY ONE-PARTY, RATHER THAN TWO-PARTY (34)	0.83	4 30 5 6 X SQ= 3.98 P = 0.028 RV YES (NO)
147	TEND TO BE THOSE WHERE THE PARTY SYSTEM IS QUALITATIVELY CLASS-ORIENTED OR MULTI-IDEOLOGICAL (23)	0.71	147	TEND TO BE THOSE WHERE THE PARTY SYSTEM IS QUALITATIVELY OTHER THAN CLASS-ORIENTED OR MULTI-IDEOLOGICAL (67)	0.85	12 11 5 62 X SQ= 19.52 P = 0. RV YES (YES)
148	ALWAYS ARE THOSE WHERE THE PARTY SYSTEM IS QUALITATIVELY OTHER THAN AFRICAN TRANSITIONAL (96)	1.00	148	LEAN LESS TOWARD BEING THOSE WHERE THE PARTY SYSTEM IS QUALITATIVELY OTHER THAN AFRICAN TRANSITIONAL (96)	0.84	0 14 20 76 X SQ= 2.30 P = 0.069 RV NO (NO)
153	ALWAYS ARE THOSE WHERE THE PARTY SYSTEM IS STABLE (42)	1.00	153	TEND TO BE THOSE WHERE THE PARTY SYSTEM IS MODERATELY STABLE OR UNSTABLE (38)	0.63	20 22 0 38 X SQ= 21.65 P = 0. RV YES (NO)
154	ALWAYS ARE THOSE WHERE THE PARTY SYSTEM IS STABLE OR MODERATELY STABLE (55)	1.00	154	TEND LESS TO BE THOSE WHERE THE PARTY SYSTEM IS STABLE OR MODERATELY STABLE (55)	0.58	20 35 0 25 X SQ= 10.26 P = 0. RV NO (NO)
157	ALWAYS ARE THOSE WHERE PERSONALISMO IS MODERATE OR NEGLIGIBLE (84)	1.00	157	TEND LESS TO BE THOSE WHERF PERSONALISMO IS MODERATE OR NEGLIGIBLE (84)	0.82	0 14 20 64 X SQ= 2.85 P = 0.039 RV NO (NO)

158	TEND TO BE THOSE WHERE PERSONALISMO IS NEGLIGIBLE (56)	0.95	158	TEND TO BE THOSE WHERE PERSONALISMO IS PRONOUNCED OR MODERATE (40)	0.51	1 39 19 37 X SQ= 12.13 P = 0. RV YES (NO)
163	ALWAYS ARE THOSE WHERE THE REGIME'S LEADERSHIP CHARISMA IS MODERATE OR NEGLIGIBLE (87)	1.00	163	LEAN LESS TOWARD BEING THOSE WHERE THE REGIME'S LEADERSHIP CHARISMA IS MODERATE OR NEGLIGIBLE (87)	0.84	0 13 20 67 X SQ= 2.44 P = 0.065 RV NO (NO)
164	TEND MORE TO BE THOSE WHERE THE REGIME'S LEADERSHIP CHARISMA IS NEGLIGIBLE (65)	0.95	164	TEND LESS TO BE THOSE WHERE THE REGIME'S LEADERSHIP CHARISMA IS NEGLIGIBLE (65)	0.58	1 33 19 46 X SQ= 8.01 P = 0.001 RV NO (NO)
168	TEND TO BE THOSE WHERE THE HORIZONTAL POWER DISTRIBUTION IS SIGNIFICANT (34)	0.73	168	TEND TO BE THOSE WHERE THE HORIZONTAL POWER DISTRIBUTION IS LIMITED OR NEGLIGIBLE (72)	0.79	16 18 6 66 X SQ= 18.77 P = 0. RV YES (NO)
169	LEAN TOWARD BEING THOSE WHERE THE HORIZONTAL POWER DISTRIBUTION IS SIGNIFICANT OR LIMITED (58)	0.73	169	LEAN TOWARD BEING THOSE WHERE THE HORIZONTAL POWER DISTRIBUTION IS NEGLIGIBLE (48)	0.50	16 42 6 42 X SQ= 2.78 P = 0.091 RV YES (NO)
170	TEND MORE TO BE THOSE WHERE THE LEGISLATIVE-EXECUTIVE STRUCTURE IS OTHER THAN PRESIDENTIAL (63)	0.90	170	TEND LESS TO BE THOSE WHERE THE LEGISLATIVE-EXECUTIVE STRUCTURE IS OTHER THAN PRESIDENTIAL (63)	0.54	2 37 19 44 X SQ= 7.76 P = 0.002 RV NO (NO)
174	TEND TO BE THOSE WHERE THE LEGISLATURE IS FULLY EFFECTIVE (28)	0.76	174	TEND TO BE THOSE WHERE THE LEGISLATURE IS PARTIALLY EFFECTIVE, LARGELY INEFFECTIVE, OR WHOLLY INEFFECTIVE (72)	0.85	16 12 5 67 X SQ= 27.67 P = 0. RV YES (YES)
175	TEND TO BE THOSE WHERE THE LEGISLATURE IS FULLY EFFECTIVE OR PARTIALLY EFFECTIVE (51)	0.76	175	TEND TO BE THOSE WHERE THE LEGISLATURE IS LARGELY INEFFECTIVE OR WHOLLY INEFFECTIVE (49)	0.56	16 35 5 44 X SQ= 5.53 P = 0.013 RV YES (NO)
178	TEND TO BE THOSE WHERE THE LEGISLATURE IS BICAMERAL (51)	0.81	178	TEND TO BE THOSE WHERE THE LEGISLATURE IS UNICAMERAL (53)	0.59	4 49 17 34 X SQ= 9.18 P = 0.001 RV YES (NO)

No.	Statement	Value	No.	Statement	Value	Statistics
179	TEND TO BE THOSE WHERE THE EXECUTIVE IS STRONG (39)	0.67	179	TEND TO BE THOSE WHERE THE EXECUTIVE IS DOMINANT (52)	0.64	7 45 14 25 X SQ= 5.12 P = 0.022 RV YES (NO)
181	TEND TO BE THOSE WHERE THE BUREAUCRACY IS MODERN, RATHER THAN SEMI-MODERN (21)	0.80	181	TEND TO BE THOSE WHERE THE BUREAUCRACY IS SEMI-MODERN, RATHER THAN MODERN (55)	0.91	16 5 4 51 X SQ= 33.76 P = 0. RV YES (YES)
183	ALWAYS ARE THOSE WHERE THE BUREAUCRACY IS TRADITIONAL, RATHER THAN POST-COLONIAL TRANSITIONAL (9)	1.00	183	LEAN TOWARD BEING THOSE WHERE THE BUREAUCRACY IS POST-COLONIAL TRANSITIONAL, RATHER THAN TRADITIONAL (25)	0.78	0 25 2 7 X SQ= 2.57 P = 0.064 RV YES (NO)
185	ALWAYS ARE THOSE WHERE PARTICIPATION BY THE MILITARY IN POLITICS IS SUPPORTIVE, RATHER THAN INTERVENTIVE (31)	1.00	185	TEND LESS TO BE THOSE WHERE PARTICIPATION BY THE MILITARY IN POLITICS IS SUPPORTIVE, RATHER THAN INTERVENTIVE (31)	0.53	0 21 7 24 X SQ= 3.71 P = 0.033 RV NO (NO)
187	TEND TO BE THOSE WHERE THE ROLE OF THE POLICE IS NOT POLITICALLY SIGNIFICANT (35)	0.68	187	TEND TO BE THOSE WHERE THE ROLE OF THE POLICE IS POLITICALLY SIGNIFICANT (66)	0.75	7 59 15 20 X SQ= 12.13 P = 0. RV YES (NO)
189	TEND LESS TO BE THOSE WHERE THE CHARACTER OF THE LEGAL SYSTEM IS OTHER THAN COMMON LAW (108)	0.77	189	TEND MORE TO BE THOSE WHERE THE CHARACTER OF THE LEGAL SYSTEM IS OTHER THAN COMMON LAW (108)	0.98	5 2 17 91 X SQ= 9.82 P = 0.003 RV NO (YES)
190	TEND LESS TO BE THOSE WHERE THE CHARACTER OF THE LEGAL SYSTEM IS CIVIL LAW, RATHER THAN COMMON LAW (32)	0.55	190	TEND MORE TO BE THOSE WHERE THE CHARACTER OF THE LEGAL SYSTEM IS CIVIL LAW, RATHER THAN COMMON LAW (32)	0.93	6 26 5 2 X SQ= 5.48 P = 0.012 RV NO (YES)
192	TEND MORE TO BE THOSE WHERE THE CHARACTER OF THE LEGAL SYSTEM IS OTHER THAN MUSLIM OR PARTLY MUSLIM (86)	0.95	192	TEND LESS TO BE THOSE WHERE THE CHARACTER OF THE LEGAL SYSTEM IS OTHER THAN MUSLIM OR PARTLY MUSLIM (86)	0.71	1 26 21 65 X SQ= 4.38 P = 0.023 RV NO (NO)
193	LEAN MORE TOWARD BEING THOSE WHERE THE CHARACTER OF THE LEGAL SYSTEM IS OTHER THAN PARTLY INDIGENOUS (86)	0.91	193	LEAN LESS TOWARD BEING THOSE WHERE THE CHARACTER OF THE LEGAL SYSTEM IS OTHER THAN PARTLY INDIGENOUS (86)	0.72	2 26 20 66 X SQ= 2.56 P = 0.095 RV NO (NO)

99 POLITIES
WHERE GOVERNMENTAL STABILITY
IS GENERALLY PRESENT AND DATES
FROM AT LEAST THE INTER-WAR PERIOD,
OR FROM THE POST-WAR PERIOD (50)

99 POLITIES
WHERE GOVERNMENTAL STABILITY
IS MODERATELY PRESENT AND DATES
FROM THE POST-WAR PERIOD,
OR IS ABSENT (36)

BOTH SUBJECT AND PREDICATE

50 IN LEFT COLUMN

AFGHANISTAN	ALBANIA	AUSTRALIA	AUSTRIA	BELGIUM	BULGARIA	CANADA	CHINA, PR	CZECHOS'KIA	DENMARK
FINLAND	FRANCE	GERMANY, E	GERMAN FR	GHANA	GUINEA	ICELAND	INDIA	IRELAND	ISRAEL
ITALY	JAPAN	KOREA, N	LIBERIA	LIBYA	LUXEMBOURG	MALAYA	MEXICO	MONGOLIA	MOROCCO
NETHERLANDS	NEW ZEALAND	NORWAY	PHILIPPINES	POLAND	PORTUGAL	RUMANIA	SA'U ARABIA	SO AFRICA	SPAIN
SWEDEN	SWITZERLAND	TUNISIA	USSR	UAR	UK	US	URUGUAY	VIETNAM, N	YUGOSLAVIA

36 IN RIGHT COLUMN

ARGENTINA	BOLIVIA	BRAZIL	BURMA	CAMBODIA	CEYLON	CHILE	COLOMBIA	CONGO(LEO)	COSTA RICA
ECUADOR	EL SALVADOR	ETHIOPIA	GREECE	GUATEMALA	HONDURAS	INDONESIA	IRAN	IRAQ	JORDAN
KOREA REP	LAOS	LEBANON	NEPAL	NICARAGUA	PAKISTAN	PANAMA	PERU	SENEGAL	SUDAN
SYRIA	THAILAND	TOGO	TURKEY	VENEZUELA	VIETNAM REP				

3 EXCLUDED BECAUSE AMBIGUOUS

HAITI HUNGARY PARAGUAY

26 EXCLUDED BECAUSE UNASCERTAINABLE

ALGERIA	BURUNDI	CAMEROUN	CEN AFR REP	CHAD	CONGO(BRA)	CUBA	CYPRUS	DAHOMEY	DOMIN REP
GABON	IVORY COAST	JAMAICA	MALAGASY R	MALI	MAURITANIA	NIGER	NIGERIA	RWANDA	SIERRE LEO
SOMALIA	TANGANYIKA	TRINIDAD	UGANDA	UPPER VOLTA	YEMEN				

	Left column			Right column		
2	TEND LESS TO BE THOSE LOCATED ELSEWHERE THAN IN WEST EUROPE, SCANDINAVIA, NORTH AMERICA, OR AUSTRALASIA (93)	0.58	2	TEND MORE TO BE THOSE LOCATED ELSEWHERE THAN IN WEST EUROPE, SCANDINAVIA, NORTH AMERICA, OR AUSTRALASIA (93)	0.97	21 1 29 35 X SQ= 14.92 P = 0. RV NO (YES)
6	TEND MORE TO BE THOSE LOCATED ELSEWHERE THAN IN THE CARIBBEAN, CENTRAL AMERICA, OR SOUTH AMERICA (93)	0.96	6	TEND LESS TO BE THOSE LOCATED ELSEWHERE THAN IN THE CARIBBEAN, CENTRAL AMERICA, OR SOUTH AMERICA (93)	0.61	2 14 48 22 X SQ= 14.60 P = 0. RV NO (YES)
19	TEND TO BE THOSE LOCATED IN NORTH AFRICA OR CENTRAL AND SOUTH AFRICA, RATHER THAN IN THE CARIBBEAN, CENTRAL AMERICA, OR SOUTH AMERICA (33)	0.80	19	TEND TO BE THOSE LOCATED IN THE CARIBBEAN, CENTRAL AMERICA, OR SOUTH AMERICA, RATHER THAN IN NORTH AFRICA OR CENTRAL AND SOUTH AFRICA (22)	0.74	8 5 2 14 X SQ= 5.62 P = 0.016 RV YES (YES)

20	LEAN TOWARD BEING THOSE LOCATED IN EAST ASIA, SOUTH ASIA, OR SOUTHEAST ASIA, RATHER THAN IN THE CARIBBEAN, CENTRAL AMERICA, OR SOUTH AMERICA (18)	0.80	20	LEAN TOWARD BEING THOSE LOCATED IN THE CARIBBEAN, CENTRAL AMERICA, OR SOUTH AMERICA, RATHER THAN IN EAST ASIA, SOUTH ASIA, OR SOUTHEAST ASIA (22)	0.58	8 10 2 14 X SQ= 2.77 P = 0.063 RV YES (NO)
26	TEND TO BE THOSE WHOSE POPULATION DENSITY IS VERY HIGH, HIGH, OR MEDIUM (48)	0.56	26	TEND TO BE THOSE WHOSE POPULATION DENSITY IS LOW (67)	0.72	28 10 22 26 X SQ= 5.66 P = 0.015 RV YES (YES)
28	TEND TO BE THOSE WHOSE POPULATION GROWTH RATE IS LOW (48)	0.60	28	TEND TO BE THOSE WHOSE POPULATION GROWTH RATE IS HIGH (62)	0.82	19 28 28 6 X SQ= 12.57 P = 0. RV YES (YES)
29	TEND TO BE THOSE WHERE THE DEGREE OF URBANIZATION IS HIGH (56)	0.80	29	TEND TO BE THOSE WHERE THE DEGREE OF URBANIZATION IS LOW (49)	0.50	35 17 9 17 X SQ= 6.26 P = 0.008 RV YES (YES)
30	TEND TO BE THOSE WHOSE AGRICULTURAL POPULATION IS MEDIUM, LOW, OR VERY LOW (57)	0.71	30	TEND TO BE THOSE WHOSE AGRICULTURAL POPULATION IS HIGH (56)	0.58	14 21 34 15 X SQ= 6.05 P = 0.013 RV YES (YES)
32	TEND LESS TO BE THOSE WHOSE GROSS NATIONAL PRODUCT IS MEDIUM, LOW, OR VERY LOW (105)	0.80	32	ALWAYS ARE THOSE WHOSE GROSS NATIONAL PRODUCT IS MEDIUM, LOW, OR VERY LOW (105)	1.00	10 0 40 36 X SQ= 6.32 P = 0.004 RV NO (NO)
33	TEND TO BE THOSE WHOSE GROSS NATIONAL PRODUCT IS VERY HIGH, HIGH, OR MEDIUM (30)	0.50	33	TEND TO BE THOSE WHOSE GROSS NATIONAL PRODUCT IS LOW OR VERY LOW (85)	0.86	25 5 25 31 X SQ= 10.48 P = 0.001 RV YES (YES)
36	TEND TO BE THOSE WHOSE PER CAPITA GROSS NATIONAL PRODUCT IS VERY HIGH, HIGH, OR MEDIUM (42)	0.64	36	TEND TO BE THOSE WHOSE PER CAPITA GROSS NATIONAL PRODUCT IS LOW OR VERY LOW (73)	0.86	32 5 18 31 X SQ= 19.45 P = 0. RV YES (YES)
37	TEND TO BE THOSE WHOSE PER CAPITA GROSS NATIONAL PRODUCT IS VERY HIGH, HIGH, MEDIUM, OR LOW (64)	0.78	37	TEND TO BE THOSE WHOSE PER CAPITA GROSS NATIONAL PRODUCT IS VERY LOW (51)	0.50	39 18 11 18 X SQ= 6.14 P = 0.011 RV YES (YES)

38	TEND LESS TO BE THOSE WHOSE INTERNATIONAL FINANCIAL STATUS IS MEDIUM, LOW, OR VERY LOW (103)	0.79	38	ALWAYS ARE THOSE WHOSE INTERNATIONAL FINANCIAL STATUS IS MEDIUM, LOW, OR VERY LOW (103)	1.00	10 0 38 36 X SQ= 6.64 P = 0.004 RV NO (NO)
39	TEND TO BE THOSE WHOSE INTERNATIONAL FINANCIAL STATUS IS VERY HIGH, HIGH, OR MEDIUM (38)	0.59	39	TEND TO BE THOSE WHOSE INTERNATIONAL FINANCIAL STATUS IS LOW OR VERY LOW (76)	0.78	29 8 20 28 X SQ= 10.08 P = 0.001 RV YES (YES)
41	TEND LESS TO BE THOSE WHOSE ECONOMIC DEVELOPMENTAL STATUS IS INTERMEDIATE, UNDERDEVELOPED, OR VERY UNDERDEVELOPED (94)	0.61	41	ALWAYS ARE THOSE WHOSE ECONOMIC DEVELOPMENTAL STATUS IS INTERMEDIATE, UNDERDEVELOPED, OR VERY UNDERDEVELOPED (94)	1.00	19 0 30 35 X SQ= 15.39 P = 0. RV NO (YES)
42	TEND TO BE THOSE WHOSE ECONOMIC DEVELOPMENTAL STATUS IS DEVELOPED OR INTERMEDIATE (36)	0.62	42	TEND TO BE THOSE WHOSE ECONOMIC DEVELOPMENTAL STATUS IS UNDERDEVELOPED OR VERY UNDERDEVELOPED (76)	0.89	30 4 18 31 X SQ= 19.77 P = 0. RV YES (YES)
43	TEND TO BE THOSE WHOSE ECONOMIC DEVELOPMENTAL STATUS IS DEVELOPED, INTERMEDIATE, OR UNDERDEVELOPED (55)	0.78	43	TEND TO BE THOSE WHOSE ECONOMIC DEVELOPMENTAL STATUS IS VERY UNDERDEVELOPED (57)	0.65	38 12 11 22 X SQ= 13.25 P = 0. RV YES (YES)
44	TEND LESS TO BE THOSE WHERE THE LITERACY RATE IS BELOW NINETY PERCENT (90)	0.52	44	ALWAYS ARE THOSE WHERE THE LITERACY RATE IS BELOW NINETY PERCENT (90)	1.00	24 0 26 36 X SQ= 21.64 P = 0. RV NO (YES)
45	TEND TO BE THOSE WHERE THE LITERACY RATE IS FIFTY PERCENT OR ABOVE (55)	0.76	45	TEND TO BE THOSE WHERE THE LITERACY RATE IS BELOW FIFTY PERCENT (54)	0.60	34 14 11 21 X SQ= 8.94 P = 0.003 RV YES (YES)
46	TEND MORE TO BE THOSE WHERE THE LITERACY RATE IS TEN PERCENT OR ABOVE (84)	0.96	46	TEND LESS TO BE THOSE WHERE THE LITERACY RATE IS TEN PERCENT OR ABOVE (84)	0.81	44 29 2 7 X SQ= 3.29 P = 0.038 RV NO (YES)
52	TEND LESS TO BE THOSE WHERE FREEDOM OF THE PRESS IS COMPLETE, INTERMITTENT, OR INTERNALLY ABSENT (82)	0.76	52	TEND MORE TO BE THOSE WHERE FREEDOM OF THE PRESS IS COMPLETE, INTERMITTENT, OR INTERNALLY ABSENT (82)	0.96	38 27 12 1 X SQ= 4.02 P = 0.026 RV NO (NO)

53	TEND LESS TO BE THOSE WHERE NEWSPAPER CIRCULATION IS LESS THAN THREE HUNDRED PER THOUSAND (101)	0.72	53	ALWAYS ARE THOSE WHERE NEWSPAPER CIRCULATION IS LESS THAN THREE HUNDRED PER THOUSAND (101)	1.00	14 0 36 36 X SQ= 10.07 P = 0. RV NO (YES)
54	TEND TO BE THOSE WHERE NEWSPAPER CIRCULATION IS ONE HUNDRED OR MORE PER THOUSAND (37)	0.60	54	TEND TO BE THOSE WHERE NEWSPAPER CIRCULATION IS LESS THAN ONE HUNDRED PER THOUSAND (76)	0.86	29 5 19 31 X SQ= 16.60 P = 0. RV YES (YES)
56	LEAN MORE TOWARD BEING THOSE WHERE THE RELIGION IS PREDOMINANTLY LITERATE (79)	0.91	56	LEAN LESS TOWARD BEING THOSE WHERE THE RELIGION IS PREDOMINANTLY LITERATE (79)	0.74	43 25 4 9 X SQ= 3.48 P = 0.063 RV NO (YES)
62	TILT LESS TOWARD BEING THOSE WHERE THE RELIGION IS CATHOLIC, RATHER THAN PROTESTANT (25)	0.71	62	ALWAYS ARE THOSE WHERE THE RELIGION IS CATHOLIC, RATHER THAN PROTESTANT (25)	1.00	5 0 12 10 X SQ= 1.92 P = 0.124 RV NO (NO)
63	TEND TO BE THOSE WHERE THE RELIGION IS PREDOMINANTLY SOME KIND OF CHRISTIAN (46)	0.61	63	TEND TO BE THOSE WHERE THE RELIGION IS PREDOMINANTLY OR PARTLY OTHER THAN CHRISTIAN (68)	0.69	30 11 19 25 X SQ= 6.64 P = 0.008 RV YES (YES)
64	LEAN MORE TOWARD BEING THOSE WHERE THE RELIGION IS CHRISTIAN, RATHER THAN MUSLIM (46)	0.83	64	LEAN LESS TOWARD BEING THOSE WHERE THE RELIGION IS CHRISTIAN, RATHER THAN MUSLIM (46)	0.58	30 11 6 8 X SQ= 3.01 P = 0.054 RV NO (YES)
67	TEND MORE TO BE THOSE THAT ARE RACIALLY HOMOGENEOUS (82)	0.90	67	TEND LESS TO BE THOSE THAT ARE RACIALLY HOMOGENEOUS (82)	0.62	44 21 5 13 X SQ= 7.71 P = 0.003 RV NO (YES)
68	LEAN TOWARD BEING THOSE THAT ARE LINGUISTICALLY HOMOGENEOUS (52)	0.57	68	LEAN TOWARD BEING THOSE THAT ARE LINGUISTICALLY WEAKLY HETEROGENEOUS OR STRONGLY HETEROGENEOUS (62)	0.64	28 13 21 23 X SQ= 2.88 P = 0.079 RV YES (YES)
70	TEND LESS TO BE THOSE THAT ARE RELIGIOUSLY, RACIALLY, OR LINGUISTICALLY HETEROGENEOUS (85)	0.66	70	TEND MORE TO BE THOSE THAT ARE RELIGIOUSLY, RACIALLY, OR LINGUISTICALLY HETEROGENEOUS (85)	0.88	16 4 31 30 X SQ= 4.14 P = 0.035 RV NO (NO)

71	LEAN LESS TOWARD BEING THOSE WHOSE DATE OF INDEPENDENCE IS AFTER 1800 (90)	0.66	71	LEAN MORE TOWARD BEING THOSE WHOSE DATE OF INDEPENDENCE IS AFTER 1800 (90)	0.86	16 5 31 30 X SQ= 3.14 P = 0.072 RV NO (NO)
74	TEND TO BE THOSE THAT ARE HISTORICALLY WESTERN (26)	0.50	74	TEND TO BE THOSE THAT ARE NOT HISTORICALLY WESTERN (87)	0.97	24 1 24 35 X SQ= 19.74 P = 0. RV YES (YES)
75	TEND TO BE THOSE THAT ARE HISTORICALLY WESTERN OR SIGNIFICANTLY WESTERNIZED (62)	0.73	75	TEND TO BE THOSE THAT ARE NOT HISTORICALLY WESTERN AND ARE NOT SIGNIFICANTLY WESTERNIZED (52)	0.50	36 18 13 18 X SQ= 3.97 P = 0.040 RV YES (YES)
76	TEND TO BE THOSE THAT ARE HISTORICALLY WESTERN, RATHER THAN HAVING BEEN SIGNIFICANTLY OR PARTIALLY WESTERNIZED THROUGH A COLONIAL RELATIONSHIP (26)	0.65	76	TEND TO BE THOSE THAT HAVE BEEN SIGNIFICANTLY OR PARTIALLY WESTERNIZED THROUGH A COLONIAL RELATIONSHIP, RATHER THAN BEING HISTORICALLY WESTERN (70)	0.97	24 1 13 30 X SQ= 24.98 P = 0. RV YES (YES)
78	TEND TO BE THOSE THAT HAVE BEEN SIGNIFICANTLY WESTERNIZED WITHOUT A COLONIAL RELATIONSHIP, RATHER THAN THROUGH SUCH A RELATIONSHIP (7)	0.55	78	TEND TO BE THOSE THAT HAVE BEEN SIGNIFICANTLY WESTERNIZED THROUGH A COLONIAL RELATIONSHIP, RATHER THAN WITHOUT SUCH A RELATIONSHIP (28)	0.94	5 16 6 1 X SQ= 6.04 P = 0.007 RV YES (YES)
79	TEND TO BE THOSE THAT WERE FORMERLY DEPENDENCIES OF BRITAIN OR FRANCE, RATHER THAN SPAIN (49)	0.88	79	TEND TO BE THOSE THAT WERE FORMERLY DEPENDENCIES OF SPAIN, RATHER THAN BRITAIN OR FRANCE (18)	0.50	15 13 2 13 X SQ= 5.04 P = 0.020 RV YES (YES)
81	TEND LESS TO BE THOSE WHERE THE TYPE OF POLITICAL MODERNIZATION IS LATER EUROPEAN OR LATER EUROPEAN DERIVED, RATHER THAN EARLY EUROPEAN OR EARLY EUROPEAN DERIVED (40)	0.65	81	ALWAYS ARE THOSE WHERE THE TYPE OF POLITICAL MODERNIZATION IS LATER EUROPEAN OR LATER EUROPEAN DERIVED, RATHER THAN EARLY EUROPEAN OR EARLY EUROPEAN DERIVED (40)	1.00	11 0 20 15 X SQ= 5.18 P = 0.009 RV NO (NO)
82	TILT TOWARD BEING THOSE WHERE THE TYPE OF POLITICAL MODERNIZATION IS EARLY OR LATER EUROPEAN OR EUROPEAN DERIVED, RATHER THAN DEVELOPED TUTELARY OR UNDEVELOPED TUTELARY (51)	0.67	82	TILT TOWARD BEING THOSE WHERE THE TYPE OF POLITICAL MODERNIZATION IS DEVELOPED TUTELARY OR UNDEVELOPED TUTELARY, RATHER THAN EARLY OR LATER EUROPEAN OR EUROPEAN DERIVED (55)	0.52	31 15 15 16 X SQ= 2.05 P = 0.105 RV YES (YES)
85	TEND TO BE THOSE WHERE THE STAGE OF POLITICAL MODERNIZATION IS ADVANCED, RATHER THAN MID- OR EARLY TRANSITIONAL (60)	0.84	85	TEND TO BE THOSE WHERE THE STAGE OF POLITICAL MODERNIZATION IS MID- OR EARLY TRANSITIONAL, RATHER THAN ADVANCED (54)	0.58	41 15 8 21 X SQ= 14.48 P = 0. RV YES (YES)

87	LEAN MORE TOWARD BEING THOSE WHOSE IDEOLOGICAL ORIENTATION IS OTHER THAN DEVELOPMENTAL (58)	0.87	87	LEAN LESS TOWARD BEING THOSE WHOSE IDEOLOGICAL ORIENTATION IS OTHER THAN DEVELOPMENTAL (58)	0.68	6 7 40 15 X SQ= 2.29 P = 0.098 RV NO (YES)
89	LEAN TOWARD BEING THOSE WHOSE IDEOLOGICAL ORIENTATION IS CONVENTIONAL (33)	0.55	89	LEAN TOWARD BEING THOSE WHOSE IDEOLOGICAL ORIENTATION IS OTHER THAN CONVENTIONAL (62)	0.71	24 8 20 20 X SQ= 3.68 P = 0.051 RV YES (YES)
92	TEND LESS TO BE THOSE WHERE THE SYSTEM STYLE IS LIMITED MOBILIZATIONAL, OR NON-MOBILIZATIONAL (93)	0.69	92	TEND MORE TO BE THOSE WHERE THE SYSTEM STYLE IS LIMITED MOBILIZATIONAL, OR NON-MOBILIZATIONAL (93)	0.94	15 2 34 34 X SQ= 6.65 P = 0.005 RV NO (YES)
95	TEND LESS TO BE THOSE WHERE THE STATUS OF THE REGIME IS CONSTITUTIONAL OR AUTHORITARIAN (95)	0.71	95	ALWAYS ARE THOSE WHERE THE STATUS OF THE REGIME IS CONSTITUTIONAL OR AUTHORITARIAN (95)	1.00	35 35 14 0 X SQ= 10.03 P = 0. RV NO (YES)
96	TEND TO BE THOSE WHERE THE STATUS OF THE REGIME IS CONSTITUTIONAL OR TOTALITARIAN (67)	0.88	96	TEND TO BE THOSE WHERE THE STATUS OF THE REGIME IS AUTHORITARIAN (23)	0.52	43 13 6 14 X SQ= 12.11 P = 0. RV YES (YES)
97	TEND TO BE THOSE WHERE THE STATUS OF THE REGIME IS TOTALITARIAN, RATHER THAN AUTHORITARIAN (16)	0.70	97	ALWAYS ARE THOSE WHERE THE STATUS OF THE REGIME IS AUTHORITARIAN, RATHER THAN TOTALITARIAN (23)	1.00	6 14 14 0 X SQ= 13.90 P = 0. RV YES (YES)
101	TEND TO BE THOSE WHERE THE REPRESENTATIVE CHARACTER OF THE REGIME IS POLYARCHIC (41)	0.57	101	TEND TO BE THOSE WHERE THE REPRESENTATIVE CHARACTER OF THE REGIME IS LIMITED POLYARCHIC, PSEUDO-POLYARCHIC, OR NON-POLYARCHIC (57)	0.74	28 7 21 20 X SQ= 5.63 P = 0.015 RV YES (NO)
106	TILT LESS TOWARD BEING THOSE WHERE THE ELECTORAL SYSTEM IS COMPETITIVE OR PARTIALLY COMPETITIVE (52)	0.58	106	TILT MORE TOWARD BEING THOSE WHERE THE ELECTORAL SYSTEM IS COMPETITIVE OR PARTIALLY COMPETITIVE (52)	0.81	25 13 18 3 X SQ= 1.80 P = 0.132 RV NO (NO)
111	TEND TO BE THOSE WHERE POLITICAL ENCULTURATION IS HIGH OR MEDIUM (53)	0.81	111	TEND TO BE THOSE WHERE POLITICAL ENCULTURATION IS LOW (42)	0.66	29 12 7 23 X SQ= 13.73 P = 0. RV YES (YES)

113	TILT MORE TOWARD BEING THOSE WHERE SECTIONALISM IS MODERATE OR NEGLIGIBLE (81)	0.79	113	TILT LESS TOWARD BEING THOSE WHERE SECTIONALISM IS MODERATE OR NEGLIGIBLE (81)	0.64	10 12 38 21 X SQ= 1.66 P = 0.136 RV NO (YES)
116	TEND TO BE THOSE WHERE INTEREST ARTICULATION BY ASSOCIATIONAL GROUPS IS SIGNIFICANT OR MODERATE (32)	0.53	116	TEND TO BE THOSE WHERE INTEREST ARTICULATION BY ASSOCIATIONAL GROUPS IS LIMITED OR NEGLIGIBLE (79)	0.80	25 7 22 28 X SQ= 7.95 P = 0.003 RV YES (YES)
119	TEND LESS TO BE THOSE WHERE INTEREST ARTICULATION BY INSTITUTIONAL GROUPS IS VERY SIGNIFICANT OR SIGNIFICANT (74)	0.60	119	TEND MORE TO BE THOSE WHERE INTEREST ARTICULATION BY INSTITUTIONAL GROUPS IS VERY SIGNIFICANT OR SIGNIFICANT (74)	0.94	30 33 20 2 X SQ= 10.89 P = 0. RV NO (YES)
120	TEND LESS TO BE THOSE WHERE INTEREST ARTICULATION BY INSTITUTIONAL GROUPS IS VERY SIGNIFICANT, SIGNIFICANT, OR MODERATE (90)	0.80	120	ALWAYS ARE THOSE WHERE INTEREST ARTICULATION BY INSTITUTIONAL GROUPS IS VERY SIGNIFICANT, SIGNIFICANT, OR MODERATE (90)	1.00	40 35 10 0 X SQ= 6.12 P = 0.004 RV NO (NO)
121	TEND TO BE THOSE WHERE INTEREST ARTICULATION BY NON-ASSOCIATIONAL GROUPS IS MODERATE, LIMITED, OR NEGLIGIBLE (61)	0.76	121	TEND TO BE THOSE WHERE INTEREST ARTICULATION BY NON-ASSOCIATIONAL GROUPS IS SIGNIFICANT (54)	0.56	12 20 38 16 X SQ= 7.62 P = 0.004 RV YES (YES)
122	TEND LESS TO BE THOSE WHERE INTEREST ARTICULATION BY NON-ASSOCIATIONAL GROUPS IS SIGNIFICANT OR MODERATE (83)	0.58	122	TEND MORE TO BE THOSE WHERE INTEREST ARTICULATION BY NON-ASSOCIATIONAL GROUPS IS SIGNIFICANT OR MODERATE (83)	0.81	29 29 21 7 X SQ= 3.88 P = 0.036 RV NO (YES)
123	TEND LESS TO BE THOSE WHERE INTEREST ARTICULATION BY NON-ASSOCIATIONAL GROUPS IS SIGNIFICANT, MODERATE, OR LIMITED (107)	0.84	123	ALWAYS ARE THOSE WHERE INTEREST ARTICULATION BY NON-ASSOCIATIONAL GROUPS IS SIGNIFICANT, MODERATE, OR LIMITED (107)	1.00	42 36 8 0 X SQ= 4.60 P = 0.019 RV NO (NO)
125	TEND TO BE THOSE WHERE INTEREST ARTICULATION BY ANOMIC GROUPS IS INFREQUENT OR VERY INFREQUENT (35)	0.62	125	TEND TO BE THOSE WHERE INTEREST ARTICULATION BY ANOMIC GROUPS IS FREQUENT OR OCCASIONAL (64)	0.91	16 30 26 3 X SQ= 19.57 P = 0. RV YES (YES)
126	TEND LESS TO BE THOSE WHERE INTEREST ARTICULATION BY ANOMIC GROUPS IS FREQUENT, OCCASIONAL, OR INFREQUENT (83)	0.62	126	ALWAYS ARE THOSE WHERE INTEREST ARTICULATION BY ANOMIC GROUPS IS FREQUENT, OCCASIONAL, OR INFREQUENT (83)	1.00	26 33 16 0 X SQ= 13.79 P = 0. RV NO (YES)

128	TILT LESS TOWARD BEING THOSE WHERE INTEREST ARTICULATION BY POLITICAL PARTIES IS SIGNIFICANT OR MODERATE (48)	0.51	128	TILT MORE TOWARD BEING THOSE WHERE INTEREST ARTICULATION BY POLITICAL PARTIES IS SIGNIFICANT OR MODERATE (48)	0.74	23 17 22 6 X SQ= 2.39 P = 0.117 RV NO (NO)
129	TEND LESS TO BE THOSE WHERE INTEREST ARTICULATION BY POLITICAL PARTIES IS SIGNIFICANT, MODERATE, OR LIMITED (56)	0.51	129	TEND MORE TO BE THOSE WHERE INTEREST ARTICULATION BY POLITICAL PARTIES IS SIGNIFICANT, MODERATE, OR LIMITED (56)	0.87	23 20 22 3 X SQ= 6.94 P = 0.004 RV NO (NO)
131	TEND TO BE THOSE WHERE INTEREST AGGREGATION BY POLITICAL PARTIES IS SIGNIFICANT OR MODERATE (30)	0.61	131	TEND TO BE THOSE WHERE INTEREST AGGREGATION BY POLITICAL PARTIES IS LIMITED OR NEGLIGIBLE (35)	0.79	17 5 11 19 X SQ= 6.87 P = 0.005 RV YES (YES)
134	TEND TO BE THOSE WHERE INTEREST AGGREGATION BY THE EXECUTIVE IS SIGNIFICANT OR MODERATE (57)	0.60	134	TEND TO BE THOSE WHERE INTEREST AGGREGATION BY THE EXECUTIVE IS LIMITED OR NEGLIGIBLE (46)	0.72	29 8 19 21 X SQ= 6.55 P = 0.009 RV YES (YES)
136	TEND LESS TO BE THOSE WHERE INTEREST AGGREGATION BY THE LEGISLATURE IS MODERATE, LIMITED, OR NEGLIGIBLE (85)	0.74	136	ALWAYS ARE THOSE WHERE INTEREST AGGREGATION BY THE LEGISLATURE IS MODERATE, LIMITED, OR NEGLIGIBLE (85)	1.00	12 0 34 26 X SQ= 6.37 P = 0.003 RV NO (NO)
137	TEND TO BE THOSE WHERE INTEREST AGGREGATION BY THE LEGISLATURE IS SIGNIFICANT OR MODERATE (29)	0.50	137	TEND TO BE THOSE WHERE INTEREST AGGREGATION BY THE LEGISLATURE IS LIMITED OR NEGLIGIBLE (68)	0.85	23 4 23 22 X SQ= 7.08 P = 0.005 RV YES (NO)
139	TEND LESS TO BE THOSE WHERE THE PARTY SYSTEM IS QUANTITATIVELY OTHER THAN ONE-PARTY (71)	0.62	139	TEND MORE TO BE THOSE WHERE THE PARTY SYSTEM IS QUANTITATIVELY OTHER THAN ONE-PARTY (71)	0.89	19 3 31 24 X SQ= 4.96 P = 0.017 RV NO (NO)
142	TEND TO BE THOSE WHERE THE PARTY SYSTEM IS QUANTITATIVELY OTHER THAN MULTI-PARTY (66)	0.74	142	TEND TO BE THOSE WHERE THE PARTY SYSTEM IS QUANTITATIVELY MULTI-PARTY (30)	0.64	12 14 34 8 X SQ= 7.37 P = 0.007 RV YES (YES)
145	TILT LESS TOWARD BEING THOSE WHERE THE PARTY SYSTEM IS QUANTITATIVELY MULTI-PARTY, RATHER THAN TWO-PARTY (30)	0.63	145	TILT MORE TOWARD BEING THOSE WHERE THE PARTY SYSTEM IS QUANTITATIVELY MULTI-PARTY, RATHER THAN TWO-PARTY (30)	0.88	12 14 7 2 X SQ= 1.57 P = 0.135 RV NO (YES)

147	LEAN LESS TOWARD BEING THOSE WHERE THE PARTY SYSTEM IS QUALITATIVELY OTHER THAN CLASS-ORIENTED OR MULTI-IDEOLOGICAL (67)	0.57	147	LEAN MORE TOWARD BEING THOSE WHERE THE PARTY SYSTEM IS QUALITATIVELY OTHER THAN CLASS-ORIENTED OR MULTI-IDEOLOGICAL (67)	0.83	17 4 23 19 X SQ= 3.09 P = 0.054 RV NO (NO)
153	TEND TO BE THOSE WHERE THE PARTY SYSTEM IS STABLE (42)	0.91	153	ALWAYS ARE THOSE WHERE THE PARTY SYSTEM IS MODERATELY STABLE OR UNSTABLE (38)	1.00	40 0 4 29 X SQ= 54.71 P = 0. RV YES (YES)
154	ALWAYS ARE THOSE WHERE THE PARTY SYSTEM IS STABLE OR MODERATELY STABLE (55)	1.00	154	TEND TO BE THOSE WHERE THE PARTY SYSTEM IS UNSTABLE (25)	0.79	44 6 0 23 X SQ= 47.34 P = 0. RV YES (YES)
158	TEND TO BE THOSE WHERE PERSONALISMO IS NEGLIGIBLE (56)	0.87	158	TEND TO BE THOSE WHERE PERSONALISMO IS PRONOUNCED OR MODERATE (40)	0.88	6 23 41 3 X SQ= 36.96 P = 0. RV YES (YES)
168	TEND TO BE THOSE WHERE THE HORIZONTAL POWER DISTRIBUTION IS SIGNIFICANT (34)	0.50	168	TEND TO BE THOSE WHERE THE HORIZONTAL POWER DISTRIBUTION IS LIMITED OR NEGLIGIBLE (72)	0.83	25 5 25 24 X SQ= 7.03 P = 0.004 RV YES (NO)
170	TEND TO BE THOSE WHERE THE LEGISLATIVE-EXECUTIVE STRUCTURE IS OTHER THAN PRESIDENTIAL (63)	0.85	170	TEND TO BE THOSE WHERE THE LEGISLATIVE-EXECUTIVE STRUCTURE IS PRESIDENTIAL (39)	0.59	7 17 41 12 X SQ= 14.35 P = 0. RV YES (YES)
172	LEAN LESS TOWARD BEING THOSE WHERE THE LEGISLATIVE-EXECUTIVE STRUCTURE IS OTHER THAN PARLIAMENTARY-ROYALIST (88)	0.75	172	LEAN MORE TOWARD BEING THOSE WHERE THE LEGISLATIVE-EXECUTIVE STRUCTURE IS OTHER THAN PARLIAMENTARY-ROYALIST (88)	0.91	12 3 36 30 X SQ= 2.31 P = 0.086 RV NO (NO)
174	TEND TO BE THOSE WHERE THE LEGISLATURE IS FULLY EFFECTIVE (28)	0.51	174	ALWAYS ARE THOSE WHERE THE LEGISLATURE IS PARTIALLY EFFECTIVE, LARGELY INEFFECTIVE, OR WHOLLY INEFFECTIVE (72)	1.00	25 0 24 25 X SQ= 17.05 P = 0. RV YES (YES)
181	TEND LESS TO BE THOSE WHERE THE BUREAUCRACY IS SEMI-MODERN, RATHER THAN MODERN (55)	0.56	181	ALWAYS ARE THOSE WHERE THE BUREAUCRACY IS SEMI-MODERN, RATHER THAN MODERN (55)	1.00	21 0 27 20 X SQ= 10.69 P = 0. RV NO (NO)

182	ALWAYS ARE THOSE WHERE THE BUREAUCRACY IS SEMI-MODERN, RATHER THAN POST-COLONIAL TRANSITIONAL (55)	1.00	182	TEND LESS TO BE THOSE WHERE THE BUREAUCRACY IS SEMI-MODERN, RATHER THAN POST-COLONIAL TRANSITIONAL (55)	0.80	27 20 0 5 X SQ= 3.89 P = 0.020 RV NO (YES)
185	TEND TO BE THOSE WHERE PARTICIPATION BY THE MILITARY IN POLITICS IS SUPPORTIVE, RATHER THAN INTERVENTIVE (31)	0.95	185	TEND TO BE THOSE WHERE PARTICIPATION BY THE MILITARY IN POLITICS IS INTERVENTIVE, RATHER THAN SUPPORTIVE (21)	0.65	1 17 20 9 X SQ= 15.59 P = 0. RV YES (YES)
187	TEND TO BE THOSE WHERE THE ROLE OF THE POLICE IS NOT POLITICALLY SIGNIFICANT (35)	0.53	187	TEND TO BE THOSE WHERE THE ROLE OF THE POLICE IS POLITICALLY SIGNIFICANT (66)	0.94	23 29 26 2 X SQ= 16.14 P = 0. RV YES (YES)
188	TEND MORE TO BE THOSE WHERE THE CHARACTER OF THE LEGAL SYSTEM IS OTHER THAN CIVIL LAW (81)	0.76	188	TEND LESS TO BE THOSE WHERE THE CHARACTER OF THE LEGAL SYSTEM IS OTHER THAN CIVIL LAW (81)	0.51	12 17 37 18 X SQ= 4.23 P = 0.035 RV NO (YES)
189	LEAN LESS TOWARD BEING THOSE WHERE THE CHARACTER OF THE LEGAL SYSTEM IS OTHER THAN COMMON LAW (108)	0.90	189	ALWAYS ARE THOSE WHERE THE CHARACTER OF THE LEGAL SYSTEM IS OTHER THAN COMMON LAW (108)	1.00	5 0 45 36 X SQ= 2.21 P = 0.072 RV NO (NO)
190	TEND LESS TO BE THOSE WHERE THE CHARACTER OF THE LEGAL SYSTEM IS CIVIL LAW, RATHER THAN COMMON LAW (32)	0.71	190	ALWAYS ARE THOSE WHERE THE CHARACTER OF THE LEGAL SYSTEM IS CIVIL LAW, RATHER THAN COMMON LAW (32)	1.00	12 17 5 0 X SQ= 3.75 P = 0.044 RV NO (YES)
194	TEND LESS TO BE THOSE THAT ARE NON-COMMUNIST (101)	0.76	194	ALWAYS ARE THOSE THAT ARE NON-COMMUNIST (101)	1.00	12 0 38 36 X SQ= 8.14 P = 0.001 RV NO (NO)

```
**************************************************************************************************

100  POLITIES                                          100  POLITIES
     WHERE GOVERNMENTAL STABILITY                           WHERE GOVERNMENTAL STABILITY
     IS GENERALLY PRESENT AND DATES FROM                    IS ABSENT  (22)
     AT LEAST THE INTER-WAR PERIOD, OR IS
     GENERALLY OR MODERATELY PRESENT, AND
     DATES FROM THE POST-WAR PERIOD  (64)

                                                                         BOTH SUBJECT AND PREDICATE
..................................................................................................

     64 IN LEFT COLUMN

AFGHANISTAN ALBANIA     AUSTRALIA   AUSTRIA     BELGIUM     BOLIVIA     BULGARIA    CAMBODIA    CANADA      CHILE
CHINA, PR   COLOMBIA    COSTA RICA  CZECHOS'KIA DENMARK     ETHIOPIA    FINLAND     FRANCE      GERMANY, E  GERMAN FR
GHANA       GREECE      GUINEA      HAITI       HUNGARY     ICELAND     INDIA       INDONESIA   IRELAND     ISRAEL
ITALY       JAPAN       KOREA, N    LIBERIA     LIBYA       LUXEMBOURG  MALAYA      MEXICO      MONGOLIA    MOROCCO
NETHERLANDS NEW ZEALAND NICARAGUA   NORWAY      PARAGUAY    PHILIPPINES POLAND      PORTUGAL    RUMANIA     SA'U ARABIA
SO AFRICA   SPAIN       SWEDEN      SWITZERLAND THAILAND    TUNISIA     USSR        UAR         UK          US
URUGUAY     VIETNAM, N  VIETNAM REP YUGOSLAVIA

     22 IN RIGHT COLUMN

ARGENTINA   BRAZIL      BURMA       CEYLON      CONGO(LEO)  ECUADOR     EL SALVADOR GUATEMALA   HONDURAS    IRAN
IRAQ        JORDAN      KOREA REP   LAOS        LEBANON     NEPAL       PAKISTAN    PANAMA      PERU        SUDAN
SYRIA       VENEZUELA

     29 EXCLUDED BECAUSE UNASCERTAINABLE

ALGERIA     BURUNDI     CAMEROUN    CEN AFR REP CHAD        CONGO(BRA)  CUBA        CYPRUS      DAHOMEY     DOMIN REP
GABON       IVORY COAST JAMAICA     MALAGASY R  MALI        MAURITANIA  NIGER       NIGERIA     RWANDA      SENEGAL
SIERRE LEO  SOMALIA     TANGANYIKA  TOGO        TRINIDAD    TURKEY      UGANDA      UPPER VOLTA YEMEN

--------------------------------------------------------------------------------------------------

  2  TEND LESS TO BE THOSE                 0.66          2  ALWAYS ARE THOSE                  1.00        22     0
     LOCATED ELSEWHERE THAN IN                              LOCATED ELSEWHERE THAN IN                     42    22
     WEST EUROPE, SCANDINAVIA,                              WEST EUROPE, SCANDINAVIA,                X SQ=    8.44
     NORTH AMERICA, OR AUSTRALASIA  (93)                    NORTH AMERICA, OR AUSTRALASIA  (93)      P =     0.001
                                                                                                     RV NO   (NO )

  6  TEND MORE TO BE THOSE                 0.86          6  TEND LESS TO BE THOSE             0.59         9     9
     LOCATED ELSEWHERE THAN IN THE                          LOCATED ELSEWHERE THAN IN THE                 55    13
     CARIBBEAN, CENTRAL AMERICA,                            CARIBBEAN, CENTRAL AMERICA,              X SQ=    5.60
     OR SOUTH AMERICA  (93)                                 OR SOUTH AMERICA  (93)                   P =     0.014
                                                                                                     RV NO   (YES)

 15  LEAN TOWARD BEING THOSE               0.75         15  LEAN TOWARD BEING THOSE           0.71         3     5
     LOCATED IN NORTH AFRICA OR                             LOCATED IN THE MIDDLE EAST, RATHER THAN        9     2
     CENTRAL AND SOUTH AFRICA, RATHER THAN                  IN NORTH AFRICA OR                       X SQ=    2.24
     IN THE MIDDLE EAST  (33)                               CENTRAL AND SOUTH AFRICA  (11)           P =     0.074
                                                                                                     RV YES  (YES)
```

19	TILT TOWARD BEING THOSE LOCATED IN NORTH AFRICA OR CENTRAL AND SOUTH AFRICA, RATHER THAN IN THE CARIBBEAN, CENTRAL AMERICA, OR SOUTH AMERICA (33)	0.50	19	TILT TOWARD BEING THOSE LOCATED IN THE CARIBBEAN, CENTRAL AMERICA, OR SOUTH AMERICA, RATHER THAN IN NORTH AFRICA OR CENTRAL AND SOUTH AFRICA (22)	0.82	9 2 9 9 X SQ= 1.74 P = 0.125 RV YES (YES)
26	TEND TO BE THOSE WHOSE POPULATION DENSITY IS VERY HIGH, HIGH, OR MEDIUM (48)	0.53	26	TEND TO BE THOSE WHOSE POPULATION DENSITY IS LOW (67)	0.73	34 6 30 16 X SQ= 3.42 P = 0.048 RV YES (NO)
28	TEND TO BE THOSE WHOSE POPULATION GROWTH RATE IS LOW (48)	0.53	28	TEND TO BE THOSE WHOSE POPULATION GROWTH RATE IS HIGH (62)	0.82	28 18 32 4 X SQ= 6.71 P = 0.006 RV YES (NO)
31	TEND LESS TO BE THOSE WHOSE AGRICULTURAL POPULATION IS HIGH OR MEDIUM (90)	0.65	31	TEND MORE TO BE THOSE WHOSE AGRICULTURAL POPULATION IS HIGH OR MEDIUM (90)	0.95	41 21 22 1 X SQ= 6.16 P = 0.005 RV NO (NO)
32	LEAN LESS TOWARD BEING THOSE WHOSE GROSS NATIONAL PRODUCT IS MEDIUM, LOW, OR VERY LOW (105)	0.84	32	ALWAYS ARE THOSE WHOSE GROSS NATIONAL PRODUCT IS MEDIUM, LOW, OR VERY LOW (105)	1.00	10 0 54 22 X SQ= 2.52 P = 0.058 RV NO (NO)
33	TEND LESS TO BE THOSE WHOSE GROSS NATIONAL PRODUCT IS LOW OR VERY LOW (85)	0.59	33	TEND MORE TO BE THOSE WHOSE GROSS NATIONAL PRODUCT IS LOW OR VERY LOW (85)	0.86	26 3 38 19 X SQ= 4.20 P = 0.021 RV NO (NO)
34	LEAN TOWARD BEING THOSE WHOSE GROSS NATIONAL PRODUCT IS VERY HIGH, HIGH, MEDIUM, OR LOW (62)	0.73	34	LEAN TOWARD BEING THOSE WHOSE GROSS NATIONAL PRODUCT IS VERY LOW (53)	0.50	47 11 17 11 X SQ= 3.10 P = 0.064 RV YES (NO)
36	TEND TO BE THOSE WHOSE PER CAPITA GROSS NATIONAL PRODUCT IS VERY HIGH, HIGH, OR MEDIUM (42)	0.55	36	TEND TO BE THOSE WHOSE PER CAPITA GROSS NATIONAL PRODUCT IS LOW OR VERY LOW (73)	0.86	35 3 29 19 X SQ= 9.58 P = 0.001 RV YES (NO)
38	LEAN LESS TOWARD BEING THOSE WHOSE INTERNATIONAL FINANCIAL STATUS IS MEDIUM, LOW, OR VERY LOW (103)	0.84	38	ALWAYS ARE THOSE WHOSE INTERNATIONAL FINANCIAL STATUS IS MEDIUM, LOW, OR VERY LOW (103)	1.00	10 0 52 22 X SQ= 2.64 P = 0.057 RV NO (NO)

39	TEND TO BE THOSE WHOSE INTERNATIONAL FINANCIAL STATUS IS VERY HIGH, HIGH, OR MEDIUM (38)	0.52	39	TEND TO BE THOSE WHOSE INTERNATIONAL FINANCIAL STATUS IS LOW OR VERY LOW (76)	0.82	33 4 30 18 X SQ= 6.43 P = 0.006 RV YES (NO)
41	TEND LESS TO BE THOSE WHOSE ECONOMIC DEVELOPMENTAL STATUS IS INTERMEDIATE, UNDERDEVELOPED, OR VERY UNDERDEVELOPED (94)	0.70	41	ALWAYS ARE THOSE WHOSE ECONOMIC DEVELOPMENTAL STATUS IS INTERMEDIATE, UNDERDEVELOPED, OR VERY UNDERDEVELOPED (94)	1.00	19 0 44 21 X SQ= 6.55 P = 0.002 RV NO (NO)
42	TEND TO BE THOSE WHOSE ECONOMIC DEVELOPMENTAL STATUS IS DEVELOPED OR INTERMEDIATE (36)	0.53	42	TEND TO BE THOSE WHOSE ECONOMIC DEVELOPMENTAL STATUS IS UNDERDEVELOPED OR VERY UNDERDEVELOPED (76)	0.90	33 2 29 19 X SQ= 10.56 P = 0. RV YES (NO)
43	TEND TO BE THOSE WHOSE ECONOMIC DEVELOPMENTAL STATUS IS DEVELOPED, INTERMEDIATE, OR UNDERDEVELOPED (55)	0.68	43	TEND TO BE THOSE WHOSE ECONOMIC DEVELOPMENTAL STATUS IS VERY UNDERDEVELOPED (57)	0.62	42 8 20 13 X SQ= 4.59 P = 0.021 RV YES (NO)
44	TEND LESS TO BE THOSE WHERE THE LITERACY RATE IS BELOW NINETY PERCENT (90)	0.61	44	ALWAYS ARE THOSE WHERE THE LITERACY RATE IS BELOW NINETY PERCENT (90)	1.00	25 0 39 22 X SQ= 10.30 P = 0. RV NO (NO)
45	TEND TO BE THOSE WHERE THE LITERACY RATE IS FIFTY PERCENT OR ABOVE (55)	0.73	45	TEND TO BE THOSE WHERE THE LITERACY RATE IS BELOW FIFTY PERCENT (54)	0.67	43 7 16 14 X SQ= 8.72 P = 0.003 RV YES (NO)
46	LEAN MORE TOWARD BEING THOSE WHERE THE LITERACY RATE IS TEN PERCENT OR ABOVE (84)	0.95	46	LEAN LESS TOWARD BEING THOSE WHERE THE LITERACY RATE IS TEN PERCENT OR ABOVE (84)	0.82	57 18 3 4 X SQ= 2.09 P = 0.079 RV NO (YES)
50	TEND LESS TO BE THOSE WHERE FREEDOM OF THE PRESS IS INTERMITTENT, INTERNALLY ABSENT, OR INTERNALLY AND EXTERNALLY ABSENT (56)	0.56	50	TEND MORE TO BE THOSE WHERE FREEDOM OF THE PRESS IS INTERMITTENT, INTERNALLY ABSENT, OR INTERNALLY AND EXTERNALLY ABSENT (56)	0.89	27 2 35 16 X SQ= 5.03 P = 0.013 RV NO (NO)
53	TEND LESS TO BE THOSE WHERE NEWSPAPER CIRCULATION IS LESS THAN THREE HUNDRED PER THOUSAND (101)	0.78	53	ALWAYS ARE THOSE WHERE NEWSPAPER CIRCULATION IS LESS THAN THREE HUNDRED PER THOUSAND (101)	1.00	14 0 50 22 X SQ= 4.26 P = 0.017 RV NO (NO)

54	TEND TO BE THOSE WHERE NEWSPAPER CIRCULATION IS ONE HUNDRED OR MORE PER THOUSAND (37)	0.53	54	TEND TO BE THOSE WHERE NEWSPAPER CIRCULATION IS LESS THAN ONE HUNDRED PER THOUSAND (76)	0.91	33 2 29 20 X SQ= 11.26 P = 0. RV YES (NO)
63	TEND TO BE THOSE WHERE THE RELIGION IS PREDOMINANTLY SOME KIND OF CHRISTIAN (46)	0.59	63	TEND TO BE THOSE WHERE THE RELIGION IS PREDOMINANTLY OR PARTLY OTHER THAN CHRISTIAN (68)	0.73	37 6 26 16 X SQ= 5.26 P = 0.014 RV YES (NO)
68	TILT TOWARD BEING THOSE THAT ARE LINGUISTICALLY HOMOGENEOUS (52)	0.57	68	TILT TOWARD BEING THOSE THAT ARE LINGUISTICALLY WEAKLY HETEROGENEOUS OR STRONGLY HETEROGENEOUS (62)	0.64	36 8 27 14 X SQ= 2.05 P = 0.137 RV YES (NO)
69	LEAN TOWARD BEING THOSE THAT ARE LINGUISTICALLY HOMOGENEOUS OR WEAKLY HETEROGENEOUS (64)	0.70	69	LEAN TOWARD BEING THOSE THAT ARE LINGUISTICALLY STRONGLY HETEROGENEOUS (50)	0.55	44 10 19 12 X SQ= 3.20 P = 0.070 RV YES (NO)
70	LEAN LESS TOWARD BEING THOSE THAT ARE RELIGIOUSLY, RACIALLY, OR LINGUISTICALLY HETEROGENEOUS (85)	0.69	70	LEAN MORE TOWARD BEING THOSE THAT ARE RELIGIOUSLY, RACIALLY, OR LINGUISTICALLY HETEROGENEOUS (85)	0.90	18 2 41 19 X SQ= 2.60 P = 0.079 RV NO (NO)
71	LEAN LESS TOWARD BEING THOSE WHOSE DATE OF INDEPENDENCE IS AFTER 1800 (90)	0.70	71	LEAN MORE TOWARD BEING THOSE WHOSE DATE OF INDEPENDENCE IS AFTER 1800 (90)	0.90	18 2 43 19 X SQ= 2.39 P = 0.081 RV NO (NO)
74	TEND LESS TO BE THOSE THAT ARE NOT HISTORICALLY WESTERN (87)	0.58	74	ALWAYS ARE THOSE THAT ARE NOT HISTORICALLY WESTERN (87)	1.00	26 0 36 22 X SQ= 11.47 P = 0. RV NO (NO)
75	TILT TOWARD BEING THOSE THAT ARE HISTORICALLY WESTERN OR SIGNIFICANTLY WESTERNIZED (62)	0.70	75	TILT TOWARD BEING THOSE THAT ARE NOT HISTORICALLY WESTERN AND ARE NOT SIGNIFICANTLY WESTERNIZED (52)	0.50	44 11 19 11 X SQ= 2.01 P = 0.122 RV YES (NO)
76	TEND TO BE THOSE THAT ARE HISTORICALLY WESTERN, RATHER THAN HAVING BEEN SIGNIFICANTLY OR PARTIALLY WESTERNIZED THROUGH A COLONIAL RELATIONSHIP (26)	0.53	76	ALWAYS ARE THOSE THAT HAVE BEEN SIGNIFICANTLY OR PARTIALLY WESTERNIZED THROUGH A COLONIAL RELATIONSHIP, RATHER THAN BEING HISTORICALLY WESTERN (70)	1.00	26 0 23 20 X SQ= 14.84 P = 0. RV YES (NO)

78	LEAN LESS TOWARD BEING THOSE THAT HAVE BEEN SIGNIFICANTLY WESTERNIZED THROUGH A COLONIAL RELATIONSHIP, RATHER THAN WITHOUT SUCH A RELATIONSHIP (28)	0.65	78	ALWAYS ARE THOSE THAT HAVE BEEN SIGNIFICANTLY WESTERNIZED THROUGH A COLONIAL RELATIONSHIP, RATHER THAN WITHOUT SUCH A RELATIONSHIP (28)	1.00	11 11 6 0 X SQ= 3.07 P = 0.055 RV NO (YES)
81	LEAN LESS TOWARD BEING THOSE WHERE THE TYPE OF POLITICAL MODERNIZATION IS LATER EUROPEAN OR LATER EUROPEAN DERIVED, RATHER THAN EARLY EUROPEAN OR EARLY EUROPEAN DERIVED (40)	0.72	81	ALWAYS ARE THOSE WHERE THE TYPE OF POLITICAL MODERNIZATION IS LATER EUROPEAN OR LATER EUROPEAN DERIVED, RATHER THAN EARLY EUROPEAN OR EARLY EUROPEAN DERIVED (40)	1.00	11 0 29 9 X SQ= 1.81 P = 0.098 RV NO (NO)
82	LEAN TOWARD BEING THOSE WHERE THE TYPE OF POLITICAL MODERNIZATION IS EARLY OR LATER EUROPEAN OR EUROPEAN DERIVED, RATHER THAN DEVELOPED TUTELARY OR UNDEVELOPED TUTELARY (51)	0.69	82	LEAN TOWARD BEING THOSE WHERE THE TYPE OF POLITICAL MODERNIZATION IS DEVELOPED TUTELARY OR UNDEVELOPED TUTELARY, RATHER THAN EARLY OR LATER EUROPEAN OR EUROPEAN DERIVED (55)	0.55	40 9 18 11 X SQ= 2.70 P = 0.066 RV YES (NO)
85	TILT TOWARD BEING THOSE WHERE THE STAGE OF POLITICAL MODERNIZATION IS ADVANCED, RATHER THAN MID- OR EARLY TRANSITIONAL (60)	0.71	85	TILT TOWARD BEING THOSE WHERE THE STAGE OF POLITICAL MODERNIZATION IS MID- OR EARLY TRANSITIONAL, RATHER THAN ADVANCED (54)	0.50	45 11 18 11 X SQ= 2.45 P = 0.115 RV YES (NO)
89	TEND TO BE THOSE WHOSE IDEOLOGICAL ORIENTATION IS CONVENTIONAL (33)	0.50	89	TEND TO BE THOSE WHOSE IDEOLOGICAL ORIENTATION IS OTHER THAN CONVENTIONAL (62)	0.80	28 3 28 12 X SQ= 3.20 P = 0.045 RV YES (NO)
92	TEND LESS TO BE THOSE WHERE THE SYSTEM STYLE IS LIMITED MOBILIZATIONAL, OR NON-MOBILIZATIONAL (93)	0.73	92	TEND MORE TO BE THOSE WHERE THE SYSTEM STYLE IS LIMITED MOBILIZATIONAL, OR NON-MOBILIZATIONAL (93)	0.95	17 1 46 21 X SQ= 3.67 P = 0.033 RV NO (NO)
93	TILT LESS TOWARD BEING THOSE WHERE THE SYSTEM STYLE IS NON-MOBILIZATIONAL (78)	0.63	93	TILT MORE TOWARD BEING THOSE WHERE THE SYSTEM STYLE IS NON-MOBILIZATIONAL (78)	0.84	23 3 40 16 X SQ= 2.02 P = 0.101 RV NO (NO)
95	TEND LESS TO BE THOSE WHERE THE STATUS OF THE REGIME IS CONSTITUTIONAL OR AUTHORITARIAN (95)	0.75	95	ALWAYS ARE THOSE WHERE THE STATUS OF THE REGIME IS CONSTITUTIONAL OR AUTHORITARIAN (95)	1.00	46 22 15 0 X SQ= 5.05 P = 0.008 RV NO (NO)
96	TEND TO BE THOSE WHERE THE STATUS OF THE REGIME IS CONSTITUTIONAL OR TOTALITARIAN (67)	0.80	96	TEND TO BE THOSE WHERE THE STATUS OF THE REGIME IS AUTHORITARIAN (23)	0.56	49 7 12 9 X SQ= 6.81 P = 0.009 RV YES (NO)

97	TEND TO BE THOSE WHERE THE STATUS OF THE REGIME IS TOTALITARIAN, RATHER THAN AUTHORITARIAN (16)	0.56	97	ALWAYS ARE THOSE WHERE THE STATUS OF THE REGIME IS AUTHORITARIAN, RATHER THAN TOTALITARIAN (23)	1.00	12 9 15 0 X SQ= 6.44 P = 0.005 RV YES (NO)
101	LEAN LESS TOWARD BEING THOSE WHERE THE REPRESENTATIVE CHARACTER OF THE REGIME IS LIMITED POLYARCHIC, PSEUDO-POLYARCHIC, OR NON-POLYARCHIC (57)	0.51	101	LEAN MORE TOWARD BEING THOSE WHERE THE REPRESENTATIVE CHARACTER OF THE REGIME IS LIMITED POLYARCHIC, PSEUDO-POLYARCHIC, OR NON-POLYARCHIC (57)	0.79	31 3 32 11 X SQ= 2.55 P = 0.077 RV NO (NO)
106	TILT LESS TOWARD BEING THOSE WHERE THE ELECTORAL SYSTEM IS COMPETITIVE OR PARTIALLY COMPETITIVE (52)	0.58	106	TILT MORE TOWARD BEING THOSE WHERE THE ELECTORAL SYSTEM IS COMPETITIVE OR PARTIALLY COMPETITIVE (52)	0.88	30 7 22 1 X SQ= 1.50 P = 0.138 RV NO (NO)
108	LEAN LESS TOWARD BEING THOSE WHERE AUTONOMOUS GROUPS ARE FULLY OR PARTIALLY TOLERATED IN POLITICS (65)	0.57	108	LEAN MORE TOWARD BEING THOSE WHERE AUTONOMOUS GROUPS ARE FULLY OR PARTIALLY TOLERATED IN POLITICS (65)	0.85	36 11 27 2 X SQ= 2.38 P = 0.071 RV NO (NO)
110	TEND LESS TO BE THOSE WHERE POLITICAL ENCULTURATION IS MEDIUM OR LOW (80)	0.72	110	ALWAYS ARE THOSE WHERE POLITICAL ENCULTURATION IS MEDIUM OR LOW (80)	1.00	13 0 34 22 X SQ= 5.80 P = 0.006 RV NO (NO)
111	TEND TO BE THOSE WHERE POLITICAL ENCULTURATION IS HIGH OR MEDIUM (53)	0.70	111	TEND TO BE THOSE WHERE POLITICAL ENCULTURATION IS LOW (42)	0.77	33 5 14 17 X SQ= 11.80 P = 0. RV YES (YES)
115	LEAN LESS TOWARD BEING THOSE WHERE INTEREST ARTICULATION BY ASSOCIATIONAL GROUPS IS MODERATE, LIMITED, OR NEGLIGIBLE (91)	0.70	115	LEAN MORE TOWARD BEING THOSE WHERE INTEREST ARTICULATION BY ASSOCIATIONAL GROUPS IS MODERATE, LIMITED, OR NEGLIGIBLE (91)	0.90	18 2 43 19 X SQ= 2.39 P = 0.081 RV NO (NO)
116	LEAN LESS TOWARD BEING THOSE WHERE INTEREST ARTICULATION BY ASSOCIATIONAL GROUPS IS LIMITED OR NEGLIGIBLE (79)	0.56	116	LEAN MORE TOWARD BEING THOSE WHERE INTEREST ARTICULATION BY ASSOCIATIONAL GROUPS IS LIMITED OR NEGLIGIBLE (79)	0.81	27 4 34 17 X SQ= 3.22 P = 0.066 RV NO (NO)
119	TEND LESS TO BE THOSE WHERE INTEREST ARTICULATION BY INSTITUTIONAL GROUPS IS VERY SIGNIFICANT OR SIGNIFICANT (74)	0.66	119	ALWAYS ARE THOSE WHERE INTEREST ARTICULATION BY INSTITUTIONAL GROUPS IS VERY SIGNIFICANT OR SIGNIFICANT (74)	1.00	42 22 22 0 X SQ= 8.44 P = 0.001 RV NO (NO)

121	TEND TO BE THOSE WHERE INTEREST ARTICULATION BY NON-ASSOCIATIONAL GROUPS IS MODERATE, LIMITED, OR NEGLIGIBLE (61)	0.72	121	TEND TO BE THOSE WHERE INTEREST ARTICULATION BY NON-ASSOCIATIONAL GROUPS IS SIGNIFICANT (54)	0.59	18 13 46 9 X SQ= 5.53 P = 0.019 RV YES (NO)
122	TEND LESS TO BE THOSE WHERE INTEREST ARTICULATION BY NON-ASSOCIATIONAL GROUPS IS SIGNIFICANT OR MODERATE (83)	0.58	122	TEND MORE TO BE THOSE WHERE INTEREST ARTICULATION BY NON-ASSOCIATIONAL GROUPS IS SIGNIFICANT OR MODERATE (83)	0.86	37 19 27 3 X SQ= 4.69 P = 0.019 RV NO (NO)
123	TILT LESS TOWARD BEING THOSE WHERE INTEREST ARTICULATION BY NON-ASSOCIATIONAL GROUPS IS SIGNIFICANT, MODERATE, OR LIMITED (107)	0.88	123	ALWAYS ARE THOSE WHERE INTEREST ARTICULATION BY NON-ASSOCIATIONAL GROUPS IS SIGNIFICANT, MODERATE, OR LIMITED (107)	1.00	56 22 8 0 X SQ= 1.73 P = 0.107 RV NO (NO)
125	TEND TO BE THOSE WHERE INTEREST ARTICULATION BY ANOMIC GROUPS IS INFREQUENT OR VERY INFREQUENT (35)	0.54	125	ALWAYS ARE THOSE WHERE INTEREST ARTICULATION BY ANOMIC GROUPS IS FREQUENT OR OCCASIONAL (64)	1.00	25 20 29 0 X SQ= 15.48 P = 0. RV YES (NO)
127	TEND TO BE THOSE WHERE INTEREST ARTICULATION BY POLITICAL PARTIES IS MODERATE, LIMITED, OR NEGLIGIBLE (72)	0.78	127	TEND TO BE THOSE WHERE INTEREST ARTICULATION BY POLITICAL PARTIES IS SIGNIFICANT (21)	0.54	12 7 42 6 X SQ= 3.72 P = 0.038 RV YES (NO)
128	TEND TO BE THOSE WHERE INTEREST ARTICULATION BY POLITICAL PARTIES IS LIMITED OR NEGLIGIBLE (45)	0.50	128	ALWAYS ARE THOSE WHERE INTEREST ARTICULATION BY POLITICAL PARTIES IS SIGNIFICANT OR MODERATE (48)	1.00	27 13 27 0 X SQ= 8.91 P = 0.001 RV YES (NO)
130	LEAN LESS TOWARD BEING THOSE WHERE INTEREST AGGREGATION BY POLITICAL PARTIES IS MODERATE, LIMITED, OR NEGLIGIBLE (71)	0.80	130	ALWAYS ARE THOSE WHERE INTEREST AGGREGATION BY POLITICAL PARTIES IS MODERATE, LIMITED, OR NEGLIGIBLE (71)	1.00	10 0 40 19 X SQ= 2.98 P = 0.052 RV NO (NO)
131	TEND TO BE THOSE WHERE INTEREST AGGREGATION BY POLITICAL PARTIES IS SIGNIFICANT OR MODERATE (30)	0.54	131	TEND TO BE THOSE WHERE INTEREST AGGREGATION BY POLITICAL PARTIES IS LIMITED OR NEGLIGIBLE (35)	0.94	19 1 16 15 X SQ= 8.71 P = 0.001 RV YES (NO)
134	TEND TO BE THOSE WHERE INTEREST AGGREGATION BY THE EXECUTIVE IS SIGNIFICANT OR MODERATE (57)	0.55	134	TEND TO BE THOSE WHERE INTEREST AGGREGATION BY THE EXECUTIVE IS LIMITED OR NEGLIGIBLE (46)	0.93	34 1 28 14 X SQ= 9.44 P = 0.001 RV YES (NO)

```
136  LEAN LESS TOWARD BEING THOSE          0.79     136  ALWAYS ARE THOSE                       1.00        12     0
     WHERE INTEREST AGGREGATION                          WHERE INTEREST AGGREGATION                         45    15
     BY THE LEGISLATURE                                  BY THE LEGISLATURE                              X SQ=   2.43
     IS MODERATE, LIMITED, OR                            IS MODERATE, LIMITED, OR                        P =    0.060
     NEGLIGIBLE  (85)                                    NEGLIGIBLE  (85)                                RV NO  (NO )

137  TEND LESS TO BE THOSE                 0.56     137  TEND MORE TO BE THOSE                  0.93        25     1
     WHERE INTEREST AGGREGATION                          WHERE INTEREST AGGREGATION                         32    14
     BY THE LEGISLATURE                                  BY THE LEGISLATURE                              X SQ=   5.60
     IS LIMITED OR NEGLIGIBLE  (68)                      IS LIMITED OR NEGLIGIBLE  (68)                  P =    0.007
                                                                                                         RV NO  (NO )

139  TEND LESS TO BE THOSE                 0.64     139  ALWAYS ARE THOSE                       1.00        22     0
     WHERE THE PARTY SYSTEM IS QUANTITATIVELY            WHERE THE PARTY SYSTEM IS QUANTITATIVELY           39    17
     OTHER THAN ONE-PARTY  (71)                          OTHER THAN ONE-PARTY  (71)                      X SQ=   6.85
                                                                                                         P =    0.002
                                                                                                         RV NO  (NO )

142  TEND TO BE THOSE                      0.75     142  TEND TO BE THOSE                       0.92        14    12
     WHERE THE PARTY SYSTEM IS QUANTITATIVELY            WHERE THE PARTY SYSTEM IS QUANTITATIVELY           43     1
     OTHER THAN MULTI-PARTY  (66)                        MULTI-PARTY  (30)                               X SQ=  18.01
                                                                                                         P =    0.
                                                                                                         RV YES (NO )

145  TILT LESS TOWARD BEING THOSE          0.64     145  TILT MORE TOWARD BEING THOSE           0.92        14    12
     WHERE THE PARTY SYSTEM IS QUANTITATIVELY            WHERE THE PARTY SYSTEM IS QUANTITATIVELY            8     1
     MULTI-PARTY, RATHER THAN                            MULTI-PARTY, RATHER THAN                        X SQ=   2.18
     TWO-PARTY  (30)                                     TWO-PARTY  (30)                                 P =    0.109
                                                                                                         RV NO  (NO )

153  TEND TO BE THOSE                      0.76     153  ALWAYS ARE THOSE                       1.00        41     0
     WHERE THE PARTY SYSTEM IS                           WHERE THE PARTY SYSTEM IS                          13    19
     STABLE  (42)                                        MODERATELY STABLE OR UNSTABLE  (38)             X SQ=  29.90
                                                                                                         P =    0.
                                                                                                         RV YES (YES)

154  TEND TO BE THOSE                      0.93     154  ALWAYS ARE THOSE                       1.00        50     0
     WHERE THE PARTY SYSTEM IS                           WHERE THE PARTY SYSTEM IS                           4    19
     STABLE OR MODERATELY STABLE  (55)                   UNSTABLE  (25)                                  X SQ=  51.63
                                                                                                         P =    0.
                                                                                                         RV YES (YES)

157  TEND TO BE THOSE                      0.90     157  TEND TO BE THOSE                       0.57         6     8
     WHERE PERSONALISMO IS                               WHERE PERSONALISMO IS                              53     6
     MODERATE OR NEGLIGIBLE  (84)                        PRONOUNCED  (14)                                X SQ=  13.22
                                                                                                         P =    0.
                                                                                                         RV YES (YES)

158  TEND TO BE THOSE                      0.73     158  TEND TO BE THOSE                       0.93        16    13
     WHERE PERSONALISMO IS                               WHERE PERSONALISMO IS                              43     1
     NEGLIGIBLE  (56)                                    PRONOUNCED OR MODERATE  (40)                    X SQ=  17.77
                                                                                                         P =    0.
                                                                                                         RV YES (NO )
```

161	TILT LESS TOWARD BEING THOSE WHERE THE POLITICAL LEADERSHIP IS ELITIST OR MODERATELY ELITIST (47)	0.54	161	TILT MORE TOWARD BEING THOSE WHERE THE POLITICAL LEADERSHIP IS ELITIST OR MODERATELY ELITIST (47)	0.79	33 11 28 3 X SQ= 1.89 P = 0.134 RV NO (NO)
163	TILT LESS TOWARD BEING THOSE WHERE THE REGIME'S LEADERSHIP CHARISMA IS MODERATE OR NEGLIGIBLE (87)	0.84	163	ALWAYS ARE THOSE WHERE THE REGIME'S LEADERSHIP CHARISMA IS MODERATE OR NEGLIGIBLE (87)	1.00	10 0 52 16 X SQ= 1.69 P = 0.112 RV NO (NO)
168	TEND LESS TO BE THOSE WHERE THE HORIZONTAL POWER DISTRIBUTION IS LIMITED OR NEGLIGIBLE (72)	0.56	168	TEND MORE TO BE THOSE WHERE THE HORIZONTAL POWER DISTRIBUTION IS LIMITED OR NEGLIGIBLE (72)	0.93	28 1 36 14 X SQ= 5.69 P = 0.007 RV NO (NO)
170	TEND TO BE THOSE WHERE THE LEGISLATIVE-EXECUTIVE STRUCTURE IS OTHER THAN PRESIDENTIAL (63)	0.74	170	TEND TO BE THOSE WHERE THE LEGISLATIVE-EXECUTIVE STRUCTURE IS PRESIDENTIAL (39)	0.56	16 9 46 7 X SQ= 4.10 P = 0.033 RV YES (NO)
174	TEND LESS TO BE THOSE WHERE THE LEGISLATURE IS PARTIALLY EFFECTIVE, LARGELY INEFFECTIVE, OR WHOLLY INEFFECTIVE (72)	0.60	174	ALWAYS ARE THOSE WHERE THE LEGISLATURE IS PARTIALLY EFFECTIVE, LARGELY INEFFECTIVE, OR WHOLLY INEFFECTIVE (72)	1.00	25 0 38 12 X SQ= 5.47 P = 0.006 RV NO (NO)
181	TEND LESS TO BE THOSE WHERE THE BUREAUCRACY IS SEMI-MODERN, RATHER THAN MODERN (55)	0.63	181	ALWAYS ARE THOSE WHERE THE BUREAUCRACY IS SEMI-MODERN, RATHER THAN MODERN (55)	1.00	21 0 36 13 X SQ= 5.20 P = 0.007 RV NO (NO)
182	ALWAYS ARE THOSE WHERE THE BUREAUCRACY IS SEMI-MODERN, RATHER THAN POST-COLONIAL TRANSITIONAL (55)	1.00	182	TEND LESS TO BE THOSE WHERE THE BUREAUCRACY IS SEMI-MODERN, RATHER THAN POST-COLONIAL TRANSITIONAL (55)	0.81	36 13 0 3 X SQ= 4.13 P = 0.025 RV NO (YES)
185	TEND TO BE THOSE WHERE PARTICIPATION BY THE MILITARY IN POLITICS IS SUPPORTIVE, RATHER THAN INTERVENTIVE (31)	0.83	185	TEND TO BE THOSE WHERE PARTICIPATION BY THE MILITARY IN POLITICS IS INTERVENTIVE, RATHER THAN SUPPORTIVE (21)	0.76	5 13 25 4 X SQ= 13.99 P = 0. RV YES (YES)
187	TEND LESS TO BE THOSE WHERE THE ROLE OF THE POLICE IS POLITICALLY SIGNIFICANT (66)	0.56	187	TEND MORE TO BE THOSE WHERE THE ROLE OF THE POLICE IS POLITICALLY SIGNIFICANT (66)	0.95	34 18 27 1 X SQ= 8.05 P = 0.002 RV NO (NO)

194	TEND LESS TO BE THOSE THAT ARE NON-COMMUNIST (101)	0.80	194	ALWAYS ARE THOSE THAT ARE NON-COMMUNIST (101)	1.00	13	0
						51	22
						X SQ=	3.80
						P =	0.018
						RV NO	(NO)

101 POLITIES
WHERE THE REPRESENTATIVE CHARACTER
OF THE REGIME IS POLYARCHIC (41)

101 POLITIES
WHERE THE REPRESENTATIVE CHARACTER
OF THE REGIME IS LIMITED POLYARCHIC,
PSEUDO-POLYARCHIC, OR
NON-POLYARCHIC (57)

PREDICATE ONLY

41 IN LEFT COLUMN

AUSTRALIA	AUSTRIA	BELGIUM	BOLIVIA	CANADA	CEYLON	COSTA RICA	CYPRUS	DENMARK	DOMIN REP
FINLAND	FRANCE	GERMAN FR	GREECE	ICELAND	INDIA	IRELAND	ISRAEL	ITALY	JAMAICA
JAPAN	LUXEMBOURG	MALAGASY R	MALAYA	MEXICO	MOROCCO	NETHERLANDS	NEW ZEALAND	NORWAY	PANAMA
PHILIPPINES	SIERRE LEO	SWEDEN	SWITZERLAND	TRINIDAD	TUNISIA	TURKEY	UK	US	URUGUAY
VENEZUELA									

57 IN RIGHT COLUMN

AFGHANISTAN	ALBANIA	ALGERIA	BRAZIL	BULGARIA	BURMA	CAMBODIA	CEN AFR REP	CHAD	CHILE
CHINA, PR	COLOMBIA	CONGO(BRA)	CUBA	CZECHOS'KIA	DAHOMEY	ECUADOR	EL SALVADOR	ETHIOPIA	GABON
GERMANY, E	GHANA	GUINEA	HAITI	HONDURAS	HUNGARY	INDONESIA	IRAN	IVORY COAST	JORDAN
KOREA, N	LAOS	LIBERIA	MALI	MAURITANIA	MONGOLIA	NEPAL	NICARAGUA	NIGER	PAKISTAN
PARAGUAY	POLAND	PORTUGAL	RUMANIA	SA'U ARABIA	SENEGAL	SO AFRICA	SPAIN	SUDAN	TANGANYIKA
THAILAND	USSR	UAR	UPPER VOLTA	VIETNAM, N	VIETNAM REP	YUGOSLAVIA			

7 EXCLUDED BECAUSE AMBIGUOUS

ARGENTINA CAMEROUN LEBANON LIBYA NIGERIA SOMALIA UGANDA

2 EXCLUDED BECAUSE UNASCERTAINED

BURUNDI RWANDA

8 EXCLUDED BECAUSE UNASCERTAINABLE

CONGO(LEO) GUATEMALA IRAQ KOREA REP PERU SYRIA TOGO YEMEN

102 POLITIES
WHERE THE REPRESENTATIVE CHARACTER
OF THE REGIME IS POLYARCHIC
OR LIMITED POLYARCHIC (59)

102 POLITIES
WHERE THE REPRESENTATIVE CHARACTER
OF THE REGIME IS PSEUDO-POLYARCHIC
OR NON-POLYARCHIC (49)

PREDICATE ONLY

59 IN LEFT COLUMN

ARGENTINA	AUSTRALIA	AUSTRIA	BELGIUM	BOLIVIA	BRAZIL	BURUNDI	CAMEROUN	CANADA	CEYLON
CHILE	COLOMBIA	CONGO(LEO)	COSTA RICA	CYPRUS	DENMARK	DOMIN REP	ECUADOR	FINLAND	FRANCE
GERMAN FR	GREECE	HONDURAS	ICELAND	INDIA	IRELAND	ISRAEL	ITALY	JAMAICA	JAPAN
LEBANON	LIBERIA	LIBYA	LUXEMBOURG	MALAGASY R	MALAYA	MEXICO	MOROCCO	NETHERLANDS	NEW ZEALAND
NIGERIA	NORWAY	PANAMA	PHILIPPINES	RWANDA	SIERRE LEO	SOMALIA	SO AFRICA	SWEDEN	SWITZERLAND
TANGANYIKA	TRINIDAD	TUNISIA	TURKEY	UGANDA	UK	US	URUGUAY	VENEZUELA	

49 IN RIGHT COLUMN

AFGHANISTAN	ALBANIA	ALGERIA	BULGARIA	BURMA	CAMBODIA	CEN AFR REP	CHAD	CHINA, PR	CONGO(BRA)
CUBA	CZECHOS'KIA	DAHOMEY	EL SALVADOR	ETHIOPIA	GABON	GERMANY, E	GHANA	GUINEA	HAITI
HUNGARY	INDONESIA	IRAN	IVORY COAST	JORDAN	KOREA, N	LAOS	MALI	MAURITANIA	MONGOLIA
NEPAL	NICARAGUA	NIGER	PAKISTAN	PARAGUAY	POLAND	PORTUGAL	RUMANIA	SA'U ARABIA	SENEGAL
SPAIN	SUDAN	THAILAND	USSR	UAR	UPPER VOLTA	VIETNAM, N	VIETNAM REP	YUGOSLAVIA	

7 EXCLUDED BECAUSE UNASCERTAINABLE

GUATEMALA	IRAQ	KOREA REP	PERU	SYRIA	TOGO	YEMEN

```
******************************************************************************************

103  POLITIES                                         103  POLITIES
     WHERE THE REPRESENTATIVE CHARACTER                    WHERE THE REPRESENTATIVE CHARACTER
     OF THE REGIME IS PSEUDO-POLYARCHIC,                   OF THE REGIME IS NON-POLYARCHIC,
     RATHER THAN                                           RATHER THAN
     NON-POLYARCHIC  (43)                                  PSEUDO-POLYARCHIC  (6)

                                                                                  SUBJECT ONLY
..........................................................................................

     43 IN LEFT COLUMN

AFGHANISTAN ALBANIA      ALGERIA     BULGARIA    CAMBODIA    CEN AFR REP CHAD        CHINA, PR   CONGO(BRA)  CUBA
CZECHOS'KIA DAHOMEY      EL SALVADOR ETHIOPIA    GABON       GERMANY, E  GHANA       GUINEA      HAITI       HUNGARY
INDONESIA   IVORY COAST  JORDAN      KOREA, N    LAOS        MALI        MAURITANIA  MONGOLIA    NICARAGUA   NIGER
PAKISTAN    PARAGUAY     POLAND      PORTUGAL    RUMANIA     SENEGAL     SPAIN       USSR        UAR         UPPER VOLTA
VIETNAM, N  VIETNAM REP  YUGOSLAVIA

     6 IN RIGHT COLUMN

BURMA       IRAN         NEPAL       SA'U ARABIA SUDAN       THAILAND

     59 EXCLUDED BECAUSE IRRELEVANT

ARGENTINA   AUSTRALIA    AUSTRIA     BELGIUM     BOLIVIA     BRAZIL      BURUNDI     CAMEROUN    CANADA      CEYLON
CHILE       COLOMBIA     CONGO(LEO)  COSTA RICA  CYPRUS      DENMARK     DOMIN REP   ECUADOR     FINLAND     FRANCE
GERMAN FR   GREECE       HONDURAS    ICELAND     INDIA       IRELAND     ISRAEL      ITALY       JAMAICA     JAPAN
LEBANON     LIBERIA      LIBYA       LUXEMBOURG  MALAGASY R  MALAYA      MEXICO      MOROCCO     NETHERLANDS NEW ZEALAND
NIGERIA     NORWAY       PANAMA      PHILIPPINES RWANDA      SIERRE LEO  SOMALIA     SO AFRICA   SWEDEN      SWITZERLAND
TANGANYIKA  TRINIDAD     TUNISIA     TURKEY      UGANDA      UK          US          URUGUAY     VENEZUELA

     7 EXCLUDED BECAUSE UNASCERTAINABLE

GUATEMALA   IRAQ         KOREA REP   PERU        SYRIA       TOGO        YEMEN

------------------------------------------------------------------------------------------

 14  LEAN MORE TOWARD BEING THOSE              0.95    14  LEAN LESS TOWARD BEING THOSE          0.67         2     2
     LOCATED ELSEWHERE THAN IN                             LOCATED ELSEWHERE THAN IN                         41     4
     THE MIDDLE EAST  (104)                                THE MIDDLE EAST  (104)                       X SQ=    2.59
                                                                                                        P =     0.068
                                                                                                        RV NO   (YES)

 15  LEAN TOWARD BEING THOSE                   0.89    15  LEAN TOWARD BEING THOSE               0.67         2     2
     LOCATED IN NORTH AFRICA OR                            LOCATED IN THE MIDDLE EAST, RATHER THAN           16     1
     CENTRAL AND SOUTH AFRICA, RATHER THAN                 IN NORTH AFRICA OR                           X SQ=    2.17
     IN THE MIDDLE EAST  (33)                              CENTRAL AND SOUTH AFRICA  (11)               P =     0.080
                                                                                                        RV YES  (YES)

 24  TEND LESS TO BE THOSE                     0.51    24  ALWAYS ARE THOSE                      1.00        22     6
     WHOSE POPULATION IS                                   WHOSE POPULATION IS                               21     0
     VERY LARGE, LARGE, OR MEDIUM  (61)                    VERY LARGE, LARGE, OR MEDIUM  (61)           X SQ=    3.33
                                                                                                        P =     0.031
                                                                                                        RV NO   (NO )
```

75	LEAN LESS TOWARD BEING THOSE THAT ARE NOT HISTORICALLY WESTERN AND ARE NOT SIGNIFICANTLY WESTERNIZED (52)	0.60	75	ALWAYS ARE THOSE THAT ARE NOT HISTORICALLY WESTERN AND ARE NOT SIGNIFICANTLY WESTERNIZED (52)	1.00	17 0 26 6 X SQ= 2.10 P = 0.080 RV NO (NO)
80	LEAN TOWARD BEING THOSE WHOSE DATE OF INDEPENDENCE IS AFTER 1914, AND THAT WERE FORMERLY DEPENDENCIES OF FRANCE, RATHER THAN BRITAIN (24)	0.81	80	ALWAYS ARE THOSE WHOSE DATE OF INDEPENDENCE IS AFTER 1914, AND THAT WERE FORMERLY DEPENDENCIES OF BRITAIN, RATHER THAN FRANCE (19)	1.00	4 2 17 0 X SQ= 2.72 P = 0.059 RV YES (NO)
91	TILT TOWARD BEING THOSE WHOSE IDEOLOGICAL ORIENTATION IS DEVELOPMENTAL, RATHER THAN TRADITIONAL (31)	0.84	91	TILT TOWARD BEING THOSE WHOSE IDEOLOGICAL ORIENTATION IS TRADITIONAL, RATHER THAN DEVELOPMENTAL (5)	0.67	16 1 3 2 X SQ= 1.47 P = 0.117 RV YES (NO)
95	LEAN LESS TOWARD BEING THOSE WHERE THE STATUS OF THE REGIME IS CONSTITUTIONAL OR AUTHORITARIAN (95)	0.61	95	ALWAYS ARE THOSE WHERE THE STATUS OF THE REGIME IS CONSTITUTIONAL OR AUTHORITARIAN (95)	1.00	25 6 16 0 X SQ= 2.02 P = 0.082 RV NO (NO)
96	TEND TO BE THOSE WHERE THE STATUS OF THE REGIME IS CONSTITUTIONAL OR TOTALITARIAN (67)	0.53	96	ALWAYS ARE THOSE WHERE THE STATUS OF THE REGIME IS AUTHORITARIAN (23)	1.00	17 0 15 6 X SQ= 3.82 P = 0.024 RV YES (NO)
97	TEND TO BE THOSE WHERE THE STATUS OF THE REGIME IS TOTALITARIAN, RATHER THAN AUTHORITARIAN (16)	0.52	97	ALWAYS ARE THOSE WHERE THE STATUS OF THE REGIME IS AUTHORITARIAN, RATHER THAN TOTALITARIAN (23)	1.00	15 6 16 0 X SQ= 3.56 P = 0.027 RV YES (NO)
99	LEAN TOWARD BEING THOSE WHERE GOVERNMENTAL STABILITY IS GENERALLY PRESENT AND DATES FROM AT LEAST THE INTER-WAR PERIOD, OR FROM THE POST-WAR PERIOD (50)	0.64	99	LEAN TOWARD BEING THOSE WHERE GOVERNMENTAL STABILITY IS MODERATELY PRESENT AND DATES FROM THE POST-WAR PERIOD, OR IS ABSENT (36)	0.83	18 1 10 5 X SQ= 2.82 P = 0.066 RV YES (NO)
100	TEND TO BE THOSE WHERE GOVERNMENTAL STABILITY IS GENERALLY PRESENT AND DATES FROM AT LEAST THE INTER-WAR PERIOD, OR IS GENERALLY OR MODERATELY PRESENT, AND DATES FROM THE POST-WAR PERIOD (64)	0.87	100	TEND TO BE THOSE WHERE GOVERNMENTAL STABILITY IS ABSENT (22)	0.67	26 2 4 4 X SQ= 5.43 P = 0.014 RV YES (YES)
113	LEAN TOWARD BEING THOSE WHERE SECTIONALISM IS MODERATE OR NEGLIGIBLE (81)	0.82	113	LEAN TOWARD BEING THOSE WHERE SECTIONALISM IS EXTREME (27)	0.60	7 3 33 2 X SQ= 2.51 P = 0.065 RV YES (NO)

121	LEAN LESS TOWARD BEING THOSE WHERE INTEREST ARTICULATION BY NON-ASSOCIATIONAL GROUPS IS SIGNIFICANT (54)	0.60	121	ALWAYS ARE THOSE WHERE INTEREST ARTICULATION BY NON-ASSOCIATIONAL GROUPS IS SIGNIFICANT (54)	1.00	26 6 17 0 X SQ= 2.10 P = 0.080 RV NO (NO)
134	LEAN LESS TOWARD BEING THOSE WHERE INTEREST AGGREGATION BY THE EXECUTIVE IS LIMITED OR NEGLIGIBLE (46)	0.57	134	ALWAYS ARE THOSE WHERE INTEREST AGGREGATION BY THE EXECUTIVE IS LIMITED OR NEGLIGIBLE (46)	1.00	17 0 23 6 X SQ= 2.43 P = 0.071 RV NO (NO)
135	LEAN TOWARD BEING THOSE WHERE INTEREST AGGREGATION BY THE EXECUTIVE IS SIGNIFICANT, MODERATE, OR LIMITED (77)	0.72	135	LEAN TOWARD BEING THOSE WHERE INTEREST AGGREGATION BY THE EXECUTIVE IS NEGLIGIBLE (13)	0.75	23 1 9 3 X SQ= 1.72 P = 0.098 RV YES (NO)
139	TEND TO BE THOSE WHERE THE PARTY SYSTEM IS QUANTITATIVELY ONE-PARTY (34)	0.73	139	ALWAYS ARE THOSE WHERE THE PARTY SYSTEM IS QUANTITATIVELY OTHER THAN ONE-PARTY (71)	1.00	30 0 11 3 X SQ= 3.94 P = 0.027 RV YES (NO)
154	TEND TO BE THOSE WHERE THE PARTY SYSTEM IS STABLE OR MODERATELY STABLE (55)	0.85	154	ALWAYS ARE THOSE WHERE THE PARTY SYSTEM IS UNSTABLE (25)	1.00	23 0 4 2 X SQ= 3.86 P = 0.037 RV YES (NO)
182	TILT TOWARD BEING THOSE WHERE THE BUREAUCRACY IS SEMI-MODERN, RATHER THAN POST-COLONIAL TRANSITIONAL (55)	0.66	182	ALWAYS ARE THOSE WHERE THE BUREAUCRACY IS POST-COLONIAL TRANSITIONAL, RATHER THAN SEMI-MODERN (25)	1.00	23 0 12 2 X SQ= 1.24 P = 0.137 RV YES (NO)
185	LEAN TOWARD BEING THOSE WHERE PARTICIPATION BY THE MILITARY IN POLITICS IS SUPPORTIVE, RATHER THAN INTERVENTIVE (31)	0.81	185	LEAN TOWARD BEING THOSE WHERE PARTICIPATION BY THE MILITARY IN POLITICS IS INTERVENTIVE, RATHER THAN SUPPORTIVE (21)	0.60	6 3 25 2 X SQ= 1.94 P = 0.088 RV YES (NO)

```
*****************************************************************************************************

104  POLITIES                                           104  POLITIES
     WHERE THE ELECTORAL SYSTEM IS                           WHERE THE ELECTORAL SYSTEM IS
     COMPETITIVE, RATHER THAN                                NON-COMPETITIVE, RATHER THAN
     NON-COMPETITIVE  (43)                                   COMPETITIVE  (30)

                                                                                        SUBJECT ONLY
.....................................................................................................

      43 IN LEFT COLUMN

ARGENTINA   AUSTRALIA   AUSTRIA     BELGIUM     BOLIVIA     BRAZIL      CANADA      CEYLON      CHILE       COSTA RICA
DENMARK     DOMIN REP   ECUADOR     FINLAND     FRANCE      GERMAN FR   GREECE      HONDURAS    ICELAND     INDIA
IRELAND     ISRAEL      ITALY       JAMAICA     JAPAN       LUXEMBOURG  MALAGASY R  MALAYA      NETHERLANDS NEW ZEALAND
NORWAY      PANAMA      PHILIPPINES SIERRE LEO  SWEDEN      SWITZERLAND TRINIDAD    TURKEY      UGANDA      UK
US          URUGUAY     VENEZUELA

      30 IN RIGHT COLUMN

ALBANIA     ALGERIA     BULGARIA    CAMBODIA    CEN AFR REP CHINA, PR   CZECHOS'KIA DAHOMEY     EL SALVADOR GABON
GERMANY, E  GHANA       GUINEA      HAITI       HUNGARY     IVORY COAST KOREA, N    LIBERIA     MALI        MAURITANIA
MONGOLIA    POLAND      PORTUGAL    RUMANIA     SPAIN       TUNISIA     USSR        VIETNAM, N  VIETNAM REP YUGOSLAVIA

      11 EXCLUDED BECAUSE AMBIGUOUS

AFGHANISTAN COLOMBIA    CYPRUS      ETHIOPIA    JORDAN      LAOS        LEBANON     LIBYA       MEXICO      PAKISTAN
SO AFRICA

      10 EXCLUDED BECAUSE IRRELEVANT

BURUNDI     CHAD        CONGO(BRA)  NICARAGUA   PARAGUAY    SA'U ARABIA SENEGAL     SOMALIA     TANGANYIKA  UPPER VOLTA

      21 EXCLUDED BECAUSE UNASCERTAINABLE

BURMA       CAMEROUN    CONGO(LEO)  CUBA        GUATEMALA   INDONESIA   IRAN        IRAQ        KOREA REP   MOROCCO
NEPAL       NIGER       NIGERIA     PERU        RWANDA      SUDAN       SYRIA       THAILAND    TOGO        UAR
YEMEN

-----------------------------------------------------------------------------------------------------

  2  TEND LESS TO BE THOSE                0.53          2  TEND MORE TO BE THOSE               0.93          20     2
     LOCATED ELSEWHERE THAN IN                             LOCATED ELSEWHERE THAN IN                         23    28
     WEST EUROPE, SCANDINAVIA,                             WEST EUROPE, SCANDINAVIA,                      X SQ=  11.50
     NORTH AMERICA, OR AUSTRALASIA  (93)                   NORTH AMERICA, OR AUSTRALASIA  (93)            P =     0.
                                                                                                          RV NO  (YES)

  3  LEAN LESS TOWARD BEING THOSE         0.74          3  LEAN MORE TOWARD BEING THOSE        0.93          11     2
     LOCATED ELSEWHERE THAN IN                             LOCATED ELSEWHERE THAN IN                         32    28
     WEST EUROPE  (102)                                    WEST EUROPE  (102)                             X SQ=   3.12
                                                                                                          P =     0.060
                                                                                                          RV NO  (NO )
```

5	ALWAYS ARE THOSE LOCATED IN SCANDINAVIA OR WEST EUROPE, RATHER THAN IN EAST EUROPE (18)	1.00	5	TEND TO BE THOSE LOCATED IN EAST EUROPE, RATHER THAN IN SCANDINAVIA OR WEST EUROPE (9)	0.82	16 2 0 9 X SQ= 16.13 P = 0. RV YES (YES)
6	TEND LESS TO BE THOSE LOCATED ELSEWHERE THAN IN THE CARIBBEAN, CENTRAL AMERICA, OR SOUTH AMERICA (93)	0.70	6	TEND MORE TO BE THOSE LOCATED ELSEWHERE THAN IN THE CARIBBEAN, CENTRAL AMERICA, OR SOUTH AMERICA (93)	0.93	13 2 30 28 X SQ= 4.65 P = 0.018 RV NO (NO)
10	TEND MORE TO BE THOSE LOCATED ELSEWHERE THAN IN NORTH AFRICA, OR CENTRAL AND SOUTH AFRICA (82)	0.93	10	TEND LESS TO BE THOSE LOCATED ELSEWHERE THAN IN NORTH AFRICA, OR CENTRAL AND SOUTH AFRICA (82)	0.63	3 11 40 19 X SQ= 8.23 P = 0.002 RV NO (YES)
15	LEAN LESS TOWARD BEING THOSE LOCATED IN NORTH AFRICA OR CENTRAL AND SOUTH AFRICA, RATHER THAN IN THE MIDDLE EAST (33)	0.60	15	ALWAYS ARE THOSE LOCATED IN NORTH AFRICA OR CENTRAL AND SOUTH AFRICA, RATHER THAN IN THE MIDDLE EAST (33)	1.00	2 0 3 11 X SQ= 2.04 P = 0.083 RV NO (YES)
19	TEND TO BE THOSE LOCATED IN THE CARIBBEAN, CENTRAL AMERICA, OR SOUTH AMERICA, RATHER THAN IN NORTH AFRICA OR CENTRAL AND SOUTH AFRICA (22)	0.81	19	TEND TO BE THOSE LOCATED IN NORTH AFRICA OR CENTRAL AND SOUTH AFRICA, RATHER THAN IN THE CARIBBEAN, CENTRAL AMERICA, OR SOUTH AMERICA (33)	0.85	3 11 13 2 X SQ= 9.96 P = 0.001 RV YES (YES)
20	TEND TO BE THOSE LOCATED IN THE CARIBBEAN, CENTRAL AMERICA, OR SOUTH AMERICA, RATHER THAN IN EAST ASIA, SOUTH ASIA, OR SOUTHEAST ASIA (22)	0.72	20	TEND TO BE THOSE LOCATED IN EAST ASIA, SOUTH ASIA, OR SOUTHEAST ASIA, RATHER THAN IN THE CARIBBEAN, CENTRAL AMERICA, OR SOUTH AMERICA (18)	0.75	5 6 13 2 X SQ= 3.31 P = 0.038 RV YES (YES)
25	LEAN LESS TOWARD BEING THOSE WHOSE POPULATION DENSITY IS MEDIUM OR LOW (98)	0.72	25	LEAN MORE TOWARD BEING THOSE WHOSE POPULATION DENSITY IS MEDIUM OR LOW (98)	0.90	12 3 31 27 X SQ= 2.46 P = 0.081 RV NO (NO)
29	TEND TO BE THOSE WHERE THE DEGREE OF URBANIZATION IS HIGH (56)	0.82	29	TEND TO BE THOSE WHERE THE DEGREE OF URBANIZATION IS LOW (49)	0.64	32 9 7 16 X SQ= 12.10 P = 0. RV YES (YES)
30	TEND TO BE THOSE WHOSE AGRICULTURAL POPULATION IS MEDIUM, LOW, OR VERY LOW (57)	0.84	30	TEND TO BE THOSE WHOSE AGRICULTURAL POPULATION IS HIGH (56)	0.69	7 20 36 9 X SQ= 18.33 P = 0. RV YES (YES)

31	TEND TO BE THOSE WHOSE AGRICULTURAL POPULATION IS LOW OR VERY LOW (24)	0.51	31	TEND TO BE THOSE WHOSE AGRICULTURAL POPULATION IS HIGH OR MEDIUM (90)	0.97	21 29 22 1 X SQ= 16.58 P = 0. RV YES (YES)
34	LEAN TOWARD BEING THOSE WHOSE GROSS NATIONAL PRODUCT IS VERY HIGH, HIGH, MEDIUM, OR LOW (62)	0.70	34	LEAN TOWARD BEING THOSE WHOSE GROSS NATIONAL PRODUCT IS VERY LOW (53)	0.53	30 14 13 16 X SQ= 3.03 P = 0.056 RV YES (YES)
36	TEND TO BE THOSE WHOSE PER CAPITA GROSS NATIONAL PRODUCT IS VERY HIGH, HIGH, OR MEDIUM (42)	0.67	36	TEND TO BE THOSE WHOSE PER CAPITA GROSS NATIONAL PRODUCT IS LOW OR VERY LOW (73)	0.70	29 9 14 21 X SQ= 8.48 P = 0.002 RV YES (YES)
37	TEND TO BE THOSE WHOSE PER CAPITA GROSS NATIONAL PRODUCT IS VERY HIGH, HIGH, MEDIUM, OR LOW (64)	0.86	37	TEND TO BE THOSE WHOSE PER CAPITA GROSS NATIONAL PRODUCT IS VERY LOW (51)	0.53	37 14 6 16 X SQ= 11.21 P = 0.001 RV YES (YES)
39	TEND TO BE THOSE WHOSE INTERNATIONAL FINANCIAL STATUS IS VERY HIGH, HIGH, OR MEDIUM (38)	0.56	39	TEND TO BE THOSE WHOSE INTERNATIONAL FINANCIAL STATUS IS LOW OR VERY LOW (76)	0.72	24 8 19 21 X SQ= 4.50 P = 0.029 RV YES (YES)
42	TEND TO BE THOSE WHOSE ECONOMIC DEVELOPMENTAL STATUS IS DEVELOPED OR INTERMEDIATE (36)	0.59	42	TEND TO BE THOSE WHOSE ECONOMIC DEVELOPMENTAL STATUS IS UNDERDEVELOPED OR VERY UNDERDEVELOPED (76)	0.73	24 8 17 22 X SQ= 5.88 P = 0.009 RV YES (YES)
43	TEND TO BE THOSE WHOSE ECONOMIC DEVELOPMENTAL STATUS IS DEVELOPED, INTERMEDIATE, OR UNDERDEVELOPED (55)	0.81	43	TEND TO BE THOSE WHOSE ECONOMIC DEVELOPMENTAL STATUS IS VERY UNDERDEVELOPED (57)	0.63	34 11 8 19 X SQ= 12.82 P = 0. RV YES (YES)
45	TEND TO BE THOSE WHERE THE LITERACY RATE IS FIFTY PERCENT OR ABOVE (55)	0.79	45	TEND TO BE THOSE WHERE THE LITERACY RATE IS BELOW FIFTY PERCENT (54)	0.52	34 12 9 13 X SQ= 5.63 P = 0.015 RV YES (YES)
50	TEND TO BE THOSE WHERE FREEDOM OF THE PRESS IS COMPLETE (43)	0.82	50	TEND TO BE THOSE WHERE FREEDOM OF THE PRESS IS INTERMITTENT, INTERNALLY ABSENT, OR INTERNALLY AND EXTERNALLY ABSENT (56)	0.92	33 2 7 23 X SQ= 31.43 P = 0. RV YES (YES)

51	TEND TO BE THOSE WHERE FREEDOM OF THE PRESS IS COMPLETE OR INTERMITTENT (60)	0.97	51	TEND TO BE THOSE WHERE FREEDOM OF THE PRESS IS INTERNALLY ABSENT, OR INTERNALLY AND EXTERNALLY ABSENT (37)	0.88	39 3 1 22 X SQ= 45.52 P = 0. RV YES (YES)
52	ALWAYS ARE THOSE WHERE FREEDOM OF THE PRESS IS COMPLETE, INTERMITTENT, OR INTERNALLY ABSENT (82)	1.00	52	TEND LESS TO BE THOSE WHERE FREEDOM OF THE PRESS IS COMPLETE, INTERMITTENT, OR INTERNALLY ABSENT (82)	0.60	41 15 0 10 X SQ= 16.34 P = 0. RV NO (YES)
54	TEND TO BE THOSE WHERE NEWSPAPER CIRCULATION IS ONE HUNDRED OR MORE PER THOUSAND (37)	0.63	54	TEND TO BE THOSE WHERE NEWSPAPER CIRCULATION IS LESS THAN ONE HUNDRED PER THOUSAND (76)	0.71	27 8 16 20 X SQ= 6.63 P = 0.007 RV YES (YES)
56	TEND MORE TO BE THOSE WHERE THE RELIGION IS PREDOMINANTLY LITERATE (79)	0.88	56	TEND LESS TO BE THOSE WHERE THE RELIGION IS PREDOMINANTLY LITERATE (79)	0.65	38 17 5 9 X SQ= 3.97 P = 0.031 RV NO (YES)
57	TEND LESS TO BE THOSE WHERE THE RELIGION IS OTHER THAN CATHOLIC (90)	0.63	57	TEND MORE TO BE THOSE WHERE THE RELIGION IS OTHER THAN CATHOLIC (90)	0.87	16 4 27 26 X SQ= 3.94 P = 0.033 RV NO (NO)
63	TEND TO BE THOSE WHERE THE RELIGION IS PREDOMINANTLY SOME KIND OF CHRISTIAN (46)	0.72	63	TEND TO BE THOSE WHERE THE RELIGION IS PREDOMINANTLY OR PARTLY OTHER THAN CHRISTIAN (68)	0.66	31 10 12 19 X SQ= 8.52 P = 0.003 RV YES (YES)
68	TEND TO BE THOSE THAT ARE LINGUISTICALLY HOMOGENEOUS (52)	0.63	68	TEND TO BE THOSE THAT ARE LINGUISTICALLY WEAKLY HETEROGENEOUS OR STRONGLY HETEROGENEOUS (62)	0.66	27 10 16 19 X SQ= 4.48 P = 0.030 RV YES (YES)
72	TEND TO BE THOSE WHOSE DATE OF INDEPENDENCE IS BEFORE 1914 (52)	0.69	72	TEND TO BE THOSE WHOSE DATE OF INDEPENDENCE IS AFTER 1914 (59)	0.64	29 10 13 18 X SQ= 6.27 P = 0.008 RV YES (YES)
75	TEND TO BE THOSE THAT ARE HISTORICALLY WESTERN OR SIGNIFICANTLY WESTERNIZED (62)	0.91	75	TEND TO BE THOSE THAT ARE NOT HISTORICALLY WESTERN AND ARE NOT SIGNIFICANTLY WESTERNIZED (52)	0.57	39 13 4 17 X SQ= 17.10 P = 0. RV YES (YES)

77	TEND TO BE THOSE THAT HAVE BEEN SIGNIFICANTLY WESTERNIZED, RATHER THAN PARTIALLY WESTERNIZED, THROUGH A COLONIAL RELATIONSHIP (28)	0.80	77	TEND TO BE THOSE THAT HAVE BEEN PARTIALLY WESTERNIZED, RATHER THAN SIGNIFICANTLY WESTERNIZED, THROUGH A COLONIAL RELATIONSHIP (41)	0.87	16 2 4 13 X SQ= 12.70 P = 0. RV YES (YES)
78	TEND TO BE THOSE THAT HAVE BEEN SIGNIFICANTLY WESTERNIZED THROUGH A COLONIAL RELATIONSHIP, RATHER THAN WITHOUT SUCH A RELATIONSHIP (28)	0.89	78	TEND TO BE THOSE THAT HAVE BEEN SIGNIFICANTLY WESTERNIZED WITHOUT A COLONIAL RELATIONSHIP, RATHER THAN THROUGH SUCH A RELATIONSHIP (7)	0.71	16 2 2 5 X SQ= 6.35 P = 0.007 RV YES (YES)
79	TEND LESS TO BE THOSE THAT WERE FORMERLY DEPENDENCIES OF BRITAIN OR FRANCE, RATHER THAN SPAIN (49)	0.58	79	TEND MORE TO BE THOSE THAT WERE FORMERLY DEPENDENCIES OF BRITAIN OR FRANCE, RATHER THAN SPAIN (49)	0.93	14 14 10 1 X SQ= 3.99 P = 0.028 RV NO (YES)
80	TEND TO BE THOSE WHOSE DATE OF INDEPENDENCE IS AFTER 1914, AND THAT WERE FORMERLY DEPENDENCIES OF BRITAIN, RATHER THAN FRANCE (19)	0.90	80	TEND TO BE THOSE WHOSE DATE OF INDEPENDENCE IS AFTER 1914, AND THAT WERE FORMERLY DEPENDENCIES OF FRANCE, RATHER THAN BRITAIN (24)	0.92	9 1 1 12 X SQ= 12.41 P = 0. RV YES (YES)
81	TEND LESS TO BE THOSE WHERE THE TYPE OF POLITICAL MODERNIZATION IS LATER EUROPEAN OR LATER EUROPEAN DERIVED, RATHER THAN EARLY EUROPEAN OR EARLY EUROPEAN DERIVED (40)	0.68	81	ALWAYS ARE THOSE WHERE THE TYPE OF POLITICAL MODERNIZATION IS LATER EUROPEAN OR LATER EUROPEAN DERIVED, RATHER THAN EARLY EUROPEAN OR EARLY EUROPEAN DERIVED (40)	1.00	10 0 21 12 X SQ= 3.40 P = 0.040 RV NO (NO)
82	TEND TO BE THOSE WHERE THE TYPE OF POLITICAL MODERNIZATION IS EARLY OR LATER EUROPEAN OR EUROPEAN DERIVED, RATHER THAN DEVELOPED TUTELARY OR UNDEVELOPED TUTELARY (51)	0.76	82	TEND TO BE THOSE WHERE THE TYPE OF POLITICAL MODERNIZATION IS DEVELOPED TUTELARY OR UNDEVELOPED TUTELARY, RATHER THAN EARLY OR LATER EUROPEAN OR EUROPEAN DERIVED (55)	0.57	31 12 10 16 X SQ= 6.27 P = 0.011 RV YES (YES)
85	TILT MORE TOWARD BEING THOSE WHERE THE STAGE OF POLITICAL MODERNIZATION IS ADVANCED, RATHER THAN MID- OR EARLY TRANSITIONAL (60)	0.72	85	TILT LESS TOWARD BEING THOSE WHERE THE STAGE OF POLITICAL MODERNIZATION IS ADVANCED, RATHER THAN MID- OR EARLY TRANSITIONAL (60)	0.53	31 16 12 14 X SQ= 1.96 P = 0.137 RV NO (YES)
86	TEND MORE TO BE THOSE WHERE THE STAGE OF POLITICAL MODERNIZATION IS ADVANCED OR MID-TRANSITIONAL, RATHER THAN EARLY TRANSITIONAL (76)	0.84	86	TEND LESS TO BE THOSE WHERE THE STAGE OF POLITICAL MODERNIZATION IS ADVANCED OR MID-TRANSITIONAL, RATHER THAN EARLY TRANSITIONAL (76)	0.60	36 18 7 12 X SQ= 4.01 P = 0.031 RV NO (YES)
87	TILT MORE TOWARD BEING THOSE WHOSE IDEOLOGICAL ORIENTATION IS OTHER THAN DEVELOPMENTAL (58)	0.82	87	TILT LESS TOWARD BEING THOSE WHOSE IDEOLOGICAL ORIENTATION IS OTHER THAN DEVELOPMENTAL (58)	0.63	6 10 28 17 X SQ= 2.01 P = 0.142 RV NO (YES)

No.	Statement	Value	No.	Statement	Value	Statistics
89	TEND TO BE THOSE WHOSE IDEOLOGICAL ORIENTATION IS CONVENTIONAL (33)	0.82	89	ALWAYS ARE THOSE WHOSE IDEOLOGICAL ORIENTATION IS OTHER THAN CONVENTIONAL (62)	1.00	28 0 6 27 X SQ= 37.85 P = 0. RV YES (YES)
92	ALWAYS ARE THOSE WHERE THE SYSTEM STYLE IS LIMITED MOBILIZATIONAL, OR NON-MOBILIZATIONAL (93)	1.00	92	TEND TO BE THOSE WHERE THE SYSTEM STYLE IS MOBILIZATIONAL (20)	0.59	0 17 43 12 X SQ= 29.83 P = 0. RV YES (YES)
93	TEND TO BE THOSE WHERE THE SYSTEM STYLE IS NON-MOBILIZATIONAL (78)	0.93	93	TEND TO BE THOSE WHERE THE SYSTEM STYLE IS MOBILIZATIONAL, OR LIMITED MOBILIZATIONAL (32)	0.72	3 21 40 8 X SQ= 30.49 P = 0. RV YES (YES)
94	ALWAYS ARE THOSE WHERE THE STATUS OF THE REGIME IS CONSTITUTIONAL (51)	1.00	94	TEND TO BE THOSE WHERE THE STATUS OF THE REGIME IS AUTHORITARIAN OR TOTALITARIAN (41)	0.92	42 2 0 24 X SQ= 55.94 P = 0. RV YES (YES)
98	LEAN LESS TOWARD BEING THOSE WHERE GOVERNMENTAL STABILITY IS GENERALLY OR MODERATELY PRESENT AND DATES FROM THE POST-WAR PERIOD, OR IS ABSENT (93)	0.65	98	LEAN MORE TOWARD BEING THOSE WHERE GOVERNMENTAL STABILITY IS GENERALLY OR MODERATELY PRESENT AND DATES FROM THE POST-WAR PERIOD, OR IS ABSENT (93)	0.87	15 4 28 26 X SQ= 3.22 P = 0.057 RV NO (NO)
100	TILT LESS TOWARD BEING THOSE WHERE GOVERNMENTAL STABILITY IS GENERALLY PRESENT AND DATES FROM AT LEAST THE INTER-WAR PERIOD, OR IS GENERALLY OR MODERATELY PRESENT, AND DATES FROM THE POST-WAR PERIOD (64)	0.81	100	TILT MORE TOWARD BEING THOSE WHERE GOVERNMENTAL STABILITY IS GENERALLY PRESENT AND DATES FROM AT LEAST THE INTER-WAR PERIOD, OR IS GENERALLY OR MODERATELY PRESENT, AND DATES FROM THE POST-WAR PERIOD (64)	0.96	29 22 7 1 X SQ= 1.59 P = 0.133 RV NO (NO)
101	TEND TO BE THOSE WHERE THE REPRESENTATIVE CHARACTER OF THE REGIME IS POLYARCHIC (41)	0.90	101	TEND TO BE THOSE WHERE THE REPRESENTATIVE CHARACTER OF THE REGIME IS LIMITED POLYARCHIC, PSEUDO-POLYARCHIC, OR NON-POLYARCHIC (57)	0.97	37 1 4 29 X SQ= 49.17 P = 0. RV YES (YES)
102	ALWAYS ARE THOSE WHERE THE REPRESENTATIVE CHARACTER OF THE REGIME IS POLYARCHIC OR LIMITED POLYARCHIC (59)	1.00	102	TEND TO BE THOSE WHERE THE REPRESENTATIVE CHARACTER OF THE REGIME IS PSEUDO-POLYARCHIC OR NON-POLYARCHIC (49)	0.93	43 2 0 28 X SQ= 61.22 P = 0. RV YES (YES)
107	TEND TO BE THOSE WHERE AUTONOMOUS GROUPS ARE FULLY TOLERATED IN POLITICS (46)	0.98	107	ALWAYS ARE THOSE WHERE AUTONOMOUS GROUPS ARE PARTIALLY TOLERATED IN POLITICS, ARE TOLERATED ONLY OUTSIDE POLITICS, OR ARE NOT TOLERATED AT ALL (65)	1.00	42 0 1 30 X SQ= 65.06 P = 0. RV YES (YES)

108	TEND TO BE THOSE WHERE AUTONOMOUS GROUPS ARE FULLY OR PARTIALLY TOLERATED IN POLITICS (65)	0.98	108	TEND TO BE THOSE WHERE AUTONOMOUS GROUPS ARE TOLERATED ONLY OUTSIDE POLITICS OR ARE NOT TOLERATED AT ALL (35)	0.86	42 4 1 25 X SQ= 49.25 P = 0. RV YES (YES)
111	TEND TO BE THOSE WHERE POLITICAL ENCULTURATION IS HIGH OR MEDIUM (53)	0.77	111	TEND TO BE THOSE WHERE POLITICAL ENCULTURATION IS LOW (42)	0.53	31 8 9 9 X SQ= 3.80 P = 0.032 RV YES (YES)
116	TEND TO BE THOSE WHERE INTEREST ARTICULATION BY ASSOCIATIONAL GROUPS IS SIGNIFICANT OR MODERATE (32)	0.67	116	ALWAYS ARE THOSE WHERE INTEREST ARTICULATION BY ASSOCIATIONAL GROUPS IS LIMITED OR NEGLIGIBLE (79)	1.00	29 0 14 27 X SQ= 28.37 P = 0. RV YES (YES)
117	TEND TO BE THOSE WHERE INTEREST ARTICULATION BY ASSOCIATIONAL GROUPS IS SIGNIFICANT, MODERATE, OR LIMITED (60)	0.91	117	TEND TO BE THOSE WHERE INTEREST ARTICULATION BY ASSOCIATIONAL GROUPS IS NEGLIGIBLE (51)	0.78	39 6 4 21 X SQ= 30.96 P = 0. RV YES (YES)
118	TEND TO BE THOSE WHERE INTEREST ARTICULATION BY INSTITUTIONAL GROUPS IS SIGNIFICANT, MODERATE, OR LIMITED (60)	0.93	118	TEND TO BE THOSE WHERE INTEREST ARTICULATION BY INSTITUTIONAL GROUPS IS VERY SIGNIFICANT (40)	0.77	3 20 39 6 X SQ= 31.89 P = 0. RV YES (YES)
119	TEND TO BE THOSE WHERE INTEREST ARTICULATION BY INSTITUTIONAL GROUPS IS MODERATE OR LIMITED (26)	0.60	119	ALWAYS ARE THOSE WHERE INTEREST ARTICULATION BY INSTITUTIONAL GROUPS IS VERY SIGNIFICANT OR SIGNIFICANT (74)	1.00	17 26 25 0 X SQ= 21.98 P = 0. RV YES (YES)
121	TEND TO BE THOSE WHERE INTEREST ARTICULATION BY NON-ASSOCIATIONAL GROUPS IS MODERATE, LIMITED, OR NEGLIGIBLE (61)	0.84	121	TEND TO BE THOSE WHERE INTEREST ARTICULATION BY NON-ASSOCIATIONAL GROUPS IS SIGNIFICANT (54)	0.50	7 15 36 15 X SQ= 8.01 P = 0.004 RV YES (YES)
122	TEND TO BE THOSE WHERE INTEREST ARTICULATION BY NON-ASSOCIATIONAL GROUPS IS LIMITED OR NEGLIGIBLE (32)	0.56	122	TEND TO BE THOSE WHERE INTEREST ARTICULATION BY NON-ASSOCIATIONAL GROUPS IS SIGNIFICANT OR MODERATE (83)	0.83	19 25 24 5 X SQ= 9.73 P = 0.001 RV YES (YES)
123	TEND LESS TO BE THOSE WHERE INTEREST ARTICULATION BY NON-ASSOCIATIONAL GROUPS IS SIGNIFICANT, MODERATE, OR LIMITED (107)	0.81	123	ALWAYS ARE THOSE WHERE INTEREST ARTICULATION BY NON-ASSOCIATIONAL GROUPS IS SIGNIFICANT, MODERATE, OR LIMITED (107)	1.00	35 30 8 0 X SQ= 4.51 P = 0.018 RV NO (NO)

125	TEND TO BE THOSE WHERE INTEREST ARTICULATION BY ANOMIC GROUPS IS INFREQUENT OR VERY INFREQUENT (35)	0.68	125	TEND TO BE THOSE WHERE INTEREST ARTICULATION BY ANOMIC GROUPS IS FREQUENT OR OCCASIONAL (64)	0.82	12 18 25 4 X SQ= 11.56 P = 0. RV YES (YES)
126	TEND LESS TO BE THOSE WHERE INTEREST ARTICULATION BY ANOMIC GROUPS IS FREQUENT, OCCASIONAL, OR INFREQUENT (83)	0.57	126	ALWAYS ARE THOSE WHERE INTEREST ARTICULATION BY ANOMIC GROUPS IS FREQUENT, OCCASIONAL, OR INFREQUENT (83)	1.00	21 22 16 0 X SQ= 10.96 P = 0. RV NO (YES)
127	TEND LESS TO BE THOSE WHERE INTEREST ARTICULATION BY POLITICAL PARTIES IS MODERATE, LIMITED, OR NEGLIGIBLE (72)	0.63	127	ALWAYS ARE THOSE WHERE INTEREST ARTICULATION BY POLITICAL PARTIES IS MODERATE, LIMITED, OR NEGLIGIBLE (72)	1.00	15 0 26 28 X SQ= 11.03 P = 0. RV NO (YES)
128	TEND TO BE THOSE WHERE INTEREST ARTICULATION BY POLITICAL PARTIES IS SIGNIFICANT OR MODERATE (48)	0.88	128	TEND TO BE THOSE WHERE INTEREST ARTICULATION BY POLITICAL PARTIES IS LIMITED OR NEGLIGIBLE (45)	0.93	36 2 5 26 X SQ= 40.55 P = 0. RV YES (YES)
129	TEND TO BE THOSE WHERE INTEREST ARTICULATION BY POLITICAL PARTIES IS SIGNIFICANT, MODERATE, OR LIMITED (56)	0.93	129	TEND TO BE THOSE WHERE INTEREST ARTICULATION BY POLITICAL PARTIES IS NEGLIGIBLE (37)	0.89	38 3 3 25 X SQ= 43.02 P = 0. RV YES (YES)
130	TILT LESS TOWARD BEING THOSE WHERE INTEREST AGGREGATION BY POLITICAL PARTIES IS MODERATE, LIMITED, OR NEGLIGIBLE (71)	0.80	130	TILT MORE TOWARD BEING THOSE WHERE INTEREST AGGREGATION BY POLITICAL PARTIES IS MODERATE, LIMITED, OR NEGLIGIBLE (71)	0.95	8 1 32 20 X SQ= 1.48 P = 0.146 RV NO (NO)
134	TEND TO BE THOSE WHERE INTEREST AGGREGATION BY THE EXECUTIVE IS SIGNIFICANT OR MODERATE (57)	0.76	134	TEND TO BE THOSE WHERE INTEREST AGGREGATION BY THE EXECUTIVE IS LIMITED OR NEGLIGIBLE (46)	0.66	32 10 10 19 X SQ= 10.68 P = 0.001 RV YES (YES)
137	TEND TO BE THOSE WHERE INTEREST AGGREGATION BY THE LEGISLATURE IS SIGNIFICANT OR MODERATE (29)	0.68	137	ALWAYS ARE THOSE WHERE INTEREST AGGREGATION BY THE LEGISLATURE IS LIMITED OR NEGLIGIBLE (68)	1.00	28 0 13 29 X SQ= 30.22 P = 0. RV YES (YES)
138	TEND TO BE THOSE WHERE INTEREST AGGREGATION BY THE LEGISLATURE IS SIGNIFICANT, MODERATE, OR LIMITED (48)	0.93	138	ALWAYS ARE THOSE WHERE INTEREST AGGREGATION BY THE LEGISLATURE IS NEGLIGIBLE (49)	1.00	39 0 3 27 X SQ= 53.95 P = 0. RV YES (YES)

139	TEND TO BE THOSE WHERE THE PARTY SYSTEM IS QUANTITATIVELY OTHER THAN ONE-PARTY (71)	0.98	139	TEND TO BE THOSE WHERE THE PARTY SYSTEM IS QUANTITATIVELY ONE-PARTY (34)	0.90	1 27 42 3 X SQ= 53.80 P = 0. RV YES (YES)
141	TEND LESS TO BE THOSE WHERE THE PARTY SYSTEM IS QUANTITATIVELY OTHER THAN TWO-PARTY (87)	0.75	141	ALWAYS ARE THOSL WHERE THE PARTY SYSTEM IS QUANTITATIVELY OTHER THAN TWO-PARTY (87)	1.00	10 0 30 30 X SQ= 6.83 P = 0.004 RV NO (YES)
142	TEND TO BE THOSE WHERE THE PARTY SYSTEM IS QUANTITATIVELY MULTI-PARTY (30)	0.55	142	TEND TO BE THOSE WHERE THE PARTY SYSTEM IS QUANTITATIVELY OTHER THAN MULTI-PARTY (66)	0.97	22 1 18 29 X SQ= 18.47 P = 0. RV YES (YES)
144	ALWAYS ARE THOSE WHERE THE PARTY SYSTEM IS QUANTITATIVELY TWO-PARTY, RATHER THAN ONE-PARTY (11)	1.00	144	ALWAYS ARE THOSE WHERE THE PARTY SYSTEM IS QUANTITATIVELY ONE-PARTY, RATHER THAN TWO-PARTY (34)	1.00	0 27 10 0 X SQ= 32.10 P = 0. RV YES (YES)
146	ALWAYS ARE THOSE WHERE THE PARTY SYSTEM IS QUALITATIVELY OTHER THAN MASS-BASED TERRITORIAL (92)	1.00	146	TEND LESS TO BE THOSE WHERE THE PARTY SYSTEM IS QUALITATIVELY OTHER THAN MASS-BASED TERRITORIAL (92)	0.81	0 5 40 22 X SQ= 5.55 P = 0.008 RV NO (YES)
147	TEND TO BE THOSE WHERE THE PARTY SYSTEM IS QUALITATIVELY CLASS-ORIENTED OR MULTI-IDEOLOGICAL (23)	0.61	147	ALWAYS ARE THOSE WHERE THE PARTY SYSTEM IS QUALITATIVELY OTHER THAN CLASS-ORIENTED OR MULTI-IDEOLOGICAL (67)	1.00	22 0 14 28 X SQ= 23.44 P = 0. RV YES (YES)
148	TILT MORE TOWARD BEING THOSE WHERE THE PARTY SYSTEM IS QUALITATIVELY OTHER THAN AFRICAN TRANSITIONAL (96)	0.95	148	TILT LESS TOWARD BEING THOSE WHERE THE PARTY SYSTEM IS QUALITATIVELY OTHER THAN AFRICAN TRANSITIONAL (96)	0.83	2 5 41 24 X SQ= 1.86 P 0.110 RV NO (YES)
153	LEAN LESS TOWARD BEING THOSE WHERE THE PARTY SYSTEM IS STABLE (42)	0.55	153	LEAN MORE TOWARD BEING THOSE WHERE THE PARTY SYSTEM IS STABLE (42)	0.78	21 18 17 5 X SQ= 2.36 P = 0.100 RV NO (NO)
154	TEND LESS TO BE THOSE WHERE THE PARTY SYSTEM IS STABLE OR MODERATELY STABLE (55)	0.71	154	TEND MORE TO BE THOSE WHERE THE PARTY SYSTEM IS STABLE OR MODERATELY STABLE (55)	0.96	27 22 11 1 X SQ= 4.04 P = 0.022 RV NO (NO)

160	TEND TO BE THOSE WHERE THE POLITICAL LEADERSHIP IS MODERATELY ELITIST OR NON-ELITIST (67)	0.95	160	TEND TO BE THOSE WHERE THE POLITICAL LEADERSHIP IS ELITIST (30)	0.63	2 17 40 10 X SQ= 25.06 P = 0. RV YES (YES)
161	TEND TO BE THOSE WHERE THE POLITICAL LEADERSHIP IS NON-ELITIST (50)	0.76	161	TEND TO BE THOSE WHERE THE POLITICAL LEADERSHIP IS ELITIST OR MODERATELY ELITIST (47)	0.78	10 21 32 6 X SQ= 17.23 P = 0. RV YES (YES)
164	TEND TO BE THOSE WHERE THE REGIME'S LEADERSHIP CHARISMA IS NEGLIGIBLE (65)	0.93	164	TEND TO BE THOSE WHERE THE REGIME'S LEADERSHIP CHARISMA IS PRONOUNCED OR MODERATE (34)	0.52	3 15 39 14 X SQ= 15.74 P = 0. RV YES (YES)
166	TEND LESS TO BE THOSE WHERE THE VERTICAL POWER DISTRIBUTION IS THAT OF FORMAL FEDERALISM OR FORMAL AND EFFECTIVE UNITARISM (99)	0.74	166	ALWAYS ARE THOSE WHERE THE VERTICAL POWER DISTRIBUTION IS THAT OF FORMAL FEDERALISM OR FORMAL AND EFFECTIVE UNITARISM (99)	1.00	11 0 32 30 X SQ= 7.15 P = 0.002 RV NO (NO)
168	TEND TO BE THOSE WHERE THE HORIZONTAL POWER DISTRIBUTION IS SIGNIFICANT (34)	0.78	168	ALWAYS ARE THOSE WHERE THE HORIZONTAL POWER DISTRIBUTION IS LIMITED OR NEGLIGIBLE (72)	1.00	32 0 9 30 X SQ= 39.53 P = 0. RV YES (YES)
169	ALWAYS ARE THOSE WHERE THE HORIZONTAL POWER DISTRIBUTION IS SIGNIFICANT OR LIMITED (58)	1.00	169	TEND TO BE THOSE WHERE THE HORIZONTAL POWER DISTRIBUTION IS NEGLIGIBLE (48)	0.90	41 3 0 27 X SQ= 55.79 P = 0. RV YES (YES)
171	TEND LESS TO BE THOSE WHERE THE LEGISLATIVE-EXECUTIVE STRUCTURE IS OTHER THAN PARLIAMENTARY-REPUBLICAN (90)	0.79	171	ALWAYS ARE THOSE WHERE THE LEGISLATIVE-EXECUTIVE STRUCTURE IS OTHER THAN PARLIAMENTARY-REPUBLICAN (90)	1.00	9 0 34 27 X SQ= 4.75 P = 0.010 RV NO (NO)
172	TEND LESS TO BE THOSE WHERE THE LEGISLATIVE-EXECUTIVE STRUCTURE IS OTHER THAN PARLIAMENTARY-ROYALIST (88)	0.58	172	ALWAYS ARE THOSE WHERE THE LEGISLATIVE-EXECUTIVE STRUCTURE IS OTHER THAN PARLIAMENTARY-ROYALIST (88)	1.00	18 0 25 28 X SQ= 13.57 P = 0. RV NO (YES)
175	ALWAYS ARE THOSE WHERE THE LEGISLATURE IS FULLY EFFECTIVE OR PARTIALLY EFFECTIVE (51)	1.00	175	ALWAYS ARE THOSE WHERE THE LEGISLATURE IS LARGELY INEFFECTIVE OR WHOLLY INEFFECTIVE (49)	1.00	42 0 0 30 X SQ= 67.94 P = 0. RV YES (YES)

178	TEND TO BE THOSE WHERE THE LEGISLATURE IS BICAMERAL (51)	0.77	178	TEND TO BE THOSE WHERE THE LEGISLATURE IS UNICAMERAL (53)	0.83	10 25 33 5 X SQ= 23.21 P = 0. RV YES (YES)
179	TEND TO BE THOSE WHERE THE EXECUTIVE IS STRONG (39)	0.92	179	ALWAYS ARE THOSE WHERE THE EXECUTIVE IS DOMINANT (52)	1.00	3 30 33 0 X SQ= 51.39 P = 0. RV YES (YES)
181	TEND TO BE THOSE WHERE THE BUREAUCRACY IS MODERN, RATHER THAN SEMI-MODERN (21)	0.50	181	ALWAYS ARE THOSE WHERE THE BUREAUCRACY IS SEMI-MODERN, RATHER THAN MODERN (55)	1.00	20 0 20 21 X SQ= 13.44 P = 0. RV YES (YES)
185	TEND TO BE THOSE WHERE PARTICIPATION BY THE MILITARY IN POLITICS IS INTERVENTIVE, RATHER THAN SUPPORTIVE (21)	0.83	185	TEND TO BE THOSE WHERE PARTICIPATION BY THE MILITARY IN POLITICS IS SUPPORTIVE, RATHER THAN INTERVENTIVE (31)	0.91	5 2 1 21 X SQ= 10.69 P = 0.001 RV YES (YES)
186	TEND TO BE THOSE WHERE PARTICIPATION BY THE MILITARY IN POLITICS IS NEUTRAL, RATHER THAN SUPPORTIVE (56)	0.97	186	TEND TO BE THOSE WHERE PARTICIPATION BY THE MILITARY IN POLITICS IS SUPPORTIVE, RATHER THAN NEUTRAL (31)	0.75	1 21 35 7 X SQ= 33.29 P = 0. RV YES (YES)
187	TEND TO BE THOSE WHERE THE ROLE OF THE POLICE IS NOT POLITICALLY SIGNIFICANT (35)	0.84	187	TEND TO BE THOSE WHERE THE ROLE OF THE POLICE IS POLITICALLY SIGNIFICANT (66)	0.97	6 28 32 1 X SQ= 39.75 P = 0. RV YES (YES)
188	TEND LESS TO BE THOSE WHERE THE CHARACTER OF THE LEGAL SYSTEM IS OTHER THAN CIVIL LAW (81)	0.51	188	TEND MORE TO BE THOSE WHERE THE CHARACTER OF THE LEGAL SYSTEM IS OTHER THAN CIVIL LAW (81)	0.86	21 4 22 25 X SQ= 7.90 P = 0.002 RV NO (YES)
189	TEND LESS TO BE THOSE WHERE THE CHARACTER OF THE LEGAL SYSTEM IS OTHER THAN COMMON LAW (108)	0.84	189	ALWAYS ARE THOSE WHERE THE CHARACTER OF THE LEGAL SYSTEM IS OTHER THAN COMMON LAW (108)	1.00	7 0 36 30 X SQ= 3.69 P = 0.037 RV NO (NO)
192	LEAN MORE TOWARD BEING THOSE WHERE THE CHARACTER OF THE LEGAL SYSTEM IS OTHER THAN MUSLIM OR PARTLY MUSLIM (86)	0.93	192	LEAN LESS TOWARD BEING THOSE WHERE THE CHARACTER OF THE LEGAL SYSTEM IS OTHER THAN MUSLIM OR PARTLY MUSLIM (86)	0.76	3 7 40 22 X SQ= 2.95 P = 0.078 RV NO (YES)

193	TEND MORE TO BE THOSE WHERE THE CHARACTER OF THE LEGAL SYSTEM IS OTHER THAN PARTLY INDIGENOUS (86)	0.93	193	TEND LESS TO BE THOSE WHERE THE CHARACTER OF THE LEGAL SYSTEM IS OTHER THAN PARTLY INDIGENOUS (86)	0.70	3 9 40 21 X SQ= 5.25 P = 0.021 RV NO (YES)
194	ALWAYS ARE THOSE THAT ARE NON-COMMUNIST (101)	1.00	194	TEND LESS TO BE THOSE THAT ARE NON-COMMUNIST (101)	0.57	0 13 43 17 X SQ= 19.81 P = 0. RV NO (YES)

105 POLITIES
WHERE THE ELECTORAL SYSTEM IS
COMPETITIVE (43)

105 POLITIES
WHERE THE ELECTORAL SYSTEM IS
PARTIALLY COMPETITIVE OR
NON-COMPETITIVE (47)

PREDICATE ONLY

43 IN LEFT COLUMN

ARGENTINA	AUSTRALIA	AUSTRIA	BELGIUM	BOLIVIA	BRAZIL	CANADA	CEYLON	CHILE	COSTA RICA
DENMARK	DOMIN REP	ECUADOR	FINLAND	FRANCE	GERMAN FR	GREECE	HONDURAS	ICELAND	INDIA
IRELAND	ISRAEL	ITALY	JAMAICA	JAPAN	LUXEMBOURG	MALAGASY R	MALAYA	NETHERLANDS	NEW ZEALAND
NORWAY	PANAMA	PHILIPPINES	SIERRE LEO	SWEDEN	SWITZERLAND	TRINIDAD	TURKEY	UGANDA	UK
US	URUGUAY	VENEZUELA							

47 IN RIGHT COLUMN

AFGHANISTAN	ALBANIA	ALGERIA	BULGARIA	BURUNDI	CAMBODIA	CEN AFR REP	CHAD	CHINA, PR	CONGO(BRA)
CZECHOS'KIA	DAHOMEY	EL SALVADOR	ETHIOPIA	GABON	GERMANY, E	GHANA	GUINEA	HAITI	HUNGARY
IVORY COAST	JORDAN	KOREA, N	LAOS	LIBERIA	LIBYA	MALI	MAURITANIA	MONGOLIA	NICARAGUA
NIGER	PAKISTAN	PARAGUAY	POLAND	PORTUGAL	RUMANIA	RWANDA	SENEGAL	SOMALIA	SPAIN
TANGANYIKA	TUNISIA	USSR	UPPER VOLTA	VIETNAM, N	VIETNAM REP	YUGOSLAVIA			

5 EXCLUDED BECAUSE AMBIGUOUS

COLOMBIA CYPRUS LEBANON MEXICO SO AFRICA

1 EXCLUDED BECAUSE IRRELEVANT

SA'U ARABIA

19 EXCLUDED BECAUSE UNASCERTAINABLE

BURMA	CAMEROUN	CONGO(LEO)	CUBA	GUATEMALA	INDONESIA	IRAN	IRAQ	KOREA REP	MOROCCO
NEPAL	NIGERIA	PERU	SUDAN	SYRIA	THAILAND	TOGO	UAR	YEMEN	

106 POLITIES
WHERE THE ELECTORAL SYSTEM IS
COMPETITIVE OR
PARTIALLY COMPETITIVE (52)

106 POLITIES
WHERE THE ELECTORAL SYSTEM IS
NON-COMPETITIVE (30)

PREDICATE ONLY

52 IN LEFT COLUMN

ARGENTINA	AUSTRALIA	AUSTRIA	BELGIUM	BOLIVIA	BRAZIL	BURUNDI	CANADA	CEYLON	CHAD
CHILE	CONGO(BRA)	COSTA RICA	DENMARK	DOMIN REP	ECUADOR	FINLAND	FRANCE	GERMAN FR	GREECE
HONDURAS	ICELAND	INDIA	IRELAND	ISRAEL	ITALY	JAMAICA	JAPAN	LUXEMBOURG	MALAGASY R
MALAYA	NETHERLANDS	NEW ZEALAND	NICARAGUA	NORWAY	PANAMA	PARAGUAY	PHILIPPINES	SENEGAL	SIERRE LEO
SOMALIA	SWEDEN	SWITZERLAND	TANGANYIKA	TRINIDAD	TURKEY	UGANDA	UK	US	UPPER VOLTA
URUGUAY	VENEZUELA								

30 IN RIGHT COLUMN

ALBANIA	ALGERIA	BULGARIA	CAMBODIA	CEN AFR REP	CHINA, PR	CZECHOS'KIA	DAHOMEY	EL SALVADOR	GABON
GERMANY, E	GHANA	GUINEA	HAITI	HUNGARY	IVORY COAST	KOREA, N	LIBERIA	MALI	MAURITANIA
MONGOLIA	POLAND	PORTUGAL	RUMANIA	SPAIN	TUNISIA	USSR	VIETNAM, N	VIETNAM REP	YUGOSLAVIA

11 EXCLUDED BECAUSE AMBIGUOUS

AFGHANISTAN	COLOMBIA	CYPRUS	ETHIOPIA	JORDAN	LAOS	LEBANON	LIBYA	MEXICO	PAKISTAN
SO AFRICA									

1 EXCLUDED BECAUSE IRRELEVANT

SA'U ARABIA

21 EXCLUDED BECAUSE UNASCERTAINABLE

BURMA	CAMEROUN	CONGO(LEO)	CUBA	GUATEMALA	INDONESIA	IRAN	IRAQ	KOREA REP	MOROCCO
NEPAL	NIGER	NIGERIA	PERU	RWANDA	SUDAN	SYRIA	THAILAND	TOGO	UAR
YEMEN									

```
107  POLITIES                                                  107  POLITIES
     WHERE AUTONOMOUS GROUPS                                        WHERE AUTONOMOUS GROUPS
     ARE FULLY TOLERATED IN POLITICS  (46)                          ARE PARTIALLY TOLERATED IN POLITICS,
                                                                    ARE TOLERATED ONLY OUTSIDE POLITICS,
                                                                    OR ARE NOT TOLERATED AT ALL  (65)

                                                                                              PREDICATE ONLY

      46 IN LEFT COLUMN

ARGENTINA   AUSTRALIA   AUSTRIA     BELGIUM     BOLIVIA     BRAZIL      CANADA      CEYLON      CHILE       COLOMBIA
COSTA RICA  CYPRUS      DENMARK     DOMIN REP   ECUADOR     FINLAND     FRANCE      GERMAN FR   GREECE      HONDURAS
ICELAND     INDIA       IRELAND     ISRAEL      ITALY       JAMAICA     JAPAN       LEBANON     LUXEMBOURG  MALAGASY R
MALAYA      NETHERLANDS NEW ZEALAND NORWAY      PANAMA      PHILIPPINES SIERRE LEO  SWEDEN      SWITZERLAND TRINIDAD
TURKEY      UGANDA      UK          US          URUGUAY     VENEZUELA

      65 IN RIGHT COLUMN

AFGHANISTAN ALBANIA     ALGERIA     BULGARIA    BURMA       BURUNDI     CAMBODIA    CAMEROUN    CEN AFR REP CHAD
CHINA, PR   CONGO(BRA)  CONGO(LEO)  CUBA        CZECHOS'KIA DAHOMEY     EL SALVADOR ETHIOPIA    GABON       GERMANY, E
GHANA       GUATEMALA   GUINEA      HAITI       HUNGARY     INDONESIA   IRAN        IRAQ        IVORY COAST JORDAN
KOREA, N    KOREA REP   LIBERIA     LIBYA       MALI        MAURITANIA  MEXICO      MONGOLIA    MOROCCO     NEPAL
NICARAGUA   NIGER       NIGERIA     PAKISTAN    PARAGUAY    POLAND      PORTUGAL    RUMANIA     RWANDA      SA'U ARABIA
SENEGAL     SOMALIA     SO AFRICA   SPAIN       SUDAN       SYRIA       TANGANYIKA  THAILAND    TUNISIA     USSR
UAR         UPPER VOLTA VIETNAM, N  VIETNAM REP YUGOSLAVIA

       1 EXCLUDED BECAUSE AMBIGUOUS

LAOS

       3 EXCLUDED BECAUSE UNASCERTAINABLE

PERU        TOGO        YEMEN
```

108 POLITIES
WHERE AUTONOMOUS GROUPS
ARE FULLY OR PARTIALLY TOLERATED
IN POLITICS (65)

108 POLITIES
WHERE AUTONOMOUS GROUPS
ARE TOLERATED ONLY OUTSIDE POLITICS
OR ARE NOT TOLERATED AT ALL (35)

PREDICATE ONLY

65 IN LEFT COLUMN

ARGENTINA	AUSTRALIA	AUSTRIA	BELGIUM	BOLIVIA	BRAZIL	BURUNDI	CAMEROUN	CANADA	CEYLON
CHAD	CHILE	COLOMBIA	CONGO(BRA)	COSTA RICA	CYPRUS	DENMARK	DOMIN REP	ECUADOR	EL SALVADOR
FINLAND	FRANCE	GABON	GERMAN FR	GREECE	HONDURAS	ICELAND	INDIA	INDONESIA	IRAN
IRELAND	ISRAEL	ITALY	JAMAICA	JAPAN	LEBANON	LIBYA	LUXEMBOURG	MALAGASY R	MALAYA
MAURITANIA	MEXICO	MOROCCO	NETHERLANDS	NEW ZEALAND	NICARAGUA	NORWAY	PANAMA	PARAGUAY	PERU
PHILIPPINES	RWANDA	SIERRE LEO	SOMALIA	SWEDEN	SWITZERLAND	TANGANYIKA	TRINIDAD	TUNISIA	TURKEY
UGANDA	UK	US	URUGUAY	VENEZUELA					

35 IN RIGHT COLUMN

AFGHANISTAN	ALBANIA	BULGARIA	CAMBODIA	CEN AFR REP	CHINA, PR	CUBA	CZECHOS'KIA	DAHOMEY	ETHIOPIA
GERMANY, E	GHANA	GUINEA	HAITI	HUNGARY	IVORY COAST	KOREA, N	LIBERIA	MALI	MONGOLIA
NEPAL	NIGER	POLAND	PORTUGAL	RUMANIA	SA'U ARABIA	SPAIN	SUDAN	THAILAND	USSR
UAR	UPPER VOLTA	VIETNAM, N	VIETNAM REP	YUGOSLAVIA					

3 EXCLUDED BECAUSE AMBIGUOUS

JORDAN LAOS SO AFRICA

12 EXCLUDED BECAUSE UNASCERTAINABLE

ALGERIA	BURMA	CONGO(LEO)	GUATEMALA	IRAQ	KOREA REP	NIGERIA	PAKISTAN	SENEGAL	SYRIA
TOGO	YEMEN								

109 POLITIES WHERE POLITICAL ENCULTURATION IS HIGH, RATHER THAN LOW (15)	109 POLITIES WHERE POLITICAL ENCULTURATION IS LOW, RATHER THAN HIGH (42)

SUBJECT ONLY

15 IN LEFT COLUMN

AUSTRALIA, AUSTRIA, COSTA RICA, DENMARK, ICELAND, IRELAND, JAPAN, LUXEMBOURG, NEW ZEALAND, NORWAY, POLAND, SWEDEN, THAILAND, TURKEY, UK

42 IN RIGHT COLUMN

ARGENTINA, BELGIUM, BOLIVIA, BRAZIL, CEN AFR REP, CHAD, CHILE, CONGO(BRA), CONGO(LEO), CYPRUS, DAHOMEY, ECUADOR, EL SALVADOR, ETHIOPIA, GABON, GUATEMALA, HAITI, INDONESIA, IRAN, IRAQ, JORDAN, KOREA REP, LAOS, LEBANON, LIBERIA, MALAYA, NEPAL, NICARAGUA, NIGER, NIGERIA, PAKISTAN, PANAMA, PERU, SA'U ARABIA, SO AFRICA, SPAIN, SYRIA, UGANDA, UPPER VOLTA, VIETNAM REP, YEMEN, YUGOSLAVIA

2 EXCLUDED BECAUSE AMBIGUOUS

SWITZERLAND, URUGUAY

38 EXCLUDED BECAUSE IRRELEVANT

AFGHANISTAN, ALGERIA, BURMA, CAMBODIA, CAMEROUN, CANADA, CEYLON, COLOMBIA, DOMIN REP, FINLAND, FRANCE, GERMAN FR, GHANA, GREECE, GUINEA, HONDURAS, INDIA, ISRAEL, ITALY, IVORY COAST, JAMAICA, LIBYA, MALAGASY R, MAURITANIA, MEXICO, MOROCCO, NETHERLANDS, PHILIPPINES, SENEGAL, SIERRE LEO, SOMALIA, SUDAN, TANGANYIKA, TOGO, TRINIDAD, TUNISIA, US, VENEZUELA

18 EXCLUDED BECAUSE UNASCERTAINED

ALBANIA, BULGARIA, BURUNDI, CHINA, PR, CUBA, CZECHOS'KIA, GERMANY, E, HUNGARY, KOREA, N, MALI, MONGOLIA, PARAGUAY, PORTUGAL, RUMANIA, RWANDA, USSR, UAR, VIETNAM, N

	Left column			Right column		
2	TEND TO BE THOSE LOCATED IN WEST EUROPE, SCANDINAVIA, NORTH AMERICA, OR AUSTRALASIA (22)	0.67	2	TEND TO BE THOSE LOCATED ELSEWHERE THAN IN WEST EUROPE, SCANDINAVIA, NORTH AMERICA, OR AUSTRALASIA (93)	0.95	10 2 5 40 X SQ= 21.90 P = 0. RV YES (YES)
3	TEND LESS TO BE THOSE LOCATED ELSEWHERE THAN IN WEST EUROPE (102)	0.73	3	TEND MORE TO BE THOSE LOCATED ELSEWHERE THAN IN WEST EUROPE (102)	0.95	4 2 11 40 X SQ= 3.55 P = 0.036 RV NO (YES)

No.	Statement	Value	No.	Statement	Value	Statistics
10	ALWAYS ARE THOSE LOCATED ELSEWHERE THAN IN NORTH AFRICA, OR CENTRAL AND SOUTH AFRICA (82)	1.00	10	TEND LESS TO BE THOSE LOCATED ELSEWHERE THAN IN NORTH AFRICA, OR CENTRAL AND SOUTH AFRICA (82)	0.69	0 13 15 29 X SQ= 4.38 P = 0.013 RV NO (NO)
26	TILT TOWARD BEING THOSE WHOSE POPULATION DENSITY IS VERY HIGH, HIGH, OR MEDIUM (48)	0.60	26	TILT TOWARD BEING THOSE WHOSE POPULATION DENSITY IS LOW (67)	0.67	9 14 6 28 X SQ= 2.25 P = 0.124 RV YES (NO)
28	TILT TOWARD BEING THOSE WHOSE POPULATION GROWTH RATE IS LOW (48)	0.60	28	TILT TOWARD BEING THOSE WHOSE POPULATION GROWTH RATE IS HIGH (62)	0.66	6 27 9 14 X SQ= 2.06 P = 0.125 RV YES (NO)
29	TEND TO BE THOSE WHERE THE DEGREE OF URBANIZATION IS HIGH (56)	0.93	29	TEND TO BE THOSE WHERE THE DEGREE OF URBANIZATION IS LOW (49)	0.60	14 16 1 24 X SQ= 10.46 P = 0.001 RV YES (NO)
30	TEND TO BE THOSE WHOSE AGRICULTURAL POPULATION IS MEDIUM, LOW, OR VERY LOW (57)	0.87	30	TEND TO BE THOSE WHOSE AGRICULTURAL POPULATION IS HIGH (56)	0.61	2 25 13 16 X SQ= 8.17 P = 0.002 RV YES (NO)
31	TEND TO BE THOSE WHOSE AGRICULTURAL POPULATION IS LOW OR VERY LOW (24)	0.60	31	TEND TO BE THOSE WHOSE AGRICULTURAL POPULATION IS HIGH OR MEDIUM (90)	0.90	6 37 9 4 X SQ= 12.86 P = 0. RV YES (YES)
32	LEAN LESS TOWARD BEING THOSE WHOSE GROSS NATIONAL PRODUCT IS MEDIUM, LOW, OR VERY LOW (105)	0.87	32	ALWAYS ARE THOSE WHOSE GROSS NATIONAL PRODUCT IS MEDIUM, LOW, OR VERY LOW (105)	1.00	2 0 13 42 X SQ= 2.53 P = 0.066 RV NO (YES)
33	TEND TO BE THOSE WHOSE GROSS NATIONAL PRODUCT IS VERY HIGH, HIGH, OR MEDIUM (30)	0.53	33	TEND TO BE THOSE WHOSE GROSS NATIONAL PRODUCT IS LOW OR VERY LOW (85)	0.81	8 8 7 34 X SQ= 4.85 P = 0.019 RV YES (YES)
34	TEND TO BE THOSE WHOSE GROSS NATIONAL PRODUCT IS VERY HIGH, HIGH, MEDIUM, OR LOW (62)	0.87	34	TEND TO BE THOSE WHOSE GROSS NATIONAL PRODUCT IS VERY LOW (53)	0.55	13 19 2 23 X SQ= 6.11 P = 0.007 RV YES (NO)

35	TEND TO BE THOSE WHOSE PER CAPITA GROSS NATIONAL PRODUCT IS VERY HIGH OR HIGH (24)	0.67	35	TEND TO BE THOSE WHOSE PER CAPITA GROSS NATIONAL PRODUCT IS MEDIUM, LOW, OR VERY LOW (91)	0.98	10 1 5 41 X SQ= 25.35 P = 0. RV YES (YES)
36	TEND TO BE THOSE WHOSE PER CAPITA GROSS NATIONAL PRODUCT IS VERY HIGH, HIGH, OR MEDIUM (42)	0.87	36	TEND TO BE THOSE WHOSE PER CAPITA GROSS NATIONAL PRODUCT IS LOW OR VERY LOW (73)	0.79	13 9 2 33 X SQ= 17.19 P = 0. RV YES (YES)
37	TEND TO BE THOSE WHOSE PER CAPITA GROSS NATIONAL PRODUCT IS VERY HIGH, HIGH, MEDIUM, OR LOW (64)	0.93	37	TEND TO BE THOSE WHOSE PER CAPITA GROSS NATIONAL PRODUCT IS VERY LOW (51)	0.55	14 19 1 23 X SQ= 8.61 P = 0.002 RV YES (NO)
38	TEND LESS TO BE THOSE WHOSE INTERNATIONAL FINANCIAL STATUS IS MEDIUM, LOW, OR VERY LOW (103)	0.80	38	ALWAYS ARE THOSE WHOSE INTERNATIONAL FINANCIAL STATUS IS MEDIUM, LOW, OR VERY LOW (103)	1.00	3 0 12 42 X SQ= 5.31 P = 0.016 RV NO (YES)
39	TEND TO BE THOSE WHOSE INTERNATIONAL FINANCIAL STATUS IS VERY HIGH, HIGH, OR MEDIUM (38)	0.67	39	TEND TO BE THOSE WHOSE INTERNATIONAL FINANCIAL STATUS IS LOW OR VERY LOW (76)	0.79	10 9 5 33 X SQ= 8.24 P = 0.003 RV YES (YES)
42	TEND TO BE THOSE WHOSE ECONOMIC DEVELOPMENTAL STATUS IS DEVELOPED OR INTERMEDIATE (36)	0.80	42	TEND TO BE THOSE WHOSE ECONOMIC DEVELOPMENTAL STATUS IS UNDERDEVELOPED OR VERY UNDERDEVELOPED (76)	0.85	12 6 3 35 X SQ= 18.62 P = 0. RV YES (YES)
43	TEND TO BE THOSE WHOSE ECONOMIC DEVELOPMENTAL STATUS IS DEVELOPED, INTERMEDIATE, OR UNDERDEVELOPED (55)	0.93	43	TEND TO BE THOSE WHOSE ECONOMIC DEVELOPMENTAL STATUS IS VERY UNDERDEVELOPED (57)	0.67	14 13 1 27 X SQ= 13.81 P = 0. RV YES (YES)
44	TEND TO BE THOSE WHERE THE LITERACY RATE IS NINETY PERCENT OR ABOVE (25)	0.80	44	TEND TO BE THOSE WHERE THE LITERACY RATE IS BELOW NINETY PERCENT (90)	0.98	12 1 3 41 X SQ= 33.54 P = 0. RV YES (YES)
45	TEND TO BE THOSE WHERE THE LITERACY RATE IS FIFTY PERCENT OR ABOVE (55)	0.93	45	TEND TO BE THOSE WHERE THE LITERACY RATE IS BELOW FIFTY PERCENT (54)	0.72	14 11 1 29 X SQ= 16.51 P = 0. RV YES (YES)

46	ALWAYS ARE THOSE WHERE THE LITERACY RATE IS TEN PERCENT OR ABOVE (84)	1.00	46	TEND LESS TO BE THOSE WHERE THE LITERACY RATE IS TEN PERCENT OR ABOVE (84)	0.67	15 28 0 14 X SQ= 4.95 P = 0.012 RV NO (NO)
50	TEND TO BE THOSE WHERE FREEDOM OF THE PRESS IS COMPLETE (43)	0.73	50	TEND TO BE THOSE WHERE FREEDOM OF THE PRESS IS INTERMITTENT, INTERNALLY ABSENT, OR INTERNALLY AND EXTERNALLY ABSENT (56)	0.74	11 9 4 25 X SQ= 7.62 P = 0.004 RV YES (YES)
53	TEND TO BE THOSE WHERE NEWSPAPER CIRCULATION IS THREE HUNDRED OR MORE PER THOUSAND (14)	0.60	53	ALWAYS ARE THOSE WHERE NEWSPAPER CIRCULATION IS LESS THAN THREE HUNDRED PER THOUSAND (101)	1.00	9 0 6 42 X SQ= 25.58 P = 0. RV YES (YES)
54	TEND TO BE THOSE WHERE NEWSPAPER CIRCULATION IS ONE HUNDRED OR MORE PER THOUSAND (37)	0.87	54	TEND TO BE THOSE WHERE NEWSPAPER CIRCULATION IS LESS THAN ONE HUNDRED PER THOUSAND (76)	0.88	13 5 2 37 X SQ= 25.24 P = 0. RV YES (YES)
55	ALWAYS ARE THOSE WHERE NEWSPAPER CIRCULATION IS TEN OR MORE PER THOUSAND (78)	1.00	55	TEND LESS TO BE THOSE WHERE NEWSPAPER CIRCULATION IS TEN OR MORE PER THOUSAND (78)	0.57	15 24 0 18 X SQ= 7.52 P = 0.001 RV NO (NO)
56	ALWAYS ARE THOSE WHERE THE RELIGION IS PREDOMINANTLY LITERATE (79)	1.00	56	TEND LESS TO BE THOSE WHERE THE RELIGION IS PREDOMINANTLY LITERATE (79)	0.52	15 21 0 19 X SQ= 8.89 P = 0.001 RV NO (NO)
62	LEAN LESS TOWARD BEING THOSE WHERE THE RELIGION IS CATHOLIC, RATHER THAN PROTESTANT (25)	0.56	62	ALWAYS ARE THOSE WHERE THE RELIGION IS CATHOLIC, RATHER THAN PROTESTANT (25)	1.00	4 0 5 8 X SQ= 2.51 P = 0.082 RV NO (YES)
63	TEND TO BE THOSE WHERE THE RELIGION IS PREDOMINANTLY SOME KIND OF CHRISTIAN (46)	0.80	63	TEND TO BE THOSE WHERE THE RELIGION IS PREDOMINANTLY OR PARTLY OTHER THAN CHRISTIAN (68)	0.79	12 9 3 33 X SQ= 13.88 P = 0. RV YES (YES)
64	TEND MORE TO BE THOSE WHERE THE RELIGION IS CHRISTIAN, RATHER THAN MUSLIM (46)	0.92	64	TEND LESS TO BE THOSE WHERE THE RELIGION IS CHRISTIAN, RATHER THAN MUSLIM (46)	0.53	12 9 1 8 X SQ= 3.72 P = 0.042 RV NO (YES)

66	LEAN TOWARD BEING THOSE THAT ARE RELIGIOUSLY HOMOGENEOUS (57)	0.73	66	LEAN TOWARD BEING THOSE THAT ARE RELIGIOUSLY HETEROGENEOUS (49)	0.59	11 16 4 23 X SQ= 3.32 P = 0.066 RV YES (NO)
67	TEND MORE TO BE THOSE THAT ARE RACIALLY HOMOGENEOUS (82)	0.93	67	TEND LESS TO BE THOSE THAT ARE RACIALLY HOMOGENEOUS (82)	0.61	14 23 1 15 X SQ= 4.05 P = 0.022 RV NO (NO)
68	TEND TO BE THOSE THAT ARE LINGUISTICALLY HOMOGENEOUS (52)	0.87	68	TEND TO BE THOSE THAT ARE LINGUISTICALLY WEAKLY HETEROGENEOUS OR STRONGLY HETEROGENEOUS (62)	0.74	13 11 2 31 X SQ= 14.19 P = 0. RV YES (YES)
69	ALWAYS ARE THOSE THAT ARE LINGUISTICALLY HOMOGENEOUS OR WEAKLY HETEROGENEOUS (64)	1.00	69	TEND TO BE THOSE THAT ARE LINGUISTICALLY STRONGLY HETEROGENEOUS (50)	0.64	15 15 0 27 X SQ= 15.83 P = 0. RV YES (YES)
70	TEND TO BE THOSE THAT ARE RELIGIOUSLY, RACIALLY, AND LINGUISTICALLY HOMOGENEOUS (21)	0.53	70	TEND TO BE THOSE THAT ARE RELIGIOUSLY, RACIALLY, OR LINGUISTICALLY HETEROGENEOUS (85)	0.95	8 2 7 36 X SQ= 13.25 P = 0. RV YES (YES)
71	TEND TO BE THOSE WHOSE DATE OF INDEPENDENCE IS BEFORE 1800 (21)	0.60	71	TEND TO BE THOSE WHOSE DATE OF INDEPENDENCE IS AFTER 1800 (90)	0.90	9 4 6 37 X SQ= 12.86 P = 0. RV YES (YES)
72	TEND TO BE THOSE WHOSE DATE OF INDEPENDENCE IS BEFORE 1914 (52)	0.80	72	TEND TO BE THOSE WHOSE DATE OF INDEPENDENCE IS AFTER 1914 (59)	0.56	12 18 3 23 X SQ= 4.39 P = 0.032 RV YES (NO)
73	ALWAYS ARE THOSE WHOSE DATE OF INDEPENDENCE IS BEFORE 1945 (65)	1.00	73	TEND LESS TO BE THOSE WHOSE DATE OF INDEPENDENCE IS BEFORE 1945 (65)	0.59	15 24 0 17 X SQ= 7.08 P = 0.002 RV NO (NO)
74	TEND TO BE THOSE THAT ARE HISTORICALLY WESTERN (26)	0.73	74	TEND TO BE THOSE THAT ARE NOT HISTORICALLY WESTERN (87)	0.95	11 2 4 39 X SQ= 25.16 P = 0. RV YES (YES)

75	TEND TO BE THOSE THAT ARE HISTORICALLY WESTERN OR SIGNIFICANTLY WESTERNIZED (62)	0.93	75	TEND TO BE THOSE THAT ARE NOT HISTORICALLY WESTERN AND ARE NOT SIGNIFICANTLY WESTERNIZED (52)	0.63	14 15 1 26 X SQ= 11.98 P = 0. RV YES (NO)
76	TEND TO BE THOSE THAT ARE HISTORICALLY WESTERN, RATHER THAN HAVING BEEN SIGNIFICANTLY OR PARTIALLY WESTERNIZED THROUGH A COLONIAL RELATIONSHIP (26)	0.92	76	TEND TO BE THOSE THAT HAVE BEEN SIGNIFICANTLY OR PARTIALLY WESTERNIZED THROUGH A COLONIAL RELATIONSHIP, RATHER THAN BEING HISTORICALLY WESTERN (70)	0.94	11 2 1 33 X SQ= 28.84 P = 0. RV YES (YES)
78	LEAN TOWARD BEING THOSE THAT HAVE BEEN SIGNIFICANTLY WESTERNIZED WITHOUT A COLONIAL RELATIONSHIP, RATHER THAN THROUGH SUCH A RELATIONSHIP (7)	0.67	78	LEAN TOWARD BEING THOSE THAT HAVE BEEN SIGNIFICANTLY WESTERNIZED THROUGH A COLONIAL RELATIONSHIP, RATHER THAN WITHOUT SUCH A RELATIONSHIP (28)	0.92	1 12 2 1 X SQ= 2.37 P = 0.071 RV YES (YES)
82	ALWAYS ARE THOSE WHERE THE TYPE OF POLITICAL MODERNIZATION IS EARLY OR LATER EUROPEAN OR EUROPEAN DERIVED, RATHER THAN DEVELOPED TUTELARY OR UNDEVELOPED TUTELARY (51)	1.00	82	TEND TO BE THOSE WHERE THE TYPE OF POLITICAL MODERNIZATION IS DEVELOPED TUTELARY OR UNDEVELOPED TUTELARY, RATHER THAN EARLY OR LATER EUROPEAN OR EUROPEAN DERIVED (55)	0.62	12 15 0 24 X SQ= 11.59 P = 0. RV YES (NO)
85	TEND TO BE THOSE WHERE THE STAGE OF POLITICAL MODERNIZATION IS ADVANCED, RATHER THAN MID- OR EARLY TRANSITIONAL (60)	0.93	85	TEND TO BE THOSE WHERE THE STAGE OF POLITICAL MODERNIZATION IS MID- OR EARLY TRANSITIONAL, RATHER THAN ADVANCED (54)	0.66	14 14 1 27 X SQ= 13.11 P = 0. RV YES (YES)
86	ALWAYS ARE THOSE WHERE THE STAGE OF POLITICAL MODERNIZATION IS ADVANCED OR MID-TRANSITIONAL, RATHER THAN EARLY TRANSITIONAL (76)	1.00	86	TEND LESS TO BE THOSE WHERE THE STAGE OF POLITICAL MODERNIZATION IS ADVANCED OR MID-TRANSITIONAL, RATHER THAN EARLY TRANSITIONAL (76)	0.59	15 24 0 17 X SQ= 7.08 P = 0.002 RV NO (NO)
87	ALWAYS ARE THOSE WHOSE IDEOLOGICAL ORIENTATION IS OTHER THAN DEVELOPMENTAL (58)	1.00	87	TEND LESS TO BE THOSE WHOSE IDEOLOGICAL ORIENTATION IS OTHER THAN DEVELOPMENTAL (58)	0.59	0 11 14 16 X SQ= 5.86 P = 0.007 RV NO (NO)
89	TEND TO BE THOSE WHOSE IDEOLOGICAL ORIENTATION IS CONVENTIONAL (33)	0.80	89	TEND TO BE THOSE WHOSE IDEOLOGICAL ORIENTATION IS OTHER THAN CONVENTIONAL (62)	0.81	12 6 3 25 X SQ= 13.17 P = 0. RV YES (YES)
94	TEND TO BE THOSE WHERE THE STATUS OF THE REGIME IS CONSTITUTIONAL (51)	0.80	94	TEND TO BE THOSE WHERE THE STATUS OF THE REGIME IS AUTHORITARIAN OR TOTALITARIAN (41)	0.61	12 11 3 17 X SQ= 4.97 P = 0.023 RV YES (YES)

96	TEND TO BE THOSE WHERE THE STATUS OF THE REGIME IS CONSTITUTIONAL OR TOTALITARIAN (67)	0.93	96	TEND TO BE THOSE WHERE THE STATUS OF THE REGIME IS AUTHORITARIAN (23)	0.50	14 13 1 13 X SQ= 6.13 P = 0.006 RV YES (YES)
97	TILT TOWARD BEING THOSE WHERE THE STATUS OF THE REGIME IS TOTALITARIAN, RATHER THAN AUTHORITARIAN (16)	0.67	97	TILT TOWARD BEING THOSE WHERE THE STATUS OF THE REGIME IS AUTHORITARIAN, RATHER THAN TOTALITARIAN (23)	0.87	1 13 2 2 X SQ= 1.61 P = 0.108 RV YES (YES)
98	TEND TO BE THOSE WHERE GOVERNMENTAL STABILITY IS GENERALLY PRESENT AND DATES AT LEAST FROM THE INTERWAR PERIOD (22)	0.60	98	TEND TO BE THOSE WHERE GOVERNMENTAL STABILITY IS GENERALLY OR MODERATELY PRESENT AND DATES FROM THE POST-WAR PERIOD, OR IS ABSENT (93)	0.88	9 5 6 37 X SQ= 11.32 P = 0.001 RV YES (YES)
99	TEND TO BE THOSE WHERE GOVERNMENTAL STABILITY IS GENERALLY PRESENT AND DATES FROM AT LEAST THE INTER-WAR PERIOD, OR FROM THE POST-WAR PERIOD (50)	0.86	99	TEND TO BE THOSE WHERE GOVERNMENTAL STABILITY IS MODERATELY PRESENT AND DATES FROM THE POST-WAR PERIOD, OR IS ABSENT (36)	0.77	12 7 2 23 X SQ= 12.70 P = 0. RV YES (YES)
100	ALWAYS ARE THOSE WHERE GOVERNMENTAL STABILITY IS GENERALLY PRESENT AND DATES FROM AT LEAST THE INTER-WAR PERIOD, OR IS GENERALLY OR MODERATELY PRESENT, AND DATES FROM THE POST-WAR PERIOD (64)	1.00	100	TEND TO BE THOSE WHERE GOVERNMENTAL STABILITY IS ABSENT (22)	0.55	13 14 0 17 X SQ= 9.42 P = 0. RV YES (NO)
101	TEND TO BE THOSE WHERE THE REPRESENTATIVE CHARACTER OF THE REGIME IS POLYARCHIC (41)	0.80	101	TEND TO BE THOSE WHERE THE REPRESENTATIVE CHARACTER OF THE REGIME IS LIMITED POLYARCHIC, PSEUDO-POLYARCHIC, OR NON-POLYARCHIC (57)	0.84	12 5 3 26 X SQ= 15.06 P = 0. RV YES (YES)
102	TEND TO BE THOSE WHERE THE REPRESENTATIVE CHARACTER OF THE REGIME IS POLYARCHIC OR LIMITED POLYARCHIC (59)	0.80	102	TEND TO BE THOSE WHERE THE REPRESENTATIVE CHARACTER OF THE REGIME IS PSEUDO-POLYARCHIC OR NON-POLYARCHIC (49)	0.58	12 15 3 21 X SQ= 4.80 P = 0.016 RV YES (NO)
105	TEND TO BE THOSE WHERE THE ELECTORAL SYSTEM IS COMPETITIVE (43)	0.92	105	TEND TO BE THOSE WHERE THE ELECTORAL SYSTEM IS PARTIALLY COMPETITIVE OR NON-COMPETITIVE (47)	0.67	12 9 1 18 X SQ= 9.99 P = 0.001 RV YES (YES)
107	TEND TO BE THOSE WHERE AUTONOMOUS GROUPS ARE FULLY TOLERATED IN POLITICS (46)	0.87	107	TEND TO BE THOSE WHERE AUTONOMOUS GROUPS ARE PARTIALLY TOLERATED IN POLITICS, ARE TOLERATED ONLY OUTSIDE POLITICS, OR ARE NOT TOLERATED AT ALL (65)	0.72	13 11 2 28 X SQ= 12.72 P = 0. RV YES (YES)

114	TEND TO BE THOSE WHERE SECTIONALISM IS NEGLIGIBLE (47)	0.80	114	TEND TO BE THOSE WHERE SECTIONALISM IS EXTREME OR MODERATE (61)	0.68	3 26 12 12 X SQ= 8.32 P = 0.002 RV YES (YES)
115	TEND TO BE THOSE WHERE INTEREST ARTICULATION BY ASSOCIATIONAL GROUPS IS SIGNIFICANT (20)	0.53	115	TEND TO BE THOSE WHERE INTEREST ARTICULATION BY ASSOCIATIONAL GROUPS IS MODERATE, LIMITED, OR NEGLIGIBLE (91)	0.92	8 3 7 36 X SQ= 11.24 P = 0.001 RV YES (YES)
116	TEND TO BE THOSE WHERE INTEREST ARTICULATION BY ASSOCIATIONAL GROUPS IS SIGNIFICANT OR MODERATE (32)	0.80	116	TEND TO BE THOSE WHERE INTEREST ARTICULATION BY ASSOCIATIONAL GROUPS IS LIMITED OR NEGLIGIBLE (79)	0.85	12 6 3 33 X SQ= 17.55 P = 0. RV YES (YES)
117	TEND TO BE THOSE WHERE INTEREST ARTICULATION BY ASSOCIATIONAL GROUPS IS SIGNIFICANT, MODERATE, OR LIMITED (60)	0.93	117	TEND TO BE THOSE WHERE INTEREST ARTICULATION BY ASSOCIATIONAL GROUPS IS NEGLIGIBLE (51)	0.54	14 18 1 21 X SQ= 8.13 P = 0.002 RV YES (NO)
118	TEND TO BE THOSE WHERE INTEREST ARTICULATION BY INSTITUTIONAL GROUPS IS SIGNIFICANT, MODERATE, OR LIMITED (60)	0.87	118	TEND TO BE THOSE WHERE INTEREST ARTICULATION BY INSTITUTIONAL GROUPS IS VERY SIGNIFICANT (40)	0.50	2 17 13 17 X SQ= 4.45 P = 0.025 RV YES (NO)
119	TEND TO BE THOSE WHERE INTEREST ARTICULATION BY INSTITUTIONAL GROUPS IS MODERATE OR LIMITED (26)	0.73	119	TEND TO BE THOSE WHERE INTEREST ARTICULATION BY INSTITUTIONAL GROUPS IS VERY SIGNIFICANT OR SIGNIFICANT (74)	0.94	4 32 11 2 X SQ= 20.96 P = 0. RV YES (YES)
120	TEND TO BE THOSE WHERE INTEREST ARITUCLATION BY INSTITUTIONAL GROUPS IS LIMITED (10)	0.53	120	ALWAYS ARE THOSE WHERE INTEREST ARTICULATION BY INSTITUTIONAL GROUPS IS VERY SIGNIFICANT, SIGNIFICANT, OR MODERATE (90)	1.00	7 34 8 0 X SQ= 17.94 P = 0. RV YES (YES)
121	TEND TO BE THOSE WHERE INTEREST ARTICULATION BY NON-ASSOCIATIONAL GROUPS IS MODERATE, LIMITED, OR NEGLIGIBLE (61)	0.93	121	TEND TO BE THOSE WHERE INTEREST ARTICULATION BY NON-ASSOCIATIONAL GROUPS IS SIGNIFICANT (54)	0.67	1 28 14 14 X SQ= 13.61 P = 0. RV YES (YES)
122	TEND TO BE THOSE WHERE INTEREST ARTICULATION BY NON-ASSOCIATIONAL GROUPS IS LIMITED OR NEGLIGIBLE (32)	0.73	122	TEND TO BE THOSE WHERE INTEREST ARTICULATION BY NON-ASSOCIATIONAL GROUPS IS SIGNIFICANT OR MODERATE (83)	0.93	4 39 11 3 X SQ= 22.68 P = 0. RV YES (YES)

123	TEND TO BE THOSE WHERE INTEREST ARTICULATION BY NON-ASSOCIATIONAL GROUPS IS NEGLIGIBLE (8)	0.53	123	ALWAYS ARE THOSE WHERE INTEREST ARTICULATION BY NON-ASSOCIATIONAL GROUPS IS SIGNIFICANT, MODERATE, OR LIMITED (107)	1.00	7 42 8 0 X SQ= 21.82 P = 0. RV YES (YES)
124	ALWAYS ARE THOSE WHERE INTEREST ARTICULATION BY ANOMIC GROUPS IS OCCASIONAL, INFREQUENT, OR VERY INFREQUENT (85)	1.00	124	TEND LESS TO BE THOSE WHERE INTEREST ARTICULATION BY ANOMIC GROUPS IS OCCASIONAL, INFREQUENT, OR VERY INFREQUENT (85)	0.77	0 9 15 30 X SQ= 2.66 P = 0.049 RV NO (NO)
125	TEND TO BE THOSE WHERE INTEREST ARTICULATION BY ANOMIC GROUPS IS INFREQUENT OR VERY INFREQUENT (35)	0.73	125	TEND TO BE THOSE WHERE INTEREST ARTICULATION BY ANOMIC GROUPS IS FREQUENT OR OCCASIONAL (64)	0.90	4 35 11 4 X SQ= 18.46 P = 0. RV YES (YES)
126	TEND TO BE THOSE WHERE INTEREST ARTICULATION BY ANOMIC GROUPS IS VERY INFREQUENT (16)	0.67	126	TEND TO BE THOSE WHERE INTEREST ARTICULATION BY ANOMIC GROUPS IS FREQUENT, OCCASIONAL, OR INFREQUENT (83)	0.97	5 38 10 1 X SQ= 23.63 P = 0. RV YES (YES)
129	LEAN MORE TOWARD BEING THOSE WHERE INTEREST ARTICULATION BY POLITICAL PARTIES IS SIGNIFICANT, MODERATE, OR LIMITED (56)	0.93	129	LEAN LESS TOWARD BEING THOSE WHERE INTEREST ARTICULATION BY POLITICAL PARTIES IS SIGNIFICANT, MODERATE, OR LIMITED (56)	0.64	13 18 1 10 X SQ= 2.60 P = 0.067 RV NO (NO)
130	TEND LESS TO BE THOSE WHERE INTEREST AGGREGATION BY POLITICAL PARTIES IS MODERATE, LIMITED, OR NEGLIGIBLE (71)	0.73	130	ALWAYS ARE THOSE WHERE INTEREST AGGREGATION BY POLITICAL PARTIES IS MODERATE, LIMITED, OR NEGLIGIBLE (71)	1.00	4 0 11 26 X SQ= 4.95 P = 0.013 RV NO (YES)
131	TILT TOWARD BEING THOSE WHERE INTEREST AGGREGATION BY POLITICAL PARTIES IS SIGNIFICANT OR MODERATE (30)	0.50	131	TILT TOWARD BEING THOSE WHERE INTEREST AGGREGATION BY POLITICAL PARTIES IS LIMITED OR NEGLIGIBLE (35)	0.77	7 5 7 17 X SQ= 1.77 P = 0.148 RV YES (YES)
132	ALWAYS ARE THOSE WHERE INTEREST AGGREGATION BY POLITICAL PARTIES IS SIGNIFICANT, MODERATE, OR LIMITED (67)	1.00	132	LEAN LESS TOWARD BEING THOSE WHERE INTEREST AGGREGATION BY POLITICAL PARTIES IS SIGNIFICANT, MODERATE, OR LIMITED (67)	0.72	13 18 0 7 X SQ= 2.79 P = 0.072 RV NO (NO)
133	ALWAYS ARE THOSE WHERE INTEREST AGGREGATION BY THE EXECUTIVE IS MODERATE, LIMITED, OR NEGLIGIBLE (76)	1.00	133	TEND LESS TO BE THOSE WHERE INTEREST AGGREGATION BY THE EXECUTIVE IS MODERATE, LIMITED, OR NEGLIGIBLE (76)	0.71	0 10 15 24 X SQ= 3.88 P = 0.021 RV NO (NO)

134	TEND TO BE THOSE WHERE INTEREST AGGREGATION BY THE EXECUTIVE IS SIGNIFICANT OR MODERATE (57)	0.87	134	TEND TO BE THOSE WHERE INTEREST AGGREGATION BY THE EXECUTIVE IS LIMITED OR NEGLIGIBLE (46)	0.66	13 11 2 21 X SQ= 9.18 P = 0.001 RV YES (YES)
137	TEND TO BE THOSE WHERE INTEREST AGGREGATION BY THE LEGISLATURE IS SIGNIFICANT OR MODERATE (29)	0.80	137	TEND TO BE THOSE WHERE INTEREST AGGREGATION BY THE LEGISLATURE IS LIMITED OR NEGLIGIBLE (68)	0.91	12 3 3 29 X SQ= 20.31 P = 0. RV YES (YES)
138	TEND TO BE THOSE WHERE INTEREST AGGREGATION BY THE LEGISLATURE IS SIGNIFICANT, MODERATE, OR LIMITED (48)	0.87	138	TEND TO BE THOSE WHERE INTEREST AGGREGATION BY THE LEGISLATURE IS NEGLIGIBLE (49)	0.58	13 13 2 18 X SQ= 6.51 P = 0.005 RV YES (YES)
141	TEND LESS TO BE THOSE WHERE THE PARTY SYSTEM IS QUANTITATIVELY OTHER THAN TWO-PARTY (87)	0.69	141	ALWAYS ARE THOSE WHERE THE PARTY SYSTEM IS QUANTITATIVELY OTHER THAN TWO-PARTY (87)	1.00	4 0 9 31 X SQ= 7.10 P = 0.005 RV NO (YES)
144	TEND TO BE THOSE WHERE THE PARTY SYSTEM IS QUANTITATIVELY TWO-PARTY, RATHER THAN ONE-PARTY (11)	0.67	144	ALWAYS ARE THOSE WHERE THE PARTY SYSTEM IS QUANTITATIVELY ONE-PARTY, RATHER THAN TWO-PARTY (34)	1.00	2 9 4 0 X SQ= 5.13 P = 0.011 RV YES (YES)
145	TEND LESS TO BE THOSE WHERE THE PARTY SYSTEM IS QUANTITATIVELY MULTI-PARTY, RATHER THAN TWO-PARTY (30)	0.60	145	ALWAYS ARE THOSE WHERE THE PARTY SYSTEM IS QUANTITATIVELY MULTI-PARTY, RATHER THAN TWO-PARTY (30)	1.00	6 15 4 0 X SQ= 4.48 P = 0.017 RV NO (YES)
147	TEND TO BE THOSE WHERE THE PARTY SYSTEM IS QUALITATIVELY CLASS-ORIENTED OR MULTI-IDEOLOGICAL (23)	0.77	147	TEND TO BE THOSE WHERE THE PARTY SYSTEM IS QUALITATIVELY OTHER THAN CLASS-ORIENTED OR MULTI-IDEOLOGICAL (67)	0.82	10 5 3 23 X SQ= 10.93 P = 0. RV YES (YES)
153	TEND TO BE THOSE WHERE THE PARTY SYSTEM IS STABLE (42)	0.86	153	TEND TO BE THOSE WHERE THE PARTY SYSTEM IS MODERATELY STABLE OR UNSTABLE (38)	0.81	12 5 2 21 X SQ= 13.85 P = 0. RV YES (YES)
154	TEND TO BE THOSE WHERE THE PARTY SYSTEM IS STABLE OR MODERATELY STABLE (55)	0.93	154	TEND TO BE THOSE WHERE THE PARTY SYSTEM IS UNSTABLE (25)	0.69	13 8 1 18 X SQ= 11.69 P = 0. RV YES (YES)

158	TEND TO BE THOSE WHERE PERSONALISMO IS NEGLIGIBLE (56)	0.87	158	TEND TO BE THOSE WHERE PERSONALISMO IS PRONOUNCED OR MODERATE (40)	0.77	2 24 13 7 X SQ= 14.39 P = 0. RV YES (YES)
161	TEND TO BE THOSE WHERE THE POLITICAL LEADERSHIP IS NON-ELITIST (50)	0.79	161	TEND TO BE THOSE WHERE THE POLITICAL LEADERSHIP IS ELITIST OR MODERATELY ELITIST (47)	0.58	3 18 11 13 X SQ= 3.83 P = 0.028 RV YES (NO)
164	TEND MORE TO BE THOSE WHERE THE REGIME'S LEADERSHIP CHARISMA IS NEGLIGIBLE (65)	0.93	164	TEND LESS TO BE THOSE WHERE THE REGIME'S LEADERSHIP CHARISMA IS NEGLIGIBLE (65)	0.59	1 13 14 19 X SQ= 4.12 P = 0.020 RV NO (NO)
168	TEND TO BE THOSE WHERE THE HORIZONTAL POWER DISTRIBUTION IS SIGNIFICANT (34)	0.80	168	TEND TO BE THOSE WHERE THE HORIZONTAL POWER DISTRIBUTION IS LIMITED OR NEGLIGIBLE (72)	0.82	12 6 3 28 X SQ= 14.83 P = 0. RV YES (YES)
169	TEND TO BE THOSE WHERE THE HORIZONTAL POWER DISTRIBUTION IS SIGNIFICANT OR LIMITED (58)	0.80	169	TEND TO BE THOSE WHERE THE HORIZONTAL POWER DISTRIBUTION IS NEGLIGIBLE (48)	0.59	12 14 3 20 X SQ= 4.84 P = 0.015 RV YES (NO)
170	ALWAYS ARE THOSE WHERE THE LEGISLATIVE-EXECUTIVE STRUCTURE IS OTHER THAN PRESIDENTIAL (63)	1.00	170	TEND TO BE THOSE WHERE THE LEGISLATIVE-EXECUTIVE STRUCTURE IS PRESIDENTIAL (39)	0.58	0 21 14 15 X SQ= 11.79 P = 0. RV YES (NO)
171	TEND LESS TO BE THOSE WHERE THE LEGISLATIVE-EXECUTIVE STRUCTURE IS OTHER THAN PARLIAMENTARY-REPUBLICAN (90)	0.71	171	TEND MORE TO BE THOSE WHERE THE LEGISLATIVE-EXECUTIVE STRUCTURE IS OTHER THAN PARLIAMENTARY-REPUBLICAN (90)	0.94	4 2 10 33 X SQ= 2.97 P = 0.048 RV NO (YES)
172	TEND TO BE THOSE WHERE THE LEGISLATIVE-EXECUTIVE STRUCTURE IS PARLIAMENTARY-ROYALIST (21)	0.57	172	TEND TO BE THOSE WHERE THE LEGISLATIVE-EXECUTIVE STRUCTURE IS OTHER THAN PARLIAMENTARY-ROYALIST (88)	0.87	8 5 6 33 X SQ= 8.34 P = 0.003 RV YES (YES)
174	TEND TO BE THOSE WHERE THE LEGISLATURE IS FULLY EFFECTIVE (28)	0.79	174	TEND TO BE THOSE WHERE THE LEGISLATURE IS PARTIALLY EFFECTIVE, LARGELY INEFFECTIVE, OR WHOLLY INEFFECTIVE (72)	0.87	11 4 3 28 X SQ= 16.46 P = 0. RV YES (YES)

175	TEND TO BE THOSE WHERE THE LEGISLATURE IS FULLY EFFECTIVE OR PARTIALLY EFFECTIVE (51)	0.86	175	TEND TO BE THOSE WHERE THE LEGISLATURE IS LARGELY INEFFECTIVE OR WHOLLY INEFFECTIVE (49)	0.59	12 13 2 19 X SQ= 6.27 P = 0.009 RV YES (NO)
178	TILT TOWARD BEING THOSE WHERE THE LEGISLATURE IS BICAMERAL (51)	0.77	178	TILT TOWARD BEING THOSE WHERE THE LEGISLATURE IS UNICAMERAL (53)	0.53	3 19 10 17 X SQ= 2.31 P = 0.104 RV YES (NO)
179	TEND TO BE THOSE WHERE THE EXECUTIVE IS STRONG (39)	0.86	179	TEND TO BE THOSE WHERE THE EXECUTIVE IS DOMINANT (52)	0.69	2 20 12 9 X SQ= 9.22 P = 0.001 RV YES (YES)
181	TEND TO BE THOSE WHERE THE BUREAUCRACY IS MODERN, RATHER THAN SEMI-MODERN (21)	0.71	181	TEND TO BE THOSE WHERE THE BUREAUCRACY IS SEMI-MODERN, RATHER THAN MODERN (55)	0.90	10 2 4 19 X SQ= 11.67 P = 0. RV YES (YES)
186	TEND MORE TO BE THOSE WHERE PARTICIPATION BY THE MILITARY IN POLITICS IS NEUTRAL, RATHER THAN SUPPORTIVE (56)	0.92	186	TEND LESS TO BE THOSE WHERE PARTICIPATION BY THE MILITARY IN POLITICS IS NEUTRAL, RATHER THAN SUPPORTIVE (56)	0.54	1 12 11 14 X SQ= 3.67 P = 0.030 RV NO (NO)
187	TEND TO BE THOSE WHERE THE ROLE OF THE POLICE IS NOT POLITICALLY SIGNIFICANT (35)	0.73	187	TEND TO BE THOSE WHERE THE ROLE OF THE POLICE IS POLITICALLY SIGNIFICANT (66)	0.91	4 31 11 3 X SQ= 18.18 P = 0. RV YES (YES)
189	TEND LESS TO BE THOSE WHERE THE CHARACTER OF THE LEGAL SYSTEM IS OTHER THAN COMMON LAW (108)	0.73	189	ALWAYS ARE THOSE WHERE THE CHARACTER OF THE LEGAL SYSTEM IS OTHER THAN COMMON LAW (108)	1.00	4 0 11 42 X SQ= 8.31 P = 0.003 RV NO (YES)
190	TEND TO BE THOSE WHERE THE CHARACTER OF THE LEGAL SYSTEM IS COMMON LAW, RATHER THAN CIVIL LAW (7)	0.50	190	ALWAYS ARE THOSE WHERE THE CHARACTER OF THE LEGAL SYSTEM IS CIVIL LAW, RATHER THAN COMMON LAW (32)	1.00	4 13 4 0 X SQ= 5.11 P = 0.012 RV YES (YES)
192	ALWAYS ARE THOSE WHERE THE CHARACTER OF THE LEGAL SYSTEM IS OTHER THAN MUSLIM OR PARTLY MUSLIM (86)	1.00	192	TEND LESS TO BE THOSE WHERE THE CHARACTER OF THE LEGAL SYSTEM IS OTHER THAN MUSLIM OR PARTLY MUSLIM (86)	0.66	0 14 15 27 X SQ= 5.13 P = 0.012 RV NO (NO)

193	ALWAYS ARE THOSE WHERE THE CHARACTER OF THE LEGAL SYSTEM IS OTHER THAN PARTLY INDIGENOUS (86)	1.00	193	TEND LESS TO BE THOSE WHERE THE CHARACTER OF THE LEGAL SYSTEM IS OTHER THAN PARTLY INDIGENOUS (86)	0.62

```
   0   16
  15   26
X SQ=   6.17
P =    0.006
RV NO  (NO )
```

```
*************************************************************************************************

110  POLITIES                                            110  POLITIES
     WHERE POLITICAL ENCULTURATION                            WHERE POLITICAL ENCULTURATION
     IS HIGH  (15)                                            IS MEDIUM OR LOW  (80)


                                                                                  PREDICATE ONLY
.................................................................................................

      15 IN LEFT COLUMN

AUSTRALIA   AUSTRIA     COSTA RICA  DENMARK     ICELAND     IRELAND     JAPAN       LUXEMBOURG  NEW ZEALAND NORWAY
POLAND      SWEDEN      THAILAND    TURKEY      UK

      80 IN RIGHT COLUMN

AFGHANISTAN ALGERIA     ARGENTINA   BELGIUM     BOLIVIA     BRAZIL      BURMA       CAMBODIA    CAMEROUN    CANADA
CEN AFR REP CEYLON      CHAD        CHILE       COLOMBIA    CONGO(BRA)  CONGO(LEO)  CYPRUS      DAHOMEY     DOMIN REP
ECUADOR     EL SALVADOR ETHIOPIA    FINLAND     FRANCE      GABON       GERMAN FR   GHANA       GREECE      GUATEMALA
GUINEA      HAITI       HONDURAS    INDIA       INDONESIA   IRAN        IRAQ        ISRAEL      ITALY       IVORY COAST
JAMAICA     JORDAN      KOREA REP   LAOS        LEBANON     LIBERIA     LIBYA       MALAGASY R  MALAYA      MAURITANIA
MEXICO      MOROCCO     NEPAL       NETHERLANDS NICARAGUA   NIGER       NIGERIA     PAKISTAN    PANAMA      PERU
PHILIPPINES SA'U ARABIA SENEGAL     SIERRE LEO  SOMALIA     SO AFRICA   SPAIN       SUDAN       SYRIA       TANGANYIKA
TOGO        TRINIDAD    TUNISIA     UGANDA      US          UPPER VOLTA VENEZUELA   VIETNAM REP YEMEN       YUGOSLAVIA

       2 EXCLUDED BECAUSE AMBIGUOUS

SWITZERLAND URUGUAY

      18 EXCLUDED BECAUSE UNASCERTAINED

ALBANIA     BULGARIA    BURUNDI     CHINA, PR   CUBA        CZECHOS'KIA GERMANY, E  HUNGARY     KOREA, N    MALI
MONGOLIA    PARAGUAY    PORTUGAL    RUMANIA     RWANDA      USSR        UAR         VIETNAM, N
-------------------------------------------------------------------------------------------------
```

111 POLITIES
WHERE POLITICAL ENCULTURATION
IS HIGH OR MEDIUM (53)

111 POLITIES
WHERE POLITICAL ENCULTURATION
IS LOW (42)

PREDICATE ONLY

53 IN LEFT COLUMN

AFGHANISTAN	ALGERIA	AUSTRALIA	AUSTRIA	BURMA	CAMBODIA	CAMEROUN	CANADA	CEYLON	COLOMBIA
COSTA RICA	DENMARK	DOMIN REP	FINLAND	FRANCE	GERMAN FR	GHANA	GREECE	GUINEA	HONDURAS
ICELAND	INDIA	IRELAND	ISRAEL	ITALY	IVORY COAST	JAMAICA	JAPAN	LIBYA	LUXEMBOURG
MALAGASY R	MAURITANIA	MEXICO	MOROCCO	NETHERLANDS	NEW ZEALAND	NORWAY	PHILIPPINES	POLAND	SENEGAL
SIERRE LEO	SOMALIA	SUDAN	SWEDEN	TANGANYIKA	THAILAND	TOGO	TRINIDAD	TUNISIA	TURKEY
UK	US	VENEZUELA							

42 IN RIGHT COLUMN

ARGENTINA	BELGIUM	BOLIVIA	BRAZIL	CEN AFR REP	CHAD	CHILE	CONGO(BRA)	CONGO(LEO)	CYPRUS
DAHOMEY	ECUADOR	EL SALVADOR	ETHIOPIA	GABON	GUATEMALA	HAITI	INDONESIA	IRAN	IRAQ
JORDAN	KOREA REP	LAOS	LEBANON	LIBERIA	MALAYA	NEPAL	NICARAGUA	NIGER	NIGERIA
PAKISTAN	PANAMA	PERU	SA'U ARABIA	SO AFRICA	SPAIN	SYRIA	UGANDA	UPPER VOLTA	VIETNAM REP
YEMEN	YUGOSLAVIA								

2 EXCLUDED BECAUSE AMBIGUOUS

SWITZERLAND URUGUAY

18 EXCLUDED BECAUSE UNASCERTAINED

ALBANIA	BULGARIA	BURUNDI	CHINA, PR	CUBA	CZECHOS'KIA	GERMANY, E	HUNGARY	KOREA, N	MALI
MONGOLIA	PARAGUAY	PORTUGAL	RUMANIA	RWANDA	USSR	UAR	VIETNAM, N		

112 POLITIES WHERE SECTIONALISM IS EXTREME, RATHER THAN NEGLIGIBLE (27)

112 POLITIES WHERE SECTIONALISM IS NEGLIGIBLE, RATHER THAN EXTREME (47)

SUBJECT ONLY

27 IN LEFT COLUMN

BELGIUM, IRAN, SUDAN, BRAZIL, IRAQ, SWITZERLAND, BURMA, JORDAN, UGANDA, CANADA, LAOS, USSR, COLOMBIA, LIBYA, UK, CONGO(LEO), MALAGASY R, YEMEN, CZECHOS'KIA, MOROCCO, YUGOSLAVIA, ECUADOR, NIGERIA, INDIA, PAKISTAN, INDONESIA, SOMALIA

47 IN RIGHT COLUMN

AUSTRIA, EL SALVADOR, ISRAEL, PANAMA, TUNISIA, BOLIVIA, FINLAND, IVORY COAST, PARAGUAY, TURKEY, BURUNDI, GERMANY, E, JAMAICA, POLAND, UAR, CEN AFR REP, GREECE, JAPAN, PORTUGAL, UPPER VOLTA, CHILE, GUATEMALA, KOREA, N, RUMANIA, VENEZUELA, CUBA, HAITI, KOREA REP, RWANDA, VIETNAM, N, COSTA RICA, HONDURAS, LUXEMBOURG, SA'U ARABIA, VIETNAM REP, CYPRUS, HUNGARY, MONGOLIA, SWEDEN, DENMARK, ICELAND, NEW ZEALAND, TANGANYIKA, DOMIN REP, IRELAND, NORWAY, TRINIDAD

3 EXCLUDED BECAUSE AMBIGUOUS

ARGENTINA, MEXICO, URUGUAY

34 EXCLUDED BECAUSE IRRELEVANT

AFGHANISTAN, CONGO(BRA), LIBERIA, SYRIA, ALBANIA, DAHOMEY, MALAYA, THAILAND, ALGERIA, ETHIOPIA, MAURITANIA, TOGO, AUSTRALIA, FRANCE, NETHERLANDS, US, BULGARIA, GABON, PERU, CAMBODIA, GERMAN FR, PHILIPPINES, CAMEROUN, GHANA, SENEGAL, CEYLON, GUINEA, SIERRE LEO, CHAD, ITALY, SO AFRICA, CHINA, PR, LEBANON, SPAIN

4 EXCLUDED BECAUSE UNASCERTAINED

MALI, NEPAL, NICARAGUA, NIGER

	Left column			Right column		
6	LEAN MORE TOWARD BEING THOSE LOCATED ELSEWHERE THAN IN THE CARIBBEAN, CENTRAL AMERICA, OR SOUTH AMERICA (93)	0.89	6	LEAN LESS TOWARD BEING THOSE LOCATED ELSEWHERE THAN IN THE CARIBBEAN, CENTRAL AMERICA, OR SOUTH AMERICA (93)	0.70	3 14 24 33 X SQ= 2.41 P = 0.087 RV NO (NO)
19	LEAN TOWARD BEING THOSE LOCATED IN NORTH AFRICA OR CENTRAL AND SOUTH AFRICA, RATHER THAN IN THE CARIBBEAN, CENTRAL AMERICA, OR SOUTH AMERICA (33)	0.73	19	LEAN TOWARD BEING THOSE LOCATED IN THE CARIBBEAN, CENTRAL AMERICA, OR SOUTH AMERICA, RATHER THAN IN NORTH AFRICA OR CENTRAL AND SOUTH AFRICA (22)	0.64	8 8 3 14 X SQ= 2.56 P = 0.071 RV YES (YES)

22	TEND TO BE THOSE WHOSE TERRITORIAL SIZE IS VERY LARGE, LARGE, OR MEDIUM (68)	0.85	22	TEND TO BE THOSE WHOSE TERRITORIAL SIZE IS SMALL (47)	0.60	23 19 4 28 X SQ= 12.23 P = 0. RV YES (YES)
24	TEND TO BE THOSE WHOSE POPULATION IS VERY LARGE, LARGE, OR MEDIUM (61)	0.70	24	TEND TO BE THOSE WHOSE POPULATION IS SMALL (54)	0.57	19 20 8 27 X SQ= 4.27 P = 0.030 RV YES (NO)
26	TILT TOWARD BEING THOSE WHOSE POPULATION DENSITY IS LOW (67)	0.67	26	TILT TOWARD BEING THOSE WHOSE POPULATION DENSITY IS VERY HIGH, HIGH, OR MEDIUM (48)	0.53	9 25 18 22 X SQ= 1.98 P = 0.146 RV YES (NO)
32	LEAN LESS TOWARD BEING THOSE WHOSE GROSS NATIONAL PRODUCT IS MEDIUM, LOW, OR VERY LOW (105)	0.85	32	LEAN MORE TOWARD BEING THOSE WHOSE GROSS NATIONAL PRODUCT IS MEDIUM, LOW, OR VERY LOW (105)	0.98	4 1 23 46 X SQ= 2.60 P = 0.056 RV NO (YES)
33	TEND LESS TO BE THOSE WHOSE GROSS NATIONAL PRODUCT IS LOW OR VERY LOW (85)	0.59	33	TEND MORE TO BE THOSE WHOSE GROSS NATIONAL PRODUCT IS LOW OR VERY LOW (85)	0.83	11 8 16 39 X SQ= 3.89 P = 0.031 RV NO (YES)
37	TEND TO BE THOSE WHOSE PER CAPITA GROSS NATIONAL PRODUCT IS VERY LOW (51)	0.59	37	TEND TO BE THOSE WHOSE PER CAPITA GROSS NATIONAL PRODUCT IS VERY HIGH, HIGH, MEDIUM, OR LOW (64)	0.70	11 33 16 14 X SQ= 5.02 P = 0.016 RV YES (YES)
38	LEAN LESS TOWARD BEING THOSE WHOSE INTERNATIONAL FINANCIAL STATUS IS MEDIUM, LOW, OR VERY LOW (103)	0.85	38	LEAN MORE TOWARD BEING THOSE WHOSE INTERNATIONAL FINANCIAL STATUS IS MEDIUM, LOW, OR VERY LOW (103)	0.98	4 1 23 45 X SQ= 2.51 P = 0.059 RV NO (YES)
45	LEAN TOWARD BEING THOSE WHERE THE LITERACY RATE IS BELOW FIFTY PERCENT (54)	0.59	45	LEAN TOWARD BEING THOSE WHERE THE LITERACY RATE IS FIFTY PERCENT OR ABOVE (55)	0.66	11 29 16 15 X SQ= 3.35 P = 0.050 RV YES (YES)
52	TILT MORE TOWARD BEING THOSE WHERE FREEDOM OF THE PRESS IS COMPLETE, INTERMITTENT, OR INTERNALLY ABSENT (82)	0.92	52	TILT LESS TOWARD BEING THOSE WHERE FREEDOM OF THE PRESS IS COMPLETE, INTERMITTENT, OR INTERNALLY ABSENT (82)	0.74	23 29 2 10 X SQ= 2.06 P = 0.106 RV NO (NO)

54	TEND MORE TO BE THOSE WHERE NEWSPAPER CIRCULATION IS LESS THAN ONE HUNDRED PER THOUSAND (76)	0.78	54	TEND LESS TO BE THOSE WHERE NEWSPAPER CIRCULATION IS LESS THAN ONE HUNDRED PER THOUSAND (76)	0.51	6 22 21 23 X SQ= 3.99 P = 0.028 RV NO (NO)
55	TEND LESS TO BE THOSE WHERE NEWSPAPER CIRCULATION IS TEN OR MORE PER THOUSAND (78)	0.59	55	TEND MORE TO BE THOSE WHERE NEWSPAPER CIRCULATION IS TEN OR MORE PER THOUSAND (78)	0.84	16 38 11 7 X SQ= 4.44 P = 0.025 RV NO (YES)
63	LEAN TOWARD BEING THOSE WHERE THE RELIGION IS PREDOMINANTLY OR PARTLY OTHER THAN CHRISTIAN (68)	0.69	63	LEAN TOWARD BEING THOSE WHERE THE RELIGION IS PREDOMINANTLY SOME KIND OF CHRISTIAN (46)	0.53	8 25 18 22 X SQ= 2.55 P = 0.087 RV YES (NO)
65	TEND TO BE THOSE WHERE THE RELIGION IS MUSLIM, RATHER THAN CATHOLIC (18)	0.75	65	TEND TO BE THOSE WHERE THE RELIGION IS CATHOLIC, RATHER THAN MUSLIM (25)	0.78	3 14 9 4 X SQ= 6.16 P = 0.008 RV YES (YES)
68	TEND TO BE THOSE THAT ARE LINGUISTICALLY WEAKLY HETEROGENEOUS OR STRONGLY HETEROGENEOUS (62)	0.70	68	TEND TO BE THOSE THAT ARE LINGUISTICALLY HOMOGENEOUS (52)	0.68	8 32 19 15 X SQ= 8.72 P = 0.002 RV YES (YES)
69	TEND TO BE THOSE THAT ARE LINGUISTICALLY STRONGLY HETEROGENEOUS (50)	0.70	69	TEND TO BE THOSE THAT ARE LINGUISTICALLY HOMOGENEOUS OR WEAKLY HETEROGENEOUS (64)	0.83	8 39 19 8 X SQ= 18.82 P = 0. RV YES (YES)
70	TEND MORE TO BE THOSE THAT ARE RELIGIOUSLY, RACIALLY, OR LINGUISTICALLY HETEROGENEOUS (85)	0.88	70	TEND LESS TO BE THOSE THAT ARE RELIGIOUSLY, RACIALLY, OR LINGUISTICALLY HETEROGENEOUS (85)	0.65	3 14 22 26 X SQ= 3.11 P = 0.047 RV NO (NO)
72	TILT TOWARD BEING THOSE WHOSE DATE OF INDEPENDENCE IS AFTER 1914 (59)	0.67	72	TILT TOWARD BEING THOSE WHOSE DATE OF INDEPENDENCE IS BEFORE 1914 (52)	0.52	9 23 18 21 X SQ= 1.72 P = 0.145 RV YES (NO)
73	TILT TOWARD BEING THOSE WHOSE DATE OF INDEPENDENCE IS AFTER 1945 (46)	0.52	73	TILT TOWARD BEING THOSE WHOSE DATE OF INDEPENDENCE IS BEFORE 1945 (65)	0.68	13 30 14 14 X SQ= 2.04 P = 0.134 RV YES (YES)

No.	Statement	Value	No.	Statement	Value	Statistics
75	TEND TO BE THOSE THAT ARE NOT HISTORICALLY WESTERN AND ARE NOT SIGNIFICANTLY WESTERNIZED (52)	0.59	75	TEND TO BE THOSE THAT ARE HISTORICALLY WESTERN OR SIGNIFICANTLY WESTERNIZED (62)	0.70	11 33 16 14 X SQ= 5.02 P = 0.016 RV YES (YES)
77	TEND TO BE THOSE THAT HAVE BEEN PARTIALLY WESTERNIZED, RATHER THAN SIGNIFICANTLY WESTERNIZED, THROUGH A COLONIAL RELATIONSHIP (41)	0.79	77	TEND TO BE THOSE THAT HAVE BEEN SIGNIFICANTLY WESTERNIZED, RATHER THAN PARTIALLY WESTERNIZED, THROUGH A COLONIAL RELATIONSHIP (28)	0.60	4 15 15 10 X SQ= 5.18 P = 0.014 RV YES (YES)
79	LEAN MORE TOWARD BEING THOSE THAT WERE FORMERLY DEPENDENCIES OF BRITAIN OR FRANCE, RATHER THAN SPAIN (49)	0.86	79	LEAN LESS TOWARD BEING THOSE THAT WERE FORMERLY DEPENDENCIES OF BRITAIN OR FRANCE, RATHER THAN SPAIN (49)	0.58	12 15 2 11 X SQ= 2.11 P = 0.090 RV NO (NO)
81	TEND LESS TO BE THOSE WHERE THE TYPE OF POLITICAL MODERNIZATION IS LATER EUROPEAN OR LATER EUROPEAN DERIVED, RATHER THAN EARLY EUROPEAN OR EARLY EUROPEAN DERIVED (40)	0.56	81	TEND MORE TO BE THOSE WHERE THE TYPE OF POLITICAL MODERNIZATION IS LATER EUROPEAN OR LATER EUROPEAN DERIVED, RATHER THAN EARLY EUROPEAN OR EARLY EUROPEAN DERIVED (40)	0.93	4 2 5 25 X SQ= 4.27 P = 0.024 RV NO (YES)
82	LEAN TOWARD BEING THOSE WHERE THE TYPE OF POLITICAL MODERNIZATION IS DEVELOPED TUTELARY OR UNDEVELOPED TUTELARY, RATHER THAN EARLY OR LATER EUROPEAN OR EUROPEAN DERIVED (55)	0.64	82	LEAN TOWARD BEING THOSE WHERE THE TYPE OF POLITICAL MODERNIZATION IS EARLY OR LATER EUROPEAN OR EUROPEAN DERIVED, RATHER THAN DEVELOPED TUTELARY OR UNDEVELOPED TUTELARY (51)	0.60	9 27 16 18 X SQ= 2.81 P = 0.080 RV YES (NO)
100	TEND LESS TO BE THOSE WHERE GOVERNMENTAL STABILITY IS GENERALLY PRESENT AND DATES FROM AT LEAST THE INTER-WAR PERIOD, OR IS GENERALLY OR MODERATELY PRESENT, AND DATES FROM THE POST-WAR PERIOD (64)	0.55	100	TEND MORE TO BE THOSE WHERE GOVERNMENTAL STABILITY IS GENERALLY PRESENT AND DATES FROM AT LEAST THE INTER-WAR PERIOD, OR IS GENERALLY OR MODERATELY PRESENT, AND DATES FROM THE POST-WAR PERIOD (64)	0.83	12 29 10 6 X SQ= 4.05 P = 0.033 RV NO (YES)
111	TILT TOWARD BEING THOSE WHERE POLITICAL ENCULTURATION IS LOW (42)	0.58	111	TILT TOWARD BEING THOSE WHERE POLITICAL ENCULTURATION IS HIGH OR MEDIUM (53)	0.66	10 23 14 12 X SQ= 2.44 P = 0.109 RV YES (YES)
121	TEND TO BE THOSE WHERE INTEREST ARTICULATION BY NON-ASSOCIATIONAL GROUPS IS SIGNIFICANT (54)	0.56	121	TEND TO BE THOSE WHERE INTEREST ARTICULATION BY NON-ASSOCIATIONAL GROUPS IS MODERATE, LIMITED, OR NEGLIGIBLE (61)	0.70	15 14 12 33 X SQ= 3.76 P = 0.047 RV YES (YES)
122	TEND MORE TO BE THOSE WHERE INTEREST ARTICULATION BY NON-ASSOCIATIONAL GROUPS IS SIGNIFICANT OR MODERATE (83)	0.89	122	TEND LESS TO BE THOSE WHERE INTEREST ARTICULATION BY NON-ASSOCIATIONAL GROUPS IS SIGNIFICANT OR MODERATE (83)	0.57	24 27 3 20 X SQ= 6.51 P = 0.005 RV NO (NO)

123	ALWAYS ARE THOSE WHERE INTEREST ARTICULATION BY NON-ASSOCIATIONAL GROUPS IS SIGNIFICANT, MODERATE, OR LIMITED (107)	1.00	123	TEND LESS TO BE THOSE WHERE INTEREST ARTICULATION BY NON-ASSOCIATIONAL GROUPS IS SIGNIFICANT, MODERATE, OR LIMITED (107)	0.85	27 40 0 7 X SQ= 2.87 P = 0.043 RV NO (NO)
125	LEAN TOWARD BEING THOSE WHERE INTEREST ARTICULATION BY ANOMIC GROUPS IS FREQUENT OR OCCASIONAL (64)	0.73	125	LEAN TOWARD BEING THOSE WHERE INTEREST ARTICULATION BY ANOMIC GROUPS IS INFREQUENT OR VERY INFREQUENT (35)	0.53	19 18 7 20 X SQ= 3.20 P = 0.070 RV YES (YES)
128	LEAN TOWARD BEING THOSE WHERE INTEREST ARTICULATION BY POLITICAL PARTIES IS SIGNIFICANT OR MODERATE (48)	0.76	128	LEAN TOWARD BEING THOSE WHERE INTEREST ARTICULATION BY POLITICAL PARTIES IS LIMITED OR NEGLIGIBLE (45)	0.50	13 21 4 21 X SQ= 2.47 P = 0.084 RV YES (NO)
134	TILT TOWARD BEING THOSE WHERE INTEREST AGGREGATION BY THE EXECUTIVE IS LIMITED OR NEGLIGIBLE (46)	0.64	134	TILT TOWARD BEING THOSE WHERE INTEREST AGGREGATION BY THE EXECUTIVE IS SIGNIFICANT OR MODERATE (57)	0.59	8 27 14 19 X SQ= 2.14 P = 0.120 RV YES (NO)
135	ALWAYS ARE THOSE WHERE INTEREST AGGREGATION BY THE EXECUTIVE IS SIGNIFICANT, MODERATE, OR LIMITED (77)	1.00	135	LEAN LESS TOWARD BEING THOSE WHERE INTEREST AGGREGATION BY THE EXECUTIVE IS SIGNIFICANT, MODERATE, OR LIMITED (77)	0.82	19 32 0 7 X SQ= 2.37 P = 0.083 RV NO (NO)
139	LEAN MORE TOWARD BEING THOSE WHERE THE PARTY SYSTEM IS QUANTITATIVELY OTHER THAN ONE-PARTY (71)	0.87	139	LEAN LESS TOWARD BEING THOSE WHERE THE PARTY SYSTEM IS QUANTITATIVELY OTHER THAN ONE-PARTY (71)	0.67	3 15 20 31 X SQ= 2.11 P = 0.092 RV NO (NO)
166	TEND LESS TO BE THOSE WHERE THE VERTICAL POWER DISTRIBUTION IS THAT OF FORMAL FEDERALISM OR FORMAL AND EFFECTIVE UNITARISM (99)	0.73	166	TEND MORE TO BE THOSE WHERE THE VERTICAL POWER DISTRIBUTION IS THAT OF FORMAL FEDERALISM OR FORMAL AND EFFECTIVE UNITARISM (99)	0.96	7 2 19 45 X SQ= 6.00 P = 0.008 RV NO (YES)
178	TEND TO BE THOSE WHERE THE LEGISLATURE IS BICAMERAL (51)	0.74	178	TEND TO BE THOSE WHERE THE LEGISLATURE IS UNICAMERAL (53)	0.64	6 27 17 15 X SQ= 7.22 P = 0.004 RV YES (YES)
182	LEAN LESS TOWARD BEING THOSE WHERE THE BUREAUCRACY IS SEMI-MODERN, RATHER THAN POST-COLONIAL TRANSITIONAL (55)	0.56	182	LEAN MORE TOWARD BEING THOSE WHERE THE BUREAUCRACY IS SEMI-MODERN, RATHER THAN POST-COLONIAL TRANSITIONAL (55)	0.83	9 29 7 6 X SQ= 2.81 P = 0.080 RV NO (YES)

192 TEND LESS TO BE THOSE 0.63
WHERE THE CHARACTER OF THE LEGAL SYSTEM
IS OTHER THAN
MUSLIM OR PARTLY MUSLIM (86)

192 TEND MORE TO BE THOSE 0.87
WHERE THE CHARACTER OF THE LEGAL SYSTEM
IS OTHER THAN
MUSLIM OR PARTLY MUSLIM (86)

10 6
17 41
X SQ= 4.61
P = 0.020
RV NO (YES)

```
*****************************************************************************************************

113  POLITIES                                          113  POLITIES
     WHERE SECTIONALISM IS                                  WHERE SECTIONALISM IS
     EXTREME  (27)                                          MODERATE OR NEGLIGIBLE  (81)


                                                                                       PREDICATE ONLY
.....................................................................................................

     27 IN LEFT COLUMN

BELGIUM     BRAZIL      BURMA       CANADA      COLOMBIA    CONGO(LEO)  CZECHOS'KIA ECUADOR     INDIA       INDONESIA
IRAN        IRAQ        JORDAN      LAOS        LIBYA       MALAGASY R  MOROCCO     NIGERIA     PAKISTAN    SOMALIA
SUDAN       SWITZERLAND UGANDA      USSR        UK          YEMEN       YUGOSLAVIA

     81 IN RIGHT COLUMN

AFGHANISTAN ALBANIA     ALGERIA     AUSTRALIA   AUSTRIA     BOLIVIA     BULGARIA    BURUNDI     CAMBODIA    CAMEROUN
CEN AFR REP CEYLON      CHAD        CHILE       CHINA, PR   CONGO(BRA)  CUBA        COSTA RICA  CYPRUS      DAHOMEY
DENMARK     DOMIN REP   EL SALVADOR ETHIOPIA    FINLAND     FRANCE      GABON       GERMANY, E  GERMAN FR   GHANA
GREECE      GUATEMALA   GUINEA      HAITI       HONDURAS    HUNGARY     ICELAND     IRELAND     ISRAEL      ITALY
IVORY COAST JAMAICA     JAPAN       KOREA, N    KOREA REP   LEBANON     LIBERIA     LUXEMBOURG  MALAYA      MAURITANIA
MONGOLIA    NETHERLANDS NEW ZEALAND NORWAY      PANAMA      PARAGUAY    PERU        PHILIPPINES POLAND      PORTUGAL
RUMANIA     RWANDA      SA'U ARABIA SENEGAL     SIERRE LEO  SO AFRICA   SPAIN       SWEDEN      SYRIA       TANGANYIKA
THAILAND    TOGO        TRINIDAD    TUNISIA     TURKEY      UAR         US          UPPER VOLTA VENEZUELA   VIETNAM, N
VIETNAM REP

      3 EXCLUDED BECAUSE AMBIGUOUS

ARGENTINA   MEXICO      URUGUAY

      4 EXCLUDED BECAUSE UNASCERTAINED

MALI        NEPAL       NICARAGUA   NIGER
-----------------------------------------------------------------------------------------------------
```

114 POLITIES WHERE SECTIONALISM IS EXTREME OR MODERATE (61)

114 POLITIES WHERE SECTIONALISM IS NEGLIGIBLE (47)

PREDICATE ONLY

61 IN LEFT COLUMN

AFGHANISTAN ALBANIA ALGERIA AUSTRALIA BELGIUM BRAZIL BULGARIA BURMA CAMBODIA CAMEROUN
CANADA CEYLON CHAD CHINA, PR COLOMBIA CONGO(BRA) CONGO(LEO) CZECHOS'KIA DAHOMEY ECUADOR
ETHIOPIA FRANCE GABON GERMAN FR GHANA GUINEA INDIA INDONESIA IRAN IRAQ
ITALY JORDAN LAOS LEBANON LIBERIA LIBYA MALAGASY R MALAYA MAURITANIA MOROCCO
NETHERLANDS NIGERIA PAKISTAN PERU PHILIPPINES SENEGAL SIERRE LEO SOMALIA SO AFRICA SPAIN
SUDAN SWITZERLAND SYRIA THAILAND TOGO UGANDA USSR UK US YEMEN
YUGOSLAVIA

47 IN RIGHT COLUMN

AUSTRIA BOLIVIA BURUNDI CEN AFR REP CHILE CUBA COSTA RICA CYPRUS DENMARK DOMIN REP
EL SALVADOR FINLAND GERMANY, E GREECE GUATEMALA HAITI HONDURAS HUNGARY ICELAND IRELAND
ISRAEL IVORY COAST JAMAICA JAPAN KOREA, N KOREA REP LUXEMBOURG MONGOLIA NEW ZEALAND NORWAY
PANAMA PARAGUAY POLAND PORTUGAL RUMANIA RWANDA SA'U ARABIA SWEDEN TANGANYIKA TRINIDAD
TUNISIA TURKEY UAR UPPER VOLTA VENEZUELA VIETNAM, N VIETNAM REP

3 EXCLUDED BECAUSE AMBIGUOUS

ARGENTINA MEXICO URUGUAY

4 EXCLUDED BECAUSE UNASCERTAINED

MALI NEPAL NICARAGUA NIGER

115 POLITIES
WHERE INTEREST ARTICULATION
BY ASSOCIATIONAL GROUPS
IS SIGNIFICANT (20)

115 POLITIES
WHERE INTEREST ARTICULATION
BY ASSOCIATIONAL GROUPS
IS MODERATE, LIMITED, OR
NEGLIGIBLE (91)

PREDICATE ONLY

20 IN LEFT COLUMN

ARGENTINA	AUSTRALIA	BELGIUM	BRAZIL	CANADA	DENMARK	FINLAND	FRANCE	GERMAN FR	ITALY
JAPAN	LUXEMBOURG	NETHERLANDS	NEW ZEALAND	NORWAY	PHILIPPINES	SWEDEN	SWITZERLAND	UK	US

91 IN RIGHT COLUMN

AFGHANISTAN	ALBANIA	ALGERIA	AUSTRIA	BOLIVIA	BULGARIA	BURMA	BURUNDI	CAMBODIA	CAMEROUN
CEN AFR REP	CEYLON	CHAD	CHILE	CHINA, PR	COLOMBIA	CONGO(BRA)	CONGO(LEO)	CUBA	COSTA RICA
CYPRUS	CZECHOS'KIA	DAHOMEY	DOMIN REP	ECUADOR	EL SALVADOR	ETHIOPIA	GABON	GERMANY, E	GHANA
GREECE	GUATEMALA	GUINEA	HAITI	HONDURAS	HUNGARY	ICELAND	INDIA	INDONESIA	IRAN
IRAQ	IRELAND	ISRAEL	IVORY COAST	JAMAICA	JORDAN	KOREA, N	LAOS	LEBANON	LIBERIA
LIBYA	MALAGASY R	MALAYA	MALI	MAURITANIA	MEXICO	MONGOLIA	MOROCCO	NEPAL	NICARAGUA
NIGER	NIGERIA	PAKISTAN	PANAMA	PARAGUAY	PERU	POLAND	RUMANIA	RWANDA	SA'U ARABIA
SENEGAL	SIERRE LEO	SOMALIA	SO AFRICA	SUDAN	SYRIA	TANGANYIKA	THAILAND	TOGO	TRINIDAD
TUNISIA	TURKEY	UGANDA	USSR	UAR	UPPER VOLTA	URUGUAY	VENEZUELA	VIETNAM, N	VIETNAM REP
YEMEN									

3 EXCLUDED BECAUSE AMBIGUOUS

PORTUGAL SPAIN YUGOSLAVIA

1 EXCLUDED BECAUSE UNASCERTAINABLE

KOREA REP

116 POLITIES WHERE INTEREST ARTICULATION BY ASSOCIATIONAL GROUPS IS SIGNIFICANT OR MODERATE (32)

116 POLITIES WHERE INTEREST ARTICULATION BY ASSOCIATIONAL GROUPS IS LIMITED OR NEGLIGIBLE (79)

BOTH SUBJECT AND PREDICATE

32 IN LEFT COLUMN

ARGENTINA	AUSTRALIA	AUSTRIA	BELGIUM	BRAZIL	CANADA	CHILE	DENMARK	FINLAND	FRANCE
GERMAN FR	GREECE	ICELAND	IRELAND	ISRAEL	ITALY	JAPAN	LEBANON	LUXEMBOURG	MEXICO
NETHERLANDS	NEW ZEALAND	NORWAY	PHILIPPINES	SO AFRICA	SWEDEN	SWITZERLAND	TURKEY	UK	US
URUGUAY	VENEZUELA								

79 IN RIGHT COLUMN

AFGHANISTAN	ALBANIA	ALGERIA	BOLIVIA	BULGARIA	BURMA	BURUNDI	CAMBODIA	CAMEROUN	CEN AFR REP
CEYLON	CHAD	CHINA, PR	COLOMBIA	CONGO(BRA)	CONGO(LEO)	CUBA	COSTA RICA	CYPRUS	CZECHOS'KIA
DAHOMEY	DOMIN REP	ECUADOR	EL SALVADOR	ETHIOPIA	GABON	GERMANY, E	GHANA	GUATEMALA	GUINEA
HAITI	HONDURAS	HUNGARY	INDIA	INDONESIA	IRAN	IRAQ	IVORY COAST	JAMAICA	JORDAN
KOREA, N	LAOS	LIBERIA	LIBYA	MALAGASY R	MALAYA	MALI	MAURITANIA	MONGOLIA	MOROCCO
NEPAL	NICARAGUA	NIGER	NIGERIA	PAKISTAN	PANAMA	PARAGUAY	PERU	POLAND	RUMANIA
RWANDA	SA'U ARABIA	SENEGAL	SIERRE LEO	SOMALIA	SUDAN	SYRIA	TANGANYIKA	THAILAND	TOGO
TRINIDAD	TUNISIA	UGANDA	USSR	UAR	UPPER VOLTA	VIETNAM, N	VIETNAM REP	YEMEN	

3 EXCLUDED BECAUSE AMBIGUOUS

PORTUGAL SPAIN YUGOSLAVIA

1 EXCLUDED BECAUSE UNASCERTAINABLE

KOREA REP

2	TEND TO BE THOSE LOCATED IN WEST EUROPE, SCANDINAVIA, NORTH AMERICA, OR AUSTRALASIA (22)	0.62	2	ALWAYS ARE THOSE LOCATED ELSEWHERE THAN IN WEST EUROPE, SCANDINAVIA, NORTH AMERICA, OR AUSTRALASIA (93)	1.00	20 0 12 79 X SQ= 56.07 P = 0. RV YES (YES)
3	TEND LESS TO BE THOSE LOCATED ELSEWHERE THAN IN WEST EUROPE (102)	0.66	3	ALWAYS ARE THOSE LOCATED ELSEWHERE THAN IN WEST EUROPE (102)	1.00	11 0 21 79 X SQ= 26.42 P = 0. RV NO (YES)
5	ALWAYS ARE THOSE LOCATED IN SCANDINAVIA OR WEST EUROPE, RATHER THAN IN EAST EUROPE (18)	1.00	5	ALWAYS ARE THOSE LOCATED IN EAST EUROPE, RATHER THAN IN SCANDINAVIA OR WEST EUROPE (9)	1.00	16 0 0 8 X SQ= 19.71 P = 0. RV YES (YES)

8	TILT MORE TOWARD BEING THOSE LOCATED ELSEWHERE THAN IN EAST ASIA SOUTH ASIA, OR SOUTHEAST ASIA (97)	0.94	8	TILT LESS TOWARD BEING THOSE LOCATED ELSEWHERE THAN IN EAST ASIA SOUTH ASIA, OR SOUTHEAST ASIA (97)	0.81	2 15 30 64 X SQ= 1.95 P = 0.144 RV NO (NO)
10	TEND MORE TO BE THOSE LOCATED ELSEWHERE THAN IN NORTH AFRICA, OR CENTRAL AND SOUTH AFRICA (82)	0.97	10	TEND LESS TO BE THOSE LOCATED ELSEWHERE THAN IN NORTH AFRICA, OR CENTRAL AND SOUTH AFRICA (82)	0.59	1 32 31 47 X SQ= 13.50 P = 0. RV NO (NO)
15	TEND TO BE THOSE LOCATED IN THE MIDDLE EAST, RATHER THAN IN NORTH AFRICA OR CENTRAL AND SOUTH AFRICA (11)	0.75	15	TEND TO BE THOSE LOCATED IN NORTH AFRICA OR CENTRAL AND SOUTH AFRICA, RATHER THAN IN THE MIDDLE EAST (33)	0.80	3 8 1 32 X SQ= 3.30 P = 0.043 RV YES (NO)
19	TEND TO BE THOSE LOCATED IN THE CARIBBEAN, CENTRAL AMERICA, OR SOUTH AMERICA, RATHER THAN IN NORTH AFRICA OR CENTRAL AND SOUTH AFRICA (22)	0.86	19	TEND TO BE THOSE LOCATED IN NORTH AFRICA OR CENTRAL AND SOUTH AFRICA, RATHER THAN IN THE CARIBBEAN, CENTRAL AMERICA, OR SOUTH AMERICA (33)	0.67	1 32 6 16 X SQ= 4.97 P = 0.013 RV YES (NO)
24	LEAN TOWARD BEING THOSE WHOSE POPULATION IS VERY LARGE, LARGE, OR MEDIUM (61)	0.66	24	LEAN TOWARD BEING THOSE WHOSE POPULATION IS SMALL (54)	0.54	21 36 11 43 X SQ= 2.91 P = 0.063 RV YES (NO)
25	TEND LESS TO BE THOSE WHOSE POPULATION DENSITY IS MEDIUM OR LOW (98)	0.72	25	TEND MORE TO BE THOSE WHOSE POPULATION DENSITY IS MEDIUM OR LOW (98)	0.91	9 7 23 72 X SQ= 5.38 P = 0.015 RV NO (YES)
28	TEND TO BE THOSE WHOSE POPULATION GROWTH RATE IS LOW (48)	0.59	28	TEND TO BE THOSE WHOSE POPULATION GROWTH RATE IS HIGH (62)	0.65	13 48 19 26 X SQ= 4.43 P = 0.032 RV YES (NO)
29	ALWAYS ARE THOSE WHERE THE DEGREE OF URBANIZATION IS HIGH (56)	1.00	29	TEND TO BE THOSE WHERE THE DEGREE OF URBANIZATION IS LOW (49)	0.68	31 23 0 48 X SQ= 36.92 P = 0. RV YES (YES)
30	TEND TO BE THOSE WHOSE AGRICULTURAL POPULATION IS MEDIUM, LOW, OR VERY LOW (57)	0.97	30	TEND TO BE THOSE WHOSE AGRICULTURAL POPULATION IS HIGH (56)	0.69	1 53 31 24 X SQ= 36.46 P = 0. RV YES (YES)

31	TEND TO BE THOSE WHOSE AGRICULTURAL POPULATION IS LOW OR VERY LOW (24)	0.69	31	TEND TO BE THOSE WHOSE AGRICULTURAL POPULATION IS HIGH OR MEDIUM (90)	0.97	10 76 22 2 X SQ= 54.46 P = 0. RV YES (YES)
33	TEND TO BE THOSE WHOSE GROSS NATIONAL PRODUCT IS VERY HIGH, HIGH, OR MEDIUM (30)	0.59	33	TEND TO BE THOSE WHOSE GROSS NATIONAL PRODUCT IS LOW OR VERY LOW (85)	0.89	19 9 13 70 X SQ= 25.31 P = 0. RV YES (YES)
34	TEND TO BE THOSE WHOSE GROSS NATIONAL PRODUCT IS VERY HIGH, HIGH, MEDIUM, OR LOW (62)	0.91	34	TEND TO BE THOSE WHOSE GROSS NATIONAL PRODUCT IS VERY LOW (53)	0.63	29 29 3 50 X SQ= 24.42 P = 0. RV YES (YES)
35	TEND TO BE THOSE WHOSE PER CAPITA GROSS NATIONAL PRODUCT IS VERY HIGH OR HIGH (24)	0.66	35	TEND TO BE THOSE WHOSE PER CAPITA GROSS NATIONAL PRODUCT IS MEDIUM, LOW, OR VERY LOW (91)	0.96	21 3 11 76 X SQ= 47.79 P = 0. RV YES (YES)
36	TEND TO BE THOSE WHOSE PER CAPITA GROSS NATIONAL PRODUCT IS VERY HIGH, HIGH, OR MEDIUM (42)	0.87	36	TEND TO BE THOSE WHOSE PER CAPITA GROSS NATIONAL PRODUCT IS LOW OR VERY LOW (73)	0.85	28 12 4 67 X SQ= 48.57 P = 0. RV YES (YES)
37	ALWAYS ARE THOSE WHOSE PER CAPITA GROSS NATIONAL PRODUCT IS VERY HIGH, HIGH, MEDIUM, OR LOW (64)	1.00	37	TEND TO BE THOSE WHOSE PER CAPITA GROSS NATIONAL PRODUCT IS VERY LOW (51)	0.63	32 29 0 50 X SQ= 34.34 P = 0. RV YES (YES)
39	TEND TO BE THOSE WHOSE INTERNATIONAL FINANCIAL STATUS IS VERY HIGH, HIGH, OR MEDIUM (38)	0.78	39	TEND TO BE THOSE WHOSE INTERNATIONAL FINANCIAL STATUS IS LOW OR VERY LOW (76)	0.86	25 11 7 67 X SQ= 39.39 P = 0. RV YES (YES)
40	TEND TO BE THOSE WHOSE INTERNATIONAL FINANCIAL STATUS IS VERY HIGH, HIGH, MEDIUM, OR LOW (71)	0.97	40	TEND TO BE THOSE WHOSE INTERNATIONAL FINANCIAL STATUS IS VERY LOW (39)	0.51	31 36 1 38 X SQ= 20.32 P = 0. RV YES (NO)
41	TEND TO BE THOSE WHOSE ECONOMIC DEVELOPMENTAL STATUS IS DEVELOPED (19)	0.53	41	TEND TO BE THOSE WHOSE ECONOMIC DEVELOPMENTAL STATUS IS INTERMEDIATE, UNDERDEVELOPED, OR VERY UNDERDEVELOPED (94)	0.96	16 3 14 76 X SQ= 33.71 P = 0. RV YES (YES)

42	TEND TO BE THOSE WHOSE ECONOMIC DEVELOPMENTAL STATUS IS DEVELOPED OR INTERMEDIATE (36)	0.87	42	TEND TO BE THOSE WHOSE ECONOMIC DEVELOPMENTAL STATUS IS UNDERDEVELOPED OR VERY UNDERDEVELOPED (76)	0.88	26 9 4 69 X SQ= 52.45 P = 0. RV YES (YES)
43	ALWAYS ARE THOSE WHOSE ECONOMIC DEVELOPMENTAL STATUS IS DEVELOPED, INTERMEDIATE, OR UNDERDEVELOPED (55)	1.00	43	TEND TO BE THOSE WHOSE ECONOMIC DEVELOPMENTAL STATUS IS VERY UNDERDEVELOPED (57)	0.73	31 21 0 56 X SQ= 43.96 P = 0. RV YES (YES)
44	TEND TO BE THOSE WHERE THE LITERACY RATE IS NINETY PERCENT OR ABOVE (25)	0.59	44	TEND TO BE THOSE WHERE THE LITERACY RATE IS BELOW NINETY PERCENT (90)	0.92	19 6 13 73 X SQ= 32.09 P = 0. RV YES (YES)
45	TEND TO BE THOSE WHERE THE LITERACY RATE IS FIFTY PERCENT OR ABOVE (55)	0.94	45	TEND TO BE THOSE WHERE THE LITERACY RATE IS BELOW FIFTY PERCENT (54)	0.70	29 22 2 52 X SQ= 33.11 P = 0. RV YES (YES)
46	ALWAYS ARE THOSE WHERE THE LITERACY RATE IS TEN PERCENT OR ABOVE (84)	1.00	46	TEND LESS TO BE THOSE WHERE THE LITERACY RATE IS TEN PERCENT OR ABOVE (84)	0.65	32 48 0 26 X SQ= 13.06 P = 0. RV NO (NO)
50	TEND TO BE THOSE WHERE FREEDOM OF THE PRESS IS COMPLETE (43)	0.74	50	TEND TO BE THOSE WHERE FREEDOM OF THE PRESS IS INTERMITTENT, INTERNALLY ABSENT, OR INTERNALLY AND EXTERNALLY ABSENT (56)	0.69	23 20 8 44 X SQ= 13.86 P = 0. RV YES (YES)
51	TEND TO BE THOSE WHERE FREEDOM OF THE PRESS IS COMPLETE OR INTERMITTENT (60)	0.97	51	TEND TO BE THOSE WHERE FREEDOM OF THE PRESS IS INTERNALLY ABSENT OR INTERNALLY AND EXTERNALLY ABSENT (37)	0.52	30 30 1 32 X SQ= 19.08 P = 0. RV YES (YES)
52	ALWAYS ARE THOSE WHERE FREEDOM OF THE PRESS IS COMPLETE, INTERMITTENT, OR INTERNALLY ABSENT (82)	1.00	52	TEND LESS TO BE THOSE WHERE FREEDOM OF THE PRESS IS COMPLETE, INTERMITTENT, OR INTERNALLY ABSENT (82)	0.74	32 46 0 16 X SQ= 8.21 P = 0.001 RV NO (NO)
54	TEND TO BE THOSE WHERE NEWSPAPER CIRCULATION IS ONE HUNDRED OR MORE PER THOUSAND (37)	0.78	54	TEND TO BE THOSE WHERE NEWSPAPER CIRCULATION IS LESS THAN ONE HUNDRED PER THOUSAND (76)	0.84	25 12 7 65 X SQ= 36.69 P = 0. RV YES (YES)

55	ALWAYS ARE THOSE WHERE NEWSPAPER CIRCULATION IS TEN OR MORE PER THOUSAND (78)	1.00	55	TEND LESS TO BE THOSE WHERE NEWSPAPER CIRCULATION IS TEN OR MORE PER THOUSAND (78)	0.55	32 42 0 35 X SQ= 19.39 P = 0. RV NO (NO)
56	TEND MORE TO BE THOSE WHERE THE RELIGION IS PREDOMINANTLY LITERATE (79)	0.97	56	TEND LESS TO BE THOSE WHERE THE RELIGION IS PREDOMINANTLY LITERATE (79)	0.60	31 45 1 30 X SQ= 13.08 P = 0. RV NO (NO)
57	TEND LESS TO BE THOSE WHERE THE RELIGION IS OTHER THAN CATHOLIC (90)	0.59	57	TEND MORE TO BE THOSE WHERE THE RELIGION IS OTHER THAN CATHOLIC (90)	0.87	13 10 19 69 X SQ= 9.21 P = 0.002 RV NO (YES)
58	TEND MORE TO BE THOSE WHERE THE RELIGION IS OTHER THAN MUSLIM (97)	0.97	58	TEND LESS TO BE THOSE WHERE THE RELIGION IS OTHER THAN MUSLIM (97)	0.78	1 17 31 62 X SQ= 4.40 P = 0.021 RV NO (NO)
62	TILT LESS TOWARD BEING THOSE WHERE THE RELIGION IS CATHOLIC, RATHER THAN PROTESTANT (25)	0.72	62	ALWAYS ARE THOSE WHERE THE RELIGION IS CATHOLIC, RATHER THAN PROTESTANT (25)	1.00	5 0 13 10 X SQ= 1.75 P = 0.128 RV NO (NO)
63	TEND TO BE THOSE WHERE THE RELIGION IS PREDOMINANTLY SOME KIND OF CHRISTIAN (46)	0.84	63	TEND TO BE THOSE WHERE THE RELIGION IS PREDOMINANTLY OR PARTLY OTHER THAN CHRISTIAN (68)	0.79	27 16 5 62 X SQ= 36.23 P = 0. RV YES (YES)
66	TILT TOWARD BEING THOSE THAT ARE RELIGIOUSLY HOMOGENEOUS (57)	0.66	66	TILT TOWARD BEING THOSE THAT ARE RELIGIOUSLY HETEROGENEOUS (49)	0.52	21 34 11 37 X SQ= 2.12 P = 0.135 RV YES (NO)
67	LEAN MORE TOWARD BEING THOSE THAT ARE RACIALLY HOMOGENEOUS (82)	0.84	67	LEAN LESS TOWARD BEING THOSE THAT ARE RACIALLY HOMOGENEOUS (82)	0.68	26 50 5 24 X SQ= 2.15 P = 0.100 RV NO (NO)
68	TEND TO BE THOSE THAT ARE LINGUISTICALLY HOMOGENEOUS (52)	0.75	68	TEND TO BE THOSE THAT ARE LINGUISTICALLY WEAKLY HETEROGENEOUS OR STRONGLY HETEROGENEOUS (62)	0.67	24 26 8 52 X SQ= 14.25 P = 0. RV YES (NO)

69	TEND TO BE THOSE THAT ARE LINGUISTICALLY HOMOGENEOUS OR WEAKLY HETEROGENEOUS (64)	0.81	69	TEND TO BE THOSE THAT ARE LINGUISTICALLY STRONGLY HETEROGENEOUS (50)	0.55	26 35 6 43 X SQ= 10.73 P = 0.001 RV YES (NO)
70	TEND LESS TO BE THOSE THAT ARE RELIGIOUSLY, RACIALLY, OR LINGUISTICALLY HETEROGENEOUS (85)	0.61	70	TEND MORE TO BE THOSE THAT ARE RELIGIOUSLY, RACIALLY, OR LINGUISTICALLY HETEROGENEOUS (85)	0.89	12 8 19 63 X SQ= 8.64 P = 0.002 RV NO (YES)
72	TEND TO BE THOSE WHOSE DATE OF INDEPENDENCE IS BEFORE 1914 (52)	0.81	72	TEND TO BE THOSE WHOSE DATE OF INDEPENDENCE IS AFTER 1914 (59)	0.68	25 25 6 52 X SQ= 18.74 P = 0. RV YES (YES)
73	TEND TO BE THOSE WHOSE DATE OF INDEPENDENCE IS BEFORE 1945 (65)	0.94	73	TEND TO BE THOSE WHOSE DATE OF INDEPENDENCE IS AFTER 1945 (46)	0.57	29 33 2 44 X SQ= 21.20 P = 0. RV YES (NO)
74	TEND TO BE THOSE THAT ARE HISTORICALLY WESTERN (26)	0.67	74	TEND TO BE THOSE THAT ARE NOT HISTORICALLY WESTERN (87)	0.95	20 4 10 75 X SQ= 44.54 P = 0. RV YES (YES)
75	ALWAYS ARE THOSE THAT ARE HISTORICALLY WESTERN OR SIGNIFICANTLY WESTERNIZED (62)	1.00	75	TEND TO BE THOSE THAT ARE NOT HISTORICALLY WESTERN AND ARE NOT SIGNIFICANTLY WESTERNIZED (52)	0.65	31 28 0 51 X SQ= 34.76 P = 0. RV YES (YES)
76	TEND TO BE THOSE THAT ARE HISTORICALLY WESTERN, RATHER THAN HAVING BEEN SIGNIFICANTLY OR PARTIALLY WESTERNIZED THROUGH A COLONIAL RELATIONSHIP (26)	0.69	76	TEND TO BE THOSE THAT HAVE BEEN SIGNIFICANTLY OR PARTIALLY WESTERNIZED THROUGH A COLONIAL RELATIONSHIP, RATHER THAN BEING HISTORICALLY WESTERN (70)	0.94	20 4 9 61 X SQ= 38.37 P = 0. RV YES (YES)
77	ALWAYS ARE THOSE THAT HAVE BEEN SIGNIFICANTLY WESTERNIZED, RATHER THAN PARTIALLY WESTERNIZED, THROUGH A COLONIAL RELATIONSHIP (28)	1.00	77	TEND TO BE THOSE THAT HAVE BEEN PARTIALLY WESTERNIZED, RATHER THAN SIGNIFICANTLY WESTERNIZED, THROUGH A COLONIAL RELATIONSHIP (41)	0.67	8 20 0 41 X SQ= 10.61 P = 0. RV YES (NO)
81	TEND LESS TO BE THOSE WHERE THE TYPE OF POLITICAL MODERNIZATION IS LATER EUROPEAN OR LATER EUROPEAN DERIVED, RATHER THAN EARLY EUROPEAN OR EARLY EUROPEAN DERIVED (40)	0.59	81	ALWAYS ARE THOSE WHERE THE TYPE OF POLITICAL MODERNIZATION IS LATER EUROPEAN OR LATER EUROPEAN DERIVED, RATHER THAN EARLY EUROPEAN OR EARLY EUROPEAN DERIVED (40)	1.00	11 0 16 21 X SQ= 8.91 P = 0.001 RV NO (YES)

82	TEND TO BE THOSE WHERE THE TYPE OF POLITICAL MODERNIZATION IS EARLY OR LATER EUROPEAN OR EUROPEAN DERIVED, RATHER THAN DEVELOPED TUTELARY OR UNDEVELOPED TUTELARY (51)	0.90	82	TEND TO BE THOSE WHERE THE TYPE OF POLITICAL MODERNIZATION IS DEVELOPED TUTELARY OR UNDEVELOPED TUTELARY, RATHER THAN EARLY OR LATER EUROPEAN OR EUROPEAN DERIVED (55)	0.71	27 21 3 51 X SQ= 29.06 P = 0. RV YES (YES)
85	TEND TO BE THOSE WHERE THE STAGE OF POLITICAL MODERNIZATION IS ADVANCED, RATHER THAN MID- OR EARLY TRANSITIONAL (60)	0.97	85	TEND TO BE THOSE WHERE THE STAGE OF POLITICAL MODERNIZATION IS MID- OR EARLY TRANSITIONAL, RATHER THAN ADVANCED (54)	0.68	31 25 1 53 X SQ= 35.60 P = 0. RV YES (YES)
86	ALWAYS ARE THOSE WHERE THE STAGE OF POLITICAL MODERNIZATION IS ADVANCED OR MID-TRANSITIONAL, RATHER THAN EARLY TRANSITIONAL (76)	1.00	86	TEND LESS TO BE THOSE WHERE THE STAGE OF POLITICAL MODERNIZATION IS ADVANCED OR MID-TRANSITIONAL, RATHER THAN EARLY TRANSITIONAL (76)	0.51	32 40 0 38 X SQ= 21.71 P = 0. RV NO (NO)
87	TEND TO BE THOSE WHOSE IDEOLOGICAL ORIENTATION IS OTHER THAN DEVELOPMENTAL (58)	0.93	87	TEND TO BE THOSE WHOSE IDEOLOGICAL ORIENTATION IS DEVELOPMENTAL (31)	0.52	2 29 28 27 X SQ= 15.35 P = 0. RV YES (YES)
89	TEND TO BE THOSE WHOSE IDEOLOGICAL ORIENTATION IS CONVENTIONAL (33)	0.93	89	TEND TO BE THOSE WHOSE IDEOLOGICAL ORIENTATION IS OTHER THAN CONVENTIONAL (62)	0.92	28 5 2 59 X SQ= 61.87 P = 0. RV YES (YES)
92	ALWAYS ARE THOSE WHERE THE SYSTEM STYLE IS LIMITED MOBILIZATIONAL, OR NON-MOBILIZATIONAL (93)	1.00	92	TEND LESS TO BE THOSE WHERE THE SYSTEM STYLE IS LIMITED MOBILIZATIONAL, OR NON-MOBILIZATIONAL (93)	0.76	0 19 32 59 X SQ= 7.79 P = 0.001 RV NO (NO)
93	TEND MORE TO BE THOSE WHERE THE SYSTEM STYLE IS NON-MOBILIZATIONAL (78)	0.94	93	TEND LESS TO BE THOSE WHERE THE SYSTEM STYLE IS NON-MOBILIZATIONAL (78)	0.62	2 29 30 47 X SQ= 9.70 P = 0. RV NO (NO)
94	ALWAYS ARE THOSE WHERE THE STATUS OF THE REGIME IS CONSTITUTIONAL (51)	1.00	94	TEND TO BE THOSE WHERE THE STATUS OF THE REGIME IS AUTHORITARIAN OR TOTALITARIAN (41)	0.64	30 21 0 37 X SQ= 30.46 P = 0. RV YES (YES)
98	TEND TO BE THOSE WHERE GOVERNMENTAL STABILITY IS GENERALLY PRESENT AND DATES AT LEAST FROM THE INTERWAR PERIOD (22)	0.50	98	TEND TO BE THOSE WHERE GOVERNMENTAL STABILITY IS GENERALLY OR MODERATELY PRESENT AND DATES FROM THE POST-WAR PERIOD, OR IS ABSENT (93)	0.95	16 4 16 75 X SQ= 28.17 P = 0. RV YES (YES)

99	TEND TO BE THOSE WHERE GOVERNMENTAL STABILITY IS GENERALLY PRESENT AND DATES FROM AT LEAST THE INTER-WAR PERIOD, OR FROM THE POST-WAR PERIOD (50)	0.78	99	TEND TO BE THOSE WHERE GOVERNMENTAL STABILITY IS MODERATELY PRESENT AND DATES FROM THE POST-WAR PERIOD, OR IS ABSENT (36)	0.56	25 22 7 28 X SQ= 7.95 P = 0.003 RV YES (YES)
101	TEND TO BE THOSE WHERE THE REPRESENTATIVE CHARACTER OF THE REGIME IS POLYARCHIC (41)	0.90	101	TEND TO BE THOSE WHERE THE REPRESENTATIVE CHARACTER OF THE REGIME IS LIMITED POLYARCHIC, PSEUDO-POLYARCHIC, OR NON-POLYARCHIC (57)	0.78	27 14 3 51 X SQ= 36.48 P = 0. RV YES (YES)
102	ALWAYS ARE THOSE WHERE THE REPRESENTATIVE CHARACTER OF THE REGIME IS POLYARCHIC OR LIMITED POLYARCHIC (59)	1.00	102	TEND TO BE THOSE WHERE THE REPRESENTATIVE CHARACTER OF THE REGIME IS PSEUDO-POLYARCHIC OR NON-POLYARCHIC (49)	0.63	32 27 0 46 X SQ= 33.37 P = 0. RV YES (YES)
105	ALWAYS ARE THOSE WHERE THE ELECTORAL SYSTEM IS COMPETITIVE (43)	1.00	105	TEND TO BE THOSE WHERE THE ELECTORAL SYSTEM IS PARTIALLY COMPETITIVE OR NON-COMPETITIVE (47)	0.76	29 14 0 44 X SQ= 41.53 P = 0. RV YES (YES)
106	ALWAYS ARE THOSE WHERE THE ELECTORAL SYSTEM IS COMPETITIVE OR PARTIALLY COMPETITIVE (52)	1.00	106	TEND TO BE THOSE WHERE THE ELECTORAL SYSTEM IS NON-COMPETITIVE (30)	0.54	29 23 0 27 X SQ= 21.45 P = 0. RV YES (YES)
107	TEND TO BE THOSE WHERE AUTONOMOUS GROUPS ARE FULLY TOLERATED IN POLITICS (46)	0.94	107	TEND TO BE THOSE WHERE AUTONOMOUS GROUPS ARE PARTIALLY TOLERATED IN POLITICS, ARE TOLERATED ONLY OUTSIDE POLITICS, OR ARE NOT TOLERATED AT ALL (65)	0.79	30 16 2 59 X SQ= 45.08 P = 0. RV YES (YES)
108	ALWAYS ARE THOSE WHERE AUTONOMOUS GROUPS ARE FULLY OR PARTIALLY TOLERATED IN POLITICS (65)	1.00	108	TEND LESS TO BE THOSE WHERE AUTONOMOUS GROUPS ARE FULLY OR PARTIALLY TOLERATED IN POLITICS (65)	0.52	31 34 0 32 X SQ= 20.29 P = 0. RV NO (NO)
111	TEND TO BE THOSE WHERE POLITICAL ENCULTURATION IS HIGH OR MEDIUM (53)	0.80	111	TEND TO BE THOSE WHERE POLITICAL ENCULTURATION IS LOW (42)	0.53	24 29 6 33 X SQ= 7.83 P = 0.003 RV YES (NO)
118	TEND TO BE THOSE WHERE INTEREST ARTICULATION BY INSTITUTIONAL GROUPS IS SIGNIFICANT, MODERATE, OR LIMITED (60)	0.91	118	TEND TO BE THOSE WHERE INTEREST ARTICULATION BY INSTITUTIONAL GROUPS IS VERY SIGNIFICANT (40)	0.52	3 33 29 31 X SQ= 14.45 P = 0. RV YES (NO)

119	TEND TO BE THOSE WHERE INTEREST ARTICULATION BY INSTITUTIONAL GROUPS IS MODERATE OR LIMITED (26)	0.69	119	TEND TO BE THOSE WHERE INTEREST ARTICULATION BY INSTITUTIONAL GROUPS IS VERY SIGNIFICANT OR SIGNIFICANT (74)	0.94	10 60 22 4 X SQ= 39.09 P = 0. RV YES (YES)
120	TEND LESS TO BE THOSE WHERE INTEREST ARTICULATION BY INSTITUTIONAL GROUPS IS VERY SIGNIFICANT, SIGNIFICANT, OR MODERATE (90)	0.69	120	ALWAYS ARE THOSE WHERE INTEREST ARTICULATION BY INSTITUTIONAL GROUPS IS VERY SIGNIFICANT, SIGNIFICANT, OR MODERATE (90)	1.00	22 64 10 0 X SQ= 19.10 P = 0. RV NO (YES)
121	TEND TO BE THOSE WHERE INTEREST ARTICULATION BY NON-ASSOCIATIONAL GROUPS IS MODERATE, LIMITED, OR NEGLIGIBLE (61)	0.91	121	TEND TO BE THOSE WHERE INTEREST ARTICULATION BY NON-ASSOCIATIONAL GROUPS IS SIGNIFICANT (54)	0.63	3 50 29 29 X SQ= 24.42 P = 0. RV YES (YES)
122	TEND TO BE THOSE WHERE INTEREST ARTICULATION BY NON-ASSOCIATIONAL GROUPS IS LIMITED OR NEGLIGIBLE (32)	0.66	122	TEND TO BE THOSE WHERE INTEREST ARTICULATION BY NON-ASSOCIATIONAL GROUPS IS SIGNIFICANT OR MODERATE (83)	0.87	11 69 21 10 X SQ= 29.17 P = 0. RV YES (YES)
123	TEND LESS TO BE THOSE WHERE INTEREST ARTICULATION BY NON-ASSOCIATIONAL GROUPS IS SIGNIFICANT, MODERATE, OR LIMITED (107)	0.75	123	ALWAYS ARE THOSE WHERE INTEREST ARTICULATION BY NON-ASSOCIATIONAL GROUPS IS SIGNIFICANT, MODERATE, OR LIMITED (107)	1.00	24 79 8 0 X SQ= 17.71 P = 0. RV NO (YES)
125	TEND TO BE THOSE WHERE INTEREST ARTICULATION BY ANOMIC GROUPS IS INFREQUENT OR VERY INFREQUENT (35)	0.79	125	TEND TO BE THOSE WHERE INTEREST ARTICULATION BY ANOMIC GROUPS IS FREQUENT OR OCCASIONAL (64)	0.82	6 55 22 12 X SQ= 29.04 P = 0. RV YES (YES)
126	TEND TO BE THOSE WHERE INTEREST ARTICULATION BY ANOMIC GROUPS IS VERY INFREQUENT (16)	0.57	126	ALWAYS ARE THOSE WHERE INTEREST ARTICULATION BY ANOMIC GROUPS IS FREQUENT, OCCASIONAL, OR INFREQUENT (83)	1.00	12 67 16 0 X SQ= 42.05 P = 0. RV YES (YES)
128	TEND TO BE THOSE WHERE INTEREST ARTICULATION BY POLITICAL PARTIES IS SIGNIFICANT OR MODERATE (48)	0.83	128	TEND TO BE THOSE WHERE INTEREST ARTICULATION BY POLITICAL PARTIES IS LIMITED OR NEGLIGIBLE (45)	0.62	25 23 5 37 X SQ= 14.51 P = 0. RV YES (YES)
129	TEND TO BE THOSE WHERE INTEREST ARTICULATION BY POLITICAL PARTIES IS SIGNIFICANT, MODERATE, OR LIMITED (56)	0.87	129	TEND TO BE THOSE WHERE INTEREST ARTICULATION BY POLITICAL PARTIES IS NEGLIGIBLE (37)	0.50	26 30 4 30 X SQ= 9.93 P = 0.001 RV YES (NO)

134	TEND TO BE THOSE WHERE INTEREST AGGREGATION BY THE EXECUTIVE IS SIGNIFICANT OR MODERATE (57)	0.86	134	TEND TO BE THOSE WHERE INTEREST AGGREGATION BY THE EXECUTIVE IS LIMITED OR NEGLIGIBLE (46)	0.54	25 32 4 38 X SQ= 12.16 P = 0. RV YES (NO)
135	ALWAYS ARE THOSE WHERE INTEREST AGGREGATION BY THE EXECUTIVE IS SIGNIFICANT, MODERATE, OR LIMITED (77)	1.00	135	TEND LESS TO BE THOSE WHERE INTEREST AGGREGATION BY THE EXECUTIVE IS SIGNIFICANT, MODERATE, OR LIMITED (77)	0.79	29 46 0 12 X SQ= 5.33 P = 0.007 RV NO (NO)
136	TEND LESS TO BE THOSE WHERE INTEREST AGGREGATION BY THE LEGISLATURE IS MODERATE, LIMITED, OR NEGLIGIBLE (85)	0.61	136	ALWAYS ARE THOSE WHERE INTEREST AGGREGATION BY THE LEGISLATURE IS MODERATE, LIMITED, OR NEGLIGIBLE (85)	1.00	12 0 19 63 X SQ= 24.59 P = 0. RV NO (YES)
137	TEND TO BE THOSE WHERE INTEREST AGGREGATION BY THE LEGISLATURE IS SIGNIFICANT OR MODERATE (29)	0.84	137	TEND TO BE THOSE WHERE INTEREST AGGREGATION BY THE LEGISLATURE IS LIMITED OR NEGLIGIBLE (68)	0.95	26 3 5 60 X SQ= 57.30 P = 0. RV YES (YES)
138	TEND TO BE THOSE WHERE INTEREST AGGREGATION BY THE LEGISLATURE IS SIGNIFICANT, MODERATE, OR LIMITED (48)	0.97	138	TEND TO BE THOSE WHERE INTEREST AGGREGATION BY THE LEGISLATURE IS NEGLIGIBLE (49)	0.72	30 18 1 47 X SQ= 37.35 P = 0. RV YES (YES)
139	ALWAYS ARE THOSE WHERE THE PARTY SYSTEM IS QUANTITATIVELY OTHER THAN ONE-PARTY (71)	1.00	139	TEND LESS TO BE THOSE WHERE THE PARTY SYSTEM IS QUANTITATIVELY OTHER THAN ONE-PARTY (71)	0.56	0 31 32 39 X SQ= 18.32 P = 0. RV NO (NO)
141	TEND LESS TO BE THOSE WHERE THE PARTY SYSTEM IS QUANTITATIVELY OTHER THAN TWO-PARTY (87)	0.76	141	TEND MORE TO BE THOSE WHERE THE PARTY SYSTEM IS QUANTITATIVELY OTHER THAN TWO-PARTY (87)	0.94	7 4 22 62 X SQ= 4.79 P = 0.031 RV NO (YES)
142	TEND TO BE THOSE WHERE THE PARTY SYSTEM IS QUANTITATIVELY MULTI-PARTY (30)	0.59	142	TEND TO BE THOSE WHERE THE PARTY SYSTEM IS QUANTITATIVELY OTHER THAN MULTI-PARTY (66)	0.80	17 13 12 51 X SQ= 11.71 P = 0.001 RV YES (YES)
144	ALWAYS ARE THOSE WHERE THE PARTY SYSTEM IS QUANTITATIVELY TWO-PARTY, RATHER THAN ONE-PARTY (11)	1.00	144	TEND TO BE THOSE WHERE THE PARTY SYSTEM IS QUANTITATIVELY ONE-PARTY, RATHER THAN TWO-PARTY (34)	0.89	0 31 7 4 X SQ= 19.31 P = 0. RV YES (YES)

146	ALWAYS ARE THOSE WHERE THE PARTY SYSTEM IS QUALITATIVELY OTHER THAN MASS-BASED TERRITORIAL (92)	1.00	146	LEAN LESS TOWARD BEING THOSE WHERE THE PARTY SYSTEM IS QUALITATIVELY OTHER THAN MASS-BASED TERRITORIAL (92)	0.88	0 8 30 58 X SQ= 2.54 P = 0.054 RV NO (NO)
147	TEND TO BE THOSE WHERE THE PARTY SYSTEM IS QUALITATIVELY CLASS-ORIENTED OR MULTI-IDEOLOGICAL (23)	0.77	147	TEND TO BE THOSE WHERE THE PARTY SYSTEM IS QUALITATIVELY OTHER THAN CLASS-ORIENTED OR MULTI-IDEOLOGICAL (67)	0.95	20 3 6 58 X SQ= 44.97 P = 0. RV YES (YES)
148	ALWAYS ARE THOSE WHERE THE PARTY SYSTEM IS QUALITATIVELY OTHER THAN AFRICAN TRANSITIONAL (96)	1.00	148	TEND LESS TO BE THOSE WHERE THE PARTY SYSTEM IS QUALITATIVELY OTHER THAN AFRICAN TRANSITIONAL (96)	0.81	0 14 32 60 X SQ= 5.42 P = 0.005 RV NO (NO)
153	TEND TO BE THOSE WHERE THE PARTY SYSTEM IS STABLE (42)	0.71	153	TEND TO BE THOSE WHERE THE PARTY SYSTEM IS MODERATELY STABLE OR UNSTABLE (38)	0.62	22 17 9 28 X SQ= 6.82 P = 0.005 RV YES (YES)
158	TEND TO BE THOSE WHERE PERSONALISMO IS NEGLIGIBLE (56)	0.75	158	TEND TO BE THOSE WHERE PERSONALISMO IS PRONOUNCED OR MODERATE (40)	0.52	8 32 24 29 X SQ= 5.38 P = 0.015 RV YES (NO)
161	TEND TO BE THOSE WHERE THE POLITICAL LEADERSHIP IS NON-ELITIST (50)	0.77	161	TEND TO BE THOSE WHERE THE POLITICAL LEADERSHIP IS ELITIST OR MODERATELY ELITIST (47)	0.59	7 37 24 26 X SQ= 9.50 P = 0.001 RV YES (NO)
164	TEND TO BE THOSE WHERE THE REGIME'S LEADERSHIP CHARISMA IS NEGLIGIBLE (65)	0.97	164	TEND TO BE THOSE WHERE THE REGIME'S LEADERSHIP CHARISMA IS PRONOUNCED OR MODERATE (34)	0.51	1 32 31 31 X SQ= 19.22 P = 0. RV YES (YES)
166	TEND LESS TO BE THOSE WHERE THE VERTICAL POWER DISTRIBUTION IS THAT OF FORMAL FEDERALISM OR FORMAL AND EFFECTIVE UNITARISM (99)	0.72	166	TEND MORE TO BE THOSE WHERE THE VERTICAL POWER DISTRIBUTION IS THAT OF FORMAL FEDERALISM OR FORMAL AND EFFECTIVE UNITARISM (99)	0.92	9 6 23 72 X SQ= 6.40 P = 0.011 RV NO (YES)
168	TEND TO BE THOSE WHERE THE HORIZONTAL POWER DISTRIBUTION IS SIGNIFICANT (34)	0.87	168	TEND TO BE THOSE WHERE THE HORIZONTAL POWER DISTRIBUTION IS LIMITED OR NEGLIGIBLE (72)	0.90	27 7 4 65 X SQ= 55.22 P = 0. RV YES (YES)

No.	Left		No.	Right		Statistics
169	ALWAYS ARE THOSE WHERE THE HORIZONTAL POWER DISTRIBUTION IS SIGNIFICANT OR LIMITED (58)	1.00	169	TEND TO BE THOSE WHERE THE HORIZONTAL POWER DISTRIBUTION IS NEGLIGIBLE (48)	0.62	31 27 0 45 X SQ= 31.91 P = 0. RV YES (YES)
170	TEND MORE TO BE THOSE WHERE THE LEGISLATIVE-EXECUTIVE STRUCTURE IS OTHER THAN PRESIDENTIAL (63)	0.78	170	TEND LESS TO BE THOSE WHERE THE LEGISLATIVE-EXECUTIVE STRUCTURE IS OTHER THAN PRESIDENTIAL (63)	0.52	7 32 25 35 X SQ= 5.04 P = 0.016 RV NO (NO)
171	TEND LESS TO BE THOSE WHERE THE LEGISLATIVE-EXECUTIVE STRUCTURE IS OTHER THAN PARLIAMENTARY-REPUBLICAN (90)	0.72	171	TEND MORE TO BE THOSE WHERE THE LEGISLATIVE-EXECUTIVE STRUCTURE IS OTHER THAN PARLIAMENTARY-REPUBLICAN (90)	0.96	9 3 23 66 X SQ= 9.64 P = 0.001 RV NO (YES)
172	TEND LESS TO BE THOSE WHERE THE LEGISLATIVE-EXECUTIVE STRUCTURE IS OTHER THAN PARLIAMENTARY-ROYALIST (88)	0.62	172	TEND MORE TO BE THOSE WHERE THE LEGISLATIVE-EXECUTIVE STRUCTURE IS OTHER THAN PARLIAMENTARY-ROYALIST (88)	0.88	12 9 20 65 X SQ= 7.50 P = 0.006 RV NO (YES)
174	TEND TO BE THOSE WHERE THE LEGISLATURE IS FULLY EFFECTIVE (28)	0.72	174	TEND TO BE THOSE WHERE THE LEGISLATURE IS PARTIALLY EFFECTIVE, LARGELY INEFFECTIVE, OR WHOLLY INEFFECTIVE (72)	0.92	23 5 9 60 X SQ= 39.95 P = 0. RV YES (YES)
175	ALWAYS ARE THOSE WHERE THE LEGISLATURE IS FULLY EFFECTIVE OR PARTIALLY EFFECTIVE (51)	1.00	175	TEND TO BE THOSE WHERE THE LEGISLATURE IS LARGELY INEFFECTIVE OR WHOLLY INEFFECTIVE (49)	0.71	32 19 0 46 X SQ= 40.28 P = 0. RV YES (YES)
178	TEND TO BE THOSE WHERE THE LEGISLATURE IS BICAMERAL (51)	0.81	178	TEND TO BE THOSE WHERE THE LEGISLATURE IS UNICAMERAL (53)	0.67	6 46 26 23 X SQ= 18.22 P = 0. RV YES (YES)
179	TEND TO BE THOSE WHERE THE EXECUTIVE IS STRONG (39)	0.89	179	TEND TO BE THOSE WHERE THE EXECUTIVE IS DOMINANT (52)	0.75	3 46 24 15 X SQ= 28.80 P = 0. RV YES (YES)
181	TEND TO BE THOSE WHERE THE BUREAUCRACY IS MODERN, RATHER THAN SEMI-MODERN (21)	0.66	181	ALWAYS ARE THOSE WHERE THE BUREAUCRACY IS SEMI-MODERN, RATHER THAN MODERN (55)	1.00	21 0 11 40 X SQ= 33.95 P = 0. RV YES (YES)

```
182  ALWAYS ARE THOSE                          1.00    182  TEND LESS TO BE THOSE                   0.62        11    40
     WHERE THE BUREAUCRACY                                  WHERE THE BUREAUCRACY                                  0    25
     IS SEMI-MODERN, RATHER THAN                            IS SEMI-MODERN, RATHER THAN                    X SQ=     4.68
     POST-COLONIAL TRANSITIONAL  (55)                       POST-COLONIAL TRANSITIONAL  (55)               P =      0.013
                                                                                                           RV NO   (NO )

186  TEND MORE TO BE THOSE                     0.93    186  TEND LESS TO BE THOSE                   0.54         2    26
     WHERE PARTICIPATION BY THE MILITARY                    WHERE PARTICIPATION BY THE MILITARY                   25    31
     IN POLITICS IS                                         IN POLITICS IS                                 X SQ=    10.38
     NEUTRAL, RATHER THAN                                   NEUTRAL, RATHER THAN                           P =       0.
     SUPPORTIVE  (56)                                       SUPPORTIVE  (56)                               RV NO   (NO )

187  TEND TO BE THOSE                          0.83    187  TEND TO BE THOSE                        0.85         5    57
     WHERE THE ROLE OF THE POLICE                           WHERE THE ROLE OF THE POLICE                          25    10
     IS NOT POLITICALLY SIGNIFICANT  (35)                   IS POLITICALLY SIGNIFICANT  (66)               X SQ=    39.13
                                                                                                           P =       0.
                                                                                                           RV YES  (YES)

188  TEND TO BE THOSE                          0.50    188  TEND TO BE THOSE                        0.82        16    14
     WHERE THE CHARACTER OF THE LEGAL SYSTEM                WHERE THE CHARACTER OF THE LEGAL SYSTEM               16    63
     IS CIVIL LAW  (32)                                     IS OTHER THAN CIVIL LAW  (81)                  X SQ=     9.93
                                                                                                           P =      0.002
                                                                                                           RV YES  (YES)

189  TEND LESS TO BE THOSE                     0.84    189  TEND MORE TO BE THOSE                   0.97         5     2
     WHERE THE CHARACTER OF THE LEGAL SYSTEM                WHERE THE CHARACTER OF THE LEGAL SYSTEM               27    77
     IS OTHER THAN COMMON LAW  (108)                        IS OTHER THAN COMMON LAW  (108)                X SQ=     4.58
                                                                                                           P =      0.020
                                                                                                           RV NO   (YES)

192  ALWAYS ARE THOSE                          1.00    192  TEND LESS TO BE THOSE                   0.65         0    27
     WHERE THE CHARACTER OF THE LEGAL SYSTEM                WHERE THE CHARACTER OF THE LEGAL SYSTEM               32    50
     IS OTHER THAN                                          IS OTHER THAN                                  X SQ=    13.09
     MUSLIM OR PARTLY MUSLIM  (86)                          MUSLIM OR PARTLY MUSLIM  (86)                  P =       0.
                                                                                                           RV NO   (NO )

193  TEND MORE TO BE THOSE                     0.97    193  TEND LESS TO BE THOSE                   0.65         1    27
     WHERE THE CHARACTER OF THE LEGAL SYSTEM                WHERE THE CHARACTER OF THE LEGAL SYSTEM               31    51
     IS OTHER THAN PARTLY INDIGENOUS  (86)                  IS OTHER THAN PARTLY INDIGENOUS  (86)          X SQ=    10.26
                                                                                                           P =       0.
                                                                                                           RV NO   (NO )

194  ALWAYS ARE THOSE                          1.00    194  TEND LESS TO BE THOSE,                  0.85         0    12
     THAT ARE NON-COMMUNIST  (101)                          THAT ARE NON-COMMUNIST  (101)                         32    66
                                                                                                           X SQ=     4.06
                                                                                                           P =      0.017
                                                                                                           RV NO   (NO )
```

117 POLITIES
WHERE INTEREST ARTICULATION
BY ASSOCIATIONAL GROUPS
IS SIGNIFICANT, MODERATE, OR
LIMITED (60)

117 POLITIES
WHERE INTEREST ARTICULATION
BY ASSOCIATIONAL GROUPS
IS NEGLIGIBLE (51)

PREDICATE ONLY

60 IN LEFT COLUMN

ALGERIA	ARGENTINA	AUSTRALIA	AUSTRIA	BELGIUM	BOLIVIA	BRAZIL	BURMA	CANADA	CEYLON
CHILE	COLOMBIA	COSTA RICA	CYPRUS	DENMARK	DOMIN REP	ECUADOR	EL SALVADOR	FINLAND	FRANCE
GERMAN FR	GREECE	GUATEMALA	HONDURAS	HUNGARY	ICELAND	INDIA	INDONESIA	IRAN	IRELAND
ISRAEL	ITALY	JAMAICA	JAPAN	LEBANON	LUXEMBOURG	MALAYA	MEXICO	MOROCCO	NETHERLANDS
NEW ZEALAND	NICARAGUA	NORWAY	PANAMA	PERU	PHILIPPINES	POLAND	SO AFRICA	SWEDEN	SWITZERLAND
SYRIA	TRINIDAD	TUNISIA	TURKEY	USSR	UAR	UK	US	URUGUAY	VENEZUELA

51 IN RIGHT COLUMN

AFGHANISTAN	ALBANIA	BULGARIA	BURUNDI	CAMBODIA	CAMEROUN	CEN AFR REP	CHAD	CHINA, PR	CONGO(BRA)
CONGO(LEO)	CUBA	CZECHOS'KIA	DAHOMEY	ETHIOPIA	GABON	GERMANY, E	GHANA	GUINEA	HAITI
IRAQ	IVORY COAST	JORDAN	KOREA, N	LAOS	LIBERIA	LIBYA	MALAGASY R	MALI	MAURITANIA
MONGOLIA	NEPAL	NIGER	NIGERIA	PAKISTAN	PARAGUAY	RUMANIA	RWANDA	SA'U ARABIA	SENEGAL
SIERRE LEO	SOMALIA	SUDAN	TANGANYIKA	THAILAND	TOGO	UGANDA	UPPER VOLTA	VIETNAM, N	VIETNAM REP
YEMEN									

3 EXCLUDED BECAUSE AMBIGUOUS

PORTUGAL SPAIN YUGOSLAVIA

1 EXCLUDED BECAUSE UNASCERTAINABLE

KOREA REP

118 POLITIES WHERE INTEREST ARTICULATION BY INSTITUTIONAL GROUPS IS VERY SIGNIFICANT (40)	118 POLITIES WHERE INTEREST ARTICULATION BY INSTITUTIONAL GROUPS IS SIGNIFICANT, MODERATE, OR LIMITED (60)

PREDICATE ONLY

40 IN LEFT COLUMN

AFGHANISTAN	ALBANIA	ALGERIA	ARGENTINA	BRAZIL	BULGARIA	BURMA	CHINA, PR	CUBA	CYPRUS
CZECHOS'KIA	ETHIOPIA	GERMANY, E	GHANA	GUINEA	HUNGARY	INDONESIA	IRAN	IRAQ	JORDAN
KOREA, N	KOREA REP	LAOS	LEBANON	LIBERIA	MONGOLIA	PAKISTAN	PARAGUAY	POLAND	PORTUGAL
RUMANIA	SPAIN	SUDAN	SYRIA	THAILAND	USSR	UAR	VIETNAM, N	VIETNAM REP	YUGOSLAVIA

60 IN RIGHT COLUMN

AUSTRALIA	AUSTRIA	BELGIUM	BOLIVIA	CAMBODIA	CANADA	CEYLON	CHILE	COLOMBIA	CONGO(LEO)
COSTA RICA	DENMARK	DOMIN REP	ECUADOR	EL SALVADOR	FINLAND	FRANCE	GERMAN FR	GREECE	GUATEMALA
HAITI	HONDURAS	ICELAND	INDIA	IRELAND	ISRAEL	ITALY	IVORY COAST	JAMAICA	JAPAN
LIBYA	LUXEMBOURG	MALAGASY R	MALAYA	MAURITANIA	MEXICO	MOROCCO	NEPAL	NETHERLANDS	NEW ZEALAND
NICARAGUA	NIGERIA	NORWAY	PANAMA	PERU	PHILIPPINES	SA'U ARABIA	SENEGAL	SO AFRICA	SWEDEN
SWITZERLAND	TANGANYIKA	TRINIDAD	TUNISIA	TURKEY	UGANDA	UK	US	URUGUAY	VENEZUELA

15 EXCLUDED BECAUSE UNASCERTAINABLE

BURUNDI	CAMEROUN	CEN AFR REP	CHAD	CONGO(BRA)	DAHOMEY	GABON	MALI	NIGER	RWANDA
SIERRE LEO	SOMALIA	TOGO	UPPER VOLTA	YEMEN					

119 POLITIES
WHERE INTEREST ARTICULATION
BY INSTITUTIONAL GROUPS
IS VERY SIGNIFICANT OR SIGNIFICANT (74)

119 POLITIES
WHERE INTEREST ARTICULATION
BY INSTITUTIONAL GROUPS
IS MODERATE OR LIMITED (26)

BOTH SUBJECT AND PREDICATE

74 IN LEFT COLUMN

AFGHANISTAN	ALBANIA	ALGERIA	ARGENTINA	BOLIVIA	BRAZIL	BULGARIA	BURMA	CAMBODIA	CEYLON
CHINA, PR	COLOMBIA	CONGO(LEO)	CUBA	CYPRUS	CZECHOS'KIA	DOMIN REP	ECUADOR	EL SALVADOR	ETHIOPIA
GERMANY, E	GHANA	GUATEMALA	GUINEA	HAITI	HONDURAS	HUNGARY	INDIA	INDONESIA	IRAN
IRAQ	ISRAEL	ITALY	IVORY COAST	JAPAN	JORDAN	KOREA, N	KOREA REP	LAOS	LEBANON
LIBERIA	LIBYA	MALAYA	MAURITANIA	MEXICO	MONGOLIA	MOROCCO	NEPAL	NICARAGUA	NIGERIA
PAKISTAN	PANAMA	PARAGUAY	PERU	POLAND	PORTUGAL	RUMANIA	SA'U ARABIA	SENEGAL	SO AFRICA
SPAIN	SUDAN	SYRIA	TANGANYIKA	THAILAND	TUNISIA	TURKEY	UGANDA	USSR	UAR
VENEZUELA	VIETNAM, N	VIETNAM REP	YUGOSLAVIA						

26 IN RIGHT COLUMN

AUSTRALIA	AUSTRIA	BELGIUM	CANADA	CHILE	COSTA RICA	DENMARK	FINLAND	FRANCE	GERMAN FR
GREECE	ICELAND	IRELAND	JAMAICA	LUXEMBOURG	MALAGASY R	NETHERLANDS	NEW ZEALAND	NORWAY	PHILIPPINES
SWEDEN	SWITZERLAND	TRINIDAD	UK	US	URUGUAY				

15 EXCLUDED BECAUSE UNASCERTAINABLE

BURUNDI	CAMEROUN	CEN AFR REP	CHAD	CONGO(BRA)	DAHOMEY	GABON	MALI	NIGER	RWANDA
SIERRE LEO	SOMALIA	TOGO	UPPER VOLTA	YEMEN					

2	TEND TO BE THOSE LOCATED ELSEWHERE THAN IN WEST EUROPE, SCANDINAVIA, NORTH AMERICA, OR AUSTRALASIA (93)	0.96	2 TEND TO BE THOSE LOCATED IN WEST EUROPE, SCANDINAVIA, NORTH AMERICA, OR AUSTRALASIA (22)	0.73	3 19 71 7 X SQ= 49.47 P = 0. RV YES (YES)
3	TEND MORE TO BE THOSE LOCATED ELSEWHERE THAN IN WEST EUROPE (102)	0.96	3 TEND LESS TO BE THOSE LOCATED ELSEWHERE THAN IN WEST EUROPE (102)	0.62	3 10 71 16 X SQ= 17.21 P = 0. RV NO (YES)
5	TEND TO BE THOSE LOCATED IN EAST EUROPE, RATHER THAN IN SCANDINAVIA OR WEST EUROPE (9)	0.75	5 ALWAYS ARE THOSE LOCATED IN SCANDINAVIA OR WEST EUROPE, RATHER THAN IN EAST EUROPE (18)	1.00	3 15 9 0 X SQ= 13.67 P = 0. RV YES (YES)

8	TEND LESS TO BE THOSE LOCATED ELSEWHERE THAN IN EAST ASIA SOUTH ASIA, OR SOUTHEAST ASIA (97)	0.77	8	TEND MORE TO BE THOSE LOCATED ELSEWHERE THAN IN EAST ASIA SOUTH ASIA, OR SOUTHEAST ASIA (97)	0.96	17 1 57 25 X SQ= 3.56 P = 0.036 RV NO (NO)
10	TEND LESS TO BE THOSE LOCATED ELSEWHERE THAN IN NORTH AFRICA, OR CENTRAL AND SOUTH AFRICA (82)	0.76	10	TEND MORE TO BE THOSE LOCATED ELSEWHERE THAN IN NORTH AFRICA, OR CENTRAL AND SOUTH AFRICA (82)	0.96	18 1 56 25 X SQ= 4.00 P = 0.021 RV NO (NO)
11	TILT LESS TOWARD BEING THOSE LOCATED ELSEWHERE THAN IN CENTRAL AND SOUTH AFRICA (87)	0.82	11	TILT MORE TOWARD BEING THOSE LOCATED ELSEWHERE THAN IN CENTRAL AND SOUTH AFRICA (87)	0.96	13 1 61 25 X SQ= 1.98 P = 0.107 RV NO (NO)
13	TEND LESS TO BE THOSE LOCATED ELSEWHERE THAN IN NORTH AFRICA OR THE MIDDLE EAST (99)	0.80	13	ALWAYS ARE THOSE LOCATED ELSEWHERE THAN IN NORTH AFRICA OR THE MIDDLE EAST (99)	1.00	15 0 59 26 X SQ= 4.71 P = 0.010 RV NO (NO)
14	LEAN LESS TOWARD BEING THOSE LOCATED ELSEWHERE THAN IN THE MIDDLE EAST (104)	0.86	14	ALWAYS ARE THOSE LOCATED ELSEWHERE THAN IN THE MIDDLE EAST (104)	1.00	10 0 64 26 X SQ= 2.55 P = 0.060 RV NO (NO)
21	TEND LESS TO BE THOSE WHOSE TERRITORIAL SIZE IS MEDIUM OR SMALL (83)	0.65	21	TEND MORE TO BE THOSE WHOSE TERRITORIAL SIZE IS MEDIUM OR SMALL (83)	0.88	26 3 48 23 X SQ= 4.12 P = 0.025 RV NO (NO)
28	TEND TO BE THOSE WHOSE POPULATION GROWTH RATE IS HIGH (62)	0.66	28	TEND TO BE THOSE WHOSE POPULATION GROWTH RATE IS LOW (48)	0.62	46 10 24 16 X SQ= 4.73 P = 0.021 RV YES (NO)
29	TEND TO BE THOSE WHERE THE DEGREE OF URBANIZATION IS LOW (49)	0.50	29	TEND TO BE THOSE WHERE THE DEGREE OF URBANIZATION IS HIGH (56)	0.96	33 23 33 1 X SQ= 13.84 P = 0. RV YES (NO)
30	TEND TO BE THOSE WHOSE AGRICULTURAL POPULATION IS HIGH (56)	0.56	30	TEND TO BE THOSE WHOSE AGRICULTURAL POPULATION IS MEDIUM, LOW, OR VERY LOW (57)	0.96	40 1 32 25 X SQ= 18.92 P = 0. RV YES (NO)

31	TEND TO BE THOSE WHOSE AGRICULTURAL POPULATION IS HIGH OR MEDIUM (90)	0.93	31	TEND TO BE THOSE WHOSE AGRICULTURAL POPULATION IS LOW OR VERY LOW (24)	0.73	68 7 5 19 X SQ= 42.25 P = 0. RV YES (YES)
32	TILT MORE TOWARD BEING THOSE WHOSE GROSS NATIONAL PRODUCT IS MEDIUM, LOW, OR VERY LOW (105)	0.93	32	TILT LESS TOWARD BEING THOSE WHOSE GROSS NATIONAL PRODUCT IS MEDIUM, LOW, OR VERY LOW (105)	0.81	5 5 69 21 X SQ= 2.08 P = 0.121 RV NO (YES)
34	TEND LESS TO BE THOSE WHOSE GROSS NATIONAL PRODUCT IS VERY HIGH, HIGH, MEDIUM, OR LOW (62)	0.55	34	TEND MORE TO BE THOSE WHOSE GROSS NATIONAL PRODUCT IS VERY HIGH, HIGH, MEDIUM, OR LOW (62)	C.81	41 21 33 5 X SQ= 4.23 P = 0.033 RV NO (NO)
35	TEND TO BE THOSE WHOSE PER CAPITA GROSS NATIONAL PRODUCT IS MEDIUM, LOW, OR VERY LOW (91)	0.92	35	TEND TO BE THOSE WHOSE PER CAPITA GROSS NATIONAL PRODUCT IS VERY HIGH OR HIGH (24)	0.69	6 18 68 8 X SQ= 36.13 P = 0. RV YES (YES)
36	TEND TO BE THOSE WHOSE PER CAPITA GROSS NATIONAL PRODUCT IS LOW OR VERY LOW (73)	0.76	36	TEND TO BE THOSE WHOSE PER CAPITA GROSS NATIONAL PRODUCT IS VERY HIGH, HIGH, OR MEDIUM (42)	0.92	18 24 56 2 X SQ= 33.77 P = 0. RV YES (YES)
39	TEND TO BE THOSE WHOSE INTERNATIONAL FINANCIAL STATUS IS LOW OR VERY LOW (76)	0.71	39	TEND TO BE THOSE WHOSE INTERNATIONAL FINANCIAL STATUS IS VERY HIGH, HIGH, OR MEDIUM (38)	0.65	21 17 52 9 X SQ= 9.38 P = 0.002 RV YES (NO)
41	TEND TO BE THOSE WHOSE ECONOMIC DEVELOPMENTAL STATUS IS INTERMEDIATE, UNDERDEVELOPED, OR VERY UNDERDEVELOPED (94)	0.94	41	TEND TO BE THOSE WHOSE ECONOMIC DEVELOPMENTAL STATUS IS CEVELOPED (19)	0.58	4 15 68 11 X SQ= 29.97 P = 0. RV YES (YES)
42	TEND TO BE THOSE WHOSE ECONOMIC DEVELOPMENTAL STATUS IS UNDERDEVELOPED OR VERY UNDERDEVELOPED (76)	0.77	42	TEND TO BE THOSE WHOSE ECONOMIC DEVELOPMENTAL STATUS IS DEVELOPED OR INTERMEDIATE (36)	0.77	16 20 55 6 X SQ= 21.85 P = 0. RV YES (YES)
43	TEND TO BE THOSE WHOSE ECONOMIC DEVELOPMENTAL STATUS IS VERY UNDERDEVELOPED (57)	0.58	43	TEND TO BE THOSE WHOSE ECONOMIC DEVELOPMENTAL STATUS IS DEVELOPED, INTERMEDIATE, OR UNDERDEVELOPED (55)	0.96	30 25 41 1 X SQ= 20.38 P = 0. RV YES (NO)

44	TEND TO BE THOSE WHERE THE LITERACY RATE IS BELOW NINETY PERCENT (90)	0.91	44	TEND TO BE THOSE WHERE THE LITERACY RATE IS NINETY PERCENT OR ABOVE (25)	0.69	7 18 67 8 X SQ= 33.54 P = 0. RV YES (YES)
45	TEND TO BE THOSE WHERE THE LITERACY RATE IS BELOW FIFTY PERCENT (54)	0.56	45	TEND TO BE THOSE WHERE THE LITERACY RATE IS FIFTY PERCENT OR ABOVE (55)	0.96	30 25 38 1 X SQ= 18.89 P = 0. RV YES (NO)
46	TEND LESS TO BE THOSE WHERE THE LITERACY RATE IS TEN PERCENT OR ABOVE (84)	0.84	46	ALWAYS ARE THOSE WHERE THE LITERACY RATE IS TEN PERCENT OR ABOVE (84)	1.00	58 26 11 0 X SQ= 3.26 P = 0.032 RV NO (NO)
50	TEND TO BE THOSE WHERE FREEDOM OF THE PRESS IS INTERMITTENT, INTERNALLY ABSENT, OR INTERNALLY AND EXTERNALLY ABSENT (56)	0.79	50	TEND TO BE THOSE WHERE FREEDOM OF THE PRESS IS COMPLETE (43)	0.92	14 23 53 2 X SQ= 35.39 P = 0. RV YES (YES)
51	TEND TO BE THOSE WHERE FREEDOM OF THE PRESS IS INTERNALLY ABSENT OR INTERNALLY AND EXTERNALLY ABSENT (37)	0.55	51	TEND TO BE THOSE WHERE FREEDOM OF THE PRESS IS COMPLETE OR INTERMITTENT (60)	0.96	30 24 36 1 X SQ= 17.16 P = 0. RV YES (NO)
54	TEND TO BE THOSE WHERE NEWSPAPER CIRCULATION IS LESS THAN ONE HUNDRED PER THOUSAND (76)	0.79	54	TEND TO BE THOSE WHERE NEWSPAPER CIRCULATION IS ONE HUNDRED OR MORE PER THOUSAND (37)	0.85	15 22 57 4 X SQ= 30.41 P = 0. RV YES (YES)
56	TEND LESS TO BE THOSE WHERE THE RELIGION IS PREDOMINANTLY LITERATE (79)	0.75	56	TEND MORE TO BE THOSE WHERE THE RELIGION IS PREDOMINANTLY LITERATE (79)	0.96	52 25 17 1 X SQ= 4.05 P = 0.020 RV NO (NO)
58	TEND LESS TO BE THOSE WHERE THE RELIGION IS OTHER THAN MUSLIM (97)	0.78	58	ALWAYS ARE THOSE WHERE THE RELIGION IS OTHER THAN MUSLIM (97)	1.00	16 0 58 26 X SQ= 5.18 P = 0.006 RV NO (NO)
62	ALWAYS ARE THOSE WHERE THE RELIGION IS CATHOLIC, RATHER THAN PROTESTANT (25)	1.00	62	TEND LESS TO BE THOSE WHERE THE RELIGION IS CATHOLIC, RATHER THAN PROTESTANT (25)	0.64	0 5 16 9 X SQ= 4.53 P = 0.014 RV NO (YES)

63	TEND TO BE THOSE WHERE THE RELIGION IS PREDOMINANTLY OR PARTLY OTHER THAN CHRISTIAN (68)	0.70	63	TEND TO BE THOSE WHERE THE RELIGION IS PREDOMINANTLY SOME KIND OF CHRISTIAN (46)	0.92	22 24 51 2 X SQ= 27.34 P = 0. RV YES (YES)
68	TEND TO BE THOSE THAT ARE LINGUISTICALLY WEAKLY HETEROGENEOUS OR STRONGLY HETEROGENEOUS (62)	0.62	68	TEND TO BE THOSE THAT ARE LINGUISTICALLY HOMOGENEOUS (52)	0.77	28 20 45 6 X SQ= 9.93 P = 0.001 RV YES (NO)
70	TEND MORE TO BE THOSE THAT ARE RELIGIOUSLY, RACIALLY, OR LINGUISTICALLY HETEROGENEOUS (85)	0.86	70	TEND LESS TO BE THOSE THAT ARE RELIGIOUSLY, RACIALLY, OR LINGUISTICALLY HETEROGENEOUS (85)	0.58	10 10 59 14 X SQ= 6.26 P = 0.009 RV NO (YES)
72	LEAN TOWARD BEING THOSE WHOSE DATE OF INDEPENDENCE IS AFTER 1914 (59)	0.52	72	LEAN TOWARD BEING THOSE WHOSE DATE OF INDEPENDENCE IS BEFORE 1914 (52)	0.72	34 18 37 7 X SQ= 3.41 P = 0.061 RV YES (NO)
74	TEND TO BE THOSE THAT ARE NOT HISTORICALLY WESTERN (87)	0.90	74	TEND TO BE THOSE THAT ARE HISTORICALLY WESTERN (26)	0.73	7 19 65 7 X SQ= 36.15 P = 0. RV YES (YES)
75	TEND LESS TO BE THOSE THAT ARE HISTORICALLY WESTERN OR SIGNIFICANTLY WESTERNIZED (62)	0.51	75	TEND MORE TO BE THOSE THAT ARE HISTORICALLY WESTERN OR SIGNIFICANTLY WESTERNIZED (62)	0.96	37 25 36 1 X SQ= 15.05 P = 0. RV NO (NO)
76	TEND TO BE THOSE THAT HAVE BEEN SIGNIFICANTLY OR PARTIALLY WESTERNIZED THROUGH A COLONIAL RELATIONSHIP, RATHER THAN BEING HISTORICALLY WESTERN (70)	0.88	76	TEND TO BE THOSE THAT ARE HISTORICALLY WESTERN, RATHER THAN HAVING BEEN SIGNIFICANTLY OR PARTIALLY WESTERNIZED THROUGH A COLONIAL RELATIONSHIP (26)	0.73	7 19 49 7 X SQ= 27.36 P = 0. RV YES (YES)
77	TILT TOWARD BEING THOSE THAT HAVE BEEN PARTIALLY WESTERNIZED, RATHER THAN SIGNIFICANTLY WESTERNIZED, THROUGH A COLONIAL RELATIONSHIP (41)	0.54	77	TILT TOWARD BEING THOSE THAT HAVE BEEN SIGNIFICANTLY WESTERNIZED, RATHER THAN PARTIALLY WESTERNIZED, THROUGH A COLONIAL RELATIONSHIP (28)	0.86	22 6 26 1 X SQ= 2.46 P = 0.101 RV YES (NO)
81	TEND MORE TO BE THOSE WHERE THE TYPE OF POLITICAL MODERNIZATION IS LATER EUROPEAN OR LATER EUROPEAN DERIVED, RATHER THAN EARLY EUROPEAN OR EARLY EUROPEAN DERIVED (40)	0.97	81	TEND LESS TO BE THOSE WHERE THE TYPE OF POLITICAL MODERNIZATION IS LATER EUROPEAN OR LATER EUROPEAN DERIVED, RATHER THAN EARLY EUROPEAN OR EARLY EUROPEAN DERIVED (40)	0.55	1 10 28 12 X SQ= 10.68 P = 0. RV NO (YES)

82	TEND TO BE THOSE WHERE THE TYPE OF POLITICAL MODERNIZATION IS DEVELOPED TUTELARY OR UNDEVELOPED TUTELARY, RATHER THAN EARLY OR LATER EUROPEAN OR EUROPEAN DERIVED (55)	0.55	82	TEND TO BE THOSE WHERE THE TYPE OF POLITICAL MODERNIZATION IS EARLY OR LATER EUROPEAN OR EUROPEAN DERIVED, RATHER THAN DEVELOPED TUTELARY OR UNDEVELOPED TUTELARY (51)	0.85	29 22 36 4 X SQ= 10.49 P = 0. RV YES (NO)
83	TILT LESS TOWARD BEING THOSE WHERE THE TYPE OF POLITICAL MODERNIZATION IS OTHER THAN NON-EUROPEAN AUTOCHTHONOUS (106)	0.88	83	ALWAYS ARE THOSE WHERE THE TYPE OF POLITICAL MODERNIZATION IS OTHER THAN NON-EUROPEAN AUTOCHTHONOUS (106)	1.00	9 0 65 26 X SQ= 2.15 P = 0.107 RV NO (NO)
85	TEND LESS TO BE THOSE WHERE THE STAGE OF POLITICAL MODERNIZATION IS ADVANCED, RATHER THAN MID- OR EARLY TRANSITIONAL (60)	0.51	85	TEND MORE TO BE THOSE WHERE THE STAGE OF POLITICAL MODERNIZATION IS ADVANCED, RATHER THAN MID- OR EARLY TRANSITIONAL (60)	0.88	37 23 36 3 X SQ= 9.93 P = 0.001 RV NO (NO)
86	TILT LESS TOWARD BEING THOSE WHERE THE STAGE OF POLITICAL MODERNIZATION IS ADVANCED OR MID-TRANSITIONAL, RATHER THAN EARLY TRANSITIONAL (76)	0.73	86	TILT MORE TOWARD BEING THOSE WHERE THE STAGE OF POLITICAL MODERNIZATION IS ADVANCED OR MID-TRANSITIONAL, RATHER THAN EARLY TRANSITIONAL (76)	0.88	53 23 20 3 X SQ= 1.89 P = 0.114 RV NO (NO)
87	TEND LESS TO BE THOSE WHOSE IDEOLOGICAL ORIENTATION IS OTHER THAN DEVELOPMENTAL (58)	0.68	87	ALWAYS ARE THOSE WHOSE IDEOLOGICAL ORIENTATION IS OTHER THAN DEVELOPMENTAL (58)	1.00	17 0 36 22 X SQ= 7.39 P = 0.002 RV NO (NO)
89	TEND TO BE THOSE WHOSE IDEOLOGICAL ORIENTATION IS OTHER THAN CONVENTIONAL (62)	0.81	89	TEND TO BE THOSE WHOSE IDEOLOGICAL ORIENTATION IS CONVENTIONAL (33)	0.96	11 22 46 1 X SQ= 36.33 P = 0. RV YES (YES)
92	TEND LESS TO BE THOSE WHERE THE SYSTEM STYLE IS LIMITED MOBILIZATIONAL, OR NON-MOBILIZATIONAL (93)	0.74	92	TEND MORE TO BE THOSE WHERE THE SYSTEM STYLE IS LIMITED MOBILIZATIONAL, OR NON-MOBILIZATIONAL (93)	0.96	19 1 54 25 X SQ= 4.56 P = 0.020 RV NO (NO)
93	TEND LESS TO BE THOSE WHERE THE SYSTEM STYLE IS NON-MOBILIZATIONAL (78)	0.56	93	TEND MORE TO BE THOSE WHERE THE SYSTEM STYLE IS NON-MOBILIZATIONAL (78)	0.96	31 1 39 25 X SQ= 12.19 P = 0. RV NO (NO)
94	TEND TO BE THOSE WHERE THE STATUS OF THE REGIME IS AUTHORITARIAN OR TOTALITARIAN (41)	0.61	94	TEND TO BE THOSE WHERE THE STATUS OF THE REGIME IS CONSTITUTIONAL (51)	0.96	25 25 39 1 X SQ= 22.15 P = 0. RV YES (YES)

95	LEAN LESS TOWARD BEING THOSE WHERE THE STATUS OF THE REGIME IS CONSTITUTIONAL OR AUTHORITARIAN (95)	0.79	95	LEAN MORE TOWARD BEING THOSE WHERE THE STATUS OF THE REGIME IS CONSTITUTIONAL OR AUTHORITARIAN (95)	0.96	56 25 15 1 X SQ= 2.97 P = 0.061 RV NO (NO)
96	TEND LESS TO BE THOSE WHERE THE STATUS OF THE REGIME IS CONSTITUTIONAL OR TOTALITARIAN (67)	0.65	96	ALWAYS ARE THOSE WHERE THE STATUS OF THE REGIME IS CONSTITUTIONAL OR TOTALITARIAN (67)	1.00	40 26 22 0 X SQ= 10.48 P = 0. RV NO (NO)
98	TEND TO BE THOSE WHERE GOVERNMENTAL STABILITY IS GENERALLY OR MODERATELY PRESENT AND DATES FROM THE POST-WAR PERIOD, OR IS ABSENT (93)	0.91	98	TEND TO BE THOSE WHERE GOVERNMENTAL STABILITY IS GENERALLY PRESENT AND DATES AT LEAST FROM THE INTERWAR PERIOD (22)	0.58	7 15 67 11 X SQ= 23.35 P = 0. RV YES (YES)
99	TEND TO BE THOSE WHERE GOVERNMENTAL STABILITY IS MODERATELY PRESENT AND DATES FROM THE POST-WAR PERIOD, OR IS ABSENT (36)	0.52	99	TEND TO BE THOSE WHERE GOVERNMENTAL STABILITY IS GENERALLY PRESENT AND DATES FROM AT LEAST THE INTER-WAR PERIOD, OR FROM THE POST-WAR PERIOD (50)	0.91	30 20 33 2 X SQ= 10.89 P = 0. RV YES (NO)
100	TEND LESS TO BE THOSE WHERE GOVERNMENTAL STABILITY IS GENERALLY PRESENT AND DATES FROM AT LEAST THE INTER-WAR PERIOD, OR IS GENERALLY OR MODERATELY PRESENT, AND DATES FROM THE POST-WAR PERIOD (64)	0.66	100	ALWAYS ARE THOSE WHERE GOVERNMENTAL STABILITY IS GENERALLY PRESENT AND DATES FROM AT LEAST THE INTER-WAR PERIOD, OR IS GENERALLY OR MODERATELY PRESENT, AND DATES FROM THE POST-WAR PERIOD (64)	1.00	42 22 22 0 X SQ= 8.44 P = 0.001 RV NO (NO)
101	TEND TO BE THOSE WHERE THE REPRESENTATIVE CHARACTER OF THE REGIME IS LIMITED POLYARCHIC, PSEUDO-POLYARCHIC, OR NON-POLYARCHIC (57)	0.75	101	TEND TO BE THOSE WHERE THE REPRESENTATIVE CHARACTER OF THE REGIME IS POLYARCHIC (41)	0.92	16 24 47 2 X SQ= 30.65 P = 0. RV YES (YES)
102	TEND TO BE THOSE WHERE THE REPRESENTATIVE CHARACTER OF THE REGIME IS PSEUDO-POLYARCHIC OR NON-POLYARCHIC (49)	0.58	102	TEND TO BE THOSE WHERE THE REPRESENTATIVE CHARACTER OF THE REGIME IS POLYARCHIC OR LIMITED POLYARCHIC (59)	0.96	29 25 40 1 X SQ= 20.40 P = 0. RV YES (NO)
105	TEND TO BE THOSE WHERE THE ELECTORAL SYSTEM IS PARTIALLY COMPETITIVE OR NON-COMPETITIVE (47)	0.68	105	ALWAYS ARE THOSE WHERE THE ELECTORAL SYSTEM IS COMPETITIVE (43)	1.00	17 25 36 0 X SQ= 28.86 P = 0. RV YES (YES)
106	TEND TO BE THOSE WHERE THE ELECTORAL SYSTEM IS NON-COMPETITIVE (30)	0.57	106	ALWAYS ARE THOSE WHERE THE ELECTORAL SYSTEM IS COMPETITIVE OR PARTIALLY COMPETITIVE (52)	1.00	20 26 26 0 X SQ= 20.62 P = 0. RV YES (YES)

107	TEND TO BE THOSE WHERE AUTONOMOUS GROUPS ARE PARTIALLY TOLERATED IN POLITICS, ARE TOLERATED ONLY OUTSIDE POLITICS, OR ARE NOT TOLERATED AT ALL (65)	0.74	107	ALWAYS ARE THOSE WHERE AUTONOMOUS GROUPS ARE FULLY TOLERATED IN POLITICS (46)	1.00	19 26 53 0 X SQ= 38.77 P = 0. RV YES (YES)
111	TEND TO BE THOSE WHERE POLITICAL ENCULTURATION IS LOW (42)	0.54	111	TEND TO BE THOSE WHERE POLITICAL ENCULTURATION IS HIGH OR MEDIUM (53)	0.92	27 22 32 2 X SQ= 13.03 P = 0. RV YES (NO)
115	TEND TO BE THOSE WHERE INTEREST ARTICULATION BY ASSOCIATIONAL GROUPS IS MODERATE, LIMITED, OR NEGLIGIBLE (91)	0.94	115	TEND TO BE THOSE WHERE INTEREST ARTICULATION BY ASSOCIATIONAL GROUPS IS SIGNIFICANT (20)	0.62	4 16 66 10 X SQ= 32.52 P = 0. RV YES (YES)
116	TEND TO BE THOSE WHERE INTEREST ARTICULATION BY ASSOCIATIONAL GROUPS IS LIMITED OR NEGLIGIBLE (79)	0.86	116	TEND TO BE THOSE WHERE INTEREST ARTICULATION BY ASSOCIATIONAL GROUPS IS SIGNIFICANT OR MODERATE (32)	0.85	10 22 60 4 X SQ= 39.09 P = 0. RV YES (YES)
117	TEND TO BE THOSE WHERE INTEREST ARTICULATION BY ASSOCIATIONAL GROUPS IS NEGLIGIBLE (51)	0.50	117	TEND TO BE THOSE WHERE INTEREST ARTICULATION BY ASSOCIATIONAL GROUPS IS SIGNIFICANT, MODERATE, OR LIMITED (60)	0.96	35 25 35 1 X SQ= 15.32 P = 0. RV YES (NO)
121	TEND TO BE THOSE WHERE INTEREST ARTICULATION BY NON-ASSOCIATIONAL GROUPS IS SIGNIFICANT (54)	0.50	121	TEND TO BE THOSE WHERE INTEREST ARTICULATION BY NON-ASSOCIATIONAL GROUPS IS MODERATE, LIMITED, OR NEGLIGIBLE (61)	0.92	37 2 37 24 X SQ= 12.75 P = 0. RV YES (NO)
122	TEND TO BE THOSE WHERE INTEREST ARTICULATION BY NON-ASSOCIATIONAL GROUPS IS SIGNIFICANT OR MODERATE (83)	0.82	122	TEND TO BE THOSE WHERE INTEREST ARTICULATION BY NON-ASSOCIATIONAL GROUPS IS LIMITED OR NEGLIGIBLE (32)	0.73	61 7 13 19 X SQ= 24.75 P = 0. RV YES (YES)
123	ALWAYS ARE THOSE WHERE INTEREST ARTICULATION BY NON-ASSOCIATIONAL GROUPS IS SIGNIFICANT, MODERATE, OR LIMITED (107)	1.00	123	TEND LESS TO BE THOSE WHERE INTEREST ARTICULATION BY NON-ASSOCIATIONAL GROUPS IS SIGNIFICANT, MODERATE, OR LIMITED (107)	0.69	74 18 0 8 X SQ= 20.75 P = 0. RV NO (YES)
125	TEND TO BE THOSE WHERE INTEREST ARTICULATION BY ANOMIC GROUPS IS FREQUENT OR OCCASIONAL (64)	0.84	125	ALWAYS ARE THOSE WHERE INTEREST ARTICULATION BY ANOMIC GROUPS IS INFREQUENT OR VERY INFREQUENT (35)	1.00	51 0 10 24 X SQ= 46.74 P = 0. RV YES (YES)

126	ALWAYS ARE THOSE WHERE INTEREST ARTICULATION BY ANOMIC GROUPS IS FREQUENT, OCCASIONAL, OR INFREQUENT (83)	1.00	126	TEND TO BE THOSE WHERE INTEREST ARTICULATION BY ANOMIC GROUPS IS VERY INFREQUENT (16)	0.67	61 8 0 16 X SQ= 45.83 P = 0. RV YES (YES)
128	TEND TO BE THOSE WHERE INTEREST ARTICULATION BY POLITICAL PARTIES IS LIMITED OR NEGLIGIBLE (45)	0.56	128	TEND TO BE THOSE WHERE INTEREST ARTICULATION BY POLITICAL PARTIES IS SIGNIFICANT OR MODERATE (48)	0.84	24 21 30 4 X SQ= 9.35 P = 0.001 RV YES (NO)
130	TEND MORE TO BE THOSE WHERE INTEREST AGGREGATION BY POLITICAL PARTIES IS MODERATE, LIMITED, OR NEGLIGIBLE (71)	0.93	130	TEND LESS TO BE THOSE WHERE INTEREST AGGREGATION BY POLITICAL PARTIES IS MODERATE, LIMITED, OR NEGLIGIBLE (71)	0.67	4 8 52 16 X SQ= 7.10 P = 0.005 RV NO (YES)
132	TEND LESS TO BE THOSE WHERE INTEREST AGGREGATION BY POLITICAL PARTIES IS SIGNIFICANT, MODERATE, OR LIMITED (67)	0.77	132	ALWAYS ARE THOSE WHERE INTEREST AGGREGATION BY POLITICAL PARTIES IS SIGNIFICANT, MODERATE, OR LIMITED (67)	1.00	30 24 9 0 X SQ= 4.71 P = 0.010 RV NO (NO)
134	TEND TO BE THOSE WHERE INTEREST AGGREGATION BY THE EXECUTIVE IS LIMITED OR NEGLIGIBLE (46)	0.66	134	TEND TO BE THOSE WHERE INTEREST AGGREGATION BY THE EXECUTIVE IS SIGNIFICANT OR MODERATE (57)	0.92	22 23 42 2 X SQ= 21.63 P = 0. RV YES (YES)
135	TEND LESS TO BE THOSE WHERE INTEREST AGGREGATION BY THE EXECUTIVE IS SIGNIFICANT, MODERATE, OR LIMITED (77)	0.75	135	ALWAYS ARE THOSE WHERE INTEREST AGGREGATION BY THE EXECUTIVE IS SIGNIFICANT, MODERATE, OR LIMITED (77)	1.00	39 25 13 0 X SQ= 5.84 P = 0.007 RV NO (NO)
137	TEND TO BE THOSE WHERE INTEREST AGGREGATION BY THE LEGISLATURE IS LIMITED OR NEGLIGIBLE (68)	0.89	137	TEND TO BE THOSE WHERE INTEREST AGGREGATION BY THE LEGISLATURE IS SIGNIFICANT OR MODERATE (29)	0.88	6 23 51 3 X SQ= 44.34 P = 0. RV YES (YES)
138	TEND TO BE THOSE WHERE INTEREST AGGREGATION BY THE LEGISLATURE IS NEGLIGIBLE (49)	0.63	138	TEND TO BE THOSE WHERE INTEREST AGGREGATION BY THE LEGISLATURE IS SIGNIFICANT, MODERATE, OR LIMITED (48)	0.96	21 25 36 1 X SQ= 23.08 P = 0. RV YES (YES)
139	TEND LESS TO BE THOSE WHERE THE PARTY SYSTEM IS QUANTITATIVELY OTHER THAN ONE-PARTY (71)	0.59	139	ALWAYS ARE THOSE WHERE THE PARTY SYSTEM IS QUANTITATIVELY OTHER THAN ONE-PARTY (71)	1.00	27 0 39 26 X SQ= 13.15 P = 0. RV NO (NO)

```
141  TEND MORE TO BE THOSE                       0.97    141  TEND LESS TO BE THOSE                       0.68         2     8
     WHERE THE PARTY SYSTEM IS QUANTITATIVELY                  WHERE THE PARTY SYSTEM IS QUANTITATIVELY               58    17
     OTHER THAN TWO-PARTY  (87)                                OTHER THAN TWO-PARTY  (87)                        X SQ=  11.35
                                                                                                                 P =    0.001
                                                                                                                 RV NO  (YES)

145  TEND MORE TO BE THOSE                       0.90    145  TEND LESS TO BE THOSE                       0.58        19    11
     WHERE THE PARTY SYSTEM IS QUANTITATIVELY                  WHERE THE PARTY SYSTEM IS QUANTITATIVELY                2     8
     MULTI-PARTY, RATHER THAN                                  MULTI-PARTY, RATHER THAN                          X SQ=   4.04
     TWO-PARTY  (30)                                           TWO-PARTY  (30)                                   P =    0.028
                                                                                                                 RV NO  (YES)

146  TILT LESS TOWARD BEING THOSE                0.87    146  ALWAYS ARE THOSE                            1.00         8     0
     WHERE THE PARTY SYSTEM IS QUALITATIVELY                   WHERE THE PARTY SYSTEM IS QUALITATIVELY                55    23
     OTHER THAN                                                OTHER THAN                                        X SQ=   1.89
     MASS-BASED TERRITORIAL  (92)                              MASS-BASED TERRITORIAL  (92)                      P =    0.102
                                                                                                                 RV NO  (NO )

147  TEND TO BE THOSE                            0.89    147  TEND TO BE THOSE                            0.77         6    17
     WHERE THE PARTY SYSTEM IS QUALITATIVELY                   WHERE THE PARTY SYSTEM IS QUALITATIVELY                48     5
     OTHER THAN                                                CLASS-ORIENTED OR MULTI-IDEOLOGICAL  (23)         X SQ=  29.36
     CLASS-ORIENTED OR MULTI-IDEOLOGICAL  (67)                                                                   P =    0.
                                                                                                                 RV YES (YES)

153  TEND TO BE THOSE                            0.57    153  TEND TO BE THOSE                            0.82        24    18
     WHERE THE PARTY SYSTEM IS                                 WHERE THE PARTY SYSTEM IS                              32     4
     MODERATELY STABLE OR UNSTABLE  (38)                       STABLE  (42)                                      X SQ=   8.14
                                                                                                                 P =    0.002
                                                                                                                 RV YES (NO )

158  TEND LESS TO BE THOSE                       0.54    158  TEND MORE TO BE THOSE                       0.85        27     4
     WHERE PERSONALISMO IS                                     WHERE PERSONALISMO IS                                  32    22
     NEGLIGIBLE  (56)                                          NEGLIGIBLE  (56)                                  X SQ=   5.94
                                                                                                                 P =    0.008
                                                                                                                 RV NO  (NO )

160  TEND TO BE THOSE                            0.51    160  ALWAYS ARE THOSE                            1.00        30     0
     WHERE THE POLITICAL LEADERSHIP IS                         WHERE THE POLITICAL LEADERSHIP IS                      29    25
     ELITIST  (30)                                             MODERATELY ELITIST OR NON-ELITIST  (67)           X SQ=  17.62
                                                                                                                 P =    0.
                                                                                                                 RV YES (NO )

161  TEND TO BE THOSE                            0.69    161  TEND TO BE THOSE                            0.84        41     4
     WHERE THE POLITICAL LEADERSHIP IS                         WHERE THE POLITICAL LEADERSHIP IS                      18    21
     ELITIST OR MODERATELY ELITIST  (47)                       NON-ELITIST  (50)                                 X SQ=  18.11
                                                                                                                 P =    0.
                                                                                                                 RV YES (YES)

164  TEND LESS TO BE THOSE                       0.65    164  TEND MORE TO BE THOSE                       0.88        22     3
     WHERE THE REGIME'S LEADERSHIP CHARISMA                    WHERE THE REGIME'S LEADERSHIP CHARISMA                 40    23
     IS NEGLIGIBLE  (65)                                       IS NEGLIGIBLE  (65)                               X SQ=   4.05
                                                                                                                 P =    0.036
                                                                                                                 RV NO  (NO )
```

168	TEND TO BE THOSE WHERE THE HORIZONTAL POWER DISTRIBUTION IS LIMITED OR NEGLIGIBLE (72)	0.83	168	TEND TO BE THOSE WHERE THE HORIZONTAL POWER DISTRIBUTION IS SIGNIFICANT (34)	0.88	11 23 55 3 X SQ= 38.24 P = 0. RV YES (YES)
169	TEND TO BE THOSE WHERE THE HORIZONTAL POWER DISTRIBUTION IS NEGLIGIBLE (48)	0.59	169	TEND TO BE THOSE WHERE THE HORIZONTAL POWER DISTRIBUTION IS SIGNIFICANT OR LIMITED (58)	0.96	27 25 39 1 X SQ= 20.97 P = 0. RV YES (NO)
170	TEND LESS TO BE THOSE WHERE THE LEGISLATIVE-EXECUTIVE STRUCTURE IS OTHER THAN PRESIDENTIAL (63)	0.58	170	TEND MORE TO BE THOSE WHERE THE LEGISLATIVE-EXECUTIVE STRUCTURE IS OTHER THAN PRESIDENTIAL (63)	0.84	27 4 37 21 X SQ= 4.34 P = 0.026 RV NO (NO)
172	TEND TO BE THOSE WHERE THE LEGISLATIVE-EXECUTIVE STRUCTURE IS OTHER THAN PARLIAMENTARY-ROYALIST (88)	0.91	172	TEND TO BE THOSE WHERE THE LEGISLATIVE-EXECUTIVE STRUCTURE IS PARLIAMENTARY-ROYALIST (21)	0.50	6 13 63 13 X SQ= 17.64 P = 0. RV YES (YES)
174	TEND TO BE THOSE WHERE THE LEGISLATURE IS PARTIALLY EFFECTIVE, LARGELY INEFFECTIVE, OR WHOLLY INEFFECTIVE (72)	0.89	174	TEND TO BE THOSE WHERE THE LEGISLATURE IS FULLY EFFECTIVE (28)	0.84	7 21 55 4 X SQ= 39.89 P = 0. RV YES (YES)
175	TEND TO BE THOSE WHERE THE LEGISLATURE IS LARGELY INEFFECTIVE OR WHOLLY INEFFECTIVE (49)	0.61	175	ALWAYS ARE THOSE WHERE THE LEGISLATURE IS FULLY EFFECTIVE OR PARTIALLY EFFECTIVE (51)	1.00	24 25 38 0 X SQ= 24.77 P = 0. RV YES (YES)
178	TEND TO BE THOSE WHERE THE LEGISLATURE IS UNICAMERAL (53)	0.54	178	TEND TO BE THOSE WHERE THE LEGISLATURE IS BICAMERAL (51)	0.84	35 4 30 21 X SQ= 9.05 P = 0.002 RV YES (NO)
179	TEND TO BE THOSE WHERE THE EXECUTIVE IS DOMINANT (52)	0.72	179	TEND TO BE THOSE WHERE THE EXECUTIVE IS STRONG (39)	0.91	44 2 17 20 X SQ= 23.52 P = 0. RV YES (YES)
181	TEND TO BE THOSE WHERE THE BUREAUCRACY IS SEMI-MODERN, RATHER THAN MODERN (55)	0.94	181	TEND TO BE THOSE WHERE THE BUREAUCRACY IS MODERN, RATHER THAN SEMI-MODERN (21)	0.72	3 18 48 7 X SQ= 33.44 P = 0. RV YES (YES)

186	TEND TO BE THOSE WHERE PARTICIPATION BY THE MILITARY IN POLITICS IS SUPPORTIVE, RATHER THAN NEUTRAL (31)	0.61	186	ALWAYS ARE THOSE WHERE PARTICIPATION BY THE MILITARY IN POLITICS IS NEUTRAL, RATHER THAN SUPPORTIVE (56)	1.00	31 0 20 24 X SQ= 22.42 P = 0. RV YES (YES)
187	TEND TO BE THOSE WHERE THE ROLE OF THE POLICE IS POLITICALLY SIGNIFICANT (66)	0.85	187	TEND TO BE THOSE WHERE THE ROLE OF THE POLICE IS NOT POLITICALLY SIGNIFICANT (35)	0.96	57 1 10 24 X SQ= 47.94 P = 0. RV YES (YES)
189	ALWAYS ARE THOSE WHERE THE CHARACTER OF THE LEGAL SYSTEM IS OTHER THAN COMMON LAW (108)	1.00	189	TEND LESS TO BE THOSE WHERE THE CHARACTER OF THE LEGAL SYSTEM IS OTHER THAN COMMON LAW (108)	0.73	0 7 74 19 X SQ= 17.49 P = 0. RV NO (YES)
190	ALWAYS ARE THOSE WHERE THE CHARACTER OF THE LEGAL SYSTEM IS CIVIL LAW, RATHER THAN COMMON LAW (32)	1.00	190	TEND LESS TO BE THOSE WHERE THE CHARACTER OF THE LEGAL SYSTEM IS CIVIL LAW, RATHER THAN COMMON LAW (32)	0.59	22 10 0 7 X SQ= 8.42 P = 0.001 RV NO (YES)
192	TEND LESS TO BE THOSE WHERE THE CHARACTER OF THE LEGAL SYSTEM IS OTHER THAN MUSLIM OR PARTLY MUSLIM (86)	0.72	192	ALWAYS ARE THOSE WHERE THE CHARACTER OF THE LEGAL SYSTEM IS OTHER THAN MUSLIM OR PARTLY MUSLIM (86)	1.00	20 0 52 26 X SQ= 7.44 P = 0.001 RV NO (NO)
193	LEAN LESS TOWARD BEING THOSE WHERE THE CHARACTER OF THE LEGAL SYSTEM IS OTHER THAN PARTLY INDIGENOUS (86)	0.81	193	LEAN MORE TOWARD BEING THOSE WHERE THE CHARACTER OF THE LEGAL SYSTEM IS OTHER THAN PARTLY INDIGENOUS (86)	0.96	14 1 59 25 X SQ= 2.41 P = 0.066 RV NO (NO)
194	TEND LESS TO BE THOSE THAT ARE NON-COMMUNIST (101)	0.82	194	ALWAYS ARE THOSE THAT ARE NON-COMMUNIST (101)	1.00	13 0 61 25 X SQ= 3.63 P = 0.020 RV NO (NO)

120 POLITIES
WHERE INTEREST ARTICULATION
BY INSTITUTIONAL GROUPS
IS VERY SIGNIFICANT, SIGNIFICANT, OR
MODERATE (90)

120 POLITIES
WHERE INTEREST ARITUCLATION
BY INSTITUTIONAL GROUPS
IS LIMITED (10)

PREDICATE ONLY

90 IN LEFT COLUMN

AFGHANISTAN	ALBANIA	ALGERIA	ARGENTINA	AUSTRIA	BELGIUM	BOLIVIA	BRAZIL	BULGARIA	BURMA
CAMBODIA	CANADA	CEYLON	CHILE	CHINA, PR	COLOMBIA	CONGO(LEO)	CUBA	COSTA RICA	CYPRUS
CZECHOS'KIA	DOMIN REP	ECUADOR	EL SALVADOR	ETHIOPIA	FRANCE	GERMANY, E	GERMAN FR	GHANA	GREECE
GUATEMALA	GUINEA	HAITI	HONDURAS	HUNGARY	INDIA	INDONESIA	IRAN	IRAQ	IRELAND
ISRAEL	ITALY	IVORY COAST	JAMAICA	JAPAN	JORDAN	KOREA, N	KOREA REP	LAOS	LEBANON
LIBERIA	LIBYA	MALAGASY R	MALAYA	MAURITANIA	MEXICO	MONGOLIA	MOROCCO	NEPAL	NETHERLANDS
NICARAGUA	NIGERIA	PAKISTAN	PANAMA	PARAGUAY	PERU	PHILIPPINES	POLAND	PORTUGAL	RUMANIA
SA'U ARABIA	SENEGAL	SO AFRICA	SPAIN	SUDAN	SWITZERLAND	SYRIA	TANGANYIKA	THAILAND	TRINIDAD
TUNISIA	TURKEY	UGANDA	USSR	UAR	URUGUAY	VENEZUELA	VIETNAM, N	VIETNAM REP	YUGOSLAVIA

10 IN RIGHT COLUMN

AUSTRALIA	DENMARK	FINLAND	ICELAND	LUXEMBOURG	NEW ZEALAND	NORWAY	SWEDEN	UK	US

15 EXCLUDED BECAUSE UNASCERTAINABLE

BURUNDI	CAMEROUN	CEN AFR REP	CHAD	CONGO(BRA)	DAHOMEY	GABON	MALI	NIGER	RWANDA
SIERRE LEO	SOMALIA	TOGO	UPPER VOLTA	YEMEN					

121 POLITIES
WHERE INTEREST ARTICULATION
BY NON-ASSOCIATIONAL GROUPS
IS SIGNIFICANT (54)

121 POLITIES
WHERE INTEREST ARTICULATION
BY NON-ASSOCIATIONAL GROUPS
IS MODERATE, LIMITED, OR
NEGLIGIBLE (61)

PREDICATE ONLY

54 IN LEFT COLUMN

```
AFGHANISTAN BURMA       BURUNDI     CAMBODIA    CAMEROUN    CEN AFR REP CEYLON      CHAD        CHINA, PR   CONGO(BRA)
CONGO(LEO)  CYPRUS      DAHOMEY     ETHIOPIA    GABON       GHANA       GUINEA      HAITI       INDIA       INDONESIA
IRAN        IRAQ        IVORY COAST JORDAN      KOREA, N    KOREA REP   LAOS        LEBANON     MALAGASY R  MALAYA
MALI        MAURITANIA  MONGOLIA    NEPAL       NIGER       NIGERIA     PAKISTAN    PHILIPPINES RWANDA      SA'U ARABIA
SENEGAL     SIERRE LEO  SOMALIA     SO AFRICA   SUDAN       SYRIA       TANGANYIKA  THAILAND    TOGO        UGANDA
UPPER VOLTA VIETNAM, N  VIETNAM REP YEMEN
```

61 IN RIGHT COLUMN

```
ALBANIA     ALGERIA     ARGENTINA   AUSTRALIA   AUSTRIA     BELGIUM     BOLIVIA     BRAZIL      BULGARIA    CANADA
CHILE       COLOMBIA    CUBA        COSTA RICA  CZECHOS'KIA DENMARK     DOMIN REP   ECUADOR     EL SALVADOR FINLAND
FRANCE      GERMANY, E  GERMAN FR   GREECE      GUATEMALA   HONDURAS    HUNGARY     ICELAND     IRELAND     ISRAEL
ITALY       JAMAICA     JAPAN       LIBERIA     LIBYA       LUXEMBOURG  MEXICO      MOROCCO     NETHERLANDS NEW ZEALAND
NICARAGUA   NORWAY      PANAMA      PARAGUAY    PERU        POLAND      PORTUGAL    RUMANIA     SPAIN       SWEDEN
SWITZERLAND TRINIDAD    TUNISIA     TURKEY      USSR        UAR         UK          US          URUGUAY     VENEZUELA
YUGOSLAVIA
```

```
******************************************************************************************************

122  POLITIES                                          122  POLITIES
     WHERE INTEREST ARTICULATION                            WHERE INTEREST ARTICULATION
     BY NON-ASSOCIATIONAL GROUPS                            BY NON-ASSOCIATIONAL GROUPS
     IS SIGNIFICANT OR MODERATE  (83)                       IS LIMITED OR NEGLIGIBLE  (32)


                                                                        BOTH SUBJECT AND PREDICATE
......................................................................................................

     83 IN LEFT COLUMN

AFGHANISTAN ALBANIA     ALGERIA     BELGIUM     BOLIVIA     BRAZIL      BURMA       BURUNDI     CAMBODIA    CAMEROUN
CANADA      CEN AFR REP CEYLON      CHAD        CHINA, PR   CONGO(BRA)  CONGO(LEO)  CYPRUS      DAHOMEY     ECUADOR
EL SALVADOR ETHIOPIA    GABON       GHANA       GUATEMALA   GUINEA      HAITI       HONDURAS    INDIA       INDONESIA
IRAN        IRAQ        ISRAEL      IVORY COAST JAMAICA     JAPAN       JORDAN      KOREA, N    KOREA REP   LAOS
LEBANON     LIBERIA     LIBYA       MALAGASY R  MALAYA      MALI        MAURITANIA  MEXICO      MONGOLIA    MOROCCO
NEPAL       NICARAGUA   NIGER       NIGERIA     PAKISTAN    PERU        PHILIPPINES POLAND      RUMANIA     RWANDA
SA'U ARABIA SENEGAL     SIERRE LEO  SOMALIA     SO AFRICA   SPAIN       SUDAN       SWITZERLAND SYRIA       TANGANYIKA
THAILAND    TOGO        TRINIDAD    TUNISIA     TURKEY      UGANDA      USSR        UAR         UPPER VOLTA VIETNAM, N
VIETNAM REP YEMEN       YUGOSLAVIA

     32 IN RIGHT COLUMN

ARGENTINA   AUSTRALIA   AUSTRIA     BULGARIA    CHILE       COLOMBIA    CUBA        COSTA RICA  CZECHOS'KIA DENMARK
DOMIN REP   FINLAND     FRANCE      GERMANY, E  GERMAN FR   GREECE      HUNGARY     ICELAND     IRELAND     ITALY
LUXEMBOURG  NETHERLANDS NEW ZEALAND NORWAY      PANAMA      PARAGUAY    PORTUGAL    SWEDEN      UK          US
URUGUAY     VENEZUELA

------------------------------------------------------------------------------------------------------

  2  TEND TO BE THOSE                        0.95        2  TEND TO BE THOSE                       0.56        4   18
     LOCATED ELSEWHERE THAN IN                              LOCATED IN WEST EUROPE, SCANDINAVIA,              79   14
     WEST EUROPE, SCANDINAVIA,                              NORTH AMERICA, OR AUSTRALASIA  (22)         X SQ=  36.23
     NORTH AMERICA, OR AUSTRALASIA  (93)                                                                P =    0.
                                                                                                        RV YES (YES)

  3  TEND MORE TO BE THOSE                   0.96        3  TEND LESS TO BE THOSE                  0.69        3   10
     LOCATED ELSEWHERE THAN IN                              LOCATED ELSEWHERE THAN IN                         80   22
     WEST EUROPE  (102)                                     WEST EUROPE  (102)                          X SQ=  14.94
                                                                                                        P =    0.
                                                                                                        RV NO  (YES)

  5  LEAN TOWARD BEING THOSE                 0.62        5  LEAN TOWARD BEING THOSE                0.79        3   15
     LOCATED IN EAST EUROPE, RATHER THAN                    LOCATED IN SCANDINAVIA OR WEST EUROPE,             5    4
     IN SCANDINAVIA OR WEST EUROPE  (9)                     RATHER THAN IN EAST EUROPE  (18)            X SQ=   2.69
                                                                                                        P =    0.072
                                                                                                        RV YES (YES)

  8  TEND LESS TO BE THOSE                   0.78        8  ALWAYS ARE THOSE                       1.00       18    0
     LOCATED ELSEWHERE THAN IN EAST ASIA                    LOCATED ELSEWHERE THAN IN EAST ASIA               65   32
     SOUTH ASIA, OR SOUTHEAST ASIA  (97)                    SOUTH ASIA, OR SOUTHEAST ASIA  (97)         X SQ=   6.67
                                                                                                        P =    0.003
                                                                                                        RV NO  (NO )
```

9	LEAN LESS TOWARD BEING THOSE LOCATED ELSEWHERE THAN IN SOUTHEAST ASIA (106)	0.89	9	ALWAYS ARE THOSE LOCATED ELSEWHERE THAN IN SOUTHEAST ASIA (106)	1.00	9 0 74 32 X SQ= 2.41 P = 0.060 RV NO (NO)
10	TEND LESS TO BE THOSE LOCATED ELSEWHERE THAN IN NORTH AFRICA, OR CENTRAL AND SOUTH AFRICA (82)	0.60	10	ALWAYS ARE THOSE LOCATED ELSEWHERE THAN IN NORTH AFRICA, OR CENTRAL AND SOUTH AFRICA (82)	1.00	33 0 50 32 X SQ= 15.95 P = 0. RV NO (NO)
13	TEND LESS TO BE THOSE LOCATED ELSEWHERE THAN IN NORTH AFRICA OR THE MIDDLE EAST (99)	0.81	13	ALWAYS ARE THOSE LOCATED ELSEWHERE THAN IN NORTH AFRICA OR THE MIDDLE EAST (99)	1.00	16 0 67 32 X SQ= 5.65 P = 0.005 RV NO (NO)
17	TEND LESS TO BE THOSE LOCATED IN THE CARIBBEAN, CENTRAL AMERICA, OR SOUTH AMERICA, RATHER THAN IN THE MIDDLE EAST (22)	0.52	17	ALWAYS ARE THOSE LOCATED IN THE CARIBBEAN, CENTRAL AMERICA, OR SOUTH AMERICA, RATHER THAN IN THE MIDDLE EAST (22)	1.00	11 0 12 10 X SQ= 5.18 P = 0.013 RV NO (NO)
19	TEND TO BE THOSE LOCATED IN NORTH AFRICA OR CENTRAL AND SOUTH AFRICA, RATHER THAN IN THE CARIBBEAN, CENTRAL AMERICA, OR SOUTH AMERICA (33)	0.73	19	ALWAYS ARE THOSE LOCATED IN THE CARIBBEAN, CENTRAL AMERICA, OR SOUTH AMERICA, RATHER THAN IN NORTH AFRICA OR CENTRAL AND SOUTH AFRICA (22)	1.00	33 0 12 10 X SQ= 15.41 P = 0. RV YES (NO)
20	TEND TO BE THOSE LOCATED IN EAST ASIA, SOUTH ASIA, OR SOUTHEAST ASIA, RATHER THAN IN THE CARIBBEAN, CENTRAL AMERICA, OR SOUTH AMERICA (18)	0.60	20	ALWAYS ARE THOSE LOCATED IN THE CARIBBEAN, CENTRAL AMERICA, OR SOUTH AMERICA, RATHER THAN IN EAST ASIA, SOUTH ASIA, OR SOUTHEAST ASIA (22)	1.00	18 0 12 10 X SQ= 8.62 P = 0.001 RV YES (NO)
22	TILT TOWARD BEING THOSE WHOSE TERRITORIAL SIZE IS VERY LARGE, LARGE, OR MEDIUM (68)	0.64	22	TILT TOWARD BEING THOSE WHOSE TERRITORIAL SIZE IS SMALL (47)	0.53	53 15 30 17 X SQ= 2.10 P = 0.138 RV YES (NO)
26	TILT TOWARD BEING THOSE WHOSE POPULATION DENSITY IS LOW (67)	0.63	26	TILT TOWARD BEING THOSE WHOSE POPULATION DENSITY IS VERY HIGH, HIGH, OR MEDIUM (48)	0.53	31 17 52 15 X SQ= 1.76 P = 0.143 RV YES (NO)
28	TEND TO BE THOSE WHOSE POPULATION GROWTH RATE IS HIGH (62)	0.65	28	TEND TO BE THOSE WHOSE POPULATION GROWTH RATE IS LOW (48)	0.66	51 11 27 21 X SQ= 7.66 P = 0.005 RV YES (NO)

29	TEND TO BE THOSE WHERE THE DEGREE OF URBANIZATION IS LOW (49)	0.64	29	ALWAYS ARE THOSE WHERE THE DEGREE OF URBANIZATION IS HIGH (56)	1.00	27 29 49 0 X SQ= 32.51 P = 0. RV YES (YES)
30	TEND TO BE THOSE WHOSE AGRICULTURAL POPULATION IS HIGH (56)	0.69	30	ALWAYS ARE THOSE WHOSE AGRICULTURAL POPULATION IS MEDIUM, LOW, OR VERY LOW (57)	1.00	56 0 25 32 X SQ= 41.14 P = 0. RV YES (YES)
31	TEND TO BE THOSE WHOSE AGRICULTURAL POPULATION IS HIGH OR MEDIUM (90)	0.93	31	TEND TO BE THOSE WHOSE AGRICULTURAL POPULATION IS LOW OR VERY LOW (24)	0.56	76 14 6 18 X SQ= 30.28 P = 0. RV YES (YES)
34	TEND TO BE THOSE WHOSE GROSS NATIONAL PRODUCT IS VERY LOW (53)	0.57	34	TEND TO BE THOSE WHOSE GROSS NATIONAL PRODUCT IS VERY HIGH, HIGH, MEDIUM, OR LOW (62)	0.81	36 26 47 6 X SQ= 11.85 P = 0. RV YES (NO)
35	TEND TO BE THOSE WHOSE PER CAPITA GROSS NATIONAL PRODUCT IS MEDIUM, LOW, OR VERY LOW (91)	0.94	35	TEND TO BE THOSE WHOSE PER CAPITA GROSS NATIONAL PRODUCT IS VERY HIGH OR HIGH (24)	0.59	5 19 78 13 X SQ= 36.64 P = 0. RV YES (YES)
36	TEND TO BE THOSE WHOSE PER CAPITA GROSS NATIONAL PRODUCT IS LOW OR VERY LOW (73)	0.81	36	TEND TO BE THOSE WHOSE PER CAPITA GROSS NATIONAL PRODUCT IS VERY HIGH, HIGH, OR MEDIUM (42)	0.81	16 26 67 6 X SQ= 35.63 P = 0. RV YES (YES)
37	TEND TO BE THOSE WHOSE PER CAPITA GROSS NATIONAL PRODUCT IS VERY LOW (51)	0.60	37	TEND TO BE THOSE WHOSE PER CAPITA GROSS NATIONAL PRODUCT IS VERY HIGH, HIGH, MEDIUM, OR LOW (64)	0.97	33 31 50 1 X SQ= 28.26 P = 0. RV YES (NO)
39	TEND TO BE THOSE WHOSE INTERNATIONAL FINANCIAL STATUS IS LOW OR VERY LOW (76)	0.77	39	TEND TO BE THOSE WHOSE INTERNATIONAL FINANCIAL STATUS IS VERY HIGH, HIGH, OR MEDIUM (38)	0.61	19 19 64 12 X SQ= 13.30 P = 0. RV YES (YES)
42	TEND TO BE THOSE WHOSE ECONOMIC DEVELOPMENTAL STATUS IS UNDERDEVELOPED OR VERY UNDERDEVELOPED (76)	0.85	42	TEND TO BE THOSE WHOSE ECONOMIC DEVELOPMENTAL STATUS IS DEVELOPED OR INTERMEDIATE (36)	0.77	12 24 69 7 X SQ= 37.47 P = 0. RV YES (YES)

43	TEND TO BE THOSE WHOSE ECONOMIC DEVELOPMENTAL STATUS IS VERY UNDERDEVELOPED (57)	0.67	43	TEND TO BE THOSE WHOSE ECONOMIC DEVELOPMENTAL STATUS IS DEVELOPED, INTERMEDIATE, OR UNDERDEVELOPED (55)	0.90	27 28 54 3 X SQ= 26.90 P = 0. RV YES (YES)
44	TEND TO BE THOSE WHERE THE LITERACY RATE IS BELOW NINETY PERCENT (90)	0.92	44	TEND TO BE THOSE WHERE THE LITERACY RATE IS NINETY PERCENT OR ABOVE (25)	0.56	7 18 76 14 X SQ= 28.29 P = 0. RV YES (YES)
45	TEND TO BE THOSE WHERE THE LITERACY RATE IS BELOW FIFTY PERCENT (54)	0.70	45	ALWAYS ARE THOSE WHERE THE LITERACY RATE IS FIFTY PERCENT OR ABOVE (55)	1.00	23 32 54 0 X SQ= 41.71 P = 0. RV YES (YES)
50	TEND TO BE THOSE WHERE FREEDOM OF THE PRESS IS INTERMITTENT, INTERNALLY ABSENT, OR INTERNALLY AND EXTERNALLY ABSENT (56)	0.68	50	TEND TO BE THOSE WHERE FREEDOM OF THE PRESS IS COMPLETE (43)	0.68	22 21 46 10 X SQ= 9.46 P = 0.002 RV YES (NO)
54	TEND TO BE THOSE WHERE NEWSPAPER CIRCULATION IS LESS THAN ONE HUNDRED PER THOUSAND (76)	0.88	54	TEND TO BE THOSE WHERE NEWSPAPER CIRCULATION IS ONE HUNDRED OR MORE PER THOUSAND (37)	0.84	10 27 71 5 X SQ= 50.82 P = 0. RV YES (YES)
55	TEND LESS TO BE THOSE WHERE NEWSPAPER CIRCULATION IS TEN OR MORE PER THOUSAND (78)	0.57	55	ALWAYS ARE THOSE WHERE NEWSPAPER CIRCULATION IS TEN OR MORE PER THOUSAND (78)	1.00	46 32 35 0 X SQ= 18.06 P = 0. RV NO (NO)
56	TEND LESS TO BE THOSE WHERE THE RELIGION IS PREDOMINANTLY LITERATE (79)	0.60	56	ALWAYS ARE THOSE WHERE THE RELIGION IS PREDOMINANTLY LITERATE (79)	1.00	47 32 31 0 X SQ= 15.80 P = 0. RV NO (NO)
57	TEND TO BE THOSE WHERE THE RELIGION IS OTHER THAN CATHOLIC (90)	0.89	57	TEND TO BE THOSE WHERE THE RELIGION IS CATHOLIC (25)	0.50	9 16 74 16 X SQ= 18.58 P = 0. RV YES (YES)
58	TEND LESS TO BE THOSE WHERE THE RELIGION IS OTHER THAN MUSLIM (97)	0.78	58	ALWAYS ARE THOSE WHERE THE RELIGION IS OTHER THAN MUSLIM (97)	1.00	18 0 65 32 X SQ= 6.67 P = 0.003 RV NO (NO)

63	TEND TO BE THOSE WHERE THE RELIGION IS PREDOMINANTLY OR PARTLY OTHER THAN CHRISTIAN (68)	0.83	63	ALWAYS ARE THOSE WHERE THE RELIGION IS PREDOMINANTLY SOME KIND OF CHRISTIAN (46)	1.00	14 32 68 0 X SQ= 62.36 P = 0. RV YES (YES)
66	TEND TO BE THOSE THAT ARE RELIGIOUSLY HETEROGENEOUS (49)	0.54	66	TEND TO BE THOSE THAT ARE RELIGIOUSLY HOMOGENEOUS (57)	0.72	34 23 40 9 X SQ= 5.04 P = 0.019 RV YES (NO)
68	TEND TO BE THOSE THAT ARE LINGUISTICALLY WEAKLY HETEROGENEOUS OR STRONGLY HETEROGENEOUS (62)	0.71	68	TEND TO BE THOSE THAT ARE LINGUISTICALLY HOMOGENEOUS (52)	0.87	24 28 58 4 X SQ= 29.16 P = 0. RV YES (YES)
69	TEND TO BE THOSE THAT ARE LINGUISTICALLY STRONGLY HETEROGENEOUS (50)	0.60	69	TEND TO BE THOSE THAT ARE LINGUISTICALLY HOMOGENEOUS OR WEAKLY HETEROGENEOUS (64)	0.97	33 31 49 1 X SQ= 27.72 P = 0. RV YES (NO)
70	TEND MORE TO BE THOSE THAT ARE RELIGIOUSLY, RACIALLY, OR LINGUISTICALLY HETEROGENEOUS (85)	0.91	70	TEND LESS TO BE THOSE THAT ARE RELIGIOUSLY, RACIALLY, OR LINGUISTICALLY HETEROGENEOUS (85)	0.53	7 14 68 16 X SQ= 16.41 P = 0. RV NO (YES)
72	TEND TO BE THOSE WHOSE DATE OF INDEPENDENCE IS AFTER 1914 (59)	0.67	72	TEND TO BE THOSE WHOSE DATE OF INDEPENDENCE IS BEFORE 1914 (52)	0.83	27 25 54 5 X SQ= 20.02 P = 0. RV YES (NO)
73	TEND TO BE THOSE WHOSE DATE OF INDEPENDENCE IS AFTER 1945 (46)	0.57	73	ALWAYS ARE THOSE WHOSE DATE OF INDEPENDENCE IS BEFORE 1945 (65)	1.00	35 30 46 0 X SQ= 26.80 P = 0. RV YES (NO)
74	TEND TO BE THOSE THAT ARE NOT HISTORICALLY WESTERN (87)	0.94	74	TEND TO BE THOSE THAT ARE HISTORICALLY WESTERN (26)	0.66	5 21 76 11 X SQ= 42.47 P = 0. RV YES (YES)
75	TEND TO BE THOSE THAT ARE NOT HISTORICALLY WESTERN AND ARE NOT SIGNIFICANTLY WESTERNIZED (52)	0.63	75	ALWAYS ARE THOSE THAT ARE HISTORICALLY WESTERN OR SIGNIFICANTLY WESTERNIZED (62)	1.00	30 32 52 0 X SQ= 34.80 P = 0. RV YES (YES)

76	TEND TO BE THOSE THAT HAVE BEEN SIGNIFICANTLY OR PARTIALLY WESTERNIZED THROUGH A COLONIAL RELATIONSHIP, RATHER THAN BEING HISTORICALLY WESTERN (70)	0.92	76	TEND TO BE THOSE THAT ARE HISTORICALLY WESTERN, RATHER THAN HAVING BEEN SIGNIFICANTLY OR PARTIALLY WESTERNIZED THROUGH A COLONIAL RELATIONSHIP (26)	0.68	5 21 60 10 X SQ= 35.35 P = 0. RV YES (YES)
77	TEND TO BE THOSE THAT HAVE BEEN PARTIALLY WESTERNIZED, RATHER THAN SIGNIFICANTLY WESTERNIZED, THROUGH A COLONIAL RELATIONSHIP (41)	0.69	77	ALWAYS ARE THOSE THAT HAVE BEEN SIGNIFICANTLY WESTERNIZED, RATHER THAN PARTIALLY WESTERNIZED, THROUGH A COLONIAL RELATIONSHIP (28)	1.00	18 10 41 0 X SQ= 14.36 P = 0. RV YES (NO)
79	TEND TO BE THOSE THAT WERE FORMERLY DEPENDENCIES OF BRITAIN OR FRANCE, RATHER THAN SPAIN (49)	0.85	79	TEND TO BE THOSE THAT WERE FORMERLY DEPENDENCIES OF SPAIN, RATHER THAN BRITAIN OR FRANCE (18)	0.71	45 4 8 10 X SQ= 15.14 P = 0. RV YES (YES)
82	TEND TO BE THOSE WHERE THE TYPE OF POLITICAL MODERNIZATION IS DEVELOPED TUTELARY OR UNDEVELOPED TUTELARY, RATHER THAN EARLY OR LATER EUROPEAN OR EUROPEAN DERIVED (55)	0.74	82	ALWAYS ARE THOSE WHERE THE TYPE OF POLITICAL MODERNIZATION IS EARLY OR LATER EUROPEAN OR EUROPEAN DERIVED, RATHER THAN DEVELOPED TUTELARY OR UNDEVELOPED TUTELARY (51)	1.00	19 32 55 0 X SQ= 46.50 P = 0. RV YES (YES)
83	LEAN LESS TOWARD BEING THOSE WHERE THE TYPE OF POLITICAL MODERNIZATION IS OTHER THAN NON-EUROPEAN AUTOCHTHONOUS (106)	0.89	83	ALWAYS ARE THOSE WHERE THE TYPE OF POLITICAL MODERNIZATION IS OTHER THAN NON-EUROPEAN AUTOCHTHONOUS (106)	1.00	9 0 74 32 X SQ= 2.41 P = 0.060 RV NO (NO)
85	TEND TO BE THOSE WHERE THE STAGE OF POLITICAL MODERNIZATION IS MID- OR EARLY TRANSITIONAL, RATHER THAN ADVANCED (54)	0.61	85	TEND TO BE THOSE WHERE THE STAGE OF POLITICAL MODERNIZATION IS ADVANCED, RATHER THAN MID- OR EARLY TRANSITIONAL (60)	0.88	32 28 50 4 X SQ= 19.79 P = 0. RV YES (NO)
86	TEND LESS TO BE THOSE WHERE THE STAGE OF POLITICAL MODERNIZATION IS ADVANCED OR MID-TRANSITIONAL, RATHER THAN EARLY TRANSITIONAL (76)	0.54	86	ALWAYS ARE THOSE WHERE THE STAGE OF POLITICAL MODERNIZATION IS ADVANCED OR MID-TRANSITIONAL, RATHER THAN EARLY TRANSITIONAL (76)	1.00	44 32 38 0 X SQ= 20.21 P = 0. RV NO (NO)
87	TEND LESS TO BE THOSE WHOSE IDEOLOGICAL ORIENTATION IS OTHER THAN DEVELOPMENTAL (58)	0.52	87	TEND MORE TO BE THOSE WHOSE IDEOLOGICAL ORIENTATION IS OTHER THAN DEVELOPMENTAL (58)	0.93	29 2 32 26 X SQ= 12.08 P = 0. RV NO (NO)
89	TEND TO BE THOSE WHOSE IDEOLOGICAL ORIENTATION IS OTHER THAN CONVENTIONAL (62)	0.82	89	TEND TO BE THOSE WHOSE IDEOLOGICAL ORIENTATION IS CONVENTIONAL (33)	0.75	12 21 55 7 X SQ= 25.93 P = 0. RV YES (YES)

94	TEND TO BE THOSE WHERE THE STATUS OF THE REGIME IS AUTHORITARIAN OR TOTALITARIAN (41)	0.56	94	TEND TO BE THOSE WHERE THE STATUS OF THE REGIME IS CONSTITUTIONAL (51)	0.77	27 24 34 7 X SQ= 7.85 P = 0.004 RV YES (NO)
96	TEND LESS TO BE THOSE WHERE THE STATUS OF THE REGIME IS CONSTITUTIONAL OR TOTALITARIAN (67)	0.63	96	TEND MORE TO BE THOSE WHERE THE STATUS OF THE REGIME IS CONSTITUTIONAL OR TOTALITARIAN (67)	0.97	37 30 22 1 X SQ= 10.67 P = 0. RV NO (NO)
97	TEND TO BE THOSE WHERE THE STATUS OF THE REGIME IS AUTHORITARIAN, RATHER THAN TOTALITARIAN (23)	0.69	97	TEND TO BE THOSE WHERE THE STATUS OF THE REGIME IS TOTALITARIAN, RATHER THAN AUTHORITARIAN (16)	0.86	22 1 10 6 X SQ= 4.97 P = 0.013 RV YES (NO)
99	TEND TO BE THOSE WHERE GOVERNMENTAL STABILITY IS MODERATELY PRESENT AND DATES FROM THE POST-WAR PERIOD, OR IS ABSENT (36)	0.50	99	TEND TO BE THOSE WHERE GOVERNMENTAL STABILITY IS GENERALLY PRESENT AND DATES FROM AT LEAST THE INTER-WAR PERIOD, OR FROM THE POST-WAR PERIOD (50)	0.75	29 21 29 7 X SQ= 3.88 P = 0.036 RV YES (NO)
101	TEND TO BE THOSE WHERE THE REPRESENTATIVE CHARACTER OF THE REGIME IS LIMITED POLYARCHIC, PSEUDO-POLYARCHIC, OR NON-POLYARCHIC (57)	0.72	101	TEND TO BE THOSE WHERE THE REPRESENTATIVE CHARACTER OF THE REGIME IS POLYARCHIC (41)	0.71	19 22 48 9 X SQ= 14.11 P = 0. RV YES (YES)
102	TEND TO BE THOSE WHERE THE REPRESENTATIVE CHARACTER OF THE REGIME IS PSEUDO-POLYARCHIC OR NON-POLYARCHIC (49)	0.55	102	TEND TO BE THOSE WHERE THE REPRESENTATIVE CHARACTER OF THE REGIME IS POLYARCHIC OR LIMITED POLYARCHIC (59)	0.78	34 25 42 7 X SQ= 8.83 P = 0.002 RV YES (NO)
105	TEND TO BE THOSE WHERE THE ELECTORAL SYSTEM IS PARTIALLY COMPETITIVE OR NON-COMPETITIVE (47)	0.68	105	TEND TO BE THOSE WHERE THE ELECTORAL SYSTEM IS COMPETITIVE (43)	0.80	19 24 41 6 X SQ= 16.84 P = 0. RV YES (YES)
107	TEND TO BE THOSE WHERE AUTONOMOUS GROUPS ARE PARTIALLY TOLERATED IN POLITICS, ARE TOLERATED ONLY OUTSIDE POLITICS, OR ARE NOT TOLERATED AT ALL (65)	0.73	107	TEND TO BE THOSE WHERE AUTONOMOUS GROUPS ARE FULLY TOLERATED IN POLITICS (46)	0.78	21 25 58 7 X SQ= 22.85 P = 0. RV YES (YES)
111	TEND TO BE THOSE WHERE POLITICAL ENCULTURATION IS LOW (42)	0.55	111	TEND TO BE THOSE WHERE POLITICAL ENCULTURATION IS HIGH OR MEDIUM (53)	0.87	32 21 39 3 X SQ= 11.43 P = 0. RV YES (NO)

114	TEND TO BE THOSE WHERE SECTIONALISM IS EXTREME OR MODERATE (61)	0.65	114	TEND TO BE THOSE WHERE SECTIONALISM IS NEGLIGIBLE (47)	0.67	51 10 27 20 X SQ= 7.80 P = 0.004 RV YES (NO)
116	TEND TO BE THOSE WHERE INTEREST ARTICULATION BY ASSOCIATIONAL GROUPS IS LIMITED OR NEGLIGIBLE (79)	0.86	116	TEND TO BE THOSE WHERE INTEREST ARTICULATION BY ASSOCIATIONAL GROUPS IS SIGNIFICANT OR MODERATE (32)	0.68	11 21 69 10 X SQ= 29.17 P = 0. RV YES (YES)
117	TEND TO BE THOSE WHERE INTEREST ARTICULATION BY ASSOCIATIONAL GROUPS IS NEGLIGIBLE (51)	0.57	117	TEND TO BE THOSE WHERE INTEREST ARTICULATION BY ASSOCIATIONAL GROUPS IS SIGNIFICANT, MODERATE, OR LIMITED (60)	0.84	34 26 46 5 X SQ= 13.78 P = 0. RV YES (NO)
119	TEND TO BE THOSE WHERE INTEREST ARTICULATION BY INSTITUTIONAL GROUPS IS VERY SIGNIFICANT OR SIGNIFICANT (74)	0.90	119	TEND TO BE THOSE WHERE INTEREST ARTICULATION BY INSTITUTIONAL GROUPS IS MODERATE OR LIMITED (26)	0.59	61 13 7 19 X SQ= 24.75 P = 0. RV YES (YES)
120	ALWAYS ARE THOSE WHERE INTEREST ARTICULATION BY INSTITUTIONAL GROUPS IS VERY SIGNIFICANT, SIGNIFICANT, OR MODERATE (90)	1.00	120	TEND LESS TO BE THOSE WHERE INTEREST ARTICULATION BY INSTITUTIONAL GROUPS IS VERY SIGNIFICANT, SIGNIFICANT, OR MODERATE (90)	0.69	68 22 0 10 X SQ= 20.27 P = 0. RV NO (YES)
125	TEND TO BE THOSE WHERE INTEREST ARTICULATION BY ANOMIC GROUPS IS FREQUENT OR OCCASIONAL (64)	0.80	125	TEND TO BE THOSE WHERE INTEREST ARTICULATION BY ANOMIC GROUPS IS INFREQUENT OR VERY INFREQUENT (35)	0.83	60 4 15 20 X SQ= 29.20 P = 0. RV YES (YES)
126	TEND TO BE THOSE WHERE INTEREST ARTICULATION BY ANOMIC GROUPS IS FREQUENT, OCCASIONAL, OR INFREQUENT (83)	0.96	126	TEND TO BE THOSE WHERE INTEREST ARTICULATION BY ANOMIC GROUPS IS VERY INFREQUENT (16)	0.54	72 11 3 13 X SQ= 30.17 P = 0. RV YES (YES)
128	TEND TO BE THOSE WHERE INTEREST ARTICULATION BY POLITICAL PARTIES IS LIMITED OR NEGLIGIBLE (45)	0.57	128	TEND TO BE THOSE WHERE INTEREST ARTICULATION BY POLITICAL PARTIES IS SIGNIFICANT OR MODERATE (48)	0.70	27 21 36 9 X SQ= 4.96 P = 0.016 RV YES (NO)
134	TILT TOWARD BEING THOSE WHERE INTEREST AGGREGATION BY THE EXECUTIVE IS LIMITED OR NEGLIGIBLE (46)	0.50	134	TILT TOWARD BEING THOSE WHERE INTEREST AGGREGATION BY THE EXECUTIVE IS SIGNIFICANT OR MODERATE (57)	0.68	36 21 36 10 X SQ= 2.09 P = 0.131 RV YES (NO)

```
137  TEND TO BE THOSE                      0.84     137  TEND TO BE THOSE                      0.62        11    18
     WHERE INTEREST AGGREGATION                          WHERE INTEREST AGGREGATION                        57    11
     BY THE LEGISLATURE                                  BY THE LEGISLATURE                            X SQ=  18.30
     IS LIMITED OR NEGLIGIBLE  (68)                      IS SIGNIFICANT OR MODERATE  (29)              P =     0.
                                                                                                       RV YES (YES)

138  TEND TO BE THOSE                      0.66     138  TEND TO BE THOSE                      0.83        23    25
     WHERE INTEREST AGGREGATION                          WHERE INTEREST AGGREGATION                        44     5
     BY THE LEGISLATURE                                  BY THE LEGISLATURE                            X SQ=  18.00
     IS NEGLIGIBLE  (49)                                 IS SIGNIFICANT, MODERATE, OR                  P =     0.
                                                         LIMITED  (48)                                 RV YES (YES)

139  LEAN LESS TOWARD BEING THOSE          0.62     139  LEAN MORE TOWARD BEING THOSE          0.81        28     6
     WHERE THE PARTY SYSTEM IS QUANTITATIVELY            WHERE THE PARTY SYSTEM IS QUANTITATIVELY          45    26
     OTHER THAN ONE-PARTY  (71)                          OTHER THAN ONE-PARTY  (71)                    X SQ=   3.06
                                                                                                       P =   0.069
                                                                                                       RV NO  (NO )

142  TEND MORE TO BE THOSE                 0.77     142  TEND LESS TO BE THOSE                 0.52        15    15
     WHERE THE PARTY SYSTEM IS QUANTITATIVELY            WHERE THE PARTY SYSTEM IS QUANTITATIVELY          50    16
     OTHER THAN MULTI-PARTY  (66)                        OTHER THAN MULTI-PARTY  (66)                  X SQ=   5.14
                                                                                                       P =   0.018
                                                                                                       RV NO  (YES)

144  TEND TO BE THOSE                      0.85     144  TEND TO BE THOSE                      0.50        28     6
     WHERE THE PARTY SYSTEM IS QUANTITATIVELY            WHERE THE PARTY SYSTEM IS QUANTITATIVELY           5     6
     ONE-PARTY, RATHER THAN                              TWO-PARTY, RATHER THAN                        X SQ=   4.05
     TWO-PARTY  (34)                                     ONE-PARTY  (11)                               P =   0.044
                                                                                                       RV YES (YES)

146  TILT LESS TOWARD BEING THOSE          0.89     146  ALWAYS ARE THOSE                      1.00         8     0
     WHERE THE PARTY SYSTEM IS QUALITATIVELY             WHERE THE PARTY SYSTEM IS QUALITATIVELY           63    29
     OTHER THAN                                          OTHER THAN                                    X SQ=   2.19
     MASS-BASED TERRITORIAL  (92)                        MASS-BASED TERRITORIAL  (92)                  P =   0.101
                                                                                                       RV NO  (NO )

147  TEND TO BE THOSE                      0.87     147  TEND TO BE THOSE                      0.54         8    15
     WHERE THE PARTY SYSTEM IS QUALITATIVELY             WHERE THE PARTY SYSTEM IS QUALITATIVELY           54    13
     OTHER THAN                                          CLASS-ORIENTED OR MULTI-IDEOLOGICAL  (23)     X SQ=  14.70
     CLASS-ORIENTED OR MULTI-IDEOLOGICAL  (67)                                                         P =     0.
                                                                                                       RV YES (YES)

148  TEND LESS TO BE THOSE                 0.82     148  ALWAYS ARE THOSE                      1.00        14     0
     WHERE THE PARTY SYSTEM IS QUALITATIVELY             WHERE THE PARTY SYSTEM IS QUALITATIVELY           64    32
     OTHER THAN                                          OTHER THAN                                    X SQ=   5.06
     AFRICAN TRANSITIONAL  (96)                          AFRICAN TRANSITIONAL  (96)                    P =   0.009
                                                                                                       RV NO  (NO )

153  LEAN TOWARD BEING THOSE               0.56     153  LEAN TOWARD BEING THOSE               0.68        23    19
     WHERE THE PARTY SYSTEM IS                           WHERE THE PARTY SYSTEM IS                         29     9
     MODERATELY STABLE OR UNSTABLE  (38)                 STABLE  (42)                                  X SQ=   3.18
                                                                                                       P =   0.061
                                                                                                       RV YES (NO )
```

158	TEND LESS TO BE THOSE WHERE PERSONALISMO IS NEGLIGIBLE (56)	0.51	158	TEND MORE TO BE THOSE WHERE PERSONALISMO IS NEGLIGIBLE (56)	0.74	32 8 33 23 X SQ= 3.82 P = 0.045 RV NO (NO)
161	LEAN TOWARD BEING THOSE WHERE THE POLITICAL LEADERSHIP IS ELITIST OR MODERATELY ELITIST (47)	0.55	161	LEAN TOWARD BEING THOSE WHERE THE POLITICAL LEADERSHIP IS NON-ELITIST (50)	0.65	36 11 30 20 X SQ= 2.35 P = 0.087 RV YES (NO)
164	TEND LESS TO BE THOSE WHERE THE REGIME'S LEADERSHIP CHARISMA IS NEGLIGIBLE (65)	0.52	164	TEND MORE TO BE THOSE WHERE THE REGIME'S LEADERSHIP CHARISMA IS NEGLIGIBLE (65)	0.94	32 2 35 30 X SQ= 14.76 P = 0. RV NO (NO)
168	TEND TO BE THOSE WHERE THE HORIZONTAL POWER DISTRIBUTION IS LIMITED OR NEGLIGIBLE (72)	0.81	168	TEND TO BE THOSE WHERE THE HORIZONTAL POWER DISTRIBUTION IS SIGNIFICANT (34)	0.65	14 20 61 11 X SQ= 19.11 P = 0. RV YES (YES)
169	TEND TO BE THOSE WHERE THE HORIZONTAL POWER DISTRIBUTION IS NEGLIGIBLE (48)	0.55	169	TEND TO BE THOSE WHERE THE HORIZONTAL POWER DISTRIBUTION IS SIGNIFICANT OR LIMITED (58)	0.77	34 24 41 7 X SQ= 7.86 P = 0.003 RV YES (NO)
174	TEND TO BE THOSE WHERE THE LEGISLATURE IS PARTIALLY EFFECTIVE, LARGELY INEFFECTIVE, OR WHOLLY INEFFECTIVE (72)	0.83	174	TEND TO BE THOSE WHERE THE LEGISLATURE IS FULLY EFFECTIVE (28)	0.53	12 16 58 14 X SQ= 11.91 P = 0. RV YES (YES)
175	TEND TO BE THOSE WHERE THE LEGISLATURE IS LARGELY INEFFECTIVE OR WHOLLY INEFFECTIVE (49)	0.61	175	TEND TO BE THOSE WHERE THE LEGISLATURE IS FULLY EFFECTIVE OR PARTIALLY EFFECTIVE (51)	0.80	27 24 43 6 X SQ= 12.81 P = 0. RV YES (NO)
178	LEAN TOWARD BEING THOSE WHERE THE LEGISLATURE IS UNICAMERAL (53)	0.58	178	LEAN TOWARD BEING THOSE WHERE THE LEGISLATURE IS BICAMERAL (51)	0.65	42 11 31 20 X SQ= 3.40 P = 0.054 RV YES (NO)
179	TEND TO BE THOSE WHERE THE EXECUTIVE IS DOMINANT (52)	0.70	179	TEND TO BE THOSE WHERE THE EXECUTIVE IS STRONG (39)	0.71	44 8 19 20 X SQ= 11.85 P = 0. RV YES (YES)

No.	Statement	Value	No.	Statement	Value	Statistics
181	TEND TO BE THOSE WHERE THE BUREAUCRACY IS SEMI-MODERN, RATHER THAN MODERN (55)	0.89	181	TEND TO BE THOSE WHERE THE BUREAUCRACY IS MODERN, RATHER THAN SEMI-MODERN (21)	0.50	5 16 39 16 X SQ= 11.97 P = 0. RV YES (YES)
182	TEND LESS TO BE THOSE WHERE THE BUREAUCRACY IS SEMI-MODERN, RATHER THAN POST-COLONIAL TRANSITIONAL (55)	0.61	182	ALWAYS ARE THOSE WHERE THE BUREAUCRACY IS SEMI-MODERN, RATHER THAN POST-COLONIAL TRANSITIONAL (55)	1.00	39 16 25 0 X SQ= 7.36 P = 0.002 RV NO (NO)
186	LEAN LESS TOWARD BEING THOSE WHERE PARTICIPATION BY THE MILITARY IN POLITICS IS NEUTRAL, RATHER THAN SUPPORTIVE (56)	0.57	186	LEAN MORE TOWARD BEING THOSE WHERE PARTICIPATION BY THE MILITARY IN POLITICS IS NEUTRAL, RATHER THAN SUPPORTIVE (56)	0.81	26 5 35 21 X SQ= 3.39 P = 0.050 RV NO (NO)
187	TEND TO BE THOSE WHERE THE ROLE OF THE POLICE IS POLITICALLY SIGNIFICANT (66)	0.77	187	TEND TO BE THOSE WHERE THE ROLE OF THE POLICE IS NOT POLITICALLY SIGNIFICANT (35)	0.64	56 10 17 18 X SQ= 13.27 P = 0. RV YES (YES)
188	TEND TO BE THOSE WHERE THE CHARACTER OF THE LEGAL SYSTEM IS OTHER THAN CIVIL LAW (81)	0.81	188	TEND TO BE THOSE WHERE THE CHARACTER OF THE LEGAL SYSTEM IS CIVIL LAW (32)	0.53	15 17 66 15 X SQ= 11.88 P = 0. RV YES (YES)
189	TEND MORE TO BE THOSE WHERE THE CHARACTER OF THE LEGAL SYSTEM IS OTHER THAN COMMON LAW (108)	0.98	189	TEND LESS TO BE THOSE WHERE THE CHARACTER OF THE LEGAL SYSTEM IS OTHER THAN COMMON LAW (108)	0.84	2 5 81 27 X SQ= 4.93 P = 0.017 RV NO (YES)
192	TEND LESS TO BE THOSE WHERE THE CHARACTER OF THE LEGAL SYSTEM IS OTHER THAN MUSLIM OR PARTLY MUSLIM (86)	0.67	192	ALWAYS ARE THOSE WHERE THE CHARACTER OF THE LEGAL SYSTEM IS OTHER THAN MUSLIM OR PARTLY MUSLIM (86)	1.00	27 0 54 32 X SQ= 12.24 P = 0. RV NO (NO)
193	TEND LESS TO BE THOSE WHERE THE CHARACTER OF THE LEGAL SYSTEM IS OTHER THAN PARTLY INDIGENOUS (86)	0.66	193	ALWAYS ARE THOSE WHERE THE CHARACTER OF THE LEGAL SYSTEM IS OTHER THAN PARTLY INDIGENOUS (86)	1.00	28 0 54 32 X SQ= 12.70 P = 0. RV NO (NO)

123 POLITIES
WHERE INTEREST ARTICULATION
BY NON-ASSOCIATIONAL GROUPS
IS SIGNIFICANT, MODERATE, OR
LIMITED (107)

123 POLITIES
WHERE INTEREST ARTICULATION
BY NON-ASSOCIATIONAL GROUPS
IS NEGLIGIBLE (8)

PREDICATE ONLY

107 IN LEFT COLUMN

AFGHANISTAN	ALBANIA	ALGERIA	ARGENTINA	BELGIUM	BOLIVIA	BRAZIL	BULGARIA	BURMA	BURUNDI
CAMBODIA	CAMEROUN	CANADA	CEN AFR REP	CEYLON	CHAD	CHILE	CHINA, PR	COLOMBIA	CONGO(BRA)
CONGO(LEO)	CUBA	COSTA RICA	CYPRUS	CZECHOS'KIA	DAHOMEY	DOMIN REP	ECUADOR	EL SALVADOR	ETHIOPIA
FINLAND	FRANCE	GABON	GERMANY, E	GERMAN FR	GHANA	GREECE	GUATEMALA	GUINEA	HAITI
HONDURAS	HUNGARY	INDIA	INDONESIA	IRAN	IRAQ	IRELAND	ISRAEL	ITALY	IVORY COAST
JAMAICA	JAPAN	JORDAN	KOREA, N	KOREA REP	LAOS	LEBANON	LIBERIA	LIBYA	MALAGASY R
MALAYA	MALI	MAURITANIA	MEXICO	MONGOLIA	MOROCCO	NEPAL	NETHERLANDS	NICARAGUA	NIGER
NIGERIA	PAKISTAN	PANAMA	PARAGUAY	PERU	PHILIPPINES	POLAND	PORTUGAL	RUMANIA	RWANDA
SA'U ARABIA	SENEGAL	SIERRE LEO	SOMALIA	SO AFRICA	SPAIN	SUDAN	SWITZERLAND	SYRIA	TANGANYIKA
THAILAND	TOGO	TRINIDAD	TUNISIA	TURKEY	UGANDA	USSR	UAR	UK	US
UPPER VOLTA	URUGUAY	VENEZUELA	VIETNAM, N	VIETNAM REP	YEMEN	YUGOSLAVIA			

8 IN RIGHT COLUMN

AUSTRALIA	AUSTRIA	DENMARK	ICELAND	LUXEMBOURG	NEW ZEALAND	NORWAY	SWEDEN

124 POLITIES
WHERE INTEREST ARTICULATION
BY ANOMIC GROUPS
IS FREQUENT (14)

124 POLITIES
WHERE INTEREST ARTICULATION
BY ANOMIC GROUPS
IS OCCASIONAL, INFREQUENT, OR
VERY INFREQUENT (85)

PREDICATE ONLY

14 IN LEFT COLUMN

CEYLON	CHINA, PR	CONGO(LEO)	GERMANY, E	IRAN	IRAQ	JORDAN	LAOS	PAKISTAN	RWANDA
SO AFRICA	SUDAN	SYRIA	VIETNAM REP						

85 IN RIGHT COLUMN

AFGHANISTAN	AUSTRALIA	AUSTRIA	BELGIUM	BOLIVIA	BRAZIL	BURMA	BURUNDI	CAMBODIA	CAMEROUN
CANADA	CEN AFR REP	CHAD	CHILE	COLOMBIA	CONGO(BRA)	COSTA RICA	CZECHOS'KIA	DAHOMEY	DENMARK
ECUADOR	EL SALVADOR	ETHIOPIA	FINLAND	GABON	GERMAN FR	GHANA	GREECE	GUATEMALA	GUINEA
HAITI	HONDURAS	ICELAND	INDIA	INDONESIA	IRELAND	ISRAEL	ITALY	IVORY COAST	JAMAICA
JAPAN	KOREA REP	LEBANON	LIBERIA	LIBYA	LUXEMBOURG	MALAGASY R	MALAYA	MALI	MAURITANIA
MEXICO	MOROCCO	NEPAL	NETHERLANDS	NEW ZEALAND	NICARAGUA	NIGER	NIGERIA	NORWAY	PANAMA
PARAGUAY	PERU	PHILIPPINES	POLAND	PORTUGAL	SA'U ARABIA	SENEGAL	SIERRE LEO	SOMALIA	SPAIN
SWEDEN	SWITZERLAND	TANGANYIKA	THAILAND	TOGO	TRINIDAD	TUNISIA	TURKEY	UGANDA	USSR
UAR	UK	UPPER VOLTA	URUGUAY	YUGOSLAVIA					

6 EXCLUDED BECAUSE AMBIGUOUS

ARGENTINA	CYPRUS	FRANCE	HUNGARY	US	VENEZUELA

6 EXCLUDED BECAUSE UNASCERTAINED

ALBANIA	BULGARIA	KOREA, N	MONGOLIA	RUMANIA	VIETNAM, N

4 EXCLUDED BECAUSE UNASCERTAINABLE

ALGERIA	CUBA	DOMIN REP	YEMEN

```
************************************************************************************************************

125  POLITIES                                           125  POLITIES
     WHERE INTEREST ARTICULATION                             WHERE INTEREST ARTICULATION
     BY ANOMIC GROUPS                                        BY ANOMIC GROUPS
     IS FREQUENT OR OCCASIONAL  (64)                         IS INFREQUENT OR VERY INFREQUENT  (35)

                                                                               BOTH SUBJECT AND PREDICATE
............................................................................................................

      64 IN LEFT COLUMN

AFGHANISTAN BOLIVIA     BRAZIL      BURMA       CAMBODIA    CAMEROUN    CEN AFR REP CEYLON      CHAD        CHINA, PR
COLOMBIA    CONGO(BRA)  CONGO(LEO)  DAHOMEY     ECUADOR     EL SALVADOR ETHIOPIA    GABON       GERMANY, E  GHANA
GUATEMALA   GUINEA      HAITI       HONDURAS    INDIA       INDONESIA   IRAN        IRAQ        IVORY COAST JAPAN
JORDAN      KOREA REP   LAOS        LEBANON     MALAYA      MALI        MAURITANIA  MEXICO      MOROCCO     NEPAL
NIGER       NIGERIA     PAKISTAN    PANAMA      PARAGUAY    PERU        POLAND      RWANDA      SA'U ARABIA SENEGAL
SIERRE LEO  SOMALIA     SO AFRICA   SPAIN       SUDAN       SYRIA       THAILAND    TOGO        TURKEY      UGANDA
USSR        UPPER VOLTA VIETNAM REP YUGOSLAVIA

      35 IN RIGHT COLUMN

AUSTRALIA   AUSTRIA     BELGIUM     BURUNDI     CANADA      CHILE       COSTA RICA  CZECHOS'KIA DENMARK     FINLAND
GERMAN FR   GREECE      ICELAND     IRELAND     ISRAEL      ITALY       JAMAICA     LIBERIA     LIBYA       LUXEMBOURG
MALAGASY R  NETHERLANDS NEW ZEALAND NICARAGUA   NORWAY      PHILIPPINES PORTUGAL    SWEDEN      SWITZERLAND TANGANYIKA
TRINIDAD    TUNISIA     UAR         UK          URUGUAY

       6 EXCLUDED BECAUSE AMBIGUOUS

ARGENTINA   CYPRUS      FRANCE      HUNGARY     US          VENEZUELA

       6 EXCLUDED BECAUSE UNASCERTAINED

ALBANIA     BULGARIA    KOREA, N    MONGOLIA    RUMANIA     VIETNAM, N

       4 EXCLUDED BECAUSE UNASCERTAINABLE

ALGERIA     CUBA        DOMIN REP   YEMEN
------------------------------------------------------------------------------------------------------------

  2  TEND TO BE THOSE                        0.98          2  TEND TO BE THOSE                      0.54           1   19
     LOCATED ELSEWHERE THAN IN                                LOCATED IN WEST EUROPE, SCANDINAVIA,                 63   16
     WEST EUROPE, SCANDINAVIA,                                NORTH AMERICA, OR AUSTRALASIA  (22)             X SQ=  35.81
     NORTH AMERICA, OR AUSTRALASIA  (93)                                                                      P =     0.
                                                                                                              RV YES (YES)

  3  TEND MORE TO BE THOSE                   0.98          3  TEND LESS TO BE THOSE                 0.69           1   11
     LOCATED ELSEWHERE THAN IN                                LOCATED ELSEWHERE THAN IN                            63   24
     WEST EUROPE  (102)                                       WEST EUROPE  (102)                              X SQ=  16.25
                                                                                                              P =     0.
                                                                                                              RV NO  (YES)
```

```
 5  TEND TO BE THOSE                      0.80     5  TEND TO BE THOSE                      0.94        1   16
    LOCATED IN EAST EUROPE, RATHER THAN               LOCATED IN SCANDINAVIA OR WEST EUROPE,           4    1
    IN SCANDINAVIA OR WEST EUROPE  (9)                RATHER THAN IN EAST EUROPE  (18)          X SQ=   8.23
                                                                                                P =    0.003
                                                                                                RV YES (YES)

 8  TEND LESS TO BE THOSE                 0.78     8  TEND MORE TO BE THOSE                 0.97       14    1
    LOCATED ELSEWHERE THAN IN EAST ASIA               LOCATED ELSEWHERE THAN IN EAST ASIA             50   34
    SOUTH ASIA, OR SOUTHEAST ASIA  (97)               SOUTH ASIA, OR SOUTHEAST ASIA  (97)       X SQ=   4.97
                                                                                                P =    0.016
                                                                                                RV NO  (NO )

10  LEAN LESS TOWARD BEING THOSE          0.61    10  LEAN MORE TOWARD BEING THOSE          0.80       25    7
    LOCATED ELSEWHERE THAN IN NORTH AFRICA,           LOCATED ELSEWHERE THAN IN NORTH AFRICA,         39   28
    OR CENTRAL AND SOUTH AFRICA  (82)                 OR CENTRAL AND SOUTH AFRICA  (82)         X SQ=   2.94
                                                                                                P =    0.072
                                                                                                RV NO  (NO )

11  TEND LESS TO BE THOSE                 0.62    11  TEND MORE TO BE THOSE                 0.89       24    4
    LOCATED ELSEWHERE THAN IN CENTRAL                 LOCATED ELSEWHERE THAN IN CENTRAL               40   31
    AND SOUTH AFRICA  (87)                            AND SOUTH AFRICA  (87)                    X SQ=   6.35
                                                                                                P =    0.006
                                                                                                RV NO  (NO )

20  LEAN TOWARD BEING THOSE               0.54    20  LEAN TOWARD BEING THOSE               0.86       14    1
    LOCATED IN EAST ASIA, SOUTH ASIA, OR              LOCATED IN THE CARIBBEAN,                       12    6
    SOUTHEAST ASIA, RATHER THAN IN                    CENTRAL AMERICA, OR SOUTH AMERICA,        X SQ=   2.07
    THE CARIBBEAN, CENTRAL AMERICA,                   RATHER THAN IN EAST ASIA, SOUTH ASIA,     P =    0.095
    OR SOUTH AMERICA  (18)                            OR SOUTHEAST ASIA  (22)                   RV YES (NO )

22  TEND TO BE THOSE                      0.70    22  TEND TO BE THOSE                      0.57       45   15
    WHOSE TERRITORIAL SIZE IS                         WHOSE TERRITORIAL SIZE IS                       19   20
    VERY LARGE, LARGE, OR MEDIUM  (68)                SMALL  (47)                               X SQ=   6.04
                                                                                                P =    0.010
                                                                                                RV YES (YES)

26  TEND TO BE THOSE                      0.69    26  TEND TO BE THOSE                      0.54       20   19
    WHOSE POPULATION DENSITY IS                       WHOSE POPULATION DENSITY IS                     44   16
    LOW  (67)                                         VERY HIGH, HIGH, OR MEDIUM  (48)          X SQ=   4.11
                                                                                                P =    0.032
                                                                                                RV YES (NO )

28  TEND TO BE THOSE                      0.68    28  TEND TO BE THOSE                      0.57       41   15
    WHOSE POPULATION GROWTH RATE                      WHOSE POPULATION GROWTH RATE                    19   20
    IS HIGH  (62)                                     IS LOW  (48)                              X SQ=   4.92
                                                                                                P =    0.018
                                                                                                RV YES (YES)

29  TEND TO BE THOSE                      0.67    29  TEND TO BE THOSE                      0.87       20   28
    WHERE THE DEGREE OF URBANIZATION                  WHERE THE DEGREE OF URBANIZATION                40    4
    IS LOW  (49)                                      IS HIGH  (56)                             X SQ=  22.42
                                                                                                P =    0.
                                                                                                RV YES (YES)
```

30	TEND TO BE THOSE WHOSE AGRICULTURAL POPULATION IS HIGH (56)	0.68	30	TEND TO BE THOSE WHOSE AGRICULTURAL POPULATION IS MEDIUM, LOW, OR VERY LOW (57)	0.80	43 7 20 28 X SQ= 19.08 P = 0. RV YES (YES)
31	TEND TO BE THOSE WHOSE AGRICULTURAL POPULATION IS HIGH OR MEDIUM (90)	0.97	31	TEND TO BE THOSE WHOSE AGRICULTURAL POPULATION IS LOW OR VERY LOW (24)	0.54	61 16 2 19 X SQ= 31.94 P = 0. RV YES (YES)
34	TEND TO BE THOSE WHOSE GROSS NATIONAL PRODUCT IS VERY LOW (53)	0.55	34	TEND TO BE THOSE WHOSE GROSS NATIONAL PRODUCT IS VERY HIGH, HIGH, MEDIUM, OR LOW (62)	0.69	29 .24 35 11 X SQ= 4.03 P = 0.035 RV YES (NO)
35	TEND TO BE THOSE WHOSE PER CAPITA GROSS NATIONAL PRODUCT IS MEDIUM, LOW, OR VERY LOW (91)	0.97	35	TEND TO BE THOSE WHOSE PER CAPITA GROSS NATIONAL PRODUCT IS VERY HIGH OR HIGH (24)	0.54	2 19 62 16 X SQ= 32.44 P = 0. RV YES (YES)
36	TEND TO BE THOSE WHOSE PER CAPITA GROSS NATIONAL PRODUCT IS LOW OR VERY LOW (73)	0.86	36	TEND TO BE THOSE WHOSE PER CAPITA GROSS NATIONAL PRODUCT IS VERY HIGH, HIGH, OR MEDIUM (42)	0.71	9 25 55 10 X SQ= 30.53 P = 0. RV YES (YES)
37	TEND TO BE THOSE WHOSE PER CAPITA GROSS NATIONAL PRODUCT IS VERY LOW (51)	0.66	37	TEND TO BE THOSE WHOSE PER CAPITA GROSS NATIONAL PRODUCT IS VERY HIGH, HIGH, MEDIUM, OR LOW (64)	0.86	22 30 42 5 X SQ= 21.90 P = 0. RV YES (YES)
39	TEND TO BE THOSE WHOSE INTERNATIONAL FINANCIAL STATUS IS LOW OR VERY LOW (76)	0.78	39	TEND TO BE THOSE WHOSE INTERNATIONAL FINANCIAL STATUS IS VERY HIGH, HIGH, OR MEDIUM (38)	0.51	14 18 49 17 X SQ= 7.45 P = 0.006 RV YES (YES)
42	TEND TO BE THOSE WHOSE ECONOMIC DEVELOPMENTAL STATUS IS UNDERDEVELOPED OR VERY UNDERDEVELOPED (76)	0.86	42	TEND TO BE THOSE WHOSE ECONOMIC DEVELOPMENTAL STATUS IS DEVELOPED OR INTERMEDIATE (36)	0.59	9 20 54 14 X SQ= 18.83 P = 0. RV YES (YES)
43	TEND TO BE THOSE WHOSE ECONOMIC DEVELOPMENTAL STATUS IS VERY UNDERDEVELOPED (57)	0.69	43	TEND TO BE THOSE WHOSE ECONOMIC DEVELOPMENTAL STATUS IS DEVELOPED, INTERMEDIATE, OR UNDERDEVELOPED (55)	0.80	19 28 43 7 X SQ= 19.89 P = 0. RV YES (YES)

45	TEND TO BE THOSE WHERE THE LITERACY RATE IS BELOW FIFTY PERCENT (54)	0.73	45	TEND TO BE THOSE WHERE THE LITERACY RATE IS FIFTY PERCENT OR ABOVE (55)	0.79	17 27 45 7 X SQ= 21.86 P = 0. RV YES (YES)
47	TEND TO BE THOSE WHERE THE LITERACY RATE IS BETWEEN FIFTY AND NINETY PERCENT, RATHER THAN NINETY PERCENT OR ABOVE (30)	0.76	47	TEND TO BE THOSE WHERE THE LITERACY RATE IS NINETY PERCENT OR ABOVE, RATHER THAN BETWEEN FIFTY AND NINETY PERCENT (25)	0.63	4 17 13 10 X SQ= 5.02 P = 0.015 RV YES (YES)
50	TEND TO BE THOSE WHERE FREEDOM OF THE PRESS IS INTERMITTENT, INTERNALLY ABSENT, OR INTERNALLY AND EXTERNALLY ABSENT (56)	0.71	50	TEND TO BE THOSE WHERE FREEDOM OF THE PRESS IS COMPLETE (43)	0.75	15 24 36 8 X SQ= 14.63 P = 0. RV YES (YES)
54	TEND TO BE THOSE WHERE NEWSPAPER CIRCULATION IS LESS THAN ONE HUNDRED PER THOUSAND (76)	0.92	54	TEND TO BE THOSE WHERE NEWSPAPER CIRCULATION IS ONE HUNDRED OR MORE PER THOUSAND (37)	0.66	5 23 59 12 X SQ= 34.60 P = 0. RV YES (YES)
56	TEND LESS TO BE THOSE WHERE THE RELIGION IS PREDOMINANTLY LITERATE (79)	0.56	56	TEND MORE TO BE THOSE WHERE THE RELIGION IS PREDOMINANTLY LITERATE (79)	0.89	34 31 27 4 X SQ= 9.52 P = 0.001 RV NO (NO)
57	LEAN MORE TOWARD BEING THOSE WHERE THE RELIGION IS OTHER THAN CATHOLIC (90)	0.86	57	LEAN LESS TOWARD BEING THOSE WHERE THE RELIGION IS OTHER THAN CATHOLIC (90)	0.69	9 11 55 24 X SQ= 3.22 P = 0.065 RV NO (YES)
62	ALWAYS ARE THOSE WHERE THE RELIGION IS CATHOLIC, RATHER THAN PROTESTANT (25)	1.00	62	TILT LESS TOWARD BEING THOSE WHERE THE RELIGION IS CATHOLIC, RATHER THAN PROTESTANT (25)	0.69	0 5 9 11 X SQ= 1.83 P = 0.123 RV NO (NO)
63	TEND TO BE THOSE WHERE THE RELIGION IS PREDOMINANTLY OR PARTLY OTHER THAN CHRISTIAN (68)	0.83	63	TEND TO BE THOSE WHERE THE RELIGION IS PREDOMINANTLY SOME KIND OF CHRISTIAN (46)	0.74	11 26 52 9 X SQ= 28.55 P = 0. RV YES (YES)
64	TEND TO BE THOSE WHERE THE RELIGION IS MUSLIM, RATHER THAN CHRISTIAN (18)	0.54	64	TEND TO BE THOSE WHERE THE RELIGION IS CHRISTIAN, RATHER THAN MUSLIM (46)	0.90	11 26 13 3 X SQ= 9.98 P = 0.001 RV YES (YES)

66	TILT TOWARD BEING THOSE THAT ARE RELIGIOUSLY HETEROGENEOUS (49)	0.55	66	TILT TOWARD BEING THOSE THAT ARE RELIGIOUSLY HOMOGENEOUS (57)	0.62	26 21 32 13 X SQ= 1.83 P = 0.135 RV YES (NO)
67	LEAN LESS TOWARD BEING THOSE THAT ARE RACIALLY HOMOGENEOUS (82)	0.66	67	LEAN MORE TOWARD BEING THOSE THAT ARE RACIALLY HOMOGENEOUS (82)	0.85	40 28 21 5 X SQ= 3.07 P = 0.055 RV NO (NO)
68	TEND TO BE THOSE THAT ARE LINGUISTICALLY WEAKLY HETEROGENEOUS OR STRONGLY HETEROGENEOUS (62)	0.75	68	TEND TO BE THOSE THAT ARE LINGUISTICALLY HOMOGENEOUS (52)	0.71	16 25 47 10 X SQ= 17.75 P = 0. RV YES (YES)
69	TEND TO BE THOSE THAT ARE LINGUISTICALLY STRONGLY HETEROGENEOUS (50)	0.63	69	TEND TO BE THOSE THAT ARE LINGUISTICALLY HOMOGENEOUS OR WEAKLY HETEROGENEOUS (64)	0.77	23 27 40 8 X SQ= 13.29 P = 0. RV YES (YES)
70	TEND MORE TO BE THOSE THAT ARE RELIGIOUSLY, RACIALLY, OR LINGUISTICALLY HETEROGENEOUS (85)	0.95	70	TEND LESS TO BE THOSE THAT ARE RELIGIOUSLY, RACIALLY, OR LINGUISTICALLY HETEROGENEOUS (85)	0.56	3 14 56 18 X SQ= 17.95 P = 0. RV NO (YES)
72	LEAN TOWARD BEING THOSE WHOSE DATE OF INDEPENDENCE IS AFTER 1914 (59)	0.63	72	LEAN TOWARD BEING THOSE WHOSE DATE OF INDEPENDENCE IS BEFORE 1914 (52)	0.59	23 20 39 14 X SQ= 3.36 P = 0.054 RV YES (NO)
73	TEND TO BE THOSE WHOSE DATE OF INDEPENDENCE IS AFTER 1945 (46)	0.53	73	TEND TO BE THOSE WHOSE DATE OF INDEPENDENCE IS BEFORE 1945 (65)	0.71	29 24 33 10 X SQ= 4.12 P = 0.032 RV YES (NO)
74	TEND TO BE THOSE THAT ARE NOT HISTORICALLY WESTERN (87)	0.95	74	TEND TO BE THOSE THAT ARE HISTORICALLY WESTERN (26)	0.59	3 20 60 14 X SQ= 32.75 P = 0. RV YES (YES)
75	TEND TO BE THOSE THAT ARE NOT HISTORICALLY WESTERN AND ARE NOT SIGNIFICANTLY WESTERNIZED (52)	0.65	75	TEND TO BE THOSE THAT ARE HISTORICALLY WESTERN OR SIGNIFICANTLY WESTERNIZED (62)	0.83	22 29 41 6 X SQ= 18.84 P = 0. RV YES (YES)

No.	Statement		No.	Statement		Statistics
76	TEND TO BE THOSE THAT HAVE BEEN SIGNIFICANTLY OR PARTIALLY WESTERNIZED THROUGH A COLONIAL RELATIONSHIP, RATHER THAN BEING HISTORICALLY WESTERN (70)	0.94	76	TEND TO BE THOSE THAT ARE HISTORICALLY WESTERN, RATHER THAN HAVING BEEN SIGNIFICANTLY OR PARTIALLY WESTERNIZED THROUGH A COLONIAL RELATIONSHIP (26)	0.61	3 20 50 13 X SQ= 28.60 P = 0. RV YES (YES)
77	TEND TO BE THOSE THAT HAVE BEEN PARTIALLY WESTERNIZED, RATHER THAN SIGNIFICANTLY WESTERNIZED, THROUGH A COLONIAL RELATIONSHIP (41)	0.71	77	TEND TO BE THOSE THAT HAVE BEEN SIGNIFICANTLY WESTERNIZED, RATHER THAN PARTIALLY WESTERNIZED, THROUGH A COLONIAL RELATIONSHIP (28)	0.62	14 8 35 5 X SQ= 3.54 P = 0.048 RV YES (NO)
80	TILT TOWARD BEING THOSE WHOSE DATE OF INDEPENDENCE IS AFTER 1914, AND THAT WERE FORMERLY DEPENDENCIES OF FRANCE, RATHER THAN BRITAIN (24)	0.63	80	TILT TOWARD BEING THOSE WHOSE DATE OF INDEPENDENCE IS AFTER 1914, AND THAT WERE FORMERLY DEPENDENCIES OF BRITAIN, RATHER THAN FRANCE (19)	0.75	12 6 20 2 X SQ= 2.28 P = 0.110 RV YES (NO)
81	LEAN MORE TOWARD BEING THOSE WHERE THE TYPE OF POLITICAL MODERNIZATION IS LATER EUROPEAN OR LATER EUROPEAN DERIVED, RATHER THAN EARLY EUROPEAN OR EARLY EUROPEAN DERIVED (40)	0.94	81	LEAN LESS TOWARD BEING THOSE WHERE THE TYPE OF POLITICAL MODERNIZATION IS LATER EUROPEAN OR LATER EUROPEAN DERIVED, RATHER THAN EARLY EUROPEAN OR EARLY EUROPEAN DERIVED (40)	0.67	1 8 16 16 X SQ= 2.92 P = 0.056 RV NO (YES)
82	TEND TO BE THOSE WHERE THE TYPE OF POLITICAL MODERNIZATION IS DEVELOPED TUTELARY OR UNDEVELOPED TUTELARY, RATHER THAN EARLY OR LATER EUROPEAN OR EUROPEAN DERIVED (55)	0.69	82	TEND TO BE THOSE WHERE THE TYPE OF POLITICAL MODERNIZATION IS EARLY OR LATER EUROPEAN OR EUROPEAN DERIVED, RATHER THAN DEVELOPED TUTELARY OR UNDEVELOPED TUTELARY (51)	0.69	17 24 38 11 X SQ= 10.76 P = 0.001 RV YES (YES)
83	TEND LESS TO BE THOSE WHERE THE TYPE OF POLITICAL MODERNIZATION IS OTHER THAN NON-EUROPEAN AUTOCHTHONOUS (106)	0.86	83	ALWAYS ARE THOSE WHERE THE TYPE OF POLITICAL MODERNIZATION IS OTHER THAN NON-EUROPEAN AUTOCHTHONOUS (106)	1.00	9 0 55 35 X SQ= 3.85 P = 0.025 RV NO (NO)
85	TEND TO BE THOSE WHERE THE STAGE OF POLITICAL MODERNIZATION IS MID- OR EARLY TRANSITIONAL, RATHER THAN ADVANCED (54)	0.68	85	TEND TO BE THOSE WHERE THE STAGE OF POLITICAL MODERNIZATION IS ADVANCED, RATHER THAN MID- OR EARLY TRANSITIONAL (60)	0.77	20 27 43 8 X SQ= 16.80 P = 0. RV YES (YES)
87	TEND LESS TO BE THOSE WHOSE IDEOLOGICAL ORIENTATION IS OTHER THAN DEVELOPMENTAL (58)	0.51	87	TEND MORE TO BE THOSE WHOSE IDEOLOGICAL ORIENTATION IS OTHER THAN DEVELOPMENTAL (58)	0.82	23 5 24 23 X SQ= 5.98 P = 0.008 RV NO (NO)
89	TEND TO BE THOSE WHOSE IDEOLOGICAL ORIENTATION IS OTHER THAN CONVENTIONAL (62)	0.85	89	TEND TO BE THOSE WHOSE IDEOLOGICAL ORIENTATION IS CONVENTIONAL (33)	0.75	8 21 44 7 X SQ= 25.47 P = 0. RV YES (YES)

No.			No.			
94	TEND TO BE THOSE WHERE THE STATUS OF THE REGIME IS AUTHORITARIAN OR TOTALITARIAN (41)	0.60	94	TEND TO BE THOSE WHERE THE STATUS OF THE REGIME IS CONSTITUTIONAL (51)	0.82	18 27 27 6 X SQ= 11.98 P = 0. RV YES (YES)
96	TEND LESS TO BE THOSE WHERE THE STATUS OF THE REGIME IS CONSTITUTIONAL OR TOTALITARIAN (67)	0.56	96	TEND MORE TO BE THOSE WHERE THE STATUS OF THE REGIME IS CONSTITUTIONAL OR TOTALITARIAN (67)	0.91	24 30 19 3 X SQ= 9.54 P = 0.001 RV NO (YES)
99	TEND TO BE THOSE WHERE GOVERNMENTAL STABILITY IS MODERATELY PRESENT AND DATES FROM THE POST-WAR PERIOD, OR IS ABSENT (36)	0.65	99	TEND TO BE THOSE WHERE GOVERNMENTAL STABILITY IS GENERALLY PRESENT AND DATES FROM AT LEAST THE INTER-WAR PERIOD, OR FROM THE POST-WAR PERIOD (50)	0.90	16 26 30 3 X SQ= 19.57 P = 0. RV YES (YES)
100	TEND LESS TO BE THOSE WHERE GOVERNMENTAL STABILITY IS GENERALLY PRESENT AND DATES FROM AT LEAST THE INTER-WAR PERIOD, OR IS GENERALLY OR MODERATELY PRESENT, AND DATES FROM THE POST-WAR PERIOD (64)	0.56	100	ALWAYS ARE THOSE WHERE GOVERNMENTAL STABILITY IS GENERALLY PRESENT AND DATES FROM AT LEAST THE INTER-WAR PERIOD, OR IS GENERALLY OR MODERATELY PRESENT, AND DATES FROM THE POST-WAR PERIOD (64)	1.00	25 29 20 0 X SQ= 15.48 P = 0. RV NO (YES)
101	TEND TO BE THOSE WHERE THE REPRESENTATIVE CHARACTER OF THE REGIME IS LIMITED POLYARCHIC, PSEUDO-POLYARCHIC, OR NON-POLYARCHIC (57)	0.80	101	TEND TO BE THOSE WHERE THE REPRESENTATIVE CHARACTER OF THE REGIME IS POLYARCHIC (41)	0.76	10 25 41 8 X SQ= 23.73 P = 0. RV YES (YES)
102	TEND TO BE THOSE WHERE THE REPRESENTATIVE CHARACTER OF THE REGIME IS PSEUDO-POLYARCHIC OR NON-POLYARCHIC (49)	0.62	102	TEND TO BE THOSE WHERE THE REPRESENTATIVE CHARACTER OF THE REGIME IS POLYARCHIC OR LIMITED POLYARCHIC (59)	0.86	22 30 36 5 X SQ= 18.33 P = 0. RV YES (YES)
105	TEND TO BE THOSE WHERE THE ELECTORAL SYSTEM IS PARTIALLY COMPETITIVE OR NON-COMPETITIVE (47)	0.72	105	TEND TO BE THOSE WHERE THE ELECTORAL SYSTEM IS COMPETITIVE (43)	0.76	12 25 31 8 X SQ= 15.25 P = 0. RV YES (YES)
106	TEND TO BE THOSE WHERE THE ELECTORAL SYSTEM IS NON-COMPETITIVE (30)	0.50	106	TEND TO BE THOSE WHERE THE ELECTORAL SYSTEM IS COMPETITIVE OR PARTIALLY COMPETITIVE (52)	0.88	18 29 18 4 X SQ= 9.70 P = 0.001 RV YES (YES)
107	TEND TO BE THOSE WHERE AUTONOMOUS GROUPS ARE PARTIALLY TOLERATED IN POLITICS, ARE TOLERATED ONLY OUTSIDE POLITICS, OR ARE NOT TOLERATED AT ALL (65)	0.77	107	TEND TO BE THOSE WHERE AUTONOMOUS GROUPS ARE FULLY TOLERATED IN POLITICS (46)	0.74	14 26 47 9 X SQ= 22.05 P = 0. RV YES (YES)

111	TEND TO BE THOSE WHERE POLITICAL ENCULTURATION IS LOW (42)	0.60	111	TEND TO BE THOSE WHERE POLITICAL ENCULTURATION IS HIGH OR MEDIUM (53)	0.86	23 25 35 4 X SQ= 15.11 P = 0. RV YES (YES)
114	TEND TO BE THOSE WHERE SECTIONALISM IS EXTREME OR MODERATE (61)	0.70	114	TEND TO BE THOSE WHERE SECTIONALISM IS NEGLIGIBLE (47)	0.61	42 13 18 20 X SQ= 7.04 P = 0.008 RV YES (YES)
116	TEND TO BE THOSE WHERE INTEREST ARTICULATION BY ASSOCIATIONAL GROUPS IS LIMITED OR NEGLIGIBLE (79)	0.90	116	TEND TO BE THOSE WHERE INTEREST ARTICULATION BY ASSOCIATIONAL GROUPS IS SIGNIFICANT OR MODERATE (32)	0.65	6 22 55 12 X SQ= 29.04 P = 0. RV YES (YES)
117	TEND TO BE THOSE WHERE INTEREST ARTICULATION BY ASSOCIATIONAL GROUPS IS NEGLIGIBLE (51)	0.61	117	TEND TO BE THOSE WHERE INTEREST ARTICULATION BY ASSOCIATIONAL GROUPS IS SIGNIFICANT, MODERATE, OR LIMITED (60)	0.82	24 28 37 6 X SQ= 14.61 P = 0. RV YES (YES)
119	ALWAYS ARE THOSE WHERE INTEREST ARTICULATION BY INSTITUTIONAL GROUPS IS VERY SIGNIFICANT OR SIGNIFICANT (74)	1.00	119	TEND TO BE THOSE WHERE INTEREST ARTICULATION BY INSTITUTIONAL GROUPS IS MODERATE OR LIMITED (26)	0.71	51 10 0 24 X SQ= 46.74 P = 0. RV YES (YES)
121	TEND TO BE THOSE WHERE INTEREST ARTICULATION BY NON-ASSOCIATIONAL GROUPS IS SIGNIFICANT (54)	0.70	121	TEND TO BE THOSE WHERE INTEREST ARTICULATION BY NON-ASSOCIATIONAL GROUPS IS MODERATE, LIMITED, OR NEGLIGIBLE (61)	0.89	45 4 19 31 X SQ= 29.07 P = 0. RV YES (YES)
122	TEND TO BE THOSE WHERE INTEREST ARTICULATION BY NON-ASSOCIATIONAL GROUPS IS SIGNIFICANT OR MODERATE (83)	0.94	122	TEND TO BE THOSE WHERE INTEREST ARTICULATION BY NON-ASSOCIATIONAL GROUPS IS LIMITED OR NEGLIGIBLE (32)	0.57	60 15 4 20 X SQ= 29.20 P = 0. RV YES (YES)
123	ALWAYS ARE THOSE WHERE INTEREST ARTICULATION BY NON-ASSOCIATIONAL GROUPS IS SIGNIFICANT, MODERATE, OR LIMITED (107)	1.00	123	TEND LESS TO BE THOSE WHERE INTEREST ARTICULATION BY NON-ASSOCIATIONAL GROUPS IS SIGNIFICANT, MODERATE, OR LIMITED (107)	0.77	64 27 0 8 X SQ= 12.99 P = 0. RV NO (YES)
128	TEND TO BE THOSE WHERE INTEREST ARTICULATION BY POLITICAL PARTIES IS LIMITED OR NEGLIGIBLE (45)	0.54	128	TEND TO BE THOSE WHERE INTEREST ARTICULATION BY POLITICAL PARTIES IS SIGNIFICANT OR MODERATE (48)	0.70	22 23 26 10 X SQ= 3.60 P = 0.042 RV YES (YES)

130	TEND MORE TO BE THOSE WHERE INTEREST AGGREGATION BY POLITICAL PARTIES IS MODERATE, LIMITED, OR NEGLIGIBLE (71)	0.95	130	TEND LESS TO BE THOSE WHERE INTEREST AGGREGATION BY POLITICAL PARTIES IS MODERATE, LIMITED, OR NEGLIGIBLE (71)	0.71	2 9 39 22 X SQ= 6.20 P = 0.007 RV NO (YES)
133	TEND LESS TO BE THOSE WHERE INTEREST AGGREGATION BY THE EXECUTIVE IS MODERATE, LIMITED, OR NEGLIGIBLE (76)	0.64	133	TEND MORE TO BE THOSE WHERE INTEREST AGGREGATION BY THE EXECUTIVE IS MODERATE, LIMITED, OR NEGLIGIBLE (76)	0.85	21 5 37 29 X SQ= 3.88 P = 0.032 RV NO (NO)
134	TEND TO BE THOSE WHERE INTEREST AGGREGATION BY THE EXECUTIVE IS LIMITED OR NEGLIGIBLE (46)	0.54	134	TEND TO BE THOSE WHERE INTEREST AGGREGATION BY THE EXECUTIVE IS SIGNIFICANT OR MODERATE (57)	0.79	26 27 30 7 X SQ= 8.19 P = 0.002 RV YES (YES)
136	ALWAYS ARE THOSE WHERE INTEREST AGGREGATION BY THE LEGISLATURE IS MODERATE, LIMITED, OR NEGLIGIBLE (85)	1.00	136	TEND LESS TO BE THOSE WHERE INTEREST AGGREGATION BY THE LEGISLATURE IS MODERATE, LIMITED, OR NEGLIGIBLE (85)	0.68	0 11 52 23 X SQ= 16.50 P = 0. RV NO (YES)
137	TEND TO BE THOSE WHERE INTEREST AGGREGATION BY THE LEGISLATURE IS LIMITED OR NEGLIGIBLE (68)	0.92	137	TEND TO BE THOSE WHERE INTEREST AGGREGATION BY THE LEGISLATURE IS SIGNIFICANT OR MODERATE (29)	0.71	4 24 48 10 X SQ= 34.23 P = 0. RV YES (YES)
138	TEND TO BE THOSE WHERE INTEREST AGGREGATION BY THE LEGISLATURE IS NEGLIGIBLE (49)	0.71	138	TEND TO BE THOSE WHERE INTEREST AGGREGATION BY THE LEGISLATURE IS SIGNIFICANT, MODERATE, OR LIMITED (48)	0.82	15 27 36 6 X SQ= 19.96 P = 0. RV YES (YES)
139	TILT LESS TOWARD BEING THOSE WHERE THE PARTY SYSTEM IS QUANTITATIVELY OTHER THAN ONE-PARTY (71)	0.66	139	TILT MORE TOWARD BEING THOSE WHERE THE PARTY SYSTEM IS QUANTITATIVELY OTHER THAN ONE-PARTY (71)	0.82	19 6 37 28 X SQ= 2.04 P = 0.145 RV NO (NO)
141	LEAN MORE TOWARD BEING THOSE WHERE THE PARTY SYSTEM IS QUANTITATIVELY OTHER THAN TWO-PARTY (87)	0.94	141	LEAN LESS TOWARD BEING THOSE WHERE THE PARTY SYSTEM IS QUANTITATIVELY OTHER THAN TWO-PARTY (87)	0.79	3 7 47 26 X SQ= 3.02 P = 0.080 RV NO (YES)
147	TEND TO BE THOSE WHERE THE PARTY SYSTEM IS QUALITATIVELY OTHER THAN CLASS-ORIENTED OR MULTI-IDEOLOGICAL (67)	0.93	147	TEND TO BE THOSE WHERE THE PARTY SYSTEM IS QUALITATIVELY CLASS-ORIENTED OR MULTI-IDEOLOGICAL (23)	0.63	3 19 43 11 X SQ= 25.80 P = 0. RV YES (YES)

No.	Statement	Proportion	No.	Statement	Proportion	Statistics
148	LEAN LESS TOWARD BEING THOSE WHERE THE PARTY SYSTEM IS QUALITATIVELY OTHER THAN AFRICAN TRANSITIONAL (96)	0.80	148	LEAN MORE TOWARD BEING THOSE WHERE THE PARTY SYSTEM IS QUALITATIVELY OTHER THAN AFRICAN TRANSITIONAL (96)	0.94	12 2 48 32 X SQ= 2.39 P = 0.077 RV NO (NO)
153	TEND TO BE THOSE WHERE THE PARTY SYSTEM IS MODERATELY STABLE OR UNSTABLE (38)	0.72	153	TEND TO BE THOSE WHERE THE PARTY SYSTEM IS STABLE (42)	0.82	11 23 29 5 X SQ= 17.55 P = 0. RV YES (YES)
154	TEND TO BE THOSE WHERE THE PARTY SYSTEM IS UNSTABLE (25)	0.52	154	TEND TO BE THOSE WHERE THE PARTY SYSTEM IS STABLE OR MODERATELY STABLE (55)	0.93	19 26 21 2 X SQ= 13.18 P = 0. RV YES (YES)
158	TEND TO BE THOSE WHERE PERSONALISMO IS PRONOUNCED OR MODERATE (40)	0.62	158	TEND TO BE THOSE WHERE PERSONALISMO IS NEGLIGIBLE (56)	0.82	31 6 19 27 X SQ= 13.73 P = 0. RV YES (YES)
161	TEND TO BE THOSE WHERE THE POLITICAL LEADERSHIP IS ELITIST OR MODERATELY ELITIST (47)	0.61	161	TEND TO BE THOSE WHERE THE POLITICAL LEADERSHIP IS NON-ELITIST (50)	0.76	31 8 20 25 X SQ= 9.34 P = 0.002 RV YES (YES)
164	TEND LESS TO BE THOSE WHERE THE REGIME'S LEADERSHIP CHARISMA IS NEGLIGIBLE (65)	0.53	164	TEND MORE TO BE THOSE WHERE THE REGIME'S LEADERSHIP CHARISMA IS NEGLIGIBLE (65)	0.82	24 6 27 28 X SQ= 6.49 P = 0.006 RV NO (YES)
168	TEND TO BE THOSE WHERE THE HORIZONTAL POWER DISTRIBUTION IS LIMITED OR NEGLIGIBLE (72)	0.89	168	TEND TO BE THOSE WHERE THE HORIZONTAL POWER DISTRIBUTION IS SIGNIFICANT (34)	0.69	6 24 51 11 X SQ= 30.66 P = 0. RV YES (YES)
169	TEND TO BE THOSE WHERE THE HORIZONTAL POWER DISTRIBUTION IS NEGLIGIBLE (48)	0.58	169	TEND TO BE THOSE WHERE THE HORIZONTAL POWER DISTRIBUTION IS SIGNIFICANT OR LIMITED (58)	0.80	24 28 33 7 X SQ= 11.18 P = 0. RV YES (YES)
170	TEND LESS TO BE THOSE WHERE THE LEGISLATIVE-EXECUTIVE STRUCTURE IS OTHER THAN PRESIDENTIAL (63)	0.52	170	TEND MORE TO BE THOSE WHERE THE LEGISLATIVE-EXECUTIVE STRUCTURE IS OTHER THAN PRESIDENTIAL (63)	0.81	27 6 29 26 X SQ= 6.34 P = 0.007 RV NO (NO)

171	TILT MORE TOWARD BEING THOSE WHERE THE LEGISLATIVE-EXECUTIVE STRUCTURE IS OTHER THAN PARLIAMENTARY-REPUBLICAN (90)	0.91	171	TILT LESS TOWARD BEING THOSE WHERE THE LEGISLATIVE-EXECUTIVE STRUCTURE IS OTHER THAN PARLIAMENTARY-REPUBLICAN (90)	0.78	5 7 51 25 X SQ= 1.90 P = 0.112 RV NO (YES)
172	TEND MORE TO BE THOSE WHERE THE LEGISLATIVE-EXECUTIVE STRUCTURE IS OTHER THAN PARLIAMENTARY-ROYALIST (88)	0.88	172	TEND LESS TO BE THOSE WHERE THE LEGISLATIVE-EXECUTIVE STRUCTURE IS OTHER THAN PARLIAMENTARY-ROYALIST (88)	0.60	7 14 52 21 X SQ= 8.47 P = 0.004 RV NO (YES)
174	TEND TO BE THOSE WHERE THE LEGISLATURE IS PARTIALLY EFFECTIVE, LARGELY INEFFECTIVE, OR WHOLLY INEFFECTIVE (72)	0.92	174	TEND TO BE THOSE WHERE THE LEGISLATURE IS FULLY EFFECTIVE (28)	0.65	4 22 48 12 X SQ= 29.04 P = 0. RV YES (YES)
175	TEND TO BE THOSE WHERE THE LEGISLATURE IS LARGELY INEFFECTIVE OR WHOLLY INEFFECTIVE (49)	0.63	175	TEND TO BE THOSE WHERE THE LEGISLATURE IS FULLY EFFECTIVE OR PARTIALLY EFFECTIVE (51)	0.76	19 26 33 8 X SQ= 11.59 P = 0. RV YES (YES)
178	TEND TO BE THOSE WHERE THE LEGISLATURE IS UNICAMERAL (53)	0.61	178	TEND TO BE THOSE WHERE THE LEGISLATURE IS BICAMERAL (51)	0.73	35 9 22 24 X SQ= 8.43 P = 0.002 RV YES (YES)
179	TEND TO BE THOSE WHERE THE EXECUTIVE IS DOMINANT (52)	0.74	179	TEND TO BE THOSE WHERE THE EXECUTIVE IS STRONG (39)	0.73	35 8 12 22 X SQ= 15.09 P = 0. RV YES (YES)
181	TEND TO BE THOSE WHERE THE BUREAUCRACY IS SEMI-MODERN, RATHER THAN MODERN (55)	0.97	181	TEND TO BE THOSE WHERE THE BUREAUCRACY IS MODERN, RATHER THAN SEMI-MODERN (21)	0.56	1 18 29 14 X SQ= 17.99 P = 0. RV YES (YES)
182	LEAN LESS TOWARD BEING THOSE WHERE THE BUREAUCRACY IS SEMI-MODERN, RATHER THAN POST-COLONIAL TRANSITIONAL (55)	0.58	182	LEAN MORE TOWARD BEING THOSE WHERE THE BUREAUCRACY IS SEMI-MODERN, RATHER THAN POST-COLONIAL TRANSITIONAL (55)	0.82	29 14 21 3 X SQ= 2.30 P = 0.086 RV NO (NO)
186	TEND LESS TO BE THOSE WHERE PARTICIPATION BY THE MILITARY IN POLITICS IS NEUTRAL, RATHER THAN SUPPORTIVE (56)	0.53	186	TEND MORE TO BE THOSE WHERE PARTICIPATION BY THE MILITARY IN POLITICS IS NEUTRAL, RATHER THAN SUPPORTIVE (56)	0.91	20 3 23 29 X SQ= 10.22 P = 0.001 RV NO (YES)

```
187  TEND TO BE THOSE                        0.86      187  TEND TO BE THOSE                        0.78        48     7
     WHERE THE ROLE OF THE POLICE                           WHERE THE ROLE OF THE POLICE                           8    25
     IS POLITICALLY SIGNIFICANT  (66)                       IS NOT POLITICALLY SIGNIFICANT  (35)              X SQ=  32.74
                                                                                                              P =    0.
                                                                                                              RV YES (YES)

189  ALWAYS ARE THOSE                        1.00      189  TEND LESS TO BE THOSE                   0.83         0     6
     WHERE THE CHARACTER OF THE LEGAL SYSTEM                WHERE THE CHARACTER OF THE LEGAL SYSTEM               64    29
     IS OTHER THAN COMMON LAW  (108)                        IS OTHER THAN COMMON LAW  (108)                   X SQ=   8.86
                                                                                                              P =    0.001
                                                                                                              RV NO  (YES)

190  ALWAYS ARE THOSE                        1.00      190  TEND LESS TO BE THOSE                   0.67        15    12
     WHERE THE CHARACTER OF THE LEGAL SYSTEM                WHERE THE CHARACTER OF THE LEGAL SYSTEM                0     6
     IS CIVIL LAW, RATHER THAN                              IS CIVIL LAW, RATHER THAN                         X SQ=   4.08
     COMMON LAW  (32)                                       COMMON LAW  (32)                                  P =    0.021
                                                                                                              RV NO  (YES)

192  TEND LESS TO BE THOSE                   0.66      192  TEND MORE TO BE THOSE                   0.89        21     4
     WHERE THE CHARACTER OF THE LEGAL SYSTEM                WHERE THE CHARACTER OF THE LEGAL SYSTEM               41    31
     IS OTHER THAN                                          IS OTHER THAN                                     X SQ=   4.78
     MUSLIM OR PARTLY MUSLIM  (86)                          MUSLIM OR PARTLY MUSLIM  (86)                     P =    0.017
                                                                                                              RV NO  (NO )

193  TEND LESS TO BE THOSE                   0.62      193  TEND MORE TO BE THOSE                   0.89        24     4
     WHERE THE CHARACTER OF THE LEGAL SYSTEM                WHERE THE CHARACTER OF THE LEGAL SYSTEM               39    31
     IS OTHER THAN PARTLY INDIGENOUS  (86)                  IS OTHER THAN PARTLY INDIGENOUS  (86)             X SQ=   6.59
                                                                                                              P =    0.005
                                                                                                              RV NO  (NO )
```

```
126  POLITIES                                      126  POLITIES
     WHERE INTEREST ARTICULATION                        WHERE INTEREST ARTICULATION
     BY ANOMIC GROUPS                                   BY ANOMIC GROUPS
     IS FREQUENT, OCCASIONAL, OR                        IS VERY INFREQUENT  (16)
     INFREQUENT  (83)

                                                                        PREDICATE ONLY

      83 IN LEFT COLUMN

AFGHANISTAN BOLIVIA     BRAZIL      BURMA       BURUNDI     CAMBODIA    CAMEROUN    CEN AFR REP CEYLON      CHAD
CHILE       CHINA, PR   COLOMBIA    CONGO(BRA)  CONGO(LEO)  COSTA RICA  CZECHOS'KIA DAHOMEY     ECUADOR     EL SALVADOR
ETHIOPIA    GABON       GERMANY, E  GHANA       GREECE      GUATEMALA   GUINEA      HAITI       HONDURAS    INDIA
INDONESIA   IRAN        IRAQ        ISRAEL      ITALY       IVORY COAST JAMAICA     JAPAN       JORDAN      KOREA REP
LAOS        LEBANON     LIBERIA     LIBYA       MALAGASY R  MALAYA      MALI        MAURITANIA  MEXICO      MOROCCO
NEPAL       NICARAGUA   NIGER       NIGERIA     PAKISTAN    PANAMA      PARAGUAY    PERU        PHILIPPINES POLAND
PORTUGAL    RWANDA      SA'U ARABIA SENEGAL     SIERRE LEO  SOMALIA     SO AFRICA   SPAIN       SUDAN       SYRIA
TANGANYIKA  THAILAND    TOGO        TRINIDAD    TUNISIA     TURKEY      UGANDA      USSR        UAR         UPPER VOLTA
URUGUAY     VIETNAM REP YUGOSLAVIA

      16 IN RIGHT COLUMN

AUSTRALIA   AUSTRIA     BELGIUM     CANADA      DENMARK     FINLAND     GERMAN FR   ICELAND     IRELAND     LUXEMBOURG
NETHERLANDS NEW ZEALAND NORWAY      SWEDEN      SWITZERLAND UK

      6 EXCLUDED BECAUSE AMBIGUOUS

ARGENTINA   CYPRUS      FRANCE      HUNGARY     US          VENEZUELA

      6 EXCLUDED BECAUSE UNASCERTAINED

ALBANIA     BULGARIA    KOREA, N    MONGOLIA    RUMANIA     VIETNAM, N

      4 EXCLUDED BECAUSE UNASCERTAINABLE

ALGERIA     CUBA        DOMIN REP   YEMEN
```

127 POLITIES
WHERE INTEREST ARTICULATION
BY POLITICAL PARTIES
IS SIGNIFICANT (21)

127 POLITIES
WHERE INTEREST ARTICULATION
BY POLITICAL PARTIES
IS MODERATE, LIMITED, OR
NEGLIGIBLE (72)

PREDICATE ONLY

21 IN LEFT COLUMN

BELGIUM	BOLIVIA	CEYLON	CHILE	CONGO(LEO)	DENMARK	FINLAND	GUATEMALA	ICELAND	ISRAEL
LEBANON	LUXEMBOURG	NETHERLANDS	NIGERIA	NORWAY	PERU	SWEDEN	SWITZERLAND	SYRIA	UGANDA
VENEZUELA									

72 IN RIGHT COLUMN

ALBANIA	ARGENTINA	AUSTRALIA	AUSTRIA	BRAZIL	BULGARIA	BURUNDI	CAMBODIA	CAMEROUN	CANADA
CEN AFR REP	CHAD	CHINA, PR	COLOMBIA	CONGO(BRA)	COSTA RICA	CZECHOS'KIA	DAHOMEY	DOMIN REP	ECUADOR
EL SALVADOR	FRANCE	GABON	GERMANY, E	GERMAN FR	GHANA	GREECE	GUINEA	HONDURAS	HUNGARY
INDIA	IRELAND	ITALY	IVORY COAST	JAMAICA	JAPAN	KOREA, N	LIBERIA	MALAGASY R	MALAYA
MALI	MAURITANIA	MEXICO	MONGOLIA	MOROCCO	NEW ZEALAND	NICARAGUA	NIGER	PANAMA	PARAGUAY
PHILIPPINES	POLAND	PORTUGAL	RUMANIA	RWANDA	SENEGAL	SIERRE LEO	SOMALIA	SPAIN	TANGANYIKA
TOGO	TRINIDAD	TUNISIA	TURKEY	USSR	UAR	UK	US	UPPER VOLTA	VIETNAM, N
VIETNAM REP	YUGOSLAVIA								

5 EXCLUDED BECAUSE AMBIGUOUS

CYPRUS	INDONESIA	LAOS	SO AFRICA	URUGUAY

5 EXCLUDED BECAUSE IRRELEVANT

AFGHANISTAN	ETHIOPIA	HAITI	LIBYA	SA'U ARABIA

12 EXCLUDED BECAUSE UNASCERTAINABLE

ALGERIA	BURMA	CUBA	IRAN	IRAQ	JORDAN	KOREA REP	NEPAL	PAKISTAN	SUDAN
THAILAND	YEMEN								

```
****************************************************************************************************

128  POLITIES                                          128  POLITIES
     WHERE INTEREST ARTICULATION                            WHERE INTEREST ARTICULATION
     BY POLITICAL PARTIES                                   BY POLITICAL PARTIES
     IS SIGNIFICANT OR MODERATE  (48)                       IS LIMITED OR NEGLIGIBLE  (45)


                                                                              BOTH SUBJECT AND PREDICATE
.....................................................................................................

      48 IN LEFT COLUMN

ARGENTINA   AUSTRALIA   AUSTRIA     BELGIUM     BOLIVIA     BRAZIL      CAMEROUN    CEYLON      CHILE       COLOMBIA
CONGO(LEO)  DENMARK     ECUADOR     EL SALVADOR FINLAND     FRANCE      GERMAN FR   GREECE      GUATEMALA   HONDURAS
ICELAND     INDIA       IRELAND     ISRAEL      ITALY       JAMAICA     JAPAN       LEBANON     LIBERIA     LUXEMBOURG
MALAGASY R  MALAYA      MOROCCO     NETHERLANDS NEW ZEALAND NIGERIA     NORWAY      PANAMA      PERU        SIERRE LEO
SOMALIA     SWEDEN      SWITZERLAND SYRIA       TRINIDAD    UGANDA      UK          VENEZUELA

      45 IN RIGHT COLUMN

ALBANIA     BULGARIA    BURUNDI     CAMBODIA    CANADA      CEN AFR REP CHAD        CHINA, PR   CONGO(BRA)  COSTA RICA
CZECHOS'KIA DAHOMEY     DOMIN REP   GABON       GERMANY, E  GHANA       GUINEA      HUNGARY     IVORY COAST KOREA, N
MALI        MAURITANIA  MEXICO      MONGOLIA    NICARAGUA   NIGER       PARAGUAY    PHILIPPINES POLAND      PORTUGAL
RUMANIA     RWANDA      SENEGAL     SPAIN       TANGANYIKA  TOGO        TUNISIA     TURKEY      USSR        UAR
US          UPPER VOLTA VIETNAM, N  VIETNAM REP YUGOSLAVIA

       5 EXCLUDED BECAUSE AMBIGUOUS

CYPRUS      INDONESIA   LAOS        SO AFRICA   URUGUAY

       5 EXCLUDED BECAUSE IRRELEVANT

AFGHANISTAN ETHIOPIA    HAITI       LIBYA       SA'U ARABIA

      12 EXCLUDED BECAUSE UNASCERTAINABLE

ALGERIA     BURMA       CUBA        IRAN        IRAQ        JORDAN      KOREA REP   NEPAL       PAKISTAN    SUDAN
THAILAND    YEMEN

-----------------------------------------------------------------------------------------------------

  2  TEND LESS TO BE THOSE                    0.62        2  TEND MORE TO BE THOSE                 0.91         18     4
     LOCATED ELSEWHERE THAN IN                               LOCATED ELSEWHERE THAN IN                          30    41
     WEST EUROPE, SCANDINAVIA,                               WEST EUROPE, SCANDINAVIA,                      X SQ=   9.00
     NORTH AMERICA, OR AUSTRALASIA  (93)                     NORTH AMERICA, OR AUSTRALASIA  (93)            P =    0.001
                                                                                                            RV NO  (YES)

  5  ALWAYS ARE THOSE                         1.00        5  TEND TO BE THOSE                      0.82         16     2
     LOCATED IN SCANDINAVIA OR WEST EUROPE,                  LOCATED IN EAST EUROPE, RATHER THAN                 0     9
     RATHER THAN IN EAST EUROPE  (18)                        IN SCANDINAVIA OR WEST EUROPE  (9)             X SQ=  16.13
                                                                                                            P =    0.
                                                                                                            RV YES (YES)
```

6	TEND LESS TO BE THOSE LOCATED ELSEWHERE THAN IN THE CARIBBEAN, CENTRAL AMERICA, OR SOUTH AMERICA (93)	0.71	6	TEND MORE TO BE THOSE LOCATED ELSEWHERE THAN IN THE CARIBBEAN, CENTRAL AMERICA, OR SOUTH AMERICA (93)	0.89	14 5 34 40 X SQ= 3.61 P = 0.040 RV NO (YES)
10	TEND MORE TO BE THOSE LOCATED ELSEWHERE THAN IN NORTH AFRICA, OR CENTRAL AND SOUTH AFRICA (82)	0.81	10	TEND LESS TO BE THOSE LOCATED ELSEWHERE THAN IN NORTH AFRICA, OR CENTRAL AND SOUTH AFRICA (82)	0.58	9 19 39 26 X SQ= 5.02 P = 0.023 RV NO (YES)
19	TEND TO BE THOSE LOCATED IN THE CARIBBEAN, CENTRAL AMERICA, OR SOUTH AMERICA, RATHER THAN IN NORTH AFRICA OR CENTRAL AND SOUTH AFRICA (22)	0.61	19	TEND TO BE THOSE LOCATED IN NORTH AFRICA OR CENTRAL AND SOUTH AFRICA, RATHER THAN IN THE CARIBBEAN, CENTRAL AMERICA, OR SOUTH AMERICA (33)	0.79	9 19 14 5 X SQ= 6.24 P = 0.008 RV YES (YES)
20	LEAN TOWARD BEING THOSE LOCATED IN THE CARIBBEAN, CENTRAL AMERICA, OR SOUTH AMERICA, RATHER THAN IN EAST ASIA, SOUTH ASIA, OR SOUTHEAST ASIA (22)	0.78	20	LEAN TOWARD BEING THOSE LOCATED IN EAST ASIA, SOUTH ASIA, OR SOUTHEAST ASIA, RATHER THAN IN THE CARIBBEAN, CENTRAL AMERICA, OR SOUTH AMERICA (18)	0.58	4 7 14 5 X SQ= 2.64 P = 0.063 RV YES (YES)
29	TEND TO BE THOSE WHERE THE DEGREE OF URBANIZATION IS HIGH (56)	0.70	29	TEND TO BE THOSE WHERE THE DEGREE OF URBANIZATION IS LOW (49)	0.55	32 17 14 21 X SQ= 4.31 P = 0.027 RV YES (YES)
30	TEND TO BE THOSE WHOSE AGRICULTURAL POPULATION IS MEDIUM, LOW, OR VERY LOW (57)	0.73	30	TEND TO BE THOSE WHOSE AGRICULTURAL POPULATION IS HIGH (56)	0.64	13 28 35 16 X SQ= 10.98 P = 0.001 RV YES (YES)
34	LEAN TOWARD BEING THOSE WHOSE GROSS NATIONAL PRODUCT IS VERY HIGH, HIGH, MEDIUM, OR LOW (62)	0.62	34	LEAN TOWARD BEING THOSE WHOSE GROSS NATIONAL PRODUCT IS VERY LOW (53)	0.56	30 20 18 25 X SQ= 2.36 P = 0.098 RV YES (YES)
36	TEND TO BE THOSE WHOSE PER CAPITA GROSS NATIONAL PRODUCT IS VERY HIGH, HIGH, OR MEDIUM (42)	0.56	36	TEND TO BE THOSE WHOSE PER CAPITA GROSS NATIONAL PRODUCT IS LOW OR VERY LOW (73)	0.73	27 12 21 33 X SQ= 7.18 P = 0.006 RV YES (YES)
37	TEND TO BE THOSE WHOSE PER CAPITA GROSS NATIONAL PRODUCT IS VERY HIGH, HIGH, MEDIUM, OR LOW (64)	0.75	37	TEND TO BE THOSE WHOSE PER CAPITA GROSS NATIONAL PRODUCT IS VERY LOW (51)	0.53	36 21 12 24 X SQ= 6.71 P = 0.006 RV YES (YES)

40	TEND MORE TO BE THOSE WHOSE INTERNATIONAL FINANCIAL STATUS IS VERY HIGH, HIGH, MEDIUM, OR LOW (71)	0.76	40	TEND LESS TO BE THOSE WHOSE INTERNATIONAL FINANCIAL STATUS IS VERY HIGH, HIGH, MEDIUM, OR LOW (71)	0.53	35 23 11 20 X SQ= 4.05 P = 0.029 RV NO (YES)
41	TEND LESS TO BE THOSE WHOSE ECONOMIC DEVELOPMENTAL STATUS IS INTERMEDIATE, UNDERDEVELOPED, OR VERY UNDERDEVELOPED (94)	0.70	41	TEND MORE TO BE THOSE WHOSE ECONOMIC DEVELOPMENTAL STATUS IS INTERMEDIATE, UNDERDEVELOPED, OR VERY UNDERDEVELOPED (94)	0.89	14 5 32 40 X SQ= 4.04 P = 0.038 RV NO (YES)
42	TEND LESS TO BE THOSE WHOSE ECONOMIC DEVELOPMENTAL STATUS IS UNDERDEVELOPED OR VERY UNDERDEVELOPED (76)	0.52	42	TEND MORE TO BE THOSE WHOSE ECONOMIC DEVELOPMENTAL STATUS IS UNDERDEVELOPED OR VERY UNDERDEVELOPED (76)	0.76	22 11 24 34 X SQ= 4.42 P = 0.029 RV NO (YES)
43	TEND TO BE THOSE WHOSE ECONOMIC DEVELOPMENTAL STATUS IS DEVELOPED, INTERMEDIATE, OR UNDERDEVELOPED (55)	0.66	43	TEND TO BE THOSE WHOSE ECONOMIC DEVELOPMENTAL STATUS IS VERY UNDERDEVELOPED (57)	0.62	31 17 16 28 X SQ= 6.23 P = 0.012 RV YES (YES)
45	TILT TOWARD BEING THOSE WHERE THE LITERACY RATE IS FIFTY PERCENT OR ABOVE (55)	0.63	45	TILT TOWARD BEING THOSE WHERE THE LITERACY RATE IS BELOW FIFTY PERCENT (54)	0.54	29 19 17 22 X SQ= 1.82 P = 0.135 RV YES (YES)
46	TEND MORE TO BE THOSE WHERE THE LITERACY RATE IS TEN PERCENT OR ABOVE (84)	0.89	46	TEND LESS TO BE THOSE WHERE THE LITERACY RATE IS TEN PERCENT OR ABOVE (84)	0.66	42 27 5 14 X SQ= 5.83 P = 0.010 RV NO (YES)
50	TEND TO BE THOSE WHERE FREEDOM OF THE PRESS IS COMPLETE (43)	0.70	50	TEND TO BE THOSE WHERE FREEDOM OF THE PRESS IS INTERMITTENT, INTERNALLY ABSENT, OR INTERNALLY AND EXTERNALLY ABSENT (56)	0.71	30 10 13 24 X SQ= 10.82 P = 0.001 RV YES (YES)
51	TEND TO BE THOSE WHERE FREEDOM OF THE PRESS IS COMPLETE OR INTERMITTENT (60)	0.98	51	TEND TO BE THOSE WHERE FREEDOM OF THE PRESS IS INTERNALLY ABSENT, OR INTERNALLY AND EXTERNALLY ABSENT (37)	0.68	42 11 1 23 X SQ= 34.78 P = 0. RV YES (YES)
52	ALWAYS ARE THOSE WHERE FREEDOM OF THE PRESS IS COMPLETE, INTERMITTENT, OR INTERNALLY ABSENT (82)	1.00	52	TEND LESS TO BE THOSE WHERE FREEDOM OF THE PRESS IS COMPLETE, INTERMITTENT, OR INTERNALLY ABSENT (82)	0.62	44 21 0 13 X SQ= 17.53 P = 0. RV NO (YES)

53	TEND LESS TO BE THOSE WHERE NEWSPAPER CIRCULATION IS LESS THAN THREE HUNDRED PER THOUSAND (101)	0.75	53	TEND MORE TO BE THOSE WHERE NEWSPAPER CIRCULATION IS LESS THAN THREE HUNDRED PER THOUSAND (101)	0.96	12 2 36 43 X SQ= 6.15 P = 0.008 RV NO (YES)
54	TEND LESS TO BE THOSE WHERE NEWSPAPER CIRCULATION IS LESS THAN ONE HUNDRED PER THOUSAND (76)	0.52	54	TEND MORE TO BE THOSE WHERE NEWSPAPER CIRCULATION IS LESS THAN ONE HUNDRED PER THOUSAND (76)	0.74	23 11 25 32 X SQ= 3.93 P = 0.032 RV NO (YES)
55	TEND MORE TO BE THOSE WHERE NEWSPAPER CIRCULATION IS TEN OR MORE PER THOUSAND (78)	0.83	55	TEND LESS TO BE THOSE WHERE NEWSPAPER CIRCULATION IS TEN OR MORE PER THOUSAND (78)	0.60	40 26 8 17 X SQ= 4.86 P = 0.019 RV NO (YES)
62	TILT LESS TOWARD BEING THOSE WHERE THE RELIGION IS CATHOLIC, RATHER THAN PROTESTANT (25)	0.74	62	ALWAYS ARE THOSE WHERE THE RELIGION IS CATHOLIC, RATHER THAN PROTESTANT (25)	1.00	5 0 14 9 X SQ= 1.37 P = 0.144 RV NO (NO)
63	TILT TOWARD BEING THOSE WHERE THE RELIGION IS PREDOMINANTLY SOME KIND OF CHRISTIAN (46)	0.56	63	TILT TOWARD BEING THOSE WHERE THE RELIGION IS PREDOMINANTLY OR PARTLY OTHER THAN CHRISTIAN (68)	0.61	27 17 21 27 X SQ= 2.19 P = 0.100 RV YES (YES)
68	TILT TOWARD BEING THOSE THAT ARE LINGUISTICALLY HOMOGENEOUS (52)	0.56	68	TILT TOWARD BEING THOSE THAT ARE LINGUISTICALLY WEAKLY HETEROGENEOUS OR STRONGLY HETEROGENEOUS (62)	0.61	27 17 21 27 X SQ= 2.19 P = 0.100 RV YES (YES)
72	TEND TO BE THOSE WHOSE DATE OF INDEPENDENCE IS BEFORE 1914 (52)	0.60	72	TEND TO BE THOSE WHOSE DATE OF INDEPENDENCE IS AFTER 1914 (59)	0.65	28 15 19 28 X SQ= 4.54 P = 0.022 RV YES (YES)
73	TEND TO BE THOSE WHOSE DATE OF INDEPENDENCE IS BEFORE 1945 (65)	0.70	73	TEND TO BE THOSE WHOSE DATE OF INDEPENDENCE IS AFTER 1945 (46)	0.53	33 20 14 23 X SQ= 4.28 P = 0.032 RV YES (YES)
74	TEND LESS TO BE THOSE THAT ARE NOT HISTORICALLY WESTERN (87)	0.62	74	TEND MORE TO BE THOSE THAT ARE NOT HISTORICALLY WESTERN (87)	0.82	18 8 29 37 X SQ= 3.82 P = 0.038 RV NO (YES)

75	TEND TO BE THOSE THAT ARE HISTORICALLY WESTERN OR SIGNIFICANTLY WESTERNIZED (62)	0.77	75	TEND TO BE THOSE THAT ARE NOT HISTORICALLY WESTERN AND ARE NOT SIGNIFICANTLY WESTERNIZED (52)	0.53	37 21 11 24 X SQ= 7.91 P = 0.003 RV YES (YES)
76	TILT LESS TOWARD BEING THOSE THAT HAVE BEEN SIGNIFICANTLY OR PARTIALLY WESTERNIZED THROUGH A COLONIAL RELATIONSHIP, RATHER THAN BEING HISTORICALLY WESTERN (70)	0.60	76	TILT MORE TOWARD BEING THOSE THAT HAVE BEEN SIGNIFICANTLY OR PARTIALLY WESTERNIZED THROUGH A COLONIAL RELATIONSHIP, RATHER THAN BEING HISTORICALLY WESTERN (70)	0.78	18 8 27 28 X SQ= 2.14 P = 0.100 RV NO (YES)
77	TEND TO BE THOSE THAT HAVE BEEN SIGNIFICANTLY WESTERNIZED, RATHER THAN PARTIALLY WESTERNIZED, THROUGH A COLONIAL RELATIONSHIP (28)	0.63	77	TEND TO BE THOSE THAT HAVE BEEN PARTIALLY WESTERNIZED, RATHER THAN SIGNIFICANTLY WESTERNIZED, THROUGH A COLONIAL RELATIONSHIP (41)	0.75	17 7 10 21 X SQ= 6.58 P = 0.007 RV YES (YES)
78	TEND MORE TO BE THOSE THAT HAVE BEEN SIGNIFICANTLY WESTERNIZED THROUGH A COLONIAL RELATIONSHIP, RATHER THAN WITHOUT SUCH A RELATIONSHIP (28)	0.94	78	TEND LESS TO BE THOSE THAT HAVE BEEN SIGNIFICANTLY WESTERNIZED THROUGH A COLONIAL RELATIONSHIP, RATHER THAN WITHOUT SUCH A RELATIONSHIP (28)	0.54	17 7 1 6 X SQ= 4.98 P = 0.025 RV NO (YES)
79	TILT LESS TOWARD BEING THOSE THAT WERE FORMERLY DEPENDENCIES OF BRITAIN OR FRANCE, RATHER THAN SPAIN (49)	0.61	79	TILT MORE TOWARD BEING THOSE THAT WERE FORMERLY DEPENDENCIES OF BRITAIN OR FRANCE, RATHER THAN SPAIN (49)	0.81	17 22 11 5 X SQ= 1.96 P = 0.138 RV NO (YES)
80	TEND TO BE THOSE WHOSE DATE OF INDEPENDENCE IS AFTER 1914, AND THAT WERE FORMERLY DEPENDENCIES OF BRITAIN, RATHER THAN FRANCE (19)	0.67	80	TEND TO BE THOSE WHOSE DATE OF INDEPENDENCE IS AFTER 1914, AND THAT WERE FORMERLY DEPENDENCIES OF FRANCE, RATHER THAN BRITAIN (24)	0.85	10 3 5 17 X SQ= 7.71 P = 0.004 RV YES (YES)
82	TEND TO BE THOSE WHERE THE TYPE OF POLITICAL MODERNIZATION IS EARLY OR LATER EUROPEAN OR EUROPEAN DERIVED, RATHER THAN DEVELOPED TUTELARY OR UNDEVELOPED TUTELARY (51)	0.64	82	TEND TO BE THOSE WHERE THE TYPE OF POLITICAL MODERNIZATION IS DEVELOPED TUTELARY OR UNDEVELOPED TUTELARY, RATHER THAN EARLY OR LATER EUROPEAN OR EUROPEAN DERIVED (55)	0.60	30 17 17 25 X SQ= 3.96 P = 0.035 RV YES (YES)
84	TILT TOWARD BEING THOSE WHERE THE TYPE OF POLITICAL MODERNIZATION IS DEVELOPED TUTELARY, RATHER THAN UNDEVELOPED TUTELARY (31)	0.59	84	TILT TOWARD BEING THOSE WHERE THE TYPE OF POLITICAL MODERNIZATION IS UNDEVELOPED TUTELARY, RATHER THAN DEVELOPED TUTELARY (24)	0.68	10 8 7 17 X SQ= 1.98 P = 0.117 RV YES (YES)
86	LEAN MORE TOWARD BEING THOSE WHERE THE STAGE OF POLITICAL MODERNIZATION IS ADVANCED OR MID-TRANSITIONAL, RATHER THAN EARLY TRANSITIONAL (76)	0.75	86	LEAN LESS TOWARD BEING THOSE WHERE THE STAGE OF POLITICAL MODERNIZATION IS ADVANCED OR MID-TRANSITIONAL, RATHER THAN EARLY TRANSITIONAL (76)	0.56	36 25 12 20 X SQ= 3.08 P = 0.054 RV NO (YES)

87	TEND TO BE THOSE WHOSE IDEOLOGICAL ORIENTATION IS OTHER THAN DEVELOPMENTAL (58)	0.76	87	TEND TO BE THOSE WHOSE IDEOLOGICAL ORIENTATION IS DEVELOPMENTAL (31)	0.50	8 20 26 20 X SQ= 4.41 P = 0.030 RV YES (YES)
89	TEND TO BE THOSE WHOSE IDEOLOGICAL ORIENTATION IS CONVENTIONAL (33)	0.71	89	TEND TO BE THOSE WHOSE IDEOLOGICAL ORIENTATION IS OTHER THAN CONVENTIONAL (62)	0.88	24 5 10 36 X SQ= 24.32 P = 0. RV YES (YES)
92	ALWAYS ARE THOSE WHERE THE SYSTEM STYLE IS LIMITED MOBILIZATIONAL, OR NON-MOBILIZATIONAL (93)	1.00	92	TEND LESS TO BE THOSE WHERE THE SYSTEM STYLE IS LIMITED MOBILIZATIONAL, OR NON-MOBILIZATIONAL (93)	0.59	0 18 48 26 X SQ= 21.88 P = 0. RV NO (YES)
93	TEND TO BE THOSE WHERE THE SYSTEM STYLE IS NON-MOBILIZATIONAL (78)	0.92	93	TEND TO BE THOSE WHERE THE SYSTEM STYLE IS MOBILIZATIONAL, OR LIMITED MOBILIZATIONAL (32)	0.55	4 24 44 20 X SQ= 21.02 P = 0. RV YES (YES)
94	TEND TO BE THOSE WHERE THE STATUS OF THE REGIME IS CONSTITUTIONAL (51)	0.95	94	TEND TO BE THOSE WHERE THE STATUS OF THE REGIME IS AUTHORITARIAN OR TOTALITARIAN (41)	0.75	39 8 2 24 X SQ= 35.54 P = 0. RV YES (YES)
95	ALWAYS ARE THOSE WHERE THE STATUS OF THE REGIME IS CONSTITUTIONAL OR AUTHORITARIAN (95)	1.00	95	TEND LESS TO BE THOSE WHERE THE STATUS OF THE REGIME IS CONSTITUTIONAL OR AUTHORITARIAN (95)	0.64	48 28 0 16 X SQ= 18.67 P = 0. RV NO (YES)
96	TEND MORE TO BE THOSE WHERE THE STATUS OF THE REGIME IS CONSTITUTIONAL OR TOTALITARIAN (67)	0.95	96	TEND LESS TO BE THOSE WHERE THE STATUS OF THE REGIME IS CONSTITUTIONAL OR TOTALITARIAN (67)	0.77	39 24 2 7 X SQ= 3.57 P = 0.033 RV NO (YES)
97	ALWAYS ARE THOSE WHERE THE STATUS OF THE REGIME IS AUTHORITARIAN, RATHER THAN TOTALITARIAN (23)	1.00	97	TILT TOWARD BEING THOSE WHERE THE STATUS OF THE REGIME IS TOTALITARIAN, RATHER THAN AUTHORITARIAN (16)	0.70	2 7 0 16 X SQ= 1.44 P = 0.120 RV YES (NO)
98	TEND LESS TO BE THOSE WHERE GOVERNMENTAL STABILITY IS GENERALLY OR MODERATELY PRESENT AND DATES FROM THE POST-WAR PERIOD, OR IS ABSENT (93)	0.71	98	TEND MORE TO BE THOSE WHERE GOVERNMENTAL STABILITY IS GENERALLY OR MODERATELY PRESENT AND DATES FROM THE POST-WAR PERIOD, OR IS ABSENT (93)	0.89	14 5 34 40 X SQ= 3.61 * P = 0.040 RV NO (YES)

99	TILT LESS TOWARD BEING THOSE WHERE GOVERNMENTAL STABILITY IS GENERALLY PRESENT AND DATES FROM AT LEAST THE INTER-WAR PERIOD, OR FROM THE POST-WAR PERIOD (50)	0.57	99	TILT MORE TOWARD BEING THOSE WHERE GOVERNMENTAL STABILITY IS GENERALLY PRESENT AND DATES FROM AT LEAST THE INTER-WAR PERIOD, OR FROM THE POST-WAR PERIOD (50)	0.79	23 22 17 6 X SQ= 2.39 P = 0.117 RV NO (NO)
100	TEND LESS TO BE THOSE WHERE GOVERNMENTAL STABILITY IS GENERALLY PRESENT AND DATES FROM AT LEAST THE INTER-WAR PERIOD, OR IS GENERALLY OR MODERATELY PRESENT, AND DATES FROM THE POST-WAR PERIOD (64)	0.67	100	ALWAYS ARE THOSE WHERE GOVERNMENTAL STABILITY IS GENERALLY PRESENT AND DATES FROM AT LEAST THE INTER-WAR PERIOD, OR IS GENERALLY OR MODERATELY PRESENT, AND DATES FROM THE POST-WAR PERIOD (64)	1.00	27 27 13 0 X SQ= 8.91 P = 0.001 RV NO (YES)
101	TEND TO BE THOSE WHERE THE REPRESENTATIVE CHARACTER OF THE REGIME IS POLYARCHIC (41)	0.82	101	TEND TO BE THOSE WHERE THE REPRESENTATIVE CHARACTER OF THE REGIME IS LIMITED POLYARCHIC, PSEUDO-POLYARCHIC, OR NON-POLYARCHIC (57)	0.83	31 7 7 35 X SQ= 31.16 P = 0. RV YES (YES)
102	TEND TO BE THOSE WHERE THE REPRESENTATIVE CHARACTER OF THE REGIME IS POLYARCHIC OR LIMITED POLYARCHIC (59)	0.98	102	TEND TO BE THOSE WHERE THE REPRESENTATIVE CHARACTER OF THE REGIME IS PSEUDO-POLYARCHIC OR NON-POLYARCHIC (49)	0.77	44 10 1 34 X SQ= 49.42 P = 0. RV YES (YES)
105	TEND TO BE THOSE WHERE THE ELECTORAL SYSTEM IS COMPETITIVE (43)	0.92	105	TEND TO BE THOSE WHERE THE ELECTORAL SYSTEM IS PARTIALLY COMPETITIVE OR NON-COMPETITIVE (47)	0.88	36 5 3 36 X SQ= 48.19 P = 0. RV YES (YES)
106	TEND TO BE THOSE WHERE THE ELECTORAL SYSTEM IS COMPETITIVE OR PARTIALLY COMPETITIVE (52)	0.95	106	TEND TO BE THOSE WHERE THE ELECTORAL SYSTEM IS NON-COMPETITIVE (30)	0.65	37 14 2 26 X SQ= 28.37 P = 0. RV YES (YES)
107	TEND TO BE THOSE WHERE AUTONOMOUS GROUPS ARE FULLY TOLERATED IN POLITICS (46)	0.81	107	TEND TO BE THOSE WHERE AUTONOMOUS GROUPS ARE PARTIALLY TOLERATED IN POLITICS, ARE TOLERATED ONLY OUTSIDE POLITICS, OR ARE NOT TOLERATED AT ALL (65)	0.86	38 6 9 38 X SQ= 38.46 P = 0. RV YES (YES)
108	TEND TO BE THOSE WHERE AUTONOMOUS GROUPS ARE FULLY OR PARTIALLY TOLERATED IN POLITICS (65)	0.98	108	TEND TO BE THOSE WHERE AUTONOMOUS GROUPS ARE TOLERATED ONLY OUTSIDE POLITICS OR ARE NOT TOLERATED AT ALL (35)	0.60	43 17 1 26 X SQ= 31.74 P = 0. RV YES (YES)
113	LEAN LESS TOWARD BEING THOSE WHERE SECTIONALISM IS MODERATE OR NEGLIGIBLE (81)	0.72	113	LEAN MORE TOWARD BEING THOSE WHERE SECTIONALISM IS MODERATE OR NEGLIGIBLE (81)	0.90	13 4 34 37 X SQ= 3.43 P = 0.056 RV NO (YES)

116	TEND TO BE THOSE WHERE INTEREST ARTICULATION BY ASSOCIATIONAL GROUPS IS SIGNIFICANT OR MODERATE (32)	0.52	116	TEND TO BE THOSE WHERE INTEREST ARTICULATION BY ASSOCIATIONAL GROUPS IS LIMITED OR NEGLIGIBLE (79)	0.88	25 5 23 37 X SQ= 14.51 P = 0. RV YES (YES)
117	TEND TO BE THOSE WHERE INTEREST ARTICULATION BY ASSOCIATIONAL GROUPS IS SIGNIFICANT, MODERATE, OR LIMITED (60)	0.83	117	TEND TO BE THOSE WHERE INTEREST ARTICULATION BY ASSOCIATIONAL GROUPS IS NEGLIGIBLE (51)	0.69	40 13 8 29 X SQ= 23.27 P = 0. RV YES (YES)
118	TEND TO BE THOSE WHERE INTEREST ARTICULATION BY INSTITUTIONAL GROUPS IS SIGNIFICANT, MODERATE, OR LIMITED (60)	0.89	118	TEND TO BE THOSE WHERE INTEREST ARTICULATION BY INSTITUTIONAL GROUPS IS VERY SIGNIFICANT (40)	0.59	5 20 40 14 X SQ= 18.24 P = 0. RV YES (YES)
121	TEND TO BE THOSE WHERE INTEREST ARTICULATION BY NON-ASSOCIATIONAL GROUPS IS MODERATE, LIMITED, OR NEGLIGIBLE (61)	0.75	121	TEND TO BE THOSE WHERE INTEREST ARTICULATION BY NON-ASSOCIATIONAL GROUPS IS SIGNIFICANT (54)	0.53	12 24 36 21 X SQ= 6.71 P = 0.006 RV YES (YES)
122	TEND LESS TO BE THOSE WHERE INTEREST ARTICULATION BY NON-ASSOCIATIONAL GROUPS IS SIGNIFICANT OR MODERATE (83)	0.56	122	TEND MORE TO BE THOSE WHERE INTEREST ARTICULATION BY NON-ASSOCIATIONAL GROUPS IS SIGNIFICANT OR MODERATE (83)	0.80	27 36 21 9 X SQ= 4.96 P = 0.016 RV NO (YES)
123	TEND LESS TO BE THOSE WHERE INTEREST ARTICULATION BY NON-ASSOCIATIONAL GROUPS IS SIGNIFICANT, MODERATE, OR LIMITED (107)	0.83	123	ALWAYS ARE THOSE WHERE INTEREST ARTICULATION BY NON-ASSOCIATIONAL GROUPS IS SIGNIFICANT, MODERATE, OR LIMITED (107)	1.00	40 45 8 0 X SQ= 6.22 P = 0.006 RV NO (YES)
125	TEND TO BE THOSE WHERE INTEREST ARTICULATION BY ANOMIC GROUPS IS INFREQUENT OR VERY INFREQUENT (35)	0.51	125	TEND TO BE THOSE WHERE INTEREST ARTICULATION BY ANOMIC GROUPS IS FREQUENT OR OCCASIONAL (64)	0.72	22 26 23 10 X SQ= 3.60 P = 0.042 RV YES (YES)
130	TILT MORE TOWARD BEING THOSE WHERE INTEREST AGGREGATION BY POLITICAL PARTIES IS MODERATE, LIMITED, OR NEGLIGIBLE (71)	0.91	130	TILT LESS TOWARD BEING THOSE WHERE INTEREST AGGREGATION BY POLITICAL PARTIES IS MODERATE, LIMITED, OR NEGLIGIBLE (71)	0.77	4 7 40 23 X SQ= 1.84 P = 0.108 RV NO (YES)
133	TEND MORE TO BE THOSE WHERE INTEREST AGGREGATION BY THE EXECUTIVE IS MODERATE, LIMITED, OR NEGLIGIBLE (76)	0.84	133	TEND LESS TO BE THOSE WHERE INTEREST AGGREGATION BY THE EXECUTIVE IS MODERATE, LIMITED, OR NEGLIGIBLE (76)	0.53	7 21 38 24 X SQ= 8.76 P = 0.003 RV NO (YES)

137	TEND TO BE THOSE WHERE INTEREST AGGREGATION BY THE LEGISLATURE IS SIGNIFICANT OR MODERATE (29)	0.55	137	TEND TO BE THOSE WHERE INTEREST AGGREGATION BY THE LEGISLATURE IS LIMITED OR NEGLIGIBLE (68)	0.91	24 4 20 39 X SQ= 18.38 P = 0. RV YES (YES)
138	TEND TO BE THOSE WHERE INTEREST AGGREGATION BY THE LEGISLATURE IS SIGNIFICANT, MODERATE, OR LIMITED (48)	0.84	138	TEND TO BE THOSE WHERE INTEREST AGGREGATION BY THE LEGISLATURE IS NEGLIGIBLE (49)	0.83	38 7 7 35 X SQ= 37.30 P = 0. RV YES (YES)
139	TEND TO BE THOSE WHERE THE PARTY SYSTEM IS QUANTITATIVELY OTHER THAN ONE-PARTY (71)	0.98	139	TEND TO BE THOSE WHERE THE PARTY SYSTEM IS QUANTITATIVELY ONE-PARTY (34)	0.72	1 31 46 12 X SQ= 44.97 P = 0. RV YES (YES)
142	TEND TO BE THOSE WHERE THE PARTY SYSTEM IS QUANTITATIVELY MULTI-PARTY (30)	0.58	142	TEND TO BE THOSE WHERE THE PARTY SYSTEM IS QUANTITATIVELY OTHER THAN MULTI-PARTY (66)	0.98	26 1 19 41 X SQ= 28.61 P = 0. RV YES (YES)
144	TEND TO BE THOSE WHERE THE PARTY SYSTEM IS QUANTITATIVELY TWO-PARTY, RATHER THAN ONE-PARTY (11)	0.89	144	TEND TO BE THOSE WHERE THE PARTY SYSTEM IS QUANTITATIVELY ONE-PARTY, RATHER THAN TWO-PARTY (34)	0.91	1 32 8 3 X SQ= 20.53 P = 0. RV YES (YES)
145	LEAN TOWARD BEING THOSE WHERE THE PARTY SYSTEM IS QUANTITATIVELY MULTI-PARTY, RATHER THAN TWO-PARTY (30)	0.76	145	LEAN TOWARD BEING THOSE WHERE THE PARTY SYSTEM IS QUANTITATIVELY TWO-PARTY, RATHER THAN MULTI-PARTY (11)	0.75	26 1 8 3 X SQ= 2.45 P = 0.065 RV YES (NO)
146	ALWAYS ARE THOSE WHERE THE PARTY SYSTEM IS QUALITATIVELY OTHER THAN MASS-BASED TERRITORIAL (92)	1.00	146	TEND LESS TO BE THOSE WHERE THE PARTY SYSTEM IS QUALITATIVELY OTHER THAN MASS-BASED TERRITORIAL (92)	0.83	0 7 43 35 X SQ= 5.76 P = 0.005 RV NO (YES)
147	TEND TO BE THOSE WHERE THE PARTY SYSTEM IS QUALITATIVELY CLASS-ORIENTED OR MULTI-IDEOLOGICAL (23)	0.59	147	ALWAYS ARE THOSE WHERE THE PARTY SYSTEM IS QUALITATIVELY OTHER THAN CLASS-ORIENTED OR MULTI-IDEOLOGICAL (67)	1.00	23 0 16 44 X SQ= 33.01 P = 0. RV YES (YES)
148	TEND MORE TO BE THOSE WHERE THE PARTY SYSTEM IS QUALITATIVELY OTHER THAN AFRICAN TRANSITIONAL (96)	0.96	148	TEND LESS TO BE THOSE WHERE THE PARTY SYSTEM IS QUALITATIVELY OTHER THAN AFRICAN TRANSITIONAL (96)	0.73	2 12 46 33 X SQ= 7.52 P = 0.003 RV NO (YES)

153	TEND TO BE THOSE WHERE THE PARTY SYSTEM IS MODERATELY STABLE OR UNSTABLE (38)	0.54	153	TEND TO BE THOSE WHERE THE PARTY SYSTEM IS STABLE (42)	0.73	19 22 22 8 X SQ= 4.13 P = 0.030 RV YES (YES)
154	TEND LESS TO BE THOSE WHERE THE PARTY SYSTEM IS STABLE OR MODERATELY STABLE (55)	0.56	154	ALWAYS ARE THOSE WHERE THE PARTY SYSTEM IS STABLE OR MODERATELY STABLE (55)	1.00	23 30 18 0 X SQ= 15.40 P = 0. RV NO (YES)
161	TEND TO BE THOSE WHERE THE POLITICAL LEADERSHIP IS NON-ELITIST (50)	0.68	161	TEND TO BE THOSE WHERE THE POLITICAL LEADERSHIP IS ELITIST OR MODERATELY ELITIST (47)	0.59	14 24 30 17 X SQ= 5.10 P = 0.017 RV YES (YES)
164	TEND TO BE THOSE WHERE THE REGIME'S LEADERSHIP CHARISMA IS NEGLIGIBLE (65)	0.89	164	TEND TO BE THOSE WHERE THE REGIME'S LEADERSHIP CHARISMA IS PRONOUNCED OR MODERATE (34)	0.54	5 22 39 19 X SQ= 15.62 P = 0. RV YES (YES)
166	TEND LESS TO BE THOSE WHERE THE VERTICAL POWER DISTRIBUTION IS THAT OF FORMAL FEDERALISM OR FORMAL AND EFFECTIVE UNITARISM (99)	0.77	166	TEND MORE TO BE THOSE WHERE THE VERTICAL POWER DISTRIBUTION IS THAT OF FORMAL FEDERALISM OR FORMAL AND EFFECTIVE UNITARISM (99)	0.96	11 2 36 43 X SQ= 5.34 P = 0.014 RV NO (YES)
168	TEND TO BE THOSE WHERE THE HORIZONTAL POWER DISTRIBUTION IS SIGNIFICANT (34)	0.58	168	TEND TO BE THOSE WHERE THE HORIZONTAL POWER DISTRIBUTION IS LIMITED OR NEGLIGIBLE (72)	0.89	25 5 18 40 X SQ= 19.60 P = 0. RV YES (YES)
169	TEND TO BE THOSE WHERE THE HORIZONTAL POWER DISTRIBUTION IS SIGNIFICANT OR LIMITED (58)	0.98	169	TEND TO BE THOSE WHERE THE HORIZONTAL POWER DISTRIBUTION IS NEGLIGIBLE (48)	0.76	42 11 1 34 X SQ= 46.22 P = 0. RV YES (YES)
170	TILT MORE TOWARD BEING THOSE WHERE THE LEGISLATIVE-EXECUTIVE STRUCTURE IS OTHER THAN PRESIDENTIAL (63)	0.70	170	TILT LESS TOWARD BEING THOSE WHERE THE LEGISLATIVE-EXECUTIVE STRUCTURE IS OTHER THAN PRESIDENTIAL (63)	0.51	14 20 32 21 X SQ= 2.34 P = 0.123 RV NO (YES)
171	TEND LESS TO BE THOSE WHERE THE LEGISLATIVE-EXECUTIVE STRUCTURE IS OTHER THAN PARLIAMENTARY-REPUBLICAN (90)	0.78	171	TEND MORE TO BE THOSE WHERE THE LEGISLATIVE-EXECUTIVE STRUCTURE IS OTHER THAN PARLIAMENTARY-REPUBLICAN (90)	0.98	10 1 36 40 X SQ= 5.67 P = 0.008 RV NO (YES)

172	TEND LESS TO BE THOSE WHERE THE LEGISLATIVE-EXECUTIVE STRUCTURE IS OTHER THAN PARLIAMENTARY-ROYALIST (88)	0.62	172	TEND MORE TO BE THOSE WHERE THE LEGISLATIVE-EXECUTIVE STRUCTURE IS OTHER THAN PARLIAMENTARY-ROYALIST (88)	0.95	18 2 30 41 X SQ= 12.42 P = 0. RV NO (YES)
174	TEND TO BE THOSE WHERE THE LEGISLATURE IS FULLY EFFECTIVE (28)	0.50	174	TEND TO BE THOSE WHERE THE LEGISLATURE IS PARTIALLY EFFECTIVE, LARGELY INEFFECTIVE, OR WHOLLY INEFFECTIVE (72)	0.93	22 3 22 39 X SQ= 17.12 P = 0. RV YES (YES)
175	TEND TO BE THOSE WHERE THE LEGISLATURE IS FULLY EFFECTIVE OR PARTIALLY EFFECTIVE (51)	0.93	175	TEND TO BE THOSE WHERE THE LEGISLATURE IS LARGELY INEFFECTIVE OR WHOLLY INEFFECTIVE (49)	0.88	41 5 3 37 X SQ= 53.84 P = 0. RV YES (YES)
176	ALWAYS ARE THOSE WHERE THE LEGISLATURE IS FULLY EFFECTIVE, PARTIALLY EFFECTIVE, OR LARGELY INEFFECTIVE (72)	1.00	176	TEND TO BE THOSE WHERE THE LEGISLATURE IS WHOLLY INEFFECTIVE (28)	0.50	44 21 0 21 X SQ= 26.46 P = 0. RV YES (YES)
178	TEND TO BE THOSE WHERE THE LEGISLATURE IS BICAMERAL (51)	0.68	178	TEND TO BE THOSE WHERE THE LEGISLATURE IS UNICAMERAL (53)	0.74	15 32 32 11 X SQ= 14.60 P = 0. RV YES (YES)
179	TEND TO BE THOSE WHERE THE EXECUTIVE IS STRONG (39)	0.84	179	TEND TO BE THOSE WHERE THE EXECUTIVE IS DOMINANT (52)	0.87	6 35 31 5 X SQ= 36.42 P = 0. RV YES (YES)
181	TEND LESS TO BE THOSE WHERE THE BUREAUCRACY IS SEMI-MODERN, RATHER THAN MODERN (55)	0.56	181	TEND MORE TO BE THOSE WHERE THE BUREAUCRACY IS SEMI-MODERN, RATHER THAN MODERN (55)	0.93	18 2 23 26 X SQ= 9.21 P = 0.001 RV NO (YES)
185	TEND TO BE THOSE WHERE PARTICIPATION BY THE MILITARY IN POLITICS IS INTERVENTIVE, RATHER THAN SUPPORTIVE (21)	0.73	185	TEND TO BE THOSE WHERE PARTICIPATION BY THE MILITARY IN POLITICS IS SUPPORTIVE, RATHER THAN INTERVENTIVE (31)	0.80	8 5 3 20 X SQ= 7.06 P = 0.006 RV YES (YES)
186	TEND TO BE THOSE WHERE PARTICIPATION BY THE MILITARY IN POLITICS IS NEUTRAL, RATHER THAN SUPPORTIVE (56)	0.92	186	TEND TO BE THOSE WHERE PARTICIPATION BY THE MILITARY IN POLITICS IS SUPPORTIVE, RATHER THAN NEUTRAL (31)	0.51	3 20 33 19 X SQ= 14.28 P = 0. RV YES (YES)

187	TEND TO BE THOSE WHERE THE ROLE OF THE POLICE IS NOT POLITICALLY SIGNIFICANT (35)	0.64	187	TEND TO BE THOSE WHERE THE ROLE OF THE POLICE IS POLITICALLY SIGNIFICANT (66)	0.84	16 32 28 6 X SQ= 17.31 P = 0. RV YES (YES)
188	TEND LESS TO BE THOSE WHERE THE CHARACTER OF THE LEGAL SYSTEM IS OTHER THAN CIVIL LAW (81)	0.56	188	TEND MORE TO BE THOSE WHERE THE CHARACTER OF THE LEGAL SYSTEM IS OTHER THAN CIVIL LAW (81)	0.84	21 7 27 37 X SQ= 7.14 P = 0.006 RV NO (YES)
189	TILT LESS TOWARD BEING THOSE WHERE THE CHARACTER OF THE LEGAL SYSTEM IS OTHER THAN COMMON LAW (108)	0.87	189	TILT MORE TOWARD BEING THOSE WHERE THE CHARACTER OF THE LEGAL SYSTEM IS OTHER THAN COMMON LAW (108)	0.98	6 1 42 44 X SQ= 2.20 P = 0.112 RV NO (YES)
192	TILT MORE TOWARD BEING THOSE WHERE THE CHARACTER OF THE LEGAL SYSTEM IS OTHER THAN MUSLIM OR PARTLY MUSLIM (86)	0.87	192	TILT LESS TOWARD BEING THOSE WHERE THE CHARACTER OF THE LEGAL SYSTEM IS OTHER THAN MUSLIM OR PARTLY MUSLIM (86)	0.73	6 12 42 32 X SQ= 2.31 P = 0.113 RV NO (YES)
193	TEND MORE TO BE THOSE WHERE THE CHARACTER OF THE LEGAL SYSTEM IS OTHER THAN PARTLY INDIGENOUS (86)	0.85	193	TEND LESS TO BE THOSE WHERE THE CHARACTER OF THE LEGAL SYSTEM IS OTHER THAN PARTLY INDIGENOUS (86)	0.62	7 17 41 28 X SQ= 5.37 P = 0.017 RV NO (YES)
194	ALWAYS ARE THOSE THAT ARE NON-COMMUNIST (101)	1.00	194	TEND LESS TO BE THOSE THAT ARE NON-COMMUNIST (101)	0.70	0 13 48 31 X SQ= 14.17 P = 0. RV NO (YES)

129 POLITIES
WHERE INTEREST ARTICULATION
BY POLITICAL PARTIES
IS SIGNIFICANT, MODERATE, OR
LIMITED (56)

129 POLITIES
WHERE INTEREST ARTICULATION
BY POLITICAL PARTIES
IS NEGLIGIBLE (37)

PREDICATE ONLY

56 IN LEFT COLUMN

ARGENTINA	AUSTRALIA	AUSTRIA	BELGIUM	BOLIVIA	BRAZIL	CAMEROUN	CEYLON	CHILE	COLOMBIA
CONGO(LEO)	COSTA RICA	DENMARK	DOMIN REP	ECUADOR	EL SALVADOR	FINLAND	FRANCE	GERMAN FR	GREECE
GUATEMALA	HONDURAS	ICELAND	INDIA	IRELAND	ISRAEL	ITALY	JAMAICA	JAPAN	LEBANON
LIBERIA	LUXEMBOURG	MALAGASY R	MALAYA	MAURITANIA	MOROCCO	NETHERLANDS	NEW ZEALAND	NICARAGUA	NIGERIA
NORWAY	PANAMA	PARAGUAY	PERU	RWANDA	SIERRE LEO	SOMALIA	SWEDEN	SWITZERLAND	SYRIA
TOGO	TRINIDAD	TURKEY	UGANDA	UK	VENEZUELA				

37 IN RIGHT COLUMN

ALBANIA	BULGARIA	BURUNDI	CAMBODIA	CANADA	CEN AFR REP	CHAD	CHINA, PR	CONGO(BRA)	CZECHOS'KIA
DAHOMEY	GABON	GERMANY, E	GHANA	GUINEA	HUNGARY	IVORY COAST	KOREA, N	MALI	MEXICO
MONGOLIA	NIGER	PHILIPPINES	POLAND	PORTUGAL	RUMANIA	SENEGAL	SPAIN	TANGANYIKA	TUNISIA
USSR	UAR	US	UPPER VOLTA	VIETNAM, N	VIETNAM REP	YUGOSLAVIA			

5 EXCLUDED BECAUSE AMBIGUOUS

CYPRUS	INDONESIA	LAOS	SO AFRICA	URUGUAY

5 EXCLUDED BECAUSE IRRELEVANT

AFGHANISTAN	ETHIOPIA	HAITI	LIBYA	SA'U ARABIA

12 EXCLUDED BECAUSE UNASCERTAINABLE

ALGERIA	BURMA	CUBA	IRAN	IRAQ	JORDAN	KOREA REP	NEPAL	PAKISTAN	SUDAN
THAILAND	YEMEN								

```
130  POLITIES                                          130  POLITIES
     WHERE INTEREST AGGREGATION                             WHERE INTEREST AGGREGATION
     BY POLITICAL PARTIES                                   BY POLITICAL PARTIES
     IS SIGNIFICANT  (12)                                   IS MODERATE, LIMITED, OR
                                                            NEGLIGIBLE  (71)

                                                                                        PREDICATE ONLY

      12 IN LEFT COLUMN

AUSTRALIA   AUSTRIA     CANADA      MEXICO      NEW ZEALAND PHILIPPINES SENEGAL     TANGANYIKA  TUNISIA     UK
US          URUGUAY

      71 IN RIGHT COLUMN

ALBANIA     ARGENTINA   BELGIUM     BOLIVIA     BRAZIL      BULGARIA    CAMBODIA    CAMEROUN    CEYLON      CHILE
CHINA, PR   CONGO(LEO)  COSTA RICA  CZECHOS'KIA DENMARK     DOMIN REP   ECUADOR     EL SALVADOR FINLAND     GERMANY, E
GERMAN FR   GUATEMALA   HONDURAS    HUNGARY     ICELAND     INDIA       IRAN        IRELAND     ISRAEL      ITALY
JAMAICA     JAPAN       JORDAN      KOREA, N    KOREA REP   LAOS        LEBANON     LIBERIA     LUXEMBOURG  MALAGASY R
MALAYA      MAURITANIA  MONGOLIA    MOROCCO     NEPAL       NETHERLANDS NICARAGUA   NIGERIA     NORWAY      PAKISTAN
PANAMA      PARAGUAY    PERU        POLAND      PORTUGAL    RUMANIA     SIERRE LEO  SOMALIA     SPAIN       SUDAN
SWEDEN      SWITZERLAND THAILAND    TRINIDAD    TURKEY      UGANDA      USSR        VENEZUELA   VIETNAM, N  VIETNAM REP
YUGOSLAVIA

       6 EXCLUDED BECAUSE AMBIGUOUS

COLOMBIA    CYPRUS      FRANCE      GREECE      INDONESIA   SO AFRICA

       5 EXCLUDED BECAUSE IRRELEVANT

AFGHANISTAN ETHIOPIA    HAITI       LIBYA       SA'U ARABIA

      14 EXCLUDED BECAUSE UNASCERTAINED

BURUNDI     CEN AFR REP CHAD        CONGO(BRA)  DAHOMEY     GABON       GHANA       GUINEA      IVORY COAST MALI
NIGER       RWANDA      UAR         UPPER VOLTA

       7 EXCLUDED BECAUSE UNASCERTAINABLE

ALGERIA     BURMA       CUBA        IRAQ        SYRIA       TOGO        YEMEN
```

```
131  POLITIES                                          131  POLITIES
     WHERE INTEREST AGGREGATION                             WHERE INTEREST AGGREGATION
     BY POLITICAL PARTIES                                   BY POLITICAL PARTIES
     IS SIGNIFICANT OR MODERATE  (30)                       IS LIMITED OR NEGLIGIBLE (35)

                                                                          BOTH SUBJECT AND PREDICATE

      30 IN LEFT COLUMN

AUSTRALIA   AUSTRIA     BOLIVIA     CAMBODIA    CAMEROUN    CANADA      GERMAN FR   INDIA       IRELAND     JAMAICA
JAPAN       LIBERIA     MALAGASY R  MALAYA      MAURITANIA  MEXICO      MOROCCO     NEW ZEALAND NIGERIA     PHILIPPINES
SENEGAL     TANGANYIKA  TRINIDAD    TUNISIA     TURKEY      UGANDA      UK          US          URUGUAY     VENEZUELA

      35 IN RIGHT COLUMN

BELGIUM     BRAZIL      CEYLON      CHILE       COSTA RICA  DENMARK     DOMIN REP   ECUADOR     EL SALVADOR FINLAND
GUATEMALA   HONDURAS    ICELAND     IRAN        ISRAEL      ITALY       JORDAN      KOREA REP   LAOS        LEBANON
LUXEMBOURG  NEPAL       NETHERLANDS NICARAGUA   NORWAY      PAKISTAN    PANAMA      PARAGUAY    PERU        SIERRE LEO
SOMALIA     SWEDEN      SWITZERLAND THAILAND    VIETNAM REP

       7 EXCLUDED BECAUSE AMBIGUOUS

ARGENTINA   COLOMBIA    CYPRUS      FRANCE      GREECE      INDONESIA   SO AFRICA

       5 EXCLUDED BECAUSE IRRELEVANT

AFGHANISTAN ETHIOPIA    HAITI       LIBYA       SA'U ARABIA

      29 EXCLUDED BECAUSE UNASCERTAINED

ALBANIA     BULGARIA    BURUNDI     CEN AFR REP CHAD        CHINA, PR   CONGO(BRA)  CZECHOS'KIA DAHOMEY     GABON
GERMANY, E  GHANA       GUINEA      HUNGARY     IVORY COAST KOREA, N    MALI        MONGOLIA    NIGER       POLAND
PORTUGAL    RUMANIA     RWANDA      SPAIN       USSR        UAR         UPPER VOLTA VIETNAM, N  YUGOSLAVIA

       9 EXCLUDED BECAUSE UNASCERTAINABLE

ALGERIA     BURMA       CONGO(LEO)  CUBA        IRAQ        SUDAN       SYRIA       TOGO        YEMEN
----------------------------------------------------------------------------------------------------------------

 10  TEND LESS TO BE THOSE                  0.67       10  TEND MORE TO BE THOSE                  0.94       10     2
     LOCATED ELSEWHERE THAN IN NORTH AFRICA,               LOCATED ELSEWHERE THAN IN NORTH AFRICA,           20    33
     OR CENTRAL AND SOUTH AFRICA  (82)                     OR CENTRAL AND SOUTH AFRICA  (82)              X SQ=   6.45
                                                                                                          P =    0.008
                                                                                                          RV NO  (YES)

 15  TEND TO BE THOSE                       0.91       15  TEND TO BE THOSE                       0.67        1     4
     LOCATED IN NORTH AFRICA OR                            LOCATED IN THE MIDDLE EAST, RATHER THAN           10     2
     CENTRAL AND SOUTH AFRICA, RATHER THAN                 IN NORTH AFRICA OR                             X SQ=   3.74
     IN THE MIDDLE EAST  (33)                              CENTRAL AND SOUTH AFRICA  (11)                 P =    0.028
                                                                                                          RV YES (YES)
```

18	LEAN TOWARD BEING THOSE LOCATED IN NORTH AFRICA OR CENTRAL AND SOUTH AFRICA, RATHER THAN IN EAST ASIA, SOUTH ASIA, OR SOUTHEAST ASIA (33)	0.67	18	LEAN TOWARD BEING THOSE LOCATED IN EAST ASIA, SOUTH ASIA, OR SOUTHEAST ASIA, RATHER THAN IN NORTH AFRICA OR CENTRAL AND SOUTH AFRICA (18)	0.78	10 2 5 7 X SQ= 2.84 P = 0.089 RV YES (YES)
19	TEND TO BE THOSE LOCATED IN NORTH AFRICA OR CENTRAL AND SOUTH AFRICA, RATHER THAN IN THE CARIBBEAN, CENTRAL AMERICA, OR SOUTH AMERICA (33)	0.62	19	TEND TO BE THOSE LOCATED IN THE CARIBBEAN, CENTRAL AMERICA, OR SOUTH AMERICA, RATHER THAN IN NORTH AFRICA OR CENTRAL AND SOUTH AFRICA (22)	0.86	10 2 6 12 X SQ= 5.36 P = 0.011 RV YES (YES)
22	TEND TO BE THOSE WHOSE TERRITORIAL SIZE IS VERY LARGE, LARGE, OR MEDIUM (68)	0.70	22	TEND TO BE THOSE WHOSE TERRITORIAL SIZE IS SMALL (47)	0.60	21 14 9 21 X SQ= 4.71 P = 0.024 RV YES (YES)
32	TEND LESS TO BE THOSE WHOSE GROSS NATIONAL PRODUCT IS MEDIUM, LOW, OR VERY LOW (105)	0.80	32	TEND MORE TO BE THOSE WHOSE GROSS NATIONAL PRODUCT IS MEDIUM, LOW, OR VERY LOW (105)	0.97	6 1 24 34 X SQ= 3.32 P = 0.042 RV NO (YES)
38	TEND LESS TO BE THOSE WHOSE INTERNATIONAL FINANCIAL STATUS IS MEDIUM, LOW, OR VERY LOW (103)	0.77	38	TEND MORE TO BE THOSE WHOSE INTERNATIONAL FINANCIAL STATUS IS MEDIUM, LOW, OR VERY LOW (103)	0.97	7 1 23 34 X SQ= 4.52 P = 0.020 RV NO (YES)
51	TILT MORE TOWARD BEING THOSE WHERE FREEDOM OF THE PRESS IS COMPLETE OR INTERMITTENT (60)	0.90	51	TILT LESS TOWARD BEING THOSE WHERE FREEDOM OF THE PRESS IS COMPLETE OR INTERMITTENT (60)	0.70	26 21 3 9 X SQ= 2.41 P = 0.104 RV NO (YES)
66	TEND TO BE THOSE THAT ARE RELIGIOUSLY HETEROGENEOUS (49)	0.57	66	TEND TO BE THOSE THAT ARE RELIGIOUSLY HOMOGENEOUS (57)	0.70	13 23 17 10 X SQ= 3.45 P = 0.044 RV YES (YES)
72	LEAN TOWARD BEING THOSE WHOSE DATE OF INDEPENDENCE IS AFTER 1914 (59)	0.55	72	LEAN TOWARD BEING THOSE WHOSE DATE OF INDEPENDENCE IS BEFORE 1914 (52)	0.68	13 23 16 11 X SQ= 2.46 P = 0.080 RV YES (YES)
73	TEND TO BE THOSE WHOSE DATE OF INDEPENDENCE IS AFTER 1945 (46)	0.52	73	TEND TO BE THOSE WHOSE DATE OF INDEPENDENCE IS BEFORE 1945 (65)	0.76	14 26 15 8 X SQ= 4.22 P = 0.035 RV YES (YES)

79	TEND TO BE THOSE THAT WERE FORMERLY DEPENDENCIES OF BRITAIN OR FRANCE, RATHER THAN SPAIN (49)	0.83	79	TEND TO BE THOSE THAT WERE FORMERLY DEPENDENCIES OF SPAIN, RATHER THAN BRITAIN OR FRANCE (18)	0.58	19 8 4 11 X SQ= 5.78 P = 0.010 RV YES (YES)
82	LEAN TOWARD BEING THOSE WHERE THE TYPE OF POLITICAL MODERNIZATION IS DEVELOPED TUTELARY OR UNDEVELOPED TUTELARY, RATHER THAN EARLY OR LATER EUROPEAN OR EUROPEAN DERIVED (55)	0.57	82	LEAN TOWARD BEING THOSE WHERE THE TYPE OF POLITICAL MODERNIZATION IS EARLY OR LATER EUROPEAN OR EUROPEAN DERIVED, RATHER THAN DEVELOPED TUTELARY OR UNDEVELOPED TUTELARY (51)	0.69	12 22 16 10 X SQ= 3.09 P = 0.067 RV YES (YES)
86	LEAN LESS TOWARD BEING THOSE WHERE THE STAGE OF POLITICAL MODERNIZATION IS ADVANCED OR MID-TRANSITIONAL, RATHER THAN EARLY TRANSITIONAL (76)	0.60	86	LEAN MORE TOWARD BEING THOSE WHERE THE STAGE OF POLITICAL MODERNIZATION IS ADVANCED OR MID-TRANSITIONAL, RATHER THAN EARLY TRANSITIONAL (76)	0.83	18 29 12 6 X SQ= 3.15 P = 0.054 RV NO (YES)
93	TILT LESS TOWARD BEING THOSE WHERE THE SYSTEM STYLE IS NON-MOBILIZATIONAL (78)	0.73	93	TILT MORE TOWARD BEING THOSE WHERE THE SYSTEM STYLE IS NON-MOBILIZATIONAL (78)	0.91	8 3 22 29 X SQ= 2.10 P = 0.101 RV NO (YES)
94	TEND MORE TO BE THOSE WHERE THE STATUS OF THE REGIME IS CONSTITUTIONAL (51)	0.93	94	TEND LESS TO BE THOSE WHERE THE STATUS OF THE REGIME IS CONSTITUTIONAL (51)	0.63	25 20 2 12 X SQ= 5.76 P = 0.012 RV NO (YES)
96	TEND MORE TO BE THOSE WHERE THE STATUS OF THE REGIME IS CONSTITUTIONAL OR TOTALITARIAN (67)	0.93	96	TEND LESS TO BE THOSE WHERE THE STATUS OF THE REGIME IS CONSTITUTIONAL OR TOTALITARIAN (67)	0.68	25 21 2 10 X SQ= 4.02 P = 0.025 RV NO (YES)
99	TEND TO BE THOSE WHERE GOVERNMENTAL STABILITY IS GENERALLY PRESENT AND DATES FROM AT LEAST THE INTER-WAR PERIOD, OR FROM THE POST-WAR PERIOD (50)	0.77	99	TEND TO BE THOSE WHERE GOVERNMENTAL STABILITY IS MODERATELY PRESENT AND DATES FROM THE POST-WAR PERIOD, OR IS ABSENT (36)	0.63	17 11 5 19 X SQ= 6.87 P = 0.005 RV YES (YES)
101	TEND TO BE THOSE WHERE THE REPRESENTATIVE CHARACTER OF THE REGIME IS POLYARCHIC (41)	0.81	101	TEND TO BE THOSE WHERE THE REPRESENTATIVE CHARACTER OF THE REGIME IS LIMITED POLYARCHIC, PSEUDO-POLYARCHIC, OR NON-POLYARCHIC (57)	0.50	22 15 5 15 X SQ= 4.88 P = 0.025 RV YES (YES)
102	TEND MORE TO BE THOSE WHERE THE REPRESENTATIVE CHARACTER OF THE REGIME IS POLYARCHIC OR LIMITED POLYARCHIC (59)	0.90	102	TEND LESS TO BE THOSE WHERE THE REPRESENTATIVE CHARACTER OF THE REGIME IS POLYARCHIC OR LIMITED POLYARCHIC (59)	0.66	27 21 3 11 X SQ= 3.96 P = 0.033 RV NO (YES)

111	TEND TO BE THOSE WHERE POLITICAL ENCULTURATION IS HIGH OR MEDIUM (53)	0.83	111	TEND TO BE THOSE WHERE POLITICAL ENCULTURATION IS LOW (42)	0.52	24 16 5 17 X SQ= 6.49 P = 0.007 RV YES (YES)
118	TEND MORE TO BE THOSE WHERE INTEREST ARTICULATION BY INSTITUTIONAL GROUPS IS SIGNIFICANT, MODERATE, OR LIMITED (60)	0.97	118	TEND LESS TO BE THOSE WHERE INTEREST ARTICULATION BY INSTITUTIONAL GROUPS IS SIGNIFICANT, MODERATE, OR LIMITED (60)	0.70	1 10 28 23 X SQ= 5.90 P = 0.007 RV NO (YES)
124	ALWAYS ARE THOSE WHERE INTEREST ARTICULATION BY ANOMIC GROUPS IS OCCASIONAL, INFREQUENT, OR VERY INFREQUENT (85)	1.00	124	TEND LESS TO BE THOSE WHERE INTEREST ARTICULATION BY ANOMIC GROUPS IS OCCASIONAL, INFREQUENT, OR VERY INFREQUENT (85)	0.82	0 6 28 28 X SQ= 3.64 P = 0.028 RV NO (YES)
127	TEND TO BE THOSE WHERE INTEREST ARTICULATION BY POLITICAL PARTIES IS MODERATE, LIMITED, OR NEGLIGIBLE (72)	0.86	127	TEND TO BE THOSE WHERE INTEREST ARTICULATION BY POLITICAL PARTIES IS SIGNIFICANT (21)	0.54	4 15 25 13 X SQ= 8.43 P = 0.002 RV YES (YES)
133	TEND LESS TO BE THOSE WHERE INTEREST AGGREGATION BY THE EXECUTIVE IS MODERATE, LIMITED, OR NEGLIGIBLE (76)	0.52	133	ALWAYS ARE THOSE WHERE INTEREST AGGREGATION BY THE EXECUTIVE IS MODERATE, LIMITED, OR NEGLIGIBLE (76)	1.00	14 0 15 32 X SQ= 17.41 P = 0. RV NO (YES)
134	TEND TO BE THOSE WHERE INTEREST AGGREGATION BY THE EXECUTIVE IS SIGNIFICANT OR MODERATE (57)	0.90	134	TEND TO BE THOSE WHERE INTEREST AGGREGATION BY THE EXECUTIVE IS LIMITED OR NEGLIGIBLE (46)	0.58	26 13 3 18 X SQ= 12.97 P = 0. RV YES (YES)
135	ALWAYS ARE THOSE WHERE INTEREST AGGREGATION BY THE EXECUTIVE IS SIGNIFICANT, MODERATE, OR LIMITED (77)	1.00	135	TEND LESS TO BE THOSE WHERE INTEREST AGGREGATION BY THE EXECUTIVE IS SIGNIFICANT, MODERATE, OR LIMITED (77)	0.77	29 24 0 7 X SQ= 5.38 P = 0.011 RV NO (YES)
136	TILT MORE TOWARD BEING THOSE WHERE INTEREST AGGREGATION BY THE LEGISLATURE IS MODERATE, LIMITED, OR NEGLIGIBLE (85)	0.90	136	TILT LESS TOWARD BEING THOSE WHERE INTEREST AGGREGATION BY THE LEGISLATURE IS MODERATE, LIMITED, OR NEGLIGIBLE (85)	0.71	3 9 26 22 X SQ= 2.21 P = 0.107 RV NO (YES)
139	TEND LESS TO BE THOSE WHERE THE PARTY SYSTEM IS QUANTITATIVELY OTHER THAN ONE-PARTY (71)	0.80	139	ALWAYS ARE THOSE WHERE THE PARTY SYSTEM IS QUANTITATIVELY OTHER THAN ONE-PARTY (71)	1.00	6 0 24 30 X SQ= 4.63 P = 0.024 RV NO (YES)

141	TEND LESS TO BE THOSE WHERE THE PARTY SYSTEM IS QUANTITATIVELY OTHER THAN TWO-PARTY (87)	0.70	141	TEND MORE TO BE THOSE WHERE THE PARTY SYSTEM IS QUANTITATIVELY OTHER THAN TWO-PARTY (87)	0.93	8 2 19 27 X SQ= 3.50 P = 0.038 RV NO (YES)
142	TEND TO BE THOSE WHERE THE PARTY SYSTEM IS QUANTITATIVELY OTHER THAN MULTI-PARTY (66)	0.85	142	TEND TO BE THOSE WHERE THE PARTY SYSTEM IS QUANTITATIVELY MULTI-PARTY (30)	0.79	4 22 22 6 X SQ= 19.10 P = 0. RV YES (YES)
145	TEND TO BE THOSE WHERE THE PARTY SYSTEM IS QUANTITATIVELY TWO-PARTY, RATHER THAN MULTI-PARTY (11)	0.67	145	TEND TO BE THOSE WHERE THE PARTY SYSTEM IS QUANTITATIVELY MULTI-PARTY, RATHER THAN TWO-PARTY (30)	0.92	4 22 8 2 X SQ= 10.82 P = 0.001 RV YES (YES)
146	LEAN LESS TOWARD BEING THOSE WHERE THE PARTY SYSTEM IS QUALITATIVELY OTHER THAN MASS-BASED TERRITORIAL (92)	0.89	146	ALWAYS ARE THOSE WHERE THE PARTY SYSTEM IS QUALITATIVELY OTHER THAN MASS-BASED TERRITORIAL (92)	1.00	3 0 24 33 X SQ= 1.87 P = 0.085 RV NO (YES)
153	TEND TO BE THOSE WHERE THE PARTY SYSTEM IS STABLE (42)	0.68	153	TEND TO BE THOSE WHERE THE PARTY SYSTEM IS MODERATELY STABLE OR UNSTABLE (38)	0.69	15 10 7 22 X SQ= 5.74 P = 0.012 RV YES (YES)
154	TEND TO BE THOSE WHERE THE PARTY SYSTEM IS STABLE OR MODERATELY STABLE (55)	0.91	154	TEND TO BE THOSE WHERE THE PARTY SYSTEM IS UNSTABLE (25)	0.56	20 14 2 18 X SQ= 10.49 P = 0. RV YES (YES)
157	TEND MORE TO BE THOSE WHERE PERSONALISMO IS MODERATE OR NEGLIGIBLE (84)	0.93	157	TEND LESS TO BE THOSE WHERE PERSONALISMO IS MODERATE OR NEGLIGIBLE (84)	0.69	2 9 28 20 X SQ= 4.28 P = 0.021 RV NO (YES)
158	LEAN MORE TOWARD BEING THOSE WHERE PERSONALISMO IS NEGLIGIBLE (56)	0.77	158	LEAN LESS TOWARD BEING THOSE WHERE PERSONALISMO IS NEGLIGIBLE (56)	0.52	7 14 23 15 X SQ= 2.99 P = 0.060 RV NO (YES)
161	TEND TO BE THOSE WHERE THE POLITICAL LEADERSHIP IS NON-ELITIST (50)	0.78	161	TEND TO BE THOSE WHERE THE POLITICAL LEADERSHIP IS ELITIST OR MODERATELY ELITIST (47)	0.60	6 18 21 12 X SQ= 6.84 P = 0.007 RV YES (YES)

164	TEND LESS TO BE THOSE WHERE THE REGIME'S LEADERSHIP CHARISMA IS NEGLIGIBLE (65)	0.68	164	TEND MORE TO BE THOSE WHERE THE REGIME'S LEADERSHIP CHARISMA IS NEGLIGIBLE (65)	0.91	9 3 19 29 X SQ= 3.52 P = 0.050 RV NO (YES)
166	TEND LESS TO BE THOSE WHERE THE VERTICAL POWER DISTRIBUTION IS THAT OF FORMAL FEDERALISM OR FORMAL AND EFFECTIVE UNITARISM (99)	0.67	166	TEND MORE TO BE THOSE WHERE THE VERTICAL POWER DISTRIBUTION IS THAT OF FORMAL FEDERALISM OR FORMAL AND EFFECTIVE UNITARISM (99)	0.91	10 3 20 32 X SQ= 4.74 P = 0.027 RV NO (YES)
176	TEND MORE TO BE THOSE WHERE THE LEGISLATURE IS FULLY EFFECTIVE, PARTIALLY EFFECTIVE, OR LARGELY INEFFECTIVE (72)	0.97	176	TEND LESS TO BE THOSE WHERE THE LEGISLATURE IS FULLY EFFECTIVE, PARTIALLY EFFECTIVE, OR LARGELY INEFFECTIVE (72)	0.79	29 22 1 6 X SQ= 2.93 P = 0.048 RV NO (YES)
178	TILT MORE TOWARD BEING THOSE WHERE THE LEGISLATURE IS BICAMERAL (51)	0.77	178	TILT LESS TOWARD BEING THOSE WHERE THE LEGISLATURE IS BICAMERAL (51)	0.56	7 14 23 18 X SQ= 2.04 P = 0.112 RV NO (YES)
182	TILT LESS TOWARD BEING THOSE WHERE THE BUREAUCRACY IS SEMI-MODERN, RATHER THAN POST-COLONIAL TRANSITIONAL (55)	0.67	182	TILT MORE TOWARD BEING THOSE WHERE THE BUREAUCRACY IS SEMI-MODERN, RATHER THAN POST-COLONIAL TRANSITIONAL (55)	0.88	14 15 7 2 X SQ= 1.37 P = 0.148 RV NO (YES)
183	TEND TO BE THOSE WHERE THE BUREAUCRACY IS POST-COLONIAL TRANSITIONAL, RATHER THAN TRADITIONAL (25)	0.87	183	TEND TO BE THOSE WHERE THE BUREAUCRACY IS TRADITIONAL, RATHER THAN POST-COLONIAL TRANSITIONAL (9)	0.71	7 2 1 5 X SQ= 3.23 P = 0.041 RV YES (YES)
187	TILT TOWARD BEING THOSE WHERE THE ROLE OF THE POLICE IS NOT POLITICALLY SIGNIFICANT (35)	0.67	187	TILT TOWARD BEING THOSE WHERE THE ROLE OF THE POLICE IS POLITICALLY SIGNIFICANT (66)	0.57	10 17 20 13 X SQ= 2.42 P = 0.119 RV YES (YES)
190	TEND TO BE THOSE WHERE THE CHARACTER OF THE LEGAL SYSTEM IS COMMON LAW, RATHER THAN CIVIL LAW (7)	0.50	190	ALWAYS ARE THOSE WHERE THE CHARACTER OF THE LEGAL SYSTEM IS CIVIL LAW, RATHER THAN COMMON LAW (32)	1.00	7 17 7 0 X SQ= 8.31 P = 0.001 RV YES (YES)

```
************************************************************************************************************
132  POLITIES                                              132  POLITIES
     WHERE INTEREST AGGREGATION                                 WHERE INTEREST AGGREGATION
     BY POLITICAL PARTIES                                       BY POLITICAL PARTIES
     IS SIGNIFICANT, MODERATE, OR                               IS NEGLIGIBLE  (9)
     LIMITED  (67)

                                                                                          PREDICATE ONLY
............................................................................................................
       67 IN LEFT COLUMN

ALGERIA      AUSTRALIA    AUSTRIA      BELGIUM      BOLIVIA      BRAZIL       BURUNDI      CAMBODIA     CAMEROUN     CANADA
CEN AFR REP  CEYLON       CHAD         CHILE        CONGO(BRA)   CONGO(LEO)   CUBA         COSTA RICA   DAHOMEY      DENMARK
DOMIN REP    ECUADOR      FINLAND      GABON        GERMAN FR    GHANA        GUINEA       HONDURAS     ICELAND      INDIA
IRELAND      ITALY        IVORY COAST  JAMAICA      JAPAN        LIBERIA      LUXEMBOURG   MALAGASY R   MALAYA       MALI
MAURITANIA   MEXICO       MOROCCO      NETHERLANDS  NEW ZEALAND  NIGER        NIGERIA      NORWAY       PHILIPPINES  RWANDA
SENEGAL      SIERRE LEO   SOMALIA      SO AFRICA    SWEDEN       SWITZERLAND  TANGANYIKA   TRINIDAD     TUNISIA      TURKEY
UGANDA       UAR          UK           US           UPPER VOLTA  URUGUAY      VENEZUELA

       9 IN RIGHT COLUMN

EL SALVADOR  GUATEMALA    ISRAEL       LAOS         LEBANON      NICARAGUA    PANAMA       PARAGUAY     VIETNAM REP

       6 EXCLUDED BECAUSE AMBIGUOUS

ARGENTINA    COLOMBIA     CYPRUS       FRANCE       GREECE       INDONESIA

       5 EXCLUDED BECAUSE IRRELEVANT

AFGHANISTAN  ETHIOPIA     HAITI        LIBYA        SA'U ARABIA

      15 EXCLUDED BECAUSE UNASCERTAINED

ALBANIA      BULGARIA     CHINA, PR    CZECHOS'KIA  GERMANY, E   HUNGARY      KOREA, N     MONGOLIA     POLAND       PORTUGAL
RUMANIA      SPAIN        USSR         VIETNAM, N   YUGOSLAVIA

      13 EXCLUDED BECAUSE UNASCERTAINABLE

BURMA        IRAN         IRAQ         JORDAN       KOREA REP    NEPAL        PAKISTAN     PERU         SUDAN        SYRIA
THAILAND     TOGO         YEMEN
------------------------------------------------------------------------------------------------------------
```

133 POLITIES
WHERE INTEREST AGGREGATION
BY THE EXECUTIVE
IS SIGNIFICANT (29)

133 POLITIES
WHERE INTEREST AGGREGATION
BY THE EXECUTIVE
IS MODERATE, LIMITED, OR
NEGLIGIBLE (76)

PREDICATE ONLY

29 IN LEFT COLUMN

BOLIVIA	BURUNDI	CAMBODIA	CAMEROUN	CEN AFR REP	CHAD	CONGO(BRA)	DAHOMEY	FRANCE	GABON
GHANA	GUINEA	INDIA	INDONESIA	IVORY COAST	MALAYA	MALI	MAURITANIA	MEXICO	MOROCCO
NIGER	PHILIPPINES	SENEGAL	TANGANYIKA	TUNISIA	UAR	US	UPPER VOLTA	VENEZUELA	

76 IN RIGHT COLUMN

AFGHANISTAN	ALBANIA	ARGENTINA	AUSTRALIA	AUSTRIA	BELGIUM	BRAZIL	BULGARIA	BURMA	CANADA
CEYLON	CHILE	CHINA, PR	COLOMBIA	CONGO(LEO)	CUBA	COSTA RICA	CZECHOS'KIA	DENMARK	DOMIN REP
ECUADOR	EL SALVADOR	ETHIOPIA	FINLAND	GERMANY, E	GERMAN FR	GREECE	GUATEMALA	HAITI	HONDURAS
HUNGARY	ICELAND	IRAN	IRELAND	ISRAEL	ITALY	JAMAICA	JAPAN	JORDAN	KOREA, N
KOREA REP	LIBERIA	LIBYA	LUXEMBOURG	MALAGASY R	MONGOLIA	NEPAL	NETHERLANDS	NEW ZEALAND	NICARAGUA
NIGERIA	NORWAY	PAKISTAN	PANAMA	PARAGUAY	POLAND	PORTUGAL	RUMANIA	RWANDA	SA'U ARABIA
SIERRE LEO	SOMALIA	SPAIN	SUDAN	SWEDEN	SWITZERLAND	THAILAND	TOGO	TRINIDAD	TURKEY
UGANDA	USSR	UK	VIETNAM, N	VIETNAM REP	YUGOSLAVIA				

5 EXCLUDED BECAUSE AMBIGUOUS

CYPRUS	LAOS	LEBANON	SO AFRICA	URUGUAY

5 EXCLUDED BECAUSE UNASCERTAINABLE

ALGERIA	IRAQ	PERU	SYRIA	YEMEN

134 POLITIES
WHERE INTEREST AGGREGATION
BY THE EXECUTIVE
IS SIGNIFICANT OR MODERATE (57)

134 POLITIES
WHERE INTEREST AGGREGATION
BY THE EXECUTIVE
IS LIMITED OR NEGLIGIBLE (46)

BOTH SUBJECT AND PREDICATE

57 IN LEFT COLUMN

AUSTRALIA	AUSTRIA	BELGIUM	BOLIVIA	BURUNDI	CAMBODIA	CAMEROUN	CANADA	CEN AFR REP	CHAD
COLOMBIA	CONGO(BRA)	COSTA RICA	DAHOMEY	DENMARK	DOMIN REP	FINLAND	FRANCE	GABON	GERMAN FR
GHANA	GREECE	GUINEA	ICELAND	INDIA	INDONESIA	IRELAND	ISRAEL	ITALY	IVORY COAST
JAMAICA	JAPAN	LUXEMBOURG	MALAGASY R	MALAYA	MALI	MAURITANIA	MEXICO	MOROCCO	NETHERLANDS
NEW ZEALAND	NIGER	NORWAY	PHILIPPINES	RWANDA	SENEGAL	SIERRE LEO	SWEDEN	TANGANYIKA	TRINIDAD
TUNISIA	TURKEY	UAR	UK	US	UPPER VOLTA	VENEZUELA			

46 IN RIGHT COLUMN

AFGHANISTAN	ALBANIA	ARGENTINA	BRAZIL	BULGARIA	BURMA	CEYLON	CHILE	CHINA, PR	CUBA
CZECHOS'KIA	ECUADOR	EL SALVADOR	ETHIOPIA	GERMANY, E	GUATEMALA	HAITI	HONDURAS	HUNGARY	IRAN
JORDAN	KOREA, N	KOREA REP	LIBERIA	LIBYA	MONGOLIA	NEPAL	NICARAGUA	NIGERIA	PANAMA
PARAGUAY	POLAND	PORTUGAL	RUMANIA	SA'U ARABIA	SOMALIA	SPAIN	SUDAN	SWITZERLAND	THAILAND
TOGO	UGANDA	USSR	VIETNAM, N	VIETNAM REP	YUGOSLAVIA				

5 EXCLUDED BECAUSE AMBIGUOUS

CYPRUS LAOS LEBANON SO AFRICA URUGUAY

7 EXCLUDED BECAUSE UNASCERTAINABLE

ALGERIA CONGO(LEO) IRAQ PAKISTAN PERU SYRIA YEMEN

2	TEND LESS TO BE THOSE LOCATED ELSEWHERE THAN IN WEST EUROPE, SCANDINAVIA, NORTH AMERICA, OR AUSTRALASIA (93)	0.67	2	TEND MORE TO BE THOSE LOCATED ELSEWHERE THAN IN WEST EUROPE, SCANDINAVIA, NORTH AMERICA, OR AUSTRALASIA (93)	0.93	19 3 38 43 X SQ= 9.36 P = 0.001 RV NO (YES)
3	TILT LESS TOWARD BEING THOSE LOCATED ELSEWHERE THAN IN WEST EUROPE (102)	0.82	3	TILT MORE TOWARD BEING THOSE LOCATED ELSEWHERE THAN IN WEST EUROPE (102)	0.93	10 3 47 43 X SQ= 1.89 P = 0.136 RV NO (NO)
5	ALWAYS ARE THOSE LOCATED IN SCANDINAVIA OR WEST EUROPE, RATHER THAN IN EAST EUROPE (18)	1.00	5	TEND TO BE THOSE LOCATED IN EAST EUROPE, RATHER THAN IN SCANDINAVIA OR WEST EUROPE (9)	0.75	15 3 0 9 X SQ= 13.67 P = 0. RV YES (YES)

6	TILT MORE TOWARD BEING THOSE LOCATED ELSEWHERE THAN IN THE CARIBBEAN, CENTRAL AMERICA, OR SOUTH AMERICA (93)	0.86	6	TILT LESS TOWARD BEING THOSE LOCATED ELSEWHERE THAN IN THE CARIBBEAN, CENTRAL AMERICA, OR SOUTH AMERICA (93)	0.74	8 12 49 34 X SQ= 1.66 P = 0.140 RV NO (YES)
10	TEND LESS TO BE THOSE LOCATED ELSEWHERE THAN IN NORTH AFRICA, OR CENTRAL AND SOUTH AFRICA (82)	0.61	10	TEND MORE TO BE THOSE LOCATED ELSEWHERE THAN IN NORTH AFRICA, OR CENTRAL AND SOUTH AFRICA (82)	0.83	22 8 35 38 X SQ= 4.57 P = 0.028 RV NO (YES)
15	TILT MORE TOWARD BEING THOSE LOCATED IN NORTH AFRICA OR CENTRAL AND SOUTH AFRICA, RATHER THAN IN THE MIDDLE EAST (33)	0.92	15	TILT LESS TOWARD BEING THOSE LOCATED IN NORTH AFRICA OR CENTRAL AND SOUTH AFRICA, RATHER THAN IN THE MIDDLE EAST (33)	0.67	2 4 22 8 X SQ= 2.02 P = 0.149 RV NO (YES)
18	TEND TO BE THOSE LOCATED IN NORTH AFRICA OR CENTRAL AND SOUTH AFRICA, RATHER THAN IN EAST ASIA, SOUTH ASIA, OR SOUTHEAST ASIA (33)	0.79	18	TEND TO BE THOSE LOCATED IN EAST ASIA, SOUTH ASIA, OR SOUTHEAST ASIA, RATHER THAN IN NORTH AFRICA OR CENTRAL AND SOUTH AFRICA (18)	0.56	22 8 6 10 X SQ= 4.22 P = 0.027 RV YES (YES)
19	TEND TO BE THOSE LOCATED IN NORTH AFRICA OR CENTRAL AND SOUTH AFRICA, RATHER THAN IN THE CARIBBEAN, CENTRAL AMERICA, OR SOUTH AMERICA (33)	0.73	19	TEND TO BE THOSE LOCATED IN THE CARIBBEAN, CENTRAL AMERICA, OR SOUTH AMERICA, RATHER THAN IN NORTH AFRICA OR CENTRAL AND SOUTH AFRICA (22)	0.60	22 8 8 12 X SQ= 4.25 P = 0.038 RV YES (YES)
24	TILT TOWARD BEING THOSE WHOSE POPULATION IS SMALL (54)	0.54	24	TILT TOWARD BEING THOSE WHOSE POPULATION IS VERY LARGE, LARGE, OR MEDIUM (61)	0.63	26 29 31 17 X SQ= 2.45 P = 0.112 RV YES (YES)
31	TEND LESS TO BE THOSE WHOSE AGRICULTURAL POPULATION IS HIGH OR MEDIUM (90)	0.68	31	TEND MORE TO BE THOSE WHOSE AGRICULTURAL POPULATION IS HIGH OR MEDIUM (90)	0.91	39 41 18 4 X SQ= 6.37 P = 0.007 RV NO (YES)
35	TEND LESS TO BE THOSE WHOSE PER CAPITA GROSS NATIONAL PRODUCT IS MEDIUM, LOW, OR VERY LOW (91)	0.65	35	TEND MORE TO BE THOSE WHOSE PER CAPITA GROSS NATIONAL PRODUCT IS MEDIUM, LOW, OR VERY LOW (91)	0.91	20 4 37 42 X SQ= 8.50 P = 0.002 RV NO (YES)
36	LEAN LESS TOWARD BEING THOSE WHOSE PER CAPITA GROSS NATIONAL PRODUCT IS LOW OR VERY LOW (73)	0.54	36	LEAN MORE TOWARD BEING THOSE WHOSE PER CAPITA GROSS NATIONAL PRODUCT IS LOW OR VERY LOW (73)	0.74	26 12 31 34 X SQ= 3.37 P = 0.064 RV NO (YES)

38	TEND LESS TO BE THOSE WHOSE INTERNATIONAL FINANCIAL STATUS IS MEDIUM, LOW, OR VERY LOW (103)	0.84	38	TEND MORE TO BE THOSE WHOSE INTERNATIONAL FINANCIAL STATUS IS MEDIUM, LOW, OR VERY LOW (103)	0.98	9 1 48 43 X SQ= 3.68 P = 0.040 RV NO (NO)
39	TILT LESS TOWARD BEING THOSE WHOSE INTERNATIONAL FINANCIAL STATUS IS LOW OR VERY LOW (76)	0.58	39	TILT MORE TOWARD BEING THOSE WHOSE INTERNATIONAL FINANCIAL STATUS IS LOW OR VERY LOW (76)	0.73	24 12 33 33 X SQ= 1.99 P = 0.144 RV NO (YES)
41	TEND LESS TO BE THOSE WHOSE ECONOMIC DEVELOPMENTAL STATUS IS INTERMEDIATE, UNDERDEVELOPED, OR VERY UNDERDEVELOPED (94)	0.73	41	TEND MORE TO BE THOSE WHOSE ECONOMIC DEVELOPMENTAL STATUS IS INTERMEDIATE, UNDERDEVELOPED, OR VERY UNDERDEVELOPED (94)	0.91	15 4 40 42 X SQ= 4.51 P = 0.022 RV NO (YES)
42	TILT LESS TOWARD BEING THOSE WHOSE ECONOMIC DEVELOPMENTAL STATUS IS UNDERDEVELOPED OR VERY UNDERDEVELOPED (76)	0.60	42	TILT MORE TOWARD BEING THOSE WHOSE ECONOMIC DEVELOPMENTAL STATUS IS UNDERDEVELOPED OR VERY UNDERDEVELOPED (76)	0.76	22 11 33 34 X SQ= 2.05 P = 0.135 RV NO (YES)
44	LEAN LESS TOWARD BEING THOSE WHERE THE LITERACY RATE IS BELOW NINETY PERCENT (90)	0.68	44	LEAN MORE TOWARD BEING THOSE WHERE THE LITERACY RATE IS BELOW NINETY PERCENT (90)	0.85	18 7 39 39 X SQ= 2.87 P = 0.066 RV NO (YES)
47	LEAN TOWARD BEING THOSE WHERE THE LITERACY RATE IS NINETY PERCENT OR ABOVE, RATHER THAN BETWEEN FIFTY AND NINETY PERCENT (25)	0.60	47	LEAN TOWARD BEING THOSE WHERE THE LITERACY RATE IS BETWEEN FIFTY AND NINETY PERCENT, RATHER THAN NINETY PERCENT OR ABOVE (30)	0.70	18 7 12 16 X SQ= 3.46 P = 0.052 RV YES (YES)
50	TEND TO BE THOSE WHERE FREEDOM OF THE PRESS IS COMPLETE (43)	0.72	50	TEND TO BE THOSE WHERE FREEDOM OF THE PRESS IS INTERMITTENT, INTERNALLY ABSENT, OR INTERNALLY AND EXTERNALLY ABSENT (56)	0.82	34 7 13 33 X SQ= 23.93 P = 0. RV YES (YES)
51	TEND TO BE THOSE WHERE FREEDOM OF THE PRESS IS COMPLETE OR INTERMITTENT (60)	0.83	51	TEND TO BE THOSE WHERE FREEDOM OF THE PRESS IS INTERNALLY ABSENT, OR INTERNALLY AND EXTERNALLY ABSENT (37)	0.67	39 13 8 27 X SQ= 20.85 P. = 0. RV YES (YES)
52	TEND MORE TO BE THOSE WHERE FREEDOM OF THE PRESS IS COMPLETE, INTERMITTENT, OR INTERNALLY ABSENT (82)	0.96	52	TEND LESS TO BE THOSE WHERE FREEDOM OF THE PRESS IS COMPLETE, INTERMITTENT, OR INTERNALLY ABSENT (82)	0.65	46 26 2 14 X SQ= 11.95 P = 0. RV NO (YES)

53	TEND LESS TO BE THOSE WHERE NEWSPAPER CIRCULATION IS LESS THAN THREE HUNDRED PER THOUSAND (101)	0.79	53	TEND MORE TO BE THOSE WHERE NEWSPAPER CIRCULATION IS LESS THAN THREE HUNDRED PER THOUSAND (101)	0.96	12 2 45 44 X SQ= 4.71 P = 0.019 RV NO (NO)
62	LEAN LESS TOWARD BEING THOSE WHERE THE RELIGION IS CATHOLIC, RATHER THAN PROTESTANT (25)	0.71	62	ALWAYS ARE THOSE WHERE THE RELIGION IS CATHOLIC, RATHER THAN PROTESTANT (25)	1.00	5 0 12 12 X SQ= 2.45 P = 0.059 RV NO (YES)
72	LEAN TOWARD BEING THOSE WHOSE DATE OF INDEPENDENCE IS AFTER 1914 (59)	0.59	72	LEAN TOWARD BEING THOSE WHOSE DATE OF INDEPENDENCE IS BEFORE 1914 (52)	0.60	23 26 33 17 X SQ= 2.92 P = 0.069 RV YES (YES)
73	TEND TO BE THOSE WHOSE DATE OF INDEPENDENCE IS AFTER 1945 (46)	0.54	73	TEND TO BE THOSE WHOSE DATE OF INDEPENDENCE IS BEFORE 1945 (65)	0.74	26 32 30 11 X SQ= 6.74 P = 0.007 RV YES (YES)
74	TEND LESS TO BE THOSE THAT ARE NOT HISTORICALLY WESTERN (87)	0.66	74	TEND MORE TO BE THOSE THAT ARE NOT HISTORICALLY WESTERN (87)	0.85	19 7 37 39 X SQ= 3.72 P = 0.040 RV NO (YES)
79	TEND TO BE THOSE THAT WERE FORMERLY DEPENDENCIES OF BRITAIN OR FRANCE, RATHER THAN SPAIN (49)	0.84	79	TEND TO BE THOSE THAT WERE FORMERLY DEPENDENCIES OF SPAIN, RATHER THAN BRITAIN OR FRANCE (18)	0.50	31 10 6 10 X SQ= 5.76 P = 0.012 RV YES (YES)
81	TEND LESS TO BE THOSE WHERE THE TYPE OF POLITICAL MODERNIZATION IS LATER EUROPEAN OR LATER EUROPEAN DERIVED, RATHER THAN EARLY EUROPEAN OR EARLY EUROPEAN DERIVED (40)	0.64	81	TEND MORE TO BE THOSE WHERE THE TYPE OF POLITICAL MODERNIZATION IS LATER EUROPEAN OR LATER EUROPEAN DERIVED, RATHER THAN EARLY EUROPEAN OR EARLY EUROPEAN DERIVED (40)	0.96	9 1 16 22 X SQ= 5.48 P = 0.011 RV NO (YES)
83	LEAN MORE TOWARD BEING THOSE WHERE THE TYPE OF POLITICAL MODERNIZATION IS OTHER THAN NON-EUROPEAN AUTOCHTHONOUS (106)	0.96	83	LEAN LESS TOWARD BEING THOSE WHERE THE TYPE OF POLITICAL MODERNIZATION IS OTHER THAN NON-EUROPEAN AUTOCHTHONOUS (106)	0.85	2 7 55 39 X SQ= 3.03 P = 0.074 RV NO (YES)
84	TEND TO BE THOSE WHERE THE TYPE OF POLITICAL MODERNIZATION IS UNDEVELOPED TUTELARY, RATHER THAN DEVELOPED TUTELARY (24)	0.63	84	TEND TO BE THOSE WHERE THE TYPE OF POLITICAL MODERNIZATION IS DEVELOPED TUTELARY, RATHER THAN UNDEVELOPED TUTELARY (31)	0.75	11 12 19 4 X SQ= 4.70 P = 0.029 RV YES (YES)

86	LEAN LESS TOWARD BEING THOSE WHERE THE STAGE OF POLITICAL MODERNIZATION IS ADVANCED OR MID-TRANSITIONAL, RATHER THAN EARLY TRANSITIONAL (76)	0.58	86	LEAN MORE TOWARD BEING THOSE WHERE THE STAGE OF POLITICAL MODERNIZATION IS ADVANCED OR MID-TRANSITIONAL, RATHER THAN EARLY TRANSITIONAL (76)	0.76	33 34 24 11 X SQ= 2.74 P = 0.092 RV NO (YES)
87	TEND TO BE THOSE WHOSE IDEOLOGICAL ORIENTATION IS DEVELOPMENTAL (31)	0.52	87	TEND TO BE THOSE WHOSE IDEOLOGICAL ORIENTATION IS OTHER THAN DEVELOPMENTAL (58)	0.90	26 3 24 27 X SQ= 12.55 P = 0. RV YES (YES)
89	TEND LESS TO BE THOSE WHOSE IDEOLOGICAL ORIENTATION IS OTHER THAN CONVENTIONAL (62)	0.54	89	TEND MORE TO BE THOSE WHOSE IDEOLOGICAL ORIENTATION IS OTHER THAN CONVENTIONAL (62)	0.84	24 5 28 27 X SQ= 6.87 P = 0.005 RV NO (NO)
91	ALWAYS ARE THOSE WHOSE IDEOLOGICAL ORIENTATION IS DEVELOPMENTAL, RATHER THAN TRADITIONAL (31)	1.00	91	TEND TO BE THOSE WHOSE IDEOLOGICAL ORIENTATION IS TRADITIONAL, RATHER THAN DEVELOPMENTAL (5)	0.63	26 3 0 5 X SQ= 14.39 P = 0. RV YES (YES)
92	TEND MORE TO BE THOSE WHERE THE SYSTEM STYLE IS LIMITED MOBILIZATIONAL, OR NON-MOBILIZATIONAL (93)	0.93	92	TEND LESS TO BE THOSE WHERE THE SYSTEM STYLE IS LIMITED MOBILIZATIONAL, OR NON-MOBILIZATIONAL (93)	0.67	4 15 53 30 X SQ= 9.82 P = 0.002 RV NO (YES)
94	TEND TO BE THOSE WHERE THE STATUS OF THE REGIME IS CONSTITUTIONAL (51)	0.84	94	TEND TO BE THOSE WHERE THE STATUS OF THE REGIME IS AUTHORITARIAN OR TOTALITARIAN (41)	0.74	37 11 7 31 X SQ= 26.91 P = 0. RV YES (YES)
95	TEND MORE TO BE THOSE WHERE THE STATUS OF THE REGIME IS CONSTITUTIONAL OR AUTHORITARIAN (95)	0.98	95	TEND LESS TO BE THOSE WHERE THE STATUS OF THE REGIME IS CONSTITUTIONAL OR AUTHORITARIAN (95)	0.66	56 29 1 15 X SQ= 17.13 P = 0. RV NO (YES)
96	TEND MORE TO BE THOSE WHERE THE STATUS OF THE REGIME IS CONSTITUTIONAL OR TOTALITARIAN (67)	0.86	96	TEND LESS TO BE THOSE WHERE THE STATUS OF THE REGIME IS CONSTITUTIONAL OR TOTALITARIAN (67)	0.65	38 26 6 14 X SQ= 4.16 P = 0.039 RV NO (YES)
97	TILT TOWARD BEING THOSE WHERE THE STATUS OF THE REGIME IS AUTHORITARIAN, RATHER THAN TOTALITARIAN (23)	0.86	97	TILT TOWARD BEING THOSE WHERE THE STATUS OF THE REGIME IS TOTALITARIAN, RATHER THAN AUTHORITARIAN (16)	0.52	6 14 1 15 X SQ= 1.86 P = 0.104 RV YES (NO)

No.	Left predicate		No.	Right predicate		Statistics
99	TEND TO BE THOSE WHERE GOVERNMENTAL STABILITY IS GENERALLY PRESENT AND DATES FROM AT LEAST THE INTER-WAR PERIOD, OR FROM THE POST-WAR PERIOD (50)	0.78	99	TEND TO BE THOSE WHERE GOVERNMENTAL STABILITY IS MODERATELY PRESENT AND DATES FROM THE POST-WAR PERIOD, OR IS ABSENT (36)	0.52	29 19 8 21 X SQ= 6.55 P = 0.009 RV YES (YES)
101	TEND TO BE THOSE WHERE THE REPRESENTATIVE CHARACTER OF THE REGIME IS POLYARCHIC (41)	0.65	101	TEND TO BE THOSE WHERE THE REPRESENTATIVE CHARACTER OF THE REGIME IS LIMITED POLYARCHIC, PSEUDO-POLYARCHIC, OR NON-POLYARCHIC (57)	0.89	35 4 19 34 X SQ= 24.74 P = 0. RV YES (YES)
102	TEND TO BE THOSE WHERE THE REPRESENTATIVE CHARACTER OF THE REGIME IS POLYARCHIC OR LIMITED POLYARCHIC (59)	0.70	102	TEND TO BE THOSE WHERE THE REPRESENTATIVE CHARACTER OF THE REGIME IS PSEUDO-POLYARCHIC OR NON-POLYARCHIC (49)	0.67	40 14 17 29 X SQ= 12.49 P = 0. RV YES (YES)
105	TEND TO BE THOSE WHERE THE ELECTORAL SYSTEM IS COMPETITIVE (43)	0.64	105	TEND TO BE THOSE WHERE THE ELECTORAL SYSTEM IS PARTIALLY COMPETITIVE OR NON-COMPETITIVE (47)	0.72	32 10 18 26 X SQ= 9.59 P = 0.001 RV YES (YES)
106	TEND TO BE THOSE WHERE THE ELECTORAL SYSTEM IS COMPETITIVE OR PARTIALLY COMPETITIVE (52)	0.80	106	TEND TO BE THOSE WHERE THE ELECTORAL SYSTEM IS NON-COMPETITIVE (30)	0.61	39 12 10 19 X SQ= 12.02 P = 0. RV YES (YES)
107	TEND TO BE THOSE WHERE AUTONOMOUS GROUPS ARE FULLY TOLERATED IN POLITICS (46)	0.60	107	TEND TO BE THOSE WHERE AUTONOMOUS GROUPS ARE PARTIALLY TOLERATED IN POLITICS, ARE TOLERATED ONLY OUTSIDE POLITICS, OR ARE NOT TOLERATED AT ALL (65)	0.80	34 9 23 36 X SQ= 14.63 P = 0. RV YES (YES)
108	TEND TO BE THOSE WHERE AUTONOMOUS GROUPS ARE FULLY OR PARTIALLY TOLERATED IN POLITICS (65)	0.82	108	TEND TO BE THOSE WHERE AUTONOMOUS GROUPS ARE TOLERATED ONLY OUTSIDE POLITICS OR ARE NOT TOLERATED AT ALL (35)	0.62	46 15 10 25 X SQ= 18.19 P = 0. RV YES (YES)
111	TEND TO BE THOSE WHERE POLITICAL ENCULTURATION IS HIGH OR MEDIUM (53)	0.79	111	TEND TO BE THOSE WHERE POLITICAL ENCULTURATION IS LOW (42)	0.68	42 10 11 21 X SQ= 16.37 P = 0. RV YES (YES)
113	LEAN MORE TOWARD BEING THOSE WHERE SECTIONALISM IS MODERATE OR NEGLIGIBLE (81)	0.85	113	LEAN LESS TOWARD BEING THOSE WHERE SECTIONALISM IS MODERATE OR NEGLIGIBLE (81)	0.67	8 14 46 29 X SQ= 3.35 P = 0.051 RV NO (YES)

117	TEND TO BE THOSE WHERE INTEREST ARTICULATION BY ASSOCIATIONAL GROUPS IS SIGNIFICANT, MODERATE, OR LIMITED (60)	0.65	117	TEND TO BE THOSE WHERE INTEREST ARTICULATION BY ASSOCIATIONAL GROUPS IS NEGLIGIBLE (51)	0.62	37 16 20 26 X SQ= 5.95 P = 0.014 RV YES (YES)
118	TEND TO BE THOSE WHERE INTEREST ARTICULATION BY INSTITUTIONAL GROUPS IS SIGNIFICANT, MODERATE, OR LIMITED (60)	0.91	118	TEND TO BE THOSE WHERE INTEREST ARTICULATION BY INSTITUTIONAL GROUPS IS VERY SIGNIFICANT (40)	0.66	4 29 41 15 X SQ= 28.61 P = 0. RV YES (YES)
119	TEND TO BE THOSE WHERE INTEREST ARTICULATION BY INSTITUTIONAL GROUPS IS MODERATE OR LIMITED (26)	0.51	119	TEND TO BE THOSE WHERE INTEREST ARTICULATION BY INSTITUTIONAL GROUPS IS VERY SIGNIFICANT OR SIGNIFICANT (74)	0.95	22 42 23 2 X SQ= 21.63 P = 0. RV YES (YES)
120	TEND LESS TO BE THOSE WHERE INTEREST ARTICULATION BY INSTITUTIONAL GROUPS IS VERY SIGNIFICANT, SIGNIFICANT, OR MODERATE (90)	0.78	120	ALWAYS ARE THOSE WHERE INTEREST ARTICULATION BY INSTITUTIONAL GROUPS IS VERY SIGNIFICANT, SIGNIFICANT, OR MODERATE (90)	1.00	35 44 10 0 X SQ= 8.90 P = 0.001 RV NO (YES)
122	TILT LESS TOWARD BEING THOSE WHERE INTEREST ARTICULATION BY NON-ASSOCIATIONAL GROUPS IS SIGNIFICANT OR MODERATE (83)	0.63	122	TILT MORE TOWARD BEING THOSE WHERE INTEREST ARTICULATION BY NON-ASSOCIATIONAL GROUPS IS SIGNIFICANT OR MODERATE (83)	0.78	36 36 21 10 X SQ= 2.09 P = 0.131 RV NO (YES)
123	TEND LESS TO BE THOSE WHERE INTEREST ARTICULATION BY NON-ASSOCIATIONAL GROUPS IS SIGNIFICANT, MODERATE, OR LIMITED (107)	0.86	123	ALWAYS ARE THOSE WHERE INTEREST ARTICULATION BY NON-ASSOCIATIONAL GROUPS IS SIGNIFICANT, MODERATE, OR LIMITED (107)	1.00	49 46 8 0 X SQ= 5.18 P = 0.008 RV NO (NO)
125	TEND TO BE THOSE WHERE INTEREST ARTICULATION BY ANOMIC GROUPS IS INFREQUENT OR VERY INFREQUENT (35)	0.51	125	TEND TO BE THOSE WHERE INTEREST ARTICULATION BY ANOMIC GROUPS IS FREQUENT OR OCCASIONAL (64)	0.81	26 30 27 7 X SQ= 8.19 P = 0.002 RV YES (YES)
130	TEND LESS TO BE THOSE WHERE INTEREST AGGREGATION BY POLITICAL PARTIES IS MODERATE, LIMITED, OR NEGLIGIBLE (71)	0.72	130	ALWAYS ARE THOSE WHERE INTEREST AGGREGATION BY POLITICAL PARTIES IS MODERATE, LIMITED, OR NEGLIGIBLE (71)	1.00	11 0 28 38 X SQ= 10.31 P = 0. RV NO (YES)
131	TEND TO BE THOSE WHERE INTEREST AGGREGATION BY POLITICAL PARTIES IS SIGNIFICANT OR MODERATE (30)	0.67	131	TEND TO BE THOSE WHERE INTEREST AGGREGATION BY POLITICAL PARTIES IS LIMITED OR NEGLIGIBLE (35)	0.86	26 3 13 18 X SQ= 12.97 P = 0. RV YES (YES)

138	TEND TO BE THOSE WHERE INTEREST AGGREGATION BY THE LEGISLATURE IS SIGNIFICANT, MODERATE, OR LIMITED (48)	0.61	138	TEND TO BE THOSE WHERE INTEREST AGGREGATION BY THE LEGISLATURE IS NEGLIGIBLE (49)	0.68	33 12 21 26 X SQ= 6.65 P = 0.006 RV YES (YES)
140	TILT LESS TOWARD BEING THOSE WHERE THE PARTY SYSTEM IS QUANTITATIVELY OTHER THAN ONE PARTY DOMINANT (87)	0.81	140	TILT MORE TOWARD BEING THOSE WHERE THE PARTY SYSTEM IS QUANTITATIVELY OTHER THAN ONE PARTY DOMINANT (87)	0.92	10 3 43 37 X SQ= 1.60 P = 0.141 RV NO (NO)
141	TEND LESS TO BE THOSE WHERE THE PARTY SYSTEM IS QUANTITATIVELY OTHER THAN TWO-PARTY (87)	0.81	141	TEND MORE TO BE THOSE WHERE THE PARTY SYSTEM IS QUANTITATIVELY OTHER THAN TWO-PARTY (87)	0.97	10 1 44 37 X SQ= 3.95 P = 0.024 RV NO (NO)
144	TEND LESS TO BE THOSE WHERE THE PARTY SYSTEM IS QUANTITATIVELY ONE-PARTY, RATHER THAN TWO-PARTY (34)	0.63	144	TEND MORE TO BE THOSE WHERE THE PARTY SYSTEM IS QUANTITATIVELY ONE-PARTY, RATHER THAN TWO-PARTY (34)	0.94	17 16 10 1 X SQ= 3.87 P = 0.031 RV NO (NO)
145	TEND LESS TO BE THOSE WHERE THE PARTY SYSTEM IS QUANTITATIVELY MULTI-PARTY, RATHER THAN TWO-PARTY (30)	0.57	145	TEND MORE TO BE THOSE WHERE THE PARTY SYSTEM IS QUANTITATIVELY MULTI-PARTY, RATHER THAN TWO-PARTY (30)	0.92	13 12 10 1 X SQ= 3.47 P = 0.031 RV NO (NO)
146	TEND LESS TO BE THOSE WHERE THE PARTY SYSTEM IS QUALITATIVELY OTHER THAN MASS-BASED TERRITORIAL (92)	0.86	146	ALWAYS ARE THOSE WHERE THE PARTY SYSTEM IS QUALITATIVELY OTHER THAN MASS-BASED TERRITORIAL (92)	1.00	7 0 44 40 X SQ= 4.17 P = 0.017 RV NO (NO)
147	TEND LESS TO BE THOSE WHERE THE PARTY SYSTEM IS QUALITATIVELY OTHER THAN CLASS-ORIENTED OR MULTI-IDEOLOGICAL (67)	0.64	147	TEND MORE TO BE THOSE WHERE THE PARTY SYSTEM IS QUALITATIVELY OTHER THAN CLASS-ORIENTED OR MULTI-IDEOLOGICAL (67)	0.88	18 4 32 30 X SQ= 4.96 P = 0.022 RV NO (NO)
148	TEND LESS TO BE THOSE WHERE THE PARTY SYSTEM IS QUALITATIVELY OTHER THAN AFRICAN TRANSITIONAL (96)	0.77	148	TEND MORE TO BE THOSE WHERE THE PARTY SYSTEM IS QUALITATIVELY OTHER THAN AFRICAN TRANSITIONAL (96)	0.98	13 1 44 40 X SQ= 6.50 P = 0.007 RV NO (NO)
154	TEND MORE TO BE THOSE WHERE THE PARTY SYSTEM IS STABLE OR MODERATELY STABLE (55)	0.89	154	TEND LESS TO BE THOSE WHERE THE PARTY SYSTEM IS STABLE OR MODERATELY STABLE (55)	0.60	33 21 4 14 X SQ= 6.69 P = 0.006 RV NO (YES)

157	TEND MORE TO BE THOSE WHERE PERSONALISMO IS MODERATE OR NEGLIGIBLE (84)	0.95	157	TEND LESS TO BE THOSE WHERE PERSONALISMO IS MODERATE OR NEGLIGIBLE (84)	0.79	3 7 53 27 X SQ= 3.55 P = 0.038 RV NO (YES)
160	ALWAYS ARE THOSE WHERE THE POLITICAL LEADERSHIP IS MODERATELY ELITIST OR NON-ELITIST (67)	1.00	160	TEND TO BE THOSE WHERE THE POLITICAL LEADERSHIP IS ELITIST (30)	0.72	0 28 53 11 X SQ= 51.36 P = 0. RV YES (YES)
161	TEND TO BE THOSE WHERE THE POLITICAL LEADERSHIP IS NON-ELITIST (50)	0.83	161	TEND TO BE THOSE WHERE THE POLITICAL LEADERSHIP IS ELITIST OR MODERATELY ELITIST (47)	0.90	9 35 44 4 X SQ= 44.80 P = 0. RV YES (YES)
164	TEND LESS TO BE THOSE WHERE THE REGIME'S LEADERSHIP CHARISMA IS NEGLIGIBLE (65)	0.56	164	TEND MORE TO BE THOSE WHERE THE REGIME'S LEADERSHIP CHARISMA IS NEGLIGIBLE (65)	0.79	24 8 31 30 X SQ= 4.13 P = 0.028 RV NO (NO)
169	TEND TO BE THOSE WHERE THE HORIZONTAL POWER DISTRIBUTION IS SIGNIFICANT OR LIMITED (58)	0.70	169	TEND TO BE THOSE WHERE THE HORIZONTAL POWER DISTRIBUTION IS NEGLIGIBLE (48)	0.67	40 14 17 28 X SQ= 11.79 P = 0. RV YES (YES)
171	TEND LESS TO BE THOSE WHERE THE LEGISLATIVE-EXECUTIVE STRUCTURE IS OTHER THAN PARLIAMENTARY-REPUBLICAN (90)	0.84	171	TEND MORE TO BE THOSE WHERE THE LEGISLATIVE-EXECUTIVE STRUCTURE IS OTHER THAN PARLIAMENTARY-REPUBLICAN (90)	0.97	9 1 46 39 X SQ= 3.37 P = 0.041 RV NO (NO)
172	TEND LESS TO BE THOSE WHERE THE LEGISLATIVE-EXECUTIVE STRUCTURE IS OTHER THAN PARLIAMENTARY-ROYALIST (88)	0.70	172	TEND MORE TO BE THOSE WHERE THE LEGISLATIVE-EXECUTIVE STRUCTURE IS OTHER THAN PARLIAMENTARY-ROYALIST (88)	0.93	17 3 39 39 X SQ= 6.60 P = 0.005 RV NO (YES)
175	TEND TO BE THOSE WHERE THE LEGISLATURE IS FULLY EFFECTIVE OR PARTIALLY EFFECTIVE (51)	0.64	175	TEND TO BE THOSE WHERE THE LEGISLATURE IS LARGELY INEFFECTIVE OR WHOLLY INEFFECTIVE (49)	0.69	35 12 20 27 X SQ= 8.59 P = 0.002 RV YES (YES)
176	TEND TO BE THOSE WHERE THE LEGISLATURE IS FULLY EFFECTIVE, PARTIALLY EFFECTIVE, OR LARGELY INEFFECTIVE (72)	0.93	176	TEND TO BE THOSE WHERE THE LEGISLATURE IS WHOLLY INEFFECTIVE (28)	0.59	51 16 4 23 X SQ= 27.32 P = 0. RV YES (YES)

179	TEND TO BE THOSE WHERE THE EXECUTIVE IS STRONG (39)	0.60	179	TEND TO BE THOSE WHERE THE EXECUTIVE IS DOMINANT (52)	0.82	21 28 31 6 X SQ= 13.11 P = 0. RV YES (YES)
181	TEND TO BE THOSE WHERE THE BUREAUCRACY IS MODERN, RATHER THAN SEMI-MODERN (21)	0.50	181	TEND TO BE THOSE WHERE THE BUREAUCRACY IS SEMI-MODERN, RATHER THAN MODERN (55)	0.97	19 1 19 31 X SQ= 16.48 P = 0. RV YES (YES)
182	TEND LESS TO BE THOSE WHERE THE BUREAUCRACY IS SEMI-MODERN, RATHER THAN POST-COLONIAL TRANSITIONAL (55)	0.53	182	TEND MORE TO BE THOSE WHERE THE BUREAUCRACY IS SEMI-MODERN, RATHER THAN POST-COLONIAL TRANSITIONAL (55)	0.84	19 31 17 6 X SQ= 6.76 P = 0.006 RV NO (YES)
183	TEND TO BE THOSE WHERE THE BUREAUCRACY IS POST-COLONIAL TRANSITIONAL, RATHER THAN TRADITIONAL (25)	0.94	183	TEND TO BE THOSE WHERE THE BUREAUCRACY IS TRADITIONAL, RATHER THAN POST-COLONIAL TRANSITIONAL (9)	0.54	17 6 1 7 X SQ= 6.84 P = 0.004 RV YES (YES)
186	TEND TO BE THOSE WHERE PARTICIPATION BY THE MILITARY IN POLITICS IS NEUTRAL, RATHER THAN SUPPORTIVE (56)	0.90	186	TEND TO BE THOSE WHERE PARTICIPATION BY THE MILITARY IN POLITICS IS SUPPORTIVE, RATHER THAN NEUTRAL (31)	0.77	5 23 47 7 X SQ= 35.11 P = 0. RV YES (YES)
187	TEND TO BE THOSE WHERE THE ROLE OF THE POLICE IS NOT POLITICALLY SIGNIFICANT (35)	0.62	187	TEND TO BE THOSE WHERE THE ROLE OF THE POLICE IS POLITICALLY SIGNIFICANT (66)	0.93	19 38 31 3 X SQ= 26.49 P = 0. RV YES (YES)
190	TEND LESS TO BE THOSE WHERE THE CHARACTER OF THE LEGAL SYSTEM IS CIVIL LAW, RATHER THAN COMMON LAW (32)	0.67	190	ALWAYS ARE THOSE WHERE THE CHARACTER OF THE LEGAL SYSTEM IS CIVIL LAW, RATHER THAN COMMON LAW (32)	1.00	14 16 7 0 X SQ= 4.58 P = 0.012 RV NO (YES)
193	TEND LESS TO BE THOSE WHERE THE CHARACTER OF THE LEGAL SYSTEM IS OTHER THAN PARTLY INDIGENOUS (86)	0.67	193	TEND MORE TO BE THOSE WHERE THE CHARACTER OF THE LEGAL SYSTEM IS OTHER THAN PARTLY INDIGENOUS (86)	0.87	19 6 38 39 X SQ= 4.41 P = 0.022 RV NO (YES)
194	ALWAYS ARE THOSE THAT ARE NON-COMMUNIST (101)	1.00	194	TEND LESS TO BE THOSE THAT ARE NON-COMMUNIST (101)	0.72	0 13 56 33 X SQ= 15.69 P = 0. RV NO (YES)

135 POLITIES WHERE INTEREST AGGREGATION BY THE EXECUTIVE IS SIGNIFICANT, MODERATE, OR LIMITED (77)	135 POLITIES WHERE INTEREST AGGREGATION BY THE EXECUTIVE IS NEGLIGIBLE (13)

PREDICATE ONLY

77 IN LEFT COLUMN

ALGERIA	ARGENTINA	AUSTRALIA	AUSTRIA	BELGIUM	BOLIVIA	BRAZIL	BURUNDI	CAMBODIA	CAMEROUN
CANADA	CEN AFR REP	CHAD	CHILE	COLOMBIA	CONGO(BRA)	COSTA RICA	DAHOMEY	DENMARK	DOMIN REP
ECUADOR	FINLAND	FRANCE	GABON	GERMAN FR	GHANA	GREECE	GUATEMALA	GUINEA	HONDURAS
ICELAND	INDIA	INDONESIA	IRAN	IRELAND	ISRAEL	ITALY	IVORY COAST	JAMAICA	JAPAN
JORDAN	LIBERIA	LIBYA	LUXEMBOURG	MALAGASY R	MALAYA	MALI	MAURITANIA	MEXICO	MOROCCO
NETHERLANDS	NEW ZEALAND	NIGER	NIGERIA	NORWAY	PAKISTAN	PANAMA	PHILIPPINES	POLAND	RWANDA
SENEGAL	SIERRE LEO	SOMALIA	SPAIN	SWEDEN	SWITZERLAND	TANGANYIKA	TRINIDAD	TUNISIA	TURKEY
UGANDA	UAR	UK	US	UPPER VOLTA	VENEZUELA	YUGOSLAVIA			

13 IN RIGHT COLUMN

AFGHANISTAN	CEYLON	EL SALVADOR	ETHIOPIA	GERMANY, E	HAITI	NEPAL	NICARAGUA	PARAGUAY	PORTUGAL
SA'U ARABIA	THAILAND	VIETNAM REP							

5 EXCLUDED BECAUSE AMBIGUOUS

CYPRUS	LAOS	LEBANON	SO AFRICA	URUGUAY

11 EXCLUDED BECAUSE UNASCERTAINED

ALBANIA	BULGARIA	CHINA, PR	CUBA	CZECHOS'KIA	HUNGARY	KOREA, N	MONGOLIA	RUMANIA	USSR
VIETNAM, N									

9 EXCLUDED BECAUSE UNASCERTAINABLE

BURMA	CONGO(LEO)	IRAQ	KOREA REP	PERU	SUDAN	SYRIA	TOGO	YEMEN

136 POLITIES
WHERE INTEREST AGGREGATION
BY THE LEGISLATURE
IS SIGNIFICANT (12)

136 POLITIES
WHERE INTEREST AGGREGATION
BY THE LEGISLATURE
IS MODERATE, LIMITED, OR
NEGLIGIBLE (85)

PREDICATE ONLY

12 IN LEFT COLUMN

BELGIUM DENMARK FINLAND IRELAND ISRAEL LUXEMBOURG NETHERLANDS NORWAY SWEDEN SWITZERLAND
US URUGUAY

85 IN RIGHT COLUMN

AFGHANISTAN ALBANIA ARGENTINA AUSTRALIA AUSTRIA BOLIVIA BRAZIL BULGARIA BURUNDI CAMBODIA
CAMEROUN CANADA CEN AFR REP CEYLON CHAD CHILE CHINA, PR CONGO(BRA) CONGO(LEO) COSTA RICA
CZECHOS'KIA DAHOMEY ECUADOR EL SALVADOR ETHIOPIA FRANCE GABON GERMANY, E GERMAN FR GHANA
GREECE GUATEMALA GUINEA HAITI HONDURAS HUNGARY ICELAND INDIA ITALY IVORY COAST
JAMAICA JAPAN JORDAN KOREA, N LAOS LEBANON LIBERIA LIBYA MALAGASY R MALAYA
MALI MAURITANIA MEXICO MONGOLIA NEPAL NEW ZEALAND NICARAGUA NIGER NIGERIA PAKISTAN
PANAMA PARAGUAY PHILIPPINES POLAND PORTUGAL RUMANIA RWANDA SENEGAL SIERRE LEO SOMALIA
SPAIN TANGANYIKA THAILAND TOGO TRINIDAD TUNISIA TURKEY UGANDA USSR UK
UPPER VOLTA VENEZUELA VIETNAM, N VIETNAM REP YUGOSLAVIA

4 EXCLUDED BECAUSE AMBIGUOUS

COLOMBIA CYPRUS INDONESIA SO AFRICA

1 EXCLUDED BECAUSE IRRELEVANT

SA'U ARABIA

13 EXCLUDED BECAUSE UNASCERTAINABLE

ALGERIA BURMA CUBA DOMIN REP IRAN IRAQ KOREA REP MOROCCO PERU SUDAN
SYRIA UAR YEMEN

```
************************************************************************************************************************

137  POLITIES                                              137  POLITIES
     WHERE INTEREST AGGREGATION                                 WHERE INTEREST AGGREGATION
     BY THE LEGISLATURE                                         BY THE LEGISLATURE
     IS SIGNIFICANT OR MODERATE  (29)                           IS LIMITED OR NEGLIGIBLE  (68)


                                                                                  BOTH SUBJECT AND PREDICATE
........................................................................................................................

      29 IN LEFT COLUMN

AUSTRALIA    AUSTRIA      BELGIUM      CANADA       CHILE        DENMARK      FINLAND      GERMAN FR    GREECE       ICELAND
INDIA        IRELAND      ISRAEL       ITALY        JAMAICA      JAPAN        LEBANON      LUXEMBOURG   NETHERLANDS  NEW ZEALAND
NORWAY       PHILIPPINES  SWEDEN       SWITZERLAND  TRINIDAD     TURKEY       UK           US           URUGUAY

      68 IN RIGHT COLUMN

AFGHANISTAN  ALBANIA      ARGENTINA    BOLIVIA      BRAZIL       BULGARIA     BURUNDI      CAMBODIA     CAMEROUN     CEN AFR REP
CEYLON       CHAD         CHINA, PR    CONGO(BRA)   CONGO(LEO)   COSTA RICA   CZECHOS'KIA  DAHOMEY      ECUADOR      EL SALVADOR
ETHIOPIA     FRANCE       GABON        GERMANY, E   GHANA        GUATEMALA    GUINEA       HAITI        HONDURAS     HUNGARY
IVORY COAST  JORDAN       KOREA, N     LAOS         LIBERIA      LIBYA        MALAGASY R   MALAYA       MALI         MAURITANIA
MEXICO       MONGOLIA     NEPAL        NICARAGUA    NIGER        NIGERIA      PAKISTAN     PANAMA       PARAGUAY     POLAND
PORTUGAL     RUMANIA      RWANDA       SENEGAL      SIERRE LEO   SOMALIA      SPAIN        TANGANYIKA   THAILAND     TOGO
TUNISIA      UGANDA       USSR         UPPER VOLTA  VENEZUELA    VIETNAM, N   VIETNAM REP  YUGOSLAVIA

       4 EXCLUDED BECAUSE AMBIGUOUS

COLOMBIA     CYPRUS       INDONESIA    SO AFRICA

       1 EXCLUDED BECAUSE IRRELEVANT

SA'U ARABIA

      13 EXCLUDED BECAUSE UNASCERTAINABLE

ALGERIA      BURMA        CUBA         DOMIN REP    IRAN         IRAQ         KOREA REP    MOROCCO      PERU         SUDAN
SYRIA        UAR          YEMEN

------------------------------------------------------------------------------------------------------------------------

  2  TEND TO BE THOSE                           0.66        2  TEND TO BE THOSE                           0.96          19    3
     LOCATED IN WEST EUROPE, SCANDINAVIA,                      LOCATED ELSEWHERE THAN IN                                10   65
     NORTH AMERICA, OR AUSTRALASIA  (22)                       WEST EUROPE, SCANDINAVIA,                            X SQ=  39.87
                                                               NORTH AMERICA, OR AUSTRALASIA  (93)                  P =     0.
                                                                                                                    RV YES (YES)

  3  TEND LESS TO BE THOSE                      0.66        3  TEND MORE TO BE THOSE                      0.96          10    3
     LOCATED ELSEWHERE THAN IN                                 LOCATED ELSEWHERE THAN IN                                19   65
     WEST EUROPE  (102)                                        WEST EUROPE  (102)                                   X SQ=  13.35
                                                                                                                    P =     0.
                                                                                                                    RV NO  (YES)
```

5	ALWAYS ARE THOSE LOCATED IN SCANDINAVIA OR WEST EUROPE, RATHER THAN IN EAST EUROPE (18)	1.00	5	TEND TO BE THOSE LOCATED IN EAST EUROPE, RATHER THAN IN SCANDINAVIA OR WEST EUROPE (9)	0.75	15 3 0 9 X SQ= 13.67 P = 0. RV YES (YES)
10	ALWAYS ARE THOSE LOCATED ELSEWHERE THAN IN NORTH AFRICA, OR CENTRAL AND SOUTH AFRICA (82)	1.00	10	TEND LESS TO BE THOSE LOCATED ELSEWHERE THAN IN NORTH AFRICA, OR CENTRAL AND SOUTH AFRICA (82)	0.59	0 28 29 40 X SQ= 14.84 P = 0. RV NO (NO)
15	ALWAYS ARE THOSE LOCATED IN THE MIDDLE EAST, RATHER THAN IN NORTH AFRICA OR CENTRAL AND SOUTH AFRICA (11)	1.00	15	TEND TO BE THOSE LOCATED IN NORTH AFRICA OR CENTRAL AND SOUTH AFRICA, RATHER THAN IN THE MIDDLE EAST (33)	0.93	3 2 0 28 X SQ= 11.93 P = 0.002 RV YES (YES)
18	ALWAYS ARE THOSE LOCATED IN EAST ASIA, SOUTH ASIA, OR SOUTHEAST ASIA, RATHER THAN IN NORTH AFRICA OR CENTRAL AND SOUTH AFRICA (18)	1.00	18	TEND TO BE THOSE LOCATED IN NORTH AFRICA OR CENTRAL AND SOUTH AFRICA, RATHER THAN IN EAST ASIA, SOUTH ASIA, OR SOUTHEAST ASIA (33)	0.70	0 28 3 12 X SQ= 3.33 P = 0.037 RV YES (NO)
19	ALWAYS ARE THOSE LOCATED IN THE CARIBBEAN, CENTRAL AMERICA, OR SOUTH AMERICA, RATHER THAN IN NORTH AFRICA OR CENTRAL AND SOUTH AFRICA (22)	1.00	19	TEND TO BE THOSE LOCATED IN NORTH AFRICA OR CENTRAL AND SOUTH AFRICA, RATHER THAN IN THE CARIBBEAN, CENTRAL AMERICA, OR SOUTH AMERICA (33)	0.67	0 28 4 14 X SQ= 4.30 P = 0.019 RV YES (NO)
26	TEND TO BE THOSE WHOSE POPULATION DENSITY IS VERY HIGH, HIGH, OR MEDIUM (48)	0.62	26	TEND TO BE THOSE WHOSE POPULATION DENSITY IS LOW (67)	0.62	18 26 11 42 X SQ= 3.75 P = 0.045 RV YES (NO)
29	TEND TO BE THOSE WHERE THE DEGREE OF URBANIZATION IS HIGH (56)	0.96	29	TEND TO BE THOSE WHERE THE DEGREE OF URBANIZATION IS LOW (49)	0.67	26 20 1 41 X SQ= 27.77 P = 0. RV YES (YES)
30	TEND TO BE THOSE WHOSE AGRICULTURAL POPULATION IS MEDIUM, LOW, OR VERY LOW (57)	0.93	30	TEND TO BE THOSE WHOSE AGRICULTURAL POPULATION IS HIGH (56)	0.66	2 44 27 23 X SQ= 25.71 P = 0. RV YES (YES)
31	TEND TO BE THOSE WHOSE AGRICULTURAL POPULATION IS LOW OR VERY LOW (24)	0.69	31	TEND TO BE THOSE WHOSE AGRICULTURAL POPULATION IS HIGH OR MEDIUM (90)	0.96	9 65 20 3 X SQ= 43.33 P = 0. RV YES (YES)

33	TEND TO BE THOSE WHOSE GROSS NATIONAL PRODUCT IS VERY HIGH, HIGH, OR MEDIUM (30)	0.52	33	TEND TO BE THOSE WHOSE GROSS NATIONAL PRODUCT IS LOW OR VERY LOW (85)	0.81	15 13 14 55 X SQ= 9.00 P = 0.003 RV YES (YES)
34	TEND TO BE THOSE WHOSE GROSS NATIONAL PRODUCT IS VERY HIGH, HIGH, MEDIUM, OR LOW (62)	0.83	34	TEND TO BE THOSE WHOSE GROSS NATIONAL PRODUCT IS VERY LOW (53)	0.62	24 26 5 42 X SQ= 14.40 P = 0. RV YES (NO)
35	TEND TO BE THOSE WHOSE PER CAPITA GROSS NATIONAL PRODUCT IS VERY HIGH OR HIGH (24)	0.66	35	TEND TO BE THOSE WHOSE PER CAPITA GROSS NATIONAL PRODUCT IS MEDIUM, LOW, OR VERY LOW (91)	0.93	19 5 10 63 X SQ= 33.88 P = 0. RV YES (YES)
36	TEND TO BE THOSE WHOSE PER CAPITA GROSS NATIONAL PRODUCT IS VERY HIGH, HIGH, OR MEDIUM (42)	0.90	36	TEND TO BE THOSE WHOSE PER CAPITA GROSS NATIONAL PRODUCT IS LOW OR VERY LOW (73)	0.79	26 14 3 54 X SQ= 37.22 P = 0. RV YES (YES)
37	TEND TO BE THOSE WHOSE PER CAPITA GROSS NATIONAL PRODUCT IS VERY HIGH, HIGH, MEDIUM, OR LOW (64)	0.97	37	TEND TO BE THOSE WHOSE PER CAPITA GROSS NATIONAL PRODUCT IS VERY LOW (51)	0.63	28 25 1 43 X SQ= 26.96 P = 0. RV YES (YES)
39	TEND TO BE THOSE WHOSE INTERNATIONAL FINANCIAL STATUS IS VERY HIGH, HIGH, OR MEDIUM (38)	0.69	39	TEND TO BE THOSE WHOSE INTERNATIONAL FINANCIAL STATUS IS LOW OR VERY LOW (76)	0.79	20 14 9 53 X SQ= 18.40 P = 0. RV YES (YES)
40	TEND TO BE THOSE WHOSE INTERNATIONAL FINANCIAL STATUS IS VERY HIGH, HIGH, MEDIUM, OR LOW (71)	0.96	40	TEND TO BE THOSE WHOSE INTERNATIONAL FINANCIAL STATUS IS VERY LOW (39)	0.53	26 31 1 35 X SQ= 17.63 P = 0. RV YES (NO)
41	TEND TO BE THOSE WHOSE ECONOMIC DEVELOPMENTAL STATUS IS DEVELOPED (19)	0.54	41	TEND TO BE THOSE WHOSE ECONOMIC DEVELOPMENTAL STATUS IS INTERMEDIATE, UNDERDEVELOPED, OR VERY UNDERDEVELOPED (94)	0.94	15 4 13 63 X SQ= 25.07 P = 0. RV YES (YES)
42	TEND TO BE THOSE WHOSE ECONOMIC DEVELOPMENTAL STATUS IS DEVELOPED OR INTERMEDIATE (36)	0.75	42	TEND TO BE THOSE WHOSE ECONOMIC DEVELOPMENTAL STATUS IS UNDERDEVELOPED OR VERY UNDERDEVELOPED (76)	0.82	21 12 7 55 X SQ= 25.93 P = 0. RV YES (YES)

<table>
<tr><td>43</td><td>ALWAYS ARE THOSE
WHOSE ECONOMIC DEVELOPMENTAL STATUS
IS DEVELOPED, INTERMEDIATE, OR
UNDERDEVELOPED (55)</td><td>1.00</td><td>43</td><td>TEND TO BE THOSE
WHOSE ECONOMIC DEVELOPMENTAL STATUS
IS VERY UNDERDEVELOPED (57)</td><td>0.69</td><td>28 21
0 47
X SQ= 35.20
P = 0.
RV YES (YES)</td></tr>
<tr><td>44</td><td>TEND TO BE THOSE
WHERE THE LITERACY RATE IS
NINETY PERCENT OR ABOVE (25)</td><td>0.62</td><td>44</td><td>TEND TO BE THOSE
WHERE THE LITERACY RATE IS
BELOW NINETY PERCENT (90)</td><td>0.90</td><td>18 7
11 61
X SQ= 25.84
P = 0.
RV YES (YES)</td></tr>
<tr><td>45</td><td>TEND TO BE THOSE
WHERE THE LITERACY RATE IS
FIFTY PERCENT OR ABOVE (55)</td><td>0.93</td><td>45</td><td>TEND TO BE THOSE
WHERE THE LITERACY RATE IS
BELOW FIFTY PERCENT (54)</td><td>0.65</td><td>26 22
2 41
X SQ= 23.83
P = 0.
RV YES (YES)</td></tr>
<tr><td>46</td><td>ALWAYS ARE THOSE
WHERE THE LITERACY RATE IS
TEN PERCENT OR ABOVE (84)</td><td>1.00</td><td>46</td><td>TEND LESS TO BE THOSE
WHERE THE LITERACY RATE IS
TEN PERCENT OR ABOVE (84)</td><td>0.64</td><td>29 41
0 23
X SQ= 11.98
P = 0.
RV NO (NO)</td></tr>
<tr><td>50</td><td>TEND TO BE THOSE
WHERE FREEDOM OF THE PRESS IS
COMPLETE (43)</td><td>0.89</td><td>50</td><td>TEND TO BE THOSE
WHERE FREEDOM OF THE PRESS IS
INTERMITTENT, INTERNALLY ABSENT, OR
INTERNALLY AND EXTERNALLY ABSENT (56)</td><td>0.74</td><td>25 14
3 39
X SQ= 26.54
P = 0.
RV YES (YES)</td></tr>
<tr><td>51</td><td>ALWAYS ARE THOSE
WHERE FREEDOM OF THE PRESS IS
COMPLETE OR INTERMITTENT (60)</td><td>1.00</td><td>51</td><td>TEND TO BE THOSE
WHERE FREEDOM OF THE PRESS IS
INTERNALLY ABSENT, OR
INTERNALLY AND EXTERNALLY ABSENT (37)</td><td>0.57</td><td>28 23
0 30
X SQ= 22.80
P = 0.
RV YES (YES)</td></tr>
<tr><td>52</td><td>ALWAYS ARE THOSE
WHERE FREEDOM OF THE PRESS IS
COMPLETE, INTERMITTENT, OR
INTERNALLY ABSENT (82)</td><td>1.00</td><td>52</td><td>TEND LESS TO BE THOSE
WHERE FREEDOM OF THE PRESS IS
COMPLETE, INTERMITTENT, OR
INTERNALLY ABSENT (82)</td><td>0.74</td><td>29 39
0 14
X SQ= 7.47
P = 0.001
RV NO (NO)</td></tr>
<tr><td>54</td><td>TEND TO BE THOSE
WHERE NEWSPAPER CIRCULATION IS
ONE HUNDRED OR MORE
PER THOUSAND (37)</td><td>0.79</td><td>54</td><td>TEND TO BE THOSE
WHERE NEWSPAPER CIRCULATION IS
LESS THAN ONE HUNDRED
PER THOUSAND (76)</td><td>0.82</td><td>23 12
6 54
X SQ= 29.78
P = 0.
RV YES (YES)</td></tr>
<tr><td>55</td><td>ALWAYS ARE THOSE
WHERE NEWSPAPER CIRCULATION IS
TEN OR MORE
PER THOUSAND (78)</td><td>1.00</td><td>55</td><td>TEND LESS TO BE THOSE
WHERE NEWSPAPER CIRCULATION IS
TEN OR MORE
PER THOUSAND (78)</td><td>0.53</td><td>29 35
0 31
X SQ= 18.14
P = 0.
RV NO (NO)</td></tr>
</table>

56	ALWAYS ARE THOSE WHERE THE RELIGION IS PREDOMINANTLY LITERATE (79)	1.00	56	TEND LESS TO BE THOSE WHERE THE RELIGION IS PREDOMINANTLY LITERATE (79)	0.56	29 36 0 28 X SQ= 16.13 P = 0. RV NO (NO)
62	TEND LESS TO BE THOSE WHERE THE RELIGION IS CATHOLIC, RATHER THAN PROTESTANT (25)	0.62	62	ALWAYS ARE THOSE WHERE THE RELIGION IS CATHOLIC, RATHER THAN PROTESTANT (25)	1.00	5 0 8 14 X SQ= 4.31 P = 0.016 RV NO (YES)
63	TEND TO BE THOSE WHERE THE RELIGION IS PREDOMINANTLY SOME KIND OF CHRISTIAN (46)	0.79	63	TEND TO BE THOSE WHERE THE RELIGION IS PREDOMINANTLY OR PARTLY OTHER THAN CHRISTIAN (68)	0.70	23 20 6 47 X SQ= 18.07 P = 0. RV YES (YES)
64	TEND MORE TO BE THOSE WHERE THE RELIGION IS CHRISTIAN, RATHER THAN MUSLIM (46)	0.96	64	TEND LESS TO BE THOSE WHERE THE RELIGION IS CHRISTIAN, RATHER THAN MUSLIM (46)	0.71	23 20 1 8 X SQ= 3.81 P = 0.028 RV NO (YES)
67	TEND MORE TO BE THOSE THAT ARE RACIALLY HOMOGENEOUS (82)	0.89	67	TEND LESS TO BE THOSE THAT ARE RACIALLY HOMOGENEOUS (82)	0.68	25 44 3 21 X SQ= 3.70 P = 0.038 RV NO (NO)
68	TEND TO BE THOSE THAT ARE LINGUISTICALLY HOMOGENEOUS (52)	0.69	68	TEND TO BE THOSE THAT ARE LINGUISTICALLY WEAKLY HETEROGENEOUS OR STRONGLY HETEROGENEOUS (62)	0.63	20 25 9 42 X SQ= 6.92 P = 0.007 RV YES (NO)
69	TEND TO BE THOSE THAT ARE LINGUISTICALLY HOMOGENEOUS OR WEAKLY HETEROGENEOUS (64)	0.79	69	TEND TO BE THOSE THAT ARE LINGUISTICALLY STRONGLY HETEROGENEOUS (50)	0.51	23 33 6 34 X SQ= 6.34 P = 0.007 RV YES (NO)
70	TEND LESS TO BE THOSE THAT ARE RELIGIOUSLY, RACIALLY, OR LINGUISTICALLY HETEROGENEOUS (85)	0.64	70	TEND MORE TO BE THOSE THAT ARE RELIGIOUSLY, RACIALLY, OR LINGUISTICALLY HETEROGENEOUS (85)	0.85	10 9 18 53 X SQ= 4.01 P = 0.048 RV NO (YES)
72	TEND TO BE THOSE WHOSE DATE OF INDEPENDENCE IS BEFORE 1914 (52)	0.68	72	TENC TO BE THOSE WHOSE DATE OF INDEPENDENCE IS AFTER 1914 (59)	0.59	19 27 9 39 X SQ= 4.69 P = 0.024 RV YES (NO)

73	TEND TO BE THOSE WHOSE DATE OF INDEPENDENCE IS BEFORE 1945 (65)	0.82	73	TEND TO BE THOSE WHOSE DATE OF INDEPENDENCE IS AFTER 1945 (46)	0.52	23 32 5 34 X SQ= 7.84 P = 0.003 RV YES (NO)
74	TEND TO BE THOSE THAT ARE HISTORICALLY WESTERN (26)	0.68	74	TEND TO BE THOSE THAT ARE NOT HISTORICALLY WESTERN (87)	0.90	19 7 9 61 X SQ= 30.43 P = 0. RV YES (YES)
75	ALWAYS ARE THOSE THAT ARE HISTORICALLY WESTERN OR SIGNIFICANTLY WESTERNIZED (62)	1.00	75	TEND TO BE THOSE THAT ARE NOT HISTORICALLY WESTERN AND ARE NOT SIGNIFICANTLY WESTERNIZED (52)	0.62	29 26 0 42 X SQ= 29.12 P = 0. RV YES (YES)
76	TEND TO BE THOSE THAT ARE HISTORICALLY WESTERN, RATHER THAN HAVING BEEN SIGNIFICANTLY OR PARTIALLY WESTERNIZED THROUGH A COLONIAL RELATIONSHIP (26)	0.73	76	TEND TO BE THOSE THAT HAVE BEEN SIGNIFICANTLY OR PARTIALLY WESTERNIZED THROUGH A COLONIAL RELATIONSHIP, RATHER THAN BEING HISTORICALLY WESTERN (70)	0.87	19 7 7 48 X SQ= 26.80 P = 0. RV YES (YES)
77	ALWAYS ARE THOSE THAT HAVE BEEN SIGNIFICANTLY WESTERNIZED, RATHER THAN PARTIALLY WESTERNIZED, THROUGH A COLONIAL RELATIONSHIP (28)	1.00	77	TEND TO BE THOSE THAT HAVE BEEN PARTIALLY WESTERNIZED, RATHER THAN SIGNIFICANTLY WESTERNIZED, THROUGH A COLONIAL RELATIONSHIP (41)	0.71	7 14 0 34 X SQ= 10.16 P = 0.001 RV YES (NO)
80	TEND TO BE THOSE WHOSE DATE OF INDEPENDENCE IS AFTER 1914, AND THAT WERE FORMERLY DEPENDENCIES OF BRITAIN, RATHER THAN FRANCE (19)	0.83	80	TEND TO BE THOSE WHOSE DATE OF INDEPENDENCE IS AFTER 1914, AND THAT WERE FORMERLY DEPENDENCIES OF FRANCE, RATHER THAN BRITAIN (24)	0.69	5 9 1 20 X SQ= 3.70 P = 0.028 RV YES (NO)
81	TEND LESS TO BE THOSE WHERE THE TYPE OF POLITICAL MODERNIZATION IS LATER EUROPEAN OR LATER EUROPEAN DERIVED, RATHER THAN EARLY EUROPEAN OR EARLY EUROPEAN DERIVED (40)	0.57	81	TEND MORE TO BE THOSE WHERE THE TYPE OF POLITICAL MODERNIZATION IS LATER EUROPEAN OR LATER EUROPEAN DERIVED, RATHER THAN EARLY EUROPEAN OR EARLY EUROPEAN DERIVED (40)	0.96	9 1 12 24 X SQ= 7.97 P = 0.003 RV NO (YES)
82	TEND TO BE THOSE WHERE THE TYPE OF POLITICAL MODERNIZATION IS EARLY OR LATER EUROPEAN OR EUROPEAN DERIVED, RATHER THAN DEVELOPED TUTELARY OR UNDEVELOPED TUTELARY (51)	0.78	82	TEND TO BE THOSE WHERE THE TYPE OF POLITICAL MODERNIZATION IS DEVELOPED TUTELARY OR UNDEVELOPED TUTELARY, RATHER THAN EARLY OR LATER EUROPEAN OR EUROPEAN DERIVED (55)	0.60	21 25 6 37 X SQ= 9.12 P = 0.001 RV YES (NO)
84	ALWAYS ARE THOSE WHERE THE TYPE OF POLITICAL MODERNIZATION IS DEVELOPED TUTELARY, RATHER THAN UNDEVELOPED TUTELARY (31)	1.00	84	TEND TO BE THOSE WHERE THE TYPE OF POLITICAL MODERNIZATION IS UNDEVELOPED TUTELARY, RATHER THAN DEVELOPED TUTELARY (24)	0.65	6 13 0 24 X SQ= 6.37 P = 0.004 RV YES (NO)

85	TEND TO BE THOSE WHERE THE STAGE OF POLITICAL MODERNIZATION IS ADVANCED, RATHER THAN MID- OR EARLY TRANSITIONAL (60)	0.93	85	TEND TO BE THOSE WHERE THE STAGE OF POLITICAL MODERNIZATION IS MID- OR EARLY TRANSITIONAL. RATHER THAN ADVANCED (54)	0.65	27 24 2 44 X SQ= 24.98 P = 0. RV YES (YES)
87	TEND MORE TO BE THOSE WHOSE IDEOLOGICAL ORIENTATION IS OTHER THAN DEVELOPMENTAL (58)	0.96	87	TEND LESS TO BE THOSE WHOSE IDEOLOGICAL ORIENTATION IS OTHER THAN DEVELOPMENTAL (58)	0.51	1 25 25 26 X SQ= 13.76 P = 0. RV NO (NO)
89	TEND TO BE THOSE WHOSE IDEOLOGICAL ORIENTATION IS CONVENTIONAL (33)	0.96	89	TEND TO BE THOSE WHOSE IDEOLOGICAL ORIENTATION IS OTHER THAN CONVENTIONAL (62)	0.93	25 4 1 50 X SQ= 56.03 P = 0. RV YES (YES)
92	ALWAYS ARE THOSE WHERE THE SYSTEM STYLE IS LIMITED MOBILIZATIONAL, OR NON-MOBILIZATIONAL (93)	1.00	92	TEND LESS TO BE THOSE WHERE THE SYSTEM STYLE IS LIMITED MOBILIZATIONAL, OR NON-MOBILIZATIONAL (93)	0.75	0 17 29 50 X SQ= 7.29 P = 0.001 RV NO (NO)
93	TEND MORE TO BE THOSE WHERE THE SYSTEM STYLE IS NON-MOBILIZATIONAL (78)	0.97	93	TEND LESS TO BE THOSE WHERE THE SYSTEM STYLE IS NON-MOBILIZATIONAL (78)	0.62	1 25 28 40 X SQ= 10.60 P = 0. RV NO (NO)
94	ALWAYS ARE THOSE WHERE THE STATUS OF THE REGIME IS CONSTITUTIONAL (51)	1.00	94	TEND TO BE THOSE WHERE THE STATUS OF THE REGIME IS AUTHORITARIAN OR TOTALITARIAN (41)	0.66	29 17 0 33 X SQ= 30.21 P = 0. RV YES (YES)
98	TEND TO BE THOSE WHERE GOVERNMENTAL STABILITY IS GENERALLY PRESENT AND DATES AT LEAST FROM THE INTERWAR PERIOD (22)	0.52	98	TEND TO BE THOSE WHERE GOVERNMENTAL STABILITY IS GENERALLY OR MODERATELY PRESENT AND DATES FROM THE POST-WAR PERIOD, OR IS ABSENT (93)	0.93	15 5 14 63 X SQ= 21.82 P = 0. RV YES (YES)
101	TEND TO BE THOSE WHERE THE REPRESENTATIVE CHARACTER OF THE REGIME IS POLYARCHIC (41)	0.96	101	TEND TO BE THOSE WHERE THE REPRESENTATIVE CHARACTER OF THE REGIME IS LIMITED POLYARCHIC, PSEUDO-POLYARCHIC, OR NON-POLYARCHIC (57)	0.82	27 10 1 47 X SQ= 44.38 P = 0. RV YES (YES)
102	ALWAYS ARE THOSE WHERE THE REPRESENTATIVE CHARACTER OF THE REGIME IS POLYARCHIC OR LIMITED POLYARCHIC (59)	1.00	102	TEND TO BE THOSE WHERE THE REPRESENTATIVE CHARACTER OF THE REGIME IS PSEUDO-POLYARCHIC OR NON-POLYARCHIC (49)	0.64	29 24 0 42 X SQ= 30.55 P = 0. RV YES (YES)

105	ALWAYS ARE THOSE WHERE THE ELECTORAL SYSTEM IS COMPETITIVE (43)	1.00	105	TEND TO BE THOSE WHERE THE ELECTORAL SYSTEM IS PARTIALLY COMPETITIVE OR NON-COMPETITIVE (47)	0.78	28 13 0 46 X SQ= 43.25 P = 0. RV YES (YES)
106	ALWAYS ARE THOSE WHERE THE ELECTORAL SYSTEM IS COMPETITIVE OR PARTIALLY COMPETITIVE (52)	1.00	106	TEND TO BE THOSE WHERE THE ELECTORAL SYSTEM IS NON-COMPETITIVE (30)	0.56	28 23 0 29 X SQ= 22.14 P = 0. RV YES (YES)
107	ALWAYS ARE THOSE WHERE AUTONOMOUS GROUPS ARE FULLY TOLERATED IN POLITICS (46)	1.00	107	TEND TO BE THOSE WHERE AUTONOMOUS GROUPS ARE PARTIALLY TOLERATED IN POLITICS, ARE TOLERATED ONLY OUTSIDE POLITICS, OR ARE NOT TOLERATED AT ALL (65)	0.79	29 14 0 52 X SQ= 47.35 P = 0. RV YES (YES)
111	TEND TO BE THOSE WHERE POLITICAL ENCULTURATION IS HIGH OR MEDIUM (53)	0.89	111	TEND TO BE THOSE WHERE POLITICAL ENCULTURATION IS LOW (42)	0.56	24 23 3 29 X SQ= 12.91 P = 0. RV YES (YES)
115	TEND TO BE THOSE WHERE INTEREST ARTICULATION BY ASSOCIATIONAL GROUPS IS SIGNIFICANT (20)	0.59	115	TEND TO BE THOSE WHERE INTEREST ARTICULATION BY ASSOCIATIONAL GROUPS IS MODERATE, LIMITED, OR NEGLIGIBLE (91)	0.95	17 3 12 62 X SQ= 31.77 P = 0. RV YES (YES)
116	TEND TO BE THOSE WHERE INTEREST ARTICULATION BY ASSOCIATIONAL GROUPS IS SIGNIFICANT OR MODERATE (32)	0.90	116	TEND TO BE THOSE WHERE INTEREST ARTICULATION BY ASSOCIATIONAL GROUPS IS LIMITED OR NEGLIGIBLE (79)	0.92	26 5 3 60 X SQ= 57.30 P = 0. RV YES (YES)
117	ALWAYS ARE THOSE WHERE INTEREST ARTICULATION BY ASSOCIATIONAL GROUPS IS SIGNIFICANT, MODERATE, OR LIMITED (60)	1.00	117	TEND TO BE THOSE WHERE INTEREST ARTICULATION BY ASSOCIATIONAL GROUPS IS NEGLIGIBLE (51)	0.71	29 19 0 46 X SQ= 37.41 P = 0. RV YES (YES)
118	TEND TO BE THOSE WHERE INTEREST ARTICULATION BY INSTITUTIONAL GROUPS IS SIGNIFICANT, MODERATE, OR LIMITED (60)	0.97	118	TEND TO BE THOSE WHERE INTEREST ARTICULATION BY INSTITUTIONAL GROUPS IS VERY SIGNIFICANT (40)	0.52	1 28 28 26 X SQ= 17.38 P = 0. RV YES (YES)
119	TEND TO BE THOSE WHERE INTEREST ARTICULATION BY INSTITUTIONAL GROUPS IS MODERATE OR LIMITED (26)	0.79	119	TEND TO BE THOSE WHERE INTEREST ARTICULATION BY INSTITUTIONAL GROUPS IS VERY SIGNIFICANT OR SIGNIFICANT (74)	0.94	6 51 23 3 X SQ= 44.34 P = 0. RV YES (YES)

120	TEND LESS TO BE THOSE WHERE INTEREST ARTICULATION BY INSTITUTIONAL GROUPS IS VERY SIGNIFICANT, SIGNIFICANT, OR MODERATE (90)	0.66	120	ALWAYS ARE THOSE WHERE INTEREST ARTICULATION BY INSTITUTIONAL GROUPS IS VERY SIGNIFICANT, SIGNIFICANT, OR MODERATE (90)	1.00	19 54 10 0 X SQ= 18.04 P = 0. RV NO (YES)
121	TEND TO BE THOSE WHERE INTEREST ARTICULATION BY NON-ASSOCIATIONAL GROUPS IS MODERATE, LIMITED, OR NEGLIGIBLE (61)	0.90	121	TEND TO BE THOSE WHERE INTEREST ARTICULATION BY NON-ASSOCIATIONAL GROUPS IS SIGNIFICANT (54)	0.59	3 40 26 28 X SQ= 17.45 P = 0. RV YES (NO)
122	TEND TO BE THOSE WHERE INTEREST ARTICULATION BY NON-ASSOCIATIONAL GROUPS IS LIMITED OR NEGLIGIBLE (32)	0.62	122	TEND TO BE THOSE WHERE INTEREST ARTICULATION BY NON-ASSOCIATIONAL GROUPS IS SIGNIFICANT OR MODERATE (83)	0.84	11 57 18 11 X SQ= 18.30 P = 0. RV YES (YES)
123	TEND LESS TO BE THOSE WHERE INTEREST ARTICULATION BY NON-ASSOCIATIONAL GROUPS IS SIGNIFICANT, MODERATE, OR LIMITED (107)	0.72	123	ALWAYS ARE THOSE WHERE INTEREST ARTICULATION BY NON-ASSOCIATIONAL GROUPS IS SIGNIFICANT, MODERATE, OR LIMITED (107)	1.00	21 68 8 0 X SQ= 16.96 P = 0. RV NO (YES)
124	ALWAYS ARE THOSE WHERE INTEREST ARTICULATION BY ANOMIC GROUPS IS OCCASIONAL, INFREQUENT, OR VERY INFREQUENT (85)	1.00	124	TEND LESS TO BE THOSE WHERE INTEREST ARTICULATION BY ANOMIC GROUPS IS OCCASIONAL, INFREQUENT, OR VERY INFREQUENT (85)	0.84	0 9 28 49 X SQ= 3.34 P = 0.028 RV NO (NO)
125	TEND TO BE THOSE WHERE INTEREST ARTICULATION BY ANOMIC GROUPS IS INFREQUENT OR VERY INFREQUENT (35)	0.86	125	TEND TO BE THOSE WHERE INTEREST ARTICULATION BY ANOMIC GROUPS IS FREQUENT OR OCCASIONAL (64)	0.83	4 48 24 10 X SQ= 34.23 P = 0. RV YES (YES)
126	TEND TO BE THOSE WHERE INTEREST ARTICULATION BY ANOMIC GROUPS IS VERY INFREQUENT (16)	0.57	126	ALWAYS ARE THOSE WHERE INTEREST ARTICULATION BY ANOMIC GROUPS IS FREQUENT, OCCASIONAL, OR INFREQUENT (83)	1.00	12 58 16 0 X SQ= 37.03 P = 0. RV YES (YES)
128	TEND TO BE THOSE WHERE INTEREST ARTICULATION BY POLITICAL PARTIES IS SIGNIFICANT OR MODERATE (48)	0.86	128	TEND TO BE THOSE WHERE INTEREST ARTICULATION BY POLITICAL PARTIES IS LIMITED OR NEGLIGIBLE (45)	0.66	24 20 4 39 X SQ= 18.38 P = 0. RV YES (YES)
129	TEND TO BE THOSE WHERE INTEREST ARTICULATION BY POLITICAL PARTIES IS SIGNIFICANT, MODERATE, OR LIMITED (56)	0.89	129	TEND TO BE THOSE WHERE INTEREST ARTICULATION BY POLITICAL PARTIES IS NEGLIGIBLE (37)	0.56	25 26 3 33 X SQ= 14.20 P = 0. RV YES (NO)

130	TEND LESS TO BE THOSE WHERE INTEREST AGGREGATION BY POLITICAL PARTIES IS MODERATE, LIMITED, OR NEGLIGIBLE (71)	0.71	130	TEND MORE TO BE THOSE WHERE INTEREST AGGREGATION BY POLITICAL PARTIES IS MODERATE, LIMITED, OR NEGLIGIBLE (71)	0.92	8 4 20 45 X SQ= 4.20 P = 0.024 RV NO (YES)
134	TEND TO BE THOSE WHERE INTEREST AGGREGATION BY THE EXECUTIVE IS SIGNIFICANT OR MODERATE (57)	0.93	134	TEND TO BE THOSE WHERE INTEREST AGGREGATION BY THE EXECUTIVE IS LIMITED OR NEGLIGIBLE (46)	0.58	25 27 2 38 X SQ= 18.21 P = 0. RV YES (NO)
135	ALWAYS ARE THOSE WHERE INTEREST AGGREGATION BY THE EXECUTIVE IS SIGNIFICANT, MODERATE, OR LIMITED (77)	1.00	135	TEND LESS TO BE THOSE WHERE INTEREST AGGREGATION BY THE EXECUTIVE IS SIGNIFICANT, MODERATE, OR LIMITED (77)	0.78	27 43 0 12 X SQ= 5.26 P = 0.007 RV NO (NO)
139	ALWAYS ARE THOSE WHERE THE PARTY SYSTEM IS QUANTITATIVELY OTHER THAN ONE-PARTY (71)	1.00	139	TEND LESS TO BE THOSE WHERE THE PARTY SYSTEM IS QUANTITATIVELY OTHER THAN ONE-PARTY (71)	0.52	0 31 29 33 X SQ= 18.95 P = 0. RV NO (NO)
141	TEND LESS TO BE THOSE WHERE THE PARTY SYSTEM IS QUANTITATIVELY OTHER THAN TWO-PARTY (87)	0.70	141	TEND MORE TO BE THOSE WHERE THE PARTY SYSTEM IS QUANTITATIVELY OTHER THAN TWO-PARTY (87)	0.97	8 2 19 60 X SQ= 10.63 P = 0.001 RV NO (YES)
142	TEND TO BE THOSE WHERE THE PARTY SYSTEM IS QUANTITATIVELY MULTI-PARTY (30)	0.52	142	TEND TO BE THOSE WHERE THE PARTY SYSTEM IS QUANTITATIVELY OTHER THAN MULTI-PARTY (66)	0.80	14 12 13 49 X SQ= 7.83 P = 0.005 RV YES (YES)
147	TEND TO BE THOSE WHERE THE PARTY SYSTEM IS QUALITATIVELY CLASS-ORIENTED OR MULTI-IDEOLOGICAL (23)	0.83	147	TEND TO BE THOSE WHERE THE PARTY SYSTEM IS QUALITATIVELY OTHER THAN CLASS-ORIENTED OR MULTI-IDEOLOGICAL (67)	0.97	20 2 4 56 X SQ= 51.19 P = 0. RV YES (YES)
148	ALWAYS ARE THOSE WHERE THE PARTY SYSTEM IS QUALITATIVELY OTHER THAN AFRICAN TRANSITIONAL (96)	1.00	148	TEND LESS TO BE THOSE WHERE THE PARTY SYSTEM IS QUALITATIVELY OTHER THAN AFRICAN TRANSITIONAL (96)	0.78	0 14 29 50 X SQ= 5.86 P = 0.004 RV NO (NO)
153	TEND TO BE THOSE WHERE THE PARTY SYSTEM IS STABLE (42)	0.81	153	TEND TO BE THOSE WHERE THE PARTY SYSTEM IS MODERATELY STABLE OR UNSTABLE (38)	0.58	21 20 5 28 X SQ= 8.91 P = 0.001 RV YES (YES)

158	TEND TO BE THOSE WHERE PERSONALISMO IS NEGLIGIBLE (56)	0.83	158	TEND TO BE THOSE WHERE PERSONALISMO IS PRONOUNCED OR MODERATE (40)	0.50	5 29 24 29 X SQ= 7.39 P = 0.005 RV YES (NO)
160	ALWAYS ARE THOSE WHERE THE POLITICAL LEADERSHIP IS MODERATELY ELITIST OR NON-ELITIST (67)	1.00	160	TEND LESS TO BE THOSE WHERE THE POLITICAL LEADERSHIP IS MODERATELY ELITIST OR NON-ELITIST (67)	0.54	0 28 28 33 X SQ= 16.68 P = 0. RV NO (NO)
161	TEND TO BE THOSE WHERE THE POLITICAL LEADERSHIP IS NON-ELITIST (50)	0.86	161	TEND TO BE THOSE WHERE THE POLITICAL LEADERSHIP IS ELITIST OR MODERATELY ELITIST (47)	0.64	4 39 24 22 X SQ= 17.01 P = 0. RV YES (YES)
163	LEAN MORE TOWARD BEING THOSE WHERE THE REGIME'S LEADERSHIP CHARISMA IS MODERATE OR NEGLIGIBLE (87)	0.97	163	LEAN LESS TOWARD BEING THOSE WHERE THE REGIME'S LEADERSHIP CHARISMA IS MODERATE OR NEGLIGIBLE (87)	0.83	1 10 28 48 X SQ= 2.20 P = 0.091 RV NO (NO)
164	TEND MORE TO BE THOSE WHERE THE REGIME'S LEADERSHIP CHARISMA IS NEGLIGIBLE (65)	0.93	164	TEND LESS TO BE THOSE WHERE THE REGIME'S LEADERSHIP CHARISMA IS NEGLIGIBLE (65)	0.53	2 27 27 31 X SQ= 11.95 P = 0. RV NO (NO)
168	TEND TO BE THOSE WHERE THE HORIZONTAL POWER DISTRIBUTION IS SIGNIFICANT (34)	0.97	168	TEND TO BE THOSE WHERE THE HORIZONTAL POWER DISTRIBUTION IS LIMITED OR NEGLIGIBLE (72)	0.97	28 2 1 63 X SQ= 76.39 P = 0. RV YES (YES)
169	ALWAYS ARE THOSE WHERE THE HORIZONTAL POWER DISTRIBUTION IS SIGNIFICANT OR LIMITED (58)	1.00	169	TEND TO BE THOSE WHERE THE HORIZONTAL POWER DISTRIBUTION IS NEGLIGIBLE (48)	0.65	29 23 0 42 X SQ= 31.31 P = 0. RV YES (YES)
170	TEND MORE TO BE THOSE WHERE THE LEGISLATIVE-EXECUTIVE STRUCTURE IS OTHER THAN PRESIDENTIAL (63)	0.90	170	TEND LESS TO BE THOSE WHERE THE LEGISLATIVE-EXECUTIVE STRUCTURE IS OTHER THAN PRESIDENTIAL (63)	0.52	3 31 26 34 X SQ= 10.55 P = 0. RV NO (NO)
171	TEND LESS TO BE THOSE WHERE THE LEGISLATIVE-EXECUTIVE STRUCTURE IS OTHER THAN PARLIAMENTARY-REPUBLICAN (90)	0.69	171	TEND MORE TO BE THOSE WHERE THE LEGISLATIVE-EXECUTIVE STRUCTURE IS OTHER THAN PARLIAMENTARY-REPUBLICAN (90)	0.97	9 2 20 63 X SQ= 12.58 P = 0. RV NO (YES)

No.	Statement	Value	No.	Statement	Value	Statistics
172	TEND LESS TO BE THOSE WHERE THE LEGISLATIVE-EXECUTIVE STRUCTURE IS OTHER THAN PARLIAMENTARY-ROYALIST (88)	0.52	172	TEND MORE TO BE THOSE WHERE THE LEGISLATIVE-EXECUTIVE STRUCTURE IS OTHER THAN PARLIAMENTARY-ROYALIST (88)	0.89	14 7 15 58 X SQ= 14.17 P = 0. RV NO (YES)
174	TEND TO BE THOSE WHERE THE LEGISLATURE IS FULLY EFFECTIVE (28)	0.86	174	TEND TO BE THOSE WHERE THE LEGISLATURE IS PARTIALLY EFFECTIVE, LARGELY INEFFECTIVE, OR WHOLLY INEFFECTIVE (72)	0.98	25 1 4 62 X SQ= 66.03 P = 0. RV YES (YES)
175	ALWAYS ARE THOSE WHERE THE LEGISLATURE IS FULLY EFFECTIVE OR PARTIALLY EFFECTIVE (51)	1.00	175	TEND TO BE THOSE WHERE THE LEGISLATURE IS LARGELY INEFFECTIVE OR WHOLLY INEFFECTIVE (49)	0.73	29 17 0 46 X SQ= 39.48 P = 0. RV YES (YES)
178	TEND TO BE THOSE WHERE THE LEGISLATURE IS BICAMERAL (51)	0.79	178	TEND TO BE THOSE WHERE THE LEGISLATURE IS UNICAMERAL (53)	0.67	6 44 23 22 X SQ= 15.29 P = 0. RV YES (YES)
179	TEND TO BE THOSE WHERE THE EXECUTIVE IS STRONG (39)	0.96	179	TEND TO BE THOSE WHERE THE EXECUTIVE IS DOMINANT (52)	0.80	1 44 24 11 X SQ= 37.31 P = 0. RV YES (YES)
181	TEND TO BE THOSE WHERE THE BUREAUCRACY IS MODERN, RATHER THAN SEMI-MODERN (21)	0.66	181	TEND TO BE THOSE WHERE THE BUREAUCRACY IS SEMI-MODERN, RATHER THAN MODERN (55)	0.97	19 1 10 36 X SQ= 27.47 P = 0. RV YES (YES)
182	ALWAYS ARE THOSE WHERE THE BUREAUCRACY IS SEMI-MODERN, RATHER THAN POST-COLONIAL TRANSITIONAL (55)	1.00	182	TEND LESS TO BE THOSE WHERE THE BUREAUCRACY IS SEMI-MODERN, RATHER THAN POST-COLONIAL TRANSITIONAL (55)	0.62	10 36 0 22 X SQ= 4.01 P = 0.024 RV NO (NO)
186	TEND TO BE THOSE WHERE PARTICIPATION BY THE MILITARY IN POLITICS IS NEUTRAL, RATHER THAN SUPPORTIVE (56)	0.96	186	TEND TO BE THOSE WHERE PARTICIPATION BY THE MILITARY IN POLITICS IS SUPPORTIVE, RATHER THAN NEUTRAL (31)	0.51	1 25 27 24 X SQ= 15.88 P = 0. RV YES (YES)
187	TEND TO BE THOSE WHERE THE ROLE OF THE POLICE IS NOT POLITICALLY SIGNIFICANT (35)	0.93	187	TEND TO BE THOSE WHERE THE ROLE OF THE POLICE IS POLITICALLY SIGNIFICANT (66)	0.85	2 50 26 9 X SQ= 44.38 P = 0. RV YES (YES)

```
189  TEND LESS TO BE THOSE                   0.76     189  ALWAYS ARE THOSE                        1.00         7    0
     WHERE THE CHARACTER OF THE LEGAL SYSTEM               WHERE THE CHARACTER OF THE LEGAL SYSTEM             22   68
     IS OTHER THAN COMMON LAW  (108)                       IS OTHER THAN COMMON LAW  (108)               X SQ=  14.27
                                                                                                         P =    0.
                                                                                                         RV NO  (YES)

190  TEND LESS TO BE THOSE                   0.61     190  ALWAYS ARE THOSE                        1.00        11   17
     WHERE THE CHARACTER OF THE LEGAL SYSTEM               WHERE THE CHARACTER OF THE LEGAL SYSTEM              7    0
     IS CIVIL LAW, RATHER THAN                             IS CIVIL LAW, RATHER THAN                     X SQ=   6.01
     COMMON LAW  (32)                                      COMMON LAW  (32)                              P =    0.008
                                                                                                         RV NO  (YES)

192  ALWAYS ARE THOSE                        1.00     192  TEND LESS TO BE THOSE                   0.73         0   18
     WHERE THE CHARACTER OF THE LEGAL SYSTEM               WHERE THE CHARACTER OF THE LEGAL SYSTEM             29   48
     IS OTHER THAN                                         IS OTHER THAN                                 X SQ=   8.06
     MUSLIM OR PARTLY MUSLIM  (86)                         MUSLIM OR PARTLY MUSLIM  (86)                 P =    0.001
                                                                                                         RV NO  (NO )

193  ALWAYS ARE THOSE                        1.00     193  TEND LESS TO BE THOSE                   0.62         0   26
     WHERE THE CHARACTER OF THE LEGAL SYSTEM               WHERE THE CHARACTER OF THE LEGAL SYSTEM             29   42
     IS OTHER THAN PARTLY INDIGENOUS  (86)                 IS OTHER THAN PARTLY INDIGENOUS  (86)         X SQ=  13.26
                                                                                                         P =    0.
                                                                                                         RV NO  (NO )

194  ALWAYS ARE THOSE                        1.00     194  TEND LESS TO BE THOSE                   0.81         0   13
     THAT ARE NON-COMMUNIST  (101)                         THAT ARE NON-COMMUNIST  (101)                       29   54
                                                                                                         X SQ=   4.96
                                                                                                         P =    0.008
                                                                                                         RV NO  (NO )
```

138 POLITIES
WHERE INTEREST AGGREGATION
BY THE LEGISLATURE
IS SIGNIFICANT, MODERATE, OR
LIMITED (48)

138 POLITIES
WHERE INTEREST AGGREGATION
BY THE LEGISLATURE
IS NEGLIGIBLE (49)

PREDICATE ONLY

48 IN LEFT COLUMN

ARGENTINA	AUSTRALIA	AUSTRIA	BELGIUM	BOLIVIA	BRAZIL	CAMEROUN	CANADA	CHILE	COLOMBIA
COSTA RICA	CYPRUS	DENMARK	DOMIN REP	ECUADOR	FINLAND	FRANCE	GERMAN FR	GREECE	GUATEMALA
HONDURAS	ICELAND	INDIA	IRELAND	ISRAEL	ITALY	JAMAICA	JAPAN	LEBANON	LIBYA
LUXEMBOURG	NETHERLANDS	NEW ZEALAND	NICARAGUA	NIGERIA	NORWAY	PANAMA	PHILIPPINES	SIERRE LEO	SWEDEN
SWITZERLAND	TRINIDAD	TURKEY	UGANDA	UK	US	URUGUAY	VENEZUELA		

49 IN RIGHT COLUMN

AFGHANISTAN	ALBANIA	BULGARIA	BURUNDI	CAMBODIA	CEN AFR REP	CEYLON	CHAD	CHINA, PR	CONGO(BRA)
CONGO(LEO)	CZECHOS'KIA	DAHOMEY	EL SALVADOR	ETHIOPIA	GABON	GERMANY, E	GHANA	GUINEA	HAITI
HUNGARY	IVORY COAST	JORDAN	KOREA, N	LAOS	LIBERIA	MALAGASY R	MALAYA	MALI	MAURITANIA
MEXICO	MONGOLIA	NEPAL	NIGER	PARAGUAY	POLAND	RUMANIA	RWANDA	SENEGAL	SOMALIA
TANGANYIKA	THAILAND	TOGO	TUNISIA	USSR	UPPER VOLTA	VIETNAM, N	VIETNAM REP	YUGOSLAVIA	

4 EXCLUDED BECAUSE AMBIGUOUS

INDONESIA PORTUGAL SO AFRICA SPAIN

1 EXCLUDED BECAUSE IRRELEVANT

SA'U ARABIA

13 EXCLUDED BECAUSE UNASCERTAINABLE

ALGERIA	BURMA	CUBA	IRAN	IRAQ	KOREA REP	MOROCCO	PAKISTAN	PERU	SUDAN
SYRIA	UAR	YEMEN							

139 POLITIES
WHERE THE PARTY SYSTEM IS QUANTITATIVELY
ONE-PARTY (34)

139 POLITIES
WHERE THE PARTY SYSTEM IS QUANTITATIVELY
OTHER THAN ONE-PARTY (71)

BOTH SUBJECT AND PREDICATE

34 IN LEFT COLUMN

ALBANIA	ALGERIA	BULGARIA	CAMBODIA	CEN AFR REP	CHAD	CHINA, PR	CUBA	CZECHOS'KIA	DAHOMEY
GABON	GERMANY, E	GHANA	GUINEA	HUNGARY	IVORY COAST	KOREA, N	LIBERIA	MALI	MAURITANIA
MONGOLIA	NIGER	POLAND	PORTUGAL	RUMANIA	SENEGAL	SPAIN	TANGANYIKA	TUNISIA	USSR
UAR	UPPER VOLTA	VIETNAM, N	YUGOSLAVIA						

71 IN RIGHT COLUMN

AFGHANISTAN	ARGENTINA	AUSTRALIA	AUSTRIA	BELGIUM	BOLIVIA	BRAZIL	BURMA	BURUNDI	CAMEROUN
CANADA	CEYLON	CHILE	COLOMBIA	CONGO(BRA)	CONGO(LEO)	COSTA RICA	CYPRUS	DENMARK	DOMIN REP
ECUADOR	EL SALVADOR	ETHIOPIA	FINLAND	FRANCE	GERMAN FR	GREECE	GUATEMALA	HAITI	HONDURAS
ICELAND	INDIA	IRELAND	ISRAEL	ITALY	JAMAICA	JAPAN	JORDAN	LAOS	LEBANON
LIBYA	LUXEMBOURG	MALAGASY R	MALAYA	MEXICO	MOROCCO	NETHERLANDS	NEW ZEALAND	NIGERIA	NORWAY
PAKISTAN	PANAMA	PARAGUAY	PERU	PHILIPPINES	RWANDA	SA'U ARABIA	SIERRE LEO	SOMALIA	SO AFRICA
SUDAN	SWEDEN	SWITZERLAND	TRINIDAD	TURKEY	UGANDA	UK	US	URUGUAY	VENEZUELA
VIETNAM REP									

2 EXCLUDED BECAUSE AMBIGUOUS

INDONESIA NICARAGUA

8 EXCLUDED BECAUSE UNASCERTAINABLE

IRAN IRAQ KOREA REP NEPAL SYRIA THAILAND TOGO YEMEN

---

2 TEND MORE TO BE THOSE LOCATED ELSEWHERE THAN IN WEST EUROPE, SCANDINAVIA, NORTH AMERICA, OR AUSTRALASIA (93) — 0.94

2 TEND LESS TO BE THOSE LOCATED ELSEWHERE THAN IN WEST EUROPE, SCANDINAVIA, NORTH AMERICA, OR AUSTRALASIA (93) — 0.72

2 20
32 51
X SQ= 5.61
P = 0.010
RV NO (NO)

5 TEND TO BE THOSE LOCATED IN EAST EUROPE, RATHER THAN IN SCANDINAVIA OR WEST EUROPE (9) — 0.82

5 ALWAYS ARE THOSE LOCATED IN SCANDINAVIA OR WEST EUROPE, RATHER THAN IN EAST EUROPE (18) — 1.00

2 16
9 0
X SQ= 16.13
P = 0.
RV YES (YES)

6 TEND MORE TO BE THOSE LOCATED ELSEWHERE THAN IN THE CARIBBEAN, CENTRAL AMERICA, OR SOUTH AMERICA (93) — 0.97

6 TEND LESS TO BE THOSE LOCATED ELSEWHERE THAN IN THE CARIBBEAN, CENTRAL AMERICA, OR SOUTH AMERICA (93) — 0.72

1 20
33 51
X SQ= 7.64
P = 0.002
RV NO (NO)

10	TEND TO BE THOSE LOCATED IN NORTH AFRICA, OR CENTRAL AND SOUTH AFRICA (33)	0.50	10	TEND TO BE THOSE LOCATED ELSEWHERE THAN IN NORTH AFRICA, OR CENTRAL AND SOUTH AFRICA (82)	0.79	17 15 17 56 X SQ= 7.73 P = 0.006 RV YES (YES)
11	TEND LESS TO BE THOSE LOCATED ELSEWHERE THAN IN CENTRAL AND SOUTH AFRICA (87)	0.59	11	TEND MORE TO BE THOSE LOCATED ELSEWHERE THAN IN CENTRAL AND SOUTH AFRICA (87)	0.82	14 13 20 58 X SQ= 5.15 P = 0.017 RV NO (YES)
14	ALWAYS ARE THOSE LOCATED ELSEWHERE THAN IN THE MIDDLE EAST (104)	1.00	14	LEAN LESS TOWARD BEING THOSE LOCATED ELSEWHERE THAN IN THE MIDDLE EAST (104)	0.90	0 7 34 64 X SQ= 2.18 P = 0.093 RV NO (NO)
15	ALWAYS ARE THOSE LOCATED IN NORTH AFRICA OR CENTRAL AND SOUTH AFRICA, RATHER THAN IN THE MIDDLE EAST (33)	1.00	15	TEND LESS TO BE THOSE LOCATED IN NORTH AFRICA OR CENTRAL AND SOUTH AFRICA, RATHER THAN IN THE MIDDLE EAST (33)	0.68	0 7 17 15 X SQ= 4.61 P = 0.012 RV NO (YES)
16	ALWAYS ARE THOSE LOCATED IN EAST ASIA, SOUTH ASIA, OR SOUTHEAST ASIA, RATHER THAN IN THE MIDDLE EAST (18)	1.00	16	TILT LESS TOWARD BEING THOSE LOCATED IN EAST ASIA, SOUTH ASIA, OR SOUTHEAST ASIA, RATHER THAN IN THE MIDDLE EAST (18)	0.56	0 7 5 9 X SQ= 1.61 P = 0.123 RV NO (NO)
19	TEND TO BE THOSE LOCATED IN NORTH AFRICA OR CENTRAL AND SOUTH AFRICA, RATHER THAN IN THE CARIBBEAN, CENTRAL AMERICA, OR SOUTH AMERICA (33)	0.94	19	TEND TO BE THOSE LOCATED IN THE CARIBBEAN, CENTRAL AMERICA, OR SOUTH AMERICA, RATHER THAN IN NORTH AFRICA OR CENTRAL AND SOUTH AFRICA (22)	0.57	17 15 1 20 X SQ= 11.15 P = 0. RV YES (YES)
20	TEND TO BE THOSE LOCATED IN EAST ASIA, SOUTH ASIA, OR SOUTHEAST ASIA, RATHER THAN IN THE CARIBBEAN, CENTRAL AMERICA, OR SOUTH AMERICA (18)	0.83	20	TEND TO BE THOSE LOCATED IN THE CARIBBEAN, CENTRAL AMERICA, OR SOUTH AMERICA, RATHER THAN IN EAST ASIA, SOUTH ASIA, OR SOUTHEAST ASIA (22)	0.69	5 9 1 20 X SQ= 3.70 P = 0.028 RV YES (NO)
25	TEND MORE TO BE THOSE WHOSE POPULATION DENSITY IS MEDIUM OR LOW (98)	0.97	25	TEND LESS TO BE THOSE WHOSE POPULATION DENSITY IS MEDIUM OR LOW (98)	0.79	1 15 33 56 X SQ= 4.56 P = 0.018 RV NO (NO)
29	TEND TO BE THOSE WHERE THE DEGREE OF URBANIZATION IS LOW (49)	0.66	29	TEND TO BE THOSE WHERE THE DEGREE OF URBANIZATION IS HIGH (56)	0.62	10 41 19 25 X SQ= 5.13 P = 0.015 RV YES (NO)

30	TEND TO BE THOSE WHOSE AGRICULTURAL POPULATION IS HIGH (56)	0.70	30	TEND TO BE THOSE WHOSE AGRICULTURAL POPULATION IS MEDIUM, LOW, OR VERY LOW (57)	0.67	23 23 10 47 X SQ= 10.87 P = 0.001 RV YES (YES)
34	LEAN TOWARD BEING THOSE WHOSE GROSS NATIONAL PRODUCT IS VERY LOW (53)	0.59	34	LEAN TOWARD BEING THOSE WHOSE GROSS NATIONAL PRODUCT IS VERY HIGH, HIGH, MEDIUM, OR LOW (62)	0.61	14 43 20 28 X SQ= 2.74 P = 0.093 RV YES (NO)
37	LEAN TOWARD BEING THOSE WHOSE PER CAPITA GROSS NATIONAL PRODUCT IS VERY LOW (51)	0.56	37	LEAN TOWARD BEING THOSE WHOSE PER CAPITA GROSS NATIONAL PRODUCT IS VERY HIGH, HIGH, MEDIUM, OR LOW (64)	0.65	15 46 19 25 X SQ= 3.23 P = 0.058 RV YES (NO)
40	TEND TO BE THOSE WHOSE INTERNATIONAL FINANCIAL STATUS IS VERY LOW (39)	0.52	40	TEND TO BE THOSE WHOSE INTERNATIONAL FINANCIAL STATUS IS VERY HIGH, HIGH, MEDIUM, OR LOW (71)	0.72	15 50 16 19 X SQ= 4.44 P = 0.025 RV YES (NO)
43	TEND TO BE THOSE WHOSE ECONOMIC DEVELOPMENTAL STATUS IS VERY UNDERDEVELOPED (57)	0.67	43	TEND TO BE THOSE WHOSE ECONOMIC DEVELOPMENTAL STATUS IS DEVELOPED, INTERMEDIATE, OR UNDERDEVELOPED (55)	0.62	11 43 22 26 X SQ= 6.41 P = 0.010 RV YES (NO)
50	TEND TO BE THOSE WHERE FREEDOM OF THE PRESS IS INTERMITTENT, INTERNALLY ABSENT, OR INTERNALLY AND EXTERNALLY ABSENT (56)	0.79	50	TEND TO BE THOSE WHERE FREEDOM OF THE PRESS IS COMPLETE (43)	0.59	6 37 22 26 X SQ= 9.38 P = 0.001 RV YES (NO)
51	TEND TO BE THOSE WHERE FREEDOM OF THE PRESS IS INTERNALLY ABSENT, OR INTERNALLY AND EXTERNALLY ABSENT (37)	0.75	51	TEND TO BE THOSE WHERE FREEDOM OF THE PRESS IS COMPLETE OR INTERMITTENT (60)	0.83	7 52 21 11 X SQ= 25.68 P = 0. RV YES (YES)
53	TEND MORE TO BE THOSE WHERE NEWSPAPER CIRCULATION IS LESS THAN THREE HUNDRED PER THOUSAND (101)	0.97	53	TEND LESS TO BE THOSE WHERE NEWSPAPER CIRCULATION IS LESS THAN THREE HUNDRED PER THOUSAND (101)	0.82	1 13 33 58 X SQ= 3.46 P = 0.033 RV NO (NO)
57	LEAN MORE TOWARD BEING THOSE WHERE THE RELIGION IS OTHER THAN CATHOLIC (90)	0.88	57	LEAN LESS TOWARD BEING THOSE WHERE THE RELIGION IS OTHER THAN CATHOLIC (90)	0.72	4 20 30 51 X SQ= 2.64 P = 0.082 RV NO (NO)

No.	Statement	Value	No.	Statement	Value	Statistics
63	LEAN MORE TOWARD BEING THOSE WHERE THE RELIGION IS PREDOMINANTLY OR PARTLY OTHER THAN CHRISTIAN (68)	0.70	63	LEAN LESS TOWARD BEING THOSE WHERE THE RELIGION IS PREDOMINANTLY OR PARTLY OTHER THAN CHRISTIAN (68)	0.51	10 35 23 36 X SQ= 2.58 P = 0.090 RV NO (NO)
67	LEAN MORE TOWARD BEING THOSE THAT ARE RACIALLY HOMOGENEOUS (82)	0.85	67	LEAN LESS TOWARD BEING THOSE THAT ARE RACIALLY HOMOGENEOUS (82)	0.66	29 44 5 23 X SQ= 3.41 P = 0.059 RV NO (NO)
68	TEND TO BE THOSE THAT ARE LINGUISTICALLY WEAKLY HETEROGENEOUS OR STRONGLY HETEROGENEOUS (62)	0.70	68	TEND TO BE THOSE THAT ARE LINGUISTICALLY HOMOGENEOUS (52)	0.55	10 39 23 32 X SQ= 4.54 P = 0.022 RV YES (NO)
69	TILT TOWARD BEING THOSE THAT ARE LINGUISTICALLY STRONGLY HETEROGENEOUS (50)	0.55	69	TILT TOWARD BEING THOSE THAT ARE LINGUISTICALLY HOMOGENEOUS OR WEAKLY HETEROGENEOUS (64)	0.62	15 44 18 27 X SQ= 1.88 P = 0.139 RV YES (NO)
72	TEND TO BE THOSE WHOSE DATE OF INDEPENDENCE IS AFTER 1914 (59)	0.72	72	TEND TO BE THOSE WHOSE DATE OF INDEPENDENCE IS BEFORE 1914 (52)	0.56	9 39 23 31 X SQ= 5.65 P = 0.011 RV YES (NO)
73	LEAN TOWARD BEING THOSE WHOSE DATE OF INDEPENDENCE IS AFTER 1945 (46)	0.56	73	LEAN TOWARD BEING THOSE WHOSE DATE OF INDEPENDENCE IS BEFORE 1945 (65)	0.63	14 44 18 26 X SQ= 2.54 P = 0.087 RV YES (NO)
75	TEND TO BE THOSE THAT ARE NOT HISTORICALLY WESTERN AND ARE NOT SIGNIFICANTLY WESTERNIZED (52)	0.59	75	TEND TO BE THOSE THAT ARE HISTORICALLY WESTERN OR SIGNIFICANTLY WESTERNIZED (62)	0.67	14 47 20 23 X SQ= 5.34 P = 0.019 RV YES (NO)
77	TEND TO BE THOSE THAT HAVE BEEN PARTIALLY WESTERNIZED, RATHER THAN SIGNIFICANTLY WESTERNIZED, THROUGH A COLONIAL RELATIONSHIP (41)	0.84	77	TEND TO BE THOSE THAT HAVE BEEN SIGNIFICANTLY WESTERNIZED, RATHER THAN PARTIALLY WESTERNIZED, THROUGH A COLONIAL RELATIONSHIP (28)	0.55	3 24 16 20 X SQ= 6.63 P = 0.005 RV YES (NO)
78	TEND TO BE THOSE THAT HAVE BEEN SIGNIFICANTLY WESTERNIZED WITHOUT A COLONIAL RELATIONSHIP, RATHER THAN THROUGH SUCH A RELATIONSHIP (7)	0.63	78	TEND TO BE THOSE THAT HAVE BEEN SIGNIFICANTLY WESTERNIZED THROUGH A COLONIAL RELATIONSHIP, RATHER THAN WITHOUT SUCH A RELATIONSHIP (28)	0.92	3 24 5 2 X SQ= 8.14 P = 0.004 RV YES (YES)

79	TEND MORE TO BE THOSE THAT WERE FORMERLY DEPENDENCIES OF BRITAIN OR FRANCE, RATHER THAN SPAIN (49)	0.95	79	TEND LESS TO BE THOSE THAT WERE FORMERLY DEPENDENCIES OF BRITAIN OR FRANCE, RATHER THAN SPAIN (49)	0.64	18 28 1 16 X SQ= 5.03 P = 0.013 RV NO (NO)
80	TEND TO BE THOSE WHOSE DATE OF INDEPENDENCE IS AFTER 1914, AND THAT WERE FORMERLY DEPENDENCIES OF FRANCE, RATHER THAN BRITAIN (24)	0.83	80	TEND TO BE THOSE WHOSE DATE OF INDEPENDENCE IS AFTER 1914, AND THAT WERE FORMERLY DEPENDENCIES OF BRITAIN, RATHER THAN FRANCE (19)	0.68	3 15 15 7 X SQ= 8.64 P = 0.002 RV YES (YES)
81	ALWAYS ARE THOSE WHERE THE TYPE OF POLITICAL MODERNIZATION IS LATER EUROPEAN OR LATER EUROPEAN DERIVED, RATHER THAN EARLY EUROPEAN OR EARLY EUROPEAN DERIVED (40)	1.00	81	LEAN LESS TOWARD BEING THOSE WHERE THE TYPE OF POLITICAL MODERNIZATION IS LATER EUROPEAN OR LATER EUROPEAN DERIVED, RATHER THAN EARLY EUROPEAN OR EARLY EUROPEAN DERIVED (40)	0.72	0 11 11 28 X SQ= 2.50 P = 0.053 RV NO (NO)
82	TEND TO BE THOSE WHERE THE TYPE OF POLITICAL MODERNIZATION IS DEVELOPED TUTELARY OR UNDEVELOPED TUTELARY, RATHER THAN EARLY OR LATER EUROPEAN OR EUROPEAN DERIVED (55)	0.66	82	TEND TO BE THOSE WHERE THE TYPE OF POLITICAL MODERNIZATION IS EARLY OR LATER EUROPEAN OR EUROPEAN DERIVED, RATHER THAN DEVELOPED TUTELARY OR UNDEVELOPED TUTELARY (51)	0.58	11 39 21 28 X SQ= 4.01 P = 0.033 RV YES (NO)
84	LEAN TOWARD BEING THOSE WHERE THE TYPE OF POLITICAL MODERNIZATION IS UNDEVELOPED TUTELARY, RATHER THAN DEVELOPED TUTELARY (24)	0.62	84	LEAN TOWARD BEING THOSE WHERE THE TYPE OF POLITICAL MODERNIZATION IS DEVELOPED TUTELARY, RATHER THAN UNDEVELOPED TUTELARY (31)	0.64	8 18 13 10 X SQ= 2.34 P = 0.088 RV YES (YES)
87	TEND TO BE THOSE WHOSE IDEOLOGICAL ORIENTATION IS DEVELOPMENTAL (31)	0.50	87	TEND TO BE THOSE WHOSE IDEOLOGICAL ORIENTATION IS OTHER THAN DEVELOPMENTAL (58)	0.75	16 13 16 39 X SQ= 4.43 P = 0.032 RV YES (YES)
89	TEND TO BE THOSE WHOSE IDEOLOGICAL ORIENTATION IS OTHER THAN CONVENTIONAL (62)	0.97	89	TEND TO BE THOSE WHOSE IDEOLOGICAL ORIENTATION IS CONVENTIONAL (33)	0.57	1 32 30 24 X SQ= 22.40 P = 0. RV YES (YES)
91	ALWAYS ARE THOSE WHOSE IDEOLOGICAL ORIENTATION IS DEVELOPMENTAL, RATHER THAN TRADITIONAL (31)	1.00	91	TILT LESS TOWARD BEING THOSE WHOSE IDEOLOGICAL ORIENTATION IS DEVELOPMENTAL, RATHER THAN TRADITIONAL (31)	0.76	16 13 0 4 X SQ= 2.36 P = 0.103 RV NO (YES)
92	TEND TO BE THOSE WHERE THE SYSTEM STYLE IS MOBILIZATIONAL (20)	0.52	92	TEND TO BE THOSE WHERE THE SYSTEM STYLE IS LIMITED MOBILIZATIONAL, OR NON-MOBILIZATIONAL (93)	0.96	17 3 16 68 X SQ= 29.46 P = 0. RV YES (YES)

93	TEND TO BE THOSE WHERE THE SYSTEM STYLE IS MOBILIZATIONAL, OR LIMITED MOBILIZATIONAL (32)	0.70	93	TEND TO BE THOSE WHERE THE SYSTEM STYLE IS NON-MOBILIZATIONAL (78)	0.89	23 8 10 62 X SQ= 33.48 P = 0. RV YES (YES)
94	TEND TO BE THOSE WHERE THE STATUS OF THE REGIME IS AUTHORITARIAN OR TOTALITARIAN (41)	0.88	94	TEND TO BE THOSE WHERE THE STATUS OF THE REGIME IS CONSTITUTIONAL (51)	0.79	3 48 22 13 X SQ= 29.97 P = 0. RV YES (YES)
95	TEND LESS TO BE THOSE WHERE THE STATUS OF THE REGIME IS CONSTITUTIONAL OR AUTHORITARIAN (95)	0.56	95	TEND MORE TO BE THOSE WHERE THE STATUS OF THE REGIME IS CONSTITUTIONAL OR AUTHORITARIAN (95)	0.99	19 67 15 1 X SQ= 28.03 P = 0. RV NO (YES)
97	TEND TO BE THOSE WHERE THE STATUS OF THE REGIME IS TOTALITARIAN, RATHER THAN AUTHORITARIAN (16)	0.68	97	TEND TO BE THOSE WHERE THE STATUS OF THE REGIME IS AUTHORITARIAN, RATHER THAN TOTALITARIAN (23)	0.91	7 10 15 1 X SQ= 8.02 P = 0.002 RV YES (YES)
98	TILT MORE TOWARD BEING THOSE WHERE GOVERNMENTAL STABILITY IS GENERALLY OR MODERATELY PRESENT AND DATES FROM THE POST-WAR PERIOD, OR IS ABSENT (93)	0.88	98	TILT LESS TOWARD BEING THOSE WHERE GOVERNMENTAL STABILITY IS GENERALLY OR MODERATELY PRESENT AND DATES FROM THE POST-WAR PERIOD, OR IS ABSENT (93)	0.75	4 18 30 53 X SQ= 1.81 P = 0.130 RV NO (NO)
99	TEND MORE TO BE THOSE WHERE GOVERNMENTAL STABILITY IS GENERALLY PRESENT AND DATES FROM AT LEAST THE INTER-WAR PERIOD, OR FROM THE POST-WAR PERIOD (50)	0.86	99	TEND LESS TO BE THOSE WHERE GOVERNMENTAL STABILITY IS GENERALLY PRESENT AND DATES FROM AT LEAST THE INTER-WAR PERIOD, OR FROM THE POST-WAR PERIOD (50)	0.56	19 31 3 24 X SQ= 4.96 P = 0.017 RV NO (NO)
100	ALWAYS ARE THOSE WHERE GOVERNMENTAL STABILITY IS GENERALLY PRESENT AND DATES FROM AT LEAST THE INTER-WAR PERIOD, OR IS GENERALLY OR MODERATELY PRESENT, AND DATES FROM THE POST-WAR PERIOD (64)	1.00	100	TEND LESS TO BE THOSE WHERE GOVERNMENTAL STABILITY IS GENERALLY PRESENT AND DATES FROM AT LEAST THE INTER-WAR PERIOD, OR IS GENERALLY OR MODERATELY PRESENT, AND DATES FROM THE POST-WAR PERIOD (64)	0.70	22 39 0 17 X SQ= 6.85 P = 0.002 RV NO (NO)
101	TEND TO BE THOSE WHERE THE REPRESENTATIVE CHARACTER OF THE REGIME IS LIMITED POLYARCHIC, PSEUDO-POLYARCHIC, OR NON-POLYARCHIC (57)	0.94	101	TEND TO BE THOSE WHERE THE REPRESENTATIVE CHARACTER OF THE REGIME IS POLYARCHIC (41)	0.66	2 39 32 20 X SQ= 29.34 P = 0. RV YES (YES)
102	TEND TO BE THOSE WHERE THE REPRESENTATIVE CHARACTER OF THE REGIME IS PSEUDO-POLYARCHIC OR NON-POLYARCHIC (49)	0.88	102	TEND TO BE THOSE WHERE THE REPRESENTATIVE CHARACTER OF THE REGIME IS POLYARCHIC OR LIMITED POLYARCHIC (59)	0.80	4 55 30 14 X SQ= 40.24 P = 0. RV YES (YES)

105	TEND TO BE THOSE WHERE THE ELECTORAL SYSTEM IS PARTIALLY COMPETITIVE OR NON-COMPETITIVE (47)	0.97	105	TEND TO BE THOSE WHERE THE ELECTORAL SYSTEM IS COMPETITIVE (43)	0.75	1 42 32 14 X SQ= 40.24 P = 0. RV YES (YES)
106	TEND TO BE THOSE WHERE THE ELECTORAL SYSTEM IS NON-COMPETITIVE (30)	0.87	106	TEND TO BE THOSE WHERE THE ELECTORAL SYSTEM IS COMPETITIVE OR PARTIALLY COMPETITIVE (52)	0.94	4 47 27 3 X SQ= 50.55 P = 0. RV YES (YES)
107	ALWAYS ARE THOSE WHERE AUTONOMOUS GROUPS ARE PARTIALLY TOLERATED IN POLITICS, ARE TOLERATED ONLY OUTSIDE POLITICS, OR ARE NOT TOLERATED AT ALL (65)	1.00	107	TEND TO BE THOSE WHERE AUTONOMOUS GROUPS ARE FULLY TOLERATED IN POLITICS (46)	0.67	0 46 34 23 X SQ= 38.31 P = 0. RV YES (YES)
108	TEND TO BE THOSE WHERE AUTONOMOUS GROUPS ARE TOLERATED ONLY OUTSIDE POLITICS OR ARE NOT TOLERATED AT ALL (35)	0.84	108	TEND TO BE THOSE WHERE AUTONOMOUS GROUPS ARE FULLY OR PARTIALLY TOLERATED IN POLITICS (65)	0.90	5 57 27 6 X SQ= 49.20 P = 0. RV YES (YES)
113	TEND MORE TO BE THOSE WHERE SECTIONALISM IS MODERATE OR NEGLIGIBLE (81)	0.91	113	TEND LESS TO BE THOSE WHERE SECTIONALISM IS MODERATE OR NEGLIGIBLE (81)	0.71	3 20 29 48 X SQ= 3.87 P = 0.040 RV NO (NO)
116	ALWAYS ARE THOSE WHERE INTEREST ARTICULATION BY ASSOCIATIONAL GROUPS IS LIMITED OR NEGLIGIBLE (79)	1.00	116	TEND LESS TO BE THOSE WHERE INTEREST ARTICULATION BY ASSOCIATIONAL GROUPS IS LIMITED OR NEGLIGIBLE (79)	0.55	0 32 31 39 X SQ= 18.32 P = 0. RV NO (NO)
117	TEND TO BE THOSE WHERE INTEREST ARTICULATION BY ASSOCIATIONAL GROUPS IS NEGLIGIBLE (51)	0.81	117	TEND TO BE THOSE WHERE INTEREST ARTICULATION BY ASSOCIATIONAL GROUPS IS SIGNIFICANT, MODERATE, OR LIMITED (60)	0.70	6 50 25 21 X SQ= 20.71 P = 0. RV YES (YES)
118	TEND TO BE THOSE WHERE INTEREST ARTICULATION BY INSTITUTIONAL GROUPS IS VERY SIGNIFICANT (40)	0.78	118	TEND TO BE THOSE WHERE INTEREST ARTICULATION BY INSTITUTIONAL GROUPS IS SIGNIFICANT, MODERATE, OR LIMITED (60)	0.80	21 13 6 52 X SQ= 24.91 P = 0. RV YES (YES)
119	ALWAYS ARE THOSE WHERE INTEREST ARTICULATION BY INSTITUTIONAL GROUPS IS VERY SIGNIFICANT OR SIGNIFICANT (74)	1.00	119	TEND LESS TO BE THOSE WHERE INTEREST ARTICULATION BY INSTITUTIONAL GROUPS IS VERY SIGNIFICANT OR SIGNIFICANT (74)	0.60	27 39 0 26 X SQ= 13.15 P = 0. RV NO (NO)

122	LEAN MORE TOWARD BEING THOSE WHERE INTEREST ARTICULATION BY NON-ASSOCIATIONAL GROUPS IS SIGNIFICANT OR MODERATE (83)	0.82	122	LEAN LESS TOWARD BEING THOSE WHERE INTEREST ARTICULATION BY NON-ASSOCIATIONAL GROUPS IS SIGNIFICANT OR MODERATE (83)	0.63	28 45 6 26 X SQ= 3.06 P = 0.069 RV NO (NO)
123	ALWAYS ARE THOSE WHERE INTEREST ARTICULATION BY NON-ASSOCIATIONAL GROUPS IS SIGNIFICANT, MODERATE, OR LIMITED (107)	1.00	123	LEAN LESS TOWARD BEING THOSE WHERE INTEREST ARTICULATION BY NON-ASSOCIATIONAL GROUPS IS SIGNIFICANT, MODERATE, OR LIMITED (107)	0.89	34 63 0 8 X SQ= 2.70 P = 0.052 RV NO (NO)
125	TILT MORE TOWARD BEING THOSE WHERE INTEREST ARTICULATION BY ANOMIC GROUPS IS FREQUENT OR OCCASIONAL (64)	0.76	125	TILT LESS TOWARD BEING THOSE WHERE INTEREST ARTICULATION BY ANOMIC GROUPS IS FREQUENT OR OCCASIONAL (64)	0.57	19 37 6 28 X SQ= 2.04 P = 0.145 RV NO (NO)
126	ALWAYS ARE THOSE WHERE INTEREST ARTICULATION BY ANOMIC GROUPS IS FREQUENT, OCCASIONAL, OR INFREQUENT (83)	1.00	126	TEND LESS TO BE THOSE WHERE INTEREST ARTICULATION BY ANOMIC GROUPS IS FREQUENT, OCCASIONAL, OR INFREQUENT (83)	0.75	25 49 0 16 X SQ= 5.90 P = 0.004 RV NO (NO)
127	ALWAYS ARE THOSE WHERE INTEREST ARTICULATION BY POLITICAL PARTIES IS MODERATE, LIMITED, OR NEGLIGIBLE (72)	1.00	127	TEND LESS TO BE THOSE WHERE INTEREST ARTICULATION BY POLITICAL PARTIES IS MODERATE, LIMITED, OR NEGLIGIBLE (72)	0.66	0 20 32 38 X SQ= 12.26 P = 0. RV NO (NO)
128	TEND TO BE THOSE WHERE INTEREST ARTICULATION BY POLITICAL PARTIES IS LIMITED OR NEGLIGIBLE (45)	0.97	128	TEND TO BE THOSE WHERE INTEREST ARTICULATION BY POLITICAL PARTIES IS SIGNIFICANT OR MODERATE (48)	0.79	1 46 31 12 X SQ= 44.97 P = 0. RV YES (YES)
129	TEND TO BE THOSE WHERE INTEREST ARTICULATION BY POLITICAL PARTIES IS NEGLIGIBLE (37)	0.94	129	TEND TO BE THOSE WHERE INTEREST ARTICULATION BY POLITICAL PARTIES IS SIGNIFICANT, MODERATE, OR LIMITED (56)	0.88	2 51 30 7 X SQ= 53.51 P = 0. RV YES (YES)
131	ALWAYS ARE THOSE WHERE INTEREST AGGREGATION BY POLITICAL PARTIES IS SIGNIFICANT OR MODERATE (30)	1.00	131	TEND TO BE THOSE WHERE INTEREST AGGREGATION BY POLITICAL PARTIES IS LIMITED OR NEGLIGIBLE (35)	0.56	6 24 0 30 X SQ= 4.63 P = 0.024 RV YES (NO)
133	TEND LESS TO BE THOSE WHERE INTEREST AGGREGATION BY THE EXECUTIVE IS MODERATE, LIMITED, OR NEGLIGIBLE (76)	0.52	133	TEND MORE TO BE THOSE WHERE INTEREST AGGREGATION BY THE EXECUTIVE IS MODERATE, LIMITED, OR NEGLIGIBLE (76)	0.82	16 12 17 53 X SQ= 8.25 P = 0.004 RV NO (YES)

138	ALWAYS ARE THOSE WHERE INTEREST AGGREGATION BY THE LEGISLATURE IS NEGLIGIBLE (49)	1.00	138	TEND TO BE THOSE WHERE INTEREST AGGREGATION BY THE LEGISLATURE IS SIGNIFICANT, MODERATE, OR LIMITED (48)	0.73	0 47 29 17 X SQ= 40.17 P = 0. RV YES (YES)
146	TEND LESS TO BE THOSE WHERE THE PARTY SYSTEM IS QUALITATIVELY OTHER THAN MASS-BASED TERRITORIAL (92)	0.76	146	ALWAYS ARE THOSE WHERE THE PARTY SYSTEM IS QUALITATIVELY OTHER THAN MASS-BASED TERRITORIAL (92)	1.00	8 0 25 60 X SQ= 12.98 P = 0. RV NO (YES)
147	ALWAYS ARE THOSE WHERE THE PARTY SYSTEM IS QUALITATIVELY OTHER THAN CLASS-ORIENTED OR MULTI-IDEOLOGICAL (67)	1.00	147	TEND LESS TO BE THOSE WHERE THE PARTY SYSTEM IS QUALITATIVELY OTHER THAN CLASS-ORIENTED OR MULTI-IDEOLOGICAL (67)	0.57	0 23 33 30 X SQ= 17.40 P = 0. RV NO (YES)
148	LEAN LESS TOWARD BEING THOSE WHERE THE PARTY SYSTEM IS QUALITATIVELY OTHER THAN AFRICAN TRANSITIONAL (96)	0.76	148	LEAN MORE TOWARD BEING THOSE WHERE THE PARTY SYSTEM IS QUALITATIVELY OTHER THAN AFRICAN TRANSITIONAL (96)	0.92	8 5 26 61 X SQ= 3.74 P = 0.055 RV NO (YES)
153	TEND TO BE THOSE WHERE THE PARTY SYSTEM IS STABLE (42)	0.79	153	TEND TO BE THOSE WHERE THE PARTY SYSTEM IS MODERATELY STABLE OR UNSTABLE (38)	0.54	19 23 5 27 X SQ= 5.98 P = 0.011 RV YES (NO)
154	ALWAYS ARE THOSE WHERE THE PARTY SYSTEM IS STABLE OR MODERATELY STABLE (55)	1.00	154	TEND LESS TO BE THOSE WHERE THE PARTY SYSTEM IS STABLE OR MODERATELY STABLE (55)	0.60	24 30 0 20 X SQ= 11.21 P = 0. RV NO (NO)
160	TEND TO BE THOSE WHERE THE POLITICAL LEADERSHIP IS ELITIST (30)	0.52	160	TEND TO BE THOSE WHERE THE POLITICAL LEADERSHIP IS MODERATELY ELITIST OR NON-ELITIST (67)	0.82	16 11 15 50 X SQ= 9.62 P = 0.001 RV YES (YES)
161	LEAN TOWARD BEING THOSE WHERE THE POLITICAL LEADERSHIP IS ELITIST OR MODERATELY ELITIST (47)	0.61	161	LEAN TOWARD BEING THOSE WHERE THE POLITICAL LEADERSHIP IS NON-ELITIST (50)	0.59	19 25 12 36 X SQ= 2.63 P = 0.080 RV YES (NO)
164	TEND TO BE THOSE WHERE THE REGIME'S LEADERSHIP CHARISMA IS PRONOUNCED OR MODERATE (34)	0.64	164	TEND TO BE THOSE WHERE THE REGIME'S LEADERSHIP CHARISMA IS NEGLIGIBLE (65)	0.81	21 12 12 50 X SQ= 16.73 P = 0. RV YES (YES)

166	ALWAYS ARE THOSE WHERE THE VERTICAL POWER DISTRIBUTION IS THAT OF FORMAL FEDERALISM OR FORMAL AND EFFECTIVE UNITARISM (99)	1.00	166	TEND LESS TO BE THOSE WHERE THE VERTICAL POWER DISTRIBUTION IS THAT OF FORMAL FEDERALISM OR FORMAL AND EFFECTIVE UNITARISM (99)	0.79	0 15 34 55 X SQ= 6.87 P = 0.002 RV NO (NO)
168	TEND TO BE THOSE WHERE THE HORIZONTAL POWER DISTRIBUTION IS LIMITED OR NEGLIGIBLE (72)	0.97	168	TEND TO BE THOSE WHERE THE HORIZONTAL POWER DISTRIBUTION IS SIGNIFICANT (34)	0.50	1 33 33 33 X SQ= 20.10 P = 0. RV YES (YES)
169	TEND TO BE THOSE WHERE THE HORIZONTAL POWER DISTRIBUTION IS NEGLIGIBLE (48)	0.91	169	TEND TO BE THOSE WHERE THE HORIZONTAL POWER DISTRIBUTION IS SIGNIFICANT OR LIMITED (58)	0.82	3 54 31 12 X SQ= 45.85 P = 0. RV YES (YES)
171	ALWAYS ARE THOSE WHERE THE LEGISLATIVE-EXECUTIVE STRUCTURE IS OTHER THAN PARLIAMENTARY-REPUBLICAN (90)	1.00	171	TEND LESS TO BE THOSE WHERE THE LEGISLATIVE-EXECUTIVE STRUCTURE IS OTHER THAN PARLIAMENTARY-REPUBLICAN (90)	0.82	0 12 30 54 X SQ= 4.68 P = 0.009 RV NO (NO)
172	ALWAYS ARE THOSE WHERE THE LEGISLATIVE-EXECUTIVE STRUCTURE IS OTHER THAN PARLIAMENTARY-ROYALIST (88)	1.00	172	TEND LESS TO BE THOSE WHERE THE LEGISLATIVE-EXECUTIVE STRUCTURE IS OTHER THAN PARLIAMENTARY-ROYALIST (88)	0.70	0 21 32 49 X SQ= 10.32 P = 0. RV NO (NO)
174	ALWAYS ARE THOSE WHERE THE LEGISLATURE IS PARTIALLY EFFECTIVE, LARGELY INEFFECTIVE, OR WHOLLY INEFFECTIVE (72)	1.00	174	TEND LESS TO BE THOSE WHERE THE LEGISLATURE IS PARTIALLY EFFECTIVE, LARGELY INEFFECTIVE, OR WHOLLY INEFFECTIVE (72)	0.55	0 28 34 34 X SQ= 19.55 P = 0. RV NO (YES)
175	TEND TO BE THOSE WHERE THE LEGISLATURE IS LARGELY INEFFECTIVE OR WHOLLY INEFFECTIVE (49)	0.97	175	TEND TO BE THOSE WHERE THE LEGISLATURE IS FULLY EFFECTIVE OR PARTIALLY EFFECTIVE (51)	0.81	1 50 33 12 X SQ= 50.17 P = 0. RV YES (YES)
176	TEND TO BE THOSE WHERE THE LEGISLATURE IS WHOLLY INEFFECTIVE (28)	0.56	176	TEND TO BE THOSE WHERE THE LEGISLATURE IS FULLY EFFECTIVE, PARTIALLY EFFECTIVE, OR LARGELY INEFFECTIVE (72)	0.89	15 55 19 7 X SQ= 19.91 P = 0. RV YES (YES)
178	TEND TO BE THOSE WHERE THE LEGISLATURE IS UNICAMERAL (53)	0.85	178	TEND TO BE THOSE WHERE THE LEGISLATURE IS BICAMERAL (51)	0.67	28 22 5 44 X SQ= 21.34 P = 0. RV YES (YES)

No.	Statement	Value	No.	Statement	Value	Statistics
179	TEND TO BE THOSE WHERE THE EXECUTIVE IS DOMINANT (52)	0.97	179	TEND TO BE THOSE WHERE THE EXECUTIVE IS STRONG (39)	0.69	31 17 1 38 X SQ= 32.98 P = 0. RV YES (YES)
181	ALWAYS ARE THOSE WHERE THE BUREAUCRACY IS SEMI-MODERN, RATHER THAN MODERN (55)	1.00	181	TEND LESS TO BE THOSE WHERE THE BUREAUCRACY IS SEMI-MODERN, RATHER THAN MODERN (55)	0.60	0 21 21 31 X SQ= 10.02 P = 0. RV NO (NO)
185	TEND TO BE THOSE WHERE PARTICIPATION BY THE MILITARY IN POLITICS IS SUPPORTIVE, RATHER THAN INTERVENTIVE (31)	0.95	185	TEND TO BE THOSE WHERE PARTICIPATION BY THE MILITARY IN POLITICS IS INTERVENTIVE, RATHER THAN SUPPORTIVE (21)	0.62	1 13 21 8 X SQ= 13.59 P = 0. RV YES (YES)
186	TEND TO BE THOSE WHERE PARTICIPATION BY THE MILITARY IN POLITICS IS SUPPORTIVE, RATHER THAN NEUTRAL (31)	0.64	186	TEND TO BE THOSE WHERE PARTICIPATION BY THE MILITARY IN POLITICS IS NEUTRAL, RATHER THAN SUPPORTIVE (56)	0.85	21 8 12 44 X SQ= 18.82 P = 0. RV YES (YES)
187	TEND TO BE THOSE WHERE THE ROLE OF THE POLICE IS POLITICALLY SIGNIFICANT (66)	0.93	187	TEND TO BE THOSE WHERE THE ROLE OF THE POLICE IS NOT POLITICALLY SIGNIFICANT (35)	0.53	27 29 2 33 X SQ= 16.01 P = 0. RV YES (NO)
188	TEND MORE TO BE THOSE WHERE THE CHARACTER OF THE LEGAL SYSTEM IS OTHER THAN CIVIL LAW (81)	0.91	188	TEND LESS TO BE THOSE WHERE THE CHARACTER OF THE LEGAL SYSTEM IS OTHER THAN CIVIL LAW (81)	0.61	3 27 30 43 X SQ= 8.07 P = 0.002 RV NO (NO)
189	ALWAYS ARE THOSE WHERE THE CHARACTER OF THE LEGAL SYSTEM IS OTHER THAN COMMON LAW (108)	1.00	189	LEAN LESS TOWARD BEING THOSE WHERE THE CHARACTER OF THE LEGAL SYSTEM IS OTHER THAN COMMON LAW (108)	0.90	0 7 34 64 X SQ= 2.18 P = 0.093 RV NO (NO)
192	TEND LESS TO BE THOSE WHERE THE CHARACTER OF THE LEGAL SYSTEM IS OTHER THAN MUSLIM OR PARTLY MUSLIM (86)	0.61	192	TEND MORE TO BE THOSE WHERE THE CHARACTER OF THE LEGAL SYSTEM IS OTHER THAN MUSLIM OR PARTLY MUSLIM (86)	0.86	13 10 20 60 X SQ= 6.77 P = 0.010 RV NO (YES)
194	TEND LESS TO BE THOSE THAT ARE NON-COMMUNIST (101)	0.62	194	ALWAYS ARE THOSE THAT ARE NON-COMMUNIST (101)	1.00	13 0 21 70 X SQ= 27.19 P = 0. RV NO (YES)

```
140  POLITIES                                        140  POLITIES
     WHERE THE PARTY SYSTEM IS QUANTITATIVELY             WHERE THE PARTY SYSTEM IS QUANTITATIVELY
     ONE PARTY DOMINANT  (13)                             OTHER THAN ONE PARTY DOMINANT  (87)

                                                                          BOTH SUBJECT AND PREDICATE

     13 IN LEFT COLUMN

BOLIVIA     BURUNDI     CAMEROUN    CONGO(BRA)  FRANCE      GREECE      INDIA       MALAGASY R  MEXICO      PARAGUAY
RWANDA      SOMALIA     VIETNAM REP

     87 IN RIGHT COLUMN

AFGHANISTAN ALBANIA     ALGERIA     ARGENTINA   AUSTRALIA   AUSTRIA     BELGIUM     BRAZIL      BULGARIA    BURMA
CAMBODIA    CANADA      CEN AFR REP CEYLON      CHAD        CHILE       CHINA, PR   COLOMBIA    CONGO(LEO)  CUBA
COSTA RICA  CYPRUS      CZECHOS'KIA DAHOMEY     DENMARK     DOMIN REP   ECUADOR     EL SALVADOR ETHIOPIA    FINLAND
GABON       GERMANY, E  GERMAN FR   GHANA       GUATEMALA   GUINEA      HAITI       HONDURAS    HUNGARY     ICELAND
IRELAND     ISRAEL      ITALY       IVORY COAST JAMAICA     JAPAN       JORDAN      KOREA, N    LAOS        LEBANON
LIBERIA     LIBYA       LUXEMBOURG  MALI        MAURITANIA  MONGOLIA    NETHERLANDS NEW ZEALAND NIGER       NIGERIA
NORWAY      PANAMA      PERU        PHILIPPINES POLAND      PORTUGAL    RUMANIA     SA'U ARABIA SENEGAL     SIERRE LEO
SPAIN       SUDAN       SWEDEN      SWITZERLAND TANGANYIKA  TRINIDAD    TUNISIA     UGANDA      USSR        UAR
UK          US          UPPER VOLTA URUGUAY     VENEZUELA   VIETNAM, N  YUGOSLAVIA

     6 EXCLUDED BECAUSE AMBIGUOUS

INDONESIA   MALAYA      MOROCCO     NICARAGUA   SO AFRICA   TURKEY

     9 EXCLUDED BECAUSE UNASCERTAINABLE

IRAN        IRAQ        KOREA REP   NEPAL       PAKISTAN    SYRIA       THAILAND    TOGO        YEMEN
-------------------------------------------------------------------------------------------------------------------

 29  TILT TOWARD BEING THOSE                 0.73       29  TILT TOWARD BEING THOSE                0.56        3   45
     WHERE THE DEGREE OF URBANIZATION                       WHERE THE DEGREE OF URBANIZATION                   8   35
     IS LOW  (49)                                           IS HIGH  (56)                                X SQ=   2.20
                                                                                                         P =    0.107
                                                                                                         RV YES (NO )

 30  LEAN TOWARD BEING THOSE                 0.69       30  LEAN TOWARD BEING THOSE                0.59        9   35
     WHOSE AGRICULTURAL POPULATION                          WHOSE AGRICULTURAL POPULATION                      4   50
     IS HIGH  (56)                                          IS MEDIUM, LOW, OR VERY LOW  (57)            X SQ=   2.54
                                                                                                         P =    0.075
                                                                                                         RV YES (NO )

 37  TEND TO BE THOSE                        0.77       37  TEND TO BE THOSE                       0.63        3   55
     WHOSE PER CAPITA GROSS NATIONAL PRODUCT                WHOSE PER CAPITA GROSS NATIONAL PRODUCT           10   32
     IS VERY LOW  (51)                                      IS VERY HIGH, HIGH, MEDIUM, OR LOW  (64)     X SQ=   5.92
                                                                                                         P =    0.013
                                                                                                         RV YES (NO )
```

40	LEAN TOWARD BEING THOSE WHOSE INTERNATIONAL FINANCIAL STATUS IS VERY LOW (39)	0.62	40	LEAN TOWARD BEING THOSE WHOSE INTERNATIONAL FINANCIAL STATUS IS VERY HIGH, HIGH, MEDIUM, OR LOW (71)	0.67	5 55 8 27 X SQ= 2.81 P = 0.064 RV YES (NO)
43	TILT TOWARD BEING THOSE WHOSE ECONOMIC DEVELOPMENTAL STATUS IS VERY UNDERDEVELOPED (57)	0.69	43	TILT TOWARD BEING THOSE WHOSE ECONOMIC DEVELOPMENTAL STATUS IS DEVELOPED, INTERMEDIATE, OR UNDERDEVELOPED (55)	0.55	4 46 9 38 X SQ= 1.72 P = 0.140 RV YES (NO)
54	TILT MORE TOWARD BEING THOSE WHERE NEWSPAPER CIRCULATION IS LESS THAN ONE HUNDRED PER THOUSAND (76)	0.85	54	TILT LESS TOWARD BEING THOSE WHERE NEWSPAPER CIRCULATION IS LESS THAN ONE HUNDRED PER THOUSAND (76)	0.59	2 35 11 50 X SQ= 2.19 P = 0.123 RV NO (NO)
84	TILT TOWARD BEING THOSE WHERE THE TYPE OF POLITICAL MODERNIZATION IS UNDEVELOPED TUTELARY, RATHER THAN DEVELOPED TUTELARY (24)	0.75	84	TILT TOWARD BEING THOSE WHERE THE TYPE OF POLITICAL MODERNIZATION IS DEVELOPED TUTELARY, RATHER THAN UNDEVELOPED TUTELARY (31)	0.55	2 21 6 17 X SQ= 1.36 P = 0.145 RV YES (NO)
85	LEAN TOWARD BEING THOSE WHERE THE STAGE OF POLITICAL MODERNIZATION IS MID- OR EARLY TRANSITIONAL, RATHER THAN ADVANCED (54)	0.69	85	LEAN TOWARD BEING THOSE WHERE THE STAGE OF POLITICAL MODERNIZATION IS ADVANCED, RATHER THAN MID- OR EARLY TRANSITIONAL (60)	0.58	4 50 9 36 X SQ= 2.40 P = 0.079 RV YES (NO)
86	TILT TOWARD BEING THOSE WHERE THE STAGE OF POLITICAL MODERNIZATION IS EARLY TRANSITIONAL, RATHER THAN ADVANCED OR MID-TRANSITIONAL (38)	0.54	86	TILT TOWARD BEING THOSE WHERE THE STAGE OF POLITICAL MODERNIZATION IS ADVANCED OR MID-TRANSITIONAL, RATHER THAN EARLY TRANSITIONAL (76)	0.70	6 60 7 26 X SQ= 1.87 P = 0.118 RV YES (NO)
87	LEAN TOWARD BEING THOSE WHOSE IDEOLOGICAL ORIENTATION IS DEVELOPMENTAL (31)	0.67	87	LEAN TOWARD BEING THOSE WHOSE IDEOLOGICAL ORIENTATION IS OTHER THAN DEVELOPMENTAL (58)	0.69	6 22 3 50 X SQ= 3.15 P = 0.058 RV YES (NO)
95	ALWAYS ARE THOSE WHERE THE STATUS OF THE REGIME IS CONSTITUTIONAL OR AUTHORITARIAN (95)	1.00	95	TILT LESS TOWARD BEING THOSE WHERE THE STATUS OF THE REGIME IS CONSTITUTIONAL OR AUTHORITARIAN (95)	0.81	12 70 0 16 X SQ= 1.48 P = 0.126 RV NO (NO)
98	ALWAYS ARE THOSE WHERE GOVERNMENTAL STABILITY IS GENERALLY OR MODERATELY PRESENT AND DATES FROM THE POST-WAR PERIOD, OR IS ABSENT (93)	1.00	98	LEAN LESS TOWARD BEING THOSE WHERE GOVERNMENTAL STABILITY IS GENERALLY OR MODERATELY PRESENT AND DATES FROM THE POST-WAR PERIOD, OR IS ABSENT (93)	0.76	0 21 13 66 X SQ= 2.65 P = 0.064 RV NO (NO)

102	TILT MORE TOWARD BEING THOSE WHERE THE REPRESENTATIVE CHARACTER OF THE REGIME IS POLYARCHIC OR LIMITED POLYARCHIC (59)	0.77	102	TILT LESS TOWARD BEING THOSE WHERE THE REPRESENTATIVE CHARACTER OF THE REGIME IS POLYARCHIC OR LIMITED POLYARCHIC (59)	0.53	10 45 3 40 X SQ= 1.75 P = 0.138 RV NO (NO)
106	LEAN MORE TOWARD BEING THOSE WHERE THE ELECTORAL SYSTEM IS COMPETITIVE OR PARTIALLY COMPETITIVE (52)	0.90	106	LEAN LESS TOWARD BEING THOSE WHERE THE ELECTORAL SYSTEM IS COMPETITIVE OR PARTIALLY COMPETITIVE (52)	0.58	9 40 1 29 X SQ= 2.57 P = 0.080 RV NO (NO)
108	TEND MORE TO BE THOSE WHERE AUTONOMOUS GROUPS ARE FULLY OR PARTIALLY TOLERATED IN POLITICS (65)	0.92	108	TEND LESS TO BE THOSE WHERE AUTONOMOUS GROUPS ARE FULLY OR PARTIALLY TOLERATED IN POLITICS (65)	0.59	12 47 1 32 X SQ= 3.90 P = 0.028 RV NO (NO)
121	TILT TOWARD BEING THOSE WHERE INTEREST ARTICULATION BY NON-ASSOCIATIONAL GROUPS IS SIGNIFICANT (54)	0.62	121	TILT TOWARD BEING THOSE WHERE INTEREST ARTICULATION BY NON-ASSOCIATIONAL GROUPS IS MODERATE, LIMITED, OR NEGLIGIBLE (61)	0.61	8 34 5 53 X SQ= 1.51 P = 0.143 RV YES (NO)
126	ALWAYS ARE THOSE WHERE INTEREST ARTICULATION BY ANOMIC GROUPS IS FREQUENT, OCCASIONAL, OR INFREQUENT (83)	1.00	126	TILT LESS TOWARD BEING THOSE WHERE INTEREST ARTICULATION BY ANOMIC GROUPS IS FREQUENT, OCCASIONAL, OR INFREQUENT (83)	0.78	12 57 0 16 X SQ= 1.96 P = 0.111 RV NO (NO)
133	TEND TO BE THOSE WHERE INTEREST AGGREGATION BY THE EXECUTIVE IS SIGNIFICANT (29)	0.54	133	TEND TO BE THOSE WHERE INTEREST AGGREGATION BY THE EXECUTIVE IS MODERATE, LIMITED, OR NEGLIGIBLE (76)	0.77	7 19 6 62 X SQ= 3.76 P = 0.041 RV YES (NO)
147	ALWAYS ARE THOSE WHERE THE PARTY SYSTEM IS QUALITATIVELY OTHER THAN CLASS-ORIENTED OR MULTI-IDEOLOGICAL (67)	1.00	147	LEAN LESS TOWARD BEING THOSE WHERE THE PARTY SYSTEM IS QUALITATIVELY OTHER THAN CLASS-ORIENTED OR MULTI-IDEOLOGICAL (67)	0.69	0 23 10 52 X SQ= 2.79 P = 0.056 RV NO (NO)
153	TEND TO BE THOSE WHERE THE PARTY SYSTEM IS MODERATELY STABLE OR UNSTABLE (38)	0.78	153	TEND TO BE THOSE WHERE THE PARTY SYSTEM IS STABLE (42)	0.64	2 39 7 22 X SQ= 4.04 P = 0.028 RV YES (NO)
160	TILT MORE TOWARD BEING THOSE WHERE THE POLITICAL LEADERSHIP IS MODERATELY ELITIST OR NON-ELITIST (67)	0.92	160	TILT LESS TOWARD BEING THOSE WHERE THE POLITICAL LEADERSHIP IS MODERATELY ELITIST OR NON-ELITIST (67)	0.67	1 25 11 51 X SQ= 1.94 P = 0.100 RV NO (NO)

169	TILT MORE TOWARD BEING THOSE WHERE THE HORIZONTAL POWER DISTRIBUTION IS SIGNIFICANT OR LIMITED (58)	0.77	169	TILT LESS TOWARD BEING THOSE WHERE THE HORIZONTAL POWER DISTRIBUTION IS SIGNIFICANT OR LIMITED (58)	0.52	10 43 3 39 X SQ= 1.82 P = 0.136 RV NO (NO)
170	TILT TOWARD BEING THOSE WHERE THE LEGISLATIVE-EXECUTIVE STRUCTURE IS PRESIDENTIAL (39)	0.62	170	TILT TOWARD BEING THOSE WHERE THE LEGISLATIVE-EXECUTIVE STRUCTURE IS OTHER THAN PRESIDENTIAL (63)	0.65	8 28 5 51 X SQ= 2.19 P = 0.123 RV YES (NO)
174	TILT MORE TOWARD BEING THOSE WHERE THE LEGISLATURE IS PARTIALLY EFFECTIVE, LARGELY INEFFECTIVE, OR WHOLLY INEFFECTIVE (72)	0.92	174	TILT LESS TOWARD BEING THOSE WHERE THE LEGISLATURE IS PARTIALLY EFFECTIVE, LARGELY INEFFECTIVE, OR WHOLLY INEFFECTIVE (72)	0.68	1 25 12 54 X SQ= 2.09 P = 0.101 RV NO (NO)
192	ALWAYS ARE THOSE WHERE THE CHARACTER OF THE LEGAL SYSTEM IS OTHFR THAN MUSLIM OR PARTLY MUSLIM (86)	1.00	192	LEAN LESS TOWARD BEING THOSE WHERE THE CHARACTER OF THE LEGAL SYSTEM IS OTHER THAN MUSLIM OR PARTLY MUSLIM (86)	0.76	0 20 13 65 X SQ= 2.53 P = 0.064 RV NO (NO)

```
****************************************************************************************************

141  POLITIES                                           141  POLITIES
     WHERE THE PARTY SYSTEM IS QUANTITATIVELY                WHERE THE PARTY SYSTEM IS QUANTITATIVELY
     TWO-PARTY  (11)                                         OTHER THAN TWO-PARTY  (87)


                                                                              BOTH SUBJECT AND PREDICATE
....................................................................................................

      11 IN LEFT COLUMN

AUSTRALIA   AUSTRIA     CANADA      COLOMBIA    HONDURAS    JAMAICA     NEW ZEALAND PHILIPPINES SIERRE LEO  UK
US

      87 IN RIGHT COLUMN

AFGHANISTAN ALBANIA     ALGERIA     ARGENTINA   BELGIUM     BOLIVIA     BRAZIL      BULGARIA    BURUNDI     CAMBODIA
CAMEROUN    CEN AFR REP CEYLON      CHAD        CHILE       CHINA, PR   CONGO(BRA)  CONGO(LEO)  COSTA RICA  CUBA
CYPRUS      CZECHOS'KIA DAHOMEY     DENMARK     DOMIN REP   ECUADOR     EL SALVADOR ETHIOPIA    FINLAND     FRANCE
GABON       GERMANY, E  GERMAN FR   GHANA       GREECE      GUATEMALA   GUINEA      HAITI       HUNGARY     ICELAND
INDIA       IRELAND     ISRAEL      ITALY       IVORY COAST JAPAN       JORDAN      KOREA, N    LAOS        LEBANON
LIBERIA     LIBYA       LUXEMBOURG  MALAGASY R  MALI        MAURITANIA  MEXICO      MONGOLIA    MOROCCO     NETHERLANDS
NIGER       NIGERIA     NORWAY      PANAMA      PARAGUAY    PERU        POLAND      PORTUGAL    RUMANIA     RWANDA
SA'U ARABIA SENEGAL     SOMALIA     SPAIN       SWEDEN      SWITZERLAND TANGANYIKA  TRINIDAD    TUNISIA     UGANDA
USSR        UAR         UPPER VOLTA VENEZUELA   VIETNAM, N  VIETNAM REP YUGOSLAVIA

      6 EXCLUDED BECAUSE AMBIGUOUS

INDONESIA   MALAYA      NICARAGUA   SO AFRICA   TURKEY      URUGUAY

      11 EXCLUDED BECAUSE UNASCERTAINABLE

BURMA       IRAN        IRAQ        KOREA REP   NEPAL       PAKISTAN    SUDAN       SYRIA       THAILAND    TOGO
YEMEN

----------------------------------------------------------------------------------------------------

  2  TEND TO BE THOSE                        0.55        2  TEND TO BE THOSE                   0.82          6   16
     LOCATED IN WEST EUROPE, SCANDINAVIA,                   LOCATED ELSEWHERE THAN IN                         5   71
     NORTH AMERICA, OR AUSTRALASIA  (22)                    WEST EUROPE, SCANDINAVIA,                  X SQ=   5.40
                                                            NORTH AMERICA, OR AUSTRALASIA  (93)        P =    0.014
                                                                                                       RV YES (NO )

 29  TILT MORE TOWARD BEING THOSE            0.80       29  TILT LESS TOWARD BEING THOSE       0.51          8   40
     WHERE THE DEGREE OF URBANIZATION                       WHERE THE DEGREE OF URBANIZATION                  2   39
     IS HIGH  (56)                                          IS HIGH  (56)                              X SQ=   2.01
                                                                                                       P =    0.100
                                                                                                       RV NO  (NO )

 31  TEND TO BE THOSE                        0.55       31  TEND TO BE THOSE                   0.81          5   70
     WHOSE AGRICULTURAL POPULATION                          WHOSE AGRICULTURAL POPULATION                     6   16
     IS LOW OR VERY LOW  (24)                               IS HIGH OR MEDIUM  (90)                    X SQ=   5.28
                                                                                                       P =    0.015
                                                                                                       RV YES (NO )
```

35	TEND TO BE THOSE WHOSE PER CAPITA GROSS NATIONAL PRODUCT IS VERY HIGH OR HIGH (24)	0.55	35	TEND TO BE THOSE WHOSE PER CAPITA GROSS NATIONAL PRODUCT IS MEDIUM, LOW, OR VERY LOW (91)	0.79	6 18 5 69 X SQ= 4.36 P = 0.023 RV YES (NO)
36	TILT TOWARD BEING THOSE WHOSE PER CAPITA GROSS NATIONAL PRODUCT IS VERY HIGH, HIGH, OR MEDIUM (42)	0.64	36	TILT TOWARD BEING THOSE WHOSE PER CAPITA GROSS NATIONAL PRODUCT IS LOW OR VERY LOW (73)	0.63	7 32 4 55 X SQ= 1.93 P = 0.108 RV YES (NO)
39	TEND TO BE THOSE WHOSE INTERNATIONAL FINANCIAL STATUS IS VERY HIGH, HIGH, OR MEDIUM (38)	0.73	39	TEND TO BE THOSE WHOSE INTERNATIONAL FINANCIAL STATUS IS LOW OR VERY LOW (76)	0.70	8 26 3 60 X SQ= 5.98 P = 0.015 RV YES (NO)
42	LEAN TOWARD BEING THOSE WHOSE ECONOMIC DEVELOPMENTAL STATUS IS DEVELOPED OR INTERMEDIATE (36)	0.64	42	LEAN TOWARD BEING THOSE WHOSE ECONOMIC DEVELOPMENTAL STATUS IS UNDERDEVELOPED OR VERY UNDERDEVELOPED (76)	0.68	7 27 4 57 X SQ= 2.94 P = 0.090 RV YES (NO)
43	LEAN TOWARD BEING THOSE WHOSE ECONOMIC DEVELOPMENTAL STATUS IS DEVELOPED, INTERMEDIATE, OR UNDERDEVELOPED (55)	0.82	43	LEAN TOWARD BEING THOSE WHOSE ECONOMIC DEVELOPMENTAL STATUS IS VERY UNDERDEVELOPED (57)	0.54	9 39 2 45 X SQ= 3.56 P = 0.051 RV YES (NO)
44	TEND TO BE THOSE WHERE THE LITERACY RATE IS NINETY PERCENT OR ABOVE (25)	0.55	44	TEND TO BE THOSE WHERE THE LITERACY RATE IS BELOW NINETY PERCENT (90)	0.78	6 19 5 68 X SQ= 3.91 P = 0.029 RV YES (NO)
45	LEAN MORE TOWARD BEING THOSE WHERE THE LITERACY RATE IS FIFTY PERCENT OR ABOVE (55)	0.82	45	LEAN LESS TOWARD BEING THOSE WHERE THE LITERACY RATE IS FIFTY PERCENT OR ABOVE (55)	0.51	9 41 2 40 X SQ= 2.65 P = 0.060 RV NO (NO)
50	ALWAYS ARE THOSE WHERE FREEDOM OF THE PRESS IS COMPLETE (43)	1.00	50	TEND TO BE THOSE WHERE FREEDOM OF THE PRESS IS INTERMITTENT, INTERNALLY ABSENT, OR INTERNALLY AND EXTERNALLY ABSENT (56)	0.58	10 31 0 43 X SQ= 9.69 P = 0. RV YES (NO)
63	TEND TO BE THOSE WHERE THE RELIGION IS PREDOMINANTLY SOME KIND OF CHRISTIAN (46)	0.91	63	TEND TO BE THOSE WHERE THE RELIGION IS PREDOMINANTLY OR PARTLY OTHER THAN CHRISTIAN (68)	0.60	10 34 1 52 X SQ= 8.42 P = 0.002 RV YES (NO)

64	ALWAYS ARE THOSE WHERE THE RELIGION IS CHRISTIAN, RATHER THAN MUSLIM (46)	1.00	64	TILT LESS TOWARD BEING THOSE WHERE THE RELIGION IS CHRISTIAN, RATHER THAN MUSLIM (46)	0.76	10 34 0 11 X SQ= 1.72 P = 0.104 RV NO (NO)
68	TILT TOWARD BEING THOSE THAT ARE LINGUISTICALLY HOMOGENEOUS (52)	0.73	68	TILT TOWARD BEING THOSE THAT ARE LINGUISTICALLY WEAKLY HETEROGENEOUS OR STRONGLY HETEROGENEOUS (62)	0.53	8 40 3 46 X SQ= 1.74 P = 0.121 RV YES (NO)
72	TILT TOWARD BEING THOSE WHOSE DATE OF INDEPENDENCE IS BEFORE 1914 (52)	0.73	72	TILT TOWARD BEING THOSE WHOSE DATE OF INDEPENDENCE IS AFTER 1914 (59)	0.56	8 37 3 47 X SQ= 2.16 P = 0.108 RV YES (NO)
74	LEAN TOWARD BEING THOSE THAT ARE HISTORICALLY WESTERN (26)	0.55	74	LEAN TOWARD BEING THOSE THAT ARE NOT HISTORICALLY WESTERN (87)	0.77	6 20 5 66 X SQ= 3.40 P = 0.063 RV YES (NO)
75	TEND MORE TO BE THOSE THAT ARE HISTORICALLY WESTERN OR SIGNIFICANTLY WESTERNIZED (62)	0.91	75	TEND LESS TO BE THOSE THAT ARE HISTORICALLY WESTERN OR SIGNIFICANTLY WESTERNIZED (62)	0.56	10 49 1 38 X SQ= 3.54 P = 0.046 RV NO (NO)
76	LEAN TOWARD BEING THOSE THAT ARE HISTORICALLY WESTERN, RATHER THAN HAVING BEEN SIGNIFICANTLY OR PARTIALLY WESTERNIZED THROUGH A COLONIAL RELATIONSHIP (26)	0.55	76	LEAN TOWARD BEING THOSE THAT HAVE BEEN SIGNIFICANTLY OR PARTIALLY WESTERNIZED THROUGH A COLONIAL RELATIONSHIP, RATHER THAN BEING HISTORICALLY WESTERN (70)	0.73	6 20 5 53 X SQ= 2.15 P = 0.087 RV YES (NO)
80	ALWAYS ARE THOSE WHOSE DATE OF INDEPENDENCE IS AFTER 1914, AND THAT WERE FORMERLY DEPENDENCIES OF BRITAIN, RATHER THAN FRANCE (19)	1.00	80	TILT TOWARD BEING THOSE WHOSE DATE OF INDEPENDENCE IS AFTER 1914, AND THAT WERE FORMERLY DEPENDENCIES OF FRANCE, RATHER THAN BRITAIN (24)	0.65	2 12 0 22 X SQ= 1.16 P = 0.144 RV YES (NO)
81	TEND TO BE THOSE WHERE THE TYPE OF POLITICAL MODERNIZATION IS EARLY EUROPEAN OR EARLY EUROPEAN DERIVED, RATHER THAN LATER EUROPEAN OR LATER EUROPEAN DERIVED (11)	0.63	81	TEND TO BE THOSE WHERE THE TYPE OF POLITICAL MODERNIZATION IS LATER EUROPEAN OR LATER EUROPEAN DERIVED, RATHER THAN EARLY EUROPEAN OR EARLY EUROPEAN DERIVED (40)	0.88	5 5 3 35 X SQ= 7.30 P = 0.006 RV YES (YES)
89	TEND TO BE THOSE WHOSE IDEOLOGICAL ORIENTATION IS CONVENTIONAL (33)	0.90	89	TEND TO BE THOSE WHOSE IDEOLOGICAL ORIENTATION IS OTHER THAN CONVENTIONAL (62)	0.70	9 21 1 49 X SQ= 11.00 P = 0. RV YES (NO)

93	ALWAYS ARE THOSE WHERE THE SYSTEM STYLE IS NON-MOBILIZATIONAL (78)	1.00	93	TEND LESS TO BE THOSE WHERE THE SYSTEM STYLE IS NON-MOBILIZATIONAL (78)	0.66	0 29 11 57 X SQ= 3.80 P = 0.031 RV NO (NO)
94	ALWAYS ARE THOSE WHERE THE STATUS OF THE REGIME IS CONSTITUTIONAL (51)	1.00	94	TEND LESS TO BE THOSE WHERE THE STATUS OF THE REGIME IS CONSTITUTIONAL (51)	0.54	11 37 0 32 X SQ= 6.68 P = 0.002 RV NO (NO)
101	TEND TO BE THOSE WHERE THE REPRESENTATIVE CHARACTER OF THE REGIME IS POLYARCHIC (41)	0.82	101	TEND TO BE THOSE WHERE THE REPRESENTATIVE CHARACTER OF THE REGIME IS LIMITED POLYARCHIC, PSEUDO-POLYARCHIC, OR NON-POLYARCHIC (57)	0.61	9 29 2 46 X SQ= 5.60 P = 0.009 RV YES (NO)
102	ALWAYS ARE THOSE WHERE THE REPRESENTATIVE CHARACTER OF THE REGIME IS POLYARCHIC OR LIMITED POLYARCHIC (59)	1.00	102	TEND LESS TO BE THOSE WHERE THE REPRESENTATIVE CHARACTER OF THE REGIME IS POLYARCHIC OR LIMITED POLYARCHIC (59)	0.52	11 44 0 41 X SQ= 7.39 P = 0.002 RV NO (NO)
105	ALWAYS ARE THOSE WHERE THE ELECTORAL SYSTEM IS COMPETITIVE (43)	1.00	105	TEND TO BE THOSE WHERE THE ELECTORAL SYSTEM IS PARTIALLY COMPETITIVE OR NON-COMPETITIVE (47)	0.60	10 30 0 45 X SQ= 10.46 P = 0. RV YES (NO)
107	ALWAYS ARE THOSE WHERE AUTONOMOUS GROUPS ARE FULLY TOLERATED IN POLITICS (46)	1.00	107	TEND TO BE THOSE WHERE AUTONOMOUS GROUPS ARE PARTIALLY TOLERATED IN POLITICS, ARE TOLERATED ONLY OUTSIDE POLITICS, OR ARE NOT TOLERATED AT ALL (65)	0.62	11 32 0 53 X SQ= 12.90 P = 0. RV YES (NO)
111	ALWAYS ARE THOSE WHERE POLITICAL ENCULTURATION IS HIGH OR MEDIUM (53)	1.00	111	TEND LESS TO BE THOSE WHERE POLITICAL ENCULTURATION IS HIGH OR MEDIUM (53)	0.54	11 37 0 31 X SQ= 6.45 P = 0.003 RV NO (NO)
115	TEND TO BE THOSE WHERE INTEREST ARTICULATION BY ASSOCIATIONAL GROUPS IS SIGNIFICANT (20)	0.55	115	TEND TO BE THOSE WHERE INTEREST ARTICULATION BY ASSOCIATIONAL GROUPS IS MODERATE, LIMITED, OR NEGLIGIBLE (91)	0.83	6 14 5 70 X SQ= 6.27 P = 0.010 RV YES (NO)
116	TEND TO BE THOSE WHERE INTEREST ARTICULATION BY ASSOCIATIONAL GROUPS IS SIGNIFICANT OR MODERATE (32)	0.64	116	TEND TO BE THOSE WHERE INTEREST ARTICULATION BY ASSOCIATIONAL GROUPS IS LIMITED OR NEGLIGIBLE (79)	0.74	7 22 4 62 X SQ= 4.79 P = 0.031 RV YES (NO)

No.	Statement	Value	No.	Statement	Value	Statistics
117	TEND TO BE THOSE WHERE INTEREST ARTICULATION BY ASSOCIATIONAL GROUPS IS SIGNIFICANT, MODERATE, OR LIMITED (60)	0.91	117	TEND TO BE THOSE WHERE INTEREST ARTICULATION BY ASSOCIATIONAL GROUPS IS NEGLIGIBLE (51)	0.51	10 41 1 43 X SQ= 5.34 P = 0.010 RV YES (NO)
118	ALWAYS ARE THOSE WHERE INTEREST ARTICULATION BY INSTITUTIONAL GROUPS IS SIGNIFICANT, MODERATE, OR LIMITED (60)	1.00	118	TEND LESS TO BE THOSE WHERE INTEREST ARTICULATION BY INSTITUTIONAL GROUPS IS SIGNIFICANT, MODERATE, OR LIMITED (60)	0.59	0 31 10 44 X SQ= 4.84 P = 0.012 RV NO (NO)
119	TEND TO BE THOSE WHERE INTEREST ARTICULATION BY INSTITUTIONAL GROUPS IS MODERATE OR LIMITED (26)	0.80	119	TEND TO BE THOSE WHERE INTEREST ARTICULATION BY INSTITUTIONAL GROUPS IS VERY SIGNIFICANT OR SIGNIFICANT (74)	0.77	2 58 8 17 X SQ= 11.35 P = 0.001 RV YES (NO)
121	TILT MORE TOWARD BEING THOSE WHERE INTEREST ARTICULATION BY NON-ASSOCIATIONAL GROUPS IS MODERATE, LIMITED, OR NEGLIGIBLE (61)	0.82	121	TILT LESS TOWARD BEING THOSE WHERE INTEREST ARTICULATION BY NON-ASSOCIATIONAL GROUPS IS MODERATE, LIMITED, OR NEGLIGIBLE (61)	0.56	2 38 9 49 X SQ= 1.68 P = 0.120 RV NO (NO)
125	LEAN TOWARD BEING THOSE WHERE INTEREST ARTICULATION BY ANOMIC GROUPS IS INFREQUENT OR VERY INFREQUENT (35)	0.70	125	LEAN TOWARD BEING THOSE WHERE INTEREST ARTICULATION BY ANOMIC GROUPS IS FREQUENT OR OCCASIONAL (64)	0.64	3 47 7 26 X SQ= 3.02 P = 0.080 RV YES (NO)
126	TEND TO BE THOSE WHERE INTEREST ARTICULATION BY ANOMIC GROUPS IS VERY INFREQUENT (16)	0.50	126	TEND TO BE THOSE WHERE INTEREST ARTICULATION BY ANOMIC GROUPS IS FREQUENT, OCCASIONAL, OR INFREQUENT (83)	0.85	5 62 5 11 X SQ= 4.83 P = 0.020 RV YES (NO)
127	ALWAYS ARE THOSE WHERE INTEREST ARTICULATION BY POLITICAL PARTIES IS MODERATE, LIMITED, OR NEGLIGIBLE (72)	1.00	127	LEAN LESS TOWARD BEING THOSE WHERE INTEREST ARTICULATION BY POLITICAL PARTIES IS MODERATE, LIMITED, OR NEGLIGIBLE (72)	0.74	0 20 11 57 X SQ= 2.37 P = 0.063 RV NO (NO)
130	TEND TO BE THOSE WHERE INTEREST AGGREGATION BY POLITICAL PARTIES IS SIGNIFICANT (12)	0.70	130	TEND TO BE THOSE WHERE INTEREST AGGREGATION BY POLITICAL PARTIES IS MODERATE, LIMITED. OR NEGLIGIBLE (71)	0.94	7 4 3 59 X SQ= 22.57 P = 0. RV YES (YES)
131	TEND TO BE THOSE WHERE INTEREST AGGREGATION BY POLITICAL PARTIES IS SIGNIFICANT OR MODERATE (30)	0.80	131	TEND TO BE THOSE WHERE INTEREST AGGREGATION BY POLITICAL PARTIES IS LIMITED OR NEGLIGIBLE (35)	0.59	8 19 2 27 X SQ= 3.50 P = 0.038 RV YES (NO)

134	TEND MORE TO BE THOSE WHERE INTEREST AGGREGATION BY THE EXECUTIVE IS SIGNIFICANT OR MODERATE (57)	0.91	134	TEND LESS TO BE THOSE WHERE INTEREST AGGREGATION BY THE EXECUTIVE IS SIGNIFICANT OR MODERATE (57)	0.54	10 44 1 37 X SQ= 3.95 P = 0.024 RV NO (NO)
137	TEND TO BE THOSE WHERE INTEREST AGGREGATION BY THE LEGISLATURE IS SIGNIFICANT OR MODERATE (29)	0.80	137	TEND TO BE THOSE WHERE INTEREST AGGREGATION BY THE LEGISLATURE IS LIMITED OR NEGLIGIBLE (68)	0.76	8 19 2 60 X SQ= 10.63 P = 0.001 RV YES (NO)
138	ALWAYS ARE THOSE WHERE INTEREST AGGREGATION BY THE LEGISLATURE IS SIGNIFICANT, MODERATE, OR LIMITED (48)	1.00	138	TEND TO BE THOSE WHERE INTEREST AGGREGATION BY THE LEGISLATURE IS NEGLIGIBLE (49)	0.57	11 34 0 45 X SQ= 10.36 P = 0. RV YES (NO)
164	LEAN MORE TOWARD BEING THOSE WHERE THE REGIME'S LEADERSHIP CHARISMA IS NEGLIGIBLE (65)	0.91	164	LEAN LESS TOWARD BEING THOSE WHERE THE REGIME'S LEADERSHIP CHARISMA IS NEGLIGIBLE (65)	0.60	1 31 10 47 X SQ= 2.72 P = 0.053 RV NO (NO)
168	TEND TO BE THOSE WHERE THE HORIZONTAL POWER DISTRIBUTION IS SIGNIFICANT (34)	0.73	168	TEND TO BE THOSE WHERE THE HORIZONTAL POWER DISTRIBUTION IS LIMITED OR NEGLIGIBLE (72)	0.73	8 22 3 61 X SQ= 7.54 P = 0.004 RV YES (NO)
169	ALWAYS ARE THOSE WHERE THE HORIZONTAL POWER DISTRIBUTION IS SIGNIFICANT OR LIMITED (58)	1.00	169	TEND LESS TO BE THOSE WHERE THE HORIZONTAL POWER DISTRIBUTION IS SIGNIFICANT OR LIMITED (58)	0.51	11 42 0 41 X SQ= 7.73 P = 0.002 RV NO (NO)
172	TEND TO BE THOSE WHERE THE LEGISLATIVE-EXECUTIVE STRUCTURE IS PARLIAMENTARY-ROYALIST (21)	0.55	172	TEND TO BE THOSE WHERE THE LEGISLATIVE-EXECUTIVE STRUCTURE IS OTHER THAN PARLIAMENTARY-ROYALIST (88)	0.83	6 14 5 70 X SQ= 6.27 P = 0.010 RV YES (NO)
174	TEND TO BE THOSE WHERE THE LEGISLATURE IS FULLY EFFECTIVE (28)	0.73	174	TEND TO BE THOSE WHERE THE LEGISLATURE IS PARTIALLY EFFECTIVE, LARGELY INEFFECTIVE, OR WHOLLY INEFFECTIVE (72)	0.79	8 17 3 64 X SQ= 10.62 P = 0.001 RV YES (NO)
175	ALWAYS ARE THOSE WHERE THE LEGISLATURE IS FULLY EFFECTIVE OR PARTIALLY EFFECTIVE (51)	1.00	175	TEND TO BE THOSE WHERE THE LEGISLATURE IS LARGELY INEFFECTIVE OR WHOLLY INEFFECTIVE (49)	0.56	11 36 0 45 X SQ= 9.84 P = 0.001 RV YES (NO)

No.	Left statement	Value	No.	Right statement	Value	Statistics
178	TILT TOWARD BEING THOSE WHERE THE LEGISLATURE IS BICAMERAL (51)	0.73	178	TILT TOWARD BEING THOSE WHERE THE LEGISLATURE IS UNICAMERAL (53)	0.55	3 46 8 37 X SQ= 2.06 P = 0.110 RV YES (NO)
179	TEND TO BE THOSE WHERE THE EXECUTIVE IS STRONG (39)	0.91	179	TEND TO BE THOSE WHERE THE EXECUTIVE IS DOMINANT (52)	0.62	1 45 10 27 X SQ= 8.96 P = 0.002 RV YES (NO)
181	LEAN TOWARD BEING THOSE WHERE THE BUREAUCRACY IS MODERN, RATHER THAN SEMI-MODERN (21)	0.60	181	LEAN TOWARD BEING THOSE WHERE THE BUREAUCRACY IS SEMI-MODERN, RATHER THAN MODERN (55)	0.76	6 14 4 45 X SQ= 3.85 P = 0.053 RV YES (NO)
186	ALWAYS ARE THOSE WHERE PARTICIPATION BY THE MILITARY IN POLITICS IS NEUTRAL, RATHER THAN SUPPORTIVE (56)	1.00	186	TEND LESS TO BE THOSE WHERE PARTICIPATION BY THE MILITARY IN POLITICS IS NEUTRAL, RATHER THAN SUPPORTIVE (56)	0.61	0 28 10 44 X SQ= 4.30 P = 0.013 RV NO (NO)
187	TEND TO BE THOSE WHERE THE ROLE OF THE POLICE IS NOT POLITICALLY SIGNIFICANT (35)	0.82	187	TEND TO BE THOSE WHERE THE ROLE OF THE POLICE IS POLITICALLY SIGNIFICANT (66)	0.67	2 49 9 24 X SQ= 7.66 P = 0.005 RV YES (NO)
189	TEND LESS TO BE THOSE WHERE THE CHARACTER OF THE LEGAL SYSTEM IS OTHER THAN COMMON LAW (108)	0.55	189	TEND MORE TO BE THOSE WHERE THE CHARACTER OF THE LEGAL SYSTEM IS OTHER THAN COMMON LAW (108)	0.98	5 2 6 85 X SQ= 21.30 P = 0. RV NO (YES)
190	TEND TO BE THOSE WHERE THE CHARACTER OF THE LEGAL SYSTEM IS COMMON LAW, RATHER THAN CIVIL LAW (7)	0.62	190	TEND TO BE THOSE WHERE THE CHARACTER OF THE LEGAL SYSTEM IS CIVIL LAW, RATHER THAN COMMON LAW (32)	0.93	3 25 5 2 X SQ= 8.52 P = 0.003 RV YES (YES)

142 POLITIES
WHERE THE PARTY SYSTEM IS QUANTITATIVELY
MULTI-PARTY (30)

142 POLITIES
WHERE THE PARTY SYSTEM IS QUANTITATIVELY
OTHER THAN MULTI-PARTY (66)

BOTH SUBJECT AND PREDICATE

30 IN LEFT COLUMN

ARGENTINA	BELGIUM	BRAZIL	CEYLON	CHILE	CONGO(LEO)	COSTA RICA	CYPRUS	DENMARK	DOMIN REP
ECUADOR	EL SALVADOR	FINLAND	GUATEMALA	ICELAND	IRELAND	ISRAEL	ITALY	LAOS	LEBANON
LUXEMBOURG	NETHERLANDS	NIGERIA	NORWAY	PANAMA	PERU	SWEDEN	SWITZERLAND	UGANDA	VENEZUELA

66 IN RIGHT COLUMN

AFGHANISTAN	ALBANIA	ALGERIA	AUSTRALIA	AUSTRIA	BOLIVIA	BULGARIA	BURUNDI	CAMBODIA	CAMEROUN
CANADA	CEN AFR REP	CHAD	CHINA, PR	COLOMBIA	CONGO(BRA)	CUBA	CZECHOS'KIA	DAHOMEY	ETHIOPIA
FRANCE	GABON	GERMANY, E	GERMAN FR	GHANA	GREECE	GUINEA	HAITI	HONDURAS	HUNGARY
INDIA	IVORY COAST	JAMAICA	JAPAN	KOREA, N	LIBERIA	LIBYA	MALAGASY R	MALI	MAURITANIA
MEXICO	MONGOLIA	NEW ZEALAND	NIGER	PARAGUAY	PHILIPPINES	POLAND	PORTUGAL	RUMANIA	RWANDA
SA'U ARABIA	SENEGAL	SIERRE LEO	SOMALIA	SPAIN	TANGANYIKA	TRINIDAD	TUNISIA	USSR	UAR
UK	US	UPPER VOLTA	VIETNAM, N	VIETNAM REP	YUGOSLAVIA				

7 EXCLUDED BECAUSE AMBIGUOUS

INDONESIA	MALAYA	MOROCCO	NICARAGUA	SO AFRICA	TURKEY	URUGUAY

12 EXCLUDED BECAUSE UNASCERTAINABLE

BURMA	IRAN	IRAQ	JORDAN	KOREA REP	NEPAL	PAKISTAN	SUDAN	SYRIA	THAILAND
TOGO	YEMEN								

	Left column			Right column		
5	ALWAYS ARE THOSE LOCATED IN SCANDINAVIA OR WEST EUROPE, RATHER THAN IN EAST EUROPE (18)	1.00	5	TEND TO BE THOSE LOCATED IN EAST EUROPE, RATHER THAN IN SCANDINAVIA OR WEST EUROPE (9)	0.56	11 7 0 9 X SQ= 6.92 P = 0.003 RV YES (YES)
6	TEND LESS TO BE THOSE LOCATED ELSEWHERE THAN IN THE CARIBBEAN, CENTRAL AMERICA, OR SOUTH AMERICA (93)	0.63	6	TEND MORE TO BE THOSE LOCATED ELSEWHERE THAN IN THE CARIBBEAN, CENTRAL AMERICA, OR SOUTH AMERICA (93)	0.86	11 9 19 57 X SQ= 5.31 P = 0.015 RV NO (YES)
10	TEND MORE TO BE THOSE LOCATED ELSEWHERE THAN IN NORTH AFRICA, OR CENTRAL AND SOUTH AFRICA (82)	0.90	10	TEND LESS TO BE THOSE LOCATED ELSEWHERE THAN IN NORTH AFRICA, OR CENTRAL AND SOUTH AFRICA (82)	0.61	3 26 27 40 X SQ= 7.12 P = 0.004 RV NO (NO)

15	TEND TO BE THOSE LOCATED IN THE MIDDLE EAST, RATHER THAN IN NORTH AFRICA OR CENTRAL AND SOUTH AFRICA (11)	0.50	15	TEND TO BE THOSE LOCATED IN NORTH AFRICA OR CENTRAL AND SOUTH AFRICA, RATHER THAN IN THE MIDDLE EAST (33)	0.93	3 2 3 26 X SQ= 4.22 P = 0.029 RV YES (YES)
19	TEND TO BE THOSE LOCATED IN THE CARIBBEAN, CENTRAL AMERICA, OR SOUTH AMERICA, RATHER THAN IN NORTH AFRICA OR CENTRAL AND SOUTH AFRICA (22)	0.79	19	TEND TO BE THOSE LOCATED IN NORTH AFRICA OR CENTRAL AND SOUTH AFRICA, RATHER THAN IN THE CARIBBEAN, CENTRAL AMERICA, OR SOUTH AMERICA (33)	0.74	3 26 11 9 X SQ= 9.48 P = 0.001 RV YES (YES)
20	LEAN TOWARD BEING THOSE LOCATED IN THE CARIBBEAN, CENTRAL AMERICA, OR SOUTH AMERICA, RATHER THAN IN EAST ASIA, SOUTH ASIA, OR SOUTHEAST ASIA (22)	0.85	20	LEAN TOWARD BEING THOSE LOCATED IN EAST ASIA, SOUTH ASIA, OR SOUTHEAST ASIA, RATHER THAN IN THE CARIBBEAN, CENTRAL AMERICA, OR SOUTH AMERICA (18)	0.50	2 9 11 9 X SQ= 2.58 P = 0.066 RV YES (YES)
22	TILT TOWARD BEING THOSE WHOSE TERRITORIAL SIZE IS SMALL (47)	0.53	22	TILT TOWARD BEING THOSE WHOSE TERRITORIAL SIZE IS VERY LARGE, LARGE, OR MEDIUM (68)	0.65	14 43 16 23 X SQ= 2.21 P = 0.117 RV YES (NO)
29	TEND TO BE THOSE WHERE THE DEGREE OF URBANIZATION IS HIGH (56)	0.72	29	TEND TO BE THOSE WHERE THE DEGREE OF URBANIZATION IS LOW (49)	0.57	21 25 8 33 X SQ= 5.54 P = 0.012 RV YES (NO)
30	TEND TO BE THOSE WHOSE AGRICULTURAL POPULATION IS MEDIUM, LOW, OR VERY LOW (57)	0.87	30	TEND TO BE THOSE WHOSE AGRICULTURAL POPULATION IS HIGH (56)	0.59	4 38 26 26 X SQ= 15.70 P = 0. RV YES (YES)
36	TEND TO BE THOSE WHOSE PER CAPITA GROSS NATIONAL PRODUCT IS VERY HIGH, HIGH, OR MEDIUM (42)	0.57	36	TEND TO BE THOSE WHOSE PER CAPITA GROSS NATIONAL PRODUCT IS LOW OR VERY LOW (73)	0.67	17 22 13 44 X SQ= 3.74 P = 0.043 RV YES (NO)
37	TEND TO BE THOSE WHOSE PER CAPITA GROSS NATIONAL PRODUCT IS VERY HIGH, HIGH, MEDIUM, OR LOW (64)	0.83	37	TEND TO BE THOSE WHOSE PER CAPITA GROSS NATIONAL PRODUCT IS VERY LOW (51)	0.52	25 32 5 34 X SQ= 8.99 P = 0.002 RV YES (NO)
40	LEAN MORE TOWARD BEING THOSE WHOSE INTERNATIONAL FINANCIAL STATUS IS VERY HIGH, HIGH, MEDIUM, OR LOW (71)	0.77	40	LEAN LESS TOWARD BEING THOSE WHOSE INTERNATIONAL FINANCIAL STATUS IS VERY HIGH, HIGH, MEDIUM, OR LOW (71)	0.56	23 34 7 27 X SQ= 2.92 P = 0.066 RV NO (NO)

```
43  TEND TO BE THOSE                          0.68      43  TEND TO BE THOSE                          0.55        19   29
    WHOSE ECONOMIC DEVELOPMENTAL STATUS                     WHOSE ECONOMIC DEVELOPMENTAL STATUS                    9   36
    IS DEVELOPED, INTERMEDIATE, OR                          IS VERY UNDERDEVELOPED  (57)                      X SQ=   3.35
    UNDERDEVELOPED  (55)                                                                                      P =    0.045
                                                                                                              RV YES (NO )

45  TEND TO BE THOSE                          0.72      45  TEND TO BE THOSE                          0.52        21   29
    WHERE THE LITERACY RATE IS                              WHERE THE LITERACY RATE IS                             8   32
    FIFTY PERCENT OR ABOVE  (55)                            BELOW FIFTY PERCENT  (54)                         X SQ=   3.97
                                                                                                              P =    0.040
                                                                                                              RV YES (NO )

46  TEND MORE TO BE THOSE                     0.90      46  TEND LESS TO BE THOSE                     0.69        27   43
    WHERE THE LITERACY RATE IS                              WHERE THE LITERACY RATE IS                             3   19
    TEN PERCENT OR ABOVE  (84)                              TEN PERCENT OR ABOVE  (84)                        X SQ=   3.67
                                                                                                              P =    0.037
                                                                                                              RV NO  (NO )

50  LEAN TOWARD BEING THOSE                   0.67      50  LEAN TOWARD BEING THOSE                   0.58        18   23
    WHERE FREEDOM OF THE PRESS IS                           WHERE FREEDOM OF THE PRESS IS                          9   32
    COMPLETE  (43)                                          INTERMITTENT, INTERNALLY ABSENT, OR               X SQ=   3.53
                                                            INTERNALLY AND EXTERNALLY ABSENT  (56)            P =    0.059
                                                                                                              RV YES (NO )

51  TEND TO BE THOSE                          0.96      51  TEND TO BE THOSE                          0.51        26   27
    WHERE FREEDOM OF THE PRESS IS                           WHERE FREEDOM OF THE PRESS IS                          1   28
    COMPLETE OR INTERMITTENT  (60)                          INTERNALLY ABSENT, OR                             X SQ=  15.65
                                                            INTERNALLY AND EXTERNALLY ABSENT  (37)            P =    0.
                                                                                                              RV YES (NO )

52  ALWAYS ARE THOSE                          1.00      52  TEND LESS TO BE THOSE                     0.73        27   41
    WHERE FREEDOM OF THE PRESS IS                           WHERE FREEDOM OF THE PRESS IS                          0   15
    COMPLETE, INTERMITTENT, OR                              COMPLETE, INTERMITTENT, OR                        X SQ=   7.11
    INTERNALLY ABSENT  (82)                                 INTERNALLY ABSENT  (82)                           P =    0.002
                                                                                                              RV NO  (NO )

54  TEND TO BE THOSE                          0.57      54  TEND TO BE THOSE                          0.70        17   19
    WHERE NEWSPAPER CIRCULATION IS                          WHERE NEWSPAPER CIRCULATION IS                        13   45
    ONE HUNDRED OR MORE                                     LESS THAN ONE HUNDRED                             X SQ=   5.20
    PER THOUSAND  (37)                                      PER THOUSAND  (76)                                P =    0.022
                                                                                                              RV YES (NO )

57  TEND LESS TO BE THOSE                     0.60      57  TEND MORE TO BE THOSE                     0.83        12   11
    WHERE THE RELIGION IS OTHER THAN                        WHERE THE RELIGION IS OTHER THAN                      18   55
    CATHOLIC  (90)                                          CATHOLIC  (90)                                    X SQ=   4.95
                                                                                                              P =    0.020
                                                                                                              RV NO  (YES)

58  ALWAYS ARE THOSE                          1.00      58  LEAN LESS TOWARD BEING THOSE              0.86         0    9
    WHERE THE RELIGION IS OTHER THAN                        WHERE THE RELIGION IS OTHER THAN                      30   57
    MUSLIM  (97)                                            MUSLIM  (97)                                      X SQ=   3.05
                                                                                                              P =    0.053
                                                                                                              RV NO  (NO )
```

62	TILT LESS TOWARD BEING THOSE WHERE THE RELIGION IS CATHOLIC, RATHER THAN PROTESTANT (25)	0.71	62	ALWAYS ARE THOSE WHERE THE RELIGION IS CATHOLIC, RATHER THAN PROTESTANT (25)	1.00	5 0 12 11 X SQ= 2.19 P = 0.125 RV NO (NO)
63	TEND TO BE THOSE WHERE THE RELIGION IS PREDOMINANTLY SOME KIND OF CHRISTIAN (46)	0.63	63	TEND TO BE THOSE WHERE THE RELIGION IS PREDOMINANTLY OR PARTLY OTHER THAN CHRISTIAN (68)	0.62	19 25 11 40 X SQ= 4.16 P = 0.028 RV YES (NO)
72	LEAN TOWARD BEING THOSE WHOSE DATE OF INDEPENDENCE IS BEFORE 1914 (52)	0.63	72	LEAN TOWARD BEING THOSE WHOSE DATE OF INDEPENDENCE IS AFTER 1914 (59)	0.59	19 26 11 37 X SQ= 3.13 P = 0.075 RV YES (NO)
75	TEND TO BE THOSE THAT ARE HISTORICALLY WESTERN OR SIGNIFICANTLY WESTERNIZED (62)	0.87	75	TEND TO BE THOSE THAT ARE NOT HISTORICALLY WESTERN AND ARE NOT SIGNIFICANTLY WESTERNIZED (52)	0.50	26 33 4 33 X SQ= 10.21 P = 0.001 RV YES (NO)
77	TEND TO BE THOSE THAT HAVE BEEN SIGNIFICANTLY WESTERNIZED, RATHER THAN PARTIALLY WESTERNIZED, THROUGH A COLONIAL RELATIONSHIP (28)	0.78	77	TEND TO BE THOSE THAT HAVE BEEN PARTIALLY WESTERNIZED, RATHER THAN SIGNIFICANTLY WESTERNIZED, THROUGH A COLONIAL RELATIONSHIP (41)	0.68	14 12 4 26 X SQ= 8.71 P = 0.002 RV YES (YES)
78	ALWAYS ARE THOSE THAT HAVE BEEN SIGNIFICANTLY WESTERNIZED THROUGH A COLONIAL RELATIONSHIP, RATHER THAN WITHOUT SUCH A RELATIONSHIP (28)	1.00	78	TEND LESS TO BE THOSE THAT HAVE BEEN SIGNIFICANTLY WESTERNIZED THROUGH A COLONIAL RELATIONSHIP, RATHER THAN WITHOUT SUCH A RELATIONSHIP (28)	0.67	14 12 0 6 X SQ= 3.76 P = 0.024 RV NO (YES)
79	TEND TO BE THOSE THAT WERE FORMERLY DEPENDENCIES OF SPAIN, RATHER THAN BRITAIN OR FRANCE (18)	0.56	79	TEND TO BE THOSE THAT WERE FORMERLY DEPENDENCIES OF BRITAIN OR FRANCE, RATHER THAN SPAIN (49)	0.84	8 31 10 6 X SQ= 7.28 P = 0.004 RV YES (YES)
80	TEND TO BE THOSE WHOSE DATE OF INDEPENDENCE IS AFTER 1914, AND THAT WERE FORMERLY DEPENDENCIES OF BRITAIN, RATHER THAN FRANCE (19)	0.75	80	TEND TO BE THOSE WHOSE DATE OF INDEPENDENCE IS AFTER 1914, AND THAT WERE FORMERLY DEPENDENCIES OF FRANCE, RATHER THAN BRITAIN (24)	0.73	6 7 2 19 X SQ= 4.12 P = 0.033 RV YES (NO)
82	TEND TO BE THOSE WHERE THE TYPE OF POLITICAL MODERNIZATION IS EARLY OR LATER EUROPEAN OR EUROPEAN DERIVED, RATHER THAN DEVELOPED TUTELARY OR UNDEVELOPED TUTELARY (51)	0.73	82	TEND TO BE THOSE WHERE THE TYPE OF POLITICAL MODERNIZATION IS DEVELOPED TUTELARY OR UNDEVELOPED TUTELARY, RATHER THAN EARLY OR LATER EUROPEAN OR EUROPEAN DERIVED (55)	0.57	22 26 8 35 X SQ= 6.43 P = 0.007 RV YES (NO)

85	TILT TOWARD BEING THOSE WHERE THE STAGE OF POLITICAL MODERNIZATION IS ADVANCED, RATHER THAN MID- OR EARLY TRANSITIONAL (60)	0.67	85	TILT TOWARD BEING THOSE WHERE THE STAGE OF POLITICAL MODERNIZATION IS MID- OR EARLY TRANSITIONAL, RATHER THAN ADVANCED (54)	0.52	20 31 10 34 X SQ= 2.26 P = 0.121 RV YES (NO)
87	LEAN MORE TOWARD BEING THOSE WHOSE IDEOLOGICAL ORIENTATION IS OTHER THAN DEVELOPMENTAL (58)	0.82	87	LEAN LESS TOWARD BEING THOSE WHOSE IDEOLOGICAL ORIENTATION IS OTHER THAN DEVELOPMENTAL (58)	0.59	4 23 18 33 X SQ= 2.71 P = 0.068 RV NO (NO)
89	TEND TO BE THOSE WHOSE IDEOLOGICAL ORIENTATION IS CONVENTIONAL (33)	0.73	89	TEND TO BE THOSE WHOSE IDEOLOGICAL ORIENTATION IS OTHER THAN CONVENTIONAL (62)	0.75	16 14 6 43 X SQ= 13.66 P = 0. RV YES (YES)
92	ALWAYS ARE THOSE WHERE THE SYSTEM STYLE IS LIMITED MOBILIZATIONAL, OR NON-MOBILIZATIONAL (93)	1.00	92	TEND LESS TO BE THOSE WHERE THE SYSTEM STYLE IS LIMITED MOBILIZATIONAL, OR NON-MOBILIZATIONAL (93)	0.71	0 19 30 46 X SQ= 9.21 P = 0.001 RV NO (NO)
94	TEND TO BE THOSE WHERE THE STATUS OF THE REGIME IS CONSTITUTIONAL (51)	0.92	94	TEND TO BE THOSE WHERE THE STATUS OF THE REGIME IS AUTHORITARIAN OR TOTALITARIAN (41)	0.56	24 23 2 29 X SQ= 14.78 P = 0. RV YES (YES)
95	ALWAYS ARE THOSE WHERE THE STATUS OF THE REGIME IS CONSTITUTIONAL OR AUTHORITARIAN (95)	1.00	95	TEND LESS TO BE THOSE WHERE THE STATUS OF THE REGIME IS CONSTITUTIONAL OR AUTHORITARIAN (95)	0.75	30 48 0 16 X SQ= 7.36 P = 0.002 RV NO (NO)
99	TEND TO BE THOSE WHERE GOVERNMENTAL STABILITY IS MODERATELY PRESENT AND DATES FROM THE POST-WAR PERIOD, OR IS ABSENT (36)	0.54	99	TEND TO BE THOSE WHERE GOVERNMENTAL STABILITY IS GENERALLY PRESENT AND DATES FROM AT LEAST THE INTER-WAR PERIOD, OR FROM THE POST-WAR PERIOD (50)	0.81	12 34 14 8 X SQ= 7.37 P = 0.007 RV YES (YES)
101	TEND TO BE THOSE WHERE THE REPRESENTATIVE CHARACTER OF THE REGIME IS POLYARCHIC (41)	0.78	101	TEND TO BE THOSE WHERE THE REPRESENTATIVE CHARACTER OF THE REGIME IS LIMITED POLYARCHIC, PSEUDO-POLYARCHIC, OR NON-POLYARCHIC (57)	0.69	18 19 5 42 X SQ= 13.19 P = 0. RV YES (NO)
102	TEND TO BE THOSE WHERE THE REPRESENTATIVE CHARACTER OF THE REGIME IS POLYARCHIC OR LIMITED POLYARCHIC (59)	0.93	102	TEND TO BE THOSE WHERE THE REPRESENTATIVE CHARACTER OF THE REGIME IS PSEUDO-POLYARCHIC OR NON-POLYARCHIC (49)	0.58	26 28 2 38 X SQ= 18.44 P = 0. RV YES (NO)

105	TEND TO BE THOSE WHERE THE ELECTORAL SYSTEM IS COMPETITIVE (43)	0.92	105	TEND TO BE THOSE WHERE THE ELECTORAL SYSTEM IS PARTIALLY COMPETITIVE OR NON-COMPETITIVE (47)	0.70	22 18 2 42 X SQ= 23.72 P = 0. RV YES (YES)
106	TEND TO BE THOSE WHERE THE ELECTORAL SYSTEM IS COMPETITIVE OR PARTIALLY COMPETITIVE (52)	0.95	106	TEND TO BE THOSE WHERE THE ELECTORAL SYSTEM IS NON-COMPETITIVE (30)	0.52	21 27 1 29 X SQ= 12.96 P = 0. RV YES (NO)
107	TEND TO BE THOSE WHERE AUTONOMOUS GROUPS ARE FULLY TOLERATED IN POLITICS (46)	0.82	107	TEND TO BE THOSE WHERE AUTONOMOUS GROUPS ARE PARTIALLY TOLERATED IN POLITICS, ARE TOLERATED ONLY OUTSIDE POLITICS, OR ARE NOT TOLERATED AT ALL (65)	0.70	23 20 5 46 X SQ= 19.25 P = 0. RV YES (YES)
111	LEAN TOWARD BEING THOSE WHERE POLITICAL ENCULTURATION IS LOW (42)	0.54	111	LEAN TOWARD BEING THOSE WHERE POLITICAL ENCULTURATION IS HIGH OR MEDIUM (53)	0.69	13 34 15 15 X SQ= 3.04 P = 0.056 RV YES (YES)
116	TEND TO BE THOSE WHERE INTEREST ARTICULATION BY ASSOCIATIONAL GROUPS IS SIGNIFICANT OR MODERATE (32)	0.57	116	TEND TO BE THOSE WHERE INTEREST ARTICULATION BY ASSOCIATIONAL GROUPS IS LIMITED OR NEGLIGIBLE (79)	0.81	17 12 13 51 X SQ= 11.71 P = 0.001 RV YES (YES)
117	TEND TO BE THOSE WHERE INTEREST ARTICULATION BY ASSOCIATIONAL GROUPS IS SIGNIFICANT, MODERATE, OR LIMITED (60)	0.83	117	TEND TO BE THOSE WHERE INTEREST ARTICULATION BY ASSOCIATIONAL GROUPS IS NEGLIGIBLE (51)	0.60	25 25 5 38 X SQ= 13.87 P = 0. RV YES (YES)
118	TEND MORE TO BE THOSE WHERE INTEREST ARTICULATION BY INSTITUTIONAL GROUPS IS SIGNIFICANT, MODERATE, OR LIMITED (60)	0.80	118	TEND LESS TO BE THOSE WHERE INTEREST ARTICULATION BY INSTITUTIONAL GROUPS IS SIGNIFICANT, MODERATE, OR LIMITED (60)	0.55	6 24 24 29 X SQ= 4.27 P = 0.032 RV NO (NO)
122	TEND TO BE THOSE WHERE INTEREST ARTICULATION BY NON-ASSOCIATIONAL GROUPS IS LIMITED OR NEGLIGIBLE (32)	0.50	122	TEND TO BE THOSE WHERE INTEREST ARTICULATION BY NON-ASSOCIATIONAL GROUPS IS SIGNIFICANT OR MODERATE (83)	0.76	15 50 15 16 X SQ= 5.14 P = 0.018 RV YES (NO)
126	TEND LESS TO BE THOSE WHERE INTEREST ARTICULATION BY ANOMIC GROUPS IS FREQUENT, OCCASIONAL, OR INFREQUENT (83)	0.60	126	TEND MORE TO BE THOSE WHERE INTEREST ARTICULATION BY ANOMIC GROUPS IS FREQUENT, OCCASIONAL, OR INFREQUENT (83)	0.89	15 50 10 6 X SQ= 7.60 P = 0.005 RV NO (YES)

127	TEND TO BE THOSE WHERE INTEREST ARTICULATION BY POLITICAL PARTIES IS SIGNIFICANT (21)	0.70	127	TEND TO BE THOSE WHERE INTEREST ARTICULATION BY POLITICAL PARTIES IS MODERATE, LIMITED, OR NEGLIGIBLE (72)	0.98	19 1 8 59 X SQ= 45.84 P = 0. RV YES (YES)
128	TEND TO BE THOSE WHERE INTEREST ARTICULATION BY POLITICAL PARTIES IS SIGNIFICANT OR MODERATE (48)	0.96	128	TEND TO BE THOSE WHERE INTEREST ARTICULATION BY POLITICAL PARTIES IS LIMITED OR NEGLIGIBLE (45)	0.68	26 19 1 41 X SQ= 28.61 P = 0. RV YES (YES)
129	ALWAYS ARE THOSE WHERE INTEREST ARTICULATION BY POLITICAL PARTIES IS SIGNIFICANT, MODERATE, OR LIMITED (56)	1.00	129	TEND TO BE THOSE WHERE INTEREST ARTICULATION BY POLITICAL PARTIES IS NEGLIGIBLE (37)	0.62	27 23 0 37 X SQ= 26.50 P = 0. RV YES (YES)
130	ALWAYS ARE THOSE WHERE INTEREST AGGREGATION BY POLITICAL PARTIES IS MODERATE, LIMITED, OR NEGLIGIBLE (71)	1.00	130	TEND LESS TO BE THOSE WHERE INTEREST AGGREGATION BY POLITICAL PARTIES IS MODERATE, LIMITED, OR NEGLIGIBLE (71)	0.74	0 11 28 32 X SQ= 6.63 P = 0.002 RV NO (NO)
131	TEND TO BE THOSE WHERE INTEREST AGGREGATION BY POLITICAL PARTIES IS LIMITED OR NEGLIGIBLE (35)	0.85	131	TEND TO BE THOSE WHERE INTEREST AGGREGATION BY POLITICAL PARTIES IS SIGNIFICANT OR MODERATE (30)	0.79	4 22 22 6 X SQ= 19.10 P = 0. RV YES (YES)
133	TEND MORE TO BE THOSE WHERE INTEREST AGGREGATION BY THE EXECUTIVE IS MODERATE, LIMITED, OR NEGLIGIBLE (76)	0.96	133	TEND LESS TO BE THOSE WHERE INTEREST AGGREGATION BY THE EXECUTIVE IS MODERATE, LIMITED, OR NEGLIGIBLE (76)	0.62	1 25 25 40 X SQ= 9.27 P = 0.001 RV NO (NO)
137	TEND TO BE THOSE WHERE INTEREST AGGREGATION BY THE LEGISLATURE IS SIGNIFICANT OR MODERATE (29)	0.54	137	TEND TO BE THOSE WHERE INTEREST AGGREGATION BY THE LEGISLATURE IS LIMITED OR NEGLIGIBLE (68)	0.79	14 13 12 49 X SQ= 7.83 P = 0.005 RV YES (YES)
138	TEND TO BE THOSE WHERE INTEREST AGGREGATION BY THE LEGISLATURE IS SIGNIFICANT, MODERATE, OR LIMITED (48)	0.86	138	TEND TO BE THOSE WHERE INTEREST AGGREGATION BY THE LEGISLATURE IS NEGLIGIBLE (49)	0.66	24 21 4 40 X SQ= 18.20 P = 0. RV YES (YES)
146	ALWAYS ARE THOSE WHERE THE PARTY SYSTEM IS QUALITATIVELY OTHER THAN MASS-BASED TERRITORIAL (92)	1.00	146	TEND LESS TO BE THOSE WHERE THE PARTY SYSTEM IS QUALITATIVELY OTHER THAN MASS-BASED TERRITORIAL (92)	0.85	0 8 30 47 X SQ= 3.26 P = 0.046 RV NO (NO)

147	TEND TO BE THOSE WHERE THE PARTY SYSTEM IS QUALITATIVELY CLASS-ORIENTED OR MULTI-IDEOLOGICAL (23)	0.56	147	TEND TO BE THOSE WHERE THE PARTY SYSTEM IS QUALITATIVELY OTHER THAN CLASS-ORIENTED OR MULTI-IDEOLOGICAL (67)	0.86	15 8 12 49 X SQ= 13.87 P = 0. RV YES (YES)
148	ALWAYS ARE THOSE WHERE THE PARTY SYSTEM IS QUALITATIVELY OTHER THAN AFRICAN TRANSITIONAL (96)	1.00	148	TEND LESS TO BE THOSE WHERE THE PARTY SYSTEM IS QUALITATIVELY OTHER THAN AFRICAN TRANSITIONAL (96)	0.79	0 13 30 48 X SQ= 5.82 P = 0.004 RV NO (NO)
153	TEND TO BE THOSE WHERE THE PARTY SYSTEM IS MODERATELY STABLE OR UNSTABLE (38)	0.58	153	TEND TO BE THOSE WHERE THE PARTY SYSTEM IS STABLE (42)	0.70	11 30 15 13 X SQ= 3.99 P = 0.042 RV YES (YES)
154	TEND TO BE THOSE WHERE THE PARTY SYSTEM IS UNSTABLE (25)	0.50	154	TEND TO BE THOSE WHERE THE PARTY SYSTEM IS STABLE OR MODERATELY STABLE (55)	0.88	13 38 13 5 X SQ= 10.46 P = 0.001 RV YES (YES)
157	LEAN LESS TOWARD BEING THOSE WHERE PERSONALISMO IS MODERATE OR NEGLIGIBLE (84)	0.79	157	LEAN MORE TOWARD BEING THOSE WHERE PERSONALISMO IS MODERATE OR NEGLIGIBLE (84)	0.93	6 4 22 57 X SQ= 2.89 P = 0.066 RV NO (YES)
161	TILT TOWARD BEING THOSE WHERE THE POLITICAL LEADERSHIP IS NON-ELITIST (50)	0.65	161	TILT TOWARD BEING THOSE WHERE THE POLITICAL LEADERSHIP IS ELITIST OR MODERATELY ELITIST (47)	0.55	9 33 17 27 X SQ= 2.26 P = 0.103 RV YES (NO)
164	ALWAYS ARE THOSE WHERE THE REGIME'S LEADERSHIP CHARISMA IS NEGLIGIBLE (65)	1.00	164	TEND LESS TO BE THOSE WHERE THE REGIME'S LEADERSHIP CHARISMA IS NEGLIGIBLE (65)	0.52	0 30 25 32 X SQ= 16.38 P = 0. RV NO (NO)
168	TEND TO BE THOSE WHERE THE HORIZONTAL POWER DISTRIBUTION IS SIGNIFICANT (34)	0.65	168	TEND TO BE THOSE WHERE THE HORIZONTAL POWER DISTRIBUTION IS LIMITED OR NEGLIGIBLE (72)	0.80	17 13 9 53 X SQ= 15.70 P = 0. RV YES (YES)
169	TEND TO BE THOSE WHERE THE HORIZONTAL POWER DISTRIBUTION IS SIGNIFICANT OR LIMITED (58)	0.96	169	TEND TO BE THOSE WHERE THE HORIZONTAL POWER DISTRIBUTION IS NEGLIGIBLE (48)	0.59	25 27 1 39 X SQ= 20.97 P = 0. RV YES (NO)

171	LEAN LESS TOWARD BEING THOSE WHERE THE LEGISLATIVE-EXECUTIVE STRUCTURE IS OTHER THAN PARLIAMENTARY-REPUBLICAN (90)	0.79	171	LEAN MORE TOWARD BEING THOSE WHERE THE LEGISLATIVE-EXECUTIVE STRUCTURE IS OTHER THAN PARLIAMENTARY-REPUBLICAN (90)	0.93	6 4 23 56 X SQ= 2.58 P = 0.072 RV NO (YES)
174	TEND TO BE THOSE WHERE THE LEGISLATURE IS FULLY EFFECTIVE (28)	0.50	174	TEND TO BE THOSE WHERE THE LEGISLATURE IS PARTIALLY EFFECTIVE, LARGELY INEFFECTIVE, OR WHOLLY INEFFECTIVE (72)	0.81	13 12 13 52 X SQ= 7.51 P = 0.004 RV YES (YES)
175	TEND TO BE THOSE WHERE THE LEGISLATURE IS FULLY EFFECTIVE OR PARTIALLY EFFECTIVE (51)	0.92	175	TEND TO BE THOSE WHERE THE LEGISLATURE IS LARGELY INEFFECTIVE OR WHOLLY INEFFECTIVE (49)	0.66	24 22 2 42 X SQ= 22.57 P = 0. RV YES (YES)
178	TEND TO BE THOSE WHERE THE LEGISLATURE IS BICAMERAL (51)	0.66	178	TEND TO BE THOSE WHERE THE LEGISLATURE IS UNICAMERAL (53)	0.62	10 39 19 24 X SQ= 4.95 P = 0.024 RV YES (NO)
179	TEND TO BE THOSE WHERE THE EXECUTIVE IS STRONG (39)	0.95	179	TEND TO BE THOSE WHERE THE EXECUTIVE IS DOMINANT (52)	0.72	1 43 20 17 X SQ= 25.43 P = 0. RV YES (YES)
181	TEND LESS TO BE THOSE WHERE THE BUREAUCRACY IS SEMI-MODERN, RATHER THAN MODERN (55)	0.54	181	TEND MORE TO BE THOSE WHERE THE BUREAUCRACY IS SEMI-MODERN, RATHER THAN MODERN (55)	0.81	12 8 14 34 X SQ= 4.45 P = 0.028 RV NO (YES)
185	TEND TO BE THOSE WHERE PARTICIPATION BY THE MILITARY IN POLITICS IS INTERVENTIVE, RATHER THAN SUPPORTIVE (21)	0.75	185	TEND TO BE THOSE WHERE PARTICIPATION BY THE MILITARY IN POLITICS IS SUPPORTIVE, RATHER THAN INTERVENTIVE (31)	0.86	6 4 2 25 X SQ= 9.01 P = 0.002 RV YES (YES)
186	TEND MORE TO BE THOSE WHERE PARTICIPATION BY THE MILITARY IN POLITICS IS NEUTRAL, RATHER THAN SUPPORTIVE (56)	0.90	186	TEND LESS TO BE THOSE WHERE PARTICIPATION BY THE MILITARY IN POLITICS IS NEUTRAL, RATHER THAN SUPPORTIVE (56)	0.58	2 25 19 34 X SQ= 6.08 P = 0.007 RV NO (NO)
187	TEND TO BE THOSE WHERE THE ROLE OF THE POLICE IS NOT POLITICALLY SIGNIFICANT (35)	0.61	187	TEND TO BE THOSE WHERE THE ROLE OF THE POLICE IS POLITICALLY SIGNIFICANT (66)	0.68	9 40 14 19 X SQ= 4.53 P = 0.024 RV YES (NO)

```
188  TEND TO BE THOSE                           0.53     188  TEND TO BE THOSE                           0.81        16    12
     WHERE THE CHARACTER OF THE LEGAL SYSTEM                  WHERE THE CHARACTER OF THE LEGAL SYSTEM                14    52
     IS CIVIL LAW  (32)                                       IS OTHER THAN CIVIL LAW  (81)                     X SQ=  10.09
                                                                                                                P =     0.001
                                                                                                                RV YES (YES)

190  LEAN MORE TOWARD BEING THOSE               0.94     190  LEAN LESS TOWARD BEING THOSE               0.67        16    12
     WHERE THE CHARACTER OF THE LEGAL SYSTEM                  WHERE THE CHARACTER OF THE LEGAL SYSTEM                 1     6
     IS CIVIL LAW, RATHER THAN                                IS CIVIL LAW, RATHER THAN                         X SQ=   2.58
     COMMON LAW  (32)                                         COMMON LAW  (32)                                  P =     0.088
                                                                                                                RV NO  (YES)

192  TEND MORE TO BE THOSE                      0.93     192  TEND LESS TO BE THOSE                      0.75         2    16
     WHERE THE CHARACTER OF THE LEGAL SYSTEM                  WHERE THE CHARACTER OF THE LEGAL SYSTEM                28    48
     IS OTHER THAN                                            IS OTHER THAN                                     X SQ=   3.33
     MUSLIM OR PARTLY MUSLIM  (86)                            MUSLIM OR PARTLY MUSLIM  (86)                     P =     0.048
                                                                                                                RV NO  (NO )

193  LEAN MORE TOWARD BEING THOSE               0.87     193  LEAN LESS TOWARD BEING THOSE               0.68         4    21
     WHERE THE CHARACTER OF THE LEGAL SYSTEM                  WHERE THE CHARACTER OF THE LEGAL SYSTEM                26    45
     IS OTHER THAN PARTLY INDIGENOUS  (86)                    IS OTHER THAN PARTLY INDIGENOUS  (86)             X SQ=   2.76
                                                                                                                P =     0.079
                                                                                                                RV NO  (NO )

194  ALWAYS ARE THOSE                           1.00     194  TEND LESS TO BE THOSE                      0.80         0    13
     THAT ARE NON-COMMUNIST  (101)                            THAT ARE NON-COMMUNIST  (101)                          30    52
                                                                                                                X SQ=   5.36
                                                                                                                P =     0.008
                                                                                                                RV NO  (NO )
```

143	POLITIES WHERE THE PARTY SYSTEM IS QUANTITATIVELY NO-PARTY (5)	143	POLITIES WHERE THE PARTY SYSTEM IS QUANTITATIVELY OTHER THAN NO-PARTY (100)

SUBJECT ONLY

5 IN LEFT COLUMN

AFGHANISTAN ETHIOPIA HAITI LIBYA SA'U ARABIA

100 IN RIGHT COLUMN

ALBANIA	ALGERIA	ARGENTINA	AUSTRALIA	AUSTRIA	BELGIUM	BOLIVIA	BRAZIL	BULGARIA	BURUNDI
CAMBODIA	CAMEROUN	CANADA	CEN AFR REP	CEYLON	CHAD	CHILE	CHINA, PR	COLOMBIA	CONGO(BRA)
CONGO(LEO)	COSTA RICA	CUBA	CYPRUS	CZECHOS'KIA	DAHOMEY	DENMARK	DOMIN REP	ECUADOR	EL SALVADOR
FINLAND	FRANCE	GABON	GERMANY, E	GERMAN FR	GHANA	GREECE	GUATEMALA	GUINEA	HONDURAS
HUNGARY	ICELAND	INDIA	IRAQ	IRELAND	ISRAEL	ITALY	IVORY COAST	JAMAICA	JAPAN
KOREA, N	LAOS	LEBANON	LIBERIA	LUXEMBOURG	MALAGASY R	MALAYA	MALI	MAURITANIA	MEXICO
MONGOLIA	MOROCCO	NETHERLANDS	NEW ZEALAND	NICARAGUA	NIGER	NIGERIA	NORWAY	PANAMA	PARAGUAY
PERU	PHILIPPINES	POLAND	PORTUGAL	RUMANIA	RWANDA	SENEGAL	SIERRE LEO	SOMALIA	SO AFRICA
SPAIN	SWEDEN	SWITZERLAND	SYRIA	TANGANYIKA	TOGO	TRINIDAD	TUNISIA	TURKEY	UGANDA
USSR	UAR	UK	US	UPPER VOLTA	URUGUAY	VENEZUELA	VIETNAM, N	VIETNAM REP	YUGOSLAVIA

1 EXCLUDED BECAUSE AMBIGUOUS

INDONESIA

9 EXCLUDED BECAUSE UNASCERTAINABLE

BURMA IRAN JORDAN KOREA REP NEPAL PAKISTAN SUDAN THAILAND YEMEN

13	TEND TO BE THOSE LOCATED IN NORTH AFRICA OR THE MIDDLE EAST (16)	0.60	13	TEND TO BE THOSE LOCATED ELSEWHERE THAN IN NORTH AFRICA OR THE MIDDLE EAST (99)	0.90	3 10 2 90 X SQ= 6.85 P = 0.013 RV YES (NO)
16	ALWAYS ARE THOSE LOCATED IN THE MIDDLE EAST, RATHER THAN IN EAST ASIA, SOUTH ASIA, OR SOUTHEAST ASIA (11)	1.00	16	TILT TOWARD BEING THOSE LOCATED IN EAST ASIA, SOUTH ASIA, OR SOUTHEAST ASIA, RATHER THAN IN THE MIDDLE EAST (18)	0.67	2 6 0 12 X SQ= 1.13 P = 0.147 RV YES (NO)
21	TILT TOWARD BEING THOSE WHOSE TERRITORIAL SIZE IS VERY LARGE OR LARGE (32)	0.60	21	TILT TOWARD BEING THOSE WHOSE TERRITORIAL SIZE IS MEDIUM OR SMALL (83)	0.75	3 25 2 75 X SQ= 1.46 P = 0.117 RV YES (NO)

30	ALWAYS ARE THOSE WHOSE AGRICULTURAL POPULATION IS HIGH (56)	1.00	30	TEND TO BE THOSE WHOSE AGRICULTURAL POPULATION IS MEDIUM, LOW, OR VERY LOW (57)	0.56	4 44 0 55 X SQ= 2.80 P = 0.044 RV YES (NO)
36	ALWAYS ARE THOSE WHOSE PER CAPITA GROSS NATIONAL PRODUCT IS LOW OR VERY LOW (73)	1.00	36	LEAN LESS TOWARD BEING THOSE WHOSE PER CAPITA GROSS NATIONAL PRODUCT IS LOW OR VERY LOW (73)	0.58	0 42 5 58 X SQ= 1.97 P = 0.082 RV NO (NO)
37	LEAN TOWARD BEING THOSE WHOSE PER CAPITA GROSS NATIONAL PRODUCT IS VERY LOW (51)	0.80	37	LEAN TOWARD BEING THOSE WHOSE PER CAPITA GROSS NATIONAL PRODUCT IS VERY HIGH, HIGH, MEDIUM, OR LOW (64)	0.63	1 63 4 37 X SQ= 2.11 P = 0.075 RV YES (NO)
45	ALWAYS ARE THOSE WHERE THE LITERACY RATE IS BELOW FIFTY PERCENT (54)	1.00	45	TEND TO BE THOSE WHERE THE LITERACY RATE IS FIFTY PERCENT OR ABOVE (55)	0.54	0 51 5 43 X SQ= 3.63 P = 0.024 RV YES (NO)
46	LEAN TOWARD BEING THOSE WHERE THE LITERACY RATE IS BELOW TEN PERCENT (26)	0.60	46	LEAN TOWARD BEING THOSE WHERE THE LITERACY RATE IS TEN PERCENT OR ABOVE (84)	0.79	2 75 3 20 X SQ= 2.17 P = 0.078 RV YES (NO)
50	ALWAYS ARE THOSE WHERE FREEDOM OF THE PRESS IS INTERMITTENT, INTERNALLY ABSENT, OR INTERNALLY AND EXTERNALLY ABSENT (56)	1.00	50	LEAN TOWARD BEING THOSE WHERE FREEDOM OF THE PRESS IS COMPLETE (43)	0.51	0 43 5 41 X SQ= 3.11 P = 0.056 RV YES (NO)
51	TEND TO BE THOSE WHERE FREEDOM OF THE PRESS IS INTERNALLY ABSENT, OR INTERNALLY AND EXTERNALLY ABSENT (37)	0.80	51	TEND TO BE THOSE WHERE FREEDOM OF THE PRESS IS COMPLETE OR INTERMITTENT (60)	0.70	1 58 4 25 X SQ= 3.29 P = 0.039 RV YES (NO)
55	TEND TO BE THOSE WHERE NEWSPAPER CIRCULATION IS LESS THAN TEN PER THOUSAND (35)	0.80	55	TEND TO BE THOSE WHERE NEWSPAPER CIRCULATION IS TEN OR MORE PER THOUSAND (78)	0.73	1 72 4 26 X SQ= 4.25 P = 0.024 RV YES (NO)
58	TEND TO BE THOSE WHERE THE RELIGION IS MUSLIM (18)	0.60	58	TEND TO BE THOSE WHERE THE RELIGION IS OTHER THAN MUSLIM (97)	0.90	3 10 2 90 X SQ= 6.85 P = 0.013 RV YES (NO)

No.	Statement		No.	Statement		
63	ALWAYS ARE THOSE WHERE THE RELIGION IS PREDOMINANTLY OR PARTLY OTHER THAN CHRISTIAN (68)	1.00	63	LEAN LESS TOWARD BEING THOSE WHERE THE RELIGION IS PREDOMINANTLY OR PARTLY OTHER THAN CHRISTIAN (68)	0.54	0 46 5 53 X SQ= 2.50 P = 0.065 RV NO (NO)
67	LEAN TOWARD BEING THOSE THAT ARE RACIALLY HETEROGENEOUS (27)	0.75	67	LEAN TOWARD BEING THOSE THAT ARE RACIALLY HOMOGENEOUS (82)	0.74	1 72 3 25 X SQ= 2.51 P = 0.064 RV YES (NO)
75	ALWAYS ARE THOSE THAT ARE NOT HISTORICALLY WESTERN AND ARE NOT SIGNIFICANTLY WESTERNIZED (52)	1.00	75	TEND TO BE THOSE THAT ARE HISTORICALLY WESTERN OR SIGNIFICANTLY WESTERNIZED (62)	0.63	0 62 5 37 X SQ= 5.37 P = 0.009 RV YES (NO)
85	ALWAYS ARE THOSE WHERE THE STAGE OF POLITICAL MODERNIZATION IS MID- OR EARLY TRANSITIONAL, RATHER THAN ADVANCED (54)	1.00	85	TEND TO BE THOSE WHERE THE STAGE OF POLITICAL MODERNIZATION IS ADVANCED, RATHER THAN MID- OR EARLY TRANSITIONAL (60)	0.56	0 56 4 44 X SQ= 2.86 P = 0.042 RV YES (NO)
91	ALWAYS ARE THOSE WHOSE IDEOLOGICAL ORIENTATION IS TRADITIONAL, RATHER THAN DEVELOPMENTAL (5)	1.00	91	ALWAYS ARE THOSE WHOSE IDEOLOGICAL ORIENTATION IS DEVELOPMENTAL, RATHER THAN TRADITIONAL (31)	1.00	0 29 3 0 X SQ= 21.31 P = 0. RV YES (YES)
94	LEAN TOWARD BEING THOSE WHERE THE STATUS OF THE REGIME IS AUTHORITARIAN OR TOTALITARIAN (41)	0.80	94	LEAN TOWARD BEING THOSE WHERE THE STATUS OF THE REGIME IS CONSTITUTIONAL (51)	0.64	1 50 4 28 X SQ= 2.22 P = 0.070 RV YES (NO)
96	TEND TO BE THOSE WHERE THE STATUS OF THE REGIME IS AUTHORITARIAN (23)	0.75	96	TEND TO BE THOSE WHERE THE STATUS OF THE REGIME IS CONSTITUTIONAL OR TOTALITARIAN (67)	0.86	1 66 3 11 X SQ= 6.02 P = 0.015 RV YES (NO)
97	ALWAYS ARE THOSE WHERE THE STATUS OF THE REGIME IS AUTHORITARIAN, RATHER THAN TOTALITARIAN (23)	1.00	97	LEAN TOWARD BEING THOSE WHERE THE STATUS OF THE REGIME IS TOTALITARIAN, RATHER THAN AUTHORITARIAN (16)	0.59	3 11 0 16 X SQ= 1.80 P = 0.090 RV YES (NO)
101	ALWAYS ARE THOSE WHERE THE REPRESENTATIVE CHARACTER OF THE REGIME IS LIMITED POLYARCHIC, PSEUDO-POLYARCHIC, OR NON-POLYARCHIC (57)	1.00	101	TILT LESS TOWARD BEING THOSE WHERE THE REPRESENTATIVE CHARACTER OF THE REGIME IS LIMITED POLYARCHIC, PSEUDO-POLYARCHIC, OR NON-POLYARCHIC (57)	0.52	0 41 4 45 X SQ= 1.84 P = 0.123 RV NO (NO)

105	ALWAYS ARE THOSE WHERE THE ELECTORAL SYSTEM IS PARTIALLY COMPETITIVE OR NON-COMPETITIVE (47)	1.00	105	TILT TOWARD BEING THOSE WHERE THE ELECTORAL SYSTEM IS COMPETITIVE (43)	0.51	0 43 4 41 X SQ= 2.22 P = 0.117 RV YES (NO)
107	ALWAYS ARE THOSE WHERE AUTONOMOUS GROUPS ARE PARTIALLY TOLERATED IN POLITICS, ARE TOLERATED ONLY OUTSIDE POLITICS, OR ARE NOT TOLERATED AT ALL (65)	1.00	107	LEAN LESS TOWARD BEING THOSE WHERE AUTONOMOUS GROUPS ARE PARTIALLY TOLERATED IN POLITICS, ARE TOLERATED ONLY OUTSIDE POLITICS, OR ARE NOT TOLERATED AT ALL (65)	0.53	0 46 5 51 X SQ= 2.62 P = 0.062 RV NO (NO)
108	TEND TO BE THOSE WHERE AUTONOMOUS GROUPS ARE TOLERATED ONLY OUTSIDE POLITICS OR ARE NOT TOLERATED AT ALL (35)	0.80	108	TEND TO BE THOSE WHERE AUTONOMOUS GROUPS ARE FULLY OR PARTIALLY TOLERATED IN POLITICS (65)	0.69	1 62 4 28 X SQ= 3.12 P = 0.043 RV YES (NO)
117	ALWAYS ARE THOSE WHERE INTEREST ARTICULATION BY ASSOCIATIONAL GROUPS IS NEGLIGIBLE (51)	1.00	117	TEND TO BE THOSE WHERE INTEREST ARTICULATION BY ASSOCIATIONAL GROUPS IS SIGNIFICANT, MODERATE, OR LIMITED (60)	0.59	0 57 5 40 X SQ= 4.49 P = 0.015 RV YES (NO)
134	ALWAYS ARE THOSE WHERE INTEREST AGGREGATION BY THE EXECUTIVE IS LIMITED OR NEGLIGIBLE (46)	1.00	134	TEND TO BE THOSE WHERE INTEREST AGGREGATION BY THE EXECUTIVE IS SIGNIFICANT OR MODERATE (57)	0.62	0 56 5 34 X SQ= 5.23 P = 0.010 RV YES (NO)
135	TEND TO BE THOSE WHERE INTEREST AGGREGATION BY THE EXECUTIVE IS NEGLIGIBLE (13)	0.80	135	TEND TO BE THOSE WHERE INTEREST AGGREGATION BY THE EXECUTIVE IS SIGNIFICANT, MODERATE, OR LIMITED (77)	0.91	1 72 4 7 X SQ= 15.13 P = 0.001 RV YES (NO)
160	LEAN TOWARD BEING THOSE WHERE THE POLITICAL LEADERSHIP IS ELITIST (30)	0.75	160	LEAN TOWARD BEING THOSE WHERE THE POLITICAL LEADERSHIP IS MODERATELY ELITIST OR NON-ELITIST (67)	0.73	3 24 1 65 X SQ= 2.27 P = 0.072 RV YES (NO)
161	ALWAYS ARE THOSE WHERE THE POLITICAL LEADERSHIP IS ELITIST OR MODERATELY ELITIST (47)	1.00	161	TEND TO BE THOSE WHERE THE POLITICAL LEADERSHIP IS NON-ELITIST (50)	0.55	4 40 0 49 X SQ= 2.71 P = 0.046 RV YES (NO)
164	TILT TOWARD BEING THOSE WHERE THE REGIME'S LEADERSHIP CHARISMA IS PRONOUNCED OR MODERATE (34)	0.75	164	TILT TOWARD BEING THOSE WHERE THE REGIME'S LEADERSHIP CHARISMA IS NEGLIGIBLE (65)	0.69	3 28 1 61 X SQ= 1.60 P = 0.106 RV YES (NO)

176	LEAN TOWARD BEING THOSE WHERE THE LEGISLATURE IS WHOLLY INEFFECTIVE (28)	0.75	176	LEAN TOWARD BEING THOSE WHERE THE LEGISLATURE IS FULLY EFFECTIVE, PARTIALLY EFFECTIVE, OR LARGELY INEFFECTIVE (72)	0.76	1 70 3 22 X SQ= 2.88 P = 0.053 RV YES (NO)
183	ALWAYS ARE THOSE WHERE THE BUREAUCRACY IS TRADITIONAL, RATHER THAN POST-COLONIAL TRANSITIONAL (9)	1.00	183	TEND TO BE THOSE WHERE THE BUREAUCRACY IS POST-COLONIAL TRANSITIONAL, RATHER THAN TRADITIONAL (25)	0.92	0 23 3 2 X SQ= 9.82 P = 0.003 RV YES (YES)
186	TILT TOWARD BEING THOSE WHERE PARTICIPATION BY THE MILITARY IN POLITICS IS SUPPORTIVE, RATHER THAN NEUTRAL (31)	0.75	186	TILT TOWARD BEING THOSE WHERE PARTICIPATION BY THE MILITARY IN POLITICS IS NEUTRAL, RATHER THAN SUPPORTIVE (56)	0.69	3 25 1 55 X SQ= 1.61 P = 0.106 RV YES (NO)

144 POLITIES
WHERE THE PARTY SYSTEM IS QUANTITATIVELY
ONE-PARTY, RATHER THAN
TWO-PARTY (34)

144 POLITIES
WHERE THE PARTY SYSTEM IS QUANTITATIVELY
TWO-PARTY, RATHER THAN
ONE-PARTY (11)

PREDICATE ONLY

34 IN LEFT COLUMN

ALBANIA, ALGERIA, BULGARIA, CAMBODIA, CEN AFR REP, CHAD, CHINA, PR, CUBA, CZECHOS'KIA, DAHOMEY, GABON, GERMANY, E, GHANA, GUINEA, HUNGARY, IVORY COAST, KOREA, N, LIBERIA, MALI, MAURITANIA, MONGOLIA, NIGER, POLAND, PORTUGAL, RUMANIA, SENEGAL, SPAIN, TANGANYIKA, TUNISIA, USSR, UAR, UPPER VOLTA, VIETNAM, N, YUGOSLAVIA

11 IN RIGHT COLUMN

AUSTRALIA, AUSTRIA, CANADA, COLOMBIA, HONDURAS, JAMAICA, NEW ZEALAND, PHILIPPINES, SIERRE LEO, UK, US

7 EXCLUDED BECAUSE AMBIGUOUS

INDONESIA, MALAYA, MOROCCO, NICARAGUA, SO AFRICA, TURKEY, URUGUAY

53 EXCLUDED BECAUSE IRRELEVANT

AFGHANISTAN, ARGENTINA, BELGIUM, BOLIVIA, BRAZIL, BURUNDI, CAMEROUN, CEYLON, CHILE, CONGO(BRA), CONGO(LEO), COSTA RICA, CYPRUS, DENMARK, DOMIN REP, ECUADOR, EL SALVADOR, ETHIOPIA, FINLAND, FRANCE, GERMAN FR, GREECE, GUATEMALA, HAITI, ICELAND, INDIA, IRELAND, ISRAEL, ITALY, JAPAN, JORDAN, LAOS, LEBANON, LIBYA, LUXEMBOURG, MALAGASY R, MEXICO, NETHERLANDS, NIGERIA, NORWAY, PAKISTAN, PANAMA, PARAGUAY, PERU, RWANDA, SA'U ARABIA, SOMALIA, SWEDEN, SWITZERLAND, TRINIDAD, UGANDA, VENEZUELA, VIETNAM REP

10 EXCLUDED BECAUSE UNASCERTAINABLE

BURMA, IRAN, IRAQ, KOREA REP, NEPAL, SUDAN, SYRIA, THAILAND, TOGO, YEMEN

145 POLITIES
WHERE THE PARTY SYSTEM IS QUANTITATIVELY
MULTI-PARTY, RATHER THAN
TWO-PARTY (30)

145 POLITIES
WHERE THE PARTY SYSTEM IS QUANTITATIVELY
TWO-PARTY, RATHER THAN
MULTI-PARTY (11)

PREDICATE ONLY

30 IN LEFT COLUMN

ARGENTINA	BELGIUM	BRAZIL	CEYLON	CHILE	CONGO(LEO)	COSTA RICA	CYPRUS	DENMARK	DOMIN REP
ECUADOR	EL SALVADOR	FINLAND	GUATEMALA	ICELAND	IRELAND	ISRAEL	ITALY	LAOS	LEBANON
LUXEMBOURG	NETHERLANDS	NIGERIA	NORWAY	PANAMA	PERU	SWEDEN	SWITZERLAND	UGANDA	VENEZUELA

11 IN RIGHT COLUMN

AUSTRALIA	AUSTRIA	CANADA	COLOMBIA	HONDURAS	JAMAICA	NEW ZEALAND	PHILIPPINES	SIERRE LEO	UK
US									

7 EXCLUDED BECAUSE AMBIGUOUS

INDONESIA	MALAYA	MOROCCO	NICARAGUA	SO AFRICA	TURKEY	URUGUAY

55 EXCLUDED BECAUSE IRRELEVANT

AFGHANISTAN	ALBANIA	ALGERIA	BOLIVIA	BULGARIA	BURUNDI	CAMBODIA	CAMEROUN	CEN AFR REP	CHAD
CHINA, PR	CONGO(BRA)	CUBA	CZECHOS'KIA	DAHOMEY	ETHIOPIA	FRANCE	GABON	GERMANY, E	GERMAN FR
GHANA	GREECE	GUINEA	HAITI	HUNGARY	INDIA	IVORY COAST	JAPAN	KOREA, N	LIBERIA
LIBYA	MALAGASY R	MALI	MAURITANIA	MEXICO	MONGOLIA	NIGER	PARAGUAY	POLAND	PORTUGAL
RUMANIA	RWANDA	SA'U ARABIA	SENEGAL	SOMALIA	SPAIN	TANGANYIKA	TRINIDAD	TUNISIA	USSR
UAR	UPPER VOLTA	VIETNAM, N	VIETNAM REP	YUGOSLAVIA					

12 EXCLUDED BECAUSE UNASCERTAINABLE

BURMA	IRAN	IRAQ	JORDAN	KOREA REP	NEPAL	PAKISTAN	SUDAN	SYRIA	THAILAND
TOGO	YEMEN								

****

146 POLITIES WHERE THE PARTY SYSTEM IS QUALITATIVELY MASS-BASED TERRITORIAL (8)	146 POLITIES WHERE THE PARTY SYSTEM IS QUALITATIVELY OTHER THAN MASS-BASED TERRITORIAL (92)

BOTH SUBJECT AND PREDICATE

....

8 IN LEFT COLUMN

ALGERIA	GHANA	GUINEA	IVORY COAST	SENEGAL	TANGANYIKA	TUNISIA	UAR

92 IN RIGHT COLUMN

ALBANIA	ARGENTINA	AUSTRALIA	AUSTRIA	BELGIUM	BOLIVIA	BRAZIL	BULGARIA	BURMA	BURUNDI
CAMEROUN	CANADA	CEN AFR REP	CEYLON	CHAD	CHILE	CHINA, PR	COLOMBIA	CONGO(BRA)	CONGO(LEO)
COSTA RICA	CYPRUS	CZECHOS'KIA	DAHOMEY	DENMARK	DOMIN REP	ECUADOR	EL SALVADOR	FINLAND	GABON
GERMANY, E	GERMAN FR	GUATEMALA	HONDURAS	HUNGARY	ICELAND	INDONESIA	IRAN	IRELAND	ISRAEL
ITALY	JAMAICA	JAPAN	JORDAN	KOREA, N	KOREA REP	LAOS	LEBANON	LIBERIA	LUXEMBOURG
MALAGASY R	MALAYA	MALI	MAURITANIA	MEXICO	MONGOLIA	NEPAL	NETHERLANDS	NEW ZEALAND	NICARAGUA
NIGER	NIGERIA	NORWAY	PAKISTAN	PANAMA	PARAGUAY	PERU	PHILIPPINES	POLAND	PORTUGAL
RUMANIA	RWANDA	SIERRE LEO	SOMALIA	SO AFRICA	SPAIN	SUDAN	SWEDEN	SWITZERLAND	THAILAND
TOGO	TRINIDAD	TURKEY	UGANDA	USSR	UK	US	UPPER VOLTA	URUGUAY	VENEZUELA
VIETNAM, N	YUGOSLAVIA								

6 EXCLUDED BECAUSE AMBIGUOUS

CAMBODIA	FRANCE	GREECE	INDIA	MOROCCO	VIETNAM REP

5 EXCLUDED BECAUSE IRRELEVANT

AFGHANISTAN	ETHIOPIA	HAITI	LIBYA	SA'U ARABIA

4 EXCLUDED BECAUSE UNASCERTAINABLE

CUBA	IRAQ	SYRIA	YEMEN

10	ALWAYS ARE THOSE LOCATED IN NORTH AFRICA, OR CENTRAL AND SOUTH AFRICA (33)	1.00	10	TEND TO BE THOSE LOCATED ELSEWHERE THAN IN NORTH AFRICA, OR CENTRAL AND SOUTH AFRICA (82)	0.76	8 22 0 70 X SQ= 16.83 P = 0. RV YES (NO)
11	TEND TO BE THOSE LOCATED IN CENTRAL AND SOUTH AFRICA (28)	0.62	11	TEND TO BE THOSE LOCATED ELSEWHERE THAN IN CENTRAL AND SOUTH AFRICA (87)	0.76	5 22 3 70 X SQ= 3.77 P = 0.032 RV YES (NO)

18	ALWAYS ARE THOSE LOCATED IN NORTH AFRICA OR CENTRAL AND SOUTH AFRICA, RATHER THAN IN EAST ASIA, SOUTH ASIA, OR SOUTHEAST ASIA (33)	1.00	18	TEND LESS TO BE THOSE LOCATED IN NORTH AFRICA OR CENTRAL AND SOUTH AFRICA, RATHER THAN IN EAST ASIA, SOUTH ASIA, OR SOUTHEAST ASIA (33)	0.59	8 22 0 15 X SQ= 3.21 P = 0.038 RV NO (NO)
19	ALWAYS ARE THOSE LOCATED IN NORTH AFRICA OR CENTRAL AND SOUTH AFRICA, RATHER THAN IN THE CARIBBEAN, CENTRAL AMERICA, OR SOUTH AMERICA (33)	1.00	19	TEND LESS TO BE THOSE LOCATED IN NORTH AFRICA OR CENTRAL AND SOUTH AFRICA, RATHER THAN IN THE CARIBBEAN, CENTRAL AMERICA, OR SOUTH AMERICA (33)	0.52	8 22 0 20 X SQ= 4.52 P = 0.015 RV NO (NO)
22	TILT MORE TOWARD BEING THOSE WHOSE TERRITORIAL SIZE IS VERY LARGE, LARGE, OR MEDIUM (68)	0.87	22	TILT LESS TOWARD BEING THOSE WHOSE TERRITORIAL SIZE IS VERY LARGE, LARGE, OR MEDIUM (68)	0.57	7 52 1 40 X SQ= 1.78 P = 0.136 RV NO (NO)
26	ALWAYS ARE THOSE WHOSE POPULATION DENSITY IS LOW (67)	1.00	26	TEND LESS TO BE THOSE WHOSE POPULATION DENSITY IS LOW (67)	0.54	0 42 8 50 X SQ= 4.56 P = 0.019 RV NO (NO)
28	TILT MORE TOWARD BEING THOSE WHOSE POPULATION GROWTH RATE IS HIGH (62)	0.86	28	TILT LESS TOWARD BEING THOSE WHOSE POPULATION GROWTH RATE IS HIGH (62)	0.54	6 48 1 41 X SQ= 1.53 P = 0.132 RV NO (NO)
29	TILT TOWARD BEING THOSE WHERE THE DEGREE OF URBANIZATION IS LOW (49)	0.75	29	TILT TOWARD BEING THOSE WHERE THE DEGREE OF URBANIZATION IS HIGH (56)	0.57	2 47 6 36 X SQ= 1.80 P = 0.137 RV YES (NO)
30	TEND TO BE THOSE WHOSE AGRICULTURAL POPULATION IS HIGH (56)	0.87	30	TEND TO BE THOSE WHOSE AGRICULTURAL POPULATION IS MEDIUM, LOW, OR VERY LOW (57)	0.58	7 38 1 53 X SQ= 4.50 P = 0.022 RV YES (NO)
33	ALWAYS ARE THOSE WHOSE GROSS NATIONAL PRODUCT IS LOW OR VERY LOW (85)	1.00	33	TILT LESS TOWARD BEING THOSE WHOSE GROSS NATIONAL PRODUCT IS LOW OR VERY LOW (85)	0.70	0 28 8 64 X SQ= 2.04 P = 0.101 RV NO (NO)
34	TILT TOWARD BEING THOSE WHOSE GROSS NATIONAL PRODUCT IS VERY LOW (53)	0.75	34	TILT TOWARD BEING THOSE WHOSE GROSS NATIONAL PRODUCT IS VERY HIGH, HIGH, MEDIUM, OR LOW (62)	0.55	2 51 6 41 X SQ= 1.65 P = 0.143 RV YES (NO)

36	ALWAYS ARE THOSE WHOSE PER CAPITA GROSS NATIONAL PRODUCT IS LOW OR VERY LOW (73)	1.00	36	TEND LESS TO BE THOSE WHOSE PER CAPITA GROSS NATIONAL PRODUCT IS LOW OR VERY LOW (73)	0.58	0 39 8 53 X SQ= 3.92 P = 0.021 RV NO (NO)
43	ALWAYS ARE THOSE WHOSE ECONOMIC DEVELOPMENTAL STATUS IS VERY UNDERDEVELOPED (57)	1.00	43	TEND TO BE THOSE WHOSE ECONOMIC DEVELOPMENTAL STATUS IS DEVELOPED, INTERMEDIATE, OR UNDERDEVELOPED (55)	0.54	0 49 8 41 X SQ= 6.67 P = 0.006 RV YES (NO)
45	ALWAYS ARE THOSE WHERE THE LITERACY RATE IS BELOW FIFTY PERCENT (54)	1.00	45	TEND TO BE THOSE WHERE THE LITERACY RATE IS FIFTY PERCENT OR ABOVE (55)	0.59	0 51 8 35 X SQ= 8.12 P = 0.001 RV YES (NO)
51	TEND TO BE THOSE WHERE FREEDOM OF THE PRESS IS INTERNALLY ABSENT, OR INTERNALLY AND EXTERNALLY ABSENT (37)	0.83	51	TEND TO BE THOSE WHERE FREEDOM OF THE PRESS IS COMPLETE OR INTERMITTENT (60)	0.68	1 54 5 25 X SQ= 4.46 P = 0.019 RV YES (NO)
54	ALWAYS ARE THOSE WHERE NEWSPAPER CIRCULATION IS LESS THAN ONE HUNDRED PER THOUSAND (76)	1.00	54	TEND LESS TO BE THOSE WHERE NEWSPAPER CIRCULATION IS LESS THAN ONE HUNDRED PER THOUSAND (76)	0.62	0 34 8 56 X SQ= 3.11 P = 0.048 RV NO (NO)
63	ALWAYS ARE THOSE WHERE THE RELIGION IS PREDOMINANTLY OR PARTLY OTHER THAN CHRISTIAN (68)	1.00	63	TEND LESS TO BE THOSE WHERE THE RELIGION IS PREDOMINANTLY OR PARTLY OTHER THAN CHRISTIAN (68)	0.53	0 43 8 48 X SQ= 4.90 P = 0.009 RV NO (NO)
64	ALWAYS ARE THOSE WHERE THE RELIGION IS MUSLIM, RATHER THAN CHRISTIAN (18)	1.00	64	TEND TO BE THOSE WHERE THE RELIGION IS CHRISTIAN, RATHER THAN MUSLIM (46)	0.86	0 43 4 7 X SQ= 12.00 P = 0.001 RV YES (NO)
69	TILT TOWARD BEING THOSE THAT ARE LINGUISTICALLY STRONGLY HETEROGENEOUS (50)	0.75	69	TILT TOWARD BEING THOSE THAT ARE LINGUISTICALLY HOMOGENEOUS OR WEAKLY HETEROGENEOUS (64)	0.57	2 52 6 39 X SQ= 1.90 P = 0.136 RV YES (NO)
73	ALWAYS ARE THOSE WHOSE DATE OF INDEPENDENCE IS AFTER 1945 (46)	1.00	73	TEND TO BE THOSE WHOSE DATE OF INDEPENDENCE IS BEFORE 1945 (65)	0.63	0 55 8 33 X SQ= 9.29 P = 0.001 RV YES (NO)

74	ALWAYS ARE THOSE THAT ARE NOT HISTORICALLY WESTERN (87)	1.00	74	TILT LESS TOWARD BEING THOSE THAT ARE NOT HISTORICALLY WESTERN (87)	0.73	0 24 8 66 X SQ= 1.57 P = 0.116 RV NO (NO)
75	LEAN TOWARD BEING THOSE THAT ARE NOT HISTORICALLY WESTERN AND ARE NOT SIGNIFICANTLY WESTERNIZED (52)	0.75	75	LEAN TOWARD BEING THOSE THAT ARE HISTORICALLY WESTERN OR SIGNIFICANTLY WESTERNIZED (62)	0.62	2 56 6 35 X SQ= 2.68 P = 0.063 RV YES (NO)
76	ALWAYS ARE THOSE THAT HAVE BEEN SIGNIFICANTLY OR PARTIALLY WESTERNIZED THROUGH A COLONIAL RELATIONSHIP, RATHER THAN BEING HISTORICALLY WESTERN (70)	1.00	76	LEAN LESS TOWARD BEING THOSE THAT HAVE BEEN SIGNIFICANTLY OR PARTIALLY WESTERNIZED THROUGH A COLONIAL RELATIONSHIP, RATHER THAN BEING HISTORICALLY WESTERN (70)	0.69	0 24 8 53 X SQ= 2.11 P = 0.099 RV NO (NO)
79	ALWAYS ARE THOSE THAT WERE FORMERLY DEPENDENCIES OF BRITAIN OR FRANCE, RATHER THAN SPAIN (49)	1.00	79	LEAN LESS TOWARD BEING THOSE THAT WERE FORMERLY DEPENDENCIES OF BRITAIN OR FRANCE, RATHER THAN SPAIN (49)	0.67	8 34 0 17 X SQ= 2.30 P = 0.090 RV NO (NO)
82	ALWAYS ARE THOSE WHERE THE TYPE OF POLITICAL MODERNIZATION IS DEVELOPED TUTELARY OR UNDEVELOPED TUTELARY, RATHER THAN EARLY OR LATER EUROPEAN OR EUROPEAN DERIVED (55)	1.00	82	TEND TO BE THOSE WHERE THE TYPE OF POLITICAL MODERNIZATION IS EARLY OR LATER EUROPEAN OR EUROPEAN DERIVED, RATHER THAN DEVELOPED TUTELARY OR UNDEVELOPED TUTELARY (51)	0.55	0 47 8 38 X SQ= 6.87 P = 0.003 RV YES (NO)
86	TILT TOWARD BEING THOSE WHERE THE STAGE OF POLITICAL MODERNIZATION IS EARLY TRANSITIONAL, RATHER THAN ADVANCED OR MID-TRANSITIONAL (38)	0.63	86	TILT TOWARD BEING THOSE WHERE THE STAGE OF POLITICAL MODERNIZATION IS ADVANCED OR MID-TRANSITIONAL, RATHER THAN EARLY TRANSITIONAL (76)	0.68	3 63 5 29 X SQ= 1.92 P = 0.117 RV YES (NO)
87	ALWAYS ARE THOSE WHOSE IDEOLOGICAL ORIENTATION IS DEVELOPMENTAL (31)	1.00	87	TEND TO BE THOSE WHOSE IDEOLOGICAL ORIENTATION IS OTHER THAN DEVELOPMENTAL (58)	0.68	8 23 0 50 X SQ= 11.56 P = 0. RV YES (NO)
93	ALWAYS ARE THOSE WHERE THE SYSTEM STYLE IS MOBILIZATIONAL, OR LIMITED MOBILIZATIONAL (32)	1.00	93	TEND TO BE THOSE WHERE THE SYSTEM STYLE IS NON-MOBILIZATIONAL (78)	0.77	8 20 0 68 X SQ= 17.62 P = 0. RV YES (NO)
96	TEND TO BE THOSE WHERE THE STATUS OF THE REGIME IS AUTHORITARIAN (23)	0.80	96	TEND TO BE THOSE WHERE THE STATUS OF THE REGIME IS CONSTITUTIONAL OR TOTALITARIAN (67)	0.80	1 60 4 15 X SQ= 6.30 P = 0.010 RV YES (NO)

97	ALWAYS ARE THOSE WHERE THE STATUS OF THE REGIME IS AUTHORITARIAN, RATHER THAN TOTALITARIAN (23)	1.00	97	TILT TOWARD BEING THOSE WHERE THE STATUS OF THE REGIME IS TOTALITARIAN, RATHER THAN AUTHORITARIAN (16)	0.50	4 15 0 15 X SQ= 1.84 P = 0.113 RV YES (NO)
101	TILT MORE TOWARD BEING THOSE WHERE THE REPRESENTATIVE CHARACTER OF THE REGIME IS LIMITED POLYARCHIC, PSEUDO-POLYARCHIC, OR NON-POLYARCHIC (57)	0.88	101	TILT LESS TOWARD BEING THOSE WHERE THE REPRESENTATIVE CHARACTER OF THE REGIME IS LIMITED POLYARCHIC, PSEUDO-POLYARCHIC, OR NON-POLYARCHIC (57)	0.54	1 36 7 43 X SQ= 2.04 P = 0.131 RV NO (NO)
102	TILT TOWARD BEING THOSE WHERE THE REPRESENTATIVE CHARACTER OF THE REGIME IS PSEUDO-POLYARCHIC OR NON-POLYARCHIC (49)	0.75	102	TILT TOWARD BEING THOSE WHERE THE REPRESENTATIVE CHARACTER OF THE REGIME IS POLYARCHIC OR LIMITED POLYARCHIC (59)	0.59	2 52 6 36 X SQ= 2.22 P = 0.132 RV YES (NO)
105	ALWAYS ARE THOSE WHERE THE ELECTORAL SYSTEM IS PARTIALLY COMPETITIVE OR NON-COMPETITIVE (47)	1.00	105	TEND TO BE THOSE WHERE THE ELECTORAL SYSTEM IS COMPETITIVE (43)	0.54	0 40 7 34 X SQ= 5.47 P = 0.012 RV YES (NO)
106	LEAN TOWARD BEING THOSE WHERE THE ELECTORAL SYSTEM IS NON-COMPETITIVE (30)	0.71	106	LEAN TOWARD BEING THOSE WHERE THE ELECTORAL SYSTEM IS COMPETITIVE OR PARTIALLY COMPETITIVE (52)	0.68	2 46 5 22 X SQ= 2.68 P = 0.091 RV YES (NO)
107	ALWAYS ARE THOSE WHERE AUTONOMOUS GROUPS ARE PARTIALLY TOLERATED IN POLITICS, ARE TOLERATED ONLY OUTSIDE POLITICS, OR ARE NOT TOLERATED AT ALL (65)	1.00	107	TEND LESS TO BE THOSE WHERE AUTONOMOUS GROUPS ARE PARTIALLY TOLERATED IN POLITICS, ARE TOLERATED ONLY OUTSIDE POLITICS, OR ARE NOT TOLERATED AT ALL (65)	0.53	0 42 8 47 X SQ= 4.87 P = 0.009 RV NO (NO)
111	ALWAYS ARE THOSE WHERE POLITICAL ENCULTURATION IS HIGH OR MEDIUM (53)	1.00	111	TEND LESS TO BE THOSE WHERE POLITICAL ENCULTURATION IS HIGH OR MEDIUM (53)	0.52	7 38 0 35 X SQ= 4.18 P = 0.016 RV NO (NO)
116	ALWAYS ARE THOSE WHERE INTEREST ARTICULATION BY ASSOCIATIONAL GROUPS IS LIMITED OR NEGLIGIBLE (79)	1.00	116	LEAN LESS TOWARD BEING THOSE WHERE INTEREST ARTICULATION BY ASSOCIATIONAL GROUPS IS LIMITED OR NEGLIGIBLE (79)	0.66	0 30 8 58 X SQ= 2.54 P = 0.054 RV NO (NO)
119	ALWAYS ARE THOSE WHERE INTEREST ARTICULATION BY INSTITUTIONAL GROUPS IS VERY SIGNIFICANT OR SIGNIFICANT (74)	1.00	119	TILT LESS TOWARD BEING THOSE WHERE INTEREST ARTICULATION BY INSTITUTIONAL GROUPS IS VERY SIGNIFICANT OR SIGNIFICANT (74)	0.71	8 55 0 23 X SQ= 1.89 P = 0.102 RV NO (NO)

No.	Subject	Value	No.	Predicate	Value	Statistics
122	ALWAYS ARE THOSE WHERE INTEREST ARTICULATION BY NON-ASSOCIATIONAL GROUPS IS SIGNIFICANT OR MODERATE (83)	1.00	122	TILT LESS TOWARD BEING THOSE WHERE INTEREST ARTICULATION BY NON-ASSOCIATIONAL GROUPS IS SIGNIFICANT OR MODERATE (83)	0.68	8 63 0 29 X SQ= 2.19 P = 0.101 RV NO (NO)
129	ALWAYS ARE THOSE WHERE INTEREST ARTICULATION BY POLITICAL PARTIES IS NEGLIGIBLE (37)	1.00	129	TEND TO BE THOSE WHERE INTEREST ARTICULATION BY POLITICAL PARTIES IS SIGNIFICANT, MODERATE, OR LIMITED (56)	0.64	0 50 7 28 X SQ= 8.41 P = 0.001 RV YES (NO)
130	ALWAYS ARE THOSE WHERE INTEREST AGGREGATION BY POLITICAL PARTIES IS SIGNIFICANT (12)	1.00	130	TEND TO BE THOSE WHERE INTEREST AGGREGATION BY POLITICAL PARTIES IS MODERATE, LIMITED, OR NEGLIGIBLE (71)	0.88	3 9 0 66 X SQ= 11.07 P = 0.003 RV YES (NO)
133	ALWAYS ARE THOSE WHERE INTEREST AGGREGATION BY THE EXECUTIVE IS SIGNIFICANT (29)	1.00	133	TEND TO BE THOSE WHERE INTEREST AGGREGATION BY THE EXECUTIVE IS MODERATE, LIMITED, OR NEGLIGIBLE (76)	0.79	7 18 0 68 X SQ= 16.76 P = 0. RV YES (NO)
138	ALWAYS ARE THOSE WHERE INTEREST AGGREGATION BY THE LEGISLATURE IS NEGLIGIBLE (49)	1.00	138	TEND TO BE THOSE WHERE INTEREST AGGREGATION BY THE LEGISLATURE IS SIGNIFICANT, MODERATE, OR LIMITED (48)	0.53	0 43 6 38 X SQ= 4.35 P = 0.026 RV YES (NO)
139	ALWAYS ARE THOSE WHERE THE PARTY SYSTEM IS QUANTITATIVELY ONE-PARTY (34)	1.00	139	TEND TO BE THOSE WHERE THE PARTY SYSTEM IS QUANTITATIVELY OTHER THAN ONE-PARTY (71)	0.71	8 25 0 60 X SQ= 12.98 P = 0. RV YES (NO)
163	TEND TO BE THOSE WHERF THE REGIME'S LEADERSHIP CHARISMA IS PRONOUNCED (13)	0.62	163	TEND TO BE THOSE WHERE THE REGIME'S LEADERSHIP CHARISMA IS MODERATE OR NEGLIGIBLE (87)	0.96	5 3 3 78 X SQ= 24.00 P = 0. RV YES (YES)
164	ALWAYS ARE THOSE WHERE THE REGIME'S LEADERSHIP CHARISMA IS PRONOUNCED OR MODERATE (34)	1.00	164	TEND TO BE THOSE WHERE THE REGIME'S LEADERSHIP CHARISMA IS NEGLIGIBLE (65)	0.77	8 18 0 62 X SQ= 17.43 P = 0. RV YES (NO)
169	ALWAYS ARE THOSE WHERE THE HORIZONTAL POWER DISTRIBUTION IS NEGLIGIBLE (48)	1.00	169	TEND TO BE THOSE WHERE THE HORIZONTAL POWER DISTRIBUTION IS SIGNIFICANT OR LIMITED (58)	0.62	0 53 8 33 X SQ= 8.94 P = 0.001 RV YES (NO)

170	ALWAYS ARE THOSE WHERE THE LEGISLATIVE-EXECUTIVE STRUCTURE IS PRESIDENTIAL (39)	1.00	170	TEND TO BE THOSE WHERE THE LEGISLATIVE-EXECUTIVE STRUCTURE IS OTHER THAN PRESIDENTIAL (63)	0.63	5 32 0 55 X SQ= 5.45 P = 0.009 RV YES (NO)
175	ALWAYS ARE THOSE WHERE THE LEGISLATURE IS LARGELY INEFFECTIVE OR WHOLLY INEFFECTIVE (49)	1.00	175	TEND TO BE THOSE WHERE THE LEGISLATURE IS FULLY EFFECTIVE OR PARTIALLY EFFECTIVE (51)	0.56	0 46 8 36 X SQ= 7.07 P = 0.002 RV YES (NO)
178	ALWAYS ARE THOSE WHERE THE LEGISLATURE IS UNICAMERAL (53)	1.00	178	TEND TO BE THOSE WHERE THE LEGISLATURE IS BICAMERAL (51)	0.51	7 43 0 44 X SQ= 4.78 P = 0.014 RV YES (NO)
179	ALWAYS ARE THOSE WHERE THE EXECUTIVE IS DOMINANT (52)	1.00	179	TEND TO BE THOSE WHERE THE EXECUTIVE IS STRONG (39)	0.53	8 34 0 38 X SQ= 6.07 P = 0.006 RV YES (NO)
188	ALWAYS ARE THOSE WHERE THE CHARACTER OF THE LEGAL SYSTEM IS OTHER THAN CIVIL LAW (81)	1.00	188	TILT LESS TOWARD BEING THOSE WHERE THE CHARACTER OF THE LEGAL SYSTEM IS OTHER THAN CIVIL LAW (81)	0.68	0 29 7 63 X SQ= 1.78 P = 0.102 RV NO (NO)
192	ALWAYS ARE THOSE WHERE THE CHARACTER OF THE LEGAL SYSTEM IS MUSLIM OR PARTLY MUSLIM (27)	1.00	192	TEND TO BE THOSE WHERE THE CHARACTER OF THE LEGAL SYSTEM IS OTHER THAN MUSLIM OR PARTLY MUSLIM (86)	0.84	7 15 0 77 X SQ= 21.74 P = 0. RV YES (NO)
193	TEND TO BE THOSE WHERE THE CHARACTER OF THE LEGAL SYSTEM IS PARTLY INDIGENOUS (28)	0.62	193	TEND TO BE THOSE WHERE THE CHARACTER OF THE LEGAL SYSTEM IS OTHER THAN PARTLY INDIGENOUS (86)	0.77	5 21 3 70 X SQ= 4.04 P = 0.028 RV YES (NO)

****

147 POLITIES WHERE THE PARTY SYSTEM IS QUALITATIVELY CLASS-ORIENTED OR MULTI-IDEOLOGICAL (23)	147 POLITIES WHERE THE PARTY SYSTEM IS QUALITATIVELY OTHER THAN CLASS-ORIENTED OR MULTI-IDEOLOGICAL (67)

BOTH SUBJECT AND PREDICATE

....

23 IN LEFT COLUMN

ARGENTINA	AUSTRALIA	AUSTRIA	BELGIUM	BRAZIL	CHILE	DENMARK	FINLAND	GERMAN FR	ICELAND
ISRAEL	ITALY	JAMAICA	JAPAN	LUXEMBOURG	NETHERLANDS	NEW ZEALAND	NORWAY	PERU	SWEDEN
SWITZERLAND	TRINIDAD	UK							

67 IN RIGHT COLUMN

ALBANIA	ALGERIA	BOLIVIA	BULGARIA	BURUNDI	CAMBODIA	CAMEROUN	CANADA	CEN AFR REP	CHAD
CHINA, PR	COLOMBIA	CONGO(BRA)	CONGO(LEO)	COSTA RICA	CUBA	CYPRUS	CZECHOS'KIA	DAHOMEY	DOMIN REP
ECUADOR	EL SALVADOR	GABON	GERMANY, E	GHANA	GUINEA	HONDURAS	HUNGARY	IVORY COAST	KOREA, N
LAOS	LEBANON	MALAGASY R	MALAYA	MALI	MAURITANIA	MEXICO	MONGOLIA	NEPAL	NICARAGUA
NIGER	NIGERIA	PANAMA	PARAGUAY	PHILIPPINES	POLAND	PORTUGAL	RUMANIA	RWANDA	SENEGAL
SIERRE LEO	SOMALIA	SPAIN	SUDAN	TANGANYIKA	THAILAND	TOGO	TUNISIA	UGANDA	USSR
UAR	US	UPPER VOLTA	VENEZUELA	VIETNAM, N	VIETNAM REP	YUGOSLAVIA			

12 EXCLUDED BECAUSE AMBIGUOUS

CEYLON	FRANCE	GREECE	GUATEMALA	INDIA	INDONESIA	IRELAND	LIBERIA	MOROCCO	SO AFRICA
TURKEY	URUGUAY								

5 EXCLUDED BECAUSE IRRELEVANT

AFGHANISTAN	ETHIOPIA	HAITI	LIBYA	SA'U ARABIA

8 EXCLUDED BECAUSE UNASCERTAINABLE

BURMA	IRAN	IRAQ	JORDAN	KOREA REP	PAKISTAN	SYRIA	YEMEN

2	TEND TO BE THOSE LOCATED IN WEST EUROPE, SCANDINAVIA, NORTH AMERICA, OR AUSTRALASIA (22)	0.65	2	TEND TO BE THOSE LOCATED ELSEWHERE THAN IN WEST EUROPE, SCANDINAVIA, NORTH AMERICA, OR AUSTRALASIA (93)	0.94	15 4 8 63 X SQ= 32.62 P = 0. RV YES (YES)
3	TEND LESS TO BE THOSE LOCATED ELSEWHERE THAN IN WEST EUROPE (102)	0.65	3	TEND MORE TO BE THOSE LOCATED ELSEWHERE THAN IN WEST EUROPE (102)	0.97	8 2 15 65 X SQ= 14.46 P = 0. RV NO (YES)

5	ALWAYS ARE THOSE LOCATED IN SCANDINAVIA OR WEST EUROPE, RATHER THAN IN EAST EUROPE (18)	1.00	5	TEND TO BE THOSE LOCATED IN EAST EUROPE, RATHER THAN IN SCANDINAVIA OR WEST EUROPE (9)	0.82	13 2 0 9 X SQ= 13.71 P = 0. RV YES (YES)
7	ALWAYS ARE THOSE LOCATED IN SOUTH AMERICA, RATHER THAN IN CENTRAL AMERICA (10)	1.00	7	TILT TOWARD BEING THOSE LOCATED IN CENTRAL AMERICA, RATHER THAN IN SOUTH AMERICA (7)	0.55	0 6 4 5 X SQ= 1.72 P = 0.103 RV YES (NO)
10	ALWAYS ARE THOSE LOCATED ELSEWHERE THAN IN NORTH AFRICA, OR CENTRAL AND SOUTH AFRICA (82)	1.00	10	TEND LESS TO BE THOSE LOCATED ELSEWHERE THAN IN NORTH AFRICA, OR CENTRAL AND SOUTH AFRICA (82)	0.58	0 28 23 39 X SQ= 12.07 P = 0. RV NO (NO)
15	ALWAYS ARE THOSE LOCATED IN THE MIDDLE EAST, RATHER THAN IN NORTH AFRICA OR CENTRAL AND SOUTH AFRICA (11)	1.00	15	LEAN TOWARD BEING THOSE LOCATED IN NORTH AFRICA OR CENTRAL AND SOUTH AFRICA, RATHER THAN IN THE MIDDLE EAST (33)	0.93	1 2 0 28 X SQ= 1.92 P = 0.097 RV YES (NO)
19	ALWAYS ARE THOSE LOCATED IN THE CARIBBEAN, CENTRAL AMERICA, OR SOUTH AMERICA, RATHER THAN IN NORTH AFRICA OR CENTRAL AND SOUTH AFRICA (22)	1.00	19	TEND TO BE THOSE LOCATED IN NORTH AFRICA OR CENTRAL AND SOUTH AFRICA, RATHER THAN IN THE CARIBBEAN, CENTRAL AMERICA, OR SOUTH AMERICA (33)	0.68	0 28 6 13 X SQ= 7.50 P = 0.003 RV YES (NO)
25	TEND LESS TO BE THOSE WHOSE POPULATION DENSITY IS MEDIUM OR LOW (98)	0.57	25	TEND MORE TO BE THOSE WHOSE POPULATION DENSITY IS MEDIUM OR LOW (98)	0.96	10 3 13 64 X SQ= 18.04 P = 0. RV NO (YES)
28	LEAN TOWARD BEING THOSE WHOSE POPULATION GROWTH RATE IS LOW (48)	0.61	28	LEAN TOWARD BEING THOSE WHOSE POPULATION GROWTH RATE IS HIGH (62)	0.60	9 38 14 25 X SQ= 2.26 P = 0.093 RV YES (NO)
29	ALWAYS ARE THOSE WHERE THE DEGREE OF URBANIZATION IS HIGH (56)	1.00	29	TEND TO BE THOSE WHERE THE DEGREE OF URBANIZATION IS LOW (49)	0.64	22 21 0 37 X SQ= 23.61 P = 0. RV YES (YES)
30	ALWAYS ARE THOSE WHOSE AGRICULTURAL POPULATION IS MEDIUM, LOW, OR VERY LOW (57)	1.00	30	TEND TO BE THOSE WHOSE AGRICULTURAL POPULATION IS HIGH (56)	0.61	0 40 23 26 X SQ= 22.93 P = 0. RV YES (NO)

```
31  TEND TO BE THOSE                          0.78       31  TEND TO BE THOSE                          0.96          5    64
    WHOSE AGRICULTURAL POPULATION                            WHOSE AGRICULTURAL POPULATION                         18     3
    IS LOW OR VERY LOW  (24)                                 IS HIGH OR MEDIUM  (90)                          X SQ=  48.06
                                                                                                              P =     0.
                                                                                                              RV YES (YES)

33  TEND TO BE THOSE                          0.57       33  TEND TO BE THOSE                          0.84         13    11
    WHOSE GROSS NATIONAL PRODUCT                             WHOSE GROSS NATIONAL PRODUCT                          10    56
    IS VERY HIGH, HIGH, OR MEDIUM  (30)                      IS LOW OR VERY LOW  (85)                         X SQ=  12.11
                                                                                                              P =     0.001
                                                                                                              RV YES (YES)

34  TEND TO BE THOSE                          0.83       34  TEND TO BE THOSE                          0.61         19    26
    WHOSE GROSS NATIONAL PRODUCT                             WHOSE GROSS NATIONAL PRODUCT                           4    41
    IS VERY HIGH, HIGH, MEDIUM,                              IS VERY LOW  (53)                                X SQ=  11.45
    OR LOW  (62)                                                                                              P =     0.001
                                                                                                              RV YES (NO )

35  TEND TO BE THOSE                          0.70       35  TEND TO BE THOSE                          0.91         16     6
    WHOSE PER CAPITA GROSS NATIONAL PRODUCT                  WHOSE PER CAPITA GROSS NATIONAL PRODUCT                7    61
    IS VERY HIGH OR HIGH  (24)                               IS MEDIUM, LOW, OR VERY LOW  (91)                X SQ=  30.85
                                                                                                              P =     0.
                                                                                                              RV YES (YES)

36  TEND TO BE THOSE                          0.91       36  TEND TO BE THOSE                          0.76         21    16
    WHOSE PER CAPITA GROSS NATIONAL PRODUCT                  WHOSE PER CAPITA GROSS NATIONAL PRODUCT                2    51
    IS VERY HIGH, HIGH, OR MEDIUM  (42)                      IS LOW OR VERY LOW  (73)                         X SQ=  29.43
                                                                                                              P =     0.
                                                                                                              RV YES (YES)

37  ALWAYS ARE THOSE                          1.00       37  TEND TO BE THOSE                          0.54         23    31
    WHOSE PER CAPITA GROSS NATIONAL PRODUCT                  WHOSE PER CAPITA GROSS NATIONAL PRODUCT                0    36
    IS VERY HIGH, HIGH, MEDIUM, OR LOW  (64)                 IS VERY LOW  (51)                                X SQ=  18.42
                                                                                                              P =     0.
                                                                                                              RV YES (NO )

39  TEND TO BE THOSE                          0.74       39  TEND TO BE THOSE                          0.77         17    15
    WHOSE INTERNATIONAL FINANCIAL STATUS                     WHOSE INTERNATIONAL FINANCIAL STATUS                   6    51
    IS VERY HIGH, HIGH, OR MEDIUM  (38)                      IS LOW OR VERY LOW  (76)                         X SQ=  17.25
                                                                                                              P =     0.
                                                                                                              RV YES (YES)

40  TEND TO BE THOSE                          0.95       40  TEND TO BE THOSE                          0.52         20    31
    WHOSE INTERNATIONAL FINANCIAL STATUS                     WHOSE INTERNATIONAL FINANCIAL STATUS                   1    33
    IS VERY HIGH, HIGH,                                      IS VERY LOW  (39)                                X SQ=  12.55
    MEDIUM, OR LOW  (71)                                                                                      P =     0.
                                                                                                              RV YES (NO )

41  TEND TO BE THOSE                          0.59       41  TEND TO BE THOSE                          0.92         13     5
    WHOSE ECONOMIC DEVELOPMENTAL STATUS                      WHOSE ECONOMIC DEVELOPMENTAL STATUS                    9    61
    IS DEVELOPED  (19)                                       IS INTERMEDIATE, UNDERDEVELOPED,                 X SQ=  23.84
                                                             OR VERY UNDERDEVELOPED  (94)                     P =     0.
                                                                                                              RV YES (YES)
```

42	TEND TO BE THOSE WHOSE ECONOMIC DEVELOPMENTAL STATUS IS DEVELOPED OR INTERMEDIATE (36)	0.82	42	TEND TO BE THOSE WHOSE ECONOMIC DEVELOPMENTAL STATUS IS UNDERDEVELOPED OR VERY UNDERDEVELOPED (76)	0.80	18 13 4 53 X SQ= 25.25 P = 0. RV YES (YES)
43	ALWAYS ARE THOSE WHOSE ECONOMIC DEVELOPMENTAL STATUS IS DEVELOPED, INTERMEDIATE, OR UNDERDEVELOPED (55)	1.00	43	TEND TO BE THOSE WHOSE ECONOMIC DEVELOPMENTAL STATUS IS VERY UNDERDEVELOPED (57)	0.68	23 21 0 44 X SQ= 28.49 P = 0. RV YES (YES)
44	TEND TO BE THOSE WHERE THE LITERACY RATE IS NINETY PERCENT OR ABOVE (25)	0.65	44	TEND TO BE THOSE WHERE THE LITERACY RATE IS BELOW NINETY PERCENT (90)	0.88	15 8 8 59 X SQ= 22.82 P = 0. RV YES (YES)
45	TEND TO BE THOSE WHERE THE LITERACY RATE IS FIFTY PERCENT OR ABOVE (55)	0.96	45	TEND TO BE THOSE WHERE THE LITERACY RATE IS BELOW FIFTY PERCENT (54)	0.60	22 25 1 37 X SQ= 18.60 P = 0. RV YES (NO)
46	ALWAYS ARE THOSE WHERE THE LITERACY RATE IS TEN PERCENT OR ABOVE (84)	1.00	46	TEND LESS TO BE THOSE WHERE THE LITERACY RATE IS TEN PERCENT OR ABOVE (84)	0.65	23 41 0 22 X SQ= 9.04 P = 0. RV NO (NO)
50	TEND TO BE THOSE WHERE FREEDOM OF THE PRESS IS COMPLETE (43)	0.83	50	TEND TO BE THOSE WHERE FREEDOM OF THE PRESS IS INTERMITTENT, INTERNALLY ABSENT, OR INTERNALLY AND EXTERNALLY ABSENT (56)	0.61	19 21 4 33 X SQ= 10.66 P = 0. RV YES (NO)
51	ALWAYS ARE THOSE WHERE FREEDOM OF THE PRESS IS COMPLETE OR INTERMITTENT (60)	1.00	51	TEND TO BE THOSE WHERE FREEDOM OF THE PRESS IS INTERNALLY ABSENT, OR INTERNALLY AND EXTERNALLY ABSENT (37)	0.52	23 26 0 28 X SQ= 16.57 P = 0. RV YES (NO)
53	TEND TO BE THOSE WHERE NEWSPAPER CIRCULATION IS THREE HUNDRED OR MORE PER THOUSAND (14)	0.52	53	TEND TO BE THOSE WHERE NEWSPAPER CIRCULATION IS LESS THAN THREE HUNDRED PER THOUSAND (101)	0.97	12 2 11 65 X SQ= 27.90 P = 0. RV YES (YES)
54	TEND TO BE THOSE WHERE NEWSPAPER CIRCULATION IS ONE HUNDRED OR MORE PER THOUSAND (37)	0.83	54	TEND TO BE THOSE WHERE NEWSPAPER CIRCULATION IS LESS THAN ONE HUNDRED PER THOUSAND (76)	0.78	19 14 4 51 X SQ= 24.49 P = 0. RV YES (YES)

55	ALWAYS ARE THOSE WHERE NEWSPAPER CIRCULATION IS TEN OR MORE PER THOUSAND (78)	1.00	55	TEND LESS TO BE THOSE WHERE NEWSPAPER CIRCULATION IS TEN OR MORE PER THOUSAND (78)	0.58	23 38 0 27 X SQ= 11.90 P = 0. RV NO (NO)
56	TEND MORE TO BE THOSE WHERE THE RELIGION IS PREDOMINANTLY LITERATE (79)	0.96	56	TEND LESS TO BE THOSE WHERE THE RELIGION IS PREDOMINANTLY LITERATE (79)	0.60	22 38 1 25 X SQ= 8.37 P = 0.001 RV NO (NO)
62	TEND LESS TO BE THOSE WHERE THE RELIGION IS CATHOLIC, RATHER THAN PROTESTANT (25)	0.58	62	ALWAYS ARE THOSE WHERE THE RELIGION IS CATHOLIC, RATHER THAN PROTESTANT (25)	1.00	5 0 7 15 X SQ= 5.16 P = 0.010 RV NO (YES)
63	TEND TO BE THOSE WHERE THE RELIGION IS PREDOMINANTLY SOME KIND OF CHRISTIAN (46)	0.83	63	TEND TO BE THOSE WHERE THE RELIGION IS PREDOMINANTLY OR PARTLY OTHER THAN CHRISTIAN (68)	0.65	19 23 4 43 X SQ= 13.75 P = 0. RV YES (NO)
64	ALWAYS ARE THOSE WHERE THE RELIGION IS CHRISTIAN, RATHER THAN MUSLIM (46)	1.00	64	LEAN LESS TOWARD BEING THOSE WHERE THE RELIGION IS CHRISTIAN, RATHER THAN MUSLIM (46)	0.79	19 23 0 6 X SQ= 2.80 P = 0.068 RV NO (NO)
68	TEND TO BE THOSE THAT ARE LINGUISTICALLY HOMOGENEOUS (52)	0.74	68	TEND TO BE THOSE THAT ARE LINGUISTICALLY WEAKLY HETEROGENEOUS OR STRONGLY HETEROGENEOUS (62)	0.62	17 25 6 41 X SQ= 7.50 P = 0.004 RV YES (NO)
69	TEND TO BE THOSE THAT ARE LINGUISTICALLY HOMOGENEOUS OR WEAKLY HETEROGENEOUS (64)	0.83	69	TEND TO BE THOSE THAT ARE LINGUISTICALLY STRONGLY HETEROGENEOUS (50)	0.50	19 33 4 33 X SQ= 6.18 P = 0.007 RV YES (NO)
70	TEND LESS TO BE THOSE THAT ARE RELIGIOUSLY, RACIALLY, OR LINGUISTICALLY HETEROGENEOUS (85)	0.64	70	TEND MORE TO BE THOSE THAT ARE RELIGIOUSLY, RACIALLY, OR LINGUISTICALLY HETEROGENEOUS (85)	0.89	8 7 14 54 X SQ= 5.19 P = 0.020 RV NO (YES)
72	TEND TO BE THOSE WHOSE DATE OF INDEPENDENCE IS BEFORE 1914 (52)	0.77	72	TEND TO BE THOSE WHOSE DATE OF INDEPENDENCE IS AFTER 1914 (59)	0.63	17 24 5 41 X SQ= 9.18 P = 0.001 RV YES (NO)

73	TEND TO BE THOSE WHOSE DATE OF INDEPENDENCE IS BEFORE 1945 (65)	0.86	73	TEND TO BE THOSE WHOSE DATE OF INDEPENDENCE IS AFTER 1945 (46)	0.54	19 30 3 35 X SQ= 9.23 P = 0.001 RV YES (NO)
74	TEND TO BE THOSE THAT ARE HISTORICALLY WESTERN (26)	0.68	74	TEND TO BE THOSE THAT ARE NOT HISTORICALLY WESTERN (87)	0.88	15 8 7 59 X SQ= 24.48 P = 0. RV YES (YES)
75	ALWAYS ARE THOSE THAT ARE HISTORICALLY WESTERN OR SIGNIFICANTLY WESTERNIZED (62)	1.00	75	TEND TO BE THOSE THAT ARE NOT HISTORICALLY WESTERN AND ARE NOT SIGNIFICANTLY WESTERNIZED (52)	0.54	23 31 0 36 X SQ= 18.42 P = 0. RV YES (NO)
76	TEND TO BE THOSE THAT ARE HISTORICALLY WESTERN, RATHER THAN HAVING BEEN SIGNIFICANTLY OR PARTIALLY WESTERNIZED THROUGH A COLONIAL RELATIONSHIP (26)	0.71	76	TEND TO BE THOSE THAT HAVE BEEN SIGNIFICANTLY OR PARTIALLY WESTERNIZED THROUGH A COLONIAL RELATIONSHIP, RATHER THAN BEING HISTORICALLY WESTERN (70)	0.86	15 8 6 49 X SQ= 21.63 P = 0. RV YES (YES)
77	ALWAYS ARE THOSE THAT HAVE BEEN SIGNIFICANTLY WESTERNIZED, RATHER THAN PARTIALLY WESTERNIZED, THROUGH A COLONIAL RELATIONSHIP (28)	1.00	77	TEND TO BE THOSE THAT HAVE BEEN PARTIALLY WESTERNIZED, RATHER THAN SIGNIFICANTLY WESTERNIZED, THROUGH A COLONIAL RELATIONSHIP (41)	0.63	6 18 0 31 X SQ= 6.32 P = 0.005 RV YES (NO)
80	ALWAYS ARE THOSE WHOSE DATE OF INDEPENDENCE IS AFTER 1914, AND THAT WERE FORMERLY DEPENDENCIES OF BRITAIN, RATHER THAN FRANCE (19)	1.00	80	TEND TO BE THOSE WHOSE DATE OF INDEPENDENCE IS AFTER 1914, AND THAT WERE FORMERLY DEPENDENCIES OF FRANCE, RATHER THAN BRITAIN (24)	0.71	3 9 0 22 X SQ= 3.32 P = 0.037 RV YES (NO)
81	TEND LESS TO BE THOSE WHERE THE TYPE OF POLITICAL MODERNIZATION IS LATER EUROPEAN OR LATER EUROPEAN DERIVED, RATHER THAN EARLY EUROPEAN OR EARLY EUROPEAN DERIVED (40)	0.63	81	TEND MORE TO BE THOSE WHERE THE TYPE OF POLITICAL MODERNIZATION IS LATER EUROPEAN OR LATER EUROPEAN DERIVED, RATHER THAN EARLY EUROPEAN OR EARLY EUROPEAN DERIVED (40)	0.92	7 2 12 23 X SQ= 3.89 P = 0.027 RV NO (YES)
82	TEND TO BE THOSE WHERE THE TYPE OF POLITICAL MODERNIZATION IS EARLY OR LATER EUROPEAN OR EUROPEAN DERIVED, RATHER THAN DEVELOPED TUTELARY OR UNDEVELOPED TUTELARY (51)	0.86	82	TEND TO BE THOSE WHERE THE TYPE OF POLITICAL MODERNIZATION IS DEVELOPED TUTELARY OR UNDEVELOPED TUTELARY, RATHER THAN EARLY OR LATER EUROPEAN OR EUROPEAN DERIVED (55)	0.60	19 25 3 38 X SQ= 12.42 P = 0. RV YES (NO)
84	ALWAYS ARE THOSE WHERE THE TYPE OF POLITICAL MODERNIZATION IS DEVELOPED TUTELARY, RATHER THAN UNDEVELOPED TUTELARY (31)	1.00	84	LEAN TOWARD BEING THOSE WHERE THE TYPE OF POLITICAL MODERNIZATION IS UNDEVELOPED TUTELARY, RATHER THAN DEVELOPED TUTELARY (24)	0.63	3 14 0 24 X SQ= 2.34 P = 0.064 RV YES (NO)

85	TEND TO BE THOSE WHERE THE STAGE OF POLITICAL MODERNIZATION IS ADVANCED, RATHER THAN MID- OR EARLY TRANSITIONAL (60)	0.87	85	TEND TO BE THOSE WHERE THE STAGE OF POLITICAL MODERNIZATION IS MID- OR EARLY TRANSITIONAL, RATHER THAN ADVANCED (54)	0.60	20 27 3 40 X SQ= 13.13 P = 0. RV YES (NO)
87	TEND TO BE THOSE WHOSE IDEOLOGICAL ORIENTATION IS OTHER THAN DEVELOPMENTAL (58)	0.94	87	TEND TO BE THOSE WHOSE IDEOLOGICAL ORIENTATION IS DEVELOPMENTAL (31)	0.51	1 29 17 28 X SQ= 9.90 P = 0.001 RV YES (NO)
89	TEND TO BE THOSE WHOSE IDEOLOGICAL ORIENTATION IS CONVENTIONAL (33)	0.94	89	TEND TO BE THOSE WHOSE IDEOLOGICAL ORIENTATION IS OTHER THAN CONVENTIONAL (62)	0.83	17 10 1 49 X SQ= 33.05 P = 0. RV YES (YES)
92	ALWAYS ARE THOSE WHERE THE SYSTEM STYLE IS LIMITED MOBILIZATIONAL, OR NON-MOBILIZATIONAL (93)	1.00	92	TEND LESS TO BE THOSE WHERE THE SYSTEM STYLE IS LIMITED MOBILIZATIONAL, OR NON-MOBILIZATIONAL (93)	0.71	0 19 23 47 X SQ= 6.79 P = 0.002 RV NO (NO)
93	TEND MORE TO BE THOSE WHERE THE SYSTEM STYLE IS NON-MOBILIZATIONAL (78)	0.96	93	TEND LESS TO BE THOSE WHERE THE SYSTEM STYLE IS NON-MOBILIZATIONAL (78)	0.57	1 28 22 37 X SQ= 9.85 P = 0.001 RV NO (NO)
94	ALWAYS ARE THOSE WHERE THE STATUS OF THE REGIME IS CONSTITUTIONAL (51)	1.00	94	TEND TO BE THOSE WHERE THE STATUS OF THE REGIME IS AUTHORITARIAN OR TOTALITARIAN (41)	0.59	21 21 0 30 X SQ= 18.83 P = 0. RV YES (YES)
98	TEND TO BE THOSE WHERE GOVERNMENTAL STABILITY IS GENERALLY PRESENT AND DATES AT LEAST FROM THE INTERWAR PERIOD (22)	0.52	98	TEND TO BE THOSE WHERE GOVERNMENTAL STABILITY IS GENERALLY OR MODERATELY PRESENT AND DATES FROM THE POST-WAR PERIOD, OR IS ABSENT (93)	0.93	12 5 11 62 X SQ= 19.52 P = 0. RV YES (YES)
101	TEND TO BE THOSE WHERE THE REPRESENTATIVE CHARACTER OF THE REGIME IS POLYARCHIC (41)	0.90	101	TEND TO BE THOSE WHERE THE REPRESENTATIVE CHARACTER OF THE REGIME IS LIMITED POLYARCHIC, PSEUDO-POLYARCHIC, OR NON-POLYARCHIC (57)	0.76	19 14 2 44 X SQ= 25.23 P = 0. RV YES (YES)
102	ALWAYS ARE THOSE WHERE THE REPRESENTATIVE CHARACTER OF THE REGIME IS POLYARCHIC OR LIMITED POLYARCHIC (59)	1.00	102	TEND TO BE THOSE WHERE THE REPRESENTATIVE CHARACTER OF THE REGIME IS PSEUDO-POLYARCHIC OR NON-POLYARCHIC (49)	0.61	22 26 0 40 X SQ= 22.06 P = 0. RV YES (NO)

105	ALWAYS ARE THOSE WHERE THE ELECTORAL SYSTEM IS COMPETITIVE (43)	1.00	105	TEND TO BE THOSE WHERE THE ELECTORAL SYSTEM IS PARTIALLY COMPETITIVE OR NON-COMPETITIVE (47)	0.74	22 14 0 40 X SQ= 31.50 P = 0. RV YES (YES)
107	ALWAYS ARE THOSE WHERE AUTONOMOUS GROUPS ARE FULLY TOLERATED IN POLITICS (46)	1.00	107	TEND TO BE THOSE WHERE AUTONOMOUS GROUPS ARE PARTIALLY TOLERATED IN POLITICS, ARE TOLERATED ONLY OUTSIDE POLITICS, OR ARE NOT TOLERATED AT ALL (65)	0.74	22 17 0 48 X SQ= 33.32 P = 0. RV YES (YES)
110	TEND LESS TO BE THOSE WHERE POLITICAL ENCULTURATION IS MEDIUM OR LOW (80)	0.55	110	TEND MORE TO BE THOSE WHERE POLITICAL ENCULTURATION IS MEDIUM OR LOW (80)	0.94	10 3 12 46 X SQ= 13.18 P = 0. RV NO (YES)
111	LEAN MORE TOWARD BEING THOSE WHERE POLITICAL ENCULTURATION IS HIGH OR MEDIUM (53)	0.77	111	LEAN LESS TOWARD BEING THOSE WHERE POLITICAL ENCULTURATION IS HIGH OR MEDIUM (53)	0.53	17 26 5 23 X SQ= 2.78 P = 0.069 RV NO (NO)
115	TEND TO BE THOSE WHERE INTEREST ARTICULATION BY ASSOCIATIONAL GROUPS IS SIGNIFICANT (20)	0.70	115	TEND TO BE THOSE WHERE INTEREST ARTICULATION BY ASSOCIATIONAL GROUPS IS MODERATE, LIMITED, OR NEGLIGIBLE (91)	0.95	16 3 7 61 X SQ= 38.01 P = 0. RV YES (YES)
116	TEND TO BE THOSE WHERE INTEREST ARTICULATION BY ASSOCIATIONAL GROUPS IS SIGNIFICANT OR MODERATE (32)	0.87	116	TEND TO BE THOSE WHERE INTEREST ARTICULATION BY ASSOCIATIONAL GROUPS IS LIMITED OR NEGLIGIBLE (79)	0.91	20 6 3 58 X SQ= 44.97 P = 0. RV YES (YES)
117	ALWAYS ARE THOSE WHERE INTEREST ARTICULATION BY ASSOCIATIONAL GROUPS IS SIGNIFICANT, MODERATE, OR LIMITED (60)	1.00	117	TEND TO BE THOSE WHERE INTEREST ARTICULATION BY ASSOCIATIONAL GROUPS IS NEGLIGIBLE (51)	0.64	23 23 0 41 X SQ= 25.36 P = 0. RV YES (YES)
118	TEND TO BE THOSE WHERE INTEREST ARTICULATION BY INSTITUTIONAL GROUPS IS SIGNIFICANT, MODERATE, OR LIMITED (60)	0.91	118	TEND TO BE THOSE WHERE INTEREST ARTICULATION BY INSTITUTIONAL GROUPS IS VERY SIGNIFICANT (40)	0.51	2 27 21 26 X SQ= 10.41 P = 0.001 RV YES (NO)
119	TEND TO BE THOSE WHERE INTEREST ARTICULATION BY INSTITUTIONAL GROUPS IS MODERATE OR LIMITED (26)	0.74	119	TEND TO BE THOSE WHERE INTEREST ARTICULATION BY INSTITUTIONAL GROUPS IS VERY SIGNIFICANT OR SIGNIFICANT (74)	0.91	6 48 17 5 X SQ= 29.36 P = 0. RV YES (YES)

No.	Statement		No.	Statement		Statistics
121	ALWAYS ARE THOSE WHERE INTEREST ARTICULATION BY NON-ASSOCIATIONAL GROUPS IS MODERATE, LIMITED, OR NEGLIGIBLE (61)	1.00	121	TEND TO BE THOSE WHERE INTEREST ARTICULATION BY NON-ASSOCIATIONAL GROUPS IS SIGNIFICANT (54)	0.57	0 38 23 29 X SQ= 20.31 P = 0. RV YES (NO)
122	TEND TO BE THOSE WHERE INTEREST ARTICULATION BY NON-ASSOCIATIONAL GROUPS IS LIMITED OR NEGLIGIBLE (32)	0.65	122	TEND TO BE THOSE WHERE INTEREST ARTICULATION BY NON-ASSOCIATIONAL GROUPS IS SIGNIFICANT OR MODERATE (83)	0.81	8 54 15 13 X SQ= 14.70 P = 0. RV YES (YES)
124	ALWAYS ARE THOSE WHERE INTEREST ARTICULATION BY ANOMIC GROUPS IS OCCASIONAL, INFREQUENT, OR VERY INFREQUENT (85)	1.00	124	TILT LESS TOWARD BEING THOSE WHERE INTEREST ARTICULATION BY ANOMIC GROUPS IS OCCASIONAL, INFREQUENT, OR VERY INFREQUENT (85)	0.87	0 7 22 47 X SQ= 1.78 P = 0.100 RV NO (NO)
125	TEND TO BE THOSE WHERE INTEREST ARTICULATION BY ANOMIC GROUPS IS INFREQUENT OR VERY INFREQUENT (35)	0.86	125	TEND TO BE THOSE WHERE INTEREST ARTICULATION BY ANOMIC GROUPS IS FREQUENT OR OCCASIONAL (64)	0.80	3 43 19 11 X SQ= 25.80 P = 0. RV YES (YES)
126	TEND TO BE THOSE WHERE INTEREST ARTICULATION BY ANOMIC GROUPS IS VERY INFREQUENT (16)	0.64	126	TEND TO BE THOSE WHERE INTEREST ARTICULATION BY ANOMIC GROUPS IS FREQUENT, OCCASIONAL, OR INFREQUENT (83)	0.98	8 53 14 1 X SQ= 33.87 P = 0. RV YES (YES)
127	TEND TO BE THOSE WHERE INTEREST ARTICULATION BY POLITICAL PARTIES IS SIGNIFICANT (21)	0.52	127	TEND TO BE THOSE WHERE INTEREST ARTICULATION BY POLITICAL PARTIES IS MODERATE, LIMITED, OR NEGLIGIBLE (72)	0.90	12 6 11 54 X SQ= 15.02 P = 0. RV YES (YES)
128	ALWAYS ARE THOSE WHERE INTEREST ARTICULATION BY POLITICAL PARTIES IS SIGNIFICANT OR MODERATE (48)	1.00	128	TEND TO BE THOSE WHERE INTEREST ARTICULATION BY POLITICAL PARTIES IS LIMITED OR NEGLIGIBLE (45)	0.73	23 16 0 44 X SQ= 33.01 P = 0. RV YES (YES)
133	ALWAYS ARE THOSE WHERE INTEREST AGGREGATION BY THE EXECUTIVE IS MODERATE, LIMITED, OR NEGLIGIBLE (76)	1.00	133	TEND LESS TO BE THOSE WHERE INTEREST AGGREGATION BY THE EXECUTIVE IS MODERATE, LIMITED, OR NEGLIGIBLE (76)	0.60	0 25 22 38 X SQ= 10.53 P = 0. RV NO (NO)
134	TEND MORE TO BE THOSE WHERE INTEREST AGGREGATION BY THE EXECUTIVE IS SIGNIFICANT OR MODERATE (57)	0.82	134	TEND LESS TO BE THOSE WHERE INTEREST AGGREGATION BY THE EXECUTIVE IS SIGNIFICANT OR MODERATE (57)	0.52	18 32 4 30 X SQ= 4.96 P = 0.022 RV NO (NO)

147/

135	ALWAYS ARE THOSE WHERE INTEREST AGGREGATION BY THE EXECUTIVE IS SIGNIFICANT, MODERATE, OR LIMITED (77)	1.00	135	LEAN LESS TOWARD BEING THOSE WHERE INTEREST AGGREGATION BY THE EXECUTIVE IS SIGNIFICANT, MODERATE, OR LIMITED (77)	0.84	22 42 0 8 X SQ= 2.51 P = 0.053 RV NO (NO)
137	TEND TO BE THOSE WHERE INTEREST AGGREGATION BY THE LEGISLATURE IS SIGNIFICANT OR MODERATE (29)	0.91	137	TEND TO BE THOSE WHERE INTEREST AGGREGATION BY THE LEGISLATURE IS LIMITED OR NEGLIGIBLE (68)	0.93	20 4 2 56 X SQ= 51.19 P = 0. RV YES (YES)
138	ALWAYS ARE THOSE WHERE INTEREST AGGREGATION BY THE LEGISLATURE IS SIGNIFICANT, MODERATE, OR LIMITED (48)	1.00	138	TEND TO BE THOSE WHERE INTEREST AGGREGATION BY THE LEGISLATURE IS NEGLIGIBLE (49)	0.70	22 18 0 43 X SQ= 29.42 P = 0. RV YES (YES)
140	ALWAYS ARE THOSE WHERE THE PARTY SYSTEM IS QUANTITATIVELY OTHER THAN ONE PARTY DOMINANT (87)	1.00	140	LEAN LESS TOWARD BEING THOSE WHERE THE PARTY SYSTEM IS QUANTITATIVELY OTHER THAN ONE PARTY DOMINANT (87)	0.84	0 10 23 52 X SQ= 2.79 P = 0.056 RV NO (NO)
142	TEND TO BE THOSE WHERE THE PARTY SYSTEM IS QUANTITATIVELY MULTI-PARTY (30)	0.65	142	TEND TO BE THOSE WHERE THE PARTY SYSTEM IS QUANTITATIVELY OTHER THAN MULTI-PARTY (66)	0.80	15 12 8 49 X SQ= 13.87 P = 0. RV YES (YES)
144	ALWAYS ARE THOSE WHERE THE PARTY SYSTEM IS QUANTITATIVELY TWO-PARTY, RATHER THAN ONE-PARTY (11)	1.00	144	TEND TO BE THOSE WHERE THE PARTY SYSTEM IS QUANTITATIVELY ONE-PARTY, RATHER THAN TWO-PARTY (34)	0.85	0 33 5 6 X SQ= 12.71 P = 0. RV YES (NO)
153	LEAN TOWARD BEING THOSE WHERE THE PARTY SYSTEM IS STABLE (42)	0.76	153	LEAN TOWARD BEING THOSE WHERE THE PARTY SYSTEM IS MODERATELY STABLE OR UNSTABLE (38)	0.51	16 22 5 23 X SQ= 3.32 P = 0.060 RV YES (NO)
157	ALWAYS ARE THOSE WHERE PERSONALISMO IS MODERATE OR NEGLIGIBLE (84)	1.00	157	LEAN LESS TOWARD BEING THOSE WHERE PERSONALISMO IS MODERATE OR NEGLIGIBLE (84)	0.84	0 10 23 52 X SQ= 2.79 P = 0.056 RV NO (NO)
160	ALWAYS ARE THOSE WHERE THE POLITICAL LEADERSHIP IS MODERATELY ELITIST OR NON-ELITIST (67)	1.00	160	TEND LESS TO BE THOSE WHERE THE POLITICAL LEADERSHIP IS MODERATELY ELITIST OR NON-ELITIST (67)	0.60	0 23 23 34 X SQ= 11.13 P = 0. RV NO (NO)

161	TEND TO BE THOSE WHERE THE POLITICAL LEADERSHIP IS NON-ELITIST (50)	0.83	161	TEND TO BE THOSE WHERE THE POLITICAL LEADERSHIP IS ELITIST OR MODERATELY ELITIST (47)	0.60	4 34 19 23 X SQ= 10.10 P = 0.001 RV YES (NO)
163	ALWAYS ARE THOSE WHERE THE REGIME'S LEADERSHIP CHARISMA IS MODERATE OR NEGLIGIBLE (87)	1.00	163	LEAN LESS TOWARD BEING THOSE WHERE THE REGIME'S LEADERSHIP CHARISMA IS MODERATE OR NEGLIGIBLE (87)	0.84	0 9 23 48 X SQ= 2.66 P = 0.053 RV NO (NO)
168	ALWAYS ARE THOSE WHERE THE HORIZONTAL POWER DISTRIBUTION IS SIGNIFICANT (34)	1.00	168	TEND TO BE THOSE WHERE THE HORIZONTAL POWER DISTRIBUTION IS LIMITED OR NEGLIGIBLE (72)	0.89	21 7 0 58 X SQ= 53.56 P = 0. RV YES (YES)
170	TEND TO BE THOSE WHERE THE LEGISLATIVE-EXECUTIVE STRUCTURE IS OTHER THAN PRESIDENTIAL (63)	0.86	170	TEND TO BE THOSE WHERE THE LEGISLATIVE-EXECUTIVE STRUCTURE IS PRESIDENTIAL (39)	0.51	3 31 19 30 X SQ= 7.77 P = 0.002 RV YES (NO)
171	TEND LESS TO BE THOSE WHERE THE LEGISLATIVE-EXECUTIVE STRUCTURE IS OTHER THAN PARLIAMENTARY-REPUBLICAN (90)	0.73	171	TEND MORE TO BE THOSE WHERE THE LEGISLATIVE-EXECUTIVE STRUCTURE IS OTHER THAN PARLIAMENTARY-REPUBLICAN (90)	0.97	6 2 16 59 X SQ= 8.11 P = 0.004 RV NO (YES)
172	TEND TO BE THOSE WHERE THE LEGISLATIVE-EXECUTIVE STRUCTURE IS PARLIAMENTARY-ROYALIST (21)	0.52	172	TEND TO BE THOSE WHERE THE LEGISLATIVE-EXECUTIVE STRUCTURE IS OTHER THAN PARLIAMENTARY-ROYALIST (88)	0.89	12 7 11 57 X SQ= 14.53 P = 0. RV YES (YES)
174	TEND TO BE THOSE WHERE THE LEGISLATURE IS FULLY EFFECTIVE (28)	0.86	174	TEND TO BE THOSE WHERE THE LEGISLATURE IS PARTIALLY EFFECTIVE, LARGELY INEFFECTIVE, OR WHOLLY INEFFECTIVE (72)	0.92	19 5 3 57 X SQ= 45.02 P = 0. RV YES (YES)
175	ALWAYS ARE THOSE WHERE THE LEGISLATURE IS FULLY EFFECTIVE OR PARTIALLY EFFECTIVE (51)	1.00	175	TEND TO BE THOSE WHERE THE LEGISLATURE IS LARGELY INEFFECTIVE OR WHOLLY INEFFECTIVE (49)	0.69	22 19 0 43 X SQ= 28.55 P = 0. RV YES (YES)
178	TEND TO BE THOSE WHERE THE LEGISLATURE IS BICAMERAL (51)	0.83	178	TEND TO BE THOSE WHERE THE LEGISLATURE IS UNICAMERAL (53)	0.71	4 44 19 18 X SQ= 17.47 P = 0. RV YES (YES)

No.	Subject	Score	No.	Predicate	Score	Statistics
179	ALWAYS ARE THOSE WHERE THE EXECUTIVE IS STRONG (39)	1.00	179	TEND TO BE THOSE WHERE THE EXECUTIVE IS DOMINANT (52)	0.70	0 39 19 17 X SQ= 24.85 P = 0. RV YES (YES)
181	TEND TO BE THOSE WHERE THE BUREAUCRACY IS MODERN, RATHER THAN SEMI-MODERN (21)	0.70	181	TEND TO BE THOSE WHERE THE BUREAUCRACY IS SEMI-MODERN, RATHER THAN MODERN (55)	0.95	16 2 7 36 X SQ= 25.47 P = 0. RV YES (YES)
182	ALWAYS ARE THOSE WHERE THE BUREAUCRACY IS SEMI-MODERN, RATHER THAN POST-COLONIAL TRANSITIONAL (55)	1.00	182	TEND LESS TO BE THOSE WHERE THE BUREAUCRACY IS SEMI-MODERN, RATHER THAN POST-COLONIAL TRANSITIONAL (55)	0.60	7 36 0 24 X SQ= 2.80 P = 0.044 RV NO (NO)
185	ALWAYS ARE THOSE WHERE PARTICIPATION BY THE MILITARY IN POLITICS IS INTERVENTIVE, RATHER THAN SUPPORTIVE (21)	1.00	185	TEND TO BE THOSE WHERE PARTICIPATION BY THE MILITARY IN POLITICS IS SUPPORTIVE, RATHER THAN INTERVENTIVE (31)	0.72	3 9 0 23 X SQ= 3.50 P = 0.034 RV YES (NO)
186	ALWAYS ARE THOSE WHERE PARTICIPATION BY THE MILITARY IN POLITICS IS NEUTRAL, RATHER THAN SUPPORTIVE (56)	1.00	186	TEND LESS TO BE THOSE WHERE PARTICIPATION BY THE MILITARY IN POLITICS IS NEUTRAL, RATHER THAN SUPPORTIVE (56)	0.56	0 23 20 29 X SQ= 11.04 P = 0. RV NO (NO)
187	TEND TO BE THOSE WHERE THE ROLE OF THE POLICE IS NOT POLITICALLY SIGNIFICANT (35)	0.95	187	TEND TO BE THOSE WHERE THE ROLE OF THE POLICE IS POLITICALLY SIGNIFICANT (66)	0.82	1 47 19 10 X SQ= 34.61 P = 0. RV YES (YES)
189	TEND LESS TO BE THOSE WHERE THE CHARACTER OF THE LEGAL SYSTEM IS OTHER THAN COMMON LAW (108)	0.78	189	TEND MORE TO BE THOSE WHERE THE CHARACTER OF THE LEGAL SYSTEM IS OTHER THAN COMMON LAW (108)	0.99	5 1 18 66 X SQ= 8.26 P = 0.004 RV NO (YES)
192	ALWAYS ARE THOSE WHERE THE CHARACTER OF THE LEGAL SYSTEM IS OTHER THAN MUSLIM OR PARTLY MUSLIM (86)	1.00	192	TEND LESS TO BE THOSE WHERE THE CHARACTER OF THE LEGAL SYSTEM IS OTHER THAN MUSLIM OR PARTLY MUSLIM (86)	0.73	0 18 23 48 X SQ= 6.26 P = 0.003 RV NO (NO)
193	ALWAYS ARE THOSE WHERE THE CHARACTER OF THE LEGAL SYSTEM IS OTHER THAN PARTLY INDIGENOUS (86)	1.00	193	TEND LESS TO BE THOSE WHERE THE CHARACTER OF THE LEGAL SYSTEM IS OTHER THAN PARTLY INDIGENOUS (86)	0.64	0 24 23 43 X SQ= 9.48 P = 0. RV NO (NO)

147/

```
194  ALWAYS ARE THOSE                         1.00        194  TEND LESS TO BE THOSE                  0.80          0    13
     THAT ARE NON-COMMUNIST   (101)                            THAT ARE NON-COMMUNIST   (101)                       23    53
                                                                                                               X SQ=    3.84
                                                                                                               P =     0.018
                                                                                                               RV NO   (NO )
```

```
****************************************************************************************************

148  POLITIES                                      148  POLITIES
     WHERE THE PARTY SYSTEM IS QUALITATIVELY            WHERE THE PARTY SYSTEM IS QUALITATIVELY
     AFRICAN TRANSITIONAL  (14)                         OTHER THAN
                                                        AFRICAN TRANSITIONAL  (96)

                                                                        BOTH SUBJECT AND PREDICATE
....................................................................................................

      14 IN LEFT COLUMN

BURUNDI     CEN AFR REP CHAD        CONGO(BRA)  DAHOMEY     GABON       MALAGASY R  MALI        MAURITANIA  NIGER
RWANDA      SIERRE LEO  TOGO        UPPER VOLTA

      96 IN RIGHT COLUMN

ALBANIA     ALGERIA     ARGENTINA   AUSTRALIA   AUSTRIA     BELGIUM     BOLIVIA     BRAZIL      BULGARIA    BURMA
CAMBODIA    CAMEROUN    CANADA      CEYLON      CHILE       CHINA, PR   COLOMBIA    CONGO(LEO)  COSTA RICA  CUBA
CYPRUS      CZECHOS'KIA DENMARK     DOMIN REP   ECUADOR     EL SALVADOR FINLAND     FRANCE      GERMANY, E  GERMAN FR
GHANA       GREECE      GUATEMALA   GUINEA      HONDURAS    HUNGARY     ICELAND     INDIA       INDONESIA   IRAN
IRAQ        IRELAND     ISRAEL      ITALY       IVORY COAST JAMAICA     JAPAN       JORDAN      KOREA, N    KOREA REP
LAOS        LEBANON     LIBERIA     LUXEMBOURG  MALAYA      MEXICO      MONGOLIA    MOROCCO     NEPAL       NETHERLANDS
NEW ZEALAND NICARAGUA   NIGERIA     NORWAY      PAKISTAN    PANAMA      PARAGUAY    PERU        PHILIPPINES POLAND
PORTUGAL    RUMANIA     SENEGAL     SOMALIA     SO AFRICA   SPAIN       SUDAN       SWEDEN      SWITZERLAND SYRIA
TANGANYIKA  THAILAND    TRINIDAD    TUNISIA     TURKEY      UGANDA      USSR        UAR         UK          US
URUGUAY     VENEZUELA   VIETNAM, N  VIETNAM REP YEMEN       YUGOSLAVIA

      5 EXCLUDED BECAUSE IRRELEVANT

AFGHANISTAN ETHIOPIA    HAITI       LIBYA       SA'U ARABIA
----------------------------------------------------------------------------------------------------

 10  ALWAYS ARE THOSE                      1.00       10  TEND TO BE THOSE                      0.82        14   17
     LOCATED IN NORTH AFRICA, OR                          LOCATED ELSEWHERE THAN IN NORTH AFRICA,            0   79
     CENTRAL AND SOUTH AFRICA  (33)                       OR CENTRAL AND SOUTH AFRICA  (82)           X SQ=  36.92
                                                                                                      P =     0.
                                                                                                      RV YES (NO )

 24  ALWAYS ARE THOSE                      1.00       24  TEND TO BE THOSE                      0.60         0   58
     WHOSE POPULATION IS                                  WHOSE POPULATION IS                           14   38
     SMALL  (54)                                          VERY LARGE, LARGE, OR MEDIUM  (61)          X SQ=  15.55
                                                                                                      P =     0.
                                                                                                      RV YES (NO )

 25  ALWAYS ARE THOSE                      1.00       25  TILT LESS TOWARD BEING THOSE          0.83         0   16
     WHOSE POPULATION DENSITY IS                          WHOSE POPULATION DENSITY IS                   14   80
     MEDIUM OR LOW  (98)                                  MEDIUM OR LOW  (98)                         X SQ=   1.55
                                                                                                      P =    0.126
                                                                                                      RV NO  (NO )
```

26	TEND MORE TO BE THOSE WHOSE POPULATION DENSITY IS LOW (67)	0.86	26	TEND LESS TO BE THOSE WHOSE POPULATION DENSITY IS LOW (67)	0.53	2 45 12 51 X SQ= 4.05 P = 0.023 RV NO (NO)
29	ALWAYS ARE THOSE WHERE THE DEGREE OF URBANIZATION IS LOW (49)	1.00	29	TEND TO BE THOSE WHERE THE DEGREE OF URBANIZATION IS HIGH (56)	0.64	0 55 14 31 X SQ= 17.40 P = 0. RV YES (NO)
30	ALWAYS ARE THOSE WHOSE AGRICULTURAL POPULATION IS HIGH (56)	1.00	30	TEND TO BE THOSE WHOSE AGRICULTURAL POPULATION IS MEDIUM, LOW, OR VERY LOW (57)	0.60	14 38 0 57 X SQ= 15.28 P = 0. RV YES (NO)
34	ALWAYS ARE THOSE WHOSE GROSS NATIONAL PRODUCT IS VERY LOW (53)	1.00	34	TEND TO BE THOSE WHOSE GROSS NATIONAL PRODUCT IS VERY HIGH, HIGH, MEDIUM, OR LOW (62)	0.62	0 60 14 36 X SQ= 16.81 P = 0. RV YES (NO)
37	ALWAYS ARE THOSE WHOSE PER CAPITA GROSS NATIONAL PRODUCT IS VERY LOW (51)	1.00	37	TEND TO BE THOSE WHOSE PER CAPITA GROSS NATIONAL PRODUCT IS VERY HIGH, HIGH, MEDIUM, OR LOW (64)	0.66	0 63 14 33 X SQ= 18.90 P = 0. RV YES (NO)
40	ALWAYS ARE THOSE WHOSE INTERNATIONAL FINANCIAL STATUS IS VERY LOW (39)	1.00	40	TEND TO BE THOSE WHOSE INTERNATIONAL FINANCIAL STATUS IS VERY HIGH, HIGH, MEDIUM, OR LOW (71)	0.75	0 68 14 23 X SQ= 26.50 P = 0. RV YES (NO)
43	ALWAYS ARE THOSE WHOSE ECONOMIC DEVELOPMENTAL STATUS IS VERY UNDERDEVELOPED (57)	1.00	43	TEND TO BE THOSE WHOSE ECONOMIC DEVELOPMENTAL STATUS IS DEVELOPED, INTERMEDIATE, OR UNDERDEVELOPED (55)	0.57	0 54 14 40 X SQ= 13.87 P = 0. RV YES (NO)
45	ALWAYS ARE THOSE WHERE THE LITERACY RATE IS BELOW FIFTY PERCENT (54)	1.00	45	TEND TO BE THOSE WHERE THE LITERACY RATE IS FIFTY PERCENT OR ABOVE (55)	0.61	0 55 14 35 X SQ= 15.79 P = 0. RV YES (NO)
46	TEND TO BE THOSE WHERE THE LITERACY RATE IS BELOW TEN PERCENT (26)	0.92	46	TEND TO BE THOSE WHERE THE LITERACY RATE IS TEN PERCENT OR ABOVE (84)	0.88	1 81 12 11 X SQ= 38.42 P = 0. RV YES (YES)

50	ALWAYS ARE THOSE WHERE FREEDOM OF THE PRESS IS COMPLETE (43)	1.00	50	TEND TO BE THOSE WHERE FREEDOM OF THE PRESS IS INTERMITTENT, INTERNALLY ABSENT, OR INTERNALLY AND EXTERNALLY ABSENT (56)	0.58	6 37 0 51 X SQ= 5.45 P = 0.007 RV YES (NO)
55	ALWAYS ARE THOSE WHERE NEWSPAPER CIRCULATION IS LESS THAN TEN PER THOUSAND (35)	1.00	55	TEND TO BE THOSE WHERE NEWSPAPER CIRCULATION IS TEN OR MORE PER THOUSAND (78)	0.82	0 77 14 17 X SQ= 36.05 P = 0. RV YES (NO)
56	TEND TO BE THOSE WHERE THE RELIGION IS PREDOMINANTLY OR PARTLY NON-LITERATE (31)	0.93	56	TEND TO BE THOSE WHERE THE RELIGION IS PREDOMINANTLY LITERATE (79)	0.82	1 75 13 16 X SQ= 30.73 P = 0. RV YES (NO)
63	ALWAYS ARE THOSE WHERE THE RELIGION IS PREDOMINANTLY OR PARTLY OTHER THAN CHRISTIAN (68)	1.00	63	TEND LESS TO BE THOSE WHERE THE RELIGION IS PREDOMINANTLY OR PARTLY OTHER THAN CHRISTIAN (68)	0.52	0 46 14 49 X SQ= 9.83 P = 0. RV NO (NO)
66	TEND TO BE THOSE THAT ARE RELIGIOUSLY HETEROGENEOUS (49)	0.91	66	TEND TO BE THOSE THAT ARE RELIGIOUSLY HOMOGENEOUS (57)	0.59	1 53 10 37 X SQ= 7.87 P = 0.002 RV YES (NO)
69	TEND TO BE THOSE THAT ARE LINGUISTICALLY STRONGLY HETEROGENEOUS (50)	0.79	69	TEND TO BE THOSE THAT ARE LINGUISTICALLY HOMOGENEOUS OR WEAKLY HETEROGENEOUS (64)	0.61	3 58 11 37 X SQ= 6.25 P = 0.008 RV YES (NO)
70	ALWAYS ARE THOSE THAT ARE RELIGIOUSLY, RACIALLY, OR LINGUISTICALLY HETEROGENEOUS (85)	1.00	70	TILT LESS TOWARD BEING THOSE THAT ARE RELIGIOUSLY, RACIALLY, OR LINGUISTICALLY HETEROGENEOUS (85)	0.78	0 20 11 70 X SQ= 1.81 P = 0.115 RV NO (NO)
73	ALWAYS ARE THOSE WHOSE DATE OF INDEPENDENCE IS AFTER 1945 (46)	1.00	73	TEND TO BE THOSE WHOSE DATE OF INDEPENDENCE IS BEFORE 1945 (65)	0.66	0 61 14 31 X SQ= 19.24 P = 0. RV YES (NO)
75	ALWAYS ARE THOSE THAT ARE NOT HISTORICALLY WESTERN AND ARE NOT SIGNIFICANTLY WESTERNIZED (52)	1.00	75	TEND TO BE THOSE THAT ARE HISTORICALLY WESTERN OR SIGNIFICANTLY WESTERNIZED (62)	0.65	0 62 14 33 X SQ= 18.61 P = 0. RV YES (NO)

76	ALWAYS ARE THOSE THAT HAVE BEEN SIGNIFICANTLY OR PARTIALLY WESTERNIZED THROUGH A COLONIAL RELATIONSHIP, RATHER THAN BEING HISTORICALLY WESTERN (70)	1.00	76	TEND LESS TO BE THOSE THAT HAVE BEEN SIGNIFICANTLY OR PARTIALLY WESTERNIZED THROUGH A COLONIAL RELATIONSHIP, RATHER THAN BEING HISTORICALLY WESTERN (70)	0.67	0 26 14 54 X SQ= 4.77 P = 0.009 RV NO (NO)
77	ALWAYS ARE THOSE THAT HAVE BEEN PARTIALLY WESTERNIZED, RATHER THAN SIGNIFICANTLY WESTERNIZED, THROUGH A COLONIAL RELATIONSHIP (41)	1.00	77	TEND TO BE THOSE THAT HAVE BEEN SIGNIFICANTLY WESTERNIZED, RATHER THAN PARTIALLY WESTERNIZED, THROUGH A COLONIAL RELATIONSHIP (28)	0.53	0 28 14 25 X SQ= 10.63 P = 0. RV YES (NO)
79	ALWAYS ARE THOSE THAT WERE FORMERLY DEPENDENCIES OF BRITAIN OR FRANCE, RATHER THAN SPAIN (49)	1.00	79	TEND LESS TO BE THOSE THAT WERE FORMERLY DEPENDENCIES OF BRITAIN OR FRANCE, RATHER THAN SPAIN (49)	0.67	12 36 0 18 X SQ= 3.95 P = 0.027 RV NO (NO)
80	TEND TO BE THOSE WHOSE DATE OF INDEPENDENCE IS AFTER 1914, AND THAT WERE FORMERLY DEPENDENCIES OF FRANCE, RATHER THAN BRITAIN (24)	0.92	80	TEND TO BE THOSE WHOSE DATE OF INDEPENDENCE IS AFTER 1914, AND THAT WERE FORMERLY DEPENDENCIES OF BRITAIN, RATHER THAN FRANCE (19)	0.58	1 18 11 13 X SQ= 6.78 P = 0.005 RV YES (NO)
82	ALWAYS ARE THOSE WHERE THE TYPE OF POLITICAL MODERNIZATION IS DEVELOPED TUTELARY OR UNDEVELOPED TUTELARY, RATHER THAN EARLY OR LATER EUROPEAN OR EUROPEAN DERIVED (55)	1.00	82	TEND TO BE THOSE WHERE THE TYPE OF POLITICAL MODERNIZATION IS EARLY OR LATER EUROPEAN OR EUROPEAN DERIVED, RATHER THAN DEVELOPED TUTELARY OR UNDEVELOPED TUTELARY (51)	0.56	0 50 14 39 X SQ= 13.12 P = 0. RV YES (NO)
84	ALWAYS ARE THOSE WHERE THE TYPE OF POLITICAL MODERNIZATION IS UNDEVELOPED TUTELARY, RATHER THAN DEVELOPED TUTELARY (24)	1.00	84	TEND TO BE THOSE WHERE THE TYPE OF POLITICAL MODERNIZATION IS DEVELOPED TUTELARY, RATHER THAN UNDEVELOPED TUTELARY (31)	0.74	0 29 14 10 X SQ= 20.09 P = 0. RV YES (YES)
86	ALWAYS ARE THOSE WHERE THE STAGE OF POLITICAL MODERNIZATION IS EARLY TRANSITIONAL, RATHER THAN ADVANCED OR MID-TRANSITIONAL (38)	1.00	86	TEND TO BE THOSE WHERE THE STAGE OF POLITICAL MODERNIZATION IS ADVANCED OR MID-TRANSITIONAL, RATHER THAN EARLY TRANSITIONAL (76)	0.76	0 73 14 23 X SQ= 28.33 P = 0. RV YES (NO)
87	ALWAYS ARE THOSE WHOSE IDEOLOGICAL ORIENTATION IS DEVELOPMENTAL (31)	1.00	87	TEND TO BE THOSE WHOSE IDEOLOGICAL ORIENTATION IS OTHER THAN DEVELOPMENTAL (58)	0.75	13 18 0 54 X SQ= 23.59 P = 0. RV YES (NO)
92	ALWAYS ARE THOSE WHERE THE SYSTEM STYLE IS LIMITED MOBILIZATIONAL, OR NON-MOBILIZATIONAL (93)	1.00	92	LEAN LESS TOWARD BEING THOSE WHERE THE SYSTEM STYLE IS LIMITED MOBILIZATIONAL, OR NON-MOBILIZATIONAL (93)	0.79	0 20 14 74 X SQ= 2.38 P = 0.068 RV NO (NO)

93	LEAN MORE TOWARD BEING THOSE WHERE THE SYSTEM STYLE IS NON-MOBILIZATIONAL (78)	0.93	93	LEAN LESS TOWARD BEING THOSE WHERE THE SYSTEM STYLE IS NON-MOBILIZATIONAL (78)	0.66	1 31 13 60 X SQ= 2.98 P = 0.059 RV NO (NO)
95	ALWAYS ARE THOSE WHERE THE STATUS OF THE REGIME IS CONSTITUTIONAL OR AUTHORITARIAN (95)	1.00	95	TILT LESS TOWARD BEING THOSE WHERE THE STATUS OF THE REGIME IS CONSTITUTIONAL OR AUTHORITARIAN (95)	0.83	14 77 0 16 X SQ= 1.64 P = 0.122 RV NO (NO)
98	ALWAYS ARE THOSE WHERE GOVERNMENTAL STABILITY IS GENERALLY OR MODERATELY PRESENT AND DATES FROM THE POST-WAR PERIOD, OR IS ABSENT (93)	1.00	98	LEAN LESS TOWARD BEING THOSE WHERE GOVERNMENTAL STABILITY IS GENERALLY OR MODERATELY PRESENT AND DATES FROM THE POST-WAR PERIOD, OR IS ABSENT (93)	0.79	0 20 14 76 X SQ= 2.30 P = 0.069 RV NO (NO)
102	LEAN TOWARD BEING THOSE WHERE THE REPRESENTATIVE CHARACTER OF THE REGIME IS PSEUDO-POLYARCHIC OR NON-POLYARCHIC (49)	0.69	102	LEAN TOWARD BEING THOSE WHERE THE REPRESENTATIVE CHARACTER OF THE REGIME IS POLYARCHIC OR LIMITED POLYARCHIC (59)	0.60	4 54 9 36 X SQ= 2.85 P = 0.071 RV YES (NO)
105	TEND TO BE THOSE WHERE THE ELECTORAL SYSTEM IS PARTIALLY COMPETITIVE OR NON-COMPETITIVE (47)	0.85	105	TEND TO BE THOSE WHERE THE ELECTORAL SYSTEM IS COMPETITIVE (43)	0.56	2 41 11 32 X SQ= 5.80 P = 0.014 RV YES (NO)
107	TEND MORE TO BE THOSE WHERE AUTONOMOUS GROUPS ARE PARTIALLY TOLERATED IN POLITICS, ARE TOLERATED ONLY OUTSIDE POLITICS, OR ARE NOT TOLERATED AT ALL (65)	0.85	107	TEND LESS TO BE THOSE WHERE AUTONOMOUS GROUPS ARE PARTIALLY TOLERATED IN POLITICS, ARE TOLERATED ONLY OUTSIDE POLITICS, OR ARE NOT TOLERATED AT ALL (65)	0.53	2 44 11 49 X SQ= 3.52 P = 0.037 RV NO (NO)
117	ALWAYS ARE THOSE WHERE INTEREST ARTICULATION BY ASSOCIATIONAL GROUPS IS NEGLIGIBLE (51)	1.00	117	TEND TO BE THOSE WHERE INTEREST ARTICULATION BY ASSOCIATIONAL GROUPS IS SIGNIFICANT, MODERATE, OR LIMITED (60)	0.65	0 60 14 32 X SQ= 18.47 P = 0. RV YES (NO)
121	ALWAYS ARE THOSE WHERE INTEREST ARTICULATION BY NON-ASSOCIATIONAL GROUPS IS SIGNIFICANT (54)	1.00	121	TEND TO BE THOSE WHERE INTEREST ARTICULATION BY NON-ASSOCIATIONAL GROUPS IS MODERATE, LIMITED, OR NEGLIGIBLE (61)	0.62	14 36 0 60 X SQ= 16.81 P = 0. RV YES (NO)
125	LEAN MORE TOWARD BEING THOSE WHERE INTEREST ARTICULATION BY ANOMIC GROUPS IS FREQUENT OR OCCASIONAL (64)	0.86	125	LEAN LESS TOWARD BEING THOSE WHERE INTEREST ARTICULATION BY ANOMIC GROUPS IS FREQUENT OR OCCASIONAL (64)	0.60	12 48 2 32 X SQ= 2.39 P = 0.077 RV NO (NO)

No.	Statement	Proportion	No.	Statement	Proportion	Statistics
126	ALWAYS ARE THOSE WHERE INTEREST ARTICULATION BY ANOMIC GROUPS IS FREQUENT, OCCASIONAL, OR INFREQUENT (83)	1.00	126	TILT LESS TOWARD BEING THOSE WHERE INTEREST ARTICULATION BY ANOMIC GROUPS IS FREQUENT, OCCASIONAL, OR INFREQUENT (83)	0.80	14 64 0 16 X SQ= 2.11 P = 0.117 RV NO (NO)
127	ALWAYS ARE THOSE WHERE INTEREST ARTICULATION BY POLITICAL PARTIES IS MODERATE, LIMITED, OR NEGLIGIBLE (72)	1.00	127	TEND LESS TO BE THOSE WHERE INTEREST ARTICULATION BY POLITICAL PARTIES IS MODERATE, LIMITED, OR NEGLIGIBLE (72)	0.73	0 21 14 58 X SQ= 3.41 P = 0.034 RV NO (NO)
128	TEND TO BE THOSE WHERE INTEREST ARTICULATION BY POLITICAL PARTIES IS LIMITED OR NEGLIGIBLE (45)	0.86	128	TEND TO BE THOSE WHERE INTEREST ARTICULATION BY POLITICAL PARTIES IS SIGNIFICANT OR MODERATE (48)	0.58	2 46 12 33 X SQ= 7.52 P = 0.003 RV YES (NO)
129	LEAN TOWARD BEING THOSE WHERE INTEREST ARTICULATION BY POLITICAL PARTIES IS NEGLIGIBLE (37)	0.64	129	LEAN TOWARD BEING THOSE WHERE INTEREST ARTICULATION BY POLITICAL PARTIES IS SIGNIFICANT, MODERATE, OR LIMITED (56)	0.65	5 51 9 28 X SQ= 3.01 P = 0.073 RV YES (NO)
133	TEND TO BE THOSE WHERE INTEREST AGGREGATION BY THE EXECUTIVE IS SIGNIFICANT (29)	0.71	133	TEND TO BE THOSE WHERE INTEREST AGGREGATION BY THE EXECUTIVE IS MODERATE, LIMITED, OR NEGLIGIBLE (76)	0.78	10 19 4 67 X SQ= 11.94 P = 0. RV YES (NO)
137	ALWAYS ARE THOSE WHERE INTEREST AGGREGATION BY THE LEGISLATURE IS LIMITED OR NEGLIGIBLE (68)	1.00	137	TEND LESS TO BE THOSE WHERE INTEREST AGGREGATION BY THE LEGISLATURE IS LIMITED OR NEGLIGIBLE (68)	0.63	0 29 14 50 X SQ= 5.86 P = 0.004 RV NO (NO)
138	TEND TO BE THOSE WHERE INTEREST AGGREGATION BY THE LEGISLATURE IS NEGLIGIBLE (49)	0.93	138	TEND TO BE THOSE WHERE INTEREST AGGREGATION BY THE LEGISLATURE IS SIGNIFICANT, MODERATE, OR LIMITED (48)	0.58	1 46 13 33 X SQ= 10.46 P = 0. RV YES (NO)
139	LEAN TOWARD BEING THOSE WHERE THE PARTY SYSTEM IS QUANTITATIVELY ONE-PARTY (34)	0.62	139	LEAN TOWARD BEING THOSE WHERE THE PARTY SYSTEM IS QUANTITATIVELY OTHER THAN ONE-PARTY (71)	0.70	8 26 5 61 X SQ= 3.74 P = 0.055 RV YES (NO)
142	ALWAYS ARE THOSE WHERE THE PARTY SYSTEM IS QUANTITATIVELY OTHER THAN MULTI-PARTY (66)	1.00	142	TEND LESS TO BE THOSE WHERE THE PARTY SYSTEM IS QUANTITATIVELY OTHER THAN MULTI-PARTY (66)	0.62	0 30 13 48 X SQ= 5.82 P = 0.004 RV NO (NO)

158	TEND TO BE THOSE WHERE PERSONALISMO IS PRONOUNCED OR MODERATE (40)	0.82	158	TEND TO BE THOSE WHERE PERSONALISMO IS NEGLIGIBLE (56)	0.64	9 31 2 54 X SQ= 6.48 P = 0.007 RV YES (NO)
160	ALWAYS ARE THOSE WHERE THE POLITICAL LEADERSHIP IS MODFRATELY ELITIST OR NON-ELITIST (67)	1.00	160	TEND LESS TO BE THOSE WHERE THE POLITICAL LEADERSHIP IS MODERATELY ELITIST OR NON-ELITIST (67)	0.66	0 27 13 53 X SQ= 4.65 P = 0.009 RV NO (NO)
161	TEND TO BE THOSE WHERE THE POLITICAL LEADERSHIP IS NON-ELITIST (50)	0.85	161	TEND TO BE THOSE WHERE THE POLITICAL LEADERSHIP IS ELITIST OR MODERATELY ELITIST (47)	0.51	2 41 11 39 X SQ= 4.43 P = 0.018 RV YES (NO)
164	TEND TO BE THOSE WHERE THE REGIME'S LEADERSHIP CHARISMA IS PRONOUNCED OR MODERATE (34)	0.82	164	TEND TO BE THOSE WHERE THE REGIME'S LEADERSHIP CHARISMA IS NEGLIGIBLE (65)	0.74	9 22 2 62 X SQ= 11.28 P = 0.001 RV YES (NO)
168	ALWAYS ARE THOSE WHERE THE HORIZONTAL POWER DISTRIBUTION IS LIMITED OR NEGLIGIBLE (72)	1.00	168	TEND LESS TO BE THOSE WHERE THE HORIZONTAL POWER DISTRIBUTION IS LIMITED OR NEGLIGIBLE (72)	0.61	0 34 14 53 X SQ= 6.59 P = 0.002 RV NO (NO)
170	TEND TO BE THOSE WHERE THE LEGISLATIVE-EXECUTIVE STRUCTURE IS PRESIDENTIAL (39)	0.69	170	TEND TO BE THOSE WHERE THE LEGISLATIVE-EXECUTIVE STRUCTURE IS OTHER THAN PRESIDENTIAL (63)	0.66	9 29 4 56 X SQ= 4.47 P = 0.029 RV YES (NO)
175	TEND TO BE THOSE WHERE THE LEGISLATURE IS LARGELY INEFFECTIVE OR WHOLLY INEFFECTIVE (49)	0.85	175	TEND TO BE THOSE WHERE THE LEGISLATURE IS FULLY EFFECTIVE OR PARTIALLY EFFECTIVE (51)	0.58	2 48 11 35 X SQ= 6.50 P = 0.006 RV YES (NO)
176	ALWAYS ARE THOSE WHERE THE LEGISLATURE IS FULLY EFFECTIVE, PARTIALLY EFFECTIVE, OR LARGELY INEFFECTIVE (72)	1.00	176	TEND LESS TO BE THOSE WHERE THE LEGISLATURE IS FULLY EFFECTIVE, PARTIALLY EFFECTIVE, OR LARGELY INEFFECTIVE (72)	0.70	13 58 0 25 X SQ= 3.85 P = 0.018 RV NO (NO)
178	TEND TO BE THOSE WHERE THE LEGISLATURE IS UNICAMERAL (53)	0.93	178	TEND TO BE THOSE WHERE THE LEGISLATURE IS BICAMERAL (51)	0.55	13 39 1 47 X SQ= 9.07 P = 0.001 RV YES (NO)

182	ALWAYS ARE THOSE WHERE THE BUREAUCRACY IS POST-COLONIAL TRANSITIONAL, RATHER THAN SEMI-MODERN (25)	1.00	182	TEND TO BE THOSE WHERE THE BUREAUCRACY IS SEMI-MODERN, RATHER THAN POST-COLONIAL TRANSITIONAL (55)	0.83	0 53 14 11 X SQ= 32.47 P = 0. RV YES (YES)
183	ALWAYS ARE THOSE WHERE THE BUREAUCRACY IS POST-COLONIAL TRANSITIONAL, RATHER THAN TRADITIONAL (25)	1.00	183	TEND LESS TO BE THOSE WHERE THE BUREAUCRACY IS POST-COLONIAL TRANSITIONAL, RATHER THAN TRADITIONAL (25)	0.65	14 11 0 6 X SQ= 4.07 P = 0.021 RV NO (YES)
186	ALWAYS ARE THOSE WHERE PARTICIPATION BY THE MILITARY IN POLITICS IS NEUTRAL, RATHER THAN SUPPORTIVE (56)	1.00	186	TEND LESS TO BE THOSE WHERE PARTICIPATION BY THE MILITARY IN POLITICS IS NEUTRAL, RATHER THAN SUPPORTIVE (56)	0.60	0 28 13 42 X SQ= 6.16 P = 0.003 RV NO (NO)
188	ALWAYS ARE THOSE WHERE THE CHARACTER OF THE LEGAL SYSTEM IS OTHER THAN CIVIL LAW (81)	1.00	188	TEND LESS TO BE THOSE WHERE THE CHARACTER OF THE LEGAL SYSTEM IS OTHER THAN CIVIL LAW (81)	0.67	0 31 14 64 X SQ= 4.88 P = 0.009 RV NO (NO)
193	TEND TO BE THOSE WHERE THE CHARACTER OF THE LEGAL SYSTEM IS PARTLY INDIGENOUS (28)	0.93	193	TEND TO BE THOSE WHERE THE CHARACTER OF THE LEGAL SYSTEM IS OTHER THAN PARTLY INDIGENOUS (86)	0.85	13 14 1 81 X SQ= 35.88 P = 0. RV YES (NO)

```
******************************************************************************************

149  POLITIES                                          149  POLITIES
     WHERE THE PARTY SYSTEM IS QUALITATIVELY                WHERE THE PARTY SYSTEM IS QUALITATIVELY
     BROADLY AGGREGATIVE, RATHER THAN                       PERSONALISTIC, SITUATIONAL, OR AD HOC,
     PERSONALISTIC, SITUATIONAL, OR AD HOC  (4)             RATHER THAN BROADLY AGGREGATIVE  (4)


                                                                                       SUBJECT ONLY
..........................................................................................

      4 IN LEFT COLUMN

CANADA      MEXICO      PHILIPPINES US

      4 IN RIGHT COLUMN

EL SALVADOR LAOS        LEBANON     PANAMA

     13 EXCLUDED BECAUSE AMBIGUOUS

CAMBODIA    CEYLON      FRANCE      GREECE      GUATEMALA   INDIA       INDONESIA   IRELAND     MOROCCO     SO AFRICA
TURKEY      URUGUAY     VIETNAM REP

     83 EXCLUDED BECAUSE IRRELEVANT

AFGHANISTAN ALBANIA     ALGERIA     ARGENTINA   AUSTRALIA   AUSTRIA     BELGIUM     BOLIVIA     BRAZIL      BULGARIA
BURUNDI     CAMEROUN    CEN AFR REP CHAD        CHILE       CHINA, PR   COLOMBIA    CONGO(BRA)  CONGO(LEO)  COSTA RICA
CUBA        CYPRUS      CZECHOS'KIA DAHOMEY     DENMARK     DOMIN REP   ECUADOR     ETHIOPIA    FINLAND     GABON
GERMANY, E  GERMAN FR   GHANA       GUINEA      HAITI       HONDURAS    HUNGARY     ICELAND     ISRAEL      ITALY
IVORY COAST JAMAICA     JAPAN       KOREA, N    LIBERIA     LIBYA       LUXEMBOURG  MALAGASY R  MALAYA      MALI
MAURITANIA  MONGOLIA    NETHERLANDS NEW ZEALAND NICARAGUA   NIGER       NIGERIA     NORWAY      PARAGUAY    PERU
POLAND      PORTUGAL    RUMANIA     RWANDA      SA'U ARABIA SENEGAL     SIERRE LEO  SOMALIA     SPAIN       SWEDEN
SWITZERLAND TANGANYIKA  TOGO        TRINIDAD    TUNISIA     UGANDA      USSR        UAR         UK          UPPER VOLTA
VENEZUELA   VIETNAM, N  YUGOSLAVIA

     11 EXCLUDED BECAUSE UNASCERTAINABLE

BURMA       IRAN        IRAQ        JORDAN      KOREA REP   NEPAL       PAKISTAN    SUDAN       SYRIA       THAILAND
YEMEN

------------------------------------------------------------------------------------------

 21  TILT TOWARD BEING THOSE               0.75      21  ALWAYS ARE THOSE                   1.00          3    0
     WHOSE TERRITORIAL SIZE IS                           WHOSE TERRITORIAL SIZE IS                       1    4
     VERY LARGE OR LARGE  (32)                           MEDIUM OR SMALL  (83)                      X SQ=   2.13
                                                                                                    P =    0.143
                                                                                                    RV YES (YES)

 22  ALWAYS ARE THOSE                      1.00      22  TILT TOWARD BEING THOSE            0.75          4    1
     WHOSE TERRITORIAL SIZE IS                           WHOSE TERRITORIAL SIZE IS                       0    3
     VERY LARGE, LARGE, OR MEDIUM  (68)                  SMALL  (47)                                X SQ=   2.13
                                                                                                    P =    0.143
                                                                                                    RV YES (YES)
```

23	ALWAYS ARE THOSE WHOSE POPULATION IS VERY LARGE OR LARGE (27)	1.00	23	ALWAYS ARE THOSE WHOSE POPULATION IS MEDIUM OR SMALL (88)	1.00	4 0 0 4 X SQ= 4.50 P = 0.029 RV YES (YES)
24	ALWAYS ARE THOSE WHOSE POPULATION IS VERY LARGE, LARGE, OR MEDIUM (61)	1.00	24	ALWAYS ARE THOSE WHOSE POPULATION IS SMALL (54)	1.00	4 0 0 4 X SQ= 4.50 P = 0.029 RV YES (YES)
33	TILT TOWARD BEING THOSE WHOSE GROSS NATIONAL PRODUCT IS VERY HIGH, HIGH, OR MEDIUM (30)	0.75	33	ALWAYS ARE THOSE WHOSE GROSS NATIONAL PRODUCT IS LOW OR VERY LOW (85)	1.00	3 0 1 4 X SQ= 2.13 P = 0.143 RV YES (YES)
34	ALWAYS ARE THOSE WHOSE GROSS NATIONAL PRODUCT IS VERY HIGH, HIGH, MEDIUM, OR LOW (62)	1.00	34	ALWAYS ARE THOSE WHOSE GROSS NATIONAL PRODUCT IS VERY LOW (53)	1.00	4 0 0 4 X SQ= 4.50 P = 0.029 RV YES (YES)
39	ALWAYS ARE THOSE WHOSE INTERNATIONAL FINANCIAL STATUS IS VERY HIGH, HIGH, OR MEDIUM (38)	1.00	39	ALWAYS ARE THOSE WHOSE INTERNATIONAL FINANCIAL STATUS IS LOW OR VERY LOW (76)	1.00	4 0 0 4 X SQ= 4.50 P = 0.029 RV YES (YES)
40	ALWAYS ARE THOSE WHOSE INTERNATIONAL FINANCIAL STATUS IS VERY HIGH, HIGH, MEDIUM, OR LOW (71)	1.00	40	TILT TOWARD BEING THOSE WHOSE INTERNATIONAL FINANCIAL STATUS IS VERY LOW (39)	0.75	4 1 0 3 X SQ= 2.13 P = 0.143 RV YES (YES)
42	TILT TOWARD BEING THOSE WHOSE ECONOMIC DEVELOPMENTAL STATUS IS DEVELOPED OR INTERMEDIATE (36)	0.75	42	ALWAYS ARE THOSE WHOSE ECONOMIC DEVELOPMENTAL STATUS IS UNDERDEVELOPED OR VERY UNDERDEVELOPED (76)	1.00	3 0 1 4 X SQ= 2.13 P = 0.143 RV YES (YES)
43	ALWAYS ARE THOSE WHOSE ECONOMIC DEVELOPMENTAL STATUS IS DEVELOPED, INTERMEDIATE, OR UNDERDEVELOPED (55)	1.00	43	ALWAYS ARE THOSE WHOSE ECONOMIC DEVELOPMENTAL STATUS IS VERY UNDERDEVELOPED (57)	1.00	4 0 0 3 X SQ= 3.51 P = 0.029 RV YES (YES)
45	ALWAYS ARE THOSE WHERE THE LITERACY RATE IS FIFTY PERCENT OR ABOVE (55)	1.00	45	TILT TOWARD BEING THOSE WHERE THE LITERACY RATE IS BELOW FIFTY PERCENT (54)	0.67	4 1 0 2 X SQ= 1.18 P = 0.143 RV YES (YES)

50	ALWAYS ARE THOSE WHERE FREEDOM OF THE PRESS IS COMPLETE (43)	1.00	50	TILT TOWARD BEING THOSE WHERE FREEDOM OF THE PRESS IS INTERMITTENT, INTERNALLY ABSENT, OR INTERNALLY AND EXTERNALLY ABSENT (56)	0.67	4 1 0 2 X SQ= 1.18 P = 0.143 RV YES (YES)
85	ALWAYS ARE THOSE WHERE THE STAGE OF POLITICAL MODERNIZATION IS ADVANCED, RATHER THAN MID- OR EARLY TRANSITIONAL (60)	1.00	85	TILT TOWARD BEING THOSE WHERE THE STAGE OF POLITICAL MODERNIZATION IS MID- OR EARLY TRANSITIONAL, RATHER THAN ADVANCED (54)	0.75	4 1 0 3 X SQ= 2.13 P = 0.143 RV YES (YES)
89	ALWAYS ARE THOSE WHOSE IDEOLOGICAL ORIENTATION IS CONVENTIONAL (33)	1.00	89	TILT TOWARD BEING THOSE WHOSE IDEOLOGICAL ORIENTATION IS OTHER THAN CONVENTIONAL (62)	0.67	4 1 0 2 X SQ= 1.18 P = 0.143 RV YES (YES)
99	ALWAYS ARE THOSE WHERE GOVERNMENTAL STABILITY IS GENERALLY PRESENT AND DATES FROM AT LEAST THE INTER-WAR PERIOD, OR FROM THE POST-WAR PERIOD (50)	1.00	99	ALWAYS ARE THOSE WHERE GOVERNMENTAL STABILITY IS MODERATELY PRESENT AND DATES FROM THE POST-WAR PERIOD, OR IS ABSENT (36)	1.00	4 0 0 4 X SQ= 4.50 P = 0.029 RV YES (YES)
100	ALWAYS ARE THOSE WHERE GOVERNMENTAL STABILITY IS GENERALLY PRESENT AND DATES FROM AT LEAST THE INTER-WAR PERIOD, OR IS GENERALLY OR MODERATELY PRESENT, AND DATES FROM THE POST-WAR PERIOD (64)	1.00	100	ALWAYS ARE THOSE WHERE GOVERNMENTAL STABILITY IS ABSENT (22)	1.00	4 0 0 4 X SQ= 4.50 P = 0.029 RV YES (YES)
101	ALWAYS ARE THOSE WHERE THE REPRESENTATIVE CHARACTER OF THE REGIME IS POLYARCHIC (41)	1.00	101	TILT TOWARD BEING THOSE WHERE THE REPRESENTATIVE CHARACTER OF THE REGIME IS LIMITED POLYARCHIC, PSEUDO-POLYARCHIC, OR NON-POLYARCHIC (57)	0.67	4 1 0 2 X SQ= 1.18 P = 0.143 RV YES (YES)
111	ALWAYS ARE THOSE WHERE POLITICAL ENCULTURATION IS HIGH OR MEDIUM (53)	1.00	111	ALWAYS ARE THOSE WHERE POLITICAL ENCULTURATION IS LOW (42)	1.00	4 0 0 4 X SQ= 4.50 P = 0.029 RV YES (YES)
115	TILT TOWARD BEING THOSE WHERE INTEREST ARTICULATION BY ASSOCIATIONAL GROUPS IS SIGNIFICANT (20)	0.75	115	ALWAYS ARE THOSE WHERE INTEREST ARTICULATION BY ASSOCIATIONAL GROUPS IS MODERATE, LIMITED, OR NEGLIGIBLE (91)	1.00	3 0 1 4 X SQ= 2.13 P = 0.143 RV YES (YES)
116	ALWAYS ARE THOSE WHERE INTEREST ARTICULATION BY ASSOCIATIONAL GROUPS IS SIGNIFICANT OR MODERATE (32)	1.00	116	TILT TOWARD BEING THOSE WHERE INTEREST ARTICULATION BY ASSOCIATIONAL GROUPS IS LIMITED OR NEGLIGIBLE (79)	0.75	4 1 0 3 X SQ= 2.13 P = 0.143 RV YES (YES)

119	TILT TOWARD BEING THOSE WHERE INTEREST ARTICULATION BY INSTITUTIONAL GROUPS IS MODERATE OR LIMITED (26)	0.75	119	ALWAYS ARE THOSE WHERE INTEREST ARTICULATION BY INSTITUTIONAL GROUPS IS VERY SIGNIFICANT OR SIGNIFICANT (74)	1.00	1 4 3 0 X SQ= 2.13 P = 0.143 RV YES (YES)
125	TILT TOWARD BEING THOSE WHERE INTEREST ARTICULATION BY ANOMIC GROUPS IS INFREQUENT OR VERY INFREQUENT (35)	0.67	125	ALWAYS ARE THOSE WHERE INTEREST ARTICULATION BY ANOMIC GROUPS IS FREQUENT OR OCCASIONAL (64)	1.00	1 4 2 0 X SQ= 1.18 P = 0.143 RV YES (YES)
128	ALWAYS ARE THOSE WHERE INTEREST ARTICULATION BY POLITICAL PARTIES IS LIMITED OR NEGLIGIBLE (45)	1.00	128	ALWAYS ARE THOSE WHERE INTEREST ARTICULATION BY POLITICAL PARTIES IS SIGNIFICANT OR MODERATE (48)	1.00	0 3 4 0 X SQ= 3.51 P = 0.029 RV YES (YES)
129	ALWAYS ARE THOSE WHERE INTEREST ARTICULATION BY POLITICAL PARTIES IS NEGLIGIBLE (37)	1.00	129	ALWAYS ARE THOSE WHERE INTEREST ARTICULATION BY POLITICAL PARTIES IS SIGNIFICANT, MODERATE, OR LIMITED (56)	1.00	0 3 4 0 X SQ= 3.51 P = 0.029 RV YES (YES)
130	ALWAYS ARE THOSE WHERE INTEREST AGGREGATION BY POLITICAL PARTIES IS SIGNIFICANT (12)	1.00	130	ALWAYS ARE THOSE WHERE INTEREST AGGREGATION BY POLITICAL PARTIES IS MODERATE, LIMITED, OR NEGLIGIBLE (71)	1.00	4 0 0 4 X SQ= 4.50 P = 0.029 RV YES (YES)
132	ALWAYS ARE THOSE WHERE INTEREST AGGREGATION BY POLITICAL PARTIES IS SIGNIFICANT, MODERATE, OR LIMITED (67)	1.00	132	ALWAYS ARE THOSE WHERE INTEREST AGGREGATION BY POLITICAL PARTIES IS NEGLIGIBLE (9)	1.00	4 0 0 4 X SQ= 4.50 P = 0.029 RV YES (YES)
134	ALWAYS ARE THOSE WHERE INTEREST AGGREGATION BY THE EXECUTIVE IS SIGNIFICANT OR MODERATE (57)	1.00	134	ALWAYS ARE THOSE WHERE INTEREST AGGREGATION BY THE EXECUTIVE IS LIMITED OR NEGLIGIBLE (46)	1.00	4 0 0 2 X SQ= 2.34 P = 0.067 RV YES (YES)
145	ALWAYS ARE THOSE WHERE THE PARTY SYSTEM IS QUANTITATIVELY TWO-PARTY, RATHER THAN MULTI-PARTY (11)	1.00	145	ALWAYS ARE THOSE WHERE THE PARTY SYSTEM IS QUANTITATIVELY MULTI-PARTY, RATHER THAN TWO-PARTY (30)	1.00	0 4 3 0 X SQ= 3.51 P = 0.029 RV YES (YES)
153	ALWAYS ARE THOSE WHERE THE PARTY SYSTEM IS STABLE (42)	1.00	153	ALWAYS ARE THOSE WHERE THE PARTY SYSTEM IS MODERATELY STABLE OR UNSTABLE (38)	1.00	4 0 0 4 X SQ= 4.50 P = 0.029 RV YES (YES)

154	ALWAYS ARE THOSE WHERE THE PARTY SYSTEM IS STABLE OR MODERATELY STABLE (55)	1.00	154	ALWAYS ARE THOSE WHERE THE PARTY SYSTEM IS UNSTABLE (25)	1.00	4 0 0 4 X SQ= 4.50 P = 0.029 RV YES (YES)
157	ALWAYS ARE THOSE WHERE PERSONALISMO IS MODERATE OR NEGLIGIBLE (84)	1.00	157	ALWAYS ARE THOSE WHERE PERSONALISMO IS PRONOUNCED (14)	1.00	0 4 4 0 X SQ= 4.50 P = 0.029 RV YES (YES)
158	ALWAYS ARE THOSE WHERE PERSONALISMO IS NEGLIGIBLE (56)	1.00	158	ALWAYS ARE THOSE WHERE PERSONALISMO IS PRONOUNCED OR MODERATE (40)	1.00	0 4 4 0 X SQ= 4.50 P = 0.029 RV YES (YES)
160	ALWAYS ARE THOSE WHERE THE POLITICAL LEADERSHIP IS MODERATELY ELITIST OR NON-ELITIST (67)	1.00	160	TILT TOWARD BEING THOSE WHERE THE POLITICAL LEADERSHIP IS ELITIST (30)	0.67	0 2 4 1 X SQ= 1.18 P = 0.143 RV YES (YES)
168	TILT TOWARD BEING THOSE WHERE THE HORIZONTAL POWER DISTRIBUTION IS SIGNIFICANT (34)	0.75	168	ALWAYS ARE THOSE WHERE THE HORIZONTAL POWER DISTRIBUTION IS LIMITED OR NEGLIGIBLE (72)	1.00	3 0 1 4 X SQ= 2.13 P = 0.143 RV YES (YES)
174	TILT TOWARD BEING THOSE WHERE THE LEGISLATURE IS FULLY EFFECTIVE (28)	0.75	174	ALWAYS ARE THOSE WHERE THE LEGISLATURE IS PARTIALLY EFFECTIVE, LARGELY INEFFECTIVE, OR WHOLLY INEFFECTIVE (72)	1.00	3 0 1 4 X SQ= 2.13 P = 0.143 RV YES (YES)
178	ALWAYS ARE THOSE WHERE THE LEGISLATURE IS BICAMERAL (51)	1.00	178	TILT TOWARD BEING THOSE WHERE THE LEGISLATURE IS UNICAMERAL (53)	0.75	0 3 4 1 X SQ= 2.13 P = 0.143 RV YES (YES)
187	ALWAYS ARE THOSE WHERE THE ROLE OF THE POLICE IS NOT POLITICALLY SIGNIFICANT (35)	1.00	187	ALWAYS ARE THOSE WHERE THE ROLE OF THE POLICE IS POLITICALLY SIGNIFICANT (66)	1.00	0 3 4 0 X SQ= 3.51 P = 0.029 RV YES (YES)

```
****************************************************************************************************

150  POLITIES                                          150  POLITIES
     WHERE THE PARTY SYSTEM IS QUALITATIVELY                WHERE THE PARTY SYSTEM IS QUALITATIVELY
     LATIN LIBERAL CONSERVATIVE, RATHER THAN                LATIN SOCIAL REVOLUTIONARY, RATHER THAN
     LATIN SOCIAL REVOLUTIONARY  (5)                        LATIN LIBERAL CONSERVATIVE  (4)

                                                                                        SUBJECT ONLY
....................................................................................................

       5 IN LEFT COLUMN

COLOMBIA    ECUADOR     HONDURAS    NICARAGUA   PARAGUAY

       4 IN RIGHT COLUMN

BOLIVIA     COSTA RICA  DOMIN REP   VENEZUELA

       1 EXCLUDED BECAUSE AMBIGUOUS

URUGUAY

     104 EXCLUDED BECAUSE IRRELEVANT

AFGHANISTAN ALBANIA     ALGERIA     ARGENTINA   AUSTRALIA   AUSTRIA     BELGIUM     BRAZIL      BULGARIA    BURMA
BURUNDI     CAMBODIA    CAMEROUN    CANADA      CEN AFR REP CEYLON      CHAD        CHILE       CHINA, PR   CONGO(BRA)
CONGO(LEO)  CYPRUS      CZECHOS'KIA DAHOMEY     DENMARK     EL SALVADOR ETHIOPIA    FINLAND     FRANCE      GABON
GERMANY, E  GERMAN FR   GHANA       GREECE      GUATEMALA   GUINEA      HAITI       HUNGARY     ICELAND     INDIA
INDONESIA   IRAN        IRAQ        IRELAND     ISRAEL      ITALY       IVORY COAST JAMAICA     JAPAN       JORDAN
KOREA, N    KOREA REP   LAOS        LEBANON     LIBERIA     LIBYA       LUXEMBOURG  MALAGASY R  MALAYA      MALI
MAURITANIA  MEXICO      MONGOLIA    MOROCCO     NEPAL       NETHERLANDS NEW ZEALAND NIGER       NIGERIA     NORWAY
PAKISTAN    PANAMA      PERU        PHILIPPINES POLAND      PORTUGAL    RUMANIA     RWANDA      SA'U ARABIA SENEGAL
SIERRE LEO  SOMALIA     SO AFRICA   SPAIN       SUDAN       SWEDEN      SWITZERLAND SYRIA       TANGANYIKA  THAILAND
TOGO        TRINIDAD    TUNISIA     TURKEY      UGANDA      USSR        UAR         UK          US          UPPER VOLTA
VIETNAM, N  VIETNAM REP YEMEN       YUGOSLAVIA

       1 EXCLUDED BECAUSE UNASCERTAINABLE

CUBA

----------------------------------------------------------------------------------------------------

 87  ALWAYS ARE THOSE                       1.00     87  TILT TOWARD BEING THOSE              0.75        0    3
     WHOSE IDEOLOGICAL ORIENTATION                       WHOSE IDEOLOGICAL ORIENTATION                    3    1
     IS OTHER THAN DEVELOPMENTAL  (58)                   IS DEVELOPMENTAL  (31)                      X SQ=   1.47
                                                                                                     P =    0.143
                                                                                                     RV YES (YES)

 89  ALWAYS ARE THOSE                       1.00     89  TILT TOWARD BEING THOSE              0.75        3    1
     WHOSE IDEOLOGICAL ORIENTATION                       WHOSE IDEOLOGICAL ORIENTATION                    0    3
     IS CONVENTIONAL  (33)                               IS OTHER THAN CONVENTIONAL  (62)            X SQ=   1.47
                                                                                                     P =    0.143
                                                                                                     RV YES (YES)
```

```
101  ALWAYS ARE THOSE                              1.00    101  ALWAYS ARE THOSE                          1.00          0     4
     WHERE THE REPRESENTATIVE CHARACTER                         WHERE THE REPRESENTATIVE CHARACTER                       5     0
     OF THE REGIME IS LIMITED POLYARCHIC,                       OF THE REGIME IS POLYARCHIC  (41)                X SQ=   5.41
     PSEUDO-POLYARCHIC, OR                                                                                   P =    0.008
     NON-POLYARCHIC  (57)                                                                                    RV YES (YES)

127  ALWAYS ARE THOSE                              1.00    127  TILT TOWARD BEING THOSE                   0.67          0     2
     WHERE INTEREST ARTICULATION                                WHERE INTEREST ARTICULATION                              5     1
     BY POLITICAL PARTIES                                       BY POLITICAL PARTIES                             X SQ=   1.60
     IS MODERATE, LIMITED, OR                                   IS SIGNIFICANT  (21)                             P =    0.107
     NEGLIGIBLE  (72)                                                                                        RV YES (YES)

131  ALWAYS ARE THOSE                              1.00    131  TILT TOWARD BEING THOSE                   0.67          0     2
     WHERE INTEREST AGGREGATION                                 WHERE INTEREST AGGREGATION                               4     1
     BY POLITICAL PARTIES                                       BY POLITICAL PARTIES                             X SQ=   1.18
     IS LIMITED OR NEGLIGIBLE (35)                              IS SIGNIFICANT OR MODERATE  (30)                 P =    0.143
                                                                                                             RV YES (YES)

160  TEND TO BE THOSE                              0.80    160  ALWAYS ARE THOSE                          1.00          4     0
     WHERE THE POLITICAL LEADERSHIP IS                          WHERE THE POLITICAL LEADERSHIP IS                        1     4
     ELITIST  (30)                                              MODERATELY ELITIST OR NON-ELITIST  (67)          X SQ=   2.98
                                                                                                             P =    0.048
                                                                                                             RV YES (YES)

161  ALWAYS ARE THOSE                              1.00    161  ALWAYS ARE THOSE                          1.00          5     0
     WHERE THE POLITICAL LEADERSHIP IS                          WHERE THE POLITICAL LEADERSHIP IS                        0     4
     ELITIST OR MODERATELY ELITIST  (47)                        NON-ELITIST  (50)                                X SQ=   5.41
                                                                                                             P =    0.008
                                                                                                             RV YES (YES)
```

151 POLITIES
LOCATED IN CENTRAL AND SOUTH AFRICA AND
WHERE THE PARTY SYSTEM IS QUALITATIVELY
REGIONAL OR REGIONAL-ETHNIC (5)

151 POLITIES
LOCATED IN CENTRAL AND SOUTH AFRICA AND
WHERE THE PARTY SYSTEM IS QUALITATIVELY
OTHER THAN
REGIONAL OR REGIONAL ETHNIC (22)

SUBJECT ONLY

5 IN LEFT COLUMN

CAMEROUN CONGO(LEO) NIGERIA SOMALIA UGANDA

22 IN RIGHT COLUMN

BURUNDI	CEN AFR REP	CHAD	CONGO(BRA)	DAHOMEY	ETHIOPIA	GABON	GHANA	GUINEA	IVORY COAST
LIBERIA	MALAGASY R	MALI	MAURITANIA	NIGER	RWANDA	SENEGAL	SIERRE LEO	SO AFRICA	TANGANYIKA
TOGO	UPPER VOLTA								

87 EXCLUDED BECAUSE IRRELEVANT

AFGHANISTAN	ALBANIA	ALGERIA	ARGENTINA	AUSTRALIA	AUSTRIA	BELGIUM	BOLIVIA	BRAZIL	BULGARIA
BURMA	CAMBODIA	CANADA	CEYLON	CHILE	CHINA, PR	COLOMBIA	COSTA RICA	CUBA	CYPRUS
CZECHOS'KIA	DENMARK	DOMIN REP	ECUADOR	EL SALVADOR	FINLAND	FRANCE	GERMANY, E	GERMAN FR	GREECE
GUATEMALA	HAITI	HONDURAS	HUNGARY	ICELAND	INDIA	INDONESIA	IRAN	IRAQ	IRELAND
ISRAEL	ITALY	JAMAICA	JAPAN	JORDAN	KOREA, N	KOREA REP	LAOS	LEBANON	LIBYA
LUXEMBOURG	MALAYA	MEXICO	MONGOLIA	MOROCCO	NEPAL	NETHERLANDS	NEW ZEALAND	NICARAGUA	NORWAY
PAKISTAN	PANAMA	PARAGUAY	PERU	PHILIPPINES	POLAND	PORTUGAL	RUMANIA	SA'U ARABIA	SPAIN
SWEDEN	SWITZERLAND	SYRIA	THAILAND	TRINIDAD	TUNISIA	TURKEY	USSR	UAR	UK
US	URUGUAY	VENEZUELA	VIETNAM, N	VIETNAM REP	YEMEN	YUGOSLAVIA			

1 EXCLUDED BECAUSE UNASCERTAINABLE

SUDAN

	Left column		Right column			
24	LEAN TOWARD BEING THOSE WHOSE POPULATION IS VERY LARGE, LARGE, OR MEDIUM (61)	0.60	LEAN TOWARD BEING THOSE WHOSE POPULATION IS SMALL (54)	0.82	3 4 2 18	X SQ= 1.85 P = 0.091 RV YES (NO)
28	ALWAYS ARE THOSE WHOSE POPULATION GROWTH RATE IS LOW (48)	1.00	TEND TO BE THOSE WHOSE POPULATION GROWTH RATE IS HIGH (62)	0.63	0 12 5 7	X SQ= 4.04 P = 0.037 RV YES (NO)
80	TILT TOWARD BEING THOSE WHOSE DATE OF INDEPENDENCE IS AFTER 1914, AND THAT WERE FORMERLY DEPENDENCIES OF BRITAIN, RATHER THAN FRANCE (19)	0.67	TILT TOWARD BEING THOSE WHOSE DATE OF INDEPENDENCE IS AFTER 1914, AND THAT WERE FORMERLY DEPENDENCIES OF FRANCE, RATHER THAN BRITAIN (24)	0.82	2 3 1 14	X SQ= 1.18 P = 0.140 RV YES (NO)

151/

No.	Left	Prop.	No.	Right	Prop.	Statistics
102	ALWAYS ARE THOSE WHERE THE REPRESENTATIVE CHARACTER OF THE REGIME IS POLYARCHIC OR LIMITED POLYARCHIC (59)	1.00	102	TEND TO BE THOSE WHERE THE REPRESENTATIVE CHARACTER OF THE REGIME IS PSEUDO-POLYARCHIC OR NON-POLYARCHIC (49)	0.67	5 7 0 14 X SQ= 4.79 P = 0.012 RV YES (NO)
113	TEND TO BE THOSE WHERE SECTIONALISM IS EXTREME (27)	0.80	113	TEND TO BE THOSE WHERE SECTIONALISM IS MODERATE OR NEGLIGIBLE (81)	0.95	4 1 1 19 X SQ= 9.77 P = 0.002 RV YES (YES)
127	TEND TO BE THOSE WHERE INTEREST ARTICULATION BY POLITICAL PARTIES IS SIGNIFICANT (21)	0.60	127	ALWAYS ARE THOSE WHERE INTEREST ARTICULATION BY POLITICAL PARTIES IS MODERATE, LIMITED, OR NEGLIGIBLE (72)	1.00	3 0 2 20 X SQ= 8.55 P = 0.004 RV YES (YES)
128	ALWAYS ARE THOSE WHERE INTEREST ARTICULATION BY POLITICAL PARTIES IS SIGNIFICANT OR MODERATE (48)	1.00	128	TEND TO BE THOSE WHERE INTEREST ARTICULATION BY POLITICAL PARTIES IS LIMITED OR NEGLIGIBLE (45)	0.85	5 3 0 17 X SQ= 9.66 P = 0.001 RV YES (YES)
133	LEAN TOWARD BEING THOSE WHERE INTEREST AGGREGATION BY THE EXECUTIVE IS MODERATE, LIMITED, OR NEGLIGIBLE (76)	0.80	133	LEAN TOWARD BEING THOSE WHERE INTEREST AGGREGATION BY THE EXECUTIVE IS SIGNIFICANT (29)	0.71	1 15 4 6 X SQ= 2.60 P = 0.055 RV YES (NO)
134	TEND TO BE THOSE WHERE INTEREST AGGREGATION BY THE EXECUTIVE IS LIMITED OR NEGLIGIBLE (46)	0.75	134	TEND TO BE THOSE WHERE INTEREST AGGREGATION BY THE EXECUTIVE IS SIGNIFICANT OR MODERATE (57)	0.86	1 18 3 3 X SQ= 3.87 P = 0.031 RV YES (YES)
138	TEND TO BE THOSE WHERE INTEREST AGGREGATION BY THE LEGISLATURE IS SIGNIFICANT, MODERATE, OR LIMITED (48)	0.60	138	TEND TO BE THOSE WHERE INTEREST AGGREGATION BY THE LEGISLATURE IS NEGLIGIBLE (49)	0.95	3 1 2 20 X SQ= 5.70 P = 0.014 RV YES (YES)
142	TEND TO BE THOSE WHERE THE PARTY SYSTEM IS QUANTITATIVELY MULTI-PARTY (30)	0.60	142	ALWAYS ARE THOSE WHERE THE PARTY SYSTEM IS QUANTITATIVELY OTHER THAN MULTI-PARTY (66)	1.00	3 0 2 20 X SQ= 8.55 P = 0.004 RV YES (YES)
154	ALWAYS ARE THOSE WHERE THE PARTY SYSTEM IS UNSTABLE (25)	1.00	154	LEAN TOWARD BEING THOSE WHERE THE PARTY SYSTEM IS STABLE OR MODERATELY STABLE (55)	0.89	0 8 2 1 X SQ= 2.81 P = 0.055 RV YES (YES)

166	TEND TO BE THOSE WHERE THE VERTICAL POWER DISTRIBUTION IS THAT OF EFFECTIVE FEDERALISM OR LIMITED FEDERALISM (15)	0.75	166	ALWAYS ARE THOSE WHERE THE VERTICAL POWER DISTRIBUTION IS THAT OF FORMAL FEDERALISM OR FORMAL AND EFFECTIVE UNITARISM (99)	1.00	3 0 1 22 X SQ= 12.03 P = 0.002 RV YES (YES)
169	ALWAYS ARE THOSE WHERE THE HORIZONTAL POWER DISTRIBUTION IS SIGNIFICANT OR LIMITED (58)	1.00	169	LEAN TOWARD BEING THOSE WHERE THE HORIZONTAL POWER DISTRIBUTION IS NEGLIGIBLE (48)	0.68	3 7 0 15 X SQ= 2.67 P = 0.052 RV YES (NO)
170	TILT TOWARD BEING THOSE WHERE THE LEGISLATIVE-EXECUTIVE STRUCTURE IS OTHER THAN PRESIDENTIAL (63)	0.80	170	TILT TOWARD BEING THOSE WHERE THE LEGISLATIVE-EXECUTIVE STRUCTURE IS PRESIDENTIAL (39)	0.70	1 14 4 6 X SQ= 2.34 P = 0.121 RV YES (NO)
171	LEAN LESS TOWARD BEING THOSE WHERE THE LEGISLATIVE-EXECUTIVE STRUCTURE IS OTHER THAN PARLIAMENTARY-REPUBLICAN (90)	0.60	171	LEAN MORE TOWARD BEING THOSE WHERE THE LEGISLATIVE-EXECUTIVE STRUCTURE IS OTHER THAN PARLIAMENTARY-REPUBLICAN (90)	0.95	2 1 3 21 X SQ= 2.22 P = 0.079 RV NO (YES)
175	TEND TO BE THOSE WHERE THE LEGISLATURE IS FULLY EFFECTIVE OR PARTIALLY EFFECTIVE (51)	0.75	175	TEND TO BE THOSE WHERE THE LEGISLATURE IS LARGELY INEFFECTIVE OR WHOLLY INEFFECTIVE (49)	0.86	3 3 1 18 X SQ= 3.87 P = 0.031 RV YES (YES)
179	ALWAYS ARE THOSE WHERE THE EXECUTIVE IS STRONG (39)	1.00	179	TEND TO BE THOSE WHERE THE EXECUTIVE IS DOMINANT (52)	0.88	0 15 2 2 X SQ= 3.91 P = 0.035 RV YES (YES)

```
152  POLITIES                                      152  POLITIES
     WHERE THE PARTY SYSTEM IS                          WHERE THE PARTY SYSTEM IS
     STABLE, RATHER THAN                                UNSTABLE, RATHER THAN
     UNSTABLE  (42)                                     STABLE  (25)

                                                                                SUBJECT ONLY

     42 IN LEFT COLUMN

ALBANIA      AUSTRALIA   AUSTRIA     BELGIUM     BULGARIA     CANADA      CHINA, PR   CZECHOS'KIA DENMARK     FINLAND
GERMANY, E   GERMAN FR   GUINEA      HUNGARY     ICELAND      INDIA       IRELAND     ISRAEL      JAPAN       KOREA, N
LIBERIA      LUXEMBOURG  MEXICO      MONGOLIA    NETHERLANDS  NEW ZEALAND NORWAY      PHILIPPINES POLAND      PORTUGAL
RUMANIA      SO AFRICA   SPAIN       SWEDEN      SWITZERLAND  TANGANYIKA  TUNISIA     USSR        UK          US
VIETNAM, N   YUGOSLAVIA

     25 IN RIGHT COLUMN

ARGENTINA    BOLIVIA     BRAZIL      CEYLON      CHILE        CONGO(LEO)  ECUADOR     EL SALVADOR GREECE      GUATEMALA
HONDURAS     IRAQ        JORDAN      KOREA REP   LAOS         LEBANON     NEPAL       PAKISTAN    PANAMA      PERU
SIERRE LEO   SOMALIA     SYRIA       THAILAND    VENEZUELA

      3 EXCLUDED BECAUSE AMBIGUOUS

COLOMBIA     INDONESIA   URUGUAY

     18 EXCLUDED BECAUSE IRRELEVANT

AFGHANISTAN  CAMBODIA    COSTA RICA  ETHIOPIA    FRANCE       GHANA       HAITI       ITALY       IVORY COAST LIBYA
MALAGASY R   MALAYA      NICARAGUA   PARAGUAY    SA'U ARABIA  SENEGAL     TURKEY      VIETNAM REP

     27 EXCLUDED BECAUSE UNASCERTAINABLE

ALGERIA      BURMA       BURUNDI     CAMEROUN    CEN AFR REP  CHAD        CONGO(BRA)  CUBA        CYPRUS      DAHOMEY
DOMIN REP    GABON       IRAN        JAMAICA     MALI         MAURITANIA  MOROCCO     NIGER       NIGERIA     RWANDA
SUDAN        TOGO        TRINIDAD    UGANDA      UAR          UPPER VOLTA YEMEN

--------------------------------------------------------------------------------------------------------------

  2  TEND LESS TO BE THOSE              0.55        2  TEND MORE TO BE THOSE              0.96         19     1
     LOCATED ELSEWHERE THAN IN                         LOCATED ELSEWHERE THAN IN                      23    24
     WEST EUROPE, SCANDINAVIA,                         WEST EUROPE, SCANDINAVIA,                X SQ=  10.83
     NORTH AMERICA, OR AUSTRALASIA  (93)               NORTH AMERICA, OR AUSTRALASIA  (93)      P =     0.
                                                                                                RV NO  (YES)

  6  TEND MORE TO BE THOSE              0.98        6  TEND LESS TO BE THOSE              0.56          1    11
     LOCATED ELSEWHERE THAN IN THE                     LOCATED ELSEWHERE THAN IN THE                  41    14
     CARIBBEAN, CENTRAL AMERICA,                       CARIBBEAN, CENTRAL AMERICA,              X SQ=  15.74
     OR SOUTH AMERICA  (93)                            OR SOUTH AMERICA  (93)                   P =     0.
                                                                                                RV NO  (YES)
```

14	LEAN MORE TOWARD BEING THOSE LOCATED ELSEWHERE THAN IN THE MIDDLE EAST (104)	0.98	14	LEAN LESS TOWARD BEING THOSE LOCATED ELSEWHERE THAN IN THE MIDDLE EAST (104)	0.84	1 4 41 21 X SQ= 2.47 P = 0.061 RV NO (YES)
19	TEND TO BE THOSE LOCATED IN NORTH AFRICA OR CENTRAL AND SOUTH AFRICA, RATHER THAN IN THE CARIBBEAN, CENTRAL AMERICA, OR SOUTH AMERICA (33)	0.83	19	TEND TO BE THOSE LOCATED IN THE CARIBBEAN, CENTRAL AMERICA, OR SOUTH AMERICA, RATHER THAN IN NORTH AFRICA OR CENTRAL AND SOUTH AFRICA (22)	0.79	5 3 1 11 X SQ= 4.37 P = 0.018 RV YES (YES)
20	TEND TO BE THOSE LOCATED IN EAST ASIA, SOUTH ASIA, OR SOUTHEAST ASIA, RATHER THAN IN THE CARIBBEAN, CENTRAL AMERICA, OR SOUTH AMERICA (18)	0.87	20	TEND TO BE THOSE LOCATED IN THE CARIBBEAN, CENTRAL AMERICA, OR SOUTH AMERICA, RATHER THAN IN EAST ASIA, SOUTH ASIA, OR SOUTHEAST ASIA (22)	0.65	7 6 1 11 X SQ= 4.03 P = 0.030 RV YES (YES)
26	TEND TO BE THOSE WHOSE POPULATION DENSITY IS VERY HIGH, HIGH, OR MEDIUM (48)	0.62	26	TEND TO BE THOSE WHOSE POPULATION DENSITY IS LOW (67)	0.68	26 8 16 17 X SQ= 4.47 P = 0.024 RV YES (YES)
28	TEND TO BE THOSE WHOSE POPULATION GROWTH RATE IS LOW (48)	0.65	28	TEND TO BE THOSE WHOSE POPULATION GROWTH RATE IS HIGH (62)	0.84	14 21 26 4 X SQ= 12.96 P = 0. RV YES (YES)
29	TEND MORE TO BE THOSE WHERE THE DEGREE OF URBANIZATION IS HIGH (56)	0.81	29	TEND LESS TO BE THOSE WHERE THE DEGREE OF URBANIZATION IS HIGH (56)	0.52	30 13 7 12 X SQ= 4.65 P = 0.024 RV NO (YES)
30	TILT MORE TOWARD BEING THOSE WHOSE AGRICULTURAL POPULATION IS MEDIUM, LOW, OR VERY LOW (57)	0.73	30	TILT LESS TOWARD BEING THOSE WHOSE AGRICULTURAL POPULATION IS MEDIUM, LOW, OR VERY LOW (57)	0.52	11 12 30 13 X SQ= 2.20 P = 0.111 RV NO (YES)
31	TEND LESS TO BE THOSE WHOSE AGRICULTURAL POPULATION IS HIGH OR MEDIUM (90)	0.57	31	TEND MORE TO BE THOSE WHOSE AGRICULTURAL POPULATION IS HIGH OR MEDIUM (90)	0.92	24 23 18 2 X SQ= 7.50 P = 0.002 RV NO (NO)
32	TEND LESS TO BE THOSE WHOSE GROSS NATIONAL PRODUCT IS MEDIUM, LOW, OR VERY LOW (105)	0.81	32	ALWAYS ARE THOSE WHOSE GROSS NATIONAL PRODUCT IS MEDIUM, LOW, OR VERY LOW (105)	1.00	8 0 34 25 X SQ= 3.75 P = 0.021 RV NO (NO)

33	TEND TO BE THOSE WHOSE GROSS NATIONAL PRODUCT IS VERY HIGH, HIGH, OR MEDIUM (30)	0.55	33	TEND TO BE THOSE WHOSE GROSS NATIONAL PRODUCT IS LOW OR VERY LOW (85)	0.88	23 3 19 22 X SQ= 10.33 P = 0.001 RV YES (YES)
34	TEND TO BE THOSE WHOSE GROSS NATIONAL PRODUCT IS VERY HIGH, HIGH, MEDIUM, OR LOW (62)	0.79	34	TEND TO BE THOSE WHOSE GROSS NATIONAL PRODUCT IS VERY LOW (53)	0.52	33 12 9 13 X SQ= 5.33 P = 0.015 RV YES (YES)
35	TEND TO BE THOSE WHOSE PER CAPITA GROSS NATIONAL PRODUCT IS VERY HIGH OR HIGH (24)	0.50	35	TEND TO BE THOSE WHOSE PER CAPITA GROSS NATIONAL PRODUCT IS MEDIUM, LOW, OR VERY LOW (91)	0.96	21 1 21 24 X SQ= 13.02 P = 0. RV YES (YES)
36	TEND TO BE THOSE WHOSE PER CAPITA GROSS NATIONAL PRODUCT IS VERY HIGH, HIGH, OR MEDIUM (42)	0.69	36	TEND TO BE THOSE WHOSE PER CAPITA GROSS NATIONAL PRODUCT IS LOW OR VERY LOW (73)	0.80	29 5 13 20 X SQ= 13.19 P = 0. RV YES (YES)
38	TEND LESS TO BE THOSE WHOSE INTERNATIONAL FINANCIAL STATUS IS MEDIUM, LOW, OR VERY LOW (103)	0.80	38	ALWAYS ARE THOSE WHOSE INTERNATIONAL FINANCIAL STATUS IS MEDIUM, LOW, OR VERY LOW (103)	1.00	8 0 32 25 X SQ= 4.00 P = 0.019 RV NO (NO)
39	TEND TO BE THOSE WHOSE INTERNATIONAL FINANCIAL STATUS IS VERY HIGH, HIGH, OR MEDIUM (38)	0.66	39	TEND TO BE THOSE WHOSE INTERNATIONAL FINANCIAL STATUS IS LOW OR VERY LOW (76)	0.80	27 5 14 20 X SQ= 11.30 P = 0. RV YES (YES)
41	TEND LESS TO BE THOSE WHOSE ECONOMIC DEVELOPMENTAL STATUS IS INTERMEDIATE, UNDERDEVELOPED, OR VERY UNDERDEVELOPED (94)	0.59	41	ALWAYS ARE THOSE WHOSE ECONOMIC DEVELOPMENTAL STATUS IS INTERMEDIATE, UNDERDEVELOPED, OR VERY UNDERDEVELOPED (94)	1.00	17 0 24 24 X SQ= 11.41 P = 0. RV NO (YES)
42	TEND TO BE THOSE WHOSE ECONOMIC DEVELOPMENTAL STATUS IS DEVELOPED OR INTERMEDIATE (36)	0.68	42	TEND TO BE THOSE WHOSE ECONOMIC DEVELOPMENTAL STATUS IS UNDERDEVELOPED OR VERY UNDERDEVELOPED (76)	0.87	28 3 13 21 X SQ= 16.72 P = 0. RV YES (YES)
43	TEND TO BE THOSE WHOSE ECONOMIC DEVELOPMENTAL STATUS IS DEVELOPED, INTERMEDIATE, OR UNDERDEVELOPED (55)	0.81	43	TEND TO BE THOSE WHOSE ECONOMIC DEVELOPMENTAL STATUS IS VERY UNDERDEVELOPED (57)	0.62	34 9 8 15 X SQ= 10.86 P = 0.001 RV YES (YES)

44	TEND TO BE THOSE WHERE THE LITERACY RATE IS NINETY PERCENT OR ABOVE (25)	0.57	44	ALWAYS ARE THOSE WHERE THE LITERACY RATE IS BELOW NINETY PERCENT (90)	1.00	24 0 18 25 X SQ= 19.84 P = 0. RV YES (YES)
45	TEND TO BE THOSE WHERE THE LITERACY RATE IS FIFTY PERCENT OR ABOVE (55)	0.86	45	TEND TO BE THOSE WHERE THE LITERACY RATE IS BELOW FIFTY PERCENT (54)	0.62	32 9 5 15 X SQ= 13.71 P = 0. RV YES (YES)
46	TEND MORE TO BE THOSE WHERE THE LITERACY RATE IS TEN PERCENT OR ABOVE (84)	0.97	46	TEND LESS TO BE THOSE WHERE THE LITERACY RATE IS TEN PERCENT OR ABOVE (84)	0.80	38 20 1 5 X SQ= 3.59 P = 0.030 RV NO (YES)
50	LEAN TOWARD BEING THOSE WHERE FREEDOM OF THE PRESS IS COMPLETE (43)	0.52	50	LEAN TOWARD BEING THOSE WHERE FREEDOM OF THE PRESS IS INTERMITTENT, INTERNALLY ABSENT, OR INTERNALLY AND EXTERNALLY ABSENT (56)	0.75	22 5 20 15 X SQ= 3.09 P = 0.057 RV YES (NO)
53	TEND LESS TO BE THOSE WHERE NEWSPAPER CIRCULATION IS LESS THAN THREE HUNDRED PER THOUSAND (101)	0.67	53	ALWAYS ARE THOSE WHERE NEWSPAPER CIRCULATION IS LESS THAN THREE HUNDRED PER THOUSAND (101)	1.00	14 0 28 25 X SQ= 8.61 P = 0.001 RV NO (NO)
54	TEND TO BE THOSE WHERE NEWSPAPER CIRCULATION IS ONE HUNDRED OR MORE PER THOUSAND (37)	0.67	54	TEND TO BE THOSE WHERE NEWSPAPER CIRCULATION IS LESS THAN ONE HUNDRED PER THOUSAND (76)	0.84	27 4 13 21 X SQ= 14.36 P = 0. RV YES (YES)
56	LEAN MORE TOWARD BEING THOSE WHERE THE RELIGION IS PREDOMINANTLY LITERATE (79)	0.90	56	LEAN LESS TOWARD BEING THOSE WHERE THE RELIGION IS PREDOMINANTLY LITERATE (79)	0.71	35 17 4 7 X SQ= 2.49 P = 0.086 RV NO (YES)
58	TEND MORE TO BE THOSE WHERE THE RELIGION IS OTHER THAN MUSLIM (97)	0.98	58	TEND LESS TO BE THOSE WHERE THE RELIGION IS OTHER THAN MUSLIM (97)	0.80	1 5 41 20 X SQ= 4.00 P = 0.024 RV NO (YES)
62	TILT LESS TOWARD BEING THOSE WHERE THE RELIGION IS CATHOLIC, RATHER THAN PROTESTANT (25)	0.64	62	ALWAYS ARE THOSE WHERE THE RELIGION IS CATHOLIC, RATHER THAN PROTESTANT (25)	1.00	5 0 9 7 X SQ= 1.61 P = 0.123 RV NO (NO)

63	TEND TO BE THOSE WHERE THE RELIGION IS PREDOMINANTLY SOME KIND OF CHRISTIAN (46)	0.68	63	TEND TO BE THOSE WHERE THE RELIGION IS PREDOMINANTLY OR PARTLY OTHER THAN CHRISTIAN (68)	0.68	28 8 13 17 X SQ= 6.85 P = 0.005 RV YES (YES)
67	TEND MORE TO BE THOSE THAT ARE RACIALLY HOMOGENEOUS (82)	0.90	67	TEND LESS TO BE THOSE THAT ARE RACIALLY HOMOGENEOUS (82)	0.61	38 14 4 9 X SQ= 6.40 P = 0.008 RV NO (YES)
71	TEND LESS TO BE THOSE WHOSE DATE OF INDEPENDENCE IS AFTER 1800 (90)	0.64	71	TEND MORE TO BE THOSE WHOSE DATE OF INDEPENDENCE IS AFTER 1800 (90)	0.92	14 2 25 22 X SQ= 4.59 P = 0.018 RV NO (NO)
74	TEND TO BE THOSE THAT ARE HISTORICALLY WESTERN (26)	0.57	74	TEND TO BE THOSE THAT ARE NOT HISTORICALLY WESTERN (87)	0.96	23 1 17 24 X SQ= 16.68 P = 0. RV YES (YES)
75	TEND MORE TO BE THOSE THAT ARE HISTORICALLY WESTERN OR SIGNIFICANTLY WESTERNIZED (62)	0.80	75	TEND LESS TO BE THOSE THAT ARE HISTORICALLY WESTERN OR SIGNIFICANTLY WESTERNIZED (62)	0.56	33 14 8 11 X SQ= 3.43 P = 0.050 RV NO (YES)
76	TEND TO BE THOSE THAT ARE HISTORICALLY WESTERN, RATHER THAN HAVING BEEN SIGNIFICANTLY OR PARTIALLY WESTERNIZED THROUGH A COLONIAL RELATIONSHIP (26)	0.74	76	TEND TO BE THOSE THAT HAVE BEEN SIGNIFICANTLY OR PARTIALLY WESTERNIZED THROUGH A COLONIAL RELATIONSHIP, RATHER THAN BEING HISTORICALLY WESTERN (70)	0.95	23 1 8 21 X SQ= 22.46 P = 0. RV YES (YES)
78	TEND TO BE THOSE THAT HAVE BEEN SIGNIFICANTLY WESTERNIZED WITHOUT A COLONIAL RELATIONSHIP, RATHER THAN THROUGH SUCH A RELATIONSHIP (7)	0.67	78	ALWAYS ARE THOSE THAT HAVE BEEN SIGNIFICANTLY WESTERNIZED THROUGH A COLONIAL RELATIONSHIP, RATHER THAN WITHOUT SUCH A RELATIONSHIP (28)	1.00	3 13 6 0 X SQ= 8.79 P = 0.001 RV YES (YES)
79	TEND TO BE THOSE THAT WERE FORMERLY DEPENDENCIES OF BRITAIN OR FRANCE, RATHER THAN SPAIN (49)	0.92	79	TEND TO BE THOSE THAT WERE FORMERLY DEPENDENCIES OF SPAIN, RATHER THAN BRITAIN OR FRANCE (18)	0.56	12 8 1 10 X SQ= 5.61 P = 0.008 RV YES (YES)
81	TEND LESS TO BE THOSE WHERE THE TYPE OF POLITICAL MODERNIZATION IS LATER EUROPEAN OR LATER EUROPEAN DERIVED, RATHER THAN EARLY EUROPEAN OR EARLY EUROPEAN DERIVED (40)	0.66	81	ALWAYS ARE THOSE WHERE THE TYPE OF POLITICAL MODERNIZATION IS LATER EUROPEAN OR LATER EUROPEAN DERIVED, RATHER THAN EARLY EUROPEAN OR EARLY EUROPEAN DERIVED (40)	1.00	10 0 19 12 X SQ= 3.76 P = 0.021 RV NO (NO)

82	LEAN MORE TOWARD BEING THOSE WHERE THE TYPE OF POLITICAL MODERNIZATION IS EARLY OR LATER EUROPEAN OR EUROPEAN DERIVED, RATHER THAN DEVELOPED TUTELARY OR UNDEVELOPED TUTELARY (51)	0.74	82	LEAN LESS TOWARD BEING THOSE WHERE THE TYPE OF POLITICAL MODERNIZATION IS EARLY OR LATER EUROPEAN OR EUROPEAN DERIVED, RATHER THAN DEVELOPED TUTELARY OR UNDEVELOPED TUTELARY (51)	0.52	29 12 10 11 X SQ= 2.27 P = 0.098 RV NO (YES)
85	TEND TO BE THOSE WHERE THE STAGE OF POLITICAL MODERNIZATION IS ADVANCED, RATHER THAN MID- OR EARLY TRANSITIONAL (60)	0.88	85	TEND TO BE THOSE WHERE THE STAGE OF POLITICAL MODERNIZATION IS MID- OR EARLY TRANSITIONAL, RATHER THAN ADVANCED (54)	0.52	37 12 5 13 X SQ= 10.86 P = 0.001 RV YES (YES)
87	TILT MORE TOWARD BEING THOSE WHOSE IDEOLOGICAL ORIENTATION IS OTHER THAN DEVELOPMENTAL (58)	0.90	87	TILT LESS TOWARD BEING THOSE WHOSE IDEOLOGICAL ORIENTATION IS OTHER THAN DEVELOPMENTAL (58)	0.69	4 5 36 11 X SQ= 2.41 P = 0.100 RV NO (YES)
89	LEAN TOWARD BEING THOSE WHOSE IDEOLOGICAL ORIENTATION IS CONVENTIONAL (33)	0.55	89	LEAN TOWARD BEING THOSE WHOSE IDEOLOGICAL ORIENTATION IS OTHER THAN CONVENTIONAL (62)	0.72	21 5 17 13 X SQ= 2.69 P = 0.085 RV YES (NO)
92	TEND LESS TO BE THOSE WHERE THE SYSTEM STYLE IS LIMITED MOBILIZATIONAL, OR NON-MOBILIZATIONAL (93)	0.66	92	ALWAYS ARE THOSE WHERE THE SYSTEM STYLE IS LIMITED MOBILIZATIONAL, OR NON-MOBILIZATIONAL (93)	1.00	14 0 27 25 X SQ= 8.89 P = 0.001 RV NO (NO)
93	TEND LESS TO BE THOSE WHERE THE SYSTEM STYLE IS NON-MOBILIZATIONAL (78)	0.59	93	TEND MORE TO BE THOSE WHERE THE SYSTEM STYLE IS NON-MOBILIZATIONAL (78)	0.91	17 2 24 20 X SQ= 5.67 P = 0.009 RV NO (NO)
95	TEND LESS TO BE THOSE WHERE THE STATUS OF THE REGIME IS CONSTITUTIONAL OR AUTHORITARIAN (95)	0.63	95	ALWAYS ARE THOSE WHERE THE STATUS OF THE REGIME IS CONSTITUTIONAL OR AUTHORITARIAN (95)	1.00	26 25 15 0 X SQ= 9.84 P = 0. RV NO (NO)
96	TEND MORE TO BE THOSE WHERE THE STATUS OF THE REGIME IS CONSTITUTIONAL OR TOTALITARIAN (67)	0.95	96	TEND LESS TO BE THOSE WHERE THE STATUS OF THE REGIME IS CONSTITUTIONAL OR TOTALITARIAN (67)	0.61	38 11 2 7 X SQ= 8.44 P = 0.002 RV NO (YES)
97	TEND TO BE THOSE WHERE THE STATUS OF THE REGIME IS TOTALITARIAN, RATHER THAN AUTHORITARIAN (16)	0.88	97	ALWAYS ARE THOSE WHERE THE STATUS OF THE REGIME IS AUTHORITARIAN, RATHER THAN TOTALITARIAN (23)	1.00	2 7 15 0 X SQ= 12.92 P = 0. RV YES (YES)

99	ALWAYS ARE THOSE WHERE GOVERNMENTAL STABILITY IS GENERALLY PRESENT AND DATES FROM AT LEAST THE INTER-WAR PERIOD, OR FROM THE POST-WAR PERIOD (50)	1.00	99	ALWAYS ARE THOSE WHERE GOVERNMENTAL STABILITY IS MODERATELY PRESENT AND DATES FROM THE POST-WAR PERIOD, OR IS ABSENT (36)	1.00	40 0 0 23 X SQ= 58.76 P = 0. RV YES (YES)
100	ALWAYS ARE THOSE WHERE GOVERNMENTAL STABILITY IS GENERALLY PRESENT AND DATES FROM AT LEAST THE INTER-WAR PERIOD, OR IS GENERALLY OR MODERATELY PRESENT, AND DATES FROM THE POST-WAR PERIOD (64)	1.00	100	TEND TO BE THOSE WHERE GOVERNMENTAL STABILITY IS ABSENT (22)	0.83	41 4 0 19 X SQ= 44.29 P = 0. RV YES (YES)
106	TEND LESS TO BE THOSE WHERE THE ELECTORAL SYSTEM IS COMPETITIVE OR PARTIALLY COMPETITIVE (52)	0.55	106	TEND MORE TO BE THOSE WHERE THE ELECTORAL SYSTEM IS COMPETITIVE OR PARTIALLY COMPETITIVE (52)	0.92	22 12 18 1 X SQ= 4.43 P = 0.019 RV NO (NO)
108	TEND LESS TO BE THOSE WHERE AUTONOMOUS GROUPS ARE FULLY OR PARTIALLY TOLERATED IN POLITICS (65)	0.59	108	TEND MORE TO BE THOSE WHERE AUTONOMOUS GROUPS ARE FULLY OR PARTIALLY TOLERATED IN POLITICS (65)	0.88	24 15 17 2 X SQ= 3.56 P = 0.034 RV NO (NO)
111	TEND TO BE THOSE WHERE POLITICAL ENCULTURATION IS HIGH OR MEDIUM (53)	0.83	111	TEND TO BE THOSE WHERE POLITICAL ENCULTURATION IS LOW (42)	0.72	24 7 5 18 X SQ= 14.30 P = 0. RV YES (YES)
116	TEND TO BE THOSE WHERE INTEREST ARTICULATION BY ASSOCIATIONAL GROUPS IS SIGNIFICANT OR MODERATE (32)	0.56	116	TEND TO BE THOSE WHERE INTEREST ARTICULATION BY ASSOCIATIONAL GROUPS IS LIMITED OR NEGLIGIBLE (79)	0.75	22 6 17 18 X SQ= 4.73 P = 0.020 RV YES (YES)
119	TEND LESS TO BE THOSE WHERE INTEREST ARTICULATION BY INSTITUTIONAL GROUPS IS VERY SIGNIFICANT OR SIGNIFICANT (74)	0.57	119	TEND MORE TO BE THOSE WHERE INTEREST ARTICULATION BY INSTITUTIONAL GROUPS IS VERY SIGNIFICANT OR SIGNIFICANT (74)	0.91	24 21 18 2 X SQ= 6.62 P = 0.005 RV NO (NO)
120	TEND LESS TO BE THOSE WHERE INTEREST ARTICULATION BY INSTITUTIONAL GROUPS IS VERY SIGNIFICANT, SIGNIFICANT, OR MODERATE (90)	0.76	120	ALWAYS ARE THOSE WHERE INTEREST ARTICULATION BY INSTITUTIONAL GROUPS IS VERY SIGNIFICANT, SIGNIFICANT, OR MODERATE (90)	1.00	32 23 10 0 X SQ= 4.77 P = 0.011 RV NO (NO)
121	TEND TO BE THOSE WHERE INTEREST ARTICULATION BY NON-ASSOCIATIONAL GROUPS IS MODERATE, LIMITED, OR NEGLIGIBLE (61)	0.79	121	TEND TO BE THOSE WHERE INTEREST ARTICULATION BY NON-ASSOCIATIONAL GROUPS IS SIGNIFICANT (54)	0.52	9 13 33 12 X SQ= 5.33 P = 0.015 RV YES (YES)

122	LEAN LESS TOWARD BEING THOSE WHERE INTEREST ARTICULATION BY NON-ASSOCIATIONAL GROUPS IS SIGNIFICANT OR MODERATE (83)	0.55	122	LEAN MORE TOWARD BEING THOSE WHERE INTEREST ARTICULATION BY NON-ASSOCIATIONAL GROUPS IS SIGNIFICANT OR MODERATE (83)	0.80	23 20 19 5 X SQ= 3.31 P = 0.064 RV NO (NO)
123	TEND LESS TO BE THOSE WHERE INTEREST ARTICULATION BY NON-ASSOCIATIONAL GROUPS IS SIGNIFICANT, MODERATE, OR LIMITED (107)	0.81	123	ALWAYS ARE THOSE WHERE INTEREST ARTICULATION BY NON-ASSOCIATIONAL GROUPS IS SIGNIFICANT, MODERATE, OR LIMITED (107)	1.00	34 25 8 0 X SQ= 3.75 P = 0.021 RV NO (NO)
125	TEND TO BE THOSE WHERE INTEREST ARTICULATION BY ANOMIC GROUPS IS INFREQUENT OR VERY INFREQUENT (35)	0.68	125	TEND TO BE THOSE WHERE INTEREST ARTICULATION BY ANOMIC GROUPS IS FREQUENT OR OCCASIONAL (64)	0.91	11 21 23 2 X SQ= 17.04 P = 0. RV YES (YES)
126	TEND LESS TO BE THOSE WHERE INTEREST ARTICULATION BY ANOMIC GROUPS IS FREQUENT, OCCASIONAL, OR INFREQUENT (83)	0.53	126	ALWAYS ARE THOSE WHERE INTEREST ARTICULATION BY ANOMIC GROUPS IS FREQUENT, OCCASIONAL, OR INFREQUENT (83)	1.00	18 23 16 0 X SQ= 12.81 P = 0. RV NO (YES)
128	TEND TO BE THOSE WHERE INTEREST ARTICULATION BY POLITICAL PARTIES IS LIMITED OR NEGLIGIBLE (45)	0.54	128	ALWAYS ARE THOSE WHERE INTEREST ARTICULATION BY POLITICAL PARTIES IS SIGNIFICANT OR MODERATE (48)	1.00	19 18 22 0 X SQ= 13.19 P = 0. RV YES (NO)
130	TEND LESS TO BE THOSE WHERE INTEREST AGGREGATION BY POLITICAL PARTIES IS MODERATE, LIMITED, OR NEGLIGIBLE (71)	0.75	130	ALWAYS ARE THOSE WHERE INTEREST AGGREGATION BY POLITICAL PARTIES IS MODERATE, LIMITED, OR NEGLIGIBLE (71)	1.00	10 0 30 22 X SQ= 4.84 P = 0.010 RV NO (NO)
131	TEND TO BE THOSE WHERE INTEREST AGGREGATION BY POLITICAL PARTIES IS SIGNIFICANT OR MODERATE (30)	0.60	131	TEND TO BE THOSE WHERE INTEREST AGGREGATION BY POLITICAL PARTIES IS LIMITED OR NEGLIGIBLE (35)	0.90	15 2 10 18 X SQ= 9.79 P = 0.001 RV YES (YES)
134	TEND TO BE THOSE WHERE INTEREST AGGREGATION BY THE EXECUTIVE IS SIGNIFICANT OR MODERATE (57)	0.59	134	TEND TO BE THOSE WHERE INTEREST AGGREGATION BY THE EXECUTIVE IS LIMITED OR NEGLIGIBLE (46)	0.78	24 4 17 14 X SQ= 5.24 P = 0.012 RV YES (NO)
136	TEND LESS TO BE THOSE WHERE INTEREST AGGREGATION BY THE LEGISLATURE IS MODERATE, LIMITED, OR NEGLIGIBLE (85)	0.73	136	ALWAYS ARE THOSE WHERE INTEREST AGGREGATION BY THE LEGISLATURE IS MODERATE, LIMITED, OR NEGLIGIBLE (85)	1.00	11 0 30 21 X SQ= 5.13 P = 0.011 RV NO (NO)

137	TEND TO BE THOSE WHERE INTEREST AGGREGATION BY THE LEGISLATURE IS SIGNIFICANT OR MODERATE (29)	0.51	137	TEND TO BE THOSE WHERE INTEREST AGGREGATION BY THE LEGISLATURE IS LIMITED OR NEGLIGIBLE (68)	0.86	21 3 20 18 X SQ= 6.50 P = 0.006 RV YES (NO)
144	LEAN TOWARD BEING THOSE WHERE THE PARTY SYSTEM IS QUANTITATIVELY ONE-PARTY, RATHER THAN TWO-PARTY (34)	0.73	144	ALWAYS ARE THOSE WHERE THE PARTY SYSTEM IS QUANTITATIVELY TWO-PARTY, RATHER THAN ONE-PARTY (11)	1.00	19 0 7 2 X SQ= 1.81 P = 0.095 RV YES (NO)
145	TILT LESS TOWARD BEING THOSE WHERE THE PARTY SYSTEM IS QUANTITATIVELY MULTI-PARTY, RATHER THAN TWO-PARTY (30)	0.61	145	TILT MORE TOWARD BEING THOSE WHERE THE PARTY SYSTEM IS QUANTITATIVELY MULTI-PARTY, RATHER THAN TWO-PARTY (30)	0.87	11 13 7 2 X SQ= 1.56 P = 0.134 RV NO (YES)
157	ALWAYS ARE THOSE WHERE PERSONALISMO IS MODERATE OR NEGLIGIBLE (84)	1.00	157	TEND TO BE THOSE WHERE PERSONALISMO IS PRONOUNCED (14)	0.50	0 10 42 10 X SQ= 21.48 P = 0. RV YES (YES)
158	TEND TO BE THOSE WHERE PERSONALISMO IS NEGLIGIBLE (56)	0.95	158	TEND TO BE THOSE WHERE PERSONALISMO IS PRONOUNCED OR MODERATE (40)	0.85	2 17 40 3 X SQ= 37.35 P = 0. RV YES (YES)
161	TILT LESS TOWARD BEING THOSE WHERE THE POLITICAL LEADERSHIP IS ELITIST OR MODERATELY ELITIST (47)	0.52	161	TILT MORE TOWARD BEING THOSE WHERE THE POLITICAL LEADERSHIP IS ELITIST OR MODERATELY ELITIST (47)	0.75	22 15 20 5 X SQ= 2.02 P = 0.105 RV NO (NO)
168	TEND TO BE THOSE WHERE THE HORIZONTAL POWER DISTRIBUTION IS SIGNIFICANT (34)	0.52	168	TEND TO BE THOSE WHERE THE HORIZONTAL POWER DISTRIBUTION IS LIMITED OR NEGLIGIBLE (72)	0.84	22 3 20 16 X SQ= 5.81 P = 0.011 RV YES (NO)
170	TEND TO BE THOSE WHERE THE LEGISLATIVE-EXECUTIVE STRUCTURE IS OTHER THAN PRESIDENTIAL (63)	0.85	170	TEND TO BE THOSE WHERE THE LEGISLATIVE-EXECUTIVE STRUCTURE IS PRESIDENTIAL (39)	0.52	6 11 35 10 X SQ= 8.14 P = 0.003 RV YES (YES)
174	TEND TO BE THOSE WHERE THE LEGISLATURE IS FULLY EFFECTIVE (28)	0.52	174	ALWAYS ARE THOSE WHERE THE LEGISLATURE IS PARTIALLY EFFECTIVE, LARGELY INEFFECTIVE, OR WHOLLY INEFFECTIVE (72)	1.00	22 0 20 18 X SQ= 12.72 P = 0. RV YES (NO)

181	TEND LESS TO BE THOSE WHERE THE BUREAUCRACY IS SEMI-MODERN, RATHER THAN MODERN (55)	0.54	181	ALWAYS ARE THOSE WHERE THE BUREAUCRACY IS SEMI-MODERN, RATHER THAN MODERN (55)	1.00	19 0 22 16 X SQ= 9.13 P = 0. RV NO (NO)
185	ALWAYS ARE THOSE WHERE PARTICIPATION BY THE MILITARY IN POLITICS IS SUPPORTIVE, RATHER THAN INTERVENTIVE (31)	1.00	185	TEND TO BE THOSE WHERE PARTICIPATION BY THE MILITARY IN POLITICS IS INTERVENTIVE, RATHER THAN SUPPORTIVE (21)	0.80	0 12 18 3 X SQ= 19.30 P = 0. RV YES (YES)
187	TEND TO BE THOSE WHERE THE ROLE OF THE POLICE IS NOT POLITICALLY SIGNIFICANT (35)	0.55	187	TEND TO BE THOSE WHERE THE ROLE OF THE POLICE IS POLITICALLY SIGNIFICANT (66)	0.86	19 18 23 3 X SQ= 7.87 P = 0.003 RV YES (NO)
188	TEND TO BE THOSE WHERE THE CHARACTER OF THE LEGAL SYSTEM IS OTHER THAN CIVIL LAW (81)	0.78	188	TEND TO BE THOSE WHERE THE CHARACTER OF THE LEGAL SYSTEM IS CIVIL LAW (32)	0.52	9 13 32 12 X SQ= 5.03 P = 0.016 RV YES (YES)
189	TILT LESS TOWARD BEING THOSE WHERE THE CHARACTER OF THE LEGAL SYSTEM IS OTHER THAN COMMON LAW (108)	0.88	189	ALWAYS ARE THOSE WHERE THE CHARACTER OF THE LEGAL SYSTEM IS OTHER THAN COMMON LAW (108)	1.00	5 0 37 25 X SQ= 1.72 P = 0.149 RV NO (NO)
190	TEND LESS TO BE THOSE WHERE THE CHARACTER OF THE LEGAL SYSTEM IS CIVIL LAW, RATHER THAN COMMON LAW (32)	0.64	190	ALWAYS ARE THOSE WHERE THE CHARACTER OF THE LEGAL SYSTEM IS CIVIL LAW, RATHER THAN COMMON LAW (32)	1.00	9 13 5 0 X SQ= 3.58 P = 0.041 RV NO (YES)
194	TEND LESS TO BE THOSE THAT ARE NON-COMMUNIST (101)	0.69	194	ALWAYS ARE THOSE THAT ARE NON-COMMUNIST (101)	1.00	13 0 29 25 X SQ= 7.72 P = 0.001 RV NO (NO)

153 POLITIES
WHERE THE PARTY SYSTEM IS
STABLE (42)

153 POLITIES
WHERE THE PARTY SYSTEM IS
MODERATELY STABLE OR UNSTABLE (38)

PREDICATE ONLY

42 IN LEFT COLUMN

ALBANIA AUSTRALIA AUSTRIA BELGIUM BULGARIA CANADA CHINA, PR CZECHOS'KIA DENMARK FINLAND
GERMANY, E GERMAN FR GUINEA HUNGARY ICELAND INDIA IRELAND ISRAEL JAPAN KOREA, N
LIBERIA LUXEMBOURG MEXICO MONGOLIA NETHERLANDS NEW ZEALAND NORWAY PHILIPPINES POLAND PORTUGAL
RUMANIA SO AFRICA SPAIN SWEDEN SWITZERLAND TANGANYIKA TUNISIA USSR UK US
VIETNAM, N YUGOSLAVIA

38 IN RIGHT COLUMN

ARGENTINA BOLIVIA BRAZIL CAMBODIA CEYLON CHILE CONGO(LEO) COSTA RICA ECUADOR EL SALVADOR
FRANCE GHANA GREECE GUATEMALA HONDURAS IRAQ ITALY IVORY COAST JORDAN KOREA REP
LAOS LEBANON MALAGASY R MALAYA NEPAL NICARAGUA PAKISTAN PANAMA PARAGUAY PERU
SENEGAL SIERRE LEO SOMALIA SYRIA THAILAND TURKEY VENEZUELA VIETNAM REP

3 EXCLUDED BECAUSE AMBIGUOUS

COLOMBIA INDONESIA URUGUAY

5 EXCLUDED BECAUSE IRRELEVANT

AFGHANISTAN ETHIOPIA HAITI LIBYA SA'U ARABIA

27 EXCLUDED BECAUSE UNASCERTAINABLE

ALGERIA BURMA BURUNDI CAMEROUN CEN AFR REP CHAD CONGO(BRA) CUBA CYPRUS DAHOMEY
DOMIN REP GABON IRAN JAMAICA MALI MAURITANIA MOROCCO NIGER NIGERIA RWANDA
SUDAN TOGO TRINIDAD UGANDA UAR UPPER VOLTA YEMEN

```
************************************************************************************************************

154  POLITIES                                      154  POLITIES
     WHERE THE PARTY SYSTEM IS                          WHERE THE PARTY SYSTEM IS
     STABLE OR MODERATELY STABLE  (55)                  UNSTABLE  (25)

                                                                                            PREDICATE ONLY
............................................................................................................

     55 IN LEFT COLUMN

ALBANIA     AUSTRALIA   AUSTRIA     BELGIUM     BULGARIA    CAMBODIA    CANADA      CHINA, PR   COSTA RICA  CZECHOS'KIA
DENMARK     FINLAND     FRANCE      GERMANY, E  GERMAN FR   GHANA       GUINEA      HUNGARY     ICELAND     INDIA
IRELAND     ISRAEL      ITALY       IVORY COAST JAPAN       KOREA, N    LIBERIA     LUXEMBOURG  MALAGASY R  MALAYA
MEXICO      MONGOLIA    NETHERLANDS NEW ZEALAND NICARAGUA   NORWAY      PARAGUAY    PHILIPPINES POLAND      PORTUGAL
RUMANIA     SENEGAL     SO AFRICA   SPAIN       SWEDEN      SWITZERLAND TANGANYIKA  TUNISIA     TURKEY      USSR
UK          US          VIETNAM, N  VIETNAM REP YUGOSLAVIA

     25 IN RIGHT COLUMN

ARGENTINA   BOLIVIA     BRAZIL      CEYLON      CHILE       CONGO(LEO)  ECUADOR     EL SALVADOR GREECE      GUATEMALA
HONDURAS    IRAQ        JORDAN      KOREA REP   LAOS        LEBANON     NEPAL       PAKISTAN    PANAMA      PERU
SIERRE LEO  SOMALIA     SYRIA       THAILAND    VENEZUELA

      3 EXCLUDED BECAUSE AMBIGUOUS

COLOMBIA    INDONESIA   URUGUAY

      5 EXCLUDED BECAUSE IRRELEVANT

AFGHANISTAN ETHIOPIA    HAITI       LIBYA       SA'U ARABIA

     27 EXCLUDED BECAUSE UNASCERTAINABLE

ALGERIA     BURMA       BURUNDI     CAMEROUN    CEN AFR REP CHAD        CONGO(BRA)  CUBA        CYPRUS      DAHOMEY
DOMIN REP   GABON       IRAN        JAMAICA     MALI        MAURITANIA  MOROCCO     NIGER       NIGERIA     RWANDA
SUDAN       TOGO        TRINIDAD    UGANDA      UAR         UPPER VOLTA YEMEN
------------------------------------------------------------------------------------------------------------
```

```
****************************************************************************************************

155  POLITIES                                          155  POLITIES
     WHERE THE PARTY SYSTEM IS                              WHERE THE PARTY SYSTEM IS
     STABLE MULTI-PARTY, RATHER THAN                        UNSTABLE MULTI-PARTY, RATHER THAN
     UNSTABLE MULTI-PARTY  (11)                             STABLE MULTI-PARTY  (13)

                                                                                     SUBJECT ONLY
....................................................................................................

     11 IN LEFT COLUMN

BELGIUM     DENMARK     FINLAND     ICELAND     IRELAND     ISRAEL      LUXEMBOURG  NETHERLANDS NORWAY      SWEDEN
SWITZERLAND

     13 IN RIGHT COLUMN

ARGENTINA   BRAZIL      CEYLON      CHILE       CONGO(LEO)  ECUADOR     EL SALVADOR GUATEMALA   LAOS        LEBANON
PANAMA      PERU        VENEZUELA

     91 EXCLUDED BECAUSE IRRELEVANT

AFGHANISTAN ALBANIA     ALGERIA     AUSTRALIA   AUSTRIA     BOLIVIA     BULGARIA    BURMA       BURUNDI     CAMBODIA
CAMEROUN    CANADA      CEN AFR REP CHAD        CHINA, PR   COLOMBIA    CONGO(BRA)  COSTA RICA  CUBA        CYPRUS
CZECHOS'KIA DAHOMEY     DOMIN REP   ETHIOPIA    FRANCE      GABON       GERMANY, E  GERMAN FR   GHANA       GREECE
GUINEA      HAITI       HONDURAS    HUNGARY     INDIA       INDONESIA   IRAN        IRAQ        ITALY       IVORY COAST
JAMAICA     JAPAN       JORDAN      KOREA, N    KOREA REP   LIBERIA     LIBYA       MALAGASY R  MALAYA      MALI
MAURITANIA  MEXICO      MONGOLIA    MOROCCO     NEPAL       NEW ZEALAND NICARAGUA   NIGER       NIGERIA     PAKISTAN
PARAGUAY    PHILIPPINES POLAND      PORTUGAL    RUMANIA     RWANDA      SA'U ARABIA SENEGAL     SIERRE LEO  SOMALIA
SO AFRICA   SPAIN       SUDAN       SYRIA       TANGANYIKA  THAILAND    TOGO        TRINIDAD    TUNISIA     TURKEY
UGANDA      USSR        UAR         UK          US          UPPER VOLTA URUGUAY     VIETNAM, N  VIETNAM REP YEMEN
YUGOSLAVIA

----------------------------------------------------------------------------------------------------

  2  TEND TO BE THOSE                     0.91       2  ALWAYS ARE THOSE                  1.00        10    0
     LOCATED IN WEST EUROPE, SCANDINAVIA,               LOCATED ELSEWHERE THAN IN                      1   13
     NORTH AMERICA, OR AUSTRALASIA  (22)                WEST EUROPE, SCANDINAVIA,               X SQ=  16.69
                                                        NORTH AMERICA, OR AUSTRALASIA  (93)     P =     0.
                                                                                                RV YES (YES)

  3  TEND LESS TO BE THOSE                0.55       3  ALWAYS ARE THOSE                  1.00         5    0
     LOCATED ELSEWHERE THAN IN                          LOCATED ELSEWHERE THAN IN                      6   13
     WEST EUROPE  (102)                                 WEST EUROPE  (102)                      X SQ=   4.96
                                                                                                P =    0.011
                                                                                                RV NO  (YES)

  6  ALWAYS ARE THOSE                     1.00       6  TEND TO BE THOSE                  0.69         0    9
     LOCATED ELSEWHERE THAN IN THE                      LOCATED IN THE CARIBBEAN,                     11    4
     CARIBBEAN, CENTRAL AMERICA,                        CENTRAL AMERICA, OR SOUTH AMERICA  (22) X SQ=   9.41
     OR SOUTH AMERICA  (93)                                                                     P =    0.001
                                                                                                RV YES (YES)
```

21	ALWAYS ARE THOSE WHOSE TERRITORIAL SIZE IS MEDIUM OR SMALL (83)	1.00	21	TEND LESS TO BE THOSE WHOSE TERRITORIAL SIZE IS MEDIUM OR SMALL (83)	0.62	0 5 11 8 X SQ= 3.27 P = 0.041 RV NO (YES)
22	TILT TOWARD BEING THOSE WHOSE TERRITORIAL SIZE IS SMALL (47)	0.73	22	TILT TOWARD BEING THOSE WHOSE TERRITORIAL SIZE IS VERY LARGE, LARGE, OR MEDIUM (68)	0.62	3 8 8 5 X SQ= 1.61 P = 0.123 RV YES (YES)
26	LEAN TOWARD BEING THOSE WHOSE POPULATION DENSITY IS VERY HIGH, HIGH, OR MEDIUM (48)	0.64	26	LEAN TOWARD BEING THOSE WHOSE POPULATION DENSITY IS LOW (67)	0.77	7 3 4 10 X SQ= 2.54 P = 0.095 RV YES (YES)
28	TEND TO BE THOSE WHOSE POPULATION GROWTH RATE IS LOW (48)	0.82	28	TEND TO BE THOSE WHOSE POPULATION GROWTH RATE IS HIGH (62)	0.85	2 11 9 2 X SQ= 8.09 P = 0.003 RV YES (YES)
29	ALWAYS ARE THOSE WHERE THE DEGREE OF URBANIZATION IS HIGH (56)	1.00	29	TEND LESS TO BE THOSE WHERE THE DEGREE OF URBANIZATION IS HIGH (56)	0.62	11 8 0 5 X SQ= 3.27 P = 0.041 RV NO (YES)
31	TEND TO BE THOSE WHOSE AGRICULTURAL POPULATION IS LOW OR VERY LOW (24)	0.82	31	TEND TO BE THOSE WHOSE AGRICULTURAL POPULATION IS HIGH OR MEDIUM (90)	0.85	2 11 9 2 X SQ= 8.09 P = 0.003 RV YES (YES)
35	ALWAYS ARE THOSE WHOSE PER CAPITA GROSS NATIONAL PRODUCT IS VERY HIGH OR HIGH (24)	1.00	35	TEND TO BE THOSE WHOSE PER CAPITA GROSS NATIONAL PRODUCT IS MEDIUM, LOW, OR VERY LOW (91)	0.92	11 1 0 12 X SQ= 16.78 P = 0. RV YES (YES)
41	TEND TO BE THOSE WHOSE ECONOMIC DEVELOPMENTAL STATUS IS DEVELOPED (19)	0.80	41	ALWAYS ARE THOSE WHOSE ECONOMIC DEVELOPMENTAL STATUS IS INTERMEDIATE, UNDERDEVELOPED, OR VERY UNDERDEVELOPED (94)	1.00	8 0 2 12 X SQ= 11.83 P = 0. RV YES (YES)
42	ALWAYS ARE THOSE WHOSE ECONOMIC DEVELOPMENTAL STATUS IS DEVELOPED OR INTERMEDIATE (36)	1.00	42	TEND TO BE THOSE WHOSE ECONOMIC DEVELOPMENTAL STATUS IS UNDERDEVELOPED OR VERY UNDERDEVELOPED (76)	0.83	10 2 0 10 X SQ= 12.10 P = 0. RV YES (YES)

43	ALWAYS ARE THOSE WHOSE ECONOMIC DEVELOPMENTAL STATUS IS DEVELOPED, INTERMEDIATE, OR UNDERDEVELOPED (55)	1.00	43	TEND TO BE THOSE WHOSE ECONOMIC DEVELOPMENTAL STATUS IS VERY UNDERDEVELOPED (57)	0.50	11 6 0 6 X SQ= 5.07 P = 0.014 RV YES (YES)
44	TEND TO BE THOSE WHERE THE LITERACY RATE IS NINETY PERCENT OR ABOVE (25)	0.91	44	ALWAYS ARE THOSE WHERE THE LITERACY RATE IS BELOW NINETY PERCENT (90)	1.00	10 0 1 13 X SQ= 16.69 P = 0. RV YES (YES)
45	ALWAYS ARE THOSE WHERE THE LITERACY RATE IS FIFTY PERCENT OR ABOVE (55)	1.00	45	TEND TO BE THOSE WHERE THE LITERACY RATE IS BELOW FIFTY PERCENT (54)	0.50	11 6 0 6 X SQ= 5.07 P = 0.014 RV YES (YES)
50	TEND TO BE THOSE WHERE FREEDOM OF THE PRESS IS COMPLETE (43)	0.91	50	TEND TO BE THOSE WHERE FREEDOM OF THE PRESS IS INTERMITTENT, INTERNALLY ABSENT, OR INTERNALLY AND EXTERNALLY ABSENT (56)	0.70	10 3 1 7 X SQ= 5.86 P = 0.008 RV YES (YES)
53	TEND TO BE THOSE WHERE NEWSPAPER CIRCULATION IS THREE HUNDRED OR MORE PER THOUSAND (14)	0.64	53	ALWAYS ARE THOSE WHERE NEWSPAPER CIRCULATION IS LESS THAN THREE HUNDRED PER THOUSAND (101)	1.00	7 0 4 13 X SQ= 8.80 P = 0.001 RV YES (YES)
54	ALWAYS ARE THOSE WHERE NEWSPAPER CIRCULATION IS ONE HUNDRED OR MORE PER THOUSAND (37)	1.00	54	TEND TO BE THOSE WHERE NEWSPAPER CIRCULATION IS LESS THAN ONE HUNDRED PER THOUSAND (76)	0.77	11 3 0 10 X SQ= 11.51 P = 0. RV YES (YES)
56	ALWAYS ARE THOSE WHERE THE RELIGION IS PREDOMINANTLY LITERATE (79)	1.00	56	TEND LESS TO BE THOSE WHERE THE RELIGION IS PREDOMINANTLY LITERATE (79)	0.62	11 8 0 5 X SQ= 3.27 P = 0.041 RV NO (YES)
62	TEND TO BE THOSE WHERE THE RELIGION IS PROTESTANT, RATHER THAN CATHOLIC (5)	0.62	62	ALWAYS ARE THOSE WHERE THE RELIGION IS CATHOLIC, RATHER THAN PROTESTANT (25)	1.00	5 0 3 6 X SQ= 3.43 P = 0.031 RV YES (YES)
63	TEND TO BE THOSE WHERE THE RELIGION IS PREDOMINANTLY SOME KIND OF CHRISTIAN (46)	0.91	63	TEND TO BE THOSE WHERE THE RELIGION IS PREDOMINANTLY OR PARTLY OTHER THAN CHRISTIAN (68)	0.54	10 6 1 7 X SQ= 3.55 P = 0.033 RV YES (YES)

No.	Left	Prop.	No.	Right	Prop.	Statistics
66	TILT TOWARD BEING THOSE THAT ARE RELIGIOUSLY HOMOGENEOUS (57)	0.82	66	TILT TOWARD BEING THOSE THAT ARE RELIGIOUSLY HETEROGENEOUS (49)	0.54	9 6 2 7 X SQ= 1.89 P = 0.105 RV YES (YES)
67	ALWAYS ARE THOSE THAT ARE RACIALLY HOMOGENEOUS (82)	1.00	67	TEND TO BE THOSE THAT ARE RACIALLY HETEROGENEOUS (27)	0.58	11 5 0 7 X SQ= 6.67 P = 0.005 RV YES (YES)
70	TEND TO BE THOSE THAT ARE RELIGIOUSLY, RACIALLY, AND LINGUISTICALLY HOMOGENEOUS (21)	0.55	70	TEND TO BE THOSE THAT ARE RELIGIOUSLY, RACIALLY, OR LINGUISTICALLY HETEROGENEOUS (85)	0.92	6 1 5 11 X SQ= 3.81 P = 0.027 RV YES (YES)
71	TEND TO BE THOSE WHOSE DATE OF INDEPENDENCE IS BEFORE 1800 (21)	0.55	71	ALWAYS ARE THOSE WHOSE DATE OF INDEPENDENCE IS AFTER 1800 (90)	1.00	6 0 5 13 X SQ= 6.77 P = 0.003 RV YES (YES)
74	ALWAYS ARE THOSE THAT ARE HISTORICALLY WESTERN (26)	1.00	74	ALWAYS ARE THOSE THAT ARE NOT HISTORICALLY WESTERN (87)	1.00	10 0 0 13 X SQ= 19.11 P = 0. RV YES (YES)
76	ALWAYS ARE THOSE THAT ARE HISTORICALLY WESTERN, RATHER THAN HAVING BEEN SIGNIFICANTLY OR PARTIALLY WESTERNIZED THROUGH A COLONIAL RELATIONSHIP (26)	1.00	76	ALWAYS ARE THOSE THAT HAVE BEEN SIGNIFICANTLY OR PARTIALLY WESTERNIZED THROUGH A COLONIAL RELATIONSHIP, RATHER THAN BEING HISTORICALLY WESTERN (70)	1.00	10 0 0 13 X SQ= 19.11 P = 0. RV YES (YES)
79	ALWAYS ARE THOSE THAT WERE FORMERLY DEPENDENCIES OF BRITAIN OR FRANCE, RATHER THAN SPAIN (49)	1.00	79	TILT TOWARD BEING THOSE THAT WERE FORMERLY DEPENDENCIES OF SPAIN, RATHER THAN BRITAIN OR FRANCE (18)	0.73	2 3 0 8 X SQ= 1.33 P = 0.128 RV YES (NO)
81	LEAN LESS TOWARD BEING THOSE WHERE THE TYPE OF POLITICAL MODERNIZATION IS LATER EUROPEAN OR LATER EUROPEAN DERIVED, RATHER THAN EARLY EUROPEAN OR EARLY EUROPEAN DERIVED (40)	0.60	81	ALWAYS ARE THOSE WHERE THE TYPE OF POLITICAL MODERNIZATION IS LATER EUROPEAN OR LATER EUROPEAN DERIVED, RATHER THAN EARLY EUROPEAN OR EARLY EUROPEAN DERIVED (40)	1.00	4 0 6 9 X SQ= 2.47 P = 0.087 RV NO (YES)
85	ALWAYS ARE THOSE WHERE THE STAGE OF POLITICAL MODERNIZATION IS ADVANCED, RATHER THAN MID- OR EARLY TRANSITIONAL (60)	1.00	85	TEND TO BE THOSE WHERE THE STAGE OF POLITICAL MODERNIZATION IS MID- OR EARLY TRANSITIONAL, RATHER THAN ADVANCED (54)	0.54	11 6 0 7 X SQ= 5.96 P = 0.006 RV YES (YES)

89	TEND TO BE THOSE WHOSE IDEOLOGICAL ORIENTATION IS CONVENTIONAL (33)	0.91	89	TEND TO BE THOSE WHOSE IDEOLOGICAL ORIENTATION IS OTHER THAN CONVENTIONAL (62)	0.57	10 3 1 4 X SQ= 2.82 P = 0.047 RV YES (YES)
98	TEND TO BE THOSE WHERE GOVERNMENTAL STABILITY IS GENERALLY PRESENT AND DATES AT LEAST FROM THE INTERWAR PERIOD (22)	0.91	98	ALWAYS ARE THOSE WHERE GOVERNMENTAL STABILITY IS GENERALLY OR MODERATELY PRESENT AND DATES FROM THE POST-WAR PERIOD, OR IS ABSENT (93)	1.00	10 0 1 13 X SQ= 16.69 P = 0. RV YES (YES)
99	ALWAYS ARE THOSE WHERE GOVERNMENTAL STABILITY IS GENERALLY PRESENT AND DATES FROM AT LEAST THE INTER-WAR PERIOD, OR FROM THE POST-WAR PERIOD (50)	1.00	99	ALWAYS ARE THOSE WHERE GOVERNMENTAL STABILITY IS MODERATELY PRESENT AND DATES FROM THE POST-WAR PERIOD, OR IS ABSENT (36)	1.00	11 0 0 13 X SQ= 20.14 P = 0. RV YES (YES)
101	ALWAYS ARE THOSE WHERE THE REPRESENTATIVE CHARACTER OF THE REGIME IS POLYARCHIC (41)	1.00	101	TEND TO BE THOSE WHERE THE REPRESENTATIVE CHARACTER OF THE REGIME IS LIMITED POLYARCHIC, PSEUDO-POLYARCHIC, OR NON-POLYARCHIC (57)	0.63	11 3 0 5 X SQ= 6.39 P = 0.005 RV YES (YES)
110	TEND TO BE THOSE WHERE POLITICAL ENCULTURATION IS HIGH (15)	0.60	110	ALWAYS ARE THOSE WHERE POLITICAL ENCULTURATION IS MEDIUM OR LOW (80)	1.00	6 0 4 13 X SQ= 7.67 P = 0.002 RV YES (YES)
111	TEND TO BE THOSE WHERE POLITICAL ENCULTURATION IS HIGH OR MEDIUM (53)	0.90	111	TEND TO BE THOSE WHERE POLITICAL ENCULTURATION IS LOW (42)	0.85	9 2 1 11 X SQ= 9.80 P = 0.001 RV YES (YES)
115	TEND TO BE THOSE WHERE INTEREST ARTICULATION BY ASSOCIATIONAL GROUPS IS SIGNIFICANT (20)	0.73	115	TEND TO BE THOSE WHERE INTEREST ARTICULATION BY ASSOCIATIONAL GROUPS IS MODERATE, LIMITED, OR NEGLIGIBLE (91)	0.85	8 2 3 11 X SQ= 5.87 P = 0.011 RV YES (YES)
116	ALWAYS ARE THOSE WHERE INTEREST ARTICULATION BY ASSOCIATIONAL GROUPS IS SIGNIFICANT OR MODERATE (32)	1.00	116	TEND TO BE THOSE WHERE INTEREST ARTICULATION BY ASSOCIATIONAL GROUPS IS LIMITED OR NEGLIGIBLE (79)	0.62	11 5 0 8 X SQ= 7.57 P = 0.002 RV YES (YES)
118	ALWAYS ARE THOSE WHERE INTEREST ARTICULATION BY INSTITUTIONAL GROUPS IS SIGNIFICANT, MODERATE, OR LIMITED (60)	1.00	118	LEAN LESS TOWARD BEING THOSE WHERE INTEREST ARTICULATION BY INSTITUTIONAL GROUPS IS SIGNIFICANT, MODERATE, OR LIMITED (60)	0.69	0 4 11 9 X SQ= 2.15 P = 0.098 RV NO (YES)

119	TEND TO BE THOSE WHERE INTEREST ARTICULATION BY INSTITUTIONAL GROUPS IS MODERATE OR LIMITED (26)	0.91	119	TEND TO BE THOSE WHERE INTEREST ARTICULATION BY INSTITUTIONAL GROUPS IS VERY SIGNIFICANT OR SIGNIFICANT (74)	0.92	1 12 10 1 X SQ= 13.44 P = 0. RV YES (YES)
120	TEND TO BE THOSE WHERE INTEREST ARITUCLATION BY INSTITUTIONAL GROUPS IS LIMITED (10)	0.55	120	ALWAYS ARE THOSE WHERE INTEREST ARTICULATION BY INSTITUTIONAL GROUPS IS VERY SIGNIFICANT, SIGNIFICANT, OR MODERATE (90)	1.00	5 13 6 0 X SQ= 6.77 P = 0.003 RV YES (YES)
121	ALWAYS ARE THOSE WHERE INTEREST ARTICULATION BY NON-ASSOCIATIONAL GROUPS IS MODERATE, LIMITED, OR NEGLIGIBLE (61)	1.00	121	LEAN LESS TOWARD BEING THOSE WHERE INTEREST ARTICULATION BY NON-ASSOCIATIONAL GROUPS IS MODERATE, LIMITED, OR NEGLIGIBLE (61)	0.69	0 4 11 9 X SQ= 2.15 P = 0.098 RV NO (YES)
122	LEAN TOWARD BEING THOSE WHERE INTEREST ARTICULATION BY NON-ASSOCIATIONAL GROUPS IS LIMITED OR NEGLIGIBLE (32)	0.73	122	LEAN TOWARD BEING THOSE WHERE INTEREST ARTICULATION BY NON-ASSOCIATIONAL GROUPS IS SIGNIFICANT OR MODERATE (83)	0.69	3 9 8 4 X SQ= 2.69 P = 0.100 RV YES (YES)
123	TEND LESS TO BE THOSE WHERE INTEREST ARTICULATION BY NON-ASSOCIATIONAL GROUPS IS SIGNIFICANT, MODERATE, OR LIMITED (107)	0.55	123	ALWAYS ARE THOSE WHERE INTEREST ARTICULATION BY NON-ASSOCIATIONAL GROUPS IS SIGNIFICANT, MODERATE, OR LIMITED (107)	1.00	6 13 5 0 X SQ= 4.96 P = 0.011 RV NO (YES)
125	ALWAYS ARE THOSE WHERE INTEREST ARTICULATION BY ANOMIC GROUPS IS INFREQUENT OR VERY INFREQUENT (35)	1.00	125	TEND TO BE THOSE WHERE INTEREST ARTICULATION BY ANOMIC GROUPS IS FREQUENT OR OCCASIONAL (64)	0.91	0 10 11 1 X SQ= 14.85 P = 0. RV YES (YES)
126	TEND TO BE THOSE WHERE INTEREST ARTICULATION BY ANOMIC GROUPS IS VERY INFREQUENT (16)	0.91	126	ALWAYS ARE THOSE WHERE INTEREST ARTICULATION BY ANOMIC GROUPS IS FREQUENT, OCCASIONAL, OR INFREQUENT (83)	1.00	1 11 10 0 X SQ= 14.85 P = 0. RV YES (YES)
132	TILT MORE TOWARD BEING THOSE WHERE INTEREST AGGREGATION BY POLITICAL PARTIES IS SIGNIFICANT, MODERATE, OR LIMITED (67)	0.91	132	TILT LESS TOWARD BEING THOSE WHERE INTEREST AGGREGATION BY POLITICAL PARTIES IS SIGNIFICANT, MODERATE, OR LIMITED (67)	0.55	10 6 1 5 X SQ= 2.06 P = 0.149 RV NO (YES)
134	TEND TO BE THOSE WHERE INTEREST AGGREGATION BY THE EXECUTIVE IS SIGNIFICANT OR MODERATE (57)	0.91	134	TEND TO BE THOSE WHERE INTEREST AGGREGATION BY THE EXECUTIVE IS LIMITED OR NEGLIGIBLE (46)	0.89	10 1 1 8 X SQ= 9.72 P = 0.001 RV YES (YES)

136	TEND TO BE THOSE WHERE INTEREST AGGREGATION BY THE LEGISLATURE IS SIGNIFICANT (12)	0.91	136	ALWAYS ARE THOSE WHERE INTEREST AGGREGATION BY THE LEGISLATURE IS MODERATE, LIMITED, OR NEGLIGIBLE (85)	1.00	10 0 1 12 X SQ= 15.78 P = 0. RV YES (YES)
137	ALWAYS ARE THOSE WHERE INTEREST AGGREGATION BY THE LEGISLATURE IS SIGNIFICANT OR MODERATE (29)	1.00	137	TEND TO BE THOSE WHERE INTEREST AGGREGATION BY THE LEGISLATURE IS LIMITED OR NEGLIGIBLE (68)	0.83	11 2 0 10 X SQ= 13.00 P = 0. RV YES (YES)
147	ALWAYS ARE THOSE WHERE THE PARTY SYSTEM IS QUALITATIVELY CLASS-ORIENTED OR MULTI-IDEOLOGICAL (23)	1.00	147	TEND TO BE THOSE WHERE THE PARTY SYSTEM IS QUALITATIVELY OTHER THAN CLASS-ORIENTED OR MULTI-IDEOLOGICAL (67)	0.64	10 4 0 7 X SQ= 6.90 P = 0.004 RV YES (YES)
158	ALWAYS ARE THOSE WHERE PERSONALISMO IS NEGLIGIBLE (56)	1.00	158	TEND TO BE THOSE WHERE PERSONALISMO IS PRONOUNCED OR MODERATE (40)	0.92	0 12 11 1 X SQ= 16.78 P = 0. RV YES (YES)
161	ALWAYS ARE THOSE WHERE THE POLITICAL LEADERSHIP IS NON-ELITIST (50)	1.00	161	TEND TO BE THOSE WHERE THE POLITICAL LEADERSHIP IS ELITIST OR MODERATELY ELITIST (47)	0.75	0 9 11 3 X SQ= 10.59 P = 0. RV YES (YES)
168	ALWAYS ARE THOSE WHERE THE HORIZONTAL POWER DISTRIBUTION IS SIGNIFICANT (34)	1.00	168	TEND TO BE THOSE WHERE THE HORIZONTAL POWER DISTRIBUTION IS LIMITED OR NEGLIGIBLE (72)	0.80	11 2 0 8 X SQ= 11.03 P = 0. RV YES (YES)
170	ALWAYS ARE THOSE WHERE THE LEGISLATIVE-EXECUTIVE STRUCTURE IS OTHER THAN PRESIDENTIAL (63)	1.00	170	TEND TO BE THOSE WHERE THE LEGISLATIVE-EXECUTIVE STRUCTURE IS PRESIDENTIAL (39)	0.67	0 8 11 4 X SQ= 8.50 P = 0.001 RV YES (YES)
172	LEAN TOWARD BEING THOSE WHERE THE LEGISLATIVE-EXECUTIVE STRUCTURE IS PARLIAMENTARY-ROYALIST (21)	0.55	172	LEAN TOWARD BEING THOSE WHERE THE LEGISLATIVE-EXECUTIVE STRUCTURE IS OTHER THAN PARLIAMENTARY-ROYALIST (88)	0.85	6 2 5 11 X SQ= 2.54 P = 0.082 RV YES (YES)
174	ALWAYS ARE THOSE WHERE THE LEGISLATURE IS FULLY EFFECTIVE (28)	1.00	174	ALWAYS ARE THOSE WHERE THE LEGISLATURE IS PARTIALLY EFFECTIVE, LARGELY INEFFECTIVE, OR WHOLLY INEFFECTIVE (72)	1.00	11 0 0 10 X SQ= 17.18 P = 0. RV YES (YES)

```
181  ALWAYS ARE THOSE                        1.00     181  ALWAYS ARE THOSE                        1.00        11     0
     WHERE THE BUREAUCRACY                                 WHERE THE BUREAUCRACY                                 0    11
     IS MODERN, RATHER THAN                                IS SEMI-MODERN, RATHER THAN                      X SQ=   18.18
     SEMI-MODERN  (21)                                     MODERN  (55)                                     P =      0.
                                                                                                            RV YES (YES)

186  ALWAYS ARE THOSE                        1.00     186  LEAN TOWARD BEING THOSE                 0.50         0     2
     WHERE PARTICIPATION BY THE MILITARY                   WHERE PARTICIPATION BY THE MILITARY                 11     2
     IN POLITICS IS                                        IN POLITICS IS                                   X SQ=    2.76
     NEUTRAL, RATHER THAN                                  SUPPORTIVE, RATHER THAN                          P =      0.057
     SUPPORTIVE  (56)                                      NEUTRAL  (31)                                    RV YES (YES)

187  ALWAYS ARE THOSE                        1.00     187  TEND TO BE THOSE                        0.89         0     8
     WHERE THE ROLE OF THE POLICE                          WHERE THE ROLE OF THE POLICE                        11     1
     IS NOT POLITICALLY SIGNIFICANT  (35)                  IS POLITICALLY SIGNIFICANT  (66)                 X SQ=   12.80
                                                                                                            P =      0.
                                                                                                            RV YES (YES)
```

156 POLITIES
WHERE PERSONALISMO IS
PRONOUNCED, RATHER THAN
NEGLIGIBLE (14)

156 POLITIES
WHERE PERSONALISMO IS
NEGLIGIBLE, RATHER THAN
PRONOUNCED (56)

SUBJECT ONLY

14 IN LEFT COLUMN

CAMBODIA, ECUADOR, EL SALVADOR, GREECE, GUATEMALA, HONDURAS, LAOS, LEBANON, NICARAGUA, PANAMA, SYRIA, THAILAND, UAR, URUGUAY

56 IN RIGHT COLUMN

ALBANIA, ALGERIA, AUSTRALIA, AUSTRIA, BELGIUM, BULGARIA, CANADA, CHINA, PR, CUBA, CZECHOS'KIA, DENMARK, FINLAND, GERMANY, E, GERMAN FR, GUINEA, HUNGARY, ICELAND, INDIA, IRELAND, ISRAEL, ITALY, JAMAICA, KOREA, N, LUXEMBOURG, MALAGASY R, MALAYA, MEXICO, MONGOLIA, MOROCCO, NETHERLANDS, NEW ZEALAND, NIGERIA, NORWAY, PARAGUAY, PHILIPPINES, POLAND, PORTUGAL, RUMANIA, SENEGAL, SIERRE LEO, SOMALIA, SO AFRICA, SPAIN, SWEDEN, SWITZERLAND, TANGANYIKA, TRINIDAD, TUNISIA, TURKEY, UGANDA, USSR, UK, US, VENEZUELA, VIETNAM, N, YUGOSLAVIA

31 EXCLUDED BECAUSE IRRELEVANT

AFGHANISTAN, ARGENTINA, BOLIVIA, BRAZIL, CAMEROUN, CEN AFR REP, CEYLON, CHAD, CHILE, COLOMBIA, CONGO(BRA), CONGO(LEO), COSTA RICA, DAHOMEY, ETHIOPIA, FRANCE, GABON, GHANA, HAITI, INDONESIA, IVORY COAST, JAPAN, LIBERIA, LIBYA, MALI, MAURITANIA, NIGER, PERU, SA'U ARABIA, UPPER VOLTA, VIETNAM REP

3 EXCLUDED BECAUSE UNASCERTAINED

BURUNDI, CYPRUS, RWANDA

11 EXCLUDED BECAUSE UNASCERTAINABLE

BURMA, DOMIN REP, IRAN, IRAQ, JORDAN, KOREA REP, NEPAL, PAKISTAN, SUDAN, TOGO, YEMEN

	Left column			Right column		
2	LEAN MORE TOWARD BEING THOSE LOCATED ELSEWHERE THAN IN WEST EUROPE, SCANDINAVIA, NORTH AMERICA, OR AUSTRALASIA (93)	0.93	2	LEAN LESS TOWARD BEING THOSE LOCATED ELSEWHERE THAN IN WEST EUROPE, SCANDINAVIA, NORTH AMERICA, OR AUSTRALASIA (93)	0.64	1 20 13 36 X SQ= 3.10 P = 0.050 RV NO (NO)
6	TEND TO BE THOSE LOCATED IN THE CARIBBEAN, CENTRAL AMERICA, OR SOUTH AMERICA (22)	0.50	6	TEND TO BE THOSE LOCATED ELSEWHERE THAN IN THE CARIBBEAN, CENTRAL AMERICA, OR SOUTH AMERICA (93)	0.89	7 6 7 50 X SQ= 8.98 P = 0.003 RV YES (YES)

15	TILT TOWARD BEING THOSE LOCATED IN THE MIDDLE EAST, RATHER THAN IN NORTH AFRICA OR CENTRAL AND SOUTH AFRICA (11)	0.67	15	TILT TOWARD BEING THOSE LOCATED IN NORTH AFRICA OR CENTRAL AND SOUTH AFRICA, RATHER THAN IN THE MIDDLE EAST (33)	0.86	2 2 1 12 X SQ= 1.42 P = 0.121 RV YES (YES)
19	TEND TO BE THOSE LOCATED IN THE CARIBBEAN, CENTRAL AMERICA, OR SOUTH AMERICA, RATHER THAN IN NORTH AFRICA OR CENTRAL AND SOUTH AFRICA (22)	0.87	19	TEND TO BE THOSE LOCATED IN NORTH AFRICA OR CENTRAL AND SOUTH AFRICA, RATHER THAN IN THE CARIBBEAN, CENTRAL AMERICA, OR SOUTH AMERICA (33)	0.67	1 12 7 6 X SQ= 4.51 P = 0.030 RV YES (YES)
22	LEAN TOWARD BEING THOSE WHOSE TERRITORIAL SIZE IS SMALL (47)	0.71	22	LEAN TOWARD BEING THOSE WHOSE TERRITORIAL SIZE IS VERY LARGE, LARGE, OR MEDIUM (68)	0.59	4 33 10 23 X SQ= 3.01 P = 0.071 RV YES (NO)
24	TEND TO BE THOSE WHOSE POPULATION IS SMALL (54)	0.79	24	TEND TO BE THOSE WHOSE POPULATION IS VERY LARGE, LARGE, OR MEDIUM (61)	0.64	3 36 11 20 X SQ= 6.69 P = 0.006 RV YES (NO)
26	LEAN TOWARD BEING THOSE WHOSE POPULATION DENSITY IS LOW (67)	0.71	26	LEAN TOWARD BEING THOSE WHOSE POPULATION DENSITY IS VERY HIGH, HIGH, OR MEDIUM (48)	0.55	4 31 10 25 X SQ= 2.23 P = 0.084 RV YES (NO)
28	LEAN TOWARD BEING THOSE WHOSE POPULATION GROWTH RATE IS HIGH (62)	0.79	28	LEAN TOWARD BEING THOSE WHOSE POPULATION GROWTH RATE IS LOW (48)	0.52	11 26 3 28 X SQ= 3.01 P = 0.069 RV YES (NO)
31	LEAN MORE TOWARD BEING THOSE WHOSE AGRICULTURAL POPULATION IS HIGH OR MEDIUM (90)	0.93	31	LEAN LESS TOWARD BEING THOSE WHOSE AGRICULTURAL POPULATION IS HIGH OR MEDIUM (90)	0.64	13 36 1 20 X SQ= 3.10 P = 0.050 RV NO (NO)
33	ALWAYS ARE THOSE WHOSE GROSS NATIONAL PRODUCT IS LOW OR VERY LOW (85)	1.00	33	TEND LESS TO BE THOSE WHOSE GROSS NATIONAL PRODUCT IS LOW OR VERY LOW (85)	0.57	0 24 14 32 X SQ= 7.33 P = 0.001 RV NO (NO)
34	TEND TO BE THOSE WHOSE GROSS NATIONAL PRODUCT IS VERY LOW (53)	0.71	34	TEND TO BE THOSE WHOSE GROSS NATIONAL PRODUCT IS VERY HIGH, HIGH, MEDIUM, OR LOW (62)	0.71	4 40 10 16 X SQ= 7.07 P = 0.005 RV YES (NO)

35	ALWAYS ARE THOSE WHOSE PER CAPITA GROSS NATIONAL PRODUCT IS MEDIUM, LOW, OR VERY LOW (91)	1.00	35	TEND LESS TO BE THOSE WHOSE PER CAPITA GROSS NATIONAL PRODUCT IS MEDIUM, LOW, OR VERY LOW (91)	0.59	0 23 14 33 X SQ= 6.80 P = 0.003 RV NO (NO)
36	TEND TO BE THOSE WHOSE PER CAPITA GROSS NATIONAL PRODUCT IS LOW OR VERY LOW (73)	0.79	36	TEND TO BE THOSE WHOSE PER CAPITA GROSS NATIONAL PRODUCT IS VERY HIGH, HIGH, OR MEDIUM (42)	0.61	3 34 11 22 X SQ= 5.45 P = 0.015 RV YES (NO)
39	TEND TO BE THOSE WHOSE INTERNATIONAL FINANCIAL STATUS IS LOW OR VERY LOW (76)	0.93	39	TEND TO BE THOSE WHOSE INTERNATIONAL FINANCIAL STATUS IS VERY HIGH, HIGH, OR MEDIUM (38)	0.53	1 29 13 26 X SQ= 7.67 P = 0.002 RV YES (NO)
41	ALWAYS ARE THOSE WHOSE ECONOMIC DEVELOPMENTAL STATUS IS INTERMEDIATE, UNDERDEVELOPED, OR VERY UNDERDEVELOPED (94)	1.00	41	TEND LESS TO BE THOSE WHOSE ECONOMIC DEVELOPMENTAL STATUS IS INTERMEDIATE, UNDERDEVELOPED, OR VERY UNDERDEVELOPED (94)	0.67	0 18 14 36 X SQ= 4.75 P = 0.014 RV NO (NO)
42	TEND TO BE THOSE WHOSE ECONOMIC DEVELOPMENTAL STATUS IS UNDERDEVELOPED OR VERY UNDERDEVELOPED (76)	0.86	42	TEND TO BE THOSE WHOSE ECONOMIC DEVELOPMENTAL STATUS IS DEVELOPED OR INTERMEDIATE (36)	0.52	2 28 12 26 X SQ= 4.93 P = 0.015 RV YES (NO)
43	TEND TO BE THOSE WHOSE ECONOMIC DEVELOPMENTAL STATUS IS VERY UNDERDEVELOPED (57)	0.77	43	TEND TO BE THOSE WHOSE ECONOMIC DEVELOPMENTAL STATUS IS DEVELOPED, INTERMEDIATE, OR UNDERDEVELOPED (55)	0.71	3 40 10 16 X SQ= 8.55 P = 0.003 RV YES (NO)
44	ALWAYS ARE THOSE WHERE THE LITERACY RATE IS BELOW NINETY PERCENT (90)	1.00	44	TEND LESS TO BE THOSE WHERE THE LITERACY RATE IS BELOW NINETY PERCENT (90)	0.59	0 23 14 33 X SQ= 6.80 P = 0.003 RV NO (NO)
45	TEND TO BE THOSE WHERE THE LITERACY RATE IS BELOW FIFTY PERCENT (54)	0.69	45	TEND TO BE THOSE WHERE THE LITERACY RATE IS FIFTY PERCENT OR ABOVE (55)	0.71	4 37 9 15 X SQ= 5.65 P = 0.011 RV YES (NO)
53	ALWAYS ARE THOSE WHERE NEWSPAPER CIRCULATION IS LESS THAN THREE HUNDRED PER THOUSAND (101)	1.00	53	LEAN LESS TOWARD BEING THOSE WHERE NEWSPAPER CIRCULATION IS LESS THAN THREE HUNDRED PER THOUSAND (101)	0.77	0 13 14 43 X SQ= 2.60 P = 0.057 RV NO (NO)

54	LEAN TOWARD BEING THOSE WHERE NEWSPAPER CIRCULATION IS LESS THAN ONE HUNDRED PER THOUSAND (76)	0.79	LEAN TOWARD BEING THOSE WHERE NEWSPAPER CIRCULATION IS ONE HUNDRED OR MORE PER THOUSAND (37)	0.52	3 28 11 26 X SQ= 3.01 P = 0.069 RV YES (NO)
74	TEND MORE TO BE THOSE THAT ARE NOT HISTORICALLY WESTERN (87)	0.93	TEND LESS TO BE THOSE THAT ARE NOT HISTORICALLY WESTERN (87)	0.56	1 24 13 30 X SQ= 5.15 P = 0.012 RV NO (NO)
76	TEND TO BE THOSE THAT HAVE BEEN SIGNIFICANTLY OR PARTIALLY WESTERNIZED THROUGH A COLONIAL RELATIONSHIP, RATHER THAN BEING HISTORICALLY WESTERN (70)	0.92	TEND TO BE THOSE THAT ARE HISTORICALLY WESTERN, RATHER THAN HAVING BEEN SIGNIFICANTLY OR PARTIALLY WESTERNIZED THROUGH A COLONIAL RELATIONSHIP (26)	0.52	1 24 12 22 X SQ= 6.49 P = 0.004 RV YES (NO)
77	TILT TOWARD BEING THOSE THAT HAVE BEEN SIGNIFICANTLY WESTERNIZED, RATHER THAN PARTIALLY WESTERNIZED, THROUGH A COLONIAL RELATIONSHIP (28)	0.75	TILT TOWARD BEING THOSE THAT HAVE BEEN PARTIALLY WESTERNIZED, RATHER THAN SIGNIFICANTLY WESTERNIZED, THROUGH A COLONIAL RELATIONSHIP (41)	0.57	9 9 3 12 X SQ= 2.02 P = 0.145 RV YES (YES)
78	ALWAYS ARE THOSE THAT HAVE BEEN SIGNIFICANTLY WESTERNIZED THROUGH A COLONIAL RELATIONSHIP, RATHER THAN WITHOUT SUCH A RELATIONSHIP (28)	1.00	LEAN LESS TOWARD BEING THOSE THAT HAVE BEEN SIGNIFICANTLY WESTERNIZED THROUGH A COLONIAL RELATIONSHIP, RATHER THAN WITHOUT SUCH A RELATIONSHIP (28)	0.60	9 9 0 6 X SQ= 2.90 P = 0.052 RV NO (YES)
79	TEND TO BE THOSE THAT WERE FORMERLY DEPENDENCIES OF SPAIN, RATHER THAN BRITAIN OR FRANCE (18)	0.58	TEND TO BE THOSE THAT WERE FORMERLY DEPENDENCIES OF BRITAIN OR FRANCE, RATHER THAN SPAIN (49)	0.85	5 22 7 4 X SQ= 5.42 P = 0.017 RV YES (YES)
84	ALWAYS ARE THOSE WHERE THE TYPE OF POLITICAL MODERNIZATION IS DEVELOPED TUTELARY, RATHER THAN UNDEVELOPED TUTELARY (31)	1.00	TILT LESS TOWARD BEING THOSE WHERE THE TYPE OF POLITICAL MODERNIZATION IS DEVELOPED TUTELARY, RATHER THAN UNDEVELOPED TUTELARY (31)	0.60	5 12 0 8 X SQ= 1.39 P = 0.140 RV NO (NO)
85	TEND TO BE THOSE WHERE THE STAGE OF POLITICAL MODERNIZATION IS MID- OR EARLY TRANSITIONAL, RATHER THAN ADVANCED (54)	0.57	TEND TO BE THOSE WHERE THE STAGE OF POLITICAL MODERNIZATION IS ADVANCED, RATHER THAN MID- OR EARLY TRANSITIONAL (60)	0.75	6 42 8 14 X SQ= 3.98 P = 0.050 RV YES (NO)
93	TILT MORE TOWARD BEING THOSE WHERE THE SYSTEM STYLE IS NON-MOBILIZATIONAL (78)	0.86	TILT LESS TOWARD BEING THOSE WHERE THE SYSTEM STYLE IS NON-MOBILIZATIONAL (78)	0.60	2 22 12 33 X SQ= 2.22 P = 0.115 RV NO (NO)

95	ALWAYS ARE THOSE WHERE THE STATUS OF THE REGIME IS CONSTITUTIONAL OR AUTHORITARIAN (95)	1.00	95	TEND LESS TO BE THOSE WHERE THE STATUS OF THE REGIME IS CONSTITUTIONAL OR AUTHORITARIAN (95)	0.71	14 39 0 16 X SQ= 3.79 P = 0.029 RV NO (NO)
96	TEND TO BE THOSE WHERE THE STATUS OF THE REGIME IS AUTHORITARIAN (23)	0.50	96	TEND TO BE THOSE WHERE THE STATUS OF THE REGIME IS CONSTITUTIONAL OR TOTALITARIAN (67)	0.94	6 49 6 3 X SQ= 12.34 P = 0.001 RV YES (YES)
97	ALWAYS ARE THOSE WHERE THE STATUS OF THE REGIME IS AUTHORITARIAN, RATHER THAN TOTALITARIAN (23)	1.00	97	TEND TO BE THOSE WHERE THE STATUS OF THE REGIME IS TOTALITARIAN, RATHER THAN AUTHORITARIAN (16)	0.84	6 3 0 16 X SQ= 10.62 P = 0. RV YES (YES)
98	ALWAYS ARE THOSE WHERE GOVERNMENTAL STABILITY IS GENERALLY OR MODERATELY PRESENT AND DATES FROM THE POST-WAR PERIOD, OR IS ABSENT (93)	1.00	98	TEND LESS TO BE THOSE WHERE GOVERNMENTAL STABILITY IS GENERALLY OR MODERATELY PRESENT AND DATES FROM THE POST-WAR PERIOD, OR IS ABSENT (93)	0.66	0 19 14 37 X SQ= 4.92 P = 0.008 RV NO (NO)
99	TEND TO BE THOSE WHERE GOVERNMENTAL STABILITY IS MODERATELY PRESENT AND DATES FROM THE POST-WAR PERIOD, OR IS ABSENT (36)	0.86	99	TEND TO BE THOSE WHERE GOVERNMENTAL STABILITY IS GENERALLY PRESENT AND DATES FROM AT LEAST THE INTER-WAR PERIOD, OR FROM THE POST-WAR PERIOD (50)	0.93	2 41 12 3 X SQ= 30.49 P = 0. RV YES (YES)
100	TEND TO BE THOSE WHERE GOVERNMENTAL STABILITY IS ABSENT (22)	0.57	100	TEND TO BE THOSE WHERE GOVERNMENTAL STABILITY IS GENERALLY PRESENT AND DATES FROM AT LEAST THE INTER-WAR PERIOD, OR IS GENERALLY OR MODERATELY PRESENT, AND DATES FROM THE POST-WAR PERIOD (64)	0.98	6 43 8 1 X SQ= 20.39 P = 0. RV YES (YES)
101	LEAN TOWARD BEING THOSE WHERE THE REPRESENTATIVE CHARACTER OF THE REGIME IS LIMITED POLYARCHIC, PSEUDO-POLYARCHIC, OR NON-POLYARCHIC (57)	0.73	101	LEAN TOWARD BEING THOSE WHERE THE REPRESENTATIVE CHARACTER OF THE REGIME IS POLYARCHIC (41)	0.58	3 31 8 22 X SQ= 2.42 P = 0.096 RV YES (NO)
111	TEND TO BE THOSE WHERE POLITICAL ENCULTURATION IS LOW (42)	0.67	111	TEND TO BE THOSE WHERE POLITICAL ENCULTURATION IS HIGH OR MEDIUM (53)	0.83	4 35 8 7 X SQ= 9.27 P = 0.002 RV YES (YES)
115	ALWAYS ARE THOSE WHERE INTEREST ARTICULATION BY ASSOCIATIONAL GROUPS IS MODERATE, LIMITED, OR NEGLIGIBLE (91)	1.00	115	TEND LESS TO BE THOSE WHERE INTEREST ARTICULATION BY ASSOCIATIONAL GROUPS IS MODERATE, LIMITED, OR NEGLIGIBLE (91)	0.70	0 16 14 37 X SQ= 4.02 P = 0.016 RV NO (NO)

No.	Statement	Value	No.	Statement	Value	Statistics
116	TILT MORE TOWARD BEING THOSE WHERE INTEREST ARTICULATION BY ASSOCIATIONAL GROUPS IS LIMITED OR NEGLIGIBLE (79)	0.79	116	TILT LESS TOWARD BEING THOSE WHERE INTEREST ARTICULATION BY ASSOCIATIONAL GROUPS IS LIMITED OR NEGLIGIBLE (79)	0.55	3 24 11 29 X SQ= 1.72 P = 0.134 RV NO (NO)
120	ALWAYS ARE THOSE WHERE INTEREST ARTICULATION BY INSTITUTIONAL GROUPS IS VERY SIGNIFICANT, SIGNIFICANT, OR MODERATE (90)	1.00	120	TILT LESS TOWARD BEING THOSE WHERE INTEREST ARTICULATION BY INSTITUTIONAL GROUPS IS VERY SIGNIFICANT, SIGNIFICANT, OR MODERATE (90)	0.81	14 44 0 10 X SQ= 1.74 P = 0.108 RV NO (NO)
125	LEAN TOWARD BEING THOSE WHERE INTEREST ARTICULATION BY ANOMIC GROUPS IS FREQUENT OR OCCASIONAL (64)	0.71	125	LEAN TOWARD BEING THOSE WHERE INTEREST ARTICULATION BY ANOMIC GROUPS IS INFREQUENT OR VERY INFREQUENT (35)	0.59	10 19 4 27 X SQ= 2.79 P = 0.068 RV YES (NO)
126	ALWAYS ARE THOSE WHERE INTEREST ARTICULATION BY ANOMIC GROUPS IS FREQUENT, OCCASIONAL, OR INFREQUENT (83)	1.00	126	TEND LESS TO BE THOSE WHERE INTEREST ARTICULATION BY ANOMIC GROUPS IS FREQUENT, OCCASIONAL, OR INFREQUENT (83)	0.65	14 30 0 16 X SQ= 4.98 P = 0.013 RV NO (NO)
131	TEND TO BE THOSE WHERE INTEREST AGGREGATION BY POLITICAL PARTIES IS LIMITED OR NEGLIGIBLE (35)	0.82	131	TEND TO BE THOSE WHERE INTEREST AGGREGATION BY POLITICAL PARTIES IS SIGNIFICANT OR MODERATE (30)	0.61	2 23 9 15 X SQ= 4.54 P = 0.018 RV YES (NO)
132	TEND TO BE THOSE WHERE INTEREST AGGREGATION BY POLITICAL PARTIES IS NEGLIGIBLE (9)	0.55	132	TEND TO BE THOSE WHERE INTEREST AGGREGATION BY POLITICAL PARTIES IS SIGNIFICANT, MODERATE, OR LIMITED (67)	0.95	5 39 6 2 X SQ= 12.84 P = 0.001 RV YES (YES)
134	LEAN TOWARD BEING THOSE WHERE INTEREST AGGREGATION BY THE EXECUTIVE IS LIMITED OR NEGLIGIBLE (46)	0.70	134	LEAN TOWARD BEING THOSE WHERE INTEREST AGGREGATION BY THE EXECUTIVE IS SIGNIFICANT OR MODERATE (57)	0.63	3 34 7 20 X SQ= 2.53 P = 0.081 RV YES (NO)
142	LEAN TOWARD BEING THOSE WHERE THE PARTY SYSTEM IS QUANTITATIVELY MULTI-PARTY (30)	0.60	142	LEAN TOWARD BEING THOSE WHERE THE PARTY SYSTEM IS QUANTITATIVELY OTHER THAN MULTI-PARTY (66)	0.71	6 15 4 37 X SQ= 2.38 P = 0.075 RV YES (NO)
147	ALWAYS ARE THOSE WHERE THE PARTY SYSTEM IS QUALITATIVELY OTHER THAN CLASS-ORIENTED OR MULTI-IDEOLOGICAL (67)	1.00	147	TEND LESS TO BE THOSE WHERE THE PARTY SYSTEM IS QUALITATIVELY OTHER THAN CLASS-ORIENTED OR MULTI-IDEOLOGICAL (67)	0.65	0 18 10 33 X SQ= 3.45 P = 0.026 RV NO (NO)

153	ALWAYS ARE THOSE WHERE THE PARTY SYSTEM IS MODERATELY STABLE OR UNSTABLE (38)	1.00	153	TEND TO BE THOSE WHERE THE PARTY SYSTEM IS STABLE (42)	0.82	0 40 12 9 X SQ= 24.95 P = 0. RV YES (YES)
154	TEND TO BE THOSE WHERE THE PARTY SYSTEM IS UNSTABLE (25)	0.83	154	TEND TO BE THOSE WHERE THE PARTY SYSTEM IS STABLE OR MODERATELY STABLE (55)	0.94	2 46 10 3 X SQ= 29.82 P = 0. RV YES (YES)
160	TEND TO BE THOSE WHERE THE POLITICAL LEADERSHIP IS ELITIST (30)	0.70	160	TEND TO BE THOSE WHERE THE POLITICAL LEADERSHIP IS MODERATELY ELITIST OR NON-ELITIST (67)	0.67	7 17 3 35 X SQ= 3.47 P = 0.037 RV YES (NO)
161	LEAN TOWARD BEING THOSE WHERE THE POLITICAL LEADERSHIP IS ELITIST OR MODERATELY ELITIST (47)	0.80	161	LEAN TOWARD BEING THOSE WHERE THE POLITICAL LEADERSHIP IS NON-ELITIST (50)	0.54	8 24 2 28 X SQ= 2.61 P = 0.083 RV YES (NO)
166	ALWAYS ARE THOSE WHERE THE VERTICAL POWER DISTRIBUTION IS THAT OF FORMAL FEDERALISM OR FORMAL AND EFFECTIVE UNITARISM (99)	1.00	166	TILT LESS TOWARD BEING THOSE WHERE THE VERTICAL POWER DISTRIBUTION IS THAT OF FORMAL FEDERALISM OR FORMAL AND EFFECTIVE UNITARISM (99)	0.82	0 10 14 46 X SQ= 1.64 P = 0.112 RV NO (NO)
168	LEAN MORE TOWARD BEING THOSE WHERE THE HORIZONTAL POWER DISTRIBUTION IS LIMITED OR NEGLIGIBLE (72)	0.85	168	LEAN LESS TOWARD BEING THOSE WHERE THE HORIZONTAL POWER DISTRIBUTION IS LIMITED OR NEGLIGIBLE (72)	0.53	2 26 11 29 X SQ= 3.20 P = 0.058 RV NO (NO)
171	ALWAYS ARE THOSE WHERE THE LEGISLATIVE-EXECUTIVE STRUCTURE IS OTHER THAN PARLIAMENTARY-REPUBLICAN (90)	1.00	171	TILT LESS TOWARD BEING THOSE WHERE THE LEGISLATIVE-EXECUTIVE STRUCTURE IS OTHER THAN PARLIAMENTARY-REPUBLICAN (90)	0.79	0 11 12 41 X SQ= 1.76 P = 0.106 RV NO (NO)
174	TEND MORE TO BE THOSE WHERE THE LEGISLATURE IS PARTIALLY EFFECTIVE, LARGELY INEFFECTIVE, OR WHOLLY INEFFECTIVE (72)	0.92	174	TEND LESS TO BE THOSE WHERE THE LEGISLATURE IS PARTIALLY EFFECTIVE, LARGELY INEFFECTIVE, OR WHOLLY INEFFECTIVE (72)	0.55	1 25 11 30 X SQ= 4.26 P = 0.021 RV NO (NO)
181	ALWAYS ARE THOSE WHERE THE BUREAUCRACY IS SEMI-MODERN, RATHER THAN MODERN (55)	1.00	181	TEND LESS TO BE THOSE WHERE THE BUREAUCRACY IS SEMI-MODERN, RATHER THAN MODERN (55)	0.58	0 20 11 28 X SQ= 5.20 P = 0.011 RV NO (NO)

183	ALWAYS ARE THOSE WHERE THE BUREAUCRACY IS TRADITIONAL, RATHER THAN POST-COLONIAL TRANSITIONAL (9)	1.00	183	ALWAYS ARE THOSE WHERE THE BUREAUCRACY IS POST-COLONIAL TRANSITIONAL, RATHER THAN TRADITIONAL (25)	1.00	0 8 3 0 X SQ= 6.54 P = 0.006 RV YES (YES)
185	TEND TO BE THOSE WHERE PARTICIPATION BY THE MILITARY IN POLITICS IS INTERVENTIVE, RATHER THAN SUPPORTIVE (21)	0.73	185	TEND TO BE THOSE WHERE PARTICIPATION BY THE MILITARY IN POLITICS IS SUPPORTIVE, RATHER THAN INTERVENTIVE (31)	0.90	8 2 3 19 X SQ= 10.64 P = 0.001 RV YES (YES)
187	TEND TO BE THOSE WHERE THE ROLE OF THE POLICE IS POLITICALLY SIGNIFICANT (66)	0.85	187	TEND TO BE THOSE WHERE THE ROLE OF THE POLICE IS NOT POLITICALLY SIGNIFICANT (35)	0.52	11 27 2 29 X SQ= 4.27 P = 0.028 RV YES (NO)
188	TEND TO BE THOSE WHERE THE CHARACTER OF THE LEGAL SYSTEM IS CIVIL LAW (32)	0.64	188	TEND TO BE THOSE WHERE THE CHARACTER OF THE LEGAL SYSTEM IS OTHER THAN CIVIL LAW (81)	0.76	9 13 5 42 X SQ= 6.72 P = 0.008 RV YES (NO)
190	ALWAYS ARE THOSE WHERE THE CHARACTER OF THE LEGAL SYSTEM IS CIVIL LAW, RATHER THAN COMMON LAW (32)	1.00	190	LEAN LESS TOWARD BEING THOSE WHERE THE CHARACTER OF THE LEGAL SYSTEM IS CIVIL LAW, RATHER THAN COMMON LAW (32)	0.65	9 13 0 7 X SQ= 2.46 P = 0.066 RV NO (NO)
194	ALWAYS ARE THOSE THAT ARE NON-COMMUNIST (101)	1.00	194	LEAN LESS TOWARD BEING THOSE THAT ARE NON-COMMUNIST (101)	0.76	0 13 14 42 X SQ= 2.68 P = 0.056 RV NO (NO)

157 POLITIES
WHERE PERSONALISMO IS
PRONOUNCED (14)

157 POLITIES
WHERE PERSONALISMO IS
MODERATE OR NEGLIGIBLE (84)

PREDICATE ONLY

14 IN LEFT COLUMN

CAMBODIA	ECUADOR	EL SALVADOR	GREECE	GUATEMALA	HONDURAS	LAOS	LEBANON	NICARAGUA	PANAMA
SYRIA	THAILAND	UAR	URUGUAY						

84 IN RIGHT COLUMN

ALBANIA	ALGERIA	ARGENTINA	AUSTRALIA	AUSTRIA	BELGIUM	BOLIVIA	BRAZIL	BULGARIA	BURUNDI
CAMEROUN	CANADA	CEN AFR REP	CEYLON	CHAD	CHILE	CHINA, PR	COLOMBIA	CONGO(BRA)	CONGO(LEO)
COSTA RICA	CUBA	CZECHOS'KIA	DAHOMEY	DENMARK	FINLAND	FRANCE	GABON	GERMANY, E	GERMAN FR
GHANA	GUINEA	HUNGARY	ICELAND	INDIA	INDONESIA	IRELAND	ISRAEL	ITALY	IVORY COAST
JAMAICA	JAPAN	KOREA, N	LIBERIA	LUXEMBOURG	MALAGASY R	MALAYA	MALI	MAURITANIA	MEXICO
MONGOLIA	MOROCCO	NETHERLANDS	NEW ZEALAND	NIGER	NIGERIA	NORWAY	PARAGUAY	PERU	PHILIPPINES
POLAND	PORTUGAL	RUMANIA	RWANDA	SENEGAL	SIERRE LEO	SOMALIA	SO AFRICA	SPAIN	SWEDEN
SWITZERLAND	TANGANYIKA	TRINIDAD	TUNISIA	TURKEY	UGANDA	USSR	UK	US	UPPER VOLTA
VENEZUELA	VIETNAM, N	VIETNAM REP	YUGOSLAVIA						

5 EXCLUDED BECAUSE IRRELEVANT

AFGHANISTAN	ETHIOPIA	HAITI	LIBYA	SA'U ARABIA

1 EXCLUDED BECAUSE UNASCERTAINED

CYPRUS

11 EXCLUDED BECAUSE UNASCERTAINABLE

BURMA	DOMIN REP	IRAN	IRAQ	JORDAN	KOREA REP	NEPAL	PAKISTAN	SUDAN	TOGO
YEMEN									

158 POLITIES
WHERE PERSONALISMO IS
PRONOUNCED OR MODERATE (40)

158 POLITIES
WHERE PERSONALISMO IS
NEGLIGIBLE (56)

PREDICATE ONLY

40 IN LEFT COLUMN

ARGENTINA	BOLIVIA	BRAZIL	CAMBODIA	CAMEROUN	CEN AFR REP	CEYLON	CHAD	CHILE	COLOMBIA
CONGO(BRA)	CONGO(LEO)	COSTA RICA	DAHOMEY	ECUADOR	EL SALVADOR	FRANCE	GABON	GHANA	GREECE
GUATEMALA	HONDURAS	INDONESIA	IVORY COAST	JAPAN	LAOS	LEBANON	LIBERIA	MALI	MAURITANIA
NICARAGUA	NIGER	PANAMA	PERU	SYRIA	THAILAND	UAR	UPPER VOLTA	URUGUAY	VIETNAM REP

56 IN RIGHT COLUMN

ALBANIA	ALGERIA	AUSTRALIA	AUSTRIA	BELGIUM	BULGARIA	CANADA	CHINA, PR	CUBA	CZECHOS'KIA
DENMARK	FINLAND	GERMANY, E	GERMAN FR	GUINEA	HUNGARY	ICELAND	INDIA	IRELAND	ISRAEL
ITALY	JAMAICA	KOREA, N	LUXEMBOURG	MALAGASY R	MALAYA	MEXICO	MONGOLIA	MOROCCO	NETHERLANDS
NEW ZEALAND	NIGERIA	NORWAY	PARAGUAY	PHILIPPINES	POLAND	PORTUGAL	RUMANIA	SENEGAL	SIERRE LEO
SOMALIA	SO AFRICA	SPAIN	SWEDEN	SWITZERLAND	TANGANYIKA	TRINIDAD	TUNISIA	TURKEY	UGANDA
USSR	UK	US	VENEZUELA	VIETNAM, N	YUGOSLAVIA				

5 EXCLUDED BECAUSE IRRELEVANT

AFGHANISTAN ETHIOPIA HAITI LIBYA SA'U ARABIA

3 EXCLUDED BECAUSE UNASCERTAINED

BURUNDI CYPRUS RWANDA

11 EXCLUDED BECAUSE UNASCERTAINABLE

BURMA DOMIN REP IRAN IRAQ JORDAN KOREA REP NEPAL PAKISTAN SUDAN TOGO
YEMEN

```
****************************************************************************************************

159  POLITIES                                          159  POLITIES
     WHERE THE POLITICAL LEADERSHIP IS                      WHERE THE POLITICAL LEADERSHIP IS
     ELITIST, RATHER THAN                                   NON-ELITIST, RATHER THAN
     NON-ELITIST  (30)                                      ELITIST  (50)

                                                                                     SUBJECT ONLY
....................................................................................................

     30 IN LEFT COLUMN

AFGHANISTAN ALBANIA     BULGARIA    CHINA, PR   CZECHOS'KIA ECUADOR     EL SALVADOR ETHIOPIA    GERMANY, E  GUATEMALA
HONDURAS    HUNGARY     JORDAN      KOREA, N    LAOS        LIBERIA     MONGOLIA    NEPAL       NICARAGUA   PARAGUAY
POLAND      PORTUGAL    RUMANIA     SA'U ARABIA SO AFRICA   SPAIN       THAILAND    USSR        VIETNAM, N  YUGOSLAVIA

     50 IN RIGHT COLUMN

ARGENTINA   AUSTRALIA   AUSTRIA     BELGIUM     BOLIVIA     BURUNDI     CAMEROUN    CANADA      CEN AFR REP CHAD
CONGO(BRA)  CONGO(LEO)  COSTA RICA  DAHOMEY     DENMARK     DOMIN REP   FINLAND     FRANCE      GABON       GERMAN FR
GREECE      ICELAND     INDIA       INDONESIA   IRELAND     ISRAEL      ITALY       IVORY COAST JAMAICA     JAPAN
LUXEMBOURG  MALAGASY R  MALAYA      MALI        MOROCCO     NETHERLANDS NEW ZEALAND NIGER       NORWAY      SENEGAL
SWEDEN      SWITZERLAND TANGANYIKA  TOGO        TRINIDAD    TUNISIA     TURKEY      UPPER VOLTA URUGUAY     VENEZUELA

     10 EXCLUDED BECAUSE AMBIGUOUS

CAMBODIA    CYPRUS      HAITI       IRAN        LEBANON     NIGERIA     PAKISTAN    RWANDA      UGANDA      UAR

     17 EXCLUDED BECAUSE IRRELEVANT

BRAZIL      CEYLON      CHILE       COLOMBIA    GHANA       GUINEA      LIBYA       MAURITANIA  MEXICO      PANAMA
PERU        PHILIPPINES SIERRE LEO  SOMALIA     UK          US          VIETNAM REP

      8 EXCLUDED BECAUSE UNASCERTAINABLE

ALGERIA     BURMA       CUBA        IRAQ        KOREA REP   SUDAN       SYRIA       YEMEN
----------------------------------------------------------------------------------------------------

  2  TEND MORE TO BE THOSE                 0.93      2  TEND LESS TO BE THOSE                 0.64         2   18
     LOCATED ELSEWHERE THAN IN                          LOCATED ELSEWHERE THAN IN                           28   32
     WEST EUROPE, SCANDINAVIA,                          WEST EUROPE, SCANDINAVIA,                       X SQ=    7.11
     NORTH AMERICA, OR AUSTRALASIA  (93)                NORTH AMERICA, OR AUSTRALASIA  (93)             P =     0.003
                                                                                                       RV NO   (NO )

  3  TILT MORE TOWARD BEING THOSE          0.93      3  TILT LESS TOWARD BEING THOSE          0.80         2   10
     LOCATED ELSEWHERE THAN IN                          LOCATED ELSEWHERE THAN IN                           28   40
     WEST EUROPE  (102)                                 WEST EUROPE  (102)                              X SQ=    1.67
                                                                                                       P =     0.122
                                                                                                       RV NO   (NO )

  5  TEND TO BE THOSE                      0.82      5  ALWAYS ARE THOSE                      1.00         2   15
     LOCATED IN EAST EUROPE, RATHER THAN                LOCATED IN SCANDINAVIA OR WEST EUROPE,              9    0
     IN SCANDINAVIA OR WEST EUROPE  (9)                 RATHER THAN IN EAST EUROPE  (18)                X SQ=   15.33
                                                                                                       P =     0.
                                                                                                       RV YES  (YES)
```

8	LEAN LESS TOWARD BEING THOSE LOCATED ELSEWHERE THAN IN EAST ASIA SOUTH ASIA, OR SOUTHEAST ASIA (97)	0.77	8	LEAN MORE TOWARD BEING THOSE LOCATED ELSEWHERE THAN IN EAST ASIA SOUTH ASIA, OR SOUTHEAST ASIA (97)	0.92	7 4 23 46 X SQ= 2.54 P = 0.090 RV NO (YES)
10	TEND MORE TO BE THOSE LOCATED ELSEWHERE THAN IN NORTH AFRICA, OR CENTRAL AND SOUTH AFRICA (82)	0.90	10	TEND LESS TO BE THOSE LOCATED ELSEWHERE THAN IN NORTH AFRICA, OR CENTRAL AND SOUTH AFRICA (82)	0.64	3 18 27 32 X SQ= 5.27 P = 0.017 RV NO (NO)
15	LEAN TOWARD BEING THOSE LOCATED IN THE MIDDLE EAST, RATHER THAN IN NORTH AFRICA OR CENTRAL AND SOUTH AFRICA (11)	0.50	15	LEAN TOWARD BEING THOSE LOCATED IN NORTH AFRICA OR CENTRAL AND SOUTH AFRICA, RATHER THAN IN THE MIDDLE EAST (33)	0.90	3 2 3 18 X SQ= 2.53 P = 0.062 RV YES (YES)
18	TEND TO BE THOSE LOCATED IN EAST ASIA, SOUTH ASIA, OR SOUTHEAST ASIA, RATHER THAN IN NORTH AFRICA OR CENTRAL AND SOUTH AFRICA (18)	0.70	18	TEND TO BE THOSE LOCATED IN NORTH AFRICA OR CENTRAL AND SOUTH AFRICA, RATHER THAN IN EAST ASIA, SOUTH ASIA, OR SOUTHEAST ASIA (33)	0.82	3 18 7 4 X SQ= 6.05 P = 0.013 RV YES (YES)
19	TILT TOWARD BEING THOSE LOCATED IN THE CARIBBEAN, CENTRAL AMERICA, OR SOUTH AMERICA, RATHER THAN IN NORTH AFRICA OR CENTRAL AND SOUTH AFRICA (22)	0.67	19	TILT TOWARD BEING THOSE LOCATED IN NORTH AFRICA OR CENTRAL AND SOUTH AFRICA, RATHER THAN IN THE CARIBBEAN, CENTRAL AMERICA, OR SOUTH AMERICA (33)	0.69	3 18 6 8 X SQ= 2.25 P = 0.112 RV YES (NO)
24	LEAN TOWARD BEING THOSE WHOSE POPULATION IS VERY LARGE, LARGE, OR MEDIUM (61)	0.63	24	LEAN TOWARD BEING THOSE WHOSE POPULATION IS SMALL (54)	0.60	19 20 11 30 X SQ= 3.21 P = 0.064 RV YES (NO)
25	TILT MORE TOWARD BEING THOSE WHOSE POPULATION DENSITY IS MEDIUM OR LOW (98)	0.93	25	TILT LESS TOWARD BEING THOSE WHOSE POPULATION DENSITY IS MEDIUM OR LOW (98)	0.80	2 10 28 40 X SQ= 1.67 P = 0.122 RV NO (NO)
31	TEND MORE TO BE THOSE WHOSE AGRICULTURAL POPULATION IS HIGH OR MEDIUM (90)	0.93	31	TEND LESS TO BE THOSE WHOSE AGRICULTURAL POPULATION IS HIGH OR MEDIUM (90)	0.62	27 31 2 19 X SQ= 7.57 P = 0.003 RV NO (NO)
36	TILT TOWARD BEING THOSE WHOSE PER CAPITA GROSS NATIONAL PRODUCT IS LOW OR VERY LOW (73)	0.67	36	TILT TOWARD BEING THOSE WHOSE PER CAPITA GROSS NATIONAL PRODUCT IS VERY HIGH, HIGH, OR MEDIUM (42)	0.52	10 26 20 24 X SQ= 1.94 P = 0.113 RV YES (NO)

41	LEAN MORE TOWARD BEING THOSE WHOSE ECONOMIC DEVELOPMENTAL STATUS IS INTERMEDIATE, UNDERDEVELOPED, OR VERY UNDERDEVELOPED (94)	0.90	41	LEAN LESS TOWARD BEING THOSE WHOSE ECONOMIC DEVELOPMENTAL STATUS IS INTERMEDIATE, UNDERDEVELOPED, OR VERY UNDERDEVELOPED (94)	0.71	3 14 27 34 X SQ= 2.93 P = 0.053 RV NO (NO)
49	TILT TOWARD BEING THOSE WHERE THE LITERACY RATE IS BETWEEN TEN AND FIFTY PERCENT, RATHER THAN BELOW TEN PERCENT (24)	0.58	49	TILT TOWARD BEING THOSE WHERE THE LITERACY RATE IS BELOW TEN PERCENT, RATHER THAN BETWEEN TEN AND FIFTY PERCENT (26)	0.71	7 6 5 15 X SQ= 1.72 P = 0.142 RV YES (YES)
50	TEND TO BE THOSE WHERE FREEDOM OF THE PRESS IS INTERMITTENT, INTERNALLY ABSENT, OR INTERNALLY AND EXTERNALLY ABSENT (56)	0.96	50	TEND TO BE THOSE WHERE FREEDOM OF THE PRESS IS COMPLETE (43)	0.75	1 30 25 10 X SQ= 29.24 P = 0. RV YES (YES)
51	TEND TO BE THOSE WHERE FREEDOM OF THE PRESS IS INTERNALLY ABSENT, OR INTERNALLY AND EXTERNALLY ABSENT (37)	0.85	51	TEND TO BE THOSE WHERE FREEDOM OF THE PRESS IS COMPLETE OR INTERMITTENT (60)	0.92	4 37 22 3 X SQ= 36.61 P = 0. RV YES (YES)
52	TEND TO BE THOSE WHERE FREEDOM OF THE PRESS IS INTERNALLY AND EXTERNALLY ABSENT (16)	0.54	52	ALWAYS ARE THOSE WHERE FREEDOM OF THE PRESS IS COMPLETE, INTERMITTENT, OR INTERNALLY ABSENT (82)	1.00	12 41 14 0 X SQ= 24.75 P = 0. RV YES (YES)
53	TEND MORE TO BE THOSE WHERE NEWSPAPER CIRCULATION IS LESS THAN THREE HUNDRED PER THOUSAND (101)	0.97	53	TEND LESS TO BE THOSE WHERE NEWSPAPER CIRCULATION IS LESS THAN THREE HUNDRED PER THOUSAND (101)	0.78	1 11 29 39 X SQ= 3.76 P = 0.026 RV NO (NO)
62	ALWAYS ARE THOSE WHERE THE RELIGION IS CATHOLIC, RATHER THAN PROTESTANT (25)	1.00	62	TILT LESS TOWARD BEING THOSE WHERE THE RELIGION IS CATHOLIC, RATHER THAN PROTESTANT (25)	0.69	0 5 7 11 X SQ= 1.26 P = 0.147 RV NO (NO)
72	LEAN TOWARD BEING THOSE WHOSE DATE OF INDEPENDENCE IS BEFORE 1914 (52)	0.68	72	LEAN TOWARD BEING THOSE WHOSE DATE OF INDEPENDENCE IS AFTER 1914 (59)	0.55	19 22 9 27 X SQ= 2.91 P = 0.061 RV YES (NO)
77	TILT TOWARD BEING THOSE THAT HAVE BEEN SIGNIFICANTLY WESTERNIZED, RATHER THAN PARTIALLY WESTERNIZED, THROUGH A COLONIAL RELATIONSHIP (28)	0.67	77	TILT TOWARD BEING THOSE THAT HAVE BEEN PARTIALLY WESTERNIZED, RATHER THAN SIGNIFICANTLY WESTERNIZED, THROUGH A COLONIAL RELATIONSHIP (41)	0.69	6 9 3 20 X SQ= 2.31 P = 0.115 RV YES (NO)

79	TEND TO BE THOSE THAT WERE FORMERLY DEPENDENCIES OF SPAIN, RATHER THAN BRITAIN OR FRANCE (18)	0.60	79	TEND TO BE THOSE THAT WERE FORMERLY DEPENDENCIES OF BRITAIN OR FRANCE, RATHER THAN SPAIN (49)	0.81	4 25 6 6 X SQ= 4.23 P = 0.040 RV YES (YES)
81	LEAN MORE TOWARD BEING THOSE WHERE THE TYPE OF POLITICAL MODERNIZATION IS LATER EUROPEAN OR LATER EUROPEAN DERIVED, RATHER THAN EARLY EUROPEAN OR EARLY EUROPEAN DERIVED (40)	0.94	81	LEAN LESS TOWARD BEING THOSE WHERE THE TYPE OF POLITICAL MODERNIZATION IS LATER EUROPEAN OR LATER EUROPEAN DERIVED, RATHER THAN EARLY EUROPEAN OR EARLY EUROPEAN DERIVED (40)	0.67	1 8 16 16 X SQ= 2.92 P = 0.056 RV NO (YES)
82	TILT TOWARD BEING THOSE WHERE THE TYPE OF POLITICAL MODERNIZATION IS EARLY OR LATER EUROPEAN OR EUROPEAN DERIVED, RATHER THAN DEVELOPED TUTELARY OR UNDEVELOPED TUTELARY (51)	0.71	82	TILT TOWARD BEING THOSE WHERE THE TYPE OF POLITICAL MODERNIZATION IS DEVELOPED TUTELARY OR UNDEVELOPED TUTELARY, RATHER THAN EARLY OR LATER EUROPEAN OR EUROPEAN DERIVED (55)	0.50	17 24 7 24 X SQ= 2.05 P = 0.130 RV YES (NO)
83	TEND LESS TO BE THOSE WHERE THE TYPE OF POLITICAL MODERNIZATION IS OTHER THAN NON-EUROPEAN AUTOCHTHONOUS (106)	0.80	83	TEND MORE TO BE THOSE WHERE THE TYPE OF POLITICAL MODERNIZATION IS OTHER THAN NON-EUROPEAN AUTOCHTHONOUS (106)	0.96	6 2 24 48 X SQ= 3.70 P = 0.047 RV NO (YES)
84	ALWAYS ARE THOSE WHERE THE TYPE OF POLITICAL MODERNIZATION IS DEVELOPED TUTELARY, RATHER THAN UNDEVELOPED TUTELARY (31)	1.00	84	TEND TO BE THOSE WHERE THE TYPE OF POLITICAL MODERNIZATION IS UNDEVELOPED TUTELARY, RATHER THAN DEVELOPED TUTELARY (24)	0.67	7 8 0 16 X SQ= 7.16 P = 0.002 RV YES (NO)
86	TEND MORE TO BE THOSE WHERE THE STAGE OF POLITICAL MODERNIZATION IS ADVANCED OR MID-TRANSITIONAL, RATHER THAN EARLY TRANSITIONAL (76)	0.86	86	TEND LESS TO BE THOSE WHERE THE STAGE OF POLITICAL MODERNIZATION IS ADVANCED OR MID-TRANSITIONAL, RATHER THAN EARLY TRANSITIONAL (76)	0.60	25 30 4 20 X SQ= 4.79 P = 0.021 RV NO (NO)
87	ALWAYS ARE THOSE WHOSE IDEOLOGICAL ORIENTATION IS OTHER THAN DEVELOPMENTAL (58)	1.00	87	TEND TO BE THOSE WHOSE IDEOLOGICAL ORIENTATION IS DEVELOPMENTAL (31)	0.50	0 22 25 22 X SQ= 16.12 P = 0. RV YES (YES)
89	TEND TO BE THOSE WHOSE IDEOLOGICAL ORIENTATION IS OTHER THAN CONVENTIONAL (62)	0.88	89	TEND TO BE THOSE WHOSE IDEOLOGICAL ORIENTATION IS CONVENTIONAL (33)	0.50	3 22 21 22 X SQ= 7.85 P = 0.003 RV YES (NO)
91	ALWAYS ARE THOSE WHOSE IDEOLOGICAL ORIENTATION IS TRADITIONAL, RATHER THAN DEVELOPMENTAL (5)	1.00	91	ALWAYS ARE THOSE WHOSE IDEOLOGICAL ORIENTATION IS DEVELOPMENTAL, RATHER THAN TRADITIONAL (31)	1.00	0 22 5 0 X SQ= 20.78 P = 0. RV YES (YES)

92	TEND LESS TO BE THOSE WHERE THE SYSTEM STYLE IS LIMITED MOBILIZATIONAL, OR NON-MOBILIZATIONAL (93)	0.55	92	ALWAYS ARE THOSE WHERE THE SYSTEM STYLE IS LIMITED MOBILIZATIONAL, OR NON-MOBILIZATIONAL (93)	1.00	13 0 16 50 X SQ= 23.67 P = 0. RV NO (YES)
94	TEND TO BE THOSE WHERE THE STATUS OF THE REGIME IS AUTHORITARIAN OR TOTALITARIAN (41)	0.93	94	TEND TO BE THOSE WHERE THE STATUS OF THE REGIME IS CONSTITUTIONAL (51)	0.94	2 33 26 2 X SQ= 44.38 P = 0. RV YES (YES)
95	TEND TO BE THOSE WHERE THE STATUS OF THE REGIME IS TOTALITARIAN (16)	0.52	95	ALWAYS ARE THOSE WHERE THE STATUS OF THE REGIME IS CONSTITUTIONAL OR AUTHORITARIAN (95)	1.00	14 50 15 0 X SQ= 28.65 P = 0. RV YES (YES)
96	TEND LESS TO BE THOSE WHERE THE STATUS OF THE REGIME IS CONSTITUTIONAL OR TOTALITARIAN (67)	0.61	96	TEND MORE TO BE THOSE WHERE THE STATUS OF THE REGIME IS CONSTITUTIONAL OR TOTALITARIAN (67)	0.94	17 33 11 2 X SQ= 8.75 P = 0.001 RV NO (YES)
101	ALWAYS ARE THOSE WHERE THE REPRESENTATIVE CHARACTER OF THE REGIME IS LIMITED POLYARCHIC, PSEUDO-POLYARCHIC, OR NON-POLYARCHIC (57)	1.00	101	TEND TO BE THOSE WHERE THE REPRESENTATIVE CHARACTER OF THE REGIME IS POLYARCHIC (41)	0.73	0 33 29 12 X SQ= 35.47 P = 0. RV YES (YES)
102	TEND TO BE THOSE WHERE THE REPRESENTATIVE CHARACTER OF THE REGIME IS PSEUDO-POLYARCHIC OR NON-POLYARCHIC (49)	0.86	102	TEND TO BE THOSE WHERE THE REPRESENTATIVE CHARACTER OF THE REGIME IS POLYARCHIC OR LIMITED POLYARCHIC (59)	0.78	4 38 25 11 X SQ= 27.29 P = 0. RV YES (YES)
105	TEND TO BE THOSE WHERE THE ELECTORAL SYSTEM IS PARTIALLY COMPETITIVE OR NON-COMPETITIVE (47)	0.92	105	TEND TO BE THOSE WHERE THE ELECTORAL SYSTEM IS COMPETITIVE (43)	0.71	2 32 23 13 X SQ= 23.16 P = 0. RV YES (YES)
106	TEND TO BE THOSE WHERE THE ELECTORAL SYSTEM IS NON-COMPETITIVE (30)	0.81	106	TEND TO BE THOSE WHERE THE ELECTORAL SYSTEM IS COMPETITIVE OR PARTIALLY COMPETITIVE (52)	0.86	4 37 17 6 X SQ= 24.68 P = 0. RV YES (YES)
107	TEND TO BE THOSE WHERE AUTONOMOUS GROUPS ARE PARTIALLY TOLERATED IN POLITICS, ARE TOLERATED ONLY OUTSIDE POLITICS, OR ARE NOT TOLERATED AT ALL (65)	0.93	107	TEND TO BE THOSE WHERE AUTONOMOUS GROUPS ARE FULLY TOLERATED IN POLITICS (46)	0.63	2 31 27 18 X SQ= 21.46 P = 0. RV YES (YES)

No.		Value	No.		Value	Statistics
108	TEND TO BE THOSE WHERE AUTONOMOUS GROUPS ARE TOLERATED ONLY OUTSIDE POLITICS OR ARE NOT TOLERATED AT ALL (35)	0.81	108	TEND TO BE THOSE WHERE AUTONOMOUS GROUPS ARE FULLY OR PARTIALLY TOLERATED IN POLITICS (65)	0.85	5 40 21 7 X SQ= 28.00 P = 0. RV YES (YES)
111	TEND TO BE THOSE WHERE POLITICAL ENCULTURATION IS LOW (42)	0.76	111	TEND TO BE THOSE WHERE POLITICAL ENCULTURATION IS HIGH OR MEDIUM (53)	0.71	4 32 13 13 X SQ= 9.60 P = 0.001 RV YES (YES)
115	ALWAYS ARE THOSE WHERE INTEREST ARTICULATION BY ASSOCIATIONAL GROUPS IS MODERATE, LIMITED, OR NEGLIGIBLE (91)	1.00	115	TEND LESS TO BE THOSE WHERE INTEREST ARTICULATION BY ASSOCIATIONAL GROUPS IS MODERATE, LIMITED, OR NEGLIGIBLE (91)	0.68	0 16 27 34 X SQ= 9.05 P = 0.001 RV NO (NO)
116	TEND MORE TO BE THOSE WHERE INTEREST ARTICULATION BY ASSOCIATIONAL GROUPS IS LIMITED OR NEGLIGIBLE (79)	0.96	116	TEND LESS TO BE THOSE WHERE INTEREST ARTICULATION BY ASSOCIATIONAL GROUPS IS LIMITED OR NEGLIGIBLE (79)	0.52	1 24 26 26 X SQ= 13.73 P = 0. RV NO (YES)
117	TEND TO BE THOSE WHERE INTEREST ARTICULATION BY ASSOCIATIONAL GROUPS IS NEGLIGIBLE (51)	0.67	117	TEND TO BE THOSE WHERE INTEREST ARTICULATION BY ASSOCIATIONAL GROUPS IS SIGNIFICANT, MODERATE, OR LIMITED (60)	0.66	9 33 18 17 X SQ= 6.29 P = 0.008 RV YES (YES)
118	TEND TO BE THOSE WHERE INTEREST ARTICULATION BY INSTITUTIONAL GROUPS IS VERY SIGNIFICANT (40)	0.73	118	TEND TO BE THOSE WHERE INTEREST ARTICULATION BY INSTITUTIONAL GROUPS IS SIGNIFICANT, MODERATE, OR LIMITED (60)	0.92	22 3 8 36 X SQ= 28.85 P = 0. RV YES (YES)
119	ALWAYS ARE THOSE WHERE INTEREST ARTICULATION BY INSTITUTIONAL GROUPS IS VERY SIGNIFICANT OR SIGNIFICANT (74)	1.00	119	TEND TO BE THOSE WHERE INTEREST ARTICULATION BY INSTITUTIONAL GROUPS IS MODERATE OR LIMITED (26)	0.54	30 18 0 21 X SQ= 20.75 P = 0. RV YES (YES)
122	LEAN MORE TOWARD BEING THOSE WHERE INTEREST ARTICULATION BY NON-ASSOCIATIONAL GROUPS IS SIGNIFICANT OR MODERATE (83)	0.80	122	LEAN LESS TOWARD BEING THOSE WHERE INTEREST ARTICULATION BY NON-ASSOCIATIONAL GROUPS IS SIGNIFICANT OR MODERATE (83)	0.60	24 30 6 20 X SQ= 2.57 P = 0.086 RV NO (NO)
123	ALWAYS ARE THOSE WHERE INTEREST ARTICULATION BY NON-ASSOCIATIONAL GROUPS IS SIGNIFICANT, MODERATE, OR LIMITED (107)	1.00	123	TEND LESS TO BE THOSE WHERE INTEREST ARTICULATION BY NON-ASSOCIATIONAL GROUPS IS SIGNIFICANT, MODERATE, OR LIMITED (107)	0.84	30 42 0 8 X SQ= 3.70 P = 0.022 RV NO (NO)

125	TEND TO BE THOSE WHERE INTEREST ARTICULATION BY ANOMIC GROUPS IS FREQUENT OR OCCASIONAL (64)	0.83	125	TEND TO BE THOSE WHERE INTEREST ARTICULATION BY ANOMIC GROUPS IS INFREQUENT OR VERY INFREQUENT (35)	0.56	19 20 4 25 X SQ= 7.57 P = 0.004 RV YES (NO)
126	ALWAYS ARE THOSE WHERE INTEREST ARTICULATION BY ANOMIC GROUPS IS FREQUENT, OCCASIONAL, OR INFREQUENT (83)	1.00	126	TEND LESS TO BE THOSE WHERE INTEREST ARTICULATION BY ANOMIC GROUPS IS FREQUENT, OCCASIONAL, OR INFREQUENT (83)	0.67	23 30 0 15 X SQ= 7.99 P = 0.001 RV NO (NO)
128	TEND TO BE THOSE WHERE INTEREST ARTICULATION BY POLITICAL PARTIES IS LIMITED OR NEGLIGIBLE (45)	0.77	128	TEND TO BE THOSE WHERE INTEREST ARTICULATION BY POLITICAL PARTIES IS SIGNIFICANT OR MODERATE (48)	0.64	5 30 17 17 X SQ= 8.55 P = 0.002 RV YES (YES)
129	TEND TO BE THOSE WHERE INTEREST ARTICULATION BY POLITICAL PARTIES IS NEGLIGIBLE (37)	0.68	129	TEND TO BE THOSE WHERE INTEREST ARTICULATION BY POLITICAL PARTIES IS SIGNIFICANT, MODERATE, OR LIMITED (56)	0.70	7 33 15 14 X SQ= 7.56 P = 0.004 RV YES (YES)
130	ALWAYS ARE THOSE WHERE INTEREST AGGREGATION BY POLITICAL PARTIES IS MODERATE, LIMITED, OR NEGLIGIBLE (71)	1.00	130	TEND LESS TO BE THOSE WHERE INTEREST AGGREGATION BY POLITICAL PARTIES IS MODERATE, LIMITED, OR NEGLIGIBLE (71)	0.77	0 8 26 27 X SQ= 4.98 P = 0.009 RV NO (NO)
131	TEND TO BE THOSE WHERE INTEREST AGGREGATION BY POLITICAL PARTIES IS LIMITED OR NEGLIGIBLE (35)	0.91	131	TEND TO BE THOSE WHERE INTEREST AGGREGATION BY POLITICAL PARTIES IS SIGNIFICANT OR MODERATE (30)	0.64	1 21 10 12 X SQ= 7.76 P = 0.004 RV YES (NO)
132	TEND TO BE THOSE WHERE INTEREST AGGREGATION BY POLITICAL PARTIES IS NEGLIGIBLE (9)	0.56	132	TEND TO BE THOSE WHERE INTEREST AGGREGATION BY POLITICAL PARTIES IS SIGNIFICANT, MODERATE, OR LIMITED (67)	0.98	4 44 5 1 X SQ= 16.54 P = 0. RV YES (YES)
134	ALWAYS ARE THOSE WHERE INTEREST AGGREGATION BY THE EXECUTIVE IS LIMITED OR NEGLIGIBLE (46)	1.00	134	TEND TO BE THOSE WHERE INTEREST AGGREGATION BY THE EXECUTIVE IS SIGNIFICANT OR MODERATE (57)	0.92	0 44 28 4 X SQ= 57.26 P = 0. RV YES (YES)
135	TEND TO BE THOSE WHERE INTEREST AGGREGATION BY THE EXECUTIVE IS NEGLIGIBLE (13)	0.56	135	ALWAYS ARE THOSE WHERE INTEREST AGGREGATION BY THE EXECUTIVE IS SIGNIFICANT, MODERATE, OR LIMITED (77)	1.00	8 46 10 0 X SQ= 26.22 P = 0. RV YES (YES)

137	ALWAYS ARE THOSE WHERE INTEREST AGGREGATION BY THE LEGISLATURE IS LIMITED OR NEGLIGIBLE (68)	1.00	137	TEND TO BE THOSE WHERE INTEREST AGGREGATION BY THE LEGISLATURE IS SIGNIFICANT OR MODERATE (29)	0.52	0 24 28 22 X SQ= 19.31 P = 0. RV YES (YES)
138	TEND TO BE THOSE WHERE INTEREST AGGREGATION BY THE LEGISLATURE IS NEGLIGIBLE (49)	0.85	138	TEND TO BE THOSE WHERE INTEREST AGGREGATION BY THE LEGISLATURE IS SIGNIFICANT, MODERATE, OR LIMITED (48)	0.64	4 30 22 17 X SQ= 13.90 P = 0. RV YES (YES)
139	TEND TO BE THOSE WHERE THE PARTY SYSTEM IS QUANTITATIVELY ONE-PARTY (34)	0.59	139	TEND TO BE THOSE WHERE THE PARTY SYSTEM IS QUANTITATIVELY OTHER THAN ONE-PARTY (71)	0.75	16 12 11 36 X SQ= 7.27 P = 0.006 RV YES (YES)
140	TILT MORE TOWARD BEING THOSE WHERE THE PARTY SYSTEM IS QUANTITATIVELY OTHER THAN ONE PARTY DOMINANT (87)	0.96	140	TILT LESS TOWARD BEING THOSE WHERE THE PARTY SYSTEM IS QUANTITATIVELY OTHER THAN ONE PARTY DOMINANT (87)	0.82	1 8 25 37 X SQ= 1.77 P = 0.141 RV NO (NO)
142	LEAN MORE TOWARD BEING THOSE WHERE THE PARTY SYSTEM IS QUANTITATIVELY OTHER THAN MULTI-PARTY (66)	0.84	142	LEAN LESS TOWARD BEING THOSE WHERE THE PARTY SYSTEM IS QUANTITATIVELY OTHER THAN MULTI-PARTY (66)	0.61	4 17 21 27 X SQ= 2.86 P = 0.061 RV NO (NO)
144	LEAN MORE TOWARD BEING THOSE WHERE THE PARTY SYSTEM IS QUANTITATIVELY ONE-PARTY, RATHER THAN TWO-PARTY (34)	0.94	144	LEAN LESS TOWARD BEING THOSE WHERE THE PARTY SYSTEM IS QUANTITATIVELY ONE-PARTY, RATHER THAN TWO-PARTY (34)	0.69	16 11 1 5 X SQ= 2.06 P = 0.085 RV NO (YES)
147	ALWAYS ARE THOSE WHERE THE PARTY SYSTEM IS QUALITATIVELY OTHER THAN CLASS-ORIENTED OR MULTI-IDEOLOGICAL (67)	1.00	147	TEND LESS TO BE THOSE WHERE THE PARTY SYSTEM IS QUALITATIVELY OTHER THAN CLASS-ORIENTED OR MULTI-IDEOLOGICAL (67)	0.55	0 19 23 23 X SQ= 12.60 P = 0. RV NO (YES)
148	ALWAYS ARE THOSE WHERE THE PARTY SYSTEM IS QUALITATIVELY OTHER THAN AFRICAN TRANSITIONAL (96)	1.00	148	TEND LESS TO BE THOSE WHERE THE PARTY SYSTEM IS QUALITATIVELY OTHER THAN AFRICAN TRANSITIONAL (96)	0.78	0 11 27 39 X SQ= 5.25 P = 0.007 RV NO (NO)
157	TEND LESS TO BE THOSE WHERE PERSONALISMO IS MODERATE OR NEGLIGIBLE (84)	0.72	157	TEND MORE TO BE THOSE WHERE PERSONALISMO IS MODERATE OR NEGLIGIBLE (84)	0.96	7 2 18 46 X SQ= 6.57 P = 0.006 RV NO (YES)

166	ALWAYS ARE THOSE WHERE THE VERTICAL POWER DISTRIBUTION IS THAT OF FORMAL FEDERALISM OR FORMAL AND EFFECTIVE UNITARISM (99)	1.00	166	TEND LESS TO BE THOSE WHERE THE VERTICAL POWER DISTRIBUTION IS THAT OF FORMAL FEDERALISM OR FORMAL AND EFFECTIVE UNITARISM (99)	0.82	0 9 30 40 X SQ= 4.53 P = 0.011 RV NO (NO)
168	TEND TO BE THOSE WHERE THE HORIZONTAL POWER DISTRIBUTION IS LIMITED OR NEGLIGIBLE (72)	0.97	168	TEND TO BE THOSE WHERE THE HORIZONTAL POWER DISTRIBUTION IS SIGNIFICANT (34)	0.56	1 27 29 21 X SQ= 20.22 P = 0. RV YES (YES)
169	TEND TO BE THOSE WHERE THE HORIZONTAL POWER DISTRIBUTION IS NEGLIGIBLE (48)	0.80	169	TEND TO BE THOSE WHERE THE HORIZONTAL POWER DISTRIBUTION IS SIGNIFICANT OR LIMITED (58)	0.73	6 35 24 13 X SQ= 18.66 P = 0. RV YES (YES)
171	LEAN MORE TOWARD BEING THOSE WHERE THE LEGISLATIVE-EXECUTIVE STRUCTURE IS OTHER THAN PARLIAMENTARY-REPUBLICAN (90)	0.96	171	LEAN LESS TOWARD BEING THOSE WHERE THE LEGISLATIVE-EXECUTIVE STRUCTURE IS OTHER THAN PARLIAMENTARY-REPUBLICAN (90)	0.80	1 10 26 40 X SQ= 2.59 P = 0.085 RV NO (NO)
172	TEND MORE TO BE THOSE WHERE THE LEGISLATIVE-EXECUTIVE STRUCTURE IS OTHER THAN PARLIAMENTARY-ROYALIST (88)	0.96	172	TEND LESS TO BE THOSE WHERE THE LEGISLATIVE-EXECUTIVE STRUCTURE IS OTHER THAN PARLIAMENTARY-ROYALIST (88)	0.70	1 15 26 35 X SQ= 5.85 P = 0.007 RV NO (NO)
175	TEND TO BE THOSE WHERE THE LEGISLATURE IS LARGELY INEFFECTIVE OR WHOLLY INEFFECTIVE (49)	0.89	175	TEND TO BE THOSE WHERE THE LEGISLATURE IS FULLY EFFECTIVE OR PARTIALLY EFFECTIVE (51)	0.70	3 33 25 14 X SQ= 22.56 P = 0. RV YES (YES)
176	TEND TO BE THOSE WHERE THE LEGISLATURE IS WHOLLY INEFFECTIVE (28)	0.79	176	ALWAYS ARE THOSE WHERE THE LEGISLATURE IS FULLY EFFECTIVE, PARTIALLY EFFECTIVE, OR LARGELY INEFFECTIVE (72)	1.00	6 47 22 0 X SQ= 48.54 P = 0. RV YES (YES)
179	TEND TO BE THOSE WHERE THE EXECUTIVE IS DOMINANT (52)	0.96	179	TEND TO BE THOSE WHERE THE EXECUTIVE IS STRONG (39)	0.66	26 14 1 27 X SQ= 23.46 P = 0. RV YES (YES)
181	TEND TO BE THOSE WHERE THE BUREAUCRACY IS SEMI-MODERN, RATHER THAN MODERN (55)	0.96	181	TEND TO BE THOSE WHERE THE BUREAUCRACY IS MODERN, RATHER THAN SEMI-MODERN (21)	0.55	1 18 22 15 X SQ= 13.08 P = 0. RV YES (YES)

```
182  ALWAYS ARE THOSE                         1.00     182  TEND TO BE THOSE                        0.52          22    15
     WHERE THE BUREAUCRACY                                  WHERE THE BUREAUCRACY                                     0    16
     IS SEMI-MODERN, RATHER THAN                            IS POST-COLONIAL TRANSITIONAL,                     X SQ=  13.91
     POST-COLONIAL TRANSITIONAL  (55)                       RATHER THAN SEMI-MODERN  (25)                      P =     0.
                                                                                                               RV YES (YES)

183  ALWAYS ARE THOSE                         1.00     183  ALWAYS ARE THOSE                        1.00           0    16
     WHERE THE BUREAUCRACY                                  WHERE THE BUREAUCRACY                                     7     0
     IS TRADITIONAL, RATHER THAN                            IS POST-COLONIAL TRANSITIONAL,                     X SQ=  18.52
     POST-COLONIAL TRANSITIONAL  (9)                        RATHER THAN TRADITIONAL  (25)                      P =     0.
                                                                                                               RV YES (YES)

185  TILT TOWARD BEING THOSE                  0.79     185  TILT TOWARD BEING THOSE                 0.60           6     3
     WHERE PARTICIPATION BY THE MILITARY                    WHERE PARTICIPATION BY THE MILITARY                      22     2
     IN POLITICS IS                                         IN POLITICS IS                                     X SQ=   1.53
     SUPPORTIVE, RATHER THAN                                INTERVENTIVE, RATHER THAN                          P =     0.111
     INTERVENTIVE  (31)                                     SUPPORTIVE  (21)                                   RV YES (NO )

186  ALWAYS ARE THOSE                         1.00     186  TEND TO BE THOSE                        0.95          22     2
     WHERE PARTICIPATION BY THE MILITARY                    WHERE PARTICIPATION BY THE MILITARY                       0    42
     IN POLITICS IS                                         IN POLITICS IS                                     X SQ=  53.70
     SUPPORTIVE, RATHER THAN                                NEUTRAL, RATHER THAN                               P =     0.
     NEUTRAL  (31)                                          SUPPORTIVE  (56)                                   RV YES (YES)

187  ALWAYS ARE THOSE                         1.00     187  TEND TO BE THOSE                        0.63          30    16
     WHERE THE ROLE OF THE POLICE                           WHERE THE ROLE OF THE POLICE                              0    27
     IS POLITICALLY SIGNIFICANT  (66)                       IS NOT POLITICALLY SIGNIFICANT  (35)               X SQ=  27.26
                                                                                                               P =     0.
                                                                                                               RV YES (YES)

192  LEAN MORE TOWARD BEING THOSE             0.93     192  LEAN LESS TOWARD BEING THOSE            0.76           2    12
     WHERE THE CHARACTER OF THE LEGAL SYSTEM                WHERE THE CHARACTER OF THE LEGAL SYSTEM                  27    38
     IS OTHER THAN                                          IS OTHER THAN                                      X SQ=   2.60
     MUSLIM OR PARTLY MUSLIM  (86)                          MUSLIM OR PARTLY MUSLIM  (86)                      P =     0.070
                                                                                                               RV NO  (NO )

193  LEAN MORE TOWARD BEING THOSE             0.87     193  LEAN LESS TOWARD BEING THOSE            0.66           4    17
     WHERE THE CHARACTER OF THE LEGAL SYSTEM                WHERE THE CHARACTER OF THE LEGAL SYSTEM                  26    33
     IS OTHER THAN PARTLY INDIGENOUS  (86)                  IS OTHER THAN PARTLY INDIGENOUS  (86)              X SQ=   3.14
                                                                                                               P =     0.065
                                                                                                               RV NO  (NO )

194  TEND LESS TO BE THOSE                    0.57     194  ALWAYS ARE THOSE                        1.00          13     0
     THAT ARE NON-COMMUNIST  (101)                          THAT ARE NON-COMMUNIST  (101)                            17    50
                                                                                                               X SQ=  22.78
                                                                                                               P =     0.
                                                                                                               RV NO  (YES)
```

****

160 POLITIES
WHERE THE POLITICAL LEADERSHIP IS
ELITIST (30)

160 POLITIES
WHERE THE POLITICAL LEADERSHIP IS
MODERATELY ELITIST OR NON-ELITIST (67)

PREDICATE ONLY

....

30 IN LEFT COLUMN

```
AFGHANISTAN ALBANIA     BULGARIA    CHINA, PR   CZECHOS'KIA ECUADOR     EL SALVADOR ETHIOPIA    GERMANY, E  GUATEMALA
HONDURAS    HUNGARY     JORDAN      KOREA, N    LAOS        LIBERIA     MONGOLIA    NEPAL       NICARAGUA   PARAGUAY
POLAND      PORTUGAL    RUMANIA     SA'U ARABIA SO AFRICA   SPAIN       THAILAND    USSR        VIETNAM, N  YUGOSLAVIA
```

67 IN RIGHT COLUMN

```
ARGENTINA   AUSTRALIA   AUSTRIA     BELGIUM     BOLIVIA     BRAZIL      BURUNDI     CAMEROUN    CANADA      CEN AFR REP
CEYLON      CHAD        CHILE       COLOMBIA    CONGO(BRA)  CONGO(LEO)  COSTA RICA  DAHOMEY     DENMARK     DOMIN REP
FINLAND     FRANCE      GABON       GERMAN FR   GHANA       GREECE      GUINEA      ICELAND     INDIA       INDONESIA
IRELAND     ISRAEL      ITALY       IVORY COAST JAMAICA     JAPAN       LIBYA       LUXEMBOURG  MALAGASY R  MALAYA
MALI        MAURITANIA  MEXICO      MOROCCO     NETHERLANDS NEW ZEALAND NIGER       NORWAY      PANAMA      PERU
PHILIPPINES SENEGAL     SIERRE LEO  SOMALIA     SWEDEN      SWITZERLAND TANGANYIKA  TOGO        TRINIDAD    TUNISIA
TURKEY      UK          US          UPPER VOLTA URUGUAY     VENEZUELA   VIETNAM REP
```

10 EXCLUDED BECAUSE AMBIGUOUS

```
CAMBODIA    CYPRUS      HAITI       IRAN        LEBANON     NIGERIA     PAKISTAN    RWANDA      UGANDA      UAR
```

8 EXCLUDED BECAUSE UNASCERTAINABLE

```
ALGERIA     BURMA       CUBA        IRAQ        KOREA REP   SUDAN       SYRIA       YEMEN
```

```
******************************************************************************************************************************

161  POLITIES                                          161  POLITIES
     WHERE THE POLITICAL LEADERSHIP IS                      WHERE THE POLITICAL LEADERSHIP IS
     ELITIST OR MODERATELY ELITIST  (47)                    NON-ELITIST  (50)


                                                                                                     PREDICATE ONLY
..............................................................................................................................

      47 IN LEFT COLUMN

AFGHANISTAN ALBANIA     BRAZIL      BULGARIA    CEYLON      CHILE       CHINA, PR   COLOMBIA    CZECHOS'KIA ECUADOR
EL SALVADOR ETHIOPIA    GERMANY, E  GHANA       GUATEMALA   GUINEA      HONDURAS    HUNGARY     JORDAN      KOREA, N
LAOS        LIBERIA     LIBYA       MAURITANIA  MEXICO      MONGOLIA    NEPAL       NICARAGUA   PANAMA      PARAGUAY
PERU        PHILIPPINES POLAND      PORTUGAL    RUMANIA     SA'U ARABIA SIERRE LEO  SOMALIA     SO AFRICA   SPAIN
THAILAND    USSR        UK          US          VIETNAM, N  VIETNAM REP YUGOSLAVIA

      50 IN RIGHT COLUMN

ARGENTINA   AUSTRALIA   AUSTRIA     BELGIUM     BOLIVIA     BURUNDI     CAMEROUN    CANADA      CEN AFR REP CHAD
CONGO(BRA)  CONGO(LEO)  COSTA RICA  DAHOMEY     DENMARK     DOMIN REP   FINLAND     FRANCE      GABON       GERMAN FR
GREECE      ICELAND     INDIA       INDONESIA   IRELAND     ISRAEL      ITALY       IVORY COAST JAMAICA     JAPAN
LUXEMBOURG  MALAGASY R  MALAYA      MALI        MOROCCO     NETHERLANDS NEW ZEALAND NIGER       NORWAY      SENEGAL
SWEDEN      SWITZERLAND TANGANYIKA  TOGO        TRINIDAD    TUNISIA     TURKEY      UPPER VOLTA URUGUAY     VENEZUELA

      10 EXCLUDED BECAUSE AMBIGUOUS

CAMBODIA    CYPRUS      HAITI       IRAN        LEBANON     NIGERIA     PAKISTAN    RWANDA      UGANDA      UAR

       8 EXCLUDED BECAUSE UNASCERTAINABLE

ALGERIA     BURMA       CUBA        IRAQ        KOREA REP   SUDAN       SYRIA       YEMEN
------------------------------------------------------------------------------------------------------------------------------
```

****

162 POLITIES WHERE THE REGIME'S LEADERSHIP CHARISMA IS PRONOUNCED, RATHER THAN NEGLIGIBLE (13)	162 POLITIES WHERE THE REGIME'S LEADERSHIP CHARISMA IS NEGLIGIBLE, RATHER THAN PRONOUNCED (65)

SUBJECT ONLY

....

13 IN LEFT COLUMN

CAMBODIA	CHINA, PR	CUBA	ETHIOPIA	FRANCE	GHANA	GUINEA	INDIA	INDONESIA	IVORY COAST
TANGANYIKA	UAR	VIETNAM, N							

65 IN RIGHT COLUMN

ALBANIA	ARGENTINA	AUSTRALIA	AUSTRIA	BELGIUM	BOLIVIA	BRAZIL	BULGARIA	CANADA	CEYLON
CHILE	COLOMBIA	COSTA RICA	CZECHOS'KIA	DENMARK	DOMIN REP	ECUADOR	EL SALVADOR	FINLAND	GERMANY, E
GERMAN FR	GREECE	GUATEMALA	HAITI	HONDURAS	HUNGARY	ICELAND	IRELAND	ISRAEL	ITALY
JAPAN	KOREA REP	LEBANON	LIBERIA	LUXEMBOURG	MALAGASY R	MALAYA	MEXICO	MONGOLIA	NETHERLANDS
NEW ZEALAND	NICARAGUA	NORWAY	PANAMA	PARAGUAY	PERU	PHILIPPINES	POLAND	PORTUGAL	RUMANIA
SIERRE LEO	SOMALIA	SO AFRICA	SPAIN	SUDAN	SWEDEN	SWITZERLAND	THAILAND	TRINIDAD	TURKEY
UK	US	URUGUAY	VENEZUELA	VIETNAM REP					

7 EXCLUDED BECAUSE AMBIGUOUS

AFGHANISTAN	CONGO(LEO)	CYPRUS	IRAN	NIGERIA	UGANDA	USSR

21 EXCLUDED BECAUSE IRRELEVANT

ALGERIA	CAMEROUN	CEN AFR REP	CHAD	CONGO(BRA)	DAHOMEY	GABON	JAMAICA	JORDAN	KOREA, N
LIBYA	MALI	MAURITANIA	MOROCCO	NIGER	PAKISTAN	SA'U ARABIA	SENEGAL	TUNISIA	UPPER VOLTA
YUGOSLAVIA									

4 EXCLUDED BECAUSE UNASCERTAINED

BURUNDI	LAOS	NEPAL	RWANDA

5 EXCLUDED BECAUSE UNASCERTAINABLE

BURMA	IRAQ	SYRIA	TOGO	YEMEN

2	LEAN MORE TOWARD BEING THOSE LOCATED ELSEWHERE THAN IN WEST EUROPE, SCANDINAVIA, NORTH AMERICA, OR AUSTRALASIA (93)	0.92	2	LEAN LESS TOWARD BEING THOSE LOCATED ELSEWHERE THAN IN WEST EUROPE, SCANDINAVIA, NORTH AMERICA, OR AUSTRALASIA (93)	0.68	1 21 12 44 X SQ= 2.14 P = 0.096 RV NO (NO)
6	TILT MORE TOWARD BEING THOSE LOCATED ELSEWHERE THAN IN THE CARIBBEAN, CENTRAL AMERICA, OR SOUTH AMERICA (93)	0.92	6	TILT LESS TOWARD BEING THOSE LOCATED ELSEWHERE THAN IN THE CARIBBEAN, CENTRAL AMERICA, OR SOUTH AMERICA (93)	0.69	1 20 12 45 X SQ= 1.88 P = 0.102 RV NO (NO)

19	TEND TO BE THOSE LOCATED IN NORTH AFRICA OR CENTRAL AND SOUTH AFRICA, RATHER THAN IN THE CARIBBEAN, CENTRAL AMERICA, OR SOUTH AMERICA (33)	0.86	19	TEND TO BE THOSE LOCATED IN THE CARIBBEAN, CENTRAL AMERICA, OR SOUTH AMERICA, RATHER THAN IN NORTH AFRICA OR CENTRAL AND SOUTH AFRICA (22)	0.77	6 6 1 20 X SQ= 6.84 P = 0.005 RV YES (YES)
20	TEND TO BE THOSE LOCATED IN EAST ASIA, SOUTH ASIA, OR SOUTHEAST ASIA, RATHER THAN IN THE CARIBBEAN, CENTRAL AMERICA, OR SOUTH AMERICA (18)	0.83	20	TEND TO BE THOSE LOCATED IN THE CARIBBEAN, CENTRAL AMERICA, OR SOUTH AMERICA, RATHER THAN IN EAST ASIA, SOUTH ASIA, OR SOUTHEAST ASIA (22)	0.71	5 8 1 20 X SQ= 4.17 P = 0.021 RV YES (NO)
22	LEAN TOWARD BEING THOSE WHOSE TERRITORIAL SIZE IS VERY LARGE, LARGE, OR MEDIUM (68)	0.77	22	LEAN TOWARD BEING THOSE WHOSE TERRITORIAL SIZE IS SMALL (47)	0.51	10 32 3 33 X SQ= 2.32 P = 0.078 RV YES (NO)
29	TEND TO BE THOSE WHERE THE DEGREE OF URBANIZATION IS LOW (49)	0.73	29	TEND TO BE THOSE WHERE THE DEGREE OF URBANIZATION IS HIGH (56)	0.76	3 44 8 14 X SQ= 7.94 P = 0.003 RV YES (NO)
30	TEND TO BE THOSE WHOSE AGRICULTURAL POPULATION IS HIGH (56)	0.75	30	TEND TO BE THOSE WHOSE AGRICULTURAL POPULATION IS MEDIUM, LOW, OR VERY LOW (57)	0.74	9 17 3 48 X SQ= 8.73 P = 0.002 RV YES (NO)
36	TEND TO BE THOSE WHOSE PER CAPITA GROSS NATIONAL PRODUCT IS LOW OR VERY LOW (73)	0.85	36	TEND TO BE THOSE WHOSE PER CAPITA GROSS NATIONAL PRODUCT IS VERY HIGH, HIGH, OR MEDIUM (42)	0.55	2 36 11 29 X SQ= 5.43 P = 0.013 RV YES (NO)
37	TEND TO BE THOSE WHOSE PER CAPITA GROSS NATIONAL PRODUCT IS VERY LOW (51)	0.77	37	TEND TO BE THOSE WHOSE PER CAPITA GROSS NATIONAL PRODUCT IS VERY HIGH, HIGH, MEDIUM, OR LOW (64)	0.80	3 52 10 13 X SQ= 14.26 P = 0. RV YES (NO)
42	TEND TO BE THOSE WHOSE ECONOMIC DEVELOPMENTAL STATUS IS UNDERDEVELOPED OR VERY UNDERDEVELOPED (76)	0.92	42	TEND TO BE THOSE WHOSE ECONOMIC DEVELOPMENTAL STATUS IS DEVELOPED OR INTERMEDIATE (36)	0.51	1 32 12 31 X SQ= 6.49 P = 0.005 RV YES (NO)
43	TEND TO BE THOSE WHOSE ECONOMIC DEVELOPMENTAL STATUS IS VERY UNDERDEVELOPED (57)	0.69	43	TEND TO BE THOSE WHOSE ECONOMIC DEVELOPMENTAL STATUS IS DEVELOPED, INTERMEDIATE, OR UNDERDEVELOPED (55)	0.68	4 43 9 20 X SQ= 4.93 P = 0.025 RV YES (NO)

45	TEND TO BE THOSE WHERE THE LITERACY RATE IS BELOW FIFTY PERCENT (54)	0.73	45	TEND TO BE THOSE WHERE THE LITERACY RATE IS FIFTY PERCENT OR ABOVE (55)	0.76	3 47 8 15 X SQ= 8.07 P = 0.003 RV YES (NO)
50	TEND TO BE THOSE WHERE FREEDOM OF THE PRESS IS INTERMITTENT, INTERNALLY ABSENT, OR INTERNALLY AND EXTERNALLY ABSENT (56)	0.83	50	TEND TO BE THOSE WHERE FREEDOM OF THE PRESS IS COMPLETE (43)	0.56	2 33 10 26 X SQ= 4.68 P = 0.024 RV YES (NO)
51	TEND TO BE THOSE WHERE FREEDOM OF THE PRESS IS INTERNALLY ABSENT, OR INTERNALLY AND EXTERNALLY ABSENT (37)	0.75	51	TEND TO BE THOSE WHERE FREEDOM OF THE PRESS IS COMPLETE OR INTERMITTENT (60)	0.71	3 42 9 17 X SQ= 7.28 P = 0.006 RV YES (NO)
53	ALWAYS ARE THOSE WHERE NEWSPAPER CIRCULATION IS LESS THAN THREE HUNDRED PER THOUSAND (101)	1.00	53	TILT LESS TOWARD BEING THOSE WHERE NEWSPAPER CIRCULATION IS LESS THAN THREE HUNDRED PER THOUSAND (101)	0.78	0 14 13 51 X SQ= 2.11 P = 0.110 RV NO (NO)
54	LEAN TOWARD BEING THOSE WHERE NEWSPAPER CIRCULATION IS LESS THAN ONE HUNDRED PER THOUSAND (76)	0.83	54	LEAN TOWARD BEING THOSE WHERE NEWSPAPER CIRCULATION IS ONE HUNDRED OR MORE PER THOUSAND (37)	0.51	2 33 10 32 X SQ= 3.48 P = 0.055 RV YES (NO)
63	TEND TO BE THOSE WHERE THE RELIGION IS PREDOMINANTLY OR PARTLY OTHER THAN CHRISTIAN (68)	0.85	63	TEND TO BE THOSE WHERE THE RELIGION IS PREDOMINANTLY SOME KIND OF CHRISTIAN (46)	0.65	2 42 11 23 X SQ= 8.77 P = 0.002 RV YES (NO)
64	TEND TO BE THOSE WHERE THE RELIGION IS MUSLIM, RATHER THAN CHRISTIAN (18)	0.50	64	TEND TO BE THOSE WHERE THE RELIGION IS CHRISTIAN, RATHER THAN MUSLIM (46)	0.95	2 42 2 2 X SQ= 4.86 P = 0.030 RV YES (YES)
68	TEND TO BE THOSE THAT ARE LINGUISTICALLY WEAKLY HETEROGENEOUS OR STRONGLY HETEROGENEOUS (62)	0.75	68	TEND TO BE THOSE THAT ARE LINGUISTICALLY HOMOGENEOUS (52)	0.62	3 40 9 25 X SQ= 4.10 P = 0.027 RV YES (NO)
69	TEND TO BE THOSE THAT ARE LINGUISTICALLY STRONGLY HETEROGENEOUS (50)	0.58	69	TEND TO BE THOSE THAT ARE LINGUISTICALLY HOMOGENEOUS OR WEAKLY HETEROGENEOUS (64)	0.75	5 49 7 16 X SQ= 4.01 P = 0.035 RV YES (NO)

72	TEND TO BE THOSE WHOSE DATE OF INDEPENDENCE IS AFTER 1914 (59)	0.69	72	TEND TO BE THOSE WHOSE DATE OF INDEPENDENCE IS BEFORE 1914 (52)	0.71	4 44 9 18 X SQ= 5.89 P = 0.010 RV YES (NO)
73	TEND TO BE THOSE WHOSE DATE OF INDEPENDENCE IS AFTER 1945 (46)	0.69	73	TEND TO BE THOSE WHOSE DATE OF INDEPENDENCE IS BEFORE 1945 (65)	0.84	4 52 9 10 X SQ= 13.34 P = 0. RV YES (NO)
74	TEND MORE TO BE THOSE THAT ARE NOT HISTORICALLY WESTERN (87)	0.92	74	TEND LESS TO BE THOSE THAT ARE NOT HISTORICALLY WESTERN (87)	0.60	1 25 12 38 X SQ= 3.58 P = 0.028 RV NO (NO)
75	TEND TO BE THOSE THAT ARE NOT HISTORICALLY WESTERN AND ARE NOT SIGNIFICANTLY WESTERNIZED (52)	0.62	75	TEND TO BE THOSE THAT ARE HISTORICALLY WESTERN OR SIGNIFICANTLY WESTERNIZED (62)	0.81	5 52 8 12 X SQ= 8.18 P = 0.003 RV YES (NO)
76	TEND MORE TO BE THOSE THAT HAVE BEEN SIGNIFICANTLY OR PARTIALLY WESTERNIZED THROUGH A COLONIAL RELATIONSHIP, RATHER THAN BEING HISTORICALLY WESTERN (70)	0.91	76	TEND LESS TO BE THOSE THAT HAVE BEEN SIGNIFICANTLY OR PARTIALLY WESTERNIZED THROUGH A COLONIAL RELATIONSHIP, RATHER THAN BEING HISTORICALLY WESTERN (70)	0.55	1 25 10 30 X SQ= 3.67 P = 0.040 RV NO (NO)
77	TEND TO BE THOSE THAT HAVE BEEN PARTIALLY WESTERNIZED, RATHER THAN SIGNIFICANTLY WESTERNIZED, THROUGH A COLONIAL RELATIONSHIP (41)	0.70	77	TEND TO BE THOSE THAT HAVE BEEN SIGNIFICANTLY WESTERNIZED, RATHER THAN PARTIALLY WESTERNIZED, THROUGH A COLONIAL RELATIONSHIP (28)	0.76	3 22 7 7 X SQ= 4.95 P = 0.019 RV YES (YES)
79	LEAN TOWARD BEING THOSE THAT WERE FORMERLY DEPENDENCIES OF BRITAIN OR FRANCE, RATHER THAN SPAIN (49)	0.89	79	LEAN TOWARD BEING THOSE THAT WERE FORMERLY DEPENDENCIES OF SPAIN, RATHER THAN BRITAIN OR FRANCE (18)	0.52	8 16 1 17 X SQ= 3.21 P = 0.055 RV YES (NO)
82	TEND TO BE THOSE WHERE THE TYPE OF POLITICAL MODERNIZATION IS DEVELOPED TUTELARY OR UNDEVELOPED TUTELARY, RATHER THAN EARLY OR LATER EUROPEAN OR EUROPEAN DERIVED (55)	0.82	82	TEND TO BE THOSE WHERE THE TYPE OF POLITICAL MODERNIZATION IS EARLY OR LATER EUROPEAN OR EUROPEAN DERIVED, RATHER THAN DEVELOPED TUTELARY OR UNDEVELOPED TUTELARY (51)	0.77	2 48 9 14 X SQ= 12.57 P = 0. RV YES (NO)
85	LEAN TOWARD BEING THOSE WHERE THE STAGE OF POLITICAL MODERNIZATION IS MID- OR EARLY TRANSITIONAL, RATHER THAN ADVANCED (54)	0.62	85	LEAN TOWARD BEING THOSE WHERE THE STAGE OF POLITICAL MODERNIZATION IS ADVANCED, RATHER THAN MID- OR EARLY TRANSITIONAL (60)	0.68	5 44 8 21 X SQ= 2.81 P = 0.062 RV YES (NO)

86	TEND TO BE THOSE WHERE THE STAGE OF POLITICAL MODERNIZATION IS EARLY TRANSITIONAL, RATHER THAN ADVANCED OR MID-TRANSITIONAL (38)	0.54	86	TEND TO BE THOSE WHERE THE STAGE OF POLITICAL MODERNIZATION IS ADVANCED OR MID-TRANSITIONAL, RATHER THAN EARLY TRANSITIONAL (76)	0.88	6 57 7 8 X SQ= 9.51 P = 0.002 RV YES (NO)
87	TEND TO BE THOSE WHOSE IDEOLOGICAL ORIENTATION IS DEVELOPMENTAL (31)	0.60	87	TEND TO BE THOSE WHOSE IDEOLOGICAL ORIENTATION IS OTHER THAN DEVELOPMENTAL (58)	0.85	6 8 4 44 X SQ= 7.17 P = 0.006 RV YES (NO)
89	TEND TO BE THOSE WHOSE IDEOLOGICAL ORIENTATION IS OTHER THAN CONVENTIONAL (62)	0.92	89	TEND TO BE THOSE WHOSE IDEOLOGICAL ORIENTATION IS CONVENTIONAL (33)	0.61	1 31 11 20 X SQ= 8.70 P = 0.001 RV YES (NO)
93	TEND TO BE THOSE WHERE THE SYSTEM STYLE IS MOBILIZATIONAL, OR LIMITED MOBILIZATIONAL (32)	0.77	93	TEND TO BE THOSE WHERE THE SYSTEM STYLE IS NON-MOBILIZATIONAL (78)	0.79	10 13 3 50 X SQ= 13.62 P = 0. RV YES (NO)
94	TEND TO BE THOSE WHERE THE STATUS OF THE REGIME IS AUTHORITARIAN OR TOTALITARIAN (41)	0.82	94	TEND TO BE THOSE WHERE THE STATUS OF THE REGIME IS CONSTITUTIONAL (51)	0.68	2 41 9 19 X SQ= 7.80 P = 0.005 RV YES (NO)
96	TEND TO BE THOSE WHERE THE STATUS OF THE REGIME IS AUTHORITARIAN (23)	0.55	96	TEND TO BE THOSE WHERE THE STATUS OF THE REGIME IS CONSTITUTIONAL OR TOTALITARIAN (67)	0.88	5 51 6 7 X SQ= 8.31 P = 0.004 RV YES (NO)
98	ALWAYS ARE THOSE WHERE GOVERNMENTAL STABILITY IS GENERALLY OR MODERATELY PRESENT AND DATES FROM THE POST-WAR PERIOD, OR IS ABSENT (93)	1.00	98	TEND LESS TO BE THOSE WHERE GOVERNMENTAL STABILITY IS GENERALLY OR MODERATELY PRESENT AND DATES FROM THE POST-WAR PERIOD, OR IS ABSENT (93)	0.71	0 19 13 46 X SQ= 3.56 P = 0.031 RV NO (NO)
101	TEND TO BE THOSE WHERE THE REPRESENTATIVE CHARACTER OF THE REGIME IS LIMITED POLYARCHIC, PSEUDO-POLYARCHIC, OR NON-POLYARCHIC (57)	0.85	101	TEND TO BE THOSE WHERE THE REPRESENTATIVE CHARACTER OF THE REGIME IS POLYARCHIC (41)	0.59	2 35 11 24 X SQ= 6.57 P = 0.005 RV YES (NO)
102	TEND TO BE THOSE WHERE THE REPRESENTATIVE CHARACTER OF THE REGIME IS PSEUDO-POLYARCHIC OR NON-POLYARCHIC (49)	0.77	102	TEND TO BE THOSE WHERE THE REPRESENTATIVE CHARACTER OF THE REGIME IS POLYARCHIC OR LIMITED POLYARCHIC (59)	0.73	3 45 10 17 X SQ= 9.38 P = 0.001 RV YES (NO)

105	TEND TO BE THOSE WHERE THE ELECTORAL SYSTEM IS PARTIALLY COMPETITIVE OR NON-COMPETITIVE (47)	0.80	105	TEND TO BE THOSE WHERE THE ELECTORAL SYSTEM IS COMPETITIVE (43)	0.70	2 39 8 17 X SQ= 6.90 P = 0.005 RV YES (NO)
106	LEAN TOWARD BEING THOSE WHERE THE ELECTORAL SYSTEM IS NON-COMPETITIVE (30)	0.60	106	LEAN TOWARD BEING THOSE WHERE THE ELECTORAL SYSTEM IS COMPETITIVE OR PARTIALLY COMPETITIVE (52)	0.75	4 41 6 14 X SQ= 3.26 P = 0.057 RV YES (NO)
107	TEND TO BE THOSE WHERE AUTONOMOUS GROUPS ARE PARTIALLY TOLERATED IN POLITICS, ARE TOLERATED ONLY OUTSIDE POLITICS, OR ARE NOT TOLERATED AT ALL (65)	0.77	107	TEND TO BE THOSE WHERE AUTONOMOUS GROUPS ARE FULLY TOLERATED IN POLITICS (46)	0.62	3 40 10 24 X SQ= 5.31 P = 0.013 RV YES (NO)
108	TEND TO BE THOSE WHERE AUTONOMOUS GROUPS ARE TOLERATED ONLY OUTSIDE POLITICS OR ARE NOT TOLERATED AT ALL (35)	0.62	108	TEND TO BE THOSE WHERE AUTONOMOUS GROUPS ARE FULLY OR PARTIALLY TOLERATED IN POLITICS (65)	0.74	5 46 8 16 X SQ= 4.77 P = 0.020 RV YES (NO)
116	TEND TO BE THOSE WHERE INTEREST ARTICULATION BY ASSOCIATIONAL GROUPS IS LIMITED OR NEGLIGIBLE (79)	0.92	116	TEND TO BE THOSE WHERE INTEREST ARTICULATION BY ASSOCIATIONAL GROUPS IS SIGNIFICANT OR MODERATE (32)	0.50	1 31 12 31 X SQ= 6.23 P = 0.005 RV YES (NO)
117	TEND TO BE THOSE WHERE INTEREST ARTICULATION BY ASSOCIATIONAL GROUPS IS NEGLIGIBLE (51)	0.62	117	TEND TO BE THOSE WHERE INTEREST ARTICULATION BY ASSOCIATIONAL GROUPS IS SIGNIFICANT, MODERATE, OR LIMITED (60)	0.74	5 46 8 16 X SQ= 4.77 P = 0.020 RV YES (NO)
121	TEND TO BE THOSE WHERE INTEREST ARTICULATION BY NON-ASSOCIATIONAL GROUPS IS SIGNIFICANT (54)	0.77	121	TEND TO BE THOSE WHERE INTEREST ARTICULATION BY NON-ASSOCIATIONAL GROUPS IS MODERATE, LIMITED, OR NEGLIGIBLE (61)	0.78	10 14 3 51 X SQ= 13.11 P = 0. RV YES (NO)
126	ALWAYS ARE THOSE WHERE INTEREST ARTICULATION BY ANOMIC GROUPS IS FREQUENT, OCCASIONAL, OR INFREQUENT (83)	1.00	126	LEAN LESS TOWARD BEING THOSE WHERE INTEREST ARTICULATION BY ANOMIC GROUPS IS FREQUENT, OCCASIONAL, OR INFREQUENT (83)	0.71	11 39 0 16 X SQ= 2.79 P = 0.053 RV NO (NO)
127	ALWAYS ARE THOSE WHERE INTEREST ARTICULATION BY POLITICAL PARTIES IS MODERATE, LIMITED, OR NEGLIGIBLE (72)	1.00	127	LEAN LESS TOWARD BEING THOSE WHERE INTEREST ARTICULATION BY POLITICAL PARTIES IS MODERATE, LIMITED, OR NEGLIGIBLE (72)	0.71	0 17 11 41 X SQ= 2.85 P = 0.054 RV NO (NO)

128	TEND TO BE THOSE WHERE INTEREST ARTICULATION BY POLITICAL PARTIES IS LIMITED OR NEGLIGIBLE (45)	0.82	128	TEND TO BE THOSE WHERE INTEREST ARTICULATION BY POLITICAL PARTIES IS SIGNIFICANT OR MODERATE (48)	0.67	2 39 9 19 X SQ= 7.31 P = 0.005 RV YES (NO)
129	TEND TO BE THOSE WHERE INTEREST ARTICULATION BY POLITICAL PARTIES IS NEGLIGIBLE (37)	0.73	129	TEND TO BE THOSE WHERE INTEREST ARTICULATION BY POLITICAL PARTIES IS SIGNIFICANT, MODERATE, OR LIMITED (56)	0.74	3 43 8 15 X SQ= 7.15 P = 0.005 RV YES (NO)
133	TEND TO BE THOSE WHERE INTEREST AGGREGATION BY THE EXECUTIVE IS SIGNIFICANT (29)	0.69	133	TEND TO BE THOSE WHERE INTEREST AGGREGATION BY THE EXECUTIVE IS MODERATE, LIMITED, OR NEGLIGIBLE (76)	0.90	9 6 4 55 X SQ= 19.86 P = 0. RV YES (YES)
138	TEND TO BE THOSE WHERE INTEREST AGGREGATION BY THE LEGISLATURE IS NEGLIGIBLE (49)	0.73	138	TEND TO BE THOSE WHERE INTEREST AGGREGATION BY THE LEGISLATURE IS SIGNIFICANT, MODERATE, OR LIMITED (48)	0.67	3 39 8 19 X SQ= 4.64 P = 0.019 RV YES (NO)
139	TEND TO BE THOSE WHERE THE PARTY SYSTEM IS QUANTITATIVELY ONE-PARTY (34)	0.67	139	TEND TO BE THOSE WHERE THE PARTY SYSTEM IS QUANTITATIVELY OTHER THAN ONE-PARTY (71)	0.81	8 12 4 50 X SQ= 9.14 P = 0.002 RV YES (NO)
142	ALWAYS ARE THOSE WHERE THE PARTY SYSTEM IS QUANTITATIVELY OTHER THAN MULTI-PARTY (66)	1.00	142	TEND LESS TO BE THOSE WHERE THE PARTY SYSTEM IS QUANTITATIVELY OTHER THAN MULTI-PARTY (66)	0.56	0 25 12 32 X SQ= 6.46 P = 0.003 RV NO (NO)
144	ALWAYS ARE THOSE WHERE THE PARTY SYSTEM IS QUANTITATIVELY ONE-PARTY, RATHER THAN TWO-PARTY (34)	1.00	144	TEND LESS TO BE THOSE WHERE THE PARTY SYSTEM IS QUANTITATIVELY ONE-PARTY, RATHER THAN TWO-PARTY (34)	0.52	9 11 0 10 X SQ= 4.46 P = 0.013 RV NO (NO)
146	TEND TO BE THOSE WHERE THE PARTY SYSTEM IS QUALITATIVELY MASS-BASED TERRITORIAL (8)	0.62	146	ALWAYS ARE THOSE WHERE THE PARTY SYSTEM IS QUALITATIVELY OTHER THAN MASS-BASED TERRITORIAL (92)	1.00	5 0 3 62 X SQ= 32.84 P = 0. RV YES (YES)
147	ALWAYS ARE THOSE WHERE THE PARTY SYSTEM IS QUALITATIVELY OTHER THAN CLASS-ORIENTED OR MULTI-IDEOLOGICAL (67)	1.00	147	TEND LESS TO BE THOSE WHERE THE PARTY SYSTEM IS QUALITATIVELY OTHER THAN CLASS-ORIENTED OR MULTI-IDEOLOGICAL (67)	0.60	0 22 9 33 X SQ= 3.86 P = 0.022 RV NO (NO)

154	ALWAYS ARE THOSE WHERE THE PARTY SYSTEM IS STABLE OR MODERATELY STABLE (55)	1.00	154	LEAN LESS TOWARD BEING THOSE WHERE THE PARTY SYSTEM IS STABLE OR MODERATELY STABLE (55)	0.69	9 41 0 18 X SQ= 2.33 P = 0.099 RV NO (NO)
168	TEND TO BE THOSE WHERE THE HORIZONTAL POWER DISTRIBUTION IS LIMITED OR NEGLIGIBLE (72)	0.92	168	TEND TO BE THOSE WHERE THE HORIZONTAL POWER DISTRIBUTION IS SIGNIFICANT (34)	0.50	1 31 12 31 X SQ= 6.23 P = 0.005 RV YES (NO)
169	TEND TO BE THOSE WHERE THE HORIZONTAL POWER DISTRIBUTION IS NEGLIGIBLE (48)	0.85	169	TEND TO BE THOSE WHERE THE HORIZONTAL POWER DISTRIBUTION IS SIGNIFICANT OR LIMITED (58)	0.74	2 46 11 16 X SQ= 13.68 P = 0. RV YES (NO)
172	ALWAYS ARE THOSE WHERE THE LEGISLATIVE-EXECUTIVE STRUCTURE IS OTHER THAN PARLIAMENTARY-ROYALIST (88)	1.00	172	LEAN LESS TOWARD BEING THOSE WHERE THE LEGISLATIVE-EXECUTIVE STRUCTURE IS OTHER THAN PARLIAMENTARY-ROYALIST (88)	0.75	0 16 12 47 X SQ= 2.51 P = 0.059 RV NO (NO)
175	TEND TO BE THOSE WHERE THE LEGISLATURE IS LARGELY INEFFECTIVE OR WHOLLY INEFFECTIVE (49)	0.83	175	TEND TO BE THOSE WHERE THE LEGISLATURE IS FULLY EFFECTIVE OR PARTIALLY EFFECTIVE (51)	0.70	2 42 10 18 X SQ= 9.83 P = 0.001 RV YES (NO)
176	TEND TO BE THOSE WHERE THE LEGISLATURE IS WHOLLY INEFFECTIVE (28)	0.58	176	TEND TO BE THOSE WHERE THE LEGISLATURE IS FULLY EFFECTIVE, PARTIALLY EFFECTIVE, OR LARGELY INEFFECTIVE (72)	0.77	5 46 7 14 X SQ= 4.36 P = 0.032 RV YES (NO)
179	ALWAYS ARE THOSE WHERE THE EXECUTIVE IS DOMINANT (52)	1.00	179	TEND TO BE THOSE WHERE THE EXECUTIVE IS STRONG (39)	0.64	13 19 0 34 X SQ= 14.73 P = 0. RV YES (NO)
185	TILT MORE TOWARD BEING THOSE WHERE PARTICIPATION BY THE MILITARY IN POLITICS IS SUPPORTIVE, RATHER THAN INTERVENTIVE (31)	0.87	185	TILT LESS TOWARD BEING THOSE WHERE PARTICIPATION BY THE MILITARY IN POLITICS IS SUPPORTIVE, RATHER THAN INTERVENTIVE (31)	0.52	1 14 7 15 X SQ= 2.01 P = 0.108 RV NO (NO)
186	TEND TO BE THOSE WHERE PARTICIPATION BY THE MILITARY IN POLITICS IS SUPPORTIVE, RATHER THAN NEUTRAL (31)	0.70	186	TEND TO BE THOSE WHERE PARTICIPATION BY THE MILITARY IN POLITICS IS NEUTRAL, RATHER THAN SUPPORTIVE (56)	0.69	7 15 3 34 X SQ= 3.95 P = 0.030 RV YES (NO)

187	TILT MORE TOWARD BEING THOSE WHERE THE ROLE OF THE POLICE IS POLITICALLY SIGNIFICANT (66) 0.77	187	TILT LESS TOWARD BEING THOSE WHERE THE ROLE OF THE POLICE IS POLITICALLY SIGNIFICANT (66) 0.51	10 30 3 29 X SQ= 1.97 P = 0.125 RV NO (NO)
188	TEND MORE TO BE THOSE WHERE THE CHARACTER OF THE LEGAL SYSTEM IS OTHER THAN CIVIL LAW (81) 0.91	188	TEND LESS TO BE THOSE WHERE THE CHARACTER OF THE LEGAL SYSTEM IS OTHER THAN CIVIL LAW (81) 0.52	1 31 10 34 X SQ= 4.28 P = 0.020 RV NO (NO)
192	TEND LESS TO BE THOSE WHERE THE CHARACTER OF THE LEGAL SYSTEM IS OTHER THAN MUSLIM OR PARTLY MUSLIM (86) 0.55	192	TEND MORE TO BE THOSE WHERE THE CHARACTER OF THE LEGAL SYSTEM IS OTHER THAN MUSLIM OR PARTLY MUSLIM (86) 0.95	5 3 6 62 X SQ= 12.61 P = 0.001 RV NO (YES)
193	TEND LESS TO BE THOSE WHERE THE CHARACTER OF THE LEGAL SYSTEM IS OTHER THAN PARTLY INDIGENOUS (86) 0.54	193	TEND MORE TO BE THOSE WHERE THE CHARACTER OF THE LEGAL SYSTEM IS OTHER THAN PARTLY INDIGENOUS (86) 0.92	6 5 7 60 X SQ= 10.24 P = 0.002 RV NO (YES)

163 POLITIES
WHERE THE REGIME'S LEADERSHIP CHARISMA
IS PRONOUNCED (13)

163 POLITIES
WHERE THE REGIME'S LEADERSHIP CHARISMA
IS MODERATE OR NEGLIGIBLE (87)

PREDICATE ONLY

13 IN LEFT COLUMN

CAMBODIA	CHINA, PR	CUBA	ETHIOPIA	FRANCE	GHANA	GUINEA	INDIA	INDONESIA	IVORY COAST
TANGANYIKA	UAR	VIETNAM, N							

87 IN RIGHT COLUMN

ALBANIA	ALGERIA	ARGENTINA	AUSTRALIA	AUSTRIA	BELGIUM	BOLIVIA	BRAZIL	BULGARIA	BURMA
CAMEROUN	CANADA	CEN AFR REP	CEYLON	CHAD	CHILE	COLOMBIA	CONGO(BRA)	COSTA RICA	CZECHOS'KIA
DAHOMEY	DENMARK	DOMIN REP	ECUADOR	EL SALVADOR	FINLAND	GABON	GERMANY, E	GERMAN FR	GREECE
GUATEMALA	HAITI	HONDURAS	HUNGARY	ICELAND	IRELAND	ISRAEL	ITALY	JAMAICA	JAPAN
JORDAN	KOREA, N	KOREA REP	LEBANON	LIBERIA	LIBYA	LUXEMBOURG	MALAGASY R	MALAYA	MALI
MAURITANIA	MEXICO	MONGOLIA	MOROCCO	NETHERLANDS	NEW ZEALAND	NICARAGUA	NIGER	NORWAY	PAKISTAN
PANAMA	PARAGUAY	PERU	PHILIPPINES	POLAND	PORTUGAL	RUMANIA	SA'U ARABIA	SENEGAL	SIERRE LEO
SOMALIA	SO AFRICA	SPAIN	SUDAN	SWEDEN	SWITZERLAND	THAILAND	TRINIDAD	TUNISIA	TURKEY
UK	US	UPPER VOLTA	URUGUAY	VENEZUELA	VIETNAM REP	YUGOSLAVIA			

7 EXCLUDED BECAUSE AMBIGUOUS

AFGHANISTAN	CONGO(LEO)	CYPRUS	IRAN	NIGERIA	UGANDA	USSR

4 EXCLUDED BECAUSE UNASCERTAINED

BURUNDI	LAOS	NEPAL	RWANDA

4 EXCLUDED BECAUSE UNASCERTAINABLE

IRAQ	SYRIA	TOGO	YEMEN

164 POLITIES
WHERE THE REGIME'S LEADERSHIP CHARISMA
IS PRONOUNCED OR MODERATE (34)

164 POLITIES
WHERE THE REGIME'S LEADERSHIP CHARISMA
IS NEGLIGIBLE (65)

PREDICATE ONLY

34 IN LEFT COLUMN

ALGERIA	CAMBODIA	CAMEROUN	CEN AFR REP	CHAD	CHINA, PR	CONGO(BRA)	CUBA	DAHOMEY	ETHIOPIA
FRANCE	GABON	GHANA	GUINEA	INDIA	INDONESIA	IVORY COAST	JAMAICA	JORDAN	KOREA, N
LIBYA	MALI	MAURITANIA	MOROCCO	NIGER	PAKISTAN	SA'U ARABIA	SENEGAL	TANGANYIKA	TUNISIA
UAR	UPPER VOLTA	VIETNAM, N	YUGOSLAVIA						

65 IN RIGHT COLUMN

ALBANIA	ARGENTINA	AUSTRALIA	AUSTRIA	BELGIUM	BOLIVIA	BRAZIL	BULGARIA	CANADA	CEYLON
CHILE	COLOMBIA	COSTA RICA	CZECHOS'KIA	DENMARK	DOMIN REP	ECUADOR	EL SALVADOR	FINLAND	GERMANY, E
GERMAN FR	GREECE	GUATEMALA	HAITI	HONDURAS	HUNGARY	ICELAND	IRELAND	ISRAEL	ITALY
JAPAN	KOREA REP	LEBANON	LIBERIA	LUXEMBOURG	MALAGASY R	MALAYA	MEXICO	MONGOLIA	NETHERLANDS
NEW ZEALAND	NICARAGUA	NORWAY	PANAMA	PARAGUAY	PERU	PHILIPPINES	POLAND	PORTUGAL	RUMANIA
SIERRE LEO	SOMALIA	SO AFRICA	SPAIN	SUDAN	SWEDEN	SWITZERLAND	THAILAND	TRINIDAD	TURKEY
UK	US	URUGUAY	VENEZUELA	VIETNAM REP					

7 EXCLUDED BECAUSE AMBIGUOUS

AFGHANISTAN CONGO(LEO) CYPRUS IRAN NIGERIA UGANDA USSR

4 EXCLUDED BECAUSE UNASCERTAINED

BURUNDI LAOS NEPAL RWANDA

5 EXCLUDED BECAUSE UNASCERTAINABLE

BURMA IRAQ SYRIA TOGO YEMEN

****

165 POLITIES
WHERE THE VERTICAL POWER DISTRIBUTION
IS THAT OF EFFECTIVE FEDERALISM (8)

165 POLITIES
WHERE THE VERTICAL POWER DISTRIBUTION
IS THAT OF LIMITED FEDERALISM,
FORMAL FEDERALISM, OR
FORMAL AND EFFECTIVE UNITARISM (106)

SUBJECT ONLY

....

8 IN LEFT COLUMN

AUSTRALIA	BRAZIL	CAMEROUN	CANADA	GERMAN FR	SWITZERLAND	UGANDA	US

106 IN RIGHT COLUMN

AFGHANISTAN	ALBANIA	ALGERIA	ARGENTINA	AUSTRIA	BELGIUM	BOLIVIA	BULGARIA	BURMA	BURUNDI
CAMBODIA	CEN AFR REP	CEYLON	CHAD	CHILE	CHINA, PR	COLOMBIA	CONGO(BRA)	COSTA RICA	CUBA
CYPRUS	CZECHOS'KIA	DAHOMEY	DENMARK	DOMIN REP	ECUADOR	EL SALVADOR	ETHIOPIA	FINLAND	FRANCE
GABON	GERMANY, E	GHANA	GREECE	GUATEMALA	GUINEA	HAITI	HONDURAS	HUNGARY	ICELAND
INDIA	INDONESIA	IRAN	IRAQ	IRELAND	ISRAEL	ITALY	IVORY COAST	JAMAICA	JAPAN
JORDAN	KOREA, N	KOREA REP	LAOS	LEBANON	LIBERIA	LIBYA	LUXEMBOURG	MALAGASY R	MALAYA
MALI	MAURITANIA	MEXICO	MONGOLIA	MOROCCO	NEPAL	NETHERLANDS	NEW ZEALAND	NICARAGUA	NIGER
NIGERIA	NORWAY	PAKISTAN	PANAMA	PARAGUAY	PERU	PHILIPPINES	POLAND	PORTUGAL	RUMANIA
RWANDA	SA'U ARABIA	SENEGAL	SIERRE LEO	SOMALIA	SO AFRICA	SPAIN	SUDAN	SWEDEN	SYRIA
TANGANYIKA	THAILAND	TOGO	TRINIDAD	TUNISIA	TURKEY	USSR	UAR	UK	UPPER VOLTA
URUGUAY	VENEZUELA	VIETNAM, N	VIETNAM REP	YEMEN	YUGOSLAVIA				

1 EXCLUDED BECAUSE UNASCERTAINABLE

CONGO(LEO)

	Left column			Right column		
2	TEND TO BE THOSE LOCATED IN WEST EUROPE, SCANDINAVIA, NORTH AMERICA, OR AUSTRALASIA (22)	0.62	2	TEND TO BE THOSE LOCATED ELSEWHERE THAN IN WEST EUROPE, SCANDINAVIA, NORTH AMERICA, OR AUSTRALASIA (93)	0.84	5 17 3 89 X SQ= 7.54 P = 0.007 RV YES (NO)
22	TILT MORE TOWARD BEING THOSE WHOSE TERRITORIAL SIZE IS VERY LARGE, LARGE, OR MEDIUM (68)	0.87	22	TILT LESS TOWARD BEING THOSE WHOSE TERRITORIAL SIZE IS VERY LARGE, LARGE, OR MEDIUM (68)	0.57	7 60 1 46 X SQ= 1.79 P = 0.137 RV NO (NO)
31	TEND TO BE THOSE WHOSE AGRICULTURAL POPULATION IS LOW OR VERY LOW (24)	0.62	31	TEND TO BE THOSE WHOSE AGRICULTURAL POPULATION IS HIGH OR MEDIUM (90)	0.82	3 86 5 19 X SQ= 6.31 P = 0.011 RV YES (NO)

33	TEND TO BE THOSE WHOSE GROSS NATIONAL PRODUCT IS VERY HIGH, HIGH, OR MEDIUM (30)	0.75	33	TEND TO BE THOSE WHOSE GROSS NATIONAL PRODUCT IS LOW OR VERY LOW (85)	0.77	6 24 2 82 X SQ= 7.99 P = 0.004 RV YES (NO)
35	TEND TO BE THOSE WHOSE PER CAPITA GROSS NATIONAL PRODUCT IS VERY HIGH OR HIGH (24)	0.62	35	TEND TO BE THOSE WHOSE PER CAPITA GROSS NATIONAL PRODUCT IS MEDIUM, LOW, OR VERY LOW (91)	0.82	5 19 3 87 X SQ= 6.41 P = 0.010 RV YES (NO)
36	TILT TOWARD BEING THOSE WHOSE PER CAPITA GROSS NATIONAL PRODUCT IS VERY HIGH, HIGH, OR MEDIUM (42)	0.62	36	TILT TOWARD BEING THOSE WHOSE PER CAPITA GROSS NATIONAL PRODUCT IS LOW OR VERY LOW (73)	0.65	5 37 3 69 X SQ= 1.39 P = 0.142 RV YES (NO)
39	TEND TO BE THOSE WHOSE INTERNATIONAL FINANCIAL STATUS IS VERY HIGH, HIGH, OR MEDIUM (38)	0.75	39	TEND TO BE THOSE WHOSE INTERNATIONAL FINANCIAL STATUS IS LOW OR VERY LOW (76)	0.70	6 32 2 73 X SQ= 4.76 P = 0.017 RV YES (NO)
41	TEND TO BE THOSE WHOSE ECONOMIC DEVELOPMENTAL STATUS IS DEVELOPED (19)	0.62	41	TEND TO BE THOSE WHOSE ECONOMIC DEVELOPMENTAL STATUS IS INTERMEDIATE, UNDERDEVELOPED, OR VERY UNDERDEVELOPED (94)	0.87	5 14 3 90 X SQ= 9.44 P = 0.003 RV YES (NO)
42	TEND TO BE THOSE WHOSE ECONOMIC DEVELOPMENTAL STATUS IS DEVELOPED OR INTERMEDIATE (36)	0.75	42	TEND TO BE THOSE WHOSE ECONOMIC DEVELOPMENTAL STATUS IS UNDERDEVELOPED OR VERY UNDERDEVELOPED (76)	0.71	6 30 2 73 X SQ= 5.19 P = 0.014 RV YES (NO)
44	TEND TO BE THOSE WHERE THE LITERACY RATE IS NINETY PERCENT OR ABOVE (25)	0.62	44	TEND TO BE THOSE WHERE THE LITERACY RATE IS BELOW NINETY PERCENT (90)	0.81	5 20 3 86 X SQ= 5.92 P = 0.012 RV YES (NO)
50	TEND TO BE THOSE WHERE FREEDOM OF THE PRESS IS COMPLETE (43)	0.87	50	TEND TO BE THOSE WHERE FREEDOM OF THE PRESS IS INTERMITTENT, INTERNALLY ABSENT, OR INTERNALLY AND EXTERNALLY ABSENT (56)	0.60	7 36 1 54 X SQ= 4.94 P = 0.020 RV YES (NO)
51	ALWAYS ARE THOSE WHERE FREEDOM OF THE PRESS IS COMPLETE OR INTERMITTENT (60)	1.00	51	TEND LESS TO BE THOSE WHERE FREEDOM OF THE PRESS IS COMPLETE OR INTERMITTENT (60)	0.58	8 51 0 37 X SQ= 3.84 P = 0.022 RV NO (NO)

54	TILT TOWARD BEING THOSE WHERE NEWSPAPER CIRCULATION IS ONE HUNDRED OR MORE PER THOUSAND (37)	0.62	54	TILT TOWARD BEING THOSE WHERE NEWSPAPER CIRCULATION IS LESS THAN ONE HUNDRED PER THOUSAND (76)	0.69	5 32 3 72 X SQ= 2.10 P = 0.113 RV YES (NO)
63	LEAN TOWARD BEING THOSE WHERE THE RELIGION IS PREDOMINANTLY SOME KIND OF CHRISTIAN (46)	0.75	63	LEAN TOWARD BEING THOSE WHERE THE RELIGION IS PREDOMINANTLY OR PARTLY OTHER THAN CHRISTIAN (68)	0.62	6 40 2 65 X SQ= 2.80 P = 0.061 RV YES (NO)
66	TEND TO BE THOSE THAT ARE RELIGIOUSLY HETEROGENEOUS (49)	0.87	66	TEND TO BE THOSE THAT ARE RELIGIOUSLY HOMOGENEOUS (57)	0.58	1 56 7 · 41 X SQ= 4.41 P = 0.022 RV YES (NO)
74	TEND TO BE THOSE THAT ARE HISTORICALLY WESTERN (26)	0.63	74	TEND TO BE THOSE THAT ARE NOT HISTORICALLY WESTERN (87)	0.80	5 21 3 83 X SQ= 5.27 P = 0.016 RV YES (NO)
76	TEND TO BE THOSE THAT ARE HISTORICALLY WESTERN, RATHER THAN HAVING BEEN SIGNIFICANTLY OR PARTIALLY WESTERNIZED THROUGH A COLONIAL RELATIONSHIP (26)	0.63	76	TEND TO BE THOSE THAT HAVE BEEN SIGNIFICANTLY OR PARTIALLY WESTERNIZED THROUGH A COLONIAL RELATIONSHIP, RATHER THAN BEING HISTORICALLY WESTERN (70)	0.76	5 21 3 66 X SQ= 3.67 P = 0.033 RV YES (NO)
81	TEND TO BE THOSE WHERE THE TYPE OF POLITICAL MODERNIZATION IS EARLY EUROPEAN OR EARLY EUROPEAN DERIVED, RATHER THAN LATER EUROPEAN OR LATER EUROPEAN DERIVED (11)	0.67	81	TEND TO BE THOSE WHERE THE TYPE OF POLITICAL MODERNIZATION IS LATER EUROPEAN OR LATER EUROPEAN DERIVED, RATHER THAN EARLY EUROPEAN OR EARLY EUROPEAN DERIVED (40)	0.84	4 7 2 38 X SQ= 5.43 P = 0.015 RV YES (NO)
89	TEND TO BE THOSE WHOSE IDEOLOGICAL ORIENTATION IS CONVENTIONAL (33)	0.83	89	TEND TO BE THOSE WHOSE IDEOLOGICAL ORIENTATION IS OTHER THAN CONVENTIONAL (62)	0.68	5 28 1 60 X SQ= 4.48 P = 0.019 RV YES (NO)
93	ALWAYS ARE THOSE WHERE THE SYSTEM STYLE IS NON-MOBILIZATIONAL (78)	1.00	93	TILT LESS TOWARD BEING THOSE WHERE THE SYSTEM STYLE IS NON-MOBILIZATIONAL (78)	0.68	0 32 8 69 X SQ= 2.22 P = 0.102 RV NO (NO)
94	ALWAYS ARE THOSE WHERE THE STATUS OF THE REGIME IS CONSTITUTIONAL (51)	1.00	94	TEND LESS TO BE THOSE WHERE THE STATUS OF THE REGIME IS CONSTITUTIONAL (51)	0.52	7 44 0 41 X SQ= 4.29 P = 0.016 RV NO (NO)

101	LEAN TOWARD BEING THOSE WHERE THE REPRESENTATIVE CHARACTER OF THE REGIME IS POLYARCHIC (41)	0.83	101	LEAN TOWARD BEING THOSE WHERE THE REPRESENTATIVE CHARACTER OF THE REGIME IS LIMITED POLYARCHIC, PSEUDO-POLYARCHIC, OR NON-POLYARCHIC (57)	0.61	5 36 1 56 X SQ= 2.89 P = 0.079 RV YES (NO)
102	ALWAYS ARE THOSE WHERE THE REPRESENTATIVE CHARACTER OF THE REGIME IS POLYARCHIC OR LIMITED POLYARCHIC (59)	1.00	102	TEND LESS TO BE THOSE WHERE THE REPRESENTATIVE CHARACTER OF THE REGIME IS POLYARCHIC OR LIMITED POLYARCHIC (59)	0.51	8 50 0 49 X SQ= 5.45 P = 0.007 RV NO (NO)
105	ALWAYS ARE THOSE WHERE THE ELECTORAL SYSTEM IS COMPETITIVE (43)	1.00	105	TEND TO BE THOSE WHERE THE ELECTORAL SYSTEM IS PARTIALLY COMPETITIVE OR NON-COMPETITIVE (47)	0.57	7 36 0 47 X SQ= 6.18 P = 0.004 RV YES (NO)
107	TEND TO BE THOSE WHERE AUTONOMOUS GROUPS ARE FULLY TOLERATED IN POLITICS (46)	0.87	107	TEND TO BE THOSE WHERE AUTONOMOUS GROUPS ARE PARTIALLY TOLERATED IN POLITICS, ARE TOLERATED ONLY OUTSIDE POLITICS, OR ARE NOT TOLERATED AT ALL (65)	0.62	7 39 1 63 X SQ= 5.51 P = 0.009 RV YES (NO)
108	ALWAYS ARE THOSE WHERE AUTONOMOUS GROUPS ARE FULLY OR PARTIALLY TOLERATED IN POLITICS (65)	1.00	108	TEND LESS TO BE THOSE WHERE AUTONOMOUS GROUPS ARE FULLY OR PARTIALLY TOLERATED IN POLITICS (65)	0.62	8 57 0 35 X SQ= 3.16 P = 0.048 RV NO (NO)
114	ALWAYS ARE THOSE WHERE SECTIONALISM IS EXTREME OR MODERATE (61)	1.00	114	TEND LESS TO BE THOSE WHERE SECTIONALISM IS EXTREME OR MODERATE (61)	0.53	8 52 0 47 X SQ= 4.98 P = 0.009 RV NO (NO)
115	TEND TO BE THOSE WHERE INTEREST ARTICULATION BY ASSOCIATIONAL GROUPS IS SIGNIFICANT (20)	0.75	115	TEND TO BE THOSE WHERE INTEREST ARTICULATION BY ASSOCIATIONAL GROUPS IS MODERATE, LIMITED, OR NEGLIGIBLE (91)	0.86	6 14 2 88 X SQ= 14.83 P = 0. RV YES (NO)
116	TEND TO BE THOSE WHERE INTEREST ARTICULATION BY ASSOCIATIONAL GROUPS IS SIGNIFICANT OR MODERATE (32)	0.75	116	TEND TO BE THOSE WHERE INTEREST ARTICULATION BY ASSOCIATIONAL GROUPS IS LIMITED OR NEGLIGIBLE (79)	0.75	6 26 2 76 X SQ= 6.58 P = 0.007 RV YES (NO)
119	TEND TO BE THOSE WHERE INTEREST ARTICULATION BY INSTITUTIONAL GROUPS IS MODERATE OR LIMITED (26)	0.71	119	TEND TO BE THOSE WHERE INTEREST ARTICULATION BY INSTITUTIONAL GROUPS IS VERY SIGNIFICANT OR SIGNIFICANT (74)	0.77	2 71 5 21 X SQ= 5.62 P = 0.013 RV YES (NO)

126	TEND TO BE THOSE WHERE INTEREST ARTICULATION BY ANOMIC GROUPS IS VERY INFREQUENT (16)	0.57	126	TEND TO BE THOSE WHERE INTEREST ARTICULATION BY ANOMIC GROUPS IS FREQUENT, OCCASIONAL, OR INFREQUENT (83)	0.87	3 79 4 12 X SQ= 6.26 P = 0.013 RV YES (NO)
131	TILT TOWARD BEING THOSE WHERE INTEREST AGGREGATION BY POLITICAL PARTIES IS SIGNIFICANT OR MODERATE (30)	0.75	131	TILT TOWARD BEING THOSE WHERE INTEREST AGGREGATION BY POLITICAL PARTIES IS LIMITED OR NEGLIGIBLE (35)	0.58	6 24 2 33 X SQ= 1.87 P = 0.130 RV YES (NO)
137	TILT TOWARD BEING THOSE WHERE INTEREST AGGREGATION BY THE LEGISLATURE IS SIGNIFICANT OR MODERATE (29)	0.62	137	TILT TOWARD BEING THOSE WHERE INTEREST AGGREGATION BY THE LEGISLATURE IS LIMITED OR NEGLIGIBLE (68)	0.73	5 24 3 64 X SQ= 2.81 P = 0.101 RV YES (NO)
138	ALWAYS ARE THOSE WHERE INTEREST AGGREGATION BY THE LEGISLATURE IS SIGNIFICANT, MODERATE, OR LIMITED (48)	1.00	138	TEND TO BE THOSE WHERE INTEREST AGGREGATION BY THE LEGISLATURE IS NEGLIGIBLE (49)	0.55	8 40 0 48 X SQ= 6.68 P = 0.006 RV YES (NO)
139	ALWAYS ARE THOSE WHERE THE PARTY SYSTEM IS QUANTITATIVELY OTHER THAN ONE-PARTY (71)	1.00	139	LEAN LESS TOWARD BEING THOSE WHERE THE PARTY SYSTEM IS QUANTITATIVELY OTHER THAN ONE-PARTY (71)	0.65	0 34 8 62 X SQ= 2.75 P = 0.051 RV NO (NO)
160	ALWAYS ARE THOSE WHERE THE POLITICAL LEADERSHIP IS MODERATELY ELITIST OR NON-ELITIST (67)	1.00	160	LEAN LESS TOWARD BEING THOSE WHERE THE POLITICAL LEADERSHIP IS MODERATELY ELITIST OR NON-ELITIST (67)	0.66	0 30 7 59 X SQ= 2.04 P = 0.094 RV NO (NO)
168	TEND TO BE THOSE WHERE THE HORIZONTAL POWER DISTRIBUTION IS SIGNIFICANT (34)	0.86	168	TEND TO BE THOSE WHERE THE HORIZONTAL POWER DISTRIBUTION IS LIMITED OR NEGLIGIBLE (72)	0.72	6 28 1 71 X SQ= 7.44 P = 0.004 RV YES (NO)
169	ALWAYS ARE THOSE WHERE THE HORIZONTAL POWER DISTRIBUTION IS SIGNIFICANT OR LIMITED (58)	1.00	169	TEND LESS TO BE THOSE WHERE THE HORIZONTAL POWER DISTRIBUTION IS SIGNIFICANT OR LIMITED (58)	0.52	7 51 0 48 X SQ= 4.40 P = 0.015 RV NO (NO)
174	TEND TO BE THOSE WHERE THE LEGISLATURE IS FULLY EFFECTIVE (28)	0.62	174	TEND TO BE THOSE WHERE THE LEGISLATURE IS PARTIALLY EFFECTIVE, LARGELY INEFFECTIVE, OR WHOLLY INEFFECTIVE (72)	0.75	5 23 3 69 X SQ= 3.44 P = 0.037 RV YES (NO)

No.			No.			
175	ALWAYS ARE THOSE WHERE THE LEGISLATURE IS FULLY EFFECTIVE OR PARTIALLY EFFECTIVE (51)	1.00	175	TEND TO BE THOSE WHERE THE LEGISLATURE IS LARGELY INEFFECTIVE OR WHOLLY INEFFECTIVE (49)	0.53	8 43 0 49 X SQ= 6.36 P = 0.006 RV YES (NO)
179	ALWAYS ARE THOSE WHERE THE EXECUTIVE IS STRONG (39)	1.00	179	TEND TO BE THOSE WHERE THE EXECUTIVE IS DOMINANT (52)	0.61	0 52 6 33 X SQ= 6.25 P = 0.005 RV YES (NO)
181	TEND TO BE THOSE WHERE THE BUREAUCRACY IS MODERN, RATHER THAN SEMI-MODERN (21)	0.83	181	TEND TO BE THOSE WHERE THE BUREAUCRACY IS SEMI-MODERN, RATHER THAN MODERN (55)	0.77	5 16 1 54 X SQ= 7.31 P = 0.005 RV YES (NO)
186	ALWAYS ARE THOSE WHERE PARTICIPATION BY THE MILITARY IN POLITICS IS NEUTRAL, RATHER THAN SUPPORTIVE (56)	1.00	186	TEND LESS TO BE THOSE WHERE PARTICIPATION BY THE MILITARY IN POLITICS IS NEUTRAL, RATHER THAN SUPPORTIVE (56)	0.61	0 31 7 49 X SQ= 2.69 P = 0.047 RV NO (NO)
187	TEND TO BE THOSE WHERE THE ROLE OF THE POLICE IS NOT POLITICALLY SIGNIFICANT (35)	0.86	187	TEND TO BE THOSE WHERE THE ROLE OF THE POLICE IS POLITICALLY SIGNIFICANT (66)	0.69	1 64 6 29 X SQ= 6.28 P = 0.007 RV YES (NO)

****

166 POLITIES WHERE THE VERTICAL POWER DISTRIBUTION IS THAT OF EFFECTIVE FEDERALISM OR LIMITED FEDERALISM (15)	166 POLITIES WHERE THE VERTICAL POWER DISTRIBUTION IS THAT OF FORMAL FEDERALISM OR FORMAL AND EFFECTIVE UNITARISM (99)

BOTH SUBJECT AND PREDICATE

....

15 IN LEFT COLUMN

ARGENTINA	AUSTRALIA	AUSTRIA	BRAZIL	CAMEROUN	CANADA	GERMAN FR	LIBYA	MALAYA	NIGERIA
PAKISTAN	SWITZERLAND	UGANDA	US	VENEZUELA					

99 IN RIGHT COLUMN

AFGHANISTAN	ALBANIA	ALGERIA	BELGIUM	BOLIVIA	BULGARIA	BURMA	BURUNDI	CAMBODIA	CEN AFR REP
CEYLON	CHAD	CHILE	CHINA, PR	COLOMBIA	CONGO(BRA)	COSTA RICA	CUBA	CYPRUS	CZECHOS'KIA
DAHOMEY	DENMARK	DOMIN REP	ECUADOR	EL SALVADOR	ETHIOPIA	FINLAND	FRANCE	GABON	GERMANY, E
GHANA	GREECE	GUATEMALA	GUINEA	HAITI	HONDURAS	HUNGARY	ICELAND	INDIA	INDONESIA
IRAN	IRAQ	IRELAND	ISRAEL	ITALY	IVORY COAST	JAMAICA	JAPAN	JORDAN	KOREA, N
KOREA REP	LAOS	LEBANON	LIBERIA	LUXEMBOURG	MALAGASY R	MALI	MAURITANIA	MEXICO	MONGOLIA
MOROCCO	NEPAL	NETHERLANDS	NEW ZEALAND	NICARAGUA	NIGER	NORWAY	PANAMA	PARAGUAY	PERU
PHILIPPINES	POLAND	PORTUGAL	RUMANIA	RWANDA	SA'U ARABIA	SENEGAL	SIERRE LEO	SOMALIA	SO AFRICA
SPAIN	SUDAN	SWEDEN	SYRIA	TANGANYIKA	THAILAND	TOGO	TRINIDAD	TUNISIA	TURKEY
USSR	UAR	UK	UPPER VOLTA	URUGUAY	VIETNAM, N	VIETNAM REP	YEMEN	YUGOSLAVIA	

1 EXCLUDED BECAUSE UNASCERTAINABLE

CONGO(LEO)

21	TEND TO BE THOSE WHOSE TERRITORIAL SIZE IS VERY LARGE OR LARGE (32)	0.60	21 TEND TO BE THOSE WHOSE TERRITORIAL SIZE IS MEDIUM OR SMALL (83)	0.78	9 22 6 77 X SQ= 7.58 P = 0.004 RV YES (NO)
24	TEND TO BE THOSE WHOSE POPULATION IS VERY LARGE, LARGE, OR MEDIUM (61)	0.80	24 TEND TO BE THOSE WHOSE POPULATION IS SMALL (54)	0.52	12 48 3 51 X SQ= 4.00 P = 0.027 RV YES (NO)
30	TEND TO BE THOSE WHOSE AGRICULTURAL POPULATION IS MEDIUM, LOW, OR VERY LOW (57)	0.80	30 TEND TO BE THOSE WHOSE AGRICULTURAL POPULATION IS HIGH (56)	0.54	3 52 12 45 X SQ= 4.60 P = 0.024 RV YES (NO)

33	TEND TO BE THOSE WHOSE GROSS NATIONAL PRODUCT IS VERY HIGH, HIGH, OR MEDIUM (30)	0.60	33	TEND TO BE THOSE WHOSE GROSS NATIONAL PRODUCT IS LOW OR VERY LOW (85)	0.79	9 21 6 78 X SQ= 8.21 P = 0.003 RV YES (NO)
34	TEND TO BE THOSE WHOSE GROSS NATIONAL PRODUCT IS VERY HIGH, HIGH, MEDIUM, OR LOW (62)	0.80	34	TEND TO BE THOSE WHOSE GROSS NATIONAL PRODUCT IS VERY LOW (53)	0.51	12 49 3 50 X SQ= 3.72 P = 0.049 RV YES (NO)
36	LEAN TOWARD BEING THOSE WHOSE PER CAPITA GROSS NATIONAL PRODUCT IS VERY HIGH, HIGH, OR MEDIUM (42)	0.60	36	LEAN TOWARD BEING THOSE WHOSE PER CAPITA GROSS NATIONAL PRODUCT IS LOW OR VERY LOW (73)	0.67	9 33 6 66 X SQ= 2.92 P = 0.082 RV YES (NO)
39	TEND TO BE THOSE WHOSE INTERNATIONAL FINANCIAL STATUS IS VERY HIGH, HIGH, OR MEDIUM (38)	0.67	39	TEND TO BE THOSE WHOSE INTERNATIONAL FINANCIAL STATUS IS LOW OR VERY LOW (76)	0.71	10 28 5 70 X SQ= 6.84 P = 0.007 RV YES (NO)
42	LEAN TOWARD BEING THOSE WHOSE ECONOMIC DEVELOPMENTAL STATUS IS DEVELOPED OR INTERMEDIATE (36)	0.57	42	LEAN TOWARD BEING THOSE WHOSE ECONOMIC DEVELOPMENTAL STATUS IS UNDERDEVELOPED OR VERY UNDERDEVELOPED (76)	0.71	8 28 6 69 X SQ= 3.27 P = 0.063 RV YES (NO)
43	TEND TO BE THOSE WHOSE ECONOMIC DEVELOPMENTAL STATUS IS DEVELOPED, INTERMEDIATE, OR UNDERDEVELOPED (55)	0.80	43	TEND TO BE THOSE WHOSE ECONOMIC DEVELOPMENTAL STATUS IS VERY UNDERDEVELOPED (57)	0.55	12 43 3 53 X SQ= 5.10 P = 0.013 RV YES (NO)
51	TEND MORE TO BE THOSE WHERE FREEDOM OF THE PRESS IS COMPLETE OR INTERMITTENT (60)	0.87	51	TEND LESS TO BE THOSE WHERE FREEDOM OF THE PRESS IS COMPLETE OR INTERMITTENT (60)	0.57	13 46 2 35 X SQ= 3.59 P = 0.041 RV NO (NO)
52	ALWAYS ARE THOSE WHERE FREEDOM OF THE PRESS IS COMPLETE, INTERMITTENT, OR INTERNALLY ABSENT (82)	1.00	52	LEAN LESS TOWARD BEING THOSE WHERE FREEDOM OF THE PRESS IS COMPLETE, INTERMITTENT, OR INTERNALLY ABSENT (82)	0.80	15 66 0 16 X SQ= 2.23 P = 0.069 RV NO (NO)
92	ALWAYS ARE THOSE WHERE THE SYSTEM STYLE IS LIMITED MOBILIZATIONAL, OR NON-MOBILIZATIONAL (93)	1.00	92	LEAN LESS TOWARD BEING THOSE WHERE THE SYSTEM STYLE IS LIMITED MOBILIZATIONAL, OR NON-MOBILIZATIONAL (93)	0.79	0 20 15 77 X SQ= 2.49 P = 0.069 RV NO (NO)

93	LEAN MORE TOWARD BEING THOSE WHERE THE SYSTEM STYLE IS NON-MOBILIZATIONAL (78)	0.93	93	LEAN LESS TOWARD BEING THOSE WHERE THE SYSTEM STYLE IS NON-MOBILIZATIONAL (78)	0.67	1 31 13 64 X SQ= 2.69 P = 0.062 RV NO (NO)
94	TEND TO BE THOSE WHERE THE STATUS OF THE REGIME IS CONSTITUTIONAL (51)	0.92	94	TEND TO BE THOSE WHERE THE STATUS OF THE REGIME IS AUTHORITARIAN OR TOTALITARIAN (41)	0.51	12 39 1 40 X SQ= 6.68 P = 0.005 RV YES (NO)
95	ALWAYS ARE THOSE WHERE THE STATUS OF THE REGIME IS CONSTITUTIONAL OR AUTHORITARIAN (95)	1.00	95	TILT LESS TOWARD BEING THOSE WHERE THE STATUS OF THE REGIME IS CONSTITUTIONAL OR AUTHORITARIAN (95)	0.83	15 79 0 16 X SQ= 1.76 P = 0.121 RV NO (NO)
101	TEND TO BE THOSE WHERE THE REPRESENTATIVE CHARACTER OF THE REGIME IS POLYARCHIC (41)	0.80	101	TEND TO BE THOSE WHERE THE REPRESENTATIVE CHARACTER OF THE REGIME IS LIMITED POLYARCHIC, PSEUDO-POLYARCHIC, OR NON-POLYARCHIC (57)	0.63	8 33 2 55 X SQ= 5.03 P = 0.015 RV YES (NO)
102	TEND TO BE THOSE WHERE THE REPRESENTATIVE CHARACTER OF THE REGIME IS POLYARCHIC OR LIMITED POLYARCHIC (59)	0.93	102	TEND TO BE THOSE WHERE THE REPRESENTATIVE CHARACTER OF THE REGIME IS PSEUDO-POLYARCHIC OR NON-POLYARCHIC (49)	0.52	14 44 1 48 X SQ= 9.00 P = 0.001 RV YES (NO)
105	TEND TO BE THOSE WHERE THE ELECTORAL SYSTEM IS COMPETITIVE (43)	0.85	105	TEND TO BE THOSE WHERE THE ELECTORAL SYSTEM IS PARTIALLY COMPETITIVE OR NON-COMPETITIVE (47)	0.58	11 32 2 45 X SQ= 6.63 P = 0.006 RV YES (NO)
106	ALWAYS ARE THOSE WHERE THE ELECTORAL SYSTEM IS COMPETITIVE OR PARTIALLY COMPETITIVE (52)	1.00	106	TEND LESS TO BE THOSE WHERE THE ELECTORAL SYSTEM IS COMPETITIVE OR PARTIALLY COMPETITIVE (52)	0.58	11 41 0 30 X SQ= 5.62 P = 0.006 RV NO (NO)
107	TEND TO BE THOSE WHERE AUTONOMOUS GROUPS ARE FULLY TOLERATED IN POLITICS (46)	0.73	107	TEND TO BE THOSE WHERE AUTONOMOUS GROUPS ARE PARTIALLY TOLERATED IN POLITICS, ARE TOLERATED ONLY OUTSIDE POLITICS, OR ARE NOT TOLERATED AT ALL (65)	0.63	11 35 4 60 X SQ= 5.67 P = 0.011 RV YES (NO)
108	ALWAYS ARE THOSE WHERE AUTONOMOUS GROUPS ARE FULLY OR PARTIALLY TOLERATED IN POLITICS (65)	1.00	108	TEND LESS TO BE THOSE WHERE AUTONOMOUS GROUPS ARE FULLY OR PARTIALLY TOLERATED IN POLITICS (65)	0.60	13 52 0 35 X SQ= 6.37 P = 0.004 RV NO (NO)

114	TEND MORE TO BE THOSE WHERE SECTIONALISM IS EXTREME OR MODERATE (61)	0.86	114	TEND LESS TO BE THOSE WHERE SECTIONALISM IS EXTREME OR MODERATE (61)	0.52	12 48 2 45 X SQ= 4.44 P = 0.020 RV NO (NO)
116	TEND TO BE THOSE WHERE INTEREST ARTICULATION BY ASSOCIATIONAL GROUPS IS SIGNIFICANT OR MODERATE (32)	0.60	116	TEND TO BE THOSE WHERE INTEREST ARTICULATION BY ASSOCIATIONAL GROUPS IS LIMITED OR NEGLIGIBLE (79)	0.76	9 23 6 72 X SQ= 6.40 P = 0.011 RV YES (NO)
118	TILT MORE TOWARD BEING THOSE WHERE INTEREST ARTICULATION BY INSTITUTIONAL GROUPS IS SIGNIFICANT, MODERATE, OR LIMITED (60)	0.79	118	TILT LESS TOWARD BEING THOSE WHERE INTEREST ARTICULATION BY INSTITUTIONAL GROUPS IS SIGNIFICANT, MODERATE, OR LIMITED (60)	0.56	3 37 11 48 X SQ= 1.61 P = 0.149 RV NO (NO)
128	TEND TO BE THOSE WHERE INTEREST ARTICULATION BY POLITICAL PARTIES IS SIGNIFICANT OR MODERATE (48)	0.85	128	TEND TO BE THOSE WHERE INTEREST ARTICULATION BY POLITICAL PARTIES IS LIMITED OR NEGLIGIBLE (45)	0.54	11 36 2 43 X SQ= 5.34 P = 0.014 RV YES (NO)
131	TEND TO BE THOSE WHERE INTEREST AGGREGATION BY POLITICAL PARTIES IS SIGNIFICANT OR MODERATE (30)	0.77	131	TEND TO BE THOSE WHERE INTEREST AGGREGATION BY POLITICAL PARTIES IS LIMITED OR NEGLIGIBLE (35)	0.62	10 20 3 32 X SQ= 4.74 P = 0.027 RV YES (NO)
135	ALWAYS ARE THOSE WHERE INTEREST AGGREGATION BY THE EXECUTIVE IS SIGNIFICANT, MODERATE, OR LIMITED (77)	1.00	135	TILT LESS TOWARD BEING THOSE WHERE INTEREST AGGREGATION BY THE EXECUTIVE IS SIGNIFICANT, MODERATE, OR LIMITED (77)	0.83	15 62 0 13 X SQ= 1.80 P = 0.115 RV NO (NO)
138	TEND TO BE THOSE WHERE INTEREST AGGREGATION BY THE LEGISLATURE IS SIGNIFICANT, MODERATE, OR LIMITED (48)	0.93	138	TEND TO BE THOSE WHERE INTEREST AGGREGATION BY THE LEGISLATURE IS NEGLIGIBLE (49)	0.57	13 35 1 47 X SQ= 10.12 P = 0.001 RV YES (NO)
139	ALWAYS ARE THOSE WHERE THE PARTY SYSTEM IS QUANTITATIVELY OTHER THAN ONE-PARTY (71)	1.00	139	TEND LESS TO BE THOSE WHERE THE PARTY SYSTEM IS QUANTITATIVELY OTHER THAN ONE-PARTY (71)	0.62	0 34 15 55 X SQ= 6.87 P = 0.002 RV NO (NO)
160	ALWAYS ARE THOSE WHERE THE POLITICAL LEADERSHIP IS MODERATELY ELITIST OR NON-ELITIST (67)	1.00	160	TEND LESS TO BE THOSE WHERE THE POLITICAL LEADERSHIP IS MODERATELY ELITIST OR NON-ELITIST (67)	0.64	0 30 12 54 X SQ= 4.68 P = 0.009 RV NO (NO)

161	TILT TOWARD BEING THOSE WHERE THE POLITICAL LEADERSHIP IS NON-ELITIST (50)	0.75	161	TILT TOWARD BEING THOSE WHERE THE POLITICAL LEADERSHIP IS ELITIST OR MODERATELY ELITIST (47)	0.52	3 44 9 40 X SQ= 2.15 P = 0.121 RV YES (NO)
168	TEND TO BE THOSE WHERE THE HORIZONTAL POWER DISTRIBUTION IS SIGNIFICANT (34)	0.62	168	TEND TO BE THOSE WHERE THE HORIZONTAL POWER DISTRIBUTION IS LIMITED OR NEGLIGIBLE (72)	0.72	8 26 5 67 X SQ= 4.46 P = 0.024 RV YES (NO)
169	TEND TO BE THOSE WHERE THE HORIZONTAL POWER DISTRIBUTION IS SIGNIFICANT OR LIMITED (58)	0.92	169	TEND TO BE THOSE WHERE THE HORIZONTAL POWER DISTRIBUTION IS NEGLIGIBLE (48)	0.51	12 46 1 47 X SQ= 6.81 P = 0.006 RV YES (NO)
175	ALWAYS ARE THOSE WHERE THE LEGISLATURE IS FULLY EFFECTIVE OR PARTIALLY EFFECTIVE (51)	1.00	175	TEND TO BE THOSE WHERE THE LEGISLATURE IS LARGELY INEFFECTIVE OR WHOLLY INEFFECTIVE (49)	0.57	14 37 0 49 X SQ= 13.44 P = 0. RV YES (NO)
178	TEND TO BE THOSE WHERE THE LEGISLATURE IS BICAMERAL (51)	0.80	178	TEND TO BE THOSE WHERE THE LEGISLATURE IS UNICAMERAL (53)	0.56	3 49 12 39 X SQ= 5.18 P = 0.013 RV YES (NO)
179	TEND TO BE THOSE WHERE THE EXECUTIVE IS STRONG (39)	0.92	179	TEND TO BE THOSE WHERE THE EXECUTIVE IS DOMINANT (52)	0.65	1 51 11 28 X SQ= 11.25 P = 0. RV YES (NO)
181	LEAN TOWARD BEING THOSE WHERE THE BUREAUCRACY IS MODERN, RATHER THAN SEMI-MODERN (21)	0.55	181	LEAN TOWARD BEING THOSE WHERE THE BUREAUCRACY IS SEMI-MODERN, RATHER THAN MODERN (55)	0.77	6 15 5 50 X SQ= 3.22 P = 0.062 RV YES (NO)
185	ALWAYS ARE THOSE WHERE PARTICIPATION BY THE MILITARY IN POLITICS IS INTERVENTIVE, RATHER THAN SUPPORTIVE (21)	1.00	185	LEAN TOWARD BEING THOSE WHERE PARTICIPATION BY THE MILITARY IN POLITICS IS SUPPORTIVE, RATHER THAN INTERVENTIVE (31)	0.63	3 18 0 31 X SQ= 2.44 P = 0.060 RV YES (NO)
186	ALWAYS ARE THOSE WHERE PARTICIPATION BY THE MILITARY IN POLITICS IS NEUTRAL, RATHER THAN SUPPORTIVE (56)	1.00	186	TEND LESS TO BE THOSE WHERE PARTICIPATION BY THE MILITARY IN POLITICS IS NEUTRAL, RATHER THAN SUPPORTIVE (56)	0.59	0 31 11 45 X SQ= 5.31 P = 0.007 RV NO (NO)

```
187  LEAN TOWARD BEING THOSE               0.62        187  LEAN TOWARD BEING THOSE           0.69         5   60
     WHERE THE ROLE OF THE POLICE                           WHERE THE ROLE OF THE POLICE                    8   27
     IS NOT POLITICALLY SIGNIFICANT  (35)                   IS POLITICALLY SIGNIFICANT  (66)            X SQ=   3.38
                                                                                                        P =    0.057
                                                                                                        RV YES (NO )
```

```
************************************************************************************************************************

167  POLITIES                                            167  POLITIES
     WHERE THE HORIZONTAL POWER DISTRIBUTION                  WHERE THE HORIZONTAL POWER DISTRIBUTION
     IS SIGNIFICANT, RATHER THAN                              IS NEGLIGIBLE, RATHER THAN
     NEGLIGIBLE  (34)                                         SIGNIFICANT  (48)


                                                                                                        SUBJECT ONLY
........................................................................................................................

      34 IN LEFT COLUMN

AUSTRALIA  AUSTRIA    BELGIUM     BRAZIL     CANADA     CHILE       COSTA RICA CYPRUS     DENMARK     DOMIN REP
FINLAND    GERMAN FR  GREECE      ICELAND    INDIA      IRELAND     ISRAEL     ITALY      JAMAICA     JAPAN
LUXEMBOURG MALAYA     NETHERLANDS NEW ZEALAND NORWAY    PHILIPPINES SO AFRICA  SWEDEN     SWITZERLAND TRINIDAD
TURKEY     UK         US          URUGUAY

      48 IN RIGHT COLUMN

AFGHANISTAN ALBANIA    ALGERIA    BULGARIA   CAMBODIA    CEN AFR REP CHAD        CHINA, PR  CONGO(BRA)  CUBA
CZECHOS'KIA DAHOMEY    ETHIOPIA   GABON      GERMANY, E  GHANA       GUINEA      HAITI      HUNGARY     INDONESIA
IRAN        IVORY COAST JORDAN    KOREA, N   LAOS        LIBERIA     MONGOLIA    NEPAL      NIGER       PAKISTAN
PARAGUAY    POLAND     PORTUGAL   RUMANIA    SA'U ARABIA SENEGAL     SPAIN       SUDAN      TANGANYIKA  THAILAND
TOGO        TUNISIA    USSR       UAR        UPPER VOLTA VIETNAM, N  VIETNAM REP YUGOSLAVIA

      24 EXCLUDED BECAUSE IRRELEVANT

BOLIVIA    BURUNDI    CAMEROUN    CEYLON     COLOMBIA   ECUADOR    EL SALVADOR FRANCE     GUATEMALA  HONDURAS
LEBANON    LIBYA      MALAGASY R  MALI       MAURITANIA MEXICO     MOROCCO     NICARAGUA  NIGERIA    PANAMA
RWANDA     SIERRE LEO SOMALIA     VENEZUELA

       1 EXCLUDED BECAUSE UNASCERTAINED

UGANDA

       8 EXCLUDED BECAUSE UNASCERTAINABLE

ARGENTINA  BURMA      CONGO(LEO)  IRAQ       KOREA REP  PERU       SYRIA      YEMEN

------------------------------------------------------------------------------------------------------------------------

  2  TEND TO BE THOSE                          0.56        2  TEND TO BE THOSE                      0.96        19     2
     LOCATED IN WEST EUROPE, SCANDINAVIA,                     LOCATED ELSEWHERE THAN IN                         15    46
     NORTH AMERICA, OR AUSTRALASIA  (22)                      WEST EUROPE, SCANDINAVIA,                      X SQ=  25.29
                                                              NORTH AMERICA, OR AUSTRALASIA  (93)            P =     0.
                                                                                                             RV YES (YES)

  3  TEND LESS TO BE THOSE                     0.71        3  TEND MORE TO BE THOSE                 0.96        10     2
     LOCATED ELSEWHERE THAN IN                                LOCATED ELSEWHERE THAN IN                         24    46
     WEST EUROPE  (102)                                       WEST EUROPE  (102)                             X SQ=   8.23
                                                                                                             P =    0.003
                                                                                                             RV NO  (YES)
```

5	ALWAYS ARE THOSE LOCATED IN SCANDINAVIA OR WEST EUROPE, RATHER THAN IN EAST EUROPE (18)	1.00	5	TEND TO BE THOSE LOCATED IN EAST EUROPE, RATHER THAN IN SCANDINAVIA OR WEST EUROPE (9)	0.82	15 2 0 9 X SQ= 15.33 P = 0. RV YES (YES)
6	LEAN LESS TOWARD BEING THOSE LOCATED ELSEWHERE THAN IN THE CARIBBEAN, CENTRAL AMERICA, OR SOUTH AMERICA (93)	0.79	6	LEAN MORE TOWARD BEING THOSE LOCATED ELSEWHERE THAN IN THE CARIBBEAN, CENTRAL AMERICA, OR SOUTH AMERICA (93)	0.94	7 3 27 45 X SQ= 2.60 P = 0.084 RV NO (YES)
10	TEND MORE TO BE THOSE LOCATED ELSEWHERE THAN IN NORTH AFRICA, OR CENTRAL AND SOUTH AFRICA (82)	0.97	10	TEND LESS TO BE THOSE LOCATED ELSEWHERE THAN IN NORTH AFRICA, OR CENTRAL AND SOUTH AFRICA (82)	0.60	1 19 33 29 X SQ= 12.57 P = 0. RV NO (YES)
15	TEND TO BE THOSE LOCATED IN THE MIDDLE EAST, RATHER THAN IN NORTH AFRICA OR CENTRAL AND SOUTH AFRICA (11)	0.75	15	TEND TO BE THOSE LOCATED IN NORTH AFRICA OR CENTRAL AND SOUTH AFRICA, RATHER THAN IN THE MIDDLE EAST (33)	0.83	3 4 1 19 X SQ= 3.27 P = 0.042 RV YES (NO)
18	TILT TOWARD BEING THOSE LOCATED IN EAST ASIA, SOUTH ASIA, OR SOUTHEAST ASIA, RATHER THAN IN NORTH AFRICA OR CENTRAL AND SOUTH AFRICA (18)	0.80	18	TILT TOWARD BEING THOSE LOCATED IN NORTH AFRICA OR CENTRAL AND SOUTH AFRICA, RATHER THAN IN EAST ASIA, SOUTH ASIA, OR SOUTHEAST ASIA (33)	0.63	1 19 4 11 X SQ= 1.75 P = 0.141 RV YES (NO)
19	TEND TO BE THOSE LOCATED IN THE CARIBBEAN, CENTRAL AMERICA, OR SOUTH AMERICA, RATHER THAN IN NORTH AFRICA OR CENTRAL AND SOUTH AFRICA (22)	0.87	19	TEND TO BE THOSE LOCATED IN NORTH AFRICA OR CENTRAL AND SOUTH AFRICA, RATHER THAN IN THE CARIBBEAN, CENTRAL AMERICA, OR SOUTH AMERICA (33)	0.86	1 19 7 3 X SQ= 11.27 P = 0. RV YES (YES)
20	TEND TO BE THOSE LOCATED IN THE CARIBBEAN, CENTRAL AMERICA, OR SOUTH AMERICA, RATHER THAN IN EAST ASIA, SOUTH ASIA, OR SOUTHEAST ASIA (22)	0.64	20	TEND TO BE THOSE LOCATED IN EAST ASIA, SOUTH ASIA, OR SOUTHEAST ASIA, RATHER THAN IN THE CARIBBEAN, CENTRAL AMERICA, OR SOUTH AMERICA (18)	0.79	4 11 7 3 X SQ= 2.98 P = 0.049 RV YES (YES)
25	TEND LESS TO BE THOSE WHOSE POPULATION DENSITY IS MEDIUM OR LOW (98)	0.68	25	TEND MORE TO BE THOSE WHOSE POPULATION DENSITY IS MEDIUM OR LOW (98)	0.96	11 2 23 46 X SQ= 9.83 P = 0.001 RV NO (YES)
29	TEND TO BE THOSE WHERE THE DEGREE OF URBANIZATION IS HIGH (56)	0.93	29	TEND TO BE THOSE WHERE THE DEGREE OF URBANIZATION IS LOW (49)	0.69	28 13 2 29 X SQ= 25.29 P = 0. RV YES (YES)

30	TEND TO BE THOSE WHOSE AGRICULTURAL POPULATION IS MEDIUM, LOW, OR VERY LOW (57)	0.94	30	TEND TO BE THOSE WHOSE AGRICULTURAL POPULATION IS HIGH (56)	0.72	2 33 32 13 X SQ= 31.83 P = 0. RV YES (YES)
31	TEND TO BE THOSE WHOSE AGRICULTURAL POPULATION IS LOW OR VERY LOW (24)	0.62	31	TEND TO BE THOSE WHOSE AGRICULTURAL POPULATION IS HIGH OR MEDIUM (90)	0.98	13 46 21 1 X SQ= 32.52 P = 0. RV YES (YES)
33	TEND TO BE THOSE WHOSE GROSS NATIONAL PRODUCT IS VERY HIGH, HIGH, OR MEDIUM (30)	0.50	33	TEND TO BE THOSE WHOSE GROSS NATIONAL PRODUCT IS LOW OR VERY LOW (85)	0.79	17 10 17 38 X SQ= 6.40 P = 0.008 RV YES (YES)
34	TEND TO BE THOSE WHOSE GROSS NATIONAL PRODUCT IS VERY HIGH, HIGH, MEDIUM, OR LOW (62)	0.79	34	TEND TO BE THOSE WHOSE GROSS NATIONAL PRODUCT IS VERY LOW (53)	0.54	27 22 7 26 X SQ= 7.99 P = 0.003 RV YES (YES)
35	TEND TO BE THOSE WHOSE PER CAPITA GROSS NATIONAL PRODUCT IS VERY HIGH OR HIGH (24)	0.56	35	TEND TO BE THOSE WHOSE PER CAPITA GROSS NATIONAL PRODUCT IS MEDIUM, LOW, OR VERY LOW (91)	0.94	19 3 15 45 X SQ= 22.51 P = 0. RV YES (YES)
36	TEND TO BE THOSE WHOSE PER CAPITA GROSS NATIONAL PRODUCT IS VERY HIGH, HIGH, OR MEDIUM (42)	0.82	36	TEND TO BE THOSE WHOSE PER CAPITA GROSS NATIONAL PRODUCT IS LOW OR VERY LOW (73)	0.79	28 10 6 38 X SQ= 27.87 P = 0. RV YES (YES)
37	TEND TO BE THOSE WHOSE PER CAPITA GROSS NATIONAL PRODUCT IS VERY HIGH, HIGH, MEDIUM, OR LOW (64)	0.97	37	TEND TO BE THOSE WHOSE PER CAPITA GROSS NATIONAL PRODUCT IS VERY LOW (51)	0.67	33 16 1 32 X SQ= 31.01 P = 0. RV YES (YES)
39	TEND TO BE THOSE WHOSE INTERNATIONAL FINANCIAL STATUS IS VERY HIGH, HIGH, OR MEDIUM (38)	0.65	39	TEND TO BE THOSE WHOSE INTERNATIONAL FINANCIAL STATUS IS LOW OR VERY LOW (76)	0.77	22 11 12 36 X SQ= 12.28 P = 0. RV YES (YES)
42	TEND TO BE THOSE WHOSE ECONOMIC DEVELOPMENTAL STATUS IS DEVELOPED OR INTERMEDIATE (36)	0.73	42	TEND TO BE THOSE WHOSE ECONOMIC DEVELOPMENTAL STATUS IS UNDERDEVELOPED OR VERY UNDERDEVELOPED (76)	0.83	24 8 9 39 X SQ= 22.80 P = 0. RV YES (YES)

43	TEND TO BE THOSE WHOSE ECONOMIC DEVELOPMENTAL STATUS IS DEVELOPED, INTERMEDIATE, OR UNDERDEVELOPED (55)	0.97	43	TEND TO BE THOSE WHOSE ECONOMIC DEVELOPMENTAL STATUS IS VERY UNDERDEVELOPED (57)	0.72	32 13 1 34 X SQ= 35.08 P = 0. RV YES (YES)
44	TEND TO BE THOSE WHERE THE LITERACY RATE IS NINETY PERCENT OR ABOVE (25)	0.53	44	TEND TO BE THOSE WHERE THE LITERACY RATE IS BELOW NINETY PERCENT (90)	0.87	18 6 16 42 X SQ= 13.83 P = 0. RV YES (YES)
45	TEND TO BE THOSE WHERE THE LITERACY RATE IS FIFTY PERCENT OR ABOVE (55)	0.88	45	TEND TO BE THOSE WHERE THE LITERACY RATE IS BELOW FIFTY PERCENT (54)	0.63	30 16 4 27 X SQ= 18.49 P = 0. RV YES (YES)
46	ALWAYS ARE THOSE WHERE THE LITERACY RATE IS TEN PERCENT OR ABOVE (84)	1.00	46	TEND LESS TO BE THOSE WHERE THE LITERACY RATE IS TEN PERCENT OR ABOVE (84)	0.62	34 28 0 17 X SQ= 14.21 P = 0. RV NO (YES)
50	TEND TO BE THOSE WHERE FREEDOM OF THE PRESS IS COMPLETE (43)	0.88	50	TEND TO BE THOSE WHERE FREEDOM OF THE PRESS IS INTERMITTENT, INTERNALLY ABSENT, OR INTERNALLY AND EXTERNALLY ABSENT (56)	0.92	29 3 4 37 X SQ= 44.24 P = 0. RV YES (YES)
51	ALWAYS ARE THOSE WHERE FREEDOM OF THE PRESS IS COMPLETE OR INTERMITTENT (60)	1.00	51	TEND TO BE THOSE WHERE FREEDOM OF THE PRESS IS INTERNALLY ABSENT, OR INTERNALLY AND EXTERNALLY ABSENT (37)	0.87	33 5 0 35 X SQ= 52.02 P = 0. RV YES (YES)
54	TEND TO BE THOSE WHERE NEWSPAPER CIRCULATION IS ONE HUNDRED OR MORE PER THOUSAND (37)	0.74	54	TEND TO BE THOSE WHERE NEWSPAPER CIRCULATION IS LESS THAN ONE HUNDRED PER THOUSAND (76)	0.80	25 9 9 37 X SQ= 21.14 P = 0. RV YES (YES)
55	ALWAYS ARE THOSE WHERE NEWSPAPER CIRCULATION IS TEN OR MORE PER THOUSAND (78)	1.00	55	TEND LESS TO BE THOSE WHERE NEWSPAPER CIRCULATION IS TEN OR MORE PER THOUSAND (78)	0.52	34 24 0 22 X SQ= 20.09 P = 0. RV NO (YES)
56	TEND MORE TO BE THOSE WHERE THE RELIGION IS PREDOMINANTLY LITERATE (79)	0.97	56	TEND LESS TO BE THOSE WHERE THE RELIGION IS PREDOMINANTLY LITERATE (79)	0.61	33 27 1 17 X SQ= 11.83 P = 0. RV NO (YES)

57	TEND LESS TO BE THOSE WHERE THE RELIGION IS OTHER THAN CATHOLIC (90)	0.68	57	TEND MORE TO BE THOSE WHERE THE RELIGION IS OTHER THAN CATHOLIC (90)	0.90	11 5 23 43 X SQ= 4.78 P = 0.022 RV NO (YES)
58	TEND MORE TO BE THOSE WHERE THE RELIGION IS OTHER THAN MUSLIM (97)	0.97	58	TEND LESS TO BE THOSE WHERE THE RELIGION IS OTHER THAN MUSLIM (97)	0.79	1 10 33 38 X SQ= 4.05 P = 0.022 RV NO (NO)
63	TEND TO BE THOSE WHERE THE RELIGION IS PREDOMINANTLY SOME KIND OF CHRISTIAN (46)	0.76	63	TEND TO BE THOSE WHERE THE RELIGION IS PREDOMINANTLY OR PARTLY OTHER THAN CHRISTIAN (68)	0.77	26 11 8 36 X SQ= 20.30 P = 0. RV YES (YES)
68	TEND TO BE THOSE THAT ARE LINGUISTICALLY HOMOGENEOUS (52)	0.65	68	TEND TO BE THOSE THAT ARE LINGUISTICALLY WEAKLY HETEROGENEOUS OR STRONGLY HETEROGENEOUS (62)	0.70	22 14 12 33 X SQ= 8.38 P = 0.003 RV YES (YES)
69	TEND TO BE THOSE THAT ARE LINGUISTICALLY HOMOGENEOUS OR WEAKLY HETEROGENEOUS (64)	0.74	69	TEND TO BE THOSE THAT ARE LINGUISTICALLY STRONGLY HETEROGENEOUS (50)	0.55	25 21 9 26 X SQ= 5.57 P = 0.013 RV YES (YES)
70	LEAN LESS TOWARD BEING THOSE THAT ARE RELIGIOUSLY, RACIALLY, OR LINGUISTICALLY HETEROGENEOUS (85)	0.67	70	LEAN MORE TOWARD BEING THOSE THAT ARE RELIGIOUSLY, RACIALLY, OR LINGUISTICALLY HETEROGENEOUS (85)	0.86	11 6 22 38 X SQ= 3.18 P = 0.053 RV NO (YES)
72	TEND TO BE THOSE WHOSE DATE OF INDEPENDENCE IS BEFORE 1914 (52)	0.70	72	TEND TO BE THOSE WHOSE DATE OF INDEPENDENCE IS AFTER 1914 (59)	0.65	23 16 10 30 X SQ= 8.03 P = 0.003 RV YES (YES)
73	TEND TO BE THOSE WHOSE DATE OF INDEPENDENCE IS BEFORE 1945 (65)	0.79	73	TEND TO BE THOSE WHOSE DATE OF INDEPENDENCE IS AFTER 1945 (46)	0.52	26 22 7 24 X SQ= 6.48 P = 0.010 RV YES (YES)
74	TEND TO BE THOSE THAT ARE HISTORICALLY WESTERN (26)	0.59	74	TEND TO BE THOSE THAT ARE NOT HISTORICALLY WESTERN (87)	0.88	19 6 13 42 X SQ= 17.52 P = 0. RV YES (YES)

75	TEND TO BE THOSE THAT ARE HISTORICALLY WESTERN OR SIGNIFICANTLY WESTERNIZED (62)	0.97	75	TEND TO BE THOSE THAT ARE NOT HISTORICALLY WESTERN AND ARE NOT SIGNIFICANTLY WESTERNIZED (52)	0.69	32 15 1 33 X SQ= 32.03 P = 0. RV YES (YES)
76	TEND TO BE THOSE THAT ARE HISTORICALLY WESTERN, RATHER THAN HAVING BEEN SIGNIFICANTLY OR PARTIALLY WESTERNIZED THROUGH A COLONIAL RELATIONSHIP (26)	0.61	76	TEND TO BE THOSE THAT HAVE BEEN SIGNIFICANTLY OR PARTIALLY WESTERNIZED THROUGH A COLONIAL RELATIONSHIP, RATHER THAN BEING HISTORICALLY WESTERN (70)	0.82	19 6 12 28 X SQ= 11.27 P = 0. RV YES (YES)
77	TEND TO BE THOSE THAT HAVE BEEN SIGNIFICANTLY WESTERNIZED, RATHER THAN PARTIALLY WESTERNIZED, THROUGH A COLONIAL RELATIONSHIP (28)	0.91	77	TEND TO BE THOSE THAT HAVE BEEN PARTIALLY WESTERNIZED, RATHER THAN SIGNIFICANTLY WESTERNIZED, THROUGH A COLONIAL RELATIONSHIP (41)	0.86	10 4 1 24 X SQ= 16.96 P = 0. RV YES (YES)
80	ALWAYS ARE THOSE WHOSE DATE OF INDEPENDENCE IS AFTER 1914, AND THAT WERE FORMERLY DEPENDENCIES OF BRITAIN, RATHER THAN FRANCE (19)	1.00	80	TEND TO BE THOSE WHOSE DATE OF INDEPENDENCE IS AFTER 1914, AND THAT WERE FORMERLY DEPENDENCIES OF FRANCE, RATHER THAN BRITAIN (24)	0.74	7 6 0 17 X SQ= 9.12 P = 0.001 RV YES (YES)
81	TEND LESS TO BE THOSE WHERE THE TYPE OF POLITICAL MODERNIZATION IS LATER EUROPEAN OR LATER EUROPEAN DERIVED, RATHER THAN EARLY EUROPEAN OR EARLY EUROPEAN DERIVED (40)	0.60	81	ALWAYS ARE THOSE WHERE THE TYPE OF POLITICAL MODERNIZATION IS LATER EUROPEAN OR LATER EUROPEAN DERIVED, RATHER THAN EARLY EUROPEAN OR EARLY EUROPEAN DERIVED (40)	1.00	10 0 15 13 X SQ= 5.15 P = 0.008 RV NO (NO)
82	TEND TO BE THOSE WHERE THE TYPE OF POLITICAL MODERNIZATION IS EARLY OR LATER EUROPEAN OR EUROPEAN DERIVED, RATHER THAN DEVELOPED TUTELARY OR UNDEVELOPED TUTELARY (51)	0.78	82	TEND TO BE THOSE WHERE THE TYPE OF POLITICAL MODERNIZATION IS DEVELOPED TUTELARY OR UNDEVELOPED TUTELARY, RATHER THAN EARLY OR LATER EUROPEAN OR EUROPEAN DERIVED (55)	0.68	25 13 7 28 X SQ= 13.71 P = 0. RV YES (YES)
84	ALWAYS ARE THOSE WHERE THE TYPE OF POLITICAL MODERNIZATION IS DEVELOPED TUTELARY, RATHER THAN UNDEVELOPED TUTELARY (31)	1.00	84	TEND LESS TO BE THOSE WHERE THE TYPE OF POLITICAL MODERNIZATION IS DEVELOPED TUTELARY, RATHER THAN UNDEVELOPED TUTELARY (31)	0.54	7 15 0 13 X SQ= 3.37 P = 0.031 RV NO (NO)
85	TEND TO BE THOSE WHERE THE STAGE OF POLITICAL MODERNIZATION IS ADVANCED, RATHER THAN MID- OR EARLY TRANSITIONAL (60)	0.85	85	TEND TO BE THOSE WHERE THE STAGE OF POLITICAL MODERNIZATION IS MID- OR EARLY TRANSITIONAL, RATHER THAN ADVANCED (54)	0.55	29 21 5 26 X SQ= 12.11 P = 0. RV YES (YES)
87	TEND MORE TO BE THOSE WHOSE IDEOLOGICAL ORIENTATION IS OTHER THAN DEVELOPMENTAL (58)	0.90	87	TEND LESS TO BE THOSE WHOSE IDEOLOGICAL ORIENTATION IS OTHER THAN DEVELOPMENTAL (58)	0.55	3 18 27 22 X SQ= 8.40 P = 0.002 RV NO (YES)

89	TEND TO BE THOSE WHOSE IDEOLOGICAL ORIENTATION IS CONVENTIONAL (33)	0.90	89	ALWAYS ARE THOSE WHOSE IDEOLOGICAL ORIENTATION IS OTHER THAN CONVENTIONAL (62)	1.00	27 0 3 44 X SQ= 58.53 P = 0. RV YES (YES)
92	ALWAYS ARE THOSE WHERE THE SYSTEM STYLE IS LIMITED MOBILIZATIONAL, OR NON-MOBILIZATIONAL (93)	1.00	92	TEND LESS TO BE THOSE WHERE THE SYSTEM STYLE IS LIMITED MOBILIZATIONAL, OR NON-MOBILIZATIONAL (93)	0.60	0 19 34 28 X SQ= 15.78 P = 0. RV NO (YES)
93	TEND TO BE THOSE WHERE THE SYSTEM STYLE IS NON-MOBILIZATIONAL (78)	0.97	93	TEND TO BE THOSE WHERE THE SYSTEM STYLE IS MOBILIZATIONAL, OR LIMITED MOBILIZATIONAL (32)	0.58	1 26 33 19 X SQ= 23.51 P = 0. RV YES (YES)
94	ALWAYS ARE THOSE WHERE THE STATUS OF THE REGIME IS CONSTITUTIONAL (51)	1.00	94	TEND TO BE THOSE WHERE THE STATUS OF THE REGIME IS AUTHORITARIAN OR TOTALITARIAN (41)	0.97	33 1 0 37 X SQ= 63.25 P = 0. RV YES (YES)
98	TEND LESS TO BE THOSE WHERE GOVERNMENTAL STABILITY IS GENERALLY OR MODERATELY PRESENT AND DATES FROM THE POST-WAR PERIOD, OR IS ABSENT (93)	0.53	98	TEND MORE TO BE THOSE WHERE GOVERNMENTAL STABILITY IS GENERALLY OR MODERATELY PRESENT AND DATES FROM THE POST-WAR PERIOD, OR IS ABSENT (93)	0.88	16 6 18 42 X SQ= 10.41 P = 0.001 RV NO (YES)
99	LEAN MORE TOWARD BEING THOSE WHERE GOVERNMENTAL STABILITY IS GENERALLY PRESENT AND DATES FROM AT LEAST THE INTER-WAR PERIOD, OR FROM THE POST-WAR PERIOD (50)	0.83	99	LEAN LESS TOWARD BEING THOSE WHERE GOVERNMENTAL STABILITY IS GENERALLY PRESENT AND DATES FROM AT LEAST THE INTER-WAR PERIOD, OR FROM THE POST-WAR PERIOD (50)	0.62	25 21 5 13 X SQ= 2.68 P = 0.093 RV NO (YES)
100	TILT MORE TOWARD BEING THOSE WHERE GOVERNMENTAL STABILITY IS GENERALLY PRESENT AND DATES FROM AT LEAST THE INTER-WAR PERIOD, OR IS GENERALLY OR MODERATELY PRESENT, AND DATES FROM THE POST-WAR PERIOD (64)	0.97	100	TILT LESS TOWARD BEING THOSE WHERE GOVERNMENTAL STABILITY IS GENERALLY PRESENT AND DATES FROM AT LEAST THE INTER-WAR PERIOD, OR IS GENERALLY OR MODERATELY PRESENT, AND DATES FROM THE POST-WAR PERIOD (64)	0.83	28 29 1 6 X SQ= 1.81 P = 0.116 RV NO (NO)
101	TEND TO BE THOSE WHERE THE REPRESENTATIVE CHARACTER OF THE REGIME IS POLYARCHIC (41)	0.91	101	TEND TO BE THOSE WHERE THE REPRESENTATIVE CHARACTER OF THE REGIME IS LIMITED POLYARCHIC, PSEUDO-POLYARCHIC, OR NON-POLYARCHIC (57)	0.98	31 1 3 46 X SQ= 61.79 P = 0. RV YES (YES)
102	ALWAYS ARE THOSE WHERE THE REPRESENTATIVE CHARACTER OF THE REGIME IS POLYARCHIC OR LIMITED POLYARCHIC (59)	1.00	102	TEND TO BE THOSE WHERE THE REPRESENTATIVE CHARACTER OF THE REGIME IS PSEUDO-POLYARCHIC OR NON-POLYARCHIC (49)	0.94	34 3 0 44 X SQ= 65.96 P = 0. RV YES (YES)

105	ALWAYS ARE THOSE WHERE THE ELECTORAL SYSTEM IS COMPETITIVE (43)	1.00	105	ALWAYS ARE THOSE WHERE THE ELECTORAL SYSTEM IS PARTIALLY COMPETITIVE OR NON-COMPETITIVE (47)	1.00	32 0 0 39 X SQ= 67.02 P = 0. RV YES (YES)
106	ALWAYS ARE THOSE WHERE THE ELECTORAL SYSTEM IS COMPETITIVE OR PARTIALLY COMPETITIVE (52)	1.00	106	TEND TO BE THOSE WHERE THE ELECTORAL SYSTEM IS NON-COMPETITIVE (30)	0.79	31 7 0 27 X SQ= 38.90 P = 0. RV YES (YES)
107	TEND TO BE THOSE WHERE AUTONOMOUS GROUPS ARE FULLY TOLERATED IN POLITICS (46)	0.94	107	TEND TO BE THOSE WHERE AUTONOMOUS GROUPS ARE PARTIALLY TOLERATED IN POLITICS, ARE TOLERATED ONLY OUTSIDE POLITICS, OR ARE NOT TOLERATED AT ALL (65)	0.98	32 1 2 45 X SQ= 64.45 P = 0. RV YES (YES)
108	TEND TO BE THOSE WHERE AUTONOMOUS GROUPS ARE FULLY OR PARTIALLY TOLERATED IN POLITICS (65)	0.97	108	TEND TO BE THOSE WHERE AUTONOMOUS GROUPS ARE TOLERATED ONLY OUTSIDE POLITICS OR ARE NOT TOLERATED AT ALL (35)	0.79	32 9 1 33 X SQ= 39.56 P = 0. RV YES (YES)
111	TEND TO BE THOSE WHERE POLITICAL ENCULTURATION IS HIGH OR MEDIUM (53)	0.81	111	TEND TO BE THOSE WHERE POLITICAL ENCULTURATION IS LOW (42)	0.59	25 14 6 20 X SQ= 8.94 P = 0.002 RV YES (YES)
115	TEND TO BE THOSE WHERE INTEREST ARTICULATION BY ASSOCIATIONAL GROUPS IS SIGNIFICANT (20)	0.53	115	ALWAYS ARE THOSE WHERE INTEREST ARTICULATION BY ASSOCIATIONAL GROUPS IS MODERATE, LIMITED, OR NEGLIGIBLE (91)	1.00	18 0 16 45 X SQ= 27.92 P = 0. RV YES (YES)
116	TEND TO BE THOSE WHERE INTEREST ARTICULATION BY ASSOCIATIONAL GROUPS IS SIGNIFICANT OR MODERATE (32)	0.79	116	ALWAYS ARE THOSE WHERE INTEREST ARTICULATION BY ASSOCIATIONAL GROUPS IS LIMITED OR NEGLIGIBLE (79)	1.00	27 0 7 45 X SQ= 50.82 P = 0. RV YES (YES)
117	TEND TO BE THOSE WHERE INTEREST ARTICULATION BY ASSOCIATIONAL GROUPS IS SIGNIFICANT, MODERATE, OR LIMITED (60)	0.97	117	TEND TO BE THOSE WHERE INTEREST ARTICULATION BY ASSOCIATIONAL GROUPS IS NEGLIGIBLE (51)	0.80	33 9 1 36 X SQ= 43.14 P = 0. RV YES (YES)
118	TEND TO BE THOSE WHERE INTEREST ARTICULATION BY INSTITUTIONAL GROUPS IS SIGNIFICANT, MODERATE, OR LIMITED (60)	0.91	118	TEND TO BE THOSE WHERE INTEREST ARTICULATION BY INSTITUTIONAL GROUPS IS VERY SIGNIFICANT (40)	0.77	3 31 31 9 X SQ= 32.19 P = 0. RV YES (YES)

119	TEND TO BE THOSE WHERE INTEREST ARTICULATION BY INSTITUTIONAL GROUPS IS MODERATE OR LIMITED (26)	0.68	119	TEND TO BE THOSE WHERE INTEREST ARTICULATION BY INSTITUTIONAL GROUPS IS VERY SIGNIFICANT OR SIGNIFICANT (74)	0.97	11 39 23 1 X SQ= 32.68 P = 0. RV YES (YES)
120	TEND LESS TO BE THOSE WHERE INTEREST ARTICULATION BY INSTITUTIONAL GROUPS IS VERY SIGNIFICANT, SIGNIFICANT, OR MODERATE (90)	0.71	120	ALWAYS ARE THOSE WHERE INTEREST ARTICULATION BY INSTITUTIONAL GROUPS IS VERY SIGNIFICANT, SIGNIFICANT, OR MODERATE (90)	1.00	24 40 10 0 X SQ= 11.20 P = 0. RV NO (YES)
121	TEND TO BE THOSE WHERE INTEREST ARTICULATION BY NON-ASSOCIATIONAL GROUPS IS MODERATE, LIMITED, OR NEGLIGIBLE (61)	0.85	121	TEND TO BE THOSE WHERE INTEREST ARTICULATION BY NON-ASSOCIATIONAL GROUPS IS SIGNIFICANT (54)	0.65	5 31 29 17 X SQ= 18.13 P = 0. RV YES (YES)
122	TEND TO BE THOSE WHERE INTEREST ARTICULATION BY NON-ASSOCIATIONAL GROUPS IS LIMITED OR NEGLIGIBLE (32)	0.59	122	TEND TO BE THOSE WHERE INTEREST ARTICULATION BY NON-ASSOCIATIONAL GROUPS IS SIGNIFICANT OR MODERATE (83)	0.85	14 41 20 7 X SQ= 15.69 P = 0. RV YES (YES)
123	TEND LESS TO BE THOSE WHERE INTEREST ARTICULATION BY NON-ASSOCIATIONAL GROUPS IS SIGNIFICANT, MODERATE, OR LIMITED (107)	0.76	123	ALWAYS ARE THOSE WHERE INTEREST ARTICULATION BY NON-ASSOCIATIONAL GROUPS IS SIGNIFICANT, MODERATE, OR LIMITED (107)	1.00	26 48 8 0 X SQ= 9.99 P = 0.001 RV NO (YES)
125	TEND TO BE THOSE WHERE INTEREST ARTICULATION BY ANOMIC GROUPS IS INFREQUENT OR VERY INFREQUENT (35)	0.80	125	TEND TO BE THOSE WHERE INTEREST ARTICULATION BY ANOMIC GROUPS IS FREQUENT OR OCCASIONAL (64)	0.82	6 33 24 7 X SQ= 24.67 P = 0. RV YES (YES)
126	TEND TO BE THOSE WHERE INTEREST ARTICULATION BY ANOMIC GROUPS IS VERY INFREQUENT (16)	0.53	126	ALWAYS ARE THOSE WHERE INTEREST ARTICULATION BY ANOMIC GROUPS IS FREQUENT, OCCASIONAL, OR INFREQUENT (83)	1.00	14 40 16 0 X SQ= 24.71 P = 0. RV YES (YES)
127	TEND LESS TO BE THOSE WHERE INTEREST ARTICULATION BY POLITICAL PARTIES IS MODERATE, LIMITED, OR NEGLIGIBLE (72)	0.63	127	ALWAYS ARE THOSE WHERE INTEREST ARTICULATION BY POLITICAL PARTIES IS MODERATE, LIMITED, OR NEGLIGIBLE (72)	1.00	11 0 19 35 X SQ= 12.95 P = 0. RV NO (YES)
128	TEND TO BE THOSE WHERE INTEREST ARTICULATION BY POLITICAL PARTIES IS SIGNIFICANT OR MODERATE (48)	0.83	128	TEND TO BE THOSE WHERE INTEREST ARTICULATION BY POLITICAL PARTIES IS LIMITED OR NEGLIGIBLE (45)	0.97	25 1 5 34 X SQ= 40.30 P = 0. RV YES (YES)

129	TEND TO BE THOSE WHERE INTEREST ARTICULATION BY POLITICAL PARTIES IS SIGNIFICANT, MODERATE, OR LIMITED (56)	0.90	129	TEND TO BE THOSE WHERE INTEREST ARTICULATION BY POLITICAL PARTIES IS NEGLIGIBLE (37)	0.89	27 4 3 31 X SQ= 36.89 P = 0. RV YES (YES)
134	TEND TO BE THOSE WHERE INTEREST AGGREGATION BY THE EXECUTIVE IS SIGNIFICANT OR MODERATE (57)	0.87	134	TEND TO BE THOSE WHERE INTEREST AGGREGATION BY THE EXECUTIVE IS LIMITED OR NEGLIGIBLE (46)	0.62	27 17 4 28 X SQ= 16.35 P = 0. RV YES (YES)
135	ALWAYS ARE THOSE WHERE INTEREST AGGREGATION BY THE EXECUTIVE IS SIGNIFICANT, MODERATE, OR LIMITED (77)	1.00	135	TEND LESS TO BE THOSE WHERE INTEREST AGGREGATION BY THE EXECUTIVE IS SIGNIFICANT, MODERATE, OR LIMITED (77)	0.71	30 25 0 10 X SQ= 8.05 P = 0.001 RV NO (YES)
137	TEND TO BE THOSE WHERE INTEREST AGGREGATION BY THE LEGISLATURE IS SIGNIFICANT OR MODERATE (29)	0.93	137	ALWAYS ARE THOSE WHERE INTEREST AGGREGATION BY THE LEGISLATURE IS LIMITED OR NEGLIGIBLE (68)	1.00	28 0 2 42 X SQ= 60.28 P = 0. RV YES (YES)
138	TEND TO BE THOSE WHERE INTEREST AGGREGATION BY THE LEGISLATURE IS SIGNIFICANT, MODERATE, OR LIMITED (48)	0.97	138	TEND TO BE THOSE WHERE INTEREST AGGREGATION BY THE LEGISLATURE IS NEGLIGIBLE (49)	0.97	31 1 1 38 X SQ= 59.40 P = 0. RV YES (YES)
139	TEND TO BE THOSE WHERE THE PARTY SYSTEM IS QUANTITATIVELY OTHER THAN ONE-PARTY (71)	0.97	139	TEND TO BE THOSE WHERE THE PARTY SYSTEM IS QUANTITATIVELY ONE-PARTY (34)	0.72	1 31 33 12 X SQ= 34.59 P = 0. RV YES (YES)
141	TEND LESS TO BE THOSE WHERE THE PARTY SYSTEM IS QUANTITATIVELY OTHER THAN TWO-PARTY (87)	0.73	141	ALWAYS ARE THOSE WHERE THE PARTY SYSTEM IS QUANTITATIVELY OTHER THAN TWO-PARTY (87)	1.00	8 0 22 41 X SQ= 9.80 P = 0.001 RV NO (YES)
142	TEND TO BE THOSE WHERE THE PARTY SYSTEM IS QUANTITATIVELY MULTI-PARTY (30)	0.57	142	TEND TO BE THOSE WHERE THE PARTY SYSTEM IS QUANTITATIVELY OTHER THAN MULTI-PARTY (66)	0.97	17 1 13 39 X SQ= 23.57 P = 0. RV YES (YES)
144	ALWAYS ARE THOSE WHERE THE PARTY SYSTEM IS QUANTITATIVELY TWO-PARTY, RATHER THAN ONE-PARTY (11)	1.00	144	ALWAYS ARE THOSE WHERE THE PARTY SYSTEM IS QUANTITATIVELY ONE-PARTY, RATHER THAN TWO-PARTY (34)	1.00	0 32 8 0 X SQ= 33.99 P = 0. RV YES (YES)

No.	Statement	Value	No.	Statement	Value	Statistics
146	ALWAYS ARE THOSE WHERE THE PARTY SYSTEM IS QUALITATIVELY OTHER THAN MASS-BASED TERRITORIAL (92)	1.00	146	TEND LESS TO BE THOSE WHERE THE PARTY SYSTEM IS QUALITATIVELY OTHER THAN MASS-BASED TERRITORIAL (92)	0.80	0 8 32 33 X SQ= 5.16 P = 0.008 RV NO (NO)
147	TEND TO BE THOSE WHERE THE PARTY SYSTEM IS QUALITATIVELY CLASS-ORIENTED OR MULTI-IDEOLOGICAL (23)	0.75	147	ALWAYS ARE THOSE WHERE THE PARTY SYSTEM IS QUALITATIVELY OTHER THAN CLASS-ORIENTED OR MULTI-IDEOLOGICAL (67)	1.00	21 0 7 39 X SQ= 39.19 P = 0. RV YES (YES)
148	ALWAYS ARE THOSE WHERE THE PARTY SYSTEM IS QUALITATIVELY OTHER THAN AFRICAN TRANSITIONAL (96)	1.00	148	TEND LESS TO BE THOSE WHERE THE PARTY SYSTEM IS QUALITATIVELY OTHER THAN AFRICAN TRANSITIONAL (96)	0.82	0 8 34 36 X SQ= 5.05 P = 0.008 RV NO (NO)
158	TEND MORE TO BE THOSE WHERE PERSONALISMO IS NEGLIGIBLE (56)	0.81	158	TEND LESS TO BE THOSE WHERE PERSONALISMO IS NEGLIGIBLE (56)	0.58	6 16 26 22 X SQ= 3.38 P = 0.042 RV NO (YES)
160	TEND TO BE THOSE WHERE THE POLITICAL LEADERSHIP IS MODERATELY ELITIST OR NON-ELITIST (67)	0.97	160	TEND TO BE THOSE WHERE THE POLITICAL LEADERSHIP IS ELITIST (30)	0.60	1 24 32 16 X SQ= 23.59 P = 0. RV YES (YES)
161	TEND TO BE THOSE WHERE THE POLITICAL LEADERSHIP IS NON-ELITIST (50)	0.82	161	TEND TO BE THOSE WHERE THE POLITICAL LEADERSHIP IS ELITIST OR MODERATELY ELITIST (47)	0.67	6 27 27 13 X SQ= 15.82 P = 0. RV YES (YES)
164	TEND TO BE THOSE WHERE THE REGIME'S LEADERSHIP CHARISMA IS NEGLIGIBLE (65)	0.94	164	TEND TO BE THOSE WHERE THE REGIME'S LEADERSHIP CHARISMA IS PRONOUNCED OR MODERATE (34)	0.62	2 26 31 16 X SQ= 22.30 P = 0. RV YES (YES)
166	TEND LESS TO BE THOSE WHERE THE VERTICAL POWER DISTRIBUTION IS THAT OF FORMAL FEDERALISM OR FORMAL AND EFFECTIVE UNITARISM (99)	0.76	166	TEND MORE TO BE THOSE WHERE THE VERTICAL POWER DISTRIBUTION IS THAT OF FORMAL FEDERALISM OR FORMAL AND EFFECTIVE UNITARISM (99)	0.98	8 1 26 47 X SQ= 7.30 P = 0.003 RV NO (YES)
170	LEAN MORE TOWARD BEING THOSE WHERE THE LEGISLATIVE-EXECUTIVE STRUCTURE IS OTHER THAN PRESIDENTIAL (63)	0.79	170	LEAN LESS TOWARD BEING THOSE WHERE THE LEGISLATIVE-EXECUTIVE STRUCTURE IS OTHER THAN PRESIDENTIAL (63)	0.59	7 17 27 24 X SQ= 2.82 P = 0.081 RV NO (YES)

171	TEND LESS TO BE THOSE WHERE THE LEGISLATIVE-EXECUTIVE STRUCTURE IS OTHER THAN PARLIAMENTARY-REPUBLICAN (90)	0.71	171	ALWAYS ARE THOSE WHERE THE LEGISLATIVE-EXECUTIVE STRUCTURE IS OTHER THAN PARLIAMENTARY-REPUBLICAN (90)	1.00	10 0 24 41 X SQ= 11.49 P = 0. RV NO (YES)
172	TEND LESS TO BE THOSE WHERE THE LEGISLATIVE-EXECUTIVE STRUCTURE IS OTHER THAN PARLIAMENTARY-ROYALIST (88)	0.56	172	TEND MORE TO BE THOSE WHERE THE LEGISLATIVE-EXECUTIVE STRUCTURE IS OTHER THAN PARLIAMENTARY-ROYALIST (88)	0.98	15 1 19 42 X SQ= 17.69 P = 0. RV NO (YES)
175	ALWAYS ARE THOSE WHERE THE LEGISLATURE IS FULLY EFFECTIVE OR PARTIALLY EFFECTIVE (51)	1.00	175	ALWAYS ARE THOSE WHERE THE LEGISLATURE IS LARGELY INEFFECTIVE OR WHOLLY INEFFECTIVE (49)	1.00	33 0 0 42 X SQ= 71.00 P = 0. RV YES (YES)
178	TEND TO BE THOSE WHERE THE LEGISLATURE IS BICAMERAL (51)	0.82	178	TEND TO BE THOSE WHERE THE LEGISLATURE IS UNICAMERAL (53)	0.77	6 33 27 10 X SQ= 23.34 P = 0. RV YES (YES)
179	TEND TO BE THOSE WHERE THE EXECUTIVE IS STRONG (39)	0.93	179	ALWAYS ARE THOSE WHERE THE EXECUTIVE IS DOMINANT (52)	1.00	2 42 28 0 X SQ= 60.28 P = 0. RV YES (YES)
181	TEND TO BE THOSE WHERE THE BUREAUCRACY IS MODERN, RATHER THAN SEMI-MODERN (21)	0.59	181	ALWAYS ARE THOSE WHERE THE BUREAUCRACY IS SEMI-MODERN, RATHER THAN MODERN (55)	1.00	20 0 14 23 X SQ= 18.34 P = 0. RV YES (YES)
182	ALWAYS ARE THOSE WHERE THE BUREAUCRACY IS SEMI-MODERN, RATHER THAN POST-COLONIAL TRANSITIONAL (55)	1.00	182	TEND LESS TO BE THOSE WHERE THE BUREAUCRACY IS SEMI-MODERN, RATHER THAN POST-COLONIAL TRANSITIONAL (55)	0.64	14 23 0 13 X SQ= 5.08 P = 0.010 RV NO (NO)
185	TILT TOWARD BEING THOSE WHERE PARTICIPATION BY THE MILITARY IN POLITICS IS INTERVENTIVE, RATHER THAN SUPPORTIVE (21)	0.67	185	TILT TOWARD BEING THOSE WHERE PARTICIPATION BY THE MILITARY IN POLITICS IS SUPPORTIVE, RATHER THAN INTERVENTIVE (31)	0.80	2 7 1 28 X SQ= 1.25 P = 0.134 RV YES (NO)
186	TEND TO BB THOSE WHERE PARTICIPATION BY THE MILITARY IN POLITICS IS NEUTRAL, RATHER THAN SUPPORTIVE (56)	0.97	186	TEND TO BE THOSE WHERE PARTICIPATION BY THE MILITARY IN POLITICS IS SUPPORTIVE, RATHER THAN NEUTRAL (31)	0.74	1 28 31 10 X SQ= 32.79 P = 0. RV YES (YES)

187	TEND TO BE THOSE WHERE THE ROLE OF THE POLICE IS NOT POLITICALLY SIGNIFICANT (35)	0.93	187	TEND TO BE THOSE WHERE THE ROLE OF THE POLICE IS POLITICALLY SIGNIFICANT (66)	0.98	2 44 27 1 X SQ= 58.12 P = 0. RV YES (YES)
188	TEND LESS TO BE THOSE WHERE THE CHARACTER OF THE LEGAL SYSTEM IS OTHER THAN CIVIL LAW (81)	0.59	188	TEND MORE TO BE THOSE WHERE THE CHARACTER OF THE LEGAL SYSTEM IS OTHER THAN CIVIL LAW (81)	0.89	14 5 20 41 X SQ= 8.31 P = 0.003 RV NO (YES)
189	TEND LESS TO BE THOSE WHERE THE CHARACTER OF THE LEGAL SYSTEM IS OTHER THAN COMMON LAW (108)	0.79	189	ALWAYS ARE THOSE WHERE THE CHARACTER OF THE LEGAL SYSTEM IS OTHER THAN COMMON LAW (108)	1.00	7 0 27 48 X SQ= 8.33 P = 0.001 RV NO (YES)
192	TEND MORE TO BE THOSE WHERE THE CHARACTER OF THE LEGAL SYSTEM IS OTHER THAN MUSLIM OR PARTLY MUSLIM (86)	0.97	192	TEND LESS TO BE THOSE WHERE THE CHARACTER OF THE LEGAL SYSTEM IS OTHER THAN MUSLIM OR PARTLY MUSLIM (86)	0.63	1 17 33 29 X SQ= 11.09 P = 0. RV NO (YES)
193	TEND MORE TO BE THOSE WHERE THE CHARACTER OF THE LEGAL SYSTEM IS OTHER THAN PARTLY INDIGENOUS (86)	0.97	193	TEND LESS TO BE THOSE WHERE THE CHARACTER OF THE LEGAL SYSTEM IS OTHER THAN PARTLY INDIGENOUS (86)	0.62	1 18 33 30 X SQ= 11.48 P = 0. RV NO (YES)
194	ALWAYS ARE THOSE THAT ARE NON-COMMUNIST (101)	1.00	194	TEND LESS TO BE THOSE THAT ARE NON-COMMUNIST (101)	0.72	0 13 34 34 X SQ= 9.24 P = 0. RV NO (YES)

168 POLITIES
WHERE THE HORIZONTAL POWER DISTRIBUTION
IS SIGNIFICANT (34)

168 POLITIES
WHERE THE HORIZONTAL POWER DISTRIBUTION
IS LIMITED OR NEGLIGIBLE (72)

PREDICATE ONLY

34 IN LEFT COLUMN

AUSTRALIA	AUSTRIA	BELGIUM	BRAZIL	CANADA	CHILE	COSTA RICA	CYPRUS	DENMARK	DOMIN REP
FINLAND	GERMAN FR	GREECE	ICELAND	INDIA	IRELAND	ISRAEL	ITALY	JAMAICA	JAPAN
LUXEMBOURG	MALAYA	NETHERLANDS	NEW ZEALAND	NORWAY	PHILIPPINES	SO AFRICA	SWEDEN	SWITZERLAND	TRINIDAD
TURKEY	UK	US	URUGUAY						

72 IN RIGHT COLUMN

AFGHANISTAN	ALBANIA	ALGERIA	BOLIVIA	BULGARIA	BURUNDI	CAMBODIA	CAMEROUN	CEN AFR REP	CEYLON
CHAD	CHINA, PR	COLOMBIA	CONGO(BRA)	CUBA	CZECHOS'KIA	DAHOMEY	ECUADOR	EL SALVADOR	ETHIOPIA
FRANCE	GABON	GERMANY, E	GHANA	GUATEMALA	GUINEA	HAITI	HONDURAS	HUNGARY	INDONESIA
IRAN	IVORY COAST	JORDAN	KOREA, N	LAOS	LEBANON	LIBERIA	LIBYA	MALAGASY R	MALI
MAURITANIA	MEXICO	MONGOLIA	MOROCCO	NEPAL	NICARAGUA	NIGER	NIGERIA	PAKISTAN	PANAMA
PARAGUAY	POLAND	PORTUGAL	RUMANIA	RWANDA	SA'U ARABIA	SENEGAL	SIERRE LEO	SOMALIA	SPAIN
SUDAN	TANGANYIKA	THAILAND	TOGO	TUNISIA	USSR	UAR	UPPER VOLTA	VENEZUELA	VIETNAM, N
VIETNAM REP	YUGOSLAVIA								

1 EXCLUDED BECAUSE UNASCERTAINED

UGANDA

8 EXCLUDED BECAUSE UNASCERTAINABLE

ARGENTINA	BURMA	CONGO(LEO)	IRAQ	KOREA REP	PERU	SYRIA	YEMEN

169 POLITIES
WHERE THE HORIZONTAL POWER DISTRIBUTION
IS SIGNIFICANT OR LIMITED (58)

169 POLITIES
WHERE THE HORIZONTAL POWER DISTRIBUTION
IS NEGLIGIBLE (48)

PREDICATE ONLY

58 IN LEFT COLUMN

AUSTRALIA	AUSTRIA	BELGIUM	BOLIVIA	BRAZIL	BURUNDI	CAMEROUN	CANADA	CEYLON	CHILE
COLOMBIA	COSTA RICA	CYPRUS	DENMARK	DOMIN REP	ECUADOR	EL SALVADOR	FINLAND	FRANCE	GERMAN FR
GREECE	GUATEMALA	HONDURAS	ICELAND	INDIA	IRELAND	ISRAEL	ITALY	JAMAICA	JAPAN
LEBANON	LIBYA	LUXEMBOURG	MALAGASY R	MALAYA	MALI	MAURITANIA	MEXICO	MOROCCO	NETHERLANDS
NEW ZEALAND	NICARAGUA	NIGERIA	NORWAY	PANAMA	PHILIPPINES	RWANDA	SIERRE LEO	SOMALIA	SO AFRICA
SWEDEN	SWITZERLAND	TRINIDAD	TURKEY	UK	US	URUGUAY	VENEZUELA		

48 IN RIGHT COLUMN

AFGHANISTAN	ALBANIA	ALGERIA	BULGARIA	CAMBODIA	CEN AFR REP	CHAD	CHINA, PR	CONGO(BRA)	CUBA
CZECHOS'KIA	DAHOMEY	ETHIOPIA	GABON	GERMANY, E	GHANA	GUINEA	HAITI	HUNGARY	INDONESIA
IRAN	IVORY COAST	JORDAN	KOREA, N	LAOS	LIBERIA	MONGOLIA	NEPAL	NIGER	PAKISTAN
PARAGUAY	POLAND	PORTUGAL	RUMANIA	SA'U ARABIA	SENEGAL	SPAIN	SUDAN	TANGANYIKA	THAILAND
TOGO	TUNISIA	USSR	UAR	UPPER VOLTA	VIETNAM, N	VIETNAM REP	YUGOSLAVIA		

1 EXCLUDED BECAUSE UNASCERTAINED

UGANDA

8 EXCLUDED BECAUSE UNASCERTAINABLE

ARGENTINA	BURMA	CONGO(LEO)	IRAQ	KOREA REP	PERU	SYRIA	YEMEN

170 POLITIES
WHERE THE LEGISLATIVE-EXECUTIVE STRUCTURE
IS PRESIDENTIAL (39)

170 POLITIES
WHERE THE LEGISLATIVE-EXECUTIVE STRUCTURE
IS OTHER THAN PRESIDENTIAL (63)

BOTH SUBJECT AND PREDICATE

39 IN LEFT COLUMN

ARGENTINA	BOLIVIA	BRAZIL	CAMEROUN	CEN AFR REP	CHILE	COLOMBIA	CONGO(BRA)	COSTA RICA	CYPRUS
DAHOMEY	DOMIN REP	ECUADOR	EL SALVADOR	GABON	GHANA	GUATEMALA	GUINEA	HAITI	HONDURAS
INDONESIA	IVORY COAST	LIBERIA	MALAGASY R	MAURITANIA	MEXICO	NICARAGUA	NIGER	PAKISTAN	PANAMA
PARAGUAY	PHILIPPINES	RWANDA	SENEGAL	TUNISIA	US	UPPER VOLTA	VENEZUELA	VIETNAM REP	

63 IN RIGHT COLUMN

AFGHANISTAN	ALBANIA	AUSTRALIA	AUSTRIA	BELGIUM	BULGARIA	BURUNDI	CAMBODIA	CANADA	CEYLON
CHAD	CHINA, PR	CONGO(LEO)	CZECHOS'KIA	DENMARK	ETHIOPIA	FINLAND	FRANCE	GERMANY, E	GERMAN FR
GREECE	HUNGARY	ICELAND	INDIA	IRAN	IRELAND	ISRAEL	ITALY	JAMAICA	JAPAN
JORDAN	KOREA, N	LAOS	LEBANON	LIBYA	LUXEMBOURG	MALAYA	MALI	MONGOLIA	MOROCCO
NEPAL	NETHERLANDS	NEW ZEALAND	NIGERIA	NORWAY	POLAND	PORTUGAL	RUMANIA	SIERRE LEO	SOMALIA
SO AFRICA	SPAIN	SWEDEN	SWITZERLAND	THAILAND	TRINIDAD	TURKEY	UGANDA	USSR	UK
URUGUAY	VIETNAM, N	YUGOSLAVIA							

1 EXCLUDED BECAUSE AMBIGUOUS

TANGANYIKA

1 EXCLUDED BECAUSE IRRELEVANT

SA'U ARABIA

11 EXCLUDED BECAUSE UNASCERTAINABLE

ALGERIA	BURMA	CUBA	IRAQ	KOREA REP	PERU	SUDAN	SYRIA	TOGO	UAR
YEMEN									

---

2 TEND MORE TO BE THOSE LOCATED ELSEWHERE THAN IN WEST EUROPE, SCANDINAVIA, NORTH AMERICA, OR AUSTRALASIA (93) — 0.97

2 TEND LESS TO BE THOSE LOCATED ELSEWHERE THAN IN WEST EUROPE, SCANDINAVIA, NORTH AMERICA, OR AUSTRALASIA (93) — 0.67

1	21
38	42
X SQ=	11.72
P =	0.
RV NO	(NO)

3 ALWAYS ARE THOSE LOCATED ELSEWHERE THAN IN WEST EUROPE (102) — 1.00

3 TEND LESS TO BE THOSE LOCATED ELSEWHERE THAN IN WEST EUROPE (102) — 0.79

0	13
39	50
X SQ=	7.46
P =	0.001
RV NO	(NO)

6	TEND LESS TO BE THOSE LOCATED ELSEWHERE THAN IN THE CARIBBEAN, CENTRAL AMERICA, OR SOUTH AMERICA (93)	0.56	6	TEND MORE TO BE THOSE LOCATED ELSEWHERE THAN IN THE CARIBBEAN, CENTRAL AMERICA, OR SOUTH AMERICA (93)	0.95	17 3 22 60 X SQ= 20.64 P = 0. RV NO (YES)
10	TEND LESS TO BE THOSE LOCATED ELSEWHERE THAN IN NORTH AFRICA, OR CENTRAL AND SOUTH AFRICA (82)	0.59	10	TEND MORE TO BE THOSE LOCATED ELSEWHERE THAN IN NORTH AFRICA, OR CENTRAL AND SOUTH AFRICA (82)	0.81	16 12 23 51 X SQ= 4.79 P = 0.022 RV NO (YES)
15	LEAN MORE TOWARD BEING THOSE LOCATED IN NORTH AFRICA OR CENTRAL AND SOUTH AFRICA, RATHER THAN IN THE MIDDLE EAST (33)	0.94	15	LEAN LESS TOWARD BEING THOSE LOCATED IN NORTH AFRICA OR CENTRAL AND SOUTH AFRICA, RATHER THAN IN THE MIDDLE EAST (33)	0.67	1 6 16 12 X SQ= 2.58 P = 0.088 RV NO (YES)
17	TEND TO BE THOSE LOCATED IN THE CARIBBEAN, CENTRAL AMERICA, OR SOUTH AMERICA, RATHER THAN IN THE MIDDLE EAST (22)	0.94	17	TEND TO BE THOSE LOCATED IN THE MIDDLE EAST, RATHER THAN IN THE CARIBBEAN, CENTRAL AMERICA, OR SOUTH AMERICA (11)	0.67	1 6 17 3 X SQ= 8.70 P = 0.002 RV YES (YES)
18	LEAN TOWARD BEING THOSE LOCATED IN NORTH AFRICA OR CENTRAL AND SOUTH AFRICA, RATHER THAN IN EAST ASIA, SOUTH ASIA, OR SOUTHEAST ASIA (33)	0.80	18	LEAN TOWARD BEING THOSE LOCATED IN EAST ASIA, SOUTH ASIA, OR SOUTHEAST ASIA, RATHER THAN IN NORTH AFRICA OR CENTRAL AND SOUTH AFRICA (18)	0.50	16 12 4 12 X SQ= 3.05 P = 0.060 RV YES (YES)
19	LEAN TOWARD BEING THOSE LOCATED IN THE CARIBBEAN, CENTRAL AMERICA, OR SOUTH AMERICA, RATHER THAN IN NORTH AFRICA OR CENTRAL AND SOUTH AFRICA (22)	0.52	19	LEAN TOWARD BEING THOSE LOCATED IN NORTH AFRICA OR CENTRAL AND SOUTH AFRICA, RATHER THAN IN THE CARIBBEAN, CENTRAL AMERICA, OR SOUTH AMERICA (33)	0.80	16 12 17 3 X SQ= 3.02 P = 0.059 RV YES (NO)
20	TEND TO BE THOSE LOCATED IN THE CARIBBEAN, CENTRAL AMERICA, OR SOUTH AMERICA, RATHER THAN IN EAST ASIA, SOUTH ASIA, OR SOUTHEAST ASIA (22)	0.81	20	TEND TO BE THOSE LOCATED IN EAST ASIA, SOUTH ASIA, OR SOUTHEAST ASIA, RATHER THAN IN THE CARIBBEAN, CENTRAL AMERICA, OR SOUTH AMERICA (18)	0.80	4 12 17 3 X SQ= 10.81 P = 0.001 RV YES (YES)
24	TEND TO BE THOSE WHOSE POPULATION IS SMALL (54)	0.69	24	TEND TO BE THOSE WHOSE POPULATION IS VERY LARGE, LARGE, OR MEDIUM (61)	0.62	12 39 27 24 X SQ= 8.14 P = 0.003 RV YES (YES)
26	TEND TO BE THOSE WHOSE POPULATION DENSITY IS LOW (67)	0.77	26	TEND TO BE THOSE WHOSE POPULATION DENSITY IS VERY HIGH, HIGH, OR MEDIUM (48)	0.59	9 37 30 26 X SQ= 10.97 P = 0. RV YES (YES)

28	TEND TO BE THOSE WHOSE POPULATION GROWTH RATE IS HIGH (62)	0.74	28	TEND TO BE THOSE WHOSE POPULATION GROWTH RATE IS LOW (48)	0.57	28 26 10 34 X SQ= 7.48 P = 0.004 RV YES (YES)
29	TEND TO BE THOSE WHERE THE DEGREE OF URBANIZATION IS LOW (49)	0.66	29	TEND TO BE THOSE WHERE THE DEGREE OF URBANIZATION IS HIGH (56)	0.67	12 38 23 19 X SQ= 7.91 P = 0.005 RV YES (YES)
30	LEAN TOWARD BEING THOSE WHOSE AGRICULTURAL POPULATION IS HIGH (56)	0.59	30	LEAN TOWARD BEING THOSE WHOSE AGRICULTURAL POPULATION IS MEDIUM, LOW, OR VERY LOW (57)	0.61	23 24 16 38 X SQ= 3.18 P = 0.065 RV YES (NO)
34	TEND TO BE THOSE WHOSE GROSS NATIONAL PRODUCT IS VERY LOW (53)	0.72	34	TEND TO BE THOSE WHOSE GROSS NATIONAL PRODUCT IS VERY HIGH, HIGH, MEDIUM, OR LOW (62)	0.68	11 43 28 20 X SQ= 13.94 P = 0. RV YES (YES)
36	TEND TO BE THOSE WHOSE PER CAPITA GROSS NATIONAL PRODUCT IS LOW OR VERY LOW (73)	0.87	36	TEND TO BE THOSE WHOSE PER CAPITA GROSS NATIONAL PRODUCT IS VERY HIGH, HIGH, OR MEDIUM (42)	0.57	5 36 34 27 X SQ= 17.88 P = 0. RV YES (YES)
40	TEND TO BE THOSE WHOSE INTERNATIONAL FINANCIAL STATUS IS VERY LOW (39)	0.56	40	TEND TO BE THOSE WHOSE INTERNATIONAL FINANCIAL STATUS IS VERY HIGH, HIGH, MEDIUM, OR LOW (71)	0.76	17 45 22 14 X SQ= 9.43 P = 0.001 RV YES (YES)
43	TEND TO BE THOSE WHOSE ECONOMIC DEVELOPMENTAL STATUS IS VERY UNDERDEVELOPED (57)	0.71	43	TEND TO BE THOSE WHOSE ECONOMIC DEVELOPMENTAL STATUS IS DEVELOPED, INTERMEDIATE, OR UNDERDEVELOPED (55)	0.65	11 40 27 22 X SQ= 10.55 P = 0.001 RV YES (YES)
45	TEND TO BE THOSE WHERE THE LITERACY RATE IS BELOW FIFTY PERCENT (54)	0.61	45	TEND TO BE THOSE WHERE THE LITERACY RATE IS FIFTY PERCENT OR ABOVE (55)	0.64	15 37 23 21 X SQ= 4.53 P = 0.023 RV YES (YES)
47	TEND TO BE THOSE WHERE THE LITERACY RATE IS BETWEEN FIFTY AND NINETY PERCENT, RATHER THAN NINETY PERCENT OR ABOVE (30)	0.93	47	TEND TO BE THOSE WHERE THE LITERACY RATE IS NINETY PERCENT OR ABOVE, RATHER THAN BETWEEN FIFTY AND NINETY PERCENT (25)	0.65	1 24 14 13 X SQ= 12.24 P = 0. RV YES (YES)

52	LEAN MORE TOWARD BEING THOSE WHERE FREEDOM OF THE PRESS IS COMPLETE, INTERMITTENT, OR INTERNALLY ABSENT (82)	0.96	52	LEAN LESS TOWARD BEING THOSE WHERE FREEDOM OF THE PRESS IS COMPLETE, INTERMITTENT, OR INTERNALLY ABSENT (82)	0.80	26 49 1 12 X SQ= 2.63 P = 0.058 RV NO (NO)
53	TEND MORE TO BE THOSE WHERE NEWSPAPER CIRCULATION IS LESS THAN THREE HUNDRED PER THOUSAND (101)	0.97	53	TEND LESS TO BE THOSE WHERE NEWSPAPER CIRCULATION IS LESS THAN THREE HUNDRED PER THOUSAND (101)	0.79	1 13 38 50 X SQ= 5.20 P = 0.009 RV NO (NO)
54	TEND MORE TO BE THOSE WHERE NEWSPAPER CIRCULATION IS LESS THAN ONE HUNDRED PER THOUSAND (76)	0.85	54	TEND LESS TO BE THOSE WHERE NEWSPAPER CIRCULATION IS LESS THAN ONE HUNDRED PER THOUSAND (76)	0.51	6 30 33 31 X SQ= 10.37 P = 0.001 RV NO (YES)
56	TEND LESS TO BE THOSE WHERE THE RELIGION IS PREDOMINANTLY LITERATE (79)	0.55	56	TEND MORE TO BE THOSE WHERE THE RELIGION IS PREDOMINANTLY LITERATE (79)	0.83	21 50 17 10 X SQ= 7.83 P = 0.005 RV NO (YES)
57	TEND LESS TO BE THOSE WHERE THE RELIGION IS OTHER THAN CATHOLIC (90)	0.64	57	TEND MORE TO BE THOSE WHERE THE RELIGION IS OTHER THAN CATHOLIC (90)	0.84	14 10 25 53 X SQ= 4.31 P = 0.030 RV NO (YES)
62	ALWAYS ARE THOSE WHERE THE RELIGION IS CATHOLIC, RATHER THAN PROTESTANT (25)	1.00	62	TEND LESS TO BE THOSE WHERE THE RELIGION IS CATHOLIC, RATHER THAN PROTESTANT (25)	0.67	0 5 14 10 X SQ= 3.54 P = 0.042 RV NO (YES)
67	TEND LESS TO BE THOSE THAT ARE RACIALLY HOMOGENEOUS (82)	0.56	67	TEND MORE TO BE THOSE THAT ARE RACIALLY HOMOGENEOUS (82)	0.85	20 53 16 9 X SQ= 9.22 P = 0.002 RV NO (YES)
70	TEND MORE TO BE THOSE THAT ARE RELIGIOUSLY, RACIALLY, OR LINGUISTICALLY HETEROGENEOUS (85)	0.91	70	TEND LESS TO BE THOSE THAT ARE RELIGIOUSLY, RACIALLY, OR LINGUISTICALLY HETEROGENEOUS (85)	0.72	3 17 32 43 X SQ= 4.07 P = 0.035 RV NO (NO)
73	LEAN TOWARD BEING THOSE WHOSE DATE OF INDEPENDENCE IS AFTER 1945 (46)	0.51	73	LEAN TOWARD BEING THOSE WHOSE DATE OF INDEPENDENCE IS BEFORE 1945 (65)	0.67	19 40 20 20 X SQ= 2.46 P = 0.095 RV YES (YES)

74	TEND MORE TO BE THOSE THAT ARE NOT HISTORICALLY WESTERN (87)	0.97	74	TEND LESS TO BE THOSE THAT ARE NOT HISTORICALLY WESTERN (87)	0.59	1 25 38 36 X SQ= 16.31 P = 0. RV NO (YES)
76	TEND TO BE THOSE THAT HAVE BEEN SIGNIFICANTLY OR PARTIALLY WESTERNIZED THROUGH A COLONIAL RELATIONSHIP, RATHER THAN BEING HISTORICALLY WESTERN (70)	0.97	76	TEND TO BE THOSE THAT ARE HISTORICALLY WESTERN, RATHER THAN HAVING BEEN SIGNIFICANTLY OR PARTIALLY WESTERNIZED THROUGH A COLONIAL RELATIONSHIP (26)	0.52	1 25 37 23 X SQ= 22.30 P = 0. RV YES (YES)
78	ALWAYS ARE THOSE THAT HAVE BEEN SIGNIFICANTLY WESTERNIZED THROUGH A COLONIAL RELATIONSHIP, RATHER THAN WITHOUT SUCH A RELATIONSHIP (28)	1.00	78	TEND TO BE THOSE THAT HAVE BEEN SIGNIFICANTLY WESTERNIZED WITHOUT A COLONIAL RELATIONSHIP, RATHER THAN THROUGH SUCH A RELATIONSHIP (7)	0.54	18 6 0 7 X SQ= 9.63 P = 0.001 RV YES (YES)
79	TEND LESS TO BE THOSE THAT WERE FORMERLY DEPENDENCIES OF BRITAIN OR FRANCE, RATHER THAN SPAIN (49)	0.56	79	TEND MORE TO BE THOSE THAT WERE FORMERLY DEPENDENCIES OF BRITAIN OR FRANCE, RATHER THAN SPAIN (49)	0.96	19 22 15 1 X SQ= 8.87 P = 0.001 RV NO (YES)
80	TEND TO BE THOSE WHOSE DATE OF INDEPENDENCE IS AFTER 1914, AND THAT WERE FORMERLY DEPENDENCIES OF FRANCE, RATHER THAN BRITAIN (24)	0.82	80	TEND TO BE THOSE WHOSE DATE OF INDEPENDENCE IS AFTER 1914, AND THAT WERE FORMERLY DEPENDENCIES OF BRITAIN, RATHER THAN FRANCE (19)	0.61	3 11 14 7 X SQ= 5.19 P = 0.015 RV YES (YES)
81	TEND MORE TO BE THOSE WHERE THE TYPE OF POLITICAL MODERNIZATION IS LATER EUROPEAN OR LATER EUROPEAN DERIVED, RATHER THAN EARLY EUROPEAN OR EARLY EUROPEAN DERIVED (40)	0.94	81	TEND LESS TO BE THOSE WHERE THE TYPE OF POLITICAL MODERNIZATION IS LATER EUROPEAN OR LATER EUROPEAN DERIVED, RATHER THAN EARLY EUROPEAN OR EARLY EUROPEAN DERIVED (40)	0.68	1 10 17 21 X SQ= 3.26 P = 0.038 RV NO (NO)
83	ALWAYS ARE THOSE WHERE THE TYPE OF POLITICAL MODERNIZATION IS OTHER THAN NON-EUROPEAN AUTOCHTHONOUS (106)	1.00	83	TEND LESS TO BE THOSE WHERE THE TYPE OF POLITICAL MODERNIZATION IS OTHER THAN NON-EUROPEAN AUTOCHTHONOUS (106)	0.86	0 9 39 54 X SQ= 4.46 P = 0.012 RV NO (NO)
84	LEAN TOWARD BEING THOSE WHERE THE TYPE OF POLITICAL MODERNIZATION IS UNDEVELOPED TUTELARY, RATHER THAN DEVELOPED TUTELARY (24)	0.67	84	LEAN TOWARD BEING THOSE WHERE THE TYPE OF POLITICAL MODERNIZATION IS DEVELOPED TUTELARY, RATHER THAN UNDEVELOPED TUTELARY (31)	0.65	7 15 14 8 X SQ= 3.28 P = 0.069 RV YES (YES)
85	TEND TO BE THOSE WHERE THE STAGE OF POLITICAL MODERNIZATION IS MID- OR EARLY TRANSITIONAL, RATHER THAN ADVANCED (54)	0.69	85	TEND TO BE THOSE WHERE THE STAGE OF POLITICAL MODERNIZATION IS ADVANCED, RATHER THAN MID- OR EARLY TRANSITIONAL (60)	0.65	12 41 27 22 X SQ= 10.03 P = 0.001 RV YES (YES)

87	TEND TO BE THOSE WHOSE IDEOLOGICAL ORIENTATION IS DEVELOPMENTAL (31)	0.62	87	TEND TO BE THOSE WHOSE IDEOLOGICAL ORIENTATION IS OTHER THAN DEVELOPMENTAL (58)	0.85	18 8 11 44 X SQ= 16.54 P = 0. RV YES (YES)
91	ALWAYS ARE THOSE WHOSE IDEOLOGICAL ORIENTATION IS DEVELOPMENTAL, RATHER THAN TRADITIONAL (31)	1.00	91	TEND LESS TO BE THOSE WHOSE IDEOLOGICAL ORIENTATION IS DEVELOPMENTAL, RATHER THAN TRADITIONAL (31)	0.67	18 8 0 4 X SQ= 4.34 P = 0.018 RV NO (YES)
92	LEAN MORE TOWARD BEING THOSE WHERE THE SYSTEM STYLE IS LIMITED MOBILIZATIONAL, OR NON-MOBILIZATIONAL (93)	0.92	92	LEAN LESS TOWARD BEING THOSE WHERE THE SYSTEM STYLE IS LIMITED MOBILIZATIONAL, OR NON-MOBILIZATIONAL (93)	0.79	3 13 36 49 X SQ= 2.25 P = 0.096 RV NO (NO)
95	ALWAYS ARE THOSE WHERE THE STATUS OF THE REGIME IS CONSTITUTIONAL OR AUTHORITARIAN (95)	1.00	95	TEND LESS TO BE THOSE WHERE THE STATUS OF THE REGIME IS CONSTITUTIONAL OR AUTHORITARIAN (95)	0.76	37 47 0 15 X SQ= 8.75 P = 0.001 RV NO (NO)
96	TEND LESS TO BE THOSE WHERE THE STATUS OF THE REGIME IS CONSTITUTIONAL OR TOTALITARIAN (67)	0.65	96	TEND MORE TO BE THOSE WHERE THE STATUS OF THE REGIME IS CONSTITUTIONAL OR TOTALITARIAN (67)	0.86	17 49 9 8 X SQ= 3.47 P = 0.042 RV NO (YES)
97	ALWAYS ARE THOSE WHERE THE STATUS OF THE REGIME IS AUTHORITARIAN, RATHER THAN TOTALITARIAN (23)	1.00	97	TEND TO BE THOSE WHERE THE STATUS OF THE REGIME IS TOTALITARIAN, RATHER THAN AUTHORITARIAN (16)	0.65	9 8 0 15 X SQ= 8.58 P = 0.001 RV YES (YES)
99	TEND TO BE THOSE WHERE GOVERNMENTAL STABILITY IS MODERATELY PRESENT AND DATES FROM THE POST-WAR PERIOD, OR IS ABSENT (36)	0.71	99	TEND TO BE THOSE WHERE GOVERNMENTAL STABILITY IS GENERALLY PRESENT AND DATES FROM AT LEAST THE INTER-WAR PERIOD, OR FROM THE POST-WAR PERIOD (50)	0.77	7 41 17 12 X SQ= 14.35 P = 0. RV YES (YES)
101	LEAN TOWARD BEING THOSE WHERE THE REPRESENTATIVE CHARACTER OF THE REGIME IS LIMITED POLYARCHIC, PSEUDO-POLYARCHIC, OR NON-POLYARCHIC (57)	0.69	101	LEAN TOWARD BEING THOSE WHERE THE REPRESENTATIVE CHARACTER OF THE REGIME IS POLYARCHIC (41)	0.54	11 30 24 26 X SQ= 3.42 P = 0.052 RV YES (NO)
110	ALWAYS ARE THOSE WHERE POLITICAL ENCULTURATION IS MEDIUM OR LOW (80)	1.00	110	TEND LESS TO BE THOSE WHERE POLITICAL ENCULTURATION IS MEDIUM OR LOW (80)	0.70	0 14 36 33 X SQ= 10.86 P = 0. RV NO (YES)

111	TEND TO BE THOSE WHERE POLITICAL ENCULTURATION IS LOW (42)	0.58	111	TEND TO BE THOSE WHERE POLITICAL ENCULTURATION IS HIGH OR MEDIUM (53)	0.68	15 32 21 15 X SQ= 4.77 P = 0.025 RV YES (YES)
115	LEAN MORE TOWARD BEING THOSE WHERE INTEREST ARTICULATION BY ASSOCIATIONAL GROUPS IS MODERATE, LIMITED, OR NEGLIGIBLE (91)	0.90	115	LEAN LESS TOWARD BEING THOSE WHERE INTEREST ARTICULATION BY ASSOCIATIONAL GROUPS IS MODERATE, LIMITED, OR NEGLIGIBLE (91)	0.73	4 16 35 44 X SQ= 3.00 P = 0.072 RV NO (NO)
116	TEND MORE TO BE THOSE WHERE INTEREST ARTICULATION BY ASSOCIATIONAL GROUPS IS LIMITED OR NEGLIGIBLE (79)	0.82	116	TEND LESS TO BE THOSE WHERE INTEREST ARTICULATION BY ASSOCIATIONAL GROUPS IS LIMITED OR NEGLIGIBLE (79)	0.58	7 25 32 35 X SQ= 5.04 P = 0.016 RV NO (NO)
119	TEND MORE TO BE THOSE WHERE INTEREST ARTICULATION BY INSTITUTIONAL GROUPS IS VERY SIGNIFICANT OR SIGNIFICANT (74)	0.87	119	TEND LESS TO BE THOSE WHERE INTEREST ARTICULATION BY INSTITUTIONAL GROUPS IS VERY SIGNIFICANT OR SIGNIFICANT (74)	0.64	27 37 4 21 X SQ= 4.34 P = 0.026 RV NO (NO)
120	LEAN MORE TOWARD BEING THOSE WHERE INTEREST ARTICULATION BY INSTITUTIONAL GROUPS IS VERY SIGNIFICANT, SIGNIFICANT, OR MODERATE (90)	0.97	120	LEAN LESS TOWARD BEING THOSE WHERE INTEREST ARTICULATION BY INSTITUTIONAL GROUPS IS VERY SIGNIFICANT, SIGNIFICANT, OR MODERATE (90)	0.84	30 49 1 9 X SQ= 1.95 P = 0.094 RV NO (NO)
123	ALWAYS ARE THOSE WHERE INTEREST ARTICULATION BY NON-ASSOCIATIONAL GROUPS IS SIGNIFICANT, MODERATE, OR LIMITED (107)	1.00	123	TEND LESS TO BE THOSE WHERE INTEREST ARTICULATION BY NON-ASSOCIATIONAL GROUPS IS SIGNIFICANT, MODERATE, OR LIMITED (107)	0.87	39 55 0 8 X SQ= 3.76 P = 0.022 RV NO (NO)
125	TEND MORE TO BE THOSE WHERE INTEREST ARTICULATION BY ANOMIC GROUPS IS FREQUENT OR OCCASIONAL (64)	0.82	125	TEND LESS TO BE THOSE WHERE INTEREST ARTICULATION BY ANOMIC GROUPS IS FREQUENT OR OCCASIONAL (64)	0.53	27 29 6 26 X SQ= 6.34 P = 0.007 RV NO (NO)
126	ALWAYS ARE THOSE WHERE INTEREST ARTICULATION BY ANOMIC GROUPS IS FREQUENT, OCCASIONAL, OR INFREQUENT (83)	1.00	126	TEND LESS TO BE THOSE WHERE INTEREST ARTICULATION BY ANOMIC GROUPS IS FREQUENT, OCCASIONAL, OR INFREQUENT (83)	0.71	33 39 0 16 X SQ= 9.86 P = 0. RV NO (NO)
128	TILT TOWARD BEING THOSE WHERE INTEREST ARTICULATION BY POLITICAL PARTIES IS LIMITED OR NEGLIGIBLE (45)	0.59	128	TILT TOWARD BEING THOSE WHERE INTEREST ARTICULATION BY POLITICAL PARTIES IS SIGNIFICANT OR MODERATE (48)	0.60	14 32 20 21 X SQ= 2.34 P = 0.123 RV YES (NO)

133	TEND TO BE THOSE WHERE INTEREST AGGREGATION BY THE EXECUTIVE IS SIGNIFICANT (29)	0.50	133	TEND TO BE THOSE WHERE INTEREST AGGREGATION BY THE EXECUTIVE IS MODERATE, LIMITED, OR NEGLIGIBLE (76)	0.86	19 8 19 51 X SQ= 13.52 P = 0. RV YES (YES)
136	LEAN MORE TOWARD BEING THOSE WHERE INTEREST AGGREGATION BY THE LEGISLATURE IS MODERATE, LIMITED, OR NEGLIGIBLE (85)	0.97	136	LEAN LESS TOWARD BEING THOSE WHERE INTEREST AGGREGATION BY THE LEGISLATURE IS MODERATE, LIMITED, OR NEGLIGIBLE (85)	0.82	1 11 33 49 X SQ= 3.34 P = 0.050 RV NO (NO)
137	TEND MORE TO BE THOSE WHERE INTEREST AGGREGATION BY THE LEGISLATURE IS LIMITED OR NEGLIGIBLE (68)	0.91	137	TEND LESS TO BE THOSE WHERE INTEREST AGGREGATION BY THE LEGISLATURE IS LIMITED OR NEGLIGIBLE (68)	0.57	3 26 31 34 X SQ= 10.55 P = 0. RV NO (NO)
140	TILT LESS TOWARD BEING THOSE WHERE THE PARTY SYSTEM IS QUANTITATIVELY OTHER THAN ONE PARTY DOMINANT (87)	0.78	140	TILT MORE TOWARD BEING THOSE WHERE THE PARTY SYSTEM IS QUANTITATIVELY OTHER THAN ONE PARTY DOMINANT (87)	0.91	8 5 28 51 X SQ= 2.19 P = 0.123 RV NO (YES)
146	TEND LESS TO BE THOSE WHERE THE PARTY SYSTEM IS QUALITATIVELY OTHER THAN MASS-BASED TERRITORIAL (92)	0.86	146	ALWAYS ARE THOSE WHERE THE PARTY SYSTEM IS QUALITATIVELY OTHER THAN MASS-BASED TERRITORIAL (92)	1.00	5 0 32 55 X SQ= 5.45 P = 0.009 RV NO (YES)
147	TEND MORE TO BE THOSE WHERE THE PARTY SYSTEM IS QUALITATIVELY OTHER THAN CLASS-ORIENTED OR MULTI-IDEOLOGICAL (67)	0.91	147	TEND LESS TO BE THOSE WHERE THE PARTY SYSTEM IS QUALITATIVELY OTHER THAN CLASS-ORIENTED OR MULTI-IDEOLOGICAL (67)	0.61	3 19 31 30 X SQ= 7.77 P = 0.002 RV NO (YES)
148	TEND LESS TO BE THOSE WHERE THE PARTY SYSTEM IS QUALITATIVELY OTHER THAN AFRICAN TRANSITIONAL (96)	0.76	148	TEND MORE TO BE THOSE WHERE THE PARTY SYSTEM IS QUALITATIVELY OTHER THAN AFRICAN TRANSITIONAL (96)	0.93	9 4 29 56 X SQ= 4.47 P = 0.029 RV NO (YES)
153	TEND TO BE THOSE WHERE THE PARTY SYSTEM IS MODERATELY STABLE OR UNSTABLE (38)	0.76	153	TEND TO BE THOSE WHERE THE PARTY SYSTEM IS STABLE (42)	0.70	6 35 19 15 X SQ= 12.43 P = 0. RV YES (YES)
158	TEND TO BE THOSE WHERE PERSONALISMO IS PRONOUNCED OR MODERATE (40)	0.74	158	TEND TO BE THOSE WHERE PERSONALISMO IS NEGLIGIBLE (56)	0.79	25 12 9 44 X SQ= 21.62 P = 0. RV YES (YES)

160	TILT MORE TOWARD BEING THOSE WHERE THE POLITICAL LEADERSHIP IS MODERATELY ELITIST OR NON-ELITIST (67)	0.80	160	TILT LESS TOWARD BEING THOSE WHERE THE POLITICAL LEADERSHIP IS MODERATELY ELITIST OR NON-ELITIST (67)	0.62	7 22 28 36 X SQ= 2.49 P = 0.105 RV NO (NO)
168	TEND MORE TO BE THOSE WHERE THE HORIZONTAL POWER DISTRIBUTION IS LIMITED OR NEGLIGIBLE (72)	0.82	168	TEND LESS TO BE THOSE WHERE THE HORIZONTAL POWER DISTRIBUTION IS LIMITED OR NEGLIGIBLE (72)	0.56	7 27 31 34 X SQ= 5.84 P = 0.010 RV NO (NO)
174	TEND MORE TO BE THOSE WHERE THE LEGISLATURE IS PARTIALLY EFFECTIVE, LARGELY INEFFECTIVE, OR WHOLLY INEFFECTIVE (72)	0.92	174	TEND LESS TO BE THOSE WHERE THE LEGISLATURE IS PARTIALLY EFFECTIVE, LARGELY INEFFECTIVE, OR WHOLLY INEFFECTIVE (72)	0.59	3 25 33 36 X SQ= 10.22 P = 0. RV NO (NO)
176	TEND MORE TO BE THOSE WHERE THE LEGISLATURE IS FULLY EFFECTIVE, PARTIALLY EFFECTIVE, OR LARGELY INEFFECTIVE (72)	0.86	176	TEND LESS TO BE THOSE WHERE THE LEGISLATURE IS FULLY EFFECTIVE, PARTIALLY EFFECTIVE, OR LARGELY INEFFECTIVE (72)	0.64	31 39 5 22 X SQ= 4.49 P = 0.020 RV NO (NO)
178	LEAN TOWARD BEING THOSE WHERE THE LEGISLATURE IS UNICAMERAL (53)	0.63	178	LEAN TOWARD BEING THOSE WHERE THE LEGISLATURE IS BICAMERAL (51)	0.58	24 26 14 36 X SQ= 3.44 P = 0.063 RV YES (NO)
181	TEND MORE TO BE THOSE WHERE THE BUREAUCRACY IS SEMI-MODERN, RATHER THAN MODERN (55)	0.96	181	TEND LESS TO BE THOSE WHERE THE BUREAUCRACY IS SEMI-MODERN, RATHER THAN MODERN (55)	0.57	1 20 23 27 X SQ= 9.47 P = 0.001 RV NO (NO)
183	ALWAYS ARE THOSE WHERE THE BUREAUCRACY IS POST-COLONIAL TRANSITIONAL, RATHER THAN TRADITIONAL (25)	1.00	183	TEND TO BE THOSE WHERE THE BUREAUCRACY IS TRADITIONAL, RATHER THAN POST-COLONIAL TRANSITIONAL (9)	0.50	12 8 0 8 X SQ= 6.13 P = 0.004 RV YES (YES)
185	TEND TO BE THOSE WHERE PARTICIPATION BY THE MILITARY IN POLITICS IS INTERVENTIVE, RATHER THAN SUPPORTIVE (21)	0.59	185	TEND TO BE THOSE WHERE PARTICIPATION BY THE MILITARY IN POLITICS IS SUPPORTIVE, RATHER THAN INTERVENTIVE (31)	0.92	10 2 7 22 X SQ= 9.94 P = 0.001 RV YES (YES)
187	TEND MORE TO BE THOSE WHERE THE ROLE OF THE POLICE IS POLITICALLY SIGNIFICANT (66)	0.83	187	TEND LESS TO BE THOSE WHERE THE ROLE OF THE POLICE IS POLITICALLY SIGNIFICANT (66)	0.51	25 30 5 29 X SQ= 7.57 P = 0.003 RV NO (NO)

	Left column			Right column		
188	TEND LESS TO BE THOSE WHERE THE CHARACTER OF THE LEGAL SYSTEM IS OTHER THAN CIVIL LAW (81)	0.55	188	TEND MORE TO BE THOSE WHERE THE CHARACTER OF THE LEGAL SYSTEM IS OTHER THAN CIVIL LAW (81)	0.77	17 14 21 48 X SQ= 4.42 P = 0.026 RV NO (YES)
190	LEAN MORE TOWARD BEING THOSE WHERE THE CHARACTER OF THE LEGAL SYSTEM IS CIVIL LAW, RATHER THAN COMMON LAW (32)	0.94	190	LEAN LESS TOWARD BEING THOSE WHERE THE CHARACTER OF THE LEGAL SYSTEM IS CIVIL LAW, RATHER THAN COMMON LAW (32)	0.70	17 14 1 6 X SQ= 2.32 P = 0.093 RV NO (YES)
193	TEND LESS TO BE THOSE WHERE THE CHARACTER OF THE LEGAL SYSTEM IS OTHER THAN PARTLY INDIGENOUS (86)	0.59	193	TEND MORE TO BE THOSE WHERE THE CHARACTER OF THE LEGAL SYSTEM IS OTHER THAN PARTLY INDIGENOUS (86)	0.84	16 10 23 53 X SQ= 6.75 P = 0.009 RV NO (YES)
194	ALWAYS ARE THOSE THAT ARE NON-COMMUNIST (101)	1.00	194	TEND LESS TO BE THOSE THAT ARE NON-COMMUNIST (101)	0.79	0 13 39 50 X SQ= 7.46 P = 0.001 RV NO (NO)

```
171  POLITIES                                          171  POLITIES
     WHERE THE LEGISLATIVE-EXECUTIVE STRUCTURE              WHERE THE LEGISLATIVE-EXECUTIVE STRUCTURE
     IS PARLIAMENTARY-REPUBLICAN  (12)                      IS OTHER THAN
                                                            PARLIAMENTARY-REPUBLICAN  (90)

                                                                        BOTH SUBJECT AND PREDICATE

     12 IN LEFT COLUMN

AUSTRIA     CONGO(LEO)  FINLAND     GERMAN FR   ICELAND     INDIA       IRELAND     ISRAEL      ITALY       SOMALIA
SO AFRICA   TURKEY

     90 IN RIGHT COLUMN

AFGHANISTAN ALBANIA     ARGENTINA   AUSTRALIA   BELGIUM     BOLIVIA     BRAZIL      BULGARIA    BURUNDI     CAMBODIA
CAMEROUN    CANADA      CEN AFR REP CEYLON      CHAD        CHILE       CHINA, PR   COLOMBIA    CONGO(BRA)  COSTA RICA
CYPRUS      CZECHOS'KIA DAHOMEY     DENMARK     DOMIN REP   ECUADOR     EL SALVADOR ETHIOPIA    FRANCE      GABON
GERMANY, E  GHANA       GREECE      GUATEMALA   GUINEA      HAITI       HONDURAS    HUNGARY     INDONESIA   IRAN
IVORY COAST JAMAICA     JAPAN       JORDAN      KOREA, N    LAOS        LEBANON     LIBERIA     LIBYA       LUXEMBOURG
MALAGASY R  MALAYA      MALI        MAURITANIA  MEXICO      MONGOLIA    MOROCCO     NEPAL       NETHERLANDS NEW ZEALAND
NICARAGUA   NIGER       NIGERIA     NORWAY      PAKISTAN    PANAMA      PARAGUAY    PHILIPPINES POLAND      RUMANIA
RWANDA      SENEGAL     SIERRE LEO  SWEDEN      SWITZERLAND TANGANYIKA  THAILAND    TOGO        TRINIDAD    TUNISIA
UGANDA      USSR        UK          US          UPPER VOLTA URUGUAY     VENEZUELA   VIETNAM, N  VIETNAM REP YUGOSLAVIA

     2 EXCLUDED BECAUSE AMBIGUOUS

PORTUGAL    SPAIN

     1 EXCLUDED BECAUSE IRRELEVANT

SA'U ARABIA

     10 EXCLUDED BECAUSE UNASCERTAINABLE

ALGERIA     BURMA       CUBA        IRAQ        KOREA REP   PERU        SUDAN       SYRIA       UAR         YEMEN

-------------------------------------------------------------------------------------------------------------------

  5  ALWAYS ARE THOSE                          1.00       5  LEAN LESS TOWARD BEING THOSE        0.53        6   10
     LOCATED IN SCANDINAVIA OR WEST EUROPE,                  LOCATED IN SCANDINAVIA OR WEST EUROPE,          0    9
     RATHER THAN IN EAST EUROPE  (18)                        RATHER THAN IN EAST EUROPE  (18)            X SQ=   2.62
                                                                                                         P =    0.057
                                                                                                         RV NO  (NO )

 17  ALWAYS ARE THOSE                          1.00      17  LEAN TOWARD BEING THOSE             0.80        2    5
     LOCATED IN THE MIDDLE EAST, RATHER THAN                 LOCATED IN THE CARIBBEAN,                       0   20
     IN THE CARIBBEAN, CENTRAL AMERICA, OR                   CENTRAL AMERICA, OR SOUTH AMERICA,          X SQ=   2.71
     SOUTH AMERICA  (11)                                     RATHER THAN IN THE MIDDLE EAST  (22)        P =    0.060
                                                                                                         RV YES (NO )
```

29	TILT TOWARD BEING THOSE WHERE THE DEGREE OF URBANIZATION IS HIGH (56)	0.75	29	TILT TOWARD BEING THOSE WHERE THE DEGREE OF URBANIZATION IS LOW (49)	0.51	9 40 3 41 X SQ= 1.82 P = 0.127 RV YES (NO)
31	TEND TO BE THOSE WHOSE AGRICULTURAL POPULATION IS LOW OR VERY LOW (24)	0.50	31	TEND TO BE THOSE WHOSE AGRICULTURAL POPULATION IS HIGH OR MEDIUM (90)	0.80	6 72 6 18 X SQ= 3.76 P = 0.032 RV YES (NO)
34	TEND TO BE THOSE WHOSE GROSS NATIONAL PRODUCT IS VERY HIGH, HIGH, MEDIUM, OR LOW (62)	0.83	34	TEND TO BE THOSE WHOSE GROSS NATIONAL PRODUCT IS VERY LOW (53)	0.53	10 42 2 48 X SQ= 4.32 P = 0.028 RV YES (NO)
35	TEND TO BE THOSE WHOSE PER CAPITA GROSS NATIONAL PRODUCT IS VERY HIGH OR HIGH (24)	0.58	35	TEND TO BE THOSE WHOSE PER CAPITA GROSS NATIONAL PRODUCT IS MEDIUM, LOW, OR VERY LOW (91)	0.81	7 17 5 73 X SQ= 7.09 P = 0.006 RV YES (NO)
36	LEAN TOWARD BEING THOSE WHOSE PER CAPITA GROSS NATIONAL PRODUCT IS VERY HIGH, HIGH, OR MEDIUM (42)	0.67	36	LEAN TOWARD BEING THOSE WHOSE PER CAPITA GROSS NATIONAL PRODUCT IS LOW OR VERY LOW (73)	0.64	8 32 4 58 X SQ= 3.09 P = 0.057 RV YES (NO)
39	TILT TOWARD BEING THOSE WHOSE INTERNATIONAL FINANCIAL STATUS IS VERY HIGH, HIGH, OR MEDIUM (38)	0.58	39	TILT TOWARD BEING THOSE WHOSE INTERNATIONAL FINANCIAL STATUS IS LOW OR VERY LOW (76)	0.67	7 29 5 60 X SQ= 2.04 P = 0.109 RV YES (NO)
42	LEAN TOWARD BEING THOSE WHOSE ECONOMIC DEVELOPMENTAL STATUS IS DEVELOPED OR INTERMEDIATE (36)	0.64	42	LEAN TOWARD BEING THOSE WHOSE ECONOMIC DEVELOPMENTAL STATUS IS UNDERDEVELOPED OR VERY UNDERDEVELOPED (76)	0.67	7 29 4 60 X SQ= 2.86 P = 0.092 RV YES (NO)
43	TEND TO BE THOSE WHOSE ECONOMIC DEVELOPMENTAL STATUS IS DEVELOPED, INTERMEDIATE, OR UNDERDEVELOPED (55)	0.83	43	TEND TO BE THOSE WHOSE ECONOMIC DEVELOPMENTAL STATUS IS VERY UNDERDEVELOPED (57)	0.56	10 39 2 49 X SQ= 4.97 P = 0.014 RV YES (NO)
51	ALWAYS ARE THOSE WHERE FREEDOM OF THE PRESS IS COMPLETE OR INTERMITTENT (60)	1.00	51	TEND LESS TO BE THOSE WHERE FREEDOM OF THE PRESS IS COMPLETE OR INTERMITTENT (60)	0.61	12 45 0 29 X SQ= 5.45 P = 0.007 RV NO (NO)

54	TILT TOWARD BEING THOSE WHERE NEWSPAPER CIRCULATION IS ONE HUNDRED OR MORE PER THOUSAND (37)	0.58	54	TILT TOWARD BEING THOSE WHERE NEWSPAPER CIRCULATION IS LESS THAN ONE HUNDRED PER THOUSAND (76)	0.67	7 29 5 59 X SQ= 1.95 P = 0.112 RV YES (NO)
66	LEAN TOWARD BEING THOSE THAT ARE RELIGIOUSLY HOMOGENEOUS (57)	0.75	66	LEAN TOWARD BEING THOSE THAT ARE RELIGIOUSLY HETEROGENEOUS (49)	0.54	9 38 3 44 X SQ= 2.39 P = 0.073 RV YES (NO)
74	TEND TO BE THOSE THAT ARE HISTORICALLY WESTERN (26)	0.60	74	TEND TO BE THOSE THAT ARE NOT HISTORICALLY WESTERN (87)	0.80	6 18 4 72 X SQ= 5.85 P = 0.012 RV YES (NO)
75	TILT MORE TOWARD BEING THOSE THAT ARE HISTORICALLY WESTERN OR SIGNIFICANTLY WESTERNIZED (62)	0.82	75	TILT LESS TOWARD BEING THOSE THAT ARE HISTORICALLY WESTERN OR SIGNIFICANTLY WESTERNIZED (62)	0.52	9 47 2 43 X SQ= 2.38 P = 0.106 RV NO (NO)
76	TEND TO BE THOSE THAT ARE HISTORICALLY WESTERN, RATHER THAN HAVING BEEN SIGNIFICANTLY OR PARTIALLY WESTERNIZED THROUGH A COLONIAL RELATIONSHIP (26)	0.60	76	TEND TO BE THOSE THAT HAVE BEEN SIGNIFICANTLY OR PARTIALLY WESTERNIZED THROUGH A COLONIAL RELATIONSHIP, RATHER THAN BEING HISTORICALLY WESTERN (70)	0.76	6 18 4 58 X SQ= 4.13 P = 0.025 RV YES (NO)
80	ALWAYS ARE THOSE WHOSE DATE OF INDEPENDENCE IS AFTER 1914, AND THAT WERE FORMERLY DEPENDENCIES OF BRITAIN, RATHER THAN FRANCE (19)	1.00	80	LEAN TOWARD BEING THOSE WHOSE DATE OF INDEPENDENCE IS AFTER 1914, AND THAT WERE FORMERLY DEPENDENCIES OF FRANCE, RATHER THAN BRITAIN (24)	0.65	3 12 0 22 X SQ= 2.48 P = 0.059 RV YES (NO)
85	TILT TOWARD BEING THOSE WHERE THE STAGE OF POLITICAL MODERNIZATION IS ADVANCED, RATHER THAN MID- OR EARLY TRANSITIONAL (60)	0.75	85	TILT TOWARD BEING THOSE WHERE THE STAGE OF POLITICAL MODERNIZATION IS MID- OR EARLY TRANSITIONAL, RATHER THAN ADVANCED (54)	0.53	9 42 3 48 X SQ= 2.36 P = 0.122 RV YES (NO)
89	TEND TO BE THOSE WHOSE IDEOLOGICAL ORIENTATION IS CONVENTIONAL (33)	0.73	89	TEND TO BE THOSE WHOSE IDEOLOGICAL ORIENTATION IS OTHER THAN CONVENTIONAL (62)	0.67	8 25 3 50 X SQ= 4.74 P = 0.019 RV YES (NO)
94	ALWAYS ARE THOSE WHERE THE STATUS OF THE REGIME IS CONSTITUTIONAL (51)	1.00	94	TEND LESS TO BE THOSE WHERE THE STATUS OF THE REGIME IS CONSTITUTIONAL (51)	0.57	9 42 0 32 X SQ= 4.64 P = 0.011 RV NO (NO)

101	TEND TO BE THOSE WHERE THE REPRESENTATIVE CHARACTER OF THE REGIME IS POLYARCHIC (41)	0.90	101	TEND TO BE THOSE WHERE THE REPRESENTATIVE CHARACTER OF THE REGIME IS LIMITED POLYARCHIC, PSEUDO-POLYARCHIC, OR NON-POLYARCHIC (57)	0.60	9 32 1 48 X SQ= 7.06 P = 0.005 RV YES (NO)
105	TEND TO BE THOSE WHERE THE ELECTORAL SYSTEM IS COMPETITIVE (43)	0.90	105	TEND TO BE THOSE WHERE THE ELECTORAL SYSTEM IS PARTIALLY COMPETITIVE OR NON-COMPETITIVE (47)	0.56	9 34 1 43 X SQ= 5.72 P = 0.007 RV YES (NO)
106	ALWAYS ARE THOSE WHERE THE ELECTORAL SYSTEM IS COMPETITIVE OR PARTIALLY COMPETITIVE (52)	1.00	106	TEND LESS TO BE THOSE WHERE THE ELECTORAL SYSTEM IS COMPETITIVE OR PARTIALLY COMPETITIVE (52)	0.60	10 41 0 27 X SQ= 4.45 P = 0.013 RV NO (NO)
107	TEND TO BE THOSE WHERE AUTONOMOUS GROUPS ARE FULLY TOLERATED IN POLITICS (46)	0.75	107	TEND TO BE THOSE WHERE AUTONOMOUS GROUPS ARE PARTIALLY TOLERATED IN POLITICS, ARE TOLERATED ONLY OUTSIDE POLITICS, OR ARE NOT TOLERATED AT ALL (65)	0.59	9 36 3 52 X SQ= 3.68 P = 0.033 RV YES (NO)
108	ALWAYS ARE THOSE WHERE AUTONOMOUS GROUPS ARE FULLY OR PARTIALLY TOLERATED IN POLITICS (65)	1.00	108	TEND LESS TO BE THOSE WHERE AUTONOMOUS GROUPS ARE FULLY OR PARTIALLY TOLERATED IN POLITICS (65)	0.64	10 53 0 30 X SQ= 3.81 P = 0.027 RV NO (NO)
111	LEAN MORE TOWARD BEING THOSE WHERE POLITICAL ENCULTURATION IS HIGH OR MEDIUM (53)	0.83	111	LEAN LESS TOWARD BEING THOSE WHERE POLITICAL ENCULTURATION IS HIGH OR MEDIUM (53)	0.54	10 39 2 33 X SQ= 2.50 P = 0.067 RV NO (NO)
116	TEND TO BE THOSE WHERE INTEREST ARTICULATION BY ASSOCIATIONAL GROUPS IS SIGNIFICANT OR MODERATE (32)	0.75	116	TEND TO BE THOSE WHERE INTEREST ARTICULATION BY ASSOCIATIONAL GROUPS IS LIMITED OR NEGLIGIBLE (79)	0.74	9 23 3 66 X SQ= 9.64 P = 0.001 RV YES (NO)
117	TEND TO BE THOSE WHERE INTEREST ARTICULATION BY ASSOCIATIONAL GROUPS IS SIGNIFICANT, MODERATE, OR LIMITED (60)	0.83	117	TEND TO BE THOSE WHERE INTEREST ARTICULATION BY ASSOCIATIONAL GROUPS IS NEGLIGIBLE (51)	0.51	10 44 2 45 X SQ= 3.62 P = 0.033 RV YES (NO)
118	ALWAYS ARE THOSE WHERE INTEREST ARTICULATION BY INSTITUTIONAL GROUPS IS SIGNIFICANT, MODERATE, OR LIMITED (60)	1.00	118	TEND LESS TO BE THOSE WHERE INTEREST ARTICULATION BY INSTITUTIONAL GROUPS IS SIGNIFICANT, MODERATE, OR LIMITED (60)	0.60	0 31 11 46 X SQ= 5.19 P = 0.007 RV NO (NO)

125	TILT TOWARD BEING THOSE WHERE INTEREST ARTICULATION BY ANOMIC GROUPS IS INFREQUENT OR VERY INFREQUENT (35)	0.58	125	TILT TOWARD BEING THOSE WHERE INTEREST ARTICULATION BY ANOMIC GROUPS IS FREQUENT OR OCCASIONAL (64)	0.67	5 51 7 25 X SQ= 1.90 P = 0.112 RV YES (NO)
128	TEND TO BE THOSE WHERE INTEREST ARTICULATION BY POLITICAL PARTIES IS SIGNIFICANT OR MODERATE (48)	0.91	128	TEND TO BE THOSE WHERE INTEREST ARTICULATION BY POLITICAL PARTIES IS LIMITED OR NEGLIGIBLE (45)	0.53	10 36 1 40 X SQ= 5.67 P = 0.008 RV YES (NO)
129	ALWAYS ARE THOSE WHERE INTEREST ARTICULATION BY POLITICAL PARTIES IS SIGNIFICANT, MODERATE, OR LIMITED (56)	1.00	129	TEND LESS TO BE THOSE WHERE INTEREST ARTICULATION BY POLITICAL PARTIES IS SIGNIFICANT, MODERATE, OR LIMITED (56)	0.55	11 42 0 34 X SQ= 6.31 P = 0.003 RV NO (NO)
134	TEND MORE TO BE THOSE WHERE INTEREST AGGREGATION BY THE EXECUTIVE IS SIGNIFICANT OR MODERATE (57)	0.90	134	TEND LESS TO BE THOSE WHERE INTEREST AGGREGATION BY THE EXECUTIVE IS SIGNIFICANT OR MODERATE (57)	0.54	9 46 1 39 X SQ= 3.37 P = 0.041 RV NO (NO)
137	TEND TO BE THOSE WHERE INTEREST AGGREGATION BY THE LEGISLATURE IS SIGNIFICANT OR MODERATE (29)	0.82	137	TEND TO BE THOSE WHERE INTEREST AGGREGATION BY THE LEGISLATURE IS LIMITED OR NEGLIGIBLE (68)	0.76	9 20 2 63 X SQ= 12.58 P = 0. RV YES (NO)
138	TEND TO BE THOSE WHERE INTEREST AGGREGATION BY THE LEGISLATURE IS SIGNIFICANT, MODERATE, OR LIMITED (48)	0.82	138	TEND TO BE THOSE WHERE INTEREST AGGREGATION BY THE LEGISLATURE IS NEGLIGIBLE (49)	0.55	9 38 2 47 X SQ= 3.99 P = 0.026 RV YES (NO)
139	ALWAYS ARE THOSE WHERE THE PARTY SYSTEM IS QUANTITATIVELY OTHER THAN ONE-PARTY (71)	1.00	139	TEND LESS TO BE THOSE WHERE THE PARTY SYSTEM IS QUANTITATIVELY OTHER THAN ONE-PARTY (71)	0.64	0 30 12 54 X SQ= 4.68 P = 0.009 RV NO (NO)
142	LEAN TOWARD BEING THOSE WHERE THE PARTY SYSTEM IS QUANTITATIVELY MULTI-PARTY (30)	0.60	142	LEAN TOWARD BEING THOSE WHERE THE PARTY SYSTEM IS QUANTITATIVELY OTHER THAN MULTI-PARTY (66)	0.71	6 23 4 56 X SQ= 2.58 P = 0.072 RV YES (NO)
147	TEND TO BE THOSE WHERE THE PARTY SYSTEM IS QUALITATIVELY CLASS-ORIENTED OR MULTI-IDEOLOGICAL (23)	0.75	147	TEND TO BE THOSE WHERE THE PARTY SYSTEM IS QUALITATIVELY OTHER THAN CLASS-ORIENTED OR MULTI-IDEOLOGICAL (67)	0.79	6 16 2 59 X SQ= 8.11 P = 0.004 RV YES (NO)

158	TEND MORE TO BE THOSE WHERE PERSONALISMO IS NEGLIGIBLE (56)	0.92	158	TEND LESS TO BE THOSE WHERE PERSONALISMO IS NEGLIGIBLE (56)	0.53	1 36 11 41 X SQ= 4.83 P = 0.012 RV NO (NO)
161	TEND TO BE THOSE WHERE THE POLITICAL LEADERSHIP IS NON-ELITIST (50)	0.83	161	TEND TO BE THOSE WHERE THE POLITICAL LEADERSHIP IS ELITIST OR MODERATELY ELITIST (47)	0.51	2 41 10 40 X SQ= 3.58 P = 0.033 RV YES (NO)
164	LEAN MORE TOWARD BEING THOSE WHERE THE REGIME'S LEADERSHIP CHARISMA IS NEGLIGIBLE (65)	0.91	164	LEAN LESS TOWARD BEING THOSE WHERE THE REGIME'S LEADERSHIP CHARISMA IS NEGLIGIBLE (65)	0.63	1 29 10 50 X SQ= 2.19 P = 0.092 RV NO (NO)
168	TEND TO BE THOSE WHERE THE HORIZONTAL POWER DISTRIBUTION IS SIGNIFICANT (34)	0.91	168	TEND TO BE THOSE WHERE THE HORIZONTAL POWER DISTRIBUTION IS LIMITED OR NEGLIGIBLE (72)	0.73	10 24 1 64 X SQ= 14.85 P = 0. RV YES (NO)
169	ALWAYS ARE THOSE WHERE THE HORIZONTAL POWER DISTRIBUTION IS SIGNIFICANT OR LIMITED (58)	1.00	169	TEND LESS TO BE THOSE WHERE THE HORIZONTAL POWER DISTRIBUTION IS SIGNIFICANT OR LIMITED (58)	0.53	11 47 0 41 X SQ= 6.93 P = 0.002 RV NO (NO)
174	TEND TO BE THOSE WHERE THE LEGISLATURE IS FULLY EFFECTIVE (28)	0.82	174	TEND TO BE THOSE WHERE THE LEGISLATURE IS PARTIALLY EFFECTIVE, LARGELY INEFFECTIVE, OR WHOLLY INEFFECTIVE (72)	0.78	9 19 2 66 X SQ= 13.92 P = 0. RV YES (NO)
175	TEND TO BE THOSE WHERE THE LEGISLATURE IS FULLY EFFECTIVE OR PARTIALLY EFFECTIVE (51)	0.91	175	TEND TO BE THOSE WHERE THE LEGISLATURE IS LARGELY INEFFECTIVE OR WHOLLY INEFFECTIVE (49)	0.52	10 41 1 44 X SQ= 5.51 P = 0.009 RV YES (NO)
176	ALWAYS ARE THOSE WHERE THE LEGISLATURE IS FULLY EFFECTIVE, PARTIALLY EFFECTIVE, OR LARGELY INEFFECTIVE (72)	1.00	176	TEND LESS TO BE THOSE WHERE THE LEGISLATURE IS FULLY EFFECTIVE, PARTIALLY EFFECTIVE, OR LARGELY INEFFECTIVE (72)	0.71	11 60 0 25 X SQ= 2.98 P = 0.035 RV NO (NO)
179	TEND TO BE THOSE WHERE THE EXECUTIVE IS STRONG (39)	0.80	179	TEND TO BE THOSE WHERE THE EXECUTIVE IS DOMINANT (52)	0.59	2 44 8 31 X SQ= 3.87 P = 0.039 RV YES (NO)

171/

181	TEND TO BE THOSE WHERE THE BUREAUCRACY IS MODERN, RATHER THAN SEMI-MODERN (21)	0.80	181	TEND TO BE THOSE WHERE THE BUREAUCRACY IS SEMI-MODERN, RATHER THAN MODERN (55)	0.78	8 13 2 46 X SQ= 10.97 P = 0.001 RV YES (NO)
187	LEAN TOWARD BEING THOSE WHERE THE ROLE OF THE POLICE IS NOT POLITICALLY SIGNIFICANT (35)	0.67	187	LEAN TOWARD BEING THOSE WHERE THE ROLE OF THE POLICE IS POLITICALLY SIGNIFICANT (66)	0.65	4 50 8 27 X SQ= 3.12 P = 0.055 RV YES (NO)
192	ALWAYS ARE THOSE WHERE THE CHARACTER OF THE LEGAL SYSTEM IS OTHER THAN MUSLIM OR PARTLY MUSLIM (86)	1.00	192	LEAN LESS TOWARD BEING THOSE WHERE THE CHARACTER OF THE LEGAL SYSTEM IS OTHER THAN MUSLIM OR PARTLY MUSLIM (86)	0.76	0 21 12 67 X SQ= 2.33 P = 0.066 RV NO (NO)

```
**********************************************************************************************************

172  POLITIES                                            172  POLITIES
     WHERE THE LEGISLATIVE-EXECUTIVE STRUCTURE                WHERE THE LEGISLATIVE-EXECUTIVE STRUCTURE
     IS PARLIAMENTARY-ROYALIST  (21)                          IS OTHER THAN PARLIAMENTARY-ROYALIST  (88)


                                                                                   BOTH SUBJECT AND PREDICATE
..........................................................................................................

       21 IN LEFT COLUMN

AUSTRALIA   BELGIUM     BURUNDI     CANADA      CEYLON      DENMARK     GREECE      JAMAICA     JAPAN       LAOS
LUXEMBOURG  MALAYA      NETHERLANDS NEW ZEALAND NIGERIA     NORWAY      SIERRE LEO  SWEDEN      TRINIDAD    UGANDA
UK

       88 IN RIGHT COLUMN

AFGHANISTAN ALBANIA     ALGERIA     ARGENTINA   AUSTRIA     BOLIVIA     BRAZIL      BULGARIA    BURMA       CAMEROUN
CEN AFR REP CHAD        CHILE       CHINA, PR   COLOMBIA    CONGO(BRA)  CONGO(LEO)  COSTA RICA  CUBA        CYPRUS
CZECHOS'KIA DAHOMEY     DOMIN REP   ECUADOR     EL SALVADOR ETHIOPIA    FINLAND     FRANCE      GABON       GERMANY, E
GERMAN FR   GHANA       GUATEMALA   GUINEA      HAITI       HONDURAS    HUNGARY     ICELAND     INDIA       INDONESIA
IRAQ        IRELAND     ISRAEL      ITALY       IVORY COAST JORDAN      KOREA, N    KOREA REP   LEBANON     LIBERIA
LIBYA       MALAGASY R  MALI        MAURITANIA  MEXICO      MONGOLIA    MOROCCO     NEPAL       NICARAGUA   NIGER
PAKISTAN    PANAMA      PARAGUAY    PERU        PHILIPPINES POLAND      PORTUGAL    RUMANIA     RWANDA      SENEGAL
SOMALIA     SO AFRICA   SUDAN       SWITZERLAND SYRIA       TANGANYIKA  TOGO        TUNISIA     TURKEY      USSR
UAR         US          UPPER VOLTA URUGUAY     VENEZUELA   VIETNAM, N  VIETNAM REP YUGOSLAVIA

       2 EXCLUDED BECAUSE AMBIGUOUS

CAMBODIA    SPAIN

       1 EXCLUDED BECAUSE IRRELEVANT

SA'U ARABIA

       3 EXCLUDED BECAUSE UNASCERTAINABLE

IRAN        THAILAND    YEMEN

----------------------------------------------------------------------------------------------------------

  2  TEND TO BE THOSE                        0.52       2  TEND TO BE THOSE                      0.89          11    10
     LOCATED IN WEST EUROPE, SCANDINAVIA,                  LOCATED ELSEWHERE THAN IN                           10    78
     NORTH AMERICA, OR AUSTRALASIA  (22)                   WEST EUROPE, SCANDINAVIA,                       X SQ=  15.80
                                                           NORTH AMERICA, OR AUSTRALASIA  (93)             P =     0.
                                                                                                           RV YES (YES)

  5  ALWAYS ARE THOSE                        1.00       5  TEND TO BE THOSE                      0.50           8     9
     LOCATED IN SCANDINAVIA OR WEST EUROPE,                LOCATED IN EAST EUROPE, RATHER THAN                  0     9
     RATHER THAN IN EAST EUROPE  (18)                      IN SCANDINAVIA OR WEST EUROPE  (9)              X SQ=   4.11
                                                                                                           P =    0.023
                                                                                                           RV YES (NO )
```

13	ALWAYS ARE THOSE LOCATED ELSEWHERE THAN IN NORTH AFRICA OR THE MIDDLE EAST (99)	1.00	13	LEAN LESS TOWARD BEING THOSE LOCATED ELSEWHERE THAN IN NORTH AFRICA OR THE MIDDLE EAST (99)	0.85	0 13 21 75 X SQ= 2.26 P = 0.069 RV NO (NO)
26	LEAN TOWARD BEING THOSE WHOSE POPULATION DENSITY IS VERY HIGH, HIGH, OR MEDIUM (48)	0.62	26	LEAN TOWARD BEING THOSE WHOSE POPULATION DENSITY IS LOW (67)	0.62	13 33 8 55 X SQ= 3.20 P = 0.051 RV YES (NO)
30	TEND TO BE THOSE WHOSE AGRICULTURAL POPULATION IS MEDIUM, LOW, OR VERY LOW (57)	0.81	30	TEND TO BE THOSE WHOSE AGRICULTURAL POPULATION IS HIGH (56)	0.55	4 48 17 39 X SQ= 7.45 P = 0.003 RV YES (NO)
31	TEND TO BE THOSE WHOSE AGRICULTURAL POPULATION IS LOW OR VERY LOW (24)	0.52	31	TEND TO BE THOSE WHOSE AGRICULTURAL POPULATION IS HIGH OR MEDIUM (90)	0.85	10 75 11 13 X SQ= 11.86 P = 0.001 RV YES (NO)
36	TEND TO BE THOSE WHOSE PER CAPITA GROSS NATIONAL PRODUCT IS VERY HIGH, HIGH, OR MEDIUM (42)	0.71	36	TEND TO BE THOSE WHOSE PER CAPITA GROSS NATIONAL PRODUCT IS LOW OR VERY LOW (73)	0.70	15 26 6 62 X SQ= 10.95 P = 0.001 RV YES (NO)
42	TEND TO BE THOSE WHOSE ECONOMIC DEVELOPMENTAL STATUS IS DEVELOPED OR INTERMEDIATE (36)	0.57	42	TEND TO BE THOSE WHOSE ECONOMIC DEVELOPMENTAL STATUS IS UNDERDEVELOPED OR VERY UNDERDEVELOPED (76)	0.72	12 24 9 62 X SQ= 5.22 P = 0.019 RV YES (NO)
43	LEAN TOWARD BEING THOSE WHOSE ECONOMIC DEVELOPMENTAL STATUS IS DEVELOPED, INTERMEDIATE, OR UNDERDEVELOPED (55)	0.71	43	LEAN TOWARD BEING THOSE WHOSE ECONOMIC DEVELOPMENTAL STATUS IS VERY UNDERDEVELOPED (57)	0.55	15 39 6 47 X SQ= 3.61 P = 0.050 RV YES (NO)
44	TEND TO BE THOSE WHERE THE LITERACY RATE IS NINETY PERCENT OR ABOVE (25)	0.52	44	TEND TO BE THOSE WHERE THE LITERACY RATE IS BELOW NINETY PERCENT (90)	0.84	11 14 10 74 X SQ= 10.78 P = 0.001 RV YES (NO)
45	LEAN TOWARD BEING THOSE WHERE THE LITERACY RATE IS FIFTY PERCENT OR ABOVE (55)	0.71	45	LEAN TOWARD BEING THOSE WHERE THE LITERACY RATE IS BELOW FIFTY PERCENT (54)	0.54	15 38 6 44 X SQ= 3.27 P = 0.051 RV YES (NO)

50	TEND TO BE THOSE WHERE FREEDOM OF THE PRESS IS COMPLETE (43)	0.89	50	TEND TO BE THOSE WHERE FREEDOM OF THE PRESS IS INTERMITTENT, INTERNALLY ABSENT, OR INTERNALLY AND EXTERNALLY ABSENT (56)	0.64	16 27 2 48 X SQ= 14.28 P = 0. RV YES (NO)
51	ALWAYS ARE THOSE WHERE FREEDOM OF THE PRESS IS COMPLETE OR INTERMITTENT (60)	1.00	51	TEND LESS TO BE THOSE WHERE FREEDOM OF THE PRESS IS COMPLETE OR INTERMITTENT (60)	0.57	18 42 0 32 X SQ= 10.10 P = 0. RV NO (NO)
54	TEND TO BE THOSE WHERE NEWSPAPER CIRCULATION IS ONE HUNDRED OR MORE PER THOUSAND (37)	0.57	54	TEND TO BE THOSE WHERE NEWSPAPER CIRCULATION IS LESS THAN ONE HUNDRED PER THOUSAND (76)	0.71	12 25 9 61 X SQ= 4.70 P = 0.021 RV YES (NO)
58	ALWAYS ARE THOSE WHERE THE RELIGION IS OTHER THAN MUSLIM (97)	1.00	58	TEND LESS TO BE THOSE WHERE THE RELIGION IS OTHER THAN MUSLIM (97)	0.83	0 15 21 73 X SQ= 2.84 P = 0.040 RV NO (NO)
62	TEND TO BE THOSE WHERE THE RELIGION IS PROTESTANT, RATHER THAN CATHOLIC (5)	0.60	62	TEND TO BE THOSE WHERE THE RELIGION IS CATHOLIC, RATHER THAN PROTESTANT (25)	0.92	3 2 2 22 X SQ= 4.54 P = 0.024 RV YES (YES)
63	TILT TOWARD BEING THOSE WHERE THE RELIGION IS PREDOMINANTLY SOME KIND OF CHRISTIAN (46)	0.57	63	TILT TOWARD BEING THOSE WHERE THE RELIGION IS PREDOMINANTLY OR PARTLY OTHER THAN CHRISTIAN (68)	0.62	12 33 9 54 X SQ= 1.84 P = 0.140 RV YES (NO)
66	TEND TO BE THOSE THAT ARE RELIGIOUSLY HETEROGENEOUS (49)	0.70	66	TEND TO BE THOSE THAT ARE RELIGIOUSLY HOMOGENEOUS (57)	0.56	6 45 14 35 X SQ= 3.42 P = 0.046 RV YES (NO)
74	TEND TO BE THOSE THAT ARE HISTORICALLY WESTERN (26)	0.52	74	TEND TO BE THOSE THAT ARE NOT HISTORICALLY WESTERN (87)	0.84	11 14 10 72 X SQ= 10.35 P = 0.001 RV YES (NO)
75	TILT MORE TOWARD BEING THOSE THAT ARE HISTORICALLY WESTERN OR SIGNIFICANTLY WESTERNIZED (62)	0.71	75	TILT LESS TOWARD BEING THOSE THAT ARE HISTORICALLY WESTERN OR SIGNIFICANTLY WESTERNIZED (62)	0.53	15 46 6 41 X SQ= 1.67 P = 0.147 RV NO (NO)

76	TEND TO BE THOSE THAT ARE HISTORICALLY WESTERN, RATHER THAN HAVING BEEN SIGNIFICANTLY OR PARTIALLY WESTERNIZED THROUGH A COLONIAL RELATIONSHIP (26)	0.55	76	TEND TO BE THOSE THAT HAVE BEEN SIGNIFICANTLY OR PARTIALLY WESTERNIZED THROUGH A COLONIAL RELATIONSHIP, RATHER THAN BEING HISTORICALLY WESTERN (70)	0.81	11 14 9 59 X SQ= 8.51 P = 0.003 RV YES (NO)
79	ALWAYS ARE THOSE THAT WERE FORMERLY DEPENDENCIES OF BRITAIN OR FRANCE, RATHER THAN SPAIN (49)	1.00	79	TEND LESS TO BE THOSE THAT WERE FORMERLY DEPENDENCIES OF BRITAIN OR FRANCE, RATHER THAN SPAIN (49)	0.67	11 37 0 18 X SQ= 3.44 P = 0.028 RV NO (NO)
80	TEND TO BE THOSE WHOSE DATE OF INDEPENDENCE IS AFTER 1914, AND THAT WERE FORMERLY DEPENDENCIES OF BRITAIN, RATHER THAN FRANCE (19)	0.88	80	TEND TO BE THOSE WHOSE DATE OF INDEPENDENCE IS AFTER 1914, AND THAT WERE FORMERLY DEPENDENCIES OF FRANCE, RATHER THAN BRITAIN (24)	0.65	7 12 1 22 X SQ= 5.17 P = 0.015 RV YES (NO)
81	TEND TO BE THOSE WHERE THE TYPE OF POLITICAL MODERNIZATION IS EARLY EUROPEAN OR EARLY EUROPEAN DERIVED, RATHER THAN LATER EUROPEAN OR LATER EUROPEAN DERIVED (11)	0.64	81	TEND TO BE THOSE WHERE THE TYPE OF POLITICAL MODERNIZATION IS LATER EUROPEAN OR LATER EUROPEAN DERIVED, RATHER THAN EARLY EUROPEAN OR EARLY EUROPEAN DERIVED (40)	0.90	7 4 4 35 X SQ= 11.31 P = 0.001 RV YES (YES)
89	TEND TO BE THOSE WHOSE IDEOLOGICAL ORIENTATION IS CONVENTIONAL (33)	0.75	89	TEND TO BE THOSE WHOSE IDEOLOGICAL ORIENTATION IS OTHER THAN CONVENTIONAL (62)	0.72	12 21 4 53 X SQ= 10.39 P = 0.001 RV YES (NO)
93	ALWAYS ARE THOSE WHERE THE SYSTEM STYLE IS NON-MOBILIZATIONAL (78)	1.00	93	TEND LESS TO BE THOSE WHERE THE SYSTEM STYLE IS NON-MOBILIZATIONAL (78)	0.64	0 31 21 54 X SQ= 9.13 P = 0. RV NO (NO)
94	TEND TO BE THOSE WHERE THE STATUS OF THE REGIME IS CONSTITUTIONAL (51)	0.95	94	TEND TO BE THOSE WHERE THE STATUS OF THE REGIME IS AUTHORITARIAN OR TOTALITARIAN (41)	0.52	19 32 1 35 X SQ= 12.29 P = 0. RV YES (NO)
95	ALWAYS ARE THOSE WHERE THE STATUS OF THE REGIME IS CONSTITUTIONAL OR AUTHORITARIAN (95)	1.00	95	TEND LESS TO BE THOSE WHERE THE STATUS OF THE REGIME IS CONSTITUTIONAL OR AUTHORITARIAN (95)	0.82	21 70 0 15 X SQ= 2.99 P = 0.038 RV NO (NO)
96	TEND MORE TO BE THOSE WHERE THE STATUS OF THE REGIME IS CONSTITUTIONAL OR TOTALITARIAN (67)	0.95	96	TEND LESS TO BE THOSE WHERE THE STATUS OF THE REGIME IS CONSTITUTIONAL OR TOTALITARIAN (67)	0.72	19 47 1 18 X SQ= 3.32 P = 0.035 RV NO (NO)

99	LEAN MORE TOWARD BEING THOSE WHERE GOVERNMENTAL STABILITY IS GENERALLY PRESENT AND DATES FROM AT LEAST THE INTER-WAR PERIOD, OR FROM THE POST-WAR PERIOD (50)	0.80	99	LEAN LESS TOWARD BEING THOSE WHERE GOVERNMENTAL STABILITY IS GENERALLY PRESENT AND DATES FROM AT LEAST THE INTER-WAR PERIOD, OR FROM THE POST-WAR PERIOD (50)	0.55	12 36 3 30 X SQ= 2.31 P = 0.086 RV NO (NO)
101	TEND TO BE THOSE WHERE THE REPRESENTATIVE CHARACTER OF THE REGIME IS POLYARCHIC (41)	0.94	101	TEND TO BE THOSE WHERE THE REPRESENTATIVE CHARACTER OF THE REGIME IS LIMITED POLYARCHIC, PSEUDO-POLYARCHIC, OR NON-POLYARCHIC (57)	0.68	17 24 1 51 X SQ= 20.50 P = 0. RV YES (NO)
102	TEND TO BE THOSE WHERE THE REPRESENTATIVE CHARACTER OF THE REGIME IS POLYARCHIC OR LIMITED POLYARCHIC (59)	0.95	102	TEND TO BE THOSE WHERE THE REPRESENTATIVE CHARACTER OF THE REGIME IS PSEUDO-POLYARCHIC OR NON-POLYARCHIC (49)	0.52	20 39 1 43 X SQ= 13.64 P = 0. RV YES (NO)
105	TEND TO BE THOSE WHERE THE ELECTORAL SYSTEM IS COMPETITIVE (43)	0.90	105	TEND TO BE THOSE WHERE THE ELECTORAL SYSTEM IS PARTIALLY COMPETITIVE OR NON-COMPETITIVE (47)	0.63	18 25 2 43 X SQ= 15.46 P = 0. RV YES (NO)
106	ALWAYS ARE THOSE WHERE THE ELECTORAL SYSTEM IS COMPETITIVE OR PARTIALLY COMPETITIVE (52)	1.00	106	TEND LESS TO BE THOSE WHERE THE ELECTORAL SYSTEM IS COMPETITIVE OR PARTIALLY COMPETITIVE (52)	0.54	19 33 0 28 X SQ= 11.48 P = 0. RV NO (NO)
107	TEND TO BE THOSE WHERE AUTONOMOUS GROUPS ARE FULLY TOLERATED IN POLITICS (46)	0.90	107	TEND TO BE THOSE WHERE AUTONOMOUS GROUPS ARE PARTIALLY TOLERATED IN POLITICS, ARE TOLERATED ONLY OUTSIDE POLITICS, OR ARE NOT TOLERATED AT ALL (65)	0.67	18 28 2 58 X SQ= 19.52 P = 0. RV YES (NO)
108	ALWAYS ARE THOSE WHERE AUTONOMOUS GROUPS ARE FULLY OR PARTIALLY TOLERATED IN POLITICS (65)	1.00	108	TEND LESS TO BE THOSE WHERE AUTONOMOUS GROUPS ARE FULLY OR PARTIALLY TOLERATED IN POLITICS (65)	0.59	19 45 0 31 X SQ= 9.72 P = 0. RV NO (NO)
110	TEND LESS TO BE THOSE WHERE POLITICAL ENCULTURATION IS MEDIUM OR LOW (80)	0.60	110	TEND MORE TO BE THOSE WHERE POLITICAL ENCULTURATION IS MEDIUM OR LOW (80)	0.91	8 6 12 63 X SQ= 9.22 P = 0.002 RV NO (YES)
111	LEAN MORE TOWARD BEING THOSE WHERE POLITICAL ENCULTURATION IS HIGH OR MEDIUM (53)	0.75	111	LEAN LESS TOWARD BEING THOSE WHERE POLITICAL ENCULTURATION IS HIGH OR MEDIUM (53)	0.52	15 36 5 33 X SQ= 2.43 P = 0.079 RV NO (NO)

No.	Left subject		No.	Right subject		Statistics
115	TEND TO BE THOSE WHERE INTEREST ARTICULATION BY ASSOCIATIONAL GROUPS IS SIGNIFICANT (20)	0.52	115	TEND TO BE THOSE WHERE INTEREST ARTICULATION BY ASSOCIATIONAL GROUPS IS MODERATE, LIMITED, OR NEGLIGIBLE (91)	0.89	11 9 10 76 X SQ= 16.58 P = 0. RV YES (YES)
116	TEND TO BE THOSE WHERE INTEREST ARTICULATION BY ASSOCIATIONAL GROUPS IS SIGNIFICANT OR MODERATE (32)	0.57	116	TEND TO BE THOSE WHERE INTEREST ARTICULATION BY ASSOCIATIONAL GROUPS IS LIMITED OR NEGLIGIBLE (79)	0.76	12 20 9 65 X SQ= 7.50 P = 0.006 RV YES (NO)
119	TEND TO BE THOSE WHERE INTEREST ARTICULATION BY INSTITUTIONAL GROUPS IS MODERATE OR LIMITED (26)	0.68	119	TEND TO BE THOSE WHERE INTEREST ARTICULATION BY INSTITUTIONAL GROUPS IS VERY SIGNIFICANT OR SIGNIFICANT (74)	0.83	6 63 13 13 X SQ= 17.64 P = 0. RV YES (YES)
123	TEND LESS TO BE THOSE WHERE INTEREST ARTICULATION BY NON-ASSOCIATIONAL GROUPS IS SIGNIFICANT, MODERATE, OR LIMITED (107)	0.71	123	TEND MORE TO BE THOSE WHERE INTEREST ARTICULATION BY NON-ASSOCIATIONAL GROUPS IS SIGNIFICANT, MODERATE, OR LIMITED (107)	0.98	15 86 6 2 X SQ= 13.59 P = 0.001 RV NO (YES)
125	TEND TO BE THOSE WHERE INTEREST ARTICULATION BY ANOMIC GROUPS IS INFREQUENT OR VERY INFREQUENT (35)	0.67	125	TEND TO BE THOSE WHERE INTEREST ARTICULATION BY ANOMIC GROUPS IS FREQUENT OR OCCASIONAL (64)	0.71	7 52 14 21 X SQ= 8.47 P = 0.004 RV YES (NO)
128	TEND TO BE THOSE WHERE INTEREST ARTICULATION BY POLITICAL PARTIES IS SIGNIFICANT OR MODERATE (48)	0.90	128	TEND TO BE THOSE WHERE INTEREST ARTICULATION BY POLITICAL PARTIES IS LIMITED OR NEGLIGIBLE (45)	0.58	18 30 2 41 X SQ= 12.42 P = 0. RV YES (NO)
134	TEND TO BE THOSE WHERE INTEREST AGGREGATION BY THE EXECUTIVE IS SIGNIFICANT OR MODERATE (57)	0.85	134	TEND TO BE THOSE WHERE INTEREST AGGREGATION BY THE EXECUTIVE IS LIMITED OR NEGLIGIBLE (46)	0.50	17 39 3 39 X SQ= 6.60 P = 0.005 RV YES (NO)
137	TEND TO BE THOSE WHERE INTEREST AGGREGATION BY THE LEGISLATURE IS SIGNIFICANT OR MODERATE (29)	0.67	137	TEND TO BE THOSE WHERE INTEREST AGGREGATION BY THE LEGISLATURE IS LIMITED OR NEGLIGIBLE (68)	0.79	14 15 7 58 X SQ= 14.17 P = 0. RV YES (NO)
138	TEND TO BE THOSE WHERE INTEREST AGGREGATION BY THE LEGISLATURE IS SIGNIFICANT, MODERATE, OR LIMITED (48)	0.81	138	TEND TO BE THOSE WHERE INTEREST AGGREGATION BY THE LEGISLATURE IS NEGLIGIBLE (49)	0.58	17 31 4 43 X SQ= 8.48 P = 0.002 RV YES (NO)

```
139  ALWAYS ARE THOSE                            1.00    139  TEND LESS TO BE THOSE                        0.60        0   32
     WHERE THE PARTY SYSTEM IS QUANTITATIVELY                 WHERE THE PARTY SYSTEM IS QUANTITATIVELY                21   49
     OTHER THAN ONE-PARTY  (71)                               OTHER THAN ONE-PARTY  (71)                          X SQ=  10.32
                                                                                                                  P =    0.
                                                                                                                  RV NO  (NO )

141  TEND LESS TO BE THOSE                       0.70    141  TEND MORE TO BE THOSE                        0.93        6    5
     WHERE THE PARTY SYSTEM IS QUANTITATIVELY                 WHERE THE PARTY SYSTEM IS QUANTITATIVELY                14   70
     OTHER THAN TWO-PARTY  (87)                               OTHER THAN TWO-PARTY  (87)                          X SQ=   6.27
                                                                                                                  P =    0.010
                                                                                                                  RV NO  (YES)

147  TEND TO BE THOSE                            0.63    147  TEND TO BE THOSE                             0.84       12   11
     WHERE THE PARTY SYSTEM IS QUALITATIVELY                  WHERE THE PARTY SYSTEM IS QUALITATIVELY                  7   57
     CLASS-ORIENTED OR MULTI-IDEOLOGICAL  (23)                OTHER THAN                                          X SQ=  14.53
                                                              CLASS-ORIENTED OR MULTI-IDEOLOGICAL  (67)           P =    0.
                                                                                                                  RV YES (YES)

158  TEND MORE TO BE THOSE                       0.80    158  TEND LESS TO BE THOSE                        0.53        4   34
     WHERE PERSONALISMO IS                                    WHERE PERSONALISMO IS                                   16   39
     NEGLIGIBLE  (56)                                         NEGLIGIBLE  (56)                                    X SQ=   3.55
                                                                                                                  P =    0.041
                                                                                                                  RV NO  (NO )

161  TEND TO BE THOSE                            0.79    161  TEND TO BE THOSE                             0.53        4   40
     WHERE THE POLITICAL LEADERSHIP IS                        WHERE THE POLITICAL LEADERSHIP IS                       15   35
     NON-ELITIST  (50)                                        ELITIST OR MODERATELY ELITIST  (47)                 X SQ=   5.11
                                                                                                                  P =    0.019
                                                                                                                  RV YES (NO )

163  ALWAYS ARE THOSE                            1.00    163  TILT LESS TOWARD BEING THOSE                 0.85        0   12
     WHERE THE REGIME'S LEADERSHIP CHARISMA                   WHERE THE REGIME'S LEADERSHIP CHARISMA                  17   67
     IS MODERATE OR NEGLIGIBLE  (87)                          IS MODERATE OR NEGLIGIBLE  (87)                     X SQ=   1.73
                                                                                                                  P =    0.117
                                                                                                                  RV NO  (NO )

168  TEND TO BE THOSE                            0.75    168  TEND TO BE THOSE                             0.77       15   19
     WHERE THE HORIZONTAL POWER DISTRIBUTION                  WHERE THE HORIZONTAL POWER DISTRIBUTION                  5   62
     IS SIGNIFICANT  (34)                                     IS LIMITED OR NEGLIGIBLE  (72)                      X SQ=  16.84
                                                                                                                  P =    0.
                                                                                                                  RV YES (NO )

169  TEND TO BE THOSE                            0.95    169  TEND TO BE THOSE                             0.52       19   39
     WHERE THE HORIZONTAL POWER DISTRIBUTION                  WHERE THE HORIZONTAL POWER DISTRIBUTION                  1   42
     IS SIGNIFICANT OR LIMITED  (58)                          IS NEGLIGIBLE  (48)                                 X SQ=  12.55
                                                                                                                  P =    0.
                                                                                                                  RV YES (NO )

170  ALWAYS ARE THOSE                            1.00    170  TEND TO BE THOSE                             0.51        0   39
     WHERE THE LEGISLATIVE-EXECUTIVE STRUCTURE                WHERE THE LEGISLATIVE-EXECUTIVE STRUCTURE               21   38
     IS OTHER THAN PRESIDENTIAL  (63)                         IS PRESIDENTIAL  (39)                               X SQ=  15.62
                                                                                                                  P =    0.
                                                                                                                  RV YES (NO )
```

174	TEND TO BE THOSE WHERE THE LEGISLATURE IS FULLY EFFECTIVE (28)	0.67	174	TEND TO BE THOSE WHERE THE LEGISLATURE IS PARTIALLY EFFECTIVE, LARGELY INEFFECTIVE, OR WHOLLY INEFFECTIVE (72)	0.82	14 14 7 62 X SQ= 16.38 P = 0. RV YES (YES)
175	TEND TO BE THOSE WHERE THE LEGISLATURE IS FULLY EFFECTIVE OR PARTIALLY EFFECTIVE (51)	0.90	175	TEND TO BE THOSE WHERE THE LEGISLATURE IS LARGELY INEFFECTIVE OR WHOLLY INEFFECTIVE (49)	0.58	19 32 2 44 X SQ= 13.56 P = 0. RV YES (NO)
178	TEND TO BE THOSE WHERE THE LEGISLATURE IS BICAMERAL (51)	0.71	178	TEND TO BE THOSE WHERE THE LEGISLATURE IS UNICAMERAL (53)	0.57	6 46 15 34 X SQ= 4.47 P = 0.026 RV YES (NO)
179	ALWAYS ARE THOSE WHERE THE EXECUTIVE IS STRONG (39)	1.00	179	TEND TO BE THOSE WHERE THE EXECUTIVE IS DOMINANT (52)	0.68	0 48 16 23 X SQ= 21.47 P = 0. RV YES (NO)
181	TEND TO BE THOSE WHERE THE BUREAUCRACY IS MODERN, RATHER THAN SEMI-MODERN (21)	0.62	181	TEND TO BE THOSE WHERE THE BUREAUCRACY IS SEMI-MODERN, RATHER THAN MODERN (55)	0.81	10 11 6 48 X SQ= 9.93 P = 0.001 RV YES (NO)
186	ALWAYS ARE THOSE WHERE PARTICIPATION BY THE MILITARY IN POLITICS IS NEUTRAL, RATHER THAN SUPPORTIVE (56)	1.00	186	TEND LESS TO BE THOSE WHERE PARTICIPATION BY THE MILITARY IN POLITICS IS NEUTRAL, RATHER THAN SUPPORTIVE (56)	0.57	0 27 20 36 X SQ= 10.83 P = 0. RV NO (NO)
187	TEND TO BE THOSE WHERE THE ROLE OF THE POLICE IS NOT POLITICALLY SIGNIFICANT (35)	0.90	187	TEND TO BE THOSE WHERE THE ROLE OF THE POLICE IS POLITICALLY SIGNIFICANT (66)	0.77	2 58 18 17 X SQ= 27.94 P = 0. RV YES (YES)
190	TEND TO BE THOSE WHERE THE CHARACTER OF THE LEGAL SYSTEM IS COMMON LAW, RATHER THAN CIVIL LAW (7)	0.56	190	TEND TO BE THOSE WHERE THE CHARACTER OF THE LEGAL SYSTEM IS CIVIL LAW, RATHER THAN COMMON LAW (32)	0.93	4 26 5 2 X SQ= 7.49 P = 0.005 RV YES (YES)
194	ALWAYS ARE THOSE THAT ARE NON-COMMUNIST (101)	1.00	194	LEAN LESS TOWARD BEING THOSE THAT ARE NON-COMMUNIST (101)	0.85	0 13 21 74 X SQ= 2.30 P = 0.068 RV NO (NO)

```
**************************************************************************************************

173  POLITIES                                            173  POLITIES
     WHERE THE LEGISLATIVE-EXECUTIVE STRUCTURE                WHERE THE LEGISLATIVE-EXECUTIVE STRUCTURE
     IS PARLIAMENTARY-REPUBLICAN OR                           IS PRESIDENTIAL-PARLIAMENTARY OR
     PARLIAMENTARY-ROYALIST, RATHER THAN                      MONARCHICAL-PARLIAMENTARY, RATHER THAN
     PRESIDENTIAL-PARLIAMENTARY OR                            PARLIAMENTARY-REPUBLICAN OR
     MONARCHICAL-PARLIAMENTARY  (33)                          PARLIAMENTARY ROYALIST  (7)

                                                                                      SUBJECT ONLY
..................................................................................................

     33 IN LEFT COLUMN

AUSTRALIA   AUSTRIA     BELGIUM     BURUNDI     CANADA      CEYLON      CONGO(LEO)  DENMARK     FINLAND     GERMAN FR
GREECE      ICELAND     INDIA       IRELAND     ISRAEL      ITALY       JAMAICA     JAPAN       LAOS        LUXEMBOURG
MALAYA      NETHERLANDS NEW ZEALAND NIGERIA     NORWAY      SIERRE LEO  SOMALIA     SO AFRICA   SWEDEN      TRINIDAD
TURKEY      UGANDA      UK

      7 IN RIGHT COLUMN

AFGHANISTAN CHAD        FRANCE      JORDAN      LEBANON     LIBYA       MOROCCO

      4 EXCLUDED BECAUSE AMBIGUOUS

CAMBODIA    PORTUGAL    SPAIN       TANGANYIKA

     58 EXCLUDED BECAUSE IRRELEVANT

ALBANIA     ARGENTINA   BOLIVIA     BRAZIL      BULGARIA    CAMEROUN    CEN AFR REP CHILE       CHINA, PR   COLOMBIA
CONGO(BRA)  COSTA RICA  CYPRUS      CZECHOS'KIA DAHOMEY     DOMIN REP   ECUADOR     EL SALVADOR ETHIOPIA    GABON
GERMANY, E  GHANA       GUATEMALA   GUINEA      HAITI       HONDURAS    HUNGARY     INDONESIA   IVORY COAST KOREA, N
LIBERIA     MALAGASY R  MALI        MAURITANIA  MEXICO      MONGOLIA    NEPAL       NICARAGUA   NIGER       PAKISTAN
PANAMA      PARAGUAY    PHILIPPINES POLAND      RUMANIA     RWANDA      SA'U ARABIA SENEGAL     SWITZERLAND TUNISIA
USSR        US          UPPER VOLTA URUGUAY     VENEZUELA   VIETNAM, N  VIETNAM REP YUGOSLAVIA

     13 EXCLUDED BECAUSE UNASCERTAINABLE

ALGERIA     BURMA       CUBA        IRAN        IRAQ        KOREA REP   PERU        SUDAN       SYRIA       THAILAND
TOGO        UAR         YEMEN
--------------------------------------------------------------------------------------------------

  2  TILT TOWARD BEING THOSE                  0.52     2  TILT TOWARD BEING THOSE              0.86       17    1
     LOCATED IN WEST EUROPE, SCANDINAVIA,                 LOCATED ELSEWHERE THAN IN                        16    6
     NORTH AMERICA, OR AUSTRALASIA  (22)                  WEST EUROPE, SCANDINAVIA,                   X SQ=   1.90
                                                          NORTH AMERICA, OR AUSTRALASIA  (93)         P =    0.105
                                                                                                      RV YES (NO )

 13  TEND TO BE THOSE                         0.94    13  TEND TO BE THOSE                     0.71        2    5
     LOCATED ELSEWHERE THAN IN NORTH AFRICA OR            LOCATED IN NORTH AFRICA OR                       31    2
     THE MIDDLE EAST  (99)                                THE MIDDLE EAST  (16)                       X SQ=  12.86
                                                                                                      P =    0.001
                                                                                                      RV YES (YES)
```

14	TEND MORE TO BE THOSE LOCATED ELSEWHERE THAN IN THE MIDDLE EAST (104)	0.94	14	TEND LESS TO BE THOSE LOCATED ELSEWHERE THAN IN THE MIDDLE EAST (104)	0.57	2 3 31 4 X SQ= 4.18 P = 0.030 RV NO (YES)
31	TILT TOWARD BEING THOSE WHOSE AGRICULTURAL POPULATION IS LOW OR VERY LOW (24)	0.52	31	TILT TOWARD BEING THOSE WHOSE AGRICULTURAL POPULATION IS HIGH OR MEDIUM (90)	0.86	16 6 17 1 X SQ= 1.90 P = 0.105 RV YES (NO)
34	LEAN TOWARD BEING THOSE WHOSE GROSS NATIONAL PRODUCT IS VERY HIGH, HIGH, MEDIUM, OR LOW (62)	0.73	34	LEAN TOWARD BEING THOSE WHOSE GROSS NATIONAL PRODUCT IS VERY LOW (53)	0.71	24 2 9 5 X SQ= 3.20 P = 0.075 RV YES (NO)
35	TILT TOWARD BEING THOSE WHOSE PER CAPITA GROSS NATIONAL PRODUCT IS VERY HIGH OR HIGH (24)	0.52	35	TILT TOWARD BEING THOSE WHOSE PER CAPITA GROSS NATIONAL PRODUCT IS MEDIUM, LOW, OR VERY LOW (91)	0.86	17 1 16 6 X SQ= 1.90 P = 0.105 RV YES (NO)
36	LEAN TOWARD BEING THOSE WHOSE PER CAPITA GROSS NATIONAL PRODUCT IS VERY HIGH, HIGH, OR MEDIUM (42)	0.70	36	LEAN TOWARD BEING THOSE WHOSE PER CAPITA GROSS NATIONAL PRODUCT IS LOW OR VERY LOW (73)	0.71	23 2 10 5 X SQ= 2.60 P = 0.081 RV YES (NO)
37	LEAN TOWARD BEING THOSE WHOSE PER CAPITA GROSS NATIONAL PRODUCT IS VERY HIGH, HIGH, MEDIUM, OR LOW (64)	0.73	37	LEAN TOWARD BEING THOSE WHOSE PER CAPITA GROSS NATIONAL PRODUCT IS VERY LOW (51)	0.71	24 2 9 5 X SQ= 3.20 P = 0.075 RV YES (NO)
39	TILT TOWARD BEING THOSE WHOSE INTERNATIONAL FINANCIAL STATUS IS VERY HIGH, HIGH, OR MEDIUM (38)	0.52	39	TILT TOWARD BEING THOSE WHOSE INTERNATIONAL FINANCIAL STATUS IS LOW OR VERY LOW (76)	0.86	17 1 16 6 X SQ= 1.90 P = 0.105 RV YES (NO)
42	TEND TO BE THOSE WHOSE ECONOMIC DEVELOPMENTAL STATUS IS DEVELOPED OR INTERMEDIATE (36)	0.59	42	TEND TO BE THOSE WHOSE ECONOMIC DEVELOPMENTAL STATUS IS UNDERDEVELOPED OR VERY UNDERDEVELOPED (76)	0.86	19 1 13 6 X SQ= 3.04 P = 0.044 RV YES (NO)
43	LEAN TOWARD BEING THOSE WHOSE ECONOMIC DEVELOPMENTAL STATUS IS DEVELOPED, INTERMEDIATE, OR UNDERDEVELOPED (55)	0.76	43	LEAN TOWARD BEING THOSE WHOSE ECONOMIC DEVELOPMENTAL STATUS IS VERY UNDERDEVELOPED (57)	0.67	25 2 8 4 X SQ= 2.53 P = 0.060 RV YES (NO)

45	LEAN TOWARD BEING THOSE WHERE THE LITERACY RATE IS FIFTY PERCENT OR ABOVE (55)	0.67	45	LEAN TOWARD BEING THOSE WHERE THE LITERACY RATE IS BELOW FIFTY PERCENT (54)	0.83	22 1 11 5 X SQ= 3.38 P = 0.064 RV YES (NO)
50	TEND TO BE THOSE WHERE FREEDOM OF THE PRESS IS COMPLETE (43)	0.80	50	TEND TO BE THOSE WHERE FREEDOM OF THE PRESS IS INTERMITTENT, INTERNALLY ABSENT, OR INTERNALLY AND EXTERNALLY ABSENT (56)	0.86	24 1 6 6 X SQ= 8.39 P = 0.002 RV YES (YES)
51	ALWAYS ARE THOSE WHERE FREEDOM OF THE PRESS IS COMPLETE OR INTERMITTENT (60)	1.00	51	TEND LESS TO BE THOSE WHERE FREEDOM OF THE PRESS IS COMPLETE OR INTERMITTENT (60)	0.71	30 5 0 2 X SQ= 4.33 P = 0.032 RV NO (YES)
54	LEAN TOWARD BEING THOSE WHERE NEWSPAPER CIRCULATION IS ONE HUNDRED OR MORE PER THOUSAND (37)	0.58	54	LEAN TOWARD BEING THOSE WHERE NEWSPAPER CIRCULATION IS LESS THAN ONE HUNDRED PER THOUSAND (76)	0.86	19 1 14 6 X SQ= 2.77 P = 0.091 RV YES (NO)
58	TEND TO BE THOSE WHERE THE RELIGION IS OTHER THAN MUSLIM (97)	0.94	58	TEND TO BE THOSE WHERE THE RELIGION IS MUSLIM (18)	0.57	2 4 31 3 X SQ= 8.15 P = 0.005 RV YES (YES)
63	LEAN TOWARD BEING THOSE WHERE THE RELIGION IS PREDOMINANTLY SOME KIND OF CHRISTIAN (46)	0.55	63	LEAN TOWARD BEING THOSE WHERE THE RELIGION IS PREDOMINANTLY OR PARTLY OTHER THAN CHRISTIAN (68)	0.86	18 1 15 6 X SQ= 2.31 P = 0.095 RV YES (NO)
64	TEND TO BE THOSE WHERE THE RELIGION IS CHRISTIAN, RATHER THAN MUSLIM (46)	0.90	64	TEND TO BE THOSE WHERE THE RELIGION IS MUSLIM, RATHER THAN CHRISTIAN (18)	0.80	18 1 2 4 X SQ= 7.25 P = 0.005 RV YES (YES)
74	LEAN TOWARD BEING THOSE THAT ARE HISTORICALLY WESTERN (26)	0.55	74	LEAN TOWARD BEING THOSE THAT ARE NOT HISTORICALLY WESTERN (87)	0.86	17 1 14 6 X SQ= 2.32 P = 0.093 RV YES (NO)
75	TEND TO BE THOSE THAT ARE HISTORICALLY WESTERN OR SIGNIFICANTLY WESTERNIZED (62)	0.75	75	TEND TO BE THOSE THAT ARE NOT HISTORICALLY WESTERN AND ARE NOT SIGNIFICANTLY WESTERNIZED (52)	0.71	24 2 8 5 X SQ= 3.68 P = 0.030 RV YES (NO)

80	TEND TO BE THOSE WHOSE DATE OF INDEPENDENCE IS AFTER 1914, AND THAT WERE FORMERLY DEPENDENCIES OF BRITAIN, RATHER THAN FRANCE (19)	0.91	80	TEND TO BE THOSE WHOSE DATE OF INDEPENDENCE IS AFTER 1914, AND THAT WERE FORMERLY DEPENDENCIES OF FRANCE, RATHER THAN BRITAIN (24)	0.75	10 1 1 3 X SQ= 3.58 P = 0.033 RV YES (YES)
82	LEAN TOWARD BEING THOSE WHERE THE TYPE OF POLITICAL MODERNIZATION IS EARLY OR LATER EUROPEAN OR EUROPEAN DERIVED, RATHER THAN DEVELOPED TUTELARY OR UNDEVELOPED TUTELARY (51)	0.58	82	LEAN TOWARD BEING THOSE WHERE THE TYPE OF POLITICAL MODERNIZATION IS DEVELOPED TUTELARY OR UNDEVELOPED TUTELARY, RATHER THAN EARLY OR LATER EUROPEAN OR EUROPEAN DERIVED (55)	0.83	18 1 13 5 X SQ= 1.99 P = 0.090 RV YES (NO)
91	ALWAYS ARE THOSE WHOSE IDEOLOGICAL ORIENTATION IS DEVELOPMENTAL, RATHER THAN TRADITIONAL (31)	1.00	91	LEAN TOWARD BEING THOSE WHOSE IDEOLOGICAL ORIENTATION IS TRADITIONAL, RATHER THAN DEVELOPMENTAL (5)	0.67	6 1 0 2 X SQ= 2.01 P = 0.083 RV YES (YES)
94	LEAN MORE TOWARD BEING THOSE WHERE THE STATUS OF THE REGIME IS CONSTITUTIONAL (51)	0.97	94	LEAN LESS TOWARD BEING THOSE WHERE THE STATUS OF THE REGIME IS CONSTITUTIONAL (51)	0.67	28 4 1 2 X SQ= 2.49 P = 0.070 RV NO (YES)
96	LEAN MORE TOWARD BEING THOSE WHERE THE STATUS OF THE REGIME IS CONSTITUTIONAL OR TOTALITARIAN (67)	0.97	96	LEAN LESS TOWARD BEING THOSE WHERE THE STATUS OF THE REGIME IS CONSTITUTIONAL OR TOTALITARIAN (67)	0.67	28 4 1 2 X SQ= 2.49 P = 0.070 RV NO (YES)
101	TEND TO BE THOSE WHERE THE REPRESENTATIVE CHARACTER OF THE REGIME IS POLYARCHIC (41)	0.93	101	TEND TO BE THOSE WHERE THE REPRESENTATIVE CHARACTER OF THE REGIME IS LIMITED POLYARCHIC, PSEUDO-POLYARCHIC, OR NON-POLYARCHIC (57)	0.60	26 2 2 3 X SQ= 5.57 P = 0.017 RV YES (YES)
105	TEND TO BE THOSE WHERE THE ELECTORAL SYSTEM IS COMPETITIVE (43)	0.90	105	TEND TO BE THOSE WHERE THE ELECTORAL SYSTEM IS PARTIALLY COMPETITIVE OR NON-COMPETITIVE (47)	0.80	27 1 3 4 X SQ= 9.11 P = 0.003 RV YES (YES)
107	TEND TO BE THOSE WHERE AUTONOMOUS GROUPS ARE FULLY TOLERATED IN POLITICS (46)	0.84	107	TEND TO BE THOSE WHERE AUTONOMOUS GROUPS ARE PARTIALLY TOLERATED IN POLITICS, ARE TOLERATED ONLY OUTSIDE POLITICS, OR ARE NOT TOLERATED AT ALL (65)	0.71	27 2 5 5 X SQ= 6.68 P = 0.007 RV YES (YES)
110	LEAN LESS TOWARD BEING THOSE WHERE POLITICAL ENCULTURATION IS MEDIUM OR LOW (80)	0.62	110	ALWAYS ARE THOSE WHERE POLITICAL ENCULTURATION IS MEDIUM OR LOW (80)	1.00	12 0 20 7 X SQ= 2.24 P = 0.077 RV NO (NO)

114	TEND LESS TO BE THOSE WHERE SECTIONALISM IS EXTREME OR MODERATE (61)	0.52	114	ALWAYS ARE THOSE WHERE SECTIONALISM IS EXTREME OR MODERATE (61)	1.00	17 7 16 0 X SQ= 3.82 P = 0.029 RV NO (NO)
116	TILT TOWARD BEING THOSE WHERE INTEREST ARTICULATION BY ASSOCIATIONAL GROUPS IS SIGNIFICANT OR MODERATE (32)	0.64	116	TILT TOWARD BEING THOSE WHERE INTEREST ARTICULATION BY ASSOCIATIONAL GROUPS IS LIMITED OR NEGLIGIBLE (79)	0.71	21 2 12 5 X SQ= 1.65 P = 0.113 RV YES (NO)
117	LEAN TOWARD BEING THOSE WHERE INTEREST ARTICULATION BY ASSOCIATIONAL GROUPS IS SIGNIFICANT, MODERATE, OR LIMITED (60)	0.79	117	LEAN TOWARD BEING THOSE WHERE INTEREST ARTICULATION BY ASSOCIATIONAL GROUPS IS NEGLIGIBLE (51)	0.57	26 3 7 4 X SQ= 2.15 P = 0.075 RV YES (NO)
118	TEND TO BE THOSE WHERE INTEREST ARTICULATION BY INSTITUTIONAL GROUPS IS SIGNIFICANT, MODERATE, OR LIMITED (60)	0.97	118	TEND TO BE THOSE WHERE INTEREST ARTICULATION BY INSTITUTIONAL GROUPS IS VERY SIGNIFICANT (40)	0.50	1 3 29 3 X SQ= 6.81 P = 0.010 RV YES (YES)
119	LEAN TOWARD BEING THOSE WHERE INTEREST ARTICULATION BY INSTITUTIONAL GROUPS IS MODERATE OR LIMITED (26)	0.60	119	LEAN TOWARD BEING THOSE WHERE INTEREST ARTICULATION BY INSTITUTIONAL GROUPS IS VERY SIGNIFICANT OR SIGNIFICANT (74)	0.83	12 5 18 1 X SQ= 2.23 P = 0.081 RV YES (NO)
125	LEAN TOWARD BEING THOSE WHERE INTEREST ARTICULATION BY ANOMIC GROUPS IS INFREQUENT OR VERY INFREQUENT (35)	0.64	125	LEAN TOWARD BEING THOSE WHERE INTEREST ARTICULATION BY ANOMIC GROUPS IS FREQUENT OR OCCASIONAL (64)	0.83	12 5 21 1 X SQ= 2.85 P = 0.068 RV YES (NO)
133	TEND TO BE THOSE WHERE INTEREST AGGREGATION BY THE EXECUTIVE IS MODERATE, LIMITED, OR NEGLIGIBLE (76)	0.90	133	TEND TO BE THOSE WHERE INTEREST AGGREGATION BY THE EXECUTIVE IS SIGNIFICANT (29)	0.50	3 3 28 3 X SQ= 3.41 P = 0.042 RV YES (YES)
134	LEAN TOWARD BEING THOSE WHERE INTEREST AGGREGATION BY THE EXECUTIVE IS SIGNIFICANT OR MODERATE (57)	0.87	134	LEAN TOWARD BEING THOSE WHERE INTEREST AGGREGATION BY THE EXECUTIVE IS LIMITED OR NEGLIGIBLE (46)	0.50	26 3 4 3 X SQ= 2.27 P = 0.073 RV YES (NO)
137	TEND TO BE THOSE WHERE INTEREST AGGREGATION BY THE LEGISLATURE IS SIGNIFICANT OR MODERATE (29)	0.72	137	TEND TO BE THOSE WHERE INTEREST AGGREGATION BY THE LEGISLATURE IS LIMITED OR NEGLIGIBLE (68)	0.83	23 1 9 5 X SQ= 4.46 P = 0.018 RV YES (NO)

138	TILT TOWARD BEING THOSE WHERE INTEREST AGGREGATION BY THE LEGISLATURE IS SIGNIFICANT, MODERATE, OR LIMITED (48)	0.81	138	TILT TOWARD BEING THOSE WHERE INTEREST AGGREGATION BY THE LEGISLATURE IS NEGLIGIBLE (49)	0.50	26 3 6 3 X SQ= 1.27 P = 0.131 RV YES (NO)
144	ALWAYS ARE THOSE WHERE THE PARTY SYSTEM IS QUANTITATIVELY TWO-PARTY, RATHER THAN ONE-PARTY (11)	1.00	144	ALWAYS ARE THOSE WHERE THE PARTY SYSTEM IS QUANTITATIVELY ONE-PARTY, RATHER THAN TWO-PARTY (34)	1.00	0 1 7 0 X SQ= 1.47 P = 0.125 RV YES (YES)
147	TILT TOWARD BEING THOSE WHERE THE PARTY SYSTEM IS QUALITATIVELY CLASS-ORIENTED OR MULTI-IDEOLOGICAL (23)	0.67	147	ALWAYS ARE THOSE WHERE THE PARTY SYSTEM IS QUALITATIVELY OTHER THAN CLASS-ORIENTED OR MULTI-IDEOLOGICAL (67)	1.00	18 0 9 2 X SQ= 1.25 P = 0.135 RV YES (NO)
153	TEND TU BE THOSE WHERE THE PARTY SYSTEM IS STABLE (42)	0.68	153	ALWAYS ARE THOSE WHERE THE PARTY SYSTEM IS MODERATELY STABLE OR UNSTABLE (38)	1.00	19 0 9 3 X SQ= 2.79 P = 0.049 RV YES (NO)
158	TEND TO BE THOSE WHERE PERSONALISMO IS NEGLIGIBLE (56)	0.84	158	TEND TO BE THOSE WHERE PERSONALISMO IS PRONOUNCED OR MODERATE (40)	0.75	5 3 27 1 X SQ= 4.22 P = 0.028 RV YES (NO)
160	TILT MORE TOWARD BEING THOSE WHERE THE POLITICAL LEADERSHIP IS MODERATELY ELITIST OR NON-ELITIST (67)	0.94	160	TILT LESS TOWARD BEING THOSE WHERE THE POLITICAL LEADERSHIP IS MODERATELY ELITIST OR NON-ELITIST (67)	0.67	2 2 29 4 X SQ= 1.50 P = 0.115 RV NO (YES)
164	TEND TO BE THOSE WHERE THE REGIME'S LEADERSHIP CHARISMA IS NEGLIGIBLE (65)	0.93	164	TEND TO BE THOSE WHERE THE REGIME'S LEADERSHIP CHARISMA IS PRONOUNCED OR MODERATE (34)	0.83	2 5 26 1 X SQ= 13.19 P = 0. RV YES (YES)
168	TEND TO BE THOSE WHERE THE HORIZONTAL POWER DISTRIBUTION IS SIGNIFICANT (34)	0.81	168	ALWAYS ARE THOSE WHERE THE HORIZONTAL POWER DISTRIBUTION IS LIMITED OR NEGLIGIBLE (72)	1.00	25 0 6 7 X SQ= 13.11 P = 0. RV YES (YES)
174	TEND TO BE THOSE WHERE THE LEGISLATURE IS FULLY EFFECTIVE (28)	0.72	174	ALWAYS ARE THOSE WHERE THE LEGISLATURE IS PARTIALLY EFFECTIVE, LARGELY INEFFECTIVE, OR WHOLLY INEFFECTIVE (72)	1.00	23 0 9 7 X SQ= 9.47 P = 0.001 RV YES (NO)

179	TEND TO BE THOSE WHERE THE EXECUTIVE IS STRONG (39)	0.92	179	TEND TO BE THOSE WHERE THE EXECUTIVE IS DOMINANT (52)	0.67	2 4 24 2 X SQ= 7.60 P = 0.006 RV YES (YES)
181	TILT TOWARD BEING THOSE WHERE THE BUREAUCRACY IS MODERN, RATHER THAN SEMI-MODERN (21)	0.69	181	TILT TOWARD BEING THOSE WHERE THE BUREAUCRACY IS SEMI-MODERN, RATHER THAN MODERN (55)	0.75	18 1 8 3 X SQ= 1.33 P = 0.126 RV YES (NO)
186	TEND TO BE THOSE WHERE PARTICIPATION BY THE MILITARY IN POLITICS IS NEUTRAL, RATHER THAN SUPPORTIVE (56)	0.97	186	TEND TO BE THOSE WHERE PARTICIPATION BY THE MILITARY IN POLITICS IS SUPPORTIVE, RATHER THAN NEUTRAL (31)	0.50	1 3 28 3 X SQ= 6.54 P = 0.011 RV YES (YES)
187	TEND TO BE THOSE WHERE THE ROLE OF THE POLICE IS NOT POLITICALLY SIGNIFICANT (35)	0.81	187	TEND TO BE THOSE WHERE THE ROLE OF THE POLICE IS POLITICALLY SIGNIFICANT (66)	0.80	6 4 26 1 X SQ= 5.41 P = 0.014 RV YES (NO)
192	TEND TO BE THOSE WHERE THE CHARACTER OF THE LEGAL SYSTEM IS OTHER THAN MUSLIM OR PARTLY MUSLIM (86)	0.88	192	TEND TO BE THOSE WHERE THE CHARACTER OF THE LEGAL SYSTEM IS MUSLIM OR PARTLY MUSLIM (27)	0.57	4 4 29 3 X SQ= 4.77 P = 0.020 RV YES (YES)

174 POLITIES WHERE THE LEGISLATURE IS FULLY EFFECTIVE (28)

174 POLITIES WHERE THE LEGISLATURE IS PARTIALLY EFFECTIVE, LARGELY INEFFECTIVE, OR WHOLLY INEFFECTIVE (72)

PREDICATE ONLY

28 IN LEFT COLUMN

AUSTRALIA	AUSTRIA	BELGIUM	CANADA	CYPRUS	DENMARK	FINLAND	GERMAN FR	ICELAND	INDIA
IRELAND	ISRAEL	ITALY	JAMAICA	JAPAN	LUXEMBOURG	MALAYA	NETHERLANDS	NEW ZEALAND	NORWAY
PHILIPPINES	SO AFRICA	SWEDEN	SWITZERLAND	TRINIDAD	UK	US	URUGUAY		

72 IN RIGHT COLUMN

AFGHANISTAN	ALBANIA	ALGERIA	ARGENTINA	BOLIVIA	BRAZIL	BULGARIA	BURUNDI	CAMBODIA	CAMEROUN
CEN AFR REP	CEYLON	CHAD	CHILE	CHINA, PR	COLOMBIA	CONGO(BRA)	COSTA RICA	CZECHOS'KIA	DAHOMEY
ECUADOR	EL SALVADOR	ETHIOPIA	FRANCE	GABON	GERMANY, E	GHANA	GREECE	GUINEA	HAITI
HONDURAS	HUNGARY	INDONESIA	IVORY COAST	JORDAN	KOREA, N	LAOS	LEBANON	LIBERIA	LIBYA
MALAGASY R	MALI	MAURITANIA	MEXICO	MONGOLIA	MOROCCO	NEPAL	NICARAGUA	NIGER	NIGERIA
PANAMA	PARAGUAY	POLAND	PORTUGAL	RUMANIA	RWANDA	SENEGAL	SIERRE LEO	SOMALIA	SPAIN
TANGANYIKA	THAILAND	TUNISIA	TURKEY	UGANDA	USSR	UAR	UPPER VOLTA	VENEZUELA	VIETNAM, N
VIETNAM REP	YUGOSLAVIA								

1 EXCLUDED BECAUSE IRRELEVANT

SA'U ARABIA

14 EXCLUDED BECAUSE UNASCERTAINABLE

BURMA	CONGO(LEO)	CUBA	DOMIN REP	GUATEMALA	IRAN	IRAQ	KOREA REP	PAKISTAN	PERU
SUDAN	SYRIA	TOGO	YEMEN						

175 POLITIES
WHERE THE LEGISLATURE IS
FULLY EFFECTIVE OR
PARTIALLY EFFECTIVE (51)

175 POLITIES
WHERE THE LEGISLATURE IS
LARGELY INEFFECTIVE OR
WHOLLY INEFFECTIVE (49)

BOTH SUBJECT AND PREDICATE

51 IN LEFT COLUMN

ARGENTINA	AUSTRALIA	AUSTRIA	BELGIUM	BOLIVIA	BRAZIL	CAMEROUN	CANADA	CEYLON	CHILE
COLOMBIA	COSTA RICA	CYPRUS	DENMARK	ECUADOR	FINLAND	FRANCE	GERMAN FR	GREECE	HONDURAS
ICELAND	INDIA	IRELAND	ISRAEL	ITALY	JAMAICA	JAPAN	LEBANON	LIBYA	LUXEMBOURG
MALAGASY R	MALAYA	MEXICO	MOROCCO	NETHERLANDS	NEW ZEALAND	NIGERIA	NORWAY	PANAMA	PHILIPPINES
SIERRE LEO	SO AFRICA	SWEDEN	SWITZERLAND	TRINIDAD	TURKEY	UGANDA	UK	US	URUGUAY
VENEZUELA									

49 IN RIGHT COLUMN

AFGHANISTAN	ALBANIA	ALGERIA	BULGARIA	BURUNDI	CAMBODIA	CEN AFR REP	CHAD	CHINA, PR	CONGO(BRA)
CZECHOS'KIA	DAHOMEY	EL SALVADOR	ETHIOPIA	GABON	GERMANY, E	GHANA	GUINEA	HAITI	HUNGARY
INDONESIA	IVORY COAST	JORDAN	KOREA, N	LAOS	LIBERIA	MALI	MAURITANIA	MONGOLIA	NEPAL
NICARAGUA	NIGER	PARAGUAY	POLAND	PORTUGAL	RUMANIA	RWANDA	SENEGAL	SOMALIA	SPAIN
TANGANYIKA	THAILAND	TUNISIA	USSR	UAR	UPPER VOLTA	VIETNAM, N	VIETNAM REP	YUGOSLAVIA	

1 EXCLUDED BECAUSE IRRELEVANT

SA'U ARABIA

14 EXCLUDED BECAUSE UNASCERTAINABLE

BURMA	CONGO(LEO)	CUBA	DOMIN REP	GUATEMALA	IRAN	IRAQ	KOREA REP	PAKISTAN	PERU
SUDAN	SYRIA	TOGO	YEMEN						

2	TEND LESS TO BE THOSE LOCATED ELSEWHERE THAN IN WEST EUROPE, SCANDINAVIA, NORTH AMERICA, OR AUSTRALASIA (93)	0.61	2	TEND MORE TO BE THOSE LOCATED ELSEWHERE THAN IN WEST EUROPE, SCANDINAVIA, NORTH AMERICA, OR AUSTRALASIA (93)	0.96	20 2 31 47 X SQ= 15.99 P = 0. RV NO (YES)
5	ALWAYS ARE THOSE LOCATED IN SCANDINAVIA OR WEST EUROPE, RATHER THAN IN EAST EUROPE (18)	1.00	5	TEND TO BE THOSE LOCATED IN EAST EUROPE, RATHER THAN IN SCANDINAVIA OR WEST EUROPE (9)	0.82	16 2 0 9 X SQ= 16.13 P = 0. RV YES (YES)
6	TEND LESS TO BE THOSE LOCATED ELSEWHERE THAN IN THE CARIBBEAN, CENTRAL AMERICA, OR SOUTH AMERICA (93)	0.73	6	TEND MORE TO BE THOSE LOCATED ELSEWHERE THAN IN THE CARIBBEAN, CENTRAL AMERICA, OR SOUTH AMERICA (93)	0.92	14 4 37 45 X SQ= 5.06 P = 0.018 RV NO (YES)

10	TEND MORE TO BE THOSE LOCATED ELSEWHERE THAN IN NORTH AFRICA, OR CENTRAL AND SOUTH AFRICA (82)	0.84	10	TEND LESS TO BE THOSE LOCATED ELSEWHERE THAN IN NORTH AFRICA, OR CENTRAL AND SOUTH AFRICA (82)	0.55	8 22 43 27 X SQ= 8.81 P = 0.002 RV NO (YES)
15	TILT LESS TOWARD BEING THOSE LOCATED IN NORTH AFRICA OR CENTRAL AND SOUTH AFRICA, RATHER THAN IN THE MIDDLE EAST (33)	0.67	15	TILT MORE TOWARD BEING THOSE LOCATED IN NORTH AFRICA OR CENTRAL AND SOUTH AFRICA, RATHER THAN IN THE MIDDLE EAST (33)	0.92	4 2 8 22 X SQ= 2.02 P = 0.149 RV NO (YES)
19	TEND TO BE THOSE LOCATED IN THE CARIBBEAN, CENTRAL AMERICA, OR SOUTH AMERICA, RATHER THAN IN NORTH AFRICA OR CENTRAL AND SOUTH AFRICA (22)	0.64	19	TEND TO BE THOSE LOCATED IN NORTH AFRICA OR CENTRAL AND SOUTH AFRICA, RATHER THAN IN THE CARIBBEAN, CENTRAL AMERICA, OR SOUTH AMERICA (33)	0.85	8 22 14 4 X SQ= 9.87 P = 0.001 RV YES (YES)
20	TEND TO BE THOSE LOCATED IN THE CARIBBEAN, CENTRAL AMERICA, OR SOUTH AMERICA, RATHER THAN IN EAST ASIA, SOUTH ASIA, OR SOUTHEAST ASIA (22)	0.74	20	TEND TO BE THOSE LOCATED IN EAST ASIA, SOUTH ASIA, OR SOUTHEAST ASIA, RATHER THAN IN THE CARIBBEAN, CENTRAL AMERICA, OR SOUTH AMERICA (18)	0.71	5 10 14 4 X SQ= 4.92 P = 0.015 RV YES (YES)
25	TEND LESS TO BE THOSE WHOSE POPULATION DENSITY IS MEDIUM OR LOW (98)	0.75	25	TEND MORE TO BE THOSE WHOSE POPULATION DENSITY IS MEDIUM OR LOW (98)	0.94	13 3 38 46 X SQ= 5.61 P = 0.013 RV NO (YES)
29	TEND TO BE THOSE WHERE THE DEGREE OF URBANIZATION IS HIGH (56)	0.79	29	TEND TO BE THOSE WHERE THE DEGREE OF URBANIZATION IS LOW (49)	0.72	37 12 10 31 X SQ= 21.38 P = 0. RV YES (YES)
30	TEND TO BE THOSE WHOSE AGRICULTURAL POPULATION IS MEDIUM, LOW, OR VERY LOW (57)	0.80	30	TEND TO BE THOSE WHOSE AGRICULTURAL POPULATION IS HIGH (56)	0.75	10 36 41 12 X SQ= 28.31 P = 0. RV YES (YES)
34	TEND TO BE THOSE WHOSE GROSS NATIONAL PRODUCT IS VERY HIGH, HIGH, MEDIUM, OR LOW (62)	0.69	34	TEND TO BE THOSE WHOSE GROSS NATIONAL PRODUCT IS VERY LOW (53)	0.63	35 18 16 31 X SQ= 8.96 P = 0.002 RV YES (YES)
36	TEND TO BE THOSE WHOSE PER CAPITA GROSS NATIONAL PRODUCT IS VERY HIGH, HIGH, OR MEDIUM (42)	0.63	36	TEND TO BE THOSE WHOSE PER CAPITA GROSS NATIONAL PRODUCT IS LOW OR VERY LOW (73)	0.82	32 9 19 40 X SQ= 18.55 P = 0. RV YES (YES)

37	TEND TO BE THOSE WHOSE PER CAPITA GROSS NATIONAL PRODUCT IS VERY HIGH, HIGH, MEDIUM, OR LOW (64)	0.80	37	TEND TO BE THOSE WHOSE PER CAPITA GROSS NATIONAL PRODUCT IS VERY LOW (51)	0.67	41 16 10 33 X SQ= 21.33 P = 0. RV YES (YES)
39	TEND TO BE THOSE WHOSE INTERNATIONAL FINANCIAL STATUS IS VERY HIGH, HIGH, OR MEDIUM (38)	0.53	39	TEND TO BE THOSE WHOSE INTERNATIONAL FINANCIAL STATUS IS LOW OR VERY LOW (76)	0.79	27 10 24 38 X SQ= 9.56 P = 0.002 RV YES (YES)
40	TEND TO BE THOSE WHOSE INTERNATIONAL FINANCIAL STATUS IS VERY HIGH, HIGH, MEDIUM, OR LOW (71)	0.78	40	TEND TO BE THOSE WHOSE INTERNATIONAL FINANCIAL STATUS IS VERY LOW (39)	0.57	38 20 11 26 X SQ= 10.20 P = 0.001 RV YES (YES)
42	TEND TO BE THOSE WHOSE ECONOMIC DEVELOPMENTAL STATUS IS DEVELOPED OR INTERMEDIATE (36)	0.57	42	TEND TO BE THOSE WHOSE ECONOMIC DEVELOPMENTAL STATUS IS UNDERDEVELOPED OR VERY UNDERDEVELOPED (76)	0.84	28 8 21 41 X SQ= 15.85 P = 0. RV YES (YES)
43	TEND TO BE THOSE WHOSE ECONOMIC DEVELOPMENTAL STATUS IS DEVELOPED, INTERMEDIATE, OR UNDERDEVELOPED (55)	0.80	43	TEND TO BE THOSE WHOSE ECONOMIC DEVELOPMENTAL STATUS IS VERY UNDERDEVELOPED (57)	0.78	39 11 10 38 X SQ= 29.77 P = 0. RV YES (YES)
45	TEND TO BE THOSE WHERE THE LITERACY RATE IS FIFTY PERCENT OR ABOVE (55)	0.72	45	TEND TO BE THOSE WHERE THE LITERACY RATE IS BELOW FIFTY PERCENT (54)	0.66	36 15 14 29 X SQ= 12.07 P = 0. RV YES (YES)
49	TEND TO BE THOSE WHERE THE LITERACY RATE IS BETWEEN TEN AND FIFTY PERCENT, RATHER THAN BELOW TEN PERCENT (24)	0.77	49	TEND TO BE THOSE WHERE THE LITERACY RATE IS BELOW TEN PERCENT, RATHER THAN BETWEEN TEN AND FIFTY PERCENT (26)	0.67	10 9 3 18 X SQ= 5.05 P = 0.017 RV YES (YES)
50	TEND TO BE THOSE WHERE FREEDOM OF THE PRESS IS COMPLETE (43)	0.75	50	TEND TO BE THOSE WHERE FREEDOM OF THE PRESS IS INTERMITTENT, INTERNALLY ABSENT, OR INTERNALLY AND EXTERNALLY ABSENT (56)	0.84	36 6 12 32 X SQ= 27.44 P = 0. RV YES (YES)
51	TEND TO BE THOSE WHERE FREEDOM OF THE PRESS IS COMPLETE OR INTERMITTENT (60)	0.98	51	TEND TO BE THOSE WHERE FREEDOM OF THE PRESS IS INTERNALLY ABSENT, OR INTERNALLY AND EXTERNALLY ABSENT (37)	0.79	47 8 1 30 X SQ= 51.07 P = 0. RV YES (YES)

52	ALWAYS ARE THOSE WHERE FREEDOM OF THE PRESS IS COMPLETE, INTERMITTENT, OR INTERNALLY ABSENT (82)	1.00	52	TEND LESS TO BE THOSE WHERE FREEDOM OF THE PRESS IS COMPLETE, INTERMITTENT, OR INTERNALLY ABSENT (82)	0.63	49 24 0 14 X SQ= 18.87 P = 0. RV NO (YES)
54	TEND TO BE THOSE WHERE NEWSPAPER CIRCULATION IS ONE HUNDRED OR MORE PER THOUSAND (37)	0.55	54	TEND TO BE THOSE WHERE NEWSPAPER CIRCULATION IS LESS THAN ONE HUNDRED PER THOUSAND (76)	0.83	28 8 23 39 X SQ= 13.52 P = 0. RV YES (YES)
56	TEND MORE TO BE THOSE WHERE THE RELIGION IS PREDOMINANTLY LITERATE (79)	0.84	56	TEND LESS TO BE THOSE WHERE THE RELIGION IS PREDOMINANTLY LITERATE (79)	0.60	43 27 8 18 X SQ= 5.98 P = 0.011 RV NO (YES)
57	TEND LESS TO BE THOSE WHERE THE RELIGION IS OTHER THAN CATHOLIC (90)	0.67	57	TEND MORE TO BE THOSE WHERE THE RELIGION IS OTHER THAN CATHOLIC (90)	0.88	17 6 34 43 X SQ= 5.14 P = 0.017 RV NO (YES)
58	LEAN MORE TOWARD BEING THOSE WHERE THE RELIGION IS OTHER THAN MUSLIM (97)	0.94	58	LEAN LESS TOWARD BEING THOSE WHERE THE RELIGION IS OTHER THAN MUSLIM (97)	0.82	3 9 48 40 X SQ= 2.60 P = 0.069 RV NO (YES)
63	TEND TO BE THOSE WHERE THE RELIGION IS PREDOMINANTLY SOME KIND OF CHRISTIAN (46)	0.63	63	TEND TO BE THOSE WHERE THE RELIGION IS PREDOMINANTLY OR PARTLY OTHER THAN CHRISTIAN (68)	0.75	32 12 19 36 X SQ= 12.78 P = 0. RV YES (YES)
65	TEND TO BE THOSE WHERE THE RELIGION IS CATHOLIC, RATHER THAN MUSLIM (25)	0.85	65	TEND TO BE THOSE WHERE THE RELIGION IS MUSLIM, RATHER THAN CATHOLIC (18)	0.60	17 6 3 9 X SQ= 5.84 P = 0.011 RV YES (YES)
68	TEND TO BE THOSE THAT ARE LINGUISTICALLY HOMOGENEOUS (52)	0.59	68	TEND TO BE THOSE THAT ARE LINGUISTICALLY WEAKLY HETEROGENEOUS OR STRONGLY HETEROGENEOUS (62)	0.65	30 17 21 31 X SQ= 4.53 P = 0.027 RV YES (YES)
69	TILT TOWARD BEING THOSE THAT ARE LINGUISTICALLY HOMOGENEOUS OR WEAKLY HETEROGENEOUS (64)	0.67	69	TILT TOWARD BEING THOSE THAT ARE LINGUISTICALLY STRONGLY HETEROGENEOUS (50)	0.50	34 24 17 24 X SQ= 2.19 P = 0.106 RV YES (YES)

72	TEND TO BE THOSE WHOSE DATE OF INDEPENDENCE IS BEFORE 1914 (52)	0.62	72	TEND TO BE THOSE WHOSE DATE OF INDEPENDENCE IS AFTER 1914 (59)	0.66	31 16 19 31 X SQ= 6.50 P = 0.008 RV YES (YES)
73	TEND TO BE THOSE WHOSE DATE OF INDEPENDENCE IS BEFORE 1945 (65)	0.70	73	TEND TO BE THOSE WHOSE DATE OF INDEPENDENCE IS AFTER 1945 (46)	0.55	35 21 15 26 X SQ= 5.37 P = 0.014 RV YES (YES)
74	TEND LESS TO BE THOSE THAT ARE NOT HISTORICALLY WESTERN (87)	0.59	74	TEND MORE TO BE THOSE THAT ARE NOT HISTORICALLY WESTERN (87)	0.88	20 6 29 43 X SQ= 8.85 P = 0.001 RV NO (YES)
75	TEND TO BE THOSE THAT ARE HISTORICALLY WESTERN OR SIGNIFICANTLY WESTERNIZED (62)	0.84	75	TEND TO BE THOSE THAT ARE NOT HISTORICALLY WESTERN AND ARE NOT SIGNIFICANTLY WESTERNIZED (52)	0.67	42 16 8 33 X SQ= 24.82 P = 0. RV YES (YES)
76	TEND LESS TO BE THOSE THAT HAVE BEEN SIGNIFICANTLY OR PARTIALLY WESTERNIZED THROUGH A COLONIAL RELATIONSHIP, RATHER THAN BEING HISTORICALLY WESTERN (70)	0.58	76	TEND MORE TO BE THOSE THAT HAVE BEEN SIGNIFICANTLY OR PARTIALLY WESTERNIZED THROUGH A COLONIAL RELATIONSHIP, RATHER THAN BEING HISTORICALLY WESTERN (70)	0.83	20 6 28 30 X SQ= 4.90 P = 0.018 RV NO (YES)
77	TEND TO BE THOSE THAT HAVE BEEN SIGNIFICANTLY WESTERNIZED, RATHER THAN PARTIALLY WESTERNIZED, THROUGH A COLONIAL RELATIONSHIP (28)	0.70	77	TEND TO BE THOSE THAT HAVE BEEN PARTIALLY WESTERNIZED, RATHER THAN SIGNIFICANTLY WESTERNIZED, THROUGH A COLONIAL RELATIONSHIP (41)	0.83	19 5 8 25 X SQ= 14.68 P = 0. RV YES (YES)
78	TEND TO BE THOSE THAT HAVE BEEN SIGNIFICANTLY WESTERNIZED THROUGH A COLONIAL RELATIONSHIP, RATHER THAN WITHOUT SUCH A RELATIONSHIP (28)	0.90	78	TEND TO BE THOSE THAT HAVE BEEN SIGNIFICANTLY WESTERNIZED WITHOUT A COLONIAL RELATIONSHIP, RATHER THAN THROUGH SUCH A RELATIONSHIP (7)	0.50	19 5 2 5 X SQ= 4.24 P = 0.022 RV YES (YES)
79	LEAN LESS TOWARD BEING THOSE THAT WERE FORMERLY DEPENDENCIES OF BRITAIN OR FRANCE, RATHER THAN SPAIN (49)	0.65	79	LEAN MORE TOWARD BEING THOSE THAT WERE FORMERLY DEPENDENCIES OF BRITAIN OR FRANCE, RATHER THAN SPAIN (49)	0.88	20 23 11 3 X SQ= 3.18 P = 0.062 RV NO (YES)
80	TEND TO BE THOSE WHOSE DATE OF INDEPENDENCE IS AFTER 1914, AND THAT WERE FORMERLY DEPENDENCIES OF BRITAIN, RATHER THAN FRANCE (19)	0.73	80	TEND TO BE THOSE WHOSE DATE OF INDEPENDENCE IS AFTER 1914, AND THAT WERE FORMERLY DEPENDENCIES OF FRANCE, RATHER THAN BRITAIN (24)	0.82	11 4 4 18 X SQ= 9.08 P = 0.002 RV YES (YES)

81	TEND LESS TO BE THOSE WHERE THE TYPE OF POLITICAL MODERNIZATION IS LATER EUROPEAN OR LATER EUROPEAN DERIVED, RATHER THAN EARLY EUROPEAN OR EARLY EUROPEAN DERIVED (40)	0.67	81	ALWAYS ARE THOSE WHERE THE TYPE OF POLITICAL MODERNIZATION IS LATER EUROPEAN OR LATER EUROPEAN DERIVED, RATHER THAN EARLY EUROPEAN OR EARLY EUROPEAN DERIVED (40)	1.00	11 0 22 14 X SQ= 4.38 P = 0.020 RV NO (NO)
82	TEND TO BE THOSE WHERE THE TYPE OF POLITICAL MODERNIZATION IS EARLY OR LATER EUROPEAN OR EUROPEAN DERIVED, RATHER THAN DEVELOPED TUTELARY OR UNDEVELOPED TUTELARY (51)	0.67	82	TEND TO BE THOSE WHERE THE TYPE OF POLITICAL MODERNIZATION IS DEVELOPED TUTELARY OR UNDEVELOPED TUTELARY, RATHER THAN EARLY OR LATER EUROPEAN OR EUROPEAN DERIVED (55)	0.67	33 14 16 29 X SQ= 9.74 P = 0.002 RV YES (YES)
84	TILT TOWARD BEING THOSE WHERE THE TYPE OF POLITICAL MODERNIZATION IS DEVELOPED TUTELARY, RATHER THAN UNDEVELOPED TUTELARY (31)	0.69	84	TILT TOWARD BEING THOSE WHERE THE TYPE OF POLITICAL MODERNIZATION IS UNDEVELOPED TUTELARY, RATHER THAN DEVELOPED TUTELARY (24)	0.59	11 12 5 17 X SQ= 2.09 P = 0.120 RV YES (NO)
85	TEND TO BE THOSE WHERE THE STAGE OF POLITICAL MODERNIZATION IS ADVANCED, RATHER THAN MID- OR EARLY TRANSITIONAL (60)	0.69	85	TEND TO BE THOSE WHERE THE STAGE OF POLITICAL MODERNIZATION IS MID- OR EARLY TRANSITIONAL, RATHER THAN ADVANCED (54)	0.63	35 18 16 31 X SQ= 8.96 P = 0.002 RV YES (YES)
87	TEND MORE TO BE THOSE WHOSE IDEOLOGICAL ORIENTATION IS OTHER THAN DEVELOPMENTAL (58)	0.85	87	TEND LESS TO BE THOSE WHOSE IDEOLOGICAL ORIENTATION IS OTHER THAN DEVELOPMENTAL (58)	0.51	6 21 33 22 X SQ= 8.90 P = 0.002 RV NO (YES)
89	TEND TO BE THOSE WHOSE IDEOLOGICAL ORIENTATION IS CONVENTIONAL (33)	0.85	89	ALWAYS ARE THOSE WHOSE IDEOLOGICAL ORIENTATION IS OTHER THAN CONVENTIONAL (62)	1.00	33 0 6 44 X SQ= 58.32 P = 0. RV YES (YES)
92	ALWAYS ARE THOSE WHERE THE SYSTEM STYLE IS LIMITED MOBILIZATIONAL, OR NON-MOBILIZATIONAL (93)	1.00	92	TEND LESS TO BE THOSE WHERE THE SYSTEM STYLE IS LIMITED MOBILIZATIONAL, OR NON-MOBILIZATIONAL (93)	0.63	0 18 51 30 X SQ= 20.92 P = 0. RV NO (YES)
93	TEND TO BE THOSE WHERE THE SYSTEM STYLE IS NON-MOBILIZATIONAL (78)	0.92	93	TEND TO BE THOSE WHERE THE SYSTEM STYLE IS MOBILIZATIONAL, OR LIMITED MOBILIZATIONAL (32)	0.53	4 25 47 22 X SQ= 22.01 P = 0. RV YES (YES)
94	ALWAYS ARE THOSE WHERE THE STATUS OF THE REGIME IS CONSTITUTIONAL (51)	1.00	94	TEND TO BE THOSE WHERE THE STATUS OF THE REGIME IS AUTHORITARIAN OR TOTALITARIAN (41)	0.94	48 2 0 34 X SQ= 72.29 P = 0. RV YES (YES)

95	ALWAYS ARE THOSE WHERE THE STATUS OF THE REGIME IS CONSTITUTIONAL OR AUTHORITARIAN (95)	1.00	95	TEND LESS TO BE THOSE WHERE THE STATUS OF THE REGIME IS CONSTITUTIONAL OR AUTHORITARIAN (95)	0.68	50 32 0 15 X SQ= 16.51 P = 0. RV NO (YES)
96	ALWAYS ARE THOSE WHERE THE STATUS OF THE REGIME IS CONSTITUTIONAL OR TOTALITARIAN (67)	1.00	96	TEND TO BE THOSE WHERE THE STATUS OF THE REGIME IS AUTHORITARIAN (23)	0.50	48 17 0 17 X SQ= 27.31 P = 0. RV YES (YES)
98	TEND LESS TO BE THOSE WHERE GOVERNMENTAL STABILITY IS GENERALLY OR MODERATELY PRESENT AND DATES FROM THE POST-WAR PERIOD, OR IS ABSENT (93)	0.69	98	TEND MORE TO BE THOSE WHERE GOVERNMENTAL STABILITY IS GENERALLY OR MODERATELY PRESENT AND DATES FROM THE POST-WAR PERIOD, OR IS ABSENT (93)	0.90	16 5 35 44 X SQ= 5.53 P = 0.013 RV NO (YES)
101	TEND TO BE THOSE WHERE THE REPRESENTATIVE CHARACTER OF THE REGIME IS POLYARCHIC (41)	0.87	101	TEND TO BE THOSE WHERE THE REPRESENTATIVE CHARACTER OF THE REGIME IS LIMITED POLYARCHIC, PSEUDO-POLYARCHIC, OR NON-POLYARCHIC (57)	0.98	39 1 6 45 X SQ= 62.54 P = 0. RV YES (YES)
102	ALWAYS ARE THOSE WHERE THE REPRESENTATIVE CHARACTER OF THE REGIME IS POLYARCHIC OR LIMITED POLYARCHIC (59)	1.00	102	TEND TO BE THOSE WHERE THE REPRESENTATIVE CHARACTER OF THE REGIME IS PSEUDO-POLYARCHIC OR NON-POLYARCHIC (49)	0.88	51 6 0 43 X SQ= 74.98 P = 0. RV YES (YES)
105	TEND TO BE THOSE WHERE THE ELECTORAL SYSTEM IS COMPETITIVE (43)	0.98	105	ALWAYS ARE THOSE WHERE THE ELECTORAL SYSTEM IS PARTIALLY COMPETITIVE OR NON-COMPETITIVE (47)	1.00	42 0 1 45 X SQ= 80.22 P = 0. RV YES (YES)
106	ALWAYS ARE THOSE WHERE THE ELECTORAL SYSTEM IS COMPETITIVE OR PARTIALLY COMPETITIVE (52)	1.00	106	TEND TO BE THOSE WHERE THE ELECTORAL SYSTEM IS NON-COMPETITIVE (30)	0.77	41 9 0 30 X SQ= 47.23 P = 0. RV YES (YES)
107	TEND TO BE THOSE WHERE AUTONOMOUS GROUPS ARE FULLY TOLERATED IN POLITICS (46)	0.86	107	ALWAYS ARE THOSE WHERE AUTONOMOUS GROUPS ARE PARTIALLY TOLERATED IN POLITICS, ARE TOLERATED ONLY OUTSIDE POLITICS, OR ARE NOT TOLERATED AT ALL (65)	1.00	44 0 7 48 X SQ= 71.09 P = 0. RV YES (YES)
108	TEND TO BE THOSE WHERE AUTONOMOUS GROUPS ARE FULLY OR PARTIALLY TOLERATED IN POLITICS (65)	0.98	108	TEND TO BE THOSE WHERE AUTONOMOUS GROUPS ARE TOLERATED ONLY OUTSIDE POLITICS OR ARE NOT TOLERATED AT ALL (35)	0.71	48 13 1 32 X SQ= 46.14 P = 0. RV YES (YES)

111	TEND TO BE THOSE WHERE POLITICAL ENCULTURATION IS HIGH OR MEDIUM (53)	0.73	111	TEND TO BE THOSE WHERE POLITICAL ENCULTURATION IS LOW (42)	0.59	35 13 13 19 X SQ= 7.05 P = 0.005 RV YES (YES)
116	TEND TO BE THOSE WHERE INTEREST ARTICULATION BY ASSOCIATIONAL GROUPS IS SIGNIFICANT OR MODERATE (32)	0.63	116	ALWAYS ARE THOSE WHERE INTEREST ARTICULATION BY ASSOCIATIONAL GROUPS IS LIMITED OR NEGLIGIBLE (79)	1.00	32 0 19 46 X SQ= 40.28 P = 0. RV YES (YES)
117	TEND TO BE THOSE WHERE INTEREST ARTICULATION BY ASSOCIATIONAL GROUPS IS SIGNIFICANT, MODERATE, OR LIMITED (60)	0.86	117	TEND TO BE THOSE WHERE INTEREST ARTICULATION BY ASSOCIATIONAL GROUPS IS NEGLIGIBLE (51)	0.80	44 9 7 37 X SQ= 40.78 P = 0. RV YES (YES)
118	TEND TO BE THOSE WHERE INTEREST ARTICULATION BY INSTITUTIONAL GROUPS IS SIGNIFICANT, MODERATE, OR LIMITED (60)	0.90	118	TEND TO BE THOSE WHERE INTEREST ARTICULATION BY INSTITUTIONAL GROUPS IS VERY SIGNIFICANT (40)	0.74	5 28 44 10 X SQ= 33.99 P = 0. RV YES (YES)
119	TEND TO BE THOSE WHERE INTEREST ARTICULATION BY INSTITUTIONAL GROUPS IS MODERATE OR LIMITED (26)	0.51	119	ALWAYS ARE THOSE WHERE INTEREST ARTICULATION BY INSTITUTIONAL GROUPS IS VERY SIGNIFICANT OR SIGNIFICANT (74)	1.00	24 38 25 0 X SQ= 24.77 P = 0. RV YES (YES)
121	TEND TO BE THOSE WHERE INTEREST ARTICULATION BY NON-ASSOCIATIONAL GROUPS IS MODERATE, LIMITED, OR NEGLIGIBLE (61)	0.76	121	TEND TO BE THOSE WHERE INTEREST ARTICULATION BY NON-ASSOCIATIONAL GROUPS IS SIGNIFICANT (54)	0.63	12 31 39 18 X SQ= 14.52 P = 0. RV YES (YES)
122	TEND LESS TO BE THOSE WHERE INTEREST ARTICULATION BY NON-ASSOCIATIONAL GROUPS IS SIGNIFICANT OR MODERATE (83)	0.53	122	TEND MORE TO BE THOSE WHERE INTEREST ARTICULATION BY NON-ASSOCIATIONAL GROUPS IS SIGNIFICANT OR MODERATE (83)	0.88	27 43 24 6 X SQ= 12.81 P = 0. RV NO (YES)
123	TEND LESS TO BE THOSE WHERE INTEREST ARTICULATION BY NON-ASSOCIATIONAL GROUPS IS SIGNIFICANT, MODERATE, OR LIMITED (107)	0.84	123	ALWAYS ARE THOSE WHERE INTEREST ARTICULATION BY NON-ASSOCIATIONAL GROUPS IS SIGNIFICANT, MODERATE, OR LIMITED (107)	1.00	43 49 8 0 X SQ= 6.36 P = 0.006 RV NO (YES)
125	TEND TO BE THOSE WHERE INTEREST ARTICULATION BY ANOMIC GROUPS IS INFREQUENT OR VERY INFREQUENT (35)	0.58	125	TEND TO BE THOSE WHERE INTEREST ARTICULATION BY ANOMIC GROUPS IS FREQUENT OR OCCASIONAL (64)	0.80	19 33 26 8 X SQ= 11.59 P = 0. RV YES (YES)

```
126  TEND LESS TO BE THOSE                 0.64     126  ALWAYS ARE THOSE                      1.00        29    41
     WHERE INTEREST ARTICULATION                         WHERE INTEREST ARTICULATION                        16     0
     BY ANOMIC GROUPS                                    BY ANOMIC GROUPS                              X SQ=  15.64
     IS FREQUENT, OCCASIONAL, OR                         IS FREQUENT, OCCASIONAL, OR                   P =     0.
     INFREQUENT  (83)                                    INFREQUENT  (83)                              RV NO  (YES)

127  TEND LESS TO BE THOSE                 0.63     127  ALWAYS ARE THOSE                      1.00        17     0
     WHERE INTEREST ARTICULATION                         WHERE INTEREST ARTICULATION                        29    40
     BY POLITICAL PARTIES                                BY POLITICAL PARTIES                          X SQ=  16.17
     IS MODERATE, LIMITED, OR                            IS MODERATE, LIMITED, OR                      P =     0.
     NEGLIGIBLE  (72)                                    NEGLIGIBLE  (72)                              RV NO  (YES)

128  TEND TO BE THOSE                      0.89     128  TEND TO BE THOSE                      0.92        41     3
     WHERE INTEREST ARTICULATION                         WHERE INTEREST ARTICULATION                         5    37
     BY POLITICAL PARTIES                                BY POLITICAL PARTIES                          X SQ=  53.84
     IS SIGNIFICANT OR MODERATE  (48)                    IS LIMITED OR NEGLIGIBLE  (45)                P =     0.
                                                                                                       RV YES (YES)

129  TEND TO BE THOSE                      0.91     129  TEND TO BE THOSE                      0.82        42     7
     WHERE INTEREST ARTICULATION                         WHERE INTEREST ARTICULATION                         4    33
     BY POLITICAL PARTIES                                BY POLITICAL PARTIES                          X SQ=  44.58
     IS SIGNIFICANT, MODERATE, OR                        IS NEGLIGIBLE  (37)                           P =     0.
     LIMITED  (56)                                                                                     RV YES (YES)

134  TEND TO BE THOSE                      0.74     134  TEND TO BE THOSE                      0.57        35    20
     WHERE INTEREST AGGREGATION                          WHERE INTEREST AGGREGATION                         12    27
     BY THE EXECUTIVE                                    BY THE EXECUTIVE                              X SQ=   8.59
     IS SIGNIFICANT OR MODERATE  (57)                    IS LIMITED OR NEGLIGIBLE  (46)                P =     0.002
                                                                                                       RV YES (YES)

137  TEND TO BE THOSE                      0.63     137  ALWAYS ARE THOSE                      1.00        29     0
     WHERE INTEREST AGGREGATION                          WHERE INTEREST AGGREGATION                         17    46
     BY THE LEGISLATURE                                  BY THE LEGISLATURE                            X SQ=  39.48
     IS SIGNIFICANT OR MODERATE  (29)                    IS LIMITED OR NEGLIGIBLE  (68)                P =     0.
                                                                                                       RV YES (YES)

138  TEND TO BE THOSE                      0.92     138  TEND TO BE THOSE                      0.98        44     1
     WHERE INTEREST AGGREGATION                          WHERE INTEREST AGGREGATION                          4    43
     BY THE LEGISLATURE                                  BY THE LEGISLATURE                            X SQ=  69.88
     IS SIGNIFICANT, MODERATE, OR                        IS NEGLIGIBLE  (49)                           P =     0.
     LIMITED  (48)                                                                                     RV YES (YES)

139  TEND TO BE THOSE                      0.98     139  TEND TO BE THOSE                      0.73         1    33
     WHERE THE PARTY SYSTEM IS QUANTITATIVELY            WHERE THE PARTY SYSTEM IS QUANTITATIVELY           50    12
     OTHER THAN ONE-PARTY  (71)                          ONE-PARTY  (34)                               X SQ=  50.17
                                                                                                       P =     0.
                                                                                                       RV YES (YES)

141  TEND LESS TO BE THOSE                 0.77     141  ALWAYS ARE THOSE                      1.00        11     0
     WHERE THE PARTY SYSTEM IS QUANTITATIVELY            WHERE THE PARTY SYSTEM IS QUANTITATIVELY           36    45
     OTHER THAN TWO-PARTY  (87)                          OTHER THAN TWO-PARTY  (87)                    X SQ=   9.84
                                                                                                       P =     0.001
                                                                                                       RV NO  (YES)
```

142	TEND TO BE THOSE WHERE THE PARTY SYSTEM IS QUANTITATIVELY MULTI-PARTY (30)	0.52	142	TEND TO BE THOSE WHERE THE PARTY SYSTEM IS QUANTITATIVELY OTHER THAN MULTI-PARTY (66)	0.95	24 2 22 42 X SQ= 22.57 P = 0. RV YES (YES)
144	ALWAYS ARE THOSE WHERE THE PARTY SYSTEM IS QUANTITATIVELY TWO-PARTY, RATHER THAN ONE-PARTY (11)	1.00	144	ALWAYS ARE THOSE WHERE THE PARTY SYSTEM IS QUANTITATIVELY ONE-PARTY, RATHER THAN TWO-PARTY (34)	1.00	0 33 11 0 X SQ= 38.83 P = 0. RV YES (YES)
146	ALWAYS ARE THOSE WHERE THE PARTY SYSTEM IS QUALITATIVELY OTHER THAN MASS-BASED TERRITORIAL (92)	1.00	146	TEND LESS TO BE THOSE WHERE THE PARTY SYSTEM IS QUALITATIVELY OTHER THAN MASS-BASED TERRITORIAL (92)	0.82	0 8 46 36 X SQ= 7.07 P = 0.002 RV NO (YES)
147	TEND TO BE THOSE WHERE THE PARTY SYSTEM IS QUALITATIVELY CLASS-ORIENTED OR MULTI-IDEOLOGICAL (23)	0.54	147	ALWAYS ARE THOSE WHERE THE PARTY SYSTEM IS QUALITATIVELY OTHER THAN CLASS-ORIENTED OR MULTI-IDEOLOGICAL (67)	1.00	22 0 19 43 X SQ= 28.55 P = 0. RV YES (YES)
148	TEND MORE TO BE THOSE WHERE THE PARTY SYSTEM IS QUALITATIVELY OTHER THAN AFRICAN TRANSITIONAL (96)	0.96	148	TEND LESS TO BE THOSE WHERE THE PARTY SYSTEM IS QUALITATIVELY OTHER THAN AFRICAN TRANSITIONAL (96)	0.76	2 11 48 35 X SQ= 6.50 P = 0.006 RV NO (YES)
160	TEND TO BE THOSE WHERE THE POLITICAL LEADERSHIP IS MODERATELY ELITIST OR NON-ELITIST (67)	0.94	160	TEND TO BE THOSE WHERE THE POLITICAL LEADERSHIP IS ELITIST (30)	0.57	3 25 44 19 X SQ= 24.82 P = 0. RV YES (YES)
161	TEND TO BE THOSE WHERE THE POLITICAL LEADERSHIP IS NON-ELITIST (50)	0.70	161	TEND TO BE THOSE WHERE THE POLITICAL LEADERSHIP IS ELITIST OR MODERATELY ELITIST (47)	0.68	14 30 33 14 X SQ= 11.92 P = 0. RV YES (YES)
164	TEND TO BE THOSE WHERE THE REGIME'S LEADERSHIP CHARISMA IS NEGLIGIBLE (65)	0.87	164	TEND TO BE THOSE WHERE THE REGIME'S LEADERSHIP CHARISMA IS PRONOUNCED OR MODERATE (34)	0.58	6 25 42 18 X SQ= 19.05 P = 0. RV YES (YES)
166	TEND LESS TO BE THOSE WHERE THE VERTICAL POWER DISTRIBUTION IS THAT OF FORMAL FEDERALISM OR FORMAL AND EFFECTIVE UNITARISM (99)	0.73	166	ALWAYS ARE THOSE WHERE THE VERTICAL POWER DISTRIBUTION IS THAT OF FORMAL FEDERALISM OR FORMAL AND EFFECTIVE UNITARISM (99)	1.00	14 0 37 49 X SQ= 13.44 P = 0. RV NO (YES)

168	TEND TO BE THOSE WHERE THE HORIZONTAL POWER DISTRIBUTION IS SIGNIFICANT (34)	0.67	168	ALWAYS ARE THOSE WHERE THE HORIZONTAL POWER DISTRIBUTION IS LIMITED OR NEGLIGIBLE (72)	1.00	33 0 16 49 X SQ= 46.78 P = 0. RV YES (YES)
169	ALWAYS ARE THOSE WHERE THE HORIZONTAL POWER DISTRIBUTION IS SIGNIFICANT OR LIMITED (58)	1.00	169	TEND TO BE THOSE WHERE THE HORIZONTAL POWER DISTRIBUTION IS NEGLIGIBLE (48)	0.86	49 7 0 42 X SQ= 70.04 P = 0. RV YES (YES)
171	TEND LESS TO BE THOSE WHERE THE LEGISLATIVE-EXECUTIVE STRUCTURE IS OTHER THAN PARLIAMENTARY-REPUBLICAN (90)	0.80	171	TEND MORE TO BE THOSE WHERE THE LEGISLATIVE-EXECUTIVE STRUCTURE IS OTHER THAN PARLIAMENTARY-REPUBLICAN (90)	0.98	10 1 41 44 X SQ= 5.51 P = 0.009 RV NO (YES)
172	TEND LESS TO BE THOSE WHERE THE LEGISLATIVE-EXECUTIVE STRUCTURE IS OTHER THAN PARLIAMENTARY-ROYALIST (88)	0.63	172	TEND MORE TO BE THOSE WHERE THE LEGISLATIVE-EXECUTIVE STRUCTURE IS OTHER THAN PARLIAMENTARY-ROYALIST (88)	0.96	19 2 32 44 X SQ= 13.56 P = 0. RV NO (YES)
178	TEND TO BE THOSE WHERE THE LEGISLATURE IS BICAMERAL (51)	0.76	178	TEND TO BE THOSE WHERE THE LEGISLATURE IS UNICAMERAL (53)	0.79	12 37 38 10 X SQ= 26.88 P = 0. RV YES (YES)
179	TEND TO BE THOSE WHERE THE EXECUTIVE IS STRONG (39)	0.86	179	ALWAYS ARE THOSE WHERE THE EXECUTIVE IS DOMINANT (52)	1.00	6 42 38 0 X SQ= 61.53 P = 0. RV YES (YES)
181	TEND LESS TO BE THOSE WHERE THE BUREAUCRACY IS SEMI-MODERN, RATHER THAN MODERN (55)	0.54	181	ALWAYS ARE THOSE WHERE THE BUREAUCRACY IS SEMI-MODERN, RATHER THAN MODERN (55)	1.00	21 0 25 24 X SQ= 13.55 P = 0. RV NO (NO)
182	TEND MORE TO BE THOSE WHERE THE BUREAUCRACY IS SEMI-MODERN, RATHER THAN POST-COLONIAL TRANSITIONAL (55)	0.83	182	TEND LESS TO BE THOSE WHERE THE BUREAUCRACY IS SEMI-MODERN, RATHER THAN POST-COLONIAL TRANSITIONAL (55)	0.60	25 24 5 16 X SQ= 3.40 P = 0.040 RV NO (YES)
185	TEND TO BE THOSE WHERE PARTICIPATION BY THE MILITARY IN POLITICS IS INTERVENTIVE, RATHER THAN SUPPORTIVE (21)	0.62	185	TEND TO BE THOSE WHERE PARTICIPATION BY THE MILITARY IN POLITICS IS SUPPORTIVE, RATHER THAN INTERVENTIVE (31)	0.81	5 6 3 26 X SQ= 4.15 P = 0.025 RV YES (NO)

186	TEND TO BE THOSE WHERE PARTICIPATION BY THE MILITARY IN POLITICS IS NEUTRAL, RATHER THAN SUPPORTIVE (56)	0.93	186	TEND TO BE THOSE WHERE PARTICIPATION BY THE MILITARY IN POLITICS IS SUPPORTIVE, RATHER THAN NEUTRAL (31)	0.65	3 26 41 14 X SQ= 28.86 P = 0. RV YES (YES)
187	TEND TO BE THOSE WHERE THE ROLE OF THE POLICE IS NOT POLITICALLY SIGNIFICANT (35)	0.73	187	TEND TO BE THOSE WHERE THE ROLE OF THE POLICE IS POLITICALLY SIGNIFICANT (66)	0.95	12 41 33 2 X SQ= 40.48 P = 0. RV YES (YES)
188	TEND LESS TO BE THOSE WHERE THE CHARACTER OF THE LEGAL SYSTEM IS OTHER THAN CIVIL LAW (81)	0.57	188	TEND MORE TO BE THOSE WHERE THE CHARACTER OF THE LEGAL SYSTEM IS OTHER THAN CIVIL LAW (81)	0.85	22 7 29 40 X SQ= 8.06 P = 0.004 RV NO (YES)
189	TEND LESS TO BE THOSE WHERE THE CHARACTER OF THE LEGAL SYSTEM IS OTHER THAN COMMON LAW (108)	0.86	189	ALWAYS ARE THOSE WHERE THE CHARACTER OF THE LEGAL SYSTEM IS OTHER THAN COMMON LAW (108)	1.00	7 0 44 49 X SQ= 5.28 P = 0.013 RV NO (YES)
192	TEND MORE TO BE THOSE WHERE THE CHARACTER OF THE LEGAL SYSTEM IS OTHER THAN MUSLIM OR PARTLY MUSLIM (86)	0.88	192	TEND LESS TO BE THOSE WHERE THE CHARACTER OF THE LEGAL SYSTEM IS OTHER THAN MUSLIM OR PARTLY MUSLIM (86)	0.68	6 15 45 32 X SQ= 4.76 P = 0.025 RV NO (YES)
193	TEND MORE TO BE THOSE WHERE THE CHARACTER OF THE LEGAL SYSTEM IS OTHER THAN PARTLY INDIGENOUS (86)	0.88	193	TEND LESS TO BE THOSE WHERE THE CHARACTER OF THE LEGAL SYSTEM IS OTHER THAN PARTLY INDIGENOUS (86)	0.59	6 20 45 29 X SQ= 9.50 P = 0.001 RV NO (YES)
194	ALWAYS ARE THOSE THAT ARE NON-COMMUNIST (101)	1.00	194	TEND LESS TO BE THOSE THAT ARE NON-COMMUNIST (101)	0.73	0 13 51 36 X SQ= 13.30 P = 0. RV NO (YES)

```
******************************************************************************************************

176  POLITIES                                          176  POLITIES
     WHERE THE LEGISLATURE IS                               WHERE THE LEGISLATURE IS
     FULLY EFFECTIVE,                                       WHOLLY INEFFECTIVE  (28)
     PARTIALLY EFFECTIVE, OR
     LARGELY INEFFECTIVE  (72)

                                                                                       PREDICATE ONLY
......................................................................................................

       72 IN LEFT COLUMN

ALGERIA     ARGENTINA   AUSTRALIA   AUSTRIA     BELGIUM     BOLIVIA     BRAZIL      BURUNDI     CAMEROUN    CANADA
CEN AFR REP CEYLON      CHAD        CHILE       COLOMBIA    CONGO(BRA)  COSTA RICA  CYPRUS      DAHOMEY     DENMARK
ECUADOR     EL SALVADOR FINLAND     FRANCE      GABON       GERMAN FR   GREECE      HONDURAS    ICELAND     INDIA
INDONESIA   IRELAND     ISRAEL      ITALY       IVORY COAST JAMAICA     JAPAN       LEBANON     LIBERIA     LIBYA
LUXEMBOURG  MALAGASY R  MALAYA      MALI        MAURITANIA  MEXICO      MOROCCO     NETHERLANDS NEW ZEALAND NICARAGUA
NIGER       NIGERIA     NORWAY      PANAMA      PHILIPPINES RWANDA      SENEGAL     SIERRE LEO  SOMALIA     SO AFRICA
SWEDEN      SWITZERLAND TANGANYIKA  TRINIDAD    TUNISIA     TURKEY      UGANDA      UK          US          UPPER VOLTA
URUGUAY     VENEZUELA

       28 IN RIGHT COLUMN

AFGHANISTAN ALBANIA     BULGARIA    CAMBODIA    CHINA, PR   CZECHOS'KIA ETHIOPIA    GERMANY, E  GHANA       GUINEA
HAITI       HUNGARY     JORDAN      KOREA, N    LAOS        MONGOLIA    NEPAL       PARAGUAY    POLAND      PORTUGAL
RUMANIA     SPAIN       THAILAND    USSR        UAR         VIETNAM, N  VIETNAM REP YUGOSLAVIA

       1 EXCLUDED BECAUSE IRRELEVANT

SA'U ARABIA

       14 EXCLUDED BECAUSE UNASCERTAINABLE

BURMA       CONGO(LEO)  CUBA        DOMIN REP   GUATEMALA   IRAN        IRAQ        KOREA REP   PAKISTAN    PERU
SUDAN       SYRIA       TOGO        YEMEN

------------------------------------------------------------------------------------------------------
```

```
******************************************************************************************

177  POLITIES                                        177  POLITIES
     WHERE THE LEGISLATURE IS                             WHERE THE LEGISLATURE IS
     PARTIALLY EFFECTIVE, RATHER THAN                     LARGELY INEFFECTIVE, RATHER THAN
     LARGELY INEFFECTIVE  (23)                            PARTIALLY EFFECTIVE  (21)

                                                                                SUBJECT ONLY
..........................................................................................

     23 IN LEFT COLUMN

ARGENTINA   BOLIVIA     BRAZIL      CAMEROUN    CEYLON      CHILE       COLOMBIA    COSTA RICA  ECUADOR     FRANCE
GREECE      HONDURAS    LEBANON     LIBYA       MALAGASY R  MEXICO      MOROCCO     NIGERIA     PANAMA      SIERRE LEO
TURKEY      UGANDA      VENEZUELA

     21 IN RIGHT COLUMN

ALGERIA     BURUNDI     CEN AFR REP CHAD        CONGO(BRA)  DAHOMEY     EL SALVADOR GABON       INDONESIA   IVORY COAST
LIBERIA     MALI        MAURITANIA  NICARAGUA   NIGER       RWANDA      SENEGAL     SOMALIA     TANGANYIKA  TUNISIA
UPPER VOLTA

     57 EXCLUDED BECAUSE IRRELEVANT

AFGHANISTAN ALBANIA     AUSTRALIA   AUSTRIA     BELGIUM     BULGARIA    CAMBODIA    CANADA      CHINA, PR   CYPRUS
CZECHOS'KIA DENMARK     ETHIOPIA    FINLAND     GERMANY, E  GERMAN FR   GHANA       GUINEA      HAITI       HUNGARY
ICELAND     INDIA       IRELAND     ISRAEL      ITALY       JAMAICA     JAPAN       JORDAN      KOREA, N    LAOS
LUXEMBOURG  MALAYA      MONGOLIA    NEPAL       NETHERLANDS NEW ZEALAND NORWAY      PARAGUAY    PHILIPPINES POLAND
PORTUGAL    RUMANIA     SA'U ARABIA SO AFRICA   SPAIN       SWEDEN      SWITZERLAND THAILAND    TRINIDAD    USSR
UAR         UK          US          URUGUAY     VIETNAM, N  VIETNAM REP YUGOSLAVIA

     14 EXCLUDED BECAUSE UNASCERTAINABLE

BURMA       CONGO(LEO)  CUBA        DOMIN REP   GUATEMALA   IRAN        IRAQ        KOREA REP   PAKISTAN    PERU
SUDAN       SYRIA       TOGO        YEMEN
------------------------------------------------------------------------------------------

  6  TEND LESS TO BE THOSE                  0.52      6  TEND MORE TO BE THOSE                 0.90       11    2
     LOCATED ELSEWHERE THAN IN THE                       LOCATED ELSEWHERE THAN IN THE                    12   19
     CARIBBEAN, CENTRAL AMERICA,                         CARIBBEAN, CENTRAL AMERICA,                 X SQ=   6.01
     OR SOUTH AMERICA  (93)                              OR SOUTH AMERICA  (93)                      P =    0.008
                                                                                                     RV NO  (YES)

 10  TEND TO BE THOSE                       0.70     10  TEND TO BE THOSE                      0.86        7   18
     LOCATED ELSEWHERE THAN IN NORTH AFRICA,             LOCATED IN NORTH AFRICA, OR                      16    3
     OR CENTRAL AND SOUTH AFRICA  (82)                   CENTRAL AND SOUTH AFRICA  (33)              X SQ=  11.51
                                                                                                     P =     0.
                                                                                                     RV YES (YES)

 15  TILT LESS TOWARD BEING THOSE           0.78     15  ALWAYS ARE THOSE                      1.00        2    0
     LOCATED IN NORTH AFRICA OR                          LOCATED IN NORTH AFRICA OR                        7   18
     CENTRAL AND SOUTH AFRICA, RATHER THAN               CENTRAL AND SOUTH AFRICA, RATHER THAN       X SQ=   1.69
     IN THE MIDDLE EAST  (33)                            IN THE MIDDLE EAST  (33)                    P =    0.103
                                                                                                     RV NO  (YES)
```

19	TEND TO BE THOSE LOCATED IN THE CARIBBEAN, CENTRAL AMERICA, OR SOUTH AMERICA, RATHER THAN IN NORTH AFRICA OR CENTRAL AND SOUTH AFRICA (22)	0.61	19	TEND TO BE THOSE LOCATED IN NORTH AFRICA OR CENTRAL AND SOUTH AFRICA, RATHER THAN IN THE CARIBBEAN, CENTRAL AMERICA, OR SOUTH AMERICA (33)	0.90	7 18 11 2 X SQ= 8.84 P = 0.002 RV YES (YES)
24	TEND TO BE THOSE WHOSE POPULATION IS VERY LARGE, LARGE, OR MEDIUM (61)	0.57	24	TEND TO BE THOSE WHOSE POPULATION IS SMALL (54)	0.86	13 3 10 18 X SQ= 6.74 P = 0.005 RV YES (YES)
29	TEND TO BE THOSE WHERE THE DEGREE OF URBANIZATION IS HIGH (56)	0.64	29	TEND TO BE THOSE WHERE THE DEGREE OF URBANIZATION IS LOW (49)	0.90	14 2 8 19 X SQ= 11.25 P = 0. RV YES (YES)
30	TEND TO BE THOSE WHOSE AGRICULTURAL POPULATION IS MEDIUM, LOW, OR VERY LOW (57)	0.61	30	TEND TO BE THOSE WHOSE AGRICULTURAL POPULATION IS HIGH (56)	0.95	9 20 14 1 X SQ= 12.98 P = 0. RV YES (YES)
34	TEND TO BE THOSE WHOSE GROSS NATIONAL PRODUCT IS VERY HIGH, HIGH, MEDIUM, OR LOW (62)	0.52	34	TEND TO BE THOSE WHOSE GROSS NATIONAL PRODUCT IS VERY LOW (53)	0.90	12 2 11 19 X SQ= 7.34 P = 0.003 RV YES (YES)
36	TEND LESS TO BE THOSE WHOSE PER CAPITA GROSS NATIONAL PRODUCT IS LOW OR VERY LOW (73)	0.74	36	ALWAYS ARE THOSE WHOSE PER CAPITA GROSS NATIONAL PRODUCT IS LOW OR VERY LOW (73)	1.00	6 0 17 21 X SQ= 4.32 P = 0.022 RV NO (YES)
37	TEND TO BE THOSE WHOSE PER CAPITA GROSS NATIONAL PRODUCT IS VERY HIGH, HIGH, MEDIUM, OR LOW (64)	0.61	37	TEND TO BE THOSE WHOSE PER CAPITA GROSS NATIONAL PRODUCT IS VERY LOW (51)	0.81	14 4 9 17 X SQ= 6.31 P = 0.007 RV YES (YES)
40	TEND TO BE THOSE WHOSE INTERNATIONAL FINANCIAL STATUS IS VERY HIGH, HIGH, MEDIUM, OR LOW (71)	0.61	40	TEND TO BE THOSE WHOSE INTERNATIONAL FINANCIAL STATUS IS VERY LOW (39)	0.85	14 3 9 17 X SQ= 7.59 P = 0.004 RV YES (YES)
42	TEND LESS TO BE THOSE WHOSE ECONOMIC DEVELOPMENTAL STATUS IS UNDERDEVELOPED OR VERY UNDERDEVELOPED (76)	0.73	42	ALWAYS ARE THOSE WHOSE ECONOMIC DEVELOPMENTAL STATUS IS UNDERDEVELOPED OR VERY UNDERDEVELOPED (76)	1.00	6 0 16 21 X SQ= 4.58 P = 0.021 RV NO (YES)

No.	Statement		No.	Statement		Statistics
43	TEND TO BE THOSE WHOSE ECONOMIC DEVELOPMENTAL STATUS IS DEVELOPED, INTERMEDIATE, OR UNDERDEVELOPED (55)	0.52	43	ALWAYS ARE THOSE WHOSE ECONOMIC DEVELOPMENTAL STATUS IS VERY UNDERDEVELOPED (57)	1.00	11 0 10 21 X SQ= 12.32 P = 0. RV YES (YES)
45	TEND TO BE THOSE WHERE THE LITERACY RATE IS FIFTY PERCENT OR ABOVE (55)	0.50	45	TEND TO BE THOSE WHERE THE LITERACY RATE IS BELOW FIFTY PERCENT (54)	0.95	11 1 11 19 X SQ= 8.31 P = 0.002 RV YES (YES)
46	TEND TO BE THOSE WHERE THE LITERACY RATE IS TEN PERCENT OR ABOVE (84)	0.86	46	TEND TO BE THOSE WHERE THE LITERACY RATE IS BELOW TEN PERCENT (26)	0.70	19 6 3 14 X SQ= 11.57 P = 0. RV YES (YES)
51	LEAN MORE TOWARD BEING THOSE WHERE FREEDOM OF THE PRESS IS COMPLETE OR INTERMITTENT (60)	0.95	51	LEAN LESS TOWARD BEING THOSE WHERE FREEDOM OF THE PRESS IS COMPLETE OR INTERMITTENT (60)	0.70	19 7 1 3 X SQ= 1.77 P = 0.095 RV NO (YES)
54	TEND LESS TO BE THOSE WHERE NEWSPAPER CIRCULATION IS LESS THAN ONE HUNDRED PER THOUSAND (76)	0.74	54	ALWAYS ARE THOSE WHERE NEWSPAPER CIRCULATION IS LESS THAN ONE HUNDRED PER THOUSAND (76)	1.00	6 0 17 21 X SQ= 4.32 P = 0.022 RV NO (YES)
55	TEND TO BE THOSE WHERE NEWSPAPER CIRCULATION IS TEN OR MORE PER THOUSAND (78)	0.74	55	TEND TO BE THOSE WHERE NEWSPAPER CIRCULATION IS LESS THAN TEN PER THOUSAND (35)	0.76	17 5 6 16 X SQ= 9.11 P = 0.002 RV YES (YES)
56	LEAN TOWARD BEING THOSE WHERE THE RELIGION IS PREDOMINANTLY LITERATE (79)	0.70	56	LEAN TOWARD BEING THOSE WHERE THE RELIGION IS PREDOMINANTLY OR PARTLY NON-LITERATE (31)	0.62	16 8 7 13 X SQ= 3.21 P = 0.068 RV YES (YES)
57	TEND LESS TO BE THOSE WHERE THE RELIGION IS OTHER THAN CATHOLIC (90)	0.57	57	TEND MORE TO BE THOSE WHERE THE RELIGION IS OTHER THAN CATHOLIC (90)	0.90	10 2 13 19 X SQ= 4.78 P = 0.017 RV NO (YES)
63	TEND LESS TO BE THOSE WHERE THE RELIGION IS PREDOMINANTLY OR PARTLY OTHER THAN CHRISTIAN (68)	0.52	63	TEND MORE TO BE THOSE WHERE THE RELIGION IS PREDOMINANTLY OR PARTLY OTHER THAN CHRISTIAN (68)	0.90	11 2 12 19 X SQ= 6.01 P = 0.008 RV NO (YES)

64	TEND TO BE THOSE WHERE THE RELIGION IS CHRISTIAN, RATHER THAN MUSLIM (46)	0.79	64	TEND TO BE THOSE WHERE THE RELIGION IS MUSLIM, RATHER THAN CHRISTIAN (18)	0.75	11 2 3 6 X SQ= 4.03 P = 0.026 RV YES (YES)
65	TEND TO BE THOSE WHERE THE RELIGION IS CATHOLIC, RATHER THAN MUSLIM (25)	0.77	65	TEND TO BE THOSE WHERE THE RELIGION IS MUSLIM, RATHER THAN CATHOLIC (18)	0.75	10 2 3 6 X SQ= 3.54 P = 0.032 RV YES (YES)
68	LEAN TOWARD BEING THOSE THAT ARE LINGUISTICALLY HOMOGENEOUS (52)	0.57	68	LEAN TOWARD BEING THOSE THAT ARE LINGUISTICALLY WEAKLY HETEROGENEOUS OR STRONGLY HETEROGENEOUS (62)	0.71	13 6 10 15 X SQ= 2.45 P = 0.076 RV YES (YES)
69	TEND TO BE THOSE THAT ARE LINGUISTICALLY HOMOGENEOUS OR WEAKLY HETEROGENEOUS (64)	0.65	69	TEND TO BE THOSE THAT ARE LINGUISTICALLY STRONGLY HETEROGENEOUS (50)	0.71	15 6 8 15 X SQ= 4.53 P = 0.019 RV YES (YES)
72	TEND TO BE THOSE WHOSE DATE OF INDEPENDENCE IS BEFORE 1914 (52)	0.61	72	TEND TO BE THOSE WHOSE DATE OF INDEPENDENCE IS AFTER 1914 (59)	0.86	14 3 9 18 X SQ= 8.18 P = 0.002 RV YES (YES)
73	TEND TO BE THOSE WHOSE DATE OF INDEPENDENCE IS BEFORE 1945 (65)	0.65	73	TEND TO BE THOSE WHOSE DATE OF INDEPENDENCE IS AFTER 1945 (46)	0.86	15 3 8 18 X SQ= 9.77 P = 0.001 RV YES (YES)
75	TEND TO BE THOSE THAT ARE HISTORICALLY WESTERN OR SIGNIFICANTLY WESTERNIZED (62)	0.70	75	TEND TO BE THOSE THAT ARE NOT HISTORICALLY WESTERN AND ARE NOT SIGNIFICANTLY WESTERNIZED (52)	0.86	16 3 7 18 X SQ= 11.51 P = 0. RV YES (YES)
77	TEND TO BE THOSE THAT HAVE BEEN SIGNIFICANTLY WESTERNIZED, RATHER THAN PARTIALLY WESTERNIZED, THROUGH A COLONIAL RELATIONSHIP (28)	0.65	77	TEND TO BE THOSE THAT HAVE BEEN PARTIALLY WESTERNIZED, RATHER THAN SIGNIFICANTLY WESTERNIZED, THROUGH A COLONIAL RELATIONSHIP (41)	0.85	13 3 7 17 X SQ= 8.44 P = 0.003 RV YES (YES)
79	TEND TO BE THOSE THAT WERE FORMERLY DEPENDENCIES OF SPAIN, RATHER THAN BRITAIN OR FRANCE (18)	0.56	79	TEND TO BE THOSE THAT WERE FORMERLY DEPENDENCIES OF BRITAIN OR FRANCE, RATHER THAN SPAIN (49)	0.88	8 14 10 2 X SQ= 5.12 P = 0.013 RV YES (YES)

80	TEND TO BE THOSE WHOSE DATE OF INDEPENDENCE IS AFTER 1914, AND THAT WERE FORMERLY DEPENDENCIES OF BRITAIN, RATHER THAN FRANCE (19)	0.50	80	TEND TO BE THOSE WHOSE DATE OF INDEPENDENCE IS AFTER 1914, AND THAT WERE FORMERLY DEPENDENCIES OF FRANCE, RATHER THAN BRITAIN (24)	0.93	4 1 4 13 X SQ= 3.16 P = 0.039 RV YES (YES)
82	TEND TO BE THOSE WHERE THE TYPE OF POLITICAL MODERNIZATION IS EARLY OR LATER EUROPEAN OR EUROPEAN DERIVED, RATHER THAN DEVELOPED TUTELARY OR UNDEVELOPED TUTELARY (51)	0.59	82	TEND TO BE THOSE WHERE THE TYPE OF POLITICAL MODERNIZATION IS DEVELOPED TUTELARY OR UNDEVELOPED TUTELARY, RATHER THAN EARLY OR LATER EUROPEAN OR EUROPEAN DERIVED (55)	0.90	13 2 9 19 X SQ= 9.54 P = 0.001 RV YES (YES)
86	TEND TO BE THOSE WHERE THE STAGE OF POLITICAL MODERNIZATION IS ADVANCED OR MID-TRANSITIONAL, RATHER THAN EARLY TRANSITIONAL (76)	0.70	86	TEND TO BE THOSE WHERE THE STAGE OF POLITICAL MODERNIZATION IS EARLY TRANSITIONAL, RATHER THAN ADVANCED OR MID-TRANSITIONAL (38)	0.81	16 4 7 17 X SQ= 9.35 P = 0.001 RV YES (YES)
87	TEND TO BE THOSE WHOSE IDEOLOGICAL ORIENTATION IS OTHER THAN DEVELOPMENTAL (58)	0.71	87	TEND TO BE THOSE WHOSE IDEOLOGICAL ORIENTATION IS DEVELOPMENTAL (31)	0.95	4 18 10 1 X SQ= 13.04 P = 0. RV YES (YES)
89	TEND TO BE THOSE WHOSE IDEOLOGICAL ORIENTATION IS CONVENTIONAL (33)	0.71	89	ALWAYS ARE THOSE WHOSE IDEOLOGICAL ORIENTATION IS OTHER THAN CONVENTIONAL (62)	1.00	10 0 4 19 X SQ= 16.24 P = 0. RV YES (YES)
94	ALWAYS ARE THOSE WHERE THE STATUS OF THE REGIME IS CONSTITUTIONAL (51)	1.00	94	TEND TO BE THOSE WHERE THE STATUS OF THE REGIME IS AUTHORITARIAN OR TOTALITARIAN (41)	0.75	21 2 0 6 X SQ= 15.55 P = 0. RV YES (YES)
96	ALWAYS ARE THOSE WHERE THE STATUS OF THE REGIME IS CONSTITUTIONAL OR TOTALITARIAN (67)	1.00	96	TEND TO BE THOSE WHERE THE STATUS OF THE REGIME IS AUTHORITARIAN (23)	0.75	21 2 0 6 X SQ= 15.55 P = 0. RV YES (YES)
101	TEND TO BE THOSE WHERE THE REPRESENTATIVE CHARACTER OF THE REGIME IS POLYARCHIC (41)	0.71	101	TEND TO BE THOSE WHERE THE REPRESENTATIVE CHARACTER OF THE REGIME IS LIMITED POLYARCHIC, PSEUDO-POLYARCHIC, OR NON-POLYARCHIC (57)	0.94	12 1 5 17 X SQ= 13.17 P = 0. RV YES (YES)
102	ALWAYS ARE THOSE WHERE THE REPRESENTATIVE CHARACTER OF THE REGIME IS POLYARCHIC OR LIMITED POLYARCHIC (59)	1.00	102	TEND TO BE THOSE WHERE THE REPRESENTATIVE CHARACTER OF THE REGIME IS PSEUDO-POLYARCHIC OR NON-POLYARCHIC (49)	0.71	23 6 0 15 X SQ= 21.85 P = 0. RV YES (YES)

No.	Statement	Value	No.	Statement	Value	Statistics
105	TEND TO BE THOSE WHERE THE ELECTORAL SYSTEM IS COMPETITIVE (43)	0.94	105	ALWAYS ARE THOSE WHERE THE ELECTORAL SYSTEM IS PARTIALLY COMPETITIVE OR NON-COMPETITIVE (47)	1.00	16 0 1 20 X SQ= 29.44 P = 0. RV YES (YES)
106	ALWAYS ARE THOSE WHERE THE ELECTORAL SYSTEM IS COMPETITIVE OR PARTIALLY COMPETITIVE (52)	1.00	106	TEND TO BE THOSE WHERE THE ELECTORAL SYSTEM IS NON-COMPETITIVE (30)	0.56	15 8 0 10 X SQ= 9.47 P = 0.001 RV YES (YES)
107	TEND TO BE THOSE WHERE AUTONOMOUS GROUPS ARE FULLY TOLERATED IN POLITICS (46)	0.74	107	ALWAYS ARE THOSE WHERE AUTONOMOUS GROUPS ARE PARTIALLY TOLERATED IN POLITICS, ARE TOLERATED ONLY OUTSIDE POLITICS, OR ARE NOT TOLERATED AT ALL (65)	1.00	17 0 6 21 X SQ= 22.27 P = 0. RV YES (YES)
113	LEAN LESS TOWARD BEING THOSE WHERE SECTIONALISM IS MODERATE OR NEGLIGIBLE (81)	0.62	113	LEAN MORE TOWARD BEING THOSE WHERE SECTIONALISM IS MODERATE OR NEGLIGIBLE (81)	0.89	8 2 13 16 X SQ= 2.42 P = 0.074 RV NO (YES)
116	TEND LESS TO BE THOSE WHERE INTEREST ARTICULATION BY ASSOCIATIONAL GROUPS IS LIMITED OR NEGLIGIBLE (79)	0.61	116	ALWAYS ARE THOSE WHERE INTEREST ARTICULATION BY ASSOCIATIONAL GROUPS IS LIMITED OR NEGLIGIBLE (79)	1.00	9 0 14 21 X SQ= 8.07 P = 0.002 RV NO (YES)
117	TEND TO BE THOSE WHERE INTEREST ARTICULATION BY ASSOCIATIONAL GROUPS IS SIGNIFICANT, MODERATE, OR LIMITED (60)	0.70	117	TEND TO BE THOSE WHERE INTEREST ARTICULATION BY ASSOCIATIONAL GROUPS IS NEGLIGIBLE (51)	0.76	16 5 7 16 X SQ= 7.47 P = 0.003 RV YES (YES)
121	TEND TO BE THOSE WHERE INTEREST ARTICULATION BY NON-ASSOCIATIONAL GROUPS IS MODERATE, LIMITED, OR NEGLIGIBLE (61)	0.70	121	TEND TO BE THOSE WHERE INTEREST ARTICULATION BY NON-ASSOCIATIONAL GROUPS IS SIGNIFICANT (54)	0.76	7 16 16 5 X SQ= 7.47 P = 0.003 RV YES (YES)
122	TEND LESS TO BE THOSE WHERE INTEREST ARTICULATION BY NON-ASSOCIATIONAL GROUPS IS SIGNIFICANT OR MODERATE (83)	0.65	122	ALWAYS ARE THOSE WHERE INTEREST ARTICULATION BY NON-ASSOCIATIONAL GROUPS IS SIGNIFICANT OR MODERATE (83)	1.00	15 21 8 0 X SQ= 6.74 P = 0.004 RV NO (YES)
127	TEND LESS TO BE THOSE WHERE INTEREST ARTICULATION BY POLITICAL PARTIES IS MODERATE, LIMITED, OR NEGLIGIBLE (72)	0.67	127	ALWAYS ARE THOSE WHERE INTEREST ARTICULATION BY POLITICAL PARTIES IS MODERATE, LIMITED, OR NEGLIGIBLE (72)	1.00	7 0 14 19 X SQ= 5.54 P = 0.009 RV NO (YES)

128	TEND TO BE THOSE WHERE INTEREST ARTICULATION BY POLITICAL PARTIES IS SIGNIFICANT OR MODERATE (48)	0.90	128	TEND TO BE THOSE WHERE INTEREST ARTICULATION BY POLITICAL PARTIES IS LIMITED OR NEGLIGIBLE (45)	0.84	19 3 2 16 X SQ= 19.57 P = 0. RV YES (YES)
129	TEND TO BE THOSE WHERE INTEREST ARTICULATION BY POLITICAL PARTIES IS SIGNIFICANT, MODERATE, OR LIMITED (56)	0.95	129	TEND TO BE THOSE WHERE INTEREST ARTICULATION BY POLITICAL PARTIES IS NEGLIGIBLE (37)	0.68	20 6 1 13 X SQ= 15.08 P = 0. RV YES (YES)
130	LEAN MORE TOWARD BEING THOSE WHERE INTEREST AGGREGATION BY POLITICAL PARTIES IS MODERATE, LIMITED, OR NEGLIGIBLE (71)	0.94	130	LEAN LESS TOWARD BEING THOSE WHERE INTEREST AGGREGATION BY POLITICAL PARTIES IS MODERATE, LIMITED, OR NEGLIGIBLE (71)	0.62	1 3 17 5 X SQ= 2.23 P = 0.072 RV NO (YES)
133	TEND TO BE THOSE WHERE INTEREST AGGREGATION BY THE EXECUTIVE IS MODERATE, LIMITED, OR NEGLIGIBLE (76)	0.73	133	TEND TO BE THOSE WHERE INTEREST AGGREGATION BY THE EXECUTIVE IS SIGNIFICANT (29)	0.75	6 15 16 5 X SQ= 7.73 P = 0.005 RV YES (YES)
134	LEAN TOWARD BEING THOSE WHERE INTEREST AGGREGATION BY THE EXECUTIVE IS LIMITED OR NEGLIGIBLE (46)	0.50	134	LEAN TOWARD BEING THOSE WHERE INTEREST AGGREGATION BY THE EXECUTIVE IS SIGNIFICANT OR MODERATE (57)	0.80	11 16 11 4 X SQ= 2.90 P = 0.058 RV YES (YES)
137	TILT LESS TOWARD BEING THOSE WHERE INTEREST AGGREGATION BY THE LEGISLATURE IS LIMITED OR NEGLIGIBLE (68)	0.80	137	ALWAYS ARE THOSE WHERE INTEREST AGGREGATION BY THE LEGISLATURE IS LIMITED OR NEGLIGIBLE (68)	1.00	4 0 16 19 X SQ= 2.34 P = 0.106 RV NO (YES)
138	TEND TO BE THOSE WHERE INTEREST AGGREGATION BY THE LEGISLATURE IS SIGNIFICANT, MODERATE, OR LIMITED (48)	0.86	138	TEND TO BE THOSE WHERE INTEREST AGGREGATION BY THE LEGISLATURE IS NEGLIGIBLE (49)	0.95	18 1 3 18 X SQ= 22.76 P = 0. RV YES (YES)
139	TEND TO BE THOSE WHERE THE PARTY SYSTEM IS QUANTITATIVELY OTHER THAN ONE-PARTY (71)	0.96	139	TEND TO BE THOSE WHERE THE PARTY SYSTEM IS QUANTITATIVELY ONE-PARTY (34)	0.74	1 14 22 5 X SQ= 18.87 P = 0. RV YES (YES)
142	TEND TO BE THOSE WHERE THE PARTY SYSTEM IS QUANTITATIVELY MULTI-PARTY (30)	0.52	142	TEND TO BE THOSE WHERE THE PARTY SYSTEM IS QUANTITATIVELY OTHER THAN MULTI-PARTY (66)	0.95	11 1 10 18 X SQ= 8.42 P = 0.002 RV YES (YES)

144	ALWAYS ARE THOSE 1.00 WHERE THE PARTY SYSTEM IS QUANTITATIVELY TWO-PARTY, RATHER THAN ONE-PARTY (11)	144	ALWAYS ARE THOSE 1.00 WHERE THE PARTY SYSTEM IS QUANTITATIVELY ONE-PARTY, RATHER THAN TWO-PARTY (34)	0 14 3 0 X SQ= 10.82 P = 0.001 RV YES (YES)
146	ALWAYS ARE THOSE 1.00 WHERE THE PARTY SYSTEM IS QUALITATIVELY OTHER THAN MASS-BASED TERRITORIAL (92)	146	TEND LESS TO BE THOSE 0.76 WHERE THE PARTY SYSTEM IS QUALITATIVELY OTHER THAN MASS-BASED TERRITORIAL (92)	0 5 19 16 X SQ= 3.22 P = 0.049 RV NO (YES)
147	LEAN LESS TOWARD BEING THOSE 0.82 WHERE THE PARTY SYSTEM IS QUALITATIVELY OTHER THAN CLASS-ORIENTED OR MULTI-IDEOLOGICAL (67)	147	ALWAYS ARE THOSE 1.00 WHERE THE PARTY SYSTEM IS QUALITATIVELY OTHER THAN CLASS-ORIENTED OR MULTI-IDEOLOGICAL (67)	3 0 14 19 X SQ= 1.71 P = 0.095 RV NO (YES)
148	TEND TO BE THOSE 0.91 WHERE THE PARTY SYSTEM IS QUALITATIVELY OTHER THAN AFRICAN TRANSITIONAL (96)	148	TEND TO BE THOSE 0.52 WHERE THE PARTY SYSTEM IS QUALITATIVELY AFRICAN TRANSITIONAL (14)	2 11 20 10 X SQ= 7.60 P = 0.003 RV YES (YES)
154	LEAN TOWARD BEING THOSE 0.71 WHERE THE PARTY SYSTEM IS UNSTABLE (25)	154	LEAN TOWARD BEING THOSE 0.75 WHERE THE PARTY SYSTEM IS STABLE OR MODERATELY STABLE (55)	5 6 12 2 X SQ= 2.92 P = 0.081 RV YES (YES)
164	TEND TO BE THOSE 0.81 WHERE THE REGIME S LEADERSHIP CHARISMA IS NEGLIGIBLE (65)	164	TEND TO BE THOSE 0.79 WHERE THE REGIME'S LEADERSHIP CHARISMA IS PRONOUNCED OR MODERATE (34)	4 15 17 4 X SQ= 12.05 P = 0. RV YES (YES)
166	TEND LESS TO BE THOSE 0.70 WHERE THE VERTICAL POWER DISTRIBUTION IS THAT OF FORMAL FEDERALISM OR FORMAL AND EFFECTIVE UNITARISM (99)	166	ALWAYS ARE THOSE 1.00 WHERE THE VERTICAL POWER DISTRIBUTION IS THAT OF FORMAL FEDERALISM OR FORMAL AND EFFECTIVE UNITARISM (99)	7 0 16 21 X SQ= 5.50 P = 0.009 RV NO (YES)
168	TEND LESS TO BE THOSE 0.76 WHERE THE HORIZONTAL POWER DISTRIBUTION IS LIMITED OR NEGLIGIBLE (72)	168	ALWAYS ARE THOSE 1.00 WHERE THE HORIZONTAL POWER DISTRIBUTION IS LIMITED OR NEGLIGIBLE (72)	5 0 16 21 X SQ= 3.63 P = 0.048 RV NO (YES)
169	ALWAYS ARE THOSE 1.00 WHERE THE HORIZONTAL POWER DISTRIBUTION IS SIGNIFICANT OR LIMITED (58)	169	TEND TO BE THOSE 0.67 WHERE THE HORIZONTAL POWER DISTRIBUTION IS NEGLIGIBLE (48)	21 7 0 14 X SQ= 18.11 P = 0. RV YES (YES)

177/

No.	Left	Value	No.	Right	Value	Statistics
178	TEND TO BE THOSE WHERE THE LEGISLATURE IS BICAMERAL (51)	0.65	178	TEND TO BE THOSE WHERE THE LEGISLATURE IS UNICAMERAL (53)	0.90	8 19 15 2 X SQ= 12.11 P = 0. RV YES (YES)
179	TEND TO BE THOSE WHERE THE EXECUTIVE IS STRONG (39)	0.78	179	ALWAYS ARE THOSE WHERE THE EXECUTIVE IS DOMINANT (52)	1.00	4 16 14 0 X SQ= 18.07 P = 0. RV YES (YES)
182	TEND TO BE THOSE WHERE THE BUREAUCRACY IS SEMI-MODERN, RATHER THAN POST-COLONIAL TRANSITIONAL (55)	0.77	182	TEND TO BE THOSE WHERE THE BUREAUCRACY IS POST-COLONIAL TRANSITIONAL, RATHER THAN SEMI-MODERN (25)	0.80	17 4 5 16 X SQ= 11.55 P = 0.001 RV YES (YES)
187	TILT LESS TOWARD BEING THOSE WHERE THE ROLE OF THE POLICE IS POLITICALLY SIGNIFICANT (66)	0.61	187	TILT MORE TOWARD BEING THOSE WHERE THE ROLE OF THE POLICE IS POLITICALLY SIGNIFICANT (66)	0.87	11 13 7 2 X SQ= 1.56 P = 0.134 RV NO (YES)
188	TEND TO BE THOSE WHERE THE CHARACTER OF THE LEGAL SYSTEM IS CIVIL LAW (32)	0.61	188	TEND TO BE THOSE WHERE THE CHARACTER OF THE LEGAL SYSTEM IS OTHER THAN CIVIL LAW (81)	0.90	14 2 9 19 X SQ= 10.39 P = 0.001 RV YES (YES)
192	TEND TO BE THOSE WHERE THE CHARACTER OF THE LEGAL SYSTEM IS OTHER THAN MUSLIM OR PARTLY MUSLIM (86)	0.78	192	TEND TO BE THOSE WHERE THE CHARACTER OF THE LEGAL SYSTEM IS MUSLIM OR PARTLY MUSLIM (27)	0.57	5 12 18 9 X SQ= 4.41 P = 0.029 RV YES (YES)
193	TEND TO BE THOSE WHERE THE CHARACTER OF THE LEGAL SYSTEM IS OTHER THAN PARTLY INDIGENOUS (86)	0.78	193	TEND TO BE THOSE WHERE THE CHARACTER OF THE LEGAL SYSTEM IS PARTLY INDIGENOUS (28)	0.71	5 15 18 6 X SQ= 9.02 P = 0.002 RV YES (YES)

178 POLITIES
WHERE THE LEGISLATURE IS UNICAMERAL (53)

178 POLITIES
WHERE THE LEGISLATURE IS BICAMERAL (51)

BOTH SUBJECT AND PREDICATE

53 IN LEFT COLUMN

ALBANIA	ALGERIA	BULGARIA	BURUNDI	CAMEROUN	CEN AFR REP	CHAD	CHINA, PR	CONGO(BRA)	CONGO(LEO)
COSTA RICA	CZECHOS'KIA	DAHOMEY	DENMARK	EL SALVADOR	FINLAND	GABON	GERMANY, E	GHANA	GREECE
GUATEMALA	GUINEA	HAITI	HONDURAS	HUNGARY	INDONESIA	ISRAEL	IVORY COAST	KOREA, N	LEBANON
MALI	MAURITANIA	MONGOLIA	NEPAL	NEW ZEALAND	NIGER	PAKISTAN	PANAMA	PARAGUAY	POLAND
RUMANIA	RWANDA	SENEGAL	SIERRE LEO	SOMALIA	SPAIN	TANGANYIKA	TOGO	TUNISIA	UGANDA
UPPER VOLTA	VIETNAM, N	VIETNAM REP							

51 IN RIGHT COLUMN

AFGHANISTAN	ARGENTINA	AUSTRALIA	AUSTRIA	BELGIUM	BOLIVIA	BRAZIL	CAMBODIA	CANADA	CEYLON
CHILE	COLOMBIA	DOMIN REP	ECUADOR	ETHIOPIA	FRANCE	GERMAN FR	ICELAND	INDIA	IRAN
IRELAND	ITALY	JAMAICA	JAPAN	JORDAN	LAOS	LIBERIA	LIBYA	LUXEMBOURG	MALAGASY R
MALAYA	MEXICO	MOROCCO	NETHERLANDS	NICARAGUA	NIGERIA	NORWAY	PERU	PHILIPPINES	PORTUGAL
SO AFRICA	SWEDEN	SWITZERLAND	TRINIDAD	TURKEY	USSR	UK	US	URUGUAY	VENEZUELA
YUGOSLAVIA									

1 EXCLUDED BECAUSE AMBIGUOUS

CYPRUS

1 EXCLUDED BECAUSE IRRELEVANT

SA'U ARABIA

9 EXCLUDED BECAUSE UNASCERTAINABLE

BURMA CUBA IRAQ KOREA REP SUDAN SYRIA THAILAND UAR YEMEN

---

2 TEND MORE TO BE THOSE LOCATED ELSEWHERE THAN IN WEST EUROPE, SCANDINAVIA, NORTH AMERICA, OR AUSTRALASIA (93) 0.91

2 TEND LESS TO BE THOSE LOCATED ELSEWHERE THAN IN WEST EUROPE, SCANDINAVIA, NORTH AMERICA, OR AUSTRALASIA (93) 0.67

5 17
48 34
X SQ= 7.53
P = 0.004
RV NO (YES)

5 TEND TO BE THOSE LOCATED IN EAST EUROPE, RATHER THAN IN SCANDINAVIA OR WEST EUROPE (9) 0.64

5 TEND TO BE THOSE LOCATED IN SCANDINAVIA OR WEST EUROPE, RATHER THAN IN EAST EUROPE (18) 0.87

4 14
7 2
X SQ= 5.54
P = 0.011
RV YES (YES)

6	LEAN MORE TOWARD BEING THOSE LOCATED ELSEWHERE THAN IN THE CARIBBEAN, CENTRAL AMERICA, OR SOUTH AMERICA (93)	0.87	6	LEAN LESS TOWARD BEING THOSE LOCATED ELSEWHERE THAN IN THE CARIBBEAN, CENTRAL AMERICA, OR SOUTH AMERICA (93)	0.73	7 14 46 37 X SQ= 2.45 P = 0.089 RV NO (YES)
7	TEND TO BE THOSE LOCATED IN CENTRAL AMERICA, RATHER THAN IN SOUTH AMERICA (7)	0.83	7	TEND TO BE THOSE LOCATED IN SOUTH AMERICA, RATHER THAN IN CENTRAL AMERICA (10)	0.82	5 2 1 9 X SQ= 4.38 P = 0.035 RV YES (YES)
10	TEND LESS TO BE THOSE LOCATED ELSEWHERE THAN IN NORTH AFRICA, OR CENTRAL AND SOUTH AFRICA (82)	0.55	10	TEND MORE TO BE THOSE LOCATED ELSEWHERE THAN IN NORTH AFRICA, OR CENTRAL AND SOUTH AFRICA (82)	0.86	24 7 29 44 X SQ= 10.91 P = 0.001 RV NO (YES)
15	LEAN MORE TOWARD BEING THOSE LOCATED IN NORTH AFRICA OR CENTRAL AND SOUTH AFRICA, RATHER THAN IN THE MIDDLE EAST (33)	0.92	15	LEAN LESS TOWARD BEING THOSE LOCATED IN NORTH AFRICA OR CENTRAL AND SOUTH AFRICA, RATHER THAN IN THE MIDDLE EAST (33)	0.64	2 4 24 7 X SQ= 2.80 P = 0.051 RV NO (YES)
19	TEND TO BE THOSE LOCATED IN NORTH AFRICA OR CENTRAL AND SOUTH AFRICA, RATHER THAN IN THE CARIBBEAN, CENTRAL AMERICA, OR SOUTH AMERICA (33)	0.77	19	TEND TO BE THOSE LOCATED IN THE CARIBBEAN, CENTRAL AMERICA, OR SOUTH AMERICA, RATHER THAN IN NORTH AFRICA OR CENTRAL AND SOUTH AFRICA (22)	0.67	24 7 7 14 X SQ= 8.36 P = 0.003 RV YES (YES)
21	TILT MORE TOWARD BEING THOSE WHOSE TERRITORIAL SIZE IS MEDIUM OR SMALL (83)	0.79	21	TILT LESS TOWARD BEING THOSE WHOSE TERRITORIAL SIZE IS MEDIUM OR SMALL (83)	0.65	11 18 42 33 X SQ= 2.06 P = 0.127 RV NO (YES)
24	TEND TO BE THOSE WHOSE POPULATION IS SMALL (54)	0.62	24	TEND TO BE THOSE WHOSE POPULATION IS VERY LARGE, LARGE, OR MEDIUM (61)	0.65	20 33 33 18 X SQ= 6.52 P = 0.007 RV YES (YES)
25	TEND MORE TO BE THOSE WHOSE POPULATION DENSITY IS MEDIUM OR LOW (98)	0.92	25	TEND LESS TO BE THOSE WHOSE POPULATION DENSITY IS MEDIUM OR LOW (98)	0.76	4 12 49 39 X SQ= 3.95 P = 0.030 RV NO (YES)
29	TEND TO BE THOSE WHERE THE DEGREE OF URBANIZATION IS LOW (49)	0.68	29	TEND TO BE THOSE WHERE THE DEGREE OF URBANIZATION IS HIGH (56)	0.77	15 36 32 11 X SQ= 17.15 P = 0. RV YES (YES)

```
30  TEND TO BE THOSE                          0.67    30  TEND TO BE THOSE                          0.73        35    14
    WHOSE AGRICULTURAL POPULATION                          WHOSE AGRICULTURAL POPULATION                         17    37
    IS HIGH  (56)                                          IS MEDIUM, LOW, OR VERY LOW  (57)               X SQ=  14.84
                                                                                                           P =     0.
                                                                                                           RV YES (YES)

34  TEND TO BE THOSE                          0.64    34  TEND TO BE THOSE                          0.71        19    36
    WHOSE GROSS NATIONAL PRODUCT                           WHOSE GROSS NATIONAL PRODUCT                          34    15
    IS VERY LOW  (53)                                      IS VERY HIGH, HIGH, MEDIUM,                     X SQ=  11.23
                                                           OR LOW  (62)                                    P =     0.
                                                                                                           RV YES (YES)

36  TEND TO BE THOSE                          0.75    36  TEND TO BE THOSE                          0.53        13    27
    WHOSE PER CAPITA GROSS NATIONAL PRODUCT                WHOSE PER CAPITA GROSS NATIONAL PRODUCT               40    24
    IS LOW OR VERY LOW  (73)                               IS VERY HIGH, HIGH, OR MEDIUM  (42)             X SQ=   7.71
                                                                                                           P =     0.005
                                                                                                           RV YES (YES)

37  TEND TO BE THOSE                          0.60    37  TEND TO BE THOSE                          0.73        21    37
    WHOSE PER CAPITA GROSS NATIONAL PRODUCT                WHOSE PER CAPITA GROSS NATIONAL PRODUCT               32    14
    IS VERY LOW  (51)                                      IS VERY HIGH, HIGH, MEDIUM, OR LOW  (64)        X SQ=  10.13
                                                                                                           P =     0.001
                                                                                                           RV YES (YES)

38  ALWAYS ARE THOSE                          1.00    38  TEND LESS TO BE THOSE                     0.80         0    10
    WHOSE INTERNATIONAL FINANCIAL STATUS                   WHOSE INTERNATIONAL FINANCIAL STATUS                  51    41
    IS MEDIUM, LOW, OR VERY LOW  (103)                     IS MEDIUM, LOW, OR VERY LOW  (103)              X SQ=   8.98
                                                                                                           P =     0.001
                                                                                                           RV NO  (YES)

39  TEND TO BE THOSE                          0.79    39  TEND TO BE THOSE                          0.51        11    26
    WHOSE INTERNATIONAL FINANCIAL STATUS                   WHOSE INTERNATIONAL FINANCIAL STATUS                  41    25
    IS LOW OR VERY LOW  (76)                               IS VERY HIGH, HIGH, OR MEDIUM  (38)             X SQ=   8.70
                                                                                                           P =     0.002
                                                                                                           RV YES (YES)

40  TEND TO BE THOSE                          0.56    40  TEND TO BE THOSE                          0.82        22    40
    WHOSE INTERNATIONAL FINANCIAL STATUS                   WHOSE INTERNATIONAL FINANCIAL STATUS                  28     9
    IS VERY LOW  (39)                                      IS VERY HIGH, HIGH,                             X SQ=  13.41
                                                           MEDIUM, OR LOW  (71)                            P =     0.
                                                                                                           RV YES (YES)

42  TEND TO BE THOSE                          0.81    42  TEND TO BE THOSE                          0.50        10    25
    WHOSE ECONOMIC DEVELOPMENTAL STATUS                    WHOSE ECONOMIC DEVELOPMENTAL STATUS                   42    25
    IS UNDERDEVELOPED OR                                   IS DEVELOPED OR INTERMEDIATE  (36)              X SQ=   9.39
    VERY UNDERDEVELOPED  (76)                                                                              P =     0.002
                                                                                                           RV YES (YES)

43  TEND TO BE THOSE                          0.73    43  TEND TO BE THOSE                          0.73        14    37
    WHOSE ECONOMIC DEVELOPMENTAL STATUS                    WHOSE ECONOMIC DEVELOPMENTAL STATUS                   37    14
    IS VERY UNDERDEVELOPED  (57)                           IS DEVELOPED, INTERMEDIATE, OR                  X SQ=  18.98
                                                           UNDERDEVELOPED  (55)                            P =     0.
                                                                                                           RV YES (YES)
```

45	TEND TO BE THOSE WHERE THE LITERACY RATE IS BELOW FIFTY PERCENT (54)	0.62	45	TEND TO BE THOSE WHERE THE LITERACY RATE IS FIFTY PERCENT OR ABOVE (55)	0.64	18 32 30 18 X SQ= 5.86 P = 0.015 RV YES (YES)
49	TEND TO BE THOSE WHERE THE LITERACY RATE IS BELOW TEN PERCENT, RATHER THAN BETWEEN TEN AND FIFTY PERCENT (26)	0.71	49	TEND TO BE THOSE WHERE THE LITERACY RATE IS BETWEEN TEN AND FIFTY PERCENT, RATHER THAN BELOW TEN PERCENT (24)	0.82	8 14 20 3 X SQ= 10.19 P = 0.001 RV YES (YES)
50	LEAN TOWARD BEING THOSE WHERE FREEDOM OF THE PRESS IS INTERMITTENT, INTERNALLY ABSENT, OR INTERNALLY AND EXTERNALLY ABSENT (56)	0.64	50	LEAN TOWARD BEING THOSE WHERE FREEDOM OF THE PRESS IS COMPLETE (43)	0.57	14 28 25 21 X SQ= 3.12 P = 0.056 RV YES (YES)
51	TEND TO BE THOSE WHERE FREEDOM OF THE PRESS IS INTERNALLY ABSENT, OR INTERNALLY AND EXTERNALLY ABSENT (37)	0.56	51	TEND TO BE THOSE WHERE FREEDOM OF THE PRESS IS COMPLETE OR INTERMITTENT (60)	0.82	17 40 22 9 X SQ= 12.16 P = 0. RV YES (YES)
53	LEAN MORE TOWARD BEING THOSE WHERE NEWSPAPER CIRCULATION IS LESS THAN THREE HUNDRED PER THOUSAND (101)	0.92	53	LEAN LESS TOWARD BEING THOSE WHERE NEWSPAPER CIRCULATION IS LESS THAN THREE HUNDRED PER THOUSAND (101)	0.80	4 10 49 41 X SQ= 2.29 P = 0.089 RV NO (YES)
55	TEND LESS TO BE THOSE WHERE NEWSPAPER CIRCULATION IS TEN OR MORE PER THOUSAND (78)	0.55	55	TEND MORE TO BE THOSE WHERE NEWSPAPER CIRCULATION IS TEN OR MORE PER THOUSAND (78)	0.82	28 42 23 9 X SQ= 7.70 P = 0.003 RV NO (YES)
56	TEND LESS TO BE THOSE WHERE THE RELIGION IS PREDOMINANTLY LITERATE (79)	0.57	56	TEND MORE TO BE THOSE WHERE THE RELIGION IS PREDOMINANTLY LITERATE (79)	0.82	28 42 21 9 X SQ= 6.41 P = 0.008 RV NO (YES)
57	TEND MORE TO BE THOSE WHERE THE RELIGION IS OTHER THAN CATHOLIC (90)	0.87	57	TEND LESS TO BE THOSE WHERE THE RELIGION IS OTHER THAN CATHOLIC (90)	0.67	7 17 46 34 X SQ= 4.85 P = 0.020 RV NO (YES)
63	TEND TO BE THOSE WHERE THE RELIGION IS PREDOMINANTLY OR PARTLY OTHER THAN CHRISTIAN (68)	0.70	63	TEND TO BE THOSE WHERE THE RELIGION IS PREDOMINANTLY SOME KIND OF CHRISTIAN (46)	0.58	16 29 37 21 X SQ= 7.00 P = 0.006 RV YES (YES)

68	LEAN TOWARD BEING THOSE THAT ARE LINGUISTICALLY WEAKLY HETEROGENEOUS OR STRONGLY HETEROGENEOUS (62)	0.63	68	LEAN TOWARD BEING THOSE THAT ARE LINGUISTICALLY HOMOGENEOUS (52)	0.55	19 28 33 23 X SQ= 2.80 P = 0.076 RV YES (YES)
72	TEND TO BE THOSE WHOSE DATE OF INDEPENDENCE IS AFTER 1914 (59)	0.69	72	TEND TO BE THOSE WHOSE DATE OF INDEPENDENCE IS BEFORE 1914 (52)	0.68	16 34 35 16 X SQ= 12.12 P = 0. RV YES (YES)
73	TEND TO BE THOSE WHOSE DATE OF INDEPENDENCE IS AFTER 1945 (46)	0.57	73	TEND TO BE THOSE WHOSE DATE OF INDEPENDENCE IS BEFORE 1945 (65)	0.74	22 37 29 13 X SQ= 8.67 P = 0.002 RV YES (YES)
74	LEAN MORE TOWARD BEING THOSE THAT ARE NOT HISTORICALLY WESTERN (87)	0.83	74	LEAN LESS TOWARD BEING THOSE THAT ARE NOT HISTORICALLY WESTERN (87)	0.66	9 17 43 33 X SQ= 2.91 P = 0.070 RV NO (YES)
75	TEND TO BE THOSE THAT ARE NOT HISTORICALLY WESTERN AND ARE NOT SIGNIFICANTLY WESTERNIZED (52)	0.60	75	TEND TO BE THOSE THAT ARE HISTORICALLY WESTERN OR SIGNIFICANTLY WESTERNIZED (62)	0.76	21 38 32 12 X SQ= 12.47 P = 0. RV YES (YES)
76	LEAN MORE TOWARD BEING THOSE THAT HAVE BEEN SIGNIFICANTLY OR PARTIALLY WESTERNIZED THROUGH A COLONIAL RELATIONSHIP, RATHER THAN BEING HISTORICALLY WESTERN (70)	0.80	76	LEAN LESS TOWARD BEING THOSE THAT HAVE BEEN SIGNIFICANTLY OR PARTIALLY WESTERNIZED THROUGH A COLONIAL RELATIONSHIP, RATHER THAN BEING HISTORICALLY WESTERN (70)	0.61	9 17 36 27 X SQ= 2.89 P = 0.065 RV NO (YES)
77	TEND TO BE THOSE THAT HAVE BEEN PARTIALLY WESTERNIZED, RATHER THAN SIGNIFICANTLY WESTERNIZED, THROUGH A COLONIAL RELATIONSHIP (41)	0.78	77	TEND TO BE THOSE THAT HAVE BEEN SIGNIFICANTLY WESTERNIZED, RATHER THAN PARTIALLY WESTERNIZED, THROUGH A COLONIAL RELATIONSHIP (28)	0.65	8 17 28 9 X SQ= 9.96 P = 0.001 RV YES (YES)
79	LEAN MORE TOWARD BEING THOSE THAT WERE FORMERLY DEPENDENCIES OF BRITAIN OR FRANCE, RATHER THAN SPAIN (49)	0.82	79	LEAN LESS TOWARD BEING THOSE THAT WERE FORMERLY DEPENDENCIES OF BRITAIN OR FRANCE, RATHER THAN SPAIN (49)	0.59	27 16 6 11 X SQ= 2.69 P = 0.084 RV NO (YES)
80	TEND TO BE THOSE WHOSE DATE OF INDEPENDENCE IS AFTER 1914, AND THAT WERE FORMERLY DEPENDENCIES OF FRANCE, RATHER THAN BRITAIN (24)	0.76	80	TEND TO BE THOSE WHOSE DATE OF INDEPENDENCE IS AFTER 1914, AND THAT WERE FORMERLY DEPENDENCIES OF BRITAIN, RATHER THAN FRANCE (19)	0.67	6 8 19 4 X SQ= 4.59 P = 0.027 RV YES (YES)

81	TEND MORE TO BE THOSE WHERE THE TYPE OF POLITICAL MODERNIZATION IS LATER EUROPEAN OR LATER EUROPEAN DERIVED, RATHER THAN EARLY EUROPEAN OR EARLY EUROPEAN DERIVED (40)	0.95	81	TEND LESS TO BE THOSE WHERE THE TYPE OF POLITICAL MODERNIZATION IS LATER EUROPEAN OR LATER EUROPEAN DERIVED, RATHER THAN EARLY EUROPEAN OR EARLY EUROPEAN DERIVED (40)	0.68	1 10 18 21 X SQ= 3.55 P = 0.035 RV NO (NO)
82	TEND TO BE THOSE WHERE THE TYPE OF POLITICAL MODERNIZATION IS DEVELOPED TUTELARY OR UNDEVELOPED TUTELARY, RATHER THAN EARLY OR LATER EUROPEAN OR EUROPEAN DERIVED (55)	0.63	82	TEND TO BE THOSE WHERE THE TYPE OF POLITICAL MODERNIZATION IS EARLY OR LATER EUROPEAN OR EUROPEAN DERIVED, RATHER THAN DEVELOPED TUTELARY OR UNDEVELOPED TUTELARY (51)	0.69	19 31 32 14 X SQ= 8.36 P = 0.002 RV YES (YES)
84	TEND TO BE THOSE WHERE THE TYPE OF POLITICAL MODERNIZATION IS UNDEVELOPED TUTELARY, RATHER THAN DEVELOPED TUTELARY (24)	0.69	84	TEND TO BE THOSE WHERE THE TYPE OF POLITICAL MODERNIZATION IS DEVELOPED TUTELARY, RATHER THAN UNDEVELOPED TUTELARY (31)	0.86	10 12 22 2 X SQ= 9.50 P = 0.001 RV YES (YES)
85	LEAN TOWARD BEING THOSE WHERE THE STAGE OF POLITICAL MODERNIZATION IS MID- OR EARLY TRANSITIONAL, RATHER THAN ADVANCED (54)	0.58	85	LEAN TOWARD BEING THOSE WHERE THE STAGE OF POLITICAL MODERNIZATION IS ADVANCED, RATHER THAN MID- OR EARLY TRANSITIONAL (60)	0.61	22 31 31 20 X SQ= 3.13 P = 0.054 RV YES (YES)
87	TEND TO BE THOSE WHOSE IDEOLOGICAL ORIENTATION IS DEVELOPMENTAL (31)	0.54	87	TEND TO BE THOSE WHOSE IDEOLOGICAL ORIENTATION IS OTHER THAN DEVELOPMENTAL (58)	0.89	25 4 21 32 X SQ= 14.68 P = 0. RV YES (YES)
89	TEND TO BE THOSE WHOSE IDEOLOGICAL ORIENTATION IS OTHER THAN CONVENTIONAL (62)	0.85	89	TEND TO BE THOSE WHOSE IDEOLOGICAL ORIENTATION IS CONVENTIONAL (33)	0.68	7 25 41 12 X SQ= 22.78 P = 0. RV YES (YES)
91	ALWAYS ARE THOSE WHOSE IDEOLOGICAL ORIENTATION IS DEVELOPMENTAL, RATHER THAN TRADITIONAL (31)	1.00	91	TEND LESS TO BE THOSE WHOSE IDEOLOGICAL ORIENTATION IS DEVELOPMENTAL, RATHER THAN TRADITIONAL (31)	0.57	25 4 0 3 X SQ= 7.32 P = 0.007 RV NO (YES)
92	TEND LESS TO BE THOSE WHERE THE SYSTEM STYLE IS LIMITED MOBILIZATIONAL, OR NON-MOBILIZATIONAL (93)	0.71	92	TEND MORE TO BE THOSE WHERE THE SYSTEM STYLE IS LIMITED MOBILIZATIONAL, OR NON-MOBILIZATIONAL (93)	0.96	15 2 37 49 X SQ= 9.87 P = 0.001 RV NO (YES)
93	TEND LESS TO BE THOSE WHERE THE SYSTEM STYLE IS NON-MOBILIZATIONAL (78)	0.56	93	TEND MORE TO BE THOSE WHERE THE SYSTEM STYLE IS NON-MOBILIZATIONAL (78)	0.88	22 6 28 45 X SQ= 11.53 P = 0. RV NO (YES)

94	TEND TO BE THOSE WHERE THE STATUS OF THE REGIME IS AUTHORITARIAN OR TOTALITARIAN (41)	0.64	94	TEND TO BE THOSE WHERE THE STATUS OF THE REGIME IS CONSTITUTIONAL (51)	0.77	13 37 23 11 X SQ= 12.68 P = 0. RV YES (YES)
95	TEND LESS TO BE THOSE WHERE THE STATUS OF THE REGIME IS CONSTITUTIONAL OR AUTHORITARIAN (95)	0.76	95	TEND MORE TO BE THOSE WHERE THE STATUS OF THE REGIME IS CONSTITUTIONAL OR AUTHORITARIAN (95)	0.94	39 47 12 3 X SQ= 4.83 P = 0.023 RV NO (YES)
97	TILT TOWARD BEING THOSE WHERE THE STATUS OF THE REGIME IS TOTALITARIAN, RATHER THAN AUTHORITARIAN (16)	0.57	97	TILT TOWARD BEING THOSE WHERE THE STATUS OF THE REGIME IS AUTHORITARIAN, RATHER THAN TOTALITARIAN (23)	0.73	9 8 12 3 X SQ= 1.53 P = 0.147 RV YES (NO)
98	TEND MORE TO BE THOSE WHERE GOVERNMENTAL STABILITY IS GENERALLY OR MODERATELY PRESENT AND DATES FROM THE POST-WAR PERIOD, OR IS ABSENT (93)	0.92	98	TEND LESS TO BE THOSE WHERE GOVERNMENTAL STABILITY IS GENERALLY OR MODERATELY PRESENT AND DATES FROM THE POST-WAR PERIOD, OR IS ABSENT (93)	0.67	4 17 49 34 X SQ= 9.18 P = 0.001 RV NO (YES)
101	TEND TO BE THOSE WHERE THE REPRESENTATIVE CHARACTER OF THE REGIME IS LIMITED POLYARCHIC, PSEUDO-POLYARCHIC, OR NON-POLYARCHIC (57)	0.80	101	TEND TO BE THOSE WHERE THE REPRESENTATIVE CHARACTER OF THE REGIME IS POLYARCHIC (41)	0.66	9 31 35 16 X SQ= 17.30 P = 0. RV YES (YES)
102	TEND TO BE THOSE WHERE THE REPRESENTATIVE CHARACTER OF THE REGIME IS PSEUDO-POLYARCHIC OR NON-POLYARCHIC (49)	0.65	102	TEND TO BE THOSE WHERE THE REPRESENTATIVE CHARACTER OF THE REGIME IS POLYARCHIC OR LIMITED POLYARCHIC (59)	0.80	18 40 33 10 X SQ= 18.85 P = 0. RV YES (YES)
105	TEND TO BE THOSE WHERE THE ELECTORAL SYSTEM IS PARTIALLY COMPETITIVE OR NON-COMPETITIVE (47)	0.78	105	TEND TO BE THOSE WHERE THE ELECTORAL SYSTEM IS COMPETITIVE (43)	0.75	10 33 36 11 X SQ= 23.48 P = 0. RV YES (YES)
106	TEND TO BE THOSE WHERE THE ELECTORAL SYSTEM IS NON-COMPETITIVE (30)	0.60	106	TEND TO BE THOSE WHERE THE ELECTORAL SYSTEM IS COMPETITIVE OR PARTIALLY COMPETITIVE (52)	0.87	17 34 25 5 X SQ= 16.97 P = 0. RV YES (YES)
107	TEND TO BE THOSE WHERE AUTONOMOUS GROUPS ARE PARTIALLY TOLERATED IN POLITICS, ARE TOLERATED ONLY OUTSIDE POLITICS, OR ARE NOT TOLERATED AT ALL (65)	0.81	107	TEND TO BE THOSE WHERE AUTONOMOUS GROUPS ARE FULLY TOLERATED IN POLITICS (46)	0.69	10 34 42 15 X SQ= 23.81 P = 0. RV YES (YES)

108	TEND TO BE THOSE WHERE AUTONOMOUS GROUPS ARE TOLERATED ONLY OUTSIDE POLITICS OR ARE NOT TOLERATED AT ALL (35)	0.51	108	TEND TO BE THOSE WHERE AUTONOMOUS GROUPS ARE FULLY OR PARTIALLY TOLERATED IN POLITICS (65)	0.85	23 40 24 7 X SQ= 12.32 P = 0. RV YES (YES)
110	TILT MORE TOWARD BEING THOSE WHERE POLITICAL ENCULTURATION IS MEDIUM OR LOW (80)	0.92	110	TILT LESS TOWARD BEING THOSE WHERE POLITICAL ENCULTURATION IS MEDIUM OR LOW (80)	0.79	3 10 35 37 X SQ= 1.96 P = 0.130 RV NO (NO)
113	TEND MORE TO BE THOSE WHERE SECTIONALISM IS MODERATE OR NEGLIGIBLE (81)	0.88	113	TEND LESS TO BE THOSE WHERE SECTIONALISM IS MODERATE OR NEGLIGIBLE (81)	0.64	6 17 44 30 X SQ= 6.55 P = 0.008 RV NO (YES)
114	TEND TO BE THOSE WHERE SECTIONALISM IS NEGLIGIBLE (47)	0.54	114	TEND TO BE THOSE WHERE SECTIONALISM IS EXTREME OR MODERATE (61)	0.68	23 32 27 15 X SQ= 3.96 P = 0.040 RV YES (YES)
116	TEND TO BE THOSE WHERE INTEREST ARTICULATION BY ASSOCIATIONAL GROUPS IS LIMITED OR NEGLIGIBLE (79)	0.88	116	TEND TO BE THOSE WHERE INTEREST ARTICULATION BY ASSOCIATIONAL GROUPS IS SIGNIFICANT OR MODERATE (32)	0.53	6 26 46 23 X SQ= 18.22 P = 0. RV YES (YES)
117	TEND TO BE THOSE WHERE INTEREST ARTICULATION BY ASSOCIATIONAL GROUPS IS NEGLIGIBLE (51)	0.71	117	TEND TO BE THOSE WHERE INTEREST ARTICULATION BY ASSOCIATIONAL GROUPS IS SIGNIFICANT, MODERATE, OR LIMITED (60)	0.82	15 40 37 9 X SQ= 26.25 P = 0. RV YES (YES)
118	TEND TO BE THOSE WHERE INTEREST ARTICULATION BY INSTITUTIONAL GROUPS IS VERY SIGNIFICANT (40)	0.54	118	TEND TO BE THOSE WHERE INTEREST ARTICULATION BY INSTITUTIONAL GROUPS IS SIGNIFICANT, MODERATE, OR LIMITED (60)	0.78	21 11 18 40 X SQ= 8.69 P = 0.002 RV YES (YES)
121	TEND TO BE THOSE WHERE INTEREST ARTICULATION BY NON-ASSOCIATIONAL GROUPS IS SIGNIFICANT (54)	0.60	121	TEND TO BE THOSE WHERE INTEREST ARTICULATION BY NON-ASSOCIATIONAL GROUPS IS MODERATE, LIMITED, OR NEGLIGIBLE (61)	0.75	32 13 21 38 X SQ= 11.50 P = 0. RV YES (YES)
125	TEND TO BE THOSE WHERE INTEREST ARTICULATION BY ANOMIC GROUPS IS FREQUENT OR OCCASIONAL (64)	0.80	125	TEND TO BE THOSE WHERE INTEREST ARTICULATION BY ANOMIC GROUPS IS INFREQUENT OR VERY INFREQUENT (35)	0.52	35 22 9 24 X SQ= 8.43 P = 0.002 RV YES (YES)

128	TEND TO BE THOSE WHERE INTEREST ARTICULATION BY POLITICAL PARTIES IS LIMITED OR NEGLIGIBLE (45)	0.68	128	TEND TO BE THOSE WHERE INTEREST ARTICULATION BY POLITICAL PARTIES IS SIGNIFICANT OR MODERATE (48)	0.74	15 32 32 11 X SQ= 14.60 P = 0. RV YES (YES)
129	TEND TO BE THOSE WHERE INTEREST ARTICULATION BY POLITICAL PARTIES IS NEGLIGIBLE (37)	0.60	129	TEND TO BE THOSE WHERE INTEREST ARTICULATION BY POLITICAL PARTIES IS SIGNIFICANT, MODERATE, OR LIMITED (56)	0.81	19 35 28 8 X SQ= 14.04 P = 0. RV YES (YES)
131	TILT TOWARD BEING THOSE WHERE INTEREST AGGREGATION BY POLITICAL PARTIES IS LIMITED OR NEGLIGIBLE (35)	0.67	131	TILT TOWARD BEING THOSE WHERE INTEREST AGGREGATION BY POLITICAL PARTIES IS SIGNIFICANT OR MODERATE (30)	0.56	7 23 14 18 X SQ= 2.04 P = 0.112 RV YES (NO)
137	TEND TO BE THOSE WHERE INTEREST AGGREGATION BY THE LEGISLATURE IS LIMITED OR NEGLIGIBLE (68)	0.88	137	TEND TO BE THOSE WHERE INTEREST AGGREGATION BY THE LEGISLATURE IS SIGNIFICANT OR MODERATE (29)	0.51	6 23 44 22 X SQ= 15.29 P = 0. RV YES (YES)
138	TEND TO BE THOSE WHERE INTEREST AGGREGATION BY THE LEGISLATURE IS NEGLIGIBLE (49)	0.75	138	TEND TO BE THOSE WHERE INTEREST AGGREGATION BY THE LEGISLATURE IS SIGNIFICANT, MODERATE, OR LIMITED (48)	0.74	12 34 36 12 X SQ= 20.57 P = 0. RV YES (YES)
139	TEND TO BE THOSE WHERE THE PARTY SYSTEM IS QUANTITATIVELY ONE-PARTY (34)	0.56	139	TEND TO BE THOSE WHERE THE PARTY SYSTEM IS QUANTITATIVELY OTHER THAN ONE-PARTY (71)	0.90	28 5 22 44 X SQ= 21.34 P = 0. RV YES (YES)
141	TILT MORE TOWARD BEING THOSE WHERE THE PARTY SYSTEM IS QUANTITATIVELY OTHER THAN TWO-PARTY (87)	0.94	141	TILT LESS TOWARD BEING THOSE WHERE THE PARTY SYSTEM IS QUANTITATIVELY OTHER THAN TWO-PARTY (87)	0.82	3 8 46 37 X SQ= 2.06 P = 0.110 RV NO (YES)
142	TEND MORE TO BE THOSE WHERE THE PARTY SYSTEM IS QUANTITATIVELY OTHER THAN MULTI-PARTY (66)	0.80	142	TEND LESS TO BE THOSE WHERE THE PARTY SYSTEM IS QUANTITATIVELY OTHER THAN MULTI-PARTY (66)	0.56	10 19 39 24 X SQ= 4.95 P = 0.024 RV NO (YES)
146	TEND LESS TO BE THOSE WHERE THE PARTY SYSTEM IS QUALITATIVELY OTHER THAN MASS-BASED TERRITORIAL (92)	0.86	146	ALWAYS ARE THOSE WHERE THE PARTY SYSTEM IS QUALITATIVELY OTHER THAN MASS-BASED TERRITORIAL (92)	1.00	7 0 43 44 X SQ= 4.78 P = 0.014 RV NO (YES)

147	TEND TO BE THOSE WHERE THE PARTY SYSTEM IS QUALITATIVELY OTHER THAN CLASS-ORIENTED OR MULTI-IDEOLOGICAL (67)	0.92	147	TEND TO BE THOSE WHERE THE PARTY SYSTEM IS QUALITATIVELY CLASS-ORIENTED OR MULTI-IDEOLOGICAL (23)	0.51	4 19 44 18 X SQ= 17.47 P = 0. RV YES (YES)
148	TEND LESS TO BE THOSE WHERE THE PARTY SYSTEM IS QUALITATIVELY OTHER THAN AFRICAN TRANSITIONAL (96)	0.75	148	TEND MORE TO BE THOSE WHERE THE PARTY SYSTEM IS QUALITATIVELY OTHER THAN AFRICAN TRANSITIONAL (96)	0.98	13 1 39 47 X SQ= 9.07 P = 0.001 RV NO (YES)
164	TEND LESS TO BE THOSE WHERE THE REGIME'S LEADERSHIP CHARISMA IS NEGLIGIBLE (65)	0.53	164	TEND MORE TO BE THOSE WHERE THE REGIME'S LEADERSHIP CHARISMA IS NEGLIGIBLE (65)	0.80	22 9 25 37 X SQ= 6.59 P = 0.008 RV NO (YES)
166	TEND MORE TO BE THOSE WHERE THE VERTICAL POWER DISTRIBUTION IS THAT OF FORMAL FEDERALISM OR FORMAL AND EFFECTIVE UNITARISM (99)	0.94	166	TEND LESS TO BE THOSE WHERE THE VERTICAL POWER DISTRIBUTION IS THAT OF FORMAL FEDERALISM OR FORMAL AND EFFECTIVE UNITARISM (99)	0.76	3 12 49 39 X SQ= 5.18 P = 0.013 RV NO (YES)
168	TEND TO BE THOSE WHERE THE HORIZONTAL POWER DISTRIBUTION IS LIMITED OR NEGLIGIBLE (72)	0.88	168	TEND TO BE THOSE WHERE THE HORIZONTAL POWER DISTRIBUTION IS SIGNIFICANT (34)	0.55	6 27 45 22 X SQ= 19.31 P = 0. RV YES (YES)
169	TEND TO BE THOSE WHERE THE HORIZONTAL POWER DISTRIBUTION IS NEGLIGIBLE (48)	0.65	169	TEND TO BE THOSE WHERE THE HORIZONTAL POWER DISTRIBUTION IS SIGNIFICANT OR LIMITED (58)	0.80	18 39 33 10 X SQ= 18.24 P = 0. RV YES (YES)
170	TEND LESS TO BE THOSE WHERE THE LEGISLATIVE-EXECUTIVE STRUCTURE IS OTHER THAN PRESIDENTIAL (63)	0.52	170	TEND MORE TO BE THOSE WHERE THE LEGISLATIVE-EXECUTIVE STRUCTURE IS OTHER THAN PRESIDENTIAL (63)	0.72	24 14 26 36 X SQ= 3.44 P = 0.043 RV NO (YES)
172	TEND MORE TO BE THOSE WHERE THE LEGISLATIVE-EXECUTIVE STRUCTURE IS OTHER THAN PARLIAMENTARY-ROYALIST (88)	0.88	172	TEND LESS TO BE THOSE WHERE THE LEGISLATIVE-EXECUTIVE STRUCTURE IS OTHER THAN PARLIAMENTARY-ROYALIST (88)	0.69	6 15 46 34 X SQ= 4.47 P = 0.026 RV NO (YES)
175	TEND TO BE THOSE WHERE THE LEGISLATURE IS LARGELY INEFFECTIVE OR WHOLLY INEFFECTIVE (49)	0.76	175	TEND TO BE THOSE WHERE THE LEGISLATURE IS FULLY EFFECTIVE OR PARTIALLY EFFECTIVE (51)	0.79	12 38 37 10 X SQ= 26.88 P = 0. RV YES (YES)

No.	Statement	Value	No.	Statement	Value	Statistics
179	TEND TO BE THOSE WHERE THE EXECUTIVE IS DOMINANT (52)	0.79	179	TEND TO BE THOSE WHERE THE EXECUTIVE IS STRONG (39)	0.66	34 15 9 29 X SQ= 16.10 P = 0. RV YES (YES)
181	TEND MORE TO BE THOSE WHERE THE BUREAUCRACY IS SEMI-MODERN, RATHER THAN MODERN (55)	0.86	181	TEND LESS TO BE THOSE WHERE THE BUREAUCRACY IS SEMI-MODERN, RATHER THAN MODERN (55)	0.60	4 17 24 26 X SQ= 4.05 P = 0.033 RV NO (NO)
182	TEND LESS TO BE THOSE WHERE THE BUREAUCRACY IS SEMI-MODERN, RATHER THAN POST-COLONIAL TRANSITIONAL (55)	0.53	182	TEND MORE TO BE THOSE WHERE THE BUREAUCRACY IS SEMI-MODERN, RATHER THAN POST-COLONIAL TRANSITIONAL (55)	0.93	24 26 21 2 X SQ= 10.73 P = 0. RV NO (YES)
183	TEND TO BE THOSE WHERE THE BUREAUCRACY IS POST-COLONIAL TRANSITIONAL, RATHER THAN TRADITIONAL (25)	0.95	183	TEND TO BE THOSE WHERE THE BUREAUCRACY IS TRADITIONAL, RATHER THAN POST-COLONIAL TRANSITIONAL (9)	0.75	21 2 1 6 X SQ= 12.58 P = 0. RV YES (YES)
186	LEAN LESS TOWARD BEING THOSE WHERE PARTICIPATION BY THE MILITARY IN POLITICS IS NEUTRAL, RATHER THAN SUPPORTIVE (56)	0.55	186	LEAN MORE TOWARD BEING THOSE WHERE PARTICIPATION BY THE MILITARY IN POLITICS IS NEUTRAL, RATHER THAN SUPPORTIVE (56)	0.74	19 11 23 32 X SQ= 2.79 P = 0.072 RV NO (YES)
187	TEND TO BE THOSE WHERE THE ROLE OF THE POLICE IS POLITICALLY SIGNIFICANT (66)	0.80	187	TEND TO BE THOSE WHERE THE ROLE OF THE POLICE IS NOT POLITICALLY SIGNIFICANT (35)	0.57	36 20 9 26 X SQ= 11.32 P = 0.001 RV YES (YES)
188	TEND MORE TO BE THOSE WHERE THE CHARACTER OF THE LEGAL SYSTEM IS OTHER THAN CIVIL LAW (81)	0.83	188	TEND LESS TO BE THOSE WHERE THE CHARACTER OF THE LEGAL SYSTEM IS OTHER THAN CIVIL LAW (81)	0.56	9 22 43 28 X SQ= 7.37 P = 0.005 RV NO (YES)
189	LEAN MORE TOWARD BEING THOSE WHERE THE CHARACTER OF THE LEGAL SYSTEM IS OTHER THAN COMMON LAW (108)	0.98	189	LEAN LESS TOWARD BEING THOSE WHERE THE CHARACTER OF THE LEGAL SYSTEM IS OTHER THAN COMMON LAW (108)	0.88	1 6 52 45 X SQ= 2.62 P = 0.058 RV NO (YES)
192	TEND LESS TO BE THOSE WHERE THE CHARACTER OF THE LEGAL SYSTEM IS OTHER THAN MUSLIM OR PARTLY MUSLIM (86)	0.69	192	TEND MORE TO BE THOSE WHERE THE CHARACTER OF THE LEGAL SYSTEM IS OTHER THAN MUSLIM OR PARTLY MUSLIM (86)	0.88	16 6 36 44 X SQ= 4.26 P = 0.030 RV NO (YES)

193	TEND LESS TO BE THOSE WHERE THE CHARACTER OF THE LEGAL SYSTEM IS OTHER THAN PARTLY INDIGENOUS (86)	0.58	193	TEND MORE TO BE THOSE WHERE THE CHARACTER OF THE LEGAL SYSTEM IS OTHER THAN PARTLY INDIGENOUS (86)	0.88	22 6 31 45 X SQ= 10.22 P = 0.001 RV NO (YES)
194	TEND LESS TO BE THOSE THAT ARE NON-COMMUNIST (101)	0.79	194	TEND MORE TO BE THOSE THAT ARE NON-COMMUNIST (101)	0.96	11 2 42 49 X SQ= 5.28 P = 0.015 RV NO (YES)

```
******************************************************************************************************************

179  POLITIES                                            179  POLITIES
     WHERE THE EXECUTIVE IS DOMINANT  (52)                    WHERE THE EXECUTIVE IS STRONG  (39)



                                                                                            BOTH SUBJECT AND PREDICATE
..................................................................................................................

      52 IN LEFT COLUMN

AFGHANISTAN ALBANIA     ALGERIA     BULGARIA    CAMBODIA    CEN AFR REP CHINA, PR   CONGO(BRA)  CUBA        CZECHOS'KIA
DAHOMEY     EL SALVADOR ETHIOPIA    FRANCE      GABON       GERMANY, E  GHANA       GUINEA      HAITI       HONDURAS
HUNGARY     INDIA       INDONESIA   IRAN        IVORY COAST JORDAN      KOREA, N    LIBERIA     MALI        MAURITANIA
MEXICO      MONGOLIA    MOROCCO     NEPAL       NICARAGUA   PAKISTAN    PARAGUAY    POLAND      PORTUGAL    RUMANIA
SA'U ARABIA SENEGAL     SO AFRICA   SPAIN       TANGANYIKA  TUNISIA     USSR        UAR         UPPER VOLTA VIETNAM, N
VIETNAM REP YUGOSLAVIA

      39 IN RIGHT COLUMN

AUSTRALIA   AUSTRIA     BELGIUM     BOLIVIA     BRAZIL      CAMEROUN    CANADA      COLOMBIA    COSTA RICA  CYPRUS
DENMARK     DOMIN REP   ECUADOR     FINLAND     GERMAN FR   ICELAND     IRELAND     ISRAEL      ITALY       JAMAICA
JAPAN       LEBANON     LIBYA       LUXEMBOURG  MALAGASY R  MALAYA      NETHERLANDS NEW ZEALAND NIGERIA     NORWAY
PANAMA      PHILIPPINES SIERRE LEO  SWEDEN      TRINIDAD    TURKEY      UK          US          VENEZUELA

       4 EXCLUDED BECAUSE AMBIGUOUS

CEYLON      CHILE       GREECE      LAOS

       2 EXCLUDED BECAUSE IRRELEVANT

SWITZERLAND URUGUAY

       6 EXCLUDED BECAUSE UNASCERTAINED

BURUNDI     CHAD        NIGER       RWANDA      SOMALIA     UGANDA

      12 EXCLUDED BECAUSE UNASCERTAINABLE

ARGENTINA   BURMA       CONGO(LEO)  GUATEMALA   IRAQ        KOREA REP   PERU        SUDAN       SYRIA       THAILAND
TOGO        YEMEN

------------------------------------------------------------------------------------------------------------------

  2  TEND MORE TO BE THOSE                     0.94          2  TEND LESS TO BE THOSE                  0.56          3   17
     LOCATED ELSEWHERE THAN IN                                  LOCATED ELSEWHERE THAN IN                           49   22
     WEST EUROPE, SCANDINAVIA,                                  WEST EUROPE, SCANDINAVIA,                        X SQ=  16.45
     NORTH AMERICA, OR AUSTRALASIA  (93)                        NORTH AMERICA, OR AUSTRALASIA  (93)              P =    0.
                                                                                                                 RV NO  (YES)

  5  TEND TO BE THOSE                          0.75          5  ALWAYS ARE THOSE                       1.00          3   13
     LOCATED IN EAST EUROPE, RATHER THAN                        LOCATED IN SCANDINAVIA OR WEST EUROPE,               9    0
     IN SCANDINAVIA OR WEST EUROPE  (9)                         RATHER THAN IN EAST EUROPE  (18)                 X SQ=  12.15
                                                                                                                 P =    0.
                                                                                                                 RV YES (YES)
```

8	TILT LESS TOWARD BEING THOSE LOCATED ELSEWHERE THAN IN EAST ASIA SOUTH ASIA, OR SOUTHEAST ASIA (97)	0.81	8	TILT MORE TOWARD BEING THOSE LOCATED ELSEWHERE THAN IN EAST ASIA SOUTH ASIA, OR SOUTHEAST ASIA (97)	0.92	10 3 42 36 X SQ= 1.57 P = 0.142 RV NO (NO)
10	TEND LESS TO BE THOSE LOCATED ELSEWHERE THAN IN NORTH AFRICA, OR CENTRAL AND SOUTH AFRICA (82)	0.63	10	TEND MORE TO BE THOSE LOCATED ELSEWHERE THAN IN NORTH AFRICA, OR CENTRAL AND SOUTH AFRICA (82)	0.87	19 5 33 34 X SQ= 5.29 P = 0.016 RV NO (YES)
19	TEND TO BE THOSE LOCATED IN NORTH AFRICA OR CENTRAL AND SOUTH AFRICA, RATHER THAN IN THE CARIBBEAN, CENTRAL AMERICA, OR SOUTH AMERICA (33)	0.73	19	TEND TO BE THOSE LOCATED IN THE CARIBBEAN, CENTRAL AMERICA, OR SOUTH AMERICA, RATHER THAN IN NORTH AFRICA OR CENTRAL AND SOUTH AFRICA (22)	0.67	19 5 7 10 X SQ= 4.66 P = 0.021 RV YES (YES)
20	LEAN TOWARD BEING THOSE LOCATED IN EAST ASIA, SOUTH ASIA, OR SOUTHEAST ASIA, RATHER THAN IN THE CARIBBEAN, CENTRAL AMERICA, OR SOUTH AMERICA (18)	0.59	20	LEAN TOWARD BEING THOSE LOCATED IN THE CARIBBEAN, CENTRAL AMERICA, OR SOUTH AMERICA, RATHER THAN IN EAST ASIA, SOUTH ASIA, OR SOUTHEAST ASIA (22)	0.77	10 3 7 10 X SQ= 2.52 P = 0.071 RV YES (YES)
25	TEND MORE TO BE THOSE WHOSE POPULATION DENSITY IS MEDIUM OR LOW (98)	0.92	25	TEND LESS TO BE THOSE WHOSE POPULATION DENSITY IS MEDIUM OR LOW (98)	0.74	4 10 48 29 X SQ= 4.22 P = 0.037 RV NO (YES)
29	TEND TO BE THOSE WHERE THE DEGREE OF URBANIZATION IS LOW (49)	0.61	29	TEND TO BE THOSE WHERE THE DEGREE OF URBANIZATION IS HIGH (56)	0.83	18 29 28 6 X SQ= 13.86 P = 0. RV YES (YES)
30	TEND TO BE THOSE WHOSE AGRICULTURAL POPULATION IS HIGH (56)	0.66	30	TEND TO BE THOSE WHOSE AGRICULTURAL POPULATION IS MEDIUM, LOW, OR VERY LOW (57)	0.85	33 6 17 33 X SQ= 20.79 P = 0. RV YES (YES)
36	TEND TO BE THOSE WHOSE PER CAPITA GROSS NATIONAL PRODUCT IS LOW OR VERY LOW (73)	0.77	36	TEND TO BE THOSE WHOSE PER CAPITA GROSS NATIONAL PRODUCT IS VERY HIGH, HIGH, OR MEDIUM (42)	0.64	12 25 40 14 X SQ= 13.89 P = 0. RV YES (YES)
37	TEND TO BE THOSE WHOSE PER CAPITA GROSS NATIONAL PRODUCT IS VERY LOW (51)	0.58	37	TEND TO BE THOSE WHOSE PER CAPITA GROSS NATIONAL PRODUCT IS VERY HIGH, HIGH, MEDIUM, OR LOW (64)	0.85	22 33 30 6 X SQ= 14.96 P = 0. RV YES (YES)

No.			No.			
39	TEND TO BE THOSE WHOSE INTERNATIONAL FINANCIAL STATUS IS LOW OR VERY LOW (76)	0.71	39	TEND TO BE THOSE WHOSE INTERNATIONAL FINANCIAL STATUS IS VERY HIGH, HIGH, OR MEDIUM (38)	0.51	15 20 36 19 X SQ= 3.58 P = 0.049 RV YES (YES)
42	TEND TO BE THOSE WHOSE ECONOMIC DEVELOPMENTAL STATUS IS UNDERDEVELOPED OR VERY UNDERDEVELOPED (76)	0.78	42	TEND TO BE THOSE WHOSE ECONOMIC DEVELOPMENTAL STATUS IS DEVELOPED OR INTERMEDIATE (36)	0.57	11 21 40 16 X SQ= 10.00 P = 0.001 RV YES (YES)
43	TEND TO BE THOSE WHOSE ECONOMIC DEVELOPMENTAL STATUS IS VERY UNDERDEVELOPED (57)	0.67	43	TEND TO BE THOSE WHOSE ECONOMIC DEVELOPMENTAL STATUS IS DEVELOPED, INTERMEDIATE, OR UNDERDEVELOPED (55)	0.81	17 30 34 7 X SQ= 17.77 P = 0. RV YES (YES)
45	TEND TO BE THOSE WHERE THE LITERACY RATE IS BELOW FIFTY PERCENT (54)	0.64	45	TEND TO BE THOSE WHERE THE LITERACY RATE IS FIFTY PERCENT OR ABOVE (55)	0.76	17 29 30 9 X SQ= 12.07 P = 0. RV YES (YES)
50	TEND TO BE THOSE WHERE FREEDOM OF THE PRESS IS INTERMITTENT, INTERNALLY ABSENT, OR INTERNALLY AND EXTERNALLY ABSENT (56)	0.88	50	TEND TO BE THOSE WHERE FREEDOM OF THE PRESS IS COMPLETE (43)	0.79	5 31 37 8 X SQ= 34.72 P = 0. RV YES (YES)
51	TEND TO BE THOSE WHERE FREEDOM OF THE PRESS IS INTERNALLY ABSENT, OR INTERNALLY AND EXTERNALLY ABSENT (37)	0.79	51	TEND TO BE THOSE WHERE FREEDOM OF THE PRESS IS COMPLETE OR INTERMITTENT (60)	0.97	9 38 33 1 X SQ= 44.90 P = 0. RV YES (YES)
52	TEND LESS TO BE THOSE WHERE FREEDOM OF THE PRESS IS COMPLETE, INTERMITTENT, OR INTERNALLY ABSENT (82)	0.62	52	ALWAYS ARE THOSE WHERE FREEDOM OF THE PRESS IS COMPLETE, INTERMITTENT, OR INTERNALLY ABSENT (82)	1.00	26 39 16 0 X SQ= 16.19 P = 0. RV NO (YES)
54	TEND TO BE THOSE WHERE NEWSPAPER CIRCULATION IS LESS THAN ONE HUNDRED PER THOUSAND (76)	0.80	54	TEND TO BE THOSE WHERE NEWSPAPER CIRCULATION IS ONE HUNDRED OR MORE PER THOUSAND (37)	0.56	10 22 40 17 X SQ= 11.08 P = 0.001 RV YES (YES)
58	TEND LESS TO BE THOSE WHERE THE RELIGION IS OTHER THAN MUSLIM (97)	0.77	58	TEND MORE TO BE THOSE WHERE THE RELIGION IS OTHER THAN MUSLIM (97)	0.95	12 2 40 37 X SQ= 4.22 P = 0.021 RV NO (NO)

No.	Subject	Value	No.	Predicate	Value	Statistics
62	ALWAYS ARE THOSE WHERE THE RELIGION IS CATHOLIC, RATHER THAN PROTESTANT (25)	1.00	62	TILT LESS TOWARD BEING THOSE WHERE THE RELIGION IS CATHOLIC, RATHER THAN PROTESTANT (25)	0.71	0 5 10 12 X SQ= 1.92 P = 0.124 RV NO (NO)
63	TEND TO BE THOSE WHERE THE RELIGION IS PREDOMINANTLY OR PARTLY OTHER THAN CHRISTIAN (68)	0.69	63	TEND TO BE THOSE WHERE THE RELIGION IS PREDOMINANTLY SOME KIND OF CHRISTIAN (46)	0.64	16 25 35 14 X SQ= 8.27 P = 0.003 RV YES (YES)
64	TEND LESS TO BE THOSE WHERE THE RELIGION IS CHRISTIAN, RATHER THAN MUSLIM (46)	0.57	64	TEND MORE TO BE THOSE WHERE THE RELIGION IS CHRISTIAN, RATHER THAN MUSLIM (46)	0.93	16 25 12 2 X SQ= 7.33 P = 0.004 RV NO (YES)
65	TEND TO BE THOSE WHERE THE RELIGION IS MUSLIM, RATHER THAN CATHOLIC (18)	0.55	65	TEND TO BE THOSE WHERE THE RELIGION IS CATHOLIC, RATHER THAN MUSLIM (25)	0.86	10 12 12 2 X SQ= 4.26 P = 0.033 RV YES (YES)
68	TEND TO BE THOSE THAT ARE LINGUISTICALLY WEAKLY HETEROGENEOUS OR STRONGLY HETEROGENEOUS (62)	0.63	68	TEND TO BE THOSE THAT ARE LINGUISTICALLY HOMOGENEOUS (52)	0.62	19 24 32 15 X SQ= 4.30 P = 0.033 RV YES (YES)
69	TEND TO BE THOSE THAT ARE LINGUISTICALLY STRONGLY HETEROGENEOUS (50)	0.51	69	TEND TO BE THOSE THAT ARE LINGUISTICALLY HOMOGENEOUS OR WEAKLY HETEROGENEOUS (64)	0.72	25 28 26 11 X SQ= 3.84 P = 0.033 RV YES (YES)
72	LEAN TOWARD BEING THOSE WHOSE DATE OF INDEPENDENCE IS AFTER 1914 (59)	0.58	72	LEAN TOWARD BEING THOSE WHOSE DATE OF INDEPENDENCE IS BEFORE 1914 (52)	0.61	21 23 29 15 X SQ= 2.27 P = 0.092 RV YES (YES)
74	TEND MORE TO BE THOSE THAT ARE NOT HISTORICALLY WESTERN (87)	0.86	74	TEND LESS TO BE THOSE THAT ARE NOT HISTORICALLY WESTERN (87)	0.55	7 17 44 21 X SQ= 9.12 P = 0.002 RV NO (YES)
75	TEND TO BE THOSE THAT ARE NOT HISTORICALLY WESTERN AND ARE NOT SIGNIFICANTLY WESTERNIZED (52)	0.59	75	TEND TO BE THOSE THAT ARE HISTORICALLY WESTERN OR SIGNIFICANTLY WESTERNIZED (62)	0.85	21 33 30 6 X SQ= 15.61 P = 0. RV YES (YES)

No.	Statement	Value	No.	Statement	Value	Statistics
76	TEND MORE TO BE THOSE THAT HAVE BEEN SIGNIFICANTLY OR PARTIALLY WESTERNIZED THROUGH A COLONIAL RELATIONSHIP, RATHER THAN BEING HISTORICALLY WESTERN (70)	0.82	76	TEND LESS TO BE THOSE THAT HAVE BEEN SIGNIFICANTLY OR PARTIALLY WESTERNIZED THROUGH A COLONIAL RELATIONSHIP, RATHER THAN BEING HISTORICALLY WESTERN (70)	0.53	7 17 32 19 X SQ= 6.09 P = 0.012 RV NO (YES)
77	TEND TO BE THOSE THAT HAVE BEEN PARTIALLY WESTERNIZED, RATHER THAN SIGNIFICANTLY WESTERNIZED, THROUGH A COLONIAL RELATIONSHIP (41)	0.71	77	TEND TO BE THOSE THAT HAVE BEEN SIGNIFICANTLY WESTERNIZED, RATHER THAN PARTIALLY WESTERNIZED, THROUGH A COLONIAL RELATIONSHIP (28)	0.68	9 13 22 6 X SQ= 5.90 P = 0.009 RV YES (YES)
80	TEND TO BE THOSE WHOSE DATE OF INDEPENDENCE IS AFTER 1914, AND THAT WERE FORMERLY DEPENDENCIES OF FRANCE, RATHER THAN BRITAIN (24)	0.73	80	TEND TO BE THOSE WHOSE DATE OF INDEPENDENCE IS AFTER 1914, AND THAT WERE FORMERLY DEPENDENCIES OF BRITAIN, RATHER THAN FRANCE (19)	0.73	6 8 16 3 X SQ= 4.48 P = 0.024 RV YES (YES)
81	TILT MORE TOWARD BEING THOSE WHERE THE TYPE OF POLITICAL MODERNIZATION IS LATER EUROPEAN OR LATER EUROPEAN DERIVED, RATHER THAN EARLY EUROPEAN OR EARLY EUROPEAN DERIVED (40)	0.89	81	TILT LESS TOWARD BEING THOSE WHERE THE TYPE OF POLITICAL MODERNIZATION IS LATER EUROPEAN OR LATER EUROPEAN DERIVED, RATHER THAN EARLY EUROPEAN OR EARLY EUROPEAN DERIVED (40)	0.68	2 8 17 17 X SQ= 1.74 P = 0.148 RV NO (YES)
82	TEND TO BE THOSE WHERE THE TYPE OF POLITICAL MODERNIZATION IS DEVELOPED TUTELARY OR UNDEVELOPED TUTELARY, RATHER THAN EARLY OR LATER EUROPEAN OR EUROPEAN DERIVED (55)	0.59	82	TEND TO BE THOSE WHERE THE TYPE OF POLITICAL MODERNIZATION IS EARLY OR LATER EUROPEAN OR EUROPEAN DERIVED, RATHER THAN DEVELOPED TUTELARY OR UNDEVELOPED TUTELARY (51)	0.68	19 25 27 12 X SQ= 4.67 P = 0.027 RV YES (YES)
85	LEAN TOWARD BEING THOSE WHERE THE STAGE OF POLITICAL MODERNIZATION IS MID- OR EARLY TRANSITIONAL, RATHER THAN ADVANCED (54)	0.53	85	LEAN TOWARD BEING THOSE WHERE THE STAGE OF POLITICAL MODERNIZATION IS ADVANCED, RATHER THAN MID- OR EARLY TRANSITIONAL (60)	0.67	24 26 27 13 X SQ= 2.69 P = 0.087 RV YES (YES)
87	TILT LESS TOWARD BEING THOSE WHOSE IDEOLOGICAL ORIENTATION IS OTHER THAN DEVELOPMENTAL (58)	0.61	87	TILT MORE TOWARD BEING THOSE WHOSE IDEOLOGICAL ORIENTATION IS OTHER THAN DEVELOPMENTAL (58)	0.78	16 7 25 25 X SQ= 1.72 P = 0.136 RV NO (YES)
89	TEND TO BE THOSE WHOSE IDEOLOGICAL ORIENTATION IS OTHER THAN CONVENTIONAL (62)	0.91	89	TEND TO BE THOSE WHOSE IDEOLOGICAL ORIENTATION IS CONVENTIONAL (33)	0.78	4 25 41 7 X SQ= 35.29 P = 0. RV YES (YES)
92	TEND LESS TO BE THOSE WHERE THE SYSTEM STYLE IS LIMITED MOBILIZATIONAL, OR NON-MOBILIZATIONAL (93)	0.63	92	ALWAYS ARE THOSE WHERE THE SYSTEM STYLE IS LIMITED MOBILIZATIONAL, OR NON-MOBILIZATIONAL (93)	1.00	19 0 32 39 X SQ= 16.25 P = 0. RV NO (YES)

93	TEND TO BE THOSE WHERE THE SYSTEM STYLE IS MOBILIZATIONAL, OR LIMITED MOBILIZATIONAL (32)	0.55	93	TEND TO BE THOSE WHERE THE SYSTEM STYLE IS NON-MOBILIZATIONAL (78)	0.92	27 3 22 36 X SQ= 19.66 P = 0. RV YES (YES)
94	TEND TO BE THOSE WHERE THE STATUS OF THE REGIME IS AUTHORITARIAN OR TOTALITARIAN (41)	0.84	94	ALWAYS ARE THOSE WHERE THE STATUS OF THE REGIME IS CONSTITUTIONAL (51)	1.00	7 38 36 0 X SQ= 53.92 P = 0. RV YES (YES)
95	TEND LESS TO BE THOSE WHERE THE STATUS OF THE REGIME IS CONSTITUTIONAL OR AUTHORITARIAN (95)	0.67	95	ALWAYS ARE THOSE WHERE THE STATUS OF THE REGIME IS CONSTITUTIONAL OR AUTHORITARIAN (95)	1.00	33 39 16 0 X SQ= 13.45 P = 0. RV NO (YES)
96	TEND LESS TO BE THOSE WHERE THE STATUS OF THE REGIME IS CONSTITUTIONAL OR TOTALITARIAN (67)	0.56	96	ALWAYS ARE THOSE WHERE THE STATUS OF THE REGIME IS CONSTITUTIONAL OR TOTALITARIAN (67)	1.00	23 38 18 0 X SQ= 19.18 P = 0. RV NO (YES)
98	TEND MORE TO BE THOSE WHERE GOVERNMENTAL STABILITY IS GENERALLY OR MODERATELY PRESENT AND DATES FROM THE POST-WAR PERIOD, OR IS ABSENT (93)	0.87	98	TEND LESS TO BE THOSE WHERE GOVERNMENTAL STABILITY IS GENERALLY OR MODERATELY PRESENT AND DATES FROM THE POST-WAR PERIOD, OR IS ABSENT (93)	0.64	7 14 45 25 X SQ= 5.12 P = 0.022 RV NO (YES)
101	TEND TO BE THOSE WHERE THE REPRESENTATIVE CHARACTER OF THE REGIME IS LIMITED POLYARCHIC, PSEUDO-POLYARCHIC, OR NON-POLYARCHIC (57)	0.90	101	TEND TO BE THOSE WHERE THE REPRESENTATIVE CHARACTER OF THE REGIME IS POLYARCHIC (41)	0.91	5 32 47 3 X SQ= 53.99 P = 0. RV YES (YES)
102	TEND TO BE THOSE WHERE THE REPRESENTATIVE CHARACTER OF THE REGIME IS PSEUDO-POLYARCHIC OR NON-POLYARCHIC (49)	0.83	102	ALWAYS ARE THOSE WHERE THE REPRESENTATIVE CHARACTER OF THE REGIME IS POLYARCHIC OR LIMITED POLYARCHIC (59)	1.00	9 39 43 0 X SQ= 57.87 P = 0. RV YES (YES)
105	TEND TO BE THOSE WHERE THE ELECTORAL SYSTEM IS PARTIALLY COMPETITIVE OR NON-COMPETITIVE (47)	0.93	105	TEND TO BE THOSE WHERE THE ELECTORAL SYSTEM IS COMPETITIVE (43)	0.97	3 33 40 1 X SQ= 58.33 P = 0. RV YES (YES)
106	TEND TO BE THOSE WHERE THE ELECTORAL SYSTEM IS NON-COMPETITIVE (30)	0.75	106	ALWAYS ARE THOSE WHERE THE ELECTORAL SYSTEM IS COMPETITIVE OR PARTIALLY COMPETITIVE (52)	1.00	10 32 30 0 X SQ= 38.11 P = 0. RV YES (YES)

No.	Left statement	Value	No.	Right statement	Value	Statistics
107	TEND TO BE THOSE WHERE AUTONOMOUS GROUPS ARE PARTIALLY TOLERATED IN POLITICS, ARE TOLERATED ONLY OUTSIDE POLITICS, OR ARE NOT TOLERATED AT ALL (65)	0.92	107	TEND TO BE THOSE WHERE AUTONOMOUS GROUPS ARE FULLY TOLERATED IN POLITICS (46)	0.90	4 35 48 4 X SQ= 57.96 P = 0. RV YES (YES)
108	TEND TO BE THOSE WHERE AUTONOMOUS GROUPS ARE TOLERATED ONLY OUTSIDE POLITICS OR ARE NOT TOLERATED AT ALL (35)	0.66	108	TEND TO BE THOSE WHERE AUTONOMOUS GROUPS ARE FULLY OR PARTIALLY TOLERATED IN POLITICS (65)	0.97	16 37 31 1 X SQ= 33.25 P = 0. RV YES (YES)
111	TEND TO BE THOSE WHERE POLITICAL ENCULTURATION IS LOW (42)	0.54	111	TEND TO BE THOSE WHERE POLITICAL ENCULTURATION IS HIGH OR MEDIUM (53)	0.76	17 29 20 9 X SQ= 6.07 P = 0.009 RV YES (YES)
116	TEND TO BE THOSE WHERE INTEREST ARTICULATION BY ASSOCIATIONAL GROUPS IS LIMITED OR NEGLIGIBLE (79)	0.94	116	TEND TO BE THOSE WHERE INTEREST ARTICULATION BY ASSOCIATIONAL GROUPS IS SIGNIFICANT OR MODERATE (32)	0.62	3 24 46 15 X SQ= 28.80 P = 0. RV YES (YES)
117	TEND TO BE THOSE WHERE INTEREST ARTICULATION BY ASSOCIATIONAL GROUPS IS NEGLIGIBLE (51)	0.65	117	TEND TO BE THOSE WHERE INTEREST ARTICULATION BY ASSOCIATIONAL GROUPS IS SIGNIFICANT, MODERATE, OR LIMITED (60)	0.85	17 33 32 6 X SQ= 20.07 P = 0. RV YES (YES)
118	TEND TO BE THOSE WHERE INTEREST ARTICULATION BY INSTITUTIONAL GROUPS IS VERY SIGNIFICANT (40)	0.61	118	TEND TO BE THOSE WHERE INTEREST ARTICULATION BY INSTITUTIONAL GROUPS IS SIGNIFICANT, MODERATE, OR LIMITED (60)	0.89	28 4 18 33 X SQ= 19.63 P = 0. RV YES (YES)
119	TEND TO BE THOSE WHERE INTEREST ARTICULATION BY INSTITUTIONAL GROUPS IS VERY SIGNIFICANT OR SIGNIFICANT (74)	0.96	119	TEND TO BE THOSE WHERE INTEREST ARTICULATION BY INSTITUTIONAL GROUPS IS MODERATE OR LIMITED (26)	0.54	44 17 2 20 X SQ= 23.52 P = 0. RV YES (YES)
120	ALWAYS ARE THOSE WHERE INTEREST ARTICULATION BY INSTITUTIONAL GROUPS IS VERY SIGNIFICANT, SIGNIFICANT, OR MODERATE (90)	1.00	120	TEND LESS TO BE THOSE WHERE INTEREST ARTICULATION BY INSTITUTIONAL GROUPS IS VERY SIGNIFICANT, SIGNIFICANT, OR MODERATE (90)	0.73	46 27 0 10 X SQ= 11.70 P = 0. RV NO (YES)
121	TEND TO BE THOSE WHERE INTEREST ARTICULATION BY NON-ASSOCIATIONAL GROUPS IS SIGNIFICANT (54)	0.56	121	TEND TO BE THOSE WHERE INTEREST ARTICULATION BY NON-ASSOCIATIONAL GROUPS IS MODERATE, LIMITED, OR NEGLIGIBLE (61)	0.79	29 8 23 31 X SQ= 10.07 P = 0.001 RV YES (YES)

122	TEND TO BE THOSE WHERE INTEREST ARTICULATION BY NON-ASSOCIATIONAL GROUPS IS SIGNIFICANT OR MODERATE (83)	0.85	122	TEND TO BE THOSE WHERE INTEREST ARTICULATION BY NON-ASSOCIATIONAL GROUPS IS LIMITED OR NEGLIGIBLE (32)	0.51	44 19 8 20 X SQ= 11.85 P = 0. RV YES (YES)
123	ALWAYS ARE THOSE WHERE INTEREST ARTICULATION BY NON-ASSOCIATIONAL GROUPS IS SIGNIFICANT, MODERATE, OR LIMITED (107)	1.00	123	TEND LESS TO BE THOSE WHERE INTEREST ARTICULATION BY NON-ASSOCIATIONAL GROUPS IS SIGNIFICANT, MODERATE, OR LIMITED (107)	0.79	52 31 0 8 X SQ= 9.28 P = 0.001 RV NO (YES)
124	TEND LESS TO BE THOSE WHERE INTEREST ARTICULATION BY ANOMIC GROUPS IS OCCASIONAL, INFREQUENT, OR VERY INFREQUENT (85)	0.84	124	ALWAYS ARE THOSE WHERE INTEREST ARTICULATION BY ANOMIC GROUPS IS OCCASIONAL, INFREQUENT, OR VERY INFREQUENT (85)	1.00	7 0 36 34 X SQ= 4.28 P = 0.016 RV NO (NO)
125	TEND TO BE THOSE WHERE INTEREST ARTICULATION BY ANOMIC GROUPS IS FREQUENT OR OCCASIONAL (64)	0.81	125	TEND TO BE THOSE WHERE INTEREST ARTICULATION BY ANOMIC GROUPS IS INFREQUENT OR VERY INFREQUENT (35)	0.65	35 12 8 22 X SQ= 15.09 P = 0. RV YES (YES)
126	ALWAYS ARE THOSE WHERE INTEREST ARTICULATION BY ANOMIC GROUPS IS FREQUENT, OCCASIONAL, OR INFREQUENT (83)	1.00	126	TEND LESS TO BE THOSE WHERE INTEREST ARTICULATION BY ANOMIC GROUPS IS FREQUENT, OCCASIONAL, OR INFREQUENT (83)	0.56	43 19 0 15 X SQ= 20.83 P = 0. RV NO (YES)
127	ALWAYS ARE THOSE WHERE INTEREST ARTICULATION BY POLITICAL PARTIES IS MODERATE, LIMITED, OR NEGLIGIBLE (72)	1.00	127	TEND LESS TO BE THOSE WHERE INTEREST ARTICULATION BY POLITICAL PARTIES IS MODERATE, LIMITED, OR NEGLIGIBLE (72)	0.64	0 13 41 23 X SQ= 15.33 P = 0. RV NO (YES)
128	TEND TO BE THOSE WHERE INTEREST ARTICULATION BY POLITICAL PARTIES IS LIMITED OR NEGLIGIBLE (45)	0.85	128	TEND TO BE THOSE WHERE INTEREST ARTICULATION BY POLITICAL PARTIES IS SIGNIFICANT OR MODERATE (48)	0.86	6 31 35 5 X SQ= 36.42 P = 0. RV YES (YES)
129	TEND TO BE THOSE WHERE INTEREST ARTICULATION BY POLITICAL PARTIES IS NEGLIGIBLE (37)	0.76	129	TEND TO BE THOSE WHERE INTEREST ARTICULATION BY POLITICAL PARTIES IS SIGNIFICANT, MODERATE, OR LIMITED (56)	0.92	10 33 31 3 X SQ= 32.51 P = 0. RV YES (YES)
134	TEND TO BE THOSE WHERE INTEREST AGGREGATION BY THE EXECUTIVE IS LIMITED OR NEGLIGIBLE (46)	0.57	134	TEND TO BE THOSE WHERE INTEREST AGGREGATION BY THE EXECUTIVE IS SIGNIFICANT OR MODERATE (57)	0.84	21 31 28 6 X SQ= 13.11 P = 0. RV YES (YES)

135	TEND LESS TO BE THOSE WHERE INTEREST AGGREGATION BY THE EXECUTIVE IS SIGNIFICANT, MODERATE, OR LIMITED (77)	0.73	135	ALWAYS ARE THOSE WHERE INTEREST AGGREGATION BY THE EXECUTIVE IS SIGNIFICANT, MODERATE, OR LIMITED (77)	1.00	30 36 11 0 X SQ= 9.18 P = 0.001 RV NO (YES)
136	ALWAYS ARE THOSE WHERE INTEREST AGGREGATION BY THE LEGISLATURE IS MODERATE, LIMITED, OR NEGLIGIBLE (85)	1.00	136	TEND LESS TO BE THOSE WHERE INTEREST AGGREGATION BY THE LEGISLATURE IS MODERATE, LIMITED, OR NEGLIGIBLE (85)	0.71	0 10 45 25 X SQ= 12.20 P = 0. RV NO (YES)
137	TEND TO BE THOSE WHERE INTEREST AGGREGATION BY THE LEGISLATURE IS LIMITED OR NEGLIGIBLE (68)	0.98	137	TEND TO BE THOSE WHERE INTEREST AGGREGATION BY THE LEGISLATURE IS SIGNIFICANT OR MODERATE (29)	0.69	1 24 44 11 X SQ= 37.31 P = 0. RV YES (YES)
138	TEND TO BE THOSE WHERE INTEREST AGGREGATION BY THE LEGISLATURE IS NEGLIGIBLE (49)	0.88	138	TEND TO BE THOSE WHERE INTEREST AGGREGATION BY THE LEGISLATURE IS SIGNIFICANT, MODERATE, OR LIMITED (48)	0.95	5 36 37 2 X SQ= 51.52 P = 0. RV YES (YES)
139	TEND TO BE THOSE WHERE THE PARTY SYSTEM IS QUANTITATIVELY ONE-PARTY (34)	0.65	139	TEND TO BE THOSE WHERE THE PARTY SYSTEM IS QUANTITATIVELY OTHER THAN ONE-PARTY (71)	0.97	31 1 17 38 X SQ= 32.98 P = 0. RV YES (YES)
141	TEND MORE TO BE THOSE WHERE THE PARTY SYSTEM IS QUANTITATIVELY OTHER THAN TWO-PARTY (87)	0.98	141	TEND LESS TO BE THOSE WHERE THE PARTY SYSTEM IS QUANTITATIVELY OTHER THAN TWO-PARTY (87)	0.73	1 10 45 27 X SQ= 8.96 P = 0.002 RV NO (YES)
142	TEND TO BE THOSE WHERE THE PARTY SYSTEM IS QUANTITATIVELY OTHER THAN MULTI-PARTY (66)	0.98	142	TEND TO BE THOSE WHERE THE PARTY SYSTEM IS QUANTITATIVELY MULTI-PARTY (30)	0.54	1 20 43 17 X SQ= 25.43 P = 0. RV YES (YES)
146	TEND LESS TO BE THOSE WHERE THE PARTY SYSTEM IS QUALITATIVELY OTHER THAN MASS-BASED TERRITORIAL (92)	0.81	146	ALWAYS ARE THOSE WHERE THE PARTY SYSTEM IS QUALITATIVELY OTHER THAN MASS-BASED TERRITORIAL (92)	1.00	8 0 34 38 X SQ= 6.07 P = 0.006 RV NO (YES)
147	ALWAYS ARE THOSE WHERE THE PARTY SYSTEM IS QUALITATIVELY OTHER THAN CLASS-ORIENTED OR MULTI-IDEOLOGICAL (67)	1.00	147	TEND TO BE THOSE WHERE THE PARTY SYSTEM IS QUALITATIVELY CLASS-ORIENTED OR MULTI-IDEOLOGICAL (23)	0.53	0 19 39 17 X SQ= 24.85 P = 0. RV YES (YES)

160	TEND TO BE THOSE WHERE THE POLITICAL LEADERSHIP IS ELITIST (30)	0.58	160	TEND TO BE THOSE WHERE THE POLITICAL LEADERSHIP IS MODERATELY ELITIST OR NON-ELITIST (67)	0.97	26 1 19 35 X SQ= 24.81 P = 0. RV YES (YES)
161	TEND TO BE THOSE WHERE THE POLITICAL LEADERSHIP IS ELITIST OR MODERATELY ELITIST (47)	0.69	161	TEND TO BE THOSE WHERE THE POLITICAL LEADERSHIP IS NON-ELITIST (50)	0.75	31 9 14 27 X SQ= 13.71 P = 0. RV YES (YES)
163	TEND LESS TO BE THOSE WHERE THE REGIME'S LEADERSHIP CHARISMA IS MODERATE OR NEGLIGIBLE (87)	0.73	163	ALWAYS ARE THOSE WHERE THE REGIME'S LEADERSHIP CHARISMA IS MODERATE OR NEGLIGIBLE (87)	1.00	13 0 35 37 X SQ= 9.83 P = 0. RV NO (YES)
164	TEND TO BE THOSE WHERE THE REGIME'S LEADERSHIP CHARISMA IS PRONOUNCED OR MODERATE (34)	0.60	164	TEND TO BE THOSE WHERE THE REGIME S LEADERSHIP CHARISMA IS NEGLIGIBLE (65)	0.92	29 3 19 34 X SQ= 22.18 P = 0. RV YES (YES)
166	TEND MORE TO BE THOSE WHERE THE VERTICAL POWER DISTRIBUTION IS THAT OF FORMAL FEDERALISM OR FORMAL AND EFFECTIVE UNITARISM (99)	0.98	166	TEND LESS TO BE THOSE WHERE THE VERTICAL POWER DISTRIBUTION IS THAT OF FORMAL FEDERALISM OR FORMAL AND EFFECTIVE UNITARISM (99)	0.72	1 11 51 28 X SQ= 11.25 P = 0. RV NO (YES)
168	TEND TO BE THOSE WHERE THE HORIZONTAL POWER DISTRIBUTION IS LIMITED OR NEGLIGIBLE (72)	0.96	168	TEND TO BE THOSE WHERE THE HORIZONTAL POWER DISTRIBUTION IS SIGNIFICANT (34)	0.72	2 28 50 11 X SQ= 43.54 P = 0. RV YES (YES)
169	TEND TO BE THOSE WHERE THE HORIZONTAL POWER DISTRIBUTION IS NEGLIGIBLE (48)	0.81	169	ALWAYS ARE THOSE WHERE THE HORIZONTAL POWER DISTRIBUTION IS SIGNIFICANT OR LIMITED (58)	1.00	10 39 42 0 X SQ= 55.30 P = 0. RV YES (YES)
171	TEND MORE TO BE THOSE WHERE THE LEGISLATIVE-EXECUTIVE STRUCTURE IS OTHER THAN PARLIAMENTARY-REPUBLICAN (90)	0.96	171	TEND LESS TO BE THOSE WHERE THE LEGISLATIVE-EXECUTIVE STRUCTURE IS OTHER THAN PARLIAMENTARY-REPUBLICAN (90)	0.79	2 8 44 31 X SQ= 3.87 P = 0.039 RV NO (YES)
172	ALWAYS ARE THOSE WHERE THE LEGISLATIVE-EXECUTIVE STRUCTURE IS OTHER THAN PARLIAMENTARY-ROYALIST (88)	1.00	172	TEND LESS TO BE THOSE WHERE THE LEGISLATIVE-EXECUTIVE STRUCTURE IS OTHER THAN PARLIAMENTARY-ROYALIST (88)	0.59	0 16 48 23 X SQ= 21.47 P = 0. RV NO (YES)

174	TEND TO BE THOSE WHERE THE LEGISLATURE IS PARTIALLY EFFECTIVE, LARGELY INEFFECTIVE, OR WHOLLY INEFFECTIVE (72)	0.96	174	TEND TO BE THOSE WHERE THE LEGISLATURE IS FULLY EFFECTIVE (28)	0.63	2 24 46 14 X SQ= 32.25 P = 0. RV YES (YES)
175	TEND TO BE THOSE WHERE THE LEGISLATURE IS LARGELY INEFFECTIVE OR WHOLLY INEFFECTIVE (49)	0.87	175	ALWAYS ARE THOSE WHERE THE LEGISLATURE IS FULLY EFFECTIVE OR PARTIALLY EFFECTIVE (51)	1.00	6 38 42 0 X SQ= 61.53 P = 0. RV YES (YES)
176	TEND TO BE THOSE WHERE THE LEGISLATURE IS WHOLLY INEFFECTIVE (28)	0.54	176	ALWAYS ARE THOSE WHERE THE LEGISLATURE IS FULLY EFFECTIVE, PARTIALLY EFFECTIVE, OR LARGELY INEFFECTIVE (72)	1.00	22 38 26 0 X SQ= 26.99 P = 0. RV YES (YES)
178	TEND TO BE THOSE WHERE THE LEGISLATURE IS UNICAMERAL	0.69 (53)	178	TEND TO BE THOSE WHERE THE LEGISLATURE IS BICAMERAL	0.76 (51)	34 9 15 29 X SQ= 16.10 P = 0. RV YES (YES)
181	TEND TO BE THOSE WHERE THE BUREAUCRACY IS SEMI-MODERN, RATHER THAN MODERN (55)	0.94	181	TEND TO BE THOSE WHERE THE BUREAUCRACY IS MODERN, RATHER THAN SEMI-MODERN (21)	0.51	2 18 29 17 X SQ= 13.69 P = 0. RV YES (YES)
185	LEAN TOWARD BEING THOSE WHERE PARTICIPATION BY THE MILITARY IN POLITICS IS SUPPORTIVE, RATHER THAN INTERVENTIVE (31)	0.81	185	LEAN TOWARD BEING THOSE WHERE PARTICIPATION BY THE MILITARY IN POLITICS IS INTERVENTIVE, RATHER THAN SUPPORTIVE (21)	0.60	7 3 29 2 X SQ= 2.03 P = 0.083 RV YES (NO)
186	TEND TO BE THOSE WHERE PARTICIPATION BY THE MILITARY IN POLITICS IS SUPPORTIVE, RATHER THAN NEUTRAL (31)	0.69	186	TEND TO BE THOSE WHERE PARTICIPATION BY THE MILITARY IN POLITICS IS NEUTRAL, RATHER THAN SUPPORTIVE (56)	0.94	29 2 13 33 X SQ= 29.26 P = 0. RV YES (YES)
187	TEND TO BE THOSE WHERE THE ROLE OF THE POLICE IS POLITICALLY SIGNIFICANT (66)	0.90	187	TEND TO BE THOSE WHERE THE ROLE OF THE POLICE IS NOT POLITICALLY SIGNIFICANT (35)	0.76	45 8 5 25 X SQ= 34.45 P = 0. RV YES (YES)
188	LEAN MORE TOWARD BEING THOSE WHERE THE CHARACTER OF THE LEGAL SYSTEM IS OTHER THAN CIVIL LAW (81)	0.82	188	LEAN LESS TOWARD BEING THOSE WHERE THE CHARACTER OF THE LEGAL SYSTEM IS OTHER THAN CIVIL LAW (81)	0.62	9 15 41 24 X SQ= 3.68 P = 0.053 RV NO (YES)

189	ALWAYS ARE THOSE WHERE THE CHARACTER OF THE LEGAL SYSTEM IS OTHER THAN COMMON LAW (108)	1.00	189	TEND LESS TO BE THOSE WHERE THE CHARACTER OF THE LEGAL SYSTEM IS OTHER THAN COMMON LAW (108)	0.82	0 7 52 32 X SQ= 7.74 P = 0.002 RV NO (YES)
190	ALWAYS ARE THOSE WHERE THE CHARACTER OF THE LEGAL SYSTEM IS CIVIL LAW, RATHER THAN COMMON LAW (32)	1.00	190	LEAN LESS TOWARD BEING THOSE WHERE THE CHARACTER OF THE LEGAL SYSTEM IS CIVIL LAW, RATHER THAN COMMON LAW (32)	0.68	9 15 0 7 X SQ= 2.10 P = 0.077 RV NO (NO)
192	TEND LESS TO BE THOSE WHERE THE CHARACTER OF THE LEGAL SYSTEM IS OTHER THAN MUSLIM OR PARTLY MUSLIM (86)	0.66	192	TEND MORE TO BE THOSE WHERE THE CHARACTER OF THE LEGAL SYSTEM IS OTHER THAN MUSLIM OR PARTLY MUSLIM (86)	0.90	17 4 33 35 X SQ= 5.60 P = 0.011 RV NO (YES)
193	TEND LESS TO BE THOSE WHERE THE CHARACTER OF THE LEGAL SYSTEM IS OTHER THAN PARTLY INDIGENOUS (86)	0.69	193	TEND MORE TO BE THOSE WHERE THE CHARACTER OF THE LEGAL SYSTEM IS OTHER THAN PARTLY INDIGENOUS (86)	0.90	16 4 36 35 X SQ= 4.34 P = 0.023 RV NO (NO)
194	TEND LESS TO BE THOSE THAT ARE NON-COMMUNIST (101)	0.75	194	ALWAYS ARE THOSE THAT ARE NON-COMMUNIST (101)	1.00	13 0 38 39 X SQ= 9.65 P = 0. RV NO (YES)

****

180	POLITIES WHERE THE BUREAUCRACY IS MODERN OR SEMI-MODERN (76)	180	POLITIES WHERE THE BUREAUCRACY IS POST-COLONIAL TRANSITIONAL OR TRADITIONAL (34)

SUBJECT ONLY

....

76 IN LEFT COLUMN

ALBANIA	ARGENTINA	AUSTRALIA	AUSTRIA	BELGIUM	BOLIVIA	BRAZIL	BULGARIA	CANADA	CEYLON
CHILE	CHINA, PR	COLOMBIA	COSTA RICA	CUBA	CYPRUS	CZECHOS'KIA	DENMARK	DOMIN REP	ECUADOR
EL SALVADOR	FINLAND	FRANCE	GERMANY, E	GERMAN FR	GHANA	GREECE	GUATEMALA	GUINEA	HAITI
HONDURAS	HUNGARY	ICELAND	INDIA	IRELAND	ISRAEL	ITALY	JAMAICA	JAPAN	KOREA, N
KOREA REP	LEBANON	LIBERIA	LIBYA	LUXEMBOURG	MALAYA	MEXICO	MONGOLIA	MOROCCO	NETHERLANDS
NEW ZEALAND	NICARAGUA	NORWAY	PANAMA	PARAGUAY	PERU	PHILIPPINES	POLAND	PORTUGAL	RUMANIA
SO AFRICA	SPAIN	SWEDEN	SWITZERLAND	SYRIA	TRINIDAD	TUNISIA	TURKEY	USSR	UAR
UK	US	URUGUAY	VENEZUELA	VIETNAM, N	YUGOSLAVIA				

34 IN RIGHT COLUMN

AFGHANISTAN	ALGERIA	BURMA	BURUNDI	CAMBODIA	CAMEROUN	CEN AFR REP	CHAD	CONGO(BRA)	CONGO(LEO)
DAHOMEY	ETHIOPIA	GABON	IRAN	IVORY COAST	JORDAN	LAOS	MALAGASY R	MALI	MAURITANIA
NEPAL	NIGER	NIGERIA	RWANDA	SA'U ARABIA	SENEGAL	SIERRE LEO	SOMALIA	SUDAN	TANGANYIKA
THAILAND	TOGO	UGANDA	UPPER VOLTA						

3 EXCLUDED BECAUSE AMBIGUOUS

INDONESIA PAKISTAN VIETNAM REP

2 EXCLUDED BECAUSE UNASCERTAINABLE

IRAQ YEMEN

2	TEND LESS TO BE THOSE LOCATED ELSEWHERE THAN IN WEST EUROPE, SCANDINAVIA, NORTH AMERICA, OR AUSTRALASIA (93)	0.71	2	ALWAYS ARE THOSE LOCATED ELSEWHERE THAN IN WEST EUROPE, SCANDINAVIA, NORTH AMERICA, OR AUSTRALASIA (93)	1.00	22 0 54 34 X SQ= 10.56 P = 0. RV NO (NO)
6	TEND LESS TO BE THOSE LOCATED ELSEWHERE THAN IN THE CARIBBEAN, CENTRAL AMERICA, OR SOUTH AMERICA (93)	0.71	6	ALWAYS ARE THOSE LOCATED ELSEWHERE THAN IN THE CARIBBEAN, CENTRAL AMERICA, OR SOUTH AMERICA (93)	1.00	22 0 54 34 X SQ= 10.56 P = 0. RV NO (NO)
10	TEND TO BE THOSE LOCATED ELSEWHERE THAN IN NORTH AFRICA, OR CENTRAL AND SOUTH AFRICA (82)	0.89	10	TEND TO BE THOSE LOCATED IN NORTH AFRICA, OR CENTRAL AND SOUTH AFRICA (33)	0.74	8 25 68 9 X SQ= 41.45 P = 0. RV YES (YES)

15	TILT LESS TOWARD BEING THOSE LOCATED IN NORTH AFRICA OR CENTRAL AND SOUTH AFRICA, RATHER THAN IN THE MIDDLE EAST (33)	0.62	15	TILT MORE TOWARD BEING THOSE LOCATED IN NORTH AFRICA OR CENTRAL AND SOUTH AFRICA, RATHER THAN IN THE MIDDLE EAST (33)	0.86	5 4 8 25 X SQ= 1.94 P = 0.107 RV NO (YES)
17	TEND TO BE THOSE LOCATED IN THE CARIBBEAN, CENTRAL AMERICA, OR SOUTH AMERICA, RATHER THAN IN THE MIDDLE EAST (22)	0.81	17	ALWAYS ARE THOSE LOCATED IN THE MIDDLE EAST, RATHER THAN IN THE CARIBBEAN, CENTRAL AMERICA, OR SOUTH AMERICA (11)	1.00	5 4 22 0 X SQ= 7.62 P = 0.004 RV YES (NO)
18	TEND TO BE THOSE LOCATED IN EAST ASIA, SOUTH ASIA, OR SOUTHEAST ASIA, RATHER THAN IN NORTH AFRICA OR CENTRAL AND SOUTH AFRICA (18)	0.56	18	TEND TO BE THOSE LOCATED IN NORTH AFRICA OR CENTRAL AND SOUTH AFRICA, RATHER THAN IN EAST ASIA, SOUTH ASIA, OR SOUTHEAST ASIA (33)	0.83	8 25 10 5 X SQ= 6.21 P = 0.009 RV YES (YES)
19	TEND TO BE THOSE LOCATED IN THE CARIBBEAN, CENTRAL AMERICA, OR SOUTH AMERICA, RATHER THAN IN NORTH AFRICA OR CENTRAL AND SOUTH AFRICA (22)	0.73	19	ALWAYS ARE THOSE LOCATED IN NORTH AFRICA OR CENTRAL AND SOUTH AFRICA, RATHER THAN IN THE CARIBBEAN, CENTRAL AMERICA, OR SOUTH AMERICA (33)	1.00	8 25 22 0 X SQ= 27.58 P = 0. RV YES (YES)
20	TEND TO BE THOSE LOCATED IN THE CARIBBEAN, CENTRAL AMERICA, OR SOUTH AMERICA, RATHER THAN IN EAST ASIA, SOUTH ASIA, OR SOUTHEAST ASIA (22)	0.69	20	ALWAYS ARE THOSE LOCATED IN EAST ASIA, SOUTH ASIA, OR SOUTHEAST ASIA, RATHER THAN IN THE CARIBBEAN, CENTRAL AMERICA, OR SOUTH AMERICA (18)	1.00	10 5 22 0 X SQ= 5.87 P = 0.007 RV YES (NO)
22	TEND TO BE THOSE WHOSE TERRITORIAL SIZE IS SMALL (47)	0.50	22	TEND TO BE THOSE WHOSE TERRITORIAL SIZE IS VERY LARGE, LARGE, OR MEDIUM (68)	0.76	38 26 38 8 X SQ= 5.72 P = 0.012 RV YES (NO)
24	LEAN TOWARD BEING THOSE WHOSE POPULATION IS VERY LARGE, LARGE, OR MEDIUM (61)	0.58	24	LEAN TOWARD BEING THOSE WHOSE POPULATION IS SMALL (54)	0.62	44 13 32 21 X SQ= 2.89 P = 0.066 RV YES (NO)
25	TEND LESS TO BE THOSE WHOSE POPULATION DENSITY IS MEDIUM OR LOW (98)	0.78	25	ALWAYS ARE THOSE WHOSE POPULATION DENSITY IS MEDIUM OR LOW (98)	1.00	17 0 59 34 X SQ= 7.36 P = 0.001 RV NO (NO)
26	TEND TO BE THOSE WHOSE POPULATION DENSITY IS VERY HIGH, HIGH, OR MEDIUM (48)	0.53	26	TEND TO BE THOSE WHOSE POPULATION DENSITY IS LOW (67)	0.85	40 5 36 29 X SQ= 12.45 P = 0. RV YES (NO)

29	TEND TO BE THOSE WHERE THE DEGREE OF URBANIZATION IS HIGH (56)	0.79	29	TEND TO BE THOSE WHERE THE DEGREE OF URBANIZATION IS LOW (49)	0.94	53 2 14 32 X SQ= 45.85 P = 0. RV YES (YES)
30	TEND TO BE THOSE WHOSE AGRICULTURAL POPULATION IS MEDIUM, LOW, OR VERY LOW (57)	0.72	30	TEND TO BE THOSE WHOSE AGRICULTURAL POPULATION IS HIGH (56)	0.94	21 31 54 2 X SQ= 37.31 P = 0. RV YES (YES)
31	TEND LESS TO BE THOSE WHOSE AGRICULTURAL POPULATION IS HIGH OR MEDIUM (90)	0.68	31	ALWAYS ARE THOSE WHOSE AGRICULTURAL POPULATION IS HIGH OR MEDIUM (90)	1.00	52 33 24 0 X SQ= 11.59 P = 0. RV NO (NO)
33	TEND LESS TO BE THOSE WHOSE GROSS NATIONAL PRODUCT IS LOW OR VERY LOW (85)	0.63	33	ALWAYS ARE THOSE WHOSE GROSS NATIONAL PRODUCT IS LOW OR VERY LOW (85)	1.00	28 0 48 34 X SQ= 14.92 P = 0. RV NO (NO)
34	TEND TO BE THOSE WHOSE GROSS NATIONAL PRODUCT IS VERY HIGH, HIGH, MEDIUM, OR LOW (62)	0.66	34	TEND TO BE THOSE WHOSE GROSS NATIONAL PRODUCT IS VERY LOW (53)	0.76	50 8 26 26 X SQ= 15.18 P = 0. RV YES (YES)
36	TEND TO BE THOSE WHOSE PER CAPITA GROSS NATIONAL PRODUCT IS VERY HIGH, HIGH, OR MEDIUM (42)	0.55	36	ALWAYS ARE THOSE WHOSE PER CAPITA GROSS NATIONAL PRODUCT IS LOW OR VERY LOW (73)	1.00	42 0 34 34 X SQ= 28.10 P = 0. RV YES (YES)
37	TEND TO BE THOSE WHOSE PER CAPITA GROSS NATIONAL PRODUCT IS VERY HIGH, HIGH, MEDIUM, OR LOW (64)	0.80	37	TEND TO BE THOSE WHOSE PER CAPITA GROSS NATIONAL PRODUCT IS VERY LOW (51)	0.94	61 2 15 32 X SQ= 50.11 P = 0. RV YES (YES)
39	TEND LESS TO BE THOSE WHOSE INTERNATIONAL FINANCIAL STATUS IS LOW OR VERY LOW (76)	0.52	39	ALWAYS ARE THOSE WHOSE INTERNATIONAL FINANCIAL STATUS IS LOW OR VERY LOW (76)	1.00	36 0 39 34 X SQ= 22.25 P = 0. RV NO (NO)
40	TEND TO BE THOSE WHOSE INTERNATIONAL FINANCIAL STATUS IS VERY HIGH, HIGH, MEDIUM, OR LOW (71)	0.79	40	TEND TO BE THOSE WHOSE INTERNATIONAL FINANCIAL STATUS IS VERY LOW (39)	0.70	57 10 15 23 X SQ= 21.33 P = 0. RV YES (YES)

42	TEND LESS TO BE THOSE WHOSE ECONOMIC DEVELOPMENTAL STATUS IS UNDERDEVELOPED OR VERY UNDERDEVELOPED (76)	0.51	42	ALWAYS ARE THOSE WHOSE ECONOMIC DEVELOPMENTAL STATUS IS UNDERDEVELOPED OR VERY UNDERDEVELOPED (76)	1.00	36 0 38 33 X SQ= 22.06 P = 0. RV NO (NO)
43	TEND TO BE THOSE WHOSE ECONOMIC DEVELOPMENTAL STATUS IS DEVELOPED, INTERMEDIATE, OR UNDERDEVELOPED (55)	0.70	43	TEND TO BE THOSE WHOSE ECONOMIC DEVELOPMENTAL STATUS IS VERY UNDERDEVELOPED (57)	0.97	52 1 22 32 X SQ= 38.63 P = 0. RV YES (YES)
44	TEND LESS TO BE THOSE WHERE THE LITERACY RATE IS BELOW NINETY PERCENT (90)	0.67	44	ALWAYS ARE THOSE WHERE THE LITERACY RATE IS BELOW NINETY PERCENT (90)	1.00	25 0 51 34 X SQ= 12.66 P = 0. RV NO (NO)
45	TEND TO BE THOSE WHERE THE LITERACY RATE IS FIFTY PERCENT OR ABOVE (55)	0.73	45	TEND TO BE THOSE WHERE THE LITERACY RATE IS BELOW FIFTY PERCENT (54)	0.94	51 2 19 32 X SQ= 38.44 P = 0. RV YES (YES)
46	ALWAYS ARE THOSE WHERE THE LITERACY RATE IS TEN PERCENT OR ABOVE (84)	1.00	46	TEND TO BE THOSE WHERE THE LITERACY RATE IS BELOW TEN PERCENT (26)	0.76	72 8 0 25 X SQ= 67.48 P = 0. RV YES (YES)
54	TEND TO BE THOSE WHERE NEWSPAPER CIRCULATION IS ONE HUNDRED OR MORE PER THOUSAND (37)	0.50	54	ALWAYS ARE THOSE WHERE NEWSPAPER CIRCULATION IS LESS THAN ONE HUNDRED PER THOUSAND (76)	1.00	37 0 37 34 X SQ= 23.69 P = 0. RV YES (NO)
55	TEND TO BE THOSE WHERE NEWSPAPER CIRCULATION IS TEN OR MORE PER THOUSAND (78)	0.96	55	TEND TO BE THOSE WHERE NEWSPAPER CIRCULATION IS LESS THAN TEN PER THOUSAND (35)	0.88	71 4 3 30 X SQ= 73.89 P = 0. RV YES (YES)
56	TEND TO BE THOSE WHERE THE RELIGION IS PREDOMINANTLY LITERATE (79)	0.87	56	TEND TO BE THOSE WHERE THE RELIGION IS PREDOMINANTLY OR PARTLY NON-LITERATE (31)	0.65	63 12 9 22 X SQ= 27.95 P = 0. RV YES (YES)
57	TEND LESS TO BE THOSE WHERE THE RELIGION IS OTHER THAN CATHOLIC (90)	0.67	57	ALWAYS ARE THOSE WHERE THE RELIGION IS OTHER THAN CATHOLIC (90)	1.00	25 0 51 34 X SQ= 12.66 P = 0. RV NO (NO)

58	TEND MORE TO BE THOSE WHERE THE RELIGION IS OTHER THAN MUSLIM (97)	0.92	58	TEND LESS TO BE THOSE WHERE THE RELIGION IS OTHER THAN MUSLIM (97)	0.76	6 8 70 26 X SQ= 3.86 P = 0.032 RV NO (YES)
63	TEND TO BE THOSE WHERE THE RELIGION IS PREDOMINANTLY SOME KIND OF CHRISTIAN (46)	0.61	63	ALWAYS ARE THOSE WHERE THE RELIGION IS PREDOMINANTLY OR PARTLY OTHER THAN CHRISTIAN (68)	1.00	46 0 29 34 X SQ= 33.61 P = 0. RV YES (YES)
64	TEND TO BE THOSE WHERE THE RELIGION IS CHRISTIAN, RATHER THAN MUSLIM (46)	0.88	64	ALWAYS ARE THOSE WHERE THE RELIGION IS MUSLIM, RATHER THAN CHRISTIAN (18)	1.00	46 0 6 8 X SQ= 25.59 P = 0. RV YES (YES)
66	TEND TO BE THOSE THAT ARE RELIGIOUSLY HOMOGENEOUS (57)	0.59	66	TEND TO BE THOSE THAT ARE RELIGIOUSLY HETEROGENEOUS (49)	0.65	42 11 29 20 X SQ= 3.94 P = 0.033 RV YES (NO)
68	TEND TO BE THOSE THAT ARE LINGUISTICALLY HOMOGENEOUS (52)	0.60	68	TEND TO BE THOSE THAT ARE LINGUISTICALLY WEAKLY HETEROGENEOUS OR STRONGLY HETEROGENEOUS (62)	0.82	45 6 30 28 X SQ= 15.20 P = 0. RV YES (NO)
69	TEND TO BE THOSE THAT ARE LINGUISTICALLY HOMOGENEOUS OR WEAKLY HETEROGENEOUS (64)	0.72	69	TEND TO BE THOSE THAT ARE LINGUISTICALLY STRONGLY HETEROGENEOUS (50)	0.76	54 8 21 26 X SQ= 20.48 P = 0. RV YES (YES)
70	TEND LESS TO BE THOSE THAT ARE RELIGIOUSLY, RACIALLY, OR LINGUISTICALLY HETEROGENEOUS (85)	0.73	70	TEND MORE TO BE THOSE THAT ARE RELIGIOUSLY, RACIALLY, OR LINGUISTICALLY HETEROGENEOUS (85)	0.93	19 2 52 28 X SQ= 4.02 P = 0.030 RV NO (NO)
72	TEND TO BE THOSE WHOSE DATE OF INDEPENDENCE IS BEFORE 1914 (52)	0.65	72	TEND TO BE THOSE WHOSE DATE OF INDEPENDENCE IS AFTER 1914 (59)	0.85	47 5 25 29 X SQ= 21.65 P = 0. RV YES (YES)
73	TEND TO BE THOSE WHOSE DATE OF INDEPENDENCE IS BEFORE 1945 (65)	0.79	73	TEND TO BE THOSE WHOSE DATE OF INDEPENDENCE IS AFTER 1945 (46)	0.82	57 6 15 28 X SQ= 33.75 P = 0. RV YES (YES)

74	TEND LESS TO BE THOSE THAT ARE NOT HISTORICALLY WESTERN (87)	0.65	74	ALWAYS ARE THOSE THAT ARE NOT HISTORICALLY WESTERN (87)	1.00	26 0 48 34 X SQ= 13.87 P = 0. RV NO (NO)
75	TEND TO BE THOSE THAT ARE HISTORICALLY WESTERN OR SIGNIFICANTLY WESTERNIZED (62)	0.81	75	TEND TO BE THOSE THAT ARE NOT HISTORICALLY WESTERN AND ARE NOT SIGNIFICANTLY WESTERNIZED (52)	0.97	61 1 14 33 X SQ= 55.46 P = 0. RV YES (YES)
76	TEND LESS TO BE THOSE THAT HAVE BEEN SIGNIFICANTLY OR PARTIALLY WESTERNIZED THROUGH A COLONIAL RELATIONSHIP, RATHER THAN BEING HISTORICALLY WESTERN (70)	0.59	76	ALWAYS ARE THOSE THAT HAVE BEEN SIGNIFICANTLY OR PARTIALLY WESTERNIZED THROUGH A COLONIAL RELATIONSHIP, RATHER THAN BEING HISTORICALLY WESTERN (70)	1.00	26 0 37 29 X SQ= 14.71 P = 0. RV NO (NO)
77	TEND TO BE THOSE THAT HAVE BEEN SIGNIFICANTLY WESTERNIZED, RATHER THAN PARTIALLY WESTERNIZED, THROUGH A COLONIAL RELATIONSHIP (28)	0.75	77	TEND TO BE THOSE THAT HAVE BEEN PARTIALLY WESTERNIZED, RATHER THAN SIGNIFICANTLY WESTERNIZED, THROUGH A COLONIAL RELATIONSHIP (41)	0.97	27 1 9 28 X SQ= 30.68 P = 0. RV YES (YES)
79	TEND LESS TO BE THOSE THAT WERE FORMERLY DEPENDENCIES OF BRITAIN OR FRANCE, RATHER THAN SPAIN (49)	0.55	79	ALWAYS ARE THOSE THAT WERE FORMERLY DEPENDENCIES OF BRITAIN OR FRANCE, RATHER THAN SPAIN (49)	1.00	22 24 18 0 X SQ= 12.88 P = 0. RV NO (YES)
80	LEAN TOWARD BEING THOSE WHOSE DATE OF INDEPENDENCE IS AFTER 1914, AND THAT WERE FORMERLY DEPENDENCIES OF BRITAIN, RATHER THAN FRANCE (19)	0.63	80	LEAN TOWARD BEING THOSE WHOSE DATE OF INDEPENDENCE IS AFTER 1914, AND THAT WERE FORMERLY DEPENDENCIES OF FRANCE, RATHER THAN BRITAIN (24)	0.71	10 7 6 17 X SQ= 3.11 P = 0.053 RV YES (YES)
82	TEND TO BE THOSE WHERE THE TYPE OF POLITICAL MODERNIZATION IS EARLY OR LATER EUROPEAN OR EUROPEAN DERIVED, RATHER THAN DEVELOPED TUTELARY OR UNDEVELOPED TUTELARY (51)	0.71	82	ALWAYS ARE THOSE WHERE THE TYPE OF POLITICAL MODERNIZATION IS DEVELOPED TUTELARY OR UNDEVELOPED TUTELARY, RATHER THAN EARLY OR LATER EUROPEAN OR EUROPEAN DERIVED (55)	1.00	51 0 21 29 X SQ= 38.71 P = 0. RV YES (YES)
83	TILT MORE TOWARD BEING THOSE WHERE THE TYPE OF POLITICAL MODERNIZATION IS OTHER THAN NON-EUROPEAN AUTOCHTHONOUS (106)	0.95	83	TILT LESS TOWARD BEING THOSE WHERE THE TYPE OF POLITICAL MODERNIZATION IS OTHER THAN NON-EUROPEAN AUTOCHTHONOUS (106)	0.85	4 5 72 29 X SQ= 1.67 P = 0.132 RV NO (YES)
84	TEND TO BE THOSE WHERE THE TYPE OF POLITICAL MODERNIZATION IS DEVELOPED TUTELARY, RATHER THAN UNDEVELOPED TUTELARY (31)	0.90	84	TEND TO BE THOSE WHERE THE TYPE OF POLITICAL MODERNIZATION IS UNDEVELOPED TUTELARY, RATHER THAN DEVELOPED TUTELARY (24)	0.76	19 7 2 22 X SQ= 18.90 P = 0. RV YES (YES)

85	TEND TO BE THOSE WHERE THE STAGE OF POLITICAL MODERNIZATION IS ADVANCED, RATHER THAN MID- OR EARLY TRANSITIONAL (60)	0.72	85	TEND TO BE THOSE WHERE THE STAGE OF POLITICAL MODERNIZATION IS MID- OR EARLY TRANSITIONAL, RATHER THAN ADVANCED (54)	0.91	55 3 21 30 X SQ= 34.51 P = 0. RV YES (YES)
86	TEND TO BE THOSE WHERE THE STAGE OF POLITICAL MODERNIZATION IS ADVANCED OR MID-TRANSITIONAL, RATHER THAN EARLY TRANSITIONAL (76)	0.88	86	TEND TO BE THOSE WHERE THE STAGE OF POLITICAL MODERNIZATION IS EARLY TRANSITIONAL, RATHER THAN ADVANCED OR MID-TRANSITIONAL (38)	0.79	67 7 9 26 X SQ= 44.28 P = 0. RV YES (YES)
87	TEND TO BE THOSE WHOSE IDEOLOGICAL ORIENTATION IS OTHER THAN DEVELOPMENTAL (58)	0.85	87	TEND TO BE THOSE WHOSE IDEOLOGICAL ORIENTATION IS DEVELOPMENTAL (31)	0.78	9 21 51 6 X SQ= 29.76 P = 0. RV YES (YES)
89	TEND TO BE THOSE WHOSE IDEOLOGICAL ORIENTATION IS CONVENTIONAL (33)	0.56	89	ALWAYS ARE THOSE WHOSE IDEOLOGICAL ORIENTATION IS OTHER THAN CONVENTIONAL (62)	1.00	33 0 26 31 X SQ= 25.02 P = 0. RV YES (YES)
92	TEND LESS TO BE THOSE WHERE THE SYSTEM STYLE IS LIMITED MOBILIZATIONAL, OR NON-MOBILIZATIONAL (93)	0.77	92	TEND MORE TO BE THOSE WHERE THE SYSTEM STYLE IS LIMITED MOBILIZATIONAL, OR NON-MOBILIZATIONAL (93)	0.94	17 2 58 32 X SQ= 3.49 P = 0.033 RV NO (NO)
94	TEND TO BE THOSE WHERE THE STATUS OF THE REGIME IS CONSTITUTIONAL (51)	0.65	94	TEND TO BE THOSE WHERE THE STATUS OF THE REGIME IS AUTHORITARIAN OR TOTALITARIAN (41)	0.72	46 5 25 13 X SQ= 6.60 P = 0.007 RV YES (NO)
95	TEND LESS TO BE THOSE WHERE THE STATUS OF THE REGIME IS CONSTITUTIONAL OR AUTHORITARIAN (95)	0.78	95	ALWAYS ARE THOSE WHERE THE STATUS OF THE REGIME IS CONSTITUTIONAL OR AUTHORITARIAN (95)	1.00	58 34 16 0 X SQ= 7.00 P = 0.002 RV NO (NO)
96	TEND TO BE THOSE WHERE THE STATUS OF THE REGIME IS CONSTITUTIONAL OR TOTALITARIAN (67)	0.89	96	TEND TO BE THOSE WHERE THE STATUS OF THE REGIME IS AUTHORITARIAN (23)	0.72	62 5 8 13 X SQ= 25.88 P = 0. RV YES (YES)
97	TEND TO BE THOSE WHERE THE STATUS OF THE REGIME IS TOTALITARIAN, RATHER THAN AUTHORITARIAN (16)	0.67	97	ALWAYS ARE THOSE WHERE THE STATUS OF THE REGIME IS AUTHORITARIAN, RATHER THAN TOTALITARIAN (23)	1.00	8 13 16 0 X SQ= 12.67 P = 0. RV YES (YES)

99	TEND TO BE THOSE WHERE GOVERNMENTAL STABILITY IS GENERALLY PRESENT AND DATES FROM AT LEAST THE INTER-WAR PERIOD, OR FROM THE POST-WAR PERIOD (50)	0.71	99	TEND TO BE THOSE WHERE GOVERNMENTAL STABILITY IS MODERATELY PRESENT AND DATES FROM THE POST-WAR PERIOD, OR IS ABSENT (36)	0.86	48 2 20 12 X SQ= 13.19 P = 0. RV YES (NO)
100	TEND TO BE THOSE WHERE GOVERNMENTAL STABILITY IS GENERALLY PRESENT AND DATES FROM AT LEAST THE INTER-WAR PERIOD, OR IS GENERALLY OR MODERATELY PRESENT, AND DATES FROM THE POST-WAR PERIOD (64)	0.81	100	TEND TO BE THOSE WHERE GOVERNMENTAL STABILITY IS ABSENT (22)	0.58	57 5 13 7 X SQ= 6.76 P = 0.007 RV YES (NO)
101	TEND TO BE THOSE WHERE THE REPRESENTATIVE CHARACTER OF THE REGIME IS POLYARCHIC (41)	0.57	101	TEND TO BE THOSE WHERE THE REPRESENTATIVE CHARACTER OF THE REGIME IS LIMITED POLYARCHIC, PSEUDO-POLYARCHIC, OR NON-POLYARCHIC (57)	0.92	39 2 30 24 X SQ= 16.42 P = 0. RV YES (NO)
102	TEND TO BE THOSE WHERE THE REPRESENTATIVE CHARACTER OF THE REGIME IS POLYARCHIC OR LIMITED POLYARCHIC (59)	0.68	102	TEND TO BE THOSE WHERE THE REPRESENTATIVE CHARACTER OF THE REGIME IS PSEUDO-POLYARCHIC OR NON-POLYARCHIC (49)	0.70	49 10 23 23 X SQ= 11.61 P = 0.001 RV YES (YES)
105	TEND TO BE THOSE WHERE THE ELECTORAL SYSTEM IS COMPETITIVE (43)	0.63	105	TEND TO BE THOSE WHERE THE ELECTORAL SYSTEM IS PARTIALLY COMPETITIVE OR NON-COMPETITIVE (47)	0.88	40 3 24 21 X SQ= 15.52 P = 0. RV YES (NO)
107	TEND TO BE THOSE WHERE AUTONOMOUS GROUPS ARE FULLY TOLERATED IN POLITICS (46)	0.57	107	TEND TO BE THOSE WHERE AUTONOMOUS GROUPS ARE PARTIALLY TOLERATED IN POLITICS, ARE TOLERATED ONLY OUTSIDE POLITICS, OR ARE NOT TOLERATED AT ALL (65)	0.91	43 3 32 29 X SQ= 19.14 P = 0. RV YES (NO)
108	LEAN TOWARD BEING THOSE WHERE AUTONOMOUS GROUPS ARE FULLY OR PARTIALLY TOLERATED IN POLITICS (65)	0.71	108	LEAN TOWARD BEING THOSE WHERE AUTONOMOUS GROUPS ARE TOLERATED ONLY OUTSIDE POLITICS OR ARE NOT TOLERATED AT ALL (35)	0.50	51 13 21 13 X SQ= 2.80 P = 0.091 RV YES (NO)
110	TEND LESS TO BE THOSE WHERE POLITICAL ENCULTURATION IS MEDIUM OR LOW (80)	0.76	110	TEND MORE TO BE THOSE WHERE POLITICAL ENCULTURATION IS MEDIUM OR LOW (80)	0.97	14 1 45 30 X SQ= 4.76 P = 0.016 RV NO (NO)
114	TEND TO BE THOSE WHERE SECTIONALISM IS NEGLIGIBLE (47)	0.54	114	TEND TO BE THOSE WHERE SECTIONALISM IS EXTREME OR MODERATE (61)	0.77	33 24 39 7 X SQ= 7.52 P = 0.005 RV YES (NO)

116	TEND LESS TO BE THOSE WHERE INTEREST ARTICULATION BY ASSOCIATIONAL GROUPS IS LIMITED OR NEGLIGIBLE (79)	0.56	116	ALWAYS ARE THOSE WHERE INTEREST ARTICULATION BY ASSOCIATIONAL GROUPS IS LIMITED OR NEGLIGIBLE (79)	1.00	32 0 40 34 X SQ= 19.59 P = 0. RV NO (NO)
117	TEND TO BE THOSE WHERE INTEREST ARTICULATION BY ASSOCIATIONAL GROUPS IS SIGNIFICANT, MODERATE, OR LIMITED (60)	0.78	117	TEND TO BE THOSE WHERE INTEREST ARTICULATION BY ASSOCIATIONAL GROUPS IS NEGLIGIBLE (51)	0.91	56 3 16 31 X SQ= 41.74 P = 0. RV YES (YES)
120	TILT LESS TOWARD BEING THOSE WHERE INTEREST ARTICULATION BY INSTITUTIONAL GROUPS IS VERY SIGNIFICANT, SIGNIFICANT, OR MODERATE (90)	0.87	120	ALWAYS ARE THOSE WHERE INTEREST ARTICULATION BY INSTITUTIONAL GROUPS IS VERY SIGNIFICANT, SIGNIFICANT, OR MODERATE (90)	1.00	66 20 10 0 X SQ= 1.70 P = 0.115 RV NO (NO)
121	TEND TO BE THOSE WHERE INTEREST ARTICULATION BY NON-ASSOCIATIONAL GROUPS IS MODERATE, LIMITED, OR NEGLIGIBLE (61)	0.79	121	TEND TO BE THOSE WHERE INTEREST ARTICULATION BY NON-ASSOCIATIONAL GROUPS IS SIGNIFICANT (54)	0.97	16 33 60 1 X SQ= 51.90 P = 0. RV YES (YES)
122	TEND LESS TO BE THOSE WHERE INTEREST ARTICULATION BY NON-ASSOCIATIONAL GROUPS IS SIGNIFICANT OR MODERATE (83)	0.58	122	ALWAYS ARE THOSE WHERE INTEREST ARTICULATION BY NON-ASSOCIATIONAL GROUPS IS SIGNIFICANT OR MODERATE (83)	1.00	44 34 32 0 X SQ= 18.20 P = 0. RV NO (NO)
125	TEND TO BE THOSE WHERE INTEREST ARTICULATION BY ANOMIC GROUPS IS INFREQUENT OR VERY INFREQUENT (35)	0.52	125	TEND TO BE THOSE WHERE INTEREST ARTICULATION BY ANOMIC GROUPS IS FREQUENT OR OCCASIONAL (64)	0.91	30 30 32 3 X SQ= 14.96 P = 0. RV YES (YES)
126	TEND LESS TO BE THOSE WHERE INTEREST ARTICULATION BY ANOMIC GROUPS IS FREQUENT, OCCASIONAL, OR INFREQUENT (83)	0.74	126	ALWAYS ARE THOSE WHERE INTEREST ARTICULATION BY ANOMIC GROUPS IS FREQUENT, OCCASIONAL, OR INFREQUENT (83)	1.00	46 33 16 0 X SQ= 8.48 P = 0.001 RV NO (NO)
128	TEND TO BE THOSE WHERE INTEREST ARTICULATION BY POLITICAL PARTIES IS SIGNIFICANT OR MODERATE (48)	0.59	128	TEND TO BE THOSE WHERE INTEREST ARTICULATION BY POLITICAL PARTIES IS LIMITED OR NEGLIGIBLE (45)	0.70	41 7 28 16 X SQ= 4.70 P = 0.029 RV YES (NO)
129	LEAN TOWARD BEING THOSE WHERE INTEREST ARTICULATION BY POLITICAL PARTIES IS SIGNIFICANT, MODERATE, OR LIMITED (56)	0.67	129	LEAN TOWARD BEING THOSE WHERE INTEREST ARTICULATION BY POLITICAL PARTIES IS NEGLIGIBLE (37)	0.57	46 10 23 13 X SQ= 2.98 P = 0.083 RV YES (NO)

133	TEND MORE TO BE THOSE WHERE INTEREST AGGREGATION BY THE EXECUTIVE IS MODERATE, LIMITED, OR NEGLIGIBLE (76)	0.81	133	TEND LESS TO BE THOSE WHERE INTEREST AGGREGATION BY THE EXECUTIVE IS MODERATE, LIMITED, OR NEGLIGIBLE (76)	0.53	13 15 57 17 X SQ= 7.47 P = 0.004 RV NO (YES)
137	TEND LESS TO BE THOSE WHERE INTEREST AGGREGATION BY THE LEGISLATURE IS LIMITED OR NEGLIGIBLE (68)	0.56	137	ALWAYS ARE THOSE WHERE INTEREST AGGREGATION BY THE LEGISLATURE IS LIMITED OR NEGLIGIBLE (68)	1.00	29 0 37 29 X SQ= 16.33 P = 0. RV NO (NO)
138	TEND TO BE THOSE WHERE INTEREST AGGREGATION BY THE LEGISLATURE IS SIGNIFICANT, MODERATE, OR LIMITED (48)	0.66	138	TEND TO BE THOSE WHERE INTEREST AGGREGATION BY THE LEGISLATURE IS NEGLIGIBLE (49)	0.86	44 4 23 25 X SQ= 19.76 P = 0. RV YES (YES)
142	TEND LESS TO BE THOSE WHERE THE PARTY SYSTEM IS QUANTITATIVELY OTHER THAN MULTI-PARTY (66)	0.62	142	TEND MORE TO BE THOSE WHERE THE PARTY SYSTEM IS QUANTITATIVELY OTHER THAN MULTI-PARTY (66)	0.85	26 4 42 23 X SQ= 3.88 P = 0.030 RV NO (NO)
144	TILT LESS TOWARD BEING THOSE WHERE THE PARTY SYSTEM IS QUANTITATIVELY ONE-PARTY, RATHER THAN TWO-PARTY (34)	0.68	144	TILT MORE TOWARD BEING THOSE WHERE THE PARTY SYSTEM IS QUANTITATIVELY ONE-PARTY, RATHER THAN TWO-PARTY (34)	0.93	21 13 10 1 X SQ= 2.07 P = 0.132 RV NO (NO)
147	TEND LESS TO BE THOSE WHERE THE PARTY SYSTEM IS QUALITATIVELY OTHER THAN CLASS-ORIENTED OR MULTI-IDEOLOGICAL (67)	0.62	147	ALWAYS ARE THOSE WHERE THE PARTY SYSTEM IS QUALITATIVELY OTHER THAN CLASS-ORIENTED OR MULTI-IDEOLOGICAL (67)	1.00	23 0 38 28 X SQ= 12.34 P = 0. RV NO (NO)
148	ALWAYS ARE THOSE WHERE THE PARTY SYSTEM IS QUALITATIVELY OTHER THAN AFRICAN TRANSITIONAL (96)	1.00	148	TEND LESS TO BE THOSE WHERE THE PARTY SYSTEM IS QUALITATIVELY OTHER THAN AFRICAN TRANSITIONAL (96)	0.55	0 14 74 17 X SQ= 34.75 P = 0. RV NO (YES)
153	TEND TO BE THOSE WHERE THE PARTY SYSTEM IS STABLE (42)	0.63	153	TEND TO BE THOSE WHERE THE PARTY SYSTEM IS MODERATELY STABLE OR UNSTABLE (38)	0.92	41 1 24 11 X SQ= 10.14 P = 0.001 RV YES (NO)
154	TEND TO BE THOSE WHERE THE PARTY SYSTEM IS STABLE OR MODERATELY STABLE (55)	0.75	154	TEND TO BE THOSE WHERE THE PARTY SYSTEM IS UNSTABLE (25)	0.58	49 5 16 7 X SQ= 4.01 P = 0.035 RV YES (NO)

158	TEND TO BE THOSE WHERE PERSONALISMO IS NEGLIGIBLE (56)	0.68	158	TEND TO BE THOSE WHERE PERSONALISMO IS PRONOUNCED OR MODERATE (40)	0.65	23 15 48 8 X SQ= 6.47 P = 0.007 RV YES (NO)
164	TEND TO BE THOSE WHERE THE REGIME'S LEADERSHIP CHARISMA IS NEGLIGIBLE (65)	0.81	164	TEND TO BE THOSE WHERE THE REGIME'S LEADERSHIP CHARISMA IS PRONOUNCED OR MODERATE (34)	0.78	14 18 59 5 X SQ= 24.88 P = 0. RV YES (YES)
168	TEND LESS TO BE THOSE WHERE THE HORIZONTAL POWER DISTRIBUTION IS LIMITED OR NEGLIGIBLE (72)	0.53	168	ALWAYS ARE THOSE WHERE THE HORIZONTAL POWER DISTRIBUTION IS LIMITED OR NEGLIGIBLE (72)	1.00	34 0 38 31 X SQ= 19.77 P = 0. RV NO (NO)
169	TEND TO BE THOSE WHERE THE HORIZONTAL POWER DISTRIBUTION IS SIGNIFICANT OR LIMITED (58)	0.68	169	TEND TO BE THOSE WHERE THE HORIZONTAL POWER DISTRIBUTION IS NEGLIGIBLE (48)	0.71	49 9 23 22 X SQ= 11.87 P = 0. RV YES (NO)
174	TEND LESS TO BE THOSE WHERE THE LEGISLATURE IS PARTIALLY EFFECTIVE, LARGELY INEFFECTIVE, OR WHOLLY INEFFECTIVE (72)	0.60	174	ALWAYS ARE THOSE WHERE THE LEGISLATURE IS PARTIALLY EFFECTIVE, LARGELY INEFFECTIVE, OR WHOLLY INEFFECTIVE (72)	1.00	28 0 42 28 X SQ= 13.78 P = 0. RV NO (NO)
175	TEND TO BE THOSE WHERE THE LEGISLATURE IS FULLY EFFECTIVE OR PARTIALLY EFFECTIVE (51)	0.66	175	TEND TO BE THOSE WHERE THE LEGISLATURE IS LARGELY INEFFECTIVE OR WHOLLY INEFFECTIVE (49)	0.82	46 5 24 23 X SQ= 16.49 P = 0. RV YES (NO)
178	TEND TO BE THOSE WHERE THE LEGISLATURE IS BICAMERAL (51)	0.61	178	TEND TO BE THOSE WHERE THE LEGISLATURE IS UNICAMERAL (53)	0.73	28 22 43 8 X SQ= 8.38 P = 0.002 RV YES (NO)
179	TEND TO BE THOSE WHERE THE EXECUTIVE IS STRONG (39)	0.53	179	TEND TO BE THOSE WHERE THE EXECUTIVE IS DOMINANT (52)	0.82	31 18 35 4 X SQ= 6.77 P = 0.006 RV YES (NO)
187	TEND LESS TO BE THOSE WHERE THE ROLE OF THE POLICE IS POLITICALLY SIGNIFICANT (66)	0.56	187	TEND MORE TO BE THOSE WHERE THE ROLE OF THE POLICE IS POLITICALLY SIGNIFICANT (66)	0.82	38 23 30 5 X SQ= 4.82 P = 0.019 RV NO (NO)

188	TEND LESS TO BE THOSE WHERE THE CHARACTER OF THE LEGAL SYSTEM IS OTHER THAN CIVIL LAW (81)	0.59	188	TEND MORE TO BE THOSE WHERE THE CHARACTER OF THE LEGAL SYSTEM IS OTHER THAN CIVIL LAW (81)	0.97	31 1 44 32 X SQ= 14.34 P = 0. RV NO (NO)
189	LEAN LESS TOWARD BEING THOSE WHERE THE CHARACTER OF THE LEGAL SYSTEM IS OTHER THAN COMMON LAW (108)	0.91	189	ALWAYS ARE THOSE WHERE THE CHARACTER OF THE LEGAL SYSTEM IS OTHER THAN COMMON LAW (108)	1.00	7 0 69 34 X SQ= 1.98 P = 0.097 RV NO (NO)
192	TEND TO BE THOSE WHERE THE CHARACTER OF THE LEGAL SYSTEM IS OTHER THAN MUSLIM OR PARTLY MUSLIM (86)	0.91	192	TEND TO BE THOSE WHERE THE CHARACTER OF THE LEGAL SYSTEM IS MUSLIM OR PARTLY MUSLIM (27)	0.52	7 17 68 16 X SQ= 21.21 P = 0. RV YES (YES)
193	TEND TO BE THOSE WHERE THE CHARACTER OF THE LEGAL SYSTEM IS OTHER THAN PARTLY INDIGENOUS (86)	0.95	193	TEND TO BE THOSE WHERE THE CHARACTER OF THE LEGAL SYSTEM IS PARTLY INDIGENOUS (28)	0.67	4 22 72 11 X SQ= 44.44 P = 0. RV YES (YES)
194	TEND LESS TO BE THOSE THAT ARE NON-COMMUNIST (101)	0.83	194	ALWAYS ARE THOSE THAT ARE NON-COMMUNIST (101)	1.00	13 0 62 34 X SQ= 5.14 P = 0.009 RV NO (NO)

181 POLITIES
WHERE THE BUREAUCRACY
IS MODERN, RATHER THAN
SEMI-MODERN (21)

181 POLITIES
WHERE THE BUREAUCRACY
IS SEMI-MODERN, RATHER THAN
MODERN (55)

PREDICATE ONLY

21 IN LEFT COLUMN

AUSTRALIA	AUSTRIA	BELGIUM	CANADA	DENMARK	FINLAND	FRANCE	GERMAN FR	ICELAND	IRELAND
ISRAEL	ITALY	LUXEMBOURG	NETHERLANDS	NEW ZEALAND	NORWAY	SO AFRICA	SWEDEN	SWITZERLAND	UK
US									

55 IN RIGHT COLUMN

ALBANIA	ARGENTINA	BOLIVIA	BRAZIL	BULGARIA	CEYLON	CHILE	CHINA, PR	COLOMBIA	COSTA RICA
CUBA	CYPRUS	CZECHOS'KIA	DOMIN REP	ECUADOR	EL SALVADOR	GERMANY, E	GHANA	GREECE	GUATEMALA
GUINEA	HAITI	HONDURAS	HUNGARY	INDIA	JAMAICA	JAPAN	KOREA, N	KOREA REP	LEBANON
LIBERIA	LIBYA	MALAYA	MEXICO	MONGOLIA	MOROCCO	NICARAGUA	PANAMA	PARAGUAY	PERU
PHILIPPINES	POLAND	PORTUGAL	RUMANIA	SPAIN	SYRIA	TRINIDAD	TUNISIA	TURKEY	USSR
UAR	URUGUAY	VENEZUELA	VIETNAM, N	YUGOSLAVIA					

3 EXCLUDED BECAUSE AMBIGUOUS

INDONESIA	PAKISTAN	VIETNAM REP

34 EXCLUDED BECAUSE IRRELEVANT

AFGHANISTAN	ALGERIA	BURMA	BURUNDI	CAMBODIA	CAMEROUN	CEN AFR REP	CHAD	CONGO(BRA)	CONGO(LEO)
DAHOMEY	ETHIOPIA	GABON	IRAN	IVORY COAST	JORDAN	LAOS	MALAGASY R	MALI	MAURITANIA
NEPAL	NIGER	NIGERIA	RWANDA	SA'U ARABIA	SENEGAL	SIERRE LEO	SOMALIA	SUDAN	TANGANYIKA
THAILAND	TOGO	UGANDA	UPPER VOLTA						

2 EXCLUDED BECAUSE UNASCERTAINABLE

IRAQ	YEMEN

182 POLITIES
WHERE THE BUREAUCRACY
IS SEMI-MODERN, RATHER THAN
POST-COLONIAL TRANSITIONAL (55)

182 POLITIES
WHERE THE BUREAUCRACY
IS POST-COLONIAL TRANSITIONAL,
RATHER THAN SEMI-MODERN (25)

PREDICATE ONLY

55 IN LEFT COLUMN

ALBANIA	ARGENTINA	BOLIVIA	BRAZIL	BULGARIA	CEYLON	CHILE	CHINA, PR	COLOMBIA	COSTA RICA
CUBA	CYPRUS	CZECHOS'KIA	DOMIN REP	ECUADOR	EL SALVADOR	GERMANY, E	GHANA	GREECE	GUATEMALA
GUINEA	HAITI	HONDURAS	HUNGARY	INDIA	JAMAICA	JAPAN	KOREA, N	KOREA REP	LEBANON
LIBERIA	LIBYA	MALAYA	MEXICO	MONGOLIA	MOROCCO	NICARAGUA	PANAMA	PARAGUAY	PERU
PHILIPPINES	POLAND	PORTUGAL	RUMANIA	SPAIN	SYRIA	TRINIDAD	TUNISIA	TURKEY	USSR
UAR	URUGUAY	VENEZUELA	VIETNAM, N	YUGOSLAVIA					

25 IN RIGHT COLUMN

ALGERIA	BURMA	BURUNDI	CAMEROUN	CEN AFR REP	CHAD	CONGO(BRA)	CONGO(LEO)	DAHOMEY	GABON
IVORY COAST	MALAGASY R	MALI	MAURITANIA	NIGER	NIGERIA	RWANDA	SENEGAL	SIERRE LEO	SOMALIA
SUDAN	TANGANYIKA	TOGO	UGANDA	UPPER VOLTA					

3 EXCLUDED BECAUSE AMBIGUOUS

INDONESIA	PAKISTAN	VIETNAM REP

30 EXCLUDED BECAUSE IRRELEVANT

AFGHANISTAN	AUSTRALIA	AUSTRIA	BELGIUM	CAMBODIA	CANADA	DENMARK	ETHIOPIA	FINLAND	FRANCE
GERMAN FR	ICELAND	IRAN	IRELAND	ISRAEL	ITALY	JORDAN	LAOS	LUXEMBOURG	NEPAL
NETHERLANDS	NEW ZEALAND	NORWAY	SA'U ARABIA	SO AFRICA	SWEDEN	SWITZERLAND	THAILAND	UK	US

2 EXCLUDED BECAUSE UNASCERTAINABLE

IRAQ	YEMEN

183 POLITIES
WHERE THE BUREAUCRACY
IS POST-COLONIAL TRANSITIONAL,
RATHER THAN TRADITIONAL (25)

183 POLITIES
WHERE THE BUREAUCRACY
IS TRADITIONAL, RATHER THAN
POST-COLONIAL TRANSITIONAL (9)

PREDICATE ONLY

25 IN LEFT COLUMN

ALGERIA, BURMA, BURUNDI, CAMEROUN, CEN AFR REP, CHAD, CONGO(BRA), CONGO(LEO), DAHOMEY, GABON, IVORY COAST, MALAGASY R, MALI, MAURITANIA, NIGER, NIGERIA, RWANDA, SENEGAL, SIERRE LEO, SOMALIA, SUDAN, TANGANYIKA, TOGO, UGANDA, UPPER VOLTA

9 IN RIGHT COLUMN

AFGHANISTAN, CAMBODIA, ETHIOPIA, IRAN, JORDAN, LAOS, NEPAL, SA'U ARABIA, THAILAND

3 EXCLUDED BECAUSE AMBIGUOUS

INDONESIA, PAKISTAN, VIETNAM REP

76 EXCLUDED BECAUSE IRRELEVANT

ALBANIA, ARGENTINA, AUSTRALIA, AUSTRIA, BELGIUM, BOLIVIA, BRAZIL, BULGARIA, CANADA, CEYLON, CHILE, CHINA, PR, COLOMBIA, COSTA RICA, CUBA, CYPRUS, CZECHOS'KIA, DENMARK, DOMIN REP, ECUADOR, EL SALVADOR, FINLAND, FRANCE, GERMANY, E, GERMAN FR, GHANA, GREECE, GUATEMALA, GUINEA, HAITI, HONDURAS, HUNGARY, ICELAND, INDIA, IRELAND, ISRAEL, ITALY, JAMAICA, JAPAN, KOREA, N, KOREA REP, LEBANON, LIBERIA, LIBYA, LUXEMBOURG, MALAYA, MEXICO, MONGOLIA, MOROCCO, NETHERLANDS, NEW ZEALAND, NICARAGUA, NORWAY, PANAMA, PARAGUAY, PERU, PHILIPPINES, POLAND, PORTUGAL, RUMANIA, SO AFRICA, SPAIN, SWEDEN, SWITZERLAND, SYRIA, TRINIDAD, TUNISIA, TURKEY, USSR, UAR, UK, US, URUGUAY, VENEZUELA, VIETNAM, N, YUGOSLAVIA

2 EXCLUDED BECAUSE UNASCERTAINABLE

IRAQ, YEMEN

184 POLITIES
WHERE PARTICIPATION BY THE MILITARY
IN POLITICS IS
INTERVENTIVE, RATHER THAN
NEUTRAL (21)

184 POLITIES
WHERE PARTICIPATION BY THE MILITARY
IN POLITICS IS
NEUTRAL, RATHER THAN
INTERVENTIVE (56)

SUBJECT ONLY

21 IN LEFT COLUMN

ARGENTINA	BRAZIL	BURMA	EL SALVADOR	GUATEMALA	HAITI	HONDURAS	IRAQ	KOREA REP	NICARAGUA
PAKISTAN	PANAMA	PARAGUAY	PERU	SUDAN	SYRIA	THAILAND	TOGO	TURKEY	UAR
YEMEN									

56 IN RIGHT COLUMN

AUSTRALIA	AUSTRIA	BELGIUM	BOLIVIA	BURUNDI	CAMEROUN	CANADA	CEN AFR REP	CEYLON	CHAD
CHILE	COLOMBIA	CONGO(BRA)	COSTA RICA	CYPRUS	DAHOMEY	DENMARK	DOMIN REP	FINLAND	GABON
GERMAN FR	GREECE	ICELAND	INDIA	IRELAND	ISRAEL	ITALY	IVORY COAST	JAMAICA	JAPAN
LIBYA	LUXEMBOURG	MALAGASY R	MALAYA	MALI	MAURITANIA	MEXICO	MOROCCO	NETHERLANDS	NEW ZEALAND
NIGER	NIGERIA	NORWAY	PHILIPPINES	RWANDA	SIERRE LEO	SWEDEN	SWITZERLAND	TANGANYIKA	TRINIDAD
TUNISIA	UGANDA	UK	US	UPPER VOLTA	URUGUAY				

7 EXCLUDED BECAUSE AMBIGUOUS

CONGO(LEO)	CUBA	FRANCE	LAOS	NEPAL	SOMALIA	VENEZUELA

31 EXCLUDED BECAUSE IRRELEVANT

AFGHANISTAN	ALBANIA	ALGERIA	BULGARIA	CAMBODIA	CHINA, PR	CZECHOS'KIA	ECUADOR	ETHIOPIA	GERMANY, E
GHANA	GUINEA	HUNGARY	INDONESIA	IRAN	JORDAN	KOREA, N	LEBANON	LIBERIA	MONGOLIA
POLAND	PORTUGAL	RUMANIA	SA'U ARABIA	SENEGAL	SO AFRICA	SPAIN	USSR	VIETNAM, N	VIETNAM REP
YUGOSLAVIA									

---

2 ALWAYS ARE THOSE LOCATED ELSEWHERE THAN IN WEST EUROPE, SCANDINAVIA, NORTH AMERICA, OR AUSTRALASIA (93) 1.00

2 TEND LESS TO BE THOSE LOCATED ELSEWHERE THAN IN WEST EUROPE, SCANDINAVIA, NORTH AMERICA, OR AUSTRALASIA (93) 0.66

0 19
21 37
X SQ= 7.72
P = 0.001
RV NO (NO)

6 TEND LESS TO BE THOSE LOCATED ELSEWHERE THAN IN THE CARIBBEAN, CENTRAL AMERICA, OR SOUTH AMERICA (93) 0.52

6 TEND MORE TO BE THOSE LOCATED ELSEWHERE THAN IN THE CARIBBEAN, CENTRAL AMERICA, OR SOUTH AMERICA (93) 0.84

10 9
11 47
X SQ= 6.57
P = 0.007
RV NO (YES)

10 LEAN MORE TOWARD BEING THOSE LOCATED ELSEWHERE THAN IN NORTH AFRICA, OR CENTRAL AND SOUTH AFRICA (82) 0.86

10 LEAN LESS TOWARD BEING THOSE LOCATED ELSEWHERE THAN IN NORTH AFRICA, OR CENTRAL AND SOUTH AFRICA (82) 0.62

3 21
18 35
X SQ= 2.83
P = 0.058
RV NO (NO)

11	TEND MORE TO BE THOSE LOCATED ELSEWHERE THAN IN CENTRAL AND SOUTH AFRICA (87)	0.90	11	TEND LESS TO BE THOSE LOCATED ELSEWHERE THAN IN CENTRAL AND SOUTH AFRICA (87)	0.68	2 18 19 38 X SQ= 2.97 P = 0.048 RV NO (NO)
14	TEND LESS TO BE THOSE LOCATED ELSEWHERE THAN IN THE MIDDLE EAST (104)	0.81	14	TEND MORE TO BE THOSE LOCATED ELSEWHERE THAN IN THE MIDDLE EAST (104)	0.96	4 2 17 54 X SQ= 3.17 P = 0.044 RV NO (YES)
15	TEND TO BE THOSE LOCATED IN THE MIDDLE EAST, RATHER THAN IN NORTH AFRICA OR CENTRAL AND SOUTH AFRICA (11)	0.57	15	TEND TO BE THOSE LOCATED IN NORTH AFRICA OR CENTRAL AND SOUTH AFRICA, RATHER THAN IN THE MIDDLE EAST (33)	0.91	4 2 3 21 X SQ= 5.14 P = 0.016 RV YES (YES)
18	LEAN TOWARD BEING THOSE LOCATED IN EAST ASIA, SOUTH ASIA, OR SOUTHEAST ASIA, RATHER THAN IN NORTH AFRICA OR CENTRAL AND SOUTH AFRICA (18)	0.57	18	LEAN TOWARD BEING THOSE LOCATED IN NORTH AFRICA OR CENTRAL AND SOUTH AFRICA, RATHER THAN IN EAST ASIA, SOUTH ASIA, OR SOUTHEAST ASIA (33)	0.81	3 21 4 5 X SQ= 2.31 P = 0.068 RV YES (NO)
19	TEND TO BE THOSE LOCATED IN THE CARIBBEAN, CENTRAL AMERICA, OR SOUTH AMERICA, RATHER THAN IN NORTH AFRICA OR CENTRAL AND SOUTH AFRICA (22)	0.77	19	TEND TO BE THOSE LOCATED IN NORTH AFRICA OR CENTRAL AND SOUTH AFRICA, RATHER THAN IN THE CARIBBEAN, CENTRAL AMERICA, OR SOUTH AMERICA (33)	0.70	3 21 10 9 X SQ= 6.31 P = 0.007 RV YES (YES)
28	TILT MORE TOWARD BEING THOSE WHOSE POPULATION GROWTH RATE IS HIGH (62)	0.75	28	TILT LESS TOWARD BEING THOSE WHOSE POPULATION GROWTH RATE IS HIGH (62)	0.52	15 29 5 27 X SQ= 2.38 P = 0.112 RV NO (NO)
30	TILT TOWARD BEING THOSE WHOSE AGRICULTURAL POPULATION IS HIGH (56)	0.62	30	TILT TOWARD BEING THOSE WHOSE AGRICULTURAL POPULATION IS MEDIUM, LOW, OR VERY LOW (57)	0.61	13 22 8 34 X SQ= 2.31 P = 0.122 RV YES (NO)
35	ALWAYS ARE THOSE WHOSE PER CAPITA GROSS NATIONAL PRODUCT IS MEDIUM, LOW, OR VERY LOW (91)	1.00	35	TEND LESS TO BE THOSE WHOSE PER CAPITA GROSS NATIONAL PRODUCT IS MEDIUM, LOW, OR VERY LOW (91)	0.66	0 19 21 37 X SQ= 7.72 P = 0.001 RV NO (NO)
36	TEND MORE TO BE THOSE WHOSE PER CAPITA GROSS NATIONAL PRODUCT IS LOW OR VERY LOW (73)	0.95	36	TEND LESS TO BE THOSE WHOSE PER CAPITA GROSS NATIONAL PRODUCT IS LOW OR VERY LOW (73)	0.52	1 27 20 29 X SQ= 10.65 P = 0. RV NO (NO)

38	ALWAYS ARE THOSE WHOSE INTERNATIONAL FINANCIAL STATUS IS MEDIUM, LOW, OR VERY LOW (103)	1.00	38	LEAN LESS TOWARD BEING THOSE WHOSE INTERNATIONAL FINANCIAL STATUS IS MEDIUM, LOW, OR VERY LOW (103)	0.86	0 8 21 48 X SQ= 1.99 P = 0.099 RV NO (NO)
41	ALWAYS ARE THOSE WHOSE ECONOMIC DEVELOPMENTAL STATUS IS INTERMEDIATE, UNDERDEVELOPED, OR VERY UNDERDEVELOPED (94)	1.00	41	TEND LESS TO BE THOSE WHOSE ECONOMIC DEVELOPMENTAL STATUS IS INTERMEDIATE, UNDERDEVELOPED, OR VERY UNDERDEVELOPED (94)	0.73	0 15 21 40 X SQ= 5.52 P = 0.008 RV NO (NO)
42	TEND MORE TO BE THOSE WHOSE ECONOMIC DEVELOPMENTAL STATUS IS UNDERDEVELOPED OR VERY UNDERDEVELOPED (76)	0.90	42	TEND LESS TO BE THOSE WHOSE ECONOMIC DEVELOPMENTAL STATUS IS UNDERDEVELOPED OR VERY UNDERDEVELOPED (76)	0.56	2 24 19 31 X SQ= 6.41 P = 0.006 RV NO (NO)
43	LEAN TOWARD BEING THOSE WHOSE ECONOMIC DEVELOPMENTAL STATUS IS VERY UNDERDEVELOPED (57)	0.67	43	LEAN TOWARD BEING THOSE WHOSE ECONOMIC DEVELOPMENTAL STATUS IS DEVELOPED, INTERMEDIATE, OR UNDERDEVELOPED (55)	0.58	7 32 14 23 X SQ= 2.83 P = 0.073 RV YES (NO)
44	ALWAYS ARE THOSE WHERE THE LITERACY RATE IS BELOW NINETY PERCENT (90)	1.00	44	TEND LESS TO BE THOSE WHERE THE LITERACY RATE IS BELOW NINETY PERCENT (90)	0.68	0 18 21 38 X SQ= 7.11 P = 0.002 RV NO (NO)
45	LEAN TOWARD BEING THOSE WHERE THE LITERACY RATE IS BELOW FIFTY PERCENT (54)	0.67	45	LEAN TOWARD BEING THOSE WHERE THE LITERACY RATE IS FIFTY PERCENT OR ABOVE (55)	0.57	7 32 14 24 X SQ= 2.58 P = 0.077 RV YES (NO)
49	TEND TO BE THOSE WHERE THE LITERACY RATE IS BETWEEN TEN AND FIFTY PERCENT, RATHER THAN BELOW TEN PERCENT (24)	0.77	49	TEND TO BE THOSE WHERE THE LITERACY RATE IS BELOW TEN PERCENT, RATHER THAN BETWEEN TEN AND FIFTY PERCENT (26)	0.68	10 7 3 15 X SQ= 4.97 P = 0.015 RV YES (YES)
50	TEND TO BE THOSE WHERE FREEDOM OF THE PRESS IS INTERMITTENT, INTERNALLY ABSENT, OR INTERNALLY AND EXTERNALLY ABSENT (56)	0.94	50	TEND TO BE THOSE WHERE FREEDOM OF THE PRESS IS COMPLETE (43)	0.87	1 40 15 6 X SQ= 31.01 P = 0. RV YES (YES)
51	TEND TO BE THOSE WHERE FREEDOM OF THE PRESS IS INTERNALLY ABSENT, OR INTERNALLY AND EXTERNALLY ABSENT (37)	0.50	51	TEND TO BE THOSE WHERE FREEDOM OF THE PRESS IS COMPLETE OR INTERMITTENT (60)	0.98	7 45 7 1 X SQ= 17.31 P = 0. RV YES (YES)

52	TEND LESS TO BE THOSE WHERE FREEDOM OF THE PRESS IS COMPLETE, INTERMITTENT, OR INTERNALLY ABSENT (82)	0.86	52	ALWAYS ARE THOSE WHERE FREEDOM OF THE PRESS IS COMPLETE, INTERMITTENT, OR INTERNALLY ABSENT (82)	1.00	12 47 2 0 X SQ= 3.17 P = 0.050 RV NO (YES)
53	ALWAYS ARE THOSE WHERE NEWSPAPER CIRCULATION IS LESS THAN THREE HUNDRED PER THOUSAND (101)	1.00	53	TEND LESS TO BE THOSE WHERE NEWSPAPER CIRCULATION IS LESS THAN THREE HUNDRED PER THOUSAND (101)	0.77	0 13 21 43 X SQ= 4.33 P = 0.015 RV NO (NO)
54	TEND MORE TO BE THOSE WHERE NEWSPAPER CIRCULATION IS LESS THAN ONE HUNDRED PER THOUSAND (76)	0.90	54	TEND LESS TO BE THOSE WHERE NEWSPAPER CIRCULATION IS LESS THAN ONE HUNDRED PER THOUSAND (76)	0.55	2 25 19 31 X SQ= 6.80 P = 0.004 RV NO (NO)
58	TEND LESS TO BE THOSE WHERE THE RELIGION IS OTHER THAN MUSLIM (97)	0.71	58	TEND MORE TO BE THOSE WHERE THE RELIGION IS OTHER THAN MUSLIM (97)	0.93	6 4 15 52 X SQ= 4.45 P = 0.021 RV NO (YES)
64	TEND LESS TO BE THOSE WHERE THE RELIGION IS CHRISTIAN, RATHER THAN MUSLIM (46)	0.54	64	TEND MORE TO BE THOSE WHERE THE RELIGION IS CHRISTIAN, RATHER THAN MUSLIM (46)	0.87	7 27 6 4 X SQ= 4.03 P = 0.043 RV NO (YES)
66	TEND TO BE THOSE THAT ARE RELIGIOUSLY HOMOGENEOUS (57)	0.75	66	TEND TO BE THOSE THAT ARE RELIGIOUSLY HETEROGENEOUS (49)	0.55	15 24 5 29 X SQ= 4.03 P = 0.035 RV YES (NO)
73	TEND TO BE THOSE WHOSE DATE OF INDEPENDENCE IS BEFORE 1945 (65)	0.75	73	TEND TO BE THOSE WHOSE DATE OF INDEPENDENCE IS AFTER 1945 (46)	0.53	15 26 5 29 X SQ= 3.50 P = 0.039 RV YES (NO)
74	ALWAYS ARE THOSE THAT ARE NOT HISTORICALLY WESTERN (87)	1.00	74	TEND LESS TO BE THOSE THAT ARE NOT HISTORICALLY WESTERN (87)	0.65	0 19 21 36 X SQ= 7.92 P = 0.001 RV NO (NO)
76	ALWAYS ARE THOSE THAT HAVE BEEN SIGNIFICANTLY OR PARTIALLY WESTERNIZED THROUGH A COLONIAL RELATIONSHIP, RATHER THAN BEING HISTORICALLY WESTERN (70)	1.00	76	TEND LESS TO BE THOSE THAT HAVE BEEN SIGNIFICANTLY OR PARTIALLY WESTERNIZED THROUGH A COLONIAL RELATIONSHIP, RATHER THAN BEING HISTORICALLY WESTERN (70)	0.65	0 19 17 35 X SQ= 6.47 P = 0.003 RV NO (NO)

	Left statement			Right statement		Statistics
79	TEND TO BE THOSE THAT WERE FORMERLY DEPENDENCIES OF SPAIN, RATHER THAN BRITAIN OR FRANCE (18)	0.50	79	TEND TO BE THOSE THAT WERE FORMERLY DEPENDENCIES OF BRITAIN OR FRANCE, RATHER THAN SPAIN (49)	0.81	8 30 8 7 X SQ= 3.90 P = 0.043 RV YES (YES)
81	ALWAYS ARE THOSE WHERE THE TYPE OF POLITICAL MODERNIZATION IS LATER EUROPEAN OR LATER EUROPEAN DERIVED, RATHER THAN EARLY EUROPEAN OR EARLY EUROPEAN DERIVED (40)	1.00	81	TEND LESS TO BE THOSE WHERE THE TYPE OF POLITICAL MODERNIZATION IS LATER EUROPEAN OR LATER EUROPEAN DERIVED, RATHER THAN EARLY EUROPEAN OR EARLY EUROPEAN DERIVED (40)	0.65	0 9 10 17 X SQ= 2.95 P = 0.039 RV NO (NO)
84	TEND TO BE THOSE WHERE THE TYPE OF POLITICAL MODERNIZATION IS DEVELOPED TUTELARY, RATHER THAN UNDEVELOPED TUTELARY (31)	0.89	84	TEND TO BE THOSE WHERE THE TYPE OF POLITICAL MODERNIZATION IS UNDEVELOPED TUTELARY, RATHER THAN DEVELOPED TUTELARY (24)	0.62	8 11 1 18 X SQ= 5.24 P = 0.019 RV YES (NO)
86	TEND MORE TO BE THOSE WHERE THE STAGE OF POLITICAL MODERNIZATION IS ADVANCED OR MID-TRANSITIONAL, RATHER THAN EARLY TRANSITIONAL (76)	0.86	86	TEND LESS TO BE THOSE WHERE THE STAGE OF POLITICAL MODERNIZATION IS ADVANCED OR MID-TRANSITIONAL, RATHER THAN EARLY TRANSITIONAL (76)	0.59	18 33 3 23 X SQ= 3.78 P = 0.032 RV NO (NO)
89	TEND TO BE THOSE WHOSE IDEOLOGICAL ORIENTATION IS OTHER THAN CONVENTIONAL (62)	0.85	89	TEND TO BE THOSE WHOSE IDEOLOGICAL ORIENTATION IS CONVENTIONAL (33)	0.57	2 27 11 20 X SQ= 5.63 P = 0.011 RV YES (NO)
92	LEAN LESS TOWARD BEING THOSE WHERE THE SYSTEM STYLE IS LIMITED MOBILIZATIONAL, OR NON-MOBILIZATIONAL (93)	0.90	92	ALWAYS ARE THOSE WHERE THE SYSTEM STYLE IS LIMITED MOBILIZATIONAL, OR NON-MOBILIZATIONAL (93)	1.00	2 0 18 56 X SQ= 2.51 P = 0.067 RV NO (YES)
94	TEND TO BE THOSE WHERE THE STATUS OF THE REGIME IS AUTHORITARIAN OR TOTALITARIAN (41)	0.71	94	TEND TO BE THOSE WHERE THE STATUS OF THE REGIME IS CONSTITUTIONAL (51)	0.98	4 43 10 1 X SQ= 28.70 P = 0. RV YES (YES)
96	TEND TO BE THOSE WHERE THE STATUS OF THE REGIME IS AUTHORITARIAN (23)	0.69	96	TEND TO BE THOSE WHERE THE STATUS OF THE REGIME IS CONSTITUTIONAL OR TOTALITARIAN (67)	0.98	4 43 9 1 X SQ= 26.64 P = 0. RV YES (YES)
98	ALWAYS ARE THOSE WHERE GOVERNMENTAL STABILITY IS GENERALLY OR MODERATELY PRESENT AND DATES FROM THE POST-WAR PERIOD, OR IS ABSENT (93)	1.00	98	TEND LESS TO BE THOSE WHERE GOVERNMENTAL STABILITY IS GENERALLY OR MODERATELY PRESENT AND DATES FROM THE POST-WAR PERIOD, OR IS ABSENT (93)	0.73	0 15 21 41 X SQ= 5.38 P = 0.008 RV NO (NO)

99	TEND TO BE THOSE WHERE GOVERNMENTAL STABILITY IS MODERATELY PRESENT AND DATES FROM THE POST-WAR PERIOD, OR IS ABSENT (36)	0.94	99	TEND TO BE THOSE WHERE GOVERNMENTAL STABILITY IS GENERALLY PRESENT AND DATES FROM AT LEAST THE INTER-WAR PERIOD, OR FROM THE POST-WAR PERIOD (50)	0.82	1 28 17 6 X SQ= 25.11 P = 0. RV YES (YES)
100	TEND TO BE THOSE WHERE GOVERNMENTAL STABILITY IS ABSENT (22)	0.72	100	TEND TO BE THOSE WHERE GOVERNMENTAL STABILITY IS GENERALLY PRESENT AND DATES FROM AT LEAST THE INTER-WAR PERIOD, OR IS GENERALLY OR MODERATELY PRESENT, AND DATES FROM THE POST-WAR PERIOD (64)	0.97	5 33 13 1 X SQ= 25.30 P = 0. RV YES (YES)
101	TEND TO BE THOSE WHERE THE REPRESENTATIVE CHARACTER OF THE REGIME IS LIMITED POLYARCHIC, PSEUDO-POLYARCHIC, OR NON-POLYARCHIC (57)	0.85	101	TEND TO BE THOSE WHERE THE REPRESENTATIVE CHARACTER OF THE REGIME IS POLYARCHIC (41)	0.74	2 37 11 13 X SQ= 12.65 P = 0. RV YES (NO)
102	TEND TO BE THOSE WHERE THE REPRESENTATIVE CHARACTER OF THE REGIME IS PSEUDO-POLYARCHIC OR NON-POLYARCHIC (49)	0.64	102	TEND TO BE THOSE WHERE THE REPRESENTATIVE CHARACTER OF THE REGIME IS POLYARCHIC OR LIMITED POLYARCHIC (59)	0.82	5 46 9 10 X SQ= 9.97 P = 0.001 RV YES (NO)
107	TEND TO BE THOSE WHERE AUTONOMOUS GROUPS ARE PARTIALLY TOLERATED IN POLITICS, ARE TOLERATED ONLY OUTSIDE POLITICS, OR ARE NOT TOLERATED AT ALL (65)	0.72	107	TEND TO BE THOSE WHERE AUTONOMOUS GROUPS ARE FULLY TOLERATED IN POLITICS (46)	0.64	5 36 13 20 X SQ= 5.94 P = 0.013 RV YES (NO)
111	TEND TO BE THOSE WHERE POLITICAL ENCULTURATION IS LOW (42)	0.68	111	TEND TO BE THOSE WHERE POLITICAL ENCULTURATION IS HIGH OR MEDIUM (53)	0.72	6 36 13 14 X SQ= 7.82 P = 0.005 RV YES (NO)
115	LEAN MORE TOWARD BEING THOSE WHERE INTEREST ARTICULATION BY ASSOCIATIONAL GROUPS IS MODERATE, LIMITED, OR NEGLIGIBLE (91)	0.90	115	LEAN LESS TOWARD BEING THOSE WHERE INTEREST ARTICULATION BY ASSOCIATIONAL GROUPS IS MODERATE, LIMITED, OR NEGLIGIBLE (91)	0.70	2 17 18 39 X SQ= 2.26 P = 0.081 RV NO (NO)
116	TEND MORE TO BE THOSE WHERE INTEREST ARTICULATION BY ASSOCIATIONAL GROUPS IS LIMITED OR NEGLIGIBLE (79)	0.85	116	TEND LESS TO BE THOSE WHERE INTEREST ARTICULATION BY ASSOCIATIONAL GROUPS IS LIMITED OR NEGLIGIBLE (79)	0.55	3 25 17 31 X SQ= 4.36 P = 0.029 RV NO (NO)
118	TEND TO BE THOSE WHERE INTEREST ARTICULATION BY INSTITUTIONAL GROUPS IS VERY SIGNIFICANT (40)	0.58	118	TEND TO BE THOSE WHERE INTEREST ARTICULATION BY INSTITUTIONAL GROUPS IS SIGNIFICANT, MODERATE, OR LIMITED (60)	0.95	11 2 8 42 X SQ= 19.92 P = 0. RV YES (YES)

119	ALWAYS ARE THOSE WHERE INTEREST ARTICULATION BY INSTITUTIONAL GROUPS IS VERY SIGNIFICANT OR SIGNIFICANT (74)	1.00	119	TEND TO BE THOSE WHERE INTEREST ARTICULATION BY INSTITUTIONAL GROUPS IS MODERATE OR LIMITED (26)	0.55	19 20 0 24 X SQ= 14.51 P = 0. RV YES (NO)
122	LEAN MORE TOWARD BEING THOSE WHERE INTEREST ARTICULATION BY NON-ASSOCIATIONAL GROUPS IS SIGNIFICANT OR MODERATE (83)	0.86	122	LEAN LESS TOWARD BEING THOSE WHERE INTEREST ARTICULATION BY NON-ASSOCIATIONAL GROUPS IS SIGNIFICANT OR MODERATE (83)	0.62	18 35 3 21 X SQ= 2.83 P = 0.058 RV NO (NO)
123	ALWAYS ARE THOSE WHERE INTEREST ARTICULATION BY NON-ASSOCIATIONAL GROUPS IS SIGNIFICANT, MODERATE, OR LIMITED (107)	1.00	123	LEAN LESS TOWARD BEING THOSE WHERE INTEREST ARTICULATION BY NON-ASSOCIATIONAL GROUPS IS SIGNIFICANT, MODERATE, OR LIMITED (107)	0.86	21 48 0 8 X SQ= 1.99 P = 0.099 RV NO (NO)
125	TEND TO BE THOSE WHERE INTEREST ARTICULATION BY ANOMIC GROUPS IS FREQUENT OR OCCASIONAL (64)	0.89	125	TEND TO BE THOSE WHERE INTEREST ARTICULATION BY ANOMIC GROUPS IS INFREQUENT OR VERY INFREQUENT (35)	0.56	17 23 2 29 X SQ= 9.81 P = 0.001 RV YES (NO)
126	ALWAYS ARE THOSE WHERE INTEREST ARTICULATION BY ANOMIC GROUPS IS FREQUENT, OCCASIONAL, OR INFREQUENT (83)	1.00	126	TEND LESS TO BE THOSE WHERE INTEREST ARTICULATION BY ANOMIC GROUPS IS FREQUENT, OCCASIONAL, OR INFREQUENT (83)	0.69	19 36 0 16 X SQ= 5.89 P = 0.004 RV NO (NO)
130	ALWAYS ARE THOSE WHERE INTEREST AGGREGATION BY POLITICAL PARTIES IS MODERATE, LIMITED, OR NEGLIGIBLE (71)	1.00	130	TEND LESS TO BE THOSE WHERE INTEREST AGGREGATION BY POLITICAL PARTIES IS MODERATE, LIMITED, OR NEGLIGIBLE (71)	0.72	0 11 14 29 X SQ= 3.29 P = 0.028 RV NO (NO)
131	TEND TO BE THOSE WHERE INTEREST AGGREGATION BY POLITICAL PARTIES IS LIMITED OR NEGLIGIBLE (35)	0.92	131	TEND TO BE THOSE WHERE INTEREST AGGREGATION BY POLITICAL PARTIES IS SIGNIFICANT OR MODERATE (30)	0.62	1 25 11 15 X SQ= 8.77 P = 0.002 RV YES (NO)
132	TEND TO BE THOSE WHERE INTEREST AGGREGATION BY POLITICAL PARTIES IS NEGLIGIBLE (9)	0.56	132	TEND TO BE THOSE WHERE INTEREST AGGREGATION BY POLITICAL PARTIES IS SIGNIFICANT, MODERATE, OR LIMITED (67)	0.98	4 51 5 1 X SQ= 19.20 P = 0. RV YES (YES)
134	TEND TO BE THOSE WHERE INTEREST AGGREGATION BY THE EXECUTIVE IS LIMITED OR NEGLIGIBLE (46)	0.87	134	TEND TO BE THOSE WHERE INTEREST AGGREGATION BY THE EXECUTIVE IS SIGNIFICANT OR MODERATE (57)	0.87	2 47 14 7 X SQ= 29.20 P = 0. RV YES (YES)

136	ALWAYS ARE THOSE WHERE INTEREST AGGREGATION BY THE LEGISLATURE IS MODERATE, LIMITED, OR NEGLIGIBLE (85)	1.00	136	LEAN LESS TOWARD BEING THOSE WHERE INTEREST AGGREGATION BY THE LEGISLATURE IS MODERATE, LIMITED, OR NEGLIGIBLE (85)	0.76	0 12 13 39 X SQ= 2.38 P = 0.059 RV NO (NO)
137	TEND TO BE THOSE WHERE INTEREST AGGREGATION BY THE LEGISLATURE IS LIMITED OR NEGLIGIBLE (68)	0.92	137	TEND TO BE THOSE WHERE INTEREST AGGREGATION BY THE LEGISLATURE IS SIGNIFICANT OR MODERATE (29)	0.53	1 27 12 24 X SQ= 6.88 P = 0.004 RV YES (NO)
148	LEAN MORE TOWARD BEING THOSE WHERE THE PARTY SYSTEM IS QUALITATIVELY OTHER THAN AFRICAN TRANSITIONAL (96)	0.95	148	LEAN LESS TOWARD BEING THOSE WHERE THE PARTY SYSTEM IS QUALITATIVELY OTHER THAN AFRICAN TRANSITIONAL (96)	0.76	1 13 19 42 X SQ= 2.24 P = 0.095 RV NO (NO)
153	ALWAYS ARE THOSE WHERE THE PARTY SYSTEM IS MODERATELY STABLE OR UNSTABLE (38)	1.00	153	TEND TO BE THOSE WHERE THE PARTY SYSTEM IS STABLE (42)	0.71	0 24 15 10 X SQ= 18.02 P = 0. RV YES (YES)
154	TEND TO BE THOSE WHERE THE PARTY SYSTEM IS UNSTABLE (25)	0.80	154	TEND TO BE THOSE WHERE THE PARTY SYSTEM IS STABLE OR MODERATELY STABLE (55)	0.85	3 29 12 5 X SQ= 16.81 P = 0. RV YES (YES)
157	TEND TO BE THOSE WHERE PERSONALISMO IS PRONOUNCED (14)	0.62	157	TEND TO BE THOSE WHERE PERSONALISMO IS MODERATE OR NEGLIGIBLE (84)	0.96	8 2 5 51 X SQ= 22.79 P = 0. RV YES (YES)
158	TEND TO BE THOSE WHERE PERSONALISMO IS PRONOUNCED OR MODERATE (40)	0.85	158	TEND TO BE THOSE WHERE PERSONALISMO IS NEGLIGIBLE (56)	0.63	11 19 2 32 X SQ= 7.53 P = 0.004 RV YES (NO)
160	TEND TO BE THOSE WHERE THE POLITICAL LEADERSHIP IS ELITIST (30)	0.50	160	ALWAYS ARE THOSE WHERE THE POLITICAL LEADERSHIP IS MODERATELY ELITIST OR NON-ELITIST (67)	1.00	6 0 6 52 X SQ= 23.11 P = 0. RV YES (YES)
161	TEND TO BE THOSE WHERE THE POLITICAL LEADERSHIP IS ELITIST OR MODERATELY ELITIST (47)	0.75	161	TEND TO BE THOSE WHERE THE POLITICAL LEADERSHIP IS NON-ELITIST (50)	0.81	9 10 3 42 X SQ= 11.98 P = 0. RV YES (NO)

164	TILT MORE TOWARD BEING THOSE WHERE THE REGIME'S LEADERSHIP CHARISMA IS NEGLIGIBLE (65)	0.87	164	TILT LESS TOWARD BEING THOSE WHERE THE REGIME'S LEADERSHIP CHARISMA IS NEGLIGIBLE (65)	0.67	2 17 14 34 X SQ= 1.68 P = 0.126 RV NO (NO)
168	TEND TO BE THOSE WHERE THE HORIZONTAL POWER DISTRIBUTION IS LIMITED OR NEGLIGIBLE (72)	0.86	168	TEND TO BE THOSE WHERE THE HORIZONTAL POWER DISTRIBUTION IS SIGNIFICANT (34)	0.56	2 31 12 24 X SQ= 6.32 P = 0.006 RV YES (NO)
169	TEND TO BE THOSE WHERE THE HORIZONTAL POWER DISTRIBUTION IS NEGLIGIBLE (48)	0.50	169	TEND TO BE THOSE WHERE THE HORIZONTAL POWER DISTRIBUTION IS SIGNIFICANT OR LIMITED (58)	0.82	7 45 7 10 X SQ= 4.49 P = 0.032 RV YES (NO)
170	TEND TO BE THOSE WHERE THE LEGISLATIVE-EXECUTIVE STRUCTURE IS PRESIDENTIAL (39)	0.83	170	TEND TO BE THOSE WHERE THE LEGISLATIVE-EXECUTIVE STRUCTURE IS OTHER THAN PRESIDENTIAL (63)	0.62	10 21 2 34 X SQ= 6.36 P = 0.009 RV YES (NO)
172	ALWAYS ARE THOSE WHERE THE LEGISLATIVE-EXECUTIVE STRUCTURE IS OTHER THAN PARLIAMENTARY-ROYALIST (88)	1.00	172	TEND LESS TO BE THOSE WHERE THE LEGISLATIVE-EXECUTIVE STRUCTURE IS OTHER THAN PARLIAMENTARY-ROYALIST (88)	0.64	0 20 19 36 X SQ= 7.52 P = 0.002 RV NO (NO)
174	ALWAYS ARE THOSE WHERE THE LEGISLATURE IS PARTIALLY EFFECTIVE, LARGELY INEFFECTIVE, OR WHOLLY INEFFECTIVE (72)	1.00	174	TEND LESS TO BE THOSE WHERE THE LEGISLATURE IS PARTIALLY EFFECTIVE, LARGELY INEFFECTIVE, OR WHOLLY INEFFECTIVE (72)	0.51	0 27 11 28 X SQ= 7.22 P = 0.002 RV NO (NO)
175	LEAN TOWARD BEING THOSE WHERE THE LEGISLATURE IS LARGELY INEFFECTIVE OR WHOLLY INEFFECTIVE (49)	0.55	175	LEAN TOWARD BEING THOSE WHERE THE LEGISLATURE IS FULLY EFFECTIVE OR PARTIALLY EFFECTIVE (51)	0.75	5 41 6 14 X SQ= 2.42 P = 0.076 RV YES (NO)
176	TEND LESS TO BE THOSE WHERE THE LEGISLATURE IS FULLY EFFECTIVE, PARTIALLY EFFECTIVE, OR LARGELY INEFFECTIVE (72)	0.64	176	ALWAYS ARE THOSE WHERE THE LEGISLATURE IS FULLY EFFECTIVE, PARTIALLY EFFECTIVE, OR LARGELY INEFFECTIVE (72)	1.00	7 55 4 0 X SQ= 15.38 P = 0. RV NO (YES)
179	TEND TO BE THOSE WHERE THE EXECUTIVE IS DOMINANT (52)	0.70	179	TEND TO BE THOSE WHERE THE EXECUTIVE IS STRONG (39)	0.72	7 13 3 33 X SQ= 4.55 P = 0.025 RV YES (NO)

```
181  ALWAYS ARE THOSE                          1.00     181  TEND TO BE THOSE                          0.50        0   19
     WHERE THE BUREAUCRACY                                   WHERE THE BUREAUCRACY                                 14   19
     IS SEMI-MODERN, RATHER THAN                             IS MODERN, RATHER THAN                            X SQ=   8.98
     MODERN  (55)                                            SEMI-MODERN  (21)                                 P =    0.001
                                                                                                               RV YES (NO )

182  TEND MORE TO BE THOSE                     0.82     182  TEND LESS TO BE THOSE                     0.51       14   19
     WHERE THE BUREAUCRACY                                   WHERE THE BUREAUCRACY                                  3   18
     IS SEMI-MODERN, RATHER THAN                             IS SEMI-MODERN, RATHER THAN                       X SQ=   3.50
     POST-COLONIAL TRANSITIONAL  (55)                        POST-COLONIAL TRANSITIONAL  (55)                  P =    0.038
                                                                                                               RV NO  (NO )

187  ALWAYS ARE THOSE                          1.00     187  TEND TO BE THOSE                          0.76       18   11
     WHERE THE ROLE OF THE POLICE                            WHERE THE ROLE OF THE POLICE                           0   34
     IS POLITICALLY SIGNIFICANT  (66)                        IS NOT POLITICALLY SIGNIFICANT  (35)              X SQ=  26.58
                                                                                                               P =    0.
                                                                                                               RV YES (YES)

188  TEND TO BE THOSE                          0.57     188  TEND TO BE THOSE                          0.73       12   15
     WHERE THE CHARACTER OF THE LEGAL SYSTEM                 WHERE THE CHARACTER OF THE LEGAL SYSTEM                9   41
     IS CIVIL LAW  (32)                                      IS OTHER THAN CIVIL LAW  (81)                     X SQ=   4.92
                                                                                                               P =    0.017
                                                                                                               RV YES (NO )

190  ALWAYS ARE THOSE                          1.00     190  TEND LESS TO BE THOSE                     0.68       12   15
     WHERE THE CHARACTER OF THE LEGAL SYSTEM                 WHERE THE CHARACTER OF THE LEGAL SYSTEM                0    7
     IS CIVIL LAW, RATHER THAN                               IS CIVIL LAW, RATHER THAN                         X SQ=   3.06
     COMMON LAW  (32)                                        COMMON LAW  (32)                                  P =    0.036
                                                                                                               RV NO  (NO )

193  TEND MORE TO BE THOSE                     0.95     193  TEND LESS TO BE THOSE                     0.70        1   17
     WHERE THE CHARACTER OF THE LEGAL SYSTEM                 WHERE THE CHARACTER OF THE LEGAL SYSTEM               19   39
     IS OTHER THAN PARTLY INDIGENOUS  (86)                   IS OTHER THAN PARTLY INDIGENOUS  (86)             X SQ=   3.93
                                                                                                               P =    0.030
                                                                                                               RV NO  (NO )
```

185 POLITIES
WHERE PARTICIPATION BY THE MILITARY
IN POLITICS IS
INTERVENTIVE, RATHER THAN
SUPPORTIVE (21)

185 POLITIES
WHERE PARTICIPATION BY THE MILITARY
IN POLITICS IS
SUPPORTIVE, RATHER THAN
INTERVENTIVE (31)

PREDICATE ONLY

21 IN LEFT COLUMN

ARGENTINA	BRAZIL	BURMA	EL SALVADOR	GUATEMALA	HAITI	HONDURAS	IRAQ	KOREA REP	NICARAGUA
PAKISTAN	PANAMA	PARAGUAY	PERU	SUDAN	SYRIA	THAILAND	TOGO	TURKEY	UAR
YEMEN									

31 IN RIGHT COLUMN

AFGHANISTAN	ALBANIA	ALGERIA	BULGARIA	CAMBODIA	CHINA, PR	CZECHOS'KIA	ECUADOR	ETHIOPIA	GERMANY, E
GHANA	GUINEA	HUNGARY	INDONESIA	IRAN	JORDAN	KOREA, N	LEBANON	LIBERIA	MONGOLIA
POLAND	PORTUGAL	RUMANIA	SA'U ARABIA	SENEGAL	SO AFRICA	SPAIN	USSR	VIETNAM, N	VIETNAM REP
YUGOSLAVIA									

7 EXCLUDED BECAUSE AMBIGUOUS

CONGO(LEO)	CUBA	FRANCE	LAOS	NEPAL	SOMALIA	VENEZUELA

56 EXCLUDED BECAUSE IRRELEVANT

AUSTRALIA	AUSTRIA	BELGIUM	BOLIVIA	BURUNDI	CAMEROUN	CANADA	CEN AFR REP	CEYLON	CHAD
CHILE	COLOMBIA	CONGO(BRA)	COSTA RICA	CYPRUS	DAHOMEY	DENMARK	DOMIN REP	FINLAND	GABON
GERMAN FR	GREECE	ICELAND	INDIA	IRELAND	ISRAEL	ITALY	IVORY COAST	JAMAICA	JAPAN
LIBYA	LUXEMBOURG	MALAGASY R	MALAYA	MALI	MAURITANIA	MEXICO	MOROCCO	NETHERLANDS	NEW ZEALAND
NIGER	NIGERIA	NORWAY	PHILIPPINES	RWANDA	SIERRE LEO	SWEDEN	SWITZERLAND	TANGANYIKA	TRINIDAD
TUNISIA	UGANDA	UK	US	UPPER VOLTA	URUGUAY				

186 POLITIES
WHERE PARTICIPATION BY THE MILITARY
IN POLITICS IS
SUPPORTIVE, RATHER THAN
NEUTRAL (31)

186 POLITIES
WHERE PARTICIPATION BY THE MILITARY
IN POLITICS IS
NEUTRAL, RATHER THAN
SUPPORTIVE (56)

PREDICATE ONLY

31 IN LEFT COLUMN

AFGHANISTAN	ALBANIA	ALGERIA	BULGARIA	CAMBODIA	CHINA, PR	CZECHOS'KIA	ECUADOR	ETHIOPIA	GERMANY, E
GHANA	GUINEA	HUNGARY	INDONESIA	IRAN	JORDAN	KOREA, N	LEBANON	LIBERIA	MONGOLIA
POLAND	PORTUGAL	RUMANIA	SA'U ARABIA	SENEGAL	SO AFRICA	SPAIN	USSR	VIETNAM, N	VIETNAM REP
YUGOSLAVIA									

56 IN RIGHT COLUMN

AUSTRALIA	AUSTRIA	BELGIUM	BOLIVIA	BURUNDI	CAMEROUN	CANADA	CEN AFR REP	CEYLON	CHAD
CHILE	COLOMBIA	CONGO(BRA)	COSTA RICA	CYPRUS	DAHOMEY	DENMARK	DOMIN REP	FINLAND	GABON
GERMAN FR	GREECE	ICELAND	INDIA	IRELAND	ISRAEL	ITALY	IVORY COAST	JAMAICA	JAPAN
LIBYA	LUXEMBOURG	MALAGASY R	MALAYA	MALI	MAURITANIA	MEXICO	MOROCCO	NETHERLANDS	NEW ZEALAND
NIGER	NIGERIA	NORWAY	PHILIPPINES	RWANDA	SIERRE LEO	SWEDEN	SWITZERLAND	TANGANYIKA	TRINIDAD
TUNISIA	UGANDA	UK	US	UPPER VOLTA	URUGUAY				

7 EXCLUDED BECAUSE AMBIGUOUS

CONGO(LEO)	CUBA	FRANCE	LAOS	NEPAL	SOMALIA	VENEZUELA

21 EXCLUDED BECAUSE IRRELEVANT

ARGENTINA	BRAZIL	BURMA	EL SALVADOR	GUATEMALA	HAITI	HONDURAS	IRAQ	KOREA REP	NICARAGUA
PAKISTAN	PANAMA	PARAGUAY	PERU	SUDAN	SYRIA	THAILAND	TOGO	TURKEY	UAR
YEMEN									

187 POLITIES
WHERE THE ROLE OF THE POLICE
IS POLITICALLY SIGNIFICANT (66)

187 POLITIES
WHERE THE ROLE OF THE POLICE
IS NOT POLITICALLY SIGNIFICANT (35)

BOTH SUBJECT AND PREDICATE

66 IN LEFT COLUMN

AFGHANISTAN	ALBANIA	ALGERIA	ARGENTINA	BOLIVIA	BULGARIA	BURMA	CAMBODIA	CAMEROUN	CEN AFR REP
CHINA, PR	COLOMBIA	CONGO(BRA)	CONGO(LEO)	CUBA	CZECHOS'KIA	DAHOMEY	ECUADOR	EL SALVADOR	ETHIOPIA
GABON	GERMANY, E	GHANA	GUATEMALA	GUINEA	HAITI	HONDURAS	HUNGARY	INDONESIA	IRAN
IRAQ	IVORY COAST	JORDAN	KOREA, N	KOREA REP	LAOS	LEBANON	LIBERIA	MONGOLIA	MOROCCO
NEPAL	NICARAGUA	NIGERIA	PAKISTAN	PARAGUAY	POLAND	PORTUGAL	RUMANIA	SA'U ARABIA	SENEGAL
SOMALIA	SO AFRICA	SPAIN	SUDAN	SYRIA	THAILAND	TOGO	TUNISIA	TURKEY	USSR
UAR	VENEZUELA	VIETNAM, N	VIETNAM REP	YEMEN	YUGOSLAVIA				

35 IN RIGHT COLUMN

AUSTRALIA	AUSTRIA	BELGIUM	CANADA	CEYLON	DENMARK	FINLAND	FRANCE	GERMAN FR	GREECE
ICELAND	INDIA	IRELAND	ISRAEL	ITALY	JAMAICA	JAPAN	LUXEMBOURG	MALAGASY R	MALAYA
MAURITANIA	MEXICO	NETHERLANDS	NEW ZEALAND	NORWAY	PHILIPPINES	SIERRE LEO	SWEDEN	SWITZERLAND	TANGANYIKA
TRINIDAD	UGANDA	UK	US	URUGUAY					

12 EXCLUDED BECAUSE UNASCERTAINED

BRAZIL	BURUNDI	CHAD	CHILE	COSTA RICA	CYPRUS	LIBYA	MALI	NIGER	PANAMA
RWANDA	UPPER VOLTA								

2 EXCLUDED BECAUSE UNASCERTAINABLE

DOMIN REP PERU

	Left column			Right column		
2	TEND TO BE THOSE LOCATED ELSEWHERE THAN IN WEST EUROPE, SCANDINAVIA, NORTH AMERICA, OR AUSTRALASIA (93)	0.97	2	TEND TO BE THOSE LOCATED IN WEST EUROPE, SCANDINAVIA, NORTH AMERICA, OR AUSTRALASIA (22)	0.57	2 20 64 15 X SQ= 36.20 P = 0. RV YES (YES)
3	TEND MORE TO BE THOSE LOCATED ELSEWHERE THAN IN WEST EUROPE (102)	0.97	3	TEND LESS TO BE THOSE LOCATED ELSEWHERE THAN IN WEST EUROPE (102)	0.69	2 11 64 24 X SQ= 14.01 P = 0. RV NO (YES)
5	TEND TO BE THOSE LOCATED IN EAST EUROPE, RATHER THAN IN SCANDINAVIA OR WEST EUROPE (9)	0.82	5	ALWAYS ARE THOSE LOCATED IN SCANDINAVIA OR WEST EUROPE, RATHER THAN IN EAST EUROPE (18)	1.00	2 16 9 0 X SQ= 16.13 P = 0. RV YES (YES)

10	LEAN LESS TOWARD BEING THOSE LOCATED ELSEWHERE THAN IN NORTH AFRICA, OR CENTRAL AND SOUTH AFRICA (82)	0.68	10	LEAN MORE TOWARD BEING THOSE LOCATED ELSEWHERE THAN IN NORTH AFRICA, OR CENTRAL AND SOUTH AFRICA (82)	0.86	21 5 45 30 X SQ= 2.82 P = 0.061 RV NO (NO)
13	TEND LESS TO BE THOSE LOCATED ELSEWHERE THAN IN NORTH AFRICA OR THE MIDDLE EAST (99)	0.80	13	TEND MORE TO BE THOSE LOCATED ELSEWHERE THAN IN NORTH AFRICA OR THE MIDDLE EAST (99)	0.97	13 1 53 34 X SQ= 4.11 P = 0.031 RV NO (NO)
14	LEAN LESS TOWARD BEING THOSE LOCATED ELSEWHERE THAN IN THE MIDDLE EAST (104)	0.86	14	LEAN MORE TOWARD BEING THOSE LOCATED ELSEWHERE THAN IN THE MIDDLE EAST (104)	0.97	9 1 57 34 X SQ= 1.89 P = 0.097 RV NO (NO)
26	LEAN TOWARD BEING THOSE WHOSE POPULATION DENSITY IS LOW (67)	0.64	26	LEAN TOWARD BEING THOSE WHOSE POPULATION DENSITY IS VERY HIGH, HIGH, OR MEDIUM (48)	0.57	24 20 42 15 X SQ= 3.22 P = 0.058 RV YES (NO)
28	TILT TOWARD BEING THOSE WHOSE POPULATION GROWTH RATE IS HIGH (62)	0.59	28	TILT TOWARD BEING THOSE WHOSE POPULATION GROWTH RATE IS LOW (48)	0.57	36 15 25 20 X SQ= 1.73 P = 0.142 RV YES (NO)
29	TEND TO BE THOSE WHERE THE DEGREE OF URBANIZATION IS LOW (49)	0.58	29	TEND TO BE THOSE WHERE THE DEGREE OF URBANIZATION IS HIGH (56)	0.78	25 25 35 7 X SQ= 9.76 P = 0.001 RV YES (YES)
30	TEND TO BE THOSE WHOSE AGRICULTURAL POPULATION IS HIGH (56)	0.67	30	TEND TO BE THOSE WHOSE AGRICULTURAL POPULATION IS MEDIUM, LOW, OR VERY LOW (57)	0.83	43 6 21 29 X SQ= 20.71 P = 0. RV YES (YES)
31	TEND TO BE THOSE WHOSE AGRICULTURAL POPULATION IS HIGH OR MEDIUM (90)	0.95	31	TEND TO BE THOSE WHOSE AGRICULTURAL POPULATION IS LOW OR VERY LOW (24)	0.57	62 15 3 20 X SQ= 32.54 P = 0. RV YES (YES)
34	TEND TO BE THOSE WHOSE GROSS NATIONAL PRODUCT IS VERY LOW (53)	0.50	34	TEND TO BE THOSE WHOSE GROSS NATIONAL PRODUCT IS VERY HIGH, HIGH, MEDIUM, OR LOW (62)	0.74	33 26 33 9 X SQ= 4.60 P = 0.021 RV YES (NO)

35	TEND TO BE THOSE WHOSE PER CAPITA GROSS NATIONAL PRODUCT IS MEDIUM, LOW, OR VERY LOW (91)	0.94	35	TEND TO BE THOSE WHOSE PER CAPITA GROSS NATIONAL PRODUCT IS VERY HIGH OR HIGH (24)	0.57	4 20 62 15 X SQ= 30.18 P = 0. RV YES (YES)
36	TEND TO BE THOSE WHOSE PER CAPITA GROSS NATIONAL PRODUCT IS LOW OR VERY LOW (73)	0.79	36	TEND TO BE THOSE WHOSE PER CAPITA GROSS NATIONAL PRODUCT IS VERY HIGH, HIGH, OR MEDIUM (42)	0.74	14 26 52 9 X SQ= 24.76 P = 0. RV YES (YES)
37	TEND TO BE THOSE WHOSE PER CAPITA GROSS NATIONAL PRODUCT IS VERY LOW (51)	0.56	37	TEND TO BE THOSE WHOSE PER CAPITA GROSS NATIONAL PRODUCT IS VERY HIGH, HIGH, MEDIUM, OR LOW (64)	0.80	29 28 37 7 X SQ= 10.67 P = 0.001 RV YES (NO)
39	TEND TO BE THOSE WHOSE INTERNATIONAL FINANCIAL STATUS IS LOW OR VERY LOW (76)	0.75	39	TEND TO BE THOSE WHOSE INTERNATIONAL FINANCIAL STATUS IS VERY HIGH, HIGH, OR MEDIUM (38)	0.57	16 20 49 15 X SQ= 9.08 P = 0.002 RV YES (YES)
42	TEND TO BE THOSE WHOSE ECONOMIC DEVELOPMENTAL STATUS IS UNDERDEVELOPED OR VERY UNDERDEVELOPED (76)	0.83	42	TEND TO BE THOSE WHOSE ECONOMIC DEVELOPMENTAL STATUS IS DEVELOPED OR INTERMEDIATE (36)	0.68	11 23 53 11 X SQ= 22.78 P = 0. RV YES (YES)
43	TEND TO BE THOSE WHOSE ECONOMIC DEVELOPMENTAL STATUS IS VERY UNDERDEVELOPED (57)	0.67	43	TEND TO BE THOSE WHOSE ECONOMIC DEVELOPMENTAL STATUS IS DEVELOPED, INTERMEDIATE, OR UNDERDEVELOPED (55)	0.83	21 29 43 6 X SQ= 20.71 P = 0. RV YES (YES)
44	TEND TO BE THOSE WHERE THE LITERACY RATE IS BELOW NINETY PERCENT (90)	0.91	44	TEND TO BE THOSE WHERE THE LITERACY RATE IS NINETY PERCENT OR ABOVE (25)	0.54	6 19 60 16 X SQ= 22.71 P = 0. RV YES (YES)
45	TEND TO BE THOSE WHERE THE LITERACY RATE IS BELOW FIFTY PERCENT (54)	0.65	45	TEND TO BE THOSE WHERE THE LITERACY RATE IS FIFTY PERCENT OR ABOVE (55)	0.80	21 28 39 7 X SQ= 16.17 P = 0. RV YES (YES)
47	TEND TO BE THOSE WHERE THE LITERACY RATE IS BETWEEN FIFTY AND NINETY PERCENT, RATHER THAN NINETY PERCENT OR ABOVE (30)	0.71	47	TEND TO BE THOSE WHERE THE LITERACY RATE IS NINETY PERCENT OR ABOVE, RATHER THAN BETWEEN FIFTY AND NINETY PERCENT (25)	0.68	6 19 15 9 X SQ= 5.92 P = 0.010 RV YES (YES)

48	LEAN TOWARD BEING THOSE WHERE THE LITERACY RATE IS BETWEEN TEN AND FIFTY PERCENT, RATHER THAN BETWEEN FIFTY AND NINETY PERCENT (24)	0.56	48	LEAN TOWARD BEING THOSE WHERE THE LITERACY RATE IS BETWEEN FIFTY AND NINETY PERCENT, RATHER THAN BETWEEN TEN AND FIFTY PERCENT (30)	0.75	15 9 19 3 X SQ= 2.27 P = 0.096 RV YES (NO)
50	TEND TO BE THOSE WHERE FREEDOM OF THE PRESS IS INTERMITTENT, INTERNALLY ABSENT, OR INTERNALLY AND EXTERNALLY ABSENT (56)	0.93	50	TEND TO BE THOSE WHERE FREEDOM OF THE PRESS IS COMPLETE (43)	0.94	4 31 51 2 X SQ= 61.11 P = 0. RV YES (YES)
51	TEND TO BE THOSE WHERE FREEDOM OF THE PRESS IS INTERNALLY ABSENT, OR INTERNALLY AND EXTERNALLY ABSENT (37)	0.70	51	ALWAYS ARE THOSE WHERE FREEDOM OF THE PRESS IS COMPLETE OR INTERMITTENT (60)	1.00	16 33 37 0 X SQ= 37.64 P = 0. RV YES (YES)
54	TEND TO BE THOSE WHERE NEWSPAPER CIRCULATION IS LESS THAN ONE HUNDRED PER THOUSAND (76)	0.84	54	TEND TO BE THOSE WHERE NEWSPAPER CIRCULATION IS ONE HUNDRED OR MORE PER THOUSAND (37)	0.66	10 23 54 12 X SQ= 23.34 P = 0. RV YES (YES)
56	TEND LESS TO BE THOSE WHERE THE RELIGION IS PREDOMINANTLY LITERATE (79)	0.67	56	TEND MORE TO BE THOSE WHERE THE RELIGION IS PREDOMINANTLY LITERATE (79)	0.89	41 31 20 4 X SQ= 4.33 P = 0.027 RV NO (NO)
58	TEND LESS TO BE THOSE WHERE THE RELIGION IS OTHER THAN MUSLIM (97)	0.76	58	TEND MORE TO BE THOSE WHERE THE RELIGION IS OTHER THAN MUSLIM (97)	0.97	16 1 50 34 X SQ= 6.02 P = 0.005 RV NO (NO)
62	ALWAYS ARE THOSE WHERE THE RELIGION IS CATHOLIC, RATHER THAN PROTESTANT (25)	1.00	62	TEND LESS TO BE THOSE WHERE THE RELIGION IS CATHOLIC, RATHER THAN PROTESTANT (25)	0.64	0 5 11 9 X SQ= 2.93 P = 0.046 RV NO (YES)
63	TEND TO BE THOSE WHERE THE RELIGION IS PREDOMINANTLY OR PARTLY OTHER THAN CHRISTIAN (68)	0.74	63	TEND TO BE THOSE WHERE THE RELIGION IS PREDOMINANTLY SOME KIND OF CHRISTIAN (46)	0.69	17 24 48 11 X SQ= 15.21 P = 0. RV YES (YES)
64	TEND LESS TO BE THOSE WHERE THE RELIGION IS CHRISTIAN, RATHER THAN MUSLIM (46)	0.52	64	TEND MORE TO BE THOSE WHERE THE RELIGION IS CHRISTIAN, RATHER THAN MUSLIM (46)	0.96	17 24 16 1 X SQ= 11.52 P = 0. RV NO (YES)

65	TEND TO BE THOSE WHERE THE RELIGION IS MUSLIM, RATHER THAN CATHOLIC (18)	0.59	TEND TO BE THOSE WHERE THE RELIGION IS CATHOLIC, RATHER THAN MUSLIM (25)	0.90	11 9 16 1 X SQ= 5.28 P = 0.010 RV YES (NO)
68	TEND TO BE THOSE THAT ARE LINGUISTICALLY WEAKLY HETEROGENEOUS OR STRONGLY HETEROGENEOUS (62)	0.63	TEND TO BE THOSE THAT ARE LINGUISTICALLY HOMOGENEOUS (52)	0.60	24 21 41 14 X SQ= 4.01 P = 0.035 RV YES (NO)
70	TEND MORE TO BE THOSE THAT ARE RELIGIOUSLY, RACIALLY, OR LINGUISTICALLY HETEROGENEOUS (85)	0.87	TEND LESS TO BE THOSE THAT ARE RELIGIOUSLY, RACIALLY, OR LINGUISTICALLY HETEROGENEOUS (85)	0.68	8 11 52 23 X SQ= 3.76 P = 0.035 RV NO (YES)
71	LEAN MORE TOWARD BEING THOSE WHOSE DATE OF INDEPENDENCE IS AFTER 1800 (90)	0.84	LEAN LESS TOWARD BEING THOSE WHOSE DATE OF INDEPENDENCE IS AFTER 1800 (90)	0.68	10 11 53 23 X SQ= 2.63 P = 0.074 RV NO (YES)
74	TEND TO BE THOSE THAT ARE NOT HISTORICALLY WESTERN (87)	0.91	TEND TO BE THOSE THAT ARE HISTORICALLY WESTERN (26)	0.59	6 20 59 14 X SQ= 25.85 P = 0. RV YES (YES)
75	TEND TO BE THOSE THAT ARE NOT HISTORICALLY WESTERN AND ARE NOT SIGNIFICANTLY WESTERNIZED (52)	0.60	TEND TO BE THOSE THAT ARE HISTORICALLY WESTERN OR SIGNIFICANTLY WESTERNIZED (62)	0.83	26 29 39 6 X SQ= 15.20 P = 0. RV YES (YES)
76	TEND TO BE THOSE THAT HAVE BEEN SIGNIFICANTLY OR PARTIALLY WESTERNIZED THROUGH A COLONIAL RELATIONSHIP, RATHER THAN BEING HISTORICALLY WESTERN (70)	0.88	TEND TO BE THOSE THAT ARE HISTORICALLY WESTERN, RATHER THAN HAVING BEEN SIGNIFICANTLY OR PARTIALLY WESTERNIZED THROUGH A COLONIAL RELATIONSHIP (26)	0.61	6 20 43 13 X SQ= 19.12 P = 0. RV YES (YES)
80	TEND TO BE THOSE WHOSE DATE OF INDEPENDENCE IS AFTER 1914, AND THAT WERE FORMERLY DEPENDENCIES OF FRANCE, RATHER THAN BRITAIN (24)	0.69	TEND TO BE THOSE WHOSE DATE OF INDEPENDENCE IS AFTER 1914, AND THAT WERE FORMERLY DEPENDENCIES OF BRITAIN, RATHER THAN FRANCE (19)	0.83	8 10 18 2 X SQ= 7.11 P = 0.004 RV YES (YES)
81	TEND MORE TO BE THOSE WHERE THE TYPE OF POLITICAL MODERNIZATION IS LATER EUROPEAN OR LATER EUROPEAN DERIVED, RATHER THAN EARLY EUROPEAN OR EARLY EUROPEAN DERIVED (40)	0.96	TEND LESS TO BE THOSE WHERE THE TYPE OF POLITICAL MODERNIZATION IS LATER EUROPEAN OR LATER EUROPEAN DERIVED, RATHER THAN EARLY EUROPEAN OR EARLY EUROPEAN DERIVED (40)	0.55	1 10 22 12 X SQ= 8.18 P = 0.002 RV NO (YES)

82	TEND TO BE THOSE WHERE THE TYPE OF POLITICAL MODERNIZATION IS DEVELOPED TUTELARY OR UNDEVELOPED TUTELARY, RATHER THAN EARLY OR LATER EUROPEAN OR EUROPEAN DERIVED (55)	0.60	82	TEND TO BE THOSE WHERE THE TYPE OF POLITICAL MODERNIZATION IS EARLY OR LATER EUROPEAN OR EUROPEAN DERIVED, RATHER THAN DEVELOPED TUTELARY OR UNDEVELOPED TUTELARY (51)	0.65	23 22 35 12 X SQ= 4.43 P = 0.030 RV YES (NO)
85	TEND TO BE THOSE WHERE THE STAGE OF POLITICAL MODERNIZATION IS MID- OR EARLY TRANSITIONAL, RATHER THAN ADVANCED (54)	0.54	85	TEND TO BE THOSE WHERE THE STAGE OF POLITICAL MODERNIZATION IS ADVANCED, RATHER THAN MID- OR EARLY TRANSITIONAL (60)	0.74	30 26 35 9 X SQ= 6.21 P = 0.011 RV YES (NO)
87	LEAN LESS TOWARD BEING THOSE WHOSE IDEOLOGICAL ORIENTATION IS OTHER THAN DEVELOPMENTAL (58)	0.62	87	LEAN MORE TOWARD BEING THOSE WHOSE IDEOLOGICAL ORIENTATION IS OTHER THAN DEVELOPMENTAL (58)	0.83	19 5 31 24 X SQ= 2.82 P = 0.076 RV NO (NO)
89	TEND TO BE THOSE WHOSE IDEOLOGICAL ORIENTATION IS OTHER THAN CONVENTIONAL (62)	0.89	89	TEND TO BE THOSE WHOSE IDEOLOGICAL ORIENTATION IS CONVENTIONAL (33)	0.83	6 24 50 5 X SQ= 40.33 P = 0. RV YES (YES)
92	TEND LESS TO BE THOSE WHERE THE SYSTEM STYLE IS LIMITED MOBILIZATIONAL, OR NON-MOBILIZATIONAL (93)	0.69	92	ALWAYS ARE THOSE WHERE THE SYSTEM STYLE IS LIMITED MOBILIZATIONAL, OR NON-MOBILIZATIONAL (93)	1.00	20 0 44 35 X SQ= 11.84 P = 0. RV NO (NO)
93	TEND LESS TO BE THOSE WHERE THE SYSTEM STYLE IS NON-MOBILIZATIONAL (78)	0.52	93	TEND MORE TO BE THOSE WHERE THE SYSTEM STYLE IS NON-MOBILIZATIONAL (78)	0.91	29 3 32 32 X SQ= 13.50 P = 0. RV NO (YES)
94	TEND TO BE THOSE WHERE THE STATUS OF THE REGIME IS AUTHORITARIAN OR TOTALITARIAN (41)	0.80	94	ALWAYS ARE THOSE WHERE THE STATUS OF THE REGIME IS CONSTITUTIONAL (51)	1.00	10 34 41 0 X SQ= 49.63 P = 0. RV YES (YES)
95	TEND LESS TO BE THOSE WHERE THE STATUS OF THE REGIME IS CONSTITUTIONAL OR AUTHORITARIAN (95)	0.74	95	ALWAYS ARE THOSE WHERE THE STATUS OF THE REGIME IS CONSTITUTIONAL OR AUTHORITARIAN (95)	1.00	46 35 16 0 X SQ= 9.02 P = 0. RV NO (NO)
96	TEND LESS TO BE THOSE WHERE THE STATUS OF THE REGIME IS CONSTITUTIONAL OR TOTALITARIAN (67)	0.53	96	ALWAYS ARE THOSE WHERE THE STATUS OF THE REGIME IS CONSTITUTIONAL OR TOTALITARIAN (67)	1.00	26 34 23 0 X SQ= 19.80 P = 0. RV NO (YES)

99	TEND TO BE THOSE WHERE GOVERNMENTAL STABILITY IS MODERATELY PRESENT AND DATES FROM THE POST-WAR PERIOD, OR IS ABSENT (36)	0.56	99	TEND TO BE THOSE WHERE GOVERNMENTAL STABILITY IS GENERALLY PRESENT AND DATES FROM AT LEAST THE INTER-WAR PERIOD, OR FROM THE POST-WAR PERIOD (50)	0.93	23 26 29 2 X SQ= 16.14 P = 0. RV YES (YES)
101	TEND TO BE THOSE WHERE THE REPRESENTATIVE CHARACTER OF THE REGIME IS LIMITED POLYARCHIC, PSEUDO-POLYARCHIC, OR NON-POLYARCHIC (57)	0.91	101	TEND TO BE THOSE WHERE THE REPRESENTATIVE CHARACTER OF THE REGIME IS POLYARCHIC (41)	0.94	5 32 49 2 X SQ= 58.22 P = 0. RV YES (YES)
102	TEND TO BE THOSE WHERE THE REPRESENTATIVE CHARACTER OF THE REGIME IS PSEUDO-POLYARCHIC OR NON-POLYARCHIC (49)	0.73	102	TEND TO BE THOSE WHERE THE REPRESENTATIVE CHARACTER OF THE REGIME IS POLYARCHIC OR LIMITED POLYARCHIC (59)	0.97	16 34 44 1 X SQ= 41.26 P = 0. RV YES (YES)
105	TEND TO BE THOSE WHERE THE ELECTORAL SYSTEM IS PARTIALLY COMPETITIVE OR NON-COMPETITIVE (47)	0.86	105	TEND TO BE THOSE WHERE THE ELECTORAL SYSTEM IS COMPETITIVE (43)	0.94	6 32 38 2 X SQ= 46.56 P = 0. RV YES (YES)
106	TEND TO BE THOSE WHERE THE ELECTORAL SYSTEM IS NON-COMPETITIVE (30)	0.70	106	TEND TO BE THOSE WHERE THE ELECTORAL SYSTEM IS COMPETITIVE OR PARTIALLY COMPETITIVE (52)	0.97	12 33 28 1 X SQ= 31.92 P = 0. RV YES (YES)
107	TEND TO BE THOSE WHERE AUTONOMOUS GROUPS ARE PARTIALLY TOLERATED IN POLITICS, ARE TOLERATED ONLY OUTSIDE POLITICS, OR ARE NOT TOLERATED AT ALL (65)	0.86	107	TEND TO BE THOSE WHERE AUTONOMOUS GROUPS ARE FULLY TOLERATED IN POLITICS (46)	0.91	9 32 54 3 X SQ= 51.90 P = 0. RV YES (YES)
108	TEND TO BE THOSE WHERE AUTONOMOUS GROUPS ARE TOLERATED ONLY OUTSIDE POLITICS OR ARE NOT TOLERATED AT ALL (35)	0.61	108	ALWAYS ARE THOSE WHERE AUTONOMOUS GROUPS ARE FULLY OR PARTIALLY TOLERATED IN POLITICS (65)	1.00	20 35 31 0 X SQ= 30.68 P = 0. RV YES (YES)
111	TEND TO BE THOSE WHERE POLITICAL ENCULTURATION IS LOW (42)	0.60	111	TEND TO BE THOSE WHERE POLITICAL ENCULTURATION IS HIGH OR MEDIUM (53)	0.91	21 30 31 3 X SQ= 19.42 P = 0. RV YES (YES)
115	TEND TO BE THOSE WHERE INTEREST ARTICULATION BY ASSOCIATIONAL GROUPS IS MODERATE, LIMITED, OR NEGLIGIBLE (91)	0.98	115	TEND TO BE THOSE WHERE INTEREST ARTICULATION BY ASSOCIATIONAL GROUPS IS SIGNIFICANT (20)	0.51	1 18 61 17 X SQ= 32.15 P = 0. RV YES (YES)

116	TEND TO BE THOSE WHERE INTEREST ARTICULATION BY ASSOCIATIONAL GROUPS IS LIMITED OR NEGLIGIBLE (79)	0.92	116	TEND TO BE THOSE WHERE INTEREST ARTICULATION BY ASSOCIATIONAL GROUPS IS SIGNIFICANT OR MODERATE (32)	0.71	5 25 57 10 X SQ= 39.13 P = 0. RV YES (YES)
117	TEND TO BE THOSE WHERE INTEREST ARTICULATION BY ASSOCIATIONAL GROUPS IS NEGLIGIBLE (51)	0.61	117	TEND TO BE THOSE WHERE INTEREST ARTICULATION BY ASSOCIATIONAL GROUPS IS SIGNIFICANT, MODERATE, OR LIMITED (60)	0.86	24 30 38 5 X SQ= 18.17 P = 0. RV YES (YES)
118	TEND TO BE THOSE WHERE INTEREST ARTICULATION BY INSTITUTIONAL GROUPS IS VERY SIGNIFICANT (40)	0.64	118	ALWAYS ARE THOSE WHERE INTEREST ARTICULATION BY INSTITUTIONAL GROUPS IS SIGNIFICANT, MODERATE, OR LIMITED (60)	1.00	37 0 21 34 X SQ= 33.68 P = 0. RV YES (YES)
119	TEND TO BE THOSE WHERE INTEREST ARTICULATION BY INSTITUTIONAL GROUPS IS VERY SIGNIFICANT OR SIGNIFICANT (74)	0.98	119	TEND TO BE THOSE WHERE INTEREST ARTICULATION BY INSTITUTIONAL GROUPS IS MODERATE OR LIMITED (26)	0.71	57 10 1 24 X SQ= 47.94 P = 0. RV YES (YES)
120	ALWAYS ARE THOSE WHERE INTEREST ARTICULATION BY INSTITUTIONAL GROUPS IS VERY SIGNIFICANT, SIGNIFICANT, OR MODERATE (90)	1.00	120	TEND LESS TO BE THOSE WHERE INTEREST ARTICULATION BY INSTITUTIONAL GROUPS IS VERY SIGNIFICANT, SIGNIFICANT, OR MODERATE (90)	0.71	58 24 0 10 X SQ= 16.22 P = 0. RV NO (YES)
121	TEND TO BE THOSE WHERE INTEREST ARTICULATION BY NON-ASSOCIATIONAL GROUPS IS SIGNIFICANT (54)	0.58	121	TEND TO BE THOSE WHERE INTEREST ARTICULATION BY NON-ASSOCIATIONAL GROUPS IS MODERATE, LIMITED, OR NEGLIGIBLE (61)	0.74	38 9 28 26 X SQ= 8.10 P = 0.003 RV YES (NO)
122	TEND TO BE THOSE WHERE INTEREST ARTICULATION BY NON-ASSOCIATIONAL GROUPS IS SIGNIFICANT OR MODERATE (83)	0.85	122	TEND TO BE THOSE WHERE INTEREST ARTICULATION BY NON-ASSOCIATIONAL GROUPS IS LIMITED OR NEGLIGIBLE (32)	0.51	56 17 10 18 X SQ= 13.27 P = 0. RV YES (YES)
123	ALWAYS ARE THOSE WHERE INTEREST ARTICULATION BY NON-ASSOCIATIONAL GROUPS IS SIGNIFICANT, MODERATE, OR LIMITED (107)	1.00	123	TEND LESS TO BE THOSE WHERE INTEREST ARTICULATION BY NON-ASSOCIATIONAL GROUPS IS SIGNIFICANT, MODERATE, OR LIMITED (107)	0.77	66 27 0 8 X SQ= 13.40 P = 0. RV NO (YES)
125	TEND TO BE THOSE WHERE INTEREST ARTICULATION BY ANOMIC GROUPS IS FREQUENT OR OCCASIONAL (64)	0.87	125	TEND TO BE THOSE WHERE INTEREST ARTICULATION BY ANOMIC GROUPS IS INFREQUENT OR VERY INFREQUENT (35)	0.76	48 8 7 25 X SQ= 32.74 P = 0. RV YES (YES)

126	ALWAYS ARE THOSE WHERE INTEREST ARTICULATION BY ANOMIC GROUPS IS FREQUENT, OCCASIONAL, OR INFREQUENT (83)	1.00	126	TEND LESS TO BE THOSE WHERE INTEREST ARTICULATION BY ANOMIC GROUPS IS FREQUENT, OCCASIONAL, OR INFREQUENT (83)	0.52	55 17 0 16 X SQ= 29.41 P = 0. RV NO (YES)
128	TEND TO BE THOSE WHERE INTEREST ARTICULATION BY POLITICAL PARTIES IS LIMITED OR NEGLIGIBLE (45)	0.67	128	TEND TO BE THOSE WHERE INTEREST ARTICULATION BY POLITICAL PARTIES IS SIGNIFICANT OR MODERATE (48)	0.82	16 28 32 6 X SQ= 17.31 P = 0. RV YES (YES)
129	TEND TO BE THOSE WHERE INTEREST ARTICULATION BY POLITICAL PARTIES IS NEGLIGIBLE (37)	0.56	129	TEND TO BE THOSE WHERE INTEREST ARTICULATION BY POLITICAL PARTIES IS SIGNIFICANT, MODERATE, OR LIMITED (56)	0.85	21 29 27 5 X SQ= 12.74 P = 0. RV YES (YES)
131	TILT TOWARD BEING THOSE WHERE INTEREST AGGREGATION BY POLITICAL PARTIES IS LIMITED OR NEGLIGIBLE (35)	0.63	131	TILT TOWARD BEING THOSE WHERE INTEREST AGGREGATION BY POLITICAL PARTIES IS SIGNIFICANT OR MODERATE (30)	0.61	10 20 17 13 X SQ= 2.42 P = 0.119 RV YES (YES)
134	TEND TO BE THOSE WHERE INTEREST AGGREGATION BY THE EXECUTIVE IS LIMITED OR NEGLIGIBLE (46)	0.67	134	TEND TO BE THOSE WHERE INTEREST AGGREGATION BY THE EXECUTIVE IS SIGNIFICANT OR MODERATE (57)	0.91	19 31 38 3 X SQ= 26.49 P = 0. RV YES (YES)
136	ALWAYS ARE THOSE WHERE INTEREST AGGREGATION BY THE LEGISLATURE IS MODERATE, LIMITED, OR NEGLIGIBLE (85)	1.00	136	TEND LESS TO BE THOSE WHERE INTEREST AGGREGATION BY THE LEGISLATURE IS MODERATE, LIMITED, OR NEGLIGIBLE (85)	0.66	0 12 52 23 X SQ= 17.90 P = 0. RV NO (YES)
137	TEND TO BE THOSE WHERE INTEREST AGGREGATION BY THE LEGISLATURE IS LIMITED OR NEGLIGIBLE (68)	0.96	137	TEND TO BE THOSE WHERE INTEREST AGGREGATION BY THE LEGISLATURE IS SIGNIFICANT OR MODERATE (29)	0.74	2 26 50 9 X SQ= 44.38 P = 0. RV YES (YES)
138	TEND TO BE THOSE WHERE INTEREST AGGREGATION BY THE LEGISLATURE IS NEGLIGIBLE (49)	0.74	138	TEND TO BE THOSE WHERE INTEREST AGGREGATION BY THE LEGISLATURE IS SIGNIFICANT, MODERATE, OR LIMITED (48)	0.83	13 29 37 6 X SQ= 24.40 P = 0. RV YES (YES)
139	TEND LESS TO BE THOSE WHERE THE PARTY SYSTEM IS QUANTITATIVELY OTHER THAN ONE-PARTY (71)	0.52	139	TEND MORE TO BE THOSE WHERE THE PARTY SYSTEM IS QUANTITATIVELY OTHER THAN ONE-PARTY (71)	0.94	27 2 29 33 X SQ= 16.01 P = 0. RV NO (YES)

141	TEND MORE TO BE THOSE WHERE THE PARTY SYSTEM IS QUANTITATIVELY OTHER THAN TWO-PARTY (87)	0.96	141	TEND LESS TO BE THOSE WHERE THE PARTY SYSTEM IS QUANTITATIVELY OTHER THAN TWO-PARTY (87)	0.73	2 9 49 24 X SQ= 7.66 P = 0.005 RV NO (YES)
142	TEND MORE TO BE THOSE WHERE THE PARTY SYSTEM IS QUANTITATIVELY OTHER THAN MULTI-PARTY (66)	0.82	142	TEND LESS TO BE THOSE WHERE THE PARTY SYSTEM IS QUANTITATIVELY OTHER THAN MULTI-PARTY (66)	0.58	9 14 40 19 X SQ= 4.53 P = 0.024 RV NO (YES)
147	TEND TO BE THOSE WHERE THE PARTY SYSTEM IS QUALITATIVELY OTHER THAN CLASS-ORIENTED OR MULTI-IDEOLOGICAL (67)	0.98	147	TEND TO BE THOSE WHERE THE PARTY SYSTEM IS QUALITATIVELY CLASS-ORIENTED OR MULTI-IDEOLOGICAL (23)	0.66	1 19 47 10 X SQ= 34.61 P = 0. RV YES (YES)
153	TEND TO BE THOSE WHERE THE PARTY SYSTEM IS MODERATELY STABLE OR UNSTABLE (38)	0.58	153	TEND TO BE THOSE WHERE THE PARTY SYSTEM IS STABLE (42)	0.77	19 23 26 7 X SQ= 7.33 P = 0.004 RV YES (YES)
157	LEAN LESS TOWARD BEING THOSE WHERE PERSONALISMO IS MODERATE OR NEGLIGIBLE (84)	0.79	157	LEAN MORE TOWARD BEING THOSE WHERE PERSONALISMO IS MODERATE OR NEGLIGIBLE (84)	0.94	11 2 41 33 X SQ= 2.80 P = 0.066 RV NO (NO)
158	TEND LESS TO BE THOSE WHERE PERSONALISMO IS NEGLIGIBLE (56)	0.52	158	TEND MORE TO BE THOSE WHERE PERSONALISMO IS NEGLIGIBLE (56)	0.83	25 6 27 29 X SQ= 7.43 P = 0.003 RV NO (YES)
160	TEND TO BE THOSE WHERE THE POLITICAL LEADERSHIP IS ELITIST (30)	0.59	160	ALWAYS ARE THOSE WHERE THE POLITICAL LEADERSHIP IS MODERATELY ELITIST OR NON-ELITIST (67)	1.00	30 0 21 34 X SQ= 28.39 P = 0. RV YES (YES)
161	TEND TO BE THOSE WHERE THE POLITICAL LEADERSHIP IS ELITIST OR MODERATELY ELITIST (47)	0.69	161	TEND TO BE THOSE WHERE THE POLITICAL LEADERSHIP IS NON-ELITIST (50)	0.79	35 7 16 27 X SQ= 16.96 P = 0. RV YES (YES)
164	TEND LESS TO BE THOSE WHERE THE REGIME'S LEADERSHIP CHARISMA IS NEGLIGIBLE (65)	0.56	164	TEND MORE TO BE THOSE WHERE THE REGIME'S LEADERSHIP CHARISMA IS NEGLIGIBLE (65)	0.85	24 5 30 29 X SQ= 7.06 P = 0.005 RV NO (NO)

166	LEAN MORE TOWARD BEING THOSE WHERE THE VERTICAL POWER DISTRIBUTION IS THAT OF FORMAL FEDERALISM OR FORMAL AND EFFECTIVE UNITARISM (99)	0.92	166	LEAN LESS TOWARD BEING THOSE WHERE THE VERTICAL POWER DISTRIBUTION IS THAT OF FORMAL FEDERALISM OR FORMAL AND EFFECTIVE UNITARISM (99)	0.77	5 8 60 27 X SQ= 3.38 P = 0.057 RV NO (YES)
168	TEND TO BE THOSE WHERE THE HORIZONTAL POWER DISTRIBUTION IS LIMITED OR NEGLIGIBLE (72)	0.97	168	TEND TO BE THOSE WHERE THE HORIZONTAL POWER DISTRIBUTION IS SIGNIFICANT (34)	0.79	2 27 57 7 X SQ= 54.60 P = 0. RV YES (YES)
169	TEND TO BE THOSE WHERE THE HORIZONTAL POWER DISTRIBUTION IS NEGLIGIBLE (48)	0.75	169	TEND TO BE THOSE WHERE THE HORIZONTAL POWER DISTRIBUTION IS SIGNIFICANT OR LIMITED (58)	0.97	15 33 44 1 X SQ= 41.50 P = 0. RV YES (YES)
170	TEND LESS TO BE THOSE WHERE THE LEGISLATIVE-EXECUTIVE STRUCTURE IS OTHER THAN PRESIDENTIAL (63)	0.55	170	TEND MORE TO BE THOSE WHERE THE LEGISLATIVE-EXECUTIVE STRUCTURE IS OTHER THAN PRESIDENTIAL (63)	0.85	25 5 30 29 X SQ= 7.57 P = 0.003 RV NO (NO)
171	LEAN MORE TOWARD BEING THOSE WHERE THE LEGISLATIVE-EXECUTIVE STRUCTURE IS OTHER THAN PARLIAMENTARY-REPUBLICAN (90)	0.93	171	LEAN LESS TOWARD BEING THOSE WHERE THE LEGISLATIVE-EXECUTIVE STRUCTURE IS OTHER THAN PARLIAMENTARY-REPUBLICAN (90)	0.77	4 8 50 27 X SQ= 3.12 P = 0.055 RV NO (YES)
172	TEND TO BE THOSE WHERE THE LEGISLATIVE-EXECUTIVE STRUCTURE IS OTHER THAN PARLIAMENTARY-ROYALIST (88)	0.97	172	TEND TO BE THOSE WHERE THE LEGISLATIVE-EXECUTIVE STRUCTURE IS PARLIAMENTARY-ROYALIST (21)	0.51	2 18 58 17 X SQ= 27.94 P = 0. RV YES (YES)
174	TEND TO BE THOSE WHERE THE LEGISLATURE IS PARTIALLY EFFECTIVE, LARGELY INEFFECTIVE, OR WHOLLY INEFFECTIVE (72)	0.98	174	TEND TO BE THOSE WHERE THE LEGISLATURE IS FULLY EFFECTIVE (28)	0.74	1 26 52 9 X SQ= 48.60 P = 0. RV YES (YES)
175	TEND TO BE THOSE WHERE THE LEGISLATURE IS LARGELY INEFFECTIVE OR WHOLLY INEFFECTIVE (49)	0.77	175	TEND TO BE THOSE WHERE THE LEGISLATURE IS FULLY EFFECTIVE OR PARTIALLY EFFECTIVE (51)	0.94	12 33 41 2 X SQ= 40.48 P = 0. RV YES (YES)
176	TEND TO BE THOSE WHERE THE LEGISLATURE IS WHOLLY INEFFECTIVE (28)	0.53	176	ALWAYS ARE THOSE WHERE THE LEGISLATURE IS FULLY EFFECTIVE, PARTIALLY EFFECTIVE, OR LARGELY INEFFECTIVE (72)	1.00	25 35 28 0 X SQ= 24.74 P = 0. RV YES (YES)

No.			No.			
178	TEND TO BE THOSE WHERE THE LEGISLATURE IS UNICAMERAL (53)	0.64	178	TEND TO BE THOSE WHERE THE LEGISLATURE IS BICAMERAL (51)	0.74	36 9 20 26 X SQ= 11.32 P = 0.001 RV YES (YES)
179	TEND TO BE THOSE WHERE THE EXECUTIVE IS DOMINANT (52)	0.85	179	TEND TO BE THOSE WHERE THE EXECUTIVE IS STRONG (39)	0.83	45 5 8 25 X SQ= 34.45 P = 0. RV YES (YES)
181	TEND TO BE THOSE WHERE THE BUREAUCRACY IS SEMI-MODERN, RATHER THAN MODERN (55)	0.97	181	TEND TO BE THOSE WHERE THE BUREAUCRACY IS MODERN, RATHER THAN SEMI-MODERN (21)	0.67	1 20 37 10 X SQ= 29.28 P = 0. RV YES (YES)
183	TILT LESS TOWARD BEING THOSE WHERE THE BUREAUCRACY IS POST-COLONIAL TRANSITIONAL, RATHER THAN TRADITIONAL (25)	0.61	183	ALWAYS ARE THOSE WHERE THE BUREAUCRACY IS POST-COLONIAL TRANSITIONAL, RATHER THAN TRADITIONAL (25)	1.00	14 5 9 0 X SQ= 1.37 P = 0.144 RV NO (NO)
186	TEND TO BE THOSE WHERE PARTICIPATION BY THE MILITARY IN POLITICS IS SUPPORTIVE, RATHER THAN NEUTRAL (31)	0.74	186	ALWAYS ARE THOSE WHERE PARTICIPATION BY THE MILITARY IN POLITICS IS NEUTRAL, RATHER THAN SUPPORTIVE (56)	1.00	31 0 11 34 X SQ= 39.38 P = 0. RV YES (YES)
189	ALWAYS ARE THOSE WHERE THE CHARACTER OF THE LEGAL SYSTEM IS OTHER THAN COMMON LAW (108)	1.00	189	TEND LESS TO BE THOSE WHERE THE CHARACTER OF THE LEGAL SYSTEM IS OTHER THAN COMMON LAW (108)	0.80	0 7 66 28 X SQ= 11.25 P = 0. RV NO (YES)
190	ALWAYS ARE THOSE WHERE THE CHARACTER OF THE LEGAL SYSTEM IS CIVIL LAW, RATHER THAN COMMON LAW (32)	1.00	190	TEND LESS TO BE THOSE WHERE THE CHARACTER OF THE LEGAL SYSTEM IS CIVIL LAW, RATHER THAN COMMON LAW (32)	0.61	15 11 0 7 X SQ= 5.26 P = 0.009 RV NO (YES)
193	LEAN LESS TOWARD BEING THOSE WHERE THE CHARACTER OF THE LEGAL SYSTEM IS OTHER THAN PARTLY INDIGENOUS (86)	0.72	193	LEAN MORE TOWARD BEING THOSE WHERE THE CHARACTER OF THE LEGAL SYSTEM IS OTHER THAN PARTLY INDIGENOUS (86)	0.89	18 4 47 31 X SQ= 2.62 P = 0.078 RV NO (NO)
194	TEND LESS TO BE THOSE THAT ARE NON-COMMUNIST (101)	0.80	194	ALWAYS ARE THOSE THAT ARE NON-COMMUNIST (101)	1.00	13 0 52 35 X SQ= 6.37 P = 0.004 RV NO (NO)

188 POLITIES
WHERE THE CHARACTER OF THE LEGAL SYSTEM
IS CIVIL LAW (32)

188 POLITIES
WHERE THE CHARACTER OF THE LEGAL SYSTEM
IS OTHER THAN CIVIL LAW (81)

BOTH SUBJECT AND PREDICATE

32 IN LEFT COLUMN

ARGENTINA	AUSTRIA	BELGIUM	BOLIVIA	BRAZIL	CHILE	COLOMBIA	COSTA RICA	DOMIN REP	ECUADOR
EL SALVADOR	FRANCE	GERMAN FR	GREECE	GUATEMALA	HAITI	HONDURAS	ITALY	LUXEMBOURG	MEXICO
NETHERLANDS	NICARAGUA	PANAMA	PARAGUAY	PERU	PORTUGAL	SPAIN	SWITZERLAND	THAILAND	TURKEY
URUGUAY	VENEZUELA								

81 IN RIGHT COLUMN

AFGHANISTAN	ALBANIA	ALGERIA	AUSTRALIA	BULGARIA	BURMA	BURUNDI	CAMBODIA	CAMEROUN	CANADA
CEN AFR REP	CEYLON	CHAD	CHINA, PR	CONGO(BRA)	CONGO(LEO)	CUBA	CYPRUS	CZECHOS'KIA	DAHOMEY
DENMARK	FINLAND	GABON	GERMANY, E	GHANA	HUNGARY	ICELAND	INDIA	INDONESIA	IRAN
IRAQ	IRELAND	ISRAEL	IVORY COAST	JAMAICA	JAPAN	JORDAN	KOREA, N	KOREA REP	LAOS
LEBANON	LIBERIA	LIBYA	MALAGASY R	MALAYA	MALI	MAURITANIA	MONGOLIA	MOROCCO	NEPAL
NEW ZEALAND	NIGER	NIGERIA	NORWAY	PAKISTAN	PHILIPPINES	POLAND	RUMANIA	RWANDA	SA'U ARABIA
SENEGAL	SIERRE LEO	SOMALIA	SO AFRICA	SUDAN	SWEDEN	SYRIA	TANGANYIKA	TOGO	TRINIDAD
TUNISIA	UGANDA	USSR	UAR	UK	US	UPPER VOLTA	VIETNAM, N	VIETNAM REP	YEMEN
YUGOSLAVIA									

2 EXCLUDED BECAUSE AMBIGUOUS

ETHIOPIA GUINEA

---

3 TEND LESS TO BE THOSE LOCATED ELSEWHERE THAN IN WEST EUROPE (102) 0.66

3 TEND MORE TO BE THOSE LOCATED ELSEWHERE THAN IN WEST EUROPE (102) 0.98

11 2
21 79
X SQ= 19.91
P = 0.
RV NO (YES)

5 ALWAYS ARE THOSE LOCATED IN SCANDINAVIA OR WEST EUROPE, RATHER THAN IN EAST EUROPE (18) 1.00

5 TEND TO BE THOSE LOCATED IN EAST EUROPE, RATHER THAN IN SCANDINAVIA OR WEST EUROPE (9) 0.56

11 7
0 9
X SQ= 6.92
P = 0.003
RV YES (YES)

6 TEND TO BE THOSE LOCATED IN THE CARIBBEAN, CENTRAL AMERICA, OR SOUTH AMERICA (22) 0.59

6 TEND TO BE THOSE LOCATED ELSEWHERE THAN IN THE CARIBBEAN, CENTRAL AMERICA, OR SOUTH AMERICA (93) 0.96

19 3
13 78
X SQ= 41.86
P = 0.
RV YES (YES)

8	TEND MORE TO BE THOSE LOCATED ELSEWHERE THAN IN EAST ASIA SOUTH ASIA, OR SOUTHEAST ASIA (97)	0.97	8	TEND LESS TO BE THOSE LOCATED ELSEWHERE THAN IN EAST ASIA SOUTH ASIA, OR SOUTHEAST ASIA (97)	0.79	1 17 31 64 X SQ= 4.21 P = 0.021 RV NO (NO)
10	ALWAYS ARE THOSE LOCATED ELSEWHERE THAN IN NORTH AFRICA, OR CENTRAL AND SOUTH AFRICA (82)	1.00	10	TEND LESS TO BE THOSE LOCATED ELSEWHERE THAN IN NORTH AFRICA, OR CENTRAL AND SOUTH AFRICA (82)	0.62	0 31 32 50 X SQ= 15.01 P = 0. RV NO (NO)
13	TEND MORE TO BE THOSE LOCATED ELSEWHERE THAN IN NORTH AFRICA OR THE MIDDLE EAST (99)	0.97	13	TEND LESS TO BE THOSE LOCATED ELSEWHERE THAN IN NORTH AFRICA OR THE MIDDLE EAST (99)	0.81	1 15 31 66 X SQ= 3.30 P = 0.038 RV NO (NO)
17	TEND TO BE THOSE LOCATED IN THE CARIBBEAN, CENTRAL AMERICA, OR SOUTH AMERICA, RATHER THAN IN THE MIDDLE EAST (22)	0.95	17	TEND TO BE THOSE LOCATED IN THE MIDDLE EAST, RATHER THAN IN THE CARIBBEAN, CENTRAL AMERICA, OR SOUTH AMERICA (11)	0.77	1 10 19 3 X SQ= 15.25 P = 0. RV YES (YES)
20	TEND TO BE THOSE LOCATED IN THE CARIBBEAN, CENTRAL AMERICA, OR SOUTH AMERICA, RATHER THAN IN EAST ASIA, SOUTH ASIA, OR SOUTHEAST ASIA (22)	0.95	20	TEND TO BE THOSE LOCATED IN EAST ASIA, SOUTH ASIA, OR SOUTHEAST ASIA, RATHER THAN IN THE CARIBBEAN, CENTRAL AMERICA, OR SOUTH AMERICA (18)	0.85	1 17 19 3 X SQ= 22.73 P = 0. RV YES (YES)
29	TEND TO BE THOSE WHERE THE DEGREE OF URBANIZATION IS HIGH (56)	0.79	29	TEND TO BE THOSE WHERE THE DEGREE OF URBANIZATION IS LOW (49)	0.55	23 33 6 41 X SQ= 8.77 P = 0.002 RV YES (NO)
30	TEND TO BE THOSE WHOSE AGRICULTURAL POPULATION IS MEDIUM, LOW, OR VERY LOW (57)	0.78	30	TEND TO BE THOSE WHOSE AGRICULTURAL POPULATION IS HIGH (56)	0.59	7 47 25 32 X SQ= 11.44 P = 0. RV YES (NO)
37	TEND TO BE THOSE WHOSE PER CAPITA GROSS NATIONAL PRODUCT IS VERY HIGH, HIGH, MEDIUM, OR LOW (64)	0.87	37	TEND TO BE THOSE WHOSE PER CAPITA GROSS NATIONAL PRODUCT IS VERY LOW (51)	0.56	28 36 4 45 X SQ= 15.61 P = 0. RV YES (NO)
43	TEND TO BE THOSE WHOSE ECONOMIC DEVELOPMENTAL STATUS IS DEVELOPED, INTERMEDIATE, OR UNDERDEVELOPED (55)	0.68	43	TEND TO BE THOSE WHOSE ECONOMIC DEVELOPMENTAL STATUS IS VERY UNDERDEVELOPED (57)	0.57	21 34 10 45 X SQ= 4.49 P = 0.022 RV YES (NO)

45	TEND TO BE THOSE WHERE THE LITERACY RATE IS FIFTY PERCENT OR ABOVE (55)	0.72	45	TEND TO BE THOSE WHERE THE LITERACY RATE IS BELOW FIFTY PERCENT (54)	0.57	23 32 9 43 X SQ= 6.54 P = 0.006 RV YES (NO)
46	ALWAYS ARE THOSE WHERE THE LITERACY RATE IS TEN PERCENT OR ABOVE (84)	1.00	46	TEND LESS TO BE THOSE WHERE THE LITERACY RATE IS TEN PERCENT OR ABOVE (84)	0.68	32 52 0 25 X SQ= 11.71 P = 0. RV NO (NO)
47	LEAN TOWARD BEING THOSE WHERE THE LITERACY RATE IS BETWEEN FIFTY AND NINETY PERCENT, RATHER THAN NINETY PERCENT OR ABOVE (30)	0.70	47	LEAN TOWARD BEING THOSE WHERE THE LITERACY RATE IS NINETY PERCENT OR ABOVE, RATHER THAN BETWEEN FIFTY AND NINETY PERCENT (25)	0.56	7 18 16 14 X SQ= 2.63 P = 0.099 RV YES (YES)
51	LEAN MORE TOWARD BEING THOSE WHERE FREEDOM OF THE PRESS IS COMPLETE OR INTERMITTENT (60)	0.78	51	LEAN LESS TOWARD BEING THOSE WHERE FREEDOM OF THE PRESS IS COMPLETE OR INTERMITTENT (60)	0.57	21 39 6 29 X SQ= 2.64 P = 0.098 RV NO (NO)
52	TEND MORE TO BE THOSE WHERE FREEDOM OF THE PRESS IS COMPLETE, INTERMITTENT, OR INTERNALLY ABSENT (82)	0.96	52	TEND LESS TO BE THOSE WHERE FREEDOM OF THE PRESS IS COMPLETE, INTERMITTENT, OR INTERNALLY ABSENT (82)	0.78	27 53 1 15 X SQ= 3.64 P = 0.034 RV NO (NO)
55	ALWAYS ARE THOSE WHERE NEWSPAPER CIRCULATION IS TEN OR MORE PER THOUSAND (78)	1.00	55	TEND LESS TO BE THOSE WHERE NEWSPAPER CIRCULATION IS TEN OR MORE PER THOUSAND (78)	0.58	32 46 0 33 X SQ= 17.08 P = 0. RV NO (NO)
56	TILT MORE TOWARD BEING THOSE WHERE THE RELIGION IS PREDOMINANTLY LITERATE (79)	0.84	56	TILT LESS TOWARD BEING THOSE WHERE THE RELIGION IS PREDOMINANTLY LITERATE (79)	0.68	27 52 5 24 X SQ= 2.16 P = 0.101 RV NO (NO)
57	TEND TO BE THOSE WHERE THE RELIGION IS CATHOLIC (25)	0.66	57	TEND TO BE THOSE WHERE THE RELIGION IS OTHER THAN CATHOLIC (90)	0.95	21 4 11 77 X SQ= 45.57 P = 0. RV YES (YES)
58	TEND MORE TO BE THOSE WHERE THE RELIGION IS OTHER THAN MUSLIM (97)	0.97	58	TEND LESS TO BE THOSE WHERE THE RELIGION IS OTHER THAN MUSLIM (97)	0.79	1 17 31 64 X SQ= 4.21 P = 0.021 RV NO (NO)

62	ALWAYS ARE THOSE WHERE THE RELIGION IS CATHOLIC, RATHER THAN PROTESTANT (25)	1.00	62	TEND TO BE THOSE WHERE THE RELIGION IS PROTESTANT, RATHER THAN CATHOLIC (5)	0.56	0 5 21 4 X SQ= 10.29 P = 0.001 RV YES (YES)
63	TEND TO BE THOSE WHERE THE RELIGION IS PREDOMINANTLY SOME KIND OF CHRISTIAN (46)	0.78	63	TEND TO BE THOSE WHERE THE RELIGION IS PREDOMINANTLY OR PARTLY OTHER THAN CHRISTIAN (68)	0.74	25 21 7 59 X SQ= 23.32 P = 0. RV YES (YES)
66	TEND TO BE THOSE THAT ARE RELIGIOUSLY HOMOGENEOUS (57)	0.75	66	TEND TO BE THOSE THAT ARE RELIGIOUSLY HETEROGENEOUS (49)	0.54	24 33 8 39 X SQ= 6.48 P = 0.010 RV YES (NO)
68	TEND TO BE THOSE THAT ARE LINGUISTICALLY HOMOGENEOUS (52)	0.69	68	TEND TO BE THOSE THAT ARE LINGUISTICALLY WEAKLY HETEROGENEOUS OR STRONGLY HETEROGENEOUS (62)	0.62	22 30 10 50 X SQ= 7.76 P = 0.003 RV YES (NO)
69	TEND TO BE THOSE THAT ARE LINGUISTICALLY HOMOGENEOUS OR WEAKLY HETEROGENEOUS (64)	0.81	69	TEND TO BE THOSE THAT ARE LINGUISTICALLY STRONGLY HETEROGENEOUS (50)	0.52	26 38 6 42 X SQ= 9.30 P = 0.001 RV YES (NO)
72	ALWAYS ARE THOSE WHOSE DATE OF INDEPENDENCE IS BEFORE 1914 (52)	1.00	72	TEND TO BE THOSE WHOSE DATE OF INDEPENDENCE IS AFTER 1914 (59)	0.74	31 20 0 58 X SQ= 46.33 P = 0. RV YES (YES)
75	TEND TO BE THOSE THAT ARE HISTORICALLY WESTERN OR SIGNIFICANTLY WESTERNIZED (62)	0.94	75	TEND TO BE THOSE THAT ARE NOT HISTORICALLY WESTERN AND ARE NOT SIGNIFICANTLY WESTERNIZED (52)	0.60	30 32 2 48 X SQ= 24.59 P = 0. RV YES (NO)
77	TEND TO BE THOSE THAT HAVE BEEN SIGNIFICANTLY WESTERNIZED, RATHER THAN PARTIALLY WESTERNIZED, THROUGH A COLONIAL RELATIONSHIP (28)	0.95	77	TEND TO BE THOSE THAT HAVE BEEN PARTIALLY WESTERNIZED, RATHER THAN SIGNIFICANTLY WESTERNIZED, THROUGH A COLONIAL RELATIONSHIP (41)	0.80	18 10 1 39 X SQ= 28.24 P = 0. RV YES (YES)
78	TEND MORE TO BE THOSE THAT HAVE BEEN SIGNIFICANTLY WESTERNIZED THROUGH A COLONIAL RELATIONSHIP, RATHER THAN WITHOUT SUCH A RELATIONSHIP (28)	0.95	78	TEND LESS TO BE THOSE THAT HAVE BEEN SIGNIFICANTLY WESTERNIZED THROUGH A COLONIAL RELATIONSHIP, RATHER THAN WITHOUT SUCH A RELATIONSHIP (28)	0.63	18 10 1 6 X SQ= 3.81 P = 0.032 RV NO (YES)

```
79  TEND TO BE THOSE                     0.94    79  TEND TO BE THOSE                     0.98       1    47
    THAT WERE FORMERLY DEPENDENCIES                  THAT WERE FORMERLY DEPENDENCIES                17     1
    OF SPAIN, RATHER THAN                            OF BRITAIN OR FRANCE,                       X SQ=  51.74
    BRITAIN OR FRANCE  (18)                          RATHER THAN SPAIN  (49)                     P =     0.
                                                                                                 RV YES (YES)

82  ALWAYS ARE THOSE                     1.00    82  TEND TO BE THOSE                     0.72      30    21
    WHERE THE TYPE OF POLITICAL MODERNIZATION        WHERE THE TYPE OF POLITICAL MODERNIZATION       0    54
    IS EARLY OR LATER EUROPEAN OR                    IS DEVELOPED TUTELARY OR                    X SQ=  41.63
    EUROPEAN DERIVED, RATHER THAN                    UNDEVELOPED TUTELARY, RATHER THAN           P =     0.
    DEVELOPED TUTELARY OR                            EARLY OR LATER EUROPEAN OR                  RV YES (YES)
    UNDEVELOPED TUTELARY  (51)                       EUROPEAN DERIVED  (55)

86  ALWAYS ARE THOSE                     1.00    86  TEND LESS TO BE THOSE                0.54      32    43
    WHERE THE STAGE OF                               WHERE THE STAGE OF                              0    37
    POLITICAL MODERNIZATION IS                       POLITICAL MODERNIZATION IS                  X SQ=  20.06
    ADVANCED OR MID-TRANSITIONAL,                    ADVANCED OR MID-TRANSITIONAL,               P =     0.
    RATHER THAN EARLY TRANSITIONAL  (76)             RATHER THAN EARLY TRANSITIONAL  (76)        RV NO  (NO )

87  TEND MORE TO BE THOSE                0.88    87  TEND LESS TO BE THOSE                0.56       3    27
    WHOSE IDEOLOGICAL ORIENTATION                    WHOSE IDEOLOGICAL ORIENTATION                  22    35
    IS OTHER THAN DEVELOPMENTAL  (58)                IS OTHER THAN DEVELOPMENTAL  (58)           X SQ=   6.51
                                                                                                 P =     0.006
                                                                                                 RV NO  (NO )

89  TEND TO BE THOSE                     0.74    89  TEND TO BE THOSE                     0.77      17    16
    WHOSE IDEOLOGICAL ORIENTATION                    WHOSE IDEOLOGICAL ORIENTATION                   6    54
    IS CONVENTIONAL  (33)                            IS OTHER THAN CONVENTIONAL  (62)            X SQ=  17.55
                                                                                                 P =     0.
                                                                                                 RV YES (YES)

92  ALWAYS ARE THOSE                     1.00    92  TEND LESS TO BE THOSE                0.76       0    19
    WHERE THE SYSTEM STYLE                           WHERE THE SYSTEM STYLE                         31    61
    IS LIMITED MOBILIZATIONAL, OR                    IS LIMITED MOBILIZATIONAL, OR               X SQ=   7.29
    NON-MOBILIZATIONAL  (93)                         NON-MOBILIZATIONAL  (93)                    P =     0.001
                                                                                                 RV NO  (NO )

93  TEND MORE TO BE THOSE                0.94    93  TEND LESS TO BE THOSE                0.62       2    29
    WHERE THE SYSTEM STYLE                           WHERE THE SYSTEM STYLE                         29    48
    IS NON-MOBILIZATIONAL  (78)                      IS NON-MOBILIZATIONAL  (78)                 X SQ=   9.05
                                                                                                 P =     0.001
                                                                                                 RV NO  (NO )

94  TEND TO BE THOSE                     0.76    94  TEND TO BE THOSE                     0.52      22    29
    WHERE THE STATUS OF THE REGIME                   WHERE THE STATUS OF THE REGIME                  7    32
    IS CONSTITUTIONAL  (51)                          IS AUTHORITARIAN OR                         X SQ=   5.32
                                                     TOTALITARIAN  (41)                          P =     0.013
                                                                                                 RV YES (NO )

95  TILT MORE TOWARD BEING THOSE         0.94    95  TILT LESS TOWARD BEING THOSE         0.82      29    64
    WHERE THE STATUS OF THE REGIME                   WHERE THE STATUS OF THE REGIME                  2    14
    IS CONSTITUTIONAL OR                             IS CONSTITUTIONAL OR                        X SQ=   1.51
    AUTHORITARIAN  (95)                              AUTHORITARIAN  (95)                         P =     0.148
                                                                                                 RV NO  (NO )
```

99	TEND TO BE THOSE WHERE GOVERNMENTAL STABILITY IS MODERATELY PRESENT AND DATES FROM THE POST-WAR PERIOD, OR IS ABSENT (36)	0.59	99	TEND TO BE THOSE WHERE GOVERNMENTAL STABILITY IS GENERALLY PRESENT AND DATES FROM AT LEAST THE INTER-WAR PERIOD, OR FROM THE POST-WAR PERIOD (50)	0.67	12 37 17 18 X SQ= 4.23 P = 0.035 RV YES (NO)
101	TEND TO BE THOSE WHERE THE REPRESENTATIVE CHARACTER OF THE REGIME IS POLYARCHIC (41)	0.59	101	TEND TO BE THOSE WHERE THE REPRESENTATIVE CHARACTER OF THE REGIME IS LIMITED POLYARCHIC, PSEUDO-POLYARCHIC, OR NON-POLYARCHIC (57)	0.64	17 24 12 43 X SQ= 3.42 P = 0.045 RV YES (NO)
102	TEND TO BE THOSE WHERE THE REPRESENTATIVE CHARACTER OF THE REGIME IS POLYARCHIC OR LIMITED POLYARCHIC (59)	0.77	102	TEND TO BE THOSE WHERE THE REPRESENTATIVE CHARACTER OF THE REGIME IS PSEUDO-POLYARCHIC OR NON-POLYARCHIC (49)	0.53	23 36 7 40 X SQ= 6.34 P = 0.009 RV YES (NO)
105	TEND TO BE THOSE WHERE THE ELECTORAL SYSTEM IS COMPETITIVE (43)	0.78	105	TEND TO BE THOSE WHERE THE ELECTORAL SYSTEM IS PARTIALLY COMPETITIVE OR NON-COMPETITIVE (47)	0.64	21 22 6 39 X SQ= 11.42 P = 0. RV YES (NO)
107	TEND TO BE THOSE WHERE AUTONOMOUS GROUPS ARE FULLY TOLERATED IN POLITICS (46)	0.68	107	TEND TO BE THOSE WHERE AUTONOMOUS GROUPS ARE PARTIALLY TOLERATED IN POLITICS, ARE TOLERATED ONLY OUTSIDE POLITICS, OR ARE NOT TOLERATED AT ALL (65)	0.68	21 25 10 53 X SQ= 10.17 P = 0.001 RV YES (NO)
114	TILT TOWARD BEING THOSE WHERE SECTIONALISM IS NEGLIGIBLE (47)	0.57	114	TILT TOWARD BEING THOSE WHERE SECTIONALISM IS EXTREME OR MODERATE (61)	0.60	12 47 16 31 X SQ= 1.87 P = 0.126 RV YES (NO)
116	TEND TO BE THOSE WHERE INTEREST ARTICULATION BY ASSOCIATIONAL GROUPS IS SIGNIFICANT OR MODERATE (32)	0.53	116	TEND TO BE THOSE WHERE INTEREST ARTICULATION BY ASSOCIATIONAL GROUPS IS LIMITED OR NEGLIGIBLE (79)	0.80	16 16 14 63 X SQ= 9.93 P = 0.002 RV YES (YES)
117	TEND TO BE THOSE WHERE INTEREST ARTICULATION BY ASSOCIATIONAL GROUPS IS SIGNIFICANT, MODERATE, OR LIMITED (60)	0.87	117	TEND TO BE THOSE WHERE INTEREST ARTICULATION BY ASSOCIATIONAL GROUPS IS NEGLIGIBLE (51)	0.57	26 34 4 45 X SQ= 15.01 P = 0. RV YES (NO)
118	TEND MORE TO BE THOSE WHERE INTEREST ARTICULATION BY INSTITUTIONAL GROUPS IS SIGNIFICANT, MODERATE, OR LIMITED (60)	0.78	118	TEND LESS TO BE THOSE WHERE INTEREST ARTICULATION BY INSTITUTIONAL GROUPS IS SIGNIFICANT, MODERATE, OR LIMITED (60)	0.53	7 31 25 35 X SQ= 4.71 P = 0.026 RV NO (NO)

121	TEND TO BE THOSE WHERE INTEREST ARTICULATION BY NON-ASSOCIATIONAL GROUPS IS MODERATE, LIMITED, OR NEGLIGIBLE (61)	0.94	121	TEND TO BE THOSE WHERE INTEREST ARTICULATION BY NON-ASSOCIATIONAL GROUPS IS SIGNIFICANT (54)	0.62	2 50 30 31 X SQ= 26.23 P = 0. RV YES (NO)
122	TEND TO BE THOSE WHERE INTEREST ARTICULATION BY NON-ASSOCIATIONAL GROUPS IS LIMITED OR NEGLIGIBLE (32)	0.53	122	TEND TO BE THOSE WHERE INTEREST ARTICULATION BY NON-ASSOCIATIONAL GROUPS IS SIGNIFICANT OR MODERATE (83)	0.81	15 66 17 15 X SQ= 11.88 P = 0. RV YES (YES)
124	ALWAYS ARE THOSE WHERE INTEREST ARTICULATION BY ANOMIC GROUPS IS OCCASIONAL, INFREQUENT, OR VERY INFREQUENT (85)	1.00	124	TEND LESS TO BE THOSE WHERE INTEREST ARTICULATION BY ANOMIC GROUPS IS OCCASIONAL, INFREQUENT, OR VERY INFREQUENT (85)	0.80	0 14 27 56 X SQ= 4.80 P = 0.009 RV NO (NO)
128	TEND TO BE THOSE WHERE INTEREST ARTICULATION BY POLITICAL PARTIES IS SIGNIFICANT OR MODERATE (48)	0.75	128	TEND TO BE THOSE WHERE INTEREST ARTICULATION BY POLITICAL PARTIES IS LIMITED OR NEGLIGIBLE (45)	0.58	21 27 7 37 X SQ= 7.14 P = 0.006 RV YES (NO)
129	TEND TO BE THOSE WHERE INTEREST ARTICULATION BY POLITICAL PARTIES IS SIGNIFICANT, MODERATE, OR LIMITED (56)	0.89	129	TEND TO BE THOSE WHERE INTEREST ARTICULATION BY POLITICAL PARTIES IS NEGLIGIBLE (37)	0.52	25 31 3 33 X SQ= 11.98 P = 0. RV YES (NO)
131	TEND TO BE THOSE WHERE INTEREST AGGREGATION BY POLITICAL PARTIES IS LIMITED OR NEGLIGIBLE (35)	0.71	131	TEND TO BE THOSE WHERE INTEREST AGGREGATION BY POLITICAL PARTIES IS SIGNIFICANT OR MODERATE (30)	0.56	7 23 17 18 X SQ= 3.40 P = 0.043 RV YES (NO)
133	LEAN MORE TOWARD BEING THOSE WHERE INTEREST AGGREGATION BY THE EXECUTIVE IS MODERATE, LIMITED, OR NEGLIGIBLE (76)	0.87	133	LEAN LESS TOWARD BEING THOSE WHERE INTEREST AGGREGATION BY THE EXECUTIVE IS MODERATE, LIMITED, OR NEGLIGIBLE (76)	0.67	4 24 26 49 X SQ= 3.17 P = 0.052 RV NO (NO)
138	TEND TO BE THOSE WHERE INTEREST AGGREGATION BY THE LEGISLATURE IS SIGNIFICANT, MODERATE, OR LIMITED (48)	0.82	138	TEND TO BE THOSE WHERE INTEREST AGGREGATION BY THE LEGISLATURE IS NEGLIGIBLE (49)	0.63	23 25 5 42 X SQ= 14.13 P = 0. RV YES (NO)
139	TEND MORE TO BE THOSE WHERE THE PARTY SYSTEM IS QUANTITATIVELY OTHER THAN ONE-PARTY (71)	0.90	139	TEND LESS TO BE THOSE WHERE THE PARTY SYSTEM IS QUANTITATIVELY OTHER THAN ONE-PARTY (71)	0.59	3 30 27 43 X SQ= 8.07 P = 0.002 RV NO (NO)

142	TEND TO BE THOSE WHERE THE PARTY SYSTEM IS QUANTITATIVELY MULTI-PARTY (30)	0.57	142	TEND TO BE THOSE WHERE THE PARTY SYSTEM IS QUANTITATIVELY OTHER THAN MULTI-PARTY (66)	0.79	16 14 12 52 X SQ= 10.09 P = 0.001 RV YES (YES)
144	LEAN TOWARD BEING THOSE WHERE THE PARTY SYSTEM IS QUANTITATIVELY TWO-PARTY, RATHER THAN ONE-PARTY (11)	0.60	144	LEAN TOWARD BEING THOSE WHERE THE PARTY SYSTEM IS QUANTITATIVELY ONE-PARTY, RATHER THAN TWO-PARTY (34)	0.79	2 31 3 8 X SQ= 1.88 P = 0.091 RV YES (NO)
146	ALWAYS ARE THOSE WHERE THE PARTY SYSTEM IS QUALITATIVELY OTHER THAN MASS-BASED TERRITORIAL (92)	1.00	146	TILT LESS TOWARD BEING THOSE WHERE THE PARTY SYSTEM IS QUALITATIVELY OTHER THAN MASS-BASED TERRITORIAL (92)	0.90	0 7 29 63 X SQ= 1.78 P = 0.102 RV NO (NO)
148	ALWAYS ARE THOSE WHERE THE PARTY SYSTEM IS QUALITATIVELY OTHER THAN AFRICAN TRANSITIONAL (96)	1.00	148	TEND LESS TO BE THOSE WHERE THE PARTY SYSTEM IS QUALITATIVELY OTHER THAN AFRICAN TRANSITIONAL (96)	0.82	0 14 31 64 X SQ= 4.88 P = 0.009 RV NO (NO)
153	TEND TO BE THOSE WHERE THE PARTY SYSTEM IS MODERATELY STABLE OR UNSTABLE (38)	0.68	153	TEND TO BE THOSE WHERE THE PARTY SYSTEM IS STABLE (42)	0.63	9 32 19 19 X SQ= 5.61 P = 0.011 RV YES (YES)
158	LEAN TOWARD BEING THOSE WHERE PERSONALISMO IS PRONOUNCED OR MODERATE (40)	0.57	158	LEAN TOWARD BEING THOSE WHERE PERSONALISMO IS NEGLIGIBLE (56)	0.65	17 23 13 42 X SQ= 2.99 P = 0.073 RV YES (NO)
163	LEAN MORE TOWARD BEING THOSE WHERE THE REGIME'S LEADERSHIP CHARISMA IS MODERATE OR NEGLIGIBLE (87)	0.97	163	LEAN LESS TOWARD BEING THOSE WHERE THE REGIME'S LEADERSHIP CHARISMA IS MODERATE OR NEGLIGIBLE (87)	0.85	1 10 31 56 X SQ= 2.04 P = 0.096 RV NO (NO)
164	TEND MORE TO BE THOSE WHERE THE REGIME'S LEADERSHIP CHARISMA IS NEGLIGIBLE (65)	0.97	164	TEND LESS TO BE THOSE WHERE THE REGIME'S LEADERSHIP CHARISMA IS NEGLIGIBLE (65)	0.52	1 31 31 34 X SQ= 17.30 P = 0. RV NO (NO)
169	TEND TO BE THOSE WHERE THE HORIZONTAL POWER DISTRIBUTION IS SIGNIFICANT OR LIMITED (58)	0.83	169	TEND TO BE THOSE WHERE THE HORIZONTAL POWER DISTRIBUTION IS NEGLIGIBLE (48)	0.55	25 33 5 41 X SQ= 11.46 P = 0. RV YES (NO)

170	TEND TO BE THOSE WHERE THE LEGISLATIVE-EXECUTIVE STRUCTURE IS PRESIDENTIAL (39)	0.55	170	TEND TO BE THOSE WHERE THE LEGISLATIVE-EXECUTIVE STRUCTURE IS OTHER THAN PRESIDENTIAL (63)	0.70	17 21 14 48 X SQ= 4.42 P = 0.026 RV YES (NO)
175	TEND TO BE THOSE WHERE THE LEGISLATURE IS FULLY EFFECTIVE OR PARTIALLY EFFECTIVE (51)	0.76	175	TEND TO BE THOSE WHERE THE LEGISLATURE IS LARGELY INEFFECTIVE OR WHOLLY INEFFECTIVE (49)	0.58	22 29 7 40 X SQ= 8.06 P = 0.004 RV YES (NO)
178	TEND TO BE THOSE WHERE THE LEGISLATURE IS BICAMERAL (51)	0.71	178	TEND TO BE THOSE WHERE THE LEGISLATURE IS UNICAMERAL (53)	0.61	9 43 22 28 X SQ= 7.37 P = 0.005 RV YES (NO)
179	LEAN TOWARD BEING THOSE WHERE THE EXECUTIVE IS STRONG (39)	0.62	179	LEAN TOWARD BEING THOSE WHERE THE EXECUTIVE IS DOMINANT (52)	0.63	9 41 15 24 X SQ= 3.68 P = 0.053 RV YES (NO)
182	ALWAYS ARE THOSE WHERE THE BUREAUCRACY IS SEMI-MODERN, RATHER THAN POST-COLONIAL TRANSITIONAL (55)	1.00	182	TEND LESS TO BE THOSE WHERE THE BUREAUCRACY IS SEMI-MODERN, RATHER THAN POST-COLONIAL TRANSITIONAL (55)	0.55	23 31 0 25 X SQ= 13.03 P = 0. RV NO (NO)
185	TEND TO BE THOSE WHERE PARTICIPATION BY THE MILITARY IN POLITICS IS INTERVENTIVE, RATHER THAN SUPPORTIVE (21)	0.80	185	TEND TO BE THOSE WHERE PARTICIPATION BY THE MILITARY IN POLITICS IS SUPPORTIVE, RATHER THAN INTERVENTIVE (31)	0.74	12 9 3 26 X SQ= 10.57 P = 0.001 RV YES (YES)
186	LEAN MORE TOWARD BEING THOSE WHERE PARTICIPATION BY THE MILITARY IN POLITICS IS NEUTRAL, RATHER THAN SUPPORTIVE (56)	0.83	186	LEAN LESS TOWARD BEING THOSE WHERE PARTICIPATION BY THE MILITARY IN POLITICS IS NEUTRAL, RATHER THAN SUPPORTIVE (56)	0.61	3 26 15 41 X SQ= 2.19 P = 0.098 RV NO (NO)
194	ALWAYS ARE THOSE THAT ARE NON-COMMUNIST (101)	1.00	194	TEND LESS TO BE THOSE THAT ARE NON-COMMUNIST (101)	0.84	0 13 32 67 X SQ= 4.41 P = 0.018 RV NO (NO)

```
189  POLITIES                                          189  POLITIES
     WHERE THE CHARACTER OF THE LEGAL SYSTEM                WHERE THE CHARACTER OF THE LEGAL SYSTEM
     IS COMMON LAW  (7)                                     IS OTHER THAN COMMON LAW  (108)

                                                                               BOTH SUBJECT AND PREDICATE

      7 IN LEFT COLUMN

AUSTRALIA   IRELAND     JAMAICA     NEW ZEALAND TRINIDAD    UK          US

     108 IN RIGHT COLUMN

AFGHANISTAN ALBANIA     ALGERIA     ARGENTINA   AUSTRIA     BELGIUM     BOLIVIA     BRAZIL      BULGARIA    BURMA
BURUNDI     CAMBODIA    CAMEROUN    CANADA      CEN AFR REP CEYLON      CHAD        CHILE       CHINA, PR   COLOMBIA
CONGO(BRA)  CONGO(LEO)  COSTA RICA  CUBA        CYPRUS      CZECHOS'KIA DAHOMEY     DENMARK     DOMIN REP   ECUADOR
EL SALVADOR ETHIOPIA    FINLAND     FRANCE      GABON       GERMANY, E  GERMAN FR   GHANA       GREECE      GUATEMALA
GUINEA      HAITI       HONDURAS    HUNGARY     ICELAND     INDIA       INDONESIA   IRAN        IRAQ        ISRAEL
ITALY       IVORY COAST JAPAN       JORDAN      KOREA, N    KOREA REP   LAOS        LEBANON     LIBERIA     LIBYA
LUXEMBOURG  MALAGASY R  MALAYA      MALI        MAURITANIA  MEXICO      MONGOLIA    MOROCCO     NEPAL       NETHERLANDS
NICARAGUA   NIGER       NIGERIA     NORWAY      PAKISTAN    PANAMA      PARAGUAY    PERU        PHILIPPINES POLAND
PORTUGAL    RUMANIA     RWANDA      SA'U ARABIA SENEGAL     SIERRE LEO  SOMALIA     SO AFRICA   SPAIN       SUDAN
SWEDEN      SWITZERLAND SYRIA       TANGANYIKA  THAILAND    TOGO        TUNISIA     TURKEY      UGANDA      USSR
UAR         UPPER VOLTA URUGUAY     VENEZUELA   VIETNAM, N  VIETNAM REP YEMEN       YUGOSLAVIA

---------------------------------------------------------------------------------------------------------------

  2  TEND TO BE THOSE                           0.71         2  TEND TO BE THOSE                          0.84         5   17
     LOCATED IN WEST EUROPE, SCANDINAVIA,                       LOCATED ELSEWHERE THAN IN                               2   91
     NORTH AMERICA, OR AUSTRALASIA  (22)                        WEST EUROPE, SCANDINAVIA,                          X SQ=   9.82
                                                                NORTH AMERICA, OR AUSTRALASIA  (93)                P =    0.003
                                                                                                                   RV YES (NO )

 10  ALWAYS ARE THOSE                           1.00        10  TILT LESS TOWARD BEING THOSE              0.69         0   33
     LOCATED ELSEWHERE THAN IN NORTH AFRICA,                    LOCATED ELSEWHERE THAN IN NORTH AFRICA,                 7   75
     OR CENTRAL AND SOUTH AFRICA  (82)                          OR CENTRAL AND SOUTH AFRICA  (82)                  X SQ=   1.69
                                                                                                                   P =    0.107
                                                                                                                   RV NO  (NO )

 29  ALWAYS ARE THOSE                           1.00        29  TEND LESS TO BE THOSE                     0.51         6   50
     WHERE THE DEGREE OF URBANIZATION                           WHERE THE DEGREE OF URBANIZATION                        0   49
     IS HIGH  (56)                                              IS HIGH  (56)                                      X SQ=   3.76
                                                                                                                   P =    0.029
                                                                                                                   RV NO  (NO )

 30  ALWAYS ARE THOSE                           1.00        30  TEND TO BE THOSE                          0.53         0   56
     WHOSE AGRICULTURAL POPULATION                              WHOSE AGRICULTURAL POPULATION                           7   50
     IS MEDIUM, LOW, OR VERY LOW  (57)                          IS HIGH  (56)                                      X SQ=   5.37
                                                                                                                   P =    0.013
                                                                                                                   RV YES (NO )
```

31	TEND TO BE THOSE WHOSE AGRICULTURAL POPULATION IS LOW OR VERY LOW (24)	0.71	31	TEND TO BE THOSE WHOSE AGRICULTURAL POPULATION IS HIGH OR MEDIUM (90)	0.82	2 88 5 19 X SQ= 8.39 P = 0.004 RV YES (NO)
35	TEND TO BE THOSE WHOSE PER CAPITA GROSS NATIONAL PRODUCT IS VERY HIGH OR HIGH (24)	0.71	35	TEND TO BE THOSE WHOSE PER CAPITA GROSS NATIONAL PRODUCT IS MEDIUM, LOW, OR VERY LOW (91)	0.82	5 19 2 89 X SQ= 8.51 P = 0.004 RV YES (NO)
36	ALWAYS ARE THOSE WHOSE PER CAPITA GROSS NATIONAL PRODUCT IS VERY HIGH, HIGH, OR MEDIUM (42)	1.00	36	TEND TO BE THOSE WHOSE PER CAPITA GROSS NATIONAL PRODUCT IS LOW OR VERY LOW (73)	0.68	7 35 0 73 X SQ= 10.20 P = 0.001 RV YES (NO)
41	TEND TO BE THOSE WHOSE ECONOMIC DEVELOPMENTAL STATUS IS DEVELOPED (19)	0.57	41	TEND TO BE THOSE WHOSE ECONOMIC DEVELOPMENTAL STATUS IS INTERMEDIATE, UNDERDEVELOPED, OR VERY UNDERDEVELOPED (94)	0.86	4 15 3 91 X SQ= 5.88 P = 0.015 RV YES (NO)
42	TEND TO BE THOSE WHOSE ECONOMIC DEVELOPMENTAL STATUS IS DEVELOPED OR INTERMEDIATE (36)	0.71	42	TEND TO BE THOSE WHOSE ECONOMIC DEVELOPMENTAL STATUS IS UNDERDEVELOPED OR VERY UNDERDEVELOPED (76)	0.70	5 31 2 74 X SQ= 3.54 P = 0.034 RV YES (NO)
43	ALWAYS ARE THOSE WHOSE ECONOMIC DEVELOPMENTAL STATUS IS DEVELOPED, INTERMEDIATE, OR UNDERDEVELOPED (55)	1.00	43	TEND TO BE THOSE WHOSE ECONOMIC DEVELOPMENTAL STATUS IS VERY UNDERDEVELOPED (57)	0.54	7 48 0 57 X SQ= 5.72 P = 0.006 RV YES (NO)
44	TEND TO BE THOSE WHERE THE LITERACY RATE IS NINETY PERCENT OR ABOVE (25)	0.71	44	TEND TO BE THOSE WHERE THE LITERACY RATE IS BELOW NINETY PERCENT (90)	0.81	5 20 2 88 X SQ= 7.93 P = 0.005 RV YES (NO)
45	ALWAYS ARE THOSE WHERE THE LITERACY RATE IS FIFTY PERCENT OR ABOVE (55)	1.00	45	TEND TO BE THOSE WHERE THE LITERACY RATE IS BELOW FIFTY PERCENT (54)	0.53	7 48 0 54 X SQ= 5.38 P = 0.013 RV YES (NO)
50	ALWAYS ARE THOSE WHERE FREEDOM OF THE PRESS IS COMPLETE (43)	1.00	50	TEND TO BE THOSE WHERE FREEDOM OF THE PRESS IS INTERMITTENT, INTERNALLY ABSENT, OR INTERNALLY AND EXTERNALLY ABSENT (56)	0.61	7 36 0 56 X SQ= 7.49 P = 0.002 RV YES (NO)

53	TEND TO BE THOSE WHERE NEWSPAPER CIRCULATION IS THREE HUNDRED OR MORE PER THOUSAND (14)	0.57	53	TEND TO BE THOSE WHERE NEWSPAPER CIRCULATION IS LESS THAN THREE HUNDRED PER THOUSAND (101)	0.91	4 10 3 98 X SQ= 9.97 P = 0.004 RV YES (NO)
54	TEND TO BE THOSE WHERE NEWSPAPER CIRCULATION IS ONE HUNDRED OR MORE PER THOUSAND (37)	0.71	54	TEND TO BE THOSE WHERE NEWSPAPER CIRCULATION IS LESS THAN ONE HUNDRED PER THOUSAND (76)	0.70	5 32 2 74 X SQ= 3.37 P = 0.037 RV YES (NO)
55	ALWAYS ARE THOSE WHERE NEWSPAPER CIRCULATION IS TEN OR MORE PER THOUSAND (78)	1.00	55	LEAN LESS TOWARD BEING THOSE WHERE NEWSPAPER CIRCULATION IS TEN OR MORE PER THOUSAND (78)	0.67	7 71 0 35 X SQ= 1.98 P = 0.097 RV NO (NO)
56	ALWAYS ARE THOSE WHERE THE RELIGION IS PREDOMINANTLY LITERATE (79)	1.00	56	TILT LESS TOWARD BEING THOSE WHERE THE RELIGION IS PREDOMINANTLY LITERATE (79)	0.70	7 72 0 31 X SQ= 1.63 P = 0.109 RV NO (NO)
63	TEND TO BE THOSE WHERE THE RELIGION IS PREDOMINANTLY SOME KIND OF CHRISTIAN (46)	0.86	63	TEND TO BE THOSE WHERE THE RELIGION IS PREDOMINANTLY OR PARTLY OTHER THAN CHRISTIAN (68)	0.63	6 40 1 67 X SQ= 4.53 P = 0.017 RV YES (NO)
66	TEND TO BE THOSE THAT ARE RELIGIOUSLY HETEROGENEOUS (49)	0.86	66	TEND TO BE THOSE THAT ARE RELIGIOUSLY HOMOGENEOUS (57)	0.57	1 56 6 43 X SQ= 3.15 P = 0.047 RV YES (NO)
68	TEND TO BE THOSE THAT ARE LINGUISTICALLY HOMOGENEOUS (52)	0.86	68	TEND TO BE THOSE THAT ARE LINGUISTICALLY WEAKLY HETEROGENEOUS OR STRONGLY HETEROGENEOUS (62)	0.57	6 46 1 61 X SQ= 3.27 P = 0.046 RV YES (NO)
69	ALWAYS ARE THOSE THAT ARE LINGUISTICALLY HOMOGENEOUS OR WEAKLY HETEROGENEOUS (64)	1.00	69	TEND LESS TO BE THOSE THAT ARE LINGUISTICALLY HOMOGENEOUS OR WEAKLY HETEROGENEOUS (64)	0.53	7 57 0 50 X SQ= 4.08 P = 0.018 RV NO (NO)
74	TEND TO BE THOSE THAT ARE HISTORICALLY WESTERN (26)	0.71	74	TEND TO BE THOSE THAT ARE NOT HISTORICALLY WESTERN (87)	0.80	5 21 2 85 X SQ= 7.18 P = 0.007 RV YES (NO)

No.	Left		No.	Right		Statistics
75	ALWAYS ARE THOSE THAT ARE HISTORICALLY WESTERN OR SIGNIFICANTLY WESTERNIZED (62)	1.00	75	TEND LESS TO BE THOSE THAT ARE HISTORICALLY WESTERN OR SIGNIFICANTLY WESTERNIZED (62)	0.51	7 55 0 52 X SQ= 4.45 P = 0.015 RV NO (NO)
76	TEND TO BE THOSE THAT ARE HISTORICALLY WESTERN, RATHER THAN HAVING BEEN SIGNIFICANTLY OR PARTIALLY WESTERNIZED THROUGH A COLONIAL RELATIONSHIP (26)	0.71	76	TEND TO BE THOSE THAT HAVE BEEN SIGNIFICANTLY OR PARTIALLY WESTERNIZED THROUGH A COLONIAL RELATIONSHIP, RATHER THAN BEING HISTORICALLY WESTERN (70)	0.76	5 21 2 68 X SQ= 5.29 P = 0.015 RV YES (NO)
80	ALWAYS ARE THOSE WHOSE DATE OF INDEPENDENCE IS AFTER 1914, AND THAT WERE FORMERLY DEPENDENCIES OF BRITAIN, RATHER THAN FRANCE (19)	1.00	80	LEAN TOWARD BEING THOSE WHOSE DATE OF INDEPENDENCE IS AFTER 1914, AND THAT WERE FORMERLY DEPENDENCIES OF FRANCE, RATHER THAN BRITAIN (24)	0.60	3 16 0 24 X SQ= 2.00 P = 0.079 RV YES (NO)
81	TEND TO BE THOSE WHERE THE TYPE OF POLITICAL MODERNIZATION IS EARLY EUROPEAN OR EARLY EUROPEAN DERIVED, RATHER THAN LATER EUROPEAN OR LATER EUROPEAN DERIVED (11)	0.80	81	TEND TO BE THOSE WHERE THE TYPE OF POLITICAL MODERNIZATION IS LATER EUROPEAN OR LATER EUROPEAN DERIVED, RATHER THAN EARLY EUROPEAN OR EARLY EUROPEAN DERIVED (40)	0.85	4 7 1 39 X SQ= 7.69 P = 0.006 RV YES (NO)
89	ALWAYS ARE THOSE WHOSE IDEOLOGICAL ORIENTATION IS CONVENTIONAL (33)	1.00	89	TEND TO BE THOSE WHOSE IDEOLOGICAL ORIENTATION IS OTHER THAN CONVENTIONAL (62)	0.69	5 28 0 62 X SQ= 7.11 P = 0.004 RV YES (NO)
93	ALWAYS ARE THOSE WHERE THE SYSTEM STYLE IS NON-MOBILIZATIONAL (78)	1.00	93	TILT LESS TOWARD BEING THOSE WHERE THE SYSTEM STYLE IS NON-MOBILIZATIONAL (78)	0.69	0 32 7 71 X SQ= 1.75 P = 0.104 RV NO (NO)
94	ALWAYS ARE THOSE WHERE THE STATUS OF THE REGIME IS CONSTITUTIONAL (51)	1.00	94	TEND LESS TO BE THOSE WHERE THE STATUS OF THE REGIME IS CONSTITUTIONAL (51)	0.52	7 44 0 41 X SQ= 4.29 P = 0.016 RV NO (NO)
98	TEND TO BE THOSE WHERE GOVERNMENTAL STABILITY IS GENERALLY PRESENT AND DATES AT LEAST FROM THE INTERWAR PERIOD (22)	0.71	98	TEND TO BE THOSE WHERE GOVERNMENTAL STABILITY IS GENERALLY OR MODERATELY PRESENT AND DATES FROM THE POST-WAR PERIOD, OR IS ABSENT (93)	0.84	5 17 2 91 X SQ= 9.82 P = 0.003 RV YES (NO)
99	ALWAYS ARE THOSE WHERE GOVERNMENTAL STABILITY IS GENERALLY PRESENT AND DATES FROM AT LEAST THE INTER-WAR PERIOD, OR FROM THE POST-WAR PERIOD (50)	1.00	99	LEAN LESS TOWARD BEING THOSE WHERE GOVERNMENTAL STABILITY IS GENERALLY PRESENT AND DATES FROM AT LEAST THE INTER-WAR PERIOD, OR FROM THE POST-WAR PERIOD (50)	0.56	5 45 0 36 X SQ= 2.21 P = 0.072 RV NO (NO)

101	ALWAYS ARE THOSE WHERE THE REPRESENTATIVE CHARACTER OF THE REGIME IS POLYARCHIC (41)	1.00	101	TEND TO BE THOSE WHERE THE REPRESENTATIVE CHARACTER OF THE REGIME IS LIMITED POLYARCHIC, PSEUDO-POLYARCHIC, OR NON-POLYARCHIC (57)	0.63	7 34 0 57 X SQ= 8.06 P = 0.002 RV YES (NO)
105	ALWAYS ARE THOSE WHERE THE ELECTORAL SYSTEM IS COMPETITIVE (43)	1.00	105	TEND TO BE THOSE WHERE THE ELECTORAL SYSTEM IS PARTIALLY COMPETITIVE OR NON-COMPETITIVE (47)	0.57	7 36 0 47 X SQ= 6.18 P = 0.004 RV YES (NO)
107	ALWAYS ARE THOSE WHERE AUTONOMOUS GROUPS ARE FULLY TOLERATED IN POLITICS (46)	1.00	107	TEND TO BE THOSE WHERE AUTONOMOUS GROUPS ARE PARTIALLY TOLERATED IN POLITICS, ARE TOLERATED ONLY OUTSIDE POLITICS, OR ARE NOT TOLERATED AT ALL (65)	0.62	7 39 0 65 X SQ= 8.14 P = 0.002 RV YES (NO)
110	TEND TO BE THOSE WHERE POLITICAL ENCULTURATION IS HIGH (15)	0.57	110	TEND TO BE THOSE WHERE POLITICAL ENCULTURATION IS MEDIUM OR LOW (80)	0.87	4 11 3 77 X SQ= 6.65 P = 0.011 RV YES (NO)
111	ALWAYS ARE THOSE WHERE POLITICAL ENCULTURATION IS HIGH OR MEDIUM (53)	1.00	111	TEND LESS TO BE THOSE WHERE POLITICAL ENCULTURATION IS HIGH OR MEDIUM (53)	0.52	7 46 0 42 X SQ= 4.21 P = 0.016 RV NO (NO)
115	TEND TO BE THOSE WHERE INTEREST ARTICULATION BY ASSOCIATIONAL GROUPS IS SIGNIFICANT (20)	0.57	115	TEND TO BE THOSE WHERE INTEREST ARTICULATION BY ASSOCIATIONAL GROUPS IS MODERATE, LIMITED, OR NEGLIGIBLE (91)	0.85	4 16 3 88 X SQ= 5.17 P = 0.019 RV YES (NO)
116	TEND TO BE THOSE WHERE INTEREST ARTICULATION BY ASSOCIATIONAL GROUPS IS SIGNIFICANT OR MODERATE (32)	0.71	116	TEND TO BE THOSE WHERE INTEREST ARTICULATION BY ASSOCIATIONAL GROUPS IS LIMITED OR NEGLIGIBLE (79)	0.74	5 27 2 77 X SQ= 4.58 P = 0.020 RV YES (NO)
117	ALWAYS ARE THOSE WHERE INTEREST ARTICULATION BY ASSOCIATIONAL GROUPS IS SIGNIFICANT, MODERATE, OR LIMITED (60)	1.00	117	TEND LESS TO BE THOSE WHERE INTEREST ARTICULATION BY ASSOCIATIONAL GROUPS IS SIGNIFICANT, MODERATE, OR LIMITED (60)	0.51	7 53 0 51 X SQ= 4.53 P = 0.015 RV NO (NO)
119	ALWAYS ARE THOSE WHERE INTEREST ARTICULATION BY INSTITUTIONAL GROUPS IS MODERATE OR LIMITED (26)	1.00	119	TEND TO BE THOSE WHERE INTEREST ARTICULATION BY INSTITUTIONAL GROUPS IS VERY SIGNIFICANT OR SIGNIFICANT (74)	0.80	0 74 7 19 X SQ= 17.49 P = 0. RV YES (NO)

120	TEND TO BE THOSE WHERE INTEREST ARITUCLATION BY INSTITUTIONAL GROUPS IS LIMITED (10)	0.57	120	TEND TO BE THOSE WHERE INTEREST ARTICULATION BY INSTITUTIONAL GROUPS IS VERY SIGNIFICANT, SIGNIFICANT, OR MODERATE (90)	0.94	3 87 4 6 X SQ= 13.38 P = 0.002 RV YES (NO)
121	ALWAYS ARE THOSE WHERE INTEREST ARTICULATION BY NON-ASSOCIATIONAL GROUPS IS MODERATE, LIMITED, OR NEGLIGIBLE (61)	1.00	121	TEND TO BE THOSE WHERE INTEREST ARTICULATION BY NON-ASSOCIATIONAL GROUPS IS SIGNIFICANT (54)	0.50	0 54 7 54 X SQ= 4.74 P = 0.014 RV YES (NO)
122	TEND TO BE THOSE WHERE INTEREST ARTICULATION BY NON-ASSOCIATIONAL GROUPS IS LIMITED OR NEGLIGIBLE (32)	0.71	122	TEND TO BE THOSE WHERE INTEREST ARTICULATION BY NON-ASSOCIATIONAL GROUPS IS SIGNIFICANT OR MODERATE (83)	0.75	2 81 5 27 X SQ= 4.93 P = 0.017 RV YES (NO)
125	ALWAYS ARE THOSE WHERE INTEREST ARTICULATION BY ANOMIC GROUPS IS INFREQUENT OR VERY INFREQUENT (35)	1.00	125	TEND TO BE THOSE WHERE INTEREST ARTICULATION BY ANOMIC GROUPS IS FREQUENT OR OCCASIONAL (64)	0.69	0 64 6 29 X SQ= 8.86 P = 0.001 RV YES (NO)
126	TEND TO BE THOSE WHERE INTEREST ARTICULATION BY ANOMIC GROUPS IS VERY INFREQUENT (16)	0.67	126	TEND TO BE THOSE WHERE INTEREST ARTICULATION BY ANOMIC GROUPS IS FREQUENT, OCCASIONAL, OR INFREQUENT (83)	0.87	2 81 4 12 X SQ= 8.38 P = 0.006 RV YES (NO)
128	TILT TOWARD BEING THOSE WHERE INTEREST ARTICULATION BY POLITICAL PARTIES IS SIGNIFICANT OR MODERATE (48)	0.86	128	TILT TOWARD BEING THOSE WHERE INTEREST ARTICULATION BY POLITICAL PARTIES IS LIMITED OR NEGLIGIBLE (45)	0.51	6 42 1 44 X SQ= 2.20 P = 0.112 RV YES (NO)
130	TEND TO BE THOSE WHERE INTEREST AGGREGATION BY POLITICAL PARTIES IS SIGNIFICANT (12)	0.57	130	TEND TO BE THOSE WHERE INTEREST AGGREGATION BY POLITICAL PARTIES IS MODERATE, LIMITED, OR NEGLIGIBLE (71)	0.89	4 8 3 68 X SQ= 7.81 P = 0.007 RV YES (NO)
131	ALWAYS ARE THOSE WHERE INTEREST AGGREGATION BY POLITICAL PARTIES IS SIGNIFICANT OR MODERATE (30)	1.00	131	TEND TO BE THOSE WHERE INTEREST AGGREGATION BY POLITICAL PARTIES IS LIMITED OR NEGLIGIBLE (35)	0.60	7 23 0 35 X SQ= 6.89 P = 0.003 RV YES (NO)
134	ALWAYS ARE THOSE WHERE INTEREST AGGREGATION BY THE EXECUTIVE IS SIGNIFICANT OR MODERATE (57)	1.00	134	TEND LESS TO BE THOSE WHERE INTEREST AGGREGATION BY THE EXECUTIVE IS SIGNIFICANT OR MODERATE (57)	0.52	7 50 0 46 X SQ= 4.28 P = 0.016 RV NO (NO)

137	ALWAYS ARE THOSE WHERE INTEREST AGGREGATION BY THE LEGISLATURE IS SIGNIFICANT OR MODERATE (29)	1.00	137	TEND TO BE THOSE WHERE INTEREST AGGREGATION BY THE LEGISLATURE IS LIMITED OR NEGLIGIBLE (68)	0.76	7 22 0 68 X SQ= 14.27 P = 0. RV YES (NO)
139	ALWAYS ARE THOSE WHERE THE PARTY SYSTEM IS QUANTITATIVELY OTHER THAN ONE-PARTY (71)	1.00	139	LEAN LESS TOWARD BEING THOSE WHERE THE PARTY SYSTEM IS QUANTITATIVELY OTHER THAN ONE-PARTY (71)	0.65	0 34 7 64 X SQ= 2.18 P = 0.093 RV NO (NO)
141	TEND TO BE THOSE WHERE THE PARTY SYSTEM IS QUANTITATIVELY TWO-PARTY (11)	0.71	141	TEND TO BE THOSE WHERE THE PARTY SYSTEM IS QUANTITATIVELY OTHER THAN TWO-PARTY (87)	0.93	5 6 2 85 X SQ= 21.30 P = 0. RV YES (NO)
145	TEND TO BE THOSE WHERE THE PARTY SYSTEM IS QUANTITATIVELY TWO-PARTY, RATHER THAN MULTI-PARTY (11)	0.83	145	TEND TO BE THOSE WHERE THE PARTY SYSTEM IS QUANTITATIVELY MULTI-PARTY, RATHER THAN TWO-PARTY (30)	0.83	1 29 5 6 X SQ= 8.31 P = 0.003 RV YES (NO)
147	TEND TO BE THOSE WHERE THE PARTY SYSTEM IS QUALITATIVELY CLASS-ORIENTED OR MULTI-IDEOLOGICAL (23)	0.83	147	TEND TO BE THOSE WHERE THE PARTY SYSTEM IS QUALITATIVELY OTHER THAN CLASS-ORIENTED OR MULTI-IDEOLOGICAL (67)	0.79	5 18 1 66 X SQ= 8.26 P = 0.004 RV YES (NO)
153	ALWAYS ARE THOSE WHERE THE PARTY SYSTEM IS STABLE (42)	1.00	153	LEAN TOWARD BEING THOSE WHERE THE PARTY SYSTEM IS MODERATELY STABLE OR UNSTABLE (38)	0.51	5 37 0 38 X SQ= 3.01 P = 0.056 RV YES (NO)
158	ALWAYS ARE THOSE WHERE PERSONALISMO IS NEGLIGIBLE (56)	1.00	158	TEND LESS TO BE THOSE WHERE PERSONALISMO IS NEGLIGIBLE (56)	0.55	0 40 7 49 X SQ= 3.70 P = 0.021 RV NO (NO)
160	ALWAYS ARE THOSE WHERE THE POLITICAL LEADERSHIP IS MODERATELY ELITIST OR NON-ELITIST (67)	1.00	160	LEAN LESS TOWARD BEING THOSE WHERE THE POLITICAL LEADERSHIP IS MODERATELY ELITIST OR NON-ELITIST (67)	0.67	0 30 7 60 X SQ= 2.00 P = 0.095 RV NO (NO)
168	ALWAYS ARE THOSE WHERE THE HORIZONTAL POWER DISTRIBUTION IS SIGNIFICANT (34)	1.00	168	TEND TO BE THOSE WHERE THE HORIZONTAL POWER DISTRIBUTION IS LIMITED OR NEGLIGIBLE (72)	0.73	7 27 0 72 X SQ= 12.71 P = 0. RV YES (NO)

172	TEND TO BE THOSE WHERE THE LEGISLATIVE-EXECUTIVE STRUCTURE IS PARLIAMENTARY-ROYALIST (21)	0.71	172	TEND TO BE THOSE WHERE THE LEGISLATIVE-EXECUTIVE STRUCTURE IS OTHER THAN PARLIAMENTARY-ROYALIST (88)	0.84	5 16 2 86 X SQ= 9.75 P = 0.003 RV YES (NO)
174	ALWAYS ARE THOSE WHERE THE LEGISLATURE IS FULLY EFFECTIVE (28)	1.00	174	TEND TO BE THOSE WHERE THE LEGISLATURE IS PARTIALLY EFFECTIVE, LARGELY INEFFECTIVE, OR WHOLLY INEFFECTIVE (72)	0.77	7 21 0 72 X SQ= 15.71 P = 0. RV YES (NO)
178	LEAN TOWARD BEING THOSE WHERE THE LEGISLATURE IS BICAMERAL (51)	0.86	178	LEAN TOWARD BEING THOSE WHERE THE LEGISLATURE IS UNICAMERAL (53)	0.54	1 52 6 45 X SQ= 2.62 P = 0.058 RV YES (NO)
179	ALWAYS ARE THOSE WHERE THE EXECUTIVE IS STRONG (39)	1.00	179	TEND TO BE THOSE WHERE THE EXECUTIVE IS DOMINANT (52)	0.62	0 52 7 32 X SQ= 7.74 P = 0.002 RV YES (NO)
181	TEND TO BE THOSE WHERE THE BUREAUCRACY IS MODERN, RATHER THAN SEMI-MODERN (21)	0.71	181	TEND TO BE THOSE WHERE THE BUREAUCRACY IS SEMI-MODERN, RATHER THAN MODERN (55)	0.77	5 16 2 53 X SQ= 5.18 P = 0.015 RV YES (NO)
186	ALWAYS ARE THOSE WHERE PARTICIPATION BY THE MILITARY IN POLITICS IS NEUTRAL, RATHER THAN SUPPORTIVE (56)	1.00	186	TEND LESS TO BE THOSE WHERE PARTICIPATION BY THE MILITARY IN POLITICS IS NEUTRAL, RATHER THAN SUPPORTIVE (56)	0.61	0 31 7 49 X SQ= 2.69 P = 0.047 RV NO (NO)
187	ALWAYS ARE THOSE WHERE THE ROLE OF THE POLICE IS NOT POLITICALLY SIGNIFICANT (35)	1.00	187	TEND TO BE THOSE WHERE THE ROLE OF THE POLICE IS POLITICALLY SIGNIFICANT (66)	0.70	0 66 7 28 X SQ= 11.25 P = 0. RV YES (NO)

```
******************************************************************************************************

190  POLITIES                                          190  POLITIES
     WHERE THE CHARACTER OF THE LEGAL SYSTEM                WHERE THE CHARACTER OF THE LEGAL SYSTEM
     IS CIVIL LAW, RATHER THAN                              IS COMMON LAW, RATHER THAN
     COMMON LAW  (32)                                       CIVIL LAW  (7)


                                                                           BOTH SUBJECT AND PREDICATE
......................................................................................................

      32 IN LEFT COLUMN

ARGENTINA   AUSTRIA     BELGIUM     BOLIVIA     BRAZIL      CHILE       COLOMBIA    COSTA RICA  DOMIN REP   ECUADOR
EL SALVADOR FRANCE      GERMAN FR   GREECE      GUATEMALA   HAITI       HONDURAS    ITALY       LUXEMBOURG  MEXICO
NETHERLANDS NICARAGUA   PANAMA      PARAGUAY    PERU        PORTUGAL    SPAIN       SWITZERLAND THAILAND    TURKEY
URUGUAY     VENEZUELA

       7 IN RIGHT COLUMN

AUSTRALIA   IRELAND     JAMAICA     NEW ZEALAND TRINICAD    UK          US

       2 EXCLUDED BECAUSE AMBIGUOUS

ETHIOPIA    GUINEA

      74 EXCLUDED BECAUSE IRRELEVANT

AFGHANISTAN ALBANIA     ALGERIA     BULGARIA    BURMA       BURUNDI     CAMBODIA    CAMEROUN    CANADA      CEN AFR REP
CEYLON      CHAD        CHINA, PR   CONGO(BRA)  CONGO(LEO)  CUBA        CYPRUS      CZECHOS'KIA DAHOMEY     DENMARK
FINLAND     GABON       GERMANY, E  GHANA       HUNGARY     ICELAND     INDIA       INDONESIA   IRAN        IRAQ
ISRAEL      IVORY COAST JAPAN       JORDAN      KOREA, N    KOREA REP   LAOS        LEBANON     LIBERIA     LIBYA
MALAGASY R  MALAYA      MALI        MAURITANIA  MONGOLIA    MOROCCO     NEPAL       NIGER       NIGERIA     NORWAY
PAKISTAN    PHILIPPINES POLAND      RUMANIA     RWANDA      SA'U ARABIA SENEGAL     SIERRE LEO  SOMALIA     SO AFRICA
SUDAN       SWEDEN      SYRIA       TANGANYIKA  TOGO        TUNISIA     UGANDA      USSR        UAR         UPPER VOLTA
VIETNAM, N  VIETNAM REP YEMEN       YUGOSLAVIA
------------------------------------------------------------------------------------------------------

  2  TILT TOWARD BEING THOSE              0.66       2  TILT TOWARD BEING THOSE              0.71        11     5
     LOCATED ELSEWHERE THAN IN                          LOCATED IN WEST EUROPE, SCANDINAVIA,             21     2
     WEST EUROPE, SCANDINAVIA,                          NORTH AMERICA, OR AUSTRALASIA  (22)         X SQ=   1.91
     NORTH AMERICA, OR AUSTRALASIA  (93)                                                            P =    0.101
                                                                                                    RV YES (NO )

 31  TILT TOWARD BEING THOSE              0.66      31  TILT TOWARD BEING THOSE              0.71        21     2
     WHOSE AGRICULTURAL POPULATION                      WHOSE AGRICULTURAL POPULATION                    11     5
     IS HIGH OR MEDIUM  (90)                            IS LOW OR VERY LOW  (24)                    X SQ=   1.91
                                                                                                    P =    0.101
                                                                                                    RV YES (NO )

 35  LEAN TOWARD BEING THOSE              0.72      35  LEAN TOWARD BEING THOSE              0.71         9     5
     WHOSE PER CAPITA GROSS NATIONAL PRODUCT            WHOSE PER CAPITA GROSS NATIONAL PRODUCT          23     2
     IS MEDIUM, LOW, OR VERY LOW  (91)                  IS VERY HIGH OR HIGH  (24)                  X SQ=   2.99
                                                                                                    P =    0.075
                                                                                                    RV YES (NO )
```

36	TEND TO BE THOSE WHOSE PER CAPITA GROSS NATIONAL PRODUCT IS LOW OR VERY LOW (73)	0.56	36	ALWAYS ARE THOSE WHOSE PER CAPITA GROSS NATIONAL PRODUCT IS VERY HIGH, HIGH, OR MEDIUM (42)	1.00	14 7 18 0 X SQ= 5.22 P = 0.010 RV YES (NO)
38	LEAN MORE TOWARD BEING THOSE WHOSE INTERNATIONAL FINANCIAL STATUS IS MEDIUM, LOW, OR VERY LOW (103)	0.91	38	LEAN LESS TOWARD BEING THOSE WHOSE INTERNATIONAL FINANCIAL STATUS IS MEDIUM, LOW, OR VERY LOW (103)	0.57	3 3 29 4 X SQ= 2.71 P = 0.059 RV NO (YES)
44	TEND TO BE THOSE WHERE THE LITERACY RATE IS BELOW NINETY PERCENT (90)	0.78	44	TEND TO BE THOSE WHERE THE LITERACY RATE IS NINETY PERCENT OR ABOVE (25)	0.71	7 5 25 2 X SQ= 4.50 P = 0.020 RV YES (NO)
47	LEAN TOWARD BEING THOSE WHERE THE LITERACY RATE IS BETWEEN FIFTY AND NINETY PERCENT, RATHER THAN NINETY PERCENT OR ABOVE (30)	0.70	47	LEAN TOWARD BEING THOSE WHERE THE LITERACY RATE IS NINETY PERCENT OR ABOVE, RATHER THAN BETWEEN FIFTY AND NINETY PERCENT (25)	0.71	7 5 16 2 X SQ= 2.24 P = 0.084 RV YES (NO)
50	TEND LESS TO BE THOSE WHERE FREEDOM OF THE PRESS IS COMPLETE (43)	0.56	50	ALWAYS ARE THOSE WHERE FREEDOM OF THE PRESS IS COMPLETE (43)	1.00	15 7 12 0 X SQ= 3.06 P = 0.036 RV NO (NO)
53	TEND TO BE THOSE WHERE NEWSPAPER CIRCULATION IS LESS THAN THREE HUNDRED PER THOUSAND (101)	0.91	53	TEND TO BE THOSE WHERE NEWSPAPER CIRCULATION IS THREE HUNDRED OR MORE PER THOUSAND (14)	0.57	3 4 29 3 X SQ= 5.95 P = 0.012 RV YES (YES)
57	TEND TO BE THOSE WHERE THE RELIGION IS CATHOLIC (25)	0.66	57	TEND TO BE THOSE WHERE THE RELIGION IS OTHER THAN CATHOLIC (90)	0.86	21 1 11 6 X SQ= 4.25 P = 0.030 RV YES (NO)
66	TEND TO BE THOSE THAT ARE RELIGIOUSLY HOMOGENEOUS (57)	0.75	66	TEND TO BE THOSE THAT ARE RELIGIOUSLY HETEROGENEOUS (49)	0.86	24 1 8 6 X SQ= 6.75 P = 0.005 RV YES (NO)
72	ALWAYS ARE THOSE WHOSE DATE OF INDEPENDENCE IS BEFORE 1914 (52)	1.00	72	TEND LESS TO BE THOSE WHOSE DATE OF INDEPENDENCE IS BEFORE 1914 (52)	0.57	31 4 0 3 X SQ= 9.13 P = 0.004 RV NO (YES)

74	TILT TOWARD BEING THOSE THAT ARE NOT HISTORICALLY WESTERN (87)	0.66	74	TILT TOWARD BEING THOSE THAT ARE HISTORICALLY WESTERN (26)	0.71	11 5 21 2 X SQ= 1.91 P = 0.101 RV YES (NO)
79	TEND TO BE THOSE THAT WERE FORMERLY DEPENDENCIES OF SPAIN, RATHER THAN BRITAIN OR FRANCE (18)	0.94	79	ALWAYS ARE THOSE THAT WERE FORMERLY DEPENDENCIES OF BRITAIN OR FRANCE, RATHER THAN SPAIN (49)	1.00	1 6 17 0 X SQ= 15.13 P = 0. RV YES (YES)
81	TEND TO BE THOSE WHERE THE TYPE OF POLITICAL MODERNIZATION IS LATER EUROPEAN OR LATER EUROPEAN DERIVED, RATHER THAN EARLY EUROPEAN OR EARLY EUROPEAN DERIVED (40)	0.83	81	TEND TO BE THOSE WHERE THE TYPE OF POLITICAL MODERNIZATION IS EARLY EUROPEAN OR EARLY EUROPEAN DERIVED, RATHER THAN LATER EUROPEAN OR LATER EUROPEAN DERIVED (11)	0.80	5 4 25 1 X SQ= 5.99 P = 0.010 RV YES (NO)
82	ALWAYS ARE THOSE WHERE THE TYPE OF POLITICAL MODERNIZATION IS EARLY OR LATER EUROPEAN OR EUROPEAN DERIVED, RATHER THAN DEVELOPED TUTELARY OR UNDEVELOPED TUTELARY (51)	1.00	82	TEND LESS TO BE THOSE WHERE THE TYPE OF POLITICAL MODERNIZATION IS EARLY OR LATER EUROPEAN OR EUROPEAN DERIVED, RATHER THAN DEVELOPED TUTELARY OR UNDEVELOPED TUTELARY (51)	0.71	30 5 0 2 X SQ= 4.33 P = 0.032 RV NO (YES)
86	ALWAYS ARE THOSE WHERE THE STAGE OF POLITICAL MODERNIZATION IS ADVANCED OR MID-TRANSITIONAL, RATHER THAN EARLY TRANSITIONAL (76)	1.00	86	TEND LESS TO BE THOSE WHERE THE STAGE OF POLITICAL MODERNIZATION IS ADVANCED OR MID-TRANSITIONAL, RATHER THAN EARLY TRANSITIONAL (76)	0.71	32 5 0 2 X SQ= 4.66 P = 0.028 RV NO (YES)
98	TEND TO BE THOSE WHERE GOVERNMENTAL STABILITY IS GENERALLY OR MODERATELY PRESENT AND DATES FROM THE POST-WAR PERIOD, OR IS ABSENT (93)	0.81	98	TEND TO BE THOSE WHERE GOVERNMENTAL STABILITY IS GENERALLY PRESENT AND DATES AT LEAST FROM THE INTERWAR PERIOD (22)	0.71	6 5 26 2 X SQ= 5.48 P = 0.012 RV YES (NO)
99	TEND TO BE THOSE WHERE GOVERNMENTAL STABILITY IS MODERATELY PRESENT AND DATES FROM THE POST-WAR PERIOD, OR IS ABSENT (36)	0.59	99	ALWAYS ARE THOSE WHERE GOVERNMENTAL STABILITY IS GENERALLY PRESENT AND DATES FROM AT LEAST THE INTER-WAR PERIOD, OR FROM THE POST-WAR PERIOD (50)	1.00	12 5 17 0 X SQ= 3.75 P = 0.044 RV YES (NO)
101	LEAN LESS TOWARD BEING THOSE WHERE THE REPRESENTATIVE CHARACTER OF THE REGIME IS POLYARCHIC (41)	0.59	101	ALWAYS ARE THOSE WHERE THE REPRESENTATIVE CHARACTER OF THE REGIME IS POLYARCHIC (41)	1.00	17 7 12 0 X SQ= 2.68 P = 0.070 RV NO (NO)
110	TEND TO BE THOSE WHERE POLITICAL ENCULTURATION IS MEDIUM OR LOW (80)	0.85	110	TEND TO BE THOSE WHERE POLITICAL ENCULTURATION IS HIGH (15)	0.57	4 4 23 3 X SQ= 3.43 P = 0.037 RV YES (YES)

119	TEND TO BE THOSE WHERE INTEREST ARTICULATION BY INSTITUTIONAL GROUPS IS VERY SIGNIFICANT OR SIGNIFICANT (74)	0.69	119	ALWAYS ARE THOSE WHERE INTEREST ARTICULATION BY INSTITUTIONAL GROUPS IS MODERATE OR LIMITED (26)	1.00	22 0 10 7 X SQ= 8.42 P = 0.001 RV YES (NO)
120	TEND TO BE THOSE WHERE INTEREST ARTICULATION BY INSTITUTIONAL GROUPS IS VERY SIGNIFICANT, SIGNIFICANT, OR MODERATE (90)	0.97	120	TEND TO BE THOSE WHERE INTEREST ARITUCLATION BY INSTITUTIONAL GROUPS IS LIMITED (10)	0.57	31 3 1 4 X SQ= 10.55 P = 0.002 RV YES (YES)
123	TILT MORE TOWARD BEING THOSE WHERE INTEREST ARTICULATION BY NON-ASSOCIATIONAL GROUPS IS SIGNIFICANT, MODERATE, OR LIMITED (107)	0.94	123	TILT LESS TOWARD BEING THOSE WHERE INTEREST ARTICULATION BY NON-ASSOCIATIONAL GROUPS IS SIGNIFICANT, MODERATE, OR LIMITED (107)	0.71	30 5 2 2 X SQ= 1.16 P = 0.141 RV NO (YES)
125	TEND TO BE THOSE WHERE INTEREST ARTICULATION BY ANOMIC GROUPS IS FREQUENT OR OCCASIONAL (64)	0.56	125	ALWAYS ARE THOSE WHERE INTEREST ARTICULATION BY ANOMIC GROUPS IS INFREQUENT OR VERY INFREQUENT (35)	1.00	15 0 12 6 X SQ= 4.08 P = 0.021 RV YES (NO)
126	LEAN TOWARD BEING THOSE WHERE INTEREST ARTICULATION BY ANOMIC GROUPS IS FREQUENT, OCCASIONAL, OR INFREQUENT (83)	0.78	126	LEAN TOWARD BEING THOSE WHERE INTEREST ARTICULATION BY ANOMIC GROUPS IS VERY INFREQUENT (16)	0.67	21 2 6 4 X SQ= 2.73 P = 0.053 RV YES (NO)
130	TEND TO BE THOSE WHERE INTEREST AGGREGATION BY POLITICAL PARTIES IS MODERATE, LIMITED, OR NEGLIGIBLE (71)	0.89	130	TEND TO BE THOSE WHERE INTEREST AGGREGATION BY POLITICAL PARTIES IS SIGNIFICANT (12)	0.57	3 4 24 3 X SQ= 4.66 P = 0.020 RV YES (YES)
131	TEND TO BE THOSE WHERE INTEREST AGGREGATION BY POLITICAL PARTIES IS LIMITED OR NEGLIGIBLE (35)	0.71	131	ALWAYS ARE THOSE WHERE INTEREST AGGREGATION BY POLITICAL PARTIES IS SIGNIFICANT OR MODERATE (30)	1.00	7 7 17 0 X SQ= 8.31 P = 0.001 RV YES (YES)
134	TEND TO BE THOSE WHERE INTEREST AGGREGATION BY THE EXECUTIVE IS LIMITED OR NEGLIGIBLE (46)	0.53	134	ALWAYS ARE THOSE WHERE INTEREST AGGREGATION BY THE EXECUTIVE IS SIGNIFICANT OR MODERATE (57)	1.00	14 7 16 0 X SQ= 4.58 P = 0.012 RV YES (NO)
137	TEND TO BE THOSE WHERE INTEREST AGGREGATION BY THE LEGISLATURE IS LIMITED OR NEGLIGIBLE (68)	0.61	137	ALWAYS ARE THOSE WHERE INTEREST AGGREGATION BY THE LEGISLATURE IS SIGNIFICANT OR MODERATE (29)	1.00	11 7 17 0 X SQ= 6.01 P = 0.008 RV YES (NO)

141	TEND TO BE THOSE WHERE THE PARTY SYSTEM IS QUANTITATIVELY OTHER THAN TWO-PARTY (87)	0.89	141	TEND TO BE THOSE WHERE THE PARTY SYSTEM IS QUANTITATIVELY TWO-PARTY (11)	0.71	3 5 25 2 X SQ= 8.52 P = 0.003 RV YES (YES)
142	LEAN TOWARD BEING THOSE WHERE THE PARTY SYSTEM IS QUANTITATIVELY MULTI-PARTY (30)	0.57	142	LEAN TOWARD BEING THOSE WHERE THE PARTY SYSTEM IS QUANTITATIVELY OTHER THAN MULTI-PARTY (66)	0.86	16 1 12 6 X SQ= 2.58 P = 0.088 RV YES (NO)
153	TEND TO BE THOSE WHERE THE PARTY SYSTEM IS MODERATELY STABLE OR UNSTABLE (38)	0.68	153	ALWAYS ARE THOSE WHERE THE PARTY SYSTEM IS STABLE (42)	1.00	9 5 19 0 X SQ= 5.46 P = 0.008 RV YES (NO)
158	TEND TO BE THOSE WHERE PERSONALISMO IS PRONOUNCED OR MODERATE (40)	0.57	158	ALWAYS ARE THOSE WHERE PERSONALISMO IS NEGLIGIBLE (56)	1.00	17 0 13 7 X SQ= 5.23 P = 0.009 RV YES (NO)
168	TEND TO BE THOSE WHERE THE HORIZONTAL POWER DISTRIBUTION IS LIMITED OR NEGLIGIBLE (72)	0.53	168	ALWAYS ARE THOSE WHERE THE HORIZONTAL POWER DISTRIBUTION IS SIGNIFICANT (34)	1.00	14 7 16 0 X SQ= 4.58 P = 0.012 RV YES (NO)
170	LEAN TOWARD BEING THOSE WHERE THE LEGISLATIVE-EXECUTIVE STRUCTURE IS PRESIDENTIAL (39)	0.55	170	LEAN TOWARD BEING THOSE WHERE THE LEGISLATIVE-EXECUTIVE STRUCTURE IS OTHER THAN PRESIDENTIAL (63)	0.86	17 1 14 6 X SQ= 2.32 P = 0.093 RV YES (NO)
172	TEND TO BE THOSE WHERE THE LEGISLATIVE-EXECUTIVE STRUCTURE IS OTHER THAN PARLIAMENTARY-ROYALIST (88)	0.87	172	TEND TO BE THOSE WHERE THE LEGISLATIVE-EXECUTIVE STRUCTURE IS PARLIAMENTARY-ROYALIST (21)	0.71	4 5 26 2 X SQ= 7.49 P = 0.005 RV YES (YES)
174	TEND TO BE THOSE WHERE THE LEGISLATURE IS PARTIALLY EFFECTIVE, LARGELY INEFFECTIVE, OR WHOLLY INEFFECTIVE (72)	0.72	174	ALWAYS ARE THOSE WHERE THE LEGISLATURE IS FULLY EFFECTIVE (28)	1.00	8 7 21 0 X SQ= 9.37 P = 0.001 RV YES (NO)
179	LEAN LESS TOWARD BEING THOSE WHERE THE EXECUTIVE IS STRONG (39)	0.62	179	ALWAYS ARE THOSE WHERE THE EXECUTIVE IS STRONG (39)	1.00	9 0 15 7 X SQ= 2.10 P = 0.077 RV NO (NO)

181	TEND TO BE THOSE WHERE THE BUREAUCRACY IS SEMI-MODERN, RATHER THAN MODERN (55)	0.74	181	TEND TO BE THOSE WHERE THE BUREAUCRACY IS MODERN, RATHER THAN SEMI-MODERN (21)	0.71	8 5 23 2 X SQ= 3.45 P = 0.034 RV YES (NO)
187	TEND TO BE THOSE WHERE THE ROLE OF THE POLICE IS POLITICALLY SIGNIFICANT (66)	0.58	187	ALWAYS ARE THOSE WHERE THE ROLE OF THE POLICE IS NOT POLITICALLY SIGNIFICANT (35)	1.00	15 0 11 7 X SQ= 5.26 P = 0.009 RV YES (NO)

	Left			Right		
191	POLITIES WHERE THE CHARACTER OF THE LEGAL SYSTEM IS MIXED CIVIL-MUSLIM-INDIGENOUS, RATHER THAN MIXED COMMON-MUSLIM-INDIGENOUS (8)		191	POLITIES WHERE THE CHARACTER OF THE LEGAL SYSTEM IS MIXED COMMON-MUSLIM-INDIGENOUS, RATHER THAN MIXED CIVIL-MUSLIM-INDIGENOUS (5)		

SUBJECT ONLY

8 IN LEFT COLUMN

CHAD DAHOMEY INDONESIA IVORY COAST MALI NIGER SENEGAL UPPER VOLTA

5 IN RIGHT COLUMN

GHANA NIGERIA SIERRE LEO TANGANYIKA UGANDA

2 EXCLUDED BECAUSE AMBIGUOUS

ETHIOPIA GUINEA

100 EXCLUDED BECAUSE IRRELEVANT

AFGHANISTAN	ALBANIA	ALGERIA	ARGENTINA	AUSTRALIA	AUSTRIA	BELGIUM	BOLIVIA	BRAZIL	BULGARIA
BURMA	BURUNDI	CAMBODIA	CAMEROUN	CANADA	CEN AFR REP	CEYLON	CHILE	CHINA, PR	COLOMBIA
CONGO(BRA)	CONGO(LEO)	COSTA RICA	CUBA	CYPRUS	CZECHOS'KIA	DENMARK	DOMIN REP	ECUADOR	EL SALVADOR
FINLAND	FRANCE	GABON	GERMANY, E	GERMAN FR	GREECE	GUATEMALA	HAITI	HONDURAS	HUNGARY
ICELAND	INDIA	IRAN	IRAQ	IRELAND	ISRAEL	ITALY	JAMAICA	JAPAN	JORDAN
KOREA, N	KOREA REP	LAOS	LEBANON	LIBERIA	LIBYA	LUXEMBOURG	MALAGASY R	MALAYA	MAURITANIA
MEXICO	MONGOLIA	MOROCCO	NEPAL	NETHERLANDS	NEW ZEALAND	NICARAGUA	NORWAY	PAKISTAN	PANAMA
PARAGUAY	PERU	PHILIPPINES	POLAND	PORTUGAL	RUMANIA	RWANDA	SA'U ARABIA	SOMALIA	SO AFRICA
SPAIN	SUDAN	SWEDEN	SWITZERLAND	SYRIA	THAILAND	TOGO	TRINIDAD	TUNISIA	TURKEY
USSR	UAR	UK	US	URUGUAY	VENEZUELA	VIETNAM, N	VIETNAM REP	YEMEN	YUGOSLAVIA

	Left			Right		
24	TEND TO BE THOSE WHOSE POPULATION IS SMALL (54)	0.87	24	TEND TO BE THOSE WHOSE POPULATION IS VERY LARGE, LARGE, OR MEDIUM (61)	0.80	1 4 7 1 X SQ= 3.41 P = 0.032 RV YES (YES)
28	LEAN TOWARD BEING THOSE WHOSE POPULATION GROWTH RATE IS HIGH (62)	0.87	28	LEAN TOWARD BEING THOSE WHOSE POPULATION GROWTH RATE IS LOW (48)	0.75	7 1 1 3 X SQ= 2.30 P = 0.067 RV YES (YES)
48	ALWAYS ARE THOSE WHERE THE LITERACY RATE IS BETWEEN FIFTY AND NINETY PERCENT, RATHER THAN BETWEEN TEN AND FIFTY PERCENT (30)	1.00	48	ALWAYS ARE THOSE WHERE THE LITERACY RATE IS BETWEEN TEN AND FIFTY PERCENT, RATHER THAN BETWEEN FIFTY AND NINETY PERCENT (24)	1.00	1 0 0 2 X SQ= 0.19 P = 0.333 RV YES (YES)

80	ALWAYS ARE THOSE WHOSE DATE OF INDEPENDENCE IS AFTER 1914, AND THAT WERE FORMERLY DEPENDENCIES OF FRANCE, RATHER THAN BRITAIN (24)	1.00	80	ALWAYS ARE THOSE WHOSE DATE OF INDEPENDENCE IS AFTER 1914, AND THAT WERE FORMERLY DEPENDENCIES OF BRITAIN, RATHER THAN FRANCE (19)	1.00	0 5 7 0 X SQ= 8.24 P = 0.001 RV YES (YES)
99	ALWAYS ARE THOSE WHERE GOVERNMENTAL STABILITY IS MODERATELY PRESENT AND DATES FROM THE POST-WAR PERIOD, OR IS ABSENT (36)	1.00	99	ALWAYS ARE THOSE WHERE GOVERNMENTAL STABILITY IS GENERALLY PRESENT AND DATES FROM AT LEAST THE INTER-WAR PERIOD, OR FROM THE POST-WAR PERIOD (50)	1.00	0 1 2 0 X SQ= 0.19 P = 0.333 RV YES (YES)
102	ALWAYS ARE THOSE WHERE THE REPRESENTATIVE CHARACTER OF THE REGIME IS PSEUDO-POLYARCHIC OR NON-POLYARCHIC (49)	1.00	102	TEND TO BE THOSE WHERE THE REPRESENTATIVE CHARACTER OF THE REGIME IS POLYARCHIC OR LIMITED POLYARCHIC (59)	0.80	0 4 8 1 X SQ= 5.87 P = 0.007 RV YES (YES)
105	ALWAYS ARE THOSE WHERE THE ELECTORAL SYSTEM IS PARTIALLY COMPETITIVE OR NON-COMPETITIVE (47)	1.00	105	TILT TOWARD BEING THOSE WHERE THE ELECTORAL SYSTEM IS COMPETITIVE (43)	0.50	0 2 7 2 X SQ= 1.58 P = 0.109 RV YES (YES)
107	ALWAYS ARE THOSE WHERE AUTONOMOUS GROUPS ARE PARTIALLY TOLERATED IN POLITICS, ARE TOLERATED ONLY OUTSIDE POLITICS, OR ARE NOT TOLERATED AT ALL (65)	1.00	107	TILT LESS TOWARD BEING THOSE WHERE AUTONOMOUS GROUPS ARE PARTIALLY TOLERATED IN POLITICS, ARE TOLERATED ONLY OUTSIDE POLITICS, OR ARE NOT TOLERATED AT ALL (65)	0.60	0 2 8 3 X SQ= 1.33 P = 0.128 RV NO (YES)
129	ALWAYS ARE THOSE WHERE INTEREST ARTICULATION BY POLITICAL PARTIES IS NEGLIGIBLE (37)	1.00	129	TEND TO BE THOSE WHERE INTEREST ARTICULATION BY POLITICAL PARTIES IS SIGNIFICANT, MODERATE, OR LIMITED (56)	0.60	0 3 7 2 X SQ= 2.86 P = 0.045 RV YES (YES)
133	ALWAYS ARE THOSE WHERE INTEREST AGGREGATION BY THE EXECUTIVE IS SIGNIFICANT (29)	1.00	133	TEND TO BE THOSE WHERE INTEREST AGGREGATION BY THE EXECUTIVE IS MODERATE, LIMITED, OR NEGLIGIBLE (76)	0.60	8 2 0 3 X SQ= 3.32 P = 0.035 RV YES (YES)
138	ALWAYS ARE THOSE WHERE INTEREST AGGREGATION BY THE LEGISLATURE IS NEGLIGIBLE (49)	1.00	138	TEND TO BE THOSE WHERE INTEREST AGGREGATION BY THE LEGISLATURE IS SIGNIFICANT, MODERATE, OR LIMITED (48)	0.60	0 3 7 2 X SQ= 2.86 P = 0.045 RV YES (YES)
139	ALWAYS ARE THOSE WHERE THE PARTY SYSTEM IS QUANTITATIVELY ONE-PARTY (34)	1.00	139	TEND TO BE THOSE WHERE THE PARTY SYSTEM IS QUANTITATIVELY OTHER THAN ONE-PARTY (71)	0.60	7 2 0 3 X SQ= 2.86 P = 0.045 RV YES (YES)

158	TEND TO BE THOSE WHERE PERSONALISMO IS PRONOUNCED OR MODERATE (40)	0.87	158	TEND TO BE THOSE WHERE PERSONALISMO IS NEGLIGIBLE (56)	0.80	7 1 1 4 X SQ= 3.41 P = 0.032 RV YES (YES)
161	ALWAYS ARE THOSE WHERE THE POLITICAL LEADERSHIP IS NON-ELITIST (50)	1.00	161	LEAN TOWARD BEING THOSE WHERE THE POLITICAL LEADERSHIP IS ELITIST OR MODERATELY ELITIST (47)	0.67	0 2 8 1 X SQ= 2.81 P = 0.055 RV YES (YES)
166	ALWAYS ARE THOSE WHERE THE VERTICAL POWER DISTRIBUTION IS THAT OF FORMAL FEDERALISM OR FORMAL AND EFFECTIVE UNITARISM (99)	1.00	166	TILT LESS TOWARD BEING THOSE WHERE THE VERTICAL POWER DISTRIBUTION IS THAT OF FORMAL FEDERALISM OR FORMAL AND EFFECTIVE UNITARISM (99)	0.60	0 2 8 3 X SQ= 1.33 P = 0.128 RV NO (YES)
172	ALWAYS ARE THOSE WHERE THE LEGISLATIVE-EXECUTIVE STRUCTURE IS OTHER THAN PARLIAMENTARY-ROYALIST (88)	1.00	172	TEND TO BE THOSE WHERE THE LEGISLATIVE-EXECUTIVE STRUCTURE IS PARLIAMENTARY-ROYALIST (21)	0.60	0 3 8 2 X SQ= 3.32 P = 0.035 RV YES (YES)
175	ALWAYS ARE THOSE WHERE THE LEGISLATURE IS LARGELY INEFFECTIVE OR WHOLLY INEFFECTIVE (49)	1.00	175	TEND TO BE THOSE WHERE THE LEGISLATURE IS FULLY EFFECTIVE OR PARTIALLY EFFECTIVE (51)	0.60	0 3 8 2 X SQ= 3.32 P = 0.035 RV YES (YES)
179	ALWAYS ARE THOSE WHERE THE EXECUTIVE IS DOMINANT (52)	1.00	179	TILT TOWARD BEING THOSE WHERE THE EXECUTIVE IS STRONG (39)	0.50	6 2 0 2 X SQ= 1.28 P = 0.133 RV YES (YES)

```
************************************************************************************************

192  POLITIES                                          192  POLITIES
     WHERE THE CHARACTER OF THE LEGAL SYSTEM                WHERE THE CHARACTER OF THE LEGAL SYSTEM
     IS MUSLIM OR PARTLY MUSLIM  (27)                       IS OTHER THAN
                                                            MUSLIM OR PARTLY MUSLIM  (86)

                                                                        BOTH SUBJECT AND PREDICATE
................................................................................................

     27 IN LEFT COLUMN

ALGERIA     CHAD        DAHOMEY     GHANA       INDONESIA   IRAN        IVORY COAST JORDAN      LIBYA       MALAYA
MALI        MAURITANIA  MOROCCO     NIGER       NIGERIA     PAKISTAN    SA'U ARABIA SENEGAL     SIERRE LEO  SUDAN
SYRIA       TANGANYIKA  TUNISIA     UGANDA      UAR         UPPER VOLTA YEMEN

     86 IN RIGHT COLUMN

AFGHANISTAN ALBANIA     ARGENTINA   AUSTRALIA   AUSTRIA     BELGIUM     BOLIVIA     BRAZIL      BULGARIA    BURMA
BURUNDI     CAMBODIA    CAMEROUN    CANADA      CEN AFR REP CEYLON      CHILE       CHINA, PR   COLOMBIA    CONGO(BRA)
CONGO(LEO)  COSTA RICA  CUBA        CYPRUS      CZECHOS'KIA DENMARK     DOMIN REP   ECUADOR     EL SALVADOR FINLAND
FRANCE      GABON       GERMANY, E  GERMAN FR   GREECE      GUATEMALA   HAITI       HONDURAS    HUNGARY     ICELAND
INDIA       IRAQ        IRELAND     ISRAEL      ITALY       JAMAICA     JAPAN       KOREA, N    KOREA REP   LAOS
LEBANON     LIBERIA     LUXEMBOURG  MALAGASY R  MEXICO      MONGOLIA    NEPAL       NETHERLANDS NEW ZEALAND NICARAGUA
NORWAY      PANAMA      PARAGUAY    PERU        PHILIPPINES POLAND      PORTUGAL    RUMANIA     RWANDA      SOMALIA
SO AFRICA   SPAIN       SWEDEN      SWITZERLAND THAILAND    TOGO        TRINIDAD    TURKEY      USSR        UK
US          URUGUAY     VENEZUELA   VIETNAM, N  VIETNAM REP YUGOSLAVIA

     2 EXCLUDED BECAUSE AMBIGUOUS

ETHIOPIA    GUINEA
------------------------------------------------------------------------------------------------

  2  ALWAYS ARE THOSE                       1.00      2  TEND LESS TO BE THOSE              0.74       0    22
     LOCATED ELSEWHERE THAN IN                           LOCATED ELSEWHERE THAN IN                    27    64
     WEST EUROPE, SCANDINAVIA,                           WEST EUROPE, SCANDINAVIA,                X SQ=   7.02
     NORTH AMERICA, OR AUSTRALASIA  (93)                 NORTH AMERICA, OR AUSTRALASIA  (93)      P =    0.002
                                                                                                  RV NO  (NO )

  6  ALWAYS ARE THOSE                       1.00      6  TEND LESS TO BE THOSE              0.74       0    22
     LOCATED ELSEWHERE THAN IN THE                       LOCATED ELSEWHERE THAN IN THE                27    64
     CARIBBEAN, CENTRAL AMERICA,                         CARIBBEAN, CENTRAL AMERICA,              X SQ=   7.02
     OR SOUTH AMERICA  (93)                              OR SOUTH AMERICA  (93)                   P =    0.002
                                                                                                  RV NO  (NO )

 10  TEND TO BE THOSE                       0.70     10  TEND TO BE THOSE                   0.86      19    12
     LOCATED IN NORTH AFRICA, OR                         LOCATED ELSEWHERE THAN IN NORTH AFRICA,       8    74
     CENTRAL AND SOUTH AFRICA  (33)                      OR CENTRAL AND SOUTH AFRICA  (82)        X SQ=  30.08
                                                                                                  P =    0.
                                                                                                  RV YES (YES)
```

13	TEND LESS TO BE THOSE LOCATED ELSEWHERE THAN IN NORTH AFRICA OR THE MIDDLE EAST (99)	0.63	13	TEND MORE TO BE THOSE LOCATED ELSEWHERE THAN IN NORTH AFRICA OR THE MIDDLE EAST (99)	0.93	10 6 17 80 X SQ= 12.90 P = 0. RV NO (YES)
17	ALWAYS ARE THOSE LOCATED IN THE MIDDLE EAST, RATHER THAN IN THE CARIBBEAN, CENTRAL AMERICA, OR SOUTH AMERICA (11)	1.00	17	TEND TO BE THOSE LOCATED IN THE CARIBBEAN, CENTRAL AMERICA, OR SOUTH AMERICA, RATHER THAN IN THE MIDDLE EAST (22)	0.79	5 6 0 22 X SQ= 8.52 P = 0.002 RV YES (NO)
18	TEND TO BE THOSE LOCATED IN NORTH AFRICA OR CENTRAL AND SOUTH AFRICA, RATHER THAN IN EAST ASIA, SOUTH ASIA, OR SOUTHEAST ASIA (33)	0.86	18	TEND TO BE THOSE LOCATED IN EAST ASIA, SOUTH ASIA, OR SOUTHEAST ASIA, RATHER THAN IN NORTH AFRICA OR CENTRAL AND SOUTH AFRICA (18)	0.56	19 12 3 15 X SQ= 7.45 P = 0.003 RV YES (YES)
20	ALWAYS ARE THOSE LOCATED IN EAST ASIA, SOUTH ASIA, OR SOUTHEAST ASIA, RATHER THAN IN THE CARIBBEAN, CENTRAL AMERICA, OR SOUTH AMERICA (18)	1.00	20	LEAN TOWARD BEING THOSE LOCATED IN THE CARIBBEAN, CENTRAL AMERICA, OR SOUTH AMERICA, RATHER THAN IN EAST ASIA, SOUTH ASIA, OR SOUTHEAST ASIA (22)	0.59	3 15 0 22 X SQ= 1.93 P = 0.083 RV YES (NO)
21	TEND TO BE THOSE WHOSE TERRITORIAL SIZE IS VERY LARGE OR LARGE (32)	0.52	21	TEND TO BE THOSE WHOSE TERRITORIAL SIZE IS MEDIUM OR SMALL (83)	0.80	14 17 13 69 X SQ= 9.08 P = 0.002 RV YES (NO)
25	ALWAYS ARE THOSE WHOSE POPULATION DENSITY IS MEDIUM OR LOW (98)	1.00	25	TEND LESS TO BE THOSE WHOSE POPULATION DENSITY IS MEDIUM OR LOW (98)	0.80	0 17 27 69 X SQ= 4.83 P = 0.011 RV NO (NO)
26	TEND TO BE THOSE WHOSE POPULATION DENSITY IS LOW (67)	0.85	26	TEND TO BE THOSE WHOSE POPULATION DENSITY IS VERY HIGH, HIGH, OR MEDIUM (48)	0.51	4 44 23 42 X SQ= 9.67 P = 0.001 RV YES (NO)
28	TEND TO BE THOSE WHOSE POPULATION GROWTH RATE IS HIGH (62)	0.77	28	TEND TO BE THOSE WHOSE POPULATION GROWTH RATE IS LOW (48)	0.51	20 41 6 42 X SQ= 5.02 P = 0.022 RV YES (NO)
29	TEND TO BE THOSE WHERE THE DEGREE OF URBANIZATION IS LOW (49)	0.73	29	TEND TO BE THOSE WHERE THE DEGREE OF URBANIZATION IS HIGH (56)	0.64	7 49 19 28 X SQ= 9.13 P = 0.001 RV YES (NO)

30	TEND TO BE THOSE WHOSE AGRICULTURAL POPULATION IS HIGH (56)	0.81	30	TEND TO BE THOSE WHOSE AGRICULTURAL POPULATION IS MEDIUM, LOW, OR VERY LOW (57)	0.61	21 33 5 52 X SQ= 12.39 P = 0. RV YES (NO)
31	ALWAYS ARE THOSE WHOSE AGRICULTURAL POPULATION IS HIGH OR MEDIUM (90)	1.00	31	TEND LESS TO BE THOSE WHOSE AGRICULTURAL POPULATION IS HIGH OR MEDIUM (90)	0.72	26 62 0 24 X SQ= 7.65 P = 0.001 RV NO (NO)
32	ALWAYS ARE THOSE WHOSE GROSS NATIONAL PRODUCT IS MEDIUM, LOW, OR VERY LOW (105)	1.00	32	TILT LESS TOWARD BEING THOSE WHOSE GROSS NATIONAL PRODUCT IS MEDIUM, LOW, OR VERY LOW (105)	0.88	0 10 27 76 X SQ= 2.15 P = 0.114 RV NO (NO)
33	TEND MORE TO BE THOSE WHOSE GROSS NATIONAL PRODUCT IS LOW OR VERY LOW (85)	0.93	33	TEND LESS TO BE THOSE WHOSE GROSS NATIONAL PRODUCT IS LOW OR VERY LOW (85)	0.67	2 28 25 58 X SQ= 5.44 P = 0.011 RV NO (NO)
34	TEND TO BE THOSE WHOSE GROSS NATIONAL PRODUCT IS VERY LOW (53)	0.67	34	TEND TO BE THOSE WHOSE GROSS NATIONAL PRODUCT IS VERY HIGH, HIGH, MEDIUM, OR LOW (62)	0.60	9 52 18 34 X SQ= 5.05 P = 0.016 RV YES (NO)
35	ALWAYS ARE THOSE WHOSE PER CAPITA GROSS NATIONAL PRODUCT IS MEDIUM, LOW, OR VERY LOW (91)	1.00	35	TEND LESS TO BE THOSE WHOSE PER CAPITA GROSS NATIONAL PRODUCT IS MEDIUM, LOW, OR VERY LOW (91)	0.72	0 24 27 62 X SQ= 7.97 P = 0.001 RV NO (NO)
36	TEND MORE TO BE THOSE WHOSE PER CAPITA GROSS NATIONAL PRODUCT IS LOW OR VERY LOW (73)	0.96	36	TEND LESS TO BE THOSE WHOSE PER CAPITA GROSS NATIONAL PRODUCT IS LOW OR VERY LOW (73)	0.52	1 41 26 45 X SQ= 15.18 P = 0. RV NO (NO)
37	TEND TO BE THOSE WHOSE PER CAPITA GROSS NATIONAL PRODUCT IS VERY LOW (51)	0.78	37	TEND TO BE THOSE WHOSE PER CAPITA GROSS NATIONAL PRODUCT IS VERY HIGH, HIGH, MEDIUM, OR LOW (64)	0.67	6 58 21 28 X SQ= 15.32 P = 0. RV YES (NO)
38	ALWAYS ARE THOSE WHOSE INTERNATIONAL FINANCIAL STATUS IS MEDIUM, LOW, OR VERY LOW (103)	1.00	38	LEAN LESS TOWARD BEING THOSE WHOSE INTERNATIONAL FINANCIAL STATUS IS MEDIUM, LOW, OR VERY LOW (103)	0.88	0 10 27 74 X SQ= 2.23 P = 0.066 RV NO (NO)

No.	Subject	Prop.	No.	Predicate	Prop.	Statistics
39	TEND MORE TO BE THOSE WHOSE INTERNATIONAL FINANCIAL STATUS IS LOW OR VERY LOW (76)	0.89	39	TEND LESS TO BE THOSE WHOSE INTERNATIONAL FINANCIAL STATUS IS LOW OR VERY LOW (76)	0.59	3 35 24 50 X SQ= 6.98 P = 0.005 RV NO (NO)
40	LEAN TOWARD BEING THOSE WHOSE INTERNATIONAL FINANCIAL STATUS IS VERY LOW (39)	0.50	40	LEAN TOWARD BEING THOSE WHOSE INTERNATIONAL FINANCIAL STATUS IS VERY HIGH, HIGH, MEDIUM, OR LOW (71)	0.70	13 57 13 25 X SQ= 2.50 P = 0.098 RV YES (NO)
42	ALWAYS ARE THOSE WHOSE ECONOMIC DEVELOPMENTAL STATUS IS UNDERDEVELOPED OR VERY UNDERDEVELOPED (76)	1.00	42	TEND LESS TO BE THOSE WHOSE ECONOMIC DEVELOPMENTAL STATUS IS UNDERDEVELOPED OR VERY UNDERDEVELOPED (76)	0.57	0 36 26 48 X SQ= 14.67 P = 0. RV NO (NO)
43	TEND TO BE THOSE WHOSE ECONOMIC DEVELOPMENTAL STATUS IS VERY UNDERDEVELOPED (57)	0.88	43	TEND TO BE THOSE WHOSE ECONOMIC DEVELOPMENTAL STATUS IS DEVELOPED, INTERMEDIATE, OR UNDERDEVELOPED (55)	0.62	3 52 23 32 X SQ= 18.18 P = 0. RV YES (NO)
44	ALWAYS ARE THOSE WHERE THE LITERACY RATE IS BELOW NINETY PERCENT (90)	1.00	44	TEND LESS TO BE THOSE WHERE THE LITERACY RATE IS BELOW NINETY PERCENT (90)	0.71	0 25 27 61 X SQ= 8.46 P = 0. RV NO (NO)
45	TEND TO BE THOSE WHERE THE LITERACY RATE IS BELOW FIFTY PERCENT (54)	0.96	45	TEND TO BE THOSE WHERE THE LITERACY RATE IS FIFTY PERCENT OR ABOVE (55)	0.67	1 54 26 26 X SQ= 30.39 P = 0. RV YES (YES)
46	TEND TO BE THOSE WHERE THE LITERACY RATE IS BELOW TEN PERCENT (26)	0.52	46	TEND TO BE THOSE WHERE THE LITERACY RATE IS TEN PERCENT OR ABOVE (84)	0.86	12 72 13 12 X SQ= 13.44 P = 0. RV YES (YES)
54	ALWAYS ARE THOSE WHERE NEWSPAPER CIRCULATION IS LESS THAN ONE HUNDRED PER THOUSAND (76)	1.00	54	TEND LESS TO BE THOSE WHERE NEWSPAPER CIRCULATION IS LESS THAN ONE HUNDRED PER THOUSAND (76)	0.56	0 37 27 47 X SQ= 15.91 P = 0. RV NO (NO)
55	TEND TO BE THOSE WHERE NEWSPAPER CIRCULATION IS LESS THAN TEN PER THOUSAND (35)	0.67	55	TEND TO BE THOSE WHERE NEWSPAPER CIRCULATION IS TEN OR MORE PER THOUSAND (78)	0.82	9 69 18 15 X SQ= 21.02 P = 0. RV YES (YES)

57	ALWAYS ARE THOSE WHERE THE RELIGION IS OTHER THAN CATHOLIC (90)	1.00	57	TEND LESS TO BE THOSE WHERE THE RELIGION IS OTHER THAN CATHOLIC (90)	0.71	0 25 27 61 X SQ= 8.46 P = 0. RV NO (NO)
58	TEND TO BE THOSE WHERE THE RELIGION IS MUSLIM (18)	0.52	58	TEND TO BE THOSE WHERE THE RELIGION IS OTHER THAN MUSLIM (97)	0.95	14 4 13 82 X SQ= 30.75 P = 0. RV YES (YES)
63	ALWAYS ARE THOSE WHERE THE RELIGION IS PREDOMINANTLY OR PARTLY OTHER THAN CHRISTIAN (68)	1.00	63	TEND TO BE THOSE WHERE THE RELIGION IS PREDOMINANTLY SOME KIND OF CHRISTIAN (46)	0.54	0 46 27 39 X SQ= 22.61 P = 0. RV YES (NO)
68	TEND TO BE THOSE THAT ARE LINGUISTICALLY WEAKLY HETEROGENEOUS OR STRONGLY HETEROGENEOUS (62)	0.78	68	TEND TO BE THOSE THAT ARE LINGUISTICALLY HOMOGENEOUS (52)	0.54	6 46 21 39 X SQ= 7.15 P = 0.004 RV YES (NO)
69	TEND TO BE THOSE THAT ARE LINGUISTICALLY STRONGLY HETEROGENEOUS (50)	0.74	69	TEND TO BE THOSE THAT ARE LINGUISTICALLY HOMOGENEOUS OR WEAKLY HETEROGENEOUS (64)	0.67	7 57 20 28 X SQ= 12.53 P = 0. RV YES (NO)
72	TEND TO BE THOSE WHOSE DATE OF INDEPENDENCE IS AFTER 1914 (59)	0.96	72	TEND TO BE THOSE WHOSE DATE OF INDEPENDENCE IS BEFORE 1914 (52)	0.61	1 50 26 32 X SQ= 24.51 P = 0. RV YES (NO)
73	TEND TO BE THOSE WHOSE DATE OF INDEPENDENCE IS AFTER 1945 (46)	0.85	73	TEND TO BE THOSE WHOSE DATE OF INDEPENDENCE IS BEFORE 1945 (65)	0.73	4 60 23 22 X SQ= 26.18 P = 0. RV YES (YES)
74	ALWAYS ARE THOSE THAT ARE NOT HISTORICALLY WESTERN (87)	1.00	74	TEND LESS TO BE THOSE THAT ARE NOT HISTORICALLY WESTERN (87)	0.69	0 26 27 58 X SQ= 9.26 P = 0. RV NO (NO)
75	TEND TO BE THOSE THAT ARE NOT HISTORICALLY WESTERN AND ARE NOT SIGNIFICANTLY WESTERNIZED (52)	0.93	75	TEND TO BE THOSE THAT ARE HISTORICALLY WESTERN OR SIGNIFICANTLY WESTERNIZED (62)	0.71	2 60 25 25 X SQ= 30.59 P = 0. RV YES (YES)

76	ALWAYS ARE THOSE THAT HAVE BEEN SIGNIFICANTLY OR PARTIALLY WESTERNIZED THROUGH A COLONIAL RELATIONSHIP, RATHER THAN BEING HISTORICALLY WESTERN (70)	1.00	76	TEND LESS TO BE THOSE THAT HAVE BEEN SIGNIFICANTLY OR PARTIALLY WESTERNIZED THROUGH A COLONIAL RELATIONSHIP, RATHER THAN BEING HISTORICALLY WESTERN (70)	0.63	0 26 25 44 X SQ= 10.98 P = 0. RV NO (NO)
77	TEND TO BE THOSE THAT HAVE BEEN PARTIALLY WESTERNIZED, RATHER THAN SIGNIFICANTLY WESTERNIZED, THROUGH A COLONIAL RELATIONSHIP (41)	0.92	77	TEND TO BE THOSE THAT HAVE BEEN SIGNIFICANTLY WESTERNIZED, RATHER THAN PARTIALLY WESTERNIZED, THROUGH A COLONIAL RELATIONSHIP (28)	0.60	2 26 23 17 X SQ= 15.86 P = 0. RV YES (YES)
79	ALWAYS ARE THOSE THAT WERE FORMERLY DEPENDENCIES OF BRITAIN OR FRANCE, RATHER THAN SPAIN (49)	1.00	79	TEND LESS TO BE THOSE THAT WERE FORMERLY DEPENDENCIES OF BRITAIN OR FRANCE, RATHER THAN SPAIN (49)	0.59	22 26 0 18 X SQ= 10.40 P = 0. RV NO (NO)
82	ALWAYS ARE THOSE WHERE THE TYPE OF POLITICAL MODERNIZATION IS DEVELOPED TUTELARY OR UNDEVELOPED TUTELARY, RATHER THAN EARLY OR LATER EUROPEAN OR EUROPEAN DERIVED (55)	1.00	82	TEND TO BE THOSE WHERE THE TYPE OF POLITICAL MODERNIZATION IS EARLY OR LATER EUROPEAN OR EUROPEAN DERIVED, RATHER THAN DEVELOPED TUTELARY OR UNDEVELOPED TUTELARY (51)	0.65	0 51 26 28 X SQ= 30.10 P = 0. RV YES (NO)
85	TEND TO BE THOSE WHERE THE STAGE OF POLITICAL MODERNIZATION IS MID- OR EARLY TRANSITIONAL, RATHER THAN ADVANCED (54)	0.69	85	TEND TO BE THOSE WHERE THE STAGE OF POLITICAL MODERNIZATION IS ADVANCED, RATHER THAN MID- OR EARLY TRANSITIONAL (60)	0.60	8 52 18 34 X SQ= 5.93 P = 0.013 RV YES (NO)
86	TEND TO BE THOSE WHERE THE STAGE OF POLITICAL MODERNIZATION IS EARLY TRANSITIONAL, RATHER THAN ADVANCED OR MID-TRANSITIONAL (38)	0.65	86	TEND TO BE THOSE WHERE THE STAGE OF POLITICAL MODERNIZATION IS ADVANCED OR MID-TRANSITIONAL, RATHER THAN EARLY TRANSITIONAL (76)	0.77	9 66 17 20 X SQ= 14.17 P = 0. RV YES (NO)
87	TEND TO BE THOSE WHOSE IDEOLOGICAL ORIENTATION IS DEVELOPMENTAL (31)	0.85	87	TEND TO BE THOSE WHOSE IDEOLOGICAL ORIENTATION IS OTHER THAN DEVELOPMENTAL (58)	0.81	17 13 3 54 X SQ= 26.50 P = 0. RV YES (YES)
89	ALWAYS ARE THOSE WHOSE IDEOLOGICAL ORIENTATION IS OTHER THAN CONVENTIONAL (62)	1.00	89	TEND LESS TO BE THOSE WHOSE IDEOLOGICAL ORIENTATION IS OTHER THAN CONVENTIONAL (62)	0.53	0 33 23 37 X SQ= 14.81 P = 0. RV NO (NO)
95	ALWAYS ARE THOSE WHERE THE STATUS OF THE REGIME IS CONSTITUTIONAL OR AUTHORITARIAN (95)	1.00	95	TEND LESS TO BE THOSE WHERE THE STATUS OF THE REGIME IS CONSTITUTIONAL OR AUTHORITARIAN (95)	0.81	26 67 0 16 X SQ= 4.44 P = 0.011 RV NO (NO)

96	TEND TO BE THOSE WHERE THE STATUS OF THE REGIME IS AUTHORITARIAN (23)	0.53	96	TEND TO BE THOSE WHERE THE STATUS OF THE REGIME IS CONSTITUTIONAL OR TOTALITARIAN (67)	0.83	8 59 9 12 X SQ= 7.92 P = 0.004 RV YES (NO)
97	ALWAYS ARE THOSE WHERE THE STATUS OF THE REGIME IS AUTHORITARIAN, RATHER THAN TOTALITARIAN (23)	1.00	97	TEND TO BE THOSE WHERE THE STATUS OF THE REGIME IS TOTALITARIAN, RATHER THAN AUTHORITARIAN (16)	0.57	9 12 0 16 X SQ= 6.88 P = 0.005 RV YES (NO)
98	TEND MORE TO BE THOSE WHERE GOVERNMENTAL STABILITY IS GENERALLY OR MODERATELY PRESENT AND DATES FROM THE POST-WAR PERIOD, OR IS ABSENT (93)	0.96	98	TEND LESS TO BE THOSE WHERE GOVERNMENTAL STABILITY IS GENERALLY OR MODERATELY PRESENT AND DATES FROM THE POST-WAR PERIOD, OR IS ABSENT (93)	0.76	1 21 26 65 X SQ= 4.38 P = 0.023 RV NO (NO)
101	TEND TO BE THOSE WHERE THE REPRESENTATIVE CHARACTER OF THE REGIME IS LIMITED POLYARCHIC, PSEUDO-POLYARCHIC, OR NON-POLYARCHIC (57)	0.82	101	TEND TO BE THOSE WHERE THE REPRESENTATIVE CHARACTER OF THE REGIME IS POLYARCHIC (41)	0.50	4 37 18 37 X SQ= 5.78 P = 0.013 RV YES (NO)
102	TEND TO BE THOSE WHERE THE REPRESENTATIVE CHARACTER OF THE REGIME IS PSEUDO-POLYARCHIC OR NON-POLYARCHIC (49)	0.68	102	TEND TO BE THOSE WHERE THE REPRESENTATIVE CHARACTER OF THE REGIME IS POLYARCHIC OR LIMITED POLYARCHIC (59)	0.63	8 51 17 30 X SQ= 6.22 P = 0.010 RV YES (NO)
105	TEND TO BE THOSE WHERE THE ELECTORAL SYSTEM IS PARTIALLY COMPETITIVE OR NON-COMPETITIVE (47)	0.83	105	TEND TO BE THOSE WHERE THE ELECTORAL SYSTEM IS COMPETITIVE (43)	0.57	3 40 15 30 X SQ= 7.84 P = 0.003 RV YES (NO)
107	TEND TO BE THOSE WHERE AUTONOMOUS GROUPS ARE PARTIALLY TOLERATED IN POLITICS, ARE TOLERATED ONLY OUTSIDE POLITICS, OR ARE NOT TOLERATED AT ALL (65)	0.88	107	TEND TO BE THOSE WHERE AUTONOMOUS GROUPS ARE FULLY TOLERATED IN POLITICS (46)	0.52	3 43 23 40 X SQ= 11.56 P = 0. RV YES (NO)
110	ALWAYS ARE THOSE WHERE POLITICAL ENCULTURATION IS MEDIUM OR LOW (80)	1.00	110	TEND LESS TO BE THOSE WHERE POLITICAL ENCULTURATION IS MEDIUM OR LOW (80)	0.78	0 15 25 53 X SQ= 5.05 P = 0.009 RV NO (NO)
114	TEND TO BE THOSE WHERE SECTIONALISM IS EXTREME OR MODERATE (61)	0.76	114	TEND TO BE THOSE WHERE SECTIONALISM IS NEGLIGIBLE (47)	0.51	19 40 6 41 X SQ= 4.46 P = 0.022 RV YES (NO)

116	ALWAYS ARE THOSE WHERE INTEREST ARTICULATION BY ASSOCIATIONAL GROUPS IS LIMITED OR NEGLIGIBLE (79)	1.00	116	TEND LESS TO BE THOSE WHERE INTEREST ARTICULATION BY ASSOCIATIONAL GROUPS IS LIMITED OR NEGLIGIBLE (79)	0.61	0 32 27 50 X SQ= 13.09 P = 0. RV NO (NO)
117	TEND TO BE THOSE WHERE INTEREST ARTICULATION BY ASSOCIATIONAL GROUPS IS NEGLIGIBLE (51)	0.70	117	TEND TO BE THOSE WHERE INTEREST ARTICULATION BY ASSOCIATIONAL GROUPS IS SIGNIFICANT, MODERATE, OR LIMITED (60)	0.63	8 52 19 30 X SQ= 8.05 P = 0.003 RV YES (NO)
119	ALWAYS ARE THOSE WHERE INTEREST ARTICULATION BY INSTITUTIONAL GROUPS IS VERY SIGNIFICANT OR SIGNIFICANT (74)	1.00	119	TEND LESS TO BE THOSE WHERE INTEREST ARTICULATION BY INSTITUTIONAL GROUPS IS VERY SIGNIFICANT OR SIGNIFICANT (74)	0.67	20 52 0 26 X SQ= 7.44 P = 0.001 RV NO (NO)
121	TEND TO BE THOSE WHERE INTEREST ARTICULATION BY NON-ASSOCIATIONAL GROUPS IS SIGNIFICANT (54)	0.81	121	TEND TO BE THOSE WHERE INTEREST ARTICULATION BY NON-ASSOCIATIONAL GROUPS IS MODERATE, LIMITED, OR NEGLIGIBLE (61)	0.65	22 30 5 56 X SQ= 16.13 P = 0. RV YES (NO)
122	ALWAYS ARE THOSE WHERE INTEREST ARTICULATION BY NON-ASSOCIATIONAL GROUPS IS SIGNIFICANT OR MODERATE (83)	1.00	122	TEND LESS TO BE THOSE WHERE INTEREST ARTICULATION BY NON-ASSOCIATIONAL GROUPS IS SIGNIFICANT OR MODERATE (83)	0.63	27 54 0 32 X SQ= 12.24 P = 0. RV NO (NO)
125	TEND MORE TO BE THOSE WHERE INTEREST ARTICULATION BY ANOMIC GROUPS IS FREQUENT OR OCCASIONAL (64)	0.84	125	TEND LESS TO BE THOSE WHERE INTEREST ARTICULATION BY ANOMIC GROUPS IS FREQUENT OR OCCASIONAL (64)	0.57	21 41 4 31 X SQ= 4.78 P = 0.017 RV NO (NO)
126	ALWAYS ARE THOSE WHERE INTEREST ARTICULATION BY ANOMIC GROUPS IS FREQUENT, OCCASIONAL, OR INFREQUENT (83)	1.00	126	TEND LESS TO BE THOSE WHERE INTEREST ARTICULATION BY ANOMIC GROUPS IS FREQUENT, OCCASIONAL, OR INFREQUENT (83)	0.78	25 56 0 16 X SQ= 5.14 P = 0.010 RV NO (NO)
128	TILT TOWARD BEING THOSE WHERE INTEREST ARTICULATION BY POLITICAL PARTIES IS LIMITED OR NEGLIGIBLE (45)	0.67	128	TILT TOWARD BEING THOSE WHERE INTEREST ARTICULATION BY POLITICAL PARTIES IS SIGNIFICANT OR MODERATE (48)	0.57	6 42 12 32 X SQ= 2.31 P = 0.113 RV YES (NO)
129	LEAN TOWARD BEING THOSE WHERE INTEREST ARTICULATION BY POLITICAL PARTIES IS NEGLIGIBLE (37)	0.61	129	LEAN TOWARD BEING THOSE WHERE INTEREST ARTICULATION BY POLITICAL PARTIES IS SIGNIFICANT, MODERATE, OR LIMITED (56)	0.66	7 49 11 25 X SQ= 3.46 P = 0.057 RV YES (NO)

132	ALWAYS ARE THOSE WHERE INTEREST AGGREGATION BY POLITICAL PARTIES IS SIGNIFICANT, MODERATE, OR LIMITED (67)	1.00	132	TILT LESS TOWARD BEING THOSE WHERE INTEREST AGGREGATION BY POLITICAL PARTIES IS SIGNIFICANT, MODERATE, OR LIMITED (67)	0.84	18 48 0 9 X SQ= 1.91 P = 0.103 RV NO (NO)
133	TEND TO BE THOSE WHERE INTEREST AGGREGATION BY THE EXECUTIVE IS SIGNIFICANT (29)	0.62	133	TEND TO BE THOSE WHERE INTEREST AGGREGATION BY THE EXECUTIVE IS MODERATE, LIMITED, OR NEGLIGIBLE (76)	0.84	15 13 9 66 X SQ= 17.46 P = 0. RV YES (YES)
137	ALWAYS ARE THOSE WHERE INTEREST AGGREGATION BY THE LEGISLATURE IS LIMITED OR NEGLIGIBLE (68)	1.00	137	TEND LESS TO BE THOSE WHERE INTEREST AGGREGATION BY THE LEGISLATURE IS LIMITED OR NEGLIGIBLE (68)	0.62	0 29 18 48 X SQ= 8.06 P = 0.001 RV NO (NO)
138	TEND TO BE THOSE WHERE INTEREST AGGREGATION BY THE LEGISLATURE IS NEGLIGIBLE (49)	0.76	138	TEND TO BE THOSE WHERE INTEREST AGGREGATION BY THE LEGISLATURE IS SIGNIFICANT, MODERATE, OR LIMITED (48)	0.56	4 44 13 34 X SQ= 4.79 P = 0.017 RV YES (NO)
139	TEND TO BE THOSE WHERE THE PARTY SYSTEM IS QUANTITATIVELY ONE-PARTY (34)	0.57	139	TEND TO BE THOSE WHERE THE PARTY SYSTEM IS QUANTITATIVELY OTHER THAN ONE-PARTY (71)	0.75	13 20 10 60 X SQ= 6.77 P = 0.010 RV YES (NO)
140	ALWAYS ARE THOSE WHERE THE PARTY SYSTEM IS QUANTITATIVELY OTHER THAN ONE PARTY DOMINANT (87)	1.00	140	LEAN LESS TOWARD BEING THOSE WHERE THE PARTY SYSTEM IS QUANTITATIVELY OTHER THAN ONE PARTY DOMINANT (87)	0.83	0 13 20 65 X SQ= 2.53 P = 0.064 RV NO (NO)
142	TEND MORE TO BE THOSE WHERE THE PARTY SYSTEM IS QUANTITATIVELY OTHER THAN MULTI-PARTY (66)	0.89	142	TEND LESS TO BE THOSE WHERE THE PARTY SYSTEM IS QUANTITATIVELY OTHER THAN MULTI-PARTY (66)	0.63	2 28 16 48 X SQ= 3.33 P = 0.048 RV NO (NO)
144	LEAN MORE TOWARD BEING THOSE WHERE THE PARTY SYSTEM IS QUANTITATIVELY ONE-PARTY, RATHER THAN TWO-PARTY (34)	0.93	144	LEAN LESS TOWARD BEING THOSE WHERE THE PARTY SYSTEM IS QUANTITATIVELY ONE-PARTY, RATHER THAN TWO-PARTY (34)	0.67	13 20 1 10 X SQ= 2.23 P = 0.076 RV NO (NO)
146	TEND LESS TO BE THOSE WHERE THE PARTY SYSTEM IS QUALITATIVELY OTHER THAN MASS-BASED TERRITORIAL (92)	0.68	146	ALWAYS ARE THOSE WHERE THE PARTY SYSTEM IS QUALITATIVELY OTHER THAN MASS-BASED TERRITORIAL (92)	1.00	7 0 15 77 X SQ= 21.74 P = 0. RV NO (YES)

147	ALWAYS ARE THOSE WHERE THE PARTY SYSTEM IS QUALITATIVELY OTHER THAN CLASS-ORIENTED OR MULTI-IDEOLOGICAL (67)	1.00	TEND LESS TO BE THOSE WHERE THE PARTY SYSTEM IS QUALITATIVELY OTHER THAN CLASS-ORIENTED OR MULTI-IDEOLOGICAL (67)	0.68	0 23 18 48 X SQ= 6.26 P = 0.003 RV NO (NO)
153	TEND TO BE THOSE WHERE THE PARTY SYSTEM IS MODERATELY STABLE OR UNSTABLE (38)	0.80	TEND TO BE THOSE WHERE THE PARTY SYSTEM IS STABLE (42)	0.57	2 39 8 30 X SQ= 3.32 P = 0.043 RV YES (NO)
160	LEAN MORE TOWARD BEING THOSE WHERE THE POLITICAL LEADERSHIP IS MODERATELY ELITIST OR NON-ELITIST (67)	0.89	LEAN LESS TOWARD BEING THOSE WHERE THE POLITICAL LEADERSHIP IS MODERATELY ELITIST OR NON-ELITIST (67)	0.65	2 27 16 50 X SQ= 2.90 P = 0.052 RV NO (NO)
164	TEND TO BE THOSE WHERE THE REGIME'S LEADERSHIP CHARISMA IS PRONOUNCED OR MODERATE (34)	0.86	TEND TO BE THOSE WHERE THE REGIME'S LEADERSHIP CHARISMA IS NEGLIGIBLE (65)	0.83	19 13 3 62 X SQ= 33.61 P = 0. RV YES (YES)
169	TEND TO BE THOSE WHERE THE HORIZONTAL POWER DISTRIBUTION IS NEGLIGIBLE (48)	0.71	TEND TO BE THOSE WHERE THE HORIZONTAL POWER DISTRIBUTION IS SIGNIFICANT OR LIMITED (58)	0.64	7 51 17 29 X SQ= 7.60 P = 0.004 RV YES (NO)
171	ALWAYS ARE THOSE WHERE THE LEGISLATIVE-EXECUTIVE STRUCTURE IS OTHER THAN PARLIAMENTARY-REPUBLICAN (90)	1.00	LEAN LESS TOWARD BEING THOSE WHERE THE LEGISLATIVE-EXECUTIVE STRUCTURE IS OTHER THAN PARLIAMENTARY-REPUBLICAN (90)	0.85	0 12 21 67 X SQ= 2.33 P = 0.066 RV NO (NO)
175	TEND TO BE THOSE WHERE THE LEGISLATURE IS LARGELY INEFFECTIVE OR WHOLLY INEFFECTIVE (49)	0.71	TEND TO BE THOSE WHERE THE LEGISLATURE IS FULLY EFFECTIVE OR PARTIALLY EFFECTIVE (51)	0.58	6 45 15 32 X SQ= 4.76 P = 0.025 RV YES (NO)
178	TEND TO BE THOSE WHERE THE LEGISLATURE IS UNICAMERAL (53)	0.73	TEND TO BE THOSE WHERE THE LEGISLATURE IS BICAMERAL (51)	0.55	16 36 6 44 X SQ= 4.26 P = 0.030 RV YES (NO)
179	TEND TO BE THOSE WHERE THE EXECUTIVE IS DOMINANT (52)	0.81	TEND TO BE THOSE WHERE THE EXECUTIVE IS STRONG (39)	0.51	17 33 4 35 X SQ= 5.60 P = 0.011 RV YES (NO)

```
181  ALWAYS ARE THOSE                       1.00    181  TILT LESS TOWARD BEING THOSE        0.69          0   21
     WHERE THE BUREAUCRACY                               WHERE THE BUREAUCRACY                            7   47
     IS SEMI-MODERN, RATHER THAN                         IS SEMI-MODERN, RATHER THAN                X SQ=   1.67
     MODERN  (55)                                        MODERN  (55)                               P =    0.105
                                                                                                    RV NO  (NO )

182  TEND TO BE THOSE                       0.67    182  TEND TO BE THOSE                    0.81          7   47
     WHERE THE BUREAUCRACY                               WHERE THE BUREAUCRACY                           14   11
     IS POST-COLONIAL TRANSITIONAL,                      IS SEMI-MODERN, RATHER THAN                X SQ=  14.09
     RATHER THAN SEMI-MODERN  (25)                       POST-COLONIAL TRANSITIONAL  (55)           P =     0.
                                                                                                    RV YES (YES)

194  ALWAYS ARE THOSE                       1.00    194  TEND LESS TO BE THOSE               0.85          0   13
     THAT ARE NON-COMMUNIST  (101)                       THAT ARE NON-COMMUNIST  (101)                   27   72
                                                                                                    X SQ=   3.30
                                                                                                    P =    0.036
                                                                                                    RV NO  (NO )
```

193 POLITIES
WHERE THE CHARACTER OF THE LEGAL SYSTEM
IS PARTLY INDIGENOUS (28)

193 POLITIES
WHERE THE CHARACTER OF THE LEGAL SYSTEM
IS OTHER THAN PARTLY INDIGENOUS (86)

BOTH SUBJECT AND PREDICATE

28 IN LEFT COLUMN

BURUNDI	CAMEROUN	CEN AFR REP	CHAD	CONGO(BRA)	CONGO(LEO)	DAHOMEY	ETHIOPIA	GABON	GHANA
GUINEA	INDONESIA	IVORY COAST	LAOS	LIBERIA	MALAGASY R	MALI	NIGER	NIGERIA	RWANDA
SENEGAL	SIERRE LEO	SO AFRICA	TANGANYIKA	TOGO	UGANDA	UPPER VOLTA	VIETNAM REP		

86 IN RIGHT COLUMN

AFGHANISTAN	ALBANIA	ALGERIA	ARGENTINA	AUSTRALIA	AUSTRIA	BELGIUM	BOLIVIA	BRAZIL	BULGARIA
CAMBODIA	CANADA	CEYLON	CHILE	CHINA, PR	COLOMBIA	COSTA RICA	CUBA	CYPRUS	CZECHOS'KIA
DENMARK	DOMIN REP	ECUADOR	EL SALVADOR	FINLAND	FRANCE	GERMANY, E	GERMAN FR	GREECE	GUATEMALA
HAITI	HONDURAS	HUNGARY	ICELAND	INDIA	IRAN	IRAQ	IRELAND	ISRAEL	ITALY
JAMAICA	JAPAN	JORDAN	KOREA, N	KOREA REP	LEBANON	LIBYA	LUXEMBOURG	MALAYA	MAURITANIA
MEXICO	MONGOLIA	MOROCCO	NEPAL	NETHERLANDS	NEW ZEALAND	NICARAGUA	NORWAY	PAKISTAN	PANAMA
PARAGUAY	PERU	PHILIPPINES	POLAND	PORTUGAL	RUMANIA	SA'U ARABIA	SOMALIA	SPAIN	SUDAN
SWEDEN	SWITZERLAND	SYRIA	THAILAND	TRINIDAD	TUNISIA	TURKEY	USSR	UAR	UK
US	URUGUAY	VENEZUELA	VIETNAM, N	YEMEN	YUGOSLAVIA				

1 EXCLUDED BECAUSE UNASCERTAINABLE

BURMA

2	ALWAYS ARE THOSE LOCATED ELSEWHERE THAN IN WEST EUROPE, SCANDINAVIA, NORTH AMERICA, OR AUSTRALASIA (93)	1.00	2	TEND LESS TO BE THOSE LOCATED ELSEWHERE THAN IN WEST EUROPE, SCANDINAVIA, NORTH AMERICA, OR AUSTRALASIA (93)	0.74	0 22 28 64 X SQ= 7.31 P = 0.002 RV NO (NO)
6	ALWAYS ARE THOSE LOCATED ELSEWHERE THAN IN THE CARIBBEAN, CENTRAL AMERICA, OR SOUTH AMERICA (93)	1.00	6	TEND LESS TO BE THOSE LOCATED ELSEWHERE THAN IN THE CARIBBEAN, CENTRAL AMERICA, OR SOUTH AMERICA (93)	0.74	0 22 28 64 X SQ= 7.31 P = 0.002 RV NO (NO)
10	TEND TO BE THOSE LOCATED IN NORTH AFRICA, OR CENTRAL AND SOUTH AFRICA (33)	0.89	10	TEND TO BE THOSE LOCATED ELSEWHERE THAN IN NORTH AFRICA, OR CENTRAL AND SOUTH AFRICA (82)	0.91	25 8 3 78 X SQ= 61.87 P = 0. RV YES (YES)

No.	Statement	Value	No.	Statement	Value	Statistics
13	ALWAYS ARE THOSE LOCATED ELSEWHERE THAN IN NORTH AFRICA OR THE MIDDLE EAST (99)	1.00	13	TEND LESS TO BE THOSE LOCATED ELSEWHERE THAN IN NORTH AFRICA OR THE MIDDLE EAST (99)	0.81	0 16 28 70 X SQ= 4.62 P = 0.011 RV NO (NO)
14	ALWAYS ARE THOSE LOCATED ELSEWHERE THAN IN THE MIDDLE EAST (104)	1.00	14	LEAN LESS TOWARD BEING THOSE LOCATED ELSEWHERE THAN IN THE MIDDLE EAST (104)	0.87	0 11 28 75 X SQ= 2.63 P = 0.063 RV NO (NO)
18	TEND TO BE THOSE LOCATED IN NORTH AFRICA OR CENTRAL AND SOUTH AFRICA, RATHER THAN IN EAST ASIA, SOUTH ASIA, OR SOUTHEAST ASIA (33)	0.89	18	TEND TO BE THOSE LOCATED IN EAST ASIA, SOUTH ASIA, OR SOUTHEAST ASIA, RATHER THAN IN NORTH AFRICA OR CENTRAL AND SOUTH AFRICA (18)	0.64	25 8 3 14 X SQ= 13.11 P = 0. RV YES (YES)
20	ALWAYS ARE THOSE LOCATED IN EAST ASIA, SOUTH ASIA, OR SOUTHEAST ASIA, RATHER THAN IN THE CARIBBEAN, CENTRAL AMERICA, OR SOUTH AMERICA (18)	1.00	20	LEAN TOWARD BEING THOSE LOCATED IN THE CARIBBEAN, CENTRAL AMERICA, OR SOUTH AMERICA, RATHER THAN IN EAST ASIA, SOUTH ASIA, OR SOUTHEAST ASIA (22)	0.61	3 14 0 22 X SQ= 2.09 P = 0.074 RV YES (NO)
22	TEND MORE TO BE THOSE WHOSE TERRITORIAL SIZE IS VERY LARGE, LARGE, OR MEDIUM (68)	0.75	22	TEND LESS TO BE THOSE WHOSE TERRITORIAL SIZE IS VERY LARGE, LARGE, OR MEDIUM (68)	0.53	21 46 7 40 X SQ= 3.20 P = 0.050 RV NO (NO)
23	TEND MORE TO BE THOSE WHOSE POPULATION IS MEDIUM OR SMALL (88)	0.93	23	TEND LESS TO BE THOSE WHOSE POPULATION IS MEDIUM OR SMALL (88)	0.72	2 24 26 62 X SQ= 4.06 P = 0.022 RV NO (NO)
24	TEND TO BE THOSE WHOSE POPULATION IS SMALL (54)	0.68	24	TEND TO BE THOSE WHOSE POPULATION IS VERY LARGE, LARGE, OR MEDIUM (61)	0.59	9 51 19 35 X SQ= 5.21 P = 0.016 RV YES (NO)
25	ALWAYS ARE THOSE WHOSE POPULATION DENSITY IS MEDIUM OR LOW (98)	1.00	25	TEND LESS TO BE THOSE WHOSE POPULATION DENSITY IS MEDIUM OR LOW (98)	0.80	0 17 28 69 X SQ= 5.04 P = 0.006 RV NO (NO)
26	TEND TO BE THOSE WHOSE POPULATION DENSITY IS LOW (67)	0.82	26	TEND TO BE THOSE WHOSE POPULATION DENSITY IS VERY HIGH, HIGH, OR MEDIUM (48)	0.50	5 43 23 43 X SQ= 7.68 P = 0.004 RV YES (NO)

No.	Statement	Value	No.	Statement	Value	Statistics
29	TEND TO BE THOSE WHERE THE DEGREE OF URBANIZATION IS LOW (49)	0.96	29	TEND TO BE THOSE WHERE THE DEGREE OF URBANIZATION IS HIGH (56)	0.71	1 55 26 22 X SQ= 34.22 P = 0. RV YES (YES)
30	TEND TO BE THOSE WHOSE AGRICULTURAL POPULATION IS HIGH (56)	0.93	30	TEND TO BE THOSE WHOSE AGRICULTURAL POPULATION IS MEDIUM, LOW, OR VERY LOW (57)	0.65	26 29 2 55 X SQ= 26.31 P = 0. RV YES (NO)
32	ALWAYS ARE THOSE WHOSE GROSS NATIONAL PRODUCT IS MEDIUM, LOW, OR VERY LOW (105)	1.00	32	LEAN LESS TOWARD BEING THOSE WHOSE GROSS NATIONAL PRODUCT IS MEDIUM, LOW, OR VERY LOW (105)	0.88	0 10 28 76 X SQ= 2.26 P = 0.066 RV NO (NO)
33	TEND MORE TO BE THOSE WHOSE GROSS NATIONAL PRODUCT IS LOW OR VERY LOW (85)	0.93	33	TEND LESS TO BE THOSE WHOSE GROSS NATIONAL PRODUCT IS LOW OR VERY LOW (85)	0.67	2 28 26 58 X SQ= 5.79 P = 0.007 RV NO (NO)
34	TEND TO BE THOSE WHOSE GROSS NATIONAL PRODUCT IS VERY LOW (53)	0.79	34	TEND TO BE THOSE WHOSE GROSS NATIONAL PRODUCT IS VERY HIGH, HIGH, MEDIUM, OR LOW (62)	0.64	6 55 22 31 X SQ= 13.69 P = 0. RV YES (NO)
35	ALWAYS ARE THOSE WHOSE PER CAPITA GROSS NATIONAL PRODUCT IS MEDIUM, LOW, OR VERY LOW (91)	1.00	35	TEND LESS TO BE THOSE WHOSE PER CAPITA GROSS NATIONAL PRODUCT IS MEDIUM, LOW, OR VERY LOW (91)	0.72	0 24 28 62 X SQ= 8.29 P = 0.001 RV NO (NO)
36	TEND MORE TO BE THOSE WHOSE PER CAPITA GROSS NATIONAL PRODUCT IS LOW OR VERY LOW (73)	0.96	36	TEND LESS TO BE THOSE WHOSE PER CAPITA GROSS NATIONAL PRODUCT IS LOW OR VERY LOW (73)	0.52	1 41 27 45 X SQ= 15.81 P = 0. RV NO (NO)
37	TEND TO BE THOSE WHOSE PER CAPITA GROSS NATIONAL PRODUCT IS VERY LOW (51)	0.96	37	TEND TO BE THOSE WHOSE PER CAPITA GROSS NATIONAL PRODUCT IS VERY HIGH, HIGH, MEDIUM, OR LOW (64)	0.73	1 63 27 23 X SQ= 38.87 P = 0. RV YES (YES)
38	ALWAYS ARE THOSE WHOSE INTERNATIONAL FINANCIAL STATUS IS MEDIUM, LOW, OR VERY LOW (103)	1.00	38	LEAN LESS TOWARD BEING THOSE WHOSE INTERNATIONAL FINANCIAL STATUS IS MEDIUM, LOW, OR VERY LOW (103)	0.88	0 10 28 74 X SQ= 2.34 P = 0.064 RV NO (NO)

39	TEND MORE TO BE THOSE WHOSE INTERNATIONAL FINANCIAL STATUS IS LOW OR VERY LOW (76)	0.93	39	TEND LESS TO BE THOSE WHOSE INTERNATIONAL FINANCIAL STATUS IS LOW OR VERY LOW (76)	0.58	2 36 26 49 X SQ= 10.17 P = 0. RV NO (NO)
40	TEND TO BE THOSE WHOSE INTERNATIONAL FINANCIAL STATUS IS VERY LOW (39)	0.71	40	TEND TO BE THOSE WHOSE INTERNATIONAL FINANCIAL STATUS IS VERY HIGH, HIGH, MEDIUM, OR LOW (71)	0.77	8 62 20 19 X SQ= 18.80 P = 0. RV YES (YES)
41	ALWAYS ARE THOSE WHOSE ECONOMIC DEVELOPMENTAL STATUS IS INTERMEDIATE, UNDERDEVELOPED, OR VERY UNDERDEVELOPED (94)	1.00	41	TEND LESS TO BE THOSE WHOSE ECONOMIC DEVELOPMENTAL STATUS IS INTERMEDIATE, UNDERDEVELOPED, OR VERY UNDERDEVELOPED (94)	0.77	0 19 28 65 X SQ= 6.11 P = 0.003 RV NO (NO)
42	TEND MORE TO BE THOSE WHOSE ECONOMIC DEVELOPMENTAL STATUS IS UNDERDEVELOPED OR VERY UNDERDEVELOPED (76)	0.96	42	TEND LESS TO BE THOSE WHOSE ECONOMIC DEVELOPMENTAL STATUS IS UNDERDEVELOPED OR VERY UNDERDEVELOPED (76)	0.58	1 35 27 48 X SQ= 12.53 P = 0. RV NO (NO)
43	TEND TO BE THOSE WHOSE ECONOMIC DEVELOPMENTAL STATUS IS VERY UNDERDEVELOPED (57)	0.96	43	TEND TO BE THOSE WHOSE ECONOMIC DEVELOPMENTAL STATUS IS DEVELOPED, INTERMEDIATE, OR UNDERDEVELOPED (55)	0.64	1 53 27 30 X SQ= 28.09 P = 0. RV YES (NO)
44	ALWAYS ARE THOSE WHERE THE LITERACY RATE IS BELOW NINETY PERCENT (90)	1.00	44	TEND LESS TO BE THOSE WHERE THE LITERACY RATE IS BELOW NINETY PERCENT (90)	0.71	0 25 28 61 X SQ= 8.80 P = 0. RV NO (NO)
45	TEND TO BE THOSE WHERE THE LITERACY RATE IS BELOW FIFTY PERCENT (54)	0.93	45	TEND TO BE THOSE WHERE THE LITERACY RATE IS FIFTY PERCENT OR ABOVE (55)	0.64	2 52 25 29 X SQ= 23.90 P = 0. RV YES (NO)
46	TEND TO BE THOSE WHERE THE LITERACY RATE IS BELOW TEN PERCENT (26)	0.74	46	TEND TO BE THOSE WHERE THE LITERACY RATE IS TEN PERCENT OR ABOVE (84)	0.93	7 76 20 6 X SQ= 46.23 P = 0. RV YES (YES)
52	ALWAYS ARE THOSE WHERE FREEDOM OF THE PRESS IS COMPLETE, INTERMITTENT, OR INTERNALLY ABSENT (82)	1.00	52	TEND LESS TO BE THOSE WHERE FREEDOM OF THE PRESS IS COMPLETE, INTERMITTENT, OR INTERNALLY ABSENT (82)	0.80	18 63 0 16 X SQ= 3.02 P = 0.037 RV NO (NO)

54	ALWAYS ARE THOSE WHERE NEWSPAPER CIRCULATION IS LESS THAN ONE HUNDRED PER THOUSAND (76)	1.00	54	TEND LESS TO BE THOSE WHERE NEWSPAPER CIRCULATION IS LESS THAN ONE HUNDRED PER THOUSAND (76)	0.56	0 37 28 47 X SQ= 16.48 P = 0. RV NO (NO)
55	TEND TO BE THOSE WHERE NEWSPAPER CIRCULATION IS LESS THAN TEN PER THOUSAND (35)	0.86	55	TEND TO BE THOSE WHERE NEWSPAPER CIRCULATION IS TEN OR MORE PER THOUSAND (78)	0.87	4 73 24 11 X SQ= 48.22 P = 0. RV YES (YES)
56	TEND TO BE THOSE WHERE THE RELIGION IS PREDOMINANTLY OR PARTLY NON-LITERATE (31)	0.93	56	TEND TO BE THOSE WHERE THE RELIGION IS PREDOMINANTLY LITERATE (79)	0.93	2 76 25 6 X SQ= 68.45 P = 0. RV YES (YES)
63	ALWAYS ARE THOSE WHERE THE RELIGION IS PREDOMINANTLY OR PARTLY OTHER THAN CHRISTIAN (68)	1.00	63	TEND TO BE THOSE WHERE THE RELIGION IS PREDOMINANTLY SOME KIND OF CHRISTIAN (46)	0.54	0 46 28 39 X SQ= 23.36 P = 0. RV YES (NO)
66	TEND TO BE THOSE THAT ARE RELIGIOUSLY HETEROGENEOUS (49)	0.92	66	TEND TO BE THOSE THAT ARE RELIGIOUSLY HOMOGENEOUS (57)	0.67	2 54 22 27 X SQ= 23.02 P = 0. RV YES (NO)
68	TEND TO BE THOSE THAT ARE LINGUISTICALLY WEAKLY HETEROGENEOUS OR STRONGLY HETEROGENEOUS (62)	0.89	68	TEND TO BE THOSE THAT ARE LINGUISTICALLY HOMOGENEOUS (52)	0.58	3 49 25 36 X SQ= 16.83 P = 0. RV YES (NO)
69	TEND TO BE THOSE THAT ARE LINGUISTICALLY STRONGLY HETEROGENEOUS (50)	0.86	69	TEND TO BE THOSE THAT ARE LINGUISTICALLY HOMOGENEOUS OR WEAKLY HETEROGENEOUS (64)	0.71	4 60 24 25 X SQ= 24.94 P = 0. RV YES (NO)
70	ALWAYS ARE THOSE THAT ARE RELIGIOUSLY, RACIALLY, OR LINGUISTICALLY HETEROGENEOUS (85)	1.00	70	TEND LESS TO BE THOSE THAT ARE RELIGIOUSLY, RACIALLY, OR LINGUISTICALLY HETEROGENEOUS (85)	0.73	0 21 25 58 X SQ= 6.76 P = 0.002 RV NO (NO)
72	TEND TO BE THOSE WHOSE DATE OF INDEPENDENCE IS AFTER 1914 (59)	0.89	72	TEND TO BE THOSE WHOSE DATE OF INDEPENDENCE IS BEFORE 1914 (52)	0.60	3 49 25 33 X SQ= 18.22 P = 0. RV YES (NO)

73	TEND TO BE THOSE WHOSE DATE OF INDEPENDENCE IS AFTER 1945 (46)	0.89	73	TEND TO BE THOSE WHOSE DATE OF INDEPENDENCE IS BEFORE 1945 (65)	0.76	3 62 25 20 X SQ= 33.73 P = 0. RV YES (YES)
75	ALWAYS ARE THOSE THAT ARE NOT HISTORICALLY WESTERN AND ARE NOT SIGNIFICANTLY WESTERNIZED (52)	1.00	75	TEND TO BE THOSE THAT ARE HISTORICALLY WESTERN OR SIGNIFICANTLY WESTERNIZED (62)	0.72	0 62 27 24 X SQ= 40.27 P = 0. RV YES (YES)
76	ALWAYS ARE THOSE THAT HAVE BEEN SIGNIFICANTLY OR PARTIALLY WESTERNIZED THROUGH A COLONIAL RELATIONSHIP, RATHER THAN BEING HISTORICALLY WESTERN (70)	1.00	76	TEND LESS TO BE THOSE THAT HAVE BEEN SIGNIFICANTLY OR PARTIALLY WESTERNIZED THROUGH A COLONIAL RELATIONSHIP, RATHER THAN BEING HISTORICALLY WESTERN (70)	0.62	0 26 26 43 X SQ= 11.66 P = 0. RV NO (NO)
77	ALWAYS ARE THOSE THAT HAVE BEEN PARTIALLY WESTERNIZED, RATHER THAN SIGNIFICANTLY WESTERNIZED, THROUGH A COLONIAL RELATIONSHIP (41)	1.00	77	TEND TO BE THOSE THAT HAVE BEEN SIGNIFICANTLY WESTERNIZED, RATHER THAN PARTIALLY WESTERNIZED, THROUGH A COLONIAL RELATIONSHIP (28)	0.65	0 28 25 15 X SQ= 25.05 P = 0. RV YES (YES)
79	ALWAYS ARE THOSE THAT WERE FORMERLY DEPENDENCIES OF BRITAIN OR FRANCE, RATHER THAN SPAIN (49)	1.00	79	TEND LESS TO BE THOSE THAT WERE FORMERLY DEPENDENCIES OF BRITAIN OR FRANCE, RATHER THAN SPAIN (49)	0.59	22 26 0 18 X SQ= 10.40 P = 0. RV NO (NO)
80	TEND TO BE THOSE WHOSE DATE OF INDEPENDENCE IS AFTER 1914, AND THAT WERE FORMERLY DEPENDENCIES OF FRANCE, RATHER THAN BRITAIN (24)	0.76	80	TEND TO BE THOSE WHOSE DATE OF INDEPENDENCE IS AFTER 1914, AND THAT WERE FORMERLY DEPENDENCIES OF BRITAIN, RATHER THAN FRANCE (19)	0.62	5 13 16 8 X SQ= 4.76 P = 0.028 RV YES (YES)
82	TEND TO BE THOSE WHERE THE TYPE OF POLITICAL MODERNIZATION IS DEVELOPED TUTELARY OR UNDEVELOPED TUTELARY, RATHER THAN EARLY OR LATER EUROPEAN OR EUROPEAN DERIVED (55)	0.96	82	TEND TO BE THOSE WHERE THE TYPE OF POLITICAL MODERNIZATION IS EARLY OR LATER EUROPEAN OR EUROPEAN DERIVED, RATHER THAN DEVELOPED TUTELARY OR UNDEVELOPED TUTELARY (51)	0.64	1 50 26 28 X SQ= 26.92 P = 0. RV YES (NO)
84	TEND TO BE THOSE WHERE THE TYPE OF POLITICAL MODERNIZATION IS UNDEVELOPED TUTELARY, RATHER THAN DEVELOPED TUTELARY (24)	0.85	84	TEND TO BE THOSE WHERE THE TYPE OF POLITICAL MODERNIZATION IS DEVELOPED TUTELARY, RATHER THAN UNDEVELOPED TUTELARY (31)	0.93	4 26 22 2 X SQ= 29.71 P = 0. RV YES (YES)
85	ALWAYS ARE THOSE WHERE THE STAGE OF POLITICAL MODERNIZATION IS MID- OR EARLY TRANSITIONAL, RATHER THAN ADVANCED (54)	1.00	85	TEND TO BE THOSE WHERE THE STAGE OF POLITICAL MODERNIZATION IS ADVANCED, RATHER THAN MID- OR EARLY TRANSITIONAL (60)	0.71	0 60 28 25 X SQ= 39.35 P = 0. RV YES (YES)

86	TEND TO BE THOSE WHERE THE STAGE OF POLITICAL MODERNIZATION IS EARLY TRANSITIONAL, RATHER THAN ADVANCED OR MID-TRANSITIONAL (38)	0.93	86	TEND TO BE THOSE WHERE THE STAGE OF POLITICAL MODERNIZATION IS ADVANCED OR MID-TRANSITIONAL, RATHER THAN EARLY TRANSITIONAL (76)	0.87	2 74 26 11 X SQ= 57.51 P = 0. RV YES (YES)
87	TEND TO BE THOSE WHOSE IDEOLOGICAL ORIENTATION IS DEVELOPMENTAL (31)	0.87	87	TEND TO BE THOSE WHOSE IDEOLOGICAL ORIENTATION IS OTHER THAN DEVELOPMENTAL (58)	0.83	20 11 3 55 X SQ= 34.09 P = 0. RV YES (YES)
89	TEND MORE TO BE THOSE WHOSE IDEOLOGICAL ORIENTATION IS OTHER THAN CONVENTIONAL (62)	0.96	89	TEND LESS TO BE THOSE WHOSE IDEOLOGICAL ORIENTATION IS OTHER THAN CONVENTIONAL (62)	0.54	1 32 23 38 X SQ= 11.78 P = 0. RV NO (NO)
91	TILT MORE TOWARD BEING THOSE WHOSE IDEOLOGICAL ORIENTATION IS DEVELOPMENTAL, RATHER THAN TRADITIONAL (31)	0.95	91	TILT LESS TOWARD BEING THOSE WHOSE IDEOLOGICAL ORIENTATION IS DEVELOPMENTAL, RATHER THAN TRADITIONAL (31)	0.73	20 11 1 4 X SQ= 1.92 P = 0.138 RV NO (YES)
94	TILT TOWARD BEING THOSE WHERE THE STATUS OF THE REGIME IS AUTHORITARIAN OR TOTALITARIAN (41)	0.67	94	TILT TOWARD BEING THOSE WHERE THE STATUS OF THE REGIME IS CONSTITUTIONAL (51)	0.59	4 47 8 32 X SQ= 1.93 P = 0.121 RV YES (NO)
95	ALWAYS ARE THOSE WHERE THE STATUS OF THE REGIME IS CONSTITUTIONAL OR AUTHORITARIAN (95)	1.00	95	TEND LESS TO BE THOSE WHERE THE STATUS OF THE REGIME IS CONSTITUTIONAL OR AUTHORITARIAN (95)	0.81	26 68 0 16 X SQ= 4.36 P = 0.012 RV NO (NO)
96	TEND TO BE THOSE WHERE THE STATUS OF THE REGIME IS AUTHORITARIAN (23)	0.64	96	TEND TO BE THOSE WHERE THE STATUS OF THE REGIME IS CONSTITUTIONAL OR TOTALITARIAN (67)	0.81	4 63 7 15 X SQ= 7.97 P = 0.004 RV YES (NO)
97	ALWAYS ARE THOSE WHERE THE STATUS OF THE REGIME IS AUTHORITARIAN, RATHER THAN TOTALITARIAN (23)	1.00	97	TEND TO BE THOSE WHERE THE STATUS OF THE REGIME IS TOTALITARIAN, RATHER THAN AUTHORITARIAN (16)	0.52	7 15 0 16 X SQ= 4.30 P = 0.014 RV YES (NO)
98	LEAN MORE TOWARD BEING THOSE WHERE GOVERNMENTAL STABILITY IS GENERALLY OR MODERATELY PRESENT AND DATES FROM THE POST-WAR PERIOD, OR IS ABSENT (93)	0.93	98	LEAN LESS TOWARD BEING THOSE WHERE GOVERNMENTAL STABILITY IS GENERALLY OR MODERATELY PRESENT AND DATES FROM THE POST-WAR PERIOD, OR IS ABSENT (93)	0.77	2 20 26 66 X SQ= 2.56 P = 0.095 RV NO (NO)

101	TEND TO BE THOSE WHERE THE REPRESENTATIVE CHARACTER OF THE REGIME IS LIMITED POLYARCHIC, PSEUDO-POLYARCHIC, OR NON-POLYARCHIC (57)	0.90	101	TEND TO BE THOSE WHERE THE REPRESENTATIVE CHARACTER OF THE REGIME IS POLYARCHIC (41)	0.51	2 39 19 37 X SQ= 10.13 P = 0. RV YES (NO)
102	TILT TOWARD BEING THOSE WHERE THE REPRESENTATIVE CHARACTER OF THE REGIME IS PSEUDO-POLYARCHIC OR NON-POLYARCHIC (49)	0.59	102	TILT TOWARD BEING THOSE WHERE THE REPRESENTATIVE CHARACTER OF THE REGIME IS POLYARCHIC OR LIMITED POLYARCHIC (59)	0.60	11 48 16 32 X SQ= 2.30 P = 0.117 RV YES (NO)
105	TEND TO BE THOSE WHERE THE ELECTORAL SYSTEM IS PARTIALLY COMPETITIVE OR NON-COMPETITIVE (47)	0.86	105	TEND TO BE THOSE WHERE THE ELECTORAL SYSTEM IS COMPETITIVE (43)	0.59	3 40 19 28 X SQ= 11.85 P = 0. RV YES (NO)
107	TEND TO BE THOSE WHERE AUTONOMOUS GROUPS ARE PARTIALLY TOLERATED IN POLITICS, ARE TOLERATED ONLY OUTSIDE POLITICS, OR ARE NOT TOLERATED AT ALL (65)	0.88	107	TEND TO BE THOSE WHERE AUTONOMOUS GROUPS ARE FULLY TOLERATED IN POLITICS (46)	0.51	3 43 23 41 X SQ= 11.25 P = 0. RV YES (NO)
108	TILT TOWARD BEING THOSE WHERE AUTONOMOUS GROUPS ARE TOLERATED ONLY OUTSIDE POLITICS OR ARE NOT TOLERATED AT ALL (35)	0.50	108	TILT TOWARD BEING THOSE WHERE AUTONOMOUS GROUPS ARE FULLY OR PARTIALLY TOLERATED IN POLITICS (65)	0.69	11 54 11 24 X SQ= 2.01 P = 0.129 RV YES (NO)
110	ALWAYS ARE THOSE WHERE POLITICAL ENCULTURATION IS MEDIUM OR LOW (80)	1.00	110	TEND LESS TO BE THOSE WHERE POLITICAL ENCULTURATION IS MEDIUM OR LOW (80)	0.78	0 15 25 54 X SQ= 4.95 P = 0.009 RV NO (NO)
111	TEND TO BE THOSE WHERE POLITICAL ENCULTURATION IS LOW (42)	0.64	111	TEND TO BE THOSE WHERE POLITICAL ENCULTURATION IS HIGH OR MEDIUM (53)	0.62	9 43 16 26 X SQ= 4.13 P = 0.034 RV YES (NO)
114	LEAN MORE TOWARD BEING THOSE WHERE SECTIONALISM IS EXTREME OR MODERATE (61)	0.73	114	LEAN LESS TOWARD BEING THOSE WHERE SECTIONALISM IS EXTREME OR MODERATE (61)	0.51	19 41 7 40 X SQ= 3.17 P = 0.068 RV NO (NO)
115	ALWAYS ARE THOSE WHERE INTEREST ARTICULATION BY ASSOCIATIONAL GROUPS IS MODERATE, LIMITED, OR NEGLIGIBLE (91)	1.00	115	TEND LESS TO BE THOSE WHERE INTEREST ARTICULATION BY ASSOCIATIONAL GROUPS IS MODERATE, LIMITED, OR NEGLIGIBLE (91)	0.76	0 20 28 62 X SQ= 6.79 P = 0.002 RV NO (NO)

116	TEND MORE TO BE THOSE WHERE INTEREST ARTICULATION BY ASSOCIATIONAL GROUPS IS LIMITED OR NEGLIGIBLE (79)	0.96	116	TEND LESS TO BE THOSE WHERE INTEREST ARTICULATION BY ASSOCIATIONAL GROUPS IS LIMITED OR NEGLIGIBLE (79)	0.62	1 31 27 51 X SQ= 10.26 P = 0. RV NO (NO)
117	TEND TO BE THOSE WHERE INTEREST ARTICULATION BY ASSOCIATIONAL GROUPS IS NEGLIGIBLE (51)	0.93	117	TEND TO BE THOSE WHERE INTEREST ARTICULATION BY ASSOCIATIONAL GROUPS IS SIGNIFICANT, MODERATE, OR LIMITED (60)	0.70	2 57 26 25 X SQ= 30.19 P = 0. RV YES (YES)
121	TEND TO BE THOSE WHERE INTEREST ARTICULATION BY NON-ASSOCIATIONAL GROUPS IS SIGNIFICANT (54)	0.96	121	TEND TO BE THOSE WHERE INTEREST ARTICULATION BY NON-ASSOCIATIONAL GROUPS IS MODERATE, LIMITED, OR NEGLIGIBLE (61)	0.70	27 26 1 60 X SQ= 34.59 P = 0. RV YES (YES)
122	ALWAYS ARE THOSE WHERE INTEREST ARTICULATION BY NON-ASSOCIATIONAL GROUPS IS SIGNIFICANT OR MODERATE (83)	1.00	122	TEND LESS TO BE THOSE WHERE INTEREST ARTICULATION BY NON-ASSOCIATIONAL GROUPS IS SIGNIFICANT OR MODERATE (83)	0.63	28 54 0 32 X SQ= 12.70 P = 0. RV NO (NO)
125	TEND MORE TO BE THOSE WHERE INTEREST ARTICULATION BY ANOMIC GROUPS IS FREQUENT OR OCCASIONAL (64)	0.86	125	TEND LESS TO BE THOSE WHERE INTEREST ARTICULATION BY ANOMIC GROUPS IS FREQUENT OR OCCASIONAL (64)	0.56	24 39 4 31 X SQ= 6.59 P = 0.005 RV NO (NO)
126	ALWAYS ARE THOSE WHERE INTEREST ARTICULATION BY ANOMIC GROUPS IS FREQUENT, OCCASIONAL, OR INFREQUENT (83)	1.00	126	TEND LESS TO BE THOSE WHERE INTEREST ARTICULATION BY ANOMIC GROUPS IS FREQUENT, OCCASIONAL, OR INFREQUENT (83)	0.77	28 54 0 16 X SQ= 6.07 P = 0.005 RV NO (NO)
128	TEND TO BE THOSE WHERE INTEREST ARTICULATION BY POLITICAL PARTIES IS LIMITED OR NEGLIGIBLE (45)	0.71	128	TEND TO BE THOSE WHERE INTEREST ARTICULATION BY POLITICAL PARTIES IS SIGNIFICANT OR MODERATE (48)	0.59	7 41 17 28 X SQ= 5.37 P = 0.017 RV YES (NO)
129	TEND TO BE THOSE WHERE INTEREST ARTICULATION BY POLITICAL PARTIES IS NEGLIGIBLE (37)	0.62	129	TEND TO BE THOSE WHERE INTEREST ARTICULATION BY POLITICAL PARTIES IS SIGNIFICANT, MODERATE, OR LIMITED (56)	0.68	9 47 15 22 X SQ= 5.75 P = 0.014 RV YES (NO)
133	TEND TO BE THOSE WHERE INTEREST AGGREGATION BY THE EXECUTIVE IS SIGNIFICANT (29)	0.62	133	TEND TO BE THOSE WHERE INTEREST AGGREGATION BY THE EXECUTIVE IS MODERATE, LIMITED, OR NEGLIGIBLE (76)	0.83	16 13 10 65 X SQ= 17.36 P = 0. RV YES (YES)

134	TEND TO BE THOSE WHERE INTEREST AGGREGATION BY THE EXECUTIVE IS SIGNIFICANT OR MODERATE (57)	0.76	134	TEND TO BE THOSE WHERE INTEREST AGGREGATION BY THE EXECUTIVE IS LIMITED OR NEGLIGIBLE (46)	0.51	19 38 6 39 X SQ= 4.41 P = 0.022 RV YES (NO)
137	ALWAYS ARE THOSE WHERE INTEREST AGGREGATION BY THE LEGISLATURE IS LIMITED OR NEGLIGIBLE (68)	1.00	137	TEND LESS TO BE THOSE WHERE INTEREST AGGREGATION BY THE LEGISLATURE IS LIMITED OR NEGLIGIBLE (68)	0.59	0 29 26 42 X SQ= 13.26 P = 0. RV NO (NO)
138	TEND TO BE THOSE WHERE INTEREST AGGREGATION BY THE LEGISLATURE IS NEGLIGIBLE (49)	0.85	138	TEND TO BE THOSE WHERE INTEREST AGGREGATION BY THE LEGISLATURE IS SIGNIFICANT, MODERATE, OR LIMITED (48)	0.62	4 44 22 27 X SQ= 14.71 P = 0. RV YES (NO)
142	LEAN MORE TOWARD BEING THOSE WHERE THE PARTY SYSTEM IS QUANTITATIVELY OTHER THAN MULTI-PARTY (66)	0.84	142	LEAN LESS TOWARD BEING THOSE WHERE THE PARTY SYSTEM IS QUANTITATIVELY OTHER THAN MULTI-PARTY (66)	0.63	4 26 21 45 X SQ= 2.76 P = 0.079 RV NO (NO)
144	TILT MORE TOWARD BEING THOSE WHERE THE PARTY SYSTEM IS QUANTITATIVELY ONE-PARTY, RATHER THAN TWO-PARTY (34)	0.93	144	TILT LESS TOWARD BEING THOSE WHERE THE PARTY SYSTEM IS QUANTITATIVELY ONE-PARTY, RATHER THAN TWO-PARTY (34)	0.68	13 21 1 10 X SQ= 2.07 P = 0.132 RV NO (NO)
146	TEND LESS TO BE THOSE WHERE THE PARTY SYSTEM IS QUALITATIVELY OTHER THAN MASS-BASED TERRITORIAL (92)	0.81	146	TEND MORE TO BE THOSE WHERE THE PARTY SYSTEM IS QUALITATIVELY OTHER THAN MASS-BASED TERRITORIAL (92)	0.96	5 3 21 70 X SQ= 4.04 P = 0.028 RV NO (YES)
147	ALWAYS ARE THOSE WHERE THE PARTY SYSTEM IS QUALITATIVELY OTHER THAN CLASS-ORIENTED OR MULTI-IDEOLOGICAL (67)	1.00	147	TEND LESS TO BE THOSE WHERE THE PARTY SYSTEM IS QUALITATIVELY OTHER THAN CLASS-ORIENTED OR MULTI-IDEOLOGICAL (67)	0.65	0 23 24 43 X SQ= 9.48 P = 0. RV NO (NO)
148	TEND LESS TO BE THOSE WHERE THE PARTY SYSTEM IS QUALITATIVELY OTHER THAN AFRICAN TRANSITIONAL (96)	0.52	148	TEND MORE TO BE THOSE WHERE THE PARTY SYSTEM IS QUALITATIVELY OTHER THAN AFRICAN TRANSITIONAL (96)	0.99	13 1 14 81 X SQ= 35.88 P = 0. RV NO (YES)
158	TEND TO BE THOSE WHERE PERSONALISMO IS PRONOUNCED OR MODERATE (40)	0.67	158	TEND TO BE THOSE WHERE PERSONALISMO IS NEGLIGIBLE (56)	0.67	16 24 8 48 X SQ= 6.91 P = 0.008 RV YES (NO)

161	LEAN TOWARD BEING THOSE WHERE THE POLITICAL LEADERSHIP IS NON-ELITIST (50)	0.68	161	LEAN TOWARD BEING THOSE WHERE THE POLITICAL LEADERSHIP IS ELITIST OR MODERATELY ELITIST (47)	0.54	8 39 17 33 X SQ= 2.82 P = 0.066 RV YES (NO)
164	TEND TO BE THOSE WHERE THE REGIME'S LEADERSHIP CHARISMA IS PRONOUNCED OR MODERATE (34)	0.76	164	TEND TO BE THOSE WHERE THE REGIME'S LEADERSHIP CHARISMA IS NEGLIGIBLE (65)	0.77	16 18 5 60 X SQ= 18.41 P = 0. RV YES (NO)
169	TEND TO BE THOSE WHERE THE HORIZONTAL POWER DISTRIBUTION IS NEGLIGIBLE (48)	0.69	169	TEND TO BE THOSE WHERE THE HORIZONTAL POWER DISTRIBUTION IS SIGNIFICANT OR LIMITED (58)	0.62	8 50 18 30 X SQ= 6.74 P = 0.006 RV YES (NO)
170	TEND TO BE THOSE WHERE THE LEGISLATIVE-EXECUTIVE STRUCTURE IS PRESIDENTIAL (39)	0.62	170	TEND TO BE THOSE WHERE THE LEGISLATIVE-EXECUTIVE STRUCTURE IS OTHER THAN PRESIDENTIAL (63)	0.70	16 23 10 53 X SQ= 6.75 P = 0.009 RV YES (NO)
175	TEND TO BE THOSE WHERE THE LEGISLATURE IS LARGELY INEFFECTIVE OR WHOLLY INEFFECTIVE (49)	0.77	175	TEND TO BE THOSE WHERE THE LEGISLATURE IS FULLY EFFECTIVE OR PARTIALLY EFFECTIVE (51)	0.61	6 45 20 29 X SQ= 9.50 P = 0.001 RV YES (NO)
178	TEND TO BE THOSE WHERE THE LEGISLATURE IS UNICAMERAL (53)	0.79	178	TEND TO BE THOSE WHERE THE LEGISLATURE IS BICAMERAL (51)	0.59	22 31 6 45 X SQ= 10.22 P = 0.001 RV YES (NO)
179	TEND MORE TO BE THOSE WHERE THE EXECUTIVE IS DOMINANT (52)	0.80	179	TEND LESS TO BE THOSE WHERE THE EXECUTIVE IS DOMINANT (52)	0.51	16 36 4 35 X SQ= 4.34 P = 0.023 RV NO (NO)
182	TEND TO BE THOSE WHERE THE BUREAUCRACY IS POST-COLONIAL TRANSITIONAL, RATHER THAN SEMI-MODERN (25)	0.87	182	TEND TO BE THOSE WHERE THE BUREAUCRACY IS SEMI-MODERN, RATHER THAN POST-COLONIAL TRANSITIONAL (55)	0.93	3 52 20 4 X SQ= 45.40 P = 0. RV YES (YES)
183	TEND TO BE THOSE WHERE THE BUREAUCRACY IS POST-COLONIAL TRANSITIONAL, RATHER THAN TRADITIONAL (25)	0.91	183	TEND TO BE THOSE WHERE THE BUREAUCRACY IS TRADITIONAL, RATHER THAN POST-COLONIAL TRANSITIONAL (9)	0.64	20 4 2 7 X SQ= 8.42 P = 0.002 RV YES (YES)

185	LEAN MORE TOWARD BEING THOSE WHERE PARTICIPATION BY THE MILITARY IN POLITICS IS SUPPORTIVE, RATHER THAN INTERVENTIVE (31)	0.89	185	LEAN LESS TOWARD BEING THOSE WHERE PARTICIPATION BY THE MILITARY IN POLITICS IS SUPPORTIVE, RATHER THAN INTERVENTIVE (31)	0.55	1 19 8 23 X SQ= 2.33 P = 0.072 RV NO (NO)
187	LEAN MORE TOWARD BEING THOSE WHERE THE ROLE OF THE POLICE IS POLITICALLY SIGNIFICANT (66)	0.82	187	LEAN LESS TOWARD BEING THOSE WHERE THE ROLE OF THE POLICE IS POLITICALLY SIGNIFICANT (66)	0.60	18 47 4 31 X SQ= 2.62 P = 0.078 RV NO (NO)
194	ALWAYS ARE THOSE THAT ARE NON-COMMUNIST (101)	1.00	194	TEND LESS TO BE THOSE THAT ARE NON-COMMUNIST (101)	0.85	0 13 28 72 X SQ= 3.45 P = 0.036 RV NO (NO)

194 POLITIES
THAT ARE COMMUNIST (13)

194 POLITIES
THAT ARE NON-COMMUNIST (101)

BOTH SUBJECT AND PREDICATE

13 IN LEFT COLUMN

ALBANIA	BULGARIA	CHINA, PR	CZECHOS'KIA	GERMANY, E	HUNGARY	KOREA, N	MONGOLIA	POLAND	RUMANIA
USSR	VIETNAM, N	YUGOSLAVIA							

101 IN RIGHT COLUMN

AFGHANISTAN	ALGERIA	ARGENTINA	AUSTRALIA	AUSTRIA	BELGIUM	BOLIVIA	BRAZIL	BURMA	BURUNDI
CAMBODIA	CAMEROUN	CANADA	CEN AFR REP	CEYLON	CHAD	CHILE	COLOMBIA	CONGO(BRA)	CONGO(LEO)
COSTA RICA	CYPRUS	DAHOMEY	DENMARK	DOMIN REP	ECUADOR	EL SALVADOR	ETHIOPIA	FINLAND	FRANCE
GABON	GERMAN FR	GHANA	GREECE	GUATEMALA	GUINEA	HAITI	HONDURAS	ICELAND	INDIA
INDONESIA	IRAN	IRAQ	IRELAND	ISRAEL	ITALY	IVORY COAST	JAMAICA	JAPAN	JORDAN
KOREA REP	LAOS	LEBANON	LIBERIA	LIBYA	LUXEMBOURG	MALAGASY R	MALAYA	MALI	MAURITANIA
MEXICO	MOROCCO	NEPAL	NETHERLANDS	NEW ZEALAND	NICARAGUA	NIGER	NIGERIA	NORWAY	PAKISTAN
PANAMA	PARAGUAY	PERU	PHILIPPINES	PORTUGAL	RWANDA	SA'U ARABIA	SENEGAL	SIERRE LEO	SOMALIA
SO AFRICA	SPAIN	SUDAN	SWEDEN	SWITZERLAND	SYRIA	TANGANYIKA	THAILAND	TOGO	TRINIDAD
TUNISIA	TURKEY	UGANDA	UAR	UK	US	UPPER VOLTA	URUGUAY	VENEZUELA	VIETNAM REP
YEMEN									

1 EXCLUDED BECAUSE IRRELEVANT

CUBA

	Left column			Right column		
24	TEND TO BE THOSE WHOSE POPULATION IS VERY LARGE, LARGE, OR MEDIUM (61)	0.85	24	TEND TO BE THOSE WHOSE POPULATION IS SMALL (54)	0.51	11 49 2 52 X SQ= 4.66 P = 0.018 RV YES (NO)
26	TEND TO BE THOSE WHOSE POPULATION DENSITY IS VERY HIGH, HIGH, OR MEDIUM (48)	0.85	26	TEND TO BE THOSE WHOSE POPULATION DENSITY IS LOW (67)	0.64	11 36 2 65 X SQ= 9.47 P = 0.002 RV YES (NO)
28	LEAN TOWARD BEING THOSE WHOSE POPULATION GROWTH RATE IS LOW (48)	0.73	28	LEAN TOWARD BEING THOSE WHOSE POPULATION GROWTH RATE IS HIGH (62)	0.59	3 58 8 40 X SQ= 2.89 P = 0.057 RV YES (NO)

33	TEND TO BE THOSE WHOSE GROSS NATIONAL PRODUCT IS VERY HIGH, HIGH, OR MEDIUM (30)	0.54	33	TEND TO BE THOSE WHOSE GROSS NATIONAL PRODUCT IS LOW OR VERY LOW (85)	0.77	7 23 6 78 X SQ= 4.24 P = 0.038 RV YES (NO)
36	LEAN TOWARD BEING THOSE WHOSE PER CAPITA GROSS NATIONAL PRODUCT IS VERY HIGH, HIGH, OR MEDIUM (42)	0.62	36	LEAN TOWARD BEING THOSE WHOSE PER CAPITA GROSS NATIONAL PRODUCT IS LOW OR VERY LOW (73)	0.67	8 33 5 68 X SQ= 3.01 P = 0.063 RV YES (NO)
39	TILT TOWARD BEING THOSE WHOSE INTERNATIONAL FINANCIAL STATUS IS VERY HIGH, HIGH, OR MEDIUM (38)	0.58	39	TILT TOWARD BEING THOSE WHOSE INTERNATIONAL FINANCIAL STATUS IS LOW OR VERY LOW (76)	0.69	7 31 5 70 X SQ= 2.54 P = 0.102 RV YES (NO)
42	TEND TO BE THOSE WHOSE ECONOMIC DEVELOPMENTAL STATUS IS DEVELOPED OR INTERMEDIATE (36)	0.62	42	TEND TO BE THOSE WHOSE ECONOMIC DEVELOPMENTAL STATUS IS UNDERDEVELOPED OR VERY UNDERDEVELOPED (76)	0.71	8 28 5 70 X SQ= 4.29 P = 0.026 RV YES (NO)
43	TILT TOWARD BEING THOSE WHOSE ECONOMIC DEVELOPMENTAL STATUS IS DEVELOPED, INTERMEDIATE, OR UNDERDEVELOPED (55)	0.69	43	TILT TOWARD BEING THOSE WHOSE ECONOMIC DEVELOPMENTAL STATUS IS VERY UNDERDEVELOPED (57)	0.54	9 45 4 53 X SQ= 1.65 P = 0.145 RV YES (NO)
45	ALWAYS ARE THOSE WHERE THE LITERACY RATE IS FIFTY PERCENT OR ABOVE (55)	1.00	45	TEND TO BE THOSE WHERE THE LITERACY RATE IS BELOW FIFTY PERCENT (54)	0.55	9 45 0 54 X SQ= 7.76 P = 0.003 RV YES (NO)
51	ALWAYS ARE THOSE WHERE FREEDOM OF THE PRESS IS INTERNALLY ABSENT, OR INTERNALLY AND EXTERNALLY ABSENT (37)	1.00	51	TEND TO BE THOSE WHERE FREEDOM OF THE PRESS IS COMPLETE OR INTERMITTENT (60)	0.72	0 60 13 23 X SQ= 22.07 P = 0. RV YES (NO)
52	TEND TO BE THOSE WHERE FREEDOM OF THE PRESS IS INTERNALLY AND EXTERNALLY ABSENT (16)	0.77	52	TEND TO BE THOSE WHERE FREEDOM OF THE PRESS IS COMPLETE, INTERMITTENT, OR INTERNALLY ABSENT (82)	0.94	3 79 10 5 X SQ= 38.12 P = 0. RV YES (YES)
54	TEND TO BE THOSE WHERE NEWSPAPER CIRCULATION IS ONE HUNDRED OR MORE PER THOUSAND (37)	0.73	54	TEND TO BE THOSE WHERE NEWSPAPER CIRCULATION IS LESS THAN ONE HUNDRED PER THOUSAND (76)	0.72	8 28 3 73 X SQ= 7.26 P = 0.005 RV YES (NO)

55	ALWAYS ARE THOSE WHERE NEWSPAPER CIRCULATION IS TEN OR MORE PER THOUSAND (78)	1.00	55	TEND LESS TO BE THOSE WHERE NEWSPAPER CIRCULATION IS TEN OR MORE PER THOUSAND (78)	0.65	11 66 0 35 X SQ= 4.05 P = 0.016 RV NO (NO)
56	ALWAYS ARE THOSE WHERE THE RELIGION IS PREDOMINANTLY LITERATE (79)	1.00	56	TEND LESS TO BE THOSE WHERE THE RELIGION IS PREDOMINANTLY LITERATE (79)	0.69	10 68 0 31 X SQ= 2.97 P = 0.035 RV NO (NO)
58	ALWAYS ARE THOSE WHERE THE RELIGION IS OTHER THAN MUSLIM (97)	1.00	58	TILT LESS TOWARD BEING THOSE WHERE THE RELIGION IS OTHER THAN MUSLIM (97)	0.82	0 18 13 83 X SQ= 1.57 P = 0.125 RV NO (NO)
67	ALWAYS ARE THOSE THAT ARE RACIALLY HOMOGENEOUS (82)	1.00	67	TEND LESS TO BE THOSE THAT ARE RACIALLY HOMOGENEOUS (82)	0.71	13 67 0 28 X SQ= 3.75 P = 0.020 RV NO (NO)
73	TEND MORE TO BE THOSE WHOSE DATE OF INDEPENDENCE IS BEFORE 1945 (65)	0.91	73	TEND LESS TO BE THOSE WHOSE DATE OF INDEPENDENCE IS BEFORE 1945 (65)	0.55	10 54 1 45 X SQ= 3.99 P = 0.024 RV NO (NO)
76	TEND TO BE THOSE THAT ARE HISTORICALLY WESTERN, RATHER THAN HAVING BEEN SIGNIFICANTLY OR PARTIALLY WESTERNIZED THROUGH A COLONIAL RELATIONSHIP (26)	0.80	76	TEND TO BE THOSE THAT HAVE BEEN SIGNIFICANTLY OR PARTIALLY WESTERNIZED THROUGH A COLONIAL RELATIONSHIP, RATHER THAN BEING HISTORICALLY WESTERN (70)	0.76	4 22 1 68 X SQ= 4.83 P = 0.019 RV YES (NO)
78	ALWAYS ARE THOSE THAT HAVE BEEN SIGNIFICANTLY WESTERNIZED WITHOUT A COLONIAL RELATIONSHIP, RATHER THAN THROUGH SUCH A RELATIONSHIP (7)	1.00	78	TEND TO BE THOSE THAT HAVE BEEN SIGNIFICANTLY WESTERNIZED THROUGH A COLONIAL RELATIONSHIP, RATHER THAN WITHOUT SUCH A RELATIONSHIP (28)	0.93	0 27 5 2 X SQ= 17.27 P = 0. RV YES (YES)
82	TILT TOWARD BEING THOSE WHERE THE TYPE OF POLITICAL MODERNIZATION IS EARLY OR LATER EUROPEAN OR EUROPEAN DERIVED, RATHER THAN DEVELOPED TUTELARY OR UNDEVELOPED TUTELARY (51)	0.73	82	TILT TOWARD BEING THOSE WHERE THE TYPE OF POLITICAL MODERNIZATION IS DEVELOPED TUTELARY OR UNDEVELOPED TUTELARY, RATHER THAN EARLY OR LATER EUROPEAN OR EUROPEAN DERIVED (55)	0.55	8 42 3 52 X SQ= 2.08 P = 0.111 RV YES (NO)
85	TEND TO BE THOSE WHERE THE STAGE OF POLITICAL MODERNIZATION IS ADVANCED, RATHER THAN MID- OR EARLY TRANSITIONAL (60)	0.92	85	TEND TO BE THOSE WHERE THE STAGE OF POLITICAL MODERNIZATION IS MID- OR EARLY TRANSITIONAL, RATHER THAN ADVANCED (54)	0.53	12 47 1 53 X SQ= 7.74 P = 0.002 RV YES (NO)

92	ALWAYS ARE THOSE WHERE THE SYSTEM STYLE IS MOBILIZATIONAL (20)	1.00	92	TEND TO BE THOSE WHERE THE SYSTEM STYLE IS LIMITED MOBILIZATIONAL, OR NON-MOBILIZATIONAL (93)	0.94	13 6 0 93 X SQ= 65.47 P = 0. RV YES (YES)
95	ALWAYS ARE THOSE WHERE THE STATUS OF THE REGIME IS TOTALITARIAN (16)	1.00	95	TEND TO BE THOSE WHERE THE STATUS OF THE REGIME IS CONSTITUTIONAL OR AUTHORITARIAN (95)	0.98	0 95 13 2 X SQ= 85.24 P = 0. RV YES (YES)
99	ALWAYS ARE THOSE WHERE GOVERNMENTAL STABILITY IS GENERALLY PRESENT AND DATES FROM AT LEAST THE INTER-WAR PERIOD, OR FROM THE POST-WAR PERIOD (50)	1.00	99	TEND LESS TO BE THOSE WHERE GOVERNMENTAL STABILITY IS GENERALLY PRESENT AND DATES FROM AT LEAST THE INTER-WAR PERIOD, OR FROM THE POST-WAR PERIOD (50)	0.51	12 38 0 36 X SQ= 8.14 P = 0.001 RV NO (NO)
102	ALWAYS ARE THOSE WHERE THE REPRESENTATIVE CHARACTER OF THE REGIME IS PSEUDO-POLYARCHIC OR NON-POLYARCHIC (49)	1.00	102	TEND TO BE THOSE WHERE THE REPRESENTATIVE CHARACTER OF THE REGIME IS POLYARCHIC OR LIMITED POLYARCHIC (59)	0.63	0 59 13 35 X SQ= 15.74 P = 0. RV YES (NO)
106	ALWAYS ARE THOSE WHERE THE ELECTORAL SYSTEM IS NON-COMPETITIVE (30)	1.00	106	TEND TO BE THOSE WHERE THE ELECTORAL SYSTEM IS COMPETITIVE OR PARTIALLY COMPETITIVE (52)	0.75	0 51 13 17 X SQ= 23.21 P = 0. RV YES (NO)
108	ALWAYS ARE THOSE WHERE AUTONOMOUS GROUPS ARE TOLERATED ONLY OUTSIDE POLITICS OR ARE NOT TOLERATED AT ALL (35)	1.00	108	TEND TO BE THOSE WHERE AUTONOMOUS GROUPS ARE FULLY OR PARTIALLY TOLERATED IN POLITICS (65)	0.74	0 64 13 22 X SQ= 24.21 P = 0. RV YES (NO)
116	ALWAYS ARE THOSE WHERE INTEREST ARTICULATION BY ASSOCIATIONAL GROUPS IS LIMITED OR NEGLIGIBLE (79)	1.00	116	TEND LESS TO BE THOSE WHERE INTEREST ARTICULATION BY ASSOCIATIONAL GROUPS IS LIMITED OR NEGLIGIBLE (79)	0.67	0 32 12 66 X SQ= 4.06 P = 0.017 RV NO (NO)
117	LEAN TOWARD BEING THOSE WHERE INTEREST ARTICULATION BY ASSOCIATIONAL GROUPS IS NEGLIGIBLE (51)	0.75	117	LEAN TOWARD BEING THOSE WHERE INTEREST ARTICULATION BY ASSOCIATIONAL GROUPS IS SIGNIFICANT, MODERATE, OR LIMITED (60)	0.57	3 56 9 42 X SQ= 3.24 P = 0.062 RV YES (NO)
118	ALWAYS ARE THOSE WHERE INTEREST ARTICULATION BY INSTITUTIONAL GROUPS IS VERY SIGNIFICANT (40)	1.00	118	TEND TO BE THOSE WHERE INTEREST ARTICULATION BY INSTITUTIONAL GROUPS IS SIGNIFICANT, MODERATE, OR LIMITED (60)	0.69	13 27 0 59 X SQ= 19.32 P = 0. RV YES (NO)

129	ALWAYS ARE THOSE WHERE INTEREST ARTICULATION BY POLITICAL PARTIES IS NEGLIGIBLE (37)	1.00	129	TEND TO BE THOSE WHERE INTEREST ARTICULATION BY POLITICAL PARTIES IS SIGNIFICANT, MODERATE, OR LIMITED (56)	0.70	0 55 13 24 X SQ= 19.70 P = 0. RV YES (NO)
134	ALWAYS ARE THOSE WHERE INTEREST AGGREGATION BY THE EXECUTIVE IS LIMITED OR NEGLIGIBLE (46)	1.00	134	TEND TO BE THOSE WHERE INTEREST AGGREGATION BY THE EXECUTIVE IS SIGNIFICANT OR MODERATE (57)	0.63	0 56 13 33 X SQ= 15.69 P = 0. RV YES (NO)
138	ALWAYS ARE THOSE WHERE INTEREST AGGREGATION BY THE LEGISLATURE IS NEGLIGIBLE (49)	1.00	138	TEND TO BE THOSE WHERE INTEREST AGGREGATION BY THE LEGISLATURE IS SIGNIFICANT, MODERATE, OR LIMITED (48)	0.57	0 47 13 36 X SQ= 12.25 P = 0. RV YES (NO)
139	ALWAYS ARE THOSE WHERE THE PARTY SYSTEM IS QUANTITATIVELY ONE-PARTY (34)	1.00	139	TEND TO BE THOSE WHERE THE PARTY SYSTEM IS QUANTITATIVELY OTHER THAN ONE-PARTY (71)	0.77	13 21 0 70 X SQ= 27.19 P = 0. RV YES (NO)
153	ALWAYS ARE THOSE WHERE THE PARTY SYSTEM IS STABLE (42)	1.00	153	TEND TO BE THOSE WHERE THE PARTY SYSTEM IS MODERATELY STABLE OR UNSTABLE (38)	0.57	13 29 0 38 X SQ= 11.86 P = 0. RV YES (NO)
158	ALWAYS ARE THOSE WHERE PERSONALISMO IS NEGLIGIBLE (56)	1.00	158	TEND LESS TO BE THOSE WHERE PERSONALISMO IS NEGLIGIBLE (56)	0.51	0 40 13 42 X SQ= 9.04 P = 0.001 RV NO (NO)
160	ALWAYS ARE THOSE WHERE THE POLITICAL LEADERSHIP IS ELITIST (30)	1.00	160	TEND TO BE THOSE WHERE THE POLITICAL LEADERSHIP IS MODERATELY ELITIST OR NON-ELITIST (67)	0.80	13 17 0 67 X SQ= 29.90 P = 0. RV YES (NO)
169	ALWAYS ARE THOSE WHERE THE HORIZONTAL POWER DISTRIBUTION IS NEGLIGIBLE (48)	1.00	169	TEND TO BE THOSE WHERE THE HORIZONTAL POWER DISTRIBUTION IS SIGNIFICANT OR LIMITED (58)	0.63	0 58 13 34 X SQ= 15.85 P = 0. RV YES (NO)
176	ALWAYS ARE THOSE WHERE THE LEGISLATURE IS WHOLLY INEFFECTIVE (28)	1.00	176	TEND TO BE THOSE WHERE THE LEGISLATURE IS FULLY EFFECTIVE, PARTIALLY EFFECTIVE, OR LARGELY INEFFECTIVE (72)	0.83	0 72 13 15 X SQ= 34.43 P = 0. RV YES (NO)

178	TEND TO BE THOSE WHERE THE LEGISLATURE IS UNICAMERAL	0.85 (53)	178	TEND TO BE THOSE WHERE THE LEGISLATURE IS BICAMERAL	0.54 (51)	11 42 2 49 X SQ= 5.28 P = 0.015 RV YES (NO)
179	ALWAYS ARE THOSE WHERE THE EXECUTIVE IS DOMINANT (52)	1.00	179	TEND TO BE THOSE WHERE THE EXECUTIVE IS STRONG (39)	0.51	13 38 0 39 X SQ= 9.65 P = 0. RV YES (NO)
181	ALWAYS ARE THOSE WHERE THE BUREAUCRACY IS SEMI-MODERN, RATHER THAN MODERN (55)	1.00	181	TEND LESS TO BE THOSE WHERE THE BUREAUCRACY IS SEMI-MODERN, RATHER THAN MODERN (55)	0.66	0 21 13 41 X SQ= 4.55 P = 0.015 RV NO (NO)
182	ALWAYS ARE THOSE WHERE THE BUREAUCRACY IS SEMI-MODERN, RATHER THAN POST-COLONIAL TRANSITIONAL (55)	1.00	182	TEND LESS TO BE THOSE WHERE THE BUREAUCRACY IS SEMI-MODERN, RATHER THAN POST-COLONIAL TRANSITIONAL (55)	0.62	13 41 0 25 X SQ= 5.56 P = 0.007 RV NO (NO)
185	ALWAYS ARE THOSE WHERE PARTICIPATION BY THE MILITARY IN POLITICS IS SUPPORTIVE, RATHER THAN INTERVENTIVE (31)	1.00	185	TEND TO BE THOSE WHERE PARTICIPATION BY THE MILITARY IN POLITICS IS INTERVENTIVE, RATHER THAN SUPPORTIVE (21)	0.54	0 21 13 18 X SQ= 9.61 P = 0.001 RV YES (NO)
186	ALWAYS ARE THOSE WHERE PARTICIPATION BY THE MILITARY IN POLITICS IS SUPPORTIVE, RATHER THAN NEUTRAL (31)	1.00	186	TEND TO BE THOSE WHERE PARTICIPATION BY THE MILITARY IN POLITICS IS NEUTRAL, RATHER THAN SUPPORTIVE (56)	0.76	13 18 0 56 X SQ= 24.41 P = 0. RV YES (NO)
187	ALWAYS ARE THOSE WHERE THE ROLE OF THE POLICE IS POLITICALLY SIGNIFICANT (66)	1.00	187	TEND LESS TO BE THOSE WHERE THE ROLE OF THE POLICE IS POLITICALLY SIGNIFICANT (66)	0.60	13 52 0 35 X SQ= 6.37 P = 0.004 RV NO (NO)

195 POLITIES
BELONGING TO THE FIFTY PERCENT THAT
HAVE PURPLE WHISKERS (58)

195 POLITIES
BELONGING TO THE FIFTY PERCENT THAT
DO NOT HAVE PURPLE WHISKERS (57)

SUBJECT ONLY

58 IN LEFT COLUMN

ALGERIA	BRAZIL	BULGARIA	CAMBODIA	CAMEROUN	CANADA	CEYLON	CHILE	CONGO(BRA)	CONGO(LEO)
COSTA RICA	CUBA	DENMARK	DOMIN REP	ECUADOR	EL SALVADOR	FINLAND	GERMANY, E	GHANA	GREECE
GUATEMALA	GUINEA	HAITI	HUNGARY	ICELAND	INDONESIA	ISRAEL	ITALY	JAMAICA	JORDAN
LAOS	LEBANON	MALAYA	MALI	MONGOLIA	NEW ZEALAND	NICARAGUA	NORWAY	PAKISTAN	PARAGUAY
PHILIPPINES	RUMANIA	RWANDA	SA'U ARABIA	SENEGAL	SPAIN	SUDAN	SWITZERLAND	SYRIA	THAILAND
TOGO	TUNISIA	TURKEY	UGANDA	URUGUAY	VENEZUELA	VIETNAM, N	YUGOSLAVIA		

57 IN RIGHT COLUMN

AFGHANISTAN	ALBANIA	ARGENTINA	AUSTRALIA	AUSTRIA	BELGIUM	BOLIVIA	BURMA	BURUNDI	CEN AFR REP
CHAD	CHINA, PR	COLOMBIA	CYPRUS	CZECHOS'KIA	DAHOMEY	ETHIOPIA	FRANCE	GABON	GERMAN FR
HONDURAS	INDIA	IRAN	IRAQ	IRELAND	IVORY COAST	JAPAN	KOREA, N	KOREA REP	LIBERIA
LIBYA	LUXEMBOURG	MALAGASY R	MAURITANIA	MEXICO	MOROCCO	NEPAL	NETHERLANDS	NIGER	NIGERIA
PANAMA	PERU	POLAND	PORTUGAL	SIERRE LEO	SOMALIA	SO AFRICA	SWEDEN	TANGANYIKA	TRINIDAD
USSR	UAR	UK	US	UPPER VOLTA	VIETNAM REP	YEMEN			

	Left column			Right column		
19	TILT TOWARD BEING THOSE LOCATED IN THE CARIBBEAN, CENTRAL AMERICA, OR SOUTH AMERICA, RATHER THAN IN NORTH AFRICA OR CENTRAL AND SOUTH AFRICA (22)	0.52	19	TILT TOWARD BEING THOSE LOCATED IN NORTH AFRICA OR CENTRAL AND SOUTH AFRICA, RATHER THAN IN THE CARIBBEAN, CENTRAL AMERICA, OR SOUTH AMERICA (33)	0.71	13 20 14 8 X SQ= 2.21 P = 0.102 RV YES (YES)
21	LEAN MORE TOWARD BEING THOSE WHOSE TERRITORIAL SIZE IS MEDIUM OR SMALL (83)	0.79	21	LEAN LESS TOWARD BEING THOSE WHOSE TERRITORIAL SIZE IS MEDIUM OR SMALL (83)	0.65	12 20 46 37 X SQ= 2.29 P = 0.099 RV NO (YES)
32	LEAN MORE TOWARD BEING THOSE WHOSE GROSS NATIONAL PRODUCT IS MEDIUM, LOW, OR VERY LOW (105)	0.97	32	LEAN LESS TOWARD BEING THOSE WHOSE GROSS NATIONAL PRODUCT IS MEDIUM, LOW, OR VERY LOW (105)	0.86	2 8 56 49 X SQ= 2.83 P = 0.053 RV NO (YES)
38	LEAN MORE TOWARD BEING THOSE WHOSE INTERNATIONAL FINANCIAL STATUS IS MEDIUM, LOW, OR VERY LOW (103)	0.96	38	LEAN LESS TOWARD BEING THOSE WHOSE INTERNATIONAL FINANCIAL STATUS IS MEDIUM, LOW, OR VERY LOW (103)	0.86	2 8 55 48 X SQ= 2.84 P = 0.053 RV NO (YES)

195/

47	TILT TOWARD BEING THOSE WHERE THE LITERACY RATE IS BETWEEN FIFTY AND NINETY PERCENT, RATHER THAN NINETY PERCENT OR ABOVE (30)	0.66	47	TILT TOWARD BEING THOSE WHERE THE LITERACY RATE IS NINETY PERCENT OR ABOVE, RATHER THAN BETWEEN FIFTY AND NINETY PERCENT (25)	0.58	10 15 19 11 X SQ= 2.12 P = 0.108 RV YES (YES)
51	LEAN LESS TOWARD BEING THOSE WHERE FREEDOM OF THE PRESS IS COMPLETE OR INTERMITTENT (60)	0.53	51	LEAN MORE TOWARD BEING THOSE WHERE FREEDOM OF THE PRESS IS COMPLETE OR INTERMITTENT (60)	0.71	26 34 23 14 X SQ= 2.54 P = 0.095 RV NO (YES)
71	TEND MORE TO BE THOSE WHOSE DATE OF INDEPENDENCE IS AFTER 1800 (90)	0.89	71	TEND LESS TO BE THOSE WHOSE DATE OF INDEPENDENCE IS AFTER 1800 (90)	0.72	6 15 51 39 X SQ= 4.31 P = 0.028 RV NO (YES)
81	TEND MORE TO BE THOSE WHERE THE TYPE OF POLITICAL MODERNIZATION IS LATER EUROPEAN OR LATER EUROPEAN DERIVED, RATHER THAN EARLY EUROPEAN OR EARLY EUROPEAN DERIVED (40)	0.89	81	TEND LESS TO BE THOSE WHERE THE TYPE OF POLITICAL MODERNIZATION IS LATER EUROPEAN OR LATER EUROPEAN DERIVED, RATHER THAN EARLY EUROPEAN OR EARLY EUROPEAN DERIVED (40)	0.65	3 8 25 15 X SQ= 3.02 P = 0.048 RV NO (YES)
83	LEAN MORE TOWARD BEING THOSE WHERE THE TYPE OF POLITICAL MODERNIZATION IS OTHER THAN NON-EUROPEAN AUTOCHTHONOUS (106)	0.97	83	LEAN LESS TOWARD BEING THOSE WHERE THE TYPE OF POLITICAL MODERNIZATION IS OTHER THAN NON-EUROPEAN AUTOCHTHONOUS (106)	0.88	2 7 56 50 X SQ= 2.01 P = 0.094 RV NO (YES)
99	TILT TOWARD BEING THOSE WHERE GOVERNMENTAL STABILITY IS MODERATELY PRESENT AND DATES FROM THE POST-WAR PERIOD, OR IS ABSENT (36)	0.50	99	TILT TOWARD BEING THOSE WHERE GOVERNMENTAL STABILITY IS GENERALLY PRESENT AND DATES FROM AT LEAST THE INTER-WAR PERIOD, OR FROM THE POST-WAR PERIOD (50)	0.67	23 27 23 13 X SQ= 2.02 P = 0.127 RV YES (YES)
127	TILT LESS TOWARD BEING THOSE WHERE INTEREST ARTICULATION BY POLITICAL PARTIES IS MODERATE, LIMITED, OR NEGLIGIBLE (72)	0.70	127	TILT MORE TOWARD BEING THOSE WHERE INTEREST ARTICULATION BY POLITICAL PARTIES IS MODERATE, LIMITED, OR NEGLIGIBLE (72)	0.85	14 7 33 39 X SQ= 2.05 P = 0.136 RV NO (YES)
131	TILT TOWARD BEING THOSE WHERE INTEREST AGGREGATION BY POLITICAL PARTIES IS LIMITED OR NEGLIGIBLE (35)	0.63	131	TILT TOWARD BEING THOSE WHERE INTEREST AGGREGATION BY POLITICAL PARTIES IS SIGNIFICANT OR MODERATE (30)	0.57	13 17 22 13 X SQ= 1.75 P = 0.140 RV YES (YES)
142	TEND LESS TO BE THOSE WHERE THE PARTY SYSTEM IS QUANTITATIVELY OTHER THAN MULTI-PARTY (66)	0.57	142	TEND MORE TO BE THOSE WHERE THE PARTY SYSTEM IS QUANTITATIVELY OTHER THAN MULTI-PARTY (66)	0.80	20 10 27 39 X SQ= 4.49 P = 0.027 RV NO (YES)

148	LEAN MORE TOWARD BEING THOSE WHERE THE PARTY SYSTEM IS QUALITATIVELY OTHER THAN AFRICAN TRANSITIONAL (96)	0.93	148	LEAN LESS TOWARD BEING THOSE WHERE THE PARTY SYSTEM IS QUALITATIVELY OTHER THAN AFRICAN TRANSITIONAL (96)	0.81	4 10 52 44 X SQ= 2.26 P = 0.091 RV NO (YES)
153	TILT TOWARD BEING THOSE WHERE THE PARTY SYSTEM IS MODERATELY STABLE OR UNSTABLE (38)	0.56	153	TILT TOWARD BEING THOSE WHERE THE PARTY SYSTEM IS STABLE (42)	0.62	19 23 24 14 X SQ= 1.91 P = 0.122 RV YES (YES)
157	TEND LESS TO BE THOSE WHERE PERSONALISMO IS MODERATE OR NEGLIGIBLE (84)	0.78	157	TEND MORE TO BE THOSE WHERE PERSONALISMO IS MODERATE OR NEGLIGIBLE (84)	0.94	11 3 40 44 X SQ= 3.45 P = 0.043 RV NO (YES)
178	LEAN TOWARD BEING THOSE WHERE THE LEGISLATURE IS UNICAMERAL (53)	0.60	178	LEAN TOWARD BEING THOSE WHERE THE LEGISLATURE IS BICAMERAL (51)	0.59	32 21 21 30 X SQ= 3.10 P = 0.077 RV YES (YES)

```
****************************************************************************************************

196  POLITIES                                              196  POLITIES
     BELONGING TO THE FORTY PERCENT THAT                        BELONGING TO THE SIXTY PERCENT THAT
     HAVE BLUE WHISKERS  (46)                                   DO NOT HAVE BLUE WHISKERS  (69)

                                                                                              SUBJECT ONLY
....................................................................................................

     46 IN LEFT COLUMN

AFGHANISTAN ALGERIA     BULGARIA    BURMA       CAMBODIA    CANADA      CEYLON      CHILE       COLOMBIA    CONGO(BRA)
DOMIN REP   EL SALVADOR GERMANY, E  GUATEMALA   GUINEA      HONDURAS    ICELAND     INDIA       IRAN        IRAQ
ISRAEL      ITALY       KOREA, N    KOREA REP   LEBANON     LUXEMBOURG  MALAGASY R  MALI        MAURITANIA  MOROCCO
NEPAL       NETHERLANDS NIGER       NIGERIA     NORWAY      PANAMA      PARAGUAY    PERU        PORTUGAL    RWANDA
SA'U ARABIA SO AFRICA   SPAIN       TRINIDAD    UPPER VOLTA URUGUAY

     69 IN RIGHT COLUMN

ALBANIA     ARGENTINA   AUSTRALIA   AUSTRIA     BELGIUM     BOLIVIA     BRAZIL      BURUNDI     CAMEROUN    CEN AFR REP
CHAD        CHINA, PR   CONGO(LEO)  COSTA RICA  CUBA        CYPRUS      CZECHOS'KIA DAHOMEY     DENMARK     ECUADOR
ETHIOPIA    FINLAND     FRANCE      GABON       GERMAN FR   GHANA       GREECE      HAITI       HUNGARY     INDONESIA
IRELAND     IVORY COAST JAMAICA     JAPAN       JORDAN      LAOS        LIBERIA     LIBYA       MALAYA      MEXICO
MONGOLIA    NEW ZEALAND NICARAGUA   PAKISTAN    PHILIPPINES POLAND      RUMANIA     SENEGAL     SIERRE LEO  SOMALIA
SUDAN       SWEDEN      SWITZERLAND SYRIA       TANGANYIKA  THAILAND    TOGO        TUNISIA     TURKEY      UGANDA
USSR        UAR         UK          US          VENEZUELA   VIETNAM, N  VIETNAM REP YEMEN       YUGOSLAVIA
----------------------------------------------------------------------------------------------------

 25  TILT LESS TOWARD BEING THOSE          0.78        25  TILT MORE TOWARD BEING THOSE          0.90         10     7
     WHOSE POPULATION DENSITY IS                           WHOSE POPULATION DENSITY IS                        36    62
     MEDIUM OR LOW  (98)                                   MEDIUM OR LOW  (98)                         X SQ=    2.10
                                                                                                       P =     0.110
                                                                                                       RV NO   (YES)

 28  TEND TO BE THOSE                      0.69        28  TEND TO BE THOSE                      0.52         31    31
     WHOSE POPULATION GROWTH RATE                          WHOSE POPULATION GROWTH RATE                       14    34
     IS HIGH  (62)                                         IS LOW  (48)                                X SQ=    4.03
                                                                                                       P =     0.033
                                                                                                       RV YES  (YES)

 33  TEND MORE TO BE THOSE                 0.85        33  TEND LESS TO BE THOSE                 0.67          7    23
     WHOSE GROSS NATIONAL PRODUCT                          WHOSE GROSS NATIONAL PRODUCT                       39    46
     IS LOW OR VERY LOW  (85)                              IS LOW OR VERY LOW  (85)                    X SQ=    3.81
                                                                                                       P =     0.033
                                                                                                       RV NO   (NO )

 39  TEND MORE TO BE THOSE                 0.80        39  TEND LESS TO BE THOSE                 0.58          9    29
     WHOSE INTERNATIONAL FINANCIAL STATUS                  WHOSE INTERNATIONAL FINANCIAL STATUS               36    40
     IS LOW OR VERY LOW  (76)                              IS LOW OR VERY LOW  (76)                    X SQ=    5.00
                                                                                                       P =     0.016
                                                                                                       RV NO   (NO )
```

44	LEAN MORE TOWARD BEING THOSE WHERE THE LITERACY RATE IS BELOW NINETY PERCENT (90)	0.87	44	LEAN LESS TOWARD BEING THOSE WHERE THE LITERACY RATE IS BELOW NINETY PERCENT (90)	0.72	6 19 40 50 X SQ= 2.61 P = 0.070 RV NO (NO)
47	LEAN TOWARD BEING THOSE WHERE THE LITERACY RATE IS BETWEEN FIFTY AND NINETY PERCENT, RATHER THAN NINETY PERCENT OR ABOVE (30)	0.71	47	LEAN TOWARD BEING THOSE WHERE THE LITERACY RATE IS NINETY PERCENT OR ABOVE, RATHER THAN BETWEEN FIFTY AND NINETY PERCENT (25)	0.56	6 19 15 15 X SQ= 2.88 P = 0.057 RV YES (YES)
77	TILT TOWARD BEING THOSE THAT HAVE BEEN SIGNIFICANTLY WESTERNIZED, RATHER THAN PARTIALLY WESTERNIZED, THROUGH A COLONIAL RELATIONSHIP (28)	0.52	77	TILT TOWARD BEING THOSE THAT HAVE BEEN PARTIALLY WESTERNIZED, RATHER THAN SIGNIFICANTLY WESTERNIZED, THROUGH A COLONIAL RELATIONSHIP (41)	0.67	15 13 14 27 X SQ= 1.84 P = 0.139 RV YES (YES)
78	LEAN MORE TOWARD BEING THOSE THAT HAVE BEEN SIGNIFICANTLY WESTERNIZED THROUGH A COLONIAL RELATIONSHIP, RATHER THAN WITHOUT SUCH A RELATIONSHIP (28)	0.94	78	LEAN LESS TOWARD BEING THOSE THAT HAVE BEEN SIGNIFICANTLY WESTERNIZED THROUGH A COLONIAL RELATIONSHIP, RATHER THAN WITHOUT SUCH A RELATIONSHIP (28)	0.68	15 13 1 6 X SQ= 2.08 P = 0.096 RV NO (YES)
100	TILT LESS TOWARD BEING THOSE WHERE GOVERNMENTAL STABILITY IS GENERALLY PRESENT AND DATES FROM AT LEAST THE INTER-WAR PERIOD, OR IS GENERALLY OR MODERATELY PRESENT, AND DATES FROM THE POST-WAR PERIOD (64)	0.66	100	TILT MORE TOWARD BEING THOSE WHERE GOVERNMENTAL STABILITY IS GENERALLY PRESENT AND DATES FROM AT LEAST THE INTER-WAR PERIOD, OR IS GENERALLY OR MODERATELY PRESENT, AND DATES FROM THE POST-WAR PERIOD (64)	0.80	23 41 12 10 X SQ= 1.64 P = 0.140 RV NO (YES)
110	LEAN MORE TOWARD BEING THOSE WHERE POLITICAL ENCULTURATION IS MEDIUM OR LOW (80)	0.92	110	LEAN LESS TOWARD BEING THOSE WHERE POLITICAL ENCULTURATION IS MEDIUM OR LOW (80)	0.79	3 12 35 45 X SQ= 2.06 P = 0.096 RV NO (NO)
131	LEAN TOWARD BEING THOSE WHERE INTEREST AGGREGATION BY POLITICAL PARTIES IS LIMITED OR NEGLIGIBLE (35)	0.68	131	LEAN TOWARD BEING THOSE WHERE INTEREST AGGREGATION BY POLITICAL PARTIES IS SIGNIFICANT OR MODERATE (30)	0.57	9 21 19 16 X SQ= 2.96 P = 0.078 RV YES (YES)
135	TILT LESS TOWARD BEING THOSE WHERE INTEREST AGGREGATION BY THE EXECUTIVE IS SIGNIFICANT, MODERATE, OR LIMITED (77)	0.78	135	TILT MORE TOWARD BEING THOSE WHERE INTEREST AGGREGATION BY THE EXECUTIVE IS SIGNIFICANT, MODERATE, OR LIMITED (77)	0.91	29 48 8 5 X SQ= 1.73 P = 0.133 RV NO (YES)
166	TEND MORE TO BE THOSE WHERE THE VERTICAL POWER DISTRIBUTION IS THAT OF FORMAL FEDERALISM OR FORMAL AND EFFECTIVE UNITARISM (99)	0.96	166	TEND LESS TO BE THOSE WHERE THE VERTICAL POWER DISTRIBUTION IS THAT OF FORMAL FEDERALISM OR FORMAL AND EFFECTIVE UNITARISM (99)	0.81	2 13 44 55 X SQ= 4.03 P = 0.025 RV NO (NO)

```
******************************************************************************************************
197  POLITIES                                        197  POLITIES
     BELONGING TO THE THIRTY PERCENT THAT                 BELONGING TO THE SEVENTY PERCENT THAT
     HAVE GREEN WHISKERS  (35)                            DO NOT HAVE GREEN WHISKERS  (80)


                                                                                        SUBJECT ONLY
......................................................................................................

     35 IN LEFT COLUMN

AFGHANISTAN ALGERIA     BRAZIL      BURUNDI     COSTA RICA  CYPRUS      GERMAN FR   GUATEMALA   HONDURAS    INDIA
IRAN        JORDAN      KOREA, N    KOREA REP   LAOS        LIBYA       MAURITANIA  MEXICO      MONGOLIA    MOROCCO
NETHERLANDS NORWAY      PAKISTAN    PHILIPPINES POLAND      SA'U ARABIA SOMALIA     SPAIN       TUNISIA     USSR
UAR         UK          VENEZUELA   VIETNAM REP YEMEN

     80 IN RIGHT COLUMN

ALBANIA     ARGENTINA   AUSTRALIA   AUSTRIA     BELGIUM     BOLIVIA     BULGARIA    BURMA       CAMBODIA    CAMEROUN
CANADA      CEN AFR REP CEYLON      CHAD        CHILE       CHINA, PR   COLOMBIA    CONGO(BRA)  CONGO(LEO)  CUBA
CZECHOS'KIA DAHOMEY     DENMARK     DOMIN REP   ECUADOR     EL SALVADOR ETHIOPIA    FINLAND     FRANCE      GABON
GERMANY, E  GHANA       GREECE      GUINEA      HAITI       HUNGARY     ICELAND     INDONESIA   IRAQ        IRELAND
ISRAEL      ITALY       IVORY COAST JAMAICA     JAPAN       LEBANON     LIBERIA     LUXEMBOURG  MALAGASY R  MALAYA
MALI        NEPAL       NEW ZEALAND NICARAGUA   NIGER       NIGERIA     PANAMA      PARAGUAY    PERU        PORTUGAL
RUMANIA     RWANDA      SENEGAL     SIERRE LEO  SO AFRICA   SUDAN       SWEDEN      SWITZERLAND SYRIA       TANGANYIKA
THAILAND    TOGO        TRINIDAD    TURKEY      UGANDA      US          UPPER VOLTA URUGUAY     VIETNAM, N  YUGOSLAVIA
------------------------------------------------------------------------------------------------------

 11  TEND MORE TO BE THOSE                  0.91          11  TEND LESS TO BE THOSE                  0.69         3   25
     LOCATED ELSEWHERE THAN IN CENTRAL                        LOCATED ELSEWHERE THAN IN CENTRAL                  32   55
     AND SOUTH AFRICA  (87)                                   AND SOUTH AFRICA  (87)                          X SQ=   5.62
                                                                                                              P =    0.009
                                                                                                              RV NO  (NO )

 13  TEND LESS TO BE THOSE                  0.69          13  TEND MORE TO BE THOSE                  0.94        11    5
     LOCATED ELSEWHERE THAN IN NORTH AFRICA OR                LOCATED ELSEWHERE THAN IN NORTH AFRICA OR          24   75
     THE MIDDLE EAST  (99)                                    THE MIDDLE EAST  (99)                           X SQ=  10.87
                                                                                                              P =    0.001
                                                                                                              RV NO  (YES)

 14  LEAN LESS TOWARD BEING THOSE           0.83          14  LEAN MORE TOWARD BEING THOSE           0.94         6    5
     LOCATED ELSEWHERE THAN IN                                LOCATED ELSEWHERE THAN IN                          29   75
     THE MIDDLE EAST  (104)                                   THE MIDDLE EAST  (104)                          X SQ=   2.20
                                                                                                              P =    0.088
                                                                                                              RV NO  (YES)

 15  TILT LESS TOWARD BEING THOSE           0.57          15  TILT MORE TOWARD BEING THOSE           0.83         6    5
     LOCATED IN NORTH AFRICA OR                               LOCATED IN NORTH AFRICA OR                          8   25
     CENTRAL AND SOUTH AFRICA, RATHER THAN                    CENTRAL AND SOUTH AFRICA, RATHER THAN           X SQ=   2.23
     IN THE MIDDLE EAST  (33)                                 IN THE MIDDLE EAST  (33)                        P =    0.132
                                                                                                              RV NO  (YES)
```

31	TILT MORE TOWARD BEING THOSE WHOSE AGRICULTURAL POPULATION IS HIGH OR MEDIUM (90)	0.88	31	TILT LESS TOWARD BEING THOSE WHOSE AGRICULTURAL POPULATION IS HIGH OR MEDIUM (90)	0.75	30 60 4 20 X SQ= 1.78 P = 0.137 RV NO (NO)
36	TILT MORE TOWARD BEING THOSE WHOSE PER CAPITA GROSS NATIONAL PRODUCT IS LOW OR VERY LOW (73)	0.74	36	TILT LESS TOWARD BEING THOSE WHOSE PER CAPITA GROSS NATIONAL PRODUCT IS LOW OR VERY LOW (73)	0.59	9 33 26 47 X SQ= 1.91 P = 0.142 RV NO (NO)
51	LEAN TOWARD BEING THOSE WHERE FREEDOM OF THE PRESS IS INTERNALLY ABSENT, OR INTERNALLY AND EXTERNALLY ABSENT (37)	0.52	51	LEAN TOWARD BEING THOSE WHERE FREEDOM OF THE PRESS IS COMPLETE OR INTERMITTENT (60)	0.68	15 45 16 21 X SQ= 2.71 P = 0.075 RV YES (NO)
56	TEND MORE TO BE THOSE WHERE THE RELIGION IS PREDOMINANTLY LITERATE (79)	0.91	56	TEND LESS TO BE THOSE WHERE THE RELIGION IS PREDOMINANTLY LITERATE (79)	0.64	29 50 3 28 X SQ= 6.63 P = 0.005 RV NO (NO)
58	TEND LESS TO BE THOSE WHERE THE RELIGION IS OTHER THAN MUSLIM (97)	0.63	58	TEND MORE TO BE THOSE WHERE THE RELIGION IS OTHER THAN MUSLIM (97)	0.94	13 5 22 75 X SQ= 15.34 P = 0. RV NO (YES)
64	TEND TO BE THOSE WHERE THE RELIGION IS MUSLIM, RATHER THAN CHRISTIAN (18)	0.52	64	TEND TO BE THOSE WHERE THE RELIGION IS CHRISTIAN, RATHER THAN MUSLIM (46)	0.87	12 34 13 5 X SQ= 9.71 P = 0.001 RV YES (YES)
66	TEND TO BE THOSE THAT ARE RELIGIOUSLY HOMOGENEOUS (57)	0.80	66	TEND TO BE THOSE THAT ARE RELIGIOUSLY HETEROGENEOUS (49)	0.57	24 33 6 43 X SQ= 10.15 P = 0.001 RV YES (NO)
68	TEND TO BE THOSE THAT ARE LINGUISTICALLY HOMOGENEOUS (52)	0.60	68	TEND TO BE THOSE THAT ARE LINGUISTICALLY WEAKLY HETEROGENEOUS OR STRONGLY HETEROGENEOUS (62)	0.61	21 31 14 48 X SQ= 3.42 P = 0.044 RV YES (NO)
84	TEND TO BE THOSE WHERE THE TYPE OF POLITICAL MODERNIZATION IS DEVELOPED TUTELARY, RATHER THAN UNDEVELOPED TUTELARY (31)	0.85	84	TEND TO BE THOSE WHERE THE TYPE OF POLITICAL MODERNIZATION IS UNDEVELOPED TUTELARY, RATHER THAN DEVELOPED TUTELARY (24)	0.60	17 14 3 21 X SQ= 8.73 P = 0.002 RV YES (YES)

85	TEND TO BE THOSE WHERE THE STAGE OF POLITICAL MODERNIZATION IS ADVANCED, RATHER THAN MID- OR EARLY TRANSITIONAL (60)	0.71	85	TEND TO BE THOSE WHERE THE STAGE OF POLITICAL MODERNIZATION IS MID- OR EARLY TRANSITIONAL, RATHER THAN ADVANCED (54)	0.55	24 36 10 44 X SQ= 5.28 P = 0.014 RV YES (NO)
91	TILT LESS TOWARD BEING THOSE WHOSE IDEOLOGICAL ORIENTATION IS DEVELOPMENTAL, RATHER THAN TRADITIONAL (31)	0.70	91	TILT MORE TOWARD BEING THOSE WHOSE IDEOLOGICAL ORIENTATION IS DEVELOPMENTAL, RATHER THAN TRADITIONAL (31)	0.92	7 24 3 2 X SQ= 1.43 P = 0.119 RV NO (YES)
107	TILT MORE TOWARD BEING THOSF WHERE AUTONOMOUS GROUPS ARE PARTIALLY TOLERATED IN POLITICS, ARE TOLERATED ONLY OUTSIDE POLITICS, OR ARE NOT TOLERATED AT ALL (65)	0.70	107	TILT LESS TOWARD BEING THOSE WHERE AUTONOMOUS GROUPS ARE PARTIALLY TOLERATED IN POLITICS, ARE TOLERATED ONLY OUTSIDE POLITICS, OR ARE NOT TOLERATED AT ALL (65)	0.54	10 36 23 42 X SQ= 1.79 P = 0.143 RV NO (NO)
118	LEAN TOWARD BEING THOSE WHERE INTEREST ARTICULATION BY INSTITUTIONAL GROUPS IS VERY SIGNIFICANT (40)	0.53	118	LEAN TOWARD BEING THOSE WHERE INTEREST ARTICULATION BY INSTITUTIONAL GROUPS IS SIGNIFICANT, MODERATE, OR LIMITED (60)	0.66	17 23 15 45 X SQ= 2.62 P = 0.082 RV YES (NO)
122	TILT MORE TOWARD BEING THOSE WHERE INTEREST ARTICULATION BY NON-ASSOCIATIONAL GROUPS IS SIGNIFICANT OR MODERATE (83)	0.83	122	TILT LESS TOWARD BEING THOSE WHERE INTEREST ARTICULATION BY NON-ASSOCIATIONAL GROUPS IS SIGNIFICANT OR MODERATE (83)	0.67	29 54 6 26 X SQ= 2.15 P = 0.115 RV NO (NO)
134	TILT TOWARD BEING THOSE WHERE INTEREST AGGREGATION BY THE EXECUTIVE IS LIMITED OR NEGLIGIBLE (46)	0.57	134	TILT TOWARD BEING THOSE WHERE INTEREST AGGREGATION BY THE EXECUTIVE IS SIGNIFICANT OR MODERATE (57)	0.60	13 44 17 29 X SQ= 1.83 P = 0.132 RV YES (NO)
161	TEND TO BE THOSE WHERE THE POLITICAL LEADERSHIP IS ELITIST OR MODERATELY ELITIST (47)	0.68	161	TEND TO BE THOSE WHERE THE POLITICAL LEADERSHIP IS NON-ELITIST (50)	0.59	19 28 9 41 X SQ= 4.89 P = 0.024 RV YES (NO)
182	TILT MORE TOWARD BEING THOSE WHERE THE BUREAUCRACY IS SEMI-MODERN, RATHER THAN POST-COLONIAL TRANSITIONAL (55)	0.83	182	TILT LESS TOWARD BEING THOSE WHERE THE BUREAUCRACY IS SEMI-MODERN, RATHER THAN POST-COLONIAL TRANSITIONAL (55)	0.63	19 36 4 21 X SQ= 2.05 P = 0.114 RV NO (NO)
183	TEND TO BE THOSE WHERE THE BUREAUCRACY IS TRADITIONAL, RATHER THAN POST-COLONIAL TRANSITIONAL (9)	0.56	183	TEND TO BE THOSE WHERE THE BUREAUCRACY IS POST-COLONIAL TRANSITIONAL, RATHER THAN TRADITIONAL (25)	0.84	4 21 5 4 X SQ= 3.48 P = 0.034 RV YES (YES)

197/

```
193  TEND MORE TO BE THOSE                   0.91        193  TEND LESS TO BE THOSE                   0.68         3   25
     WHERE THE CHARACTER OF THE LEGAL SYSTEM                  WHERE THE CHARACTER OF THE LEGAL SYSTEM               32   54
     IS OTHER THAN PARTLY INDIGENOUS  (86)                    IS OTHER THAN PARTLY INDIGENOUS  (86)            X SQ=   5.78
                                                                                                               P =    0.009
                                                                                                               RV NO  (NO )
```

198 POLITIES
BELONGING TO THE TWENTY PERCENT THAT
HAVE PINK WHISKERS (23)

198 POLITIES
BELONGING TO THE EIGHTY PERCENT THAT
DO NOT HAVE PINK WHISKERS (92)

SUBJECT ONLY

23 IN LEFT COLUMN

AFGHANISTAN, AUSTRALIA, COSTA RICA, CUBA, DAHOMEY, FRANCE, GREECE, ISRAEL, JAPAN, KOREA, N, KOREA REP, LAOS, MONGOLIA, NIGER, PERU, SA'U ARABIA, SIERRE LEO, SUDAN, TANGANYIKA, TRINIDAD, TUNISIA, TURKEY, VIETNAM REP

92 IN RIGHT COLUMN

ALBANIA, ALGERIA, ARGENTINA, AUSTRIA, BELGIUM, BOLIVIA, BRAZIL, BULGARIA, BURMA, BURUNDI, CAMBODIA, CAMEROUN, CANADA, CEN AFR REP, CEYLON, CHAD, CHILE, CHINA, PR, COLOMBIA, CONGO(BRA), CONGO(LEO), CYPRUS, CZECHOS'KIA, DENMARK, DOMIN REP, ECUADOR, EL SALVADOR, ETHIOPIA, FINLAND, GABON, GERMANY, E, GERMAN FR, GHANA, GUATEMALA, GUINEA, HAITI, HONDURAS, HUNGARY, ICELAND, INDIA, INDONESIA, IRAN, IRAQ, IRELAND, ITALY, IVORY COAST, JAMAICA, JORDAN, LEBANON, LIBERIA, LIBYA, LUXEMBOURG, MALAGASY R, MALAYA, MALI, MAURITANIA, MEXICO, MOROCCO, NEPAL, NETHERLANDS, NEW ZEALAND, NICARAGUA, NIGERIA, NORWAY, PAKISTAN, PANAMA, PARAGUAY, PHILIPPINES, POLAND, PORTUGAL, RUMANIA, RWANDA, SENEGAL, SOMALIA, SO AFRICA, SPAIN, SWEDEN, SWITZERLAND, SYRIA, THAILAND, TOGO, UGANDA, USSR, UAR, UK, US, UPPER VOLTA, URUGUAY, VENEZUELA, VIETNAM, N, YEMEN, YUGOSLAVIA

---

39 LEAN MORE TOWARD BEING THOSE WHOSE INTERNATIONAL FINANCIAL STATUS IS LOW OR VERY LOW (76) 0.83

39 LEAN LESS TOWARD BEING THOSE WHOSE INTERNATIONAL FINANCIAL STATUS IS LOW OR VERY LOW (76) 0.63

4 34
19 57
X SQ= 2.46
P = 0.085
RV NO (NO)

63 LEAN MORE TOWARD BEING THOSE WHERE THE RELIGION IS PREDOMINANTLY OR PARTLY OTHER THAN CHRISTIAN (68) 0.78

63 LEAN LESS TOWARD BEING THOSE WHERE THE RELIGION IS PREDOMINANTLY OR PARTLY OTHER THAN CHRISTIAN (68) 0.55

5 41
18 50
X SQ= 3.23
P = 0.057
RV NO (NO)

82 LEAN TOWARD BEING THOSE WHERE THE TYPE OF POLITICAL MODERNIZATION IS DEVELOPED TUTELARY OR UNDEVELOPED TUTELARY, RATHER THAN EARLY OR LATER EUROPEAN OR EUROPEAN DERIVED (55) 0.70

82 LEAN TOWARD BEING THOSE WHERE THE TYPE OF POLITICAL MODERNIZATION IS EARLY OR LATER EUROPEAN OR EUROPEAN DERIVED, RATHER THAN DEVELOPED TUTELARY OR UNDEVELOPED TUTELARY (51) 0.52

6 45
14 41
X SQ= 2.41
P = 0.086
RV YES (NO)

114 LEAN TOWARD BEING THOSE WHERE SECTIONALISM IS NEGLIGIBLE (47) 0.64

114 LEAN TOWARD BEING THOSE WHERE SECTIONALISM IS EXTREME OR MODERATE (61) 0.62

8 53
14 33
X SQ= 3.58
P = 0.052
RV YES (NO)

199 POLITIES
BELONGING TO THE TEN PERCENT THAT
HAVE YELLOW WHISKERS (12)

199 POLITIES
BELONGING TO THE NINETY PERCENT
THAT DO NOT HAVE YELLOW WHISKERS (103)

SUBJECT ONLY

12 IN LEFT COLUMN

AUSTRIA, BELGIUM, CAMEROUN, CHAD, CONGO(BRA), CONGO(LEO), CZECHOS'KIA, DAHOMEY, ISRAEL, PARAGUAY, SOMALIA, UPPER VOLTA

103 IN RIGHT COLUMN

AFGHANISTAN, ALBANIA, ALGERIA, ARGENTINA, AUSTRALIA, BOLIVIA, BRAZIL, BULGARIA, BURMA, BURUNDI, CAMBODIA, CANADA, CEN AFR REP, CEYLON, CHILE, CHINA, PR, COLOMBIA, COSTA RICA, CUBA, CYPRUS, DENMARK, DOMIN REP, ECUADOR, EL SALVADOR, ETHIOPIA, FINLAND, FRANCE, GABON, GERMANY, E, GERMAN FR, GHANA, GREECE, GUATEMALA, GUINEA, HAITI, HONDURAS, HUNGARY, ICELAND, INDIA, INDONESIA, IRAN, IRAQ, IRELAND, ITALY, IVORY COAST, JAMAICA, JAPAN, JORDAN, KOREA, N, KOREA REP, LAOS, LEBANON, LIBERIA, LIBYA, LUXEMBOURG, MALAGASY R, MALAYA, MALI, MAURITANIA, MEXICO, MONGOLIA, MOROCCO, NEPAL, NETHERLANDS, NEW ZEALAND, NICARAGUA, NIGER, NIGERIA, NORWAY, PAKISTAN, PANAMA, PERU, PHILIPPINES, POLAND, PORTUGAL, RUMANIA, RWANDA, SA'U ARABIA, SENEGAL, SIERRE LEO, SO AFRICA, SPAIN, SUDAN, SWEDEN, SWITZERLAND, SYRIA, TANGANYIKA, THAILAND, TOGO, TRINIDAD, TUNISIA, TURKEY, UGANDA, USSR, UAR, UK, US, URUGUAY, VENEZUELA, VIETNAM, N, VIETNAM REP, YEMEN, YUGOSLAVIA

---

10 TEND TO BE THOSE LOCATED IN NORTH AFRICA, OR CENTRAL AND SOUTH AFRICA (33) — 0.58

10 TEND TO BE THOSE LOCATED ELSEWHERE THAN IN NORTH AFRICA, OR CENTRAL AND SOUTH AFRICA (82) — 0.75

7 26
5 77
X SQ= 4.25
P = 0.037
RV YES (NO)

18 ALWAYS ARE THOSE LOCATED IN NORTH AFRICA OR CENTRAL AND SOUTH AFRICA, RATHER THAN IN EAST ASIA, SOUTH ASIA, OR SOUTHEAST ASIA (33) — 1.00

18 TEND LESS TO BE THOSE LOCATED IN NORTH AFRICA OR CENTRAL AND SOUTH AFRICA, RATHER THAN IN EAST ASIA, SOUTH ASIA, OR SOUTHEAST ASIA (33) — 0.59

7 26
0 18
X SQ= 2.82
P = 0.042
RV NO (NO)

19 TILT MORE TOWARD BEING THOSE LOCATED IN NORTH AFRICA OR CENTRAL AND SOUTH AFRICA, RATHER THAN IN THE CARIBBEAN, CENTRAL AMERICA, OR SOUTH AMERICA (33) — 0.87

19 TILT LESS TOWARD BEING THOSE LOCATED IN NORTH AFRICA OR CENTRAL AND SOUTH AFRICA, RATHER THAN IN THE CARIBBEAN, CENTRAL AMERICA, OR SOUTH AMERICA (33) — 0.55

7 26
1 21
X SQ= 1.76
P = 0.126
RV NO (NO)

23 ALWAYS ARE THOSE WHOSE POPULATION IS MEDIUM OR SMALL (88) — 1.00

23 LEAN LESS TOWARD BEING THOSE WHOSE POPULATION IS MEDIUM OR SMALL (88) — 0.74

0 27
12 76
X SQ= 2.78
P = 0.066
RV NO (NO)

No.	Left statement	Value	No.	Right statement	Value	Statistics
37	TILT TOWARD BEING THOSE WHOSE PER CAPITA GROSS NATIONAL PRODUCT IS VERY LOW (51)	0.67	37	TILT TOWARD BEING THOSE WHOSE PER CAPITA GROSS NATIONAL PRODUCT IS VERY HIGH, HIGH, MEDIUM, OR LOW (64)	0.58	4 60 8 43 X SQ= 1.79 P = 0.129 RV YES (NO)
40	TILT TOWARD BEING THOSE WHOSE INTERNATIONAL FINANCIAL STATUS IS VERY LOW (39)	0.58	40	TILT TOWARD BEING THOSE WHOSE INTERNATIONAL FINANCIAL STATUS IS VERY HIGH, HIGH, MEDIUM, OR LOW (71)	0.67	5 66 7 32 X SQ= 2.06 P = 0.110 RV YES (NO)
46	TEND TO BE THOSE WHERE THE LITERACY RATE IS BELOW TEN PERCENT (26)	0.58	46	TEND TO BE THOSE WHERE THE LITERACY RATE IS TEN PERCENT OR ABOVE (84)	0.81	5 79 7 19 X SQ= 6.96 P = 0.007 RV YES (NO)
49	ALWAYS ARE THOSE WHERE THE LITERACY RATE IS BELOW TEN PERCENT, RATHER THAN BETWEEN TEN AND FIFTY PERCENT (26)	1.00	49	TEND TO BE THOSE WHERE THE LITERACY RATE IS BETWEEN TEN AND FIFTY PERCENT, RATHER THAN BELOW TEN PERCENT (24)	0.56	0 24 7 19 X SQ= 5.44 P = 0.010 RV YES (NO)
55	TEND TO BE THOSE WHERE NEWSPAPER CIRCULATION IS LESS THAN TEN PER THOUSAND (35)	0.58	55	TEND TO BE THOSE WHERE NEWSPAPER CIRCULATION IS TEN OR MORE PER THOUSAND (78)	0.72	5 73 7 28 X SQ= 3.38 P = 0.046 RV YES (NO)
56	LEAN TOWARD BEING THOSE WHERE THE RELIGION IS PREDOMINANTLY OR PARTLY NON-LITERATE (31)	0.50	56	LEAN TOWARD BEING THOSE WHERE THE RELIGION IS PREDOMINANTLY LITERATE (79)	0.74	6 73 6 25 X SQ= 2.07 P = 0.093 RV YES (NO)
69	TEND TO BE THOSE THAT ARE LINGUISTICALLY STRONGLY HETEROGENEOUS (50)	0.75	69	TEND TO BE THOSE THAT ARE LINGUISTICALLY HOMOGENEOUS OR WEAKLY HETEROGENEOUS (64)	0.60	3 61 9 41 X SQ= 3.96 P = 0.030 RV YES (NO)
73	TILT TOWARD BEING THOSE WHOSE DATE OF INDEPENDENCE IS AFTER 1945 (46)	0.67	73	TILT TOWARD BEING THOSE WHOSE DATE OF INDEPENDENCE IS BEFORE 1945 (65)	0.62	4 61 8 38 X SQ= 2.46 P = 0.118 RV YES (NO)
77	TILT MORE TOWARD BEING THOSE THAT HAVE BEEN PARTIALLY WESTERNIZED, RATHER THAN SIGNIFICANTLY WESTERNIZED, THROUGH A COLONIAL RELATIONSHIP (41)	0.88	77	TILT LESS TOWARD BEING THOSE THAT HAVE BEEN PARTIALLY WESTERNIZED, RATHER THAN SIGNIFICANTLY WESTERNIZED, THROUGH A COLONIAL RELATIONSHIP (41)	0.56	1 27 7 34 X SQ= 1.79 P = 0.130 RV NO (NO)

84	TEND TO BE THOSE WHERE THE TYPE OF POLITICAL MODERNIZATION IS UNDEVELOPED TUTELARY, RATHER THAN DEVELOPED TUTELARY (24)	0.88	84	TEND TO BE THOSE WHERE THE TYPE OF POLITICAL MODERNIZATION IS DEVELOPED TUTELARY, RATHER THAN UNDEVELOPED TUTELARY (31)	0.64	1 30 7 17 X SQ= 5.39 P = 0.016 RV YES (NO)
86	TILT TOWARD BEING THOSE WHERE THE STAGE OF POLITICAL MODERNIZATION IS EARLY TRANSITIONAL, RATHER THAN ADVANCED OR MID-TRANSITIONAL (38)	0.58	86	TILT TOWARD BEING THOSE WHERE THE STAGE OF POLITICAL MODERNIZATION IS ADVANCED OR MID-TRANSITIONAL, RATHER THAN EARLY TRANSITIONAL (76)	0.70	5 71 7 31 X SQ= 2.62 P = 0.101 RV YES (NO)
87	TEND TO BE THOSE WHOSE IDEOLOGICAL ORIENTATION IS DEVELOPMENTAL (31)	0.73	87	TEND TO BE THOSE WHOSE IDEOLOGICAL ORIENTATION IS OTHER THAN DEVELOPMENTAL (58)	0.71	8 23 3 55 X SQ= 6.15 P = 0.014 RV YES (NO)
117	LEAN TOWARD BEING THOSE WHERE INTEREST ARTICULATION BY ASSOCIATIONAL GROUPS IS NEGLIGIBLE (51)	0.75	117	LEAN TOWARD BEING THOSE WHERE INTEREST ARTICULATION BY ASSOCIATIONAL GROUPS IS SIGNIFICANT, MODERATE, OR LIMITED (60)	0.58	3 57 9 42 X SQ= 3.36 P = 0.062 RV YES (NO)
161	TILT TOWARD BEING THOSE WHERE THE POLITICAL LEADERSHIP IS NON-ELITIST (50)	0.75	161	TILT TOWARD BEING THOSE WHERE THE POLITICAL LEADERSHIP IS ELITIST OR MODERATELY ELITIST (47)	0.52	3 44 9 41 X SQ= 2.04 P = 0.123 RV YES (NO)
178	TEND TO BE THOSE WHERE THE LEGISLATURE IS UNICAMERAL (53)	0.83	178	TEND TO BE THOSE WHERE THE LEGISLATURE IS BICAMERAL (51)	0.53	10 43 2 49 X SQ= 4.32 P = 0.029 RV YES (NO)
181	TILT TOWARD BEING THOSE WHERE THE BUREAUCRACY IS MODERN, RATHER THAN SEMI-MODERN (21)	0.60	181	TILT TOWARD BEING THOSE WHERE THE BUREAUCRACY IS SEMI-MODERN, RATHER THAN MODERN (55)	0.75	3 18 2 53 X SQ= 1.34 P = 0.126 RV YES (NO)
182	TEND TO BE THOSE WHERE THE BUREAUCRACY IS POST-COLONIAL TRANSITIONAL, RATHER THAN SEMI-MODERN (25)	0.78	182	TEND TO BE THOSE WHERE THE BUREAUCRACY IS SEMI-MODERN, RATHER THAN POST-COLONIAL TRANSITIONAL (55)	0.75	2 53 7 18 X SQ= 7.92 P = 0.003 RV YES (NO)
186	TILT MORE TOWARD BEING THOSE WHERE PARTICIPATION BY THE MILITARY IN POLITICS IS NEUTRAL, RATHER THAN SUPPORTIVE (56)	0.89	186	TILT LESS TOWARD BEING THOSE WHERE PARTICIPATION BY THE MILITARY IN POLITICS IS NEUTRAL, RATHER THAN SUPPORTIVE (56)	0.62	1 30 8 48 X SQ= 1.57 P = 0.149 RV NO (NO)

Appendices

A. Raw Characteristic Code Sheet

```
                        10        20        30        40        50        60
AFGHANISTAN    KGGHBBAIIHHHHHKABGAH4GBIGBAG3GBBHAAB44HHJ444A3HGJHBAHBAZG21112
ALBANIA        JHHGABAIHIHBHGRBAABH4BAAAGBGGH5BHAB5H55HAAAGAGHGPHAABBAYA22222
ALGERIA        PBGHABAHH5HGGGKAAGHGBHABAB6GG6BBGAB66666AB6G6BHG6GAAGBAPG11122
ARGENTINA      SBBHBAGGGGBBBBBAAABGGBA3G3H3AAG3AAG3B3GGIKGBGGB6ABB6BAAAG22222
AUSTRALIA      AAGHAAHGABAAAAHBAABAAAAHGAAAAAABAHHHBABBHKAGGGAAIABBAGBGG22212
AUSTRIA        ZHGGBAGGBGBAABBAAAAA4BAHGABAAAAGBGHHBABBHKAGGGBAGABBAGBAG22221
BELGIUM        ZHGABAHGAGAAABBAAGBA4AAHGAAAAAGAAGBHAGBAIKAGGGHAIABBAGBAG22221
BOLIVIA        SBHHABAIIIHGBGXBBGBGGBBBBAGAAAGGGBBBABAGBRGBGGHBABBBBGAAG22222
BRAZIL         SABHAABGHGBBBGBABABGHBA3GAHBAAGAAABBBGGGIKGBBGAAABBBBA5AG12122
BULGARIA       JHGGBABHGHBBHBGAABBB4BAAAGBGGH5BHAG5H55HAAAGAGHGPHAABBAYA11222
BURMA          YGBHBBAHIHGBBGJAAGHIAHG3ABHH66BAGAAB6666666666G66666GAA6G21222
BURUNDI        HHHGBBAIIIHH5H55AAHIHIGBG565BB5GH6AGH5AHBS65G5HBIGA5GG5BG22122
CAMBODIA       YHHHBBAIIIHGGHJAABHIBHG3BBGGGGBBHBABHBAHA3BA3AHG3HBAHBAZG11222
CAMEROUN       HGHHBBAIIIHHAHZBAGHIBIGBG6636BBBH6ABBBAGBG6BGBABABABGGABG12221
CANADA         QABHAAHBABAAABHBAGBAAAAHGAAAAABAAGBHHABBHJAGGGAAIABBAGBIG11222
CEN AFR REP    HGHHBBAIIIHH5HS3AGHIBIGBGB6GGGGGH6ABH5AHAS6BGBHGAGAAGGABG22222
CEYLON         XHGBABBHIHHB6GQBAGHGAHG3GAHAAABBGBAAAGHHI3GBBGHBIBB3BGBZG11222
CHAD           HBHHABAIIIHHAHYBBGHIBIGBG56GBBGBH6ABH5AHAS6BGBHGBGA5GG5RG22221
CHILE          SGGHAAGHGGGBABBA3ABGGBAHGAGBAAGGBGGGAGGBIKGBBGHAABB3BG5AG11222
CHINA, PR      IAAGA5ABI5G5HG55A3AB4GAAAGBGGH5BHAAAH55HAAAGAAHGPHAABBAYA22222
COLOMBIA       SBGHAABHHGBBAGBABABGGBBHGAGB3ABAGBGBB3B3HQ3BBGHBABBBBGAAG21222
CONGO(BRA)     HGHHABAIIIHH5HXBAGHIBIGBG56GBBGBH6ABH5AHBS6BGBHGAGAAGGABG11221
CONGO(LEO)     HBGHBBAHIHHHBHXBAGHIHIGBG6H666GAHBAAA66HIGGBG366G6A6G3ABG12221
COSTA RICA     GHHHA3BIHI3BABBAAABGGBAHGAGAAAAGGGGGGGBGIRBBGGHAABABBG5AG12112
CUBA           BHGGAABHGHGBHBBABABGGBA3AG6G6H5GHAG66656A66G6AHG666AB3A6B12212
CYPRUS         KHHGBBBIGIBBABRBAGHGAHAHGA6A3AGGGAA33333IH6533HAAA3BBG5ZG22122
CZECHOS'KIA    JHGGBABGBGAAHBHBAGGA4BAAAGBGGH5AHAGGH55HAAAGAGHGPHAABBAYA22221
DAHOMEY        HHHHBBAIIIHH5HZBAGHIBIGBG56GGGBH6ABH5AHAS6BGBHGAGAAGGARG22211
DENMARK        RHHGBAGGBGAAAAAAAAAA4BAHGAAAAAAGAHHHAGBAIKAGGGHAIAABAGBXG12222
DOMIN REP      BHHGAABIHHHBAGBABABGGBBBGA6AAABGGBG6GGB6IR66GGHAA6BBBG6AG11222
ECUADOR        SGHHAABIHHGGAGXBBGBGGBBHGAHBAAGAGBBBBGGGIQGAAGHBABBBBBAAG12222
EL SALVADOR    GHHBABBIHIHG5GBABABGGBBGGBHGGBGGGBBBBHHHIPGAAGHBAGAABAAAG11222
ETHIOPIA       HBGH5BAHIHHHGHZBBGAH4GBIGBGG3GGBHAAB44HHJ444AAHGKHBAHBA3G22222
FINLAND        RGHHBABHBGAAAAAAABGA4BAHGAAAAABGAHGHAGBAIKAGGGHAGAABAGBXG12222
FRANCE         ZGBGBAGBABAABBBAAAAA4AAHGABAAABBAGG3B3AGB3BBGAHBBBBAA3BAG22212
GABON          HGHHBBAIIIHH5HXBAGHIBIGBG56GGBGBH6ABH5AHAS6BGBHGAGAAGGABG22222
GERMANY, E     JHGBBAGGB5AAHAHBAA3A4BAAAGBGGH5GHAGAH5HHAAAGAGHGPHAABBAYA11222
GERMAN FR      ZGBBBAHBBBAAAAHBAA3A4BAHGABAAABBAGGHBBBBGKAGGGAAGABBAGBAG22122
GHANA          HGGH5BAIIHHGGGZBAGHIAIGBABBGGGBBHAABH5AHABBBBAHGAHAABBASG12222
GREECE         ZHGGBABHGHBB3BGAAABA4BAHGAGAAABGBGGGB3BBB3GAGGHAIBA3BGBAG12212
GUATEMALA      GHHHABAIHHHG5GXBBGBGGBA3G6H666GGGBBBAHGGI3GAAGHBA6A6BAAAG11122
GUINEA         HGHHABAIIIH5GHYBAGHIBIGBABBGGGBBHAABH5AHABAGBAHGAHAABBA3G11222
HAITI          BHHBBBAIIIHGGGXBBABIBBBGG33GGGGGHBAB44HHJ4443GHGAHAABAAAG12222
HONDURAS       GHHHABAIHIHG5GBABABGGBBHGAHBAABGGBBBBGGHQGAAGHBABAABAAAG21122
HUNGARY        JHGGBABHGGBAHBHBAAGA4BAAAG3GGG5GGAG3H55HAAAGAGHGPHAABBAYA12222
ICELAND        RHHHAAGIBIBAAAAAAAGA4BAHGAAAAAAGBHHHAGBBIKAGGGHAGABBAGBXG11222
INDIA          XBABABABIBGGAGIAAGHGAHA3GABAAABAGBABBBABB3AGGAGAGABABGBZG21122
INDONESIA      YBBGABAGIGHBGGKAAGHIHHGBBBGG6BGAGAAB33A3333BGAHGAGAA3BARG12222
IRAN           KBBHAAAHIHHGGHKAAGAI4GB3GBHH6BGAGAAA66G666633HG66BAHBAPG21122
IRAQ           KGGHAAAHHHGG6GKAAGGIAHAGG6H666GAHAAA666666G666H666666AAZG21222
IRELAND        ZHHGBABHBHBAABBAAAGAABAHGAAAAAAGBGGHBBBAI3AGGGHAGABBAGBGG22222
ISRAEL         KHHGAAHHBH3BBBPAAGH3AHABBABAAABGBBBGAHBAIKAGGGHAGAABAGBZG11211
ITALY          ZGBBBAGBBBABABBAAABA4BAHGABAAABBABGGBGBBIKBGGGHAGABBAGBAG11222
IVORY COAST    HGHHABAIIIHH5HZBAGHIBIGBB36GGGBGHBABH5AHABBBGAHGAGAAGGARG22222
JAMAICA        BHHBAABIG5GBAGHBBAHGAHG3GA6AAABGGGBGBBBBHK6GGBHAIABBBGBGG12222
```

A. Raw Characteristic Code Sheet (*cont.*)

```
                      10        20        30        40        50        60
JAPAN         IGBABABBGBBAAAQBAAAB4GAHGABAAAAGABBBBBBBGKABGGHAIABBBGBZG22212
JORDAN        KHHHAABIIIHGHGKAAAHIAHAIGBHG33GAHAAA66GH66G6ABHGJHBAHBAPG12122
KOREA, N      IHGG55AII5H5H555AA3H4HAAAGBGGH5GHAA5H55HAAAGABHGPHAABBAYA21112
KOREA REP     IHBAAAAHIHHBGG33AA3H4HA66BH666GG6AAB666666G66GH66666BAAZG21112
LAOS          YGHHABAIIIHHBHYBAGHIBHGGGBHG33GAHAAA3H3HIPGAA5HGIHB3H3ABG12112
LEBANON       KHHBAABIGH35BGRBAAGGBHAHGAH33AGBBAABAH3BIPGA3GHBBBABBBAZG11222
LIBERIA       HHHHBBAIIIH5BHXBAGBH4HG3GBABGGGBHABGBBGHA3ABAGHGAGBABBAHG22222
LIBYA         PBHHAAAIIIGGBHKAAAHIHHG3GAB33BBAHBBG44GGJ444BBBBJBBBBG5PG22122
LUXEMBOURG    ZHHBBAGIAHAAAABAAAAA4AAHGAAAAAAGAHHHAGBAIKAGGGHAIABBAGBAG21222
MALAGASY R    HGHHABAIIIHGAHXB3AHIBIG3GA6AAABAHGAGBBBHBSBGGGGBABBBGGBBG21222
MALAYA        YHGGA3BHGHGGAGQBBGHIAHGBGABAAAGBGBABBBAH3HBGGGBAIABBBGBQG12222
MALI          HBHHABAIIIHHAHYBBGHIBIGBG56GGG55H6ABH5AHAS6BGBHBHGAAGG5RG11222
MAURITANIA    HBHHABAIIIH5AHKABGHIBIGBBA6GGBBBHBABGBAHAS6BBBHBAGAAGGBPG21122
MEXICO        GBBHAABGHGBBAGBABABGGBAHGABA3BB3BBBBHAAHBJAGBGGBABBABGBAG22122
MONGOLIA      IBHHABAIIIH5HBJAAAGH4HAAAGBGGH5GHAA5H55HAAAGAGHGPHAABBAYA12112
MOROCCO       PGGHAAAHIHH5BGKAAGHIBHA3BABA6BBAGBBBBBA6336GGBHBJBBABGAPG21122
NEPAL         XHGGABAIIIHHGHQB3GAJ4GG36BHH6GG5HBAB66HH66G6A5HGKHAAH3AZG21222
NETHERLANDS   ZHGABAGGBGAAABHBAAAA4AAHGAAAAABBAGGHAGBAIKAGGGHAIABBAGBAG21122
NEW ZEALAND   AGHHAAGHAGAAAAHBAABAAAAHGAAAAAAGAHHHBABBHKAGGGHAIAABAGBGG12222
NICARAGUA     GHHHAAAIHIHG5GBABABGGBB3GBGGBBG5GBBGGHHG3QBAAGHBAGBABAAAG12222
NIGER         HBHHABAIIIHHAHYBBGHIBIGBG56G6GG5H6ABH5AHAS6BGBHGAGA5GG5RG21212
NIGERIA       HBBGBBBHIHHGBHZBAGHIAIG3GA6366GAHBABABGGIG6G33BBIBBBGGASG21222
NORWAY        RGHHBAGHAGAAAAAAAAAA4BAHGAAAAAAGAHHHAGBAIKAGGGHAIABBAGBXG11122
PAKISTAN      XBBGABBGIGGGGHKAAGHIAHA36BHG36GAHAAA666666G63BBGA6AA3AAQG12122
PANAMA        GHHHAABIHIHBABBABBBGGBB3GAHAAAGGGBGBBHGGIPGABGHBABABBA5AG21222
PARAGUAY      SGHHA3BIIIHBHGBA3ABGGBB3GB3GBB5GHAGBGHHHBQBGAGHGAHAABAAAG11221
PERU          SBGHAABHHHGGBGXBBGBGGBB6G6H666GBGBBBA666IKGBBGH666B6BA6AG21212
PHILIPPINES   YGBGA3BHHGGBAGBAAGHGHHAHGABAAABBAGAGHAABHJAGBGHAAABBBGBIG12122
POLAND        JGBGBABGGGBAGBBAAAGA4BAAAGBGGGAGGABBH5GHAAAGAGHGPHAABBAYA22122
PORTUGAL      ZHGGB3BHHHGBGGBAAAAA4BA3GGAGGG5G3AGGH5H3AIAGAGHG3HBABBAAG21222
RUMANIA       JGBGBAAGGGBAHBGAABBB4BAAAGBGGH5GHAB5H55HAAAGAGHGPHAABBAYA12222
RWANDA        HHHGBBAIIIHH5H55AAHIHIGBG5656B5GH6AAG5BHBS6535HBAGA5GG5BG11222
SA'U ARABIA   KBGHAB5HHH3HHHKA5AGJ4HHIGBAH4GGGHBAB44H4J444ABHG444AHBAKG11112
SENEGAL       HGHHABAIIHHH5HKAAGHIBIGBB66GB6BBHBABHAAHABBGGBHGAGAAGBARG12222
SIERRE LEO    HHHHABAIIIHHAHZBAGHIAIGBGA6AAABBH6ABBGBGHSGGBGHBIBABGGBSG22212
SOMALIA       HGHHBBAIIIHHAHKAAAHIHIGBG563BBBAH6ABBGGHBGGGBGHBGGA5G3AZG22121
SO AFRICA     HBGHAAGGGGBGBGXBBGB3AABHG3AB33GBBBAA333333AGAGHAGABAABAJG21222
SPAIN         ZGBGBABGGGGBGGBAABAA4BA33GAGGGGB3ABBH5G3AIAGAGHG3HAABBAAG11122
SUDAN         HBGHBBAIIHHHGHYBBGHIAHABBBHH6GBAHAAA66666666GHG6666GAAQG12212
SWEDEN        RGGHBAGGAGAAAAAAAAAA4BAHGAAAAAAGAHHHAGBAIKAGGGHAIABBAGBXG22222
SWITZERLAND   ZHHBBAGGAGAAAAHBAGAA4AAHGAAAAA3AAGBHAGGAIKAGGGAAQABGAGBAG12222
SYRIA         KHHHAAAIHHH5BGKAABGIBHAGG6H666GBGAAAA66666GA66H66666BAAPG12222
TANGANYIKA    HBGHBBAIIIHHAHZBAGHIAIGBB56BBBBGHBAGHAAHABAGGAHG3GAAGGBSG22212
THAILAND      YGBGABAHIHHBGGJAABAH4GBIGBGH6GABHAAB66HH66GAAGHG6H66HAAAG12222
TOGO          HHHH5BAIIIHH6HXBAGHIBIGBG66666BBH6ABG66H6S66G6HG66A6GAABG12222
TRINIDAD      BHHBA3GIG5GBAGRBBBHGAHG3GA6AAABGGGBGBBBBGK6GGGHAIABBBGBGG21212
TUNISIA       PHHHAAAIHHHGGGKAAAHIBHABBABAGBBGGBBGHAAHABAGGBHGAGAABGAPG12112
TURKEY        KBBHAAAGHGGGBGKAABAB4GAHGA6AAAAGBBBBGBBB33BGGGHAGBBBBAAAG12212
UGANDA        HGGHBBAIIIHHAHXBAGHIAIG3GA63AAGAHBABABGGIG6G33A5IBA5GGBSG12222
USSR          JAAH5ABABAAAGB55AGAB4GAAAGAGGH5AGABBH55HAAAGA3GGPHBABBAYA22122
UAR           PBBHAABHHGHGHGKAAAHGAHABABBG6G5GGABGH5A6AB6A3AHG6H6ABAAPG22122
UK            ZGBBBAHBABAAAAHBAAAA4AAHGAAAAAAAAHGHBABBHKAGBGHAIABBAGBGG22122
US            QAAHBAHAAAAAAAHBBAAAAAAHGAAAAABBAHG3HAAAHJAGBGAAAABBAGBGG22222
UPPER VOLTA   HGHHABAIIIHH5HYBAGHIBIGBG56GBGGGH6ABH5AHAS6BGBHGAGAAGG5RG21221
URUGUAY       SHHHBAGHGHBBABBAAABGGBAHGABAAA33BGGG3A3A333AGGHAQABGBGBAG11222
VENEZUELA     SBGHAABHBG3BGGBABABGGBABBAHAAABGBBG3ABAGIRGGGGBBABBBB3AAG12122
VIETNAM, N    YHGGB55HI5H5H555ABHIBHGAAGBGGH5GHAA5H55HAAAGAAHGPHAABBAYA12222
VIETNAM REP   YHGGA3AHIHHBGG33ABHIBHG3A3GGGGGHAAAHHHHB3BBBGHGAHAA3BABG22112
YEMEN         KGHHBBAIIIHH6HKA5AG64HG666666GAH6A666666666H666666AAKG22122
YUGOSLAVIA    JGBGBBAGGGBBGGHBAGGB4BAAAGBGGGGA3ABBH5GHAAAGABGGPHBABBAYA12222
```

B. Finished Characteristic Code Sheet

	10, 20, 30	40, 50, 60, 70	80, 90, 100
AFGHANISTAN	22224242222211111444212122222112222	22221222222442222222121222422212222	11122444444414212221222212111122132
ALBANIA	12222242222222444444222221112112222	21222222211214222221122222424421112	21122444442124112222411221221122122
ALGERIA	22224242212112244114112122212112212	21225222221421221221121222422211222	22221211124221111222111212126622122
ARGENTINA	22224122222222442422111122221222112	11211211211214211211112222211111111	21121211242124113332322313322231411
AUSTRALIA	21224242222222444444112122211222111	11111111111144111111122222411421112	21111144141124112212422111411111411
AUSTRIA	21121242222222444444222121121222111	11211211111144111211112222211111111	11111144442124112212422111421111411
BELGIUM	21121242222222444444222111221222111	11211111111144111211112222211111222	21111144441124112212422111411111411
BOLIVIA	22224122222222442422112222212112222	22222222221421211221222122424422222	21121211242124211222121111422111411
BRAZIL	22224122222222442422111122211212112	21211211211214211221122222211112112	21121211442124113332322111422221411
BULGARIA	12222242222222444444222121121212212	11221211211214222111222222411411212	21121442442124112222411221221122122
BURMA	22224241122222424241211122222112212	22221221211214211221122222424411222	22222224114221223322311212122222266
BURUNDI	22224242211222244114222221122112222	22222222222442555222222522424451115	22222224444222221222122515526651442
CAMBODIA	22224241122222424241222222222112222	22222222221421221222122222424411212	22222224124221223223321212122122122
CAMEROUN	22224242211222244114212222222112222	22222222222442111222222221424421222	22222224124222221222122616626631466
CANADA	21224242222222444444111122211221111	11111111111144111211122222411421222	21111144141124112212422111411111411
CEN AFR REP	22224242211222244114212222222112222	22222222222442555222222222424431222	22222224124222221222122212126622122
CEYLON	22224241222222424241222111212212212	22221222211214666221122222424421222	22221211114221223332322111422211411
CHAD	22224242211222244114112222212112222	22222222222442111222222221242442222	22222224124222221222122515526622142
CHILE	22224122222222442422212122211222212	11211221211214111211112222211113113	21121211242124112212422111422121411
CHINA, PR	22224241222222424241111121115111112	22511221251555222221322252424451335	11121442444414112222411221221122122
COLOMBIA	22224122222222442422112122211212212	21211211211214111221112222211112112	21121211242124212212422111422121433
CONGO(BRA)	22224242211222244114212222212112222	22222222222442555222222122424421222	22222224124222221222122515526622142
CONGO(LEO)	22224242211222244114112122222112212	22221222222442211222222122424421222	22222224444222221222122616622261466
COSTA RICA	22224112222222442422222222213212222	21222223211214111211112222211111111	21121211242124112212422111422111411
CUBA	22224142222222442422222121111212212	11221221211214222111122222211112112	21121211242124113222311221226622166
CYPRUS	22224242222211111444222221122212222	11222211211214111211112222424421222	22221211114221112212422111426611433
CZECHOS'KIA	12222242222222444444222121121212111	11211111111144222211122222411421222	22111144442124112222411221221122122
DAHOMEY	22224242211222244114222222222112222	22222222222442555222222221424421222	22222224124222221222122515526622122
DENMARK	21211242222222444444222221121222111	11211111111144111111122222111411111	11111144442124112212422111411111411
DOMIN REP	22224142222222442422222221111212222	21221222211214111221112222211112112	21121211242124211222122111426611411
ECUADOR	22224122222222442422212222211212222	21221221221421111221222122424422222	21121211242124212212422111422221411
EL SALVADOR	22224112222222442422222211212212222	21222222221421555221112222211112112	21121211242124212122422212122222122
ETHIOPIA	22224242211222244114112122252112212	22221222222442212222222221424422222	11122444444414212221222212122122132
FINLAND	21211242222222444444212222221212211	11211111111144111111122222111411212	22111144442124112212422111411111411
FRANCE	21121242222222444444211121121221111	11111111111144211211112222211111111	11111144441124112212422111421111411
GABON	22224242211222244114212222222112222	22222222222442555222222122424421222	22222224124222221222122515526622122
GERMANY, E	12222242222222444444222111221222111	11551111111144222111122222411421112	33311144442124112222411221221122122
GERMAN FR	21121242222222444444211111221221111	11111111111144111111122222411421112	33311144442124112212422111421111411
GHANA	22224242211222244114212122252112222	22221222221421221221222221424421222	22222224114222221222111212121122122
GREECE	21121242222222444444222121121212212	11221211211214331211122222411411111	21111144442124112212422111422111411
GUATEMALA	22224112222222442422222222212112222	21221222221421555221222122424422222	21121211242124113332322616622266666
GUINEA	22224242211222244114212222212112222	22222222225455221222222212424421222	22222224124222221222111212121122122
HAITI	22224142222222442422222211222112222	22222222221421221221222122424422112	21122224142124212122422233323122122
HONDURAS	22224112222222442422222222212112222	21222222221421555221112222211112112	21121211242124212212422111422221411
HUNGARY	12222242222222444444222121121212212	11211211111144222211122222411421112	22111144442124112222411221223122122
ICELAND	21211242222222444444222222211222221	11222211111144111111122222111411111	22111144442124112212422111411111411
INDIA	22224241222222424241111111212111112	22111221221421111221122222424411222	22221211114221113232322111421111411
INDONESIA	22224241122222424241111121112112112	22211222211214221221121222422211222	22222224444221221222121212122122166
IRAN	22224242222211111444111122211112212	22221222221421221222121222422211222	11122224444414213223322212122222266
IRAQ	22224242222211111444212122211112212	21221221221421266221121222422211222	22122224114221112122422616622266666
IRELAND	21121242222222444444222221121212211	11221211111144111211112222211111111	22111144112124112212422111411111411
ISRAEL	22224242222211111444222221111222211	11221331211214211211122222424411222	22231333114221111222121111421111411
ITALY	21121242222222444444211111221221111	11111111211214111211112222211111111	21111144442124112212422111421111411
IVORY COAST	22224242211222244114212222212112222	22222222222442555222222221424421222	22222224124222221222121313326622122
JAMAICA	22224142222222442422222211211212222	11225221211214111221122222411422112	22221211114221223232322111426611411

B. Finished Characteristic Code Sheet (*cont.*)

Country			
JAPAN	22224241222222424241211111221211112	11111211111144111111112222242442112	111214424444141122124221114211111411
JORDAN	22224242222211111444222222211212222	22222222221421222211212224222111111	222222241142211122212222121222221 32
KOREA, N	22224241222222424241222121155112222	22225222251555222255322252424451115	333224444442211122224112212211 22122
KOREA REP	22224241222222424241221111211112212	22221222211214221221322233424431113	333224444442211166626262121222666 66
LAOS	22224241122222424241212222121112222	22222222222442211222222212424421222	222222241242212221224222121222221 32
LEBANON	22224242222211111444222211211212222	11221223251555211221122222424421112	221212111242211122124221114222314 33
LIBERIA	22224242211222244114222222222112222	22222222251555211222222122424421222	211224444442212232323222121111214 22
LIBYA	22224242212112244114112222211112222	22222221221421211222121222422211111	222222244442212232333221114211314 32
LUXEMBOURG	21121242222222444444222211221222221	11221111111144111111112222211111111	111111444411241122124221114111114 11
MALAGASY R	22224242211222244114212222212112222	22222222221421111222222122424423112	222222241242222232323221114266114 11
MALAYA	22224241122222424241222121113212212	11221221221421111221122222424422222	222222241142212212221221114211114 11
MALI	22224242211222244114112222212112222	22222222222442111222222212424422222	222222241242222212221225155266221 22
MAURITANIA	22224242211222244114112222212112222	22222222225455111222121222422212222	222222241242222212221211114266221 22
MEXICO	22224112222222442422111222211212112	21211211211214111221112222211112112	211212112421241122124221114211114 33
MONGOLIA	22224241222222424241112222212112222	22222222255555222211122222424411111	221224444442211122224112212211221 22
MOROCCO	22224242212112244114212122221112212	22221222225455211221121222422211222	222222241242211132333211114211114 66
NEPAL	22224241222222424241222121112112222	22222222222442212212222122424423222	111224444444142232233262121222222 66
NETHERLANDS	21121242222222444444222111221222111	11211111111144111211122222411421112	111111444411241122124221114111114 11
NEW ZEALAND	21224242222222444444212222211222211	11211111111144111111122222411421112	211111441411241122124221114111114 11
NICARAGUA	22224112222222442422222222211112222	21222222221421555211122222111112112	211212112421241213332422212122121 42
NIGER	22224242211222244114112222212112222	22222222222442111222222212424422222	222222241242222122122251552666221 62
NIGERIA	22224242211222244114111121122212212	22221222221421211222222221424421222	222222241142222232333221114266314 66
NORWAY	21211242222222444444212222221222211	11211111111144111111122222111411111	111111444421241122124221114111114 11
PAKISTAN	22224241222222424241111121112212112	22211221221421221222121222422211222	222222241142211133223262121222221 32
PANAMA	22224112222222442422222222211212222	21222222211214111211112222211112212	211212112421241213332422111422211 411
PARAGUAY	22224122222222442422212222213212222	22222222211214222211122222111113113	211212112421241213332422212123122 142
PERU	22224122222222442422112122211212212	21221221221421211221222122424422222	211212112421241666262261662266666 6
PHILIPPINES	22224241122222424241211121113212212	21211221211214111221112222211111222	222212114442211122124221114211114 11
POLAND	12222242222222444444211121121212112	11211211111144221211112222211111111	221111444421241122224112212211221 22
PORTUGAL	21121242222222444444222121123212212	21221221211214221221112222211111111	111111444421241123324222121112212 2
RUMANIA	12222242222222444444211121121112112	11211211111144222211122222411411212	211214424421241122224112212211221 22
RWANDA	22224242211222244114222221122112222	22222222222442555222222522424451115	222222244442222212221225155266514 62
SA'U ARABIA	22224242222211111444112122212552212	21221233222442222221212224222215115	221224444442214422122222121111222 44
SENEGAL	22224242211222244114212222212112222	22221222222442555222121222422211222	222222241242222172212161662262214 2
SIERRE LEO	22224242211222244114222222212112222	22222222222442111222222221424421222	222222241142222212221211114266114 11
SOMALIA	22224242211222244114212222222112222	22222222222442111222121222422211111	222222244442222212221225155266314 42
SO AFRICA	22224242211222244114112122211222112	11211211221421211221222122424422222	211332331411242122124223333111214 33
SPAIN	21121242222222444444211121121212112	11211221211214221221112222211111212	111111444421241123324332212111221 22
SUDAN	22224242211222244114112122222112222	22221222222442221222222212424422222	222222241142211112221212121222222 66
SWEDEN	21211242222222444444212122221222111	11211111111144111111122222111411111	111111444421241122124221114111114 11
SWITZERLAND	21121242222222444444222211221222111	11211111111144111111122222411421222	111111444411241122124221114111114 11
SYRIA	22224242222211111444222222211112222	21221222221455211221212224222211212	221222241242211211224226166222666 66
TANGANYIKA	22224242211222244114112122222112222	22222222222442111222222221424421222	222222241142222122212151552662144 2
THAILAND	22224241122222424241211121112112212	22221222211214221221122222424411212	111224444444142122212222121221222 66
TOGO	22224242211222244114222222252112222	22222222222442666222222122424421222	222222241242222122212261662266666 6
TRINIDAD	22224142222222442422222221121322222	11225221211214111221122222424422212	222212111142212232323221114266114 11
TUNISIA	22224242212112244114222222211112222	21221222221421221221121222422211111	222222241242211122212211114211114 22
TURKEY	22224242222211111444111122211112112	21211221221421211221121222422211212	111214424444141122124221114226114 11
UGANDA	22224242211222244114212122222112222	22222222222442111222222122424421222	222222241142222232333221114266314 11
USSR	12222242222222444444111122251211111	11111111111144221211122222454451222	111214424444141122224112212111221 22
UAR	22224242212112244114111222211212212	21211222221421222211212224222211111	222212111142211112221112121211221 66
UK	21121242222222444444211112221221111	11111111111144111111122222411421112	111111444411241122124221114111114 11
US	21224242222222444444111122221221111	11111111111144111111122222411422112	111111441411241122124221114111114 11
UPPER VOLTA	22224242211222244114212222212112222	22222222222442555222222212424421222	222222241242222122212251552662214 2
URUGUAY	22224122222222442422222222221222212	11221211211214111211112222211111111	211212112421241122124221114211114 11
VENEZUELA	22224122222222442422112122211212211	11211331211214221221112222211112112	211212112421241112221211114222114 11
VIETNAM, N	22224241122222424241222121125512212	22225222255555222255322255424451212	222222241242212222224112212211221 22
VIETNAM REP	22224241122222424241222121113112212	22221222211214221221322233424431212	222222241242212233223112333221221 22
YEMEN	22224242222211111444212222222112222	22222222222442266222121222422215115	221224444442212266266666666266666 66
YUGOSLAVIA	12222242222222444444211121122112112	11211211211214221221122222411421222	221214424421241122224112212211221 22

B. Finished Characteristic Code Sheet (*cont.*)

	110 · 120 · 130 · 140	150 · 160 · 170	180 · 190
AFGHANISTAN	32242142122211111211444444222222 2	221444444444444444111333222222222222	24212442421122442221112
ALBANIA	22255542122211121155522225522522212	222142224441114222111222222222222422	24111214421122442212222 2
ALGERIA	22642142122111121166666666166166612	22214122444666422266642122222662622	12112421421122441221112 2
ARGENTINA	11122233311111122133321123322122122	21241212444222242122222221666122421	11261214114112142222222 2
AUSTRALIA	11111142111122222222221111121121122	12222212444111422222222221111221111	14221144242221242222212
AUSTRIA	11111122221122122222221111121121122	12222212444111422222222221111212111	14221144242212142222222 1
BELGIUM	11122211111122121122211122121111122	21241212444111122222222222111221111	14221144242212142222222 1
BOLIVIA	11122222222121121121111121111122121	22244222424222442122222222421122421	11221214242112142222222 2
BRAZIL	11122211111111121121121122122122122	21241212444222242142122211111122421	11221214114512142221212 2
BULGARIA	22255542122211122155522225522522212	22214222444111422211122222222222422	24111214421122442211122 2
BURMA	62642111122111111121166666627666622	66666262644666466666666262266666266 6	66662421114122442622122 2
BURUNDI	12155522222266611122122225511112222 1	22244221442666452522255522421221122	12152421242522442122212 2
CAMBODIA	22242142122221111121122221111122212	22214322344421411133311122222223322	24212442421122442221122 2
CAMEROUN	62142142122266611121121121111122121	22244222441666442122242111421122421	11122421242122442121222 1
CANADA	11142111111122121122222211121121122	12222222144111422222222211111221111	14221144242222442221122 2
CEN AFR REP	22222222222266611121122255111122212	22214221442666442122242122222122422	12112421242122442122222 2
CEYLON	11142142122121111111111122122222222	21241232344222242142122222421221121	11231214242222442221122 2
CHAD	12122242122266611121122255111122212	22214221442666442122242122222222222	12152421242522411122222 1
CHILE	11122222221122122122111122122121122	21241212444222242142122222111122421	11231214242512142221122 2
CHINA, PR	22255542122211111111122225522522212	22214222444111422211111222222222422	24111214421122442212222 2
COLOMBIA	31142111122121122121121133321133122	12222222414333442142122222421122421	11221214242112142222122 2
CONGO(BRA)	12122242122266611121122255111122221	22244221442666442122242122222122422	12112421242122442121122 1
CONGO(LEO)	62622211122221111111111126126622222	21241222441222242122233366666212166	66162421333122442121222 1
CUBA	62255522222211122166666666122566612	21241222424442144212222222111122421	11121214242512142221211 2
COSTA RICA	11111122222122122122122122121122122	22214622464666422266611122222662666	66611214333122442241221 2
CYPRUS	31122222222111111133333333333333122	21241222444666455533333322111122411	14321214242522442222212 2
CZECHOS'KIA	22255511122211122122122225522522212	22214222444111422211122222222222422	24111214421122442212222 1
DAHOMEY	22222242122266611121122255111122212	22214221442666442122242122222122422	12112421242122411122221 1
DENMARK	11111122211122222222211122121111122	21241212444111122222222222111221111	14121144242222442221222 2
DOMIN REP	11142122222121122166622122121166122	21241222424666466622222222111122466	66221214242612142221122 2
ECUADOR	11122211122121121121121122122122122	21241222414222211111122222421122421	11221214421112142221222 2
EL SALVADOR	22122222222121121121121122222222222	21241222244222211111122222421122422	12111214114112142221122 2
ETHIOPIA	32222242122211111121144444422222222	22144444442444444411111122222222422	24212442421132333122222 2
FINLAND	11142122211122222122211122121111122	21241212444111122222222222111212111	14121144242222442221222 2
FRANCE	11142142111122122133321133311122121	22244332344421442122211122421222221	11211144333212142222221 2
GABON	22122242122266611121122255111122212	22214221442666442122242122222122422	12112421242122442122222 2
GERMANY, E	22255522222211122111222255222222212	22214222444111422211122222222222422	24111214421122442211122 2
GERMAN FR	11142142111122122122221121121121122	22244212444111422222222211111212111	14221144242212142222212 2
GHANA	22242142122211111121122255111122212	22214122442421442142111122222122422	24111214421122421121222 2
GREECE	11142122221122122122121133321121121	22244332344222411122222222111221121	11131214242212142221221 2
GUATEMALA	62622222222121121121111122222122122	21241232344222211111122222421122466	66161214114112142221112 2
GUINEA	22242142122211111121122255111122212	22214122442111422242111122222122422	24111214421132333121122 2
HAITI	22222222222221111121144444422222222	22144444444444444433322222222122422	24111214114112142221222 2
HONDURAS	11142122222121121121121122122122122	12222222414222411111122222421122421	11111214114112142222112 2
HUNGARY	22255522222111122133322225522522212	22214222444111422211122222222222422	24111214421122442211222 2
ICELAND	11111122221122222222211122121121122	21241212444111122222222222111212111	14221144242222442221122 2
INDIA	11142111122121111121121121111121121	22244332344111422222211122111212111	14211214242222442222112 2
INDONESIA	62122211122111111121133333331113333 3	33333232344333442122211122222122422	12113333421122411121222 2
IRAN	62122211122111111111666226221666666	66666262644666466633333322222226666	66212442421122441222112 2
IRAQ	62622211122211111111666666666666666	66266662644222466666666622666662666	66666666114122442222122 2
IRELAND	11111122221122122122221121121111122	21241232344111222222222222111212111	14221144242221242222222 2
ISRAEL	11142122221121121122111122221111122	21241212444111222222222222111212111	14121144242222442221121 1
ITALY	11142142111121122122121122121121122	21241212444421422222222222111212111	14221144242212142221122 2
IVORY COAST	22242122222221111121122255111122212	22214122442421442122211122222122422	12112421242122411122222 2
JAMAICA	11142122222122121122121121121121122	12222212444666422222242122111221111	14221214242221242221222 2

B. Finished Characteristic Code Sheet (*cont.*)

	110 120 130 140	150 160 170	180 190
JAPAN	11111122211121121121121121121121122	22244212444111442122222222111221111	142212142422224422222212
JORDAN	32322211122211111111166622622122222	26646262644222466611142122222222222	242124424211224412212122
KOREA, N	2225552222221111155522225522522212	22214222444111422211142122222222422	241112144211224422121112
KOREA REP	62622222226661111121166622622666666	66666262644222466666662222266666266	666612141141224422221112
LAOS	33322211122211111111133322233322222	21241222244222211111155522222221122	242324423331224421212112
LEBANON	3112224212111111121111122233321122	21241222244222211133322222421222221	111212144211224422211222
LIBERIA	22222242122211121122121121122122212	22214232442111442111222222222122422	122112144211224421222222
LIBYA	32142111122221121122144444422122122	22144444444444444442142121421222221	112212142425224412222122
LUXEMBOURG	11111122211122222222211122121111122	21241212444111122222222222111221111	142211442422121422221222
MALAGASY R	11142111122222111122121121121122221	22244221442421422222222222421122421	112224212422224421221222
MALAYA	11122242122121111121121121111122223	33233222444421422222222221111221111	142212142422224412212222
MALI	22255555222666111211222551111122212	22214221442666442122242122421222422	121124212425224111211222
MAURITANIA	22142142122221111121122121111122212	22214221442666442142142122421122422	121124212422224412221122
MEXICO	32142133321121121121122211111122221	22244222144111422242122222421122421	112112142422121422222122
MONGOLIA	2225552222221111155522225522522212	22214222444111422211122222222222422	241112144211224422112112
MOROCCO	62142111122121121121121121111166623	23233332344666422222242122421222221	112112142421224412221122
NEPAL	62222255522221111121166622622222266	66666222644222466611155522222222422	241124423331224422221222
NETHERLANDS	11142142111122122122211122121111122	21241212444111122222222222111221111	142211442422121422221122
NEW ZEALAND	11111122211122222222211111121121122	12222212444111422222222222111221111	141211442422212422212222
NICARAGUA	12122255522121121122122122222222133	33233222414421411111122222421122422	122112141141121422212222
NIGER	62222255522266611121122255111122212	22214221442666442122242122222122422	121524212425224111221212
NIGERIA	62622211122221111121111121122122122	21241222441666422233333321421221121	112224212421224211221222
NORWAY	11111122211122222222211122121111122	21241212444111122222222222111221111	142211442422224422211122
PAKISTAN	32622211122211111111166622626122626	66646262644222466633342121222122466	661133331141224412212122
PANAMA	11122222222121122121121122222122122	21241222244222211142122222421122421	111212141145121422221222
PARAGUAY	12155522222211122121122122222222221	22244222414421422211122222222122422	241112141141121422211221
PERU	66122242122121121121111122666666622	21241212444222242142122222666662666	662612141146121422221212
PHILIPPINES	11142142111122111122122211111121122	12222222144111422242122222111122411	142212142422224422212122
POLAND	22211122222111121121122225522122212	22214222444111422211122222222222422	241112144211224422122122
PORTUGAL	22255522233311122122122225522222312	22214222444111422211122222222232322	242112144211121422221222
RUMANIA	22255522222211121155522225522522212	22214222444111422211122222222222422	241112144211224422112222
RWANDA	62155522222266611111122155121122221	22244221442666452533355224211122422	121524212425224421211222
SA•U ARABIA	42222222222221111121144444422244422	22144444444444444411142122224444444	444124424211224412211112
SENEGAL	12642142122221111121122211111122212	22214122442421422222242122222122422	121124214211224111212222
SIERRE LEO	11142142122266611121121122121122122	12222221442222422242122222421221121	111224212422224211222212
SOMALIA	12142111122266611121121122122122221	22244222441222422242122222421212122	121524213331224422222121
SO AFRICA	32322242121121111111133333133333323	33233232342111422211122222111212111	142111444211224421221222
SPAIN	22222242133311121121122225522122312	22214222444111422211122222222233322	241112144211121422211122
SUDAN	62242111122211111111166626622666622	66666222646666466666622222222662666	666624211141224412212212
SWEDEN	11111122211122222222211122121111122	21241212444111122222222222111221111	142211442422224422222222
SWITZERLAND	11133311111122121122211122122111122	21241212444111122222222211111222411	142411442422121422212222
SYRIA	62622242122111111111111116666666666	66266662644222411166666622666662666	666612141141224412212222
TANGANYIKA	12142122222221111221222111111122212	22214122442111422222211122222322322	121124212422224211222212
THAILAND	62211142122211111121166622622222266	66666222644222411111122222222226622	246624421141121422212222
TOGO	66642142122266611121122166622622266	66266221442666466622266622222622666	661624211141224421212222
TRINIDAD	11142122222122121122121121121121122	22244212444666422222222222111221111	142212142422212422221212
TUNISIA	22142122222121121122122211111122212	22214122444111422222242122222122422	121112142421224412212112
TURKEY	11111122221121121121122121121121123	33233232344421422222222222111212121	112212141141121422212212
UGANDA	11122211122221111121111121122122122	21241222441666422233333311555221121	111524212422224211212222
USSR	22255511122111121121122225522522212	22214222444111422211133322222222422	242112144211224422122122
UAR	62255522222111121122122225511166612	22214122444666411133311122222662622	246112141141224412222122
UK	11111111111122222122222111112112112 2	12222212444111422242122222111221111	142211442422212422222122
US	11142142111122222133322211111111122	12222222144111422242122211111122411	142211442422212422222222
UPPER VOLTA	12222222222226661112112225511112221 2	22214221442666442122242122222122422	121124212425224111221221
URUGUAY	11133333321122122122133311133311122	33233232334333411122222222111222411	142412142422121422211222
VENEZUELA	11142122221121122133311121111122122	21241222424222222222222221421122421	112212143331121422212122
VIETNAM, N	2225552222221111155522225522522212	22214222444111422211111122222222422	241112144211224422112222
VIETNAM REP	22222222222221111111112222222222221	22244322344421442142122222222122422	241133334211224421222112
YEMEN	66622211122266611166666666666666666	66666662644666466666666622666666666	666666661141224412222122
YUGOSLAVIA	22222211133311121121122225522122212	22214222444111422211142122222222422	242112144211224422112222

C. Check List of Deletions

ALL HAND WINNOWED PREDICATE DELETIONS INCLUDED

CELL ORDER IS TOP LEFT, TOP RIGHT, BOTTOM LEFT, BOTTOM RIGHT

SO = 'SUBJECT ONLY' DELETION

RD = REDUNDANT DELETION

1	5	0	18	9	0	RD
1	10	0	33	9	73	RD
1	11	0	28	9	78	RD
1	27	7	24	2	82	SO
1	34	8	54	1	52	RD
1	43	8	47	1	56	RD
1	46	9	75	0	26	RD
1	50	0	43	9	47	RD
1	78	0	28	4	3	RD
1	86	9	67	0	38	RD
1	93	9	23	0	78	RD
1	94	0	51	9	32	RD
1	96	9	58	0	23	RD
1	97	0	23	9	7	RD
1	100	9	55	0	22	RD
1	101	0	41	9	48	RD
1	104	0	43	9	21	SO
1	105	0	43	9	38	RD
1	107	0	46	9	56	RD
1	119	9	65	0	26	RD
1	127	0	21	9	63	RD
1	128	0	48	9	36	RD
1	133	0	29	9	67	RD
1	137	0	29	9	59	RD
1	142	0	30	9	57	RD
1	144	9	25	0	11	RD
1	152	9	33	0	25	SO
1	154	9	46	0	25	RD
1	159	9	21	0	50	SO
1	161	9	38	0	50	RD
1	167	0	34	9	39	SO
1	168	0	34	9	63	RD
1	174	0	28	9	63	RD
1	175	0	51	9	40	RD
1	180	9	67	0	34	SO
1	188	0	32	9	72	RD
1	192	0	27	9	77	RD
1	193	0	28	9	77	RD
2	3	13	0	9	93	RD
2	4	5	0	17	93	SO
2	5	18	0	0	9	RD
2	6	0	22	22	71	RD
2	8	0	18	22	75	RD
2	10	0	33	22	60	RD
2	11	0	28	22	65	RD
2	13	0	16	22	77	RD

2	14	0	11	22	82	RD
2	31	5	85	17	7	RD
2	32	6	4	16	89	RD
2	38	7	3	15	88	RD
2	40	21	50	1	38	RD
2	46	22	62	0	26	RD
2	47	18	7	4	26	RD
2	48	4	26	0	24	RD
2	58	0	18	22	75	RD
2	59	0	13	22	78	SO
2	64	22	24	0	18	RD
2	65	8	17	0	18	RD
2	75	22	40	0	52	RD
2	76	22	4	0	70	RD
2	82	22	29	0	55	RD
2	86	22	54	0	38	RD
2	87	0	31	22	36	RD
2	92	0	20	21	72	RD
2	104	20	23	2	28	SO
2	108	20	45	2	33	RD
2	109	10	5	2	40	SO
2	117	20	40	0	51	RD
2	118	2	38	20	40	RD
2	123	14	93	8	0	RD
2	129	18	38	4	33	RD
2	133	2	27	20	56	RD
2	136	10	2	12	73	RD
2	139	2	32	20	51	RD
2	141	6	5	16	71	RD
2	148	0	14	22	74	RD
2	152	19	23	1	24	SO
2	154	21	34	1	24	RD
2	155	10	1	0	13	SO
2	156	1	13	20	36	SO
2	159	2	28	18	32	SO
2	160	2	28	20	47	RD
2	162	1	12	21	44	SO
2	165	5	3	17	89	SO
2	167	19	15	2	46	SO
2	173	17	16	1	6	SO
2	176	20	52	2	26	RD
2	180	22	54	0	34	SO
2	184	0	21	19	37	SO
2	192	0	27	22	64	RD
3	2	13	9	0	93	RD
3	5	13	5	0	9	RD

3	6	0	22	13	80	RD
3	8	0	18	13	84	RD
3	10	0	33	13	69	RD
3	11	0	28	13	74	RD
3	46	13	71	0	26	RD
3	47	9	16	4	26	RD
3	48	4	26	0	24	RD
3	57	8	17	5	85	RD
3	58	0	18	13	84	RD
3	64	13	33	0	18	RD
3	65	8	17	0	18	RD
3	74	13	13	0	87	RD
3	75	13	49	0	52	RD
3	76	13	13	0	70	RD
3	81	6	5	7	33	RD
3	82	13	38	0	55	RD
3	86	13	63	0	38	RD
3	87	0	31	13	45	RD
3	92	0	20	12	81	RD
3	102	11	48	2	47	RD
3	104	11	32	2	28	SO
3	106	11	41	2	28	RD
3	108	11	54	2	33	RD
3	109	4	11	2	40	SO
3	117	11	49	0	51	RD
3	129	11	45	2	35	RD
3	152	10	32	1	24	SO
3	154	12	43	1	24	RD
3	155	5	6	0	13	SO
3	159	2	28	10	40	SO
3	167	10	24	2	46	SO
3	169	11	47	2	46	RD
3	180	13	63	0	34	SO
3	184	0	21	10	46	SO
3	192	0	27	13	73	RD
4	2	5	17	0	93	RD
4	5	5	13	0	9	RD
4	36	5	37	0	73	RD
4	37	5	59	0	51	RD
4	43	5	50	0	57	RD
4	45	5	50	0	54	RD
4	47	5	20	0	30	RD
4	54	5	32	0	76	RD
4	74	5	21	0	87	RD
4	75	5	57	0	52	RD
4	76	5	21	0	70	RD

4	82	5	46	0	55	RD
4	99	5	45	0	36	RD
4	102	5	54	0	49	RD
4	104	5	38	0	30	SO
4	109	4	11	0	42	SO
4	117	5	55	0	51	RD
4	118	0	40	5	55	RD
4	119	0	74	5	21	RD
4	121	0	54	5	56	RD
4	125	0	64	5	30	RD
4	128	5	43	0	45	RD
4	129	5	51	0	37	RD
4	138	5	43	0	49	RD
4	152	5	37	0	25	SO
4	155	5	6	0	13	SO
4	167	5	29	0	48	SO
4	169	5	53	0	48	RD
4	175	5	46	0	49	RD
5	1	0	9	18	0	SO
5	2	18	0	0	9	RD
5	3	13	0	5	9	RD
5	4	5	0	13	9	SO
5	27	7	7	11	2	SO
5	92	0	9	17	0	RD
5	104	16	0	2	9	SO
5	120	11	9	7	0	RD
5	159	2	9	15	0	SO
5	167	15	0	2	9	SO
6	2	0	22	22	71	RD
6	3	0	13	22	80	RD
6	8	0	18	22	75	RD
6	10	0	33	22	60	RD
6	11	0	28	22	65	RD
6	13	0	16	22	77	RD
6	14	0	11	22	82	RD
6	17	0	11	22	0	RD
6	19	0	33	22	0	RD
6	20	0	18	22	0	RD
6	27	2	29	20	64	SO
6	35	1	23	21	70	RD
6	47	0	25	14	16	RD
6	49	8	16	0	26	RD
6	58	0	18	22	75	RD
6	63	16	30	6	62	RD
6	64	16	30	0	18	RD
6	65	15	10	0	18	RD

6	74	0	26	22	65	RD
6	104	13	30	2	28	SO
6	106	15	37	2	28	RD
6	108	19	46	2	33	RD
6	112	3	24	14	33	SO
6	129	18	38	1	36	RD
6	139	1	33	20	51	RD
6	148	0	14	21	75	RD
6	152	1	41	11	14	SO
6	155	0	11	9	4	SO
6	156	7	7	6	50	SO
6	162	1	12	20	45	SO
6	167	7	27	3	45	SO
6	176	16	56	2	26	RD
6	177	11	12	2	19	SO
6	180	22	54	0	34	SO
6	184	10	11	9	47	SO
6	192	0	27	22	64	RD
8	2	0	22	18	75	RD
8	3	0	13	18	84	RD
8	6	0	22	18	75	RD
8	9	9	0	9	97	RD
8	10	0	33	18	64	RD
8	11	0	28	18	69	RD
8	13	0	16	18	81	RD
8	16	0	11	18	0	RD
8	18	0	33	18	0	RD
8	20	18	0	0	22	RD
8	27	10	21	8	76	SO
8	50	4	39	13	43	RD
8	57	1	24	17	73	RD
8	59	0	13	18	82	SO
8	74	0	26	18	69	RD
8	76	0	26	11	59	RD
8	119	17	57	1	25	RD
8	125	14	50	1	34	RD
8	148	0	14	18	78	RD
8	159	7	23	4	46	SO
9	8	9	9	0	97	RD
9	10	0	33	9	73	RD
9	11	0	28	9	78	RD
9	16	0	11	9	9	RD
9	18	0	33	9	9	RD
9	20	9	9	0	22	RD
9	27	6	25	3	81	SO
9	47	0	25	5	25	RD
9	74	0	26	9	78	RD
9	77	1	27	7	34	RD
10	1	0	9	33	73	SO
10	2	0	22	33	60	RD
10	3	0	13	33	69	RD
10	6	0	22	33	60	RD
10	8	0	18	33	64	RD
10	9	0	9	33	73	RD
10	11	28	0	5	82	RD
10	12	5	0	28	82	SO
10	14	0	11	33	71	RD
10	15	0	11	33	0	RD
10	18	33	0	0	18	RD
10	19	33	0	0	22	RD
10	23	2	25	31	57	RD

10	27	3	28	30	54	SO
10	32	0	10	33	72	RD
10	33	1	29	32	53	RD
10	38	0	10	33	70	RD
10	39	2	36	31	45	RD
10	41	0	19	33	61	RD
10	42	1	35	32	44	RD
10	44	0	25	33	57	RD
10	48	0	30	8	16	RD
10	49	8	16	21	5	RD
10	53	0	14	33	68	RD
10	57	0	25	33	57	RD
10	59	8	5	23	77	SO
10	60	6	1	27	76	SO
10	61	8	0	25	79	SO
10	63	0	46	33	35	RD
10	65	0	25	8	10	RD
10	71	1	20	32	58	RD
10	74	0	26	32	55	RD
10	75	2	60	30	22	RD
10	89	1	32	26	36	RD
10	91	25	6	1	4	RD
10	104	3	40	11	19	SO
10	109	0	15	13	29	SO
10	123	33	74	0	8	RD
10	127	3	18	25	47	RD
10	135	28	49	1	12	RD
10	136	0	12	28	57	RD
10	142	3	27	26	40	RD
10	144	17	17	1	10	RD
10	157	1	13	28	56	RD
10	159	3	27	18	32	SO
10	160	3	27	24	43	RD
10	167	1	33	19	29	SO
10	168	1	33	30	42	RD
10	177	7	16	18	3	SO
10	180	8	68	25	9	SO
10	184	3	18	21	35	SO
10	199	7	5	26	77	SO
11	1	0	9	28	78	SO
11	2	0	22	28	65	RD
11	3	0	13	28	74	RD
11	6	0	22	28	65	RD
11	8	0	18	28	69	RD
11	9	0	9	28	78	RD
11	10	28	5	0	82	RD
11	13	0	16	28	71	RD
11	14	0	11	28	76	RD
11	15	0	11	28	5	RD
11	18	28	5	0	18	RD
11	19	28	5	0	22	RD
11	23	1	26	27	61	RD
11	27	3	28	25	59	SO
11	31	27	63	1	23	RD
11	32	0	10	28	77	RD
11	33	1	29	27	58	RD
11	35	0	24	28	63	RD
11	36	1	41	27	46	RD
11	38	0	10	28	75	RD
11	39	1	37	27	49	RD
11	41	0	19	28	66	RD

11	42	1	35	27	49	RD
11	44	0	25	28	62	RD
11	48	0	30	4	20	RD
11	49	4	20	21	5	RD
11	53	0	14	28	73	RD
11	57	0	25	28	62	RD
11	59	8	5	18	82	SO
11	60	6	1	22	81	SO
11	61	8	0	20	84	SO
11	63	0	46	28	40	RD
11	65	0	25	3	15	RD
11	71	1	20	27	63	RD
11	74	0	26	27	60	RD
11	75	0	62	27	25	RD
11	89	1	32	23	39	RD
11	91	22	9	1	4	RD
11	104	3	40	9	21	SO
11	109	0	15	13	29	SO
11	115	0	20	28	63	RD
11	116	1	31	27	52	RD
11	135	23	54	1	12	RD
11	136	0	12	26	59	RD
11	142	3	27	22	44	RD
11	144	14	20	1	10	RD
11	159	3	27	16	34	SO
11	160	3	27	21	46	RD
11	167	1	33	16	32	SO
11	168	1	33	25	47	RD
11	177	5	18	16	5	SO
11	180	4	72	24	10	SO
11	184	2	19	18	38	SO
11	197	3	32	25	55	SO
11	199	7	5	21	82	SO
12	10	5	28	0	82	RD
12	13	5	11	0	99	RD
12	18	5	28	0	18	RD
12	19	5	28	0	22	RD
12	48	0	30	4	20	RD
12	63	0	46	5	63	RD
12	64	0	46	5	13	RD
12	65	0	25	5	13	RD
12	72	0	52	5	54	RD
12	173	0	33	2	5	SO
12	197	5	30	0	80	SO
13	2	0	22	16	77	RD
13	6	0	22	16	77	RD
13	8	0	18	16	81	RD
13	11	0	28	16	71	RD
13	12	5	0	11	99	SO
13	14	11	0	5	99	RD
13	15	11	0	5	28	RD
13	16	11	0	0	18	RD
13	17	11	0	0	22	RD
13	18	5	28	0	18	RD
13	19	5	28	0	22	RD
13	48	2	28	8	16	RD
13	49	8	16	3	23	RD
13	57	0	25	16	74	RD
13	65	0	25	13	5	RD
13	74	0	26	15	72	RD
13	86	13	63	2	36	RD

13	90	3	2	9	93	SO
13	143	3	2	10	90	SO
13	173	2	31	5	2	SO
13	188	1	31	15	66	RD
13	197	11	24	5	75	SO
14	2	0	22	11	82	RD
14	6	0	22	11	82	RD
14	10	0	33	11	71	RD
14	11	0	28	11	76	RD
14	13	11	5	0	99	RD
14	15	11	0	0	33	RD
14	16	11	0	0	18	RD
14	17	11	0	0	22	RD
14	57	0	25	11	79	RD
14	64	0	46	8	10	RD
14	65	0	25	8	10	RD
14	74	0	26	10	77	RD
14	90	3	2	6	96	SO
14	103	2	41	2	4	SO
14	152	1	41	4	21	SO
14	173	2	31	3	4	SO
14	184	4	17	2	54	SO
14	197	6	29	5	75	SO
15	10	0	33	11	0	RD
15	11	0	28	11	5	RD
15	13	11	5	0	28	RD
15	14	11	0	0	33	RD
15	36	3	1	8	32	RD
15	59	0	8	11	23	SO
15	61	0	8	11	25	SO
15	71	3	1	8	32	RD
15	88	2	0	8	33	SO
15	90	3	1	6	28	SO
15	103	2	16	2	1	SO
15	104	2	3	0	11	SO
15	116	3	1	8	32	RD
15	135	4	28	2	1	RD
15	156	2	1	2	12	SO
15	159	3	3	2	18	SO
15	167	3	1	4	19	SO
15	174	2	1	4	29	RD
15	177	2	7	0	18	SO
15	180	5	8	4	25	SO
15	184	4	3	2	21	SO
15	197	6	8	5	25	SO
16	8	0	18	11	0	RD
16	9	0	9	11	9	RD
16	13	11	0	0	18	RD
16	14	11	0	0	18	RD
16	27	2	10	9	8	SO
16	90	3	1	6	15	SO
16	143	2	0	6	12	SO
17	6	0	22	11	0	RD
17	13	11	0	0	22	RD
17	14	11	0	0	22	RD
17	49	4	8	3	0	RD
17	59	0	5	11	17	SO
17	64	0	16	8	0	RD
17	65	0	15	8	0	RD
17	71	3	0	8	22	RD
17	90	3	0	6	22	SO

17	180	5	22	4	0	SO
18	8	0	18	33	0	RD
18	9	0	9	33	9	RD
18	10	33	0	0	18	RD
18	11	28	0	5	18	RD
18	12	5	0	28	18	SO
18	13	5	0	28	18	RD
18	27	3	10	30	8	SO
18	32	0	3	33	15	RD
18	33	1	5	32	13	RD
18	38	0	2	33	15	RD
18	39	2	6	31	12	RD
18	48	0	7	8	4	RD
18	52	22	13	1	4	RD
18	59	8	0	23	18	SO
18	61	8	0	25	15	SO
18	135	28	7	1	4	RD
18	148	14	0	17	18	RD
18	159	3	7	18	4	SO
18	167	1	4	19	11	SO
18	180	8	10	25	5	SO
18	184	3	4	21	5	SO
18	199	7	0	26	18	SO
19	6	0	22	33	0	RD
19	10	33	0	0	22	RD
19	11	28	0	5	22	RD
19	12	5	0	28	22	SO
19	13	5	0	28	22	RD
19	31	32	18	1	4	RD
19	36	1	7	32	15	RD
19	39	2	6	31	16	RD
19	42	1	5	32	16	RD
19	48	0	14	8	8	RD
19	49	8	8	21	0	RD
19	60	6	0	27	22	SO
19	61	8	0	25	22	SO
19	64	0	16	8	0	RD
19	65	0	15	8	0	RD
19	85	5	9	28	13	RD
19	88	0	2	33	13	SO
19	101	4	9	20	10	RD
19	104	3	13	11	2	SO
19	108	15	19	12	2	RD
19	112	8	3	8	14	SO
19	116	1	6	32	16	RD
19	135	28	15	1	4	RD
19	148	14	0	17	21	RD
19	152	5	1	3	11	SO
19	153	5	1	7	14	RD
19	156	1	7	12	6	SO
19	159	3	6	18	8	SO
19	160	3	6	24	14	RD
19	162	6	1	6	20	SO
19	163	6	1	21	21	RD
19	167	1	7	19	3	SO
19	168	1	7	30	13	RD
19	177	7	11	18	2	SO
19	180	8	22	25	0	SO
19	184	3	10	21	9	SO
19	195	13	14	20	8	SO
19	199	7	1	26	21	SO

20	6	0	22	18	0	RD
20	8	18	0	0	22	RD
20	9	9	0	9	22	RD
20	27	10	2	8	20	SO
20	32	3	0	15	22	RD
20	36	2	7	16	15	RD
20	59	0	5	18	17	SO
20	64	1	16	2	0	RD
20	71	4	0	12	22	RD
20	92	6	1	12	21	RD
20	104	5	13	6	2	SO
20	152	7	1	6	11	SO
20	153	7	1	9	14	RD
20	162	5	1	8	20	SO
20	167	4	7	11	3	SO
20	180	10	22	5	0	SO
21	12	3	2	29	81	SO
21	22	32	36	0	47	RD
21	23	14	13	18	70	RD
21	25	1	16	31	67	RD
21	27	4	27	28	56	SO
21	48	6	24	11	13	RD
21	112	12	15	7	40	SO
21	143	3	2	25	75	SO
21	149	3	1	0	4	SO
21	155	0	11	5	8	SO
21	195	12	46	20	37	SO
22	27	11	20	57	27	SO
22	60	7	0	60	43	SO
22	112	23	4	19	28	SO
22	149	4	0	1	3	SO
22	155	3	8	8	5	SO
22	156	4	10	33	23	SO
22	162	10	3	32	33	SO
22	165	7	1	60	46	SO
22	180	38	38	26	8	SO
23	11	1	27	26	61	RD
23	24	27	34	0	54	RD
23	32	10	0	17	88	RD
23	38	9	1	17	86	RD
23	49	6	18	0	26	RD
23	59	0	13	27	73	SO
23	86	24	52	3	35	RD
23	112	11	16	6	41	SO
23	149	4	0	0	4	SO
23	177	6	17	1	20	SO
23	180	21	55	4	30	SO
23	199	0	12	27	76	SO
24	1	8	1	53	53	SO
24	27	23	8	38	46	SO
24	59	4	9	57	43	SO
24	60	1	6	55	48	SO
24	103	22	21	6	0	SO
24	112	19	8	20	27	SO
24	149	4	0	0	4	SO
24	151	3	2	4	18	SO
24	156	3	11	36	20	SO
24	159	19	11	20	30	SO
24	177	13	10	3	18	SO
24	180	44	32	13	21	SO
24	191	1	7	4	1	SO

25	27	0	31	17	67	SO
25	104	12	31	3	27	SO
25	159	2	28	10	40	SO
25	167	11	23	2	46	SO
25	180	17	59	0	34	SO
26	1	8	1	40	66	SO
26	11	3	25	45	42	RD
26	12	0	5	48	62	SO
26	21	5	27	43	40	RD
26	23	17	10	31	57	RD
26	25	17	0	31	67	RD
26	27	31	0	17	67	SO
26	32	7	3	41	64	RD
26	33	20	10	28	57	RD
26	35	14	10	34	57	RD
26	41	11	8	36	58	RD
26	42	21	15	26	50	RD
26	44	16	9	32	58	RD
26	46	44	40	3	23	RD
26	48	19	11	6	18	RD
26	59	1	12	45	55	SO
26	60	0	7	43	60	SO
26	61	1	7	44	60	SO
26	64	26	20	2	16	RD
26	65	13	12	2	16	RD
26	86	37	39	11	27	RD
26	100	34	30	6	16	RD
26	109	9	6	14	28	SO
26	112	9	18	25	22	SO
26	115	11	9	33	58	RD
26	152	26	16	8	17	SO
26	154	30	25	8	17	RD
26	155	7	4	3	10	SO
26	156	4	10	31	25	SO
26	180	40	36	5	29	SO
27	1	7	2	24	82	SO
27	9	6	3	25	81	RD
27	11	3	25	28	59	RD
27	21	4	28	27	56	RD
27	26	31	17	0	67	RD
27	46	27	57	3	23	RD
27	48	14	16	3	21	RD
27	52	20	62	9	7	RD
27	59	0	13	29	71	SO
27	95	17	78	13	3	RD
27	96	24	43	4	19	RD
27	104	9	34	13	17	SO
27	112	5	22	17	30	SO
27	152	17	25	4	21	SO
27	154	20	35	4	21	RD
27	155	3	8	0	13	SO
27	159	14	16	10	40	SO
27	165	0	8	31	75	SO
27	180	23	53	5	29	SO
28	1	1	7	61	41	SO
28	12	5	0	57	48	SO
28	27	12	18	50	30	SO
28	33	10	19	52	29	RD
28	35	6	17	56	31	RD
28	41	3	15	57	33	RD
28	44	4	20	58	28	RD

28	47	4	20	20	10	RD
28	53	3	11	59	37	RD
28	60	6	1	53	46	SO
28	64	19	27	15	3	RD
28	98	6	15	56	33	RD
28	100	28	32	18	4	RD
28	109	6	9	27	14	SO
28	126	52	27	4	12	RD
28	151	0	5	12	7	SO
28	152	14	26	21	4	SO
28	154	23	29	21	4	RD
28	155	2	9	11	2	SO
28	156	11	3	26	28	SO
28	157	11	3	42	39	RD
28	184	15	5	29	27	SO
28	191	7	1	1	3	SO
28	196	31	14	31	34	SO
29	3	12	0	44	49	RD
29	4	5	0	51	49	SO
29	11	1	27	55	22	RD
29	23	18	7	38	42	RD
29	32	8	1	48	48	RD
29	33	25	4	31	45	RD
29	38	9	1	46	48	RD
29	41	19	0	35	49	RD
29	47	25	0	17	7	RD
29	48	17	7	11	12	RD
29	49	11	12	0	26	RD
29	57	19	2	37	47	RD
29	59	3	10	53	37	SO
29	60	0	7	55	42	SO
29	61	0	8	55	41	SO
29	64	39	3	9	9	RD
29	65	19	2	9	9	RD
29	71	15	4	38	45	RD
29	95	46	44	9	3	RD
29	98	18	3	38	46	RD
29	104	32	7	9	16	SO
29	109	14	1	16	24	SO
29	110	14	1	33	43	RD
29	115	19	0	35	48	RD
29	123	48	49	8	0	RD
29	126	32	44	16	0	RD
29	136	12	0	34	42	RD
29	152	30	7	13	12	SO
29	155	11	0	8	5	SO
29	162	3	8	44	14	SO
29	163	3	8	49	30	RD
29	167	28	2	13	29	SO
29	174	23	2	26	39	RD
29	177	14	8	2	19	SO
29	180	53	14	2	32	SO
30	3	0	13	56	44	RD
30	4	0	5	56	52	SO
30	25	3	14	53	43	RD
30	27	11	19	45	38	SO
30	31	56	33	0	24	RD
30	32	2	8	54	49	RD
30	33	6	24	50	33	RD
30	38	1	9	54	47	RD
30	44	1	24	55	33	RD

30	46	27	57	25	0	RD
30	59	10	3	44	54	SO
30	60	7	0	45	57	SO
30	61	7	1	47	56	SO
30	64	4	42	14	3	RD
30	65	2	23	14	3	RD
30	71	7	14	47	41	RD
30	98	2	19	54	38	RD
30	104	7	36	20	9	SO
30	109	2	13	25	16	SO
30	110	2	13	46	33	RD
30	119	40	32	1	25	RD
30	120	41	47	0	10	RD
30	123	56	49	0	8	RD
30	127	5	16	36	35	RD
30	143	4	0	44	55	SO
30	144	23	10	2	9	RD
30	152	11	30	12	13	SO
30	162	9	3	17	48	SO
30	163	9	3	35	51	RD
30	167	2	32	33	13	SO
30	177	9	14	20	1	SO
30	180	21	54	31	2	SO
30	184	13	8	22	34	SO
31	4	1	4	89	20	SO
31	104	21	22	29	1	SO
31	109	6	9	37	4	SO
31	152	24	18	23	2	SO
31	155	2	9	11	2	SO
31	156	13	1	36	20	SO
31	159	27	2	31	19	SO
31	162	12	1	42	23	SO
31	165	3	5	86	19	SO
31	167	13	21	46	1	SO
31	173	16	17	6	1	SO
31	180	52	24	33	0	SO
31	184	20	1	36	20	SO
32	109	2	13	0	42	SO
32	112	4	23	1	46	SO
32	152	8	34	0	25	SO
32	167	7	27	2	46	SO
32	180	10	66	0	34	SO
33	1	6	3	24	82	SO
33	3	9	4	21	81	RD
33	11	1	27	29	58	RD
33	25	9	8	21	77	RD
33	32	10	0	20	85	RD
33	34	30	32	0	53	RD
33	47	18	7	7	23	RD
33	49	4	20	0	26	RD
33	53	9	5	21	80	RD
33	55	29	49	1	34	RD
33	61	0	8	30	74	SO
33	64	22	24	3	15	RD
33	71	13	8	15	75	RD
33	86	29	47	1	37	RD
33	98	12	10	18	75	RD
33	100	26	38	3	19	RD
33	109	8	7	8	34	SO
33	112	11	16	8	39	SO
33	149	3	1	0	4	SO

33	152	23	19	3	22	SO
33	154	26	29	3	22	RD
33	156	0	14	24	32	SO
33	165	6	2	24	82	SO
33	167	17	17	10	38	SO
33	180	28	48	0	34	SO
33	196	7	39	23	46	SO
34	1	8	1	54	52	SO
34	27	24	7	38	46	SO
34	59	3	10	59	41	SO
34	60	0	7	58	45	SO
34	61	2	6	57	47	SO
34	104	30	13	14	16	SO
34	109	13	2	19	23	SO
34	149	4	0	0	4	SO
34	152	33	9	12	13	SO
34	156	4	10	40	16	SO
34	167	27	7	22	26	SO
34	173	24	9	2	5	SO
34	177	12	11	2	19	SO
34	180	50	26	8	26	SO
35	4	5	0	19	91	SO
35	59	0	13	24	76	SO
35	104	21	22	3	27	SO
35	109	10	5	1	41	SO
35	152	21	21	1	24	SO
35	155	11	0	1	12	SO
35	156	0	14	23	33	SO
35	159	3	27	19	31	SO
35	162	1	12	22	43	SO
35	165	5	3	19	87	SO
35	167	19	15	3	45	SO
35	173	17	16	1	6	SO
35	180	24	52	0	34	SO
35	184	0	21	19	37	SO
36	1	8	1	34	72	SO
36	3	12	1	30	72	RD
36	4	5	0	37	73	SO
36	11	1	27	41	46	RD
36	25	12	5	30	68	RD
36	27	16	15	26	58	SO
36	32	8	2	34	71	RD
36	35	24	0	18	73	RD
36	37	42	22	0	51	RD
36	38	9	1	32	71	RD
36	47	25	0	14	16	RD
36	48	14	16	2	22	RD
36	59	1	12	41	59	SO
36	60	0	7	42	61	SO
36	61	0	8	42	62	SO
36	64	34	12	0	18	RD
36	65	13	12	0	18	RD
36	71	13	8	27	63	RD
36	86	39	37	3	35	RD
36	88	0	5	40	55	SO
36	98	18	4	24	69	RD
36	100	35	29	3	19	RD
36	104	29	14	9	21	SO
36	106	30	22	9	21	RD
36	108	32	33	9	26	RD
36	109	13	2	9	33	SO

36	110	13	2	21	59	RD
36	112	7	20	22	25	SO
36	115	18	2	22	69	RD
36	118	12	28	30	30	RD
36	124	2	12	32	53	RD
36	127	13	8	26	46	RD
36	129	28	28	11	26	RD
36	143	0	5	42	58	SO
36	144	10	24	7	4	RD
36	152	29	13	5	20	SO
36	154	32	23	5	20	RD
36	155	11	0	4	9	SO
36	156	3	11	34	22	SO
36	157	3	11	38	46	RD
36	159	10	20	26	24	SO
36	162	2	11	36	29	SO
36	163	2	11	38	49	RD
36	165	5	3	37	69	SO
36	167	28	6	10	38	SO
36	173	23	10	2	5	SO
36	177	6	17	0	21	SO
36	180	42	34	0	34	SO
36	184	1	20	27	29	SO
36	197	9	26	33	47	SO
37	1	9	0	55	51	SO
37	4	5	0	59	51	SO
37	59	4	9	60	40	SO
37	60	0	7	64	39	SO
37	61	0	8	64	40	SO
37	104	37	6	14	16	SO
37	109	14	1	19	23	SO
37	112	11	16	33	14	SO
37	143	1	4	63	37	SO
37	152	34	8	14	11	SO
37	162	3	10	52	13	SO
37	167	33	1	16	32	SO
37	173	24	9	2	5	SO
37	177	14	9	4	17	SO
37	180	61	15	2	32	SO
38	104	9	34	1	27	SO
38	109	3	12	0	42	SO
38	112	4	23	1	45	SO
38	152	8	32	0	25	SO
38	167	8	26	1	45	SO
38	180	10	64	0	34	SO
38	184	0	21	8	48	SO
39	1	6	2	32	74	SO
39	3	9	4	29	72	RD
39	4	4	1	34	75	SO
39	11	1	27	37	49	RD
39	14	1	10	37	66	RD
39	31	20	70	18	5	RD
39	38	10	0	27	76	RD
39	40	38	32	0	39	RD
39	41	17	1	20	74	RD
39	47	21	3	11	19	RD
39	49	5	19	0	26	RD
39	53	11	2	27	74	RD
39	55	37	40	1	34	RD
39	57	13	12	25	64	RD
39	59	1	12	37	62	SO

39	60	0	7	37	65	SO
39	61	0	8	38	65	SO
39	64	29	16	4	14	RD
39	65	13	12	4	14	RD
39	71	14	7	23	67	RD
39	86	37	38	1	37	RD
39	98	15	7	23	69	RD
39	100	33	30	4	18	RD
39	104	24	19	8	21	SO
39	109	10	5	9	33	SO
39	110	10	5	21	59	RD
39	123	32	74	6	2	RD
39	126	19	63	13	3	RD
39	136	8	4	26	58	RD
39	149	4	0	0	4	SO
39	152	27	14	5	20	SO
39	154	30	24	5	20	RD
39	156	1	13	29	26	SO
39	157	1	13	36	47	RD
39	165	6	2	32	73	SO
39	167	22	12	11	36	SO
39	173	17	16	1	6	SO
39	177	8	15	1	20	SO
39	180	36	39	0	34	SO
39	196	9	36	29	40	SO
39	198	4	19	34	57	SO
40	27	24	5	47	34	SO
40	59	5	8	66	29	SO
40	60	1	6	67	33	SO
40	61	3	5	66	34	SO
40	104	33	8	14	13	SO
40	109	14	1	23	19	SO
40	149	4	0	1	3	SO
40	152	34	6	16	9	SO
40	167	29	3	25	20	SO
40	177	14	9	3	17	SO
40	180	57	15	10	23	SO
41	4	4	1	15	93	SO
41	11	0	28	19	66	RD
41	32	7	3	12	91	RD
41	38	8	2	10	91	RD
41	42	19	17	0	76	RD
41	43	19	34	0	57	RD
41	46	19	63	0	26	RD
41	47	18	7	1	27	RD
41	51	16	43	3	33	RD
41	55	19	57	0	35	RD
41	59	0	13	19	79	SO
41	64	18	27	0	18	RD
41	65	4	20	0	18	RD
41	69	14	49	5	44	RD
41	76	18	8	0	69	RD
41	86	19	55	0	38	RD
41	100	19	44	0	21	RD
41	104	16	25	3	27	SO
41	106	16	34	3	27	RD
41	108	16	47	3	32	RD
41	109	7	8	1	41	SO
41	118	3	37	16	42	RD
41	120	10	78	9	1	RD
41	123	13	92	6	2	RD

41	136	9	2	10	74	RD
41	152	17	24	0	24	SO
41	155	8	2	0	12	SO
41	156	0	14	18	36	SO
41	159	3	27	14	34	SO
41	160	3	27	16	49	RD
41	165	5	3	14	90	SO
41	167	15	18	3	45	SO
41	180	19	55	0	34	SO
41	184	0	21	15	40	SO
42	1	8	1	28	75	SO
42	4	5	0	31	76	SO
42	59	1	12	35	62	SO
42	60	0	7	36	64	SO
42	61	0	8	36	65	SO
42	104	24	17	8	22	SO
42	109	12	3	6	35	SO
42	149	3	1	0	4	SO
42	152	28	13	3	21	SO
42	155	10	0	2	10	SO
42	156	2	12	28	26	SO
42	162	1	12	32	31	SO
42	165	6	2	30	73	SO
42	167	24	9	8	39	SO
42	173	19	13	1	6	SO
42	177	6	16	0	21	SO
42	180	36	38	0	33	SO
42	184	2	19	24	31	SO
43	1	8	1	47	56	SO
43	3	13	0	42	57	RD
43	4	5	0	50	57	SO
43	23	21	6	34	51	RD
43	25	12	4	43	53	RD
43	27	20	11	35	46	SO
43	35	24	0	31	57	RD
43	41	19	0	34	57	RD
43	42	36	0	17	57	RD
43	47	25	0	20	9	RD
43	48	20	9	9	15	RD
43	49	9	15	0	25	RD
43	57	18	6	37	51	RD
43	58	4	13	51	44	RD
43	59	3	10	52	45	SO
43	60	0	7	54	46	SO
43	61	0	8	55	46	SO
43	65	18	6	4	13	RD
43	71	16	5	37	50	RD
43	90	0	4	54	46	SO
43	95	42	50	12	4	RD
43	98	19	2	36	55	RD
43	100	42	20	8	13	RD
43	104	34	8	11	19	SO
43	108	40	24	11	22	RD
43	109	14	1	13	27	SO
43	110	14	1	31	47	RD
43	136	12	0	37	47	RD
43	144	12	22	9	2	RD
43	149	4	0	0	3	SO
43	152	34	8	9	15	SO
43	154	38	16	9	15	RD
43	155	11	0	6	6	SO
43	156	3	10	40	16	SO
43	157	3	10	47	36	RD
43	162	4	9	43	20	SO
43	163	4	9	48	36	RD
43	167	32	1	13	34	SO
43	173	25	8	2	4	SO
43	177	11	10	0	21	SO
43	180	52	22	1	32	SO
43	184	7	14	32	23	SO
44	1	6	3	19	87	SO
44	4	5	0	20	90	SO
44	59	0	13	25	75	SO
44	104	19	24	6	24	SO
44	109	12	3	1	41	SO
44	152	24	18	0	25	SO
44	155	10	1	0	13	SO
44	156	0	14	23	33	SO
44	162	1	12	23	42	SO
44	165	5	3	20	86	SO
44	167	18	16	6	42	SO
44	180	25	51	0	34	SO
44	184	0	21	18	38	SO
45	1	9	0	46	54	SO
45	3	13	0	42	54	RD
45	4	5	0	50	54	SO
45	11	0	27	55	27	RD
45	12	0	5	55	49	SO
45	15	2	8	0	32	RD
45	18	0	32	8	6	RD
45	19	0	32	14	8	RD
45	23	20	6	35	48	RD
45	25	13	3	42	51	RD
45	27	22	6	33	48	SO
45	31	32	52	23	1	RD
45	32	8	1	47	53	RD
45	33	25	4	30	50	RD
45	38	9	1	45	53	RD
45	44	25	0	30	54	RD
45	46	55	25	0	26	RD
45	48	30	0	0	24	RD
45	53	14	0	41	54	RD
45	57	22	3	33	51	RD
45	58	1	17	54	37	RD
45	59	0	12	55	40	SO
45	60	0	7	53	47	SO
45	61	0	8	53	46	SO
45	65	22	3	1	17	RD
45	71	15	5	37	49	RD
45	88	0	5	45	44	SO
45	98	18	3	37	51	RD
45	100	43	16	7	14	RD
45	104	34	9	12	13	SO
45	106	35	17	12	13	RD
45	109	14	1	11	29	SO
45	110	14	1	29	49	RD
45	112	11	16	29	15	SO
45	123	47	54	8	0	RD
45	124	3	10	41	42	RD
45	126	28	52	16	0	RD
45	143	0	5	51	43	SO
45	149	4	0	1	2	SO
45	152	32	5	9	15	SO
45	155	11	0	6	6	SO
45	156	4	9	37	15	SO
45	157	4	9	47	32	RD
45	162	3	8	47	15	SO
45	163	3	8	50	33	RD
45	167	30	4	16	27	SO
45	173	22	11	1	5	SO
45	174	25	3	26	40	RD
45	177	11	11	1	19	SO
45	180	51	19	2	32	SO
45	184	7	14	32	24	SO
46	1	9	0	75	26	SO
46	27	27	3	57	23	SO
46	60	0	6	80	20	SO
46	61	2	6	80	20	SO
46	90	2	3	78	20	SO
46	104	41	2	21	5	SO
46	109	15	0	28	14	SO
46	143	2	3	75	20	SO
46	152	38	1	20	5	SO
46	167	34	0	28	17	SO
46	177	19	3	6	14	SO
46	180	72	0	8	25	SO
47	4	5	0	20	30	SO
47	109	12	2	1	10	SO
47	152	24	8	0	9	SO
47	155	10	1	0	6	SO
47	156	0	4	23	14	SO
47	165	5	1	20	29	SO
47	184	0	7	18	14	SO
48	12	0	4	30	20	SO
48	27	14	3	16	21	SO
48	59	0	7	30	17	SO
48	112	5	9	15	8	SO
48	180	26	16	2	6	SO
49	12	4	0	20	26	SO
49	60	0	6	24	20	SO
49	159	7	5	6	15	SO
49	167	4	0	9	17	SO
49	177	7	3	4	14	SO
49	180	16	0	6	25	SO
49	184	10	3	7	15	SO
50	1	0	9	43	47	SO
50	3	9	3	34	53	RD
50	4	5	0	38	56	SO
50	12	0	5	43	51	SO
50	14	1	10	42	46	RD
50	27	7	21	36	35	SO
50	31	24	50	19	5	RD
50	35	18	6	25	50	RD
50	41	15	4	28	50	RD
50	44	18	7	25	49	RD
50	51	43	17	0	37	RD
50	52	43	38	0	16	RD
50	53	13	1	30	55	RD
50	64	27	15	2	15	RD
50	65	13	9	2	15	RD
50	74	18	7	25	47	RD
50	88	0	4	42	41	SO
50	90	0	5	42	44	SO
50	95	43	36	0	16	RD
50	96	37	27	0	20	RD
50	104	33	7	2	23	SO
50	109	11	4	9	25	SO
50	110	11	4	28	38	RD
50	115	17	3	26	49	RD
50	120	27	55	10	0	RD
50	136	11	1	28	41	RD
50	143	0	5	43	41	SO
50	144	5	23	10	0	RD
50	145	18	9	10	0	RD
50	149	4	0	1	2	SO
50	152	22	20	5	15	SO
50	154	26	26	5	15	RD
50	155	10	1	3	7	SO
50	159	1	25	30	10	SO
50	162	2	10	33	26	SO
50	165	7	1	36	54	SO
50	167	29	4	3	37	SO
50	173	24	6	1	6	SO
50	184	1	15	40	6	SO
50	189	7	0	36	56	RD
51	1	0	9	60	28	SO
51	27	10	18	50	19	SO
51	59	8	1	52	36	SO
51	90	0	5	56	29	SO
51	104	39	1	3	22	SO
51	143	1	4	58	25	SO
51	159	4	22	37	3	SO
51	162	3	9	42	17	SO
51	165	8	0	51	37	SO
51	167	33	0	5	35	SO
51	173	30	0	5	2	SO
51	177	19	1	7	3	SO
51	184	7	7	45	1	SO
52	1	3	6	79	10	SO
52	27	20	9	62	7	SO
52	90	2	3	73	13	SO
52	104	41	0	15	10	SO
52	112	23	2	29	10	SO
52	152	32	10	19	1	SO
52	159	12	14	41	0	SO
52	167	34	0	24	16	SO
52	173	31	0	5	2	SO
52	184	12	2	47	0	SO
53	4	5	0	9	101	SO
53	27	1	30	13	71	SO
53	104	13	30	1	29	SO
53	109	9	6	0	42	SO
53	152	14	28	0	25	SO
53	155	7	4	0	13	SO
53	156	0	14	13	43	SO
53	159	1	29	11	39	SO
53	162	0	13	14	51	SO
53	167	13	21	1	47	SO
53	180	14	62	0	34	SO
53	184	0	21	13	43	SO
54	1	7	2	30	74	SO
54	4	5	0	32	76	SO
54	15	2	9	0	33	RD
54	18	0	33	2	14	RD

54	19	0	33	6	16	RD
54	21	6	26	31	50	RD
54	25	9	8	28	68	RD
54	32	8	2	29	74	RD
54	38	9	1	27	74	RD
54	40	32	39	5	34	RD
54	46	36	47	0	26	RD
54	47	25	0	11	19	RD
54	48	11	19	0	24	RD
54	51	27	33	9	26	RD
54	53	14	0	23	76	RD
54	55	37	41	0	35	RD
54	59	0	13	37	61	SO
54	60	0	7	37	66	SO
54	61	0	8	37	66	SO
54	64	32	14	0	18	RD
54	65	13	12	0	18	RD
54	71	12	9	23	66	RD
54	98	16	6	21	70	RD
54	100	33	29	2	20	RD
54	104	27	16	8	20	SO
54	106	27	25	8	20	RD
54	109	13	2	5	37	SO
54	110	13	2	14	66	RD
54	112	6	21	22	23	SO
54	115	18	2	19	70	RD
54	124	1	13	27	58	RD
54	127	11	10	23	47	RD
54	152	27	13	4	21	SO
54	154	30	23	4	21	RD
54	155	11	0	3	10	SO
54	156	3	11	28	26	SO
54	162	2	10	33	32	SO
54	165	5	3	32	72	SO
54	167	25	9	9	37	SO
54	173	19	14	1	6	SO
54	177	6	17	0	21	SO
54	180	37	37	0	34	SO
54	184	2	19	25	31	SO
55	1	9	0	69	35	SO
55	27	24	5	54	30	SO
55	59	6	7	72	26	SO
55	60	0	7	75	28	SO
55	61	1	7	75	28	SO
55	90	2	3	75	25	SO
55	104	40	3	19	9	SO
55	109	15	0	24	18	SO
55	112	16	11	38	7	SO
55	143	1	4	72	26	SO
55	152	37	3	19	6	SO
55	167	34	0	24	22	SO
55	177	17	6	5	16	SO
55	180	71	3	4	30	SO
56	1	9	0	70	31	SO
56	3	13	0	66	31	RD
56	11	3	25	76	6	RD
56	14	11	0	68	31	RD
56	23	24	1	55	30	RD
56	27	24	3	55	28	SO
56	31	55	30	23	1	RD
56	44	25	0	54	31	RD

56	48	28	0	15	9	RD
56	49	15	9	6	20	RD
56	53	14	0	65	31	RD
56	59	0	13	79	16	SO
56	60	0	7	79	24	SO
56	61	0	8	79	23	SO
56	71	19	1	58	30	RD
56	98	20	2	59	29	RD
56	104	38	5	17	9	SO
56	109	15	0	21	19	SO
56	110	15	0	50	28	RD
56	123	71	31	8	0	RD
56	136	12	0	53	28	RD
56	152	35	4	17	7	SO
56	155	11	0	8	5	SO
56	167	33	1	27	17	SO
56	168	33	1	41	27	RD
56	174	27	1	43	25	RD
56	177	16	7	8	13	SO
56	180	63	9	12	22	SO
56	197	29	3	50	28	SO
56	199	6	6	73	25	SO
57	11	0	28	25	62	RD
57	17	0	11	15	7	RD
57	19	0	33	15	7	RD
57	49	3	21	0	26	RD
57	56	25	54	0	31	RD
57	58	0	18	25	72	RD
57	59	0	13	25	75	SO
57	62	0	5	25	0	RD
57	63	25	21	0	68	RD
57	64	25	21	0	18	RD
57	65	25	0	0	18	RD
57	66	25	32	0	49	RD
57	77	16	12	0	41	RD
57	104	16	27	4	26	SO
57	106	18	34	4	26	RD
57	108	21	44	4	31	RD
57	112	3	24	14	33	SO
57	167	11	23	5	43	SO
57	177	10	13	2	19	SO
57	180	25	51	0	34	SO
58	3	0	13	18	84	RD
58	12	5	0	13	97	SO
58	17	8	3	0	22	RD
58	19	8	25	0	22	RD
58	27	2	29	16	68	SO
58	35	0	24	18	73	RD
58	41	0	19	18	76	RD
58	48	1	29	9	15	RD
58	53	0	14	18	83	RD
58	56	18	61	0	31	RD
58	57	0	25	18	72	RD
58	59	0	13	18	82	SO
58	63	0	46	18	50	RD
58	64	0	46	18	0	RD
58	65	0	25	18	0	RD
58	66	18	39	0	49	RD
58	90	3	2	11	91	SO
58	112	9	18	4	43	SO
58	143	3	2	10	90	SO

58	152	1	41	5	20	SO
58	167	1	33	10	38	SO
58	168	1	33	14	58	RD
58	173	2	31	4	3	SO
58	180	6	70	8	26	SO
58	184	6	15	4	52	SO
58	197	13	22	5	75	SO
59	11	8	18	5	82	RD
59	18	8	23	0	18	RD
59	20	0	18	5	17	RD
59	27	0	29	13	71	SO
59	39	1	37	12	62	RD
59	44	0	25	13	75	RD
59	48	0	30	7	17	RD
59	56	0	79	13	16	RD
59	57	0	25	13	75	RD
59	58	0	18	13	82	RD
59	63	0	46	13	53	RD
59	66	0	57	13	36	RD
59	68	2	48	11	51	RD
59	109	0	15	11	31	SO
59	152	2	40	5	20	SO
59	155	0	11	4	9	SO
59	197	1	33	12	67	SO
60	11	6	22	1	81	RD
60	36	0	42	7	61	RD
60	42	0	36	7	64	RD
60	45	0	53	7	47	RD
60	49	0	24	6	20	RD
60	54	0	37	7	66	RD
60	56	0	79	7	24	RD
60	63	0	46	7	56	RD
60	66	0	57	7	42	RD
60	68	0	50	7	53	RD
60	72	0	51	7	50	RD
60	94	0	51	3	33	RD
60	101	0	41	7	46	RD
60	104	0	43	2	24	SO
60	116	0	32	7	68	RD
60	122	7	71	0	32	RD
60	128	0	48	5	36	RD
60	167	0	34	6	38	SO
60	177	0	23	4	17	SO
60	180	1	71	6	28	SO
60	191	4	4	0	5	SO
61	11	8	20	0	84	RD
61	15	0	11	8	25	RD
61	18	8	25	0	15	RD
61	19	8	25	0	22	RD
61	36	0	42	8	62	RD
61	42	0	36	8	65	RD
61	56	0	79	8	23	RD
61	63	0	46	8	57	RD
61	66	0	57	8	41	RD
61	68	0	51	8	52	RD
61	116	0	32	8	69	RD
61	122	8	72	0	32	RD
61	162	4	8	1	62	SO
61	167	0	34	5	41	SO
61	180	1	73	7	27	SO
61	191	2	6	4	1	SO

61	196	1	44	7	60	SO
61	197	0	33	8	71	SO
62	4	5	0	0	25	SO
62	36	5	13	0	12	RD
62	47	5	6	0	16	RD
62	54	5	13	0	12	RD
62	57	0	25	5	0	RD
62	99	5	12	0	10	RD
62	109	4	5	0	8	SO
62	119	0	16	5	9	RD
62	125	0	9	5	11	RD
62	128	5	14	0	9	RD
62	152	5	9	0	7	SO
62	155	5	3	0	6	SO
62	159	0	7	5	11	SO
63	1	7	1	39	67	SO
63	3	13	0	33	68	RD
63	4	5	0	41	68	SO
63	11	0	28	46	40	RD
63	12	0	5	46	63	SO
63	14	0	11	46	57	RD
63	17	0	11	16	6	RD
63	19	0	33	16	6	RD
63	20	1	17	16	6	RD
63	21	8	23	38	45	RD
63	25	10	7	36	61	RD
63	31	25	64	21	3	RD
63	33	22	7	24	61	RD
63	35	22	1	24	67	RD
63	38	7	2	38	65	RD
63	47	23	1	20	10	RD
63	48	20	10	3	21	RD
63	49	3	21	0	26	RD
63	51	30	30	12	24	RD
63	53	13	1	33	67	RD
63	56	46	32	0	31	RD
63	58	0	18	46	50	RD
63	59	0	13	46	53	SO
63	60	0	7	46	56	SO
63	61	0	8	46	57	SO
63	64	46	0	0	18	RD
63	65	25	0	0	18	RD
63	71	12	8	32	58	RD
63	90	0	5	46	55	SO
63	98	17	4	29	64	RD
63	100	37	26	6	16	RD
63	104	31	12	10	19	SO
63	108	36	29	10	24	RD
63	109	12	3	9	33	SO
63	110	12	3	24	56	RD
63	112	8	18	25	22	SO
63	115	19	1	24	66	RD
63	118	13	26	33	27	RD
63	120	36	53	10	0	RD
63	123	38	68	8	0	RD
63	124	1	13	36	48	RD
63	126	21	61	16	0	RD
63	129	31	25	13	23	RD
63	136	11	1	32	52	RD
63	143	0	5	46	53	SO
63	144	10	23	10	1	RD

63	152	28	13	8	17	SO
63	154	33	21	8	17	RD
63	155	10	1	6	7	SO
63	162	2	11	42	23	SO
63	163	2	11	44	43	RD
63	165	6	2	40	65	SO
63	167	26	8	11	36	SO
63	173	18	15	1	6	SO
63	174	21	7	23	48	RD
63	177	11	12	2	19	SO
63	180	46	29	0	34	SO
63	198	5	18	41	50	SO
64	1	7	0	39	18	SO
64	3	13	0	33	18	RD
64	11	0	3	46	15	RD
64	12	0	5	46	13	SO
64	17	0	8	16	0	RD
64	19	0	8	16	0	RD
64	27	16	2	30	16	SO
64	31	25	17	21	0	RD
64	33	22	3	24	15	RD
64	35	22	0	24	18	RD
64	41	18	0	27	18	RD
64	48	20	1	3	9	RD
64	53	13	0	33	18	RD
64	57	25	0	21	18	RD
64	58	0	18	46	0	RD
64	63	46	0	0	18	RD
64	65	25	0	0	18	RD
64	66	33	18	13	0	RD
64	90	0	3	46	11	SO
64	98	17	2	29	16	RD
64	104	31	1	10	3	SO
64	109	12	1	9	8	SO
64	112	8	9	25	4	SO
64	120	36	16	10	0	RD
64	123	38	18	8	0	RD
64	124	1	5	36	11	RD
64	143	0	3	46	10	SO
64	152	28	1	8	5	SO
64	162	2	2	42	2	SO
64	167	26	1	11	10	SO
64	173	18	2	1	4	SO
64	177	11	3	2	6	SO
64	180	46	6	0	8	SO
64	184	7	6	27	4	SO
64	197	12	13	34	5	SO
65	12	0	5	25	13	SO
65	27	9	2	16	16	SO
65	90	0	3	25	11	SO
65	104	16	1	4	3	SO
65	112	3	9	14	4	SO
65	143	0	3	25	10	SO
65	162	2	2	23	2	SO
65	167	11	1	5	10	SO
65	177	10	3	2	6	SO
65	180	25	6	0	8	SO
66	4	5	0	52	49	SO
66	11	3	22	54	27	RD
66	12	5	0	52	49	SO
66	27	17	8	40	41	SO

66	46	49	30	5	18	RD
66	59	0	13	57	36	SO
66	60	0	7	57	42	SO
66	61	0	8	57	41	SO
66	70	21	0	32	48	RD
66	109	11	4	16	23	SO
66	110	11	4	38	39	RD
66	116	21	11	34	37	RD
66	122	34	40	23	9	RD
66	136	9	3	35	42	RD
66	155	9	2	6	7	SO
66	165	1	7	56	41	SO
66	180	42	29	11	20	SO
66	184	15	5	24	29	SO
66	197	24	6	33	43	SO
67	1	9	0	71	29	SO
67	27	27	3	53	26	SO
67	33	26	4	54	25	RD
67	44	24	1	56	28	RD
67	59	5	7	73	22	SO
67	60	3	4	72	25	SO
67	109	14	1	23	15	SO
67	143	1	3	72	25	SO
67	152	38	4	14	9	SO
67	155	11	0	5	7	SO
68	3	10	3	42	59	RD
68	11	4	24	48	38	RD
68	25	12	5	40	57	RD
68	31	33	56	18	6	RD
68	35	17	7	35	55	RD
68	41	13	6	38	55	RD
68	44	18	7	34	55	RD
68	46	46	37	5	21	RD
68	53	12	2	40	60	RD
68	55	44	33	7	28	RD
68	59	2	11	48	51	SO
68	60	0	7	50	53	SO
68	61	0	8	51	52	SO
68	64	35	11	7	11	RD
68	65	21	4	7	11	RD
68	69	52	12	0	50	RD
68	100	36	27	8	14	RD
68	104	27	16	10	19	SO
68	106	31	21	10	19	RD
68	108	38	27	11	23	RD
68	109	13	2	11	31	SO
68	110	13	2	27	53	RD
68	112	8	19	32	15	SO
68	113	8	19	40	40	RD
68	115	15	5	35	55	RD
68	116	24	8	26	52	RD
68	126	29	53	12	4	RD
68	137	20	9	25	42	RD
68	144	10	23	8	3	RD
68	162	3	9	40	25	SO
68	167	22	12	14	33	SO
68	168	22	12	27	44	RD
68	174	17	11	30	41	RD
68	177	13	10	6	15	SO
68	180	45	30	6	28	SO
68	197	21	14	31	48	SO

69	4	5	0	59	50	SO
69	59	2	11	60	39	SO
69	60	0	7	60	43	SO
69	61	0	8	61	42	SO
69	109	15	0	15	27	SO
69	112	8	19	39	8	SO
69	162	5	7	49	16	SO
69	167	25	9	21	26	SO
69	177	15	8	6	15	SO
69	180	54	21	8	26	SO
70	4	4	1	17	83	SO
70	12	3	2	18	82	SO
70	37	17	45	4	39	RD
70	46	19	59	1	21	RD
70	55	19	55	2	28	RD
70	59	0	12	21	72	SO
70	66	21	32	0	48	RD
70	67	21	54	0	29	RD
70	68	21	22	0	62	RD
70	69	21	34	0	50	RD
70	86	19	52	2	32	RD
70	109	8	7	2	36	SO
70	111	14	38	2	36	RD
70	112	3	22	14	26	SO
70	117	15	44	5	38	RD
70	123	15	82	6	2	RD
70	155	6	5	1	11	SO
70	167	11	22	6	38	SO
70	180	19	52	2	28	SO
71	4	3	2	18	88	SO
71	11	1	27	20	63	RD
71	17	3	8	0	22	RD
71	32	6	3	15	87	RD
71	37	15	47	6	43	RD
71	38	5	4	15	86	RD
71	40	20	48	1	38	RD
71	41	10	7	11	81	RD
71	47	12	11	3	26	RD
71	53	8	4	13	86	RD
71	59	0	13	21	75	SO
71	72	21	31	0	59	RD
71	73	21	44	0	46	RD
71	86	20	52	1	37	RD
71	90	3	2	16	82	SO
71	99	16	31	5	30	RD
71	100	18	43	2	19	RD
71	109	9	6	4	37	SO
71	117	14	45	5	44	RD
71	123	16	87	5	3	RD
71	136	7	5	13	69	RD
71	152	14	25	2	22	SO
71	154	16	36	2	22	RD
71	155	6	5	0	13	SO
71	195	6	51	15	39	SO
72	3	11	1	41	58	RD
72	12	0	5	52	54	SO
72	23	16	9	36	50	RD
72	31	33	55	19	3	RD
72	32	8	1	44	58	RD
72	33	22	6	30	53	RD
72	34	36	23	16	36	RD

72	35	17	5	35	54	RD
72	38	8	1	43	58	RD
72	41	15	2	36	56	RD
72	44	17	6	35	53	RD
72	46	49	31	3	23	RD
72	53	10	2	42	57	RD
72	57	22	3	30	56	RD
72	58	3	15	49	44	RD
72	60	0	7	51	50	SO
72	61	1	7	51	50	SO
72	64	35	9	3	15	RD
72	65	22	3	3	15	RD
72	71	21	0	31	59	RD
72	73	52	13	0	46	RD
72	74	18	6	33	52	RD
72	104	29	13	10	18	SO
72	106	31	20	10	18	RD
72	109	12	3	18	23	SO
72	112	9	18	23	21	SO
72	115	17	2	33	56	RD
72	118	16	21	36	23	RD
72	122	27	54	25	5	RD
72	123	45	58	7	1	RD
72	133	5	24	44	28	RD
72	135	34	42	9	3	RD
72	136	9	3	37	45	RD
72	137	19	9	27	39	RD
72	144	9	23	8	3	RD
72	159	19	9	22	27	SO
72	162	4	9	44	18	SO
72	167	23	10	16	30	SO
72	168	23	10	27	43	RD
72	174	17	10	30	40	RD
72	177	14	9	3	18	SO
72	180	47	25	5	29	SO
72	199	3	9	49	50	SO
73	1	8	0	57	46	SO
73	3	12	0	53	46	RD
73	4	5	0	60	46	SO
73	9	1	8	64	38	RD
73	11	3	25	62	21	RD
73	12	0	5	65	41	SO
73	31	44	44	20	2	RD
73	32	8	1	57	45	RD
73	33	25	3	40	43	RD
73	35	21	1	44	45	RD
73	38	8	1	56	45	RD
73	39	32	5	33	41	RD
73	47	23	0	20	9	RD
73	49	12	12	5	21	RD
73	60	0	7	64	37	SO
73	61	1	7	64	37	SO
73	64	42	2	7	11	RD
73	65	24	1	7	11	RD
73	71	21	0	44	46	RD
73	72	52	0	13	46	RD
73	104	32	10	15	13	SO
73	109	15	0	24	17	SO
73	112	13	14	30	14	SO
73	115	18	1	44	45	RD
73	116	29	2	33	44	RD

73	123	57	46	8	0	RD
73	127	16	5	37	32	RD
73	135	39	37	10	2	RD
73	136	11	1	44	38	RD
73	137	23	5	32	34	RD
73	159	25	3	25	24	SO
73	160	25	3	33	33	RD
73	162	4	9	52	10	SO
73	163	4	9	54	29	RD
73	167	26	7	22	24	SO
73	168	26	7	34	36	RD
73	174	20	7	36	34	RD
73	177	15	8	3	18	SO
73	180	57	15	6	28	SO
73	184	15	5	26	29	SO
73	199	4	8	61	38	SO
74	3	13	0	13	87	RD
74	4	5	0	21	87	SO
74	9	0	9	26	78	RD
74	11	0	27	26	60	RD
74	14	0	10	26	77	RD
74	32	6	4	20	83	RD
74	38	7	3	18	83	RD
74	40	25	44	1	38	RD
74	46	26	56	0	26	RD
74	47	22	3	4	25	RD
74	48	4	25	0	23	RD
74	51	19	39	6	31	RD
74	58	0	18	26	69	RD
74	59	0	12	26	73	SO
74	64	26	20	0	18	RD
74	65	9	16	0	18	RD
74	75	26	35	0	52	RD
74	76	26	0	0	69	RD
74	86	26	48	0	38	RD
74	104	20	22	6	24	SO
74	106	20	31	6	24	RD
74	109	11	4	2	39	SO
74	118	6	34	20	38	RD
74	133	2	27	24	51	RD
74	136	10	1	16	69	RD
74	148	0	14	26	68	RD
74	152	23	17	1	24	SO
74	154	25	28	1	24	RD
74	155	10	0	0	13	SO
74	156	1	13	24	30	SO
74	157	1	13	25	57	RD
74	162	1	12	25	38	SO
74	165	5	3	21	83	SO
74	167	19	13	6	42	SO
74	173	17	14	1	6	SO
74	180	26	48	0	34	SO
74	184	0	21	19	36	SO
75	1	8	1	54	51	SO
75	3	13	0	49	52	RD
75	4	5	0	57	52	SO
75	9	1	8	61	44	RD
75	23	20	7	42	45	RD
75	25	15	2	47	50	RD
75	41	19	0	41	52	RD
75	47	25	0	24	6	RD

75	48	24	6	11	12	RD
75	49	11	12	0	26	RD
75	57	25	0	37	52	RD
75	58	3	15	59	37	RD
75	59	4	8	58	42	SO
75	60	0	7	61	41	SO
75	61	0	8	62	41	SO
75	64	46	0	3	15	RD
75	65	25	0	3	15	RD
75	71	16	5	44	45	RD
75	74	26	0	35	52	RD
75	76	26	0	28	41	RD
75	77	28	0	0	41	RD
75	90	0	5	62	39	SO
75	98	18	3	44	49	RD
75	100	44	19	11	11	RD
75	103	17	26	0	6	SO
75	104	39	4	13	17	SO
75	109	14	1	15	26	SO
75	110	14	1	35	44	RD
75	112	11	16	33	14	SO
75	113	11	16	47	33	RD
75	123	54	52	8	0	RD
75	124	3	10	48	37	RD
75	127	17	4	41	31	RD
75	143	0	5	62	37	SO
75	144	14	20	10	1	RD
75	148	0	14	62	33	RD
75	152	33	8	14	11	SO
75	162	5	8	52	12	SO
75	163	5	8	55	31	RD
75	167	32	1	15	33	SO
75	173	24	8	2	5	SO
75	174	26	1	32	40	RD
75	177	16	7	3	18	SO
75	180	61	14	1	33	SO
76	1	4	0	22	70	SO
76	4	5	0	21	70	SO
76	59	0	12	26	56	SO
76	109	11	1	2	33	SO
76	152	23	8	1	21	SO
76	155	10	0	0	13	SO
76	156	1	12	24	22	SO
76	162	1	10	25	30	SO
76	165	5	3	21	66	SO
76	167	19	12	6	28	SO
76	180	26	37	0	29	SO
76	184	0	17	19	35	SO
77	60	0	7	28	32	SO
77	61	0	7	28	32	SO
77	104	16	4	2	13	SO
77	112	4	15	15	10	SO
77	156	9	3	9	12	SO
77	159	6	3	9	20	SO
77	162	3	7	22	7	SO
77	167	10	1	4	24	SO
77	177	13	7	3	17	SO
77	180	27	9	1	28	SO
78	1	0	4	28	3	SO
78	27	4	4	24	3	SO
78	32	1	3	27	4	RD

78	38	1	2	27	4	RD
78	51	18	2	5	5	RD
78	104	16	2	2	5	SO
78	109	1	2	12	1	SO
78	152	3	6	13	0	SO
78	156	9	0	9	6	SO
78	196	15	1	13	6	SO
79	19	26	0	3	18	RD
79	20	9	0	3	18	RD
79	48	7	11	14	7	RD
79	65	1	14	10	0	RD
79	104	14	10	14	1	SO
79	106	19	12	14	1	RD
79	108	24	16	14	1	RD
79	112	12	2	15	11	SO
79	132	35	10	4	5	RD
79	152	12	1	8	10	SO
79	153	12	1	15	13	RD
79	155	2	0	3	8	SO
79	156	5	7	22	4	SO
79	157	5	7	36	10	RD
79	159	4	6	25	6	SO
79	160	4	6	32	11	RD
79	162	8	1	16	17	SO
79	177	8	10	14	2	SO
79	180	22	18	24	0	SO
79	184	8	8	30	7	SO
80	27	6	2	13	22	SO
80	45	7	1	12	21	RD
80	49	8	4	4	13	RD
80	60	1	6	18	16	SO
80	103	4	17	2	0	SO
80	104	9	1	1	12	SO
80	137	5	1	9	20	RD
80	151	2	1	3	14	SO
80	167	7	0	6	17	SO
80	173	10	1	1	3	SO
80	177	4	4	1	13	SO
80	180	10	6	7	17	SO
80	191	0	7	5	0	SO
81	3	6	7	5	33	RD
81	19	1	0	0	20	RD
81	27	1	15	10	25	SO
81	32	4	2	7	38	RD
81	38	5	2	6	37	RD
81	43	11	29	0	10	RD
81	47	10	13	0	19	RD
81	54	10	22	1	18	RD
81	74	10	16	0	24	RD
81	76	10	16	1	20	RD
81	92	0	9	11	30	RD
81	95	10	28	0	11	RD
81	100	11	29	0	9	RD
81	104	10	21	0	12	SO
81	106	10	23	0	12	RD
81	112	4	5	2	25	SO
81	144	0	11	5	3	RD
81	152	10	19	0	12	SO
81	155	4	6	0	9	SO
81	159	1	16	8	16	SO
81	160	1	16	10	22	RD

81	165	4	2	7	38	SO
81	167	10	15	0	13	SO
81	176	11	24	0	12	RD
81	184	0	10	9	17	SO
81	195	3	25	8	15	SO
82	1	8	0	43	55	SO
82	3	13	0	38	55	RD
82	4	5	0	46	55	SO
82	9	0	8	51	47	RD
82	12	0	5	51	50	SO
82	14	0	8	51	47	RD
82	17	0	8	20	2	RD
82	20	0	14	20	2	RD
82	31	29	52	22	2	RD
82	32	6	1	45	54	RD
82	33	23	3	28	52	RD
82	35	22	1	29	54	RD
82	38	7	1	43	54	RD
82	47	23	0	19	10	RD
82	48	19	10	9	13	RD
82	49	9	13	0	23	RD
82	60	0	7	51	44	SO
82	61	0	7	51	45	SO
82	64	44	2	0	15	RD
82	65	24	1	0	15	RD
82	74	26	0	24	54	RD
82	75	48	10	2	45	RD
82	76	26	0	21	48	RD
82	77	19	9	1	39	RD
82	98	18	2	33	53	RD
82	100	40	18	9	11	RD
82	104	31	10	12	16	SO
82	106	33	17	12	16	RD
82	108	37	25	12	17	RD
82	109	12	0	15	24	SO
82	112	9	16	27	18	SO
82	115	18	1	30	53	RD
82	123	43	55	8	0	RD
82	124	2	10	39	39	RD
82	136	11	1	35	42	RD
82	137	21	6	25	37	RD
82	144	11	21	8	3	RD
82	152	29	10	12	11	SO
82	159	17	7	24	24	SO
82	162	2	9	48	14	SO
82	163	2	9	49	35	RD
82	167	25	7	13	28	SO
82	173	18	13	1	5	SO
82	174	20	7	27	38	RD
82	177	13	9	2	19	SO
82	180	51	21	0	29	SO
82	198	6	14	45	41	SO
83	22	8	60	1	46	RD
83	50	1	42	8	48	RD
83	57	0	25	9	81	RD
83	64	0	46	3	15	RD
83	65	0	25	3	15	RD
83	72	9	43	0	59	RD
83	73	9	56	0	46	RD
83	74	0	26	9	78	RD
83	78	0	28	4	3	RD

83	90	3	2	4	98	SO
83	91	0	31	3	2	RD
83	159	6	24	2	48	SO
83	180	4	72	5	29	SO
83	195	2	56	7	50	SO
84	9	8	0	23	24	RD
84	11	2	24	29	0	RD
84	12	5	0	26	24	SO
84	14	8	0	23	24	RD
84	23	7	1	24	23	RD
84	49	10	3	4	19	RD
84	59	1	6	30	16	SO
84	61	0	7	28	17	SO
84	68	11	4	20	20	RD
84	75	10	0	21	24	RD
84	77	9	0	15	24	RD
84	156	5	0	12	8	SO
84	159	7	0	8	16	SO
84	167	7	0	15	13	SO
84	176	15	20	8	2	RD
84	180	19	2	7	22	SO
84	184	8	1	11	18	SO
84	197	17	3	14	21	SO
84	199	1	7	30	17	SO
85	1	9	0	51	54	SO
85	3	13	0	47	54	RD
85	4	5	0	55	54	SO
85	11	1	27	59	27	RD
85	23	22	5	38	49	RD
85	25	12	5	48	49	RD
85	27	21	10	39	44	SO
85	31	38	52	22	2	RD
85	47	25	0	19	11	RD
85	48	19	11	9	15	RD
85	49	9	15	1	24	RD
85	57	18	7	42	47	RD
85	59	1	12	59	40	SO
85	60	1	6	56	46	SO
85	61	0	8	59	44	SO
85	64	38	8	9	8	RD
85	71	16	5	40	49	RD
85	86	60	16	0	38	RD
85	98	18	3	42	51	RD
85	100	45	18	11	11	RD
85	104	31	12	16	14	SO
85	109	14	1	14	27	SO
85	110	14	1	31	48	RD
85	143	0	4	56	44	SO
85	149	4	0	1	3	SO
85	152	37	5	12	13	SO
85	154	41	14	12	13	RD
85	155	11	0	6	7	SO
85	156	6	8	42	14	SO
85	162	5	8	44	21	SO
85	167	29	5	21	26	SO
85	174	24	4	29	43	RD
85	177	11	12	2	19	SO
85	180	55	21	3	30	SO
85	197	24	10	36	44	SO
86	1	9	0	67	38	SO
86	3	13	0	63	38	RD

86	11	3	25	73	13	RD
86	23	24	3	52	35	RD
86	31	53	37	23	1	RD
86	41	19	0	55	38	RD
86	47	25	0	24	6	RD
86	49	18	6	3	22	RD
86	53	14	0	62	38	RD
86	59	6	7	70	29	SO
86	60	1	6	72	30	SO
86	61	1	7	74	29	SO
86	64	45	1	11	6	RD
86	65	25	0	11	6	RD
86	71	20	1	52	37	RD
86	85	60	0	16	38	RD
86	98	20	1	56	37	RD
86	104	36	7	18	12	SO
86	109	15	0	24	17	SO
86	115	20	0	52	38	RD
86	123	68	38	8	0	RD
86	127	17	4	44	28	RD
86	157	12	2	53	31	RD
86	159	25	4	30	20	SO
86	162	6	7	57	8	SO
86	163	6	7	64	22	RD
86	167	31	3	27	20	SO
86	168	31	3	40	31	RD
86	174	25	3	41	31	RD
86	177	16	7	4	17	SO
86	180	67	9	7	26	SO
86	184	18	3	33	23	SO
86	197	27	7	49	31	SO
86	199	5	7	71	31	SO
87	1	0	9	31	49	SO
87	3	0	13	31	45	RD
87	11	22	2	9	56	RD
87	12	3	0	28	58	SO
87	23	2	17	29	41	RD
87	31	30	36	1	21	RD
87	35	2	22	29	36	RD
87	41	0	19	29	39	RD
87	47	0	25	4	15	RD
87	49	6	8	19	4	RD
87	57	2	17	29	41	RD
87	59	5	3	24	55	SO
87	60	6	1	25	54	SO
87	61	6	1	25	56	SO
87	64	2	37	7	6	RD
87	65	2	17	7	6	RD
87	89	0	33	31	23	RD
87	91	31	0	0	5	RD
87	104	6	28	10	17	SO
87	109	0	14	11	16	SO
87	120	17	48	0	10	RD
87	136	1	11	25	40	RD
87	137	1	25	25	26	RD
87	150	0	3	3	1	SO
87	152	4	36	5	11	SO
87	157	1	9	27	42	RD
87	159	0	25	22	22	SO
87	162	6	4	8	44	SO
87	163	6	4	21	48	RD

87	167	3	27	18	22	SO
87	168	3	27	27	29	RD
87	177	4	10	18	1	SO
87	180	9	51	21	6	SO
87	199	8	3	23	55	SO
88	94	0	48	3	30	RD
88	134	0	57	2	32	RD
88	149	0	4	2	1	SO
88	152	0	40	4	13	SO
88	153	0	40	4	23	RD
88	155	0	11	2	5	SO
88	156	3	8	0	53	SO
88	158	3	29	0	53	RD
88	159	2	23	0	49	SO
88	184	4	6	0	55	SO
89	1	0	9	33	53	SO
89	4	5	0	28	62	SO
89	11	1	23	32	39	RD
89	32	7	2	26	60	RD
89	38	8	1	25	59	RD
89	40	29	31	4	28	RD
89	41	16	3	17	57	RD
89	47	19	6	9	11	RD
89	49	4	13	0	25	RD
89	52	32	34	0	15	RD
89	53	13	1	20	61	RD
89	57	13	5	20	57	RD
89	58	1	15	32	47	RD
89	60	0	7	33	51	SO
89	61	0	7	33	53	SO
89	64	27	11	1	15	RD
89	65	13	5	1	15	RD
89	71	12	7	20	53	RD
89	87	0	31	33	23	RD
89	92	0	20	33	41	RD
89	95	32	45	0	14	RD
89	96	32	22	0	19	RD
89	98	16	3	17	59	RD
89	100	28	28	3	12	RD
89	104	28	6	0	27	SO
89	106	27	14	0	27	RD
89	109	12	3	6	25	SO
89	110	12	3	18	44	RD
89	124	1	12	28	39	RD
89	127	11	5	18	41	RD
89	133	4	23	25	34	RD
89	136	11	1	18	50	RD
89	149	4	0	1	2	SO
89	150	3	0	1	3	SO
89	152	21	17	5	13	SO
89	154	25	23	5	13	RD
89	155	10	1	3	4	SO
89	159	3	21	22	22	SO
89	160	3	21	28	28	RD
89	162	1	11	31	20	SO
89	163	1	11	31	39	RD
89	165	5	1	28	60	SO
89	167	27	3	0	44	SO
89	176	33	25	0	25	RD
89	177	10	4	0	19	SO
89	180	33	26	0	31	SO

89	184	2	11	27	20	SO
90	14	3	6	2	96	RD
90	50	0	42	5	44	RD
90	64	0	46	3	11	RD
90	65	0	25	3	11	RD
90	91	0	31	5	0	RD
90	94	0	47	5	33	RD
90	97	5	15	0	16	RD
90	101	0	40	5	49	RD
90	107	0	45	5	54	RD
90	143	3	1	0	96	SO
90	152	0	42	2	22	SO
90	159	5	24	0	49	SO
90	161	5	40	0	49	RD
90	167	0	34	5	40	SO
90	173	0	31	2	3	SO
90	175	0	47	4	43	RD
90	180	0	74	5	24	SO
90	197	3	28	2	74	SO
91	11	22	1	9	4	RD
91	14	1	3	30	2	RD
91	50	10	0	11	5	RD
91	87	31	0	0	5	RD
91	90	0	5	31	0	SO
91	103	16	3	1	2	SO
91	133	22	0	8	5	RD
91	143	0	3	29	0	SO
91	159	0	5	22	0	SO
91	173	6	0	1	2	SO
91	197	7	3	24	2	SO
92	1	9	0	11	93	SO
92	3	0	12	20	81	RD
92	27	12	18	8	75	SO
92	49	3	21	0	25	RD
92	59	0	13	20	78	SO
92	93	20	12	0	78	RD
92	98	1	20	19	73	RD
92	100	17	46	1	21	RD
92	101	0	41	20	36	RD
92	104	0	43	17	12	SO
92	115	0	20	19	71	RD
92	127	0	21	18	53	RD
92	136	0	12	17	67	RD
92	144	18	15	0	11	RD
92	152	14	27	0	25	SO
92	159	13	16	0	50	SO
92	167	0	34	19	28	SO
92	168	0	34	19	52	RD
92	174	0	28	18	53	RD
92	180	17	58	2	32	SO
92	184	2	18	0	56	SO
93	1	9	0	23	78	SO
93	3	0	12	32	66	RD
93	12	4	1	28	77	SO
93	31	30	55	2	22	RD
93	52	18	60	12	4	RD
93	57	3	21	29	57	RD
93	59	1	12	31	64	SO
93	71	2	17	28	60	RD
93	92	20	0	12	78	RD
93	104	3	40	21	8	SO

93	127	3	18	25	46	RD
93	137	1	28	25	40	RD
93	144	24	9	0	11	RD
93	152	17	24	2	20	SO
93	154	22	32	2	20	RD
93	156	2	12	22	33	SO
93	159	13	15	9	41	SO
93	162	10	3	13	50	SO
93	163	10	3	21	63	RD
93	165	0	8	32	69	SO
93	167	1	33	26	19	SO
93	168	1	33	30	39	RD
93	174	1	27	28	42	RD
94	1	0	9	51	32	SO
94	3	11	2	40	39	RD
94	4	5	0	46	41	SO
94	27	11	18	40	23	SO
94	31	30	39	21	1	RD
94	35	21	3	30	38	RD
94	41	16	3	33	38	RD
94	44	19	6	32	35	RD
94	53	13	1	38	40	RD
94	55	45	27	6	12	RD
94	57	17	7	34	34	RD
94	60	0	3	51	33	SO
94	65	17	7	5	8	RD
94	88	0	3	48	30	SO
94	90	0	5	47	33	SO
94	95	51	23	0	16	RD
94	96	51	16	0	23	RD
94	104	42	0	2	24	SO
94	109	12	3	11	17	SO
94	124	1	8	44	25	RD
94	132	43	8	3	5	RD
94	135	46	15	1	12	RD
94	136	12	0	34	33	RD
94	143	1	4	50	28	SO
94	144	2	23	11	0	RD
94	159	2	26	33	2	SO
94	162	2	9	41	19	SO
94	165	7	0	44	41	SO
94	167	33	0	1	37	SO
94	173	28	1	4	2	SO
94	177	21	0	2	6	SO
94	180	46	25	5	13	SO
94	184	4	10	43	1	SO
95	1	0	9	95	7	SO
95	11	27	0	68	16	RD
95	27	17	13	78	3	SO
95	40	57	12	35	2	RD
95	48	23	6	22	0	RD
95	65	21	4	17	0	RD
95	73	49	13	44	1	RD
95	75	50	12	45	4	RD
95	86	59	15	35	1	RD
95	94	51	0	23	16	RD
95	96	51	16	23	0	RD
95	97	23	0	0	16	RD
95	100	46	15	22	0	RD
95	103	25	16	6	0	SO
95	104	43	0	13	15	SO
95	107	45	1	47	15	RD
95	119	56	15	25	1	RD
95	136	12	0	67	16	RD
95	152	26	15	25	0	SO
95	154	38	15	25	0	RD
95	156	14	0	39	16	SO
95	157	14	0	66	16	RD
95	159	14	15	50	0	SO
95	167	33	0	30	16	SO
95	168	33	0	54	16	RD
95	174	27	0	55	15	RD
95	180	58	16	34	0	SO
96	1	9	0	58	23	SO
96	3	13	0	54	23	RD
96	11	5	6	62	17	RD
96	27	24	4	43	19	SO
96	33	26	2	41	21	RD
96	35	24	0	43	23	RD
96	41	19	0	46	23	RD
96	47	25	0	23	5	RD
96	48	23	5	10	9	RD
96	53	14	0	53	23	RD
96	60	0	3	64	19	SO
96	64	42	3	5	8	RD
96	65	21	3	5	8	RD
96	88	0	2	62	16	SO
96	90	0	5	63	15	SO
96	94	51	0	16	23	RD
96	95	51	23	16	0	RD
96	97	0	23	16	0	RD
96	100	49	12	7	9	RD
96	103	17	15	0	6	SO
96	104	42	0	17	7	SO
96	109	14	1	13	13	SO
96	110	14	1	38	20	RD
96	115	19	0	45	22	RD
96	120	56	22	10	0	RD
96	122	37	22	30	1	RD
96	124	3	5	51	17	RD
96	132	44	7	3	4	RD
96	135	50	11	3	8	RD
96	136	12	0	50	15	RD
96	143	1	3	66	11	SO
96	152	38	2	11	7	SO
96	156	6	6	49	3	SO
96	157	6	6	58	7	RD
96	159	17	11	33	2	SO
96	162	5	6	51	7	SO
96	163	5	6	58	13	RD
96	167	33	0	17	19	SO
96	173	28	1	4	2	SO
96	174	27	0	38	17	RD
96	177	21	0	2	6	SO
96	180	62	8	5	13	SO
96	184	4	9	43	1	SO
97	1	0	9	23	7	SO
97	11	6	0	17	16	RD
97	27	4	13	19	3	SO
97	35	0	3	23	13	RD
97	40	12	12	10	2	RD
97	41	0	3	23	13	RD
97	47	0	6	5	6	RD
97	48	5	6	9	0	RD
97	58	8	0	15	16	RD
97	65	3	4	8	0	RD
97	86	13	15	9	1	RD
97	90	5	0	15	16	SO
97	95	23	0	0	16	RD
97	96	0	16	23	0	RD
97	100	12	15	9	0	RD
97	103	15	16	6	0	SO
97	109	1	2	13	2	SO
97	122	22	10	1	6	RD
97	143	3	0	11	16	SO
97	152	2	15	7	0	SO
97	156	6	0	3	16	SO
97	157	6	0	7	16	RD
97	180	8	16	13	0	SO
98	4	5	0	17	93	SO
98	40	20	51	2	37	RD
98	47	16	9	2	28	RD
98	51	17	43	5	32	RD
98	55	19	59	3	32	RD
98	86	20	56	1	37	RD
98	99	22	28	0	36	RD
98	100	22	42	0	22	RD
98	104	15	28	4	26	SO
98	109	9	6	5	37	SO
98	118	5	35	17	43	RD
98	123	15	92	7	1	RD
98	152	20	22	0	25	SO
98	155	10	1	0	13	SO
98	156	0	14	19	37	SO
98	162	0	13	19	46	SO
98	167	16	18	6	42	SO
98	180	20	56	2	32	SO
98	184	0	21	15	41	SO
99	1	8	0	42	36	SO
99	3	12	1	38	35	RD
99	4	5	0	45	36	SO
99	12	4	0	46	36	SO
99	27	18	6	32	30	SO
99	31	28	34	21	2	RD
99	34	39	19	11	17	RD
99	35	23	1	27	35	RD
99	40	42	25	6	11	RD
99	47	24	0	10	14	RD
99	50	24	5	26	23	RD
99	55	43	26	5	10	RD
99	59	2	6	48	30	SO
99	86	43	26	6	10	RD
99	88	0	4	48	21	SO
99	98	22	0	28	36	RD
99	100	50	11	0	22	RD
99	103	18	10	1	5	SO
99	109	12	2	7	23	SO
99	110	12	2	24	33	RD
99	115	18	2	29	33	RD
99	124	3	10	39	23	RD
99	130	10	1	32	26	RD
99	132	31	12	1	7	RD
99	135	35	18	4	7	RD
99	149	4	0	0	4	SO
99	152	40	0	0	23	SO
99	155	11	0	0	13	SO
99	156	2	12	41	3	SO
99	157	2	12	45	14	RD
99	167	25	5	21	13	SO
99	180	48	20	2	12	SO
99	184	1	17	28	6	SO
99	195	23	23	27	13	SO
100	1	9	0	55	22	SO
100	3	13	0	51	22	RD
100	14	3	5	61	17	RD
100	27	23	2	41	20	SO
100	35	23	1	41	21	RD
100	47	25	0	18	7	RD
100	88	1	4	58	9	SO
100	98	22	0	42	22	RD
100	99	50	0	11	22	RD
100	103	26	4	2	4	SO
100	104	29	7	22	1	SO
100	109	13	0	14	17	SO
100	112	12	10	29	6	SO
100	120	54	22	10	0	RD
100	124	4	9	50	11	RD
100	126	38	20	16	0	RD
100	129	29	13	25	0	RD
100	132	35	6	4	5	RD
100	149	4	0	0	4	SO
100	152	41	0	4	19	SO
100	155	11	0	1	12	SO
100	156	6	8	43	1	SO
100	167	28	1	29	6	SO
100	180	57	13	5	7	SO
100	184	5	13	33	1	SO
100	196	23	12	41	10	SO
101	1	0	9	41	48	SO
101	4	5	0	36	57	SO
101	60	0	7	41	46	SO
101	90	0	5	40	49	SO
101	104	37	4	1	29	SO
101	109	12	3	5	26	SO
101	143	0	4	41	45	SO
101	149	4	0	1	2	SO
101	150	0	5	4	0	SO
101	155	11	0	3	5	SO
101	156	3	8	31	22	SO
101	159	0	29	33	12	SO
101	162	2	11	35	24	SO
101	165	5	1	36	56	SO
101	167	31	3	1	46	SO
101	173	26	2	2	3	SO
101	177	12	5	1	17	SO
101	180	39	30	2	24	SO
101	184	2	11	37	13	SO
102	1	0	9	59	40	SO
102	4	5	0	54	49	SO
102	27	13	18	46	31	SO
102	60	0	7	59	38	SO
102	88	0	3	55	39	SO
102	90	0	5	55	41	SO
102	104	43	0	2	28	SO

102	151	5	0	7	14	SO
102	159	4	25	38	11	SO
102	162	3	10	45	17	SO
102	165	8	0	50	49	SO
102	167	34	0	3	44	SO
102	173	32	1	4	3	SO
102	177	23	0	6	15	SO
102	180	49	23	10	23	SO
102	184	5	9	46	10	SO
102	191	0	8	4	1	SO
103	112	7	3	17	1	SO
103	133	16	0	25	6	RD
103	143	3	1	37	0	SO
103	152	16	0	4	2	SO
103	180	23	0	17	6	SO
103	184	6	3	10	0	SO
104	1	0	9	43	21	SO
104	4	5	0	38	30	SO
104	11	3	9	40	21	RD
104	27	9	13	34	17	SO
104	35	21	3	22	27	RD
104	38	9	1	34	27	RD
104	40	33	14	8	13	RD
104	41	16	3	25	27	RD
104	44	19	6	24	24	RD
104	46	41	21	2	5	RD
104	53	13	1	30	29	RD
104	55	40	19	3	9	RD
104	60	0	2	43	24	SO
104	64	31	10	1	3	RD
104	65	16	4	1	3	RD
104	73	32	15	10	13	RD
104	74	20	6	22	24	RD
104	95	43	13	0	15	RD
104	96	42	17	0	7	RD
104	105	43	0	0	30	RD
104	106	42	0	0	30	RD
104	109	12	1	9	9	SO
104	110	12	1	28	16	RD
104	115	20	0	23	27	RD
104	120	32	26	10	0	RD
104	124	1	3	36	19	RD
104	133	7	10	35	19	RD
104	135	40	15	1	5	RD
104	136	12	0	29	29	RD
104	152	21	18	11	1	SO
104	159	2	17	32	6	SO
104	162	2	6	39	14	SO
104	163	2	6	40	23	RD
104	165	7	0	36	30	SO
104	167	32	0	0	27	SO
104	174	26	0	16	30	RD
104	176	42	10	0	20	RD
104	177	16	0	0	10	SO
104	180	40	21	3	8	SO
105	1	0	9	43	38	SO
105	4	5	0	38	47	SO
105	60	0	6	43	37	SO
105	104	43	0	0	30	SO
105	109	12	1	9	18	SO
105	143	0	4	43	41	SO

105	159	2	23	32	13	SO
105	162	2	8	39	17	SO
105	165	7	0	36	47	SO
105	167	32	0	0	39	SO
105	173	27	3	1	4	SO
105	177	16	1	0	20	SO
105	180	40	24	3	21	SO
105	191	0	7	2	2	SO
106	1	0	9	52	21	SO
106	12	0	2	52	28	SO
106	27	11	13	41	17	SO
106	88	0	2	46	25	SO
106	104	42	0	0	30	SO
106	109	13	1	13	9	SO
106	152	22	18	12	1	SO
106	159	4	17	37	6	SO
106	162	4	6	41	14	SO
106	165	7	0	45	30	SO
106	167	31	0	7	27	SO
106	173	29	0	2	0	SO
106	177	15	0	8	10	SO
107	1	0	9	46	56	SO
107	4	5	0	41	65	SO
107	12	0	5	46	60	SO
107	60	0	6	46	54	SO
107	61	1	7	45	55	SO
107	88	0	4	42	52	SO
107	90	0	5	45	54	SO
107	104	42	1	0	30	SO
107	109	13	2	11	28	SO
107	143	0	5	46	51	SO
107	159	2	27	31	18	SO
107	162	3	10	40	24	SO
107	165	7	1	39	63	SO
107	167	32	2	1	45	SO
107	173	27	5	2	5	SO
107	177	17	6	0	21	SO
107	180	43	32	3	29	SO
107	184	5	13	36	20	SO
107	191	0	8	2	3	SO
108	1	0	9	65	26	SO
108	27	14	15	51	20	SO
108	60	1	5	64	26	SO
108	90	0	4	61	29	SO
108	104	42	1	4	25	SO
108	109	13	2	19	12	SO
108	143	1	4	62	28	SO
108	152	24	17	15	2	SO
108	155	11	0	10	0	SO
108	159	5	21	40	7	SO
108	162	5	8	46	16	SO
108	165	8	0	57	35	SO
108	167	32	1	9	33	SO
108	177	21	1	12	7	SO
108	180	51	21	13	13	SO
109	4	4	0	11	42	SO
109	11	0	13	15	29	RD
109	40	14	23	1	19	RD
109	41	7	1	8	41	RD
109	47	12	1	2	10	RD
109	59	0	11	15	31	SO

109	104	12	9	1	9	SO
109	106	13	13	1	9	RD
109	108	13	19	2	12	RD
109	110	15	0	0	42	RD
109	111	15	0	0	42	RD
109	112	1	14	12	12	SO
109	113	1	14	14	24	RD
109	136	5	1	10	31	RD
109	152	12	5	1	18	SO
109	155	6	1	0	11	SO
109	156	1	8	13	7	SO
109	159	2	13	11	13	SO
109	160	2	13	12	18	RD
109	167	12	6	3	20	SO
109	173	12	7	0	3	SO
109	180	14	21	1	16	SO
109	184	2	13	11	14	SO
110	4	4	1	11	79	SO
110	59	0	13	15	67	SO
110	104	12	28	1	16	SO
110	109	15	0	0	42	SO
110	112	1	23	12	23	SO
110	152	12	17	1	24	SO
110	155	6	4	0	13	SO
110	167	12	19	3	31	SO
110	173	12	20	0	7	SO
110	180	14	45	1	30	SO
111	4	5	0	48	42	SO
111	12	4	0	49	42	SO
111	59	2	11	51	31	SO
111	88	0	5	51	26	SO
111	104	31	9	8	9	SO
111	109	15	0	0	42	SO
111	112	10	14	23	12	SO
111	149	4	0	0	4	SO
111	152	24	5	7	18	SO
111	155	9	1	2	11	SO
111	156	4	8	35	7	SO
111	159	4	13	32	13	SO
111	167	25	6	14	20	SO
111	184	6	13	36	14	SO
112	21	12	7	15	40	RD
112	23	11	6	16	41	RD
112	27	5	17	22	30	SO
112	36	7	22	20	25	RD
112	48	5	15	9	8	RD
112	57	3	14	24	33	RD
112	58	9	4	18	43	RD
112	64	8	25	9	4	RD
112	103	7	17	3	1	SO
112	109	1	12	14	12	SO
112	110	1	12	23	23	RD
112	113	27	0	0	47	RD
112	114	27	0	0	47	RD
112	124	7	3	19	35	RD
112	151	4	0	1	6	SO
112	165	4	0	22	47	SO
112	173	9	16	3	0	SO
112	180	13	39	10	7	SO
112	198	2	14	25	33	SO
113	103	7	33	3	2	SO

113	109	1	14	14	24	SO
113	112	27	0	0	47	SO
113	151	4	1	1	19	SO
113	177	8	13	2	16	SO
114	4	0	5	61	42	SO
114	27	13	17	48	30	SO
114	59	10	3	51	42	SO
114	109	3	12	26	12	SO
114	112	27	0	0	47	SO
114	165	8	0	52	47	SO
114	173	17	16	7	0	SO
114	180	33	39	24	7	SO
115	4	4	1	16	90	SO
115	59	0	13	20	76	SO
115	104	20	23	0	27	SO
115	109	8	7	3	36	SO
115	149	3	1	0	4	SO
115	152	16	23	2	22	SO
115	155	8	3	2	11	SO
115	156	0	14	16	37	SO
115	159	0	27	16	34	SO
115	162	1	12	19	43	SO
115	165	6	2	14	88	SO
115	167	18	16	0	45	SO
115	180	20	52	0	34	SO
115	184	2	18	17	39	SO
116	1	0	8	32	71	SO
116	4	5	0	27	79	SO
116	11	1	27	31	52	RD
116	32	7	3	25	76	RD
116	38	8	2	24	75	RD
116	47	19	6	10	16	RD
116	48	10	16	2	22	RD
116	53	13	1	19	78	RD
116	59	1	12	31	65	SO
116	60	0	7	32	68	SO
116	61	0	8	32	69	SO
116	64	27	16	1	17	RD
116	65	13	10	1	17	RD
116	71	12	7	19	70	RD
116	95	31	63	0	13	RD
116	96	30	34	0	22	RD
116	100	27	34	4	17	RD
116	104	29	14	0	27	SO
116	109	12	3	6	33	SO
116	110	12	3	18	59	RD
116	115	20	0	12	79	RD
116	117	32	28	0	51	RD
116	124	1	13	27	54	RD
116	127	13	8	17	52	RD
116	130	9	3	20	47	RD
116	133	5	24	24	48	RD
116	149	4	0	1	3	SO
116	152	22	17	6	18	SO
116	154	25	27	6	18	RD
116	155	11	0	5	8	SO
116	156	3	11	24	29	SO
116	159	1	26	24	26	SO
116	160	1	26	30	37	RD
116	162	1	12	31	31	SO
116	163	1	12	31	52	RD

116	165	6	2	26	76	SO
116	167	27	7	0	45	SO
116	173	21	12	2	5	SO
116	176	32	40	0	25	RD
116	177	9	14	0	21	SO
116	180	32	40	0	34	SO
116	184	3	17	25	31	SO
117	4	5	0	55	51	SO
117	60	0	7	60	40	SO
117	61	0	8	60	41	SO
117	90	0	5	58	40	SO
117	104	39	4	6	21	SO
117	109	14	1	18	21	SO
117	143	0	5	57	40	SO
117	159	9	18	33	17	SO
117	162	5	8	46	16	SO
117	167	33	1	9	36	SO
117	173	26	7	3	4	SO
117	177	16	7	5	16	SO
117	180	56	16	3	31	SO
118	1	9	0	31	60	SO
118	4	0	5	40	55	SO
118	27	17	12	23	48	SO
118	59	1	9	39	51	SO
118	60	3	0	32	60	SO
118	104	3	39	20	6	SO
118	109	2	13	17	17	SO
118	155	0	11	4	9	SO
118	159	22	8	3	36	SO
118	167	3	31	31	9	SO
118	173	1	29	3	3	SO
118	184	11	8	2	42	SO
119	1	9	0	65	26	SO
119	4	0	5	74	21	SO
119	37	39	25	35	1	RD
119	38	4	6	68	20	RD
119	40	49	22	22	2	RD
119	47	7	18	23	7	RD
119	48	23	7	23	1	RD
119	52	51	25	15	1	RD
119	53	2	12	72	14	RD
119	55	53	25	19	1	RD
119	64	22	24	16	0	RD
119	65	16	9	16	0	RD
119	69	38	22	35	4	RD
119	73	43	21	28	4	RD
119	104	17	25	26	0	SO
119	108	31	26	30	0	RD
119	109	4	11	32	2	SO
119	110	4	11	55	13	RD
119	118	40	0	34	26	RD
119	120	74	16	0	10	RD
119	124	13	0	48	24	RD
119	129	29	22	25	3	RD
119	136	1	11	56	15	RD
119	144	26	1	2	8	RD
119	149	1	3	4	0	SO
119	152	24	18	21	2	SO
119	154	35	20	21	2	RD
119	155	1	10	12	1	SO
119	156	12	2	32	22	SO

119	159	30	0	18	21	SO
119	165	2	5	71	21	SO
119	167	11	23	39	1	SO
119	173	12	18	5	1	SO
119	176	34	25	28	0	RD
119	180	51	25	19	1	SO
119	184	19	0	20	24	SO
119	197	27	5	47	21	SO
120	4	0	5	90	5	SO
120	104	32	10	26	0	SO
120	109	7	8	34	0	SO
120	152	32	10	23	0	SO
120	155	5	6	13	0	SO
120	156	14	0	44	10	SO
120	159	30	0	31	8	SO
120	167	24	10	40	0	SO
120	180	66	10	20	0	SO
120	184	19	0	34	10	SO
121	1	0	9	54	52	SO
121	4	0	5	54	56	SO
121	12	0	5	54	56	SO
121	60	7	0	42	61	SO
121	61	8	0	43	61	SO
121	90	5	0	43	59	SO
121	103	26	17	6	0	SO
121	104	7	36	15	15	SO
121	109	1	14	28	14	SO
121	112	15	12	14	33	SO
121	152	9	33	13	12	SO
121	155	0	11	4	9	SO
121	162	10	3	14	51	SO
121	167	5	29	31	17	SO
121	177	7	16	16	5	SO
121	180	16	60	33	1	SO
122	4	0	5	83	27	SO
122	11	28	0	55	32	RD
122	14	11	0	72	32	RD
122	21	27	5	56	27	RD
122	32	5	5	78	27	RD
122	38	4	6	78	25	RD
122	40	43	28	35	4	RD
122	41	4	15	78	16	RD
122	46	52	32	26	0	RD
122	47	7	18	16	14	RD
122	48	16	14	24	0	RD
122	51	37	23	29	8	RD
122	53	2	12	81	20	RD
122	59	13	0	68	32	SO
122	60	7	0	71	32	SO
122	61	8	0	72	32	SO
122	64	14	32	18	0	RD
122	65	9	16	18	0	RD
122	98	9	13	74	19	RD
122	100	37	27	19	3	RD
122	104	19	24	25	5	SO
122	106	27	25	25	5	RD
122	108	39	26	29	6	RD
122	109	4	11	39	3	SO
122	110	4	11	67	13	RD
122	112	24	3	27	20	SO
122	113	24	3	54	27	RD

122	115	6	14	74	17	RD
122	118	32	8	36	24	RD
122	121	54	0	29	32	RD
122	123	83	24	0	8	RD
122	129	32	24	31	6	RD
122	133	26	3	48	28	RD
122	136	3	9	65	20	RD
122	152	23	19	20	5	SO
122	154	32	23	20	5	RD
122	155	3	8	9	4	SO
122	159	24	6	30	20	SO
122	160	24	6	42	25	RD
122	162	11	2	35	30	SO
122	167	14	20	41	7	SO
122	177	15	8	21	0	SO
122	180	44	32	34	0	SO
122	184	18	3	35	21	SO
122	197	29	6	54	26	SO
123	4	1	4	106	4	SO
123	104	35	8	30	0	SO
123	109	7	8	42	0	SO
123	112	27	0	40	7	SO
123	152	34	8	25	0	SO
123	155	6	5	13	0	SO
123	159	30	0	42	8	SO
123	167	26	8	48	0	SO
123	180	68	8	34	0	SO
123	184	21	0	48	8	SO
124	88	3	2	8	73	SO
124	104	1	36	3	19	SO
124	109	0	15	9	30	SO
124	112	7	19	3	35	SO
124	152	3	31	7	16	SO
124	159	5	18	1	44	SO
124	167	1	29	8	32	SO
124	184	4	15	2	50	SO
125	4	0	5	64	30	SO
125	12	1	3	63	32	SO
125	21	22	5	42	30	RD
125	38	3	5	59	30	RD
125	40	37	26	27	7	RD
125	41	2	15	62	19	RD
125	44	4	17	60	18	RD
125	46	38	33	23	2	RD
125	51	27	27	23	5	RD
125	53	2	11	62	24	RD
125	55	35	30	29	5	RD
125	59	11	2	52	32	SO
125	60	7	0	54	35	SO
125	65	9	11	13	3	RD
125	86	34	28	29	7	RD
125	98	5	16	59	19	RD
125	104	12	25	18	4	SO
125	108	28	31	23	4	RD
125	109	4	11	35	4	SO
125	110	4	11	54	18	RD
125	112	19	7	18	20	SO
125	115	2	15	59	19	RD
125	118	25	4	26	30	RD
125	120	51	25	0	9	RD
125	124	14	0	50	35	RD

125	126	64	19	0	16	RD
125	129	27	25	21	8	RD
125	135	40	31	11	2	RD
125	144	19	7	3	7	RD
125	149	1	2	4	0	SO
125	152	11	23	21	2	SO
125	155	0	11	10	1	SO
125	156	10	4	19	27	SO
125	159	19	4	20	25	SO
125	160	19	4	32	29	RD
125	167	6	24	33	7	SO
125	173	12	21	5	1	SO
125	176	34	31	18	3	RD
125	180	30	32	30	3	SO
125	184	17	2	23	29	SO
126	4	0	5	83	11	SO
126	59	13	0	68	16	SO
126	104	21	16	22	0	SO
126	109	5	10	38	1	SO
126	152	18	16	23	0	SO
126	155	1	10	11	0	SO
126	156	14	0	30	16	SO
126	159	23	0	30	15	SO
126	162	11	0	39	16	SO
126	165	3	4	79	12	SO
126	167	14	16	40	0	SO
126	173	18	15	6	0	SO
126	180	46	16	33	0	SO
126	184	19	0	36	16	SO
127	1	0	9	21	63	SO
127	4	5	0	16	72	SO
127	104	15	26	0	28	SO
127	150	0	5	2	1	SO
127	151	3	2	0	20	SO
127	159	1	21	13	34	SO
127	162	0	11	17	41	SO
127	167	11	19	0	35	SO
127	177	7	14	0	19	SO
128	1	0	9	48	36	SO
128	3	11	2	37	43	RD
128	4	5	0	43	45	SO
128	11	8	17	40	28	RD
128	25	14	1	34	44	RD
128	27	8	18	40	27	SO
128	31	29	42	19	3	RD
128	35	19	5	29	40	RD
128	44	17	8	31	37	RD
128	49	10	6	5	14	RD
128	60	0	5	48	36	SO
128	104	36	5	2	26	SO
128	112	13	4	21	21	SO
128	115	17	3	31	39	RD
128	119	24	30	21	4	RD
128	120	36	33	9	1	RD
128	126	30	35	15	1	RD
128	127	21	0	27	45	RD
128	129	48	8	0	37	RD
128	136	10	1	34	42	RD
128	149	0	4	3	0	SO
128	151	5	0	3	17	SO
128	152	19	22	18	0	SO

128	159	5	17	30	17	SO
128	160	5	17	39	24	RD
128	162	2	9	39	19	SO
128	163	2	9	42	32	RD
128	167	25	5	1	34	SO
128	177	19	2	3	16	SO
128	180	41	28	7	16	SO
128	191	0	7	3	2	SO
129	1	0	9	56	28	SO
129	4	5	0	51	37	SO
129	27	11	15	45	22	SO
129	60	0	5	56	28	SO
129	104	38	3	3	25	SO
129	109	13	1	18	10	SO
129	149	0	4	3	0	SO
129	151	5	0	6	14	SO
129	152	19	22	18	0	SO
129	159	7	15	33	14	SO
129	162	3	8	43	15	SO
129	167	27	3	4	31	SO
129	177	20	1	6	13	SO
129	180	46	23	10	13	SO
129	191	0	7	3	2	SO
130	104	8	32	1	20	SO
130	109	4	11	0	26	SO
130	149	4	0	0	4	SO
130	152	10	30	0	22	SO
130	159	0	26	8	27	SO
130	177	1	17	3	5	SO
130	184	0	14	11	29	SO
131	4	0	5	30	30	SO
131	11	8	2	22	33	RD
131	21	11	4	19	31	RD
131	100	19	16	1	15	RD
131	109	7	7	5	17	SO
131	129	21	27	8	1	RD
131	130	12	0	18	35	RD
131	132	30	19	0	9	RD
131	149	4	0	0	4	SO
131	150	0	4	2	1	SO
131	152	15	10	2	18	SO
131	156	2	9	23	15	SO
131	159	1	10	21	12	SO
131	160	1	10	26	20	RD
131	165	6	2	24	33	SO
131	184	1	11	25	15	SO
131	188	7	17	23	18	RD
131	189	7	0	23	35	RD
131	195	13	22	17	13	SO
131	196	9	19	21	16	SO
132	88	0	2	65	2	SO
132	109	13	0	18	7	SO
132	149	4	0	0	4	SO
132	152	26	1	10	5	SO
132	155	10	1	6	5	SO
132	156	5	6	39	2	SO
132	159	4	5	44	1	SO
132	184	4	5	51	1	SO
133	1	0	9	29	67	SO
133	12	3	1	26	75	SO
133	27	5	25	24	51	SO

133	60	5	1	24	70	SO
133	61	5	3	24	70	SO
133	103	16	25	0	6	SO
133	104	7	35	10	19	SO
133	109	0	15	10	24	SO
133	151	1	4	15	6	SO
133	159	0	28	21	28	SO
133	162	9	4	6	55	SO
133	167	4	27	16	30	SO
133	173	3	28	3	3	SO
133	177	6	16	15	5	SO
133	180	13	57	15	17	SO
133	184	1	16	21	33	SO
133	191	8	0	2	3	SO
134	1	0	9	57	37	SO
134	4	5	0	52	46	SO
134	11	19	7	38	39	RD
134	88	0	2	57	32	SO
134	90	0	5	55	36	SO
134	100	34	28	1	14	RD
134	103	17	23	0	6	SO
134	104	32	10	10	19	SO
134	109	13	2	11	21	SO
134	110	13	2	40	29	RD
134	112	8	14	27	19	SO
134	115	17	3	40	39	RD
134	116	25	4	32	38	RD
134	124	1	7	52	30	RD
134	126	38	36	15	1	RD
134	132	52	11	1	6	RD
134	133	29	0	28	46	RD
134	135	57	18	0	13	RD
134	136	10	1	42	39	RD
134	137	25	2	27	38	RD
134	143	0	5	56	34	SO
134	149	4	0	0	2	SO
134	151	1	3	18	3	SO
134	152	24	17	4	14	SO
134	155	10	1	1	8	SO
134	156	3	7	34	20	SO
134	159	0	28	44	4	SO
134	162	10	3	31	30	SO
134	167	27	4	17	28	SO
134	168	27	4	30	38	RD
134	173	26	4	3	3	SO
134	174	24	1	31	38	RD
134	177	11	11	16	4	SO
134	184	2	14	47	7	SO
134	189	7	0	50	46	RD
134	191	8	0	3	2	SO
134	197	13	17	44	29	SO
135	88	0	2	71	6	SO
135	90	1	4	70	8	SO
135	103	23	9	1	3	SO
135	104	40	1	15	5	SO
135	112	19	0	32	7	SO
135	143	1	4	72	7	SO
135	159	8	10	46	0	SO
135	167	30	0	25	10	SO
135	184	8	5	52	1	SO
136	4	4	1	8	84	SO

136	104	12	29	0	29	SO
136	109	5	10	1	31	SO
136	152	11	30	0	21	SO
136	155	10	1	0	12	SO
136	159	0	28	11	35	SO
136	167	12	18	0	42	SO
136	180	12	54	0	29	SO
136	184	0	13	12	39	SO
137	1	0	9	29	59	SO
137	4	5	0	24	68	SO
137	11	0	26	29	42	RD
137	25	12	4	17	64	RD
137	32	7	3	22	65	RD
137	38	8	2	21	64	RD
137	47	18	7	8	15	RD
137	48	8	15	2	16	RD
137	53	13	1	16	67	RD
137	59	0	11	29	55	SO
137	61	0	8	29	58	SO
137	71	11	9	17	57	RD
137	86	27	35	2	33	RD
137	95	29	50	0	16	RD
137	96	29	33	0	15	RD
137	99	23	23	4	22	RD
137	100	25	32	1	14	RD
137	104	28	13	0	29	SO
137	108	29	29	0	31	RD
137	109	12	3	3	29	SO
137	110	12	3	15	49	RD
137	127	12	7	16	52	RD
137	133	3	23	24	44	RD
137	136	12	0	17	68	RD
137	138	29	16	0	49	RD
137	144	0	32	8	2	RD
137	152	21	20	3	18	SO
137	154	23	30	3	18	RD
137	155	11	0	2	10	SO
137	159	0	28	24	22	SO
137	162	1	10	27	31	SO
137	165	5	3	24	64	SO
137	167	28	2	0	42	SO
137	173	23	9	1	5	SO
137	176	29	36	0	27	RD
137	177	4	16	0	19	SO
137	180	29	37	0	29	SO
137	184	1	12	27	24	SO
138	1	0	9	48	40	SO
138	4	5	0	43	49	SO
138	60	0	6	48	39	SO
138	90	0	4	45	43	SO
138	104	39	3	0	27	SO
138	109	13	2	13	18	SO
138	151	3	2	1	20	SO
138	155	11	0	8	4	SO
138	159	4	22	30	17	SO
138	162	3	8	39	19	SO
138	165	8	0	40	48	SO
138	167	31	1	1	38	SO
138	173	26	6	3	3	SO
138	177	18	3	1	18	SO
138	180	44	23	4	25	SO

138	191	0	7	3	2	SO
139	1	9	0	25	71	SO
139	31	33	47	1	23	RD
139	35	3	21	31	50	RD
139	36	9	33	25	38	RD
139	41	3	16	31	53	RD
139	42	8	28	26	40	RD
139	52	17	59	11	5	RD
139	60	5	2	26	68	SO
139	103	30	11	0	3	SO
139	104	1	42	27	3	SO
139	112	3	20	15	31	SO
139	115	0	20	31	51	RD
139	120	27	55	0	10	RD
139	132	19	48	0	8	RD
139	136	0	12	31	50	RD
139	137	0	29	31	33	RD
139	140	0	13	34	53	RD
139	141	0	11	34	53	RD
139	142	1	29	33	33	RD
139	144	33	1	0	11	RD
139	151	0	5	14	7	SO
139	152	19	23	0	20	SO
139	159	16	11	12	36	SO
139	162	8	4	12	50	SO
139	163	8	4	25	59	RD
139	165	0	8	34	62	SO
139	167	1	33	31	12	SO
139	177	1	22	14	5	SO
139	191	7	0	2	3	SO
140	36	2	38	11	49	RD
140	134	10	43	3	37	RD
140	139	0	34	13	53	RD
140	142	0	30	13	53	RD
140	159	1	25	8	37	SO
141	30	2	41	9	44	RD
141	37	10	47	1	40	RD
141	51	10	44	0	30	RD
141	92	0	19	11	67	RD
141	96	11	53	0	14	RD
141	104	10	30	0	30	SO
141	106	10	38	0	30	RD
141	108	11	48	0	32	RD
141	109	4	9	0	31	SO
141	139	0	34	11	53	RD
141	142	0	30	11	55	RD
141	144	0	34	11	0	RD
141	145	0	30	11	0	RD
141	149	3	1	0	4	SO
141	167	8	22	0	41	SO
141	176	11	55	0	26	RD
142	1	0	9	30	57	SO
142	4	5	0	25	66	SO
142	11	3	22	27	44	RD
142	31	18	55	12	10	RD
142	35	13	11	17	55	RD
142	55	26	39	4	25	RD
142	64	19	25	0	9	RD
142	65	12	11	0	9	RD
142	73	23	32	7	31	RD
142	86	25	38	5	27	RD

142	93	2	26	28	39	RD
142	100	14	43	12	1	RD
142	104	22	18	1	29	SO
142	108	25	33	1	31	RD
142	115	11	9	19	54	RD
142	121	7	32	23	34	RD
142	123	25	63	5	3	RD
142	132	21	41	6	2	RD
142	136	10	1	16	61	RD
142	139	1	33	29	33	RD
142	140	0	13	30	53	RD
142	141	0	11	30	55	RD
142	145	30	0	0	11	RD
142	149	0	4	4	0	SO
142	151	3	2	0	20	SO
142	152	11	30	13	5	SO
142	156	6	4	15	37	SO
142	159	4	21	17	27	SO
142	160	4	21	22	39	RD
142	162	0	12	25	32	SO
142	163	0	12	25	50	RD
142	167	17	13	1	39	SO
142	176	25	40	1	24	RD
142	177	11	10	1	18	SO
142	180	26	42	4	23	SO
142	195	20	27	10	39	SO
143	64	0	46	3	10	RD
143	65	0	25	3	10	RD
143	90	3	0	1	96	SO
143	103	3	37	1	0	SO
143	159	3	24	0	49	SO
143	167	0	34	4	37	SO
143	173	0	33	2	4	SO
143	180	2	73	3	25	SO
143	197	3	27	2	73	SO
144	1	9	0	25	11	SO
144	104	0	10	27	0	SO
144	109	2	4	9	0	SO
144	152	19	7	0	2	SO
144	159	16	1	11	5	SO
144	162	9	0	11	10	SO
144	165	0	3	34	8	SO
144	167	0	8	32	0	SO
144	173	0	7	1	0	SO
144	177	0	3	14	0	SO
144	180	21	10	13	1	SO
145	109	6	4	15	0	SO
145	149	0	3	4	0	SO
145	152	11	7	13	2	SO
146	12	3	0	5	92	SO
146	27	0	27	8	65	SO
146	42	0	34	8	56	RD
146	48	0	27	4	15	RD
146	65	0	23	4	7	RD
146	72	0	46	8	42	RD
146	89	0	31	8	44	RD
146	104	0	40	5	22	SO
146	109	0	14	0	35	SO
146	128	0	43	7	35	RD
146	131	3	24	0	33	RD
146	134	7	44	0	40	RD

146	142	0	30	8	47	RD
146	144	8	24	0	11	RD
146	147	0	23	8	56	RD
146	162	5	3	0	62	SO
146	167	0	32	8	33	SO
146	168	0	32	8	54	RD
146	174	0	27	8	55	RD
146	177	0	19	5	16	SO
147	1	0	9	23	58	SO
147	4	5	0	18	67	SO
147	11	0	25	23	42	RD
147	27	3	23	20	44	SO
147	38	5	3	18	62	RD
147	47	15	8	7	17	RD
147	52	23	41	0	13	RD
147	71	9	7	13	58	RD
147	86	21	37	2	30	RD
147	95	23	50	0	16	RD
147	96	21	37	0	13	RD
147	99	17	23	4	19	RD
147	104	22	14	0	28	SO
147	106	22	23	0	28	RD
147	108	23	31	0	30	RD
147	109	10	3	5	23	SO
147	120	14	52	9	1	RD
147	123	15	67	8	0	RD
147	129	23	23	0	37	RD
147	136	9	1	13	59	RD
147	139	0	33	23	30	RD
147	146	0	8	23	56	RD
147	148	0	14	23	53	RD
147	155	10	0	4	7	SO
147	156	0	10	18	33	SO
147	158	5	27	18	33	RD
147	159	0	23	19	23	SO
147	162	0	9	22	33	SO
147	164	1	24	22	33	RD
147	167	21	7	0	39	SO
147	169	21	26	0	39	RD
147	173	18	9	0	2	SO
147	176	22	38	0	24	RD
147	177	3	14	0	19	SO
147	180	23	38	0	28	SO
148	2	0	22	14	74	RD
148	6	0	21	14	75	RD
148	8	0	18	14	78	RD
148	11	14	13	0	83	RD
148	15	0	9	14	17	RD
148	18	14	17	0	18	RD
148	19	14	17	0	21	RD
148	23	0	27	14	69	RD
148	31	14	72	0	24	RD
148	33	0	30	14	66	RD
148	35	0	24	14	72	RD
148	36	0	42	14	54	RD
148	39	0	38	14	57	RD
148	41	0	19	14	75	RD
148	42	0	36	14	58	RD
148	44	0	25	14	71	RD
148	49	1	21	12	11	RD
148	51	6	53	0	33	RD

148	54	0	37	14	57	RD
148	57	0	25	14	71	RD
148	60	4	3	10	88	SO
148	68	3	46	11	49	RD
148	71	0	19	14	73	RD
148	72	0	49	14	43	RD
148	74	0	26	14	68	RD
148	85	0	60	14	36	RD
148	89	0	33	13	45	RD
148	101	2	39	9	44	RD
148	104	2	41	5	24	SO
148	115	0	20	14	72	RD
148	116	0	32	14	60	RD
148	122	14	64	0	32	RD
148	134	13	44	1	40	RD
148	147	0	23	14	53	RD
148	151	0	5	14	7	SO
148	159	0	27	11	39	SO
148	167	0	34	8	36	SO
148	174	0	28	13	55	RD
148	177	2	20	11	10	SO
148	180	0	74	14	17	SO
148	184	1	19	13	42	SO
148	195	4	52	10	44	SO
149	88	0	2	4	1	SO
149	131	4	0	0	4	RD
149	141	3	0	1	4	RD
149	142	0	4	4	0	RD
149	152	4	0	0	4	SO
149	156	0	4	4	0	SO
149	184	0	2	4	0	SO
150	159	4	0	0	4	SO
150	184	3	0	1	3	SO
151	112	4	1	0	6	SO
151	129	5	6	0	14	RD
151	139	0	14	5	7	RD
151	148	0	14	5	7	RD
151	152	0	4	2	1	SO
151	165	2	0	2	22	SO
151	177	3	2	1	15	SO
151	199	3	4	2	18	SO
152	1	9	0	33	25	SO
152	3	10	1	32	24	RD
152	4	5	0	37	25	SO
152	27	17	4	25	21	SO
152	37	34	14	8	11	RD
152	40	34	16	6	9	RD
152	47	24	0	8	9	RD
152	52	32	19	10	1	RD
152	55	37	19	3	6	RD
152	59	2	5	40	20	SO
152	64	28	8	1	5	RD
152	88	0	4	40	13	SO
152	90	0	2	42	22	SO
152	98	20	0	22	25	RD
152	103	16	4	0	2	SO
152	104	21	11	18	1	SO
152	109	12	1	5	18	SO
152	110	12	1	17	24	RD
152	115	16	2	23	22	RD
152	124	3	7	31	16	RD

152	129	19	18	22	0	RD
152	132	26	10	1	5	RD
152	139	19	0	23	20	RD
152	142	11	13	30	5	RD
152	149	4	0	0	4	SO
152	151	0	2	4	1	SO
152	153	42	0	0	25	RD
152	154	42	0	0	25	RD
152	155	11	0	0	13	SO
152	156	0	10	40	3	SO
152	173	19	6	0	2	SO
152	177	1	12	3	2	SO
152	180	41	16	1	7	SO
152	184	0	12	24	5	SO
153	1	9	0	33	38	SO
153	4	5	0	37	38	SO
153	27	17	7	25	31	SO
153	59	2	6	40	32	SO
153	88	0	4	40	23	SO
153	104	21	17	18	5	SO
153	109	12	2	5	21	SO
153	149	4	0	0	4	SO
153	152	42	0	0	25	SO
153	155	11	0	0	13	SO
153	156	0	12	40	9	SO
153	173	19	9	0	3	SO
153	177	1	16	3	5	SO
153	180	41	24	1	11	SO
153	184	0	15	24	10	SO
154	1	9	0	46	25	SO
154	27	20	4	35	21	SO
154	59	3	5	52	20	SO
154	88	0	4	50	13	SO
154	90	0	2	54	22	SO
154	103	23	4	0	2	SO
154	104	27	11	22	1	SO
154	109	13	1	8	18	SO
154	149	4	0	0	4	SO
154	151	0	2	8	1	SO
154	152	42	0	0	25	SO
154	155	11	0	0	13	SO
154	156	2	10	46	3	SO
154	162	9	0	41	18	SO
154	177	5	12	6	2	SO
154	180	49	16	5	7	SO
154	184	3	12	29	5	SO
155	4	5	0	6	13	SO
155	27	3	0	8	13	SO
155	36	11	4	0	9	RD
155	47	10	0	1	6	RD
155	59	0	4	11	9	SO
155	88	0	2	11	5	SO
155	100	11	1	0	12	RD
155	109	6	0	1	11	SO
155	138	11	8	0	4	RD
155	152	11	0	0	13	SO
155	153	11	0	0	13	RD
155	154	11	0	0	13	RD
155	156	0	6	11	1	SO
155	157	0	6	11	7	RD
155	159	0	4	11	3	SO

155	160	0	4	11	8	RD
155	184	0	6	11	2	SO
156	47	0	23	4	14	RD
156	88	3	0	8	53	SO
156	109	1	13	8	7	SO
156	119	12	32	2	22	RD
156	149	0	4	4	0	SO
156	152	0	40	10	3	SO
156	155	0	11	6	1	SO
156	157	14	0	0	56	RD
156	158	14	0	0	56	RD
156	159	7	17	2	28	SO
156	184	8	2	2	32	SO
156	195	11	29	3	27	SO
157	88	3	0	8	76	SO
157	90	1	0	12	81	SO
157	149	0	4	4	0	SO
157	152	0	42	10	10	SO
157	155	0	11	6	7	SO
157	156	14	0	0	56	SO
157	159	7	18	2	46	SO
157	184	8	5	2	51	SO
158	1	0	9	40	47	SO
158	4	0	5	40	51	SO
158	27	5	20	35	36	SO
158	59	8	3	32	53	SO
158	60	5	1	34	52	SO
158	88	3	0	29	53	SO
158	109	2	13	24	7	SO
158	149	0	4	4	0	SO
158	152	2	40	17	3	SO
158	155	0	11	12	1	SO
158	156	14	0	0	56	SO
158	167	6	26	16	22	SO
158	173	5	27	3	1	SO
158	180	23	48	15	8	SO
158	184	11	2	19	32	SO
158	191	7	1	1	4	SO
159	1	9	0	21	50	SO
159	11	3	16	27	34	RD
159	27	14	10	16	40	SO
159	35	3	19	27	31	RD
159	73	25	25	3	24	RD
159	88	2	0	23	49	SO
159	90	5	0	24	49	SO
159	93	13	9	15	41	RD
159	104	2	32	17	6	SO
159	109	2	11	13	13	SO
159	120	30	31	0	8	RD
159	124	5	1	18	44	RD
159	127	1	13	21	34	RD
159	133	0	21	28	28	RD
159	136	0	11	28	35	RD
159	143	3	0	24	49	SO
159	150	4	0	0	4	SO
159	155	0	11	4	3	SO
159	156	7	2	17	28	SO
159	160	30	0	0	50	RD
159	161	30	0	0	50	RD
159	167	1	27	24	13	SO
159	173	2	25	2	3	SO
159	174	1	23	27	24	RD
159	184	6	3	0	42	SO
159	197	11	9	19	41	SO
160	1	9	0	21	67	SO
160	27	14	12	16	55	SO
160	88	2	0	23	61	SO
160	90	5	0	24	65	SO
160	104	2	40	17	10	SO
160	109	2	12	13	18	SO
160	143	3	1	24	65	SO
160	149	0	4	2	1	SO
160	150	4	1	0	4	SO
160	155	0	11	4	8	SO
160	156	7	3	17	35	SO
160	159	30	0	0	50	SO
160	165	0	7	30	59	SO
160	167	1	32	24	16	SO
160	173	2	29	2	4	SO
160	184	6	6	0	52	SO
161	1	9	0	38	50	SO
161	4	0	5	47	45	SO
161	90	5	0	40	49	SO
161	104	10	32	21	6	SO
161	109	3	11	18	13	SO
161	143	4	0	40	49	SO
161	150	5	0	0	4	SO
161	152	22	20	15	5	SO
161	155	0	11	9	3	SO
161	156	8	2	24	28	SO
161	159	30	0	0	50	SO
161	167	6	27	27	13	SO
161	184	9	3	10	42	SO
161	191	0	8	2	1	SO
162	31	12	42	1	23	RD
162	35	1	22	12	43	RD
162	44	1	23	12	42	RD
162	61	4	1	8	62	SO
162	65	2	23	2	2	RD
162	104	2	39	6	14	SO
162	115	1	19	12	43	RD
162	122	11	35	2	30	RD
162	134	10	31	3	30	RD
162	137	1	27	10	31	RD
162	163	13	0	0	65	RD
162	164	13	0	0	65	RD
162	167	1	31	11	16	SO
162	173	1	26	1	1	SO
162	174	1	25	11	35	RD
162	177	1	17	3	4	SO
163	61	4	3	8	82	SO
163	104	2	40	6	23	SO
163	162	13	0	0	65	SO
163	167	1	32	11	31	SO
164	12	5	0	29	65	SO
164	60	5	1	26	62	SO
164	61	6	1	27	62	SO
164	90	3	1	28	64	SO
164	104	3	39	15	14	SO
164	109	1	14	13	19	SO
164	143	3	1	28	61	SO
164	162	13	0	0	65	SO
164	167	2	31	26	16	SO
164	173	2	26	5	1	SO
164	177	4	17	15	4	SO
164	180	14	59	18	5	SO
164	184	2	14	17	34	SO
165	27	0	31	8	75	SO
165	47	5	20	1	29	RD
165	104	7	36	0	30	SO
165	106	7	45	0	30	RD
165	112	4	22	0	47	SO
165	144	0	34	3	8	RD
165	151	2	2	0	22	SO
165	166	8	7	0	99	RD
165	167	6	28	0	48	SO
165	176	8	64	0	28	RD
165	196	1	45	7	61	SO
166	22	12	55	3	44	RD
166	104	11	32	0	30	SO
166	112	7	19	2	45	SO
166	129	11	44	2	35	RD
166	144	0	34	4	7	RD
166	151	3	1	0	22	SO
166	156	0	14	10	46	SO
166	159	0	30	9	40	SO
166	165	8	0	7	99	SO
166	167	8	26	1	47	SO
166	176	14	58	0	28	RD
166	177	7	16	0	21	SO
166	191	0	8	2	3	SO
166	196	2	44	13	55	SO
167	1	0	9	34	39	SO
167	4	5	0	29	48	SO
167	11	1	16	33	32	RD
167	32	7	2	27	46	RD
167	38	8	1	26	45	RD
167	40	29	25	3	20	RD
167	41	15	3	18	45	RD
167	49	4	9	0	17	RD
167	52	34	24	0	16	RD
167	53	13	1	21	47	RD
167	60	0	6	34	38	SO
167	61	0	5	34	41	SO
167	64	26	11	1	10	RD
167	65	11	5	1	10	RD
167	86	31	27	3	20	RD
167	90	0	5	34	40	SO
167	95	33	30	0	16	RD
167	96	33	17	0	19	RD
167	104	32	0	0	27	SO
167	109	12	3	6	20	SO
167	110	12	3	19	31	RD
167	124	1	8	29	32	RD
167	133	4	16	27	30	RD
167	136	12	0	18	42	RD
167	143	0	4	34	37	SO
167	159	1	24	27	13	SO
167	162	1	11	31	16	SO
167	163	1	11	32	31	RD
167	165	6	0	28	48	SO
167	168	34	0	0	48	RD
167	169	34	0	0	48	RD
167	173	25	1	0	3	SO
167	174	28	0	5	42	RD
167	176	33	14	0	28	RD
167	177	5	0	0	14	SO
167	180	34	23	0	22	SO
167	184	2	7	31	10	SO
168	1	0	9	34	63	SO
168	4	5	0	29	72	SO
168	60	0	7	34	61	SO
168	61	0	8	34	62	SO
168	104	32	9	0	30	SO
168	109	12	3	6	28	SO
168	149	3	1	0	4	SO
168	152	22	20	3	16	SO
168	155	11	0	2	8	SO
168	156	2	11	26	29	SO
168	159	1	29	27	21	SO
168	162	1	12	31	31	SO
168	165	6	1	28	71	SO
168	167	34	0	0	48	SO
168	173	25	6	0	7	SO
168	177	5	16	0	21	SO
168	180	34	38	0	31	SO
168	184	2	12	31	24	SO
169	1	0	9	58	39	SO
169	4	5	0	53	48	SO
169	27	13	18	45	30	SO
169	60	1	6	57	38	SO
169	90	0	5	55	40	SO
169	104	41	0	3	27	SO
169	109	12	3	14	20	SO
169	151	3	0	7	15	SO
169	159	6	24	35	13	SO
169	162	2	11	46	16	SO
169	165	7	0	51	48	SO
169	167	34	0	0	48	SO
169	173	30	1	4	3	SO
169	177	21	0	7	14	SO
169	180	49	23	9	22	SO
169	184	7	7	45	10	SO
170	1	0	9	39	54	SO
170	11	15	10	24	53	RD
170	25	2	14	37	49	RD
170	27	7	23	32	40	SO
170	31	36	42	3	21	RD
170	32	1	9	38	54	RD
170	33	6	24	33	39	RD
170	35	2	22	37	41	RD
170	38	1	9	38	52	RD
170	39	10	27	29	35	RD
170	41	1	18	37	44	RD
170	42	6	30	32	32	RD
170	44	1	24	38	39	RD
170	59	8	3	30	59	SO
170	71	1	20	38	40	RD
170	90	0	4	39	52	SO
170	98	2	19	37	44	RD
170	100	16	46	9	7	RD
170	109	0	14	21	15	SO
170	127	4	15	30	38	RD
170	151	1	4	14	6	SO

170	152	6	35	11	10	SO
170	154	14	40	11	10	RD
170	155	0	11	8	4	SO
170	167	7	27	17	24	SO
170	171	0	12	39	49	RD
170	172	0	21	39	38	RD
170	184	10	2	21	34	SO
171	6	0	20	12	70	RD
171	37	9	46	3	44	RD
171	40	10	50	2	36	RD
171	102	12	47	0	41	RD
171	104	9	34	0	27	SO
171	109	4	10	2	33	SO
171	151	2	3	1	21	SO
171	156	0	12	11	41	SO
171	159	1	26	10	40	SO
171	160	1	26	11	55	RD
171	167	10	24	0	41	SO
171	170	0	39	12	49	RD
171	172	0	21	12	66	RD
171	173	12	21	0	7	SO
172	4	3	2	18	86	SO
172	41	10	9	11	77	RD
172	47	11	14	4	24	RD
172	52	19	59	0	15	RD
172	53	8	6	13	82	RD
172	64	12	33	0	15	RD
172	92	0	20	21	68	RD
172	104	18	25	0	28	SO
172	109	8	6	5	33	SO
172	117	16	43	5	42	RD
172	118	1	36	18	40	RD
172	120	12	73	7	3	RD
172	126	11	67	10	6	RD
172	129	18	38	2	33	RD
172	133	2	26	18	54	RD
172	144	0	32	6	5	RD
172	155	6	5	2	11	SO
172	159	1	26	15	35	SO
172	160	1	26	18	49	RD
172	162	0	12	16	47	SO
172	164	1	31	16	47	RD
172	167	15	19	1	42	SO
172	171	0	12	21	66	RD
172	173	21	12	0	7	SO
172	176	20	52	1	24	RD
172	184	0	19	20	36	SO
172	189	5	2	16	86	RD
172	191	0	8	3	2	SO
173	12	0	2	33	5	SO
173	52	31	5	0	2	RD
173	90	0	2	31	3	SO
173	102	32	4	1	3	RD
173	109	12	0	7	3	SO
173	112	9	3	16	0	SO
173	126	18	6	15	0	RD
173	143	0	2	33	4	SO
173	152	19	0	6	2	SO
173	159	2	2	25	3	SO
173	162	1	1	26	1	SO
173	167	25	0	1	3	SO

173	169	30	4	1	3	RD
173	171	12	0	21	7	RD
173	172	21	0	12	7	RD
173	175	29	4	3	3	RD
173	176	31	5	1	2	RD
174	1	0	9	28	63	SO
174	3	9	4	19	68	SO
174	4	5	0	23	72	SO
174	5	14	4	0	9	SO
174	10	1	29	27	43	SO
174	11	1	24	27	48	SO
174	15	2	4	1	29	SO
174	18	1	29	4	11	SO
174	25	11	5	17	67	SO
174	26	18	26	10	46	SO
174	28	11	41	17	27	SO
174	29	23	26	2	39	SO
174	30	1	45	27	26	SO
174	31	8	68	20	4	SO
174	32	7	3	21	69	SO
174	33	15	14	13	58	SO
174	34	23	30	5	42	SO
174	35	19	5	9	67	SO
174	36	26	15	2	57	SO
174	37	27	30	1	42	SO
174	38	8	2	20	68	SO
174	39	19	18	9	53	SO
174	40	24	34	2	35	SO
174	41	15	4	12	67	SO
174	42	22	14	5	57	SO
174	43	28	22	0	48	SO
174	44	18	7	10	65	SO
174	45	25	26	3	40	SO
174	46	28	46	0	21	SO
174	47	18	7	7	19	SO
174	49	3	16	0	21	SO
174	50	26	16	2	42	SO
174	51	28	27	0	31	SO
174	52	28	45	0	14	SO
174	53	13	1	15	71	SO
174	54	22	14	6	56	SO
174	55	28	42	0	28	SO
174	56	27	43	1	25	SO
174	58	0	12	28	60	SO
174	61	0	8	28	62	SO
174	62	5	0	7	16	SO
174	63	21	23	7	48	SO
174	64	21	23	0	12	SO
174	65	7	16	0	12	SO
174	68	17	30	11	41	SO
174	72	17	30	10	40	SO
174	73	20	36	7	34	SO
174	74	18	8	8	64	SO
174	75	26	32	1	40	SO
174	76	18	8	8	50	SO
174	77	6	18	1	32	SO
174	80	7	8	0	22	SO
174	81	10	1	10	26	SO
174	82	20	27	7	38	SO
174	84	7	16	0	22	SO
174	85	24	29	4	43	SO

174	86	25	41	3	31	SO
174	87	2	25	23	32	SO
174	89	23	10	2	48	SO
174	92	0	18	28	53	SO
174	93	1	28	27	42	SO
174	94	27	23	0	34	SO
174	95	27	55	0	15	SO
174	96	27	38	0	17	SO
174	98	16	5	12	67	SO
174	99	25	24	0	25	SO
174	100	25	38	0	12	SO
174	101	27	13	1	50	SO
174	102	28	29	0	43	SO
174	104	26	16	0	30	SO
174	105	26	16	0	46	SO
174	106	26	24	0	30	SO
174	107	27	17	1	54	SO
174	108	27	34	0	33	SO
174	109	11	3	4	28	SO
174	110	11	3	15	51	SO
174	111	22	26	4	28	SO
174	115	17	3	11	66	SO
174	116	23	9	5	60	SO
174	117	28	25	0	44	SO
174	118	1	32	27	27	SO
174	119	7	55	21	4	SO
174	120	18	59	10	0	SO
174	121	5	38	23	34	SO
174	122	12	58	16	14	SO
174	123	20	72	8	0	SO
174	125	4	48	22	12	SO
174	126	10	60	16	0	SO
174	127	10	7	15	54	SO
174	128	22	22	3	39	SO
174	129	22	27	3	34	SO
174	130	8	4	18	44	SO
174	132	26	38	1	7	SO
174	133	4	25	21	44	SO
174	134	24	31	1	38	SO
174	135	25	47	0	12	SO
174	136	12	0	14	66	SO
174	137	25	4	1	62	SO
174	138	26	19	1	46	SO
174	139	0	34	28	34	SO
174	140	1	12	25	54	SO
174	141	8	3	17	64	SO
174	142	13	13	12	52	SO
174	144	0	33	8	3	SO
174	146	0	8	27	55	SO
174	147	19	3	5	57	SO
174	148	0	13	28	55	SO
174	149	3	1	0	4	SO
174	152	22	20	0	18	SO
174	153	22	20	2	29	SO
174	154	24	31	0	18	SO
174	155	11	0	0	10	SO
174	156	1	11	25	30	SO
174	157	1	11	26	55	SO
174	158	2	34	25	30	SO
174	159	1	27	23	24	SO
174	160	1	27	26	37	SO

174	161	4	40	23	24	SO
174	162	1	11	25	35	SO
174	163	1	11	26	53	SO
174	164	2	29	25	35	SO
174	165	5	3	23	69	SO
174	167	28	5	0	42	SO
174	168	28	5	0	65	SO
174	169	28	28	0	42	SO
174	170	3	33	25	36	SO
174	171	9	2	19	66	SO
174	172	14	7	14	62	SO
174	173	23	9	0	7	SO
174	175	28	23	0	49	SO
174	176	28	44	0	28	SO
174	178	4	45	23	25	SO
174	179	2	46	24	14	SO
174	180	28	42	0	28	SO
174	181	20	1	8	41	SO
174	182	8	41	0	21	SO
174	184	0	11	27	28	SO
174	186	1	28	27	28	SO
174	187	1	52	26	9	SO
174	189	7	0	21	72	SO
174	190	8	21	7	0	SO
174	192	1	20	27	50	SO
174	193	1	25	27	47	SO
174	194	0	13	28	59	SO
175	1	0	9	51	40	SO
175	3	11	2	40	47	RD
175	4	5	0	46	49	SO
175	11	6	19	45	30	RD
175	27	10	18	41	31	SO
175	31	28	48	23	1	RD
175	32	8	2	43	47	RD
175	33	20	9	31	40	RD
175	35	21	3	30	46	RD
175	38	9	1	42	46	RD
175	41	16	3	33	46	RD
175	44	19	6	32	43	RD
175	46	47	27	3	18	RD
175	53	13	1	38	48	RD
175	55	45	25	6	22	RD
175	60	0	6	51	39	SO
175	64	32	12	3	9	RD
175	86	41	25	10	24	RD
175	88	0	3	47	41	SO
175	90	0	4	47	43	SO
175	104	42	0	0	30	SO
175	109	12	2	13	19	SO
175	110	12	2	36	30	RD
175	115	20	0	31	46	RD
175	120	39	38	10	0	RD
175	124	2	6	43	35	RD
175	133	10	19	37	28	RD
175	135	45	27	1	11	RD
175	136	12	0	34	46	RD
175	151	3	1	3	18	SO
175	159	3	25	33	14	SO
175	162	2	10	42	18	SO
175	163	2	10	46	33	RD
175	165	8	0	43	49	SO

175	167	33	0	0	42	SO
175	173	29	3	4	3	SO
175	174	28	0	23	49	RD
175	176	51	21	0	28	RD
175	177	23	0	0	21	SO
175	180	46	24	5	23	SO
175	184	5	6	41	14	SO
175	191	0	8	3	2	SO
176	1	0	9	72	19	SO
176	2	20	2	52	26	SO
176	5	16	2	0	9	SO
176	6	16	2	56	26	SO
176	8	6	9	66	19	SO
176	9	3	5	69	23	SO
176	10	26	4	46	24	SO
176	11	22	3	50	25	SO
176	18	26	4	6	9	SO
176	20	6	9	16	2	SO
176	24	31	20	41	8	SO
176	25	14	2	58	26	SO
176	26	27	17	45	11	SO
176	27	13	15	59	13	SO
176	31	49	27	23	1	SO
176	35	21	3	51	25	SO
176	37	45	12	27	16	SO
176	50	42	0	16	28	SO
176	51	54	1	4	27	SO
176	52	59	14	0	14	SO
176	53	13	1	59	27	SO
176	67	49	23	21	3	SO
176	78	22	2	2	5	SO
176	81	11	0	24	12	SO
176	83	2	6	70	22	SO
176	84	15	8	20	2	SO
176	87	24	3	34	21	SO
176	89	33	0	25	25	SO
176	90	0	4	68	22	SO
176	91	24	3	0	4	SO
176	92	1	17	71	10	SO
176	93	11	18	61	8	SO
176	94	50	0	6	28	SO
176	95	71	11	0	15	SO
176	96	50	15	6	11	SO
176	97	6	11	0	15	SO
176	101	40	0	23	28	SO
176	102	57	0	15	28	SO
176	104	42	0	10	20	SO
176	105	42	0	21	25	SO
176	106	49	1	10	20	SO
176	107	44	0	28	27	SO
176	108	60	1	8	25	SO
176	115	20	0	52	25	SO
176	116	32	0	40	25	SO
176	117	49	4	23	21	SO
176	118	8	25	51	3	SO
176	119	34	28	25	0	SO
176	120	49	28	10	0	SO
176	123	64	28	8	0	SO
176	124	3	5	62	16	SO
176	125	34	18	31	3	SO
176	126	49	21	16	0	SO

176	127	17	0	48	21	SO
176	128	44	0	21	21	SO
176	129	48	1	17	20	SO
176	130	12	0	40	22	SO
176	131	29	1	22	6	SO
176	133	25	4	42	23	SO
176	134	51	4	16	23	SO
176	135	64	8	3	9	SO
176	136	12	0	53	27	SO
176	137	29	0	36	27	SO
176	138	45	0	22	25	SO
176	139	15	19	55	7	SO
176	141	11	0	55	26	SO
176	142	25	1	40	24	SO
176	143	1	3	70	22	SO
176	144	14	19	11	0	SO
176	147	22	0	38	24	SO
176	148	13	0	58	25	SO
176	158	30	6	38	17	SO
176	159	6	22	47	0	SO
176	160	6	22	60	3	SO
176	161	19	25	47	0	SO
176	162	5	7	46	14	SO
176	163	5	7	62	17	SO
176	165	8	0	64	28	SO
176	166	14	0	58	28	SO
176	167	33	0	14	28	SO
176	168	33	0	37	28	SO
176	169	56	0	14	28	SO
176	170	31	5	39	22	SO
176	171	11	0	60	25	SO
176	172	20	1	52	24	SO
176	173	31	1	5	2	SO
176	174	28	0	44	28	SO
176	175	51	0	21	28	SO
176	178	31	18	40	8	SO
176	179	22	26	38	0	SO
176	181	21	0	29	20	SO
176	182	29	20	21	0	SO
176	183	21	0	0	7	SO
176	184	7	4	55	0	SO
176	185	7	4	7	22	SO
176	186	7	22	55	0	SO
176	187	25	28	35	0	SO
176	189	7	0	65	28	SO
176	194	0	13	72	15	SO
177	11	5	16	18	5	RD
177	23	6	1	17	20	RD
177	39	8	1	15	20	RD
177	49	7	4	3	14	RD
177	60	0	4	23	17	SO
177	85	11	2	12	19	RD
177	104	16	0	0	10	SO
177	108	21	12	1	7	RD
177	151	3	1	2	15	SO
177	152	1	3	12	2	SO
177	153	1	3	16	5	RD
177	162	1	3	17	4	SO
177	167	5	0	0	14	SO
177	175	23	0	0	21	RD
177	180	18	4	5	16	SO

177	191	0	8	3	1	SO
177	199	1	5	22	16	SO
178	3	2	11	51	40	RD
178	11	22	5	31	46	RD
178	23	6	17	47	34	RD
178	27	19	9	34	42	SO
178	31	49	31	4	20	RD
178	32	1	9	52	42	RD
178	33	9	21	44	30	RD
178	35	6	18	47	33	RD
178	41	5	14	47	36	RD
178	44	8	17	45	34	RD
178	46	29	47	20	3	RD
178	52	29	47	11	2	RD
178	60	5	1	44	50	SO
178	71	4	16	47	34	RD
178	86	27	41	26	10	RD
178	90	0	3	51	43	SO
178	104	10	33	25	5	SO
178	109	3	10	19	17	SO
178	112	6	17	27	15	SO
178	115	3	17	49	32	RD
178	119	35	30	^	21	RD
178	122	42	31	11	20	RD
178	126	41	33	3	13	RD
178	127	7	13	40	30	RD
178	132	28	37	7	2	RD
178	136	3	9	47	36	RD
178	144	27	5	3	8	RD
178	149	0	4	3	1	SO
178	167	6	27	33	10	SO
178	174	4	23	45	25	RD
178	176	31	40	18	8	RD
178	177	8	15	19	2	SO
178	180	28	43	22	8	SO
178	195	32	21	21	30	SO
178	199	10	2	43	49	SO
179	1	9	0	43	39	SO
179	3	3	8	49	31	RD
179	4	0	5	52	34	SO
179	11	15	4	37	35	RD
179	31	48	22	3	17	RD
179	35	4	19	48	20	RD
179	38	3	7	47	32	RD
179	40	28	28	21	9	RD
179	41	4	14	48	23	RD
179	44	7	17	45	22	RD
179	46	34	37	13	2	RD
179	53	1	12	51	27	RD
179	55	31	34	19	5	RD
179	73	27	27	23	11	RD
179	90	4	0	44	37	SO
179	104	3	33	30	0	SO
179	109	2	12	20	9	SO
179	110	2	12	35	26	RD
179	115	1	17	48	22	RD
179	133	20	6	30	31	RD
179	144	32	0	1	10	RD
179	151	0	2	15	2	SO
179	159	26	1	14	27	SO
179	162	13	0	19	34	SO

179	165	0	6	52	33	SO
179	167	2	28	42	0	SO
179	173	2	24	4	2	SO
179	177	4	14	16	0	SO
179	180	31	35	18	4	SO
179	184	7	3	13	33	SO
179	191	6	0	2	2	SO
180	1	9	0	67	34	SO
180	3	13	0	63	34	RD
180	11	4	24	72	10	RD
180	23	21	4	55	30	RD
180	27	23	5	53	29	SO
180	32	10	0	66	34	RD
180	35	24	0	52	34	RD
180	38	10	0	64	34	RD
180	41	19	0	55	34	RD
180	48	26	2	16	6	RD
180	49	16	6	0	25	RD
180	53	14	0	62	34	RD
180	60	1	6	71	28	SO
180	61	1	7	73	27	SO
180	65	25	0	6	8	RD
180	90	0	5	74	24	SO
180	98	20	2	56	32	RD
180	103	23	17	0	6	SO
180	104	40	3	21	8	SO
180	109	14	1	21	16	SO
180	112	13	10	39	7	SO
180	115	20	0	52	34	RD
180	119	51	19	25	1	RD
180	123	68	34	8	0	RD
180	136	12	0	54	29	RD
180	143	2	3	73	25	SO
180	152	41	1	16	7	SO
180	167	34	0	23	22	SO
180	177	18	5	4	16	SO
180	182	55	0	0	25	RD
180	199	5	7	71	27	SO
181	1	0	9	21	46	SO
181	2	19	3	2	52	SO
181	3	10	3	11	52	SO
181	4	5	0	16	55	SO
181	5	15	3	0	9	SO
181	6	0	22	21	33	SO
181	8	0	10	21	45	SO
181	28	6	35	15	17	SO
181	29	21	32	0	14	SO
181	30	0	21	21	33	SO
181	31	2	50	19	5	SO
181	32	6	4	15	51	SO
181	33	14	14	7	41	SO
181	34	19	31	2	24	SO
181	35	20	4	1	51	SO
181	36	21	21	0	34	SO
181	37	21	40	0	15	SO
181	38	7	3	14	50	SO
181	39	17	19	4	35	SO
181	40	20	37	1	14	SO
181	41	16	3	4	51	SO
181	42	20	16	0	38	SO
181	43	21	31	0	22	SO

181	44	18	7	3	48	SO
181	45	20	31	1	18	SO
181	47	18	7	2	24	SO
181	50	18	15	3	34	SO
181	51	21	25	0	24	SO
181	52	21	37	0	13	SO
181	53	12	2	9	53	SO
181	54	20	17	1	36	SO
181	62	5	0	6	19	SO
181	63	19	27	2	27	SO
181	64	19	27	0	6	SO
181	67	19	35	2	18	SO
181	71	10	6	10	46	SO
181	73	19	38	1	14	SO
181	74	19	7	0	48	SO
181	75	20	41	0	14	SO
181	76	19	7	1	36	SO
181	79	7	15	0	18	SO
181	81	11	0	9	31	SO
181	82	20	31	1	20	SO
181	85	20	35	1	20	SO
181	86	21	46	0	9	SO
181	87	1	8	20	31	SO
181	89	20	13	1	25	SO
181	92	0	17	21	37	SO
181	93	1	21	20	32	SO
181	94	20	26	0	25	SO
181	95	20	38	0	16	SO
181	96	20	42	0	8	SO
181	98	16	4	5	51	SO
181	99	21	27	0	20	SO
181	100	21	36	0	13	SO
181	101	20	19	1	29	SO
181	102	21	28	0	23	SO
181	104	20	20	0	21	SO
181	105	20	20	0	24	SO
181	106	20	22	0	21	SO
181	107	20	23	1	31	SO
181	108	20	31	0	21	SO
181	109	10	4	2	19	SO
181	110	10	4	10	35	SO
181	111	18	20	2	19	SO
181	115	16	4	5	47	SO
181	116	21	11	0	40	SO
181	117	21	35	0	16	SO
181	118	0	27	21	28	SO
181	119	3	48	18	7	SO
181	120	11	55	10	0	SO
181	121	1	15	20	40	SO
181	122	5	39	16	16	SO
181	123	13	55	8	0	SO
181	125	1	29	18	14	SO
181	126	3	43	16	0	SO
181	127	10	8	10	41	SO
181	128	18	23	2	26	SO
181	129	18	28	2	21	SO
181	130	6	4	13	41	SO
181	134	19	19	1	31	SO
181	135	20	31	0	7	SO
181	136	11	1	9	45	SO
181	137	19	10	1	36	SO
181	138	20	24	0	23	SO
181	139	0	21	21	31	SO
181	141	6	4	14	45	SO
181	142	12	14	8	34	SO
181	144	0	21	6	4	SO
181	147	16	7	2	36	SO
181	152	19	22	0	16	SO
181	153	19	22	2	22	SO
181	154	21	28	0	16	SO
181	155	11	0	0	11	SO
181	156	0	11	20	28	SO
181	157	0	11	21	39	SO
181	158	1	22	20	28	SO
181	159	1	22	18	15	SO
181	160	1	22	20	26	SO
181	161	3	33	18	15	SO
181	164	1	13	20	39	SO
181	165	5	1	16	54	SO
181	166	6	5	15	50	SO
181	167	20	14	0	23	SO
181	168	20	14	1	37	SO
181	169	21	28	0	23	SO
181	170	1	23	20	27	SO
181	171	8	2	13	46	SO
181	172	10	6	11	48	SO
181	173	18	8	1	3	SO
181	174	20	8	1	41	SO
181	175	21	25	0	24	SO
181	176	21	29	0	20	SO
181	178	4	24	17	26	SO
181	179	2	29	18	17	SO
181	184	0	14	19	19	SO
181	186	1	20	19	19	SO
181	187	1	37	20	10	SO
181	189	5	2	16	53	SO
181	190	8	23	5	2	SO
181	192	0	7	21	47	SO
181	194	0	13	21	41	SO
181	199	3	2	18	53	SO
182	1	9	0	46	25	SO
182	6	22	0	33	25	SO
182	10	7	24	48	1	SO
182	11	3	23	52	2	SO
182	15	4	0	7	24	SO
182	18	7	24	10	1	SO
182	19	7	24	22	0	SO
182	22	26	20	29	5	SO
182	23	15	2	40	23	SO
182	24	32	7	23	18	SO
182	25	10	0	45	25	SO
182	26	28	3	27	22	SO
182	27	18	3	37	22	SO
182	28	35	11	17	13	SO
182	29	32	0	14	25	SO
182	30	21	24	33	1	SO
182	33	14	0	41	25	SO
182	34	31	4	24	21	SO
182	36	21	0	34	25	SO
182	37	40	1	15	24	SO
182	39	19	0	35	25	SO
182	40	37	5	14	19	SO
182	42	16	0	38	25	SO
182	43	31	1	22	24	SO
182	44	7	0	48	25	SO
182	45	31	1	18	24	SO
182	46	51	4	0	20	SO
182	49	15	3	0	20	SO
182	50	15	10	34	5	SO
182	51	25	13	24	2	SO
182	52	37	15	13	0	SO
182	54	17	0	36	25	SO
182	55	50	2	3	23	SO
182	56	43	5	8	20	SO
182	57	19	0	36	25	SO
182	60	1	5	50	20	SO
182	61	1	6	52	19	SO
182	63	27	0	27	25	SO
182	64	27	0	6	4	SO
182	65	19	0	6	4	SO
182	66	30	5	20	17	SO
182	68	30	4	24	21	SO
182	69	38	4	16	21	SO
182	72	31	0	21	25	SO
182	73	38	0	14	25	SO
182	74	7	0	48	25	SO
182	75	41	1	14	24	SO
182	76	7	0	36	25	SO
182	77	27	1	9	24	SO
182	79	15	21	18	0	SO
182	82	31	0	20	25	SO
182	84	18	3	2	22	SO
182	85	35	2	20	23	SO
182	86	46	2	9	23	SO
182	87	8	21	31	0	SO
182	89	13	0	25	22	SO
182	92	17	2	37	23	SO
182	95	38	25	16	0	SO
182	97	8	4	16	0	SO
182	99	27	0	20	5	SO
182	100	36	0	13	3	SO
182	101	19	2	29	15	SO
182	103	23	12	0	2	SO
182	105	20	3	24	16	SO
182	107	23	3	31	21	SO
182	112	9	7	29	6	SO
182	114	22	17	29	6	SO
182	116	11	0	40	25	SO
182	117	35	2	16	23	SO
182	121	15	24	40	1	SO
182	122	39	25	16	0	SO
182	125	29	21	14	3	SO
182	131	14	7	15	2	SO
182	132	25	22	6	0	SO
182	133	11	14	39	10	SO
182	134	19	17	31	6	SO
182	135	31	21	7	0	SO
182	137	10	0	36	22	SO
182	138	24	4	23	18	SO
182	147	7	0	36	24	SO
182	148	0	14	53	11	SO
182	153	22	1	22	6	SO
182	157	11	0	39	22	SO
182	159	22	0	15	16	SO
182	160	22	0	26	19	SO
182	161	33	3	15	16	SO
182	164	13	14	39	4	SO
182	167	14	0	23	13	SO
182	168	14	0	37	22	SO
182	174	8	0	41	21	SO
182	175	25	5	24	16	SO
182	176	29	21	20	0	SO
182	177	17	5	4	16	SO
182	178	24	21	26	2	SO
182	180	55	0	0	25	SO
182	184	14	3	19	18	SO
182	186	20	2	19	18	SO
182	188	23	0	31	25	SO
182	192	7	14	47	11	SO
182	193	3	20	52	4	SO
182	194	13	0	41	25	SO
182	197	19	4	36	21	SO
182	199	2	7	53	18	SO
183	8	1	4	24	5	SO
183	9	1	3	24	6	SO
183	10	24	1	1	8	SO
183	11	23	1	2	8	SO
183	13	1	4	24	5	SO
183	14	0	4	25	5	SO
183	15	0	4	24	1	SO
183	18	24	1	1	4	SO
183	24	7	6	18	3	SO
183	29	0	2	25	7	SO
183	40	5	5	19	4	SO
183	50	10	0	5	9	SO
183	51	13	1	2	8	SO
183	52	15	6	0	3	SO
183	56	5	7	20	2	SO
183	66	5	6	17	3	SO
183	71	0	5	25	4	SO
183	72	0	5	25	4	SO
183	73	0	6	25	3	SO
183	83	0	5	25	4	SO
183	84	3	4	22	0	SO
183	86	2	5	23	3	SO
183	87	21	0	0	6	SO
183	90	0	5	23	1	SO
183	91	21	0	0	5	SO
183	94	5	0	4	9	SO
183	96	5	0	4	9	SO
183	98	0	2	25	7	SO
183	102	10	0	14	9	SO
183	108	12	1	7	6	SO
183	131	7	1	2	5	SO
183	132	22	1	0	1	SO
183	133	14	1	10	7	SO
183	134	17	1	6	7	SO
183	135	21	3	0	5	SO
183	143	0	3	23	2	SO
183	148	14	0	11	6	SO
183	156	0	3	8	0	SO
183	157	0	3	22	0	SO
183	159	0	7	16	0	SO
183	160	0	7	19	0	SO

183	161	3	7	16	0	SO
183	169	9	0	13	9	SO
183	170	12	0	8	8	SO
183	176	21	0	0	7	SO
183	178	21	1	2	6	SO
183	186	2	6	18	0	SO
183	187	14	9	5	0	SO
183	193	20	2	4	7	SO
183	197	4	5	21	4	SO
184	3	0	10	21	46	RD
184	31	20	36	1	20	RD
184	47	0	18	7	14	RD
184	88	4	0	6	55	SO
184	103	6	10	3	0	SO
184	109	2	11	13	14	SO
184	120	19	34	0	10	RD
184	124	4	2	15	50	RD
184	133	1	21	16	33	RD
184	135	8	52	5	1	RD
184	149	0	4	2	0	SO
184	150	3	1	0	3	SO
184	152	0	24	12	5	SO
184	155	0	11	6	2	SO
184	156	8	2	2	32	SO
184	159	6	0	3	42	SO
184	167	2	31	7	10	SO
185	1	0	9	21	22	SO
185	6	10	1	11	30	SO
185	17	4	5	10	1	SO
185	19	3	7	10	1	SO
185	20	4	7	10	1	SO
185	26	5	16	16	15	SO
185	27	2	14	19	17	SO
185	28	15	14	5	13	SO
185	36	1	11	20	20	SO
185	42	2	9	19	21	SO
185	44	0	6	21	25	SO
185	47	0	6	7	7	SO
185	51	7	4	7	26	SO
185	52	12	17	2	13	SO
185	57	7	3	14	28	SO
185	67	10	26	9	4	SO
185	74	0	6	21	24	SO
185	76	0	6	17	13	SO
185	77	10	3	7	9	SO
185	78	10	3	1	5	SO
185	79	8	10	8	1	SO
185	80	5	2	2	7	SO
185	88	4	0	6	28	SO
185	92	2	17	18	13	SO
185	93	3	20	15	10	SO
185	94	4	2	10	27	SO
185	95	19	14	0	15	SO
185	96	4	17	9	11	SO
185	97	9	11	0	15	SO
185	98	0	7	21	24	SO
185	99	1	20	17	9	SO
185	100	5	25	13	4	SO
185	101	2	0	11	30	SO
185	102	5	4	9	27	SO
185	103	6	25	3	2	SO

185	104	5	1	2	21	SO
185	105	5	1	5	25	SO
185	106	7	2	2	21	SO
185	107	5	2	13	29	SO
185	108	9	4	4	23	SO
185	117	12	9	8	19	SO
185	118	11	26	8	5	SO
185	127	3	1	10	22	SO
185	128	8	3	5	20	SO
185	129	12	3	1	20	SO
185	138	7	2	5	22	SO
185	139	1	21	13	8	SO
185	142	6	2	4	25	SO
185	144	1	21	1	0	SO
185	147	3	0	9	23	SO
185	152	0	18	12	3	SO
185	153	0	18	15	7	SO
185	154	3	22	12	3	SO
185	156	8	3	2	19	SO
185	157	8	3	5	23	SO
185	158	11	7	2	19	SO
185	159	6	22	3	2	SO
185	160	6	22	6	5	SO
185	162	1	7	14	15	SO
185	163	1	7	16	21	SO
185	164	2	13	14	15	SO
185	166	3	0	18	31	SO
185	167	2	1	7	28	SO
185	169	7	3	7	28	SO
185	170	10	7	2	22	SO
185	175	5	3	6	26	SO
185	176	7	7	4	22	SO
185	179	7	29	3	2	SO
185	188	12	3	9	26	SO
185	193	1	8	19	23	SO
185	194	0	13	21	18	SO
186	1	9	0	22	56	SO
186	2	2	19	29	37	SO
186	5	2	15	9	0	SO
186	6	1	9	30	47	SO
186	8	7	5	24	51	SO
186	14	5	2	26	54	SO
186	15	5	2	7	21	SO
186	17	5	2	1	9	SO
186	18	7	21	7	5	SO
186	20	7	5	1	9	SO
186	24	22	23	9	33	SO
186	25	2	12	29	44	SO
186	27	14	12	17	44	SO
186	30	17	22	12	34	SO
186	31	28	36	2	20	SO
186	35	3	19	28	37	SO
186	40	22	32	6	22	SO
186	41	3	15	28	40	SO
186	49	7	7	4	15	SO
186	50	1	40	29	6	SO
186	51	4	45	26	1	SO
186	52	17	47	13	0	SO
186	53	1	13	30	43	SO
186	58	7	4	24	52	SO
186	63	9	27	21	29	SO

186	64	9	27	7	4	SO
186	65	3	12	7	4	SO
186	73	20	26	9	29	SO
186	78	3	13	5	1	SO
186	81	1	9	11	17	SO
186	83	5	1	26	55	SO
186	84	11	11	3	18	SO
186	87	5	20	22	27	SO
186	89	3	27	25	20	SO
186	90	4	0	25	52	SO
186	91	5	20	4	0	SO
186	92	17	0	13	56	SO
186	93	20	7	10	49	SO
186	94	2	43	27	1	SO
186	95	14	56	15	0	SO
186	96	17	43	11	1	SO
186	101	0	37	30	13	SO
186	102	4	46	27	10	SO
186	104	1	35	21	7	SO
186	105	1	35	25	15	SO
186	106	2	39	21	7	SO
186	107	2	36	29	20	SO
186	108	4	48	23	7	SO
186	109	1	11	12	14	SO
186	111	7	36	12	14	SO
186	115	0	17	28	39	SO
186	116	2	25	26	31	SO
186	117	9	36	19	20	SO
186	118	26	2	5	42	SO
186	119	31	20	0	24	SO
186	120	31	34	0	10	SO
186	122	26	35	5	21	SO
186	123	31	48	0	8	SO
186	124	6	2	17	50	SO
186	125	20	23	3	29	SO
186	126	23	36	0	16	SO
186	127	1	15	22	37	SO
186	128	3	33	20	19	SO
186	129	3	36	20	16	SO
186	130	1	11	22	29	SO
186	132	8	51	2	1	SO
186	133	5	21	23	33	SO
186	134	5	47	23	7	SO
186	135	13	52	6	1	SO
186	136	0	12	26	39	SO
186	137	1	27	25	24	SO
186	138	2	36	22	18	SO
186	139	21	12	8	44	SO
186	141	0	10	28	44	SO
186	142	2	19	25	34	SO
186	143	3	1	25	55	SO
186	144	21	11	0	10	SO
186	147	0	20	23	29	SO
186	148	0	13	28	42	SO
186	155	0	11	2	2	SO
186	159	22	0	2	42	SO
186	160	22	0	5	52	SO
186	161	25	10	2	42	SO
186	162	7	3	15	34	SO
186	163	7	3	21	48	SO
186	165	0	7	31	49	SO

186	166	0	11	31	45	SO
186	167	1	31	28	10	SO
186	168	1	31	30	24	SO
186	169	3	45	28	10	SO
186	172	0	20	27	36	SO
186	173	1	28	3	3	SO
186	174	1	27	28	28	SO
186	175	3	41	26	14	SO
186	176	7	55	22	0	SO
186	178	19	23	11	32	SO
186	179	29	13	2	33	SO
186	181	1	19	20	19	SO
186	182	20	19	2	18	SO
186	183	2	18	6	0	SO
186	187	31	11	0	34	SO
186	188	3	15	26	41	SO
186	189	0	7	31	49	SO
186	194	13	0	18	56	SO
186	199	1	8	30	48	SO
187	1	9	0	57	35	SO
187	4	0	5	66	30	SO
187	25	5	12	61	23	RD
187	32	2	8	64	27	RD
187	33	13	16	53	19	RD
187	38	1	9	63	26	RD
187	40	40	27	23	6	RD
187	41	3	16	62	18	RD
187	46	45	31	17	3	RD
187	52	37	34	16	0	RD
187	53	1	13	65	22	RD
187	55	41	30	23	5	RD
187	98	7	15	59	20	RD
187	100	34	27	18	1	RD
187	104	6	32	28	1	SO
187	109	4	11	31	3	SO
187	110	4	11	48	22	RD
187	124	12	1	43	32	RD
187	127	7	12	41	22	RD
187	130	2	10	43	23	RD
187	132	25	32	7	1	RD
187	135	33	33	12	1	RD
187	144	28	2	2	9	RD
187	149	0	4	3	0	SO
187	152	19	23	18	3	SO
187	154	27	27	18	3	RD
187	155	0	11	8	1	SO
187	156	11	2	27	29	SO
187	159	30	0	16	27	SO
187	162	10	3	30	29	SO
187	165	1	6	64	29	SO
187	167	2	27	44	1	SO
187	173	6	26	4	1	SO
187	177	11	7	13	2	SO
187	180	38	30	23	5	SO
187	184	18	0	11	34	SO
188	1	0	9	32	72	SO
188	11	0	26	32	55	RD
188	19	0	31	19	3	RD
188	49	9	15	0	25	RD
188	61	0	7	32	71	SO
188	64	25	21	1	17	RD

188	65	21	4	1	17	RD
188	73	31	33	0	45	RD
188	104	21	22	4	25	SO
188	106	22	30	4	25	RD
188	108	26	39	5	28	RD
188	132	18	48	5	4	RD
188	152	9	32	13	12	SO
188	154	15	39	13	12	RD
188	156	9	5	13	42	SO
188	157	9	5	21	62	RD
188	162	1	10	31	34	SO
188	167	14	20	5	41	SO
188	177	14	9	2	19	SO
188	180	31	44	1	32	SO
188	184	12	9	15	41	SO
188	189	0	7	32	74	RD
188	190	32	0	0	7	RD
188	192	0	27	32	54	RD
188	193	0	26	32	54	RD
189	37	7	57	0	51	RD
189	51	7	53	0	37	RD
189	102	7	52	0	49	RD
189	104	7	36	0	30	SO
189	106	7	45	0	30	RD
189	108	7	58	0	35	RD
189	109	4	11	0	42	SO
189	118	0	40	7	53	RD
189	138	7	41	0	49	RD
189	144	0	34	5	6	RD
189	152	5	37	0	25	SO
189	167	7	27	0	48	SO
189	169	7	51	0	48	RD
189	175	7	44	0	49	RD
189	176	7	65	0	28	RD
189	180	7	69	0	34	SO
189	188	0	32	7	74	RD
189	190	0	32	7	0	RD
190	73	31	5	0	2	RD
190	109	4	4	13	0	SO
190	111	14	7	13	0	RD
190	145	16	1	3	5	RD
190	152	9	5	13	0	SO
190	154	15	5	13	0	RD
190	156	9	0	13	7	SO
190	184	12	0	15	7	SO
190	188	32	0	0	7	RD
190	189	0	7	32	0	RD
191	60	4	0	4	5	SO
191	61	2	4	6	1	SO
191	128	0	3	7	2	RD
191	134	8	3	0	2	RD
191	177	0	3	8	1	SO
192	1	0	9	27	77	SO
192	3	0	13	27	73	RD
192	11	14	12	13	74	RD
192	12	5	0	22	86	SO
192	19	19	12	0	22	RD
192	22	21	45	6	41	RD
192	27	4	27	23	59	SO
192	41	0	19	27	65	RD
192	48	1	29	10	14	RD
192	53	0	14	27	72	RD
192	60	5	1	22	80	SO
192	61	6	1	21	82	SO
192	64	0	46	14	4	RD
192	65	0	25	14	4	RD
192	71	1	19	26	63	RD
192	104	3	40	7	22	SO
192	109	0	15	14	27	SO
192	112	10	17	6	41	SO
192	115	0	20	27	62	RD
192	120	20	68	0	10	RD
192	136	0	12	18	65	RD
192	159	2	27	12	38	SO
192	162	5	6	3	62	SO
192	167	1	33	17	29	SO
192	168	1	33	23	47	RD
192	173	4	29	4	3	SO
192	174	1	27	20	50	RD
192	177	5	18	12	9	SO
192	180	7	68	17	16	SO
192	188	0	32	27	54	RD
193	1	0	9	28	77	SO
193	3	0	13	28	73	RD
193	11	25	3	3	83	RD
193	15	0	11	25	8	RD
193	19	25	8	0	22	RD
193	31	27	62	1	23	RD
193	49	4	20	20	6	RD
193	53	0	14	28	72	RD
193	57	0	25	28	61	RD
193	59	8	5	18	81	SO
193	60	6	1	21	81	SO
193	61	8	0	19	84	SO
193	64	0	46	2	16	RD
193	71	1	20	27	62	RD
193	74	0	26	27	59	RD
193	104	3	40	9	21	SO
193	109	0	15	16	26	SO
193	119	14	59	1	25	RD
193	123	28	78	0	8	RD
193	136	0	12	26	59	RD
193	157	1	13	25	59	RD
193	159	4	26	17	33	SO
193	160	4	26	21	46	RD
193	162	6	7	5	60	SO
193	167	1	33	18	30	SO
193	168	1	33	25	47	RD
193	174	1	27	25	47	RD
193	177	5	18	15	6	SO
193	180	4	72	22	11	SO
193	184	1	19	17	39	SO
193	188	0	32	26	54	RD
193	197	3	32	25	54	SO
194	1	9	0	4	101	SO
194	2	0	22	13	79	RD
194	5	0	18	9	0	RD
194	6	0	21	13	80	RD
194	10	0	33	13	68	RD
194	11	0	28	13	73	RD
194	16	0	11	4	14	RD
194	18	0	33	4	14	RD
194	20	4	14	0	21	RD
194	27	10	20	3	81	SO
194	34	10	51	3	50	RD
194	46	11	72	0	26	RD
194	50	0	43	13	42	RD
194	64	7	38	0	18	RD
194	86	12	63	1	37	RD
194	87	0	31	13	45	RD
194	89	0	33	13	48	RD
194	93	13	18	0	78	RD
194	94	0	51	13	27	RD
194	96	13	53	0	23	RD
194	97	0	23	13	2	RD
194	100	13	51	0	22	RD
194	101	0	41	13	43	RD
194	104	0	43	13	17	SO
194	105	0	43	13	34	RD
194	107	0	45	13	52	RD
194	115	0	20	12	78	RD
194	119	13	61	0	25	RD
194	127	0	21	13	58	RD
194	128	0	48	13	31	RD
194	133	0	29	13	62	RD
194	137	0	29	13	54	RD
194	142	0	30	13	52	RD
194	144	13	20	0	11	RD
194	147	0	23	13	53	RD
194	152	13	29	0	25	SO
194	154	13	42	0	25	RD
194	156	0	14	13	42	SO
194	159	13	17	0	50	SO
194	161	13	34	0	50	RD
194	167	0	34	13	34	SO
194	168	0	34	13	58	RD
194	170	0	39	13	50	RD
194	172	0	21	13	74	RD
194	174	0	28	13	59	RD
194	175	0	51	13	36	RD
194	180	13	62	0	34	SO
194	188	0	32	13	67	RD
194	192	0	27	13	72	RD
194	193	0	28	13	72	RD
195	156	11	3	29	27	SO
196	61	1	7	44	60	SO
196	130	2	10	31	40	RD
196	165	1	7	45	61	SO
197	12	5	0	30	80	SO
197	59	1	12	33	67	SO
197	61	0	8	33	71	SO
197	65	8	17	13	5	RD
197	86	27	49	7	31	RD
197	90	3	2	28	74	SO
197	119	27	47	5	21	RD
197	143	3	2	27	73	SO
197	159	11	19	9	41	SO
197	199	1	11	34	69	RD
198	112	2	25	14	33	SO
198	113	2	25	20	61	RD
199	11	7	21	5	82	RD
199	72	3	49	9	50	RD
199	151	3	2	4	18	SO
199	177	1	22	5	16	SO
199	180	5	71	7	27	SO
199	197	1	34	11	69	RD